Relative atomic masses (A_r) of the elements with $Z \leq 112$

Element	Symbol	Atomic number	Relative atomic mass*, A_r
Actinium	Ac	89	[227]
Aluminium	Al	13	26.982
Americium	Am	95	[243]
Antimony	Sb	51	121.76
Argon	Ar	18	39.948
Arsenic	As	33	74.922
Astatine	At	85	[210]
Barium	Ba	56	137.33
Berkelium	Bk	97	[247]
Beryllium	Be	4	9.0122
Bismuth	Bi	83	208.98
Bohrium	Bh	107	[267]
Boron	B	5	10.81
Bromine	Br	35	79.904
Cadmium	Cd	48	112.41
Caesium	Cs	55	132.91
Calcium	Ca	20	40.078
Californium	Cf	98	[251]
Carbon	C	6	12.011
Cerium	Ce	58	140.12
Chlorine	Cl	17	35.45
Chromium	Cr	24	51.996
Cobalt	Co	27	58.933
Copernicium	Cn	112	[285]
Copper	Cu	29	63.546
Curium	Cm	96	[247]
Darmstadtium	Ds	110	[281]
Dubnium	Db	105	[268]
Dysprosium	Dy	66	162.50
Einsteinium	Es	99	[252]
Erbium	Er	68	167.26
Europium	Eu	63	151.96
Fermium	Fm	100	[257]
Fluorine	F	9	18.998
Francium	Fr	87	[223]
Gadolinium	Gd	64	157.25
Gallium	Ga	31	69.723
Germanium	Ge	32	72.63
Gold	Au	79	196.97
Hafnium	Hf	72	178.49
Hassium	Hs	108	[270]
Helium	He	2	4.0026
Holmium	Ho	67	164.93
Hydrogen	H	1	1.008
Indium	In	49	114.82
Iodine	I	53	126.90
Iridium	Ir	77	192.22
Iron	Fe	26	55.845
Krypton	Kr	36	83.798
Lanthanum	La	57	138.91
Lawrencium	Lr	103	[262]
Lead	Pb	82	207.2
Lithium	Li	3	6.94
Lutetium	Lu	71	174.97
Magnesium	Mg	12	24.305
Manganese	Mn	25	54.938
Meitnerium	Mt	109	[278]
Mendelevium	Md	101	[258]

Element	Symbol	Atomic number	Relative atomic mass*, A_r
Mercury	Hg	80	200.59
Molybdenum	Mo	42	95.95
Neodymium	Nd	60	144.24
Neon	Ne	10	20.180
Neptunium	Np	93	[237]
Nickel	Ni	28	58.693
Niobium	Nb	41	92.906
Nitrogen	N	7	14.007
Nobelium	No	102	[259]
Osmium	Os	76	190.23
Oxygen	O	8	15.999
Palladium	Pd	46	106.42
Phosphorus	P	15	30.974
Platinum	Pt	78	195.08
Plutonium	Pu	94	[244]
Polonium	Po	84	[209]
Potassium	K	19	39.098
Praseodymium	Pr	59	140.91
Promethium	Pm	61	[145]
Protactinium	Pa	91	231.04
Radium	Ra	88	[226]
Radon	Rn	86	[222]
Rhenium	Re	75	186.21
Rhodium	Rh	45	102.91
Roentgenium	Rg	111	[281]
Rubidium	Rb	37	85.468
Ruthenium	Ru	44	101.07
Rutherfordium	Rf	104	[263]
Samarium	Sm	62	150.36
Scandium	Sc	21	44.956
Seaborgium	Sg	106	[271]
Selenium	Se	34	78.971
Silicon	Si	14	28.085
Silver	Ag	47	107.87
Sodium	Na	11	22.990
Strontium	Sr	38	87.62
Sulfur	S	16	32.06
Tantalum	Ta	73	180.95
Technetium	Tc	43	[98]
Tellurium	Te	52	127.60
Terbium	Tb	65	158.93
Thallium	Tl	81	204.38
Thorium	Th	90	232.04
Thulium	Tm	69	168.93
Tin	Sn	50	118.71
Titanium	Ti	22	47.867
Tungsten	W	74	183.84
Uranium	U	92	238.03
Vanadium	V	23	50.942
Xenon	Xe	54	131.29
Ytterbium	Yb	70	173.05
Yttrium	Y	39	88.906
Zinc	Zn	30	65.38
Zirconium	Zr	40	91.224

* Relative atomic masses are quoted to five significant figures where known. In a few cases (e.g. Pb) this is not possible because A_r depends on the source of the element, so is variable.

[ ] denotes a radioactive element, the number given is the mass number of the most stable isotope.

Chemistry^3

INTRODUCING INORGANIC, ORGANIC, AND PHYSICAL CHEMISTRY

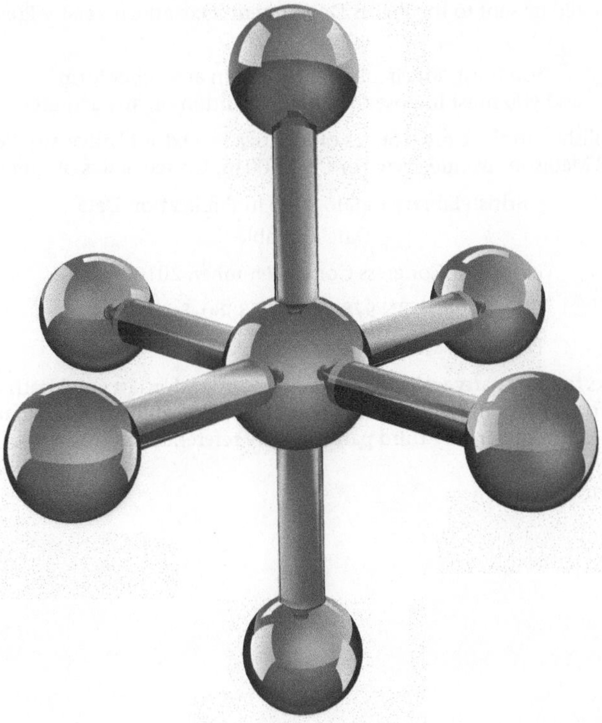

ANDREW BURROWS | JOHN HOLMAN | ANDREW PARSONS
GWEN PILLING | GARETH PRICE

Third Edition

OXFORD
UNIVERSITY PRESS

OXFORD
UNIVERSITY PRESS

Great Clarendon Street, Oxford, OX2 6DP,
United Kingdom

Oxford University Press is a department of the University of Oxford.
It furthers the University's objective of excellence in research, scholarship,
and education by publishing worldwide. Oxford is a registered trade mark of
Oxford University Press in the UK and in certain other countries

First edition 2009
Second edition 2013
Impression: 3

Published in the United States of America by Oxford University Press
198 Madison Avenue, New York, NY 10016, United States of America

British Library Cataloguing in Publication Data
Data available

Library of Congress Control Number: 2016956736

ISBN 978–0–19–873380–5

Printed by CPI Group (UK) Ltd, Croydon CR0 4YY

Summary of contents

Contents

8 Gases 342

Breathing under water

9 Reaction kinetics 382

Methane in the troposphere

10 Molecular spectroscopy 448

Searching for life on Mars

11 Analytical chemistry 512

Drugs, sport, and analytical chemistry

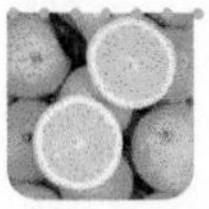

12 Molecular characterization 556

Using isotope ratios to analyse orange juice

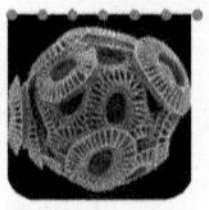

A message to readers

Why *Chemistry*^{3}?

This book covers the three main branches of chemistry—physical, inorganic, and organic—in broadly equal proportions. By covering all three branches in a single volume, the book is able to show the connections between them, reflecting the way that modern chemical research is carried out.

This book is for students starting to study chemistry at university. Our author team includes specialists in physical chemistry (Gareth Price), inorganic chemistry (Andrew Burrows), and organic chemistry (Andrew Parsons). The team also has two specialists in chemical education (John Holman, and for the two prior editions Gwen Pilling) who have taught chemistry at both school and university level, and know how to smooth the transition between the two.

Chemistry is a rapidly developing subject, and with this author team we have been able not only to reflect chemistry as it impacts the world today, but also to draw on research on how students learn chemistry best.

University courses vary quite widely in their content, and we have thoroughly researched different courses to make sure we cover the majority of what is taught. The opening chapter, *Fundamentals*, reviews key concepts that may have been encountered prior to university; it can be dipped into to check on principles that may not have been understood the first time round, or used to build an awareness of fundamental concepts that may not be familiar already. The remaining 27 chapters can individually be identified as physical, inorganic, or organic, but there are extensive cross-references to make the connections between the three branches, and we have been careful to use consistent terminology throughout.

Our approach to organic chemistry is mechanistic: we believe students learn better if they understand the underlying mechanisms, rather than just memorizing reactions of the functional groups.

Understanding how students learn

As an author team, we understand the parts of chemistry that students find hardest when starting at university, and we've taken care to build on what students know from their pre-university studies. We have drawn on our own experience of teaching chemistry, to guide students through difficult concepts and to tackle misconceptions. The book is accessible, but that does not mean it cuts corners—the treatment is as rigorous and comprehensive as any introductory book you will find.

We know that readers may want to dip in to find the information they are looking for, rather than read the whole chapter in one go, so we have ensured each chapter is presented as short, self-contained sections. We explain new ideas carefully, breaking them down into stages and using colour wherever possible. Many students learn best through pictures, and we have used diagrams to complement and, where appropriate, to replace words. We have used annotations in equations to show what each component represents and in diagrams to explain the processes at work.

We appreciate that for many students, mathematics can be a barrier to understanding chemistry. We have not avoided the mathematical challenges: where necessary, we have put the more difficult mathematics in boxes so that students can miss it out on the first reading, and come back to it later. Where students need help with basic maths, there are mathematical hints in the margin, linked to a 'Maths toolkit' at the end of the book.

Chemistry is everywhere in our lives, and this book uses everyday contexts—in the chapter opener features, photographs, and context boxes—to illustrate the applications of chemical principles and theory. We want you to enjoy this book for the way it brings chemistry alive and shows you the power and elegance of chemistry to explain the natural world. And we want it to give you a flying start to your university studies.

Good luck!

Andrew Burrows John Holman Andrew Parsons Gwen Pilling Gareth Price

New to this edition

We've been pleased with the great response to the first and second editions of our one-volume text for first year undergraduates, and in this new edition we have responded to lecturers' and students' comments by adding some new digital content, as well as updating the text throughout. The third edition has:

- New online resources including screencasts which pull together theory and practical aspects of organic chemistry.
- 200 new figures included to better illustrate and communicate key concepts.
- An additional 50 end-of-chapter problems, with outline answers, including some more challenging questions. The questions are linked clearly to relevant sections of the chapter.
- The first chapter has been thoroughly revised to provide a clear and accessible bridge between school and university level study.
- Updated examples of applications to illustrate chemical principles.
- More detailed explanation in places to further clarify, for example, some of the more challenging reaction mechanisms.
- Inclusion of a greater number of photos/images to emphasize/link to key points in the text.
- Expanded use of colour coding to link the text to diagrams more clearly.

In addition, we've added new content that students and lecturers asked for, including:

- Expanded use of Frontier Molecular Orbital Theory to understand the reactivity of organic compounds.

Getting the most from *Chemistry*[3]

This book is designed to help you understand and learn. Here are some of the ways in which the book will introduce and explain new information and ideas, and then help you to consolidate and put them into practice.

How to use this book to . . . see the bigger picture

The book builds on pre-university studies to help you to establish a set of chemical principles that can then be developed through further studies. It is important to form an understanding of the subject as a whole. To help you do this, the book includes frequent cross-references, which highlight the connections between topics within a chapter and between the branches of chemistry covered in different chapters. Each **cross-reference** gives a specific section and page reference and explains the relationship between topics.

> HIO_4 and HIO_3 are called hypervalent compounds because the iodine atoms in these compounds have more than eight electrons in the valence shell. Hypervalency is discussed in Section 5.5 (p.245).

Chapters are carefully structured to **build on** previous chapters progressively. Topic lists at the start of each chapter give the sections from earlier in the book that you should be familiar with before reading the chapter.

> **This chapter builds on the following topics:**
>
> - Hydrocarbons Section 2.5, p.79
> - Functional groups containing one or more heteroatoms Section 2.6, p.90
> - Functional groups containing carbonyl groups Section 2.7, p.98
> - Introduction to molecular spectroscopy Section 10.1, p.450
> - Molecular energies and spectroscopy Section 10.2, p.453

How to use this book to . . . place chemistry in context

The chemistry presented in the book has fascinating applications. Each chapter starts with an eye-catching and informative **chapter opening page** that presents such an application.

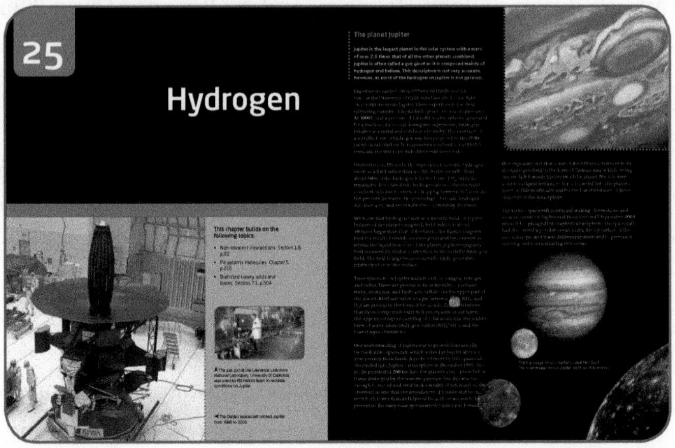

25
Hydrogen
The planet Jupiter

Throughout each chapter, **boxes** and **photographs** show some of the everyday and extraordinary applications of key chemical principles, and how they impact on society.

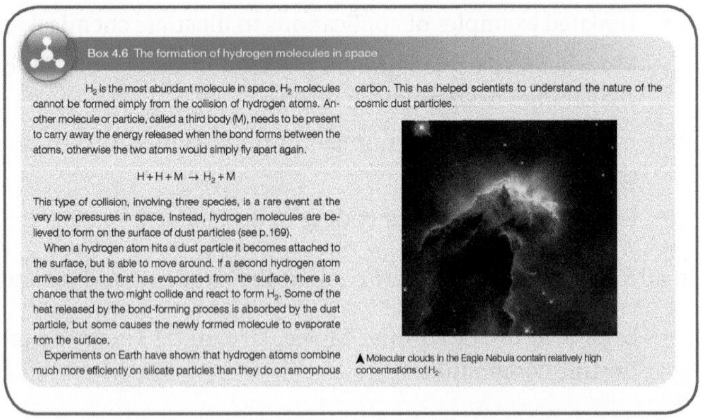

Box 4.6 The formation of hydrogen molecules in space

H_2 is the most abundant molecule in space. H_2 molecules cannot be formed simply from the collision of hydrogen atoms. Another molecule or particle, called a third body (M), needs to be present to carry away the energy released when the bond forms between the atoms, otherwise the two atoms would simply fly apart again.

$$H + H + M \rightarrow H_2 + M$$

This type of collision, involving three species, is a rare event at the very low pressures in space. Instead, hydrogen molecules are believed to form on the surface of dust particles (see p.169).

When a hydrogen atom hits a dust particle it becomes attached to the surface, but is able to move around. If a second hydrogen atom arrives before the first has evaporated from the surface, there is a chance that the two might collide and react to form H_2. Some of the heat released by the bond-forming process is absorbed by the dust particle, but some causes the newly formed molecule to evaporate from the surface.

Experiments on Earth have shown that hydrogen atoms combine much more efficiently on silicate particles than they do on amorphous carbon. This has helped scientists to understand the nature of the cosmic dust particles.

▲ Molecular clouds in the Eagle Nebula contain relatively high concentrations of H_2.

How to use this book to . . . think like a chemist

When studying chemistry at university, it is essential that you get to grips with its language. The book contains a number of different features to help you do this. New terms appear in bold where they are first introduced, and are clearly defined.

The language of chemistry draws heavily on **symbols and abbreviations.** One of the challenges of studying chemistry is to know what they all represent—particularly when the same letter or symbol is used to mean different things. Appendix 1 (p.1331) lists all the abbreviations and symbols used in the book, together with their meaning(s). When a new symbol is used, an **information note** in the margin explains what it represents. These information notes also offer summaries of important information and useful reminders.

> **i** Just as $pH = -\log_{10}[H^+(aq)]$,
> $pOH = -\log_{10}[OH^-(aq)]$,
> $pNa = -\log_{10}[Na^+(aq)]$,
> $pF = -\log_{10}[F^-(aq)]$, etc.

When key equations are introduced, colour-coded annotations point out what each symbol represents.

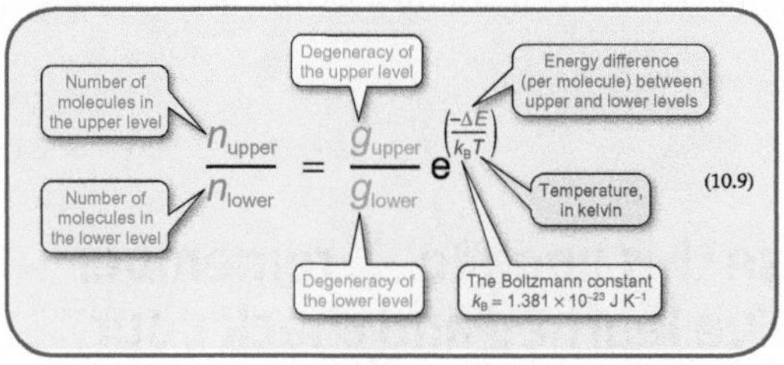
Number of molecules in the upper level
Number of molecules in the lower level
Degeneracy of the upper level
Degeneracy of the lower level
Energy difference (per molecule) between upper and lower levels
Temperature, in kelvin
The Boltzmann constant $k_B = 1.381 \times 10^{-23}$ J K^{-1}
$\frac{n_{upper}}{n_{lower}} = \frac{g_{upper}}{g_{lower}} e^{\left(\frac{-\Delta E}{k_B T}\right)}$
(10.9)

Worked examples illustrate important aspects of the topics being discussed and provide an opportunity to practise how to use equations and the theoretical concepts presented. Worked examples begin with a problem and then provide a strategy for solving it, broken down into clear steps. Each worked example ends with a question for you to test your understanding.

> **Worked example 14.9** Using Gibbs energy changes of formation to calculate Gibbs energy changes of reaction
>
> Using data from Appendix 7 (p.1352), calculate the standard Gibbs energy change, $\Delta_r G^{\ominus}_{298}$, for each of the following reactions and comment on whether the reactions are likely to occur.
>
> (a) $N_2O_4(g) \rightarrow 2NO_2(g)$
>
> (b) $3Fe_2O_3(s) + 2NH_3(g) \rightarrow 6FeO(s) + 3H_2O(l) + N_2(g)$
>
> **Strategy**
>
> Use Equation 14.18 directly to calculate $\Delta_r G^{\ominus}_{298}$. The reaction is spontaneous if $\Delta G < 0$.
>
> **Solution**
>
> *Use Equation 14.18 and substitute in the values of $\Delta_f G^{\ominus}_{298}$ for the reactants and products.*
>
> (b) $\Delta_r G^{\ominus}_{298} = [6\Delta_f G^{\ominus}_{298}(FeO(s)) + 3\Delta_f G^{\ominus}_{298}(H_2O(l)) + \Delta_f G^{\ominus}_{298}(N_2(g))] - [3\Delta_f G^{\ominus}_{298}(Fe_2O_3(s)) + 2\Delta_f G^{\ominus}_{298}(NH_3(g))]$
> $= [6\times(-244.1 \text{ kJ mol}^{-1}) + 3\times(-237.1 \text{ kJ mol}^{-1}) + 0] - [3\times(-742.2 \text{ kJ mol}^{-1}) + 2\times(-16.5 \text{ kJ mol}^{-1})]$
> $= [-2175.9 \text{ kJ mol}^{-1}] - [-2259.6 \text{ kJ mol}^{-1}]$
> $= +83.7 \text{ kJ mol}^{-1}$
>
> In each case $\Delta_r G^{\ominus}_{298} > 0$, so neither of these reactions is spontaneous under these conditions.

How to use this book to ... get to grips with maths

For many, the mathematics involved in chemistry is a challenge. As authors and educators, we understand this and have provided support to help you get to grips with the essential mathematical element of chemistry. **Maths information notes** give short reminders of the mathematical procedure being used.

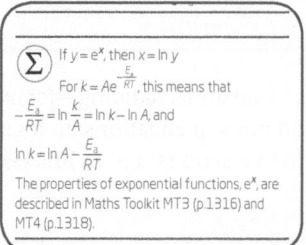

> **Σ** If $y = e^x$, then $x = \ln y$
>
> For $k = Ae^{-\frac{E_a}{RT}}$, this means that
>
> $-\frac{E_a}{RT} = \ln\frac{k}{A} = \ln k - \ln A$, and
>
> $\ln k = \ln A - \frac{E_a}{RT}$
>
> The properties of exponential functions, e^x, are described in Maths Toolkit MT3 (p.1316) and MT4 (p.1318).

Where a more extensive reminder is needed, there is a cross-reference to the **Maths Toolkit** at the end of the book. This gives fuller explanations of the key areas of maths you need in your study of chemistry.

> **MT1 Working with numbers**
>
> The system of numbers that we use is based on whole numbers (called integer numbers), which may be positive, negative, or zero. For example, the atomic number of an element, Z, the number of protons in the nucleus, is an integer. For uranium, U, $Z = 92$.
>
> Of course, not everything you encounter takes an integer value. Think about measuring a quantity of reactant needed for a reaction. It is unlikely that a whole number of kilograms or grams will be needed so it is necessary to subdivide integers. This is done by using fractions or, more commonly in scientific work, the decimal system.
>
> By convention, the '−' sign before a negative number is *always* included but the '+' sign is often omitted before a positive number. No sign before a number means it is positive.

Another challenge of chemistry is getting the **units** right in calculations. We colour units red in worked examples and where they are first introduced or discussed in the chapter. Using the right units throughout the different stages in a calculation can be difficult; we use strikethroughs to show how units are cancelled to help you track units at each step of a calculation.

> **Solution**
>
> Use Equation 3.2
>
> $E = h\nu$ (the Planck constant $h = 6.626 \times 10^{-34}$ J s) (3.2)
>
> $E = (6.626 \times 10^{-34} \text{ J } \cancel{s}) \times (4.41 \times 10^{14} \cancel{s^{-1}}) = 2.92 \times 10^{-19}$ J
>
> *Question*
>
> What is the energy of a r
> of 909 kHz?

How to use this book to . . . visualize chemistry

Diagrams are used wherever possible to explain chemical principles. We include **annotations** to show the processes involved or the sequence of a reaction.

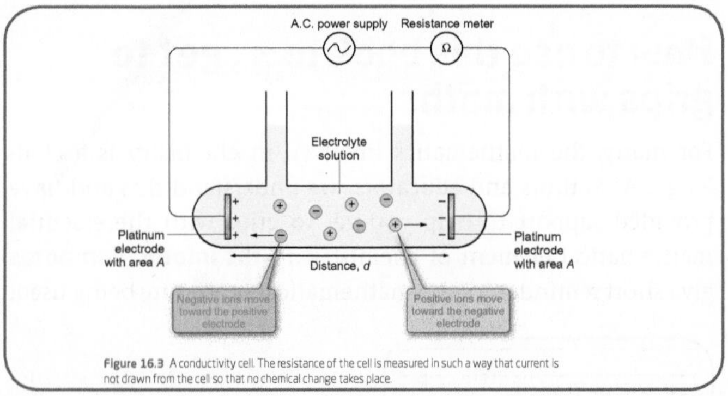
A.C. power supply
Resistance meter
Ω
Electrolyte solution
Platinum electrode with area A
Distance, d
Platinum electrode with area A
Negative ions move toward the positive electrode
Positive ions move toward the negative electrode
Figure 16.3 A conductivity cell. The resistance of the cell is measured in such a way that current is not drawn from the cell so that no chemical change takes place.

Colour is used within the text itself to aid understanding—for example, to differentiate numbers and units in equations, to highlight oxidizing and reducing groups in reactions, and to link text and diagrams.

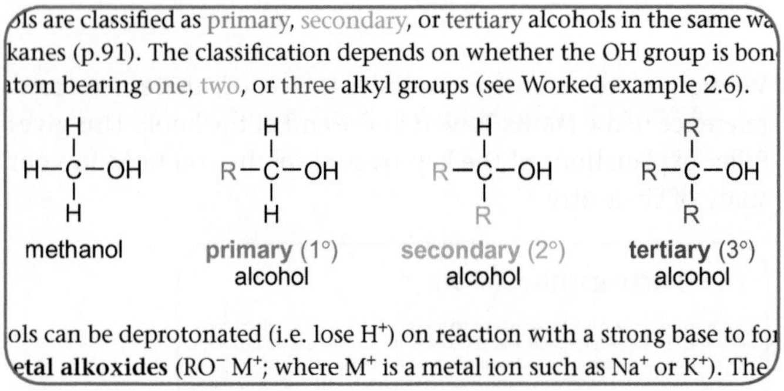
ols are classified as primary, secondary, or tertiary alcohols in the same wa
kanes (p.91). The classification depends on whether the OH group is bon
tom bearing one, two, or three alkyl groups (see Worked example 2.6).
methanol
primary (1°) alcohol
secondary (2°) alcohol
tertiary (3°) alcohol
ols can be deprotonated (i.e. lose H^+) on reaction with a strong base to fo
etal alkoxides ($RO^- M^+$; where M^+ is a metal ion such as Na^+ or K^+). The

To complement the two-dimensional representations of molecular structures and reactions in the book, numerous links are provided to online interactive three-dimensional (3D) structures and animated reactions. On **ChemTube3D**, structures can be rotated, zoomed, and have different properties displayed. Reaction animations take you through sequences step-by-step from beginning to end, with 3D curly arrows indicating reaction mechanisms. An icon in the margin indicates relevant resources are available on ChemTube3D, and these should be made use of to improve your comprehension of the shape of molecules and appreciation of its importance.

This icon indicates that related interactive resources are available through ChemTube3D. A full explanation of how to get to these resources is given in a blue panel at the beginning of chapters that refer to the website.

Screencasts have been filmed for the third edition which pull together theory and practical aspects of organic chemistry. These are linked to specific sections.

When you see this icon follow the link provided to visit the Online Resource Centre and view the related screencast.

How to use this book to . . . take learning further

A selection of **interactive activities** has been created to help you develop and test your understanding. These include: relating atomic spectra and energy levels, acid–base titrations, atomic orbitals and molecular orbitals, exploring close-packed structures, and naming organic compounds. A margin icon will direct you to relevant activities in the book's Online Resource Centre.

When you see this icon, go to the Online Resource Centre to complete an activity and explore the concepts being discussed. www.oxfordtextbooks.co.uk/orc/burrows3e/

All screencasts and videos are also available in the Online Resource Centre.

www.oup.com/uk/orc/chemistry/burrows3e/01student/video/

How to use this book to . . . remember what you've learned and check your understanding

It is impossible to remember everything you read in a textbook. Equally, it can be difficult to know what the most important pieces of information are—the things you really should remember above all else. To help you, we've pulled out **information notes** in the margin, which give short summaries of key concepts and information.

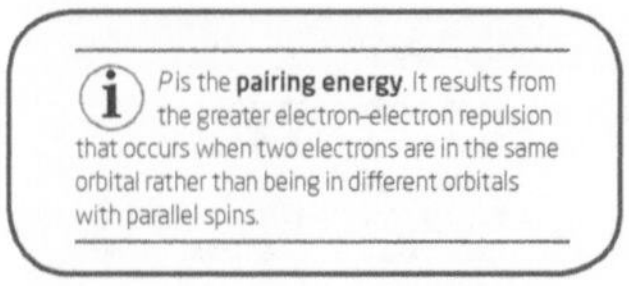
P is the **pairing energy**. It results from the greater electron–electron repulsion that occurs when two electrons are in the same orbital rather than being in different orbitals with parallel spins.

We also provide **summaries** of the key ideas, concepts, and equations that are discussed in each section. These summaries also direct you to related end-of-chapter questions.

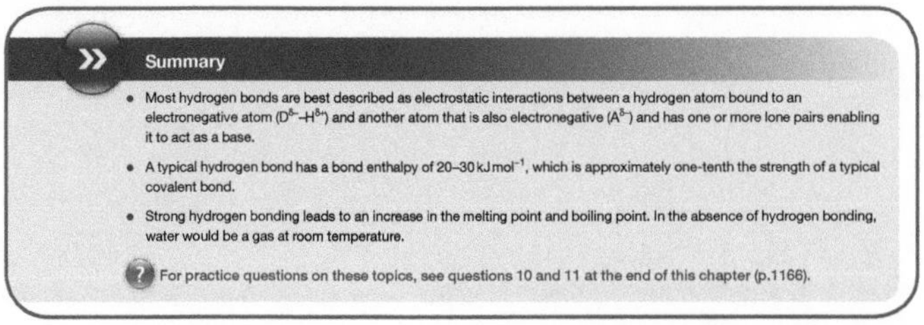

Summary

- Most hydrogen bonds are best described as electrostatic interactions between a hydrogen atom bound to an electronegative atom ($D^{\delta-}$–$H^{\delta+}$) and another atom that is also electronegative ($A^{\delta-}$) and has one or more lone pairs enabling it to act as a base.
- A typical hydrogen bond has a bond enthalpy of 20–30 kJ mol^{-1}, which is approximately one-tenth the strength of a typical covalent bond.
- Strong hydrogen bonding leads to an increase in the melting point and boiling point. In the absence of hydrogen bonding, water would be a gas at room temperature.

For practice questions on these topics, see questions 10 and 11 at the end of this chapter (p.1166).

At the end of each chapter, a **concept review** provides a comprehensive list of what you should be able to do by the end of the chapter, and **key equations** are summarized in a table. There is a highly detailed **index** to help you find everything you need in the book.

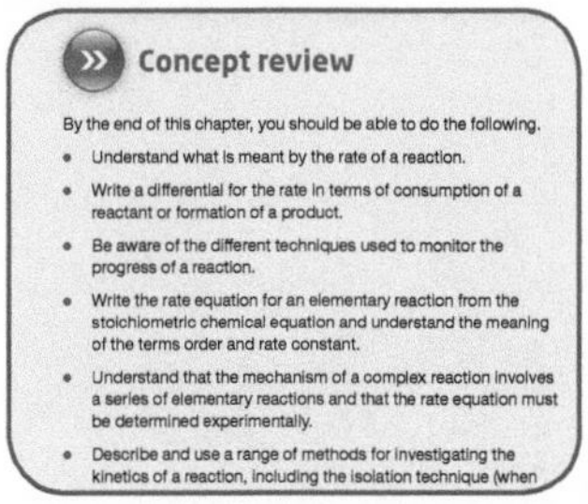

Concept review

By the end of this chapter, you should be able to do the following.

- Understand what is meant by the rate of a reaction.
- Write a differential for the rate in terms of consumption of a reactant or formation of a product.
- Be aware of the different techniques used to monitor the progress of a reaction.
- Write the rate equation for an elementary reaction from the stoichiometric chemical equation and understand the meaning of the terms order and rate constant.
- Understand that the mechanism of a complex reaction involves a series of elementary reactions and that the rate equation must be determined experimentally.
- Describe and use a range of methods for investigating the kinetics of a reaction, including the isolation technique (when

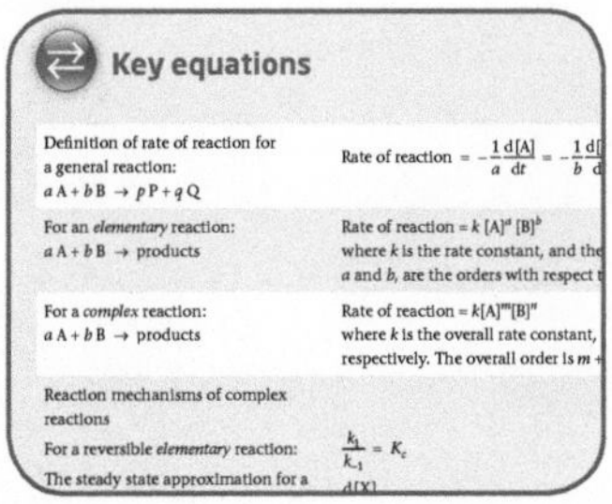

Key equations

Definition of rate of reaction for a general reaction: $a\,A + b\,B \rightarrow p\,P + q\,Q$	Rate of reaction $= -\frac{1}{a}\frac{d[A]}{dt} = -\frac{1}{b}\frac{d[}{d}$
For an *elementary* reaction: $a\,A + b\,B \rightarrow$ products	Rate of reaction $= k\,[A]^a\,[B]^b$ where k is the rate constant, and the a and b, are the orders with respect
For a *complex* reaction: $a\,A + b\,B \rightarrow$ products	Rate of reaction $= k[A]^m[B]^n$ where k is the overall rate constant, respectively. The overall order is m +
Reaction mechanisms of complex reactions	
For a reversible *elementary* reaction:	$\frac{k_1}{k_{-1}} = K_c$
The steady state approximation for a	

There are a number of screencasts available in the accompanying Online Resource Centre in which authors take you through selected examples, problems, and challenging concepts, just as your tutor might do. There are also numerous opportunities in the book and in the accompanying Online Resource Centre for you to apply your knowledge and test your understanding.

Each worked example ends with a question which allows you to practise the skill that is explored in the example itself. Some boxes contain questions to encourage you to apply the chemical principles to a contextual situation. The **end-of-chapter questions** range from simple exercises to questions that bring together several ideas in the chapter and require more advanced problem-solving skills. We have marked the more challenging questions with an asterisk.

Cross-references at the end of questions allow you to review the relevant section of text, and you can check your understanding using the numerical **answers** at the end of the book. Fully worked answers to the questions are available in a **solutions manual** in the Online Resource Centre.

In addition to the materials already mentioned, the Online Resource Centre also contains:

- 3D rotatable structures for all numbered molecular structures.
- Chapter learning outcomes, summaries, and key equations.
- Multiple-choice questions for students to test themselves.
- Figures in electronic format to facilitate lecture presentations—free online for all registered adopting lecturers.
- Test bank—free online for all registered adopting lecturers.

  A bank of multiple-choice questions, which can be downloaded into your assessment software and customized. Answers to all the questions are accompanied by feedback and references to the book so students can use the questions for formative assessment. Feedback can be turned off to enable summative assessment. Lecturers can integrate the test bank with their own assessment tools to grade and monitor students' progress. The test bank is available in Respondus and QTI XML formats.

www.oxfordtextbooks.co.uk/orc/burrows3e/

Acknowledgements

First and foremost, we would like to thank our families for their patience and support, without which this book would not have been possible. In particular, we thank Julie and Luke (ADB), Sue and Thomas (AFP), Christine and Alison (GJP), Mike (GMP), and Wendy (JSH). We would also like to thank our colleagues at the Universities of Bath and York for their friendship and guidance, and in particular acknowledge the contributions of Professor Robin Perutz, Professor Guy Dodson, Professor Duncan Bruce, Professor Lucy Carpenter, Dr John Moore, Professor Richard Taylor (Department of Chemistry, University of York), Professor Dwayne Heard, Professor Mike Pilling, and Dr Annette Taylor (School of Chemistry, University of Leeds). Our thanks also go to Sandie Dann (Loughborough University), Bridgette Duncombe (Imperial College London), Jason Eames (University of Hull), and David Smith (University of Bristol), who have written the test bank and solutions manual on the web site, Heather Powell, and Luke Scrivens (University of York) for help with checking answers, and Nick Greeves (University of Liverpool) who created the ChemTube3D resources. We would also like to thank the undergraduate students who provided detailed feedback and suggestions which helped us refine the text. Producing this book has been a long and complex process, and it would have been impossible without the dedication, support, and encouragement of Oxford University Press, and especially Ruth Hughes for the first edition, Dewi Jackson for the second edition, and Nicola Hartley for the third edition.

We would like to thank those who gave their time and expertise to review chapters, who include but are not limited to:

Arno Kraft, Heriot-Watt University
Angelo Amoroso, Cardiff University
David L. Andrews, University of East Anglia
Helen Aspinall, University of Liverpool
James Barker, University of Kingston
Inder Bhamra, University of Liverpool
Jason Birkett, Nottingham Trent University
Laurence Boyle, University of Kent
Alan Brailsford, Drug Control Centre, King's College London
Christopher Bray, Queen Mary, University of London
Craig Butts, University of Bristol
Andy Cammidge, University of East Anglia
Lucy Carpenter, University of York
Mike Casey, University College Dublin
Oliver Choroba, Gordonstoun School
Steve Christie, Loughborough University
David Cole-Hamilton, University of St Andrews
David Collison, The University of Manchester
David Cowan, Drug Control Centre, King's College London
Joe Crayston, University of St Andrews
David Crowther, Sheffield Hallam University
Paul Cullis, University of Leicester
Tony Curtis, Keele University
Edmund Cussen, University of Strathclyde
Sandie Dann, Loughborough University
Robert Dryfe, The University of Manchester
Jeroen van Duijneveldt, University of Bristol
Bridgette J. Duncombe, Imperial College London
Jason Eames, University of Hull
Andrew Ellis, University of Leicester
Ashleigh Fletcher, University of Strathclyde
Fiona Gray, University of St Andrews
David Grayson, Trinity College Dublin
Pete Griffiths, Cardiff University
Martin Grossel, University of Southampton
Anthony Harriman, Newcastle University
Laurence Harwood, University of Reading
Anthony Haynes, University of Sheffield
Richard Henchman, The University of Manchester
Richard Henderson, Newcastle University
Mike Hewlins, Cardiff University
Bob Hill, University of Glasgow
Mike Hird, University of Hull
Peter Hollins, University of Reading
Hon Wai Lam, University of Edinburgh
Richard Jones, Keele University
Simon Jones, University of Sheffield
Simon Lancaster, University of East Anglia
Dónal Leech, National University of Ireland, Galway
Ally Lewis, University of York
Peter Licence, University of Nottingham
David Littlejohn, University of Strathclyde
Paul Low, Durham University
Mike Lyons, Trinity College Dublin
Mary Masson, University of Aberdeen
Sarah Masters, University of Edinburgh
Joe McDouall, The University of Manchester
Chris McErlean, The University of Sydney
David McGarvey, Keele University
Richard Nichols, University of Liverpool
Roger Nix, Queen Mary, University of London

Michael North, Newcastle University
Elizabeth Page, University of Reading
John Parkinson, University of Strathclyde
Douglas Philip, University of St Andrews
Sheena Radford, University of Leeds
David Rice, University of Reading
Peter Rutledge, The University of Sydney
Angela Savage, formerly of National University of Ireland, Galway
Patrick Steel, Durham University
John Storey, University of Aberdeen
James Sullivan, University College Dublin
Jane Thomas-Oates, University of York
Anthony D. Walmsley, University of Hull
Darren Walsh, University of Nottingham
Mike Watkinson, Queen Mary, University of London
George Weaver, Loughborough University
Michael J. Went, University of Kent
David Worrall, Loughborough University
Eckart Wrede, Durham University
Rossana Wright, University of Nottingham

1

Fundamentals

▼ MRSA bacteria can be detected by growing on a dish containing mannitol agar medium. Yellow areas show the presence of MRSA.

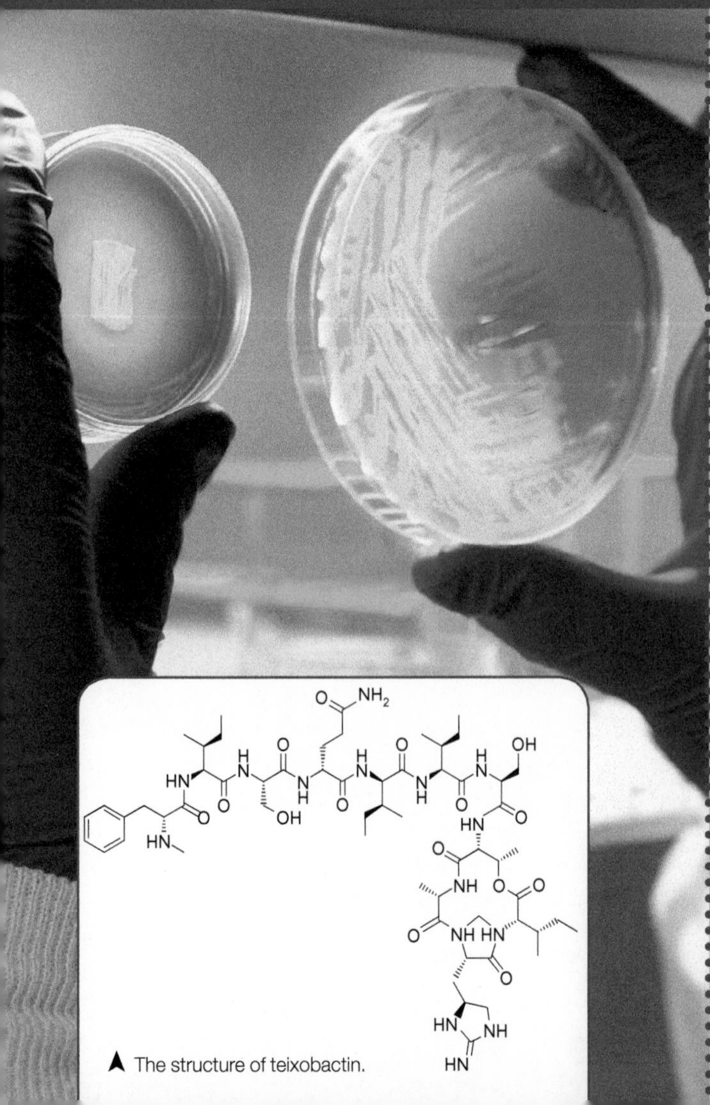

▲ The structure of teixobactin.

Making substances: keeping ahead in the fight against bacteria

Antibiotic-resistant bacteria are one of the most serious threats to world health. We have become used to the idea that bacterial diseases and infections, like tuberculosis (TB) and septicaemia, can be relatively easily cured with antibiotics, though in Victorian times they would have been deadly.

But bacteria are fighting back. New strains of bacteria such as methicillin-resistant *Staphylococcus aureus* (MRSA) have evolved and are threatening hospital patients. MRSA bacteria are present on the skin, nose, and throat of many healthy people where they seem to cause few problems. It is when MRSA gets into the lungs of frail or seriously ill patients that the infections become life-threatening.

MRSA is not the only antibiotic-resistant bacterium. In fact, antibiotic resistance is spreading faster than new antibiotics can be introduced. This is a challenge for biomedical science across the world, and chemistry will lead the way in finding the solution.

One promising discovery is teixobactin, whose discovery was reported in the journal *Nature* in 2015. Teixobactin treats many common bacterial infections including TB and *Clostridium difficile* (a diarrhoea-causing bacterium that resists antibiotics). It was isolated from bacteria found in the soil—the bacteria produce the antibiotic in order to fight off other bacteria. But extracting the teixobactin from the soil bacteria is difficult because, like many soil bacteria, it cannot be cultured in the laboratory. Chemists from Northeastern University in Boston found a way to use a 'lab on a chip' to grow the bacteria in the soil and then isolate the antibiotic. This method could pave the way for the discovery of other new antibiotics to add to the medical armoury.

What do chemists do?

Chemists work in a variety of places—not just in research laboratories. Basic research is very important, but you will find chemists working in many areas of daily life. Figure 1.1 has some examples.

Carrying out measurements and mathematical modelling: studying the atmosphere

Most of the world's population lives in cities, and cities can be very polluted places. Chemists are finding ways to measure the concentration of pollutants such as hydrocarbons and nitrogen oxides on a street-by-street basis. These high-resolution measurements help city planners to decide better policies for preventing health damage by pollution.

On a global scale, an understanding of the chemical reactions that take place in the atmosphere is needed to tackle the environmental problems associated with pollution. Atmospheric chemists measure the concentrations of substances in the air at a range of locations near the ground, or at different altitudes using balloons, aircraft, and satellites. The results are analysed to look for patterns in behaviour and their causes. To make sense of the thousands of reactions that take place in the atmosphere, individual reactions are studied in detail in laboratory experiments. The way that chemicals in the air interact to form photochemical smogs is studied using smog-simulation chambers—huge reactors in which air, containing carefully controlled mixtures of pollutants, is exposed to sunlight.

Some atmospheric chemists work as computer modellers, combining the vast amount of information from experimental studies to produce models that are used to make predictions and to inform government policies on air pollution locally, nationally, and internationally.

▲ The European smog-simulation chamber in Valencia, Spain.

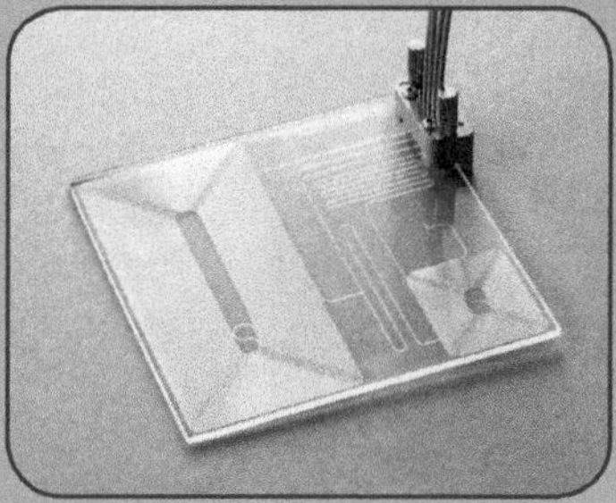

▲ This 'lab on a chip' carries a miniature gas chromatograph (see Section 11.3, p.528). It is used to analyse air samples to assess their quality. There are actually two gas chromatography columns arranged in series: the silvery lines are the columns the gases flow through.

▼ **Figure 1.1** Some examples of the work that chemists do.

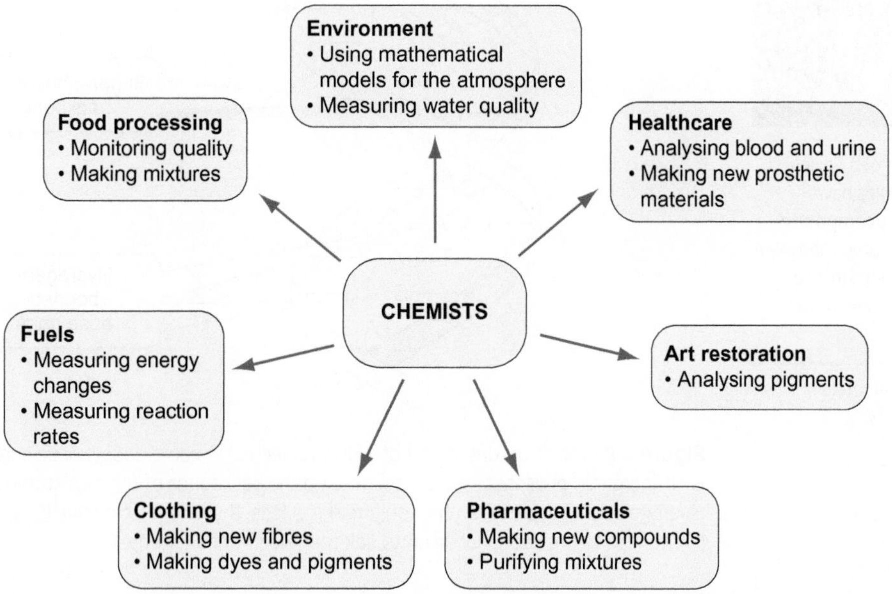
Environment
• Using mathematical models for the atmosphere
• Measuring water quality
Food processing
• Monitoring quality
• Making mixtures
Healthcare
• Analysing blood and urine
• Making new prosthetic materials
CHEMISTS
Fuels
• Measuring energy changes
• Measuring reaction rates
Art restoration
• Analysing pigments
Clothing
• Making new fibres
• Making dyes and pigments
Pharmaceuticals
• Making new compounds
• Purifying mixtures

1.1 Chemistry: the central science

The traditional division of chemistry into physical, inorganic, and organic is an arbitrary one and the majority of chemists work across these divides. Most real problems also require chemists to interact with scientists in other disciplines. For example, chemists, physicists, mathematicians, and meteorologists work together on the vital problem of predicting the extent of future global warming and its effect on climate.

In the area of materials, inorganic chemists, materials scientists, physicists, and engineers all work together on the design and production of new materials, such as solid-state oxide- and nitride-based substances for use in mobile phones. The development of organic light-emitting diodes (OLEDs) is revolutionizing the technology of display screens with the production of thin, light-weight monitors for use in computers, television sets, and the miniature display screens on mobile phones.

Advances in polymer chemistry, and in the biosciences and medicine, have resulted in the successful transplant of a synthetic windpipe into a cancer patient.

Many of the devices that we use every day—including mobile phones—depend on batteries. Electric cars offer a low pollution future, but a big limitation is the size and cost of their batteries. Chemistry is at the heart of research to find new, lighter, low-cost batteries that are quickly charged.

In the field of renewable energy, dye-sensitized solar cells are being developed that will provide a cheaper alternative to silicon-based products. Light-harvesting dyes containing ruthenium are adsorbed onto a thin film of titanium dioxide (TiO_2) nanoparticles in contact with a redox electrolyte. The dye molecules absorb solar radiation shining on them. The resulting excited dye molecules pass on their energy by transferring electrons to the titanium dioxide, so generating an electric current. The system is arranged so that the electrolyte supplies electrons back to the dye so that the cycle can be repeated over and over again.

Many molecules in biological systems, such as DNA (the carrier of the genetic code), ATP (the source of energy in cells), and chlorophyll (the pigment that captures the energy used to drive photosynthesis), contain inorganic as well as organic components. Figure 1.2 shows the structure of part of a DNA molecule. To explain how such molecules work, biochemists need to understand ideas from physical chemistry such as thermodynamics, kinetics, and electrochemistry. Chemists, biochemists, and biologists collaborate to provide a complete picture. Box 1.1 describes how chemistry is central to understanding the role of DNA and underpins genomics.

Li-ion BATTERY

Lithium-ion batteries power many devices, including some electric cars. Chemists are looking for alternatives to lithium-ion batteries that are safer and more quickly charged. A promising possibility is the aluminium-ion battery: aluminium is cheaper than lithium and less flammable, and ionizing aluminium releases three electrons compared with lithium's one.

The Swiss photochemist Michael Gratzel invented the dye-sensitized solar cell.

This synthetic windpipe is made of a flexible polymer in which tiny nanoparticles have been incorporated (called a **nanocomposite** material). The material is highly porous and stem cells from the patient are embedded in it to prevent rejection of the newly grown tissue.

Nanoparticles have diameters from 1 nm to 100 nm (where 1 nm = 1×10^{-9} m).

The hydrogen bonding in DNA is discussed in Box 25.5 (p.1158).

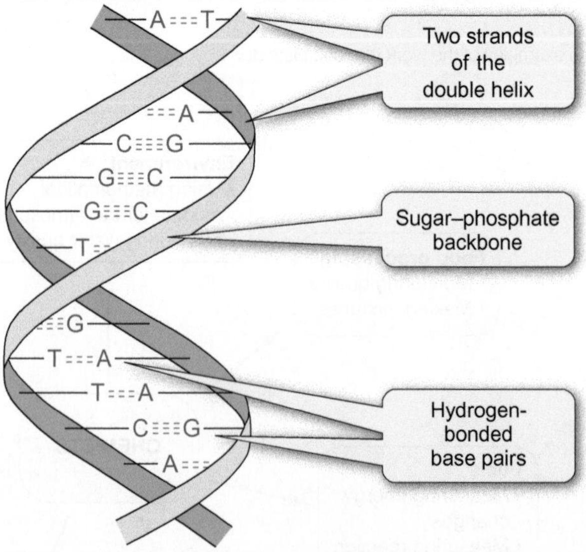
A T
A
C G
G C
G C
T
G
T A
T A
C G
A
Two strands of the double helix
Sugar–phosphate backbone
Hydrogen-bonded base pairs

Figure 1.2 The structure of part of a DNA molecule. 'Organic' deoxyribose sugar units alternate with 'inorganic' phosphate groups to make up the backbones of the two strands of the double helix. Hydrogen bonding between the four organic bases, thymine (T), cytosine (C), adenine (A), and guanine (G), bonded to the deoxyribose units, link the two strands together.

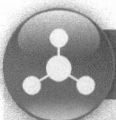

Box 1.1 The hundred thousand genome project

The full set of genes for an organism is called its **genome**. Mapping out the complete sequence of base pairs comprising a genome, and how these are organized into genes, is one of the aims of **genomics**—and it's all chemistry.

An international project, the Human Genome Project, was set up in 1990 to map out the 3 billion (3×10^9) base pairs in a human genome. Working out the sequence of bases involved a huge database of information and took 16 years to complete at a total cost of nearly $3 billion. In 2016, thanks to advances in the technology of genome sequencing, a human genome could be mapped in about 24 hours at a cost of under $1000.

How genome sequencing works

The extraordinary progress in genome sequencing is down to interdisciplinary collaboration with chemistry at its centre. Sequencing technology is changing all the time, but the methods all involve chemistry and computing.

One method is shown in Figure 1—it is called 'sequencing by synthesis' and it is fully automated in a sequencing machine. A complementary strand of DNA is built on to a single strand of the original DNA whose sequence you want to identify. Chemists have designed a system that emits a flash of light when the correct base joins the growing complementary strand. The four bases (A, T, G, and C) are added one after the other, and when the correct one is added, you get a flash of light. A camera, linked to a computer, records the flash, and then the machine moves on to the next base in the DNA chain and the process is repeated. By knowing the sequence of bases being joined to the complementary strand, we can then deduce the sequence of our original DNA strand. Massive computing power has made it possible to automate the process in a very rapid sequencing machine. It is this combination of chemistry with computing power that has driven the dramatic increase in sequencing speed and the reduction of cost.

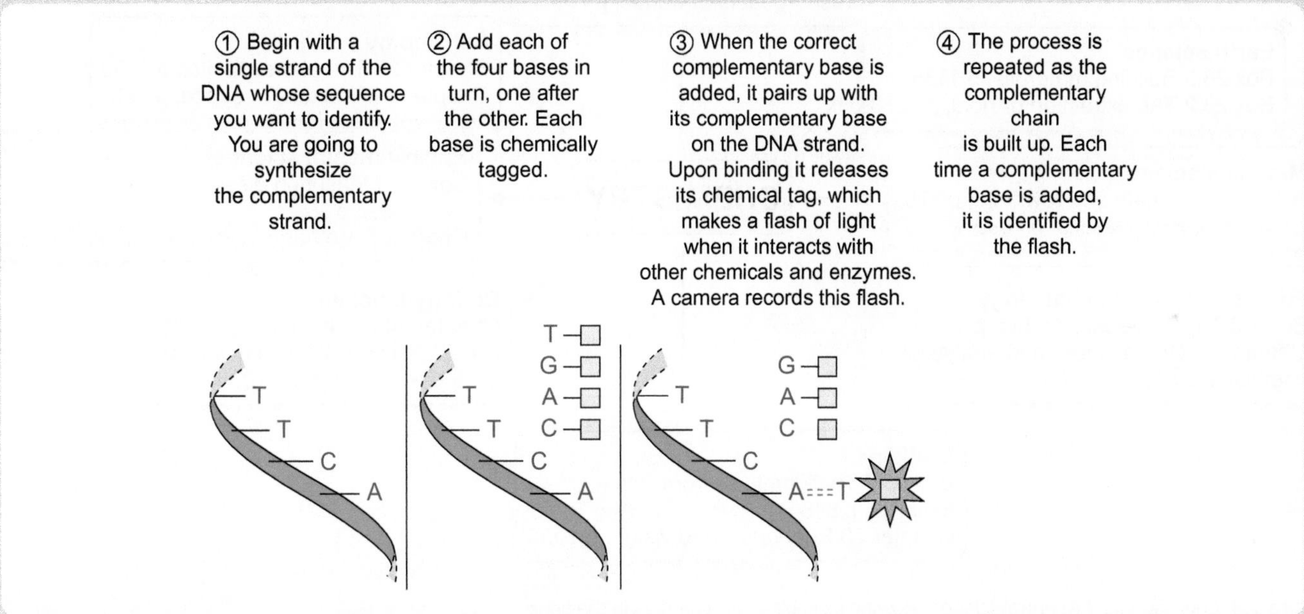

① Begin with a single strand of the DNA whose sequence you want to identify. You are going to synthesize the complementary strand.
② Add each of the four bases in turn, one after the other. Each base is chemically tagged.
③ When the correct complementary base is added, it pairs up with its complementary base on the DNA strand. Upon binding it releases its chemical tag, which makes a flash of light when it interacts with other chemicals and enzymes. A camera records this flash.
④ The process is repeated as the complementary chain is built up. Each time a complementary base is added, it is identified by the flash.
T
T
C
A
T
G
A
C
A:::T

▲ One method for sequencing DNA by synthesizing the complementary strand.

Hundreds of thousands of genomes

As genome sequencing becomes faster and cheaper, scientists are realizing the enormous potential of genomics. Apart from identical twins, every person has their own unique genome, but the differences between individuals' genomes are very small—each human shares about 99.9% of their genome with everyone else. But the differences are interesting and tell us a lot. If scientists look closely at the genomes of people who suffer from particular diseases, they can learn about the influence of genetics on disease, and even develop therapeutic strategies that are tailored to individuals carrying particular genes.

The logo of the original Human Genome Project shows its interdisciplinary nature. ➤

This is the thinking behind the 100 000 Genomes Project, a research effort which will sequence 100 000 whole genomes from patients of the British National Health Service by 2017. The project will focus on patients with rare diseases, and on patients with cancer. ➜

→

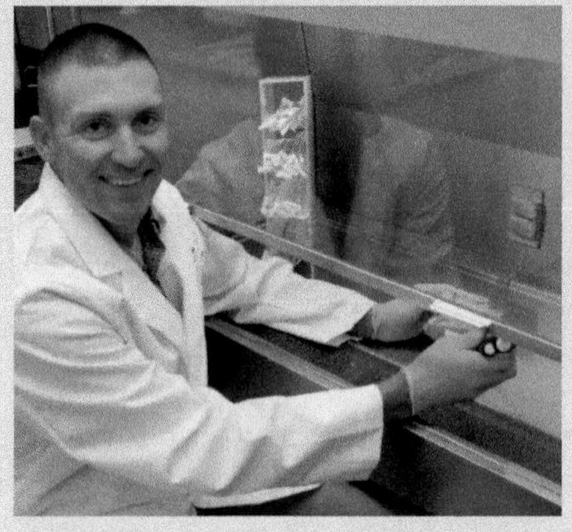

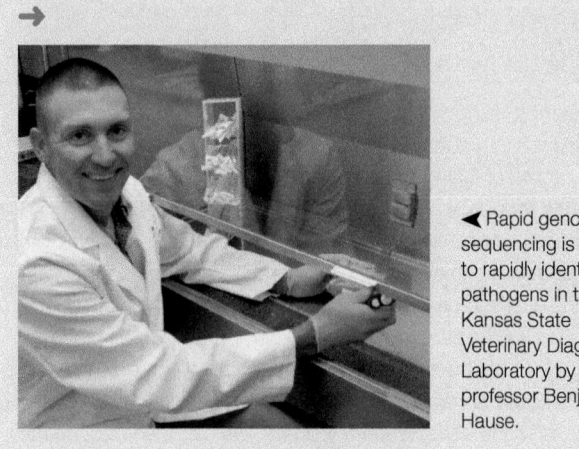

◄ Rapid genome sequencing is used to rapidly identify pathogens in the Kansas State Veterinary Diagnostic Laboratory by assistant professor Benjamin Hause.

Cancerous tumours start to grow when the DNA in normal cells changes in particular ways. 50 000 of the genomes in the 100 000 Genomes Project will be from people suffering from cancer. Each one of 25 000 cancer sufferers will give two samples of DNA: one from normal cells and one from a tumour. By comparing the DNA of tumours with the DNA of normal cells, scientists can find out more about what genetic changes might have stimulated the development of the tumours—and how to treat them. New treatments for cancer could be just one of the many valuable outcomes from the 100 000 Genomes Project.

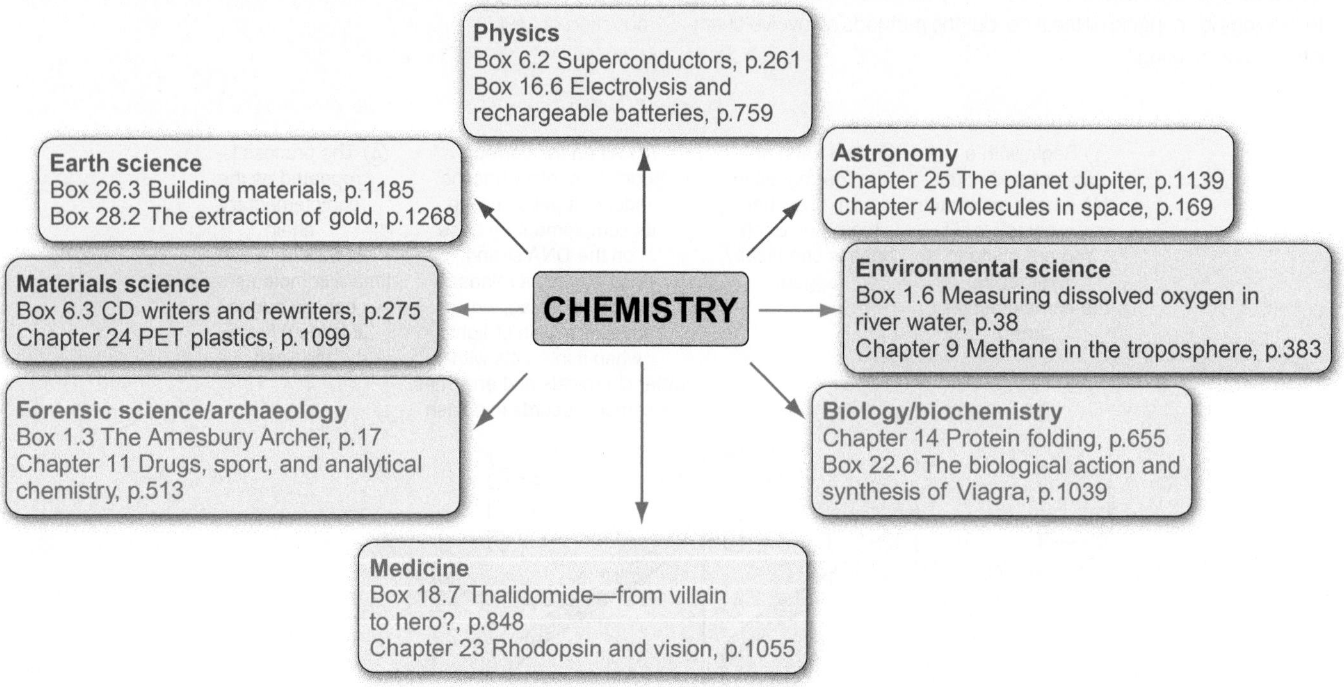
Physics
Box 6.2 Superconductors, p.261
Box 16.6 Electrolysis and rechargeable batteries, p.759
Earth science
Box 26.3 Building materials, p.1185
Box 28.2 The extraction of gold, p.1268
Astronomy
Chapter 25 The planet Jupiter, p.1139
Chapter 4 Molecules in space, p.169
Materials science
Box 6.3 CD writers and rewriters, p.275
Chapter 24 PET plastics, p.1099
CHEMISTRY
Environmental science
Box 1.6 Measuring dissolved oxygen in river water, p.38
Chapter 9 Methane in the troposphere, p.383
Forensic science/archaeology
Box 1.3 The Amesbury Archer, p.17
Chapter 11 Drugs, sport, and analytical chemistry, p.513
Biology/biochemistry
Chapter 14 Protein folding, p.655
Box 22.6 The biological action and synthesis of Viagra, p.1039
Medicine
Box 18.7 Thalidomide—from villain to hero?, p.848
Chapter 23 Rhodopsin and vision, p.1055

Figure 1.3 Chemistry as the central science, showing examples from chapters in the book.

In this book, we have highlighted the role of chemistry as the central science by including examples of applications of chemistry in many related fields. Some of these are shown in Figure 1.3.

1.2 Measurement, units, and nomenclature

Chemical knowledge is constantly increasing. New compounds are being synthesized at a phenomenal rate and data on the behaviour of chemical systems are being collected and analysed in research laboratories around the world. To allow chemists to communicate and compare their work effectively, there has to be an internationally agreed set of rules for them to work by. One of the roles of the International Union of Pure and Applied Chemistry (IUPAC) is to make recommendations for the naming of chemical compounds, the use of units and symbols, and the use of chemical terminology.

When scientists make measurements, they report each physical quantity (mass, volume, pressure, etc.) as a number and a defined unit multiplied together.

For example,

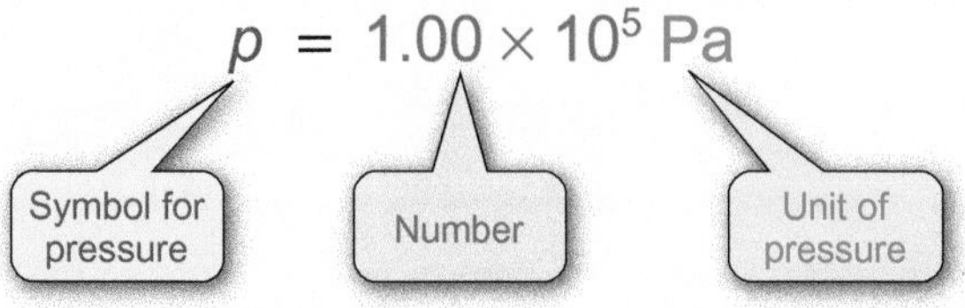
p = 1.00 × 10^5 Pa
Symbol for pressure
Number
Unit of pressure

In general,

$$\text{physical quantity} = \text{number} \times \text{unit}$$

SI units

SI units (Système International d'Unités) are an internationally accepted system of metric units. The system is based on seven **base units**, of which six are commonly used in chemistry. These are shown in Table 1.1. Note that the symbol for the physical quantity is printed in italics. The *names* of units that are named after scientists, such as the kelvin, start with a lower-case letter rather than a capital letter, though the *symbol* for the unit *does* start with an upper-case (capital) letter. Thus, the symbol for kelvin is K, not k.

All other units can be derived from the base units and are called **derived units**. Most of the commonly used derived units have their own name. Some of these are listed in Table 1.2.

A unit can be modified by a prefix, which denotes multiplication or division by a power of ten. Table 1.3 shows the prefixes commonly used in chemistry. So, for example, $1\,\text{kg} = 1 \times 10^3\,\text{g} = 1000\,\text{g}$, and $1\,\mu\text{m} = 1 \times 10^{-6}\,\text{m}$. To avoid confusion over duplication of symbols, the symbols for mega, giga, and tera are in upper-case letters, and the symbol for micro is μ (mu, Greek letter m).

The International Union of Pure and Applied Chemistry (IUPAC) is a non-governmental body that advances aspects of the chemical sciences worldwide.

Length

Chemists deal with matter on a macroscopic scale in the laboratory, but explain its behaviour in terms of atoms and molecules. This requires a wide range of distances (see Figure 1.4). You will need to become familiar with the multiplication prefixes in Table 1.3 used to describe lengths on atomic and molecular scales.

A typical C–H bond length in a hydrocarbon is 0.000000000109 m. Using scientific notation, it can be written as $1.09 \times 10^{-10}\,\text{m}$, though this is still rather cumbersome.

Always keep the number and the unit of a physical quantity together. One is meaningless without the other. For example, what is meant by the statement: 'The volume of the box is 17'?
It could be $17\,\text{cm}^3$, $17\,\text{dm}^3$, $17\,\text{m}^3$, or any other unit of volume.

Table 1.1 SI base quantities and base units used in chemistry

SI base quantity	Symbol for quantity	Name of SI unit	Symbol for SI unit
Mass	m	kilogram	kg
Length	l	metre	m
Time	t	second	s
Temperature	T	kelvin	K
Amount of substance	n	mole	mol
Electric current	I	ampere	A

Table 1.2 SI derived units commonly used in chemistry

Physical quantity	Name of SI unit	Symbol for SI unit	Written in terms of SI base units
Frequency	hertz	Hz	s^{-1}
Force	newton	N	$kg\,m\,s^{-2}$
Pressure	pascal	Pa	$kg\,m^{-1}s^{-2}$ ($= N\,m^{-2} = J\,m^{-3}$)
Energy, heat, work	joule	J	$kg\,m^{2}s^{-2}$ ($= N\,m = Pa\,m^{3}$)
Celsius temperature*	degree Celsius	°C	K
Electric charge	coulomb	C	A s
Electric potential	volt	V	$kg\,m^{2}s^{-3}A^{-1}$ ($= J\,C^{-1}$)

*The Celsius temperature, θ, is defined by: $\theta/°C = T/K - 273.15$ (where T is the temperature on the kelvin scale).

Table 1.3 Some SI multiplication prefixes

Multiplication factor	10		10^{3}	10^{6}	10^{9}	10^{12}	
Name	deca		kilo	mega	giga	tera	
Abbreviation	da		k	M	G	T	
Multiplication factor	10^{-1}	10^{-2}	10^{-3}	10^{-6}	10^{-9}	10^{-12}	10^{-15}
Name	deci	centi	milli	micro	nano	pico	femto
Abbreviation	d	c	m	μ	n	p	f

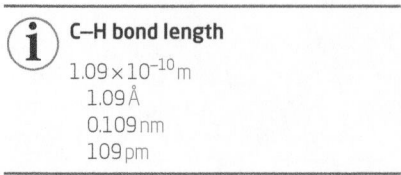

Useful conversions

1 m = 100 cm
= 1000 mm
= 1×10^{6} μm
= 1×10^{9} nm
= 1×10^{12} pm

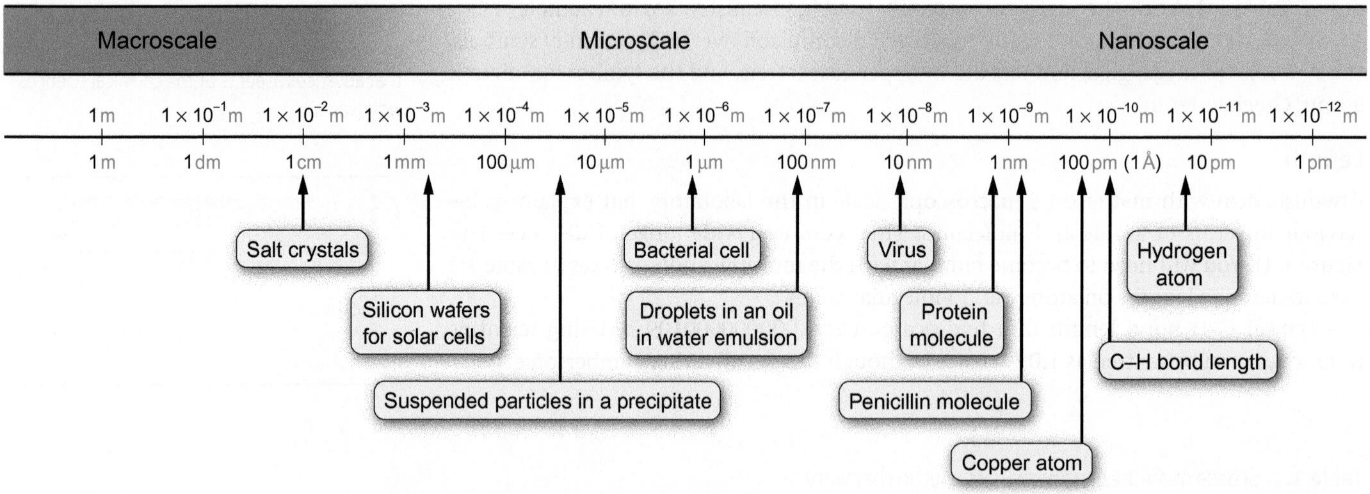
Macroscale
Microscale
Nanoscale
1 m
1 × 10^-1 m
1 × 10^-2 m
1 × 10^-3 m
1 × 10^-4 m
1 × 10^-5 m
1 × 10^-6 m
1 × 10^-7 m
1 × 10^-8 m
1 × 10^-9 m
1 × 10^-10 m
1 × 10^-11 m
1 × 10^-12 m
1 m
1 dm
1 cm
1 mm
100 μm
10 μm
1 μm
100 nm
10 nm
1 nm
100 pm (1 Å)
10 pm
1 pm
Salt crystals
Silicon wafers for solar cells
Suspended particles in a precipitate
Bacterial cell
Droplets in an oil in water emulsion
Virus
Protein molecule
Penicillin molecule
Copper atom
Hydrogen atom
C–H bond length

Figure 1.4 Units of distance from macroscale to microscale and nanoscale.

C–H bond length

1.09×10^{-10} m
1.09 Å
0.109 nm
109 pm

The value of the bond length becomes more manageable if you express it in nanometres (0.109 nm) or in picometres (109 pm). Most bond lengths are roughly the same order of magnitude and a convenient, non-SI, unit is the angstrom, Å, (where $1\,Å = 1 \times 10^{-10}$ m), because most bonds are between 1 Å and 3 Å in length. So the C–H bond length can be written as 1.09 Å. The angstrom is less used these days in favour of SI units, but you will see all three units, nanometres, picometres, and angstroms, for atomic and molecular distances in chemistry textbooks. This book uses mainly picometres.

Temperature

Two temperature scales are commonly used in chemistry. The Celsius temperature scale is the one used in everyday life. Temperatures are reported in the weather forecast in degrees Celsius; pure water normally boils at 100 °C and freezes at 0 °C. Temperatures, however, can go much lower than this. The lowest temperature possible is –273.15 °C—known as **absolute** zero on the kelvin scale. This is the scale adopted for the international standard in science. It is often called the **thermodynamic scale** or, sometimes, the **absolute scale** of temperature.

The symbol for temperatures in kelvin is T. The symbol θ (Greek theta) is used for temperatures in degrees Celsius. To convert between the two scales, you need to use the expression below.

Scientific notation (sometimes called **standard form**) is a convenient way of writing very large, or very small, numbers. The number is written as a number between 1 and 10 (expressed as a decimal) multiplied by an appropriate power of 10. (See Maths Toolkit, MT1, p.1304.)

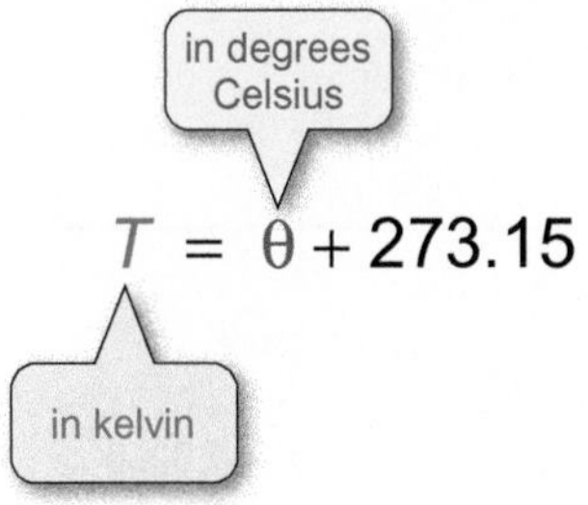
in degrees
Celsius
T = θ + 273.15
in kelvin

So, to obtain a temperature in kelvin, you add 273.15 to the temperature in degrees Celsius. (The 273.15 is exact, but often this degree of accuracy is not warranted by the data and it is enough to add 273.) Another way of writing this relationship is

$$T/\mathrm{K} = \theta/^{\circ}\mathrm{C} + 273.15 \quad (1.1)$$

You can see in Chapter 13 why temperatures in kelvin are called thermodynamic temperatures.

Note that the degree symbol (°) is not used with temperatures in kelvin. T is a product of a number and the unit K, so T/K is a pure number; similarly, $\theta/^{\circ}\mathrm{C}$ is a pure number. This means that Equation 1.1 is a relationship between pure numbers. The magnitude of a degree Celsius is the same as that of a kelvin and so temperature *differences* on the two scales are numerically the same.

The units of any physical quantity can be derived from the expression for the quantity in terms of SI base quantities. You just substitute the base units for these quantities into the expression. For example, for the volume of the rectangular reaction chamber shown in the margin

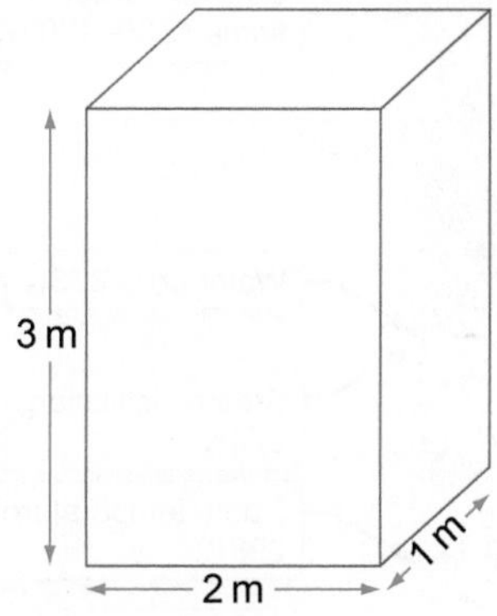
3 m
2 m
1 m

Rectangular reaction chamber.

$$\begin{aligned}\text{volume} &= \text{width} \times \text{depth} \times \text{height}\\ &= (2\,\mathrm{m}) \times (1\,\mathrm{m}) \times (3\,\mathrm{m})\\ &= 6\,(\mathrm{m} \times \mathrm{m} \times \mathrm{m}) = 6\,\mathrm{m}^3\end{aligned}$$

So, the SI unit of volume is m^3. However, $1\,\mathrm{m}^3$ is a large volume for most laboratory chemistry, where volumes are usually measured in dm^3 or cm^3.

The type of analysis used above in working out the unit of volume is called **dimensional analysis** (or sometimes **quantity calculus**, though it has nothing to do with calculus). Worked example 1.1 illustrates the use of dimensional analysis in working out derived units.

Some non-SI units, such as the litre (L) for volumes (instead of m^3) and the atmosphere (atm) or bar (bar) for pressures (instead of Pa), are still used and you will need to be familiar with these.

Worked example 1.1 Derived units

Work out the SI derived unit for density.

$$\text{Density} = \frac{\text{mass}}{\text{volume}}$$

Strategy

Insert SI base units into the expression for density.

Solution

$$\text{SI unit of density} = \frac{\text{SI unit of mass}}{\text{SI unit of volume}}$$

$$= \frac{\text{kg}}{\text{m}^3} = \text{kg m}^{-3}$$

(Note, however, that densities are often quoted in g cm^{-3}.)

Question

Work out the SI derived unit for molar mass (the mass of one mole of a substance).

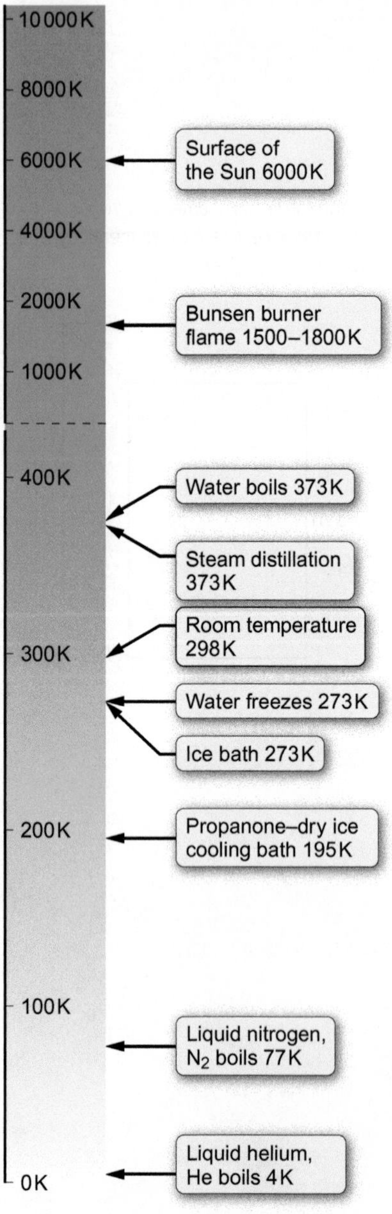
10 000 K
8000 K
6000 K
4000 K
2000 K
1000 K
400 K
300 K
200 K
100 K
0 K
Surface of the Sun 6000 K
Bunsen burner flame 1500–1800 K
Water boils 373 K
Steam distillation 373 K
Room temperature 298 K
Water freezes 273 K
Ice bath 273 K
Propanone–dry ice cooling bath 195 K
Liquid nitrogen, N2 boils 77 K
Liquid helium, He boils 4 K

The thermodynamic temperature scale.

Useful conversions

Volume

$1\,\text{L} = 1\,\text{dm}^3$
$= 1000\,\text{cm}^3$
$= 1 \times 10^{-3}\,\text{m}^3$

$1\,\text{mL (millilitre)} = 1 \times 10^{-3}\,\text{L} = 1\,\text{cm}^3$

$1\,\mu\text{L (microlitre)} = 1 \times 10^{-6}\,\text{L} = 0.001\,\text{cm}^3$

Pressure

$1\,\text{atm} = 1.01325 \times 10^5\,\text{Pa}$

$1\,\text{bar} = 1.00000 \times 10^5\,\text{Pa}$

Carrying out calculations

Below are five rules for carrying out calculations successfully. Dimensional analysis is very important and will prevent many of the common errors with units.

- ALWAYS set out your working clearly to avoid careless errors.
- ALWAYS include both the *number* and the *unit* of physical quantities *in your working*—not just in the answer.
- Units must be consistent (e.g. all volumes must be in the same units, so if volumes are in m^3, density must be in kg m^{-3}).
- Convert quantities to the correct units *before* you substitute them into an expression.
- Make sure you quote the answer to the appropriate number of significant figures (if you are multiplying or dividing, the significant figures in the answer should be the same as the smallest number of significant figures in the data you have used in the calculation). See the sample calculation below.

In this book, the units in worked examples are printed in red and any cancelling out of units is shown using diagonal strikethroughs. This is illustrated in the sample calculation below. Get used to always setting out calculations like this. It takes a little longer, but it allows you to spot errors in units that may result in an incorrect answer, and ensures that you have the correct units in your answer.

The sample calculation below uses the ideal gas equation. Don't worry if you have not met this before. It was chosen because it contains a variety of units. All the information you need is in the example.

Sample calculation

Use the ideal gas equation, $pV = nRT$, to work out the volume (V) of 1.00 mol (n) of a gas at temperature $T = 298\,\text{K}$ and pressure $p = 1.00 \times 10^5\,\text{Pa}$. The gas constant R has a value of $8.314\,\text{J K}^{-1}\,\text{mol}^{-1}$.

Rearrange the equation to give an expression for the volume, V, of the gas.

$$V = \frac{nRT}{p}$$

Substitute values for n, R, T, and p in the equation. Remember to include units.

$$V = \frac{(1.00\,\text{mol}) \times (8.316\,\text{J}\,\text{K}^{-1}\,\text{mol}^{-1}) \times (298\,\text{K})}{1.00 \times 10^{5}\,\text{Pa}}$$

Check that the units are consistent. (If they are not, convert to consistent units.) In this case, use Table 1.2 (p.8) to write joules in terms of pascals and metres (or pascals in terms of joules and metres).

$$V = \frac{(1.00\,\text{mol}) \times (8.314\,\text{Pa}\,\text{m}^{3}\,\text{K}^{-1}\,\text{mol}^{-1}) \times (298\,\text{K})}{1.00 \times 10^{5}\,\text{Pa}}$$

Cancel out the units to give the units for volume and work out the answer.

$$V = \frac{(1.00\,\cancel{\text{mol}}) \times (8.314\,\cancel{\text{Pa}}\,\text{m}^{3}\,\cancel{\text{K}^{-1}}\,\cancel{\text{mol}^{-1}}) \times (298\,\cancel{\text{K}})}{1.00 \times 10^{5}\,\cancel{\text{Pa}}}$$

$$= 0.248\,\text{m}^{3}$$

The answer is quoted to 3 significant figures to correspond with the smallest number of significant figures in the data.

You can find help with working with numbers in the Maths Toolkit, MT1 (p.1304). Rearranging equations is discussed in MT2 (p.1312).

You can find out more about the ideal gas equation and this type of calculation in Sections 8.1 (p.8) and 8.2 (p.345).

Nomenclature

In the early days of chemistry, names of compounds were often based on the source, or on a property, of the compound. For example:

- limestone (source of calcium carbonate);
- acetic acid (Latin: *acetum*, vinegar);
- putrescine (smell of rotting animal flesh).

Many natural products are still named from their sources. For example, the name penicillin comes from the name of the mould, *Penicillium notatum*, that produces it (Figure 1.5). Unfortunately, these **common names** (sometimes called **trivial names**) do not help you to work out the structure of the compound. To do this, IUPAC has devised a systematic method for naming organic and inorganic compounds. The rules for naming compounds are called **nomenclature**.

The rules for naming inorganic compounds are discussed in Section 1.4 (p.21). Chapter 2 shows you how to work out systematic **IUPAC names** for organic compounds. Practising chemists, however, tend to use a mixture of common names and IUPAC names, so you do need to be familiar with both. For example, you probably know the compound phenylethene by its systematic name, rather than by its common name styrene, but you are likely to know the addition polymer formed from it as polystyrene, rather than poly(phenylethene).

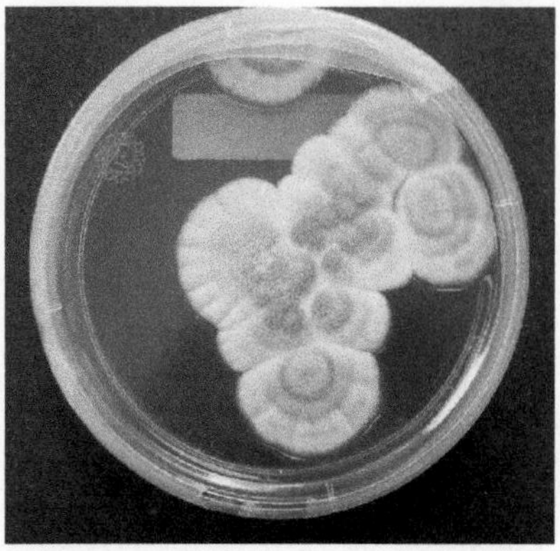

Figure 1.5 Penicillin gets its name from the mould *Penicillium notatum.* Its antibiotic action was first observed by Alexander Fleming in 1928.

In chemistry, a **natural product** is a chemical compound or substance produced by a living organism.

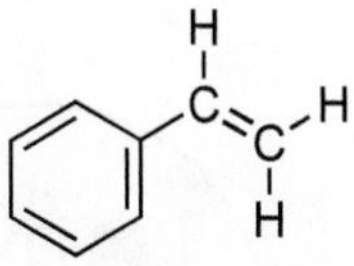

Common name: styrene
IUPAC name: phenylethene

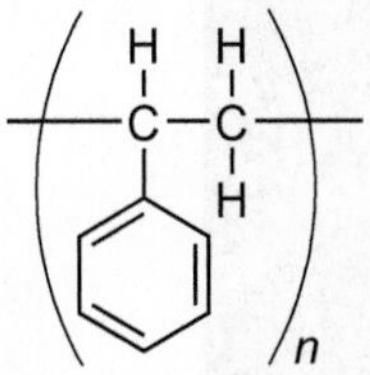

(where n is a very large number)

Common name: polystyrene
IUPAC name: poly(phenylethene)

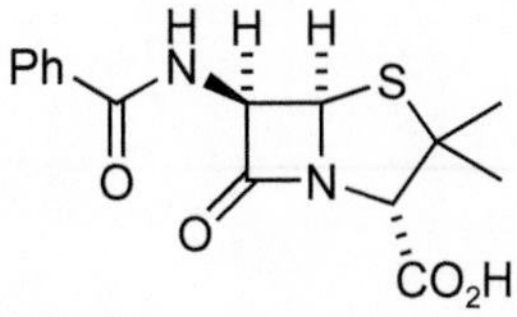

The name penicillin G is used for this important antibiotic. This is much easier to remember than the IUPAC name.

Common name: penicillin G
IUPAC name: (2*S*,5*R*,6*R*)-6-(benzamido)-3,3-dimethyl-7-oxo-4-thia-1-aza-bicyclo[3.2.0]heptane-2-carboxylic acid

The IUPAC rules for naming organic compounds are discussed in Sections 2.4–2.8 (pp.79–109). A list of frequently used common names is included at the end of Chapter 2 (p.109). The naming of inorganic compounds is discussed in Section 1.4 (p.21).

Conventions for drawing the structures of organic compounds are discussed in Section 2.2 (p.73).

In this book, both common and systematic names are generally given when a compound is first introduced. For simple compounds, such as ethene and ethanoic acid, the IUPAC names (rather than ethylene and acetic acid, respectively) are used subsequently. For more complex compounds, IUPAC names can be cumbersome so common names are used where these are available. In the case of penicillin, for example, the common name is certainly more convenient.

Summary

- SI units are based on seven base units; other units are derived from these and are called derived units.
- Bond lengths are conveniently measured in picometres (1 pm = 1×10^{-12} m).
- Temperature can be measured in degrees Celsius or in kelvin. The kelvin scale is the international standard in science and is called the thermodynamic temperature scale or the absolute temperature scale.
- Many compounds have both a common name and a systematic IUPAC name.

? For practice questions on these topics, see questions 1–4 at the end of this chapter (p.66).

1.3 Atoms and the mole

Sodium is a soft, silvery metal made up of sodium atoms. Chlorine is a green gas with a characteristic, choking smell. The gas is made up of chlorine atoms bound together in pairs as Cl_2 molecules. Sodium metal and chlorine gas are both very reactive elements and are toxic to humans. Yet, when they react together (Figure 1.6), they produce white crystals of sodium chloride, common salt—an essential component of our diet.

The above reaction illustrates the difference between elements and compounds. Sodium metal and chlorine gas are **elements** because they contain only one kind of atom (see margin note). Sodium chloride is a compound. **Compounds** are made up of atoms of more than one element. The atoms in compounds are bonded together in molecules or in network structures. The bonding in network structures can be ionic (as in the case of sodium chloride) or covalent.

i Atoms of an element all have the same number of protons in the nucleus. They may not be identical as some may have different numbers of neutrons. Isotopes of elements are discussed on p.14.

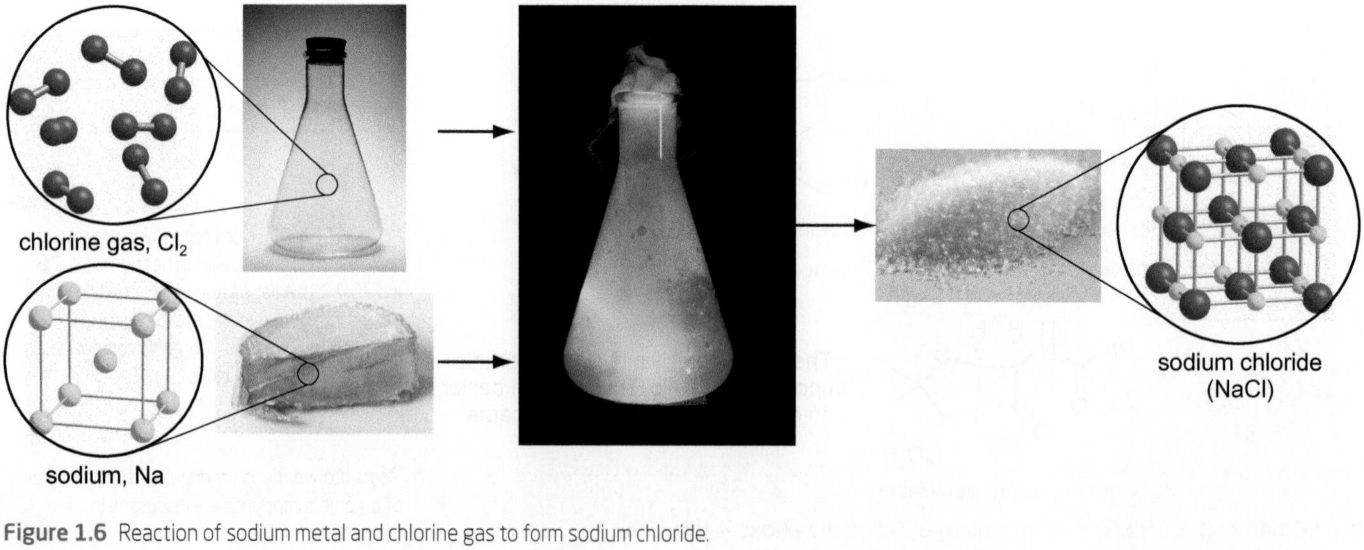
chlorine gas, Cl_2
sodium, Na
sodium chloride
(NaCl)

Figure 1.6 Reaction of sodium metal and chlorine gas to form sodium chloride.

The properties of a compound bear no resemblance to those of the constituent elements—an idea that is often poorly understood by non-chemists. Important issues, such as the presence of mercury in some injections given to babies, are frequently misreported in the media. In this case, the toxic properties of mercury metal (the *element*) were commonly quoted, when a *compound* of mercury (ethylmercury) was actually being used as the preservative in the injection. Similarly, many people on 'low sodium' diets may not realize that they are cutting down on sodium *ions* not sodium *atoms*.

'Table salt contains the deadly poisons chlorine and cyanide!'—or does it? It is important to know whether an ingredient is an element or a compound and exactly how the atoms or ions are bound. Cl^- ions are safe but Cl_2 is not; $[Fe(CN)_6]^{3-}$ is added as a drying agent and is safe, but free CN^- ions are not

Structure of the atom

A simple classical model of the atom is shown in Figure 1.7. A tiny dense positively charged **nucleus** is surrounded by negatively charged **electrons**. The nucleus is made up of positively charged **protons** and (except for hydrogen atoms) uncharged **neutrons**, held together by a strong nuclear force. This force overcomes the electrostatic repulsion between the positively charged protons. Overall, an atom is electrically neutral. Properties of the electron, proton, and neutron are summarized in Table 1.4.

During the twentieth century other subatomic particles, such as neutrinos, muons, and quarks, were discovered. These particles tell physicists about the fundamental nature of matter, but their existence has little impact on chemistry. This is because chemical reactivity is largely related to the arrangements and movements of electrons.

In the classical model of atomic structure, the orbiting electrons are arranged in shells around the nucleus. Electrons in the outermost shell (**valence electrons**) are usually the ones involved in chemical reactions. This simple model is sufficient to explain many observations. It does, however, have severe limitations, and Chapters 3 and 4 elaborate on this model using the ideas of quantum mechanics.

Elements differ from each other by the number of protons they contain in the nucleus. This number of protons is the **atomic number** (proton number) of an element, and it is given the symbol Z.

You can find all the known chemical elements in the Periodic Table inside the cover of this book. Covalent molecules are discussed in Chapters 4 and 5, and network structures in Chapter 6.

There are four fundamental forces in nature—electromagnetic, weak nuclear, strong nuclear, and gravitational.

You can read about the historical development of ideas about atomic structure in Section 3.1 (p.114).

Atomic number (Z)

This is the number of protons in the nucleus of an atom and is characteristic of an element. For example:

Hydrogen	$Z=1$
Helium	$Z=2$
Carbon	$Z=6$
Uranium	$Z=92$

Elements are arranged in the Periodic Table in order of increasing atomic number (see Section 3.6, p.143). Values of Z for all the elements are given on the inside front cover.

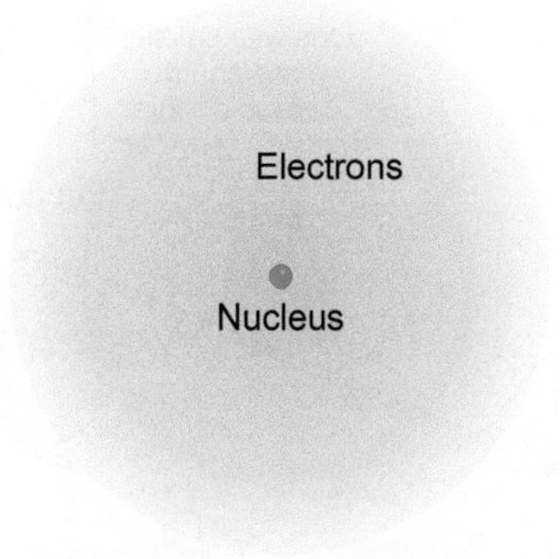
Electrons
Nucleus

Figure 1.7 A simple classical model of the atom. Negatively charged electrons surround a positively charged nucleus, so overall the atom is electrically neutral.

Table 1.4 Properties of subatomic particles

Particle	Relative mass	Relative charge	Location in atom
Electron, e	1	−1	Around nucleus
Proton, p	1836	+1	In the nucleus
Neutron, n	1839	0	In the nucleus

Isotopes

Although the number of protons is fixed for an element, the number of neutrons can vary. Atoms of the same element with different numbers of neutrons in their nuclei are known as **isotopes**. The existence of isotopes was predicted by Frederick Soddy in 1911, and later demonstrated by Francis Aston by the effect of magnetic fields on ions. This led Aston to the invention of the mass spectrometer. Box 1.2 describes how a simple mass spectrometer works.

The **mass number** of an isotope is the total number of protons and neutrons in the nucleus. Hydrogen has three isotopes, with mass numbers of 1, 2, and 3. About 99% of hydrogen atoms have nuclei containing just one proton. This isotope of hydrogen is the only isotope of any element not to have any neutrons in its nucleus. About 1% of the hydrogen atoms have nuclei that contain a neutron in addition to the proton, so have a mass number of 2. There is also a radioactive isotope of hydrogen, atoms of which have two neutrons in their nuclei and, therefore, a mass number of 3. The isotopes of hydrogen are unusual in that they are given special names and symbols. The isotope with a mass number of 2 is called deuterium (D) and that with mass number 3 is called tritium (T).

Isotopes are represented by atomic symbols in which the **atomic number** is given as a subscript, and the mass number as a superscript, both written before the chemical symbol

Box 1.2 Mass spectrometry

A good way to learn the principles of mass spectrometry is to look at the magnetic mass spectrometer shown in Figure 1—though as we explain later, most modern mass spectrometers are 'time-of-flight' machines.

In the basic spectrometer in Figure 1, a substance is placed in a vacuum and vaporized. The sample is then bombarded with electrons to form ions, which are accelerated using an electric field. The accelerated ions pass through a magnetic field, and the extent to

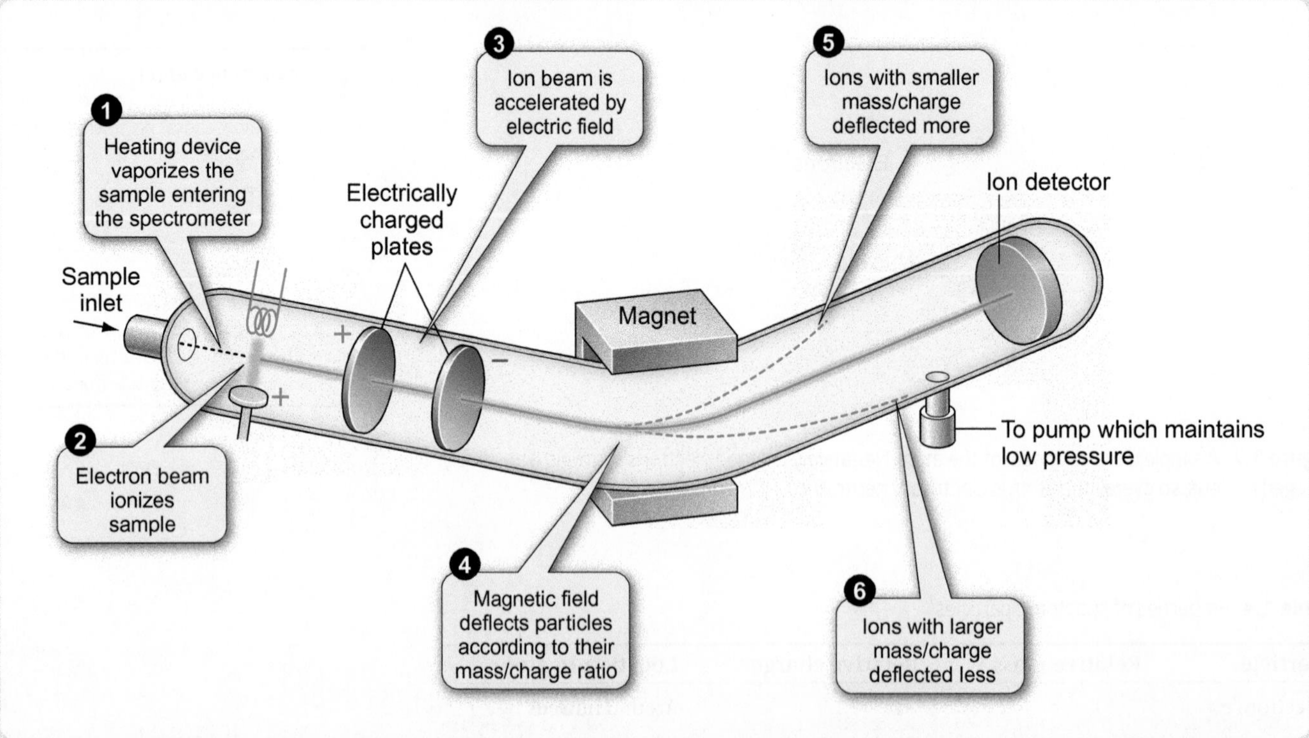

1 Heating device vaporizes the sample entering the spectrometer
Sample inlet
2 Electron beam ionizes sample
Electrically charged plates
3 Ion beam is accelerated by electric field
Magnet
4 Magnetic field deflects particles according to their mass/charge ratio
5 Ions with smaller mass/charge deflected more
Ion detector
To pump which maintains low pressure
6 Ions with larger mass/charge deflected less

▲ **Figure 1** Diagram of a simple mass spectrometer.

→

→

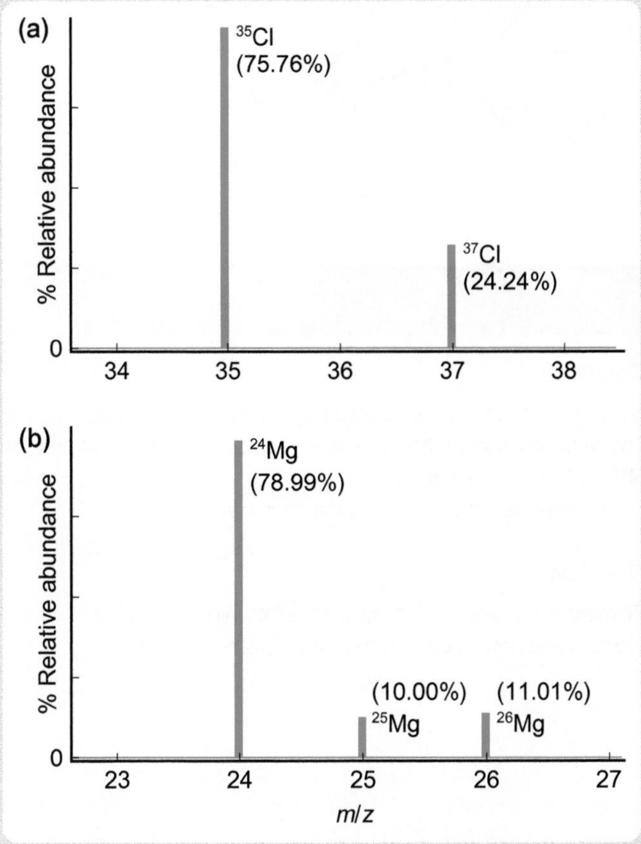
(a)
% Relative abundance
^{35}Cl
(75.76%)
^{37}Cl
(24.24%)
0
34
35
36
37
38
(b)
% Relative abundance
^{24}Mg
(78.99%)
(10.00%)
^{25}Mg
(11.01%)
^{26}Mg
0
23
24
25
26
27
m/z

▲ **Figure 2** Mass spectra of (a) chlorine and (b) magnesium, showing the presence of isotopes.

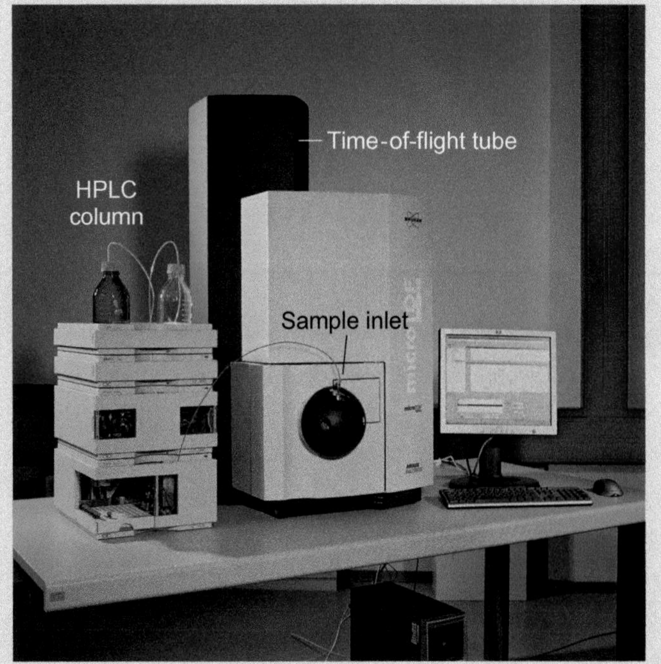
Time-of-flight tube
HPLC
column
Sample inlet

▲ **Figure 3** A time-of-flight mass spectrometer linked to a high-performance liquid chromatography (HPLC) column.

which they are deflected depends on *m/z*, their mass to charge ratio—the greater the mass to charge ratio, the smaller the deflection. The ions that strike the detector produce an electric current. The magnetic field is gradually increased so that ions with different *m/z* values are targeted at the detector. The detector signal is recorded as the magnetic field is varied producing a **mass spectrum**. (Alternatively, the magnetic field can be kept fixed and the accelerating electric field varied.) The mass spectra for atomic chlorine and magnesium in Figure 2 show their isotopic compositions.

The mass spectra of molecules are more complex than those of atoms, because the ion formed from the molecule (the **molecular ion**) can break into fragments, giving rise to characteristic peaks in the spectrum. The mass spectrum of a molecule, therefore, gives information about both its molecular mass and its structure.

Most modern mass spectrometers operate in a different way that does not require a magnetic field. In a **time-of-flight mass spectrometer** (Figure 3), ions are accelerated by an electric field and the time it takes for the ions to reach a detector is measured. The time taken depends on *m/z*—heavier ions move more slowly—so all ions with the same *m/z* value arrive at the detector at the same time.

Complex mixtures can be analysed by linking a mass spectrometer to a gas chromatograph or a high-performance liquid chromatography column. The components of the mixture are first separated by chromatography and then fed directly into the mass spectrometer for identification. Examples of these techniques are described in Box 11.1 (p.522) and Section 11.3 (p.528). Section 12.1 (p.558) discusses the use of mass spectrometry in determining the structures of organic compounds.

Question

Look at the mass spectrum of magnesium in Figure 2(b). How many protons and neutrons are there in the nuclei of each of the three isotopes?

for the element. Normal hydrogen is therefore represented as $^{1}_{1}H$, deuterium as $^{2}_{1}H$, and tritium as $^{3}_{1}H$. Since the atomic number is constant for a particular element and can be worked out from the chemical symbol, the subscript is often omitted. These symbols then become ^{1}H, ^{2}H, and ^{3}H, respectively. They are sometimes written as hydrogen-1, hydrogen-2, and hydrogen-3. You can practise writing and using atomic symbols in Worked example 1.2.

Atomic symbol for tritium

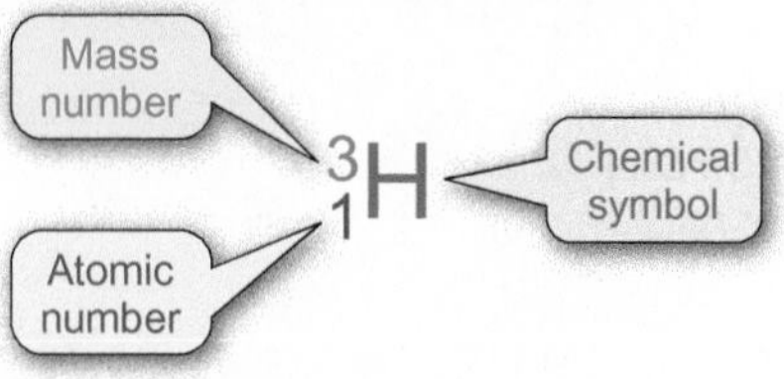
Mass number
3
1
H
Chemical symbol
Atomic number

Worked example 1.2 Atomic symbols

An isotope of molybdenum has 54 neutrons. What is its atomic symbol?

Strategy

Use the table on the inside front cover to determine the chemical symbol for molybdenum and its atomic number, *Z*.

Add the number of protons to the number of neutrons to give the mass number.

Solution

Molybdenum (Mo) has an atomic number of 42 so it has 42 protons. The mass number of the isotope is given by the number of protons (42) plus the number of neutrons (54).

The mass number is 96, and the atomic symbol is $^{96}_{42}Mo$.

Question

Element X has the atomic symbol $^{203}_{90}X$. What is the name of the element and how many neutrons are there in the nucleus?

Box 8.4 (p.367) explains how the difference in physical properties of two isotopes is used in the enrichment of uranium.

Atomic mass units and daltons
IUPAC defines an atomic mass unit (u) as $\frac{1}{12}$ the mass of an atom of ^{12}C. The atomic mass unit has a value of 1.661×10^{-27} kg, which corresponds closely to the mass of a proton or neutron. More recently, these units have been called daltons (Da); they are used particularly by biochemists. So, 1 Da = 1 u, and the mass of one ^{12}C atom is 12 Da. (Note that atomic mass units and daltons are non-SI units.)

Sometimes the Avogadro constant is given the symbol *L*.

An **integer** is a whole number.

The term **relative molar mass** is often used for M_r as this applies to both molecular and non-molecular substances.

The different isotopes of an element normally react chemically in an identical manner. However, since their masses are different, their physical properties show some variation. These differences are usually small, but for the isotopes of hydrogen they are more marked, because of the large proportionate difference between the masses of the isotopes. Box 1.3 describes an example of isotope analysis in an archaeological investigation.

Relative atomic mass and moles of atoms

The mass of a hydrogen atom is very small, approximately 1.674×10^{-27} kg. This is not a practical value to use to compare atomic masses, so instead a relative scale is used. The standard in the relative atomic mass scale is an atom of carbon-12 (^{12}C) for which the **relative atomic mass (A_r)** is defined as exactly 12 (i.e. 12.0000 . . ., etc.). All atomic masses are quoted relative to this value. Note that A_r values have no units.

Exactly 12 g of ^{12}C contain 6.022×10^{23} atoms. The relative atomic mass weighed out in grams of any element contains the same number of atoms—whatever the element. This number of atoms is called a **mole of atoms**. The number of atoms in one mole is called the **Avogadro constant**, N_A, and has a value of 6.022×10^{23} mol^{-1}.

The relative atomic mass of *naturally occurring* carbon is not exactly 12 but 12.011. This is because naturally occurring samples of carbon contain small quantities of ^{13}C (carbon-13) in addition to ^{12}C.

The relative atomic masses of many elements are close to an integer as they contain mainly one isotope. However, there are exceptions. The relative atomic mass of chlorine is 35.453 due to the presence of ^{35}Cl (abundance 75.76%) and ^{37}Cl (abundance 24.24%). Worked example 1.3 shows how the relative atomic mass is calculated.

The masses of individual isotopes relative to ^{12}C (**relative isotopic masses**) are not exactly whole numbers. Some values for selected elements are given in Table 1.5.

Box 1.3 The Amesbury Archer

The grave of a man known as the Amesbury Archer was discovered near to Stonehenge in southern England in 2002. This grave dates from the early Bronze Age and is the richest grave from this time ever found in Britain. The wealth of the objects found there, including the country's first known gold objects, indicate that the archer was a man of status. Oxygen isotope analysis of the man's dental enamel gave an insight into his origin.

Most of the oxygen that goes into the formation of tooth and bone comes from water, which is ultimately derived from rain or snow. Although the most abundant isotope of oxygen is ^{16}O, there are two other stable isotopes, ^{17}O and ^{18}O, present in minor amounts. The ratio of the isotopes depends on a number of environmental factors such as temperature and altitude. Drinking water in warm climates has a higher ratio of ^{18}O:^{16}O than that in cold climates. Determination of this ^{18}O:^{16}O ratio in teeth provides a means of determining where a person might have lived at the time their teeth formed.

Oxygen isotope analysis revealed a relatively low ^{18}O:^{16}O ratio for the Amesbury Archer, indicating that the man came from a colder climate region than was found in Britain at the time. He most likely came from somewhere in the Alps region, probably Switzerland, Austria, or Germany.

Box 3.9 on p.161 describes how the Amesbury Archer was dated using radiocarbon dating.

Question

How do the three isotopes of oxygen vary in the number of protons, neutrons, and electrons present?

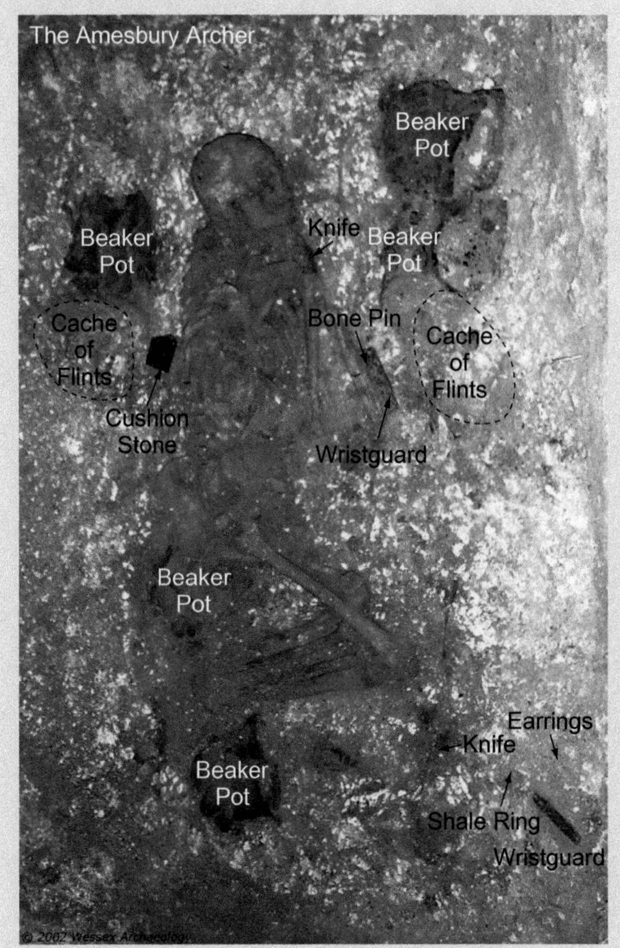
The Amesbury Archer
Beaker Pot
Beaker Pot
Knife
Beaker Pot
Cache of Flints
Bone Pin
Cache of Flints
Cushion Stone
Wristguard
Beaker Pot
Earrings
Knife
Beaker Pot
Shale Ring
Wristguard
© 2002 Wessex Archaeology

The grave of the Amesbury Archer dating from around 2300 BC. It was discovered in 2002. ➤

Worked example 1.3 Relative atomic mass of magnesium

Use the information from the mass spectrum in Figure 2(b) in Box 1.2 (p.14) to work out the relative atomic mass, A_r, of naturally occurring magnesium. (Assume that the relative mass of ^{24}Mg is exactly 24, that of ^{25}Mg is 25, and that of ^{26}Mg is 26.)

Strategy

Use the percentage abundances in the mass spectrum to work out the average relative mass of 100 atoms. Hence find the average mass of one atom (the 'weighted average').

Solution

Average relative mass of 100 atoms of magnesium

$$= (24 \times 78.99) + (25 \times 10.00) + (26 \times 11.01)$$

$$= 1895.76 + 250.00 + 286.26 = 2432.02$$

Average relative mass of one atom of magnesium, A_r(Mg)

$$= 24.32$$

(*Note*. The calculations in this worked example are simplified because the isotopic masses are not exactly whole numbers (see Table 1.5).)

Question

Naturally occurring bromine contains ^{79}Br (abundance 50.69%) and ^{81}Br (abundance 49.31%). Calculate the relative atomic mass of naturally occurring bromine.

Table 1.5 Relative masses and natural abundances of isotopes of selected elements

Element	Isotope	Relative isotopic mass	Abundance / %
H	^{1}H	1.0078	99.9885
	^{2}H	2.0141	0.0115
C	^{12}C	12 (exactly, by definition)	98.93
	^{13}C	13.0034	1.07
N	^{14}N	14.0031	99.64
	^{15}N	15.0001	0.36
O	^{16}O	15.9949	99.757
	^{17}O	16.9991	0.038
	^{18}O	17.9992	0.205
F	^{19}F	18.9984	100
Mg	^{24}Mg	23.9850	78.99
	^{25}Mg	24.9858	10.00
	^{26}Mg	25.9826	11.01
P	^{31}P	30.9738	100
Cl	^{35}Cl	34.9689	75.76
	^{37}Cl	36.9659	24.24
Br	^{79}Br	78.9183	50.69
	^{81}Br	80.9163	49.31
I	^{127}I	126.9045	100

Relative formula mass

In the same way that relative atomic mass is used to compare the masses of atoms of elements, **relative formula mass** is used to compare the formula masses of compounds. Where the formula represents a discrete molecule (such as CH_4 or H_2SO_4), the relative formula mass is called the **relative molecular mass**. Both relative formula mass and relative molecular mass are given the symbol $\boldsymbol{M_r}$.

The relative formula mass of a substance is worked out by first writing down the formula of the substance and then adding together the relative atomic masses of each of the atoms in the formula. For example,

$$M_r(H_2SO_4) = (1.01 \times 2) + (32.07 \times 1) + (16.00 \times 4) = 98.09$$

Use the correct symbol

- M_r is the symbol for the relative formula mass (or relative molecular mass) of a substance relative to an atom of ^{12}C. No units.
- M is the symbol for the molar mass of a substance. Units: g mol^{-1} (or kg mol^{-1}).
- m is the symbol for the actual mass of one atom or molecule or one formula unit of a substance. Units: g (or kg).

Amount of substance and molar mass

The unit of **amount of substance** is the **mole** (symbol mol). A mole of a substance always contains 6.022×10^{23} (N_A) entities. These entities may be atoms, molecules, or groups of ions in a formula unit, so it is important to indicate whether you are dealing with a mole of atoms, a mole of molecules, etc. The easiest way to do this is to give the formula of the entity the mole refers to. For example, 'a mole of chlorine (Cl_2)' or 'a mole of atomic chlorine (Cl)'. A mole of Cl_2 molecules contains twice as many atoms as a mole of Cl atoms.

The mole

A **mole** (1 mol) of a substance is the **amount of substance** that contains as many formula units (atoms, molecules, groups of ions, etc.) as there are in exactly 12 g of ^{12}C. The number of formula units in a mole is 6.022×10^{23} mol^{-1}, N_A, the **Avogadro constant**.

The mole is important because chemists need to know when the amounts of different compounds are the same, not in terms of their masses but in terms of the number of molecules each contains, since chemical reactions take place between molecules in specific stoichiometric ratios.

Chemical amounts are defined so that the mass of one mole (the **molar mass**, *M*) is equal to the relative atomic mass, or the relative formula mass, in grams (or kg). Thus, the molar mass of ^{12}C is exactly $12\,g\,mol^{-1}$ ($0.012\,kg\,mol^{-1}$) and the molar mass of H_2SO_4 is $98.09\,g\,mol^{-1}$ ($0.09809\,kg\,mol^{-1}$). In this book, molar mass is quoted in $g\,mol^{-1}$. In general,

$$\text{amount (in mol)} = \frac{\text{mass (in g)}}{\text{molar mass (in g mol}^{-1})} \qquad (1.2)$$

Equation 1.2 is used in Worked example 1.4 to calculate the amount in moles in a given mass of glucose.

Stoichiometry
The relationship between the amounts of reactants and products is called the stoichiometry of the reaction. A balanced chemical equation shows the stoichiometry of a reaction.

Millimoles
For small quantities of a substance, you may see the chemical amount quoted in millimoles (mmol), where $1\,mmol = 1 \times 10^{-3}\,mol$.

Worked example 1.4 Amount of substance

How many moles of molecules are contained in 25.4 g of glucose ($C_6H_{12}O_6$)?

Strategy

Work out the relative formula mass (M_r) of glucose.

The molar mass of glucose is the relative formula mass in grams.

Use Equation 1.2 to work out the amount in moles.

Solution

$$M_r\,(C_6H_{12}O_6) = (12.01 \times 6) + (1.01 \times 12) + (16.00 \times 6) = 180.18$$

From Equation 1.2

$$\text{amount (in mol)} = \frac{\text{mass (in g)}}{\text{molar mass (in g mol}^{-1})} \qquad (1.2)$$

$$= \frac{25.4\,\cancel{g}}{180.18\,\cancel{g}\,mol^{-1}} = 0.141\,mol$$

Question

What mass of sodium chloride contains 5.82 mol of NaCl?

Chemical formulae

- The **empirical formula** of a substance tells you the *ratio* of the numbers of different types of atom in the substance.
- A **molecular formula** tells you the actual number of different types of atom in a molecule.
- A **structural formula** shows how the atoms in a molecule are bonded together. This is sometimes also called a **displayed formula.**
- **Skeletal formulae** use lines to represent the carbon framework in an organic compound.

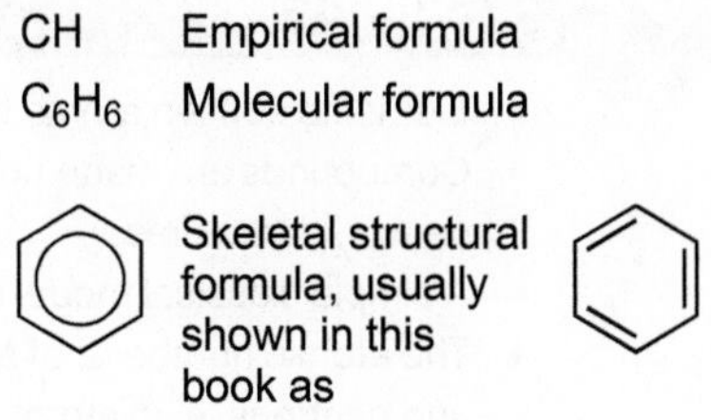
CH Empirical formula
C6H6 Molecular formula
Skeletal structural formula, usually shown in this book as
Chemical formulae for benzene.

Worked example 1.5 shows how the empirical formula of a compound is calculated from its composition by mass. The formula of a compound, such as sodium chloride (NaCl), which has an extended network structure, is an empirical formula. For a molecular compound, such as benzene, you can write both an empirical formula (CH) and a molecular formula (C_6H_6).

You can find more about drawing the structural formulae of organic compounds in Section 2.2 (p.73).

Worked example 1.5 Empirical formulae and molecular formulae

An organic compound contains 52.2% carbon, 13.1% hydrogen, and 34.7% oxygen by mass.

(a) Work out its empirical formula.

(b) Mass spectrometry shows the compound has a relative molecular mass of 46. What is its molecular formula?

Strategy

(a) You are given the mass composition in percentages, so write down the masses of C, H, and O in 100 g of the compound. Set out your working in columns as shown below.

Work out the number of moles of atoms of C, H, and O in these masses.

Find the simplest ratio of moles of atoms of C:H:O by dividing each amount by the smallest value. The ratio gives the empirical formula.

(b) Work out the relative formula mass corresponding to the empirical formula and compare this with the relative formula mass of the compound obtained from mass spectrometry.

Solution

(a)

	C	H	O
% Mass	52.2%	13.1%	34.7%
Mass in g in 100 g compound	52.2 g	13.1 g	34.7 g
Amount (in mol)	$\frac{52.2\ \cancel{g}}{12.01\ \cancel{g}\ mol^{-1}}$ = 4.35 mol	$\frac{13.1\ \cancel{g}}{1.01\ \cancel{g}\ mol^{-1}}$ = 13.0 mol	$\frac{34.7\ \cancel{g}}{16.0\ \cancel{g}\ mol^{-1}}$ = 2.17 mol
Simplest ratio (divide by smallest)	$\frac{4.35\ \cancel{mol}}{2.17\ \cancel{mol}} = 2.00$	$\frac{13.0\ \cancel{mol}}{2.17\ \cancel{mol}} = 5.99$	$\frac{2.17\ \cancel{mol}}{2.17\ \cancel{mol}} = 1.00$

Ratio of moles of atoms C:H:O is 2 : 6 : 1.

Empirical formula is C_2H_6O.

(b) $M_r(C_2H_6O) = (12.0 \times 2) + (1.01 \times 6) + (16.0 \times 1) = 46$. This corresponds to the M_r obtained from mass spectrometry, so the molecular formula of the compound is also C_2H_6O.

Question

A compound was found to contain 1.18% hydrogen, 42.00% chlorine, and 56.82% oxygen by mass. What is its empirical formula?

Summary

- Elements contain atoms that all have the same number of protons.
- Compounds are made up of atoms of more than one element; the atoms are bonded together in molecules or in extended network structures.
- A simple classical model of the atom consists of a tiny dense positively charged nucleus surrounded by electrons.
- The atomic number Z of an element is the number of protons in an atom. The mass number is the total number of protons and neutrons in an atom.
- Many elements have more than one isotope. Isotopes are atoms of an element that contain different numbers of neutrons in their nuclei. An isotope can be represented by an atomic symbol showing the mass number and atomic number.
- A mole (1 mol) of a substance is the amount of substance that contains as many formula units (atoms, molecules, groups of ions, etc.) as there are in exactly 12 g of ^{12}C. The number of formula units in a mole is 6.022×10^{23} mol^{-1}, N_A, the Avogadro constant.
- The empirical formula of a substance gives the *ratio* of the numbers of different types of atom in the substance; the molecular formula gives the actual number of different types of atom in a molecule. The structural formula shows how the atoms in a molecule are bonded together.

For practice questions on these topics, see questions 5–8 at the end of this chapter (p.66).

1.4 Chemical equations

Chemical equations summarize what happens in a chemical reaction. They tell you:

- what substances react;
- what products are formed;
- the relative amounts of the substances involved.

If an equation is balanced, you can use it to work out how much of each reactant is needed to react exactly, and how much of each product to expect if the reaction goes to completion.

When you balance an equation, you are really doing an atom count (or a count of ions and electrons if it is an ionic equation; see p.26). Atoms are neither created nor destroyed in a chemical change—simply rearranged into new molecules or new network structures. This is one of the fundamental laws of chemistry and is known as the **law of conservation of mass**. A balanced chemical equation is a sort of balance sheet of where each atom starts and where it ends up. There must be the same number of atoms at the end of the reaction as there was at the start.

The law of conservation of mass

The French chemist, Antoine Lavoisier, noted in 1774 that, if nothing is allowed to enter or leave a reaction vessel, the *total* mass is the same after a chemical reaction has taken place as it was before the reaction.

Balancing equations

The first step in constructing a balanced equation is to write down the correct formula for each of the substances involved. If the substance is molecular, you need the molecular formula (e.g. water H_2O, methane CH_4, ethanol C_2H_5OH or C_2H_6O). If the compound is ionic, you can work out the empirical formula from the formulae of the ions involved (see Table 1.6). For example, calcium chloride contains Ca^{2+} ions and Cl^- ions, so the formula of calcium chloride is $CaCl_2$. Note that an ion made up from several atoms is written in brackets when there is more than one in the formula of the compound. For example, ammonium sulfate contains NH_4^+ ions and SO_4^{2-} ions, so the formula of ammonium sulfate is $(NH_4)_2SO_4$. The subscripts in the formulae cannot be changed. Atoms or ions combine in fixed ratios to form compounds, so water is always H_2O, calcium chloride is always $CaCl_2$, etc. Different numbers would create different substances.

Ethanol burns in air or oxygen to form carbon dioxide and water. The unbalanced equation looks like this.

$$C_2H_5OH + O_2 \rightarrow CO_2 + H_2O \quad \textit{unbalanced equation}$$

Brandy contains ethanol which burns in oxygen with a blue flame.

The equation is **unbalanced** because there are different numbers of each type of atom on either side. For example, there are two C atoms on the left but only one C atom on the right.

To balance the equation, you need to insert numbers *in front* of the formulae, so that there are equal numbers of each type of atom on either side of the equation. These numbers tell you the number of *formula units* involved in the reaction. For the reaction of ethanol with oxygen, you can do this in three steps, balancing the numbers of C, H, and O atoms in turn.

A **balanced equation** is one with equal numbers of each type of atom on either side of the equation. The charge must also be same on each side. This is particularly important in redox equations where electrons are added to balance the charges (see p.27).

Step 1 Balance the number of C atoms by inserting a 2 before CO_2 (i.e. two molecules of CO_2).

$$C_2H_5OH + O_2 \rightarrow 2CO_2 + H_2O$$

Step 2 Balance the number of H atoms by inserting a 3 before H_2O (i.e. three molecules of H_2O).

$$C_2H_5OH + O_2 \rightarrow 2CO_2 + 3H_2O$$

Step 3 Balance the number of O atoms by inserting a 3 before O_2 (i.e. three molecules of O_2). Remember there is an O atom in C_2H_5OH, so there are a total of seven O atoms on either side of the equation.

$$C_2H_5OH + 3O_2 \rightarrow 2CO_2 + 3H_2O$$

Finally Check you have equal numbers of each type of atom on either side.

In older books, sulfur is spelt 'sulphur'. The spelling with 'f' is recommended by IUPAC.

Table 1.6 Names and formulae of some common ions. For metallic elements where there is more than one common ion, the oxidation state is given (e.g. iron(II) and iron(III)). For anions, you may have previously used the alternative names in brackets, though these names are no longer recommended by IUPAC (e.g. see Figure 27.28 in Section 27.6, p.1244).

Positive ions (cations)		Negative ions (anions)	
Charge 1+		Charge 1−	
H^+	hydrogen	H^-	hydride
Li^+	lithium	F^-	fluoride
Na^+	sodium	Cl^-	chloride
K^+	potassium	Br^-	bromide
Cu^+	copper(I)	I^-	iodide
Ag^+	silver(I)	OH^-	hydroxide
NH_4^+	ammonium	N_3^-	azide
Charge 2+		NO_2^-	nitrite (nitrate(III))
Mg^{2+}	magnesium	NO_3^-	nitrate (nitrate(V))
Ca^{2+}	calcium	CN^-	cyanide
Sr^{2+}	strontium	OCN^-	cyanate
Ba^{2+}	barium	SCN^-	thiocyanate
Mn^{2+}	manganese(II)	HCO_3^-	hydrogencarbonate
Fe^{2+}	iron(II)	HSO_4^-	hydrogensulfate
Co^{2+}	cobalt(II)	$H_2PO_4^-$	dihydrogenphosphate
Ni^{2+}	nickel(II)	$CH_3CO_2^-$	ethanoate
Cu^{2+}	copper(II)	ClO^-	hypochlorite (chlorate(I))
Zn^{2+}	zinc	ClO_2^-	chlorite (chlorate(III))
Cd^{2+}	cadmium(II)	ClO_3^-	chlorate (chlorate(V))
Sn^{2+}	tin(II)	ClO_4^-	perchlorate (chlorate(VII))
Pb^{2+}	lead(II)	Charge 2−	
Charge 3+		O^{2-}	oxide
Al^{3+}	aluminium	S^{2-}	sulfide
Fe^{3+}	iron(III)	CO_3^{2-}	carbonate
Cr^{3+}	chromium(III)	$C_2O_4^{2-}$	ethanedioate
		SO_3^{2-}	sulfite (sulfate(IV))
		SO_4^{2-}	sulfate (sulfate(VI))
		$S_2O_3^{2-}$	thiosulfate
		Charge 3−	
		N^{3-}	nitride
		PO_4^{3-}	phosphate

State symbols are often included in an equation to show whether each substance is a gas, a liquid, or a solid, or is present in aqueous solution. For the reactants and products at room temperature, the balanced equation for the combustion of ethanol then becomes

$$C_2H_5OH\,(l) + 3O_2\,(g) \rightarrow 2CO_2\,(g) + 3H_2O\,(l) \qquad (1.3)$$

Equation 1.3 tells you that one molecule of C_2H_5OH reacts with exactly three molecules of O_2 to form two molecules of CO_2 and three molecules of H_2O. Because a mole of each substance contains the same number of molecules (N_A), the equation also tells you that 1 mol of C_2H_5OH reacts exactly with 3 mol of O_2 to form 2 mol of CO_2 and 3 mol of H_2O. The amounts in moles allow you to work out the *masses* that react exactly (see Worked example 1.6).

	C_2H_5OH (l)	+ $3O_2$ (g)	→	$2CO_2$ (g)	+	$3H_2O$ (l)
Molecules:	1 molecule	3 molecules		2 molecules		3 molecules
Moles:	1 mol	3 mol		2 mol		3 mol
Masses:	46 g	3 × 32 g		2 × 44 g		3 × 18 g

The relationship between the amounts of reactants and products is called the **stoichiometry** of the reaction. The numbers before the formula units used to balance the equation are the **stoichiometric coefficients**.

State symbols
State symbols are included in chemical equations to show the physical state (see Section 1.7, p.47) of the reactants and products.
(g) gas
(l) liquid
(s) solid
(aq) aqueous solution (solution in water)

A balanced equation is sometimes called a **stoichiometric** equation.

Worked example 1.6 Using a balanced equation to work out reacting masses

Use the balanced equation for the combustion of ethanol to work out the mass of carbon dioxide produced when 25 g of ethanol are burned in a plentiful supply of air.

Strategy

Write the balanced equation for the reaction (Equation 1.3).

Under the equation, state what the equation tells you about the amounts in moles of the substances you are interested in.

Change the amounts in moles to masses in grams.

Scale the masses to the ones in the question.

Solution

$$C_2H_5OH\,(l) + 3O_2\,(g) \rightarrow 2CO_2\,(g) + 3H_2O\,(l) \qquad (1.3)$$

C_2H_5OH	CO_2
1 mol	2 mol
46 g	(2 × 44 g) = 88 g
1 g	$\left(\frac{88}{46}\right)$ g
25 g	$\left(\frac{88 \times 25}{46}\right)$ g = 48 g

So, 48 g of CO_2 are produced when 25 g of ethanol are burned.

Question

What mass of oxygen is needed to react exactly with 25 g of ethanol?

Types of chemical reaction

Chemical reactions can be classified in a number of ways. Below are four common types of chemical reaction that you need to be able to recognize.

1. **Acid–base reactions** involve the transfer of a proton (H^+ ion); see Sections 7.1 and 7.2 (pp.304–318). In Section 7.8 (p.336) this definition is broadened to include other types of acids and bases.
2. **Redox reactions** involve the transfer of electrons (see p.27).
3. **Precipitation reactions** involve the formation of a solid product when two solutions are mixed (see p.26).
4. **Complexation reactions** involve the formation of a complex ion (or an uncharged complex) in which a central metal ion is surrounded by electron-donating ligands; see Section 28.3 (p.1265).

Chemical equilibrium is discussed in Section 1.9 (p.56). These ideas are taken further in Chapter 15 (p.694).

If a reaction is carried out with other than stoichiometric amounts of reactants, so that one or more of the reactants is in excess, the yield is calculated from the amount of the **limiting reactant**—the one that is completely used up in the reaction.

Working out the yield of a reaction

The maximum mass of product that can be obtained from a chemical reaction can be worked out from the balanced equation and is called the **theoretical yield**. Some reactions go to completion and the mass of product that is actually obtained in the laboratory or in a chemical plant is very close to this. Often though, the **actual yield** is less than the theoretical yield. Some loss of product can occur during isolation from the reaction mixture and purification. There may be competing **side reactions** leading to other products. In addition, some reactions do not go to completion, but reach a state of equilibrium in which both reactants and products are present.

The **percentage yield** of a reaction tells you what percentage of the theoretical yield was actually obtained.

$$\text{Percentage yield} = \frac{\text{actual yield}}{\text{theoretical yield}} \times 100\% \qquad (1.4)$$

It is usual to quote percentage yields as integers (see Worked example 1.7).

In a synthesis involving several steps, a few steps with low yields can have a disastrous effect on the overall yield. Suppose a synthesis involves three steps, each of which gives a 90% yield of product.

$$\text{A} \xrightarrow{90\%} \text{B} \xrightarrow{90\%} \text{C} \xrightarrow{90\%} \text{D}$$

Worked example 1.7 Percentage yield of a reaction

Suppose you prepare a sample of aspirin by heating 10 g salicylic acid (2-hydroxybenzoic acid) with an excess of ethanoic anhydride. You obtain 6.2 g of pure aspirin. What is the percentage yield of the reaction?

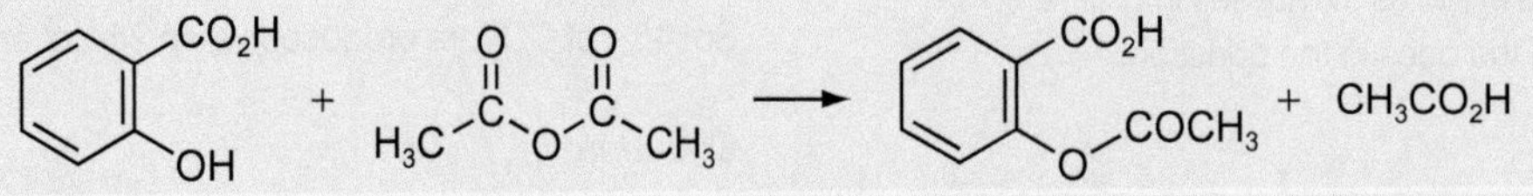
CO2H
OH
+
O
O
H3C
C
O
C
CH3
CO2H
COCH3
O
+ CH3CO2H

Strategy

Use the balanced equation above to work out the theoretical yield from 10 g salicylic acid.

Use Equation 1.4 to work out the percentage yield.

Solution

1 mol of salicylic acid gives 1 mol of aspirin.

138 g of salicylic acid gives 180 g of aspirin.

1 g of salicylic acid gives $\frac{180}{138}$ g of aspirin.

10 g of salicylic acid gives $\left(\frac{180 \times 10}{138}\right)$ g = 13.0 g of aspirin.

From Equation 1.4

$$\text{percentage yield} = \frac{\text{actual yield}}{\text{theoretical yield}} \times 100\% \qquad (1.4)$$

$$= \frac{6.2\ \cancel{g}}{13.0\ \cancel{g}} \times 100\% = 48\%$$

Question

In the Haber process, nitrogen reacts with hydrogen to form ammonia:

$$N_2(g) + 3H_2(g) \rightleftharpoons 2NH_3(g)$$

The reaction does not go to completion and an equilibrium mixture of reactants and products is formed.

To investigate the equilibrium, 1.00 mol of nitrogen and 3.00 mol of hydrogen are sealed in a container at 100 atm pressure and 400 °C. The equilibrium mixture contains 8.57 g of ammonia. What is the percentage yield of the reaction?

The starting compound for each step is the product of the previous step. So the overall yield for the sequence A → D is:

$$\frac{90}{100} \times \frac{90}{100} \times \frac{90}{100} = \frac{729000}{100000} = \frac{72.9}{100} \times 100 \text{ or } 73\%$$

If, however, each step resulted in a 30% conversion to the required product, the overall yield after three stages would only be:

$$\frac{30}{100} \times \frac{30}{100} \times \frac{30}{100} = \frac{27000}{100000} = \frac{2.7}{100} \times 100 \text{ or } 3\%$$

The percentage conversion to a required product tells you nothing about the amount of *waste* generated by the process—something that is particularly important in the chemical industry. Calculating the atom efficiency as in Box 1.4 helps in understanding the amount of waste in a reaction.

Although a high percentage yield is desirable in an industrial process, sometimes, as in the case of the Haber process in Worked example 1.7, it is more economical to accept a lower yield if this means the rate of the reaction is higher. The reactants are separated from the product and recycled through the reactor.

Box 1.4 Atom efficiency and green chemistry

Chemists make use of the law of conservation of mass to work out the **atom efficiency** of a reaction. This is a way of assessing how efficiently a reaction makes use of the reactant atoms and is a good indicator of the *waste* generated by the reaction. Such considerations are becoming increasingly important as the chemical industry seeks to develop more economical and more sustainable manufacturing processes.

Atom efficiency is calculated from the stoichiometric equation for the overall process.

$$\text{Atom efficiency} = \frac{\text{molar mass of desired product}}{\text{sum of molar masses of reactants}} \times 100\% \quad (1.5)$$

Working out the atom efficiency of alternative production routes provides a convenient way of comparing economic and environmental factors. For example, at one time phenol was manufactured by sulfonating benzene with concentrated sulfuric acid and then treating the product with sodium hydroxide. The overall equation for the reaction is

C_6H_6 +	H_2SO_4 +	2NaOH →	C_6H_5OH +	$Na_2SO_3 + 2H_2O$
benzene			phenol	sodium sulfite
1 mol	1 mol	2 mol	1 mol	1 mol
78.1 g	98.1 g	2 × 40.0 g	94.1 g	126 g

Atom efficiency

$$= \frac{94.1\,\text{g mol}^{-1}}{78.1\,\text{g mol}^{-1} + 98.1\,\text{g mol}^{-1} + (2 \times 40.0\,\text{g mol}^{-1})} \times 100\%$$

$$= \frac{94.1\,\cancel{\text{g}}\,\cancel{\text{mol}^{-1}}}{256.2\,\cancel{\text{g}}\,\cancel{\text{mol}^{-1}}} \times 100\%$$

$= 36.7\%$

This means that well under half the mass of the reactants ends up in the desired product. The rest is waste—unless uses can be found for the 126 g of sodium sulfite produced for every 94.1 g of phenol.

The calculation of atom efficiency assumes that all the reactants are converted to the products, as shown by the equation (i.e. there is 100% yield of the products). In practice, the yield was only about 88%, which brought the atom efficiency down to 32.3%.

Some sodium sulfite is used in the wood pulp and paper industry, but most of it is waste.

This is one of the reasons why the sulfonation process has now been superseded by the **cumene process**. In the first stage of the cumene process, benzene reacts with propene to form 2-phenylpropane (cumene). The cumene is then treated with oxygen to form an unstable hydroperoxide, which breaks down to give phenol and propanone (acetone). (The conventions for drawing benzene rings are discussed in Section 22.1 (p.1004). The mechanism for this reaction is described in Box 22.2 (p.1007).)

▼ Ski and snowboard wear is usually made of nylon. Phenol is one of the feedstocks used in the production of nylon.

→

→

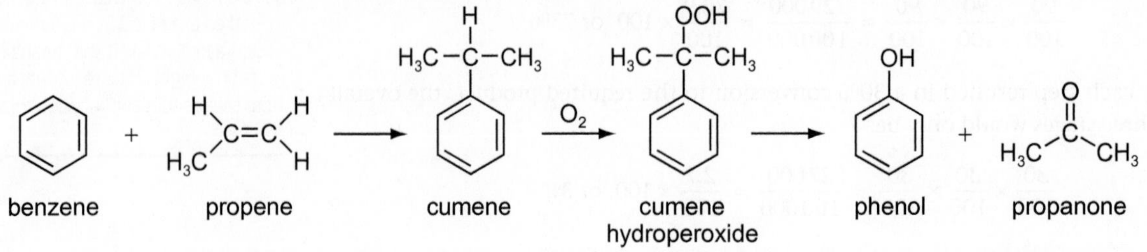

benzene
propene
cumene
O_2
cumene hydroperoxide
phenol
propanone

▲ The cumene process for the manufacture of phenol.

The overall equation for the three-stage process is

$$\underset{1\,\text{mol}}{C_6H_6} + \underset{1\,\text{mol}}{C_3H_5OH} + O_2 \rightarrow \underset{1\,\text{mol}}{C_6H_5OH} + \underset{1\,\text{mol}}{CH_3COCH_3}$$

Propanone is a valuable product of the reaction and is used to make plastics and industrial solvents. None is wasted, so the atom efficiency of the process would be 100% if the conversion to products were complete. In practice, the atom efficiency is less than this because the obtained yield is not 100%.

Question

Demand for propanone is rising at a lower rate than the demand for phenol. Calculate the atom efficiency of the cumene process if half the propanone is considered as waste and the overall conversion to products is 90%.

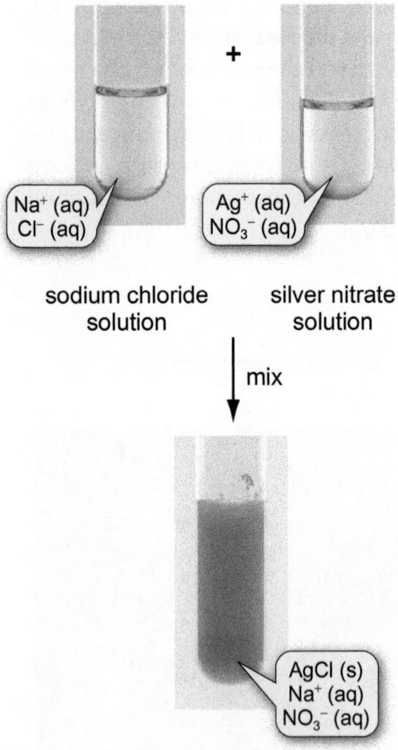

Na^+ (aq)
Cl^- (aq)
+
Ag^+ (aq)
NO_3^- (aq)
sodium chloride solution
silver nitrate solution
mix
AgCl (s)
Na^+ (aq)
NO_3^- (aq)

Figure 1.8 Precipitation of silver chloride.

Ionic equations

For reactions involving ionic compounds, the balanced equation is often shortened to include only the ions that take part in the reaction. For example, if you add a solution of silver nitrate to a solution of sodium chloride, a white precipitate of silver chloride forms (see Figure 1.8). (The white precipitate of silver chloride darkens on exposure to air.)

$$AgNO_3\,(aq) + NaCl\,(aq) \rightarrow AgCl\,(s) + NaNO_3\,(aq)$$

In a dilute solution of an ionic compound in water, the ions are surrounded by water molecules and behave independently of each other. So, a more accurate way to write the equation is

$$\underbrace{Ag^+(aq) + NO_3^-(aq)}_{\text{silver nitrate solution}} + \underbrace{Na^+(aq) + Cl^-(aq)}_{\text{sodium chloride solution}} \rightarrow \underbrace{AgCl(s)}_{\text{white precipitate of silver chloride}} + \underbrace{Na^+(aq) + NO_3^-(aq)}_{\text{sodium nitrate solution}}$$

The Na^+ ions and NO_3^- ions are present in solution before and after the reaction. They do not take part in the reaction and so are called **spectator ions**. They can be left out of the equation, which then becomes

$$Ag^+\,(aq) + Cl^-\,(aq) \rightarrow AgCl\,(s)$$

This type of equation, which excludes spectator ions and shows only the ions actually taking part in the reaction, is called an ionic equation. Note the importance of the state symbols, which clearly indicate that the above equation represents a precipitation reaction. An ionic equation can include non-ionic substances, such as metals, or covalent molecules, such as H_2O (l) or CO_2 (g). State symbols are also useful for identifying these in the equation. You can practise writing ionic equations in Worked example 1.8.

ℹ Always check when writing and balancing an ionic equation that the overall charge on each side of the equation is the same.

Worked example 1.8 Ionic equations

Construct an ionic equation for the neutralization of hydrochloric acid by sodium hydroxide solution.

Strategy

Write out a full, balanced equation for the reaction showing the ions present in the reactants and products. Identify any non-ionic substances and include state symbols in the equation.

Cross out the spectator ions that do not take part in the reaction.

Include the remaining ions and any non-ionic substances in the ionic equation.

Solution

The full, balanced equation for the reaction is

$$\underset{\text{hydrochloric acid}}{H^+(aq) + Cl^-(aq)} + \underset{\text{sodium hydroxide solution}}{Na^+(aq) + OH^-(aq)} \rightarrow \underset{\text{water (covalent molecule)}}{H_2O\,(l)} + \underset{\text{sodium chloride solution}}{Na^+(aq) + Cl^-(aq)}$$

Crossing out spectator ions

$$H^+(aq) + \cancel{Cl^-}(aq) + \cancel{Na^+}(aq) + OH^-(aq) \rightarrow H_2O\,(l) + \cancel{Na^+}(aq) + \cancel{Cl^-}(aq)$$

the ionic equation becomes

$$H^+(aq) + OH^-(aq) \rightarrow H_2O\,(l)$$

This ionic equation is the same for all neutralization reactions, whatever the acid and base used.

Question

Construct an ionic equation for the reaction of hydrofluoric acid (HF) with a solution of sodium carbonate (Na_2CO_3).

Oxidation and reduction

When a piece of magnesium burns in air (Figure 1.9), the magnesium gains oxygen and is oxidized. The product, magnesium oxide, is an ionic solid containing Mg^{2+} ions and O^{2-} ions.

$$Mg\,(s) + \tfrac{1}{2}O_2\,(g) \rightarrow MgO\,(s)$$

If you look at what is happening to the Mg atoms and to the O atoms in the reaction, you can describe the reaction by two **half equations**:

$$Mg \rightarrow Mg^{2+} + 2e^- \qquad \text{oxidation}$$

$$\tfrac{1}{2}O_2 + 2e^- \rightarrow O^{2-} \qquad \text{reduction}$$

The magnesium loses electrons in one half reaction and oxygen gains electrons in the other. Another way of looking at **oxidation** is to say that oxidation is loss of electrons. **Reduction**, then, occurs when electrons are gained. So, in this reaction, the oxygen is being reduced. Reduction and oxidation occur simultaneously when magnesium burns in air. It is a reduction–oxidation, or **redox**, reaction.

Half equations express two contributions to an overall redox reaction. (If you add the two half equations above together and cancel the $2e^-$, you get the overall equation for the reaction.) Remember the half equations are just a schematic representation—the electrons are not actually free in the reaction mixture.

These definitions of oxidation and reduction allow reactions to be classed as redox reactions even if they do not involve oxygen. For example, the burning of magnesium in chlorine closely resembles the reaction of magnesium with oxygen. Magnesium chloride is an ionic solid containing Mg^{2+} and Cl^- ions.

$$Mg\,(s) + Cl_2\,(g) \rightarrow MgCl_2\,(s)$$

$$Mg \rightarrow Mg^{2+} + 2e^- \qquad \text{oxidation}$$

$$Cl_2 + 2e^- \rightarrow 2Cl^- \qquad \text{reduction}$$

The magnesium atoms lose electrons and are oxidized. The chlorine atoms gain electrons and are reduced.

In the above reactions, O_2 and Cl_2 remove electrons from magnesium atoms and are called **oxidizing agents**. Similarly, the Mg atoms give electrons to O or Cl atoms and are **reducing agents**.

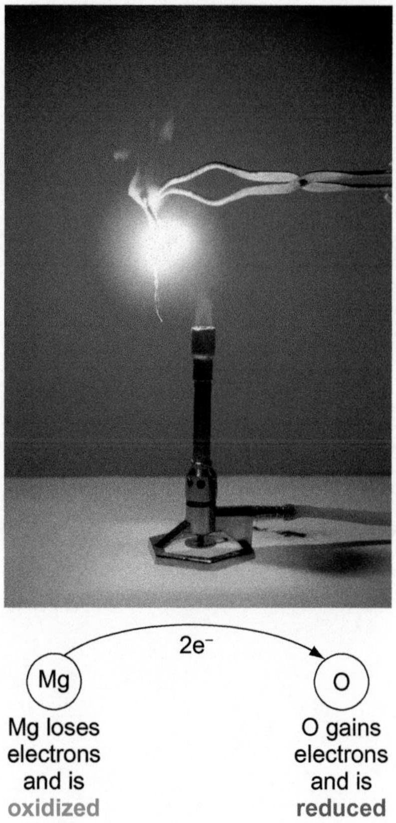
2e⁻
Mg
O
Mg loses electrons and is oxidized
O gains electrons and is reduced

Figure 1.9 The burning of magnesium in air is a redox reaction.

The structures of ionic solids are discussed in Sections 6.4 (p.277) and 6.5 (p.286).

The following mnemonic is helpful in remembering the definitions of oxidation and reduction.

OILRIG

Oxidation Is Loss of electrons; Reduction Is Gain of electrons.

Oxidation is loss of electrons. An **oxidizing agent** removes electrons from something else.
Reduction is gain of electrons. A **reducing agent** gives electrons to something else.

An element is oxidized when its oxidation state increases.
An element is reduced when its oxidation state decreases.

Oxidation states

Sometimes when a redox reaction involves non-ionic compounds, it is not possible to break the reaction down into half reactions involving electron transfer. For example, when hydrogen burns in air, water is formed

$$H_2(g) + \tfrac{1}{2}O_2(g) \rightarrow H_2O\ (l)$$

The equation is similar to the magnesium–oxygen one. Hydrogen is gaining oxygen and is being oxidized, but the product, water, is a covalent molecule.

To extend the idea of redox to reactions like this, chemists use the idea of **oxidation states** (sometimes called **oxidation numbers**). The atoms of each element in a substance are assigned oxidation states to show how much they are oxidized or reduced.

Rules for assigning oxidation states

1. The oxidation state of atoms in a pure element is zero.
2. In an uncharged compound, the sum of all the oxidation states is zero.
3. For simple ions containing only one element, such as Na^+ or Cl^-, the oxidation state is the same as the charge on the ion.
4. For ions containing more than one element, the sum of all the oxidation states is equal to the charge on the ion.
5. In compounds and ions, some elements have oxidation states that rarely change. These are:

   F: −1 (always);
   O: −2 (except in O_2^{2-}, O_2^-, OF_2, O_2F_2, and some metal oxides);
   H: +1 (except when combined with metals as H^-);
   Cl: −1 (except when combined with O and F);
   Group 1 metals: +1;
   Group 2 metals: +2.

Note that oxidation states always have a *sign* and a *number*. (The sign is written *before* the number to avoid confusion with charges.)

Applying the rules–some examples

Water (H_2O):

- Hydrogen is in oxidation state +1 (*rule 5*); oxygen is in oxidation state −2 (*rule 5*).
- There are two hydrogen atoms and one oxygen atom, so, for the compound, the sum of the oxidation states is $2 \times (+1) + (-2) = 0$ (*rule 2*).

Sulfate ion (SO_4^{2-}):

- Oxygen is in oxidation state −2 (*rule 5*).
- There are four O atoms, so the total contribution of the O atoms to the oxidation state of the ion is $4 \times (-2) = -8$.
- The sum of the oxidation states in the ion is −2 (the charge on the ion) (*rule 4*), so the oxidation state of S in the ion must be +6

Dichromate ion ($Cr_2O_7^{2-}$):

- Oxygen is in oxidation state −2 (*rule 5*).
- There are seven O atoms, so the total contribution of the O atoms to the oxidation state of the ion is $7 \times (-2) = -14$.
- The sum of the oxidation states in the ion is −2 (the charge on the ion) (*rule 4*), so the combined oxidation state of the two Cr atoms in the ion must be +12. So each Cr has an oxidation state of +6.

Oxidation states and naming inorganic compounds

Most inorganic compounds have systematic names (see Section 1.2, p.6) that tell you which elements are present, in what combination, and in what oxidation states.

The naming of organic compounds is covered in Chapter 2.

Cations

Cations containing a single element have the same name as the element. For example, H^+ is the hydrogen ion and Na^+ is the sodium ion. When an element can form more than one ion, the oxidation state of the element is given in brackets as a Roman numeral after the element name. Iron can form Fe^{2+} and Fe^{3+} ions, and these are called the iron(II) ion and the iron(III) ion, respectively (see Table 1.6 on p.22). Note that there is no space between the element name and the oxidation state. Examples of cations containing more than one atom are given in Table 1.7.

An alternative approach to naming cations is to place the charge on the ion in brackets. Using this method, the ions are called the iron(2+) ion and the iron(3+) ion, respectively.

Table 1.7 Examples of polyatomic cations

Formula	Name
O_2^+	dioxygen ion
H_3O^+	oxonium ion
NH_4^+	ammonium ion
NMe_4^+	tetramethylammonium ion

Anions

Anions containing a single element have the same stem as the element name, but the ending changes to -ide. For example, Cl^- is the chloride ion and O^{2-} is the oxide ion. There are a large number of polyatomic anions, and examples of these are given in Table 1.6 (p.22).

The endings -ate and -ite usually indicate that the ions contain oxygen, and these ions are called oxoanions. The thiocyanate ion (SCN^-) has an -ate ending even though it contains no oxygen. The thio- prefix shows that a sulfur atom has replaced the oxygen atom in cyanate (OCN^-).

Some elements form more than one oxoanion. For example, Table 1.8 shows how the four oxyanions of chlorine are named. The names give an indication of the numbers of oxygen atoms present:

- -ate anions contain more oxygen than -ite anions;
- the per- prefix implies more oxygen than the -ate anion;
- the hypo- prefix implies less oxygen than the -ite anion.

Table 1.8 Naming the oxoanions of chlorine

Formula	Name	Oxidation state of chlorine
ClO_4^-	perchlorate	+7
ClO_3^-	chlorate	+5
ClO_2^-	chlorite	+3
ClO^-	hypochlorite	+1

As with cations, the charge on an anion can be placed in brackets at the end of the name. This is useful when more than one anion is possible. For example, the superoxide ion (O_2^-) is also known as the dioxide(1−) ion.

Alternatively, longer systematic names can be used to unambiguously provide the formula. For example, the IUPAC name for the perchlorate ion (ClO_4^-) is tetraoxidochlorate(1−) (see Figure 27.28 in Section 27.6, p.1244). Using this nomenclature, the sulfate ion (SO_4^{2-}) is tetraoxidosulfate(2−). In this book, you will see the simplest names, such as perchlorate, sulfate, sulfite, nitrate, and nitrite.

Compounds

The name of an ionic compound consists of the cation name followed by the anion name. In this way, NaCl is sodium chloride, $MgBr_2$ is magnesium bromide, $(NH_4)_2SO_4$ is ammonium sulfate, and $FeCl_2$ is iron(II) chloride.

Covalently bonded binary compounds are named in a similar way to ionic compounds, with the more electropositive element given first with its element name, followed by the more electronegative atom with its element name changed to use the anionic -ide ending. For example, HCl is hydrogen chloride and H_2S is hydrogen sulfide. When the elements can combine in more than one way, the compounds are distinguished by indicating the numbers of atoms of the elements present in the formula using the prefixes shown in Table 1.9 (p.30). For example, SO_2 is sulfur dioxide and SO_3 is sulfur trioxide. Normally, when only one atom of an element is present, the prefix 'mono-' is omitted. An exception to this is CO, which is called carbon monoxide. Sometimes prefixes are needed for both elements so, for example, N_2O_5 is dinitrogen pentoxide. Some further examples are given in Worked example 1.9.

A **binary compound** is a compound containing only two different elements.

Electronegativity is the power of an atom in a molecule to attract electrons to itself. An atom with low electronegativity is said to be electropositive. See Section 4.3 (p.177).

Table 1.9 Prefixes indicating the numbers of atoms present

Number of atoms	Prefix	Example*
1	mono-	carbon monoxide (CO)
2	di-	carbon dioxide (CO_2)
3	tri-	sulfur trioxide (SO_3)
4	tetra-	silicon tetrachloride ($SiCl_4$)
5	penta-	dinitrogen pentoxide (N_2O_5)
6	hexa-	sulfur hexafluoride (SF_6)
7	hepta-	iodine heptafluoride (IF_7)
8	octa-	triuranium octaoxide (U_3O_8)
9	nona-	tetraphosphorus nonasulfide (P_4S_9)
10	deca-	disulfur decafluoride (S_2F_{10})

*When the element name begins with a- or o-, the a- or o- of the prefix is usually dropped to make pronunciation easier. For example, it is pentoxide not pentaoxide.

Worked example 1.9 Naming inorganic compounds

What are the systematic names for (a) BrF_5, (b) CaI_2, and (c) $Fe(NO_3)_2$?

Strategy

For binary compounds, change the name of the more electronegative element so that it ends in -ide. For ionic compounds, indicate the oxidation state using Roman numerals if more than one oxidation state is possible. For covalently bonded compounds, use prefixes to indicate the numbers of atoms in the formula if the elements can combine in different combinations.

Solution

(a) Bromine and fluorine can combine in more than one way, so the numbers of atoms in the formula need to be identified. Fluorine is more electronegative, so the compound is bromine pentafluoride.

(b) Calcium only forms the +2 oxidation state. The compound is calcium iodide.

(c) Iron forms more than one type of ion, so the oxidation state needs to be indicated. The anion is nitrate (NO_3^-) so the compound is iron(II) nitrate.

Question

What are the systematic names for (a) SeF_4 and (b) $Fe(ClO_4)_3$?

Changes of oxidation state

You can use oxidation states to find out what has been oxidized and what has been reduced in a reaction. For example, consider the reaction between chlorine and iodide ions.

$$Cl_2(aq) + 2I^-(aq) \rightarrow 2Cl^-(aq) + I_2(aq)$$

Half reactions:

$$2I^-(aq) \rightarrow I_2(aq) + 2e^- \quad \text{oxidation}$$

$$Cl_2(aq) + 2e^- \rightarrow 2Cl^-(aq) \quad \text{reduction}$$

In terms of electron transfer, chlorine is reduced and iodine is oxidized. Now look at the oxidation states:

$$Cl_2\,(aq) + 2\,I^-\,(aq) \rightarrow 2\,Cl^-\,(aq) + I_2\,(aq)$$

Oxidation state of I: −1 (×2) → 0

Oxidation state of Cl: 0 → −1 (×2)

- The oxidation state of iodine increases from −1 to 0, and iodine has been oxidized.
- The oxidation state of chlorine decreases from 0 to −1, and chlorine has been reduced.

Note that the overall increase in oxidation states balances the overall decrease. Another example is given in Worked example 1.10.

Worked example 1.10 Changes of oxidation state

When concentrated sulfuric acid is added to solid potassium bromide, red-brown fumes of bromine vapour are seen and sulfur dioxide gas is also given off. The ionic equation is shown below. Assign oxidation states to the elements in each of the reactants and products in the equation. Use these values to decide what has been oxidized and what reduced.

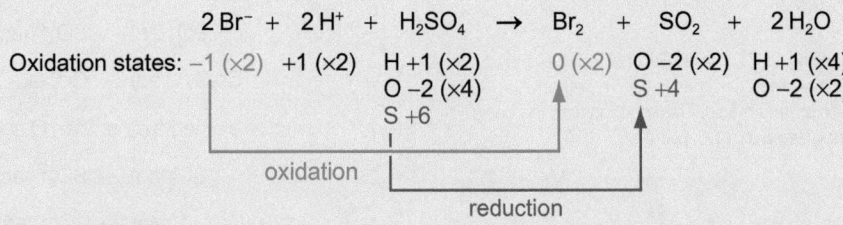

$$2\,Br^- + 2\,H^+ + H_2SO_4 \rightarrow Br_2 + SO_2 + 2\,H_2O$$

▲ Addition of concentrated sulfuric acid to potassium bromide. Red-brown fumes of bromine vapour (Br_2) are produced.

Strategy

Use the rules on p.28 to assign oxidation states for each element in the equation.

Write the values under the reactants and products in the equation.

Elements for which the oxidation state increases have been oxidized in the reaction; elements for which the oxidation state decreases have been reduced.

Solution

	$2\,Br^-$	+ $2\,H^+$	+ H_2SO_4	→ Br_2	+ SO_2	+ $2\,H_2O$
Oxidation states:	−1 (×2)	+1 (×2)	H +1 (×2) O −2 (×4) S +6	0 (×2)	O −2 (×2) S +4	H +1 (×4) O −2 (×2)

oxidation: Br −1 (×2) → 0 (×2)

reduction: S +6 → S +4

- H is +1 in reactants and products, so no redox.
- O is −2 in reactants and products, so no redox.
- Br increases from −1 in 2 Br^- to 0 in Br_2, so it has been oxidized.
- S decreases from +6 in H_2SO_4 to +4 in SO_2, so it has been reduced.

Question

Use oxidation states to decide what has been oxidized and what reduced in the reaction of sodium with water.

$$2\,Na\,(s) + 2\,H_2O\,(l) \rightarrow 2\,NaOH\,(aq) + H_2\,(g)$$

Constructing and balancing redox equations from half equations

Constructing an overall equation from the two half equations can be tricky at first and requires some practice. Here are the steps to follow. You can practise using them in Worked example 1.11.

When combining half equations to give an overall equation, make sure that the number of electrons released in oxidation is the same as the number of electrons gained in reduction.

Redox reactions are important in electrochemistry. Section 16.3 (p.735) shows how you can predict whether one compound will oxidize or reduce another. Redox reactions of *p*-block elements and their compounds are discussed in Chapter 27 and those of *d*-block elements and their compounds in Chapter 28.

Visit the Online Resource Centre to view screencast 1.1 which walks you through Worked example 1.11 to illustrate constructing and balancing a redox equation from half equations.

Step 1 Identify what is being oxidized and what reduced, and then write the two unbalanced half equations for oxidation and reduction.

Step 2 Balance all the elements in the half equations except O and H.

Step 3 Balance O by adding H_2O. Then, in *acidic solution*, balance H by adding H^+; in *alkaline solution*, balance H by adding H_2O to the side of the half equation that needs H (just *one* H_2O for each H needed) whilst adding OH^- to the other side. Then cancel out any surplus water.

Step 4 Balance electric charges in the half equations by adding electrons to the left-hand side for reduction and to the right-hand side for oxidation.

Step 5 Multiply the half equations, if necessary, by factors so that each equation transfers the same number of electrons.

Step 6 Add the two half equations together so that the numbers of electrons transferred cancel out.

Step 7 Simplify the overall equation, if necessary, by cancelling species that appear on both sides.

Step 8 Check that atoms and charges on each side of the equation balance. Check that you have no electrons left in the overall equation.

Worked example 1.11 Constructing and balancing a redox equation from half equations

Sulfur dioxide gas turns an acidified solution of potassium dichromate from orange to blue-green. The SO_2 is oxidized to SO_4^{2-} ions. The orange $Cr_2O_7^{2-}$ ions are reduced to blue-green Cr^{3+} ions. Construct half equations and a balanced overall equation for the reaction.

▼ Another redox reaction involving $Cr_2O_7^{2-}$ ions. Ethanol reduces the orange $Cr_2O_7^{2-}$ ions in (a) to blue-green Cr^{3+} ions in (b). The ethanol is oxidized to ethanal and then to ethanoic acid. This reaction was the basis of early breath tests for ethanol (See Box 11.6, p.542).

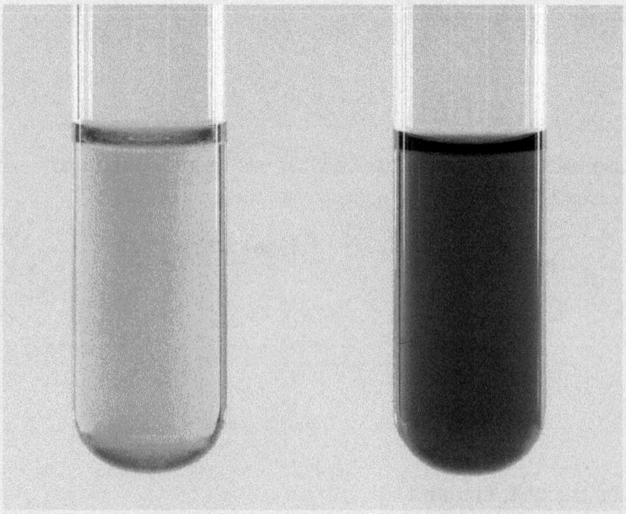

Strategy

Start by writing unbalanced half equations for the oxidation of SO_2 and the reduction of $Cr_2O_7^{2-}$ ions. Then follow through the steps 1–8 above.

Solution

Step 1 Unbalanced half equations.

$$SO_2(aq) \rightarrow SO_4^{2-}(aq) \quad \text{oxidation}$$

$$Cr_2O_7^{2-}(aq) \rightarrow Cr^{3+}(aq) \quad \text{reduction}$$

Step 2 Balance elements other than O and H.

$$SO_2(aq) \rightarrow SO_4^{2-}(aq) \quad \text{oxidation}$$

$$Cr_2O_7^{2-}(aq) \rightarrow 2Cr^{3+}(aq) \quad \text{reduction}$$

Step 3 Balance O and H by adding H_2O and H^+ ions (in acidic solution).

$$SO_2(aq) + 2H_2O(l) \rightarrow SO_4^{2-}(aq) \quad \text{oxidation}$$

$$SO_2(aq) + 2H_2O(l) \rightarrow SO_4^{2-}(aq) + 4H^+(aq)$$

$$Cr_2O_7^{2-}(aq) \rightarrow 2Cr^{3+}(aq) + 7H_2O(l) \quad \text{reduction}$$

$$Cr_2O_7^{2-}(aq) + 14H^+(aq) \rightarrow 2Cr^{3+}(aq) + 7H_2O(l)$$

Step 4 Balance electric charge by adding electrons.

$$SO_2(aq) + 2H_2O(l) \rightarrow SO_4^{2-}(aq) + 4H^+(aq) + 2e^- \quad \text{oxidation}$$

$$Cr_2O_7^{2-}(aq) + 14H^+(aq) + 6e^- \rightarrow 2Cr^{3+}(aq) + 7H_2O(l) \quad \text{reduction}$$

→

→

Step 5 Multiply the oxidation half equation × 3, so that the number of electrons transferred is the same in both half equations.

$$3SO_2(aq) + 6H_2O(l) \rightarrow 3SO_4^{2-}(aq) + 12H^+(aq) + 6e^- \quad \text{oxidation}$$

$$Cr_2O_7^{2-}(aq) + 14H^+(aq) + 6e^- \rightarrow 2Cr^{3+}(aq) + 7H_2O(l) \quad \text{reduction}$$

Step 6 Add the two half equations and cancel out the electrons.

$$3SO_2(aq) + 6H_2O(l) \rightarrow 3SO_4^{2-}(aq) + 12H^+(aq) + \cancel{6e^-} \quad \text{oxidation}$$

$$Cr_2O_7^{2-}(aq) + 14H^+(aq) + \cancel{6e^-} \rightarrow 2Cr^{3+}(aq) + 7H_2O(l) \quad \text{reduction}$$

$$Cr_2O_7^{2-}(aq) + 14H^+(aq) + 3SO_2(aq) + 6H_2O(l) \rightarrow 3SO_4^{2-}(aq) + 12H^+(aq) + 2Cr^{3+}(aq) + 7H_2O(l)$$

Steps 7 and 8 Simplify the equation by cancelling out H^+(aq) ions and H_2O (l) molecules. Atoms and charges on each side of the equation balance and no free electrons are left over.

$$Cr_2O_7^{2-}(aq) + 2H^+(aq) + 3SO_2(aq) \rightarrow 3SO_4^{2-}(aq) + 2Cr^{3+}(aq) + H_2O(l)$$

Question

Fe^{2+} ions are oxidized to Fe^{3+} by MnO_4^- ions in acidic solution. The MnO_4^- ions are reduced to Mn^{2+} ions. Construct half equations and a balanced overall equation for the reaction.

Redox in organic reactions: oxidation levels

Carbon has nine oxidation states ranging from −4 in CH_4 to +4 in CO_2 and CCl_4. For most organic molecules, however, it is not helpful to assign oxidation states to individual atoms. What, for example, are the oxidation states of carbon and hydrogen in ethane (C_2H_6) and in propane (C_3H_8)? Organic chemists use a system of oxidation levels based on the extent of oxidation or reduction of *the carbon atom that is part of, or attached to, the functional group*. Thus, alcohols and halogenoalkanes are at the same oxidation level because they can be interconverted without using oxidizing or reducing agents.

To convert an alcohol into an aldehyde or a ketone requires an oxidizing agent, so aldehydes and ketones are at a higher oxidation level than alcohols. Aldehydes can be oxidized to carboxylic acids, so carboxylic acids are at a higher oxidation level still. For example,

$$CH_3CH_2\text{–}OH \underset{OH^-}{\overset{HBr}{\rightleftharpoons}} CH_3CH_2\text{–}Br$$

No redox—reactant and product at same oxidation level

$$CH_3CH_2\text{–}OH \underset{\substack{NaBH_4 \\ \text{reducing agent}}}{\overset{\substack{\text{oxidizing agent} \\ Cr_2O_7^{2-},\ H^+}}{\rightleftharpoons}} CH_3\text{–}CHO$$

Redox reaction—product at a different oxidation level from the reactant

A **functional group**, such as –Cl and –OH, on the hydrocarbon skeleton of an organic molecule gives the molecule characteristic properties; see Section 2.3 (p.77).

You can read about the rules for assigning oxidation levels in organic compounds in Section 19.2 (p.888). The oxidation of alcohols is discussed in Box 23.2 (p.1064). There is a list of common oxidizing and reducing agents in organic chemistry in Appendix 4 (p.1338).

Summary

- A balanced equation is one with equal numbers of each type of atom on either side of the equation. It is a balance sheet of where each atom starts and where it ends up. For an ionic equation, the total charge on either side of the equation must be the same.
- The relationship between the amounts of reactants and products is called the stoichiometry of the reaction. It allows you to calculate the masses that react exactly and the masses of the products formed.
- The percentage yield of a reaction tells you what percentage of the theoretical yield is actually obtained.
- An ionic equation shows only the ions that take part in the reaction, plus any non-ionic substances involved. Spectator ions are not included.

→

- Oxidation is loss of electrons or increase in oxidation state. Reduction is gain of electrons or decrease in oxidation state.
- Redox reactions can be formally represented by two half equations, one for oxidation and the other for reduction.
- Systematic names for inorganic compounds tell you which elements are present, in what combination, and in what oxidation states.
- For organic compounds, oxidation levels are more useful than oxidation states.

For practice questions on these topics, see questions 9–15 at the end of this chapter (p.66).

1.5 Working out how much you have

Concentrations of solutions

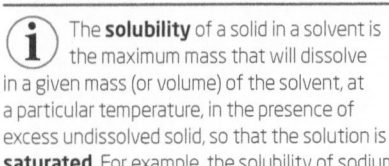

The **solubility** of a solid in a solvent is the maximum mass that will dissolve in a given mass (or volume) of the solvent, at a particular temperature, in the presence of excess undissolved solid, so that the solution is **saturated**. For example, the solubility of sodium chloride at 298 K is 36.0 g per 100 g of water.

A solution with a concentration of 1 mol dm^{-3} is sometimes called a one molar solution, abbreviated to 1 M.

Many chemical reactions take place in solution. The substance dissolved is called the **solute**, and the liquid it dissolves in the **solvent**. The **concentration** of a solution tells you how much of the solute is dissolved in a particular volume of the solution.

Concentrations can be expressed in various units depending on the circumstances. Sometimes units of grams per dm^3 (grams per litre) are used. A solution containing 20 g of sodium chloride dissolved in 1 dm^{-3} of solution has a concentration of 20 g dm^{-3}. You may sometimes see the solubility of a substance quoted in grams per 100 g of solvent.

Usually, however, chemists are more interested in the *chemical amount* of substance present rather than the mass. The amount in moles tells you the number of formula units present and relates to the chemical equation for a reaction. So, the preferred units for measuring concentrations are mol dm^{-3}.

The molarity of a solution is the amount (in moles) of solute dissolved per dm^3 of *solution* (not solvent). The units are mol dm^{-3}.

To convert g dm^{-3} to mol dm^{-3}, you need to know the molar mass of the solute.

$$\text{Concentration (mol dm}^{-3}) = \frac{\text{concentration (g dm}^{-3})}{\text{molar mass (g mol}^{-1})} \qquad (1.6)$$

The use of Equation 1.6 is illustrated in Worked example 1.12. Box 1.5 describes an alternative way of expressing very low concentrations.

Box 1.5 Measuring low concentrations: parts per million

It is often more convenient to express low concentrations in parts per million (ppm) or even parts per billion (ppb; where 1 billion = 1×10^9). If you use these units, it is important to be clear about what they refer to.

Parts per million by mass

This unit is often used for low concentrations of an impurity in a liquid or a solid. For example, the concentration of a pesticide in a ground water sample may be quoted as 1 ppm (by mass). This means that 1 g of the water sample contains 1×10^{-6} g (1 μg) of the pesticide.

Concentration in ppm (by mass)

$$= \frac{\text{mass of component}}{\text{total mass of liquid or solid}} \times 10^6 \text{ ppm} \qquad (1.7)$$

Parts per million by volume

For gases, ppm by volume is used. For example, the concentration of carbon dioxide in the atmosphere in 2011 was 392 ppm (by volume). To understand what this means it helps to remember that, for an ideal gas, the volume is proportional to the number of moles (or molecules). Thus, one mole of air contains 392×10^{-6} mol (392 μmol) of CO_2. In terms of molecules, one million molecules of air contain 392 molecules of CO_2.

→

▲ The Mauna Loa Observatory in Hawaii has been recording atmospheric CO_2 concentrations since 1958, when the concentration of CO_2 was 316 ppm by volume (see Box 10.5, p.485). In 2011, the CO_2 concentration was 392 ppm by volume.

Concentration in ppm (by volume)

$$= \frac{\text{mass of component}}{\text{total moles of gas}} \times 10^6 \text{ ppm} \qquad (1.8)$$

Percentage composition

392 ppm (by volume) is the same as 0.0392% by volume.

Question

The label on a bottle of mineral water says it contains '27 mg calcium per litre'.

(a) What is the concentration of Ca^{2+} ions in ppm (by mass) and the percentage by mass of Ca^{2+} ions?

(b) What is the concentration of Ca^{2+} ions in $mol\,dm^{-3}$?

Worked example 1.12 Units of concentration

2.54 g of sodium chloride were dissolved in water and the solution made up to 100 cm^3. What is the concentration of the solution in $mol\,dm^{-3}$?

Strategy

Work out the concentration of the solution in $g\,dm^{-3}$. (Remember: $1\,dm^3 = 1000\,cm^3$.)
Use Equation 1.6 to convert $g\,dm^{-3}$ to $mol\,dm^{-3}$.

Solution

100 cm^3 contain 2.54 g of NaCl, so 1000 cm^3 contain 25.4 g of NaCl.

$$M_r(\text{NaCl}) = 22.99 + 35.45 = 58.44$$

From Equation 1.6,

$$\text{concentration } (mol\,dm^{-3}) = \frac{\text{concentration } (g\,dm^{-3})}{\text{molar mass } (g\,mol^{-3})} \qquad (1.6)$$

$$= \frac{25.4\,\cancel{g}\,dm^{-3}}{58.44\,\cancel{g}\,mol^{-1}}$$

$$= 0.435\,mol\,dm^{-3}$$

Question

43.7 g of anhydrous copper(II) sulfate ($CuSO_4$) were dissolved in water and the solution was made up to 2.00 dm^3. What is the molarity of the solution?

If you know the concentration (c) of a solution, you can work out the amount (n) of solute in a particular volume (V).

$$\underset{\text{mol}}{\text{Amount }(n)} = \underset{\text{mol dm}^{-3}}{\text{concentration }(c)} \times \underset{\text{dm}^3}{\text{volume of solution }(V)}$$

So,

$$n = c \times V \qquad (1.9)$$

If you use this equation, remember to convert the volume of solution into dm^3.

Volumetric analysis: titrations

A solution whose concentration is known exactly is called a **standard solution**.

For a chemical reaction taking place between two solutions, if you know the stoichiometric equation for the reaction and the concentration of one of the reactants, you can carry out a **titration** to find the concentration of the other reactant. This involves adding small amounts of one solution from a burette to a known volume of the other solution in a conical flask until the reaction is just complete.

The procedure is called **volumetric analysis** because it involves making accurate measurements of volumes of solutions. The **equivalence point** is the point in a titration when the amount of added reagent has reacted exactly with the solution in the flask. An **indicator** is usually added so that a colour change indicates when this has happened. The point at which the indicator changes colour is the **end-point** of the titration.

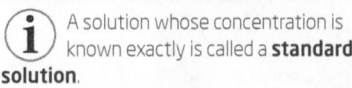

Worked example 1.13 Finding how much solute is in a given volume of solution

What mass of sodium carbonate (Na_2CO_3) must be dissolved in water to give 75.0 cm^3 of a solution with a concentration of 2.05 mol dm^{-3}?

Strategy

(a) Work out the amount of Na_2CO_3 in 75.0 cm^3 of the 2.05 mol dm^{-3} solution. You can do this from first principles or by using Equation 1.9 (but remember to convert the volume of the solution to dm^3). (1 dm^3 = 1000 cm^3.)

(b) Use Equation 1.2 (p.19) to convert the amount in moles to the mass in grams.

Solution

(a) Working out the amount of Na_2CO_3 from first principles:

1000 cm^3 (1 dm^3) of solution contain 2.05 mol of Na_2CO_3

1 cm^3 of solution contains $\frac{2.05}{1000}$ mol of Na_2CO_3

75.0 cm^3 of solution contain $\frac{2.05 \times 75.0}{1000}$ mol = 0.1538 mol of Na_2CO_3

or,

Working out the amount of Na_2CO_3 using Equation 1.9:

$$n = c \times V \qquad (1.9)$$

$$\text{amount of } Na_2CO_3 = 2.05\text{ mol } \cancel{dm^{-3}} \times 0.0750 \cancel{dm^3}$$
$$= 0.1538\text{ mol}$$

(b) $M_r(Na_2CO_3) = (2 \times 22.99) + 12.01 + (3 \times 16.00) = 106.0$

Use Equation 1.2,

$$\text{amount (in mol)} = \frac{\text{mass (in g)}}{\text{molar mass (in g mol}^{-1})} \qquad (1.2)$$

and rearrange to give an expression for the mass.

$$\text{Mass of } Na_2CO_3 \text{ needed} = \text{amount} \times \text{molar mass}$$
$$= 0.1538 \cancel{mol} \times 106.0\text{ g } \cancel{mol^{-1}}$$
$$= 16.3\text{ g}$$

Question

(a) A solution of sodium carbonate has a concentration of 0.157 mol dm^{-3}. What volume of solution contains exactly 1.00 g of sodium carbonate?

(b) What mass of sodium thiosulfate crystals ($Na_2S_2O_3.5H_2O$) must be dissolved in 250 cm^3 of water to give a solution with a concentration of 0.0750 mol dm^{-3}?

Figure 1.10 shows a typical titration apparatus that you may have used in the laboratory. The unknown concentration is calculated by following the steps in Worked example 1.14. An automated titration apparatus used for routine titrations in industry is shown in Figure 1.11. It monitors the pH of the solution as the titration proceeds and detects the equivalence point by responding to a rapid change in pH that occurs at that point.

The pH changes during acid–base titrations are explained in Section 7.4 (p.322).

Visit the Online Resource Centre to view screencast 1.2 which walks you through Worked example 1.14 to illustrate the stages in carrying out the calculation for a typical titration.

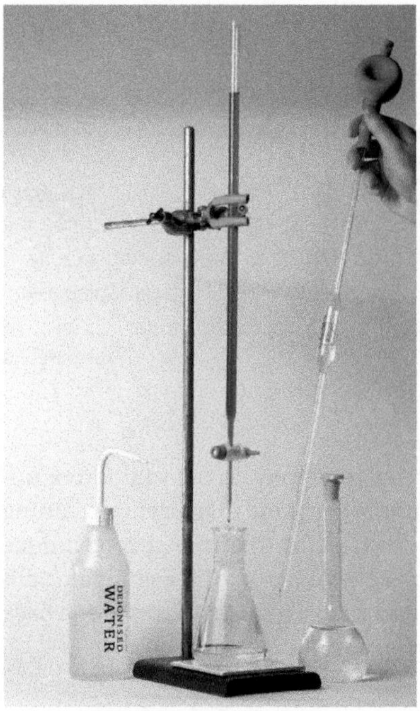
DEIONISED
WATER

Figure 1.10 Typical apparatus for a titration, showing a burette, pipette, and a volumetric flask used for making up a standard solution.

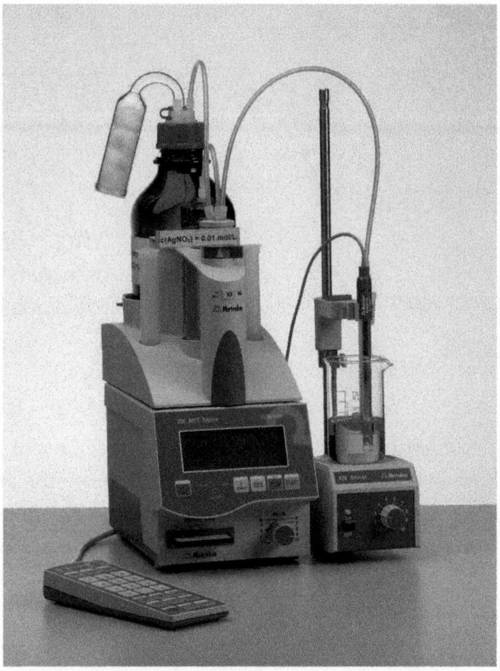

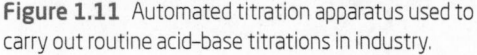

Figure 1.11 Automated titration apparatus used to carry out routine acid–base titrations in industry.

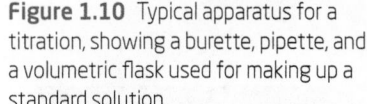

Worked example 1.14 An acid–base titration

A standard solution of sulfuric acid can be used to find the concentration of a solution of sodium hydroxide, whose concentration is unknown. In a titration, 0.100 mol dm^{-3} H_2SO_4 was added from a burette to 25.0 cm^3 of NaOH containing a few drops of phenolphthalein as the indicator. The solution changed from pink to colourless after the addition of 21.4 cm^3 of the sulfuric acid. What is the concentration of the sodium hydroxide solution?

Strategy

The steps below can be adapted for any titration calculation.

Step 1 Write down the stoichiometric equation for the reaction.

Step 2 State what the equation tells you about the amounts of the substances you are interested in.

Step 3 Use the concentration of the standard solution to work out the number of moles that have reacted.

Step 4 From your answer to Step 2, find the number of moles of the other reactant present in the volume of solution used. This is the amount that reacted exactly.

Step 5 Convert this to the amount in 1 dm^3 of solution to give the concentration in mol dm^{-3}.

Solution

Step 1 The stoichiometric equation for the reaction is

$$H_2SO_4(aq) + 2\,NaOH(aq) \rightarrow Na_2SO_4(aq) + 2\,H_2O(l)$$

Step 2 The equation tells you that 1 mol of H_2SO_4 reacts exactly with 2 mol of NaOH.

Step 3 21.4 cm^3 of 0.100 mol dm^{-3} H_2SO_4 were added.

Working out the amount of H_2SO_4 added from first principles:

1 000 cm^3 of H_2SO_4(aq) contain 0.100 mol of H_2SO_4

→

$1\ cm^3$ of $H_2SO_4(aq)$ contains $\frac{0.100}{1000}$ mol of H_2SO_4

$21.4\ cm^3$ of $H_2SO_4(aq)$ contain $\frac{0.100 \times 21.4}{1000}$ mol

$= 2.14 \times 10^{-3}$ mol H_2SO_4

or,

Working out the amount of H_2SO_4 added using Equation 1.9 (p.36):

$$n = c \times V \qquad (1.9)$$

$$\text{Amount of } H_2SO_4 = 0.100\ \text{mol}\ \cancel{\text{dm}^{-3}} \times 0.0214\ \cancel{\text{dm}^3}$$

$$= 2.14 \times 10^{-3}\ \text{mol}$$

(When using Equation 1.9, remember to convert the volume of the solution to dm^3.)

Step 4 From Step 2,

amount of NaOH to react exactly

$= 2 \times (2.14 \times 10^{-3}$ mol)

This was contained in $25.0\ cm^3$ of solution.

Step 5 Concentration of NaOH solution

$$= \frac{2 \times (2.14 \times 10^{-3}\ \text{mol})}{0.025\ \text{dm}^3}$$

$= 0.171$ mol dm^{-3}

Question

$14.8\ cm^3$ of 0.00105 mol dm^{-3} nitric acid ($HNO_3(aq)$) reacted exactly with $9.85\ cm^3$ of calcium hydroxide solution ($Ca(OH)_2(aq)$). Calculate the concentration of the calcium hydroxide solution in (a) mol dm^{-3} and (b) in g dm^{-3}.

Sometimes it is necessary to add an *excess* of a standard reagent to the solution of unknown concentration and then determine the amount of the standard reagent remaining after the reaction by a **back titration** using a second standard solution. This technique is used when the rate of reaction is slow, making the determination of the end-point difficult, or when the solution of unknown concentration is unstable. An example of a back titration in the analysis of river water is described in Box 1.6.

Box 1.6 Measuring dissolved oxygen in river water

In the UK, analysis of the dissolved oxygen in river water is carried out routinely by the water industry. A 'healthy' river has dissolved oxygen concentrations ranging from around 5 mg dm^{-3} to 10 mg dm^{-3}. (The EC standard for a salmon river requires 50% of samples to be at least 9 mg dm^{-3} and never less than 6 mg dm^{-3}.) In heavily polluted water, oxygen is used up by aerobic bacteria decomposing the organic material in the river. The concentration of oxygen falls and the river becomes lifeless.

Measurements are taken on site using electronic probes, which are pre-calibrated in the laboratory at regular intervals.

One method used to pre-calibrate the electronic probes involves performing a **Winkler titration**. This is a back titration that takes place in three stages.

Stage 1 The water sample is treated with a solution of manganese(II) sulfate ($MnSO_4(aq)$) followed by an alkaline solution of potassium iodide (KI (aq)). Both these solutions are present in excess to ensure that all the dissolved oxygen reacts. The following reaction takes place.

$$Mn^{2+}(aq) + 2OH^-(aq) \rightarrow Mn(OH)_2(s)$$

Adequate dissolved oxygen is essential for good water quality. Analysis is carried out on site using an electronic probe.

→ A precipitate of manganese(II) hydroxide forms which is oxidized by the dissolved oxygen to form manganese(III) hydroxide.

$$2Mn(OH)_2(s) + \tfrac{1}{2}O_2(aq) + H_2O(l) \rightarrow 2Mn(OH)_3(s) \qquad (1.10)$$

Stage 2 The precipitate of manganese(III) hydroxide is allowed to settle and the solution is acidified with sulfuric acid. The flask is swirled gently until all the precipitate dissolves. In the acidified solution, the liberated $Mn^{3+}(aq)$ ions are rapidly reduced back to $Mn^{2+}(aq)$ by the $I^-(aq)$ ions.

$$2Mn(OH)_3(s) + 2I^-(aq) + 6H^+(aq) \rightarrow 2Mn^{2+}(aq) + I_2(aq) + 6H_2O(l) \qquad (1.11)$$

Stage 3 The mixture is transferred to a conical flask and the $I_2(aq)$ is titrated with a standard solution of sodium thiosulfate in the presence of a starch indicator. The blue colour changes to a pale yellow colour at the end-point.

$$I_2(aq) + 2S_2O_3^{2-}(aq) \rightarrow 2I^-(aq) + S_4O_6^{2-}(aq) \qquad (1.12)$$

Question

A Winkler titration was performed on a 150 cm^3 sample of river water. 12.2 cm^3 of a 0.0100 mol dm^{-3} sodium thiosulfate solution were required.

(a) Calculate the concentration of the dissolved oxygen in mg dm^{-3}.

(b) Suggest why a back titration is used in this determination, rather than a direct titration of O_2 against $Mn^{2+}(aq)$.

Gravimetric analysis

Analytical methods that are based on accurate measurements of mass rather than volume are known as gravimetric analyses. One technique involves reacting the substance being analysed in solution so that a precipitate of known composition is produced. The precipitate is then separated by filtration, washed, dried, and accurately weighed.

For example, one way of determining the calcium content of natural waters is to treat a sample of water with an excess of ethanedioic acid (oxalic acid, $H_2C_2O_4$) followed by a solution of ammonia.

Measurement of mass is one of the simplest but most accurate measurements that can be made in the laboratory.

$$\underset{\text{ethanedioic acid}}{\begin{matrix}CO_2H\\|\\CO_2H\end{matrix}}(aq) + 2NH_3(aq); \qquad \underset{\text{ethanedioate ions}}{\begin{matrix}CO_2^-\\|\\CO_2^-\end{matrix}}(aq) + 2NH_4^+(aq)$$

The ammonia reacts with the ethanedioic acid to produce ethanedioate ions and ensures that all the calcium ions in the water sample are precipitated as calcium ethanedioate (calcium oxalate)

$$Ca^{2+}(aq) + C_2O_4^{2-}(aq) \rightarrow \underset{\text{calcium ethanedioate}}{CaC_2O_4(s)} \qquad (1.13)$$

The precipitate of calcium ethanedioate is collected in a weighed filtering crucible (see Figure 1.12), dried, and heated strongly in air. The calcium ethanedioate is converted to calcium oxide

$$CaC_2O_4(s) \rightarrow CaO(s) + CO(g) + CO_2(g) \qquad (1.14)$$

When the crucible is cool, the crucible and remaining solid are weighed. Then the crucible and contents are heated, cooled, and weighed again, to check that the conversion to calcium oxide is complete. Hence, the mass of CaO produced from the calcium ions in a given volume of water can be found. Worked example 1.15 shows how the results are calculated.

PYREX
ENGLAND

Figure 1.12 A sintered glass weighing crucible.

Worked example 1.15 Gravimetric analysis

In an analytical laboratory, a 250.0 cm^3 sample of river water was treated with excess ethanedioic acid and ammonia solution to precipitate all the Ca^{2+} ions present as CaC_2O_4. The precipitate was filtered, washed, dried, and heated strongly in a crucible whose empty mass was 24.3782 g. The mass of the crucible and CaO was 24.5186 g. Calculate the concentration of Ca^{2+} ions in the water in g per 100 cm^3 of water.

Strategy

Calculate the mass of CaO formed.
Calculate the amount of CaO formed.
Use Equations 1.13 and 1.14 to work out the stoichiometric relationship between the Ca^{2+} ions in the water and the CaO produced.
Work out the amount in moles of Ca^{2+} ions in the 250.0 cm^3 water sample.
Convert this to g per 100.0 cm^3 of water.

Solution

$$\text{Mass of CaO formed} = 24.5186\,\text{g} - 24.3782\,\text{g} = 0.1404\,\text{g}$$

$$M_r(\text{CaO}) = 40.08 + 16.00 = 56.08$$

From Equation 1.2 (p.19),

$$\text{amount of CaO formed} = \frac{0.1404\,\cancel{\text{g}}}{56.08\,\cancel{\text{g}}\,\text{mol}^{-1}} = 2.5036 \times 10^{-3}\,\text{mol}$$

From Equations 1.13 and 1.14, 1 mol of CaO is formed from 1 mol of Ca^{2+} ions in the water sample.

250.0 cm^3 of the water sample contain 2.5036×10^{-3} mol of Ca^{2+}(aq)

100.0 cm^3 of the water sample contain

$$\frac{(2.5036 \times 10^{-3}\,\text{mol}) \times 100.0\,\cancel{\text{cm}^3}}{250.0\,\cancel{\text{cm}^3}} = 1.0014 \times 10^{-3}\,\text{mol of } Ca^{2+}(\text{aq})$$

Question

A 4.500 g sample of an oil containing the pesticide DDT ($C_{14}H_9Cl_5$) was heated with sodium in alcohol to liberate all the chlorine present as Cl^- ions. The mixture was treated with silver nitrate solution and 1.509 g of solid AgCl were recovered. Calculate the percentage (by mass) of DDT in the sample.

Summary

- The molarity of a solution is the amount (in mol) of solute dissolved in 1 dm^3 of solution.
- The amount (in mol) of solute (n) dissolved in a volume (V) of solution (in dm^{-3}) is given by:
  $n = c \times V$, where c is the concentration of the solution in mol dm^{-3}.
- Volumetric analysis involves accurate measurements of volumes of solutions.
- Gravimetric analysis involves accurate measurements of mass.

For practice questions on these topics, see questions 16 and 17 at the end of this chapter (pp.66–7).

1.6 Energy changes in chemical reactions

Transfer of energy

Energy is the capacity to do work. **Work** is done when motion occurs against a force. For example, when you lift up a book, you are doing work against the force of gravity and this requires energy. The plunger in the syringe in Figure 1.13 is pushed back against gravity and against the atmosphere so that work is done.

The transfer of energy is a familiar occurrence in everyday life. When a fossil fuel, such as coal or oil, is burned in a power station, chemical energy is converted to thermal energy, which is used to heat water. The steam generated produces motion in a turbine and the mechanical energy of the rotating turbine is converted to electrical energy. The electrical energy is delivered to your home where it can produce motion in an electric motor, or heat in an electric fire, or light from a lamp.

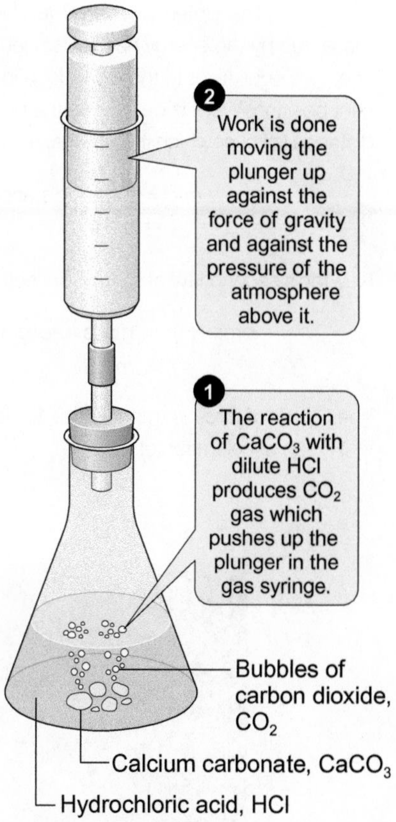
2
Work is done moving the plunger up against the force of gravity and against the pressure of the atmosphere above it.
1
The reaction of $CaCO_3$ with dilute HCl produces CO_2 gas which pushes up the plunger in the gas syringe.
Bubbles of carbon dioxide, CO_2
Calcium carbonate, $CaCO_3$
Hydrochloric acid, HCl

Figure 1.13 A simple chemical reaction that produces motion. When the reaction is carried out in an open flask, work is done pushing back the atmosphere.

The unit for work, and hence for energy, is the joule, J. The work done is one joule when a force of one newton acts over a distance of one metre:

$$1\,J = 1\,N\,m = 1\,(kg\,m\,s^{-2})\,m = 1\,kg\,m^2\,s^{-2}$$

(Newtons and joules are both derived SI units; see Table 1.2 (p.8).)

Think about the bungee jumper in Figure 1 in Box 1.7 (p.42). The height of the platform gives the jumper a large potential energy. When the jumper dives off the platform, the potential energy is converted into kinetic energy. Towards the end of the fall, the cord starts to stretch and slows down the descent. As the cord stretches, the kinetic energy of the jumper is converted into potential energy—this time stored in the stretched out molecules of the cord. Potential energy and kinetic energy are discussed in more detail in Box 1.7.

Thermochemistry

Energy changes are a characteristic feature of chemical reactions. Many chemical reactions give out energy and some take energy in. Usually the energy is in the form of heat or work, but reactions (such as the one in Figure 1.14) can also emit light or sound. In an electrochemical cell, a chemical reaction is arranged so that chemical energy is converted directly to electrical energy; see Section 16.3 (p.735).

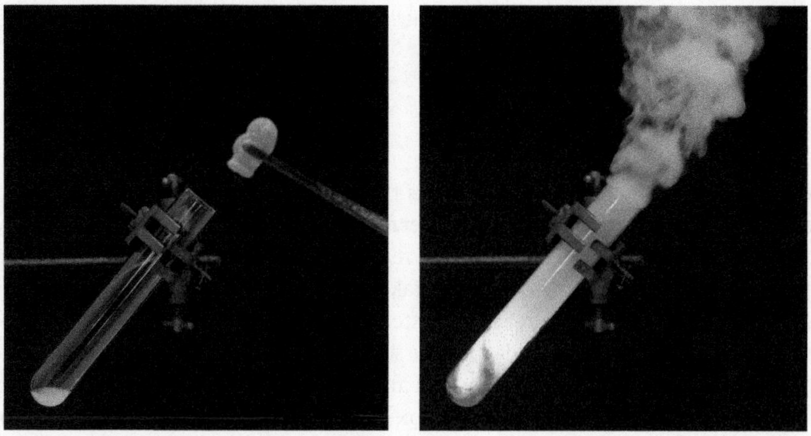

Figure 1.14 The screaming jelly baby. Potassium chlorate is heated strongly until it decomposes and gives off oxygen. When the jelly baby is added, the sugar it contains is oxidized to carbon dioxide and water in a violent reaction. Chemical energy is converted to heat, light, and sound, and work is done by the expanding gases pushing against the atmosphere. A similar, but less violent, reaction takes place in your body when you metabolize glucose.

Box 1.7 Potential energy and kinetic energy

The potential energy (E_{PE}) stored in an object can be measured by the energy needed to get it to its current position. This energy is equivalent to the work done in moving the object to its present position. Work is defined as the product of the force (F) and the distance (d) over which it operates, as shown in Equation 1.15

$$\text{work} = \text{force} \times \text{distance} = F \times d \quad (1.15)$$

Newton's second law of motion defines force as

$$\text{force } (F) = \text{mass} \times \text{acceleration} = m \times a$$

The SI unit of force is the newton. 1 N is the force that gives a mass of 1 kg an acceleration of 1 m s^{-2}.

Figure 1 Transfer of potential energy to kinetic energy during a bungee jump.

For the bungee jumper in Figure 1, the force acting is that of gravity—given from Newton's second law by the mass (m) of the person multiplied by the acceleration due to gravity, g = 9.81 m s^{-2} so that

$$F = m \times g$$

Combining this with Equation 1.15, the potential energy of a bungee jumper standing on the platform at a height h above the surface of the Earth is given by Equation 1.16.

$$\text{Potential energy} = E_{PE} = mgh \quad (1.16)$$

The kinetic energy (E_{KE}) of an object depends on its mass, m, and the speed, v, at which it is moving.

$$\text{Kinetic energy} = E_{KE} = \tfrac{1}{2}mv^2 \quad (1.17)$$

Later in this book, you will meet the terms potential energy and kinetic energy applied to atoms and molecules. The potential energy of an atom or molecule is related to its position. The forces acting arise from interactions between atoms in a molecule, or between atoms or ions in a network structure, or between molecules (see Section 17.3, p.783). The kinetic energy is related to motion, such as the random motion of molecules in gases (see Section 8.4, p.358). In Section 10.5 (p.476), there is a discussion of the interconversion of kinetic and potential energy during the vibration of a chemical bond.

Questions

(a) Calculate the minimum energy needed for a bungee jumper weighing 65.0 kg to climb from the ground to a platform 35.0 m high.

(b) Calculate the kinetic energy of a nitrogen molecule travelling at 500 m s^{-1}.

A reaction that gives out energy and heats the surroundings (the air, the reaction flask, the car engine, etc.) is described as **exothermic**. The reacting system loses energy; the products end up with less energy than the reactants had—but the surroundings end up with more and get hotter. A reaction that takes in energy and cools the surroundings is **endothermic**. In this case, the reacting system gains energy; the products end up with more energy than the reactants had—but the surroundings end up with less and get cooler.

The heat transferred between the reaction and the surroundings in an open container is the **enthalpy change** for the reaction (see Box 1.8). The enthalpy change for a reaction can be shown on an **enthalpy level diagram** (see Figures 1.15 and 1.16, which show enthalpy level diagrams for an exothermic reaction and for an endothermic reaction, respectively).

There is no way to measure the enthalpy, H, of a substance. All we can do is measure the *change* in enthalpy, ΔH, when a reaction occurs.

$$\text{Change in enthalpy} = \Delta H = H(\text{products}) - H(\text{reactants})$$

The relationship between heat and work is known as **thermodynamics** and is discussed in Chapters 13 and 14. Thermodynamics is fundamental to much of chemistry and biochemistry.

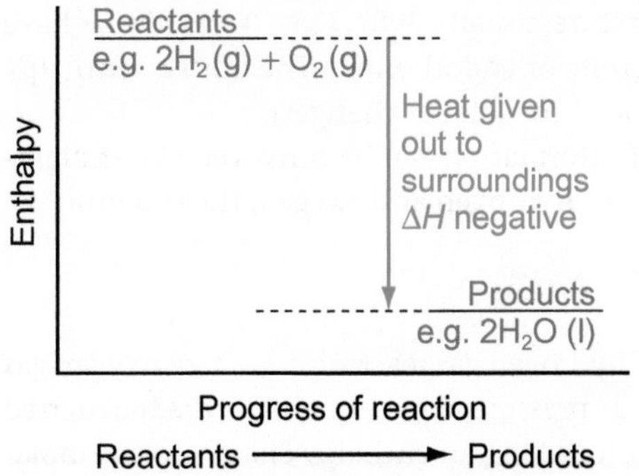
Reactants
e.g. 2H2 (g) + O2 (g)
Heat given out to surroundings
ΔH negative
Products
e.g. 2H2O (l)
Enthalpy
Progress of reaction
Reactants → Products

Figure 1.15 Enthalpy level diagram for an exothermic reaction, for example, the reaction of hydrogen and oxygen to form water: $2\,H_2(g) + O_2(g) \rightarrow 2\,H_2O(l)$.

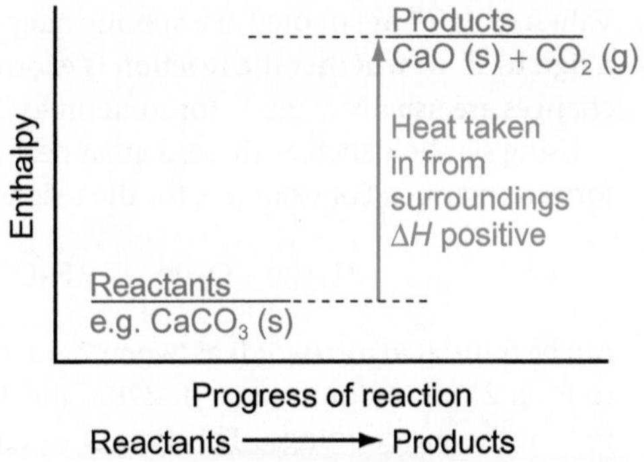
Products
CaO (s) + CO2 (g)
Heat taken in from surroundings
ΔH positive
Reactants
e.g. CaCO3 (s)
Enthalpy
Progress of reaction
Reactants → Products

Figure 1.16 Enthalpy level diagram for an endothermic reaction, for example, the decomposition of calcium carbonate: $CaCO_3(s) \rightarrow CaO(s) + CO_2(g)$.

The Greek symbol Δ
Δ (Delta) is used to mean 'a change in'. For any property or quantity, *X*,

$$\Delta X = X(\text{after the change}) - X(\text{before the change}) = X(\text{final}) - X(\text{initial})$$

Δ is commonly used in chemistry to represent a change in *X* during a chemical reaction, so

$$\Delta X = X(\text{products}) - X(\text{reactants})$$

Box 1.8 Enthalpy and internal energy

Enthalpy change is the heat transferred between a reaction and the surroundings at *constant pressure*. This applies to many reactions carried out in the laboratory in open containers. If you seal the container, so that the reaction takes place at *constant volume*, the heat transferred is called the **internal energy change.**

If the reaction involves a change in the number of moles of gas present, the internal energy change is different from the enthalpy change. Suppose, for example, that a gas is given off from an exothermic reaction taking place in an open container. The gas does work pushing back the atmosphere. Some of the chemical energy from the reaction is used to do work—the rest is converted to heat. If the same reaction takes place in a sealed container, no work is done against the atmosphere and all the chemical energy is converted to heat. You can find out more about enthalpy and internal energy in Section 13.5 (p.636).

For an exothermic change, the value of ΔH is *negative*. This is because, from the point of view of the chemical system, energy is *lost* to the surroundings. Conversely, for an endothermic change, the value of ΔH is *positive*, because the system *gains* energy from the surroundings.

Enthalpy change for a chemical reaction is denoted by the symbol $\Delta_r H$. For common types of reactions, other subscripts may be used, for example, $\Delta_c H$ for the enthalpy change of a combustion reaction (see Section 13.3, p.622).

You should get into the habit of *always* showing the sign when you write a ΔH value, whether it is positive or negative.

Standard conditions and thermochemical equations

The value of $\Delta_r H$ depends on the temperature and pressure at which the reaction occurs, as well as on the physical state of the components. For example, if water is a reactant or product, the value of $\Delta_r H$ depends on whether the water is liquid or steam.

A set of **standard conditions** (denoted by the superscript symbol $^{\ominus}$) is defined to allow comparisons between values. Standard conditions are:

- a pressure of 1 bar (1.00×10^5 Pa = 100 kPa);
- the reactant and products in their standard states (pure compound at 1 bar pressure or concentration for a solution of exactly 1 mol dm^{-3}).

The standard enthalpy change of reaction, $\Delta_r H^{\ominus}_{298}$, is the enthalpy change for a reaction at 1 bar and 298 K (25 °C) with all components in their standard states.

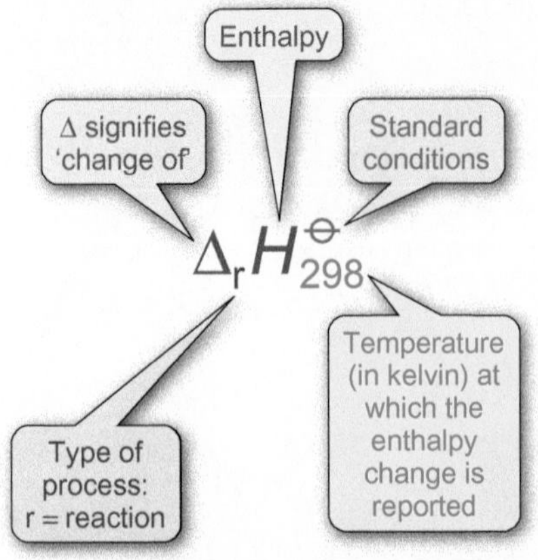
Enthalpy
Δ signifies 'change of'
Standard conditions
$\Delta_r H^{\ominus}_{298}$
Temperature (in kelvin) at which the enthalpy change is reported
Type of process: r = reaction

Values of $\Delta_r H^{\ominus}$ are quoted at a specific temperature, usually 298 K (25 °C) and *always* have a sign to show whether the reaction is exothermic or endothermic. The units of enthalpy changes are usually kJ mol^{-1} (or sometimes J mol^{-1} for smaller changes).

Using symbols such as these, a great deal of information can be conveyed in a straightforward manner. For example, for the reaction of hydrogen and oxygen, the equation

$$2H_2(g) + O_2(g) \rightarrow 2H_2O(l) \quad \Delta_r H^{\ominus}_{298} = -572\,\text{kJ mol}^{-1}$$

can be translated into words as 'when 2 mol of hydrogen gas react with 1 mol of oxygen gas to form 2 mol of liquid water at 298 K and 1 bar pressure, 572 kJ of energy are transferred to heat the surroundings'. This type of equation linking an enthalpy change to the molar amounts in a balanced equation is known as a **thermochemical equation.**

Note that the enthalpy change for the reverse reaction (under the same conditions of temperature and pressure) has the same magnitude but the opposite sign.

$$\Delta_r H(\text{forward reaction}) = -\Delta_r H(\text{backward reaction}) \qquad (1.18)$$

Note too that, in this context, the units kJ mol^{-1} mean kJ per molar amounts as stated in the equation. If you use 1 mol of hydrogen, rather than 2 mol, then $\frac{1}{2} \times 572$ kJ = 286 kJ of energy are transferred. So,

$$H_2(g) + \tfrac{1}{2}O_2(g) \rightarrow H_2O\,(l) \qquad {}_tH^{\ominus}_{298} = -286\,\text{kJ mol}^{-1}$$

For the thermal decomposition of calcium carbonate in Figure 1.16 (p.43)

$$CaCO_3(s) \rightarrow CaO(s) + CO_2(g) \qquad {}_tH^{\ominus}_{298} = -286\,\text{kJ mol}^{-1}$$

You can practise using a thermochemical equation to calculate the heat transferred in a reaction in Worked example 1.16 and in Box 1.9.

Worked example 1.16 Using thermochemical equations

A portable camp stove burns propane fuel. The thermochemical equation for the combustion of propane is

$$C_3H_8(g) + 5O_2(g) \rightarrow 3CO_2(g) + 4H_2O(l)$$
$$\Delta_f H^{\ominus}_{298} = -2220\,\text{kJ mol}^{-1}$$

How much heat energy is supplied when 500 g of propane burn in a plentiful supply of air so that combustion is complete?

Strategy

Use the thermochemical equation to write down the heat energy released when 1 mol of propane burns.

Use Equation 1.2 to work out the amount in moles of propane burned:

$$\text{amount (in mol)} = \frac{\text{mass (in g)}}{\text{molar mass (in g mol}^{-1})} \qquad (1.2)$$

Find the energy released when this amount of propane burns.

Solution

From the thermochemical equation, 2220 kJ of energy are supplied when 1 mol propane reacts with oxygen.

$M_r\,(C_3H_8) = 44.1$

$$500\,\text{g propane contain} = \frac{500\,\cancel{g}}{44.1\,\cancel{g}\,\text{mol}^{-1}} = 11.3\,\text{mol}$$

$$\text{Heat supplied by burning 500 g propane} = 2220\,\text{kJ}\,\cancel{\text{mol}^{-1}} \times 11.3\,\cancel{\text{mol}}$$
$$= 25\,200\,\text{kJ}$$

(Note that the answer does not have a sign because the direction of heat transfer is indicated by saying 'heat supplied'. When you quote a value for an *enthalpy change*, ΔH, however, *you must always include a sign.*)

Question

When 1 mol of pentane burns in a plentiful supply of air, 3537 kJ of energy are transferred as heat to the surroundings at 298 K. Write a balanced thermochemical equation for the reaction, and use it to work out the enthalpy change when 500 g of pentane are burned.

Box 1.9 Butane hair stylers

In cordless butane hair stylers, heat is generated by the oxidation of butane by air in the presence of a catalyst.

$$C_4H_{10}(g) + 6\tfrac{1}{2}O_2(g) \rightarrow 4CO_2(g) + 5H_2O(l)$$
$$\Delta_r H^{\ominus}_{298} = -2878\ \mathrm{kJ\,mol^{-1}}$$

The catalyst is often platinum powder, which is dispersed on an aluminium oxide support. The butane is supplied as a liquid in a small pressurized container. When the container is pierced, the butane gas released passes over the catalyst and oxidizes exothermically. When the hair styler is first switched on, the catalyst is heated by a battery, but the temperature quickly rises as the reaction takes place and no further external heating is needed.

Question

The label on a 'butane' refill for a hair styler states that it contains 14 g of 'isobutane'. The formula of 'isobutane' is C_4H_{10}.

(a) Assuming the fuel burns completely, according to the thermochemical equation above, calculate the heat energy supplied from 14 g of 'isobutane'.

(b) Suggest a structural formula (see Section 1.3, p.12) for 'isobutane' and give its systematic IUPAC name.

(You can read about structural isomers and naming hydrocarbons in Section 2.5, p.79. You will be able to calculate enthalpy changes under temperatures other than 298 K after studying Section 13.4, p.633.)

A cordless hair styler uses liquid butane as the fuel. ➤

Where does the energy come from?

Chemical reactions involve breaking and making chemical bonds. Bonds break in the reactants, the atoms rearrange, and new bonds form in the products. The energy changes in chemical reactions come from the energy changes when bonds are broken and made.

A chemical bond arises from an electrostatic attraction between the electrons and nuclei of the atoms or ions forming the bond. Breaking a bond requires an input of energy to overcome this attraction and separate the atoms or ions. When a bond is formed, energy is released.

- Bond breaking is an endothermic process.
- Bond formation is an exothermic process.

The difference in energy between the bond-breaking and bond-forming processes in a chemical reaction determines whether the overall change is exothermic or endothermic. Thus the reaction of hydrogen with oxygen is exothermic because more energy is released when the new bonds are formed in H_2O than was taken in to break the bonds in H_2 and O_2 (Figure 1.17)

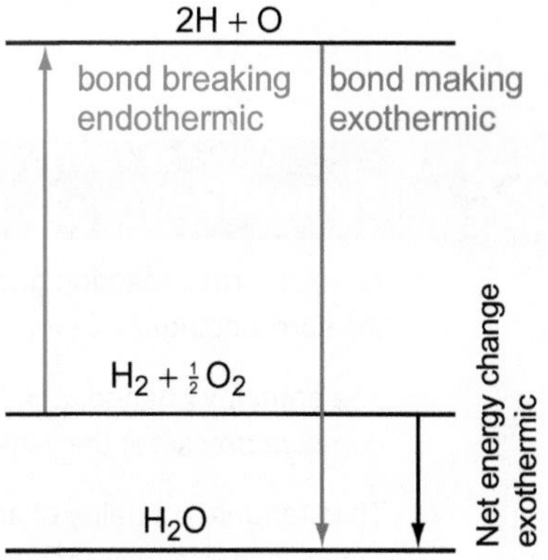

2H + O
bond breaking endothermic
bond making exothermic
H2 + ½O2
H2O
Net energy change exothermic

Figure 1.17 Bond breaking and bond making in the reaction of H_2 with O_2.

Enthalpy profiles for exothermic and endothermic processes

Figure 1.18 shows the **enthalpy profile** for an exothermic reaction. The **progress of reaction** as reactants are converted to products is plotted on the *x*-axis. Enthalpy is plotted on the *y*-axis. Figure 1.19 shows the enthalpy profile for an endothermic reaction.

Progress of reaction is sometimes called the **extent of reaction** or **the reaction coordinate**. See Sections 9.1 (p.384) and 9.7 (p.425).

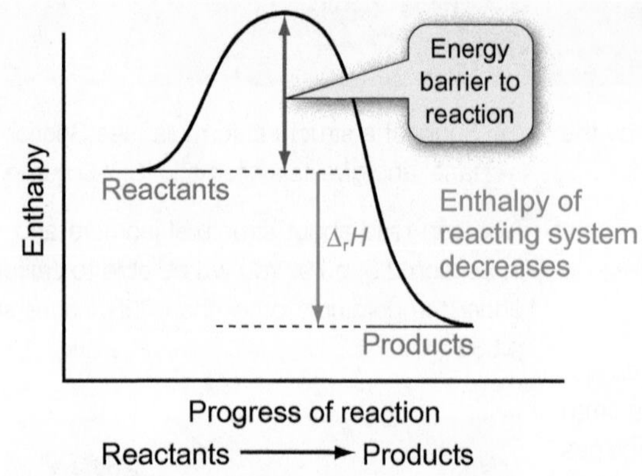
Energy barrier to reaction
Enthalpy
Reactants
$\Delta_r H$
Enthalpy of reacting system decreases
Products
Progress of reaction
Reactants ⟶ Products

Figure 1.18 Enthalpy profile for an exothermic reaction.

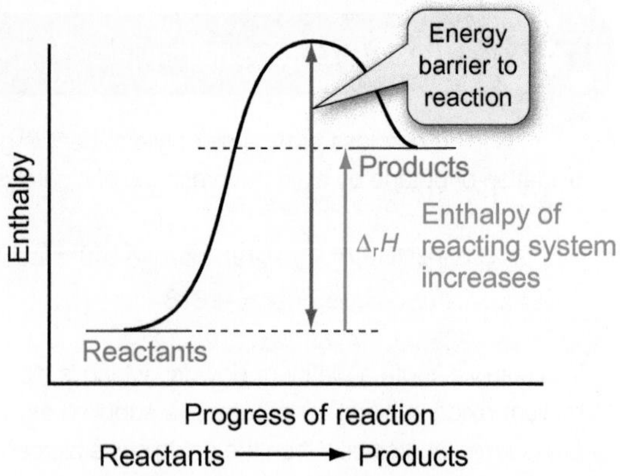
Energy barrier to reaction
Enthalpy
Products
$\Delta_r H$
Enthalpy of reacting system increases
Reactants
Progress of reaction
Reactants ⟶ Products

Figure 1.19 Enthalpy profile for an endothermic reaction.

The energy barrier is called the activation enthalpy. It arises because energy is needed to start the bond-breaking process. This is why some reactions need heating to get them started. When you use a match to ignite methane or a spark to ignite petrol vapour in a car engine, you are supplying the energy that is needed to break bonds. Once the reaction gets underway, the energy released from making the new bonds supplies energy for further bond breaking. For some reactions with a small energy barrier, there is enough energy in the surroundings at room temperature to get them started and no heating is necessary. Other reactions, such as the endothermic decomposition of calcium carbonate, need continuous heating.

The height of the energy barrier is related to the *rate* of the reaction. It is the energy that molecules must possess before they can react. The greater the proportion of molecules with this energy, the faster the reaction takes place. This is why raising the temperature increases the rate of reaction—it increases the proportion of molecules with enough energy to overcome the energy barrier.

Rates of reaction (kinetics) are discussed in Chapter 9. Section 9.7 (p.425) looks at energy barriers and the effect of temperature on the rate of a reaction.

Summary

- An exothermic reaction gives out energy and heats the surroundings; an endothermic reaction takes in energy and cools the surroundings.
- The enthalpy change, ΔH, for a reaction is the heat transferred between the reacting system and the surroundings *at constant pressure*; the value of ΔH is negative for an exothermic reaction and positive for an endothermic reaction.
- The standard enthalpy change of reaction, $\Delta_r H^{\circ}_{298}$, is the enthalpy change for a reaction at 1 bar and 298 K (25 °C) with all components in their standard states.
- A thermochemical equation links an enthalpy change to the molar amounts in a balanced equation for the reaction.
- Bond breaking is an endothermic process; bond formation is an exothermic process.

For a practice question on these topics, see question 18 at the end of this chapter (p.67).

1.7 States of matter and phase changes

At room temperature and pressure, oxygen is a gas, water is a liquid, and copper is a solid. The different physical states (gas, liquid, and solid) are called **states of matter**.

- A **solid** is a rigid form of matter. It has a shape and occupies a fixed volume (at a particular temperature and pressure).
- A **liquid** is a fluid form of matter. It occupies a fixed volume (at a particular temperature and pressure) but has no fixed shape. It has a well-defined, horizontal surface and, below this surface, it takes up the shape of the container.
- A **gas** is also a fluid form of matter. It spreads out to fill the space containing it and takes up the shape of the container.

Solids and liquids do not expand very much when you heat them, but gases expand a lot. Gases can be compressed by increasing the pressure, whereas liquids and solids cannot easily be compressed except at high pressures.

These *macroscopic* properties of the states of matter can be explained on a *molecular scale* in terms of the arrangement and motion of the molecules. This is called the **kinetic–molecular model** for the structure of matter and is summarized in Figure 1.20. Note that the term 'molecule' is used rather loosely here to mean the basic units making up the matter, which may be molecules, atoms, or ions.

A solid is rigid because the molecules cannot easily move past one another due to strong interactions between them. The molecules do vibrate, though, about fixed average positions. The vibrations become more vigorous as the temperature is raised, until the molecules have enough energy to move from their fixed positions and the solid melts to form a liquid. The molecules in a liquid are still very close together but they are able to move around in a restricted way. There is relatively little 'free space' in a liquid and collisions are very frequent. On further heating, the molecules gain more energy and move around faster. Eventually, they have enough energy to overcome completely the attractions of the other molecules in the liquid and escape from the surface to form a gas. (Remember that, when a molecular solid changes from solid → liquid → gas, it is only the intermolecular attractions *between* the molecules that are broken, not the covalent bonds *within* the molecules—see Section 1.8 (p.52).)

Fluids
The molecules in a fluid have enough energy to overcome the forces between molecules and move past one another so that the substance flows. A liquid flows under the influence of gravity from one container to another. A gas spreads out to fill the space containing it.

Gases and vapours
The term 'vapour' is often used for the gaseous state of a substance that is normally a solid or a liquid at room temperature. For example, chemists talk about water vapour and sodium vapour lamps, but oxygen gas and carbon dioxide gas. This definition is refined in Section 8.6 (p.372).

Solids and liquids
Structurally, liquids are much closer to solids than to gases. A typical liquid contains only about 3% more empty space than a close-packed solid. Solids and liquids are known as condensed phases.

The structure and properties of solids are discussed in Chapter 6. The behaviour of gases is covered in Chapter 8, and that of liquids and solutions in Chapter 17.

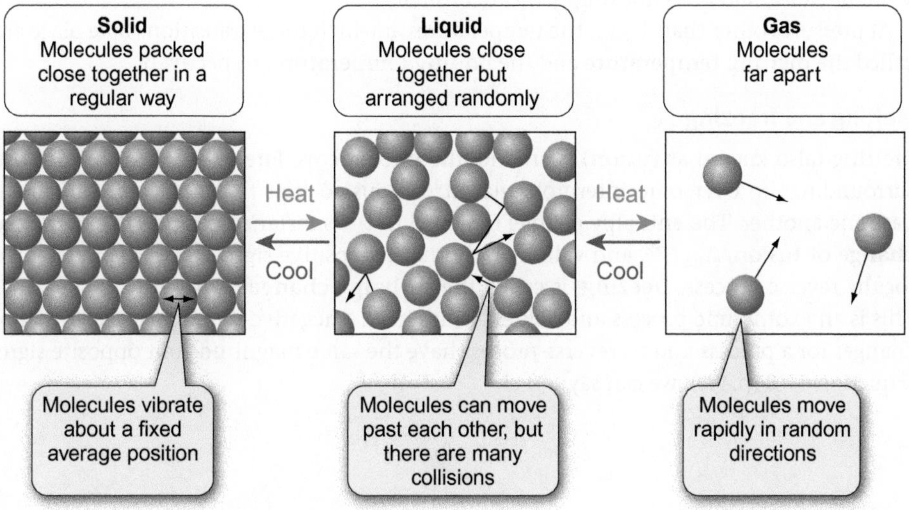
Solid
Molecules packed close together in a regular way
Liquid
Molecules close together but arranged randomly
Gas
Molecules far apart
Heat
Cool
Heat
Cool
Molecules vibrate about a fixed average position
Molecules can move past each other, but there are many collisions
Molecules move rapidly in random directions

Figure 1.20 Kinetic–molecular model for the structure of matter.

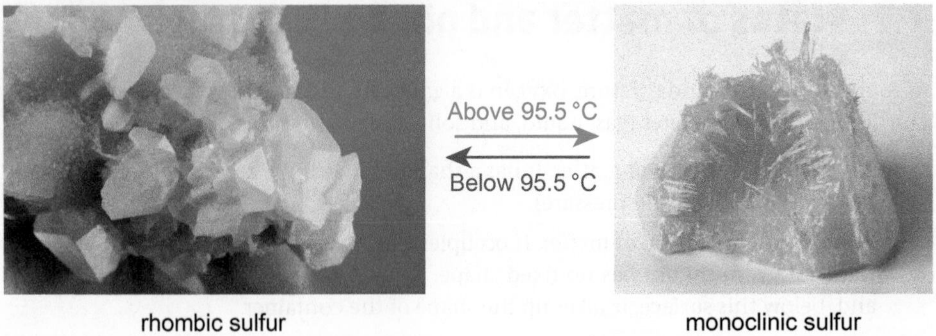
Above 95.5 °C
Below 95.5 °C
rhombic sulfur
monoclinic sulfur

Figure 1.21 The two solid phases (allotropes) of sulfur. The transition temperature is 95.5 °C at atmospheric pressure.

Phase changes

A **phase** is a form of matter that is uniform throughout, both in chemical composition and its physical state.

The three states of matter are different **phases**. A transition from one phase to another is called a **phase change** or **phase transition**. The melting of ice to form liquid water is an example of a phase change. Within the solid state, there may be more than one phase. Sulfur can exist as two different solid phases, rhombic and monoclinic (Figure 1.21). The sulfur molecules are packed differently in the two forms to give different crystal structures. At atmospheric pressure, rhombic sulfur changes into monoclinic sulfur at 95.5 °C. Below this **transition temperature**, rhombic sulfur is the more stable phase. Above the transition temperature, monoclinic sulfur is the more stable phase. In practice, the transition from one solid phase to the other has a high energy barrier (see Section 1.6, p.41) and is relatively slow.

In contrast, the transitions between the three states of matter usually take place quickly. At 1 atm pressure, the temperature at which:

- a solid changes into a liquid is called its **normal melting point (m.p.** or T_m);
- a liquid changes into a gas is called its **normal boiling point (b.p.** or T_b).

Values for m.p. and b.p. are usually quoted in °C rather than in kelvin; T_m and T_b are *always* quoted in kelvin. Note that 'normal' here has a precise scientific sense, meaning 1 atm pressure, and does not simply mean 'ordinary'.

At pressures other than 1 atm, the temperatures at which these transitions take place are called the **melting temperature** and the **boiling temperature**, respectively.

Melting and freezing

The $^{\ominus}$ superscript signifies that these are standard enthalpy changes; see Section 1.6 (p.41).

Melting (also known as **fusion**) is an endothermic process. Energy is taken in from the surroundings to overcome intermolecular attractions so that the molecules can move past one another. The enthalpy change for 1 mol of the substance is called the **enthalpy change of fusion**, $\Delta_{fus}H^{\ominus}$, and values always have a positive sign. The enthalpy change for the reverse process, **freezing**, is called the **enthalpy change of freezing**, $\Delta_{freezing}H^{\ominus}$. This is an exothermic process and values always have a negative sign. Since the enthalpy changes for a process and its reverse process have the same magnitude but opposite signs (Equation 1.18 (p.44)), we can say:

$$\Delta_{fus}H^{\ominus} = -\Delta_{freezing}H^{\ominus}$$

Vaporization and condensation

Vaporization is an endothermic process taking in energy from the surroundings. Energy is needed to overcome intermolecular attractions in the liquid and so separate the molecules. The enthalpy change for 1 mol of the substance is called the **enthalpy change of vaporization**, $\Delta_{vap}H^{\ominus}$, and values always have a positive sign. The enthalpy change for the reverse process, condensation, is called the **enthalpy change of condensation**, $\Delta_{cond}H^{\ominus}$, and values always have a negative sign. When a gas condenses to form a liquid, intermolecular attractions are re-formed. So, condensation is an exothermic process and heat is released into the surroundings.

$$\Delta_{vap}H^{\ominus} = -\Delta_{cond}H^{\ominus}$$

Evaporation and **boiling** both involve molecules in the liquid overcoming the attraction of neighbouring molecules and escaping into the vapour phase. However, there are differences between the two processes. Evaporation happens only at the surface of the liquid. It can take place at any temperature and is relatively slow. At the same time, molecules in the vapour are colliding with the liquid surface and condensing. After a time, if the liquid is in a sealed container, evaporation and condensation take place at the same rate and the system reaches equilibrium (see Section 1.9). The **vapour pressure** of a liquid is the pressure of the vapour in equilibrium with the liquid at a particular temperature.

A liquid boils when its vapour pressure is equal to the external pressure above the liquid. Boiling is a much more rapid process than evaporation and takes place at a fixed temperature (for a given pressure). Bubbles of gas form in the body of the liquid and molecules escape from inside the liquid as well as from the surface.

Figure 1.22 Iodine crystals sublime on heating to give purple iodine vapour. Reverse sublimation takes place when the vapour cools and iodine crystals are deposited around the mouth of the tube.

Sublimation and reverse sublimation

When crystals of iodine are heated *gently*, purple iodine vapour is given off from the solid (Figure 1.22). No liquid iodine is formed. As the vapour reaches the cooler part of the tube, solid iodine is deposited. The conversion of a solid directly to vapour is called **sublimation**. The enthalpy change for 1 mol of the substance is called the **enthalpy change of sublimation**, $\Delta_{sub}H^{\ominus}$, and values always have a positive sign. The reverse process is known as reverse sublimation and the enthalpy change is called the **enthalpy change of reverse sublimation**, $\Delta_{reverse\ sub}H^{\ominus}$, and values always have a negative sign.

Reverse sublimation is sometimes called **solid deposition** or **vapour deposition**.

$$\Delta_{sub}H^{\ominus} = -\Delta_{reverse\ sub}H^{\ominus}$$

The conditions of temperature and pressure under which two phases exist in equilibrium can be plotted on a graph to give a **phase diagram**. Phase diagrams are discussed in Section 17.1 (p.766). You can find more about enthalpy changes of state in Section 13.2 (p.619).

Figure 1.23 shows the transitions between the three states of matter and the **enthalpy changes of state** involved. The phase changes of water and the enthalpy changes involved are discussed in Box 1.10.

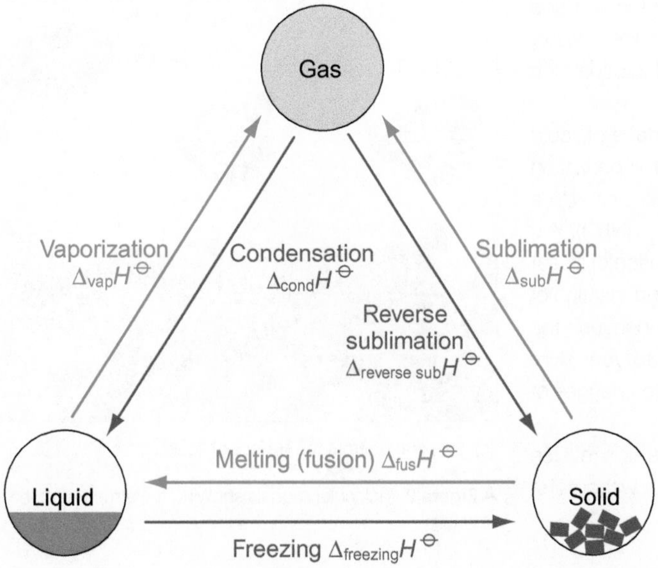
Gas
Vaporization
$\Delta_{vap}H^{\ominus}$
Condensation
$\Delta_{cond}H^{\ominus}$
Sublimation
$\Delta_{sub}H^{\ominus}$
Reverse sublimation
$\Delta_{reverse\ sub}H^{\ominus}$
Melting (fusion) $\Delta_{fus}H^{\ominus}$
Liquid
Solid
Freezing $\Delta_{freezing}H^{\ominus}$

Figure 1.23 Transitions between the three states of matter and the enthalpy changes involved. (Red arrows represent endothermic phase changes and blue arrows exothermic phase changes.)

Box 1.10 Phase changes of water

The phase changes in the water cycle are central to life on Earth. They have a controlling influence on the climate—the formation of dew, rain, snow, and frost and the evaporation of water from rivers and the oceans are all phase changes of water. The enthalpy changes of state accompanying these changes affect the temperature of the atmosphere and the surface of the Earth and, together with ocean currents, are responsible for moving huge quantities of energy around the planet. The Gulf Stream, for example, carries warm water from the Caribbean north-eastwards and warms the seas to the west of Great Britain and Ireland.

The phase changes are accompanied by changes in the structure of water. In ice, each water molecule forms hydrogen bonds to four other water molecules (Figure 1). (Hydrogen bonds are discussed in Section 1.8, p.52.)

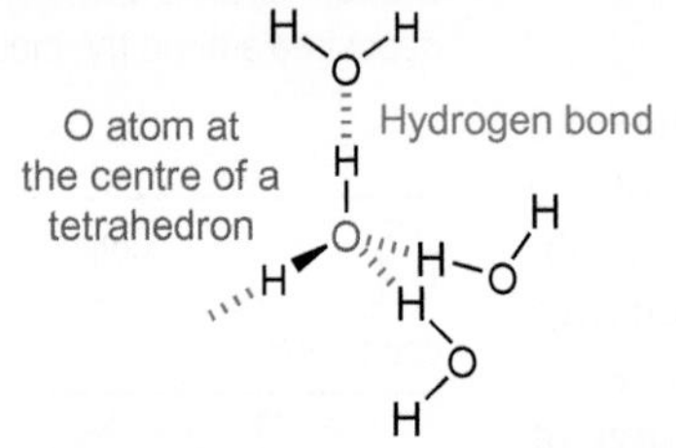
O atom at
the centre of a
tetrahedron
Hydrogen bond

Figure 1 Hydrogen bonding in ice.

The result is the open, regular structure shown in Figure 2(a). For ice to melt, the molecules must have enough energy to overcome the attraction of the hydrogen bonds and break away from their neighbours. Hydrogen bonding is still present in liquid water (Figure 2(b)) but it is more transient—the molecules in liquid water are tumbling past each other and hydrogen bonds are constantly forming and breaking. At 0 °C, the molecules are closer together in liquid water than in ice, so the water is denser than the ice—and ice floats on water.

Figure 2(c) shows molecules escaping from the surface of liquid water to form water vapour. The enthalpy change of vaporization of water is very high (compared to the values for molecules with a similar molar mass, such as methane (CH_4) and ammonia (NH_3)). You appreciate this when you sweat in hot weather. Evaporation of sweat produces a cooling effect on your skin. Conversely, condensation of water is an exothermic process. Steam scalds badly because the energy released as it condenses to water is passed to your skin. Figure 3 shows an enthalpy level diagram for the phase changes of water.

The formation of frost directly from water vapour in the air is reverse sublimation (Figure 4). When the frost vanishes as vapour without first melting, it is by sublimation.

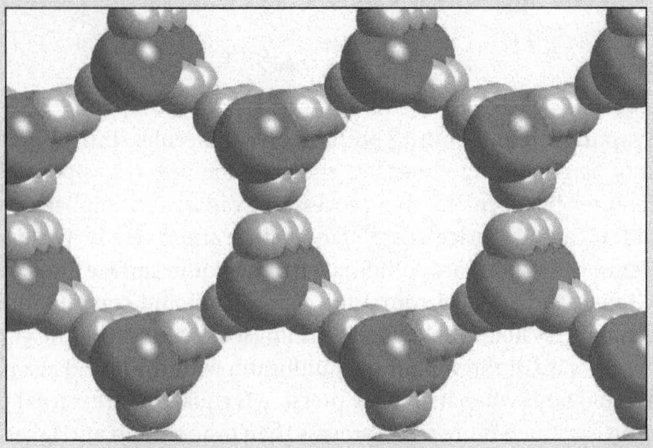

(a) The open structure of ice

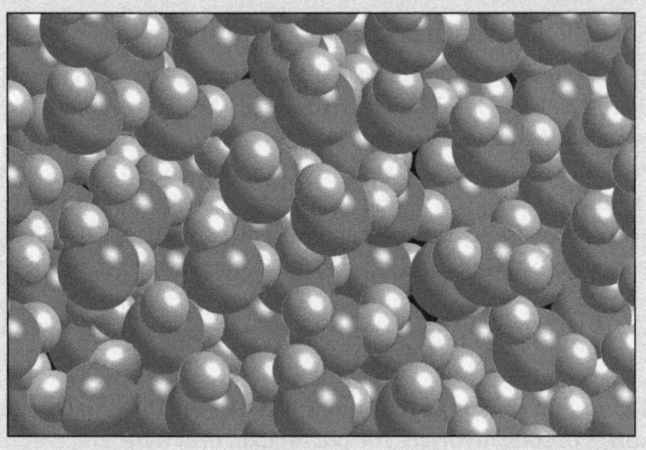

(b) The denser structure of liquid water

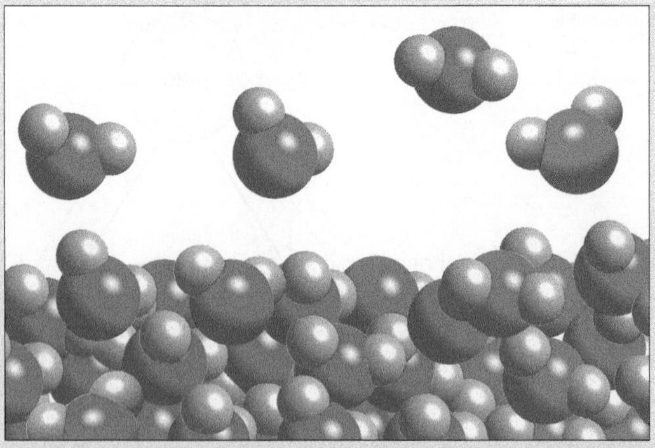

(c) Evaporation from the surface of liquid water

Figure 2 Molecular models showing the structure of the three phases of water.

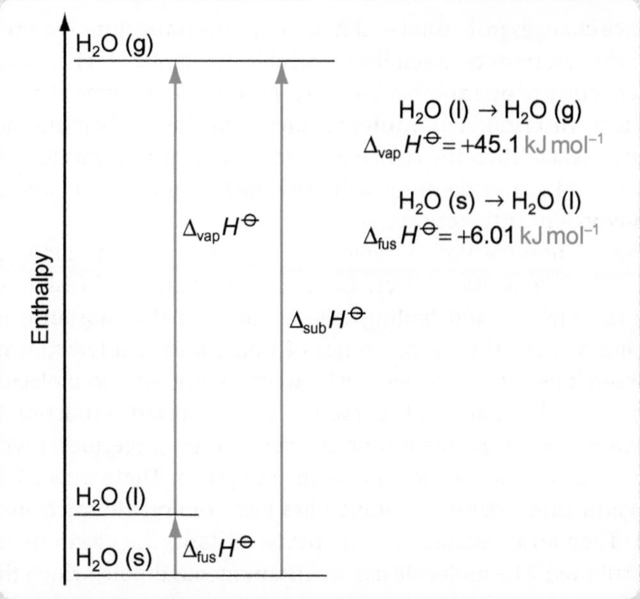
Enthalpy
H_2O (g)
H_2O (l) → H_2O (g)
$\Delta_{vap}H^{\ominus}$ = +45.1 kJ mol^{-1}
H_2O (s) → H_2O (l)
$\Delta_{fus}H^{\ominus}$ = +6.01 kJ mol^{-1}
$\Delta_{vap}H^{\ominus}$
$\Delta_{sub}H^{\ominus}$
H_2O (l)
H_2O (s)
$\Delta_{fus}H^{\ominus}$

▲ **Figure 3** Enthalpy level diagram for the phase changes of water. (The values have been adjusted to 0 °C so only enthalpy changes associated with phase changes are involved.)

▲ **Figure 4** Light, feathery hoar frost forms by reverse sublimation of water vapour from cold, relatively dry air. Frost formed from the freezing of dew (liquid water) is denser and more icy.

Questions

(a) In warm countries, water can be kept cool by storing it in unglazed earthenware pots. Explain the process involved.

(b) Suggest why wet clothes dry on a washing line but not in a plastic bag.

(c) Using the data in Figure 3, write down values at 0 °C for:
(i) $\Delta_{sub}H^{\ominus}(H_2O)$; (ii) $\Delta_{cond}H^{\ominus}(H_2O)$; (iii) $\Delta_{freezing}H^{\ominus}(H_2O)$.

(d) Suggest why the value of $\Delta_{vap}H^{\ominus}(H_2O)$ is very much more endothermic than the value of $\Delta_{fus}H^{\ominus}(H_2O)$.

Summary

- There are three common states of matter: solid; liquid; and gas.
- The three states of matter are phases. Within the solid state, there may be more than one phase.
- The kinetic–molecular model for the structure of matter explains macroscopic properties in terms of the arrangement and motion of the molecules.
- Transitions between states of matter are called phase changes.
- The enthalpy changes accompanying phase changes are called enthalpy changes of state. Values refer to 1 mol of the substance.
- Melting (fusion), vaporization, and sublimation all involve the breaking of intermolecular attractions and are endothermic processes. The reverse processes—freezing, condensation, and reverse sublimation, respectively—involve the formation of intermolecular interactions and are exothermic processes.
- $\Delta_{fus}H^{\ominus} = -\Delta_{freezing}H^{\ominus}$
  $\Delta_{vap}H^{\ominus} = -\Delta_{cond}H^{\ominus}$
  $\Delta_{sub}H^{\ominus} = -\Delta_{reverse\ sub}H^{\ominus}$

For a practice question on Sections 1.7 and 1.8, see question 19 at the end of this chapter (p.67).

1.8 Non-covalent interactions

When a molecular substance changes from solid → liquid → gas, the transitions require an input of energy to overcome attractions between the molecules. The molecules in a gas are far apart and move independently of one another but, even here, there are some attractive forces between the molecules. **Intermolecular interactions** (sometimes called intermolecular forces) are very much weaker than the covalent bonds holding the atoms together within the molecule (Figure 1.24). They do not involve the sharing of electron pairs and are sometimes called **non-covalent interactions.**

There are different types of non-covalent interactions depending on the molecules involved. The strength of the interactions affects the enthalpy changes of fusion and vaporization (and hence the melting and boiling temperatures) of the substance (see Section 1.7). Table 1.10 summarizes the various types of non-covalent interactions. It includes interactions between ions and molecules and between ionic groups in molecules.

The attraction between molecules is always the result of an electrostatic attraction between opposite charges. One type of interaction occurs between *all* molecules, including the atoms in monatomic gases, such as helium, neon, and argon. These interactions are called **London dispersion interactions** (or sometimes just **London interactions** or **dispersion interactions**). They arise because, at a particular instant, the electrons in a molecule are unevenly distributed. The molecule has an instantaneous dipole, which then induces a dipole in an adjacent molecule. (A longer, but more descriptive, name for these interactions is **instantaneous dipole–induced dipole interactions.**) Figure 1.25 shows how these interactions arise between two isolated neon atoms—but you can draw similar diagrams for any pair of atoms or molecules.

When a molecule has a permanent dipole, adjacent molecules align themselves to maximize the attractions between positive and negative charges. Such **dipole–dipole interactions** are more marked in the solid state where molecules have fixed positions (Figure 1.26) but are also present in liquids and gases. In addition, there will be London dispersion forces between the molecules. The total attraction between two molecules is the sum of the different types of attractions possible.

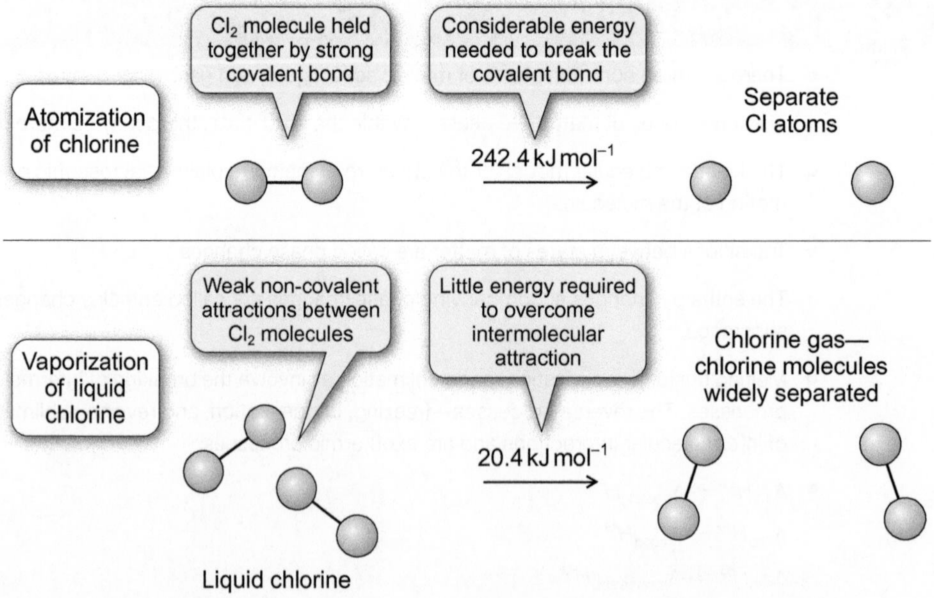
Atomization of chlorine
Cl_2 molecule held together by strong covalent bond
Considerable energy needed to break the covalent bond
242.4 kJ mol^{-1}
Separate Cl atoms
Vaporization of liquid chlorine
Weak non-covalent attractions between Cl_2 molecules
Little energy required to overcome intermolecular attraction
20.4 kJ mol^{-1}
Chlorine gas—chlorine molecules widely separated
Liquid chlorine

Figure 1.24 Covalent bonds within a molecule are much stronger than the attractions between molecules.

Non-covalent interactions This term is normally used to include all attractions other than covalent, ionic, and metallic bonding. When the attraction is between two molecules, it is called an **intermolecular interaction**.

Previously, you may have used the term **van der Waals interactions** (or **van der Waals forces**) to describe London dispersion interactions. You may also see this term in some textbooks. However, the van der Waals term is also used to include all the different types of interactions and is best avoided; see Sections 8.6 (p.372) and 17.3 (p.783).

Polar molecules and molecular dipoles
A polar molecule is one that has a permanent separation of charge and, hence, a dipole moment. For example, the molecule HCl has a dipole moment because the Cl atom is more electronegative than the H atom and the covalent bond is polarized.

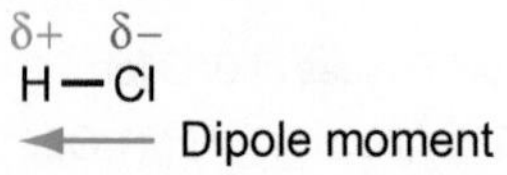
δ+ δ−
H—Cl
Dipole moment

Similarly, the C=O bond in methanal, HCHO, is polarized and the methanal molecule has a dipole moment.

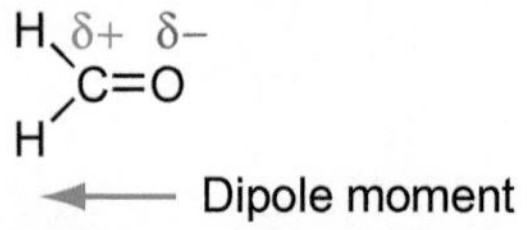
δ+ δ−
C=O
Dipole moment

Electronegativity is discussed in Section 4.3 (p.177). There is more about polar molecules and molecular dipoles in Section 5.3 (p.235). Polarization in organic molecules is discussed in Section 19.1 (p.863).

Table 1.10 Types of non-covalent interactions. For comparison, the energies of covalent bonds lie in the range 150 kJ mol^{-1} to 1000 kJ mol^{-1}.

Interaction	Acts between	For example	Typical energy / kJ mol^{-1}
London dispersion	All types of molecules	Ne, H_2, CH_4, Cl_2, HCl	~ 5 (in gas phase)
Dipole–dipole	Polar molecules	HCl	2 (in gas phase)
Dipole–induced dipole	A polar molecule and a molecule that may or may not be polar	Between HCl and Cl_2 or between HCl and CH_4	< 2 (in gas phase)
Ion–dipole	An ion and a polar molecule	A metal ion dissolved in water	15
Hydrogen bond	A molecule containing the electronegative atoms O, N, or F bound to H and a molecule containing O, N, or F. The H atom forms a link between the two electronegative atoms	H_2O, NH_3, HF	10–40*
Ion–ion	Ionic groups	Between $-NH_3^+$ and $-CO_2^-$ groups in proteins	250

* The hydrogen bond in the ion, $[HF_2]^-$ is particularly strong, 165 kJ mol^{-1}; see Section 25.3 (p.1156).

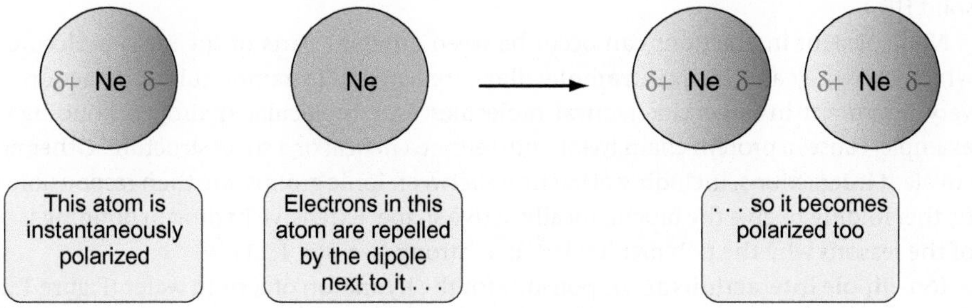
δ+ Ne δ−
Ne
δ+ Ne δ−
δ+ Ne δ−
This atom is instantaneously polarized
Electrons in this atom are repelled by the dipole next to it . . .
. . . so it becomes polarized too

Figure 1.25 London dispersion interactions between neon atoms.

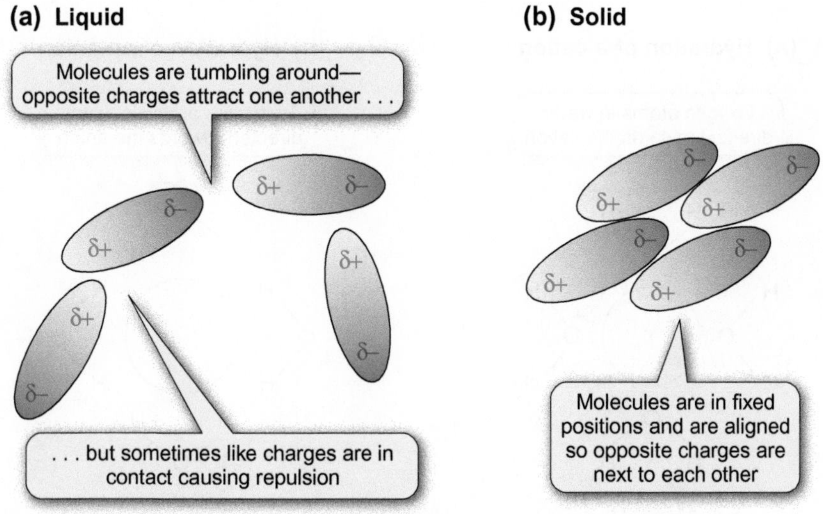
(a) Liquid
Molecules are tumbling around—opposite charges attract one another . . .
δ+
δ−
. . . but sometimes like charges are in contact causing repulsion
(b) Solid
Molecules are in fixed positions and are aligned so opposite charges are next to each other

Figure 1.26 Dipole–dipole interactions between polar molecules in (a) the liquid state and (b) the solid state.

Hydrogen bonding is discussed in more detail in Section 25.3 (p.1156) where the definition is broadened to include examples of hydrogen bonding involving atoms other than O, N, and F.

Intra means within a single molecule; **inter** means between separate molecules.

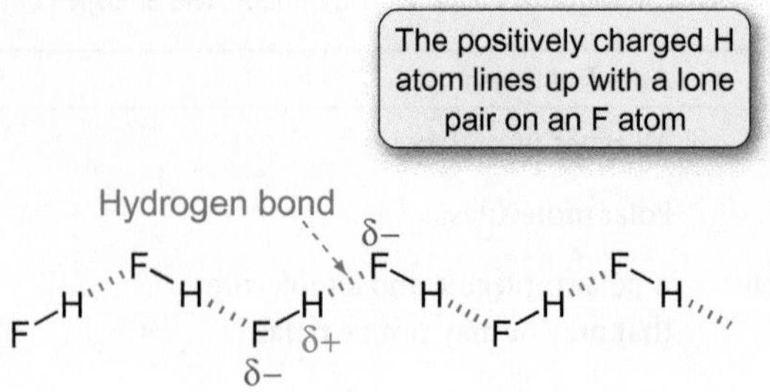
The positively charged H atom lines up with a lone pair on an F atom
Hydrogen bond
δ−
δ+
δ−

Figure 1.27 Hydrogen bonding in HF.

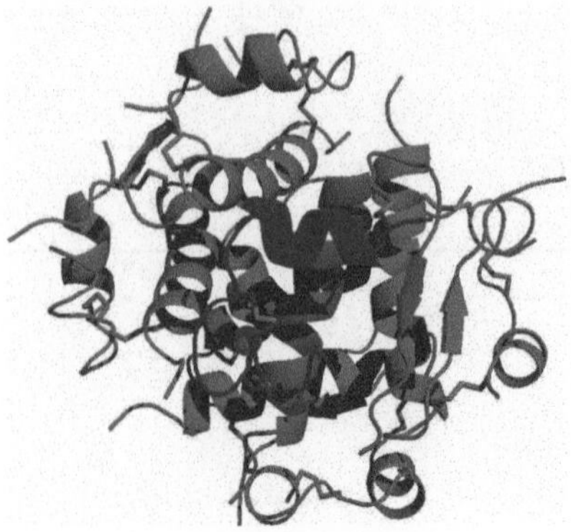

Ribbon structure of the monomer of the protein, insulin. Non-covalent interactions cause protein chains to fold into the biochemically active shape.

A molecule with a permanent dipole, such as HCl, can *induce* a dipole in an adjacent nonpolar molecule. The positive end of the permanent dipole attracts electrons in the non-polar molecule, making it temporarily polar. The result is called a **dipole–induced dipole interaction**. This type of interaction is important in mixtures of molecules (e.g. between HCl and Cl_2 molecules).

Hydrogen bonding is an extreme form of dipole–dipole interaction. A hydrogen atom attached to an electronegative atom, O, N, or F, is attracted to an electronegative atom on an adjacent molecule. The hydrogen atom carries a partial positive charge and forms a link between the two electronegative atoms. Figure 1.27 shows hydrogen bonding in solid HF.

Non-covalent interactions can occur between different parts of the same molecule, in which case they are called **intramolecular** interactions. Intramolecular interactions are very important in large biochemical molecules. Intramolecular hydrogen bonding, for example, causes a protein chain to fold into either a helical or a sheet structure. Other non-covalent interactions, including attractions between ionic groups, are then responsible for further folding to give the biochemically active shape. Extensive hydrogen bonding is one of the reasons why the polymer Kevlar® is so strong (see Box 1.11).

Ion–dipole interactions are responsible for the hydration of ions in water (Figure 1.28). The attachment of water molecules to a central ion is known as **hydration**. The interactions are particularly strong for small ions with high charges, such as Al^{3+} ions, which are strongly hydrated in water. Other polar solvents behave in a similar way. The general process is called **solvation**.

You can find more about intermolecular interactions in Sections 8.6 (p.372 gases), 17.3 (p.783, liquids and solutions), and 6.1 (p.257, solids). Hydrogen bonding is discussed in detail in Section 25.3 (p.1156). Hydrogen bonding governs the structure and reactions of many organic compounds, for example, see Section 23.3 (p.1082).

The hydration of ions affects the behaviour of ions in aqueous solution, for example; see Section 16.2 (p.730). Solvation effects in organic chemistry are discussed in Chapter 19 (pp.881 and 889). Energy changes on solvation are described in Section 17.4 (p.789). The hydration of Group 1 ions is discussed in Section 26.3 (p.1180).

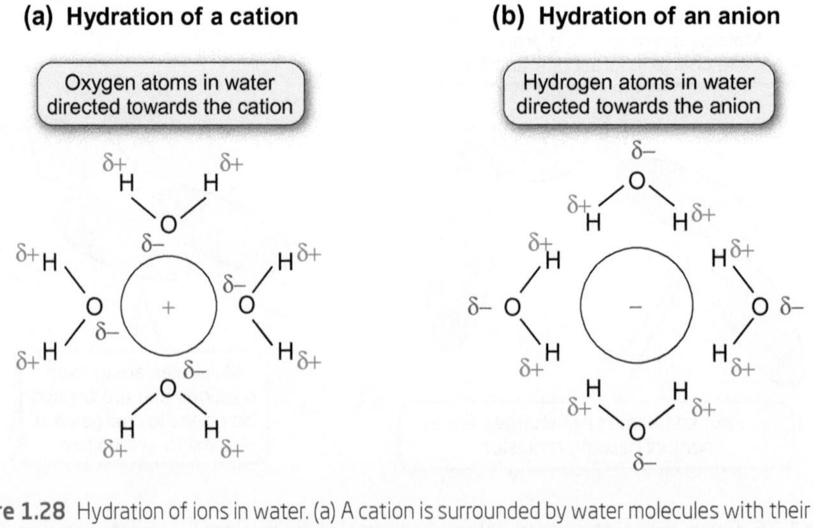
(a) Hydration of a cation
Oxygen atoms in water directed towards the cation
(b) Hydration of an anion
Hydrogen atoms in water directed towards the anion

Figure 1.28 Hydration of ions in water. (a) A cation is surrounded by water molecules with their oxygen atoms directed towards the ion. (b) An anion is surrounded by water molecules with hydrogen atoms directed toward the ion. (*Both* hydrogen atoms in each water molecule will not always be directed towards the ion.)

Box 1.11 Why is Kevlar® so strong?

The polymer Kevlar® was first made in 1965 by Stephanie Kwolek while working in the DuPont Laboratories in the USA. Since then uses for the polymer have multiplied and it is now used in a wide variety of applications ranging from bulletproof vests to tennis racquets and other sporting equipment.

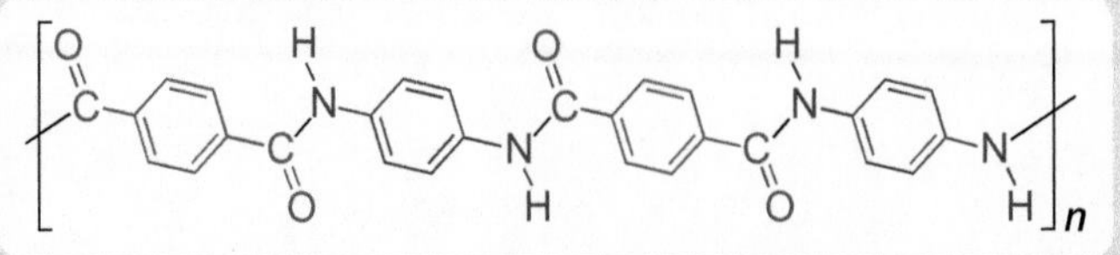

▲ Kevlar®

▲ The combination of light weight and strength in Kevlar® means that it is used for sporting equipment.

The polymer is lightweight, flexible, and resistant to heat, fire, and chemicals. Ropes and cables made from Kevlar® are five times stronger than steel on an equal weight basis, so small diameter ropes can be used to moor large ships. The secret of Kevlar®'s remarkable properties lies in its highly ordered, crystalline structure.

There are three main reasons for the strength of Kevlar®. First, the benzene rings, linked by amide groups, give each polymer chain a rigidity due to limited bond rotation. Second, strong hydrogen bonding *between* the chains fixes them into position and prevents individual chains from slipping past one another. Finally, when Kevlar® is spun into fibres, the polymer chains orientate themselves along the fibre axis with hydrogen bonds linking adjacent chains. The flat sheets of linked chains are stacked along the radii giving a highly ordered, crystalline structure (Figure 1). The polymer is processed in the liquid crystal phase (see Box 17.1, p.773)—the result is an extremely strong fibre.

Question

In terms of intermolecular attractions, suggest why it is possible to make strong ropes out of Kevlar® or nylon, but not from polythene (poly(ethene)).

▼ Figure 1 The crystalline structure of Kevlar®

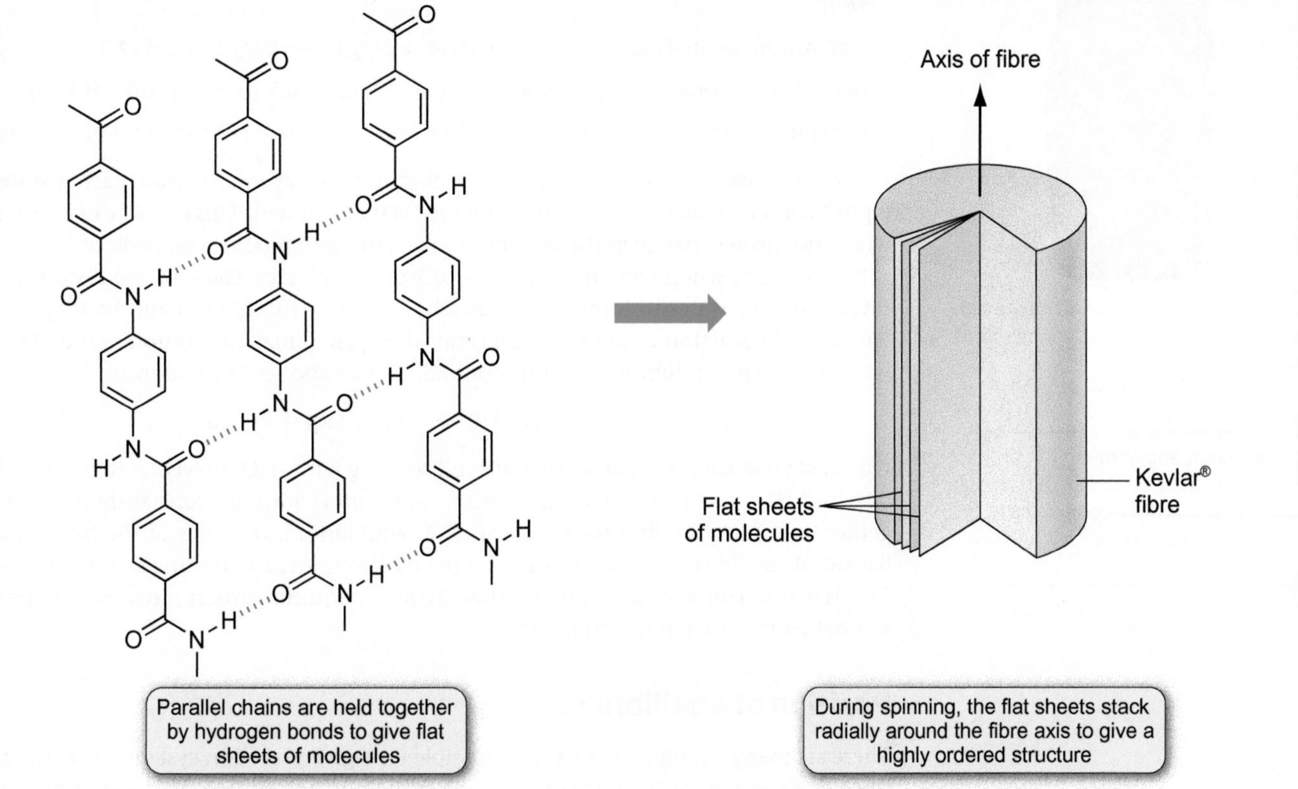
Axis of fibre
Flat sheets of molecules
Kevlar® fibre
Parallel chains are held together by hydrogen bonds to give flat sheets of molecules
During spinning, the flat sheets stack radially around the fibre axis to give a highly ordered structure

Summary

- Non-covalent interactions include interactions between molecules, between ions and molecules, and between ionic groups within molecules.
- London dispersion interactions are present in all molecules and in monatomic gases. They arise from the electrostatic attraction between an instantaneous dipole in a molecule and an induced dipole in an adjacent molecule.
- Other types of interactions include:
  - dipole–dipole interactions;
  - dipole–induced dipole interactions;
  - ion–dipole interactions;
  - hydrogen bonds;
  - ion–ion interactions.
- These types of interactions are summarized in Table 1.10 (p.53).

? For a practice question on Sections 1.7 and 1.8, see question 19 at the end of this chapter (p.67).

1.9 Chemical equilibrium: how far has a reaction gone?

Dynamic equilibrium

perrier

A dynamic equilibrium is established between carbon dioxide and water in a stoppered bottle of sparkling water.

Some reactions *go to completion* and are represented by a single arrow in the chemical equation. This tells you that virtually all the reactants are converted into products. For example,

combustion of methane	$CH_4\,(g) + 2\,O_2\,(g) \rightarrow CO_2\,(g) + 2\,H_2O\,(l)$
neutralization of an acid by a base	$HCl\,(aq) + NaOH\,(aq) \rightarrow NaCl\,(aq) + H_2O\,(l)$
precipitation of silver chloride	$AgNO_3\,(aq) + NaCl\,(aq) \rightarrow AgCl\,(s) + NaNO_3\,(aq)$

Many other reactions do not go to completion in this way, but instead reach a **state of equilibrium** in which both reactants and products are present. This is a state of balance in which no further change in the amounts of reactants and products takes place.

For example, when carbon dioxide gas dissolves in water, some of the dissolved gas reacts with the water to form hydrogencarbonate ions ($HCO_3^-\,(aq)$) and hydrogen ions ($H^+\,(aq)$). The reaction comes to equilibrium when just a small proportion of the $CO_2\,(aq)$ has reacted. The equilibrium mixture contains all four species in solution.

$$CO_2\,(aq) + H_2O\,(l) \rightleftharpoons HCO_3^-\,(aq) + H^+\,(aq)$$

On a *molecular scale*, CO_2 molecules are still reacting with H_2O molecules to form HCO_3^- ions and H^+ ions. At the same time, HCO_3^- ions and H^+ ions are constantly reacting together to reform CO_2 and H_2O molecules. At equilibrium, the rates of the forward and backward reactions are equal, so that, on the *macroscopic scale*, nothing *seems* to happen. This is why chemical equilibrium is called **dynamic equilibrium**. It is represented by the symbol $\rightleftharpoons$ in the chemical equation.

> **i Dynamic equilibrium**
> The amounts of reactants and products at equilibrium are constant because the forward reaction is taking place at the same rate as the reverse reaction.

Position of equilibrium

There are many equilibrium mixtures possible for a given reaction system, depending on the concentrations of the substances mixed at the start and the reaction conditions, such as temperature and pressure. Chemists often talk about the **position of equilibrium**. This means one particular set of equilibrium concentrations for the reaction. If the conditions, or one of the concentrations, are changed, the system is no longer in equilibrium and the concentrations of reactants and products change until a new position of equilibrium is reached. If this change results in the formation of more products, you say that the position of equilibrium *moves to the right* (meaning towards the products on the right-hand side of the equation). Conversely, if the change results in the formation of more reactants, you say that the position of equilibrium *moves to the left* (meaning towards the left-hand side of the equation).

The position of equilibrium can be altered by changing the:

- concentrations of reacting substances (in solution);
- pressures of reacting gases;
- temperature.

Henri Le Chatelier studied data from many equilibrium reactions and, in 1888, proposed a series of rules enabling him to make qualitative predictions about the effects of changes on an equilibrium. The rules are summarized in **Le Chatelier's principle** which states that:

when an external change is made to a system in dynamic equilibrium, the system responds to minimize the effect of the change.

Le Chatelier's principle is a useful way of *remembering* how an equilibrium system responds to external changes, but it is does not *explain* these changes. For explanations of the way equilibrium position changes, see Section 15.5 (page 713).

Applying Le Chatelier's principle

Changing the concentration

Yellow iron(III) ions (Fe^{3+}(aq)) react with colourless thiocyanate ions (SCN^-(aq)) to form deep red $[Fe(SCN)]^{2+}$(aq) ions. The reaction is reversible and an equilibrium is set up

$$\underset{\text{yellow}}{Fe^{3+}(aq)} + \underset{\text{colourless}}{SCN^-(aq)} \rightleftharpoons \underset{\text{deep red}}{[FeSCN]^{2+}(aq)}$$

The intensity of the red colour of the solution is a good indication of the concentration of $[Fe(SCN)]^{2+}$(aq) in the mixture. If the concentration of either Fe^{3+}(aq) or SCN^-(aq) is increased, the solution goes a darker red, showing the position of equilibrium has moved to the right. This is in accordance with Le Chatelier's principle because moving to the right reduces the concentrations of the reactants and minimizes the effect of the imposed change. The concentration of Fe^{3+}(aq) can be reduced by adding ammonium chloride, as chloride ions react with Fe^{3+}(aq) to form $[FeCl_4]^-$(aq) ions. If you do this, the red colour of the solution becomes paler, indicating that the position of equilibrium has moved to the left.

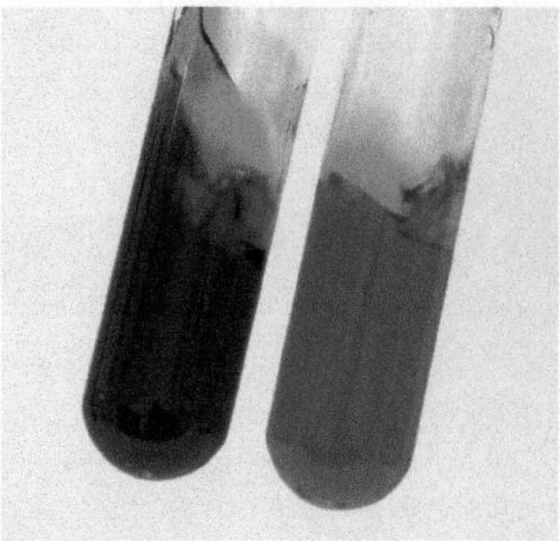

A solution containing $[Fe(SCN)]^{2+}$(aq) is a deep red colour (tube on left). On addition of NH_4Cl, the solution becomes paler (tube on the right).

Preparative chemistry is all about turning reactants into products, and chemists usually want to shift the position of equilibrium as far to the right as possible when dealing with reversible reactions. One way to do this is to remove one of the products from the reaction vessel, so that the reaction is constantly being shifted towards the product side. This is simple if one of the products is a gas. For example, when limestone is heated in an open container and the carbon dioxide is allowed to escape, the reaction goes to completion. If the reaction is carried out in a sealed container, however, the system comes to equilibrium.

$$CaCO_3(s) \rightleftharpoons CaO(s) + CO_2(g)$$

In Box 1.12, the effects of changes in concentration on a series of related equilibria are discussed.

Changing the pressure

Many important industrial processes involve reversible reactions that take place in the gas phase. Increasing the pressure at which the reaction is carried out moves the position of equilibrium towards the side of the equation with fewer gas molecules, as this reduces the pressure and minimizes the effect of the imposed change.

For example, ammonia is made industrially from nitrogen and hydrogen by the Haber process.

$$N_2(g) + 3H_2(g) \rightleftharpoons 2NH_3(g)$$

The Haber process is sometimes referred to as the Haber–Bosch process. The synthetic method was developed by Fritz Haber in 1909. The process was then scaled up by Carl Bosch, a chemical engineer employed by BASF (Badische Anilin und Soda Fabrik) near Mannheim in Germany. The first industrial plant went into production in 1913.

There are 4 molecules of gaseous reactants on the left-hand side of the equation, but only 2 molecules of gaseous product on the right-hand side. So, an increase in pressure causes the position of equilibrium to shift towards the right and increases the yield of ammonia. For this reason, the Haber process is carried out at high pressure, between 25 atm and 150 atm.

Note that pressure only has an effect on the position of equilibrium when there is a change in the number of gaseous molecules during the reaction. For a reaction such as

$$H_2(g) + I_2(g) \rightleftharpoons 2\,HI(g)$$

where there is the same number of gaseous molecules on each side of the equation, a change of pressure does not affect the position of equilibrium.

By convention, when a ΔH value is quoted alongside the equation for a reversible reaction, the value given refers to the forward reaction, that is, left to right as written.

Visit the Online Resource Centre to view video clip 1.1 which illustrates the effect of temperature on the NO_2/N_2O_4 equilibrium.

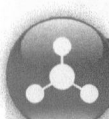

Visit the Online Resource Centre to view video clip 1.2 which uses Le Chatelier's principle to explain how the NO_2/N_2O_4 equilibrium responds to changes of pressure.

Changing the temperature

The enthalpy changes for a reaction and its reverse reaction have the same magnitude but opposite signs (Equation 1.18, p.44). If the temperature is increased, the position of equilibrium moves in the direction of the endothermic change because this lowers the temperature and minimizes the effect of the imposed change.

For example, the dark brown gas, nitrogen dioxide (NO_2), is in equilibrium with its colourless dimer, dinitrogen tetroxide (N_2O_4). (See Chapter 15 (p.696) for photographs of this reaction.) The forward reaction is exothermic.

$$\underset{\text{brown}}{2\,NO_2(g)} \rightleftharpoons \underset{\text{colourless}}{N_2O_4(g)} \quad \Delta_r H^{\ominus}_{298} = -57.0\,\text{kJ mol}^{-1}$$

If a sealed glass container of the brown equilibrium mixture is placed in boiling water, the brown colour deepens, because the position of equilibrium moves in the direction of the endothermic change towards the reactant. When the container is cooled in ice, the gas mixture turns almost colourless as the position of equilibrium moves in the direction of the exothermic change towards the product.

Box 1.12 Connecting equilibria and cave chemistry

In many systems, particularly naturally occurring systems, two or more equilibria are linked together, so that the product of the first equilibrium is the reactant in the next, and so on. You can use Le Chatelier's principle to predict the effect of changing the concentration of a substance, in one equilibrium reaction, on concentrations in the other linked equilibria.

When carbon dioxide in the air comes into contact with water, the equilibria in Equations 1.19 and 1.20 are established.

$$CO_2(g) \rightleftharpoons CO_2(aq) \qquad (1.19)$$

$$CO_2(aq) + H_2O(l) \rightleftharpoons HCO_3^-(aq) + H^+(aq) \qquad (1.20)$$

As a result, rainwater is slightly acidic. As rainwater slowly percolates through limestone rocks, a third equilibrium (Equation 1.21) becomes involved.

$$CaCO_3(s) + H^+(aq) \rightleftharpoons Ca^{2+}(aq) + HCO_3^-(aq) \qquad (1.21)$$

Combining these three equilibria, the overall reaction is

$$CaCO_3(s) + CO_2(g) + H_2O(l) \rightleftharpoons Ca^{2+}(aq) + 2HCO_3^-(aq) \qquad (1.22)$$

The reversible reaction in Equation 1.22 is responsible for the spectacular caves and potholes found in limestone country. The direction in which the reaction proceeds depends on the conditions. The concentration of carbon dioxide in the air in contact with the water seeping through the limestone rocks is 10–40 times higher than the normal atmospheric concentration.

This is because water percolating through the ground dissolves carbon dioxide from decomposing organic matter. The calcium carbonate in the rocks dissolves—creating fissures and caves.

The temperature inside the caves is approximately the same as in the limestone rocks, but the concentration of carbon dioxide is now similar to that in the atmosphere. When the percolating water drips from the roof of a cave, carbon dioxide escapes from solution into the surrounding air and solid calcium carbonate precipitates out from the solution. Stalactites made from the precipitated calcium carbonate slowly grow down from the ceiling. Similarly, stalagmites grow up from the cave floor at the point where the drops land.

Questions

(a) Use Le Chatelier's principle and Equation 1.22 to explain both the formation of caves and the presence of stalactites and stalagmites in limestone areas.

(b) The concentration of carbon dioxide in the atmosphere is currently increasing from year to year. What effect does this increase in carbon dioxide concentration have on the acidity of rainwater?

→

▲ Stalactites and stalagmites in an underground cave.

Equilibrium constant K_c

When a reaction reaches equilibrium, the concentrations of reactants and products are related. For the reaction of hydrogen and iodine to produce hydrogen iodide

$$H_2(g) + I_2(g) \rightleftharpoons 2HI(g)$$

Table 1.11 shows the equilibrium concentrations of $H_2(g)$, $I_2(g)$, and HI (g) for different initial reaction mixtures at 730 K. In the first three experiments, mixtures of hydrogen and iodine were placed in sealed reaction vessels and allowed to come to equilibrium. In the final two experiments, hydrogen iodide alone was sealed in the reaction vessel.

The experiments show that, for this reaction, at 730 K, the ratio

$$\frac{[HI(g)]^2}{[H_2(g)]\times[I_2(g)]}$$

is constant at about 46.7. The constant is called the **equilibrium constant, K_c**.

The multiplication sign between the square brackets is often omitted.

Table 1.11 Initial and equilibrium concentrations for the reaction $H_2(g) + I_2(g) \rightleftharpoons 2HI(g)$ at 730 K

Experiment	Initial concentrations / mol dm^{-3}			Equilibrium concentrations / mol dm^{-3}			K_c
	$[H_2(g)]$	$[I_2(g)]$	$[HI(g)]$	$[H_2(g)]$	$[I_2(g)]$	$[HI(g)]$	
1	2.40×10^{-2}	1.38×10^{-2}	0	1.14×10^{-2}	$0.1\text{–}2 \times 10^{-2}$	2.52×10^{-2}	46.4
2	2.44×10^{-2}	1.98×10^{-2}	0	0.77×10^{-2}	0.31×10^{-2}	3.34×10^{-2}	46.7
3	2.46×10^{-2}	1.76×10^{-2}	0	0.92×10^{-2}	0.22×10^{-2}	3.08×10^{-2}	46.9
4	0	0	3.04×10^{-2}	0.345×10^{-2}	0.345×10^{-2}	2.35×10^{-2}	46.9
5	0	0	7.58×10^{-2}	0.86×10^{-2}	0.86×10^{-2}	5.86×10^{-2}	46.4

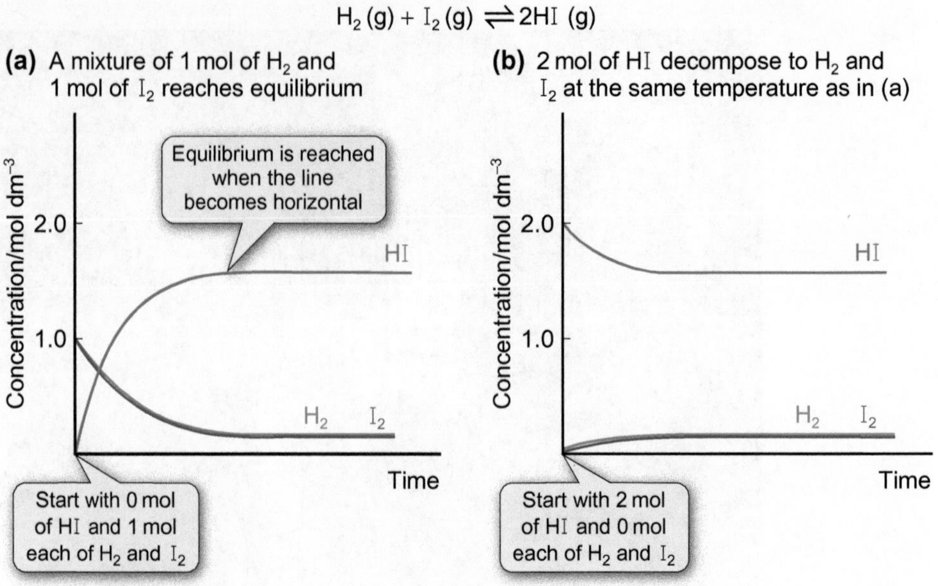

H2 (g) + I2 (g) ⇌ 2HI (g)
(a) A mixture of 1 mol of H2 and 1 mol of I2 reaches equilibrium
(b) 2 mol of HI decompose to H2 and I2 at the same temperature as in (a)
Equilibrium is reached when the line becomes horizontal
Concentration/mol dm−3
2.0
1.0
HI
H2
I2
Time
Start with 0 mol of HI and 1 mol each of H2 and I2
Start with 2 mol of HI and 0 mol each of H2 and I2

Figure 1.29 Equilibrium is reached in a sealed vessel from either (a) a mixture of hydrogen and iodine, or from (b) pure hydrogen iodide.

The **square brackets** in the expression for the equilibrium constant mean the concentration in mol dm^{-3}. Do not confuse these with the square brackets used for complex ions, such as $[FeCl_4]^-$. **Equilibrium concentrations** (the concentrations when the reaction has reached equilibrium) are sometimes indicated by a subscript 'eqm' after the square brackets $[]_{eqm}$, though this can become quite cumbersome and the subscripts are often omitted.

Visit the Online Resource Centre to view screencast 1.3 which discusses the use of equilibrium constants in terms of concentrations and in terms of partial pressures.

Values of K_c are constant for a particular temperature, so you should always state the temperature when you quote an equilibrium constant.

$$K_c = \frac{[\mathrm{HI(g)}]^2}{[\mathrm{H_2(g)}]\times[\mathrm{I_2(g)}]} = 46.7 \text{ at } 730\,\mathrm{K}$$

Under the same conditions, the same position of equilibrium is reached whether the equilibrium is approached from the reactants or from the products. In fact, once a system is at equilibrium, it is impossible to tell whether the equilibrium was arrived at by starting with the reactants or with the products. Figure 1.29 shows how the concentrations of reactants and products change as the reaction approaches equilibrium.

For the general reaction

$$a\,\mathrm{A} + b\,\mathrm{B} \rightleftharpoons c\,\mathrm{C} + d\,\mathrm{D}$$

the equilibrium constant, K_c, for the reaction is given by the expression:

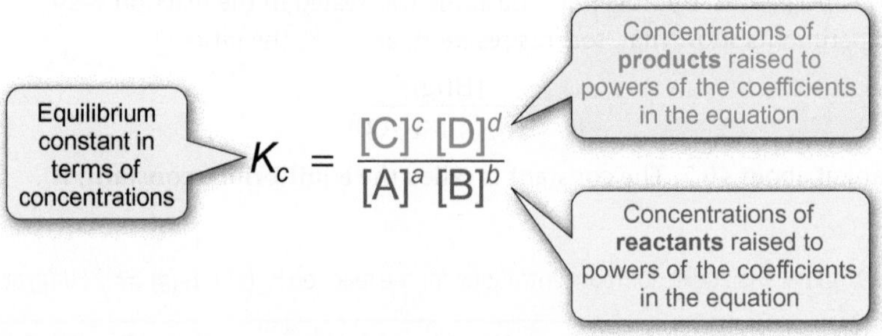

Equilibrium constant in terms of concentrations
Kc = [C]c [D]d / [A]a [B]b
Concentrations of products raised to powers of the coefficients in the equation
Concentrations of reactants raised to powers of the coefficients in the equation

Units of K_c

The units of K_c vary depending on the reaction.

Example 1

$$\mathrm{H_2(g) + I_2(g) \rightleftharpoons 2HI(g)}$$

$$K_c = \frac{[\mathrm{HI(g)}]^2}{[\mathrm{H_2(g)}][\mathrm{I_2(g)}]}$$

Units of K_c are given by $\dfrac{(\mathrm{mol\,dm^{-3}})^2}{(\mathrm{mol\,dm^{-3}})(\mathrm{mol\,dm^{-3}})}$.

So, for this reaction, K_c has no units, since the units on the top and bottom of the expression cancel out.

Example 2

$$N_2(g) + 3H_2(g) \rightleftharpoons 2NH_3(g)$$

$$K_c = \frac{[NH_3(g)]^2}{[N_2(g)][H_2(g)]^3}$$

$$\text{Units of } K_c \text{ are given by } \frac{(\text{mol dm}^{-3})^2}{(\text{mol dm}^{-3})(\text{mol dm}^{-3})^3} = \text{mol}^{-2}\,\text{dm}^6$$

Example 3

$$2N_2(g) + 6H_2(g) \rightleftharpoons 4NH_3(g)$$

$$K_c = \frac{[NH_3(g)]^4}{[N_2(g)]^2[H_2(g)]^6}$$

$$\text{Units of } K_c \text{ are given by } \frac{(\text{mol dm}^{-3})^4}{(\text{mol dm}^{-3})^2(\text{mol dm}^{-3})^6} = \text{mol}^{-4}\,\text{dm}^{12}$$

You can see that the units for K_c vary from reaction to reaction and must be worked out from the equation for the reaction and the expression for K_c. Examples 2 and 3 show why, whenever you quote a value for K_c, you must always give the stoichiometric equation for the reaction it refers to. You should also give the temperature at which the measurements were made. Worked example 1.17 illustrates the use of the equilibrium constant, K_c.

Worked example 1.17 Equilibrium constants in terms of concentrations

The ester, ethyl ethanoate, can be prepared by reacting the carboxylic acid, ethanoic acid, with the alcohol, ethanol. The reaction is reversible and comes to equilibrium

$$CH_3CO_2H\,(l) + C_2H_5OH\,(l) \rightleftharpoons CH_3CO_2C_2H_5\,(l) + H_2O\,(l)$$

(a) Write an expression for K_c for the reaction and comment on the units of K_c.

(b) The table below shows equilibrium concentrations of the reactants and products at 373 K.

	Equilibrium concentration/mol dm^{-3}
ethanoic acid	0.0342
ethanol	7.033
ethyl ethanoate	0.981
water	0.981

Calculate the equilibrium constant for the reaction at 373 K.

Strategy

(a) Use Equation 1.23 to write the expression for K_c and then use this expression to work out the units of K_c.

(b) Substitute the equilibrium concentrations into the expression for K_c to find its value at 373 K.

Solution

(a) Using Equation 1.23,

$$K_c = \frac{[CH_3CO_2C_2H_5(l)]\,[H_2O\,(l)]}{[CH_3CO_2H(l)]\,[C_2H_5OH(l)]}$$

$$\text{Units of } K_c \text{ are given by } \frac{(\cancel{\text{mol dm}^{-3}})(\cancel{\text{mol dm}^{-3}})}{(\cancel{\text{mol dm}^{-3}})(\cancel{\text{mol dm}^{-3}})}$$

so K_c for this reaction has no units.

(b)
$$K_c = \frac{(0.981\,\cancel{\text{mol dm}^{-3}}) \times (0.981\,\cancel{\text{mol dm}^{-3}})}{(0.0342\,\cancel{\text{mol dm}^{-3}}) \times (7.033\,\cancel{\text{mol dm}^{-3}})}$$
$$= 4.00$$

Question

An equimolar mixture of ethanoic acid and ethanol was heated at 373 K. At equilibrium, the concentration of ethanoic acid was found to be 0.820 mol dm^{-3}. Calculate the concentration of ethyl ethanoate in the equilibrium mixture. (*Hint*. Use the stoichiometric equation for the reaction to determine the concentration of ethanol in the equilibrium mixture.)

Equilibrium constant K_p

For a reaction taking place in the gas phase, it is often more convenient to write the equilibrium constant in terms of partial pressures of the gases, rather than in terms of their concentrations.

Partial pressures are covered in more detail in Section 8.3 (p.355).

If you assume that each gas in the equilibrium mixture behaves independently of the other gases present, the total pressure of the mixture, p_{total}, is equal to the sum of the partial pressures of the gases present. The **partial pressure**, p, of a gas in a mixture is the pressure the gas would exert if it were present alone.

For the reaction of nitrogen and hydrogen to form ammonia

$$N_2(g) + 3H_2(g) \rightleftharpoons 2NH_3(g)$$

$$p_{total} = p_{N_2} + p_{H_2} + p_{NH_3}$$

At constant temperature, the partial pressure of each gas is proportional to its concentration (see Equation 1.25 below), so you can write an expression for an equilibrium constant, K_p

$$K_p = \frac{p_{NH_3}^{\,2}}{p_{N_2} p_{H_2}^{\,3}}$$

For this reaction, the units of K_p are $\frac{\text{Pa}^2}{\text{Pa Pa}^3} = \text{Pa}^{-2}$ (if the pressures are measured in pascals),

or $\frac{\text{atm}^2}{\text{atm atm}^3} = \text{atm}^{-2}$ (if the pressures are measured in atmospheres).

For the general reaction

$$a\,\text{A} + b\,\text{B} \rightleftharpoons c\,\text{C} + d\,\text{D}$$

the equilibrium constant, K_p, is given by the expression

Equilibrium constant in terms of partial pressures

$$K_p = \frac{p_C^{\,c}\, p_D^{\,d}}{p_A^{\,a}\, p_B^{\,b}} \qquad (1.24)$$

Partial pressures of **products** raised to powers of the coefficients in the equation

Partial pressures of **reactants** raised to powers of the coefficients in the equation

For gaseous reactions, either K_p or K_c can be used. As with K_c, the **units** of K_p vary from reaction to reaction and need to be worked out from each K_p expression. Worked example 1.18 illustrates the use of the equilibrium constant, K_p.

Relationship between concentration and partial pressure of a gas

The ideal gas equation is discussed in Sections 8.1 (p.345) and 8.2 (p.349).

You can use the ideal gas equation, $pV = nRT$, to work out the relationship between concentration and partial pressure of a gas, where p = partial pressure of the gas, V = volume, n = amount in moles, T = temperature, and R (the gas constant) = 8.314 J K^{-1}mol^{-1}. The concentration of the gas is given by n/V.

$$\text{For an ideal gas, } p = \frac{nRT}{V} = \text{concentration} \times RT \qquad (1.25)$$

At constant T, $p \propto$ concentration.

Relationship between K_p and K_c

In general, $K_p = K_c(RT)^{\Delta n}$, where Δn is the change in the number of moles of gas on going from reactants to products and R is the gas constant. For the ammonia reaction,

$$\Delta n = 2\,\text{mol} - 4\,\text{mol} = -2\,\text{mol},\ \text{so}\ K_p = K_c(RT)^{-2}$$

You can derive this relationship by substituting expressions for the partial pressures in terms of concentrations using Equation 1.25 into the expression for K_p (e.g. $p_{NH_3} = [NH_3] \times RT$).

Worked example 1.18 Equilibrium constants in terms of partial pressures

A 1 : 3 molar mixture of N_2 and H_2 is heated at 673 K and 50 atm so that it comes to equilibrium. The equilibrium mixture contains 15.3% NH_3 (by volume). Calculate a value for K_p for the reaction at this temperature.

Strategy

Write the balanced equation for the reaction.

Work out the equilibrium percentages of N_2 and H_2. (They react in the ratio 1 : 3, so the unreacted gases will still be present in this ratio at equilibrium.)

Use the equilibrium percentages to work out the partial pressure of each gas present at equilibrium. (Note that the equilibrium percentages *by volume* are also the equilibrium percentages *by amount in moles*.)

Substitute these values into the expression for K_p.

Solution

$$N_2(g) + 3H_2(g) \rightleftharpoons 2NH_3(g)$$

At equilibrium, the percentage of NH_3 (by volume) is 15.3% (0.153 as a fraction). The remaining 84.7% must be N_2 and H_2 in the ratio 1 : 3.

$$\text{Equilibrium \% of } N_2 = \tfrac{1}{4} \times 84.7\% = 21.18\%\ (0.2118 \text{ as a decimal})$$

$$\text{Equilibrium \% of } H_2 = \tfrac{3}{4} \times 84.7\% = 63.53\%\ (0.6353 \text{ as a decimal})$$

$$\text{Total pressure} = 50\,\text{atm} = p_{N_2} + p_{H_2} + p_{NH_3}$$

$$p_{N_2} = 0.2118 \times 50\,\text{atm} = 10.59\,\text{atm}$$

$$p_{H_2} = 0.6352 \times 50\,\text{atm} = 31.77\,\text{atm}$$

$$p_{NH_3} = 0.153 \times 50\,\text{atm} = 7.65\,\text{atm}$$

Substituting these values into the expression for K_p (Equation 1.24)

$$K_p = \frac{p_{NH_3}^{\ 2}}{p_{N_2}\,p_{H_2}^{\ 3}} = \frac{(7.65\,\text{atm})^2}{(10.59\,\text{atm}) \times (31.77\,\text{atm})^3}$$

$$= 1.7 \times 10^{-4}\,\text{atm}^{-2}\ (\text{at } 673\,\text{K})$$

(Note that p_{N_2} = (mole fraction of N_2) × p_{total}; see Section 8.3 (p.355). The answer is given to 2 significant figures to correspond with the smallest number of significant figures in the data.)

Question

A second investigation was carried out at the same temperature (673 K) but at a higher pressure. At equilibrium, the partial pressure of N_2 was 18.7 atm and the partial pressure of H_2 was 56.1 atm.

(a) Calculate the partial pressure of NH_3 in the equilibrium mixture.

(b) What was (i) the total pressure for the second investigation and (ii) the percentage of NH_3 in the equilibrium mixture?

Values of equilibrium constants vary enormously (see Table 1.12). Note, in particular, the effect on the value of K_p of expressing the value in pascals rather than atmospheres ($1\,\text{atm} = 1.01 \times 10^5\,\text{Pa}$). It is vital to quote units alongside values of K_c and K_p.

All reactions can be considered as equilibrium reactions. Even reactions that seem to go to completion actually have a very small amount of reactant left in equilibrium with the products. In some cases, the position of equilibrium is so far towards the products it is impossible to detect any reactants left in the reaction mixture. Such reactions have very large equilibrium constants. Similarly, a reaction that is observed not to go at all may be regarded as having a vanishingly small equilibrium constant.

You can find out more about chemical equilibria in Section 15.1 (p.697), where a *thermodynamic equilibrium constant*, *K*, is defined in terms of *activities*, rather than concentrations or partial pressures. This equilibrium constant is used in thermodynamics and has the advantage of not having units.

Table 1.12 Some values of K_c and K_p for gaseous reactions at 298 K

Reaction	K_c		K_p			
	Value	Units	Value	Units	Value	Units
$2H_2(g) + O_2(g) \rightleftharpoons 2H_2O(g)$	3.4×10^{81}	$mol^{-1}\,dm^3$	1.4×10^{75}	Pa^{-1}	1.4×10^{80}	atm^{-1}
$N_2(g) + 3H_2(g) \rightleftharpoons 2NH_3(g)$	3.6×10^8	$mol^{-2}\,dm^6$	5.8×10^{-5}	Pa^{-2}	6.0×10^5	atm^{-2}
$H_2(g) + I_2(g) \rightleftharpoons 2HI(g)$	620	—	620	—	620	—
$2NO_2(g) \rightleftharpoons N_2O_4(g)$	170	$mol^{-1}\,dm^3$	6.7×10^{-5}	Pa^{-1}	6.8	atm^{-1}
$N_2(g) + O_2(g) \rightleftharpoons 2NO(g)$	4.6×10^{-31}	—	4.6×10^{-31}	—	4.6×10^{-31}	—

Summary

- A chemical equilibrium is a dynamic equilibrium in which the forward and reverse reactions are taking place at the same rate, so concentrations remain constant.
- Le Chatelier's principle summarizes the effects of external changes on an equilibrium: when an external change is made to a system in dynamic equilibrium, the system responds to minimize the effect of the change.
- Equilibrium constants can be expressed in terms of concentrations, K_c, or, for gaseous reactions, in terms of partial pressures, K_p, where $K_p = K_c(RT)^{\Delta n}$.
- For a general reaction: $a\text{A} + b\text{B} \rightleftharpoons c\text{C} + d\text{D}$, $K_c = \dfrac{[\text{C}]^c\,[\text{D}]^d}{[\text{A}]^a\,[\text{B}]^b}$ and $K_p = \dfrac{p_\text{C}^{\,c}\,p_\text{D}^{\,d}}{p_\text{A}^{\,a}\,p_\text{B}^{\,b}}$

Table 1.13 summarizes the effect of changing conditions on the composition of equilibrium mixtures and on equilibrium constants. The important thing to remember is that K_c and K_p are constant unless the temperature changes. Note that catalysts do not affect the position of equilibrium or the equilibrium constant. They increase the *rates* of both the forward and reverse reactions so that the same position of equilibrium is reached—but it is reached more quickly.

Table 1.13 The effect of changing conditions on the composition of equilibrium mixtures and equilibrium constants

Change in	Composition	K_c or K_p
Concentration	Changes	Unchanged
Partial pressure	Changes	Unchanged
Total pressure	May change	Unchanged
Temperature	Changes	Changes
Catalyst	Unchanged	Unchanged

? For practice questions on these topics, see questions 20 and 21 at the end of this chapter (p.67).

Concept review

By the end of this chapter, you should be able to do the following.

- Understand and be able to use IUPAC base units and derived units.
- Describe the structure of an atom in terms of protons, neutrons, and electrons.
- Understand the terms mass number and atomic number and write atomic symbols for chemical elements.
- Describe how a simple mass spectrometer can be used to show the isotopic composition of an element.
- Understand and use the terms: relative atomic mass (A_r); relative formula (or molecular) mass (M_r); Avogadro constant (N_A); the mole (mol) as a unit of amount of substance.
- Work out the empirical formula of a compound from its elemental composition and understand the relationship between empirical formula and molecular formula.
- Write a balanced equation for a reaction, including state symbols, and use it to calculate reacting masses.
- Work out the yield of a chemical reaction.
- Write ionic equations.
- Understand and use the terms redox, oxidation and reduction, oxidizing agent, and reducing agent.
- Write half equations for a redox reaction and use them to construct a balanced overall equation for the reaction.
- Assign oxidation states and use changes of oxidation state to decide what has been oxidized and what reduced in a redox reaction.
- Perform calculations involving concentrations of compounds in solution.
- Understand the techniques of volumetric analysis and gravimetric analysis and be able to carry out relevant calculations.
- Understand what is meant by the terms: exothermic; endothermic; enthalpy change of a reaction; standard conditions; and thermochemical equation.
- Draw an enthalpy level diagram and relate enthalpy changes to bond-breaking and bond-formation processes.
- Describe the states of matter and phase changes in terms of a kinetic–molecular model of matter.
- Recognize different types of non-covalent (intermolecular) interactions.
- Understand that chemical equilibrium is a dynamic equilibrium and use Le Chatelier's principle to predict the effect of changes on the position of equilibriums.
- Write expressions for the equilibrium constants, K_c and K_p, and carry out simple calculations.

Key equations

Temperature in kelvin	$T/\mathrm{K} = \theta/^{\circ}\mathrm{C} + 273.15$	(1.1)
Chemical amount	$\text{amount (in mol)} = \dfrac{\text{mass (in g)}}{\text{molar mass(in g mol}^{-1})}$	(1.2)
Percentage yield	$\text{percentage yield} = \dfrac{\text{actual yield}}{\text{theoretical yield}} \times 100\%$	(1.4)
Molarity of a solution	$\text{Concentration (mol dm}^{-3}) = \dfrac{\text{concentration (g dm}^{-3})}{\text{molar mass (g mol}^{-1})}$	(1.6)
The relationship between concentration (c) in mol dm^{-3}, amount in moles (n), and the volume of solution (V) in dm^3	$n = c \times V$	(1.9)
Potential energy	$\text{potential energy} = E_{PE} = mgh$	(1.16)
Kinetic energy	$\text{kinetic energy} = E_{KE} = \frac{1}{2}mv^2$	(1.17)
Thermochemical equations	$\Delta_r H(\text{forward reaction}) = -\Delta_r H(\text{backward reaction})$	(1.18)
Equilibrium constant for the reaction: $a\,\mathrm{A} + b\,\mathrm{B} \rightleftharpoons c\,\mathrm{C} + d\,\mathrm{D}$	In terms of concentrations: $K_c = \dfrac{[\mathrm{C}]^c[\mathrm{D}]^d}{[\mathrm{A}]^a[\mathrm{B}]^b}$	(1.23)
	In terms of partial pressures: $K_p = \dfrac{p_C^c\, p_D^d}{p_A^a\, p_B^b}$	(1.24)

Questions

More challenging questions are marked with an asterisk *.

1 The C–C bond length in a crystal of diamond is 0.154 nm. What is this distance in (a) metres, (b) picometres, (c) angstroms? (Section 1.2)

2 Oxygen gas liquefies at −183.0 °C and freezes at −218.4 °C. Work out its melting point, T_m, and boiling point, T_b, in kelvin. (Section 1.2)

3 A sealed flask holds 10 dm^3 of gas. What is this volume in (a) cm^3, (b) m^3, (c) litres? (Section 1.2)

4 What is the SI derived unit for the speed of a molecule? (Section 1.2)

5 How many moles of atoms are contained in the following masses: (a) 22.0 g of magnesium; (b) 43.2 g of chlorine; (c) 126 mg of gold; (d) 1.00 kg of mercury? (Section 1.3)

6 Calculate the amount of each substance contained in the following masses: (a) 89.2 g of carbon dioxide (CO_2); (b) 43.2 g of chlorine (Cl_2); (c) 0.48 kg of calcium hydroxide ($Ca(OH)_2$); (d) 25 tonnes of water, H_2O (1 tonne = 1×10^6 g). (Section 1.3)

7 What is the mass (in g) of (a) 5.46 mol of CuO; (b) 0.107 mol of $KMnO_4$; (c) 2.85 mmol of C_2H_5OH; (d) 1.95 μmol of HCN? (Section 1.3)

8 The structure of succinic acid is shown below. (Section 1.3)

CH_2CO_2H
|
CH_2CO_2H succinic acid

(a) Write down the molecular formula of succinic acid and work out its molar mass.

(b) What is the empirical formula of succinic acid?

(c) What is the percentage of carbon in succinic acid?

(d) Calculate the amount of succinic acid in a 0.125 g sample of the pure acid.

(e) How many molecules of succinic acid are present in the 0.125 g sample?

(f) How many carbon atoms are present in the 0.125 g sample?

9 Write down the formulae for sodium azide and sodium nitride. Use these to explain the difference between the N_3^- ion and the N^{3-} ion. (Section 1.4)

10 The fertilizer, ammonium nitrate, is made by reacting ammonia with nitric acid. (Section 1.4)

(a) Write a balanced equation, with state symbols, for the reaction of ammonia gas with nitric acid to form a solution of ammonium nitrate.

(b) Rewrite the equation to show the ions present in the reactants and products, and hence write an ionic equation for the reaction.

(c) What *type* of chemical reaction does your equation represent?

Nitric acid is manufactured from ammonia. The first stage in this process involves burning ammonia in oxygen on the surface of a platinum gauze catalyst. The products are NO and H_2O.

(d) Construct a balanced equation for the burning of NH_3 in O_2.

(e) Calculate the maximum mass of NO that could be obtained by burning 1.00 kg of NH_3 and the mass of oxygen required.

(f) In practice, 1.45 kg of NO were obtained from the reaction in (e). What was the percentage yield of the reaction?

11 Pentanoic acid ($C_4H_9CO_2H$) can be synthesized in three steps from butan-1-ol (C_4H_9OH) as shown below (Section 1.4):

$$\underset{\text{butan-1-ol}}{C_4H_9OH} \xrightarrow{65\%} \underset{\text{1-chlorobutane}}{C_4H_9Cl} \xrightarrow{85\%} \underset{\text{pentanenitrile}}{C_4H_9CN} \xrightarrow{67\%} \underset{\text{pentanoic acid}}{C_4H_9CO_2H}$$

(a) What is the overall yield for the conversion of butan-1-ol to pentanoic acid?

(b) If you carried out the synthesis starting with 20.0 g of butan-1-ol, what mass of pentanoic acid would you obtain?

(c) Even when there are no side reactions in an organic reaction, a 100% yield of product is rarely obtained. Suggest reasons why this might be.

12 A stream running out from a copper mine contains a dilute solution of copper sulfate. As it passes over an iron grid, copper metal deposits on the grid. (Section 1.4)

(a) Write a balanced equation, with state symbols, for the reaction taking place.

(b) Write an ionic equation for the reaction.

(c) Assign oxidation states to the elements in each of the reactants and products in the equation in (b). Use these values to decide what has been oxidized and what reduced.

13 (a) What are the systematic names for (i) CS_2; (ii) Cl_2O_7; (iii) XeF_6; (iv) $(NH_4)_2SO_4$; (v) $CrCl_3$; (vi) KIO_4?

(b) Write the formula of each of the following compounds: (i) sodium sulfite; (ii) barium carbonate; (iii) iron(II) chloride; (iv) sodium thiosulfate; (v) diiodine pentoxide; (vi) dinitrogen oxide. (Section 1.4)

14 In most compounds, H has an oxidation state of +1 and O has an oxidation state of −2. The following compounds are exceptions to this rule. Assume each metal has the oxidation state of its most common ion and that F has an oxidation state of −1. Find the oxidation state of H or O in each compound: (a) KO_2; (b) Na_2O_2; (c) MgH_2; (d) $LiAlH_4$; (e) OF_2. (Section 1.4)

15* Sodium chromate (Na_2CrO_4) can be prepared by oxidizing a chromium(III) salt with sodium peroxide (Na_2O_2) in alkaline solution. The Cr^{3+} ions are oxidized to CrO_4^{2-} ions. The O_2^{2-} ions are reduced to OH^- ions. Construct half equations and a balanced overall equation for the reaction. (Section 1.4)

16* To prepare a very dilute solution, it is more accurate to make up a more concentrated standard solution, and carry out a series of successive dilutions, than to weigh out a very small mass of the solute.

A solution was made by dissolving 0.587 g of $KMnO_4$ in dilute sulfuric acid and making the volume of solution up to 1 dm^3 in a volumetric flask. 10.0 cm^3 of this solution were transferred to a second 1 dm^3 volumetric flask and diluted to the mark with water. The dilution process was then repeated once, that is, 10.0 cm^3 of this solution were transferred to a 1 dm^3 volumetric flask and diluted to the mark with water. (Section 1.5)

(a) What mass (in mg) of $KMnO_4$ would you have had to weigh out to make 500 cm^3 of a solution with the same concentration as the final dilute solution?

(b) What is the concentration of the final dilute $KMnO_4$ solution in mol dm^{-3}?

17* The most common type of kidney stone is formed from calcium ethanedioate (CaC_2O_4) which precipitates out in the urinary tract when concentrations of Ca^{2+} ions and $C_2O_4^{2-}$ ions become too high. Magnesium ions are known to inhibit the formation of kidney stones. To analyse the concentrations of Ca^{2+}(aq) and Mg^{2+}(aq) in a sample of urine, the ions were precipitated as ethanedioates and the mixed precipitate of CaC_2O_4 and MgC_2O_4 was analysed by gravimetric analysis.

The solid ethanedioates were first heated to form a mixture of $CaCO_3$ and MgO. The mass of this mixture was 0.0433 g. This solid was then heated more strongly to give a mixture of CaO and MgO. The mass of the solid residue from this process was 0.0285 g. What was the mass of the Ca^{2+} ions in the original sample of urine? (Section 1.5)

18 The standard enthalpy change of combustion of heptane, C_7H_{16}, at 298 K, is -4817 kJ mol^{-1}. (Section 1.6)

(a) Write a thermochemical equation for the complete combustion of heptane to carbon dioxide and water.

(b) What is the enthalpy change when 50 g of heptane are burned?

(c) What mass of heptane would be needed to provide 100 MJ of energy?

19* (a) List the non-covalent interactions present in liquid water. Which is responsible for the strongest interactions between the molecules?

(b) Explain why the value of $\Delta_{vap}H^{\ominus}(H_2O)$ is unusually high for a molecule of its size.

(c) In a storm, 3 cm of rain falls on the city of Leeds, which has an area of approximately 500 km^2. Estimate the energy released as heat when this quantity of water condenses from vapour to form rain. (Density of water is 1.00 g cm^{-3}; $\Delta_{vap}H^{\ominus}(H_2O) = +40.7$ kJ mol^{-1} at 298 K.)

(d) The output from a large 2000 MW power station is 2000 MJ s^{-1}. How long would it take the power station to deliver the same quantity of energy as was released by the condensation of the rain in (c)? (Sections 1.7 and 1.8)

20* Nitrogen dioxide gas is heated in a sealed container at 700 K until the system comes to equilibrium. The nitrogen dioxide dissociates into nitrogen monoxide and oxygen in an endothermic process (Section 1.9):

$$2NO_2(g) \rightleftharpoons 2NO(g) + O_2(g)$$

The equilibrium constant at 700 K is 2.78×10^{-2} mol dm^{-3}.

(a) Write an expression for K_c.

(b) State how the position of equilibrium would be affected by:

(i) an increase in temperature
(ii) an increase in the total pressure.

(c) At equilibrium at 700 K, the concentration of nitrogen monoxide was found to be 0.017 mol dm^{-3}. What was the concentration of nitrogen dioxide in the equilibrium mixture?

21* One stage in the manufacture of methanol from methane involves the conversion of synthesis gas (a mixture of CO and H_2) to methanol. The conversion is carried out over a catalyst at a temperature of around 500 K and a pressure of 100 atm. (Section 1.9)

$$CO(g) + 2H_2(g) \rightleftharpoons CH_3OH(g) \quad \Delta H = -90.7 \text{ kJ mol}^{-1}$$

(a) Write an expression for K_p for the reaction.

(b) At 500 K and 100 atm pressure, an equilibrium mixture contains 42% CH_3OH and 48% CO. Calculate a value for K_p at this temperature.

(c) Use Le Chatelier's principle to predict what would happen to the percentage of methanol in the mixture if: (i) the temperature increases; (ii) the pressure increases; (iii) hydrogen is added at constant temperature and pressure.

2

The language of organic chemistry

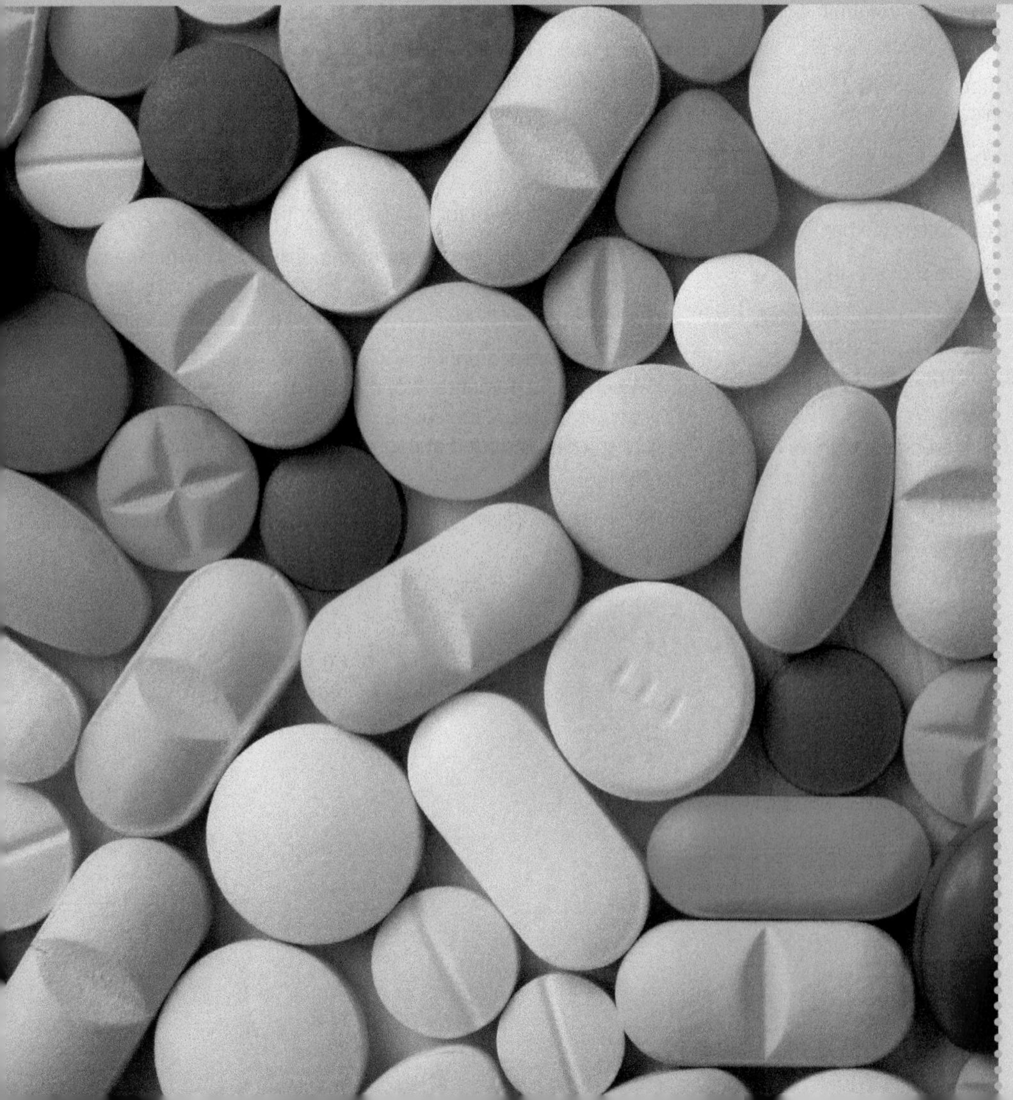

This chapter builds on the following topics:

- **Nomenclature** Section 1.2, p.6
- **Chemical formulae** Section 1.3, p.12
- **Formal charge** Section 5.1, p.219
- **Valence shell electron pair repulsion theory** Section 5.2, p.223
- **Valence bond theory for polyatomic molecules** Section 5.4, p.236
- **Resonance** Section 5.5, p.243

◄ A tablet is a mixture of the active compound (the drug) and inactive substances called excipients (such as binders, glidants, and lubricants), which improve the administration or absorption of the medicine. The English artist Damien Hirst has turned tablets into art objects by displaying them in steel medicine cabinets—in 2007, one of the cabinets sold at auction for £9.65 million, which broke the European record for work by a living artist.

Designer medicines for treating high blood pressure: an ACE approach

High blood pressure is dangerous because it forces the heart to work harder and, if this continues for a long time, the heart and arteries may cease to function. This can lead to stroke, heart attack, heart failure, or kidney failure.

A group of compounds called ACE inhibitors provides one way of treating high blood pressure. They work by inhibiting the activity of an enzyme, called **a**ngiotensin-**c**onverting **e**nzyme (ACE). ACE catalyses a reaction that produces a compound that increases blood pressure. Hence, stopping the enzyme working is a way of preventing high blood pressure.

▼ Captopril has three functional groups, which are found in compounds called thiols, amides, and carboxylic acids.

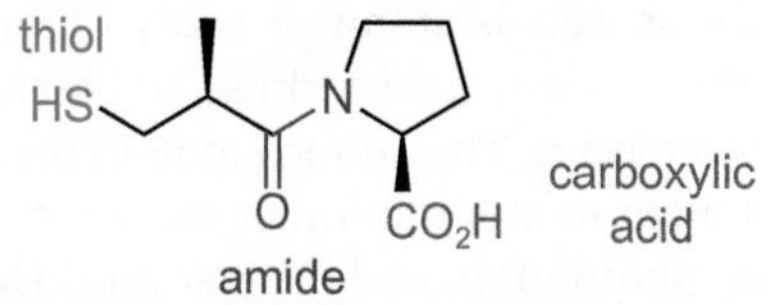
thiol
HS
N
O
CO_2H
carboxylic acid
amide

To inhibit ACE, medicinal chemists have designed molecules that fit inside the active site of the enzyme—this is where the molecule that produces high blood pressure is made. The first of these designer molecules was captopril and this fits tightly in the active site of the enzyme because of its shape and because the three functional groups interact with the functional groups on the surface of the enzyme (see figure below). Changing the type of functional groups in captopril, and also their position in the molecule, influences how tightly the molecule binds in the active site. To design effective ACE inhibitors it is important to recognize the important role the functional groups play. Hence, chemists must be able to identify functional groups, and these are introduced in Section 2.3 (p.77). Chemists also need to be able to draw molecules correctly. For example, in the structure of captopril shown above, the black lines represent the carbon framework and the wedge-shaped bonds indicate the three-dimensional structure. The functional groups are shown using colours. Drawing structures is discussed in Section 2.2 (p.73).

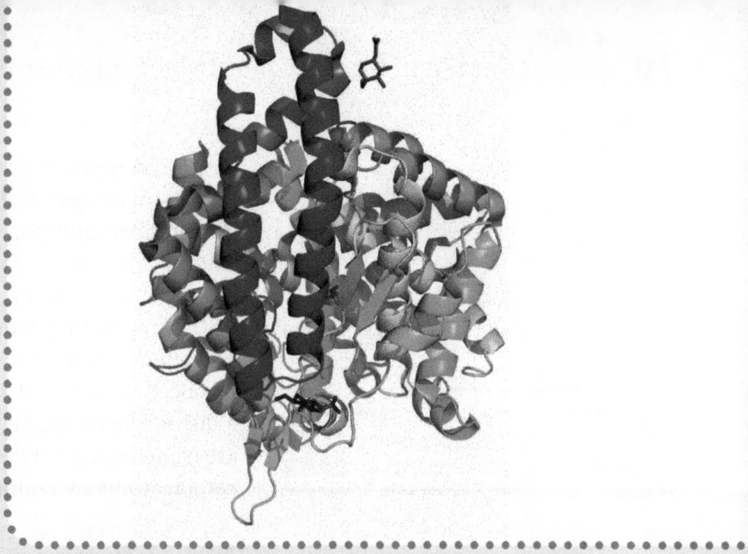

▲ The crystal structure of human ACE bound to captopril was published in 2004. In this ribbon diagram, coiled ribbons represent the folding of the protein chain.

What's in a name?

Although captopril has a single structure it is known by more than one name (see the box below). Organic chemists use a systematic approach to give an accurate name for the compound that is based on the structure. From this name, the chemical structure of captopril can be deduced so it is important that chemists can use this language, which is introduced in Section 2.4.

- Generic name: captopril
- Systematic name: 1-[(3-mercapto-2-methyl)-propanoyl]pyrrolidine-2-carboxylic acid
- Brand name: Capoten®

Unfortunately, the systematic name is very long and much too complicated for general use by doctors and the general public. Pharmaceutical companies use shorter and more distinctive names to identify and publicize their medicines. The brand (trade) name, Capoten®, distinguishes this medicine from other products and can only be used by the owner of the registered trademark. The owner uses this name for the lifetime of a patent, but when the patent expires other pharmaceutical companies can also market the same medicine under the generic name, which is captopril.

▼ A model of how captopril interacts with different groups in the active site of the enzyme. In the body, since the pH is 7–8, the solution is slightly alkaline, which means that in many captopril molecules, the thiol and carboxylic acid groups are deprotonated (i.e. they have lost H^+).

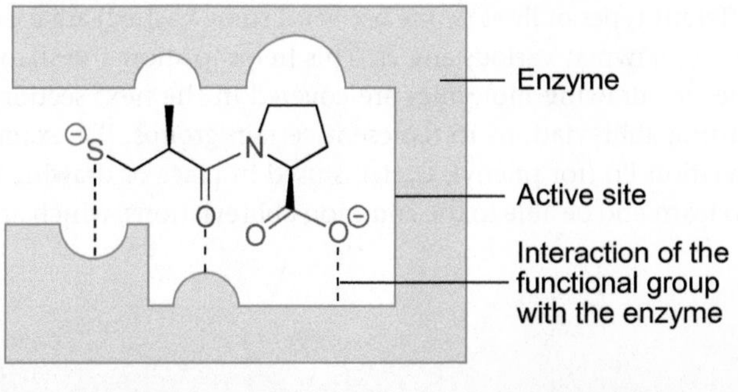
S
N
O
O
O
Enzyme
Active site
Interaction of the functional group with the enzyme

▼ The discovery of captopril began with the isolation of bradykinin (Figure 2.1, p.71) from the venom of the poisonous Brazilian viper (*Bothrops jararaca*)—the venom catastrophically lowers blood pressure in its prey.

Organic compounds contain carbon together with one or more other elements, such as hydrogen, oxygen, and nitrogen. There are a tremendous number of organic compounds. The number easily surpasses the total number of all other compounds that do not contain carbon, and organic compounds show a wide range of useful properties. This chapter begins by discussing the importance of organic chemistry, which is the chemistry of most carbon-containing compounds. Each organic compound has a unique molecular structure, and guidelines for drawing the structures of organic compounds in different ways are introduced in Section 2.2. Related to its molecular structure, an organic compound also has a distinctive name. An important aim of this chapter is to show how organic compounds are named using a systematic approach. This is a chemical language that identifies how the carbon atoms are arranged and also locates any functional groups within the molecule.

Metal (and ammonium) carbonates, hydrogencarbonates, cyanates, and isocyanates are classed as inorganic compounds.

2.1 Why are organic compounds important?

A covalent bond is formed when a pair of electrons is shared between two atoms. Section 4.2 (p.173).

Organic chemistry concentrates on the study of compounds that contain carbon. It might seem odd to have a major branch of chemistry based on only one element, so what is special about carbon?

The answer lies in sharing electrons to form **covalent bonds**. The carbon atom is able to share electrons with other carbon atoms as well as atoms of many other elements. Importantly, carbon atoms can bond together to form chains, rings, and branches. Such structures form a **carbon framework** for the molecule. The carbon atoms in the framework are often attached to hydrogen atoms and often to atoms of other elements (called **heteroatoms**), such as oxygen, nitrogen, sulfur, phosphorus, and the halogens. Heteroatoms are often part of a **functional group** on the carbon skeleton and give the molecule characteristic properties (Section 2.3, p.77). Carbon is unique in being able to make such a wide range of strong covalent bonds. This means that compounds containing carbon are by far the greatest in number and variety. At present, around 16 million organic compounds are known and the number is rising!

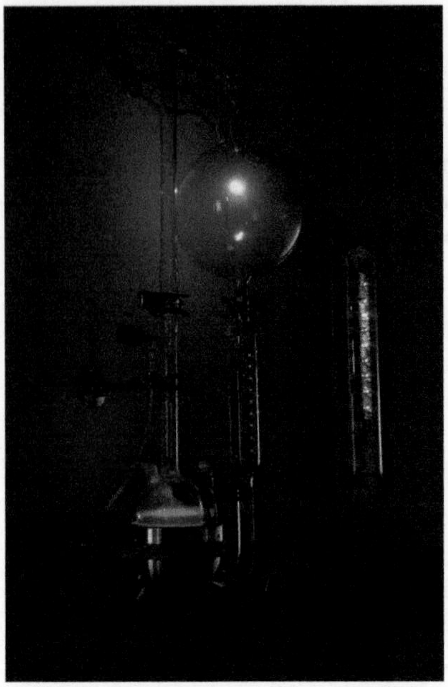

To study the origins of life, in the Miller–Urey experiment, an electrical spark was used to imitate lightning.

Most compounds in living matter are made up of organic molecules (or molecules with an organic part) and these molecules are central to life on this planet. In one theory for the origin of life on Earth, the organic gas methane (CH_4) is believed to be the source of carbon in other organic molecules. Methane is thought to have been a major component of the primordial atmosphere of early Earth along with ammonia, water, carbon dioxide, and hydrogen. In the presence of high-energy radiation these compounds can combine to form more complex molecules including sugars and amino acids. These building blocks react further to produce protein and deoxyribonucleic acid (DNA) molecules, which may have been vital for the first self-replicating systems. This theory was supported by Stanley L. Miller and Harold C. Urey who, in 1953, showed that amino acids and hydroxy acids are formed from a mixture of methane, ammonia, water vapour, and hydrogen gases in the presence of a spark discharge (used to represent lightning flashes on early Earth).

In 2008, the brown residues collected from the experiments in the 1950s were re-examined using modern spectroscopic techniques. Whereas Miller and Urey detected only five amino acids, the more recent analysis showed small amounts of nine additional amino acids.

Organic compounds play such an important part of our everyday lives that it makes sense for us to understand them. Our bodies contain around 18% carbon by weight and this is distributed in various organic compounds. Examples include DNA, which contains our genetic information, proteins that catalyse all of the reactions in our bodies, and sugars that are oxidized in respiration to provide energy. The presence of these and many other compounds is vital for our existence.

Naturally occurring compounds have a wide range of structures and biological activities. To understand how a molecule produces a biological effect it is important to know its shape (Box 18.1, p.822).

Chemists are interested in the structures of naturally occurring compounds (**natural products**) to help them understand the roles of these compounds. The structures can be complex as illustrated by those for cholesterol, prostaglandin $PGF_{2\alpha}$, and bradykinin in Figure 2.1. You will see that different types of lines (some bold and some hashed) are used in the structures and the lines are drawn at various angles. This helps to show the shape of the molecules and guidelines for drawing molecules are covered in the next section. Structures can also be drawn using abbreviations to represent certain groups. For example, for bradykinin, the abbreviation Ph (for phenyl, C_6H_5) is used in place of drawing a benzene ring. It is important to learn and be able to use common abbreviations, which are listed in Table 2.4 (p.108).

A **natural product** is a compound that is synthesized biologically by a living organism.

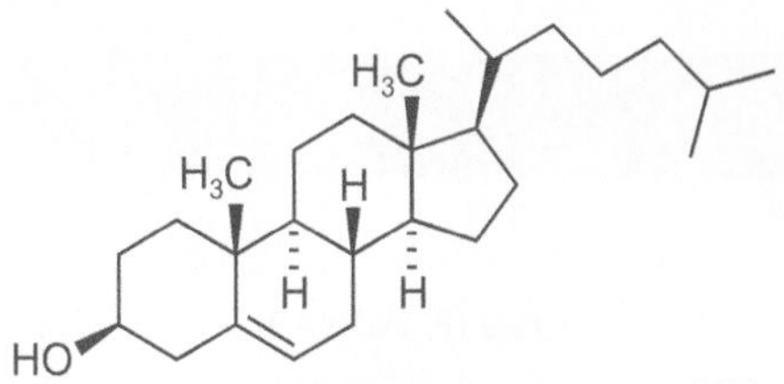

cholesterol

Cholesterol is present in all animal cells and is an important part of cell membranes. It acts as a chemical messenger (a hormone) but high concentrations of cholesterol are a primary cause of heart disease

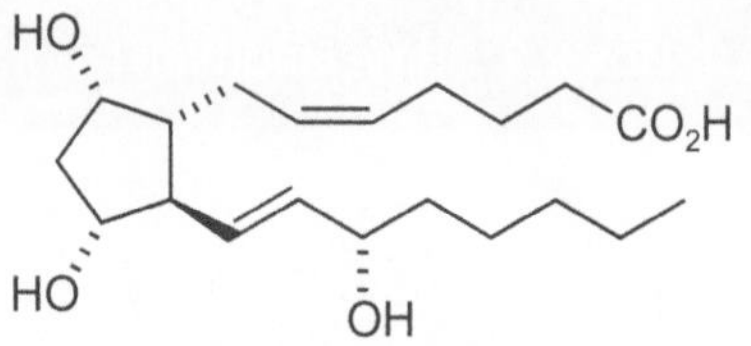

prostaglandin $PGF_{2\alpha}$

Prostaglandins are a family of compounds found in all body tissues. They regulate a number of physiological responses including blood pressure, blood clotting, and the induction of labour

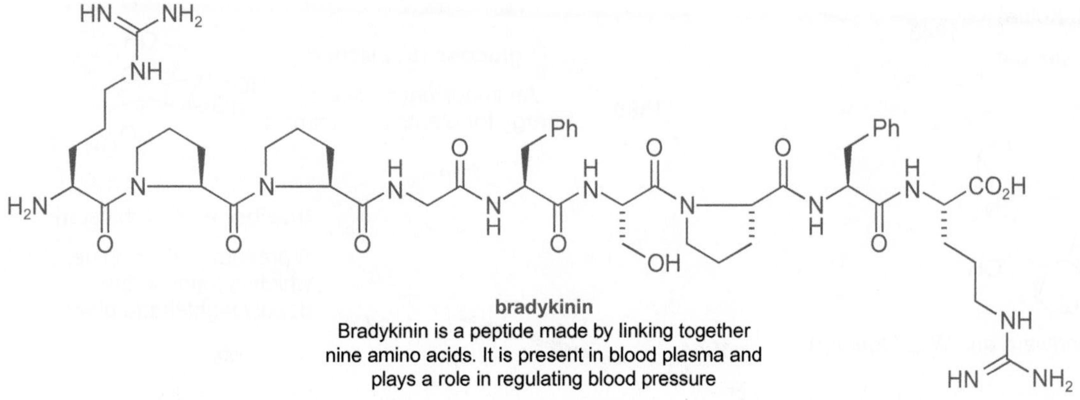

bradykinin

Bradykinin is a peptide made by linking together nine amino acids. It is present in blood plasma and plays a role in regulating blood pressure

Figure 2.1 The structures and roles of some organic compounds found in the body.

Originally, organic chemistry focused on the study of molecules of life by the extraction, purification, and analysis of substances from plants and animals (Box 2.1). Today, however, organic chemists have broadened their horizons and are also interested in **synthetic compounds**, which are not found in nature but are made in the laboratory. For example, our clothes contain organic molecules but, whereas cotton and silk are natural fibres, polyesters and nylons are synthetic. Perfumes, deodorants, and toothpastes all contain synthetic compounds, as do medicines, pesticides, food, paintings, glues, camera film, carpeting, and many other items that have significantly improved the quality of our lives.

The name synthetic comes from the word **synthesis**, which describes the process of making molecules, usually by joining together small molecules to form new, larger, and more complicated products.

Box 2.1 Friedrich Wöhler: first synthesis of a naturally occurring organic compound

Friedrich Wöhler (1800–1882) was the first person to make a naturally occurring organic compound in the laboratory. In 1828, he heated an aqueous solution of ammonium cyanate (an inorganic compound) and accidentally prepared urea, which is excreted in the urine of mammals (an average human excretes 30 g a day). This important discovery dispelled the 'vitalism' theory that argued that a 'vital force' was required for the synthesis of an organic compound.

But some may say that the theory is still alive in the minds of people who argue that industrially produced fertilizers, such as ammonium nitrate (NH_4NO_3), are alien to the environment. They advocate the use of 'organic' fertilizers as the natural way to replenish the land. (By 'organic' they mean 'grown without industrially produced fertilizers, pesticides, or herbicides', not 'organic' as the term is used in chemistry.) So-called 'organic' fertilizers, such as manure, provide plants with the same essential nutrients as industrially produced fertilizers. For example, both manure and ammonium nitrate supply plants with NO_3^- ions—the ions from both sources are identical.

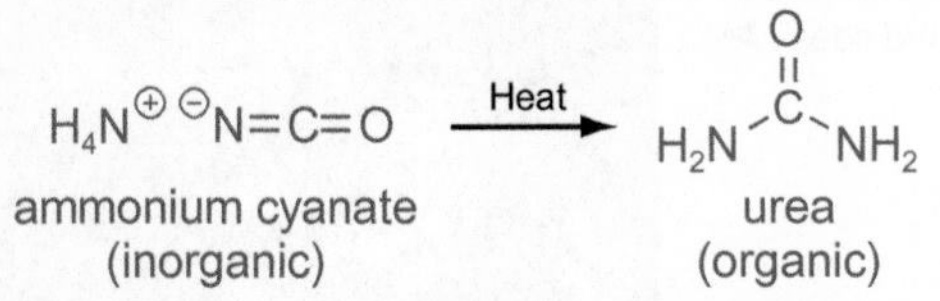
Heat
ammonium cyanate
(inorganic)
urea
(organic)

▲ Ammonium cyanate reacts to form urea.

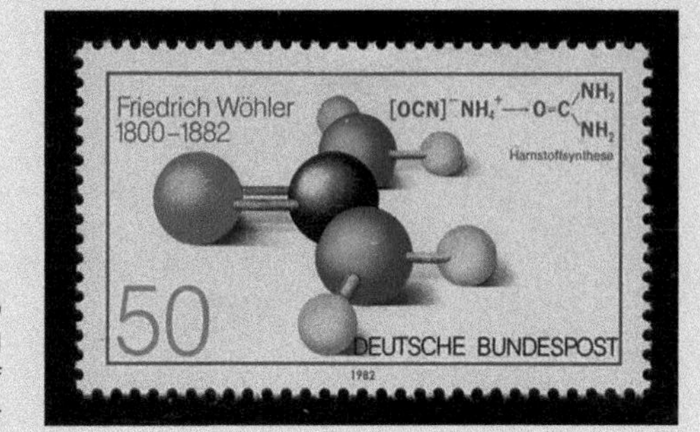
Friedrich Wöhler
1800–1882
Harnstoffsynthese
50
DEUTSCHE BUNDESPOST

This stamp commemorated the centennial of Wöhler's death. ▶

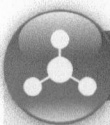

Box 2.2 Some landmark laboratory syntheses of natural products

1828

urea (F. Wöhler)

Occurs in the urine of all mammals

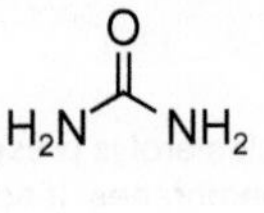

O
H_2N
NH_2

1845

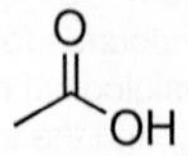

O
OH

ethanoic acid (H. Kolbe)

The acid found in vinegar

1890

glucose (E. Fischer)

An important source of energy for plants and animals

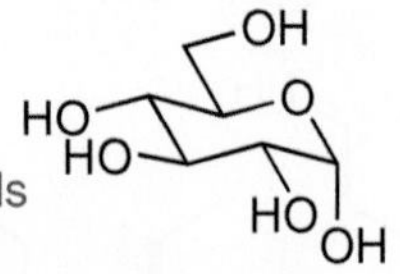

OH
HO
HO
O
HO
OH

1917

tropinone (R. Robinson)

A precursor of atropine, which is found in the deadly nightshade plant

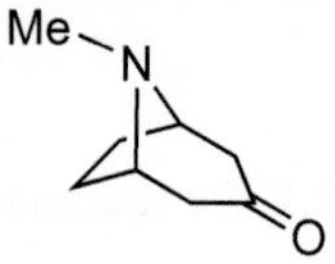

Me
N
O

1944

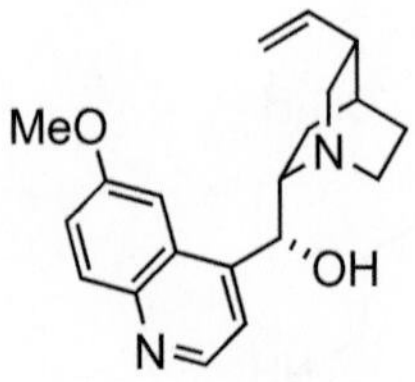

MeO
N
OH
N

quinine (R.B. Woodward and W.E. Doering)

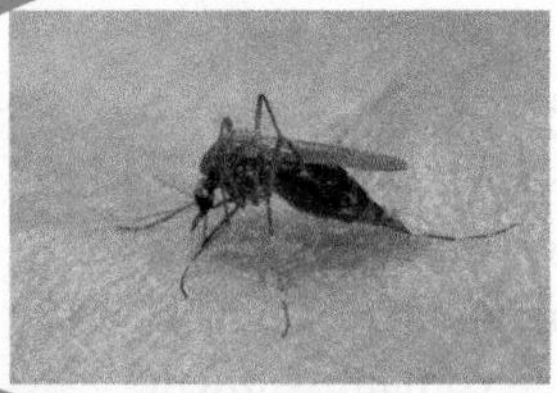

◀ Quinine, from the bark of the cinchona tree, has been used to treat and prevent malaria for over 300 years. Malaria kills more than a million people each year and up to 500 million people are infected with the malaria parasites that are passed on by mosquitoes. The bitter taste of antimalarial quinine tonic led British colonials in India to mix it with gin, to create the gin and tonic cocktail.

1973

vitamin B_{12} (R.B. Woodward and A. Eschenmoser)

Required for the production of red blood cells. Me is shorthand for the methyl group, CH_3

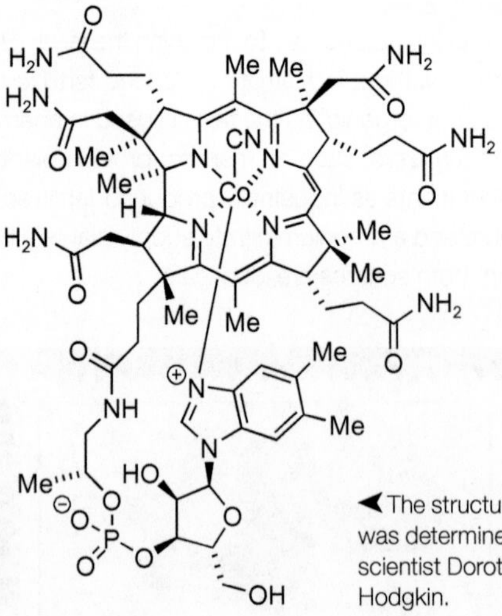

H_2N
H_2N
NH_2
Me Me
CN
Co
N
NH_2
H_2N
Me
NH_2
NH
Me
HO
N
O
P
OH

◀ The structure of vitamin B_{12} was determined by the British scientist Dorothy Crowfoot Hodgkin.

1995

brevetoxin B (K.C. Nicolaou)

A neurotoxin produced by algae

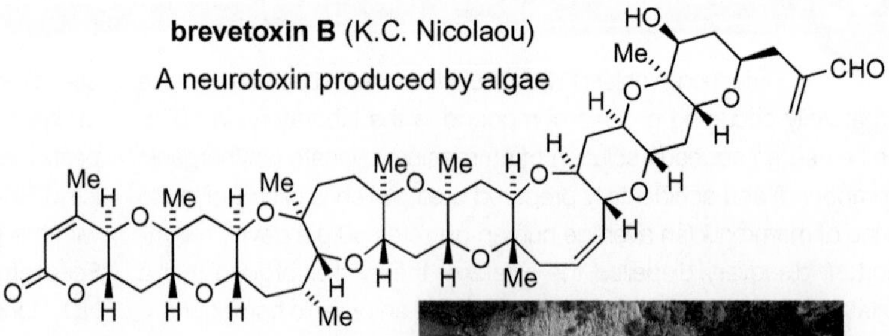

HO
Me
CHO
H
O
Me

Brevetoxin B is secreted by harmful algal blooms that proliferate during 'red tides'. ▶

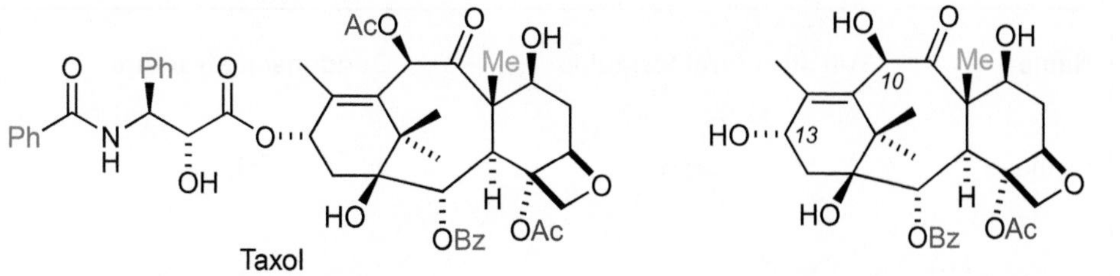
AcO
O
OH
Me
O
Ph
O
Ph
N
H
O
OH
HO
H
ŌBz
ŌAc
HO
O
OH
Me
10
HO
13
HO
H
ŌBz
ŌAc

Taxol

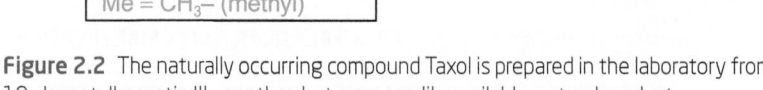

Bz = C_6H_5CO– (benzoyl)
Ac = CH_3CO– (acetyl)
Ph = C_6H_5– (phenyl)
Me = CH_3– (methyl)

10-deacetylbaccatin III

Similar to Taxol but lacks the acetyl group at position 10 and the ester side chain at position 13

Figure 2.2 The naturally occurring compound Taxol is prepared in the laboratory from 10-deacetylbaccatin III—another, but more readily available, natural product.

The anti-cancer drug Taxol is isolated from the bark of the Pacific Yew tree (*Taxus brevifolia*). However, sacrificing a 100 year-old tree provides only about 300 mg of Taxol, which is equivalent to a single dose for a cancer patient. This is unsustainable, so a laboratory synthesis was developed starting from 10-deacetylbaccatin III, harvested from the needles of the much quicker growing European Yew tree (*Taxus baccata*), which is a renewable source (shown above).

Part of the chemical industry is concerned with the large-scale preparation of compounds found in nature. Natural products are often produced more cheaply in the laboratory than by isolation from the natural source. This is especially the case with compounds such as Taxol (see Figure 2.2), which are formed in only small quantities in nature. Remember that synthetic natural products are identical to the same compounds isolated from nature (Box 2.2).

Unfortunately, some synthetic organic compounds can also cause environmental problems when released into the atmosphere. Chlorofluorocarbons (CFCs), once used as organic refrigerants and aerosol propellants, have contributed to both ozone depletion and global warming. Today, solutions are being developed for such problems and chemists also play a key part in reducing the amount of chemical waste that is produced as well as developing methods for reusing and recycling organic products.

Common replacements for CFCs (like CF_2Cl_2) are hydrofluorocarbons, such as CH_2FCF_3 and $H_2C{=}CFCF_3$ (Box 27.8, p.1241).

Summary

- Organic chemistry is the study of carbon compounds.
- Organic molecules are central to life.
- Organic chemists prepare synthetic compounds in the laboratory.

2.2 Drawing organic compounds

Although chemists from different countries may not understand each other's spoken words, chemical formulae are universal. It is important to know what atoms are present in a molecule and, apart from the smallest of molecules, how the atoms are arranged. Several different conventions are used for showing chemical structures, each providing different levels of detail. One way of representing the structures of small molecules is by using **dot and cross structures** (sometimes called **electron dot** or **Lewis structures**). These structures show the number of outer electrons in each atom and how these are paired to form covalent bonds and lone pairs. This is illustrated in the margin for methane (CH_4), where the four outer electrons of carbon (shown by crosses) are shared with the electrons from four hydrogen atoms (shown by dots) to form four covalent bonds.

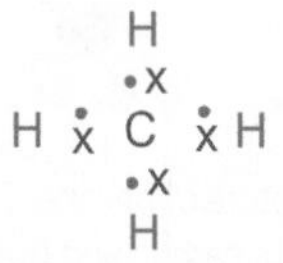
H
H C H
H

Dot and cross structure of methane (CH_4).

Lewis structures are discussed in Section 4.2 (p.173).

For larger molecules, showing all of the outer electrons is cumbersome and simpler forms are used. In a **full structural formula** (sometimes called a **displayed formula**), a line is drawn between two atoms to show a shared pair of electrons. Two or three lines represent double or triple bonds, respectively. As shown in Figure 2.3 for propanone (CH_3COCH_3) and

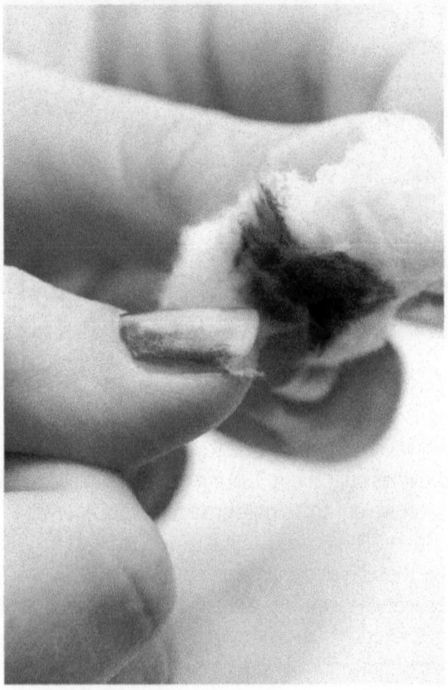

Propanone, or acetone, is an important organic solvent used in cleaning agents such as nail polish remover.

Name	Full structural formulae	Condensed structure
methane	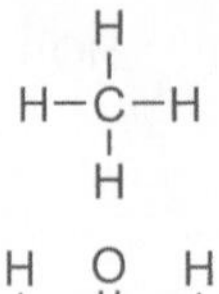	CH_4
propanone	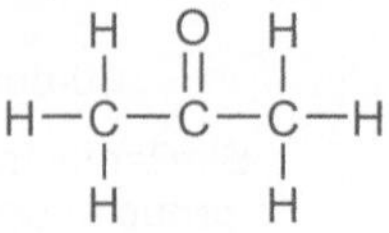	CH_3COCH_3 or MeCOMe
propan-2-ol	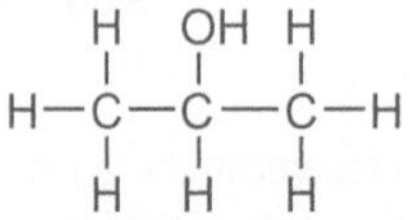	$CH_3CH(OH)CH_3$ or MeCH(OH)Me

Figure 2.3 Full structural formulae and condensed structures.

propan-2-ol ($CH_3CH(OH)CH_3$), the lone pairs of electrons on oxygen (or any other heteroatom) are not usually included unless they are needed to illustrate a reaction mechanism.

In **condensed structural formulae**, all of the atoms are shown but some, if not all, of the bonds are omitted. Lone pairs are omitted and substituents can be enclosed in brackets, as shown for propan-2-ol in Figure 2.3. Common alkyl and aryl group substituents and other functional groups are often shown in abbreviated form. For example, 'Me' is used to represent a methyl group (rather than CH_3) and 'Ph' for a phenyl group (rather than C_6H_5). Further use of these and other abbreviations is discussed in Section 2.3 (p.77). Note also that CO_2H is used for a carboxylic acid (rather than COOH).

The ultimate abbreviation of organic structures uses lines to represent the carbon framework. In these **skeletal structures**, each line segment is understood to have a carbon atom at each end unless another atom, such as oxygen or nitrogen, is shown. Hydrogen atoms attached to oxygen or nitrogen atoms are shown but not those attached to carbon. You must remember that the correct number of hydrogen atoms (the number required to fill the valence shell of each carbon) is present. As in full structural formulae, two or three lines represent multiple bonds. Importantly, chains of carbon atoms containing C–C and C=C bonds are shown in a zigzag manner to indicate the approximate shape of the molecule. Some skeletal structures and the corresponding condensed structures are shown in Figure 2.4.

Carbon atoms with four single bonds are sp^3 hybridized and the four bonds point to the corners of a tetrahedron. The bond angles around each carbon atom are approximately 109.5° and this is represented in two dimensions by the zigzag skeletal structures in Figure 2.4 (though note that the bond angles in the skeletal structures are actually 120°). Double bonds between carbon atoms (which are sp^2 hybridized) have bonds angles of 120° so double bonds are usually incorporated into the zigzag carbon chain of a skeletal structure—but remember that the presence of a double bond will disrupt the regular structure of a saturated carbon

Name	Condensed structure	Skeletal structure
butane	$CH_3CH_2CH_2CH_3$	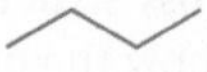
butan-1-ol	$CH_3CH_2CH_2CH_2OH$	
but-3-en-2-one	$CH_3COCH{=}CH_2$	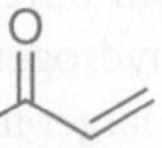
but-2-yne	$CH_3C{\equiv}CCH_3$	

Figure 2.4 Condensed structures and skeletal structures.

A reaction mechanism gives a step-by-step picture of how reactants are converted into products. It shows which bonds break and which bonds form, and in what order. (For an introduction to organic reaction mechanisms, see Chapter 19, p.860.)

A **heteroatom** is an atom that is not carbon or hydrogen.

Alkyl group substituents (e.g. CH_3) contain only carbon and hydrogen atoms, which are joined together by C–C and C–H single bonds (Box 2.3, p.81).

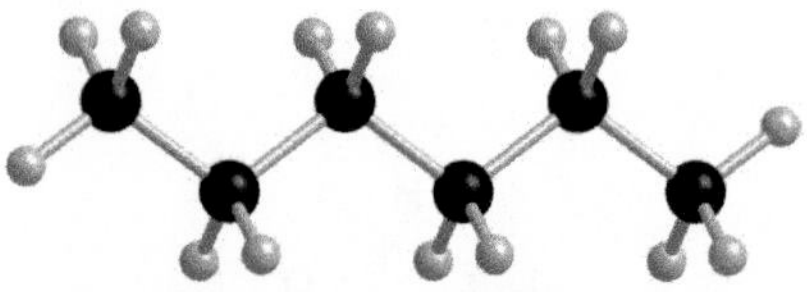

The zigzag structure of hexane (C_6H_{14}; black spheres represent carbon and the grey spheres represent hydrogen). The bond angles are all 109.5°.

Hybridization describes how different atomic orbitals of similar energies combine to form a set of equivalent hybrid orbitals. In methane (CH_4), the carbon $2s$ and $2p$ orbitals are hybridized to form four equivalent sp^3 orbitals. The sp^3 orbitals combine with the $1s$ orbitals of four hydrogen atoms to form the four covalent bonds. (You can find more about hybridization in Section 5.4, p.236.)

chain. However, triple bonds between carbon atoms (which are *sp* hybridized) have bond angles of 180° so a linear structure is shown. Skeletal structures are commonly used, particularly for large molecules, because they are easily drawn, the functional groups are clearly shown, and the structures of carbon skeletons are simplified.

Benzene (C_6H_6) is a special case. For the skeletal structure, all of the carbon and hydrogen atoms of benzene are omitted and a circle inside a six-membered ring is used to represent the six delocalized electrons. Alternatively, a **Kekulé structure** shows alternating single and double carbon–carbon bonds within a hexagon. To illustrate clearly the mechanisms of reactions at the benzene ring a single Kekulé structure is preferred (and will be used throughout this book), as this provides an accurate accounting of electrons.

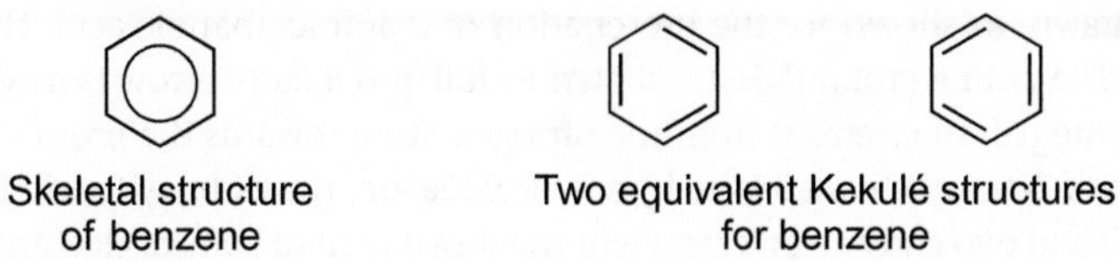
Skeletal structure of benzene
Two equivalent Kekulé structures for benzene

T+

Benzene is a carcinogen that can cause cancer in humans (Box 21.4, p.988). Until the late 1970s, many hardware shops stocked benzene for general-purpose use, as a solvent, and it was commonly found in school and university laboratories.

Organic molecules are three-dimensional, so showing chemical structures on paper (in two dimensions) can give a misleading picture of what the molecule actually looks like. This has led to the use of the **hashed–wedged line notation**, which provides a simple way of drawing organic molecules to indicate their three-dimensional nature. This is shown below for methane, which has a tetrahedral shape.

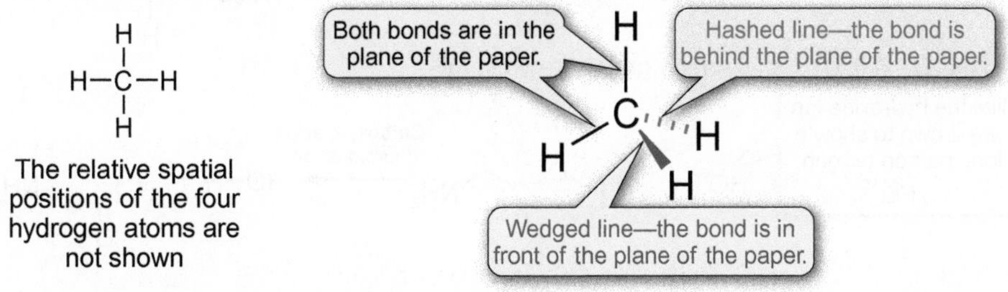
The relative spatial positions of the four hydrogen atoms are not shown
Both bonds are in the plane of the paper.
Hashed line—the bond is behind the plane of the paper.
Wedged line—the bond is in front of the plane of the paper.

A molecular model of methane (CH_4).

To represent the three-dimensional structure of organic molecules with carbon chains, the zigzag carbon skeleton is drawn first and hashed or wedged lines are added to represent the two remaining bonds on tetrahedral carbon atoms. This is shown in Figure 2.5 for 2-hydroxybutanoic acid, $CH_3CH_2CH(OH)CO_2H$ (and also for captopril on p.69). In Figure 2.5, the three-dimensional structure is not shown in structure (a), but for structure (b) the hashed line shows the bond that is pointing behind the page, while the wedged line shows the bond that points out of the page towards the reader. These lines are drawn pointing in opposite directions so as to further indicate that the carbon atom is tetrahedral. Variations of this notation are also common. For example, in structure (c), the wedged line to the hydrogen atom is not shown and the hashed line to the OH group points directly up (with respect to the zigzag chain). Although this does not represent the shape of the molecule quite as well, it is a neater representation.

The number 2 carbon atom of 2-hydroxybutanoic acid has four different substituents and is called an asymmetric carbon atom. This means that the molecule is chiral and two isomers are possible. For carbon atom number 2, one isomer has the OH group pointing

Visit the Online Resource Centre to view screencast 2.1 which walks you through different ways of drawing organic compounds (including use of hashed and wedged lines) focusing on skeletal structures.

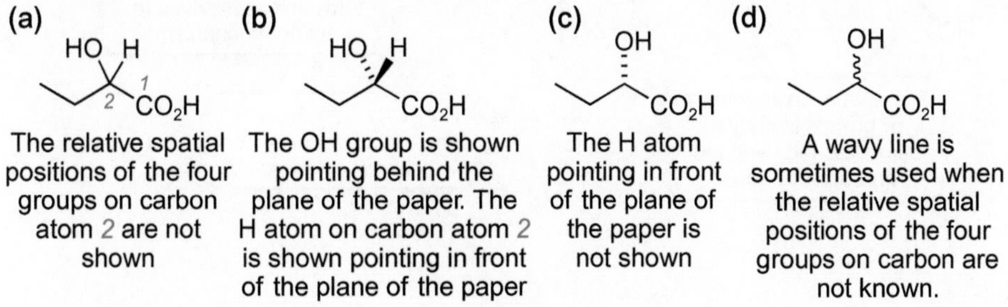
(a) The relative spatial positions of the four groups on carbon atom 2 are not shown
(b) The OH group is shown pointing behind the plane of the paper. The H atom on carbon atom 2 is shown pointing in front of the plane of the paper
(c) The H atom pointing in front of the plane of the paper is not shown
(d) A wavy line is sometimes used when the relative spatial positions of the four groups on carbon are not known.

Figure 2.5 (a)–(d) Different representations of 2-hydroxybutanoic acid.

You will sometimes see a wavy line also used in structures such as (d), where, for half of the molecules, the OH group points behind the plane of the paper, and for the other half, the OH group points in front of the plane of the paper.

Sometimes the hashed line is drawn with lines of different length.

A molecule is chiral if it cannot be superimposed on its mirror image. The two non-superimposable mirror images are different compounds and are called **enantiomers** (Section 18.4, p.838).

in front of the plane of the paper and the H pointing behind, while the second isomer has the OH group pointing behind the plane of the paper and the H pointing in front. A wavy line usually indicates the presence of an equal mixture of chiral isomers (structure (d)).

As you can see, organic molecules can be represented in a number of different ways and the choice of structure depends on the context. In general, skeletal structures are preferred (and used throughout this book) with important parts of the molecule drawn in full with hashed and wedged lines as appropriate. For chemical reactions, the most helpful structures show, in detail, only those parts of the molecules involved in the reaction. The reacting functional groups are drawn in full, and this includes showing lone pairs on any heteroatoms (e.g. on any O or N atoms), while other parts of the molecule are simplified. This allows clear reaction mechanisms to be drawn, as shown for the protonation of 2-aminoethanoic acid, $HO_2CCH_2NH_2$, in Figure 2.6. The amine group (NH_2) is drawn in full and a curly arrow is used to show the movement of the pair of electrons from the nitrogen atom towards the proton. Similarly, for deprotonation of 2-aminoethanoic acid by a hydroxide ion, the carboxylic acid group (CO_2H) is drawn in full and two curly arrows represent the movement of pairs of electrons.

2-Aminoethanoic acid is also called glycine. Glycine is produced in the body and has the simplest structure of the naturally occurring amino acids. Amino acids, as the name suggests, contain both an amine group and a carboxylic acid group (Box 2.9, p.104).

A proton is a hydrogen ion (H^+) and deprotonation is the removal of H^+ from a molecule.

Using curly arrows to represent mechanisms is an important part of organic chemistry (see Section 19.1, p.863).

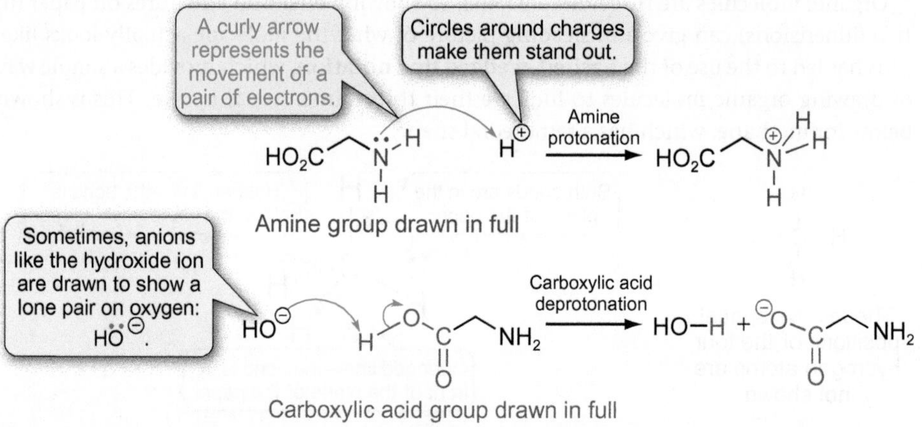
A curly arrow represents the movement of a pair of electrons.
Circles around charges make them stand out.
Amine protonation
Amine group drawn in full
Sometimes, anions like the hydroxide ion are drawn to show a lone pair on oxygen:
Carboxylic acid deprotonation
Carboxylic acid group drawn in full

Figure 2.6 Protonation and deprotonation of 2-aminoethanoic acid.

Summary

- Organic compounds are drawn using full structural formulae, condensed structures, and skeletal structures.
- The three-dimensional structure of a molecule is represented using hashed–wedged notation.
- Skeletal structures are generally preferred with important parts of the molecule drawn in full.

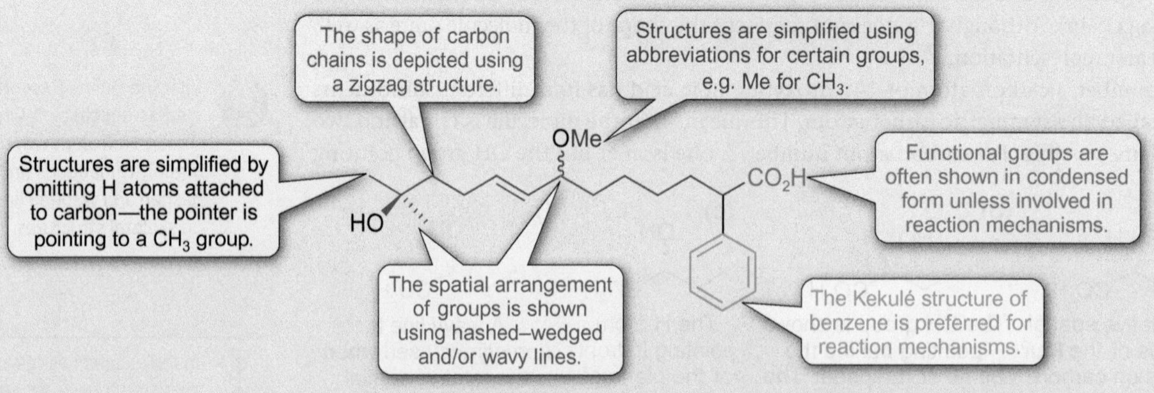
The shape of carbon chains is depicted using a zigzag structure.
Structures are simplified using abbreviations for certain groups, e.g. Me for CH3.
Structures are simplified by omitting H atoms attached to carbon—the pointer is pointing to a CH3 group.
Functional groups are often shown in condensed form unless involved in reaction mechanisms.
The spatial arrangement of groups is shown using hashed–wedged and/or wavy lines.
The Kekulé structure of benzene is preferred for reaction mechanisms.

For practice questions on these topics, see questions 1–3 at the end of this chapter (p.109).

2.3 Carbon frameworks and functional groups

When faced with the large number of organic compounds, how do you start to study their chemistry? One way is to group together structurally related compounds according to their **functional groups**. Functional groups are groups of atoms that give the molecule characteristic chemical properties. An example of a functional group is the hydroxyl group (–OH). Compounds that have this group attached to a hydrocarbon chain or ring are called **alcohols**. Alcohols are an example of a **class of compounds**.

The **carbon framework** attached to the functional group usually does not significantly affect the chemical properties—the C–C and C–H bonds in the framework are difficult to break because they are strong, non-polar covalent bonds. This means that compounds with similar structures, and the same functional groups, undergo the same types of reactions. A general notation is often used to represent the structure of a class of compounds. For alcohols this is R–OH, where R represents the carbon framework.

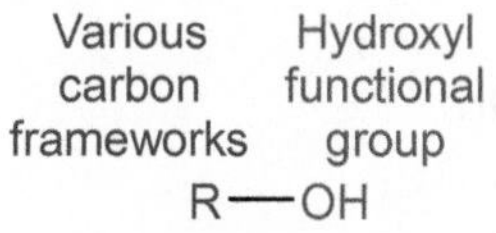
Various carbon frameworks
Hydroxyl functional group
R—OH
General structure of an alcohol

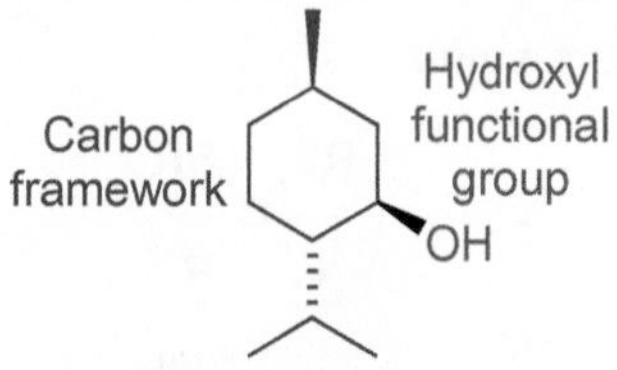
Carbon framework
Hydroxyl functional group
OH
menthol

Isolated from peppermint oil. It is a local mild anaesthetic. As it imparts a tingling sensation to the skin, it is used in aftershave lotions

Peppermint oils are obtained from various species of the *Mentha* plant, including *Mentha piperita*. The oils are widely used for flavouring, cosmetic, and medicinal purposes, including relieving the symptoms of coughs and colds.

In organic structures, R represents the carbon framework, typically an alkyl group (Box 2.3, p.81).

Menthol is a chiral molecule (Section 18.4, p.838)—the (–)-enantiomer of menthol is found in nature and is drawn here.

The carbon framework of organic molecules can be thought of as being derived from **alkanes**, the simplest class of hydrocarbons, which are made up of C–C and C–H single bonds. For example, ethane (CH_3CH_3) has six C–H bonds and one C–C bond. When a C–H bond in ethane is replaced by a C–OH bond, this gives ethanol (CH_3CH_2OH), which is an alcohol. When the C–C bond in ethane is replaced by a C=C bond, this gives ethene ($H_2C{=}CH_2$). The C=C bond is a functional group characteristic of a class of compounds called **alkenes**. As is typical for a class of compounds, most alkenes undergo the same types of reactions, which take place at or near the C=C bond.

To understand the chemical reactions of organic molecules, you need to be able to recognize functional groups. To help with this, a summary of the most important classes of compounds and functional groups is shown in Figure 2.7. This is organized according to the types of heteroatoms and multiple bonds that are part of the functional groups. The letter R is used to represent the structure of the carbon framework although, where indicated, R can also represent an H atom. When carbon frameworks are attached to the benzene ring, these compounds are called **substituted benzenes**.

Alkenes react with numerous electrophiles in electrophilic addition reactions (Section 21.3, p.970)

Hydrocarbons are compounds that contain only carbon and hydrogen.

Summary

- Organic compounds are classified by their functional groups.
- A functional group is a group of atoms in a compound that is responsible for the characteristic chemical reactions of the compound.

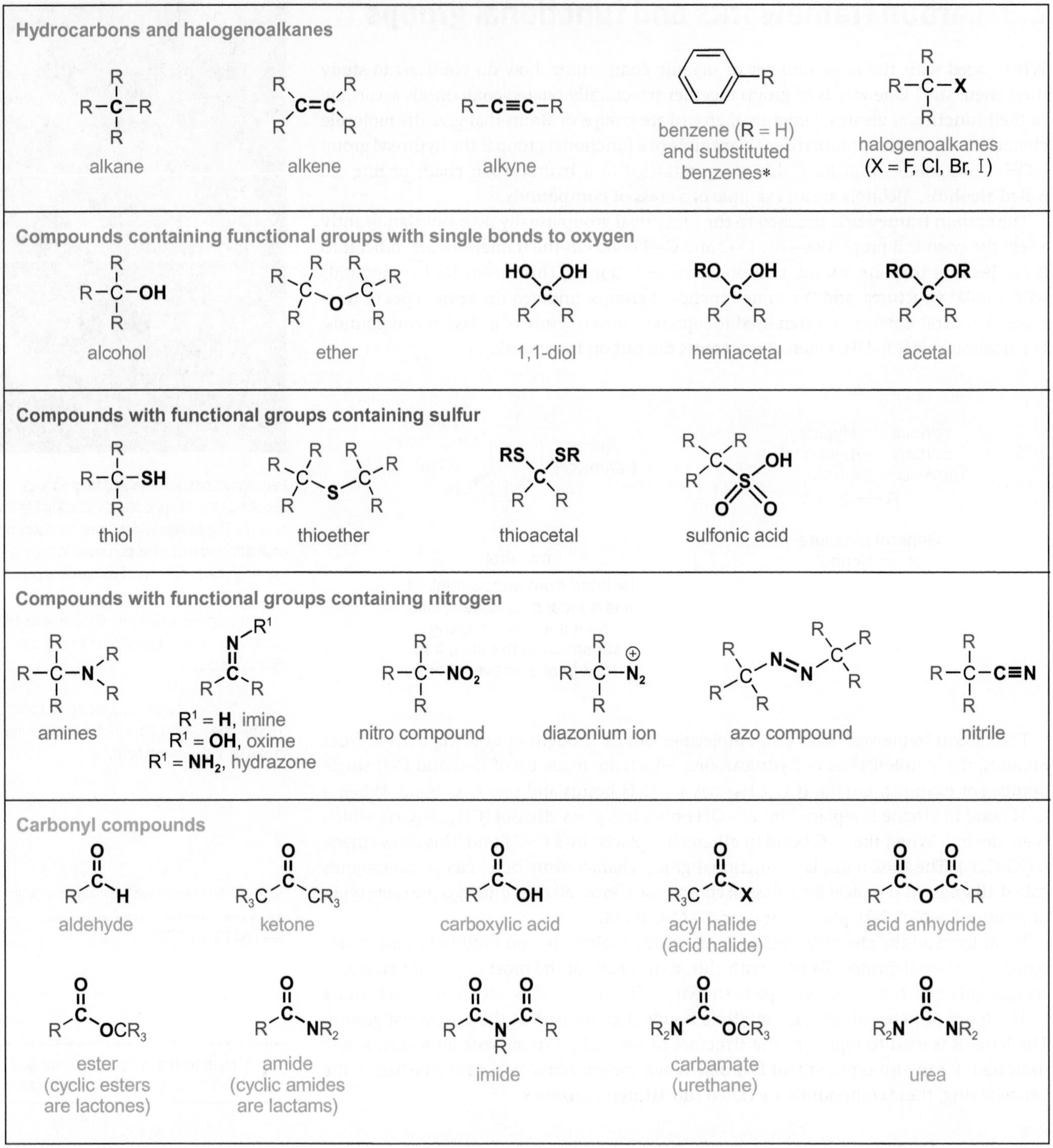
Hydrocarbons and halogenoalkanes
alkane
alkene
alkyne
benzene (R = H) and substituted benzenes*
halogenoalkanes (X = F, Cl, Br, I)
Compounds containing functional groups with single bonds to oxygen
alcohol
ether
1,1-diol
hemiacetal
acetal
Compounds with functional groups containing sulfur
thiol
thioether
thioacetal
sulfonic acid
Compounds with functional groups containing nitrogen
amines
R1 = H, imine
R1 = OH, oxime
R1 = NH2, hydrazone
nitro compound
diazonium ion
azo compound
nitrile
Carbonyl compounds
aldehyde
ketone
carboxylic acid
acyl halide (acid halide)
acid anhydride
ester (cyclic esters are lactones)
amide (cyclic amides are lactams)
imide
carbamate (urethane)
urea

Figure 2.7 Classes of organic compounds and functional groups. Atoms that make up the functional group for each class of compounds, apart from benzene and substituted benzenes, are shown in **bold** in black; R (in purple) represents a carbon framework or H.

* Substituted benzenes can have more than one carbon framework attached to the six-membered ring.

2.4 Naming organic compounds

Starting from the early days of organic chemistry, the names of new compounds have often been based on their source or use. For example, some carboxylic acids (RCO_2H) are named after their source (see Table 2.1). Unfortunately, these common (trivial) names do not help us to work out the structure of the compound; for this, you need to use the systematic method developed by IUPAC.

Alkanes form the basis of the IUPAC naming system and compounds are named as derivatives of the parent alkane. The next section shows how alkanes are named using the IUPAC rules. Sections on naming other hydrocarbons and then compounds with functional groups containing at least one heteroatom follow. IUPAC names are generally used throughout the book. However, common names are also given for some frequently used small molecules and you need to be familiar with these as well as the IUPAC names (see Table 2.5, p.109). Common names are also useful for complex molecules, where the IUPAC name is very long (e.g. captopril on p.69).

The IUPAC naming system is introduced in Section 1.2 (p.6).

A compound can have more than one name, but a name must specify only one compound.

Butanoic acid has a very unpleasant smell—it is a major cause of the odours of human vomit and flatulence. Perhaps unsurprisingly, it has been used as a component in stink bombs.

Table 2.1 Systematic and common names of some carboxylic acids

Structure	IUPAC name	Common name	Natural source (Latin name)
HCO_2H	methanoic acid	formic acid	ant (*formica*)
CH_3CO_2H	ethanoic acid	acetic acid	vinegar (*acetum*)
$CH_3CH_2CH_2CO_2H$	butanoic acid	butyric acid	butter (*butyrum*)
$CH_3CH_2CH_2CH_2CO_2H$	pentanoic acid	valeric acid	a flowering plant (*valeriana*)
$CH_3CH_2CH_2CH_2CH_2CO_2H$	hexanoic acid	caproic acid	goat (*caper*)

2.5 Hydrocarbons

Carbon has four bonding (valence) electrons in its outer shell and so tends to form a total of four bonds. If all four bonds are to hydrogen, then this gives methane (CH_4), which is the smallest member of the alkane family. Linking carbon atoms using single bonds forms open chains of different lengths. These compounds are commonly referred to as **aliphatic alkanes** (see Figure 2.8). Carbon atoms linked in one continuous chain are called **straight-chain alkanes**, whereas **branched alkanes** are molecules containing one or more branching points. In contrast to aliphatic alkanes, **alicyclic** alkanes are cyclic molecules that have a ring (closed chain) of carbon atoms.

An important characteristic of all alkanes is that they only contain single bonds and each compound contains the maximum number of hydrogen atoms. These compounds are called **saturated** hydrocarbons. Aliphatic alkanes have the general molecular formula C_nH_{2n+2} (where n is 1, 2, 3, 4, etc.) while alicyclic alkanes have the general formula C_nH_{2n} (where n is 3, 4, 5, etc.). Compounds with double or triple bonds, such as alkenes and alkynes, are **unsaturated** compounds because they contain fewer hydrogen atoms per carbon. Unsaturated compounds can be aliphatic or alicyclic, while unsaturated molecules containing a benzene ring are called **aromatic** hydrocarbons.

Alkanes are used as fuels, for generating heat (using natural gas and heating oil) and mechanical energy (from petrol). Recently, propane, or a mixture of butane and propane, has been used as a fuel for Olympic torches. The London 2012 Olympic Torch features 8000 small cut-out circles, representing 8000 inspirational people who carried it on its journey (around the UK) before the start of the games.

Alkanes

The structures and names of the first 12 straight-chain alkanes are shown in Table 2.2. Each member differs from the next by a CH_2 (methylene) group. A series of compounds related in this way is called a **homologous series** (*homos* is Greek for 'the same as'). Each

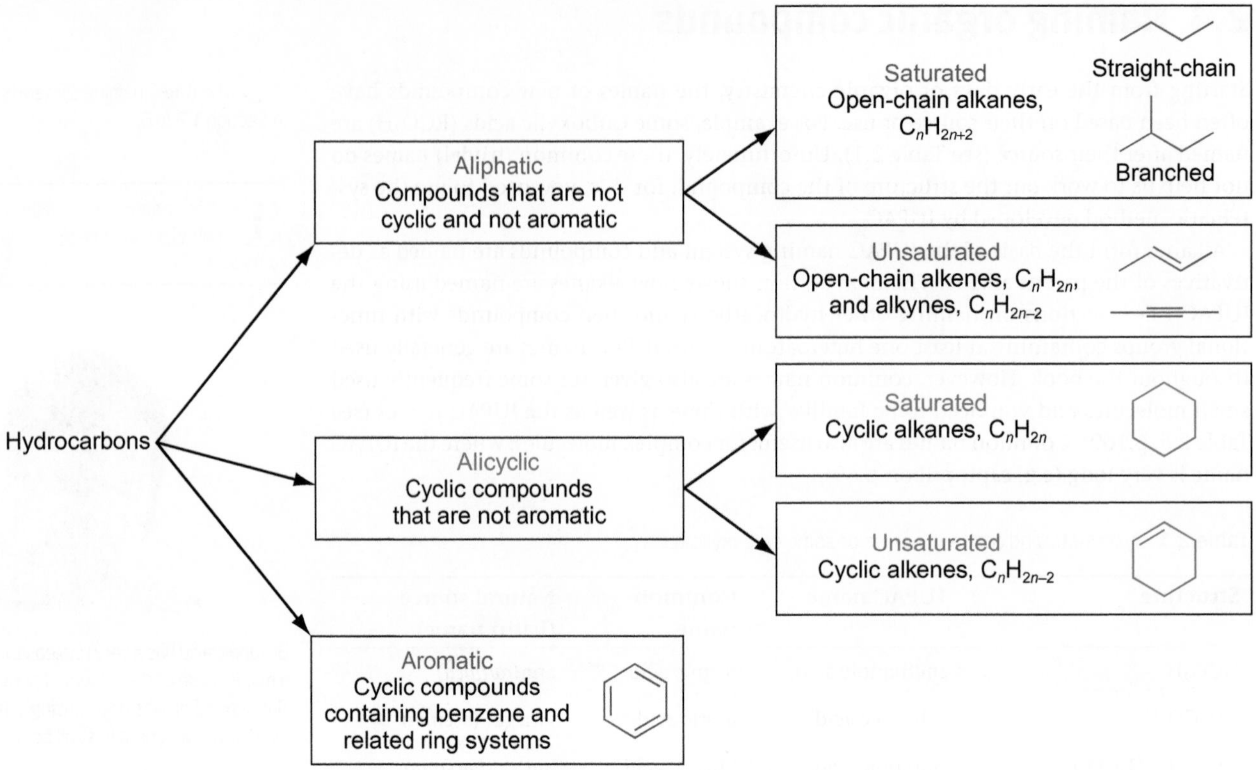
Hydrocarbons
Aliphatic
Compounds that are not cyclic and not aromatic
Alicyclic
Cyclic compounds that are not aromatic
Aromatic
Cyclic compounds containing benzene and related ring systems
Saturated
Open-chain alkanes, C_nH_{2n+2}
Straight-chain
Branched
Unsaturated
Open-chain alkenes, C_nH_{2n}, and alkynes, C_nH_{2n-2}
Saturated
Cyclic alkanes, C_nH_{2n}
Unsaturated
Cyclic alkenes, C_nH_{2n-2}

Figure 2.8 Different types of hydrocarbons.

Sometimes the names of straight-chain alkanes are given the prefix '*n*-' (where *n* means normal). For example, butane ($CH_3CH_2CH_2CH_3$) is sometimes called *n*-butane.

The suffix for alkanes is -ane.

Table 2.2 Names, molecular formulae, and structures of the first 12 straight-chain alkanes

Number of carbons	Name	Molecular formula	Condensed structure
1	methane	CH_4	CH_4
2	ethane	C_2H_6	CH_3CH_3
3	propane	C_3H_8	$CH_3CH_2CH_3$
4	butane	C_4H_{10}	$CH_3CH_2CH_2CH_3$
5	pentane	C_5H_{12}	$CH_3CH_2CH_2CH_2CH_3$
6	hexane	C_6H_{14}	$CH_3CH_2CH_2CH_2CH_2CH_3$
7	heptane	C_7H_{16}	$CH_3CH_2CH_2CH_2CH_2CH_2CH_3$
8	octane	C_8H_{18}	$CH_3CH_2CH_2CH_2CH_2CH_2CH_2CH_3$
9	nonane	C_9H_{20}	$CH_3CH_2CH_2CH_2CH_2CH_2CH_2CH_2CH_3$
10	decane	$C_{10}H_{22}$	$CH_3CH_2CH_2CH_2CH_2CH_2CH_2CH_2CH_2CH_3$
11	undecane	$C_{11}H_{24}$	$CH_3CH_2CH_2CH_2CH_2CH_2CH_2CH_2CH_2CH_2CH_3$
12	dodecane	$C_{12}H_{26}$	$CH_3CH_2CH_2CH_2CH_2CH_2CH_2CH_2CH_2CH_2CH_2CH_3$

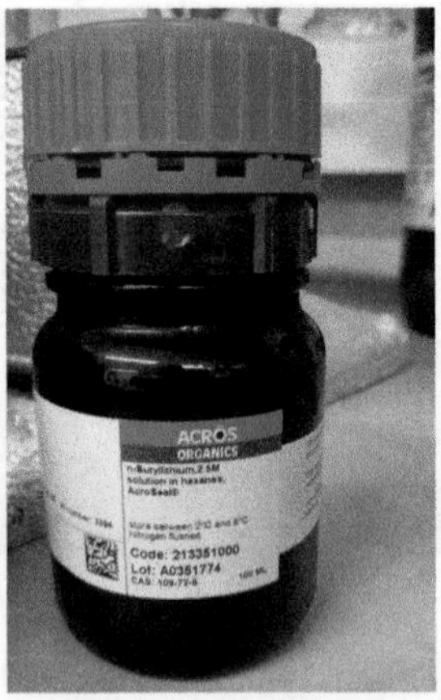
ACROS
ORGANICS

Alkanes are fairly unreactive and they are used as solvents in the laboratory. For example, reactive organometallic compounds, such as *n*-butyllithium (BuLi), are available commercially as solutions in alkanes, such as pentane, hexane, or heptane (Section 26.10, p.1195).

member of a homologous series is known as a **homologue** and homologues undergo essentially the same types of chemical reactions.

The IUPAC name of a molecule is made up of three parts: the **suffix**; the **parent chain**; and the **prefix** (see Figure 2.9). As shown in Table 2.2, all alkane names end with -ane and this ending is the suffix used for all saturated hydrocarbons. Whereas the suffix tells you

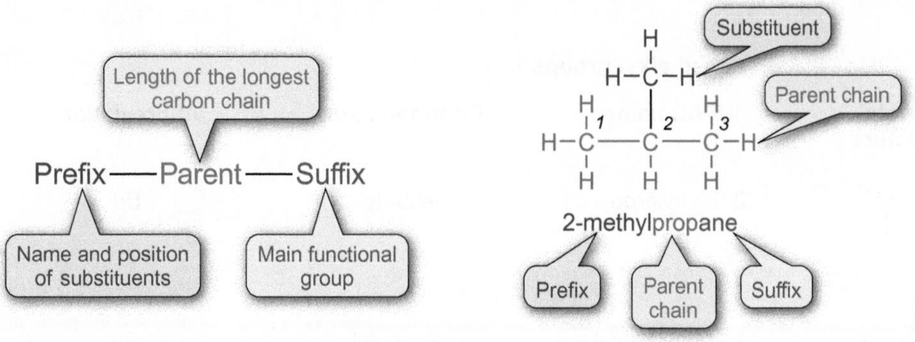
Length of the longest carbon chain
Prefix—Parent—Suffix
Name and position of substituents
Main functional group
Substituent
Parent chain
2-methylpropane
Prefix
Parent chain
Suffix

Figure 2.9 Naming compounds using the IUPAC system.

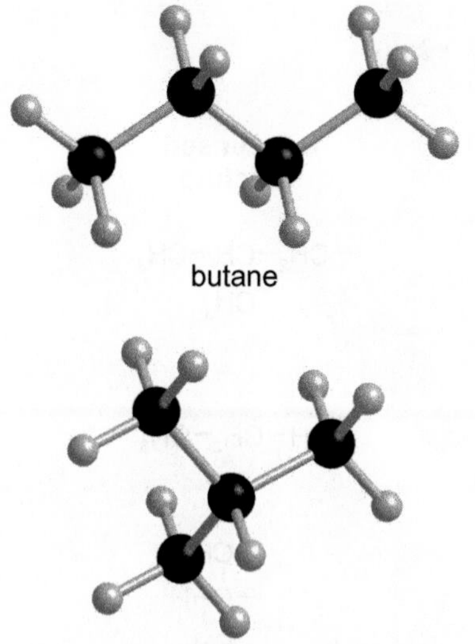
butane
2-methylpropane

Models showing the different shapes of butane ($CH_3CH_2CH_2CH_3$) and 2-methylpropane ($CH_3CH(CH_3)CH_3$)

the major functional group of the molecule (i.e. -ane for alkane), the parent chain tells you the length of the longest carbon chain (i.e. 'eth' for two carbon atoms, 'pent' for five carbon atoms). For naming straight-chain alkanes this is all you need consider, but for branched alkanes (or for molecules that contain more than one functional group) a prefix is needed.

Branched-chain alkanes are formed when a carbon atom forms bonds to three or four other carbon atoms (rather than to two other carbon atoms in a straight chain). The simplest example is 2-methylpropane shown in Figure 2.9. As the parent chain contains three carbons (numbered *1–3*) the molecule is a derivative of propane. The CH_3 substituent is bonded to the number 2 carbon atom, so the prefix 2-methyl is used. (For this example, the 2- is not strictly needed because the CH_3 group can only be joined to the number 2 carbon in methylpropane.) Names for substituent groups such as CH_3 are given in Box 2.3.

Butane ($CH_3CH_2CH_2CH_3$) and 2-methylpropane ($CH_3CH(CH_3)CH_3$) both have the same molecular formula (C_4H_{10}) but differ in the order in which the atoms are connected together. They are **structural isomers**. Note that butane and 2-methylpropane are different compounds and so have different physical properties, such as boiling points.

Structural isomers (sometimes called constitutional isomers) are compounds that have the same molecular formula but differ in the order the atoms are bonded together. Different types of isomers are discussed in detail in Section 18.1 (p.816).

Box 2.3 Alkyl group substituents

Groups, such as CH_3, that are attached to the main carbon chain are called **substituents**. The methyl group is an example of an **alkyl group** because it is saturated and contains only carbon and hydrogen atoms. Removing a hydrogen atom from an alkane gives an alkyl group, so the name of an alkyl group is derived from the corresponding alkane. The 'ane' in alkane is replaced by 'yl' and so methyl (CH_3) is derived from methane (CH_4).

Alkyl groups can also contain branched chains as illustrated by the four substituents in the figure below. For convenience, these groups are often known by their shorter common names rather than IUPAC names. The 2-methylpropyl substituent is often called the isobutyl group and the abbreviation ^{i}Bu is used in chemical structures. The use of 'iso' comes from the common name for alkanes that have an otherwise unbranched chain terminating in a $-CH(CH_3)_2$ group: the common name for 2-methylpropane ($H_3C-CH(CH_3)_2$) is isobutane as this is a structural isomer of butane. The use of superscripts s and t in ^{s}Bu and ^{t}Bu, respectively, relates to the point of attachment of the alkyl group. The ^{s}Bu group is attached by a secondary (*sec*-) carbon atom and the ^{t}Bu group by a tertiary (*tert*-) carbon atom (this is discussed later in this section on p.83).

Straight-chain alkyl groups

Condensed structure	IUPAC name	Abbreviation
$-CH_3$	methyl	Me
$-CH_2CH_3$	ethyl	Et
$-CH_2CH_2CH_3$	propyl	Pr
$-CH_2CH_2CH_2CH_3$	butyl	Bu

▲ Names of straight-chain alkyl groups.

→

Branched alkyl groups

Condensed structure	Skeletal structure	IUPAC name	Common name	Abbreviation
$-CH_2-CH-CH_3$ CH_3	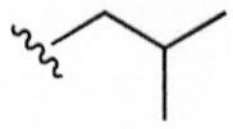	2-methylpropyl	isobutyl	^{i}Bu
CH_3 $-CH-CH_2-CH_3$	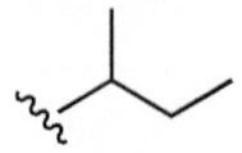	1-methylpropyl	*sec*-butyl	^{s}Bu
CH_3 $-C-H$ CH_3	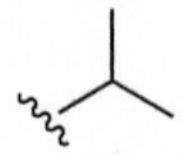	1-methylethyl	isopropyl	^{i}Pr
CH_3 $-C-CH_3$ CH_3	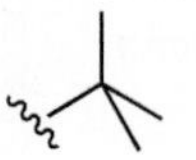	1,1-dimethylethyl	*tert*-butyl	^{t}Bu

▲Names of branched alkyl groups. The wavy line (∿∿∿) is used to indicate an incomplete structure. It shows the position at which the group is attached to the rest of the molecule.

When drawing organic structures, the symbol R is commonly used to show an alkyl group—this means that a methyl, ethyl, or another alkyl group is present within the molecule. For example, RCH_2OH is used to represent a series of alcohols that have different alkyl groups attached to $-CH_2OH$.

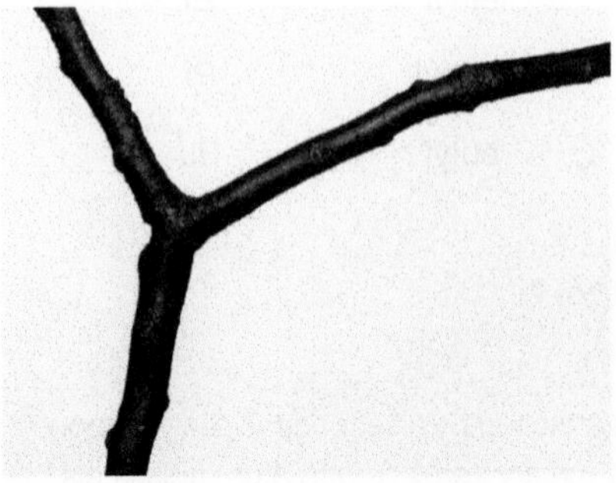

Branched alkanes have the carbon atoms arranged like branches in a tree.

Rules for naming branched alkanes

As the alkane carbon chain length increases so does the number of possible structural isomers. For example, for the hydrocarbon $C_{30}H_{62}$, over 4 billion structural isomers are possible! The IUPAC names of these and other alkanes can be determined using the following five-step sequence.

- **Step 1** Find the name of the longest continuous carbon chain.
- **Step 2** Identify and name the substituents attached to this chain.
- **Step 3** Number the chain consecutively, starting at the end nearest a branch point.
- **Step 4** Designate the location of each substituent group by an appropriate number and name.
- **Step 5** Write the complete name, by listing substituent groups in alphabetical order (e.g. *b*utyl comes before *m*ethyl). The prefixes di-, tri-, and tetra-, used to show the presence of several groups of the same kind, are not generally considered when writing in alphabetical order (e.g. *m*ethyl comes before di*p*ropyl and *e*thyl comes before di*m*ethyl).

These rules are used to name the three following aliphatic alkanes and Worked example 2.1 provides further practice in naming branched alkanes.

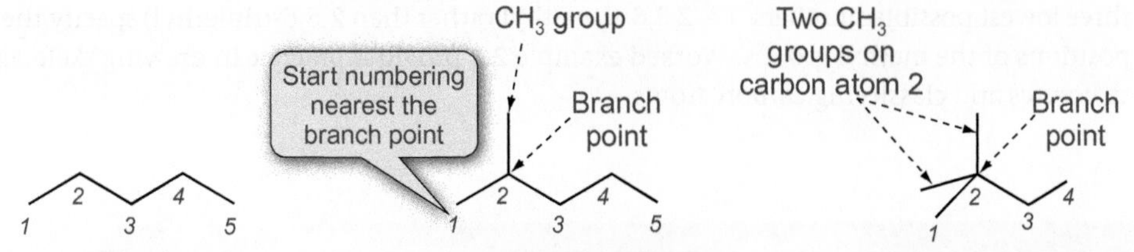

CH3 group
Two CH3 groups on carbon atom 2
Start numbering nearest the branch point
Branch point
Branch point

pentane
Five carbon chain with no substituents

2-methylpentane
Five carbon chain with a methyl group on carbon atom 2

2,2-dimethylbutane
Four carbon chain with two methyl groups on carbon atom 2

Worked example 2.1 Naming alkanes

Give the IUPAC name of the following alkane.

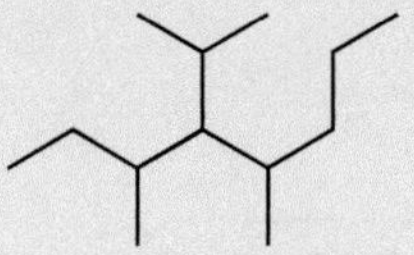

Strategy

Use the rules on p.82.

Find the longest carbon chain and then identify and name the substituents attached to the chain.

Number the chain, starting at the end nearest a branch point, and designate the position of each substituent by a number.

Write the complete name, listing groups in alphabetical order but remembering that prefixes are not generally considered when writing in alphabetical order.

Solution

The longest continuous chain has eight carbons so the compound is a derivative of octane.

An isopropyl (–$CH(CH_3)_2$) group and two methyl (–CH_3) groups are attached to the main chain.

The carbon chain is numbered from left to right because carbon atom 3 is the first branch point (see below). The methyl groups are at positions 3 and 5, while the isopropyl group is at position 4.

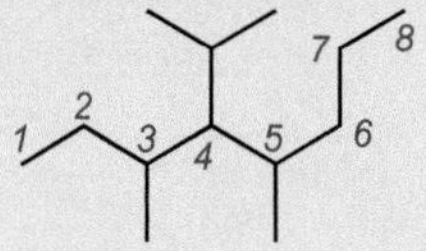

The name of the compound is 4-isopropyl-3,5-dimethyloctane (as isopropyl comes before methyl in the alphabet; the prefix di is used to show the presence of two methyl groups).

Question

Give the IUPAC name of the following compound.

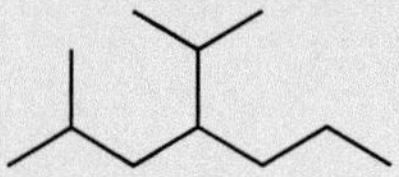

The structures of pentane, 2-methylpentane, and 2,2-dimethylbutane above show that a carbon atom can be bonded to one, two, three, or even four other carbon atoms. The carbon atoms are classified as **primary** (1°), **secondary** (2°), **tertiary** (3°), or **quaternary** (4°) carbon atoms. A primary carbon atom has one bond to another carbon atom, whereas secondary, tertiary, and quaternary carbon atoms have two, three, and four bonds to other carbon atoms, respectively.

The four different types of carbon atoms present in 2,2,5-trimethylhexane are highlighted in Figure 2.10. Notice that, in naming this compound, the numbering of the main carbon chain starts from the end nearest the quaternary carbon (C). This ensures that the

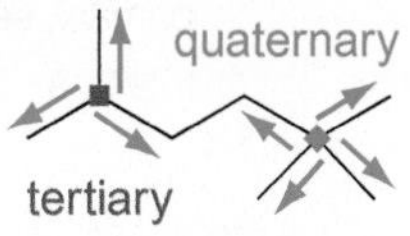

quaternary
tertiary

One helpful way to determine whether a carbon atom is 1°, 2°, 3°, or 4° is to draw an arrow (→) from the carbon in question, to each carbon that it is bonded to. The number of arrows identifies the carbon as 1° (one arrow), 2° (two arrows), 3° (three arrows), or 4° (four arrows).

Green = primary (1°) carbon
Orange = secondary (2°) carbon
Purple = tertiary (3°) carbon
Blue = quaternary (4°) carbon

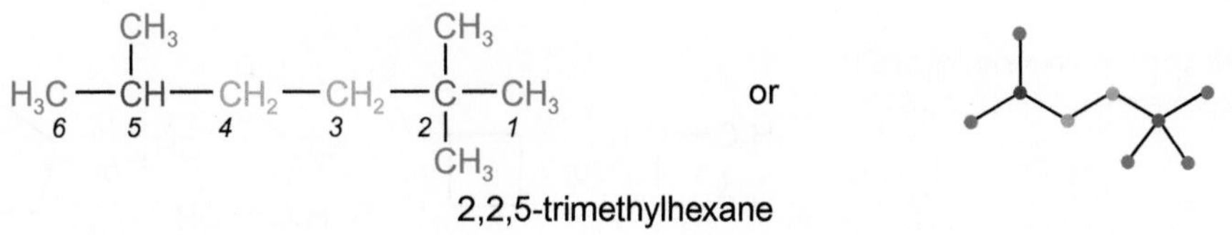

or
2,2,5-trimethylhexane

Figure 2.10 Primary, secondary, tertiary, and quaternary carbon atoms.

three lowest possible numbers (i.e. 2,2,5-trimethyl rather than 2,5,5-trimethyl) specify the positions of the methyl groups. Worked example 2.2 provides practice in drawing skeletal structures and classifying carbon atoms.

Worked example 2.2 Skeletal structures and classifying carbon atoms

Draw the skeletal structure of 5-ethyl-2,4,4-trimethyl-7-propylundecane and label each of the carbon atoms as primary, secondary, tertiary, or quaternary.

Strategy

Start by identifying the parent chain (see Table 2.2, p.80). Draw the chain as a zigzag structure and number the carbons.

Identify the substituents (Box 2.3, p.81) and use the numbers in the name of the compound to determine their positions.

Count the number of carbon atoms attached to each carbon to decide whether it is primary, secondary, tertiary, or quaternary.

Solution

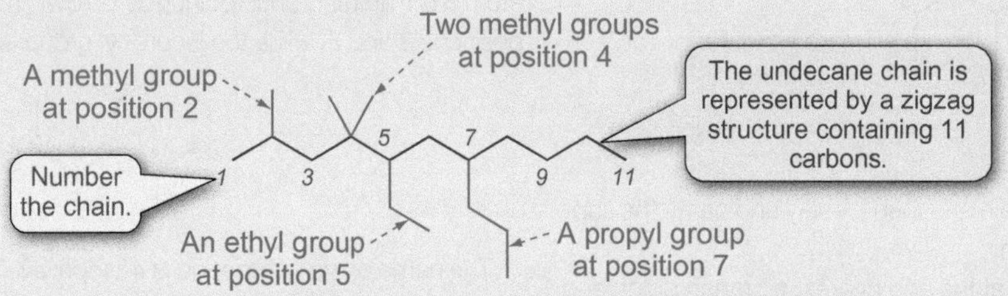
Two methyl groups at position 4
A methyl group at position 2
The undecane chain is represented by a zigzag structure containing 11 carbons.
Number the chain.
1
3
5
7
9
11
An ethyl group at position 5
A propyl group at position 7

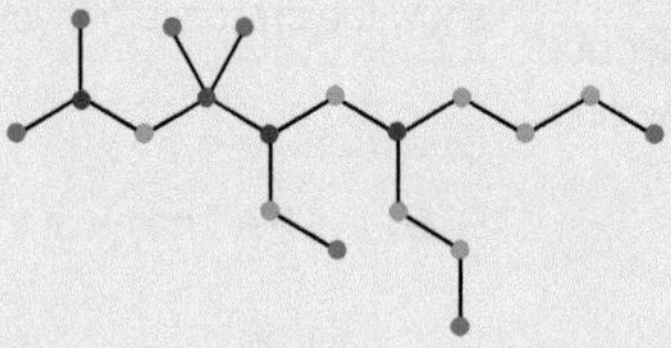

Green = primary (1°) carbon
Orange = secondary (2°) carbon
Purple = tertiary (3°) carbon
Blue = quaternary (4°) carbon

Question

Draw the skeletal structure of 4-*tert*-butyl-3,3-dimethylheptane and label each of the carbon atoms as primary, secondary, tertiary, or quaternary.

The three-dimensional arrangement of the carbon and hydrogen atoms in cycloalkanes is discussed in Section 18.2, p.819.

So far, we have only considered aliphatic alkanes, but carbon atoms can also be part of a ring. Cyclic alkanes are called **alicyclic alkanes** or **cycloalkanes**. Cycloalkanes of ring sizes from 3 upwards are known, although 5- and 6-membered rings are of particular importance because these rings are formed the most easily (Section 18.2, p.826). Cycloalkanes have the general molecular formula C_nH_{2n} and so have two fewer hydrogen atoms than aliphatic alkanes with the same number of carbon atoms.

In steroids, three cyclohexane rings are linked to a cyclopentane ring (Box 18.3, p.830).

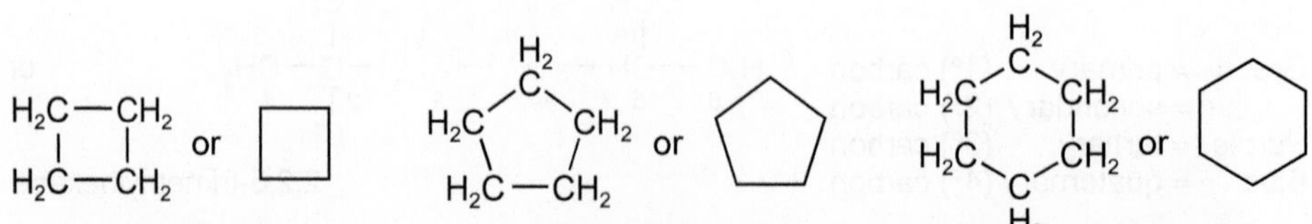

cyclobutane cyclopentane cyclohexane

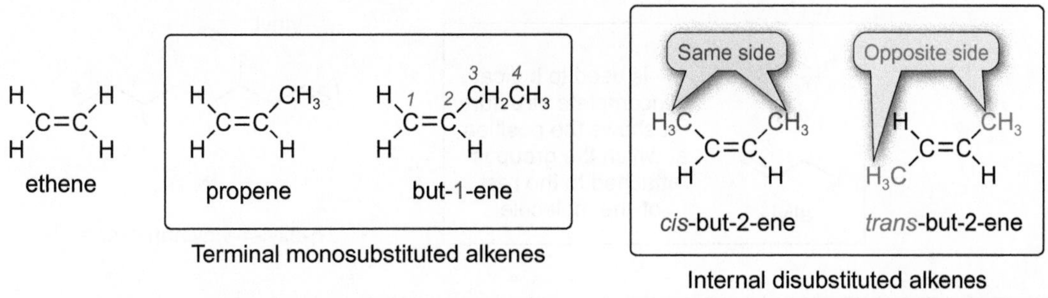
ethene
propene
but-1-ene
Terminal monosubstituted alkenes
Same side
Opposite side
cis-but-2-ene
trans-but-2-ene
Internal disubstituted alkenes

Figure 2.11 Naming alkenes.

Alkenes and alkynes

Alkenes contain carbon–carbon double bonds (C=C) and are said to be unsaturated. They have the general formula C_nH_{2n} (where $n = 2, 3$, etc.).

For alkenes, the ending -ene is used to show the presence of a C=C bond, but other than this the names are similar to those of alkanes. The simplest alkene is ethene (C_2H_4) and this is followed by propene (C_3H_6) and butene (C_4H_8), and these have two, three, and four carbon atoms, respectively.

Figure 2.11 shows how alkenes are named. For butene, the C=C bond could be at the end or in the middle of the chain. When the double bond is at the end of the chain, the alkene is called a **terminal alkene** and the number 1 is included before -ene to indicate the position of the double bond, that is, but-1-ene. Numbering of the chain starts from the end nearest the double bond and the lowest numbered carbon atom within the C=C bond is used to indicate the position. For but-1-ene there is only one alkyl group attached to the C=C bond so this is a **monosubstituted alkene**.

For all alkenes, the carbon atoms of the double bond and the atoms attached to these carbons all lie in the same plane and rotation about the C=C bond is not possible at room temperature. As a result, some alkenes can exist in two isomeric forms. But-2-ene ($CH_3CH{=}CHCH_3$) is an example of an **internal disubstituted alkene** and, because each carbon atom in the C=C bond has two different groups attached to it (i.e. CH_3 and H), two different isomers can be drawn, as shown in Figure 2.11. These compounds, which have a different arrangement of the CH_3 and H groups in space, are called *cis*- and ***trans*-isomers**. The prefix *cis*- indicates that the two CH_3 groups are on the *same side* of the double bond, while the prefix *trans*- shows that the CH_3 substituents are on the *opposite side* of the double bond. The two molecules are **stereoisomers** (see Worked example 2.3).

For the alicyclic alkenes, cyclopentene (5-ring) and cyclohexene (6-ring) shown below, the double bonds must be *cis*- (otherwise the molecules would be too strained, see Section 18.2, p.826) so the prefixes are not usually written. Also, you do not need to number the cyclopentene and cyclohexene rings because there is only one possible structure that can be drawn for each case. However, numbering is required when there are substituents on the ring. For both aliphatic and alicyclic alkenes, the carbon atoms of the double bonds are numbered so that the smallest possible numbers indicate the position of alkyl substituents. For alicyclic alkenes, the numbering always starts at the C=C bond.

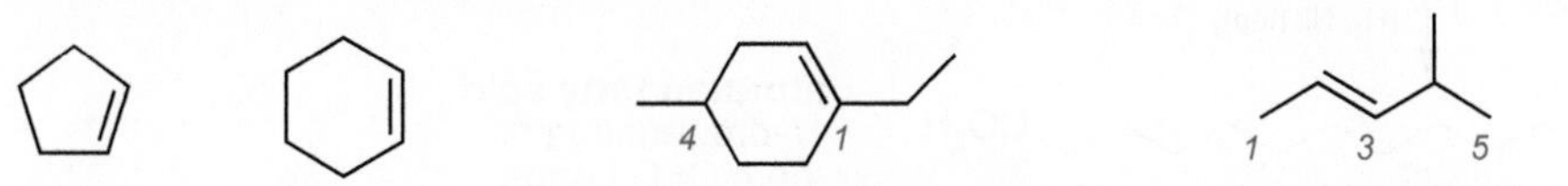
4
1
1
3
5

cyclopentene cyclohexene 1-ethyl-4-methylcyclohexene (not 2-ethyl-5-methylcyclohexene) 4-methylpent-2-ene (not 2-methylpent-4-ene)

Common names are often used for two prevalent substituents that contain a C=C double bond. These are **vinyl** for $-CH{=}CH_2$ (the IUPAC name is ethenyl) and **allyl** for $-CH_2-CH{=}CH_2$ (the IUPAC name is prop-2-enyl).

The suffix for alkenes is -ene.

The position of the C=C bond may also be written at the beginning of the name, for example, 2-butene rather than but-2-ene.

The number of alkyl or aryl groups attached to the C=C bond determines whether the alkene is a monosubstituted ($RCH{=}CH_2$), disubstituted ($R_2C{=}CH_2$ or $RCH{=}CHR$), trisubstituted ($R_2C{=}CHR$), or tetrasubstituted ($R_2C{=}CR_2$) alkene.

The C=C double bond in alkenes undergoes a large number of addition reactions. In these reactions, addition of a molecule across the C=C bond produces a saturated product, as discussed in Section 21.3 (p.970).

A molecular model of propene ($H_3C{-}CH{=}CH_2$). In this computer generated ball and stick model, a single stick represents the C=C double bond.

Stereoisomerism is discussed in Section 18.1 (p.817). The ***Z*- and *E*-notation** is a more general method of naming *cis*- and *trans*-alkene isomers and is explained on p.834.

The three-dimensional arrangement of the carbon and hydrogen atoms in cyclopentene and cyclohexene is discussed in Section 21.3, p.976.

Some alkenes can exist as stereoisomers, called *E*- or *Z*-isomers (Section 18.3, p.833). The *E*-isomer of 4-methylpent-2-ene is drawn here.

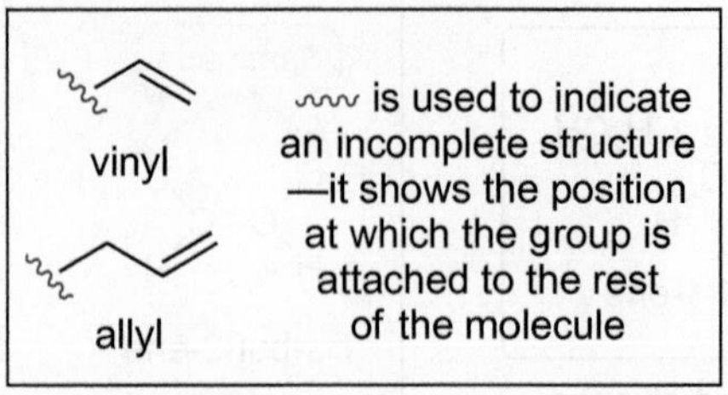
vinyl
allyl
∿∿ is used to indicate an incomplete structure —it shows the position at which the group is attached to the rest of the molecule

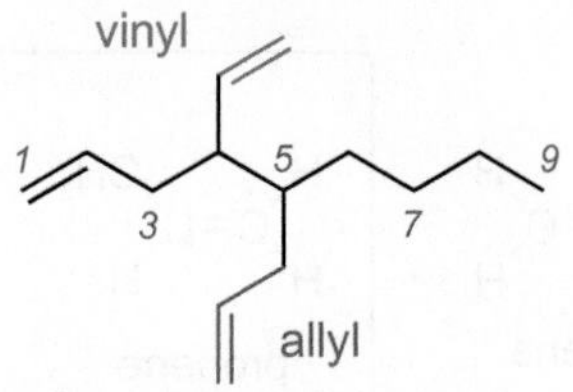
vinyl
1
3
5
7
9
allyl
5-allyl-4-vinylnon-1-ene

Box 2.4 Is butter healthier than margarine?

▲ Several scientific studies have found that when toast is dropped from a table, it falls butter-side down most of the time—one study won the Ig Nobel prize in 1996.

It now appears that neither butter nor margarine is particularly good for our health. Butter is a fat made from saturated fatty acids, such as stearic acid, which is associated with high concentrations of cholesterol in the blood and an increased risk of heart disease. Stearic acid contains an unbranched alkyl chain with a carboxylic acid group (CO_2H) at one end. In the past, margarines have been promoted as a healthier alternative to butter because they contain unsaturated fatty acids, such as linoleic acid. Linoleic acid is a polyunsaturated fatty acid because it contains more than one C=C bond.

The C=C bonds of naturally occurring unsaturated fatty acids are usually the *cis*-isomers, but margarines also contain *trans*-fatty acids, which are formed during the manufacture of margarine from vegetable oils. The process involves partial hydrogenation of some of the C=C bonds in naturally occurring *cis*-fatty acids to form some saturated fatty acids and improves the texture and lifespan of the margarine. Recent research suggests that *trans*-fatty acids can also raise cholesterol concentrations in the blood. The different effects of *cis*- and *trans*-unsaturated fatty acids in the body presumably reflect their different shapes. Margarines with low concentrations of *trans*-fatty acids are now generally recommended as being healthier and in January 2008, the British Retail Consortium announced that major UK retailers had ceased adding *trans*-fatty acids to their own products.

▼ Benecol contains very few *trans*-fatty acids and does not contain hydrogenated vegetable oils.

Benecol
Proven to Reduce Cholesterol
Benecol
Light Spread

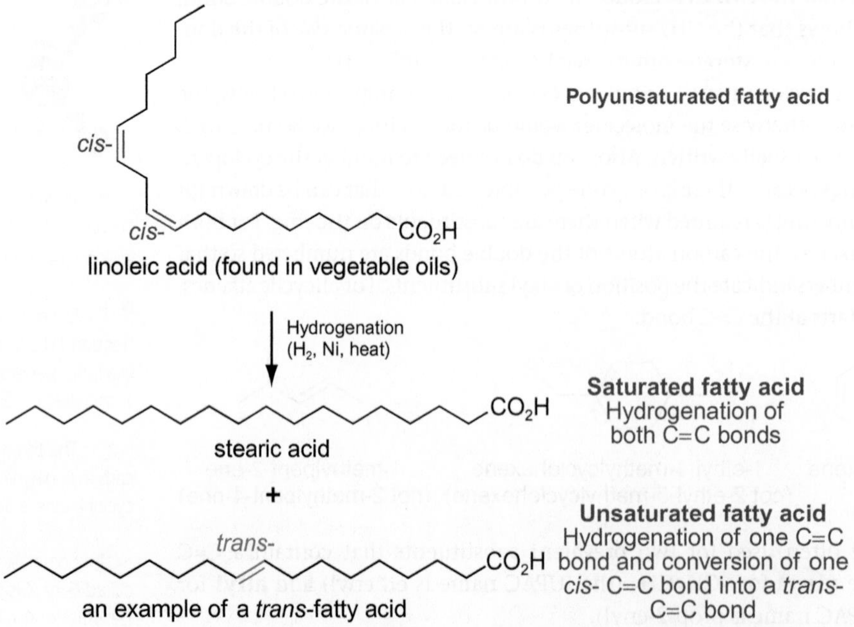
Polyunsaturated fatty acid
cis-
cis-
CO_2H
linoleic acid (found in vegetable oils)
Hydrogenation
(H_2, Ni, heat)
CO_2H
Saturated fatty acid
Hydrogenation of both C=C bonds
stearic acid
+
Unsaturated fatty acid
Hydrogenation of one C=C bond and conversion of one cis- C=C bond into a trans- C=C bond
trans-
CO_2H
an example of a trans-fatty acid

Adjacent C=C bonds | All C=C bonds separated by one C–C bond | C=C bonds separated by more than one C–C bond

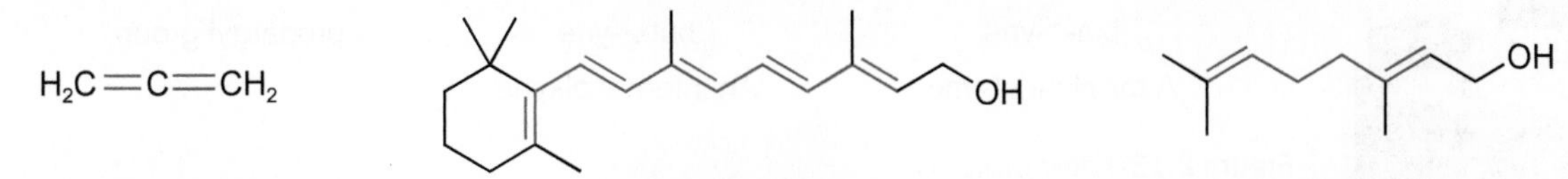
H₂C=C=CH₂
OH
OH

allene (propadiene)
The simplest cumulated diene

vitamin A (retinol)
A pentaene with five conjugated double bonds

geraniol
An isolated diene present in rose, lemongrass, and other flowers

Figure 2.12 Molecules containing more than one C=C bond.

Dienes contain two C=C bonds, **trienes** contain three, and if a molecule contains many C=C bonds it is called a **polyene**. Figure 2.12 shows some examples. **Cumulated** double bonds are next to one another, as in allene; double bonds in **conjugated** alkenes are separated by *one* single bond; and those in **isolated** alkenes by *more than one* single bond.

When naming a diene, the letter 'a' is inserted between the parent name (indicating the longest continuous carbon chain) and the word diene, for example, butadiene, not butdiene, as shown below. As usual, the chain is numbered in the direction that gives the alkyl substituents the lowest possible numbers.

2-Methylbuta-1,3-diene, or isoprene, is released into the atmosphere in huge quantities from vegetation (Box 9.3, p.400).

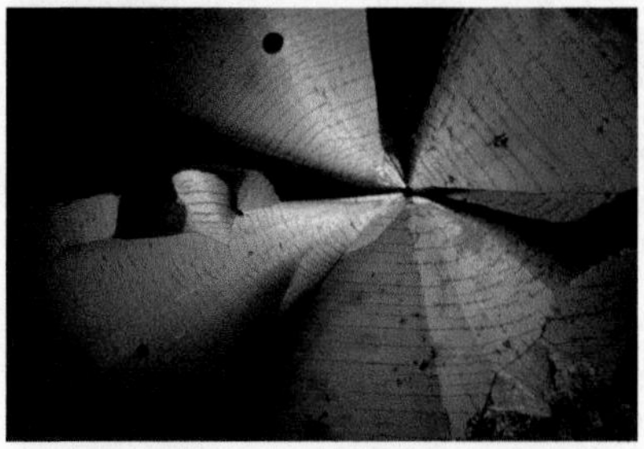

Crystals of vitamin A observed under polarized light by a microscope (the light is diffracted by the crystals to produce the colourful pattern).

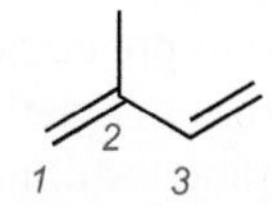
2
1
3

2-methylbuta-1,3-diene

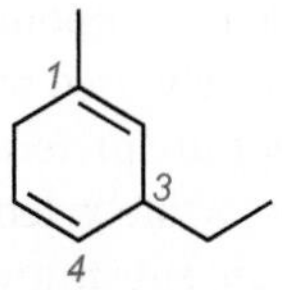
1
3
4

3-ethyl-1-methylcyclohexa-1,4-diene

Worked example 2.3 Naming alkenes

Draw the *trans*,*trans*- and *cis*,*trans*-isomers of hexa-2,4-diene.

Strategy

Draw the parent chain as a zigzag structure and number the carbons.

Draw the two double bonds at the appropriate positions.

For the *trans*,*trans*-isomer, the alkyl groups should be on the opposite sides of both double bonds.

For the *cis*,*trans*-isomer, the C=C bond at the 2-position should have the alkyl groups on the same side, whereas the C=C bond at the 4-position should have alkyl groups on the opposite sides of the double bond.

Solution

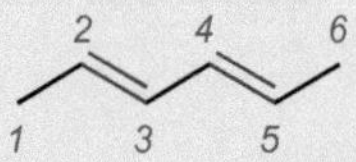
2
4
6
1
3
5

trans,*trans*-hexa-2,4-diene

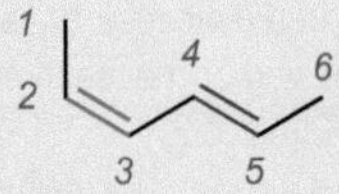
1
4
6
2
3
5

cis,*trans*-hexa-2,4-diene

Question

Draw the *cis*,*cis*- and *cis*,*trans*-isomers of octa-3,5-diene.

Alkynes are unsaturated compounds that contain C≡C triple bonds. They have the general formula, C_nH_{2n-2} (where $n = 2, 3$, etc.). For alkynes, the ending -yne is used to show the presence of a triple bond but, other than this, the names are similar to those of alkenes.

The simplest alkyne is ethyne, C_2H_2 (which has the common name acetylene), and this is followed by propyne and butyne, and these have two, three, and four carbon atoms, respectively. Figure 2.13 shows how alkynes are named. But-1-yne is an example of a

The suffix for alkynes is -yne.

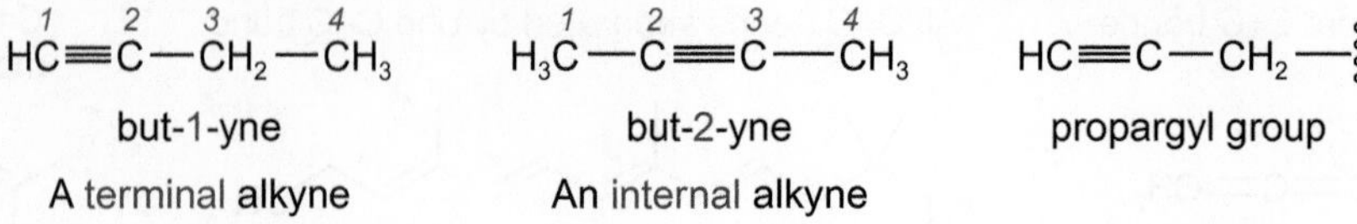
1 2 3 4
HC≡C—CH₂—CH₃
but-1-yne
A terminal alkyne
1 2 3 4
H₃C—C≡C—CH₃
but-2-yne
An internal alkyne
HC≡C—CH₂—
propargyl group

Figure 2.13 Naming alkynes.

Ethyne (acetylene) is used for oxyacetylene gas welding and cutting because of the extremely high temperature of the flame (over 3000 °C).

Like alkenes, alkynes undergo addition reactions. These reactions occur in two stages: first to give a C=C double bond and then to give a saturated product as discussed in Section 21.5 (p.994).

terminal alkyne and but-2-yne is an internal alkyne. The numbering of the carbon atoms starts from the end closest to the triple bond and the lower-numbered carbon atom within the alkyne bond is used to indicate the position. One alkyne-containing substituent that is often given a common name is the propargyl group (the IUPAC name is prop-2-ynyl), which has the triple bond at the end of a three-carbon chain.

Benzene and arenes

Aromatic compounds such as benzene are classed as arenes and the symbol 'Ar' is often used to represent an aryl group, just as R is used for an alkyl group. Benzene has the formula C_6H_6 and it is better to show the benzene ring as three conjugated double bonds (the Kekulé structure) in a hexagon of carbon atoms because it allows you to keep track of the movement of electron pairs in reaction mechanisms (see Section 2.2, p.73).

Benzene rings containing an alkyl substituent can be named as alkyl-substituted benzenes or as phenyl-substituted alkanes as shown in Figure 2.14. The **phenyl group** is C_6H_5–, while the related group ($C_6H_5CH_2$–) is called the **benzyl group**. Common names are often used for benzene compounds bearing different substituents. For example, the name toluene (for the widely used solvent) is generally used in preference to methylbenzene, styrene and cumene are preferred to ethenylbenzene (or phenylethene) and isopropylbenzene, respectively, while xylenes is normally used to describe dimethylbenzenes.

When benzene rings are substituted, there are rules for describing the positions of the substituents. For benzene rings containing two alkyl substituents, three isomers are possible. The numbers 1,2-, 1,3-, or 1,4- are used to designate the positions of the groups. Alternatively, the names *ortho-*, *meta-*, or *para-* (abbreviated *o-*, *m-*, or *p-*, respectively) are used to show the positions of the groups: *o-* for 1,2-, *m-* for 1,3-, and *p-* for 1,4-disubstituted benzenes (see margin). If the two substituents are different, the names are listed in alphabetical order before the ending 'benzene'. Benzene rings containing three substituents are named similarly, although numbers must be used since the use of *o-*, *m-*, or *p-* is not applicable (see Worked example 2.4). Alternatively, compounds can be named by considering benzene as a phenyl substituent on an alkane chain.

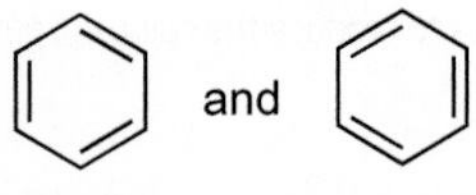
and

Kekulé structures of benzene

Benzene and related aromatic compounds undergo **substitution** reactions in which another group replaces a hydrogen atom on the aromatic ring. In the process, the aromatic ring is retained at the end of the reaction, as described in Section 22.2 (p.1013).

'Ph' stands for phenyl (C_6H_5–), *not* phenol (C_6H_5OH). The abbreviation 'Ar' stands for an aryl group.

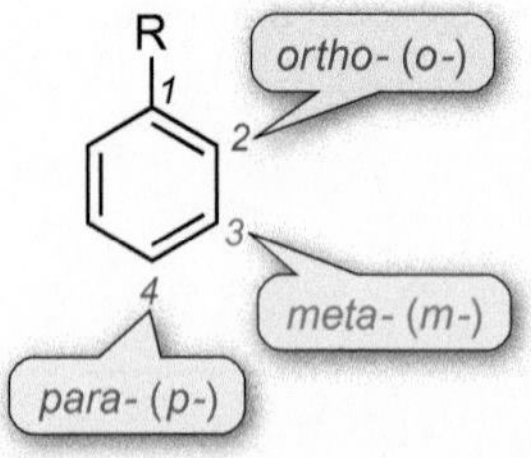
R
1
2
3
4
ortho- (o-)
meta- (m-)
para- (p-)

In industry, styrene is used to make polystyrene (Section 1.2, p.6 and Box 20.2, p.923), while cumene is converted into phenol (Box 1.4, p.25 and Box 22.2, p.1007).

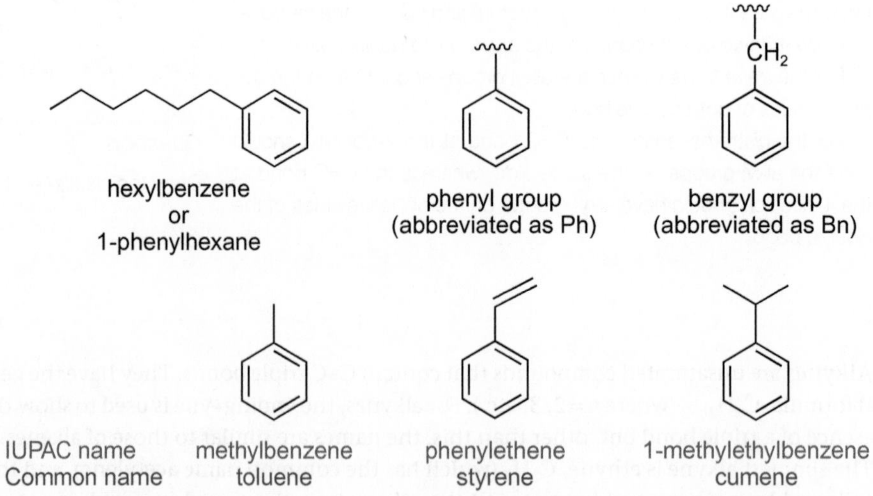
hexylbenzene
or
1-phenylhexane
phenyl group
(abbreviated as Ph)
CH₂
benzyl group
(abbreviated as Bn)
IUPAC name
Common name
methylbenzene
toluene
phenylethene
styrene
1-methylethylbenzene
cumene

Figure 2.14 Naming arenes.

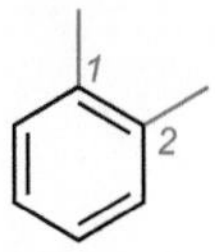
1
2

1,2-dimethylbenzene
(Common name: *o*-xylene)

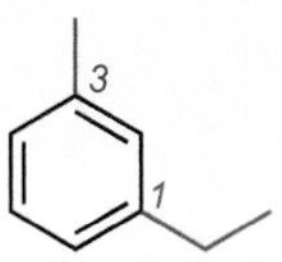
3
1

1-ethyl-3-methylbenzene

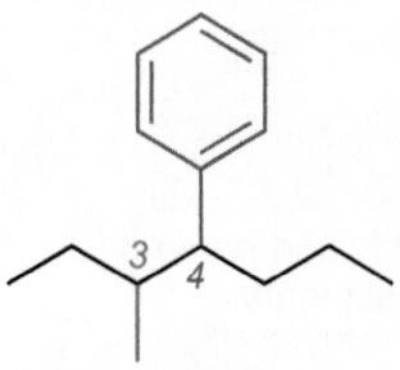
3
4

(3-methylheptan-4-yl)benzene
or
3-methyl-4-phenylheptane

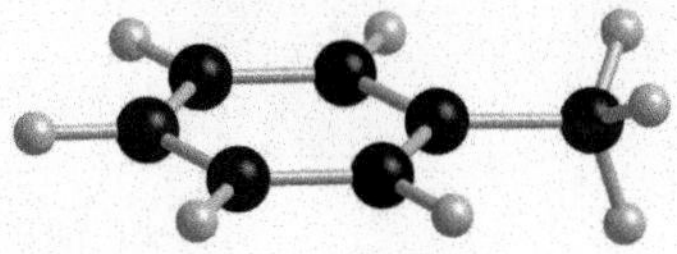

A molecular model of toluene, $C_6H_5CH_3$ (a common solvent for reactions of organic molecules).

Worked example 2.4 Naming substituted benzenes

Draw the skeletal structure of 4-ethyl-2-isopropyl-1-methylbenzene.

Strategy

Identify the structure of the three substituents on the benzene ring: ethyl is $-CH_2CH_3$; isopropyl is $-CH(CH_3)_2$; and methyl is $-CH_3$.

Draw the Kekulé structure of the benzene ring and a connecting straight line to represent the methyl group. The isopropyl group is then included at position 2, with respect to the methyl group, and an ethyl group at position 4.

Solution

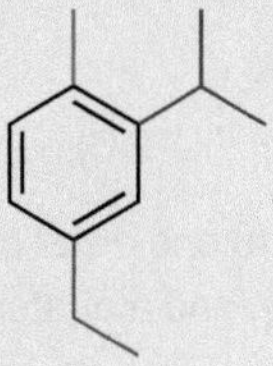

4-ethyl-2-isopropyl-1-methylbenzene

Question

Draw the skeletal structure of 2-methyl-3-phenylpent-2-ene.

There are also a number of compounds that contain more than one benzene ring. When two benzene rings are joined by a single bond, the molecule is called **biphenyl**. If two or more benzene rings share two carbon atoms, the molecules are called **polynuclear** aromatic hydrocarbons. Common examples include naphthalene, anthracene, and phenanthrene.

Polynuclear aromatic hydrocarbons, such as naphthalene ($C_{10}H_8$), have a planar shape.

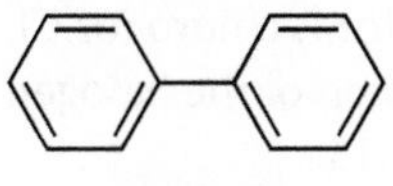

biphenyl

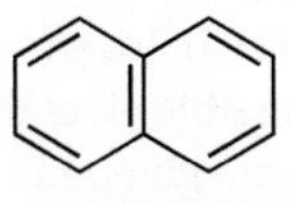

naphthalene (moth repellent and insecticide)

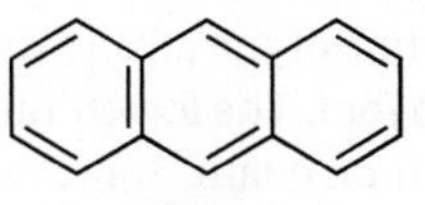

anthracene

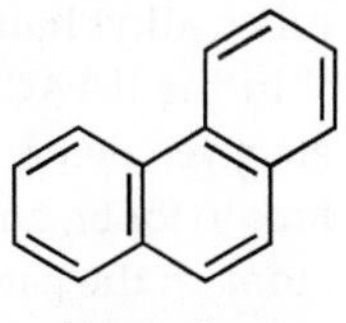

phenanthrene

Replacing one or more hydrogen atoms in biphenyl, with chlorine, forms polychlorinated biphenyls (PCBs). PCBs are very stable and break down very slowly, so they can remain in the environment for a long time (Box 11.3, p.537).

Summary

- Organic compounds are classified by their functional groups, which are groups of atoms that give the compound characteristic chemical properties.
- The IUPAC names of organic compounds are determined from the longest continuous carbon chain, called the parent chain. A prefix is added before the parent name to indicate any substituents. These are listed in alphabetical order and their positions given by numbers. A functional group suffix is included after the parent name to indicate the major functional group: -ane for alkane; -ene for alkene; and -yne for alkyne.

→

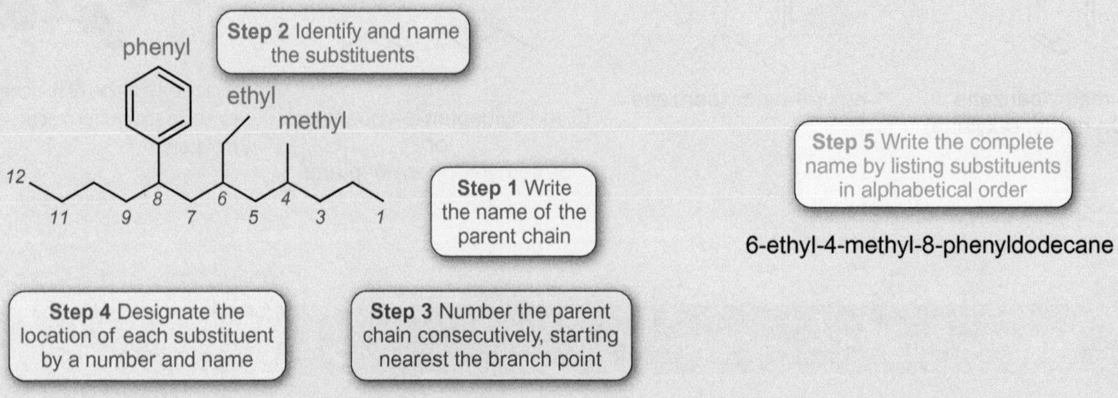
Step 2 Identify and name the substituents
phenyl
ethyl
methyl
12
11
9
8
7
6
5
4
3
1
Step 1 Write the name of the parent chain
Step 5 Write the complete name by listing substituents in alphabetical order
6-ethyl-4-methyl-8-phenyldodecane
Step 4 Designate the location of each substituent by a number and name
Step 3 Number the parent chain consecutively, starting nearest the branch point

- A compound can have a systematic (IUPAC) name and common names, but a name must specify only one compound.
- A primary carbon atom is bonded to one other carbon atom, a secondary carbon is bonded to two carbons, a tertiary carbon is bonded to three carbons, and a quaternary carbon is bonded to four carbons.

For practice questions on these topics, see questions 1–2 at the end of this chapter (p.109).

2.6 Functional groups containing one or more heteroatoms

Halogenoalkanes and related compounds

Halogenoalkanes: R–X

The C–X bonds in halogenoalkanes (where X = Cl, Br, or I) are relatively weak, which means that these bonds are easily broken in **substitution** and **elimination** reactions. Substitution reactions lead to the halogen atom being replaced by another group whilst elimination reactions result in the loss of HX (Section 20.3, p.930 and Section 20.4, p.943, respectively).

A molecular model of 2-chlorobutane, $CH_3CHClCH_2CH_3$. The green sphere represents chlorine—the use of green comes from the CPK colouring convention, and more recently, from popular software products, that assigns bright colours to each element.

Halogenoalkanes have the molecular formula RX, where R is an alkyl group (e.g. methyl, ethyl, or propyl) and X is fluorine, chlorine, bromine, or iodine. These are commonly called **alkyl halides**.

In the IUPAC system, halogenoalkanes are named as substituted alkanes and prefixes are given before the name of the alkane to define the halogen—fluoro for F, chloro for Cl, bromo for Br, and iodo for I. The lowest number possible gives the position of the halogen atom on the parent carbon chain. Some examples are given in Figure 2.15.

The common names of alkyl halides consist of the name of the alkyl group together with a separate word to define the halogen atom—fluoride for F, chloride for Cl, bromide for Br, and iodide for I. For example, bromomethane is commonly known as methyl bromide.

Alkyl halides are classified as primary, secondary, or tertiary halides depending on whether the halogen is bonded to a primary, secondary, or tertiary carbon (see p.83). A primary alkyl halide has one alkyl group bonded to the carbon bearing the halogen, while secondary and tertiary halides have two and three alkyl groups, respectively (see Worked example 2.5).

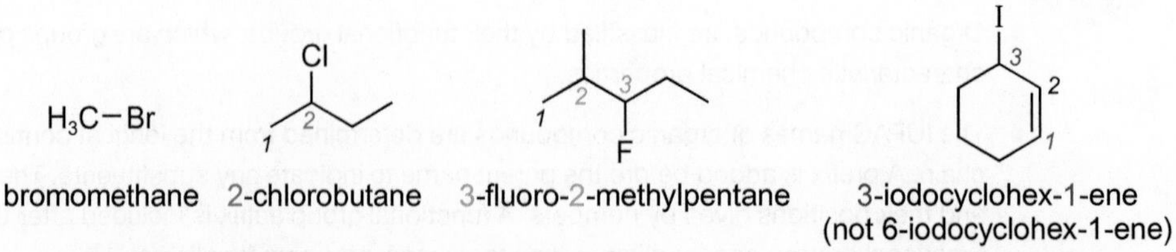
H3C–Br
bromomethane
Cl
2-chlorobutane
F
3-fluoro-2-methylpentane
I
3-iodocyclohex-1-ene
(not 6-iodocyclohex-1-ene)

Figure 2.15 Naming halogenoalkanes.

methyl halide | **primary** (1°) alkyl halide | **secondary** (2°) alkyl halide | **tertiary** (3°) alkyl halide

When the halogen atom is bonded directly to an aromatic ring, such as a benzene ring, the compounds are commonly called **aryl halides**. When a halogen atom is attached directly to a C=C bond, the compounds are commonly called **vinyl halides. Benzyl** and **allyl halides** have a CH_2X group attached to a benzene ring and the H_2C=CH group, respectively.

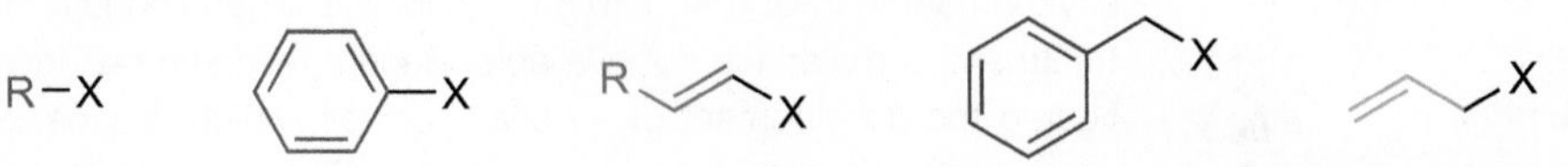

alkyl halide | aryl halide | vinyl halide | benzyl halide | allyl halide

Whereas allyl bromide (H_2C=$CHCH_2Br$) and benzyl bromide ($PhCH_2Br$) undergo nucleophilic substitution reactions, vinyl bromide (or bromoethene; H_2C=CHBr) and phenyl bromide (or bromobenzene; PhBr) do not (Section 20.3, p.930).

Worked example 2.5 Naming halogenoalkenes

Draw the skeletal structure of 4-bromo-5-methylhex-1-ene. Is the halogen bonded to a primary, secondary, or tertiary carbon atom?

Strategy

Draw a zigzag structure of hexane and number the chain.

Draw a double bond between carbon atoms 1 and 2.

Include a bromine substituent at position 4 and a methyl group at position 5.

Count the number of alkyl groups attached to the carbon atom bearing the bromine to determine if the carbon is primary, secondary, or tertiary.

Solution

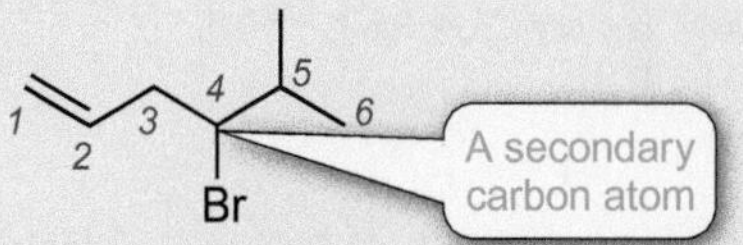
1
2
3
4
5
6
Br
A secondary carbon atom

4-bromo-5-methylhex-1-ene

Question

Draw the skeletal structure of 5,6-dichloro-2-methyloct-2-ene. Are the chlorine atoms bonded to primary, secondary, or tertiary carbon atoms?

Box 2.5 The DDT dilemma

A number of organic compounds containing chlorine have been used as insecticides. They include dichlorodiphenyltrichloroethane (DDT), which has successfully been used to kill insects that carry diseases such as typhus and malaria. However, DDT is a very stable molecule, which means that it remains in the environment for many years. It is also fat-soluble and accumulates in the fatty tissues of many animals, fish, and birds that lie towards the end of food chains. In the 1940s the population of many birds of prey dropped dramatically because their eggs did not survive to hatching due to fragile eggshells. The high concentrations of DDT found in female birds may stop the calcium supply for egg formation. As a result, there is an ongoing anti-DDT campaign by environmental activists that discourages the use of DDT to control malaria. Some countries have stopped using DDT altogether, although these nations have experienced an increase in the incidence of malaria. It has been claimed that somewhere on the Earth, on average every 12 seconds, a child dies of DDT-preventable malaria.

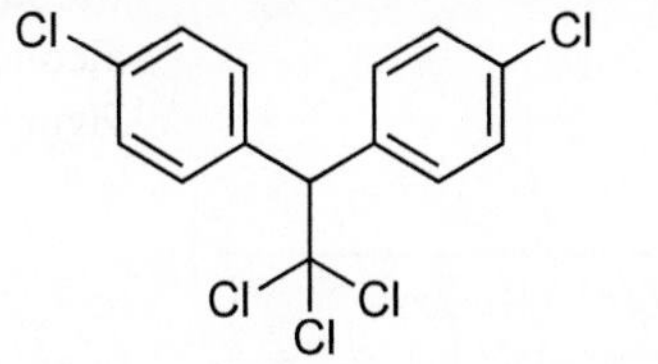

DDT

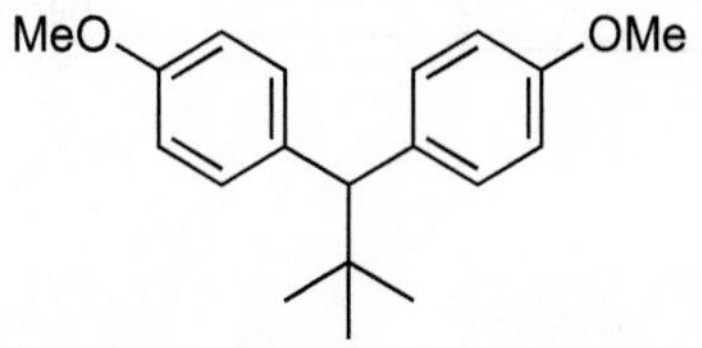

1,1-di(4-methoxyphenyl)-2,2-dimethylpropane

▲ DDT and 1,1-di(4-methoxyphenyl)-2,2-dimethylpropane

➜

DDT

▲ DDT was used extensively in the 1940s to 1960s to eradicate the insects that carried typhus and malaria; it was also used as an agricultural insecticide. The photo shows the first public test of an insecticidal fogging machine on a beach in New York State in 1945.

One solution to the problem is to use molecules that have similar structures (called **analogues**) to that of DDT, and so could prevent malaria, but are more easily broken down in the environment to give non-hazardous products. With this in mind, chemists have prepared similar structures and have found that the biological activity of DDT is related to the size and shape of the molecule. Molecules similar to DDT, such as 1,1-di(4-methoxyphenyl)-2,2-dimethylpropane, which have methyl (CH_3) or methoxy groups (CH_3O) in place of the chlorines, can produce similar biological effects. Unfortunately, this analogue is even more persistent in the environment than DDT and, to date, a viable DDT analogue for the treatment of malaria is not available. Recent research has investigated the development of malaria vaccines, which is especially novel as there are currently no vaccines available that treat parasitic infections.

Question

Some insects have developed a resistance to DDT by producing enzymes that catalyse the conversion of DDT into a related compound called 1,1-dichloro-2,2-di(4-chlorophenyl)ethene (DDE). DDE is not an insecticide as this compound is flatter than DDT and the change in shape alters the toxicity. Draw a skeletal structure for DDE.

Halogenoalkanes containing more than one halogen atom include tetrachloromethane (CCl_4; common name carbon tetrachloride) and trichloromethane ($CHCl_3$; common name chloroform), which are useful organic solvents. However, their use declined after they were shown to be toxic to humans. In comparison, dichloromethane (CH_2Cl_2; common name methylene chloride) is a less toxic solvent that is widely used in chemistry laboratories, and in paint strippers and degreasers.

Alcohols:
R–OH
The suffix for alcohols is -ol.

Alcohols, ethers, and related compounds

Alcohols and phenols

All alcohols contain the **hydroxyl (–OH) group** and have the general formula ROH. In the IUPAC system, the 'e' (at the end of the name) of the parent alkane is replaced by the suffix '-ol' (e.g. CH_3CH_2OH is called ethanol). A number before '-ol' in the name of the alcohol indicates the position of the OH group on the parent carbon chain (see Figure 2.16). The common names of alcohols are made up of the name of the alkyl group

The position of the OH group is sometimes shown at the front of the name of the alcohol, for example, 1-propanol rather than propan-1-ol (for $CH_3CH_2CH_2OH$).

Phenol is an important starting material for the preparation of pharmaceuticals, weedkillers, and synthetic resins. It is prepared from benzene using the cumene hydroperoxide process (Box 22.2, p.1007).

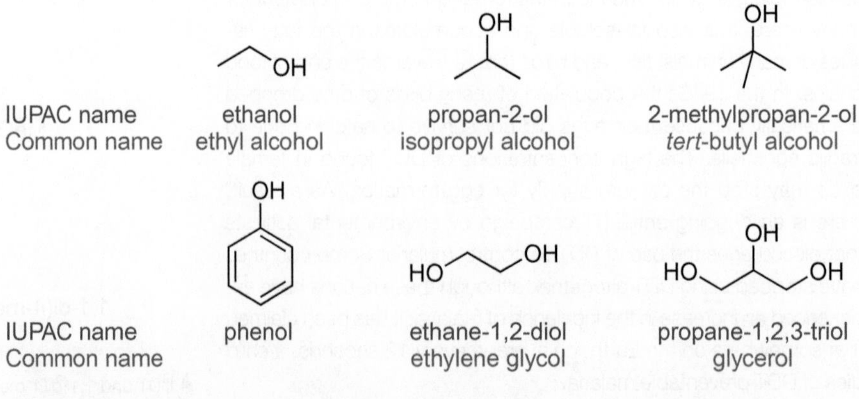
OH
OH
OH
IUPAC name
Common name
ethanol
ethyl alcohol
propan-2-ol
isopropyl alcohol
2-methylpropan-2-ol
tert-butyl alcohol
OH
HO
OH
OH
HO
OH
IUPAC name
Common name
phenol
ethane-1,2-diol
ethylene glycol
propane-1,2,3-triol
glycerol

Figure 2.16 Naming alcohols and phenols.

followed by the word 'alcohol', for example, ethyl alcohol rather than ethanol, isopropyl alcohol rather than propan-2-ol (for $CH_3CHOHCH_3$), and *tert*-butyl alcohol rather than 2-methylpropan-2-ol (for $(CH_3)_3COH$). Sometimes, the IUPAC and common names are merged together so that isopropyl alcohol and *tert*-butyl alcohol are often called isopropanol and *tert*-butanol, respectively.

When the OH group is attached to a benzene ring this compound is called phenol (C_6H_5OH; sometimes written as PhOH). It is the simplest member of a class of hydroxy benzene derivatives (called phenols), which all contain an OH group attached to a benzene ring.

Compounds with two or three OH groups are called diols and triols, respectively, and these are named *without* dropping the final 'e' of the appropriate alkane. Industrially important examples include ethane-1,2-diol (used as an antifreeze in car radiators) and propane-1,2,3-triol (obtained as a by-product from the conversion of fats and oils into soaps; see Box 24.6, p.1120), which are commonly called ethylene glycol and glycerol, respectively (Figure 2.16).

Alcohols are classified as primary, secondary, or tertiary alcohols in the same way as halogenoalkanes (p.91). The classification depends on whether the OH group is bonded to a carbon atom bearing one, two, or three alkyl groups (see Worked example 2.6).

H
H−C−OH
H
H
R−C−OH
H
H
R−C−OH
R
R
R−C−OH
R

methanol | primary (1°) alcohol | secondary (2°) alcohol | tertiary (3°) alcohol

Glycerol ($HOCH_2CH(OH)CH_2OH$), which is found in some food supplement bars, may improve the performance of endurance athletes and bodybuilders by reducing dehydration in the body.

Alcohols can be deprotonated (i.e. lose H^+) on reaction with a strong base to form salts, called **metal alkoxides** ($RO^- M^+$; where M^+ is a metal ion such as Na^+ or K^+). The name of the salt is given by the name of the metal cation, followed by the name of the alkoxide anion (RO^-) as a separate word. Alkoxide ions are named after the alcohol by changing the '-anol' ending to '-oxide'. For example, $CH_3CH_2O^- Na^+$ is sodium ethoxide and $(CH_3)_3CO^- K^+$ is commonly called potassium *tert*-butoxide:

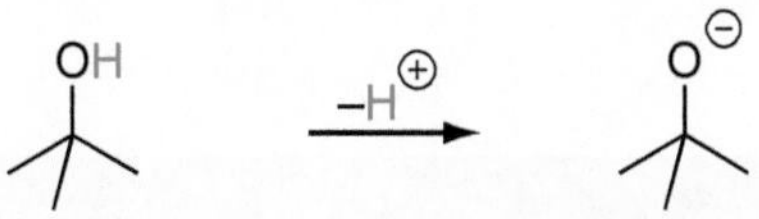
OH
−H⊕
O⊖

IUPAC name	2-methylpropan-2-ol	2-methyl-2-propoxide
Common name	*tert*-butanol	*tert*-butoxide

Alcohols are important in synthesis because they participate in a wide range of reactions. For example, the OH group in alcohols is replaced in substitution reactions (Section 20.2, p.925), and oxidation of alcohols forms aldehydes (RCHO), ketones (RCOR), or carboxylic acids (RCO_2H) (Box 23.2, p.1064).

Metal alkoxides ($RO^- M^+$) are useful bases in elimination reactions (Section 20.4, p.943).

Circles are often drawn around negative and positive charges to make them stand out.

Worked example 2.6 Naming and classifying alcohols

Draw the skeletal structure of 3-*tert*-butyl-5-isopropylcyclohexanol. Is this a primary, secondary, or tertiary alcohol?

Strategy

Draw the skeletal structure of cyclohexane and number the ring.

Include an OH group at position 1, a *tert*-butyl group at position 3, and an isopropyl group at position 5.

Count the number of alkyl groups attached to the carbon atom bearing the OH group to determine if the alcohol is primary, secondary, or tertiary.

Solution

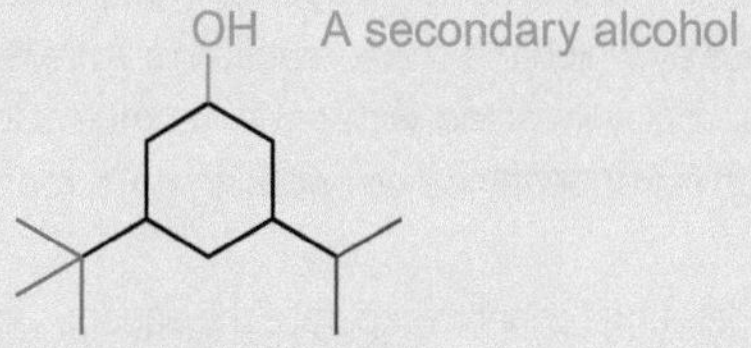
OH
A secondary alcohol

3-*tert*-butyl-5-isopropylcyclohexanol

Question

Draw the skeletal structure of 1-methyl-2-vinylcyclopentanol. Is this a primary, secondary, or tertiary alcohol?

Ethers are relatively inert compounds and are frequently used as solvents in which to carry out organic reactions (Box 2.6, below).

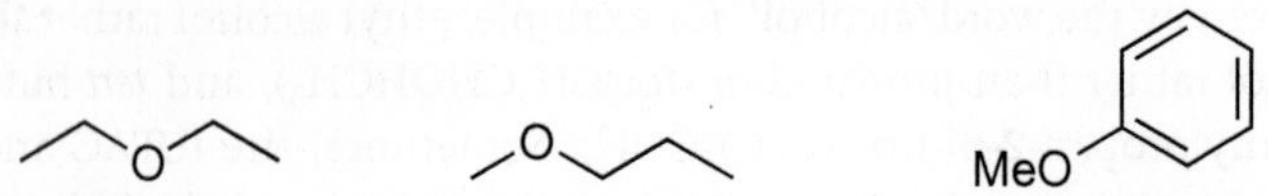

IUPAC name	ethoxyethane	1-methoxypropane	methoxybenzene
Common name	diethyl ether	methyl propyl ether	methyl phenyl ether or anisole

Figure 2.17 Naming ethers.

Ethers

Ethers
R–O–R

Ethers are compounds in which an oxygen atom is bonded to two alkyl or aryl groups. They have the general formula ROR and the two R groups can be the same or different—these are known as symmetrical or unsymmetrical ethers, respectively. In the IUPAC system, ethers are named as substituted alkanes and the –OR group is called an **alkoxy group**. For example, $-OCH_3$ is a methoxy group and $-OCH_2CH_3$ is an ethoxy group. The common names of ethers are formed by writing the names of the alkyl groups in alphabetical order, followed by the word 'ether'. Figure 2.17 shows some examples.

Heterocyclic compounds are cyclic organic compounds in which one or more carbon atoms are replaced by a heteroatom such as oxygen, nitrogen, or sulfur.

Cyclic ethers are known, including 3-membered ring epoxides (oxiranes) and also 5- and 6-membered ring ethers, such as tetrahydrofuran, tetrahydropyran, and 1,4-dioxane (see below). These molecules, which have at least one oxygen atom as part of the ring, are collectively called **heterocyclic** ethers. Aromatic heterocyclic compounds, which have a delocalized ring of electrons, are also known. The most important of these is furan in which a lone pair of electrons on the oxygen atom is part of the delocalized electron system (Section 22.1, p.1009).

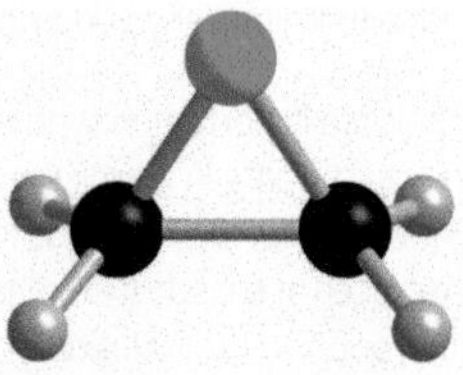

A molecular model of ethylene oxide, $(CH_2)_2O$ (the red sphere represents oxygen).

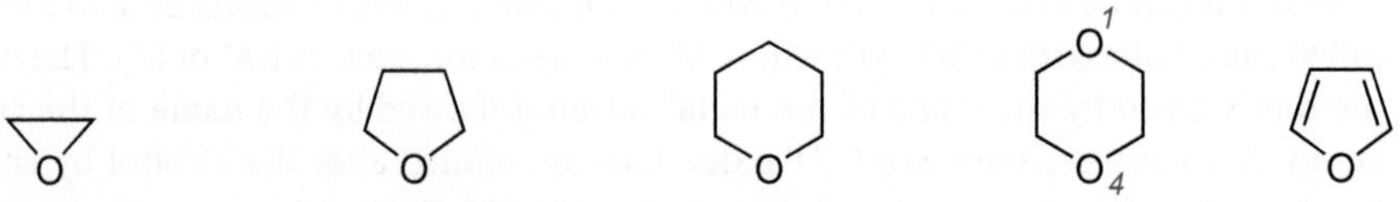

ethylene oxide, an epoxide (oxirane) | tetrahydrofuran (abbreviated THF) | tetrahydropyran | 1,4-dioxane | furan

Box 2.6 An organic work-up

Diethyl ether ($CH_3CH_2OCH_2CH_3$) and tetrahydrofuran ($(CH_2)_4O$) are important solvents for organic reactions—many organic compounds are soluble in these low-boiling liquids. At the end of an organic reaction, the crude product, in a water-immiscible organic solvent, such as diethyl ether, is often washed by shaking it with water or aqueous acid in a separating funnel. This can help to purify the product by extracting ionic by-products into the aqueous phase. The organic product is then isolated by evaporation of the organic solvent. The process by which the crude reaction mixture is converted into the crude product by washing with aqueous media followed by solvent evaporation is frequently called 'working up' a reaction.

Using a separating funnel to separate the aqueous layer in an organic work-up from the organic product dissolved in an organic solvent. ➤

Other related compounds

Compounds with two C–O bonds from the *same* carbon atom are *not* ethers. These compounds can be **1,1-diols** (sometimes called hydrates), **hemiacetals**, or **acetals**, depending on whether the oxygen substituents are hydroxyl (HO–) groups or alkoxyl (RO–) groups.

When an alkoxyl group is directly attached to an alkene double bond, the functional group is an **enol ether**. This name arises from **enols**, which have an OH group directly bonded to the C=C bond.

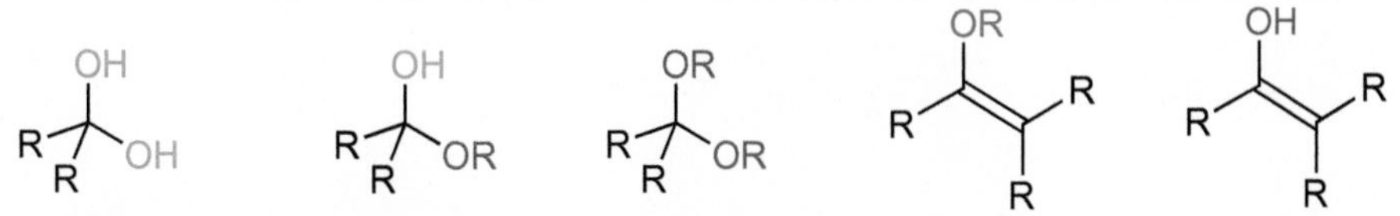

1,1-diol (hydrate) hemiacetal acetal enol ether enol

As oxygen and sulfur are in the same group in the Periodic Table, it is probably not surprising that related functional groups containing sulfur (in place of oxygen) are known. The sulfur analogue of an alcohol, with the general formula RSH, is a **thiol** (see below). For ethers, the sulfur version is called a **thioether** (sometimes called a sulfide), while replacing the two oxygen atoms in an acetal with sulfur produces a **thioacetal**. Cyclic thioethers and thioacetals are known, as are sulfur-containing aromatic heterocycles such as thiophene. The prefix thio is therefore used to name a compound that has sulfur in place of oxygen.

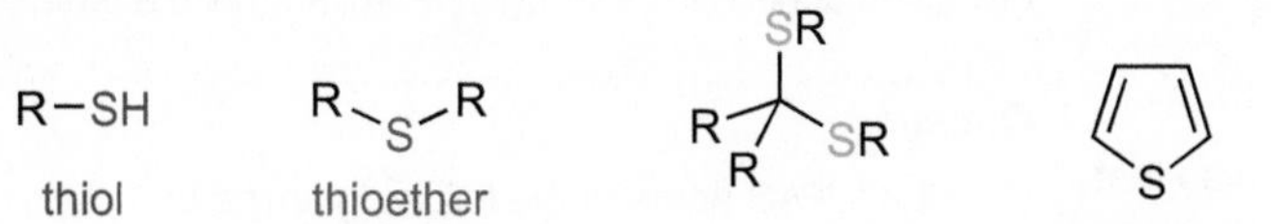

thiol thioether (sulfide) thioacetal thiophene

Amines and related compounds

Replacing the hydrogen atoms on ammonia (NH_3) with one, two, or three alkyl or aryl groups produces **amines**. Aliphatic amines have the nitrogen bonded to alkyl groups, whereas aromatic amines have the nitrogen directly bonded to one or more aryl groups.

The number of alkyl or aryl groups attached to *nitrogen* determines whether the amine is a primary, secondary, or tertiary amine. When nitrogen is bonded to four alkyl groups, the nitrogen has a positive charge. The resulting ion is called a quaternary ammonium ion.

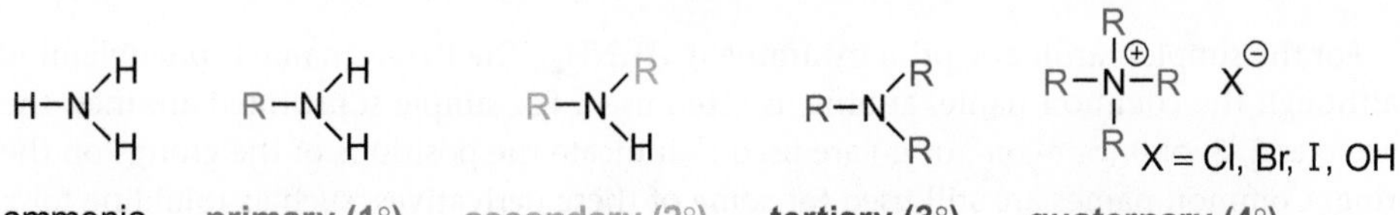
X = Cl, Br, I, OH

ammonia primary (1°) amine secondary (2°) amine tertiary (3°) amine quaternary (4°) ammonium salt

In the IUPAC system, amines are called alkanamines where the 'e' at the end of the name of the parent alkane is replaced by the suffix '-amine'. The lowest possible number indicates the position of nitrogen on the parent alkane chain, which is written in the middle of the name alkanamine. For secondary and tertiary amines (R_2NH and R_3N), the name of the alkyl or aryl group(s) on nitrogen is preceded by the letter '*N*' (in italics)—this shows that the group is bonded to nitrogen and not to carbon. The names of all substituents, whether on carbon or nitrogen, are listed in alphabetical order. Some examples are shown in Figure 2.18 (and Worked example 2.7).

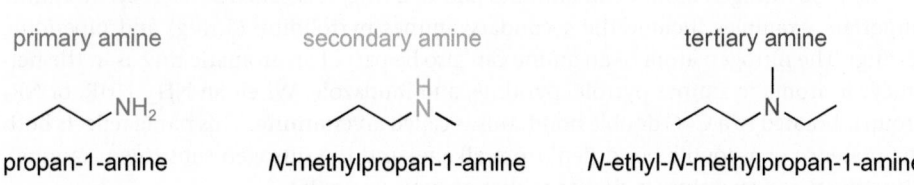

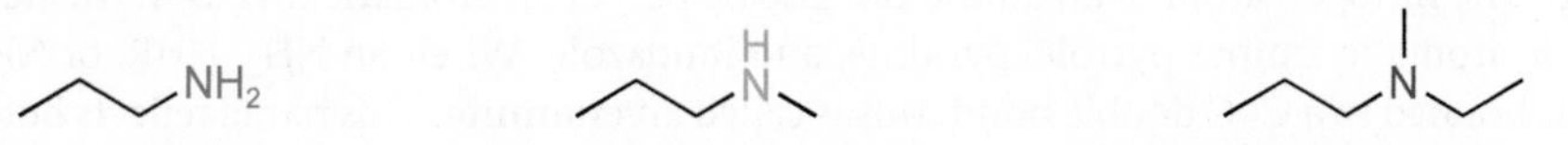
primary amine
secondary amine
tertiary amine

propan-1-amine *N*-methylpropan-1-amine *N*-ethyl-*N*-methylpropan-1-amine

Figure 2.18 Naming amines.

Enols are in equilibrium with carbonyls, which have a C=O double bond, e.g. $R_2C{=}CH{-}OH \rightleftharpoons R_2CH{-}CH{=}O$. The keto–enol equilibrium is discussed in Section 23.3, p.1082.

The pungent odours of onions, garlic, and skunks are due to volatile (low-boiling) thiols. Ethanethiol (CH_3CH_2SH) is detected at extremely low concentrations by the human nose. This property has been exploited in industry where a small amount of ethanethiol is added to natural gas. Natural gas is odourless and highly flammable, so a gas leak could go undetected and lead to an explosion. Adding a pungent thiol means that gas leaks can easily be detected and subsequently repaired.

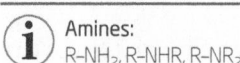

Amines: $R{-}NH_2$, R–NHR, $R{-}NR_2$.

This is different from the convention for halogenoalkanes (p.91) and alcohols (p.93)—these are determined as primary, secondary, or tertiary on the basis of the number of alkyl groups attached to the *carbon* atom bearing the halogen atom or hydroxyl group.

The suffix for amines is -amine.

When named as a substituent on a carbon chain, the $-NH_2$ group is called the amino group. For example, $HOCH_2CH_2NH_2$ is called 2-aminoethanol.

The position of the amine group is sometimes shown at the front of the name, for example, 1-propanamine rather than propan-1-amine (for $CH_3CH_2CH_2NH_2$).

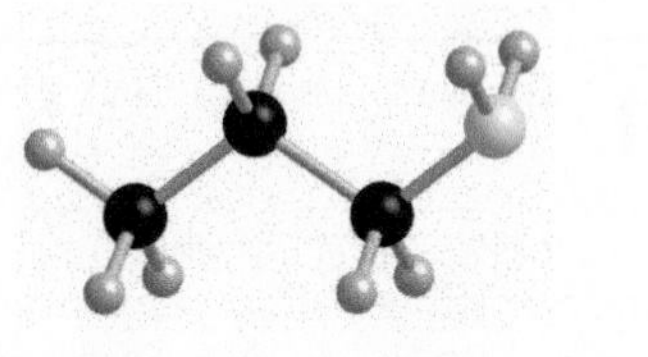

A molecular model of propan-1-amine, $CH_3CH_2CH_2NH_2$ (the blue sphere represents nitrogen).

The common name for an amine is obtained by listing the names of the alkyl groups attached to nitrogen (in alphabetical order) followed by the suffix 'amine', for example, propylamine rather than propan-1-amine. Unlike alcohols (ROH), ethers (ROR), and halogenoalkanes (RX), the common names of amines are written as a single word.

Worked example 2.7 Naming and classifying amines

Give the IUPAC name of the following amine and classify the amine as primary, secondary, or tertiary.

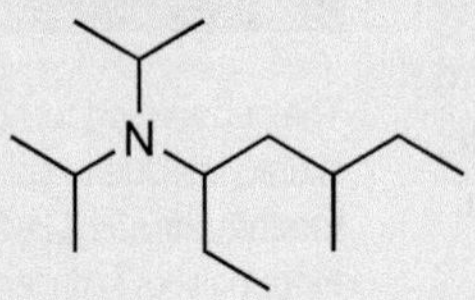
N

Strategy

Identify the longest continuous chain and number the chain so that the nitrogen substituent has the lowest possible number.

Name the longest carbon chain as an alkanamine with the position of the nitrogen group written in the middle of the name.

Identify the alkyl substituent on the alkanamine chain and its position.

Name the two other alkyl groups on nitrogen; this name should be preceded by *N,N*-.

Give the full name, listing the substituents in alphabetical order.

Count the number of alkyl groups attached to the nitrogen atom to determine if the amine is primary, secondary, or tertiary.

Solution

N,N-Diisopropyl-5-methylheptan-3-amine. There are three alkyl groups attached to the nitrogen atom, so this is a tertiary amine.

Question

Give the IUPAC name of the following amine and classify the amine as primary, secondary, or tertiary.

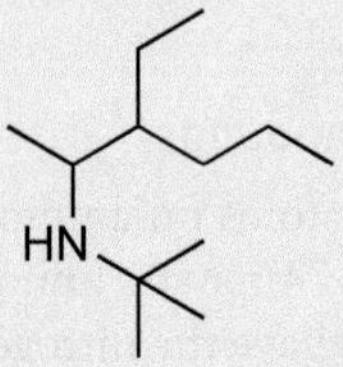
HN

For the simplest aromatic primary amine ($C_6H_5NH_2$), the IUPAC name is phenylamine although the common name, aniline, is often used. For simple substituted anilines, the prefixes 2, 3, or 4 (or *o*-, *m*-, or *p*-) are used to indicate the positions of the groups on the ring. Common names are still used for some of these derivatives, such as toluidine for a methyl-substituted aniline.

Aniline (phenylamine, $PhNH_2$) is used to prepare azo dyes such as Aniline Yellow (Chapter 22 opening page, p.1003).

In industry, aniline is produced from benzene using a two-step process—first, nitration of benzene (PhH) forms nitrobenzene ($PhNO_2$), and this is followed by reduction of the nitro substituent (Section 22.4, p.1040).

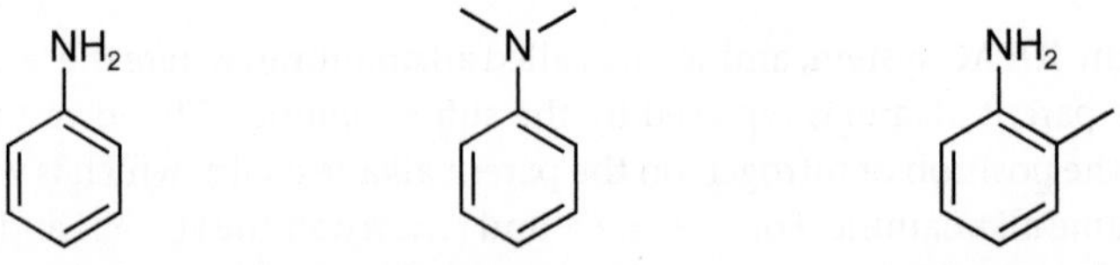
NH2
N
NH2

IUPAC name	phenylamine	*N,N*-dimethylphenylamine	2-methylphenylamine
Common name	aniline	*N,N*-dimethylaniline	2-methylaniline or *o*-toluidine

Enamines have an NH_2, NHR, or NR_2 group attached to a C=C bond. These compounds are often very reactive and difficult to isolate. Enamines with a hydrogen atom on nitrogen are in equilibrium with imines, e.g. $R_2C{=}CH{-}NHR \rightleftharpoons R_2CH{-}CH{=}NR$ (Section 23.2, p.1078).

When the nitrogen atom of an amine is part of a ring, it is called a heterocyclic amine. Important examples include the secondary amines pyrrolidine (5-ring) and piperidine (6-ring). The nitrogen atom of an amine can also be part of an aromatic ring as in the heterocyclic aromatic amines pyrrole, pyridine, and imidazole. When an NH_2, NHR, or NR_2 group is bonded to a C=C double bond, this is called an **enamine**. This name reflects both the presence of a double bond ('en' from alkene) and the nitrogen substituent (amine). The structure of enamines is similar to that of enols (see p.95).

Heterocyclic aliphatic amines Heterocyclic aromatic amines

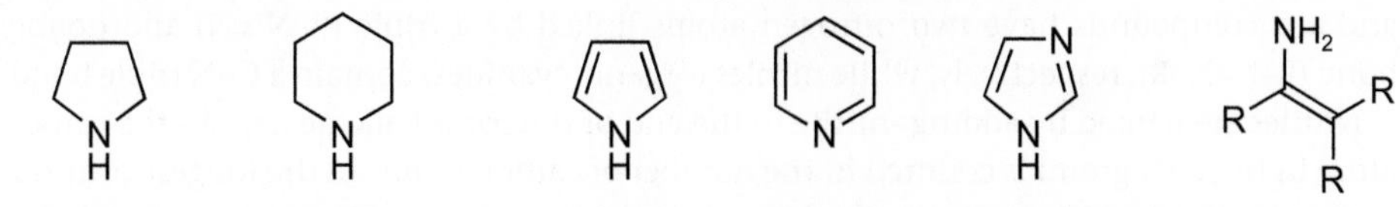

pyrrolidine piperidine pyrrole pyridine imidazole enamine

Another important functional group containing nitrogen is the nitro group. **Nitro compounds** have the general formula RNO_2, where R is an alkyl or aryl group. The structure of the nitro group is best represented as shown in the illustration below, as this allows reaction mechanisms to be easily drawn.

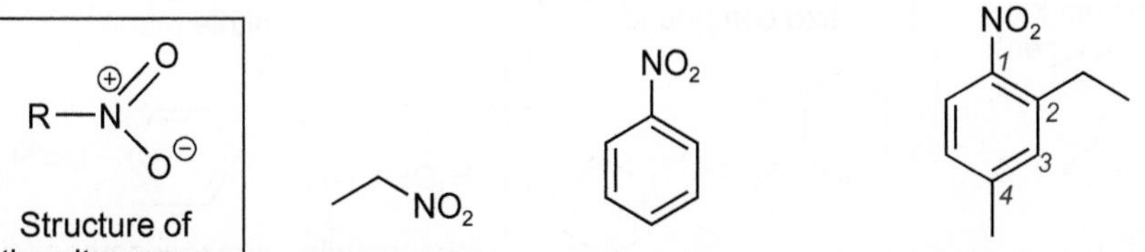

Structure of the nitro group nitroethane nitrobenzene 2-ethyl-4-methyl-1-nitrobenzene

In the IUPAC system, the prefix 'nitro' is used to indicate the presence of the NO_2 group on an alkane chain or benzene ring. For substituted nitrobenzenes, the lowest possible numbers are used to indicate the positions of other substituents, which are listed in alphabetical order.

The structure of the nitro group is one of the trickier ones to remember, so it is worth practising drawing it.

Pyridine (C_5H_5N) is a common base for organic reactions (see, for example, Chapter 20.2, p.927)

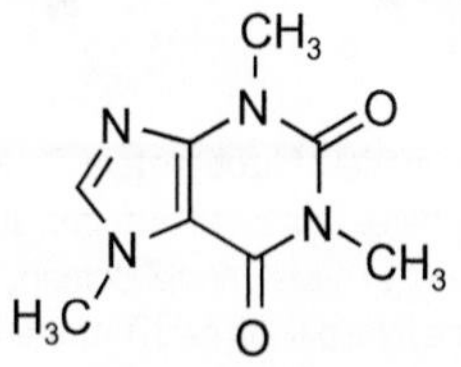

caffeine

Caffeine ($C_8H_{10}N_4O_2$) belongs to a family of heterocycles called purines, which have an imidazole ring fused to another nitrogen-containing heterocycle, called a pyrimidine (Box 5.1, p.218). Drinking beverages containing caffeine, such as coffee, can reduce physical fatigue and prevent or treat drowsiness. For healthy adults with no medical issues, it is generally agreed upon that up to 0.4 g of caffeine can be consumed daily without any adverse effects. Caffeine can be extracted from coffee beans using supercritical carbon dioxide (Chapter 17 opening page, p.765).

Box 2.7 Pain-killing alkaloids

Alkaloids are basic, nitrogen-containing natural products that are isolated from plants or animals. The structures of these compounds can be complex and many exhibit potent biological effects. Indeed, around 30% of all medicines on the market today are based on alkaloids, including the tertiary amines morphine and codeine.

Morphine is a powerful painkiller (analgesic) but is highly addictive. These properties are strongly influenced by the OH group on the benzene ring. Changing the phenolic OH group on morphine to OMe gives codeine, which is much less addictive but is about one-tenth as effective as a painkiller. It is mostly used as an anti-cough agent.

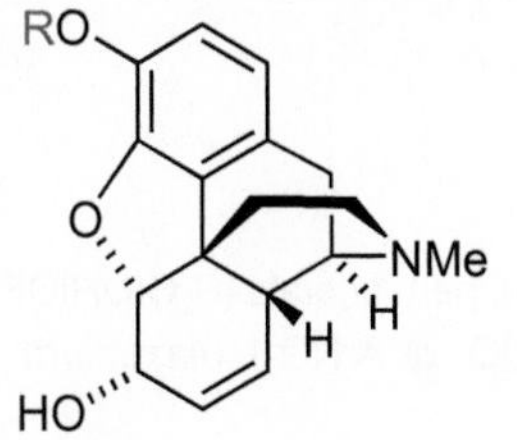

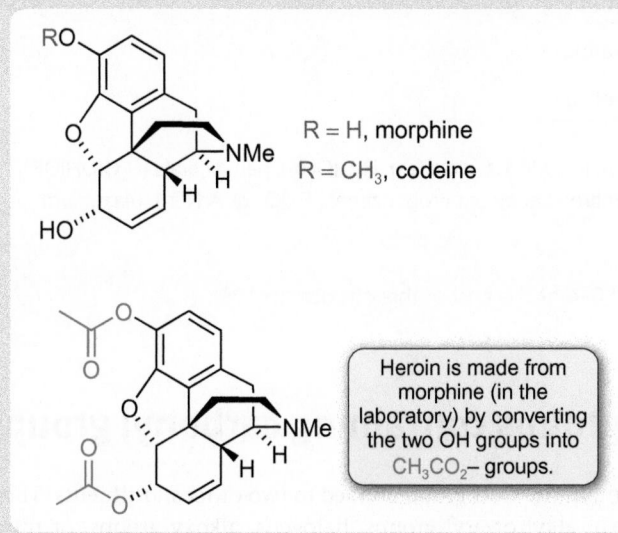

R = H, morphine
$R = CH_3$, codeine

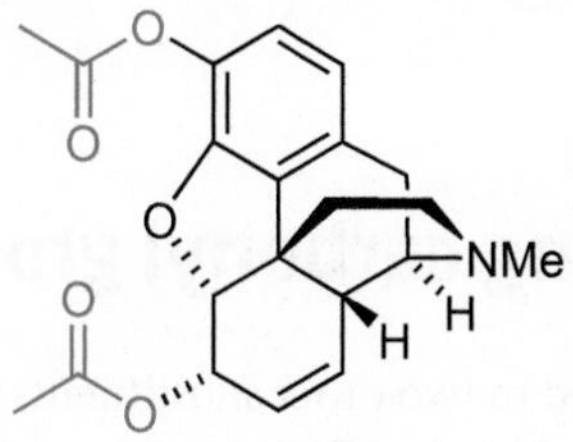

Heroin is made from morphine (in the laboratory) by converting the two OH groups into CH_3CO_2– groups.

▲ Morphine is found in opium, which is the name of the latex produced within the seeds of the opium poppy (*Papaver somniferum*).

Amides contain a nitrogen atom along with a C=O group (e.g. RCONHR) and are considered in Section 2.7 (p.104).

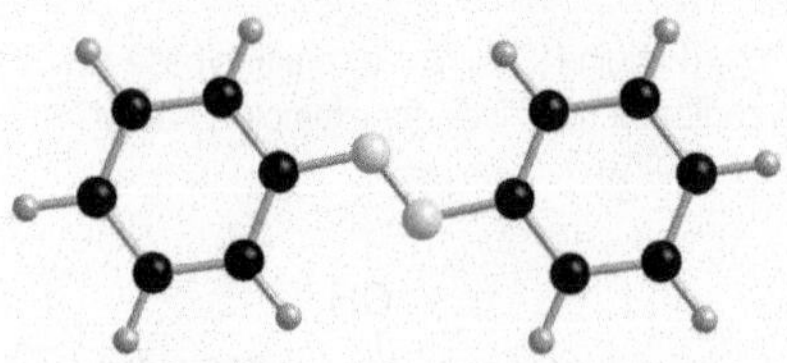

A molecular model of azobenzene, Ph–N=N–Ph (the blue spheres represent nitrogen atoms). Azobenzene can have *E* and *Z* isomers, just like alkenes—the lone pair on each nitrogen atom acts as a substituent (Section 22.4, p.1043). The more stable *E*-isomer is shown here. (In this computer generated ball and stick model, a single stick represents the N=N double bond.)

Diazonium salts can be converted into synthetic dyes, called azo dyes (Chapter 22 opening page, p.1003)

Diazonium ions, **azo compounds**, and **nitriles** also contain nitrogen. Diazonium ions and azo compounds have two nitrogen atoms linked by a triple (R–N$^+$≡N) and double bond (R–N=N–R), respectively, while nitriles (organic cyanides) contain a C≡N triple bond.

Nitriles are named by adding -nitrile to the end of the parent alkane name—the carbon atom in the C≡N group is counted in the number of carbon atoms in the longest continuous chain. If the CN group is attached to a ring, then the ending 'carbonitrile' is added to the name of the parent ring. There are two types of common name in use: often nitriles (RCN) are named after the corresponding carboxylic acid (RCO_2H; Section 2.7, p.101), by replacing -ic or -oic with -onitrile, or they are called alkyl cyanides. So, common names for ethanenitrile (CH_3CN) are ethanonitrile (sometimes acetonitrile) and methyl cyanide.

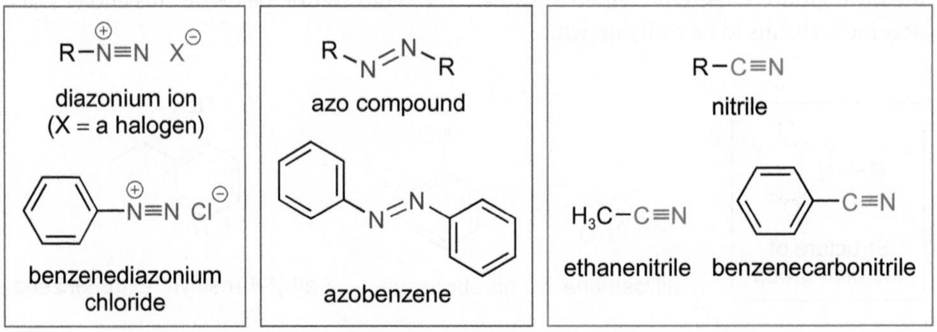
R–N≡N X
diazonium ion
(X = a halogen)
N≡N Cl
benzenediazonium chloride
R–N=N–R
azo compound
N=N
azobenzene
R–C≡N
nitrile
H3C–C≡N
ethanenitrile
C≡N
benzenecarbonitrile

Summary

- Important functional groups containing a single heteroatom include halogenoalkanes (RX), alcohols (ROH), phenols (e.g. PhOH), ethers (ROR), enols (e.g. RCH=CH–OH), thiols (RSH), amines (e.g. RNH_2), enamines (e.g. RCH=CH–NH_2), and nitriles (RCN).
- The general formulae and IUPAC names of the most important functional groups containing one heteroatom are shown in the table.

Compound class	General formula	IUPAC name
Halogenoalkane	RX	haloalkane
Alcohol	ROH	alkanol
Ether	ROR	alkoxyalkane
Thiol	RSH	alkanethiol
Amine	RNH_2, R_2NH, R_3N	alkanamine
Nitrile	RCN	alkanenitrile

- Related functional groups containing more than one heteroatom include 1,1-diols (e.g. $R_2C(OH)_2$), hemiacetals ($R_2C(OH)OR$), acetals ($R_2C(OR)_2$), thioacetals ($R_2C(SR)_2$), nitro compounds (nitroalkanes or nitrobenzenes, RNO_2 or $ArNO_2$), diazonium ions (RN_2^+), and azo compounds (RN=NR).

? For practice questions on these topics, see questions 3–4 at the end of this chapter (p.109).

2.7 Functional groups containing carbonyl groups

Carbonyl compounds contain the C=O group bonded to two other substituents. These substituents can be hydrogen, alkyl or aryl groups, halogens, alkoxy groups, or nitrogen-based groups. For example, compounds containing acyl groups, such as ethanoyl

(CH_3CO–) and benzoyl (PhCO–) groups, bonded to various substituents (e.g. H, Cl, CH_3, OCH_3, and NH_2) are all known. As a consequence, carbonyl compounds represent an extremely large class of compounds and the C=O group is probably the most important functional group in organic chemistry. This can be seen from the large number of commercially and biologically important molecules that contain a C=O bond and the fact that C=O groups play a vital role in the synthesis of organic molecules in the laboratory.

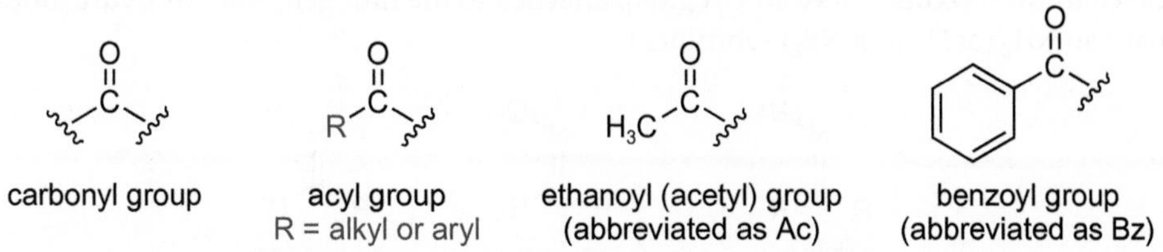
carbonyl group
acyl group
R = alkyl or aryl
ethanoyl (acetyl) group
(abbreviated as Ac)
benzoyl group
(abbreviated as Bz)

As oxygen is more electronegative than carbon, the electrons in the C=O bond are not equally shared ($^{\delta+}C{=}O^{\delta-}$). It is the polar nature of the C=O bond that lies at the heart of the rich chemistry of carbonyl compounds (discussed in Chapters 23 and 24). As there are so many different carbonyl functional groups, the naming of these compounds is divided into two sections—the first discusses aldehydes, ketones, and similar compounds, while the second covers carboxylic acids and related compounds.

Aldehydes, ketones, and related compounds

An **aldehyde** (RCHO) contains a carbonyl group bonded to at least one hydrogen atom, whereas a **ketone** (RCOR) has a carbonyl group bonded to two alkyl or aryl groups.

In the IUPAC system, an aldehyde is named by removing the 'e' at the end of the name of the parent alkane and replacing this by 'al' (as shown in Figure 2.19). As the C=O group of an aldehyde (RCHO) is always at the end of the carbon chain, the position of the C=O group does not have to be specified. When the –CHO group is attached to a ring, the molecule is named by adding the suffix -carbaldehyde to the name of the ring. However, common names are also widely used for a number of aldehydes.

In the IUPAC system, a ketone is named by removing the 'e' from the end of the name of the parent alkane and replacing this by 'one' (as shown in Figure 2.20). The smallest ketone is propanone (CH_3COCH_3), an important organic solvent often known by its common name, acetone. When naming larger ketones, the parent chain is numbered so that the carbonyl carbon atom has the lowest number. The number is given in the middle of

Whereas Bz stands for benzoyl (PhCO), Bn stands for benzyl ($PhCH_2$).

Electronegativity is introduced in Section 4.3 (p.177). Polar bonds are discussed in Section 5.3 (p.235).

Aldehyde: R–CHO.
Ketone: R–CO–R.

The chemistry of aldehydes and ketones is discussed in Chapter 23 (p.1054).

The suffix for aldehydes is -al (or -carbaldehyde when –CHO is attached to a ring). The suffix for ketones is -one.

A molecular model of propanone, CH_3COCH_3 (the red sphere represents oxygen). In this computer generated ball and stick model a single stick represents the C=O double bond (Box 5.1, p.218).

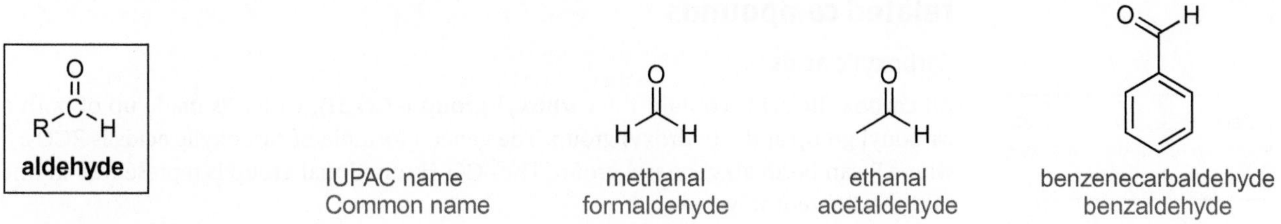
aldehyde
IUPAC name
Common name
methanal
formaldehyde
ethanal
acetaldehyde
benzenecarbaldehyde
benzaldehyde

Figure 2.19 Naming aldehydes.

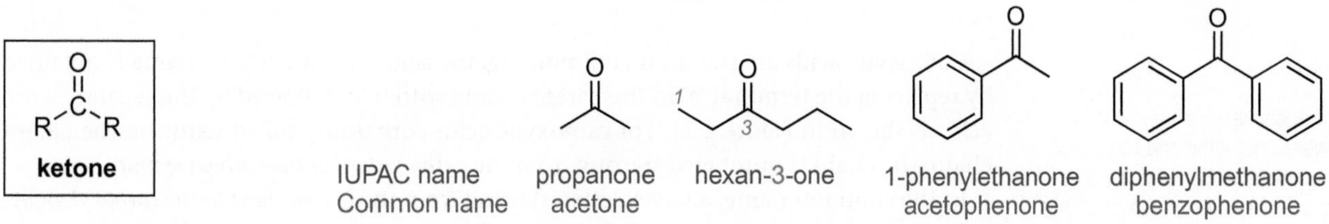
ketone
IUPAC name
Common name
propanone
acetone
hexan-3-one
1-phenylethanone
acetophenone
diphenylmethanone
benzophenone

Figure 2.20 Naming ketones.

The position of the C=O group is sometimes shown at the front of the name, for example, 3-hexanone rather than hexan-3-one (for $CH_3CH_2COCH_2CH_2CH_3$).

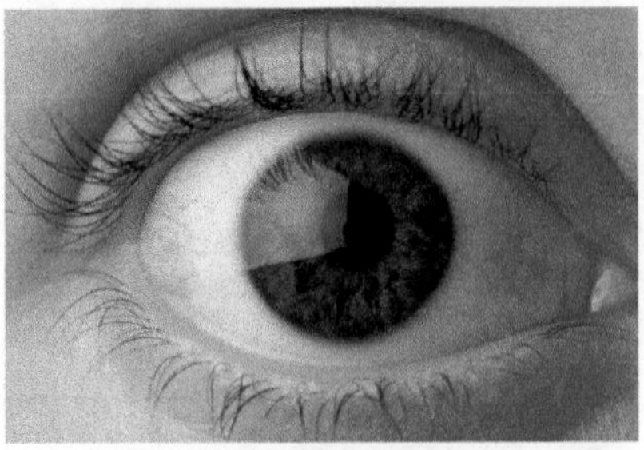

An imine, called rhodopsin, plays a crucial role in vision; see Chapter 23 opening page (p.1055)

the name before '-one' (e.g. hexan-3-one), as illustrated in Worked example 2.8. For aromatic ketones, the common names acetophenone ($PhCOCH_3$) and benzophenone (PhCOPh) are widely used.

Related compounds containing a C=NR rather than a C=O double bond also exist. The names of these functional groups depend on the nature of the R substituent attached to nitrogen. When this is a hydrogen atom or an alkyl or aryl group, these molecules are called **imines.** **Oximes** have an OH group attached to the nitrogen, whereas **hydrazones** have an NH_2 (or NHR or NR_2) substituent.

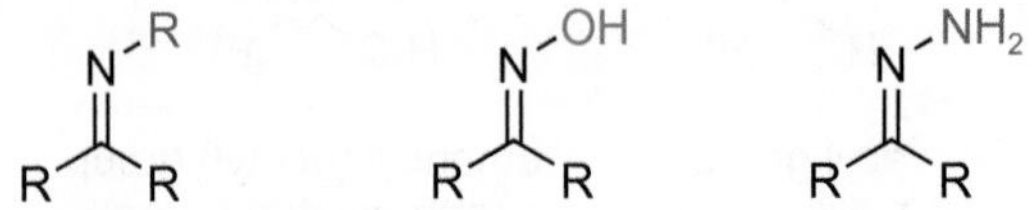

N
R
OH
NH2

imine
R = H, alkyl, aryl

oxime

hydrazone

Worked example 2.8 Drawing and naming ketones

Draw the skeletal structure of 4-benzyl-3-methyloctan-2-one.

Strategy

Draw the zigzag structure of octane and number the chain.

Include the carbonyl at position 2, a methyl group at position 3, and a benzyl group at position 4.

Solution

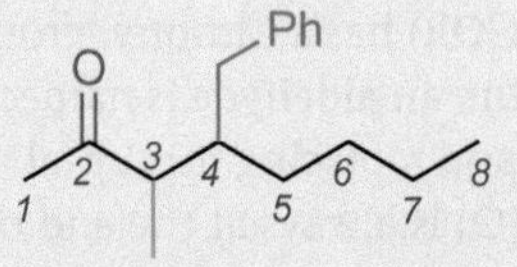

Ph
O
1
2
3
4
5
6
7
8

4-benzyl-3-methyloctan-2-one

Question

Draw the skeletal structure of 2-allyl-3,3-dimethylhexanal.

Carboxylic acids, acyl halides, acid anhydrides, esters, and related compounds

Carboxylic acids

Carboxylic acid: R–CO_2H. The suffix for carboxylic acids is -oic acid (or carboxylic acid when –COOH is attached to a ring).

All **carboxylic** acids contain the **carboxyl group (–CO_2H)**, which is made up of both a carbonyl group and a hydroxyl group. The general formula of carboxylic acids is RCO_2H, where R can be an alkyl or aryl group. The –CO_2H functional group is represented in one of three different ways:

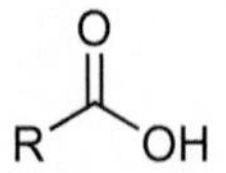

O
R
OH

or RCO_2H or RCOOH

The chemistry of carboxylic acids and related compounds is discussed in Chapter 24 (p.1098).

Carboxylic acids are widely found in nature so they were among the first classes of compounds to be isolated and studied by chemists. Many common names are still used, which are derived from the name of the source (see Section 2.4, p.79).

Carboxylic acids are the most common organic acids and the IUPAC name is obtained by replacing the terminal 'e' in the parent alkane with 'oic' followed by the separate word acid, as shown in Figure 2.21. For carboxylic acids containing substituents on the parent chain, the chain is numbered starting from the *carbon atom of the carboxyl group*.

In the common name, a Greek letter, starting from the *carbon next to the carboxyl group*, gives the position of substituents. The carbon atom next to the carboxyl group, or indeed next to any carbonyl group, is called the **α carbon** (alpha-carbon).

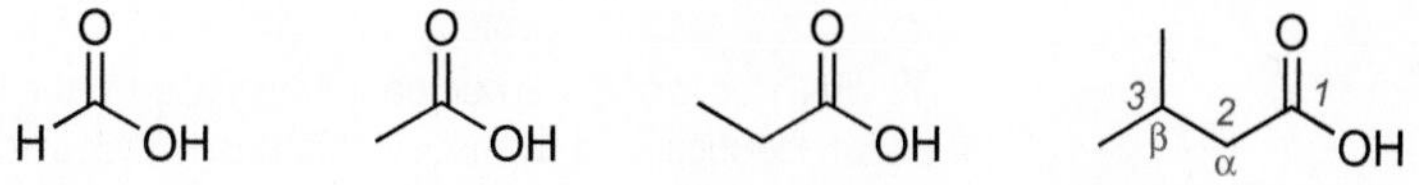

IUPAC name	methanoic acid	ethanoic acid	propanoic acid	3-methylbutanoic acid
Common name	formic acid	acetic acid	propionic acid	β-methylbutyric acid

Figure 2.21 Naming carboxylic acids.

Ants use methanoic acid (HCO_2H; formic acid) to make a stinging bite. Methanoic acid used to be prepared by boiling ants in a pan.

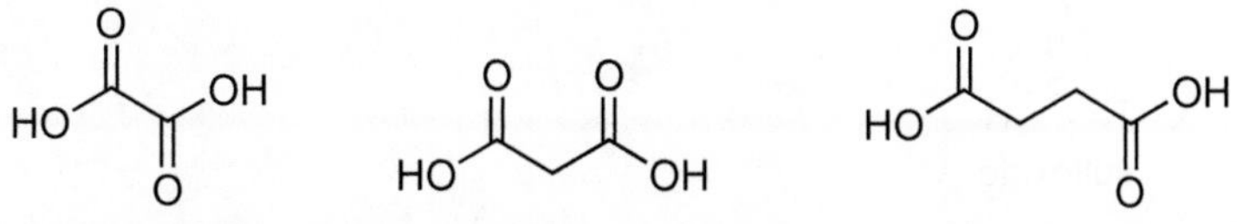

IUPAC name	ethanedioic acid	propanedioic acid	butanedioic acid
Common name	oxalic acid	malonic acid	succinic acid

Figure 2.22 Naming dicarboxylic acids.

To name dicarboxylic acids using IUPAC nomenclature, the suffix -dioic acid is added to the name of the parent alkane. As both carboxyl groups must be at the ends of the alkane chain, numbers are not required to indicate the positions of these groups. Once again, many dicarboxylic acids are known by their common names as shown in Figure 2.22.

When the carboxyl group is attached to a ring, the suffix -carboxylic acid is given after the name of the parent alkane and numbers are used to indicate the positions of any substituents on the ring. However, for the simplest aromatic carboxylic acid ($PhCO_2H$), the common name benzoic acid is generally used in place of the systematic name benzenecarboxylic acid.

Dicarboxylic acids are used to prepare copolymers, such as polyesters and polyamides, including nylon-6,6 (p.1136). The conjugate bases of dicarboxylic acids, such as the oxalate ion $^{-}O_2C\text{–}CO_2^{-}$, are excellent bidentate ligands for many transition metal ions (Section 28.3, p.1267).

4-methylcyclohexanecarboxylic acid   benzoic acid

Reaction of benzoic acid with sodium hydroxide forms sodium benzoate, $PhCO_2^{-}Na^{+}$ (and water). Sodium benzoate is a widely used food preservative, present in carbonated drinks and fruit juices, with the E number E211.

When carboxylic acids are treated with a base, such as NaOH, they lose a proton to form **carboxylate salts**, as shown below. The name of these salts is derived from the name of the cation, followed by the name of the carboxylate anion (RCO_2^{-}) as a separate word. Carboxylate ions are named after the carboxylic acid by changing the '-ic' ending of the acid to '-ate'.

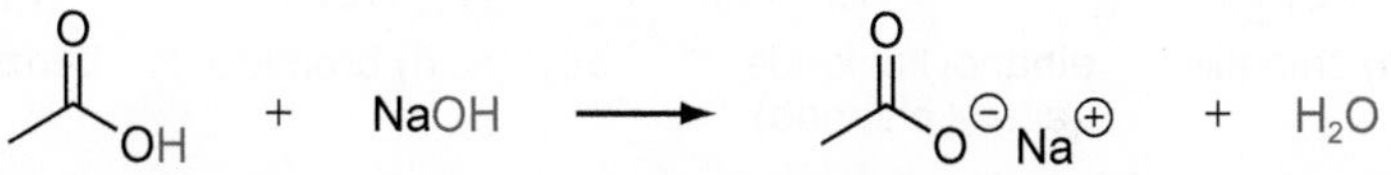

ethanoic acid   sodium ethanoate

Box 2.8 Sulfonic acids and related compounds

Sulfonic acids have the general formula RSO_3H. Like carboxylic acids, these molecules are easily deprotonated to form salts (e.g. RSO_3^{-} Na^{+}). The SO_3^{-} groups increase the solubility of organic compounds in water and this property has been exploited in synthetic detergents and dyes (including food colourings). One example is Allura Red (E129), a red food colouring.

Sulfonic acids can also be converted into related functional groups such as sulfonyl chlorides, sulfonamides, and sulfonate esters. For comparison, the structures of sulfones and sulfoxides are also shown. In these compounds, the oxidation state (Section 1.4, p.28) of sulfur is +2 (sulfone) and 0 (sulfoxide), rather than +4 as in sulfonic acids, sulfonyl chlorides, sulfonamides, and sulfonate esters. →

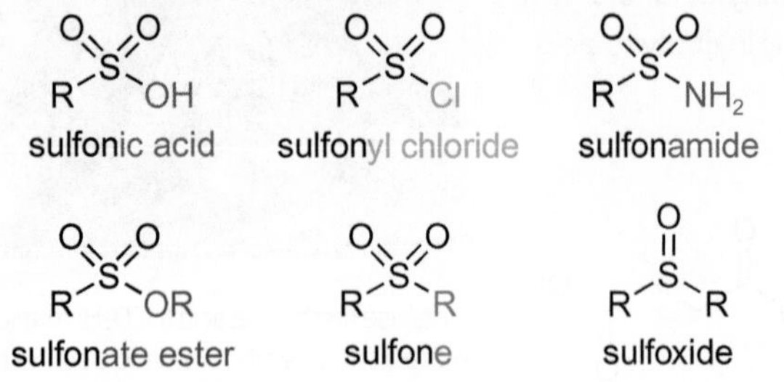
sulfonic acid
sulfonyl chloride
sulfonamide
sulfonate ester
sulfone
sulfoxide

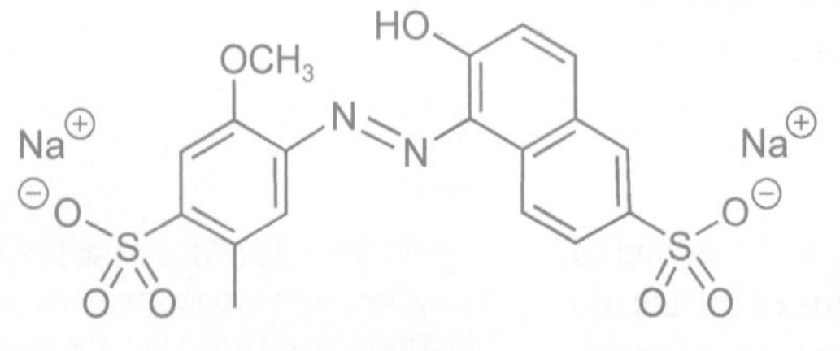

Allura Red is a food colouring containing two sodium sulfonate groups

▼ Allura Red is one of a number of dyes used in the food, pharmaceutical, and cosmetic industries. It is also used as a stain to highlight parts of a cell or biological tissue in order to aid viewing by optical microscopes. In the photo, a white blood cell is stained purple, allowing it to be easily distinguished from the smaller red blood cells.

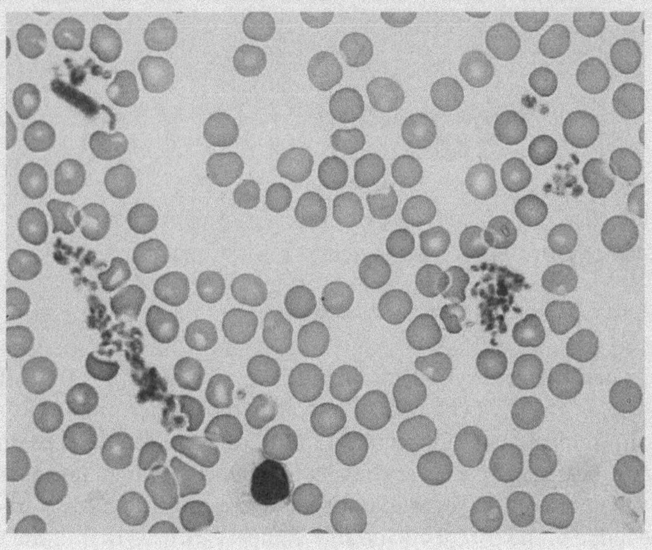

Carboxylic acid derivatives

There are a large number of compounds structurally related to carboxylic acids in which the OH group is replaced by another substituent. Important examples are acyl (acid) halides, acid anhydrides, esters, and amides, which can all be prepared from carboxylic acids. Hence these functional groups are commonly grouped together as carboxylic acid derivatives.

Acyl halides (acid halides) have a halogen atom in place of the OH group of a carboxylic acid. They have the general formula RCOX, where X is fluorine, chlorine, bromine, or iodine; the most common examples are acyl chlorides (RCOCl) and acyl bromides (RCOBr). These compounds are generally named by replacing 'ic' from the name of the carboxylic acid with 'yl chloride' or 'yl bromide', respectively, as shown in Figure 2.23.

Acyl halide: R–COX.
The suffixes for acyl chlorides and acyl bromides are -oyl chloride and -oyl bromide, respectively.

The preparation of acyl halides (RCOCl), esters (RCO_2R), and amides (e.g. $RCONH_2$) from carboxylic acids (RCO_2H) is described in Section 24.2 (p.1108).

Acid anhydride: R–CO_2CO–R.
The suffix for acid anhydrides is -oic anhydride.

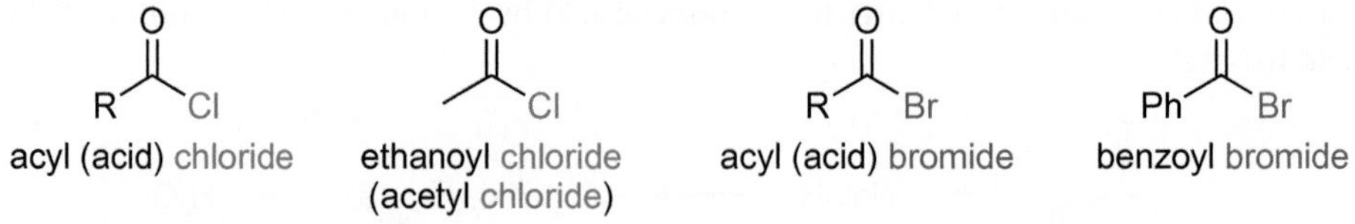
acyl (acid) chloride
ethanoyl chloride
(acetyl chloride)
acyl (acid) bromide
benzoyl bromide

Figure 2.23 Naming acyl halides.

For **acid anhydrides**, the OH group of a carboxylic acid is replaced by OCOR. In effect, two acyl groups (RCO) are linked together by a single oxygen atom. These compounds are prepared by removing water from two carboxylic acid molecules.

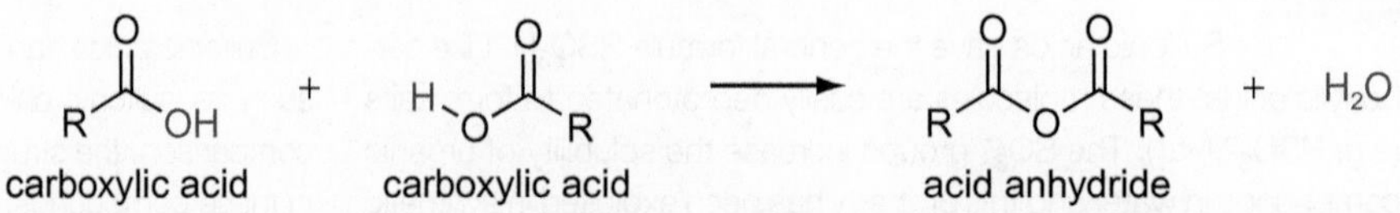
carboxylic acid
carboxylic acid
acid anhydride
H_2O

Formation of acid anhydrides from carboxylic acids requires a **dehydrating agent**, such as phosphorus(V) oxide (or phosphorus pentoxide, P_4O_{10}), Section 27.4, p.1227. The dehydrating agent removes the water from the reaction mixture as soon as it is formed and this forces the reaction to produce more of the acid anhydride (Le Chatelier's principle, Section 1.9, p.57).

If the two carboxylic acids are the same, a **symmetrical anhydride** is formed. **Mixed anhydrides** are prepared from two different carboxylic acids. Symmetrical anhydrides are named by replacing the word 'acid' in the name of the carboxylic acid from which it was formed with 'anhydride'. The names of mixed anhydrides are obtained from the names

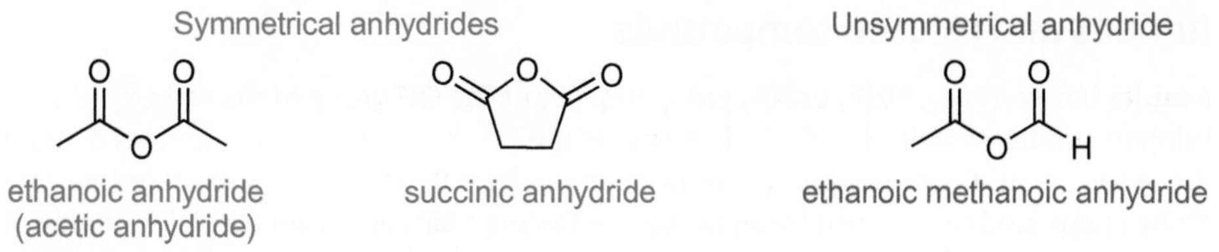
Symmetrical anhydrides
Unsymmetrical anhydride
ethanoic anhydride
(acetic anhydride)
succinic anhydride
ethanoic methanoic anhydride

Figure 2.24 Naming symmetrical and unsymmetrical anhydrides.

of both acids (given in alphabetical order) followed by anhydride (see Figure 2.24). Loss of water from dicarboxylic acids can also give 5- and 6-membered cyclic anhydrides, which are named after the parent dicarboxylic acid, for example, succinic anhydride from succinic acid ($HO_2CCH_2CH_2CO_2H$).

Replacing the OH group of a carboxylic acid with an alkoxy (OR) or aryloxy (OAr) group produces **esters**. These have the general formula RCO_2R (RCOOR) and the IUPAC and common names of esters are derived from the parent carboxylic acids (RCO_2H). The name of the alkyl or aryl group bonded to oxygen is given first, followed by the name of the acid (as a separate word) with the ending '-ic acid' replaced by '-ate', as shown in Figure 2.25 (and Worked example 2.9).

Cyclic esters are called **lactones**. The simplest lactones are known by their common names, which are derived by replacing '-oic acid' from the name of the parent carboxylic acid with the suffix '-olactone'. A Greek letter is used to indicate the different carbon atoms in the ring. Whereas 4-membered lactones are called β-lactones (because the β-carbon atom forms the C–O link), 5- and 6-membered ring lactones are called γ- and δ-lactones, respectively; see Figure 2.26.

The characteristic odours and tastes of fruits are partly due to low-molecular-mass esters. In bananas, the sweet fruity odour of 3-methylbutyl ethanoate ($CH_3CO_2CH_2CH_2CH(CH_3)_2$) attracts numerous feeders from fruit flies to fruit bats and monkeys.

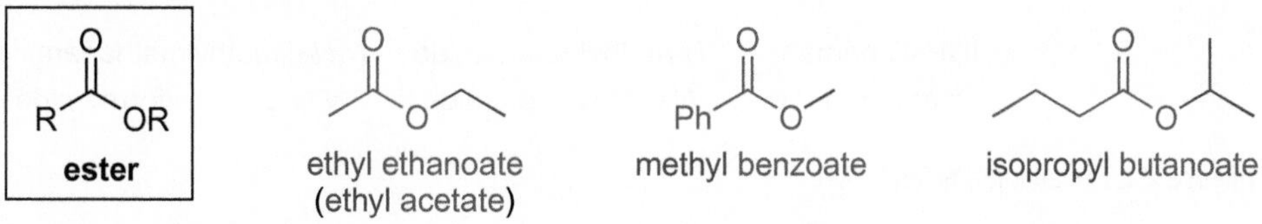
ester
ethyl ethanoate
(ethyl acetate)
methyl benzoate
isopropyl butanoate

Figure 2.25 Naming esters.

Ester: $R–CO_2R$.
The suffix for esters is -oate.

The suffix for lactones is -olactone.

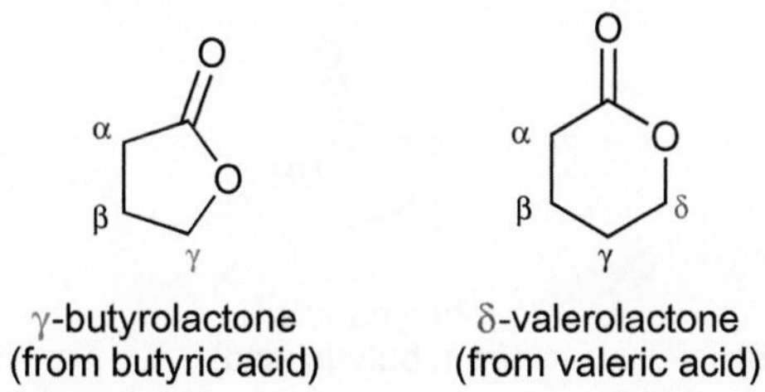
α
β
γ
δ
γ-butyrolactone
(from butyric acid)
δ-valerolactone
(from valeric acid)

Figure 2.26 Naming lactones.

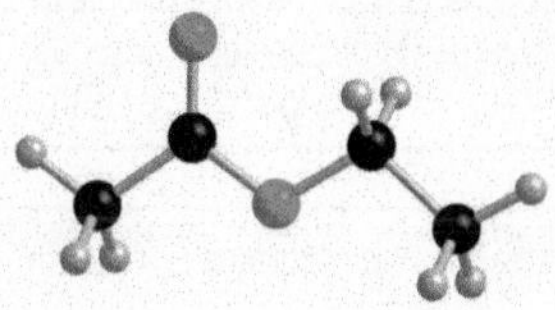

A molecular model of ethyl ethanoate, $CH_3CO_2CH_2CH_3$ (a common solvent for reactions of organic molecules). In this computer generated ball and stick model a single stick represents the C=O double bond (Box 5.1, p.218).

Worked example 2.9 Drawing and naming esters

Draw the skeletal structure of phenyl 2-methylbutanoate.

Strategy

Draw the zigzag structure of butane and number the chain.

At position 1, draw a carbonyl double bond and also a single bond to oxygen.

Include a phenyl group on the singly bonded oxygen (at position 1) and a methyl group at position 2.

Solution

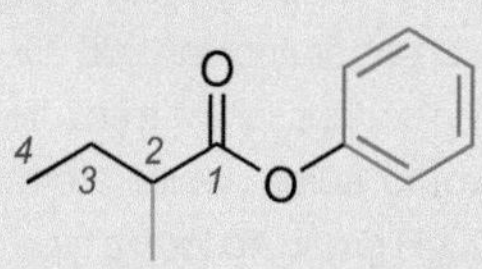

phenyl 2-methylbutanoate

Question

Draw the skeletal structure of isopropyl benzoate.

Reaction of γ-butyrolactone (GBL), see Figure 2.26, with water forms γ-hydroxybutyric acid (GHB, 4-hydroxybutanoic acid, $HOCH_2CH_2CH_2CO_2H$), which has been identified as a date rape drug. In the body, GHB produces euphoric and sedative effects; it is colourless and odourless and it has been reported that flavoured drinks can mask its salty taste.

Amide: R–$CONH_2$, R–CONHR, R–$CONR_2$. The suffix for amides is -amide.

Ethanamide (CH_3CONH_2) has been detected near the centre of our Milky Way Galaxy, some 26,000 light years from Earth. As ethanamide has an amide bond, similar to the bond linking amino acids in proteins (Box 2.9 below), this finding supports the theory that organic molecules that can lead to life (on Earth) can form in space.

N,N-Dimethylformamide (abbreviated DMF), $HCONMe_2$, is a common solvent for chemical reactions. It is an example of a polar aprotic solvent—it contains a polar group, but no O–H or N–H bonds (Section 20.3, p.937).

The suffix for lactams is -olactam.

The presence of a β-lactam ring is crucial for the antibiotic activity of penicillins such as penicillin G (Section 1.2, p.11)

Amides and related compounds

Amides have an NH_2, NHR, or NR_2 group in place of the OH group of the carboxylic acid. Primary amides contain the –$CONH_2$ group, while secondary and tertiary amides contain the –CONHR and –$CONR_2$ groups, respectively (where R can be an alkyl or aryl group). These compounds are named by replacing the ending '-oic acid' from the IUPAC name of the carboxylic acid (or '-ic acid' from the common name) with '-amide'. Some examples are given in Figure 2.27. The simplest aromatic primary amide is benzenecarboxamide, $PhCONH_2$ (or benzamide), the name being derived from benzenecarboxylic acid, $PhCO_2H$ (or benzoic acid). For secondary and tertiary amides, the name of the alkyl or aryl groups attached to nitrogen is given first and the letter '*N*' (in italics) is included to show that these groups are bonded to nitrogen.

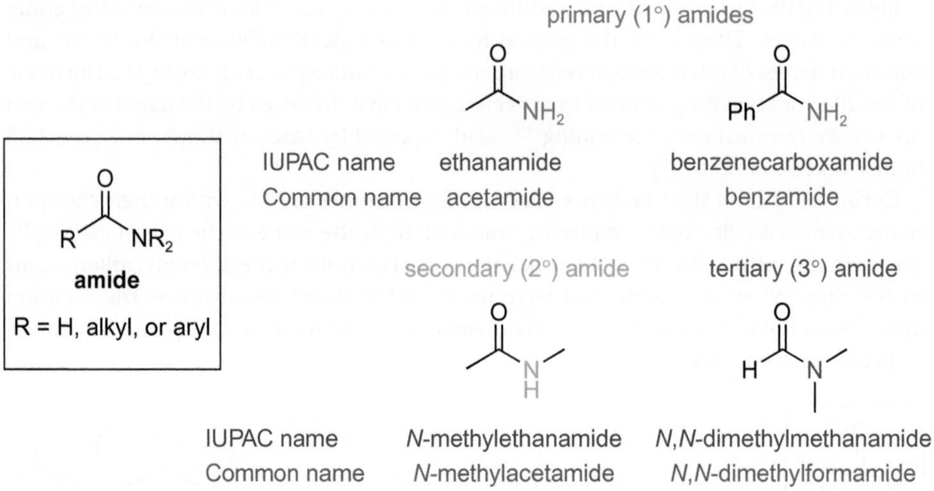
primary (1°) amides
amide
R = H, alkyl, or aryl
IUPAC name
Common name
ethanamide
acetamide
benzenecarboxamide
benzamide
secondary (2°) amide
tertiary (3°) amide
IUPAC name
Common name
N-methylethanamide
N-methylacetamide
N,N-dimethylmethanamide
N,N-dimethylformamide

Figure 2.27 Naming amides.

Cyclic amides are called **lactams**. As shown in Figure 2.28, their common names are similar to those of lactones, except that *-lactam* is used in place of *-lactone*.

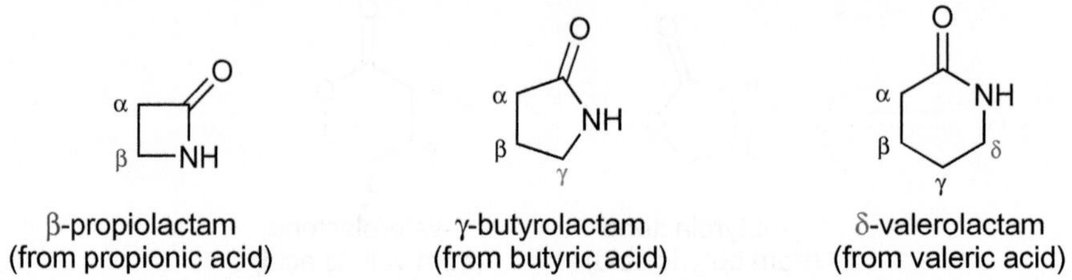
β-propiolactam
(from propionic acid)
γ-butyrolactam
(from butyric acid)
δ-valerolactam
(from valeric acid)

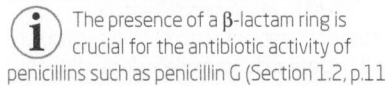

Figure 2.28 Naming lactams.

Box 2.9 An artificial sweetener

Aspartame is an artificial sweetener that was discovered by chance in an American laboratory in 1965. Dr James Schlatter was preparing a compound in connection with an anti-ulcer project. He was linking together two **amino acids**, which are molecules that contain both an –NH_2 and a –CO_2H group, to form a type of compound called a dipeptide. On one occasion, Schlatter was reacting together the amino acids aspartic acid and phenylalanine. When cleaning the laboratory glassware, he accidentally tasted a powder on his fingers and found this to have a sugary taste. He quickly established that the powder was the dipeptide formed when one of the –CO_2H groups on aspartic acid reacts with the –NH_2 group on phenylalanine, to form an amide bond, and the other –CO_2H group on aspartic acid reacts to form a –CO_2Me group. It was not until 14 years later that the dipeptide was approved for sale in France, where it was named Canderel—after the French words for candy (**candi**) and bilberries (**airelles**), which have a sweet flavour. One teaspoonful of Canderel has less than 2 calories, which has helped to ensure that, in the UK alone, people get through around 4.6 million servings a day.

→

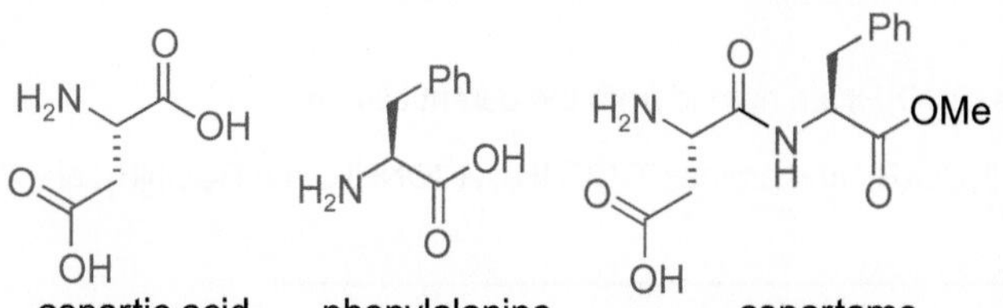

aspartic acid phenylalanine aspartame

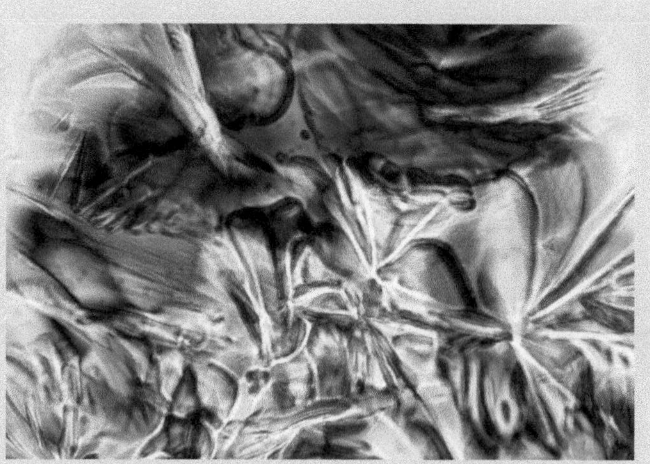

A photomicrograph of a crystal of aspartame under polarized light (the light is diffracted by the crystals to produce the colourful pattern). ➤

There are a number of other functional groups that are related to amides. These include **imides**, which have two acyl groups bonded to a nitrogen atom, **carbamates** (urethanes), and also **ureas**.

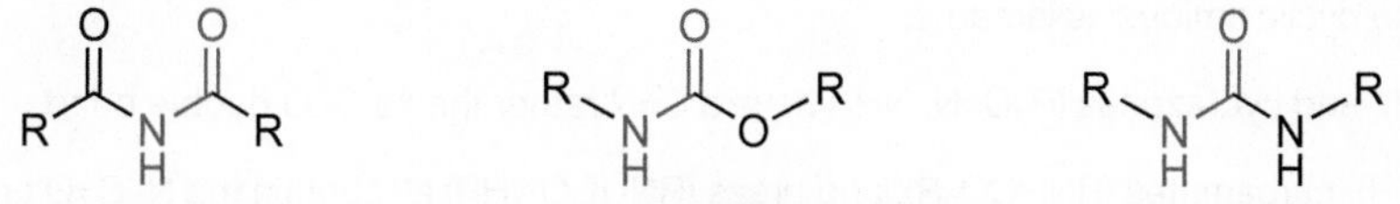

imide carbamate (urethane) urea

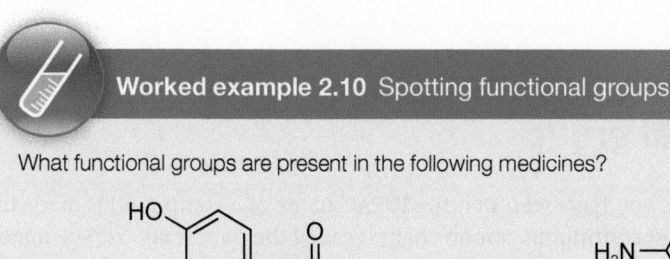

Urea (H_2NCONH_2) was the first organic compound to be synthesized from wholly inorganic starting materials (Box 2.1, p.71).

In multistep syntheses, amines are often temporarily protected from reacting with electrophiles by conversion into carbamates (Worked example 24.2, p.1117).

Worked example 2.10 Spotting functional groups

What functional groups are present in the following medicines?

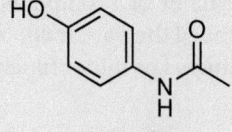

paracetamol — Widely used to relieve pain and fever

sulfanilamide — An antibacterial agent

Solution

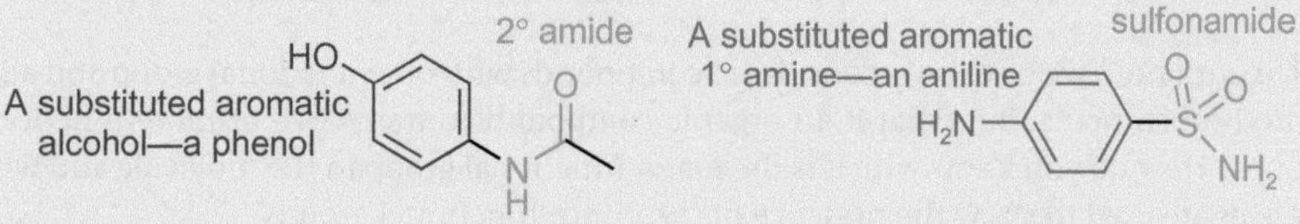
2° amide
A substituted aromatic
1° amine—an aniline
sulfonamide
A substituted aromatic
alcohol—a phenol

Question

Name all five functional groups in aspartame (see Box 2.9, p.104).

Visit the Online Resource Centre to view screencast 2.2. This screencast explains the importance of being able to identify the groups of atoms in organic compounds that give them their chemical properties, using examples from nature and medicine.

Summary

- All carbonyl compounds contain the C=O group.
- Aldehydes (RCHO), ketones (RCOR), and carboxylic acids (RCO_2H) are named from the parent alkane.
- Acyl halides (RCOX), acid anhydrides (RCO_2COR), esters (RCO_2R), and amides ($RCONH_2$, RCONHR, and $RCONR_2$) are named from the parent carboxylic acid.

Compound class	General formula	IUPAC name
Aldehyde	RCHO	alkanal
Ketone	RCOR	alkanone
Carboxylic acid	RCO_2H	alkanoic acid
Acyl chloride or bromide	RCOCl or RCOBr	alkanoyl chloride or bromide
Acid anhydride	$(RCO)_2O$	alkanoic anhydride
Ester	RCO_2R	alkyl alkanoate
Amide	$RCONR_2$, RCONHR, $RCONH_2$	alkanamide

- Cyclic esters are known as lactones and cyclic amides as lactams.
- Imines (R_2C=N–R), oximes (R_2C=N–OH), and hydrazones (R_2C=N–NH_2) have a C=N rather than a C=O double bond.
- Amides ($RCONH_2$), imides (RCONHCOR), carbamates ($RNHCO_2R$), and ureas (RNHCONHR) all contain the N–C=O group.

? For practice questions on these topics, see questions 3–4 at the end of this chapter (p.109).

2.8 Naming compounds with more than one functional group

The name prefix comes from a combination of the Latin words 'pre' meaning before and 'fix' meaning attach.

Amines are classified as primary (RNH_2), secondary (R_2NH), or tertiary (R_3N), based on the number of alkyl or aryl groups attached to nitrogen (Section 2.6, p.95). Similarly, amides are classified as primary ($RCONH_2$), secondary (RCONHR), or tertiary ($RCONR_2$) (Section 2.7, p.104).

Sequentially replacing the hydrogen atoms in ethene (H_2C=CH_2) with alkyl or aryl groups gives monosubstituted (RCH=CH_2), disubstituted (R_2C=CH_2 or RCH=CHR), trisubstituted (R_2C=CHR), and tetrasubstituted (R_2C=CR_2) alkenes (Section 2.5, p.85). Similarly, alkynes are classified as monosubstituted (RC≡CH) or disubstituted (RC≡CR).

In the previous sections, you have seen that the IUPAC name of a compound is made up of three parts. The longest continuous carbon chain is called the **parent chain**. The major functional group is identified in the **suffix**. Any substituents or minor functional groups are listed in the **prefix**.

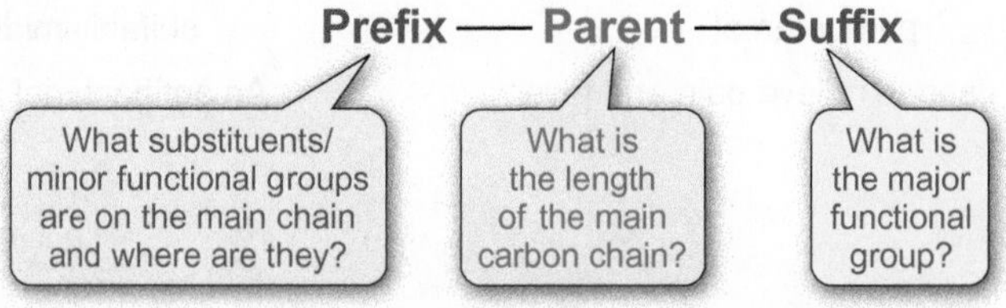
Prefix — Parent — Suffix
What substituents/ minor functional groups are on the main chain and where are they?
What is the length of the main carbon chain?
What is the major functional group?

Up until now, we have only considered compounds with one functional group and alkyl or aryl substituents. But what if an organic compound contains two or more functional groups? How do you know which is the major functional group in the molecule and what prefixes are used to show the presence of the minor functional groups?

In order to derive names of multifunctional compounds, the functional groups must be prioritized. The IUPAC system has produced an order of precedence of functional groups, which is shown in Table 2.3. The carboxylic acid functional group (–CO_2H) is given the highest priority, whilst the halo functional groups (–F, –Cl, –Br, or –I) of the halogenoalkanes have the lowest priority.

Table 2.3 The order of precedence of common functional groups

	Compound class (functional group)	Suffix (if the group is higher in precedence)	Prefix (if the group is lower in precedence)
Highest priority	Carboxylic acid (RCO_2H)	-oic acid	
↓	Ester (RCO_2R)	-oate	alkoxycarbonyl-
	Amide (e.g. $RCONH_2$)	-amide	amido-
	Nitrile (RCN)	-nitrile	cyano-
	Aldehyde (RCHO)	-al	oxo-
	Ketone (RCOR)	-one	oxo-
	Alcohol (ROH)	-ol	hydroxy-
	Amine (e.g. RNH_2)	-amine	amino-
	Alkene (e.g. RCH=CHR)	-ene	alkenyl-
	Alkyne (e.g. RC≡CR)	-yne	alkynyl-
	Alkane (RH)	-ane	alkyl-
	Ether (ROR)		alkoxy-
Lowest priority	Halogenoalkane (RX)		halo-

Once you have identified the functional groups in a molecule, use Table 2.3 to prioritize the groups. For the major functional group the suffix name is used, while prefix names are used to identify the presence of minor functional groups. This is illustrated in Figure 2.29.

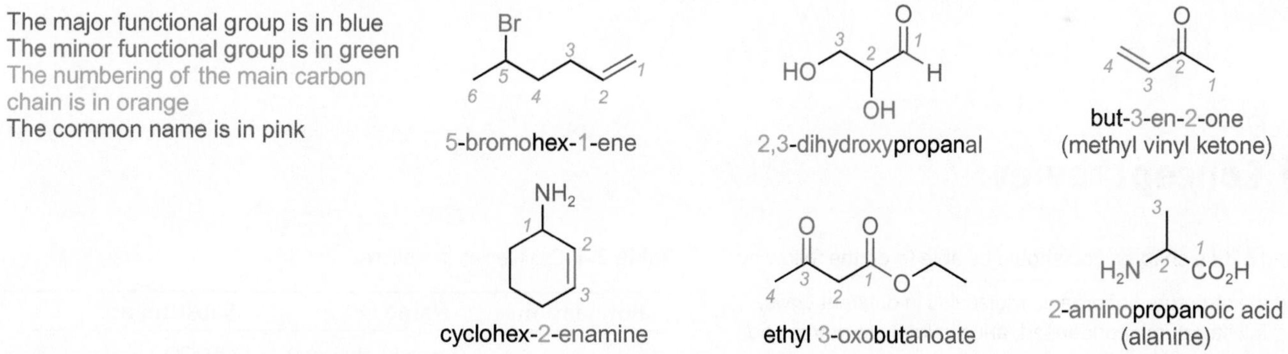
The major functional group is in blue
The minor functional group is in green
The numbering of the main carbon chain is in orange
The common name is in pink
5-bromohex-1-ene
2,3-dihydroxypropanal
but-3-en-2-one
(methyl vinyl ketone)
cyclohex-2-enamine
ethyl 3-oxobutanoate
2-aminopropanoic acid
(alanine)

Figure 2.29 Naming polyfunctional molecules.

To name substituted benzenes, the prefixes for the functional groups are listed in alphabetical order followed by the word benzene (as shown in Figure 2.30). For disubstituted benzenes, which have two substituents, the substituent of lower alphabetical order is given the number 1 and the ring is numbered so that the second substituent has the lowest possible number. Where common names are used (e.g. toluene and benzoic acid), the group associated with the common name (e.g. $-CH_3$ for toluene and $-CO_2H$ for benzoic acid) is given the number-1 position. Finally, when there are more than two substituents, the ring is numbered so that the lowest possible numbers are used and the substituents are listed in alphabetical order.

To test your knowledge of naming organic compounds, see the drag and drop activity on the Online Resource Centre.

The principal functional group determines the numbering.
Cl
NO_2
OH
Cl
NO_2
NH_2
OCH_3
OCH_3
CO_2H
Br
OH

1-chloro-3-nitrobenzene
(Substituents are numbered in alphabetical order)

3-chloro-4-nitrophenol
(The OH is the principal functional group, which is assigned to be on carbon atom 1)

2,4-dimethoxyaniline
(Or 2,4-dimethoxybenzenamine; the NH_2 is the principal functional group and it is assigned to be carbon atom 1)

3-bromo-4-hydroxybenzoic acid
($-CO_2H$ has a higher priority than –OH or –Br; so benzoic acid is preferred to phenol or bromobenzene)

Figure 2.30 Naming polysubstituted benzenes.

Summary

Organic compounds with more than one functional group are named as follows.

Step 1 Find the longest carbon chain and name this as the parent alkane.

Step 2 Identify the major functional group and replace -ane with a suffix (see Table 2.3).

Step 3 Number the chain starting nearest the major functional group.

Step 4 Identify any substituents, including minor functional groups on the chain and their number. The names and numbers are given in the prefix in alphabetical order.

? For practice questions on this topic, see questions 5–7 at the end of this chapter (p.109).

Concept review

By the end of this chapter, you should be able to do the following.

- Draw the structures of organic molecules in different ways, using full structural, condensed, and skeletal structures and hashed–wedged notation.
- Recognize and use common abbreviations in chemical structures (Table 2.4). This includes the use of R and Ar to represent alkyl and aryl groups, respectively.
- Recognize important organic functional groups (see Section 2.3, p.78).
- Name organic compounds using IUPAC nomenclature.
- Know the common names for a number of well-known organic solvents and reagents that have been accepted into the IUPAC system (Table 2.5).
- Classify carbon atoms as primary, secondary, tertiary, or quaternary.

Table 2.4 Common abbreviations

Abbreviation	Name	Substituent
Ac	acetyl (ethanoyl)	$CH_3CO–$
Bn	benzyl	$PhCH_2–$
Bu	butyl	$CH_3CH_2CH_2CH_2–$
^{i}Bu	isobutyl	$(CH_3)_2CHCH_2–$
^{t}Bu	*tert*-butyl	$(CH_3)_3C–$
Bz	benzoyl	$C_6H_5CO–$
Et	ethyl	$CH_3CH_2–$
Me	methyl	$CH_3–$
Ph	phenyl	$C_6H_5–$
Pr	propyl	$CH_3CH_2CH_2–$
^{i}Pr	isopropyl	$(CH_3)_2CH–$

Table 2.5 Names and condensed structures for some well-known compounds

Common name	IUPAC name	Condensed structure
Acetaldehyde	ethanal	CH_3CHO
Acetic acid	ethanoic acid	CH_3CO_2H
Acetone	propanone	CH_3COCH_3
Acetonitrile	ethanenitrile	CH_3CN
Acetyl chloride	ethanoyl chloride	CH_3COCl
Acetylene	ethyne	$HC{\equiv}CH$
Aniline	phenylamine	$C_6H_5NH_2$
Anisole	methoxybenzene	$C_6H_5OCH_3$
Benzaldehyde	benzenecarbaldehyde	C_6H_5CHO
Benzoic acid	benzenecarboxylic acid	$C_6H_5CO_2H$
Chloroform	trichloromethane	$CHCl_3$
Dimethylformamide	*N,N*-dimethylmethanamide	$(H_3C)_2NCHO$
Ethyl acetate	ethyl ethanoate	$CH_3CO_2CH_2CH_3$
Formaldehyde	methanal	HCHO
Phenol	hydroxybenzene	C_6H_5OH
Toluene	methylbenzene	$C_6H_5CH_3$
ortho-Xylene	1,2-dimethylbenzene	$1,2\text{-}(CH_3)_2C_6H_4$

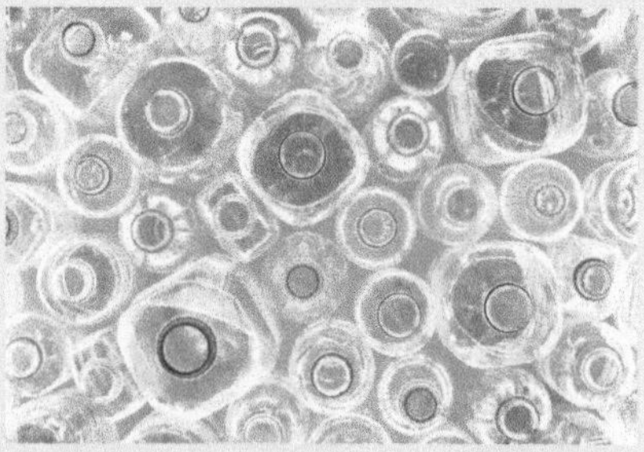

1,3-Dimethylbenzene and 1,4-dimethylbenzene are commonly called *meta*-xylene and *para*-xylene, respectively. In industry, oxidation of *para*-xylene is used to form terephthalic acid ($1,4\text{-}C_6H_4(CO_2H)_2$), which can be converted into the polyester PET. PET is used to make clothing and plastic bottles (Chapter 24 opening page, pp.1098–1099).

Questions

More challenging questions are indicated by an asterisk *.

1. Draw skeletal structures of the following compounds: (a) 2,3,5-trimethylhexane; (b) 5-ethyl-6-isopropyl-3-methylnonane. (Section 2.5)
2. Draw skeletal structures and give the names of the five structural isomers of the hydrocarbon with molecular formula C_6H_{14}. (Section 2.5)
3. Draw skeletal structures of the following compounds (Sections 2.6 and 2.7):
   (a) $PhCH(OH)CH_2CH(NH_2)CH_3$;
   (b) $EtCOCH(Me)CO_2Me$;
   (c) $Me_2C{=}CHCH_2CH_2C(CH_3){=}CHCH_2OH$;
   (d) 5-(3-nitrophenyl)-5-oxopentanoic acid.
4. Name all of the functional groups in the following molecules. (Sections 2.6 and 2.7)

Enalaprilat is an ACE inhibitor (Chapter 2 opening page, pp.68–69).

The synthesis and biological action of salbutamol is discussed in Box 22.8 (p.1049).

(a)

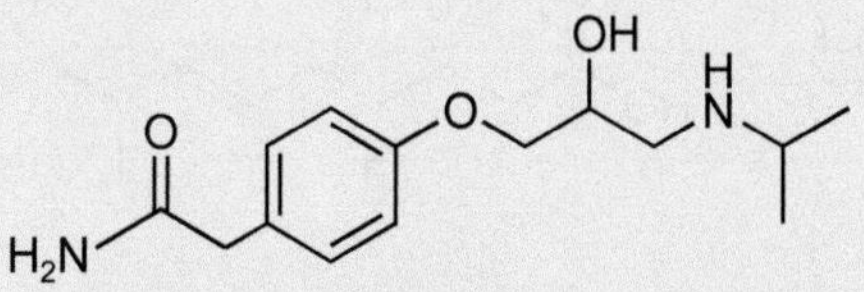

OH
H
N
O
O
H2N

atenolol

Used for the treatment and prevention of heart disease

(b)

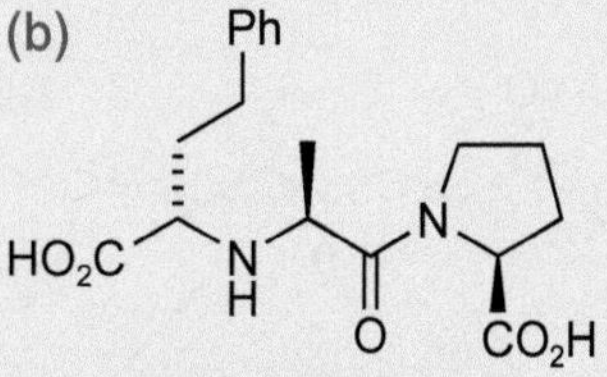

Ph
HO2C
N
H
N
O
CO2H

enalaprilat

Used for the treatment of high blood pressure (hypertension)

(c)

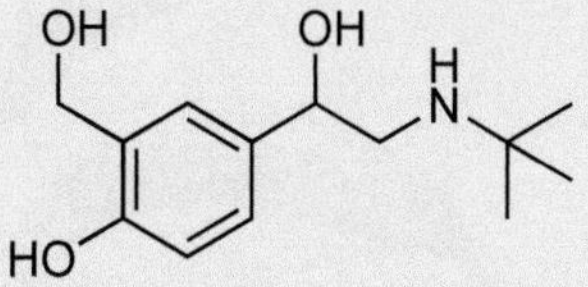

OH
OH
H
N
HO

salbutamol

Used for the treatment of asthma

5. Give IUPAC names for the following compounds (Section 2.8):
   (a) $BrCH_2CH_2CH_2CHMe_2$; (b) $H_2C{=}CHCH_2OH$; (c) $H_2NCH_2CH_2CO_2Me$.

6. The following questions relate to the synthesis of the local anaesthetic ambucaine, starting from 3-hydroxybenzoic acid, which is shown in the scheme below. (Sections 2.2–2.8)

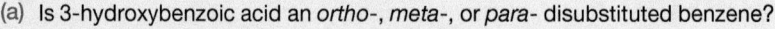

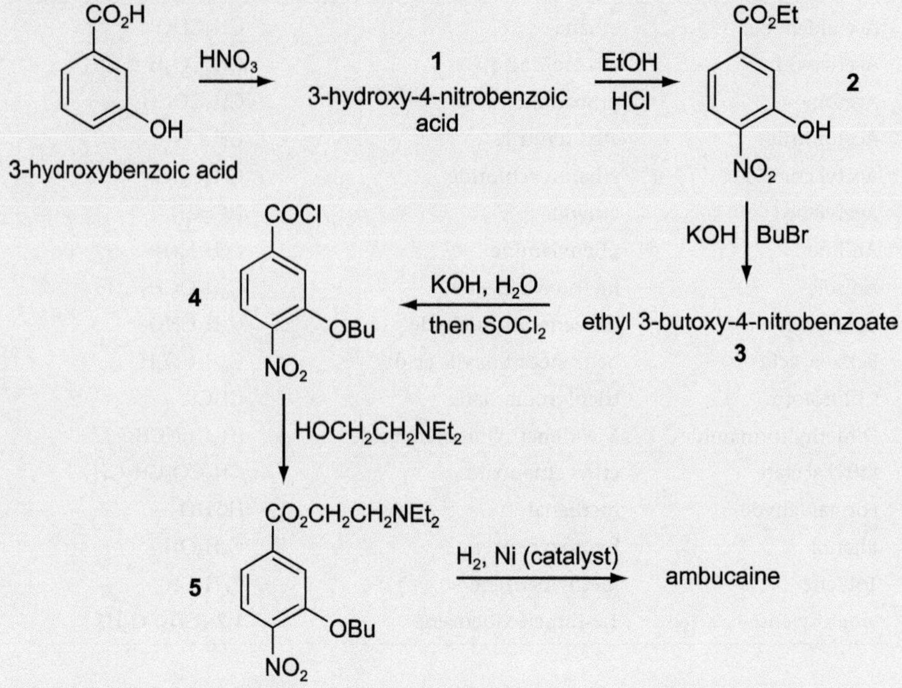

CO_2H
OH
3-hydroxybenzoic acid
HNO_3
1
3-hydroxy-4-nitrobenzoic acid
EtOH
HCl
CO_2Et
OH
NO_2
2
KOH | BuBr
ethyl 3-butoxy-4-nitrobenzoate
3
KOH, H_2O
then $SOCl_2$
COCl
OBu
NO_2
4
$HOCH_2CH_2NEt_2$
$CO_2CH_2CH_2NEt_2$
OBu
NO_2
5
H_2, Ni (catalyst)
ambucaine

2-(diethylamino)ethyl 3-butoxy-4-nitrobenzoate

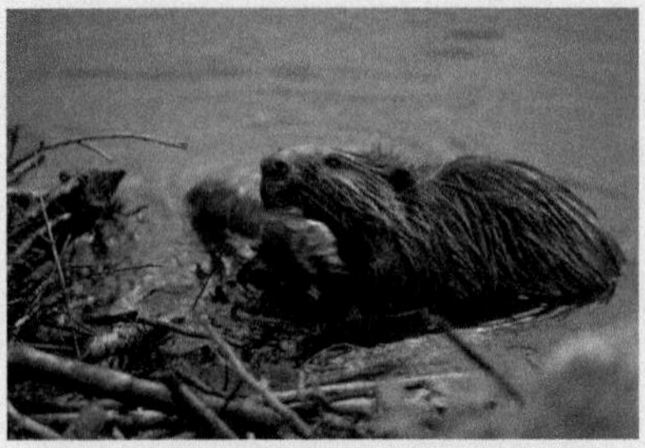

3-Hydroxybenzoic acid is one of at least 50 compounds present in beaver castoreum. Castoreum is a secretion beavers use to mark their territories—due to its strong smell it has long been used in perfumery and as a flavour ingredient in food.

(a) Is 3-hydroxybenzoic acid an *ortho*-, *meta*-, or *para*- disubstituted benzene?

(b) Draw skeletal structures for compounds **1** and **3**.

(c) Give the IUPAC names for compounds **2** and **4**.

(d) Name all of the functional groups in compound **5**.

(e) Given that the $-NO_2$ group of compound **5** is reduced to an $-NH_2$ group, on reaction with H_2 and Ni (catalyst), draw a skeletal structure for ambucaine.

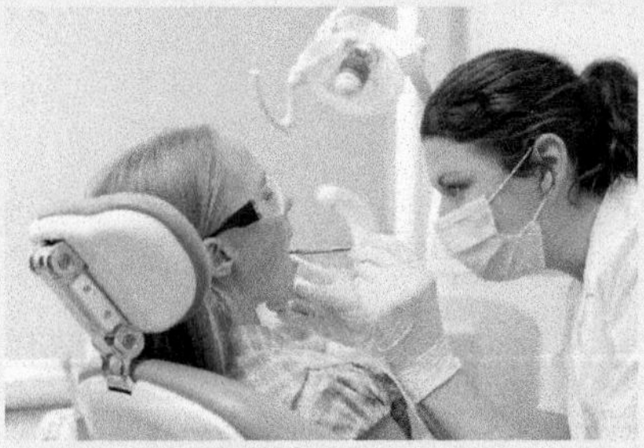

A local anaesthetic causes a loss of sensation to the area to which it is applied. For example, it allows patients to undergo dental procedures with reduced pain and distress.

7. The hot spicy flavour of chilli peppers of the *Capsicum* family is due mainly to a compound called capsaicin. The following questions relate to the synthesis of capsaicin shown below. (Sections 2.2–2.8)

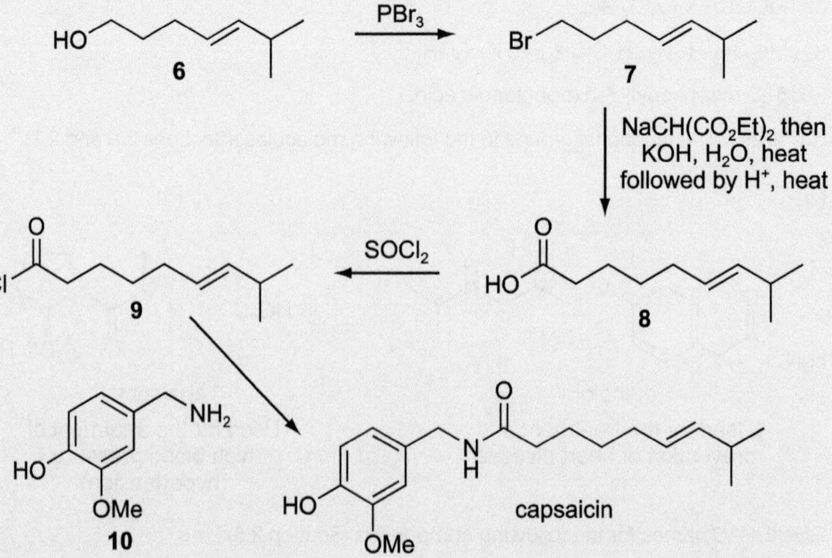

HO
6
PBr_3
Br
7
$NaCH(CO_2Et)_2$ then
KOH, H_2O, heat
followed by H^+, heat
O
HO
8
$SOCl_2$
O
Cl
9
NH_2
HO
OMe
10
O
N
H
HO
OMe
capsaicin

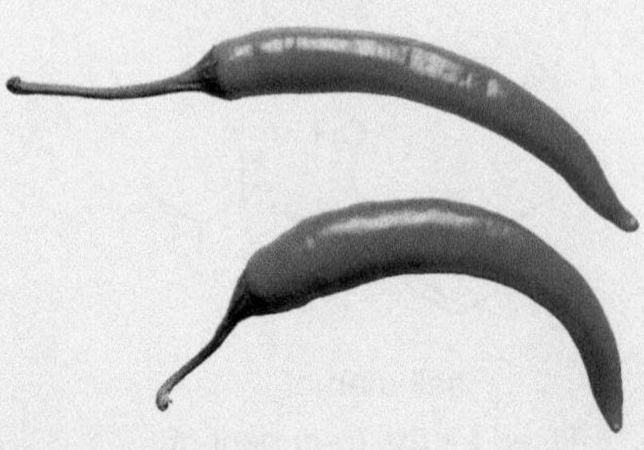

The fiery flavour of chilli peppers is due mainly to capsaicin. The chilli sauce '16 Million Reserve' is made of pure capsaicin and it is 8000 times more fiery than Tabasco sauce.

(a) Draw a skeletal structure of the stereoisomer of alkene **6**.

(b) In compound **7**, is the bromine bonded to a primary, secondary, or tertiary carbon atom?

(c) For the synthesis of compound **8**, $NaCH(CO_2Et)_2$ is prepared from $CH_2(CO_2Et)_2$, which has the common name diethyl malonate. Draw a skeletal structure of diethyl malonate and give the IUPAC name.

(d) Draw the structure of compound **8** and indicate the α and β carbon atoms.

(e) Give the IUPAC name for compound **10**.

(f) Name all of the functional groups in capsaicin.

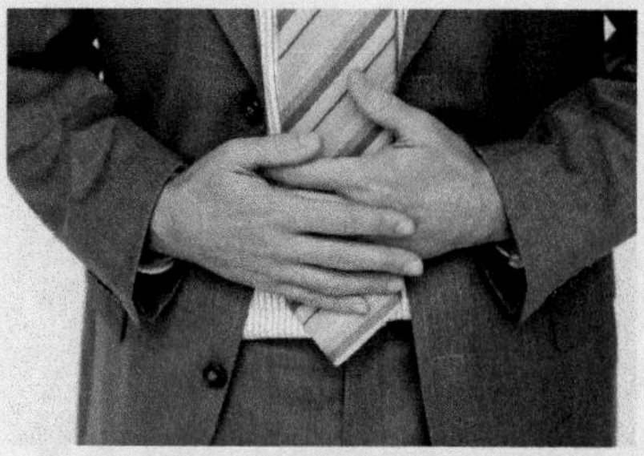

Heartburn is a painful burning sensation, usually associated with regurgitation of gastric acid—acid produced in the stomach eats away at the lining of the oesophagus and causes pain.

8.* Omeprazole is a medicine, first marketed by AstraZeneca, to treat heartburn, acid indigestion, and ulcers in the stomach. The following questions relate to the synthesis of omeprazole shown below. (Sections 2.5–2.8)

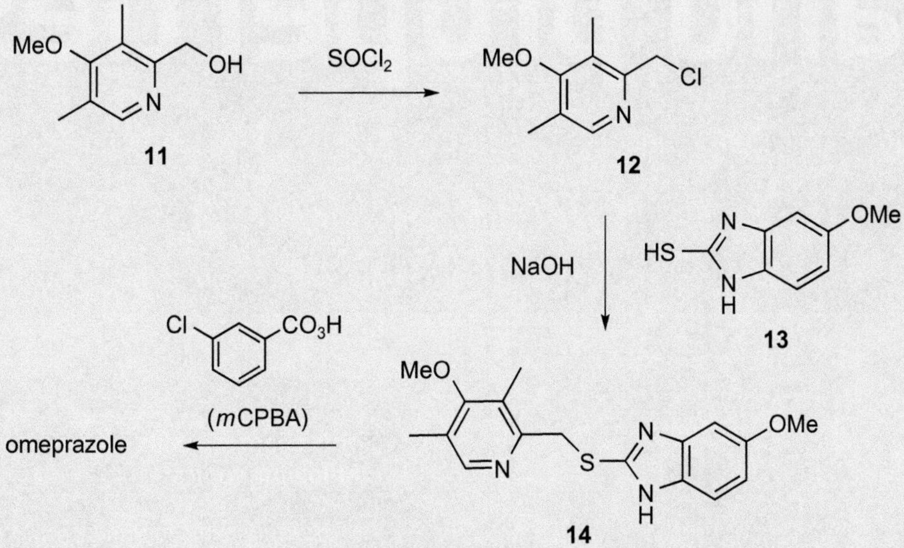
MeO
OH
SOCl2
MeO
Cl
N
N
NaOH
HS
N
OMe
N
H
Cl
CO3H
(mCPBA)
omeprazole
MeO
N
S
N
OMe
N
H

11 **12** **13** **14**

(a) Is the alcohol in compound **11** a primary, secondary, or tertiary alcohol?

(b) Reaction of **11** with $SOCl_2$, to give **12**, requires a solvent such as toluene. Draw a skeletal structure for toluene.

(c) What is the name of the heterocyclic aromatic ring in compound **11**?

(d) Give the IUPAC name for compound **12**.

(e) What is the name of the functional group in compound **13**, that reacts with **12**, to form **14**?

(f) The abbreviation *m*CPBA stands for *meta*-chloroperbenzoic acid. What does the name '*meta*-' indicate?

(g) Reaction of **14** with *m*CPBA leads to oxidation of the thioether (sulfide) to give a sulfoxide. Draw a skeletal structure for omeprazole.

LITHIUM BATTERY
CR2032
3V

$SOCl_2$ (thionyl chloride) is a common chlorinating agent used in organic synthesis (Section 20.2, p.927). It is also a component of lithium-thionyl chloride batteries where $SOCl_2$ acts as the cathode (Chapter 16 opening page, pp.926–927).

3

Atomic structure and properties

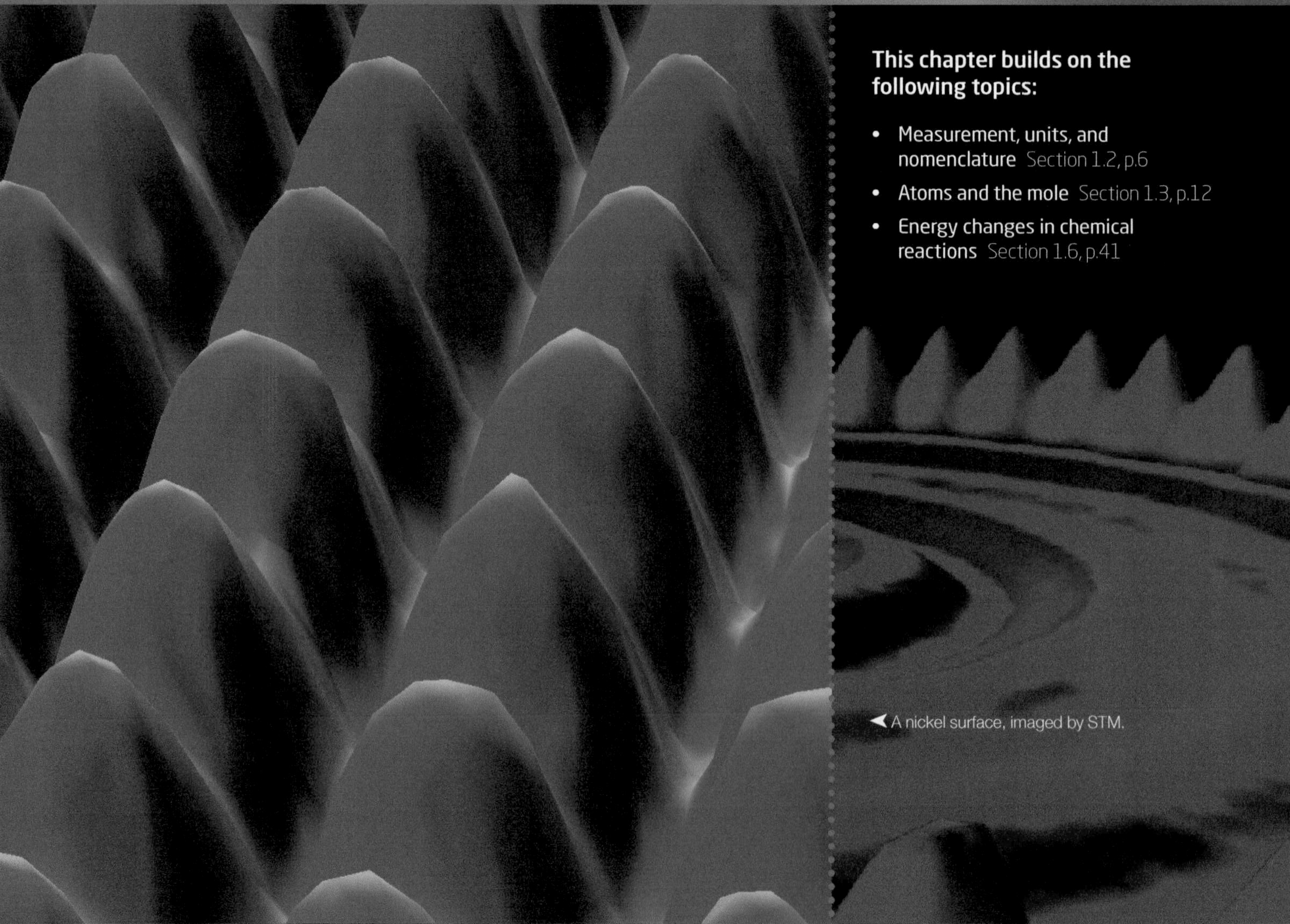

This chapter builds on the following topics:

- Measurement, units, and nomenclature Section 1.2, p.6
- Atoms and the mole Section 1.3, p.12
- Energy changes in chemical reactions Section 1.6, p.41

◀ A nickel surface, imaged by STM.

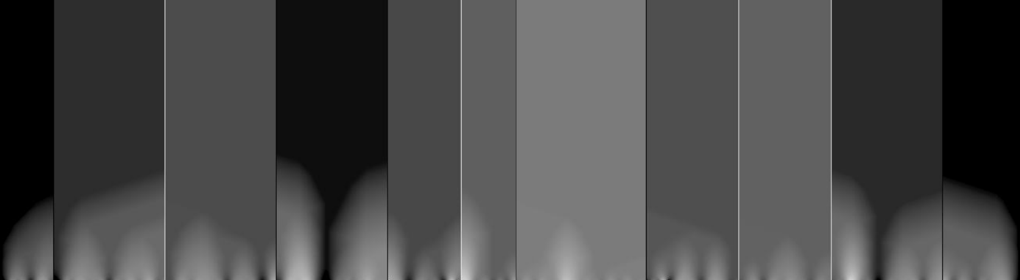

It is not possible to 'see' atoms in the conventional way using light, no matter how high the magnification. This is because light behaves as a wave, and an optical microscope only forms a clear image if the object you want to see is larger than the wavelength of light (400 nm–700 nm). Recently, clever techniques have been developed that allow optical microscopes to see objects of several nanometres dimensions. Individual atoms are, however, too small to be seen this way.

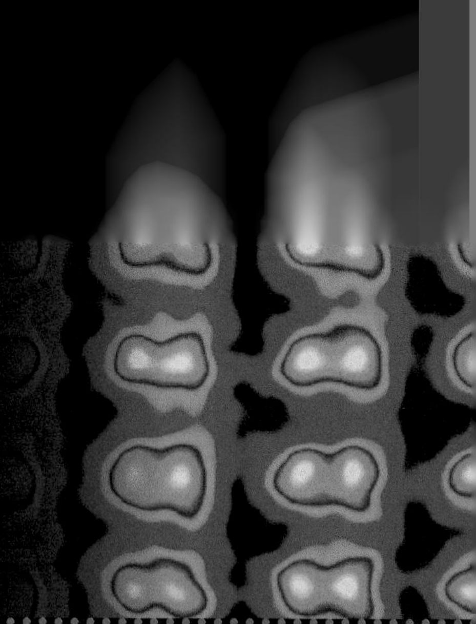

▲ Each spot in this TEM image represents a column of silicon atoms. The columns are aligned in pairs, and each pair is separated by only 78 pm (1 pm = 1×10^{-12} m).

First developed in 1981, a technique called scanning tunnelling microscopy (STM) allows the atoms on a surface to be visualized. In STM, a fine metal tip is scanned systematically across the surface of a sample. If the tip is brought to within about 1 nm (1×10^{-9} m) of the surface, electrons in the atoms of the metal tip can jump across the gap to the atoms in the surface. This movement of electrons can only be explained by quantum mechanics, and is called *quantum mechanical tunnelling*. The tunnelling of the electrons causes a small current to flow between the tip and the surface. The size of the current is related to the number of atoms tunnelling, and it is very sensitive to changes in the distance between the tip and the surface. A computer monitors the current, and moves the tip up and down to keep the current constant. The path of the tip then forms a topographical map of the surface. The pictures generated by STM suggest atoms have solid surfaces. This isn't really true, and what appears as a solid surface is really a cloud of electrons.

As well as visualizing atoms, it is possible to use STM to manipulate individual atoms. Researchers at IBM discovered that if they brought the tip very close to an atom, the atom stuck to it and could be dragged across the surface and deposited elsewhere. This enabled them to build up images on the surface.

Transmission electron microscopy (TEM) is another method of visualizing very small objects. Electrons have wave-like properties (see Section 3.4), and the first electron microscope using this principle was built in 1931 by scientists in Germany. Their instrument was not able to see atoms, mainly because of imperfections in the way the electrons were focused. Nowadays, technology is able to correct for this and, using a variation of TEM, researchers have formed images of silicon surfaces in which they have observed columns of atoms less than 0.1 nm apart.

▼ The 'quantum corral' consists of 48 iron atoms that have been arranged in a circle on a copper surface by STM.

The idea of the atom was first put forward by the ancient Greek philosophers Democritus (shown here on a bank note) and Leucippus in the fifth century BC. They argued that if an object were divided into smaller and smaller parts eventually a fundamental building block would be reached. These building blocks, or atoms, were thought to be indivisible and indestructible. It was over 2000 years before experimental evidence was obtained to support the existence of atoms.

Atoms are of fundamental importance in chemistry. All matter is made up from atoms and ions that are derived from them, and understanding the structure of the atom is key to appreciating how bonds are formed. This is important because much of chemistry is concerned with the making and breaking of bonds.

The story of how scientists came to understand atomic structure stretches over 2000 years, but many of the key developments occurred during the first part of the twentieth century. This work and the introduction of **quantum mechanics** led to a revolution in science, and quantum theory is regarded by many people as the most important scientific breakthrough of all time. Many of the ideas of quantum mechanics that you will meet in this chapter may seem strange, especially if you have not met them before. This is because the familiar ways in which the world works no longer apply when considering objects on the atomic scale. However, the appreciation of the structure of the atom that you will gain from this chapter will underpin much of what is to come in this book.

In this chapter, you will look at atomic spectroscopy alongside the structure of the atom. Atomic spectroscopy provided much of the evidence for quantum theory, and it is helpful to see how the evidence led to the development of the theory. To understand spectroscopy you need to look at the nature of electromagnetic radiation, which will also be important later on in the book.

3.1 The classical picture of the atom

In order to appreciate modern theories of atomic structure it is helpful to see them in an historical context. What led scientists to develop the theories? What was the evidence for them? In this section the historical development of atomic theory is charted, and some of the key ideas are introduced.

The atomic concept

The English chemist John Dalton is usually regarded as the father of modern atomic theory. Chemists had shown that when two reagents react together they always do so in the same mass ratio. For example, 16 g of oxygen combine with 2 g of hydrogen to form water, whereas 8 g of oxygen combine with 1 g of hydrogen. Chemists had also demonstrated that the total mass of a sealed vessel and its contents is the same before and after a reaction. This is a fundamental law of chemistry, known as the **law of conservation of mass**. In the early 1800s Dalton used the concept of atoms to explain these observations.

The law of conservation of mass means that you can write a balanced equation for a chemical reaction; see Section 1.4 (p.21).

Dalton argued that all matter is composed of solid and indivisible atoms. He said that there are as many different types of atom as there are elements, with the atoms of different elements having different masses. He explained chemical reactions as changing the ways in which the atoms are grouped together. Since atoms cannot be created or destroyed in a reaction, the overall mass of the products must equal that of the reactants. This explained the law of conservation of mass. Dalton also argued that, in a given compound, the atoms are always present in the same ratio, which explains why reagents always react together in the same mass ratio.

Dalton's theory was important because, for the first time in chemistry, it enabled macroscopic behaviour to be understood in terms of microscopic structure.

Subatomic particles

A simple classical structure of the atom with a dense positively charged nucleus surrounded by negatively charged electrons is described in Section 1.3 (p.13).

Atoms are not as indivisible as Dalton believed. The idea that atoms have internal structure was first indicated by experiments towards the end of the nineteenth century and at

Online Support. The icon in the margin indicates that accompanying interactive resources are provided online to help your understanding: just type www.chemtube3d.com/burrows/123 into your browser, replacing 123 with the number of the page where you see the icon. For pages linking to more than one resource, type 123-1, 123-2, etc. (replacing 123 with the page number) for access to successive links.

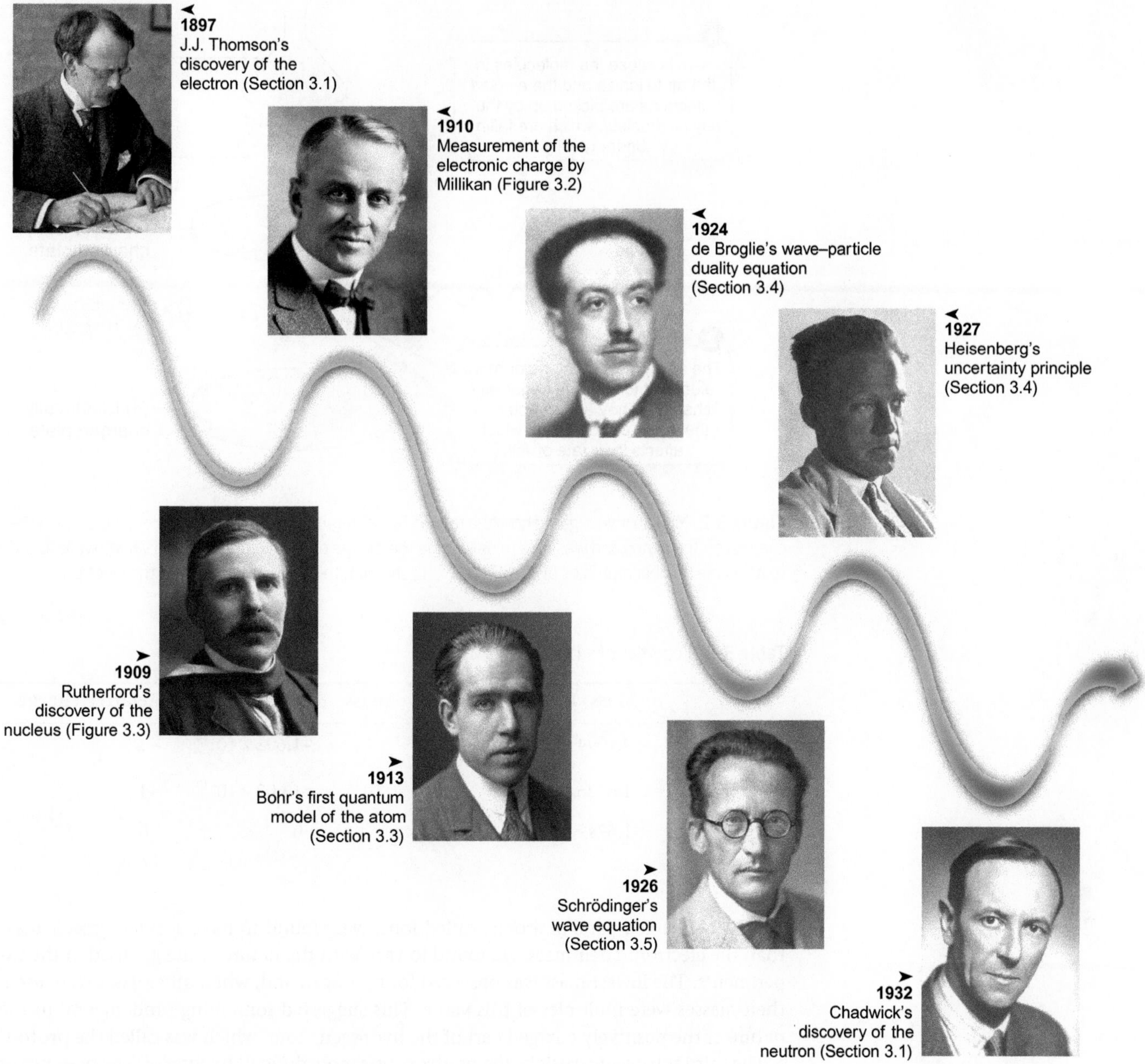
1897
J.J. Thomson's discovery of the electron (Section 3.1)
1910
Measurement of the electronic charge by Millikan (Figure 3.2)
1924
de Broglie's wave–particle duality equation (Section 3.4)
1927
Heisenberg's uncertainty principle (Section 3.4)
1909
Rutherford's discovery of the nucleus (Figure 3.3)
1913
Bohr's first quantum model of the atom (Section 3.3)
1926
Schrödinger's wave equation (Section 3.5)
1932
Chadwick's discovery of the neutron (Section 3.1)

Figure 3.1 A timeline showing important developments in understanding the structure of the atom.

the start of the twentieth century. A timeline showing the most significant discoveries and theories of atomic structure, along with the major contributors, is shown in Figure 3.1.

The first indication that atoms had internal structure was the discovery of the electron. When a high voltage electric current was passed through a gas at low pressure, negatively charged particles were observed to travel between the electrodes. These so-called *cathode rays* were the same whatever gas was used, and we now know them as **electrons**. J.J. Thomson applied electric and magnetic fields to a beam of electrons and used the deviations from a straight line to calculate the ratio of charge to mass for the particles.

Thirteen years later, the American scientist Robert Millikan was able to determine the charge on the electron as -1.60×10^{-19} C by observing the rate of fall of charged oil droplets using the apparatus shown schematically in Figure 3.2. Since Thomson had determined the charge to mass ratio for the electron, knowing its charge enabled the electronic mass to be determined.

The coulomb (symbol, C) is the SI unit for electric charge. See Table 1.2 (p.8).

Since atoms are electrically neutral, the demonstration that negatively charged electrons are present in all matter implied that positively charged species are also present. These

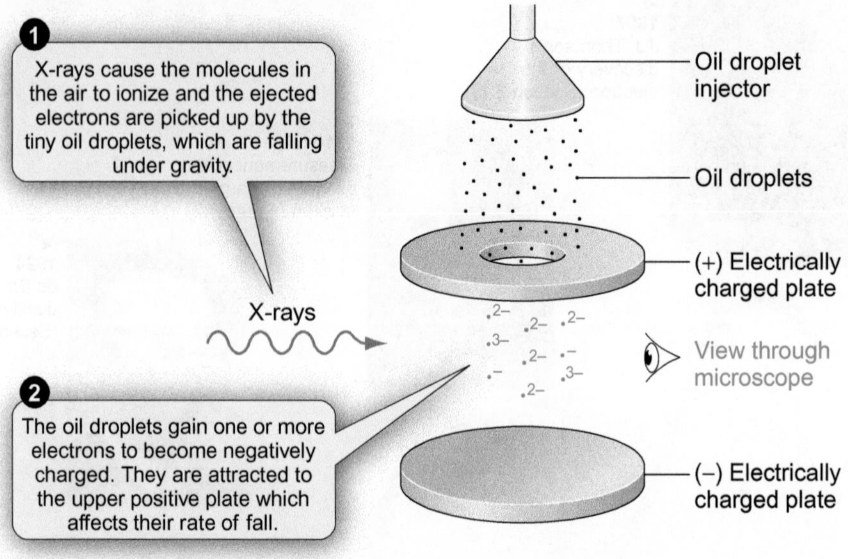
1
X-rays cause the molecules in the air to ionize and the ejected electrons are picked up by the tiny oil droplets, which are falling under gravity.
X-rays
2
The oil droplets gain one or more electrons to become negatively charged. They are attracted to the upper positive plate which affects their rate of fall.
Oil droplet injector
Oil droplets
(+) Electrically charged plate
View through microscope
(–) Electrically charged plate

Figure 3.2 Millikan investigated how the rate of fall of negatively charged oil droplets varied with the charges on the plates and was able to determine the charge on each droplet. All the values were found to be whole number multiples of -1.60×10^{-19} C, showing this to be the charge on the electron.

Table 3.1 Properties of subatomic particles

	Mass / kg	Relative mass	Charge / C	Relative charge
Electron, e	9.1094×10^{-31}	1	-1.602×10^{-19}	−1
Proton, p	1.6726×10^{-27}	1836	$+1.602 \times 10^{-19}$	+1
Neutron, n	1.6749×10^{-27}	1839	0	0

positively charged parts of atoms, called ions, were found to have a much greater mass than the electron. Their mass was found to vary with the nature of the gas used in the experiments. The lowest mass was observed for hydrogen and, when other gases were used, their masses were multiples of this value. This suggested something fundamental in the nature of the positively charged part of the hydrogen atom, which was called the **proton**.

The third subatomic particle, the **neutron**, was more difficult to observe. The presence of these electrically neutral particles was predicted on the basis of charge to mass ratios in ions but they were not observed until 1932. Neutrons were finally identified by James Chadwick as the particles emitted on bombarding beryllium and boron atoms with α-particles, which are positively charged particles that we now know consist of two protons and two neutrons. Properties of the electron, proton, and neutron are given in Table 3.1.

α-particles are helium nuclei. α-emission is described in Section 3.8 (p.159).

First ideas of atomic structure

Although scientists realized that atoms contained smaller building blocks, they had no idea of how they were arranged. The first evidence for this was obtained by the New Zealand-born scientist Ernest Rutherford.

Rutherford, then working in Manchester, asked two of his co-workers—Hans Geiger and Ernest Marsden—to shoot α-particles at a thin metal film, only a few atoms thick. This experiment is shown schematically in Figure 3.3. Most of the positively charged particles passed through the film with only minor deflections. However, a very small fraction—about one in 20000—were deflected by angles greater than 90° due to repulsive

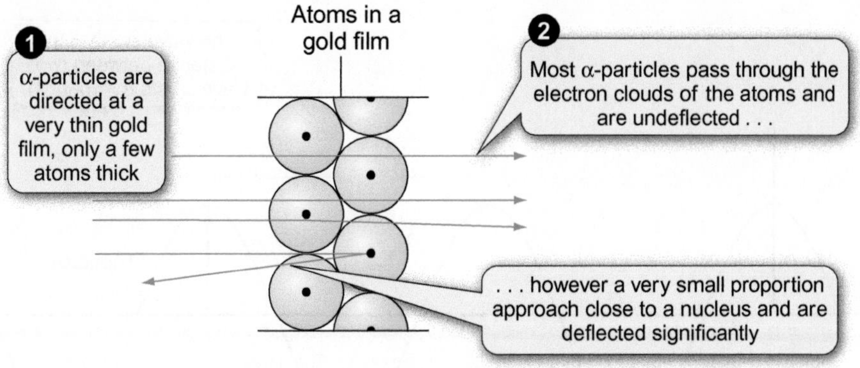
Atoms in a gold film
1
α-particles are directed at a very thin gold film, only a few atoms thick
2
Most α-particles pass through the electron clouds of the atoms and are undeflected . . .
. . . however a very small proportion approach close to a nucleus and are deflected significantly

Figure 3.3 The tiny number of deviations observed in the Geiger–Marsden experiment demonstrated that the size of the nucleus is very much smaller than the size of the atom. If a gold nucleus could be expanded to 10 cm in diameter, the diameter of the atom would be approximately 2.6 km.

interactions with a positively charged **nucleus** with some bouncing back the way they had come.

This was surprising since before this experiment most scientists believed the atom to consist of negatively charged electrons embedded in a positively charged sphere. This was sometimes called the 'plum pudding model'. The experimental results were not consistent with this theory. Instead the tiny number of deviations suggested to Rutherford an atomic structure consisting of a very small dense centre of positive charge, the nucleus, surrounded by a large volume of mainly empty space containing the electrons. We now know that the nucleus contains protons and neutrons, which are held together by a strong nuclear force. This force overcomes the electrical repulsion between the positively charged protons.

The structure of the atom suggested by Rutherford's experiment is shown in Figure 1.7 (p.13).

Summary

- Dalton's atomic theory in the early 1800s explained the law of conservation of mass and explained why reagents always react together in the same mass ratio.
- The first evidence for the internal structure of the atom was obtained in the early twentieth century. Experiments led to a model of the atom consisting of a tiny dense positively charged nucleus surrounded by electrons.

3.2 Electromagnetic radiation and quantization

One of the best ways of probing the structure of atoms is by studying their interaction with electromagnetic radiation. Atoms can absorb electromagnetic radiation, or be made to emit it. The study of these absorptions and emissions is known as **spectroscopy**, and **atomic spectroscopy** led to dramatic insights into atomic structure. Before you look at these processes, it is important to understand the nature of electromagnetic radiation.

Atomic spectroscopy gives information about the structure of atoms. The absorption and emission of radiation by molecules is called **molecular spectroscopy** and is discussed in Chapters 10 and 12.

Electromagnetic radiation

Electromagnetic radiation is a form of energy consisting of oscillating electric and magnetic fields. These oscillations travel through space at a speed c which is equal to $2.998 \times 10^8\,\mathrm{m\,s^{-1}}$, known as the **speed of light**. Visible light is a form of electromagnetic radiation, as are radio waves, microwaves, and X-rays. These types of electromagnetic radiation each have a characteristic **wavelength**, λ. The wavelength is related to the **frequency**, ν, by Equation 3.1

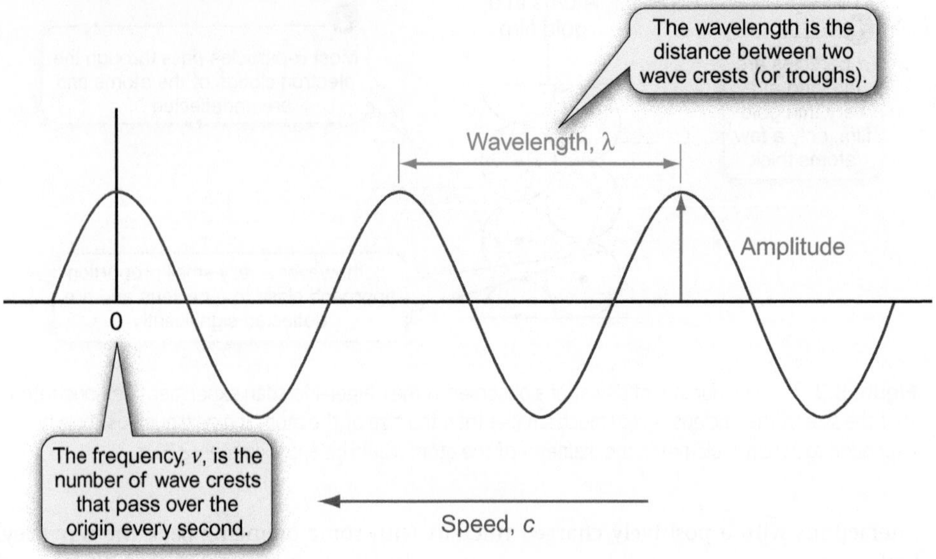
The wavelength is the distance between two wave crests (or troughs).
Wavelength, λ
Amplitude
0
The frequency, ν, is the number of wave crests that pass over the origin every second.
Speed, c

Figure 3.4 When the wave is travelling at speed c, the frequency, ν, is equal to the number of wave crests that pass the origin (0) per second.

Greek symbols

λ lambda (wavelength)
ν nu (frequency)

$1\,s^{-1} = 1\,Hz$

Visit the Online Resource Centre to view screencast 3.1 which walks you through the relationship between wavelength and frequency.

You can find help with handling and interconverting units in Section 1.2 (p.6). Information on rearranging equations is given in Maths Toolkit MT2 (p.1313).

$$c = \lambda\nu \tag{3.1}$$

The wavelength, frequency, and amplitude of a wave are shown in Figure 3.4.

Wavelengths are normally quoted in metres (m) (or nm when discussing visible radiation). Frequencies are given in s^{-1}. This is such a common unit that it is given its own name and symbol, hertz (Hz). Worked example 3.1 provides practice at interconverting wavelength and frequency for electromagnetic radiation.

Worked example 3.1 Interconverting wavelength and frequency

What is the frequency of red light with a wavelength of 680 nm?

Strategy

Use Equation 3.1, $c = \lambda\nu$, and rearrange it to find the frequency, ν. The value of ν will be in hertz.

Solution

Rearranging Equation 3.1 by dividing both sides by λ gives

$$\nu = \frac{c}{\lambda}$$

$c = 2.998 \times 10^8\,m\,s^{-1}$
$\lambda = 680\,nm = 680 \times 10^{-9}\,m$

Putting the values into the equation

$$\nu = \frac{2.998 \times 10^8\,\cancel{m}\,s^{-1}}{680 \times 10^{-9}\,\cancel{m}}$$

$$= 4.41 \times 10^{14}\,s^{-1} = 4.41 \times 10^{14}\,Hz$$

Question

Radio 5 *live* in the UK broadcasts at 909 kHz. What wavelength does this correspond to?

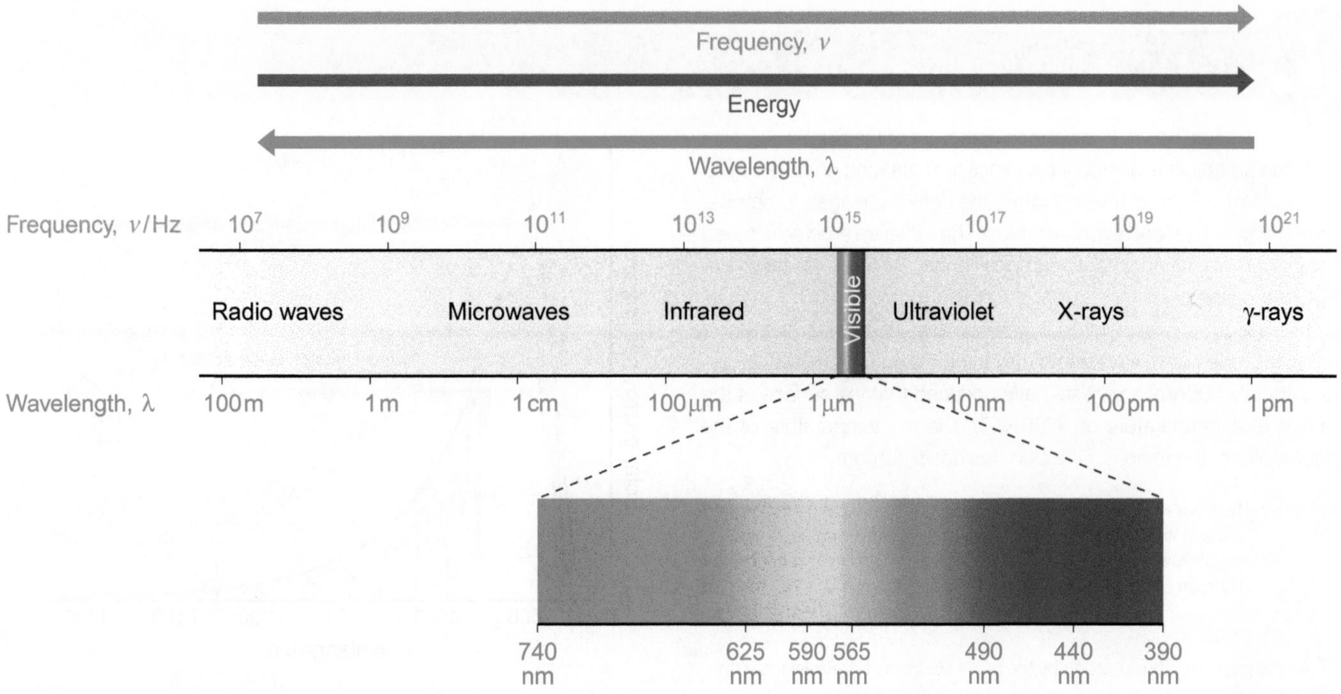
Frequency, ν
Energy
Wavelength, λ
Frequency, ν/Hz
10^7 10^9 10^11 10^13 10^15 10^17 10^19 10^21
Radio waves
Microwaves
Infrared
Visible
Ultraviolet
X-rays
γ-rays
Wavelength, λ
100 m 1 m 1 cm 100 μm 1 μm 10 nm 100 pm 1 pm
740 nm 625 nm 590 nm 565 nm 490 nm 440 nm 390 nm

Figure 3.5 The electromagnetic spectrum.

The different types of electromagnetic radiation are shown on the **electromagnetic spectrum** in Figure 3.5. Human eyes are only receptive to a relatively small range of wavelengths, so visible light forms just the part of the electromagnetic spectrum with wavelengths between approximately 390 nm and 740 nm (see Table 3.2). The colour of visible light depends on its wavelength, with red light at the top end of the wavelength range (low frequency) and violet light at the lower end (high frequency).

The concept of electromagnetic radiation is so common in physics and chemistry that it is often just referred to as radiation. So, for example, infrared radiation refers to electromagnetic radiation in the infrared part of the spectrum.

Most of the regions of the electromagnetic spectrum are useful in chemistry, and can give information about different atomic or molecular processes. Some of the most important of these are summarized in Table 3.3, and you will meet these types of spectroscopy in later chapters.

For common SI prefixes see Table 1.3 (p.8).

Do not confuse electromagnetic radiation with the ionizing radiation emitted by some materials in the form of α- and β-particles. Ionizing radiation (radioactivity) is described in Section 3.8 (p.159).

Table 3.2 The colours of visible light

Colour	Wavelength range / nm
Red	740–625
Orange	625–590
Yellow	590–565
Green	565–490
Blue	490–440
Violet	440–390

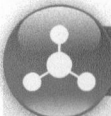

Box 3.1 Radiation from the Sun

How do we know the temperature of the Sun?

When an object is heated it becomes 'red hot' and emits red light. As the temperature is raised further, the colour changes to orange, then yellow, white, and eventually blue. The radiation emitted is called 'black body radiation' and its composition depends on the temperature of the object.

The spectrum of sunlight shows how the intensity of the Sun's radiation varies with wavelength. From the maximum intensity at approximately 500 nm, scientists have deduced that the surface of the Sun is at a temperature of 5780 K. This is the temperature of the *photosphere*, the region of the Sun that emits sunlight.

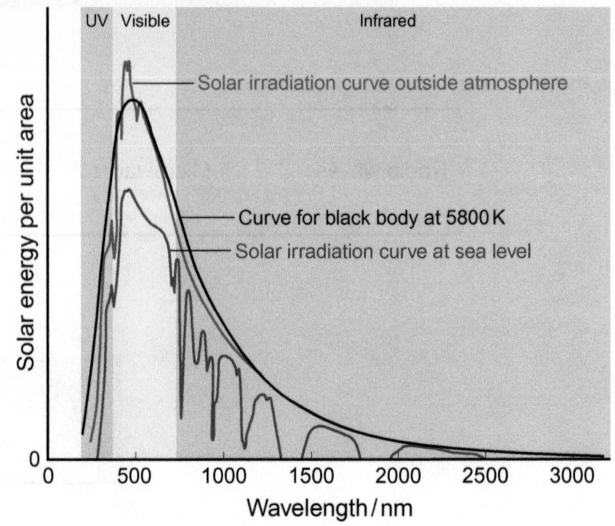

UV
Visible
Infrared
Solar irradiation curve outside atmosphere
Curve for black body at 5800 K
Solar irradiation curve at sea level
Solar energy per unit area
0
0
500
1000
1500
2000
2500
3000
Wavelength / nm

The solar irradiation curve at the top of the Earth's atmosphere and at sea level together with a black body radiation curve for a body at 5800 K. *Atmosphere Climate: An Earth System Perspective* by Thomas E. Graedel and Paul J. Crutzen © 1993 by W.H. Freeman and Company. Used with permission.

An image of the Sun recorded by the SOHO Extreme Ultraviolet Imaging Telescope.

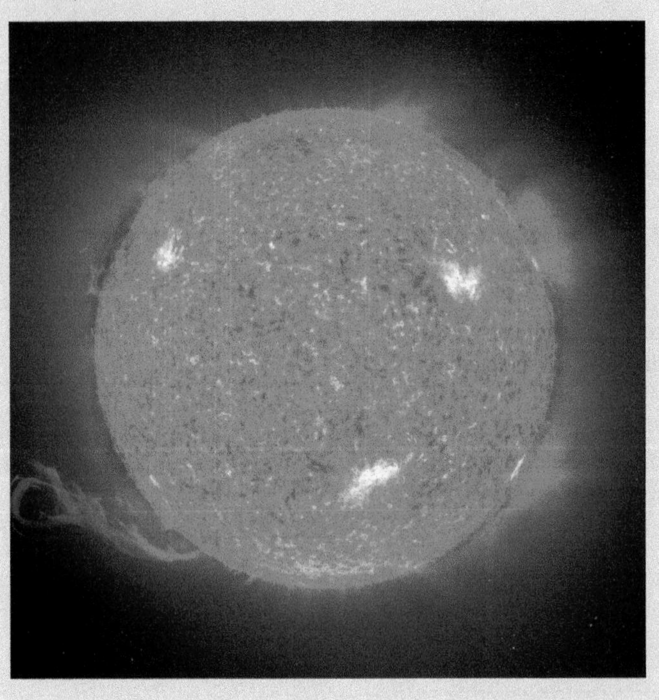

The plots above show the spectrum of sunlight reaching the edge of the Earth's atmosphere and the spectrum of sunlight at sea level, together with the spectrum of black body radiation for an object at 5800 K. About 10% of the solar radiation energy is ultraviolet, 45% is visible light, and 45% is infrared. Infrared radiation has a wavelength greater than 740 nm and you experience it as heat. Ultraviolet radiation has a wavelength of less than 390 nm and is responsible for sunburn and skin cancers.

The electromagnetic spectrum of sunlight received at the Earth's surface differs from that emitted by the Sun because some of the radiation is absorbed in the atmosphere. Ultraviolet radiation is absorbed in the stratosphere (upper atmosphere) by O_2 and ozone (O_3). The mechanism for this and the effect of chlorofluorocarbons (CFCs) on ozone is explored in Box 27.6 (p.1233). Infrared radiation is absorbed in the troposphere (lower atmosphere) by greenhouse gases such as CO_2 and H_2O, and this is discussed in Box 27.2 (p.1218).

Question

Why does the Sun appear yellow?

Quantization

In the previous section, electromagnetic radiation was considered as a wave, with a characteristic wavelength and frequency. In some situations, the behaviour of light is easier to explain by thinking of it, not as waves, but as particles. In 1900 the German physicist Max Planck proposed that electromagnetic radiation could only be emitted or absorbed in packets or quanta of radiation, which were later called photons. The energy of a photon is proportional to its frequency (Equation 3.2)

$$E = h\nu \qquad (3.2)$$

where h is known as the Planck constant, and has a value of 6.626×10^{-34} J s.

Quantization and the 'ultraviolet catastrophe'

When classical physics was applied to the emissions from 'black bodies', a problem arose. It successfully predicted the intensities of the emissions at low frequencies, but vastly overestimated the high-frequency emissions. This is known as the 'ultraviolet catastrophe'. Planck's idea of quantization arose as a solution to this problem.

Table 3.3 Examples of spectroscopy

Electromagnetic radiation used	Type of spectroscopy	Transition involved	Section
Radio waves	Nuclear magnetic resonance (NMR)	Between nuclear spin levels in a magnetic field	10.7, 12.3
Microwaves	Electron spin resonance (ESR)	Between electronic spin levels in a magnetic field	Box 3.6, 10.7
Microwaves	Rotational	Between molecular rotational states	10.4
Infrared	Infrared (IR)	Between molecular vibrational states	10.5, 12.2
Visible, ultraviolet	Ultraviolet–visible (UV-VIS)	Between molecular electronic energy levels	10.6, 11.4, 28.6
Infrared, visible, ultraviolet, X-rays	Atomic	Between atomic electronic energy levels	3.3, Boxes 3.2 and 3.3, 11.5

Look at Figure 3.5 (p.119). Ultraviolet radiation has a higher frequency than visible light. Since the energy of a photon is proportional to its frequency, each ultraviolet photon has more energy than a visible photon. This explains why the ultraviolet radiation emitted by the Sun is more harmful than the visible radiation.

Worked example 3.2 provides practice at converting frequency to energy.

Worked example 3.2 Converting frequency to energy

What is the energy of a mole of red light photons with a frequency of 4.41×10^{14} Hz?

Strategy

Use Equation 3.2 to determine the energy of one photon. Multiply this value by the Avogadro constant to find the energy of a mole of photons. (The number of entities in a mole is given by the Avogadro constant, N_A, see Section 1.3, p.16.)

Solution

Use Equation 3.2

$$E = h\nu \text{ (the Planck constant } h = 6.626 \times 10^{-34}\text{ J s)} \quad (3.2)$$

$$E = (6.626 \times 10^{-34}\text{ J s}) \times (4.41 \times 10^{14}\text{ s}^{-1}) = 2.92 \times 10^{-19}\text{ J}$$

This is the energy of one photon.

Multiply by the Avogadro constant ($N_A = 6.022 \times 10^{23}\text{ mol}^{-1}$) to find the energy of a mole of photons. So

$$E = (2.92 \times 10^{-19}\text{ J}) \times (6.022 \times 10^{23}\text{ mol}^{-1})$$

$$= 176\,000\text{ J mol}^{-1}$$

$$= 176\text{ kJ mol}^{-1}$$

Question

What is the energy of a mole of radio wave photons with a frequency of 909 kHz?

Quantum theory allows us to picture a ray of light as a stream of photons, with the energy related to the frequency of the electromagnetic radiation and the intensity related to the number of photons present. Good evidence for this approach comes from the photoelectric effect.

The photoelectric effect

When ultraviolet radiation strikes a metal surface, electrons are ejected. This is called the **photoelectric effect** (Figure 3.6). Electrons are only ejected when the frequency of the

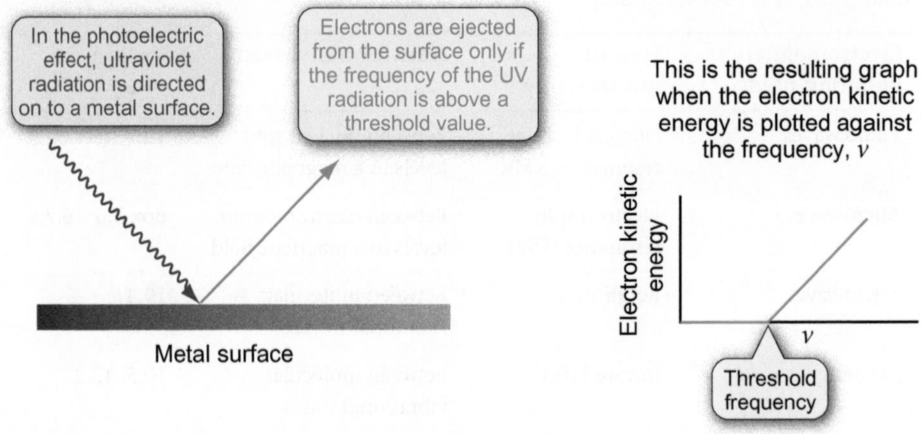
In the photoelectric effect, ultraviolet radiation is directed on to a metal surface.
Electrons are ejected from the surface only if the frequency of the UV radiation is above a threshold value.
Metal surface
This is the resulting graph when the electron kinetic energy is plotted against the frequency, ν
Electron kinetic energy
ν
Threshold frequency

Figure 3.6 The photoelectric effect. The threshold frequency of the incident radiation is characteristic of the metal.

Image intensifiers, used in many night vision devices, rely on the photoelectric effect. Photons from a low light source hit a surface that emits electrons. These are accelerated to a plate which emits multiple electrons for each one striking it. These emitted electrons strike a green phosphor which emits photons that are viewable through an eyepiece.

Greek symbols
Φ phi (upper-case) (work function)

Kinetic energy E_{KE} is introduced in Section 1.6 (p.41).

ultraviolet radiation is above a certain threshold, which is specific to the metal. However, once this threshold is passed electrons are ejected, regardless of the intensity of the radiation, though at high intensity more electrons are emitted. These observations were impossible to rationalize using classical physics.

In 1905, Albert Einstein explained the photoelectric effect using a quantum approach. Einstein reasoned that electrons can only be ejected from a surface if incoming photons transfer a minimum value of energy to atoms on the metal surface. If a photon does not have enough energy, an electron will not be ejected regardless of the intensity of the radiation, as none of the individual photons has enough energy to eject the electron. Once above the threshold, the excess energy of the photon is converted into the kinetic energy of the ejected electron. This is summarized in Equation 3.3.

$$h\nu = \Phi + E_{KE} \tag{3.3}$$

where Φ is the minimum energy required to remove the electron, known as the **work function**, and E_{KE} is the kinetic energy of the ejected electron. An example of the use of Equation 3.3 is given in Worked example 3.3.

Worked example 3.3 The photoelectric effect

When sodium is bombarded with ultraviolet radiation of wavelength 475 nm, electrons with kinetic energy of 30.0 kJ mol⁻¹ are ejected. What is the work function of sodium (in J)?

Strategy

Use Equation 3.3 to find the work function Φ

$$h\nu = \Phi + E_{KE} \tag{3.3}$$

Equation 3.3 refers to a single photon, so the kinetic energy given needs to be converted from a molar quantity to a value for a single electron.

Solution

First convert λ into ν using Equation 3.1

$$c = \lambda\nu \tag{3.1}$$

$$\nu = \frac{c}{\lambda} = \frac{2.998\times10^{8}\,\cancel{\text{m}}\,\text{s}^{-1}}{475\times10^{-9}\,\cancel{\text{m}}}$$

$$= 6.31\times10^{14}\,\text{s}^{-1}$$

Now find $h\nu$ for a single photon using Equation 3.2

$$E = h\nu \tag{3.2}$$

$$= (6.626\times10^{-34}\,\text{J}\,\cancel{\text{s}})\times(6.31\times10^{14}\,\cancel{\text{s}^{-1}})$$

$$= 4.18\times10^{-19}\,\text{J}$$

→

→ Next, divide the electron kinetic energy by the Avogadro constant to find the kinetic energy of a single electron.

$$E_{KE} = \frac{30.0 \times 10^3 \text{ J mol}^{-1}}{6.022 \times 10^{23} \text{ mol}^{-1}}$$

$$= 4.98 \times 10^{-20} \text{ J}$$

Rearrange Equation 3.3 to determine Φ

$$\Phi = h\nu - E_{KE}$$

So

$$\Phi = (4.18 \times 10^{-19} \text{ J}) - (4.98 \times 10^{-20} \text{ J})$$

$$= 3.68 \times 10^{-19} \text{ J}$$

Question

Calculate the kinetic energy of the ejected electrons (in kJ mol^{-1}) if ultraviolet radiation of wavelength 450 nm is used in this experiment.

Wave–particle duality of light

The photoelectric effect shows how electromagnetic radiation can behave as particles. There is considerable evidence, however, such as the observation of diffraction, that electromagnetic radiation behaves as a wave. In the early part of the nineteenth century, Thomas Young demonstrated that when light passes through two closely spaced slits as shown in Figure 3.7, each slit gives rise to a circular wave and these waves interfere with each other to give a diffraction pattern consisting of a series of bright and dark lines.

When the peak of the first wave coincides with the peak of the second wave **constructive interference** occurs, and the amplitudes of the two waves add together as shown in Figure 3.8(a). In contrast, when the peak of the first wave coincides with the trough of the second wave **destructive interference** results, and the two waves cancel each other out as shown in Figure 3.8(b). The series of lines in Figure 3.7 results from constructive and destructive interference of the two circular waves generated by the slits.

So, some experiments provide evidence of light consisting of waves, whereas others provide evidence of light consisting of particles. Light is described as having **wave–particle duality**. The wave–particle duality of light is counterintuitive, but it is a fundamental part of modern science. We treat light as waves when it is useful to do so, and as particles when it is useful to do so.

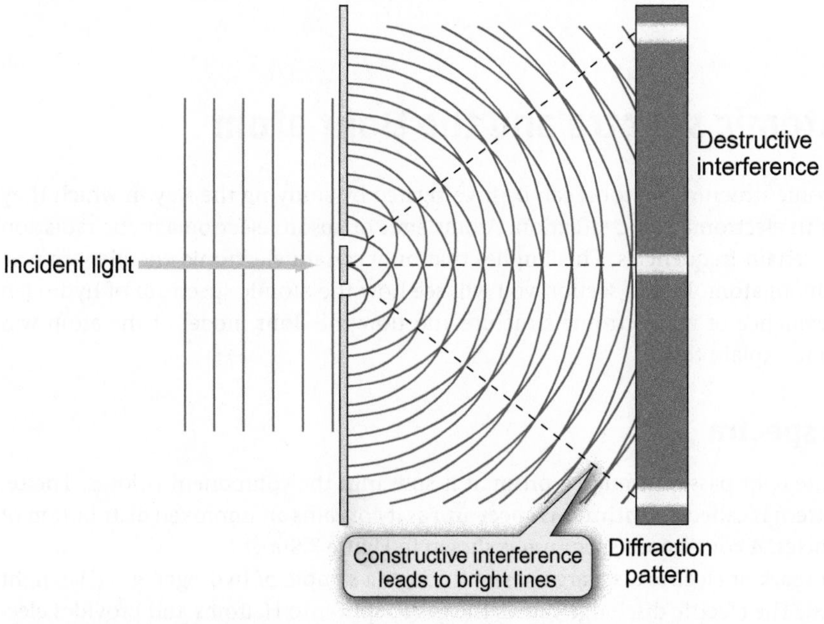
Destructive interference
Incident light
Constructive interference leads to bright lines
Diffraction pattern

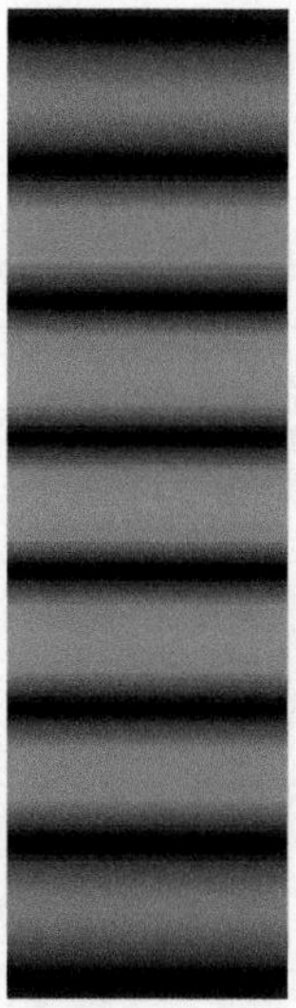

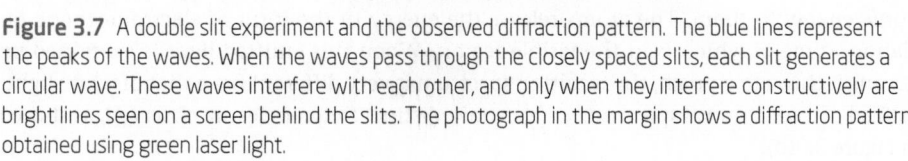

Figure 3.7 A double slit experiment and the observed diffraction pattern. The blue lines represent the peaks of the waves. When the waves pass through the closely spaced slits, each slit generates a circular wave. These waves interfere with each other, and only when they interfere constructively are bright lines seen on a screen behind the slits. The photograph in the margin shows a diffraction pattern obtained using green laser light.

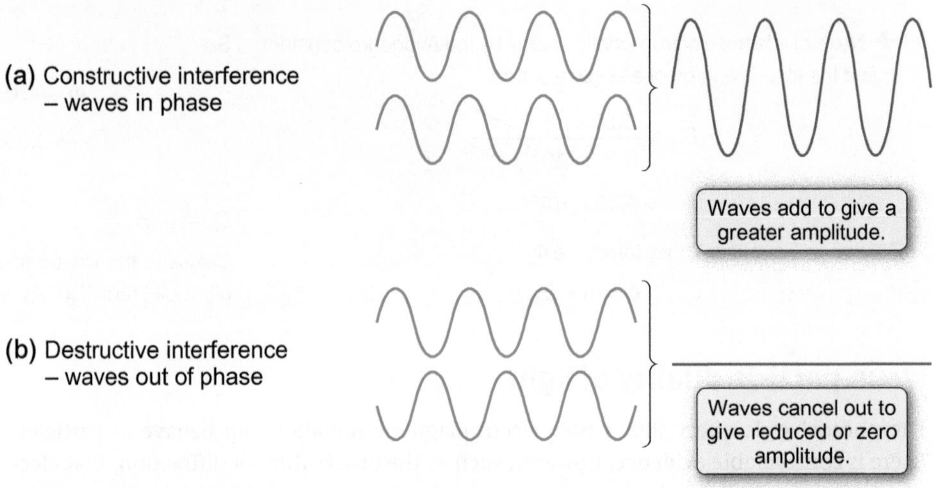
(a) Constructive interference
– waves in phase
Waves add to give a greater amplitude.
(b) Destructive interference
– waves out of phase
Waves cancel out to give reduced or zero amplitude.

Figure 3.8 (a) Constructive interference (waves in phase) and (b) destructive interference (waves out of phase).

Summary

- Electromagnetic radiation is a form of energy, consisting of oscillating electric and magnetic fields that travel at the speed of light, *c*.
- Electromagnetic radiation is quantized into photons. The energy of a photon, *E*, is equal to $E = h\nu$.
- In some circumstances electromagnetic radiation behaves as a wave; in others it behaves as particles.

For practice questions on these topics, see questions 1 and 2 at the end of this chapter (p.165).

This waterfall, Skogafoss in Iceland, has its own rainbow. The water droplets in the fine mist generated by the waterfall act as tiny prisms, splitting white light into its constituent colours.

3.3 Atomic spectra and the Bohr atom

The electronic structure of atoms can be investigated by studying the way in which they interact with electromagnetic radiation. Atoms emit or absorb electromagnetic radiation with only certain frequencies. This implies that only certain electronic energies are possible within an atom. In this section you will see how the atomic spectrum of hydrogen provides evidence of its electronic structure and how the Bohr model of the atom was developed to explain this.

Atomic spectra

When white light passes through a prism, it is split into the component colours. The resulting pattern is called a **continuous spectrum** as it contains an unbroken distribution of all frequencies. A continuous spectrum is shown in Figure 3.9(a).

When a spark or electric discharge passes through a sample of hydrogen gas (H_2), light is produced. The electric discharge causes the gas to split into H atoms and provides electronic energy to them. The atoms release the energy in the form of electromagnetic radiation, some of which is in the visible region. When this emitted light is examined by passing it through a prism, the resulting spectrum is very different from that of white light. Instead of a continuous spectrum, only certain distinct frequencies are observed, as shown in Figure 3.9(b).

(a) Continuous spectrum
Frequency, ν
In a continuous spectrum, all frequencies are emitted.
(b) Atomic spectrum of hydrogen
Frequency, ν
In an atomic spectrum, only discrete frequencies are emitted. The spectrum is different for each element and characteristic of that element.

Figure 3.9 (a) The continuous spectrum of white light and (b) the atomic spectrum of hydrogen.

Since each frequency corresponds to a particular energy, the nature of this spectrum implies that the atom can only lose energy in certain fixed amounts. This in turn leads to the conclusion that the electron in a hydrogen atom can only exist at particular **energy levels** within the atom. In other words, the electronic energy of an atom is **quantized**, and the lines in the spectrum correspond to transitions between the energy levels, as shown in Equation 3.4

$$h\nu = E_2 - E_1 \tag{3.4}$$

where E_1 represents the energy of the lower energy level and E_2 that of the upper energy level. By examining the frequencies of the spectral lines you can determine the relative energies of the electronic energy levels within the atom. The energy levels suggested by the atomic spectrum of hydrogen are shown in Figure 3.10. Each transition corresponds to a line in the spectrum.

The activity at the Online Resource Centre allows you to investigate the relationship between the atomic spectrum of hydrogen and the electronic energy levels.

The first person to identify a pattern in the atomic spectrum of hydrogen was the Swiss mathematician Johann Balmer. He demonstrated that the frequencies of the lines in the visible part of the electromagnetic spectrum all fitted the expression in Equation 3.5.

$$\nu \propto \frac{1}{4} - \frac{1}{n^2} \qquad n = 3, 4, 5, \ldots \tag{3.5}$$

The symbol '∝' in Equation 3.5 stands for 'proportional to'. Quantities are proportional if, when one is changed, the other changes in a corresponding fashion. See Maths Toolkit MT1 (p.1312).

These lines are now called the **Balmer series**. As n increases, the frequencies of the lines get closer and closer together. This fits in with the observed spectrum in which the lines converge (see Figure 3.9(b)). The integer n is used to number the energy levels, as shown in Figure 3.10.

Further series of lines were subsequently discovered in the ultraviolet and infrared regions of the electromagnetic spectrum. The Swedish scientist Johannes Rydberg demonstrated that the frequencies of all of the lines in the atomic spectrum of hydrogen are given by Equation 3.6

$$\nu = R_H\left[\frac{1}{n_1^2} - \frac{1}{n_2^2}\right] \quad \text{Rydberg equation} \tag{3.6}$$

where n_1 and n_2 are both integers, $n_2 > n_1$, and R_H is the Rydberg constant, which has a value of 3.29×10^{15} Hz. When $n_1 = 1$, the transitions correspond to the **Lyman series** of lines, which is observed in the ultraviolet region. When $n_1 = 2$, this becomes Balmer's expression and describes the lines in the visible region. Higher values of n_1 correspond to transitions in the infrared part of the electromagnetic spectrum. Details of the first five series of lines are summarized in Table 3.4, and examples of the use of Equation 3.6 are given in Worked example 3.4.

Equation 3.6 is an example of an **empirical** expression, as it is based on experimental observations rather than theory.

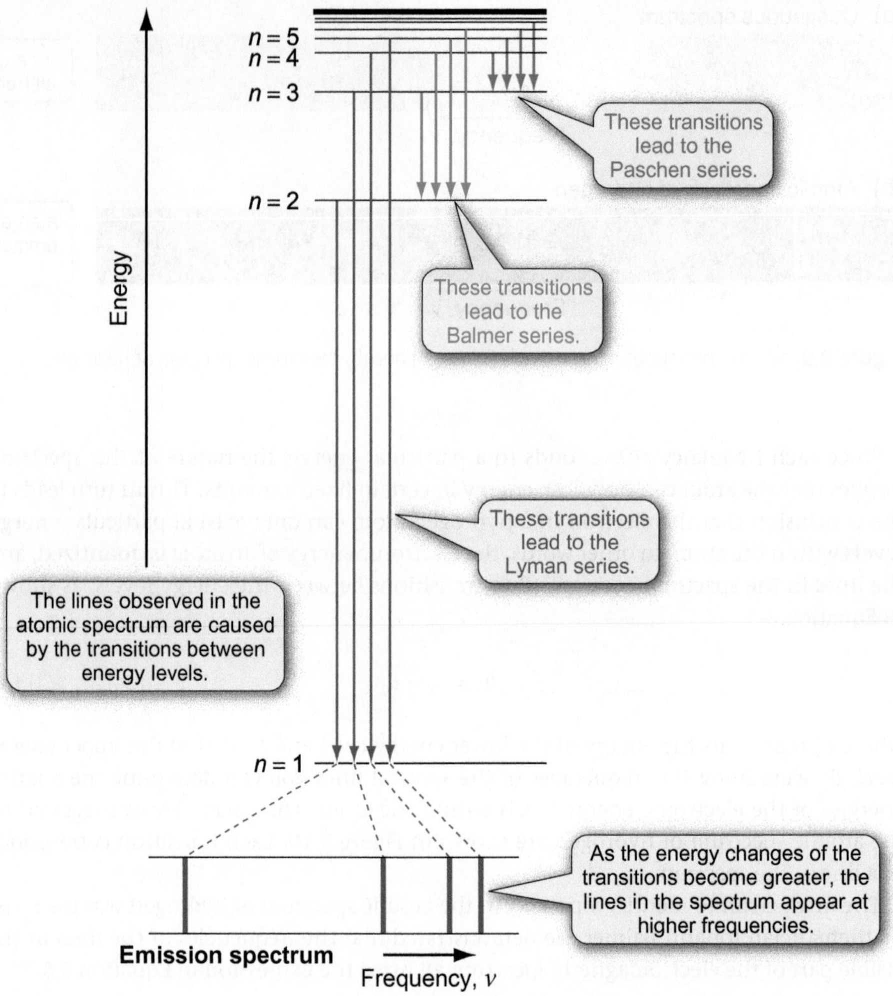

n = 5
n = 4
n = 3
n = 2
n = 1
Energy
These transitions lead to the Paschen series.
These transitions lead to the Balmer series.
These transitions lead to the Lyman series.
The lines observed in the atomic spectrum are caused by the transitions between energy levels.
As the energy changes of the transitions become greater, the lines in the spectrum appear at higher frequencies.
Emission spectrum
Frequency, ν

Figure 3.10 The electronic energy levels in the hydrogen atom and the transitions between them. Each transition corresponds to a line in the atomic emission spectrum of hydrogen. The difference in energy between two levels is equal to the energy of the resultant spectral line. The part of the emission spectrum shown is the Lyman series, in the ultraviolet region, which results from transitions down to the $n = 1$ energy level.

Table 3.4 The atomic spectrum of hydrogen

Series	Region of the electromagnetic spectrum	n_1	n_2
Lyman	Ultraviolet	1	2, 3, 4, . . .
Balmer	Visible	2	3, 4, 5, . . .
Paschen	Infrared	3	4, 5, 6, . . .
Brackett	Infrared	4	5, 6, 7, . . .
Pfund	Infrared	5	6, 7, 8, . . .

Worked example 3.4 The atomic spectrum of hydrogen

For the Paschen series of spectral lines, $n_1 = 3$ and the series lies in the infrared part of the electromagnetic spectrum. What are the frequencies of the first three lines in this series?

Strategy

Use the Rydberg equation, Equation 3.6

$$\nu = R_H\left[\frac{1}{n_1^2} - \frac{1}{n_2^2}\right]$$

$$(R_H = 3.29 \times 10^{15}\,\text{Hz},\ n_1 = 3)$$

n_2 must be greater than n_1 so, for the first three lines in the series, $n_2 = 4$, 5, and 6.

Solution

For the first line

$$\nu = R_H\left[\frac{1}{3^2} - \frac{1}{4^2}\right] = R_H\left[\frac{1}{9} - \frac{1}{16}\right]$$

$$= R_H \times 0.0486 = 1.60 \times 10^{14}\,\text{Hz}$$

For the second line

$$\nu = R_H\left[\frac{1}{3^2} - \frac{1}{5^2}\right] = R_H\left[\frac{1}{9} - \frac{1}{25}\right]$$

$$= R_H \times 0.0711$$

$$= 2.34 \times 10^{14}\,\text{Hz}$$

For the third line

$$\nu = R_H\left[\frac{1}{3^2} - \frac{1}{6^2}\right] = R_H\left[\frac{1}{9} - \frac{1}{36}\right]$$

$$= R_H \times 0.0833$$

$$= 2.74 \times 10^{14}\,\text{Hz}$$

Question

What are the frequencies for the first three lines of the Brackett series, for which $n_1 = 4$?

Ionization energies from atomic spectra

The ionization energy of hydrogen is the minimum energy required to completely remove the electron from an atom. From the atomic spectrum of hydrogen in Figure 3.9 (p.125), you can see that the lines in a particular series get closer together with increasing n_2. The lines eventually converge to a continuum at $n_2 = \infty$, as shown in Figure 3.11, when the electron is no longer part of the atom. Worked example 3.5 uses Equation 3.6 to calculate the frequency of radiation needed to remove an electron completely. The electron is originally in the lowest energy level ($n_1 = 1$). When it has been removed from the atom $n_2 = \infty$.

Trends in ionization energies are described in Section 3.7 (p.154).

Worked example 3.5 The ionization energy of hydrogen

What is the ionization energy of hydrogen in kJ mol^{-1}?

Strategy

Use the Rydberg equation (Equation 3.6) to work out the frequency of the line corresponding to the transition from $n_1 = 1$ to $n_2 = \infty$ (infinity).

Then use Equation 3.2 to convert the frequency into the energy of the transition for a single atom.

Finally, convert the energy of the transition for one atom into that for a mole of atoms.

Solution

Use Equation 3.6

$$\nu = R_H\left[\frac{1}{n_1^2} - \frac{1}{n_2^2}\right] \quad (R_H = 3.29 \times 10^{15}\,\text{Hz}) \qquad (3.6)$$

$n_1 = 1$ and $n_2 = \infty$, so

$$\nu = R_H\left[\frac{1}{1^2} - \frac{1}{\infty^2}\right]$$

Since $\infty^2 = \infty$, and $1/\infty = 0$.

$$\nu = R_H[1 - 0] = R_H = 3.29 \times 10^{15}\,\text{Hz}$$

Use Equation 3.2

$$E = h\nu \ (h = 6.626 \times 10^{-34}\,\text{J s}) \qquad (3.2)$$

$$= (6.626 \times 10^{-34}\,\text{J s}) \times (3.29 \times 10^{15}\,\text{s}^{-1})$$

$$2.18 \times 10^{-18}\,\text{J} = 2.18 \times 10^{-21}\,\text{kJ}$$

This is the energy required to completely remove an electron from one atom. For a mole of atoms, multiply by the Avogadro constant to get the value in kJ mol^{-1}

$$E = (2.18 \times 10^{-21}\,\text{kJ}) \times (6.022 \times 10^{23}\,\text{mol}^{-1})$$

$$= 1310\,\text{kJ mol}^{-1} \text{ (to 3 significant figures)}$$

Question

What is the ionization energy for an excited state of hydrogen in which the electron has already been promoted to the $n = 2$ level?

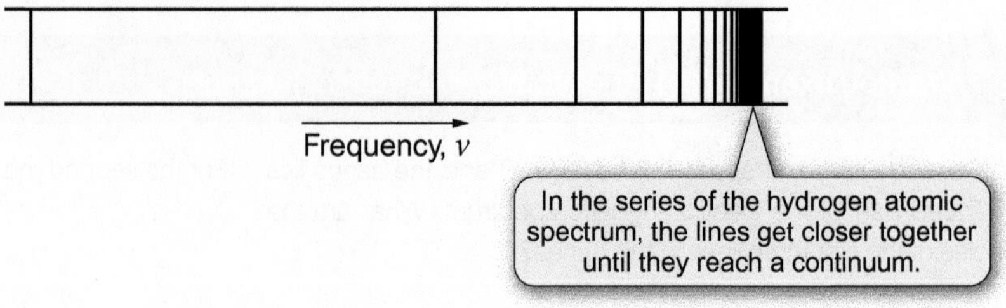

Frequency, ν
In the series of the hydrogen atomic spectrum, the lines get closer together until they reach a continuum.

Figure 3.11 Spectral lines converging to a continuum in the Lyman series.

The spectra you have seen so far are examples of **emission spectra**. To obtain these, atoms are excited to higher electronic energy levels and, as they return to a lower energy level, they emit energy as electromagnetic radiation. Since only certain frequencies are emitted, emission spectra consist of bright lines against a dark background.

An alternative way of looking at these transitions is through **absorption spectra**. In this case the samples are irradiated with a continuous spectrum of electromagnetic radiation. Photons with energies that correspond to transitions between the lowest state, called the **ground state**, and a higher state (an **excited state**) will be absorbed. In an absorption spectrum these transitions appear as dark lines against a bright background. The emission and absorption spectra for hydrogen are shown in Figure 3.12. The lines have the same frequencies in both spectra as the transitions are between the same energy levels.

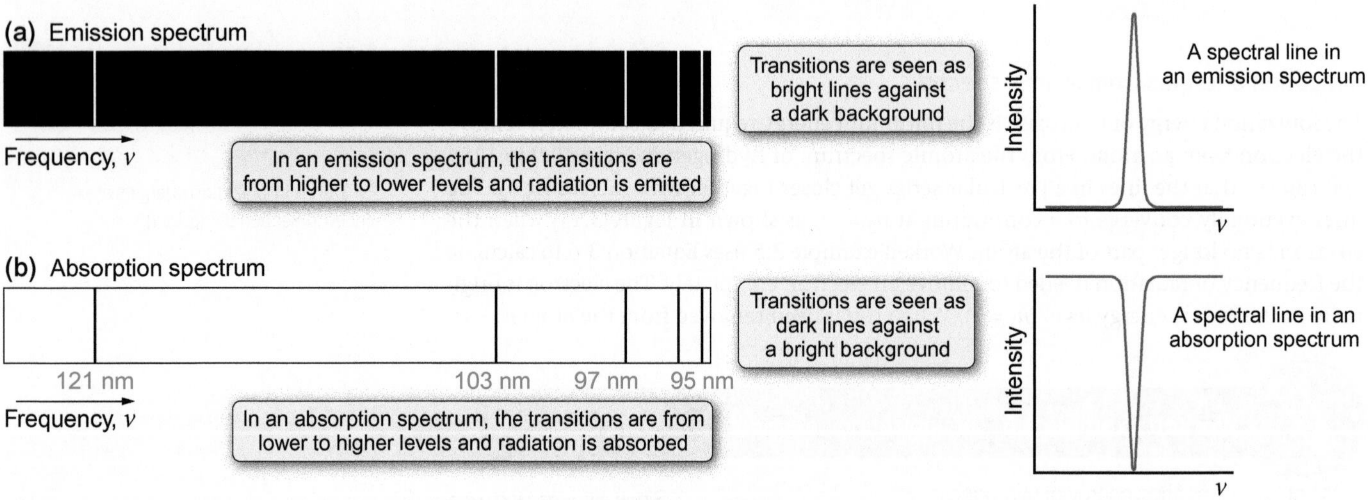

(a) Emission spectrum
Frequency, ν
In an emission spectrum, the transitions are from higher to lower levels and radiation is emitted
Transitions are seen as bright lines against a dark background
Intensity
A spectral line in an emission spectrum
ν
(b) Absorption spectrum
121 nm
103 nm
97 nm
95 nm
Frequency, ν
In an absorption spectrum, the transitions are from lower to higher levels and radiation is absorbed
Transitions are seen as dark lines against a bright background
Intensity
A spectral line in an absorption spectrum
ν

Figure 3.12 (a) Emission spectrum and (b) absorption spectrum for the hydrogen atom.

Box 3.2 Lighting up the sky

The gunpowder used in fireworks is a mixture of a fuel, such as powdered carbon and sulfur, and an oxidizing agent, such as potassium perchlorate ($KClO_4$) or potassium nitrate (KNO_3). Potassium salts are preferred to sodium salts because they are less hygroscopic (they do not absorb water on storage). Also the intense yellow light emitted by sodium salts can mask other colours.

Metals and metal compounds are added to the gunpowder to produce coloured light when the gunpowder burns:

- intense white light from burning magnesium, aluminium, or titanium;
- yellow light from sodium salts, often Na_3AlF_6 which is not hygroscopic;
- orange light from calcium salts, such as $CaCl_2$;
- red light from strontium salts, such as $SrCO_3$;
- green light from barium salts, such as $Ba(NO_3)_2$, together with a chloride source:
- blue light from copper salts, such as $CuCO_3$. Blue is the most difficult colour to produce.

→

→ You can see the same colours if you carry out flame tests on the metal salts in the laboratory. At the high temperature of the flame, electrons in the metal atoms are promoted to higher energy levels. The colours arise from these energetically excited atoms or ions. When the electrons return to the ground state, the atoms emit light of characteristic frequencies. The atomic emission spectra of sodium, strontium, and barium are shown below, together with the colours of their flame tests.

The technique of atomic emission spectroscopy makes use of these principles to analyse samples for the presence of elements. For example, a sample of river water might be analysed for pollutants, or the composition of steel monitored during manufacture. The spectrum for each element is unique and characteristic as the spacings between atomic energy levels are different for each element. The *intensity* of the light emitted at particular frequencies is measured and used to determine the concentrations of metals in the sample. (The use of atomic absorption and emission spectra in analytical chemistry is discussed more fully in Section 11.5, p.547.)

▲ Metal salts are used in fireworks to produce the different colours.

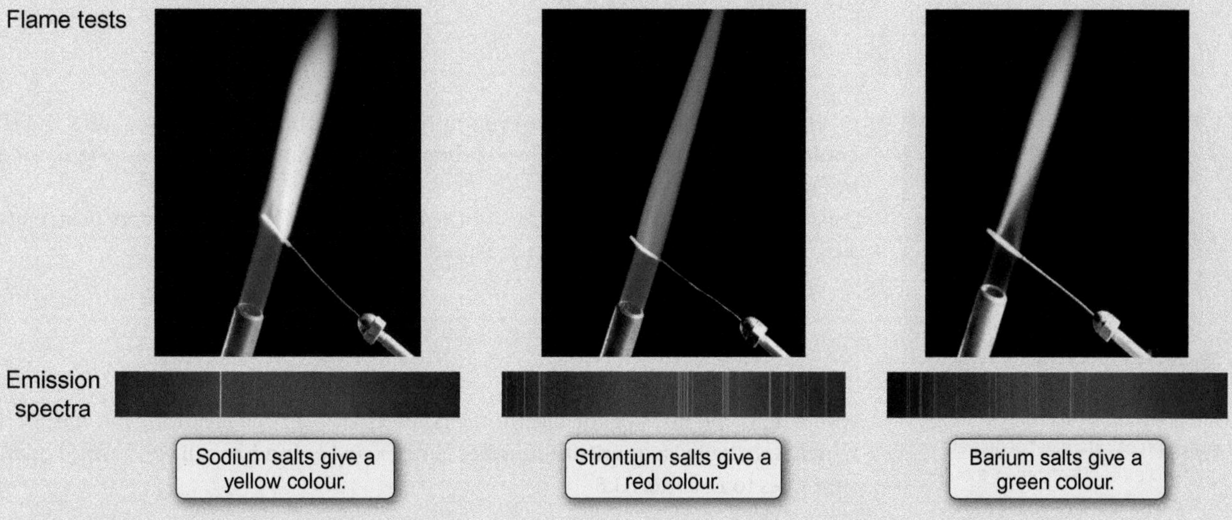
Flame tests
Emission spectra
Sodium salts give a yellow colour.
Strontium salts give a red colour.
Barium salts give a green colour.

▲ The atomic emission spectra of sodium, strontium, and barium, together with the colours of their flame tests.

Question

Why do excited atoms only emit radiation of certain frequencies?

The Bohr atom

Atomic spectroscopy demonstrated that the energy of an electron in an atom is quantized. In other words, the atom cannot have an arbitrary value of electronic energy—only certain values are allowed. The first theoretical model of the atom to account for the spectroscopic data was proposed by the Danish physicist Niels Bohr in 1913.

Bohr suggested that in hydrogen the electron moves around the nucleus in a fixed **orbit**, rather like a planet moving around the Sun, and that the energy associated with each orbit has a fixed value. By absorbing energy, an electron could move from one orbit to another further out from the nucleus, and by emitting energy it could move to an

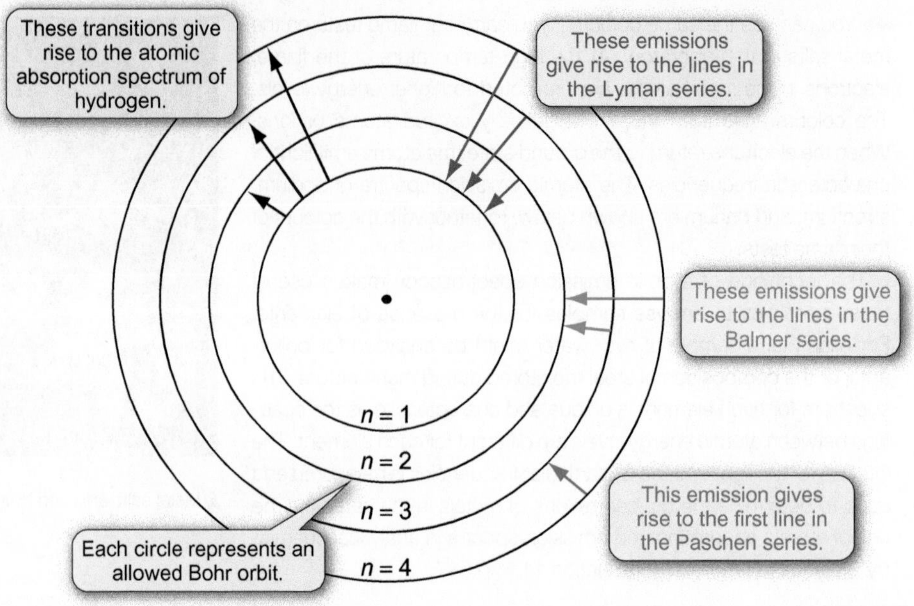
These transitions give rise to the atomic absorption spectrum of hydrogen.
These emissions give rise to the lines in the Lyman series.
These emissions give rise to the lines in the Balmer series.
n = 1
n = 2
n = 3
n = 4
This emission gives rise to the first line in the Paschen series.
Each circle represents an allowed Bohr orbit.

Figure 3.13 The Bohr atom. Bohr suggested that transitions of electrons between the orbits give rise to the atomic spectrum.

Although the quantization of energy is not intuitive from everyday life, most people are familiar with the idea of quantization, though maybe not by name. In football, for example, results are quantized, with the quanta of football being goals. It is not possible for a team to score a fraction of a goal or win $1\frac{1}{2}-1$.

In Equation 3.8, the constant k is related to the constants in Equation 3.7 by $k = 2\pi^2 m_e e^4/h^2$.

The lowest energy state is that with the most negative value of energy. This is called the ground state.

orbit closer to the nucleus. However, only certain orbits were allowed, and the electron could not lie between them. These orbits, and the transition between them, are shown in Figure 3.13.

Bohr used classical mechanics to work out an expression for the energy of an electron in a hydrogen atom. This is shown in Equation 3.7

$$E = -\frac{2\pi^2 m_e e^4}{n^2 h^2} \tag{3.7}$$

where m_e is the mass of the electron, e is the charge on the electron, and n is an integer, which Bohr called a **quantum number**. Since π, m_e, e, and h are all constants, Equation 3.7 simplifies to Equation 3.8

$$E = -\frac{k}{n^2} \tag{3.8}$$

and the electronic energy is determined only by the quantum number n.

As a line in the spectrum arises from a transition between two energy levels, this equation is consistent with Rydberg's empirical expression (Equation 3.6, p.125). The lowest energy (most negative) state is for $n = 1$. In the Bohr model this represents the orbit closest to the nucleus. When $n = 2$ or higher the atom is in an excited state, and further from the nucleus.

Despite the success of this model in explaining the atomic spectrum of hydrogen, Bohr was unable to extend it to account for the spectrum of any other atom. Nor was he able to explain why only certain orbits were allowed. Indeed, according to classical physics, the orbiting electron should emit electromagnetic radiation and collapse into the nucleus. Although the Bohr model of the atom was soon superseded, two key concepts were retained in later theories. The quantization of energy levels and quantum numbers remain important elements of current atomic theories but, before looking at these, you need to know more about the nature of electrons, and this is discussed in Section 3.4.

Box 3.3 The composition of stars

The observed solar spectrum is not quite continuous. In 1841, the German optician Joseph von Fraunhofer noted dark lines in the otherwise continuous spectrum of sunlight. He counted over 600 lines in the visible spectrum but he did not understand what caused them.

Fraunhofer's lines are the absorption spectra of elements present in the Sun's atmosphere. Light emitted by the photosphere of the Sun (see Box 3.1, p.120) passes through an atmosphere of gases that absorb characteristic frequencies. Elements present in these gases can be identified by matching the Fraunhofer lines with the atomic spectra of known elements.

Several scientists made accurate measurements of the solar emission spectrum during the eclipse of 1868. The two yellow sodium lines were clearly visible, along with a third line at 588 nm that did not correspond to the spectrum of any known element. It was proposed that a new element was present in the gases surrounding the Sun, and the element was named helium from the Greek word *helios* meaning 'sun'. Almost 30 years later, helium was discovered on Earth as a product of radioactive decay of the uranium ore cleveite.

▲ The galaxy M81 in the constellation of Ursa Major. Light from this spiral galaxy takes 12 million years to reach the Earth and is red-shifted, indicating that M81 is moving away from our galaxy.

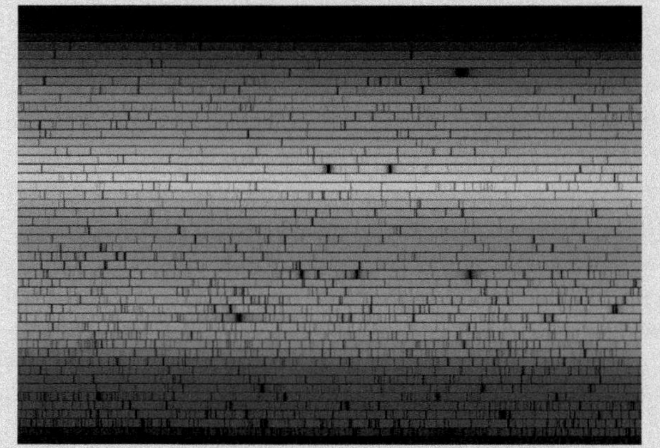

▲ Visible light emitted from the Sun, with the shortest wavelength (around 400 nm) at the bottom left, and wavelength increasing from left to right along each strip, and from bottom to top. The dark Fraunhofer lines are caused by the emitted light being absorbed by gases in the Sun's atmosphere.

When Vesto Slipher studied the light from distant galaxies in the 1920s, he found that the positions of the lines in the hydrogen spectrum were slightly altered. The lines were clearly not due to another element, since their wavelengths all related to those of the Balmer series by the same ratio, but all of the lines were at a longer wavelength than expected. Such an increase in wavelength is known as a *red shift*. The observed red shift is due to the *Doppler effect*—when an object emitting light is moving away from an observer, the wavelength appears to increase. Measurements on light from most other galaxies showed that their light was also red-shifted. This led to the conclusion that the Universe is expanding.

Question

How would Slipher's observation have changed if the Universe was contracting?

Summary

- Spectroscopic measurements show that atoms can only absorb or emit certain frequencies of electromagnetic radiation.
- Atomic spectra demonstrate that the electronic energies of atoms are quantized, with only certain values allowed.
- The Bohr model was the first quantum model of the atom.

For practice questions on these topics, see questions 3 and 4 at the end of this chapter (p.165).

3.4 The nature of the electron

Light can behave as a wave under some circumstances and as a particle under others (see Section 3.2, p.123). This wave–particle duality is not restricted to light and is a fundamental property of all matter. The wave properties of electrons are important in understanding modern theories of the structure of the atom. In this section the nature of the electron is explored in more detail.

The wave properties of matter

The wave properties of matter were first proposed by the French scientist Louis de Broglie in 1924. He suggested that all matter has wave properties associated with it, and that the wavelength is inversely proportional to the mass (m) and the velocity (v) of the matter, as shown in Equation 3.9

$$\lambda = \frac{h}{mv} \qquad (3.9)$$

where h is the Planck constant.

At first sight this seems a bizarre suggestion, as everyday objects such as cars or books show no evidence at all for wave behaviour. However, the tiny value of the Planck constant in this equation means that these objects have wavelengths that are so small that their wave properties are unnoticeable. For example, a rifle bullet of mass 5 g travelling at $1000\,\mathrm{m\,s^{-1}}$ has an associated wavelength of $1 \times 10^{-34}\,\mathrm{m}$. This is far too small to be detectable since the typical distance between atoms in a molecule is approximately $1 \times 10^{-10}\,\mathrm{m}$, around 10^{24} times larger. The wave character of everyday objects is not important but, for much less massive objects like subatomic particles, it cannot be ignored.

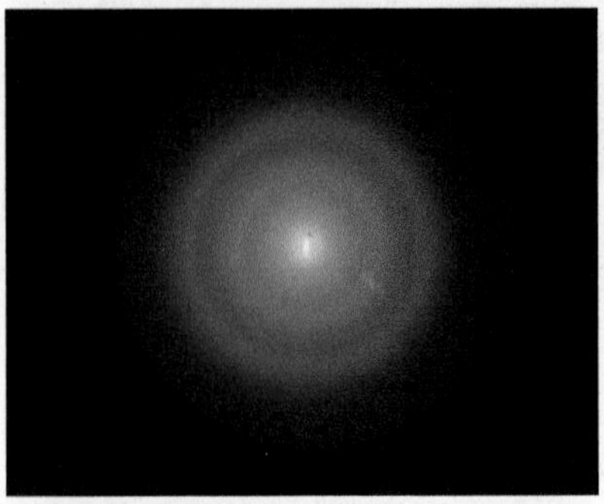

Figure 3.14 An electron diffraction pattern, arising from firing a beam of electrons at a thin sheet of graphite. The electrons pass through and hit a luminescent screen, producing the patterns of rings associated with diffraction. This demonstrates that electrons can behave as waves.

Constructive and destructive interference are described in Section 3.2 (p.123).

The first evidence for the wave properties of matter was obtained in 1925 in independent experiments carried out by Clinton Davisson and Lester Germer in the USA and by G.P. Thomson in Britain. Davisson and Germer directed a beam of electrons at a crystal of nickel and observed a diffraction pattern. The wavelength of the electron at the velocity used in the experiment is similar to the separation between the nickel atoms in the crystal, so the crystal acts as a diffraction grating. The electron diffraction pattern observed for graphite is shown in Figure 3.14. The light and dark rings arise from constructive and destructive interference, respectively.

An example of the use of the de Broglie equation, Equation 3.9, is given in Worked example 3.6.

Worked example 3.6 Using the de Broglie equation

What is the wavelength associated with an electron travelling at one-tenth of the speed of light? (The mass of an electron is given in Table 3.1, p.116.)

Strategy

Use Equation 3.9

$$\lambda = \frac{h}{mv} \quad (h = 6.626 \times 10^{-34}\,\mathrm{J\,s}) \qquad (3.9)$$

Remember, $1\,\mathrm{J} = 1\,\mathrm{kg\,m^2\,s^{-2}}$ (see Table 1.2, p.8). When using SI units, masses are measured in kg (Section 1.2, p.7).

Solution

For the electron

$$m = 9.109 \times 10^{-31}\,\mathrm{kg}$$

and $v = 0.1 \times c = 2.998 \times 10^{7}\,\mathrm{m\,s^{-1}}$

From Equation 3.9:

$$\lambda = \frac{6.626 \times 10^{-34}\,\cancel{\mathrm{kg}}\,\mathrm{m}^{\cancel{2}}\,\cancel{\mathrm{s}^{-2}}\,\cancel{\mathrm{s}}}{(9.109 \times 10^{-31}\,\cancel{\mathrm{kg}}) \times (2.998 \times 10^{7}\,\cancel{\mathrm{m}}\,\cancel{\mathrm{s}^{-1}})}$$

$$= 2.426 \times 10^{-11}\,\mathrm{m}$$

Question

What is the wavelength of a neutron travelling at a tenth of the speed of light? The mass of a neutron is given in Table 3.1 (p.116).

Electron diffraction and neutron diffraction are both used to determine the structures of molecules. Diffraction occurs because the wavelength of the particle is of a similar order of magnitude to the interatomic separations.

The significance of electron diffraction in electron microscopy is described on p.113 and its use to measure bond lengths is described in Box 4.1 (p.172).

Wave-particle duality of electrons

Like photons, electrons show wave–particle duality. In some instances they act as waves, in others they act as particles. The counterintuitive nature of the quantum world is summed up by the fact that the father and son scientists J.J. and G.P. Thomson both won Nobel Prizes in Physics. J.J. Thomson's prize in 1906 was for showing that electrons are particles. G.P. Thomson's prize in 1937 was for determining that electrons are waves. Both were right . . .

The Heisenberg uncertainty principle

Since objects such as electrons have wave properties, it is impossible to know their precise position and momentum simultaneously. This concept was developed by the German physicist Werner Heisenberg in the 1920s and is described as the **Heisenberg uncertainty principle**.

The Heisenberg uncertainty principle is quantified by Equation 3.10

$$\Delta p \Delta q \geq \frac{h}{4\pi} \qquad (3.10)$$

where Δp is the uncertainty in momentum and Δq is the uncertainty in position. An example of the use of Equation 3.10 is given in Worked example 3.7.

Momentum (p) is equal to mass (m) multiplied by velocity (v), i.e. $p = mv$.

Worked example 3.7 The Heisenberg uncertainty principle

The velocity of a rifle bullet of mass of exactly 5.000 g is known to within $1 \times 10^{-6}\,\mathrm{m\,s^{-1}}$. Calculate the uncertainty in its position.

Strategy

Since momentum (p) = mass (m) × velocity (v), you can calculate the uncertainty in the momentum, Δp, using $\Delta p = m\Delta v$.

Then, rearrange Equation 3.10, and use this to calculate Δq, the uncertainty in the position.

Solution

$$\Delta p = m\Delta v$$
$$= (5.000 \times 10^{-3}\,\mathrm{kg}) \times (1 \times 10^{-6}\,\mathrm{m\,s^{-1}})$$
$$= 5.000 \times 10^{-9}\,\mathrm{kg\,m\,s^{-1}}$$

Rearranging Equation 3.10,

$$\Delta q \geq \frac{h}{4\pi\Delta p}$$

($h = 6.626 \times 10^{-34}\,\mathrm{kg\,m^2\,s^{-1}}$, see Worked example 3.6)

$$\geq \frac{6.626 \times 10^{-34}\,\cancel{\mathrm{kg}}\,\mathrm{m}^{\cancel{2}}\,\cancel{\mathrm{s}^{-1}}}{4 \times 3.1416 \times (5.000 \times 10^{-9}\,\cancel{\mathrm{kg}}\,\cancel{\mathrm{m}}\,\cancel{\mathrm{s}^{-1}})}$$

$$\geq 1.055 \times 10^{-26}\,\mathrm{m}$$

Question

Calculate the uncertainty in the position of an electron whose velocity is known to within $1 \times 10^{-6}\,\mathrm{m\,s^{-1}}$.

As with the wavelength of a rifle bullet (p.132), the uncertainty in the position of a bullet in flight is too small to be significant. The position may be difficult to measure, but it is possible. For an electron, with much lower mass, this is not the case. One of the consequences of this is that it is not possible to talk about the precise location of an electron in an atom. Instead you need to consider the probabilities of the electrons being in certain volumes of space. This idea is developed in more detail in Section 3.5 (p.134).

Heisenberg also developed the first theory of quantum mechanics. However, the mathematics involved in his matrix mechanics was so complicated that this approach made little impact compared with wave mechanics. Heisenberg's particle-based approach and Schrödinger's wave-based approach (Section 3.5) were later shown to be equivalent.

Summary

- All matter has a wavelength associated with it, and very small objects such as electrons can behave as either particles or waves.
- The Heisenberg uncertainty principle states that it is impossible to know *both* the precise position and momentum of an object. For everyday objects the uncertainty is small enough to be insignificant, but for atoms and subatomic particles the uncertainty cannot be ignored.

? For a practice question on these topics, see question 5 at the end of this chapter (p.165).

3.5 Wavefunctions and atomic orbitals

Once the wave–particle duality of electrons had been established, this could be incorporated into mathematical models of the atom. The most successful approach to this was initiated by the Austrian physicist Erwin Schrödinger, who developed **wave mechanics**.

Wavefunctions and the Schrödinger equation

Schrödinger introduced the **wavefunction**, denoted by the Greek letter ψ (psi), as a function with a value that varies with position. The wavefunction for an electron can be calculated using the **Schrödinger equation**. This is a partial differential equation, whose form is shown in Equation 3.11.

$$\underbrace{-\frac{h^2}{8\pi^2 m}\left(\frac{\partial^2}{\partial x^2}+\frac{\partial^2}{\partial y^2}+\frac{\partial^2}{\partial z^2}\right)\psi}_{\text{kinetic energy}} + \underbrace{E_{PE}\psi}_{\substack{\text{potential}\\ \text{energy}}} = \underbrace{E\psi}_{\substack{\text{total}\\ \text{energy}}} \qquad (3.11)$$

Greek symbols
ψ psi (wavefunction, the amplitude of the electron wave)
π pi (3.14159 to 6 significant figures)
τ tau (volume)

E is the total energy associated with the wavefunction ψ, m is the mass of the electron, and E_{PE} is its potential energy, which depends on its position.

Potential energy E_{PE} is introduced in Section 1.6 (p.41).

The $\partial^2/\partial x^2$, $\partial^2/\partial y^2$, and $\partial^2/\partial z^2$ terms are partial differentials. A partial differential shows how a function depends on one variable when several are changing. Here, the differentials show how the wavefunction, ψ, changes with distance from the nucleus. $\partial^2\psi/\partial x^2$ describes how ψ changes along the x-axis at constant values of y and z; $\partial^2\psi/\partial y^2$ describes how ψ changes along the y-axis at constant values of x and z.

Differentials are mathematical functions arising from calculus and are used to describe changing conditions. See Maths Toolkit MT6 (p.1324).

Fortunately, you do not have to solve this equation in order to use the results that arise from it. It is only possible to solve the Schrödinger equation exactly for two-body problems, which in the case of atoms means systems that contain a nucleus and a single electron. So, the only neutral atom for which the equation can be solved is hydrogen, though solutions are also possible for single-electron ions such as He^+ and Li^{2+}. Despite this, there are a number of good assumptions that allow the solutions obtained for hydrogen to be adapted for other atoms.

While the wavefunction contains detailed information on the behaviour of the electron, it is not directly measurable and does not have a physical interpretation. The German physicist Max Born suggested how wavefunctions could be related to a measurable property. In the **Born interpretation**, the square of the wavefunction ψ^2 is proportional to the probability of finding the electron within a small volume of space dτ. Therefore, in a region where ψ^2 is large, the probability of finding the electron is high, and in a region where ψ^2 is small the probability of finding the electron is low.

dτ represents a very small volume of space—an infinitesimally small volume. This allows the use of mathematical methods, known as calculus, described in the Maths Toolkit MT6 (p.1324) and MT7 (p.1327).

Since the Schrödinger equation treats electrons as waves, the electrons are spread out (delocalized) rather than located in one place at a given time. Instead of the fixed orbits of the Bohr model, when using quantum mechanics you need to think in terms of probabilities. The probability per unit volume is equivalent to **electron density**.

Solutions to the Schrödinger equation

Schrödinger was able to solve his equation in 1927 and show that only certain energies are possible for the electron in a hydrogen atom. More specifically, he showed that these energy levels were given by Equation 3.12

$$E_n = \frac{hR}{n^2} \tag{3.12}$$

Do not confuse the constant R in Equation 3.12, with the gas constant; see Section 8.1 (p.349).

where

$$R = \frac{m_e e^4}{8h^3\varepsilon_0^2} \text{ and } n = 1, 2, 3, \ldots$$

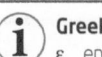

Greek symbols
ε epsilon (permittivity)

R is a constant: m_e is the mass of the electron, e the charge on the electron, h the Planck constant, and ε_0 is the vacuum permittivity, a fundamental constant with value $8.854 \times 10^{-12}\,C^2 J^{-1} m^{-1}$. These energy levels fit exactly with those experimentally determined from atomic spectra.

Compare Equation 3.12 with the Rydberg equation in Equation 3.6 (p.125) and the Bohr equation in Equation 3.8 (p.130).

In a further justification for Schrödinger's theory, the calculated value of the constant R is 3.29×10^{15} Hz, and matches the experimental value for the Rydberg constant. As in the Bohr atom, the integer n is called a quantum number. Since a further two quantum numbers also emerge from the Schrödinger equation, n is referred to as the **principal quantum number**. The value of n can be any integer from 1 to ∞, though $n = 1$ to $n = 7$ are the most important chemically.

The reason for the particular importance of these values of n and l will become apparent in Section 3.6 (p.143) when you look at the structures of many-electron atoms.

The second quantum number l is known as the **secondary quantum number**, though it is also referred to as the angular, azimuthal, or orbital quantum number. The value of l can be any whole number from 0 to $(n - 1)$, though only values of 0, 1, 2, and 3 are important chemically.

The secondary quantum number l takes values 0 to $(n-1)$. The magnetic quantum number m_l takes values $-l$ to $+l$.

The third quantum number m_l is known as the **magnetic quantum number**. The value of m_l can be any whole number from $-l$ to $+l$.

Together these three quantum numbers characterize an **atomic orbital**. An atomic orbital is the region of space defined by a wavefunction.

An atomic orbital is the region of space defined by a wavefunction.

Quantum numbers provide important information about the properties of particular atomic orbitals. For the hydrogen atom, n tells you about the energy of the orbital, l tells you about the shape of the orbital, and m_l tells you about the orientation of the orbital. The limitations on the values of l and m_l described above tell you about the numbers and types of orbitals possible.

When describing orbitals the value of l is represented by a letter.

	$l = 0$	$l = 1$	$l = 2$	$l = 3$
type of orbital	s	p	d	f

For $n = 1$, there can only be one value of l, $l = 0$, and one value of m_l, $m_l = 0$. The orbital with these three values of n, l, and m_l is known as the 1s orbital.

For $n = 1$,	$l = 0$	$m_l = 0$	1s orbital

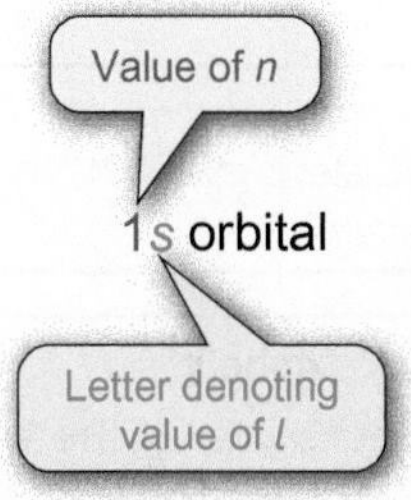
Value of n
1s orbital
Letter denoting value of l

For $n = 2$, both $l = 0$ and $l = 1$ are possible, which means that both 2s and 2p orbitals are possible. When $l = 0$, m_l can only be 0, so there is one 2s orbital. However, when $l = 1$, there are three possible values of m_l, which are −1, 0, and +1. This means that there are three distinct 2p orbitals.

For $n = 2$	$l = 0$	$m_l = 0$	one 2s orbital
	$l = 1$	$m_l = -1, 0, +1$	three 2p orbitals

All the orbitals with the same value of n are said to be in the same **shell**, whereas those with the same values of both n and l are said to be in the same **subshell**. There are four orbitals in the $n = 2$ shell, one of which is in the s subshell and three in the p subshell.

For $n = 3$, there are nine possible orbitals—one 3s, three 3p, and five 3d orbitals. The quantum numbers for these are shown overleaf.

Visit the Online Resource Centre to view screencast 3.2 which walks you through solutions to the Schrödinger equation and the values of the quantum numbers.

Principal quantum number	Secondary quantum number	Magnetic quantum number	Orbital defined by these quantum numbers
n	l $0 \le l \le (n-1)$	m_l $-l \le m_l \le +l$	
3	0	0	3*s* orbital (×1)
	1	−1 0 +1	3*p* orbitals (×3)
	2	−2 −1 0 +1 +2	3*d* orbitals (×5)

The use of these ideas to calculate the number of atomic orbitals with $n = 4$ is shown in Worked example 3.8.

Worked example 3.8 Atomic orbitals

How many atomic orbitals are possible in the shell where $n = 4$?

Strategy

Use the definitions above to work out what values of l and m_l are permitted for $n = 4$.

Solution

l can take a value from 0 to $(n - 1)$ so, for $n = 4$, $l = 0, 1, 2,$ or 3.

For $l = 0$, $m_l = 0$,
so there is one 4*s* orbital.

For $l = 1$, $m_l = -1, 0,$ or $+1$,
so there are three 4*p* orbitals.

For $l = 2$, $m_l = -2, -1, 0, +1,$ or $+2$,
so there are five 4*d* orbitals.

For $l = 3$, $m_l = -3, -2, -1, 0, +1, +2,$ or $+3$,
so there are seven 4*f* orbitals.

This gives a total of 1 + 3 + 5 + 7 = 16 atomic orbitals.

Question

What atomic orbital has the quantum numbers $n = 6$ and $l = 2$?

Radial and angular wavefunctions

More information on spherical coordinates is given in Maths Toolkit MT5 (p.1323).

Greek symbols
θ theta (angular coordinate)
ϕ phi (angular coordinate)

The wavefunction ψ is a mathematical function that varies with position. This variation means ψ is a function of the Cartesian coordinates *x*, *y*, and *z*, and this is written in shorthand as ψ(*x*, *y*, *z*). Because atoms are spherical, it is more useful to represent the wavefunction as a function of one distance, *r*, and two angles, θ and ϕ. The parameters *r*, θ, and ϕ are known as **spherical coordinates**. Cartesian coordinates and spherical coordinates are equally valid ways of specifying the position of a point with respect to an origin, as shown in Figure 3.15.

By writing the wavefunction as a function of *r*, θ, and ϕ, it is possible to break it into a radial part that just depends on *r* and an angular part that just depends on θ and ϕ. This is shown in Equation 3.13.

$$\psi(x, y, z) = \psi(r, \theta, \phi) = \underbrace{R(r)}_{\text{radial wavefunction}} \times \underbrace{Y(\theta, \phi)}_{\text{angular wavefunction}} \qquad (3.13)$$

R(*r*) is called the **radial wavefunction**, and contains information about what happens to the wavefunction as the distance from the nucleus increases. Y(θ, ϕ) is called the **angular wavefunction**, and contains information about the shape of the orbital.

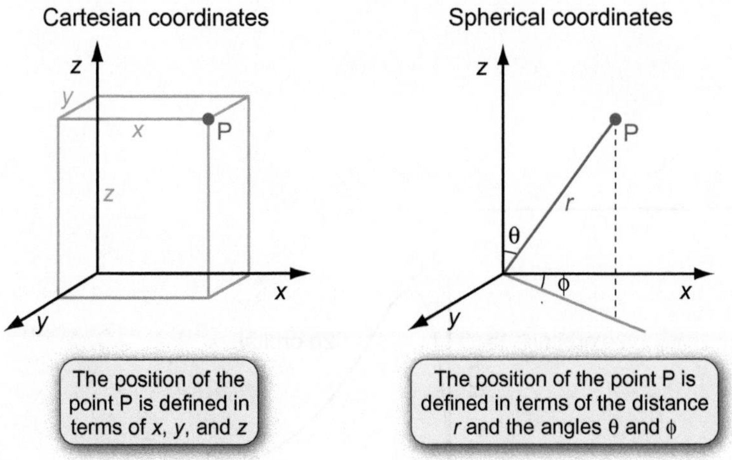

Cartesian coordinates
Spherical coordinates
z
y
x
P
z
x
y
z
P
r
θ
ϕ
x
y
The position of the point P is defined in terms of x, y, and z
The position of the point P is defined in terms of the distance r and the angles θ and ϕ

Figure 3.15 Cartesian and spherical coordinates both define the position of a point, P, with respect to an origin.

The wavefunction derived from the Schrödinger equation for the 1*s* orbital is shown in Equation 3.14, and is separated into the radial and angular parts.

$$\psi(r,\theta,\phi) = \mathrm{R}(r) \times \mathrm{Y}(\theta,\phi) = \left(\frac{4}{a_0^3}\right)^{1/2} \exp\left(\frac{-r}{a_0}\right) \times \left(\frac{1}{4\pi}\right)^{1/2} \tag{3.14}$$

$$= \left(\frac{1}{\pi a_0^3}\right)^{1/2} \exp\left(\frac{-r}{a_0}\right)$$

An expression in the form exp(x) or, alternatively, e^x, is called an exponential. More information on exponential functions is given in Maths Toolkit MT3 (p.1317). An object to the power of $\frac{1}{2}$ is the same as the square root of that object, i.e. $x^{1/2} = \sqrt{x}$.

The quantity a_0 is known as the **Bohr radius** and is defined as in Equation 3.15.

$$a_0 = \frac{\varepsilon_0 h^2}{\pi m_e e^2} = 52.9 \text{ pm} \tag{3.15}$$

The Bohr radius represents the distance from the nucleus to the electron in the Bohr model of the hydrogen atom, and is the most probable distance for finding the 1*s* electron from the nucleus in the quantum mechanical approach.

Figure 3.16(a) shows a plot of the radial wavefunction, R(r), against the distance from the nucleus, r, for the 1*s* orbital in a hydrogen atom. Although the exponential decay tends towards zero, R(r) only reaches zero at $r = \infty$.

The radial wavefunctions for the 2*s* and 2*p* orbitals in a hydrogen atom are shown in Figure 3.16(b). The radial wavefunction for the 2*s* orbital is more complicated than that for the 1*s* orbital. R(r) starts with a positive value, goes through zero becoming negative, and then tends towards zero at high values of r.

A point at which a wavefunction is zero is called a **node**, and when this occurs in the radial wavefunction it is called a **radial node**. For the 2*s* orbital the node has a spherical shape since, at this particular value of r, the wavefunction is zero regardless of the direction from the nucleus.

The radial wavefunction for the 2*p* orbital is zero at the nucleus, increases to a maximum value, then decreases towards zero at high values of r. In a *p* orbital there is a node at the nucleus, which is present in both the radial and angular wavefunctions.

A wavefunction can be positive or negative. This sign is not related to the charge on the electron which is always negative.

Figure 3.16(c) shows the radial wavefunctions for the 3*s*, 3*p*, and 3*d* orbitals. The 3*s* orbital has two radial nodes and the 3*p* orbital one radial node for $r > 0$. Generally, the first time an orbital type occurs (*s*, *p*, *d*, etc.) it has no radial node for $r > 0$, the second time it has one, the third time it has two, and so on. The 1*s*, 2*p*, 3*d*, and 4*f* orbitals are the first of their types, and all have no radial nodes for $r > 0$. In contrast, the 2*s*, 3*p*, 4*d*, and 5*f* orbitals have one radial node for $r > 0$ and the 3*s*, 4*p*, 5*d*, and 6*f* orbitals have two.

The 3*p* orbital has one radial node in addition to a node at the nucleus. This is best expressed by saying it has one radial node for $r > 0$.

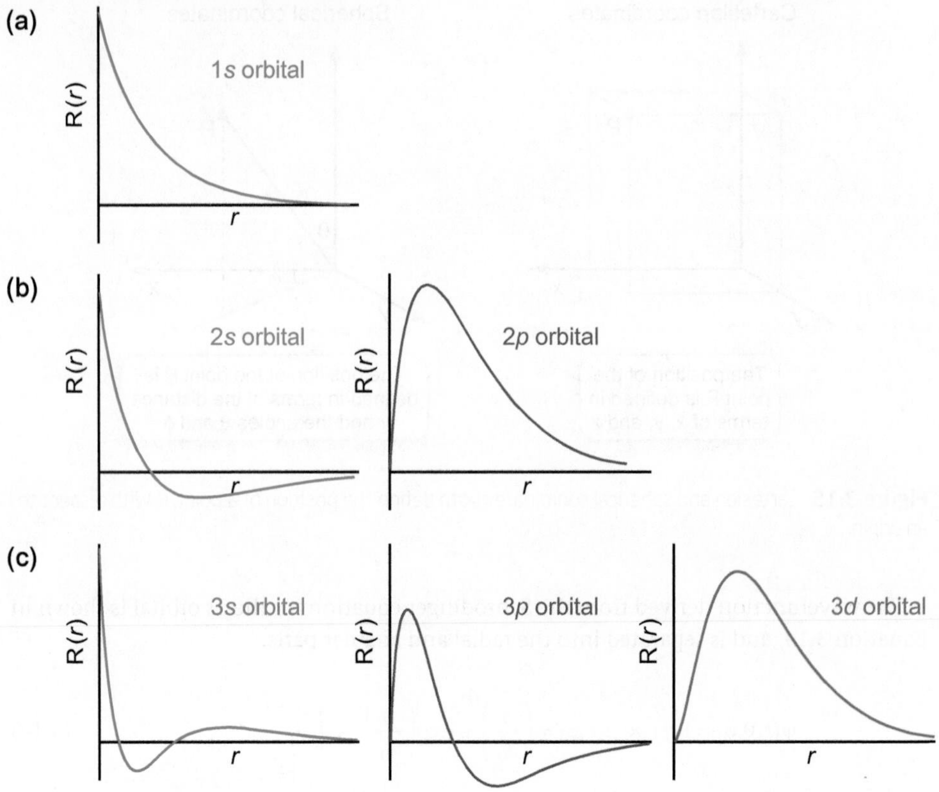
(a)
1s orbital
R(r)
r
(b)
2s orbital
R(r)
r
2p orbital
R(r)
r
(c)
3s orbital
R(r)
r
3p orbital
R(r)
r
3d orbital
R(r)
r

Figure 3.16 Variation of the radial wavefunction R(r) with r for atomic orbitals in a hydrogen atom (a) $n = 1$, (b) $n = 2$, and (c) $n = 3$. (These plots are not all the same scale.)

In summary:

- an s orbital has $(n - 1)$ radial nodes;
- a p orbital has $(n - 2)$ radial nodes for $r > 0$;
- a d orbital has $(n - 3)$ radial nodes for $r > 0$;
- an f orbital has $(n - 4)$ radial nodes for $r > 0$.

Worked example 3.9 uses these principles to show how the radial wavefunction for the 5s orbital varies with distance from the nucleus.

Worked example 3.9 Radial wavefunctions

How does the radial wavefunction for the 5s orbital vary with the distance from the nucleus?

Strategy

In order to describe the radial wavefunction for a particular atomic orbital, you need to know three things: the value of R(r) at $r = 0$; the value of R(r) at $r = \infty$; and the number of radial nodes present.

Solution

An s orbital has no node at the nucleus, which means that, at $r = 0$, R(r) > 0. (The only kind of orbital for which R(r) ≠ 0 at the nucleus is an s orbital.)

At $r = \infty$, R(r) = 0, and this is true for all orbitals.

An s orbital has $(n - 1)$ radial nodes, so for the 5s orbital there are four radial nodes.

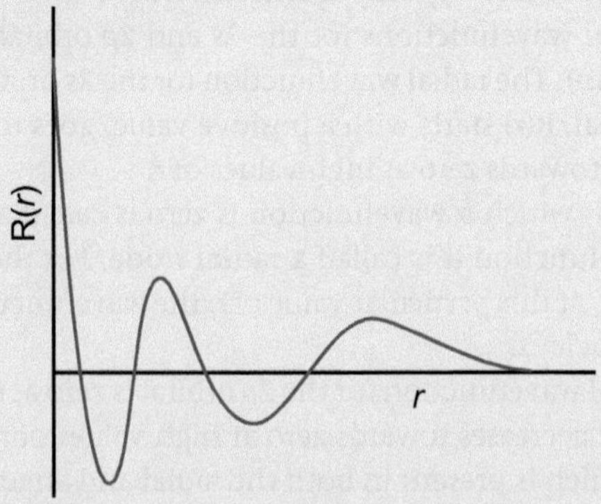
R(r)
r

Question

How many radial nodes does a 7s orbital have?

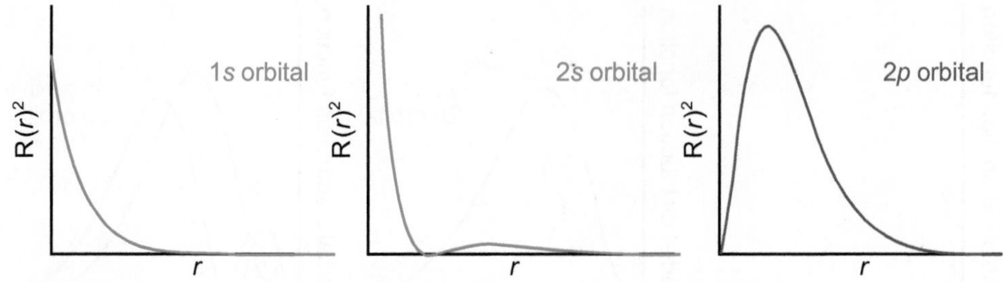
1s orbital
2s orbital
2p orbital
R(r)²
r

Figure 3.17 Variation of $R(r)^2$ with r for atomic orbitals with $n=1$ and 2 in a hydrogen atom. Although $R(r)$ can be positive or negative for the 2*s* orbital, depending on r, $R(r)^2$ is positive for all values of r.

According to the Born interpretation of the wavefunction, $\psi^2 d\tau$ represents the probability of finding an electron within a small volume $d\tau$. Therefore the function $R(r)^2 d\tau$ (sometimes written $R^2(r)d\tau$) gives you the probability of finding an electron within the volume $d\tau$ at a distance r from the nucleus. This is equivalent to the electron density at that point.

The Born interpretation was introduced on p.134.

Plots of $R(r)^2$ against r are shown for the 1*s*, 2*s*, and 2*p* orbitals in Figure 3.17. Note that $R(r)^2$ is always greater than or equal to zero—probability cannot be negative.

For an electron in an atom, the probability of finding it somewhere in the whole of space is 1, since it must be somewhere. This can be expressed mathematically as an integral as in Equation 3.16.

$$\int_{r=0}^{r=\infty} \psi^2 \, d\tau = 1 \qquad (3.16)$$

When represented in this manner wavefunctions are said to be **normalized**, and are sometimes described as **single-electron wavefunctions**.

Σ The square of a negative number is positive, so, although the wavefunction, ψ, can be positive or negative, ψ^2 is always positive.

Σ On a graph, the integral of a function gives the area under a curve that describes the function. The expression in Equation 3.16 tells you that the total area under each of the curves in Figure 3.16 from $r=0$ to $r=\infty$ must equal one, i.e. the total probability of finding the electron is 1 since it must exist somewhere. Integration is described in Maths Toolkit MT7 (p.1327).

Radial distribution functions

Although plots of $R(r)^2$ against r can be helpful, often it is more useful to know the most probable distance of finding an electron from the nucleus. This is obtained by plotting the **radial distribution function** against r. The **radial distribution function** is defined in Equation 3.17.

$$\text{Radial distribution function} = 4\pi r^2 R(r)^2 \qquad (3.17)$$

The radial distribution function is related to the probability of finding an electron in a spherical shell of radius r and thickness dr. On first sight this might be expected to be the same as $R(r)^2$, but that isn't the case. The radial distribution function is different from $R(r)^2$ because, the farther you go away from the nucleus, the greater the number of points there are at that distance. Figure 3.18 shows two spheres of different size. Imagine a thin shell of radius dr at the surface of each sphere. The volume of this shell is $4\pi r^2 dr$ and it is greater for the larger sphere since r is larger. To get the probability of finding the electron at a distance between r and $(r + dr)$ from the nucleus it is essential to take the increase in volume with increasing r into account. The probability of finding an electron in a thin spherical shell of radius r and thickness dr is given by $4\pi r^2 R(r)^2 dr$.

Plots of the radial distribution functions for the atomic orbitals with $n = 1$, 2, and 3 are shown in Figure 3.19. Multiplying $R(r)^2$ by $4\pi r^2$ has two main effects on the functions. Firstly, the maxima are pushed further from the nuclei since $4\pi r^2$ increases with r. Secondly, the radial distribution function for all orbitals is zero at $r = 0$, since multiplying $R(r)^2$ by zero equals zero.

The maximum in a radial distribution function plot represents the most probable distance for the electron from the nucleus. For the hydrogen 1*s* orbital, this is equivalent to the Bohr radius in the Bohr model of the atom.

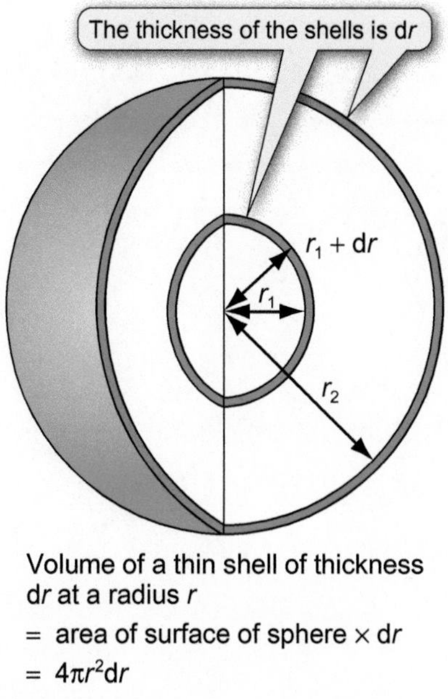
The thickness of the shells is dr
r₁ + dr
r₁
r₂
Volume of a thin shell of thickness dr at a radius r
= area of surface of sphere × dr
= 4πr²dr

Figure 3.18 The volume of the thin shell at radius r_2 is greater than that at r_1.

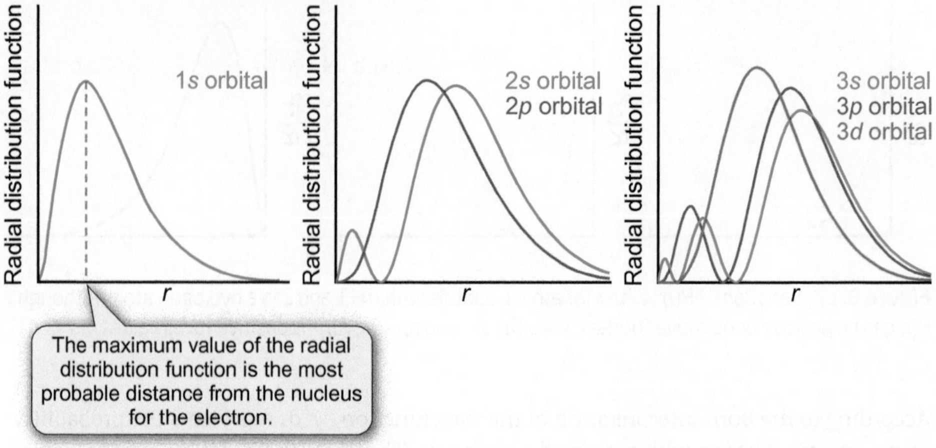
Radial distribution function
1s orbital
r
Radial distribution function
2s orbital
2p orbital
r
Radial distribution function
3s orbital
3p orbital
3d orbital
r
The maximum value of the radial distribution function is the most probable distance from the nucleus for the electron.

Figure 3.19 Variation of the radial distribution function with r for atomic orbitals with $n = 1, 2,$ and 3 in a hydrogen atom.

The shapes of atomic orbitals

The shape of an atomic orbital is related to its angular wavefunction, $Y(\theta, \phi)$. Neither of the angles θ or ϕ appears in the angular wavefunction for a hydrogen 1*s* orbital (Equation 3.14, p.137). This means that for the 1*s* orbital the angular wavefunction is a constant and the wavefunction is the same in all directions from the nucleus. Consequently, the 1*s* orbital is spherical. The same applies for all other *s* orbitals, and they differ only in terms of their size and the number of radial nodes they possess.

The shapes of orbitals are shown using **boundary surfaces**. From the plot of $R(r)$ against r for a 1*s* orbital in Figure 3.16(a) (p.138), you can see that the radial wavefunction $R(r)$ only reaches zero at $r = \infty$. This means it is not possible to define a region that contains the entire wavefunction. A boundary surface gets around this problem by containing a certain proportion of the wavefunction, typically 95%. As such, the boundary surface represents the shape of the orbital. The boundary surface for the 1*s* orbital is spherical. Boundary surfaces for the 1*s*, 2*s*, and 3*s* orbitals are shown in Figure 3.20. As n increases the orbitals get larger and consequently more diffuse.

Boundary surfaces normally show the sign (called the **phase**) of the wavefunction at the surface. For an *s* orbital, the phase of the wavefunction is the same over the whole of the boundary surface. The phase of the wavefunction can be shown either by using + and – labels, or (as we do in this book) by using different colours.

The angular wavefunctions for the three 2*p* orbitals are dependent on θ and/or ϕ. For example, the angular wavefunction for the $2p_z$ orbital is proportional to $\cos\theta$. This means

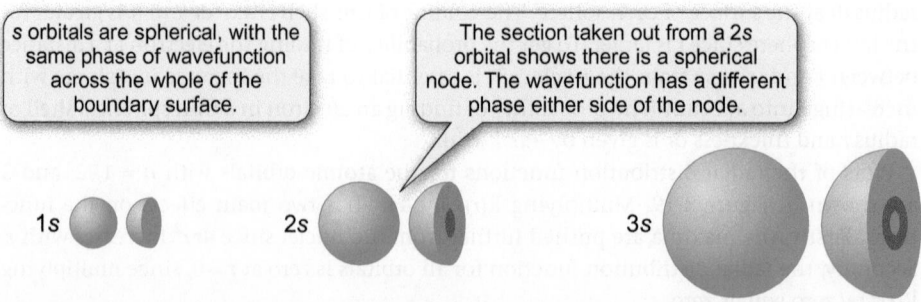

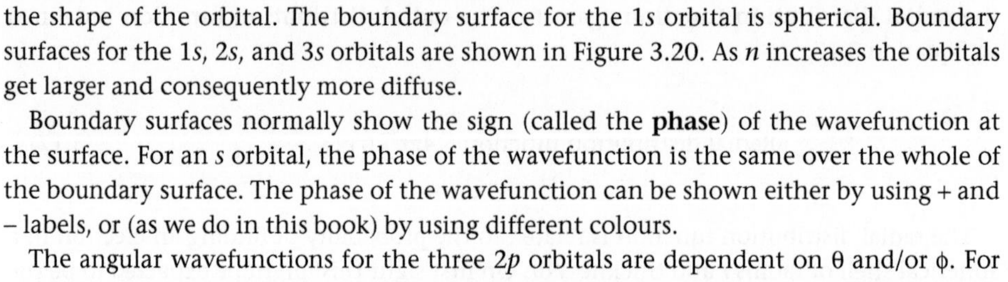
s orbitals are spherical, with the same phase of wavefunction across the whole of the boundary surface.
The section taken out from a 2s orbital shows there is a spherical node. The wavefunction has a different phase either side of the node.
1s
2s
3s

Figure 3.20 Boundary surfaces of the hydrogen 1*s*, 2*s*, and 3*s* orbitals. The cutaway sections to the right of each boundary surface show the radial nodes for the 2*s* and 3*s* orbitals.

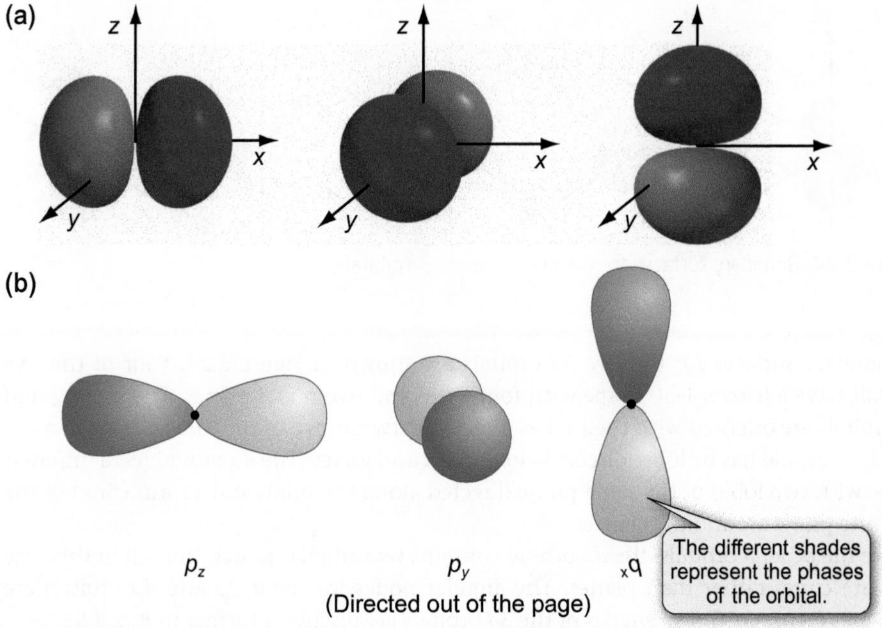
(a)
z
x
y
(b)
p_z
p_y
p_x
(Directed out of the page)
The different shades represent the phases of the orbital.

Figure 3.21 (a) Boundary surfaces of the three hydrogen $2p$ orbitals. The $2p_y$ orbital has the same shape as the $2p_x$ and $2p_z$ orbitals and is shown coming out of the page. (b) Commonly used depictions of the three hydrogen $2p$ orbitals.

that the wavefunction depends on the direction from the nucleus. In addition, in some directions $\cos\theta$ will be positive while in others $\cos\theta$ will be negative.

The shapes of the three $2p$ orbitals are shown in Figure 3.21(a). The three orbitals have the same shape, but differ in their orientations. The $2p_x$ orbital is aligned along the x-axis, the $2p_y$ orbital along the y-axis, and the $2p_z$ orbital along the z-axis. Since the angular wavefunction depends on $\cos\theta$, the boundary surface for a p orbital contains both positive and negative phases. The $2p$ orbitals are often depicted in the simplified form shown in Figure 3.21(b). This emphasizes their directionality, which is important in understanding many chemical reactions, as you will see later in the book.

Sine and cosine functions of an angle are explained in Maths Toolkit MT5 (p.1322).

A p orbital contains a **nodal plane** on which ψ is zero. Any point on the nodal plane has $\psi = 0$. The nodal plane for a $2p$ orbital is shown in Figure 3.22. Since this nodal plane arises from the angular wavefunction it is called an **angular node**.

p orbitals with higher values of n have one or more radial nodes in addition to the angular node. In contrast to s orbitals, this affects their boundary surfaces, though these still have the same general shape as the $2p$ orbitals. The boundary surface for a $3p$ orbital is shown in Figure 3.23.

Interactive 3D animation of *s*, *p*, and *d* orbitals

Box 3.4 The *p* orbitals and complex numbers

How do the p orbitals shown in Figure 3.21 relate to values of m_l?

The p orbitals that are normally used in chemistry are not those that come directly from the Schrödinger equation. This is because the solutions to the Schrödinger equation for $l > 0$ contain complex numbers. Complex numbers contain both real and imaginary parts. The imaginary parts are multiples of i, where i is $(-1)^{1/2}$, the square root of -1. It is possible to take combinations of these complex number solutions so that the imaginary parts cancel out, leaving only real numbers. It is these real solutions that are shown in Figure 3.21.

For the p orbitals

$$p_x = p_{+1} + p_{-1}$$
$$p_y = p_{+1} - p_{-1}$$
$$p_z = p_0$$

where p_{+1}, p_0, and p_{-1} represent the solutions of the Schrödinger equation with $l = 1$ and $m_l = +1$, 0, and -1, respectively.

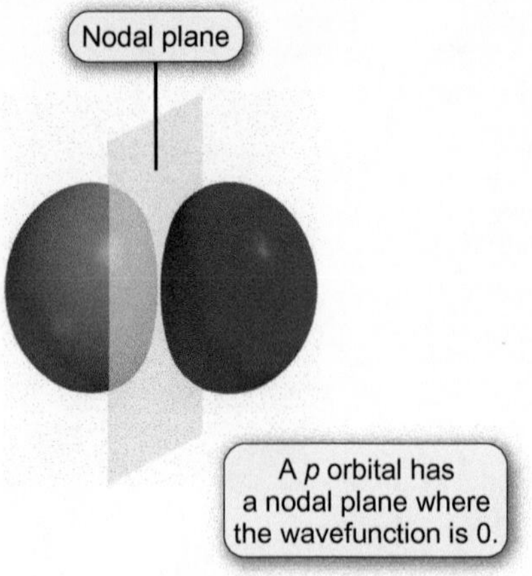
Nodal plane
A p orbital has a nodal plane where the wavefunction is 0.

Figure 3.22 The boundary surface for a hydrogen $2p$ orbital showing the position of the nodal plane.

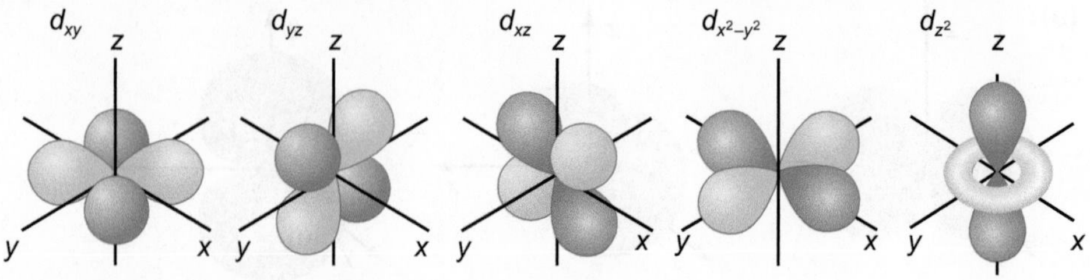
d_{xy}
d_{yz}
d_{xz}
$d_{x^2-y^2}$
d_{z^2}
z
x
y

Figure 3.24 Boundary surfaces for the five hydrogen $3d$ orbitals.

Boundary surfaces for the five $3d$ orbitals are shown in Figure 3.24. Four of the five orbitals have a 'clover-leaf' shape with four lobes and two nodal planes. The d_{xy}, d_{xz}, and d_{yz} orbitals are oriented with their lobes directed between two of the three Cartesian axes. The $d_{x^2-y^2}$ orbital has its lobes directed along the x and y axes. The d_{z^2} orbital has a different shape with two lobes of the same phase directed along the z-axis and a torus (ring) of the opposite phase around the centre.

Like the other d orbitals, the d_{z^2} orbital contains two angular nodes, though in this case they are cones rather than planes. The angular nodes for the $d_{x^2-y^2}$ and d_{z^2} orbitals are shown in Figure 3.25. The shapes of the $3d$ orbitals are discussed further in Box 3.5.

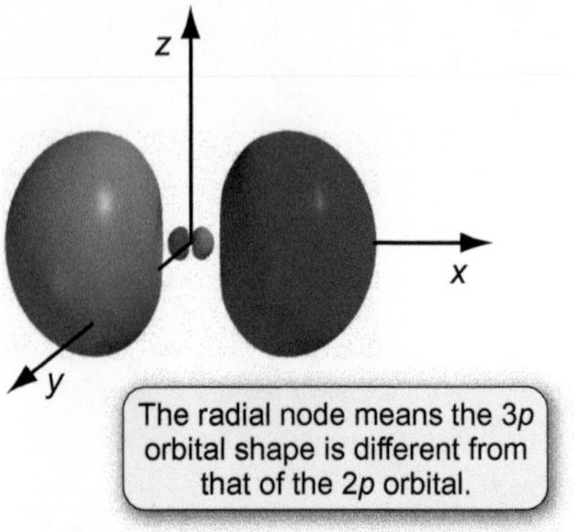
z
x
y
The radial node means the $3p$ orbital shape is different from that of the $2p$ orbital.

Figure 3.23 The boundary surface for the hydrogen $3p_x$ orbital.

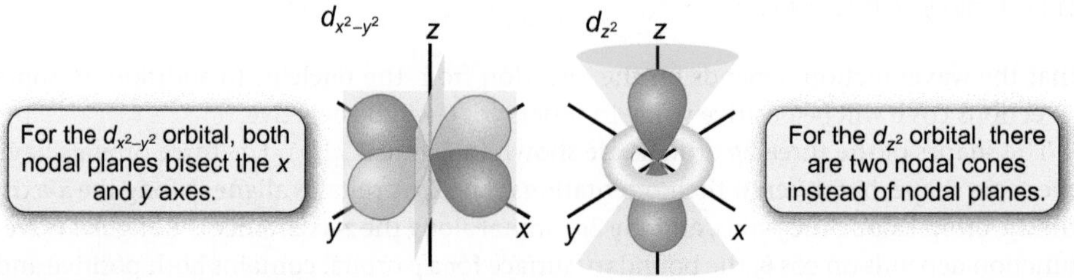
$d_{x^2-y^2}$
d_{z^2}
z
x
y
For the $d_{x^2-y^2}$ orbital, both nodal planes bisect the x and y axes.
For the d_{z^2} orbital, there are two nodal cones instead of nodal planes.

Figure 3.25 Boundary surfaces for the hydrogen $d_{x^2-y^2}$ and d_{z^2} orbitals showing the positions of the angular nodes.

Box 3.5 The shapes of the *d* orbitals

Why does the d_{z^2} orbital have a different shape to the other four d orbitals?

The Schrödinger equation provides six solutions for $l = 2$, but there are only five d orbitals, corresponding to the five allowed values of m_l. Rather than throw away a valid answer, two of the solutions (those corresponding to the $d_{z^2-x^2}$ and $d_{z^2-y^2}$ orbitals) are combined together to give the d_{z^2} orbital.

▼ The unusual shape of the d_{z^2} orbital results from it being a combination of the $d_{z^2-x^2}$ and $d_{z^2-y^2}$ orbitals.

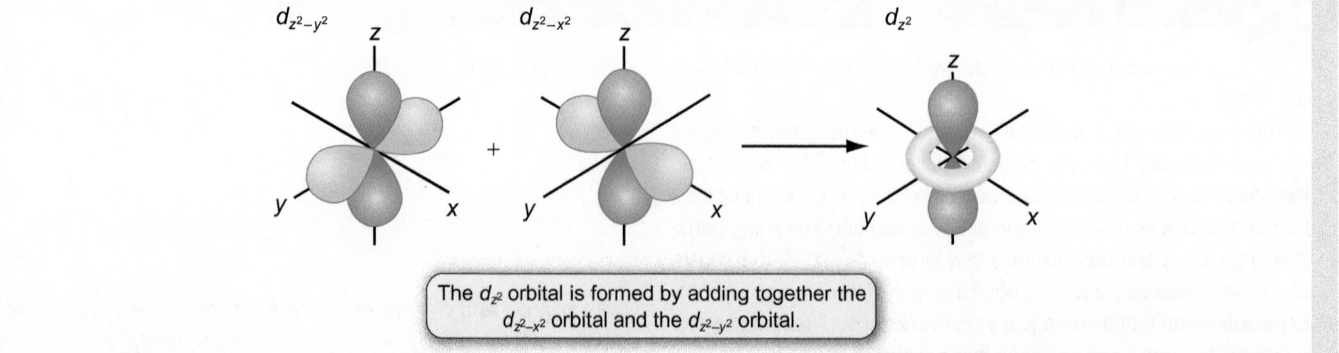
$d_{z^2-y^2}$
$d_{z^2-x^2}$
d_{z^2}
z
x
y
+
The d_{z^2} orbital is formed by adding together the $d_{z^2-x^2}$ orbital and the $d_{z^2-y^2}$ orbital.

Summary

- In wave mechanics, the electron in the hydrogen atom is represented by a wavefunction ψ that is obtained from the Schrödinger equation. The region of space defined by a wavefunction is called an atomic orbital.
- It is impossible to know simultaneously the precise location and momentum of an electron in an atom, though the probability of finding the electron within a volume $d\tau$ is related to ψ^2. This gives the electron density.
- The wavefunction ψ can be divided into a radial part, $R(r)$, and an angular part, $Y(\theta, \phi)$. $R(r)$ tells you how the atomic orbital changes with distance from the nucleus. $Y(\theta, \phi)$ tells you about the shape of the atomic orbital.
- Atomic orbitals are characterized by three quantum numbers, n, l, and m_l.
  - The principal quantum number n can be any positive integer.
  - The secondary quantum number l can be any whole number between 0 and $(n - 1)$.
  - The magnetic quantum number m_l can be any whole number from $-l$ to $+l$.
- The shape of an atomic orbital depends on the value of l.
  - *s* orbitals ($l = 0$) are spherical.
  - *p* orbitals ($l = 1$) have a dumb-bell shape with one angular node.
  - *d* orbitals ($l = 2$) have two angular nodes.

? For practice questions on these topics, see questions 6–10 at the end of this chapter (p.165).

3.6 Many-electron atoms

It is only possible to solve the Schrödinger equation for atoms with one electron. At first sight this is worrying, since it appears to limit the use of the equation to hydrogen and ions such as He^+ and Li^{2+}. Atoms other than hydrogen have more than one electron and these are known as **many-electron atoms**. In this section you will see how approximations are used to allow the solutions to the Schrödinger equation to be applied to many-electron atoms and how their atomic orbitals are occupied.

Greek symbols

Ψ psi (upper case): wavefunction for an *N*-electron atom

ψ psi (lower case): single electron wavefunction

Atomic orbitals in many-electron atoms

Although the Schrödinger equation cannot be solved exactly for many-electron atoms, there are assumptions that allow the results for hydrogen to be used as a starting point for other atoms. The key assumption is known as the **orbital approximation**. This states that the wavefunction for an *N*-electron atom, Ψ, is equal to the product of *N* single-electron wavefunctions. This is shown mathematically in Equation 3.18.

$$\Psi(1, 2, 3, \ldots N) = \psi(1) \times \psi(2) \times \psi(3) \times \ldots \times \psi(N) \qquad (3.18)$$

Since single-electron wavefunctions are the solutions to the Schrödinger equation for hydrogen, the orbital approximation allows you to use the hydrogen atomic orbitals as the basis for explaining the structure of atoms with more than one electron.

While you can use the hydrogen atomic orbitals for other atoms, there is an important difference. For hydrogen, the energy of an orbital depends only on the principal quantum number, *n*. For example, the 2*s* and 2*p* orbitals for hydrogen have the same energy as each other, and are said to be **degenerate**. Similarly, the 3*s*, 3*p*, and 3*d* orbitals are degenerate.

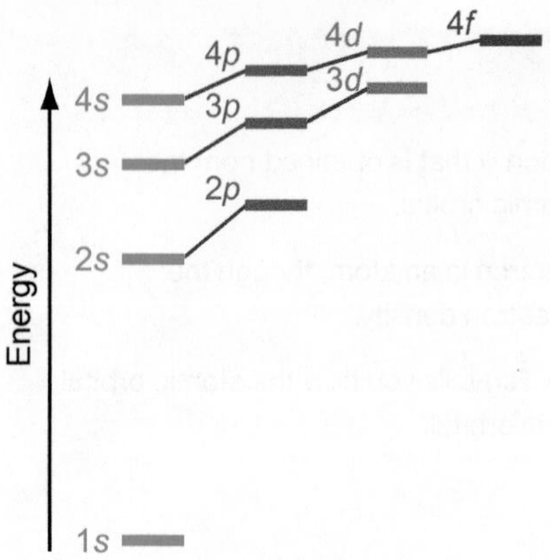

Energy
4s
3s
2s
1s
4p
3p
2p
4d
3d
4f

Figure 3.26 The order of atomic orbital energies for many-electron atoms. The energy of an orbital does not depend solely on *n* as it does for hydrogen, but is also affected by *l*. For $Z \geq 21$, the 3*d* orbitals lie below the 4*s* orbital in energy.

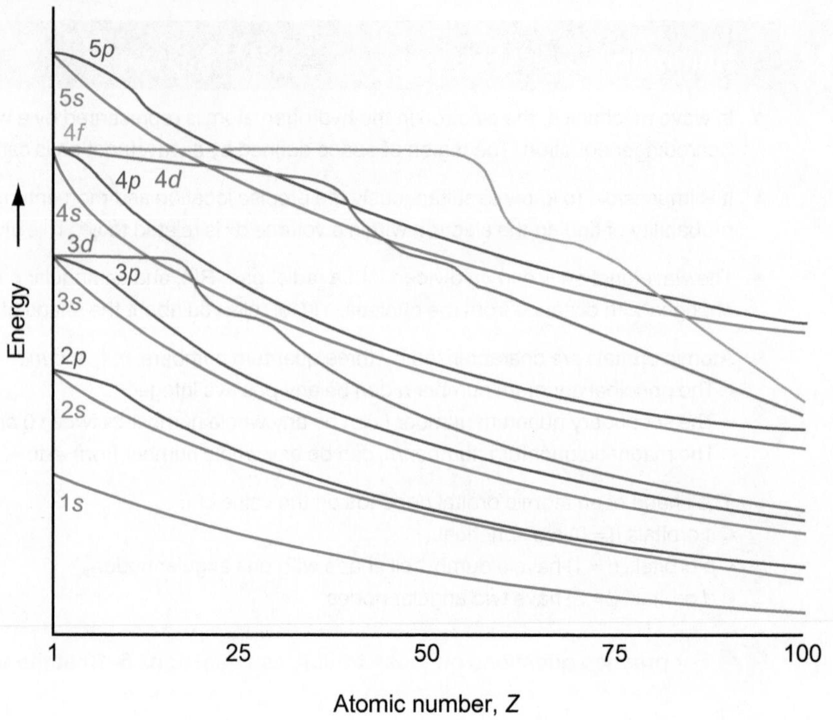

Energy
5p
5s
4f
4p 4d
4s
3d
3p
3s
2p
2s
1s
1
25
50
75
100
Atomic number, Z

Figure 3.27 Variation of the atomic orbital energies with atomic number, *Z*, for many-electron atoms.

The reason for the non-degeneracy of *s*, *p*, and *d* orbitals in many-electron atoms is described on p.150.

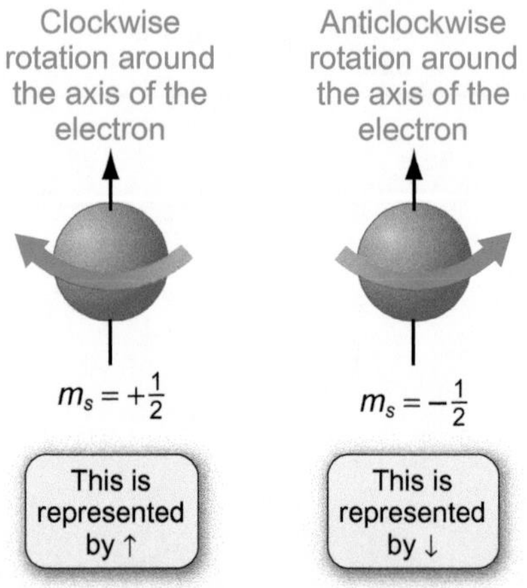

Clockwise rotation around the axis of the electron
Anticlockwise rotation around the axis of the electron
$m_s = +\frac{1}{2}$
$m_s = -\frac{1}{2}$
This is represented by ↑
This is represented by ↓

Figure 3.28 The two electron magnetic spin quantum numbers, $m_s = +\frac{1}{2}$ and $-\frac{1}{2}$, can be represented by clockwise and anticlockwise rotation about the axis of the electron.

The relationship between *s* and m_s is similar to that between *l* and m_l.

This is not the case for many-electron atoms where the lower the value of the secondary quantum number *l*, the lower in energy the orbital.

This means that the 2*s* orbital is of lower energy than the 2*p* orbital. Similarly, the 3*s* orbital is lower in energy than the 3*p* orbital, which in turn is lower than the 3*d* orbital. The relative energies for the orbitals with $n = 1$–4 in a many-electron atom are shown in Figure 3.26. As a consequence of the dependency on *l*, the 4*s* orbital is lower in energy than the 3*d* orbital for some elements.

Although it is tempting to use a single order for the energies of individual atomic orbitals, this would be misleading. This is because the relative energies of the atomic orbitals depend on *Z*. For example, the 4*s* orbital is lower in energy than the 3*d* orbital for K ($Z = 19$) and Ca ($Z = 20$), but is higher in energy for Sc ($Z = 21$) and heavier elements. This is shown in Figure 3.27.

Electron spin

So far you have met three quantum numbers, *n*, *l*, and m_l, which together define an atomic orbital. There is a fourth quantum number, m_s, that relates to the spin of the electron in the orbital.

Spin is a fundamental property of subatomic particles. If you imagine an electron as a sphere, you can think of spin as angular momentum resulting from the electron spinning on its axis in the same way as the Earth spins on its axis as it revolves around the Sun. An electron has a **spin quantum number** *s* of $\frac{1}{2}$. This is an intrinsic value for the electron, and it cannot be changed. In an atom, the electron has two possible spin states, corresponding to clockwise and anticlockwise rotation, as shown in Figure 3.28. The **spin magnetic quantum number** m_s represents the direction of spin of the electron, and it can take a value of $+\frac{1}{2}$ or $-\frac{1}{2}$. These two spin values are often represented as ↑ and ↓, respectively.

Box 3.6 Electron spin resonance spectroscopy

Electron spin resonance (ESR) spectroscopy, also known as electron paramagnetic resonance (EPR) spectroscopy, is a technique in which electron spin is detected and measured. ESR spectroscopy only detects radicals, which are molecules containing at least one unpaired electron. As you will see in Chapter 4, most molecules have all their electrons paired, which means they are invisible in ESR spectra as the electron spins of the individual electrons cancel each other out. This makes ESR spectroscopy very powerful as it allows radicals to be detected in the presence of many other substances.

Ionizing radiation (Section 3.8) produces radicals, and ESR analysis of the teeth of people affected by the Chernobyl nuclear reactor disaster in 1986 has been used to assess the radiation dose that they received. Another use of ESR spectroscopy is in determining whether a food has been irradiated. Irradiation is a process in which food is treated with ionizing radiation in order to kill micro-organisms and prolong shelf lives. In the European Union, irradiation is only permitted for certain foodstuffs, and irradiated food must be clearly labelled. Irradiation leads to an increase in the number of radicals in solid and dry parts of the food, such as crystalline sugar or cellulose. These radicals can be detected by ESR spectroscopy. ESR spectroscopy is also discussed in Section 10.7 (p.496).

The irradiated strawberries (left) were exposed to radiation after they were picked, which killed any micro-organisms. Irradiation can prolong their shelf life by 14 days. ➤

The aufbau principle

Aufbau is German for 'building up', and the **aufbau principle** involves building up the electronic structure of an atom by filling the lowest energy orbitals first. This method can be used to determine the **electronic configuration** for a particular atom.

$Z = 1$ (hydrogen)

The lowest energy state (ground state) of hydrogen is when the single electron is in the 1*s* orbital. The electronic configuration for hydrogen is written as $1s^1$ with the superscript showing the number of electrons in the orbital. This electron can have a spin magnetic quantum number (m_s) of either $+\frac{1}{2}$ or $-\frac{1}{2}$. Both these states normally have the same energy.

H $1s^1$

The electronic configuration of a many-electron atom is a way of representing which orbitals the electrons occupy in the ground state.

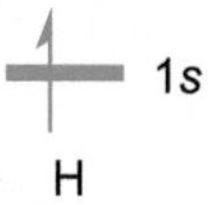
1s
H

$Z = 2$ (helium)

A helium atom contains two electrons, and both of these are in the 1*s* orbital, giving an electronic configuration of $1s^2$. However, the two electrons have different values of m_s. In 1925 the Austrian scientist Wolfgang Pauli proposed that no two electrons in an atom could have the same four quantum numbers. This is now known as the **Pauli exclusion principle**. Since any particular orbital is defined by the three quantum numbers *n*, *l*, and m_l, a consequence of the Pauli exclusion principle is that a maximum of two electrons can reside in any orbital. Furthermore, when there are two electrons in an orbital, as in the helium 1*s* orbital, these electrons must have different values of m_s. In cases such as this, the spins are said to be paired.

The **Pauli exclusion principle** states no two electrons in the same atom can have the same four quantum numbers.

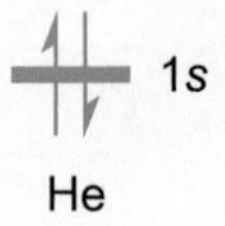
1s
He

Since there is only one orbital with the principal quantum number $n = 1$, there can be only two electrons in this shell. One of these electrons has the set of quantum numbers $n = 1, l = 0, m_l = 0, m_s = +\frac{1}{2}$, while the other has $n = 1, l = 0, m_l = 0, m_s = -\frac{1}{2}$.

$$\text{He} \qquad 1s^2$$

$Z = 3$ (lithium)

A lithium atom contains three electrons. It is not possible for all of these to be in the 1*s* orbital as this would violate the Pauli exclusion principle. The third electron must go into the next lowest energy available orbital, which is the 2*s* orbital. The electronic configuration of lithium is therefore written

$$\text{Li} \qquad 1s^2\ 2s^1$$

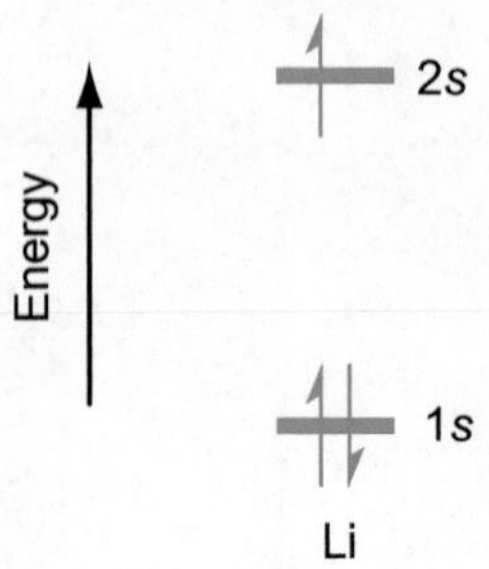
2s
Energy
1s
Li

The 1*s* and 2*s* electrons in lithium are chemically very different. The electron in the 2*s* orbital is readily lost to form the Li^+ ion, while the two electrons in the 1*s* orbital are not involved in any reactions. Generally, only electrons in the highest energy orbitals are involved in chemistry. These electrons are known as the **valence electrons**. In contrast, the electrons in the lower energy orbitals, typically those with lower values of *n*, are known as **core electrons.**

Since the filled $n = 1$ shell corresponds to the electronic configuration of helium, the electronic configuration of lithium is also written

$$\text{Li} \qquad [\text{He}]\ 2s^1$$

This 'shorthand' notation, using the filled shell electronic configuration of a Group 18 element, becomes more useful with higher values of *Z*.

$Z = 4$ (beryllium)

For the next element, beryllium, the additional electron goes into the 2*s* orbital. The two electrons in this orbital have their spins paired.

$$\text{Be} \qquad 1s^2\ 2s^2 \qquad \text{or} \qquad [\text{He}]\ 2s^2$$

$Z = 5$ (boron)

The 2*s* orbital is now filled, and the extra electron in boron goes into the next available orbital. This is one of the three 2*p* orbitals, all of which are degenerate. The electronic configuration of boron is

$$\text{B} \qquad 1s^2\ 2s^2\ 2p^1 \qquad \text{or} \qquad [\text{He}]\ 2s^2\ 2p^1$$

$Z = 6$–10 (carbon, nitrogen, oxygen, fluorine, and neon)

A total of six electrons can go into the three 2*p* orbitals, so the electronic configurations of the next five elements—carbon, nitrogen, oxygen, fluorine, and neon—are as follows.

C	$1s^2\ 2s^2\ 2p^2$	or	$[\text{He}]\ 2s^2\ 2p^2$
N	$1s^2\ 2s^2\ 2p^3$	or	$[\text{He}]\ 2s^2\ 2p^3$
O	$1s^2\ 2s^2\ 2p^4$	or	$[\text{He}]\ 2s^2\ 2p^4$
F	$1s^2\ 2s^2\ 2p^5$	or	$[\text{He}]\ 2s^2\ 2p^5$
Ne	$1s^2\ 2s^2\ 2p^6$	or	$[\text{He}]\ 2s^2\ 2p^6$

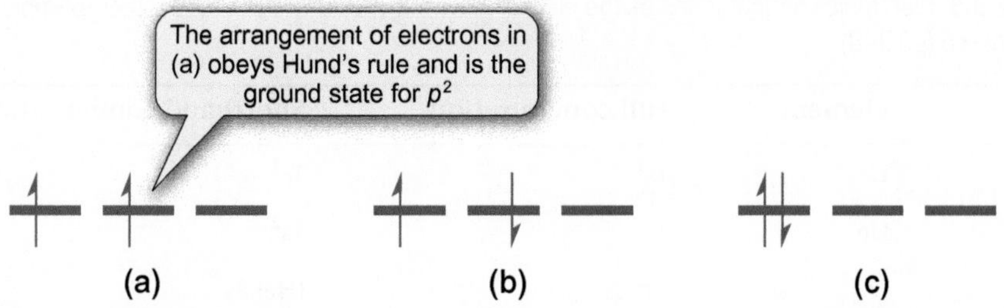
The arrangement of electrons in (a) obeys Hund's rule and is the ground state for p^2
(a)
(b)
(c)

Figure 3.29 Arrangements (a), (b), and (c) represent three possible ways of arranging two electrons in the *p* orbitals. Using Hund's rule, arrangement (a) has the lowest energy as it has the greatest number of parallel electrons.

Electronic configurations written in this way are very convenient and widely used, but they do not tell the whole story. Figure 3.29 shows three possible arrangements for two electrons in the 2*p* orbitals, none of which break the Pauli exclusion principle. All of these arrangements are allowed, but they do not all have the same energy. Which of these is the ground state?

The ground state is the arrangement with the maximum number of parallel electrons, a principle that is known as **Hund's rule** after the German scientist Friedrich Hund who originally proposed it. There are two reasons why Hund's rule applies. Firstly, parallel electrons give rise to less electron–electron repulsion than paired electrons. Secondly, there is an increase in stability from having pairs of parallel electrons, something that is known as **exchange energy** (see Box 3.7). These two factors lead to arrangement (a) in Figure 3.29 being of lower energy than either arrangement (b) or (c).

Hund's rule states that, when filling a set of degenerate orbitals, electrons are added with parallel spins to different orbitals rather than pairing two electrons in one orbital.

Box 3.7 Exchange energy

Exchange energy is a quantum mechanical effect that arises due to parallel electrons being indistinguishable and interchangeable, and it leads to stabilization of configurations that contain parallel electrons. The more pairs of parallel electrons there are present in an atom, the greater the exchange energy. For example, a p^2 configuration obeying Hund's rule has one pair of parallel electrons, whereas a p^3 configuration has three pairs of parallel electrons, in the p_x and p_y orbitals, p_x and p_z orbitals, and p_y and p_z orbitals, respectively.

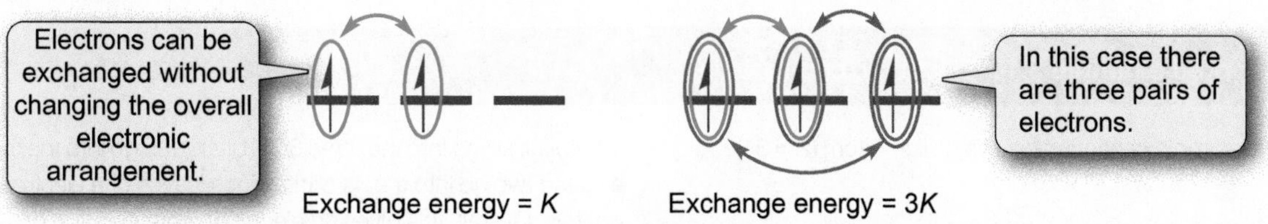
Electrons can be exchanged without changing the overall electronic arrangement.
Exchange energy = K
Exchange energy = $3K$
In this case there are three pairs of electrons.

High values of exchange energy occur for half-filled shells (p^3, d^5, and f^7 configurations), and this explains why these arrangements are relatively stable.

Question

Given that the exchange energy for a pair of parallel electrons is K, calculate the exchange energy for the electronic configuration $3d^4$.

The electronic configuration of oxygen is shown in Figure 3.30. In this representation, the energy increases (becomes less negative) up the *y*-axis. The electronic configurations for elements with $Z = 1$ to $Z = 20$ are summarized in Table 3.5.

The aufbau principle, used in combination with the Pauli exclusion principle and Hund's rule, allows the electronic configuration of any element to be predicted, and Worked example 3.10 provides an example of this.

Visit the Online Resource Centre to view screencast 3.3 which walks you through how to determine the electronic configuration of an atom.

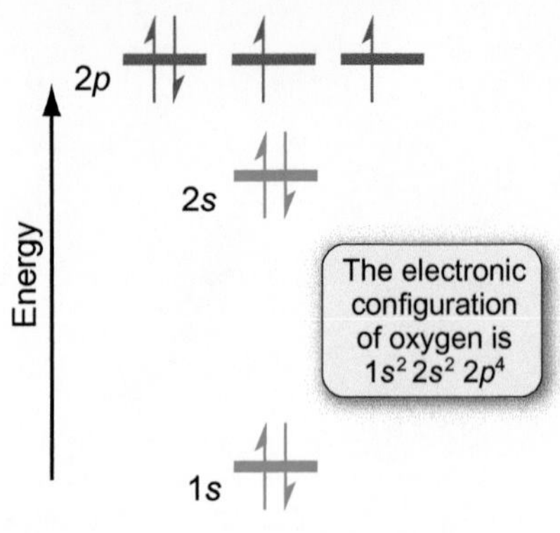
2p
2s
1s
Energy
The electronic configuration of oxygen is $1s^2\,2s^2\,2p^4$

Figure 3.30 The electronic configuration of a ground state oxygen atom. The atom has two unpaired electrons.

Table 3.5 Electronic configurations of the elements with $Z = 1$ to $Z = 20$. For further examples see Appendix 6 (p.1348).

Z	Element	Full configuration	Shorthand configuration
1	H	$1s^1$	$1s^1$
2	He	$1s^2$	$1s^2$
3	Li	$1s^2\,2s^1$	[He] $2s^1$
4	Be	$1s^2\,2s^2$	[He] $2s^2$
5	B	$1s^2\,2s^2\,2p^1$	[He] $2s^2\,2p^1$
6	C	$1s^2\,2s^2\,2p^2$	[He] $2s^2\,2p^2$
7	N	$1s^2\,2s^2\,2p^3$	[He] $2s^2\,2p^3$
8	O	$1s^2\,2s^2\,2p^4$	[He] $2s^2\,2p^4$
9	F	$1s^2\,2s^2\,2p^5$	[He] $2s^2\,2p^5$
10	Ne	$1s^2\,2s^2\,2p^6$	[He] $2s^2\,2p^6$
11	Na	$1s^2\,2s^2\,2p^6\,3s^1$	[Ne] $3s^1$
12	Mg	$1s^2\,2s^2\,2p^6\,3s^2$	[Ne] $3s^2$
13	Al	$1s^2\,2s^2\,2p^6\,3s^2\,3p^1$	[Ne] $3s^2\,3p^1$
14	Si	$1s^2\,2s^2\,2p^6\,3s^2\,3p^2$	[Ne] $3s^2\,3p^2$
15	P	$1s^2\,2s^2\,2p^6\,3s^2\,3p^3$	[Ne] $3s^2\,3p^3$
16	S	$1s^2\,2s^2\,2p^6\,3s^2\,3p^4$	[Ne] $3s^2\,3p^4$
17	Cl	$1s^2\,2s^2\,2p^6\,3s^2\,3p^5$	[Ne] $3s^2\,3p^5$
18	Ar	$1s^2\,2s^2\,2p^6\,3s^2\,3p^6$	[Ne] $3s^2\,3p^6$
19	K	$1s^2\,2s^2\,2p^6\,3s^2\,3p^6\,4s^1$	[Ar] $4s^1$
20	Ca	$1s^2\,2s^2\,2p^6\,3s^2\,3p^6\,4s^2$	[Ar] $4s^2$

Worked example 3.10 Electronic configurations

What is the electronic configuration of a gallium atom ($Z = 31$)?

Strategy

Use Figure 3.27 (p.144) to determine the order in which the orbitals are filled. Then use the aufbau principle to put the electrons into the orbitals, starting from the lowest energy orbital.

Solution

From its atomic number, Z, you know that gallium has 31 electrons. Using the aufbau principle, two electrons go into the 1*s* orbital, two go into the 2*s* orbital, six go into the three 2*p* orbitals, two go into the 3*s* orbital, six go into the three 3*p* orbitals, ten go into the five 3*d* orbitals, and two go into the 4*s* orbital. This leaves one electron for the 4*p* orbitals. Overall, this gives an electronic configuration of

$$\text{Ga}\quad 1s^2\,2s^2\,2p^6\,3s^2\,3p^6\,3d^{10}\,4s^2\,4p^1 \quad \text{or} \quad [\text{Ar}]\,3d^{10}\,4s^2\,4p^1$$

Question

What is the electronic configuration of tellurium ($Z = 52$)? Give both the full electronic configuration and the version using the Group 18 element shorthand.

Anomalies in electronic configuration

If you know the relative energies of the atomic orbitals, you can make predictions about the electronic configurations of elements. Some elements, however, have unexpected electronic configurations. These tend to result from interactions between electrons, such as electron–electron repulsions and exchange energy (see Box 3.7).

For an atom of scandium ($Z = 21$), the $3d$ orbitals are lower in energy than the $4s$ orbital, so we might expect an electronic configuration of [Ar] $3d^3$. Instead, scandium has the observed electronic configuration of [Ar] $3d^1\ 4s^2$. Indeed, all of the first row *d*-block elements have electronic configurations in which the $4s$ orbital is occupied, despite it being higher in energy that the $3d$ orbitals. Why is this? Calculations show that for these elements it is favourable to occupy the higher energy $4s$ orbital as this leads to a reduction in electron–electron repulsions. These repulsions are greater when all the electrons are in the *d* orbitals, since these are more compact than the $4s$ orbital. The reduction in electron–electron repulsion provided by occupying the $4s$ orbital overcomes the cost of occupying this higher energy orbital and leads to a lower overall energy for the atom. When an atom of a *d* block element is ionized, a $4s$ electron is lost rather than a $3d$ electron. Ionization removes the electron in the highest energy orbital (see Section 28.1, p.1258).

Chromium ($Z = 24$) has the observed electronic configuration [Ar] $3d^5\ 4s^1$, as opposed to the expected electronic configuration [Ar] $3d^4\ 4s^2$. This can be explained using exchange energy (Box 3.7). The five parallel *d* electrons in the $3d^5\ 4s^1$ configuration give rise to a very high value of exchange energy. For the $3d^4\ 4s^2$ configuration with only four parallel *d* electrons, the exchange energy is considerably lower, and the difference in exchange energy between the two configurations is of higher magnitude than the difference in electron–electron repulsions.

The structure of the Periodic Table

When the elements are arranged in order of increasing atomic number, periodic variations in their properties are observed. The form of the Periodic Table can be understood in terms of the electronic configurations of the elements.

Box 3.8 Atomic numbers and the Periodic Table

Periodic variations in the properties of the elements when ranked in order of their atomic weights (what we now call relative atomic masses) were first noticed in the nineteenth century, before the structure of the atom and the concept of atomic number were known. The most important contribution was made by the Russian chemist Dmitri Mendeleev, who arranged elements in order of their atomic weights so that elements within the same vertical column have similar properties.

Using his chemical knowledge, Mendeleev noticed when the properties of the elements didn't fit the position implied by their atomic weights. He reasoned that a number of elements had yet to be discovered, so he left spaces for them and predicted their properties. When germanium was discovered in 1886, 15 years after Mendeleev's prediction, it was found to have very similar properties to those he had envisaged.

Mendeleev also recognized that, based on its chemical properties, tellurium should go before iodine, and not after it as was suggested by the atomic weights. He assumed that the atomic weight of tellurium had been measured incorrectly and placed it before iodine. Modern measurements confirm that A_r for tellurium is greater than A_r for iodine. Tellurium lies before iodine in the modern Periodic Table (where elements are arranged in order of their

Д. И. Менделеев в домашнем кабинете.
С фот. Ф. И. Блумбаха. 1904 г.

Dmitri Mendeleev constructed the first Periodic Table in 1869. ➤

atomic number rather than relative atomic mass) because it has the lower atomic number.

How were atomic numbers experimentally determined? When fast-moving electrons hit a metal target in an evacuated tube, the target was observed to emit X-rays. An incoming electron causes ionization of a core electron in an atom of the target, and an X-ray is emitted by a valence electron dropping down to the vacancy in the lower energy level.

Henry Moseley was one of the scientists working with Rutherford in Manchester in the early part of the twentieth century. In 1913, he showed that the frequency of the emitted X-rays is proportional to Z^2, where Z is the atomic number of the element in the target. He was able to account for all of the atomic numbers between aluminium ($Z = 13$) and gold ($Z = 79$), and show that there were four missing elements, at $Z = 43$, 61, 72, and 75. The elements with atomic numbers 43 (technetium) and 61 (promethium) are radioactive, and do not occur naturally. The elements with atomic numbers 72 and 75 are hafnium and rhenium, respectively, and these were both discovered in the 1920s. Moseley did not live to see this, as he was killed in the First World War in 1915 aged only 27.

Question

Moseley showed that copper emits X-rays with $\lambda = 1.549 \times 10^{-10}$ m. Calculate the energy of a photon with this wavelength.

For a given value of the principal quantum number *n*, orbital energies increase with increasing values of the secondary quantum number *l*.

You saw earlier how the energies of the atomic orbitals determine the order in which they are filled. You can therefore use this to justify the form taken by the Periodic Table, which is shown in two different ways in Figure 3.31. In the standard representation, given in Figure 3.31(a), the lanthanide and actinide elements in the *f* block are shown separated from the other elements at the bottom of the Periodic Table. In Figure 3.31(b) these elements are shown in their correct position. The Periodic Table can be divided vertically to give **groups**, and the standard way of numbering these is from 1 to 18 (older methods ignore the *d* elements and number from 1 to 8). The Periodic Table can also be divided horizontally to give **periods**.

There are four regions within the Periodic Table as shown by the different colour blocks in Figure 3.31(a). There is a block of two columns of elements on the left (Groups 1 and 2), a block of six columns on the right (Groups 13 to 18), a block of ten columns between these (Groups 3 to 12), and two rows of fourteen elements at the bottom (the lanthanides and actinides). The number of columns in each block (2, 6, 10, and 14) correspond to the number of electrons needed to fill sets of *s*, *p*, *d*, and *f* orbitals. The blocks of the Periodic Table associated with filling these orbitals are referred to as the s block, *p* block, *d* block, and *f* block, respectively. The Periodic Table, showing all the elements together with their atomic numbers and relative atomic masses, is shown on the inside front cover.

The *d* block doesn't start until the fourth period, after the 4*s* orbital has been filled, as the 3*d* orbitals are higher in energy than the 4*s* orbital for potassium and calcium. Similarly, the *f* block, when fitted into its proper position, does not start until the sixth period, after the 6*s* orbital has been filled.

Shielding and penetration

For the hydrogen atom, the energies of the atomic orbitals are related only to the principal quantum number, *n*, and not to the secondary quantum number, *l*. So, for hydrogen, the 2*s* and 2*p* orbitals are degenerate. This is not the case for many-electron atoms where the 2*s* orbital has a lower energy than the 2*p* orbital. Why is there this difference?

In an atom with more than one electron, there are electron–electron repulsions in addition to the attractive interactions between the nucleus and the electrons. These electron–electron repulsions serve to **shield** the electrons from the full nuclear charge. In lithium, for example, the two electrons in the 1*s* orbital lie largely between the nucleus and the 2*s* electron. This means that the charge felt by the 2*s* electron, known as the **effective nuclear charge**, Z_{eff}, is considerably less than the actual nuclear charge of +3.

A 2*p* electron is more effectively shielded than a 2*s* electron in a many-electron atom. This can be understood by considering the radial distribution functions for the two

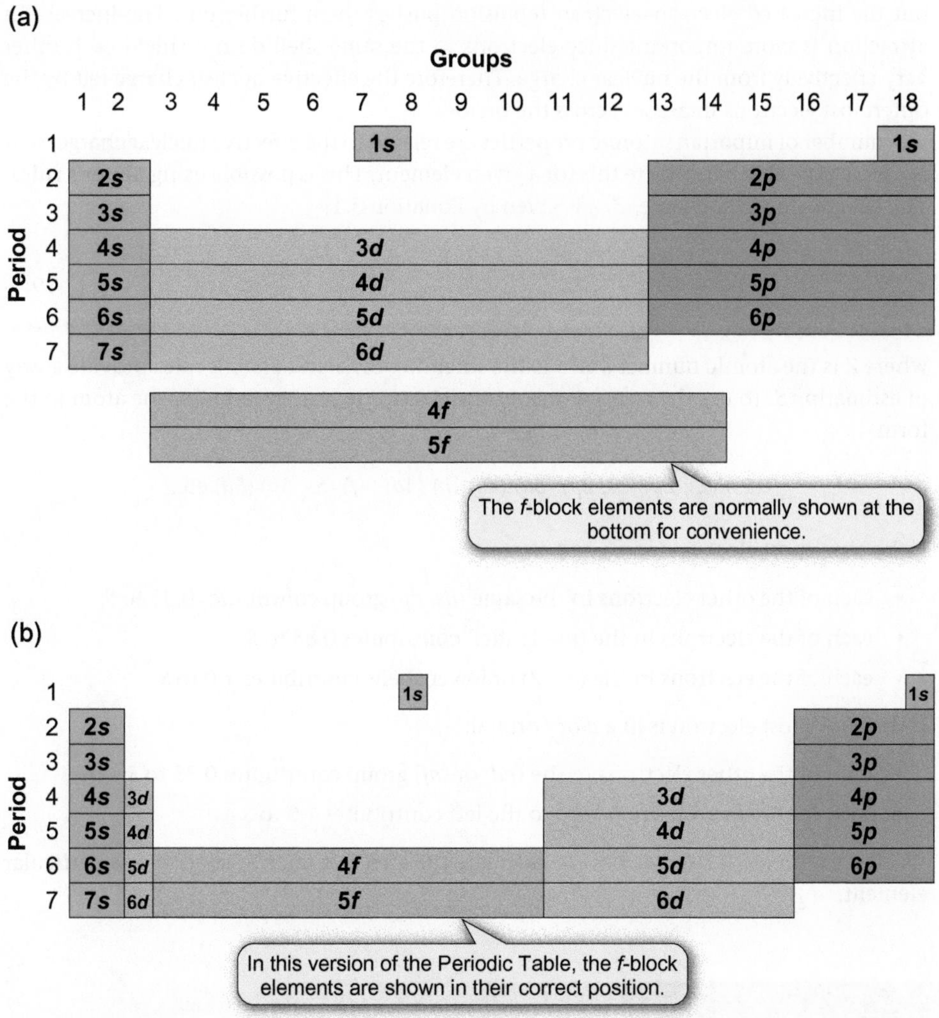
(a)
Groups
1 2 3 4 5 6 7 8 9 10 11 12 13 14 15 16 17 18
Period
1s
1s
2s
3s
4s
5s
6s
7s
3d
4d
5d
6d
2p
3p
4p
5p
6p
4f
5f
The f-block elements are normally shown at the bottom for convenience.
(b)
In this version of the Periodic Table, the f-block elements are shown in their correct position.

Figure 3.31 The structure of the Periodic Table showing the orbitals that are filled across the periods. (a) The 'standard' version; (b) the 'proper' version, with the *f*-block elements inserted into their rightful position. In both versions, the *s* block is shown in red, the *p* block in blue, the *d* block in green, and the *f* block in purple. Hydrogen is not assigned to a particular group (see Chapter 25).

orbitals. Figure 3.32 shows the 1*s*, 2*s*, and 2*p* radial distribution functions on the same scale. The 2*s* orbital has a radial node whereas the 2*p* orbital does not. The presence of this radial node means that there is an area of electron density relatively close to the nucleus for the 2*s* electron. This electron density is largely absent for a 2*p* electron. The 2*s* electron is said to **penetrate** the 1*s* electrons and thus a 2*s* electron feels a higher effective nuclear charge than a 2*p* electron. This means that the 2*s* orbital is stabilized more than the 2*p* orbital. Generally, *s* electrons are more penetrating than *p* electrons, and *p* electrons are more penetrating than *d* electrons due to the shapes of their radial distribution functions.

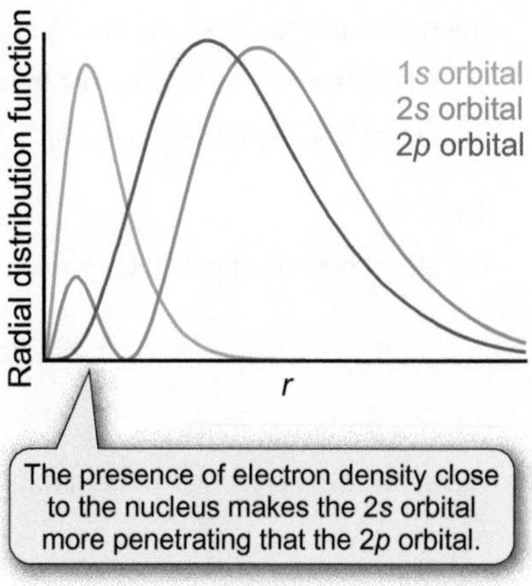
Radial distribution function
r
1s orbital
2s orbital
2p orbital
The presence of electron density close to the nucleus makes the 2s orbital more penetrating that the 2p orbital.

Figure 3.32 The radial distribution functions for the 1*s*, 2*s*, and 2*p* orbitals. The region of electron density close to the nucleus for the 2*s* orbital means this orbital is more penetrating than the 2*p* orbital.

Effective nuclear charge and Slater's rules

The 2*s* and 2*p* orbitals both become more stable (lower in energy) from left to right across the second period, from lithium to neon. Increasing the atomic number by one leads to one more proton in the nucleus and one more electron in the next available orbital with the lowest energy. The increased nuclear charge pulls the orbitals closer to the nucleus, but the increased electron–electron repulsion pushes them further out. The increase in attraction is more important since electrons in the same shell do not shield each other very effectively from the nuclear charge. Therefore the effective nuclear charge felt by the outermost electrons increases across the period.

A number of important atomic properties are related to the effective nuclear charge, so it is useful to be able to estimate this for a given element. This is possible using **Slater's rules**. The effective nuclear charge, Z_{eff}, is given by Equation 3.19

$$Z_{eff} = Z - S \qquad (3.19)$$

where Z is the atomic number and S is the shielding constant. Slater's rules provide a way of estimating S. To use the rules, write out the electronic configuration of the atom in the form

(1*s*) (2*s*, 2*p*) (3*s*, 3*p*) (3*d*) (4*s*, 4*p*) (4*d*) (4*f*) (5*s*, 5*p*) (5*d*) etc.

If the outermost electron is an *s* or *p* electron:

- each of the other electrons in the same (*ns*, *np*) group contributes 0.35 to S;
- each of the electrons in the ($n-1$) shell contributes 0.85 to S;
- each of the electrons in the ($n-2$) or lower shells contributes 1.0 to S.

If the outermost electron is in a *d* or *f* orbital:

- each of the other electrons in the (*nd*) or (*nf*) group contributes 0.35 to S;
- each of the electrons in groups to the left contributes 1.0 to S.

Worked example 3.11 shows how to calculate the effective nuclear charge for a particular element.

Worked example 3.11 Effective nuclear charge

What is the effective nuclear charge for fluorine ($Z = 9$)?

Strategy

The first step is to determine the electronic configuration.

Then use Slater's rules to calculate the shielding constant, S.

Finally, use Equation 3.19 to calculate Z_{eff}.

Solution

The electronic configuration for fluorine, written in the form above is

$$(1s^2)\ (2s^2, 2p^5)$$

Each of the electrons in the (2*s*, 2*p*) group contributes 0.35 to S. There are seven electrons in this group, but this includes the electron being considered. An electron cannot shield itself, so this group contributes 6×0.35.

Each of the electrons in the ($1s^2$) group contributes 0.85, that is, 2×0.85.

$$S = (6 \times 0.35) + (2 \times 0.85) = 3.8$$

Using Equation 3.19

$$Z_{eff} = Z - S$$
$$= 9 - 3.8 = 5.2$$

Question

What are the effective nuclear charges for (a) chlorine ($Z = 17$) and (b) bromine ($Z = 35$)?

The effective nuclear charges for elements with $Z \leq 18$, calculated using Slater's rules, are shown in Figure 3.33. Slater's rules do not take into account the stability resulting from maximizing the numbers of parallel electrons, nor do they distinguish between *s* and *p* electrons. Despite these simplifications, they are useful in explaining the *general* trends in atomic properties such as ionization energies.

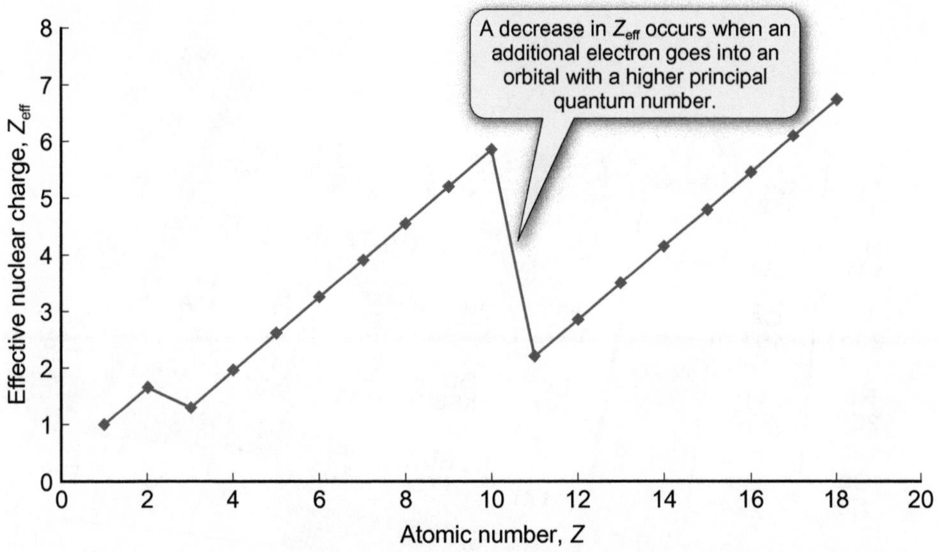
A decrease in Z_{eff} occurs when an additional electron goes into an orbital with a higher principal quantum number.
Effective nuclear charge, Z_{eff}
Atomic number, Z
0
1
2
3
4
5
6
7
8
0
2
4
6
8
10
12
14
16
18
20

Figure 3.33 The variation in the effective nuclear charge (Z_{eff}) for the elements hydrogen to argon. Values are calculated using Slater's rules.

Summary

- The energies of the atomic orbitals for hydrogen depend only on n, but for other atoms electron–electron repulsions mean that orbitals with different values of l have different orbital energies. The lower the value of l, the lower the orbital energy.
- Each atomic orbital can hold two electrons, one with the spin magnetic quantum number $m_s = +\frac{1}{2}$, the other with $m_s = -\frac{1}{2}$.
- In a many-electron atom, the orbitals are filled using the aufbau principle, with the lowest energy orbital filled first, in combination with the Pauli exclusion principle, which states that no two electrons in an atom can have the same four quantum numbers.
- Orbitals of the same energy are filled using Hund's rule, with the number of parallel electrons maximized.
- Slater's rules allow the effective nuclear charge for an electron in an atom to be calculated.

For practice questions on these topics, see questions 11–16 at the end of this chapter (p.165–166).

3.7 Atomic properties and periodicity

Patterns in the properties of the elements were fundamental in the construction of the Periodic Table, long before the electronic basis of it was understood. Trends in chemical and physical properties across periods or down groups are the basis of **periodicity**. This section examines trends in atomic and ionic radii, followed by ionization energies and electron gain energies.

Atomic and ionic radii

It is difficult to measure the size of an individual atom because of the wave nature of the electrons. The electron density doesn't simply stop at a certain distance from the nucleus, but instead fades away gradually. However, when atoms pack together in solids, or form bonds in molecules, their nuclei are generally found at characteristic distances from each other.

The **atomic radius** of an element is defined as half of the distance between the nuclei of neighbouring atoms in the pure element. If the element is a metal, this distance is also known as the **metallic radius**. If the element is a non-metal (excluding Group 18), the

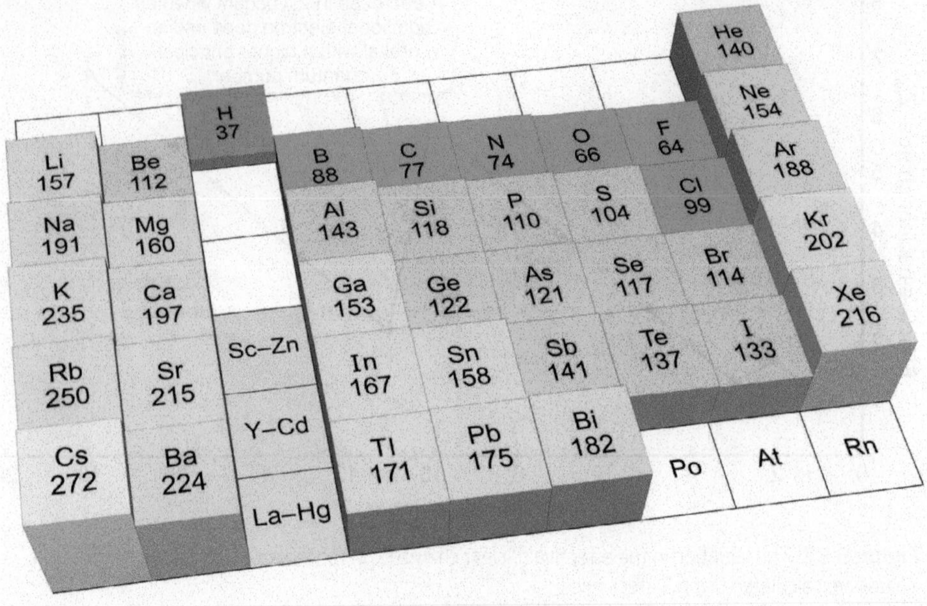
He
140
Ne
154
H
37
Li
157
Be
112
B
88
C
77
N
74
O
66
F
64
Ar
188
Na
191
Mg
160
Al
143
Si
118
P
110
S
104
Cl
99
Kr
202
K
235
Ca
197
Ga
153
Ge
122
As
121
Se
117
Br
114
Xe
216
Rb
250
Sr
215
Sc–Zn
In
167
Sn
158
Sb
141
Te
137
I
133
Cs
272
Ba
224
Y–Cd
Tl
171
Pb
175
Bi
182
Po
At
Rn
La–Hg

Figure 3.34 The atomic radii of the *s*-block and *p*-block elements. Values are in pm.

distance is known as the **covalent radius.** The variation of the atomic radius with the atomic number for the main group elements, those with *s*- or *p*-valence electrons, is shown in Figure 3.34. The observed decrease in size across the periods from Li to F and from Na to Cl is consistent with the increase in effective nuclear charge shown in Figure 3.33. As Z_{eff} increases, the electrons are pulled closer to the nucleus and consequently the atomic radius decreases.

This trend does not continue into the Group 18 elements. These elements are monatomic gases; they do not form molecules so they do not have covalent radii. Therefore, the atomic radii plotted for these elements are the **van der Waals radii**. The van der Waals radius for an atom is half the distance between non-bonded nuclei of neighbouring atoms in the structure of the solid. Since there are no bonds between the atoms in these structures, van der Waals radii are considerably larger than covalent radii.

The effective nuclear charge, as calculated using Slater's rules (p.151), decreases slightly down a group for the first few elements before reaching a plateau. However, atomic radii generally increase down groups. As a group is descended, the outermost electron lies in a larger orbital, for which the maximum in the radial distribution function lies further from the nucleus. In Group 1, for example, the outermost electron for lithium is in the 2*s* orbital, for sodium in the 3*s* orbital, and for potassium in the 4*s* orbital. This is the main factor in explaining the general increase in atomic radii down a group.

Ionic radii are discussed in more detail in Section 6.5 (p.286). Atomic and ionic radii for selected elements are given in Appendix 8 (p.1363).

The **ionic radius** of an element can be determined from the internuclear distance in ionic solids. For example, the distance between the centres of Na^+ and Cl^- ions in sodium chloride is the sum of the cationic radius of sodium and the anionic radius of chlorine. Since electrons are lost when atoms are converted into cations, the cationic radius of an element is smaller than its atomic radius. Conversely, when electrons are added to atoms to convert them into anions there is greater electron–electron repulsion, and anionic radii are larger than atomic radii. Both cationic and anionic radii follow the same trends as atomic radii: the radii decrease across a period and increase down a group.

Ionization energies

The **ionization energy** of an element is defined as the energy change when an electron is removed from an atom of the element in the gas phase. For the element X the ionization energy (I_1) is the energy change for the process in Equation 3.20

$$X(g) \rightarrow X^+(g) + e^-(g) \qquad (I_1) \qquad (3.20)$$

Values are normally given in kJ mol^{-1} (i.e. the energy change when each atom in a mole of atoms of the element loses one electron). Since, for most elements, there are still electrons remaining in the X^+ ion, the process in Equation 3.20 is known as the **first ionization energy**. Each element has Z successive ionization energies, where Z is the atomic number of that element. For example, the second ionization energy (I_2) for X is the energy required for the process in Equation 3.21

$$X^+(g) \rightarrow X^{2+}(g) + e^-(g) \qquad (I_2) \tag{3.21}$$

All ionization reactions are *endothermic*—they require an input of energy to occur, so all have positive values. Each successive ionization energy for an element is always greater than the previous one, as the electron is being removed from a progressively more positively charged ion.

The first eight ionization energies for the elements with $Z = 1$ to $Z = 20$ are given in Table 3.6, and Figure 3.35 shows the first ionization energy, I_1, for the *s*-block and *p*-block elements. There is a general increase in I_1 from left to right across each period which is consistent with the increases in effective nuclear charge. This is shown graphically in Figure 3.36 for the elements of the second and third periods. The electrons are held more tightly going from lithium to neon, and from sodium to argon, so it takes more energy to remove them.

Table 3.6 Ionization energies (kJ mol^{-1}) for the elements with $Z = 1$ to $Z = 20$. For a particular atom, the red line represents removal of an electron from a lower shell, hence the large jump in the observed value of I

Element	I_1	I_2	I_3	I_4	I_5	I_6	I_7	I_8
H	1312							
He	2372	5251						
Li	520	7298	11815					
Be	900	1757	14849	21007				
B	801	2427	3660	25026	32827			
C	1086	2353	4620	6223	37831	47277		
N	1402	2856	4578	7475	9445	53267	64360	
O	1314	3389	5300	7469	10990	13327	71331	84078
F	1681	3374	6050	8408	11023	15164	17868	92038
Ne	2081	3952	6122	9371	12177	15238	19999	23070
Na	496	4562	6910	9543	13354	16613	20117	25496
Mg	738	1451	7733	10543	13630	18020	21711	25661
Al	578	1817	2745	11577	14842	18379	23326	27466
Si	787	1577	3232	4356	16091	19806	23780	29287
P	1012	1907	2914	4964	6274	21267	25431	29872
S	1000	2252	3357	4556	7004	8496	27107	31720
Cl	1251	2298	3822	5159	6540	9362	11018	33604
Ar	1521	2666	3931	5771	7238	8781	11995	13842
K	419	3052	4420	5877	7975	9590	11343	14944
Ca	590	1145	4912	6491	8153	10496	12270	14206

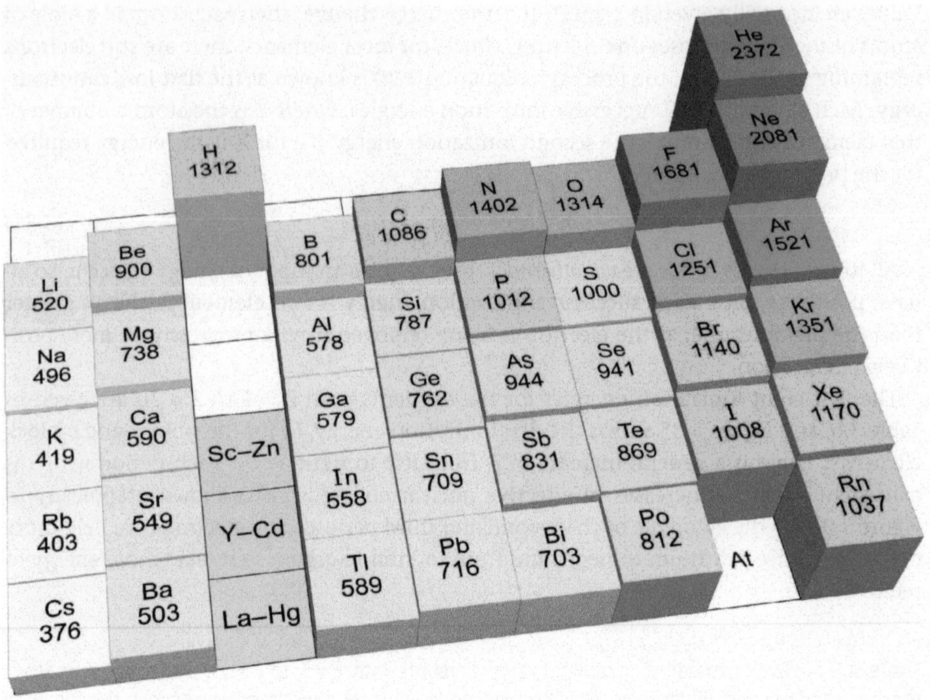
He 2372
H 1312
N 1402
O 1314
F 1681
Ne 2081
C 1086
B 801
Be 900
Li 520
Ar 1521
Cl 1251
S 1000
P 1012
Si 787
Al 578
Mg 738
Na 496
Kr 1351
Br 1140
Se 941
As 944
Ge 762
Ga 579
Ca 590
K 419
Xe 1170
I 1008
Te 869
Sb 831
Sn 709
In 558
Sc–Zn
Sr 549
Rb 403
Rn 1037
Po 812
Bi 703
Pb 716
Tl 589
Y–Cd
La–Hg
Ba 503
Cs 376
At

Figure 3.35 The first ionization energies for the *s*-block and *p*-block elements. Values are in kJ mol^{-1}.

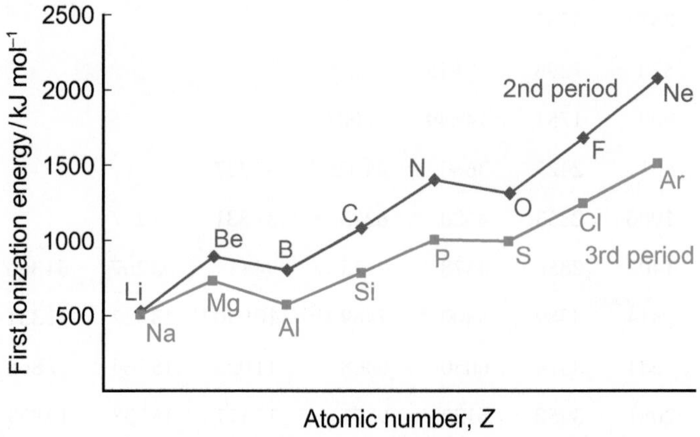
2500
2000
1500
1000
500
First ionization energy/kJ mol⁻¹
Atomic number, Z
2nd period
3rd period
Li
Be
B
C
N
O
F
Ne
Na
Mg
Al
Si
P
S
Cl
Ar

Figure 3.36 The variation of the first ionization energies for the elements of the second and third periods.

The ionization energy of a hydrogen atom is discussed in terms of the energy levels involved in Section 3.3 (p.126).

The increasing trend in I_1 across a period is not completely smooth. For the elements of the second period, I_1 decreases from beryllium to boron, and also decreases from nitrogen to oxygen. Both of these exceptions are explained by a closer look at the electronic configurations. The valence electronic configuration of boron is $2s^2\,2p^1$, so the electron lost on ionization is a $2p$ electron. The valence electronic configuration of beryllium is $2s^2$, which means in this case a $2s$ electron is lost on ionization. The $2p$ electron is in a higher energy orbital than the $2s$ electron, so it takes less energy to remove the boron $2p$ electron, despite the increase in Z_{eff} from beryllium to boron.

The valence electronic configuration for oxygen is $2s^2\,2p^4$, so one of the paired $2p$ electrons is removed on ionization. In contrast, the valence electronic configuration for nitrogen is $2s^2\,2p^3$, so one of the parallel electrons is lost. The electron–electron repulsion

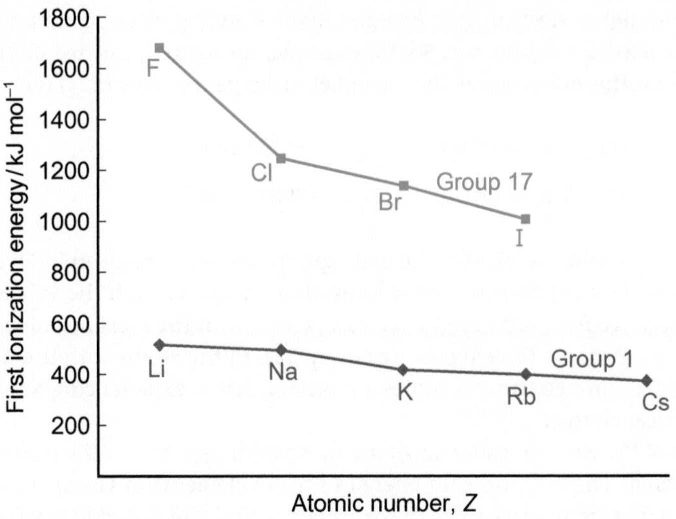
1800
1600
1400
1200
1000
800
600
400
200
First ionization energy / kJ mol⁻¹
F
Cl
Br
I
Group 17
Li
Na
K
Rb
Cs
Group 1
Atomic number, Z

Figure 3.37 The variation of the first ionization energies for the elements of Groups 1 and 17.

is greater between the paired electrons in oxygen as they are in the same orbital, which means the reduction in electron–electron repulsion on ionization is greater for oxygen than it is for nitrogen. This outweighs the increase in Z_{eff} from nitrogen to oxygen. In addition, there is no loss of exchange energy on ionizing an oxygen atom since both O and O^+ contain three parallel $2p$ electrons. In contrast, there is a loss of exchange energy on ionizing a nitrogen atom since a N^+ ion contains only two parallel $2p$ electrons whereas a N atom contains three parallel $2p$ electrons.

The ionization energy generally decreases down a group because the electron is being removed from an atomic orbital that is further away from the nucleus. This is shown graphically for the Group 1 and Group 17 elements in Figure 3.37.

The general trends in ionization energy data have implications for reactivity. The first ionization energies for the Group 1 elements are low, so the *ns* electron is easy to remove from these elements. This accounts for the strongly reducing nature of the alkali metals, all of which reduce water to hydrogen with formation of M^+. The second ionization energy for these metals is invariably very large (Table 3.6) and M^{2+} ions are not observed. The second ionization involves loss of an electron from the lower energy *p* orbital.

For the Group 2 metals the second ionization energy, I_2, while higher than I_1, is relatively low, whereas I_3 is very high. Thus these metals readily form M^{2+} ions, but do not form M^{3+} ions. Looking at Table 3.6, you can see that there is generally a large jump between the *N*th and $(N + 1)$th ionization energy for any element in Group *N*.

The chemistry of the Group 1 and Group 2 elements is described in Section 26.1 (p.1170) and Section 26.6 (p.1183).

In thermodynamic calculations, you will often need to use an enthalpy change of ionization rather than an ionization energy, i.e. $\Delta_i H(1)$ rather than I_1. The differences between these values are small and only important in accurate work. See Sections 6.5 (p.292) and 13.5 (p.642).

Electron gain energies

The **electron gain energy** (E_{eg}) of an element is defined as the energy change that occurs when an electron is attached to an atom in the gas phase. For the element X the electron gain energy is the energy change associated with the process in Equation 3.22

$$X(g) + e^- \rightarrow X^-(g) \quad (E_{eg}) \tag{3.22}$$

Values are normally given in kJ mol^{-1} (i.e. the energy change when each atom in a mole of atoms of the element gains one electron). Further electrons can be added to X^-, so the process in Equation 3.22 is also known as the first electron gain energy of X (E_{eg1}). *Adding an electron to an atom can be either an exothermic or endothermic process*, so values for the first electron gain energy can be positive or negative. An exothermic process is associated with a negative energy change. Hence fluorine, for example, has an electron gain energy of –328 kJ mol^{-1} which means that attachment of an electron to a fluorine atom releases energy.

Second and higher electron gain energies involve adding electrons to anions, so these processes are *always* endothermic. So, for example, for oxygen, the first electron gain energy (E_{eg1}) is exothermic whereas the second electron gain energy (E_{eg2}) is endothermic.

$$O\ (g) + e^- \rightarrow O^-\ (g) \qquad E_{eg1} = -141\ kJ\ mol^{-1}$$

$$O^-\ (g) + e^- \rightarrow O^{2-}\ (g) \qquad E_{eg2} = +798\ kJ\ mol^{-1}$$

Electron affinity (E_a) is a commonly used term that is related to electron gain energy. IUPAC defines electron affinity as the energy **released** on attachment of an electron to an atom. So, in contrast to standard thermodynamic conventions, exothermic processes have positive values of electron affinity and endothermic processes negative values. This means that

$$E_a = -E_{eg}$$

Care is needed in using electron affinity values, as not all sources use the general convention described above. In this book we use electron gain energies instead of electron affinities to avoid ambiguity.

The first electron gain energies for the main group elements are given in Figure 3.38. The trends are less clear cut than those for ionization energy, though the values become increasingly negative from left to right across a period, with the most negative values at the top right of the Periodic Table (ignoring Group 18). In the atoms with the most negative values, the incoming electron occupies a *p* orbital that is experiencing a relatively large effective nuclear charge.

The value of the electron gain energy for an element depends on the energy of the next available orbital. The electron gain energies for the elements of Group 18 are not accurately known, but are positive since the *p* orbitals are full and the additional electron must occupy a much higher energy *s* orbital.

From Figure 3.38 you can see that the electron gain energy becomes more positive from carbon to nitrogen. Why is the addition of an electron to nitrogen an endothermic process, while addition of an electron to carbon is an exothermic process? This can be explained by considering the electronic configurations of the atoms involved. Carbon has a valence shell configuration of $2s^2\ 2p^2$, so the additional electron goes into the unoccupied *p* orbital and leads to an increase in exchange energy. In contrast, nitrogen has the valence shell configuration $2s^2\ 2p^3$ so the incoming electron needs to be paired with an electron already present, and does not lead to an increase in exchange energy. Paired electrons experience more electron–electron repulsion than parallel electrons, so electron–electron repulsion is greater for N^- than for C^-. This outweighs the increase in effective nuclear charge from carbon to nitrogen, and ensures that addition of an electron to nitrogen is unfavourable.

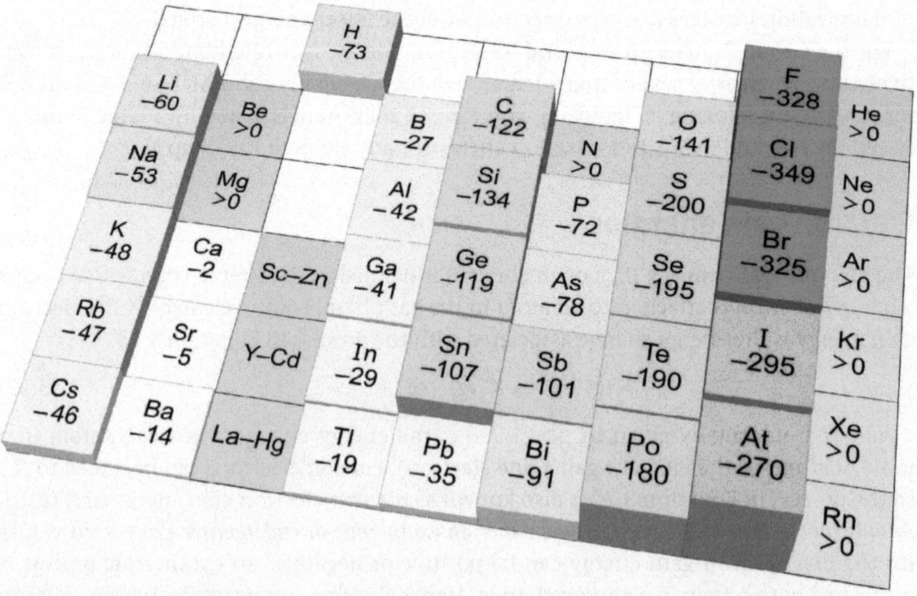

H -73
Li -60
Be >0
B -27
C -122
N >0
O -141
F -328
He >0
Na -53
Mg >0
Al -42
Si -134
P -72
S -200
Cl -349
Ne >0
K -48
Ca -2
Sc–Zn
Ga -41
Ge -119
As -78
Se -195
Br -325
Ar >0
Rb -47
Sr -5
Y–Cd
In -29
Sn -107
Sb -101
Te -190
I -295
Kr >0
Cs -46
Ba -14
La–Hg
Tl -19
Pb -35
Bi -91
Po -180
At -270
Xe >0
Rn >0

Figure 3.38 The first electron gain energies for the *s*-block and *p*-block elements. For the elements with endothermic values, E_{eg1} is not accurately known. Values are in kJ mol^{-1}.

Summary

- Atomic radii tend to decrease from left to right across a period of the Periodic Table, whereas first ionization energies tend to increase across a period. This is due to the increase in the effective nuclear charge on crossing a period.
- Atomic radii tend to increase and first ionization energies tend to decrease down groups of the Periodic Table. This is because the outermost electrons are in atomic orbitals that are located farther away from the nuclei.
- The electron gain energy is the energy change when an electron is added to an atom in the gas phase. First electron gain energies are most negative for elements in the top right of the Periodic Table (excluding Group 18), though the trends are less clear cut than for ionization energies.

? For practice questions on these topics, see questions 17 and 18 at the end of this chapter (p.166).

3.8 Nuclear chemistry

The chemical reactions of the elements and their compounds are related to their electronic configurations. Electrons can be shared between atoms, giving rise to covalent bonding (Chapter 4 and 5), or donated from one atom to another, giving rise to ionic bonding (Chapter 6). In both these processes the nuclei remain unchanged. The alchemist's goal of transmutation—the transformation of base metals, such as lead, into gold—is not possible since the energies involved in chemical reactions are far too small to give changes in nuclear structure.

However, there are a number of processes that do involve nuclear rearrangements. The final part of this chapter briefly considers two aspects of nuclear chemistry—radioactivity and the synthesis of artificial elements.

Radioactivity

Radioactivity (ionizing radiation) was discovered by the French scientist Henri Becquerel in 1896. Becquerel stored photographic plates in the same drawer as a sample of a uranium ore, and was surprised to discover that the plates darkened in a similar manner to observations with the recently discovered X-rays. The new rays did not appear to be related to any external source of energy, so Becquerel reasoned they must be coming from the ore.

The Polish-born scientist Marie Sklodowska Curie, together with her French husband Pierre Curie, spent many years studying the radioactivity produced by ores of uranium and thorium. They found that a uranium ore called pitchblende gave off far more radiation than would be expected, given the amount of uranium present. The Curies searched for the source of this additional radiation and eventually isolated two previously unknown, highly radioactive elements, which they named radium and polonium. These elements were later shown to be the products of the radioactive decay of uranium atoms.

Marie Sklodowska Curie was the first woman to be awarded a Nobel Prize, and the first person to receive two, one for physics and one for chemistry. A lifetime working with radioactive compounds means that even now her laboratory notebooks have to be stored in a lead-lined case.

Ernest Rutherford identified three different types of ionizing radiation given off by radioactive elements, which he termed α-, β-, and γ-radiation. The first two of these were found to consist of charged particles. **α-radiation** consists of positively charged particles comprising two protons and two neutrons. Thus α-particles are identical to helium nuclei. **β-radiation** consists of negatively charged particles, which were later recognized as electrons. **γ-radiation** is a type of electromagnetic radiation with a very high frequency (Figure 3.5, p.119), so consists of very high energy photons. The properties of α-, β-, and γ-radiation are given in Table 3.7. Other types of ionizing radiation have since been identified, but α-, β-, and γ-radiation remain the most important.

i Rutherford used α-particles to probe the structure of the atom (see Section 3.1, p.116).

The radiation emitted by radioactive elements such as uranium is a result of nuclear decay. Although most nuclei are stable, some are not. If a nucleus decays with emission of an α-particle, it loses two protons and two neutrons. As a result, the atomic number

Table 3.7 Properties of ionizing radiation

Type of radiation	Composition	Symbol	Relative mass	Relative charge	Properties
α-radiation	He nuclei	${}^{4}_{2}He$	7350	+2	Poorly penetrating (stopped by a few cm of air) but very harmful if inhaled or ingested
β-radiation	Electrons	${}^{0}_{-1}e$	1	−1	Moderately penetrating (stopped by a few mm of aluminium)
γ-radiation	High energy electromagnetic radiation	$h\nu$	0	0	Very penetrating (stopped by a few cm of lead)

of the element is reduced by two and its mass number is reduced by four. An example of α-emission is the decay of ${}^{238}U$

$$ {}^{238}_{92}U \rightarrow {}^{234}_{90}Th + {}^{4}_{2}He $$

Energy is also emitted since the combined masses of the thorium nuclei and helium nuclei are slightly less than the mass of the uranium nucleus. The lost mass is converted into energy according to Einstein's famous equation (Equation 3.23)

$$ E = mc^2 \tag{3.23} $$

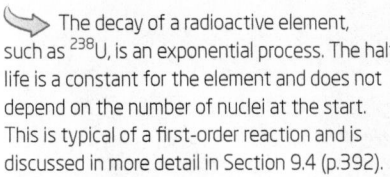

The decay of a radioactive element, such as ${}^{238}U$, is an exponential process. The half life is a constant for the element and does not depend on the number of nuclei at the start. This is typical of a first-order reaction and is discussed in more detail in Section 9.4 (p.392).

The rate of radioactive decay is normally given in terms of a **half life**, $t_{1/2}$. This is the time required for half of the nuclei within a sample to decay. The half life for the decay of ${}^{238}U$ is 4.5×10^9 years. The very long half life explains why uranium is still naturally present on the Earth, despite being unstable.

β-emission involves loss of an electron. There are no electrons present in nuclei, so how can nuclear decay lead to β-emission? For this to happen, a neutron within the nucleus is converted into a proton and an electron. When an element emits a β-particle, the mass number is unchanged but the atomic number *increases* by one. An example of β-emission is the decay of ${}^{14}C$, a naturally occurring isotope of carbon that is used in radiocarbon dating

$$ {}^{14}_{6}C \rightarrow {}^{14}_{7}N + {}^{0}_{-1}e $$

Worked example 3.12 shows how to work out the product from a radioactive decay sequence.

Worked example 3.12 Radioactive decay

What is the product when ${}^{234}_{90}Th$ undergoes a series of two β-emissions followed by two α-emissions?

Strategy

Determine the product of each of the successive reactions, bearing in mind the change in atomic number and mass number caused by α-emission and β-emission.

Solution

The following four nuclear transformations are involved.

i) ${}^{234}_{90}Th \rightarrow {}^{234}_{91}Pa + {}^{0}_{-1}e$ β-emission

ii) ${}^{234}_{91}Pa \rightarrow {}^{234}_{92}U + {}^{0}_{-1}e$ β-emission

iii) ${}^{234}_{92}U \rightarrow {}^{230}_{90}Th + {}^{4}_{2}He$ α-emission

iv) ${}^{230}_{90}Th \rightarrow {}^{226}_{88}Ra + {}^{4}_{2}He$ α-emission

The product of the four transformations is therefore ${}^{226}_{88}Ra$.

Question

${}^{226}_{88}Ra$ is also unstable, and following a series of nine more radioactive intermediates it is eventually converted into the stable ${}^{206}_{82}Pb$ isotope. What combination of emissions occur in the latter part of this process to convert ${}^{214}_{83}Bi$ into ${}^{206}_{82}Pb$?

Box 3.9 Dating the past

The grave of the Bronze Age man known as the Amesbury Archer was discovered in 2002. Box 1.3 (p.17) describes how oxygen isotope analysis was used to suggest that he probably originated from somewhere in the Alps. How do we know that the man was alive in the early Bronze Age?

The age of organic archaeological finds can be determined using radiocarbon dating. Naturally occurring carbon contains 98.9% ^{12}C, 1.1% ^{13}C, and a tiny amount (~1×10^{-10}%) of ^{14}C. ^{14}C is radioactive and has a half life of 5730 years. The ^{14}C in all living tissue decays very slowly, but is constantly replenished through eating and breathing. Once the Archer died he was no longer able to renew the ^{14}C in his body. At this point the amount of ^{14}C in his body started to decrease exponentially with a constant half life. By determining how much ^{14}C was still present, the age of the Archer was determined. Samples from the Amesbury Archer were dated twice to ensure consistency. Both sets of results suggested he was alive sometime between 2400 BC and 2200 BC.

▲ Radiocarbon dating has enabled the dates of the reigns of Egyptian pharaohs such as Khafre to be established more accurately.

The amount of ^{14}C present in a sample is often measured using accelerator mass spectrometry (AMS) as this requires only a few milligrams of sample. AMS works on a similar principle to 'normal' mass spectrometry (see Box 1.2, p.14). After the sample has been chemically prepared, it is ionized, and the ions are accelerated and separated on the basis of their m/z ratio. Very high potentials are used to accelerate the ions, which makes them easier to identify. The detector allows individual atoms to be counted. The age is calculated by comparing the ^{14}C:^{13}C ratio detected with those measured on standards that have been accurately dated by dendrochronology (dating based on the analysis of patterns of tree rings). A team of researchers based in Oxford recently used radiocarbon measurements from AMS to analyse plant samples from the funerary offerings of Egyptian pharaohs. They compared their results with historical records that were based on reign lengths and astronomical observations. This comparison suggested that the New Kingdom in Egypt began between 1570 BC and 1544 BC, earlier than many previous historical estimates.

Question

In what year will the amount of ^{14}C be half its value when the Amesbury Archer was alive? Assume he died in 2300 BC.

Where does the γ-radiation emitted in radioactive decay come from? Just as the transitions between the electronic energy levels in an atom give rise to the emission of photons in the X-ray, ultraviolet, visible, or infrared regions of the electromagnetic spectrum, transitions between energy levels associated with the nucleus give rise to γ-emissions. Sometimes the nuclei produced following α- or β-emission are in excited states. If an excited state is metastable, it may exist for a considerable time before emission of a γ-photon returns it to the ground state.

> ℹ A metastable state is an excited state that has a significant lifetime.

An example of this is γ-emission from an excited metastable state of ^{99}Tc (denoted ^{99m}Tc), which itself is formed from the β-emission of ^{99}Mo

$$^{99}_{42}\text{Mo} \rightarrow {}^{99\text{m}}_{43}\text{Tc} + {}^{0}_{-1}\text{e}$$

$$^{99\text{m}}_{43}\text{Tc} \rightarrow {}^{99}_{43}\text{Tc} + h\nu$$

The use of ^{99m}Tc in medical imaging is described in Box 28.1 (p.1257).

γ-emission often accompanies α- or β-emission.

The transuranic elements

The naturally occurring element with the highest atomic number is uranium (Z = 92). However, from the Periodic Table on the inside front cover of the book you can see that elements with higher values of Z are also known. These are called the **transuranic elements**.

The role of IUPAC is described in Section 1.2 (p.7).

The particle accelerator used at Darmstadt to discover elements 107 to 111.

All transuranic elements are synthetic, as are a few of the lighter elements such as technetium ($Z = 43$). The transuranic elements currently recognized by IUPAC are given in Table 3.8 along with their year of discovery and details of their longest lived isotope. All are radioactive, and their stability generally decreases with increasing atomic number. The elements with atomic numbers 113, 115, and 117–118 were recognized by IUPAC in December 2015, and their discovery has completed the seventh period of the Periodic Table.

Much of the pioneering work on the synthesis of transuranic elements was carried out by Glenn Seaborg and co-workers at Berkeley in the USA. Indeed, Seaborg was the only person ever to have an element named after him while he was alive. Recent discoveries have also been made in Darmstadt, Germany, and in Dubna, Russia. The heaviest elements are made using particle accelerators. In these experiments, small nuclei are accelerated at very high energies into targets containing larger nuclei so that the two nuclei fuse to form

Table 3.8 The transuranic elements

	Z	Year of synthesis	Longest lived isotope	Half life
Neptunium (Np)	93	1940	^{237}Np	2.37×10^6 years
Plutonium (Pu)	94	1940	^{244}Pu	8.08×10^7 years
Americium (Am)	95	1944	^{243}Am	7370 years
Curium (Cm)	96	1944	^{247}Cm	1.56×10^7 years
Berkelium (Bk)	97	1949	^{247}Bk	1380 years
Californium (Cf)	98	1950	^{251}Cf	898 years
Einsteinium (Es)	99	1952	^{252}Es	472 days
Fermium (Fm)	100	1952	^{257}Fm	101 days
Mendelevium (Md)	101	1955	^{258}Md	52 days
Nobelium (No)	102	1958	^{259}No	58 min
Lawrencium (Lr)	103	1961	^{262}Lr	4 h
Rutherfordium (Rf)	104	1964	^{263}Rf	10 min
Dubnium (Db)	105	1967	^{268}Db	32 h
Seaborgium (Sg)	106	1974	^{271}Sg	2.4 min
Bohrium (Bh)	107	1981	^{267}Bh	17 s
Hassium (Hs)	108	1984	^{270}Hs	22 s
Meitnerium (Mt)	109	1982	^{278}Mt	8 s
Darmstadtium (Ds)	110	1994	^{281}Ds	10 s
Roentgenium (Rg)	111	1994	^{281}Rg	26 s
Copernicium (Cn)	112	1996	^{285}Cn	30 s
Nihonium (Nh)	113	2004	^{286}Nh	20 s
Flerovium (Fl)	114	1999	^{289}Fl	2.1 s
Moscovium (Mc)	115	2003	^{289}Mc	220 ms
Livermorium (Lv)	116	2000	^{293}Lv	53 ms
Tennessine (Ts)	117	2010	^{294}Ts	51 ms
Oganesson (Og)	118	2005	^{294}Og	0.9 ms

a new element. For example, rutherfordium was first produced by bombarding a plutonium-242 target with neon-22 nuclei

$$^{242}_{94}\mathrm{Pu} + ^{22}_{10}\mathrm{Ne} \rightarrow ^{260}_{104}\mathrm{Rf} + 4\,^{1}_{0}\mathrm{n}$$

Some of the transuranic elements are technologically very important. For example, plutonium is used to power satellites and americium is used in smoke detectors (Box 3.10).

Box 3.10 Smoke detectors

Smoke detectors save thousands of lives every year, and it is recommended that all homes have one on each floor. A smoke detector consists of two parts: a sensor to sense the smoke and an alarm to wake people up in case of fire. The most commonly used domestic smoke detectors use an ionization chamber and a source of ionizing radiation to detect smoke. Inside the smoke detector is a very small quantity (around 0.0002 g) of the synthetic element americium. The isotope ^{241}Am is used as it has a long half life (432 years), and is a good source of α-particles.

The ionization chamber consists of two plates with a voltage across them, along with the ^{241}Am source. The α-particles generated by the americium ionize the oxygen and nitrogen molecules in the air. The positive ions and liberated electrons are each attracted to the plate of opposite charge and the resultant small current is detected by the electronic circuitry. When smoke enters the ionization chamber, the smoke particles attach to the ions and neutralize them. The circuitry senses the drop in current between the plates that results from this and sets off the alarm.

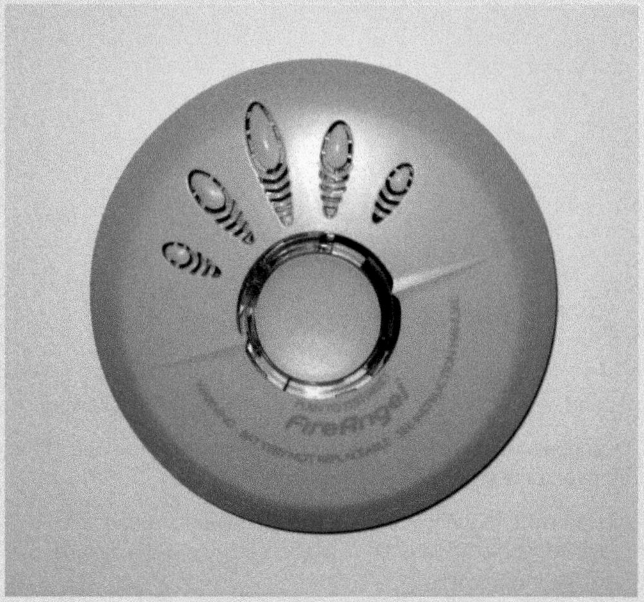
FireAngel
A smoke detector.

Question

Write the nuclear equation for the α-decay of an atom of ^{241}Am.

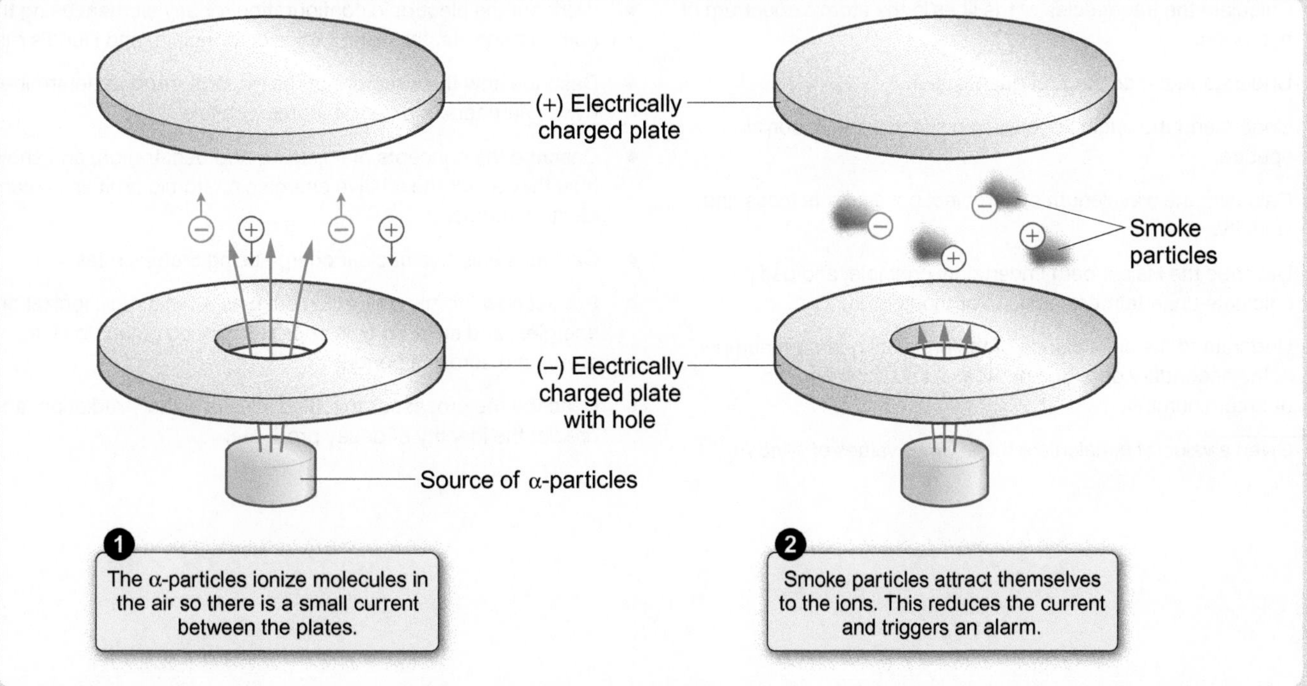
(+) Electrically charged plate
Smoke particles
(–) Electrically charged plate with hole
Source of α-particles
1
The α-particles ionize molecules in the air so there is a small current between the plates.
2
Smoke particles attract themselves to the ions. This reduces the current and triggers an alarm.
How a smoke detector works.

Summary

- There are three main types of ionizing radiation: α-, β-, and γ-radiation. α- and β-radiation consist of charged particles, whereas γ-radiation consists of high-energy photons.
- Elements with atomic number greater than 92 are not naturally occurring and are called transuranic elements. They can be made by colliding nuclei together with very high energies.

For practice questions on these topics, see questions 19 and 20 at the end of this chapter (p.166).

Concept review

By the end of this chapter, you should be able to do the following.

- Describe experiments that helped determine the structure of the atom and the charge on the electron.
- Interconvert frequency and wavelength for electromagnetic radiation, and calculate the energy associated with photons from the frequency.
- Describe experiments that helped show wave–particle duality of photons and electrons.
- Calculate the frequencies of the lines in the atomic spectrum of hydrogen.
- Understand the concept of quantization.
- Understand the origins of emission spectra and absorption spectra.
- Calculate the wavelength of an object based on its mass and velocity.
- Describe the Heisenberg uncertainty principle, and use it to calculate uncertainty in position or momentum.
- Understand the significance of the principal quantum number, *n*, the secondary quantum number, *l*, and the magnetic quantum number, m_l.
- Given a value of *n*, calculate the allowed values of *l* and m_l.
- Describe the significance of ψ, ψ^2, and the radial distribution function.
- Understand that the wavefunction can be divided into a radial part, $R(r)$, and an angular part, $Y(\theta, \phi)$.
- Sketch plots of the radial wavefunction and the radial distribution function for any atomic orbital.
- Sketch the boundary surface for any atomic orbital and understand what it means.
- Work out the electronic configuration for any element using the aufbau principle, the Pauli exclusion principle, and Hund's rule.
- Describe how the structure of the Periodic Table is determined by the numbers of different atomic orbitals.
- Describe the concepts of shielding and penetration, and show how they affect the relative energies of atomic orbitals in many-electron atoms.
- Calculate effective nuclear charge using Slater's rules.
- Predict how atomic properties such as atomic radii, ionization energies, and electron gain energies vary on going down a group and across a period.
- Describe the processes involved in α-, β-, and γ-radiation, and predict the identity of decay products.

Key equations

Relationship between electromagnetic radiation wavelength and frequency	$c = \lambda\nu$	(3.1)
Energy of a photon	$E = h\nu$	(3.2)
Rydberg equation	$\nu = R_H\left[\frac{1}{n_1^2} - \frac{1}{n_2^2}\right]$	(3.6)
de Broglie equation	$\lambda = \frac{h}{mv}$	(3.9)
Radial and angular wavefunction	$\psi(x, y, z) = \psi(r, \theta, \phi) = R(r) \times Y(\theta, \phi)$	(3.13)
Radial distribution function	Radial distribution function $= 4\pi r^2 R(r)^2$	(3.17)
Effective nuclear charge	$Z_{eff} = Z - S$	(3.19)

Questions

More challenging questions are indicated by an asterisk *.

1 What is the energy (in $kJ\,mol^{-1}$) of X-ray photons with a wavelength of 100 pm? (Section 3.2)

2 The Cl–Cl bond in Cl_2 has a bond energy of $242\,kJ\,mol^{-1}$. Assuming that absorption of photons of this energy will break the bond, what is the minimum frequency of electromagnetic radiation that is required? What part of the electromagnetic spectrum does this correspond to? (Section 3.2)

3 What is the wavelength of light for a line in the atomic spectrum of hydrogen for which $n_1 = 2$ and $n_2 = 4$? What part of the electromagnetic spectrum does this correspond to? (Section 3.3)

4 Calculate the wavelengths for the first three lines in the Lyman series. (Section 3.3)

5 What is the wavelength of a helium atom with a velocity of $1.00 \times 10^3\,m\,s^{-1}$? (Section 3.4)

6 Which of the following sets of quantum numbers are allowed? What atomic orbitals do the allowed combinations correspond to? (Section 3.5)

(a) $n = 2, l = 2, m_l = 2$

(b) $n = 5, l = 3, m_l = -2$

(c) $n = 3, l = -1, m_l = 1$

(d) $n = 2, l = 1, m_l = 1$

(e) $n = 4, l = 0, m_l = 1$

7* How many orbitals are possible for $n = 5$? Identify the orbital types giving the number of each. (Section 3.5)

8 The value of m_l for a particular orbital is –2. What are the smallest possible values for n and l? (Section 3.5)

9 For the following atomic orbitals, give the values of the quantum numbers n and l. In each case indicate what values for m_l are allowed. (Section 3.5)

(a) 2*s*

(b) 5*f*

(c) 6*p*

10* Sketch radial wavefunctions, radial distribution functions, and boundary diagrams for 6*s* and 5*p* electrons. (Section 3.5)

11 Explain why the 2*s* orbital is at a lower energy than the 2*p* orbital for lithium, whereas the 2*s* and 2*p* orbitals have the same energy for Li^{2+}. (Section 3.6)

12 Of the following arrangements of *p* electrons, which represents the ground state, which are excited states and which are impossible? (Section 3.6)

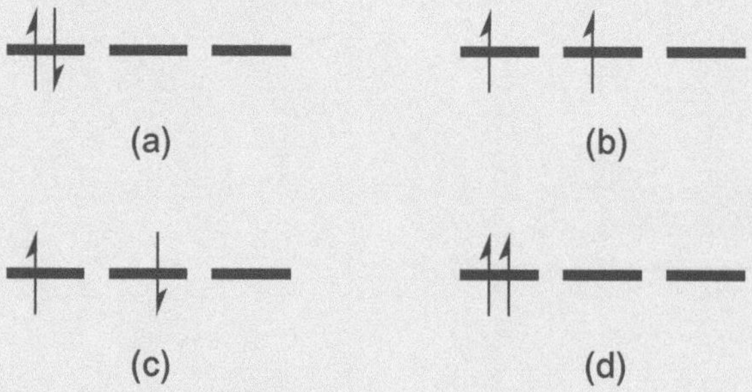
(a)
(b)
(c)
(d)

13 Using the Group 18 element shorthand, give the electronic configurations for the following elements or ions: (a) arsenic; (b) cobalt; (c) holmium; (d) bromide ion. How many unpaired electrons are there in each case? (Section 3.6)

14* Which elements would you expect to have the following electronic configurations: (a) [Ar] $4s^2$; (b) [Ne] $3s^2$ $3p^5$; (c) [Kr] $4d^7$ $5s^2$? The actual configuration for the element in (c) is [Kr] $4d^8$ $5s^1$. Suggest a reason for this. (Section 3.6)

15 Considering the number of parallel electron pairs, calculate the exchange energy for the d^5 and f^7 configurations. Assume that the exchange energy for a pair of parallel electrons is *K*. (Section 3.6)

16 Use Slater's rules to calculate the effective nuclear charge for phosphorus and for arsenic. Do the values account for the difference in atomic radius for these elements (Figure 3.34, p.154)? (Section 3.6)

17 For each of the following pairs of elements, which has the larger first ionization energy: (a) sodium and magnesium; (b) magnesium and aluminium; (c) magnesium and calcium? (Section 3.7)

18* Account for the following trends in ionization energy. (Section 3.7)

(a) The increase from oxygen to fluorine.

(b) The decrease from nitrogen to phosphorus.

(c) The decrease from phosphorus to sulfur.

19 ^{210}Pb decays into ^{206}Pb in a pathway involving two β-emissions followed by an α-emission. What are the intermediate isotopes? (Section 3.8)

20 Identify the transuranic element isotopes X, Y, and Z that were prepared by the following nuclear reactions. (Section 3.8)

(a) $^{241}\text{Am} + {}^{4}\text{He} \rightarrow \text{X} + 2\,{}^{1}\text{n}$

(b) $^{208}\text{Pb} + {}^{64}\text{Ni} \rightarrow \text{Y} + {}^{1}\text{n}$

(c) $^{249}\text{Cf} + {}^{18}\text{O} \rightarrow \text{Z} + 4\,{}^{1}\text{n}$

4

Diatomic molecules

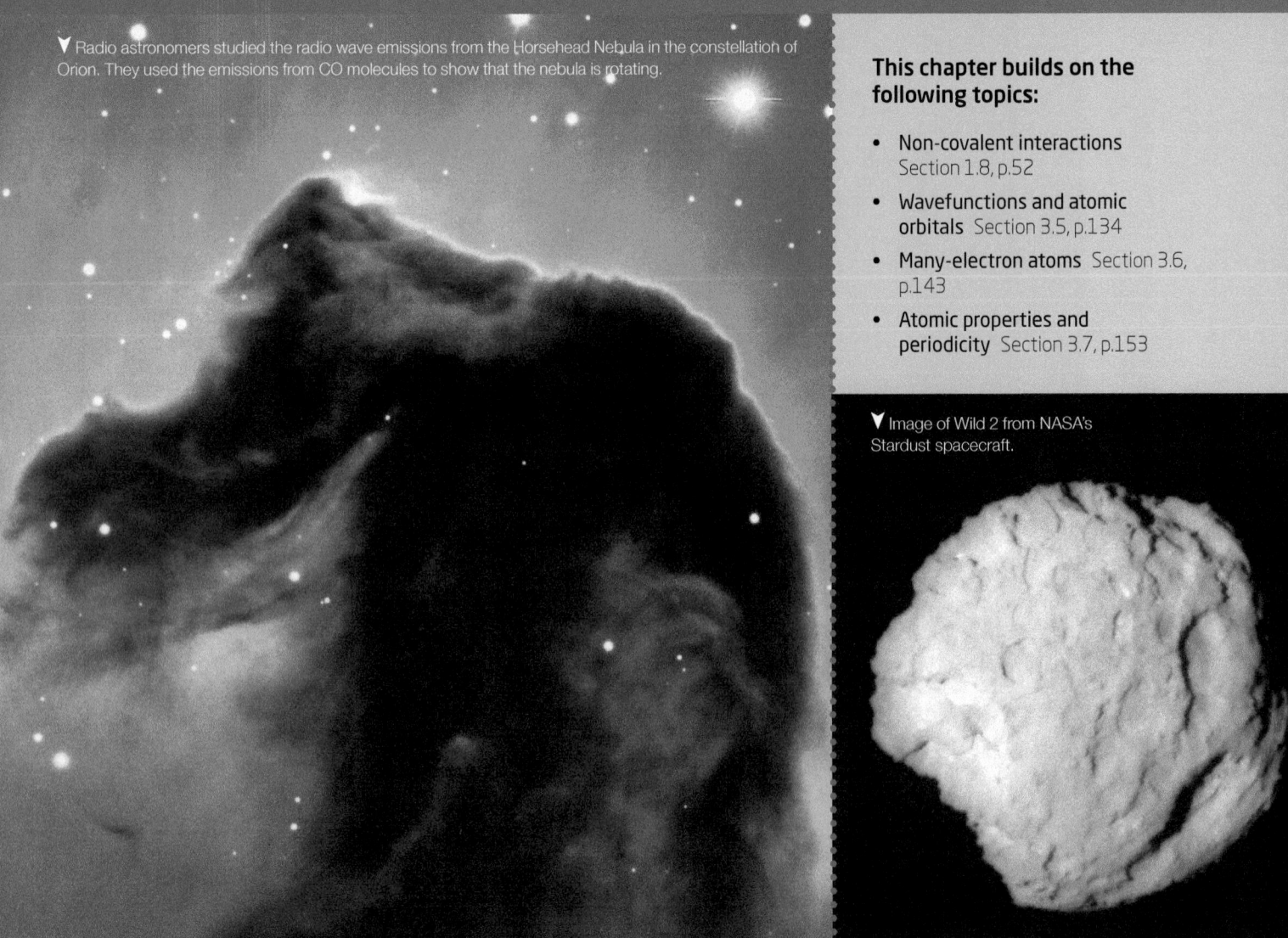

Radio astronomers studied the radio wave emissions from the Horsehead Nebula in the constellation of Orion. They used the emissions from CO molecules to show that the nebula is rotating.

This chapter builds on the following topics:

- **Non-covalent interactions** Section 1.8, p.52
- **Wavefunctions and atomic orbitals** Section 3.5, p.134
- **Many-electron atoms** Section 3.6, p.143
- **Atomic properties and periodicity** Section 3.7, p.153

Image of Wild 2 from NASA's Stardust spacecraft.

Molecules in space

Molecules form in the cool regions of space when individual atoms or ions collide and make covalent bonds with one another. Molecules cannot form in stars because the bonds would not survive the high temperatures there. At the temperatures in stars, the atoms do not even retain their electrons, and matter exists as a plasma of ionized atoms and unbound electrons.

Over 100 different molecules have been identified by radio astronomers in dense molecular clouds, such as those in the Horsehead Nebula. Some of the molecules that have been detected are shown in the table below. Many of these molecules are too reactive to be isolated on Earth.

Diatomic molecules		Polyatomic molecules	
H_2	C_2	H_2O	CH_3OH
OH	CN	HCN	CH_3CHO
CO	NO	CH_4	CH_3CH_2OH
SO	SiO	NH_3	$HC_{11}N$

Dense molecular clouds are only dense by astronomical standards. Even the densest gas clouds are about 100 billion times thinner than the Earth's atmosphere. This means that collisions between atoms or molecules are few and far between, and reactions only take place very slowly. Astrochemists believe that most reactions take place on the surface of cosmic dust particles—tiny particles of carbon, silica (SiO_2), iron, and other substances. Molecules with simple structures such as H_2O, NH_3, and CH_4 freeze onto the surface of the particles and are brought into close proximity. Some scientists believe that such molecules were the building blocks that reacted together to make the molecules that formed the basis of life on Earth. The energy for the reactions was supplied by high-energy electromagnetic radiation from nearby stars.

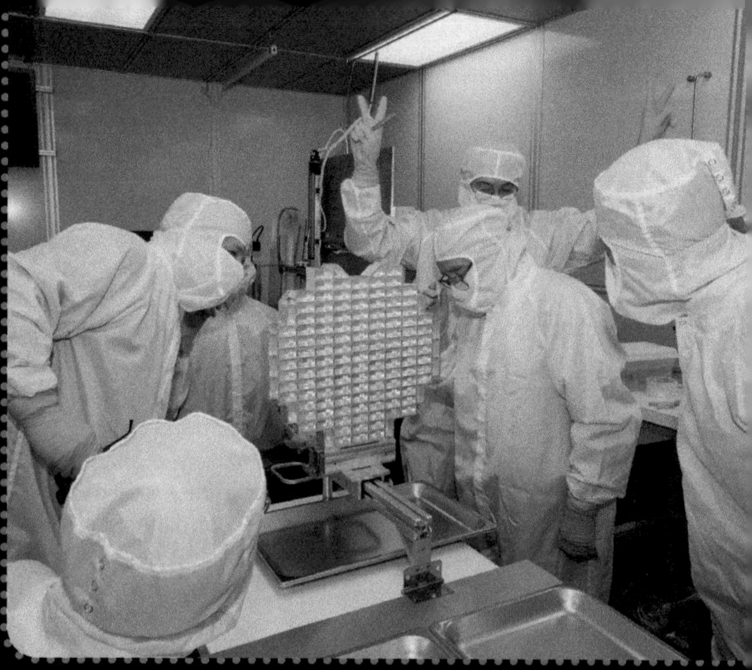

The Stardust sample capsule is analysed in a clean room to prevent contamination of the comet samples.

How do we know about the presence of molecules so far away? The molecular clouds are very cold, with temperatures of only just above absolute zero, so the electromagnetic radiation emitted by the molecules is of very long wavelength ($\lambda \sim 1$ mm) in the radio wave part of the electromagnetic spectrum. Fortunately, this radiation is not absorbed in the Earth's atmosphere and can be detected by radio telescopes on Earth. Radio telescopes can be tuned to a frequency that is characteristic of a particular molecule, such as CO, and then scan a region of sky, mapping out the distribution and intensity of emission at this frequency.

The Stardust mission reached the comet Wild 2 in January 2004, and collected particles from a dense gas cloud within 250 km of the comet's core. Stardust ejected a capsule that returned to Earth in January 2006 with these samples, and their analysis revealed an abundance of silicates. Many of these structures are known to form at high temperatures, which suggests that material was ejected from the inner solar system to the colder outer solar system where comets are believed to originate. The samples also contained organic molecules including the amino acid glycine ($NH_2CH_2CO_2H$), suggesting that at least one amino acid was delivered to Earth before life began.

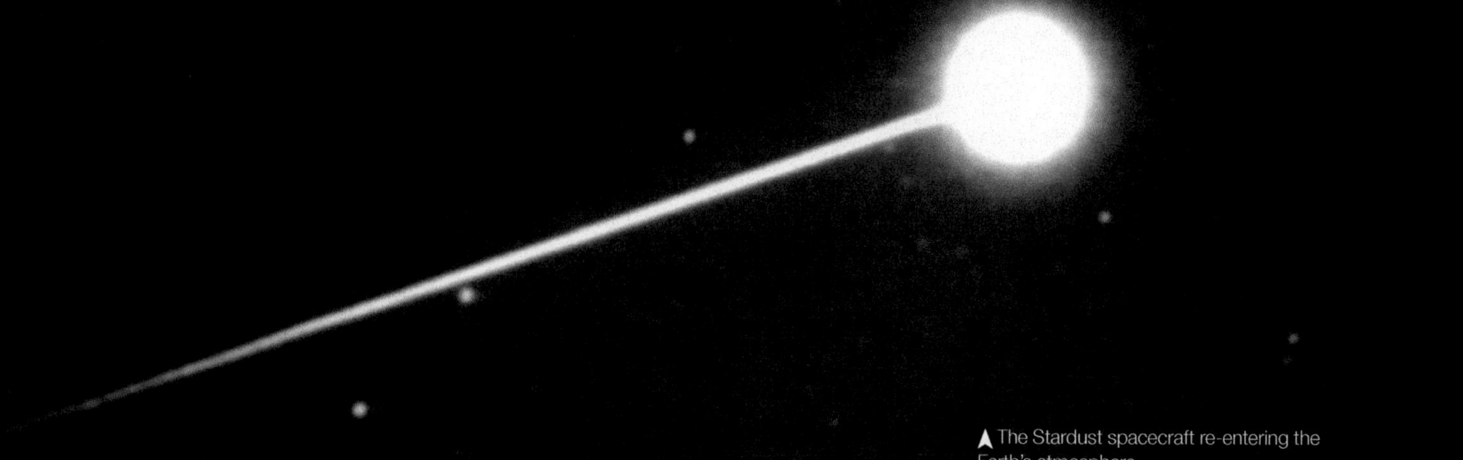

The Stardust spacecraft re-entering the Earth's atmosphere.

Chemical reactions involve the making and breaking of bonds. To understand how and when reactions occur, you need to know about the nature of bonds, and how they are formed and broken. The two most familiar types of bonding are covalent and ionic. In **covalent bonds**, electrons are shared between atoms and the increased electron density between the nuclei holds them together. In **ionic bonds** electrons are transferred from one atom to another, and the compound is held together by attractions between the oppositely charged ions.

This chapter is concerned with covalent bonding. It covers the bonding in **diatomic molecules**, molecules containing only two atoms, and considers three different approaches to bonding. The first of these is the familiar explanation for covalent bonding in terms of shared pairs of electrons, which was first proposed by Gilbert Lewis. This theory, though still useful, has a number of limitations, so two more detailed approaches—valence bond theory and molecular orbital theory—are also introduced.

A diatomic molecule contains only two atoms.

Ionic and covalent bonding. When sea water freezes the salt is expelled from the ice forming 'saltcicles'—icicles made from sodium chloride (NaCl). The water molecules in ice contain covalent bonds, whereas the bonding in NaCl is ionic.

It is helpful to think of bonding as a spectrum, with covalent and ionic bonding at the two extremes, as shown in Figure 4.1. The atoms of one element attract electrons to a different extent to the atoms of another element. This means that covalent bonds between atoms of different elements are normally **polar**, with one of the atoms in the bond carrying a partial positive charge (δ+) and the other atom carrying a partial negative charge (δ–). As a result, there is an ionic contribution to the covalent bonding. Both valence bond theory and molecular orbital theory are able to allow for these ionic contributions to the bonding.

This chapter introduces these theories of bonding by looking at diatomic molecules. The following chapter (Chapter 5) examines covalent bonding in polyatomic molecules, where more than two atoms are present. Chapter 6 looks at extended structures, and also develops the theory of ionic bonding.

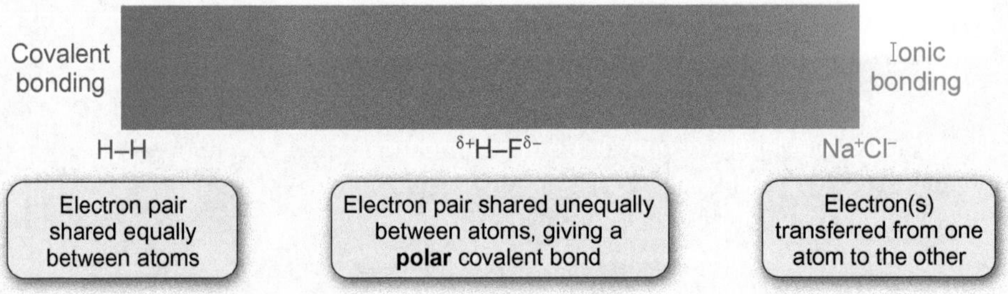
Covalent bonding
Ionic bonding
H–H
$^{\delta+}H–F^{\delta-}$
Na^+Cl^-
Electron pair shared equally between atoms
Electron pair shared unequally between atoms, giving a **polar** covalent bond
Electron(s) transferred from one atom to the other

Figure 4.1 The types of bonding can be thought of as a spectrum, with covalent bonding at one extreme and ionic bonding at the other. Another way of showing different contributions to bonding is shown in Section 6.7 (p.298).

4.1 Features of diatomic molecules

Many of the elements that exist as gases at room temperature consist of diatomic molecules rather than single atoms. Oxygen (O_2), nitrogen (N_2), chlorine (Cl_2), and hydrogen (H_2) are just four examples. Stanislao Cannizzaro identified this phenomenon in 1858 and, building on earlier work by fellow Italian chemist Amedeo Avogadro, he established the difference between what we now know as relative atomic masses and relative molecular masses. Although Cannizzaro recognized that the gases existed as diatomic molecules, he could not explain this. This is understandable, as a theory of covalent bonding is impossible without knowing about electrons, and the electron was not discovered until 1897 (Section 3.1, p.114). Cannizzaro's observation remained unexplained for almost 60 years.

All diatomic molecules have a characteristic bond length and bond dissociation enthalpy. The **bond length** is defined as the mean distance between the centres of two atoms in a molecule. For a **homonuclear diatomic**, such as H_2, in which both atoms are identical, the bond length is twice the value of the atomic radius.

Atomic radii are given in Figure 3.34 (p.154).

Online Support. The icon in the margin indicates that accompanying interactive resources are provided online to help your understanding: just type www.chemtube3d.com/burrows/123 into your browser, replacing 123 with the number of the page where you see the icon. For pages linking to more than one resource, type 123-1, 123-2, etc. (replacing 123 with the page number) for access to successive links.

Table 4.1 Bond dissociation enthalpies, *D*, and bond lengths for selected homonuclear diatomic molecules

	D/kJ mol^{-1}	Bond length/pm
H_2	+435.8	74.1
Li_2	+105.0	267.3
Na_2	+74.8	307.9
K_2	+57.0	390.5
N_2	+944.9	109.8
P_2	+489.1	189.3
O_2	+498.5	120.7
S_2	+430.0	188.9
F_2	+158.7	141.2
Cl_2	+242.4	198.8
Br_2	+193.9	228.1
I_2	+152.3	266.6

The **bond dissociation enthalpy** is a measure of the bond strength, and is defined as the standard enthalpy change for the reaction in which the bond in question is broken. It is given the symbol *D*, and is always endothermic. For the H_2 molecule, the bond dissociation enthalpy is the enthalpy change for the following reaction at 298 K

$$H_2(g) \rightarrow 2H(g) \quad D(\text{H–H}) = +436\,\text{kJ mol}^{-1}$$

The bond lengths and bond dissociation enthalpies for a selection of homonuclear diatomic molecules are given in Table 4.1.

Figure 4.2 shows how the potential energy of a simple diatomic molecule such as H_2 varies with the distance between the atoms. The minimum of the potential energy curve corresponds to the equilibrium bond length and bond strength. The deeper the minimum of the curve, the stronger the bond between the atoms.

For polyatomic molecules, mean bond enthalpies are generally more useful than specific bond dissociation enthalpies. See Section 13.3 (p.630) for more details.

Bond dissociation enthalpies are often called bond dissociation energies. Although the two have similar values, they are not exactly the same. The difference between energy and enthalpy is explained in Sections 6.5 (p.292) and 13.5 (p.642).

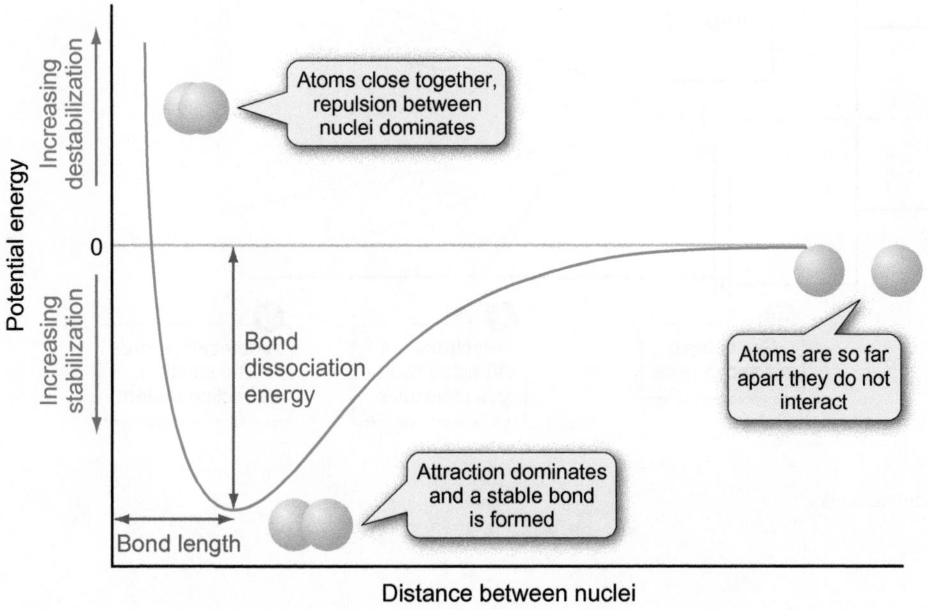
Potential energy
Increasing destabilization
Increasing stabilization
0
Atoms close together, repulsion between nuclei dominates
Bond dissociation energy
Atoms are so far apart they do not interact
Attraction dominates and a stable bond is formed
Bond length
Distance between nuclei

Figure 4.2 The potential energy of two hydrogen atoms varies with the distance between them. The minimum in the potential energy curve is the most stable situation.

Figure 4.2 ignores zero point energy, which is introduced in Box 10.1 (p.454).

Box 4.1 How can we measure bond lengths?

Bond lengths can be measured using diffraction techniques. Diffraction is a type of interference between waves, and it occurs when the wavelength is similar to the size of the object involved. To obtain bond lengths from a diffraction experiment, the wavelength used must be similar to the distances between atoms, typically 100–300 pm. Diffraction experiments were historically important in establishing that both light and electrons can act as waves (see Sections 3.2, p.123 and 3.4, p.132).

The best technique for studying molecules in the gas phase is electron diffraction. This is because the wavelength of a beam of electrons is of the same order as the distance between the atoms in the molecules. The electron beam wavelength can be tuned to the value required, as the wavelength depends on the velocity of the electrons (see Section 3.4, p.132).

Electron diffraction patterns consist of diffuse, concentric circles, which are recorded on photographic plates or by electron counters (see Figure 3.14, p.132). A mathematical analysis of the relationship between the intensities and the scattering angles leads to values for the distances between atoms.

The bond lengths in simple molecules can also be obtained from microwave spectroscopy, which looks at transitions between rotational energy levels (see Section 10.4, p.472).

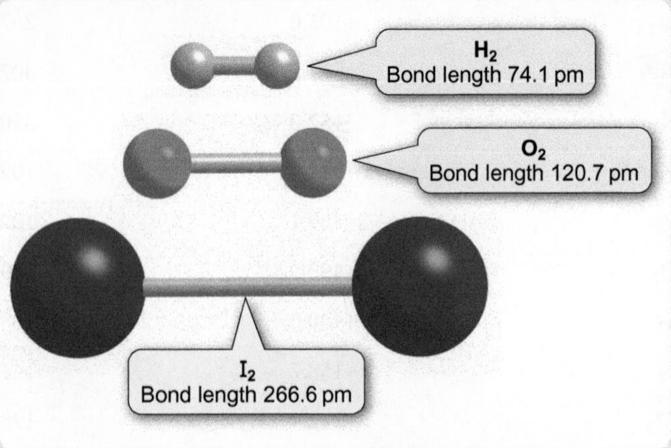

H_2
Bond length 74.1 pm
O_2
Bond length 120.7 pm
I_2
Bond length 266.6 pm

▲ Examples of diatomic molecules with their bond lengths.

Question

Diffraction of electromagnetic radiation is also used to determine structures, and for crystalline solids X-ray diffraction is commonly used (see Box 6.4, p.279). Suggest why these experiments use X-rays and not ultraviolet radiation or γ-rays.

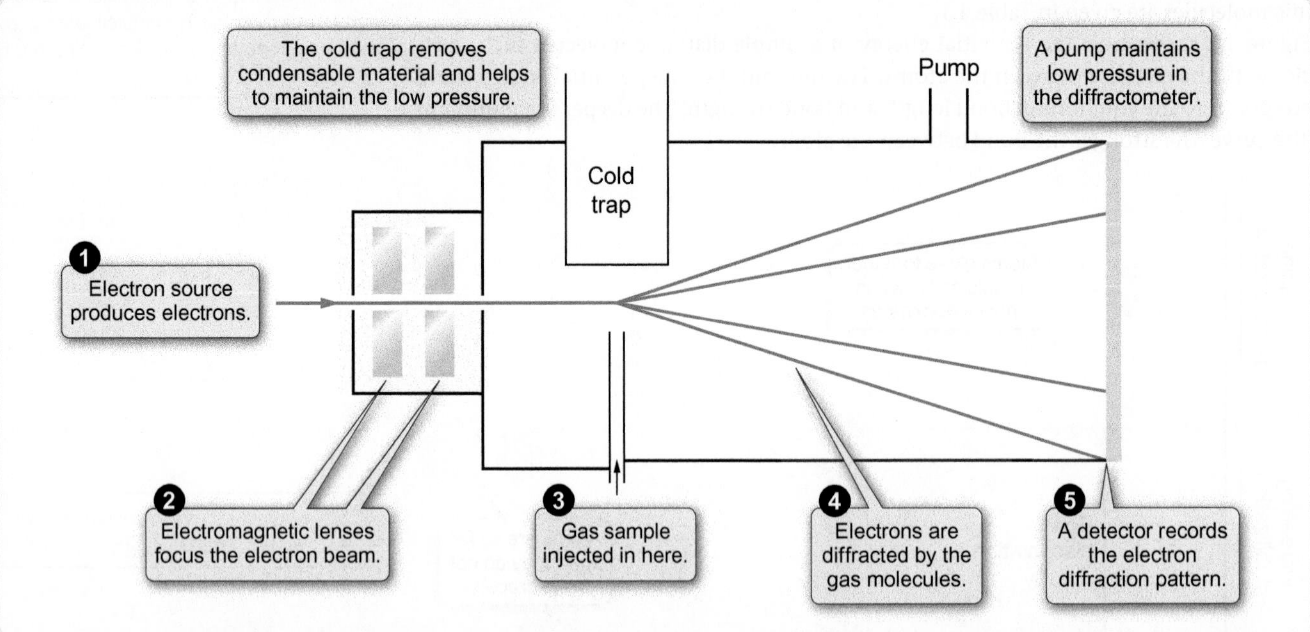

The cold trap removes condensable material and helps to maintain the low pressure.
Pump
A pump maintains low pressure in the diffractometer.
Cold trap
1
Electron source produces electrons.
2
Electromagnetic lenses focus the electron beam.
3
Gas sample injected in here.
4
Electrons are diffracted by the gas molecules.
5
A detector records the electron diffraction pattern.

▲ A schematic diagram of how an electron diffractometer works.

Summary

- Diatomic molecules contain only two atoms.
- The bond length in a diatomic molecule is defined as the mean distance between the centres of the two atoms.
- The bond dissociation enthalpy is a measure of the bond strength, and is defined as the standard enthalpy change of the reaction, at a specified temperature, in which the bond is broken.

4.2 The Lewis model

The bonding in many molecules can be explained by the tendency of the atoms to achieve the electronic configuration of the closest noble gas (Group 18 element). This is sometimes called the **octet rule**, since for most atoms eight electrons in the outer shell corresponds to the filled *s* and *p* orbitals ($ns^2\ np^6$). Hydrogen is an exception to this, since there is no $1p$ orbital. Hydrogen reaches the noble gas configuration of helium with just two electrons in the outer shell ($1s^2$).

All the orbitals with the same value of *n* are said to be in the same shell. See Section 3.5 (p.135) for more details.

The first model of covalent bonding was introduced by the American scientist Gilbert Lewis in 1916. Lewis recognized that it was possible for atoms in a molecule to obey the octet rule by sharing electrons. Electron sharing can be represented using dot and cross diagrams with different symbols representing the electrons on each atom. Figure 4.3 shows how electron sharing in F_2, O_2, and N_2 gives rise to single, double, and triple bonds, respectively. Electrons are always shared in pairs, and in a shorthand representation, a line between atoms is used to denote a shared pair of electrons. Electron pairs not involved in bonds are called **lone pairs**.

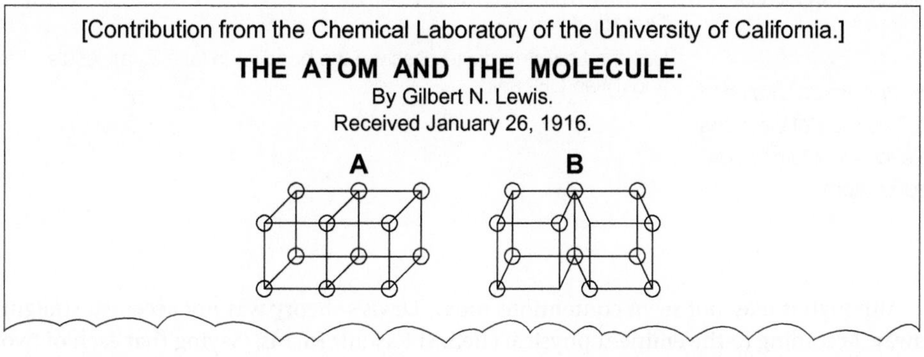
[Contribution from the Chemical Laboratory of the University of California.]

THE ATOM AND THE MOLECULE.

By Gilbert N. Lewis.

Received January 26, 1916.

A
B

The birth of a theory that is familiar to every student of chemistry. Lewis first published his theory of covalent bonding in 1916, though his attempts to show this with cubic atoms sharing edges was criticized for mimicking art—Picasso had just made cubism popular. Reprinted with permission from the Journal of the American Chemical Society.

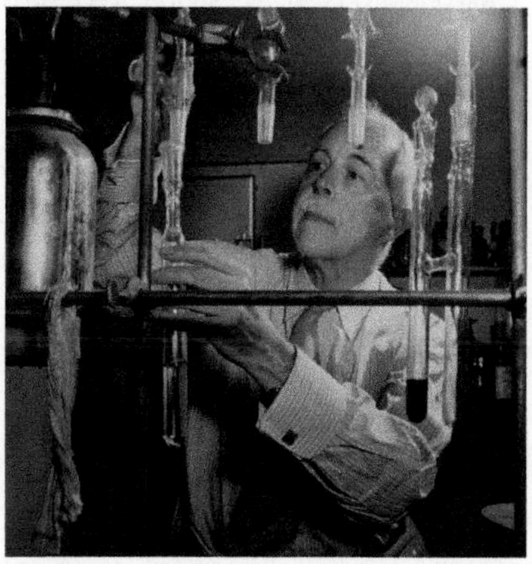

Gilbert Lewis was the first person to develop the concept of electron sharing in molecules.

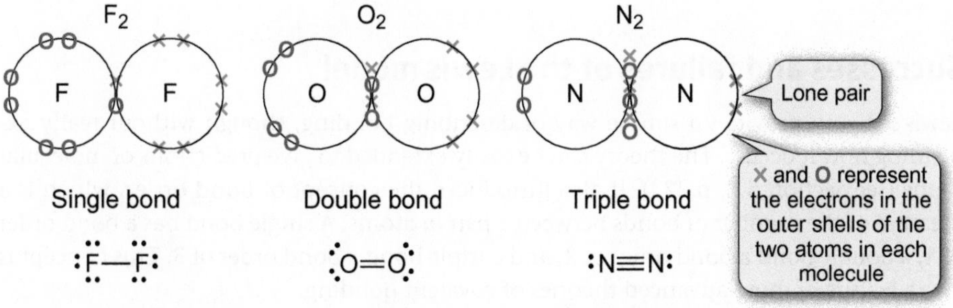
F2
O2
N2
F
F
O
O
N
N
Lone pair
x and O represent the electrons in the outer shells of the two atoms in each molecule
Single bond
Double bond
Triple bond
:F—F:
O=O
:N≡N:

Figure 4.3 Dot and cross diagrams showing the bonding in F_2, O_2, and N_2, together with simplified representations. The sharing of electrons leads to each atom attaining a noble gas configuration.

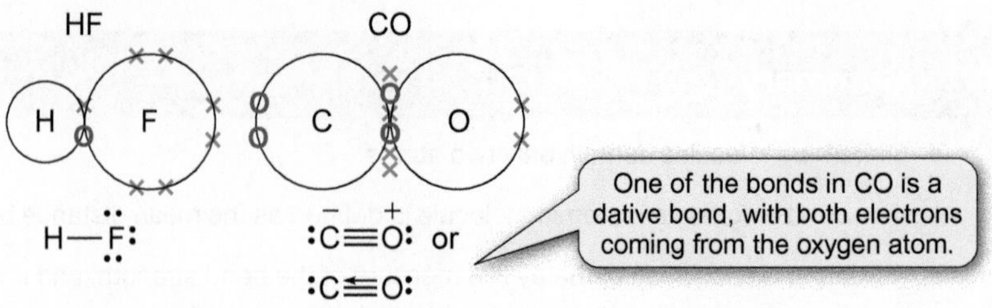
HF
CO
H
F
C
O
H—F:
:C≡O: or
:C≡O:
One of the bonds in CO is a dative bond, with both electrons coming from the oxygen atom.

Figure 4.4 Dot and cross diagrams showing the bonding in HF and CO, together with simplified representations.

A **heteronuclear diatomic** is a diatomic molecule containing two different atoms. Figure 4.4 shows how the Lewis model accounts for bonding in the heteronuclear diatomic molecules HF and CO. You can practise drawing a dot and cross diagram in Worked example 4.1.

Worked example 4.1 The Lewis model

Sulfur exists in the gas phase as S_2 molecules. Use the Lewis model to describe the bonding in a S_2 molecule.

Strategy

Work out the electronic configuration of sulfur, and draw a diagram with electrons shared so that the atoms both obey the octet rule.

Solution

Sulfur has the electronic configuration $1s^2\ 2s^2\ 2p^6\ 3s^2\ 3p^4$, so each sulfur atom has six electrons in its outer shell. Two pairs of electrons are shared between the sulfur atoms to give each a noble gas configuration. There are two lone pairs on each sulfur atom.

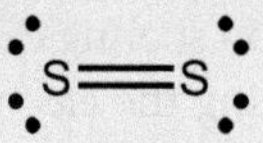
S═S

The bond between the sulfur atoms is therefore a double bond.

Question

Use the Lewis model to describe the bonding in (a) a P_2 molecule and (b) an OH^- anion.

Although it may not seem contentious today, Lewis's theory was not accepted straight away. According to the eminent physical chemist Kasimir Fajans, 'Saying that each of two atoms can attain closed electron shells by sharing a pair of electrons is equivalent to saying that a husband and wife, by having a total of two dollars in a joint account and each having six dollars in individual bank accounts, have got eight dollars apiece.' The name 'covalent bond' was introduced by Irving Langmuir, who helped popularize the theory in the early 1920s.

Successes and failures of the Lewis model

> The bond order is the number of bonds between two atoms.

Lewis's theory provides a simple way of describing bonding, though without really explaining how it occurs. The theory can be easily extended to give predictions of molecular shape (see Section 5.2, p.223). It also introduced the concept of **bond order**, which is a measure of the number of bonds between a pair of atoms. A single bond has a bond order of 1, a double bond a bond order of 2, and a triple bond a bond order of 3. This concept is also a feature of more advanced theories of covalent bonding.

Although the Lewis model is successful in explaining the structures of many molecules, it does not always work. There are many stable molecules in which the atoms do not have filled octets. For example, if nitrogen monoxide (nitric oxide, NO) is drawn with a double bond, the nitrogen atom has only seven valence electrons and so is **electron deficient**. Alternatively, if it is drawn with a triple bond, the oxygen atom has nine valence electrons. This gives an **expanded octet**, which is not possible for a second row element such as oxygen. It is not possible to draw a Lewis structure for NO in which both nitrogen and oxygen have complete octets.

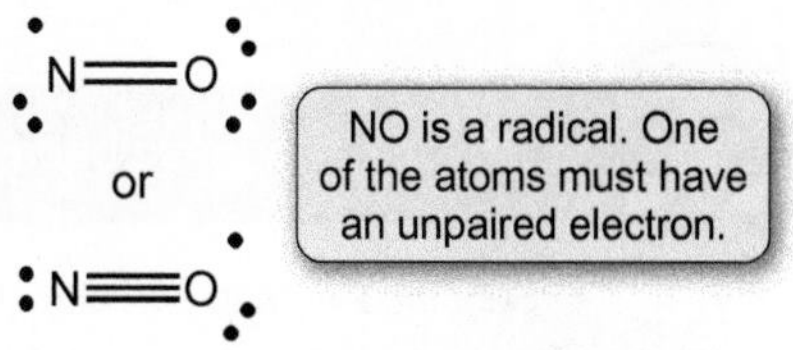
N=O
or
N≡O
NO is a radical. One of the atoms must have an unpaired electron.

Although not present in NO, there are many stable compounds that do contain expanded octets. In phosphorus pentafluoride (PF_5), for example, the phosphorus atom contributes five electrons and the fluorine atoms contribute one each to the bonding, so there are ten electrons around the central phosphorus atom. Atoms with expanded octets are described as **hypervalent**.

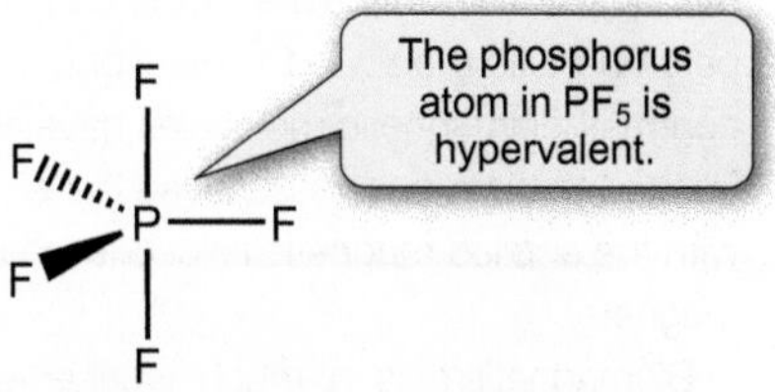
F
F
P
F
F
F
The phosphorus atom in PF5 is hypervalent.

The Lewis model also fails to explain the bonding in many transition metal compounds, such as *cis*-$[PtCl_2(NH_3)_2]$, which is used in chemotherapy for some types of cancer.

On closer inspection, there are problems with the way in which the Lewis model describes bonding in simple molecules such as oxygen. The double-bonded structure of O_2 derived from the Lewis model is shown in Figure 4.3 (p.173), but this representation cannot explain why liquid oxygen is attracted to a magnet, as shown in Figure 4.5. In the Lewis structure of O_2 all the electrons are paired, so it is expected to be diamagnetic (see Box 4.2). In reality, oxygen is paramagnetic, which suggests it contains unpaired electrons.

Transition metal compounds are detailed in Chapter 28. The use of *cis*-$[PtCl_2(NH_3)_2]$ is described in Box 28.3 (p.1275).

Paramagnetic compounds are weakly attracted to magnets; diamagnetic compounds are weakly repelled from magnets.

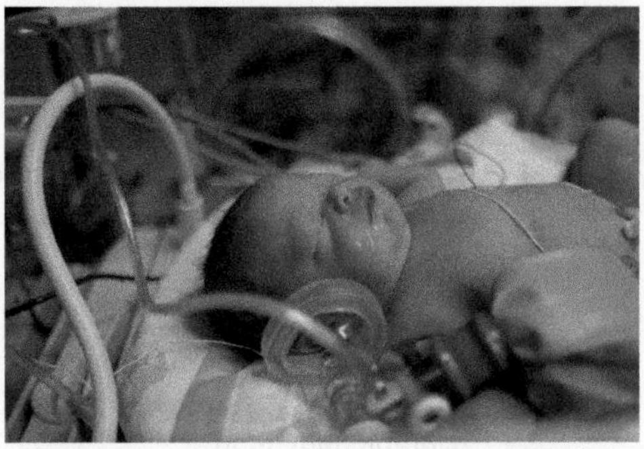

The paramagnetism of O_2 is used to monitor the oxygen content of air, for example in incubators for premature babies, by measuring the magnetism of the gas.

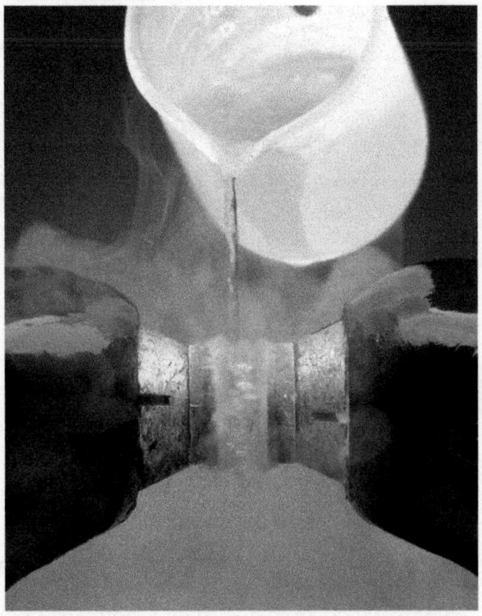

Figure 4.5 Liquid oxygen is poured between the poles of a powerful electromagnet. The liquid oxygen is attracted to the magnet and hangs in mid-air between the poles of the magnet as it evaporates.

Box 4.2 Magnetic behaviour

Substances that are attracted to magnets are called **paramagnetic**, and those that are repelled from magnets are called **diamagnetic**. Paramagnetism is caused by unpaired electrons, so paramagnetic molecules contain one or more unpaired electrons. In contrast, diamagnetic molecules have all their electron spins paired. Paramagnetism occurs because the electron spins are able to align with the applied magnetic field. In essence, the spins behave as tiny magnets.

Diamagnetism is a much weaker effect than paramagnetism. Diamagnetism is found in all materials, but is only observed in the absence of other types of magnetism.

Paramagnetism and diamagnetism can be measured using an apparatus known as a **Gouy balance**. When the magnetic field is applied, a paramagnetic compound will move down into the field, whereas a diamagnetic material will move up out of the field.

▲ Although diamagnetism is a much weaker effect than paramagnetism, it can be observed spectacularly with a very powerful magnet. This photograph shows a ball of water, about 5 cm in diameter, suspended in the bore of an electromagnet.

Question

Which of the atoms and ions Na, Ne, and O^{2-} are paramagnetic?

▼ Magnetism can be measured using a Gouy balance. The magnetic nature of the sample is determined by detecting the extent to which it is pulled into or pushed out of the magnetic field.

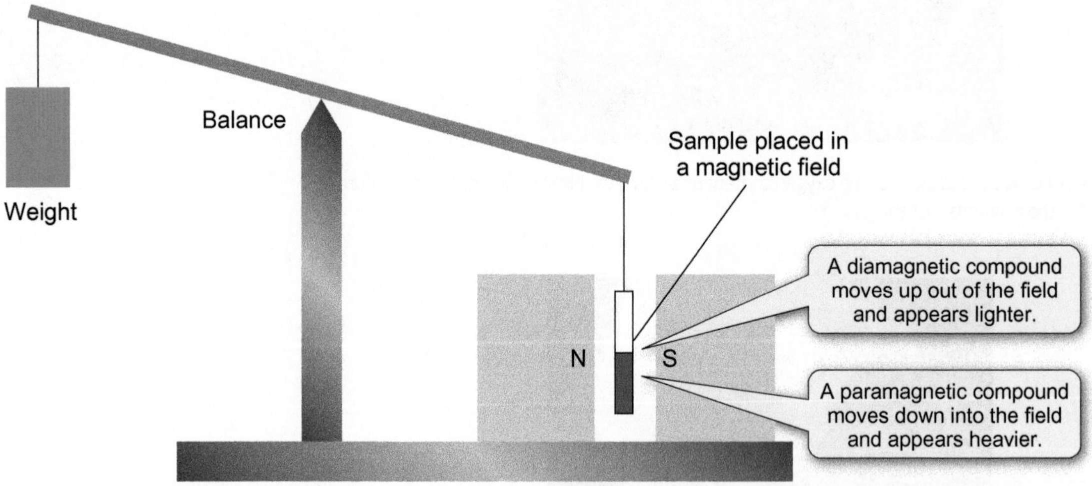
Balance
Weight
Sample placed in a magnetic field
N
S
A diamagnetic compound moves up out of the field and appears lighter.
A paramagnetic compound moves down into the field and appears heavier.

Summary

- The Lewis model was the first theory to explain bonding through the sharing of electrons.
- The bond order is the number of bonds between a pair of atoms.
- There are many stable molecules in which the atoms do not have filled octets, or have expanded octets.
- The Lewis model cannot explain the paramagnetism of oxygen, which indicates that O_2 contains unpaired electrons.

For a practice question on these topics, see question 1 at the end of this chapter (p.213).

4.3 Electronegativity

To understand more advanced theories of bonding, you need to be aware of the concept of electronegativity. **Electronegativity**, χ, is the power of an atom in a molecule to attract electrons to itself. In a heteronuclear diatomic, the shared electrons in a covalent bond are pulled towards the more electronegative atom.

Although electronegativity is a simple concept, it is not directly measurable. There are several commonly used scales for electronegativity that attempt to quantify it. The most commonly used of these was introduced by Linus Pauling and is now known as Pauling electronegativity.

Greek symbols
χ chi (electronegativity)

An atom with low electronegativity is said to be **electropositive**.

Pauling electronegativity, χ^P

Pauling noticed that the bond dissociation enthalpies for heteronuclear diatomics of the general formula XY were always greater than the average of the bond dissociation enthalpies of the homonuclear diatomics X_2 and Y_2. For example, the bond dissociation enthalpies for H_2 and F_2 are +435.8 kJ mol^{-1} and +158.7 kJ mol^{-1}, respectively, so the average of these is (435.8 kJ mol^{-1} + 158.7 kJ mol^{-1})/2 = +297.3 kJ mol^{-1}. The actual value of the bond dissociation enthalpy for HF is +569.7 kJ mol^{-1}, which is much higher than this.

Pauling reasoned that the difference between the experimental value of the bond dissociation enthalpy and that calculated by working out the average of the homonuclear bond dissociation enthalpies was a measure of the ionic character of the XY bond. He defined **electronegativity**, χ^P, where P denotes the Pauling scale, according to Equation 4.1.

Difference in electronegativity between X and Y

$$= |\chi^P(X) - \chi^P(Y)| = (\Delta D)^{\frac{1}{2}} \quad (\Delta D \text{ in eV}) \tag{4.1}$$

The symbol || in $|\chi^P(X) - \chi^P(Y)|$ means 'take the positive value of the difference'. Also, $(\Delta D)^{1/2} = \sqrt{\Delta D}$.

ΔD is the difference between the observed value for the bond dissociation enthalpy of the heteronuclear molecule XY and the average of the bond dissociation enthalpies for X_2 and Y_2, as shown in Equation 4.2.

$$\Delta D = D_{(X-Y)} - \frac{D_{(X-X)} + D_{(Y-Y)}}{2} \tag{4.2}$$

Pauling used values of ΔD in electronvolts (eV) and defined the electronegativity for fluorine as $\chi^P = 4$. A more convenient expression for χ^P, in which values of ΔD in kJ mol^{-1} are used, is shown in Equation 4.3.

1 eV = 96.485 kJ mol^{-1}. The currently accepted electronegativity for fluorine is 3.98.

$$|\chi^P(X) - \chi^P(Y)| = 0.102(\Delta D)^{\frac{1}{2}} \quad (\Delta D \text{ in kJ mol}^{-1}) \tag{4.3}$$

Figure 4.6 gives values for the Pauling electronegativities of the *s*- and *p*-block elements. Although, from Equations 4.1 and 4.3, you might expect that the units of χ^P are (eV)$^{1/2}$ or (kJ mol^{-1})$^{1/2}$, electronegativities are normally treated as having no units. The values in Figure 4.6 show that electronegativity generally *increases across a period* and *decreases down a group*.

The bond dissociation enthalpies of a vast number of compounds have now been measured, and used to calculate electronegativities. These values show that the electronegativity of an atom is not a constant but depends on its oxidation state. When an atom has two or more possible oxidation states, the electronegativity increases with increasing oxidation state. For example, χ^P for lead(IV) is greater than χ^P for lead(II). This makes sense as an electron-poor lead(IV) centre attracts electrons to a greater extent than a lead(II) centre which has two more electrons. However, the differences in electronegativity between different oxidation states are generally small, so the values of electronegativity given in Figure 4.6 can be used with a good degree of accuracy.

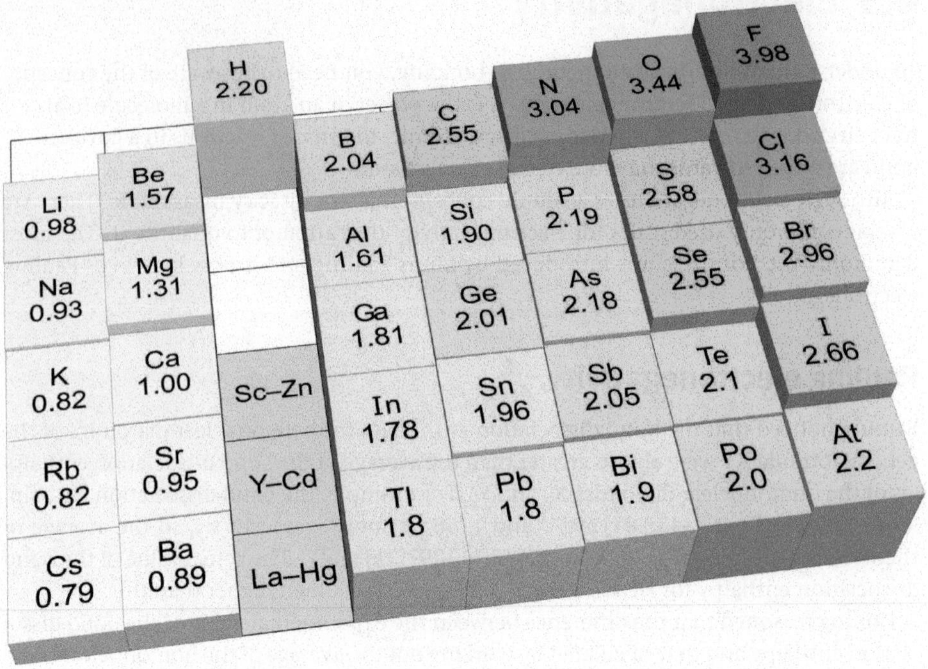
H
2.20
Li
0.98
Be
1.57
B
2.04
C
2.55
N
3.04
O
3.44
F
3.98
Na
0.93
Mg
1.31
Al
1.61
Si
1.90
P
2.19
S
2.58
Cl
3.16
K
0.82
Ca
1.00
Sc–Zn
Ga
1.81
Ge
2.01
As
2.18
Se
2.55
Br
2.96
Rb
0.82
Sr
0.95
Y–Cd
In
1.78
Sn
1.96
Sb
2.05
Te
2.1
I
2.66
Cs
0.79
Ba
0.89
La–Hg
Tl
1.8
Pb
1.8
Bi
1.9
Po
2.0
At
2.2

Figure 4.6 Pauling electronegativity (χ^P) values for the *s*- and *p*-block elements. Although Pauling originally used a value of 4 for fluorine, the value now used is slightly lower to make the electronegativities more self-consistent.

Electronegativity and bond polarity

Dipoles and polar molecules are described in more detail in Section 5.3 (p.235). The consequences of polarity on the reactivity of both organic and inorganic compounds are seen throughout the book.

If you know the electronegativities of the two atoms within a diatomic molecule, you can predict the way in which the bond is polarized. For example, the electronegativities of hydrogen and fluorine are 2.20 and 3.98, respectively. Since fluorine has the higher electronegativity, the fluorine atom attracts the electrons in the HF bond to itself more than the hydrogen atom does. This means that the HF bond is polarized with the hydrogen atom carrying a partial positive charge ($\delta+$) and the fluorine atom a partial negative charge ($\delta-$).

$$\overset{\delta+}{\mathrm{H}}\text{—}\overset{\delta-}{\mathrm{F}}$$

Summary

- Electronegativity is the power of an atom in a molecule to attract electrons to itself.
- The most commonly used scale for electronegativity is known as Pauling electronegativity, χ^P.
- Electronegativity generally increases across a period and decreases down a group.

For practice questions on these topics, see questions 2–4 at the end of this chapter (p.213).

4.4 Valence bond theory and molecular orbital theory

There are two theories of bonding, more advanced than the Lewis model, that give a better understanding of the covalent bond and allow problems such as those mentioned in Section 4.2 to be addressed. These approaches are valence bond theory and molecular orbital theory. These theories were both developed in the first half of the twentieth century, and for much of this time they were in competition. Valence bond (VB) theory was conceived by Linus Pauling who won the Nobel Prize in 1954 for this work. Molecular orbital (MO) theory was developed by Friedrich Hund and Robert Mulliken, and Mulliken received the Nobel Prize in 1966.

The Schrödinger equation makes it possible to describe electrons. However, the equation can only be solved exactly for a two-body system such as the hydrogen atom, which consists of a proton and an electron. This means it is impossible to solve the Schrödinger equation for any molecule, since there are at least two nuclei in addition to one or more electrons. The valence bond and molecular orbital theories differ in the approximations that they make to overcome this problem, though one approximation is common to both.

The Schrödinger equation is described in Section 3.5 (p.134).

Since nuclei are much heavier than electrons, they move more slowly. The **Born–Oppenheimer approximation** allows the nuclei to be regarded as being stationary, with only the electrons moving. This allows the Schrödinger equation to be solved just for the wavefunctions of the electrons.

The **valence bond** approach is closely linked to Lewis's idea of bonding, with shared pairs of electrons forming the bonds between atoms. A bond arises from the interaction between the atomic orbitals on two atoms to give a bonding orbital that is located between the two atoms. This bonding orbital contains two electrons.

In the **molecular orbital** approach, the atomic orbitals of the atoms in a molecule combine to form a set of molecular orbitals that are spread out over the whole molecule. These orbitals are said to be delocalized. The molecular orbitals are a property of the molecule as a whole, and the electrons in the molecule are distributed within these orbitals.

Why do we need two theories? Both approaches provide ways of calculating the wavefunction for a molecule and generally give good descriptions of the bonding. We tend to use the theories for different situations: valence bond theory is more useful in some circumstances, whereas molecular orbital theory is more useful in others.

Valence bond theory is very good for providing a qualitative picture of the bonding, and is the simpler of the two theories to extend from diatomic to polyatomic molecules. It is particularly useful for considering a series of molecules with one part in common. Organic compounds with the same functional groups are a good example of this, and this explains why valence bond theory is widely used in organic chemistry. It is also useful in the interpretation of infrared spectra of large molecules, as the peaks in these spectra are due to the vibrations of particular bonds between atoms.

Most modern computational chemistry makes use of molecular orbital theory as it is more quantitative. In addition, it is particularly useful for looking at excited states of molecules and for interpreting electronic spectra. It also provides a much simpler explanation of the paramagnetism of O_2.

The paramagnetism of O_2 was introduced in Section 4.2 (p.175).

The valence bond approach to diatomic molecules is described in Section 4.5. Much of the strength of valence bond theory comes from its use in polyatomic molecules, so the application of the theory is looked at in more detail in Chapter 5. The remainder of this chapter is then spent considering molecular orbital theory.

Valence bond theory is described in more detail in Section 5.4 (p.236).

Summary

- Valence bond theory and molecular orbital theory are theories of bonding involving the interactions between atomic orbitals.
- Valence bond theory is good for providing qualitative pictures of bonding, and is especially useful in organic chemistry.
- Molecular orbital theory provides a more quantitative picture, and is very useful for small molecules.

4.5 Valence bond theory

Valence bond theory was the first quantum mechanical theory of bonding. In this section the theory is developed for the homonuclear diatomic molecules H_2 and N_2, and then extended to the heteronuclear diatomic molecule HF.

Valence bond treatment of H_2

The valence bond approach to the bonding in H_2 involves considering how the 1*s* orbitals on two hydrogen atoms interact together to give a wavefunction for the molecule as a whole. The interactions between the 1*s* orbitals give rise to a σ-bonding orbital, which lies directly between the two hydrogen atoms. Calculations show that, for an accurate description of the bonding in H_2, the ionic form $H^+ H^-$ needs to be considered as well as the covalent form. The real structure is an average of the covalent and ionic forms. This leads to one of the most important concepts in valence bond theory—resonance. The structure of hydrogen can be represented as

$$\text{H–H} \leftrightarrow \text{H}^+\text{H}^- \leftrightarrow \text{H}^-\text{H}^+$$

with the double-headed arrow used to indicate resonance. The resonance forms are shown separately to give a complete picture of the character of the actual bond. The three resonance forms of hydrogen do not have independent existence and *the resonance forms do not exist in rapid equilibrium with each other*. There is only one wavefunction for H_2 which has covalent and ionic contributions to it.

Resonance is used in valence bond theory when no single Lewis-type structure accurately represents the observed bonding in a molecule. The actual structure is represented by an average of several **resonance forms**, and this is indicated in structural formulae using double-headed arrows.

The resonance forms do not contribute equally to the character of the bond, and the bonding in H_2 is a weighted average of the different resonance forms. The covalent resonance form makes a larger contribution to the molecular wavefunction than the ionic resonance forms. The two ionic resonance forms are equally important, so overall H_2 does not have a permanent dipole, as it would if one of these were more important than the other.

The mathematical basis for the valence bond description of H_2 is given in Box 4.3.

Box 4.3 Molecular wavefunctions for H_2

In a molecule, the wavefunctions of atoms are denoted by the Greek letter ϕ (phi) and wavefunctions of molecules by the Greek letter ψ (psi). The two hydrogen atoms are labelled H_A and H_B, and these have the wavefunctions ϕ_A and ϕ_B, respectively. If the atom H_A is associated with electron 1 and the atom H_B associated with electron 2, and the atoms are too far apart to interact, the wavefunction for the two atoms together can be described by the expression in Equation 4.4.

$$\psi = \phi_{A(1)}\phi_{B(2)} \qquad (4.4)$$

In order for there to be an attractive interaction between the atoms, the electrons must have their spins in opposite directions, but otherwise electrons 1 and 2 are indistinguishable. This means that when the atoms are brought close enough together to interact, it is just as likely that electron 1 will be associated with H_B and electron 2 with H_A as it is that electron 1 is with H_A and electron 2 is with H_B. This can be shown mathematically by adding a second term to Equation 4.4 to give Equation 4.5.

$$\psi = \phi_{A(1)}\phi_{B(2)} + \phi_{A(2)}\phi_{B(1)} \qquad (4.5)$$

When Equation 4.5 is used in calculations it gives a predicted bond length for H_2 of 87 pm and a predicted bond dissociation enthalpy of +303 kJ mol^{-1}. In reality, the bond is shorter and stronger than this, so Equation 4.5 underestimates the strength of the bonding between the hydrogen atoms in H_2.

One assumption made by Equation 4.5 is that each hydrogen atom always has an electron associated with it. If both electrons are associated with the atom H_A, this gives an ionic form of hydrogen, $H_A^- H_B^+$. Similarly, if both electrons are associated with H_B, this gives the ionic form $H_A^+ H_B^-$. These two possibilities are taken into account by including two additional terms to Equation 4.5 to give Equation 4.6. The expression for the molecular wavefunction, ψ, then becomes

$$\psi = \phi_{A(1)}\phi_{B(2)} + \phi_{A(2)}\phi_{B(1)} + \lambda\phi_{A(1)}\phi_{A(2)} + \lambda\phi_{B(1)}\phi_{B(2)} \qquad (4.6)$$

λ is a constant and, for H_2, λ has a value of less than one. This is because the ionic terms involve greater electron–electron repulsions than the covalent terms. Inclusion of the ionic terms with λ approximately 0.25 brings the calculated values of the bond length and bond dissociation enthalpy much closer to those observed experimentally (see Table 4.1, p.171).

Question

Why is the electron–electron repulsion greater for the ionic resonance forms of H_2 than for the covalent form?

Valence bond treatment of N_2

The electronic configuration of a nitrogen atom is $1s^2\,2s^2\,2p^3$, with the three 2*p* electrons all unpaired. N_2 can be treated in a similar way to H_2, though in this case it is interactions between the 2*p* orbitals that are important rather than those between the 1*s* orbitals. Since

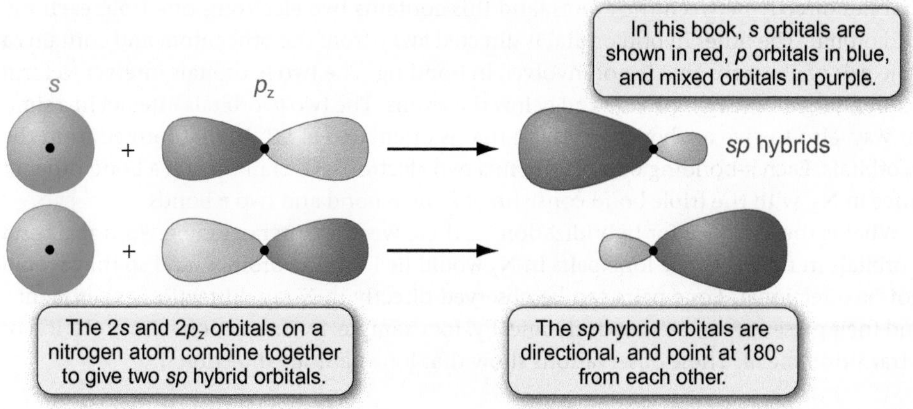
In this book, s orbitals are shown in red, p orbitals in blue, and mixed orbitals in purple.
s
p_z
sp hybrids
The 2s and $2p_z$ orbitals on a nitrogen atom combine together to give two sp hybrid orbitals.
The sp hybrid orbitals are directional, and point at 180° from each other.

Figure 4.7 *sp* hybridization involves mixing of the *s* and p_z orbitals on an atom to form two *sp* hybrid orbitals. The black dots represent the positions of the nuclei.

there are three *p* orbitals on each atom, each containing a single electron, three covalent bonds are formed between the two atoms to give the noble gas configuration.

Even taking into account ionic contributions, this approach does not give a good picture of the bonding in N_2. To understand the bonding in this and other molecules, another key concept of valence bond theory needs to be introduced. This is **hybridization**, which is a way of mixing the atomic orbitals on an atom as it interacts with another atom.

The nitrogen atoms in N_2 are ***sp*** **hybridized**. This means the *s* and p_z orbitals on each nitrogen atom interact to form two *sp* hybrid orbitals as shown in Figure 4.7. Since the p_z orbital is directional, these two *sp* hybrid orbitals are also directional, with one pointing towards the other nitrogen atom and the other directed 180° from it. The original *s* and p_z orbitals contained a total of three electrons between them. These are distributed between the two *sp* hybrid orbitals, so that one contains two electrons, and the other contains one electron. The unhybridized p_x and p_y orbitals on each nitrogen atom contain one electron each.

Mixing the s and p_z orbitals in this way gives two *sp* hybrid orbitals that are directional and so form stronger bonds when they interact with other orbitals.

The interactions between the orbitals of the two nitrogen atoms are shown in Figure 4.8. Two *sp* hybrid orbitals, one from each of the atoms, interact to form a σ-bonding orbital

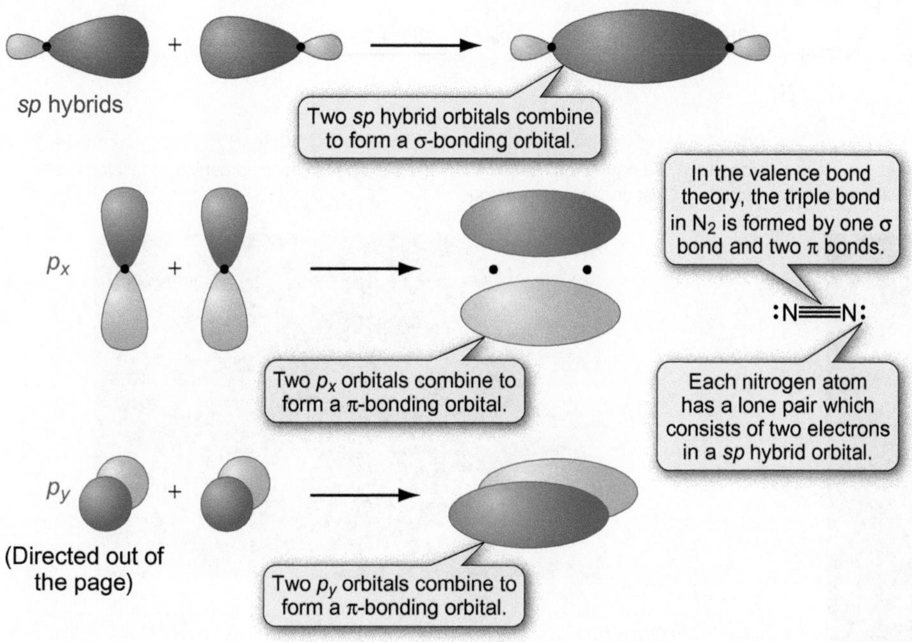
sp hybrids
Two sp hybrid orbitals combine to form a σ-bonding orbital.
In the valence bond theory, the triple bond in N_2 is formed by one σ bond and two π bonds.
p_x
:N≡N:
Two p_x orbitals combine to form a π-bonding orbital.
Each nitrogen atom has a lone pair which consists of two electrons in a sp hybrid orbital.
p_y
(Directed out of the page)
Two p_y orbitals combine to form a π-bonding orbital.

Figure 4.8 Valence bond theory predicts the triple bond in N_2 to consist of one σ bond and two π bonds.

that lies directly between the atoms, and this contains two electrons, one from each hybrid orbital. The other hybrid orbital is directed away from the other atom and contains a lone pair of electrons that is not involved in bonding. The two p_x orbitals interact to form a π-bonding orbital lying above and below the atoms. The two p_y orbitals interact in a similar way, also to give a π-bonding orbital that is orientated at 90° to that formed from the p_x orbitals. Each π-bonding orbital contains two electrons. Overall there is a bond order of three in N_2, with the triple bond consisting of one σ bond and two π bonds.

You will meet hybridization in greater detail in Chapter 5 when you look at the bonding in polyatomic molecules. See Section 5.4 (p.236).

What is the evidence for hybridization? If there were no interaction between the s and p orbitals in nitrogen, the lone pairs in N_2 would lie in the 2s orbitals, and so they would not be directional. Lone pairs can be observed directly in X-ray diffraction experiments, and their presence can be shown chemically, for example, by interaction with an H^+ ion or a transition metal. These observations show that lone pairs are directional.

Box 4.4 Making an unreactive molecule react

The strength of the nitrogen–nitrogen triple bond lies at the heart of the chemistry of N_2. With a bond dissociation enthalpy of +944.8 kJ mol^{-1}, it is the strongest homonuclear bond, and this makes N_2 very unreactive. This inertness can be exploited when working with very reactive compounds. Many compounds react violently with oxygen or moisture, which means they cannot be used in air. It is possible to work with these compounds by excluding air, and manipulating them under nitrogen instead. This can be done using a glove box. Materials are taken into the glove box through a port that has an inner and outer door. The outer door is closed, and the port evacuated using a vacuum pump before being filled with nitrogen from a cylinder. This evacuation–filling procedure is repeated several times to ensure all traces of O_2 are removed before opening the inner door.

Although the triple bond in N_2 is very strong, there are ways in which it can be *activated* so that it will react. The Haber process for converting N_2 and H_2 into ammonia (NH_3) uses an iron catalyst at a temperature of between 400°C and 500°C and a pressure of between 25 atm and 150 atm. N_2 is adsorbed onto the surface of the catalyst where it forms weak bonds with the surface iron atoms. The interaction between N_2 and the iron surface weakens the N≡N bond.

Biological reduction of N_2 to form NH_3 is known as **nitrogen fixation**. This process is carried out by bacteria and related *Rhizobium* micro-organisms that live in the root nodules of leguminous plants such as clover and peas. Nitrogen fixation requires an enzyme called nitrogenase, which is made up of two proteins, one containing iron, the other containing iron and molybdenum. The iron protein transfers electrons to the iron–molybdenum protein, which donates these electrons to N_2, reducing it sequentially to N_2H_2, N_2H_4, and finally to two molecules of NH_3. One of the goals of research in transition metal chemistry is to make compounds that can mimic this process. This would enable ammonia to be prepared more cheaply than is currently possible.

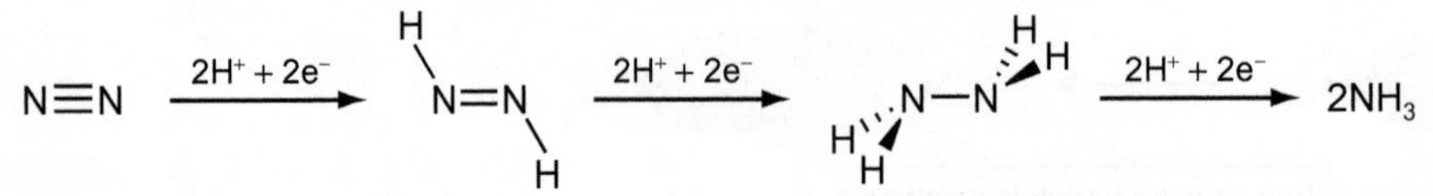
N≡N
2H⁺ + 2e⁻
N=N
H
H
2H⁺ + 2e⁻
N—N
H
H
H
H
2H⁺ + 2e⁻
2NH₃

▼ Using a glove box allows manipulations to be carried out under a nitrogen atmosphere. This is useful for compounds that react with oxygen or moisture.

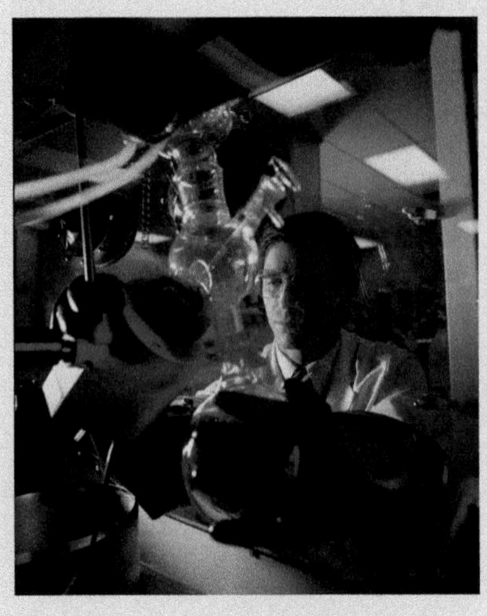

▼ Leguminous plants such as peas contain micro-organisms in their roots that are able to fix nitrogen.

Question

Given the high bond dissociation enthalpy of the N≡N bond, suggest why the reaction between N_2 and H_2 to give ammonia is exothermic.

Valence bond treatment of heteronuclear diatomics

The bonding in a heteronuclear diatomic molecule can be explained in a similar way to the bonding in a homonuclear diatomic such as H_2. The bonding in HF, for example, can be represented by the three resonance forms below

$$\text{H–F} \leftrightarrow \text{H}^+\,\text{F}^- \leftrightarrow \text{H}^-\,\text{F}^+$$

There is one major difference between the treatment of a homonuclear diatomic such as H_2 and a heteronuclear diatomic such as HF. In H_2 the two ionic contributions (for H^+ H^- and H^- H^+) are equal, whereas for HF this is not the case. Since fluorine is more electronegative than hydrogen, the contribution from the H^+ F^- resonance form is far more important than that from the H^- F^+ resonance form. This means that the electrons of the covalent bond are closer to the fluorine atom than they are to the hydrogen atom. Valence bond theory uses the different contributions from resonance forms to explain the polarity of molecules such as HF.

Defining the hydrogen 1*s* and fluorine $2p_z$ wavefunctions as ϕ_H and ϕ_F, respectively

$$\psi = \phi_{H(1)}\phi_{F(2)} + \phi_{H(2)}\phi_{F(1)} + x\phi_{F(1)}\phi_{F(2)} + y\phi_{H(1)}\phi_{H(2)}$$

where x and y are constants, and $x >> y$.

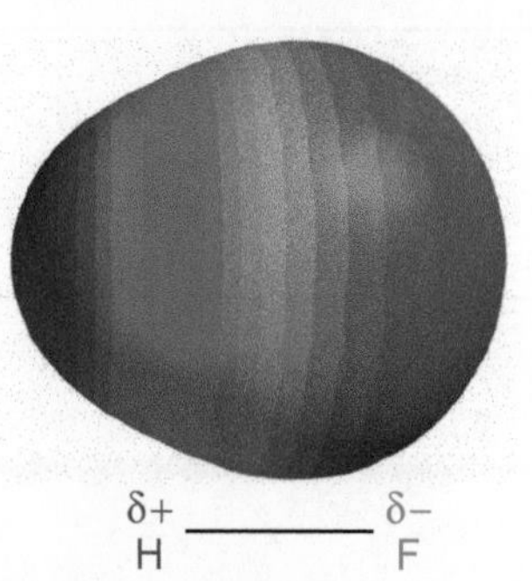
δ+
δ−
H
F
An electrostatic potential map for hydrogen fluoride, HF. The colours represent the charge distribution, with red areas having an abundance of electrons and blue areas a relative absence of electrons. This shows the polarity of the molecule.

Worked example 4.2 Valence bond treatment of LiH

Lithium hydride is an ionic solid under normal conditions of temperature and pressure, but it exists in the gas phase as LiH molecules. Draw three resonance forms to describe the bonding in the LiH molecule. Which of the ionic forms is more important?

Strategy

Draw out resonance forms analogous to those for HF. Use the relative values of electronegativity for lithium and hydrogen to determine which ionic form is more important.

Solution

$$\text{Li–H} \leftrightarrow \text{Li}^+\text{H}^- \leftrightarrow \text{Li}^-\text{H}^+$$

From Figure 4.6 (p.178), χ^P for hydrogen is 2.20 and χ^P for lithium is 0.98. Hydrogen is more electronegative, so the ionic form Li^+ H^- is more important than the ionic form Li^- H^+.

Question

For the molecule ICl, which of the ionic forms would you expect to contribute more to the molecular wavefunction?

Summary

- Valence bond theory uses resonance to describe covalent and ionic contributions to a molecular wavefunction.
- Double-headed arrows are used to show the different resonance forms that contribute to the wavefunction. The bonding is a weighted average of the resonance forms.
- Hybridization is the mixing of atomic orbitals on an atom as it interacts with another atom.
- For a heteronuclear diatomic, the resonance form with the negative charge on the more electronegative atom makes a greater contribution to the molecular wavefunction than the resonance form with the negative charge on the more electropositive atom. This accounts for the polarity of the bond.

? For practice questions on the topics in Sections 4.4 and 4.5, see questions 5 and 6 at the end of this chapter (p.213).

4.6 Molecular orbital theory

> **Greek symbols**
> ϕ (phi) an atomic orbital
> ψ (psi) a molecular orbital.
> Some books use the symbol ψ for both atomic and molecular orbitals.

Molecular orbital theory involves the formation of molecular orbitals (MOs) that are spread out over the molecule. An atomic orbital (AO) is a single-electron wavefunction on an atom (see Section 3.5, p.135) and, by analogy, a molecular orbital is a single-electron wavefunction on a molecule. As with valence bond theory, the Greek letter ϕ (phi) is used to identify an atomic orbital. In this case ψ represents a molecular orbital.

In addition to the Born–Oppenheimer approximation (p.179), there are two further important approximations that are made in molecular orbital theory—the orbital approximation and the linear combination of atomic orbitals.

The orbital approximation

The use of the orbital approximation in many-electron atoms is described in Section 3.6 (p.143).

For an atom, this approximation allows the overall wavefunction for an atom containing N electrons to be written as a product of N single-electron wavefunctions. This is useful, as single-electron wavefunctions can be calculated from the Schrödinger equation, whereas wavefunctions for more than one electron cannot.

The orbital approximation is also valid for molecules, so the wavefunction of an N-electron molecule Ψ is written as the product of N single-electron wavefunctions, as shown in Equation 4.7.

$$\Psi(1,2,3,\ldots N) = \psi(1)\times\psi(2)\times\psi(3)\times\ldots\times\psi(N) \tag{4.7}$$

The single-electron wavefunctions in molecules, ψ, are called **molecular orbitals**. The orbital approximation assumes that each electron moves in an average potential that incorporates the interactions with the nuclei and the other electrons in the molecule.

The linear combination of atomic orbitals

Constructive and destructive interference of waves is described in Section 3.2 (p.123).

The molecular orbitals are formed through an approach called the **linear combination of atomic orbitals (LCAO)**. As atomic orbitals are single-electron wavefunctions, they have wave character, and so the atomic orbitals on one atom interfere (interact) with those on another atom. This interference can occur both constructively and destructively.

Constructive interference results from the in-phase combination of the wavefunctions, and corresponds mathematically to adding the wavefunctions together. Destructive interference results from the out-of-phase combination of the wavefunctions, and is given mathematically by subtracting one wavefunction from the other. The LCAO approach allows molecular orbitals to be formed from *the addition or subtraction of atomic orbitals*. In the next section you will see how the linear combination of two hydrogen 1*s* orbitals is used to explain the bonding in H_2.

Summary

- There are two important approximations used in molecular orbital theory: these are the orbital approximation and the formation of molecular orbitals through the linear combination of atomic orbitals (LCAO).
- Using LCAO, molecular orbitals are formed from the addition or subtraction of atomic orbitals.

4.7 Molecular orbitals in hydrogen (H_2)

The *in-phase combination* is given by the sum of the atomic orbitals. If the two atoms in a hydrogen molecule are labelled H_A and H_B, then $\phi_{1s}(H_A)$ is the wavefunction of the 1*s* atomic orbital on H_A and $\phi_{1s}(H_B)$ is the wavefunction of the 1*s* atomic orbital on H_B.

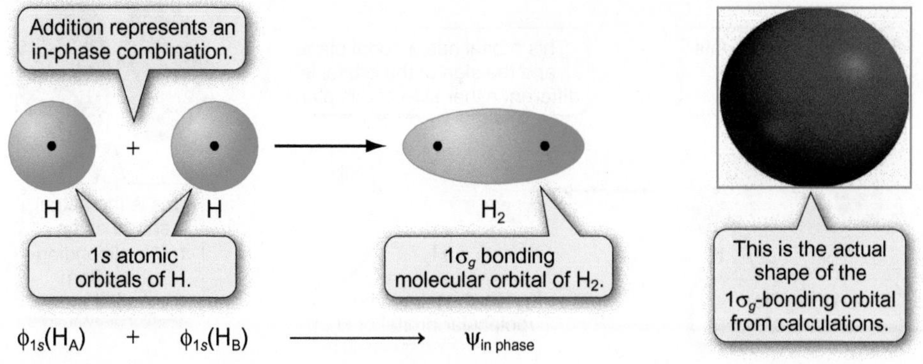
Addition represents an in-phase combination.
+
H
H
H_2
1s atomic orbitals of H.
$1\sigma_g$ bonding molecular orbital of H_2.
This is the actual shape of the $1\sigma_g$-bonding orbital from calculations.
$\phi_{1s}(H_A)$ + $\phi_{1s}(H_B)$
$\psi_{\text{in phase}}$

Figure 4.9 The boundary surface for the in-phase combination of two hydrogen 1*s* orbitals. The black dots represent the positions of the nuclei. The boundary surface contains 95% of the wavefunction.

The in-phase combination of these orbitals gives a molecular orbital, $\psi_{\text{in phase}}$, as shown by Equation 4.8

$$\psi_{\text{in phase}} = \phi_{1s}(H_A) + \phi_{1s}(H_B) \tag{4.8}$$

This combination is shown as a boundary surface in Figure 4.9. This type of figure is similar to the boundary surfaces for atomic orbitals in Chapter 3 (p.140) and gives a good representation of the shape of a molecular orbital. As with atomic orbitals, the molecular wavefunction tends towards zero with increasing distance from the nuclei, but only reaches zero at $r = \infty$. Boundary surfaces are normally drawn to contain 95% of the wavefunction. The orbital is usually shown more elongated than the actual calculated shape in order to emphasize its directionality.

For the in-phase combination of atomic orbitals, the wavefunction is greater between the nuclei than it was before the atoms interacted. The increased electron density between the nuclei holds the positive nuclei together—this means that a bond is formed. The molecular orbital arising from the in-phase combination of atomic orbitals is called a **bonding orbital**.

> ℹ The in-phase combination of atomic orbitals is given by addition, and the out-of-phase combination is given by subtraction.

The *out-of-phase combination* of the atomic orbitals is given by the difference between the atomic orbitals, as shown in Equation 4.9

$$\psi_{\text{out of phase}} = \phi_{1s}(H_A) - \phi_{1s}(H_B) \tag{4.9}$$

The boundary surface for the out-of-phase combination of hydrogen atomic orbitals is shown in Figure 4.10. In contrast to the in-phase combination, the wavefunction does not have the same phase over the entire surface. There is a nodal plane between the nuclei, and the phase of the wavefunction is different on either side of this. As with a *p* orbital, the wavefunction at any point on the nodal plane is zero, which means that the electron density, given by ψ^2, on this plane is also zero. Again, the orbital is often shown in an elongated manner to emphasize the directionality.

> ℹ As with atomic orbitals, the phases of wavefunctions can be represented by positive and negative signs or, as in this book, by different shading. Positive parts of the wavefunction are shown darker, whereas negative parts of the wavefunction are lighter.

The magnitude of the wavefunction for the out-of-phase combination is less between the nuclei than if there was no interaction between the atoms. The molecular orbital arising from the out-of-phase combination of atomic orbitals is called an **antibonding orbital**.

The combination of two hydrogen 1*s* atomic orbitals gives rise to two molecular orbitals, a bonding orbital and an antibonding orbital. Generally, *the number of molecular orbitals is always equal to the number of atomic orbitals from which they were formed.*

> ℹ The linear combination of two 1*s* atomic orbitals gives two molecular orbitals, one bonding and the other antibonding. Generally, the linear combination of *n* atomic orbitals gives *n* molecular orbitals.

Figure 4.11 shows the linear combinations of atomic orbitals graphically, using radial wavefunction plots. Box 4.5 (p.188) explains how the expressions for the bonding and antibonding orbitals in Equations 4.8 and 4.9 relate to electron density.

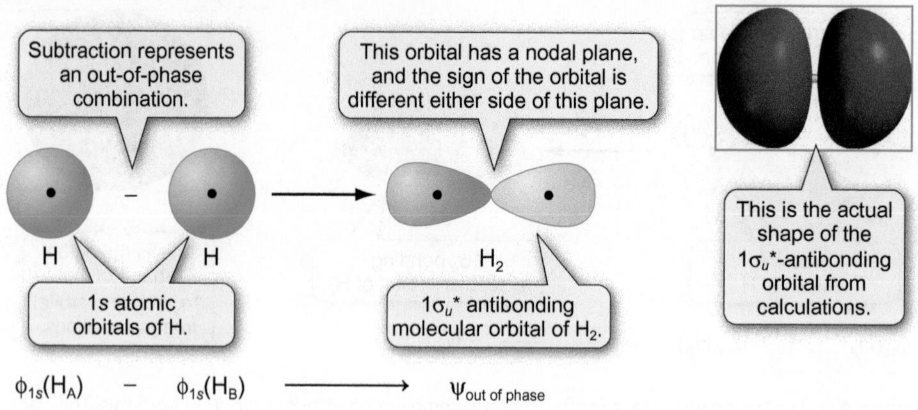
Subtraction represents an out-of-phase combination.
This orbital has a nodal plane, and the sign of the orbital is different either side of this plane.
This is the actual shape of the $1\sigma_u$*-antibonding orbital from calculations.
H
H
H_2
1s atomic orbitals of H.
$1\sigma_u$* antibonding molecular orbital of H_2.
$\phi_{1s}(H_A)$ − $\phi_{1s}(H_B)$ → $\psi_{\text{out of phase}}$

Figure 4.10 The boundary surface for the out-of-phase combination of two hydrogen 1*s* orbitals.

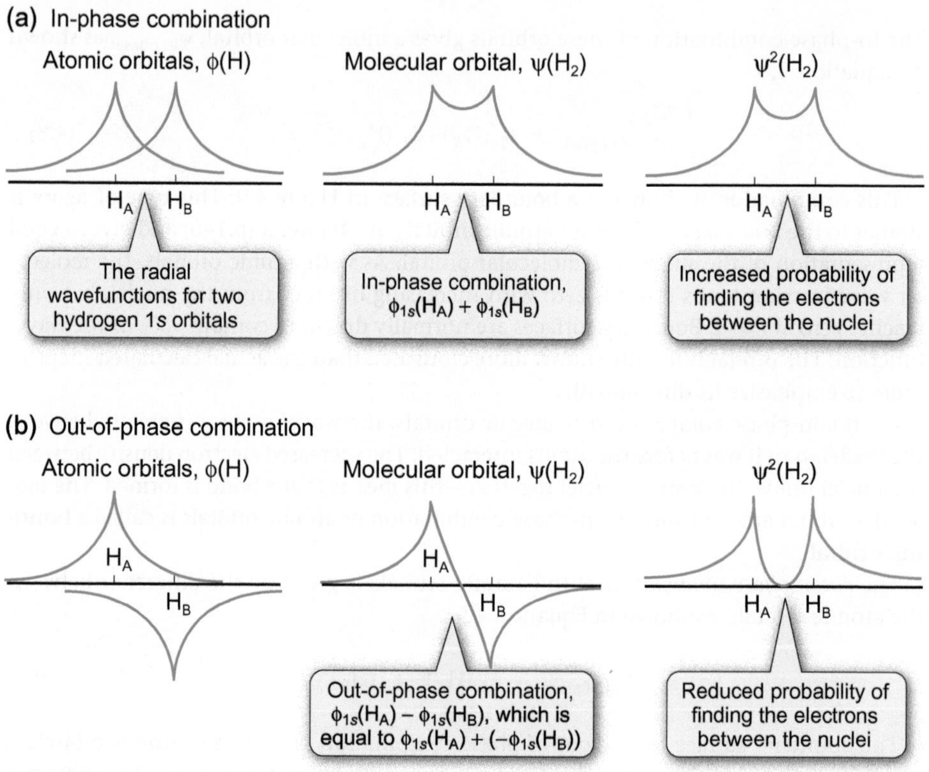
(a) In-phase combination
Atomic orbitals, ϕ(H)
Molecular orbital, $\psi(H_2)$
$\psi^2(H_2)$
H_A H_B
The radial wavefunctions for two hydrogen 1s orbitals
In-phase combination, $\phi_{1s}(H_A) + \phi_{1s}(H_B)$
Increased probability of finding the electrons between the nuclei
(b) Out-of-phase combination
Atomic orbitals, ϕ(H)
Molecular orbital, $\psi(H_2)$
$\psi^2(H_2)$
H_A H_B
Out-of-phase combination, $\phi_{1s}(H_A) - \phi_{1s}(H_B)$, which is equal to $\phi_{1s}(H_A) + (-\phi_{1s}(H_B))$
Reduced probability of finding the electrons between the nuclei

Figure 4.11 The radial wavefunctions for two hydrogen 1*s* orbitals combined in (a) an in-phase manner and (b) an out-of-phase manner.

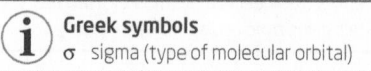

ℹ Radial wavefunction plots for the hydrogen atom are shown in Section 3.5 (p.138). The plots in Figure 4.11 extend in both directions away from the nuclei.

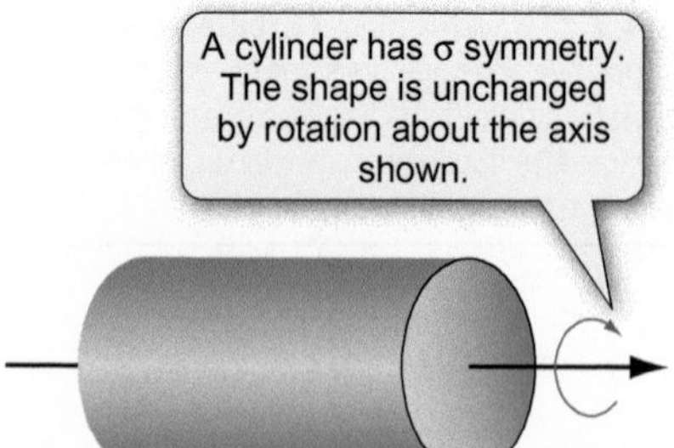
A cylinder has σ symmetry. The shape is unchanged by rotation about the axis shown.

Figure 4.12 A σ orbital, like the cylinder shown, is unchanged by rotating through any angle about the internuclear axis.

ℹ **Greek symbols**
σ sigma (type of molecular orbital)

Orbital labels

In Figures 4.9 and 4.10, the bonding and antibonding orbitals in H_2 have the labels $1\sigma_g$ and $1\sigma_u$*, respectively. The first part of each label (1) shows that these are the lowest energy σ_g and σ_u orbitals in the molecule. The second two parts—σ and either *g* or *u*—tell you about the symmetry of the orbital. An antibonding orbital is often also given a superscripted asterisk, so the $1\sigma_u$ orbital is labelled as the $1\sigma_u$* orbital. Asterisks tend only to be used for simple molecules, as in more complicated cases the bonding or antibonding nature of an orbital might be difficult to determine.

Both the bonding and antibonding orbitals in H_2 have cylindrical symmetry (Figure 4.12), and any orbital with this type of symmetry is called a σ orbital. If you rotate a σ orbital about the internuclear axis through any angle, the orbital does not change its appearance.

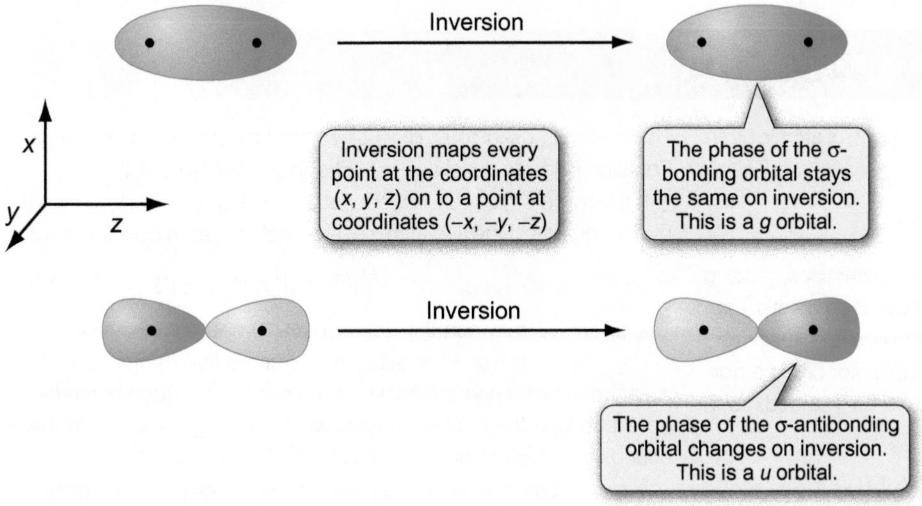
Inversion
x
y
z
Inversion maps every point at the coordinates (x, y, z) on to a point at coordinates (−x, −y, −z)
The phase of the σ-bonding orbital stays the same on inversion. This is a g orbital.
Inversion
The phase of the σ-antibonding orbital changes on inversion. This is a u orbital.

Figure 4.13 On inversion, the phase (sign) of a *g* orbital stays the same, but that of a *u* orbital changes.

Inversion is a mathematical operation under which every point at coordinates (*x*, *y*, *z*) moves to the coordinates (−*x*, −*y*, −*z*).

The symmetry label for an orbital contains information on its cylindrical symmetry (e.g. σ or π) and its parity (i.e. *g* or *u*).

Rotatable figures of the H_2 molecular orbitals.

The *g* and *u* labels are called **parity labels**, and these refer to the symmetry of the orbital with respect to the mathematical operation of inversion. When an orbital is inverted, every point at coordinates (*x*, *y*, *z*) is mapped on to a point at coordinates (−*x*, −*y*, −*z*), as shown in Figure 4.13.

If the orbital looks identical after inversion it is described as a *g* type orbital, from the German word *gerade* meaning even. If the orbital keeps the same shape but changes its phase it is described as a *u* type orbital, from the German word *ungerade* meaning odd.

Figure 4.13 shows that the σ-bonding orbital is unchanged on inversion, that is, it is σ_g, whereas the σ-antibonding orbital changes phase on inversion, that is, it is σ_u.

Worked example 4.3 Parity labels

Symmetry labels are not restricted to molecules, and can be applied to any shape. What parity labels (i.e. *g* or *u*) do the following shapes have?

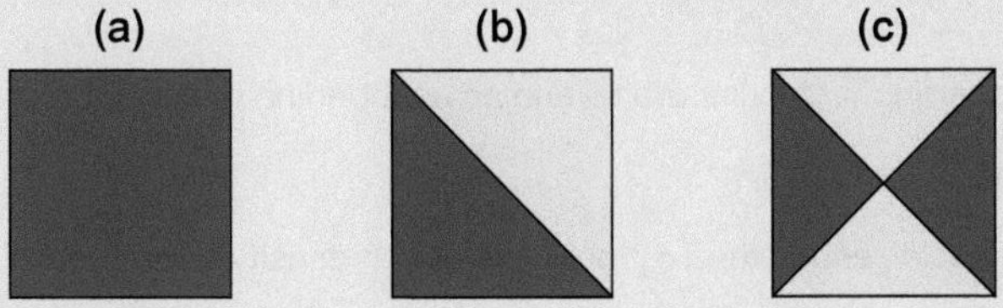
(a)
(b)
(c)

Strategy

Draw the shapes again after mapping all the points on to their inverse. The shapes are two-dimensional, so when they are inverted the point (*x*, *y*) maps on to (−*x*, −*y*). If the shape and arrangement of colours are identical after inversion, the object has *g* symmetry. If the colours have swapped positions, the object has *u* symmetry.

Solution

After inversion, the shapes are as follows.

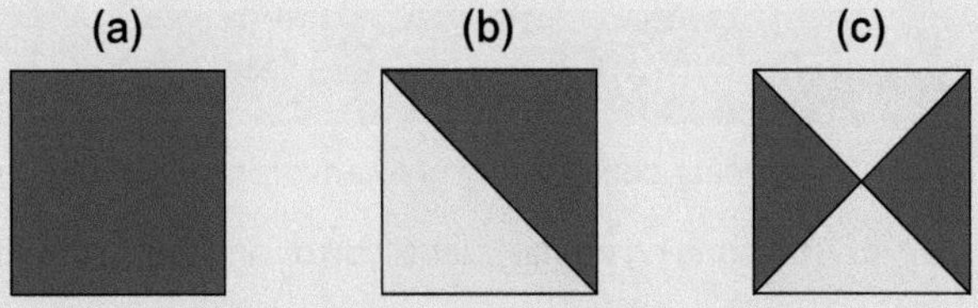
(a)
(b)
(c)

(a) and (c) have their colours in the same position as they were before the inversion, so have *g* symmetry. For (b), the colours have swapped positions, so this has *u* symmetry.

Question

What parity label does this hexagon have?

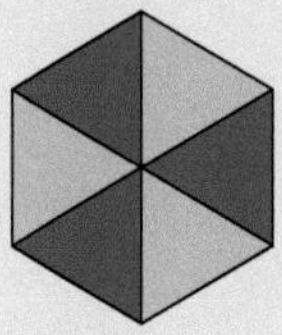

Box 4.5 Linear combinations of atomic orbitals

In Equation 4.8, the in-phase combination of atomic orbitals is given by

$$\psi_{\text{in phase}} = \phi_{1s}(H_A) + \phi_{1s}(H_B)$$

Since atomic orbitals are single-electron wavefunctions, adding these together gives a wavefunction with a magnitude greater than one. However, a molecular orbital is defined as a single-electron wavefunction, so for this to be correct a scaling factor called a **normalization constant** (often written N) needs to be included, as in Equation 4.10.

$$\psi_{\text{in phase}} = N[\phi_{1s}(H_A) + \phi_{1s}(H_B)] \quad (4.10)$$

N is a constant, with a value of less than one. In the case of H_2, $N = 0.5^{1/2}$

$$\psi_{\text{in phase}} = 0.5^{1/2} \times [\phi_{1s}(H_A) + \phi_{1s}(H_B)] \quad (4.11)$$

The electron density in a molecular orbital is given by ψ^2, so the electron density in the in-phase combination of atomic orbitals is found by squaring Equation 4.11.

$$\begin{aligned}\psi^2 &= [(0.5^{1/2}) \times [\phi_{1s}(H_A) + \phi_{1s}(H_B)]]^2 \\ &= (0.5^{1/2})^2 \times [\phi_{1s}(H_A) + \phi_{1s}(H_B)]^2 \\ &= 0.5\,[\phi_{1s}(H_A) + \phi_{1s}(H_B)] \times [\phi_{1s}(H_A) + \phi_{1s}(H_B)] \\ &= 0.5\,[\phi_{1s}(H_A)\phi_{1s}(H_A) + \phi_{1s}(H_A)\phi_{1s}(H_B) + \phi_{1s}(H_B)\phi_{1s}(H_A) \\ &\quad + \phi_{1s}(H_B)\phi_{1s}(H_B)] \\ &= 0.5\,[\{\phi_{1s}(H_A)\}^2 + 2\{\phi_{1s}(H_A)\phi_{1s}(H_B)\} + \{\phi_{1s}(H_B)\}^2] \quad (4.12)\end{aligned}$$

The methods used to multiply algebraic functions such as that in Equation 4.11 are described in Maths Toolkit MT2 (p.1314).

If the two atomic orbitals were not interacting, ψ^2 would be given by the squares of the two atomic wavefunctions, as in Equation 4.13

$$\psi^2_{\text{no interaction}} = 0.5[\{\phi_{1s}(H_A)\}^2 + \{\phi_{1s}(H_B)\}^2] \quad (4.13)$$

The difference between the Equations 4.12 and 4.13 is $+2\{\phi_{1s}(H_A)\phi_{1s}(H_B)\}$. This has a positive value, which means that electron density is enhanced between the nuclei as a result of constructive interference between the atomic orbitals, which is in agreement with the boundary surface of the bonding orbital in Figure 4.9 (p.185).

A similar process can be applied for the out-of-phase combination, using the same normalization constant.

$$\psi_{\text{out of phase}} = 0.5^{1/2} \times [\phi_{1s}(H_A) - \phi_{1s}(H_B)] \quad (4.14)$$

As before, the electron density is given by ψ^2

$$\begin{aligned}\psi^2 &= [(0.5^{1/2}) \times [\phi_{1s}(H_A) - \phi_{1s}(H_B)]]^2 \\ &= 0.5\,[\phi_{1s}(H_A) - \phi_{1s}(H_B)] \times [\phi_{1s}(H_A) - \phi_{1s}(H_B)] \\ &= 0.5\,[\phi_{1s}(H_A)\phi_{1s}(H_A) - \phi_{1s}(H_A)\phi_{1s}(H_B) \\ &\quad - \phi_{1s}(H_B)\phi_{1s}(H_A) + \phi_{1s}(H_B)\phi_{1s}(H_B)] \\ &= 0.5\,[\{\phi_{1s}(H_A)\}^2 - 2\{\phi_{1s}(H_A)\phi_{1s}(H_B)\} + \{\phi_{1s}(H_B)\}^2] \quad (4.15)\end{aligned}$$

The difference between Equations 4.15 and 4.13 is the term $-2\{\phi_{1s}(H_A)\phi_{1s}(H_B)\}$, which is negative in Equation 4.15. This means that electron density is reduced between the nuclei as a result of destructive interference between the atomic orbitals, which is in agreement with the boundary surface of the antibonding orbital in Figure 4.10 (p.186).

Summary

- The linear combination of two hydrogen 1s orbitals gives a bonding molecular orbital and an antibonding molecular orbital.
- The bonding orbital is labelled $1\sigma_g$ and the antibonding orbital is labelled $1\sigma_u{}^*$.
- In the $1\sigma_g$ orbital, electron density is enhanced between the nuclei, and in the $1\sigma_u{}^*$ orbital, electron density is reduced between the nuclei.

4.8 Molecular orbital energy level diagrams

You can show the relative energies of the σ_g and $\sigma_u{}^*$ orbitals on a **molecular orbital energy level diagram**. An example is shown in Figure 4.14. In this type of diagram, the atomic orbitals of the two atoms are shown on the left- and right-hand sides and the resultant

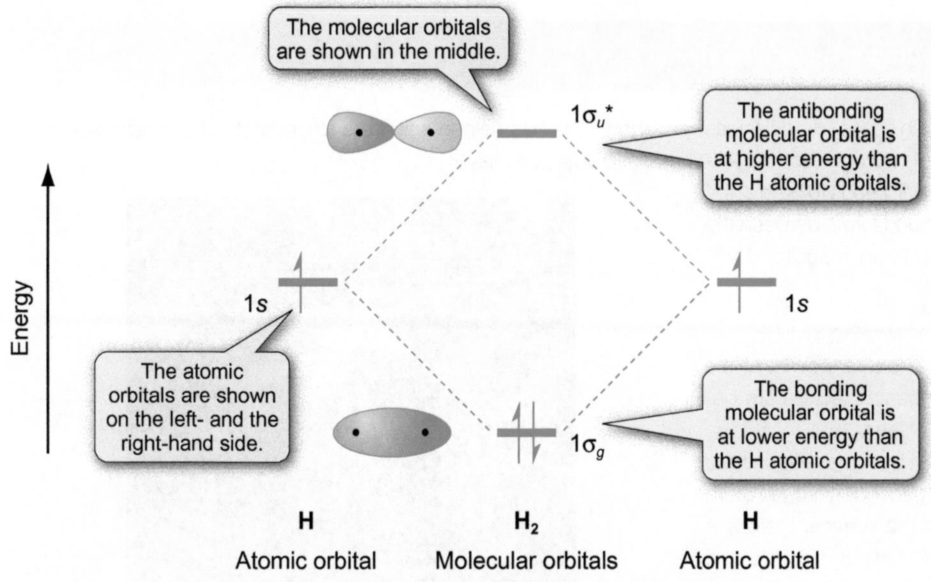
The molecular orbitals are shown in the middle.
1σu*
The antibonding molecular orbital is at higher energy than the H atomic orbitals.
Energy
1s
1s
The atomic orbitals are shown on the left- and the right-hand side.
1σg
The bonding molecular orbital is at lower energy than the H atomic orbitals.
H
H2
H
Atomic orbital
Molecular orbitals
Atomic orbital

Figure 4.14 The molecular orbital energy level diagram for H_2. H_2 has a filled bonding orbital and an empty antibonding orbital so the bond order in H_2 is 1.

molecular orbitals are placed in the centre. The *y*-axis represents increasing energy. Dotted lines are shown connecting the molecular orbitals with the atomic orbitals that contribute to them. Electrons are placed into the molecular orbitals using the **aufbau principle**, with the **Pauli exclusion principle** applying in the same way as it does for atomic orbitals. Each molecular orbital can contain a maximum of two electrons, with their spins in opposite directions.

The filling of atomic orbitals is described in Section 3.6 (p.145).

The molecular orbital energy level diagram for hydrogen (H_2) is shown in Figure 4.14. The $1\sigma_g$ bonding molecular orbital is lower in energy than the atomic orbitals, and the $1\sigma_u{}^*$ antibonding molecular orbital is higher in energy than the atomic orbitals. In Figure 4.14, the amount by which the $1\sigma_g$ orbital decreases in energy is shown as equal to the amount by which the $1\sigma_u{}^*$ orbital increases in energy. This is not strictly true, as experimental measurements show that the increase in energy of the $1\sigma_u{}^*$ orbital from the contributing atomic orbitals has a greater magnitude than the decrease in energy of the $1\sigma_g$ orbital from the atomic orbitals. This is because the approximations used (Section 4.6) are not completely accurate.

There are two electrons in H_2, one from each hydrogen atom, and these are both in the $1\sigma_g$ orbital with their spins paired. The **bond order** is a measure of the stabilization resulting from the formation of the molecule, and in the Lewis model it was defined as the number of electron pairs shared between two atoms. In molecular orbital theory, the bond order is given by the expression in Equation 4.16. An orbital containing a single electron counts as a half-filled orbital

$$\text{bond order} = \frac{\text{number of filled}}{\text{bonding orbitals}} - \frac{\text{number of filled}}{\text{antibonding orbitals}} \qquad (4.16)$$

For H_2 there is one filled bonding orbital ($1\sigma_g$) and no filled antibonding orbitals, as the $1\sigma_u{}^*$ orbital is empty. The bond order in H_2 is $(1 - 0) = 1$. Molecular orbital theory predicts a single bond between the two hydrogen atoms, which is the same as that predicted by both the Lewis model and valence bond theory.

Box 4.6 The formation of hydrogen molecules in space

H_2 is the most abundant molecule in space. H_2 molecules cannot be formed simply from the collision of hydrogen atoms. Another molecule or particle, called a third body (M), needs to be present to carry away the energy released when the bond forms between the atoms, otherwise the two atoms would simply fly apart again.

$$H + H + M \rightarrow H_2 + M$$

This type of collision, involving three species, is a rare event at the very low pressures in space. Instead, hydrogen molecules are believed to form on the surface of dust particles (see p.169).

When a hydrogen atom hits a dust particle it becomes attached to the surface, but is able to move around. If a second hydrogen atom arrives before the first has evaporated from the surface, there is a chance that the two might collide and react to form H_2. Some of the heat released by the bond-forming process is absorbed by the dust particle, but some causes the newly formed molecule to evaporate from the surface.

Experiments on Earth have shown that hydrogen atoms combine much more efficiently on silicate particles than they do on amorphous carbon. This has helped scientists to understand the nature of the cosmic dust particles.

▲ Molecular clouds in the Eagle Nebula contain relatively high concentrations of H_2.

Molecular hydrogen ions

The ion H_2^+ has been detected in the gas phase, and there is also some evidence for H_2^-. The existence of these ions can be explained using the molecular orbital energy level diagrams shown in Figure 4.15. The molecular orbitals are identical to those for H_2—the two 1*s* atomic orbitals combine to give σ_g and σ_u* molecular orbitals. The only difference is the number of electrons present, as H_2^+ has only one electron and H_2^- has three. As a result, the bond in both ions is weaker than that in H_2. You can see this most clearly by working out the bond orders for the two ions. A half-filled orbital contributes $\frac{1}{2}$ to the bond order.

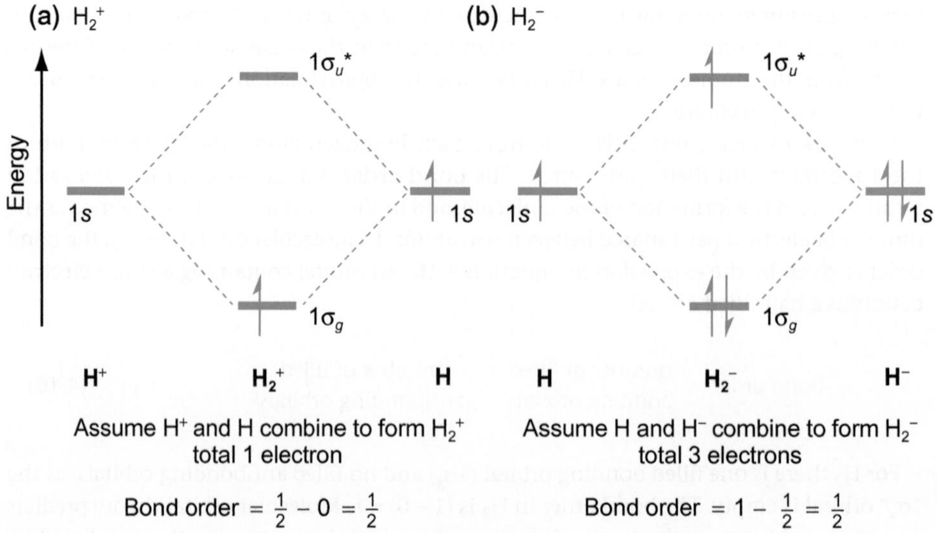
(a) H2+
(b) H2−
Energy
1σu*
1s
1s
1σg
1σu*
1s
1s
1σg
H+
H2+
H
H
H2−
H−
Assume H+ and H combine to form H2+
total 1 electron
Assume H and H− combine to form H2−
total 3 electrons
Bond order = 1/2 − 0 = 1/2
Bond order = 1 − 1/2 = 1/2

Figure 4.15 The molecular orbital energy level diagrams for (a) H_2^+ and (b) H_2^-.

Table 4.2 Properties of H_2^+ and H_2

	Bond order	D/kJ mol^{-1}	Bond length/pm	Magnetic properties
H_2^+	$\frac{1}{2}$	+256	105.2	Paramagnetic
H_2	1	+435.8	74.1	Diamagnetic

The bond dissociation enthalpies and bond lengths for H_2^+ and H_2 are given in Table 4.2. Generally, for the same pair of atoms, *as the bond order increases, the bond dissociation enthalpy increases and the bond length decreases*. Although the value is not well established, the bond dissociation enthalpy for H_2^- is lower than that for H_2^+, despite the bond orders for the two ions being identical. This is because the greater electron–electron repulsion present in H_2^- weakens the bond. Both H_2^+ and H_2^- are paramagnetic since both ions contain an unpaired electron.

The importance of H_2^+ for the magnetic field of the planet Jupiter is described in Chapter 25 (p.1139).

Helium molecules (He_2)

The 1*s* orbitals on two helium atoms can also interact to form a σ_g bonding orbital and a σ_u* antibonding orbital. Why then doesn't He_2 exist? This is easy to understand using the molecular orbital energy level diagram in Figure 4.16. The molecular orbitals arise from the in-phase and out-of-phase combinations of 1*s* orbitals, so are the same as those in H_2. Each helium atom contributes two electrons, so there would be four electrons in the He_2 molecule. This means that both the $1\sigma_g$ and $1\sigma_u$* orbitals would be filled. The bond order in He_2 is given by (1 – 1) = 0. This means that there is no stabilization on forming a bond. Generally, *a molecule can only exist if the bond order between the atoms is greater than zero*.

Although He_2 does not exist, the He_2^+ ion has been detected in the gas phase, and shown to have a bond dissociation enthalpy of +297 kJ mol^{-1}. Since He_2^+ has one electron less than He_2, there is only one electron in the $1\sigma_u$* orbital for this ion. He_2^+ has a bond order of $\left(1-\frac{1}{2}\right)=\frac{1}{2}$.

Lithium molecules (Li_2)

2*s* orbitals interact in a similar manner to 1*s* orbitals. Two 2*s* orbitals combine to give a $2\sigma_g$ bonding molecular orbital and a $2\sigma_u$* antibonding molecular orbital. The molecular orbital energy level diagram for Li_2 is shown in Figure 4.17. Figure 4.17 shows both the

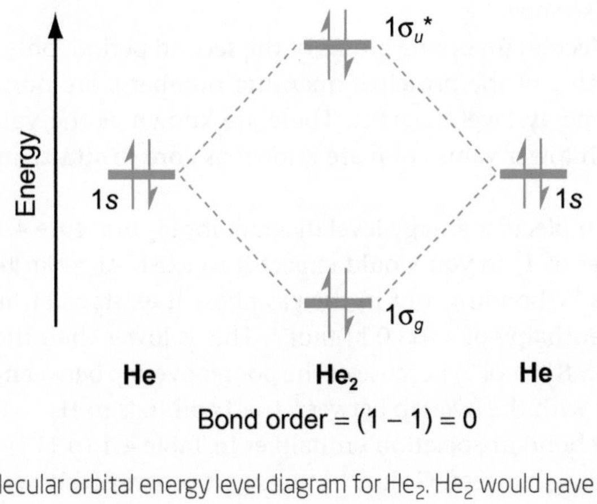

$1\sigma_u$*
Energy
1s
1s
$1\sigma_g$
He
He_2
He
Bond order = (1 – 1) = 0

Figure 4.16 The molecular orbital energy level diagram for He_2. He_2 would have a bond order of zero, so it does not exist.

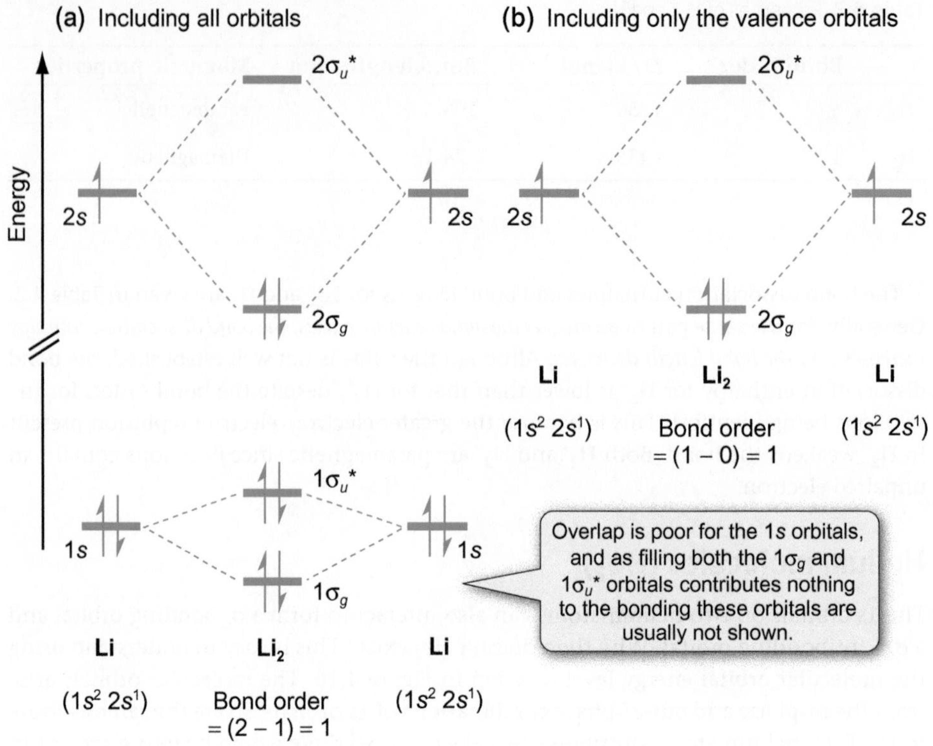

(a) Including all orbitals
(b) Including only the valence orbitals
Energy
2σu*
2s
2σg
1σu*
1s
1σg
Li
Li2
(1s2 2s1)
Bond order = (2 – 1) = 1
Bond order = (1 – 0) = 1
Overlap is poor for the 1s orbitals, and as filling both the 1σg and 1σu* orbitals contributes nothing to the bonding these orbitals are usually not shown.

Figure 4.17 The molecular orbital energy level diagram for Li_2, which has a bond order of 1. In (a) all of the molecular orbitals are shown, but in (b) the 1*s* orbitals and their resultant molecular orbitals are omitted as they contribute nothing to the bonding.

full molecular orbital energy level diagram for Li_2 and a simplified version that omits the 1*s* orbitals and their resultant molecular orbitals. The occupation of both the $1\sigma_g$ and $1\sigma_u$* orbitals leads to a net bond order of zero, so these orbitals contribute nothing to the bonding.

Another reason for not including these orbitals in the energy level diagram is that in a molecule such as Li_2, in which the 2*s* orbitals are interacting together, the distance between the nuclei is much greater than it is in H_2. The bond length in Li_2 is 267.3 pm, whereas that in H_2 is 74.1 pm. This means that in Li_2 the smaller 1*s* orbitals will not interact very much as the vast majority of their wavefunctions lie closer to the nuclei. This is shown in Figure 4.18. The degree of interaction between atomic orbitals is known as **overlap**. When the overlap is large, the interaction between the orbitals is large, but when the overlap is small the interaction is also small.

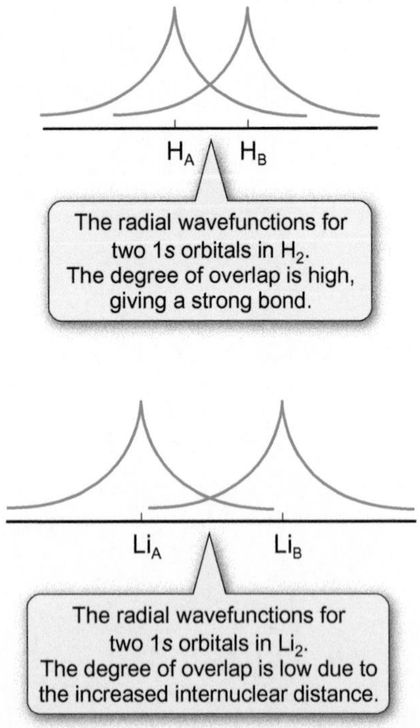

HA
HB
The radial wavefunctions for two 1s orbitals in H2. The degree of overlap is high, giving a strong bond.
LiA
LiB
The radial wavefunctions for two 1s orbitals in Li2. The degree of overlap is low due to the increased internuclear distance.

Figure 4.18 The interactions between 1*s* orbitals in H_2 and Li_2.

For diatomic molecules involving atoms of the second period, only the atomic orbitals with the highest value of the principal quantum number *n* are normally shown in the molecular orbital energy level diagram. These are known as the **valence orbitals**. Any atomic orbitals with lower values of *n* are known as **core orbitals**, and these are usually omitted.

Returning to the molecular energy level diagram for Li_2 in Figure 4.17, you can see that Li_2 has a bond order of 1, so you would expect it to exist. At room temperature lithium is a solid with metallic bonding, but in the gas phase it exists as Li_2 molecules and has a bond dissociation enthalpy of +105.0 kJ mol^{-1}. This is lower than the bond dissociation enthalpy of H_2 (+435.8 kJ mol^{-1}) because of the poorer overlap between the diffuse 2*s* orbitals when compared with the overlap between the 1*s* orbitals in H_2.

The values of the bond dissociation enthalpies in Table 4.1 (p.171) for Li_2, Na_2, and K_2 show that the bonding between Group 1 atoms gets progressively weaker as the group is descended. This is due to the increasingly diffuse nature of the *ns* orbitals as *n* increases.

Worked example 4.4 The Be_2 molecule

Would you expect Be_2 to exist?

Strategy

Construct a molecular orbital energy level diagram for Be_2 and determine its bond order. If the bond order is greater than zero, you would expect Be_2 to exist.

Solution

Beryllium has the electronic configuration $1s^2\ 2s^2$. Interactions between the core 1*s* orbitals can be ignored, so the molecular orbital energy level diagram for Be_2 involves the interactions between the two Be 2*s* orbitals to form a $2\sigma_g$ bonding orbital and a $2\sigma_u{}^*$ antibonding orbital. There are four electrons to put into these molecular orbitals.

The bond order is given by the number of filled bonding orbitals minus the number of filled antibonding orbitals, so the bond order for Be_2 is (1 – 1) = 0. Therefore, you would not expect Be_2 to exist.

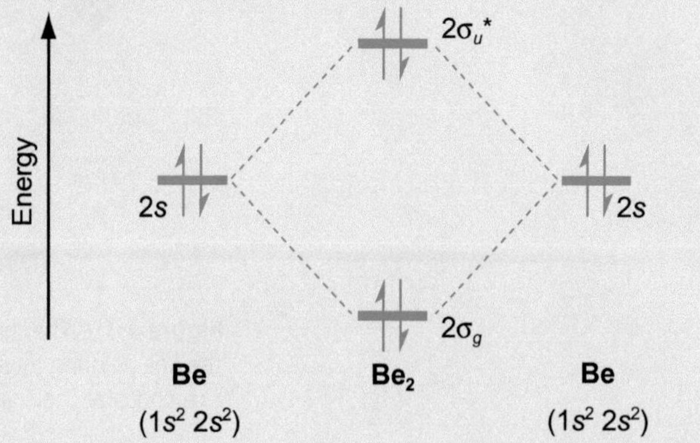

Energy
$2\sigma_u{}^*$
2s
2s
$2\sigma_g$
Be
Be_2
Be
($1s^2\ 2s^2$)
($1s^2\ 2s^2$)

Question

Would you expect $Be_2{}^+$ to exist?

Summary

- Electrons are placed into the molecular orbitals in an energy level diagram using the aufbau principle, in accordance with the Pauli exclusion principle.
- The bond order in a diatomic molecule is equal to the number of filled bonding orbitals minus the number of filled antibonding orbitals, with a single electron in an orbital contributing $\frac{1}{2}$.
- A molecule can exist if the bond order is greater than zero.
- When the degree of overlap between atomic orbitals is large, the interaction between them is large.
- Molecular orbital energy level diagrams can be used to predict whether molecules or ions exist.

For practice questions on the topic in Sections 4.6–4.8, see questions 7–10 at the end of this chapter (pp.213–214).

4.9 Linear combinations of *p* orbitals

Linear combinations of the 1*s* or 2*s* orbitals explain the existence of H_2 and Li_2, and the non-existence of He_2 and Be_2. To explain the bonding in the diatomic molecules of the other second period elements (B_2, C_2, N_2, O_2, and F_2), you also need to consider the interactions between the 2*p* orbitals.

Since *p* orbitals are directional, the p_x, p_y, and p_z orbitals do not all interact in the same way. For a general diatomic molecule, X_2, the *z*-axis is normally defined as running along the X–X bond. This means that the p_z orbitals on the two X atoms are directed towards each other, whereas the p_x and p_y orbitals on each atom are oriented at 90° to the direction of the other atom, as shown in Figure 4.19.

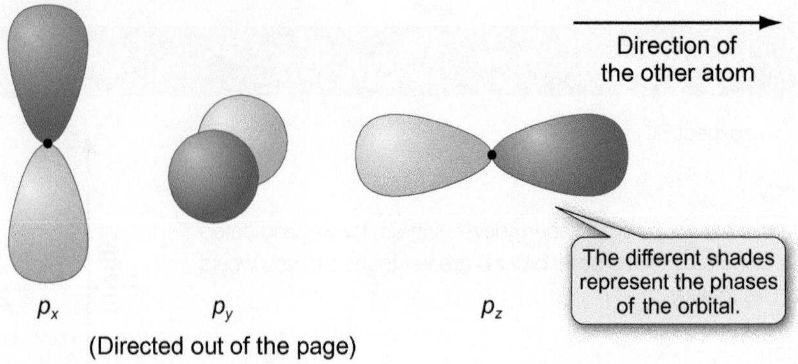
Direction of
the other atom
The different shades
represent the phases
of the orbital.
p_x
p_y
(Directed out of the page)
p_z

Figure 4.19 The three p orbitals point in different directions. In a diatomic molecule, the p_z orbital is the one pointing towards the other atom. In this chapter, the simplified forms of the p orbitals are used to emphasize their directionality (see Section 3.5, p.141).

Boundary surfaces showing the in-phase and out-of-phase combinations of two p_z orbitals are shown in Figure 4.20. The in-phase combination leads to a σ_g bonding orbital, with increased electron density between the nuclei. In contrast, the out-of-phase combination leads to a $\sigma_u{}^*$ antibonding orbital, with reduced electron density between the nuclei.

The p_x and p_y orbitals are perpendicular to the X–X axis. Boundary surfaces showing the in-phase and out-of-phase combinations of two p_x orbitals are shown in Figure 4.21. The molecular orbitals derived from the p_y orbitals are identical in shape, but rotated by 90° around the internuclear axis and so go into and out of the page.

The in-phase interaction of two p_x orbitals gives rise to a bonding orbital, with enhanced electron density between the nuclei. This orbital does not have cylindrical symmetry, as a rotation of 180° around the X–X axis maps the orbital onto itself, but with a change in

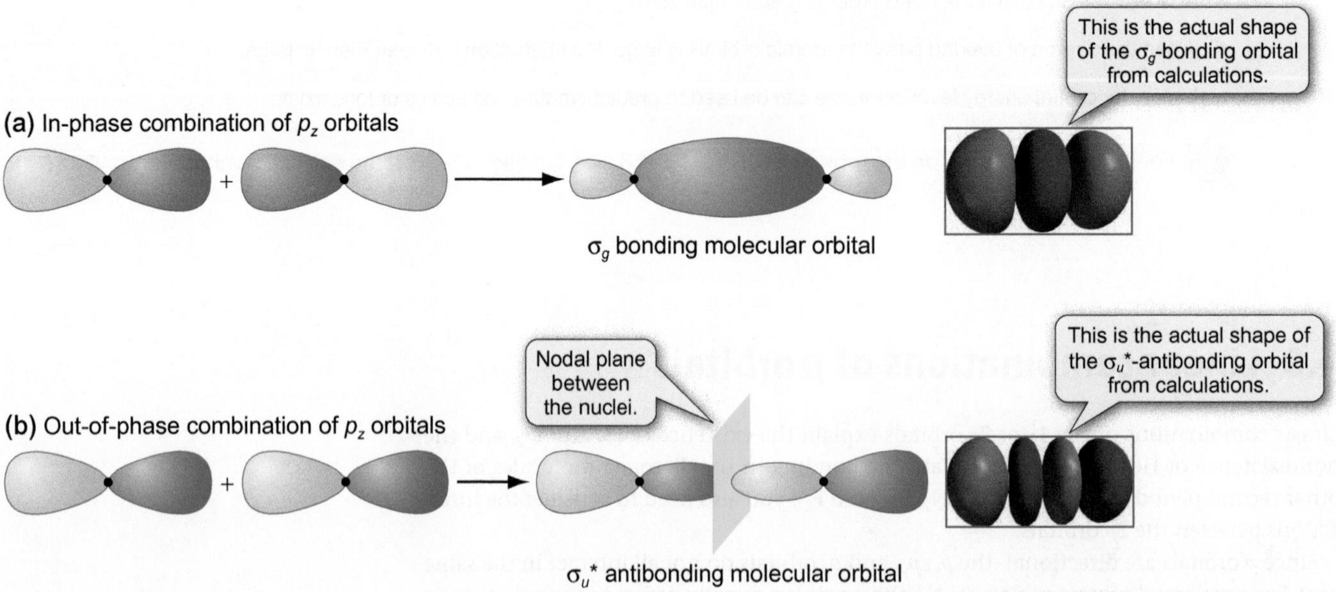
This is the actual shape
of the σ_g-bonding orbital
from calculations.
(a) In-phase combination of p_z orbitals
+
σ_g bonding molecular orbital
Nodal plane
between
the nuclei.
This is the actual shape of
the $\sigma_u{}^*$-antibonding orbital
from calculations.
(b) Out-of-phase combination of p_z orbitals
+
$\sigma_u{}^*$ antibonding molecular orbital

Figure 4.20 Boundary surfaces for the combination of two p_z atomic orbitals. (a) The in-phase combination gives a σ_g bonding molecular orbital and (b) the out-of-phase combination gives a $\sigma_u{}^*$ antibonding molecular orbital.

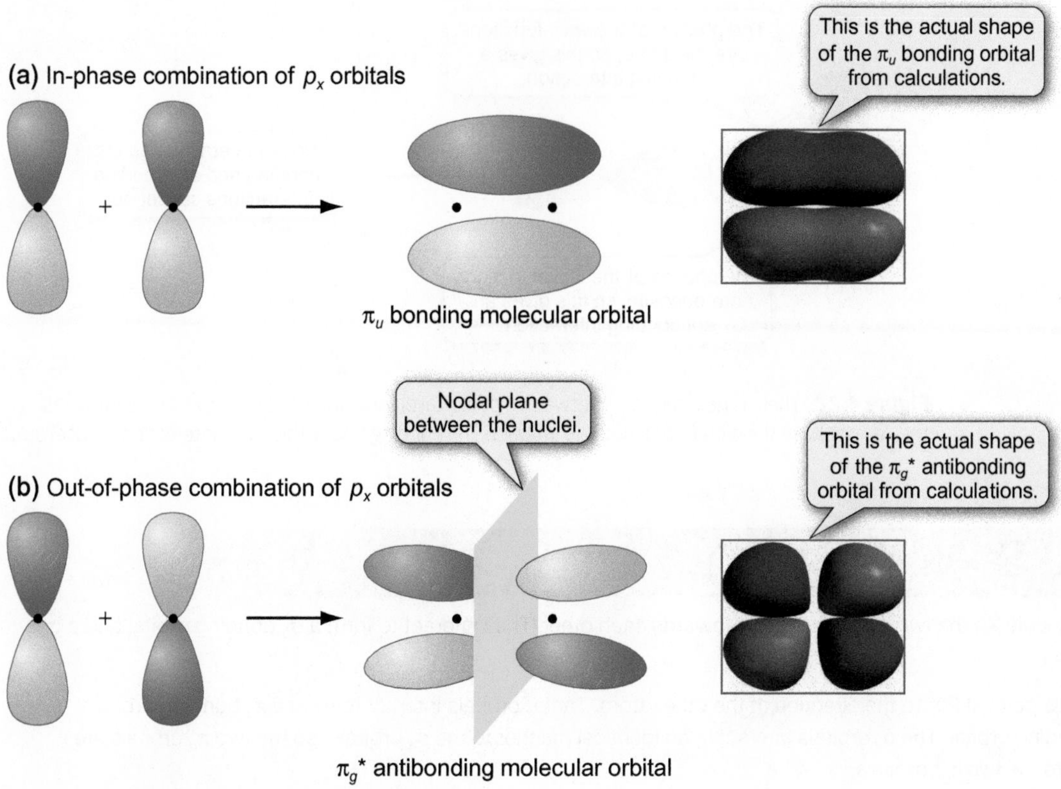
(a) In-phase combination of p_x orbitals
This is the actual shape of the π_u bonding orbital from calculations.
+
π_u bonding molecular orbital
Nodal plane between the nuclei.
(b) Out-of-phase combination of p_x orbitals
This is the actual shape of the $\pi_g{}^*$ antibonding orbital from calculations.
+
$\pi_g{}^*$ antibonding molecular orbital

Figure 4.21 Boundary surfaces for the combination of two p_x atomic orbitals. (a) The in-phase combination gives a π_u bonding molecular orbital and (b) the out-of-phase combination gives a $\pi_g{}^*$ antibonding molecular orbital.

phase. This is called a π orbital, and the symmetry label is π_u as, on inversion, the phase of the orbital changes. The out-of-phase interaction of two p_x orbitals gives an antibonding orbital, with reduced electron density between the nuclei. This orbital is also a π orbital, but it does not change its phase on inversion, so the antibonding orbital has the symmetry label $\pi_g{}^*$.

Greek symbols
π pi (type of molecular orbital)

For π orbitals in homonuclear diatomic molecules, the π_u is bonding and the π_g is antibonding. This is opposite to the labelling of the σ orbitals.

The two p_y orbitals interact in an identical manner to the two p_x orbitals to give a bonding π_u orbital and an antibonding $\pi_g{}^*$ orbital. The two pairs of π_u bonding and $\pi_g{}^*$ antibonding orbitals are degenerate.

The interactions between p_x orbitals and between p_y orbitals (**π-overlap**) are generally less effective than the interactions between p_z orbitals (**σ-overlap**), as the atomic orbitals are not pointing directly at each other. This means that the π orbitals are stabilized and destabilized relative to the atomic orbitals by a lesser degree than the σ orbitals in molecular orbital energy level diagrams, as you will see in the next section. Also, π-overlap is more distance sensitive than σ-overlap, with the bond dissociation enthalpy decreasing faster as the internuclear separation increases. This explains why multiple bonds are more common in molecules of second period elements than they are for heavier elements where bond lengths are longer. For example, elemental nitrogen exists as N_2 molecules with N≡N triple bonds, but phosphorus exists in several forms (allotropes), each of which contains P–P single bonds.

For more details on nitrogen and phosphorus chemistry see Section 27.4 (p.1223).

Figures 4.20 and 4.21 show how p_x, p_y, and p_z orbitals interact to give σ and π molecular orbitals. But why is there no interaction between the p_x orbital on one atom and the p_z orbital on the other? As you can see from Figure 4.22, any bonding interactions are counterbalanced by equal and opposite antibonding interactions. This means there is no net interaction. The orbitals are said to have the wrong symmetry to interact with each other.

The activity on the Online Resource Centre allows you to investigate how atomic orbitals interact to give molecular orbitals.

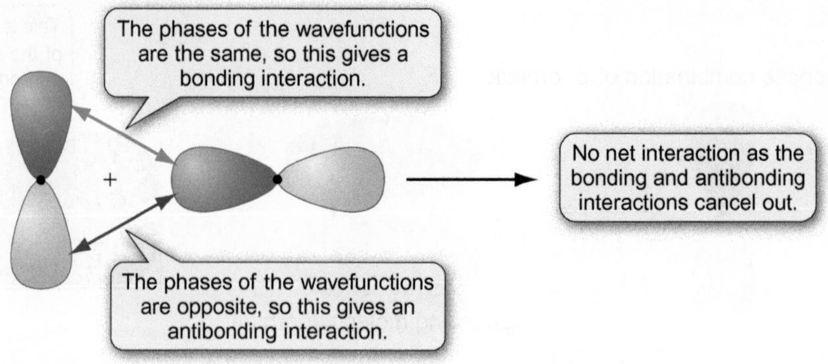
The phases of the wavefunctions are the same, so this gives a bonding interaction.
+
No net interaction as the bonding and antibonding interactions cancel out.
The phases of the wavefunctions are opposite, so this gives an antibonding interaction.

Figure 4.22 There is no interaction between a p_x orbital on one atom and a p_z orbital on another as they do not have the correct symmetry to interact. The bonding and antibonding interactions cancel out.

Summary

- In a diatomic molecule X_2, the two p_z orbitals point towards each other. They interact to form a σ_g bonding orbital and a $\sigma_u{}^*$ antibonding orbital.
- The two p_x orbitals point at 90° to the direction of the other atom. The p_x orbitals interact to form a π_u bonding orbital and a $\pi_g{}^*$ antibonding orbital. The p_y orbitals interact in an identical manner to the p_x orbitals, so the two π_u orbitals are degenerate, as are the two $\pi_g{}^*$ orbitals.
- π-overlap is weaker than σ-overlap and more sensitive to distance.
- Atomic orbitals only interact to give molecular orbitals if they have the correct symmetry.

4.10 Bonding in fluorine (F_2) and oxygen (O_2)

Figure 4.23 shows the molecular orbital energy level diagram for fluorine (F_2). (The overlap of the 1s atomic orbitals is poor and does not contribute to the bonding.) The 2s atomic orbitals interact as in Li_2 to give a $2\sigma_g$ bonding and a $2\sigma_u{}^*$ antibonding orbital. The 2p orbitals combine as described in Section 4.9 to give σ_g and π_u bonding orbitals and $\sigma_u{}^*$ and $\pi_g{}^*$ antibonding orbitals, with the σ_g stabilized more than the π_u, and the $\sigma_u{}^*$ destabilized more than the $\pi_g{}^*$. The σ_g and $\sigma_u{}^*$ orbitals derived from the $2p_z$ orbitals are the third set of orbitals with these symmetry labels in the molecule, so they are labelled $3\sigma_g$ and $3\sigma_u{}^*$, respectively. The π_u and $\pi_g{}^*$ are the first types of these orbitals in the molecule, so they are labelled $1\pi_u$ and $1\pi_g{}^*$.

Each fluorine atom has the valence electronic configuration $2s^2\ 2p^5$, so there are a total of $(2 \times 7) = 14$ electrons to put into the molecular orbitals. There are four filled bonding orbitals ($2\sigma_g$, $3\sigma_g$, and two $1\pi_u$) and three filled antibonding orbitals ($2\sigma_u{}^*$ and two $1\pi_g{}^*$). This gives a bond order for F_2 of $(4 - 3) = 1$, in agreement with the Lewis model (Figure 4.3, p.173). The $1\pi_g{}^*$ orbitals are the **highest occupied molecular orbitals** (HOMOs) and the $3\sigma_u{}^*$ orbital is the **lowest unoccupied molecular orbital** (LUMO). HOMO and LUMO are commonly used terms in molecular orbital theory and together are sometimes called **frontier orbitals**. The way a molecule reacts is often related to the nature and energies of the frontier orbitals.

Oxygen is fluorine's left-hand neighbour in the second period, and has a valence electronic configuration of $2s^2\ 2p^4$. The arrangement of molecular orbitals is the same as for F_2, but for O_2 there are a total of $(2 \times 6) = 12$ electrons to accommodate. The molecular orbital energy level diagram for O_2 is shown in Figure 4.24. Filling the orbitals using the

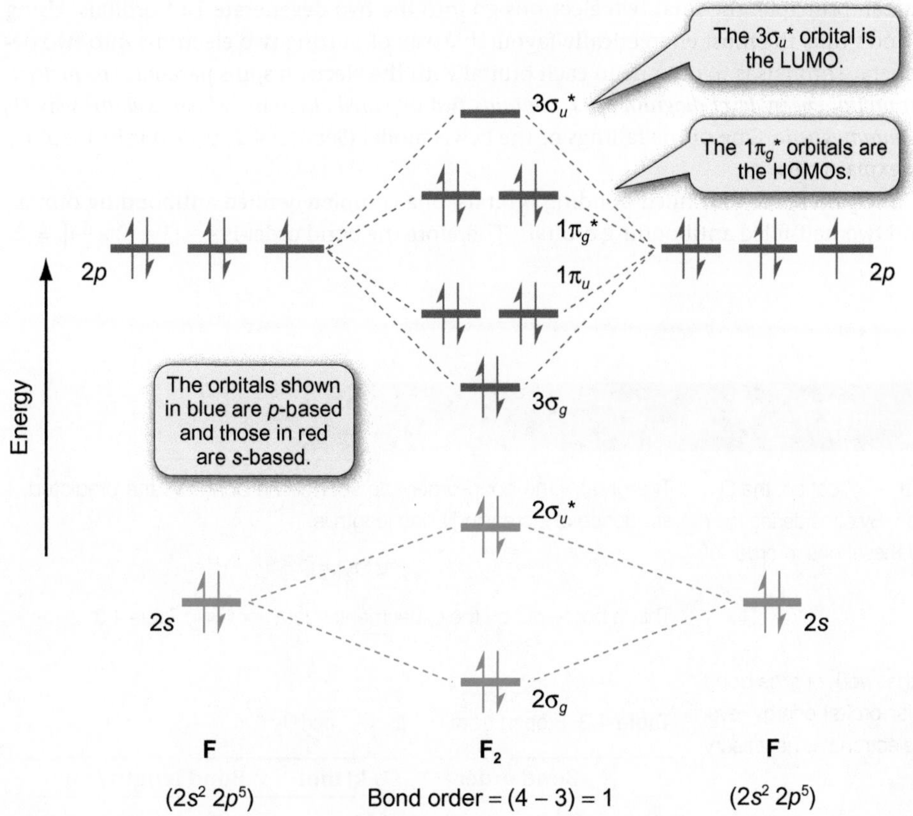

The $3\sigma_u^*$ orbital is the LUMO.
The $1\pi_g^*$ orbitals are the HOMOs.
The orbitals shown in blue are *p*-based and those in red are *s*-based.
$3\sigma_u^*$
$1\pi_g^*$
$1\pi_u$
$3\sigma_g$
$2\sigma_u^*$
$2\sigma_g$
2*p*
2*p*
2*s*
2*s*
Energy
F
F_2
F
($2s^2\ 2p^5$)
Bond order = (4 – 3) = 1
($2s^2\ 2p^5$)

Figure 4.23 The molecular orbital energy level diagram for F_2.

Interactive 3D animation of F_2 molecular orbitals.

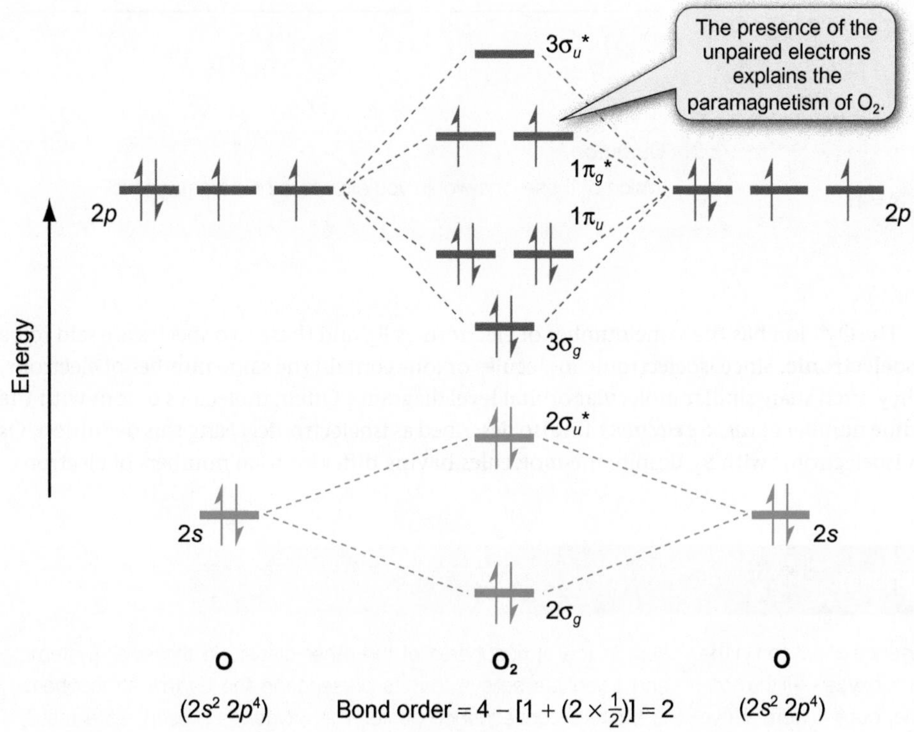

The presence of the unpaired electrons explains the paramagnetism of O_2.
$3\sigma_u^*$
$1\pi_g^*$
$1\pi_u$
$3\sigma_g$
$2\sigma_u^*$
$2\sigma_g$
2*p*
2*p*
2*s*
2*s*
Energy
O
O_2
O
($2s^2\ 2p^4$)
Bond order = $4 - [1 + (2 \times \frac{1}{2})] = 2$
($2s^2\ 2p^4$)

Figure 4.24 The molecular orbital energy level diagram for O_2.

Visit the Online Resource Centre to view screencast 4.1 which walks you through how to construct the MO energy level diagram for O_2.

Hund's rule is described in Section 3.6 (p.147).

There is a more reactive form of O_2 which is called singlet oxygen. Singlet oxygen has a higher energy than 'normal' oxygen, and it is formed when the two $\pi_g{}^*$ electrons are paired in the same orbital.

aufbau principle, the final two electrons go into the two degenerate $1\pi_g{}^*$ orbitals. Using Hund's rule, the most energetically favourable way of putting two electrons into two degenerate orbitals is to put one in each orbital with the electron spins parallel. *The molecular orbital energy level diagram of O_2 contains two unpaired electrons, which explains why O_2 is paramagnetic.* One of the failings of the Lewis model (Section 4.2, p.175) is its inability to explain this.

In O_2 there are four filled bonding orbitals, one completely filled antibonding orbital, and two half-filled antibonding orbitals. Therefore the bond order is $4 - [1 + (2 \times \frac{1}{2})] = 2$.

Worked example 4.5 $O_2{}^+$, $O_2{}^-$, and $O_2{}^{2-}$

It is possible to isolate compounds containing the $O_2{}^+$ cation, the $O_2{}^-$ (superoxide) anion, and the $O_2{}^{2-}$ (peroxide) anion. By considering their molecular orbital energy level diagrams, place these ions in order of increasing O–O bond length.

Strategy

To predict the relative values of the bond lengths, work out the bond orders for the different ions. Use the molecular orbital energy level diagram in Figure 4.24, adding or taking away electrons as necessary to form the ions.

Solution

$O_2{}^+$ has one fewer electron than O_2. Take an electron from the highest occupied molecular orbital, which is the antibonding $1\pi_g{}^*$ orbital.

The bond order in $O_2{}^+$ is $(4 - 1\frac{1}{2}) = 2\frac{1}{2}$.

$O_2{}^-$ has one more electron than O_2. Add an electron to the $1\pi_g{}^*$ orbital.

The bond order in $O_2{}^-$ is $(4 - 2\frac{1}{2}) = 1\frac{1}{2}$.

$O_2{}^{2-}$ has one more electron than $O_2{}^-$. Add this electron to the $1\pi_g{}^*$ orbital on $O_2{}^-$.

The bond order in $O_2{}^{2-}$ is $(4 - 3) = 1$.

The greater the bond order, the shorter the bond, so the predicted sequence of increasing bond length is:

$$O_2{}^+ < O_2{}^- < O_2{}^{2-}$$

This is borne out by the experimental data shown in Table 4.3.

Table 4.3 Properties of $O_2{}^+$, O_2, $O_2{}^-$, and $O_2{}^{2-}$

	Bond order	D/kJ mol^{-1}	Bond length/pm
$O_2{}^+$	$2\frac{1}{2}$	+643	112
O_2	2	+498.5	120.7
$O_2{}^-$	$1\frac{1}{2}$	+360	128
$O_2{}^{2-}$	1	+155	149

Question

Which of these ions would you expect to be paramagnetic?

Molecules or ions with the same number of electrons are called isoelectronic.

The $O_2{}^{2-}$ ion has the same number of electrons as F_2, and these two species are said to be **isoelectronic**. Since isoelectronic molecules or ions contain the same number of electrons, they often share similar molecular orbital level diagrams. Often, molecules or ions with the same number of *valence electrons* are also described as isoelectronic. Using this definition, O_2 is isoelectronic with S_2, despite the molecules having different total numbers of electrons.

Box 4.7 Oxygen in the atmosphere

Most life on Earth requires the presence of oxygen in the atmosphere for respiration. Dry air contains 21% oxygen—if the content were less than 17% we could not breathe, but if it were above 25% organic matter would be highly flammable. Oxygen is not present in the atmospheres of the other planets in the solar system, and scientists believe that its presence in the Earth's atmosphere is a direct consequence of life. The oxygen is present as a result of photosynthesis in plants and algae, but scientists believe →

→

that cyanobacteria were responsible for the initial increase in oxygen in the atmosphere, about 2300 million years ago. For the past 400 million years, the percentage of oxygen has been relatively constant.

Opinions are divided on what the Earth's atmosphere looked like in the past. While initially hydrogen was likely to have been present, there is evidence that the atmosphere 3–4 billion years ago consisted mainly of nitrogen, carbon dioxide, and water vapour. Much of the water condensed to form the oceans, while the carbon dioxide dissolved in these oceans and precipitated out as carbonate minerals such as limestone.

Evidence suggests that life evolved in the seas. Life was only possible on land when there was enough O_2 in the atmosphere (about 10%) for O_3 to start forming an ozone layer that protected land plants from ultraviolet radiation from the Sun (see Box 27.6, p.1233). The first land animals appeared about 300 million years ago, by which time the oxygen concentration was at the current level.

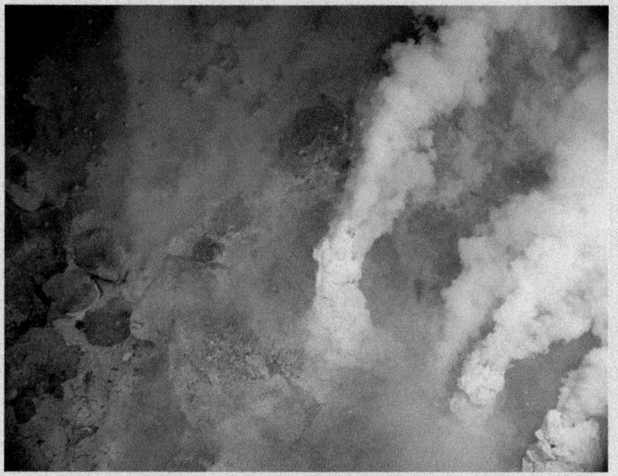

▲ Hydrothermal vents on the ocean floor release sulfurous mineral-rich fluid at extremely high temperatures. Archaeobacteria use the minerals during metabolism, and provide the basis for an ecosystem independent of sunlight. Archaeobacteria are thought to have evolved up to 3.5 billion years ago before the Earth's atmosphere contained oxygen.

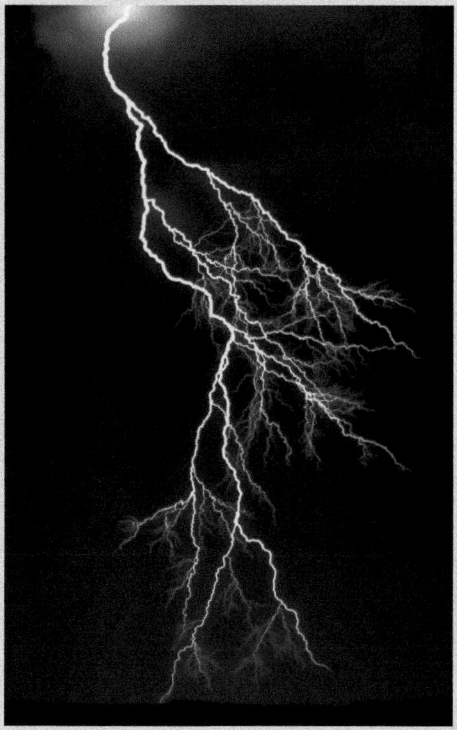

▲ Experiments have shown that sparking mixtures of methane, ammonia, and water can lead to amino acids. Some scientists think that lightning on the early Earth was important in the origin of life.

Question

Suggest why H_2 is no longer an important component of the atmosphere.

» Summary

- Molecular orbital energy level diagrams can be used to explain trends in bond lengths and bond dissociation enthalpies.
- Unlike the Lewis model, molecular orbital theory is able to explain why O_2 is paramagnetic.
- In a molecule, the two most important orbitals are usually the highest occupied molecular orbital (HOMO) and the lowest unoccupied molecular orbital (LUMO).
- Isoelectronic species have the same number of electrons.

? **For practice questions on the topic in Sections 4.9 and 4.10, see questions 11–13 at the end of this chapter (p.214).**

4.11 s-p mixing

The method used for F_2 and O_2 in Section 4.10 is easily extended to the other second period diatomics B_2, C_2, and N_2, but there is an additional complication. Photoelectron spectroscopy experiments show that for these molecules the $3\sigma_g$ orbital lies *higher* in energy than the $1\pi_u$ orbitals. This does not fit in with what you have seen about the relative strength of σ and π interactions, so what is happening?

Box 4.8 Measuring the energies of molecular orbitals

How are molecular orbital energies determined? They can be calculated using computational methods or measured experimentally using photoelectron spectroscopy (PES), a technique that is related to the photoelectric effect (Section 3.2, p.121).

In PES, a sample is bombarded with photons of known energy ($h\nu$) and the kinetic energy of the expelled electrons (E_{KE}) is measured. Assuming that there is no loss of energy due to the rearrangement of electrons in the ion formed (Koopman's rule), this leads to Equation 4.17

$$h\nu = I + E_{KE} \tag{4.17}$$

In this case the ionization energy, I, is the energy required to remove a single electron from a particular molecular orbital, so it provides a measure of the orbital energy. Since ν is known and E_{KE} is measured, Equation 4.17 allows I to be calculated.

Ultraviolet radiation is used to measure the orbital energies of the valence electrons (UV-PES), and the higher energy X-ray radiation is used for the core electrons (X-PES). The UV photoelectron spectrum for O_2 is shown below.

Each ejected electron is observed as a series of lines rather than a single peak. This is because ionization usually leads to cations in a range of vibrational energy states. The gaps between the lines relate to the vibrational frequency of the ion, which in turn is related to its bond order. By measuring the vibrational frequencies, the bonding and antibonding orbitals can be distinguished. Vibrational spectroscopy is described in Section 10.5 (p.476).

Question

In a PES experiment, a sample of nitrogen was irradiated with photons of frequency 5.13×10^{15} Hz. Ejected electrons with kinetic energy 8.97×10^{-19} J were observed. Calculate the orbital energy (in kJ mol^{-1}) of these electrons in N_2.

▼ The UV photoelectron spectrum of O_2. Adapted from Ebsworth, Rankin and Cradock, 'Structural Methods in Inorganic Chemistry', 2nd ed., p.274, Fig. 6.15a (Blackwell Scientific Publications, 1991).

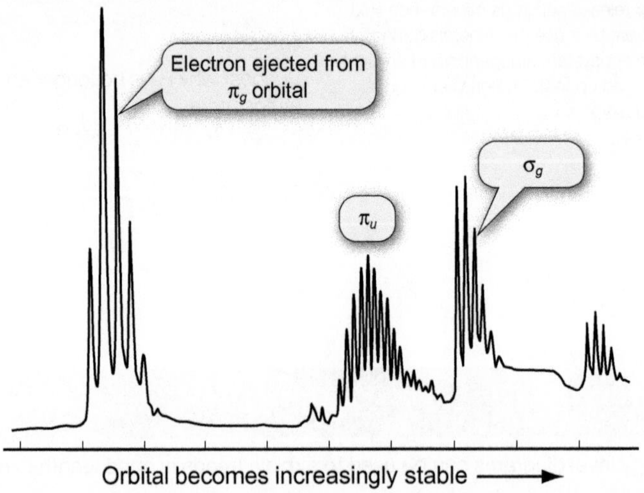
Electron ejected from
π_g orbital
π_u
σ_g
Orbital becomes increasingly stable

To explain the bonding in F_2 and O_2, you can assume that the interactions between the *s* orbitals and the interactions between the *p* orbitals are completely independent. This is not really the case, and generally orbitals of the same symmetry interact together. This means that the $2\sigma_g$ orbital derived from the 2*s* orbitals interacts with the $3\sigma_g$ orbital derived from the $2p_z$ orbitals. Similarly the *s*- and *p*-based σ_u orbitals interact together. These

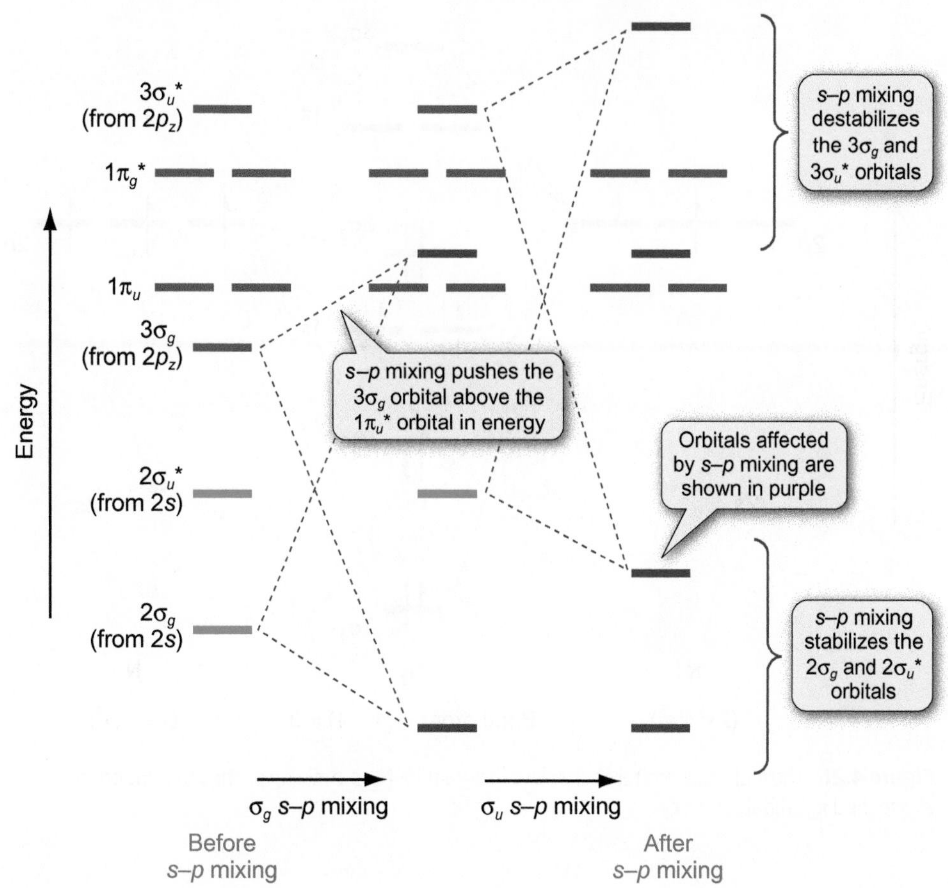

$3\sigma_u^*$ (from $2p_z$)
$1\pi_g^*$
$1\pi_u$
$3\sigma_g$ (from $2p_z$)
$2\sigma_u^*$ (from $2s$)
$2\sigma_g$ (from $2s$)
Energy
s–p mixing destabilizes the $3\sigma_g$ and $3\sigma_u^*$ orbitals
s–p mixing pushes the $3\sigma_g$ orbital above the $1\pi_u^*$ orbital in energy
Orbitals affected by s–p mixing are shown in purple
s–p mixing stabilizes the $2\sigma_g$ and $2\sigma_u^*$ orbitals
σ_g s–p mixing
σ_u s–p mixing
Before s–p mixing
After s–p mixing

Figure 4.25 The effects of *s-p* mixing on the energies of the molecular orbitals for N_2. Note that this diagram shows the molecular orbitals only—the atomic orbitals are omitted. Mixing of the σ_g and σ_u^* orbitals is shown separately for clarity. The π_u and π_g^* orbitals are unaffected by *s-p* mixing.

interactions are shown for N_2 in Figure 4.25. The degree of interaction between the orbitals depends on the energy gap between them. Orbitals that are close together in energy give a large interaction, whereas those that are far apart in energy give a small interaction.

The interaction between σ orbitals derived from the *s* and *p* atomic orbitals is known as ***s–p* mixing**. *s–p* mixing leads to the stabilization of the $2\sigma_g$ and $2\sigma_u^*$ orbitals and the destabilization of the $3\sigma_g$ and $3\sigma_u^*$ orbitals. The destabilization of the $3\sigma_g$ orbital is particularly important for N_2 as this pushes the $3\sigma_g$ orbital above the $1\pi_u$ orbitals in energy. The π orbitals are not affected by *s–p* mixing as they are of the wrong symmetry to interact with the *s*-based σ orbitals.

s-p mixing in molecular orbital theory fulfils a similar role to that of hybridization in valence bond theory.

Visit the Online Resource Centre to view screencast 4.2 which walks you through the origins of *s-p* mixing.

Bonding in N_2

The molecular orbital energy level diagram for N_2 is shown in Figure 4.26. Each nitrogen atom has the electronic configuration $2s^2\,2p^3$, so there are $(2 \times 5) = 10$ electrons to put into the molecular orbitals. The *s–p* mixing is shown by additional dotted lines connecting the *s* and *p* atomic orbitals with all of the σ_g and σ_u^* molecular orbitals. Although each *s* and p_z atomic orbital now contributes to more molecular orbitals than before, the total number of molecular orbitals is still equal to the total number of atomic orbitals. There are four filled bonding orbitals and one filled antibonding orbital in the molecular orbital energy level diagram for N_2, so the bond order is $(4 - 1) = 3$, in agreement with the Lewis model. Both valence bond and molecular orbital theories involve the 2*s* and 2*p* atomic orbitals in describing the bonding in N_2: valence bond theory uses hybridization whereas molecular orbital theory uses *s–p* mixing.

The valence bond description of the bonding in N_2 is described in Section 4.5 (p.180).

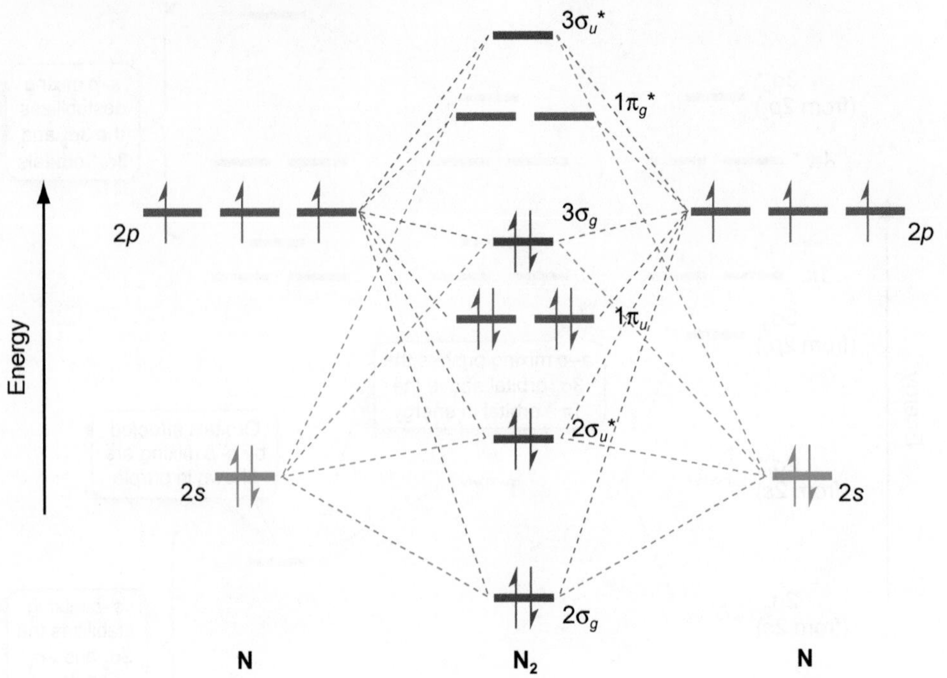

$3\sigma_u^*$
$1\pi_g^*$
$3\sigma_g$
$1\pi_u$
$2\sigma_u^*$
$2\sigma_g$
2p
2p
2s
2s
Energy

N | N_2 | N

$(2s^2\ 2p^3)$ | Bond order = (4 − 1) = 3 | $(2s^2\ 2p^3)$

Figure 4.26 The molecular orbital energy level diagram for N_2. *s–p* mixing pushes the $3\sigma_g$ orbital above the $1\pi_u$ orbital in energy.

Table 4.4 Properties of N_2^+, N_2, N_2^-, and N_2^{2-}

	Bond order	D/kJ mol^{-1}	Bond length/pm
N_2^+	$2\frac{1}{2}$	+840	112
N_2	3	+944.9	109.8
N_2^-	$2\frac{1}{2}$	+765	119
N_2^{2-}	2	Unknown	122

Bond dissociation enthalpy values and bond lengths for N_2, N_2^+, N_2^-, and N_2^{2-} are given in Table 4.4. The bond dissociation enthalpy of N_2^+ is less than that of N_2 and the bond length is greater. This is in contrast to observations for O_2^+ and O_2 (Table 4.3, p.198), as O_2^+ has a higher bond dissociation enthalpy and a shorter bond length than O_2. The reason for this difference can be seen from the molecular orbital energy level diagrams for O_2 and N_2, in Figures 4.24 and 4.26. To form O_2^+ from O_2, an antibonding electron is removed, so the bond order increases. In contrast, the electron removed from N_2 to form N_2^+ is a bonding electron, so the bond order in the cation is lower.

The *s–p* mixing described for N_2 also occurs for O_2 and F_2, but it is weaker due to the greater energy difference between the 2*s* and 2*p* orbitals in these atoms. The *s–p* mixing is not large enough to change the ordering of the orbitals in O_2 and F_2, and is usually ignored. The energy difference between the 2*s* and 2*p* orbitals increases across the second period from lithium to neon. Both 2*s* and 2*p* orbitals are stabilized by the increase in effective nuclear charge across the period, but the 2*s* orbital is stabilized to a greater extent as it is more penetrating than the 2*p* orbital.

See Section 3.6 (p.150) for a fuller discussion of effective nuclear charges and penetration.

Box 4.9 The colours of the polar lights

The aurora are shimmering, coloured swathes of light in the sky that are often seen in polar regions. The aurora borealis (northern lights) and the aurora australis (southern lights) are formed by electrons interacting with atoms and molecules high in the Earth's atmosphere. The electrons come from the Sun, and are emitted along with protons from the Sun's surface in solar flares. This high velocity stream of charged particles is known as the solar wind. The charged particles are trapped by the Earth's magnetic field and directed into the atmosphere around the magnetic poles.

At high altitudes, the N_2 and O_2 molecules in the atmosphere are largely dissociated into atoms as a result of interactions with high-energy ultraviolet radiation emitted from the Sun.

$$O_2(g) \xrightarrow{h\nu} 2O(g)$$
$$N_2(g) \xrightarrow{h\nu} 2N(g)$$

The electrons from the solar wind collide with the atoms and molecules in the atmosphere, exciting them by promoting electrons into higher energy orbitals. Each excited atom or molecule emits light of a characteristic wavelength as it returns to the ground state. At high altitudes (250 km), oxygen atoms emit a red colour with a wavelength of 630 nm when excited by low energy electrons. All-red aurorae are very rare, as this colour is often masked by emissions lower in the atmosphere. The most common aurora colour is yellow-green, which is emitted by oxygen atoms at lower altitudes (100–200 km) when excited by higher energy electrons. This emission is observed at a wavelength of 558 nm.

At altitudes between 60 km and 90 km, the colours are mainly due to N_2 molecules and N_2^+ ions. Excited N_2 molecules emit in the red part of the spectrum, with wavelengths between 661 nm and 686 nm. These colours are often seen on the lower edges and borders of the aurora. Excited N_2^+ ions emit blue and purple light, with wavelengths of 470 nm and 391 nm.

Question

Are the photons emitted by excited N_2^+ ions in aurorae of higher or lower energy than those emitted by excited N_2 molecules?

The aurora borealis (northern lights).

Bonding in B_2 and C_2

For both B_2 and C_2, *s–p* mixing is high, and the energy levels of the molecular orbitals occur in the same order as in N_2. Molecular orbital energy level diagrams for B_2 and C_2 are shown in Figure 4.27. B_2 has a bond order of one, whereas C_2 has a bond order of two.

The double bond in C_2 is made up of two π bonds. It is different from the double bond in ethene (C_2H_4) which contains one σ bond and one π bond. See Section 5.4 (p.239) for more details.

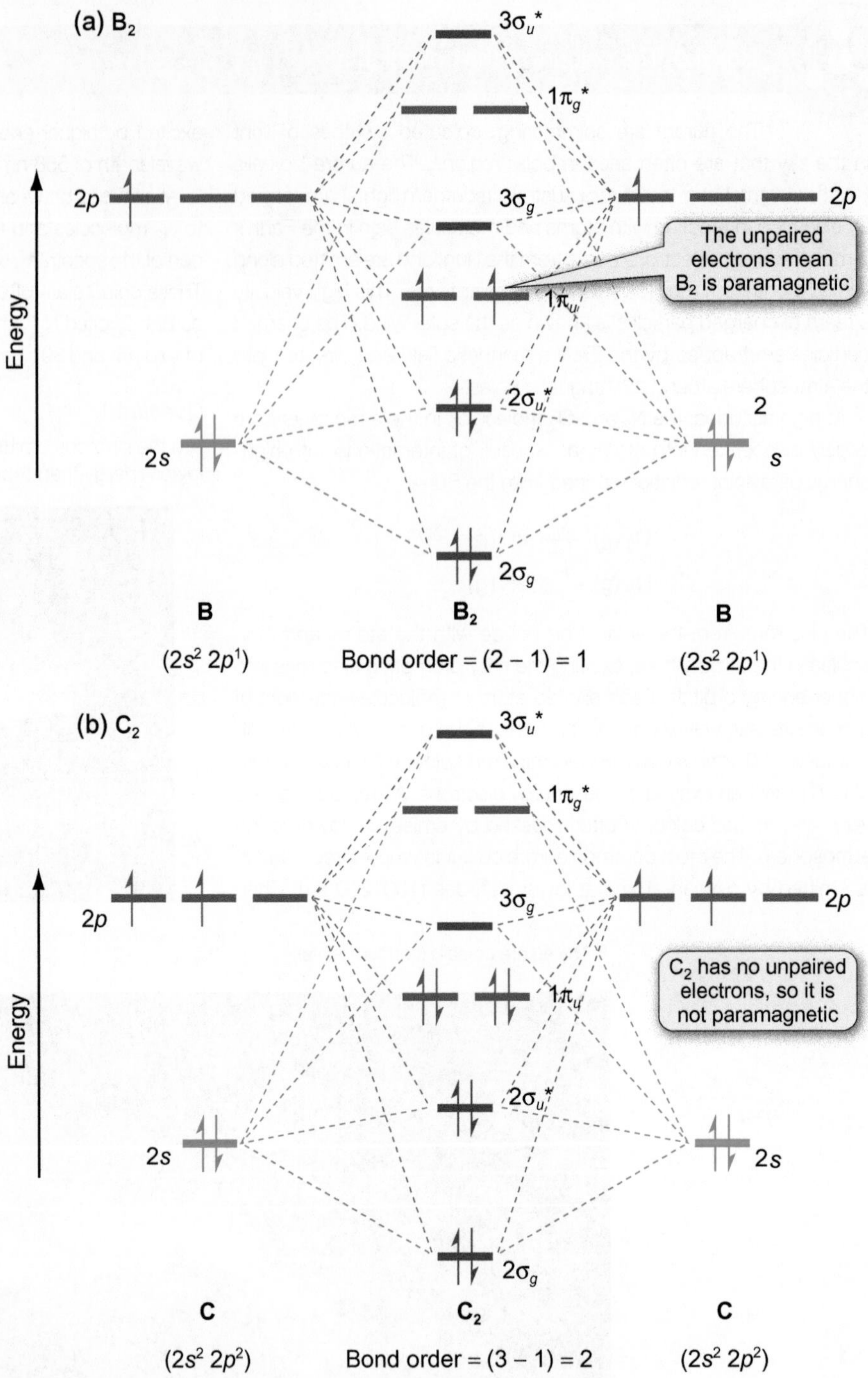
(a) B_2
$3\sigma_u{}^*$
$1\pi_g{}^*$
2p
$3\sigma_g$
2p
The unpaired electrons mean B_2 is paramagnetic
$1\pi_u$
Energy
$2\sigma_u{}^*$
2
2s
s
$2\sigma_g$
B
B_2
B
$(2s^2\ 2p^1)$
Bond order = (2 – 1) = 1
$(2s^2\ 2p^1)$
(b) C_2
$3\sigma_u{}^*$
$1\pi_g{}^*$
$3\sigma_g$
2p
2p
C_2 has no unpaired electrons, so it is not paramagnetic
$1\pi_u$
Energy
$2\sigma_u{}^*$
2s
2s
$2\sigma_g$
C
C_2
C
$(2s^2\ 2p^2)$
Bond order = (3 – 1) = 2
$(2s^2\ 2p^2)$

Figure 4.27 The molecular orbital energy level diagrams for (a) B_2 and (b) C_2.

In both B_2 and C_2, the filled $2\sigma_g$ bonding orbital and filled $2\sigma_u{}^*$ antibonding orbital essentially cancel each other out, since their net contribution to the bond order is zero. This makes C_2 unusual, in that the two bonds are both π bonds.

Magnetic behaviour is described in Box 4.2 (p.176).

B_2 has two unpaired electrons so it is predicted to be paramagnetic, whereas C_2 has only paired electrons so it is predicted to be diamagnetic, but not paramagnetic. Both B_2 and C_2 have been observed in the gas phase, and their magnetic properties match these predictions. This provides good evidence for *s–p* mixing as, if it did not occur, B_2 would be expected to be diamagnetic, not paramagnetic, and C_2 would be expected to be paramagnetic.

The relative energies of the molecular orbitals for all of the second period homonuclear diatomics are summarized in Figure 4.28.

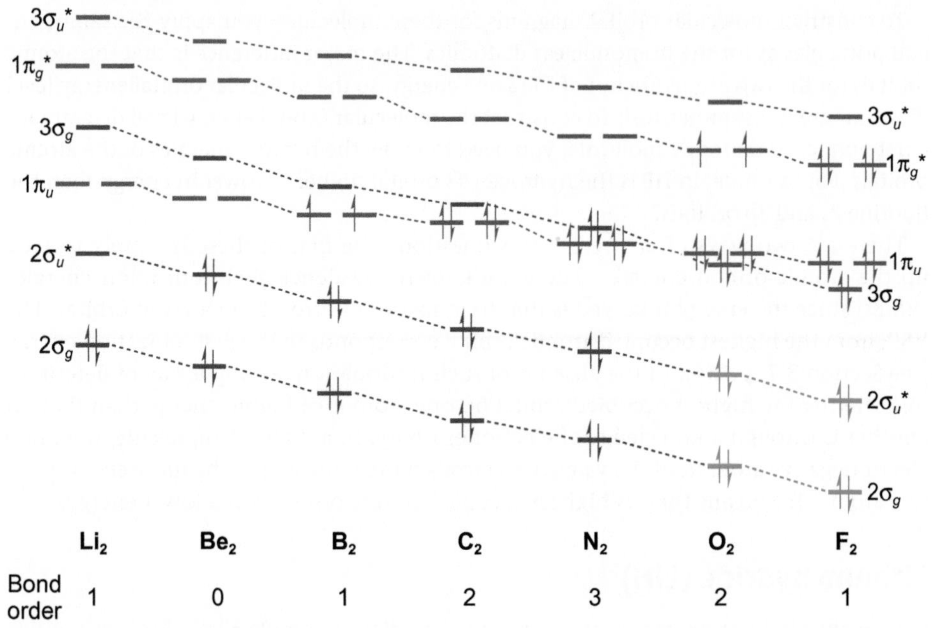

$3\sigma_u^*$
$1\pi_g^*$
$3\sigma_g$
$1\pi_u$
$2\sigma_u^*$
$2\sigma_g$
$3\sigma_u^*$
$1\pi_g^*$
$1\pi_u$
$3\sigma_g$
$2\sigma_u^*$
$2\sigma_g$
Li_2
Be_2
B_2
C_2
N_2
O_2
F_2
Bond order
1
0
1
2
3
2
1

Figure 4.28 The relative energies of the molecular orbitals for the homonuclear diatomics formed by the second row elements.

Summary

- Molecular orbitals of the same symmetry interact through *s–p* mixing.
- *s–p* mixing is important for B_2, C_2, and N_2. The molecular orbital energy level order is different from that expected if you assume independent overlap of the *s* orbitals and *p* orbitals.
- *s–p* mixing is less important for O_2 and F_2. The molecular orbital energy levels are in the expected order.

For practice questions on these topics, see questions 14–16 at the end of this chapter (p.214).

4.12 Heteronuclear diatomics

Heteronuclear diatomic molecules, such as gaseous lithium hydride (LiH) and hydrogen fluoride (HF), contain two *different* atoms. Bond lengths and bond dissociation enthalpies for selected heteronuclear diatomic molecules are given in Table 4.5.

Table 4.5 Bond dissociation enthalpies, *D*, and bond lengths for selected heteronuclear diatomic molecules

	D / kJ mol^{-1}	Bond length / pm
HF	+569.7	91.7
HCl	+431.4	127.5
HBr	+366.2	141.5
HI	+298.3	160.9
CO	+1076.6	112.8
NO	+630.6	115.1

To construct molecular orbital diagrams for these molecules, you apply the same general principles as for the homonuclear diatomics. The major difference is that the atomic orbitals for the two atoms are not of the same energy, so the molecular orbital energy level diagrams are not symmetrical. To construct the molecular orbital energy level diagram for a heteronuclear diatomic molecule, you need to know the relative energies of the atomic orbitals. For example, in HF is the hydrogen 1*s* orbital higher or lower in energy than the fluorine 2*s* and 2*p* orbitals?

There are two ways of answering this question. The first of these is simply to look up the atomic orbital energies. These are known as valence shell ionization energies (VSIEs) since the energy involved is that to ionize an electron from a given orbital. The VSIE from the highest occupied atomic orbital corresponds to the first ionization energy (see Section 3.7, p.154). In the absence of such information, a simple way of determining whether the highest occupied orbital on one atom is of higher energy than that on another is through a knowledge of electronegativity. In a diatomic molecule, the more electronegative atom holds its valence electrons more tightly than the more electropositive atom. This means that its highest occupied atomic orbital is at a lower energy.

Electronegativity is described in Section 4.3 (p.177).

Lithium hydride (LiH)

Lithium hydride is an ionic solid under normal conditions of temperature and pressure, but it exists in the gas phase as LiH molecules. You can see the valence bond treatment of the bonding in gaseous LiH in Worked example 4.2 (p.183).

One of the simplest examples of a heteronuclear diatomic molecule is lithium hydride (LiH). Lithium has the electronic configuration $1s^2\,2s^1$. The filled 1*s* orbital is a core orbital (Section 4.8) and does not interact with the hydrogen 1*s* orbital. The interactions to consider are those between the lithium 2*s* orbital and the hydrogen 1*s* orbital. These orbitals have the correct symmetry to interact with each other, and combine to give a σ bonding orbital and a σ* antibonding orbital. The boundary surfaces for these orbitals are shown in Figure 4.29, together with the molecular orbital energy level diagram for LiH.

The σ-bonding orbital is distorted towards the electronegative hydrogen atom whereas the σ*-antibonding orbital is distorted towards the electropositive lithium atom. Neither orbital has an inversion centre, so the *g* and *u* labels are not used.

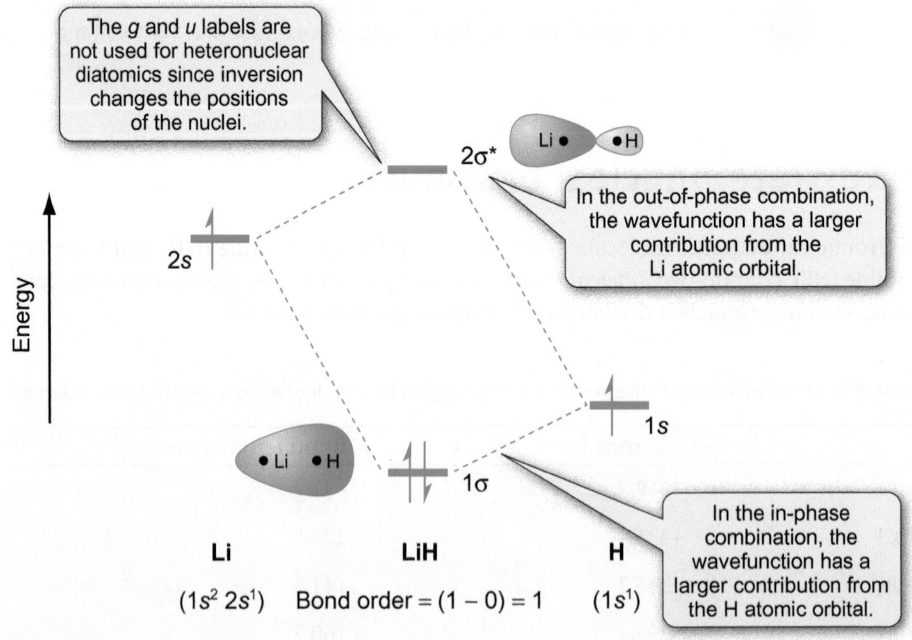
The g and u labels are not used for heteronuclear diatomics since inversion changes the positions of the nuclei.
2σ*
Li
H
In the out-of-phase combination, the wavefunction has a larger contribution from the Li atomic orbital.
Energy
2s
1s
1σ
In the in-phase combination, the wavefunction has a larger contribution from the H atomic orbital.
Li
LiH
H
(1s² 2s¹)
Bond order = (1 – 0) = 1
(1s¹)

Figure 4.29 The molecular orbital energy level diagram for LiH, together with boundary surfaces for the molecular orbitals.

In the molecular orbital energy level diagram, the 1*s* orbital of hydrogen is lower in energy than the 2*s* orbital of lithium. Hydrogen is more electronegative than lithium, meaning it is harder to remove an electron from hydrogen. Both the hydrogen 1*s* orbital and the lithium 2*s* orbital contribute to the 1σ bonding orbital, but the molecular orbital lies closer in energy to the hydrogen 1*s* orbital than the lithium 2*s* orbital. This means that it is closer to the hydrogen 1*s* orbital in character, and the contribution from this atomic orbital is larger. Similarly, the 2σ* antibonding orbital is closer in energy to the lithium 2*s* orbital, so this atomic orbital provides the greater contribution to it. This is shown mathematically in Box 4.10. Overall, the bonding orbital is filled, and the antibonding orbital is empty, and the bond order in LiH is (1 – 0) = 1.

The greater contribution of the hydrogen 1*s* orbital to the filled 1σ molecular orbital provides an explanation for the polarity of the Li–H bond in which the hydrogen has a partial negative charge (δ–) and the lithium a partial positive charge (δ+). LiH has a bond dissociation enthalpy of +238.0 kJ mol^{-1} and a bond length of 159.5 pm. The bond is weaker and longer than the H–H bond in H_2 due to the poorer overlap between the 1*s* and 2*s* orbitals.

Box 4.10 Linear combinations of atomic orbitals in LiH

For a homonuclear diatomic molecule such as H_2, the atomic orbitals from the two atoms contribute equally to the molecular orbitals. For a heteronuclear diatomic molecule this is not the case, and the electron density in a bonding orbital is distorted towards the more electronegative atom. In a linear combination expression this is shown by giving different weights to the contributions of the two atomic orbitals. For LiH, the bonding molecular orbital (1σ) is given by an in-phase combination of the atomic orbitals, but with unequal weightings, as shown in Equation 4.18

$$\psi_{\text{in phase}} = N[\phi_{\text{H}}(1s) + \lambda\phi_{\text{Li}}(2s)] \qquad (4.18)$$

Since the bonding orbital is closer in energy and character to the hydrogen 1*s* orbital than it is to the lithium 2*s* orbital, the contribution of the hydrogen atomic orbital is greater. This is taken into account in Equation 4.18 by introducing the constant λ, which has a value of less than one. *N* is the normalization constant, which has a value of less than one to make $\psi_{\text{in phase}}$ a single-electron wavefunction.

The linear combination expression for the antibonding orbital (2σ*) is given by the out-of-phase combination of atomic orbitals, as given in Equation 4.19

$$\psi_{\text{out of phase}} = N[\lambda\phi_{\text{H}}(1s) - \phi_{\text{Li}}(2s)] \qquad (4.19)$$

In this case, the contribution is greater from the lithium atomic orbital, so the constant λ is placed before the hydrogen orbital. Again, λ has a value of less than one.

Question

Write a linear combination expression for the bonding orbital in the molecule NaLi.

Hydrogen fluoride (HF)

The first step in constructing a molecular orbital energy level diagram for a heteronuclear diatomic is to write out the electronic configuration of each atom. If you do not know the valence shell ionization energies, you can work out the relative energies of the atomic orbitals by comparing the electronegativities of the atoms. In HF, fluorine is more electronegative than hydrogen. This means that the fluorine 2*p* orbitals are lower in energy than the hydrogen 1*s* orbital. It may seem odd that a 2*p* orbital is lower in energy than a 1*s* orbital, but the far greater nuclear charge present in fluorine pulls the atomic orbitals closer to the nucleus and makes the electrons more difficult to remove.

Which orbitals interact? This is largely a question of symmetry. The fluorine p_z orbital points towards the hydrogen 1*s* orbital, so the two orbitals interact to form bonding and antibonding orbitals, as shown by the boundary surfaces in Figure 4.30. The in-phase combination leads to a σ-bonding orbital, and the out-of-phase combination gives a σ*-antibonding orbital. The bonding orbital has more fluorine character, and the antibonding orbital has more hydrogen character.

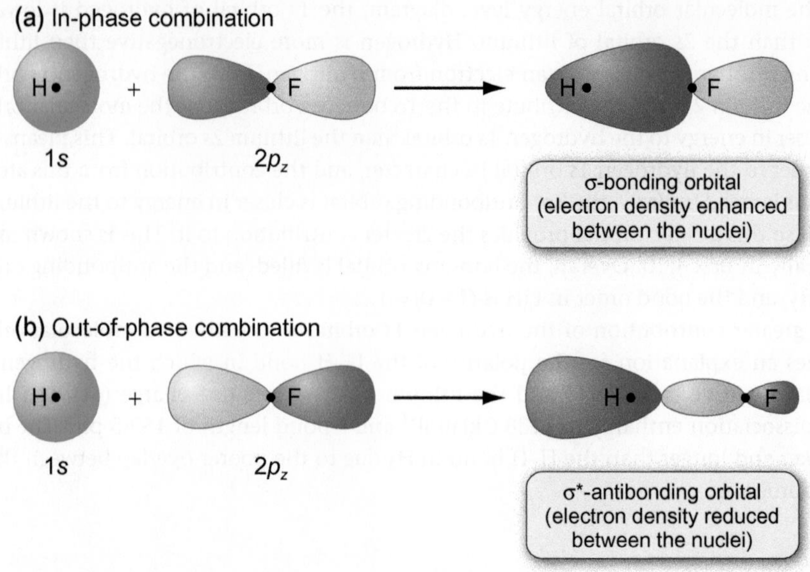
(a) In-phase combination
H•
+
F
H•
F
1s
$2p_z$
σ-bonding orbital
(electron density enhanced
between the nuclei)
(b) Out-of-phase combination
H•
+
F
H•
F
1s
$2p_z$
σ*-antibonding orbital
(electron density reduced
between the nuclei)

Figure 4.30 The interaction between a hydrogen 1*s* orbital and a fluorine 2p_z orbital leads to (a) a σ-bonding orbital and (b) a σ*-antibonding orbital.

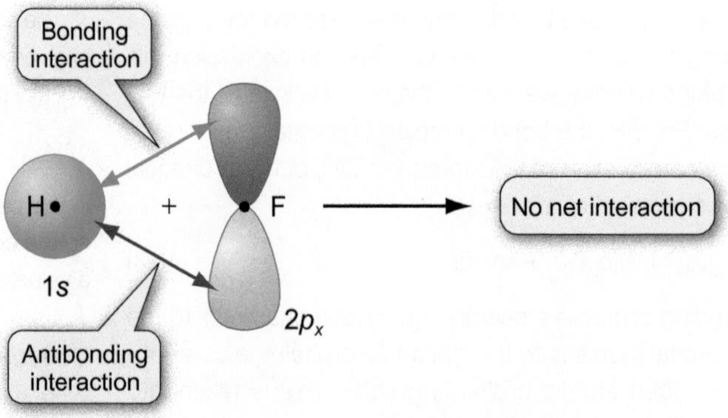
Bonding
interaction
H•
+
F
No net interaction
1s
$2p_x$
Antibonding
interaction

Figure 4.31 There is no net interaction between a hydrogen 1*s* orbital and a fluorine 2p_x orbital.

The activity on the Online Resource Centre allows you to investigate how atomic orbitals interact to give molecular orbitals.

The fluorine p_x and p_y orbitals do not have the correct symmetry to interact with the hydrogen 1*s* orbital, as shown in Figure 4.31. Since there are no other hydrogen atomic orbitals for these orbitals to interact with, they remain unaltered in the molecule. They are called **non-bonding orbitals**, and take the molecular orbital label 1π.

Finally, there is the fluorine 2*s* orbital. This has the correct symmetry to interact with the hydrogen 1*s* orbital, but there is a very large energy gap between these two atomic orbitals. The larger the energy separation between orbitals, the smaller the interaction between them. When the energy gap is very large, as in this case, there is essentially no interaction at all. To a first approximation, the hydrogen 1*s* orbital and the fluorine 2*s* orbital do not interact, and the latter acts as a non-bonding σ orbital.

The molecular orbital energy level diagram for HF is shown in Figure 4.32. The σ orbitals are labelled sequentially as 1σ, 2σ, 3σ, and 4σ*, though the 1σ orbital is derived from the fluorine 1*s* orbital and is not shown on the diagram. The 1π and 2σ non-bonding orbitals lie *at the same energy* as the atomic orbitals they are derived from. There are eight valence electrons to place into the molecular orbital diagram, one from hydrogen and seven from

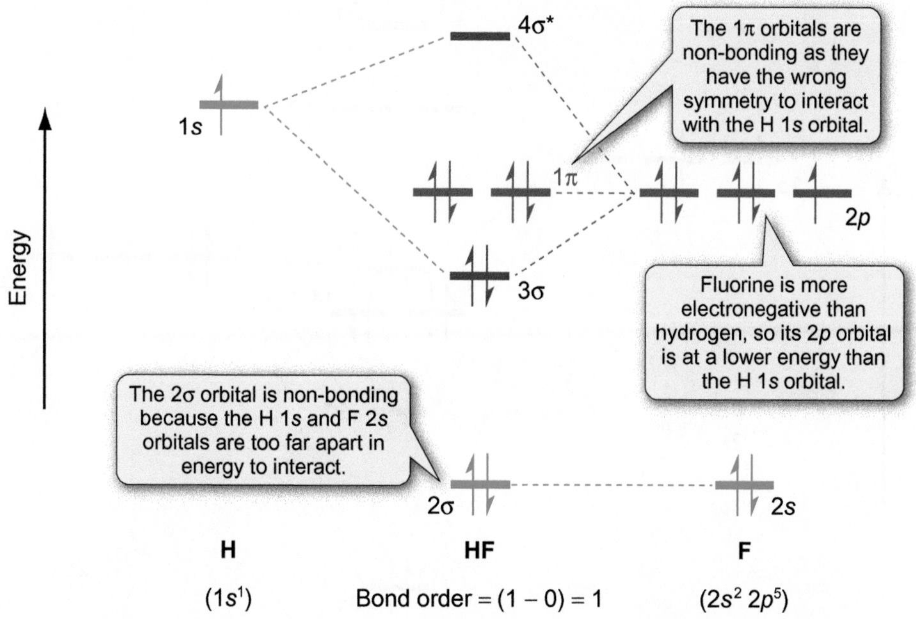
Energy
1s
4σ*
1π
3σ
2σ
2p
2s
The 1π orbitals are non-bonding as they have the wrong symmetry to interact with the H 1s orbital.
Fluorine is more electronegative than hydrogen, so its 2p orbital is at a lower energy than the H 1s orbital.
The 2σ orbital is non-bonding because the H 1s and F 2s orbitals are too far apart in energy to interact.
H
HF
F
(1s¹)
Bond order = (1 – 0) = 1
(2s² 2p⁵)

Figure 4.32 The molecular orbital energy level diagram for HF. The 1σ molecular orbital is essentially the F 1s atomic orbital and is not shown.

Visit the Online Resource Centre to view screencast 4.3 which walks you through how to construct the MO energy level diagram for HF.

fluorine. This means the 2σ, 3σ, and 1π orbitals are filled while the 4σ* orbital is empty. There is one filled bonding orbital (3σ) and there are no filled antibonding orbitals, so the bond order in HF is (1 – 0) = 1. The presence of filled non-bonding orbitals has no effect on the bond order. The bond order calculated from the molecular orbital energy level diagram is the same as that calculated from the Lewis model and from valence bond theory.

Interactive 3D animation of HF molecular orbitals.

Worked example 4.6 Equations for the molecular orbitals in HF

Write mathematical expressions for the wavefunctions of the 3σ, 4σ*, and 1π orbitals in HF.

Strategy

Follow the LCAO method used for LiH in Box 4.10. In each case, identify the atomic orbitals that contribute to the molecular orbital and decide which of these makes the greater contribution. Use a constant, λ, to show this in the equations, and state whether λ is greater than or less than 1.

Solution

The **3σ orbital** is formed from a bonding combination of the fluorine $2p_z$ orbital and the hydrogen 1*s* orbital, with a greater contribution from the fluorine orbital. Therefore

$$\psi = N[\phi_F(2p_z) + \lambda\phi_H(1s)] \quad \text{where } \lambda < 1.$$

The **4σ* orbital** is formed from an antibonding combination of the fluorine $2p_z$ orbital and the hydrogen 1*s* orbital, with a greater contribution from the hydrogen orbital. Therefore

$$\psi^* = N[\lambda\phi_F(2p_z) - \phi_H(1s)] \quad \text{where } \lambda < 1.$$

The **1π orbitals** are just the unchanged p_x and p_y fluorine atomic orbitals. Therefore

$$\psi = \phi_F(2p_x) \quad \text{and} \quad \psi = \phi_F(2p_y)$$

No normalization constant is required in this case, as the atomic orbitals are already single-electron wavefunctions.

Question

Write an expression for the wavefunction of the bonding orbital in a molecule of LiF.

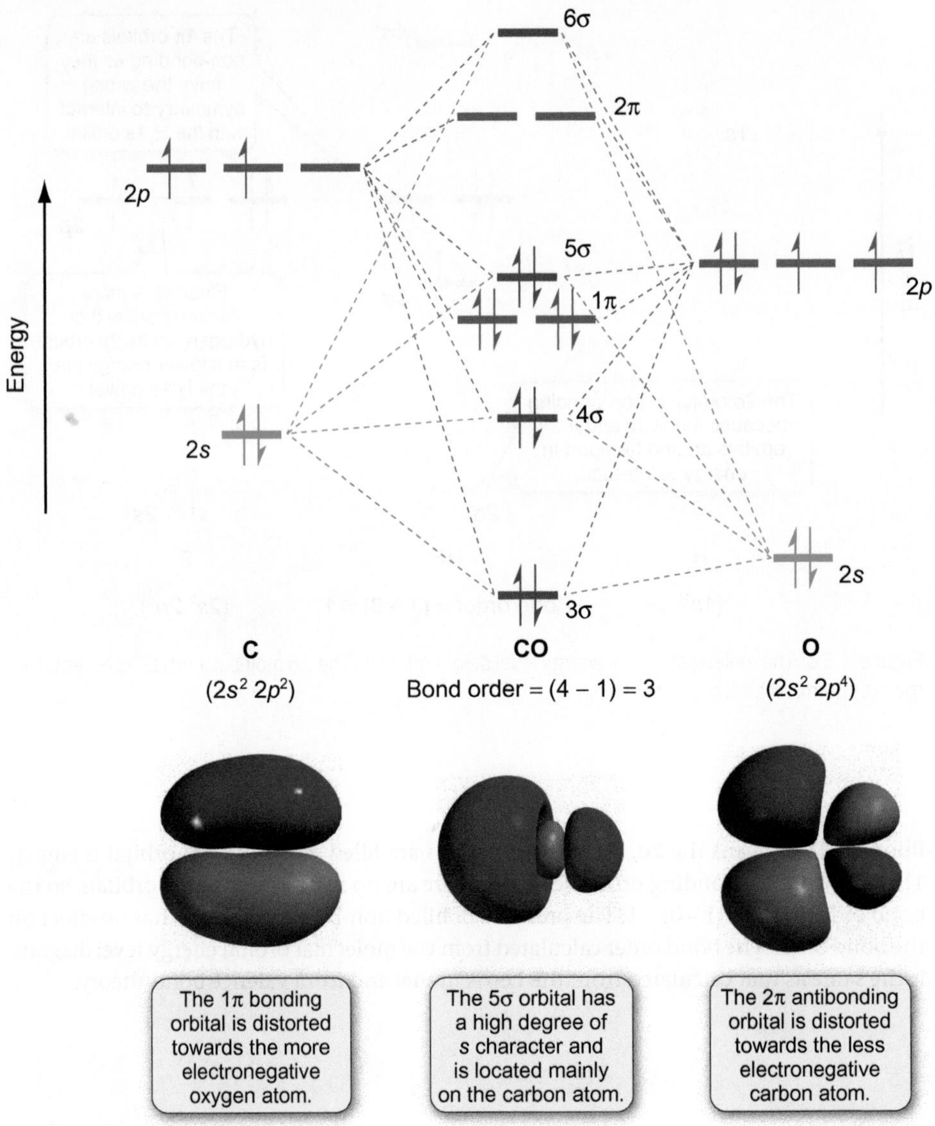

Energy
6σ
2π
2p
5σ
1π
2p
4σ
2s
2s
3σ
C
(2s² 2p²)
CO
Bond order = (4 – 1) = 3
O
(2s² 2p⁴)
The 1π bonding orbital is distorted towards the more electronegative oxygen atom.
The 5σ orbital has a high degree of s character and is located mainly on the carbon atom.
The 2π antibonding orbital is distorted towards the less electronegative carbon atom.

Figure 4.33 The molecular orbital energy level diagram for CO together with calculated boundary surfaces for the 1π, 5σ, and 2π orbitals.

Carbon monoxide (CO)

The molecular orbital energy level diagram for carbon monoxide is shown in Figure 4.33. Oxygen is more electronegative than carbon, so the oxygen $2p$ orbitals are lower in energy than the carbon $2p$ orbitals. The oxygen $2s$ orbital interacts with the carbon $2s$ orbital to give σ bonding and antibonding molecular orbitals. The set of three oxygen $2p$ orbitals interacts with the three carbon $2p$ orbitals giving σ bonding and antibonding molecular orbitals (from the p_z orbitals) and π-bonding and antibonding molecular orbitals (from the p_x and p_y orbitals). s–p mixing is generally large in heteronuclear diatomics due to the relative closeness in energy of some of the atomic orbitals, in this case the carbon $2s$ and the oxygen $2p_z$ orbitals.

There are four filled bonding orbitals and one filled antibonding orbital, so the bond order in CO is (4 – 1) = 3. This is in agreement with the Lewis model, though in the Lewis description one of the bonds is a dative bond, with both electrons from the same atom (Figure 4.4, p.174).

i Without s–p mixing the 4σ orbital in CO is antibonding and the 5σ orbital is bonding. s–p mixing stabilizes the antibonding orbital, making it more bonding in character, and destabilizes the bonding orbital, making it more antibonding. In some treatments of the bonding in CO, the 4σ orbital is regarded as bonding whereas the 5σ is regarded as antibonding.

Carbon monoxide is toxic due to its interaction with the iron centres in haemoglobin (see Box 28.7, p.1298).

Worked example 4.7 Bond lengths in NO, NO^+, and NO^-

Use molecular orbital theory to explain the trend in the observed bond lengths for NO, NO^+, and NO^-.

NO	NO^+	NO^-
115 pm	106 pm	124 pm

Strategy

Construct a molecular orbital energy level diagram for nitrogen monoxide, NO; then determine bond orders for NO, NO^+, and NO^-.

Solution

The molecular orbital energy level diagram for NO contains a similar arrangement of orbitals as in CO, though less *s*–*p* mixing means the 5σ orbital lies very close to the 1π orbitals in energy and in some treatments is shown above it. For NO, there is one more electron to accommodate than for CO.

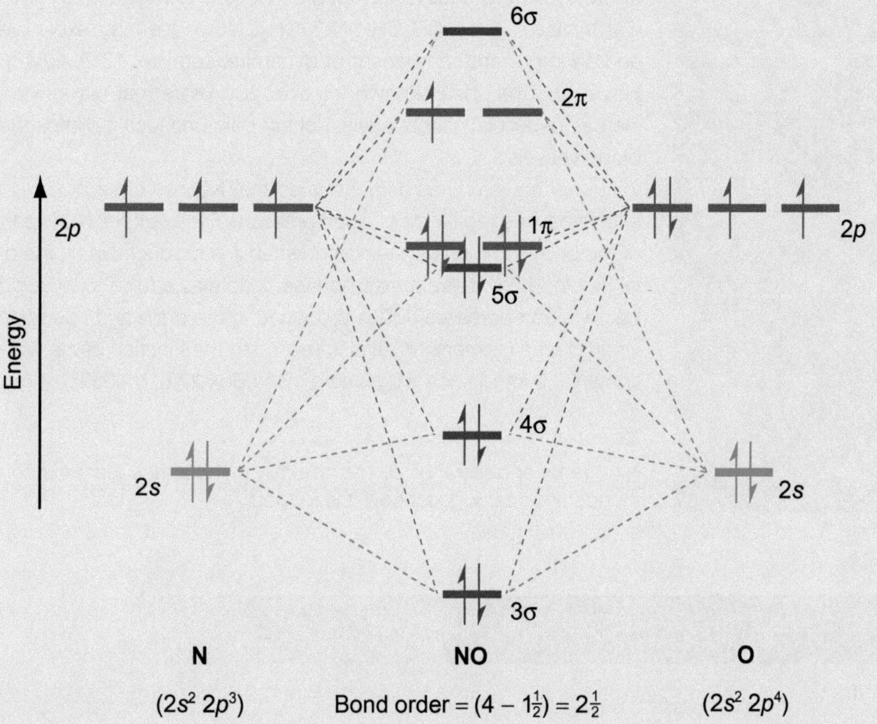
6σ
2π
1π
5σ
4σ
3σ
2p
2p
2s
2s
Energy
N
NO
O
($2s^2\ 2p^3$)
Bond order = $(4 - 1\frac{1}{2}) = 2\frac{1}{2}$
($2s^2\ 2p^4$)

NO has four filled bonding orbitals and $1\frac{1}{2}$ filled antibonding orbitals, giving a bond order of $(4 - 1\frac{1}{2}) = 2\frac{1}{2}$. For NO^+, the $2\pi^*$ electron is removed, so there is one less antibonding electron and the bond order is $(4 - 1) = 3$. For NO^- an additional electron is added to the $2\pi^*$ orbital, so the bond order is reduced to $(4 - 2) = 2$. The trend in bond lengths agrees with that expected from the bond orders, that is, NO^+ has the shortest (strongest) bond and NO^- has the longest (weakest) bond.

Question

What common anion is isoelectronic with NO^+?

Box 4.11 Using nitrogen monoxide to send biological signals

In 1986, a surprising discovery was made about nitrogen monoxide (NO). The American scientists Bob Furchgott, Lou Ignarro, and Ferid Murad discovered that NO acts as a signal molecule in biological systems. The compound penetrates membranes and regulates the function of cells. This was the first time that a gas had been observed to act as a biological signal, and the discovery led to Furchgott, Ignarro, and Murad receiving the Nobel Prize for Medicine in 1998.

▼ A coloured scanning electron microscopy (SEM) photo of blood vessels.

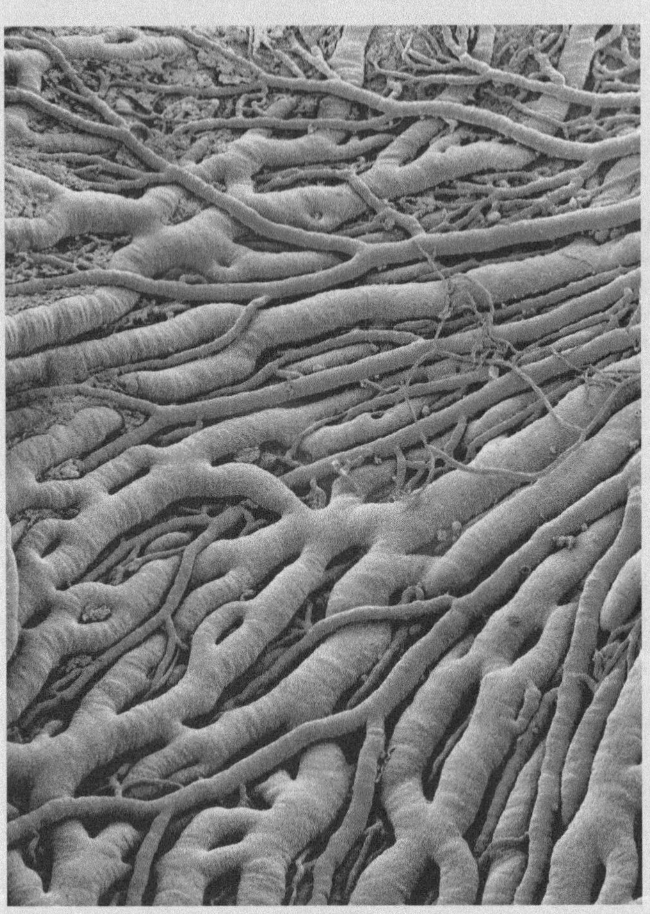

The discovery began with research on how the muscles surrounding blood vessels relax or contract to regulate the amount of blood flowing through the vessels. The organic cation acetylcholine ($MeOC(O)CH_2CH_2NMe_3^+$) was known to be a messenger that could trigger the relaxation of these muscles. Furchgott showed that the acetylcholine acted on surface cells inside the blood vessels (known as *endothelium* cells) rather than on the muscles themselves. This implied that the endothelium cells were releasing an unknown signal molecule that made the muscles relax. Furchgott and Ignarro showed that this molecule was NO.

Murad showed that nitroglycerine, commonly prescribed by doctors to treat angina, is a source of NO. Nitroglycerine (glyceryl trinitrate, $O_2NOCH_2CH(ONO_2)CH_2ONO_2$) is a high explosive and an important component of dynamite (see Box 13.7, p.643). However, it has been known for over 100 years that nitroglycerine has beneficial effects against chest pain because it dilates the blood vessels.

Further research has demonstrated that NO has other biological effects beyond regulation of blood pressure. For example, NO is part of the body's immune system and, when it is produced in a type of white blood cell called a *macrophage*, it acts as a toxin to invading bacteria and parasites. NO is produced in the brain and has been implicated in memory. NO is also related to the function of the well-known anti-impotence drug Viagra® (see Box 22.6, p.1039).

Question

Nitrogen monoxide is produced naturally in the atmosphere during thunderstorms. Suggest how it is formed.

Summary

- Atomic orbitals combine together if:
  - (a) they have the correct symmetry, *and*
  - (b) there is efficient overlap, *and*
  - (c) if the orbitals are relatively close together in energy.
- The relative energies of the atomic orbitals of the contributing atoms in a heteronuclear diatomic can be predicted using electronegativities.
- In a linear combination expression, the different atomic orbitals contribute unequally and the bonding orbital lies closer in energy to the atomic orbital that contributes more to it.
- Heteronuclear diatomics can have non-bonding orbitals, in addition to bonding and antibonding orbitals.

For practice questions on these topics, see questions 17–19 at the end of this chapter (p.214).

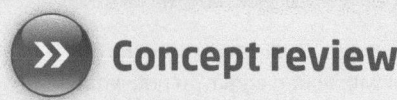

Concept review

By the end of this chapter, you should be able to do the following.

- Describe the basis of the Lewis, valence bond, and molecular orbital theories of bonding and understand the similarities and differences between them.
- Draw Lewis structures for molecules and determine their bond orders.
- Understand the meaning of the valence bond terms resonance and hybridization.
- Draw the resonance forms for homonuclear and heteronuclear diatomic molecules.
- Understand the concept of electronegativity, and use the definition of χ^P to predict bond dissociation enthalpies for heteronuclear diatomic molecules.
- Draw boundary surfaces to show how molecular orbitals are formed from the linear combination of atomic orbitals (LCAO).
- Assign symmetry labels to orbitals using σ, π, g, and u.
- Recognize that π-overlap is both less efficient and more distance-sensitive than σ-overlap.
- Construct molecular orbital energy level diagrams for homonuclear diatomics and understand why there is a crossover in the order of the energy levels between N_2 and O_2 in the second period.
- Construct molecular orbital energy level diagrams for heteronuclear diatomics and understand why the degree of interaction between atomic orbitals depends on the energy gap between them.
- Interpret molecular orbital energy level diagrams and extract from them chemical information such as bond order and magnetic properties.
- Understand the basis of photoelectron spectroscopy and how it can be used to measure molecular orbital energy levels.

Key equations

The in-phase linear combinations of atomic orbitals	$\psi_{\text{in phase}} = \phi_{1s}(H_A) + \phi_{1s}(H_B)$	(4.8)
The out-of-phase linear combinations of atomic orbitals	$\psi_{\text{out of phase}} = \phi_{1s}(H_A) - \phi_{1s}(H_B)$	(4.9)
Definition of bond order	bond order = $\dfrac{\text{number of filled}}{\text{bonding orbitals}} - \dfrac{\text{number of filled}}{\text{antibonding orbitals}}$	(4.16)

Questions

More challenging questions are indicated by an asterisk *.

1. Draw Lewis structures for the following molecules and ions, and in each case identify the bond order. (Section 4.2)

   Cl_2 Se_2 HBr ClO^-

2. Use the following information to predict the electronegativity of iodine (Section 4.3)

   $\chi^P(H) = 2.20$ $D(H{-}H) = 436\ kJ\ mol^{-1}$
   $D(I{-}I) = 152\ kJ\ mol^{-1}$ $D(H{-}I) = 298\ kJ\ mol^{-1}$

3. Use the data in Table 4.1 (p.171) and Figure 4.6 (p.178) to estimate a value for the bond dissociation enthalpy of ClBr. (Section 4.3)

4.* An alternative scale of electronegativity to that of Pauling was proposed by Robert Mulliken, who argued that an electronegative atom was likely to have both a high ionization energy, as it would not readily lose electrons, and a high negative electron gain energy, as it would be energetically favourable for it to gain electrons. Mulliken defined electronegativity as an average of these terms, as shown below.

$$\chi^M = \tfrac{1}{2}(I_1 - E_{eg1})$$

Use the data in Figures 3.35 (p.156) and 3.38 (p.158) to calculate χ_M for the Group 17 elements, and comment on the trend observed.

5. Draw one covalent and two ionic resonance forms that contribute to the bonding in ClF in the valence bond approach. Which of the ionic forms is more important? Why? (Sections 4.4, 4.5)

6. Use valence bond theory to describe the bonding in the cyanide anion, CN^-. What orbitals interact to form the bonds between the atoms? (Sections 4.4, 4.5)

7.* H_2 absorbs ultraviolet radiation of the wavelength 109 nm. What is the origin of this absorption and what energy (in $kJ\ mol^{-1}$) does it correspond to? Why does absorbing ultraviolet radiation of this wavelength cause the H_2 molecules to split into atoms? (Sections 4.6–4.8)

8. Use the molecular orbital energy level diagram for Li_2 (Figure 4.17, p.192) to work out the bond orders of Li_2, Li_2^+, and Li_2^-. Which of these are paramagnetic? (Sections 4.6–4.8)

9. Use the linear combination of atomic orbitals approach to write expressions for the wavefunctions for the in-phase and out-of-phase combinations of two lithium 2*s* orbitals. Which molecular orbitals in Li_2 do these combinations correspond to? (Sections 4.6–4.8)

10. The mass spectrum of a sample of beryllium contains an intense peak at *m/z* 9 and a less intense peak at *m/z* 18. Identify the species involved and use molecular orbital theory to support the existence of the species giving rise to the peak of lower intensity. (Sections 4.6–4.8)

11. Sketch boundary surface diagrams for the HOMO and LUMO in F_2. (Sections 4.9–4.10)

12. Draw a molecular orbital energy level diagram for Ne_2. Would you expect this molecule to exist? (Sections 4.9–4.10)

13. Which of the following pairs of molecules or ions are isoelectronic with each other? (Sections 4.9–4.10)

(a) CO and NO^-

(b) CN^- and NO^+

(c) N_2^- and O_2^-

14. Draw a labelled molecular orbital energy level diagram for the acetylide dianion C_2^{2-} and use it to explain why the bond length in C_2^{2-} (119 pm) is less than that in C_2 (124 pm). (Section 4.11)

15. Give examples of neutral homonuclear and heteronuclear diatomic molecules that are isoelectronic with C_2^{2-}. (Section 4.11)

16.* Construct a labelled molecular orbital energy level diagram for Si_2, stating any assumptions you have made. Measurements suggest that Si_2 is diamagnetic in the gas phase. Is this consistent with your diagram? If not, what changes could be made to the order of the molecular orbitals? How could these be justified? (Section 4.11)

17.* Draw a labelled MO diagram for the •OH radical. The bonding orbital is described by the wavefunction

$$\psi = N[\lambda\phi(H_{1s}) + \phi(O_{2p_z})]$$

Comment on the magnitude of λ, and any assumptions that have been made in obtaining this expression. What is the bond order in •OH and OH^-? (Section 4.12)

18. Draw a labelled molecular orbital energy level diagram for the cyanide anion CN^-. What is the bond order in CN^-? How would the bond length and magnetic behaviour of neutral CN differ from those observed for CN^-? (Section 4.12)

19. What are the symmetry labels for the following molecular orbitals in the molecules A_2 and AB? (Section 4.12)

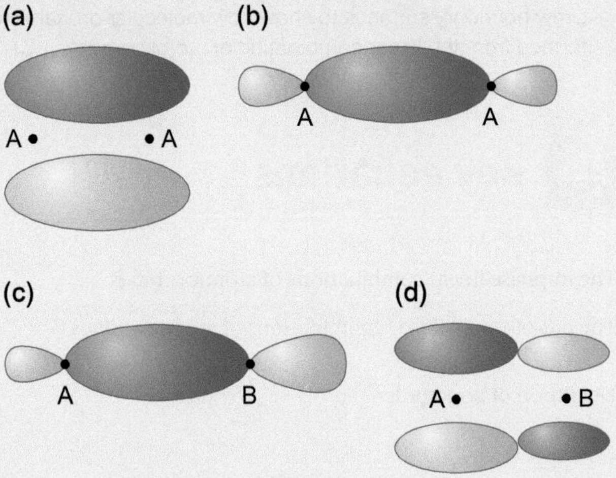
(a)
(b)
A
A
A
A
(c)
(d)
A
B
A
B

Is A or B more electronegative?

5

Polyatomic molecules

This chapter builds on the following topics:

- **Many-electron atoms** Section 3.6, p.143
- **Atomic properties and periodicity** Section 3.7, p.153
- **Diatomic molecules** Chapter 4, p.168

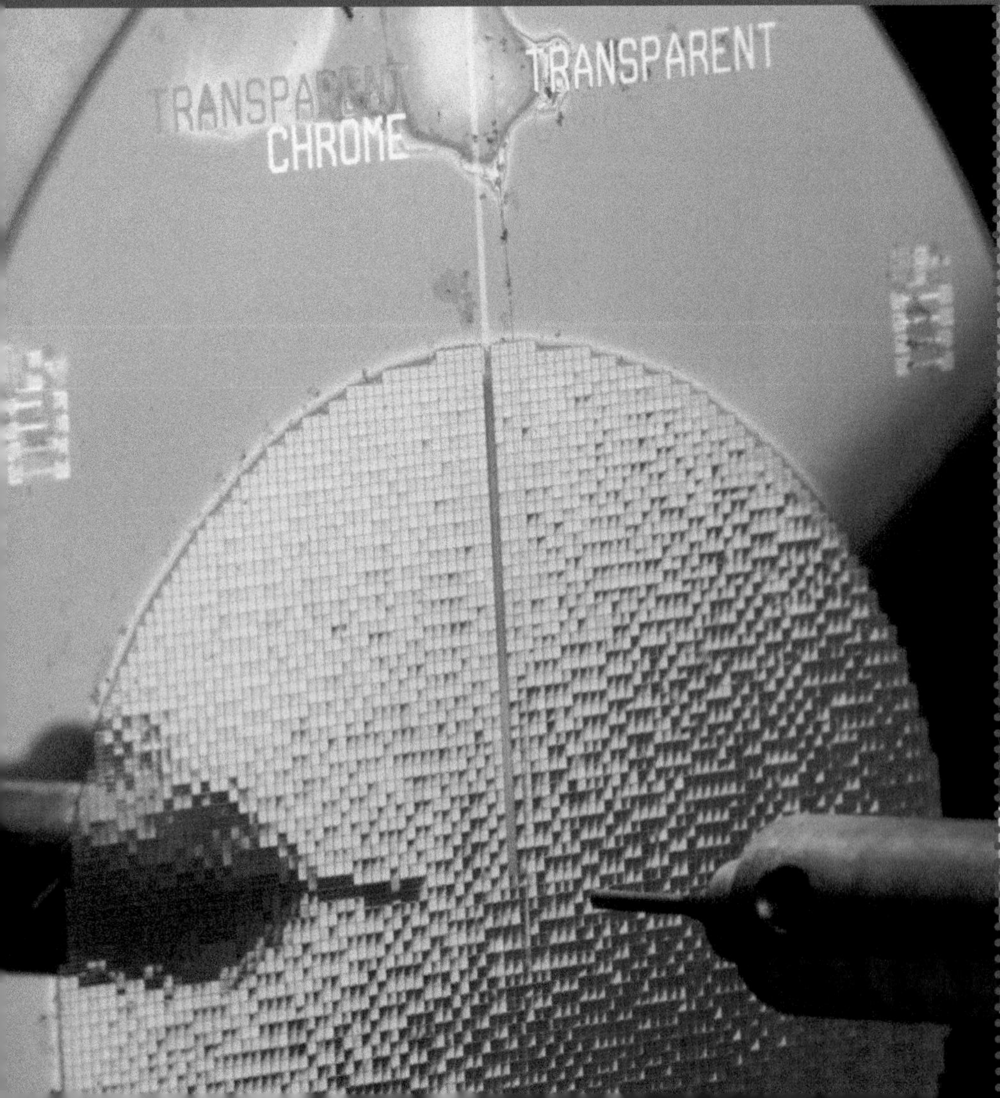
TRANSPARENT
CHROME

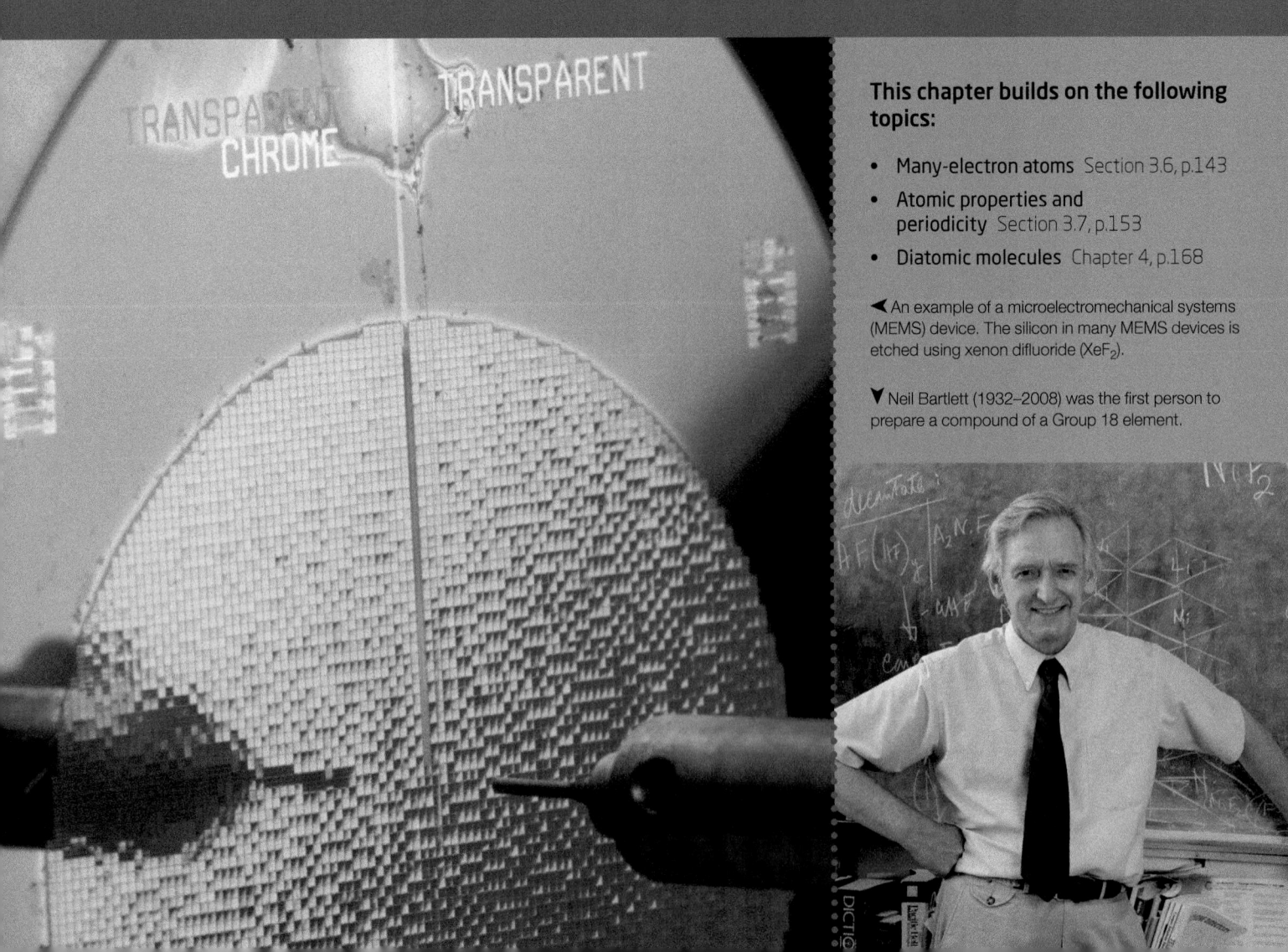

◀ An example of a microelectromechanical systems (MEMS) device. The silicon in many MEMS devices is etched using xenon difluoride (XeF_2).

▼ Neil Bartlett (1932–2008) was the first person to prepare a compound of a Group 18 element.

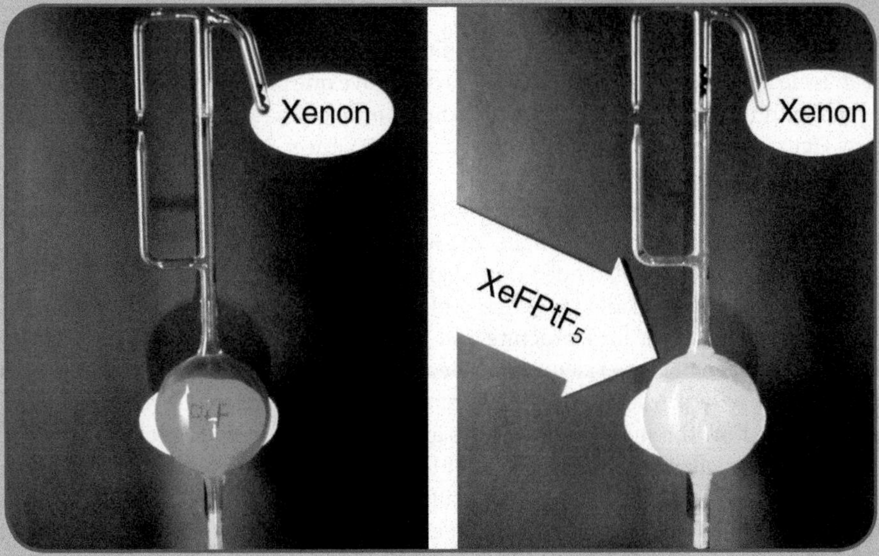
Xenon
Xenon
XeFPtF5

The first example of a compound of a Group 18 element. Mixing the colourless gas xenon with the red gas PtF_6 gave a yellow solid. The label for the product reflects the original formulation for this—it is now known to be a mixture of products including $[XeF][PtF_6]$.

Xenon compounds

The Group 18 elements were once known as the inert gases. Their $ns^2\ np^6$ electronic configuration makes them unreactive and, until the 1960s, scientists believed that the filled octet prevented them from forming any compounds. Indeed, in 1924, Friedrich Paneth, an expert on isotopes, stated that 'the unreactivity of the noble gas elements belongs to the surest of experimental results'.

The first person to prove Paneth wrong was Neil Bartlett in 1962. He had made the compound $O_2^+PtF_6^-$ and, while preparing an undergraduate lecture at the University of British Columbia, Canada, he realized that the first ionization enthalpy of xenon (+1170 kJ mol^{-1}) was very similar to that of O_2 (+1177 kJ mol^{-1}). He reasoned that if PtF_6 was a strong enough oxidizing agent to remove an electron from O_2, it should also remove an electron from xenon. He reacted PtF_6 vapour with xenon and obtained a yellow solid to which he assigned the formula $XePtF_6$. The actual identity of the reaction product was unclear for many years, but chemists have subsequently discovered that the major crystalline compound is $[XeF]^+[PtF_6]^-$, with other compounds containing the $[XeF]^+$ cation also present.

Since the 1960s, many other examples of xenon compounds have been made including the fluorides XeF_2 and XeF_4 and the oxide XeO_3. These are examples of polyatomic molecules, and their shapes are described in this chapter along with a way of explaining the bonding in XeF_2. The chemistry of xenon compounds is described in more detail in Chapter 27.

XeF_2 is used as a fluorinating agent in organic chemistry. For example, it reacts with uracil to form 5-fluorouracil, which is used in chemotherapy. Over the past few years, XeF_2 has been used as a selective etchant in the emerging MEMS (microelectromechanical systems) area of the semiconductor industry. MEMS are micromachines. In addition to the standard electronic components present on an integrated circuit, MEMS can also contain sensors and mechanical elements. This allows MEMS to detect something about the environment then respond to it. One common use for MEMS is as sensors in airbags, where they are able to detect rapid deceleration and then respond to it by closing a contact to make the airbag inflate.

One of the techniques used to make MEMS uses selective etching away of parts of a silicon wafer. The surface is patterned with a thin layer of unreactive silicon dioxide and then exposed to XeF_2. This reacts with the exposed silicon

$$2\,XeF_2(g) + Si(s) \rightarrow 2\,Xe(g) + SiF_4(g)$$

etching it away. XeF_2 is particularly useful as an etchant because both products from the reaction with silicon are gases, so do not contaminate the surfaces.

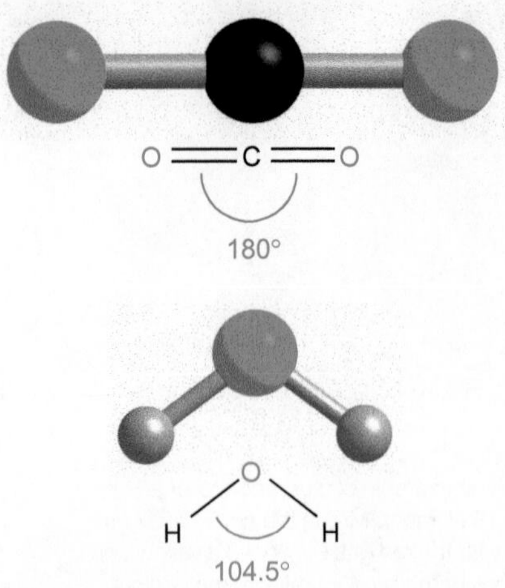
O
C
O
180°
O
H
H
104.5°

Figure 5.1 The triatomic molecules CO_2 and H_2O have different shapes. CO_2 is linear and H_2O is bent.

The properties of a covalent compound are influenced by the shape of its molecules. This is not an issue for diatomic molecules, as these only have one possible shape. However, for molecules with three or more atoms—**polyatomic molecules**—different shapes are possible. As Figure 5.1 shows, a triatomic molecule can be linear, like carbon dioxide (CO_2), or bent, like water (H_2O).

In this chapter you will learn how to predict the shapes of polyatomic molecules and how to explain the bonding within them. You will see how the three bonding theories developed in Chapter 4 can be extended from diatomic to polyatomic molecules. The Lewis model, valence bond theory, and molecular orbital theory will be looked at in turn, as each provides a different insight into structure and bonding. Sections 5.1 and 5.2 show how the Lewis model develops into valence shell electron pair repulsion (VSEPR) theory, which is an excellent way of predicting the shapes of molecules. In Section 5.3 you will look at polarity, and see how the shapes of molecules affect their properties. Sections 5.4 and 5.5 extend the ideas of valence bond theory to explain the bonding in polyatomic molecules. Finally, Sections 5.6 and 5.7 show how molecular orbital theory can be extended to polyatomic molecules. This is particularly useful in explaining the bonding in compounds that do not obey the octet rule, as the other theories are less informative for these cases.

Box 5.1 Representing structures

Valence bond theory and molecular orbital theory are quantum mechanical approaches to bonding. In both theories, bonds are formed by the interaction between the orbitals on atoms. Since electrons do not have fixed positions, atomic orbitals and molecular orbitals are represented by boundary surfaces, which contain a proportion (often 95%) of the electron density (see Sections 3.5 (p.140) and 4.7 (p.184)). It is also possible to represent the electron density in a polyatomic molecule in this way, as shown in Figure 1 for caffeine. While this type of figure can be useful to see where the regions of high electron density are in a molecule, it is difficult to see what is going on in terms of bonding, so molecules are not normally shown in this way.

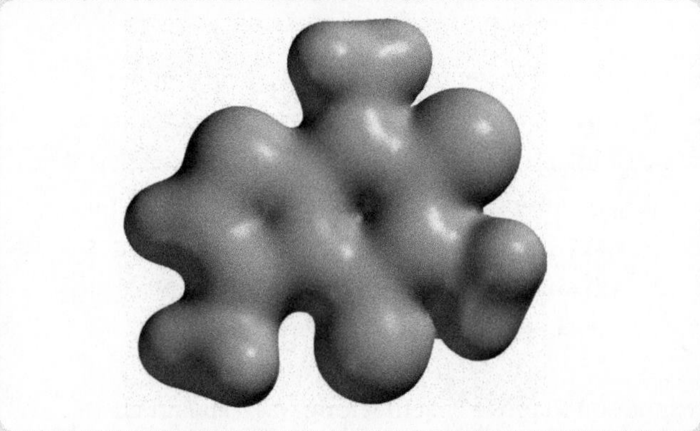

▲ **Figure 1** A boundary surface for caffeine, showing 95% of the electron density in the molecule.

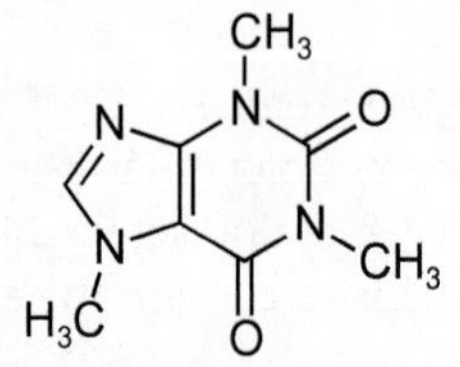

▲ Caffeine.

A very common way of representing molecules is to use a **ball and stick model**, as shown for caffeine in Figure 2. In this case, the atoms are shown as balls and the bonds are shown as sticks. The colour of the ball indicates the type of atom and in this book, for example, carbon atoms are black, oxygen atoms are red, and nitrogen atoms are blue. The size of the ball reflects the size of the atom in question. The sticks indicate connectivity, so a double bond appears the same

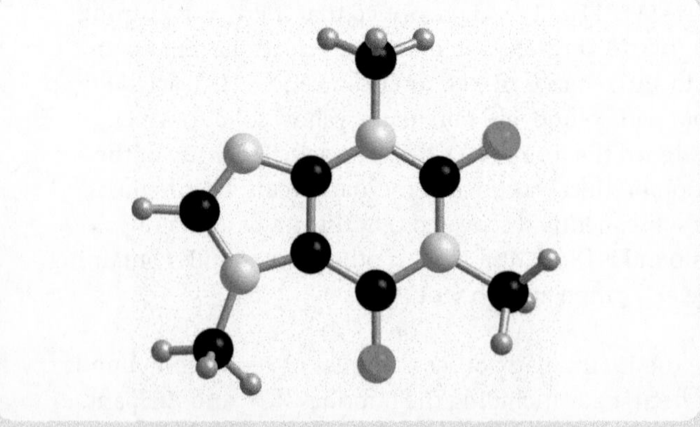

▲ **Figure 2** A ball and stick representation of the structure of caffeine. →

Online Support. The icon in the margin indicates that accompanying interactive resources are provided online to help your understanding: just type www.chemtube3d.com/burrows/123 into your browser, replacing 123 with the number of the page where you see the icon. For pages linking to more than one resource, type 123-1, 123-2, etc. (replacing 123 with the page number) for access to successive links.

as a single bond. In this book, ball and stick models are often used to show molecules alongside the structural formulae. Ball and stick models are also used for ionic structures in Chapter 6, though in these cases the sticks represent closest contacts rather than bonds. Ball and stick models are useful for seeing the spatial arrangements of the atoms and bonds.

Although ball and stick models are commonly used, it is not always easy to see the shape of a molecule using these. Another representation is the **space-filling model**. In this the atoms are shown much larger than they are in a ball and stick model, typically using their van der Waals radii (see Section 3.7, p.153). The space-filling model for caffeine is shown in Figure 3. In a space-filling model, neighbouring atoms are shown in contact.

▲ **Figure 3** A space-filling representation of the structure of caffeine.

5.1 The Lewis model

You can draw Lewis structures for polyatomic molecules using the same procedures as those for diatomic molecules (Section 4.2, p.173). The atoms within a molecule share electrons through the formation of covalent bonds to form energetically more favourable structures. For many atoms, this involves the formation of filled octets (filled outer shells or valence shells) and the atoms are said to be obeying the **octet rule**. This is equivalent to the atoms having filled *s* and *p* valence orbitals. The non-bonding electrons on each atom are present as **lone pairs**. The Lewis structures shown in Figure 5.2 show how some atoms in the second period achieve filled octets through forming covalent bonds.

The second period elements, from lithium to neon, never have more than eight electrons in their valence orbitals. However, there are molecules of these elements for which there are *less* than eight electrons in the valence orbitals of one of the atoms, and in these cases the atom is said to be **electron poor**. Examples include the nitrogen atom in NO_2 and the boron atom in BH_3 and these molecules are shown in Figure 5.3. Compounds containing an atom with an incomplete octet tend to be very reactive.

ⓘ The octet rule states that atoms combine so that in a molecule each atom has eight electrons in its valence shell.

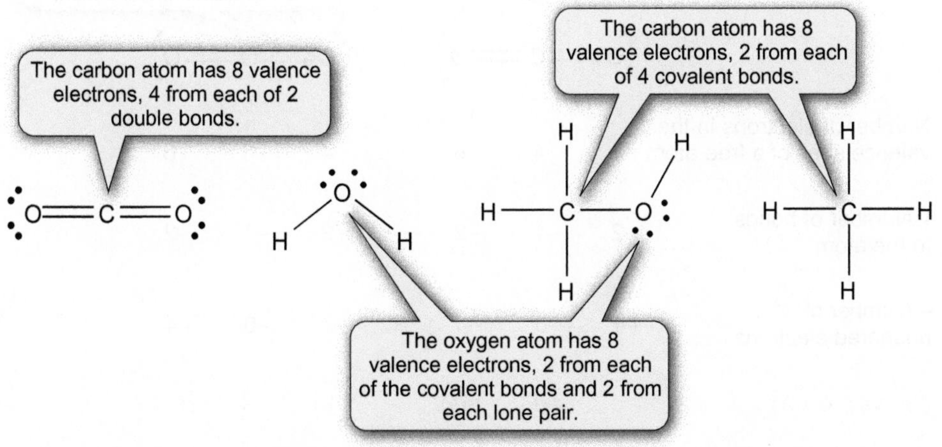

The carbon atom has 8 valence electrons, 4 from each of 2 double bonds.
The carbon atom has 8 valence electrons, 2 from each of 4 covalent bonds.
The oxygen atom has 8 valence electrons, 2 from each of the covalent bonds and 2 from each lone pair.

Figure 5.2 Lewis structures showing the covalent bonds and the lone pairs in CO_2, H_2O, CH_3OH, and CH_4. Each non-hydrogen atom has a filled octet of valence electrons.

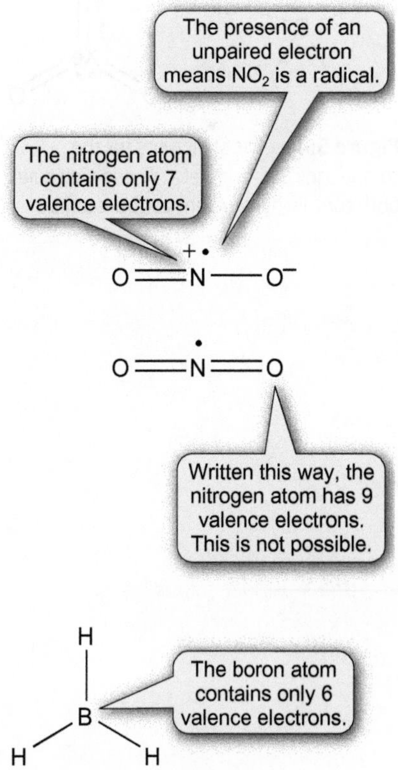

The presence of an unpaired electron means NO2 is a radical.
The nitrogen atom contains only 7 valence electrons.
Written this way, the nitrogen atom has 9 valence electrons. This is not possible.
The boron atom contains only 6 valence electrons.

Figure 5.3 Lewis structures for the compounds NO_2 and BH_3, which contain an electron-poor atom.

The reasons why second period elements do not form hypervalent compounds are discussed in Section 5.5 (p.245).

Treatment of hypervalent compounds by valence bond theory is described in Section 5.5 (p.245) and by molecular orbital theory in Section 5.7 (p.249).

Elements in the third period and beyond can break the octet rule. In compounds such as PCl_5 and XeO_3 (Figure 5.4), the central atom has more than eight electrons in its outer shell. These atoms with expanded octets are said to be **hypervalent**, as are the molecules formed. Valence bond theory and molecular orbital theory explain hypervalency in different ways as you will see later in the chapter.

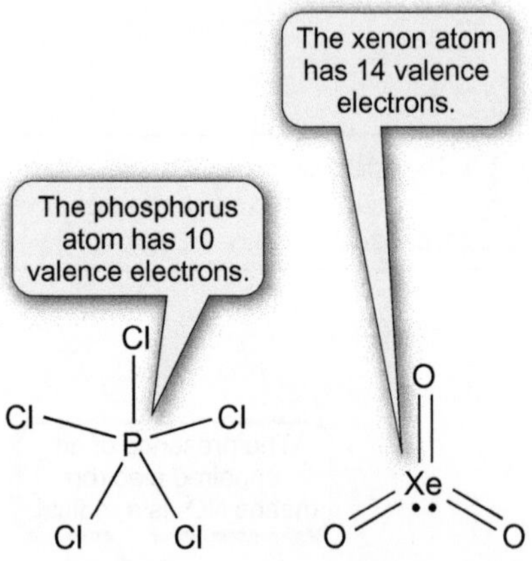
The xenon atom has 14 valence electrons.
The phosphorus atom has 10 valence electrons.
Cl
Cl
Cl
P
Cl
Cl
O
Xe
O
O

Figure 5.4 Lewis structures for the compounds PCl_5 and XeO_3. These molecules both contain hypervalent central atoms.

Formal charge

For many polyatomic molecules, you can draw several structures in which the atoms are connected in different ways. For example, the atoms in CO_2 could be connected in the order OCO or COO. How can you predict which of these is the actual structure? One way is to look at the extent to which the atoms have gained or lost electrons in each arrangement. This is assessed by working out the formal charge on each atom in the proposed structure. The most likely structure is that in which the formal charges are lowest.

The **formal charge** on an atom in a molecule is defined as the difference between the number of valence electrons in the free atom and the number of electrons assigned to that atom in a Lewis structure. The electrons within a covalent bond are divided equally between the atoms. This gives the expression in Equation 5.1

$$\text{Formal charge} = (\text{number of electrons in the valence shell of the free atom}) - (\text{number of bonds to the atom}) - (\text{number of unshared electrons}) \quad (5.1)$$

Here's an example. In carbon dioxide (Figure 5.5) there are four valence electrons associated with the carbon atom ($2s^2\,2p^2$) and six with each oxygen atom ($2s^2\,2p^4$). For the arrangement OCO, sharing the electrons from each of the double bonds between the contributing atoms gives four electrons to the carbon atom and two to each of the oxygen atoms. Each oxygen atom also has four unshared electrons. Using Equation 5.1, the formal charge on each atom is zero.

The other possible arrangement of atoms for CO_2 is also shown in Figure 5.5. In this case, one of the oxygen atoms is in the centre, giving COO. For this structure, the bonds to the carbon atom are both **dative bonds**, with the two electrons in the dative bond donated from the same atom, in this case the central oxygen atom. Dative bonds are often shown in Lewis structures as arrows.

For COO, the central oxygen atom has only four electrons associated with it and no unshared electrons, so it has a formal charge of +2. The carbon atom has six electrons associated with it, two more than it started with, so its formal charge is −2. The sum of formal

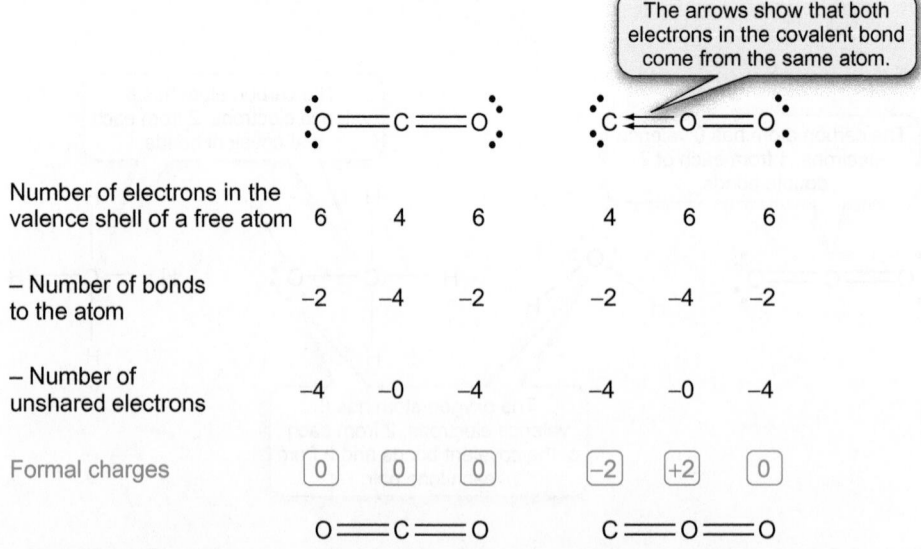

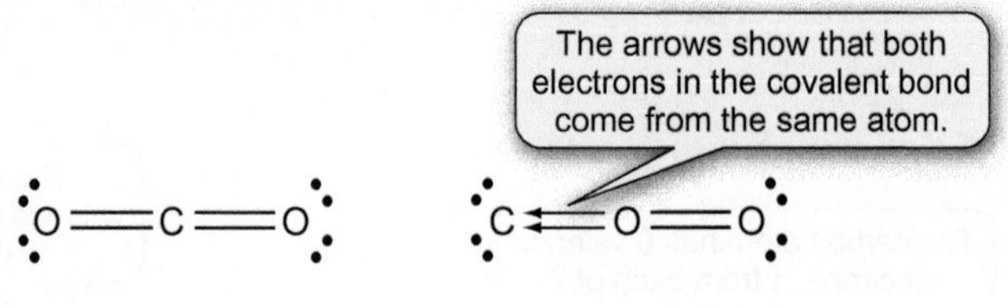
The arrows show that both electrons in the covalent bond come from the same atom.
O
C
O
C
O
O

	O	C	O	C	O	O
Number of electrons in the valence shell of a free atom	6	4	6	4	6	6
– Number of bonds to the atom	−2	−4	−2	−2	−4	−2
– Number of unshared electrons	−4	−0	−4	−4	−0	−4
Formal charges	0	0	0	−2	+2	0

O=C=O C=O=O

Figure 5.5 Formal charges for the atoms in two possible structures for carbon dioxide.

charges is equal to the overall charge, so for a neutral molecule this must be zero. Generally speaking, molecules adopt structures in which the formal charges on the atoms are as low as possible. This helps explain why OCO is favoured over COO as a structure for carbon dioxide. Worked example 5.1 shows how to work out the formal charges on two possible structures for N_2O, and then use the results to predict which structure is favoured.

Worked example 5.1 Formal charges in N_2O

Does dinitrogen oxide (N_2O) adopt a structure with a central nitrogen atom (NNO) or a central oxygen atom (NON)?

Strategy

Draw Lewis structures for both possibilities, showing the lone pairs and any dative bonds. Use Equation 5.1 to determine the formal charges for the atoms in both structures. The structure with the lower formal charges will be the one adopted.

Solution

The formal charges are smaller overall for the formula with the central nitrogen atom, so this is the structure adopted by N_2O.

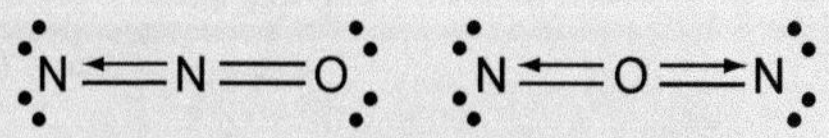

	N	N	O	N	O	N
Number of electrons in the valence shell of a free atom	5	5	6	5	6	5
– Number of bonds to the atom	−2	−4	−2	−2	−4	−2
– Number of unshared electrons	−4	−0	−4	−4	−0	−4
Formal charges	−1	+1	0	−1	+2	−1

N═N═O N═O═N

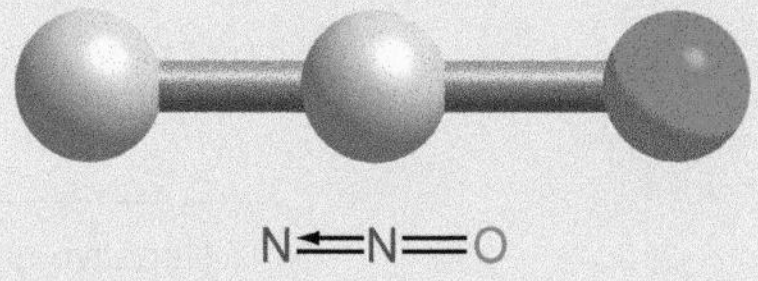

Question

Use formal charges to decide whether oxygen difluoride adopts a structure with a central oxygen atom (FOF) or a central fluorine atom (FFO).

Box 5.2 N_2O: from laughing gas to dragsters

The gas dinitrogen oxide (nitrous oxide, N_2O) was discovered in the 1770s by the English chemist and clergyman Joseph Priestley. When he investigated the physiological effects of N_2O, Humphrey Davy noted that it caused giggling, disorientation, and hallucinations in people inhaling it. Davy coined the term 'laughing gas' for nitrous oxide, and during the first half of the nineteenth century the main use for N_2O was recreational. People paid money at travelling shows and carnivals to inhale a minute's worth of the gas.

Davy noted that N_2O had anaesthetic properties, but it wasn't until the 1840s that the gas was first used as a dental and medical painkiller. Initially it was less popular than rival anaesthetics such as ether (diethyl ether, $CH_3CH_2OCH_2CH_2$) and chloroform ($CHCl_3$), since these are more potent and easier to use. However, both of these have side effects, and ether is highly flammable, which was another drawback to its use. In contrast, pure nitrous oxide is a safe but mild anaesthetic, and is still used today as a painkiller in dentistry and childbirth.

N_2O is unreactive at room temperature, but on heating it decomposes exothermically into N_2 and O_2.

$$2N_2O(g) \xrightarrow{300^\circ C} 2N_2(g) + O_2(g)$$

If N_2O is injected into the combustion chamber of an engine, the power output increases. This is because when N_2O decomposes three moles of gas are formed from two moles of N_2O, and the

LAUGHING GAS

▲ In the nineteenth century many people, including the poet Samuel Coleridge, inhaled N_2O for fun. N_2O is often formed along with other more toxic nitrogen oxides so inhaling it under these conditions is potentially dangerous.

➜

▲ Dragsters often use N_2O to provide rapid acceleration.

increase in pressure provides an extra boost to the pistons. The increased oxygen content of the air following decomposition also allows for more efficient fuel combustion.

N_2O can be injected into the fuel lines of racing cars to give more power to the engine. It is introduced into the engine's air intake manifold as a liquid where it vaporizes, increasing the pressure and lowering the intake temperature through its latent heat of vaporization. It is not feasible to run a car continuously using N_2O, so it is mainly used in short, quick events such as drag racing where maximum acceleration is important.

Question

Calculate the percentage by mass of oxygen in N_2O. How does this compare to the percentage of oxygen in air?

Resonance forms are described in Section 4.5 (p.180).

When negative charges are present in a molecule, the lowest energy structure is the one with these charges on the most electronegative atoms. The two structures in Figure 5.6 are resonance forms for the thiocyanate ion (SCN^-). Both have formal charges of −1 on one of the end atoms. Since nitrogen is more electronegative than sulfur, you would expect Structure A to be favoured. Indeed, bond length measurements suggest it makes a larger contribution to the structure of SCN^- than Structure B.

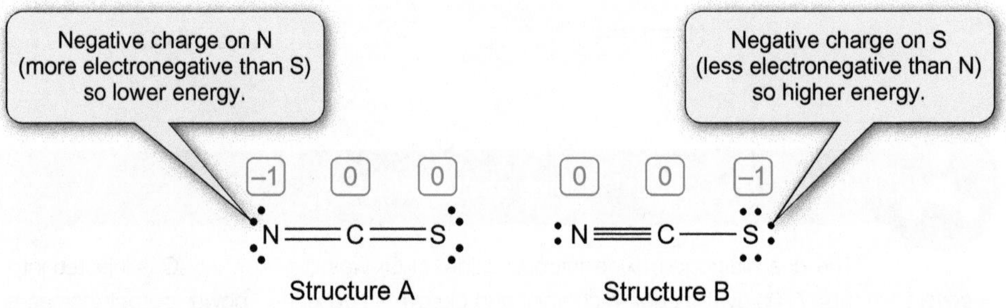
Negative charge on N (more electronegative than S) so lower energy.
Negative charge on S (less electronegative than N) so higher energy.
−1
0
0
0
0
−1
N
C
S
N
C
S
Structure A
Structure B

Figure 5.6 Two resonance forms for the thiocyanate anion. Structure A is more important as the negative charge is located on the more electronegative atom.

Summary

- The second period elements, from lithium to neon, generally obey the octet rule and never have more than eight electrons in their valence orbitals. Molecules with fewer than eight valence electrons are said to be electron poor and are generally very reactive. Examples of electron-poor molecules include NO_2 and BH_3.
- Larger atoms can exhibit hypervalency, in which the octet is expanded and there are more than eight electrons in the valence orbitals. Examples of hypervalent molecules include PCl_5 and XeO_3.
- Possible structures of a molecule can be assessed by calculating formal charges for the atoms. This allows the most likely structure to be predicted.

For practice questions on these topics, see questions 1 and 2 at the end of this chapter (p.253).

5.2 Valence shell electron pair repulsion theory

Valence shell electron pair repulsion theory, normally abbreviated to **VSEPR theory**, was developed by Nevil Sidgwick and Herbert Powell in 1940. It is a simple but powerful way of predicting the shapes of molecules. As the name suggests, it is only the valence electrons (typically *ns* and *np*) that are involved in determining the molecular shape.

There are three main assumptions in VSEPR theory, the first two of which come directly from the Lewis model.

1. The atoms in a molecule are held together by pairs of electrons known as **bonding pairs**.
2. Some atoms within a molecule may have pairs of electrons that are not involved in bonding. These are called **lone pairs**.
3. As the electron pairs are negatively charged, they repel each other. On each atom the electron pairs adopt positions as far apart from one another as possible.

The shapes adopted by molecules with a given number of electron pairs around the central atom can be found using geometry. The shapes of molecules of the general formula AX_n, containing a central A atom with *n* X atoms around it, are shown in Figure 5.7. For these molecules, the electron pairs are all bonding pairs.

Using VSEPR theory for structures based on tetrahedra and octahedra

Worked example 5.1 (p.221) shows how to work out the formal charges on the atoms in N_2O to determine whether nitrogen or oxygen is the central atom.

The first step in working out the shape of a molecule is to identify which is the central atom. This is often straightforward given the formula or the known valencies for the atoms involved but, if you don't know which atom is in the centre, you can work it out by determining the formal charges on the atoms in the different possibilities.

Once you know which is the central atom, you need to calculate the number of electron pairs around it. This is best done by drawing out the Lewis structure to show the covalent bonds and lone pairs.

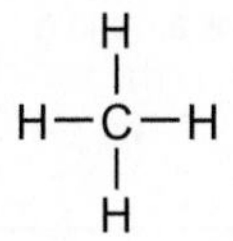

1 Lewis structure of methane (CH_4).

For example, in methane (CH_4, (**1**)), the central atom is carbon, and it forms covalent bonds to four hydrogen atoms. The electronic configuration of carbon is $1s^2\ 2s^2\ 2p^2$, so there are four electrons in the carbon valence shell. Each hydrogen atom contributes one electron through formation of a covalent bond, so there are a total of eight electrons around the carbon atom. This gives four electron pairs, all of which are bonding pairs. These electron pairs lie as far apart from each other as possible, so the CH_4 molecule is

CH_4	
Electrons from C	4
Electrons from 4 H	4
Total no. electrons around C	8
4 electron pairs around C	
4 bonding pairs and no lone pairs	
Tetrahedral	

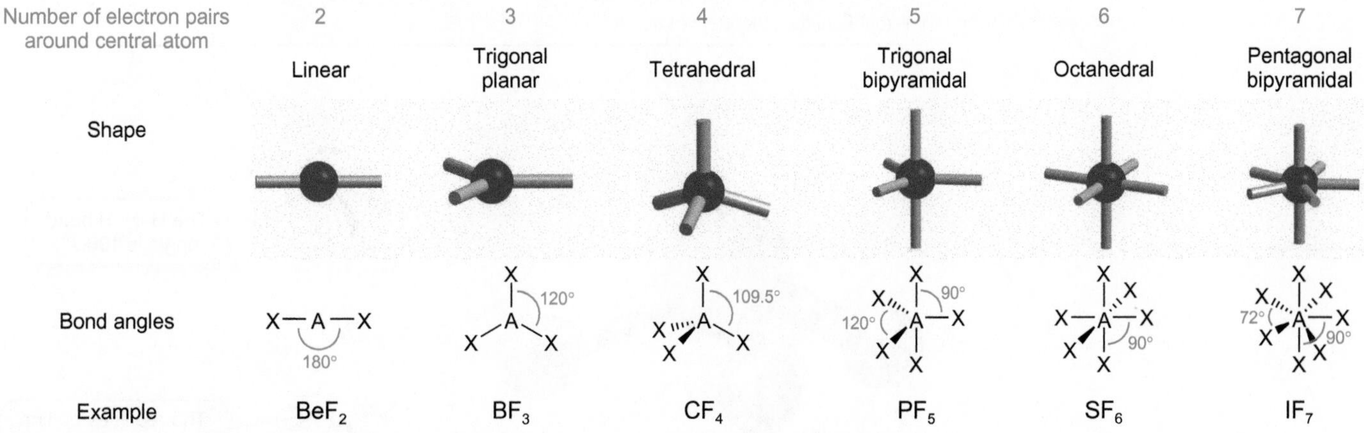
Number of electron pairs around central atom
2
3
4
5
6
7
Linear
Trigonal planar
Tetrahedral
Trigonal bipyramidal
Octahedral
Pentagonal bipyramidal
Shape
Bond angles
180°
120°
109.5°
90°
120°
90°
72°
90°
Example
BeF2
BF3
CF4
PF5
SF6
IF7

Figure 5.7 The geometries adopted by molecules of the general formula AX_n ($n = 2–7$) in which all of the electron pairs are bonding pairs. The use of wedges and hashes to show bonds coming out of and going into the page is described in Section 2.2 (p.75).

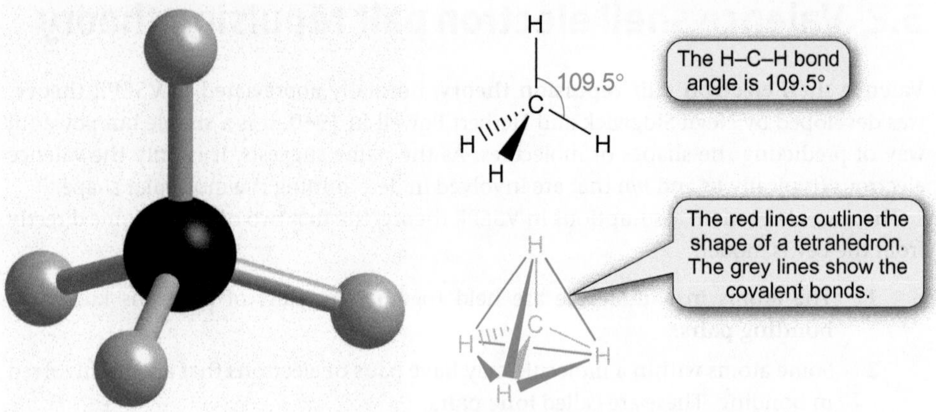
H
109.5°
C
H
H
H
The H–C–H bond angle is 109.5°.
The red lines outline the shape of a tetrahedron. The grey lines show the covalent bonds.

Figure 5.8 The methane molecule is tetrahedral. The four hydrogen atoms are positioned at the vertices of a regular tetrahedron, with the carbon atom in the centre.

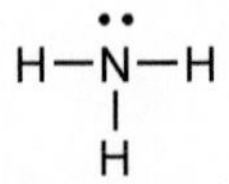
H—N—H
H

2 Lewis structure of ammonia (NH_3).

NH_3	
Electrons from N	5
Electrons from 3 H	3
Total no. electrons around N	**8**
4 electron pairs around N	
3 bonding pairs and 1 lone pair	
Trigonal pyramidal	

Lone pairs are important in deciding the shape of a molecule, but they are ignored in the description of the shape of a molecule.

tetrahedral in shape (Figure 5.7). The methane molecule is shown in Figure 5.8. The bond angles in methane are all 109.5°, which is the regular tetrahedral angle.

In ammonia (NH_3) the central atom is nitrogen, which has an electronic configuration $1s^2\ 2s^2\ 2p^3$. This means there are five electrons in the nitrogen valence shell. There are three hydrogen atoms bonded to the nitrogen, each of which forms a covalent bond and contributes one electron. This gives a total of eight electrons, so there are four electron pairs in NH_3, the same as in CH_4. There is, however, one important difference. In CH_4, all the electron pairs are bonding pairs. In NH_3 (**2**), only three of the electron pairs are bonding—the fourth is a lone pair. The molecular shape of NH_3 is based on a tetrahedron (Figure 5.7) since there are four electron pairs, but it is *only the atoms that count in the description of the shape*. The shape of the NH_3 molecule is described as *trigonal pyramidal* (sometimes abbreviated simply to pyramidal). The molecule is shown in Figure 5.9.

A lone pair occupies more space than a bonding pair, since it lies closer to the central atom. This means that the repulsion between a lone pair and a bonding pair is greater than that between two bonding pairs. In NH_3, the lone pair repels the electron pairs in the N–H bonds more than these bonding pairs repel each other. As a result the H–N–H angles are 106.7°, which is less than the value for a regular tetrahedron.

Worked example 5.2 shows how to use VSEPR theory to work out the shape of water molecules.

A lone pair (lp) takes up more space around the central atom than a bonding pair (bp). Generally, the relative strengths of repulsions between electron pairs are in the order lp–lp > lp–bp > bp–bp.

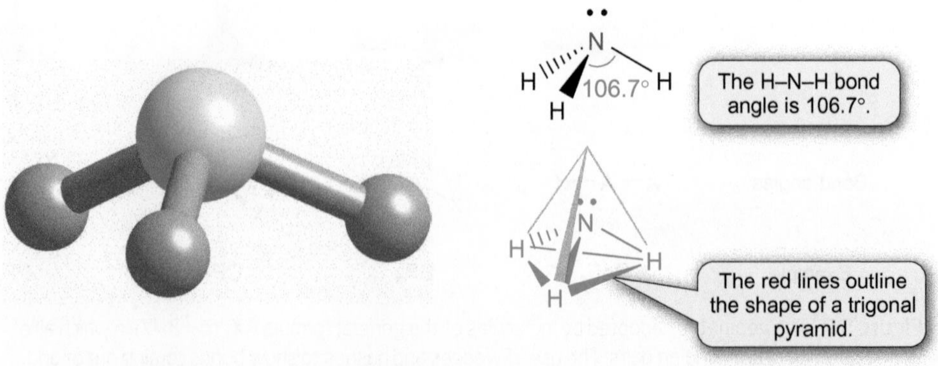
N
H
106.7°
H
H
The H–N–H bond angle is 106.7°.
N
H
H
H
The red lines outline the shape of a trigonal pyramid.

Figure 5.9 Ammonia adopts a trigonal pyramidal structure.

Worked example 5.2 The shape of water molecules

What shape is a water molecule?

Strategy

Using VSEPR theory, work out the number of electron pairs around the central oxygen atom, and use Figure 5.7 (p.223) to work out the shape that the molecule is based on. Decide how many of the electron pairs are bonding pairs and how many are lone pairs, and use this to describe the shape of the molecule, commenting on the H–O–H bond angle.

Solution

The Lewis structure of H_2O is shown below.

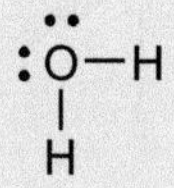

Oxygen has the electronic configuration $1s^2\ 2s^2\ 2p^4$, so it has six electrons in its valence shell. Each of the two hydrogen atoms forms a covalent bond to the oxygen atom and contributes one electron. This gives a total of eight electrons, or four electron pairs around the oxygen atom.

As there are four electron pairs around the oxygen atom, the shape of a water molecule is based on a tetrahedron (Figure 5.7), but with two lone pairs. The water molecule is shown below, and is described as being *bent*.

Lone pair–lone pair repulsion is greater than lone pair–bonding pair repulsion, so the H–O–H angle in water is less than the H–N–H angles in NH_3. The experimental value for the bond angle in water is 104.5°.

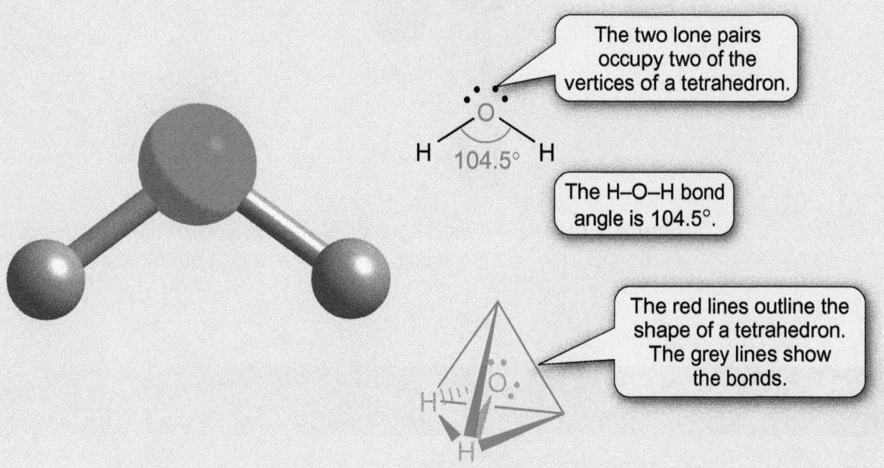
The two lone pairs occupy two of the vertices of a tetrahedron.
104.5°
The H–O–H bond angle is 104.5°.
The red lines outline the shape of a tetrahedron. The grey lines show the bonds.

Question

What is the shape of SiF_4?

Finding the shapes of ions

When working out the shape of an ion, you also need to include the charge in the calculation. For the hexafluorophosphate anion (PF_6^-), the central atom is phosphorus. Phosphorus has a valence shell electronic configuration of $3s^2\ 3p^3$, giving five valence electrons. There are six fluorine atoms, each of which contributes one electron through a covalent bond. As shown in Figure 5.10, the phosphorus atom has a formal charge of −1, so an extra electron must be added to the overall number. This gives a total of 12 electrons around the phosphorus atom, so there are six electron pairs. The shape of the PF_6^- anion is based on an octahedron (Figure 5.7, p.223). Since all of the electron pairs are bonding pairs, the shape of the ion is octahedral, as shown in Figure 5.11.

Worked example 5.3 shows how to use VSEPR theory to work out the shape of a cation.

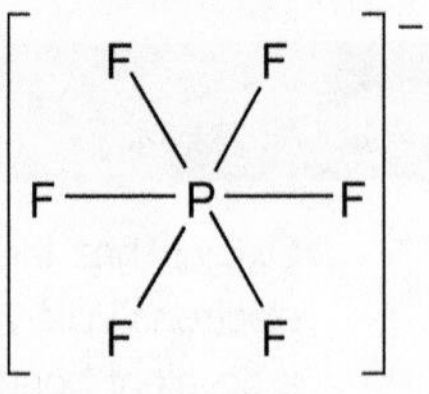
F
F
F
P
F
F
F

Number of electrons in the valence shell of a free P atom	5
– Number of bonds to the P atom	–6
– Number of unshared electrons	–0
Formal charge on P	–1

Lewis structure of PF_6^-

Figure 5.10 Working out the formal charge on the P atom in PF_6^-.

PF_6^-	
Electrons from P	5
Electrons from 6 F	6
Electrons from charge	1
Total no. electrons around P	**12**
6 electron pairs around P	
6 bonding pairs and no lone pairs	
Octahedral	

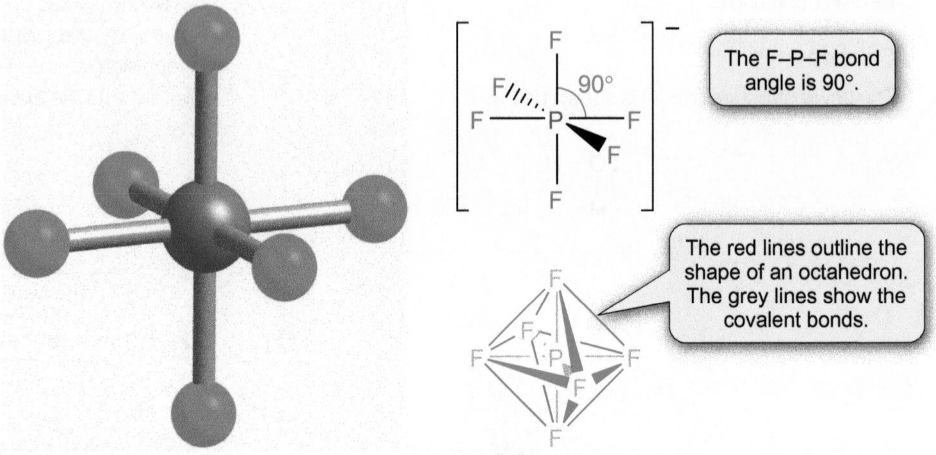
90°
The F–P–F bond angle is 90°.
The red lines outline the shape of an octahedron. The grey lines show the covalent bonds.

Figure 5.11 The PF_6^- ion is octahedral in shape. Each of the fluorine atoms sits on one of the six vertices of a regular octahedron with the phosphorus atom in the centre.

Worked example 5.3 Using VSEPR theory with ions

Use VSEPR theory to predict the shape of the XeF_5^+ ion.

Strategy

Work out the number of electron pairs around the central xenon atom and decide which of these are bonding pairs and which are lone pairs. Remember to account for the positive charge (work out the formal charge on Xe). Use Figure 5.7 (p.223) to decide on the shape by minimizing the repulsion between the electron pairs.

Solution

The Lewis structure of XeF_5^+ is given below.

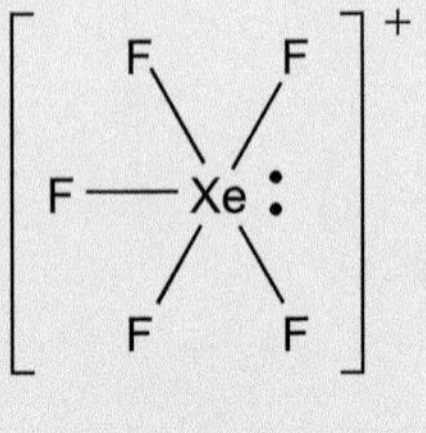
F
F
F
Xe
F
F

Number of electrons in the valence shell of a free Xe atom	8
– Number of bonds to the Xe atom	–5
– Number of unshared electrons	–2
Formal charge on Xe	+1

Lewis structure of XeF_5^+

Xenon has the valence shell electronic configuration of $5s^2\,5p^6$, so it has eight electrons in its valence shell. Each of the fluorine atoms forms a covalent bond to the xenon and donates one electron. The xenon atom has a formal charge of +1, so one electron needs to be *subtracted* to give the total number of electrons.

Electrons from Xe	8
Electrons from 5 F	5
Electrons from charge	–1
Total no. electrons around Xe	**12**

so there are six electron pairs around Xe.

There are six electron pairs, so the shape of XeF_5^+ is based on an octahedron. Of these six electron pairs, five are bonding pairs, which means that there must be one lone pair. The structure of XeF_5^+ is shown below, and the shape is described as *square pyramidal*. The F–Xe–F angles are less than 90° as the lone pair–bonding pair repulsions are greater than the bonding pair–bonding pair repulsions.

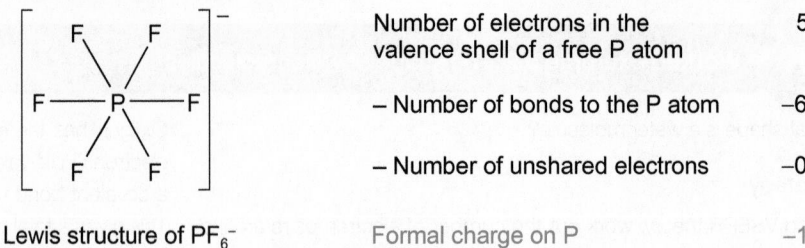

→

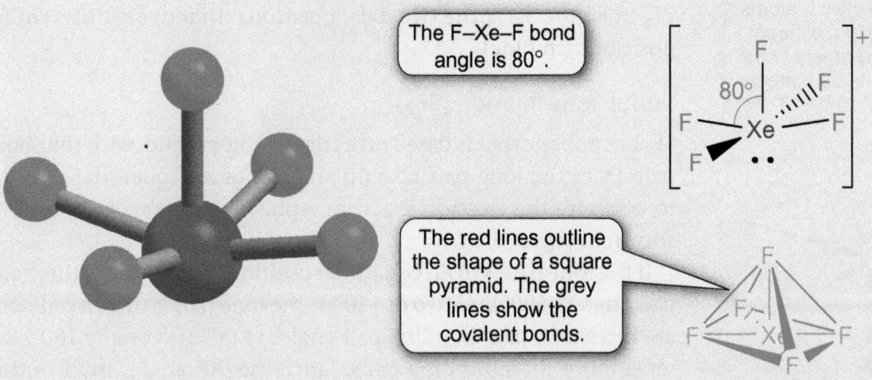
The F–Xe–F bond angle is 80°.
80°
Xe
F
+
The red lines outline the shape of a square pyramid. The grey lines show the covalent bonds.

Question

What shape is the ICl_4^- ion based on? Decide the position(s) you would expect any lone pairs to occupy, and thus predict the shape of ICl_4^-.

Summary

- VSEPR theory is used to predict the shapes of molecules by looking at the repulsions between electron pairs in the valence shell on the central atom in a molecule.
- Pairs of electrons are arranged to minimize the repulsions between them. For 2–7 pairs of electrons around a central atom, the regular geometries are:

  $n = 2$, linear;
  $n = 3$, trigonal planar;
  $n = 4$, tetrahedral;
  $n = 5$, trigonal bipyramidal;
  $n = 6$, octahedral;
  $n = 7$, pentagonal bipyramidal.

- The electron pairs include bonding pairs and lone pairs. The shape of the molecule is described by the positions of the bonding pairs only.
- A molecule with four electron pairs is tetrahedral if there are no lone pairs, trigonal pyramidal if there is one lone pair, and bent if there are two lone pairs.
- A molecule with six electron pairs is octahedral if there are no lone pairs, square pyramidal if there is one lone pair, and square planar if there are two lone pairs (see Figure 5.17, p.230).
- Lone pairs take up more space than bonding pairs, so the strength of repulsions decreases in the order:

  lp–lp > lp–bp > bp–bp (lp = lone pair; bp = bonding pair)

? For a practice question on this topic, see question 3 at the end of this chapter (p.253).

Structures based on the trigonal bipyramid

So far you have seen examples of structures based on tetrahedra and octahedra. Molecules with five electron pairs around the central atom are based on a trigonal bipyramid (see Figure 5.7, p.223). In this case, there is an additional complication as not all

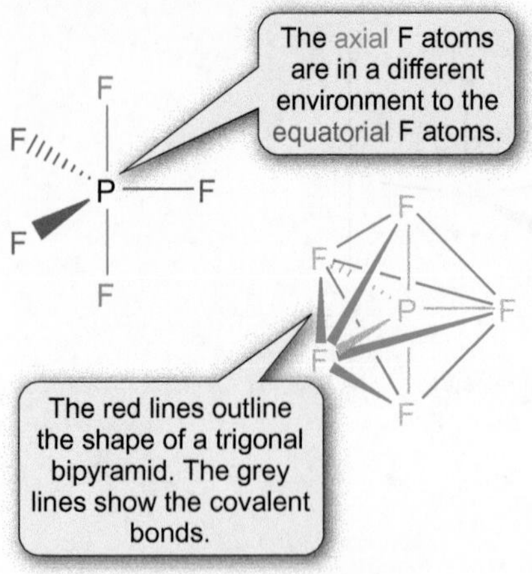

The axial F atoms are in a different environment to the equatorial F atoms.
The red lines outline the shape of a trigonal bipyramid. The grey lines show the covalent bonds.

Figure 5.12 In a trigonal bipyramid, the axial and equatorial positions are not the same.

SF_4	
Electrons from S	6
Electrons from 4 F	4
Total no. electrons around S	10
5 electron pairs around S	
4 bonding pairs and 1 lone pair	

Visit the Online Resource Centre to view screencast 5.1 which walks you through the use of VSEPR theory to predict the shape of an SF_4 molecule.

the vertices in a trigonal bipyramid are equivalent. As you can see from the structure of PF_5 in Figure 5.12, the two *axial* positions (in red) are different from the three *equatorial* positions (in blue).

Sulfur tetrafluoride (SF_4)

SF_4 has a shape that is based on a trigonal bipyramid, with four bonding pairs and one lone pair. Does the lone pair take up an axial or an equatorial position? To decide, you need to compare the electron–electron repulsions for the two possible geometries, which are shown in Figure 5.13.

If the lone pair is in an equatorial position (Structure A), there are two lone pair–bonding pair angles of 90° and two of 120°. If the lone pair is in an axial position (Structure B), there are three lone pair–bonding pair angles of 90° and one of 180°. Remember that lone pairs repel more than bonding pairs. Since the 90° angles involve the largest repulsions, the most favourable position for the lone pair is in one of the equatorial positions (Structure A), and SF_4 adopts the *disphenoidal* geometry (sometimes called a sawhorse geometry) shown in Figure 5.14.

Since the lone pair takes up more space than the bonding pairs, both the axial and the equatorial fluorine atoms are bent back from their 'ideal' positions. This is reflected in the observed bond angles, 173° and 102°, which are both less than the regular trigonal bipyramid angles of 180° and 120° (see Figure 5.7).

Chlorine trifluoride (ClF_3)

Chlorine trifluoride (ClF_3) also has a structure based on a trigonal bipyramid, though in this case there are three bonding pairs and two lone pairs. There are three possible geometries for ClF_3, as shown in Figure 5.15, and these differ in the positions occupied by the lone pairs:

- the lone pairs could both go into equatorial sites giving a *T-shaped* molecule;
- the lone pairs could both go into axial sites giving a *trigonal planar* molecule;
- one lone pair could go into an axial site and one in an equatorial site giving a *pyramidal* molecule.

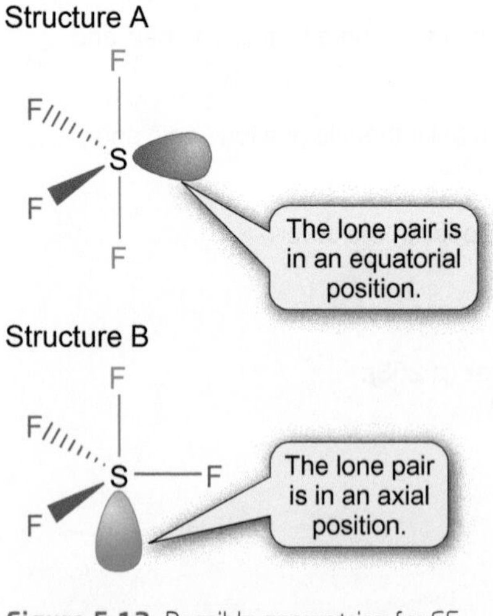

Structure A
The lone pair is in an equatorial position.
Structure B
The lone pair is in an axial position.

Figure 5.13 Possible geometries for SF_4, for which there are 4 bonding pairs and 1 lone pair around the sulfur atom.

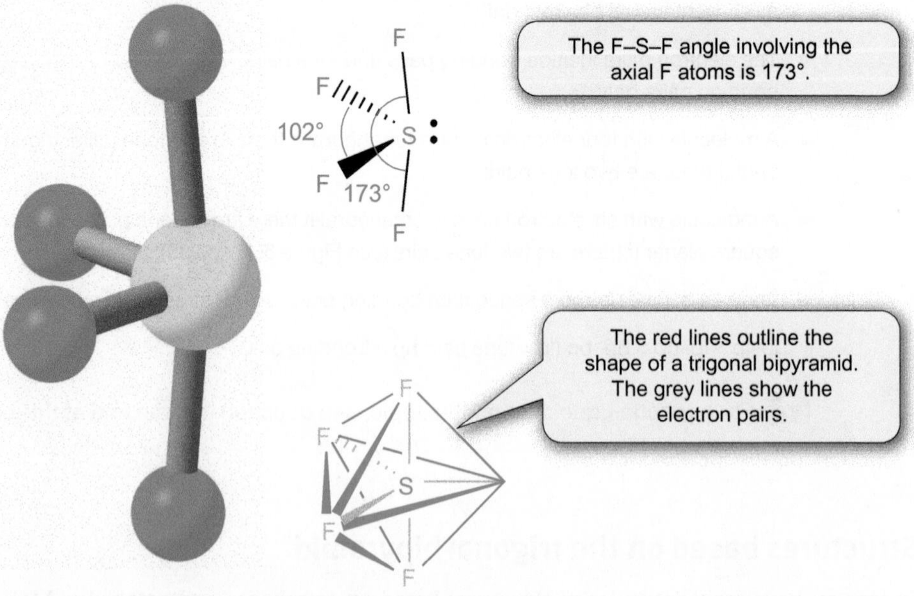

102°
173°
The F–S–F angle involving the axial F atoms is 173°.
The red lines outline the shape of a trigonal bipyramid. The grey lines show the electron pairs.

Figure 5.14 SF_4 has a disphenoidal geometry in which one of the equatorial vertices of a trigonal bipyramid is occupied by a lone pair.

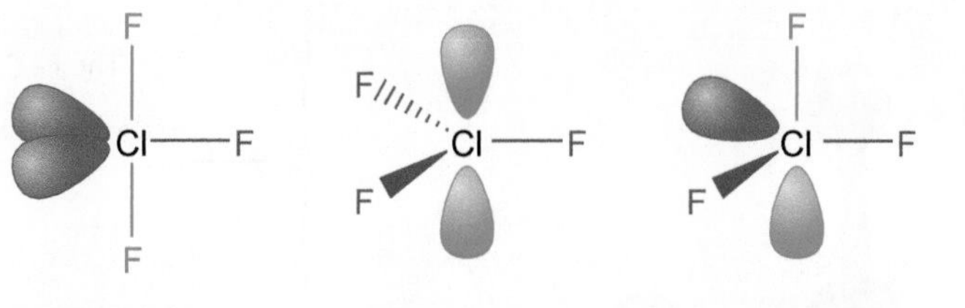
F
Cl
F
F
F
Cl
F
F
F
Cl
F
F

Shape of molecule:	T-shaped	Trigonal planar	Pyramidal
90° repulsions	2 bp–bp, 4 bp–lp	6 bp–lp	2 bp–bp, 3 bp–lp, 1 lp–lp
120° repulsions	2 bp–lp, 1 lp–lp	3 bp–bp	1 bp–bp, 2 bp–lp
180° repulsions	1 bp–bp	1 lp–lp	1 bp–lp

Figure 5.15 The three possible geometries for ClF_3. In each geometry, two of the five positions of a trigonal bipyramid are occupied by lone pairs. bp = bonding pair; lp = lone pair.

ClF_3	
Electrons from Cl	7
Electrons from 3 F	3
Total no. electrons around Cl	**10**
5 electron pairs around Cl	
3 bonding pairs and 2 lone pairs	

Since lone pair–lone pair repulsions are greater than lone pair–bonding pair or bonding pair–bonding pair repulsions, it is tempting to suggest the molecule is trigonal planar. In this geometry the lone pairs are as far apart as possible and the lone pair–lone pair repulsion would be minimized. However, this is not the shape observed for ClF_3.

To understand why, you need to take into account all of the repulsions between electron pairs. In a trigonal bipyramid there are three different bond angles: 90°, 120°, and 180°. As with SF_4, it is the repulsions with 90° angles that are the most unfavourable, and it is these that dictate the shape adopted by the molecule. The repulsions associated with the three structures are shown in Figure 5.15. If you look at the 90° repulsions, you can see that the T-shape has the smallest number of repulsions involving lone pairs, so this is the shape observed for ClF_3 (Figure 5.16).

The shapes of molecules such as ClF_3 are determined experimentally using diffraction techniques. See Boxes 4.1 (p.172) and 6.4 (p.279) for details.

Box 5.3 The fluorinating ability of ClF_3 and BrF_3

Chlorine trifluoride (ClF_3) is an extremely reactive gas that causes most combustible materials to ignite. Violent reactions, including explosions, can occur when gaseous ClF_3 is in contact with water, ice, or silicon-containing compounds such as sand and glass, or even asbestos. It also reacts violently with metals and metal oxides, especially if these are in powder form. For these reasons, ClF_3 was investigated for its use in incendiary bombs during the Second World War.

ClF_3 is one of the strongest fluorinating agents known, and this has led to it being used as a cleaning agent in the semiconductor industry and in the reprocessing of spent fuel rods from nuclear reactors. During the nuclear reaction, some of the uranium is converted into fission products, many of which are lanthanides (metals with partly filled 4*f* orbitals). Other uranium nuclei absorb neutrons and are converted, by β-emission, into plutonium

$$^{238}_{92}U + n \rightarrow {}^{239}_{92}U \xrightarrow{\beta\text{-emission}} {}^{239}_{93}Np \xrightarrow{\beta\text{-emission}} {}^{239}_{94}Pu$$

The challenge is to separate unreacted uranium from these fission products. On reaction with ClF_3, the metals are fluorinated. Uranium is converted into UF_6, plutonium into PuF_4, and the lanthanides into trifluorides. UF_6 is volatile, so it is easily separated from the other metal fluorides, which are solids with high melting points.

Bromine trifluoride (BrF_3) has a similar structure to ClF_3. It is used as a 'chemical cutter' in the petroleum industry. When steel tubing needs cutting within an oil well, a capsule is lowered into the tube, then BrF_3 is sprayed in a ring on the inside of the tube. The reaction converts the iron into iron(III) fluoride, cutting through the tubing.

▲ Steel tubes cut from the inside using the 'chemical cutter' BrF_3.

Question

As is shown in Figure 5.16, the bond angles for ClF_3 are less than 90°. Explain why the molecule is distorted from a regular T-shaped geometry in this manner.

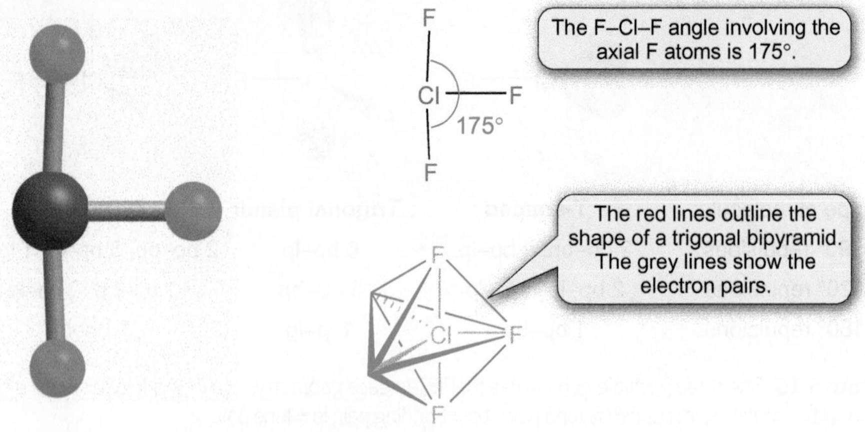
F
Cl
F
175°
F
The F–Cl–F angle involving the axial F atoms is 175°.
The red lines outline the shape of a trigonal bipyramid. The grey lines show the electron pairs.
F
Cl
F
F

Figure 5.16 The T-shaped geometry adopted by ClF_3.

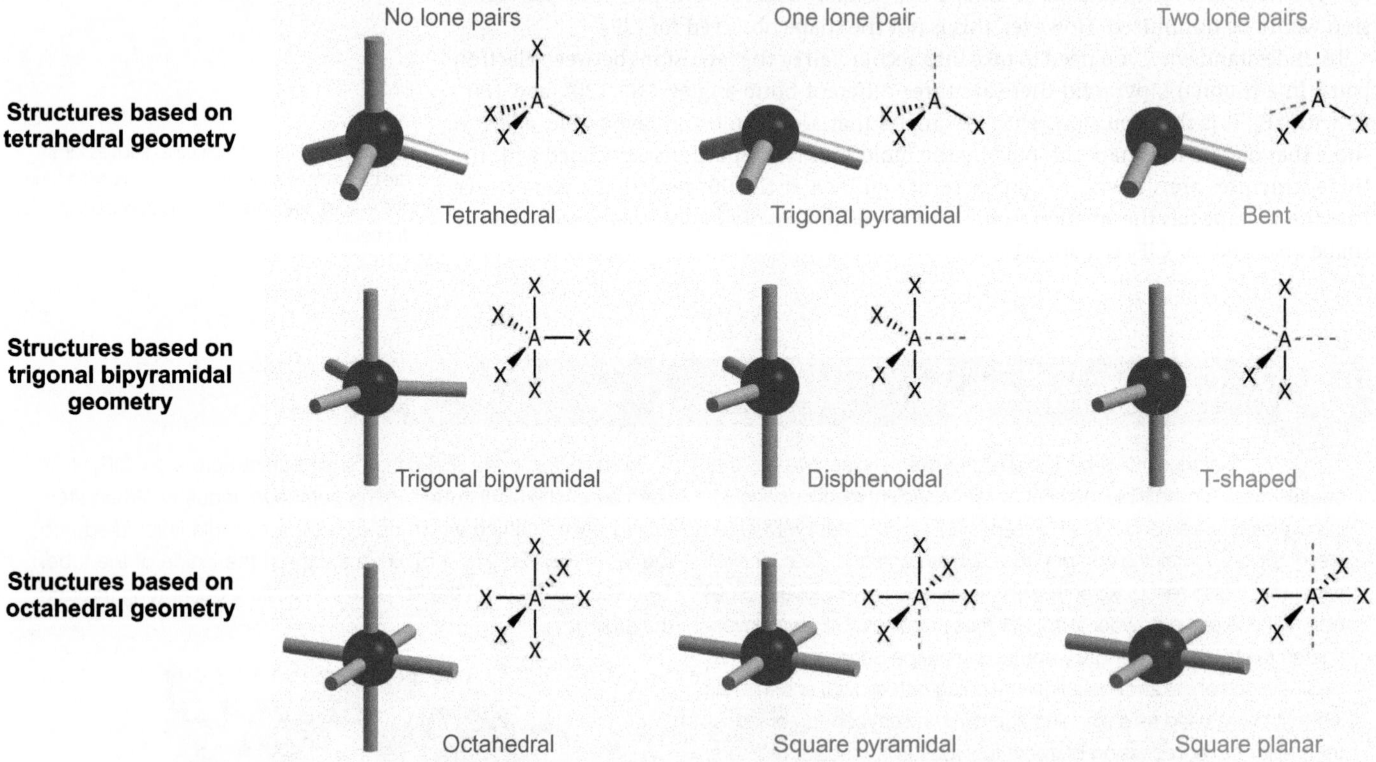
No lone pairs
One lone pair
Two lone pairs
Structures based on tetrahedral geometry
Tetrahedral
Trigonal pyramidal
Bent
Structures based on trigonal bipyramidal geometry
Trigonal bipyramidal
Disphenoidal
T-shaped
Structures based on octahedral geometry
Octahedral
Square pyramidal
Square planar

Figure 5.17 Geometries adopted by molecules based on the tetrahedral, trigonal bipyramidal, and octahedral geometries containing zero, one, or two lone pairs. For those containing lone pairs, the observed shape generally is distorted from that shown because of the greater repulsive effect of the lone pairs. The only exception is square planar, as in that case the effects from the two lone pairs cancel each other out.

The shapes of molecules based on the tetrahedron, trigonal bipyramid, and octahedron with zero, one, or two lone pairs are summarized in Figure 5.17. The observed bond angles for molecules with some of these geometries deviate from those in the regular geometries because the lone pairs occupy more space than the bonding pairs.

> (i) π bonds point in the same direction as σ bonds and so do not contribute to molecular shape. See Sections 4.5 (p.181) and 4.9 (p.193) for more details on π bonding.

Compounds containing multiple bonds

A double bond normally consists of one σ bond and one π bond, whereas a triple bond contains one σ bond and two π bonds. In terms of the shape of a molecule, the electrons

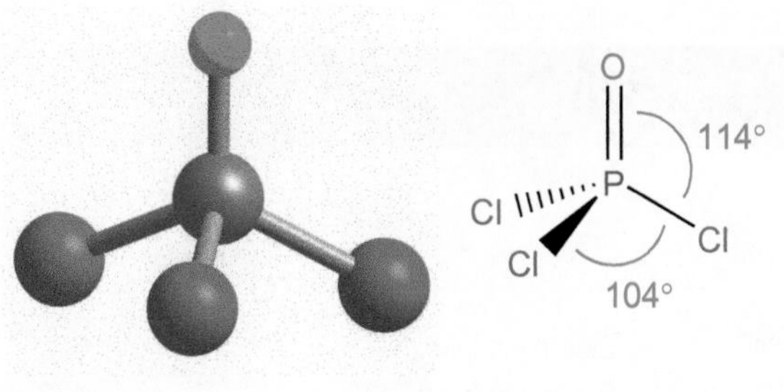
O
P
Cl
Cl
Cl
114°
104°

Figure 5.18 The shape of $POCl_3$.

in a double bond or triple bond occupy only one position round the central atom. This means that the shape is dictated by the σ bond framework, so you can treat single, double, and triple bonds all in the same general way.

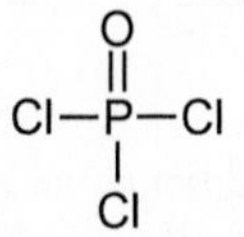

3 Lewis structure of phosphorus oxytrichloride ($POCl_3$).

Phosphorus oxytrichloride ($POCl_3$)

Phosphorus has the valence electronic configuration $3s^2\ 3p^3$, so the central phosphorus atom contributes five electrons. The chlorine atoms donate one electron each to form single bonds to phosphorus and the oxygen atom donates two electrons to form a double bond. There are a total of five electron pairs around the phosphorus atom, but one of these is a π-electron pair, so it does not contribute to the basic shape. This leaves four σ-electron pairs, all of which are bonding pairs. $POCl_3$ (**3**) is therefore tetrahedral.

The double bond takes up more space than a single bond and, as it consists of two electron pairs, it repels the other electron pairs more than a single bond. This leads to distortions in the tetrahedron, as shown in Figure 5.18, with the O–P–Cl angles wider than the Cl–P–Cl angles.

Visit the Online Resource Centre to view screencast 5.2 which walks you through the use of VSEPR theory to work out the shape of a $POCl_3$ molecule.

The carbonate ion (CO_3^{2-})

The Lewis structure of the carbonate ion (CO_3^{2-}, **4**) contains two C–O single bonds and one C=O double bond. As you will see in Section 5.5, this is not an accurate description of the bonding in the carbonate ion as experimental measurements show that all three bond lengths are the same. However, this does not affect the results from VSEPR theory because the shape is determined only by the σ-bonded framework. There are four electrons from the central carbon atom and two electrons from the oxygen atom that is doubly bonded to the carbon. Each of the singly bonded oxygen atoms contributes one electron. Since the ionic charge is located on these oxygen atoms (and not on the central carbon atom), it does not need to be included separately in the calculation.

This gives a total of eight electrons around the carbon atom, so there are four electron pairs. Of these, one is a π pair, leaving three σ pairs of electrons, which are the ones that contribute to the shape. VSEPR therefore predicts that the carbonate ion has a trigonal planar shape. This is the shape observed for the ion, as shown in Figure 5.19.

Worked example 5.4 shows how to use VSEPR theory to work out the shape of the nitrate ion.

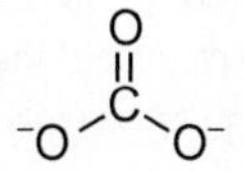

4 Lewis structure of the carbonate ion (CO_3^{2-}).

In ions containing double or triple bonds the charges are usually located on the outer atoms, giving a formal charge of 0 to the central atom.

$POCl_3$	
Electrons from P	5
Electrons from 3 Cl	3
Electrons from O	2
Total no. electrons around P	**10**
5 electron pairs around P, but one is a π pair	
4 σ-electron pairs	
4 bonding pairs and no lone pairs	
Tetrahedral	

CO_3^{2-}	
Electrons from C	4
Electrons from 3 O	4 (2 + 1 + 1)
Electrons from charge	0
Total no. electrons around C	**8**
4 electron pairs around C, but one is a π pair	
3 σ-electron pairs	
3 bonding pairs and no lone pairs	
Trigonal planar	

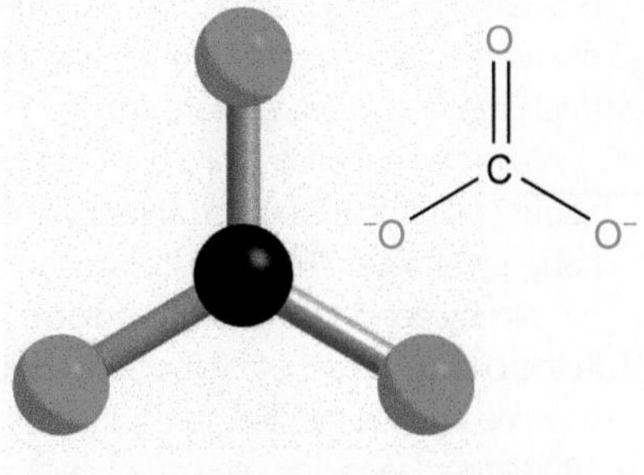

Figure 5.19 The trigonal planar geometry adopted by the carbonate ion. Measurements show that the three carbon–oxygen bond lengths are identical and the bond angles are all 120°.

Worked example 5.4 The nitrate ion

Use VSEPR theory to predict the shape of the nitrate ion (NO_3^-).

Strategy

Draw a Lewis structure for the nitrate ion using a combination of single and double bonds; then count up the electrons around the central nitrogen atom. Since nitrogen is a second period element you cannot have more than 8 electrons around this atom. (Remember to include the charge on the nitrogen atom.) Decide which of the electron pairs are σ-electron pairs, π-electron pairs, and lone pairs. Remember, π-electron pairs do not contribute to the shape.

Solution

The Lewis structure for nitrate contains two single N–O bonds and one double N=O bond. The nitrogen atom has a formal charge of +1, and two oxygen atoms carry negative charges.

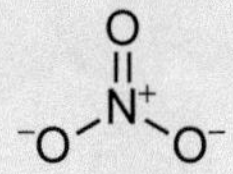

Electrons from N	5
Electrons from 3O	4 (2 + 1 + 1)
Electrons from charge	−1
Total no. electrons around N	8

4 electron pairs around N of which one is a π-electron pair

3 σ-electron pairs

3 bonding pairs and no lone pairs

Nitrate is trigonal planar.

Question

Use VSEPR theory to predict the shape of the phosphate ion (PO_4^{3-}).

Box 5.4 Nitrates in water

Many rivers, estuaries, and ground water reservoirs contain concentrations of the nitrate ion (NO_3^-) that are higher than those occurring naturally. The most important sources of additional nitrates in water are agriculture, sewage effluent, and acid rain (which contains dissolved NO_2).

Nitrate ions are essential for the growth of healthy plants, and farmers often add fertilizers to soil to replace those ions taken up by previous crops. This can be in the form of inorganic fertilizers, such as ammonium nitrate, or organic manures. Nitrate compounds are very soluble in water and can enter the water system either as run-off or from percolation through the soil into ground water. Run-off generally occurs in areas of impervious rock where the water table lies just below the surface. In areas of porous rock, such as chalk and limestone, water is carried further down towards the water table. This process can be as slow as one metre per year, so the concentrations of nitrate ions present in ground water today are due to the agricultural practices of many years ago.

Are raised concentrations of nitrate ions in water anything to worry about? Nitrates have been linked with methaemoglobinaemia (blue baby syndrome). This is a rare condition in young babies that is caused by conversion of haemoglobin—the oxygen carrying protein in the blood—into methaemoglobin. Nitrite ions (NO_2^-), made when bacteria convert nitrate into nitrite, are known to facilitate this. In the 1940s there was an outbreak of cases of methaemoglobinaemia in the USA amongst babies that had been fed bottled milk made using well water that was shown to contain high concentrations of both nitrate ions and bacteria. However, the condition is now much rarer, despite nitrate concentrations not decreasing, so the significance of nitrate in this outbreak is still unclear.

Over recent years, biologists have noticed that amphibian populations across the world are decreasing and frog deformities such as missing limbs are increasing. Although fungal infections have been

▲ Nitrates and phosphates cause a rapid increase in populations of algae, and these form mats that block light to the plants beneath. The dying plants are broken down by bacteria in the water, which deprives the water of oxygen. This is called eutrophication.

→

The Oregon spotted frog is three to four times more vulnerable to nitrate and nitrite than other frog species, and populations of this species have decreased more than those of other amphibians in the north-west of the USA.

shown to be a factor in this, there is unlikely to be one simple explanation for the declining populations. Research has shown that frog populations are vulnerable to increased concentrations of nitrate and nitrite ions, especially at the larval stages. One explanation may be that high nitrate concentrations in water lead to an increase in the population of parasitic flatworms, and it is these that cause the deformities.

Question

Use VSEPR theory to predict the shape of the nitrite ion.

Radicals

If you attempt to work out the shape of NO_2 (**5**) using VSEPR theory you encounter a problem. The calculation gives a total of $3\frac{1}{2}$ electron pairs of which $2\frac{1}{2}$ are σ-electron pairs that contribute to the shape.

Since there are two σ bonding pairs there must be only half a lone pair. How is this possible? Half a lone pair implies a single unpaired electron, so NO_2 is a radical (see Section 5.1, p.219). The single electron has to go somewhere and it occupies the same position as a lone pair would. This suggests NO_2 has a bent structure. Measurements confirm this and show that the O–N–O angle is 134.1°. This angle is wide for a structure based on trigonal planar geometry, as the single electron repels the bonding pairs much less than a lone pair would.

If NO_2 is reduced to the nitrite ion (NO_2^-), an additional electron is added, completing the lone pair. This leads to a decrease in the bond angle to 115.4° as electron–electron repulsion involving the lone pair is much greater than that involving the single electron (Figure 5.20).

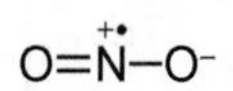

5 Lewis structure of nitrogen dioxide (NO_2).

NO_2	
Electrons from N	5
Electrons from 2 O	3 (1 + 2)
Charge on N (+1)	−1
Total no. electrons around N	**7**
$3\frac{1}{2}$ electron pairs around N, but one is a π pair	
$2\frac{1}{2}$ σ-electron pairs	
2 bonding pairs and half a lone pair	

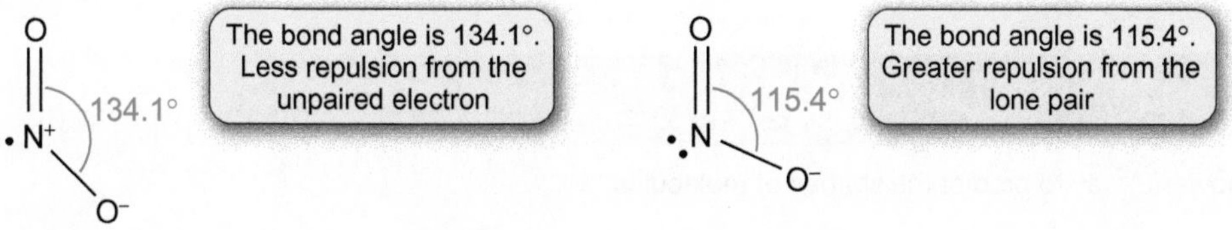

(a)
O
134.1°
N+
O−
The bond angle is 134.1°. Less repulsion from the unpaired electron
(b)
O
115.4°
N
O−
The bond angle is 115.4°. Greater repulsion from the lone pair

Figure 5.20 The shapes of (a) NO_2 and (b) NO_2^-.

Molecules with more than one centre

All the examples considered so far contain a central atom bonded to a number of peripheral atoms. What happens for a molecule such as hydrazine (N_2H_4) (**6**) for which there is no single central atom? In this case you can use VSEPR theory for both of the nitrogen atoms. Each nitrogen atom has five valence electrons, and receives two electrons from the hydrogen atoms and one electron from the N–N bond. This gives four electron pairs around each nitrogen atom, one of which is a lone pair, so the shape around both nitrogen atoms is trigonal pyramidal.

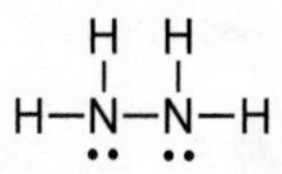

6 Lewis structure of hydrazine (N_2H_4).

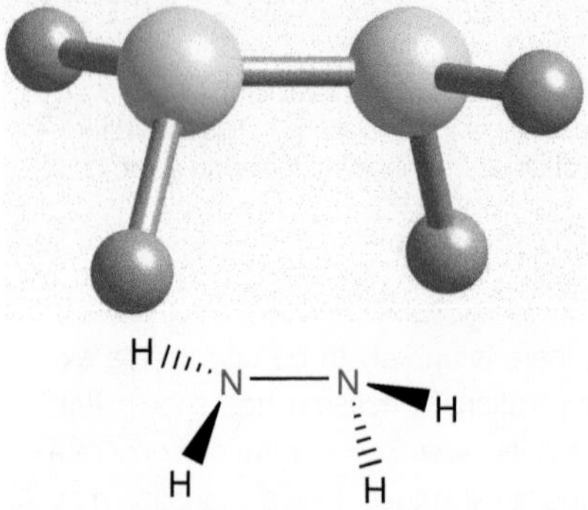
H
N
N
H
H
H

Figure 5.21 The lowest energy conformation of hydrazine (N_2H_4).

More details on the conformations of organic molecules such as C_2H_6 are given in Section 18.2 (p.819).

Molecules such as hydrazine can adopt several shapes (conformations) as rotation around the central N–N bond is a relatively low energy process and so occurs easily. Figure 5.21 shows the lowest energy conformation for hydrazine.

Limitations of VSEPR theory

VSEPR theory accurately predicts the shapes of many molecules, but it has limitations and it is important to know when it doesn't work. The theory does not apply to *d*-block metal complexes, as the shapes of these compounds are generally determined by the energies and electronic occupation of the *d* orbitals (see Section 28.4, p.1279).

There are also cases in which the assumptions made are not correct. For example, VSEPR theory predicts that the anion $TeBr_6^{2-}$ (7) has a shape based on a *pentagonal bipyramid* with one site containing a lone pair (Figure 5.7, p.223).

However, this is not the shape observed. Crystallographic measurements show that the ion is octahedral, as shown in Figure 5.22. This is explained by the lone pair occupying an *s* orbital. This electron pair is described as stereochemically inactive as it does not influence the geometry. The assumption of VSEPR theory that all of the valence electrons contribute to the shape is incorrect in this case, and also for related ions such as $SeCl_6^{2-}$ and $TeCl_6^{2-}$ and compounds like XeF_6.

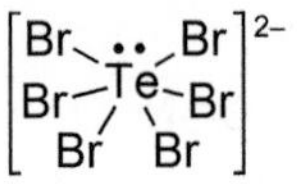
Br
Br
Te
Br
Br
Br
Br
2–

7 Lewis structure of $TeBr_6^{2-}$.

$TeBr_6^{2-}$	
Electrons from Te	6
Electrons from 6 Br	6
Electrons from charge	2
Total no. electrons around Te	**14**
7 electron pairs around Te	
6 bonding pairs and 1 lone pair	

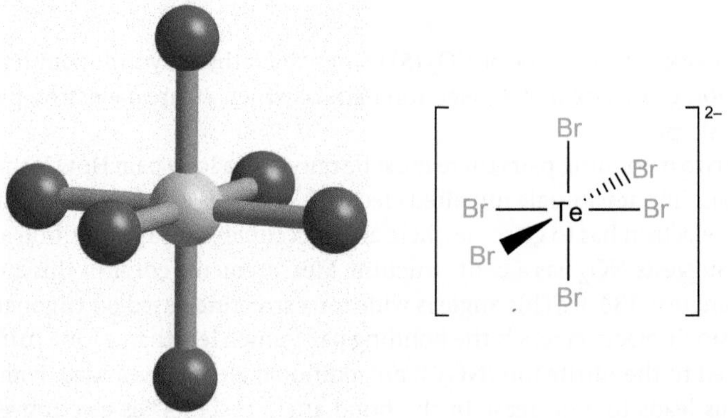
Br
Br
Br
Te
Br
Br
Br
2–

Figure 5.22 The $TeBr_6^{2-}$ anion is octahedral despite having seven valence electron pairs.

Summary

- VSEPR theory is a simple but powerful way to predict the shapes of molecules.
- The electron pairs take up positions as far away from each other as possible to minimize electron–electron repulsions.
- Lone pairs take up more space around an atom than bonding pairs. This means that lone pair–lone pair repulsion is greater than lone pair–bonding pair repulsion, which in turn is greater than bonding pair–bonding pair repulsion.
- π electrons do not contribute to the molecular shape, but double and triple bonds take up more space than single bonds. Consequently, the repulsion is greater between the electrons in a double (or triple) bond and the electrons in a single bond, than between the electrons in two single bonds.
- VSEPR theory does not work for transition metal compounds and for a few exceptions such as $TeBr_6^{2-}$ for which the assumptions made in the theory are incorrect.

For practice questions on these topics, see questions 3–6 at the end of this chapter (p.253).

5.3 Bond polarity and polar molecules

A heteronuclear diatomic molecule always has an electric dipole moment because of the difference in electronegativity between the two atoms. Chlorine is more electronegative than hydrogen, so the H–Cl bond is polar and the molecule has a dipole with the chlorine atom carrying a partial negative charge, written δ–, and the hydrogen atom carrying a partial positive charge, written δ+.

The dipoles in heteronuclear diatomic molecules are described in Section 1.8 (p.52).

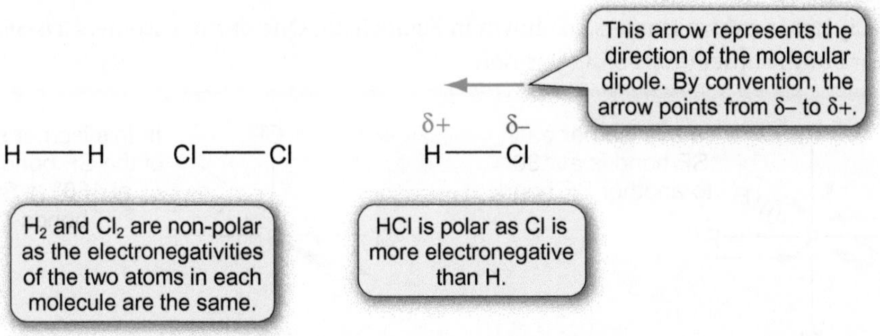
H—H
Cl—Cl
δ+ δ–
H—Cl
This arrow represents the direction of the molecular dipole. By convention, the arrow points from δ– to δ+.
H$_2$ and Cl$_2$ are non-polar as the electronegativities of the two atoms in each molecule are the same.
HCl is polar as Cl is more electronegative than H.

Selected electronegativities

F 3.98
O 3.44
Cl 3.16
N 3.04
C 2.55
H 2.20

Values for all of the *s*- and *p*-block elements are given in Figure 4.6 (p. 178).

Any molecule with a non-zero dipole moment is **polar**. The units of dipole moment are debyes, D.

For polyatomic molecules things are a little more complicated, as the dipoles from polar bonds can cancel each other out. This means that not all molecules with polar bonds have a dipole moment. In carbon dioxide, the two C=O bonds are both polar, as oxygen is more electronegative than carbon. However, the linear shape of the CO_2 molecule means that the dipole moments from these two bonds cancel each other out as shown in Figure 5.23, so consequently CO_2 is non-polar.

Dipoles are vector quantities—they have both direction and magnitude. The length of the arrow is used to indicate the magnitude of the dipole. (Velocity is another vector quantity, see Box 8.2, p.360.)

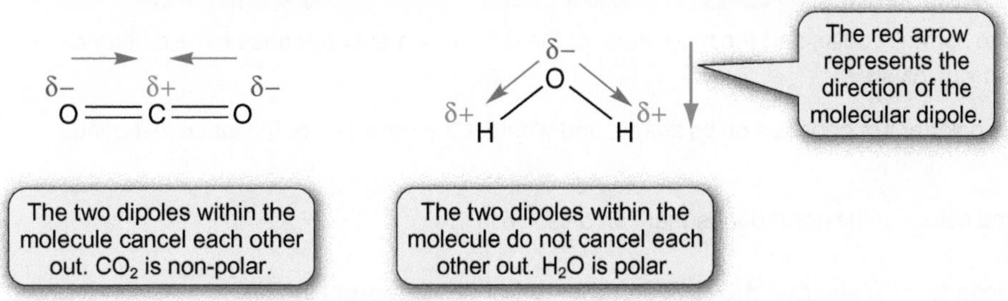
δ– δ+ δ–
O=C=O
δ–
O
δ+ H H δ+
The red arrow represents the direction of the molecular dipole.
The two dipoles within the molecule cancel each other out. CO$_2$ is non-polar.
The two dipoles within the molecule do not cancel each other out. H$_2$O is polar.

Figure 5.23 Both CO_2 and H_2O contain polar bonds, but only H_2O is a polar molecule.

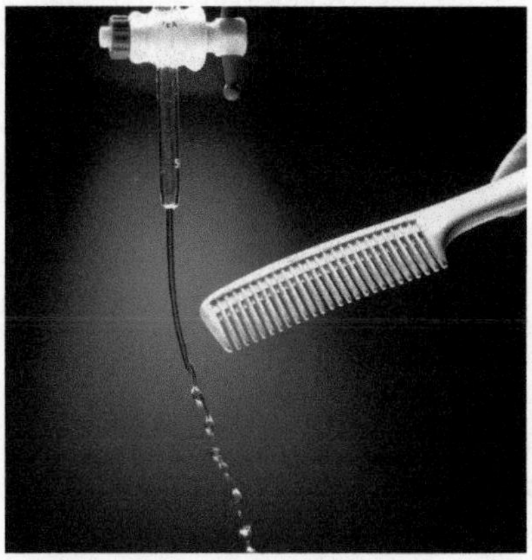

A stream of water can be deflected by a charge because the molecules are polar. The polar water molecules are being attracted towards a comb which is carrying an electric charge.

The two O–H bonds in water are also polar. In contrast to CO_2, the 104.5° bond angle in water means that the two dipoles do not cancel each other out, so overall the water molecule is polar. Polarity is one of the most important properties of a solvent, and it helps dictate the types of compound that the solvent is likely to dissolve. The high dipole moment of H_2O molecules makes water a good solvent for ionic compounds such as sodium chloride.

Molecules of the general formula AX_n with no lone pairs are always non-polar, as all of the dipoles cancel out. Therefore CH_4, PF_5, and SF_6 all have zero dipole moments. In contrast, molecules with one lone pair, such as NH_3 and SF_4, are always polar, as the dipoles cannot cancel out completely. For molecules with two or more lone pairs, the positions of these determine whether the molecule is polar. If a molecule has two lone pairs at 180° from each other, as in the square planar XeF_4 molecule, the molecule is non-polar. If the angle between the two lone pairs is less than 180° as in ClF_3, the molecule is polar.

Interactive 3D rotatable figures of H_2O, CH_4, PF_5, SF_6, NH_3, SF_4, XeF_4, and ClF_3.

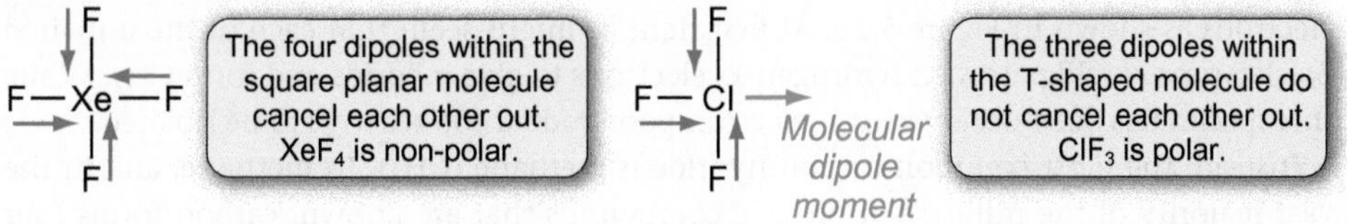
F—Xe—F
The four dipoles within the square planar molecule cancel each other out. XeF$_4$ is non-polar.
F—Cl
Molecular dipole moment
The three dipoles within the T-shaped molecule do not cancel each other out. ClF$_3$ is polar.

Interactive 3D rotatable figure of $CBrCl_3$.

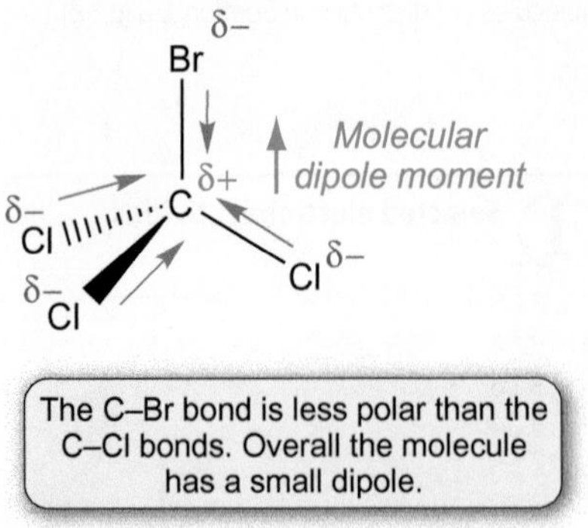
δ−
Br
Molecular
dipole moment
δ+
C
δ−
Cl
δ−
Cl
Cl
δ−
The C–Br bond is less polar than the C–Cl bonds. Overall the molecule has a small dipole.

Figure 5.24 $CBrCl_3$ is polar because the dipoles of the three C–Cl bonds are not fully cancelled by the smaller dipole of the C–Br bond.

Molecules with more than one type of atom bonded to the central atom can be polar, but are not necessarily so. In the tetrahedral molecule $CBrCl_3$, each of the bonds is polar, but the dipole of the C–Br bond is smaller than the dipoles of the C–Cl bonds as the electronegativity difference between carbon and bromine is less than that between carbon and chlorine. Figure 5.24 shows that $CBrCl_3$ has a small dipole moment, as the three large C–Cl dipoles are not completely cancelled by the smaller C–Br dipole. This is in contrast to CCl_4 which is also tetrahedral, though in this case the four C–Cl dipoles cancel completely and the molecule is non-polar.

SF_4Cl_2 can form two isomers, as shown in Figure 5.25. One of these isomers (*cis*-SF_4Cl_2) is polar but the other (*trans*-SF_4Cl_2) is not.

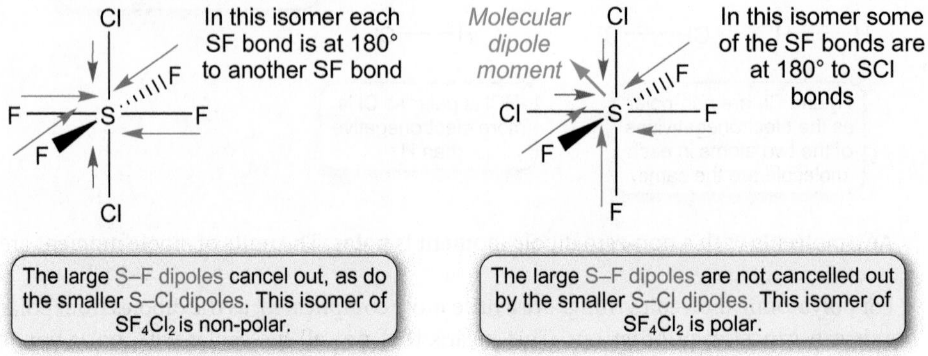
Cl
In this isomer each SF bond is at 180° to another SF bond
F
S
F
F
F
Cl
Molecular dipole moment
Cl
In this isomer some of the SF bonds are at 180° to SCl bonds
F
Cl
S
F
F
F
The large S–F dipoles cancel out, as do the smaller S–Cl dipoles. This isomer of SF_4Cl_2 is non-polar.
The large S–F dipoles are not cancelled out by the smaller S–Cl dipoles. This isomer of SF_4Cl_2 is polar.

Figure 5.25 SF_4Cl_2 forms two isomers, one of which is polar and the other of which is non-polar.

Summary

- Bonds between different atoms are normally polar, and the magnitude of the dipole moment depends on the difference in electronegativity between the atoms involved.
- Whether a polyatomic molecule is polar or not depends on its shape, and whether the polarities of the individual bonds cancel out.
- The polarity of a solvent affects the nature of the compounds that can dissolve in it.

For practice questions on these topics, see questions 7–9 at the end of this chapter (p.253).

Valence bond theory and molecular orbital theory are introduced in Section 4.4 (p.178) for diatomic molecules.

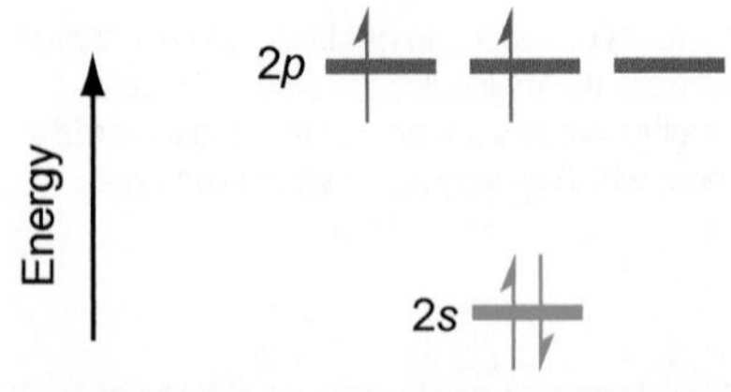
Energy
2p
2s

Figure 5.26 The ground electronic state of a carbon atom.

CH_2 is a short-lived and highly reactive molecule. It is formed in combustion reactions and is important in the formation of soot and condensed aromatic ring systems.

5.4 Valence bond theory for polyatomic molecules

While VSEPR theory is a powerful way of predicting the shapes of molecules, it does not explain *how* the bonding occurs. For this you need to use either valence bond (VB) theory or molecular orbital (MO) theory. VB theory is easier than MO theory to extend to polyatomic molecules, so this is discussed first.

In VB theory, all the bonds are localized between pairs of atoms. This makes the theory particularly useful for organic molecules, as the similarities between compounds with the same functional groups are easy to see.

The electronic configuration of carbon is $1s^2\,2s^2\,2p^2$, so a carbon atom has two unpaired electrons as shown in Figure 5.26. At first sight, it might seem that each of the unpaired 2*p* electrons would pair with hydrogen 1*s* electrons to give σ bonds and form CH_2. While this species has been observed, it is electron poor and far too reactive to be isolated.

Instead, the most common carbon hydride is methane (CH_4). In methane, and in the vast majority of the millions of organic compounds that are known, carbon forms four bonds, not two. This means that the 2*s* electrons must be involved in the bonding in addition to the 2*p* electrons. Bond measurements show that the four C–H bonds in methane are equivalent. This would not be the case if two were formed using the carbon 2*p* electrons and two were formed using the 2*s* electrons.

In VB theory, the atoms in molecules can change the energies and shapes of their atomic orbitals to increase the opportunities for bonding with other atoms. This process is known as hybridization. Hybridization involves interactions between atomic orbitals on the same atom during the formation of bonds with another atom.

Hybridization is introduced in Section 4.5 (p.180) when discussing the bonding in N_2.

sp^3 hybridization

The carbon atom in methane is said to be sp^3 hybridized. **sp^3 hybridization** in carbon involves the interaction between the 2*s* orbital and the three 2*p* orbitals to form four sp^3 hybrid orbitals, as shown in Figure 5.27. Each hybrid orbital contains one electron.

Interactive 3D methane (sp^3) hybrid orbital.

The four sp^3 hybrid orbitals are identical in energy. Each has a node at the nucleus and contains two lobes, one of which is much larger than the other. This makes the orbitals directional, and they point at 109.5° from each other, towards the vertices of a tetrahedron with the carbon nucleus in the centre. In methane, each hybrid orbital interacts with a 1*s* orbital on a hydrogen atom to form a σ-bond.

These interactions give four identical covalent C–H σ bonds, as shown in Figure 5.28. The overlap between the hybrid orbitals and the hydrogen atomic orbitals is large because of the directional nature of the hybrid orbitals, so the bonds are stronger than if they were formed with unhybridized orbitals. Although there is an energetic cost to promote the *s* electrons into the hybrid orbitals, this is outweighed by the gain in energy on forming four bonds rather than two.

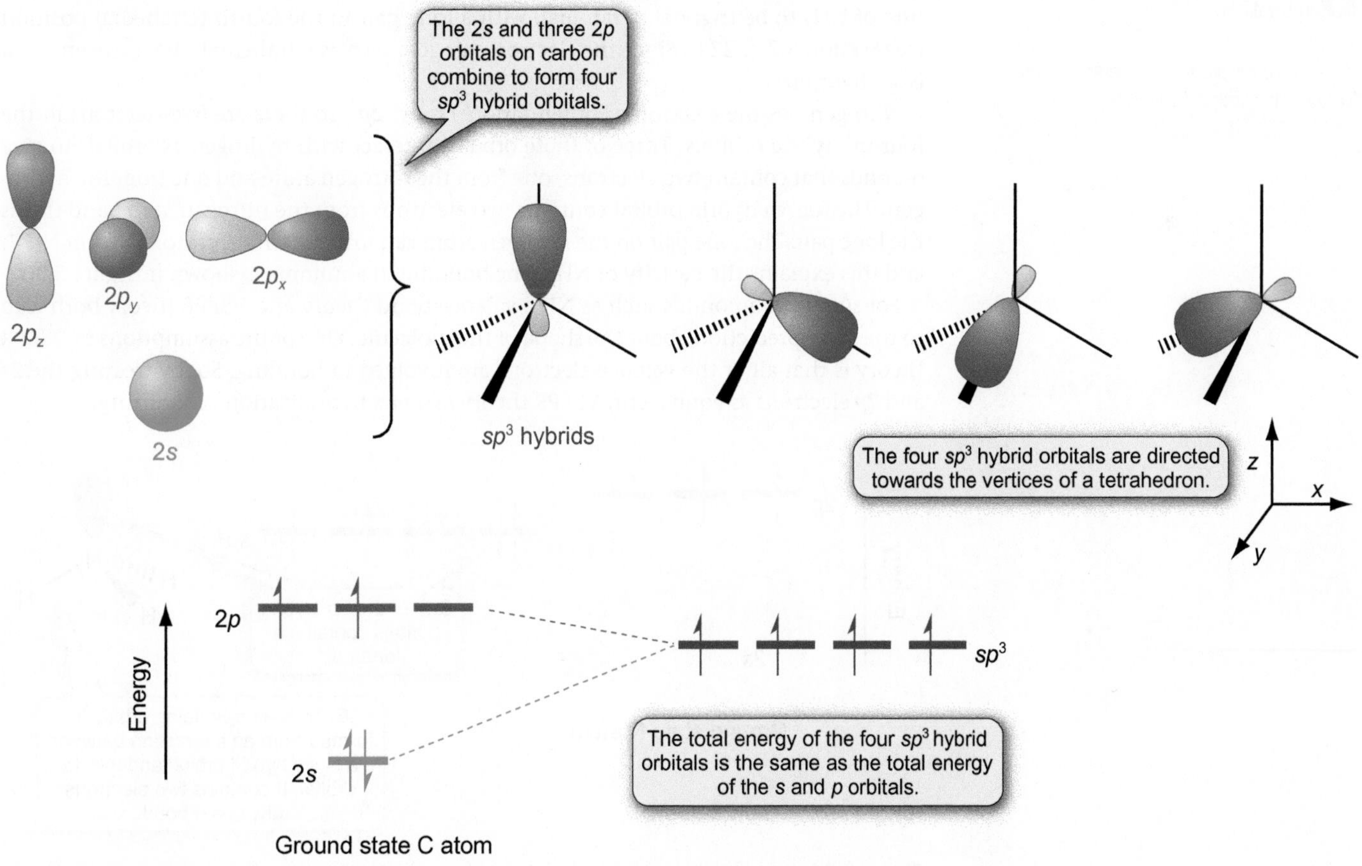
The 2s and three 2p orbitals on carbon combine to form four sp3 hybrid orbitals.
2pz
2py
2px
2s
sp3 hybrids
The four sp3 hybrid orbitals are directed towards the vertices of a tetrahedron.
z
x
y
Energy
2p
2s
sp3
The total energy of the four sp3 hybrid orbitals is the same as the total energy of the s and p orbitals.
Ground state C atom

Figure 5.27 sp^3 hybridization for carbon. The four sp^3 hybrid orbitals have one lobe larger than the other and are directed from the centre of a tetrahedron to the four vertices.

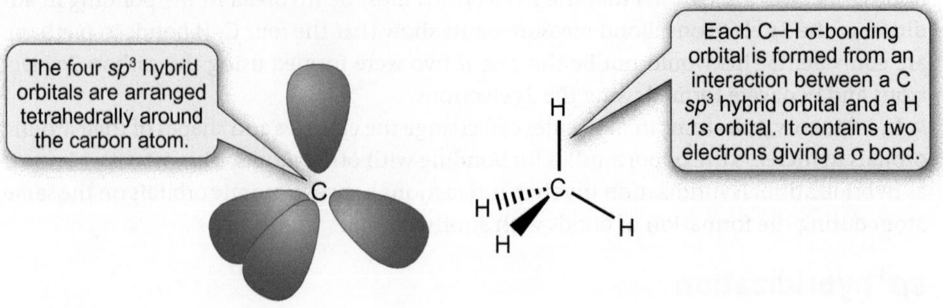
The four sp^3 hybrid orbitals are arranged tetrahedrally around the carbon atom.
Each C–H σ-bonding orbital is formed from an interaction between a C sp^3 hybrid orbital and a H 1s orbital. It contains two electrons giving a σ bond.
C
H
H
H
H
C

Figure 5.28 Formation of each of the four σ bonds in methane arises from the interaction between a carbon sp^3 hybrid orbital and a hydrogen 1s orbital.

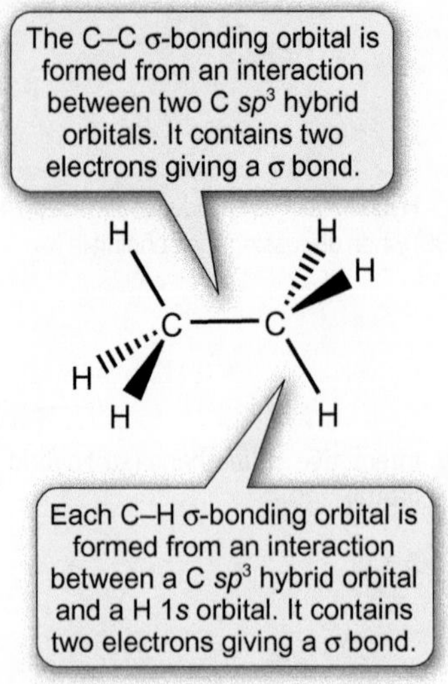
The C–C σ-bonding orbital is formed from an interaction between two C sp^3 hybrid orbitals. It contains two electrons giving a σ bond.
H
H
H
C
C
H
H
H
Each C–H σ-bonding orbital is formed from an interaction between a C sp^3 hybrid orbital and a H 1s orbital. It contains two electrons giving a σ bond.

Figure 5.29 The bonding in ethane. Both carbon atoms are in tetrahedral environments, which is consistent with sp^3 hybridization.

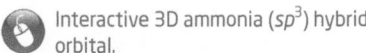

The conformations of ethane are described in Section 18.2 (p.820).

Interactive 3D ammonia (sp^3) hybrid orbital.

The basic properties of ammonia are discussed in Section 7.1 (p.306)

It is commonly said that methane is tetrahedral *because* the carbon atom is sp^3 hybridized. This is not true. CH_4 is tetrahedral as this geometry minimizes the repulsions between the electron pairs in the four C–H bonds while maximizing the strength of each bond. Tetrahedral geometry is consistent with sp^3 hybridization of the central atom, but not caused by it.

You can use a very similar picture to explain the bonding in other alkanes. Ethane (C_2H_6) contains two tetrahedral carbon atoms, linked through a C–C bond. Both of the carbon atoms are sp^3 hybridized. The interactions involved in forming the bonds in ethane are summarized in Figure 5.29.

Ammonia

sp^3 hybridization is not restricted to organic molecules. VSEPR theory predicts the structure of NH_3 to be trigonal pyramidal, with a lone pair in the fourth tetrahedral position (see Section 5.2, p.224). Since the electron pair geometry is tetrahedral, the nitrogen atom is sp^3 hybridized.

Nitrogen has the electronic configuration $1s^2\ 2s^2\ 2p^3$, so there are five electrons in the four sp^3 hybrid orbitals. Three of these orbitals interact with hydrogen 1s orbitals to give σ-bonds that contain two electrons, one from the nitrogen atom and one from the hydrogen. The fourth hybrid orbital contains two electrons from the nitrogen atom, and this is the lone pair. The lone pair on the nitrogen atom can interact with a proton to form NH_4^+, and this explains the basicity of NH_3. The bonding in ammonia is shown in Figure 5.30.

For simple compounds such as NH_3, valence bond theory and VSEPR theory both lead to the same prediction about the shape of the molecule. One of the assumptions of VSEPR theory is that all of the valence electrons are involved in bonding. So, by treating the 2s and 2p electrons as equivalent, VSEPR theory *assumes* hybridization is occurring.

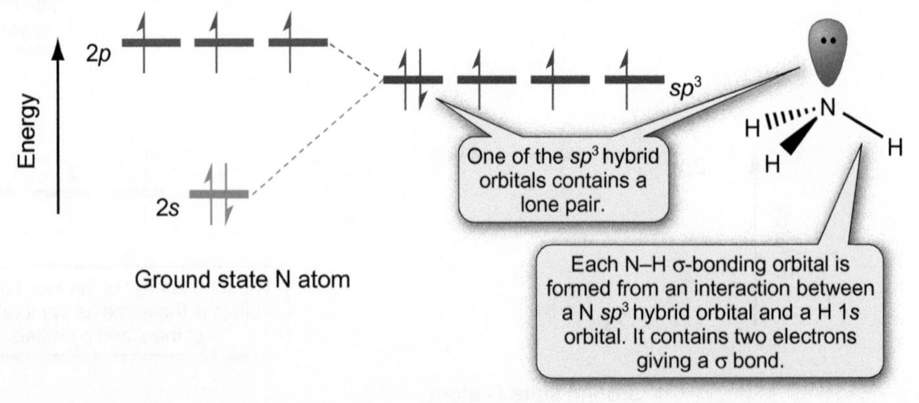
Energy
2p
2s
sp3
Ground state N atom
One of the sp^3 hybrid orbitals contains a lone pair.
H
N
H
H
Each N–H σ-bonding orbital is formed from an interaction between a N sp^3 hybrid orbital and a H 1s orbital. It contains two electrons giving a σ bond.

Figure 5.30 In ammonia there are three σ bonds formed from the interactions between nitrogen sp^3 hybrid orbitals with hydrogen 1s orbitals. The fourth sp^3 hybrid orbital contains a lone pair of electrons.

You might expect the shape of phosphine (PH_3) to be identical to that of NH_3. Phosphorus is below nitrogen in Group 15 so it has the same number of valence electrons ($3s^2\ 3p^3$). However, experimental measurements show that the H–P–H bond angle in phosphine is 93.8°, much closer to 90° than to a tetrahedral angle of 109.5°. Furthermore, PH_3 is much less basic than NH_3, which suggests that the lone pair on PH_3 is not available to interact with a proton to form PH_4^+. These observations can be explained by assuming that the lone pair is in the 3*s* orbital, while the electrons involved in forming the P–H bonds are simply those in the unhybridized 3*p* orbitals. The experimental evidence suggests that hybridization is not occurring in PH_3.

The chemistry of PH_3 is described in Section 25.2 (p.1150).

sp^2 hybridization

Ethene (C_2H_4, **8**) is an alkene and has a carbon–carbon double bond. Experimental measurements show that the geometry around the carbon atoms in ethene is approximately trigonal planar. This geometry is consistent with ***sp*² hybridization**. The 2*s*, $2p_x$, and $2p_y$ orbitals on each atom combine together to give three identical sp^2 hybrid orbitals orientated in a plane at 120° to each other. The p_z orbitals are not involved in this process and remain unhybridized, as shown in Figure 5.31. The sp^2 hybrid orbitals on the two carbon atoms each contain a single electron as do the p_z orbitals.

sp^2 hybrid orbitals on the two carbon atoms interact together to give a C–C σ-bond that contains two electrons. The two remaining sp^2 hybrid orbitals on each carbon atom interact with hydrogen 1*s* orbitals to give C–H σ-bonds that each contain two electrons. These interactions lead to the planar σ-bond framework of the ethene molecule. The *unhybridized* p_z orbitals interact to give a π-bond. This contains two electrons and has electron density above and below the molecular plane, as shown in Figure 5.32.

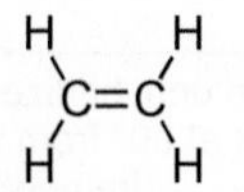
H H
C=C
H H

8 Lewis structure of ethene (C_2H_4).

Interactive 3D ethene (sp^2) hybrid orbital.

The chemistry of alkenes is described in Chapter 21 (p.962).

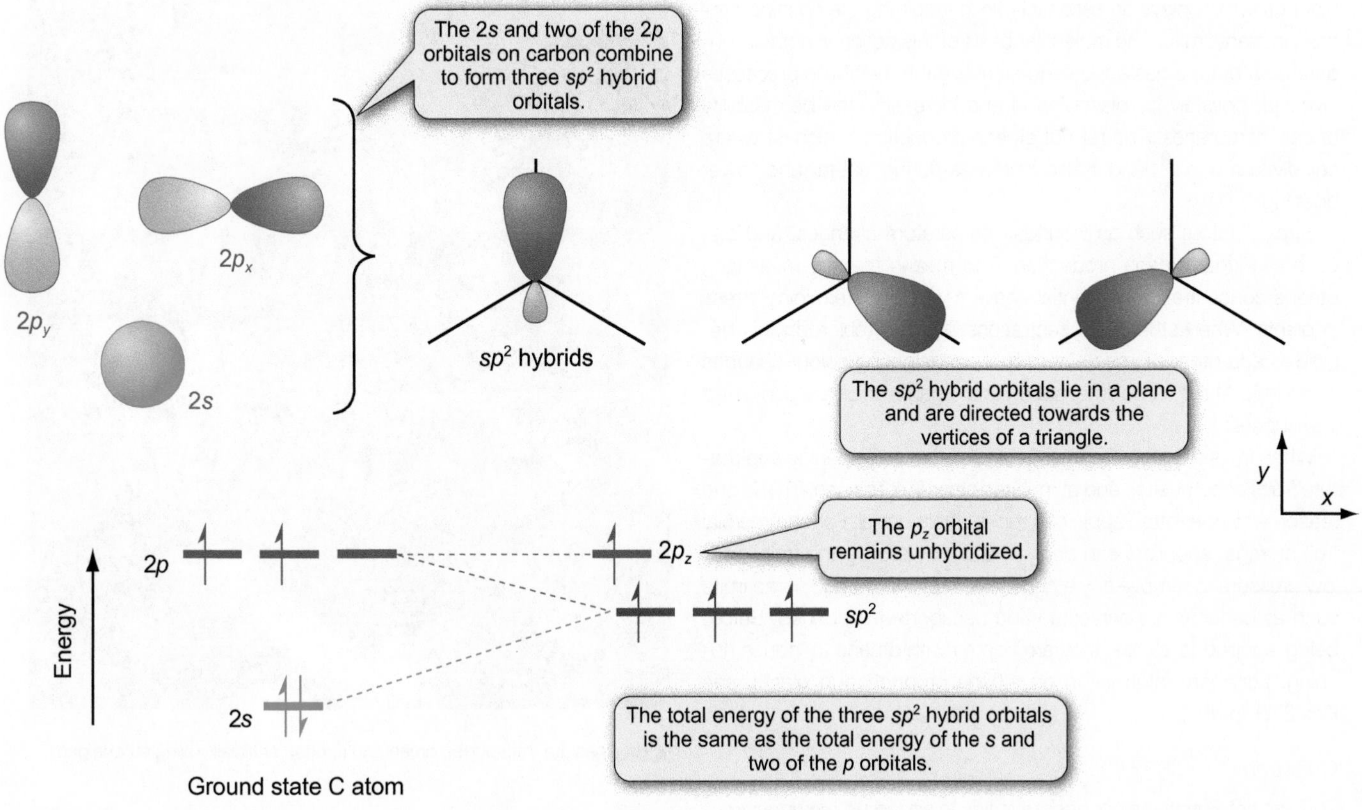
The 2s and two of the 2p orbitals on carbon combine to form three sp² hybrid orbitals.
2p_y
2p_x
2s
sp² hybrids
The sp² hybrid orbitals lie in a plane and are directed towards the vertices of a triangle.
y
x
Energy
2p
2p_z
The p_z orbital remains unhybridized.
sp²
2s
The total energy of the three sp² hybrid orbitals is the same as the total energy of the s and two of the p orbitals.
Ground state C atom

Figure 5.31 sp^2 hybridization in carbon. The three sp^2 hybrid orbitals are directed from the centre of a triangle to the three vertices.

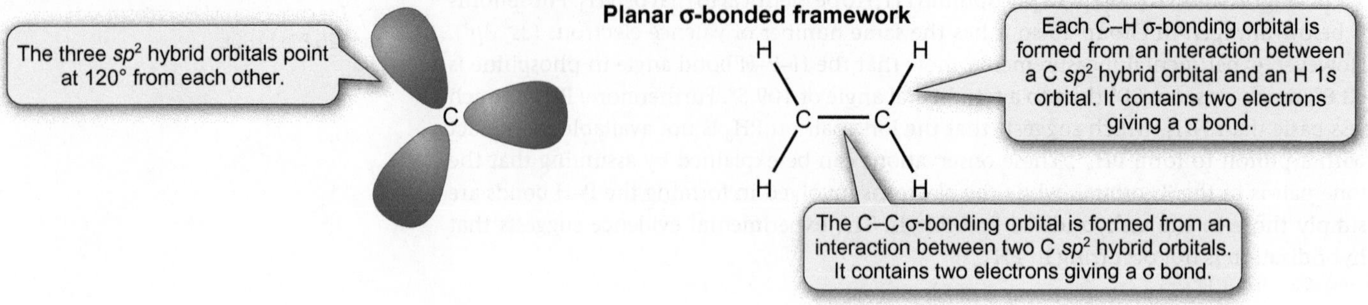
Planar σ-bonded framework
The three sp^2 hybrid orbitals point at 120° from each other.
C
H
H
C
C
H
H
Each C–H σ-bonding orbital is formed from an interaction between a C sp^2 hybrid orbital and an H 1s orbital. It contains two electrons giving a σ bond.
The C–C σ-bonding orbital is formed from an interaction between two C sp^2 hybrid orbitals. It contains two electrons giving a σ bond.

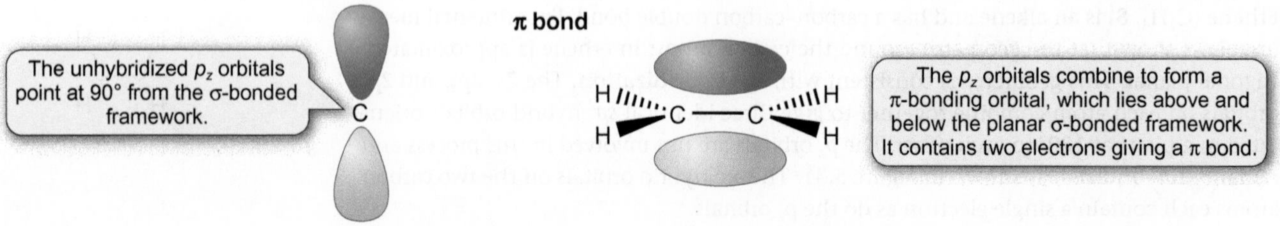
π bond
The unhybridized p_z orbitals point at 90° from the σ-bonded framework.
C
H
H
C
C
H
H
The p_z orbitals combine to form a π-bonding orbital, which lies above and below the planar σ-bonded framework. It contains two electrons giving a π bond.

Figure 5.32 The bonding in ethene. The C=C double bond consists of a σ bond and a π bond.

Box 5.5 Ethene and the ripening of fruit

Ethene is produced naturally by plants and is important in plant growth. It plays an essential role in triggering the ripening process in many fruits. The molecular basis of this action is not fully understood, but the gas appears to stimulate the metabolic processes involved, possibly by dissolving in and increasing the permeability of cell membranes. The rate of ethene production is highest where cell division occurs, and it also increases during leaf fall and flower opening.

Stress factors such as wounds, temperature changes, and disease all trigger ethene production. This means that, by measuring ethene concentrations, scientists have a way of determining stress in plants. When ethene concentrations are high, chlorophyll is degraded and other pigments are produced, causing colour changes in the fruit. At the same time, starch and organic acids are converted into sugars.

When fruits are stored, the ethene released from ripening and mature fruits accumulates and stimulates other fruit to ripen. This is one reason why one rotten apple can ruin a whole barrel. For commercial fruit storage, suppliers can delay ripening by storing the fruit under low pressure to remove any ethene released. In contrast, some fruits such as bananas are harvested and transported unripened. Before being shipped to stores, they are treated with ethene to induce ripening. For more information on ethene production in plants, see Box 21.1 (p.965).

▲ Bananas are transported green and ripened artificially using ethene gas.

Question

The ancient Egyptians cut gashes in figs to speed up the ripening process. How does this work?

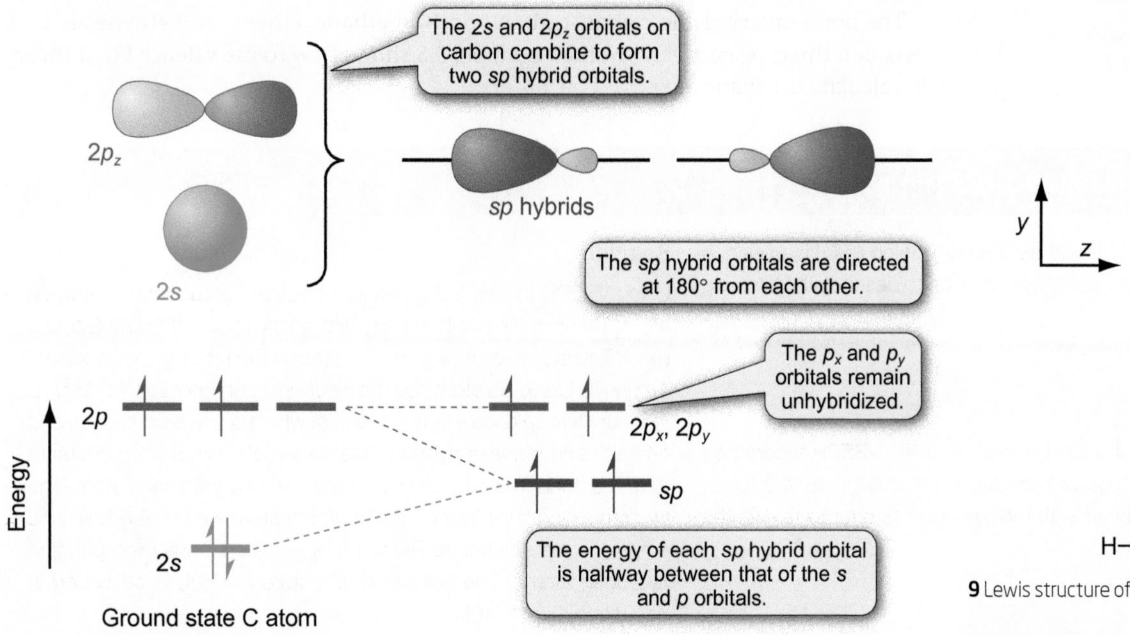
The 2s and 2pz orbitals on carbon combine to form two sp hybrid orbitals.
2pz
2s
sp hybrids
y
z
The sp hybrid orbitals are directed at 180° from each other.
The px and py orbitals remain unhybridized.
2p
2px, 2py
sp
Energy
2s
The energy of each sp hybrid orbital is halfway between that of the s and p orbitals.
Ground state C atom

Figure 5.33 *sp* hybridization for carbon. The two *sp* hybrid orbitals are directed at 180° from each other.

H–C≡C–H

9 Lewis structure of ethyne (C_2H_2).

Interactive 3D ethyne (sp) hybrid orbital.

The valence bond approach to the bonding in N_2 is described in Section 4.5 (p.180).

The chemistry of alkynes is described in Chapter 21 (p.962).

The three *p* orbitals are degenerate in an isolated carbon atom. By convention, the p_z orbital is always the 'odd' one when this degeneracy is lost. For sp^2 hybridization the p_z is the only unhybridized *p* orbital, and for *sp* hybridization it is the only *p* orbital involved in hybridization.

sp hybridization

Ethyne (C_2H_2, **9**) is an alkyne and has a carbon–carbon triple bond. Experimental measurements show that the geometry around the carbon atoms in ethyne is linear. This geometry is consistent with ***sp* hybridization** in a similar manner to the VB treatment of N_2. The 2*s* and $2p_z$ orbitals on each carbon atom hybridize to form two *sp* hybrid orbitals, which are directed at 180° to each other as shown in Figure 5.33.

The interaction between *sp* hybrid orbitals on the two carbon atoms leads to formation of a C–C σ-bond, and interactions between the remaining *sp* hybrid orbital on each carbon atom and a 1*s* orbital on a hydrogen atom give C–H σ-bonds. This leaves unhybridized $2p_x$ and $2p_y$ orbitals on each atom. These interact as shown in Figure 5.34 to give two π-bonds, arranged at 90° to each other. The triple bond in ethyne consists of one σ bond and two π bonds.

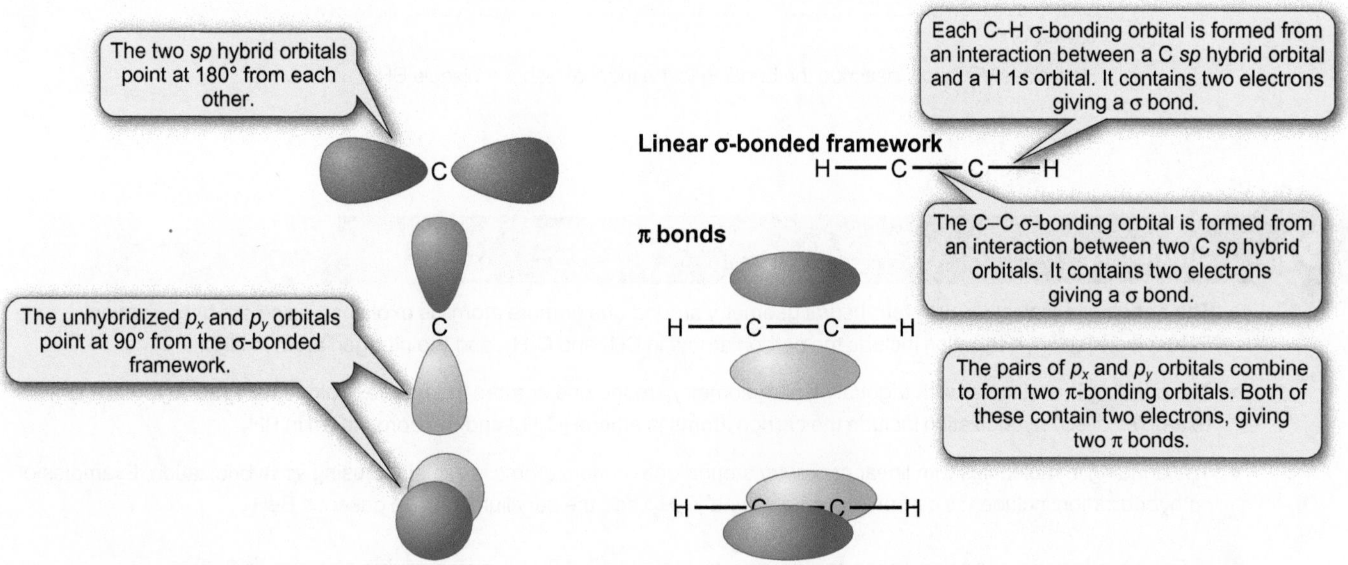
The two sp hybrid orbitals point at 180° from each other.
C
C
The unhybridized px and py orbitals point at 90° from the σ-bonded framework.
Each C–H σ-bonding orbital is formed from an interaction between a C sp hybrid orbital and a H 1s orbital. It contains two electrons giving a σ bond.
Linear σ-bonded framework
H—C—C—H
π bonds
The C–C σ-bonding orbital is formed from an interaction between two C sp hybrid orbitals. It contains two electrons giving a σ bond.
H—C—C—H
The pairs of px and py orbitals combine to form two π-bonding orbitals. Both of these contain two electrons, giving two π bonds.
H—C—C—H

Figure 5.34 The bonding in ethyne. The C≡C triple bond comprises a σ bond and two π bonds.

For definitions of bond order see Sections 4.2 (p.174) and 4.8 (p.189).

The bond orders of the carbon–carbon bonds in ethane, ethene, and ethyne are one, two, and three, respectively. Worked example 5.5 shows how to use valence bond theory to calculate the shape of BeH_2.

Worked example 5.5 Bonding in BeH_2

In the gas phase, BeH_2 exists as discrete molecules. Use valence bond theory and the concept of hybridization to describe the bonding in the BeH_2 molecule.

Strategy

Determine the shape of a BeH_2 molecule using VSEPR theory and from this deduce the type of hybridization that is present. Interact the beryllium hybrid orbitals with the hydrogen 1*s* orbitals to give the bonds in BeH_2.

Solution

Using VSEPR theory, BeH_2 has four electrons around the central Be atom. These make two bonding pairs, so the shape of the molecule is linear. This is consistent with the Be atom in BeH_2 being *sp* hybridized. An unhybridized Be atom has the electronic configuration $1s^2\,2s^2$, so there are two valence electrons, one of which is placed in each hybrid orbital. Each of these hybrid orbitals interacts with a H 1*s* orbital to form a σ bond that contains two electrons, one from each atom. The two remaining *p* orbitals on the beryllium atom are unhybridized and empty. (In the solid state BeH_2 forms polymers containing bridging hydrogen atoms. The solid state structure of BeH_2 is described in Section 25.2 (p.1146).)

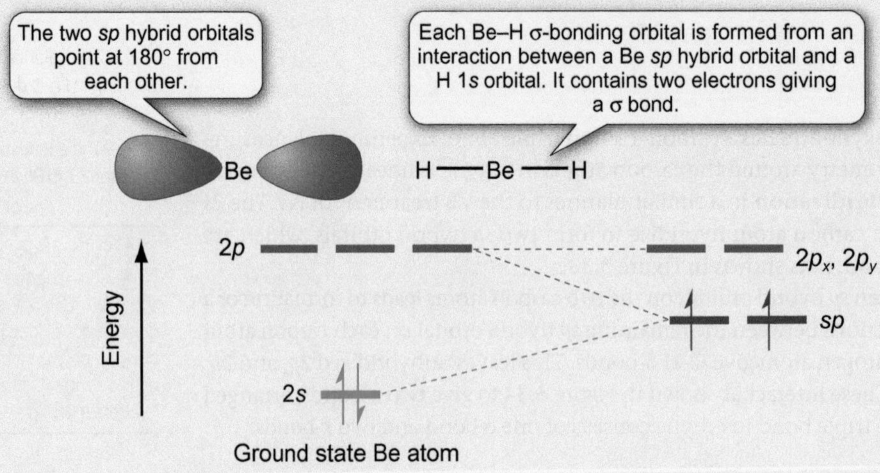
The two sp hybrid orbitals point at 180° from each other.
Each Be–H σ-bonding orbital is formed from an interaction between a Be sp hybrid orbital and a H 1s orbital. It contains two electrons giving a σ bond.
Be
H—Be—H
2p
2px, 2py
sp
Energy
2s
Ground state Be atom

Question

How does VB theory describe the bonding in the trigonal planar molecule BH_3?

Summary

- The bonding in molecules with tetrahedral geometry around one or more atoms is explained using sp^3 hybridization. Examples of sp^3 hybridization include the carbon atoms in CH_4 and C_2H_6, and the nitrogen atom in NH_3.
- The bonding in molecules with trigonal planar geometry around one or more atoms is explained using sp^2 hybridization. Examples of sp^2 hybridization include the carbon atoms in ethene (C_2H_4) and the boron atom in BH_3.
- The bonding in molecules with linear geometry around one or more atoms is explained using *sp* hybridization. Examples of *sp* hybridization include the carbon atoms in ethyne (C_2H_2) and the beryllium atom in gaseous BeH_2.

For practice questions on these topics, see questions 10–12 at the end of this chapter (p.253).

5.5 Resonance

Another important concept in valence bond theory is resonance. As with diatomic molecules, it is possible to get a better description of the bonding in a polyatomic molecule by including ionic resonance forms as well as covalent ones. In this way, the bonding in the BeH_2 molecule can be represented by the resonance forms that are shown in Figure 5.35.

Resonance is introduced in Section 4.5 (p.180) where ionic structures are included in the valence bond description of H_2.

Resonance forms are shown with a double-headed arrow between them. The 'real' structure is a weighted average of the different resonance forms and it is lower in energy than any single contributing resonance form. The resonance forms do not all contribute equally to the molecular wavefunction, and for the BeH_2 molecule, the covalent form contributes the most. Do the ionic forms contribute equally? To answer this, you need to consider the electronegativities of beryllium (χ^P 1.57) and hydrogen (χ^P 2.20). This tells you that the Be–H bonds are polarized with the hydrogen atoms δ– and the beryllium atom δ+. In turn, this means that the ionic resonance forms in which the hydrogen atom is negatively charged contribute more to the molecular wavefunction than those in which the hydrogen atom is positively charged. Resonance forms with a high charge on one atom, such as those in Figure 5.35 without any covalent bonds, usually contribute little to the molecular wavefunction.

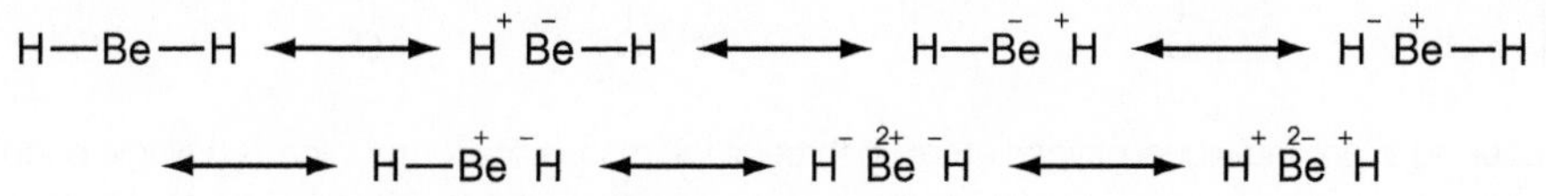

Figure 5.35 The resonance forms for the linear BeH_2 molecule.

Resonance in anions

Resonance is particularly useful for ions that contain combinations of single and double bonds to oxygen atoms. For example, the Lewis structure for the carbonate ion contains one double bond and two single bonds (see Section 5.2, p.231). Measurements show that the three carbon–oxygen bonds are identical with bond lengths of 129 pm, which is between a typical C–O single bond (142 pm) and a typical C=O double bond (121 pm). Valence bond theory explains the regular trigonal planar structure of carbonate through the existence of three resonance forms.

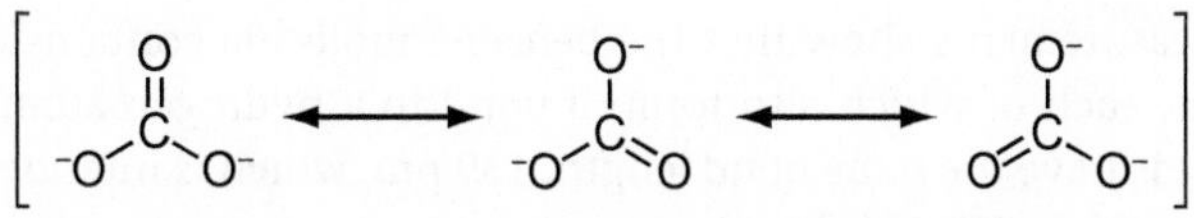

In all of the resonance forms the central carbon atom is sp^2 hybridized. Each oxygen atom has an average negative charge of two-thirds $\left(\frac{(-1)+(-1)+0}{3} = -\frac{2}{3}\right)$ and the bond order of each C–O bond is $1\frac{1}{3}$ $\left(\frac{1+1+2}{3} = \frac{4}{3}\right)$.

Worked example 5.6 uses the resonance structures for the sulfate ion to work out the average charge on each oxygen atom and the S–O bond order.

None of the resonance forms has independent existence. The real structure is an average of the resonance forms, and there is not an equilibrium or oscillation between them. As an analogy, consider a mule which is a hybrid between a donkey and a horse. A mule is always a mule; it is not a donkey some of the time and a horse some of the time.

Mule = [Horse ⟷ Donkey]

Worked example 5.6 Resonance in the SO_4^{2-} ion

The four sulfur–oxygen bonds in the sulfate ion, SO_4^{2-}, have been shown experimentally to have equal bond lengths. Draw resonance structures for SO_4^{2-} to account for this, and determine the average S–O bond order and the average charge on each oxygen atom.

Strategy

Draw a simple Lewis structure of the sulfate ion with two S–O single bonds to negatively charged oxygen atoms and two S=O double bonds to neutral oxygen atoms. Draw resonance forms to show that the four bonds are equivalent.

Solution

In each of the resonance forms, two of the four sulfur–oxygen bonds are double bonds. There are six resonance forms all of equal importance.

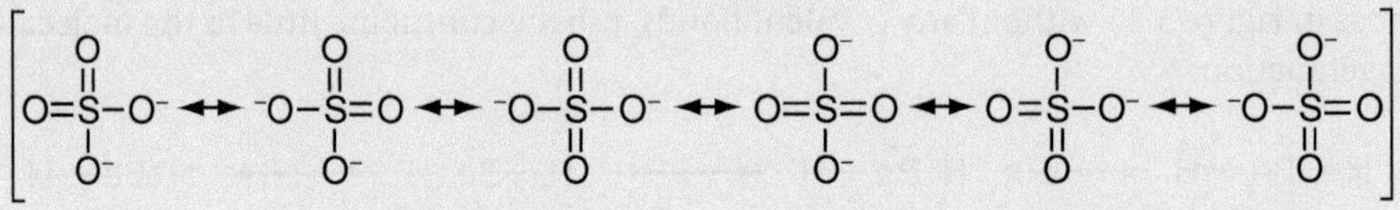

Looking at any individual oxygen atom, in three of the resonance forms it forms a single bond and in three it forms a double bond. The average S–O bond order is therefore $1\frac{1}{2}$. Looking again at an oxygen atom, in three of the resonance forms it carries a negative charge and in three it is neutral. This means that on average each oxygen atom carries half a negative charge $(-\frac{1}{2})$.

Question

The three bonds in the sulfite ion, SO_3^{2-}, are all identical. Draw out the resonance forms for sulfite and determine the average S–O bond order and the average charge on each oxygen atom.

Benzene (C_6H_6)

Experimental measurements show that the benzene molecule contains a planar ring of six carbon atoms, each of which also forms a bond to a hydrogen atom. All of the carbon– carbon bonds have the same bond length, 139 pm, which is intermediate between a carbon–carbon single and double bond.

Each of the carbon atoms in benzene has a trigonal planar geometry. This means that, in the VB description, the carbon atoms are sp^2 hybridized. Interactions between the carbon sp^2 hybrid orbitals lead to the 6-membered ring. Interactions between the carbon sp^2 hybrid orbitals and hydrogen 1*s* orbitals complete the planar σ bonded framework (Figure 5.36).

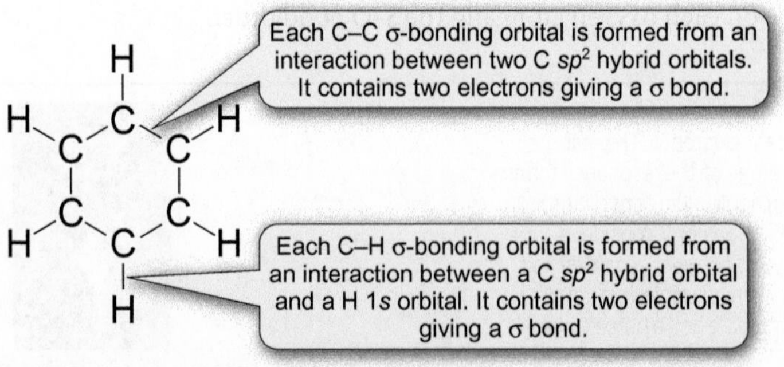
Each C–C σ-bonding orbital is formed from an interaction between two C sp² hybrid orbitals. It contains two electrons giving a σ bond.
Each C–H σ-bonding orbital is formed from an interaction between a C sp² hybrid orbital and a H 1s orbital. It contains two electrons giving a σ bond.

Figure 5.36 The planar σ-bonded framework in benzene.

Interactive 3D benzene (sp^2) hybrid orbital.

This leaves the six unhybridized $2p_z$ orbitals, one on each carbon atom, each containing a single electron. Although these could interact to give a structure containing alternating single and double bonds this does not fit in with the experimental observations on the C–C bond lengths, which are all identical. The equivalence of the carbon–carbon bonds in benzene is shown in VB theory by drawing two resonance forms.

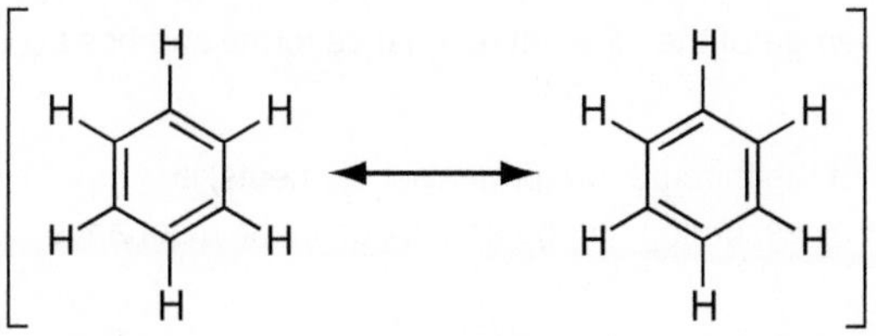

In the VB description of benzene, the 'real' structure is an average of these two resonance forms.

In benzene, the π electrons are said to be **delocalized** (spread out) over the benzene ring. The delocalized structure is more stable than the hypothetical structure with three isolated double bonds and, because of this stability, benzene is described as **aromatic**.

Aromaticity is a special kind of resonance, and the bonding in benzene and other aromatic compounds is described in more detail in Section 22.1 (p. 1004).

Hypervalency

Hybridization (sp^3, sp^2, or sp) can be used to explain the bonding in all compounds in which the central atom is an element of the second period. This is because these elements cannot contain more than eight electrons in their valence shell. Even an ion such as nitrate (NO_3^-), which at first sight appears to contradict this, has only eight electrons in the valence shell since the nitrogen atom has a formal charge of +1 (see Worked example 5.4, p.232).

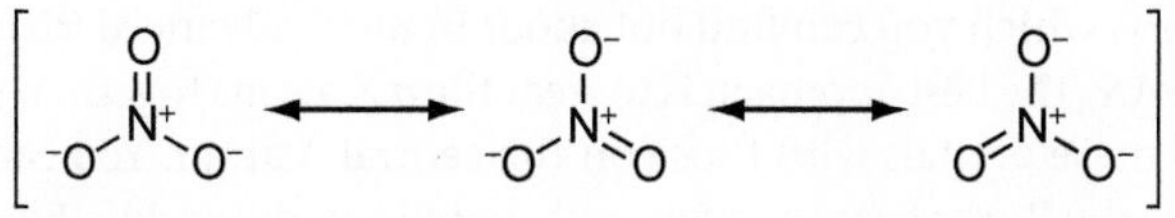

There are two explanations for why these elements always obey the octet rule. Firstly, they are relatively small, so putting five or more electron pairs around them is not possible as the electron pairs would be very close together and repel each other too much. Secondly, the extra electrons would need to go into an empty orbital, and these are all too high in energy to be occupied for these elements.

For elements in the third period, neither of these factors is so important. These elements are larger, and have unoccupied orbitals that may be accessible. There are many examples of hypervalent compounds containing elements of the third period or heavier elements, and you can see the shapes of some of these molecules and ions (PF_6^-, SF_4, ClF_3) earlier in this chapter.

A possible way to account for hypervalency in VB theory is to include d orbitals in the hybridization scheme. To explain the bonding in the trigonal bipyramidal PF_5 molecule, the phosphorus atom would be dsp^3 hybridized, so that the hybridization process involves the $3d_{z^2}$ orbital in addition to the $3s$ and three $3p$ orbitals. In the octahedral SF_6 molecule the sulfur atom would be d^2sp^3 hybridized, with both the $d_{x^2-y^2}$ and d_{z^2} orbitals involved in the hybridization. Detailed calculations, however, suggest that the d orbitals are not very important in the bonding of p-block compounds, even for elements of the third period or lower in the Periodic Table.

Hypervalency can be represented in VB theory without breaking the octet rule, though only by using charge separation. For example, each of the five resonance forms of PF_5 shown below contains only eight electrons around the central phosphorus atom.

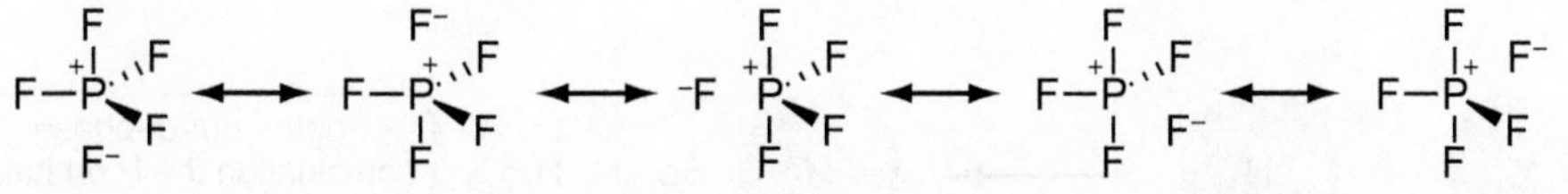

Molecular orbital theory, in the next section, provides a way of looking at the bonding in hypervalent compounds without invoking the d orbitals or using charge separation.

Summary

- Resonance is used to explain the bonding in molecules with bonds that appear to be different in a Lewis structure but are found to be the same experimentally.
- Resonance forms do not exist independently. The actual structure is an *average* of the different resonance forms and has a lower energy than any individual resonance form.
- In valence bond theory, hypervalency in compounds containing elements of the third period (or heavier elements) is explained using hybridization schemes involving one or more *d* orbitals or by assuming charge separation in the resonance forms.

For practice questions on these topics, see questions 13–15 at the end of this chapter (p.253).

5.6 A molecular orbital approach to the bonding in polyatomic molecules

The importance of symmetry in determining which atomic orbitals interact to form molecular orbitals is described in Section 4.9 (p.195).

In a diatomic molecule, the atomic orbitals can only combine to give molecular orbitals if they have the correct symmetry. The same applies for polyatomic molecules. The main difficulty in extending molecular orbital theory to polyatomic molecules lies in knowing which orbitals have the correct symmetry to interact. Generally, this can be worked out using group theory, which you can find out about in more advanced texts.

For a molecule AX_n the best approach is to treat the n X atoms together, then interact the combinations of these orbitals with those on the central A atom. You can get the general idea by looking at the linear triatomic molecule beryllium dihydride (BeH_2).

Molecular orbital approach to beryllium dihydride (BeH_2)

In the solid state, BeH_2 has a polymeric structure with bridging hydrides, as described in Section 25.2 (p.1146).

The MO approach to the bonding in the BeH_2 molecule involves two stages. In the first stage, you combine the orbitals on the two hydrogen atoms together. In the second stage you decide how these combinations of orbitals interact with the orbitals on the central beryllium atom.

Stage 1: Analysis of the hydrogen 1*s* orbitals

There are two hydrogen 1*s* orbitals in BeH_2, just as in H_2, so both in-phase and out-of-phase combinations of these are possible. These two combinations are similar in symmetry to the σ_g and σ_u orbitals in H_2, but there is one important difference. In BeH_2 the hydrogen atoms are much further apart, since they have a beryllium atom lying between them. This means that the overlap between the hydrogen atomic orbitals is very small, and as a result the energy difference between the in-phase combination and the out-of-phase combination is negligible. These two combinations of hydrogen 1*s* orbitals are shown in Figure 5.37.

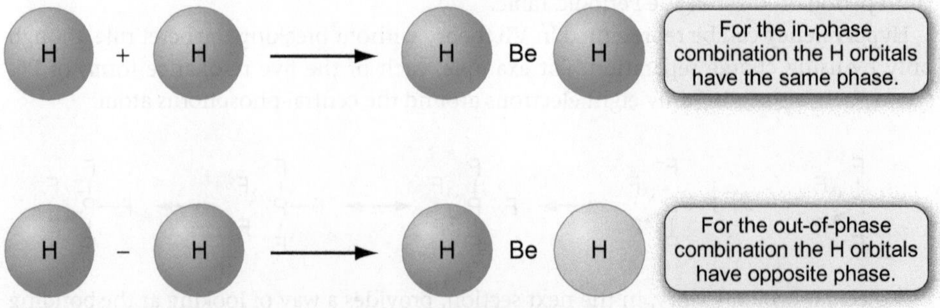
H
+
H
H
Be
H
For the in-phase combination the H orbitals have the same phase.
H
–
H
H
Be
H
For the out-of-phase combination the H orbitals have opposite phase.

Figure 5.37 The in-phase and out-of-phase combinations of hydrogen 1*s* orbitals in BeH_2.

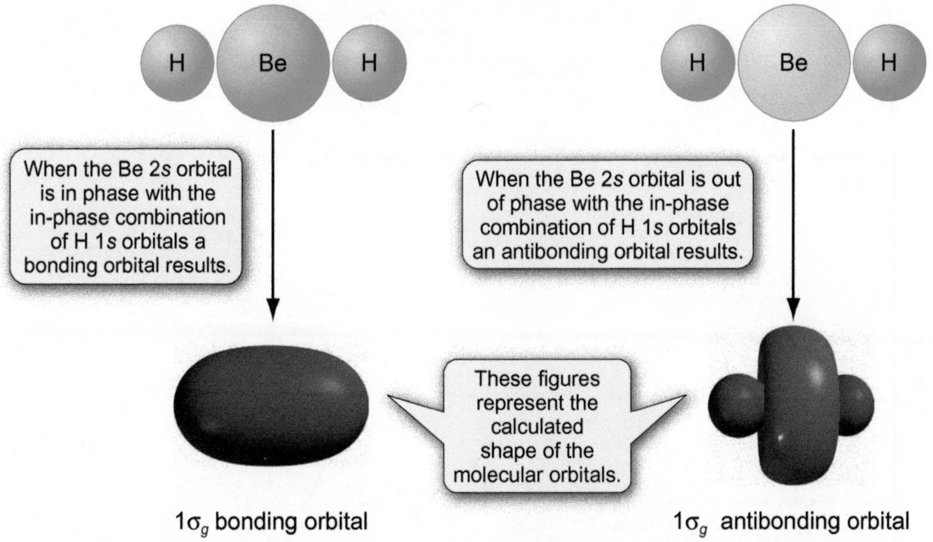
H
Be
H
When the Be 2s orbital is in phase with the in-phase combination of H 1s orbitals a bonding orbital results.
When the Be 2s orbital is out of phase with the in-phase combination of H 1s orbitals an antibonding orbital results.
These figures represent the calculated shape of the molecular orbitals.
$1\sigma_g$ bonding orbital
$1\sigma_g$ antibonding orbital

Figure 5.38 The interaction between the in-phase combination of hydrogen 1*s* orbitals and the Be 2*s* orbital gives a bonding orbital and an antibonding orbital.

Stage 2: Interactions of the hydrogen 1*s* orbital combinations with the beryllium orbitals

Beryllium has an electronic configuration of $1s^2\ 2s^2$ but, since the *s*–*p* gap in beryllium is relatively small, you need to consider interactions with both the beryllium 2*s* and 2*p* orbitals.

The variation in the *s*–*p* separation on going across the second period is described in Section 4.11 (p.200).

The *g* and *u* parity labels are defined in Section 4.7 (p.186).

The in-phase combination of hydrogen orbitals has the correct symmetry to interact with the beryllium 2*s* orbital. This interaction can itself be in-phase, leading to a bonding orbital, or out-of-phase leading to an antibonding orbital. These combinations are shown in Figure 5.38. Both of these molecular orbitals have cylindrical symmetry, so they are σ orbitals and both remain unaltered on inversion. Consequently the bonding and antibonding orbitals *both* have the symmetry label σ_g.

The out-of-phase combination of hydrogen orbitals does not have the correct symmetry to interact with the beryllium 2*s* orbital. It can, however, interact with the beryllium $2p_z$ orbital. This interaction can also be in phase, leading to a bonding orbital, or out of phase leading to an antibonding orbital. These combinations are shown in Figure 5.39. Again

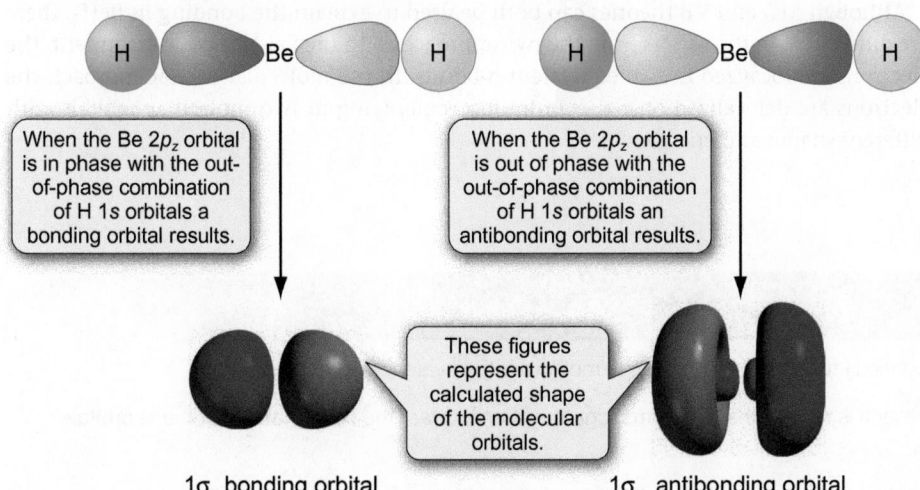

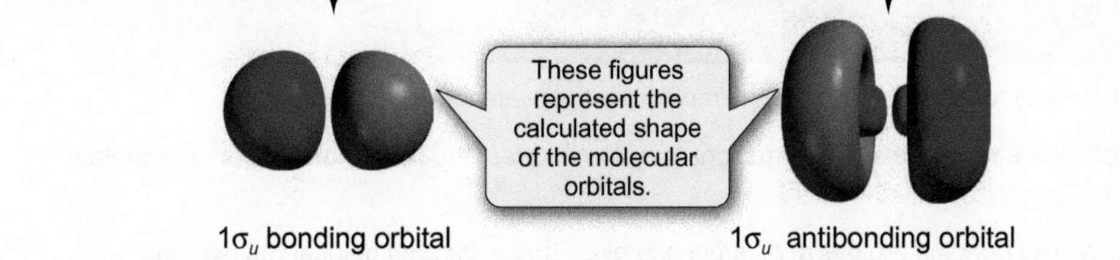
H
Be
H
When the Be $2p_z$ orbital is in phase with the out-of-phase combination of H 1s orbitals a bonding orbital results.
When the Be $2p_z$ orbital is out of phase with the out-of-phase combination of H 1s orbitals an antibonding orbital results.
These figures represent the calculated shape of the molecular orbitals.
$1\sigma_u$ bonding orbital
$1\sigma_u$ antibonding orbital

Figure 5.39 The interaction between the out-of-phase combination of hydrogen 1*s* orbitals and the Be $2p_z$ orbital gives a bonding orbital and an antibonding orbital.

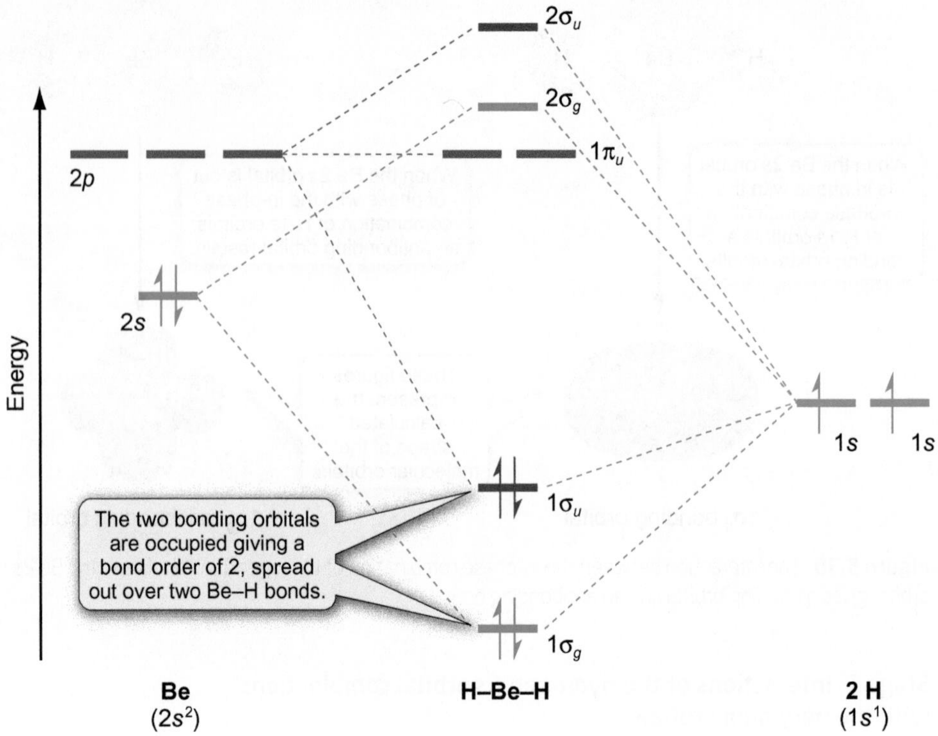

2σu
2σg
1πu
2p
2s
Energy
1s
1s
1σu
The two bonding orbitals are occupied giving a bond order of 2, spread out over two Be–H bonds.
1σg
Be
(2s²)
H–Be–H
2 H
(1s¹)

Figure 5.40 The molecular orbital energy level diagram for gaseous BeH_2.

both molecular orbitals have cylindrical symmetry, though in this case both the bonding and antibonding orbitals change phase on inversion. This means that both orbitals have the symmetry label σ_u.

The beryllium $2p_x$ and $2p_y$ orbitals cannot interact with either combination of hydrogen orbitals, so these become non-bonding molecular orbitals.

The molecular orbital energy level diagram for BeH_2 is shown in Figure 5.40. Both of the bonding orbitals are filled, whereas the non-bonding and antibonding orbitals are empty. The two filled bonding orbitals are of different energies, and are delocalized over the whole of the molecule. This means that each of the individual Be–H bonds has electron density from the two bonding orbitals. The two Be–H bonds are therefore equivalent.

Although MO and VB theories can both be used to explain the bonding in BeH_2, there are differences in the results from the two approaches. In the valence bond treatment, the electrons are localized in two equivalent σ-bonds. In the molecular orbital approach the electrons are delocalized over the entire molecule, lying in two molecular orbitals with different shapes and energies.

Compare the molecular orbital treatment of the bonding in BeH_2 described here to the valence bond approach described in Worked example 5.5 (p.242).

The advantages and disadvantages of MO and VB theories are described in Section 4.4 (p.178).

Summary

- Molecular orbital theory is more difficult to extend to polyatomic molecules than valence bond theory.
- For a molecule AX_n the best approach is to treat the n X atoms together, then interact the combinations of these orbitals with those on the central A atom.
- The linear molecule BeH_2 has two filled bonding orbitals of different energies. These are both delocalized over the molecule so that the two Be–H bonds are equivalent.

5.7 Partial molecular orbital schemes

In many cases you can get a good understanding of the bonding in a molecule using a **partial molecular orbital scheme**. In this case only the key orbitals are included. These schemes and the resultant **partial MO energy level diagrams** are simpler to construct than full MO energy level diagrams. They are particularly useful in describing the bonding in hypervalent and electron poor compounds that do not obey the octet rule.

Xenon difluoride (XeF_2)

VSEPR theory predicts that xenon difluoride (XeF_2, **10**) is a linear molecule.

The xenon atom in XeF_2 is hypervalent, as it has ten valence electrons around it. VB theory can explain this using dsp^3 hybridization involving the d_{z^2} orbital. However, theoretical studies suggest the *d* orbitals are too high in energy to be involved in the bonding of *p*-block compounds, even for a fifth period element like xenon (see Section 5.5, p.245). How can the hypervalency be explained without invoking *d* orbitals?

F–Xe–F

10 Lewis structure of xenon difluoride (XeF_2).

XeF_2	
Electrons from Xe	8
Electrons from 2 F	2
Total no. electrons around Xe	**10**
5 electron pairs around Xe	
2 bonding pairs and 3 lone pairs	

You can construct a partial MO energy level scheme for XeF_2 by just using the p_z orbitals on the three atoms, where the *z*-axis is defined as the F–Xe–F vector. This approach is similar to that used to construct the MO energy level diagram for BeH_2. Firstly, the fluorine orbitals are treated together, and then the resulting combinations are interacted with the orbitals on the central xenon atom.

The first step involves taking combinations of the fluorine p_z orbitals as shown in Figure 5.41. There are two combinations—in phase and out of phase—and in F_2 these would correspond to the σ_g and σ_u orbitals, respectively. In XeF_2, the presence of the central xenon atom means there is a much greater distance between the fluorine atoms than in F_2. Consequently, the overlap between the fluorine orbitals in XeF_2 is negligible, and the in-phase and out-of-phase combinations are of approximately the same energy.

How do these combinations of fluorine p_z orbitals interact with the xenon p_z orbital? The in-phase combination has the wrong symmetry to interact with the xenon p_z orbital, as shown in Figure 5.42(a). Any bonding interaction with the left-hand fluorine atom is counterbalanced by an antibonding interaction with the right-hand fluorine atom. This combination of fluorine orbitals does not interact with the xenon atom and is therefore non-bonding.

In contrast, the out-of-phase combination has the correct symmetry to interact with the Xe p_z orbital, as shown in Figure 5.42(b). This gives rise to both a bonding and an antibonding orbital, depending on whether the xenon orbital is in phase or out of phase with the fluorine orbitals.

Overall, the three p_z orbitals (on Xe and the two F atoms) interact to give a bonding orbital, a non-bonding orbital, and an antibonding orbital. Both the bonding and antibonding combinations have the symmetry label σ_u, whereas the non-bonding orbital (which is just the in-phase combination of fluorine orbitals) has the symmetry label σ_g. The relative energies of these orbitals are shown on the partial MO energy level diagram in Figure 5.43.

Since there are two electrons in the xenon p_z orbital and one each in the fluorine p_z orbitals, there are four electrons to place into the molecular orbitals. This means that the bonding and non-bonding orbitals are filled but the antibonding orbital is empty. This bonding pattern is described as a **3-centre 4-electron bond**. The extra two electrons in the expanded octet are located on the fluorine atoms as the filled σ_g non-bonding orbital has no xenon contribution. This enables molecular orbital theory to explain hypervalency without expanding the octet of the central atom.

ⓘ Some books call these interactions 3-centre 2-electron bonds instead of 3-centre 4-electron bonds as only two of the four electrons contribute to the bonding.

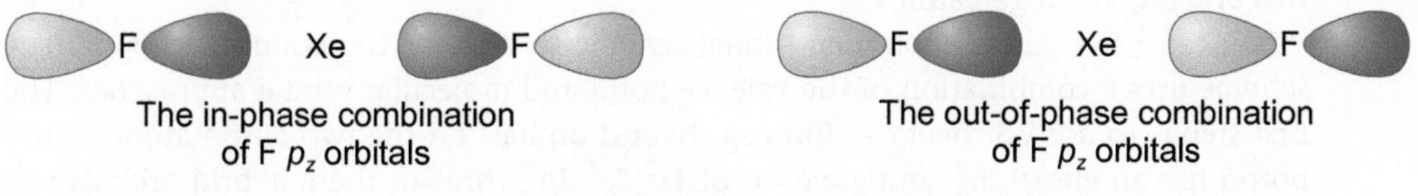
F
Xe
F
The in-phase combination of F p_z orbitals
F
Xe
F
The out-of-phase combination of F p_z orbitals

Figure 5.41 The in-phase and out-of-phase combinations of fluorine $2p_z$ orbitals in XeF_2.

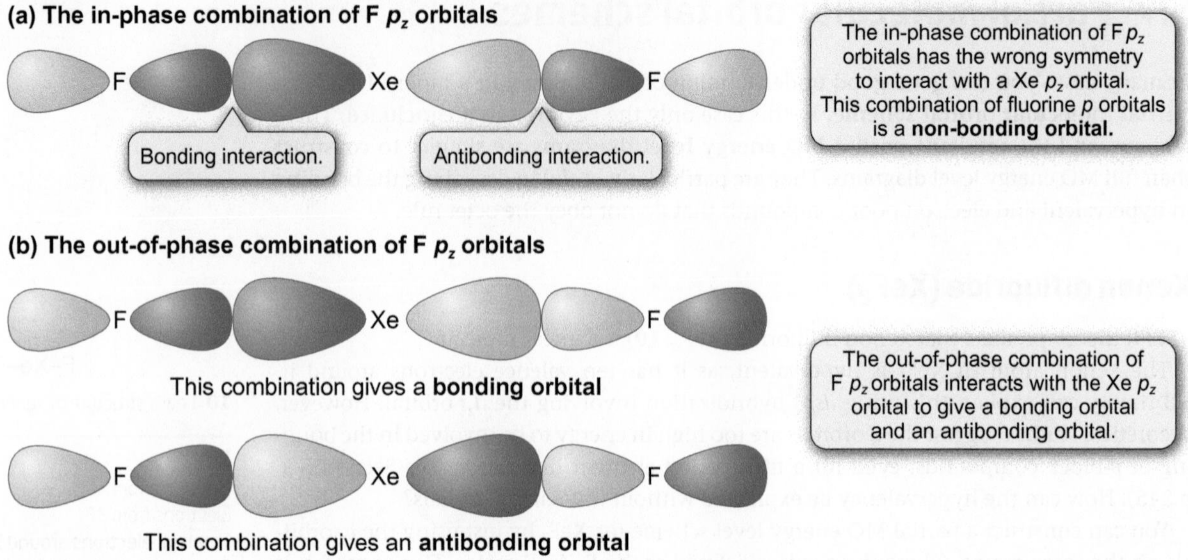
(a) The in-phase combination of F p_z orbitals
F
Xe
F
Bonding interaction.
Antibonding interaction.
The in-phase combination of F p_z orbitals has the wrong symmetry to interact with a Xe p_z orbital. This combination of fluorine p orbitals is a **non-bonding orbital.**
(b) The out-of-phase combination of F p_z orbitals
This combination gives a **bonding orbital**
The out-of-phase combination of F p_z orbitals interacts with the Xe p_z orbital to give a bonding orbital and an antibonding orbital.
This combination gives an **antibonding orbital**

Figure 5.42 The interactions between the combinations of fluorine $2p_z$ orbitals and the Xe $5p_z$ orbital. (a) The in-phase combination of F $2p_z$ orbitals and (b) the out-of-phase combination of F $2p_z$ orbitals.

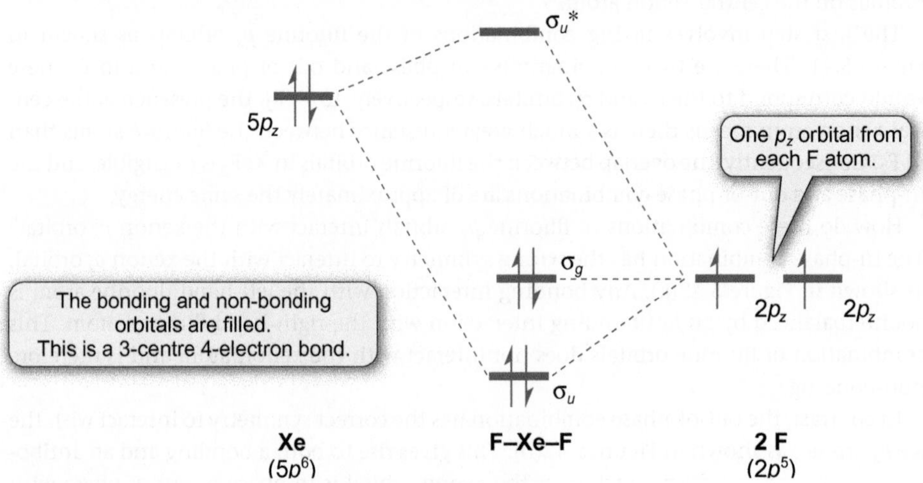
σ_u*
$5p_z$
One p_z orbital from each F atom.
σ_g
$2p_z$
$2p_z$
The bonding and non-bonding orbitals are filled. This is a 3-centre 4-electron bond.
σ_u
Xe
($5p^6$)
F–Xe–F
2 F
($2p^5$)

Figure 5.43 A partial molecular orbital energy level diagram for XeF_2.

Visit the Online Resource Centre to view screencast 5.3 which walks you through the bonding in a XeF_2 molecule.

Since there is only one filled bonding orbital, the overall bond order is one, but this is spread over two Xe–F bonds. The bond order for each individual bond is therefore $\frac{1}{2}$. 'Normal' bonds are 2-centre 2-electron bonds, so a 3-centre 4-electron bond provides a means of incorporating additional electrons within a molecule.

Diborane (B_2H_6)

Borane (BH_3) has only six electrons in the boron valence shell, so by VSEPR theory you would predict it to be trigonal planar. At room temperature, borane doesn't exist as simple BH_3 molecules, but it forms a dimer, diborane, with the formula B_2H_6 (**11**). Diborane has two bridging hydrogen atoms.

You can use a partial molecular orbital scheme to describe the bonding in B_2H_6. The scheme uses a combination of the valence bond and molecular orbital approaches. The first step is to use VB theory to form sp^3 hybrid orbitals on the two boron atoms. Since boron has an electronic configuration of $1s^2\ 2s^2\ 2p^1$, three of these hybrid orbitals will contain an electron while the fourth will be empty. Two of the hybrid orbitals interact

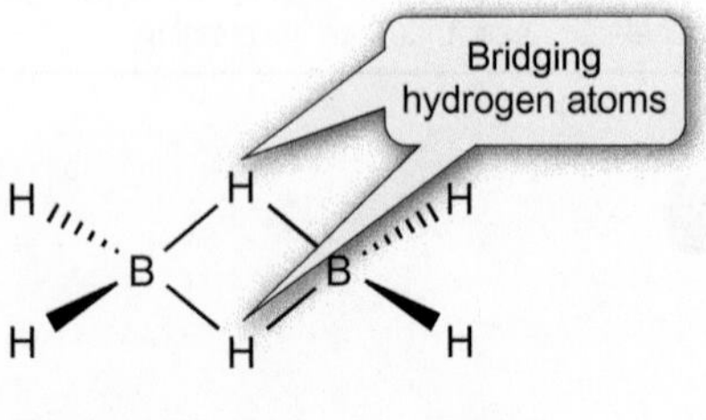
Bridging hydrogen atoms
H
B
H
H
B
H
H
H

11 The structure of diborane (B_2H_6).

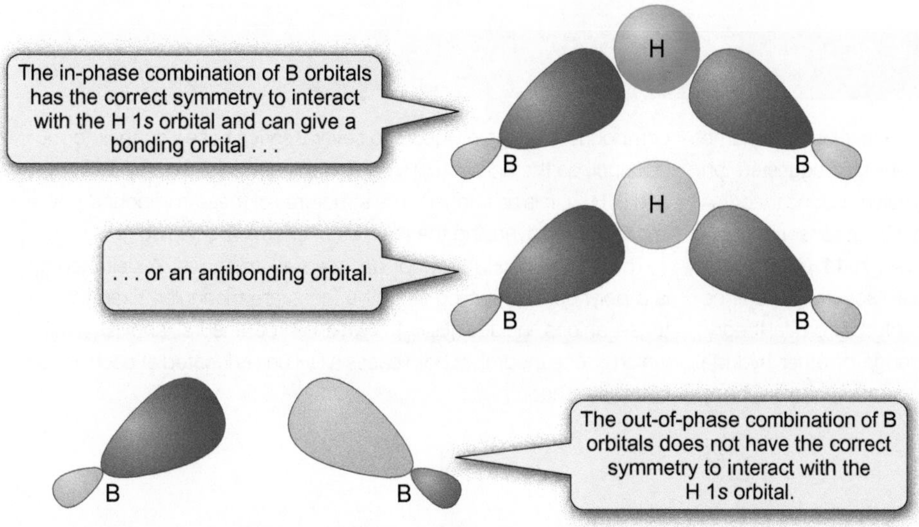
The in-phase combination of B orbitals has the correct symmetry to interact with the H 1s orbital and can give a bonding orbital . . .
H
B
B
. . . or an antibonding orbital.
H
B
B
B
B
The out-of-phase combination of B orbitals does not have the correct symmetry to interact with the H 1s orbital.

Figure 5.44 The in-phase and out-of-phase combinations of boron sp^3 hybrid orbitals in B_2H_6 and their interactions with the H 1*s* orbital.

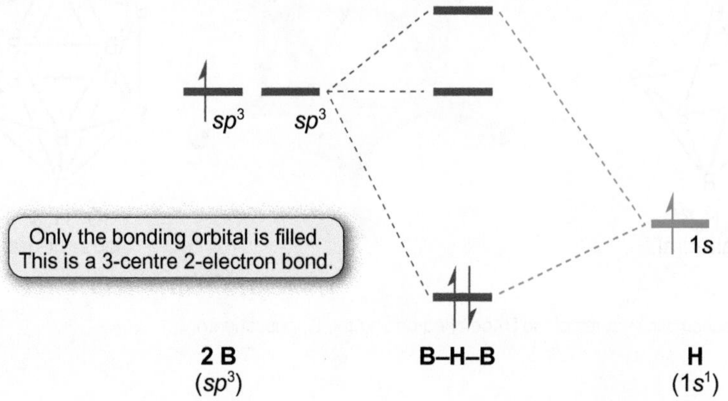
sp3
sp3
Only the bonding orbital is filled. This is a 3-centre 2-electron bond.
1s
2 B
(sp3)
B–H–B
H
(1s1)

Figure 5.45 A partial molecular orbital energy level diagram for the B-H-B bonding in B_2H_6.

with hydrogen 1*s* orbitals to form σ bonds. These contain two electrons, one from the boron atom and one from the hydrogen atom, and are normal 2-centre 2-electron bonds.

This leaves two sp^3 hybrid orbitals on each boron atom, one of which contains a single electron while the other is empty. Both of these hybrid orbitals are involved in three-centre bonding with bridging hydrogen atoms. Using MO theory, the first step is to look at the combinations of the sp^3 hybrid orbitals on the two boron atoms. These boron orbitals can be either in phase or out of phase with each other, as shown in Figure 5.44. As with the fluorine orbitals in XeF_2, the two combinations of boron orbitals are of approximately the same energy as the overlap is small.

The hydrogen 1*s* orbital interacts with the in-phase combination of boron orbitals to give a bonding orbital and an antibonding orbital. However, the 1*s* orbital cannot interact with the out-of-phase combination of boron orbitals since it does not have the correct symmetry. Overall, the three orbitals combine to give a bonding orbital, a non-bonding orbital, and an antibonding orbital, in a similar manner to those in XeF_2. The difference between the 3-centre bonds in B_2H_6 and XeF_2 lies in the occupation of the orbitals, as shown in Figure 5.45.

In B_2H_6 there are only two electrons to put into the scheme, one electron from a boron atom and one from the hydrogen atom. This means that the bonding orbital is filled, while the non-bonding and antibonding orbitals are empty. The overall interaction between the three atoms is a **3-centre 2-electron bond**. There are two of these 3-centre 2-electron bonds in B_2H_6. 3-centre 2-electron bonding is used to describe the bonding in many electron-poor compounds such as the boron hydride clusters described in Box 5.6. Compounds containing 3-centre 2-electron bonds are said to be **electron deficient**.

Visit the Online Resource Centre to view screencast 5.4 which walks you through the bonding in a B_2H_6 molecule.

Box 5.6 Boron hydrides

The structure of B_2H_6 was matter of speculation for some time. The great American chemist, Linus Pauling, suggested an ethane-like structure, and supported this using valence bond theory. A second year undergraduate student at Oxford, Christopher Longuet-Higgins, proposed the bridged structure shown in **11** (p.250) and showed how this was consistent with the infrared spectrum. Later he was the first to rationalize this structure using molecular orbital theory.

In addition to B_2H_6, boron forms a large range of other hydride compounds that adopt cage-like structures called *clusters*. These compounds can be divided into several series based on their formulae, such as the anions $[B_nH_n]^{2-}$, and the neutral compounds B_nH_{n+4} and B_nH_{n+6} (*n* is an integer). The structures of these compounds were important in advancing the theories of chemical bonding.

The anions $[B_nH_n]^{2-}$ adopt *deltahedral* structures. A deltahedron is a polyhedron in which all of the faces are triangular. Examples of these anions include $[B_6H_6]^{2-}$ which is octahedral and $[B_{12}H_{12}]^{2-}$ which is icosahedral. In both cases a BH unit is located at each vertex of the deltahedron.

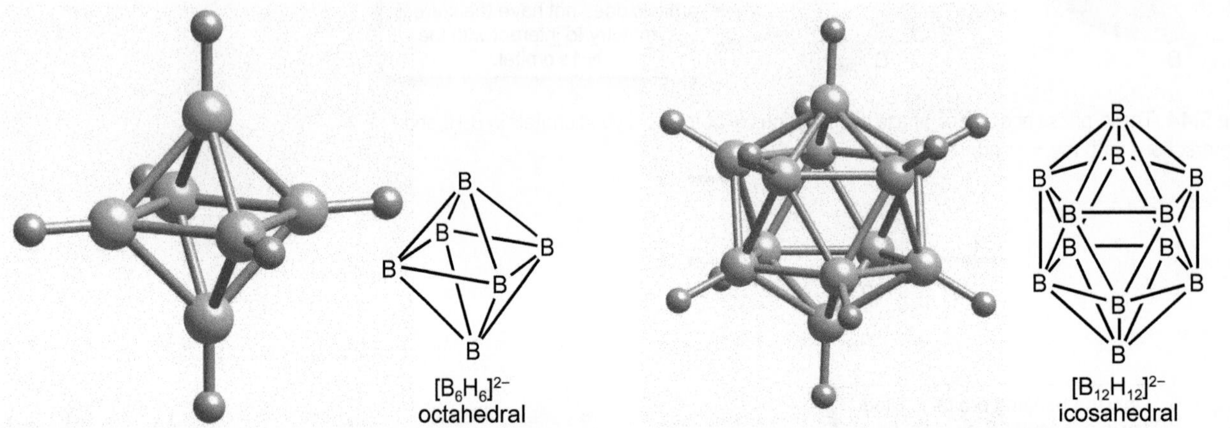
$[B_6H_6]^{2-}$
octahedral
$[B_{12}H_{12}]^{2-}$
icosahedral

▲ The structures of $[B_6H_6]^{2-}$ and $[B_{12}H_{12}]^{2-}$, together with schematic pictures of the B_6 octahedron and the B_{12} icosahedron.

Summary

- Partial molecular orbital energy level diagrams are a useful way of describing the bonding in molecules that break the octet rule.
- The bonding in hypervalent compounds such as XeF_2 can be described using 3-centre 4-electron bonds that consist of a filled bonding orbital and a filled non-bonding orbital.
- The bonding in electron-deficient compounds such as B_2H_6 can be described using 3-centre 2-electron bonds that consist of a filled bonding orbital and an empty non-bonding orbital.

For practice questions on these topics, see questions 16 and 17 at the end of this chapter (p.253).

Concept review

By the end of this chapter, you should be able to do the following.

- Work out the formal charges on the atoms in a molecule or ion and use these to predict which of several possible structures a compound is likely to adopt.
- Predict the shapes of molecules and ions using VSEPR theory.
- Understand the limitations of VSEPR theory.
- Understand the bonding in a compound by predicting the hybridization of the central atom(s), and describe how the hybrid orbitals interact with the orbitals on other atoms.

- Use resonance forms to describe the bonding in compounds that contain bonds that appear to be different in the Lewis structure but are found to be the same experimentally.
- Describe the bonding in a linear triatomic molecule such as BeH_2 using MO theory.
- Construct partial MO energy level diagrams to explain the bonding in hypervalent compounds such as XeF_2, and describe the bonding in terms of a 3-centre 4-electron bond.
- Construct partial MO energy level diagrams to explain the bonding in electron-deficient compounds such as B_2H_6, and describe the bonding in terms of a 3-centre 2-electron bond.

? Questions

More challenging questions are indicated by an asterisk *.

1. Draw Lewis structures for the following compounds.

BF_3 SiH_4 CS_2 PCl_3 PF_5 KrF_2

For each molecule, indicate whether the central atom obeys the octet rule, is hypervalent, or is electron poor. (Section 5.1)

2. Three possible resonance forms for the cyanate anion are shown below. Work out the formal charge on each of the atoms in these three structures. Which is the most important resonance form? (Section 5.1)

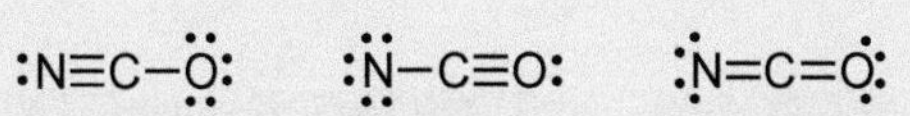

Structure A Structure B Structure C

3. Use VSEPR theory to predict shapes for the following molecules and ions: (a) BF_3; (b) XeF_5^-; (c) SbF_5; (d) I_3^-; (e) ICl_4^-; (f) SF_6. (Section 5.2)

4. Use VSEPR theory to predict shapes for the following molecules: (a) SO_2; (b) SO_3; (c) F_3SN; (d) O_3; (e) $XeOF_4$; (f) N_2H_2; (g) SeF_4. (Section 5.2)

5. Use VSEPR theory to predict shapes for the following oxoanions: (a) ClO_3^-; (b) ClO_4^-; (c) ClO_2^-; (d) $S_2O_3^{2-}$ (see below). (Section 5.2)

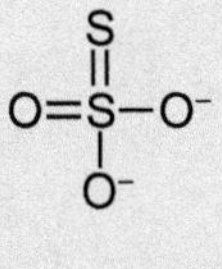

6. Place the following species in order of increasing bond angle: (a) NO_2; (b) NO_2^+; (c) NO_2^-. (Section 5.2)

7. Which of the molecules and ions in Question 3 have dipoles? (Section 5.3)

8. Three isomers of difluoroethene ($C_2H_2F_2$) are shown below. Which of these compounds are polar? (Section 5.3)

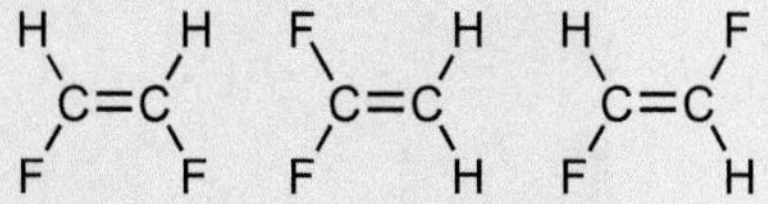

9.* The dipole moments for some phosphorus halides are given below. Account for the values, and predict the direction of the dipole where present. (Section 5.3)

PF_3, 1.03 D PCl_3, 0.56 D PF_5, 0.0 D

10 Identify the type of hybridization present in the non-hydrogen atoms of the following molecules. (Section 5.4)

(a) 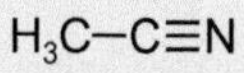 (b) 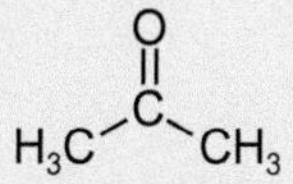 (c) 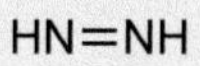

11 The structure of caffeine is shown in Box 5.1 (p.218). Identify the hybridization present in all of the non-hydrogen atoms in this molecule. (Section 5.4)

12* The bond angle in H_2S is 92°, much less than that in H_2O. Suggest a reason for this. Would you expect H_2S to be a stronger or weaker base than H_2O? (Section 5.4)

13 The four Cl–O bonds in the perchlorate anion (ClO_4^-) have identical bond lengths. Draw resonance forms to explain this. What is the average charge on each oxygen atom? (Section 5.5)

14 In the ethanoate anion ($CH_3CO_2^-$) the C–O bonds have been shown experimentally to have the same bond length (125 pm). This differs from ethanoic acid (CH_3CO_2H), which contains two different carbon-oxygen distances (122 pm, 132 pm). Draw resonance forms for the ethanoate anion consistent with this observation and determine the average C–O bond order. (Section 5.5)

15 In both the sulfate (SO_4^{2-}) and sulfite (SO_3^{2-}) anions, the S–O bond lengths are all the same. Draw resonance forms for the ions to support this. By considering the average S–O bond order in each case, predict which ion has the shorter S–O bond lengths. (Section 5.5)

16* The Xe–F bond length in XeF_2 is 198 pm, whereas the equivalent distance in the XeF^+ ion is 187 pm. Rationalize this difference by considering the bonding in the two species. (Section 5.7)

17* For the linear anion ICl_2^-, draw in-phase and out-of-phase combinations of the chlorine p_z orbitals. Determine how these interact with the iodine p_z orbital and use this information to construct a partial MO energy level diagram for ICl_2^-. What is the bond order of each I–Cl bond? (Section 5.7)

6

Solids

This chapter builds on the following topics:

- **Energy changes in chemical reactions** Section 1.6, p.41
- **Atomic properties and periodicity** Section 3.7, p.153
- **Valence bond theory for polyatomic molecules** Section 5.4, p.236

◀ The structure of a synthetic zeolite (ZSM-5) that is used industrially to catalyse hydrocarbon isomerization reactions. It has a high Si:Al ratio, and contains strongly acidic H^+ ions. Notice the open structure, showing pores and channels.

Zeolites

In 1756, Baron Axel Cronstedt, a Swedish amateur geologist, discovered a mineral that bubbled and gave off clouds of steam when heated strongly. He called this strange mineral 'zeolite', which is Greek for 'boiling stone'. Many different zeolites are now known, some naturally occurring and some synthetically produced. Zeolites have transformed industrial chemistry because of their unusual structures and properties, which are responsible for their use as catalysts, drying agents, and ion exchangers.

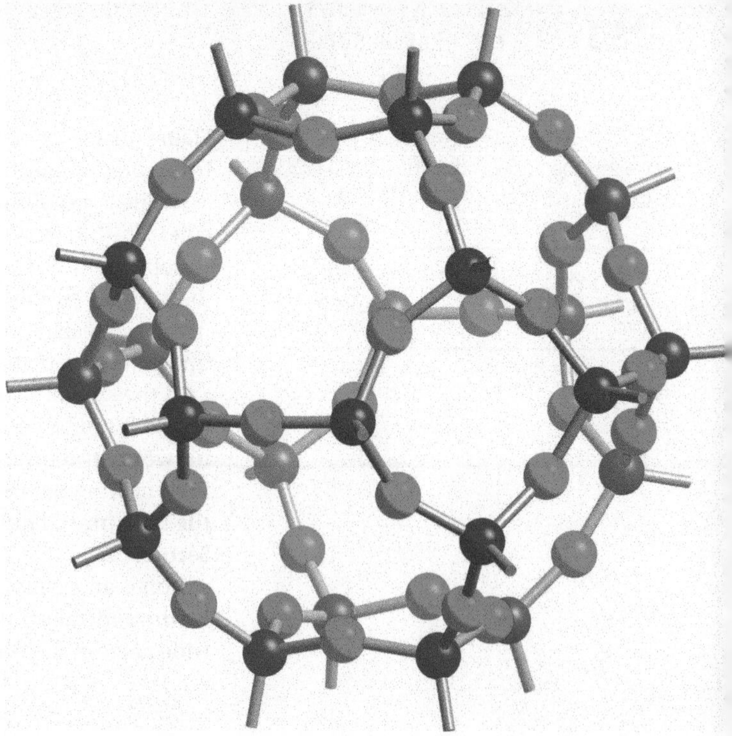

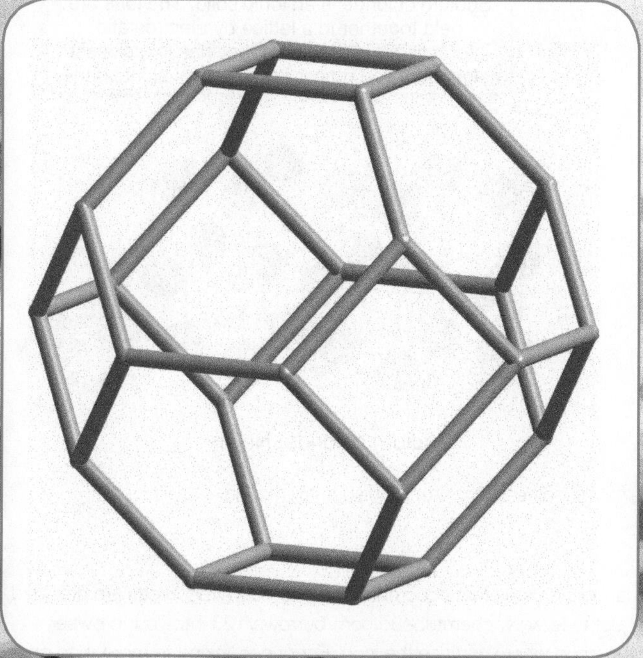

▲ The sodalite cage occurs in many zeolite structures. It is shown with the atoms above and in a simplified form below, with the vertices representing the Si/Al atoms and the oxygen atoms lying along each edge. The simplified form is a polyhedron containing six 4-membered rings and eight 6-membered rings.

Zeolites are *aluminosilicates* and have the general formula $M_xAl_xSi_{1-x}O_2 \cdot yH_2O$ where M^+ is either a metal ion or H^+, x has a value between 0 and 1, and y is the number of water molecules in the formula unit. There is a wide range of possible structures, but all have in common the linking together of SiO_4 and AlO_4 tetrahedra through their corner oxygen atoms to form covalent network structures. Many zeolites have open structures containing extensive networks of interlocking pores and channels. The water molecules lie in the pores and are released on heating, which explains the bubbling and steam observed by Cronstedt. The porous nature of zeolites and the presence of replaceable M^+ ions on their internal surfaces are the key factors in their applications.

When zeolites are used as *catalysts*, M^+ is normally H^+, and the zeolite acts as a strong solid acid. The cracking of crude oil—breaking long-chain alkane molecules into smaller, more useful ones—and the conversion of alcohols to hydrocarbons are two examples of industrial processes that use zeolites as catalysts. In the *methanol to gasoline process*, a stream of methanol gas passes over the catalyst and a dehydration–polymerization reaction takes place inside the pores. Analysis of the products shows a sharp cut-off at fractions with 11 carbon atoms, as this is the largest hydrocarbon that can fit in the pores. Zeolite catalysts are said to be 'shape-selective' as only molecules of specific sizes and shapes can fit into the cavities and react.

Zeolites that have been heated to remove the water molecules can re-absorb water into the pores, so they are commonly used as drying agents. In laboratories, you may have seen zeolites labelled as *molecular sieves*. They can absorb a wide range of other molecules as well as water. Zeolites are used as oil absorbents, and their ability to absorb ammonia and organic molecules explains why they are used in cat litter.

Another major use of zeolites is for *ion exchange*. In this case, M^+ is normally Na^+. The most common application for these ion-exchange properties is in washing powders to enable the powder to work well in hard water areas. The role of the zeolites is to remove Ca^{2+} and Mg^{2+} ions, which would otherwise precipitate in the form of salts from solution. Instead, the Ca^{2+} and Mg^{2+} ions are retained in the zeolites, releasing Na^+ ions into solution. A similar ion-exchange process occurs when zeolites are used in the treatment of nuclear waste. In this case they remove radioactive $^{90}Sr^{2+}$ ions.

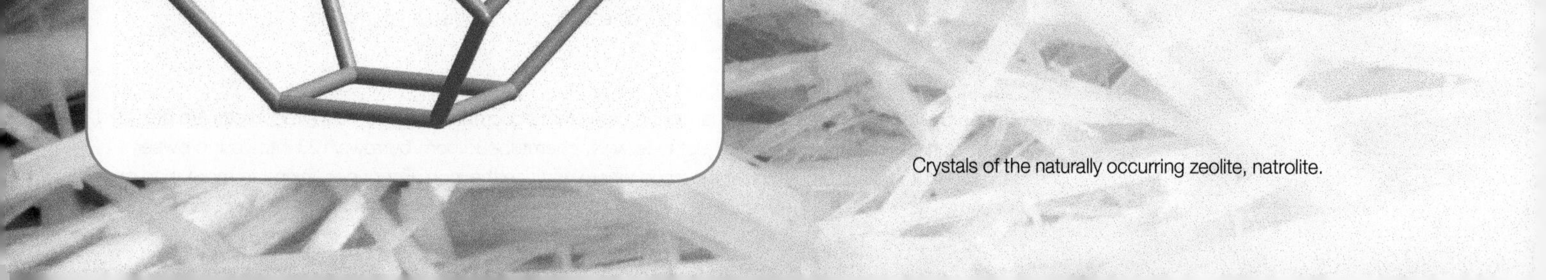

Crystals of the naturally occurring zeolite, natrolite.

The origins of intermolecular interactions are described in Sections 1.8 (p.52) and 17.3 (p.783).

Sublimation is the change of a solid directly to a vapour without passing through the liquid state. See Section 1.7 (p.49).

Figure 6.1 shows four examples of different types of solid state structures—iodine, silicon, iron, and sodium chloride.

Iodine is a **molecular solid.** Covalent bonds between the iodine atoms lead to I_2 molecules. The interactions *between* these molecules are much weaker than the covalent bonds, so solid iodine easily sublimes to form iodine vapour. If you gently warm a few crystals of iodine in a test tube, you will see the purple iodine vapour above the solid. Sublimation involves breaking only the intermolecular interactions; the I–I covalent bonds remain intact in the vapour, which contains I_2 molecules.

In the structure of silicon, each atom is in a tetrahedral geometry and forms covalent bonds to four other silicon atoms. This leads to an infinite structure known as a **covalent network structure.** Covalent network structures like that of silicon typically have very high melting points and boiling points, as covalent bonds need to be broken to convert the solid into a liquid or gas. Covalent network structures are described in more detail in Section 6.1.

Iron is an example of a **metal**. In a metal, the valence electrons are delocalized, which is why a metal conducts electricity when a potential difference is applied. Metals have a wide range of melting points. Most are solids at room temperature, although one metallic

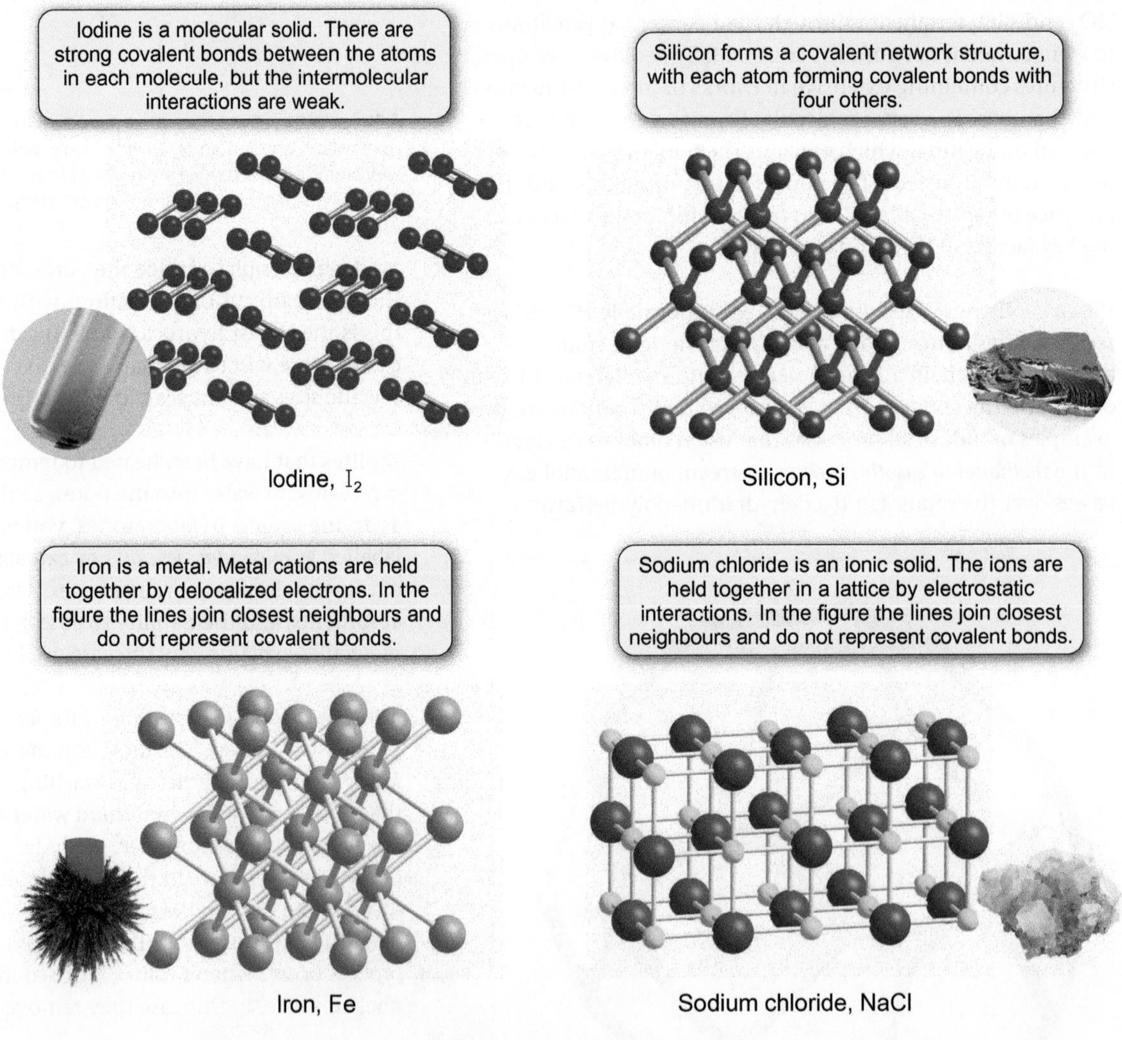
Iodine is a molecular solid. There are strong covalent bonds between the atoms in each molecule, but the intermolecular interactions are weak.
Silicon forms a covalent network structure, with each atom forming covalent bonds with four others.
Iodine, I_2
Silicon, Si
Iron is a metal. Metal cations are held together by delocalized electrons. In the figure the lines join closest neighbours and do not represent covalent bonds.
Sodium chloride is an ionic solid. The ions are held together in a lattice by electrostatic interactions. In the figure the lines join closest neighbours and do not represent covalent bonds.
Iron, Fe
Sodium chloride, NaCl

Figure 6.1 The structures of the solids iodine, silicon, iron, and sodium chloride.

Online Support. The icon in the margin indicates that accompanying interactive resources are provided online to help your understanding: just type www.chemtube3d.com/burrows/123 into your browser, replacing 123 with the number of the page where you see the icon. For pages linking to more than one resource, type 123-1, 123-2, etc. (replacing 123 with the page number) for access to successive links.

Table 6.1 Properties of substances with different types of bonding

	Examples	Electrical conductivity	Melting point and boiling point	Physical properties	Solubility
Molecular structures	Iodine (I_2) Water (H_2O) Carbon dioxide (CO_2)	Insulators	Low	Gases, liquids, or solids. If solid, soft and often brittle	Some compounds dissolve in water. Many dissolve in organic solvents
Covalent network structures	Carbon (C, both diamond and graphite) Silicon dioxide (SiO_2) Boron nitride (BN)	Normally insulators	Very high	Hard and brittle solids	Generally insoluble in water and organic solvents
Metals	Iron (Fe) Copper (Cu) Gold (Au)	Good conductors when solid or liquid	Generally high	Hard solids, but malleable and ductile	Insoluble in water (though some react) and organic solvents
Ionic compounds	Sodium chloride (NaCl) Magnesium oxide (MgO) Calcium fluoride (CaF_2)	Conduct when molten or in aqueous solution	High	Hard and brittle solids	Some compounds dissolve in water. Insoluble in organic solvents

element—mercury—is a liquid at 25 °C. The structure and bonding in metals is described in Sections 6.2 and 6.3.

Sodium chloride (rock salt, NaCl) is an **ionic solid.** When sodium metal reacts with chlorine gas a redox process occurs, and electrons are transferred from sodium to chlorine to give Na^+Cl^-. **Electrostatic interactions** (also known as Coulombic interactions) between the oppositely charged ions hold an ionic solid together. Ionic compounds generally have high melting points and boiling points. They conduct electricity in solution and the liquid phase, but not normally in the solid state. Ionic bonding is described in more detail in Sections 6.5 and 6.6.

The formation of NaCl from Na and Cl_2 is illustrated in Figure 1.6 (p.12).

Some of the characteristic properties of molecular solids, covalent network structures, metals, and ionic solids are summarized in Table 6.1. In this chapter, you will look more closely at covalent, metallic, and ionic bonding in solid state structures. This is important as the properties of solid state materials depend on their structures and bonding. Weaker intermolecular interactions are described in Chapter 17.

The structures shown in Figure 6.1 are all examples of **crystalline** materials. Crystalline compounds have very regular structures and, as you will see later in this chapter, these can often be built up from relatively small and simple repeating units. For example, all of the atoms in a piece of silicon have the same tetrahedral geometry, and lie at the same distance from their neighbours. A crystalline solid has **long-range order** as it is possible to predict the positions of every atom from those of only a few. A regular extended structure is called a **lattice.**

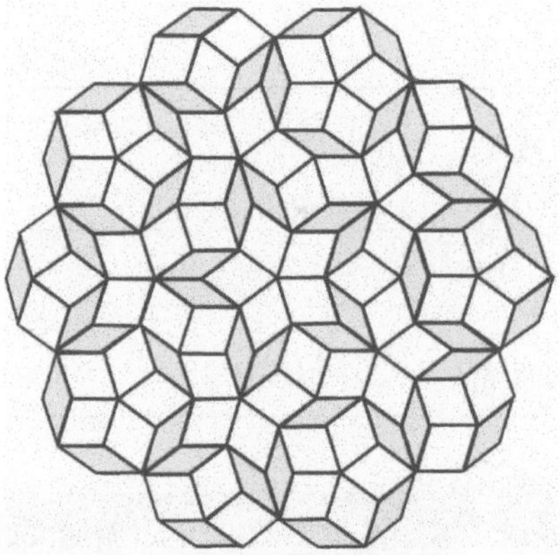

There are examples of structures which are ordered but do not have long-range order, in a similar manner to the Penrose tiling pattern shown. These structures, such as the alloy $Al_{63}Cu_{24}Fe_{13}$, are known as **quasicrystals,** and the Israeli scientist Dan Schectman was awarded the Nobel Prize in Chemistry in 2011 for his discovery of quasicrystals.

6.1 Covalent network structures

Covalent network structures are infinite structures in which the atoms are linked by covalent bonds. This type of structure is found in many non-metal elements and their compounds. Examples include the elements carbon, boron, and phosphorus, and the compounds boron nitride (BN) and silicon dioxide (SiO_2).

Chapter 27 includes a discussion of the structures of carbon (p.1216), boron (p.1210), phosphorus (p.1224), and boron nitride (p.1215). Silicon dioxide is described on p.259.

Allotropy

Some elements can form more than one structure. Diamond, graphite, and buckminsterfullerene (C_{60}) are all forms of carbon. Different structural forms of the same *element* are known as **allotropes.** The structures of these three allotropes of carbon are shown in Figure 6.2.

Interactive 3D rotatable figures of diamond and graphite.

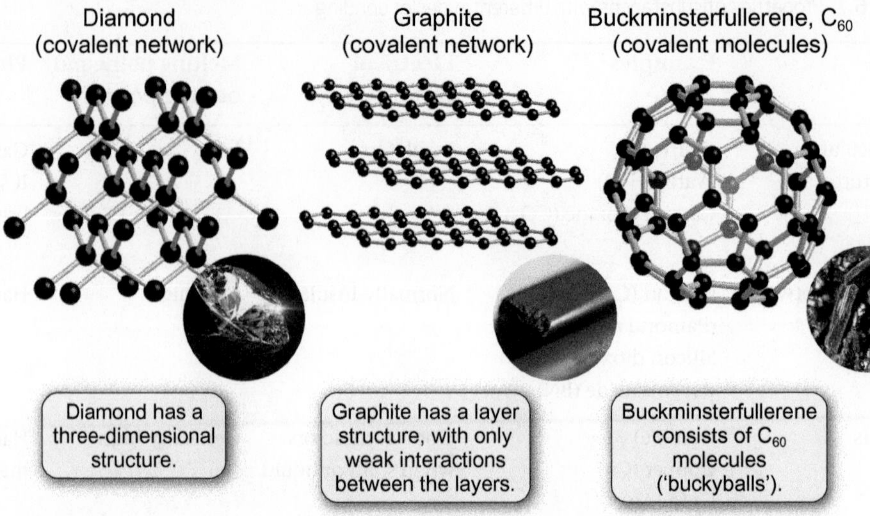
Diamond
(covalent network)
Graphite
(covalent network)
Buckminsterfullerene, C_{60}
(covalent molecules)
Diamond has a three-dimensional structure.
Graphite has a layer structure, with only weak interactions between the layers.
Buckminsterfullerene consists of C_{60} molecules ('buckyballs').

Figure 6.2 Diamond, graphite, and buckminsterfullerene are all allotropes of carbon.

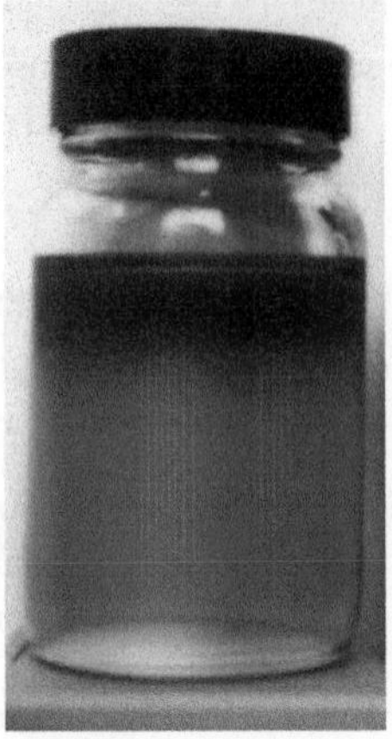

Buckminsterfullerene dissolves in benzene to form a red solution. The fact that it is soluble is good evidence for its molecular nature.

Hybridization of carbon is described in Section 5.4 (p.236).

Unlike diamond, which is an excellent electrical insulator, graphite is a good conductor of electricity. This is because the delocalized π system forms a partially filled band, so electrons are able to move within it. This means that conduction is much higher within the plane of the layers than between layers. The origin of bands is described in Section 6.3 (p.274).

Diamond and graphite both form covalent network structures, whereas buckminsterfullerene consists of discrete C_{60} molecules that form a molecular solid. Diamond and graphite have very different properties, and these can be explained by looking at the structures. In diamond each carbon atom has a tetrahedral environment. The C–C bonds are formed from the interaction of sp^3 hybrid orbitals on the carbon atoms. Any distortion of a diamond involves deforming or breaking very strong C–C bonds, so diamond is one of the hardest materials known.

Graphite forms a layer structure, with strong C–C bonds *within* the layers, but relatively weak interactions *between* the layers. The carbon atoms are in a trigonal planar environment, which is consistent with sp^2 hybridization. Interactions between the sp^2 hybrid orbitals lead to the C–C bonds. The unhybridized p orbitals interact to form a π system that is delocalized over the whole layer. The interactions between the layers are weakened in the presence of water molecules so that it is easy to cleave graphite, and for this reason graphite is used as a lubricant and as the 'lead' in pencils. When you write with a pencil, you are breaking the interlayer attractions and leaving graphite layers on the paper. Pencil marks are easy to remove from paper with a rubber because the layers do not bind strongly to the paper. In the absence of molecules such as H_2O, the interactions between the layers are stronger, and as a consequence graphite cannot be used as a lubricant on satellites.

Box 6.1 Graphene, nanotubes, and nanotechnology

Graphene and nanotubes are other, more recently discovered, allotropes of carbon. Graphene consists of single sheets of graphite. Although scientists had been attempting to make graphene for many years, it was first prepared successfully by the Russian-born scientists Andre Geim and Konstantin Novoselov in 2004. This work, carried out at the University of Manchester, UK, led to them receiving the Nobel Prize in Physics in 2010. Nanotubes were discovered 13 years earlier by the Japanese scientist Sumio Iijima. Nanotubes consist of graphene sheets that are rolled up and connected into cylinders. The diameter of the cylinders is typically 5 nm–15 nm, which is why they are called nanotubes. Nanotubes can be open at the ends, but are usually capped by combinations of 5- and 6-membered rings of carbon atoms, which are similar to half buckminsterfullerene molecules. Nanotubes can either have a single cylindrical wall, or have multiple walls.

Both graphene and nanotubes have interesting properties. Like graphite, they are electrical conductors. Indeed, graphene has very high electrical conductivity, with the electrons moving faster than in any other material, and could find application in high speed transistors. Graphene is transparent, and may in the future be used in transparent flexible touchscreens. Nanotubes have been used as ultra-thin molecular wires, and researchers at NASA have reported a method for producing integrated circuits using nanotubes instead of copper to connect the components together.

Other potential applications of nanotubes arise from their hollow nature. Scientists have inserted molecules of biological interest into nanotubes. This leads to the possibility of these materials being used as tiny needles, able to inject drug molecules into a single cell. This is an example of **nanotechnology**, which is a loose term used to describe any technological development on the nanometre scale. →

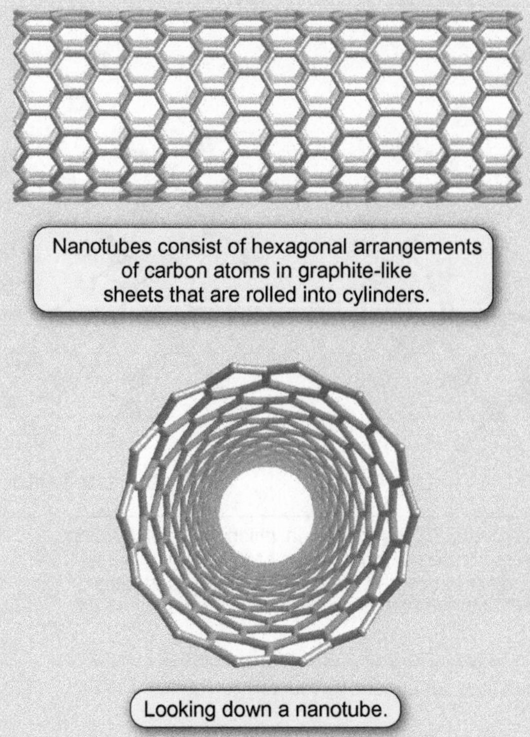
Nanotubes consist of hexagonal arrangements of carbon atoms in graphite-like sheets that are rolled into cylinders.
Looking down a nanotube.

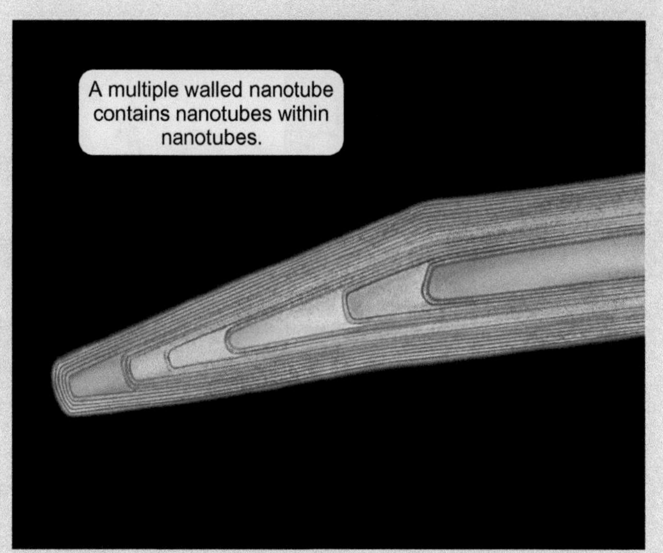
A multiple walled nanotube contains nanotubes within nanotubes.

▲ A coloured electron microscope image of a multiple wall nanotube.

◄ The structure of an open-ended nanotube.

Question

The photograph of a nanotube shown above was obtained using a form of electron microscopy. Suggest why it is not possible to see nanotubes using a conventional optical microscope.

Tin was used in medieval Europe to make cathedral organ pipes but, in very cold winters, wart-like structures were observed on the surface of the metal. This was called 'tin pest'. At the time, tin pest was blamed on the devil, but we now know that, below 13 °C, tin slowly transforms from a metallic allotrope (β-tin) to a non-metallic allotrope (α-tin). Since the conversion involves an increase in volume of 27%, the metal crumbles into powder as it changes phase. Nowadays, tin pest is prevented by alloying tin with antimony or bismuth.

Allotropy is very common, with many *p*-block elements, such as sulfur, phosphorus, and tin, having more than one crystal structure. Each different structural form of an element or compound is considered to be a different **phase** (see Section 1.7, p.47). The conversion of one phase to another is called a **phase transition.**

The structures of the *p*-block elements are described in Chapter 27 and the allotropes of sulfur are shown in Figure 1.21 (p.48).

Silicon dioxide

Silicon dioxide (silica, SiO_2) forms covalent network structures containing Si–O single bonds. There are a large number of different structures, most of which contain SiO_4 tetrahedra, but they differ in the way in which these are linked together through the bridging oxygen atoms. Different crystal structures of the same *compound* are known as **polymorphs.** The structures of two SiO_2 polymorphs—quartz and β-cristobalite—are shown in Figure 6.3.

Quartz is a polymorph of SiO_2.

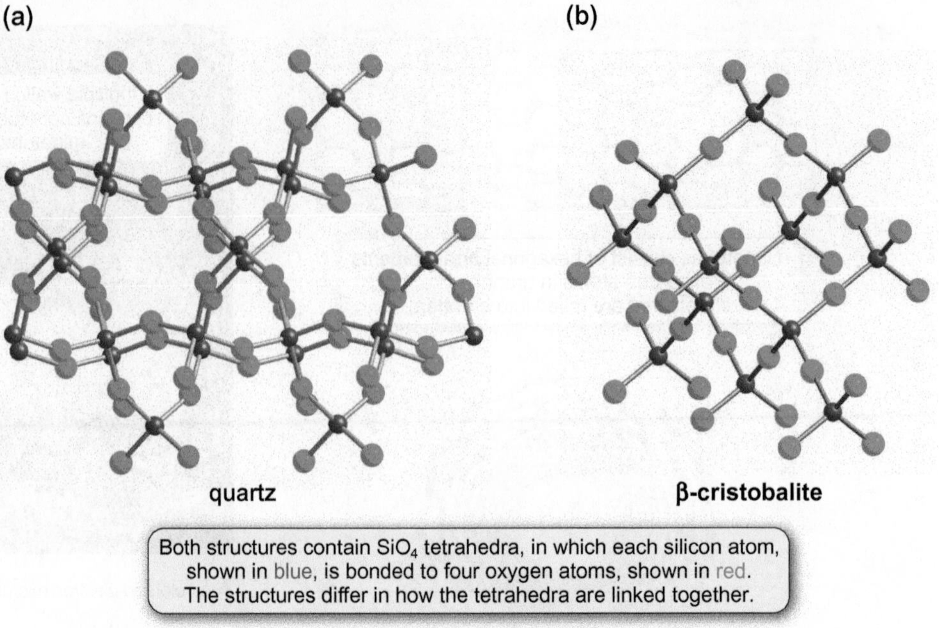
(a)
(b)
quartz
β-cristobalite
Both structures contain SiO_4 tetrahedra, in which each silicon atom, shown in blue, is bonded to four oxygen atoms, shown in red. The structures differ in how the tetrahedra are linked together.

Figure 6.3 The structures of (a) quartz and (b) β-cristobalite. These are polymorphs of SiO_2, with the silicon atoms all in tetrahedral geometries surrounded by four oxygen atoms.

Quartz is the most common form of SiO_2, and also the most common mineral on the surface of the Earth. It is found in most geological environments and is a component of almost every type of rock. β-cristobalite is formed at high temperature and pressure, and has a structure related to that of diamond. If all of the SiO_4 tetrahedra in Figure 6.3(b) were replaced by carbon atoms, the diamond structure shown in Figure 6.2 (p.258) would result.

Ceramics and glasses

The word 'ceramic' comes from the Greek term for 'pottery', so the term is commonly used to describe materials fashioned from clay at room temperature and hardened by heat. More generally, ceramics are compounds formed between metallic and non-metallic elements, often oxides, and typically have covalent network structures, often with ions enclosed in the network. Examples of ceramics include structural clay products, such as bricks, porcelain, and cements. Most ceramics are crystalline materials characterized by hardness, wear-resistance, and brittleness. Ceramics are usually good thermal and electrical insulators.

Laboratory glassware is normally made from borosilicate glass, such as Pyrex™, as this does not expand much when heated or cooled.

Sand is made up from small grains of quartz. When sand is heated until it melts and then cooled rapidly it forms an amorphous structure known as a glass. Unlike the structure of quartz in Figure 6.3, **amorphous** solids do not have long-range order, so they are structurally similar to liquids. Although each silicon atom in glass is in a tetrahedral environment, the distances between the silicon atoms are not all the same, so you cannot predict the position of one atom from that of another.

The glass formed from cooling quartz has excellent optical properties. It is used to make lenses, but, unfortunately, this glass is very brittle so easily broken. For most uses, sodium carbonate is added to the molten silica before it cools. This gives a tougher material, and it is this glass that is used to make bottles, jars, and windows. Addition of boric acid (H_3BO_3) leads to boron atoms being included into the glass. The borosilicate glasses that result have very low thermal expansions, so they do not expand very much when heated or cooled. This means that borosilicate glass is less likely to crack when its temperature changes quickly. Most laboratory glassware is made from borosilicate glass, which is better known as Pyrex™.

Box 6.2 Superconductors

The conductivity of a metal generally decreases as the temperature is lowered. In 1911, the Dutch physicist Heike Kamerlingh Onnes made a surprising discovery. He found that, at very low temperatures, some metals lost *all* their electrical resistance. He called this property **superconductivity**. In a superconductor, electrical current flows indefinitely, with no loss of energy. The temperature below which a metal becomes superconducting is known as its critical temperature. The metal with the highest known critical temperature is a niobium–germanium alloy, for which the critical temperature is 23 K (−250 °C). Metals such as this are superconducting in liquid helium. The range of applications for superconducting metals is limited to situations where the high cost of cooling is outweighed by energy saving. Examples include coils for producing high magnetic fields in NMR spectrometers and magnetic resonance imaging (MRI) instruments in hospitals.

▲ The Meissner effect. The small magnet is floating above a liquid nitrogen cooled sample of the superconductor $YBa_2Cu_3O_7$.

In the 1950s, John Bardeen, Leon Cooper, and Robert Schrieffer proposed an explanation for superconductivity. They suggested that, when superconductors conduct electricity, a passing electron attracts the nuclei of the atoms in the lattice, causing a slight ripple as the atom is displaced towards the electron. The local distortion of the lattice is rich in positive charge and a second electron is indirectly attracted to the first to form a 'Cooper pair'. These Cooper pairs travel through the metal lattice together, and are less affected by vibrations in the lattice than single electrons—hence the lack of resistance.

In 1986, a new class of high temperature superconductors was discovered. These ceramic materials include $YBa_2Cu_3O_7$, which has a critical temperature of 93 K (−180 °C). Unlike the superconducting metals, this material is superconducting in liquid nitrogen, which has a boiling point of 77 K (−196 °C) and is much cheaper than liquid helium. $YBa_2Cu_3O_7$ has a complex layer structure with CuO_2 sheets separated by 'charge reservoirs', and the mechanism of the superconductivity in $YBa_2Cu_3O_7$ and related compounds is still not fully understood.

▲ An experimental Japanese train running on a superconducting magnetic levitation track. One of these trains reached a record speed of 603 km/h in 2015. Superconducting magnets, cooled by liquid helium and liquid nitrogen, are on board the vehicle.

Like other ceramics, $YBa_2Cu_3O_7$ is brittle so it is difficult to form into wires. Recently superconducting power cables based on a Bi–Sr–Ca–Cu–O ceramic have been developed and shown to transmit three times the electricity of conventional copper cables.

▼ Interactions between the electrons and the lattice effectively bind two electrons into a Cooper pair, whose electrons travel through the lattice together.

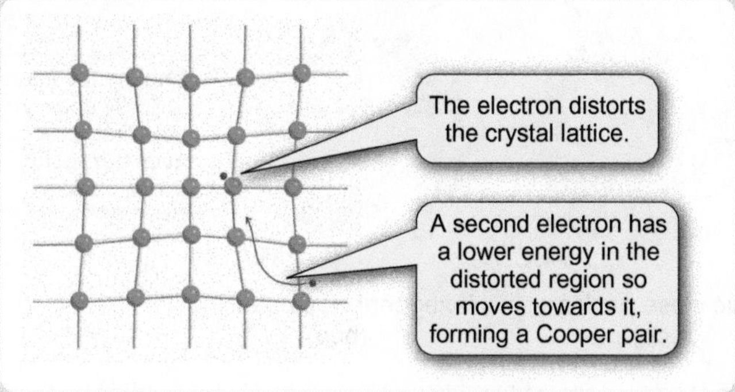
The electron distorts the crystal lattice.
A second electron has a lower energy in the distorted region so moves towards it, forming a Cooper pair.

Another important property of superconductors is their capacity to exclude a magnetic field below their critical temperature. These materials are so strongly non-magnetic (diamagnetic) that they move away from (or repel) a magnetic field. This means that a superconductor can be used to levitate a magnet or vice versa. This is known as the **Meissner effect** after its discoverer Walter Meissner. Research on transport systems that use magnetic levitation and superconductors as a means of reducing friction is currently underway.

Question

A major goal of superconductor research is to prepare a material that is superconducting at room temperature. Why would such a material be important?

Summary

- Covalent network structures are infinite structures in which the atoms are linked by covalent bonds.
- Different structural forms of the same *element* are known as allotropes. Graphite, diamond, buckminsterfullerene (C_{60}), nanotubes, and graphene are all allotropes of carbon.
- Different structural forms of the same *compound* are known as polymorphs. Quartz and β-cristobalite are two polymorphs of SiO_2.
- Each different structural form of an element or compound is a different phase. The conversion of one phase to another is called a phase transition.

▲The plums on this stack are close packed.

Close packing was identified as the most efficient way of packing spheres by the astronomer Johannes Kepler in 1611. Although widely accepted, Kepler's proposal was only proved mathematically in 1998 by Thomas Hales. His proof needed 250 pages of logic.

Visit the Online Resource Centre view screencast 6.1 which walks you through the concept of close packing.

6.2 Structures based on the packing of spheres

Close packing

In solid state structures, it is often helpful to think of the atoms or ions as being simple hard spheres. You cannot place spheres in a box so that all of the space is occupied. There are always some gaps between the spheres. The most efficient way for spheres to pack is that in which this empty space is minimized. This type of packing is called **close packing**, and many solids adopt structures that are based on close packing.

The best way to understand close packing is to start with a single layer of spheres. The most efficient way for these spheres to pack is shown in Figure 6.4(a). Packing in this way

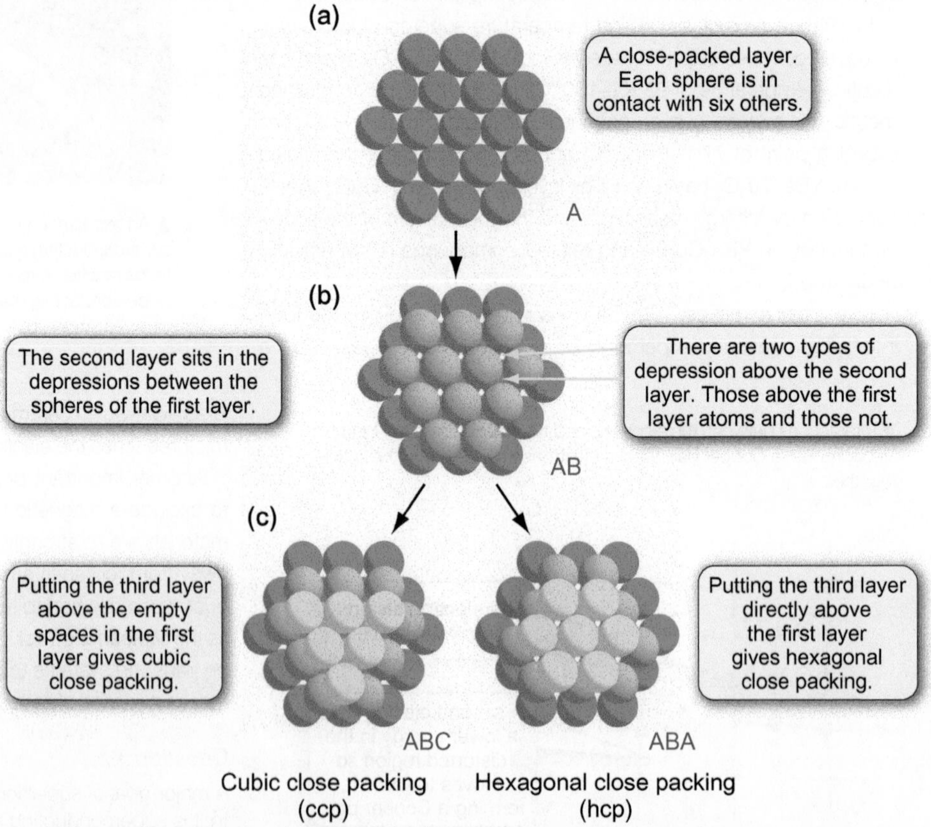
(a)
A close-packed layer. Each sphere is in contact with six others.
A
(b)
The second layer sits in the depressions between the spheres of the first layer.
There are two types of depression above the second layer. Those above the first layer atoms and those not.
AB
(c)
Putting the third layer above the empty spaces in the first layer gives cubic close packing.
Putting the third layer directly above the first layer gives hexagonal close packing.
ABC
ABA
Cubic close packing (ccp)
Hexagonal close packing (hcp)

Figure 6.4 The close packing of spheres can lead to either cubic close packing or hexagonal close packing.

enables each sphere to touch six neighbouring spheres in the layer. The next step is to place a second layer of spheres on top of this layer. To minimize the space between the layers, the new spheres sit in the depressions between the spheres of the first layer. This leads to the arrangement in Figure 6.4(b).

When it comes to placing a third row, there is a complication. This is because not all of the depressions between the second layer spheres are identical. One set of depressions lies directly above the first layer spheres, whereas the other set does not. This means there are two possible positions for the third layer, as shown in Figure 6.4(c). The two arrangements can be described by the positions of the spheres in each layer as ABA and ABC, respectively. In the ABC arrangement the third layer lies in a different position from the first layer, when viewed from above. In the ABA arrangement, the third layer is directly over the first. These two arrangements repeat to give the two most important examples of close packing. The ABCABC arrangement is known as **cubic close packing (ccp)** and the ABABAB arrangement as **hexagonal close packing (hcp)**.

Cubic close packing (ccp) has the repeat pattern ABCABC . . ., whereas hexagonal close packing (hcp) has the repeat pattern ABABAB . . . . Using models will help you to understand the difference between ccp and hcp.

Interactive 3D cubic close packing.

There are many other possible close-packing arrangements such as ABCAB and ABCB, but ccp and hcp are the most common, and the most important in chemistry. The majority of metals adopt one of the two main close-packing arrangements. Aluminium, copper, and nickel are examples of metals that form cubic close-packed structures, and magnesium, cobalt, and zinc are examples of metals that form hexagonal close-packed structures. Worked example 6.1 gives an example of recognizing if a particular layer arrangement is close packed or not.

Worked example 6.1 Close packing

Is the structure with the layer arrangement AABBAABB close packed?

Strategy

Determine whether the atoms in each layer lie in the depressions between the atoms in the layer below. If so, the structure is close packed. If not, it is not close packed.

Solution

The AABBAABB packing arrangement contains layers in which the atoms lie directly on top of the atoms in the layer beneath (AA and BB) rather than in the depressions. These layers are not close packed, so the structure AABBAABB is not close packed.

Question

Is the ABACABAC structure close packed?

In a solid state structure, the **coordination number** is defined as the number of nearest neighbours an atom has. Cubic and hexagonal close-packed structures both have coordination numbers of 12. There are six nearest neighbours in the same row, three in the row above, and three in the row below. This is shown in the expanded views of the structures in Figure 6.5.

Cubic structures

Where does the term 'cubic' in cubic close packing come from? The cubic close-packed structure is usually drawn in a different orientation from that in Figure 6.5 to emphasize the cubic symmetry. This is shown is Figure 6.6. Figure 6.6(a) shows a space-filling model, with the atoms in contact with each other. The close-packed layers are emphasized in this figure by showing them in different colours. In Figure 6.6(b) the atoms are shown not touching each other. Although this sort of figure can be misleading as the atoms are really in contact, you will meet many figures like this as they allow you to see more clearly how the atoms are arranged.

Space-filling models are described in Box 5.1 (p.218).

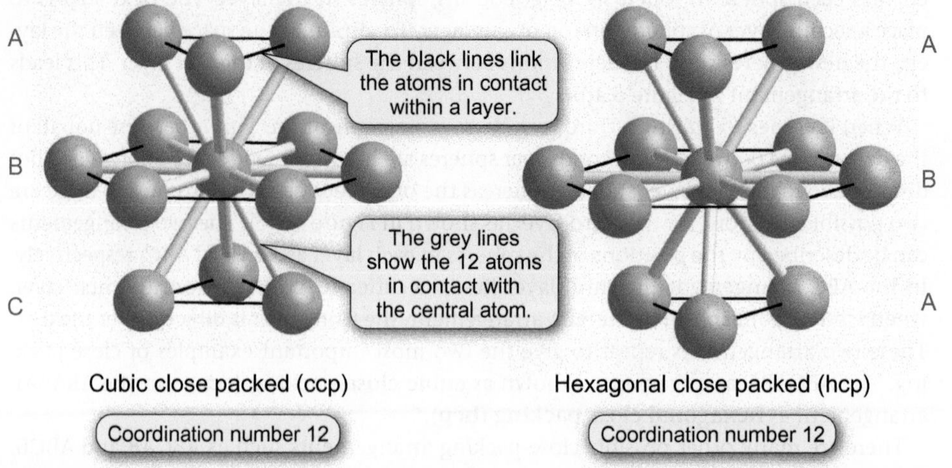
A
B
C
The black lines link the atoms in contact within a layer.
The grey lines show the 12 atoms in contact with the central atom.
A
B
A
Cubic close packed (ccp)
Hexagonal close packed (hcp)
Coordination number 12
Coordination number 12

Figure 6.5 The atoms in cubic and hexagonal close-packed structures have coordination numbers of 12. In this figure the atoms are not shown in contact in order to see their geometries more clearly.

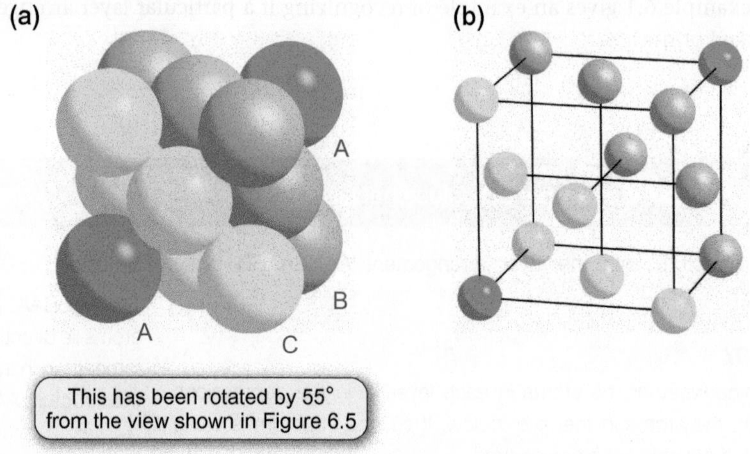
(a)
(b)
A
B
A
C
This has been rotated by 55° from the view shown in Figure 6.5

Figure 6.6 Cubic close-packed structures are usually shown in this orientation to emphasize the cubic symmetry; (a) shows the location of the close-packed layers, whereas (b) shows the cubic geometry. Remember when looking at figures like this that the atoms are really in contact with each other.

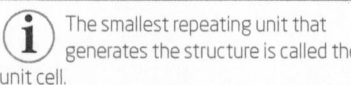

Interactive 3D body-centred cubic and primitive cubic structures.

The cubic close-packed structure is also known as **face-centred cubic (fcc)**, as there are atoms on each of the faces of the cube, in addition to each of the eight vertices. The face-centred cubic structure is shown in Figure 6.7 along with **body-centred cubic (bcc)** and **primitive cubic** structures. In the primitive cubic case, there are atoms only on the vertices of the cube, whereas for the body-centred cubic structure there is also an atom in the centre of the cube. Primitive and body-centred cubic structures are not close packed, so both contain more empty space than the ccp or hcp structures. The coordination number in the body-centred cubic structure is 8, while that in the primitive cubic structure is 6.

Although not as common as the close-packed structures, some metals form body-centred cubic structures. Examples include iron, sodium, and potassium. There is only one example of a metal forming a primitive cubic structure, and that is the radioactive element polonium.

The smallest repeating unit that generates the structure is called the unit cell.

The parts of the three cubic structures shown in Figure 6.7 are known as unit cells. A **unit cell** is the smallest possible repeating unit for a structure. The way in which identical unit cells for the primitive cubic structure stack to give the overall structure is shown in Figure 6.8. The positions of the metal atoms in such structures are called **lattice points**. More complicated structures can be based on these lattices but with ions or molecules on the lattice points instead of atoms.

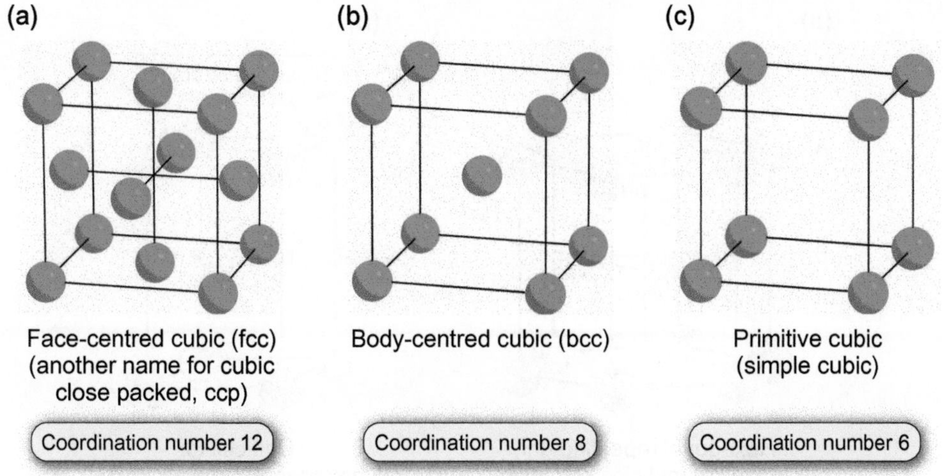
(a)
(b)
(c)
Face-centred cubic (fcc)
(another name for cubic
close packed, ccp)
Body-centred cubic (bcc)
Primitive cubic
(simple cubic)
Coordination number 12
Coordination number 8
Coordination number 6

Figure 6.7 The unit cells of important cubic lattices. (a) Face-centred cubic (fcc), (b) body-centred cubic (bcc), and (c) primitive cubic.

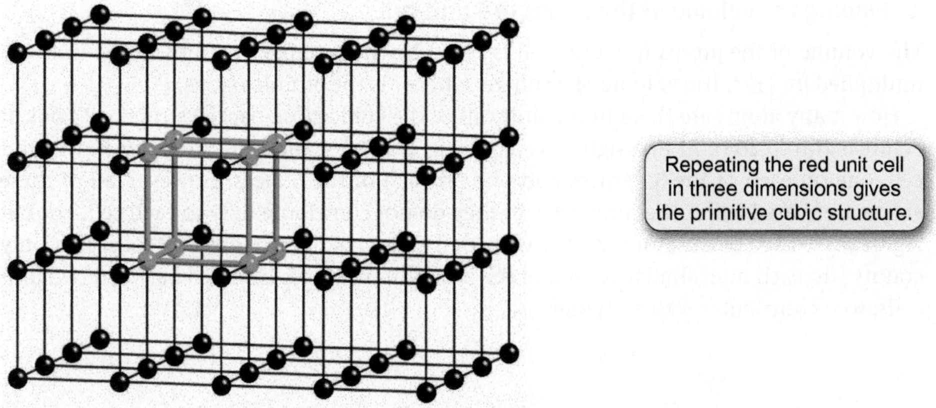
Repeating the red unit cell
in three dimensions gives
the primitive cubic structure.

Figure 6.8 The unit cell is the smallest repeating unit in a crystal structure. Repeating the red unit cell in three dimensions gives the primitive cubic structure.

Hexagonal structures

Part of the hexagonal close-packed structure is shown in Figure 6.9(a). This is a repeating unit, but it is not the unit cell as it is not the smallest repeating unit. The unit cell for the hcp structure is shown in Figure 6.9(b). Unlike the ccp unit cell, the angles in the hcp unit cell are not all 90°.

Interactive 3D hexagonal close-packed structure.

Packing efficiency

Close-packed structures are the most efficient ways possible to pack spheres. The **packing efficiency** is the percentage of the structure that is filled by the atoms. Calculations of packing efficiencies enable you to compare the packing in different structures.

The packing efficiency for a structure is given by the expression in Equation 6.1

$$\text{packing efficiency} = \frac{\text{volume of the atoms in a unit cell}}{\text{total volume of a unit cell}} \times 100 \qquad (6.1)$$

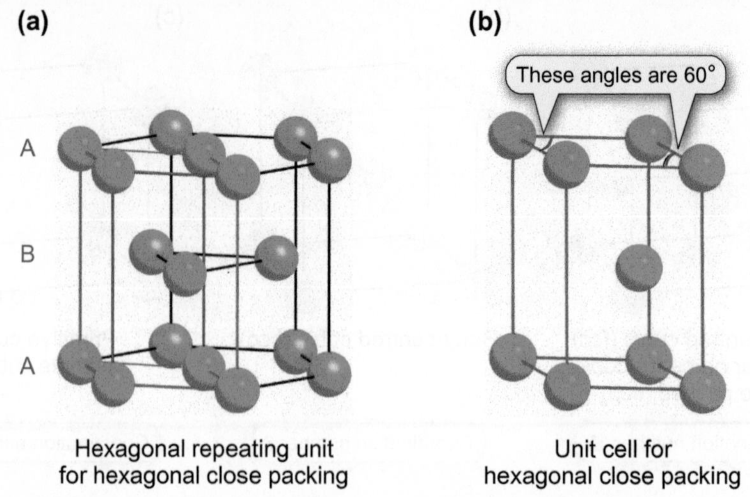
(a)
(b)
These angles are 60°
A
B
A
Hexagonal repeating unit for hexagonal close packing
Unit cell for hexagonal close packing

Figure 6.9 The hexagonal close-packed structure. (a) The hexagonal repeating unit. (b) The unit cell.

The volume of a sphere with radius r is $\frac{4}{3}\pi r^3$. The geometry of objects such as spheres is described in Maths Toolkit MT5 (p.1321).

1. Finding the volume of the atoms in a unit cell

The volume of the atoms in a unit cell is given by the number of atoms in the unit cell multiplied by $\frac{4}{3}\pi r^3$, the volume of a sphere, where r is the atomic radius.

How many atoms are there in the unit cell of the cubic close-packed structure? Look at Figure 6.7(a) (p.265). At first sight it is tempting to suggest there are 14 atoms in the unit cell, one on each of the 8 vertices and one on each of the 6 faces. However, all of these atoms are shared with other unit cells, so they do not contribute fully to each cell. Look at Figure 6.10. Each of the atoms on the vertices is shared between eight unit cells, so it only counts $\frac{1}{8}$ to each one. Similarly, each of the atoms on the faces is shared between two unit cells, so it contributes $\frac{1}{2}$ to each one.

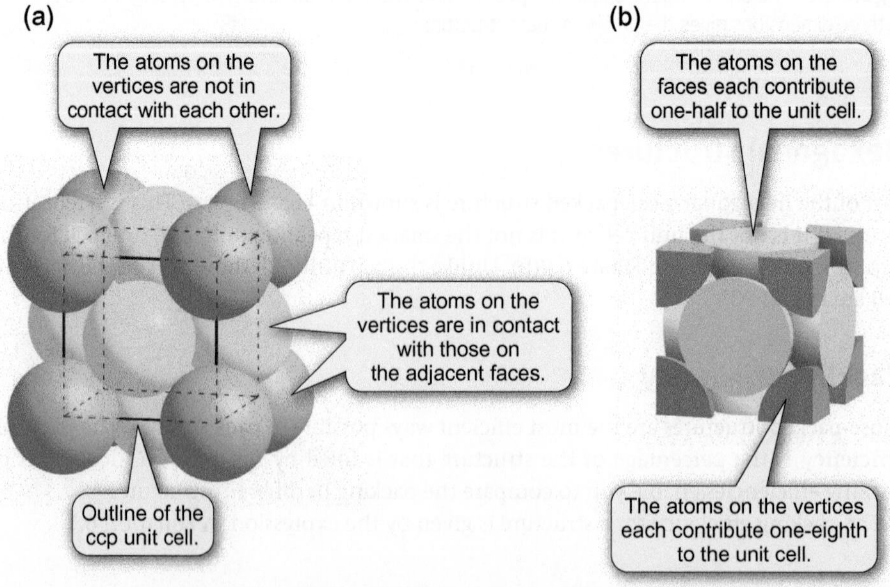
(a)
(b)
The atoms on the vertices are not in contact with each other.
The atoms on the vertices are in contact with those on the adjacent faces.
Outline of the ccp unit cell.
The atoms on the faces each contribute one-half to the unit cell.
The atoms on the vertices each contribute one-eighth to the unit cell.

Figure 6.10 None of the atoms in the cubic close-packed unit cell lies completely within one unit cell—all are shared with neighbouring unit cells. (a) shows all the atoms contributing to a unit cell and (b) shows just the parts of the atoms that lie within the unit cell.

Position of atom	Number of atoms in this position	Share of the atom belonging to the unit cell	Total number of atoms contributing to the unit cell
Vertex	8	$\frac{1}{8}$	$8\times\frac{1}{8}$
Face	6	$\frac{1}{2}$	$6\times\frac{1}{2}$

The total number of atoms, N, in a cubic close-packed unit cell is

$$N = \left(8\times\tfrac{1}{8}\right) + \left(6\times\tfrac{1}{2}\right) = 4$$

so the total volume of the atoms in a unit cell of a ccp structure is equal to $4\times\frac{4}{3}\pi r^3 = \frac{16}{3}\pi r^3$.

2. Finding the total volume of a unit cell

Look at the unit cell in Figure 6.10. In a cube, all of the sides are of equal length, so the volume is given by l^3, where l is the length of a side of the unit cell. The atoms on the vertices are not in contact with each other. An atom on a vertex touches an atom at the centre of a face, so the distance between their centres is $2r$. Looking at just one face of the cube, as in Figure 6.11, the distance between opposite vertices is $4r$. Knowing this, you can calculate the distance l using Pythagoras's theorem.

Applying Pythagoras's theorem to the blue triangle in Figure 6.11

$$\begin{aligned} l^2 + l^2 &= (4r)^2 \\ 2l^2 &= 16r^2 \\ l^2 &= 8r^2 \\ l &= 8^{1/2}r \end{aligned}$$

The volume of the unit cell is l^3, so

$$\begin{aligned} l^3 &= (8^{1/2}r)^3 \\ &= (8^{1/2})^3 \times (r)^3 \\ &= 8^{3/2}r^3 \end{aligned}$$

3. Working out the packing efficiency

Putting the expressions for the volume of the atoms in the unit cell and the total volume of the unit cell into Equation 6.1 (p.265) gives an expression for the packing efficiency. The radii cancel out.

$$\begin{aligned} \text{Packing efficiency} &= \frac{\frac{16}{3}\pi r^3}{8^{3/2} r^3}\times 100 \\ &= \frac{\frac{16}{3}\pi}{8^{3/2}}\times 100 \quad (\pi = 3.1416) \\ &= 74.0\% \end{aligned}$$

Worked example 6.2 provides another example of a packing efficiency calculation.

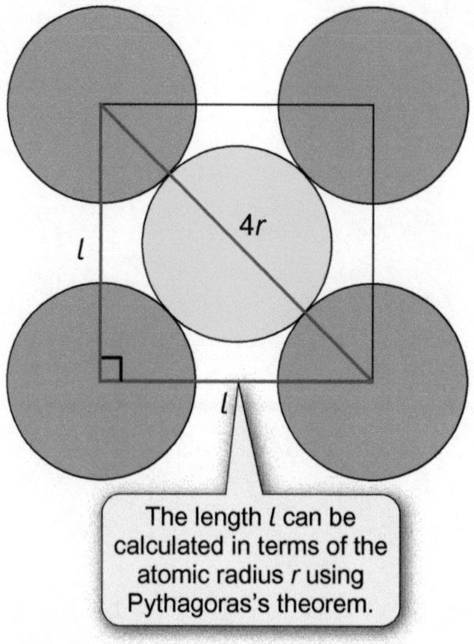
4r
l
l
The length l can be calculated in terms of the atomic radius r using Pythagoras's theorem.

Figure 6.11 One face of a cubic close-packed unit cell.

Σ For a right-angled triangle with sides of lengths a, b, and c, the length of the longest side (the hypotenuse) c is given by Pythagoras's theorem, $a^2 + b^2 = c^2$. See Maths Toolkit MT5 (p.1322).

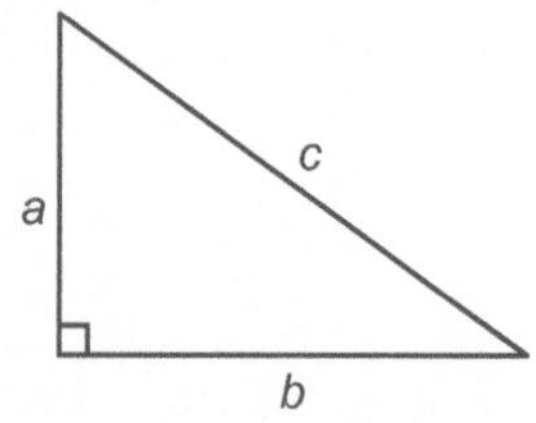
a
b
c

Σ The rules on multiplying indices are described in Maths Toolkit MT1 (p.1309).
$(8^{1/2})^3 = 8^{1/2\times3} = 8^{3/2}$.

Visit the Online Resource Centre to view screencast 6.2 which walks you through how to calculate the packing efficiency of a cubic close-packed structure.

Worked example 6.2 Packing efficiency

What is the packing efficiency of the body-centred cubic (bcc) structure?

Strategy

1 Find the volume of the atoms in a unit cell using Figure 6.7(b) (p.265) to work out the number of atoms in the unit cell.
2 Find the total volume of the unit cell. Use Pythagoras's theorem to determine the unit cell side length.
3 Use Equation 6.1 (p.265) to work out the packing efficiency. →

→ Solution

1. Finding the volume of the atoms in a unit cell

Of the nine atoms shown in Figure 6.7(b), the eight on the vertices are each shared between eight unit cells, so only contribute $\frac{1}{8}$ to each cell. The atom in the centre is fully within the unit cell, so it contributes 1.

Position of atom contributing	Number of atoms in this position	Share of the atom belonging to the unit cell	Total number of atoms to the unit cell
Vertex	8	$\frac{1}{8}$	$8 \times \frac{1}{8}$
Centre	1	1	1×1

The number of atoms, N, in the bcc unit cell is

$$N = \left(8 \times \tfrac{1}{8}\right) + (1 \times 1) = 2$$

so the total volume of the atoms is $2 \times \frac{4}{3}\pi r^3 = \frac{8}{3}\pi r^3$

2. Finding the total volume of a unit cell

In a bcc unit cell, the vertex atoms are in contact with the central atom, which means the distance from a corner of the unit cell to the centre is $2r$. The distance between opposite vertices in the cube is therefore $4r$, as shown in blue below.

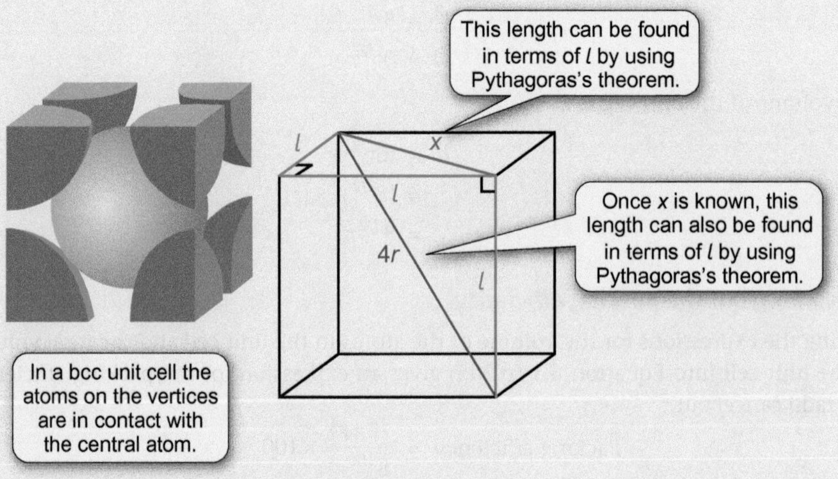
This length can be found in terms of l by using Pythagoras's theorem.
l
x
l
$4r$
l
Once x is known, this length can also be found in terms of l by using Pythagoras's theorem.
In a bcc unit cell the atoms on the vertices are in contact with the central atom.

Applying Pythagoras's theorem to the triangle on the top face of the unit cell, the green length x is given by

$$x^2 = l^2 + l^2 = 2l^2$$
$$x = (2^{1/2})l$$

Applying the theorem to the second triangle (with green, blue, and red sides) gives the relationship between l and r

$$(4r)^2 = x^2 + l^2$$
$$16r^2 = ((2^{1/2})l)^2 + l^2$$
$$= 2l^2 + l^2 = 3l^2$$

Rearranging gives:

$$l^2 = \tfrac{16}{3}r^3$$
$$l = \left(\tfrac{16}{3}\right)^{1/2} r$$

→

→ The volume of the cell is l^3, so

$$l^3 = \left(\left(\tfrac{16}{3}\right)^{1/2} r\right)^3$$
$$= \left(\tfrac{16}{3}\right)^{3/2} r^3$$

3. Working out the packing efficiency

Putting the expressions for the volume of the atoms and the total volume of the unit cell into Equation 6.1 gives

$$\text{packing efficiency} = \frac{\text{volume of the atoms in a unit cell}}{\text{total volume of a unit cell}} \times 100$$
$$= \frac{2 \times \frac{4}{3}\pi \cancel{r^3}}{\left(\frac{16}{3}\right)^{3/2} \cancel{r^3}} \times 100$$
$$= \frac{\frac{8}{3}\pi}{\left(\frac{16}{3}\right)^{3/2}} \times 100$$
$$= 68.0\%$$

Question

Show that the packing efficiency of the primitive cubic structure in Figure 6.7(c) (p.265) is 52.4%.

Table 6.2 Common structural types based on the packing of spheres

Structural type	Abbreviation	Number of atoms in the unit cell	Packing efficiency / %
Cubic close packed	ccp (or fcc)	4	74.0
Hexagonal close packed	hcp	2	74.0
Body-centred cubic	bcc	2	68.0
Primitive cubic	—	1	52.4

Key points of the common structural types that are based on the packing of spheres are summarized in Table 6.2.

Cell projection diagrams

Although unit cell diagrams such as those in Figure 6.7 (p.265) clearly show the positions of the atoms in a lattice, they can be difficult to draw for more complicated structures. For this reason, cell projection diagrams are often used instead. A **cell projection diagram** is a two-dimensional representation of the unit cell, looking at it from above. The *x*- and *y*-coordinates are shown on the projection as normal, but the *z*-coordinate is given as a number, between 0 and 1 for each atom, with 0 representing the bottom of the unit cell and 1 the top. This number can either be written next to each atom or, as we have done here, the atoms can be colour-coded, with the *z*-coordinates given in a key. The cell projection diagram for the cubic close-packed structure is shown in Figure 6.12. For the six atoms on the faces, four have a *z*-coordinate of $\frac{1}{2}$. This means that the atoms are halfway between the bottom of the unit cell ($z = 0$) and the top of the unit cell ($z = 1$). The atoms on the bottom and top faces have *z*-coordinates of 0 and 1, respectively. The colours have no significance other than to identify specific atoms.

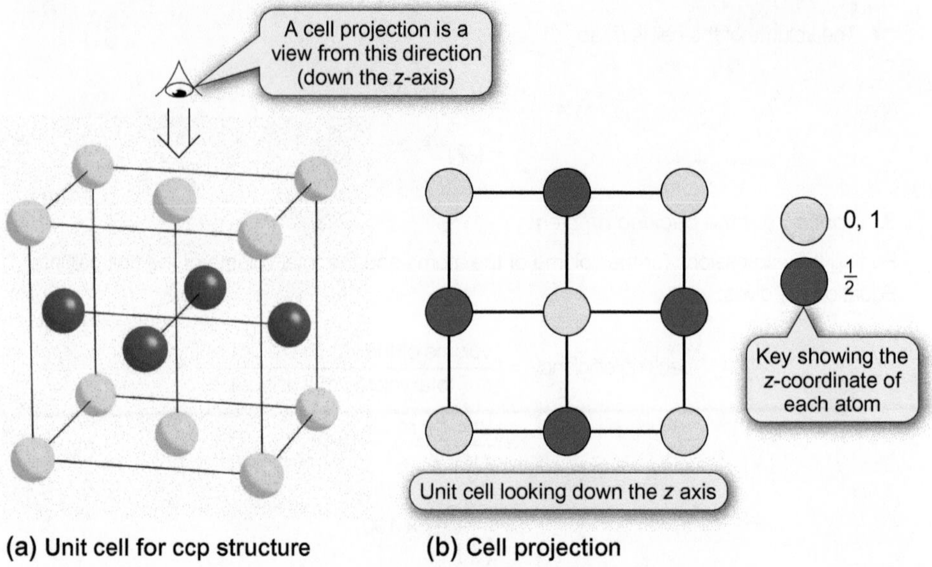
A cell projection is a view from this direction (down the z-axis)
0, 1
1/2
Key showing the z-coordinate of each atom
Unit cell looking down the z axis
(a) Unit cell for ccp structure
(b) Cell projection

Figure 6.12 (a) The unit cell for the cubic close-packed (ccp) structure. (b) A cell projection diagram for the ccp structure. The different coloured atoms have different *z*-coordinates.

Worked example 6.3 provides another example of constructing a cell projection diagram.

Worked example 6.3 Cell projection diagrams

Draw a cell projection diagram for a unit cell in the body-centred cubic structure.

Strategy

The unit cell for a body-centred cubic structure is shown in Figure 6.7(b) (p.265). Draw the unit cell in two dimensions, looking at it from the top (down the *z*-axis).

Work out the *z*-coordinate for each of the atoms, and show this on your diagram by using a different colour for atoms with a *z*-coordinate of $\frac{1}{2}$ than for those with *z*-coordinates of 0 and 1.

Solution

The atoms on the vertices have *z*-coordinates of 0 and 1. The atom in the centre of the cube has a *z*-coordinate of $\frac{1}{2}$. The cell projection diagram looks like this

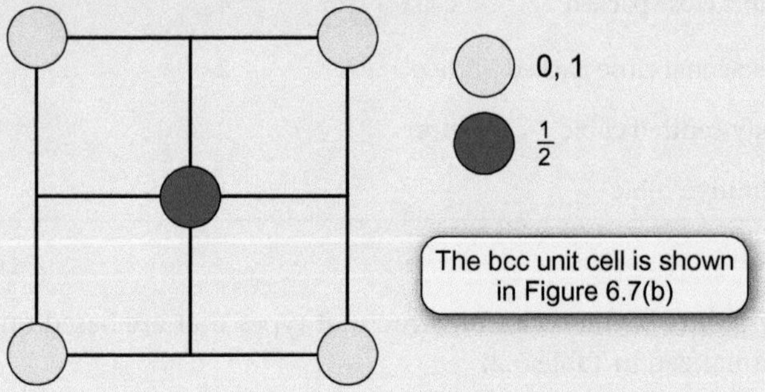
0, 1
1/2
The bcc unit cell is shown in Figure 6.7(b)

Question

Draw a cell projection diagram for the primitive cubic unit cell.

Interstitial sites

If the atoms take up 74.0% of the space in a close-packed structure, what happens to the remaining 26.0%? Spheres cannot pack without leaving some free space, and this free space is very important in the structures of compounds that are based on close packing but with other atoms or ions in the gaps (see Section 6.4, p.277).

The gaps between the atoms are called **interstitial sites**. The sites are defined by their geometries, and examples of possible geometries are shown in Figure 6.13.

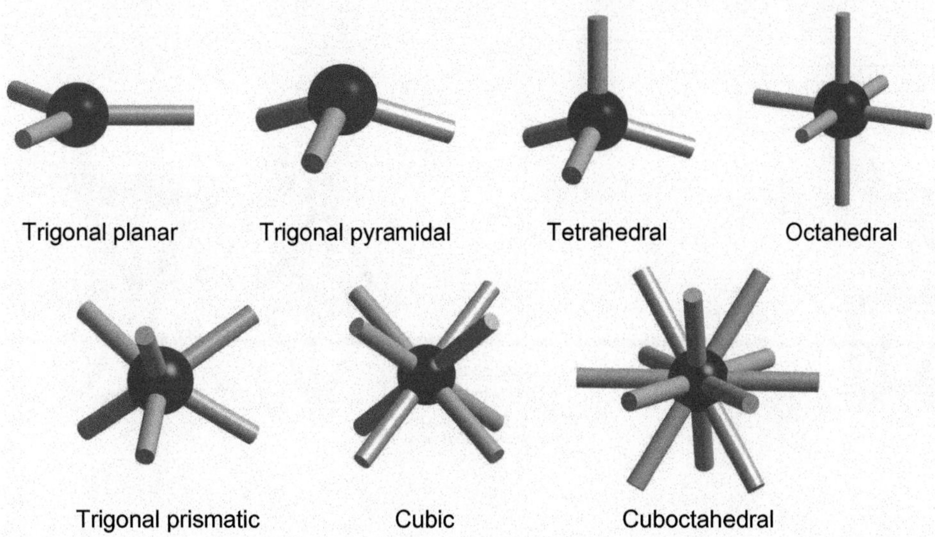
Trigonal planar
Trigonal pyramidal
Tetrahedral
Octahedral
Trigonal prismatic
Cubic
Cuboctahedral

Figure 6.13 Examples of geometries that can be observed at interstitial sites.

For a close-packed structure there are two types of interstitial sites—**octahedral sites** and **tetrahedral sites**—as shown in Figure 6.14. For each atom in a close-packed structure there is one octahedral site and two tetrahedral sites. The octahedral sites lie between a triangle of atoms in the row above and another triangle of atoms in the row below. The tetrahedral sites lie between a triangle of atoms in one row and a single atom in the other.

By drawing the interstitial sites as polyhedra as in Figure 6.15, their positions are seen more clearly. In this figure the atoms are shown much smaller than their real sizes so that the octahedra and tetrahedra are easier to see.

As you will see in Section 6.4, the structures of many ionic compounds are *based* on the close-packed structures described in this section, but with some of the ions located in the interstitial sites.

You might find it hard to visualize the shapes and positions of the interstitial sites. If this is so, you will find it very useful to look at or build models of the structures.

Visit the Online Resource Centre to view screencast 6.3 which walks you through how to identify the interstitial sites in the cubic close-packed structure.

Density

Density, ρ, is defined as mass per unit volume (Equation 6.2).

$$\rho = \frac{m}{V} \tag{6.2}$$

Greek letters:
ρ (rho) density

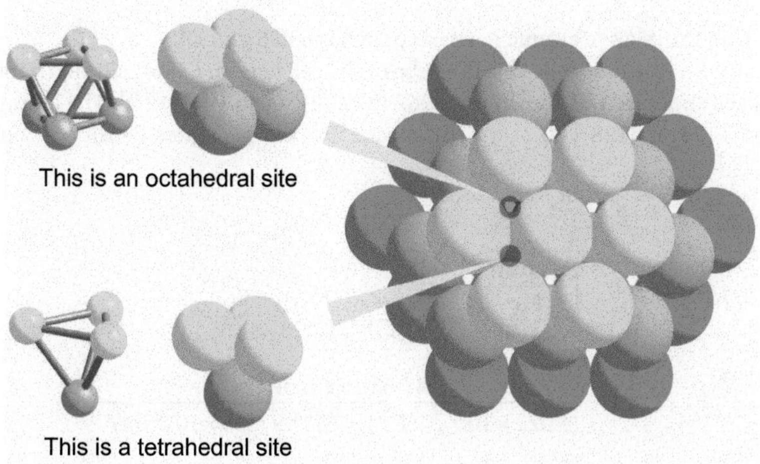
This is an octahedral site
This is a tetrahedral site

Figure 6.14 Examples of the octahedral and tetrahedral interstitial sites in a close-packed structure.

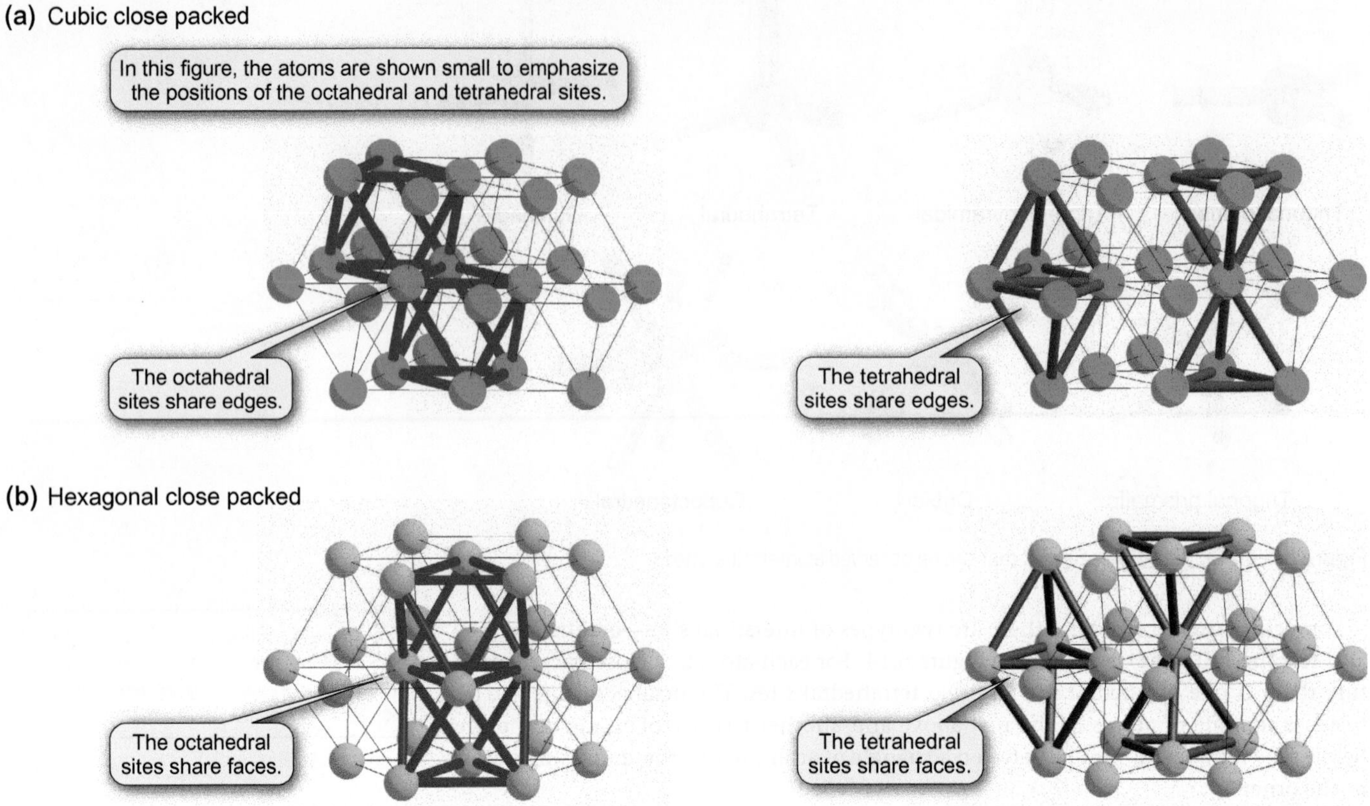
(a) Cubic close packed
In this figure, the atoms are shown small to emphasize the positions of the octahedral and tetrahedral sites.
The octahedral sites share edges.
The tetrahedral sites share edges.
(b) Hexagonal close packed
The octahedral sites share faces.
The tetrahedral sites share faces.

Figure 6.15 The octahedral and tetrahedral interstitial sites in the cubic close-packed and hexagonal close-packed structures shown as polyhedra.

where m is mass and V is volume. The density of a substance is normally reported in the units of $\mathrm{g\,cm^{-3}}$.

If you know the contents and the dimensions of a unit cell, you can use this to calculate the density for that substance. For a unit cell, m is the total mass of the atoms in the unit cell, which is given by the sum of the molar masses of the atoms within the unit cell divided by the Avogadro constant, N_A. Copper, for example, has a cubic close-packed structure. This means that there are four copper atoms in the unit cell (see p.266), so m is given by:

$$m = \frac{4 \times A(\mathrm{Cu})}{N_A}$$

where $A(\mathrm{Cu})$, the relative atomic mass of copper, is $63.55\,\mathrm{g\,mol^{-1}}$.

The volume of the unit cell, V, is the cube of the cell length, l. The cell length for a cubic close-packed structure was shown on p.267 to be $(8)^{1/2}r$, so $V = (8)^{3/2}r^3$. The atomic radius of copper is 128 pm (128×10^{-12} m). This means that the calculated density of copper is:

$$\rho = \frac{m}{V}$$

$$= \frac{\left(\frac{4A(\mathrm{Cu})}{N_A}\right)}{(8)^{3/2}r^3} = \frac{4A(\mathrm{Cu})}{N_A(8)^{3/2}r^3}$$

$$= \frac{4 \times 63.55\,\mathrm{g}\,\cancel{\mathrm{mol^{-1}}}}{(6.022 \times 10^{23}\,\cancel{\mathrm{mol^{-1}}}) \times (8)^{3/2} \times (128 \times 10^{-12}\,\mathrm{m})^3}$$

$$= 8.90 \times 10^6\,\mathrm{g\,m^{-3}}$$

$$= 8.90\,\mathrm{g\,cm^{-3}}$$

This calculated density is very close to the experimentally determined value of the density for copper, which is $8.96\,g\,cm^{-3}$.

Summary

- The most efficient manner in which atoms can pack is called close packing. There are two important types of close packing. Cubic close packing (ccp) has a layer repeat ABCABC . . . and hexagonal close packing (hcp) has a layer repeat of ABABAB . . .
- A unit cell is the smallest possible repeating unit for a structure.
- The cubic close-packed structure is also known as face-centred cubic (fcc).
- Body-centred cubic (bcc) and primitive cubic structures can also be formed, but they have lower packing efficiencies.
- Packing efficiencies for structures can be calculated by dividing the volume of the atoms in a unit cell by the total volume of the unit cell. The value is usually quoted as a percentage.
- Close-packed structures contain octahedral and tetrahedral interstitial sites.
- The density of a substance can be calculated by dividing the total mass of its unit cell by the volume of its unit cell.

? For practice questions on these topics, see questions 2–4 at the end of this chapter (p.300).

6.3 Metallic bonding

What holds the atoms together in a metal? Metals typically have low ionization enthalpies, so they are able to lose their valence electrons relatively easily. The **free electron model** of bonding in a metal considers the lattice as being made up of metal ions, surrounded by a 'sea' of delocalized electrons (Figure 6.16). The metal ions can vibrate, but are otherwise fixed into position. In contrast, the valence electrons are free to move. Normally the electrons move in random directions but, when a potential difference is applied, the electrons move from high to low potential, which gives rise to a current. This explains how metals can conduct electricity.

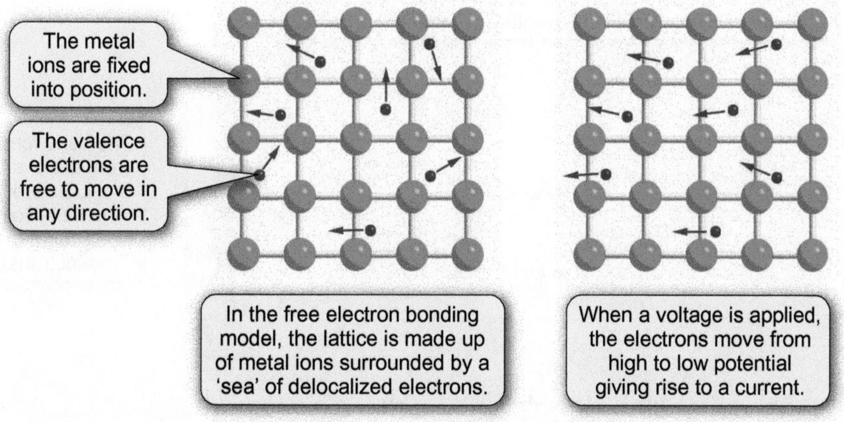
The metal ions are fixed into position.
The valence electrons are free to move in any direction.
In the free electron bonding model, the lattice is made up of metal ions surrounded by a 'sea' of delocalized electrons.
When a voltage is applied, the electrons move from high to low potential giving rise to a current.

Figure 6.16 In the free electron model of bonding in metals, there is a lattice of metal ions with the valence electrons free to move. (The lines between the ions are to emphasize their fixed positions—they do not represent covalent bonds. The ions are in fact touching one another.)

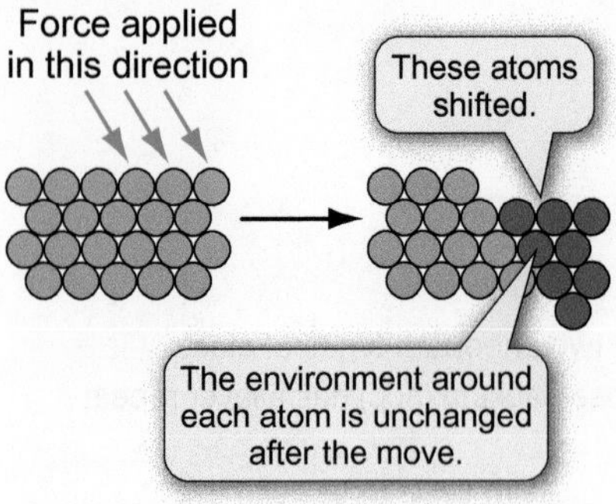
Force applied
in this direction
These atoms
shifted.
The environment around
each atom is unchanged
after the move.

Figure 6.17 The origin of the malleability of metals.

Metals are also **malleable**, meaning they can be hammered into sheets, and **ductile**, meaning they can be drawn into wires. These properties arise because one plane of metal ions can slip past another when a force is applied. This is shown in Figure 6.17. The environment of each ion is unchanged following the movement.

The free electron model explains why metals conduct electricity, but it doesn't explain how the conductivity changes with temperature or why some materials are semiconductors. **Band theory** is a more advanced approach to metallic bonding that has its basis in molecular orbital theory.

Band theory

In Section 4.8 (p.191) the bonding between two lithium atoms to form Li_2 is described in terms of molecular orbital theory. The two 2*s* orbitals combine to give a filled σ_g orbital and an empty $\sigma_u{}^*$ orbital, so Li_2 has a bond order of one. In solid lithium, there are many millions of atoms, each of which has a 2*s* orbital. These atomic orbitals all interact together to give many millions of molecular orbitals, all with very similar energies. The energy gaps between these molecular orbitals are so small that the orbitals overlap to give a **band**, as shown in Figure 6.18.

As each 2*s* orbital is half-filled, the band is also half-filled. It is known as a **valence band** because it contains the valence shell electrons. When a potential difference is applied to a sample of lithium, the electrons are able to move within the band, which is delocalized over all of the metal atoms, so the metal conducts electricity.

When a metal atom has more than one type of atomic orbital in its valence shell, each forms a band through interactions with the atomic orbitals on the other atoms. In magnesium, for example, there is a band from the interactions of the 3*s* orbitals and a band from the interactions of the 3*p* orbitals. Magnesium has the electronic configuration [Ne] $3s^2$, so the 3*s* band is filled and the 3*p* band is empty. As the 3*s* and 3*p* atomic orbitals are relatively close together in energy, the two bands overlap to give a single partially filled valence band. This explains how magnesium can conduct electricity, despite each atom having a filled 3*s* orbital.

To conduct electricity a material must have a partially filled band. Such a band is called a **conduction band**. Compounds that have filled bands may be insulators or semiconductors. The type of behaviour adopted depends on the **band gap**, which is the energy separation between the top of the filled band and the bottom of the next available empty band, as shown in Figure 6.19. If the band gap is small, some of the electrons from the filled band will have enough thermal energy to occupy the empty band. Being partially filled,

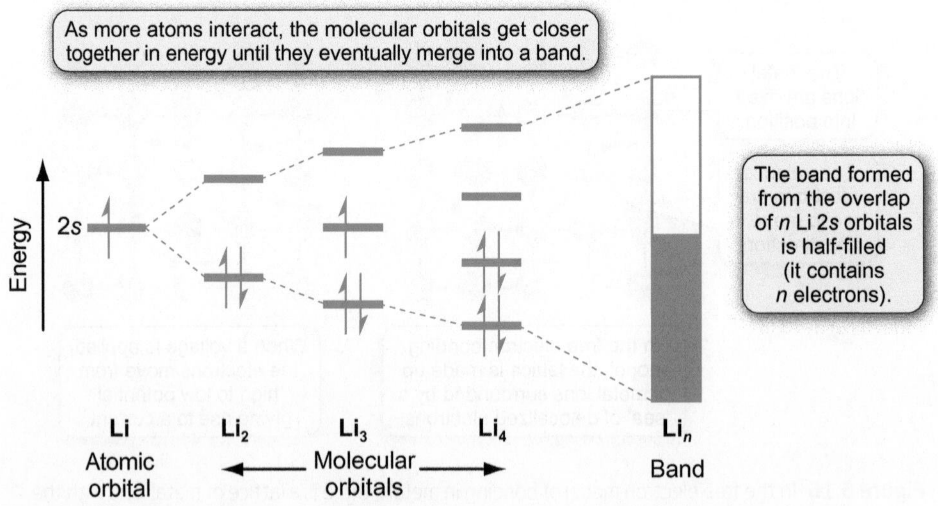
As more atoms interact, the molecular orbitals get closer together in energy until they eventually merge into a band.
Energy
2s
The band formed from the overlap of n Li 2s orbitals is half-filled (it contains n electrons).
Li
Li2
Li3
Li4
Lin
Atomic orbital
Molecular orbitals
Band

Figure 6.18 The origin of the valence band in lithium metal.

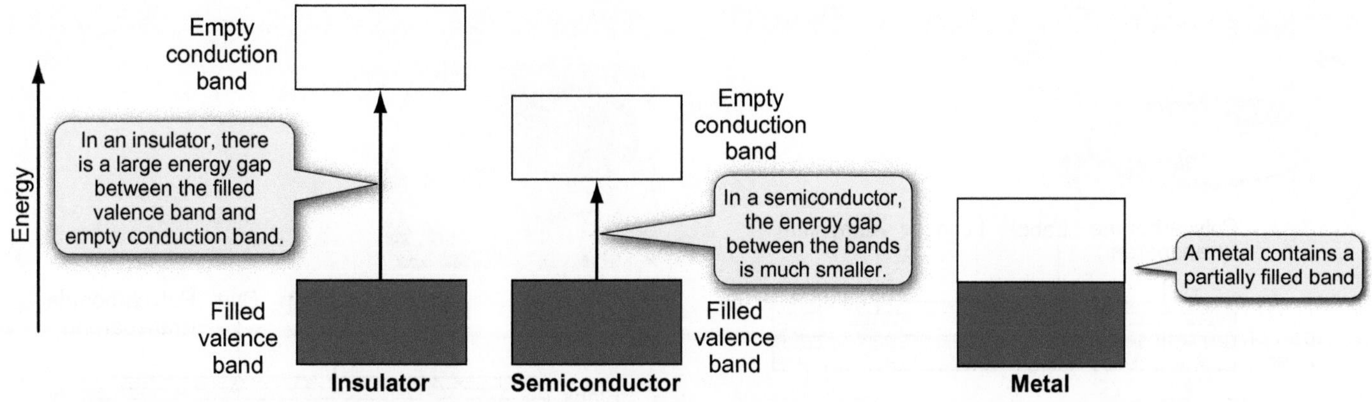
Energy
Empty conduction band
In an insulator, there is a large energy gap between the filled valence band and empty conduction band.
Filled valence band
Insulator
Empty conduction band
In a semiconductor, the energy gap between the bands is much smaller.
Filled valence band
Semiconductor
A metal contains a partially filled band
Metal

Figure 6.19 Insulators and semiconductors both have band gaps between the filled valence band and the empty conduction band. In contrast, metals have partially filled bands.

this becomes a conduction band, and the material acts as a semiconductor. Its conductivity increases with temperature, because at higher temperatures there are more electrons in the conduction band. If the band gap is large, the conduction band remains unpopulated, and the material is an insulator. Generally, as the band gap decreases, the conductivity of a material increases. For the Group 14 elements, diamond has a high band gap, so is an insulator; silicon and germanium have smaller band gaps, so are semiconductors, whereas tin and lead have no band gap, so are conductors.

A **semiconductor** has electrical conductivity between that of a conductor and that of an insulator. The resistance of a semiconductor decreases at higher temperatures or if there are impurities.

Alloys

An alloy is a metallic material that contains two or more metals or a metal and a non-metal. Alloys are far more widely used commercially than pure metallic elements as they are often stronger and less susceptible to corrosion. From a structural perspective, there are two main types of alloy: solid solutions and intermetallic compounds. A **solid solution** is a mixture of the metals and the relative proportions of the metals can be varied. For example, bronze is a solid solution of copper and tin in which the proportion of tin can vary from 1% to 40%. In contrast, an **intermetallic compound** has a fixed composition, such as the sodium–mercury amalgam $NaHg_2$.

Box 6.3 CD writers and rewriters

The information on a compact disc (CD) is stored in digital form as a series of 0s and 1s. The 0s and 1s are represented by bumps and flat areas, respectively, on the reflective surface of the CD. The CD contains a continuous spiral track, running from the centre to the edge, with a width of 0.5 μm and a length of approximately 5 km. When light from a laser passes over a flat area, it is reflected back and detected by a sensor. When it passes over a bump, the laser beam is scattered and so is not reflected back. In a commercially produced CD (Figure 1), the bumps are imprinted onto discs made from polycarbonate that is coated with aluminium to create the reflective surface. To protect the metal surface it is coated with a layer of lacquer, onto which the label is printed.

This process of imprinting and coating discs is not suitable for home use, and recordable CDs (CD-Rs) work by a different process. CD-Rs contain a smooth reflective layer of aluminium under a layer of a photosensitive dye. On a blank CD-R disc (Figure 2a), the dye layer is *translucent*, allowing light to pass through to the reflective metal layer. This means that the light is reflected back and detected by the sensor. When the dye is heated, it turns *opaque*, forming non-reflecting areas (Figure 2b). The laser reading the CD is only reflected back to the sensor when the dye is translucent, so the flat areas and bumps of a normal CD are replaced by reflective and non-reflective surfaces. To record on to the CD, the CD writer uses a second laser, the 'write laser', which is more powerful than the 'read laser'. The write layer turns the dye opaque by heating it to 600 °C—hence the term 'burning' a CD. Each manufacturer uses a different dye, and each dye has its own recording and lifetime characteristics.

CD-Rs are cheap and compatible with most CD players, but they can only be 'written' on once. Rewritable CDs (CD-RWs) can be erased and rewritten. CD-RWs rely on a phase-change material that

→

→

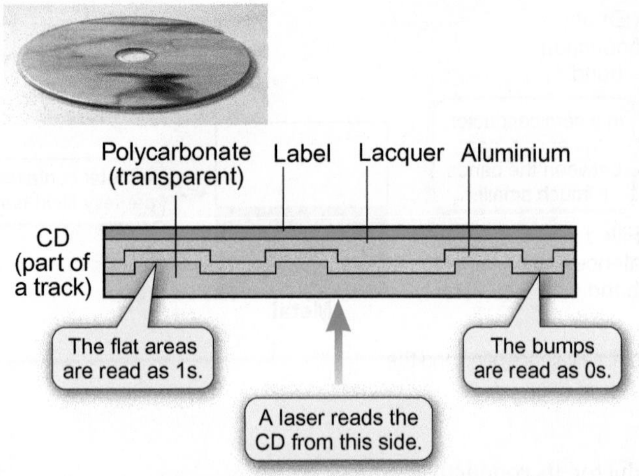
Polycarbonate (transparent)
Label
Lacquer
Aluminium
CD (part of a track)
The flat areas are read as 1s.
The bumps are read as 0s.
A laser reads the CD from this side.

▲ **Figure 1** Cross-section of a commercially produced compact disc (CD).

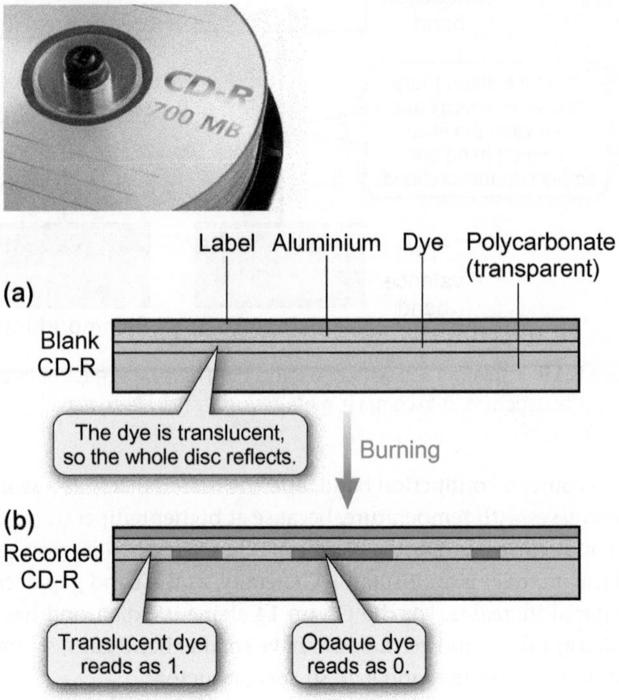
Label
Aluminium
Dye
Polycarbonate (transparent)
(a)
Blank CD-R
The dye is translucent, so the whole disc reflects.
Burning
(b)
Recorded CD-R
Translucent dye reads as 1.
Opaque dye reads as 0.

▲ **Figure 2** Cross-section of a recordable compact disc (CD-R).

is an alloy of silver, indium, antimony, and tellurium. At room temperature, the Ag–In–Sb–Te alloy exists in a crystalline form that is translucent (Figure 3a). If this is heated to 600 °C with the write laser, the alloy undergoes a phase change, losing crystallinity. The resulting amorphous form of the alloy is opaque, and the material remains in this state when it is cooled (Figure 3b). The opaque spots serve the same purpose as the bumps on a commercial CD, and represent 0s. The spots that remain crystalline represent 1s.

A CD rewriter contains a third laser in addition to the high-intensity write laser and the low-intensity read laser. The intermediate intensity erase laser heats the Ag–In–Sb–Te alloy to 200 °C. By holding the material at this temperature, the alloy is restored to its crystalline state, thus erasing all the encoded 0s. The disk can now be rewritten. Although more flexible than CD-Rs, CD-RWs are less reflective and as a result they often cannot be read by older CD players. The main use of CD-RWs is for data storage.

Question

Detailed studies on heat-treated samples of Ag–In–Sb–Te alloys suggest several phases are formed, including one that can be described as a face-centred cubic arrangement of tellurium with a 1 : 1 mixture of silver and antimony occupying the octahedral sites. What is the formula of this phase?

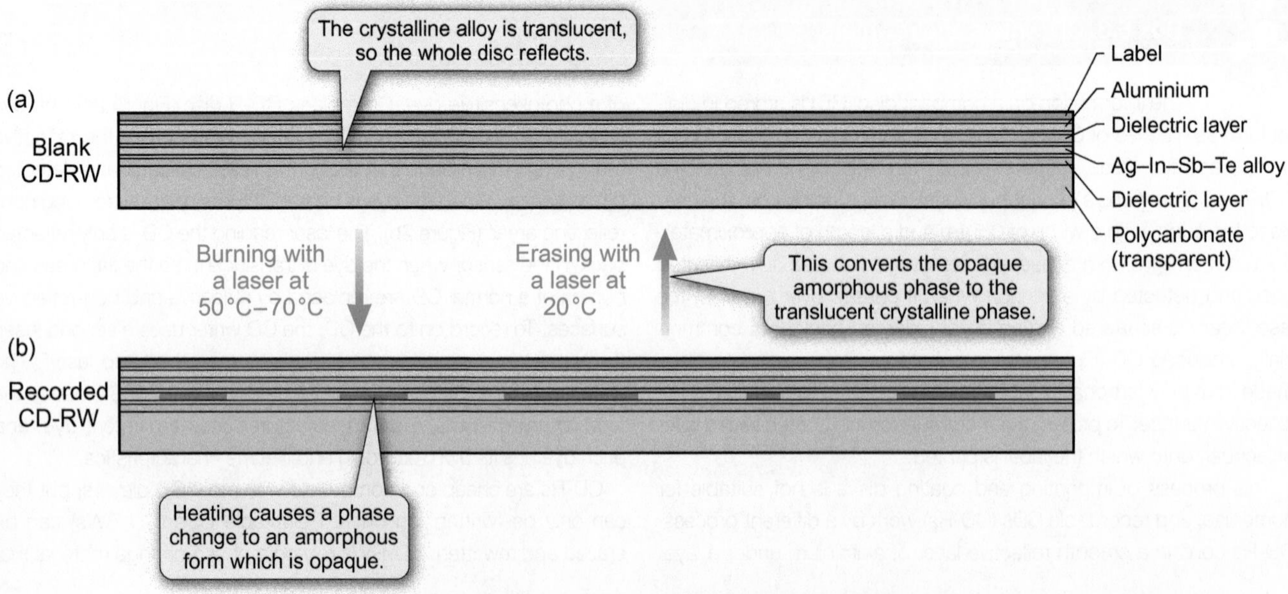
The crystalline alloy is translucent, so the whole disc reflects.
(a)
Blank CD-RW
Label
Aluminium
Dielectric layer
Ag–In–Sb–Te alloy
Dielectric layer
Polycarbonate (transparent)
Burning with a laser at 50°C–70°C
Erasing with a laser at 20°C
This converts the opaque amorphous phase to the translucent crystalline phase.
(b)
Recorded CD-RW
Heating causes a phase change to an amorphous form which is opaque.

▲ **Figure 3** Cross-section of a rewritable compact disc (CD-RW).

Solid solutions are the most widely used type of alloy, and these can be divided into two classes. In a **substitutional alloy**, atoms of one metal are exchanged for those of another metal. For example, up to 40% of the atoms in the ccp lattice of copper can be substituted by zinc atoms giving the alloy *brass*. Substitutional alloys form when the two metals have similar atomic radii. The atomic radius of copper is 128 pm and that of zinc 137 pm, so the zinc atoms are able to sit in positions that would be normally be occupied by copper atoms without major distortions to the lattice.

In an **interstitial alloy**, atoms of one element are present in the interstitial sites of a metal lattice. Steel is an alloy of iron containing up to 2% carbon. At high temperatures, steel forms the phase *austenite* in which the small carbon atoms occupy interstitial sites in a cubic close-packed iron lattice.

Both substitutional and interstitial alloys tend to be harder and stronger than the elemental metals. This is because the presence of different-sized atoms makes it more difficult for the planes of atoms to slide over one another. However, the presence of very large atoms such as bismuth can drastically alter the packing, and lead to soft materials with low melting points.

Summary

- The bonding in metals and their conductivity can be explained using band theory, which is derived from molecular orbital theory.
- Alloys are metallic materials containing a mixture of two or more metals, or a metal and a non-metal.
- Substitutional alloys, such as brass, have structures in which the atoms of one element are substituted by those of another.
- Interstitial alloys, such as the austenite phase of steel, have structures in which one element lies in the interstitial sites of another.

For a practice question on this topic, see question 5 at the end of this chapter (p.300).

6.4 Structures of compounds

Binary solids contain two elements. The structures of many binary solids can be explained using a close-packed arrangement of one of the elements with the other located in the interstitial sites. Many ionic compounds can be described in this way, though this type of description is also useful for some covalent network structures. In this section, you will look at the most important structural types, which take their names from typical examples.

The sodium chloride structure is also called the rock salt structure. It is just as valid to describe the NaCl structure as having a cubic close-packed arrangement of Na^+ ions with Cl^- ions in the octahedral sites, though the larger ion is normally used to define the close-packed structure.

Structures based on close packing

Sodium chloride (NaCl)

The unit cell for **sodium chloride** is shown in Figure 6.20, together with a cell projection diagram. There are two coordination numbers for a binary solid, one for each type of atom or ion. Both the Na^+ and Cl^- ions have coordination numbers of 6 and have octahedral geometries.

Interactive 3D NaCl structure.

Compare the unit cell for NaCl in Figure 6.20 with that for the cubic close-packed structure in Figure 6.7(a) (p.265). The Cl^- anions occupy the same positions as the metal atoms, whereas the Na^+ cations occupy the same positions as the octahedral sites (see Figure 6.14). The NaCl structure is therefore a cubic close-packed lattice of chloride ions with sodium ions in the octahedral sites.

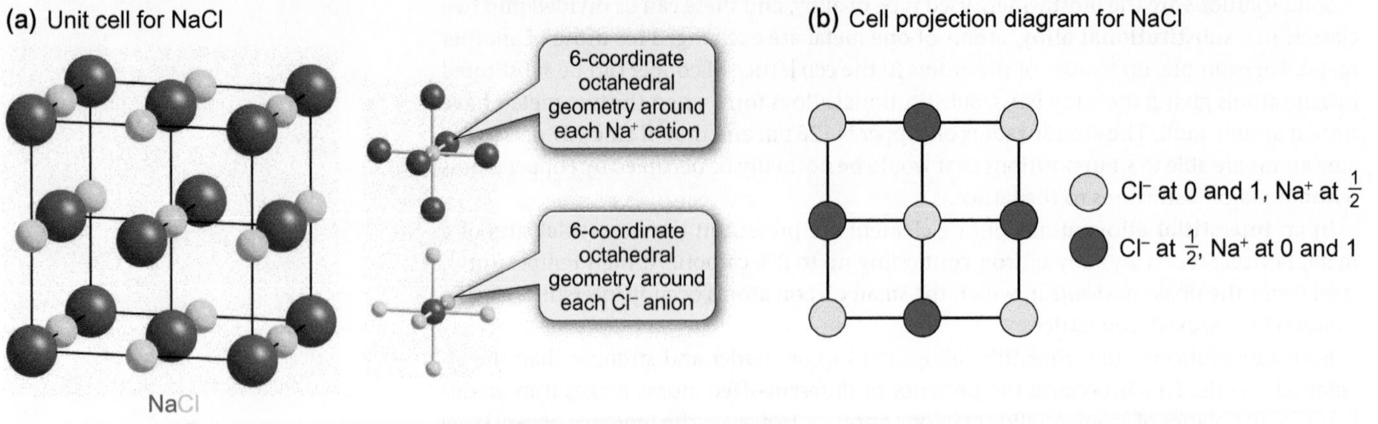
(a) Unit cell for NaCl
6-coordinate octahedral geometry around each Na⁺ cation
6-coordinate octahedral geometry around each Cl⁻ anion
NaCl
(b) Cell projection diagram for NaCl
Cl⁻ at 0 and 1, Na⁺ at 1/2
Cl⁻ at 1/2, Na⁺ at 0 and 1

Figure 6.20 (a) The unit cell and (b) cell projection diagram for sodium chloride (NaCl).

Interactive 3D fluorite and anti-fluorite structures.

Crystals of fluorite (CaF_2).

The NaCl structure is very common and is adopted by many other compounds. Most Group 1 metal halides have this structure, as do many oxides such as magnesium oxide (MgO). The structure is not limited to compounds with monatomic cations and anions. For example, iron disulfide (FeS_2, iron pyrites) adopts the NaCl structure, with $[S_2]^{2-}$ ions taking the place of the Cl^- ions and Fe^{2+} ions taking the place of the Na^+ ions.

Fluorite (CaF_2)

The structure of **fluorite** (CaF_2) is shown in Figure 6.21 along with its cell projection diagram. Fluorite has a structure based on a cubic close-packed arrangement of Ca^{2+} cations, with the F^- anions occupying the tetrahedral sites. Since there are two tetrahedral sites per atom position, this leads to the observed 1:2 stoichiometry. Each Ca^{2+} ion has a coordination number of 8 and is in a cubic environment. Each F^- ion has a coordination number of 4 and is in a tetrahedral environment. The unit cell contains 4 Ca^{2+} ions and 8 F^- ions. Many other compounds adopt this structure including zirconium(IV) oxide (ZrO_2) and barium chloride ($BaCl_2$).

The related structure in which there is a cubic close-packed arrangement of anions, with the cations in the tetrahedral sites, is called the **antifluorite** structure. Examples of compounds adopting the antifluorite structure include sodium oxide (Na_2O) and silver(I) sulfide (Ag_2S).

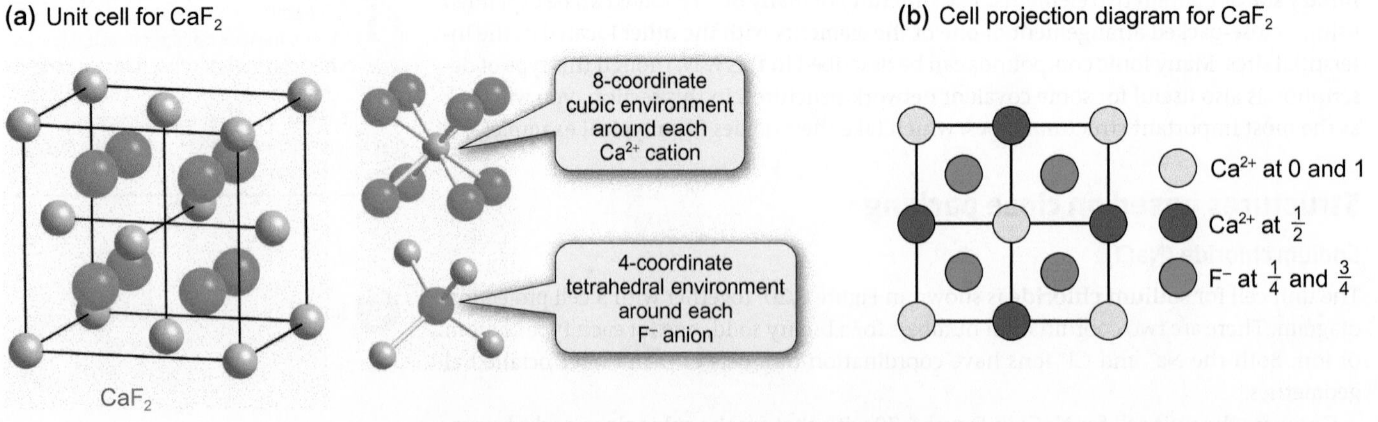
(a) Unit cell for CaF₂
8-coordinate cubic environment around each Ca²⁺ cation
4-coordinate tetrahedral environment around each F⁻ anion
CaF₂
(b) Cell projection diagram for CaF₂
Ca²⁺ at 0 and 1
Ca²⁺ at 1/2
F⁻ at 1/4 and 3/4

Figure 6.21 (a) The unit cell and (b) cell projection diagram for fluorite (CaF_2).

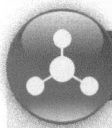

Box 6.4 X-ray crystallography

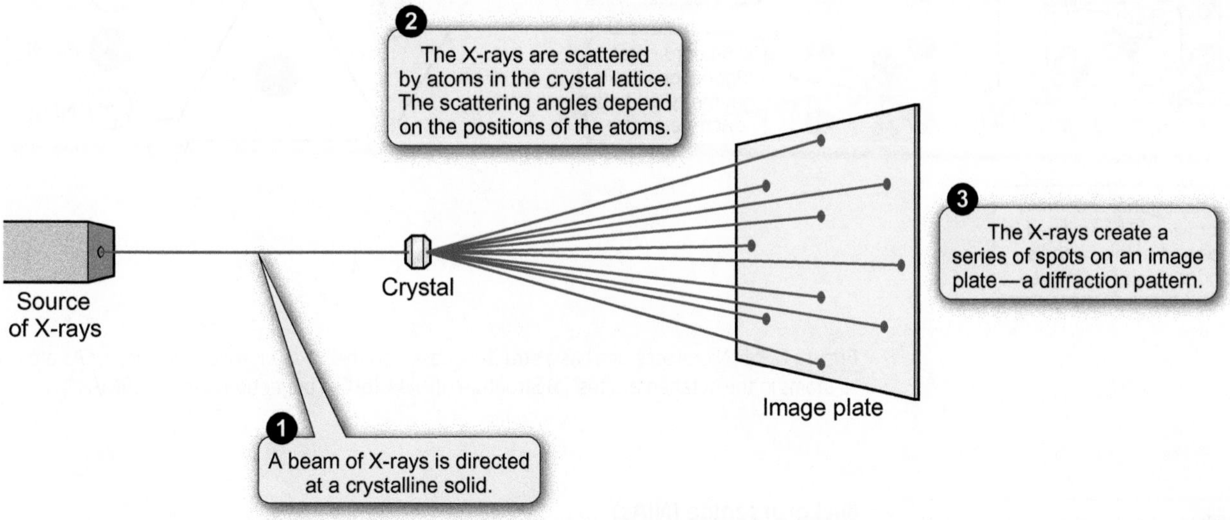
2
The X-rays are scattered by atoms in the crystal lattice. The scattering angles depend on the positions of the atoms.
Source of X-rays
Crystal
3
The X-rays create a series of spots on an image plate—a diffraction pattern.
Image plate
1
A beam of X-rays is directed at a crystalline solid.

▲ A typical diffraction pattern for a single crystal.

How do chemists determine the structures of solids? One very powerful method for determining the structure of a crystalline solid is X-ray crystallography. This technique relies on the fact that the spacing between atoms (100 pm–300 pm) is similar in magnitude to the wavelength of X-rays.

The electron clouds of the atoms in the crystal scatter the X-rays, with the degree of scattering proportional to the atomic number, Z. The scattering pattern from a single molecule is too weak to be detected. In a crystal, the individual atoms, ions, or molecules all lie in identical positions, so they scatter the X-rays in an identical manner. This means that the scattering pattern for each atom, ion, or molecule is amplified and able to be detected. In a modern instrument, the scattering leads to a **diffraction pattern** that is observed as a series of spots on an image plate. From this diffraction pattern, the electron density in the crystal can be reconstructed using a computer. This allows the bond length and bond angle data to be determined. However, since the scattering is proportional to Z, it can be difficult to locate hydrogen atoms, especially if they are close to much heavier atoms.

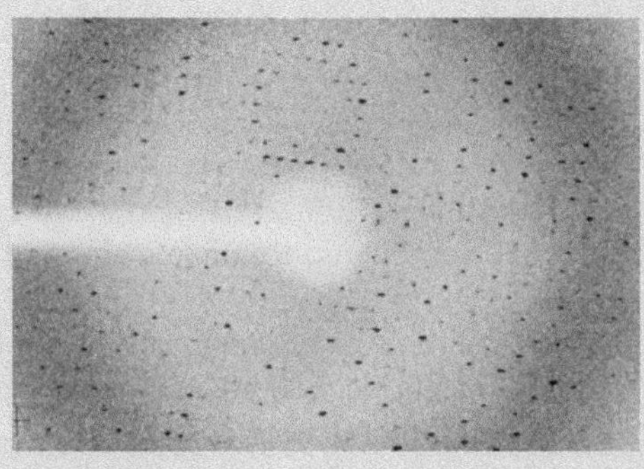

▲ The single crystal diffraction experiment using a modern detector.

The biggest problem with X-ray crystallography often lies in preparing the crystals for the experiment. Although crystals of only 0.1 mm × 0.1 mm × 0.1 mm are needed, for some compounds these are difficult to grow. Proteins are especially troublesome, as their crystals often diffract very poorly and can decompose in the X-ray beam. Despite these difficulties, thanks to patient experimental work, X-ray crystallography has been the key technique for finding the structure of proteins.

X-ray crystallography was used to obtain most of the structures shown in this book. Historically, X-ray crystallography was important in providing the first evidence of the double helical structure of DNA (see Box 25.5, p.1158) and the structures of many biologically important molecules.

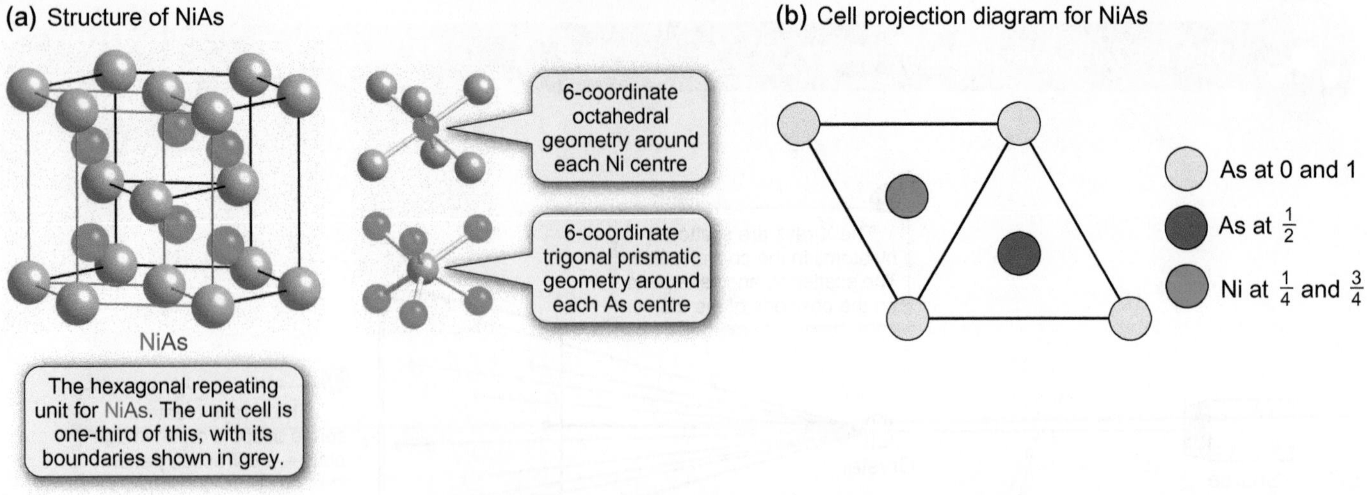

(a) Structure of NiAs
(b) Cell projection diagram for NiAs
6-coordinate octahedral geometry around each Ni centre
6-coordinate trigonal prismatic geometry around each As centre
NiAs
The hexagonal repeating unit for NiAs. The unit cell is one-third of this, with its boundaries shown in grey.
As at 0 and 1
As at $\frac{1}{2}$
Ni at $\frac{1}{4}$ and $\frac{3}{4}$

Figure 6.22 Nickel arsenide has a structure based on the hexagonal close packing of As atoms, with Ni atoms in the octahedral sites. (a) Structure of NiAs. (b) Cell projection diagram of NiAs.

Compounds adopting the NiAs structure tend to have a high degree of covalent character. See Section 6.6 (p.296).

Interactive 3D NiAs structure.

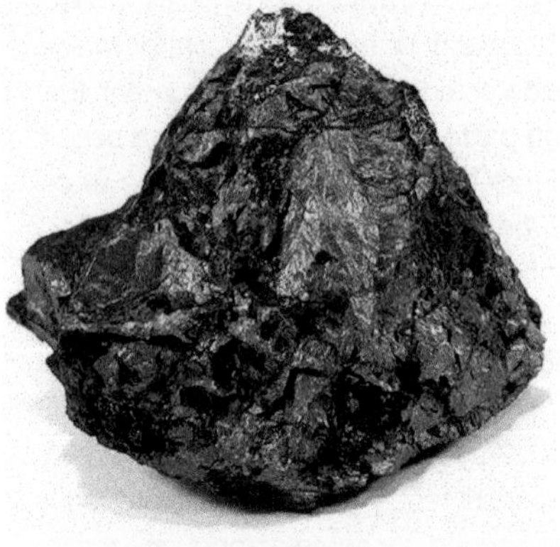

Nickeline (NiAs) occurs naturally as large clumps, rather than single crystals.

Black crystals of sphalerite, mixed with some quartz crystals.

Nickel arsenide (NiAs)

Nickel arsenide (NiAs), also known as nickeline, has a structure based on a hexagonal close-packed arrangement of arsenic atoms, with nickel atoms in the octahedral sites. The NiAs structure, together with the cell projection diagram, is shown in Figure 6.22. Both nickel and arsenic atoms have coordination numbers of 6, with the geometry around the nickel atoms octahedral and that around the arsenic atoms trigonal prismatic (see Figure 6.13, p.271). The nickel arsenide structure is adopted by many transition metal compounds containing Group 14–16 anions. Examples include cobalt(II) selenide (CoSe) and nickel(II) telluride (NiTe).

The NiAs structure is an hcp analogue of the ccp-based NaCl structure, with the metal atoms occupying all of the octahedral sites. There is no equivalent hcp analogue of the fluorite or antifluorite structure in which one set of ions occupies all of the tetrahedral sites. This is because the tetrahedral sites in an hcp structure are closer together than those in the ccp arrangement (see Figure 6.15, p.272). If all of the tetrahedral sites in an hcp structure were occupied, repulsion between the ions of the same charge would destabilize the lattice.

Sphalerite and wurtzite (ZnS)

Sphalerite (also known as zinc blende) and **wurtzite** are polymorphs of zinc sulfide (ZnS). Both have close-packed structures based on S^{2-} anions, with Zn^{2+} ions in half of the tetrahedral sites, which are alternatingly filled and empty. The sphalerite structure is based on cubic close packing, and is shown in Figure 6.23(a). The wurtzite structure is based on hexagonal close packing, and is shown in Figure 6.23(b).

In both cases the Zn^{2+} and S^{2-} ions have coordination numbers of 4 and tetrahedral geometries.

Copper(I) chloride (CuCl) and silver(I) iodide (AgI) adopt sphalerite structures. Zinc oxide (ZnO), silicon carbide (SiC), and boron nitride (BN) all adopt wurtzite structures, though the bonding in SiC and BN is covalent rather than ionic.

Cadmium chloride ($CdCl_2$) and cadmium iodide (CdI_2)

Cadmium chloride ($CdCl_2$) and cadmium iodide (CdI_2) adopt structures in which the anions are close packed and the cations occupy half of the octahedral sites. These structures are shown in Figure 6.24. The $CdCl_2$ structure is based on cubic close packing and the CdI_2 structure based on hexagonal close packing.

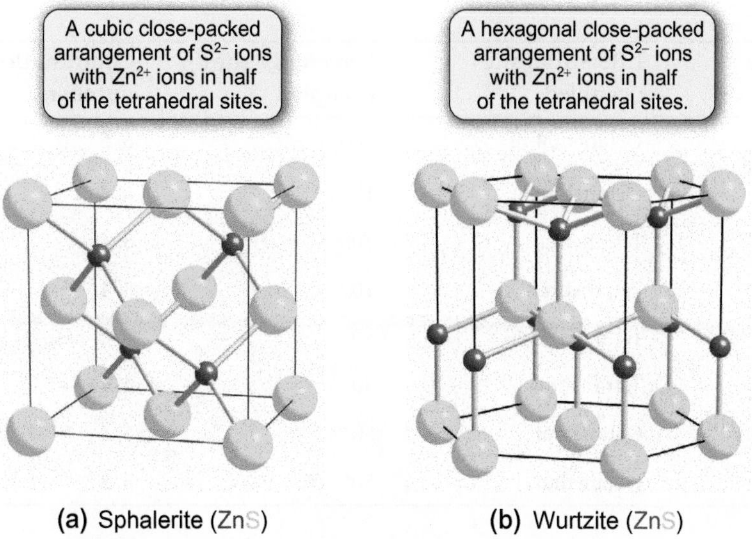
A cubic close-packed arrangement of S^{2-} ions with Zn^{2+} ions in half of the tetrahedral sites.
A hexagonal close-packed arrangement of S^{2-} ions with Zn^{2+} ions in half of the tetrahedral sites.
(a) Sphalerite (ZnS)
(b) Wurtzite (ZnS)

Figure 6.23 The structures of the sphalerite and wurtzite polymorphs of zinc sulfide (ZnS).

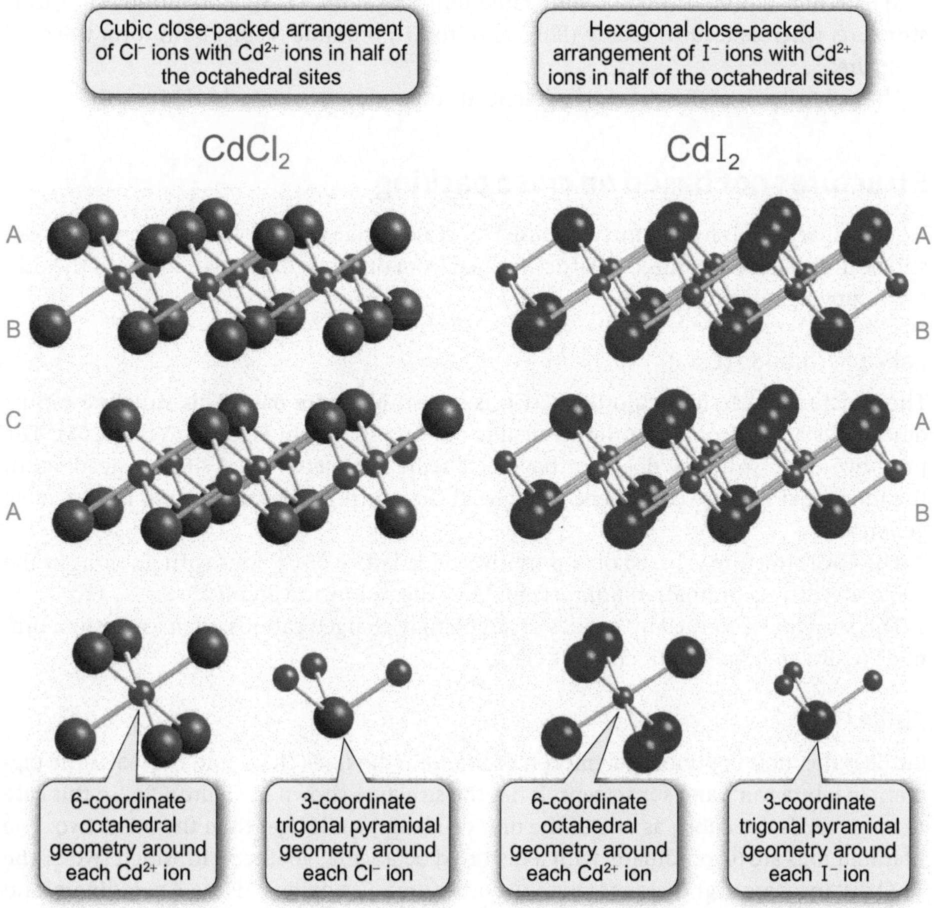
Cubic close-packed arrangement of Cl^- ions with Cd^{2+} ions in half of the octahedral sites
Hexagonal close-packed arrangement of I^- ions with Cd^{2+} ions in half of the octahedral sites
$CdCl_2$
CdI_2
A
B
C
A
A
B
A
B
6-coordinate octahedral geometry around each Cd^{2+} ion
3-coordinate trigonal pyramidal geometry around each Cl^- ion
6-coordinate octahedral geometry around each Cd^{2+} ion
3-coordinate trigonal pyramidal geometry around each I^- ion

Figure 6.24 The layer structures of cadmium chloride ($CdCl_2$) and cadmium iodide (CdI_2).

Gallium nitride, GaN, has the wurtzite structure, and is arguably the most important semiconductor material to be discovered since silicon. One of the uses of GaN is in blue light-emitting diodes (LEDs). The invention of blue LEDs enabled white light to be produced in a more energy efficient way than was previously possible, and led to the Nobel Prize in Physics in 2014 for the Japanese scientists Isamu Akasaki, Hiroshi Amano, and Sunji Nakamura.

Interactive 3D sphalerite and wurtzite structures.

Interactive 3D CdI_2 and $CdCl_2$ structures.

Table 6.3 Structures based on close packing

Crystal structure	Type of close packing	Interstitial sites occupied	Percentage sites occupied	Coordination numbers
Sodium chloride (NaCl)	Cubic	Octahedral	100	6, 6
Nickel arsenide (NiAs)	Hexagonal	Octahedral	100	6, 6
Fluorite (CaF_2)	Cubic	Tetrahedral	100	8, 4
not observed	Hexagonal	Tetrahedral	100	8, 4
Sphalerite (ZnS)	Cubic	Tetrahedral	50	4, 4
Wurtzite (ZnS)	Hexagonal	Tetrahedral	50	4, 4
Cadmium chloride ($CdCl_2$)	Cubic	Octahedral	50*	6, 3
Cadmium iodide (CdI_2)	Hexagonal	Octahedral	50*	6, 3

* Alternate layers filled and empty.

In both cases the Cd^{2+} cations fill alternate layers of octahedral sites, which means that both $CdCl_2$ and CdI_2 form layer structures. The Cd^{2+} ions have a coordination number of 6, and possess octahedral geometry. The halide ions have a coordination number of 3 and trigonal pyramidal geometry.

Magnesium chloride ($MgCl_2$) and tantalum(IV) sulfide (TaS_2) both form the $CdCl_2$ structure, whereas lead(II) iodide (PbI_2) and iron(II) bromide ($FeBr_2$) both form the CdI_2 structure.

The key features of the structures described above are summarized in Table 6.3.

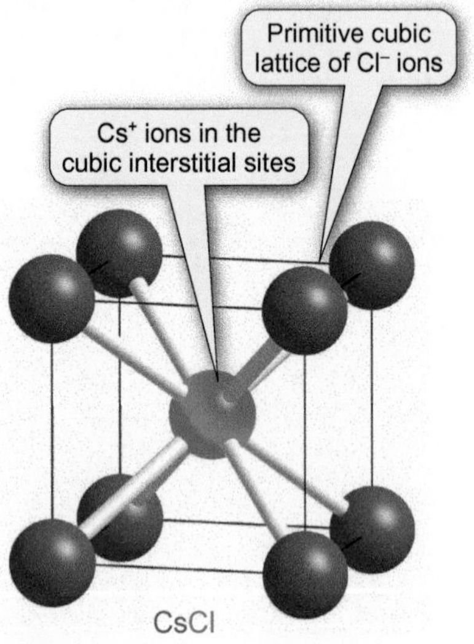
Primitive cubic lattice of Cl^- ions
Cs^+ ions in the cubic interstitial sites
CsCl

Figure 6.25 The unit cell for CsCl.

Interactive 3D CsCl structure.

If all of the ions in the CsCl structure were identical, the structure would be body-centred cubic. However, it is wrong to describe the CsCl structure as body-centred cubic, as in that structure the vertex and centre positions must be the same.

Interactive 3D rutile structure.

Structures not based on close packing

Not all binary solids form structures based on close packing. However, their structures can still be understood in terms of a lattice formed from one type of atom or ion with the others occupying interstitial sites.

Caesium chloride (CsCl)

The structure of caesium chloride (CsCl) is shown in Figure 6.25. This structure can be described starting from the primitive cubic packing shown in Figure 6.7(c) (p.265). The primitive cubic structure does not have octahedral and tetrahedral sites. Instead, all of the interstitial sites are equivalent. They lie at the centre of the cube and so have a cubic geometry.

The CsCl structure is based on a primitive cubic lattice of Cl^- ions with Cs^+ ions in the cubic sites. The coordination numbers for both the cation and anion are 8.

This structure is formed by halides of large singly charged cations, such as caesium bromide (CsBr) and thallium(I) chloride (TlCl).

Rutile (TiO_2)

Rutile is the most common polymorph of titanium dioxide (TiO_2), and is used as the pigment in white paint and sunscreens. It has the structure shown in Figure 6.26. In this case the unit cell is not cubic, as one of the unit cell lengths is longer than the other two. The titanium ions are 6-coordinate, with a distorted octahedral geometry in which two of the Ti–O distances are slightly longer than the other four. The oxide ions are 3-coordinate, and adopt trigonal planar geometries.

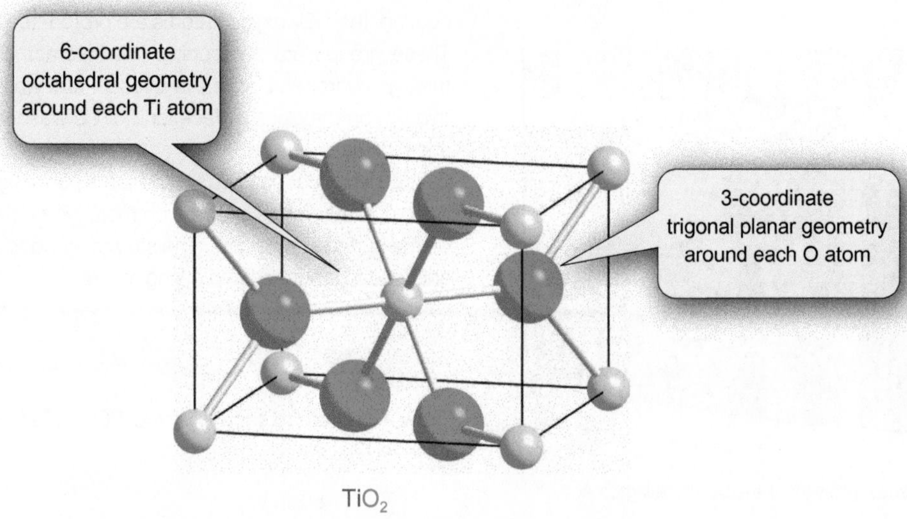
6-coordinate
octahedral geometry
around each Ti atom
3-coordinate
trigonal planar geometry
around each O atom
TiO2

Figure 6.26 The unit cell for rutile (TiO_2).

Box 6.5 Self-cleaning windows

Pilkington Activ™, launched in 2001, was the first type of glass to be produced that has the ability to clean itself. The glass contains a virtually transparent coating of titanium dioxide, 15 nm thick, which is deposited during the manufacturing process. The TiO_2 coating is durable, since it is bonded to the glass surface, and has two functions that allow it to act as a self-cleaning glass.

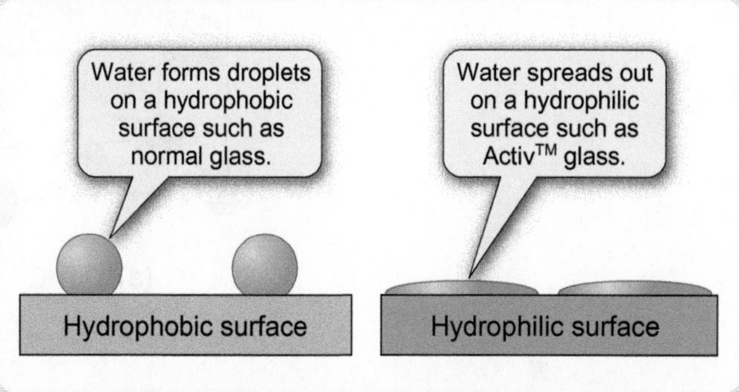
Water forms droplets on a hydrophobic surface such as normal glass.
Water spreads out on a hydrophilic surface such as Activ™ glass.
Hydrophobic surface
Hydrophilic surface

Firstly, TiO_2 absorbs ultraviolet photons from sunlight. On absorption of a photon, an electron is promoted from the filled valence band into the empty conduction band. The promoted electron is then able to interact with oxygen adsorbed on the surface to produce a superoxide ion (O_2^-)

$$O_2 + e^- \rightarrow O_2^-$$

Once it has been activated in this way, the photoactive form of TiO_2 acts as an oxidizing agent by accepting electrons into the vacancies in the valence band. It obtains these electrons by oxidation of water, converting H_2O into very reactive hydroxyl radicals ($OH^\bullet$)

$$H_2O \rightarrow OH^\bullet + H^+ + e^-$$

The hydroxyl radicals and superoxide ions are both strong oxidizing agents and are able to oxidize most of the organic molecules present in dirt, converting them eventually to CO_2 and H_2O. This means that photoactivation of TiO_2 provides a means for getting rid of most organic dirt.

The second way in which the TiO_2 coating leads to self-cleaning relates to interactions with water molecules. On the surface of the

▲ The manufacture of sheet glass within a float glass chamber.

→

→

▲ Comparison of regular sheet glass (on the left) and Pilkington Activ™ (on the right) in the rain.

coating, the TiO_2 oxygen atoms are protonated, forming OH groups. These groups are hydrophilic, and interact with water molecules through hydrogen bonds. As a result, rain water spreads out into a thin film on the glass surface, and dirt is washed off the window in a sheet of water.

This behaviour contrasts with that of normal glass, which is hydrophobic. Water forms droplets on most types of glass, and these run off the surface in streams. The streams tend to concentrate the dirt, and lead to smudges and drying marks.

Question

Draw a diagram to show how the TiO_2 surface interacts with water molecules.

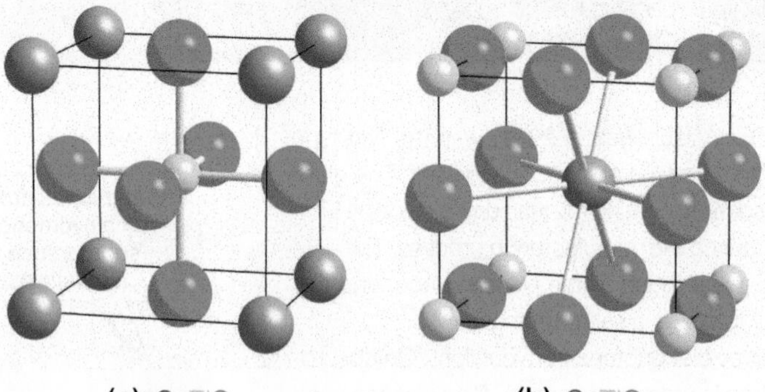

(a) $CaTiO_3$ (b) $CaTiO_3$

The two unit cells give the same structure when repeated in three dimensions.

Figure 6.27 Two equivalent unit cells for perovskite ($CaTiO_3$).

Interactive 3D perovskite structure.

Perovskite ($CaTiO_3$)

Perovskite is calcium titanate ($CaTiO_3$). The perovskite structure is very common, and many compounds with this structure have interesting electronic properties that have led to uses in solar cells, sensors, and lasers. Two unit cells for perovskite are shown in Figure 6.27. The two unit cells are equivalent, as they both repeat to give the same overall structure.

In the representation shown in Figure 6.27(a), the structure can be described as a primitive cubic packing of Ca^{2+} cations, with O^{2-} anions on the faces and a Ti^{4+} cation in the centre. This agrees with the $CaTiO_3$ formula, since the unit cell contains the ions shown below.

Ion type	Number of atoms in the unit cell	Share of the atom belonging to the unit cell	Total number of atoms contributing to the unit cell	Number of atoms in the formula
Ca^{2+}	8	$\frac{1}{8}$	$8 \times \frac{1}{8}$	1
Ti^{4+}	1	1	1×1	1
O^{2-}	6	$\frac{1}{2}$	$6 \times \frac{1}{2}$	3

The representation shown in Figure 6.27(b) is described in Worked example 6.4.

Worked example 6.4 Unit cells and formulae

Describe the unit cell shown in Figure 6.27(b) in terms of a primitive cubic packing of Ti^{4+} ions, stating the positions of the other ions. Show that this unit cell agrees with the $CaTiO_3$ formula.

Strategy

Look at the positions of the O^{2-} and Ca^{2+} ions. Count the total number of each of the three different ions in the unit cell. An ion lying on a vertex is shared between eight unit cells, so it contributes $\frac{1}{8}$ to each one. An ion on an edge is shared between four unit cells, so it contributes $\frac{1}{4}$ to each one. An ion on a face is shared between two unit cells, so it contributes $\frac{1}{2}$ to each one.

Solution

The unit cell can be described as a primitive cubic packing of Ti^{4+} cations, with O^{2-} anions on the edges and a Ca^{2+} cation in the centre. This gives the following ions in the unit cell which agrees with the $CaTiO_3$ formula.

Ion type	Number of atoms in the unit cell	Share of the atom belonging to the unit cell	Total number of atoms contributing to the unit cell	Number of atoms in the formula
Ca^{2+}	1	1	1×1	1
Ti^{4+}	8	$\frac{1}{8}$	$8 \times \frac{1}{8}$	1
O^{2-}	12	$\frac{1}{4}$	$12 \times \frac{1}{4}$	3

Question

The unit cell below is for an oxide of rhenium with the rhenium atoms in green and the oxygen atoms in red. What is the formula of the compound?

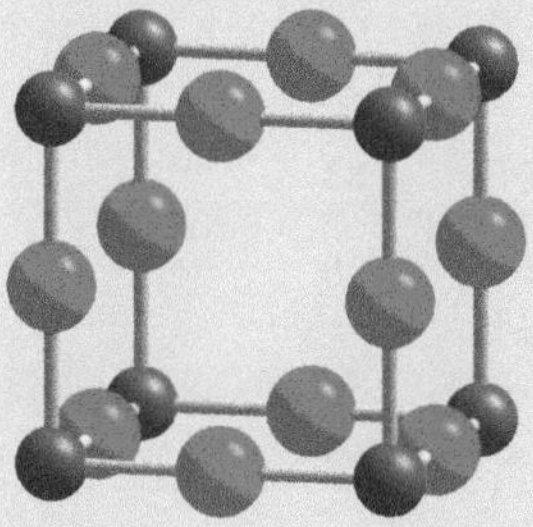

Summary

- The structures of many binary compounds can be described using a close-packed lattice of one atom or ion, with the other type occupying the octahedral or tetrahedral interstitial sites.
- The NaCl structure has a cubic close-packed array of chloride ions with the Na^+ ions occupying the octahedral sites whereas the CaF_2 structure has a cubic close-packed array of calcium ions with F^- ions occupying the tetrahedral sites.
- Structures can also be based on lattices that are not close packed. An example of this is the structure of CsCl, which consists of a primitive cubic lattice of Cl^- ions with Cs^+ ions in the cubic sites.

? For practice questions on these topics, see questions 6–11 at the end of this chapter (p.301).

6.5 The ionic model

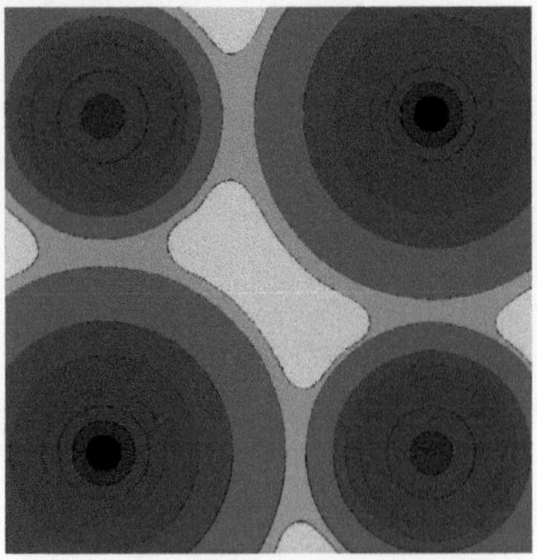

Figure 6.28 An electron density map for part of the structure of sodium chloride. The darker the colour, the higher the electron density.

More ionic radii are given in Appendix 8 (p.1363). These data show how the ionic radii change with coordination number.

p-block compounds are discussed in Chapter 27, and *d*-block compounds are discussed in Chapter 28.

Ionic solids are held together by electrostatic interactions between the cations and anions. In the ionic model, the ions are assumed to be hard spheres with fixed sizes. The radii of the spheres are known as **ionic radii.** The distance between the centres of two ions in an ionic solid can be measured accurately using X-ray crystallography. This distance is the sum of the cationic and anionic radii. It is, however, difficult to determine the values of the individual ionic radii from this distance. As you can see from Figure 6.28, it is hard to tell at exactly what point the electron density of the cation stops and that of the anion starts.

The values of ionic radii given in Tables 6.4 and 6.5 were determined using X-ray crystallography, and apply to the ions in 6-coordinate crystal structures. The point of minimum electron density between the nuclei was used to define where the cation stops and the anion starts. In reality, the radius of a particular cation or anion varies with the compound it is in, and differences in coordination number lead to significant changes. For this reason, values of ionic radii need to be used with some care.

Like atomic radii (Section 3.7, p.153), ionic radii increase down a group with the increase in the principal quantum number. Most *s*- and *p*-block ions have a Group 18 element configuration, though some *p*-block ions such as Sn^{2+} and Pb^{2+} retain two *s* electrons. Transition metal ions may have a number of *d* electrons. The changes in ionic radii going down the group for the Group 1 cations and the Group 17 anions are shown in Figure 6.29.

For ions of the same electronic configuration (called isoelectronic ions), the greater the nuclear charge, the smaller the ion. This is shown by the ionic radii for the isoelectronic

Table 6.4 Ionic radii for common cations*

M^+	Ionic radius / pm	M^{2+}	Ionic radius / pm	M^{3+}	Ionic radius / pm
Li^+	76	Mg^{2+}	72	Al^{3+}	54
Na^+	102	Ca^{2+}	100	Ti^{3+}	67
K^+	138	Sr^{2+}	118	Cr^{3+}	62
Rb^+	152	Ba^{2+}	135	Fe^{3+}	55
Cs^+	167	Mn^{2+}	83		
Ag^+	129	Fe^{2+}	78		
Tl^+	150	Ni^{2+}	69		
		Cu^{2+}	73		
		Zn^{2+}	74		

* For 6-coordinate structures. *d*-block metal ions are in high spin configurations (see Section 28.5, p.1284).

Table 6.5 Ionic radii for common anions*

M^-	Ionic radius / pm	M^{2-}	Ionic radius / pm	M^{3-}	Ionic radius / pm
F^-	133	O^{2-}	140	N^{3-}	171
Cl^-	181	S^{2-}	184		
Br^-	196	Se^{2-}	198		
I^-	220	Te^{2-}	221		

* For 6-coordinate structures.

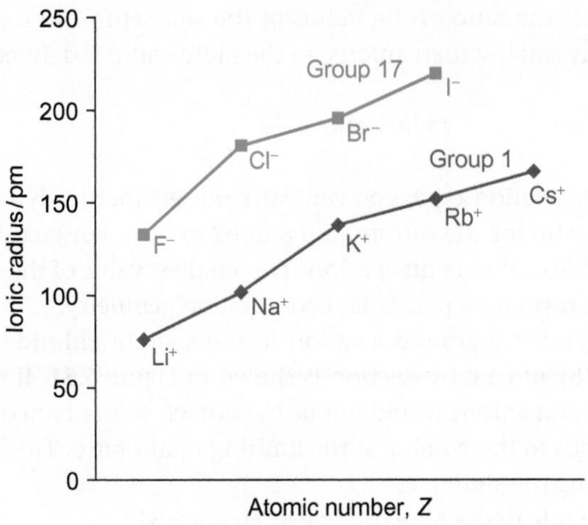
Group 17
Group 1
I⁻
Br⁻
Cl⁻
F⁻
Cs⁺
Rb⁺
K⁺
Na⁺
Li⁺
Ionic radius/pm
Atomic number, Z
250
200
150
100
50
0

Figure 6.29 The ionic radii for the Group 1 cations and the Group 17 anions.

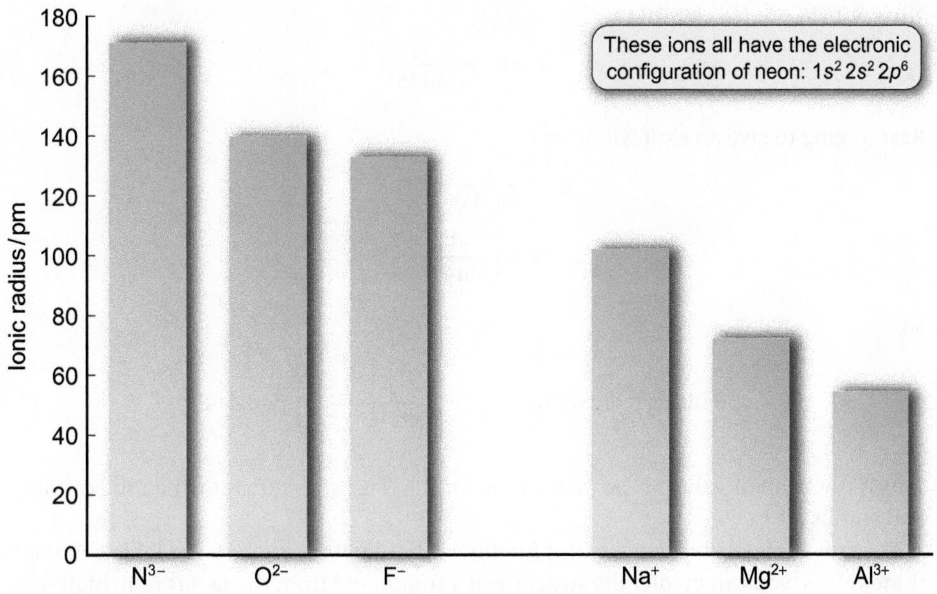
These ions all have the electronic configuration of neon: 1s² 2s² 2p⁶
Ionic radius/pm
180
160
140
120
100
80
60
40
20
0
N³⁻
O²⁻
F⁻
Na⁺
Mg²⁺
Al³⁺

Figure 6.30 The ionic radii of the isoelectronic ions Al^{3+}, Mg^{2+}, Na^+, F^-, O^{2-}, and N^{3-}.

ions N^{3-}, O^{2-}, F^-, Na^+, Mg^{2+}, and Al^{3+}, which are plotted in Figure 6.30. The increase in electron–electron repulsions is another factor in the large size of the anions.

Prediction of structures

Section 6.4 describes a number of different structures. Is it possible to predict which of these structures a particular compound will adopt? To some extent, the answer is yes. To begin with, you need to return to the idea of ions as charged, fixed-sized hard spheres.

In an ionic solid, the ions are assumed to be in contact with their nearest neighbours that have the opposite charge. This means that the distance between the centres of neighbouring ions is equal to the sum of the ionic radii, $r_+ + r_-$, where r_+ is the cationic radius and r_- is the anionic radius. The most stable structure is the one that maximizes the cation–anion contacts. In other words, the compound will adopt the structure with the maximum coordination number, subject to the cations and anions being in contact with each other.

The **radius ratio** is the ratio of the radius of the smaller ion to that of the larger ion. Cations are normally smaller than anions, so the radius ratio is defined as in Equation 6.3

$$\text{radius ratio} = \frac{r_+}{r_-} \tag{6.3}$$

The limiting radius ratio is the smallest value of the radius ratio for a particular coordination type.

For a particular coordination type, you can work out geometrically the smallest possible value of the radius ratio for the cations and anions to be in contact. If the radius ratio is any smaller, the cation–anion contact is lost. The smallest value of the radius ratio is called the **limiting radius ratio** for a particular coordination geometry.

The octahedral geometry around a cation in the sodium chloride structure is shown in Figure 6.20 (p.278) and a cross-section is shown in Figure 6.31. If the cation were any smaller, the cations and anions would not be in contact, so the radii of the ions shown in the figure corresponds to them being at the limiting radius ratio. The limiting radius ratio can be found by using the sine rule.

Applying the sine rule to the blue triangle in Figure 6.31

$$\frac{r_+ + r_-}{\sin 90°} = \frac{r_-}{\sin 45°}$$

Since sin 90° = 1, this becomes

$$r_+ + r_- = \frac{r_-}{\sin 45°}$$

Rearranging to give an expression for r_+

$$r_+ = \frac{r_-}{\sin 45°} - r_-$$
$$= r_-\left(\frac{1}{\sin 45°} - 1\right)$$

So

$$\text{limiting radius ratio} = \frac{r_+}{r_-} = \left(\frac{1}{\sin 45°} - 1\right) = 0.414$$

This is the smallest value of the radius ratio for the NaCl structure for which the cations and anions are in contact.

The ranges of possible radius ratios for different geometries are given in Table 6.6. Using these values, you can predict the structure of a compound from the radius ratio of its ions. This is known as the **radius ratio rule**.

The limiting radius ratio for tetrahedral geometry is calculated in Worked example 6.5.

For a right-angled triangle, the sine of an angle is defined as the length of the side opposite to the angle in question divided by the length of the hypotenuse. Trigonometric functions are described in Maths Toolkit MT5 (p.1322).

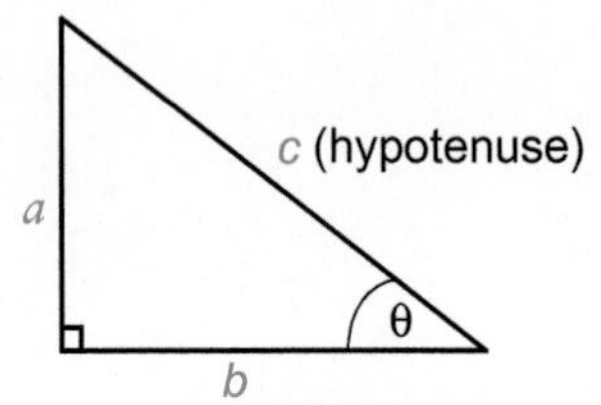

c (hypotenuse)
a
b
θ

$$\sin\theta = \frac{\text{opposite}}{\text{hypotenuse}} = \frac{a}{c}$$

The **sine rule** states that, for *any* triangle with sides of lengths *a*, *b*, and *c*, and angles of *A*, *B*, and *C*,

$$\frac{a}{\sin A} = \frac{b}{\sin B} = \frac{c}{\sin C}$$

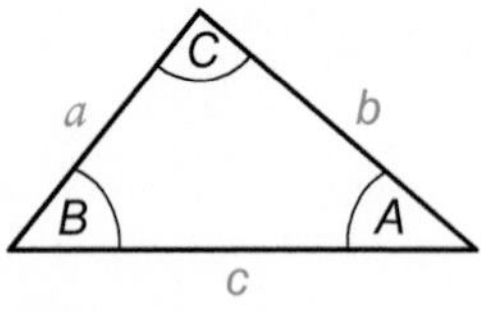

C
a
b
B
A
c

Visit the Online Resource Centre to view screencast 6.4 which walks you through how to calculate the limiting radius ratio.

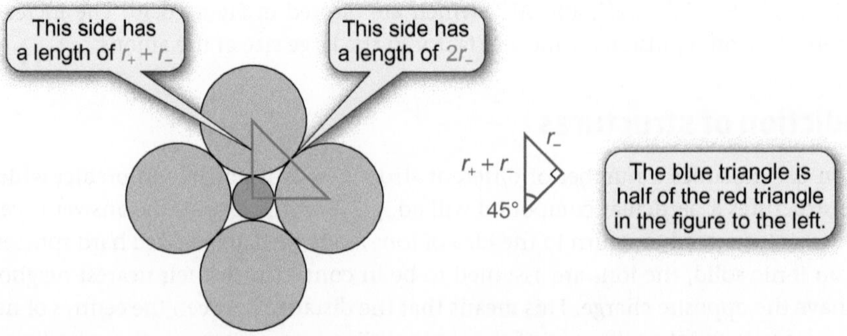

This side has a length of r+ + r−
This side has a length of 2r−
r−
r+ + r−
45°
The blue triangle is half of the red triangle in the figure to the left.

Figure 6.31 The geometry around a Na^+ cation in the structure of NaCl. The two other Cl^- ions, one going into the page and the other coming out of the page, have been omitted for clarity.

Table 6.6 Structural predictions from the radius ratio rule

Radius ratio	Coordination number	Geometry around cation
< 0.155	2	Linear
0.155–0.225	3	Trigonal
0.225–0.414	4	Tetrahedral
0.414–0.732	6	Octahedral
0.732–1.000	8	Cubic
> 1.000	12	Cuboctahedral

Worked example 6.5 Limiting radius ratios

What is the limiting radius ratio for tetrahedral geometry around a positive ion?

Strategy

Draw out part of a tetrahedron so that the anions on the vertices are in contact. The cation sits at the centre of the tetrahedron.

Construct a right-angled triangle from the centre of the tetrahedron with lengths in terms of r_+ and r_-.

Use the sine rule to work out the limiting radius ratio.

Solution

Using the sine rule

$$\frac{r_+ + r_-}{\sin 90°} = \frac{r_-}{\sin 54.8°}$$

54.8° is half the tetrahedral angle of 109.5°.

Since sin 90° = 1, this becomes

$$r_+ + r_- = \frac{r_-}{\sin 54.8°}$$

Rearranging to give an expression for r_+

$$r_+ = \frac{r_-}{\sin 54.8°} - r_-$$

$$= r_-\left(\frac{1}{\sin 54.8°} - 1\right)$$

$$\text{Limiting radius ratio} = \frac{r_+}{r_-} = \left(\frac{1}{\sin 54.8°} - 1\right)$$

$$= 0.225$$

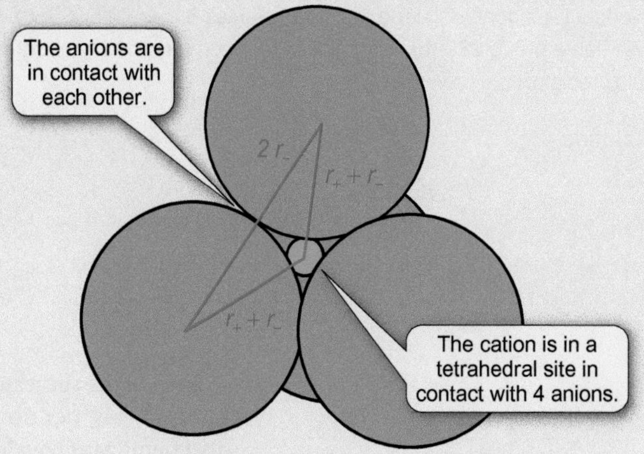
The anions are in contact with each other.
2r−
r+ + r−
r+ + r−
The cation is in a tetrahedral site in contact with 4 anions.

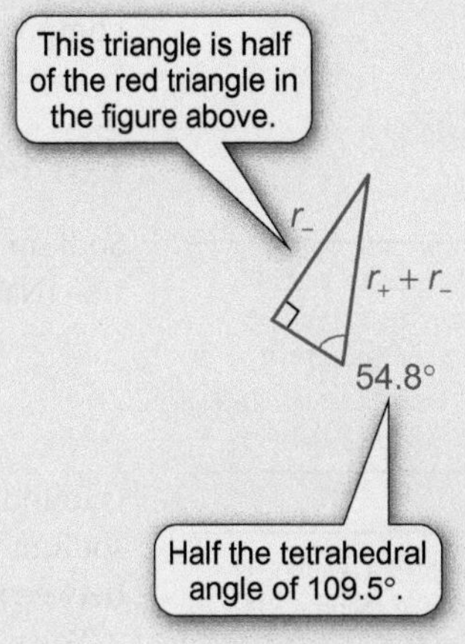
This triangle is half of the red triangle in the figure above.
r−
r+ + r−
54.8°
Half the tetrahedral angle of 109.5°.

Question

What is the limiting radius ratio for trigonal planar geometry around a positive ion? Comment on its value.

Does the radius ratio rule work? Ionic radii for common cations and anions are given in Tables 6.4 and 6.5. For NaCl the radius ratio is

$$\frac{r_+}{r_-} = \frac{r(\mathrm{Na^+})}{r(\mathrm{Cl^-})} = \frac{102\,\mathrm{pm}}{181\,\mathrm{pm}} = 0.56$$

This is within the range for 6-coordination and octahedral geometry, which is consistent with the structure adopted by NaCl. In this case, the theoretical prediction of the radius ratio rule corresponds with experimental observation. Another example is described in Worked example 6.6.

Worked example 6.6 The radius ratio rule

What is the radius ratio for the ions in CsCl? Does the structure of CsCl obey the radius ratio rule?

Strategy

Work out r_+/r_- for CsCl using the data in Tables 6.4 and 6.5 (p.286). Use Table 6.6 (p.289) to determine what geometry the radius ratio rule predicts.

Solution

$$\frac{r_+}{r_-} = \frac{r(\mathrm{Cs^+})}{r(\mathrm{Cl^-})} = \frac{167\,\not{\mathrm{pm}}}{181\,\not{\mathrm{pm}}} = 0.92$$

This is within the range for 8-coordination and cubic geometry, which is consistent with the observed structure for CsCl (primitive cubic lattice of Cl^- ions with Cs^+ ions in the centres of the cubes—see Figure 6.25, p.282). CsCl obeys the radius ratio rule.

Question

Magnesium oxide, MgO, adopts the sodium chloride structure. Does MgO obey the radius ratio rule?

Despite the successes here, the radius ratio rule does not always work. Ions are not really hard spheres, nor do they have fixed radii. In many so-called ionic compounds there are also significant covalent interactions, which can have an effect on the structure adopted. However, the radius ratio rule provides a quick and easy way to predict the structure for a compound, and it has a reasonable success rate.

Born-Haber cycles and lattice enthalpies

> The enthalpy change for a reaction between elements in their standard states to give one mole of a compound in its standard state is known as an enthalpy change of formation. See Section 13.3 (p.624).

> Hess's law states that, if a reaction is carried out in a series of steps, $\Delta H^\ominus$ for the reaction will be equal to the sum of the enthalpy changes for the individual steps. This means that the overall enthalpy change for a process is independent of the number of steps or the nature of the path by which the reaction is carried out, provided the starting and finishing states are the same in each case. For more details see Section 13.3 (p.623).

Sodium metal reacts violently with chlorine gas to give the ionic compound sodium chloride (Na^+Cl^-). The standard enthalpy change for the reaction

$$\mathrm{Na\,(s)} + \tfrac{1}{2}\,\mathrm{Cl_2\,(g)} \rightarrow \mathrm{NaCl\,(s)}$$

is found experimentally to be $-411\,\mathrm{kJ\,mol^{-1}}$. This is the **enthalpy change of formation** of sodium chloride [$\Delta_f H^\ominus$(NaCl)] since one mole of sodium chloride is formed in a reaction between the elements in their standard states.

You can imagine this reaction occurring in five discrete steps.

Step 1 Conversion of sodium metal into the gas phase atoms (atomization)

$\mathrm{Na\,(s)} \rightarrow \mathrm{Na\,(g)}$ $\Delta_a H^\ominus$(Na) is the enthalpy change of atomization for Na

Step 2 Cleavage of the Cl–Cl bond to convert Cl_2 molecules into Cl atoms

$\tfrac{1}{2}\,\mathrm{Cl_2\,(g)} \rightarrow \mathrm{Cl\,(g)}$ $\Delta_a H^\ominus$(Cl) is the enthalpy change of atomization for Cl

Step 3 Ionization of sodium to form Na^+

$Na(g) \rightarrow Na^+(g) + e^-$ $\Delta_i H(1)^{\ominus}(Na)$ is the first ionization enthalpy for Na

Step 4 Addition of an electron to chlorine to form Cl^-

$Cl(g) + e^- \rightarrow Cl^-(g)$ $\Delta_{eg} H(1)^{\ominus}(Cl)$ is the electron gain enthalpy for Cl

Step 5 Combination of the gaseous ions to form the solid state structure

$Na^+(g) + Cl^-(g) \rightarrow NaCl(s)$ $-\Delta_{latt} H^{\ominus}(NaCl)$ is minus the lattice enthalpy for NaCl

Put together, these enthalpy changes form an enthalpy cycle called a **Born–Haber cycle.** The Born–Haber cycle for NaCl is shown in Figure 6.32.

The reaction between sodium and chlorine does not proceed through the steps in this precise sequence, but you can use Hess's law to calculate the enthalpy change for any of the individual steps, provided the others are known.

The enthalpy change with the largest magnitude in the Born–Haber cycle for formation of NaCl is the formation of the ionic lattice from the gaseous ions. This is minus the lattice enthalpy of NaCl, as the lattice enthalpy is the enthalpy change for the conversion of one mole of the ionic solid into the gaseous ions, that is

$$NaCl(s) \rightarrow Na^+(g) + Cl^-(g) \quad \Delta_{latt} H^{\ominus}(NaCl)$$

$\Delta_{latt} H^{\ominus}(NaCl)$ is strongly endothermic due to the electrostatic interactions between the ions in the solid, and the $-\Delta_{latt} H^{\ominus}(NaCl)$ term in the Born–Haber cycle ensures that the formation of sodium chloride from sodium and chlorine is an exothermic reaction.

Generally, the **lattice enthalpy** for a compound A_xB_y is the enthalpy change for the process in Equation 6.4

$$A_xB_y(s) \rightarrow xA^{y+}(g) + yB^{x-}(g) \quad \Delta_{latt} H^{\ominus}(A_xB_y) \qquad (6.4)$$

You can apply Hess's law to the Born–Haber cycle in Figure 6.32 to work out a value for the lattice enthalpy. The enthalpy change of formation of NaCl is given by the expression

$$\Delta_f H^{\ominus}(NaCl) = \Delta_a H^{\ominus}(Na) + \Delta_i H(1)^{\ominus}(Na) + \Delta_a H^{\ominus}(Cl) + \Delta_{eg} H(1)^{\ominus}(Cl) - \Delta_{latt} H^{\ominus}(NaCl)$$

The **enthalpy change of atomization** for Na is identical to the enthalpy change of sublimation for Na. The enthalpy change of atomization for a diatomic molecule such as Cl_2 is equal to half of the Cl–Cl bond dissociation enthalpy.

Magnitude means the numerical value without the sign.

The lattice enthalpy is defined as the enthalpy change that occurs when a mole of an ionic solid is converted into the gaseous ions. Some books define lattice enthalpy as the enthalpy change when the gaseous ions are converted *into* one mole of an ionic solid. This gives the same number, but with the opposite sign.

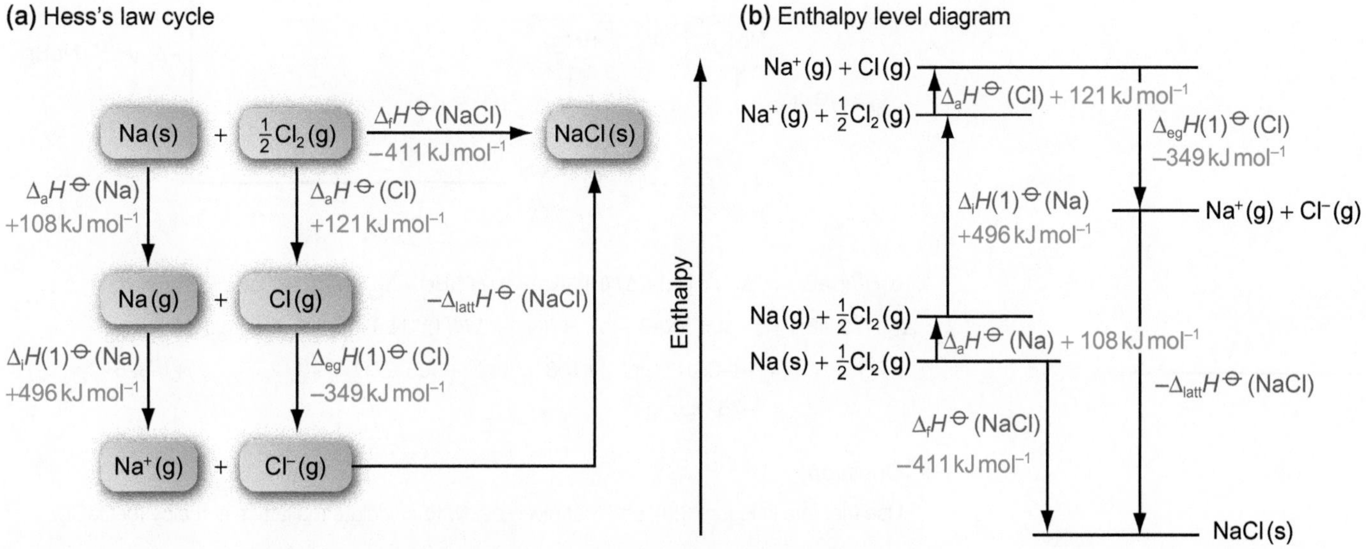
(a) Hess's law cycle
Na (s)
+
½Cl₂ (g)
ΔfH⊖ (NaCl)
−411 kJ mol⁻¹
NaCl (s)
ΔaH⊖ (Na)
+108 kJ mol⁻¹
ΔaH⊖ (Cl)
+121 kJ mol⁻¹
Na (g)
Cl (g)
−ΔlattH⊖ (NaCl)
ΔiH(1)⊖ (Na)
+496 kJ mol⁻¹
ΔegH(1)⊖ (Cl)
−349 kJ mol⁻¹
Na⁺ (g)
Cl⁻ (g)
(b) Enthalpy level diagram
Na⁺ (g) + Cl (g)
Na⁺ (g) + ½Cl₂ (g)
ΔaH⊖ (Cl) + 121 kJ mol⁻¹
ΔegH(1)⊖ (Cl)
−349 kJ mol⁻¹
ΔiH(1)⊖ (Na)
+496 kJ mol⁻¹
Na⁺ (g) + Cl⁻ (g)
Enthalpy
Na (g) + ½Cl₂ (g)
ΔaH⊖ (Na) + 108 kJ mol⁻¹
Na (s) + ½Cl₂ (g)
−ΔlattH⊖ (NaCl)
ΔfH⊖ (NaCl)
−411 kJ mol⁻¹
NaCl (s)

Figure 6.32 A Born–Haber cycle for the enthalpy change of formation of NaCl. This Born–Haber cycle is shown in two forms: (a) emphasizes the different components involved and shows the enthalpy changes between them, (b) shows the magnitudes of the enthalpy changes to scale.

In a Born–Haber cycle, all of the terms are enthalpy changes. Sometimes internal energy changes (ΔU) are used instead of enthalpy changes in Born–Haber cycles as the values are often more readily available. The differences between ΔH and ΔU are usually small, so they are not important except in very accurate work. The relationship between ΔH and ΔU is discussed further in Section 13.5 (p.642).

By rearranging this equation, you can obtain an expression for the lattice enthalpy for NaCl

$$\Delta_{latt}H^{\ominus}(\text{NaCl}) = -\Delta_f H^{\ominus}(\text{NaCl}) + \Delta_a H^{\ominus}(\text{Na}) + \Delta_i H(1)^{\ominus}(\text{Na}) + \Delta_a H^{\ominus}(\text{Cl}) + \Delta_{eg} H(1)^{\ominus}(\text{Cl})$$

Using the values from Figure 6.32

$$\begin{aligned}\Delta_{latt}H^{\ominus}(\text{NaCl}) &= -(-411\,\text{kJ mol}^{-1}) + 108\,\text{kJ mol}^{-1} + 496\,\text{kJ mol}^{-1} \\ &\quad + 121\,\text{kJ mol}^{-1} + (-349\,\text{kJ mol}^{-1}) \\ &= +787\,\text{kJ mol}^{-1}\end{aligned}$$

The Born–Haber cycle for sodium bromide is constructed in Worked example 6.7 and used to calculate the lattice enthalpy for this compound.

Worked example 6.7 Born–Haber cycles

Use the data in Figure 6.32 (p.291) and those given below to calculate the lattice enthalpy for NaBr.

$$\begin{aligned}\Delta_f H^{\ominus}(\text{NaBr}) &= -361\,\text{kJ mol}^{-1} \\ \Delta_a H^{\ominus}(\text{Br}) &= +112\,\text{kJ mol}^{-1} \\ \Delta_{eg} H(1)^{\ominus}(\text{Br}) &= -325\,\text{kJ mol}^{-1}\end{aligned}$$

Strategy

Draw a Born–Haber cycle for the formation of NaBr. Take care over the signs of the enthalpy changes when you write in the values.

Solution

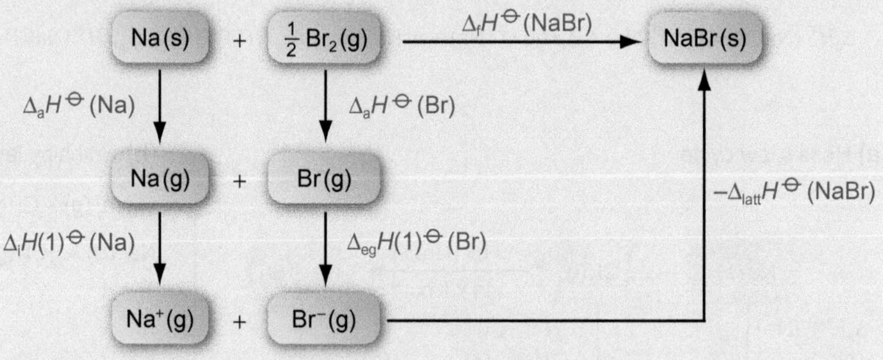
Na(s)
+
1/2 Br2(g)
ΔfH⦵(NaBr)
NaBr(s)
ΔaH⦵ (Na)
ΔaH⦵ (Br)
Na(g)
+
Br(g)
−ΔlattH⦵ (NaBr)
ΔiH(1)⦵ (Na)
ΔegH(1)⦵ (Br)
Na+(g)
+
Br−(g)

$$\begin{aligned}\Delta_f H^{\ominus}(\text{NaBr}) &= \Delta_a H^{\ominus}(\text{Na}) + \Delta_i H(1)^{\ominus}(\text{Na}) + \Delta_a H^{\ominus}(\text{Br}) + \Delta_{eg} H(1)^{\ominus}(\text{Br}) - \Delta_{latt} H^{\ominus}(\text{NaBr}) \\ \Delta_{latt} H^{\ominus}(\text{NaBr}) &= -\Delta_f H^{\ominus}(\text{NaBr}) + \Delta_a H^{\ominus}(\text{Na}) + \Delta_i H(1)^{\ominus}(\text{Na}) + \Delta_a H^{\ominus}(\text{Br}) + \Delta_{eg} H(1)^{\ominus}(\text{Br}) \\ &= -(-361\,\text{kJ mol}^{-1}) + 108\,\text{kJ mol}^{-1} + 496\,\text{kJ mol}^{-1} + 112\,\text{kJ mol}^{-1} + (-325\,\text{kJ mol}^{-1}) \\ &= +752\,\text{kJ mol}^{-1}\end{aligned}$$

Question

Use the data in Figure 6.32 and that given below to calculate the lattice enthalpy for $CaCl_2$.

$$\Delta_f H^{\ominus}(\text{CaCl}) = -796\,\text{kJ mol}^{-1} \qquad \Delta_i H(1)^{\ominus}(\text{Ca}) = +590\,\text{kJ mol}^{-1}$$

$$\Delta_a H^{\ominus}(\text{Ca}) = +178\,\text{kJ mol}^{-1} \qquad \Delta_i H(2)^{\ominus}(\text{Ca}) = +1145\,\text{kJ mol}^{-1}$$

Summary

- Coordination numbers and structures of ionic compounds can be predicted using the radius ratio rule, though this is not always reliable as it assumes ions are hard spheres with fixed radii.
- Hess's law enables enthalpy changes to be determined by using Born–Haber cycles.
- The lattice enthalpy for the compound A_xB_y is the enthalpy change for the process

$$A_xB_y\,(s) \rightarrow x\,A^{y+}\,(g) + y\,B^{x-}\,(g)$$

- In a Born–Haber cycle, the enthalpy change with the largest magnitude is usually the lattice enthalpy.

? For practice questions on these topics, see questions 12–13 at the end of this chapter (p.301).

6.6 Calculating lattice energy

Born–Haber cycles allow values for lattice enthalpies to be determined from experimental data. It is also useful to be able to calculate lattice enthalpies without needing to know other thermodynamic data. In the ionic model, the energy change when the gaseous ions come together to form an ionic solid comes from the electrostatic interactions between the ions. By considering all the electrostatic interactions present in an ionic solid, a theoretical value for the lattice energy can be obtained.

The calculations described in this section give internal energy changes (ΔU) rather than enthalpy changes (ΔH). Section 13.5 describes how to convert between internal energies and enthalpies but, since the differences are relatively small, in this section the difference is ignored.

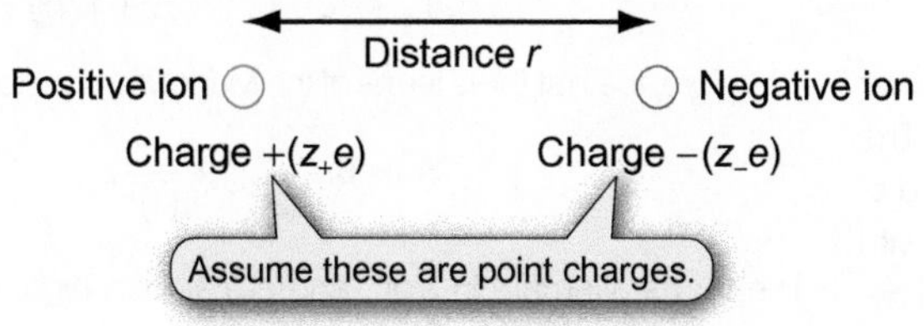
Distance r
Positive ion
Negative ion
Charge $+(z_+e)$
Charge $-(z_-e)$
Assume these are point charges.

The lattice energy is the difference in potential energy between the ions in the solid lattice and the ions widely separated as a gas. The change in internal energy (at 0 K) when two ions, of charges $+z_+$ and $-z_-$, are brought from an infinite distance to a distance r is given by Equation 6.5.

$$\Delta U = \frac{-z_+z_-e^2}{4\pi\varepsilon_0 r} \quad (6.5)$$

where z_+ and z_- are positive integers, equal to the charges on the ions, e is the charge on the electron (1.6022×10^{-19} C), r is the distance between the ions (in metres), and ε_0 is the permittivity of a vacuum. The latter is a fundamental constant, with the value $8.8542 \times 10^{-12}\,C^2J^{-1}m^{-1}$. This expression assumes that the ions are point charges.

In a crystal, there are many millions of interactions between pairs of ions. Those between cations and anions are attractive, whereas those between ions of the same charge are repulsive. To account for all of these interactions, a term known as the **Madelung constant,** A, is introduced, giving Equation 6.6. This equation also includes the Avogadro constant so that ΔU becomes a molar quantity

$$\Delta U = \frac{-AN_Az_+z_-e^2}{4\pi\varepsilon_0 r} \quad (6.6)$$

Box 6.6 Determining the Madelung constant

The unit cell for the NaCl structure is shown in Figure 6.20 (p.278). This is shown again below, though in this case ions at different distances from the red central sodium ion are shown in different colours.

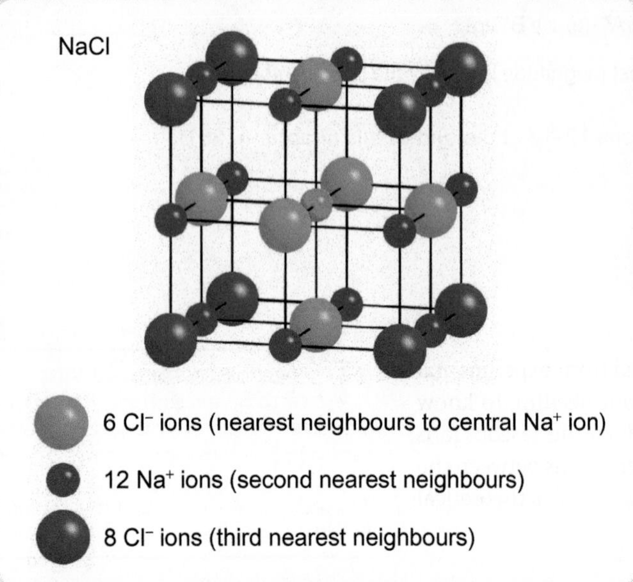
NaCl
6 Cl^- ions (nearest neighbours to central Na^+ ion)
12 Na^+ ions (second nearest neighbours)
8 Cl^- ions (third nearest neighbours)

Start with the red Na^+ ion in the centre of the unit cell and consider the interaction of this ion with all the other ions in the structure. The nearest neighbours are the six green chloride ions, which lie at a distance r away. Using Equation 6.5, the interactions of these ions with the central Na^+ ion gives an *attractive* energy ΔU_1, which is given by

$$\Delta U_1 = 6 \times \frac{-(1 \times 1)e^2}{4\pi\varepsilon_0 r} = -6 \times \frac{e^2}{4\pi\varepsilon_0 r}$$

The next nearest neighbours are the sodium ions coloured blue. There are twelve of these ions, each on an edge of the unit cell, and they lie at a distance of $(2^{1/2})r$ away from the central Na^+ ion. (You can show the distance is $(2^{1/2})r$ using Pythagoras's theorem.)

The interactions between the central Na^+ ion and these Na^+ ions give a *repulsive* energy ΔU_2, which is given by

$$\Delta U_2 = 12 \times \frac{+(1 \times 1)e^2}{4\pi\varepsilon_0 (2)^{1/2} r} = \frac{12}{(2)^{1/2}} \times \frac{e^2}{4\pi\varepsilon_0 r}$$

The next set of nearest ions are the chloride ions, coloured purple. There are eight of these ions, each sitting on a vertex of the unit cell, and they lie at a distance of $(3^{1/2})r$ from the central atom. Together these give an attractive energy ΔU_3, which is given by

$$\Delta U_3 = 8 \times \frac{-(1 \times 1)e^2}{4\pi\varepsilon_0 (3)^{1/2} r} = \frac{-8}{(3)^{1/2}} \times \frac{e^2}{4\pi\varepsilon_0 r}$$

The three energy terms ΔU_1, ΔU_2, and ΔU_3 account for all of the ions in a unit cell, but the interactions do not stop there. Other terms are calculated in a similar way using neighbouring unit cells.

The total energy ΔU is given by the sum of all the terms, ΔU_n, from $n = 1$ to $n = \infty$. The energy is given by the expression in Equation 6.8, where only the first three terms, ΔU_1, ΔU_2, and ΔU_3, have been written out explicitly

$$\begin{aligned}\Delta U &= \Delta U_1 + \Delta U_2 + \Delta U_3 + \ldots \qquad (6.8)\\ &= \frac{-6e^2}{4\pi\varepsilon_0 r} + \frac{+12e^2}{4\pi\varepsilon_0 (2)^{1/2} r} + \frac{-8e^2}{4\pi\varepsilon_0 (3)^{1/2} r} + \ldots \\ &= \left(6 - \frac{12}{(2)^{1/2}} + \frac{8}{(3)^{1/2}} - \ldots\right)\left(\frac{-e^2}{4\pi\varepsilon_0 r}\right)\end{aligned}$$

Therefore, the first three terms of the Madelung constant, A, for NaCl are

$$A = 6 - \frac{12}{(2)^{1/2}} + \frac{8}{(3)^{1/2}}$$

The terms alternate in sign, and are all of a high, though generally decreasing, magnitude. A large number of terms need to be included to get to the actual value for the Madelung constant, which is 1.7476 for the NaCl structure.

Question

Draw a fragment of the NaCl structure in which a unit cell is extended out from the centre, and use this to identify the next set of closest ions to the central Na^+ ion. Use the number of these ions and their distance from the central Na^+ ion to calculate the fourth term in the Madelung constant for NaCl.

Table 6.7 Values of Madelung constants for common structural types

Crystal structure	A
Sodium chloride (NaCl)	1.7476
Fluorite (CaF_2)	2.5194
Sphalerite (ZnS)	1.6381
Wurtzite (ZnS)	1.6413
Rutile (TiO_2)	2.385
Caesium chloride (CsCl)	1.7627

The lattice energy is the negative of this potential energy, that is,

$$\text{Lattice energy} = \Delta_{\text{latt}}U = \frac{AN_A z_+ z_- e^2}{4\pi\varepsilon_0 r} \qquad (6.7)$$

The value of the Madelung constant depends on the structure of the compound. For the sodium chloride structure, A is 1.7476, and values of A for other structures are given in Table 6.7. The calculation of the Madelung constant for sodium chloride is described in Box 6.6.

The expression in Equation 6.7 would be accurate if the ions were all point charges. This is not true in reality, which means that additional short-range forces between the ions also need to be included. These forces arise from the repulsions between overlapping electron clouds, and rise steeply with decreasing r.

Table 6.8 Values of the Born exponent n for different ions*

Electronic configuration of ion	Examples of cations	Examples of anions	n
[He]	Li^+, Be^{2+}	H^-	5
[Ne]	Na^+, Mg^{2+}, Al^{3+}	F^-, O^{2-}, N^{3-}	7
[Ar] or [Ar] $3d^{10}$	K^+, Ca^{2+}, Zn^{2+}	Cl^-, S^{2-}	9
[Kr] or [Kr] $3d^{10}$	Rb^+, Sr^{2+}, Cd^{2+}	Br^-, Se^{2-}	10
[Xe] or [Xe] $3d^{10}$	Cs^+, Ba^{2+}, Hg^{2+}	I^-, Te^{2-}	12

* For a compound XY, the overall value of n is given by the average of the values for the cation and the anion.

The repulsive energy term, E_{rep}, is given by Equation 6.9

$$E_{rep} = \frac{B}{r^n} \tag{6.9}$$

where both B and n are constants. The constant n is called the **Born exponent**. When the repulsive energy has been taken into account, Equation 6.7 is converted into Equation 6.10, which is known as the **Born–Landé equation**

$e = 1.6022 \times 10^{-19}$ C,
$\varepsilon_0 = 8.8542 \times 10^{-12}$ C^2 J^{-1} m^{-1}

$$\Delta_{latt}U = \frac{AN_A z_+ z_- e^2}{4\pi\varepsilon_0 r}\left(1 - \frac{1}{n}\right) \tag{6.10}$$

The Born exponent n can be obtained experimentally from the compressibility of the solid, which relates to how the lattice energy changes with applied pressure. An approximate value of the Born exponent for a particular compound can be predicted from the electronic configuration of the ions. Values of n for different ions are given in Table 6.8.

When the two ions in a compound have the same value of n, that is also the value for the compound. For example, n is 7 for both Na^+ and F^-, so, for NaF, $n = 7$. When the ions do not have the same Born exponents, the compound takes the average value of the ions. The Na^+ ion has $n = 7$, and the Cl^- ion $n = 9$ so, for NaCl,

$$n = \frac{7+9}{2} = 8$$

Using the Born–Landé equation

You can use the Born–Landé equation (Equation 6.10) to calculate the theoretical lattice energy of NaCl.

For NaCl, $r = 281.5$ pm, $z_+ = 1$, $z_- = 1$, $A = 1.7476$, and $n = 8$.

$$\begin{aligned}\Delta_{latt}U &= \frac{(1.7476) \times (6.02217 \times 10^{23}\ \text{mol}^{-1}) \times (1) \times (1) \times (1.6022 \times 10^{-19}\ \cancel{\text{C}})^2}{4 \times \pi \times (8.8542 \times 10^{-12}\ \cancel{\text{C}^2}\ \text{J}^{-1}\ \cancel{\text{m}^{-1}}) \times (281.5 \times 10^{-12}\ \cancel{\text{m}})} \times \left(1 - \frac{1}{8}\right)\\ &= +754700\ \text{J mol}^{-1}\\ &= +754.7\ \text{kJ mol}^{-1}\end{aligned}$$

This compares with the value of +787 kJ mol^{-1} for the lattice enthalpy that was calculated from the experimental data using a Born–Haber cycle on p.292. The agreement between the theoretical value given by the Born–Landé equation and the experimental value is relatively good. This suggests the theory behind the Born–Landé equation is a good way of describing ionic bonding, in this case at least.

The Born–Landé equation is used in Worked example 6.8 to calculate the lattice enthalpy of magnesium oxide.

Worked example 6.8 Using the Born–Landé equation

Use the Born–Landé equation to calculate the lattice energy of MgO, which adopts the NaCl structure with a Mg–O distance of 210.9 pm.

Strategy

The Born–Landé equation is given in Equation 6.10.

Determine the Born exponent, n, for MgO from the values for the ions given in Table 6.8. Look up the value for the Madelung constant for MgO in Table 6.7 (p.294).

Solution

$$\Delta_{\text{latt}}U = \frac{AN_A z_+ z_- e^2}{4\pi\varepsilon_0 r}\left(1-\frac{1}{n}\right) \tag{6.10}$$

Mg^{2+} and O^{2-} both have the electronic configuration [Ne]. From Table 6.8, both have Born exponents $n = 7$, so this is also the value for MgO.

For MgO (from Table 6.7), $A = 1.7476$, $z_+ = 2$, and $z_- = 2$. Putting these values into Equation 6.9

$$\Delta_{\text{latt}}U = \frac{(1.7476)\times(6.02217\times10^{23}\,\text{mol}^{-1})\times(2)\times(2)\times(1.6022\times10^{-19}\,\cancel{\text{C}})^2}{4\times\pi\times(8.8542\times10^{-12}\,\cancel{\text{C}^2}\,\text{J}^{-1}\,\cancel{\text{m}}^{-1})\times(210.9\times10^{-12}\,\cancel{\text{m}})}\times\left(1-\frac{1}{7}\right)$$

$$= +3.947\times10^{6}\ \text{J mol}^{-1}$$

$$= +3947\ \text{kJ mol}^{-1}$$

This has a much larger magnitude than the lattice energy of NaCl. This difference is mainly due to the higher charges on the ions in MgO.

Question

Calculate the lattice energy for CsCl given that the closest Cs–Cl distance is 348.1 pm.

Deviations from the ionic model

For NaCl, the calculated lattice energy from the Born–Landé equation is close to the experimental value derived from a Born–Haber cycle. This suggests that the ionic model is a good description for the bonding in NaCl. However, no compound is completely ionic. Even in NaCl, the positive charge on the Na^+ cations distorts the spherical electron clouds of the Cl^- anions, leading to some degree of covalent character.

Atoms or ions that are easily distorted are said to be **polarizable**. Large anions are more polarizable than small anions as the outer electron density is further from the nucleus and so less controlled by it. In contrast, small cations with high charges such as Be^{2+} and Al^{3+} are strongly **polarizing** and tend to induce distortions in anions. Such cations have a high charge density. Compounds containing small highly charged cations and large polarizable anions tend to have the highest degrees of polarization. This leads to high covalent character. The polarization of the anions is shown in Figure 6.33.

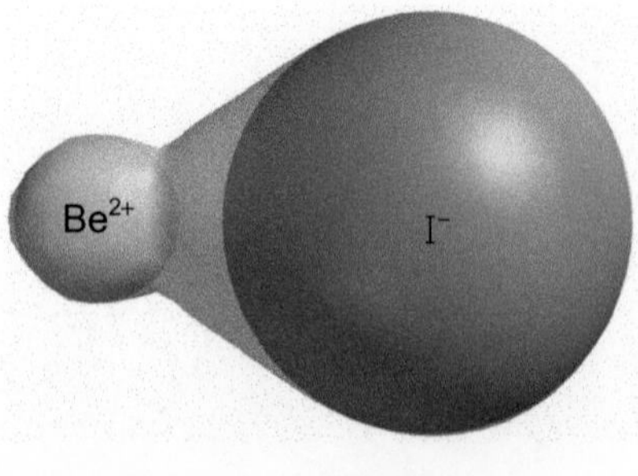
Be2+
I−

Figure 6.33 The small Be^{2+} cation polarizes the large I^- anion, pulling electron density, shown as a boundary surface in grey, towards itself.

For ionic compounds with a large covalent character, the Born–Landé equation underestimates the lattice energy, and the actual values obtained from a Born–Haber cycle are considerably higher. This means that the Born–Landé equation can be used as a way of estimating the extent to which a compound is ionic. The closer the value from the Born–Landé equation is to that from the Born–Haber cycle, the more ionic the compound is. For example, the calculated lattice energy from the Born–Landé equation for silver(I) fluoride is within 30 kJ mol^{-1} of the value determined experimentally using a Born–Haber cycle. In contrast, the calculated lattice energy from the Born–Landé equation for silver(I) iodide is almost 200 kJ mol^{-1} lower than that determined using a Born–Haber cycle. These differences reflect the more polarizable nature of the large iodide anions.

The Kapustinskii equation

The Born–Landé equation gives good predictions of lattice energy for many ionic compounds. There is, however, a drawback in using it, as both the Madelung constant and interionic distance r require a detailed knowledge of the structure, which may not be available. The Russian chemist Anatoly Kapustinskii recognized that, when the Madelung constant A was divided by the number of ions in the formula unit of the compound (v), it gave very similar values for a wide variety of structures. The value of A/v increases slightly with coordination number, but so does the interionic separation, and these two factors largely cancel out. By using an average value of the Born exponent, n, combining together all of the constants to give a single constant k, and assuming that the interionic distance is the sum of the two ionic radii ($r_+ + r_-$), the Born–Landé equation can be simplified into the expression given in Equation 6.11. This is known as the **Kapustinskii equation**

$$\Delta_{latt}U = \frac{kvz_+z_-}{r_+ + r_-} \qquad (6.11)$$

where z_+ and z_- are positive integers giving the charges on the ions, r_+ and r_- are the ionic radii in pm, v is the number of ions in the formula unit, and k is a constant with a value of 107 900 pm kJ mol^{-1}. Using non-SI units for k allows values for the ionic radii to be used in pm, without having to convert them into metres. The lattice energy, $\Delta_{latt}U$, is then given directly in kJ mol^{-1}. An example of the use of the Kapustinskii equation is given in Worked example 6.9.

Table 6.9 Thermochemical radii for common cations and anions

Ion	Ionic radius / pm
NH_4^+	136
OH^-	152
O_2^{2-}	167
CN^-	187
NO_3^-	200
ClO_4^-	225
CO_3^{2-}	189
SO_4^{2-}	218
PO_4^{3-}	230

Worked example 6.9 Using the Kapustinskii equation

Use the Kapustinskii equation to estimate the lattice energy for NaCl, and compare this value with that calculated from the Born–Landé equation.

Strategy

The Kapustinskii equation is given in Equation 6.11.

Use the ionic radius data from Tables 6.4 and 6.5 (p.286).

The Born–Landé equation is used to calculate the lattice energy for NaCl on p.295.

Solution

For NaCl, there are two ions in the formula unit (Na^+ and Cl^-) so $v = 2$.

Both ions have single charges, so $z_+ = 1$ and $z_- = 1$.

From Table 6.4, $r_+ = 102$ pm and from Table 6.5, $r_- = 181$ pm.

Putting these values into Equation 6.10 gives

$$\Delta_{latt}U = \frac{(107\,900\ \cancel{pm}\ kJ\,mol^{-1}) \times 2 \times 1 \times 1}{(102\ \cancel{pm}) + (181\ \cancel{pm})}$$

$$= +763\ kJ\,mol^{-1}$$

This compares with the value of +754.7 kJ mol^{-1} from the Born–Landé equation, a difference of just over 1%.

Question

Use the Kapustinskii equation to estimate the lattice energy for MgO.

As well as providing a simple means of estimating lattice energies, the Kapustinskii equation allows the ionic radii for polyatomic ions such as carbonate and sulfate to be estimated from the lattice energies of their compounds. These values are often called **thermochemical radii** because of the way in which they have been calculated. Thermochemical radii for some common polyatomic cations and anions are given in Table 6.9.

The Kapustinskii equation is very useful in predicting trends in the behaviour of ionic compounds. You will use the Kapustinskii equation for this in Chapter 26.

Summary

- Lattice energies can be calculated using the Born–Landé equation.
- Compounds containing polarizable anions and polarizing cations have high covalent character.
- Compounds with a large degree of covalent character have higher lattice energies than anticipated from the Born–Landé equation.
- In the absence of full information, good estimates of lattice energies can be made using the Kapustinskii equation.

For practice questions on these topics, see questions 14–17 at the end of this chapter (p.301).

6.7 Predicting bond types

How can the type of bonding in a binary compound be predicted? One of the key indicators of the bonding is the electronegativities of the elements involved.

For a binary compound, the greater the difference in electronegativity between the elements, the greater the ionic character of the bond. When the electronegativities for both elements are high, the compound tends to be covalent. Since high electronegativities are associated with *p*-block elements, *p*-block binary compounds such as ClF_3 and CO_2 are normally covalent, though there are exceptions for some of the heavier elements in low oxidation states, such as the ionic compound lead(II) fluoride, PbF_2.

When the electronegativities of the atoms are both low, the substance will be an alloy, with metallic bonding.

This information is summarized on the bond-type triangle in Figure 6.34. This type of representation was first developed in the 1930s and is sometimes referred to as a

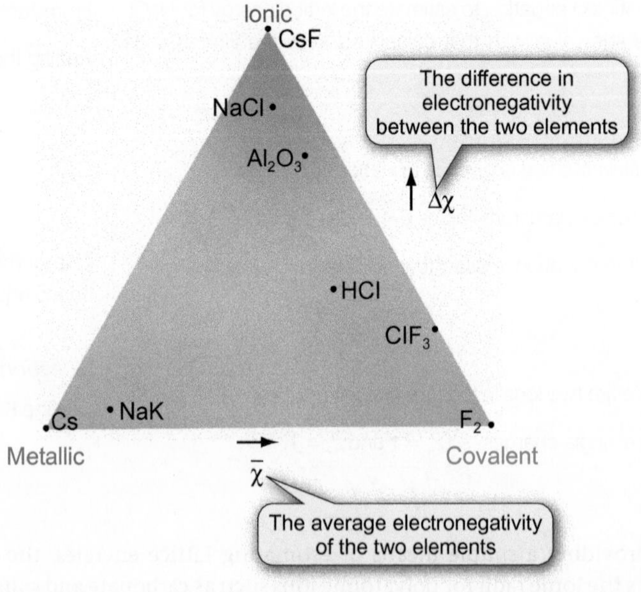

Ionic
CsF
NaCl
Al2O3
The difference in electronegativity between the two elements
Δχ
HCl
ClF3
Cs
NaK
F2
Metallic
Covalent
χ̄
The average electronegativity of the two elements

Figure 6.34 A van Arkel–Ketelaar triangle. For a binary compound, the type of bonding can be predicted by plotting the difference in electronegativity for the two atoms against their average electronegativity.

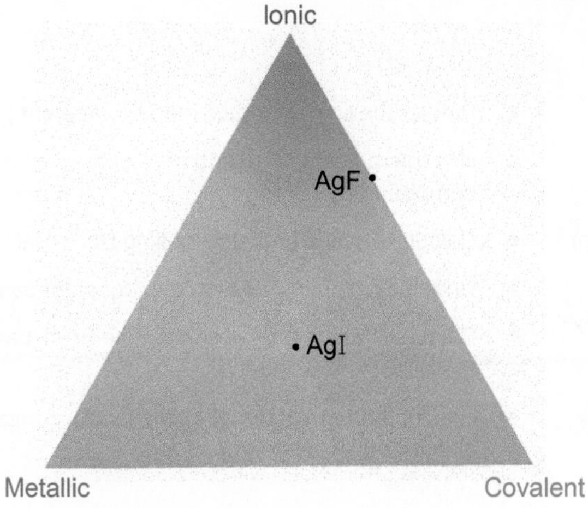
Ionic
AgF
AgI
Metallic
Covalent

Figure 6.35 A van Arkel–Ketelaar triangle showing the positions of AgF and AgI.

van Arkel–Ketelaar triangle, after two of its pioneers. In a van Arkel–Ketelaar triangle, the three corners represent the extremes of metallic, ionic, and covalent bonding, with caesium (Cs), caesium fluoride (CsF), and fluorine (F_2) in these corner positions. The average electronegativity of the elements, $\bar{\chi}$, is plotted on the *x*-axis and the difference in electronegativity, $\Delta\chi$, is plotted on the *y*-axis. For the molecule AB

$$\bar{\chi} = \frac{\chi_A + \chi_B}{2}$$

and

$$\Delta\chi = |\chi_A - \chi_B|$$

Most compounds lie in between these extremes, and their positions in the van Arkel–Ketelaar triangle give a good indication of their bonding. So, for example, NaCl lies near to the ionic corner so is predominantly ionic whereas ClF_3 lies near to the covalent corner, so is best described as covalent.

By comparing the lattice enthalpy calculated from a Born–Haber cycle with the lattice energy calculated from the Born–Landé equation or Kapustinskii equation, you can get a good idea of the importance of the covalent contribution to bonding in an ionic solid (Section 6.6, p.296). For example, the measured lattice enthalpy for AgF is close to that calculated from the Born–Landé equation, suggesting that the ionic model is a good description of bonding in this case. AgF lies near to the ionic corner of the triangle, consistent with this. There is a much poorer match between the measured and calculated lattice enthalpies for AgI, suggesting significant covalent interactions. Consistent with this, AgI lies further from the ionic corner of the van Arkel–Ketelaar triangle, and closer to the covalent corner. This is shown in Figure 6.35.

Summary

- The type of bonding present in a binary compound can be predicted by considering the electronegativities of the elements involved.
- This can be shown on a van Arkel–Ketelaar triangle, which plots the average electronegativity of the elements against the difference in their electronegativities.

For a practice question on this topic, see question 18 at the end of this chapter (p.301).

Concept review

By the end of this chapter, you should be able to do the following.

- Use the terms allotrope, polymorph, and unit cell, giving examples of each.
- Understand the differences between cubic close packing (ccp) and hexagonal close packing (hcp).
- Draw unit cells and cell projection diagrams for ccp, hcp, body-centred cubic (bcc), and primitive cubic structures.
- Calculate packing efficiencies and densities.
- Describe the bonding in metals using the free electron model and band theory.
- Describe the structures of binary compounds based on the packing of one type of ion or atom with the other type occupying interstitial sites.
- Predict the limiting radius ratio for different geometries.
- Use the radius ratio rule to predict the structures of ionic compounds.
- Calculate lattice enthalpies using Born–Haber cycles.
- Understand how the Madelung constant can be calculated.
- Calculate lattice energies using the Born–Landé equation and the Kapustinskii equation.
- Predict the type of bonding in a binary compound from the electronegativities of the atoms.

Key equations

packing efficiency	$\text{packing efficiency} = \dfrac{\text{volume of the atoms in a unit cell}}{\text{total volume of a unit cell}} \times 100$	(6.1)
radius ratio	$\text{radius ratio} = \dfrac{r_+}{r_-}$	(6.2)
Density	$\rho = \dfrac{m}{V}$	(6.3)
lattice enthalpy	$A_xB_y(s) \rightarrow xA^{y+}(g) + yB^{x-}(g)\ [\Delta_{latt}H^{\ominus}(A_xB_y)]$	(6.4)
Born–Landé equation	$\Delta_{latt}U = \dfrac{AN_A z_+ z_- e^2}{4\pi\varepsilon_0 r}\left(1-\dfrac{1}{n}\right)$	(6.10)
Kapustinskii equation	$\Delta_{latt}U = \dfrac{kvz_+z_-}{r_+ + r_-}$	(6.11)

Questions

More challenging questions are indicated by an asterisk *.

1. From the properties in the table below, describe whether each of the elements shown exists as a molecular structure, covalent network structure, or a metal. (Section 6.0)

	T_m/K	T_b/K	Conductivity
Boron	2348	4270	Insulator
Phosphorus (white)	317	554	Insulator
Lead	601	2022	Conductor

2. Draw the unit cell and a cell projection diagram for the hexagonal close-packed structure. How many atoms does the unit cell contain? Indicate the positions of the octahedral and tetrahedral sites. (Section 6.2)

3.* The body-centred cubic structure contains octahedral interstitial sites. Draw the unit cell and a cell projection diagram and mark the position of these, verifying the total number of octahedral sites per unit cell is six. (Section 6.2)

4. Nickel adopts the cubic close-packed structure. By considering the contents and the volume of the unit cell, use the data below to estimate the density of nickel (in g cm^{-3}). (Section 6.2)

Ar(Ni) = 58.693, atomic radius of Ni 125 pm

5. The UK £1 coin is made from a nickel–brass alloy, containing 70% copper, 5.5% nickel, and 24.5% zinc. Is this likely to be a substitutional alloy or an interstitial alloy? (Section 6.3)

6. Draw the cell projection diagram for the rutile structure, shown in Figure 6.26 (p.283). (Section 6.4)

7. The unit cell for an oxide of copper is shown below. Use the figure to determine the formula of this compound. Draw a cell projection diagram and describe the structure in terms of filling interstitial sites. (Section 6.4)

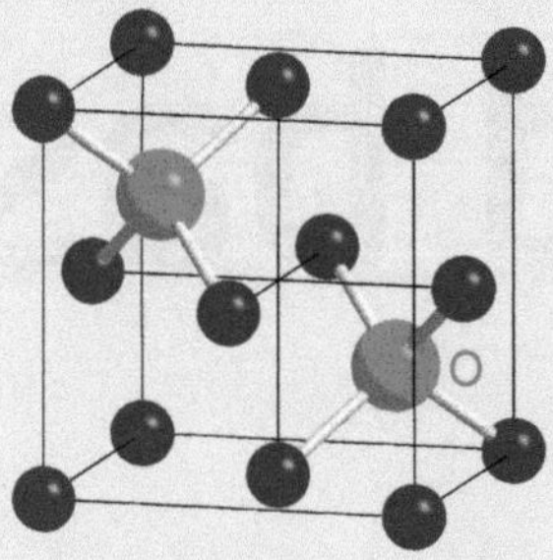
O
Cu

8. Lithium bismuthide (Li_3Bi) adopts a structure based on a cubic close-packed lattice of Bi^{3-} anions with Li^+ cations occupying the octahedral *and* tetrahedral sites. Draw a cell projection diagram for Li_3Bi. (Section 6.4)

9. Cadmium iodide forms the structure shown in Figure 6.24 (p.281), in which the iodide ions are hexagonally close packed and the Cd^{2+} ions occupy half the octahedral sites. Draw a cell projection diagram for cadmium iodide. (Section 6.4)

10. A mixed oxidation state oxide of iron has a structure based on a cubic close-packed arrangement of oxide ions with Fe^{2+} ions in one-eighth of the tetrahedral sites and Fe^{3+} ions in half of the octahedral sites. What is the formula of the compound? (Section 6.4)

11. Molybdenum sulfide, MoS_2, has a structure containing hexagonal layers of sulfide ions in a AABBAABB arrangement with molybdenum ions occupying interstitial sites between layers of the same type (i.e. AA or BB), as shown below.

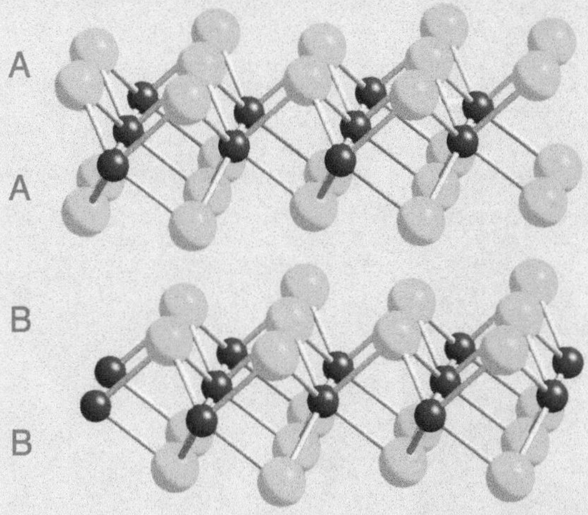
A
A
B
B

(a) Is the structure of MoS_2 based on close packing?

(b) What are the coordination numbers and coordination geometries of the cations and anions?

(c) Suggest a reason why MoS_2 can be used as a solid lubricant.

12. Do the following compounds obey the radius ratio rule? The observed structure is given in square brackets. (Section 6.5)

(a) copper(I) chloride [sphalerite structure];

(b) thallium(I) chloride [caesium chloride structure];

(c) caesium iodide [caesium chloride structure];

(d) potassium fluoride [sodium chloride structure];

(e) zinc sulfide [sphalerite structure].

13. Construct a Born–Haber cycle for the formation of MgO (s) from Mg (s) and O_2 (g) and use the information below to calculate the lattice enthalpy of MgO. (Section 6.5)

$\Delta_f H^\ominus$(MgO)	−602 kJ mol^{-1}
$\Delta_i H(1)^\ominus$(Mg)	+738 kJ mol^{-1}
$\Delta_i H(2)^\ominus$(Mg)	+1451 kJ mol^{-1}
$\Delta_a H^\ominus$(Mg)	+147 kJ mol^{-1}
$\Delta_a H^\ominus$(O)	+249 kJ mol^{-1}
$\Delta_{eg} H(1)^\ominus$(O)	−141 kJ mol^{-1}
$\Delta_{eg} H(2)^\ominus$(O)	+798 kJ mol^{-1}

14. Use the Born–Landé equation to calculate the lattice energy of KF, which adopts the sodium chloride structure with a K–F distance of 268.4 pm. (Section 6.6)

15. Use the Kapustinskii equation and the data in Tables 6.4 and 6.5 (p.286) to estimate the lattice energies for KF, KCl, and KBr. Suggest a reason for the trend you observe. (Section 6.6)

16. Use the Kapustinskii equation and the data in Tables 6.4, 6.5 (p.286), and 6.9 (p.297) to estimate the lattice energies for $CaCl_2$ and $CaSO_4$. (Section 6.6)

17.*Use the data below to construct Born–Haber cycles for lithium chloride and silver chloride, and from those work out the lattice enthalpies for both compounds. Use the ionic radii for Li^+, Ag^+, and Cl^- given in Tables 6.4 and 6.5 (p.286) to estimate the lattice energies for lithium chloride and silver chloride using the Kapustinskii equation. Compare the values you have obtained from these two approaches, and comment on the differences between them. (Section 6.6)

$\Delta H_a^\ominus$(Li) = 159 kJ mol^{-1}, $\Delta H_a^\ominus$(Ag) = 285 kJ mol^{-1},
$\Delta H_a^\ominus$(Cl) = 121 kJ mol^{-1}

$\Delta H_i(1)^\ominus$(Li) = 520 kJ mol^{-1}, $\Delta H_i(1)^\ominus$(Ag) = 731 kJ mol^{-1},
$\Delta H_{eg}(1)^\ominus$(Cl) = −349 kJ mol^{-1}

$\Delta H_f^\ominus$(LiCl) = −408 kJ mol^{-1}, $\Delta H_f^\ominus$(AgCl) = −127 kJ mol^{-1}

18. Draw a van Arkel–Ketelaar triangle, and use the electronegativities of the elements to estimate the positions of KF and PbI_2. Use this to estimate the type of bonding in the two compounds. (Section 6.7)

7

Acids and bases

▼ The blue colour in cornflowers is caused by a metal compound of cyanidin.

This chapter builds on the following topics:

- **Chemical equilibrium—how far has a reaction gone?** Section 1.9, p.56
- **Valence shell electron pair repulsion theory** Section 5.2, p.223
- **Bond polarity and polar molecules** Section 5.3, p.235

▼ The colour of hydrangeas depends on the pH of the soil.

Acids and bases in the garden

Keen gardeners know that the colour of hydrangeas varies with the pH of the soil. In acidic soil the flowers are blue, whereas in neutral or slightly alkaline soil the flowers are pink. The hydrangea is not simply acting as an acid–base indicator though, because the pH within the blue and pink flowers is identical. The key to the colour of the flowers is the presence of aluminium ions, which are needed to give the blue colour. The pH of the soil is important in dictating the mobility of the aluminium ions.

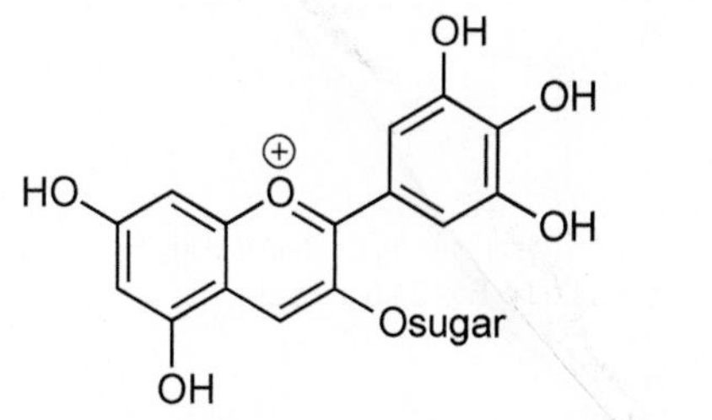

OH
OH
HO
O
OH
Osugar
OH

Delphinidin

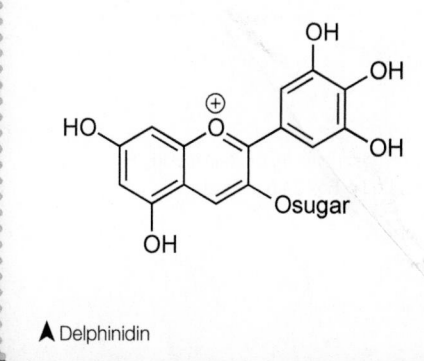

The Al^{3+} ions are released from aluminium silicates in clay and acid soils. The concentrations of the aluminium species present depend on the pH, with the following equilibria involved

$$[Al(H_2O)_6]^{3+}(aq) + OH^-(aq) \rightleftharpoons [Al(H_2O)_5(OH)]^{2+}(aq) + H_2O\,(l)$$

$$[Al(H_2O)_5(OH)]^{2+}(aq) + OH^-(aq) \rightleftharpoons [Al(H_2O)_4(OH)_2]^{+}(aq) + H_2O\,(l)$$

$$[Al(H_2O)_4(OH)_2]^{+}(aq) + OH^-(aq) \rightleftharpoons Al(OH)_3\,(s) + 4H_2O\,(l)$$

$$Al(OH)_3\,(s) + OH^-(aq) \rightleftharpoons [Al(OH)_4]^-(aq)$$

When the pH is less than 7, the aluminium is present as soluble $[Al(H_2O)_6]^{3+}$, $[Al(H_2O)_5(OH)]^{2+}$, and $[Al(H_2O)_4(OH)_2]^+$ ions, whereas at the higher pHs present in alkaline soils, it is mainly in the form of insoluble $Al(OH)_3$. (The pH scale is described in Section 7.2, and acid–base indicators are described in Section 7.5.)

How do the aluminium ions influence the hydrangea colour? The main pigment in hydrangea flowers is a derivative of delphinidin, which is one of a class of naturally occurring compounds called anthocyanins. Delphinidin is normally pink, but conversion into an aluminium compound causes a colour change to blue.

As long as aluminium ions are present in the soil, you can control the colour of hydrangeas by changing the soil pH. To turn blue hydrangeas pink, you need to add base, and this is normally done with lime (calcium oxide, CaO). Calcium oxide reacts with water to form calcium hydroxide, which is a source of OH^- ions

$$CaO\,(s) + H_2O\,(l) \rightarrow Ca(OH)_2\,(s)$$

Hydroxide ions push the four aluminium equilibria towards the right, immobilizing the aluminium ions as insoluble $Al(OH)_3$, which keeps the hydrangeas pink. To turn pink hydrangeas blue, you need to decrease the pH of the soil. This is normally done by adding aluminium sulfate, which has the additional benefit of raising the concentration of Al^{3+} ions. Aqueous solutions of aluminium ions are acidic due to hydrolysis (reaction with water) of the hexaaqua ion

$$[Al(H_2O)_6]^{3+}(aq) + H_2O\,(l) \rightleftharpoons [Al(H_2O)_5(OH)]^{2+}(aq) + H_3O^+(aq)$$

This type of hydrolysis reaction to give an acid solution is discussed in Box 7.3 (p.312) and, for aluminium, in Section 27.2 (p.1213).

Anthocyanins are very common molecules in nature. As another example, cyanidin contributes to the colours of many fruits, including blackberries, strawberries, cherries, and red apples, and is also present in red wine. It is responsible for the red colour of roses and the blue colour of cornflowers. The colour of cyanidin depends on the pH.

As with hydrangeas, the difference in colour between red roses and blue cornflowers is related to the presence of metal ions. The blue compound in cornflowers has been identified as a compound of cyanidin with iron, magnesium, and calcium whereas the cyanidin in rose petals is not coordinated to a metal ion.

The colour of cyanidin depends on the pH.

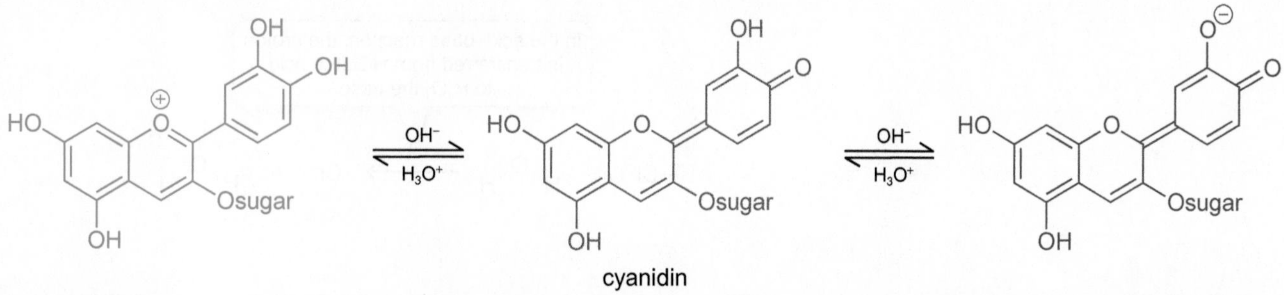

OH
OH
HO
O
Osugar
OH
OH⁻
H₃O⁺
OH
O
HO
O
Osugar
OH
cyanidin
OH⁻
H₃O⁺
O
O
HO
O
Osugar
OH

The chemistry of soap making is described in Box 24.6 (p.1120).

Acids and bases have been known since ancient times. Arabic alchemists, such as Jabir ibn Hayyan (often Latinized to Geben), prepared hydrochloric acid and nitric acid in the 700s. The combination of nitric acid and hydrochloric acid known as *aqua regia* was particularly interesting to the alchemists as it dissolves gold. Bases have an even longer history, with ashes containing sodium hydroxide used to make soaps as long ago as 2800 BC.

As early as the seventeenth century it was recognized that acids and bases were in some way opposites of each other. Towards the end of the nineteenth century, the Swedish chemist Svante Arrhenius proposed the first chemical definitions of these terms. He defined an acid as a substance that ionizes in water to give H^+ ions and anions, and a base as a substance that ionizes in water to give hydroxide ions (OH^-) and cations. A base that is soluble in water is called an **alkali**, and basic aqueous solutions are known as **alkaline** solutions.

An acid reacts with a base in a **neutralization** reaction to produce water

$$\underset{\text{from acid}}{H^+(aq)} + \underset{\text{from base}}{OH^-(aq)} \rightarrow \underset{\text{water}}{H_2O(l)}$$

Ionic equations and spectator ions are discussed in Section 1.4 (p.26).

The anion from the acid and the cation from the alkali remain in solution. They are sometimes called **spectator ions**, because they take no part in the reaction. When water is evaporated from the solution, these ions come together to form an ionic compound, known as a **salt**. For example

$$\underset{\text{acid}}{HCl(aq)} + \underset{\text{base}}{NaOH(aq)} \rightarrow \underset{\text{salt}}{NaCl(aq)} + \underset{\text{water}}{H_2O(l)}$$

- **Arrhenius acid**: a substance that ionizes in water to give H^+.
- **Arrhenius base**: a substance that ionizes in water to give OH^-.

While Arrhenius's definitions were a major breakthrough at the time, they have a number of limitations. For example, they only apply in aqueous solution, but acid–base reactions can occur in other solvents too. Also, it is difficult to see how the familiar base ammonia (NH_3) can function, as it does not contain an OH group.

In this chapter, two broader definitions of acids and bases are described. Section 7.1 introduces Brønsted–Lowry theory, in which an acid is defined as a proton donor and a base is defined as a proton acceptor. Section 7.8 introduces the Lewis theory, in which an acid is defined as an electron pair acceptor and a base is defined as an electron pair donor. This is a wider definition—all Brønsted–Lowry acids are Lewis acids, but not all Lewis acids are Brønsted–Lowry acids.

7.1 Brønsted-Lowry acids and bases

- A **Brønsted-Lowry acid** is a proton (H^+) donor.
- A **Brønsted-Lowry base** is a proton (H^+) acceptor.

A more general theory of acids and bases than that of Arrhenius was proposed independently in 1923 by Johannes Brønsted in Denmark and Martin Lowry in Britain, and it is now usually known as the **Brønsted–Lowry theory**. According to the Brønsted–Lowry theory, an acid donates H^+ in a chemical reaction and the substance that accepts the H^+ is a base. Since a hydrogen atom consists of only a proton and an electron, an H^+ ion is simply a proton. This means that a Brønsted–Lowry acid is a proton donor and a Brønsted–Lowry base is a proton acceptor. An **acid–base reaction** is one in which a proton is transferred from an acid to a base.

Hydrogen chloride is a gas and contains covalent HCl molecules. When hydrogen chloride gas dissolves in water, the following reaction takes place

$$\underset{\text{acid}}{HCl(g)} + \underset{\text{base}}{H_2O(l)} \rightarrow \underset{\text{chloride ion}}{Cl^-(aq)} + \underset{\text{oxonium ion}}{H_3O^+(aq)}$$

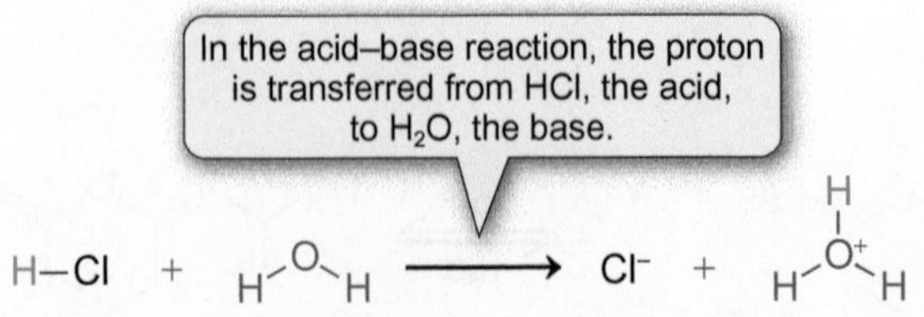
In the acid–base reaction, the proton is transferred from HCl, the acid, to H2O, the base.
H–Cl
Cl⁻

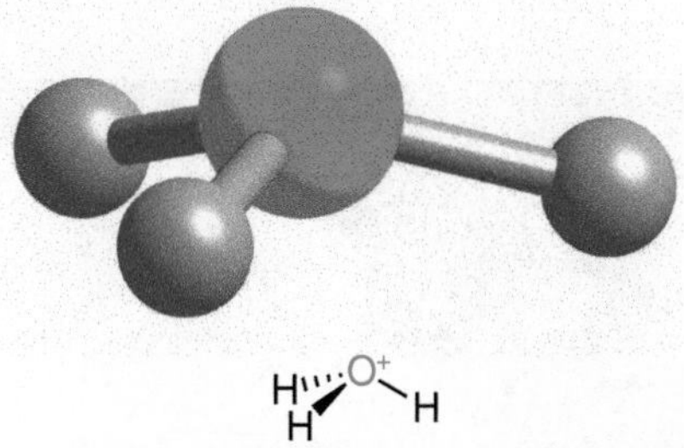

Figure 7.1 The oxonium ion, H_3O^+. The shape of the ion is trigonal pyramidal, as predicted by VSEPR theory (see Section 5.2, p.223).

The corrosion on this limestone figure is caused by acid rain. Rainwater is naturally acidic (pH ~ 5) because CO_2 in the atmosphere dissolves in water to give carbonic acid, which then ionizes, releasing H_3O^+ ions (see Box 1.12, p.58).

This is an acid–base reaction in which a proton is transferred from HCl to H_2O. In this reaction, H_2O behaves as a base.

In water, the H^+ ion forms a strong interaction with a water molecule, and the product is best represented as H_3O^+, the **oxonium ion**, which is shown in Figure 7.1. The H_3O^+(aq) ion gives rise to the familiar acidic properties of all aqueous solutions of acids. The oxonium ion interacts strongly with water molecules, and is always hydrated when water is the solvent.

Hydration, and the more general process of solvation, is discussed in Box 7.1.

The oxonium ion is also known as the hydroxonium ion and the hydronium ion, though the latter term is not recommended by IUPAC.

Box 7.1 Solvation

When a compound dissolves, the individual ions or molecules interact with the solvent molecules. This process is called **solvation**, and the ions or molecules are said to be solvated. Hydration is a specific type of solvation with water as the solvent. The hydration of ions that occurs when an ionic compound dissolves in water is described in Section 1.8 (p.54).

An ion or molecule in solution is surrounded by solvent molecules with which it forms non-covalent interactions. These solvent molecules are known as the **solvation shell**. For ions, the greater the charge density, the larger the solvation shell. Small, highly charged ions like Al^{3+} typically have more solvent molecules associated with them than large, low charged ions such as Cs^+.

Solvation shells can be several molecules thick. For example, in acidic solutions, each H_3O^+ ion forms hydrogen bonds with three water molecules to give $[(H_3O)(H_2O)_3]^+$, which is sometimes written as $[H_9O_4]^+$. These three water molecules comprise the *first solvation shell* of H_3O^+. A second group of water molecules forms hydrogen bonds with the water molecules in the first shell, and these form the *second solvation shell*. Since the solvent molecules are in constant motion, solvation shells are not static, and they continually change their composition with hydrogen bonds breaking and forming. For this reason, hydration in a chemical formula is just represented by (aq).

Typically, polar molecules interact strongly with polar solvents through dipole–dipole interactions and/or hydrogen bonds. This means that polar molecules tend to dissolve in polar solvents. Common non-aqueous polar solvents include:

- ethanenitrile (acetonitrile, CH_3CN);
- propanone (acetone, CH_3COCH_3);
- methanol (CH_3OH);

▼ Solvation of a metal cation with water. The water molecules in the first solvation shell coordinate to the metal ion and have a high degree of order. The second solvation shell is less ordered and more dynamic.

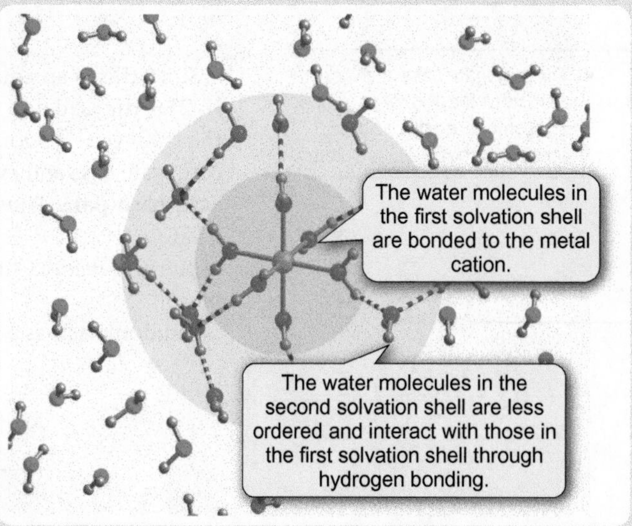
The water molecules in the first solvation shell are bonded to the metal cation.
The water molecules in the second solvation shell are less ordered and interact with those in the first solvation shell through hydrogen bonding.

→

Sugar (sucrose) dissolves in water such as the hot, aqueous solution of a cup of tea. Strong hydrogen bonds between the sucrose and water molecules make solvation in water favourable.

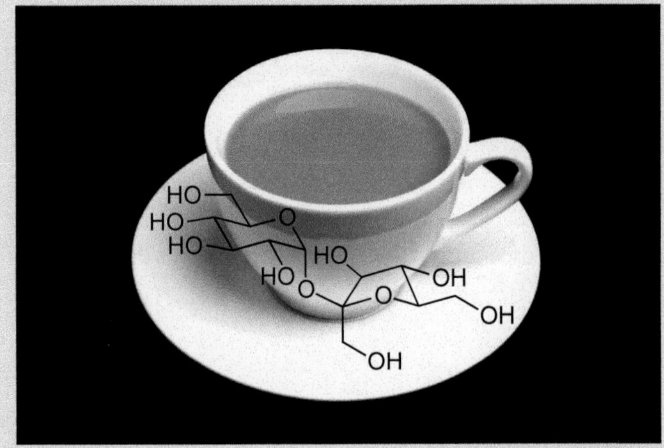

- ethanol (CH_3CH_2OH);
- dimethyl sulfoxide (DMSO; $(CH_3)_2SO$);
- liquid ammonia (NH_3).

Methanol, ethanol, and water all contain O–H groups that can form hydrogen bonds and also act as a source of protons in reactions. These solvents are known as **protic solvents.** In ethanenitrile, propanone, and dimethyl sulfoxide there are no hydrogen atoms that form hydrogen bonds or that can be transferred to another molecule. These solvents are known as **aprotic solvents.** Dipoles and polar molecules are described in Section 5.3 (p.235) and the uses of polar protic solvents and polar aprotic solvents in organic substitution reactions are discussed in Section 20.3 (p.936). Protic solvents are needed to solvate anions (see Figure 1.28, p.54).

Question

Suggest why sucrose does not dissolve in hexane.

The gas phase reaction between HCl and NH_3 is also an acid–base reaction

$$\underset{\text{acid}}{HCl\,(g)} + \underset{\text{base}}{NH_3\,(g)} \rightarrow NH_4^+Cl^-\,(s)$$

In this reaction NH_3 is acting as a Brønsted–Lowry base (but not an Arrhenius base) by accepting a proton from HCl. This is an example of an acid–base reaction that does not involve water.

The reaction between ethanoic acid, CH_3CO_2H, and water is an equilibrium in which CH_3CO_2H is the acid and H_2O is the base

$$\underset{\text{acid}}{CH_3CO_2H\,(aq)} + \underset{\text{base}}{H_2O\,(l)} \rightleftharpoons \underset{\text{conjugate base}}{CH_3CO_2^-\,(aq)} + \underset{\text{conjugate acid}}{H_3O^+\,(aq)}$$

- The **conjugate base** of an acid consists of the acid minus a proton.
- The **conjugate acid** of a base consists of the base plus a proton.

Since this is an equilibrium, the reverse reaction also occurs. For the reverse reaction, H_3O^+ is the acid and the ethanoate ion ($CH_3CO_2^-$) is the base. $CH_3CO_2^-$ is described as the conjugate base of CH_3CO_2H, and H_3O^+ is the conjugate acid of H_2O. Every acid has a conjugate base and every base has a conjugate acid. Together, they are called a **conjugate acid–base pair.** Thus, there are two conjugate acid–base pairs in the above equation, $CH_3CO_2H\,(aq)/CH_3CO_2^-\,(aq)$ and $H_3O^+\,(aq)/H_2O\,(l)$. The conjugate base of an acid consists of the acid *minus* a single proton. The conjugate acid of a base consists of the base *plus* a proton.

Ammonia acts as a Brønsted–Lowry base in water through the equilibrium

$$\underset{\text{acid}}{H_2O\,(l)} + \underset{\text{base}}{NH_3\,(aq)} \rightleftharpoons \underset{\text{conjugate base}}{OH^-\,(aq)} + \underset{\text{conjugate acid}}{NH_4^+\,(aq)}$$

In this reaction, OH^- is the conjugate base of water and NH_4^+ is the conjugate acid of ammonia. The conjugate acid–base pairs are $H_2O\,(l)/OH^-\,(aq)$ and $NH_4^+\,(aq)/NH_3\,(aq)$.

Worked example 7.1 Conjugate acids and conjugate bases

For the proton transfer equilibrium of HCN with water, identify the conjugate acid–base pairs involved.

Strategy

Write out the equation for the equilibrium reaction between HCN and H_2O. For the forward reaction, decide which reactant is acting as an acid and which as a base. The product formed by loss of a proton from the acid is its conjugate base. The product formed by addition of a proton to the base is its conjugate acid.

Solution

In the forward reaction, HCN is the acid and H_2O the base. For the reverse reaction, CN^- acts as the base and H_3O^+ as the acid.

$$\underset{\text{acid}}{HCN(aq)} + \underset{\text{base}}{H_2O(l)} \rightleftharpoons \underset{\text{conjugate base}}{CN^-(aq)} + \underset{\text{conjugate acid}}{H_3O^+(aq)}$$

$CN^-(aq)$ is the conjugate base of HCN(aq) and $H_3O^+(aq)$ is the conjugate acid of $H_2O(l)$. So, the conjugate acid–base pairs involved in the equilibrium are $HCN(aq)/CN^-(aq)$ and $H_3O^+(aq)/H_2O(l)$.

Question

Pure ethanoic acid (CH_3CO_2H) reacts with liquid ammonia (NH_3) to form ions in a proton transfer equilibrium. Write an equation for the reaction and identify the conjugate acid–base pairs involved. (Omit state symbols.)

Box 7.2 Acids, alkalis, and human tissue

Acids and alkalis have a corrosive effect on human tissue, with the effect depending on the concentration.

Dilute acids and alkalis both hydrolyse the peptide links in proteins such as those in skin and eyes, eventually breaking the proteins down into soluble amino acids (see Section 19.2, p.895). However, alkalis are more effective at this than acids, and as a result alkalis are more corrosive on the skin than acids at a given concentration. Compare the following.

	Concentration/ $mol\,dm^{-3}$	Hazard label required
Sodium hydroxide solution	0.05	Irritant
	0.5	Corrosive
Sulfuric acid	0.5	Irritant
	1.5	Corrosive

John George Haigh, the notorious acid bath murderer.

Concentrated acids have a more extensive chemical effect than dilute acids on organic material. This was exploited by the acid bath murderer John Haigh, who between 1944 and 1949 disposed of at least six people by dissolving their bodies in concentrated sulfuric acid (H_2SO_4). By protonating all OH and NH_2 groups in proteins, concentrated sulfuric acid initiates a series of reactions that eventually remove all oxygen, hydrogen, and nitrogen atoms, so the final product is carbon. When investigators in the Haigh case sifted through the acid sludge they found several incriminating objects including gallstones and a pair of dentures. Gallstones are made of cholesterol, which does not react with acids. By matching the dentures with casts that had been made previously by a dentist, the investigators were able to identify one of the victims.

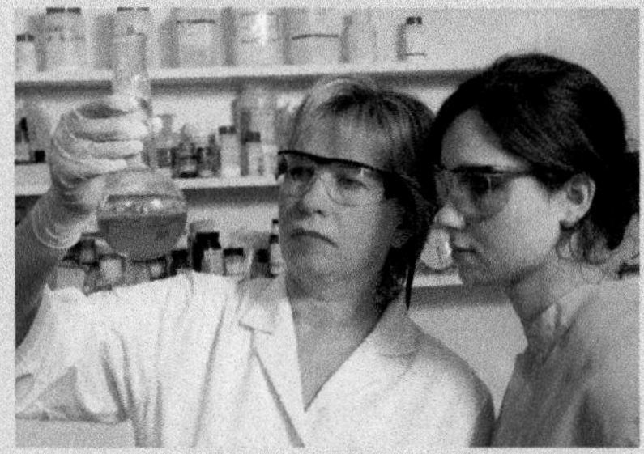

Students in a chemical laboratory wearing safety glasses.

The parts of the body most vulnerable to acids and alkalis are the eyes. Unlike the skin, whose surface consists of dead cells, the cornea of the eye is made of living cells, with no protection from the hydrolysing effect of acids and bases. This is why eye protection is so important when working with even the most dilute acids and alkalis.

Question

What mass of sulfuric acid is present in $500\,cm^3$ of a solution with the concentration $2\,mol\,dm^{-3}$ of SO_4^{2-}?

Summary

- According to the Brønsted–Lowry definition, an acid is a proton (H^+) donor and a base is a proton acceptor. An acid–base reaction involves a proton transfer.
- The oxonium ion (H_3O^+(aq)) is responsible for the acidic properties of aqueous solutions of acids.
- An acid–base reaction involves conjugate acid–base pairs. The conjugate base of an acid consists of the acid minus a proton, and the conjugate acid of a base consists of the base plus a proton.

For a practice question on these topics, see question 1 at the end of this chapter (p.339).

7.2 The strengths of acids and bases

The stronger an acid, the weaker its conjugate base. The stronger a base, the weaker its conjugate acid.

Strong or concentrated?

- **Concentration** is a measure of the amount of substance in a given volume of solution.
- **Strength** is a measure of the extent to which an acid can donate H^+, or the extent to which a base can accept H^+.

This means you can have a concentrated solution of a weak acid and a dilute solution of a strong acid.

Acids vary widely in strength. Some are powerful proton donors and are described as **strong acids**. Similarly, a **strong base** has a high tendency to accept protons. If an acid has a strong tendency to donate protons, its conjugate base will have a weak tendency to accept them and will be a weak base. Conversely, the conjugate acid of a strong base will be a weak acid. This is illustrated in Table 7.1, which shows some examples of acid–base conjugate pairs and their relative strengths.

Water can act as an acid or a base, depending on the reaction. With a stronger acid than itself, water acts as a base. For example, in the reaction with methanoic acid (HCO_2H(aq)), water *accepts* a proton to form H_3O^+(aq)

$$\underset{\text{acid}}{HCO_2H(aq)} + \underset{\text{base}}{H_2O(l)} \rightleftharpoons \underset{\text{conjugate base}}{HCO_2^-(aq)} + \underset{\text{conjugate acid}}{H_3O^+(aq)}$$

With a stronger base than itself, water acts as an acid. In the reaction with ammonia, H_2O *donates* a proton to form NH_4^+(aq) and OH^-(aq)

$$\underset{\text{acid}}{H_2O(l)} + \underset{\text{base}}{NH_3(aq)} \rightleftharpoons \underset{\text{conjugate base}}{OH^-(aq)} + \underset{\text{conjugate acid}}{NH_4^+(aq)}$$

Table 7.1 The relative strengths of acids and their conjugate bases

	Conjugate acid		Conjugate base	
Strongest acid	$HClO_4(aq) + H_2O(l)$ perchloric acid	$\rightleftharpoons$	$H_3O^+(aq) + ClO_4^-(aq)$ perchlorate ion	Weakest base
	$HCl(aq) + H_2O(l)$ hydrochloric acid	$\rightleftharpoons$	$H_3O^+(aq) + Cl^-(aq)$ chloride ion	
	$H_2SO_4(aq) + H_2O(l)$ sulfuric acid	$\rightleftharpoons$	$H_3O^+(aq) + HSO_4^-(aq)$ hydrogensulfate ion	
Increasing acid strength ↑	$H_3O^+(aq) + H_2O(l)$ oxonium ion	$\rightleftharpoons$	$H_3O^+(aq) + H_2O(l)$ water	Increasing base strength ↓
	$CH_3CO_2H(aq) + H_2O(l)$ ethanoic acid	$\rightleftharpoons$	$H_3O^+(aq) + CH_3CO_2^-(aq)$ ethanoate ion	
	$NH_4^+(aq) + H_2O(l)$ ammonium ion	$\rightleftharpoons$	$H_3O^+(aq) + NH_3(aq)$ ammonia	
	$H_2O(l) + H_2O(l)$ water	$\rightleftharpoons$	$H_3O^+(aq) + OH^-(aq)$ hydroxide ion	
Weakest acid	$C_2H_5OH(aq) + H_2O(l)$ ethanol	$\rightleftharpoons$	$H_3O^+(aq) + C_2H_5O^-(aq)$ ethoxide ion	Strongest base

Strength of acids

Strong acids have a high tendency to donate protons. In aqueous solution, their reaction with water to produce H_3O^+(aq) essentially goes to completion and the acid is fully ionized. Hydrochloric acid (HCl) and nitric acid (HNO_3) are examples of strong acids

$$HCl\,(aq) + H_2O\,(l) \rightarrow Cl^-\,(aq) + H_3O^+\,(aq)$$

$$HNO_3\,(aq) + H_2O\,(l) \rightarrow NO_3^-\,(aq) + H_3O^+\,(aq)$$

For weak acids, such as ethanoic acid (CH_3CO_2H), the reaction with water is incomplete. The tendency of the acid to donate protons to a base is much weaker, and so the reaction comes to equilibrium when only a small proportion of the acid molecules have reacted to form ions.

The strength of an acid, HA, in aqueous solution is related to the equilibrium constant for the following reaction

$$HA\,(aq) + H_2O\,(l) \rightleftharpoons A^-\,(aq) + H_3O^+\,(aq)$$

$$K_c = \frac{[H_3O^+(aq)][A^-(aq)]}{[HA\,(aq)][H_2O\,(l)]} \qquad (7.1)$$

The equilibrium constant K_c is introduced in Section 1.9 (p.59).

The term $[H_2O\,(l)]$ is essentially constant because the reaction takes place in dilute solution. A new constant, K_a, is defined which incorporates $[H_2O\,(l)]$, as shown in Equation 7.2

$$K_a = \frac{[H_3O^+(aq)][A^-(aq)]}{[HA\,(aq)]} \qquad \text{units of } K_a \text{ are mol dm}^{-3} \qquad (7.2)$$

This equilibrium constant, K_a, is called the **acidity constant** (or sometimes the acid dissociation constant).

Table 7.2 shows the acidity constants for some common acids in water at 298 K (25 °C). For a strong acid such as perchloric acid ($HClO_4$), the position of the equilibrium lies far over to the right-hand side, so K_a is very high and the acid is completely in the form of the ions in aqueous solution. For a weak acid, such as hydrocyanic acid (HCN), only a very small proportion of the molecules are in the form of ions, so K_a is much smaller.

Since many K_a values are either very small or very large numbers, it is more convenient to quote them as pK_a values, where

$$pK_a = -\log_{10} K_a \qquad (7.3)$$

The higher the value of K_a, the stronger the acid, and the lower pK_a (see Table 7.2). Strong acids like hydrochloric acid have negative pK_a values.

Strictly speaking, Equation 7.3 uses a slightly different version of K_a to that defined in Equation 7.2. This is because K_a has units of mol dm^{-3}, and it is only possible to take a logarithm of a number with no units. Equation 7.3 uses the thermodynamic equilibrium constant, K, which involves activities rather than concentrations of H_3O^+, A^-, and HA. The activity, a_A, of a species A in solution is defined by Equation 7.4

$$a_A = \frac{[A]}{[A]^{\ominus}} \qquad (7.4)$$

where $[A]^{\ominus}$ = 1 mol dm^{-3}. This means that the activity has the same numerical value as the concentration, but has no units.

Since strong acids are completely ionized in water, they appear to be of equal strengths. This is called the **levelling effect**. To compare the strengths of these acids, a solvent that is a weaker proton acceptor than water, such as diethyl ether, must be used. In this solvent, ionization is not complete, even for strong acids, and this allows the acidity constants to be measured. However, changing the solvent will also affect the acidity constant. For example, the pK_a for HCl is –7 in water, but 8.9 in ethanenitrile.

Dissociation is an old chemical term that means 'breaking up' or 'moving apart'. It can be used to mean the breaking of a chemical bond, the ionization of a molecule, or the reaction of an acid or base with water. In this book, we use the term dissociation to mean the homolytic breaking of a specific bond in a molecule.

Logarithms

A logarithmic scale is a convenient way of bringing numbers of widely differing magnitudes on to the same scale. In general, for a number y, the logarithm 'to the base 10' is given by x, where

$$y = 10^x \text{ and } \log_{10}(y) = x$$

- When $y = 10$, $y = 10^1$ $\log_{10}(10) = 1$
- When $y = 1$, $y = 10^0$ $\log_{10}(1) = 0$
- When $y = 0.1$, $y = 10^{-1}$ $\log_{10}(0.1) = -1$

You can find out more about logarithms in Maths Toolkit MT3 (p.1316).

Activities are defined in Section 14.6 (p.686), and thermodynamic equilibrium constants are described in more detail in Section 15.1 (p.698).

The reasons for the trend in pK_a for the hydrohalic acids, HF, HCl, HBr, and HI, are discussed in Section 25.2 (p.1153). Trends in pK_a for organic acids are described in Section 19.2 (p.896).

Table 7.2 Acidity constants (K_a) and values of pK_a for some common acids at 298 K

Acid		K_a/mol dm^{-3}	pK_a	
Hydriodic acid	HI	1×10^{10}	−10	Strongest acid
Perchloric acid	$HClO_4$	1×10^{10}	−10	
Hydrobromic acid	HBr	1×10^{9}	−9	Strong acids
Hydrochloric acid	HCl	1×10^{7}	−7	
Sulfuric acid	H_2SO_4	1×10^{3}	−3	
Nitric acid	HNO_3	25	−1.4	
Trichloroethanoic acid	CCl_3CO_2H	2.2×10^{-1}	0.66	
Chlorous acid	$HClO_2$	1.1×10^{-2}	1.94	
Hydrofluoric acid	HF	6.3×10^{-4}	3.20	
Nitrous acid	HNO_2	5.6×10^{-4}	3.25	
Methanoic acid	HCO_2H	1.8×10^{-4}	3.75	
Benzoic acid	$C_6H_5CO_2H$	6.3×10^{-5}	4.20	Weak acids
Ethanoic acid	CH_3CO_2H	1.7×10^{-5}	4.76	
Propanoic acid	$CH_3CH_2CO_2H$	1.3×10^{-5}	4.87	
Carbonic acid	H_2CO_3	4.5×10^{-7}	6.35	
Hypochlorous acid	HOCl	4.0×10^{-8}	7.40	
Hydrocyanic acid	HCN	6.2×10^{-10}	9.21	
Phenol	C_6H_5OH	1.0×10^{-10}	9.99	Weakest acid

The pH scale

In a similar way to the definition of pK_a, it is also convenient to express concentrations of $H_3O^+(aq)$ ions on a logarithmic scale. This is the **pH scale**, which was devised at the beginning of the twentieth century by the Danish biochemist Søren Sørensen. pH is defined in Equation 7.5

- An **acidic** solution is defined as one in which $[H^+(aq)] > [OH^-(aq)]$.
- An **alkaline** solution is defined as one in which $[H^+(aq)] < [OH^-(aq)]$.

$$pH = -\log_{10}[H_3O^+(aq)] \qquad (7.5)$$

pH measures the concentration of H_3O^+ ions in a solution. As with the definition of pK_a, Equation 7.5 should strictly use the activity of H_3O^+ rather than the concentration.

Converting back from a value for $x = \log_{10} y$ to a value of y is called taking an **antilog**. $y = 10^x$, so $[H_3O^+(aq)] = 10^{-pH}$. Make sure you know how to work out antilogs on your calculator.

Figure 7.2 shows the relationship between pH and $[H_3O^+(aq)]$, and Figure 7.3 shows the pH of a range of commonly occurring solutions.

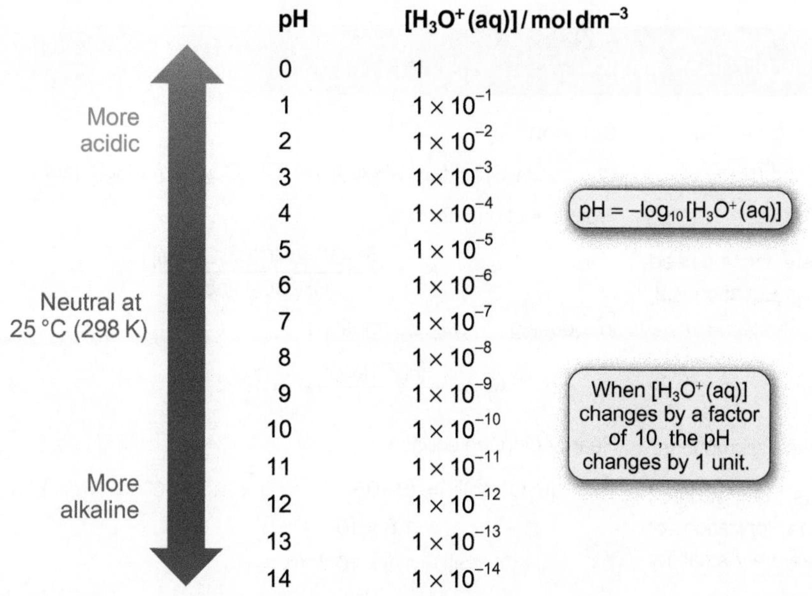
pH
$[H_3O^+(aq)]$ / mol dm^{-3}
More acidic
Neutral at 25 °C (298 K)
More alkaline
0 1
1 1×10^{-1}
2 1×10^{-2}
3 1×10^{-3}
4 1×10^{-4}
5 1×10^{-5}
6 1×10^{-6}
7 1×10^{-7}
8 1×10^{-8}
9 1×10^{-9}
10 1×10^{-10}
11 1×10^{-11}
12 1×10^{-12}
13 1×10^{-13}
14 1×10^{-14}
$pH = -\log_{10}[H_3O^+(aq)]$
When $[H_3O^+(aq)]$ changes by a factor of 10, the pH changes by 1 unit.

Figure 7.2 The pH scale was developed by the microbiologist Søren Sørensen while working at the Carlsberg Laboratory in Denmark. Note that pH values of less than 0 and greater than 14 are possible.

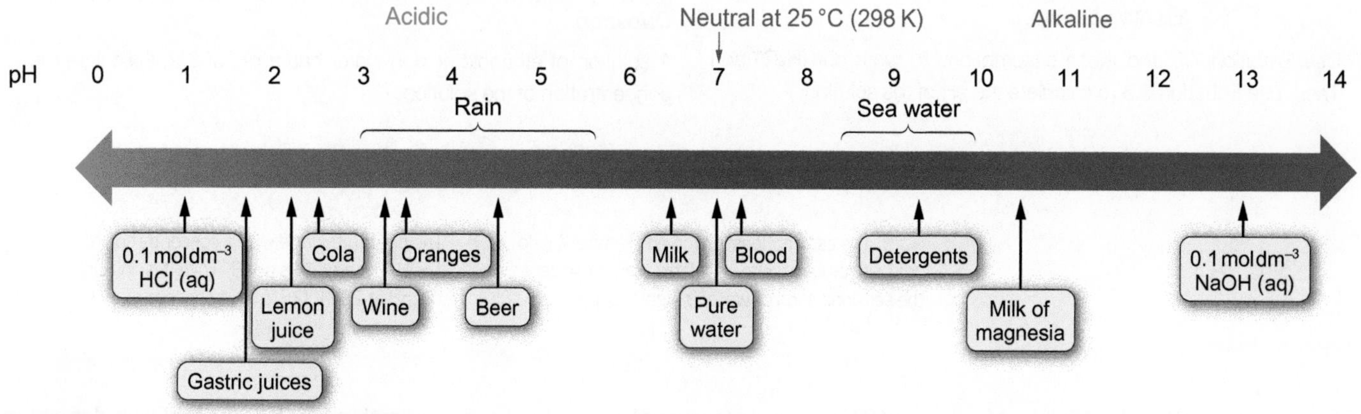
Acidic
Neutral at 25 °C (298 K)
Alkaline
pH 0 1 2 3 4 5 6 7 8 9 10 11 12 13 14
Rain
Sea water
0.1 mol dm^{-3} HCl (aq)
Gastric juices
Lemon juice
Cola
Wine
Oranges
Beer
Milk
Pure water
Blood
Detergents
Milk of magnesia
0.1 mol dm^{-3} NaOH (aq)

Figure 7.3 The pH of some common solutions.

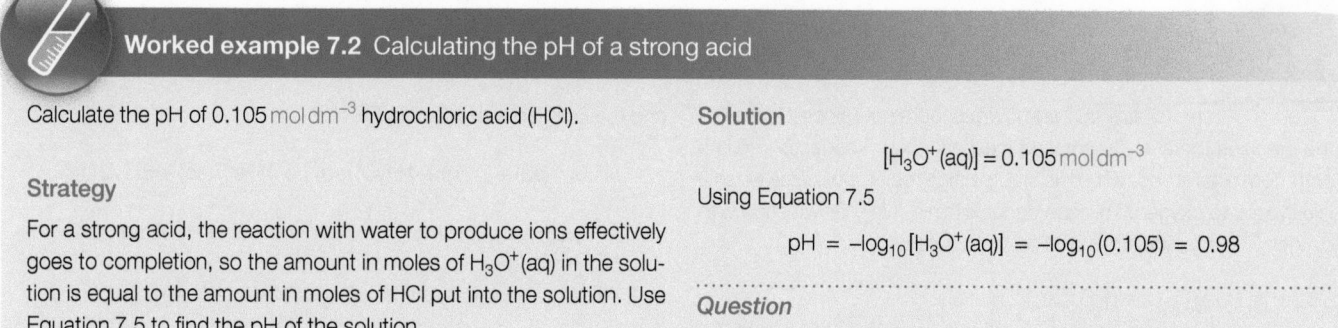

Worked example 7.2 Calculating the pH of a strong acid

Calculate the pH of 0.105 mol dm^{-3} hydrochloric acid (HCl).

Strategy

For a strong acid, the reaction with water to produce ions effectively goes to completion, so the amount in moles of $H_3O^+(aq)$ in the solution is equal to the amount in moles of HCl put into the solution. Use Equation 7.5 to find the pH of the solution.

Solution

$$[H_3O^+(aq)] = 0.105 \text{ mol dm}^{-3}$$

Using Equation 7.5

$$pH = -\log_{10}[H_3O^+(aq)] = -\log_{10}(0.105) = 0.98$$

Question

Calculate the pH of 0.245 mol dm^{-3} nitric acid (HNO_3).

Worked example 7.3 Calculating the pH of a weak acid

Calculate the pH of 0.105 mol dm^{-3} ethanoic acid (CH_3CO_2H) at 298 K. (K_a for ethanoic acid is 1.7×10^{-5} mol dm^{-3} at 298 K.)

Strategy

For a weak acid, the reaction with water is incomplete and you need to work out $[H_3O^+(aq)]$ from the expression for K_a (Equation 7.2) on p.309.

$$K_a = \frac{[H_3O^+(aq)][A^-(aq)]}{[HA(aq)]} \tag{7.2}$$

For a dilute solution of a weak acid you can make two assumptions.

Assumption 1 $[H_3O^+(aq)] = [A^-(aq)]$. This neglects the $[H_3O^+(aq)]$ present in the solution from the ionization of water (see p.313), but this is usually very small by comparison.

Assumption 2 $[HA(aq)]$ at equilibrium is equal to the concentration of HA used to make the solution. This neglects the small proportion of HA molecules that ionize. The error introduced by making this assumption is small and for many purposes the pH obtained is sufficiently accurate.

Use Equation 7.2 and these assumptions to work out $[H_3O^+(aq)]$. Then, use Equation 7.5 to calculate the pH of the solution.

Solution

$$CH_3CO_2H(aq) + H_2O(l) \rightleftharpoons CH_3CO_2^-(aq) + H_3O^+(aq)$$

From Equation 7.2,

$$K_a = \frac{[H_3O^+(aq)][CH_3CO_2^-(aq)]}{[CH_3CO_2H(aq)]}$$

Introducing the two assumptions

$$K_a = \frac{[H_3O^+(aq)]^2}{0.105\,\text{mol dm}^{-3}} = 1.7 \times 10^{-5}\,\text{mol dm}^{-3}$$

Rearranging the equation

$$[H_3O^+(aq)]^2 = (0.105\,\text{mol dm}^{-3}) \times (1.7 \times 10^{-5}\,\text{mol dm}^{-3})$$
$$= 1.8 \times 10^{-6}\,\text{mol}^2\,\text{dm}^{-6}$$
$$[H_3O^+(aq)] = 1.3 \times 10^{-3}\,\text{mol dm}^{-3}$$

Using Equation 7.5,

$$pH = -\log_{10}[H_3O^+(aq)] = -\log_{10}(1.3 \times 10^{-3}) = 2.9 \text{ (at 298 K)}$$

In this case $[H_3O^+] = 1.3 \times 10^{-3}$ mol dm^{-3}. This is very much less than [HA] ([HA] = 0.105 mol dm^{-3}).

Question

A solution of ethanoic acid in water has a pH of 3.2. Calculate the concentration of the solution.

The assumptions in Worked example 7.3 do not hold for very dilute solutions, as in these cases the self-ionization of water is significant (see p.513).

The pH of nitric acid with a concentration of 1×10^{-8} mol dm^{-3} is not 8. Since it is an acid, the pH at 298 K (25 °C) must be less than 7.

Although pOH is not as commonly used as pH, the definition helps provide a short cut in a number of calculations. For example, see Section 7.4 (p.323).

pOH is a convenient way to describe the concentration of $[OH^-]$ ions. pOH is defined in an analogous way to pH, as shown in Equation 7.6

$$pOH = -\log_{10}[OH^-(aq)] \tag{7.6}$$

Box 7.3 Acidic water in disused mines

Iron pyrites (FeS_2) is known as fool's gold since it looks like the precious metal and confused inexperienced prospectors. It is a common mineral and, while it is stable inside rocks, it oxidizes when it is exposed to oxygen. This can occur within a mine. On reaction with oxygen, the disulfide ion is oxidized to sulfate

$$2FeS_2(s) + 7O_2(g) + 6H_2O(l) \rightarrow 2Fe^{2+}(aq) + 4SO_4^{2-}(aq) + 4H_3O^+(aq)$$

and the Fe^{2+} ion is oxidized to Fe^{3+}

$$4Fe^{2+}(aq) + O_2(g) + 4H_3O^+(aq) \rightarrow 4Fe^{3+}(aq) + 6H_2O(l)$$

The $Fe^{3+}(aq)$ ions hydrolyse in solution to release $H_3O^+(aq)$ ions

$$\underset{\text{acid}}{[Fe(H_2O)_6]^{3+}(aq)} + \underset{\text{base}}{H_2O(l)} \rightleftharpoons \underset{\text{conjugate base}}{[Fe(H_2O)_5(OH)]^{2+}(aq)} + \underset{\text{conjugate acid}}{H_3O^+(aq)}$$

→

▲ The Richmond Mine, at Iron Mountain, California contains some of the most acidic mine-water ever reported.

$$[Fe(H_2O)_5(OH)]^{2+}(aq) + H_2O(l) \rightleftharpoons [Fe(H_2O)_4(OH)_2]^{+}(aq) + H_3O^{+}(aq)$$

acid, base, conjugate base, conjugate acid

$$[Fe(H_2O)_4(OH)_2]^{+}(aq) + H_2O(l) \rightleftharpoons [Fe(H_2O)_3(OH)_3](s) + H_3O^{+}(aq)$$

acid, base, conjugate base, conjugate acid

As a result of this oxidation and hydrolysis process, the water in mines can become very acidic. The high temperature in some mines causes water to evaporate, which leads to further concentration of the acid. The pH of groundwater in an abandoned mine in California has been measured as −3.6.

As the acidic water drains from mines it becomes more dilute, and this causes the $Fe^{3+}(aq)$ ions to hydrolyse further, shifting the positions of the equilibria above to the right-hand sides. The iron(III) is precipitated as hydrated $Fe(OH)_3$.

In addition to iron, drainage water from mines normally carries high concentrations of ions of other metals such as manganese, aluminium, zinc, cadmium, and lead. These all cause environmental problems. Acidic water draining from mines must be treated to remove the metal ions and to increase the pH of the water. Artificial waterfalls are often introduced to increase turbulence and speed up the precipitation of $Fe(OH)_3$. Bases such as CaO, $Ca(OH)_2$, or NaOH are sometimes added to neutralize the solutions and precipitate the other dissolved metal ions as hydroxides or oxides.

A 'natural' approach that is growing in popularity is the use of reed beds. Both aerobic and anaerobic wetlands play a role, though the precise mechanisms are not well understood. Iron is precipitated as $Fe(OH)_3$ in aerobic reed beds, whereas heavy metal ions, such as those of cadmium and lead, are precipitated as sulfides in anaerobic wetlands, so immobilizing them. Anaerobic wetlands have a layer of organic matter that contains sulfate-reducing bacteria. The reduction of sulfate to H_2S produces HCO_3^- ions from the organic matter, which neutralize the H_3O^+ ions. Although much slower than chemical treatment, reed beds work without day-to-day management and provide good habitats for wildlife.

▲ Reed beds can be used to remove metal ions from water.

Question

What is the concentration of H_3O^+ in a solution of pH −3.6?

Ionization of water

Although water contains mostly covalent H_2O molecules, it does ionize to a very small extent. When water ionizes, some molecules behave as acids and some as bases

$$H_2O(l) + H_2O(l) \rightleftharpoons OH^-(aq) + H_3O^+(aq)$$

acid, base, conjugate base, conjugate acid

The equilibrium constant for the reaction is given by

$$K_c = \frac{[H_3O^+(aq)][OH^-(aq)]}{[H_2O(l)]^2}$$

The term $[H_2O(l)]^2$ is effectively constant because only a very small proportion of water molecules are ionized, so a new constant, K_w, is defined as in Equation 7.7

$$K_w = [H_3O^+(aq)][OH^-(aq)] \quad \text{units of } K_w \text{ are mol}^2\,\text{dm}^{-6} \tag{7.7}$$

An older name for the self-ionization constant is the *ionic product*.

K_w is called the **self-ionization constant** and, at 298 K, $K_w = 1.00 \times 10^{-14}\,\text{mol}^2\,\text{dm}^{-6}$.

In pure water

$$[H_3O^+(aq)] = [OH^-(aq)]$$

so

$$K_w = [H_3O^+(aq)]^2 = 1.00 \times 10^{-14}\,\text{mol}^2\,\text{dm}^{-6}$$

Therefore

$$[H_3O^+(aq)] = 1.00 \times 10^{-7}\,\text{mol}\,\text{dm}^{-3} \text{ and pH} = 7.00$$

Pure water has pH 7 at 298 K (25 °C). The ionization of water is endothermic so, by Le Chatelier's principle (see Section 1.9, p.56), the value of K_w increases with increasing temperature. Thus, the pH of pure water is less than 7 above 298 K and greater than 7 below 298 K. Some values are shown in Table 7.3.

Pure water is neutral regardless of the temperature. It is the position of 'neutral' on the pH scale that changes with temperature. The concentrations of the $H_3O^+(aq)$ and $OH^-(aq)$ ions change with temperature, but at all temperatures the concentrations of $H_3O^+(aq)$ and $OH^-(aq)$ are equal to one another. However, the increased concentration of H_3O^+ ions in water at higher temperatures can lead to problems, such as corrosion in boilers.

Taking logarithms of both sides of Equation 7.7 gives

$$\log_{10}K_w = \log_{10}[H_3O^+(aq)] + \log_{10}[OH^-(aq)]$$

Working with logarithms
Two useful expressions:

$$\log_{10}(a \times b) = \log_{10}a + \log_{10}b$$

$$\log_{10}\left(\frac{a}{b}\right) = \log_{10}a - \log_{10}b$$

You can find out more about working with logarithms in Maths Toolkit MT3 (p.1316).

Substituting the value of K_w for water at 298 K and rearranging gives Equation 7.8, which relates pH and pOH

$$\text{pH} + \text{pOH} = 14.00 \tag{7.8}$$

Table 7.3 The effect of temperature on K_w and the pH of pure water

Temperature / °C	Temperature / K	K_w / mol^2 dm^{-6}	pH
0	273	1.5×10^{-15}	7.41
10	283	3.0×10^{-15}	7.26
20	293	6.8×10^{-15}	7.08
25	298	1.0×10^{-14}	7.00
30	303	1.5×10^{-14}	6.91
40	313	3.0×10^{-14}	6.76
50	323	5.5×10^{-14}	6.63
60	333	9.5×10^{-14}	6.51

Strengths of bases

In the same way as you derived an expression for K_a for acids, you can derive an expression for a constant K_b for the proton transfer equilibrium for a base in water. Take, for example, the reaction of ammonia with water

$$NH_3\,(aq) + H_2O\,(l) \rightleftharpoons NH_4^+(aq) + OH^-(aq)$$

For this reaction

$$K_c = \frac{[NH_4^+(aq)][OH^-(aq)]}{[NH_3\,(aq)][H_2O\,(l)]}$$

and since $[H_2O\,(l)]$ is effectively constant

$$K_b = \frac{[NH_4^+(aq)][OH^-(aq)]}{[NH_3\,(aq)]}$$

The constant, K_b, is called a **basicity constant** (or sometimes a base dissociation constant). Like K_a, it is defined so that it incorporates the constant $[H_2O\,(l)]$ term from the expression for K_c. The experimental value of K_b for ammonia in water at 298 K (25 °C) is 1.8×10^{-5} mol dm^{-3}. The low value tells you that only a small proportion of the dissolved NH_3 molecules are present as NH_4^+ ions.

In general, for a base, B, in water

$$B\,(aq) + H_2O\,(l) \rightleftharpoons BH^+(aq) + OH^-(aq) \qquad (7.9)$$

$$K_b = \frac{[BH^+(aq)][OH^-(aq)]}{[B\,(aq)]} \qquad (7.10)$$

and

$$pK_b = -\log_{10} K_b \qquad (7.11)$$

The higher the value of K_b, the stronger the base, and the lower pK_b.

For many purposes it is more convenient to describe the strength of a base in terms of the K_a of the conjugate acid. For example, $NH_4^+(aq)$ is the conjugate acid of $NH_3\,(aq)$ so you need to consider the following reaction

$$\underset{\text{conjugate acid}}{NH_4^+(aq)} + \underset{\text{conjugate base}}{H_2O\,(l)} \rightleftharpoons \underset{\text{base}}{NH_3\,(aq)} + \underset{\text{acid}}{H_3O^+(aq)}$$

$$K_a = \frac{[H_3O^+(aq)][NH_3\,(aq)]}{[NH_4^+(aq)]}$$

Table 7.4 shows acidity constants for the conjugate acids of some common bases at 298 K (25 °C). The stronger a base, the weaker is its conjugate acid. So a strong base has a low value of K_a and a high value of pK_a.

Relationship between K_a and K_b for a conjugate acid-base pair

The following expressions highlight the link between K_a and K_b for a conjugate acid–base pair. For ammonia in water, multiplication of the two equilibrium constants in Equations 7.2 and 7.10 gives

$$K_a \times K_b = \frac{[H_3O^+(aq)]\cancel{[NH_3(aq)]}}{\cancel{[NH_4^+(aq)]}} \times \frac{\cancel{[NH_4^+(aq)]}[OH^-(aq)]}{\cancel{[NH_3(aq)]}}$$

$$= [H_3O^+(aq)][OH^-(aq)] = K_w$$

So

$$K_a \times K_b = K_w \qquad (7.12)$$

Since aqueous ammonia solutions contain ammonium cations and hydroxide anions in addition to NH_3 molecules, they are often labelled as 'ammonium hydroxide'. This 'compound' cannot be isolated pure. Removal of water from the solution pulls the position of equilibrium to the left and eventually gaseous ammonia is the only product.

When chemists talk about the K_a or the pK_a of a base they are referring to the conjugate acid of the base.

Table 7.4 Acidity constants (K_a) and values of pK_a for the conjugate acids of some weak bases at 298 K (25 °C). The pK_b of the base is also shown

Base		Conjugate acid	K_a / mol dm^{-3}	pK_a	pK_b	
Urea	$CO(NH_2)_2$	$CO(NH_2)(NH_3)^+$	7.8×10^{-1}	0.11	13.89	Weakest base
Phenylamine	$C_6H_5NH_2$	$C_6H_5NH_3^+$	1.3×10^{-5}	4.87	9.13	
Pyridine	C_5H_5N	$C_5H_5NH^+$	5.9×10^{-6}	5.23	8.77	
Hydrazine	NH_2NH_2	$NH_2NH_3^+$	7.9×10^{-9}	8.1	5.9	
Ammonia	NH_3	NH_4^+	5.6×10^{-10}	9.25	4.75	
Trimethylamine	$(CH_3)_3N$	$(CH_3)_3NH^+$	1.6×10^{-10}	9.80	4.20	
Methylamine	CH_3NH_2	$CH_3NH_3^+$	2.2×10^{-11}	10.66	3.34	
Triethylamine	$(CH_2CH_3)_3N$	$(CH_2CH_3)_3NH^+$	1.8×10^{-11}	10.75	3.25	Strongest base

Taking logarithms of both sides gives

$$\log_{10}(K_a \times K_b) = \log_{10} K_w$$

$$\log_{10} K_a + \log_{10} K_b = \log_{10} K_w$$

Since pK_a = $-\log_{10} K_a$ and pK_b = $-\log_{10} K_b$, multiplication throughout by -1 gives

$$pK_a + pK_b = pK_w \qquad (7.13)$$

where pK_w = $-\log_{10} K_w$ = 14.00 at 298 K.

Equation 7.13 quantifies the relationship between the strength of an acid and that of its conjugate base described on p.308. If an acid has a large K_a, then its conjugate base has a small K_b, and vice versa.

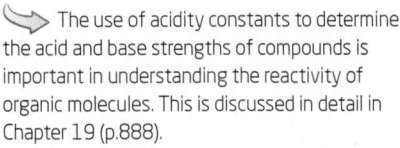

The use of acidity constants to determine the acid and base strengths of compounds is important in understanding the reactivity of organic molecules. This is discussed in detail in Chapter 19 (p.888).

Worked example 7.4 Strengths of acids and bases

Use Tables 7.1 (p.308), 7.2 (p.310), and 7.4 above to decide the following.

(a) Which in the following pairs is the stronger acid:

(i) HCO_2H or $C_6H_5CO_2H$?

(ii) NH_4^+ or $CH_3NH_3^+$?

(b) Which in the following pairs is the stronger base:

(i) CN^- or OCl^-?

(ii) CH_3NH_2 or $N(CH_3)_3$?

Strategy

The greater the K_a (or the smaller pK_a) of a weak acid, the stronger the acid.

The stronger the acid, the weaker its conjugate base.

For (b), use Table 7.1 to help you write out the conjugate acid–base pair for each species.

Use the K_a values in Tables 7.2 and 7.4 to help you decide relative strengths.

Solution

(a) (i) $K_a(HCO_2H) > K_a(C_6H_5CO_2H)$, so HCO_2H is the stronger acid.

(ii) $K_a(NH_4^+) > K_a(CH_3NH_3^+)$, so NH_4^+ is the stronger acid.

(b) (i) $K_a(HOCl) > K_a(HCN)$, so HOCl is the stronger acid. The stronger acid has the weaker conjugate base, so CN^- is a stronger base than OCl^-.

(ii) $K_a[(CH_3)_3NH^+] > K_a(CH_3NH_3^+)$, so $(CH_3)_3NH^+$ is the stronger acid. The stronger acid has the weaker conjugate base, so CH_3NH_2 is a stronger base than $N(CH_3)_3$.

Question

Use Table 7.2 to predict which is the stronger base, F^- ions or NO_3^- ions.

The position of equilibrium in an acid-base reaction

By looking at pK_a values, you can predict in which direction an acid–base reaction will go. For example, take the following equilibrium involving ethanoic acid and trichloroethanoic acid

$$\underset{\text{acid}}{CCl_3CO_2H\,(aq)} + \underset{\text{base}}{CH_3CO_2^-\,(aq)} \rightleftharpoons \underset{\text{conjugate base}}{CCl_3CO_2^-\,(aq)} + \underset{\text{conjugate acid}}{CH_3CO_2H\,(aq)}$$

From the data in Table 7.2 (p.310), CCl_3CO_2H is a stronger acid than CH_3CO_2H as it has a lower pK_a value. This means it is better at transferring a proton, so the equilibrium in the reaction above lies to the right-hand side.

Using pK_a values to calculate the equilibrium constant in an acid-base reaction

For a general reaction between an acid (HA) and a base (B)

$$HA + B \rightleftharpoons BH^+$$

you can use the values of $K_a(HA)$ and $K_a(BH^+)$ to determine the equilibrium constant, K_c.

Given that

$$K_a(HA) = \frac{[H_3O^+(aq)][A^-(aq)]}{[HA(aq)]} \quad \text{and} \quad K_a(BH^+) = \frac{[H_3O^+(aq)][B(aq)]}{[BH^+(aq)]}$$

$$\frac{K_a(HA)}{K_a(BH^+)} = \frac{\cancel{[H_3O^+(aq)]}[A^-(aq)]}{[HA(aq)]} \times \frac{[BH^+(aq)]}{\cancel{[H_3O^+(aq)]}[B(aq)]}$$

$$= \frac{[A^-(aq)][BH^+(aq)]}{[HA(aq)][B(aq)]}$$

$$= K_c$$

$$K_c = \frac{K_a(HA)}{K_a(BH^+)}, \text{ so talking logarithms of both sides}$$

$$\log_{10}K_c = \log_{10}K_a(HA) - \log_{10}K_a(BH^+)$$

$$pK_c = pK_a(HA) - pK_a(BH^+) \qquad (7.14)$$

Visit the Online Resource Centre to view screencast 7.1 which walks you through how to calculate the equilibrium constant for an acid-base reaction.

Equation 7.14 is used in Section 19.2 (p.895), to predict the extent of organic reactions.

Box 7.4 Controlling pH in a swimming pool

Swimming pools are potential breeding grounds for bacteria, so disinfecting them is essential for the health of the swimmers. The most common method for this is chlorination. When chlorine dissolves in water, it reacts to form hypochlorous acid (HOCl) which is the active bactericide

$$Cl_2\,(aq) + 2\,H_2O\,(l) \rightleftharpoons H_3O^+\,(aq) + Cl^-\,(aq) + HOCl\,(aq)$$

HOCl is a weak acid, so the equilibrium

$$HOCl\,(aq) + H_2O\,(l) \rightleftharpoons OCl^-\,(aq) + H_3O^+\,(aq)$$

lies on the left-hand side. If the pH is increased, H_3O^+ ions are removed and the equilibrium shifts to the right. This converts more HOCl into OCl^-, which is a less effective bactericide because the negative charge on the ion prevents it from passing through the bacteria cell walls.

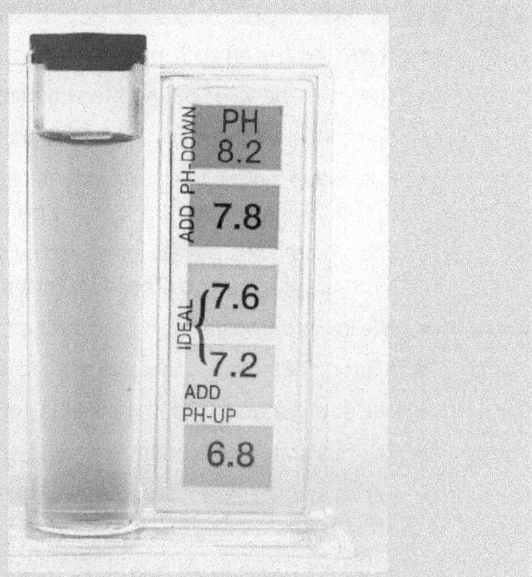
PH 8.2
ADD PH-DOWN
7.8
IDEAL
7.6
7.2
ADD PH-UP
6.8

A tester kit uses an indicator to measure the pH of swimming pools. Indicators are described in Section 7.5 (p.327).

→ To maximize the concentration of HOCl, the pH must be maintained between 4 and 6. Unfortunately, under these acidic conditions, the water would attack the concrete of the pool, not to mention the swimmers and their swimwear. A compromise pH of between 7.3 and 7.4 is chosen to protect the pool and maintain as much of the HOCl as possible in the active form. The pH is adjusted by adding sodium carbonate to neutralize excess acid. Pools are typically buffered by using a sodium hydrogencarbonate system. Buffering is necessary as swimmers can introduce alkalis from their sweat, and acids from urine (see Section 7.3, p.319, for an explanation of how buffers work).

The original source of chlorine was from cylinders of the gas, but this is potentially hazardous. Nowadays the source is usually trichlorocyanuric acid (trichlor, $C_3N_3Cl_3O_3$) which reacts with water to form HOCl.

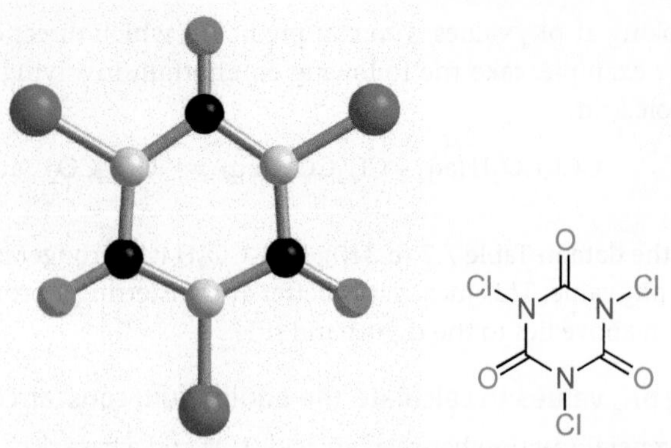

▲ Trichlorocyanuric acid (trichlor).

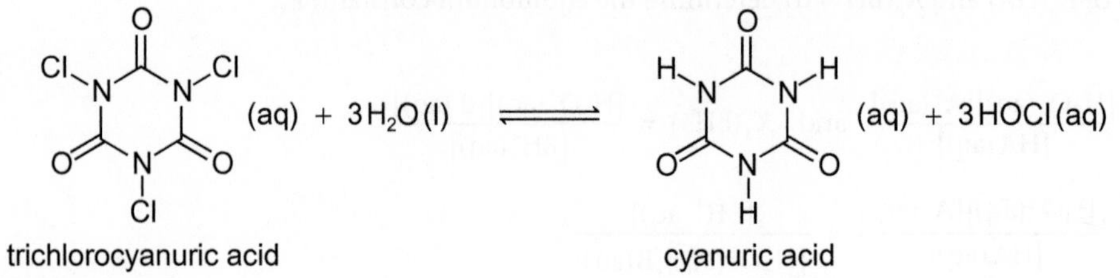
(aq) + 3H2O(l)
(aq) + 3HOCl(aq)
trichlorocyanuric acid
cyanuric acid

Trichlor is a solid and, although it needs to be handled with gloves, it is much easier to use than chlorine gas.

Trichlor also acts as a 'chlorine buffer'. In outdoor pools as much as 90% of the HOCl can be lost due to photolytic decomposition in sunlight

$$2\,HOCl(aq) \xrightarrow{h\nu} 2\,HCl(aq) + O_2(g)$$

The equilibria between trichlor, cyanuric acid, and the partially chlorinated cyanuric acids enable much of the chlorine to be present in compounds that do not decompose in sunlight. These equilibria adjust to the loss of HOCl by moving to the right and generating more of this active molecule.

Question

The pK_a of HOCl is 7.40. Calculate the pH of a 0.100 mol dm^{-3} HOCl solution, explaining any assumptions you make.

Summary

- The strength of an acid is measured by its acidity constant, K_a. For convenience, this is often quoted as pK_a, where p$K_a = -\log_{10}K_a$.
- The lower the value of pK_a, the stronger the acid, and the greater the value of K_a.
- The concentration of $[H_3O^+(aq)]$ in solution is given by the pH, which is defined as pH $= -\log_{10}[H_3O^+(aq)]$.
- Pure water ionizes to a small extent, and this is quantified by the self-ionization constant of water, K_w. At 298 K, $K_w = 1.00 \times 10^{-14}$ mol^2 dm^{-6}, so the pH of pure water at 298 K is 7.00.
- The strength of a base is measured by its basicity constant, K_b, though it is often more convenient to provide the K_a value for the conjugate acid. These are related since $K_a \times K_b = K_w$.
- The lower the value of pK_b, the stronger the base, and the greater the value of K_b.
- A strong base has a low value of K_a for its conjugate acid, and a high value of pK_a.
- The stronger an acid, the weaker its conjugate base. The stronger a base, the weaker its conjugate acid.
- The position of equilibrium in an acid–base reaction depends on the relative strengths of the acids (or bases).
- For an acid–base reaction, $HA + B \rightleftharpoons A^- + BH^+$, the equilibrium constant for the reaction is found using pK_c = pK_a(HA) – pK_a(BH^+).

? For practice questions on these topics, see questions 2–6 at the end of this chapter (p.339).

7.3 Buffer solutions

Controlling the pH is important in living systems as many reactions must take place within a narrow range of pH values. It can be important too in many chemical or biochemical processes in industry, and medical and cosmetic products are often 'buffered' so that the pH accurately matches the part of the body where it is applied. A **buffer solution** resists changes of pH. The pH stays approximately constant even when small amounts of acids and alkalis are added. Furthermore, its pH does not change when the solution is diluted.

Buffer solutions are usually made either from a weak acid and one of its salts (e.g. CH_3CO_2H and CH_3CO_2Na) or from a weak base and one of its salts (e.g. NH_3 and NH_4Cl). In other words, a buffer solution contains a weak acid and its conjugate base, or a weak base and its conjugate acid. By choosing suitable conjugate acid–base pairs and adjusting their concentrations, you can make buffer solutions with almost any desired pH.

The action of a buffer solution made up from a weak acid, HA (aq), and its conjugate base, A^-(aq) ions, depends on the equilibrium:

$$HA(aq) + H_2O(l) \rightleftharpoons A^-(aq) + H_3O^+(aq) \qquad (7.1)$$

A typical buffer of this sort is made by dissolving equimolar amounts of ethanoic acid and sodium ethanoate (the salt) in water.

You can make two good approximations about the species present in the equilibrium.

Assumption 1 *All the A^-(aq) ions come from the salt.* HA (aq) supplies very few A^-(aq) ions in comparison with the salt, which is an ionic compound and so totally in the form of ions.

Assumption 2 *Almost all the HA molecules put into the buffer solution remain unchanged.* You make the same assumption when calculating the pH of a weak acid (see Worked example 7.3), but it is a better approximation for a buffer solution because the high concentration of A^-(aq) ions from the salt pushes the position of equilibrium even further to the left.

When a small amount of an acidic substance is added to the buffer solution, the extra H_3O^+(aq) ions disturb the equilibrium. By Le Chatelier's principle (see Section 1.9, p.57), some A^-(aq) ions from the salt react with H_3O^+(aq) ions and the position of equilibrium moves to the left to minimize the effect of the change. A significant fall in pH is prevented.

If a small amount of alkali is added, the OH^-(aq) ions react with H_3O^+(aq) ions in the buffer solution, disturbing the equilibrium. The weak acid HA (aq) reacts with H_2O (l) and the position of equilibrium moves to the right to produce more H_3O^+(aq) ions and minimize the effect of the change. A significant rise in pH is prevented.

The presence of both the weak acid and its salt are necessary for the buffer solution to work. There must be plenty of HA (aq) to act as a *source* of extra H_3O^+(aq) ions when they are needed, and plenty of A^-(aq) ions to act as a *sink* for any extra H_3O^+(aq) ions that have been added. The presence of HA (aq) means that this type of buffer solution has an acidic pH. How a buffer solution keeps the pH of a solution balanced can be thought of as a see-saw.

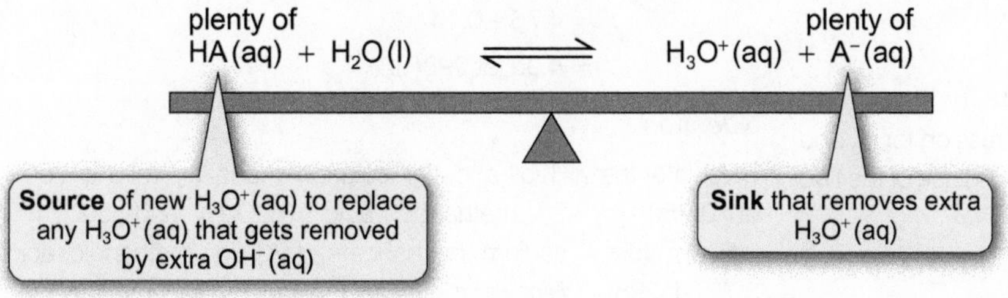
plenty of
HA(aq) + H₂O(l)
plenty of
H₃O⁺(aq) + A⁻(aq)
Source of new H₃O⁺(aq) to replace any H₃O⁺(aq) that gets removed by extra OH⁻(aq)
Sink that removes extra H₃O⁺(aq)

An example of a buffer solution with an alkaline pH is ammonia solution plus ammonium chloride. In this case, the weak base, NH_3(aq), acts as the proton sink and the NH_4^+(aq) ions from the salt act as the proton source.

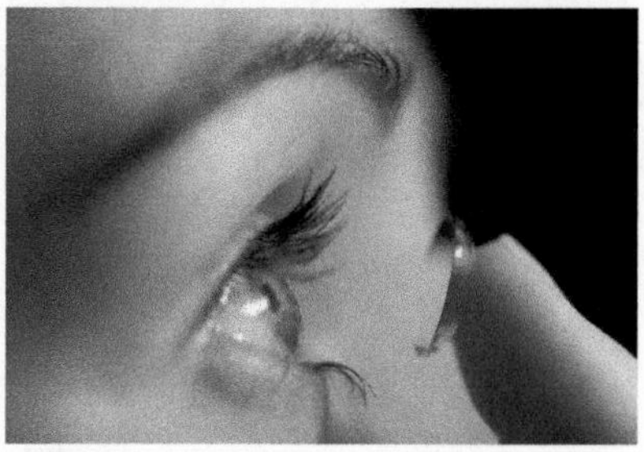

Contact lenses are cleaned in a solution that is buffered to match the pH of the surface of the eye. A boric acid/borate buffer is commonly used.

A buffer solution contains a weak acid and its conjugate base, or a weak base and its conjugate acid.

Calculations with buffer solutions

Reorganizing Equation 7.2 (p.309) gives

$$K_a = \frac{[H_3O^+(aq)][A^-(aq)]}{[HA(aq)]} = [H_3O^+(aq)] \times \frac{[A^-(aq)]}{[HA(aq)]}$$

If you then make use of **assumptions 1 and 2** above, this becomes Equation 7.15

$$K_a = [H_3O^+(aq)] \times \frac{[\text{conjugate base}]}{[\text{acid}]} \qquad (7.15)$$

where [conjugate base] and [acid] are the concentrations of the salt and acid used to prepare the buffer solution.

Taking logarithms of both sides (see p.309) and multiplying through by −1 gives

$$-\log_{10} K_a = -\log_{10}[H_3O^+(aq)] - \log_{10}\frac{[\text{conjugate base}]}{\text{acid}}$$

Rearranging the equation, this becomes Equation 7.16

$$\text{pH} = \text{p}K_a + \log_{10}\frac{[\text{conjugate base}]}{[\text{acid}]} \qquad (7.16)$$

Equation 7.16 is called the **Henderson–Hasselbalch equation.** It shows that the pH of the buffer solution depends on *two* factors:

- the value of K_a, which provides a 'coarse tuning' of the buffer pH. The choice of the weak acid or base determines which *region* of the pH range the buffer is in;
- the ratio of [conjugate base]/[acid], which provides a 'fine tuning' of the buffer pH. Changing the ratio from about 3 : 1 to about 1 : 3 changes the $[H_3O^+(aq)]$ by a factor of 9, which alters the pH by approximately 1 unit. The ratio should not be too far outside this range; otherwise there will not be enough HA (aq) or A^- (aq) for the buffer to act as a source or a sink.

Visit the Online Resource Centre to view screencast 7.2 which walks you through the derivation of the Henderson-Hasselbalch equation.

If you look at Equation 7.16, you can see why the pH of a buffer solution is not affected by dilution. When you add water, the concentrations of the acid and the conjugate base are reduced equally, so that the *ratio* of their concentrations remains the same and the pH is unchanged.

Worked example 7.5 Buffer solutions

Calculate the pH of a buffer solution containing 0.150 mol dm^{-3} ethanoic acid and 0.205 mol dm^{-3} sodium ethanoate. (pK_a for ethanoic acid is 4.76 at 298 K.)

Strategy

Either, substitute values directly into Equation 7.16, *or*, if you have trouble remembering equations, start from the expression for K_a and rearrange to give an expression for $[H_3O^+(aq)]$, making use of the two assumptions above.

Solution

Using Equation 7.16

$$\text{pH} = \text{p}K_a + \log_{10}\frac{(0.205\ \text{mol dm}^{-3})}{(0.150\ \text{mol dm}^{-3})} \qquad (7.16)$$

$$= 4.76 + 0.14$$

$$= 4.90 \text{ (at 298 K).}$$

Question

Calculate the pH of a buffer solution made by mixing 250 cm^3 of 0.250 mol dm^{-3} methanoic acid (HCO_2H) and 500 cm^3 of 0.500 mol dm^{-3} sodium methanoate. (pK_a for methanoic acid is 3.75 at 298 K.) *Hint.* In this case the concentrations of acid and conjugate base in the buffer are different from those in the solutions added to make it because they dilute one another when they are mixed. Remember to take this dilution into account.

Box 7.5 Buffering in the blood

Human blood has a narrow pH range of 7.35–7.45, which must be maintained for metabolic processes to function properly. To keep the pH in this range requires a delicate balance between the concentrations of the conjugate acid–base pairs making up the buffer system. The main buffer is a carbonic acid/hydrogencarbonate system, which involves the following three equilibria. (These are the same equilibria discussed in Box 1.12, p.58, in relation to cave chemistry, but to explain the buffer action Equation 1.20 is broken down into two stages.)

$$CO_2(g) \rightleftharpoons CO_2(aq)$$

$$CO_2(aq) + H_2O(l) \rightleftharpoons H_2CO_3(aq)$$

$$\underset{\text{acid}}{H_2CO_3(aq)} + \underset{\text{base}}{H_2O(l)} \rightleftharpoons \underset{\text{conjugate base}}{HCO_3^-(aq)} + \underset{\text{conjugate acid}}{H_3O^+(aq)}$$

Carbonic acid (H_2CO_3) is a weak acid and $HCO_3^-(aq)$ is its conjugate base. At the temperature in the human body, the pK_a for carbonic acid is 6.3, which is about 1 pH unit outside the pH range for blood. However, the normal concentration of $CO_2(g)$ in the lungs maintains a ratio of $HCO_3^-(aq)/H_2CO_3(aq)$ in blood plasma of about 13 : 1. Substituting these values into Equation 7.16 gives a buffer pH of 7.4.

$$pH = pK_a + \log_{10}\frac{[HCO_3^-(aq)]}{[H_2CO_3(aq)]}$$

$$pH = 6.3 + \log_{10}(13) = 7.4$$

The carbonic acid concentration in the blood is largely controlled by breathing and respiration. When you breathe more quickly, the amount of CO_2 exhaled increases and your system becomes deficient in $CO_2(g)$. Hydrogencarbonate ion concentration is largely controlled by excretion in urine.

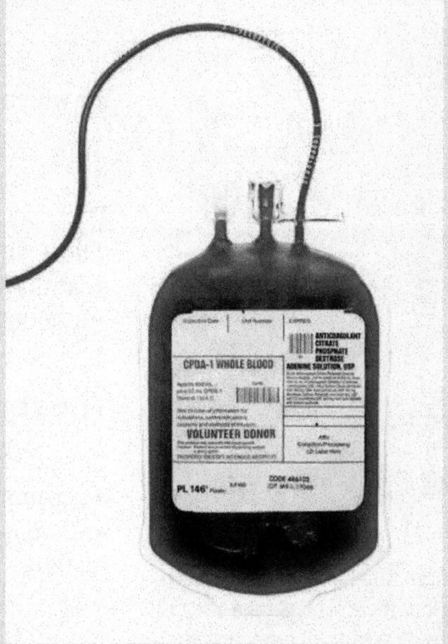
CPDA-1 WHOLE BLOOD
VOLUNTEER DONOR
ANTICOAGULANT
CITRATE
PHOSPHATE
DEXTROSE
ADENINE SOLUTION, USP

▲ Blood contains an efficient buffer system that keeps the pH around 7.4.

Question

If blood pH rises above 7.45, a potentially life-threatening condition called *alkalosis* can result. This can happen in patients who are hyperventilating from severe anxiety, or in climbers suffering from oxygen deficiency at high altitude. Although no longer recommended first aid, one way to treat alkalosis is to get the patient to breathe into a bag so that the exhaled CO_2 is re-inhaled. Use the above equilibria to explain how this brings down blood pH.

Summary

- A buffer solution is resistant to changes in pH on addition of small quantities of acid or base, and on dilution.
- Buffers consist of either a weak acid and its salt or a weak base and its salt.
- The pH of a buffer solution can be calculated using the Henderson–Hasselbalch equation (Equation 7.16).

For a practice question on this topic, see question 7 at the end of this chapter (p.339).

7.4 pH changes in acid–base titrations

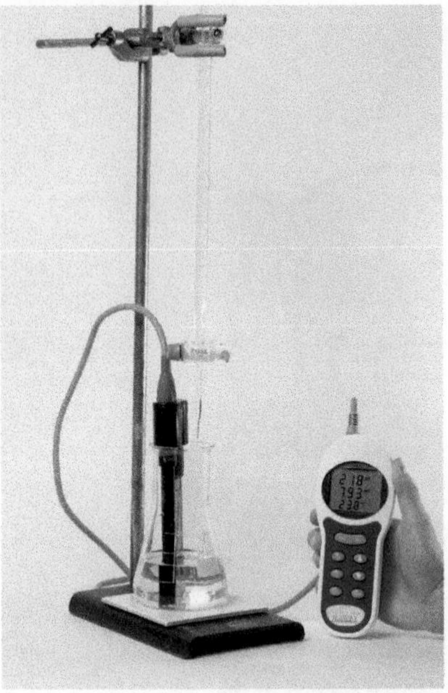

In a typical acid–base titration, the alkali is added from the burette into the solution of the acid in a conical flask. The progress of the reaction can be followed with a pH meter.

Acid–base titrations are described in Section 1.5 (p.36). The use of a pH meter is explained in Section 11.2 (p.524), and indicators are discussed in Section 7.5 (p.327).

An acid–base reaction in aqueous solution is a neutralization reaction

$$\underset{\text{acid}}{HX\,(aq)} + \underset{\text{base}}{MOH\,(aq)} \rightarrow \underset{\text{salt}}{MX\,(aq)} + \underset{\text{water}}{H_2O\,(l)}$$

for which the ionic reaction is

$$H_3O^+(aq) + OH^-(aq) \rightarrow 2H_2O\,(l)$$

These reactions are carried out quantitatively in titrations. Normally, indicators are used to detect when neutralization has occurred, but the progress of the reaction can be followed quantitatively by using a pH meter to measure the pH of the solution in the conical flask as the titration proceeds.

The reaction between a strong acid and a strong base

Consider the reaction between $20.0\,cm^3$ of $0.100\,mol\,dm^{-3}$ hydrochloric acid solution and $0.100\,mol\,dm^{-3}$ aqueous sodium hydroxide. In the titration, controlled amounts of the sodium hydroxide solution are added from a burette to the solution of the acid in a conical flask

$$HCl\,(aq) + NaOH\,(aq) \rightarrow NaCl\,(aq) + H_2O\,(l)$$

Before any of the sodium hydroxide is added, the pH of the solution is

$$\begin{aligned} pH &= -\log_{10}[H_3O^+(aq)] \\ &= -\log_{10}(0.100) \\ &= 1.00 \end{aligned}$$

The reaction will be complete when $20.0\,cm^3$ of the sodium hydroxide solution have been added. At this point, the solution in the conical flask contains sodium chloride solution. $[H_3O^+(aq)] = [OH^-(aq)]$, and the solution is neutral, pH = 7.00.

How does the pH of the solution change during the titration before neutralization?

As the titration proceeds, the $OH^-(aq)$ ions from the added NaOH react with some of the $H_3O^+(aq)$ ions from the acid. The concentration of $H_3O^+(aq)$ goes down, and the pH rises. Because 1 mol $H_3O^+(aq)$ reacts with 1 mol $OH^-(aq)$, the pH of the mixture at any point in the titration before neutralization can be calculated from the expression

$$[H_3O^+] = \frac{(\text{original amount in moles of acid}) - (\text{amount in moles of base added})}{(\text{volume of acid}) + (\text{volume of base added})} \qquad (7.17)$$

You can use either Equation 7.17 or Equation 7.18 to work out the pH at any point in a titration before neutralization.

You can work out the amounts in moles of acid and base from first principles as shown in Worked example 1.13 (p.36) or by using Equation 1.9, $n = c \times V$, where n is the amount in moles, c is the concentration in $mol\,dm^{-3}$, and V is the volume of the solution in dm^3. Using Equation 1.9,

$$[H_3O^+] = \frac{(\text{conc. of acid} \times \text{volume of acid}) - (\text{conc. of base} \times \text{volume of base added})}{(\text{volume of acid}) + (\text{volume of base added})} \qquad (7.18)$$

In this case

$$[H_3O^+] = \frac{(0.100\,mol\,dm^{-3} \times 0.0200\,\cancel{dm^3}) - (0.100\,mol\,dm^{-3} \times x\,\cancel{dm^3})}{(0.0200 + x)\,\cancel{dm^3}}$$

where x is the volume of added NaOH. Remember that volumes are given in dm^3 not cm^3.

$$[H_3O^+] = \frac{0.00200 - 0.100x}{0.0200 + x} \text{mol dm}^{-3}$$

Using Equation 7.17

$$pH = -\log_{10}\left(\frac{0.00200 - 0.100x}{0.0200 + x}\right)$$

Calculating the pH for some different values of x

$$x = 10.0\,\text{cm}^3 = 0.0100\,\text{dm}^3 \qquad pH = 1.48$$
$$x = 19.0\,\text{cm}^3 = 0.0190\,\text{dm}^3 \qquad pH = 2.59$$
$$x = 19.9\,\text{cm}^3 = 0.0199\,\text{dm}^3 \qquad pH = 3.60$$
$$x = 19.99\,\text{cm}^3 = 0.01999\,\text{dm}^3 \qquad pH = 4.60$$

When $20.0\,\text{cm}^3$ of NaOH solution has been added, all of the acid has reacted. This is known as the **equivalence point** of the titration. For the titration between a strong acid and a strong base the equivalence point occurs at pH 7.00.

How does the pH of the solution change after neutralization?

To work out the pH of the solution past the equivalence point, you need to calculate the concentration of OH^-, then use the relationship between pOH and pH (Equation 7.8, p.314) to calculate the pH.

For example, if $21.0\,\text{cm}^3$ sodium hydroxide is added, $20.0\,\text{cm}^3$ is used to neutralize the hydrochloric acid, leaving $1.0\,\text{cm}^3$ of the $0.100\,\text{mol dm}^{-3}$ solution, in a total volume of $41.0\,\text{cm}^3$. The concentration of OH^- is given by

$$[OH^-] = \frac{(0.100\,\text{mol dm}^{-3} \times 0.0010\,\cancel{\text{dm}^3})}{0.0410\,\cancel{\text{dm}^3}}$$
$$= 2.44 \times 10^{-3}\,\text{mol dm}^{-3}$$

Visit the Online Resource Centre to view screencast 7.3 which walks you through the way pH changes in a titration.

so

$$pOH = -\log_{10}(2.44 \times 10^{-3}) = 2.6$$

Using Equation 7.8

$$pH + pOH = 14.00 \text{ (at 298 K)}$$
$$pH = 14.00 - pOH = 11.4$$

Titration curve for a strong acid and a strong base

A plot showing the variation of pH with the amount of added base is called a **titration curve**, and the titration curve for the reaction between a strong acid and a strong base is shown in Figure 7.4. The general shape of this curve is identical for all titrations of strong acids with strong bases. The pH is low at the start of the titration, and it changes slowly with added base until near the equivalence point when it rises sharply. This rise continues beyond the equivalence point to a high pH, after which the rate of increase in pH becomes much slower.

The equivalence point occurs at the steepest part of the curve, which for a strong acid and a strong base is at pH 7.00. If a less concentrated acid is used, the starting pH is higher, and consequently the steep portion of the curve is shorter. For example, if a $0.010\,\text{mol dm}^{-3}$ solution of HCl were used, the starting pH would be

$$pH = -\log_{10}(0.010) = 2.0$$

as shown in Figure 7.5. The equivalence point is still at pH 7.00, regardless of the concentrations of the acid and base used.

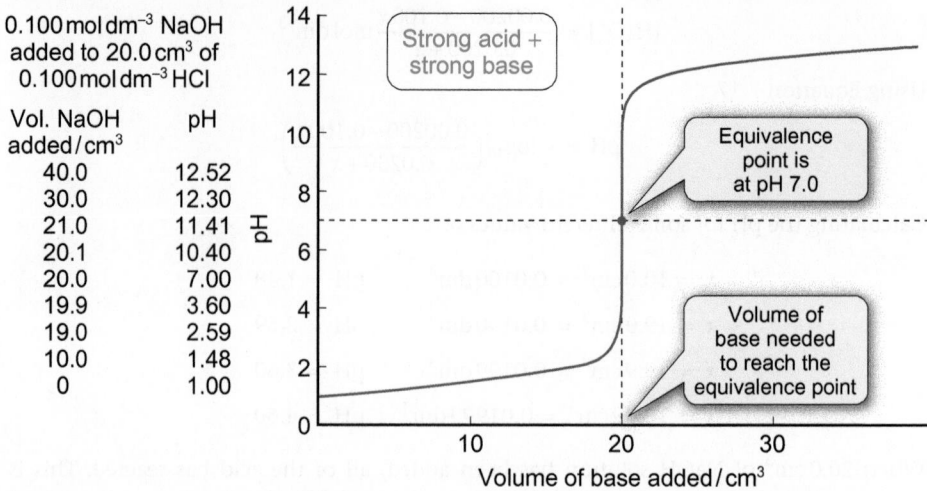

0.100 mol dm⁻³ NaOH added to 20.0 cm³ of 0.100 mol dm⁻³ HCl

Vol. NaOH added/cm³	pH
40.0	12.52
30.0	12.30
21.0	11.41
20.1	10.40
20.0	7.00
19.9	3.60
19.0	2.59
10.0	1.48
0	1.00

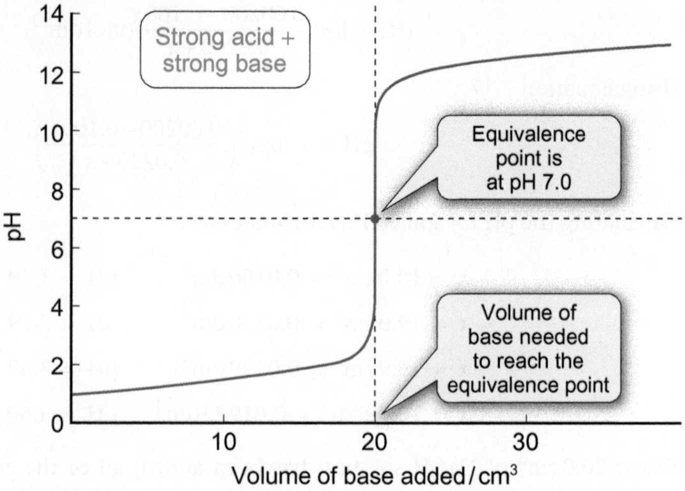
Strong acid + strong base
Equivalence point is at pH 7.0
Volume of base needed to reach the equivalence point
pH
Volume of base added/cm³

Figure 7.4 The titration curve for the addition of a strong base to a strong acid. In this case, 20.0 cm^3 of 0.100 mol dm^{-3} hydrochloric acid is titrated with a solution of 0.100 mol dm^{-3} NaOH.

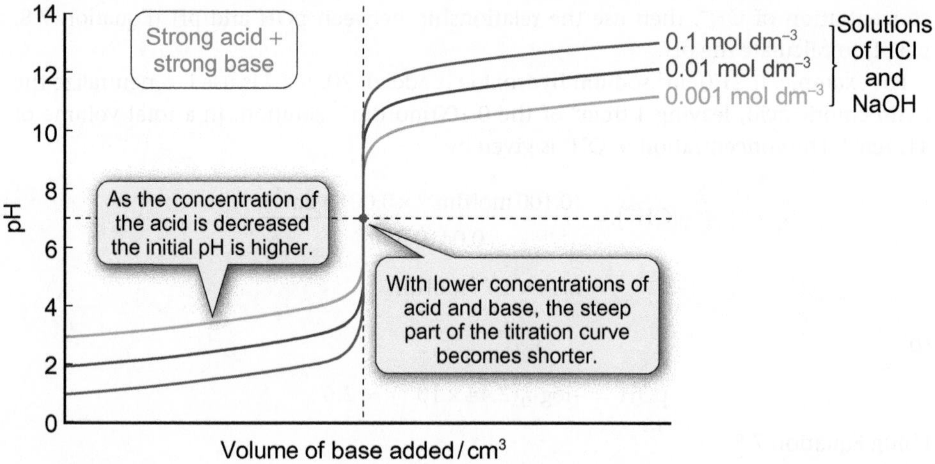
Strong acid + strong base
0.1 mol dm⁻³
0.01 mol dm⁻³
0.001 mol dm⁻³
Solutions of HCl and NaOH
As the concentration of the acid is decreased the initial pH is higher.
With lower concentrations of acid and base, the steep part of the titration curve becomes shorter.
pH
Volume of base added/cm³

Figure 7.5 The titration curve for the addition of a strong base to a strong acid, showing the effect of the concentrations of the acid and base used.

Worked example 7.6 Calculating the pH in a titration reaction

20.0 cm^3 of 0.240 mol dm^{-3} nitric acid is transferred to a conical flask and titrated against 0.400 mol dm^{-3} sodium hydroxide solution. What is the pH of the solution when 8.0 cm^3 of the base has been added?

Strategy

Work out the volume of NaOH (aq) needed to reach the equivalence point.

If the reaction has not reached the equivalence point, use Equation 7.17 or Equation 7.18 to calculate $[H_3O^+]$, and then Equation 7.5 (p.310) to work out the pH.

If the reaction has passed the equivalence point, calculate $[OH]^-$, and use the relationship between pH and pOH (Equation 7.8, p.314) to work out the pH.

Remember to convert volumes of solutions from cm^3 to dm^3 to use in Equation 7.17 or Equation 7.18.

→

→ Solution

$$HNO_3(aq) + NaOH(aq) \rightarrow NaNO_3(aq) + H_2O(l)$$

20.0 cm^3 of 0.240 mol dm^{-3} HNO_3 contain $\frac{0.240\,\text{mol} \times 20.0\,\cancel{cm^3}}{1000\,\cancel{cm^3}} = 4.80 \times 10^{-3}$ mol

Since 1 mol HNO_3(aq) reacts with 1 mol NaOH(aq), 4.80×10^{-3} mol of NaOH are needed to reach the equivalence point.

0.400 mol of NaOH are contained in 1000 cm^3.

4.80×10^{-3} mol of NaOH are contained in $\frac{4.80 \times 10^{-3}\,\cancel{mol} \times 1000\,cm^3}{0.400\,\cancel{mol}} = 12.0\,cm^3$ of solution.

Volume of NaOH need to reach the equivalence point is 12.0 cm^3.

Addition of 8.0 cm^3 NaOH is before the equivalence point. Using Equation 7.18

$$[H_3O^+] = \frac{(\text{conc. of acid} \times \text{volume of acid}) - (\text{conc. of base} \times \text{volume of base added})}{(\text{volume of acid}) + (\text{volume of base added})} \quad (7.18)$$

$$[H_3O^+] = \frac{(0.240\,\text{mol}\,dm^{-3} \times 0.0200\,\cancel{dm^3}) - (0.400\,\text{mol}\,dm^{-3} \times 0.0080\,\cancel{dm^3})}{(0.0200 + 0.0080)\,\cancel{dm^3}}$$

$$= 0.0571\,\text{mol}\,dm^{-3}$$

$$pH = -\log_{10}(0.0571) = 1.24$$

Question

What is the pH of the solution in the conical flask when 11.9 cm^3 of NaOH have been added?

The reaction between a weak acid and a strong base

When a weak acid is titrated against a strong base, the same volume of base is needed to reach the equivalence point as would be the case if the acid were a strong acid of the same concentration. This is because 1 mol of the base reacts with 1 mol of the acid, whether the acid is strong or weak. After the equivalence point, the titration curve is the same shape as for a titration between a strong acid and a strong base. There are, however, differences in the shape of the titration curve before neutralization, as seen in Figure 7.6.

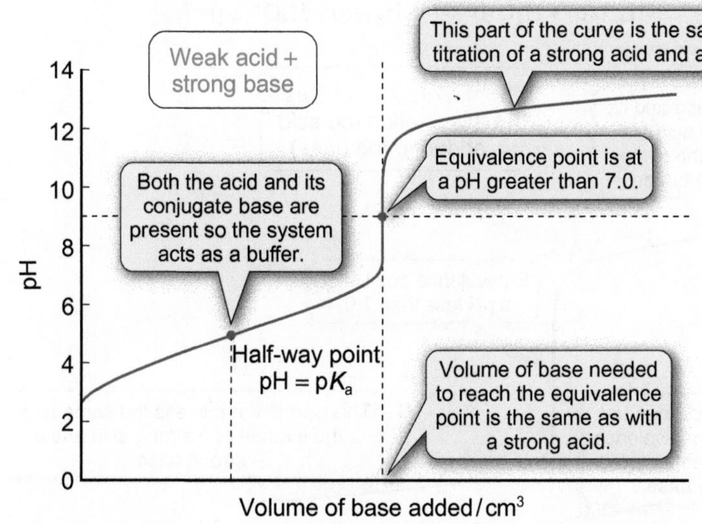
Weak acid + strong base
This part of the curve is the same as for the titration of a strong acid and a strong base.
Equivalence point is at a pH greater than 7.0.
Both the acid and its conjugate base are present so the system acts as a buffer.
Half-way point pH = pKa
Volume of base needed to reach the equivalence point is the same as with a strong acid.
pH
14
12
10
8
6
4
2
0
Volume of base added/cm3

Figure 7.6 The titration curve for the addition of a strong base to a weak acid, with the same concentrations as in Figure 7.4.

This equilibrium is important in the action of buffers. See Section 7.3 (p.319) for more details.

To understand these differences, consider the reaction between ethanoic acid and sodium hydroxide. Ethanoic acid is present in solution in equilibrium with ethanoate ions

$$CH_3CO_2H\,(aq) + H_2O\,(l) \rightleftharpoons CH_3CO_2^-\,(aq) + H_3O^+\,(aq)$$

Since the acid is only partially ionized, the initial pH is higher than for a titration with a strong acid. Both the acid and anion are present, so the system acts as a buffer when the base is added. The base removes H_3O^+, but this shifts the equilibrium above to the right generating more H_3O^+. This means that the pH changes slowly before neutralization.

Another important difference between this titration curve and that for a strong acid–strong base is the pH at the equivalence point. For this titration the equivalence point is at a pH higher than 7. This is because the salt (sodium ethanoate) hydrolyses (reacts with water), so the neutralized solution is alkaline

Salts of weak acids and strong bases are alkaline in aqueous solution because of hydrolysis.

$$CH_3CO_2^-\,(aq) + H_2O\,(l) \rightleftharpoons CH_3CO_2H\,(aq) + OH^-\,(aq)$$

For ethanoic acid

$$K_a = \frac{[CH_3CO_2^-(aq)][H_3O^+(aq)]}{[CH_3CO_2H(aq)]}$$

At the point of half-neutralization

$$[CH_3CO_2^-(aq)] = [CH_3CO_2H\,(aq)]$$

So

$$K_a = [H_3O^+(aq)]$$

and pK_a = pH when the acid is half neutralized.

The implications of this relationship for organic reactivity are explored in Section 19.2 (p.888).

The reaction between a strong acid and a weak base

Salts of strong acids and weak bases are acidic in aqueous solution because of hydrolysis.

For a reaction between a weak acid and a weak base there is no well-defined end-point, so titrations are not useful.

The principles behind the reaction between a strong acid and a weak base are similar to those in the reaction between a weak acid and a strong base. A typical example of a strong acid–weak base reaction is that between hydrochloric acid and ammonia, and a titration curve for this reaction is shown in Figure 7.7.

In this case, hydrochloric acid is added from the burette to the ammonia solution in the conical flask, so the curve is a different shape to those in Figures 7.4–7.6. In this case, the pH starts at a high value and finishes at a low value. The solution before the equivalence point contains both $NH_3\,(aq)$ and $NH_4^+(aq)$, so it acts as a buffer, resisting change in pH with addition of acid. The equivalence point comes at a pH lower than 7 due to hydrolysis of the ammonium chloride product

$$NH_4^+(aq) + H_2O\,(l) \rightleftharpoons NH_3\,(aq) + H_3O^+\,(aq)$$

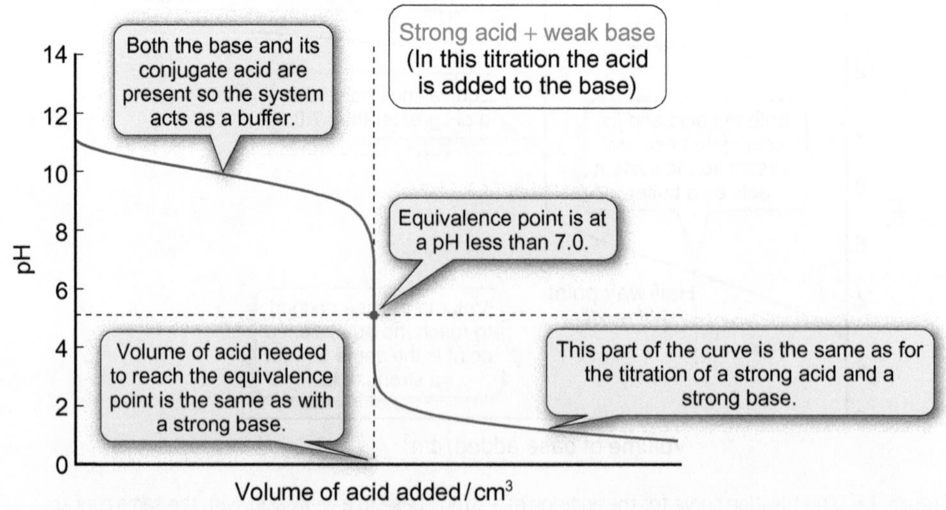
Both the base and its conjugate acid are present so the system acts as a buffer.
Strong acid + weak base (In this titration the acid is added to the base)
Equivalence point is at a pH less than 7.0.
Volume of acid needed to reach the equivalence point is the same as with a strong base.
This part of the curve is the same as for the titration of a strong acid and a strong base.
pH
14
12
10
8
6
4
2
0
Volume of acid added / cm³

Figure 7.7 The titration curve for the addition of a strong acid to a weak base.

Summary

- Acid–base reactions in solution can be carried out quantitatively using titrations.
- A titration curve shows how the pH of the solution changes during the titration.
- For the addition of a strong base to a strong acid, the pH starts low and increases slowly until very close to the equivalence point. It then rises sharply through the equivalence point before increasing slowly again at high pH. The equivalence point is at pH 7.00.
- For the addition of a strong base to a weak acid, the pH starts higher, and initially rises slowly as the system acts as a buffer. The equivalence point is at pH > 7.00 due to hydrolysis of the salt. At half-neutralization, pH = pK_a.
- For the addition of a strong acid to a weak base, the pH initially decreases slowly as the system acts as a buffer. The equivalence point is at pH < 7.00 due to hydrolysis of the salt.

For a practice question on this topic, see question 8 at the end of this chapter (p.339).

7.5 Indicators

While it is possible to follow the course of an acid–base titration using a pH meter to measure $[H_3O^+]$, it is usually easier and more convenient to use an **indicator**. The **end-point** of the reaction is the point at which the indicator changes colour. With a good choice of indicator, the end-point is the same as the equivalence point, so the change in colour of the indicator shows the equivalence point of the reaction.

In acid–base titrations, an indicator is used to determine the equivalence point through a colour change. The indicator is chosen so its colour change corresponds to the equivalence point.

An acid–base indicator is normally a weak acid or a weak base that has different colours in its neutral (uncharged) and ionic forms. A weak acid is partially ionized in solution. Writing the indicator as HIn, the following equilibrium occurs

$$\underset{\text{acid}}{\text{HIn (aq)}} + H_2O\,(l) \rightleftharpoons \underset{\text{conjugate base}}{\text{In}^-\text{(aq)}} + H_3O^+\,(aq)$$

The acidity constant, K_{in}, for the indicator is given by

$$K_{in} = \frac{[H_3O^+][In^-]}{[HIn]}$$

Taking logarithms of both sides and then rearranging gives

$$pH = pK_{in} + \log_{10}\frac{[In^-]}{[HIn]} \quad (7.19)$$

which is analogous to the Henderson–Hasselbalch equation (Equation 7.16, p.320).

You would expect the indicator to change colour when the conversion of HIn to In^- is half complete, that is, when $[HIn] = [In^-]$. At this point, from Equation 7.19

$$pH = pK_{in} + \log_{10}(1)$$

$$pH = pK_{in}$$

Σ $\log_{10}(1) = 0$

An indicator does not change colour instantaneously. Generally you can only see the colour of the neutral (or ionic) form if there is a 10 : 1 excess of that compound. For example, an indicator that is red in the neutral form and blue in the ionic form will appear completely red when $[HIn] > 10 \times [In^-]$ and completely blue when $[In^-] > 10 \times [HIn]$. This means the colour change typically occurs over 2 pH units. Titrations involving a strong acid or base (or both) involve a sharp pH change at the equivalence point (Figures 7.4–7.7), so the colour change of the indicator occurs with a very small addition of base or acid.

Since an indicator is a weak acid or base, it is important to use only a few drops. If too much is added, the indicator will affect the pH of the solution.

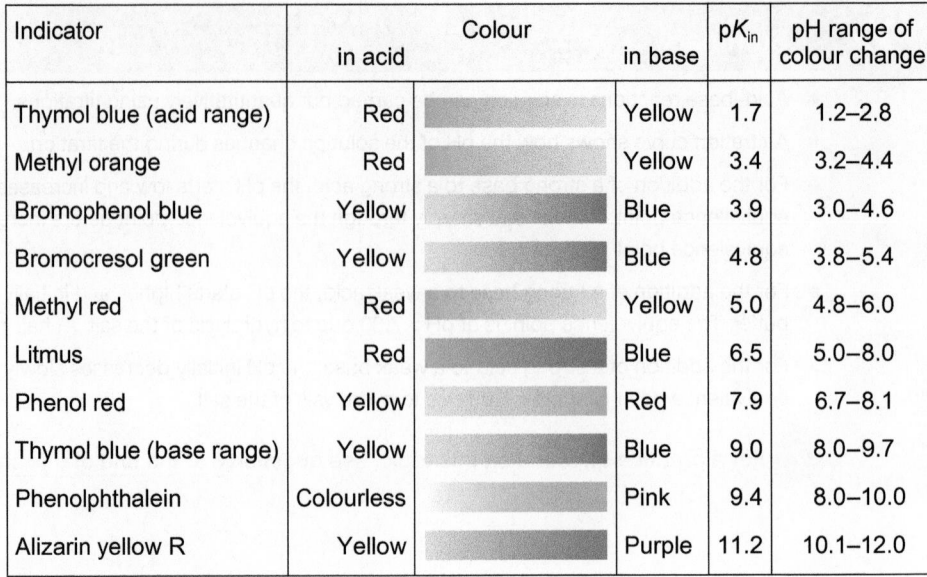

Indicator	Colour			pK_{in}	pH range of colour change
	in acid		in base		
Thymol blue (acid range)	Red		Yellow	1.7	1.2–2.8
Methyl orange	Red		Yellow	3.4	3.2–4.4
Bromophenol blue	Yellow		Blue	3.9	3.0–4.6
Bromocresol green	Yellow		Blue	4.8	3.8–5.4
Methyl red	Red		Yellow	5.0	4.8–6.0
Litmus	Red		Blue	6.5	5.0–8.0
Phenol red	Yellow		Red	7.9	6.7–8.1
Thymol blue (base range)	Yellow		Blue	9.0	8.0–9.7
Phenolphthalein	Colourless		Pink	9.4	8.0–10.0
Alizarin yellow R	Yellow		Purple	11.2	10.1–12.0

Figure 7.8 Examples of commonly used indicators for acid–base titrations.

Choosing an indicator

The pH at which an indicator changes colour depends on the value of pK_{in}, which varies from indicator to indicator. A selection of common acid–base indicators is given in Figure 7.8, together with pH ranges over which they change colour.

The choice of indicator for a titration depends on the strengths of the acid and base that are being used. For a titration between a strong acid and a strong base, such as that between hydrochloric acid and sodium hydroxide, the equivalence point is at pH 7, so an indicator with pK_{in} of around 7 is needed. From Figure 7.8, phenol red would be a good choice, as it changes colour over the pH range 6.7 to 8.1. This is shown on the titration curve in Figure 7.9(a). A good indicator for a particular titration changes colour with the addition of 1 or 2 drops of acid or base, and this corresponds to the steepest part of the titration curve.

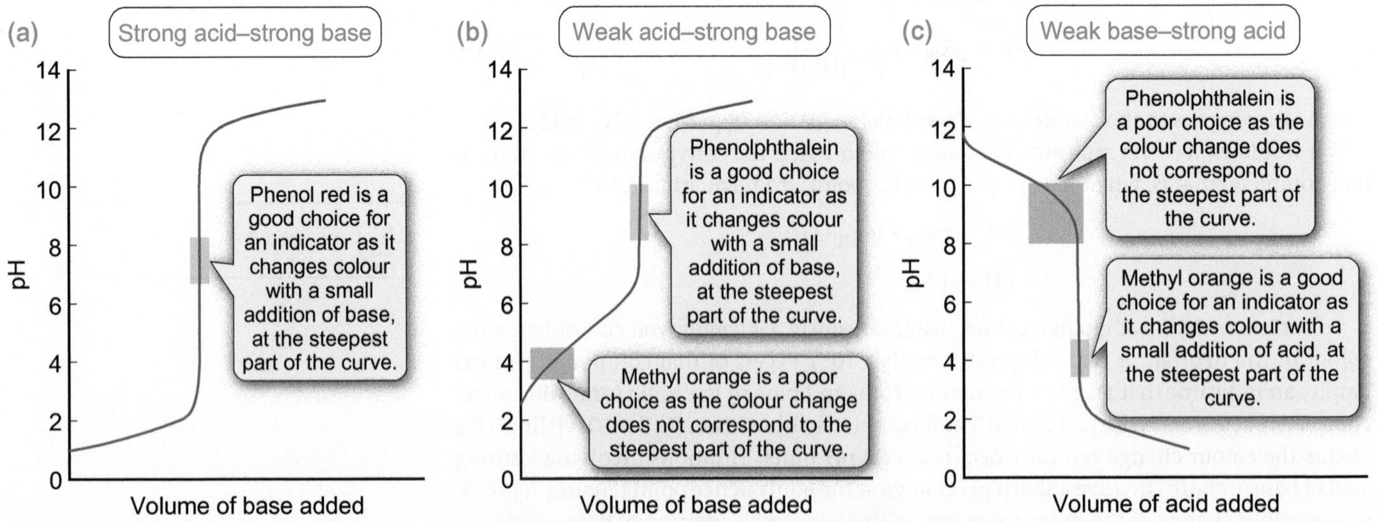
(a)
Strong acid–strong base
Phenol red is a good choice for an indicator as it changes colour with a small addition of base, at the steepest part of the curve.
pH
Volume of base added
(b)
Weak acid–strong base
Phenolphthalein is a good choice for an indicator as it changes colour with a small addition of base, at the steepest part of the curve.
Methyl orange is a poor choice as the colour change does not correspond to the steepest part of the curve.
pH
Volume of base added
(c)
Weak base–strong acid
Phenolphthalein is a poor choice as the colour change does not correspond to the steepest part of the curve.
Methyl orange is a good choice for an indicator as it changes colour with a small addition of acid, at the steepest part of the curve.
pH
Volume of acid added

Figure 7.9 The indicator chosen for a titration must have a colour change range that corresponds with the steep part of the titration curve.

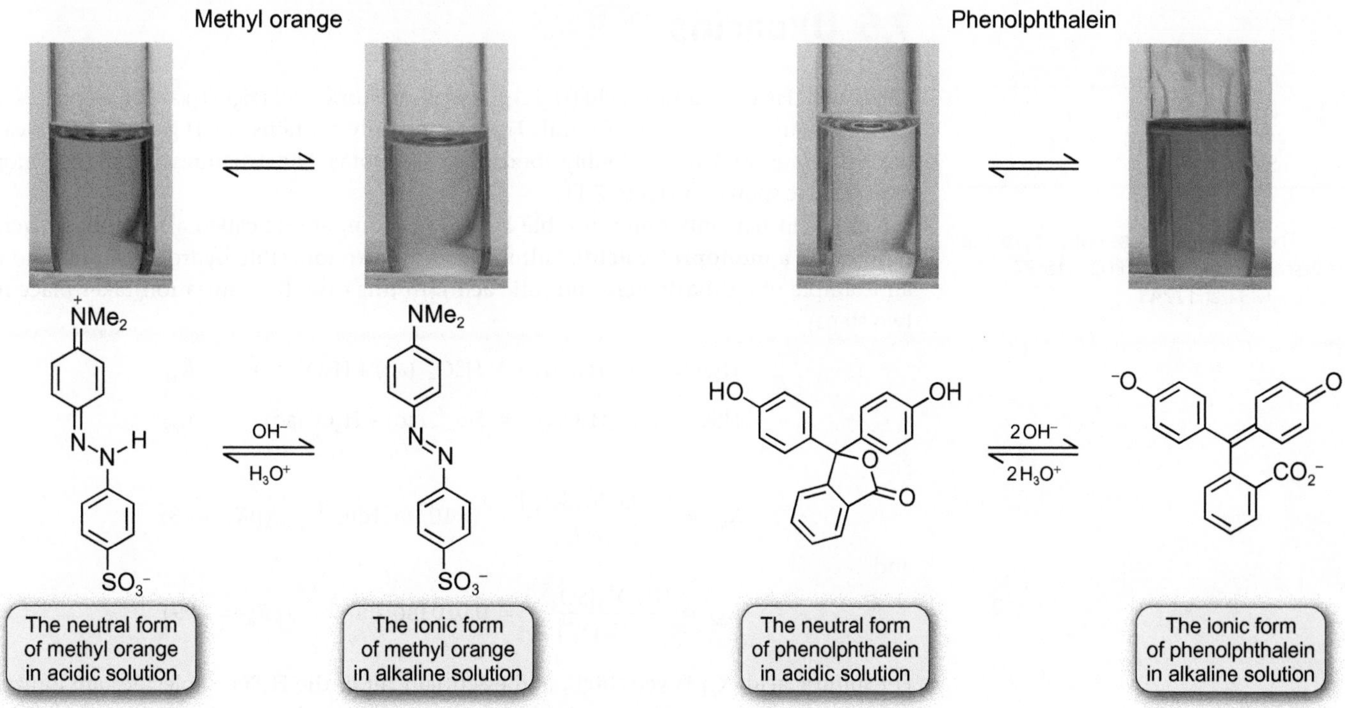
Methyl orange
Phenolphthalein
OH⁻
H₃O⁺
2OH⁻
2H₃O⁺
The neutral form of methyl orange in acidic solution
The ionic form of methyl orange in alkaline solution
The neutral form of phenolphthalein in acidic solution
The ionic form of phenolphthalein in alkaline solution

Figure 7.10 From left to right, methyl orange in acidic solution, methyl orange in alkaline solution, phenolphthalein in acidic solution, and phenolphthalein in alkaline solution. Note that 'neutral' here means 'no overall charge'.

For a titration between ethanoic acid (a weak acid) and sodium hydroxide (a strong base) the equivalence point is at a higher pH. In this case, an indicator such as phenolphthalein which has a higher pK_{in} gives the best results. As Figure 7.9(b) shows, phenolphthalein changes colour at the equivalence point with a small amount of added base. Methyl orange would be a much poorer choice in this case since a large addition of base would be needed to observe the colour change and the colour change does not correspond to the equivalence point. For the titration between hydrochloric acid and ammonia, an indicator with a lower pK_{in}, such as methyl orange, is used, whereas phenolphthalein would not be suitable. The colour changes involved with these indicators are shown in Figure 7.10, together with the structures of the methyl orange and phenolphthalein in their neutral and ionic forms.

Litmus and universal indicator change colour over a wider range of pH units than most indicators, so are best used to give a general idea of whether a solution is acidic or basic rather than to determine the end-point in a titration. Universal indicator is a mixture of indicators, each with a different pK_{in}, so the colour varies over a wide pH range.

The activity at the Online Resource Centre allows you to see the effect of the indicator on the end-point of a titration.

Summary

- Acid–base titrations can be followed easily by adding an indicator. An acid–base indicator is normally a combination of a weak acid with its conjugate base, or a weak base with its conjugate acid. The neutral and ionic forms have different colours.
- An indicator is chosen so that pK_{in} = pH at the equivalence point of a titration.
- The equivalence point occurs on the steep part of a titration curve. Indicators typically change colour over ~2 pH units, and at the steep part of a titration curve this occurs with addition of 1 or 2 drops of acid or base.

For a practice question on this topic, see question 9 at the end of this chapter (p.339).

7.6 Oxoacids

> An oxoacid H_mXO_n normally has m OH groups and $(n-m)$ doubly bonded oxygen atoms.

> The structures and chemistry of oxoacids are described in more detail in Chapter 27, pp.1225, 1237, and 1243.

Nitric acid (HNO_3), sulfuric acid (H_2SO_4), and phosphoric acid (H_3PO_4) are all examples of **oxoacids**. An oxoacid of the formula H_mXO_n normally contains m OH groups, which can ionize in water, and $(n-m)$ doubly bonded oxygen atoms. The structures of some common oxoacids are shown in Figure 7.11.

Nitric acid has only one ionizable hydrogen atom, and is called a **monobasic acid** (also called a **monoprotic acid**). Sulfuric acid has two ionizable hydrogen atoms and is an example of a **dibasic acid** (diprotic acid). In this case, the ionization takes place in two steps

$$H_2SO_4\,(aq) + H_2O\,(l) \rightleftharpoons HSO_4^-\,(aq) + H_3O^+\,(aq) \qquad K_{a1}$$

$$HSO_4^-\,(aq) + H_2O\,(l) \rightleftharpoons SO_4^{2-}\,(aq) + H_3O^+\,(aq) \qquad K_{a2}$$

So

$$K_{a1} = \frac{[H_3O^+][HSO_4^-]}{[H_2SO_4]} = 1\times10^3\ \text{mol dm}^{-3} \qquad (pK_{a1} = -3)$$

and

$$K_{a2} = \frac{[H_3O^+][SO_4^{2-}]}{[HSO_4^-]} = 0.010\ \text{mol dm}^{-3} \qquad (pK_{a2} = 1.99)$$

For sulfuric acid, K_{a1} is very high, and essentially all of the H_2SO_4 molecules are ionized in aqueous solution. K_{a2} is considerably lower as the second proton is lost from the HSO_4^- anion rather than from a neutral molecule. Removing H^+ from a negatively charged ion is harder than removing it from a neutral molecule. This means that both HSO_4^- and SO_4^{2-} ions are present in sulfuric acid solutions.

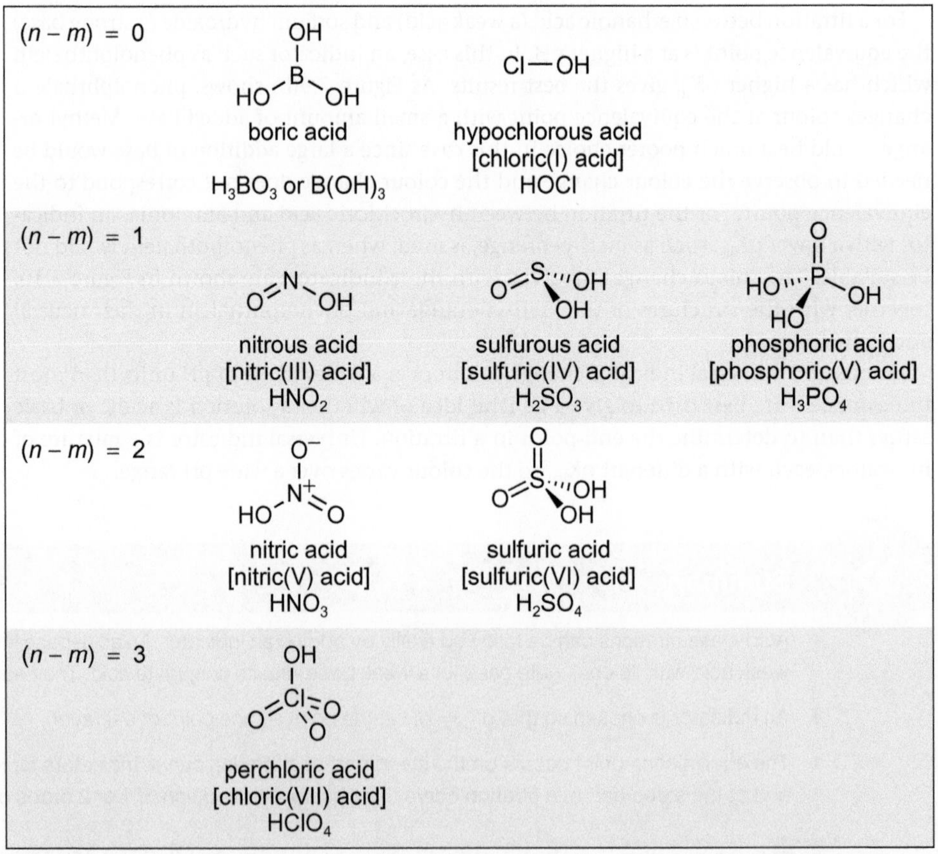

(n – m) = 0
boric acid
H3BO3 or B(OH)3
hypochlorous acid
[chloric(I) acid]
HOCl
(n – m) = 1
nitrous acid
[nitric(III) acid]
HNO2
sulfurous acid
[sulfuric(IV) acid]
H2SO3
phosphoric acid
[phosphoric(V) acid]
H3PO4
(n – m) = 2
nitric acid
[nitric(V) acid]
HNO3
sulfuric acid
[sulfuric(VI) acid]
H2SO4
(n – m) = 3
perchloric acid
[chloric(VII) acid]
HClO4

Figure 7.11 The structures of some common oxoacids of the general formula H_mXO_n.

Phosphoric acid is a **tribasic acid** (triprotic acid)

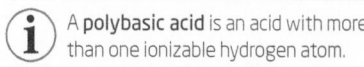

A **polybasic acid** is an acid with more than one ionizable hydrogen atom.

$$H_3PO_4\,(aq) + H_2O\,(l) \rightleftharpoons H_2PO_4^-\,(aq) + H_3O^+\,(aq) \qquad K_{a1}$$

$$H_2PO_4^-\,(aq) + H_2O\,(l) \rightleftharpoons HPO_4^{2-}\,(aq) + H_3O^+\,(aq) \qquad K_{a2}$$

$$HPO_4^{2-}\,(aq) + H_2O\,(l) \rightleftharpoons PO_4^{3-}\,(aq) + H_3O^+\,(aq) \qquad K_{a3}$$

For phosphoric acid:

$$K_{a1} = \frac{[H_3O^+][H_2PO_4^-]}{[H_3PO_4]} = 6.9 \times 10^{-3}\,\text{mol dm}^{-3} \qquad (pK_{a1} = 2.16)$$

$$K_{a2} = \frac{[H_3O^+][HPO_4^{2-}]}{[H_2PO_4^-]} = 6.2 \times 10^{-8}\,\text{mol dm}^{-3} \qquad (pK_{a2} = 7.21)$$

$$K_{a3} = \frac{[H_3O^+][PO_4^{3-}]}{[HPO_4^{2-}]} = 4.8 \times 10^{-13}\,\text{mol dm}^{-3} \qquad (pK_{a3} = 12.32)$$

Food-grade phosphoric acid is used as an acidifying agent, for example, to give cola drinks their tangy taste.

As for sulfuric acid, each successive value of K_a is lower as the proton is more difficult to remove. This is because it is removed from a species with increasing negative charge.

The strengths of oxoacids

The pK_a values for some common oxoacids are given in Table 7.5.

Table 7.5 pK_a values for some common oxoacids

$(n-m)=0$		$(n-m)=1$		$(n-m)=2$		$(n-m)=3$	
HOCl	7.40	HNO_2	3.25	HNO_3	−1.4	$HClO_4$	−10
HOBr	8.55	$HClO_2$	1.94	$HClO_3$	−1		
HOI	10.0	H_2SO_3	1.85, 7.2	HIO_3	0.78		
H_4SiO_4	9.9	H_3PO_4	2.16, 7.21, 12.32	H_2SO_4	−3, 1.99		
H_3PO_3	1.3	H_5IO_6	3.3, 6.7	H_2SeO_4	<0, 2.1		

From the data in Table 7.5, it can be seen that the value of pK_{a1} depends on $(n-m)$ as below.

$(n-m)$	Approximate pK_a
0	7.4–10.0
1	1.8–3.3
2	−3–0.8
3	< −8

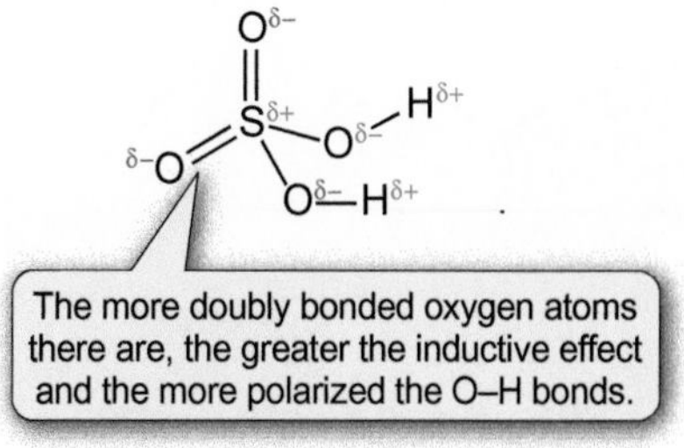
The more doubly bonded oxygen atoms there are, the greater the inductive effect and the more polarized the O–H bonds.

There are two reasons for the importance of $(n-m)$. Firstly, there is an inductive effect (−I), with the electronegative oxygen atoms drawing electrons away from hydrogen atoms in the O–H bonds. This weakens the O–H bonds, so that a proton is more easily removed. The more doubly bonded oxygen atoms that are present, the greater this inductive effect, and the weaker the O–H bond.

Inductive effects in organic chemistry are described in Section 19.1 (p.870).

The second effect involves the conjugate base. The more resonance forms there are, the greater the delocalization of the electron, so the more stable the anion. For example, in the hydrogensulfate ion (HSO_4^-) the negative charge is delocalized on three oxygen atoms, whereas in the hydrogensulfite ion (HSO_3^-) it is delocalized on only two oxygen atoms. The resonance forms for the ions are shown below.

Resonance forms are introduced in Section 4.5 (p.180).

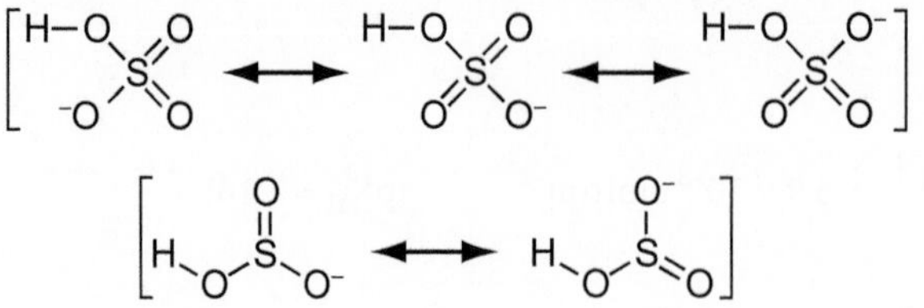

The greater delocalization of the negative charge in HSO_4^- compared with HSO_3^- is a factor in the greater acidity of H_2SO_4 compared to H_2SO_3

The effects of charge delocalization in organic chemistry are described in Sections 19.1 (p.873) and 19.2 (p.889).

Protonation of the conjugate base—the reverse reaction in the acid–base equilibrium—is easier for HSO_3^- as the average charge on each oxygen atom is greater.

In Table 7.5 there appears to be an exception to these generalities. The pK_{a1} value for phosphonic acid (H_3PO_3), 1.3, does not fit in with the other acids for which $(n - m) = 0$, but is closer to the pK_{a1} value for acids with $(n - m) = 1$. This suggests there are only two rather than three ionizable hydrogen atoms in H_3PO_3. In the structure of H_3PO_3, one of the three hydrogen atoms is bonded directly to the phosphorus atom. The P–H bond is not very polar, so it is difficult to ionize this proton. H_3PO_3 is therefore a dibasic acid. In a similar manner, phosphinic acid (H_3PO_2) is a monobasic acid with a pK_a value of 2.0. In this acid, two of the three hydrogen atoms are bonded to the phosphorus atom, so only one is ionizable.

H3PO3 contains two ionizable hydrogen atoms.
H3PO2 contains one ionizable hydrogen atom.

phosphonic acid [phosphoric(III) acid] phosphinic acid [phosphoric(I) acid]

Although not as large as the effect of changing $(n - m)$, increasing the electronegativity of the central atom removes some electron density from the OH bonds. For this reason H_2SO_4 is a slightly stronger acid than H_2SeO_4.

The reaction between a dibasic acid and a strong base

For a dibasic acid, two moles of base are needed for full neutralization of one mole of acid. The two neutralization equilibria have significantly different pK_a values, so you can treat the two reactions with base independently.

A titration curve for the addition of sodium hydroxide to a weak dibasic acid is shown in Figure 7.12. The titration curve has two separate equivalence points on it, and both of these can be detected by repeating the reaction using different indicators.

Speciation curves

For a weak acid, the concentrations of the acid HA and the anion A^- in the solution vary with pH. The concentration of A^- is raised by increasing the pH of the solution as this neutralizes H_3O^+. This pushes the position of equilibrium to the right-hand side

$$HA\,(aq) + H_2O\,(l) \rightleftharpoons A^-\,(aq) + H_3O^+\,(aq)$$

In contrast, decreasing the pH pushes the position of equilibrium to the left-hand side, giving a higher concentration of HA.

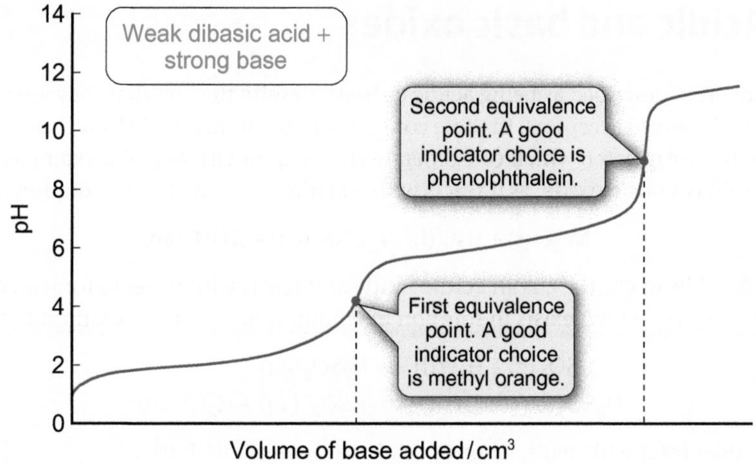
Weak dibasic acid + strong base
Second equivalence point. A good indicator choice is phenolphthalein.
First equivalence point. A good indicator choice is methyl orange.
pH
14
12
10
8
6
4
2
0
Volume of base added / cm³

Figure 7.12 The titration curve for the addition of a strong base to a weak dibasic acid.

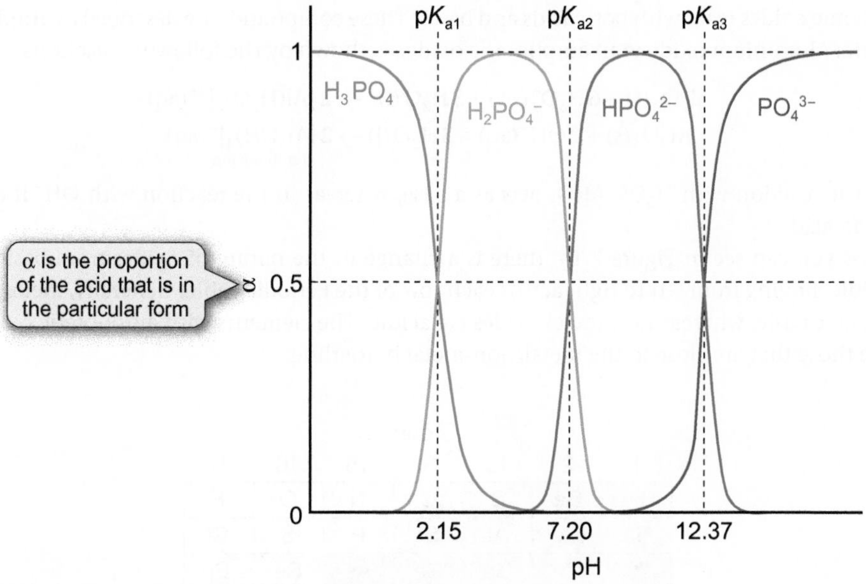
pKa1
pKa2
pKa3
H₃PO₄
H₂PO₄⁻
HPO₄²⁻
PO₄³⁻
α is the proportion of the acid that is in the particular form
α
1
0.5
0
2.15
7.20
12.37
pH

Figure 7.13 The speciation curve for phosphoric acid (H_3PO_4).

For a polybasic acid, there are several possible species present in solution. For example, solutions of phosphoric acid can contain H_3PO_4, $H_2PO_4^-$, HPO_4^{2-}, and PO_4^{3-}. The variation of concentration of these species with pH is shown on a **speciation curve** (Figure 7.13). At low pH (below pK_{a1}), H_3PO_4 (red) is the most important species, but as the pH is increased $H_2PO_4^-$ (orange), HPO_4^{2-} (green), and finally PO_4^{3-} (blue) become in turn the dominant species.

Summary

- For an acid with more than one ionizable hydrogen atom (a polybasic acid), successive values of pK_a increase, as it is easier to remove a proton from a neutral molecule than from an anion.
- Oxoacids with the general formula H_mXO_n increase in strength with increasing value of $(n - m)$.
- The variation of the concentrations of the acid and associated anions with pH is shown using speciation curves.

For practice questions on these topics, see questions 10–12 at the end of this chapter (p.340).

7.7 Acidic and basic oxides

Oxides are often described as being acidic or basic, despite the fact that they are not obvious proton donors or acceptors. In these compounds, the terms 'acidic' and 'basic' relate to their reaction with water, which can generate $H_3O^+(aq)$ or $OH^-(aq)$. For example, sodium oxide (Na_2O) is a **basic oxide**, as it reacts with water to form an alkaline solution of NaOH

$$Na_2O\,(s) + H_2O\,(l) \rightarrow 2\,Na^+(aq) + 2\,OH^-(aq)$$

Sulfur dioxide, in contrast, is an **acidic oxide** as it reacts with water to form an oxoacid, in this case H_2SO_3, which exists in solution in equilibrium with the oxoanion and H_3O^+

$$SO_2\,(g) + H_2O\,(l) \rightarrow H_2SO_3\,(aq)$$
$$H_2SO_3\,(aq) + H_2O\,(l) \rightleftharpoons HSO_3^-\,(aq) + H_3O^+(aq)$$

Basic oxides react with acids, whereas acidic oxides react with bases

$$Na_2O\,(s) + 2\,HCl\,(g) \rightarrow 2\,NaCl\,(s) + H_2O\,(l)$$
$$SO_2\,(g) + OH^-(aq) \rightarrow HSO_3^-\,(aq)$$

Acidic, basic, and amphoteric oxides are described in more detail in Chapter 26 (p.1186), and Chapter 27 (p.1206).

Some oxides react with both acids and bases. These compounds are described as **amphoteric**. Aluminium oxide is an amphoteric oxide, as shown by the following reactions

$$Al_2O_3\,(s) + 6\,H_3O^+(aq) + 3\,H_2O\,(l) \rightarrow 2\,[Al(H_2O)_6]^{3+}(aq)$$
$$Al_2O_3\,(s) + 2\,OH^-(aq) + 3\,H_2O\,(l) \rightarrow 2\,\underbrace{[Al(OH)_4]^-(aq)}_{\text{aluminate ion}}$$

In the reaction with H_3O^+, Al_2O_3 acts as a base, whereas in the reaction with OH^- it acts as an acid.

As you can see in Figure 7.14, there is a change in the nature of oxides from basic to acidic moving from left to right across each row of the Periodic Table. Generally, metal oxides are basic, whereas non-metal oxides are acidic. The elements with amphoteric oxides are those that are close to the metal/non-metal borderline.

Group

1	2	13	14	15	16	17
Li	Be	B	C	N	O	F
Na	Mg	Al	Si	P	S	Cl
K	Ca	Ga	Ge	As	Se	Br
Rb	Sr	In	Sn	Sb	Te	I
Cs	Ba	Tl	Pb	Bi	Po	At

Key

	Acidic oxides
	Amphoteric or neutral oxides
	Basic oxides

Figure 7.14 The positions of acidic, basic, and amphoteric oxides in the Periodic Table.

Box 7.6 Cooking with acids and bases

Cooking involves chemical reactions, and many of these make use of acids and bases. Acids are particularly common in food, and taste sour or sharp. Citrus fruits such as lemons contain citric acid, whose structure is shown below, and apples contain malic acid, also shown below. Many cola drinks contain phosphoric acid (H_3PO_4). Vinegar is a dilute solution of ethanoic acid (CH_3CO_2H).

A very common type of reaction in cooking involves the formation of carbon dioxide, which is used to make bread and cakes rise.

→

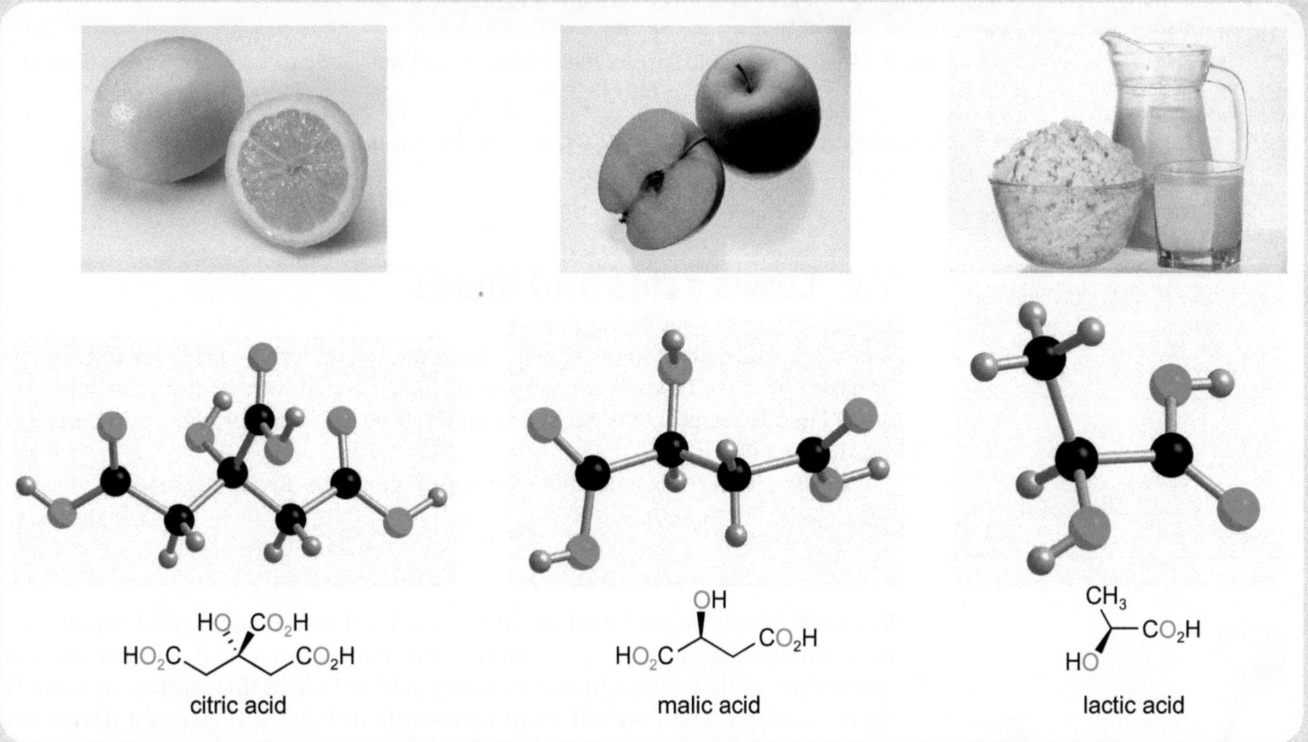
HO CO2H
HO2C CO2H
citric acid
OH
HO2C CO2H
malic acid
CH3
CO2H
HO
lactic acid

▲ Some acids are naturally present in foods. Malic acid and lactic acid contain chiral carbon atoms. See Section 18.4 (p.838) for more details.

Carbon dioxide is often formed by fermentation, which involves the conversion of carbohydrates into ethanol and CO_2, promoted by the addition of yeast. Another way to generate CO_2 is to use acid–base chemistry. A common method involves the reaction between sodium hydrogencarbonate ($NaHCO_3$), also known as *baking soda*, and lactic acid, which is present in milk products such as yoghurt

$$CH_3CH(OH)CO_2H(aq) + HCO_3^-(aq) \rightarrow CH_3CH(OH)CO_2^-(aq) + H_2O(l) + CO_2(g)$$

If the ingredients do not themselves provide an acid, *baking powder* can be used instead of baking soda. Self-raising flour contains baking powder. Baking powder contains a combination of sodium dihydrogenphosphate (NaH_2PO_4) and sodium hydrogencarbonate. These compounds do not react together in the solid state, but when water is added $H_2PO_4^-$ produces H_3O^+ ions, which react with HCO_3^- to form CO_2

$$H_2PO_4^-(aq) + H_2O(l) \rightleftharpoons HPO_4^{2-}(aq) + H_3O^+(aq)$$

$$H_3O^+(aq) + HCO_3^-(aq) \rightarrow H_2O(l) + CO_2(g)$$

Although most chefs would probably not think of themselves as chemists, Heston Blumenthal, proprietor of the Fat Duck restaurant in Berkshire, England, collaborates with academic chemists and has pioneered a molecular approach to cooking. His green tea sour mousse was designed to cleanse the palate, and the flavour ingredients—green tea, lime, and vodka—all have a role to play.

▲ Chef Heston Blumenthal (right) after receiving an Honorary Fellowship from The Royal Society of Chemistry for his creative applications of science to cooking.

Polyphenols in the green tea together with citric acid from the lime increase salivation whereas the ethanol in the vodka helps disperse fat molecules in the mouth.

Question

Predict the form of the titration curve for the reaction between malic acid and sodium hydroxide solution.

Summary

- Oxides are often described as acidic or basic on the basis of their reactions with water, or their reactions with bases and acids.
- Compounds that react with both acids and bases, such as Al_2O_3, are described as amphoteric.

7.8 Lewis acids and bases

- A Lewis acid is an electron pair acceptor.
- A Lewis base is an electron pair donor.

A dative bond is a covalent bond in which the two electrons originate from the same atom.

A more general definition of acids and bases than that given by Brønsted and Lowry was proposed by G. N. Lewis in the same year, 1923, though it was not routinely used until much later. A **Lewis acid** is defined as an electron pair acceptor and a **Lewis base** as an electron pair donor.

For example, in the reaction between boron trifluoride (BF_3) and ammonia

$$\underset{\text{Lewis acid}}{BF_3} + \underset{\text{Lewis base}}{NH_3} \rightarrow \underset{\text{adduct}}{F_3\overset{-}{B}\text{–}\overset{+}{N}H_3}$$

BF_3 *accepts* the nitrogen lone pair into a vacant orbital on boron, so it acts as a Lewis acid. Ammonia *donates* the lone pair into the boron orbital, so it acts as a Lewis base. Neither molecule acts as a Brønsted–Lowry acid or base in this reaction as there is no proton transfer. The covalent bond between boron and nitrogen is a **dative bond**, with both electrons coming from the Lewis base. Dative bonds are often shown as arrows indicating the direction of the electron donation ($F_3B \leftarrow NH_3$). The product is called an **adduct**. Group 13 compounds such as BF_3 are good Lewis acids because they only have six valence electrons, and addition of two electrons from the Lewis base completes the octet.

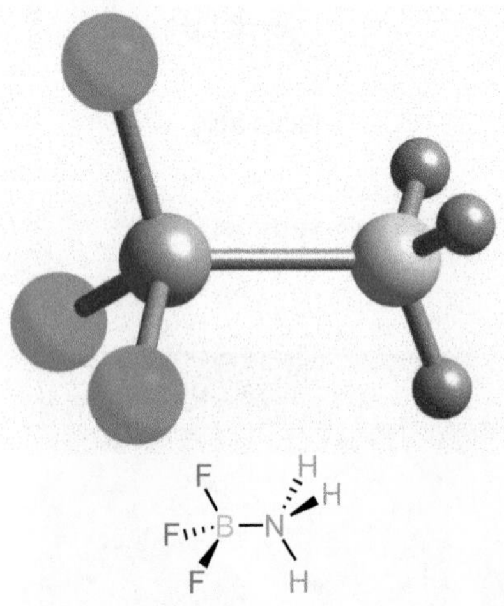

The adduct formed by boron trifluoride and ammonia.

Other *p*-block compounds can also act as Lewis acids, but only if the central atom is in the third period or lower of the Periodic Table. So, SiF_4 acts as a Lewis acid in the reaction with fluoride ions

$$\underset{\text{Lewis acid}}{SiF_4} + \underset{\text{Lewis base}}{2F^-} \rightarrow \underset{\text{adduct}}{SiF_6^{2-}}$$

Similarly, many other *p*-block halides such as AsF_3, PF_5, SF_4, and BrF_3 all react with fluoride ions in Lewis acid–Lewis base reactions. In contrast, CF_4 does not react with F^- because carbon is unable to expand its octet to accept the electron pair.

The chemistry of the boron halides in described in Section 27.2 (p.1213).

The reasons why second period elements do not form hypervalent compounds are discussed in Section 5.5 (p.245).

Metal cations are Lewis acidic, and Lewis acid–Lewis base interactions involving metal ions are the basis of **coordination chemistry**. For example, the Ni^{2+} ion forms Lewis acid–Lewis base interactions with six water molecules to form the $[Ni(H_2O)_6]^{2+}$ ion. When a molecule or anion interacts in this way with a metal ion it is called a **ligand**, and it is said to **coordinate** to the metal ion. In coordination chemistry, the Lewis acids (the metal ions) and the Lewis bases (the ligands) are divided into different classes, depending on their properties. Hard and soft acids and bases are described in Section 28.3 (p.1277).

The reactions between Lewis acids and Lewis bases are a key feature of *p*-block chemistry, and these are described in more detail in Chapter 27. The interactions between Lewis acids and Lewis bases in coordination chemistry are described in Sections 26.4 (p.1181) and 26.8 (p.1192) for the *s*-block metals and in Chapter 28 for the *d*-block metals.

Molecules with complete octets are also able to act as Lewis acids if they have multiple bonds and can accommodate an additional pair of electrons through rearrangement of their electrons. For example, CO_2 acts as a Lewis acid on reaction with hydroxide

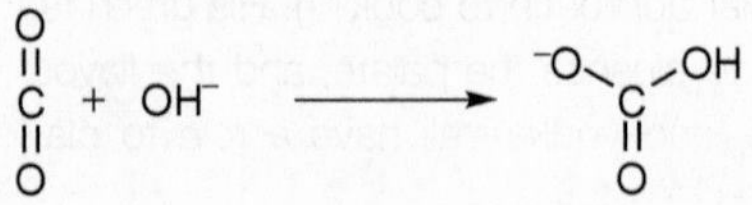

Reactions of Lewis acids and Lewis bases

Lewis acids react with Lewis bases to form adducts. They can also undergo displacement reactions, in which one Lewis base (or Lewis acid) is displaced by another from an adduct. The general reaction, in which A is a Lewis acid and both B and B′ are Lewis bases is

$$\overset{-}{A}-\overset{+}{B} + B' \rightarrow \overset{-}{A}-\overset{+}{B'} + B$$

Pyridine is a stronger Lewis base than diethyl ether, so pyridine displaces diethyl ether from the adduct formed with BF_3

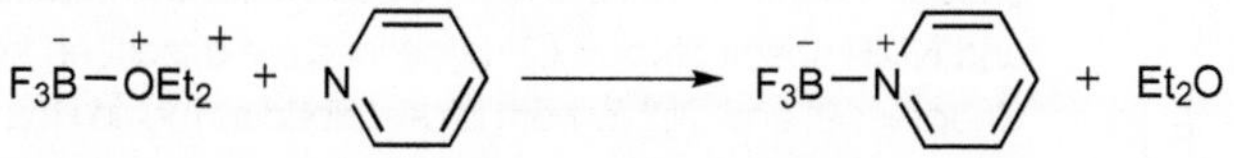
F3B—OEt2
+
N
F3B—N
+ Et2O

Generally, an ion or molecule needs to contain a lone pair to act as a Lewis base. Lewis basicity tends to decrease across a period so, for example, ammonia is a stronger Lewis base than water. In addition, Lewis basicity tends to decrease down a group. Thus ammonia is a stronger Lewis base than phosphine, PH_3.

The Lewis definitions of acid and base are more general than the Brønsted–Lowry definitions. A proton is an electron-pair acceptor, so a Brønsted–Lowry acid is just one type of Lewis acid. Similarly, a Brønsted–Lowry base is an example of a Lewis base that uses an electron pair to bind specifically to a proton as opposed to another molecule or ion. When H^+ accepts a lone pair from H_2O to form H_3O^+, H^+ is acting as a Lewis acid and H_2O as a Lewis base.

The terms Lewis acid and Lewis base are thermodynamic terms and relate to equilibria and the products of reactions. When considering kinetics and reaction mechanisms it is more normal to refer to an electron pair donor as a **nucleophile** and an electron pair acceptor as an **electrophile**.

Nucleophiles and electrophiles are discussed in Section 19.1, p.879.

Worked example 7.7 Lewis acids and bases

Identify the Lewis acids and Lewis bases in the following reactions.

(a) $AsF_5 + F^- \rightarrow AsF_6^-$

(b) $Cu^{2+} + 6NH_3 \rightarrow [Cu(NH_3)_6]^{2+}$

Strategy

Identify whether either reactant can interact with an electron pair and so has the ability to act as a Lewis acid, or whether it has an electron pair it can donate, so allowing it to act as a Lewis base.

Solution

(a) AsF_5 is trigonal bipyramidal (use VSEPR theory; Section 5.2, p.223) so it does not have a lone pair on the arsenic. Arsenic is a third period element so it can form hypervalent compounds (see Section 5.5, p.245). AsF_5 is itself hypervalent, but can act as a Lewis acid, accepting a lone pair. F^- has lone pairs so it can act as a Lewis base.

(b) The Cu^{2+} ion acts as a Lewis acid. NH_3 has a lone pair so it acts as a Lewis base.

Question

Identify the Lewis acid and the Lewis base in the reaction between boron triiodide, BI_3, and trimethylamine, NMe_3. What is the product of the reaction?

Box 7.7 Superacids

An acid that is stronger than 100% sulfuric acid is called a superacid. Superacids were developed by the Hungarian–American chemist George Olah, who won the Nobel Prize for his work in 1994. A superacid is formed by dissolving a powerful Lewis acid such as antimony pentafluoride (SbF_5) in a Brønsted–Lowry acid such as hydrofluoric acid (HF) or fluorosulfuric acid (HSO_3F). The combination of SbF_5 and HSO_3F is known as *magic acid* because it can dissolve candle wax. Magic acid is 1×10^{18} times stronger than 100% sulfuric acid.

→

Superacids are so strong because the Brønsted–Lowry acid donates a lone pair to the Lewis acid. The H–X bond is broken, and the proton becomes attached to another molecule of the Brønsted–Lowry acid. This leads to high concentrations of cations such as H_2F^+ and $H_2SO_3F^+$ in which the proton is extremely weakly bound.

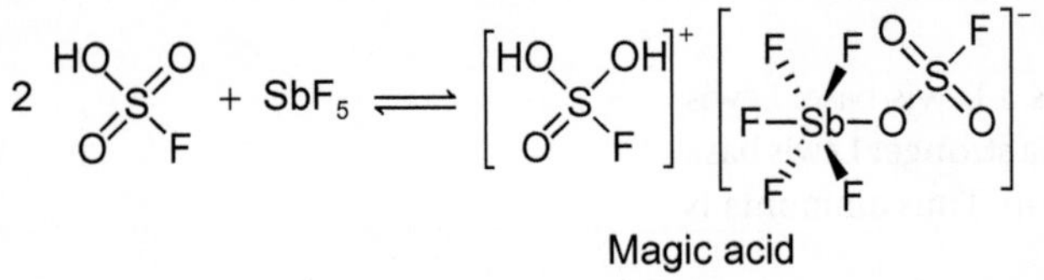
2
HO
O
S
O
F
+ SbF5
HO
OH
S
O
F
F
F
Sb
O
S
O
F
F
F
Magic acid

Superacids are strong enough to protonate hydrocarbons giving carbocations such as CH_3^+ and CH_5^+. Protonation is the first step in dissolving candle wax, which is made from long-chain hydrocarbons such as $C_{30}H_{62}$.

Carbocations are intermediates in many organic reactions, but it was only through the use of superacids that they could be produced in large enough quantities to be directly observed and characterized using IR and NMR spectroscopy. Carbocations are described in more detail in Chapter 19, and Olah's work is described further in Box 19.3, p.868.

Question

Use VSEPR theory to predict the shapes of the SbF_5 molecule and the SbF_6^- anion. Refer to Chapter 5 for the background to VSEPR theory if you need help.

Summary

- According to the Lewis theory, an acid is an electron-pair acceptor and a base is an electron-pair donor. These are generally referred to as Lewis acids and Lewis bases.
- The Lewis definition of acids and bases is broader than the Brønsted–Lowry definition. Brønsted–Lowry acids and bases are particular examples of Lewis acids and bases.

? For practice questions on this topic, see questions 13–14 at the end of this chapter (p.340).

Concept review

By the end of this chapter, you should be able to do the following.

- Know the Arrhenius, Brønsted–Lowry, and Lewis definitions of acids and bases.
- Identify the conjugate acid and conjugate base for an acid–base reaction.
- Calculate the pH of a solution of a strong or weak acid and a strong or weak base.
- Manipulate logarithms to calculate pH from $[H_3O^+(aq)]$ and pK_a from K_a and carry out the reverse processes.
- Calculate K_b and K_a for a base.
- Determine the position of an acid–base equilibrium using K_a and K_b.
- Calculate the pH of a buffer solution and suggest an appropriate buffer solution for a given pH.
- Determine the pH at any point in the titration between a strong acid and a strong base.
- Sketch the titration curve for a titration between a strong acid and a strong base, a weak acid and a strong base, and a strong base and a weak acid.
- Suggest an appropriate indicator for a given acid–base reaction.
- Understand the factors involved in influencing the value of pK_{a1} for an oxoacid.
- Identify Lewis acids and Lewis bases.

Key equations

definition of K_a	$K_a = \frac{[H_3O^+(aq)][A^-(aq)]}{[HA(aq)]}$ (units of K_a mol dm^{-3})	(7.2)
definition of pK_a	$pK_a = -\log_{10} K_a$	(7.3)
definition of pH	$pH = -\log_{10}[H_3O^+(aq)]$	(7.5)
the self-ionization constant of water	$K_w = [H_3O^+(aq)][OH^-(aq)]$	(7.7)
definition of K_b	$K_b = \frac{[BH^+(aq)][OH^-(aq)]}{[B(aq)]}$ (units of K_b mol dm^{-3})	(7.10)
Henderson–Hasselbalch equation	$pH = pK_a + \log_{10}\frac{[\text{conjugate base}]}{[\text{acid}]}$	(7.16)

Questions

More challenging questions are indicated by an asterisk *.

1 Identify the acid, base, conjugate acid, and conjugate base in the following equilibria. (Section 7.1)

(a) $HCO_2H(aq) + H_2O(l) \rightleftharpoons HCO_2^-(aq) + H_3O^+(aq)$

(b) $CH_3CH_2NH_2(aq) + H_2O(l) \rightleftharpoons CH_3CH_2NH_3^+(aq) + OH^-(aq)$

(c) $H_2SO_4 + CH_3CH_2OH \rightleftharpoons HSO_4^- + CH_3CH_2OH_2^+$

2 Aspirin is acetylsalicylic acid which has a pK_a of 3.49. (Section 7.2)

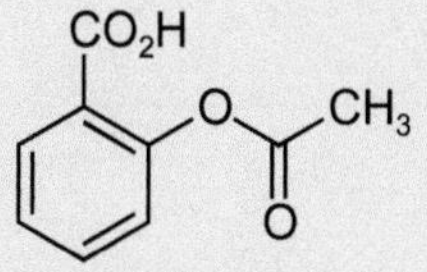

(a) Calculate the pH of a 0.10 mol dm^{-3} solution of this acid.

(b) Would you expect the sodium salt of acetylsalicylic acid to be acidic, neutral, or basic in solution? Explain your reasoning.

3 K_w for water at 40 °C is 3.0×10^{-14} mol^2 dm^{-6}. Is pH 7 neutral, acidic, or basic for an aqueous solution at this temperature? (Section 7.2)

4 Calculate the following (Section 7.2):

(a) $[H_3O^+(aq)]$ in a solution of HCl at pH 1.5;

(b) $[OH^-(aq)]$ in a solution of KOH at pH 12.8.

5 Hydroxylamine (NH_2OH) is a weak base. A 0.20 mol dm^{-3} solution of hydroxylamine has a pH of 9.7. (Section 7.2)

(a) Calculate K_b and pK_b for hydroxylamine.

(b) What is the pK_a of NH_3OH^+?

6 Methylamine (CH_3NH_2) is a base with pK_b 3.34. Identify the conjugate acid of methylamine, and determine the pK_a for this at 298 K. (Section 7.2)

7* The pH inside most cells is maintained at around 7.4 by a phosphate buffer made up from the $H_2PO_4^-(aq)$ ion and its conjugate base, $HPO_4^{2-}(aq)$. The pK_a of the acid $H_2PO_4^-(aq)$ is 7.2. (Section 7.3)

$$H_2PO_4^-(aq) + H_2O(l) \rightleftharpoons H_3O^+(aq) + HPO_4^{2-}(aq)$$

(a) Write an expression for K_a for $H_2PO_4^-(aq)$.

(b) Calculate the ratio of $[HPO_4^{2-}(aq)]/[H_2PO_4^-(aq)]$ needed to give a pH of 7.4 in the cell.

(c) A typical value for the total phosphate concentration in a cell, $[H_2PO_4^-(aq)] + [HPO_4^{2-}(aq)]$, is 0.020 mol dm^{-3}. Calculate typical values of $[HPO_4^{2-}(aq)]$ and $[H_2PO_4^-(aq)]$ inside a cell.

8 In a titration, a solution of 0.200 mol dm^{-3} NaOH solution is added from a burette to 30.0 cm^3 of 0.100 mol dm^{-3} HNO_3 solution in a conical flask. Calculate the pH of the solution (Section 7.4):

(a) before any NaOH is added;

(b) after addition of 5.0 cm^3 of the NaOH solution;

(c) after addition of 10.0 cm^3 of the NaOH solution;

(d) at the equivalence point;

(e) after addition of 20.0 cm^3 of the NaOH solution.

9 For the following titrations, select which of the two indicators given is more appropriate. (Section 7.5)

(a) Propanoic acid titrated against sodium hydroxide solution using either phenolphthalein (pK_{in} 9.4) or methyl orange (pK_{in} 3.4).

(b) Hydrochloric acid titrated against methylamine solution using either thymol blue (pK_{in} 9.0) or bromophenol blue (pK_{in} 3.9).

10 Estimate pK_{a1} for the following oxoacids: (a) $HBrO_3$; (b) H_2SeO_3; (c) H_2CO_3. The actual values of pK_{a1} and pK_{a2} for H_2CO_3 are 6.35 and 10.33, respectively. Suggest a reason for the large difference between these and your calculated values. (Section 7.6)

11 The structure of diphosphoric acid ($H_4P_2O_7$) is shown below. Its four pK_a values are 1.0, 1.8, 6.6, and 9.6. Suggest reasons for the relative values. (Section 7.6)

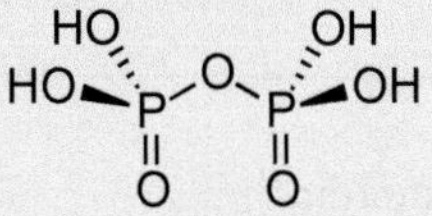

12 Citric acid (see Box 7.6, p.334) has the speciation curve shown below in water at 298 K. Use this to estimate values for pK_{a1}, pK_{a2}, and pK_{a3}. (Section 7.6)

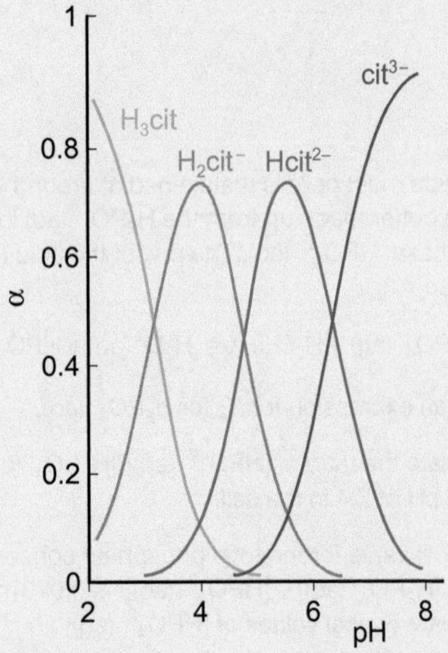

1
0.8
0.6
0.4
0.2
0
α
H3cit
H2cit⁻
Hcit2⁻
cit3⁻
2
4
6
8
pH

13 Identify the Lewis acid and Lewis base in the following reactions. (Section 7.8)

(a) $I_2 + I^- \rightarrow I_3^-$

(b) $NH_3 + HBr \rightarrow NH_4Br$

(c) $SiF_4 + 2py \rightarrow SiF_4(py)_2$

(d) $CO_2 + OH^- \rightarrow HCO_3^-$

Which, if any, of the reactions are also Brønsted–Lowry acid–base reactions.

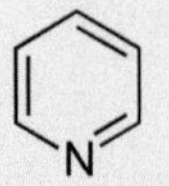

py = pyridine

14 Sulfur dioxide can act as either a Lewis acid or a Lewis base. Suggest how it reacts with (a) NMe_3 and (b) SbF_5, in each case identifying the product and stating whether SO_2 is acting as a Lewis acid or a Lewis base. (Section 7.8)

8

Gases

This chapter builds on the following topics:

- **Measurement, units, and nomenclature** Section 1.2, p.6
- **Atoms and the mole** Section 1.3, p.12
- **Chemical equations** Section 1.4, p.21
- **Energy changes in chemical reactions** Section 1.6, p.41
- **States of matter and phase changes** Section 1.7, p.47
- **Non-covalent interactions** Section 1.8, p.52

◄ A SCUBA cylinder and regulator for underwater breathing. The regulator valve at the top of the cylinder controls the pressure of the air leaving the cylinder.

Breathing under water

SCUBA stands for Self-Contained Underwater Breathing Apparatus. When using SCUBA equipment, a diver breathes from a cylinder containing highly compressed air. A regulator valve at the top of the cylinder reduces the pressure of the air as it leaves the cylinder so that the air the diver breathes has the same pressure as the surrounding water. At a depth of 10 m, the air in the diver's lungs is at a pressure of 2 atm; at 20 m the pressure is 3 atm, and so on. At these higher pressures, each breath taken contains more molecules than at the surface. The gas laws (Section 8.1) describe how the pressure, volume, and number of moles of gas are interrelated.

The change in pressure with depth has some important consequences and divers have to take precautions to avoid a number of hazards. Since the pressure of the gases in the lungs increases with depth, the partial pressures of nitrogen and oxygen also increase. (Partial pressures are introduced in Section 8.3.) Although it is necessary for life, oxygen can be toxic in too high concentrations.

Oxygen and nitrogen in the inhaled gas mixture pass into the bloodstream through the lungs. The amount of gas that dissolves at a given temperature depends on the partial pressure of the gas and therefore on the depth. It also depends on the solubility constant in water or blood, which is higher for N_2 than for O_2. (You can read about the solubility of gases in liquids in Section 17.4, p.789.) If the diver stays underwater for extended periods of time, quite large quantities of nitrogen dissolve in the bloodstream. This can cause *nitrogen narcosis*, which produces similar effects to alcohol intoxication. Divers become disoriented and lose spatial awareness so the onset is very dangerous. In extreme cases, divers have been unable to find the surface and have drowned.

If the diver surfaces too quickly, the dissolved gases come out of solution as bubbles. As the pressure reduces, the volume of the bubbles increases and this causes *decompression sickness*, sometimes known as 'the bends'. To avoid these problems deep-sea divers breathe a mixture of oxygen and helium (helium is less soluble in blood than nitrogen) and either return to the surface very slowly or enter decompression chambers, which slowly return them to atmospheric pressure.

The deeper the dive, the higher the pressure and the greater the number of molecules of gas in a given volume. Breathing too high a concentration of oxygen can be dangerous so, for deep dives below 150 m, the proportion of oxygen in the helium–oxygen mixture is reduced to below 10%. For very deep dives, the proportion of oxygen can be as low as 1%. At normal pressure, the proportion of oxygen in air is about 20%. So if the pressure is 20 times greater than normal, the proportion of oxygen in the mixture needs to be one twentieth of its normal value (1%) in order to get the same number of dissolved molecules in the blood.

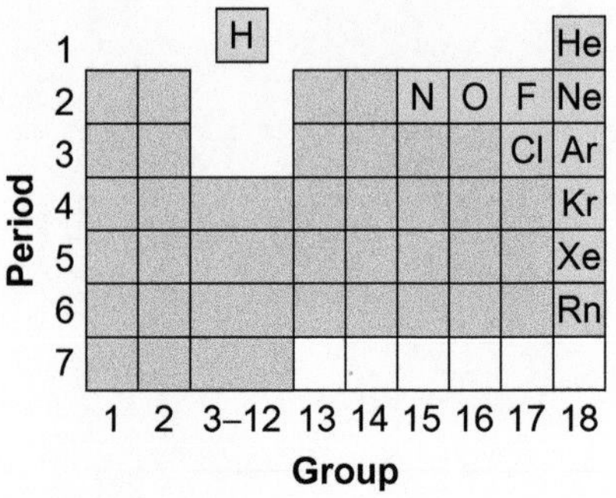
H
He
N O F Ne
Cl Ar
Kr
Xe
Rn
Period
1 2 3 4 5 6 7
1 2 3–12 13 14 15 16 17 18
Group

Figure 8.1 The elements that are gases at room temperature are shaded blue in the Periodic Table. H_2, N_2, O_2, F_2, and Cl_2 are diatomic (two atoms covalently bound in each molecule). The gases in Group 18 are monatomic (exist as isolated atoms).

Eleven *elements* exist as gases under normal conditions. Apart from hydrogen, these elements lie to the right of the Periodic Table, as shown in Figure 8.1. They include the monatomic gases of Group 18, such as helium (He) and neon (Ne), and diatomic gases, such as hydrogen (H_2), nitrogen (N_2), and oxygen (O_2).

Many more gaseous *compounds* exist, ranging from the commonplace such as carbon monoxide (CO) and carbon dioxide (CO_2), through to the less common but familiar substances such as hydrogen sulfide (H_2S), sulfur dioxide (SO_2), and butane (C_4H_{10}).

Why are chemists interested in gases? In historical terms, the study of gases involved some of the earliest and most important scientific measurements and helped to develop many ideas of atomic theory and stoichiometry. In more modern terms, many reactions that are important in the chemical industry involve reactions between gases. These include the oxidation of sulfur dioxide (SO_2) in the production of sulfuric acid (H_2SO_4), the Haber process for the production of ammonia (NH_3), and many reactions in the petrochemicals industry.

In addition, there are a number of atmospheric environmental problems facing society, including ozone depletion in the stratosphere (upper atmosphere), atmospheric pollution in the troposphere (the lower atmosphere), and the effect of increasing concentrations of greenhouse gases on global warming. Tackling these problems needs a good understanding of gas phase chemistry.

Another reason for studying gases is that it is relatively straightforward to explain their behaviour at a molecular level. Even small volumes of gases contain enormous numbers of molecules but their behaviour can be described by some quite simple laws. One of the aims of this chapter is to show how the observable properties of gases arise from the behaviour of the individual molecules in the gas. A final reason for discussing gases in some detail is that it illustrates some of the different ways in which ideas and models in chemistry, and science in general, are developed.

The quantitative relationship between the amounts of reactants and products is called the stoichiometry of the reaction (see Section 1.4, p.21).

The properties of solids, liquids, and gases are discussed in Sections 1.7 (p.47) and 17.1 (p.766).

A **fluid** is a substance that is capable of flow, that is, one where the molecules are not fixed in position relative to each other. All gases and liquids are fluids.

So, think about what you already know about gases. Gases are highly mobile and flow from regions of high pressure to lower pressure—think about a windy day or opening a can of carbonated drink. One definition of a gas is that it is a fluid that expands to fill its container. Another characteristic is that all gases mix easily with other gases. Think about what happens if you open a bottle of perfume in a room. Volatile components readily evaporate to form gases that mix with the gases in the air until they reach the same concentration throughout the room. This happens because the molecules in a gas are free to move relative to each other (characteristic of a fluid). The molecules interact with each other only weakly and are relatively far apart, unlike the molecules in liquids and solids (Figure 8.2).

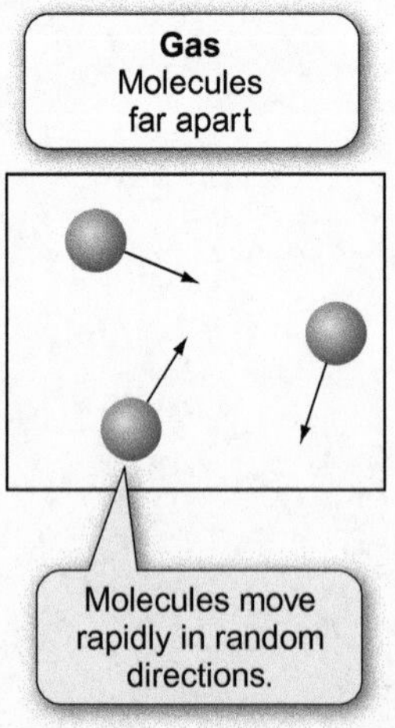
Gas
Molecules far apart
Molecules move rapidly in random directions.

Figure 8.2 In a gas, the molecules are further apart than in a solid or liquid, and they move at random.

By the laws of the game, a football must be 68–70 cm in circumference and must be inflated to a pressure of 0.6–1.1 atm. A typical football therefore has a volume of 5.8 dm^3 and contains around 0.24 mol of gas, which is equivalent to 1.4×10^{23} molecules. To describe the behaviour of this enormous number of molecules with some simple equations is one of the triumphs of physical chemistry.

8.1 The gas laws: an empirical approach

An illustration of the variable physical properties, p, V, and n, of a gas. Changing the temperature, T, will change p and/or V.

Some of the very earliest experiments in physical chemistry involved the study of gases. Although carried out over three hundred years ago, these experiments provide a good example of the *empirical* approach to understanding chemical systems. In this approach large numbers of experimental observations are made, then the observations are correlated to derive laws that describe and explain the behaviour of systems in general terms. In Section 8.4, you will see how a *theoretical* approach can be used to derive the same laws. In this approach, a model is constructed to describe the behaviour of the molecules in the system and is used to develop a theory for the behaviour of gases. The test of the theoretical model is how well it describes the experimental results and how well it can predict the properties of gases.

What properties of a gas can be measured? Think about blowing up a balloon. As you blow more air into the balloon (i.e. as the amount of gas in moles, n, increases), the pressure, p, increases. The volume, V, of the balloon also increases. If you change the temperature, T, then the pressure and/or volume will change as a result. So, how are these quantities related?

Boyle's law, Charles's law, and Avogadro's law

Boyle's law

In 1661, Robert Boyle performed experiments to investigate how the volume of a gas changed when the pressure changed. He found that:

At constant temperature, the volume of a fixed amount of gas is reduced in proportion as the pressure increases.

For example, if he *doubled* the pressure, then the volume occupied by the gas *halved*. This is illustrated in Figure 8.3.

This relationship between p and V, at constant temperature T, and for a fixed amount of gas, can be expressed mathematically by the equation

$$p \propto \frac{1}{V} \tag{8.1}$$

The volume is **inversely proportional** to the pressure and vice versa

$$V \propto \frac{1}{p}$$

The symbol $\propto$ means 'proportional to'. This means that, if the term on the left-hand side of an equation is increased, the term on the right-hand side increases in the same ratio (Maths Toolkit MT1, p.1304).

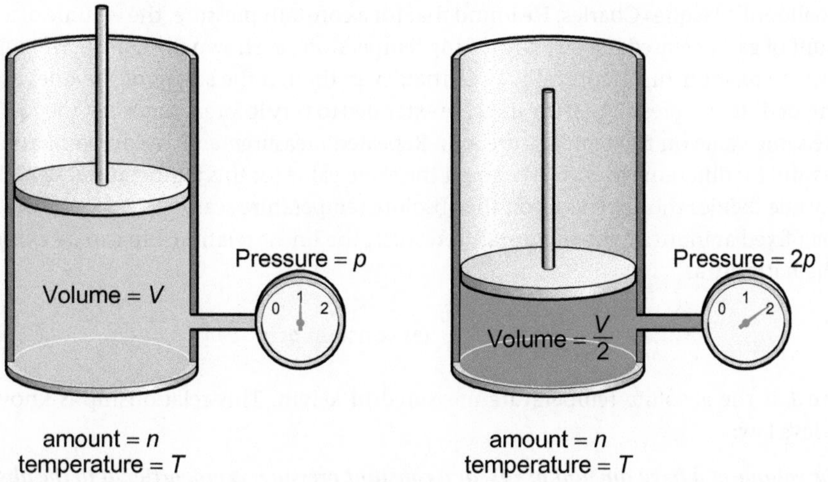
Pressure = p
Volume = V
0 1 2
amount = n
temperature = T
Pressure = 2p
Volume = V/2
0 1 2
amount = n
temperature = T

Figure 8.3 Boyle's law: if the volume of a fixed amount of gas is halved at constant temperature, the gas pressure doubles.

The SI unit of pressure is the pascal, Pa; see Section 1.2 (p.6). The units of pressure are discussed in Section 8.2 (p.349).

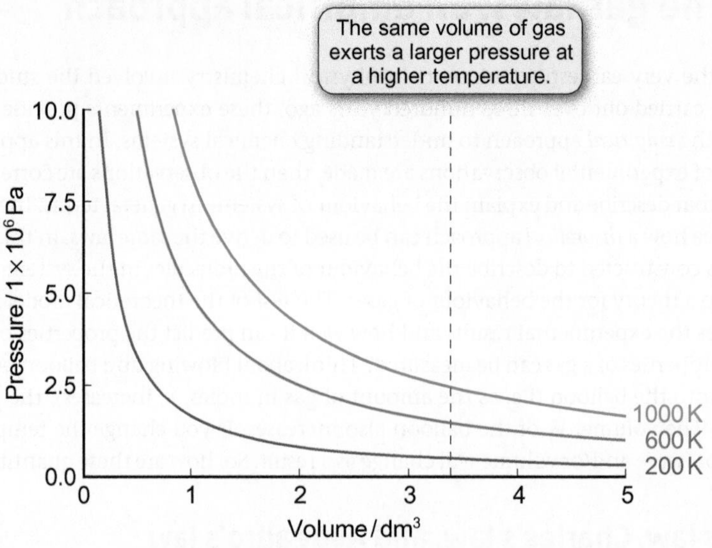
The same volume of gas exerts a larger pressure at a higher temperature.
10.0
7.5
5.0
2.5
0.0
Pressure/1 × 10^6 Pa
0
1
2
3
4
5
Volume/dm^3
1000 K
600 K
200 K

Figure 8.4 The relationship between pressure and volume of 1 mol of a gas at constant temperature.

The term **isotherm** comes from the Greek for 'constant temperature'.

A graphical view of this relationship is given in Figure 8.4. The plots are known as ***p–V* isotherms**. An alternative way of expressing the relationship is

$$p = \text{constant} \times \frac{1}{V} \text{(at constant } T, n)$$

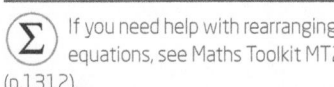

If you need help with rearranging equations, see Maths Toolkit MT2 (p.1312).

The constant is known as the proportionality constant. This last equation can be rearranged to give

$$pV = \text{constant (at constant } T, n) \tag{8.2}$$

The relationship between pressure and volume is known as **Boyle's** law after the Irish scientist Robert Boyle who published it in 1662. It gives a quantitative description of the familiar experience that, when you compress a gas by reducing its volume (e.g. in a bicycle pump), the pressure exerted by the gas increases.

Charles's law

The kelvin scale of temperature is described in Section 1.2 (p.6). Temperatures measured in kelvin are also known as absolute or thermodynamic temperatures.

In 1787, the temperature dependence of gas properties was studied by the French scientist and balloonist Jacques Charles. He found that for a constant pressure, the volume of a fixed amount of gas increased linearly with rising temperature, as shown in Figure 8.5. Figure 8.6 shows the relationship graphically: note that, even though the slopes of the lines are different for different pressures, if the lines are extended to very low temperatures, they all meet at the same value on the temperature axis. Repeated measurements for different amounts of gas and for different gases all give about the same value for this temperature, –273.15 °C. This value defines the zero point on the absolute temperature scale, –273.15 °C = 0 K.

For a fixed amount of gas at constant pressure, the linear relationship can be expressed mathematically as

$$V \propto T \quad \text{(at constant } p, n) \tag{8.3}$$

From Charles's measurements, –273.15 °C is the temperature at which the volume of a gas would reach zero. Of course, zero volume has no physical significance but the Belfast-born physicist William Thomson (1824–1907), later called Lord Kelvin, realized that this indicated a temperature at which all the energy would be lost from a system. This is the lowest possible temperature that could be reached—hence it is referred to as 'absolute zero'. The unit of 'absolute temperature' is named after Lord Kelvin.

where T is the absolute temperature measured in kelvin. This relationship is known as Charles's law:

The volume of a fixed amount of gas, at a constant pressure, is proportional to the absolute temperature.

If the absolute temperature doubles, the volume occupied by the gas doubles.

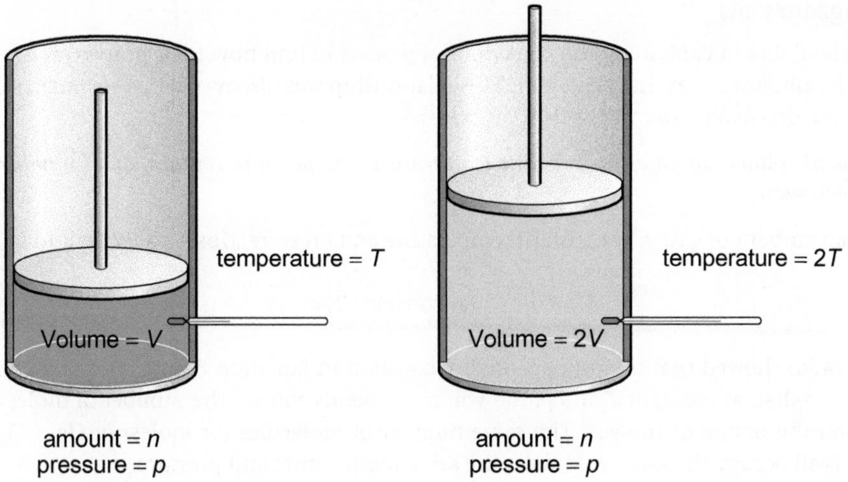

temperature = T
Volume = V
amount = n
pressure = p
temperature = $2T$
Volume = $2V$
amount = n
pressure = p

Figure 8.5 Charles's law: if the temperature of a fixed amount of gas is doubled at constant pressure, the volume doubles.

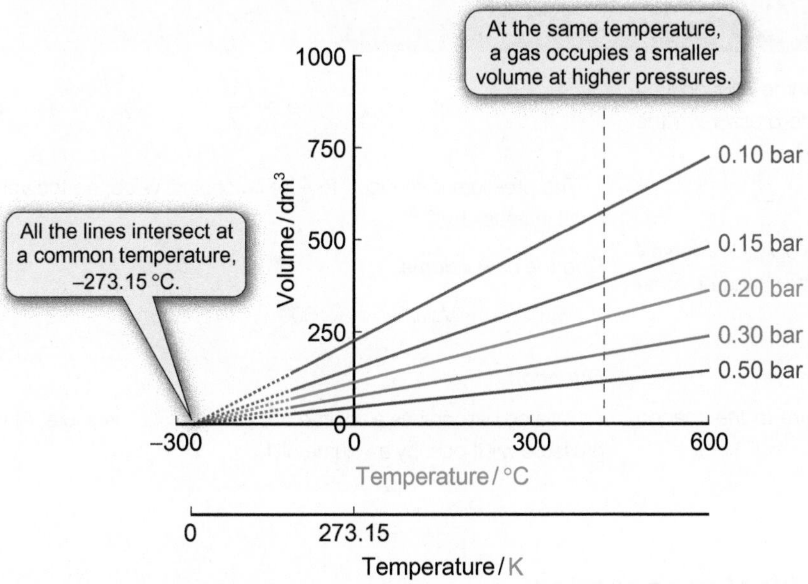

At the same temperature, a gas occupies a smaller volume at higher pressures.
All the lines intersect at a common temperature, −273.15 °C.
Volume / dm^3
1000
750
500
250
0
0.10 bar
0.15 bar
0.20 bar
0.30 bar
0.50 bar
−300
0
300
600
Temperature / °C
0
273.15
Temperature / K

Figure 8.6 The relationship between the volume and temperature for 1 mol of a gas at constant pressure. The lines represent experiments conducted at different pressures.

1 bar = 1×10^5 Pa
The bar is a derived SI unit (see Section 1.2, p.6, and Section 8.2, p.349).

Jacques Charles was renowned as a balloonist as well as a scientist. In 1783 he launched the world's first hydrogen-filled balloon in Paris. It landed 21 km away in the village of Gonesse, where terrified villagers attacked it with pitchforks.

Avogadro's law

The final link in explaining the behaviour of gases is to find how their properties depend on the amount of gas that is present. This relationship was discovered by the Italian scientist Amadeo Avogadro in 1811. His law states:

> ***Equal volumes of gases at constant temperature and pressure contain equal numbers of molecules.***

For an amount of gas, n, at constant temperature and pressure, this is equivalent to saying

$$V \propto n \quad \text{(at constant } T, p) \qquad (8.4)$$

Avogadro showed that the proportionality constant in Equation 8.4 was the same for all gases so that, at constant T and p, the volume depends *only* on the number of moles and *not* on the nature of the gas. The same number of molecules (or moles) of He, CO_2, or C_4H_{10} all occupy the same volume at the same temperature and pressure.

Worked example 8.1 Using Boyle's law

A balloon occupies a volume of 600 cm^3. What volume will it occupy if the surrounding pressure is reduced so that the pressure in the balloon is reduced to one-third of its starting value?

Strategy

From Boyle's law (Equation 8.1) you know that the volume is inversely proportional to pressure. You can therefore relate the change in pressure to the change in volume.

Solution

Use Equation 8.1 to relate the change of pressure to the change in volume.

$$p \propto \frac{1}{V} \qquad (8.1)$$

The pressure is reduced to $\frac{1}{3}$ of its original value, so the volume must increase by 3 times.

Find the new volume.

$$\text{Volume} = 3 \times 600\,\text{cm}^3 = 1800\,\text{cm}^3$$

Question

An inflated balloon has a volume of 1 dm^3 at 1 atm pressure. At what pressure will it occupy a volume of 1.5 dm^3?

The ideal gas equation

The three experimental laws of Boyle, Charles, and Avogadro link together the four variable properties of a gas, n, p, V, and T. They show that the volume occupied by a gas is directly proportional to the amount in moles and to the absolute temperature. It is also inversely proportional to the pressure. Collecting together the information from the previous section

Boyle's law:	$V \propto \frac{1}{p}$	(at constant T, n)
Charles's law:	$V \propto T$	(at constant p, n)
Avogadro's law:	$V \propto n$	(at constant T, p)

Combining all three of these observations gives

$$V \propto \frac{nT}{p}$$

from which it follows that

$$pV \propto nT$$

The proportionality in this equation is equivalent to multiplying by a single constant (which is the value of all the proportionality constants from the separate laws collected together). This is usually expressed as:

$$pV = nRT \text{ where } R \text{ is a constant} \quad (8.5)$$

R is known as the gas constant and has the same value for all ideal gases. The value of R in SI units is **8.314 J K^{-1} mol^{-1}**. It turns out that this value is much more fundamental in chemistry than simply relating the properties of gases (see Section 8.4, p.358).

Each of the laws described above is an approximation. They were obeyed within the limits of accuracy that Boyle, Charles, and Avogadro were able to achieve in their measurements. Later, more accurate experiments showed that the laws are not obeyed exactly under all conditions. A gas whose properties exactly obey Equation 8.5 is called an **ideal gas** (sometimes called a **perfect gas**). Equation 8.5 is, therefore, known as the **ideal gas equation**. It links all of the properties required to define the state of a gas and so is called an **equation of state**. In fact, for many gases, such as N_2, O_2, and He, around atmospheric pressure and temperature, the ideal gas equation describes their behaviour reasonably well and gases are usually assumed to behave ideally unless they are at high pressures or very low temperatures.

The gas constant is represented by the symbol R.

R = 8.314 J K^{-1} mol^{-1} (SI units)
R = 0.08206 dm^3 atm K^{-1} mol^{-1}

Summary

- The three experimentally observed laws of gas behaviour can be combined to give the *ideal gas equation*

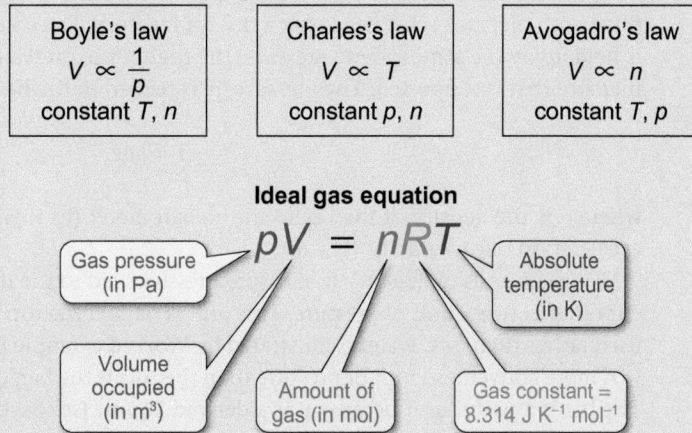

Boyle's law	Charles's law	Avogadro's law
$V \propto \frac{1}{p}$	$V \propto T$	$V \propto n$
constant T, n	constant p, n	constant T, p

Ideal gas equation

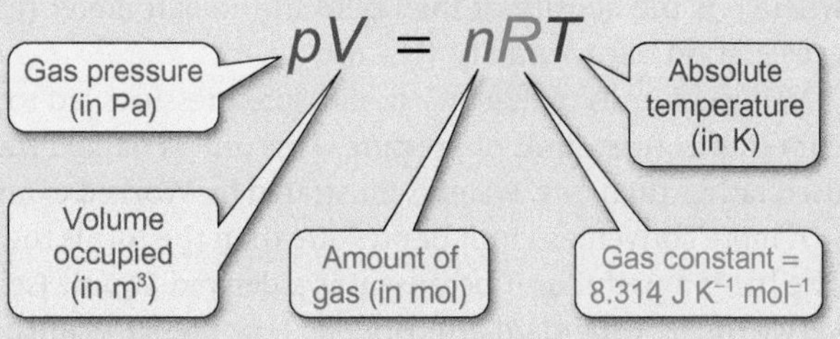
pV = nRT
Gas pressure (in Pa)
Volume occupied (in m³)
Amount of gas (in mol)
Gas constant = 8.314 J K⁻¹ mol⁻¹
Absolute temperature (in K)

8.2 Using the ideal gas equation

The ideal gas equation (Equation 8.5) relates the observable properties of a gas and allows you to predict the effect of changing conditions on gas behaviour.

Use of SI units and some guidelines for carrying out calculations are given in Section 1.2 (p.6).

Pressure

An important matter when performing chemical calculations is to keep track of the units of the various quantities involved. This is particularly important with gas pressures, because a number of different units of pressure are in common use.

CHANGE
RAIN
FAIR
29
30
28
31
INS Hg

Barometers are useful in forecasting the weather. This American garden barometer gives the pressure in millibars ('Mbs') and inches of mercury ('INS Hg').

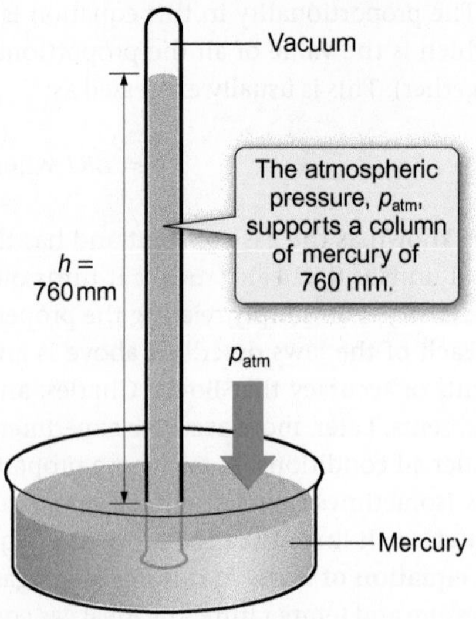
Vacuum
The atmospheric pressure, p_{atm}, supports a column of mercury of 760 mm.
h = 760 mm
p_{atm}
Mercury

Figure 8.7 The principle behind pressure measurement using a mercury barometer. The atmospheric pressure supports the column of mercury.

The SI unit of pressure is the pascal, Pa. 1 Pa is the pressure when a force of 1 N acts on an area of 1 m^2 so that 1 Pa = 1 N m^{-2}. However, 1 Pa is a rather small quantity and not a very convenient unit to use; for example, atmospheric pressure is about 100000 Pa. The kilopascal (1 kPa = 1000 Pa) is often used.

Historically, pressures were measured by determining the height of a column of mercury, as illustrated in Figure 8.7, which shows a simple mercury barometer. A glass tube filled with mercury is inverted over a dish of mercury. The column of mercury in the tube is held up by the atmospheric pressure; the higher the pressure, the longer the column of mercury that is supported. The pressure, p, is related to the height, h, of the column by:

$$P = h\rho g \tag{8.6}$$

where ρ is the density of the liquid in the barometer (in this case, mercury) and g is the acceleration due to gravity, 9.81 m s^{-2}.

Using columns of mercury to measure pressures led to the use of the millimetre of mercury (mmHg) as a unit of pressure. This unit is named the torr. The reason that mercury is used rather than, say, water is illustrated by Worked example 8.2.

A more convenient unit of pressure than the torr is the bar, defined as exactly 1×10^5 Pa. The bar is a useful unit because it is a derived SI unit (see Section 1.2, p.6) whose value is very close to 1 standard atmosphere. Low pressures are often quoted in millibar (1 mbar = 1×10^{-3} bar), while high pressures are measured in kilobar (1 kbar = 1×10^3 bar). A standard pressure of one atmosphere, 1 atm, is defined as 760 Torr. Some of these pressure units and their conversion factors are summarized in Table 8.1.

Some useful SI relationships

1 Pa = 1 N m^{-2}
= 1 kg m^{-1} s^{-2}
= 1 J m^{-3}

Table 8.1 Interrelationship of units for measuring gas pressure

Unit name	Symbol	Value
pascal	Pa	1 N m^{-2}
bar	bar	1×10^5 Pa = 100 kPa
torr	Torr	1 mmHg = 133.32 Pa
standard atmosphere	atm	1.013 bar = 101 325 Pa = 1.01325×10^5 Pa = 760 Torr

Worked example 8.2 Using columns of liquid to measure pressures

Around room temperature, the density of mercury is 13.6 $g\,cm^{-3}$ and that of water is 1.0 $g\,cm^{-3}$. Calculate the height of a column of each liquid that would be supported by a pressure of 1 atm and suggest why mercury is used in barometers rather than water.

Strategy

Substitute the values given into Equation 8.6, which gives the relationship between pressure and column height. Work in SI units and remember to keep the units consistent.

Solution

Convert the values given for the density and pressure to SI units.

The SI unit of density is $kg\,m^{-3}$. Since 1 kg = 1×10^{-3} g and $1\,m^3 = 1\times10^6\,cm^3$

$$1\,g\,cm^{-3} = 1\frac{g}{cm^3} = \frac{1\times10^{-3}\,kg}{1\times10^{-6}\,m^3} = 1\times10^3\,kg\,m^{-3}$$

So

$$\rho_{Hg} = 13.6\times10^3\,kg\,m^{-3} = 13600\,kg\,m^{-3} \text{ and}$$

$$\rho_{H_2O} = 1.0\times10^3\,kg\,m^{-3} = 1000\,kg\,m^{-3}$$

Use Equation 8.6 to find the column height, h.

Rearrange Equation 8.6 ($p = h\rho g$) to give an expression for h

$$h = \frac{p}{\rho g} \quad (g = 9.81\,m\,s^{-2})$$

From Table 8.1, 1 atm = 101 325 Pa

For mercury

$$h_{Hg} = \frac{101\,325\,Pa}{13600\,kg\,m^{-3}\times9.81\,m\,s^{-2}} = \frac{101\,325\,\cancel{kg}\,m^{-1}\,\cancel{s^{-2}}}{13600\,\cancel{kg}\,m^{-3}\times9.81\,m\,\cancel{s^{-2}}} = 0.759\,m$$

(When cancelling units, remember that 1 Pa = 1 $N\,m^{-2}$ = 1 $kg\,m^{-1}\,s^{-2}$.)

For water

$$h_{H_2O} = \frac{101\,325\,Pa}{1000\,kg\,m^{-3}\times9.81\,m\,s^{-2}} = 10.3\,m$$

This shows why mercury is used in a barometer. A mercury barometer has a column 0.759 m high, but, if water were to be used, the barometer would need to be over 10 m high—as high as a house and rather too large for convenience!

Question

The pressure at the top of a high mountain is 0.4 atm. Calculate the height of the mercury column in a barometer at the top of the mountain.

Calculations using the ideal gas equation

For a fixed amount of gas, if you know any two quantities out of pressure, volume, and temperature, then the third can be calculated using the ideal gas equation

$$pV = nRT \quad (8.5)$$

Alternatively, if you know the pressure, volume, and temperature of a gas, you can calculate the amount of gas present. This is illustrated in Worked example 8.3.

The effect of changing any of the conditions can also be calculated using an alternative form of the ideal gas equation

For a fixed amount of gas $$\frac{pV}{T} = nR = \text{constant}$$

So, for a fixed amount of gas under two different sets of conditions, denoted by the subscripts 1 and 2

$$\frac{p_1V_1}{T_1} = \frac{p_2V_2}{T_2} \quad (8.7)$$

Worked example 8.4 (p.352) illustrates how this equation is used.

Visit the Online Resource Centre to view screencast 8.1 which explains the importance of getting the units right when you use the ideal gas law.

When using the ideal gas equation, remember to convert the quantities given into SI units, that is, pressure in Pa, volume in m^3, temperature in K.

When using Equation 8.7, there is no need to convert p and V to SI units, because the units cancel out, though units of p and V must be the same on both sides of the equation. However, you *must* convert temperatures to kelvin.

Worked example 8.3 Using the ideal gas equation to calculate the amount of gas

A small cylinder of oxygen gas has a volume of 5.0 dm^3. The cylinder is stored at 20 °C and the pressure inside is 56 bar. Assuming oxygen behaves as an ideal gas under these conditions, calculate the mass of O_2 in the cylinder.

Strategy

You know the pressure, volume, and temperature of the gas so you can use the ideal gas equation to calculate the amount in moles, n. Rearrange the ideal gas equation to give an expression for n. Find the amount in moles, then multiply this by the molar mass to give the mass of gas.

First, the pressure, volume, and temperature must be converted to SI units.

Solution

Convert the given values to SI units.

$$1\,\text{bar} = 1\times10^5\,\text{Pa, so } 56\,\text{bar} = 56\times10^5\,\text{Pa} = 5.6\times10^6\,\text{Pa}$$

$$1\,\text{dm}^3 = 1\times10^{-3}\,\text{m}^3\text{, so } 5.0\,\text{dm}^3 = 5.0\times10^{-3}\,\text{m}^3$$

$$20\,°\text{C} = (20+273)\,\text{K} = 293\,\text{K}$$

Rearrange the ideal gas equation (Equation 8.5) to find the number of moles of O_2.

$$pV = nRT \qquad (R = 8.314\,\text{J K}^{-1}\,\text{mol}^{-1})$$

$$n = \frac{pV}{RT} = \frac{(5.6\times10^6\,\text{Pa})\times(5.0\times10^{-3}\,\text{m}^3)}{8.314\,\text{J K}^{-1}\,\text{mol}^{-1}\times293\,\text{K}} \qquad (1\,\text{Pa} = 1\,\text{J m}^{-3})$$

$$= \frac{(5.6\times10^6\,\cancel{\text{J}}\,\cancel{\text{m}^{-3}})\times(5.0\times10^{-3}\,\cancel{\text{m}^3})}{8.314\,\cancel{\text{J}}\,\cancel{\text{K}^{-1}}\,\text{mol}^{-1}\times293\,\cancel{\text{K}}} = 11.5\,\text{mol}$$

Thus, the cylinder contains 11.5 mol of gas.

Use the molar mass to find the mass of O_2.

$$M_r(O_2) = 32$$

$$\text{Mass} = n\times M \quad \text{where } M \text{ is the molar mass.}$$

$$\text{Mass of } O_2 = (11.5\,\cancel{\text{mol}})\times(32\,\text{g}\,\cancel{\text{mol}^{-1}})$$

$$= 370\,\text{g}$$

Question

A gas cylinder contains 10 kg of oxygen at a pressure of 20 bar at 25 °C. Calculate the volume of the cylinder, in m^3.

You can find more on interconverting units in Section 1.2 (p.6).

In Worked example 8.3, the units of all the quantities were converted to SI units so that the gas constant, R, has the value of 8.314 J K^{-1} mol^{-1}. The ideal gas constant can be expressed in different units, for example, $R = 0.08206$ dm^3 atm mol^{-1} K^{-1}, and this can be useful if pressures are measured in atmospheres. However, it is a better general strategy to get used to always using SI units.

Worked example 8.4 The effect of changing conditions

Calculate the pressure inside the cylinder described in Worked example 8.3 when the temperature increases to 30 °C.

Strategy

The volume of the gas is constant and the cylinder contains a fixed amount of gas. You can, therefore, use the ideal gas equation written for two sets of conditions, as in Equation 8.7. Substitution of the given values allows you to find the new pressure.

Solution

Write the ideal gas equation for two sets of conditions.

$$\frac{p_1V_1}{T_1} = \frac{p_2V_2}{T_2} \qquad (8.7)$$

Since V is constant, Equation 8.7 simplifies to

$$\frac{p_1}{T_1} = \frac{p_2}{T_2}$$

where p_1 and T_1 are the original conditions (56 bar and 20 °C) and p_2 and T_2 the values under the new conditions (30 °C and the new pressure). Remember, you must convert °C to K.

Substitute in the given data.

$$\frac{56\,\text{bar}}{293\,\text{K}} = \frac{p_2}{303\,\text{K}}$$

Rearrange and solve to find p_2.

$$p_2 = \frac{56\,\text{bar}\times303\,\cancel{\text{K}}}{293\,\cancel{\text{K}}} = 58\,\text{bar}$$

Question

A container fitted with a piston contains gas at 1 atm and 25 °C. Calculate the pressure if the piston is moved so as to halve the volume. (Assume the temperature is constant.)

Box 8.1 Car air bags

Most cars are fitted with a *supplementary restraint system*, more commonly known as an air bag. This is designed to inflate within a few milliseconds in a collision to prevent an occupant hitting the steering wheel or front windscreen. The diagram shows a basic design of an air bag system.

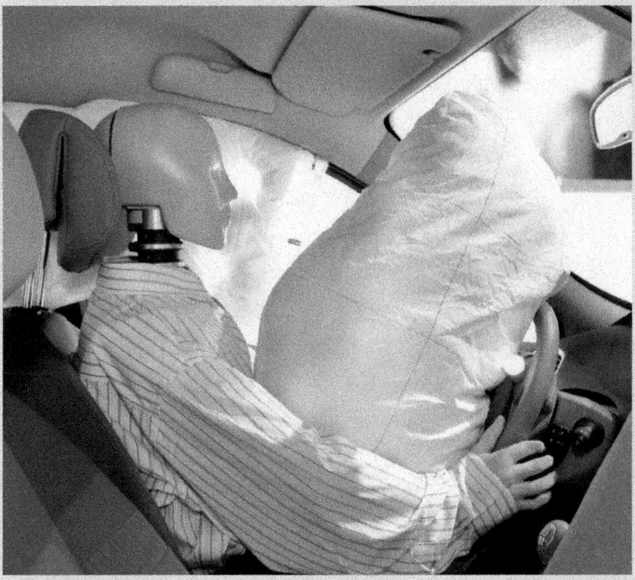

▲ An air bag system in a simulated car crash. In a collision, a chemical reaction is initiated that rapidly produces a large volume of gas to inflate the bag.

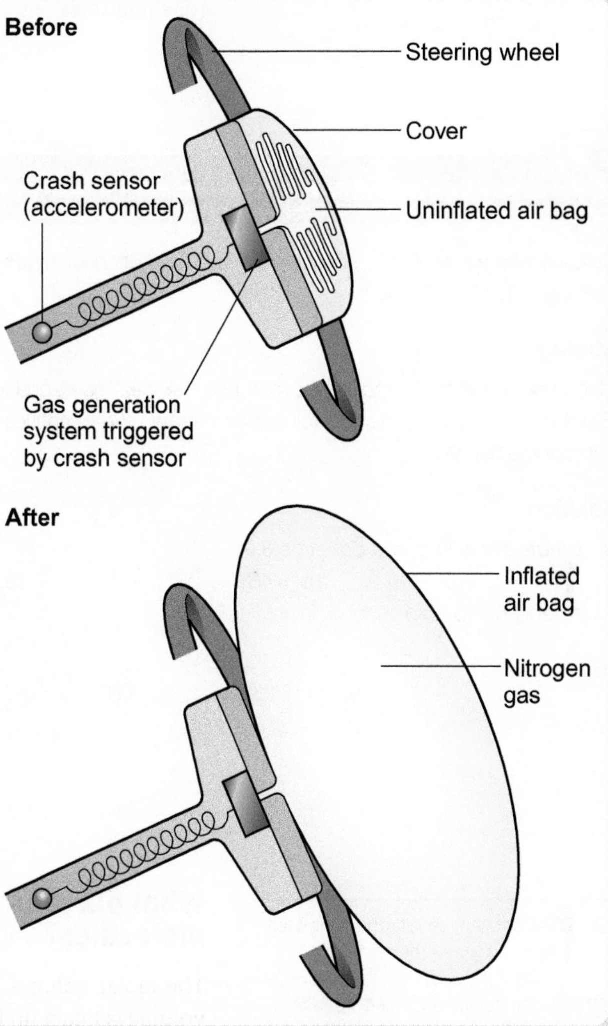
Before
Steering wheel
Cover
Crash sensor (accelerometer)
Uninflated air bag
Gas generation system triggered by crash sensor
After
Inflated air bag
Nitrogen gas

On impact, an *accelerometer* detects that the car has slowed down violently and sends an electrical signal, triggering a chemical reaction that produces a gas to inflate the folded bag. There are several gas generation systems in use, but a common one uses the decomposition of sodium azide (NaN_3). The reaction is extremely fast and releases a large volume of unreactive nitrogen gas

$$2\,NaN_3\,(s) \rightarrow 2\,Na\,(s) + 3\,N_2\,(g)$$

The other product of the decomposition is sodium, a highly reactive metal. To make the products safe, the NaN_3 is mixed with potassium nitrate (KNO_3) and silica (SiO_2). As the NaN_3 decomposes, the sodium metal formed is oxidized by the potassium nitrate to sodium oxide (Na_2O), which combines with the silica to make sodium silicate (Na_2SiO_3), a safe unreactive powder.

Question

An air bag needs 60 dm^3 of gas to inflate it at 298 K. Calculate:

(a) the amount in moles of N_2 gas at 298 K required to inflate the air bag to a pressure of 1 atm;

(b) the amount in moles of NaN_3 required to produce this amount of gas;

(c) the mass of NaN_3 required.

Standard conditions and molar volumes

The ideal gas equation can be used to calculate the volume of one mole of a gas under different conditions. The local surrounding (**ambient**) conditions vary depending on the location (since pressure depends on altitude) and the weather conditions. Therefore, a standard set of conditions needs to be defined. The IUPAC definition of **standard ambient temperature and pressure, SATP**, is **298.15 K** (25 °C) **and 1 bar**.

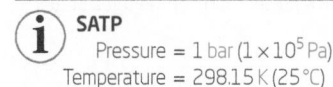

SATP
Pressure = 1 bar (1×10^5 Pa)
Temperature = 298.15 K (25 °C)

At SATP, the molar volume of an ideal gas is $0.0248\,m^3$ ($24.8\,dm^3$).

Under these conditions, one mole of an ideal gas has a volume of $0.0248\,m^3$ ($24.8\,dm^3$) (see Worked example 8.5). This value applies to *all* gases and vapours, irrespective of formula or molar mass assuming they behave ideally. The volume occupied by one mole of a substance is called the **molar volume** and is given the symbol V_m. It has units of $m^3\,mol^{-1}$ (or sometimes $dm^3\,mol^{-1}$).

Worked example 8.5 The molar volume of an ideal gas

Calculate the volume (in m^3) occupied by 1.00 mol of an ideal gas at 298 K and 1.00 atm pressure.

Strategy

Since you know n, T, and p, you can use the ideal gas equation (Equation 8.5) to calculate the volume by inserting the quantities in appropriate SI units.

Solution

Substitute the values into Equation 8.5.

$$pV = nRT \qquad (8.5)$$

From Table 8.1 (p.350)

$$1.00\,atm = 101\,325\,Pa$$
$$= 101\,325\,J\,m^{-3} \qquad (1\,Pa = 1\,J\,m^{-3})$$

Rearrange Equation 8.5 to give an expression for V.

$$V = \frac{nRT}{p}$$

$$= \frac{1.00\,\cancel{mol} \times 8.314\,\cancel{J}\,\cancel{K}^{-1}\,\cancel{mol}^{-1} \times 298\,\cancel{K}}{101325\,\cancel{J}\,m^{-3}}$$

$$= 0.0245\,m^3$$

1.00 mol of a gas occupies $0.0245\,m^3$ at 298 K and 1 atm. (Note that the value of V_m at 1 atm is very close to that at 1 bar.)

Question

Calculate the molar volume (in m^3) of an ideal gas at 1.00 atm and exactly 0 °C.

STP A different set of conditions used to be employed as the standard for gases. This was 1 atm and 0 °C, often known as **STP** (standard temperature and pressure). Under these conditions, the molar volume of an ideal gas is $22.4\,dm^3$.

What percentage of the volume of a gas is taken up by its molecules?

The molar volume, V_m, is the total volume occupied by the gas—but what part of this volume is taken up by the actual molecules?

In liquid water, H_2O (l), the molecules are close together and almost touching. To a first approximation, the molar volume of H_2O (l) equal to the actual volume of the molecules. The density of H_2O (l) at 25 °C is $1.0\,g\,cm^{-3}$. The molar mass of H_2O is $18.0\,g\,mol^{-1}$, so one mole of H_2O (l) occupies a volume of about $18\,cm^3$.

Now imagine that one mole of H_2O (l) becomes vapour, H_2O (g), at SATP. It now occupies $24.8\,dm^3$ ($24\,800\,cm^3$). The calculation in the last paragraph estimates that the actual volume of the molecules is about $18\,cm^3$, so the percentage of the total volume taken up by the molecules in the H_2O (g) can be calculated.

$$\text{Percentage of total volume taken up by the molecules} = \frac{18\,cm^3}{24\,800\,cm^3} \times 100$$
$$= 0.07\%$$

Figure 1.19 in Section 1.7 (p.46) shows a molecular representation of the three states of matter.

So, when one mole of H_2O (l) evaporates to form H_2O (g), of the $24\,800\,cm^3$ that it occupies, the actual molecules take up only about $18\,cm^3$, only ≈0.07% of the total volume. The remainder is free space.

This calculation assumes that H_2O (g) behaves ideally—which is a good approximation at 1 bar pressure. For other gases, the precise percentage will vary but the principle will be the same—*the atoms or molecules comprise only a very small fraction of the total volume of a gas*. However, they are moving around so fast and freely that they occupy the whole of the container (Section 8.5, p.362).

The terms *gas* and *vapour* have been used here. The distinction is that 'gas' is usually used to describe compounds that are gases at room temperature, while 'vapour' describes the gas phase of compounds that are liquids or solids at room temperature—such as water. More precise definitions are given in Section 8.6 (p.372). However, the two terms are often used interchangeably.

Summary

- If any three of n, p, V, and T are known, the fourth can be calculated using the ideal gas equation $pV = nRT$.
- For a fixed amount of gas, the effect of changing the conditions can be calculated using $\frac{p_1V_1}{T_1} = \frac{p_2V_2}{T_2}$.
- Standard conditions for a gas at SATP (standard ambient temperature and pressure) are defined as 298.15 K (25 °C) and 1 bar.
- The molar volume, V_m, of an ideal gas at SATP is 0.0248 $m^3\,mol^{-1}$ (24.8 $dm^3\,mol^{-1}$), although the molecules take up only a small fraction (<0.1%) of this volume.

For practice questions on these topics, see questions 1, 4–9, 11 and 14 at the end of this chapter (p.378).

8.3 Mixtures of gases

The discussion so far has focused on pure gases. However, since the features of an ideal gas do not depend on the nature of the gas, the ideas can be extended to mixtures of gases. There is plenty of space between the molecules in a gas. This means that, when gases mix together, the molecules can easily intermingle.

Figure 8.8 shows three containers of equal volume. The first contains a gas exerting a pressure p_A, the second a different gas exerting a pressure p_B. In the third container, the two gases mixed together occupy the same volume. What is the pressure of the mixture?

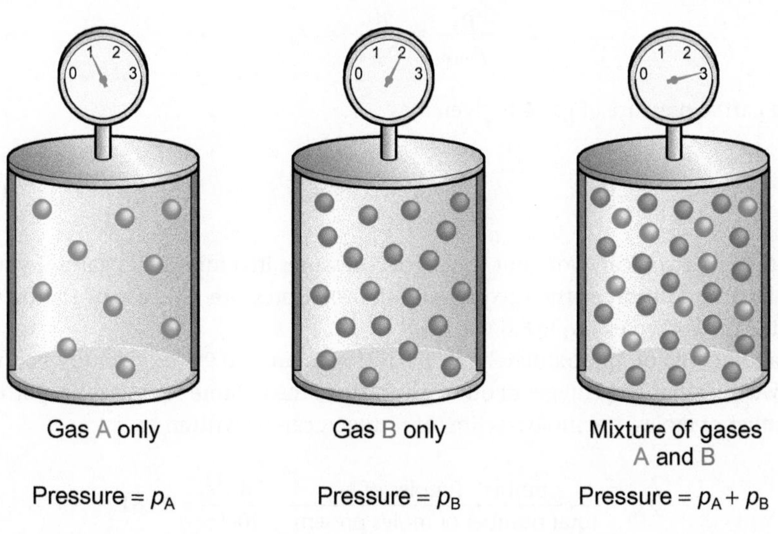
Gas A only
Pressure = p_A
Gas B only
Pressure = p_B
Mixture of gases A and B
Pressure = $p_A + p_B$

Figure 8.8 The pressure of the gas mixture is the sum of the individual partial pressures—this is an example of Dalton's law.

Given the large distances between them, the molecules do not interact so each gas in the mixture exerts the same pressure as if it were the only substance in the container. Therefore, the total pressure is simply the sum of the two individual pressures. This is expressed in **Dalton's law**:

The total pressure exerted by a mixture of gases is the sum of the partial pressures of each individual gas.

The **partial pressure** is the pressure that would be exerted if the gas were alone in the container.

In general, for a mixture of gases A, B, C, ..., Dalton's law says that the total pressure, p_{total} is given by:

The symbol Σ (sigma) means 'the sum of' (see Maths Toolkit MT1, p.1304).

$$p_{\text{total}} = p_A + p_B + p_C + \ldots = \sum_i p_i \quad (8.8)$$

where p_i represents the partial pressure of any component, i, in the mixture.

It is easy to measure the total pressure exerted by a mixture of gases with a barometer or pressure gauge. However, it is not possible to measure the contribution to this total of each individual gas. To describe the proportion of each component in a mixture the **mole fraction**, x, of the component is used. This is given by:

The **mole fraction**, x_A, of component A in a mixture gives the fraction of the total number of moles that is A. Although the mole fraction is defined here in terms of gases, it can also be used to define the proportions of components in liquid or solid mixtures (Section 17.4, p.789).

$$\text{mole fraction of A} = x_A = \frac{\text{number of moles of A}}{\text{total number of moles present}} = \frac{n_A}{n_{\text{total}}} \quad (8.9)$$

For a mixture containing 1 mol of nitrogen and 3 mol of hydrogen

$$\text{mole fraction of nitrogen:} \quad x_{N_2} = \frac{1\,\cancel{\text{mol}}}{2\,\cancel{\text{mol}}} = 0.25$$

$$\text{mole fraction of hydrogen:} \quad x_{H_2} = \frac{3\,\cancel{\text{mol}}}{4\,\cancel{\text{mol}}} = 0.75$$

Note that the sum of the mole fractions for all the components in a mixture is 1. Note also that mole fraction has no units, because it is a ratio.

For an ideal gas at constant V and T, $p_A \propto n_A$. So, for a component A in a mixture of gases

$$\frac{p_A}{p_{\text{total}}} = \frac{n_A}{n_{\text{total}}} = x_A$$

and the partial pressure of gas A is given by

$$p_A = x_A\, p_{\text{total}} \quad (8.10)$$

This means that you can work out the partial pressures in a mixture of gases if you know the molar composition of the mixture and the total pressure. The use of Dalton's law is illustrated in Worked example 8.6.

Dry air consists of approximately 78.1% nitrogen and 20.9% oxygen (by volume) together with smaller percentages of other gases. Since the volume of a gas is proportional to the number of moles, the mole fraction of nitrogen can be written as

$$x_{N_2} = \frac{\text{number of moles of } N_2}{\text{total number of moles present}} = \frac{78.1\%}{100\%} = 0.781$$

so that, in air at 1 atm pressure, the partial pressure of nitrogen is $0.781 \times 1\,\text{atm} = 0.781\,\text{atm}$.

Worked example 8.6 Partial pressures of gases

When delivered to a customer, natural gas consists of almost pure (>99%) methane. However, when the unpurified natural gas comes from the gas field, it is a mixture of several gases. The composition varies with its origin, but a typical composition by volume is:

methane (CH_4): 80%
ethane (C_2H_6): 6%
propane (C_3H_8): 3%
butane (C_4H_{10}): 4%
carbon dioxide (CO_2): 2%
nitrogen (N_2): 2%
hydrogen sulfide (H_2S): 3%

If the gas comes out of the ground at a pressure of 5.07×10^5 Pa (5 atm), calculate the partial pressure (in Pa) of methane before purification.

Strategy

Use Dalton's law to calculate the partial pressure of methane from its mole fraction and the total pressure.

Solution

For ideal gases at constant T and p, $n \propto V$ so the values given also represent the *molar* percentages. (The molar percentage is the mole fraction multiplied by 100.)

Find the mole fraction of methane, x_{CH_4}, from the percentage composition.

Take 100 mol of gas; this contains 80 mol of CH_4.

$$x_{CH_4} = \frac{80\ \cancel{mol}}{100\ \cancel{mol}} = 0.8$$

Use Equation 8.10 to find the partial pressure of methane, p_{CH_4}

$$p_A = x_A p_{total} \quad (8.10)$$

The total pressure is 5.07×10^5 Pa.

$$p_{CH_4} = 0.8 \times 5.07 \times 10^5\ \text{Pa}$$

$$= 4 \times 10^5\ \text{Pa} = 400\ \text{kPa} \quad (1\ \text{kPa} = 1 \times 10^3\ \text{Pa})$$

Question

Calculate the partial pressure of ethane in the gas at the same total pressure. Give your answer in atm.

A flare from a natural gas field. Natural gas is a mixture of gases, with methane being the major component.

Summary

- Dalton's law states that the total pressure exerted by a mixture of gases is the sum of the partial pressures of each individual gas

$$p_{total} = p_A + p_B + p_C + \ldots = \sum_i p_i$$

- Mole fraction of gas A in a mixture $= \dfrac{\text{number of moles of A}}{\text{total number of moles present}} = \dfrac{n_A}{n_{total}}$
- Partial pressure of gas A $= p_A = x_A p_{total}$

For practice questions on these topics, see questions 6, 7, 10 and 22 at the end of this chapter (p.378).

The Scottish physicist James Clerk Maxwell as an undergraduate at Cambridge. With other scientists, Maxwell developed the kinetic theory of gases, but is perhaps better known for his contributions to electromagnetism.

8.4 Kinetic molecular theory and the gas laws

The ideal gas equation was obtained using an empirical approach by combining laws derived from experimental measurements (Section 8.1, p.345). Although it allows you to look at what happens when the conditions are changed, the approach does not tell you what is happening to the molecules in the gas. Chemists need to understand processes on a *molecular* level.

A common approach in science is to set up a **model**. The first stage in developing a model is to define the basic properties of the system and then make some assumptions about its behaviour. You can use models to derive equations and develop a theory that describes the situation under study in mathematical terms—then test whether the theory matches the experimental observations. The real test of a model is that it can be used to make *predictions* that can be tested experimentally.

The **kinetic molecular theory of gases** (often called more simply the **kinetic theory**) is a good example of such an approach. The theory is based on a simple model that describes gas behaviour in terms of the movement of molecules. It was developed during the nineteenth century by a number of scientists including Ludwig Boltzmann and James Clerk Maxwell.

Assumptions of the kinetic model of gases

The basis of the kinetic model is that *a gas consists of molecules that are in constant random motion in all directions throughout the container*. As shown in Figure 8.9, the molecules collide with each other and with the walls of the container. The steady pressure exerted by the gas on the container walls is explained in terms of the collisions of the gas molecules with the walls.

If an object—including a molecule—is moving it must possess **kinetic energy**. For a molecule of mass m, moving with a speed s, the kinetic energy, E_{KE}, is given by

$$E_{KE} = \tfrac{1}{2}ms^2 \qquad (8.11)$$

Kinetic energy is introduced in Section 1.6 (p.41).

Thus, a molecule with a higher mass has greater kinetic energy than one with a lower mass moving at the same speed. Alternatively, for two molecules of the same mass, the one moving faster has higher kinetic energy. The mean kinetic energy, $\bar{E}_{KE}$, of the molecules in a gas is directly proportional to the absolute temperature, so the molecules move faster when the gas is hotter.

$$\bar{E}_{KE} \propto T \qquad (T \text{ in kelvin}) \qquad (8.12)$$

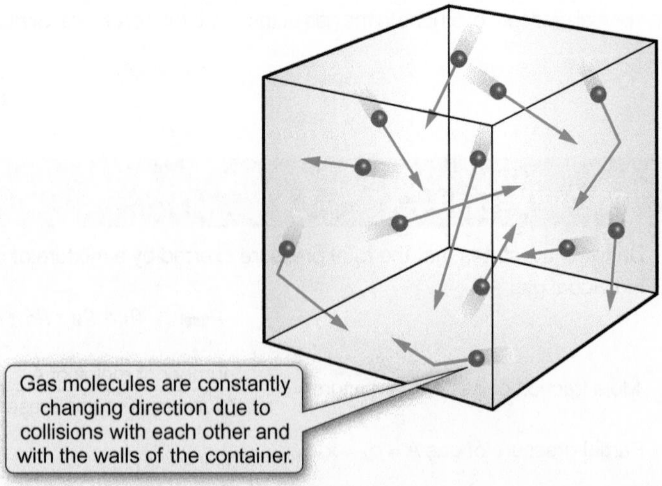
Gas molecules are constantly changing direction due to collisions with each other and with the walls of the container.

Figure 8.9 In a gas, the molecules are in continuous random motion in all directions, and continually collide with each other and with the walls of the container.

In developing the model, there are three major assumptions.

- The gas molecules have negligible size compared with the container—they are treated as *point masses*.
- Collisions between the molecules and the container walls are *elastic* so that there is no energy loss or gain on collision.
- The molecules do not interact with each other.

Are these assumptions reasonable? The size of the molecules is very much smaller than the total volume occupied by the gas (Section 8.2, p.349). The third assumption is perhaps the most questionable since molecules *do* interact. However, these interactions are significant only when the molecules are very close together. At the relatively large intermolecular distances in a gas, the assumption of negligible interactions is quite reasonable.

The first and last bullet points in the assumptions for the kinetic model are the same as the properties assumed for an ideal gas (see Section 8.6, p.372).

Interactions between molecules are introduced in Section 1.8 (p.52) and discussed further in Section 17.3 (p.783).

Using the kinetic model of gases to derive the ideal gas equation

So, how are these assumptions incorporated into a theory that explains the observable properties of a gas?

Pressure is exerted as a result of molecules colliding with the walls of the container. The pressure depends on how many molecules hit the wall of the container per second and the force they exert on the wall when they hit it, which depends on how fast they are moving.

The number of collisions is influenced by the concentration of gas molecules, that is, the number of moles in a given volume. The speed of the molecules depends on the temperature. This means that you might expect a relation between the volume, temperature, pressure and number of moles of a gas taking into account the three assumptions listed above.

By considering these relationships, it can be shown that the gas pressure is related to an average value for the speed of the molecules, c, by the expression

$$p = \frac{1}{3}\left(\frac{nN_A}{V}\right)mc^2 \qquad (8.13a)$$

where m is the mass of a molecule and N_A is the Avogadro constant. The average speed, c, of the molecules used in Equation 8.13a is the **root mean square (rms) speed**. It is the square root of the mean of the squares of the individual speeds. Box 8.2 shows how Equation 8.13a is derived by considering the change of momentum as molecules collide with the walls of the container.

Rearranging Equation 8.13a in the form

$$pV = \tfrac{1}{3}nN_A mc^2 \qquad (8.13b)$$

gives an equation that is beginning to look similar to the ideal gas equation. However, it does not have a temperature term. The temperature can be introduced by making use of Equation 8.11, $E_{KE} = \frac{1}{2}ms^2$. This is the kinetic energy of an individual molecule. The mean kinetic energy per molecule, $\bar{E}_{KE}$, taking into account all the molecules in a gas, is then

$$\bar{E}_{KE} = \tfrac{1}{2}mc^2 \qquad (8.14)$$

where

$$c^2 = \frac{s_1^2 + s_2^2 + \cdots + s_N^2}{N} \text{ and } c = \left(\frac{s_1^2 + s_2^2 + \cdots + s_N^2}{N}\right)^{1/2} = \text{rms speed}$$

To explain the meaning of root mean square speed, consider a simple example. If a sample contains a total of five molecules moving with speeds s_1, s_2, s_3, s_4, and s_5 then the rms speed is

$$c = \left(\frac{s_1^2 + s_2^2 + s_3^2 + s_4^2 + s_5^2}{5}\right)^{1/2}$$

The symbol $x^{1/2}$ means the same as $\sqrt{x}$; they are alternative ways of expressing the square root of x (Maths Toolkit MT1, p.1304).

From Equation 8.14, $mc^2 = 2\bar{E}_{KE}$. Substituting for mc^2 in Equation 8.13b

$$pV = \tfrac{1}{3}nN_A mc^2 = \tfrac{1}{3}nN_A(2 \times \bar{E}_{KE})$$

Since $\bar{E}_{KE} \propto T$ (Equation 8.12) and N_A is constant, this can be written as:

$$pV = \text{constant} \times nT$$

Visit the Online Resource Centre to view screencast 8.2 which explains how Equation 8.13b links the kinetic theory to the empirical ideal gas equation.

This derivation of the equation shows why the gas constant, R, is fundamental in chemistry. It describes the relationship between energy and temperature.

which has the same form as the empirical relationship for an ideal gas, $pV = nRT$ (Equation 8.5). So, *this theoretical approach produces the same result as the empirical approach.*

Thus, the ideal gas equation can be derived either from *experimental* observations or from a *theoretical* model. The theoretical derivation relies on the gas obeying the three assumptions in the model and predicts that it should follow Equation 8.5. A reasonable conclusion is, therefore, that an ideal gas is one that satisfies the three assumptions of the kinetic model.

Box 8.2 Calculating the pressure of a gas from the kinetic theory

This box explains how Equation 8.13a is derived from the kinetic theory. The pressure exerted by a gas is due to the molecules exerting a force on the container when they collide with its walls.

Pressure is defined as the force exerted per unit area. In a gas, the force is equal to the *rate of change of momentum* of the molecules when they collide with the container walls. To calculate the pressure you need to know how many molecules collide with the wall in one second and how much change of momentum takes place with each collision.

The momentum of a moving object is given by

$$\text{momentum} = \text{mass} \times \text{velocity} = mv$$

Velocity is a **vector** quantity—it has both direction and magnitude. So values of velocity have a sign that depends on the direction of motion. (In contrast, speed is a **scalar** quantity and has only magnitude.) A velocity, v, can be resolved into the sum of three components v_x, v_y, and v_z in the x-, y-, and z-directions, as shown in Figure 1.

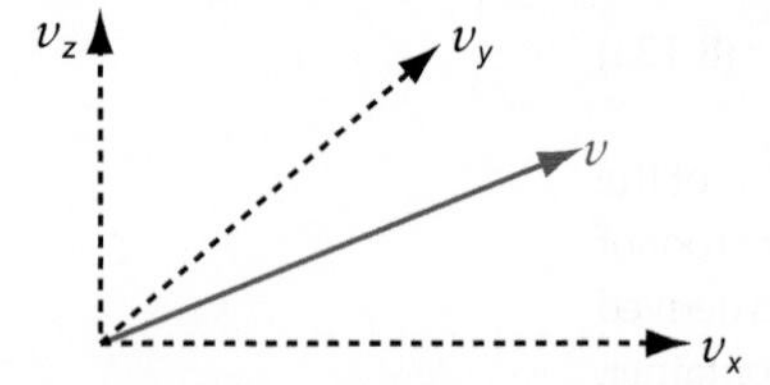

v_z
v_y
v
v_x

Figure 1 The velocity, v, in any direction can be resolved into components along the x-, y-, and z-axes.

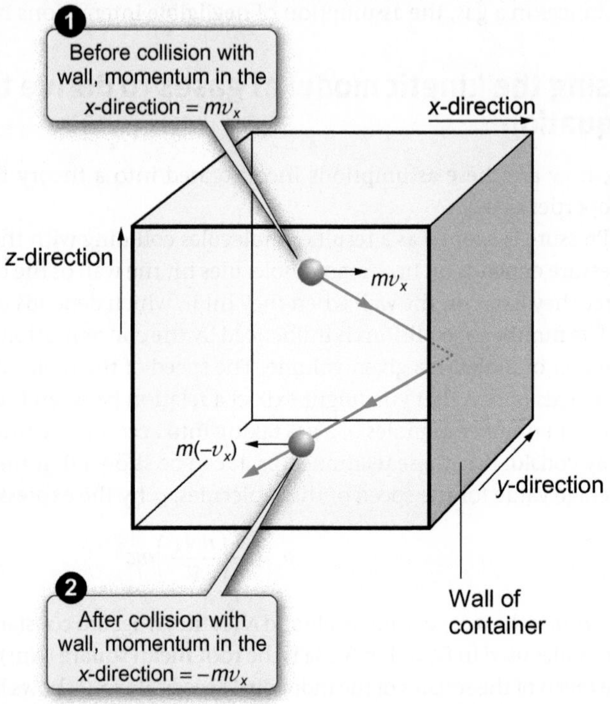

1
Before collision with wall, momentum in the x-direction = mv_x
x-direction
z-direction
mv_x
$m(-v_x)$
y-direction
2
After collision with wall, momentum in the x-direction = $-mv_x$
Wall of container

Figure 2 The change in momentum in the x-direction of a molecule on colliding with the wall of the container.

There will be gas molecules moving in all directions. To simplify the calculations, concentrate on the momentum changes that occur in only one direction—say the x-direction—and on one of the walls of the container, as shown in Figure 2. Momentum, like velocity, is a vector quantity.

If v_x is the velocity component of a molecule in the x-direction, then the momentum in the x-direction before the collision is mv_x. Since the collision is elastic, after the collision, the momentum must be $-mv_x$, that is, the same magnitude but in the opposite direction. Thus, the change of momentum per collision is given by

$$\begin{aligned}\text{change of momentum} &= (\text{momentum before collision}) \\ &\quad - (\text{momentum after collision}) \\ &= (mv_x) - (-mv_x) \\ &= 2mv_x \qquad (8.15)\end{aligned}$$

(As far as collisions with this wall are concerned, the momentum in the y- and z-directions is unchanged on collision.)

Next, you need to calculate the number of molecules colliding with the wall in one second. Assume, for a moment, that *all* the molecules in the gas have the same velocity, v_x, in the x-direction.

A velocity of v_x, measured in $m\,s^{-1}$, means that in one second a molecule will travel a distance v_x metres in the x-direction. Thus, if the molecule is within a distance v_x of the wall, it will hit the wall during this 1 s interval; if it is further away, it will not. For a container with a wall of area A, all the molecules that are moving toward the wall that are contained within a volume Av_x will hit, as shown in Figure 3.

If the gas contains N molecules per unit volume, then

$$\text{number of molecules hitting the wall per second} = \tfrac{1}{2}NAv_x \qquad (8.16)$$

The factor of $\frac{1}{2}$ is introduced because, on average, only half the molecules will be heading towards the wall, while the other half will be heading in the opposite direction.

Combining Equations 8.15 and 8.16, you can obtain an expression for the total change of momentum in one second, that is, for the rate of change of momentum. →

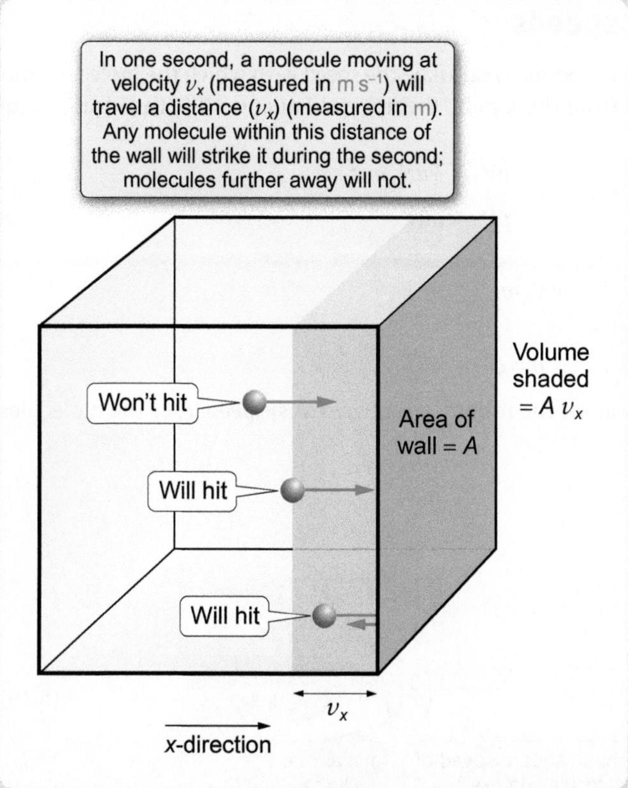
In one second, a molecule moving at velocity v_x (measured in m s⁻¹) will travel a distance (v_x) (measured in m). Any molecule within this distance of the wall will strike it during the second; molecules further away will not.
Volume shaded = $A\,v_x$
Won't hit
Area of wall = A
Will hit
Will hit
v_x
x-direction

▲ **Figure 3** How many molecules will hit the wall in one second? Only molecules within the shaded volume will hit the wall during the one-second interval and therefore contribute to the pressure during this time.

$$\text{rate of change of momentum} = \text{number of molecules colliding with the wall in one second} \times \text{change of momentum on each collision}$$

$$= (\tfrac{1}{2}NAv_x)\times(2mv_x) = NAmv_x^2$$

This expression assumes that all molecules are moving with the same velocity v_x in the x-direction. In fact there is a range of velocities, so you need to replace v_x^2 by $\overline{v_x^2}$, which is its value averaged over all the gas molecules. Then

$$\text{the rate of change of momentum} = NAm\overline{v_x^2} \qquad (8.17)$$

Equation 8.17 gives the force on the wall caused by all the collisions in the x-direction. Since pressure = force ÷ area, the pressure on the wall is obtained by dividing by the area, A, of the wall

$$p = \frac{NAm\overline{v_x^2}}{A} = NAm\overline{v_x^2} \qquad (8.18)$$

Since molecules are moving randomly in all directions, it is possible to obtain an expression for $\overline{v_x^2}$ in terms of the total mean square velocity, $\overline{v^2}$, where

$$\overline{v^2} = \overline{v_x^2} + \overline{v_y^2} + \overline{v_z^2}$$

In any realistic sample of gas, there will be a very large number of molecules moving randomly in all directions, so that

$$\overline{v_x^2} = \overline{v_y^2} = \overline{v_z^2} = \tfrac{1}{3}\overline{v^2}$$

The square of a vector quantity is always positive (i.e. it is a scalar quantity) so $\overline{v^2} = c^2$, where c is the root mean square speed defined on p.359. This means that

$$\overline{v_x^2} = \tfrac{1}{3}c^2$$

Remember that, for the purposes of this derivation, N is the number of molecules per unit volume. It is more convenient to think about the numbers of moles, n, of the gas so, as an alternative, you can write

$$N = \frac{nN_A}{V}$$

where n is the number of moles of gas in the total volume of the container, V, and N_A is the Avogadro constant.

Incorporating these last two changes into Equation 8.18 leads to

$$p = \frac{1}{3}\left(\frac{nN_A}{V}\right)mc^2 \qquad (8.13a)$$

Summary

- The ideal gas equation can be derived using the kinetic theory of gases, making the assumptions that:
  - the gas molecules have negligible size;
  - collisions are *elastic* so there is no energy loss;
  - the molecules do not interact with each other.
- Using the kinetic theory of gases and these assumptions, the pressure of a gas, p, can be calculated by finding the rate of change of momentum of molecules colliding with the walls of the container.

$$p = \frac{1}{3}\left(\frac{nN_A}{V}\right)mc^2$$

where m is the mass of a molecule and c is the root mean square (rms) speed. V is the volume of gas, n is the amount in moles, and N_A is Avogadro's constant.

? For practice questions on these topics, see questions 25 and 26 at the end of this chapter (p.380).

8.5 The speeds of molecules in a gas

Average molecular speeds

The kinetic theory of gases allows you to calculate the speed with which the molecules move.

Equation 8.13b, derived from the kinetic theory, and Equation 8.5, the ideal gas equation, can be combined

$$pV = \tfrac{1}{3}nN_A mc^2 \quad (8.13b)$$

$$pV = nRT \quad (8.5)$$

so that

$$\tfrac{1}{3}nN_A mc^2 = nRT$$

This can be rearranged to give

$$mN_A c^2 = 3RT$$

which leads to an expression for the root mean square (rms) speed, c, of the molecules in a gas

$$c^2 = \frac{3RT}{mN_A}$$

and

$$c = \left(\frac{3RT}{M}\right)^{1/2} \quad (8.19)$$

Gas constant = 8.314 J K^{-1} mol^{-1}

temperature (in K)

root mean square speed of molecules (in m s^{-1})

molar mass (in kg mol^{-1})

> ℹ The symbol c is also used for the speed of light. It is usually obvious which meaning is being used, but be careful not to confuse them.

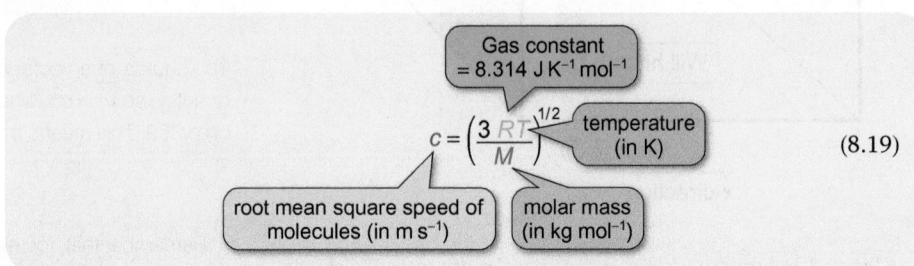

Remember that m refers to the mass of an individual molecule (e.g. for neon (Ne) $m = 3.35 \times 10^{-26}$ kg). Multiplying by N_A (6.023×10^{23} mol^{-1}) gives the molar mass, M (for Ne, $M = 0.02018$ kg mol^{-1} = 20.18 g mol^{-1}).

So, how fast do the molecules move? Equation 8.19 is important since it allows you to calculate c for a particular gas at a particular temperature. An illustration is given in Worked example 8.7. The rms speed for a nitrogen molecule at around room temperature is about 500 m s^{-1}. This means that, if there were no collisions, the molecule would travel a distance equivalent to the length of about five football pitches in one second. An alternative way of looking at this is to calculate the speed in more familiar units. A speed of 515 m s^{-1} is equivalent to 1150 miles per hour, comparable with the speed of a supersonic jet aircraft or the muzzle velocity of a bullet from a high-powered rifle.

> @ Visit the Online Resource Centre to view screencast 8.3 which explains the significance of Equation 8.19 and the way molecular speed is linked to temperature and molar mass.

> ℹ The rms speed, $\bar{c}$, is close in value to the mean speed (see Figure 8.11).

Worked example 8.7 How fast do molecules move?

Calculate the root mean square molecular speed in nitrogen gas at 298 K.

Strategy

You can use Equation 8.19 directly. To be consistent, SI units must be used throughout: temperatures in kelvin and molar masses in kilograms (not grams) per mole.

Solution

Find the molar mass, M, of a nitrogen molecule.

$A_r(\mathrm{N}) = 14.0$, so $M(\mathrm{N_2}) = 28.0\ \mathrm{g\,mol^{-1}} = 28.0 \times 10^{-3}\ \mathrm{kg\,mol^{-1}}$

Use Equation 8.19 to find the root mean square speed, c.

$$c = \left(\frac{3RT}{M}\right)^{1/2} = \left(\frac{3 \times 8.314\ \mathrm{J\,K^{-1}\,mol^{-1}} \times 298\ \mathrm{K}}{28.0 \times 10^{-3}\ \mathrm{kg\,mol^{-1}}}\right)^{1/2}$$

$$= \left(\frac{3 \times 8.314\ \cancel{\mathrm{kg}}\,\mathrm{m^2\,s^{-2}}\,\cancel{\mathrm{K}}^{-1}\,\cancel{\mathrm{mol}}^{-1} \times 298\ \cancel{\mathrm{K}}}{28.0 \times 10^{-3}\ \cancel{\mathrm{kg}}\ \cancel{\mathrm{mol}}^{-1}}\right)^{1/2} \quad (1\ \mathrm{J} = 1\ \mathrm{kg\,m^2\,s^{-2}})$$

$$= 515\ \mathrm{m\,s^{-1}}$$

Question

Calculate the root mean square speeds of helium molecules and carbon dioxide molecules at 15 °C.

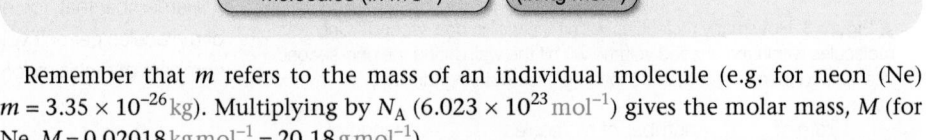

Table 8.2 Root mean square molecular speeds for gases at 20 °C

Gas	c/m s^{-1}	Gas	c/m s^{-1}
Bromine (Br_2)	213	Nitrogen (N_2)	511
Sulfur trioxide (SO_3)	302	Carbon dioxide (CO_2)	407
Butane (C_4H_{10})	355	Water (H_2O)	637
Chlorine (Cl_2)	320	Ammonia (NH_3)	655
Argon (Ar)	327	Helium (He)	1352
Oxygen (O_2)	478	Hydrogen (H_2)	2704

Equation 8.19 shows that the rms speed is higher at higher temperatures and is lower for molecules with higher mass. For example, for oxygen molecules (M = 32.0 g mol^{-1}), the rms speed, c, is 461 m s^{-1} at 0 °C and 539 m s^{-1} at 100 °C. For radon gas (M = 222 g mol^{-1}) the corresponding values are 175 m s^{-1} at 0 °C and 205 m s^{-1} at 100 °C. Some further examples are given in Table 8.2 for gases at 20 °C. Try using Equation 8.19 to explain the trends in the values shown.

The distribution of molecular speeds

At any instant, some molecules will be moving faster than the speeds calculated from Equation 8.19 while others will be moving more slowly. In a sample of gas, there is a *distribution* of speeds. The distribution of speeds has the form shown in Figure 8.10. This was shown by James Clerk Maxwell and later by Ludwig Boltzmann and is, therefore, known as the **Maxwell–Boltzmann distribution.**

The maximum value in the curves represents *the most probable* value of the speed, that is, the one that occurs most often (see Figure 8.11). The distribution of speeds is influenced by the molar mass, M, of the gas and its temperature, T, as you can see in Figure 8.12. For gases with high molar mass, there is a narrow spread of speeds distributed around a relatively low speed. Gases with lower molar masses have a significant fraction of their molecules moving at higher speeds. In gases at low temperatures, most molecules move with speeds close to this most probable value. At higher temperatures there is a much broader spread of speeds.

ⓘ You will see the name Boltzmann in several places in this book in connection with reaction kinetics (Sections 9.7, p.425, and 9.8, p.432), with spectroscopy (Section 10.3, p.458), and with entropy (Section 14.1, p.656).

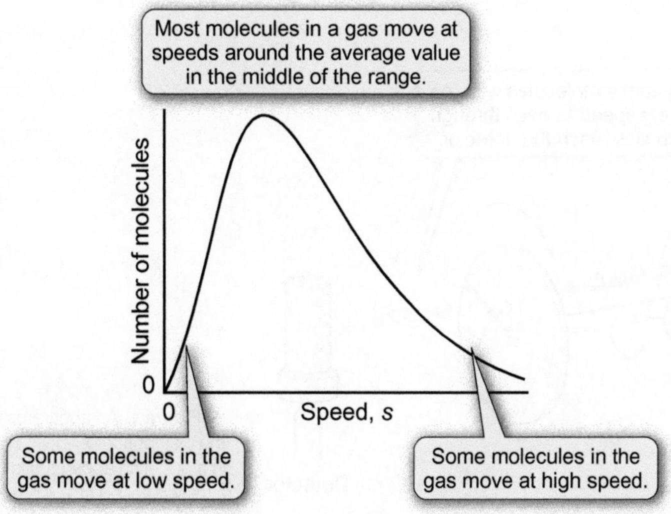
Most molecules in a gas move at speeds around the average value in the middle of the range.
Number of molecules
0
0
Speed, s
Some molecules in the gas move at low speed.
Some molecules in the gas move at high speed.

Figure 8.10 The Maxwell–Boltzmann distribution of molecular speeds.

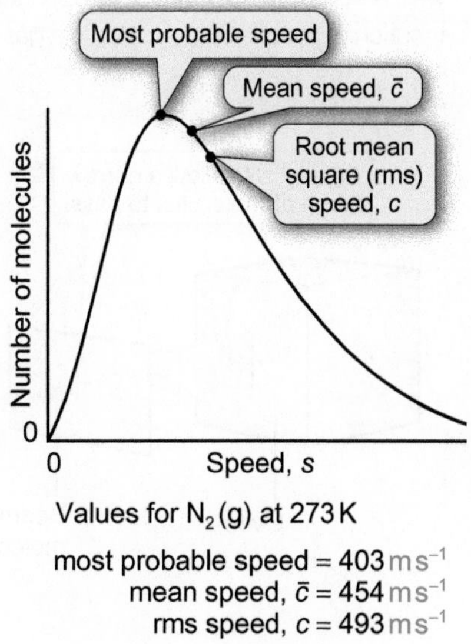
Most probable speed
Mean speed, c̄
Root mean square (rms) speed, c
Number of molecules
0
0
Speed, s
Values for N₂(g) at 273 K
most probable speed = 403 m s⁻¹
mean speed, c̄ = 454 m s⁻¹
rms speed, c = 493 m s⁻¹

Figure 8.11 Three different ways of quoting the 'average' speed of molecules in a gas.

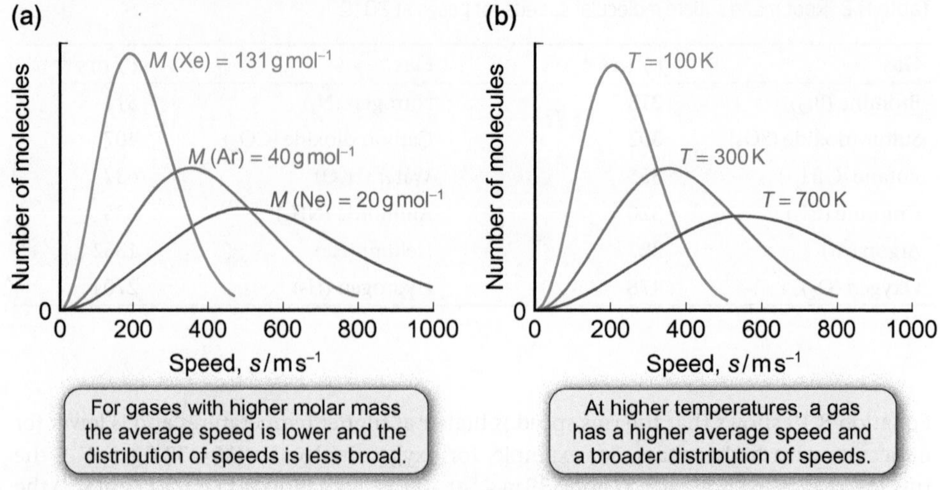
(a)
M (Xe) = 131 g mol⁻¹
M (Ar) = 40 g mol⁻¹
M (Ne) = 20 g mol⁻¹
Number of molecules
Speed, s/m s⁻¹
For gases with higher molar mass the average speed is lower and the distribution of speeds is less broad.
(b)
T = 100 K
T = 300 K
T = 700 K
Number of molecules
Speed, s/m s⁻¹
At higher temperatures, a gas has a higher average speed and a broader distribution of speeds.

Figure 8.12 The effect of molar mass and temperature on the distribution of molecular speeds. (a) The effect of molar mass for three Group 18 gases at 300 K. (b) The effect of temperature for $M = 40\ g\ mol^{-1}$.

In this discussion of molecular speeds, it is important to remember that the distribution describes the whole sample of a gas. Provided the conditions remain the same, the *distribution* of speeds through the whole sample remains constant. However, for an individual molecule, the speed and direction of movement continually change as a result of collisions with the walls of the container and with other molecules.

Box 8.3 describes an experiment to measure the distribution of speeds in a gas.

Box 8.3 Measuring the distribution of speeds in a gas

When the Maxwell–Boltzmann distribution was derived mathematically, there was no way of confirming its validity. However, there is now evidence that it matches experimental observations very well. A schematic diagram of an apparatus used to measure the distribution of speeds of molecules in a gas is shown in Figure 1.

A sample of gas is contained in a heated oven, from which a narrow beam of molecules can escape through a small hole into an evacuated chamber. The beam is directed through parallel slits, towards two rotating discs, separated by a distance d, each of which contains a narrow slit. The discs are rotated at the same speed but

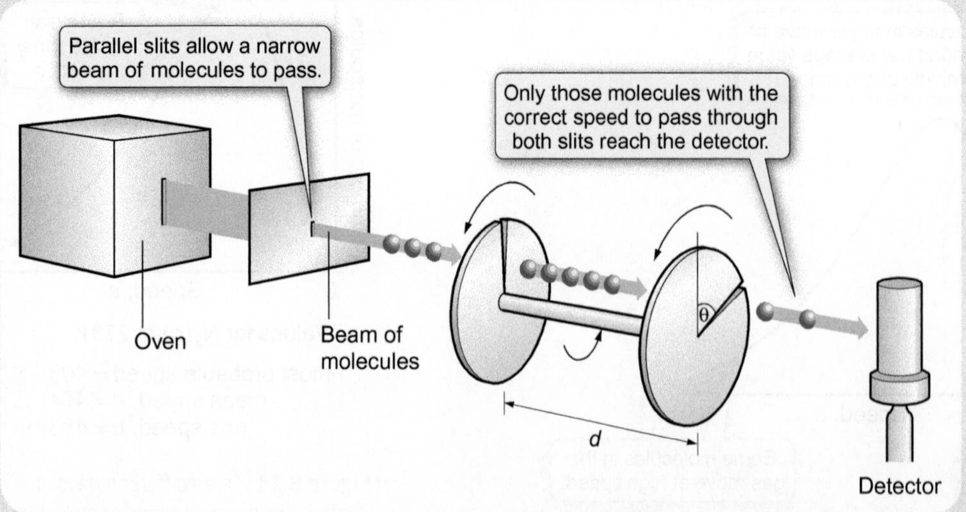
Parallel slits allow a narrow beam of molecules to pass.
Only those molecules with the correct speed to pass through both slits reach the detector.
Oven
Beam of molecules
θ
d
Detector

Figure 1 An apparatus to determine the distribution of speeds of molecules in a gas. The apparatus is contained in a sealed vessel at low pressure.

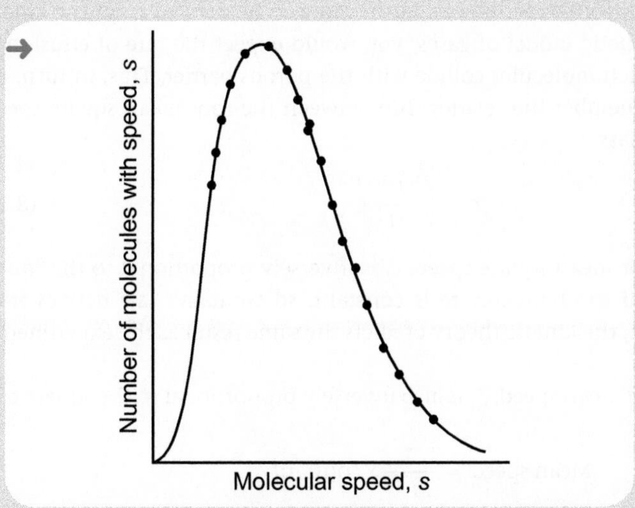
Number of molecules with speed, s
Molecular speed, s

▲ **Figure 2** A typical result. The plot shows the number of molecules reaching the detector against speed.

with the slits at slightly different angles. The angle between the two slits is θ.

After passing through the first slit, only those gas molecules with a particular speed will reach the second disc in time to pass through its slit and be detected. The number of molecules reaching the detector is measured and the speed of the molecules is calculated from the distance, d, between the discs and the rate of rotation. By changing the angle, θ, between the two slits, different speeds can be selected. A plot of the number of molecules reaching the detector against the speed can then be constructed as in Figure 2.

The experiments have provided experimental confirmation of the theoretical Maxwell–Boltzmann distribution.

Question

Why must the apparatus be contained in a sealed vessel at low pressure?

Effusion and diffusion

Further experimental evidence to support the kinetic theory comes from the study of effusion. This is the process by which gas molecules pass through a small hole such as a pore in a membrane, as illustrated in Figure 8.13.

In 1833, Thomas Graham studied the effusion of different gases and found that:

At a given temperature and gas pressure, the rate of effusion (i.e. the number of molecules passing through the hole per second) is inversely proportional to the square root of the molar mass.

This is known as **Graham's law**.

$$\text{Rate of effusion} \propto \frac{1}{M^{1/2}} \qquad (8.20)$$

In effusion experiments, for Graham's law to be obeyed, the hole must be of a similar size to the distance molecules travel between collisions—called the mean free path (see p.369). For nitrogen gas at SATP, the mean free path is 6.7×10^{-8} m.

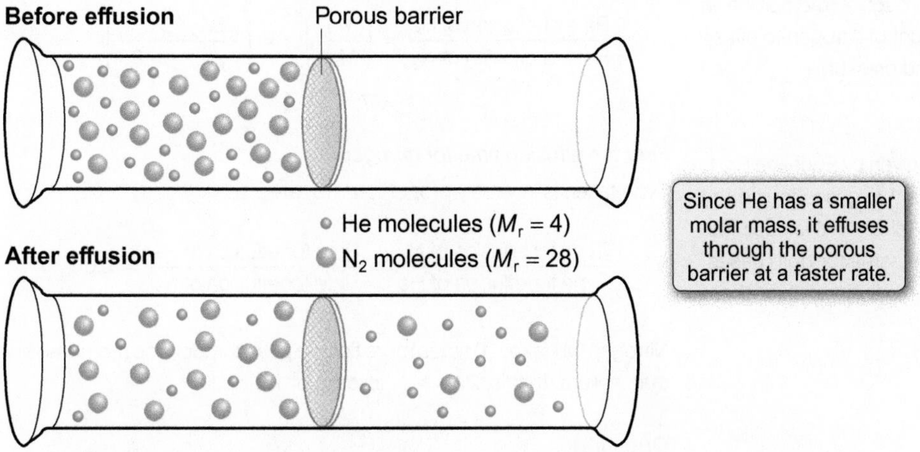
Before effusion
Porous barrier
He molecules ($M_r = 4$)
N_2 molecules ($M_r = 28$)
After effusion
Since He has a smaller molar mass, it effuses through the porous barrier at a faster rate.

Figure 8.13 A schematic representation of effusion. The rate at which gas molecules move through the porous barrier depends inversely on the square root of their molar mass as shown by Graham's law. Thus, more helium passes through in a certain time than nitrogen.

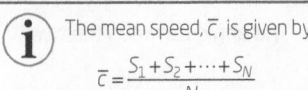

The mean speed, $\bar{c}$, is given by

$$\bar{c} = \frac{S_1 + S_2 + \cdots + S_N}{N}$$

Helium balloons are often made from plastic coated with a very thin film of metal to reduce the rate at which helium effuses out. Why is this more important with a helium-filled balloon than with an air-filled balloon?

Graham's law is derived from experiments, but it can also be derived from the kinetic theory. In terms of the kinetic model of gases, you would expect the rate of effusion to depend on the rate at which molecules collide with the porous barrier. This, in turn, depends on their speed. Remember the relationship between the root mean square speed, temperature, and molar mass

$$c = \left(\frac{3RT}{M}\right)^{1/2} \quad (8.19)$$

This shows that the root mean square speed, c, is inversely proportional to the square root of the molar mass, if the temperature is constant, so Graham's law derives from Equation 8.19. Once again, the kinetic theory predicts the same result as the experimental law.

It can be shown that the mean speed, $\bar{c}$, is also inversely proportional to the square root of the molar mass

$$\text{Mean speed} = \frac{1}{M^{1/2}} \times \text{constant}$$

For a mixture of two gases, A and B, with molar masses, M_A and M_B, the above equation will apply to each gas

$$\frac{\text{Mean speed of A}}{\text{Mean speed of B}} = \frac{\text{constant}/M_A^{1/2}}{\text{constant}/M_B^{1/2}} = \left(\frac{M_B}{M_A}\right)^{1/2}$$

Therefore,

$$\frac{\text{Rate of effusion of A}}{\text{Rate of effusion of B}} = \left(\frac{M_B}{M_A}\right)^{1/2} \quad (8.21)$$

Gases with different molar masses will, therefore, effuse at different rates. A gas with a low molar mass (such as helium) effuses faster than a gas with a higher molar mass (such as nitrogen) as shown in Worked example 8.8. This explains why a balloon filled with helium collapses faster than one filled with air.

A practical example of the use of effusion as a separation technique for gases is described in Box 8.4.

Worked example 8.8 Relative rates of effusion

A certain amount of helium gas passes through a membrane in 1.0 hour. How long will it take the same amount of nitrogen to effuse under the same conditions of temperature and pressure?

Strategy

The relative rates of effusion of two gases are given by Equation 8.21. Substitute values for the molar masses of the two gases to find the rates of effusion and use this to find the time taken. In this case, the units are unimportant, as long as they are the same for both gases, since they cancel. However it is good practice always to include them in the calculation.

Solution

Find the relative rates of effusion.

$$\frac{\text{Rate of effusion of A}}{\text{Rate of effusion of B}} = \left(\frac{M_B}{M_A}\right)^{1/2} \quad (8.21)$$

$$\frac{\text{Rate of effusion of He}}{\text{Rate of effusion of } N_2} = \left(\frac{M_{N_2}}{M_{He}}\right)^{1/2} = \left(\frac{28.0\,\cancel{g\,mol^{-1}}}{4.0\,\cancel{g\,mol^{-1}}}\right)^{1/2}$$

$$= 7^{1/2} = 2.6$$

Find the effusion time for nitrogen.

Time taken is inversely proportional to rate

$$\frac{\text{Time for effusion of } N_2}{\text{Time for effusion of He}} = \frac{\text{Rate for effusion of He}}{\text{Rate for effusion of } N_2} = 2.6$$

Nitrogen will take 2.6 times more time to pass through the membrane than helium, that is, 2.6 hours instead of 1.0 hour.

Question

Calculate the relative rates of effusion of oxygen, carbon dioxide, and nitrogen through a porous film at a fixed temperature.

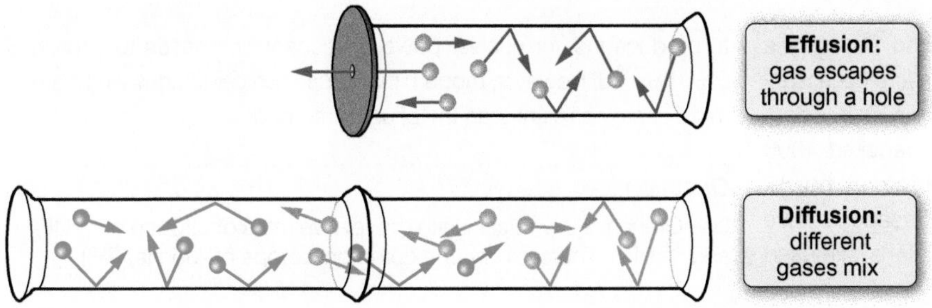
Effusion:
gas escapes through a hole
Diffusion:
different gases mix

Figure 8.14 Effusion and diffusion. Both processes arise from the random motion of gas molecules. Effusion is the movement of gases through a small hole, or porous barrier, while diffusion involves mixing of gases.

Diffusion of Br_2. Bromine vapour evaporates from the liquid in the bottom of the jar and diffuses to mix with the air.

The process of diffusion also depends on the movement of molecules. Figure 8.14 shows how diffusion differs from effusion. **Diffusion** occurs when two (or more) gases come into contact and mix. There are large spaces between the molecules, so they can easily mix. The only impediment to mixing is the collisions between the molecules and this means that mixing does not occur instantaneously but takes some time. For example, if you open a bottle of perfume it takes some time for the volatile components to spread around the room. The time taken for gases to mix depends on the speed of the molecules and on the frequency of intermolecular collisions. It thus depends on the temperature, molar masses, and the pressure of the gas—the higher the pressure, the more frequent the collisions.

Box 8.4 Enriching uranium: a practical application of effusion

Natural uranium consists of 99.3% of the ^{238}U isotope and 0.7% ^{235}U (with very small amounts of some other isotopes). For the production of atomic weapons at the end of the Second World War (1939–1945), and later for some types of nuclear power generation, material containing a much higher percentage of ^{235}U was needed. The process of increasing the proportion of ^{235}U is called enriching.

Since they are isotopes, the chemistry of ^{238}U and ^{235}U is identical. This means that any separation must be based on physical properties. Although uranium is a metal, it readily forms volatile uranium hexafluoride (UF_6). This is a liquid at room temperature but boils at 56 °C.

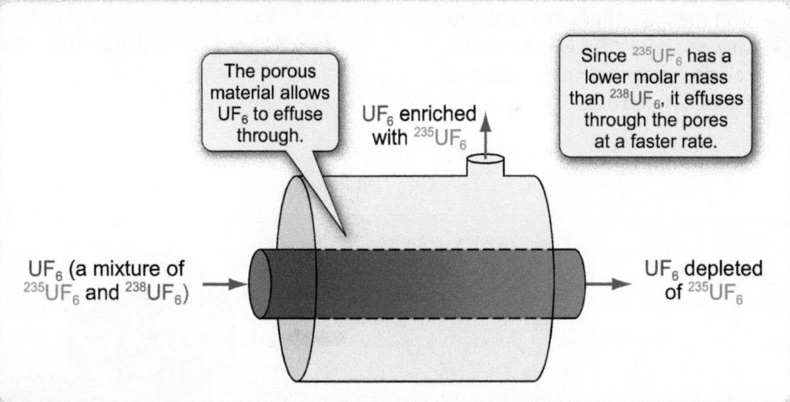
The porous material allows UF_6 to effuse through.
UF_6 enriched with $^{235}UF_6$
Since $^{235}UF_6$ has a lower molar mass than $^{238}UF_6$, it effuses through the pores at a faster rate.
UF_6 (a mixture of $^{235}UF_6$ and $^{238}UF_6$)
UF_6 depleted of $^{235}UF_6$

Figure 2 Effusion is used to enrich UF_6 with $^{235}UF_6$. The difference in masses is small so many separation steps are needed to separate $^{235}UF_6$ from the main isotope $^{238}UF_6$.

Figure 1 The uranium separation plant in Oak Ridge, USA.

As part of the secret Manhattan Project to develop an atomic bomb, a huge separation plant was built in the USA in 1943, at Oak Ridge, Tennessee. The plant used effusion as the basis of the separation method, as shown in Figure 2. Vaporized UF_6 produced from natural uranium passed through a tube whose walls were made of a porous ceramic. A low pressure was maintained outside the tube and the gas effused through the walls.

After passing through the porous barrier, the UF_6 gas was slightly enriched in the lighter isotope since, considering the molar →

➜ masses, $^{235}UF_6$ passes through slightly faster than $^{238}UF_6$. The small difference meant that many separation stages were necessary to obtain a realistic amount of pure ^{235}U. The plant at Oak Ridge, which was half a mile long and covered 43 acres, required 4000 stages to produce the 90% pure ^{235}U necessary for an atomic bomb. Because of the difficulty in production, most fission weapons since that time have used plutonium. Uranium enrichment is still used to produce fuel for nuclear power stations which needs to contain 3%–4% ^{235}U. However, modern plants use gas centrifuges which are more energy efficient than the effusion method.

Question

Use Graham's law to calculate the relative rates of effusion of $^{238}UF_6$ and $^{235}UF_6$. (There is only one common isotope of fluorine, ^{19}F.)

Collisions between molecules

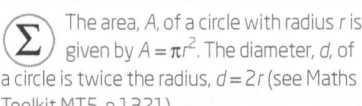

Collision theory is applied to the kinetics of reactions in Section 9.8 (p.432).

Chemical reactions occur when molecules collide with each other. Is it possible to use the results from kinetic theory to estimate how often collisions between molecules take place in gases?

To do this, one of the assumptions of the ideal gas model has to be questioned. If the molecules are simply point masses with no size there would be no collisions between molecules at all. Intuitively, you would expect that for large molecules there would be more chance of a collision taking place than for small molecules. But how exactly does the molecular size affect the chance of a collision?

Collision cross-section

The area, A, of a circle with radius r is given by $A = \pi r^2$. The diameter, d, of a circle is twice the radius, $d = 2r$ (see Maths Toolkit MT5, p.1321).

Visit the Online Resource Centre to view screencast 8.4 which explains collision cross-section by talking you through Figure 8.15.

Think about two molecules, A and B, approaching each other. The molecules are of the same type, and so are the same size. As shown in Figure 8.15, they will collide if the centre of one (say molecule B) comes within a distance of two radii—one diameter—of molecule A. This area is the *target* presented by A. The area of the target for molecule B to hit molecule A is, therefore, a circle with an area of πd^2, where d is the diameter of the molecule. This area is known as the collision cross-section, σ (see Figure 8.16).

$$\text{Collision cross-section, } \sigma = \text{area of target presented} = \pi d^2$$

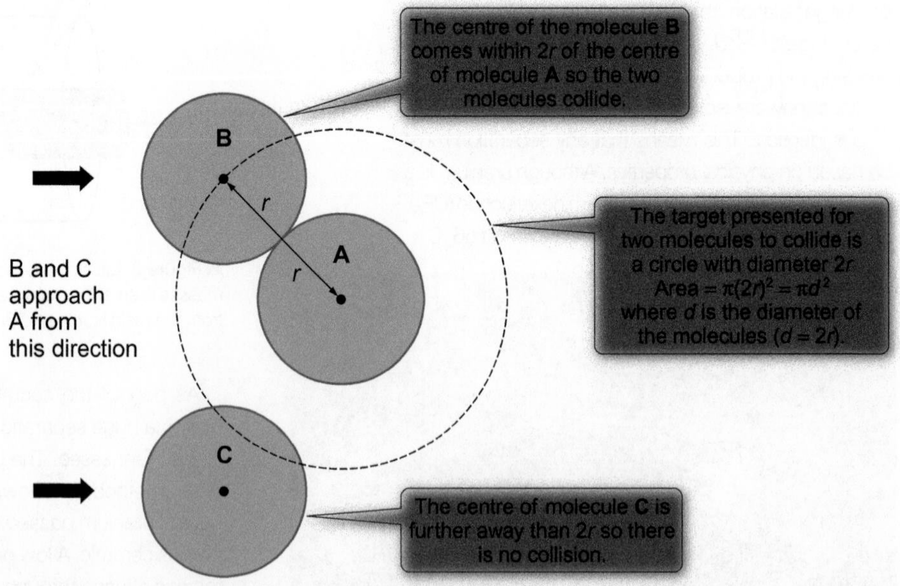
The centre of the molecule B comes within 2r of the centre of molecule A so the two molecules collide.
B
r
A
r
B and C approach A from this direction
The target presented for two molecules to collide is a circle with diameter 2r Area = π(2r)² = πd² where d is the diameter of the molecules (d = 2r).
C
The centre of molecule C is further away than 2r so there is no collision.

Figure 8.15 Two molecules will collide if their centres come within two molecular radii. (For simplicity, assume molecule **A** is stationary and molecules **B** and **C** are approaching in the direction of the arrows. Only spherical molecules are shown but the principle can be extended to other shapes.)

Table 8.3 Collision cross-sections of molecules ($1\ \text{nm}^2 = 1 \times 10^{-18}\ \text{m}^2$)

Gas	σ / nm^2
Helium (He)	0.21
Hydrogen (H_2)	0.27
Argon (Ar)	0.36
Oxygen (O_2)	0.40
Nitrogen (N_2)	0.43
Carbon dioxide (CO_2)	0.52
Sulfur dioxide (SO_2)	0.58
Benzene (C_6H_6)	0.88
Chlorine (Cl_2)	0.93

σ **Collision cross-section.** The target area presented for another molecule to collide with. Units: m^2.

λ **Mean free path.** The mean distance travelled between collisions. Units: m.

Z **Collision frequency.** The mean number of collisions per second. Units: s^{-1}.

Figure 8.16 Key quantities associated with molecular collisions.

The collision cross-section is related to the size of the molecules and some values for common gases are given in Table 8.3.

Collision frequency and mean free path

The mean number of collisions that a molecule undergoes per second is known as the **collision frequency**, Z (see Figure 8.16).

The number of collisions that a molecule will undergo in 1 s depends on the distance that it moves and the number of gas molecules per unit volume (i.e. the pressure). In a given time period, a molecule will collide with any other molecule whose centre lies within a cylinder with cross-sectional area, σ.

Using kinetic theory to develop this idea, it can be shown that Z is related to the collision cross-section, σ, and the mean speed, $\bar{c}$, by the expression

$$Z = \sqrt{2}\,N_A(\bar{c}\sigma)\times\left(\frac{p}{RT}\right) \tag{8.22}$$

Equation 8.22 is derived by considering the distance a molecule will travel in 1 s and the number of molecules in a cylinder of this length with cross-sectional area, σ. You can find the derivation in a more advanced textbook.

The use of Equation 8.22 is illustrated in Worked example 8.9, which shows that, on average, a nitrogen molecule undergoes around 7 billion collisions every second at SATP.

SATP refers to 1 bar (1×10^5 Pa) pressure and 298.15 K (25 °C).

Knowing the average speed at which molecules move and the number of collisions they undergo, the distance travelled between collisions can be calculated. This is called the **mean free path**, λ (see Figure 8.16)

$$\lambda = \frac{\text{distance travelled in 1 second}}{\text{number of collisions in 1 second}} = \frac{\bar{c}}{Z}$$

Using Equation 8.22 to substitute for Z

$$\lambda = \frac{\bar{c}}{\sqrt{2}\,N_A(\bar{c}\sigma)\times\left(\frac{p}{RT}\right)} = \frac{\not{\bar{c}}RT}{\sqrt{2}\,N_A(\not{\bar{c}}\sigma)p}$$

$$\lambda = \frac{RT}{\sqrt{2}\,N_A\sigma p} \tag{8.23}$$

Visit the Online Resource Centre to view screencast 8.5 which explains mean free path and walks you through Equation 8.23.

The mean free path, collision frequency, and mean speed are related by

$$\lambda \times Z = \bar{c}.$$

Worked example 8.9 How often do molecules collide?

For nitrogen gas at SATP, calculate:

(a) the collision frequency, Z;

(b) the mean free path, λ, of the N_2 molecules.

(For N_2, $\sigma = 0.43\,\text{nm}^2$ and the mean speed, $\bar{c} = 475\,\text{m s}^{-1}$.)

Strategy

(a) Use Equation 8.22 to find the collision frequency. To do this, you need to convert σ to units of m^2 and to use SI units for p, R, and T.

(b) Use Equation 8.23 to find the mean free path (or use the relationship $\lambda = \bar{c}/Z$).

Solution

(a) ***Substitute the values into Equation 8.22 to find the collision frequency, Z.***

$$Z = \sqrt{2}N_A(\bar{c}\sigma)\times\left(\frac{p}{RT}\right) \qquad (8.22)$$

$$\sigma = (1\,\text{nm}^2 = 1\times10^{-18}\,\text{m}^2)$$

$$Z = \sqrt{2}\times(6.02\times10^{23}\,\cancel{\text{mol}^{-1}})\times475\,\text{ms}^{-1}\times(0.43\times10^{-18}\,\text{m}^2)\times\left(\frac{1\times10^5\,\text{Pa}}{8.314\,\text{J}\,\cancel{\text{K}^{-1}}\,\cancel{\text{mol}^{-1}}\times298\,\cancel{\text{K}}}\right)$$

$$= 7.0\times10^9\,\text{s}^{-1} \qquad (1\,\text{Pa} = 1\,\text{J}\,\text{m}^{-3})$$

(b) ***Use Equation 8.23 to find the mean free path, λ.***

$$\lambda = \frac{RT}{\sqrt{2}N_A\sigma p} \qquad (8.23)$$

$$\lambda = \frac{8.314\,\cancel{\text{J}}\,\cancel{\text{K}^{-1}}\,\cancel{\text{mol}^{-1}}\times298\,\cancel{\text{K}}}{\sqrt{2}\times(6.02\times10^{23}\,\cancel{\text{mol}^{-1}})\times(0.43\times10^{-18}\,\text{m}^2)\times(1\times10^5\,\cancel{\text{J}}\,\text{m}^{-3})}$$

$$= 6.7\times10^{-8}\,\text{m}$$

Question

Calculate the collision frequency in hydrogen gas at SATP and account for the difference from that for nitrogen. (For H_2, $\sigma = 0.27\,\text{nm}^2$ and the mean speed $\bar{c} = 1780\,\text{m s}^{-1}$ at 298 K.)

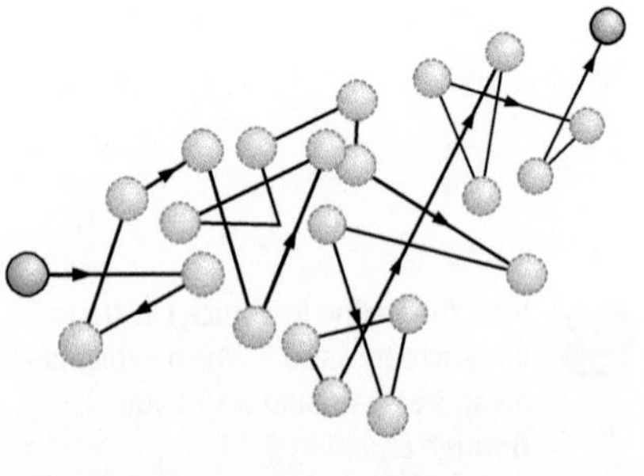

Figure 8.17 The 'random walk' of a gas molecule. An individual gas molecule undergoes many random changes of direction as a result of collisions with other molecules and with the container.

As shown in Worked example 8.9, the mean free path, λ, for N_2 at SATP is 6.7×10^{-8} m, so a nitrogen molecule travels approximately 67 nm between collisions. Compare this with the molecular size of approximately 0.3 nm. It is a characteristic of gases that their molecules move a large distance between collisions compared with their size. This is very different from the situation in solids and liquids.

In a gas at atmospheric pressure, collisions occur on average every 0.1 ns (i.e. at a rate of $1\times10^{10}\,\text{s}^{-1}$). This means that, over any observable time in a real container, gas molecules undergo a huge number of collisions. The molecules move in all directions at random and undergo many changes in direction, as shown in Figure 8.17. This is the origin of the random motion of dust or other particles suspended in a gas, known as **Brownian motion**. The dust particles themselves have no intrinsic motion but they move and change direction under the constant random bombardment of gas molecules.

Under normal pressures, gas molecules undergo many more collisions with each other than they do with the walls of the container. This means that, despite the molecules moving with the speed of a rifle bullet, gases diffuse quite slowly, because they have a long way to travel on their random route.

Equation 8.23 shows how the mean free path, λ, depends on conditions. For example, the value of λ decreases at higher pressure. At a pressure of 2 atm, a nitrogen molecule travels on average about 33 nm before colliding with another molecule. At a pressure of 1 atm, the molecule travels twice this distance. This is because there are more molecules in a given volume at higher pressure. At low pressures (such as those inside a mass spectrometer) the mean free path is several metres (see Worked example 8.10).

In interstellar space the pressure is so low (around 1×10^{-18} atm) that collisions are very rare. Chemical reactions are therefore very slow and the timescale for products to be formed is immense. (Reactions in interstellar space are discussed in Box 4.6, p.190.)

Collision frequencies are also important in reaction kinetics (Section 9.8, p.432). Z gives the number of collisions that occur in 1 second. If every collision resulted in a reaction, Z would represent the rate of the gas phase reaction. However, this is not the case, because most collisions do not result in a reaction.

Worked example 8.10 Molecular collisions at low pressures

A vacuum pump maintains the air pressure inside a mass spectrometer at 1×10^{-8} bar. Calculate the mean free path of the air molecules at 298 K under these conditions. (You can assume that a nitrogen molecule is representative of air.)

Strategy

Use Equation 8.23 (p.369) to calculate the mean free path. Remember to use SI units.

Solution

Use Equation 8.23 to find the mean free path, λ.

$$\lambda = \frac{RT}{\sqrt{2}\, N_A \sigma p} \qquad (8.23)$$

From Table 8.3 (p.369), the collision cross-section for N_2

$$\sigma = 0.43\,\text{nm}^2 = 0.43 \times 10^{-18}\,\text{m}^2 \quad (1\,\text{nm}^2 = 1 \times 10^{-18}\,\text{m}^2)$$

$$p = 1 \times 10^{-8}\,\text{bar} = 1 \times 10^{-3}\,\text{Pa} \quad (1\,\text{bar} = 1 \times 10^{5}\,\text{Pa} = 1 \times 10^{5}\,\text{J m}^{-3})$$

$$\lambda = \frac{8.314\,\cancel{\text{J}}\,\cancel{\text{K}^{-1}}\,\cancel{\text{mol}^{-1}} \times 298\,\cancel{\text{K}}}{\sqrt{2} \times (6.02 \times 10^{23}\,\cancel{\text{mol}^{-1}}) \times (0.43 \times 10^{-18}\,\text{m}^2) \times (1 \times 10^{-3}\,\cancel{\text{J}}\,\text{m}^{-3})} = 6.8\,\text{m}$$

This shows that it is unlikely that a molecule will undergo a collision with another gas molecule in the path length of the spectrometer, which is less than 1 metre (see Box 1.2, p.14). This is important to prevent interference with the mass spectrum being measured.

Question

Calculate the pressure at which the mean free path of N_2 is 10 cm at 298 K.

Summary

- Gas molecules move at high speeds that depend on the temperature and their molar mass. For example, N_2 molecules at 298 K have an average speed of about 500 m s^{-1}.
- The distribution of molecular speeds is described by the Maxwell–Boltzmann distribution. The maximum value in the curve represents the most probable value of the speed.
- The rate of effusion of a gas is proportional to $1/M^{1/2}$ (Graham's law).
- Characteristic properties of molecules in a gas include:
  - the *collision cross-section*, σ, the target area presented for another molecule to collide with;
  - the *collision frequency*, Z, the mean number of collisions per second;
  - the *mean free path*, λ, the mean distance travelled between collisions.
- Gas molecules at atmospheric pressure undergo collisions at a rate of about 1×10^{10} s^{-1} and have mean free paths many times larger than the size of the molecules.

➜

- There are two different ways of representing the average speed of a molecule. (The numerical values are quite similar.)
  - Root mean square (rms) speed, c. The expressions for the mean kinetic energy ($\bar{E}_{KE} = \frac{1}{2}mc^2$) and for pV ($pV = \frac{1}{3}nMc^2$) depend on the mean square of the molecular speeds

$$c = \left(\frac{s_1^2 + s_2^2 + \cdots s_N^2}{N}\right)^{1/2} = \text{root mean square speed}$$

  - Mean speed, $\bar{c}$. Effusion, collision frequency, and mean free path depend on the mean speed.

$$\bar{c} = \frac{s_1 + s_2 + \cdots + s_N}{N}$$

For practice questions on these topics, see questions 11, 12, 15–19, 24, 25, 28–30 at the end of this chapter (p.378).

8.6 Real gases

How well does the ideal gas equation work?

For many gases (such as N_2, O_2, and He) at around atmospheric pressure, the ideal gas equation describes their behaviour very well. However, this is not the case under all conditions so the equation needs to be tested more rigorously by comparison with experimental data. The behaviour of other gases (such as CO_2 and butane) deviates slightly from the ideal model behaviour even at atmospheric pressure. So, how might the ideal gas equation be tested?

An ideal gas obeys Boyle's law (Equation 8.1, p.345) under all conditions. If the volume of a fixed amount of gas is measured as a function of pressure at constant temperature, a plot of p versus $(1/V)$ will give a straight line if the gas obeys Boyle's law. Figure 8.18 shows that there is very good agreement between Boyle's law and experimental data at low pressures but deviations from ideal behaviour occur as the pressure increases.

An alternative approach is to plot p versus V. Plots of pressure versus volume at fixed temperatures are called ***p*–*V* isotherms**. Examples for three temperatures are shown in Figure 8.19, where the solid lines are predicted from the ideal gas equation and the points are experimental measurements. There is again quite a good fit of the experimental results to ideal behaviour, particularly at higher temperatures and lower pressures. However, at low temperatures or high pressures, there are marked deviations from ideal behaviour.

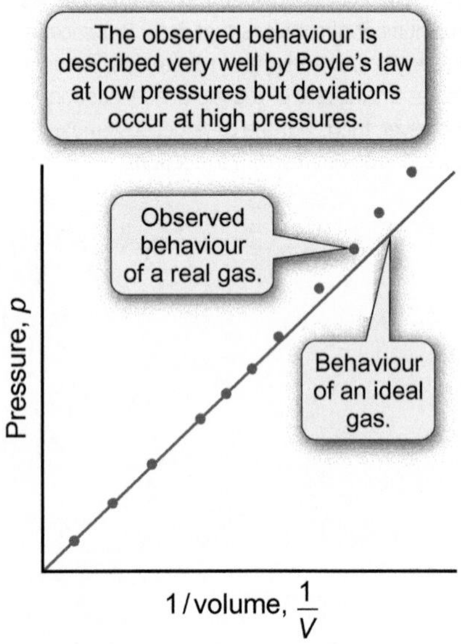
The observed behaviour is described very well by Boyle's law at low pressures but deviations occur at high pressures.
Observed behaviour of a real gas.
Behaviour of an ideal gas.
Pressure, p
1/volume, 1/V

Figure 8.18 Comparison of the behaviour of a real gas (blue points) with that of an ideal gas (blue line).

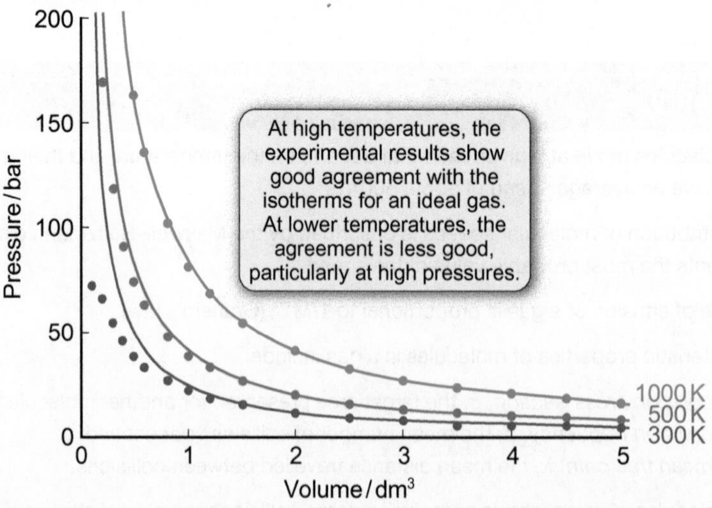
At high temperatures, the experimental results show good agreement with the isotherms for an ideal gas. At lower temperatures, the agreement is less good, particularly at high pressures.
Pressure/bar
200
150
100
50
0
0
1
2
3
4
5
Volume/dm³
1000 K
500 K
300 K

Figure 8.19 *p*–*V* isotherms for gases. The points show experimental measurements; the solid lines show the behaviour of an ideal gas.

You can read more about plotting functions in terms of straight line graphs in Maths Toolkit MT4 (p.1318).

Isotherms are lines connecting measurements at the same temperatures. A related concept is that of isobars—lines of constant pressure—which are often indicated on weather maps.

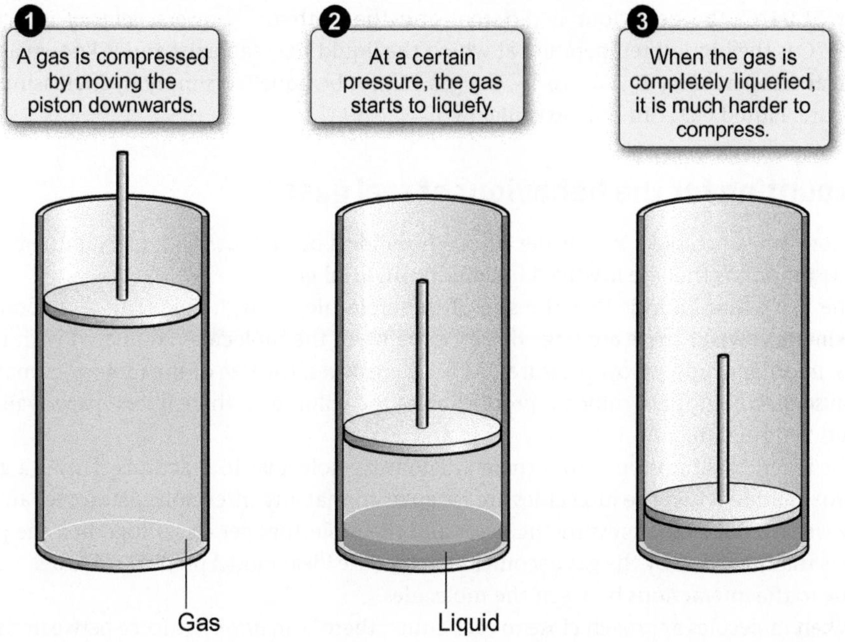
1
A gas is compressed by moving the piston downwards.
2
At a certain pressure, the gas starts to liquefy.
3
When the gas is completely liquefied, it is much harder to compress.
Gas
Liquid

Figure 8.20 As pressure is applied to a gas, the attractive intermolecular forces become important and the gas condenses to a liquid.

Panna cotta and liquid nitrogen.

The plots in Figures 8.18 and 8.19 suggest that the pressure of a sample of gas can be increased indefinitely with little change in gas behaviour. However, this does not happen in reality because, if the pressure is increased enough, especially at low temperatures, the gas will liquefy, as shown in Figure 8.20.

So, how does liquefaction affect the appearance of isotherms for a real gas? This is illustrated in Figure 8.21. At low pressures, the isotherm looks like that for an ideal gas. However, as the pressure increases, a value is reached where the gas starts to liquefy and a marked decrease in volume occurs with no change in gas pressure. This shows up as a flat 'step' in the graph. When all the gas has turned to liquid, the curve is *much* steeper since it takes very large pressure changes to significantly change the volume of a liquid—it is much less *compressible*.

Some experimentally measured isotherms for carbon dioxide are shown in Figure 8.22 and demonstrate the behaviour of a real gas at different temperatures. At 50 °C, the isotherm looks like that for an ideal gas. At this temperature, the compound exists as a gas at any pressure. At the lower temperature of 20 °C, the flattened portion of the isotherm shows that liquefaction occurs when the pressure reaches about 60 atm.

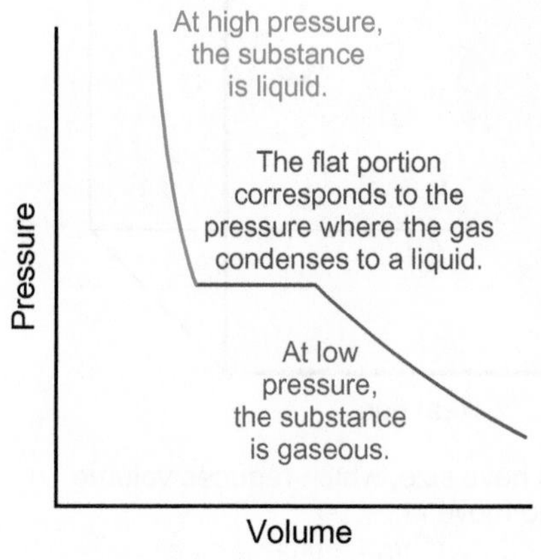
At high pressure, the substance is liquid.
The flat portion corresponds to the pressure where the gas condenses to a liquid.
At low pressure, the substance is gaseous.
Pressure
Volume

Figure 8.21 A *p*–*V* isotherm showing condensation of a real gas.

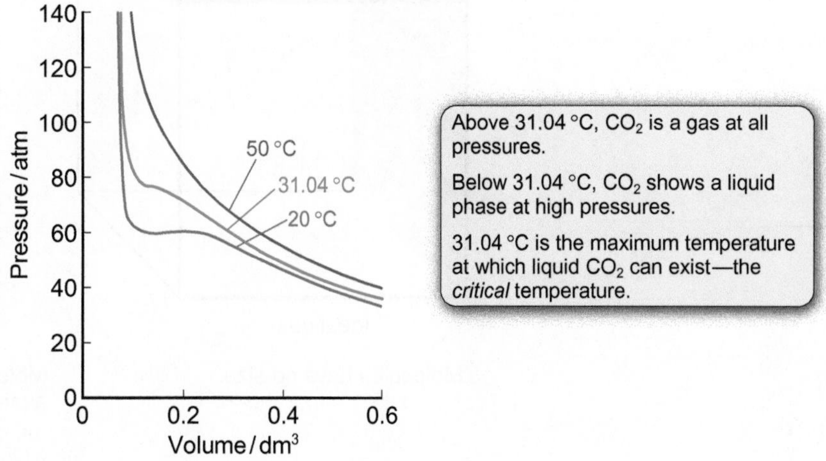
140
120
100
80
60
40
20
0
Pressure / atm
0
0.2
0.4
0.6
Volume / dm³
50 °C
31.04 °C
20 °C
Above 31.04 °C, CO₂ is a gas at all pressures.
Below 31.04 °C, CO₂ shows a liquid phase at high pressures.
31.04 °C is the maximum temperature at which liquid CO₂ can exist—the critical temperature.

Figure 8.22 Experimental *p*–*V* isotherms for 1 mol of carbon dioxide.

At 31.04 °C, the behaviour is different and the isotherm shows a *point of inflection*. 31.04 °C is the highest temperature at which the liquid CO_2 can exist and is known as the **critical temperature**, T_C. Above T_C, the gas cannot be liquefied simply by increasing the pressure. Liquid CO_2 can only exist below T_C.

A gas is sometimes defined as a substance at a higher temperature than its T_C, whereas a vapour is a gaseous form of matter at a temperature below its T_C. However, this does not apply in the case of CO_2 which is called a gas, not a vapour. The terms 'gas' and 'vapour' are often used interchangeably.

There is further discussion of critical temperatures in Section 17.1 (p.766).

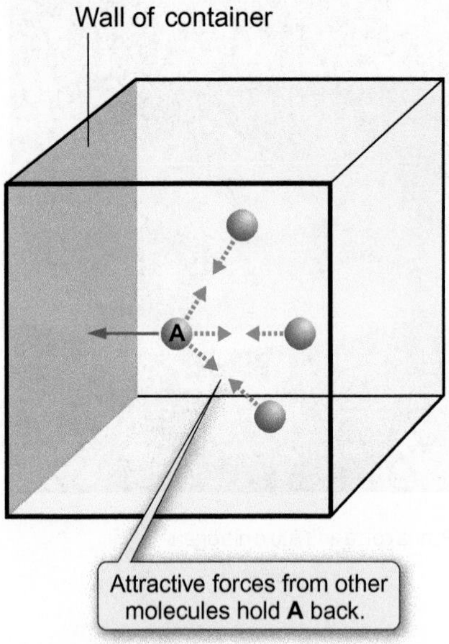
Wall of container
A
Attractive forces from other molecules hold **A** back.

Figure 8.23 The speed with which molecule **A** hits the wall is reduced by the attraction of the other molecules. This means that the gas exerts a lower pressure than that predicted for an ideal gas.

Accounting for the behaviour of real gases

So, how can we account for the deviations from ideal behaviour? Let's re-examine two of the assumptions that are involved in defining an ideal gas.

The first assumption is that the size of the molecules is negligible. This is a good approximation when there are large distances between the molecules compared with their sizes, in other words, at low pressures. At high pressures, the size of the molecules matters because they occupy a significant proportion of the volume, so there is less space available to move around in.

The second assumption is that there are no intermolecular interactions. This is a good approximation when the molecules are far apart so that any intermolecular forces are felt very weakly. But as the pressure increases and the molecules get closer together, the pressure actually exerted by the gas becomes less than the ideal model predicts. This behaviour is due to the interactions between the molecules.

When molecules approach close to each other, there is an *attractive* force between them, even for non-polar molecules, such as Ar or N_2. These attractive forces are responsible for holding the molecules together in a liquid. In the gas phase, the attractive forces hold back the molecules a little, so they collide with the walls with slightly less force (Figure 8.23). The assumption of no intermolecular interactions is most realistic under conditions of low pressures and high temperatures. At low pressures, the molecules are relatively far apart so that intermolecular attractions are relatively insignificant. At high temperatures, the kinetic energy of the molecules is high enough to make the energy of attraction insignificant.

The reasons for deviation from ideal behaviour are summarized in Figure 8.24. Molecules in which the intermolecular attractions are stronger show larger deviations from ideal behaviour under a given set of conditions.

The ideal gas equation is best viewed as an approximation. It describes gas behaviour under some conditions but it is not a perfect description of experimental behaviour. However, it is very useful because it closely describes experimental observations for many gases under most commonly encountered situations. It is only under conditions where the

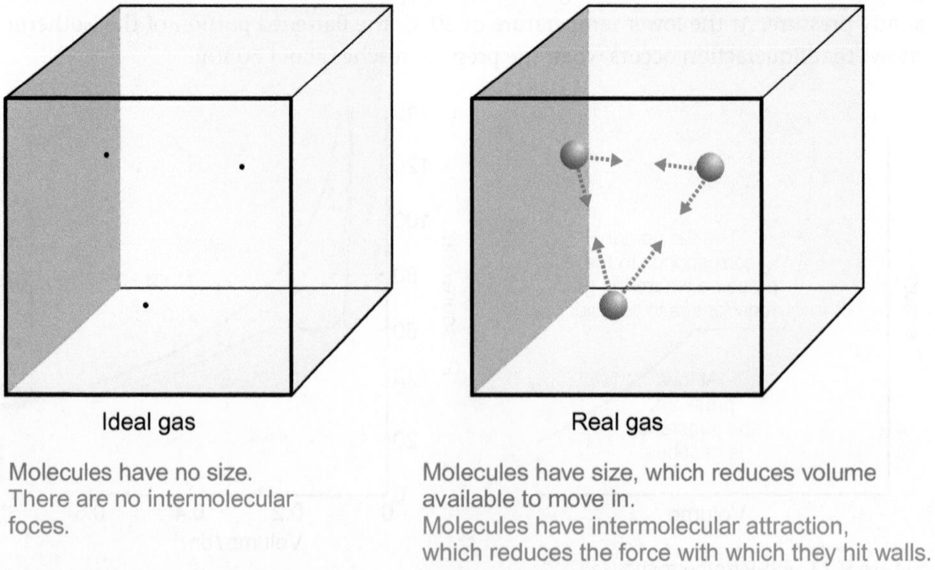
Ideal gas
Real gas
Molecules have no size.
There are no intermolecular foces.
Molecules have size, which reduces volume available to move in.
Molecules have intermolecular attraction, which reduces the force with which they hit walls.

Figure 8.24 A summary of the differences between ideal and real gases.

molecules are relatively close together, or moving so slowly that their size and interactions become significant, that marked deviations occur.

Another way of expressing the ideal gas equation is as a **limiting law**

$$\lim_{p\to 0}(pV) = nRT \tag{8.24}$$

This indicates that the value in brackets equals the value on the right-hand side of the equation in the limit as $p \to 0$. In other words, the closer the value of p is to zero, the better the equation is obeyed by real systems.

The van der Waals equation for gases

The first successful attempt to account for the behaviour of real gases was introduced by the Dutch physicist Johannes van der Waals in 1873. He started with the ideal gas equation and introduced some additional terms to account for the behaviour of real gases.

Intermolecular attractions are introduced in Section 1.8 (p.52). They are discussed further in Section 17.3 (p.783).

The first correction accounts for the attractions between the molecules. These mean that, in a real gas, the pressure is somewhat less than in the ideal case. All the surrounding molecules attract the one about to strike the wall, slowing it down so that when it hits the wall it exerts a slightly smaller force (see Figure 8.23). Thus, van der Waals suggested that a correction factor should be added to the actual value of p to make up for the slightly reduced pressure, and so obtain the ideal value

Another 'ideal' model is used in Section 17.4 (p.789) which describes ideal solutions. Again, this is an idealized description. It explains the behaviour of liquid mixtures under some circumstances.

$$p + a\left(\frac{n}{V}\right)^2$$

where a is a constant related to the strength of the attraction between the molecules in the particular gas. Some values of a are shown in Table 8.4. The $a(n/V)^2$ term represents the difference between the measured pressure, p, and the ideal value. So, adding the $a(n/V)^2$ factor 'tops up' the real (measured) pressure so it equals the ideal value.

The second correction concerns the finite size of the molecules. Because the molecules have a small volume of their own, the volume in which molecules can move is less than the total volume of the container. So, van der Waals subtracted a correction factor from the actual value of V to allow for this reduction. This takes the form $(V - nb)$, where the value of b is a constant related to the volume of the molecules. Some values of b are shown in Table 8.4.

Table 8.4 Some van der Waals constants for gases

	SI units		Non-SI units	
	a^* / Pa m^6 mol^{-2}	b / 10^{-5} m^3 mol^{-1}	a / atm dm^6 mol^{-2}	b / dm^3 mol^{-1}
Nitrogen	0.137	3.9	1.35	0.039
Oxygen	0.138	3.2	1.36	0.032
Carbon dioxide	0.364	4.3	3.59	0.043
Carbon monoxide	0.147	3.9	1.45	0.039
Helium	0.003	2.4	0.034	0.024
Hydrogen	0.025	2.7	0.244	0.027
Argon	0.137	3.2	1.35	0.032
Water	0.228	4.3	2.25	0.043

* Since 1 Pa = 1 N m^{-2} and 1 J = 1 N m, it follows that 1 Pa = 1 J m^{-3}. Alternative SI units for a are, therefore, J m^3 mol^{-2}.

When these two corrections are introduced into the ideal gas equation the result is Equation 8.25. Equation 8.25 is the van der Waals equation of state.

Visit the Online Resource Centre to view screencast 8.6 which walks you through the van der Waals equation (Equation 8.25).

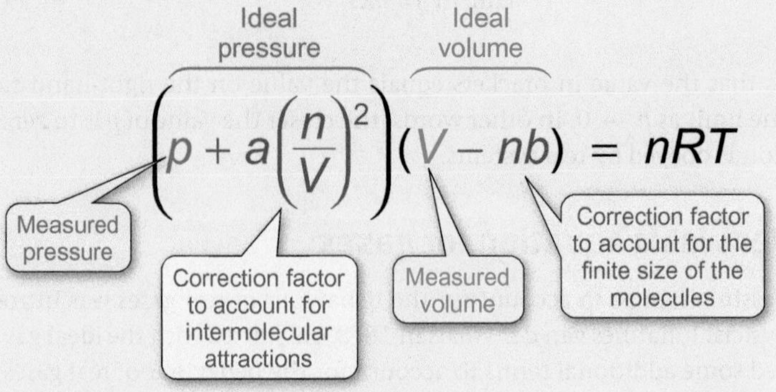
Ideal pressure
Ideal volume
Measured pressure
Correction factor to account for intermolecular attractions
Measured volume
Correction factor to account for the finite size of the molecules

$$\left(p + a\left(\frac{n}{V}\right)^2\right)(V - nb) = nRT \qquad (8.25)$$

Equation 8.25 is the van der Waals equation of state.

Although the origin of the *a* and *b* factors can be explained, their values cannot be calculated theoretically. The values shown in Table 8.4 have been determined empirically by fitting experimental measurements to the equation. The values of *a* and *b* in the table are given both in SI units and in some commonly used non-SI units.

You can see how to use the van der Waals equation in Worked example 8.11.

The van der Waals equation gives a better description of gas behaviour than the ideal gas equation, but it is still not exact. There are several other equations of state for real gases and liquids, all of which rely on giving a better account of the approximations involved in the ideal model. This type of work is particularly important for chemical engineers who have to design and operate chemical plants to run chemical reactions and processes in the gas phase on a very large scale. An accurate knowledge of gas behaviour is needed to operate the plant efficiently and safely.

Worked example 8.11 Using the van der Waals equation

A laboratory gas cylinder filled with nitrogen has a volume of 0.150 m^3 and contains 35.0 kg of gas when delivered. Calculate the pressure in bar inside the cylinder at 293 K (20 °C): (a) using the ideal gas equation; and (b) using the van der Waals equation. (The van der Waals constants for nitrogen are $a = 0.137$ Pa m^6 mol^{-2} and $b = 3.9 \times 10^{-5}$ m^3 mol^{-1}. $R = 8.314$ J K^{-1} mol^{-1}.)

Strategy

First, calculate the number of moles in 35.0 kg of nitrogen. Then rearrange the ideal gas equation (Equation 8.5) and the van der Waals equation (Equation 8.25) to give expressions for pressure, *p*, and substitute in the appropriate data.

Solution

Find the amount in moles of N_2 in the cylinder.

$$M_r(N_2) = 28.0$$

$$35.0\,\text{kg of N}_2 \text{ contain } \frac{35.0 \times 10^3\,\text{g}}{28\,\text{g mol}^{-1}} = 1250\,\text{mol}$$

→

→ (a) *Rearrange the ideal gas equation, Equation 8.5, to give the pressure, p.*

$$p = \frac{nRT}{V} = \frac{1250\,\cancel{\text{mol}} \times 8.314\,\text{J}\,\cancel{\text{K}^{-1}}\,\cancel{\text{mol}^{-1}} \times 293\,\cancel{\text{K}}}{0.150\,\text{m}^3}$$

$$p = 2.031 \times 10^7\ \text{J m}^{-3}$$

$$= 2.031 \times 10^7\ \text{Pa} \quad (1\,\text{Pa} = 1\,\text{J m}^{-3})$$

$$= 203\,\text{bar} \quad (1\,\text{bar} = 1 \times 10^5\,\text{Pa})$$

(b) *Rearrange the van der Waals equation, Equation 8.25, to give the pressure, p.*

$$\left(p + a\left(\frac{n}{V}\right)^2\right)(V - nb) = nRT \qquad (8.25)$$

$$p + a\left(\frac{n}{V}\right)^2 = \frac{nRT}{(V - nb)}$$

$$p = \frac{nRT}{(V - nb)} - a\left(\frac{n}{V}\right)^2$$

For nitrogen, $a = 0.137\ \text{Pa m}^6\,\text{mol}^{-2} = 0.137\ \text{J m}^3\,\text{mol}^{-2}$ and $b = 3.9 \times 10^{-5}\ \text{m}^3\,\text{mol}^{-1}$

$$p = \frac{1250\,\cancel{\text{mol}} \times 8.314\,\text{J}\,\cancel{\text{mol}^{-1}}\,\cancel{\text{K}^{-1}} \times 293\,\cancel{\text{K}}}{0.150\,\text{m}^3 - [1250\,\cancel{\text{mol}} \times (3.9 \times 10^{-5}\,\text{m}^3\,\cancel{\text{mol}^{-1}})]} - (0.317\,\text{J m}^3\,\cancel{\text{mol}^{-2}}) \times \left(\frac{1250\,\cancel{\text{mol}}}{0.150\,\text{m}^3}\right)^2$$

$$= \frac{3.045 \times 10^6\ \text{J}}{0.1013\,\text{m}^3} - 9.51 \times 10^6\ \text{J m}^{-3}$$

$$= 2.055 \times 10^7\ \text{Pa}$$

$$= 206\,\text{bar}$$

Question

For the nitrogen gas in this example, work out the percentage reduction in V due to the term nb and compare it to the percentage increase in p due to the $a(n/V)^2$ term. Which correction factor makes the larger change?

Summary

- For many gases, the ideal gas equation describes experimental behaviour well at low pressures and high temperatures but deviations occur at high pressures and low temperatures.
- A more realistic description of gas behaviour is given by the van der Waals equation.
- The van der Waals equation is derived by introducing into the ideal gas equation terms to account for the intermolecular attractions, $a(n/V)^2$, and the finite molecular size, nb

$$\left(p + a\left(\frac{n}{V}\right)^2\right)(V - nb) = nRT$$

? For practice questions on this topic, see questions 21, 23, 27, 31–33 at the end of this chapter (p.378).

Concept review

By the end of this chapter, you should be able to do the following.

- Describe the experimental evidence for the relation between n, p, V, and T, in the ideal gas equation.
- Use the ideal gas equation to solve problems relating to changing the conditions of gases.
- Apply Dalton's law of partial pressures to mixtures of gases.
- Describe the assumptions underlying the kinetic theory of gases.
- Calculate the average (root mean square) speed of molecules in a gas and account for its dependence on the temperature and the molar mass of the gas.
- Explain the difference between the root mean square speed, c, and the mean speed, $\bar{c}$.
- Describe the Maxwell–Boltzmann distribution of molecular speeds and its dependence on the temperature and the molar mass of the gas.
- Explain what is meant by effusion and apply Graham's law.
- Calculate the frequency of collisions and mean free path between molecules in a gas.
- Account for deviations of real gases from ideal behaviour.
- Use the van der Waals equation to describe the behaviour of real gases.
- Understand the role of models in explaining the behaviour of chemical systems and the difference between an empirical approach and a theoretical approach.

Key equations

The ideal gas law	$pV = nRT$	(8.5)
Partial pressures and Dalton's law	$p_{\text{total}} = p_{\text{A}} + p_{\text{B}} + p_{\text{C}} +$	(8.8)
	$p_{\text{A}} = x_{\text{A}} p_{\text{total}}$	(8.10)
	$x_A = \dfrac{n_{\text{A}}}{n_{\text{total}}}$	(8.9)
p–V relationship derived from the kinetic theory	$pV = \frac{1}{3} nN_{\text{A}} mc^2$	(8.13b)
Root mean square speed of gas molecules	$c = \left(\dfrac{3RT}{M}\right)^{1/2}$	(8.19)
Graham's law of effusion	Rate of effusion $\propto \dfrac{1}{M^{1/2}}$	(8.20)
Mean free path	$\lambda = \dfrac{RT}{\sqrt{2} N_{\text{A}} \sigma p}$	(8.23)
van der Waals equation	$\left(p + a\left(\dfrac{n}{2}\right)^2\right)(V - nb) = nRT$	(8.25)

Questions

More challenging questions are indicated by an asterisk *.

Note: For some questions, data will be needed from tables within the chapter and from the Periodic Table on the inside front cover.

1 3.036 g of a gas occupy a volume of 426 cm^3 at 273 K and 1.00 atm pressure. Calculate the molar mass of the gas. (Section 8.2)

2 At 27 °C and 1.0 atm pressure, the density of a gaseous hydrocarbon is 1.22 g dm^{-3}. What is the hydrocarbon? (Section 8.2)

3 A sample of gas has a volume of 346 cm^3 at 25 °C when the pressure is 1.00 atm. What volume will it occupy if the conditions are changed to 35 °C and 1.25 atm? (Section 8.2)

4 The air that we breathe is about 21% oxygen, by volume. Exhaled air contains about 14% oxygen. The absorption of oxygen within the lungs takes place in tiny spherical compartments called alveoli which have a diameter of the order of 0.1 mm. Estimate the number of oxygen molecules absorbed in one breath in each of the alveoli. (Assume that body temperature is 37 °C.) (Section 8.2)

5 Incandescent light bulbs, which have been phased out in favour of less energy-consuming lighting, are filled with an inert gas to prevent the filament from burning. Find the mass of argon needed to fill a 75.0 cm^3 light bulb to a pressure of 1.05 atm at 25.0 °C. (Section 8.2)

6 A vessel of volume 50.0 dm^3 contains 2.50 mol of argon and 1.20 mol of nitrogen at 273.15 K.

(i) Calculate the partial pressure in bar of each gas.

(ii) Calculate the total pressure in bar.

(iii) How many additional moles of nitrogen must be pumped into the vessel in order to raise the pressure to 5 bar? (Sections 8.2 and 8.3)

7* A mixture of methane and ethane is in a 500 cm^3 container at 298.15 K. The pressure is 1.25 bar and the mass of gas is 0.530 g. Find the mass of methane in the mixture. (Sections 8.2 and 8.3)

8* Two bulbs A and B, with volumes $V_A = 1.00\,dm^3$ and $V_B = 5.00\,dm^3$, are connected via a tap. The volume of the connecting tubing is negligible. Bulb A contains gas at a pressure of 6.00 bar while bulb B contains a vacuum.

(a) The temperature of the whole apparatus is maintained at 298 K. If the tap is opened, calculate the pressure of gas in the system after opening the tap.

(b) The tap is closed and bulb B is then immersed in an oil bath at a temperature of 423 K while the temperature of bulb A is maintained at 298 K. Calculate the resulting pressures in each bulb.

(c) The tap is opened again. What is the final pressure and the number of moles of gas in each bulb? (Section 8.2)

9* Divers' 'bends' are caused by the formation of bubbles of nitrogen in blood as the solubility reduces when the diver returns to the surface. The solubility of nitrogen in water at 1.00 atm pressure is 13.0 $mg\,kg^{-1}$ at body temperature of 37 °C and increases linearly with pressure. In water, the pressure increases at the rate of 1.00 atm per 10 m depth.

Estimate the volume of gas that comes out of solution when a diver who has 4.5 kg of blood rapidly ascends from a depth of 50 m of water to the surface. Assume the solubility of nitrogen in blood is the same as in water. (Section 8.2)

10 A mixture of nitrogen and carbon dioxide contains 38.4% N_2 by mass. What is the mole fraction of nitrogen in the mixture? If the total pressure is 1.2 atm, what is the partial pressure of each gas in Pa? (Section 8.3)

11 How much faster is the rate of effusion of helium than that of carbon dioxide, when both gases are at the same temperature? (Section 8.5)

12 Two identical flasks contain nitrogen gas at the same pressure. Each has an identical pin-hole. One flask is kept at 25 °C while the other is heated to 125 °C. Calculate the relative rates of effusion of nitrogen from the two flasks. (Section 8.5)

13 An evacuated flask is filled with dry air and weighed. The same flask is filled at the same temperature to the same pressure with moist air on a humid day. Will it weigh more, less, or the same? Explain your answer. (Section 8.2)

14 The equation for the complete combustion of methane is

$$CH_4(g) + 2O_2(g) \rightarrow CO_2(g) + 2H_2O(g)$$

What volume of oxygen at SATP is needed to react exactly with 10 g of methane? (Section 8.2)

15 The average (root mean square) speed of an oxygen molecule is 425 $m\,s^{-1}$ at 0 °C. Calculate the average speed at 100 °C. (Section 8.5)

16 A balloon is filled with nitrogen and placed in a box containing helium. What will happen, assuming the balloon is porous to both gases? (Section 8.5)

17 The atmospheric pressure and temperature both fall as the altitude increases. The total pressure at a height of 20 km is about 5% of that at the Earth's surface (i.e. at sea level) and the temperature is about –55 °C. Estimate the collision frequency and the mean free path of oxygen molecules under these conditions. (For oxygen at this temperature, the mean speed, $\bar{c} = 380\,m\,s^{-1}$.) (Section 8.5)

18 Argon has a collision cross-section of 0.36 nm^2. At 0 °C, what pressure is needed so that its mean free path becomes equal to the diameter of the molecules? (Section 8.5)

19 Suggest why there is little hydrogen or helium in the atmosphere around the Earth or Mars although these gases form the major constituent of the atmosphere of huge planets such as Jupiter or Saturn. (Section 8.5)

20 The atmosphere of a spacecraft with volume 27 m^3 consists of 80% helium and 20% oxygen by volume. The gases continually escape by effusion through small leaks in the walls. The leak amounts to 1000 Pa per day. The temperature inside the spacecraft is 20 °C. What masses of helium and oxygen must be carried to replace the gas that leaks during a 10-day mission? (Section 8.5)

21 (a) State (i) the ideal gas equation (ii) the van der Waals equation of state. Explain how the additional terms in the van der Waals equation account for the actual behaviour of real gases.

(b) Without performing any numerical calculations, show that, in the limit of high temperatures and low pressures, the van der Waals and ideal gas equations are identical. (Section 8.6)

22 Dry air with the following composition is used to fill a SCUBA cylinder for a dive. (Section 8.6)

Gas	Composition of dry air by volume
N_2	78%
O_2	21%
Ar	1%

(a) At 10 m depth, a diver experiences an external pressure of 2 atm. Write an expression for the total pressure of the air in terms of the partial pressures of N_2, O_2, and Ar.

(b) What is the molar percentage of oxygen in the air inhaled at 2 atm?

(c) What is the partial pressure of O_2 in air inhaled at: (i) 1 atm; (ii) 2 atm?

(d) How does the number of molecules of O_2 inhaled per breath at 10 m depth compare with the number inhaled per breath at sea level? Suggest why some deep-sea divers dive with a gas mixture containing 10% oxygen.

23 A 10 dm^3 SCUBA cylinder is filled with air to a pressure of 300 atm at a temperature of 20 °C (293 K). (Section 8.6)

(a) Calculate the amount in moles of gas in the cylinder, assuming the air behaves as an ideal gas.

(b) When the diver jumps into cold water at 278 K, the pressure gauge shows an alarming drop in pressure. Explain the reason why and calculate the new pressure inside the cylinder.

(c) In fact, the compressed gases do not behave as ideal gases. Explain why. Use the van der Waals equation (for air, $a = 0.137\ \mathrm{Pa\,m^6\,mol^{-2}}$ and $b = 3.7 \times 10^{-5}\ \mathrm{m^3\,mol^{-1}}$) to show that the amount of air in the cylinder is 115 mol. In view of your answer to part (a) above, what are the implications of this for divers?

24* The space between stars in the universe contains a mixture of gases at very low pressure as well as some dust. The gas is mainly hydrogen with a little helium, but many other molecules have also been detected in tiny amounts (see Chapter 4, p.168). For a helium atom, the collision cross-section $\sigma = 0.21\ \mathrm{nm^2}$. The conditions in interstellar space are around $T = 10$ K and $p = 1 \times 10^{-18}$ atm. (Section 8.5)

(a) Calculate the mean free path, λ, of helium atoms at SATP.

(b) Estimate the value of λ for a helium atom in interstellar space. (*Hint*. For a given gas, $\lambda \propto \frac{T}{p}$.)

(c) Estimate the time between collisions for helium atoms in interstellar space. (For the atoms at 10 K, $\bar{c} = 230\ \mathrm{m\,s^{-1}}$)

25 Sketch graphs to show how the distribution of molecular speeds differs between (Section 8.5):

(a) helium at 100 K and helium at 300 K;

(b) helium at 100 K and xenon at 100 K.

26 Using the kinetic theory of gases, it can be shown that (Section 8.4)

$$pV = \tfrac{1}{3}nN_A mc^2$$

(a) Explain each of the terms in this equation.

(b) Show how this equation is equivalent to the ideal gas equation $pV = nRT$.

(c) Using these equations as examples, explain the difference between an empirical approach and a theoretical approach to modelling a chemical system.

27 Consider the following gases at SATP (Section 8.6):

argon, krypton, nitrogen, methane, hydrogen chloride, chlorine, carbon dioxide, helium.

(a) Which gas would be expected to most closely follow ideal behaviour?

(b) Which gas would deviate most from ideal behaviour?

(c) Which gas would have the highest and lowest root mean square speeds?

(d) Which gas would effuse most slowly?

28 The molecular speeds in a sample of 100 molecules are distributed as follows (Section 8.5):

Number of molecules	10	20	40	15	10	5
Speed/$\mathrm{m\,s^{-1}}$	60	80	100	120	140	160

(a) What is the most probable speed?

(b) Calculate the mean speed of the molecules in the sample.

(c) Calculate the rms speed of the molecules in the sample.

29 *The density of nitrogen gas in a container at 300 K and 1.0 bar pressure is 1.25 g dm^{-3} (Section 8.5).

(a) Calculate the rms speed of the molecules.

(b) At what the temperature will the rms speed be twice as fast?

30* A sample of gaseous uranium hexafluoride, UF_6, is held at a temperature of 300 K and a pressure of 0.1 mbar. The collision diameter of UF_6 is 0.40 nm. (Section 8.5)

(a) What is the rms speed of the molecules?

(b) Estimate the collision frequency and mean free path under these conditions.

31 (a) Explain how the fact that gases such as nitrogen or carbon dioxide can be liquefied by applying high pressures shows that the ideal gas equation can only be an approximation.

(b) Why is a lower pressure needed to liquefy CO_2 than for N_2? (Section 8.6)

32 Predict which of the following substances will have (a) the highest and (b) the lowest values of the van der Waals constants *a* and *b*.

N_2, He, NO_2 (Section 8.6)

33 Calculate the temperature at which 20.0 mol of helium would exert a pressure of 120 atm in a 10.0 dm^3 cylinder, using (a) the ideal gas equation and (b) the van der Waals equation. For He, $a = 0.034\ \mathrm{dm^6\,atm\,mol^{-2}}$ and $b = 0.024\ \mathrm{dm^3\,mol^{-1}}$. (Section 8.6)

9

Reaction kinetics

▼ Methane concentrations in the troposphere showing their variation with time and latitude. (The colours show the concentration ranges indicated on the vertical axis.) The concentration is higher in the Northern Hemisphere than it is in the Southern Hemisphere, because more methane is emitted in the Northern Hemisphere and the transfer of air across the Equator is slow. Note that the peaks in the Northern Hemisphere coincide with the troughs in the Southern Hemisphere. © NOAA

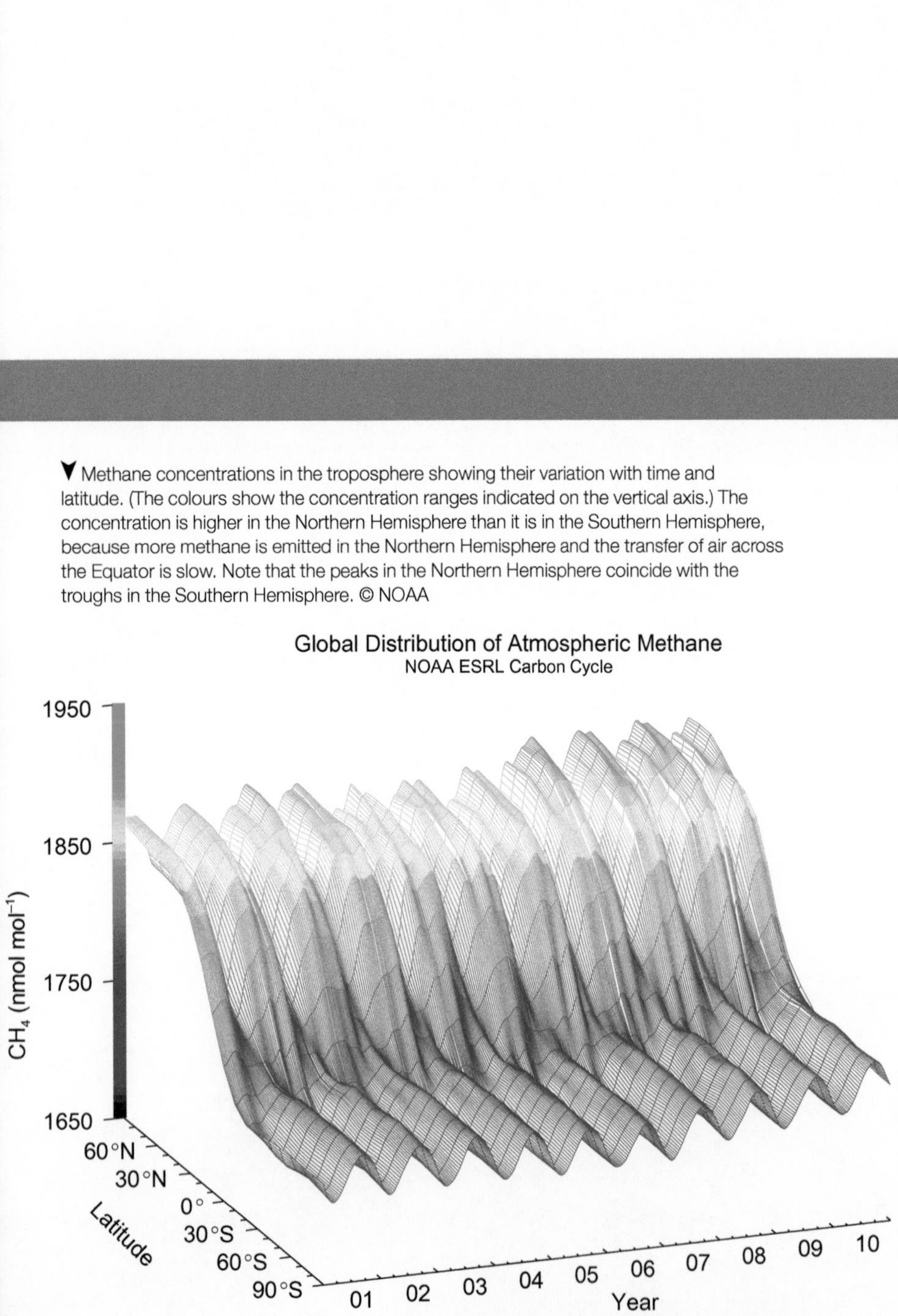
Global Distribution of Atmospheric Methane
NOAA ESRL Carbon Cycle
CH_4 (nmol mol^{-1})
1950
1850
1750
1650
Latitude
60°N
30°N
0°
30°S
60°S
90°S
Year
01
02
03
04
05
06
07
08
09
10

Three-dimensional representation of the latitudinal distribution of atmospheric methane in the marine boundary layer. Data from the Carton Cycle cooperative air sampling network were used. The surface represents data smoothed in time and latitude. Contact: Dr. Ed Dlugokencky, NOAA ESRL Carbon Cycle, Boulder, Colorado, (303) 497–6228, ed.dlugokencky@noaa.gov, http://wwe.esrl.noaa.gov/gmd/ccgg/.

This chapter builds on the following topics:

- Measurement, units, and nomenclature Section 1.2, p.6
- Amount of substance and molar mass Section 1.3, p.12
- Balancing equations Section 1.4, p.21
- Concentrations of solutions Section 1.5, p.34
- Enthalpy profiles for exothermic and endothermic processes Section 1.6, p.41
- Chemical equilibrium: how far has a reaction gone? Section 1.9, p.56
- Kinetic molecular theory and the gas laws Section 8.4, p.358
- The distribution of molecular speeds Section 8.5, p.362

Increased concentrations of methane in the troposphere are thought to be due to melting permafrost in regions of Arctic tundra. ➤

Methane in the troposphere

Methane absorbs infrared radiation and is an important greenhouse gas. To investigate the link between increased concentrations of greenhouse gases and climate change, scientists need information about the processes that release methane into the atmosphere and about what happens to the methane once there. In particular, they need to know *how quickly* the methane is removed and what is its *average lifetime* in the troposphere (the lower atmosphere). (Box 9.3 on p.400 discusses the significance of the average lifetime of emitted methane.)

Methane is emitted from a variety of natural and human-related sources. It is produced when bacteria living in airless (anaerobic) places break down carbohydrates. The biggest emissions come from wetlands, ruminant animals (such as cattle), and biomass burning. The average concentration of methane in the troposphere in 2010 was 1.79 ppm (by volume). This is a substantial increase from its pre-industrial (1750–1800) concentration of around 0.7 ppm.

The main way methane is removed from the troposphere is by reaction with hydroxyl radicals, $^{\bullet}OH$. The first step of this process is

$$CH_4 + {}^{\bullet}OH \rightarrow H_2O + CH_3{}^{\bullet}$$

in which the $^{\bullet}OH$ radical removes a hydrogen atom from methane to form a molecule of water and a $CH_3{}^{\bullet}$ radical. (You can read about the formation of radicals in Section 19.1, p.863.) The rate of the reaction of methane with $^{\bullet}OH$ is crucial in determining the lifetime of methane molecules in the troposphere. The reaction is a simple elementary reaction (Section 9.4) and the rate of reaction is governed by the following rate equation

$$\text{rate of reaction} = k[CH_4][{}^{\bullet}OH]$$

where k is the rate constant. The value of the rate constant, k, is related to how fast the reaction proceeds and is obtained from laboratory experiments.

In the troposphere, the concentration of methane is approximately constant (at least on a day-to-day basis—it is increasing but only slowly), which means methane must be removed at roughly the same rate as it is being emitted. So,

$$\begin{aligned}\text{the average rate of emission} &\\ &= \text{the average rate of removal}\\ &= k[CH_4][{}^{\bullet}OH]\end{aligned}$$

where $[CH_4]$ and $[{}^{\bullet}OH]$ are the average concentrations of CH_4 and $^{\bullet}OH$, respectively.

Rearranging this equation, the average concentration of methane in the troposphere is given by

$$[CH_4] = \frac{\text{average rate of emission}}{k[{}^{\bullet}OH]}$$

so that the concentration of methane depends inversely on the concentration of $^{\bullet}OH$ radicals.

The three-dimensional figure opposite shows methane concentrations in the troposphere over a 10 year period. There is a seasonal variation because the reactions that generate $^{\bullet}OH$ radicals from water require sunlight, so concentrations of $^{\bullet}OH$ are higher in the summer months. Methane is, therefore, removed more rapidly in the summer and its concentration is lower.

You can see that methane concentrations have been rising sharply since 2007, after a decade or so of holding fairly steady. There is concern that this may be due to the melting of permafrost in regions of Arctic tundra, which releases previously trapped methane into the atmosphere.

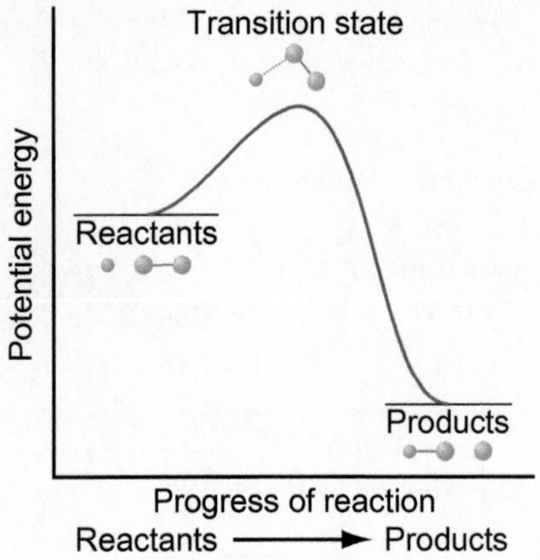
Transition state
Potential energy
Reactants
Products
Progress of reaction
Reactants ⟶ Products

Figure 9.1 A general reaction profile showing the progress of an elementary reaction from reactants to products, via a transient, high-energy transition state.

Enthalpy profiles for exothermic and endothermic reactions are shown in Section 1.6 (p.41). Energy and enthalpy profiles are often called **reaction profiles**.

Paul Crutzen, Mario Molina, and F. Sherwood Rowland were awarded the Nobel Prize in Chemistry in 1995 for their work concerning the formation and removal of ozone in the stratosphere. You can read about some of the reactions involved in the ozone layer and the role of CFCs in ozone depletion in Box 27.6 (p.1233).

Helium-filled balloons are sent up into the stratosphere carrying instruments to measure ozone concentrations, which can be used to study the kinetics of reactions in the stratosphere. This balloon is being launched near McMurdo Station, in Antarctica.

Chemistry is about reactions, so a knowledge of the rates at which they occur is of central importance. The study of rates of reactions is called **reaction kinetics**.

Kinetic studies provide information about the **mechanisms** of reactions—the detailed routes by which they take place. Most of the reactions you encounter are **complex reactions** in which an overall reaction takes place by a series of simple steps, called **elementary reactions**. The elementary reactions combine together to give the overall reaction. In this chapter, elementary reactions are discussed first in Section 9.4, followed by complex reactions in Section 9.5.

9.1 Why study reaction kinetics?

The study of reaction kinetics falls into two major areas. The first is a search for fundamental information about the interactions and energy changes that take place at a molecular level during an elementary reaction. The progress of a reaction can be charted by constructing an energy profile such as the one shown in Figure 9.1. At the highest energy on the curve is a transient structure that can either return to the reactants, or go on to form the products. This is the **transition state** for the reaction. Detailed information about the structure of transition states contributes to an understanding of how the reactants are transformed into the products. Theories of how reactions take place are discussed in Section 9.8.

The second major area of reaction kinetics involves the study of complex reactions, and the elementary reactions that make them up. Experimental studies on a reaction in solution might lead to the determination of a rate equation, which shows how the rate of a reaction depends on the concentrations of the reactants, and to a value of the rate constant for the reaction. Such experiments have provided much of the evidence for the mechanisms of chemical reactions, and examples of this are described in Section 9.6.

Rate constants are related to how fast a reaction proceeds at a particular temperature and compilations of rate constants are widely used in applications of gas phase reaction kinetics. Rate constants are the key to deciding on the relative importance of elementary reactions in a system of interconnected reactions, such as exists in the atmosphere. A knowledge of rate constants of elementary reactions allowed, for example, the mechanism for ozone depletion in the stratosphere (upper atmosphere) to be worked out by Paul Crutzen.

In industry, the rates at which reactions occur are linked to profitability and safety. Understanding the factors that govern the rates, particularly temperature (Section 9.7), is essential. In the pharmaceutical industry, a knowledge of kinetics can help chemists maximize the formation of a desired product and minimize competing side-reactions.

An understanding of the role of catalysts in speeding up reactions, including the action of enzymes in living organisms, comes from kinetic studies. This is discussed in Section 9.9. First, though, this chapter starts by defining what is meant by the rate of a reaction (Section 9.2) and looks at ways of monitoring concentrations as a reaction proceeds (Section 9.3).

9.2 What is meant by the rate of a reaction?

The **rate of a reaction** is the rate at which reactants are converted into products. You can measure this by monitoring either the consumption of a reactant or the formation of a product.

Iron powder burning—a chemical reaction with a fast rate.

Iron rusting—a chemical reaction with a slow rate.

Unlike in a car, which has a speedometer to measure speed, reaction rate cannot be measured directly. Instead, you measure the concentration of a reactant or product at different times—just as you could measure the speed of a car by measuring the distance travelled after different times.

Consider first a simple reaction in which a single reactant is converted into a single product, for example, in an isomerization reaction

$$\text{reactant} \rightarrow \text{product}$$

Figure 9.2 plots the concentration of the *reactant* against time. It shows how the rate of the reaction changes as the reaction proceeds and is called a **kinetic profile**. The curve is steepest at the start, when $t = 0$, and becomes progressively less steep as the reaction proceeds. At the end of the reaction, the plot becomes horizontal because the concentration of the reactant is no longer changing. The steepness of the curve tells you how fast the reaction is taking place. The rate of reaction at a particular instant can be found by drawing a tangent to the curve at that time and measuring the gradient of the tangent. The rate at the start of the reaction, when $t = 0$, is called the **initial rate** of the reaction.

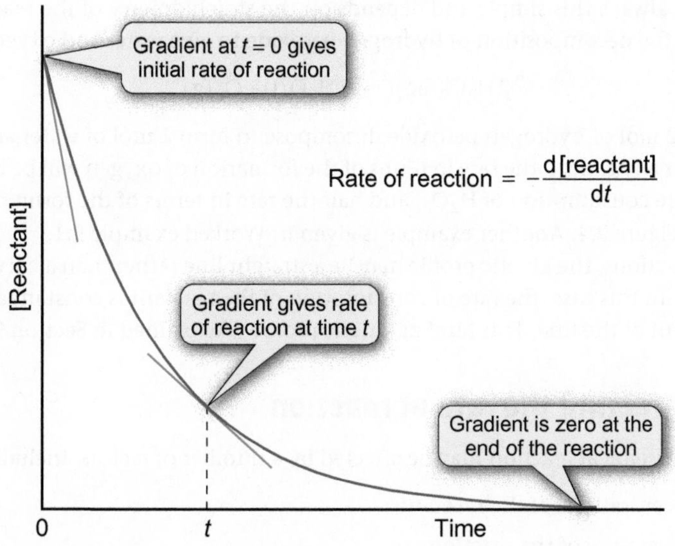

Gradient at t = 0 gives initial rate of reaction
Rate of reaction = − d[reactant]/dt
[Reactant]
Gradient gives rate of reaction at time t
Gradient is zero at the end of the reaction
0
t
Time

Figure 9.2 The concentration of the *reactant* plotted against time.

Drawing tangents and measuring gradients

A tangent is a straight line drawn so that it just touches the curve at the point of interest. It has the same slope (gradient) as the curve at that point.

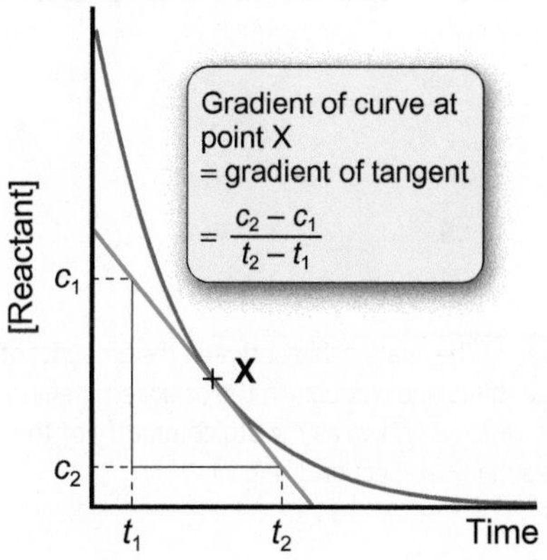

Gradient of curve at point X
= gradient of tangent
= (c2 − c1)/(t2 − t1)
[Reactant]
c1
c2
X
t1
t2
Time

In this case, the gradient has a negative value because the concentration decreases with time (c_2 is smaller than c_1). You can find more information about gradients in Maths Toolkit, MT4 (p.1318).

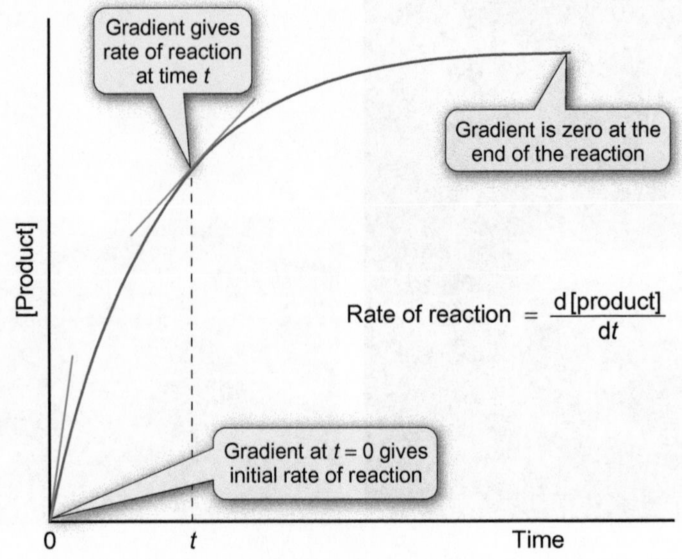
Gradient gives
rate of reaction
at time t
Gradient is zero at the
end of the reaction
[Product]
Rate of reaction = d[product]/dt
Gradient at t = 0 gives
initial rate of reaction
0
t
Time

Figure 9.3 The concentration of the *product* plotted against time.

The gradient of a plot of [A] against t is the **differential** of [A] with respect to t and is denoted by $\frac{d[A]}{dt}$. If you need help with differentials, look in Maths Toolkit, MT6 (p.1324).

You can express the gradient at a particular time as the differential, $\frac{d[\text{reactant}]}{dt}$. The value of the gradient is negative in Figure 9.2, because the reactant is being used up, so the rate of reaction $= -\frac{d[\text{reactant}]}{dt}$. The minus sign is included because rates are always quoted as positive quantities.

Now look at the plot of the concentration of the *product* against time in Figure 9.3. This time the gradient of the curve has positive values because the product is being formed in the reaction. At any time, t, the rate of formation of the product in this reaction is the same as the rate of consumption of the reactant, so

$$\text{rate of reaction} = \frac{d[\text{product}]}{dt} = -\frac{d[\text{reactant}]}{dt}$$

The relationship between the amounts of reactants and products in the balanced chemical equation is known as the **stoichiometry** of the reaction; see Section 1.4 (p.21).

The relationship between the rate of removal of reactant and the rate of formation of product isn't always this simple and depends on the stoichiometry of the reaction. Take, for example, the decomposition of hydrogen peroxide to give water and oxygen.

$$2H_2O_2(aq) \rightarrow 2H_2O(l) + O_2(g)$$

The more rigorous definition of the rate of a reaction in Equation 9.4 on p.390 takes into account the stoichiometry of the reaction and ensures that the rate of reaction has the same value whether given in terms of consumption of a reactant or the formation of a product.

In this case, 2 mol of hydrogen peroxide decompose to form 2 mol of water and 1 mol of oxygen. So, at any time, t, the rate in terms of the formation of oxygen will be half the rate in terms of the consumption of H_2O_2, and half the rate in terms of the formation of H_2O, as shown in Figure 9.4. Another example is given in Worked example 9.1.

In some reactions, the kinetic profile may be a straight line rather than a curve as shown in Figure 9.5. In this case, the rate of consumption of the reactant is constant and is given by the gradient of the line. This kind of kinetic profile is explained in Section 9.5 (p.405).

Factors affecting the rate of reaction

The rate of a chemical reaction may be affected by a number of factors, including:

- the concentrations of the reactants;
- the temperature of the reaction;
- the intensity of radiation (if the reaction involves radiation such as light);

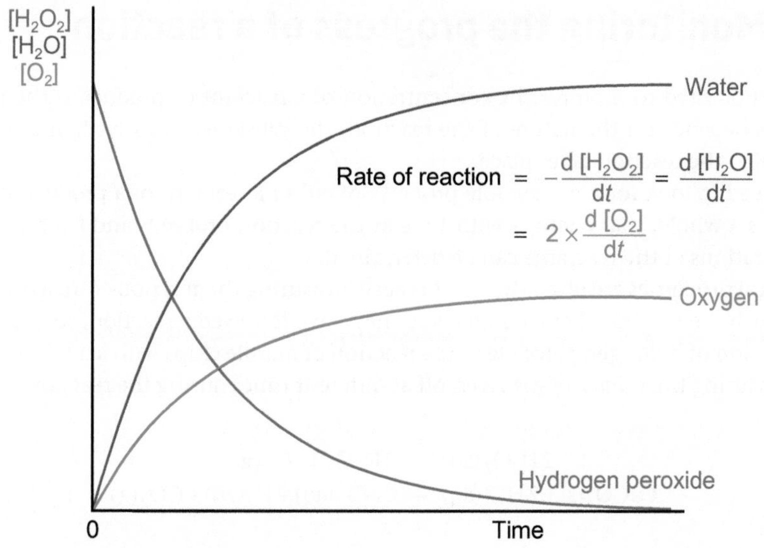
[H$_2$O$_2$]
[H$_2$O]
[O$_2$]
Water
Rate of reaction $= -\frac{d[H_2O_2]}{dt} = \frac{d[H_2O]}{dt}$
$= 2 \times \frac{d[O_2]}{dt}$
Oxygen
Hydrogen peroxide
0
Time

Figure 9.4 The decomposition of hydrogen peroxide.

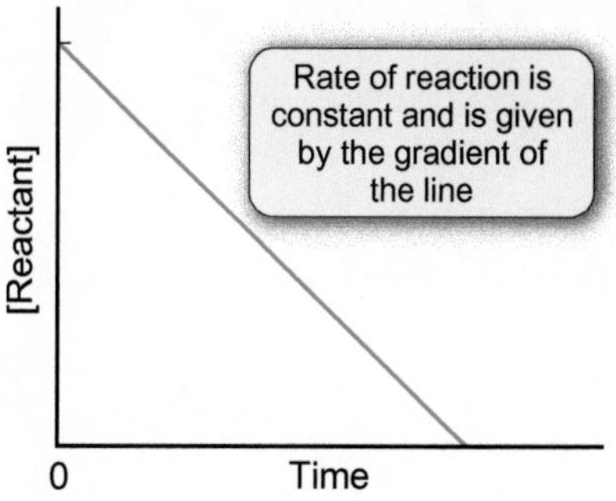
Rate of reaction is constant and is given by the gradient of the line
[Reactant]
0
Time

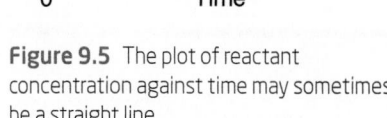

Figure 9.5 The plot of reactant concentration against time may sometimes be a straight line.

- the particle size of a solid (related to the surface area of the solid particles);
- the polarity of a solvent (for reactions in solution involving ions);
- the presence of a catalyst.

The next four sections (Sections 9.3–9.6) are concerned with the effects of concentration.

Units of rate of reaction

Concentrations are normally measured in $mol\,dm^{-3}$, so the units for rate of reaction are usually $mol\,dm^{-3}\,s^{-1}$.

Worked example 9.1 The rate of a reaction

For the following reaction, the initial rate of formation of I_2(aq) was found to be. $2.5 \times 10^{-3}\,mol\,dm^{-3}\,s^{-1}$.

$$S_2O_8^{2-}(aq) + 2I^-(aq) \rightarrow 2SO_4^{2-}(aq) + I_2(aq)$$

What was:

(a) the initial rate of formation of SO_4^{2-}(aq);

(b) the initial rate of consumption of $S_2O_8^{2-}$(aq)?

Strategy

Use the stoichiometric equation to find the relationship between the number of moles of I_2(aq) formed and the number of moles of SO_4^{2-}(aq) formed in the reaction, and the number of moles of $S_2O_8^{2-}$(aq) consumed. This gives the relation between the rates of reaction.

Solution

1 mol of $S_2O_8^{2-}$(aq) reacts to form 2 mol of SO_4^{2-}(aq) and 1 mol of I_2(aq).

(a) The initial rate of formation of SO_4^{2-}(aq)

$= 2 \times$ initial rate of formation I_2(aq)

$= 5.0 \times 10^{-3}\,mol\,dm^{-3}\,s^{-1}$.

(b) The initial rate of consumption of $S_2O_8^{2-}$(aq)

$=$ initial rate of formation I_2(aq)

$= 2.5 \times 10^{-3}\,mol\,dm^{-3}\,s^{-1}$.

Question

For the reaction between nitrogen and hydrogen

$$N_2(g) + 3H_2(g) \rightarrow 2NH_3(g)$$

the rate of formation of ammonia was measured as $10\,mmol\,dm^{-3}\,s^{-1}$. What was the rate of consumption of hydrogen?

9.3 Monitoring the progress of a reaction

The method used to monitor the concentration of a reactant or product as the reaction proceeds depends on the nature of the reaction, the substances involved, and the speed with which the reaction takes place.

You need to look for a measurable property of either a reactant or a product, or of the system as a whole, that changes with time as the reaction proceeds and from which the concentrations of the reactants can be determined.

Physical properties are often the best because measuring them is non-intrusive and does not disturb the reaction. For example, you may have followed a reaction, such as the decomposition of hydrogen peroxide or the reaction of marble chips with acid, by collecting and measuring the *volume of gas* given off at different times during the reaction.

Marble is a form of calcium carbonate, $CaCO_3$.

$$2H_2O_2(aq) \rightarrow 2H_2O(l) + O_2(g)$$
$$CaCO_3(s) + 2HCl(aq) \rightarrow CaCl_2(aq) + H_2O(l) + CO_2(g)$$

Alternatively, you may have carried out the reaction in an open conical flask on a balance and measured the *change in mass* as the reaction proceeded. Figure 9.6 shows some typical results for these types of experiments.

For reactions taking place in the gas phase (in a closed container of constant volume), it is often possible to follow the reaction by monitoring the *change in pressure* as the reaction proceeds.

The relationship between gas pressures, gas volumes, and the amount in moles is described in Section 8.1 (p.345).

Spectroscopic techniques, including spectrophotometric analysis, are described in Section 11.4 (p.542). The theory underlying the techniques can be found in Chapter 10. NMR spectroscopy is discussed in Sections 10.7 (p.496) and 12.3 (p.578).

Spectroscopic methods are useful where a reactant or product has a strong absorption in an accessible part of the spectrum. The intensity of absorption can then be used to measure the concentration of that substance. When the reaction involves a coloured compound, a *spectrophotometer* (sometimes called a colorimeter) can be conveniently used to monitor the concentration of that substance. For example, the reaction of purple $[MnO_4]^-$ ions with hydrogen peroxide in acidic solution can be monitored in this way

$$2[MnO_4]^-(aq) + 5H_2O_2(aq) + 6H^+(aq) \rightarrow 2Mn^{2+}(aq) + 5O_2(g) + 8H_2O(l)$$

Sometimes it is possible to follow a reaction using *NMR spectroscopy* if there is a characteristic signal that changes during the reaction. For example, in the esterification of trifluoroethanoic acid (CF_3CO_2H) with benzyl alcohol ($C_6H_5CH_2OH$)

$$CF_3CO_2H + C_6H_5CH_2OH \rightarrow CF_3CO_2CH_2C_6H_5 + H_2O$$

the consumption of benzyl alcohol can be monitored by the change in the signal from the CH_2 hydrogens in the ^1H NMR spectrum.

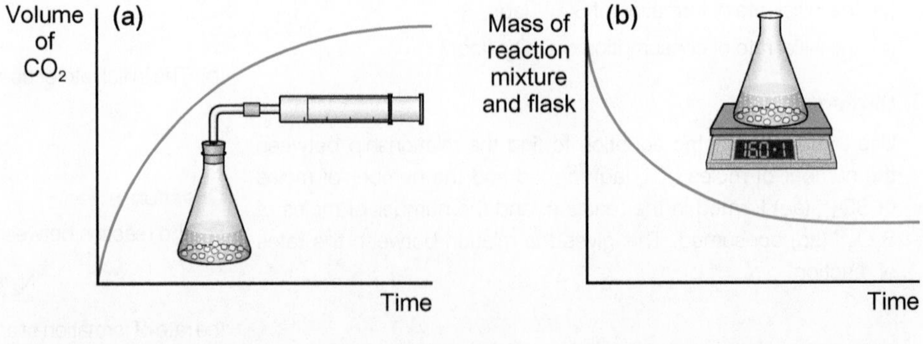
Volume of CO2
(a)
Time
Mass of reaction mixture and flask
(b)
Time

Figure 9.6 Monitoring the reaction of marble chips with dilute hydrochloric acid by (a) measuring the volume of CO_2 given off, and (b) measuring the mass of the reaction mixture and flask as the reaction proceeds.

Spectroscopic techniques based on the use of *flow tubes and lasers* have allowed faster and faster reactions to be monitored. These techniques are discussed at the end of Section 9.4 (p.402) and in Box 9.5 (p.406).

If there is a change in the number or type of ions present in the solution, their concentration can be monitored by measuring the *conductivity* of the solution. For example, the hydrolysis of 2-chloro-2-methylpropane generates ionic products

$$(CH_3)_3CCl(aq) + H_2O(l) \rightarrow (CH_3)_3COH(aq) + H^+(aq) + Cl^-(aq)$$

Reactions in which the concentration of $H^+(aq)$ ions changes can be measured by using a glass electrode to follow the *pH of the solution* during the reaction. This would provide another way of following this hydrolysis reaction.

For reactions of chiral compounds, where there is a *change in optical activity*, a polarimeter can be used to measure the rotation of the plane of polarized light as a function of time. An early example of the use of this method, first carried out in 1850, was to follow the rate of the acid-catalysed conversion of sucrose into a mixture of glucose and fructose

$$\underset{\text{D-(+)-sucrose}}{C_{12}H_{22}O_{11}(aq)} + H_2O(l) \xrightarrow{H^+} \underset{\text{D-(+)-glucose}}{C_6H_{12}O_6(aq)} + \underset{\text{D-(−)-fructose}}{C_6H_{12}O_6(aq)}$$

Since D-(−)-fructose has a larger specific rotation than D-(+)-glucose, the resulting mixture is laevorotatory (rotates plane-polarized light in an anticlockwise direction).

Titrations are sometimes used to measure concentrations in rate studies, but you need to withdraw a small sample to carry out the titration and this disturbs the reaction. It is also a slow method and there needs to be some way of quenching (stopping or slowing down) the reaction as soon as you withdraw the sample.

The key to success in all the methods is to ensure that there is rapid mixing at the start (compared to the time the reaction takes) and that the temperature is kept constant. The rates of many reactions are dependent on temperature so, if the temperature varies, your results will be meaningless. This is particularly important for exothermic or endothermic reactions where there are heat changes during the reaction. There must be good thermal contact between the reaction vessel and the surroundings (e.g. a water bath) to keep the temperature constant.

Measurement of the conductivity of ionic solutions is discussed in Section 16.2 (p.730). The use of a glass electrode to measure pH is described in Section 11.2 (p.523). The definition of pH and calculations involving pH are in Section 7.2 (p.308).

You can find out about chiral compounds and the use of a polarimeter to measure optical activity in Section 18.4 (p.838). The d/l nomenclature for sugars is explained in Box 18.6 (p.842).

9.4 Elementary reactions

Rate equations

Elementary reactions are single-step reactions that involve one or two molecules or atoms. For example

$$H^\bullet + Cl_2 \rightarrow HCl + Cl^\bullet \quad (9.1)$$

There are two important features of elementary reactions. The first is that the equation represents the actual changes that take place at a molecular level during the reaction—in the case above, this involves the collision of an H atom with a Cl_2 molecule to produce a molecule of HCl and a Cl atom. For a complex reaction, which is made up of a series of elementary reactions, the chemical equation just summarizes the overall stoichiometry of the reaction. It doesn't tell you about the changes that take place at a molecular level.

The second feature of an elementary reaction is that you can use the chemical equation to write a **rate equation** for the reaction. A rate equation expresses the rate of the reaction at a particular instant in terms of concentrations of the reactants at that instant.

For the reaction in Equation 9.1, the rate equation is

$$\text{rate of reaction} = k[H^\bullet][Cl_2]$$

where k is the **rate constant** for the reaction. The value of k is characteristic of the reaction and is constant for a particular temperature. The rate constant is related to how fast the reaction proceeds at that temperature. Figure 9.7 shows plots of the concentration of reactant against time for different temperatures, that is, for different values of k.

In Equation 9.1, the hydrogen and chlorine atoms, $H^\bullet$ and $Cl^\bullet$, are both radicals.

A **radical** is an atom or molecule containing an unpaired electron; see Section 5.1 (p.219) and Section 19.1 (p.863).

Fast reactions involving a radical reactant (and their reverse reactions) are generally single-step **elementary reactions**. A **complex reaction** consists of a series of elementary steps (see Section 9.4). You should always assume that a reaction is complex unless there is evidence otherwise.

A rate equation is sometimes called a **rate law**.

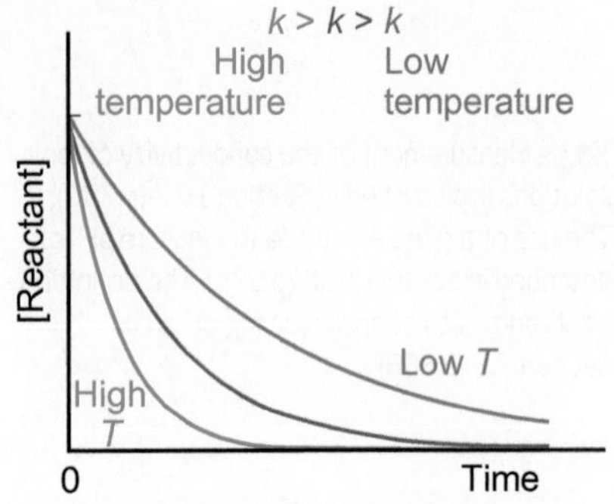
k > k > k
High temperature
Low temperature
[Reactant]
Low T
High T
0
Time

Figure 9.7 Plots of the concentration of a reactant against time at three different temperatures, that is, for different values of the rate constant, *k*.

For the elementary reaction

$$C_2H_6 \rightarrow 2CH_3^{\bullet} \qquad (9.2)$$

the rate equation is

$$\text{rate of reaction} = k[C_2H_6]$$

Defining the rate of a reaction

For the reaction of $H^{\bullet}$ with Cl_2 in Equation 9.1, the rate of reaction has the same value whether you consider the rate of consumption of a reactant or the rate of formation of a product. For the reaction of ethane to form two methyl radicals in Equation 9.2, this is not the case. The rate of formation of methyl radicals is twice the rate of consumption of ethane (see Section 9.2, p.384). In the reverse process, shown in Equation 9.3

$$2CH_3^{\bullet} \rightarrow C_2H_6 \qquad (9.3)$$

the rate of consumption of methyl radicals is twice the rate of formation of ethane.

It would be helpful to have a definition for the rate of a reaction that gives the same value of the rate whether you are monitoring the rate of consumption of any of the reactants or the rate of formation of any of the products. This is done as follows. For a general reaction

$$aA + bB \rightarrow pP + qQ$$

$$\text{rate of reaction} = -\frac{1}{a}\frac{d[A]}{dt} = -\frac{1}{b}\frac{d[B]}{dt} = \frac{1}{p}\frac{d[P]}{dt} = \frac{1}{q}\frac{d[Q]}{dt} \qquad (9.4)$$

Expressing the rate of consumption of a reactant or the rate of formation of a product as a differential is described in Section 9.2 (p.384).

This corresponds to the expressions for the rates given in Section 9.2 (p.384), except that the terms $\frac{1}{a}, \frac{1}{b}$, etc. have been added to take account of the stoichiometry of the reaction.

This definition is used to write the differential rate equations for elementary reactions in the rest of the chapter.

Reaction order

In general, for an **elementary reaction**

$$aA + bB \rightarrow \text{products}$$

the rate equation takes the form

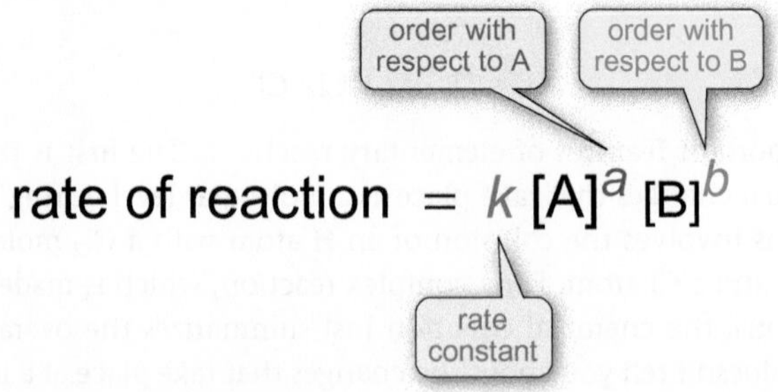
order with respect to A
order with respect to B
rate of reaction = k [A]^a [B]^b
rate constant

The **overall order** of the reaction is given by $a + b$.

Why do rate equations take this form? If the elementary reaction involves two molecules A and B, these molecules must collide in order to react. The rate of collision of A and B depends on how many A and B molecules are present in a given volume—in other words, on their concentrations. There is more about collision theory of reaction in Section 9.8 (p.432).

- For Equation 9.1, the reaction is first order with respect to $H^{\bullet}$ and first order with respect to $[Cl_2]$. Overall, the reaction is second order ($a + b = 2$).
- For Equation 9.2, the reaction is first order with respect to $[C_2H_6]$. Overall, the reaction is first order. (There is only one reactant, so no substance B in the rate equation.)

The **order of a reaction** with respect to a reactant (in an elementary reaction) gives the power to which the concentration of that substance is raised in the rate equation.

The number 1 is assumed, rather than written, in the rate equation, as is the case for stoichiometric coefficients in chemical equations.

- For Equation 9.3, the reaction is second order with respect to $[CH_3^\bullet]$. Overall, the reaction is second order.

In practice, elementary reactions involve either one or two reactants, and *a* and *b* can only take the values 1 or 2. So, the overall order of an elementary reaction can only be first order or second order.

Differential rate equations

A rate equation where the rate of consumption of a reactant, or the rate of formation of a product, is written as a differential (see Section 9.2, p.384) is known as a **differential rate equation.** Using the definition for the rate of reaction in Equation 9.4, you can now write differential rate equations for the reactions in Equations 9.1–9.3. These are shown in Table 9.1, expressed in terms of both the consumption of reactants and the formation of products.

For example, for

$$2\,CH_3^\bullet \rightarrow C_2H_6 \quad (9.3)$$

$$\text{rate of reaction} = -\frac{1}{2}\frac{d[CH_3^\bullet]}{dt} = \frac{d[C_2H_6]}{dt} = k[CH_3^\bullet]$$

So

$$\frac{d[CH_3^\bullet]}{dt} = 2 \times k[CH_3^\bullet]^2$$

You can practise writing differential rate equations in Worked example 9.2.

Units of *k*

Rates of reaction are expressed in $mol\,dm^{-3}\,s^{-1}$ (see Section 9.2, p.384). The units of *k* depend on the order of the reaction. For a first order reaction

$$\text{rate} = k[A]$$

so $k = \text{rate}/[A]$, and the units of *k* are given by

$$\frac{\cancel{mol\,dm^{-3}}\,s^{-1}}{\cancel{mol\,dm^{-3}}} = s^{-1}$$

For a second order reaction

$$\text{rate} = k[A]^2$$

so $k = \text{rate}/[A]^2$ and the units of *k* are given by

$$\frac{\cancel{mol\,dm^{-3}}\,s^{-1}}{(\cancel{mol\,dm^{-3}})(mol\,dm^{-3})} = dm^3\,mol^{-1}\,s^{-1}$$

Table 9.1 Rate equations expressed in terms of differentials showing the relationship between the rate of consumption of reactants and the rate of formation of products

Reaction	Rate of consumption of reactants	Rate of formation of products
(9.1) $H^\bullet + Cl_2 \rightarrow HCl + Cl^\bullet$	$-\frac{d[H^\bullet]}{dt} = -\frac{d[Cl_2]}{dt} = k[H^\bullet][Cl_2]$	$\frac{d[HCl]}{dt} = \frac{d[Cl^\bullet]}{dt} = k[H^\bullet][Cl_2]$
(9.2) $C_2H_6 \rightarrow 2\,CH_3^\bullet$	$-\frac{d[C_2H_6]}{dt} = k[C_2H_6]$	$\frac{d[CH_3^\bullet]}{dt} = 2 \times k[C_2H_6]$
(9.3) $2\,CH_3^\bullet \rightarrow C_2H_6$	$-\frac{d[CH_3^\bullet]}{dt} = 2 \times k[CH_3^\bullet]^2$	$\frac{d[ClH_6]}{dt} = k[CH_3^\bullet]^2$

Σ Equations involving differentials are called **differential equations**, see Maths Toolkit MT6 (p.1324).

Σ **Differentiation and integration**
When you differentiate an expression for a variable, *y*, with respect to a second variable, *x*, the result (a differential $\frac{dy}{dx}$) tells you the rate at which *y* changes with respect to *x*. Integration is the opposite of differentiation. In this case, you start with an expression describing the rate of change, that is, the gradient of a plot (a curve or a straight line) and integrate it to find an expression for the curve or straight line itself.

Radioactive decay is a first order process. The rate depends only on the concentration of the radioactive isotope.

Worked example 9.2 Writing differential rate equations for elementary reactions

The following elementary reaction between ozone and oxygen atoms occurs in the atmosphere.

$$O_3 + O \rightarrow 2O_2$$

(a) Write down an expression for the rate of the reaction.

(b) What is the order of the reaction with respect to (i) O_3, and (ii) O? What is the overall order of the reaction?

(c) Use differential expressions, as in Table 9.1 (p.391), to write rate equations in terms of the rates of consumption of reactants and the rate of formation of the product.

(Note that both O_2 and O atoms have two unpaired electrons and are *biradicals*; see Section 4.10, p.196. The dots are omitted here for simplicity.)

Strategy

The reaction is an elementary process, so you can use the chemical equation to write the rate equation. The stoichiometric coefficients in the chemical equation give the orders of reaction with respect to each reactant. Use Equation 9.4 and Table 9.1 to help you write the rate equation in terms of differentials.

Solution

(a) Rate of reaction $= k[O_3][O]$.

(b) The reaction is first order with respect to O_3 and first order with respect to O. The overall order of the reaction is 1 + 1 = 2.

(c) From Equation 9.4

$$\text{rate of reaction} = -\frac{d[O_3]}{dt} = \frac{1}{2}\frac{d[O_2]}{dt}$$

Combining this equation with the rate equation in (a)

$$\text{rate of consumption of reactants: } -\frac{d[O_3]}{dt} = -\frac{d[O]}{dt} = -k[O_3][O]$$

$$\text{rate of formation of products: } \frac{d[O_2]}{dt} = 2 \times k[O_3][O]$$

Question

The alkaline hydrolysis of bromomethane is an elementary reaction

$$CH_3Br + OH^- \rightarrow CH_3OH + Br^-$$

(a) Write a rate equation for the reaction in terms of the differential for consumption of CH_3Br.

(b) What is the overall order of the reaction?

Integrated rate equations

An **integrated rate equation** gives the concentration of a reactant as a function of time.

Rate equations tell you the rate of reaction at a particular instant during the reaction. However, you can't measure rates of reaction directly—experimental data usually consist of measurements of concentration of a reactant or product at different times. It is useful to have an expression that shows how the concentration varies with time. This information comes from the **integrated rate equation** for the reaction.

The integrated rate equation is derived mathematically from the differential rate equation for the reaction. You can see how this is done for a first order elementary reaction in Box 9.1. The resulting equation, and how it is used, are described below.

Visit the Online Resource Centre to view screencast 9.1 which explains the difference between differential and integrated rate equations and walks you through Box 9.1 to demonstrate how the integrated rate equation for a first order reaction is derived from the differential rate equation for the reaction.

A first order elementary reaction

For a first order elementary reaction

$$A \rightarrow \text{products}$$

the integrated rate equation (Equation 9.6a) is an expression linking the concentration of the reactant A at a particular time, $[A]_t$, to the time, t, the concentration of A at the start of the reaction, $[A]_0$, and the rate constant, k.

$$\ln[A]_t = \ln[A]_0 - kt \qquad (9.6a)$$

$$y = mx + c$$
$$\ln[A]_t = -kt + \ln[A]_0$$

So, if you know the concentration of A at the start and a value for the rate constant, the integrated rate equation allows you to work out the concentration of a reactant or product at any time during the reaction.

Alternatively, if you can measure the concentrations during the reaction, the integrated rate equation can be used to find a value for the rate constant, k. This is easily done because Equation 9.6a corresponds to the equation for a straight line ($y = mx + c$), so a plot of

$\ln[A]_t$ against t is a straight line with a gradient $-k$ (Figure 9.8). The plot also helps you to determine the order of the reaction. If the experimental results plotted in this way lead to a straight line, then the reaction is first order.

Equations 9.6b and 9.6c are two useful alternative ways of writing Equation 9.6a (see Box 9.1). Equation 9.6c is an equation for **exponential decay** and tells you about the shape of the plot of [A] against time. A common feature of all first order reactions is that the concentration of the reactant decays exponentially with time, as shown in Figure 9.9.

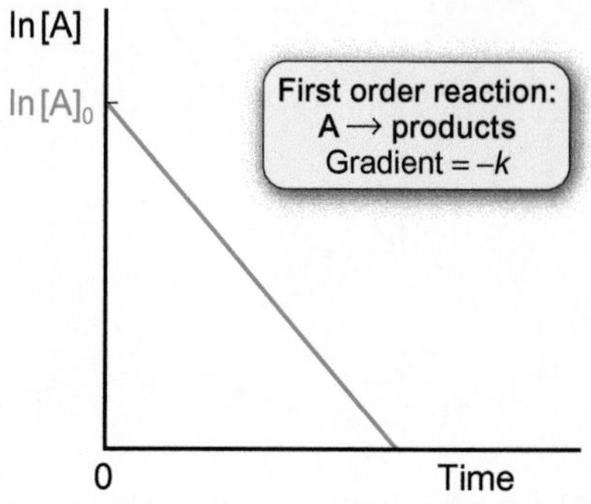

ln [A]
ln [A]0
First order reaction:
A ⟶ products
Gradient = –k
0
Time

Figure 9.8 For the first order reaction, A → products, a plot of ln [A] against t is a straight line. The rate constant k can be found from the gradient.

Σ Straight line graphs

The equation for a straight line is

$$y = mx + c$$

You can find help with plotting and interpreting straight line graphs in Maths Toolkit MT4 (p.1318).

Σ Natural logarithms

'ln' means 'logarithm to the base e', where e = 2.7183.

So, $\ln x = \log_e x$ and is known as a natural logarithm. Natural logarithms are explained in Maths Toolkit MT3 (p.1316).

Box 9.1 Deriving the integrated rate equation for a first order reaction

From Equation 9.4 (p.390) the differential rate equation for the first order elementary reaction

$$A \rightarrow \text{products}$$

is given by

$$\text{rate of reaction} = -\frac{d[A]}{dt} = k[A]$$

To solve this differential equation to obtain a relationship between [A] and t, first separate these two variables so that [A] is on one side of the equation and t is on the other.

$$\frac{d[A]}{[A]} = -k\,dt$$

Now integrate both sides of the equation from $[A] = [A]_0$ at $t = 0$, to $[A] = [A]_t$ at time t. (*Note*. Go to Maths Toolkit MT7 (p.1327) if you need help with carrying out integration.)

$$\int_{[A]_0}^{[A]_t} \frac{d[A]}{[A]} = -k\int_0^t dt$$

To do this you need to use the standard integral

$$\int_{x_1}^{x_2} \frac{dx}{x} = [\ln x]_{x_1}^{x_2} = \ln x_2 - \ln x_1$$

Then

$$[\ln[A]]_{[A]_0}^{[A]_t} = -k[t]_0^t$$

$$\ln[A]_t - \ln[A]_0 = -k(t-0)$$

which can be rearranged to give

$$\ln[A]_t = \ln[A]_0 - kt \qquad (9.6a)$$

Using the expression, $\ln\frac{a}{b} = \ln a - \ln b$, Equation 9.6a can be written as

$$\ln\frac{[A]_t}{[A]_0} = -kt \qquad (9.6b)$$

A third way of writing this equation makes use of the fact that $\ln y = x$ can be written as $y = e^x$, so Equation 9.6b becomes

$$\frac{[A]_t}{[A]_0} = e^{-kt}$$

so that

$$[A]_t = [A]_0\, e^{-kt} \qquad (9.6c)$$

Equation 9.6c has the form of an **exponential decay**, which is an important feature of first order reactions. (The properties of exponential functions, e^x (also written exp(x)), are described in Maths Toolkit MT3 (p.1316) and MT4 (p.1318).)

A second order elementary reaction

For a second order elementary reaction

$$A + A \rightarrow \text{products}$$

the integrated rate equation is given by

$$\frac{1}{[A]_t} = \frac{1}{[A]_0} + 2kt \qquad (9.7b)$$

ⓘ

$$y = mx + c$$

$$\frac{1}{[A]_t} = 2kt + \frac{1}{[A]_0}$$

You can see how this is derived from the differential rate equation in Box 9.2. In this case, a plot of $\frac{1}{[A]_t}$ against t is a straight line with gradient $2k$ (Figure 9.10). Equations 9.7a and 9.7c are alternative ways of writing Equation 9.7b (see Box 9.2).

Figure 9.11 shows plots of [A] against t for a first order and for a second order reaction with the same initial rate. Note that, for the second order reaction, [A] approaches zero more slowly than for the first order reaction.

Worked example 9.3 illustrates the use of integrated rate equations.

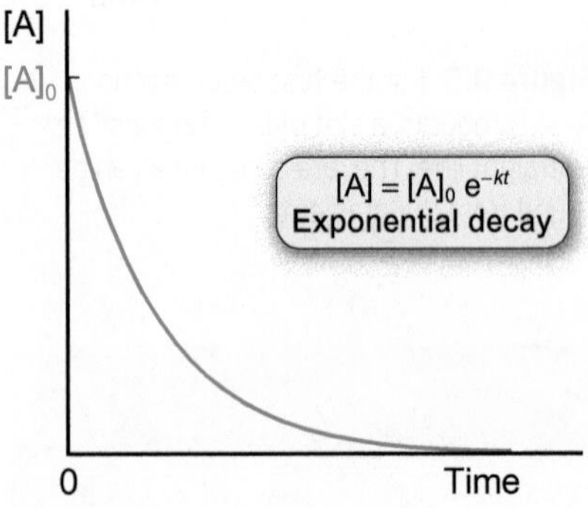
[A]
$[A]_0$
$[A] = [A]_0 e^{-kt}$
Exponential decay
0
Time

Figure 9.9 For the first order reaction, A $\rightarrow$ products, a plot of [A] against t is an exponential decay curve.

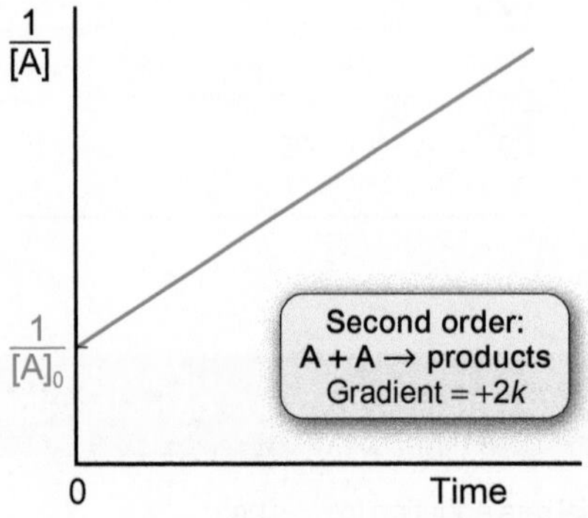
$\frac{1}{[A]}$
$\frac{1}{[A]_0}$
Second order:
A + A → products
Gradient = +2k
0
Time

Figure 9.10 For a second order reaction, A + A $\rightarrow$ products, a plot of 1/[A] against t is a straight line. The rate constant k can be found from the gradient.

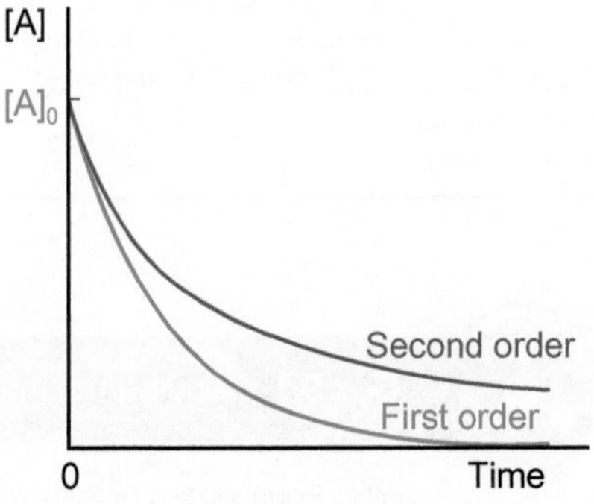
[A]
$[A]_0$
Second order
First order
0
Time

Figure 9.11 A comparison of decay curves for a first order and a second order reaction with the same initial decay rate.

Box 9.2 Deriving the integrated rate equation for a second order reaction

This follows the general procedure given for deriving the integrated rate equation for a first order reaction in Box 9.1.

From Equation 9.4 (p.390) the differential rate equation for the second order elementary reaction

$$A + A \rightarrow \text{products}$$

is given by

$$\text{rate of reaction} = -\frac{1}{2}\frac{d[A]}{dt} = k[A]^2$$

so

$$-\frac{d[A]}{dt} = 2 \times k[A]^2$$

Separating the variables, the equation can be rewritten as

$$\frac{d[A]}{[A]^2} = -2k\,dt$$

To obtain a relationship between [A] and t, integrate both sides of the equation from $[A] = [A]_0$ at $t = 0$, to $[A] = [A]_t$ at time t.

$$\int_{[A]_0}^{[A]_t} \frac{d[A]}{[A]^2} = -2k\int_0^t dt$$

(Note that $2k$ is constant, so goes outside the integral.)

To do this you need to use the standard integral

$$\int_{x_1}^{x_2} \frac{dx}{x^2} = \left[-\frac{1}{x}\right]_{x_1}^{x_2} = \left(-\frac{1}{x_2}\right) - \left(-\frac{1}{x_1}\right) = -\frac{1}{x_2} + \frac{1}{x_1}$$

Then

$$\left[-\frac{1}{[A]}\right]_{[A]_0}^{[A]_t} = -2k[t]_0^t$$

$$-\frac{1}{[A]_t} + \frac{1}{[A]_0} = -2k(t - 0)$$

which can be rearranged to give

$$\frac{1}{[A]_t} + \frac{1}{[A]_0} = 2kt \qquad (9.7a)$$

Here are two useful alternative forms of Equation 9.7a

$$\frac{1}{[A]_t} = \frac{1}{[A]_0} 2kt \qquad (9.7b)$$

$$[A]_t = \frac{[A]_0}{1 + 2kt[A]_0} \qquad (9.7c)$$

Equation 9.7c allows you to predict the concentration of A at any time after the start of the reaction.

Worked example 9.3 Writing integrated rate equations

Write an integrated rate equation for the decomposition of ethane to give two methyl radicals in Equation 9.2 (p.390). Explain how a value of the rate constant k can be found from experimental measurements of concentration of ethane at different times during the course of the reaction.

$$C_2H_6 \rightarrow 2CH_3^{\bullet} \quad (9.2)$$

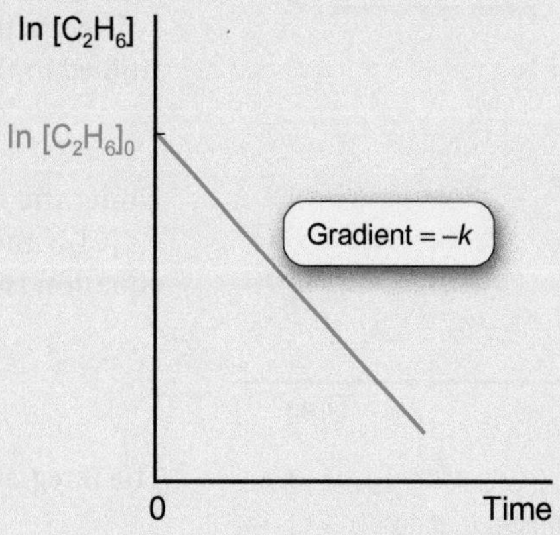
ln [C2H6]
ln [C2H6]0
Gradient = –k
0
Time

Strategy

The reaction is a first order elementary reaction, so the integrated rate equations are given by Equations 9.6(a–c) in Box 9.1.

Solution

Use Equation 9.2 to write the differential equation.

$$-\frac{d[C_2H_6]}{dt} = k[C_2H_6]$$

Use Equation 9.6a on p.390 to write the integrated rate equation.

$$\ln[C_2H_6]_t = \ln[C_2H_6]_0 - kt$$

This is the most useful form of the integrated rate equation in this case because it is the equation of a straight line that can be used to find the rate constant. A plot of $\ln[C_2H_6]$ against t is a straight line, with gradient equal to $-k$.

Question

Write an integrated rate equation for reaction of two methyl radicals to form ethane in Equation 9.3 (p.390). Explain how a value of the rate constant k can be found from experimental measurements of concentration of the methyl radical at different times during the course of the reaction.

$$2CH_3^{\bullet} \rightarrow C_2H_6 \quad (9.3)$$

Pseudo-first order reactions

In elementary reactions where there is more than one reactant, such as A + B → products, the concentrations of *both* A and B change during the reaction and affect the rate. A useful method used to study the kinetics involves isolating one reactant at a time by having the other reactant in large excess. Suppose B is in large excess: this means that the concentration of B is effectively constant throughout the reaction and any changes in the rate are due to changes in the concentration of A alone.

> This is called the **isolation technique**. Its use in the study of the kinetics of complex reactions is described in Section 9.5 (p.405).

$$\text{rate of reaction} = k[A][B]$$

If [A] << [B], you can assume that [B] hardly changes in the reaction and is equal to its initial value $[B]_0$. Then

> For example, if $[B]_0 = 1.00\ \text{mol dm}^{-3}$ and $[A]_0 = 0.01\ \text{mol dm}^{-3}$, then the concentration of B at the end of the reaction when A is all used up is $(1.00 - 0.01)\ \text{mol dm}^{-3} = 0.99\ \text{mol dm}^{-3}$. So, [B] is almost constant.

$$\text{rate of reaction} = k'[A], \quad \text{where } k' = k[B]_0$$

The differential rate equation can now be written as a first order process

$$-\frac{d[A]}{dt} = k'[A]$$

Using Equation 9.6a (p.392) to write the integrated rate equation

$$\ln[A]_t = \ln[A]_0 - k't \quad (9.8a)$$

so substituting for k'

$$\ln[A]_t = \ln[A]_0 - k[B]_0 t \quad (9.8b)$$

> $y = mx + c$
> $\ln[A]_t = -k[B_0]t + \ln[A]_0$

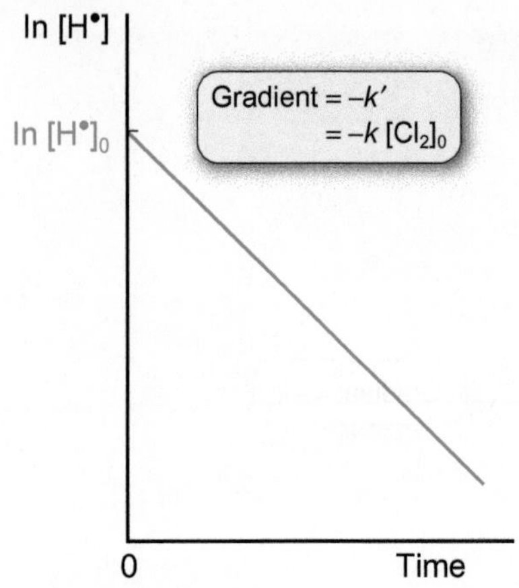
ln [H•]
ln [H•]0
Gradient = −k′
= −k [Cl2]0
0
Time

Figure 9.12 Plot of ln $[H^\bullet]$ against t for the pseudo-first order reaction, $H^\bullet + Cl_2 \rightarrow HCl + Cl^\bullet$, when Cl_2 is in large excess.

In the presence of a large excess of B, the reaction has all the appearance of a first order reaction. The reaction is said to show **pseudo-first order** kinetics under these conditions and k' is called the **effective rate constant** for a fixed concentration of B, $[B]_0$. A plot of ln[A] against t is a straight line with gradient $-k[B]_0$.

The reaction of chlorine molecules with hydrogen atoms in Equation 9.1 (p.389) can be studied in this way.

$$H^\bullet + Cl_2 \rightarrow HCl + Cl^\bullet \quad (9.1)$$

Under the reaction conditions, $H^\bullet$ is present in only very small concentrations, so $[H^\bullet] << [Cl_2]$, and you can assume $[Cl_2] = [Cl_2]_0$ throughout the reaction. The differential rate equation is

$$-\frac{d[H]}{dt} = k'[H^\bullet], \quad \text{where } k' = k[Cl_2]_0$$

The integrated rate equation is then

$$\ln[H^\bullet] = \ln[H^\bullet]_0 - k[Cl_2]_0 t$$

Figure 9.12 shows a plot of ln $[H^\bullet]$ against time.

Half lives

> The **half life** of a reactant is the time taken for its concentration to fall to half its initial value.

An important characteristic of a first order reaction is the **half life**, $t_{1/2}$, of the reactant. This is the time taken for the concentration of the reactant to fall to half its initial value. You can use half lives to decide whether a reaction is first order, and to get a value for the rate constant, k. For a first order reaction, A → products, you can find the half life of A by substituting $[A]_t = [A]_0/2$ at $t = t_{1/2}$ into the integrated rate equation.

Using the integrated rate equation in the form of Equation 9.6b (Box 9.1, p.393)

$$\ln\frac{[A]_t}{[A]_0} = -kt \quad (9.6b)$$

For $[A]_t = [A]_0/2$

$$\ln\frac{[A]_0}{2[A]_0} = -kt_{1/2}$$

Rearranging this to give an expression for $t_{1/2}$

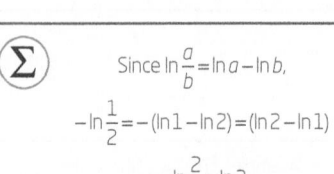

> Since $\ln\frac{a}{b} = \ln a - \ln b$,
>
> $-\ln\frac{1}{2} = -(\ln 1 - \ln 2) = (\ln 2 - \ln 1)$
>
> $= \ln\frac{2}{1} = \ln 2$

$$t_{1/2} = \frac{-\ln\frac{1}{2}}{k}$$

$$t_{1/2} = \frac{\ln 2}{k} \quad (9.9)$$

Since $\ln 2 = 0.693$

$$t_{1/2} = \frac{0.693}{k}$$

> The half life provides a useful indication of the rate constant for a first order reaction. A reaction with a large rate constant has a short half life.

The key point is that, since Equation 9.9 does not contain the term $[A]_0$,

the half life of the reactant in a first order reaction is independent of its initial concentration, so $t_{1/2}$ is constant through the course of the reaction.

If the concentration of A at any time, t, is $[A]_t$, it will fall to $\frac{1}{2}[A]_t$ after an interval of $t_{1/2}$.

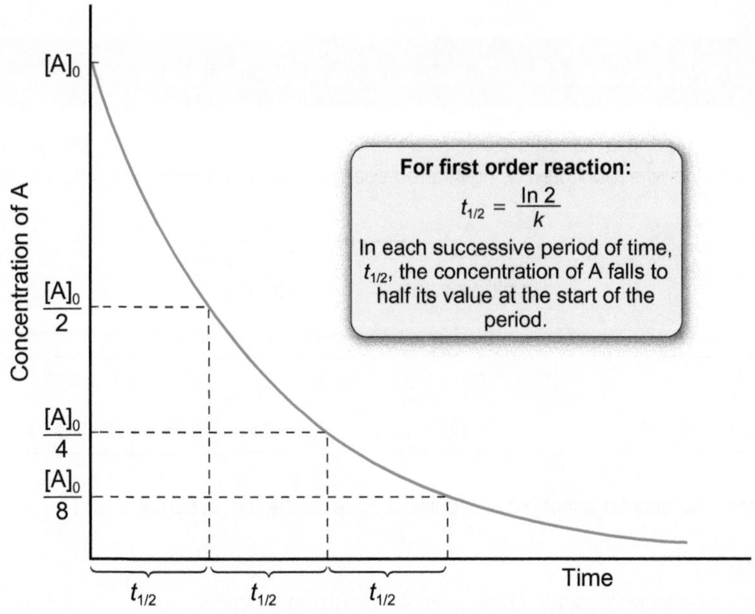
[A]0
Concentration of A
[A]0/2
[A]0/4
[A]0/8
For first order reaction:
t1/2 = ln 2/k
In each successive period of time, t1/2, the concentration of A falls to half its value at the start of the period.
t1/2
t1/2
t1/2
Time

Figure 9.13 For a first order reaction, A → products, the half life of A is independent of its initial concentration.

Figure 9.13 shows three successive half lives, when the concentration of A falls from $[A]_0 \rightarrow \frac{[A]_0}{2} \rightarrow \frac{[A]_0}{4} \rightarrow \frac{[A]_0}{8}$. After n successive half lives, the concentration of A will be $\frac{[A]_0}{2^n}$.

In contrast to a first order reaction, the half life of a second order reaction *does* depend on the initial concentration of the reactant. For the second order reaction

$$A + A \rightarrow \text{products}$$

you can find the half life of A by substituting $[A]_t = [A]_0/2$ at $t = t_{1/2}$ into the integrated rate equation.

Using the integrated rate equation in the form of Equation 9.7a (Box 9.2, p.394)

$$\frac{1}{[A]_t} - \frac{1}{[A]_0} = 2kt \qquad (9.7a)$$

For $[A]_t = [A]_0/2$

$$\frac{2}{[A]_0} - \frac{1}{[A]_0} = \frac{1}{[A]_0} 2kt_{1/2}$$

Rearranging this to give an expression for $t_{1/2}$

$$t_{1/2} = \frac{1}{2k[A]_0} \qquad (9.10)$$

In this case, the half life is inversely proportional to the concentration of the reactant.

A constant half life indicates a first order reaction. If you inspect a set of data of concentration against time for a reaction and see that the initial concentration falls to half its value in a certain time as in Figure 9.13, and that another concentration falls to half its value in the same time, the reaction is likely to be first order. To be sure, you would need to plot a graph of ln[A] against t to show that this gives a straight line.

In Worked example 9.4, you can use both the half life method and the integrated rate equation to find the order of a reaction and a value for the rate constant. The significance of the lifetimes of molecules in the atmosphere is discussed in Box 9.3.

Visit the Online Resource Centre to view screencast 9.2 which walks you through Figure 9.13 and explains how to find the half life of a reaction.

Photochromic sunglasses darken or lighten in response to light intensity. This is achieved by incorporating a light-sensitive dye into the lenses. The fading of the dye in the presence of light is a first order process. Manufacturers publish half lives of the different coloured dyes used.

The decay of a radioactive isotope follows an exponential curve and is a first order process. Its half life is a characteristic property of a radioactive isotope. Examples are discussed in various parts of the book: ^{14}C in Box 3.9 (p.161), ^{3}H (tritium) in Box 25.7 (p.1162), ^{40}K in Box 27.9 (p.1249), and ^{98}Tc in Box 28.1 (p.1257).

In fact, *any* fractional life remains constant during a first order reaction. So, for example, if you measure the 'quarter' life, $t_{1/4}$, the time for the concentration of a reactant to fall to a quarter of its initial value, you find it is independent of the initial concentration and remains constant through the course of the reaction.

The concentration of cyclopropane at $t = 0$ is 1.50×10^{-3} mol dm^{-3}. So, [cyclopropane] has been divided by 10^{-3} mol dm^{-3} to give the numbers in the table. This is a common way of tabulating data. Remember: quantity = number × unit, so number = quantity/unit (see Section 1.2, p.6).

Worked example 9.4 Isomerization of cyclopropane

When cyclopropane is heated to 750 K in a closed container, it isomerizes to form propene. The reaction was monitored using infrared spectroscopy and the following data obtained.

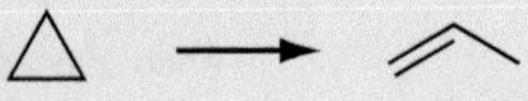

cyclopropane propene

t/min	0	5.0	10	20	30	40	50	60
[cyclopropane]/10^{-3} mol dm^{-3}	1.50	1.23	1.01	0.68	0.46	0.31	0.21	0.14

Show that the reaction is first order and find a value for the rate constant, k, at 750 K.

Strategy

This worked example illustrates two ways of tackling the problem.

1. Using half lives. Plot a graph of [cyclopropane] against time and measure successive half lives. If the reaction is first order, the half life will be independent of the concentration of cyclopropane. Then use $t_{1/2} = \frac{\ln 2}{k}$ (Equation 9.9) to find a value for the rate constant.
2. Using an integrated rate equation. Use the integrated rate equation for a first order reaction: $\ln[A]_t = \ln[A]_0 - kt$ (Equation 9.6b) and plot a graph of ln[cyclopropane]$_t$ against t. (Remember ln = log$_e$.) For a first order reaction, the plot is a straight line with gradient $-k$.

Solution

1. Using half lives

Plot a graph of [cyclopropane] against t, as shown below. Measure at least three half lives.

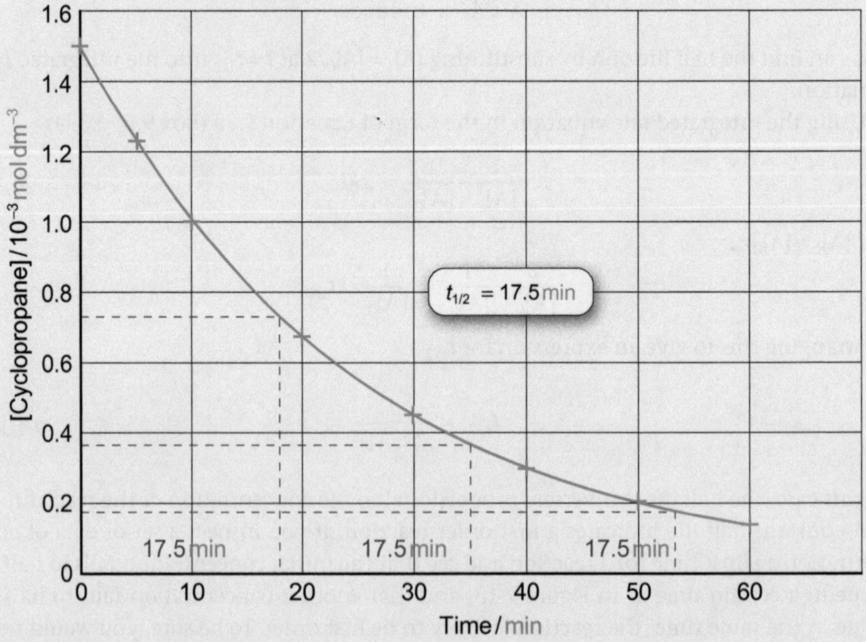
[Cyclopropane]/10⁻³ mol dm⁻³
1.6
1.4
1.2
1.0
0.8
0.6
0.4
0.2
0
10
20
30
40
50
60
Time/min
t1/2 = 17.5 min
17.5 min
17.5 min
17.5 min

The measured half lives have a constant value showing that the reaction is first order.
To find k, substitute an average value for the half life into Equation 9.9

$$t_{1/2} = \frac{\ln 2}{k} \qquad (9.9)$$

$$\text{so } k = \frac{0.693}{17.5\,\text{min}} = 0.040\,\text{min}^{-1} = 6.6 \times 10^{-4}\,\text{s}^{-1} \quad \left(1\,\text{min}^{-1} = \frac{1}{60}\text{s}^{-1}\right)$$

2. Using an integrated rate equation

The data give the following plot for ln [cyclopropane] against *t*.

t/min	0	5.0	10	20	30	40	50	60
ln ([cyclopropane]/10^{-3} mol dm^{-3})	0.405	0.207	0.001	−0.386	−0.777	−1.171	−1.561	−1.966

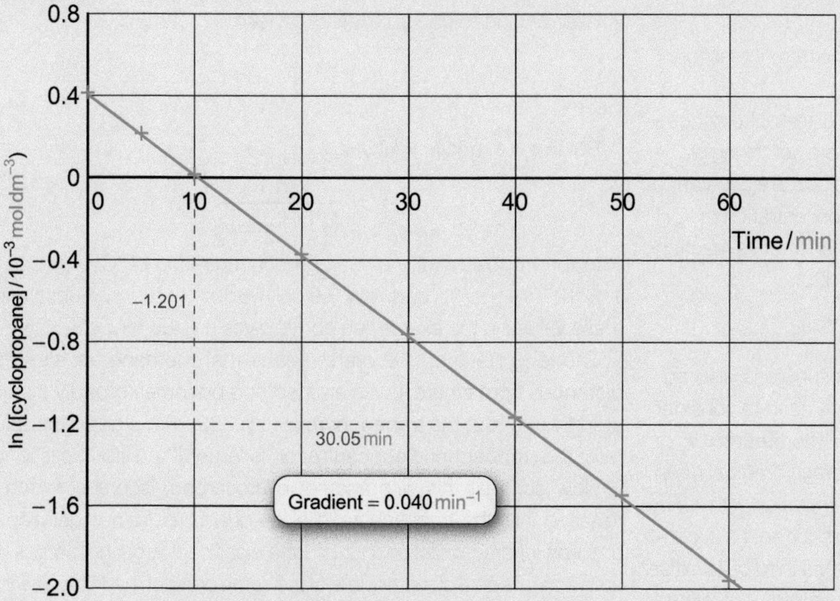

0.8
0.4
0
−0.4
−0.8
−1.2
−1.6
−2.0
0
10
20
30
40
50
60
Time/min
ln ([cyclopropane]/10−3 mol dm−3)
−1.201
30.05 min
Gradient = 0.040 min−1

Note that [cyclopropane] is divided by its units so that you are finding the logarithm of a number (rather than a quantity with units). Here the units of [cyclopropane] are 10^{-3} mol dm^{-3}. You could have plotted the graph using units of mol dm^{-3} for [cyclopropane] and obtained the same gradient. This is because, when you find the gradient, you are subtracting two values of ln [cyclopropane]:

$$\ln a_1 - \ln a_2 = \ln\frac{a_1}{a_2} \text{ and, } \ln 10^{-3}a_1 - 10^{-3}a_2 = \ln\frac{a_1}{a_2}$$

See Maths note on p.385 about finding the gradient of a straight line.

The plot is a straight line showing that the reaction is first order.

The gradient of the line = −0.040 min^{-1}, so

$$k = 0.040\,\text{min}^{-1} = 6.6 \times 10^{-4}\,\text{s}^{-1}$$

which agrees with the value from the half life method.

Generally speaking, it is preferable to use the integrated rate equation and plot a straight-line graph to determine *k*, rather than use the half life method, because it is difficult to find half lives accurately from experimental data.

Question

The reaction of methyl radicals to form ethane was investigated using the flash photolysis technique described on p.402. The following results were obtained at 295 K.

t/10^{-4} s	0	3.00	6.00	10.0	15.0	20.0	25.0	30.0
[$CH_3^\bullet$]/10^{-8} mol dm^{-3}	1.50	1.10	0.93	0.74	0.55	0.45	0.41	0.35

Show that the reaction is second order and find a value for the rate constant at 295 K.

Box 9.3 Atmospheric lifetime of methane

The impact of methane on climate change (see chapter opener, p.382) depends on the amount emitted from the Earth's surface, the length of time it stays in the troposphere, and its ability to absorb infrared radiation. This box is about calculating the lifetime of methane in the troposphere. Methane is removed from the troposphere by reaction with $^{\bullet}OH$ radicals. These radicals act as scavengers and are responsible for removing many of the compounds we emit

$$CH_4 + {}^{\bullet}OH \rightarrow CH_3^{\bullet} + H_2O$$

This is a second order elementary reaction and the rate of reaction = $k[CH4][^{\bullet}OH]$.

The kinetics of reactions in the troposphere are different from those in laboratory experiments, because the reactions are not taking place in a closed container. In the troposphere, $^{\bullet}OH$ radicals are constantly being produced and removed. As a result, the concentration of $^{\bullet}OH$ is approximately constant, $[^{\bullet}OH]_{constant}$, and you can treat its reaction with CH4 as a pseudo-first order reaction (see p.395).

$$\text{Rate of reaction} = k'[CH_4], \text{ where } k' = k[^{\bullet}OH]_{constant}$$

The total concentration of methane in the troposphere is also approximately constant, but to work out a lifetime you need to consider the fate of specific emitted methane molecules. The **lifetime**, τ, of methane molecules in the troposphere is the *average time* between emission of a molecule of CH_4 and its removal by reaction with $^{\bullet}OH$.

By this definition, τ is equal to the time it takes for an initial concentration $[CH_4]_0$ to fall to $[CH_4]_0/e$—approximately a third of its initial value, while in the presence of a constant $[^{\bullet}OH]$. (The lifetime is related to the probability of a molecule still being present after emission. This falls exponentially with time, which explains the presence of 'e' in the expression for τ.) On the decay curve below, compare τ with the half life, $t_{1/2}$, which is the time it takes for the concentration to fall to half of its initial value.

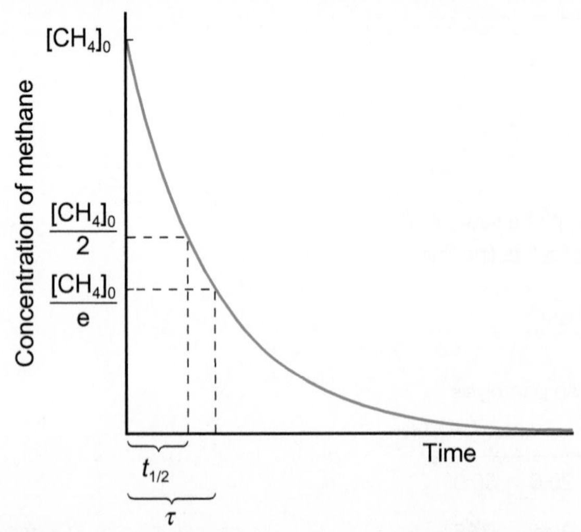
$[CH_4]_0$
Concentration of methane
$\frac{[CH_4]_0}{2}$
$\frac{[CH_4]_0}{e}$
Time
$t_{1/2}$
τ

▲ Decay curve of emitted methane showing lifetime, τ, and half life, $t_{1/2}$ ($e = 2.7183$)

Using the integrated rate equation for a first order reaction in the form of Equation 9.6b (Box 9.1, p.393)

$$\ln\frac{[CH_4]_t}{[CH_4]_0} = -k't \tag{9.6b}$$

When $t = \tau$, $[CH_4]_t = [CH_4]_0/e$, so

$$\ln\frac{[CH_4]_0}{e[CH_4]_0} = -k'\tau, \quad \text{so } \ln\frac{1}{e} = -k'\tau$$

Rearranging to give an expression for τ

$$\tau = \frac{-\ln\frac{1}{e}}{k'} = \frac{\ln e}{k'}$$

But $\ln e = 1$, and $k' = k[^{\bullet}OH]_{constant}$, so

$$\tau = \frac{1}{k[^{\bullet}OH]_{constant}}$$

In the troposphere, the average concentration of $^{\bullet}OH$ radicals is $1 \times 10^{-15}\,mol\,dm^{-3}$ and $k = 3.9 \times 106\,dm^3\,mol^{-1}\,s^{-1}$. Substituting these values in the expression above gives a value for τ of ~8 years.

The long lifetime of 8 years means that methane travels large distances from where it was emitted and becomes broadly distributed throughout the troposphere (see p.382). For a given emission rate, the tropospheric concentration is larger if the lifetime is long.

Now contrast this with another hydrocarbon, isoprene, which is released into the troposphere in huge quantities from vegetation. It absorbs infrared radiation, but is not a significant greenhouse gas. Its emission rate on a global scale is of the same order of magnitude as that of methane, but it has a short lifetime (τ about 1 hour) so it is only found close to its emission source and its tropospheric concentration is small.

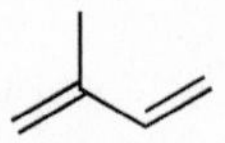

isoprene
[2-methylbuta-1,3-diene]

▲ Tropical rainforests are a source of isoprene.

Question

Calculate a value for the half life (in years) of methane under the conditions in the troposphere.

Kinetic techniques used to study elementary reactions

Elementary reactions often involve atoms and other radicals and occur very rapidly, so that special techniques are needed to study them. Two of the most widely used are **flow techniques** and **flash photolysis**. Both methods are designed to produce well-mixed reactants on a very short timescale. The concentration of one of the reactants is then measured after a short time interval, often using a spectroscopic technique. A series of experiments is carried out, varying the time interval in each experiment, to give values of concentration at different times after the start of the reaction. The conditions are usually chosen so that one reactant is in large excess so that the reaction follows pseudo-first order kinetics (see p.395).

Another method for studying fast reactions uses a stopped-flow technique and is described for complex reactions in Box 9.5 (p.406).

The discharge flow method

This is commonly used to study the reactions of atoms with molecules, such as the reaction of chlorine radicals with ozone, an important reaction in the stratosphere.

$$Cl^{\bullet} + O_3 \rightarrow ClO^{\bullet} + O_2$$

The chlorine radicals are formed by flowing chlorine gas (Cl_2), diluted in N_2, though a microwave discharge, which causes some of the Cl_2 molecules to dissociate to form $Cl^{\bullet}$ radicals. The experimental setup is shown in Figure 9.14.

O_3 (also diluted in N_2) is then introduced through an injector and mixes rapidly with the $Cl^{\bullet}/Cl_2/N_2$ mixture already flowing through the apparatus. The whole mixture then flows down a tube at a constant speed, u. The $Cl^{\bullet}$ radicals are monitored by a detector at a distance x along the tube, which corresponds to the time ($t = x/u$) since the reactants were mixed and the reaction started. The experiment is then repeated and the time varied by changing the position of the injector. In this way, the concentration of $Cl^{\bullet}$ radicals can be measured for a series of reaction times.

The conditions are arranged so that $[Cl^{\bullet}] \ll [O_3]$ and the reaction shows pseudo-first order kinetics. Using Equations 9.8(a) and (b) on p.395

$$\ln [Cl^{\bullet}] = \ln [Cl^{\bullet}]_0 - k't \text{ where } t = x/u \text{ and } k' = k[O_3]_0$$

A plot of $\ln [Cl^{\bullet}]$ against t gives a straight line with a gradient of $-k'$ (the pseudo-first order rate constant). The whole set of experiments is then repeated for different $[O_3]_0$. The value of k' is plotted against $[O_3]_0$, which gives a straight line with gradient k (the overall, second order rate constant for the reaction).

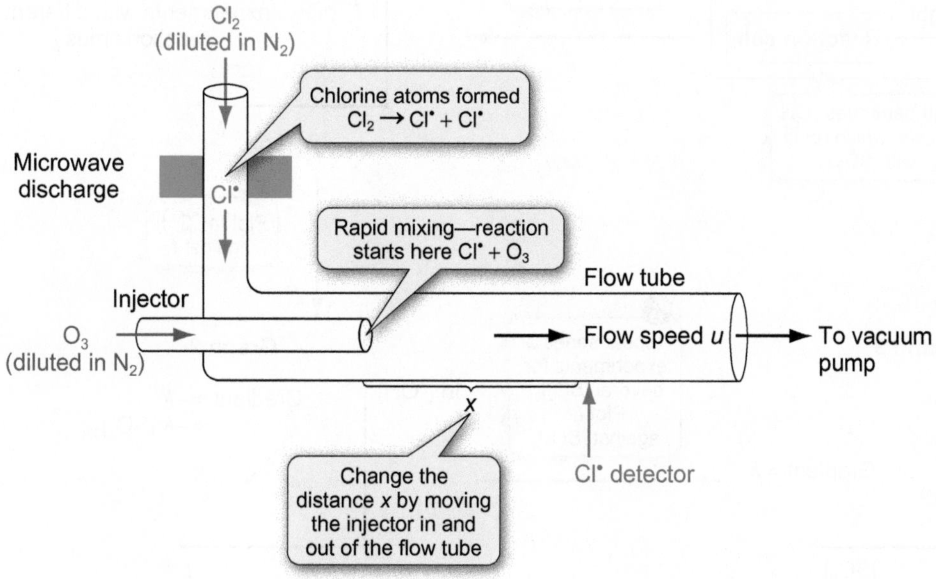
Cl2 (diluted in N2)
Chlorine atoms formed Cl2 → Cl• + Cl•
Microwave discharge
Cl•
Rapid mixing—reaction starts here Cl• + O3
Flow tube
Injector
O3 (diluted in N2)
Flow speed u
To vacuum pump
x
Cl• detector
Change the distance x by moving the injector in and out of the flow tube

Figure 9.14 The discharge flow method for investigating fast reactions.

Flash photolysis

Photolysis is the breaking of bonds in a molecule by the absorption of light.

The thermodynamics of the formation of vapour trails (contrails) from jet aircraft is discussed in Box 15.4 (p.707).

There is a wide variety of flash photolysis techniques using different energy sources and with different methods of monitoring the reaction. The technique can also be adapted to study reactions in solution.

Visit the Online Resource Centre to view screencast 9.3 walks you through Figure 9.15 to explain the stages in a flash photolysis experiment to find the rate constant for the reaction of ${}^{\bullet}OH$ radicals with SO_2.

In one example of this technique, a brief intense flash of light from a laser is used to produce radicals very rapidly by photolysis of a precursor molecule. The radicals generated react with a second reactant already in the reaction cell. The concentration of the radical rises rapidly during the flash and then decays as it reacts. Its concentration is monitored in a series of experiments as a function of time.

For example, the reaction of ${}^{\bullet}OH$ radicals with SO_2 molecules has been studied in this way. (This reaction is important in the formation of vapour trails from jet aircraft.)

$$ {}^{\bullet}OH + SO_2 \rightarrow HOSO_2{}^{\bullet} $$

In this case, the reaction cell initially contains a mixture of H_2O_2 and SO_2. A laser operating at 248 nm photolyses the H_2O_2 molecules to produce ${}^{\bullet}OH$ radicals

$$ H_2O_2 + h\nu \rightarrow 2{}^{\bullet}OH $$

The conditions are arranged so that $[{}^{\bullet}OH] << [SO_2]$ and the reaction shows pseudo-first order kinetics. The concentration of hydroxyl radicals is measured spectroscopically. The sequence of events in the investigation is shown in Figure 9.15.

A series of experiments is carried out with several delay times after the flash, using a constant value of $[SO_2]_0$. The results (Graph 1) give a kinetic profile of $[{}^{\bullet}OH]$ against time. A first order plot of $\ln[{}^{\bullet}OH]$ against time (Graph 2) is a straight line with gradient $-k'$ (the pseudo-first order rate constant). The whole set of experiments is then repeated for different $[SO_2]_0$ and the overall rate constant found from a plot of k' against $[SO_2]_0$ (Graph 3) as described on p.401 for the $Cl^{\bullet} + O_3$ reaction.

The length of the initial laser pulse must be significantly shorter than the timescale of the reaction being studied. Gas phase reactions of radicals occur on microsecond and milli second timescales and are studied using lasers with pulse lengths ~10 ns (nanosecond, 1×10^{-9} s). Faster reactions, such as the reactions involved in photosynthesis, require lasers with shorter pulse lengths ~1 ps (picosecond, 1×10^{-12} s). Even shorter laser pulses on the femtosecond timescale (1 fs = 1×10^{-15} s) have been used to observe the vibrations and

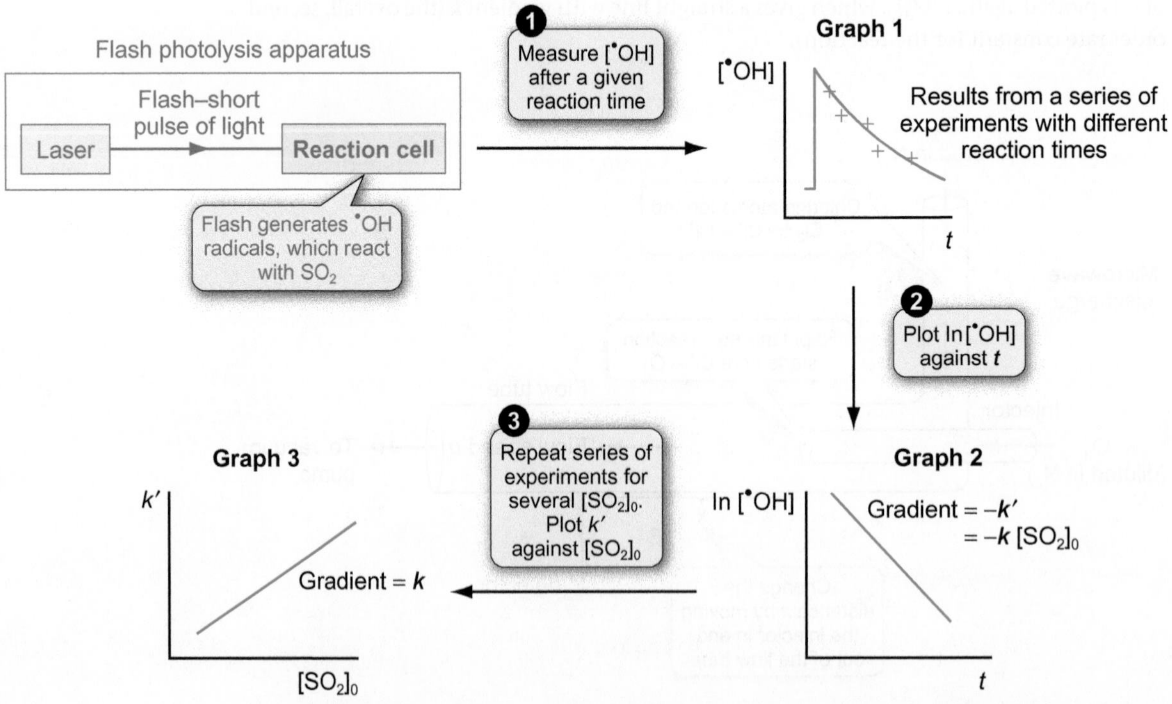

Flash photolysis apparatus
Laser
Flash–short pulse of light
Reaction cell
Flash generates •OH radicals, which react with SO2
1
Measure [•OH] after a given reaction time
Graph 1
[•OH]
Results from a series of experiments with different reaction times
t
2
Plot ln[•OH] against t
Graph 2
ln [•OH]
Gradient = –k′ = –k [SO2]0
t
3
Repeat series of experiments for several [SO2]0. Plot k′ against [SO2]0
Graph 3
k′
Gradient = k
[SO2]0

Figure 9.15 Stages in a flash photolysis experiment, for example, the reaction of ${}^{\bullet}OH$ radicals with SO_2.

Femtosecond lasers are being used to investigate the rapid transfer of solar energy between molecules in photosynthesis.

The Egyptian born chemist, Ahmed Zewail, was awarded the Nobel Prize in 1999 for his development of femtochemistry, the probing of chemical reactions on the femtosecond (1 fs = 1×10^{-15} s) timescale. His technique has been called the world's fastest camera.

rotations of chemical bonds, and their rearrangements, as product molecules are formed from the reactants. The laser is effectively 'freezing' the molecules in time and taking a snapshot. This new area of chemistry is called **femtochemistry**.

Box 9.4 describes one of the earliest uses of flash photolysis and the significance of the results almost 50 years later.

Femtochemistry allows chemists to observe the transition state for a reaction (see Section 9.8, p.434).

Box 9.4 Using flash photolysis to monitor $ClO^{\bullet}$ radicals

Ronald Norrish and George Porter were awarded the Nobel Prize in Chemistry in 1967 for their invention of flash photolysis in the late 1940s. One of the earliest publications using the new technique described the photochemical reaction between Cl_2 and O_2 and the observation of an absorption spectrum due to the short-lived intermediate, $ClO^{\bullet}$. Porter then used this spectrum to monitor the concentration of $ClO^{\bullet}$ at different times to study the kinetics of its decay.

The absorption spectrum was obtained by first firing a photolytic flash lamp (causes photolysis) to produce $ClO^{\bullet}$ radicals and then firing a second flash lamp a measured time delay after the first. Light from the second flash passed through the reaction cell and then through a spectrometer. The spectrum was recorded on a photographic plate. A series of experiments was conducted with different time delays between the two flashes.

Figure 1 is a photograph from Porter's 1952 paper and shows a series of spectra recorded at different times after the photolytic flash. Each spectrum contains a series of bands corresponding to the vibrational structure of the absorption spectrum in the region of 270 nm. The strength of the $ClO^{\bullet}$ absorption decreases with time as the $ClO^{\bullet}$ reacts. The concentration of $ClO^{\bullet}$ was determined from the fraction of light absorbed at a specific wavelength (see Section 10.3, p.458).

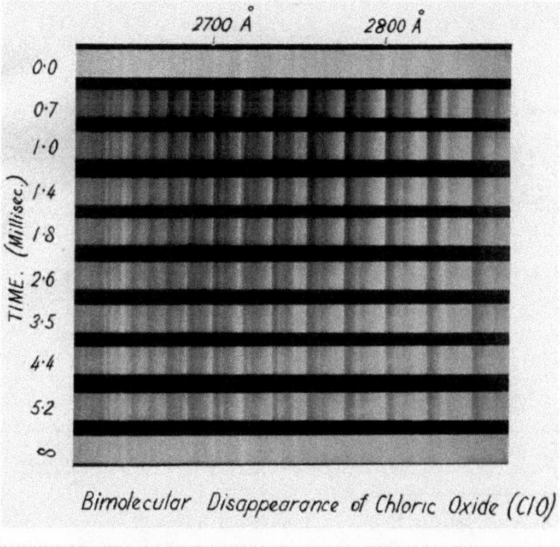
2700 Å
2800 Å
TIME. (Millisec.)
0·0
0·7
1·0
1·4
1·8
2·6
3·5
4·4
5·2
∞
Bimolecular Disappearance of Chloric Oxide (ClO)

Figure 1 $ClO^{\bullet}$ spectra recorded at different times after the photolytic flash.

By plotting $1/[ClO^{\bullet}]$ against time and obtaining a straight line, Porter showed that the decay of $ClO^{\bullet}$ is a second order process and obtained the rate constant for the reaction

$$ClO^{\bullet} + ClO^{\bullet} \rightarrow \text{products}$$

→

➜ Fifty years later, interest in this reaction was reawakened and new studies, based essentially on the same technique as that used by Porter, were conducted using a modern instrument at the Jet Propulsion Laboratory in California. The impetus for the new investigation was the observation of the stratospheric ozone 'hole' by Joe Farman and his colleagues of the British Antarctic Survey at Halley Bay in Antarctica. $ClO^{\bullet}$ is involved in a catalytic cycle leading to ozone depletion (see Boxes 27.6, p.1233 and 27.8, p.1241).

However, the catalytic cycle cannot explain the dramatic reduction in ozone in the Antarctic spring when the concentration of oxygen atoms is very low. (In the cycle, $ClO^{\bullet}$ radicals react with O to regenerate $Cl^{\bullet}$ radicals.) An alternative mechanism for forming $Cl^{\bullet}$ from $ClO^{\bullet}$ was needed and the second order reaction of $ClO^{\bullet}$, first investigated by Porter, provided a possible route to $Cl^{\bullet}$, via its dimer, ClOOCl

$$ClO^{\bullet} + ClO^{\bullet} \rightarrow ClOOCl$$

$$ClOOCl + h\nu \rightarrow Cl^{\bullet} + ClOO^{\bullet}$$

$$ClOO^{\bullet} \rightarrow Cl^{\bullet} + O_2$$

Experiments (like the one in Figure 2) were designed to measure the yields of all the possible products from the reaction $ClO^{\bullet} + ClO^{\bullet}$. The results showed that dimer formation is the major route to the formation of $Cl^{\bullet}$ radicals from $ClO^{\bullet}$ in the Antarctic spring.

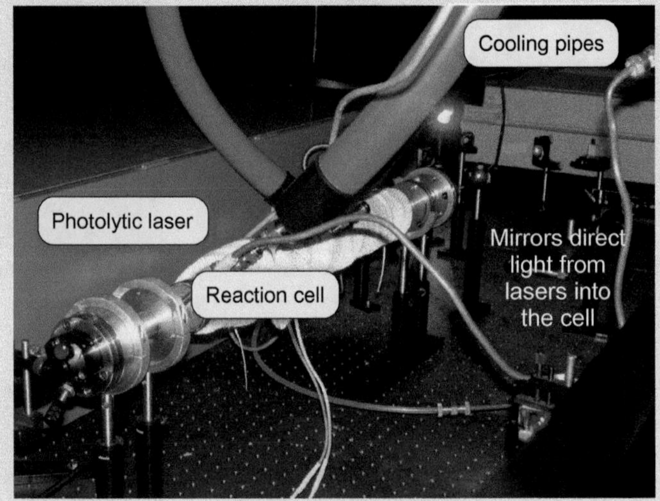
Cooling pipes
Photolytic laser
Mirrors direct light from lasers into the cell
Reaction cell

Figure 1 A flash photolysis apparatus set up to monitor the decay of radicals. The photolytic laser is on the left. The probe laser (which fires the second flash) is off to the right. Mirrors direct the light from the two lasers into the reaction cell. The red light being reflected by the mirrors is from the probe laser.

Question

Write a differential rate equation and an integrated rate equation for the formation of the $ClO^{\bullet}$ dimer from $ClO^{\bullet}$.

Summary

- An elementary reaction is a single step reaction involving one or two molecules or atoms.
- For a general reaction $aA + bB \rightarrow pP + qQ$

$$\text{rate of reaction} = -\frac{1}{a}\frac{d[A]}{dt} = -\frac{1}{b}\frac{d[B]}{dt} = \frac{1}{p}\frac{d[P]}{dt} = \frac{1}{q}\frac{d[Q]}{dt}$$

- The rate equation for an elementary reaction can be written directly from the chemical equation for the reaction.
- A rate equation in which the rate of consumption of a reactant, or the rate of formation of a product, is written as a differential is called a differential rate equation.
- An integrated rate equation gives the concentration of a reactant as a function of time.
- The half life of a reactant is the time taken for its concentration to fall to half its initial value.
- Table 9.2 summarizes the equations for the first order, second order, and pseudo-first order elementary reactions that you have met in Section 9.4.

Table 9.2 A summary of equations for elementary reactions

Type of reaction	Order of reaction	Differential rate equation	Integrated rate equation	Half life, $t_{1/2}$
$A \rightarrow$ products	First order	$-\frac{d[A]}{dt} = k[A]$	$\ln[A]_t = \ln[A]_0 - kt$	$\frac{\ln 2}{k}$
$A + A \rightarrow$ products	Second order	$-\frac{d[A]}{dt} = 2k[A]^2$	$\frac{1}{[A]_t} = \frac{1}{[A]_0} + 2kt$	$\frac{1}{2k[A]_0}$
$A + B \rightarrow$ products where $[A] << [B]$	Pseudo-first order	$-\frac{d[A]}{dt} = k'[A]$ where $k' = k[B]_0$	$\ln[A]_t = \ln[A]_0 - k't = \ln[A]_0 - kt[B]_0$	$\frac{\ln 2}{k'} = \frac{\ln 2}{k[B]_0}$

? **For practice questions on the topics in Sections 9.1–9.4, see questions 2, 3, 6, 8–13, and 15–17 at the end of this chapter (pp.444–445).**

9.5 Complex reactions: experimental methods

Most of the reactions you will meet are not simple elementary reactions. They are **complex reactions** in which the mechanism involves a series of elementary reactions. For a complex reaction, you cannot write down the overall rate equation directly from the stoichiometric equation for the reaction. The rate equation *must* be determined from experiments to investigate how the rate depends on the concentration of each of the reactants.

For a general reaction in which A and B are reactants

$$a\text{A} + b\text{B} \rightarrow \text{products}$$

the overall rate equation is given by an equation of the form

$$\text{rate of reaction} = k[\text{A}]^m[\text{B}]^n \qquad (9.11)$$

where k is the **overall rate constant** for the reaction and m and n are the **orders of the reaction**, with respect to A and B, respectively. The orders often have values of 0, 1, or 2, but sometimes can take higher, or even fractional, values. The **overall order** of the reaction is $(m + n)$.

Note that the stoichiometric coefficients in the chemical equation, a and b, do not appear in the overall rate equation. It may sometimes turn out that m and n take the values of a and b, *but you cannot assume this*.

Sometimes a substance can appear in the rate equation even though it does not appear among the reactants in the stoichiometric equation. This might be a product of the reaction or a catalyst. If the rate of the overall reaction is found experimentally to be unaffected by the concentration of a reactant, the reaction is said to be **zero order** with respect to that reactant ($m = 0$) and it does not appear in the overall rate equation.

Here are some examples of complex reactions.

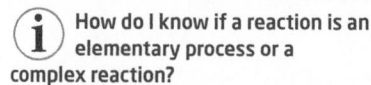

How do I know if a reaction is an elementary process or a complex reaction?
Fast reactions involving a radical reactant (and their reverse reactions) are generally elementary reactions. You should always assume that a reaction is a complex reaction unless you read otherwise. Reactions of the type A + B + C or A + 2 B must be complex because the probability of more than two reactant molecules colliding simultaneously is very low.

Remember For a complex reaction, the rate equation and the reaction order **must** be determined from experiments to investigate how reaction rate depends on the concentration of each of the reactants. You cannot predict them by simply looking at the stoichiometric equation for the reaction.

Visit the Online Resource Centre to view screencast 9.4 which explains the difference between elementary and complex reactions.

Example 1 The reaction of iodide ions (I^-) with peroxodisulfate ions ($S_2O_8^{2-}$)

The equation for the reaction is

$$S_2O_8^{2-}(aq) + 2I^-(aq) \rightarrow 2SO_4^{2-}(aq) + I_2(aq)$$

Experiments show that that the rate equation for the reaction is

$$\text{rate of reaction} = k[S_2O_8^{2-}(aq)][I^-(aq)]$$

so this reaction is first order with respect to $S_2O_8^{2-}$, first order with respect to I^-, and second order overall.

Example 2 The reaction of propanone (CH_3COCH_3) with iodine (I_2)

The equation for the reaction is

$$CH_3COCH_3(aq) + I_2(aq) \rightarrow CH_3COCH_2I(aq) + H^+(aq) + I^-(aq)$$

The mechanism of the acid-catalysed reaction of propanone with halogens is discussed in Section 23.3 (p.1082).

The reaction is catalysed by acids.

Experiments show that the rate equation for the reaction is

$$\text{rate of reaction} = k[CH_3COCH_3(aq)][H^+(aq)]$$

so this reaction is first order with respect to both CH_3COCH_3 and H^+. Note that the catalyst, H^+, appears in the rate equation even though it is not used up in the reaction. In contrast, I_2 does not appear in the rate equation, even though it is a reactant. The reaction is zero order with respect to I_2. The overall order of the reaction is two.

Example 3 The reaction between hydrogen (H_2) and iodine (I_2)

The equation for the reaction is

$$H_2(g) + I_2(g) \rightarrow 2HI(g)$$

Experiments show that the rate equation for the reaction is

$$\text{rate of reaction} = k[H_2(g)][I_2(g)]$$

so this reaction is first order with respect to H_2 and first order with respect to I_2, and second order overall. (Further experiments show that this is not an elementary reaction, but a complex reaction involving I atoms.)

Example 4 The reaction between hydrogen (H_2) and bromine (Br_2)

The equation for the reaction is

$$H_2(g) + Br_2(g) \rightarrow 2HBr(g)$$

Experiments show that the rate equation for the reaction is

$$\text{rate of reaction} = k[H_2(g)][Br_2(g)]^{1/2}$$

so this reaction is first order with respect to H_2, but has an order of $\frac{1}{2}$ with respect to bromine. The overall order is $\frac{3}{2}$.

Compare this with Example 3. Similar looking reactions do not always have the same kinetics.

The experimentally determined orders for complex reactions give chemists an insight into the **mechanisms** by which reactions occur (see Section 9.6, p.416). The mechanism of the reaction $H_2 + Br_2$ involves a series of four elementary reactions and is described in more detail in Section 9.6 (p.423).

Choosing an experimental method

To investigate the kinetics of a complex reaction, you need to determine how the rate of the reaction depends on the concentration of each of the reactants—and also on the concentration of any catalyst involved. From these experiments you can find the order of the reaction with respect to each of these components, then construct an overall rate equation and find the overall rate constant for the reaction.

The first step is to decide on an experimental method for monitoring the concentration of one of the reactants (or a product) as a function of time. This is discussed in Section 9.3. Box 9.5 describes a method for studying fast reactions in solution.

Having decided how you are going to monitor the reaction, you then need to think about how to carry out the experiments and analyse the results. There is a range of procedures available. Which one to choose depends on the particular reaction, what information you require, and the accuracy needed.

In all kinetics experiments, it is important to carry out the reactions at a constant temperature (e.g. in a thermostatically controlled water bath).

Box 9.5 The stopped-flow technique

The stopped-flow technique is used for studying fast reactions in solution. It is essentially a way of rapidly mixing the reactants.

Two solutions, containing the separate reactants, are contained in syringes, which are mechanically driven to force the solutions into the mixing chamber and then down a short tube and into a third syringe. The solutions enter the mixing chamber turbulently and this promotes rapid mixing. After a short time, the plunger in the third syringe hits a stop and the flow is immediately halted. The concentration of one of the reactants, or of a product, is then monitored as a function of time using, for example, absorption spectroscopy, and a kinetic profile is determined.

The kinetics of the reaction (order and rate constant) are determined by measuring kinetic profiles for a range of initial concentrations. Typical reaction timescales are ~20 ms.

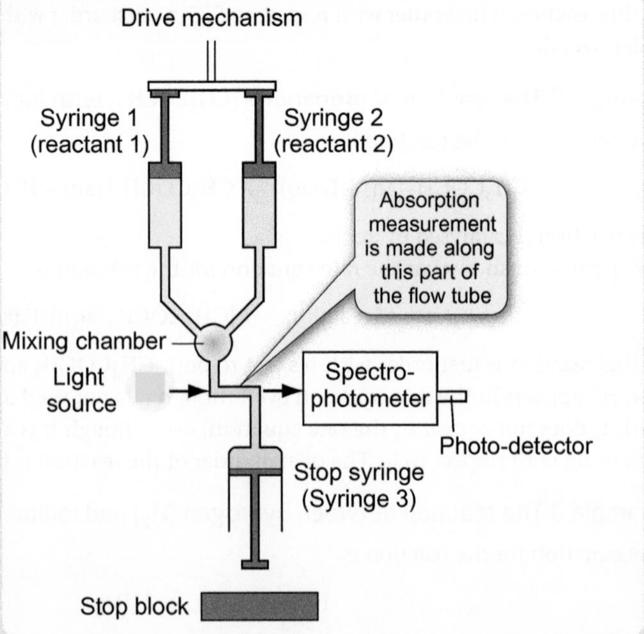
Drive mechanism
Syringe 1
(reactant 1)
Syringe 2
(reactant 2)
Absorption measurement is made along this part of the flow tube
Mixing chamber
Light source
Spectro-photometer
Photo-detector
Stop syringe
(Syringe 3)
Stop block

Diagram of a stopped-flow apparatus.

The isolation technique

The most common type of reaction involves more than one reactant. The trouble is that, as the reaction proceeds, the concentrations of *all* the reactants decrease and affect the overall rate of the reaction. This problem is overcome by isolating one reactant at a time, by having all the other reactants in large excess. As a result, their concentrations can be assumed to be constant throughout the reaction and the effect of a single reactant on the rate of reaction can be studied. This is sometimes called the **isolation technique.**

For a reaction of the type, A + B → products, where the rate of reaction = $k[A]^m[B]^n$, you would need to conduct two sets of experiments. In the first set, the concentration of B is in large excess, so that the order with respect to A can be determined. Rate of reaction = $k'_A[A]^m$, where k'_A is an effective rate constant for the reaction of A, and $k'_A = k[B]_0^n$.

In a second set of experiments, the concentration of A is in large excess, and the order with respect to B is determined. Rate of reaction = $k'_B[B]^n$, where k'_B is an effective rate constant for the reaction of B, and $k'_B = k[A]_0^m$.

The next task is to decide what method to use in each set of experiments to determine the order with respect to a single reactant. There are a number of options:

- drawing tangents to the concentration–time curve;
- initial rates method;
- using integrated rate equations;
- using half lives.

The **isolation technique** is introduced in Section 9.4 (p.395) where pseudo-first order elementary reactions are discussed. It is used in conjunction with one of the methods listed below (except the initial rates method where it is unnecessary).

Drawing tangents to the concentration-time curve

First, the readings of concentration of the reactant A against time are used to plot a kinetic profile. If the plot is a straight line, the rate of reaction is independent of the concentration of A and the reaction is zero order with respect to A. In this case, rate of reaction = k and the rate constant can be found from the gradient of the straight line.

More likely, the kinetic profile will be a curve of the type shown in Figure 9.16. Finding the rate of reaction at different values of [A] involves drawing tangents to the curve at different points and calculating the gradient of the tangent at each point. The gradient of each tangent is a measure of the rate of the reaction at that concentration of A.

Zero order reactions may seem odd. How can the rate of reaction be independent of one of the reactants? This is explained in Section 9.6, p.417.

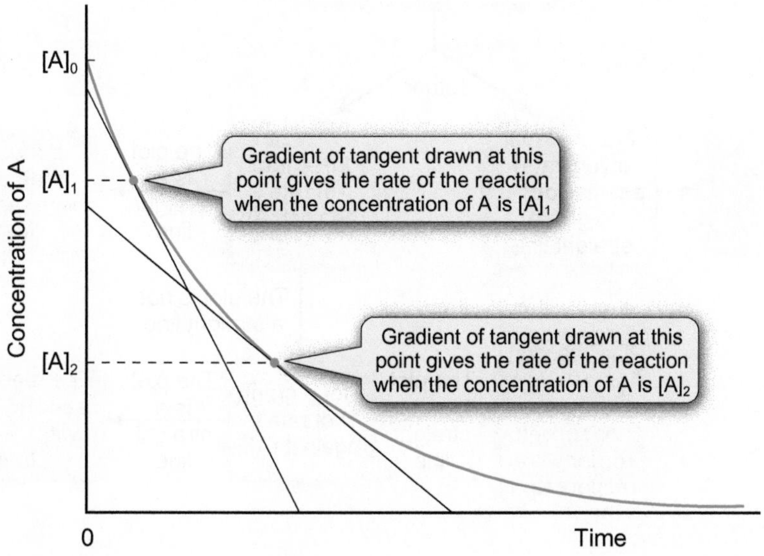
[A]0
[A]1
[A]2
Concentration of A
Gradient of tangent drawn at this point gives the rate of the reaction when the concentration of A is [A]1
Gradient of tangent drawn at this point gives the rate of the reaction when the concentration of A is [A]2
0
Time

Figure 9.16 Drawing tangents to a concentration–time curve.

- You can then use the values of the rates of reaction determined from the tangents to plot a graph of the rate against [A]. If this graph is a straight line, the rate of reaction is directly proportional to [A] and the reaction is pseudo-first order with respect to A.
- If the plot is a curve, the reaction is not first order. You cannot, however, assume it must then be second order with respect to A—it might be third order, or fractional. You must now plot a graph of the rate against $[A]^2$. If this plot is a straight line, the reaction is second order with respect to A. If this plot is also a curve, the reaction is neither first nor second order with respect to A.

Σ $\log ab = \log a + \log b$ and $\log a^b = b \log a$
Further help with logarithms can be found in Maths Toolkit MT3 (p.1316).

A less laborious and more general method of analysing the data involves plotting the *logarithms* of the rate and the concentration (called a **log–log plot**).

$$\text{rate of reaction} = k'[A]^m$$

Taking logarithms (either $\log_{10}$ or $\log_e$) of each side of the equation gives $\log(\text{rate}) = \log(k'[A]^m)$, so that

$$\log(\text{rate}) = m\log[A] + \log k' \tag{9.12}$$

Σ $y = mx + c$
$\log(\text{rate}) = m\log[A] + \log k'$
(here m is the order with respect to A)

This equation has the form of a straight line, $y = mx + c$, where m is the gradient and c the intercept. A plot of log(rate) against log[A] is *always* a straight line, whatever the order of the reaction with respect to A. The order is found from the gradient of the line and the rate constant is found from the intercept. The sequence of reasoning is summarized in Figure 9.17.

It is difficult to draw accurate tangents to a curve—particularly if there is some scatter in the experimental data that do not fit on to a smooth curve. To help, there are software packages that allow you to feed in the concentration–time data to obtain the gradient as

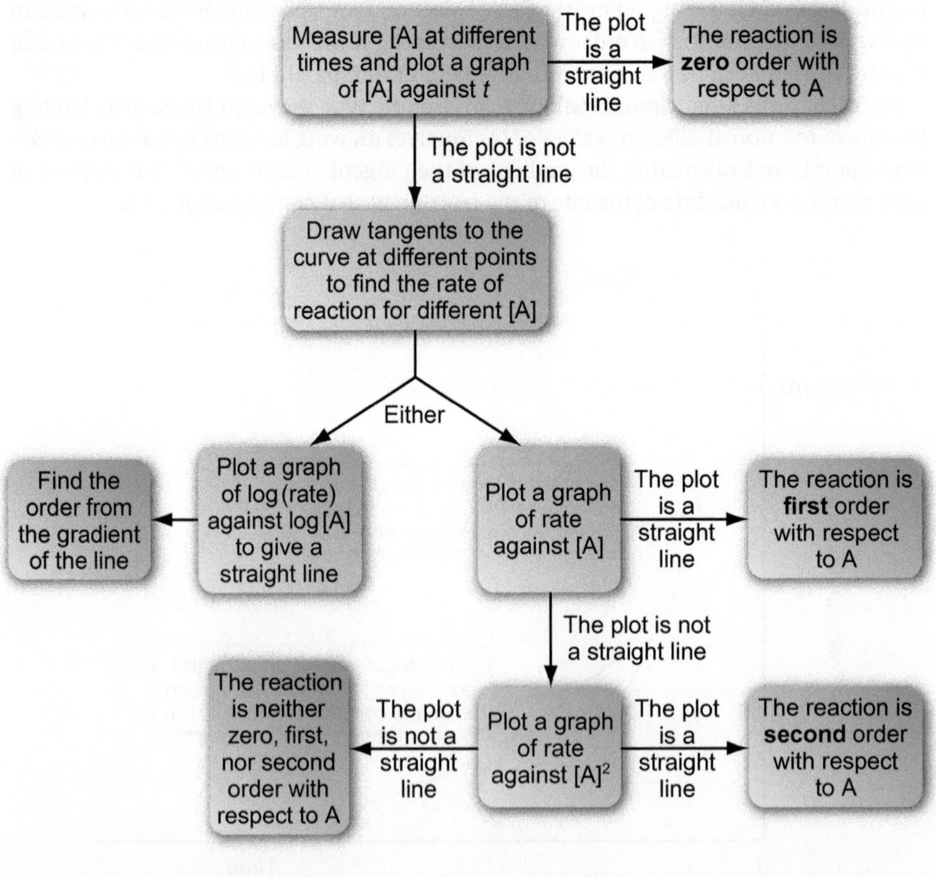

Measure [A] at different times and plot a graph of [A] against t
The plot is a straight line
The reaction is zero order with respect to A
The plot is not a straight line
Draw tangents to the curve at different points to find the rate of reaction for different [A]
Either
Find the order from the gradient of the line
Plot a graph of log (rate) against log [A] to give a straight line
Plot a graph of rate against [A]
The plot is a straight line
The reaction is first order with respect to A
The plot is not a straight line
The reaction is neither zero, first, nor second order with respect to A
The plot is not a straight line
Plot a graph of rate against [A]²
The plot is a straight line
The reaction is second order with respect to A

Figure 9.17 A summary of the sequence of reasoning to determine the order with respect to A by drawing tangents to a concentration–time curve and subsequent graphical methods.

a function of concentration. Even so, the method is not used in accurate work. Log–log plots are quite good for determining reaction orders, but tend to give large errors on the values of rate constants.

Initial rates method

The initial rates method provides a more accurate variant of the previous method. As before, readings of concentration of a reactant (or product) at different times are used to plot a kinetic profile, but this time only one tangent is drawn to the curve at $t = 0$. The gradient of the tangent gives the rate of reaction at the start of the reaction. The experiment is then repeated several times using different initial concentrations.

For example, to investigate the decomposition of dinitrogen pentoxide

$$2\,N_2O_5\,(g) \rightarrow 4\,NO_2\,(g) + O_2\,(g)$$

a series of separate experiments is carried out, each with a different initial concentration of N_2O_5. In each experiment, the concentration of N_2O_5 is monitored as the reaction proceeds and a graph of $[N_2O_5]$ against time is plotted to give a kinetic profile. For each graph, only the tangent at $t = 0$ is drawn, so the reaction need only be followed long enough to allow an accurate tangent to be drawn at this point. The gradients of the tangents give the initial rate of the reaction for different initial concentrations of N_2O_5.

The way to process and interpret the initial rate data to assign a reaction order follows the reasoning in Figure 9.17. You can often get a preliminary idea of the orders by inspecting the tabulated data as described in Worked example 9.5. The most reliable method is to plot a graph of log(initial rate) versus log(initial concentration) for each reactant (see p.408).

Figure 9.18(a) shows the tangents drawn to find the initial rates in five separate experiments. Figure 9.18(b) shows a plot of initial rate against initial concentration. This is a straight line, showing that the rate is proportional to $[N_2O_5]$ and the reaction is first order.

The initial rate method still has the disadvantage of drawing a tangent to a curve, and the tangent at $t = 0$ is particularly difficult to draw accurately. This problem can be overcome by measuring $\Delta[N_2O_5]$, the amount of reactant consumed in a given time Δt, instead of drawing the tangent. Provided $\Delta[N_2O_5]$ is less than about 10% of $[N_2O_5]_0$, $\dfrac{\Delta[N_2O_5]}{\Delta t}$ is a good approximation to the initial rate. This is illustrated in Figure 9.19.

The **initial rate** of a reaction is the rate of change in concentration of a reactant at the instant the reaction starts.

For a reaction involving more than one reactant, the isolation technique is not necessary when using the initial rates method, since the concentrations of all the reactants are approximately constant (at their initial values) for the time interval of the initial rate.

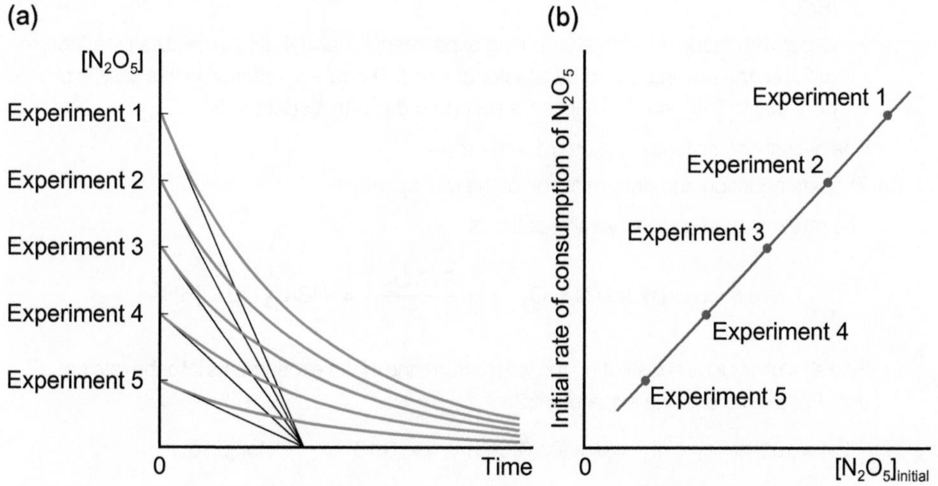

(a)
[N2O5]
Experiment 1
Experiment 2
Experiment 3
Experiment 4
Experiment 5
0
Time
(b)
Initial rate of consumption of N2O5
Experiment 1
Experiment 2
Experiment 3
Experiment 4
Experiment 5
0
[N2O5]initial

Figure 9.18 (a) The initial rate of reaction for each experiment (1–5) is found by drawing a tangent to the curve at the start of the reaction. (b) A plot of the five initial rates obtained in (a) against the initial concentration of N_2O_5 for each experiment.

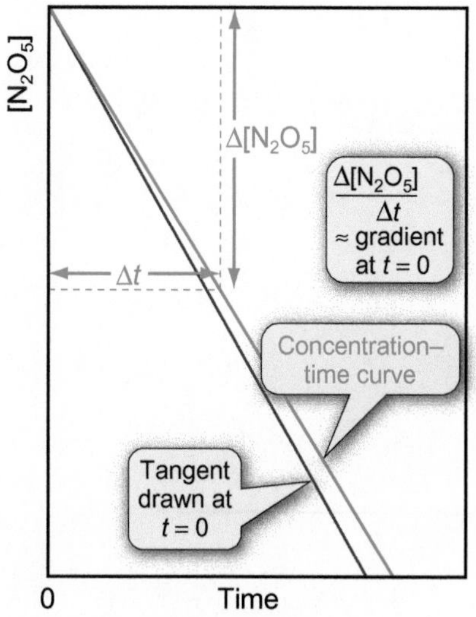

[N2O5]
Δ[N2O5]
Δ[N2O5] / Δt ≈ gradient at t = 0
Δt
Concentration–time curve
Tangent drawn at t = 0
0
Time

Figure 9.19 This is the first part of one of the curves in Figure 9.18(a), near $t = 0$. It shows $\Delta[N_2O_5]$ for a 10% change in concentration. The error introduced by approximating the gradient to $[N_2O_5]/\Delta t$ is 4.8%.

Br^- (aq) and BrO_3^- (aq) are both colourless, but in aqueous solution they form coloured Br_2 (aq). The colour change can be used to monitor the rate of the reaction.

Worked example 9.5 Using the initial rate method to investigate the reaction between bromate ions and bromide ions

Bromate ions (BrO_3^-) react with bromide ions (Br^-) in acid solution to form bromine (Br_2)

$$BrO_3^-(aq) + 5\,Br^-(aq) + 6\,H^+(aq) \rightarrow 3\,Br_2(aq) + 3\,H_2O(l)$$

The concentration of BrO_3^- (aq) was monitored at 298 K in four separate experiments, each with different initial concentrations of the reactants as shown in the following table.

	Initial concentration / mol dm^{-3}			
Experiment	[BrO_3^-(aq)]	[Br^-(aq)]	[H^+(aq)]	Initial rate / 10^{-3} mol dm^{-3} s^{-1}
1	0.10	0.10	0.10	1.2
2	0.20	0.10	0.10	2.4
3	0.10	0.30	0.10	3.6
4	0.20	0.10	0.20	9.6

(a) Determine the order of reaction with respect to each reactant and give the overall order of reaction.

(b) Write the overall rate equation for the reaction and find a value of the rate constant, k, at 298 K.

Strategy

By inspecting the data, find how the initial rate varies for each reactant as its concentration changes. To isolate the effect of each reactant, compare experiments that differ in the concentration of one substance at a time.

Solution

(a) Finding the orders of reaction.

- Order with respect to BrO_3^-. Comparing experiments 1 and 2, [BrO_3^-] doubles and the rate doubles, so the reaction is first order with respect to BrO_3^-.
- Order with respect to Br^-. Comparing experiments 1 and 3, [Br^-] increases by a factor of 3 and the rate increases by a factor of 3, so the reaction is first order with respect to [Br^-].
- Order with respect to H^+. Comparing experiments 2 and 4, [H^+] increases by a factor of 2, but the rate increased by a factor of 4 (2^2). The rate of reaction in this case is proportional to $[H^+]^2$, so the reaction is second order with respect to H^+.

The overall order of the reaction = 1 + 1 + 2 = 4.

(b) The rate equation and determination of the rate constant.

The rate equation for the overall reaction is

$$\text{rate of consumption of } BrO_3^- = -\frac{E[BrO_3^-]}{dt} = k[BrO_3^-][Br^-][H^+]^2$$

Find a value for k by substituting the values from one of the experiments into the rate equation. For example, using experiment 2

$$2.4 \times 10^{-3}\,\text{mol dm}^{-3}\,\text{s}^{-1} = k \times (0.20\,\text{mol dm}^{-3}) \times (0.10\,\text{mol dm}^{-3}) \times (0.10\,\text{mol dm}^{-3})^2$$

$$k = 12\,\text{dm}^9\,\text{mol}^{-3}\,\text{s}^{-1}$$

The average value of k calculated from the four experiments is 12 dm^9 mol^{-3} s^{-1}.

→

Question

The acid-catalysed bromination of propanone was investigated using the initial rate method at 298 K

$$CH_3COCH_3\,(aq) + Br_2\,(aq) \xrightarrow{H^+} CH_3COCH_2Br\,(aq) + Br^-\,(aq) + H^+\,(aq)$$

The concentration of Br_2 was monitored in five separate experiments. By inspection of the following data, determine the order of the reaction with respect to each of the substances in the table.

	Initial concentration / mol dm^{-3}			
Experiment	[CH_3COCH_3 (aq)]	[Br_2 (aq)]	[H^+ (aq)]	Initial rate / 10^{-5} mol dm^{-3} s^{-1}
1	0.30	0.05	0.05	5.60
2	0.30	0.10	0.05	5.60
3	0.30	0.05	0.10	11.1
4	0.40	0.05	0.20	30.5
5	0.40	0.05	0.05	7.55

Using integrated rate equations

Integrated rate equations provide the most accurate method of determining reaction orders and rate constants, and are widely used in experimental work in chemical kinetics. The integrated rate equations derived in Section 9.4 for first and second order elementary reactions can be applied to complex reactions, when considering the order with respect to a single reactant. This is achieved using the isolation technique described in Section 9.5 (p.407), in which all reactants except the one under investigation are in excess. The rate constant in the integrated rate equation is then an effective rate constant and the order determined is called a **pseudo-order.**

For a pseudo-first order reaction, the integrated rate equation is

$$\ln[A]_t = \ln[A]_0 - k't \text{ (where } k' \text{ is the effective rate constant)} \qquad \text{(from 9.6a)}$$

so a plot of ln[A] against t is a straight line with a gradient of $-k'$.

For a second order reaction, the integrated rate equation is

$$\frac{1}{[A]_t} = \frac{1}{[A]_0} + 2k't \qquad \text{(from 9.7b)}$$

so a plot of $\frac{1}{[A]}$ against t is a straight line with a gradient of $+2k'$.

The integrated rate equation for a zero order reaction is derived in Box 9.6 and is given by

$$[A]_t = [A]_0 - k't \qquad (9.13)$$

so a plot of $[A]_t$ against t is a straight line with a gradient of $-k'$. Box 9.7 describes an example of a zero order reaction.

The advantage of integrated rate equations is they allow you to plot your experimental readings of concentration and time directly as straight lines, without having to calculate the rate of reaction from the gradient of a tangent to the concentration–time plot. (They also allow you to make good use of all of your data as it becomes harder to draw tangents towards the end of a reaction.)

Worked example 9.6 illustrates the use of integrated rate equations in the analysis of the kinetics of a complex reaction.

Integrated rate equations are derived and explained for first and second order elementary reactions in Section 9.4 (p.389). The same equations apply for complex reactions when investigating the order with respect to a single reactant.

ⓘ The '2' in Equation 9.7b arises from the stoichiometric equation for the reaction (see Section 9.4, p.390 and p.393). When using Equation 9.7b for a complex reaction, check whether the '2' is appropriate for the particular reaction being studied.

ⓘ $y = mx + c$

$[A]_t = -k't + [A]_0$

ⓘ For a zero order reaction, rate = k, so the units of k are the same as the units of rate: mol dm^{-3} s^{-1}.

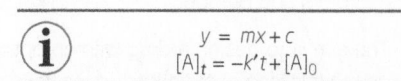

Box 9.6 Deriving the integrated rate equation for a zero order reaction

This follows the general procedure given for deriving the integrated rate equation for a first order reaction in Box 9.1 (p.393).

The differential rate equation for a pseudo-zero order reaction

$$A \rightarrow \text{products}$$

is given by

$$\text{rate of reaction} = -\frac{d[A]}{dt}$$
$$= k'$$

Separating the variables, the equation can be rewritten as

$$d[A] = -k'dt$$

Integrate both sides of the equation from $[A] = [A]_0$ at $t = 0$, to $[A] = [A]_t$ at time t.

$$\int_{[A]_0}^{[A]_t} d[A] = -k'\int_0^t dt$$

This gives

$$[[A]_t]_{[A]_0}^{[A]_t} = -k'[t]_0^t$$
$$([A]_t - [A]_0) = -k'(t-0)$$

which can be rearranged to

$$[A]_t = [A]_0 - k't \quad (9.13)$$

Box 9.7 Pharmacokinetics

The study of how a drug reaches its target and what happens to it in the body is known as **pharmacokinetics**. This information is important in the pharmaceutical industry when a new drug is being developed. There is no point in perfecting the structure of a new compound to bind to a specific receptor site, if the compound has no chance of reaching it.

There are four stages to consider:

- **drug absorption**: this is the route by which the drug reaches the bloodstream from the point of administration;
- **drug distribution**: once absorbed, the drug is distributed around the body in the blood supply and then more slowly from there to the tissues and organs;
- **drug metabolism**: this is the breaking down of the drug in the body due to the action of enzymes;
- **drug excretion**: the main route for excretion is via the kidneys and bladder.

The aim is to design a drug with an optimum lifetime in the body—one that is long enough for it to reach its target and be effective, but short enough to ensure it does not build up in the body.

The pharmacokinetics of ethanol have been studied and the results used to set the legal limits for alcohol in the blood and urine. Ethanol is rapidly absorbed into the bloodstream from the stomach and small intestine, and is quickly distributed to all parts of the body. Its metabolism in the liver is catalysed by a dehydrogenase enzyme (see Box 23.3 (p.1066)) and the degradation process follows zero order kinetics with respect to ethanol. This is because all the active sites in the enzyme molecules become filled with ethanol molecules.

▲ The degradation of ethanol in the body follows zero order kinetics.

The slow step in the degradation takes place in the active sites and the kinetics of this step governs the overall kinetics. The concentration of active sites with bound ethanol molecules is constant, so the rate of reaction is independent of the concentration of unbound ethanol. (The kinetics of enzyme-catalysed reactions is explained more fully in Box 9.8, p.440.)

Question

The breakdown in the body of the chemotherapy drug, cisplatin, is found to follow first order kinetics. The rate constant at body temperature is $1.87 \times 10^{-3}\,\text{min}^{-1}$. The concentration of the drug in the body of a cancer patient is $5.16 \times 10^{-4}\,\text{mol dm}^{-3}$. What will the concentration be after 24 hours?

Using half lives

The expressions for half lives, $t_{1/2}$, derived in Section 9.4 for first order and second order elementary reactions can be applied to complex reactions when considering the order with respect to a single reactant. The most useful half life is the one for first order reactions, which is a constant, and so independent of the concentration of A

$$t_{1/2} = \frac{\ln 2}{k} \tag{9.9}$$

Half lives for first and second order elementary reactions are discussed in Section 9.4 (p.396). The same expressions for half lives apply for complex reactions when investigating the order with respect to a single reactant (see Table 9.2, p.404).

Worked example 9.6 Using integrated rate equations to investigate the reaction of iodine with hex-1-ene

Hex-1-ene reacts with I_2 in an addition reaction to give 1,2-diiodohexane. (The addition of halogens to alkenes is discussed in Section 21.3, p.970.)

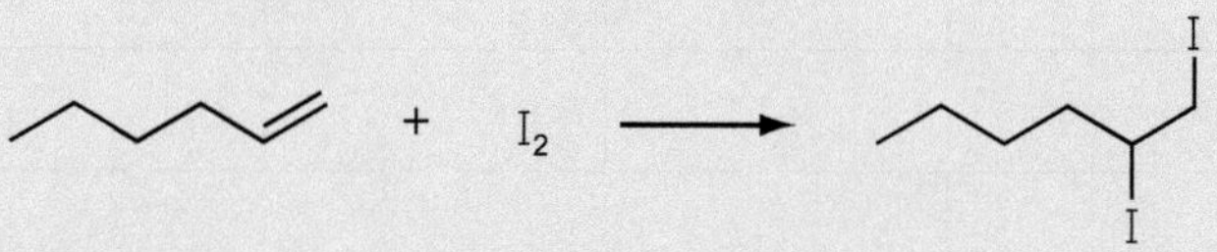

+
I2
I
I

The reaction was carried out in solution at 298 K using a large excess of hex-1-ene. Use the data in the following table to find the order of the reaction with respect to I_2.

t/s	0	500	1500	2500	3500	4500	5500	6500	7500
$[I_2]/10^{-3}$ mol dm^{-3}	20.0	17.5	14.1	11.7	10.1	8.9	7.9	7.1	6.5

Strategy

The aim of this worked example is to demonstrate the use of integrated equations in finding the order of the reaction with respect to I_2.

Assuming no other substances are involved, the overall rate equation for the reaction can be written as

$$\text{rate of reaction} = \frac{d[I_2]}{dt} = k[I_2]^m[\text{hex-1-ene}]^n$$

Since hex-1-ene is in large excess, [hex-1-ene] can be assumed to be constant throughout the reaction, so

$$-\frac{d[I_2]}{dt} = k'[I_2]^m$$

where k' is an effective rate constant.

To find m, use integrated rate equations for zero, first, and second order reactions to plot $[I_2]$ against t, $\ln[I_2]$ against t, and $1/[I_2]$ against t, respectively. Whichever plot gives a straight line corresponds to the integrated rate equation for the reaction.

Solution

- To test for *zero order kinetics*, use the integrated rate equation

$$[A]_t = [A]_0 - k't \tag{9.13}$$

Use the data in the question to plot a graph of $[I_2]$ against time. →

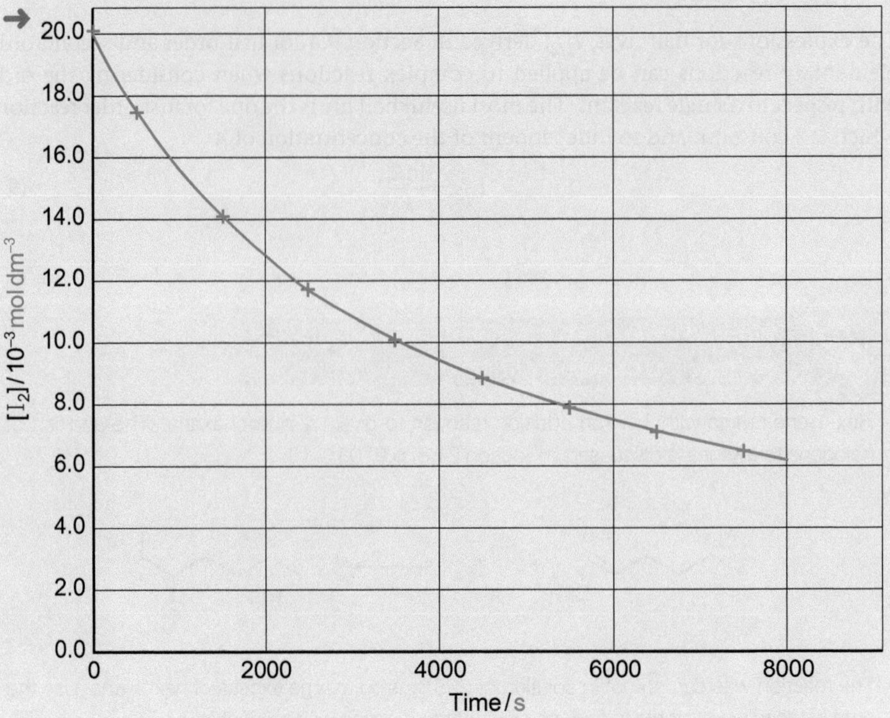

20.0
18.0
16.0
14.0
12.0
10.0
8.0
6.0
4.0
2.0
0.0
[I2]/10−3 mol dm−3
0
2000
4000
6000
8000
Time / s

The plot is not a straight line, so the reaction is not zero order with respect to I_2.

- To test for *first order kinetics*, use the integrated rate equation

$$\ln[A]_t = \ln[A]_0 - k't \qquad (9.6a)$$

and plot a graph of $\ln[I_2]$ against time.

t/s	0	500	1500	2500	3500	4500	5500	6500	7500
$\ln([I_2]/\text{mol dm}^{-3}$	−3.91	−4.05	−4.26	−4.45	−4.60	−4.72	−4.84	−4.95	−5.04

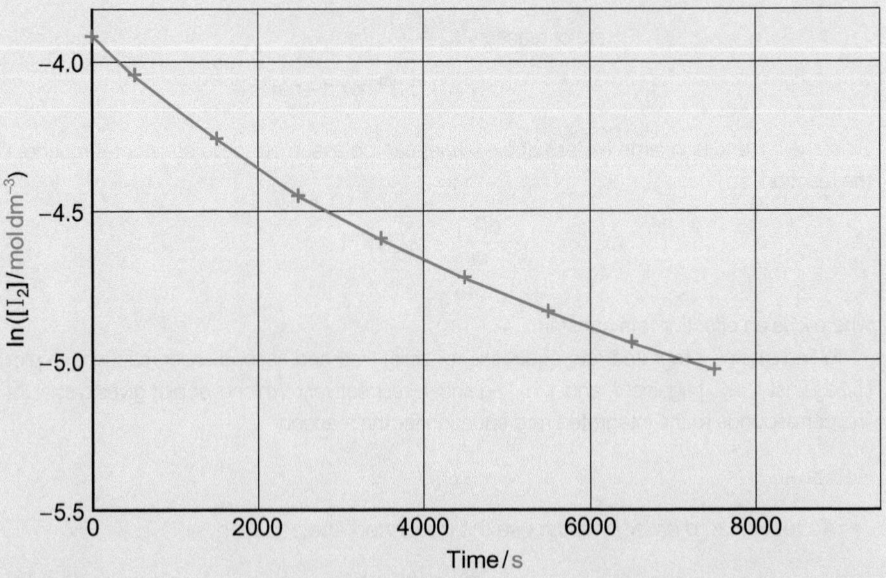

−4.0
−4.5
−5.0
−5.5
ln([I2]/mol dm−3)
0
2000
4000
6000
8000
Time / s

The plot is not a straight line, so the reaction is not first order with respect to I_2.

The logarithm of $([I_2]/\text{mol dm}^{-3})$ is plotted here. You could have plotted the logarithm of $[I_2]/10^{-3}\text{ mol dm}^{-3}$ and this would also give a curve (see the maths note in Worked example 9.4 (p.398)). A logarithm has no units, but you should always indicate on the y-axis the units you have used for the concentration in calculating the logarithm.

- To test for *second order kinetics*, use the integrated rate equation

$$\frac{1}{[A]_t} = \frac{1}{[A]_0} + k't \qquad \text{(from 9.7b)}$$

and plot a graph of $\frac{1}{[I_2]}$ against time.

(The factor of 2 is omitted in Equation 9.7b because the stoichiometry for I_2 is 1; see Section 9.4, pp.390–394.)

t/s	0	500	1500	2500	3500	4500	5500	6500	7500
$\frac{1}{[I_2]}$/ dm^3 mol^{-1}	50.0	57.1	70.9	85.5	100	112	127	141	154

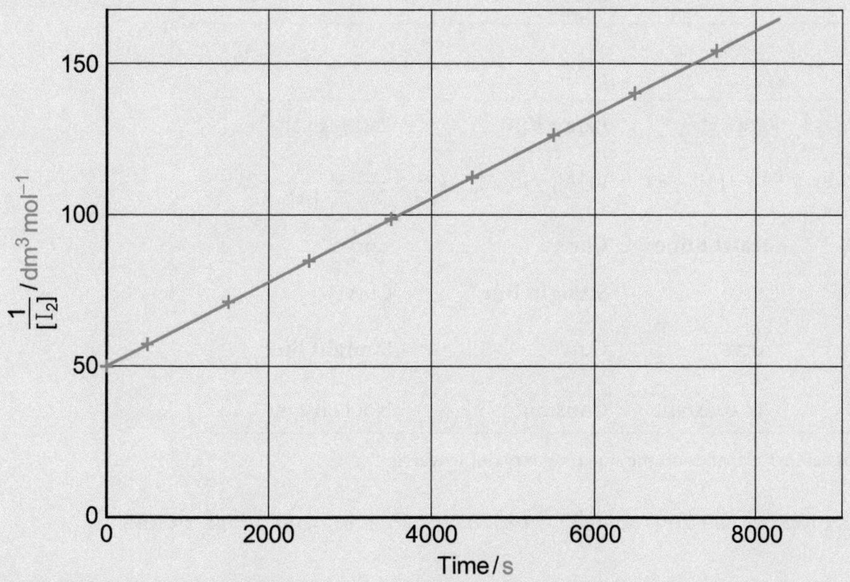

150
100
50
0
0
2000
4000
6000
8000
Time/s

This plot is linear and confirms that the reaction is second order with respect to I_2. The pseudo-second order rate equation is then

$$-\frac{d[I_2]}{dt} = k'[I_2]^2$$

Question

Suggest an alternative way of showing that the reaction is not first order with respect to I_2 that would have saved you plotting ln[I_2] against time.

Summary

- A complex reaction takes place by a series of elementary reactions.
- The overall rate equation of a complex reaction cannot be predicted from the stoichiometric equation. It *must* be determined experimentally.
- For the complex reaction, $aA + bB \rightarrow$ products:

$$\text{rate of reaction} = k[A]^m[B]^n$$

where k is the overall rate constant, and m and n are the orders with respect to A and B, respectively. The overall order of the reaction is $m + n$.

- The rate of a reaction can also depend on the concentration of a substance that does not appear as a reactant in the chemical equation, such as a catalyst.
- Orders of complex reactions often have values of 0, 1, or 2, but sometimes can take higher, or fractional, values.
- For reactions involving more than one reactant, the effect of the concentration of each reactant on the rate of reaction is studied using the isolation technique for each reactant in turn.
- Experimental methods for investigating the kinetics of complex reactions include: drawing tangents to a concentration–time curve; the initial rate method; using integrated rate equations; using half lives.
- Integrated rate equations provide the most accurate and reliable method for determining reaction orders and rate constants. Table 9.3 summarizes the characteristics for reaction orders 0, 1, and 2.

Table 9.3 Using integrated rate equations to study complex reactions

	Order		
	0	1	2
Rate equation	Rate = k'	Rate = $k'[\text{A}]$	Rate = $k'[\text{A}]^2$
Integrated rate equation	$[\text{A}]_t = [\text{A}]_0 - k't$	$\ln[\text{A}]_t = \ln[\text{A}]_0 - k't$	$\frac{1}{[\text{A}]_t} = \frac{1}{[\text{A}]_0} + 2k't$ *
Plot of [A] vs. t	**Straight line**	Curve	Curve
Plot of ln [A] vs. t	Curve	**Straight line**	Curve
Plot of $\frac{1}{[\text{A}]}$ vs. t	Curve	Curve	**Straight line**
Half life, $t_{1/2}$	Not constant	**Constant**	Not constant

* The multiplication factor before k' depends on the stoichiometry of the reaction.

? For practice questions on these topics, see questions 5, 7, 14, 18 and 28 at the end of this chapter (pp.444–446).

9.6 Complex reactions: reaction mechanisms

The series of elementary steps that make up a complex reaction is known as the mechanism of the reaction. Each step has its own rate constant—some steps may take place quickly; others more slowly. Sometimes there are very large differences between the timescales of the individual steps. For example, the timescale for the overall reaction may be minutes or hours, whereas an individual elementary step may take place on a timescale of only microseconds or nanoseconds.

Working out organic reaction mechanisms from kinetic studies

The mechanisms of nucleophilic substitution reactions of halogenoalkanes are discussed in Section 20.3 (p.930).

Kinetic studies of the type described in this section are the main source of evidence for the mechanisms of many organic reactions. For example, the kinetics of nucleophilic substitution reactions of halogenoalkanes were extensively studied by Sir Christopher Ingold and co-workers at University College, London, in the 1930s.

$$\text{R–X} + \text{Nu}^- \rightarrow \text{R–Nu} + \text{X}^-$$

where X = Cl, Br, and I and Nu^- is a nucleophile, such as OH^-.

The researchers showed that there are two extreme types of kinetics for these reactions. In some cases, the rate of nucleophilic substitution was found to be first order with respect to the halogenoalkane, but zero order with respect to the nucleophile—that is, the rate of reaction does not depend on the concentration of the nucleophile. The overall order of the reaction is then one. The alkaline hydrolysis of 2-bromo-2-methylpropane (*tert*-butyl bromide) shows this type of kinetics

$$(CH_3)_3C\text{–}Br + OH^- \rightarrow (CH_3)_3C\text{–}OH + Br^-$$

$$\text{rate of reaction} = k[(CH_3)_3CBr]$$

In other reactions, the rate of nucleophilic substitution was found to be first order with respect to the halogenoalkane and first order with respect to the nucleophile. The overall order of the reaction is then two. The alkaline hydrolysis of bromomethane shows this type of kinetics

$$CH_3\text{–}Br + OH^- \rightarrow CH_3\text{–}OH + Br^-$$

$$\text{rate of reaction} = k[CH_3Br][OH^-]$$

For many halogenoalkanes, the kinetics are more complex, suggesting that the reactions are taking place by mechanisms that lie somewhere between the two.

So, how can these experimental observations be linked to a reaction mechanism in each case? Why is the rate of hydrolysis of $(CH_3)_3C$–Br independent of the concentration of OH^- ions? This was explained by proposing the two-step mechanism shown below. Each step is an elementary reaction with its own rate constant. Step 1 proceeds much more slowly than step 2.

$$\text{Step 1:}\quad (CH_3)_3C\text{–}Br \xrightarrow[k_1]{\text{slow}} (CH_3)_3C^+ + Br^- \quad \text{Rate of reaction} = k_1[(CH_3)_3C-Br]$$

$$\text{Step 2:}\quad (CH_3)_3C^+ + OH^- \xrightarrow[k_2]{\text{fast}} (CH_3)_3C\text{–}OH \quad \text{Rate of reaction} = k_2[(CH_3)_3C^+][OH^-]$$

Step 1 acts as a bottleneck for the reaction—rather like an exit gate acting as a bottleneck for a crowd pushing its way out of a stadium. It doesn't matter how fast people move away from the gate once they are through it—the rate of emptying the stadium is governed by the rate people pass through the gate. The rate of step 1 determines the rate of the overall reaction, so step 1 is called the **rate-determining step**. The rate equation for step 1 becomes the rate equation for the overall reaction. The rate-determining step in this case involves only a single molecule of the reactant and is said to be **unimolecular**. This type of mechanism is known as an **S_N1 mechanism** (substitution, nucleophilic, unimolecular).

Explaining zero order
The idea of a rate-determining step explains why it is possible to have a rate equation that is zero order for a particular reactant. If the reactant is not involved in the rate-determining step, its concentration won't affect the rate of the reaction. But the reactant is still involved in the reaction and gets used up.

In contrast, the hydrolysis of bromomethane is second order. To explain this, a one-step mechanism was proposed

$$CH_3\text{–}Br + OH^- \rightarrow CH_3\text{–}OH + Br^-$$

This is an elementary reaction and the rate equation is given by

$$\text{rate of reaction} = k[CH_3Br][OH^-]$$

In the reaction, the OH^- ion starts to form a bond with the carbon atom in bromomethane at the same time as the bond between the carbon atom and bromine starts to lengthen and break. The transition from reactants to products takes place smoothly in a single step (called a **concerted** process).

A reaction in which the reactants are converted to products in a single step is said to be a **concerted reaction**.

In this case, the mechanism for the reaction involves a single step, which is rate-determining. Since this step involves two molecules, the reaction is said to be **bimolecular**. This type of mechanism is known as an **S_N2 mechanism** (substitution, nucleophilic, bimolecular).

The **molecularity** of a reaction is the number of molecules taking part in the rate-determining step of the reaction. It applies only to elementary reactions. The idea of molecularity is developed further in Section 9.8 (p.432).

Once a mechanism has been proposed for an organic reaction, chemists look carefully at the features of the reaction, particularly any stereochemical implications, and design experiments to test the theory. A mechanism can never be proved absolutely. A valid mechanism must consist of one or more elementary reactions, the sum of which gives the overall equation for the reaction. It must correctly predict the experimentally determined rate equation and be consistent with all related kinetic experimental data and stereochemical observations.

The detailed mechanisms of these nucleophilic substitution reactions, and the stereochemical implications, are described in Section 20.3 (p.930). The effect of solvent polarity on the rate of reaction is also discussed.

The two important points to note at this stage are, firstly, the link between kinetics experiments and theoretical reaction mechanisms and, secondly, the idea of a slow, rate-determining step governing the kinetics of a complex reaction. These ideas are taken further in Section 9.8.

Kinetics of reversible reactions

All the examples in the previous sections involve reactions that essentially go to completion in one direction. For many reactions, this is not the case and the reaction comes to an equilibrium position. As the reaction approaches equilibrium, the forward reaction is opposed by the reverse (back) reaction.

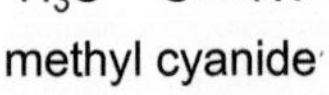

methyl cyanide

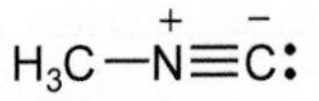

methyl **iso**cyanide

The IUPAC name for methyl cyanide is ethanenitrile; see Section 2.6 (p.98). It is a polar aprotic solvent often known by its common name, acetonitrile; see Section 20.3 (p.930).

For example, in the isomerization of methyl isocyanide to methyl cyanide

$$CH_3-NC \underset{k_{-1}}{\overset{k_1}{\rightleftharpoons}} CH_3-CN$$

the forward and reverse reactions are first order elementary reactions and k_1 and k_{-1} are the rate constants for the forward and reverse reactions, respectively.

As soon as CH_3CN is formed in the reaction, some of it reacts to reform CH_3NC, so the measured rate of reaction involves not only the rate of consumption of CH_3NC, but also the rate of formation of CH_3NC. The rate equation is given by

$$-\frac{d[CH_3NC]}{dt} = k_1[CH_3NC] - k_{-1}[CH_3CN]$$

At equilibrium, the net rate of consumption of CH_3NC is zero, because the rates of the forward and reverse reactions are equal. So, at equilibrium

$$k_1[CH_3NC]_{eqm} = k_{-1}[CH_3CN]_{eqm}$$

Rearranging this equation

$$\frac{k_1}{k_{-1}} = \frac{[CH_3CN]_{eqm}}{[CH_3CN]_{eqm}}$$

The right-hand side of this equation corresponds to the equilibrium constant, K_c, for the reaction, so the ratio of the rate constants for the forward and reverse reactions is equal to the equilibrium constant for the reaction. (Remember the idea of dynamic equilibrium.)

$$\frac{k_1}{k_{-1}} = K_c \qquad (9.14)$$

Dynamic equilibrium and basic ideas about chemical equilibrium and K_c are discussed in Section 1.9 (p.59). The thermodynamics of equilibrium is developed in Chapter 15.

This is an important relationship, which links kinetics and equilibrium for an elementary reaction. It is a general expression for an elementary reaction and does not just apply to first order reactions. (But remember, it does not apply to a complex reaction involving several elementary steps, where the equilibrium constant is written for the overall reaction.) If the forward rate constant is much larger than the reverse rate constant, then $K_c >> 1$. The position of equilibrium is well over to the right and the products predominate in the equilibrium mixture. If the opposite is true, then $K_c << 1$ and the reactants predominate in the equilibrium mixture.

Kinetics of parallel reactions

Sometimes more than one set of products can be obtained from the same reactants. For example, the reaction of a hydrogen atom with an oxygen molecule can proceed by two independent elementary reactions (these reactions take place when hydrogen burns in oxygen)

Both O_2 and O have two unpaired electrons and are biradicals. The dots have been omitted here for simplicity.

$$H^\bullet + O_2 \xrightarrow{k_1} HO^\bullet + O$$

$$H^\bullet + O_2 \xrightarrow{k_2} HO_2^\bullet$$

The reactions take place in parallel, and the ratio of products formed from each reaction depends on the relative magnitudes of the rate constants, k_1 and k_2. It is analogous to the situation in Figure 9.20 on p.419 when the water can leave the tank by two exit pipes of differing diameters. The ratio of the volumes of water leaving by each pipe depends on the relative cross-sectional areas of the pipes.

The rate of consumption of H$^\bullet$ atoms is the sum of the rates of the two individual reactions

$$-\frac{d[H^\bullet]}{dt} = k_1[H^\bullet][O_2]+k_2[H^\bullet][O_2] = (k_1+k_2)[H^\bullet][O_2]$$

$$\text{The rate of the overall reaction} = -\frac{d[H^\bullet]}{dt} = k[H^\bullet][O_2]$$

where k is the overall rate constant for the consumption of H$^\bullet$ atoms, so

$$k = (k_1+k_2)$$

Under pseudo-first order conditions when $[H^\bullet] \ll [O_2]$ (see Section 9.4, pp.395 and 407)

$$-\frac{d[H^\bullet]}{dt} = k[H^\bullet][O_2]_0 = k'[H^\bullet], \text{ where } k' = (k_1+k_2)[O_2]_0$$

and $[O_2]_0$ is the initial concentration of O_2.

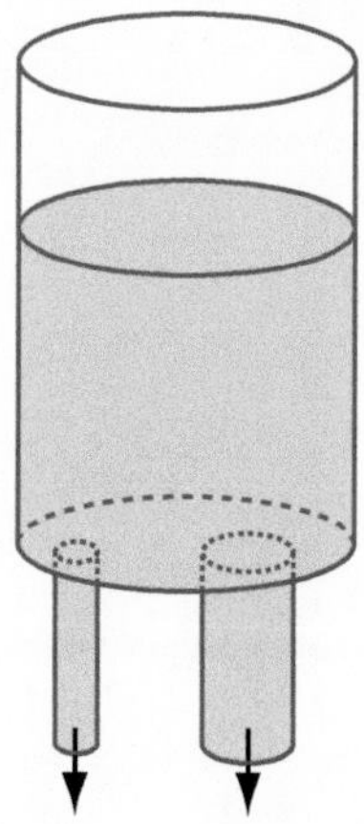

Figure 9.20 Water flowing out of a tank with two exit pipes is analogous to a reaction mechanism containing two parallel reactions.

Kinetics of consecutive reactions

Many chemical reactions occur in a sequence—the product of one reaction becomes the reactant for the next reaction in the sequence, and so on. This is true for many reactions in the atmosphere, for many biochemical processes that proceed by a cascade of reactions, and for many industrial processes.

You can see an example of a biochemical cascade process in Box 19.2 (p.864).

Earlier in this section, you saw that complex reactions take place by a series of elementary steps. How can the kinetics of such a sequence of consecutive reactions be understood?

Consider the sequence of first order elementary reactions:

$$\underset{\text{reactant}}{A} \xrightarrow{k_1} \underset{\text{intermediate}}{B} \xrightarrow{k_2} \underset{\text{product}}{C}$$

The first step is a first order reaction and the concentration of A decays exponentially, but the behaviour of B and C depends on the relative magnitudes of the rate constants k_1 and k_2.

Exponential decay in a first order reaction is described in Section 9.4 (p.393).

- If k_1 is much larger than k_2 (i.e. if the first step is much faster than the second step), there is a considerable time lag between the loss of A and the formation of C. The first reaction takes place quickly and the intermediate B builds up to a concentration comparable to the starting concentration of A, before reacting slowly to form C.
- If k_2 is much larger than k_1 (i.e. if the first step is much slower than the second step), B reacts rapidly once it is formed so its concentration is always small. The formation of C then follows closely the decay of A. The concentration of B is always small, so the rate of change of the concentration, $\frac{d[B]}{dt}$, is also small.

These two situations are illustrated in Figures 9.21(a) and (b). In each case, the rate constants, k_1 and k_2, differ by a factor of 5. In (a) $k_1 = 5k_2$ and in (b) $k_2 = 5k_1$. The dashed curves in (b) show what happens if B is even more reactive and $k_2 = 50k_1$. The maximum concentration reached by B is then very small, as is the rate of change of B. In general, if B is a highly reactive species, such as a reactive radical, then [B] and $\frac{d[B]}{dt}$ are very small. This is the basis of the steady state approximation.

Steady state approximation

One way to simplify the mathematical analysis of a series of reactions involving highly reactive species, such as the intermediate B above, is to make use of the **steady state approximation**. This states that the rate of change of concentration of a reactive species, X, in a sequence of consecutive reactions is approximately zero. X gets used up as fast as it is formed, so its concentration stays constant—it reaches a **steady state**.

$$\frac{d[X]}{dt} = 0$$

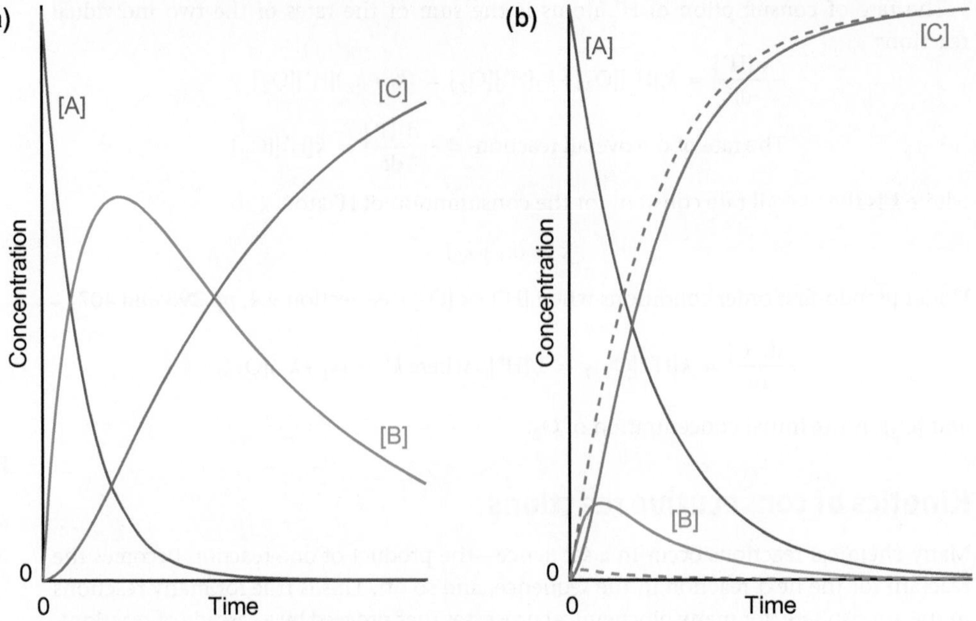
(a)
[A]
[C]
[B]
Concentration
0
0
Time
(b)
[A]
[C]
[B]
Concentration
0
0
Time

Figure 9.21 Plots showing the concentrations of A, B, and C in the sequence of reactions A → B → C. (a) $k_1 = 5k_2$. (b) $k_2 = 5k_1$ (solid lines) and $k_2 = 50k_1$ (dashed lines).

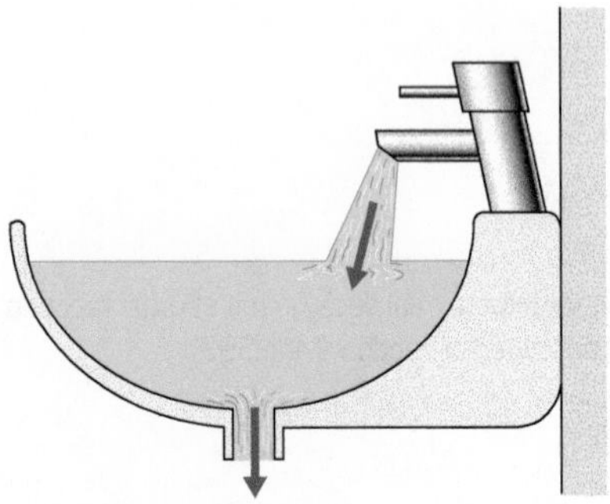

Figure 9.22 One example of a steady state situation.

You can practise using the steady state approximation in Worked example 9.7 (p.422). It is used in the next subsection to explain the kinetics of a series of reactions involving a reversible step and is applied to the reaction of H_2 and Br_2 on p.423.

The reactions in a natural gas flame reach a steady state. Gas and air flow into the flame and the combustion products move away from the flame. The concentration of radical intermediates in the flame is approximately constant.

The steady state approximation is usually a very good approximation for reactions involving radicals because many radicals have high reactivity. When a steady state is reached, [X] is constant and

$$\frac{d[X]}{dt} = (\text{rate of formation of X} - \text{rate of consumption of X}) = 0 \qquad (9.15)$$

It is rather like the situation in Figure 9.22, where water is running into a washbasin with the plug out of the waste pipe. Before long, it gets to a point where water is running out as fast as it is running in, and the level of water in the basin stays the same. If you turned the tap on more, the level of the water would rise, but that would make the water run out faster because of the higher pressure. Before long it would reach a steady state again, but with a higher water level in the basin.

Equilibrium and steady state

At equilibrium, the concentrations of reactants and products are constant because the rates of the forward and reverse reactions in the equilibrium are the same. A series of reactions reaches a steady state when the rate of formation of an intermediate is the same as its rate of removal—but these two processes are not the reverse of one another and they don't come to equilibrium.

A mechanism with a reversible step

Sometimes in a sequence of consecutive reactions, some of the steps are reversible. Consider the following reaction sequence in which the first step is reversible.

$$A \underset{k_{-1}}{\overset{k_1}{\rightleftharpoons}} B$$

$$B \xrightarrow{k_2} C$$

There are two types of kinetics depending on the relative magnitudes of the rate constants.

(a) Pre-equilibrium

If the reaction from B → C is relatively slow, then A and B have time to equilibrate. This is known as a **pre-equilibrium**. Then

$$\frac{[B]}{[A]} = K_c$$

where K_c is the equilibrium constant for the first step. Using Equation 9.14 (p.418)

$$\frac{k_1}{k_{-1}} = K_c \tag{9.14}$$

But $K_c = \frac{[B]}{[A]}$, so you can write an expression for [B] in terms of the rate constants for the forward and reverse reactions

$$[B] = K_c[A] = \frac{[k_1]}{[k_{-1}]}[A]$$

The overall rate of reaction is the rate of consumption of A, which is equal to the rate of formation of the product C. The conversion of B to C takes place slowly and is the rate-determining step. The rate equation for this step is the rate equation for the overall reaction

$$\text{rate of reaction} = -\frac{d[A]}{dt} = \frac{d[C]}{dt} = k_2[B] = k_2\frac{k_1}{k_{-1}}[A] \tag{9.16}$$

In fact, B is being removed from the pre-equilibrium, but very slowly compared to the rate at which the equilibrium is established—so the approximation that the first step comes to equilibrium is a good one.

(b) Short-lived intermediate

If B is very short-lived (either k_{-1} or k_2 is much larger than k_1), then the steady state approximation can be applied to B (Equation 9.15, p.420).

When a steady state is reached

$$\frac{d[B]}{dt} = (\text{rate of formation of B} - \text{rate of consumption of B}) = 0$$

$$\frac{d[B]}{dt} = k_1[A] - (k_{-1}[B] + k_2[B]) = 0$$

$$\text{so, } k_1[A] = (k_{-1} + k_2)[B]$$

Rearranging to give an expression for [B]

$$[B] = \frac{k_1[A]}{(k_{-1} + k_2)}$$

and

$$\text{rate of reaction} = \frac{d[C]}{dt} = k_2[B] = \frac{k_1k_2[A]}{(k_{-1}+k_2)} \tag{9.17}$$

There are two extreme cases.

- If k_{-1} is much larger than k_2 (i.e. the conversion of B back to A is much faster than the conversion of B to C), k_2 can be omitted from the bottom line in Equation 9.17 and then

$$\frac{d[C]}{dt} = k\frac{k_1}{k_{-1}}[A]$$

This is the same result as in Equation 9.16. A pre-equilibrium occurs that then leads to formation of C. Because k_{-1} is much larger than k_1, the concentration of B in this pre-equilibrium is very small.

In this case $k_{-1} >> k_1$ and $k_{-1} >> k_2$.

Visit the Online Resource Centre to view screencast 9.5 which walks you through Worked example 9.7 to illustrate the use of the steady state approximation in deriving a rate equation to test a proposed mechanism.

In this case $k_2 >> k_1$ and $k_2 >> k_{-1}$.

O_2 has two unpaired electrons and is a biradical. NO and NO_2 both have an unpaired electron and are radicals. The dots have been omitted in Worked example 9.7 for simplicity.

- If k_2 is much larger than k_{-1} (i.e. the conversion of B to C is much faster than the conversion of B back to A), k_{-1} can be omitted from the bottom line in Equation 9.17 and then

$$\frac{d[C]}{dt} = k_1[A]$$

which means that the conversion of A → B is the rate-determining step and equilibrium is not established.

You can use the steady state approximation to derive a rate equation in Worked example 9.7.

Worked example 9.7 The steady state approximation: deriving a rate equation to test a proposed reaction mechanism

A sample of the colourless gas, NO, rapidly turns brown when exposed to air due to the formation of NO_2. The overall reaction is

$$2\,NO\,(g) + O_2\,(g) \rightarrow 2\,NO_2\,(g)$$

Kinetic studies show that the reaction is second order with respect to NO and first order with respect to O_2. The following mechanism has been proposed

$$2\,NO \underset{k_{-1}}{\overset{k_1}{\rightleftharpoons}} N_2O_2$$

$$N_2O_2 + O_2 \xrightarrow{k_2} 2\,NO_2$$

in which N_2O_2 is a short-lived intermediate. Show that the proposed mechanism is compatible with the observed kinetics.

Strategy

To test whether the proposed mechanism is plausible, work out an expression for the overall rate equation that follows from the mechanism and compare it with the experimentally determined rate equation. (Remember that the steps in the proposed mechanism are elementary reactions.)

Step 1 Write down the experimentally determined rate equation.

Step 2 The intermediate, N_2O_2, is short lived so you can apply the steady state approximation (Equation 9.15, p.420) to find an expression for $[N_2O_2]$.

Step 3 Use this to work out a rate equation from the mechanism in terms of [NO] and $[O_2]$, and compare this with the experimentally determined rate equation.

Step 4 Think about the relative magnitude of the rate constants for the individual steps and decide which reaction is the rate-determining step and so governs the overall kinetics.

Solution

Step 1 Experimentally determined rate equation.

$$\text{Rate of reaction} = k[NO]^2[O_2]$$

Step 2 Using the steady state approximation to find an expression for $[N_2O_2]$.

From Equation 9.15 (p.420) when a steady state is reached

$$\frac{d[N_2O_2]}{dt} = (\text{rate of formation } N_2O_2 - \text{rate of consumption of } N_2O_2) = 0$$

$$\frac{d[N_2O_2]}{dt} = k_1[NO]^2 - (k_{-1}[N_2O_2] + k_2[N_2O_2][O_2]) = 0$$

$$\text{so, } k_1[NO]^2 = (k_{-1} + k_2[O_2]) \times [N_2O_2]$$

Rearranging to give an expression for $[N_2O_2]$

$$[N_2O_2] = \frac{k_1[NO]^2}{(k_{-1} + k_2[O_2])} \qquad (9.18)$$

Step 3 Deriving a rate equation from the mechanism.

From Equation 9.4 (p.390) for the rate of reaction

$$\text{rate of reaction} = -\frac{1}{2}\frac{d[NO]}{dt} = \frac{1}{2}\frac{d[NO_2]}{dt} = k_2[N_2O_2][O_2]$$

so the rate of formation of NO_2 is given by the rate equation

$$\frac{d[NO_2]}{dt} = 2k_2[N_2O_2][O_2]$$

Substituting the expression for $[N_2O_2]$ in Equation 9.18 into the above equation

$$\frac{d[NO_2]}{dt} = 2k_2 \times \frac{k_1[NO]^2}{(k_{-1} + k_2[O_2])} \times [O_2]$$

$$= \frac{2k_1k_2[NO]^2[O_2]}{(k_{-1} + k_2[O_2])} \qquad (9.19)$$

This is the overall rate equation for the reaction in terms of measurable concentrations. It does not, however, fit the observed kinetics, which show that the reaction is second order with respect to NO and first order with respect to O_2.

Step 4 Think about the relative magnitude of the rate constants.

Compare the relative magnitudes of k_{-1} and $k_2[O_2]$ in Equation 9.19. If $k_{-1} >> k_2[O_2]$, the rate equation becomes

$$\frac{d[NO_2]}{dt} = \frac{2k_1k_2}{k_{-1}}[NO]^2[O_2]$$

which is now consistent with the observed kinetics (step 1). This would be the case if the second step is the rate-determining step for the reaction. The reaction is sufficiently slow that it does not disturb the pre-equilibrium between NO and N_2O_2, which is established quickly.

Note that this does not *prove* that the mechanism is correct—merely that it is consistent with the experimental observations.

Question

(a) Show that the proposed mechanism is consistent with the overall equation for the reaction.

(b) Give the molecularity of each of the three elementary reactions in the proposed mechanism.

(c) Suggest why the reaction: $2NO + O_2 \rightarrow 2NO_2$ is unlikely to be an elementary reaction.

(d) An alternative mechanism that has been considered is

$NO + NO \rightarrow NO_2 + N$ slow

$N + O_2 \rightarrow NO_2$ fast

Explain why this mechanism can be discounted.

Chain reactions

Radical chain reactions involve a sequence of reactions in which a radical reacts and then is regenerated in subsequent reactions. The overall reaction of hydrogen and bromine provides a good example of a chain reaction.

$$H_2(g) + Br_2(g) \rightarrow 2HBr(g)$$

The first step involves the dissociation of Br_2 to form bromine radicals. This may be brought about by heat (**thermal dissociation**) or light (**photodissociation**)

1 $Br_2 \xrightarrow{k_1} 2Br^{\bullet}$

This is followed by a pair of reactions in which HBr is formed and $Br^{\bullet}$ radicals are regenerated

2 $Br^{\bullet} + H_2 \xrightarrow{k_2} H^{\bullet} + HBr$

3 $H^{\bullet} + Br_2 \xrightarrow{k_3} Br^{\bullet} + HBr$

The fourth step is the recombination of $Br^{\bullet}$ radicals to form Br_2.

4 $Br^{\bullet} + Br^{\bullet} \xrightarrow{k_4} Br_2$

The reaction sequence can be represented diagrammatically as follows

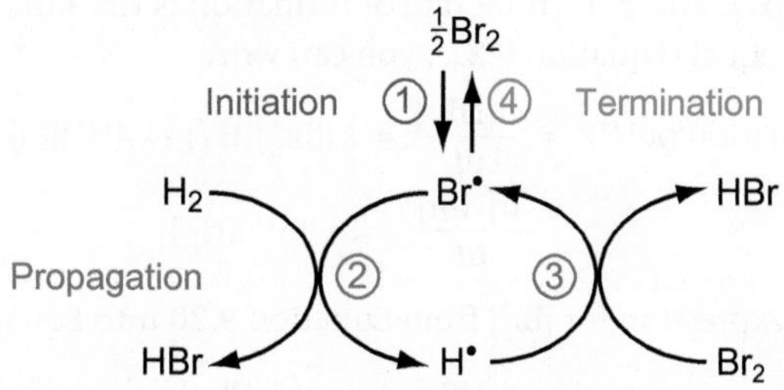
½Br2
Initiation ① ④ Termination
H2
Br•
HBr
Propagation ② ③
HBr
H•
Br2

- Step 1 is called an **initiation reaction**—it leads to the net formation of radicals.
- Step 4 is a **termination reaction** and removes radicals from the system.
- Steps 2 and 3 are **propagation reactions** and keep the reaction going. They involve the replacement of one radical by another but do not change the overall concentration of radicals.

The experimentally determined rate equation obtained from kinetic studies is

$$\text{rate of reaction} = k[H_2(g)][Br_2(g)]^{1/2}$$

where k is the overall rate constant (see Section 9.5, p.405).

How can this be explained in terms of the mechanism for the reaction? How is it possible to have a fractional reaction order? The intermediate radicals in the mechanism proposed above are all very reactive and short lived, so you can use the steady state approximation

In the atmosphere, the mechanism for the reaction of NO with O_2 described in Worked example 9.7 occurs only very slowly, because [NO] is so small and the rate depends on $[NO]^2$. NO from car exhausts, for example, is converted to NO_2 much more quickly by other processes, such as reaction with O_3.

You might think that the reaction $H^{\bullet} + H^{\bullet} \rightarrow H_2$ would also be a termination reaction. In fact, $[H^{\bullet}] << [Br^{\bullet}]$ so this reaction proceeds very much more slowly than the recombination of $Br^{\bullet}$ radicals to from Br_2 and has no significant effect.

(Equation 9.15 on p.420). There are three stages in the analysis and these are summarized below. The first stage gives an expression for the concentration of $Br^\bullet$ radicals. The second stage determines the relationship between the rates of the two propagation reactions. These expressions are then combined in the third stage to give a rate equation for the formation of HBr, which can be compared with the experimentally determined rate equation above.

Deriving a rate equation for the $H_2 + Br_2$ chain reaction

Stage 1 Finding an expression for $[Br^\bullet]$

Using the steady state approximation, you can say that when a steady state is reached the total radical concentration will be constant. The propagation reactions do not affect this concentration (they just convert one radical into another), so the rate of the initiation reaction (step 1: forms $Br^\bullet$) and the rate of the termination reaction (step 4: removes $Br^\bullet$) must be equal, and

$$2k_1[Br_2] = 2k_4[Br^\bullet]^2$$

(The '2' on each side of the equation is there because there are 2 $Br^\bullet$ involved in step 1 and step 4.) Rearranging this equation to give an expression for $[Br^\bullet]$

$$[Br^\bullet] = \left(\frac{k_1[Br_2]}{k_4}\right)^{1/2} \tag{9.20}$$

Stage 2 Finding the relationship between the rates of the propagation steps

In practice, the rates of the propagation reactions (steps 2 and 3) are much greater than the rates of the initiation and termination reactions (step 1 and step 4). As a result, the reaction cycles through the propagation reactions many times for each initiation and termination reaction—hence the name **chain reaction**. In order to maintain a steady state in radical concentrations, the rates of these two propagation reactions must be equal. If, for example, step 2 were faster than step 3, $[H^\bullet]$ would rise and $[Br^\bullet]$ would decline. So

$$\text{rate of step 2} = \text{rate of step 3}$$

$$k_2[Br^\bullet][H_2] = k_3[H^\bullet][Br_2] \tag{9.21}$$

Stage 3 The rate of formation of HBr

HBr is formed in steps 2 and 3. Its total rate of formation is the sum of the rates of these steps. Since they are equal (Equation 9.21), you can write

$$\text{rate of formation of HBr} = \frac{d[HBr]}{dt} = k_2[Br^\bullet][H_2] + k_3[H^\bullet][Br_2], \text{and}$$

$$\frac{d[HBr]}{dt} = 2k_2[Br^\bullet][H_2] \tag{9.22}$$

Now, substitute the expression for $[Br^\bullet]$ from Equation 9.20 into Equation 9.22:

$$\frac{d[HBr]}{dt} = 2k_2\left(\frac{k_1[Br_2]}{k_4}\right)^{1/2}[H_2]$$

so that

$$\frac{d[HBr]}{dt} = 2k_2\left(\frac{k_1}{k_4}\right)^{1/2}[Br_2]^{1/2}[H_2] \tag{9.23}$$

Equation 9.23 is the rate equation from the proposed reaction mechanism. It gives the order with respect to Br_2 as 0.5 and the order with respect to H_2 as 1. The overall order is 1.5. This agrees with the experimentally determined rate equation. The rate constant k is given by

$$k = 2k_2\left(\frac{k_1}{k_4}\right)^{1/2}$$

The kinetics of many chain reactions can be analysed in this way and often give fractional orders.

The reaction of H_2 and O_2, initiated by a spark, is a chain reaction. The mechanism includes a branching reaction where one radical reacts to form two radicals. This leads to an exponential increase in the concentration of radicals, and in the rate of the reaction. The whole process occurs very rapidly—and the system explodes.

Chain reactions involving organic radicals are discussed in Section 19.2 (p.864) and Section 20.2 (p.921).

Summary

- The series of elementary steps that make up a complex reaction is known as the mechanism of the reaction.
- A valid mechanism must correctly predict the experimentally determined rate equation and be consistent with all related kinetic experimental data and stereochemical observations. The sum of the elementary steps must give the overall equation for the reaction.
- Kinetic studies are the main source of evidence for the mechanism of many organic reactions.
- The rate equation for the slow rate-determining step is the rate equation for the overall reaction.
- For an equilibrium reaction where the forward and reverse reactions are elementary processes, the ratio of the rate constants for the forward and reverse reactions is equal to the equilibrium constant for the reaction.

$$\frac{k_1}{k_{-1}} = K_c$$

- The steady state approximation is used to simplify analysis of the kinetics of consecutive reactions involving a highly reactive intermediate, B. It is assumed that the concentration of B is very small, and that the rate of change of the concentration of B is zero.

For practice questions on these topics, see questions 23–26, 29, and 30 at the end of this chapter (pp.445–446).

9.7 Effect of temperature on the rate of a reaction

The Arrhenius equation

The rates of most chemical reactions increase as the temperature is raised. Applications of this are all around you. Food is cooked to speed up desirable reactions between chemical compounds in the food, whereas food is refrigerated to slow down chemical reactions that lead to decomposition and decay.

The rate constant, k, changes with temperature and determines the way in which the rate of reaction varies with temperature. This behaviour is summarized by the **Arrhenius equation**

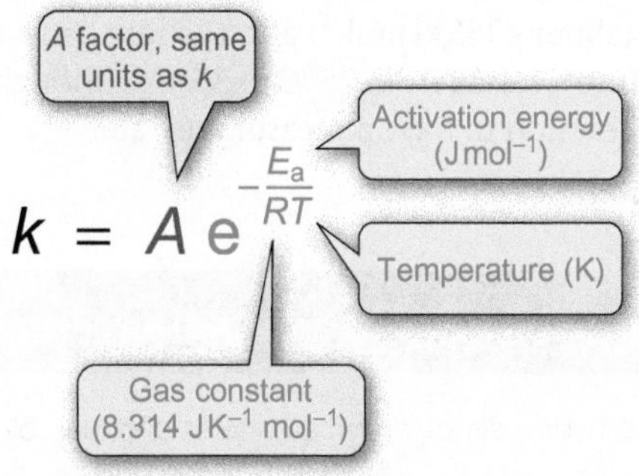
A factor, same units as k
Activation energy (J mol⁻¹)
k = A e^(−Ea/RT)
Temperature (K)
Gas constant (8.314 J K⁻¹ mol⁻¹)

The constant A is sometimes called the Arrhenius factor or the pre-exponential factor.

Note that the presence of the gas constant does not mean the equation only applies to gases. It applies equally well to reactions in solution. Figure 9.23 shows how the value of k increases exponentially with temperature.

Taking logarithms of both sides gives an alternative form of the Arrhenius equation

$$\ln k = \ln A - \frac{E_a}{R}\left(\frac{1}{T}\right) \qquad (9.24b)$$

The rate constant, k, and hence the reaction rate, increase with temperature and decrease with increasing E_a.

If $y = e^x$, then $x = \ln y$

For $k = Ae^{-\frac{E_a}{RT}}$, this means that

$-\frac{E_a}{RT} = \ln\frac{k}{A} = \ln k - \ln A$, and

$\ln k = \ln A - \frac{E_a}{RT}$

The properties of exponential functions, e^x, are described in Maths Toolkit MT3 (p.1316) and MT4 (p.1318).

Towards the end of the nineteenth century, the Swedish chemist, Svante Arrhenius, examined kinetic data from many reactions and noticed that a plot of the logarithm of the rate constant, ln k, against $1/T$ is a straight line. This led to the relationship in Equations 9.24(a and b), which became known as the **Arrhenius equation**.

$$y = mx + c$$

$$\ln k = -\frac{E_a}{RT}\left(\frac{1}{T}\right) + \ln A$$

Cooking involves chemical reactions and the rate of cooking depends critically on temperature.

When you use an Arrhenius plot of ln *k* against $1/T$ to find a value of *A*, remember that the intercept must be with the *y*-axis where $\frac{1}{T}=0$.

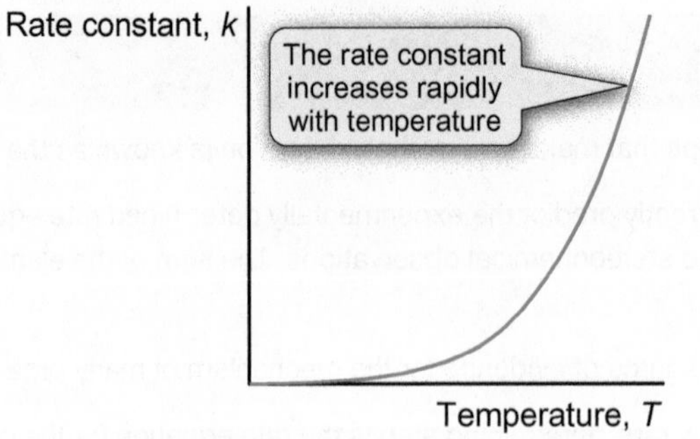

Rate constant, *k*
The rate constant increases rapidly with temperature
Temperature, *T*

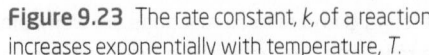

Figure 9.23 The rate constant, *k*, of a reaction increases exponentially with temperature, *T*.

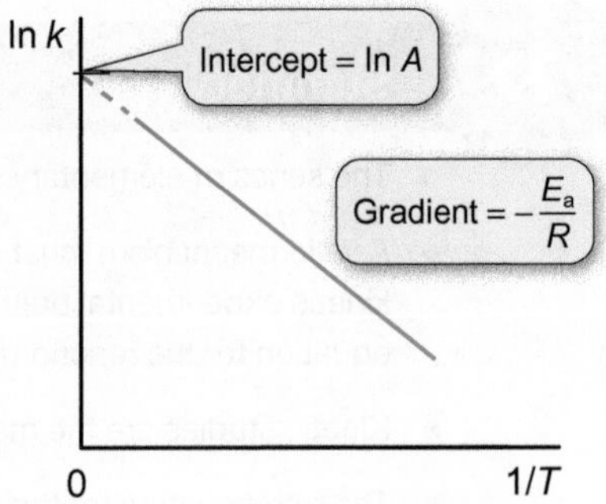

ln *k*
Intercept = ln *A*
Gradient = $-\frac{E_a}{R}$
0
1/*T*

Figure 9.24 An Arrhenius plot of ln *k* against $1/T$. The activation energy can be found from the gradient and a value of *A* from the intercept at $1/T=0$.

Be careful to be accurate when using the terms **rate** and **rate constant.** When all concentrations are 1 mol dm^{-3}, the two quantities have the same numerical value—but the units are different. For other concentrations, they have different numerical values. *k* is constant for a given temperature, whereas the rate usually varies throughout the reaction.

As a rough rule of thumb (taking into account a range of activation energies), the rates of reaction for many organic reactions in solution increase by a factor of 2 to 4 when the temperature is raised from 25 °C to 35 °C.

Equation 9.24b is useful because it is the equation of a straight line ($y = mx + c$). A plot of ln *k* against $\frac{1}{T}$ (Figure 9.24) is a straight line with gradient $-\frac{E_a}{R}$ and intercept ln *A* (see Worked example 9.8).

The value of the activation energy, E_a, determines how sensitive the reaction is to changes of temperature. A high value of E_a corresponds to a reaction that is very sensitive to changes of temperature, whereas a low value of E_a corresponds to a reaction that is less sensitive to changes of temperature.

An often quoted rule of thumb is that the rate of a reaction increases by a factor of two for a 10 K rise in temperature. This applies in only limited circumstances and should be treated with caution. For some reactions of organic compounds in solution, where the activation energy is about +50 kJ mol^{-1}, the rate of the reaction does double when the temperature is raised from 25 °C to 35 °C. It is not a general rule, though, and does not apply to other temperature ranges or to reactions with different activation energies.

Reactions in the gas phase have a wide range of sensitivities to temperature. Some, with low activation energies, are only weakly sensitive to temperature. Radical recombination reactions in the gas phase, such as the reaction of two methyl radicals to give ethane, have zero activation energy (see Table 9.4, p.431) and their rate constants are independent of temperature. The reverse reaction, the reaction of ethane to form two methyl radicals, has a high activation energy (about +365 kJ mol^{-1}) and the rate constant increases by a factor of 100 when the temperature is raised from 25 °C to 35 °C, although much higher temperatures are required for the reaction to have a measurable rate.

Worked example 9.8 Finding an activation energy

The rate constant for the hydrolysis of bromoethane in alkaline solution was measured at different temperatures.

$$C_2H_5Br(aq) + OH^-(aq) \rightarrow C_2H_5OH(aq) + Br^-(aq)$$

Use the results in the following table to find a value for the activation energy of the reaction.

Temperature/°C	25	28	31	34	37	40	43	46
$k/10^{-5}\,dm^3\,mol^{-1}\,s^{-1}$	8.5	13	19	25	37	51	70	96

→

→

Strategy

Use the Arrhenius equation in the form of Equation 9.24b

$$\ln k = \ln A - \frac{E_a}{R}\left(\frac{1}{T}\right) \quad (9.24b)$$

Plot a graph of $\ln k$ against $1/T$ (T must be in K).

The gradient of the line $-\dfrac{E_a}{R}$.

Solution

Tabulate the data you need to plot the graph of $\ln k$ against $1/T$. Take care to include the correct units.

Temperature/°C	25	28	31	34	37	40	43	46
Temperature/K	298	301	304	307	310	313	316	319
$(1/T)/10^{-3}\,K^{-1}$	3.36	3.32	3.29	3.26	3.23	3.19	3.16	3.13
$k/10^{-5}\,dm^3\,mol^{-1}\,s^{-1}$	8.5	13	19	25	37	51	70	96
$\ln(k/10^{-5}\,dm^3\,mol^{-1}\,s^{-1})$	2.14	2.56	2.94	3.22	3.61	3.93	4.25	4.56

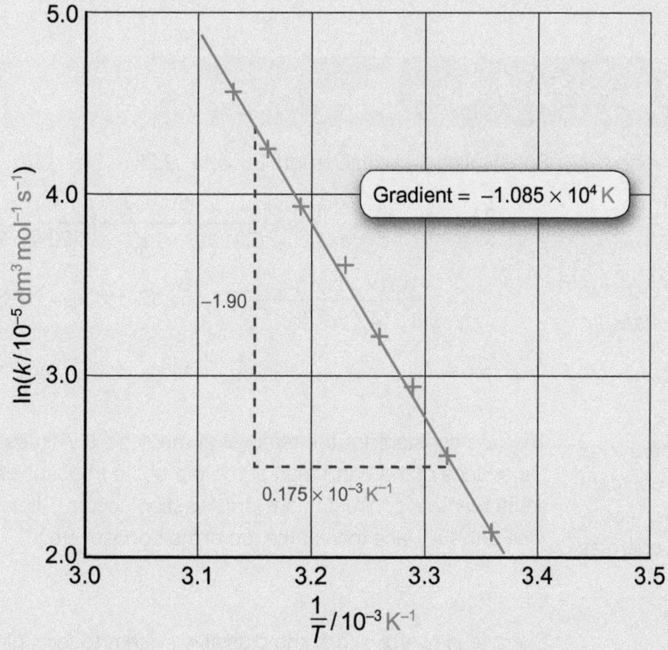

5.0
4.0
3.0
2.0
3.0
3.1
3.2
3.3
3.4
3.5
$\ln(k/10^{-5}\,dm^3\,mol^{-1}\,s^{-1})$
$\frac{1}{T}/10^{-3}\,K^{-1}$
Gradient = -1.085×10^4 K
−1.90
$0.175 \times 10^{-3}\,K^{-1}$

The graph is a straight line.

$$\text{Gradient} = -1.085 \times 10^4\,K = -\frac{E_a}{R}$$

So, $E_a = -(R \times \text{gradient}) = -(8.314\,J\,K^{-1}\,mol^{-1} \times -1.085 \times 10^4\,K) = 9.02 \times 10^4\,J\,mol^{-1}$

Activation energy $E_a = +90\,kJ\,mol^{-1}$

Question

The decomposition of a drug in humans was found to be a first order process. The activation energy for the decomposition is $+95\,kJ\,mol^{-1}$ and $A = 5 \times 10^{10}\,s^{-1}$.

(a) What is the rate constant for the decomposition at blood temperature, 310 K?

(b) How long will it take the concentration of the drug in the blood to fall to half its initial value?

Σ See the Maths notes in Worked example 9.4 (p.398) and Worked example 9.6 (p.413) about plotting logarithm values.

Σ See Maths note on p. 385 about finding the gradient of a straight line.

If you know the value of the rate constant, k_1, at a particular temperature, T_1, you can use the Arrhenius equation to calculate the value of the rate constant, k_2, at any other temperature, T_2.

Using Equation 9.24b (p.425)

$$\ln k_1 = \ln A - \frac{E_a}{R}\left(\frac{1}{T_1}\right)$$

You can write a corresponding equation for T_2

$$\ln k_2 = \ln A - \frac{E_a}{R}\left(\frac{1}{T_2}\right)$$

Subtracting the first equation from the second gives

$$\ln k_2 - \ln k_1 = \ln\frac{k_2}{k_1} = \frac{E_a}{R}\left(\frac{1}{T_1} - \frac{1}{T_2}\right) \quad (9.25)$$

Equation 9.25 allows you to find the ratio of two rate constants at different temperatures, as illustrated in Worked example 9.9.

Worked example 9.9 Comparing rate constants at different temperatures

The main mechanism for removing methane from the atmosphere is the reaction with hydroxyl radicals

$$CH_4 + {}^{\bullet}OH \rightarrow CH_3^{\bullet} + H_2O$$

The activation energy for the reaction is +14.1 kJ mol^{-1}. Calculate the ratio of the rate constants for the reaction at the Earth's surface (at 295 K) and at the top of the troposphere (at 220 K).

Strategy

Use Equation 9.25 to find a value for the ratio k_2/k_1, where k_2 is the rate constant at the Earth's surface and k_1 is the rate constant at the top of the troposphere. When using the Arrhenius equation, temperatures should always be in K and activation energies in J mol^{-1}.

Solution

$T_2 = 295\,\text{K}$ and $T_1 = 220\,\text{K}$.

Substituting these values into Equation 9.25

$$\ln\frac{k_2}{k_1} = \frac{E_a}{R}\left(\frac{1}{T_1} - \frac{1}{T_2}\right) = \frac{+14.1\times10^3\,\text{J mol}^{-1}}{8.314\,\text{J K}^{-1}\,\text{mol}^{-1}}\left(\frac{1}{220\,\text{K}} - \frac{1}{295\,\text{K}}\right)$$

$$= \frac{+14.1\times10^3\,\cancel{\text{J}}\,\cancel{\text{mol}^{-1}}}{8.314\,\cancel{\text{J}}\,\cancel{\text{K}^{-1}}\,\cancel{\text{mol}^{-1}}}(1.156\times10^{-3}\,\cancel{\text{K}^{-1}}) = 1.96$$

$$\frac{k_2}{k_1} = 7.1$$

The rate constant for the removal of methane is 7 times greater at the surface of the Earth than at the top of the troposphere. (For the same reactant concentrations, the reaction occurs 7 times faster at the Earth's surface than at the top of the troposphere.)

Question

Sucrose is hydrolysed in the digestive system to form glucose and fructose. The activation energy for the reaction is +108 kJ mol^{-1}. At 298 K, the rate constant is 1.85×10^{-4} dm^3 mol^{-1} s^{-1}. Calculate the rate constant for the reaction at body temperature, 37 °C (310 K).

Energy profiles

The temperature dependence of a reaction can be interpreted through an **energy profile** of the reaction. This is a plot of potential energy, E_{PE}, as the reaction takes place. Figure 9.25 shows the energy profile for the elementary reaction

$$H^{\bullet} + F_2 \rightarrow HF + F^{\bullet} \quad (9.26)$$

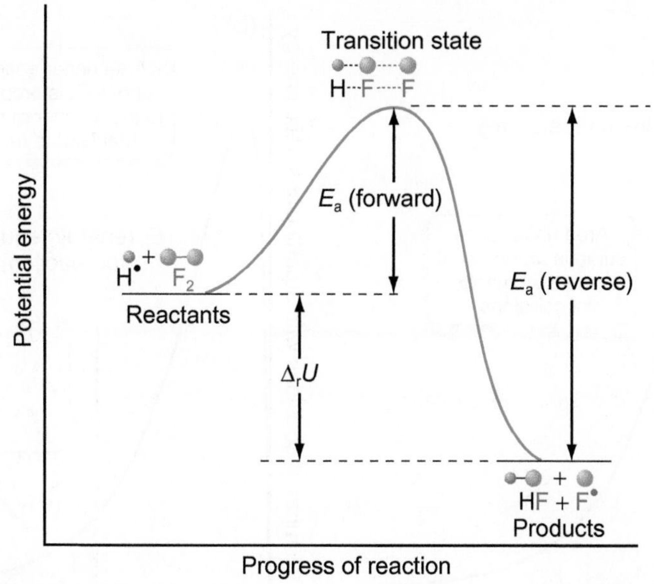
Transition state
H···F····F
E_a (forward)
E_a (reverse)
H• F2
Reactants
$\Delta_r U$
HF + F•
Products
Potential energy
Progress of reaction

Figure 9.25 Energy profile for the elementary reaction $H^{\bullet} + F_2 \rightarrow HF + F^{\bullet}$ (not drawn to scale).

The *x*-axis follows the **progress of the reaction** from reactants to products. (It is sometimes called the reaction coordinate or extent of reaction.) The reaction is exothermic, so the products are at lower potential energy than the reactants. But, for the reaction to proceed, an energy barrier has first to be overcome because of the need to rearrange the electrons in the system.

The highest point on the energy profile is called the **transition state**. The difference in energy between the reactants and the transition state is the **activation energy**, E_a. In Figure 9.25, the activation energy for the forward reaction is labelled E_a (forward). The activation energy for the reverse reaction, E_a (reverse), is larger than E_a (forward) because the reverse reaction is endothermic. $\Delta_r U$ is the energy change for the reaction.

The activation energy for the forward reaction is +10 kJ mol^{-1}, which is much less than the energy needed to break the bond in F_2 (+158 kJ mol^{-1}). This is because, in the transition state, the F–F bond is not fully broken and the H–F bond is already partly formed.

> The energy change of reaction $\Delta_r U$ is the change in internal energy (see Section 13.5, p.636). The value of $\Delta_r U$ is usually very close to the value of the enthalpy change of reaction, $\Delta_r H$.

Distribution of energies

The energies of reactant molecules are spread over a wide range and the distribution of energies among the molecules depends on the temperature.

In a gas, the distribution of molecular speeds (*s*) is described by the Maxwell–Boltzmann distribution. The shape of the distribution of kinetic energies of molecules is very similar, since E_{KE} is proportional to s^2. Figure 9.26(a) plots the number of collisions with a particular kinetic energy at a temperature T_1 (blue curve) and at a higher temperature T_2 (red curve). As the temperature increases, more molecules move at higher speeds and have higher kinetic energies. As a result, more collisions occur with higher kinetic energies. There is still a spread of energies at T_2, but a greater fraction of collisions have higher energies.

> The Maxwell–Boltzmann distribution of molecular speeds in a gas is described in Section 8.5 (p.362). Figure 8.12(b) shows the distribution of speeds at different temperatures.

Consider now an elementary reaction, A + B → products, taking place in the gas phase. The activation energy, E_a, for the reaction is shown by the dotted line in Figure 9.26(b). The dotted line shows the minimum kinetic energy along the line of approach needed in a collision for reaction to occur. The shaded area under the curve to the right of E_a gives the number of collisions with enough energy to react. This area is greater at the higher temperature.

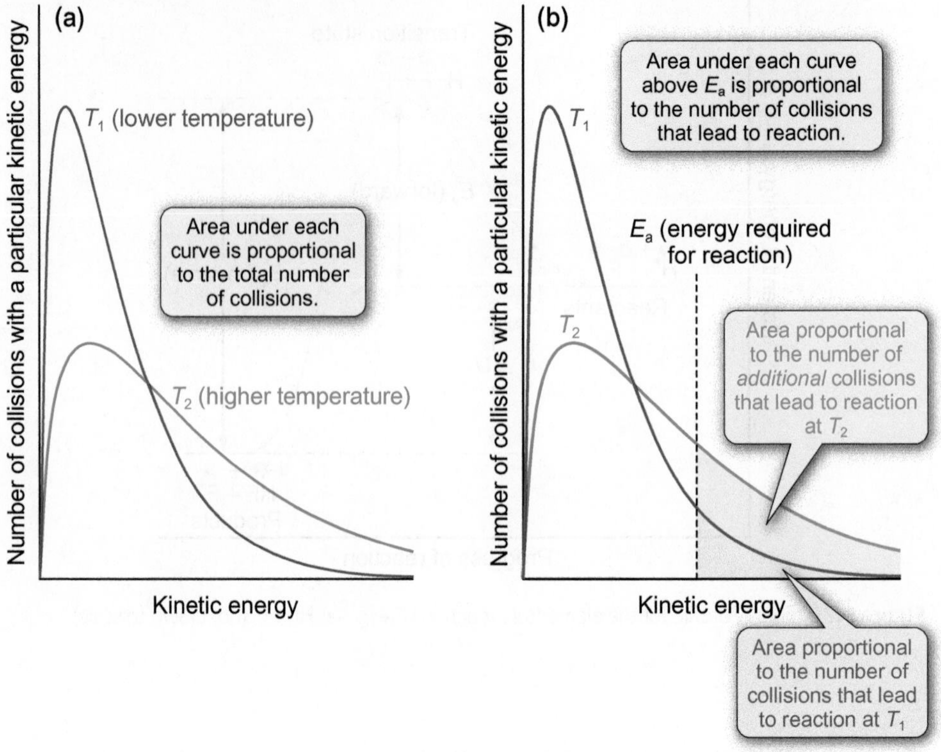
(a)
Number of collisions with a particular kinetic energy
T1 (lower temperature)
Area under each curve is proportional to the total number of collisions.
T2 (higher temperature)
Kinetic energy
(b)
Number of collisions with a particular kinetic energy
Area under each curve above Ea is proportional to the number of collisions that lead to reaction.
T1
Ea (energy required for reaction)
T2
Area proportional to the number of additional collisions that lead to reaction at T2
Kinetic energy
Area proportional to the number of collisions that lead to reaction at T1

Figure 9.26 (a) Number of collisions with a particular kinetic energy at temperature, T_1, and at a higher temperature, T_2. (b) The number of collisions with enough energy to react is greater at the higher temperature.

Look again at the form of the Arrhenius equation (Equation 9.24a, p.430). The rate constant increases with temperature because, at a higher temperature, more colliding molecules have sufficient energy to overcome the energy barrier. The term $e^{-\frac{E_a}{RT}}$ gives the fraction of collisions that have the required energy. As T increases, E_a/RT becomes smaller and approaches zero (so that $e^{-\frac{E_a}{RT}}$ approaches 1). If $e^{-\frac{E_a}{RT}} = 1$, all the collisions would have enough energy to react and $k = A$.

Fraction of collisions with the minimum energy for reaction

$$k = A\,e^{-\frac{E_a}{RT}} \qquad (9.24a)$$

Rate constant if all the collisions had enough energy to overcome the energy barrier

Table 9.4 shows values of A factors and activation energies for some first and second order elementary reactions. Note the radical recombination reaction of two methyl radicals, which has zero activation energy. There is no energy barrier to reaction, so the rate of the reaction is independent of temperature.

Table 9.4 *A factors and activation energies for some first and second order elementary reactions*

Elementary reaction	A	E_a / kJ mol^{-1}
First order (gas phase)	/ s^{-1}	
$CH_3NC \rightarrow CH_3CN$	5.0×10^{13}	+160
cyclopropane → propene	1.6×10^{15}	+272
$C_6H_5CH_2CH_3 \rightarrow C_6H_5CH_2^{\bullet} + CH_3^{\bullet}$	7.1×10^{15}	+314
$C_2H_6 \rightarrow 2\,CH_3^{\bullet}$	2.0×10^{17}	+365
Second order (gas phase)	/ dm^3 mol^{-1} s^{-1}	
$CH_3^{\bullet} + CH_3^{\bullet} \rightarrow C_2H_6$	3.6×10^{10}	0
$Cl^{\bullet} + O_3 \rightarrow ClO^{\bullet} + O_2$	1.7×10^{10}	+2
$H^{\bullet} + F_2 \rightarrow HF + F^{\bullet}$	9×10^{9}	+10
$HF + F^{\bullet} \rightarrow H^{\bullet} + F_2$	1.3×10^{10}	+422
Second order (aqueous solution)	/ dm^3 mol^{-1} s^{-1}	
$C_2H_5Br + OH^- \rightarrow C_2H_5OH + Br^-$	4.3×10^{11}	+90

Activation energy for a complex reaction

The discussion of the effect of temperature so far refers only to elementary reactions, but the rates of complex reactions also depend on temperature. If the kinetics of the reactions is governed by a rate-determining step (such as in the S_N1 hydrolysis of a halogenoalkane; see Section 9.6, p.416), the Arrhenius equation applies to the rate constant for that step.

For many complex reactions, however, the rate constants of several reactions contribute to the overall rate constant. Very often, the temperature dependence of the overall rate constant does follow the Arrhenius equation and a plot of $\ln k$ against $1/T$ gives a straight line with gradient $-E_a/R$. E_a in this case is called an **effective activation energy** and its interpretation depends on the reaction being studied.

Summary

- The way the rate constant varies with temperature is described by the Arrhenius equation

$$k = Ae^{-\frac{E_a}{RT}}$$

- A plot of $\ln k$ against $1/T$ is a straight line with a gradient $-E_a/R$.
- The activation energy, E_a, is the energy barrier reactants must overcome in order to form products.
- The rate constant of a reaction increases with temperature because a greater fraction of reactant molecules has enough energy to pass over the energy barrier.
- An energy profile shows how the potential energy of the system changes as the reaction progresses.

? For practice questions on these topics, see questions 1, and 19–22 at the end of this chapter (p.443 and p.445).

9.8 Theories of reactions

Theories of elementary reactions provide a framework for understanding how reactions take place. A reaction theory is based on a **model** of how the reaction occurs at a molecular level. The model is then described in mathematical terms so that predictions can be made about the reaction. One way to assess the value of a theory—and in turn the quality of the model on which it is based—is to compare the theoretical estimate of a rate constant with the experimental value. These predicted rate constants are often used in situations where experimental measurements are difficult to make, or they may be used to help design an experiment. There are two commonly used theories of reaction.

- **Collision theory** is based on a simple model in which the reactant molecules behave like hard spheres. Collisions between the spheres may result in reaction. A calculated collision frequency is used to estimate the *A* factor in the Arrhenius equation for the reaction.
- **Transition state theory** is a more sophisticated and detailed theory that assumes the reactants form a transition state—a short-lived complex at the maximum on the energy profile. The *A* factor is then calculated, based on an assumed structure for the transition state.

To obtain a theoretical estimate of the rate constant, the calculated *A* factor is combined with a value for E_a calculated using quantum mechanics.

In this chapter, the theories are used to estimate the *A* factor for a reaction, which can then be compared with the value of *A* obtained experimentally from an Arrhenius plot as described in Section 9.7 (p.425). Before looking at these theories in detail, you need to understand what is meant by the molecularity of an elementary reaction.

Molecularity

The idea of the molecularity of an elementary reaction is introduced in Section 9.6 (p.416), where the S_N1 (unimolecular) and S_N2 (bimolecular) mechanisms for the hydrolysis of halogenoalkanes are discussed.

The kinetics of the methyl isocyanide isomerization are discussed in Section 9.6 (p.416).

The **molecularity** of an elementary reaction (see Section 9.6, p.416) can now be more accurately defined as the number of reactant molecules involved in the transition state.

In a **unimolecular** reaction, the transition state involves just one molecule of the reactant. A bond in the reactant molecule may break to form two product molecules, or the reactant molecule may rearrange during the reaction to form the product, as in the isomerization of methyl isocyanide

$$CH_3\text{–}NC \rightleftharpoons CH_3\text{–}CN$$

The transition state involves one molecule of CH_3NC in a high-energy state in which the CH_3–N bond is partly broken and the new CH_3–C bond is partly formed. For this to happen, the NC group partly rotates

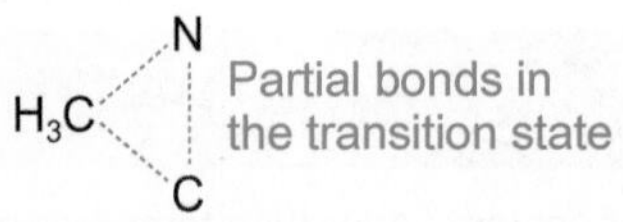
N
H3C
C
Partial bonds in the transition state

In a **bimolecular** reaction, the transition state involves two reactant molecules. In some cases, the two molecules come together to form a single larger molecule. In other cases, as the two molecules come together, bonds break and new bonds form to give two new product molecules. For example, in the reaction between $H^{\bullet}$ and F_2 (Equation 9.26, p.428 and Figure 9.25, p.429), the transition state involves one atom of $H^{\bullet}$ and one molecule of F_2 in a high-energy state in which the H–F bond is partly formed and the F–F bond partly broken.

$$\underset{\text{reactant}}{H^{\bullet} + F_2} \rightarrow \underset{\text{intermediate}}{H\cdots F\cdots F} \rightarrow \underset{\text{product}}{HF + F^{\bullet}}$$

It is important not to confuse *molecularity* with the *order* of a reaction. A bimolecular reaction is second order, but the origins of the terms are different. Reaction orders are experimentally determined from the observed kinetics of the reaction. The molecularity is a theoretical interpretation of the kinetics to explain the way in which the reaction occurs. Molecularities apply only to the elementary steps of a complex reaction, whereas the overall order of the reaction may take a quite different value.

Collision theory

In the model for collision theory, for a reaction to occur two molecules must collide with a certain minimum kinetic energy along their line of approach. The energy profiles used in the collision theory represent the potential energy of the two molecules as they approach and then collide and react to form the products. These energy profiles are discussed in Section 9.7 (see, for example, Figure 9.25, p.429, for the $H^{\bullet} + F_2$ reaction). For a reaction to occur, colliding molecules must have sufficient energy to overcome the energy barrier, E_a.

Consider a simple bimolecular reaction in the gas phase between two reactants A and B. Collision theory uses the kinetic theory of gases developed in Section 8.4 (p.358) to find an expression for the A factor in the Arrhenius equation, which is related to the number of collisions between A and B molecules per second. This expression for the A factor shown below

$$A = \sigma\left(\frac{8k_BT}{\pi\mu}\right)^{1/2} N_A$$

depends on the collision cross-section, σ, which is related to the size of the molecules and their average relative speed. (k_B is the Boltzmann constant. μ is called the reduced mass and is equal to $\frac{m_A m_B}{m_A + m_B}$, where m_A and m_B are the masses of the reactant molecules. N_A is the Avogadro constant.)

Speeds of molecules, collision cross-section, and the collision frequency, Z, (for collisions between molecules in a pure gas), are introduced in Section 8.5 (p.362). The Boltzmann constant k_B is the gas constant R divided by the Avogadro constant and is discussed in Section 10.3 (p.458), when discussing the population of energy levels in spectroscopy.

But not all the collisions result in reaction, so to obtain an expression for the rate constant, k, the expression above is multiplied by the fraction of the collisions, $e^{-\frac{E_a}{RT}}$, that have enough energy to surmount the energy barrier, E_a, and react. So,

$$k = \sigma\left(\frac{8k_BT}{\pi\mu}\right)^{1/2} N_A e^{-\frac{E_a}{RT}}$$

For most reactions, this gives a value of the rate constant that is too large, sometimes by several orders of magnitude. This is explained by assuming that not all collisions with enough energy to overcome the energy barrier actually go on to form products. Better agreement is obtained if a probability factor, P, is introduced, where the value of P is less than one

$$k = P\sigma\left(\frac{8k_BT}{\pi\mu}\right)^{1/2} N_A e^{-\frac{E_a}{RT}} = Ae^{\frac{E_a}{RT}}$$

Comparing this expression with the Arrhenius equation, you can see that

$$A = P\sigma\left(\frac{8k_BT}{\pi\mu}\right)^{1/2} N_A \quad (9.27)$$

P is sometimes interpreted as a **steric factor**—the reactants need to collide in a specific orientation to react. Figure 9.27 shows the gas phase reaction

$$CH_3I + K^{\bullet} \rightarrow CH_3^{\bullet} + KI$$

Reaction can only take place if the potassium atom ($K^{\bullet}$) approaches in the direction of the iodine atom in the CH_3I molecule. This reduces the effective cross-section from that expected using a hard sphere model for collisions.

It is difficult, though, to predict P to give reliable estimates for rate constants. Collision theory is widely applied to simple reactions, such as those between atoms and diatomic molecules, but not to more complicated reactions. Transition theory, discussed below, provides a more detailed theoretical treatment.

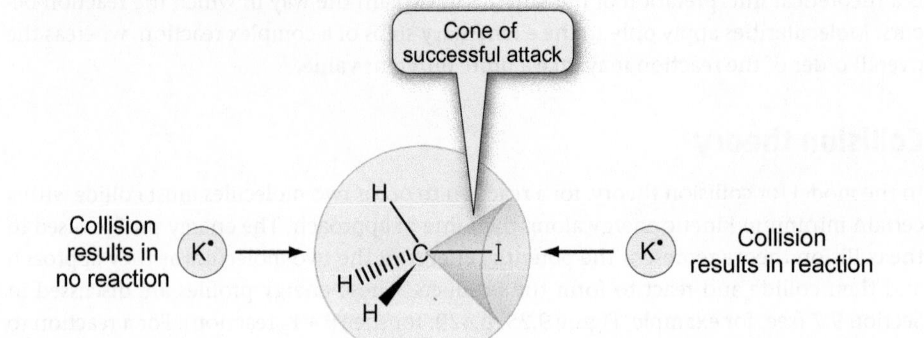

Cone of successful attack
Collision results in no reaction
K•
H
H
C
I
H
K•
Collision results in reaction

Figure 9.27 In the gas phase reaction between a K atom and a molecule of CH_3I, only those collisions in which K approaches CH_3I along a direction that lies inside the cone lead to reaction—even though the energy of collisions in other orientations may exceed the activation energy.

Transition state theory

In the model for the transition state theory, the reaction pathway from reactants to products passes through a transition state at the highest point of the energy profile. This is illustrated in Figure 9.28 for a bimolecular reaction with a rate constant k

$$\mathrm{A + BC \rightarrow AB + C}\ (\text{rate constant } k)$$

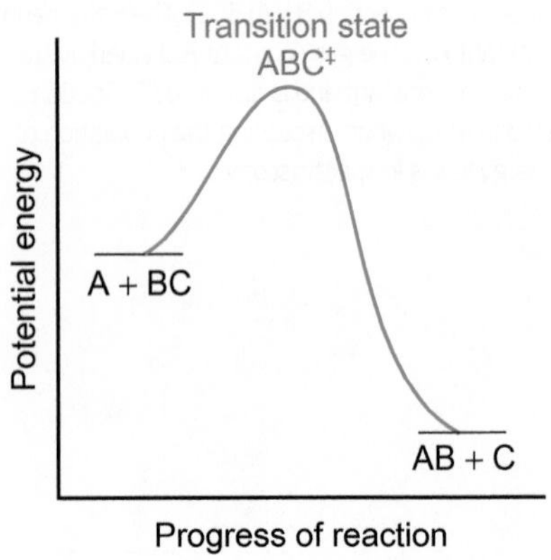

Transition state
ABC‡
Potential energy
A + BC
AB + C
Progress of reaction

Figure 9.28 Energy profile for the reaction A + BC → AB + C.

Transition state theory assumes that a pre-equilibrium (see Section 9.6, p.416) is established between the reactants and the transition state, $ABC^{\ddagger}$, which then goes on to form the products

$$\mathrm{A + BC} \overset{K^{\ddagger}}{\rightleftharpoons} \mathrm{ABC}^{\ddagger} \overset{k^{\ddagger}}{\longrightarrow} \mathrm{AB + C}$$

where $K^{\ddagger}$ is the equilibrium constant for the reversible initial step to form the transition state and $k^{\ddagger}$ is the rate constant for the formation of products from the transition state.

To obtain an expression for the rate constant, k, for the overall reaction, the theory uses the ideas developed on p.420 for a series of elementary steps involving a pre-equilibrium, together with an estimate of the rate constant, $k^{\ddagger}$. The resulting equation is

$$k = \frac{k_B T}{h} K^{\ddagger} \qquad (9.28)$$

where k_B is the Boltzmann constant (see p.433) and h is the Planck constant.

This is an important expression because it links k with thermodynamic functions for the formation of the transition state. $K^{\ddagger}$ is the equilibrium constant for the formation of the transition state and is related to the corresponding Gibbs energy change by the equation

$$\Delta^{\ddagger}G = -RT \ln K^{\ddagger}$$

where $\Delta^{\ddagger}G$ is the Gibbs energy change for the formation of the transition state from the reactants and is called the **Gibbs energy of activation.**

Rearranging the equation to give an expression for $K^{\ddagger}$

$$\ln K^{\ddagger} = -\frac{\Delta^{\ddagger}G}{RT} \text{ and } K^{\ddagger} = e^{-\frac{\Delta^{\ddagger}G}{RT}} \qquad (\text{If } \ln y = x, \text{then } y = e^{x})$$

Equation 9.28 can now be rewritten as

$$k = \frac{k_B T}{h} e^{-\frac{\Delta^{\ddagger}G}{RT}}$$

This expression can be further extended, since $\Delta^{\ddagger}G = \Delta^{\ddagger}H - T\Delta^{\ddagger}S$, where $\Delta^{\ddagger}H$ is the **enthalpy of activation** and $\Delta^{\ddagger}S$ is the **entropy of activation**. Using the mathematical relationship,

The symbol ‡ is used as a superscript to denote the transition state and also the quantities (such as $\Delta^{\ddagger}G$ and $K^{\ddagger}$) involved in the formation and reaction of the transition state.

The Planck constant, h, is introduced in Section 3.2 (p.117). It appears in Equation 9.28 because the expression is derived by considering the energy levels associated with $ABC^{\ddagger}$.

For a reaction, the equilibrium constant, K, is related to the standard Gibbs energy change of reaction, $\Delta_r G^{\ominus}$, by the equation

$\Delta_r G^{\ominus} = -RT \ln K$ (see Section 15.3, p.705)

and hence to the standard enthalpy change, $\Delta_r H^{\ominus}$, and standard entropy change, $\Delta_r S^{\ominus}$

$\Delta_r G^{\ominus} = \Delta_r H^{\ominus} - T\Delta_r S^{\ominus}$ (see Section 14.5, p.675).

$e^{a+b} = e^a \times e^b$, you can obtain an expression for the overall rate constant in terms of $\Delta^{\ddagger}H$ and $\Delta^{\ddagger}S$. Comparing this with the Arrhenius equation

$$k = \frac{k_B T}{h} e^{\frac{\Delta^{\ddagger}S}{R}} e^{-\frac{\Delta^{\ddagger}H}{RT}} = A e^{-\frac{E_a}{RT}}$$

you can see that the term in blue corresponds to the A factor in the Arrhenius equation and $\Delta^{\ddagger}H$ corresponds to E_a.

$$A = \frac{k_B T}{h} e^{\frac{\Delta^{\ddagger}S}{R}} \qquad (9.29)$$

Equation 9.29 shows the importance of entropy changes during the formation of the transition state, since these affect the value of the rate constant and hence the rate of the reaction.

The next section describes the use of $\Delta^{\ddagger}G$ in plotting Gibbs energy profiles for reactions.

Visit the Online Resource Centre to view screencast 9.6 which discusses transition state theory and the importance of Gibbs energy profiles.

Gibbs energy profiles for reactions

An alternative way to follow the progress of a reaction is to plot Gibbs energy, G, as the reaction takes place (rather than the potential energy as in Section 9.7). This is called a **Gibbs energy profile**. Transition state theory links the Gibbs energy of activation, $\Delta^{\ddagger}G$, directly to the rate constant, k, for the reaction—the greater the value of $\Delta^{\ddagger}G$, the smaller the value of the rate constant and the slower the reaction. As shown in the previous section, the value of $\Delta^{\ddagger}G$ contains a contribution both from the enthalpy of activation, $\Delta^{\ddagger}H$, and the entropy of activation, $\Delta^{\ddagger}S$. Calculated values of $\Delta^{\ddagger}S$, using Equation 9.29 and experimental A factors, provide information about the structures of transition states.

Figure 9.29 shows the Gibbs energy profile for a concerted single step reaction, such as the alkaline hydrolysis of bromomethane, which proceeds by an S_N2 mechanism (see Section 9.6, p.417). There is a single maximum corresponding to the transition state for the reaction. The Gibbs energy change for the reaction, $\Delta_r G$, corresponds to the difference in Gibbs energy between the reactants and products.

For a complex reaction involving two steps, such as the alkaline hydrolysis of 2-bromo-2-methylpropane, which proceeds by an S_N1 mechanism (see Section 9.6, p.417), the Gibbs energy profile shows two maxima corresponding to the transition states for the two steps, and a minimum corresponding to the formation of the intermediate. In the example in Figure 9.30 (p.436), $k_1 \ll k_2$, so that $\Delta^{\ddagger}G_1$, the Gibbs energy of activation for the formation of the transition state in step 1, is much larger than $\Delta^{\ddagger}G_2$, the Gibbs energy of activation for the formation of the transition state in step 2.

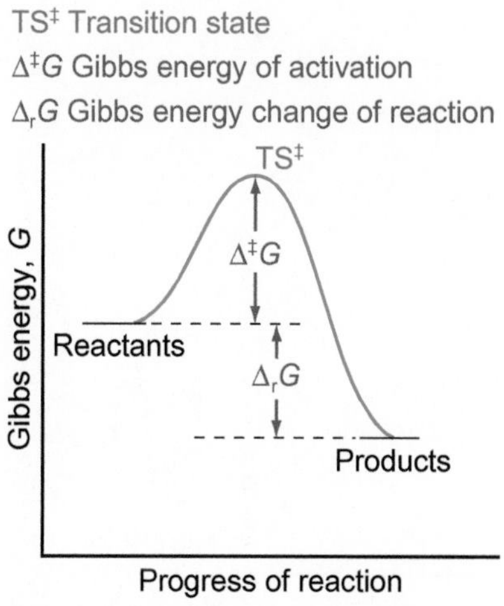
TS‡ Transition state
Δ‡G Gibbs energy of activation
ΔrG Gibbs energy change of reaction
TS‡
Δ‡G
Reactants
ΔrG
Products
Gibbs energy, G
Progress of reaction

Figure 9.29 Gibbs energy profile for a single step reaction, such as the alkaline hydrolysis of bromomethane.

The mechanism of S_N2 reactions is discussed in Section 20.3. Figure 20.10 (p.930) shows a Gibbs energy profile for an S_N2 reaction.

The mechanism of S_N1 reactions is discussed in Section 20.3. Figure 20.15 (p.934) shows a Gibbs energy profile for an S_N1 reaction.

An **intermediate** in a reaction can be detected spectroscopically, but cannot usually be isolated. It occurs at a minimum in the Gibbs energy profile. A **transition state** occurs at a maximum in the Gibbs energy profile. It is very difficult to detect and cannot be isolated.

Take care not to confuse the Gibbs energy of activation, $\Delta^{\ddagger}G$, with the Gibbs energy change for the overall reaction, $\Delta_r G$. The Gibbs energy change for a reaction is related to the spontaneity of the reaction. For a reaction to be spontaneous, the value of $\Delta_r G$ must be negative. Not all spontaneous reactions take place in practice, however. In cases where the Gibbs energy of activation is very large, the rate constant may be too small for significant reaction to occur under normal conditions.

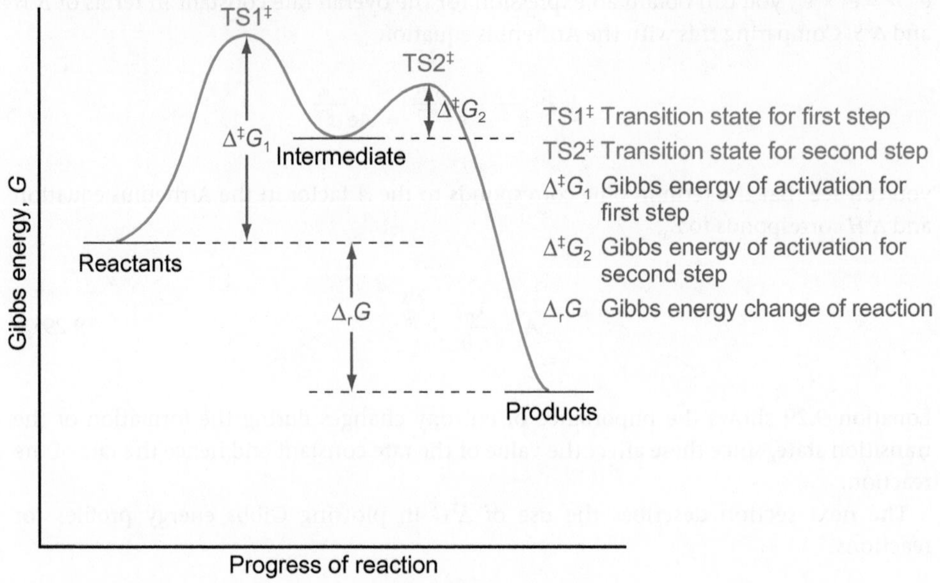
TS1‡
TS2‡
Δ‡G2
Δ‡G1
Intermediate
Reactants
ΔrG
Products
Gibbs energy, G
Progress of reaction
TS1‡ Transition state for first step
TS2‡ Transition state for second step
Δ‡G1 Gibbs energy of activation for first step
Δ‡G2 Gibbs energy of activation for second step
ΔrG Gibbs energy change of reaction

Figure 9.30 Gibbs energy profile for a two-step reaction, such as the alkaline hydrolysis of 2-bromo-2-methylpropane.

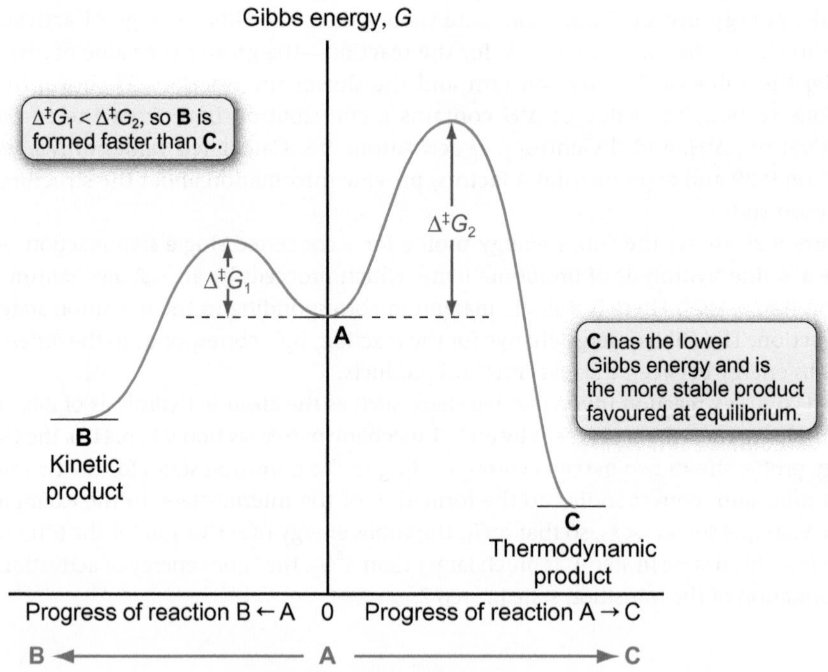
Gibbs energy, G
Δ‡G1 < Δ‡G2, so B is formed faster than C.
Δ‡G2
Δ‡G1
A
C has the lower Gibbs energy and is the more stable product favoured at equilibrium.
B
Kinetic product
C
Thermodynamic product
Progress of reaction B ← A 0 Progress of reaction A → C
B A C

Figure 9.31 Kinetic and thermodynamic products.

Competing reactions

The kinetics of parallel reactions are discussed in Section 9.6 (p.418).

When two reactions take place independently in parallel, each reaction has its own Gibbs energy profile. For example, Figure 9.31 shows the Gibbs energy profiles for the parallel reactions

$$A \rightleftharpoons B \text{ and } A \rightleftharpoons C$$

The reaction $A \rightarrow B$ has the lower $\Delta^{\ddagger}G$, but the product C has the lower Gibbs energy and is favoured at equilibrium. The reaction $A \rightarrow B$ will occur more rapidly than $A \rightarrow C$, so product B will form initially. B is known as the **kinetic product**. For longer reaction times, since both reactions are reversible, the system comes to equilibrium to form C as the major product. C is known as the **thermodynamic product**. This process takes longer because the rate constant for A → C is smaller than that for A → B.

Gibbs energy profiles are used extensively in organic chemistry and biochemistry to illustrate the mechanism of reactions. They can be particularly useful when considering competing reactions, as they can help chemists choose the best reaction conditions to maximize a desired reaction and minimize unwanted reactions.

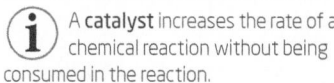

For examples of the use of Gibbs energy profiles in organic chemistry, see Sections 19.1 (p.867) and 20.3 (p.930).

Summary

- The molecularity of a reaction is the number of reactant molecules involved in forming the transition state.
- Collision theory is most useful for bimolecular reactions in the gas phase. It relates the *A* factor in the Arrhenius equation to the frequency of collisions taking place with the correct orientation for reaction.
- Transition state theory shows that the rate constant, *k*, depends on the Gibbs energy of activation, $\Delta^{\ddagger}G$. The *A* factor is related to $\Delta^{\ddagger}S$ and the activation energy, E_a, to $\Delta^{\ddagger}H$.
- Gibbs energy profiles are used extensively in organic chemistry and biochemistry to illustrate the mechanisms of reactions.

For practice questions on these topics, see questions 4, 27, and 30 at the end of this chapter (pp.444–447).

9.9 Catalysis

A **catalyst** is a substance that increases the rate of a chemical reaction without being consumed in the reaction. It does not appear in the overall equation for the reaction. Some substances slow down chemical reactions and these are often called negative catalysts or inhibitors.

A **catalyst** increases the rate of a chemical reaction without being consumed in the reaction.

The catalyst does not affect the *amount* of product formed—only the *rate* at which it is formed. Only a small amount of catalyst is usually needed. It is not changed chemically in the reaction and can be recovered at the end of the reaction. Sometimes a catalyst may be changed *physically*. For example, the surface of a solid catalyst may crumble or become roughened. This suggests that the catalyst is taking part in the reaction but is being regenerated.

Figure 9.32 shows Gibbs energy profiles for an uncatalysed reaction and for the same reaction in the presence of a catalyst. The catalyst acts by providing an alternative pathway for the reaction in which the rate-determining step has a lower Gibbs energy of activation than that of the uncatalysed reaction. At a given temperature, the rate constant is greater for the catalysed reaction, so the reaction is faster. Note that in Figure 9.32 the uncatalysed reaction proceeds through a single transition state, whereas the mechanism for the catalysed reaction involves the formation of an intermediate. Table 9.5 compares the effective activation energies for some catalysed and uncatalysed reactions.

For a complex reaction made up of a number of elementary reactions, E_a for the overall reaction is called the effective activation energy; see Section 9.7 (p.431).

There are two types of catalytic reaction.

- In **heterogeneous catalysis**, the catalyst is in a different physical state from that of the reactants. Many industrial processes involve this type of catalysis, in which a mixture of gases or liquids is passed over a solid catalyst, such as a *d*-block metal or one of its compounds. An advantage of heterogeneous catalysis is that the products (and any unreacted starting materials) are easily separated from the catalyst.

Solid heterogeneous catalysts are often deposited on structures designed to maximize their surface area.

- In **homogeneous catalysis**, the catalyst is in the same physical state as the reactants. Many organic acid–base catalysed reactions involve this type of catalysis. Enzyme reactions in cells take place in aqueous solution and involve homogeneous catalysis.

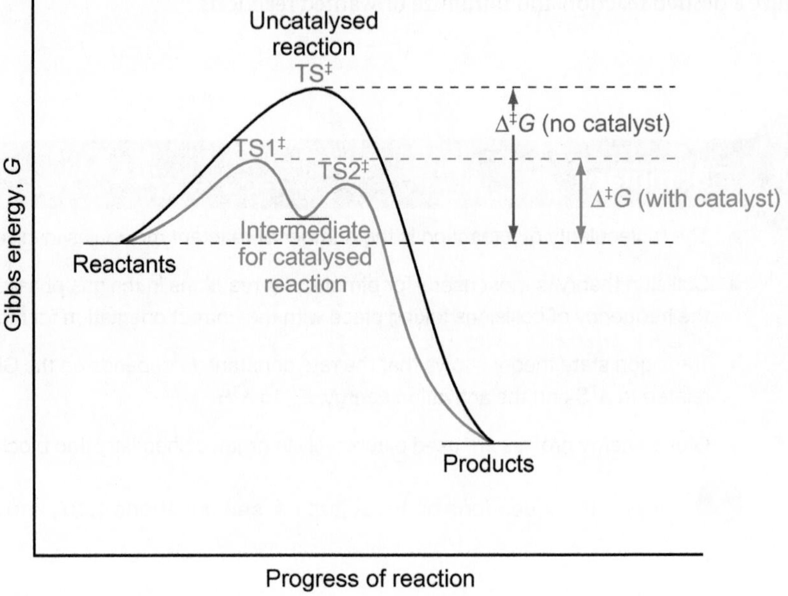
Uncatalysed reaction
TS‡
Δ‡G (no catalyst)
TS1‡
TS2‡
Δ‡G (with catalyst)
Intermediate for catalysed reaction
Reactants
Products
Gibbs energy, G
Progress of reaction

Figure 9.32 Gibbs energy profiles for a catalysed and an uncatalysed reaction.

Table 9.5 Effective activation energies of catalysed and uncatalysed reactions

Reaction	Catalyst	E_a/kJ mol^{-1}
$2NH_3(g) \rightarrow N_2(g) + 3H_2(g)$	none	+350
	tungsten	+162
$2N_2O(g) \rightarrow 2N_2(g) + O_2(g)$	none	+245
	gold	+121
$2HI(g) \rightarrow H_2(g) + I_2(g)$	none	+184
	gold	+105
$2H_2O_2(aq) \rightarrow 2H_2O(l) + O_2(aq)$	none	+75
	platinum	+49
	catalase (enzyme)	+23

Enzyme-catalysed reactions

Enzymes catalyse a wide range of reactions in your body. These reactions take place in dilute solution at 37 °C. Without the presence of enzymes, the reactions would take place much too slowly to sustain life. An enzyme is normally a protein molecule, and is specific for a particular reaction.

The enzyme interacts with the reactant, called the **substrate**, at a specific location in the enzyme called the **active site**. The three-dimensional shape of the active site fits the shape

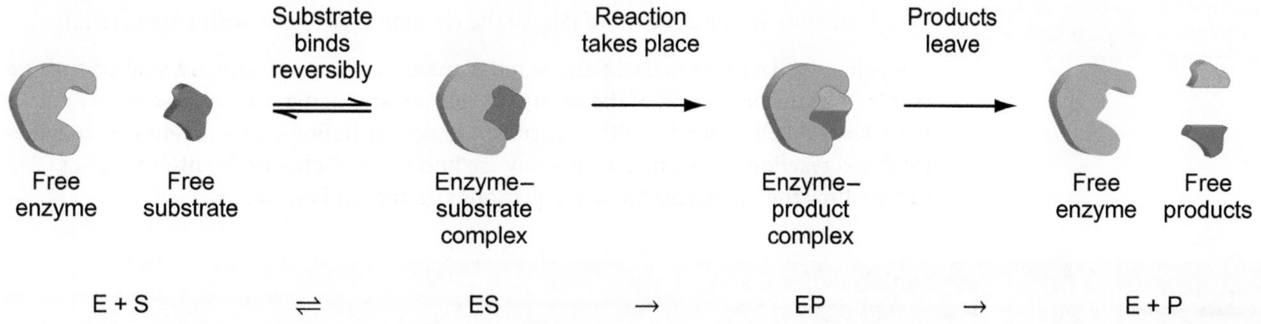
Substrate binds reversibly
Reaction takes place
Products leave
Free enzyme
Free substrate
Enzyme–substrate complex
Enzyme–product complex
Free enzyme
Free products
E + S ⇌ ES → EP → E + P

Figure 9.33 The simple 'lock and key' model for enzyme catalysis.

of the substrate. The substrate becomes bound to the active site by non-covalent interactions that depend very specifically on the three-dimensional shapes of the enzyme and substrate. The reaction then takes place in the active site. The products are less strongly bound and fit the active site less well, so they are released from the enzyme, which is then free to react with another molecule of the substrate. The whole process, sometimes called the 'lock and key' model, is summarized in Figure 9.33.

Figure 9.34 shows the Gibbs energy profile for this simple model of enzyme catalysis.

What happens, of course, is more complex. Scientists believe that in many cases, the substrate is not quite a perfect fit and must alter its shape to fit into the active site. This means that both the substrate and the active site are in strained arrangements, which can help the reaction to occur.

Experimental studies of enzyme-catalysed reactions are often carried out by measuring the initial rate of formation of the product in a solution containing a low concentration of the enzyme. A series of experiments is carried out with different initial concentrations of the substrate, $[S]_0$, to find out how the initial rate depends on [S]. Another series of experiments in which $[S]_0$ is constant and the initial enzyme concentration, $[E]_0$, is varied gives the dependence of the rate on [E].

Kinetic experiments such as these showed that enzyme-catalysed reactions have the following two characteristics:

- For a given initial enzyme concentration, and *low values* of $[S]_0$, the rate of product formation is directly proportional to $[S]_0$, so the reaction is *first order* with respect to [S].

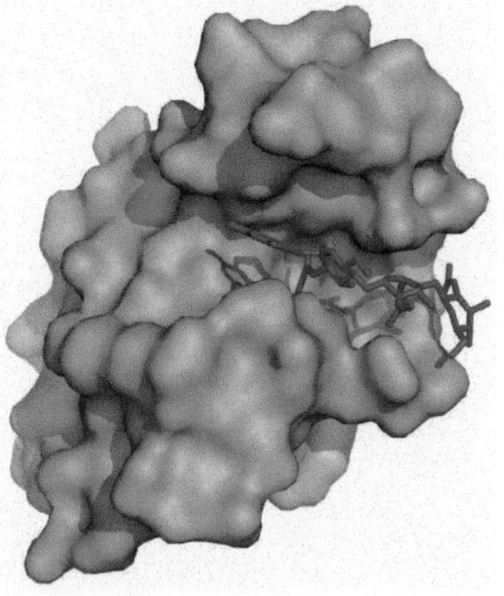

Computer graphics representation of a lysozyme molecule, an important catalytic enzyme that is capable of cleaving sugar structures in biological systems. In this image the protein is shown in green-blue and the substrate in purple. The classical 'lock and key' description of enzyme action is clearly illustrated in this image by the insertion of the substrate into the active site.

The initial rate method of investigating the kinetics of a reaction is described in Section 9.5 (p.409).

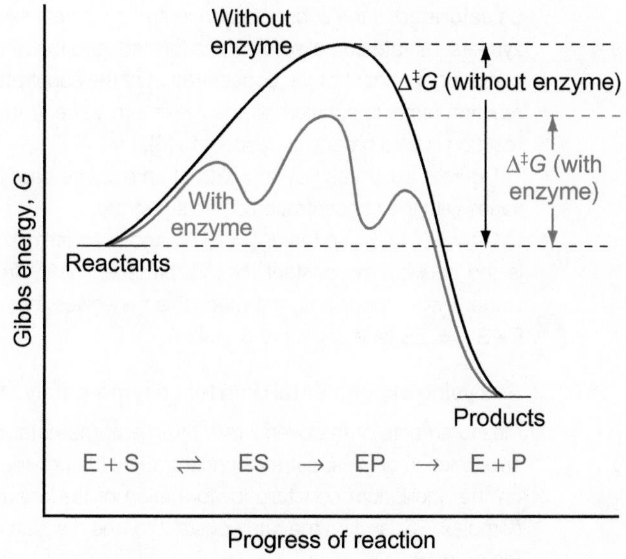
Without enzyme
$\Delta^{\ddagger}G$ (without enzyme)
$\Delta^{\ddagger}G$ (with enzyme)
With enzyme
Reactants
Products
Gibbs energy, G
E + S ⇌ ES → EP → E + P
Progress of reaction

Figure 9.34 Gibbs energy profile for an enzyme-catalysed reaction.

Zero order reactions are discussed in Section 9.5 (Boxes 9.6 and 9.7, p.412).

- For a given initial enzyme concentration, and *high values* of $[S]_0$, **the rate of product formation is independent of $[S]_0$, so the reaction is *zero order* with respect to [S].**

Biochemists Leonor Michaelis and Maud Menten, working in Germany, studied enzyme reactions at the beginning of the twentieth century and derived an expression to explain their kinetic behaviour for different substrate concentrations. The kinetics of enzyme-catalysed reactions explained in this way are known as **Michaelis–Menten kinetics**. You can see how the Michaelis–Menten equation is derived in Box 9.8.

Box 9.8 The Michaelis–Menten mechanism

A quantitative analysis of the kinetics of enzyme-catalysed reactions uses the steady state approximation and other ideas developed in Section 9.6 (p.416) for complex reactions.

The enzyme–product complex, EP, is short-lived and only reacts to give E + P, so the reaction scheme in Figure 9.33 can be simplified to

$$E + S \underset{k_{-1}}{\overset{k_1}{\rightleftharpoons}} ES \xrightarrow{k_2} E + P$$

Applying the steady state approximation (Equation 9.15, p.420) to the concentration of the enzyme–substrate complex, ES

$$\frac{d[ES]}{dt} = (\text{rate of formation of ES} - \text{rate of consumption of ES}) = 0$$

$$= k_1[E][S] - k_{-1}[ES] - k_2[ES]$$

$$\frac{d[ES]}{dt} = k_1[E][S] - (k_{-1} + k_2)[ES] = 0 \qquad (9.30)$$

The concentration of the substrate is usually much larger than that of the enzyme, so that [S] is effectively constant in a particular experiment.

It is not possible to measure [ES] but the total enzyme concentration $[E]_0$ is known. (The free enzyme concentration, [E], in Equation 9.30 is less than $[E]_0$, because some of the enzyme is bound up in the enzyme–substrate complex.) So

$$[E]_0 = [E] + [ES] \quad \text{and} \quad [E] = [E]_0 - [ES]$$

Substituting for [E] in Equation 9.30 gives

$$\frac{d[ES]}{dt} = k_1[E]_0[S] - k_1[ES][S] - (k_{-1} + k_2)[ES] = 0$$

Rearranging this equation

$$[ES] = \frac{k_1[E]_0[S]}{(k_1[S] + k_{-1} + k_2)}$$

The rate of formation of the product, P, is given by

$$\frac{d[P]}{dt} = k_2[ES] = \frac{k_1k_2[E]_0[S]}{(k_1[S] + k_{-1} + k_2)}$$

Dividing the numerator and the denominator by k_1 gives

$$\text{rate of formation of product} = \frac{k_2[E]_0[S]}{[S] + \left(\frac{k_{-1} + k_2}{k_1}\right)}$$

This can be written as

$$\text{rate of the enzyme-catalysed reaction} = \frac{k_2[E]_0[S]}{[S] + K_M} \qquad (9.31)$$

$$\text{where } K_M = \frac{k_{-1} + k_2}{k_1}$$

Equation 9.31 is called the **Michaelis–Menten equation** and K_M the **Michaelis constant**. You can distinguish two extreme situations.

(a) Reactions with low substrate concentrations

When [S] is very small, $K_M >> [S]$, so that

$$\text{rate of the reaction} = \frac{k_2[E]_0[S]}{K_M}$$

The rate of reaction increases linearly with [S], that is, it is first order with respect to [S].

(b) Reactions with large substrate concentrations

When [S] is large, $[S] >> K_M$, so that

$$\text{rate of reaction} = k_2[E]_0$$

The rate of reaction is now independent of [S]. Under these conditions, all of the enzyme molecules have substrates attached at any moment during the reaction and $[ES] = [E]_0$. The enzyme is said to be **saturated**. If the substrate concentration increases, no more enzyme–substrate complexes can be formed, and the rate of the reaction is independent of the concentration of the substrate. The rate of reaction remains constant at the maximum value, $(\text{rate})_{max}$, and the reaction is zero order with respect to [S].

Figure 1 illustrates how the rate of an enzyme-catalysed reaction varies with the concentration of the substrate.

Often $k_{-1} >> k_2$, so that $K_M = k_{-1}/k_1$ and is equal to $1/K_c$, where K_c is the equilibrium constant for the formation of ES from E and S. Under these conditions, the mechanism involves a pre-equilibrium, $E + S \rightleftharpoons ES$ (see Section 9.6, p.421).

Analysing experimental data for enzyme-catalysed reactions

Measurements of the overall rate of an enzyme-catalysed reaction, as a function of the substrate concentration, [S], provides information on the equilibrium constant for formation of the enzyme-substrate complex, ES, and on the rate constant for the reaction of ES to form the product, P.

→

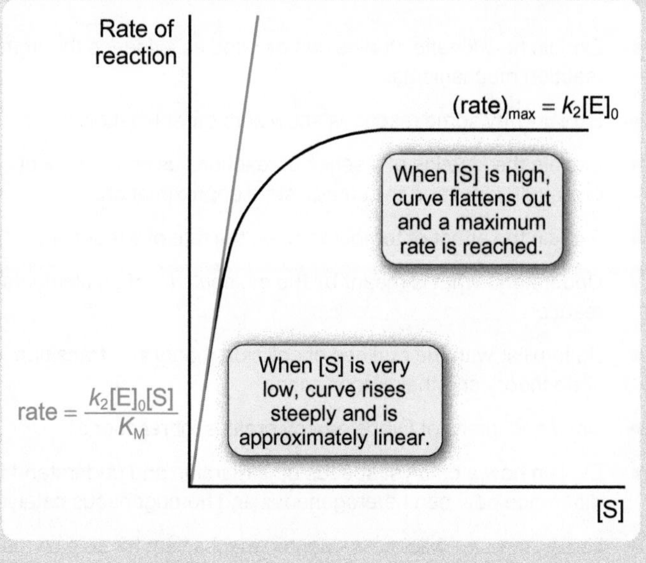

Rate of reaction
(rate)max = k2[E]0
When [S] is high, curve flattens out and a maximum rate is reached.
When [S] is very low, curve rises steeply and is approximately linear.
rate = k2[E]0[S] / KM
[S]

▲ **Figure 1** The dependence of the rate of an enzyme-catalysed reaction on the concentration of the substrate, [S]. (The plot is obtained from a series of experiments with different values of [S].)

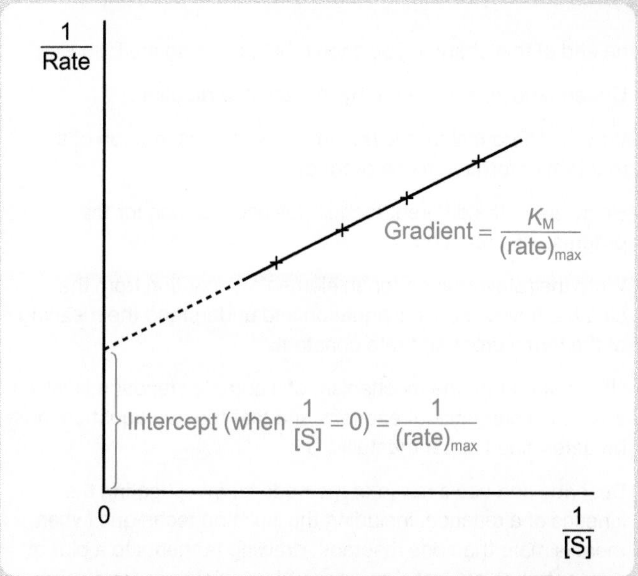

1 / Rate
Gradient = KM / (rate)max
Intercept (when 1/[S] = 0) = 1/(rate)max
0
1 / [S]

▲ **Figure 2** A Lineweaver–Burk plot of 1/(rate) against 1/[S] allows the Michaelis constant, K_M, and the maximum rate of reaction, $(\text{rate})_{max}$, to be found.

It is always a good idea to use a straight line plot to analyse experimental data. This can be done here by taking the reciprocal of the rate from Equation 9.31.

$$\frac{1}{\text{rate}} = \frac{1}{k_2[E]_0} + \frac{K_M}{k_2[E]_0[S]} \qquad (9.32)$$

$$= \frac{1}{(\text{rate})_{max}} + \frac{K_M}{(\text{rate})_{max}[S]}$$

where the maximum rate, $(\text{rate})_{max} = k_2[E]_0$

A plot of 1/rate against 1/[S] gives a straight line. The intercept on the *y*-axis (where 1/[S] = 0) is $1/(\text{rate})_{max}$ and the gradient of the line is $K_M/(\text{rate})_{max}$. The ratio of the gradient to the intercept is K_M. This is called a **Lineweaver–Burk plot** and is illustrated in Figure 2. A value for k_2 can also be found, since $[E]_0$ is known and

$$k_2 = \frac{(\text{rate})_{max}}{[E]_0}$$

Question

The following results were obtained for the enzyme-catalysed hydrolysis of ATP at 20 °C.

The initial concentration of the enzyme, ATPase, was 1.5×10^{-8} mol dm^{-3}.

$[ATP]_0$/µmol dm^{-3}	0.5	1.0	2.0	4.0
Initial rate/µmol dm^{-3} s^{-1}	0.54	0.82	1.11	1.36

Determine the maximum rate, $(\text{rate})_{max}$, and the Michaelis constant, K_M, for the reaction.

Summary

- A catalyst increases the rate of a chemical reaction without being consumed in the reaction.
- Catalysts act by providing an alternative pathway for reaction in which the rate-determining step has a lower Gibbs energy of activation than the uncatalysed reaction.
- The Michaelis–Menten equation is used to analyse the kinetics of enzyme-catalysed reactions.

For a practice question on these topics, see question 31 at the end of this chapter (p.447).

Concept review

By the end of this chapter, you should be able to do the following.

- Understand what is meant by the rate of a reaction.
- Write a differential for the rate in terms of consumption of a reactant or formation of a product.
- Be aware of the different techniques used to monitor the progress of a reaction.
- Write the rate equation for an elementary reaction from the stoichiometric chemical equation and understand the meaning of the terms order and rate constant.
- Understand that the mechanism of a complex reaction involves a series of elementary reactions and that the rate equation must be determined experimentally.
- Describe and use a range of methods for investigating the kinetics of a reaction, including the isolation technique (when there is more than one reactant), drawing tangents to a plot of concentration against time, measuring initial rates, and using integrated rate equations and half lives.
- Explain how kinetic studies can be used as evidence for organic reaction mechanisms.
- Explain why some reactions show zero order kinetics.
- Explain the kinetics of a series of reactions using the idea of pre-equilibrium and the steady state approximation.
- Explain the effect of temperature on the rate of a reaction.
- Understand what is meant by the molecularity of an elementary reaction.
- Be familiar with the outlines of collision theory and transition state theory and their importance.
- Draw and interpret Gibbs energy profiles for reactions.
- Explain how a catalyst speeds up a reaction and understand the difference between heterogeneous and homogeneous catalysis.
- Understand the Michaelis–Menten mechanism for an enzyme-catalysed reaction, and use a Lineweaver–Burk plot to analyse experimental data.

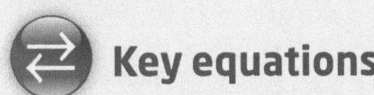

Key equations

Definition of rate of reaction for a general reaction: $a\,A + b\,B \rightarrow p\,P + q\,Q$	Rate of reaction $= -\frac{1}{a}\frac{d[A]}{dt} = -\frac{1}{b}\frac{d[B]}{dt} = \frac{1}{p}\frac{d[P]}{dt} = \frac{1}{q}\frac{d[Q]}{dt}$	(9.4)
For an *elementary* reaction: $a\,A + b\,B \rightarrow$ products	Rate of reaction $= k\,[A]^a\,[B]^b$ where k is the rate constant, and the stoichiometric coefficients in the chemical equation, a and b, are the orders with respect to A and B, respectively. The overall order is $a + b$.	(9.5)
For a *complex* reaction: $a\,A + b\,B \rightarrow$ products	Rate of reaction $= k[A]^m[B]^n$ where k is the overall rate constant, and m and n are the orders with respect to A and B, respectively. The overall order is $m + n$.	(9.11)
Reaction mechanisms of complex reactions		
For a reversible *elementary* reaction:	$\frac{k_1}{k_{-1}} = K_c$	(9.14)
The steady state approximation for a reactive intermediate X:	$\frac{d[X]}{dt} =$ (rate of formation of X − rate of consumption of X) $= 0$	(9.15)
The Arrhenius equation	$k = Ae^{-\frac{E_a}{RT}}$	(9.24a)
	$\ln k = \ln A - \frac{E_a}{R}\left(\frac{1}{T}\right)$	(9.24b)
	$\ln k_2 - \ln K_1 = \ln\frac{k_2}{k_1} = \frac{E_a}{R}\left(\frac{1}{T_1} - \frac{1}{T_2}\right)$	(9.25)
Michaelis–Menten equation	Rate of an enzyme-catalysed reaction $= \frac{k_2[E]_0[S]}{[S] + K_M}$ where $K_M = \frac{k_{-1} + k_2}{k_1}$	(9.31)

Table 9.2 A summary of equations for elementary reactions

Type of reaction	Order of reaction	Differential rate equation	Integrated rate equation	Half life, $t_{1/2}$
A → products	First order	$-\dfrac{d[A]}{dt} = k[A]$	$\ln [A]_t = \ln [A]_0 - kt$	$\dfrac{\ln 2}{k}$
A + A → products	Second order	$-\dfrac{d[A]}{dt} = 2k[A]^2$	$\dfrac{1}{[A]_t} = \dfrac{1}{[A]_0} + 2kt$	$\dfrac{1}{2k[A]_0}$
A + B → products where [A] <<[B]	Pseudo-first order	$-\dfrac{d[A]}{dt} = k'[A]$ where $k' = k[B]_0$	$\ln [A]_t = \ln [A]_0 - k't = \ln [A]_0 - kt[B]_0$	$\dfrac{\ln 2}{k'} = \dfrac{\ln 2}{k[B]_0}$

Table 9.3 Using integrated rate equations to study complex reactions **

	Order		
	0	1	2
Rate equation	Rate = k'	Rate = $k'[A]$	Rate = $k'[A]_2$
Integrated rate equation	$[A]_t = [A]_0 - k't$	$\ln[A]_t = \ln[A]_0 - k't$	$\dfrac{1}{[A]_t} = \dfrac{1}{[A]_0} + 2k't$ *
Plot of [A] vs. t	**Straight line**	Curve	Curve
Plot of ln[A] vs. t	Curve	**Straight line**	Curve
Plot of $\dfrac{1}{[A]}$ vs. t	Curve	Curve	**Straight line**
Half life, $t_{1/2}$	Not constant	**Constant**	Not constant

* The multiplication factor before k' depends on the stoichiometry of the reaction.
** The order of a complex reaction may be higher than 2 or have a fractional value.

Questions

More challenging questions are indicated by an asterisk *.

Note: Radical dots are not shown on stable radicals such as O_2, NO, and NO_2 unless relevant to the reaction.

1 The energy profiles A–D represent four different reactions. All the diagrams are drawn to the same scale. (Sections 9.1 and 9.7)

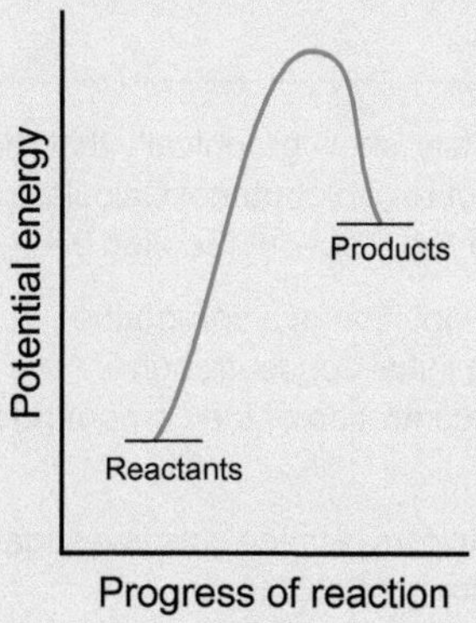
Potential energy
Products
Reactants
Progress of reaction
A

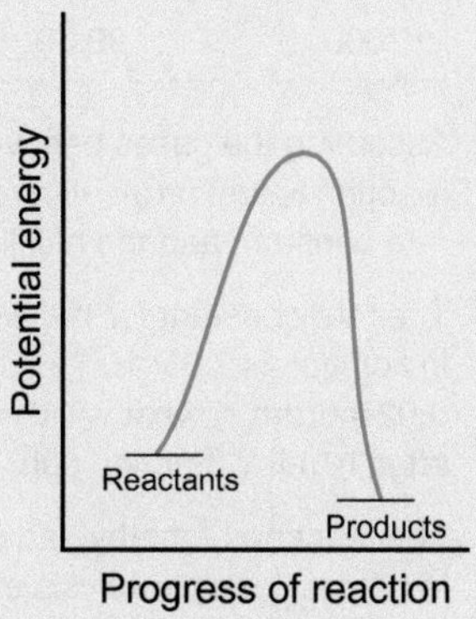
Potential energy
Reactants
Products
Progress of reaction
B

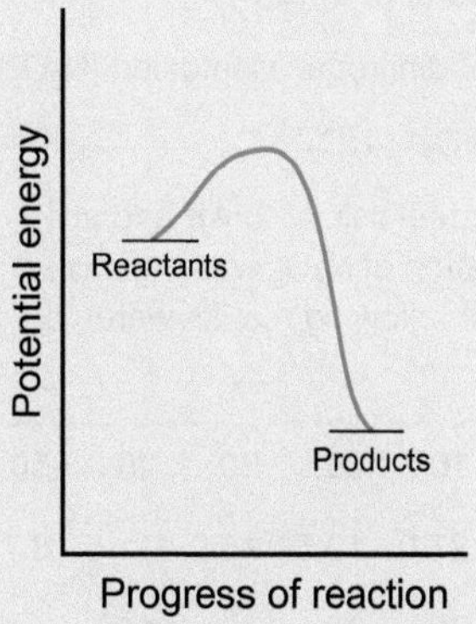
Potential energy
Reactants
Products
Progress of reaction
C

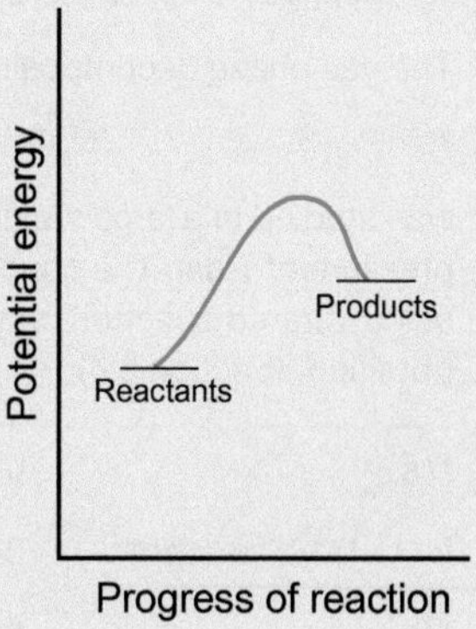
Potential energy
Products
Reactants
Progress of reaction
D

Which of the energy profiles A–D represents

(a) the most exothermic reaction;

(b) the most endothermic reaction;

(c) the reaction with the largest activation energy;

(d) the reaction with the smallest activation energy?

2 The oxidation of ammonia in air is catalysed by platinum metal

$$4NH_3(g) + 5O_2(g) \rightarrow 4NO(g) + 6H_2O(g)$$

Write an expression for the rate of reaction in terms of differentials for the consumption of the reactants and formation of the products. (Sections 9.2 and 9.4)

3 Under certain experimental conditions, the rate of the following reaction is 5.86×10^{-6} mol dm^{-3} s^{-1}

$$2N_2O(g) \rightarrow 2N_2(g) + O_2(g)$$

Calculate values for $\frac{d[N_2O]}{dt}$, $\frac{d[N_2]}{dt}$, and $\frac{d[O_2]}{dt}$ under these conditions. (Sections 9.2 and 9.4)

4 Write the rate equation for the following elementary reactions and give the molecularity for each reaction. (Sections 9.4 and 9.8)

(a) $Cl^{\bullet} + O_3 \rightarrow ClO^{\bullet} + O_2$

(b) $CH_3N_2CH_3 \rightarrow 2CH_3^{\bullet} + N_2$

(c) $2Cl^{\bullet} \rightarrow Cl_2$

(d) $NO_2^{\bullet} + F_2 \rightarrow NO_2F + F^{\bullet}$

5 For the complex reaction of NO and H_2

$$2NO(g) + 2H_2(g) \rightarrow N_2(g) + 2H_2O(g)$$

the rate equation is given by

rate of reaction $= k[NO]^2[H_2]$ (Section 9.5)

(a) What are the orders of the reaction with respect to NO and H_2?

(b) What is the overall order of the reaction?

(c) What will happen to the rate of reaction if:

(i) $[H_2]$ is doubled;

(ii) $[H_2]$ is halved;

(iii) [NO] is doubled;

(iv) [NO] is increased by a factor of three?

6 The gas phase decomposition of dinitrogen pentoxide (N_2O_5)

$$N_2O_5 \rightarrow NO_2^{\bullet} + NO_3^{\bullet}$$

was studied in a large stainless steel cell, at 294 K and a pressure of 1 bar. The concentration of N_2O_5 was monitored using infrared spectroscopy. The following results were obtained. (Section 9.4)

t/s	0	10	20	30	40	50
$[N_2O_5]/10^{-9}$ mol dm^{-3}	34.0	27.0	19.5	15.0	11.5	8.7
t/s	60	70	80	90	100	
$[N_2O_5]/10^{-9}$ mol dm^{-3}	6.6	5.1	3.9	2.9	2.2	

Show that the reaction is first order and find a value for the rate constant at 294 K.

7 The data below were obtained for the decomposition of difluorine oxide (F_2O) at 298 K. (Sections 9.4 and 9.5)

$$2F_2O(g) \rightarrow 2F_2(g) + O_2(g)$$

t/s	0	60	120	180	240	300	360	420
$[F_2O]/10^{-3}$ mol dm^{-3}	7.2	5.5	4.6	3.8	3.3	2.9	2.6	2.4

Verify that this is a second order reaction and determine the rate constant at 298 K.

8 From the data provided in the table, deduce the rate equation and the value of the rate constant for the following reaction. (Section 9.4)

$$CH_3COCH_3(aq) + Br_2(aq) + H^+(aq) \rightarrow CH_3COCH_2Br(aq) + 2H^+(aq) + Br^-(aq)$$

	Initial concentration of CH_3COCH_3 /mol dm^{-3}	Initial concentration of Br_2 /mol dm^{-3}	Initial concentration of H^+ /mol dm^{-3}	Initial rate of reaction /mol dm^{-3} s^{-1}
1	1.00	1.00	1.00	4.0×10^{-3}
2	2.00	1.00	1.00	8.0×10^{-3}
3	2.00	2.00	1.00	8.0×10^{-3}
4	1.00	1.00	2.00	8.0×10^{-3}

9 For a particular first order reaction, half of the reactant is used up after 15 s. What fraction of the reactant will remain after 1 min? (Section 9.4)

10 Cyclobutane decomposes to form ethane according to:

$$C_4H_8(g) \rightarrow 2C_2H_4(g)$$

A quantity of cyclobutane was sealed in a container and exerted a pressure of 53.30 kPa at 700 K. The pressure changed during the reaction as follows. At the end of the reaction, the pressure was 106.60 kPa.

Time/s	Total pressure/kPa
0	53.30
2000	64.53
4000	73.59
6000	80.53
8000	85.99
10 000	90.39

Assuming the gases behave ideally, show graphically that the reaction is first order with respect to cyclobutane. Calculate the rate constant and the half life for the reaction. (Section 9.4)

11 The rate constant for the decomposition of a compound in solution is 2.0×10^{-4} s^{-1}. If the initial concentration is 0.02 mol dm^{-3}, what will the concentration of the compound be after 10 min? (Section 9.4)

12 The reaction of methyl radicals to form ethane was investigated in a laser flash photolysis experiment at 300 K.

$$CH_3^{\bullet} + CH_3^{\bullet} \rightarrow C_2H_6$$

The rate constant for this reaction at 300 K is 3.7×10^{10} dm^3 mol^{-1} s^{-1}. The concentration of methyl radicals, $[CH_3^\bullet]$, at time $t = 0$ was 1.70×10^{-8} mol dm^{-3}. Calculate a value for $[CH_3^\bullet]$ at $t = 1.00 \times 10^{-3}$ s. (Section 9.4)

13 The acid-catalysed hydrolysis of sucrose shows first order kinetics. The half life for the reaction at room temperature was found to be 190 min. Calculate the rate constant for the reaction under these conditions. (Section 9.4)

14 The reaction between $H_2PO_4^-$ and OH^- was investigated at 298 K using the initial rate method

$$H_2PO_4^-(aq) + OH^-(aq) \rightarrow HPO_4^{2-}(aq) + H_2O(aq)$$

The following results were obtained. (Section 9.5)

	Initial rate/10^{-3} mol dm^{-3} min^{-1}	$[OH^-]_0/10^{-3}$ mol dm^{-3}	$[H_2PO_4^-]_0/10^{-3}$ mol dm^{-3}
Experiment 1	2.0	0.40	3.0
Experiment 2	3.7	0.55	3.0
Experiment 3	7.1	0.75	3.0

(a) Plot a log–log graph to determine the order of reaction with respect to OH^-(aq).

(b) What further experiments would you need to do to find the order with respect to $H_2PO_4^-$?

15 The addition of bromine to propene is an elementary reaction with a rate constant, *k*

$$CH_2{=}CHCH_3 + Br_2 \rightarrow CH_2BrCHBrCH_3$$

Kinetic studies were carried out at 298 K using excess Br_2. For $[Br_2]_0 = 0.20$ mol dm^{-3}, the pseudo-first order rate constant, k', for the reaction was found to be 900 s^{-1}. What is the value of *k* at 298 K? (Section 9.4)

16* The elementary reaction between ethanal (CH_3CHO) and $^\bullet OH$ radicals

$$CH_3CHO + {}^\bullet OH \rightarrow CH_3CO^\bullet + H_2O$$

was studied at 298 K, by laser flash photolysis, using laser-induced fluorescence to detect $^\bullet OH$ as a function of time. The ethanal concentration was 3.3×10^{-7} mol dm^{-3} and this concentration was much higher than the concentration of $^\bullet OH$ radicals. The following data for the concentration of $^\bullet OH$ radicals, relative to their concentration at zero time, were obtained. (Section 9.4)

$t/10^{-3}$ s	0	0.2	0.4	0.6	0.8	1.0	1.2	1.5
$[{}^\bullet OH]/[{}^\bullet OH]_0$	1	0.55	0.31	0.16	0.09	0.05	0.03	0.01

(a) Write a rate equation for the reaction.

(b) Show that, under the above conditions, the reaction follows pseudo-first order kinetics.

(c) Determine the pseudo-first order rate constant, k'.

17* The investigation described in Question 16 was repeated a number of times, each time using a different concentration of ethanal, $[CH_3CHO]_0$. A value for k' was found in each case. The following results were obtained.

$[\text{ethanal}]_0/10^{-7}$ mol dm^{-3}	1.2	2.4	4.0	5.1
$k'/10^3$ s^{-1}	1.12	2.10	3.65	4.50

Confirm that the reaction is first order with respect to ethanal and determine the second order rate constant for the reaction. (Incorporate the value of k' determined in Question 16 into your analysis.) (Section 9.4)

18 The decomposition of ammonia on a platinum surface at 856 °C

$$2NH_3(g) \rightarrow N_2(g) + 3H_2(g)$$

shows the following dependence of the concentration of ammonia gas on time.

t/s	0	200	400	600	800	1000	1200
$[NH_3]/10^{-3}$ mol dm^{-3}	2.10	1.85	1.47	1.23	0.86	0.57	0.34

Find the order of the reaction and a value for the rate constant at 856 °C. Suggest an explanation for the order you obtain. (Section 9.5)

19 Rate constants at a series of temperatures were obtained for the decomposition of azomethane

$$CH_3N_2CH_3 \rightarrow 2CH_3^\bullet + N_2$$

T/K	523	541	560	576	593
$k/10^{-6}$ s^{-1}	1.8	15	60	160	950

Use the data in the table to find the activation energy, E_a, for the reaction. (Section 9.7)

20 The hydrolysis of hydrogencarbonate ions in water at a high pH:

$$HCO_3^-(aq) + OH^-(aq) \rightarrow CO_3^{2-}(aq) + H_2O(l)$$

follows the rate law: Rate $= k\,[HCO_3^-]$ where $k = 0.01$ s^{-1} at 25 °C.

(a) What is the overall order of the reaction?

(b) What is the half-life of HCO_3^- if the initial concentration of HCO_3^-, $[HCO_3^-]_o = 0.001$ mol dm^{-3}?

(c) What is the rate constant of the reaction at 350 K if the activation energy is 10.0 kJ mol^{-1}?

(d) The hydrolysis reaction is exothermic. Sketch a curve to show how the energy of the system changes in going from reactants to products. Indicate the activation energy for the forward and reverse reactions and the overall change in energy for the reaction. (Sections 9.4 and 9.7)

21 Calculate the activation energy for a reaction in which the rate constant is 5 times faster at 50 °C than at 20 °C. (Section 9.7)

22 The *A* factor for the reaction of methane with hydroxyl radicals is 1.11×10^9 dm^3 mol^{-1} s^{-1}.

$$CH_4 + {}^\bullet OH \rightarrow CH_3^\bullet + H_2O$$

The activation energy for the reaction is +14.1 kJ mol^{-1}. (Section 9.7)

(a) Calculate the rate constant for the reaction at 220 K, which corresponds to a region close to the top of the troposphere.

(b) Compare the value you obtain in (a) with the value for the rate constant at 300 K, when $k = 3.9 \times 10^6\,\text{dm}^3\,\text{mol}^{-1}\,\text{s}^{-1}$. This corresponds to a region nearer the Earth's surface, see Box 9.3 (p.400). Comment on the difference between the values for the rate constant at the two temperatures.

23 The following mechanism has been suggested for the gas phase oxidation of hydrogen bromide

$$HBr + O_2 \rightarrow HOOBr$$

$$HOOBr + HBr \rightarrow 2\,HOBr$$

$$HOBr + HBr \rightarrow H_2O + Br_2$$

No HOBr is found in the final products. Experimentally, the overall reaction is found to be first order in HBr and in O_2. (Section 9.6)

(a) Write a balanced stoichiometric equation for the reaction.

(b) Show that the above mechanism is consistent with the observed orders of reaction.

(c) Which step is likely to be the rate-determining step?

24 The mechanism for the formation of a DNA double helix from two strands A and B is as follows. (Section 9.6)

$$\text{strand A} + \text{strand B} \overset{\text{fast}}{\rightleftharpoons} \text{unstable helix}$$

$$\text{unstable helix} \overset{\text{slow}}{\rightleftharpoons} \text{stable double helix}$$

(a) Experiments show that the overall reaction is first order with respect to strand A and first order with respect to strand B. Write the equation for the overall reaction.

(b) Write the rate equation for the overall reaction.

(c) Assuming the processes involved are elementary reactions, derive an expression for the rate constant for the overall reaction in terms of the rate constants for the individual steps.

25 The reaction of methane with hydrogen atoms is an elementary process.

$$CH_4 + H^{\bullet} \rightleftharpoons CH_3^{\bullet} + H_2$$

The rate constant for the forward reaction at 1 000 K is $1.6 \times 10^8\,\text{dm}^3\,\text{mol}^{-1}\,\text{s}^{-1}$. The equilibrium constant, K_c, at the same temperature is 19.8. Calculate the rate constant for the reverse reaction at 1000 K. (Section 9.6)

26* Molecules move much more slowly in solution than in the gas phase. The progress of a molecule is frequently stopped, and the direction of motion changed, on collision with solvent molecules.

A reaction in solution between two reactants, A and B, can be described by a model involving three processes. In the first, the reactants diffuse towards one another (rate constant, k_d). When they encounter one another they form AB, called an *encounter complex*, and stay together for ~10^{-10} s, trapped in a *solvent cage*. The separation of A and B by leaving the solvent cage is described by a rate constant k_{-d}. Alternatively, the encounter complex can react, to form the products, with a rate constant k_r. The overall mechanism is

$$A + B \underset{k_{-d}}{\overset{k_d}{\rightleftharpoons}} \underset{\text{encounter comlex}}{AB} \overset{k_r}{\rightleftharpoons} \text{product}$$

The individual steps can be treated as elementary reactions, and the steady state approximation can be applied to AB, since it is so short lived. (Section 9.6)

(a) What is the order of each of the three steps in the mechanism and what are the units of the rate constants for each step?

(b) Show that the rate of forming the products is given by

$$\text{rate of reaction} = \frac{k_d k_r}{k_{-d} + k_r}[A][B]$$

and write down an expression for the overall rate constant, k.

(c) Simplify this expression for a case in which:

(i) reaction to form the products is much faster than diffusion of A and B from the solvent cage;

(ii) diffusion of A and B from the solvent cage is much faster than reaction to form the products.

In each case, state which is the rate-determining step.

27 An exothermic reaction proceeds by the following three-step mechanism

$$A \overset{\text{fast}}{\rightleftharpoons} B \overset{\text{slow}}{\rightleftharpoons} C \overset{\text{fast}}{\rightleftharpoons} D$$

The first step occurs rapidly and a pre-equilibrium is established. The second step is slow and is the rate-determining step. (Section 9.8)

(a) Draw the shape of the Gibbs energy profile for the reaction.

(b) How would the Gibbs energy profile differ if the third step were the rate-determining step of the reaction?

28 After intravenous injection of a drug to treat hypertension (high blood pressure), the blood plasma of the patient was analysed for the remaining drug at various times after the injection. (Section 9.5)

t / min	50	100	150	200	250	300	400	500
[drug]/10^{-9} g cm^{-3}	650	445	304	208	142	97	45	21

(a) Is the removal of the drug in the body a first or a second order process?

(b) Calculate the rate constant, k, and the half life, $t_{1/2}$, for the process.

(c) An essential part of drug development is achieving an optimum value of $t_{1/2}$ for effective operation and elimination of the drug from the bloodstream. What would be the possible problems if $t_{1/2}$ were too short or too long?

29* The reaction between hydrogen and iodine is a complex reaction.

$$H_2 + I_2 \rightarrow 2HI$$

Kinetics experiments show that the reaction is first order with respect to H_2 and first order with respect to I_2. The following mechanism has been proposed

1 $I_2 \underset{k_{-1}}{\overset{k_1}{\rightleftharpoons}} 2I^•$ Equilibrium constant K_1

2 $H_2 + I^• \underset{k_{-2}}{\overset{k_2}{\rightleftharpoons}} H_2I^•$ Equilibrium constant K_2

3 $H_2I^• + I^• \xrightarrow{k_3} 2HI$

where each step is an elementary reaction. Reaction 3 is the rate-determining step and both reactions 1 and 2 form pre-equilibria. (Section 9.6)

(a) Assuming reaction 1 is at equilibrium, obtain an expression for $[I^•]$ in terms of $[I_2]$ and the rate constants k_1 and k_{-1}.

(b) Assuming reaction 2 is at equilibrium, obtain an expression for $[H_2I^•]$ in terms of $[H_2]$, $[I^•]$, and the rate constants k_2 and k_{-2}.

(c) Write down the rate equation for reaction 3 and substitute in this the expressions for $[I^•]$ and $[H_2I^•]$ you obtained in (a) and (b).

(d) Compare the rate equation you have derived from the reaction mechanism with that found experimentally and write an expression for the rate constant, k, for the overall reaction.

30* The dissociation of propane in which a C–C bond breaks to form a methyl radical and an ethyl radical is a *unimolecular* reaction.

$$C_3H_8 \rightarrow CH_3^• + C_2H_5^•$$

The rate of formation of $CH_3^•$ (and of $C_2H_5^•$) is given by

$$\frac{d[CH_3^•]}{dt} = k_{overall}[C_3H_8], \text{ where } k_{overall} \text{ is the unimolecular rate}$$

constant for the reaction. (Sections 9.6 and 9.8)

(a) Explain what is meant by the term '*unimolecular*'.

The propane molecules obtain sufficient energy to dissociate by colliding with other molecules, M, where M may be an unreactive gas such as nitrogen. The mechanism for this process can be written as

$$C_3H_8 + M \underset{k_{-1}}{\overset{k_1}{\rightleftharpoons}} C_3H_8^* + M$$

$$C_3H_8^* \xrightarrow{k_2} CH_3^• + C_2H_5^•$$

where $C_3H_8^*$ is a propane molecule in a high energy state, which has sufficient energy to dissociate.

(b) By applying the steady state approximation to $C_3H_8^*$, derive an expression for $[C_3H_8^*]$. Since the rate of formation of CH_3^* is equal to $k_2[C_3H_8^*]$, show that

$$k_{overall} = \frac{k_1k_2[M]}{k_{-1}[M]+k_2}$$

(c) Show that:

(i) when [M] is very large, $k_{overall} \approx \frac{k_1k_2}{k_{-1}}$

(ii) when [M] is very small, $k_{overall} \approx k_1[M]$

(d) What are the rate-determining stages in the mechanism under each of the conditions in **(c)(i)** and **(c)(ii)**? Sketch a graph to show the dependence of $k_{overall}$ on [M].

(e) It is always better to find a linear expression to analyse experimental data. Show that

$$\frac{1}{k_{overall}} = \frac{k_{-1}}{k_1k_2} + \frac{1}{k_1[M]}$$

A plot of $1/k_{overall}$ (y-axis) against $1/[M]$ (x-axis) is a straight line. What are the gradient and the intercept on the y-axis at $1/[M] = 0$?

31 At *low substrate concentrations*, the initial rate of an enzyme catalysed reaction was found to be directly proportional to the initial substrate concentration, $[S]_0$, and directly proportional to the initial enzyme concentration, $[E]_0$. (Section 9.9)

(a) Outline the series of experiments that led to these results.

(b) Write a rate equation for the reaction under these conditions and explain the observed kinetics in terms of the mechanism for the reaction. What is the rate-determining step under these conditions?

At *much higher substrate concentrations*, the initial rate was found to be constant and independent of the initial substrate concentrations. The initial enzyme concentration was the same in each experiment.

(c) Write a rate equation for the reaction under these conditions and explain the observed kinetics in terms of the mechanism for the reaction. What is the rate-determining step under these conditions?

10

Molecular spectroscopy

➤ Infrared spectra of rocks on the surface of Mars. The brown spectrum was obtained from a spectrometer on an orbiting satellite, Mars Global Surveyor (MGS). The red spectrum was measured by the miniature thermal emission spectrometer (mini-TES) on the Mars Rover, and shows the higher sensitivity allowed by putting the spectrometer on the surface of the planet. The CO_2 absorption has been removed from the satellite spectrum.

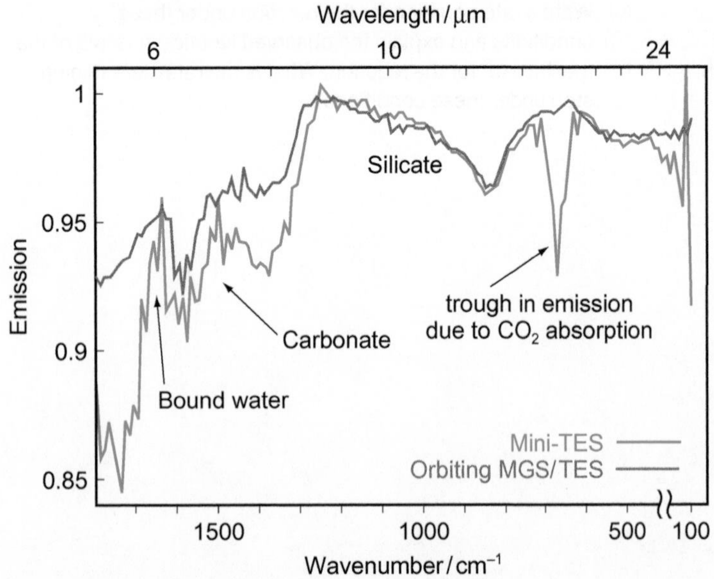

Wavelength / μm
6
10
24
1
0.95
0.9
0.85
Emission
Silicate
Carbonate
Bound water
trough in emission due to CO_2 absorption
Mini-TES
Orbiting MGS/TES
1500
1000
500
100
Wavenumber / cm^{-1}

This chapter builds on the following topics:

- Measurement, units, and nomenclature Section 1.2, p.6
- Atoms and the mole Section 1.3, p.12
- Energy changes in chemical reactions Section 1.6, p.41
- Non-covalent interactions Section 1.8, p.52
- Electromagnetic radiation and quantization Section 3.2, p.117
- Atomic spectra and the Bohr atom Section 3.3, p.124
- The nature of the electron Section 3.4, p.132
- Wavefunctions and atomic orbitals Section 3.5, p.134
- Electron spin Section 3.6, p.143
- Features of diatomic molecules Section 4.1, p.170
- Molecular orbital theory Section 4.6, p.184

One of the Mars Rovers exploring the Martian surface. The thermal emission spectrometer is located on the neck at the top of the Rover. ➤

Searching for life on Mars

Is there life on Mars? The question has long fascinated people. Mars is the planet most like Earth in the solar system and has been studied from orbiting satellites and, more recently, from robot vehicles moving on the surface of the planet. A vital part of any Mars mission is to look for evidence for the presence of water—or for signs that water was once present. On Earth, life occurs wherever water is found, so it is a good indicator of whether life exists—or has ever existed—on Mars. Even if no water is detected, some minerals, such as clays and carbonates, only form in the presence of water so their detection would indicate that there once was water on the planet.

NASA launched two missions in July 2003 that landed 'Mars Rovers' on the surface in January 2004. These have sent back a huge amount of information from their on-board instruments. One of the instruments mounted on the neck of the Rover—a *Miniature Thermal Emission Spectrometer* (mini-TES)—monitors infrared radiation emitted from the soil and rocks on the surface. Infrared radiation is emitted as a result of the vibrations of chemical bonds (see Section 10.5, p.476). The energy of the radiation emitted corresponds to transitions between quantized vibrational energy levels in the compounds (see Section 10.2, p.453) as they return from an excited state to the lower energy ground state.

The figure opposite shows an example of the infrared spectra obtained. The brown spectrum was obtained from an orbiting satellite and the red spectrum, which shows much greater sensitivity, was obtained from a Mars Rover. The spectra are emission spectra (see Section 10.3, p.458), and the emission peaks *rise upwards* in the spectrum. The horizontal axis of the spectrum records the *wavenumber* (in cm^{-1}) of the radiation emitted (wavenumber = 1/wavelength), which is proportional to the energy of the transition. The wavenumbers of the emission peaks are characteristic of the vibration of particular chemical bonds and parts of the spectrum can be assigned to particular types of rocks.

The emission around 1640 cm^{-1} is characteristic of water bound within the crystal structure of a transparent silicate rock, such as a zeolite. (The structure of zeolites is discussed in Chapter 6, p.255) The broad emission around 1480 cm^{-1} is thought to be due to carbonate rocks. The emission pattern around 840 cm^{-1} is best matched by transparent silicate or zeolite rocks, or crystalline SiO_2. The large *trough* in the emission at 667 cm^{-1} is due to absorption of infrared radiation by carbon dioxide in the thin Martian atmosphere, which has about 0.7% of the atmospheric pressure on the Earth. (Compare the position of this absorption with the absorption spectrum of carbon dioxide in Figure 1 in Box 10.5 on p.485.)

The presence of carbonates and the water bound in silicate rocks suggests that water vapour was once present in the Martian atmosphere. These are only preliminary findings, however, and much more evidence is needed before scientists can be certain.

10.1 Introduction to molecular spectroscopy

Atomic spectroscopy is introduced in Section 3.3 (p.124). Some applications of atomic spectroscopy are described in Section 11.5 (p.547).

The use of spectroscopic techniques in the structure determination of organic molecules is described in Chapter 12. Section 11.4 (p.542) describes some examples of quantitative analysis using molecular spectroscopy.

The quantization of energies in atoms is described in Section 3.2 (p.117).

Molecular spectroscopy uses the interaction of electromagnetic radiation with matter to obtain information about molecules. It includes a range of techniques and methods which can reveal a wealth of information about molecules and their reactions.

Why is spectroscopy important to a chemist? Much of what is known about molecules has been discovered through spectroscopic methods. These methods can be used to confirm the *identity* of compounds as well as for quantitative analysis—measuring *how much* of a compound is present. A wide range of molecular properties, including the precise arrangement of the atoms, the bond lengths and angles, as well as the strength of the bonds, can be determined. Information on dynamic properties—how the atoms move within molecules—can also be found.

As in atoms, molecular energies are quantized and so can only take certain discrete values. Spectroscopy involves the transition of molecules between these discrete energy levels; indeed, if energy levels were not quantized, spectroscopic methods would not exist! Another application of spectroscopy is the testing and confirmation of theories that describe energy quantization.

Figure 10.1 shows the infrared spectrum of acetylsalicylic acid—more commonly known as aspirin. The spectrum consists of a series of peaks (called **absorption bands**) with varying intensities. In infrared spectra, the peaks represent absorption of radiation due to different bond vibrations that take place in the molecule. For example, the large peaks around 1700 cm^{-1} and 1750 cm^{-1} correspond to absorption of radiation due to the stretching and compressing of the two C=O carbonyl bonds.

So, what information is needed to interpret this spectrum? What properties of the molecule influence the vibration of its bonds? Why does the C=O bond in aspirin absorb around 1700 cm^{-1}? To what energy does 1700 cm^{-1} correspond? Why are some peaks large and others small? This chapter answers these types of questions for several different types of spectroscopy and describes the information that each type of spectroscopy can give.

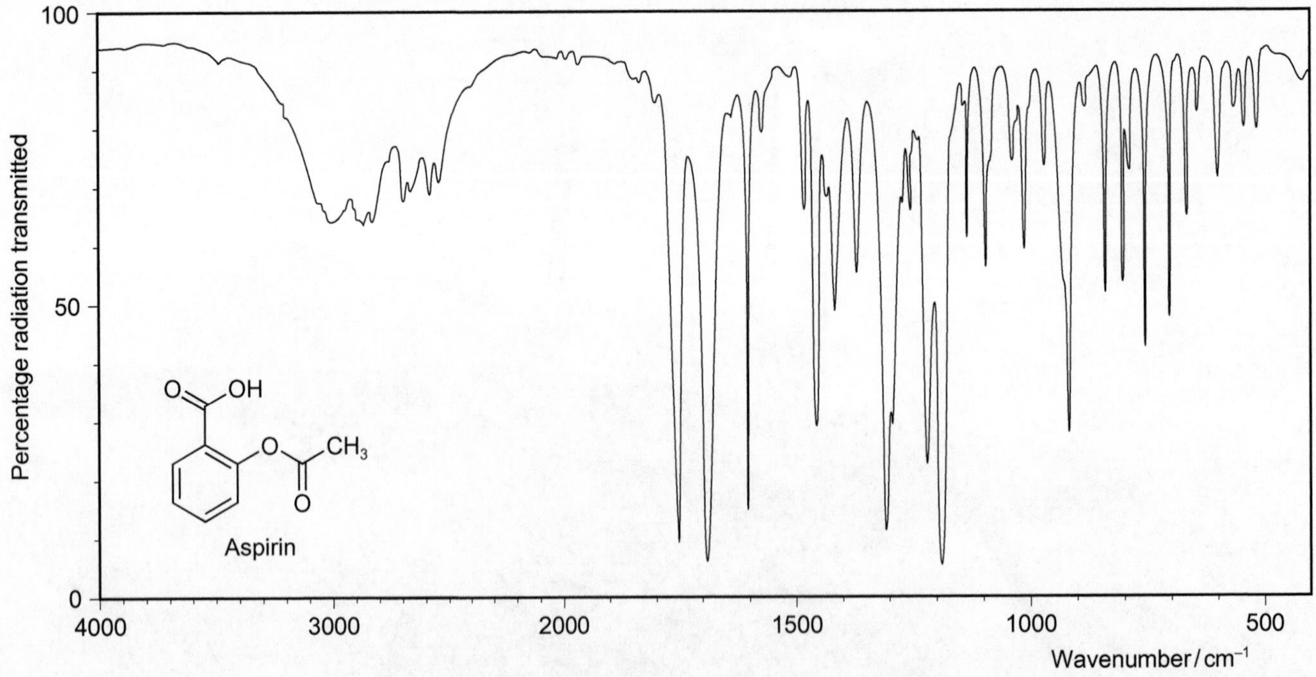
100
50
0
Percentage radiation transmitted
4000
3000
2000
1500
1000
500
Wavenumber / cm^{-1}
O
OH
O
CH$_3$
O
Aspirin

Figure 10.1 The infrared spectrum of aspirin. The sample was prepared by grinding powdered aspirin with potassium bromide and compressing the mixture into a thin disc. KBr has no absorptions in the region in which aspirin absorbs.

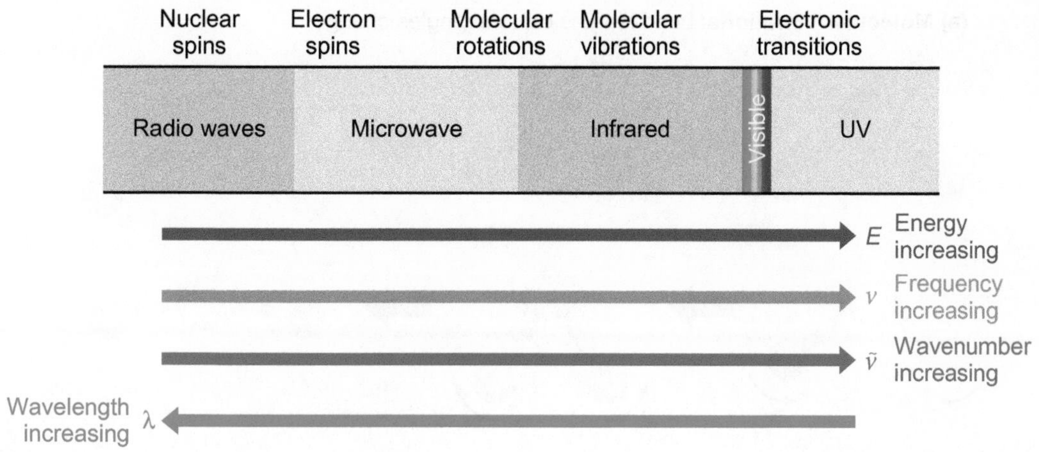
Nuclear spins
Electron spins
Molecular rotations
Molecular vibrations
Electronic transitions
Radio waves
Microwave
Infrared
Visible
UV
E Energy increasing
ν Frequency increasing
ṽ Wavenumber increasing
Wavelength increasing λ

Figure 10.2 Different processes involve transitions of different energies, so that the spectra appear in different regions of the electromagnetic spectrum. The complete electromagnetic spectrum is shown in Appendix 11 (p.1371).

Electromagnetic radiation

Understanding the nature of electromagnetic (EM) radiation is fundamental to understanding spectroscopy. Electromagnetic radiation consists of an oscillating electric and magnetic field that carries energy through space at the speed of light, c. The wavelength of the radiation, λ, is related to its frequency, ν, by

$$c = \lambda\nu \quad (3.1)$$

The link between the energy, E, of a photon of radiation and its frequency, ν, is

$$E = h\nu \quad (3.2)$$

where h is the Planck constant. The energy of the photon is therefore proportional to its frequency and inversely proportional to its wavelength since $\nu = c/\lambda$.

You can find a fuller account of the nature of electromagnetic radiation in Section 3.2 (p.117).

A molecule can absorb a photon and be raised from one molecular energy level to another. The energy of the photon must correspond exactly to the difference in energy between the initial and final energy levels. The spacing between the energy levels depends on the type of molecular transition. Different types of molecular transition involve absorption of energy from different parts of the electromagnetic spectrum, as shown in Figure 10.2.

Figure 10.3 A prism splits white light into the different colours of the visible spectrum.

The electromagnetic spectrum spans a vast range of energies, ranging from $1 \times 10^{-3}\,\mathrm{J\,mol^{-1}}$ at the radiofrequency end of the spectrum to $>1 \times 10^{9}\,\mathrm{J\,mol^{-1}}$ for γ-rays. All of these types of electromagnetic radiation can interact with molecules but the interactions involve different processes within the molecule. The energies of γ-rays and X-rays are much higher than those associated with chemical bonds and are involved in changes that occur within atomic *nuclei*. Low-energy ultraviolet and visible light (Figure 10.3) have energies that correspond to transitions between electronic energy levels. Moving along the spectrum, infrared radiation has energy equivalent to the transitions involved in vibrations of chemical bonds (Figure 10.4(a)). Next comes microwave radiation, which corresponds to the energies of the transitions involved when molecules rotate (Figure 10.4(b)). At the low-energy end of the electromagnetic spectrum are radio waves, which correspond to the small energy differences associated with electron spin or nuclear spin transitions when molecules are placed in a magnetic field.

The full range of the electromagnetic spectrum is shown in Appendix 11 (p.1371).

In discussing the various types of spectroscopy described in this chapter, it is useful to bear in mind the approximate energies involved in these transitions and these are summarized in Table 10.1. The energy levels within molecules that give rise to these transitions can be calculated in principle by solving the Schrödinger equation for an appropriate model for the type of change. This is explained in the next section.

The energy of radiation is inversely proportional to its wavelength. Therefore, vibrational spectra occur at longer wavelengths than electronic spectra. Rotational spectra are at even longer wavelengths.

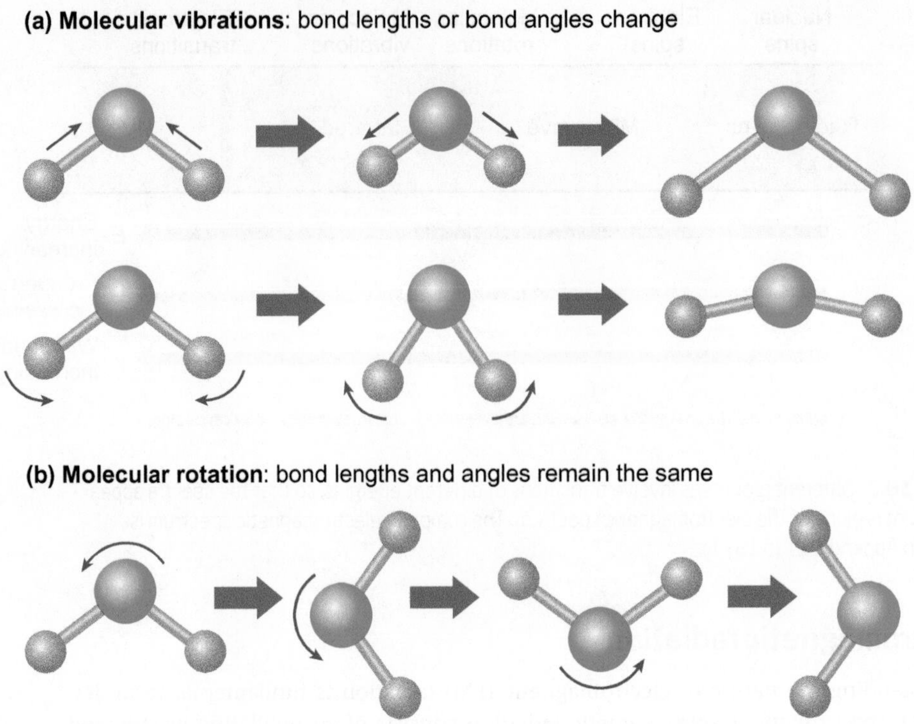
(a) Molecular vibrations: bond lengths or bond angles change
(b) Molecular rotation: bond lengths and angles remain the same

Figure 10.4 (a) Molecular vibrations and (b) molecular rotations involve quantized energies that give rise to spectra.

Table 10.1 The energies of molecular transitions

Molecular transition	Approximate energy range / kJ mol^{-1}	Type of radiation
Electronic transitions	50–1000	ultraviolet and visible
Bond vibrations	1–50	infrared
Molecular rotation	0.01–1	microwaves
Electron spin	5×10^{-4}–0.01	microwaves
Nuclear spin	1×10^{-6}–1×10^{-4}	radio waves

Summary

- Spectroscopy involves the interaction and exchange of electromagnetic radiation with matter.
- Molecular spectroscopy can be used to identify compounds, to measure how much of a compound is present, and to determine molecular properties.
- Absorption of energy can only take place when the energy of the radiation exactly matches the difference between molecular energy levels.
- The electromagnetic spectrum includes radiation with a large range of energies that can interact with various processes within molecules.

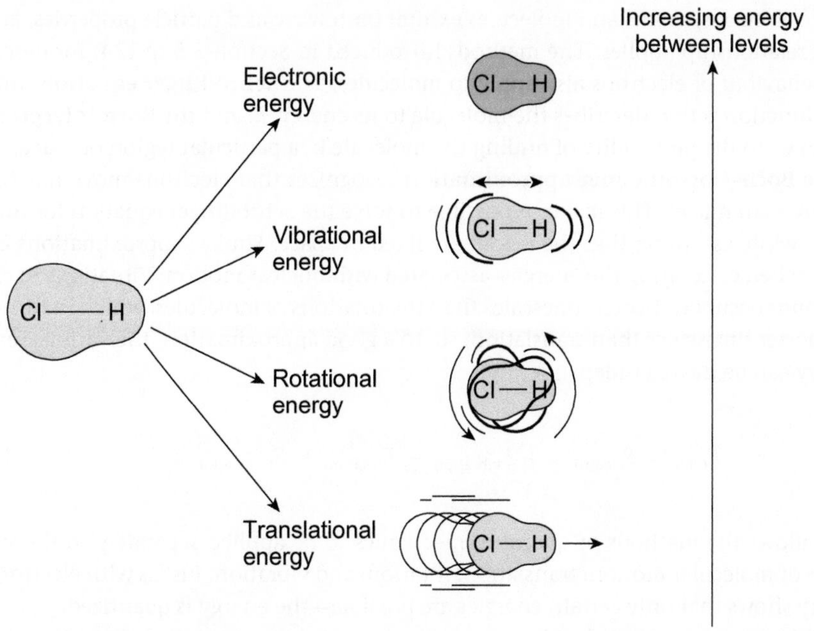
Increasing energy between levels
Electronic energy
Vibrational energy
Rotational energy
Translational energy
Cl—H

Figure 10.5 An HCl molecule has energy associated with different aspects of its motion and behaviour of its electrons. Changes in all these forms of energy are quantized.

10.2 Molecular energies and spectroscopy

Atomic spectra arise from transitions between **electronic** energy levels. Calculating electronic energy levels in atoms is done by deriving an expression for the wavefunction that describes the electron and solving for the energy using the Schrödinger equation. The method can be adapted for calculating the electronic energy levels in molecules.

Using the Schrödinger equation to find energy levels in atoms is described in Section 3.5 (p.134). The quantization of energy is introduced in Section 3.2 (p.117).

Other types of energy in molecules are associated with the behaviour of the atomic nuclei and their electrons (Figure 10.5). The movement of the molecule as a whole through space is associated with **translational** energy. The molecule can rotate, giving rise to **rotational** energy, and the atoms within the molecule can move relative to one another, as if they were on springs, giving **vibrational** energy. All these forms of energy are **quantized** and there are energy levels associated with each of them. The molecule cannot take up energy continuously, but can only do so in discrete amounts, leading to transitions between the energy levels.

This section uses some of the general principles of quantum theory introduced in Chapter 3 to show how the quantization of *molecular energy* arises. Central to this is the use of **models** to describe the different types of motion, which allow the calculation of molecular energy levels.

Quantization of molecular energy

The momentum of a particle is the product of its mass, m, and its velocity, v. Momentum = mv.

The wave–particle duality of the electron is discussed in Sections 3.2 (p.123) and 3.4 (p.133). The **de Broglie equation** encompasses both wave and particle properties, by relating the momentum, mv, of the electron to its wavelength λ

$$\lambda = \frac{h}{mv} \tag{3.9}$$

The de Broglie equation is introduced in Section 3.4 (p.132). The Schrödinger equation and the Born interpretation are introduced in Section 3.5 (p.134).

where h is the Planck constant.

Like electrons, atoms and molecules exhibit both wave and particle properties, and the same relationship applies. The methods introduced in Section 3.5 (p.134) for describing the behaviour of electrons also apply to molecules. The **Schrödinger equation** links the wavefunction ψ that describes the molecule to its energy, E, and the **Born interpretation** relates ψ^2 to the probability of finding the molecule in a particular region of space.

The **Born–Oppenheimer approximation** recognizes that electrons move much more rapidly than nuclei. This makes it possible to solve the Schrödinger equation for the electrons, while assuming that the nuclei are fixed in space. Similar approximations can be made when calculating the energies associated with nuclear motion. Vibrations in chemical bonds occur on shorter timescales than the rotations of molecules, which in turn occur on shorter timescales than translation, so, to a good approximation, the various forms of energy can be treated independently.

$$E_{\text{total}} = E_{\text{electronic}} + E_{\text{vibration}} + E_{\text{rotation}} + E_{\text{translation}} \tag{10.1}$$

This allows the methods of quantum mechanics to be applied separately to the various forms of molecular motion: translation; rotation; and vibration. Just as with electrons, the theory shows that only certain energies are possible—the energy is quantized.

While solution of the Schrödinger equation is necessary for a full analysis, the essential principles governing the quantization of molecular energies can be understood by considering the de Broglie equation. An example of how this can be done is given in Box 10.1. This uses the de Broglie equation to demonstrate how quantization arises and the nature of the wavefunctions for a simple model system, a 'particle moving in a one-dimensional box'.

This is the simplest model possible for motion of a molecule and can be extended into three dimensions to describe translations (see p.456).

Box 10.1 Particle in a one-dimensional box

Imagine a particle of mass m, for example, a helium atom, that can only move back and forth on one axis, the x-axis as in Figure 1. This is described as motion in 'a one-dimensional box'. The length of the box is l and the potential energy, E_{PE}, is zero inside the box and infinite outside. This means that the particle can move along the x-axis between $x = 0$ and $x = l$ but can never get out of the box due to the infinite energy barrier.

In order to find expressions for the energies of the particle, some properties of the wavefunction that describes its motion need to be defined.

Remember that ψ^2 is related to the probability of finding the particle (see Section 3.5, p.134). This has two consequences:

- since E_{PE} is infinite outside, the particle cannot leave the box. As a result, the probability of finding the particle outside the box is zero, so here ψ^2 (and hence ψ) is 0.
- because ψ is related to the probability of finding the particle, its value cannot change discontinuously. ψ is zero just outside the box, so it must also be zero at the sides of the box, that is, $\psi = 0$ at $x = 0$ and $x = l$.

With these properties, a relationship for the energy of the particle can be found. The total energy of the particle, E, is the sum of its potential energy and kinetic energy (see Section 1.6, p.41). The potential energy is zero inside the box, so that E is equal to the kinetic energy of the particle due to its motion (translation). From the expression for kinetic energy (Equation 1.17, p.42)

$$E = E_{KE} = \frac{mv^2}{2}$$

where v is the velocity of the particle. Rewriting this equation to give an expression for the momentum

$$v^2 = \frac{2E}{m} \quad \text{so } v = \left(\frac{2E}{m}\right)^{1/2} \quad \text{and momentum, } mv = (2mE)^{1/2}$$

Substituting this expression for momentum into the de Broglie equation (Equation 3.9, p.132)

$$\lambda = \frac{h}{mv} = \frac{h}{(2mE)^{1/2}}$$

→

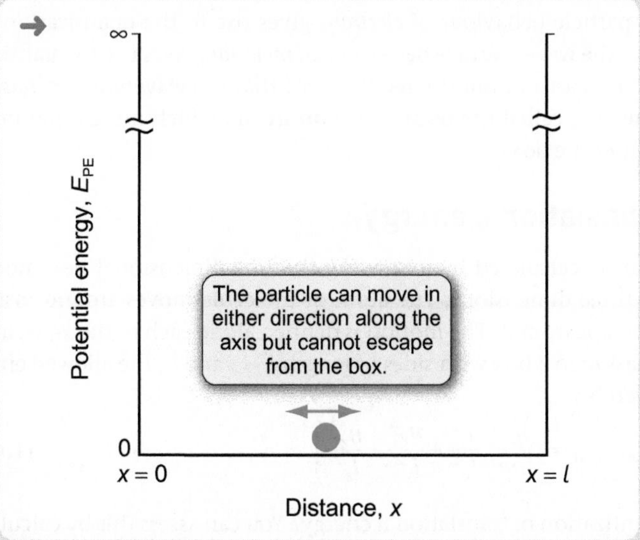

∞
Potential energy, E_{PE}
The particle can move in either direction along the axis but cannot escape from the box.
0
$x = 0$
$x = l$
Distance, x

▲ **Figure 1** A particle in a one-dimensional box.

and rearranging the resulting equation to give an expression for E leads to a relationship between the energy of the wavefunction and the wavelength, λ

$$E = \frac{h^2}{2m\lambda^2} \qquad (10.2)$$

The wavelength of the wave must be such that it fits the length of the box so that its value is zero at $x = 0$ and $x = l$. Figure 2 shows that this means that the length of the box must correspond to an integral number of half wavelengths.

The first allowed energy level occurs at the energy where exactly half a wave (in green) fits into the box, so $l = \frac{1}{2}\lambda$, that is, $\lambda = 2l$. This is the longest wavelength that can satisfy $\psi = 0$ at $x = 0$ and $x = l$. The energy associated with this wavefunction—the lowest allowed energy—can be found from Equation 10.2

$$E = \frac{h^2}{2m \times (2l)^2} = \frac{h^2}{8ml^2} \qquad (10.3)$$

The next energy level occurs at the energy where a whole wave (in blue) fits into the box, so that $\lambda = l$. Here,

$$E = \frac{h^2}{2ml^2}$$

which you can rewrite as

$$E = \frac{4h^2}{8ml^2} = \frac{(2)^2h^2}{8ml^2}$$

to show the relationship to the energy of the first level in Equation 10.3.

The next energy level occurs where three half-waves (in pink) fit into the box so $l = \frac{3}{2}\lambda$ and

$$E = \frac{h^2}{2m \times \left(\frac{2}{3}l\right)^2} = \frac{9h^2}{8ml^2} = \frac{(3)^2h^2}{8ml^2}$$

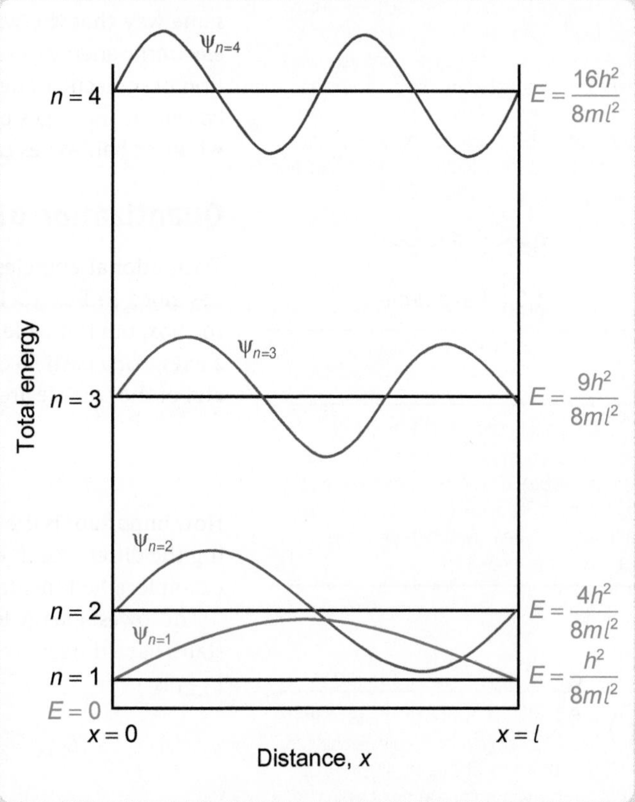

$\psi_{n=4}$
$n = 4$
$E = \frac{16h^2}{8ml^2}$
$\psi_{n=3}$
Total energy
$n = 3$
$E = \frac{9h^2}{8ml^2}$
$\psi_{n=2}$
$n = 2$
$E = \frac{4h^2}{8ml^2}$
$\psi_{n=1}$
$n = 1$
$E = \frac{h^2}{8ml^2}$
$E = 0$
$x = 0$
$x = l$
Distance, x

▲ **Figure 2** Energy levels and wavefunctions for a particle in a one-dimensional box.

The pattern continues as more half-wavelengths fit into the box giving the general expression for the allowed energy levels

$$E_n = \frac{n^2h^2}{8ml^2} \qquad (10.4)$$

where n is a **translational quantum number** that can take values $n = 1, 2, 3, \ldots \infty$. (Note that this is different from the principal quantum number n used for atomic orbitals in Chapter 3.) Each energy level is associated with a wavefunction, ψ, with the appropriate wavelength. The energy levels and their wavefunctions are shown in Figure 2.

Equation 10.2 tells you that, when $E = 0$, λ is infinite and the wavefunction is a straight line. Since $\psi = 0$ at the edges of the box, when $E = 0$ both ψ and ψ^2 must be zero everywhere within the box, so the probability of finding the particle with zero kinetic energy is also zero. Thus, this energy is not allowed and n cannot take a value of 0. The **zero point energy** of the particle—its energy at $T = 0$ K when it occupies the lowest energy level—cannot be zero. The lowest energy level is that with $n = 1$. The occurrence of a zero point energy is quite common in quantum systems although it does not arise in classical mechanics. All forms of linear motion have zero point energy.

Question

Write an expression for the energy level, E, when $n = 6$ and sketch the shape of the wavefunction.

One application of the particle in a box model is to calculate how electronic energies in conjugated hydrocarbons vary with the length of the molecule. The particle is an electron and the box is the π electron system. Conjugated systems such as hexatriene contain alternating single and double bonds; see Section 2.5 (p.79) and Section 19.1 (p.876).

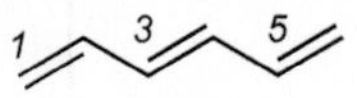

1,3,5-hexatriene

The results from the treatment of the 'particle in a box' in Box 10.1 show that, in the same way that the wave–particle behaviour of *electrons* gives rise to the quantization of electronic energy in atoms, the wave–particle behaviour of *molecules* gives rise to quantization of molecular energy. This arises from the requirement that the wavefunction has to be zero at the edges of the box, called the **boundary condition**, which means that only whole or half waves can fit in the box.

Quantization of translational energy

Compare Equation 10.5 (three dimensions) with Equation 10.4 (one dimension) in Box 10.1. The quantum numbers, n_x, n_y, and n_z, in Equation 10.5 determine the wavelengths of the wavefunctions in the directions of the three axes.

Translational energies can be calculated by extending the one-dimensional box model described in Box 10.1 to three dimensions. The atom or molecule moves around inside the box, but is unable to escape from it. The motion is defined along each of the *x*-, *y*-, and *z*-axes. For a particle of mass m in a box with sides of length l_x, l_y, and l_z, the allowed energies of the particle are given by

$$E_{\text{translation}} = \frac{h^2}{8m}\left(\frac{n_x^{\,2}}{l_x^{\,2}} + \frac{n_y^{\,2}}{l_y^{\,2}} + \frac{n_z^{\,2}}{l_z^{\,2}}\right) \qquad (10.5)$$

How important is the quantization of translational energy? You can assess this by calculating the difference in energy between the lowest and the first excited energy levels in, for example, a helium atom in a cubic box of side 1 m.

The lowest energy level has $n_x = n_y = n_z = 1$. (The level is *triply degenerate* because all three states have the same energy.) From Equation 10.5, the corresponding translational energy, $E_{1,1,1}$ is

In Equation 10.5, m is the mass of the individual atom or molecule. For helium, $A_r = 4.00$ so the mass of 1 mol is 4.00 g (4.00×10^{-3} kg). The mass of one atom of helium is found by dividing this by the Avogadro constant.

$$\begin{aligned} E_{1,1,1} &= \frac{h^2}{8m}\left(\frac{1}{l_x^{\,2}} + \frac{1}{l_y^{\,2}} + \frac{1}{l_z^{\,2}}\right) \\ &= \frac{(6.626 \times 10^{-34}\,\text{J s})^2}{8 \times (4.00 \times 10^{-3}\,\text{kg mol}^{-1})/6.022 \times 10^{23}\,\text{mol}^{-1}} \times \left(\frac{3}{1\,\text{m}^2}\right) \\ &= 2.48 \times 10^{-41}\,\text{J} \end{aligned}$$

Energy states with the same energy are described as **degenerate**; see, for example, Section 3.6 (p.143).

The first excited level has n_x (or n_y or n_z) = 2. Using Equation 10.5 shows that $E_{2,1,1}$ (= $E_{1,2,1}$ = $E_{1,1,2}$) = 4.96×10^{-41} J. So the energy difference between the lowest level and the first excited level is (4.96×10^{-41} J–2.48×10^{-41} J) = 2.48×10^{-41} J per molecule which corresponds to 1.50×10^{-17} J mol^{-1}.

This energy difference is extremely small compared with the average molar kinetic energy of helium atoms at 298 K, which is 3720 J mol^{-1}. Thus, the spacing of the levels is so small that it is not possible to observe transitions between translational energy levels using spectroscopic methods. The translational energy levels can in practice be treated as continuous—translational motion is adequately described by classical mechanics.

Quantization of rotational, vibrational, and electronic energy

Like translational energies, the energies of other molecular processes, such as rotations and vibrations, are quantized. For each type of motion, a model can be developed to describe the molecular process involved and the model can be used to find the allowed energy levels. With some modification of the shape of the box, the particle in a box model described in Box 10.1 (p.454) can be used to describe vibrational motion. A related model in which a particle moves on a ring can be extended to describe rotation. You will meet these different types of models in detail in more advanced courses on quantum mechanics. In each case, the model shows that the energy changes involved are quantized.

For rotational, vibrational, and electronic energy, the differences between the energy levels are such that transitions between them can be observed and this gives rise to spectra. Different processes involve transitions of different energies, so that the spectra appear in different regions of the electromagnetic spectrum (see Table 10.1, p.452 and Figure 10.2, p.451). Electronic, vibrational, and rotational transitions happen simultaneously, and the different energy levels 'stack' up on each other, as shown in Figure 10.6.

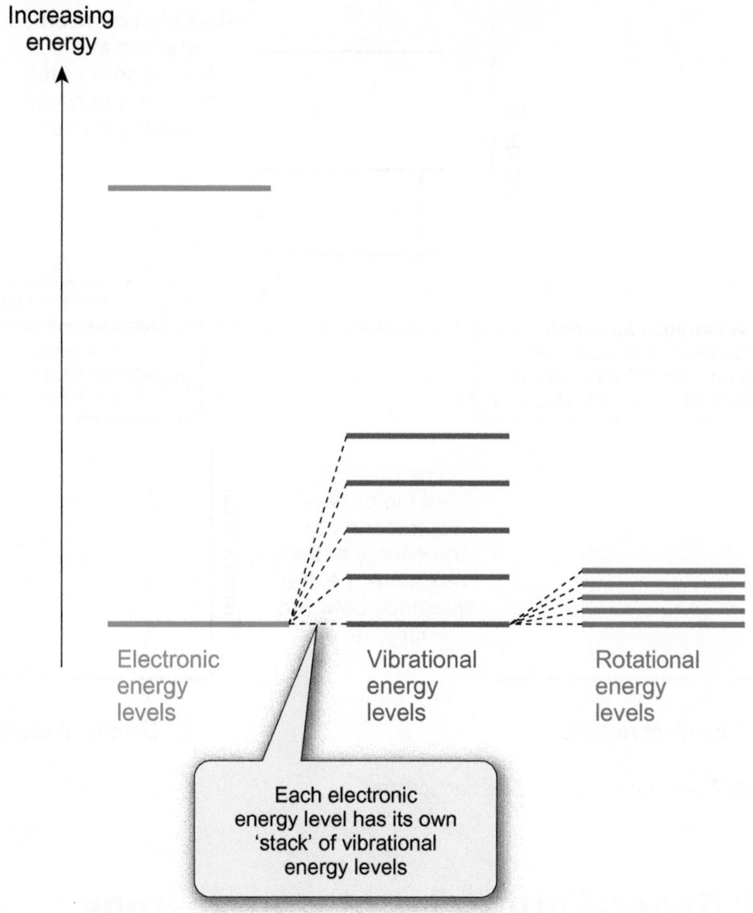
Increasing energy
Electronic energy levels
Vibrational energy levels
Rotational energy levels
Each electronic energy level has its own 'stack' of vibrational energy levels

Figure 10.6 Each electronic energy level has within it separate vibrational and rotational energy levels. Translational energy levels also exist, but are not shown here. Note that the levels are not to scale.

Summary

- Motion in molecules results in translational, rotational, and vibrational energies, all of which are quantized.
- The Schrödinger equation can be applied to molecules as well as to atoms.
- The Born–Oppenheimer approximation allows the different types of molecular energy to be treated independently

$$E_{\text{total}} = E_{\text{electronic}} + E_{\text{vibration}} + E_{\text{rotation}} + E_{\text{translation}}$$

- Molecules can be treated as both waves and particles. Applying the de Broglie equation to a molecule moving in a box is a simple model that demonstrates the quantization of energy.
- Transitions between energy levels involve discrete quantities of energy and give rise to spectra.
- Translational energy levels are too close together for transitions between them to be observed spectroscopically, but rotational and vibrational energy levels are spaced far enough apart to be the basis of important spectroscopic techniques.
- Because of the different spacings between the energy levels for different types of molecular energy, the spectra appear in different parts of the electromagnetic spectrum.

For practice questions on these topics, see questions 1–5 at the end of this chapter (p.510).

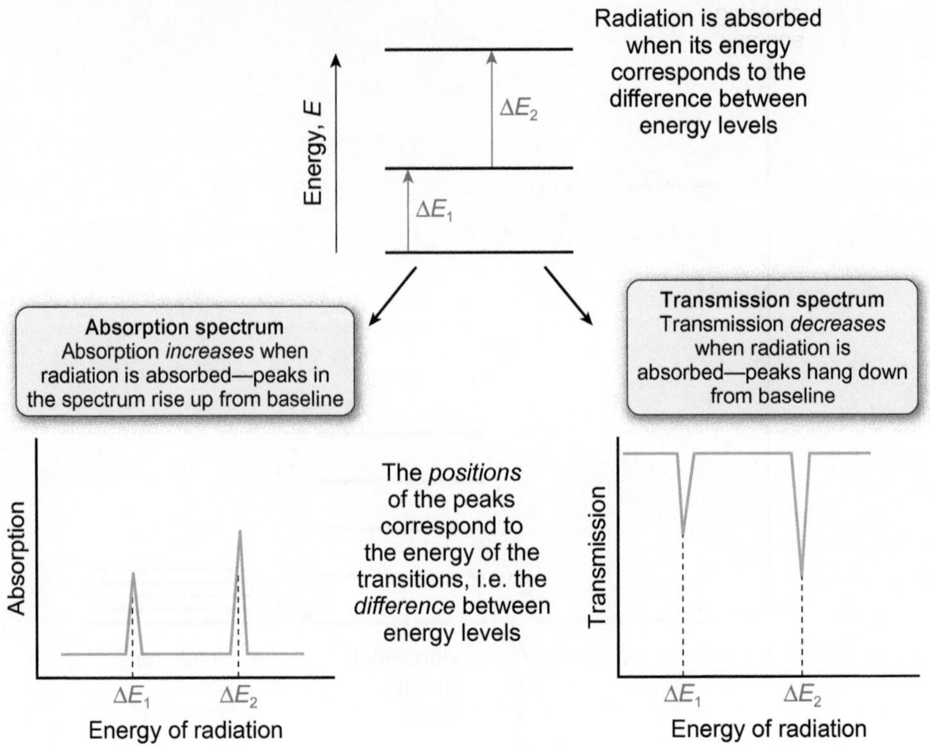
Radiation is absorbed when its energy corresponds to the difference between energy levels
Energy, E
ΔE2
ΔE1
Absorption spectrum
Absorption increases when radiation is absorbed—peaks in the spectrum rise up from baseline
Transmission spectrum
Transmission decreases when radiation is absorbed—peaks hang down from baseline
Absorption
ΔE1 ΔE2
Energy of radiation
The positions of the peaks correspond to the energy of the transitions, i.e. the difference between energy levels
Transmission
ΔE1 ΔE2
Energy of radiation

Figure 10.7 Absorption and transmission spectra.

10.3 General principles of spectroscopy

Absorption and emission spectra

All forms of spectroscopy have some basic features in common. In a basic experiment, the sample under study is irradiated with electromagnetic radiation, the energy of which is gradually changed. Whenever the energy exactly corresponds to a gap between energy levels within the sample, the radiation may be absorbed. This is known as **absorption spectroscopy**. The lowest energy state of the molecule is known as the **ground state** and absorption of radiation increases the energy of the molecule and promotes it to an **excited state**.

The spectrum obtained can be displayed in two ways, as illustrated in Figure 10.7. In an **absorption spectrum**, the amount of radiation absorbed is plotted against the energy of the radiation. The peaks rise up from a base line in the spectrum. An alternative is to plot a **transmission spectrum**, which measures the proportion of energy transmitted through the sample. If no absorption takes place, 100% of the radiation is transmitted through the sample. When absorption takes place, a *downward* peak is produced since the amount of radiation transmitted goes down. For example, the infrared spectrum for aspirin shown in Figure 10.1 (p.450) is plotted as a transmission spectrum.

Another form of spectroscopy is **emission spectroscopy**. Here, radiation is emitted when a species in an excited state falls back to the ground state. The sample may be heated or irradiated to achieve the excited states or subjected to an electric discharge. The peaks in the spectrum *rise upwards* from the baseline. The infrared spectra obtained by the Mars Rover (p.449) are emission spectra.

Raman spectroscopy (Section 10.5, p.488) involves *scattering* rather than absorption or emission of light. Radiation is directed onto a sample, and some of the radiation that is scattered has a different frequency from the incident radiation. The difference in frequency provides information about the energy levels of the sample.

ℹ It is important to remember that the radiation absorbed by the molecule corresponds to *differences* in energy levels, not to the actual energies of the levels.

Absorption and emission atomic spectra are introduced in Section 3.3 (p.124). Some applications in analytical chemistry are described in Section 11.5 (p.547).

ℹ The colours observed when some metal compounds are placed in a flame are due to emission of light from electronic excited states (see Box 3.2, p.128). The excitation is caused by the high temperatures. This principle also forms the basis of atomic emission spectrometry (see Section 11.5, p.547).

ℹ **Scattering** involves an instantaneous collision between a photon and a molecule in which there is a transfer of energy so that the molecule gains or loses energy. The resulting radiation is scattered from the molecule in all directions. Its energy may be lower, higher, or the same as the incident radiation.

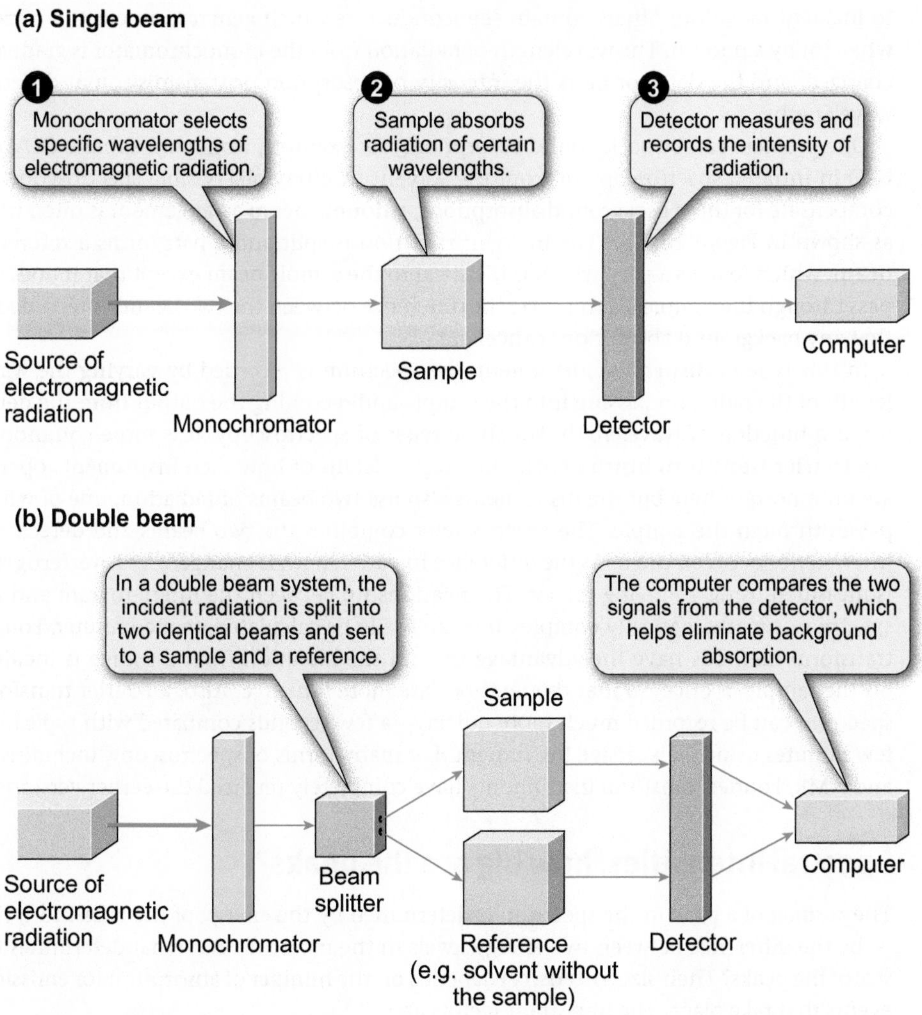
(a) Single beam
1
Monochromator selects specific wavelengths of electromagnetic radiation.
2
Sample absorbs radiation of certain wavelengths.
3
Detector measures and records the intensity of radiation.
Source of electromagnetic radiation
Monochromator
Sample
Detector
Computer
(b) Double beam
In a double beam system, the incident radiation is split into two identical beams and sent to a sample and a reference.
The computer compares the two signals from the detector, which helps eliminate background absorption.
Sample
Source of electromagnetic radiation
Monochromator
Beam splitter
Reference (e.g. solvent without the sample)
Detector
Computer

Figure 10.8 The general features of an absorption spectrometer. (a) A single beam and (b) a double beam instrument.

Experimental methods

Electronics and computer analysis mean that many users simply treat a spectrometer as a 'black box' into which a sample is placed and out of which a spectrum emerges. However, some appreciation of what happens inside a spectrometer helps in understanding the principles of spectroscopy.

Although each form of spectroscopy needs different equipment and the spectrometers often look very different, they have some general features in common. These are indicated in the diagram in Figure 10.8, which illustrates absorption spectroscopy.

Figure 10.8(a) shows a simple single beam spectrometer. A source of radiation that emits in the appropriate region of the spectrum is needed. The source emits a range of wavelengths and a wavelength selector is included to select specific wavelengths. This is known as a **monochromator** because it selects a single wavelength at a time. One type of monochromator is a diffraction grating, which, like a prism (see Figure 10.3, p.451), splits the radiation into its component wavelengths, so that only a narrow band of wavelengths is incident on the sample.

The radiation passes through the sample held in a suitable container and then on to a detector that produces an electric signal in proportion to the intensity of radiation coming out of the sample. Some detectors rely on sensitive detection of the heat generated due

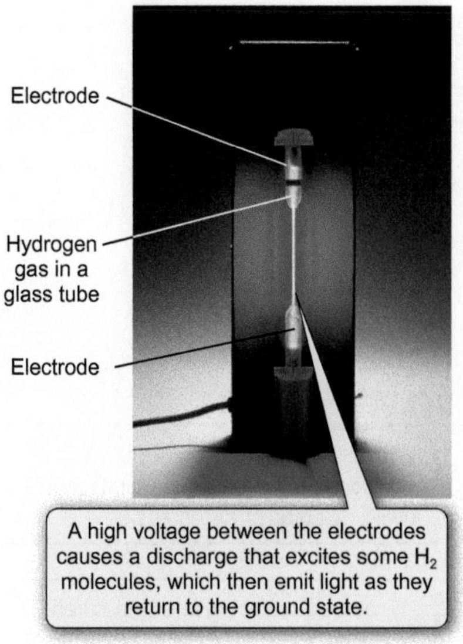
Electrode
Hydrogen gas in a glass tube
Electrode
A high voltage between the electrodes causes a discharge that excites some H_2 molecules, which then emit light as they return to the ground state.

H_2 gas emission. Even simple molecules such as hydrogen emit light when excited in a strong electric field.

to incident radiation; others contain semiconductors which generate an electric current when hit by a photon. The wavelength of radiation from the monochromator is gradually changed, and the detector plots the intensity of absorption or transmission at different wavelengths.

Visit the Online Resource Centre to view video clip 10.1 which illustrates the use of an absorption spectrometer.

One problem can be background absorptions, for example, from atmospheric CO_2 and H_2O in infrared spectroscopy or from the solvent in ultraviolet–visible spectroscopy. To compensate for these background absorptions, **a double beam** arrangement is often used, as shown in Figure 10.8(b). The incident radiation is split and a part forms a **reference beam**, which follows a path which is identical to the sample beam except that it does not pass through the sample. In this case, the difference between the two beams is recorded so that any background absorptions cancel out.

In a **dispersive spectrometer**, the incident radiation is split into its component wavelengths, allowing the wavelength of the radiation incident on the sample to be varied.

In this type of **dispersive** instrument, the spectrum is recorded by varying the wavelength of the radiation passing into the sample and recording the output from the detector as a function of wavelength. For some types of spectroscopy, it is more common to use **Fourier transform** instruments. The precise details of how such instruments operate are unimportant here but the instruments also use two beams of radiation, one of which passes through the sample. The spectrometer combines the two beams and detects the interference between them. As the difference in path length is changed, an **interferogram** is measured using a moving mirror. The relationship between the interferogram and the spectrum is mathematically complex but can easily be calculated on a computer. Fourier transform methods have the advantage that all the radiation from a source is incident on the sample at once, so that the methods are more sensitive. Also, a Fourier transform spectrum can be recorded much more quickly—a few seconds compared with typically a few minutes using a dispersive instrument. For many forms of spectroscopy, including IR and NMR, Fourier transform instruments have completely replaced the earlier versions.

Fourier transform NMR spectroscopy is discussed in Section 12.3 (p.578).

Spectral intensities: how big are the peaks?

The *position* of a peak in the spectrum is determined by the energy of the transition, that is, by the difference between two energy levels in the molecule. But what determines the *size* of the peaks? Their size (**intensity**) depends on the number of absorption (or emission) events that take place. The important factors are:

- the concentration of the sample;
- the distance the radiation travels in the sample (called the **path length**);
- how many molecules within the sample are in the correct energy state to absorb or emit at this frequency; and
- how likely it is that the transition will occur.

These factors are now discussed in turn.

The Beer-Lambert law

The effect on the spectrum of the amount of material is most easily seen by considering a compound in solution, as illustrated in Figure 10.9.

When radiation with an intensity I_0 is incident on a sample in solution with a concentration c, some radiation is absorbed and the rest emerges with a lower intensity, I_t. The **transmittance**, T, of the solution is defined by

Note that the transmittance, T, and absorbance, A, do not have units since they involve the ratio of two intensities.

$$T = \frac{I_t}{I_0} \qquad (10.6)$$

A has no units; c is usually measured in mol dm^{-3} and l in cm. The units of ε are often given as dm^3 mol^{-1} cm^{-1}. The SI units for ε are m^2 mol^{-1}.

The relationship between transmittance and concentration is not linear so that a second measure, the **absorbance**, A, of the solution is defined as

$$A = -\log_{10}T$$

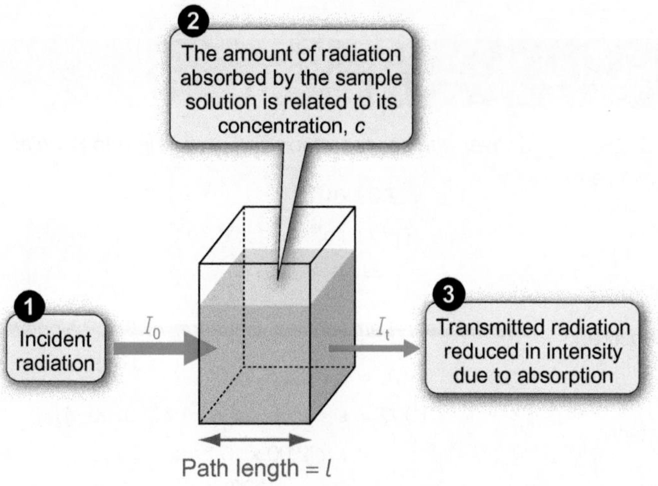
2
The amount of radiation absorbed by the sample solution is related to its concentration, c
1
Incident radiation
I_0
I_t
3
Transmitted radiation reduced in intensity due to absorption
Path length = l

Figure 10.9 The transmittance, T, of a substance in solution depends on the concentration and is defined in terms of the intensity of the incident radiation, I_0, and the intensity of the transmitted radiation, I_t.

This is usually expressed as

$$A = \log_{10}\left(\frac{I_0}{I_t}\right) \quad (10.7)$$

Using the absorbance, rather than the transmittance, has the advantage that, for dilute solutions, A is proportional to concentration, as expressed by the **Beer–Lambert law**

$$A = \varepsilon \times c \times l \quad (10.8)$$

where l is the path length in the solution through which the radiation passes. The **molar absorption coefficient**, ε, is a measure of how effectively the molecules absorb radiation. A high value of ε indicates that a large amount of radiation is absorbed at a given concentration.

Summarizing Equations 10.7 and 10.8:

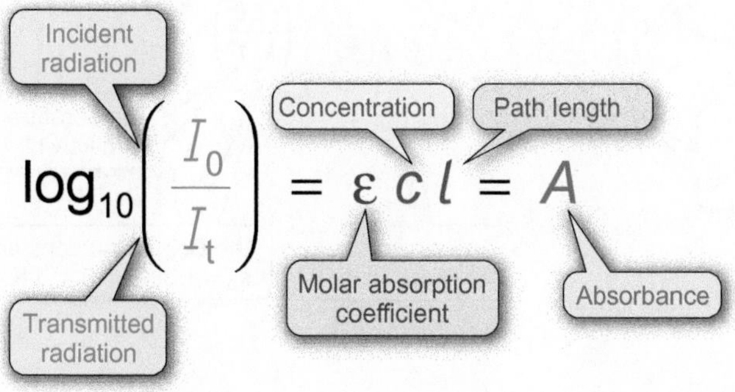
Incident radiation
Concentration
Path length
$\log_{10}\left(\frac{I_0}{I_t}\right) = \varepsilon c l = A$
Molar absorption coefficient
Absorbance
Transmitted radiation

Worked example 10.1 illustrates how the Beer–Lambert law can be used to calculate a value for ε.

Equation 10.8 shows a linear relation between the absorbance and concentration. The Beer–Lambert law is the basis of using spectroscopic methods for quantitative analysis (see Section 11.4, p.542).

Older names for ε and A are the *extinction coefficient* and the *optical density*, respectively.

Although the Beer–Lambert law is usually used with solutions, it can be applied more generally. For samples in the gas phase, the concentration can be measured in terms of the gas pressure.

Worked example 10.1 Using the Beer–Lambert law

A 2.00×10^{-3} mol dm^{-3} solution of a compound absorbs 50.0% of the light of a certain wavelength passing through 1.00 cm of a sample. Calculate the molar absorption coefficient, ε, at this wavelength. Quote your answer in SI units.

Strategy

The incident intensity is 100%. 50% of the incident light is absorbed by the solution, so 50% is transmitted. So you can calculate the absorbance using Equation 10.7 and then use the Beer–Lambert law (Equation 10.8) to find ε. Be careful to keep the units consistent.

Solution

Calculate the absorbance of the solution using Equation 10.7.

$$A = \log_{10}\left(\frac{I_0}{I_t}\right) \qquad (10.7)$$

$$= \log_{10}\left(\frac{100\%}{50\%}\right)$$

$$= 0.301$$

Convert the concentration and path length to SI units.

$$c = 2.00 \times 10^{-3}\ \text{mol dm}^{-3} = 2.00\ \text{mol m}^{-3}$$

$$(1\ \text{mol dm}^{-3} = 1 \times 10^{3}\ \text{mol m}^{-3})$$

$$l = 1.00\ \text{cm} = 0.0100\ \text{m}$$

Use the Beer–Lambert law (Equation 10.8) to calculate ε.

$$A = \varepsilon \times c \times l \qquad (10.8)$$

$$0.301 = \varepsilon \times (2.00\ \text{mol m}^{-3}) \times (0.0100\ \text{m})$$

$$= \varepsilon \times (2.00 \times 10^{-2}\ \text{mol m}^{-2})$$

$$\varepsilon = \frac{0.301}{(2.00 \times 10^{-2}\ \text{mol m}^{-2})}$$

$$= 15.1\ \text{m}^2\ \text{mol}^{-1}$$

Question

What concentration of the solution is required to absorb 35% of the light at the same wavelength in an identical cell?

Population of energy levels

Visit the Online Resource Centre to view screencast 10.1 which walks you through the use of the equation 10.9.

Another factor that determines the intensity of spectral peaks (see p.460) is the **population** of the energy levels, that is, how many molecules are in each level. Absorption of a photon results in the **promotion** of a molecule from a lower level to a higher level. The more molecules there are in the lower energy level, the more likely it is that a photon will be absorbed and a molecule promoted.

The populations in the energy levels are described by the **Boltzmann distribution** (Equation 10.9). If two energy levels are separated by an energy difference ΔE, as shown in Figure 10.10, then the number of molecules in the upper level, n_{upper}, is related to the number in the lower level, n_{lower}, by

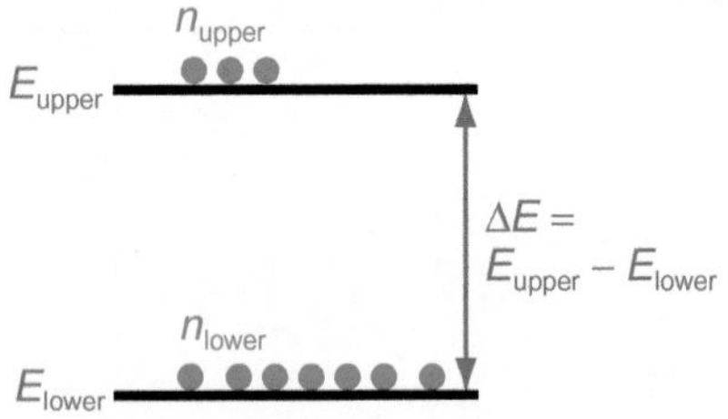

n upper
E upper
ΔE = E upper − E lower
n lower
E lower

Figure 10.10 The Boltzmann distribution. The population of the upper energy level depends on the energy difference between the upper and lower levels. (In this illustration, the upper and lower levels have the same degeneracy.)

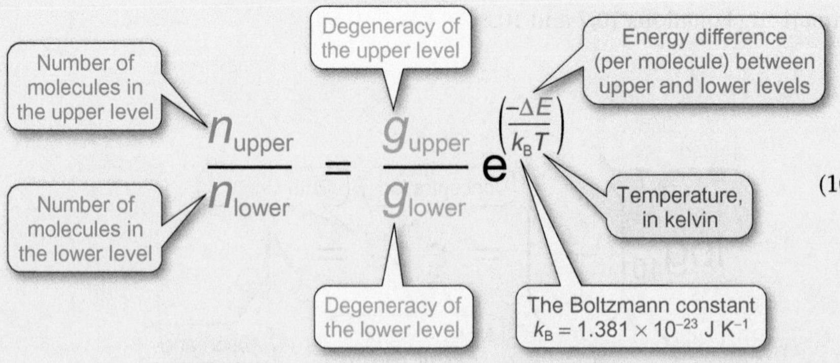

Number of molecules in the upper level
Degeneracy of the upper level
Energy difference (per molecule) between upper and lower levels
Number of molecules in the lower level
Temperature, in kelvin
Degeneracy of the lower level
The Boltzmann constant $k_B = 1.381 \times 10^{-23}$ J K^{-1}

$$\frac{n_{\text{upper}}}{n_{\text{lower}}} = \frac{g_{\text{upper}}}{g_{\text{lower}}} e^{\left(\frac{-\Delta E}{k_B T}\right)} \qquad (10.9)$$

Equation 10.9 shows that, if energy levels are closely spaced so that ΔE is small (much less than k_BT), the value of the exponential term is close to one (because $e^0 = 1$). This means that the ratio of the populations is approximately equal to the ratio of the degeneracies. If the degeneracies of both states are one, the populations, n_{upper} and n_{lower}, are similar.

On the other hand, if the energy levels are widely spaced so that ΔE is large (ΔE is much greater than k_BT), the ratio of the populations is now close to zero and very few molecules are in the upper state. This is illustrated in Worked example 10.2. Equation 10.9 also shows that the population of higher levels increases as the temperature increases.

For an absorption peak, the population of the *lower* energy level in the transition affects the intensity of the peak. For an emission peak, the intensity depends on the population of the *upper* level.

Equation 10.9 applies to any two energy levels, not just an excited state and the ground state. The degeneracy of a level is the number of states with the same energy.

The Boltzmann constant, k_B, is related to two other fundamental constants. It is the gas constant, R, expressed per molecule rather than per mole, that is, R divided by the Avogadro constant. $k_B = R/N_A = (8.314\,\text{J K}^{-1}\,\text{mol}^{-1}/6.022\times10^{23}\,\text{mol}^{-1}) = 1.381\times10^{-23}\,\text{J K}^{-1}$.

The properties of exponential functions, e^x (also written exp(x)), are described in Maths Toolkit, MT3 (p.1316). and MT4 (p.1318).

Worked example 10.2 Using the Boltzmann distribution

Calculate the relative populations at 298 K between two energy levels, each with a degeneracy of one, separated by: (a) 200 kJ mol^{-1}; (b) 5000 J mol^{-1}; and (c) 1.00 J mol^{-1}.

Strategy

You can use the Boltzmann distribution (Equation 10.9) directly since you are given the degeneracies and the energy difference between the two levels. You will need to convert the values for the energy differences from energy per mole to energy per molecule.

Solution

Use Equation 10.9 to find the relative populations.

$$\frac{n_{\text{upper}}}{n_{\text{lower}}} = \frac{g_{\text{upper}}}{n_{\text{lower}}}e^{\left(\frac{-\Delta E}{k_BT}\right)} \qquad (10.9)$$

(a) An energy difference of 200 kJ mol^{-1} = 200 000 J mol^{-1}. Divide by N_A (where $N_A = 6.022\times10^{23}$ mol^{-1}) to convert to an energy per molecule, and put g_{upper} and g_{lower} each equal to 1.

$$\frac{n_{\text{upper}}}{n_{\text{lower}}} = \frac{1}{1}e^{\left(\frac{-(200\,000\,\text{J mol}^{-1})/(6.022\times10^{23}\,\text{mol}^{-1})}{(1.381\times10^{-23}\,\text{J K}^{-1})\times298\,\text{K}}\right)} = 8.44\times10^{-36}$$

$$n_{\text{upper}} = 8.44\times10^{-36}\times n_{\text{lower}}$$

Effectively *all* of the molecules are in the lower energy state with very few in the upper energy state.

(An energy difference of 200 kJ mol^{-1} corresponds to the transitions in electronic spectroscopy (Section 10.6, p.491). Radiation of this energy is visible (yellow–orange) light, λ = 598 nm.)

(b) Using the same method for an energy difference of 5000 J mol^{-1}

$$\frac{n_{\text{upper}}}{n_{\text{lower}}} = \frac{1}{1}e^{\left(\frac{-(5000\,\text{J mol}^{-1})/(6.022\times10^{23}\,\text{mol}^{-1})}{(1.381\times10^{-23}\,\text{J K}^{-1})\times298\,\text{K}}\right)} = 0.13$$

$$n_{\text{upper}} = 0.13\times n_{\text{lower}}$$

Here most of the molecules are in the lower energy state. (An energy difference of 5000 J mol^{-1} corresponds to the transitions in vibrational spectroscopy (Section 10.5, p.476). Radiation of this energy is in the infrared part of the EM spectrum, $\lambda = 2.39\times10^{-5}$ m.)

(c) Using the same method for an energy difference of 1.00 J mol^{-1}

$$\frac{n_{\text{upper}}}{n_{\text{lower}}} = \frac{1}{1}e^{\left(\frac{-(1.00\,\text{J mol}^{-1})/(6.022\times10^{23}\,\text{mol}^{-1})}{(1.381\times10^{-23}\,\text{J K}^{-1})\times298\,\text{K}}\right)} = 0.9995$$

$$n_{\text{upper}} = 0.9995\times n_{\text{lower}}$$

The higher and lower energy states are about equally populated ($n_{\text{upper}} \approx n_{\text{lower}}$).

(An energy difference of 1 J mol^{-1} corresponds to the transitions in spin resonance techniques (Section 10.7, p.496). Radiation of this energy corresponds to radio waves, λ = 0.12 m.)

Question

Calculate the relative populations of two energy levels, each with a degeneracy of one, separated by 5.0 kJ mol^{-1} at: (a) 200 K; (b) 300 K; (c) 500 K. What is the effect of temperature on the relative populations in the two levels?

Selection rules

Most molecules have a large number of energy levels. In principle, therefore, a very large number of transitions are possible. However, not all of the possible transitions give rise to absorptions (or emissions) in the spectra. Only transitions between certain energy levels can take place; these are **allowed** transitions. Others are **forbidden**—there is a very low probability of a transition and almost no energy absorption or emission takes place, so the intensity of the absorption or emission peak is close to zero.

Wavefunctions and symmetry are introduced in Section 3.5 (p.134).

Each form of spectroscopy has **selection rules** that tell you which transitions are allowed. *Gross selection rules* say what properties of the molecule are needed for it to absorb radiation; *specific selection rules* say which levels transitions are allowed between. You will meet the selection rules for each type of spectroscopy in the following sections.

Box 10.2 Lasers

A good illustration of the effects of quantization and the population of energy levels comes from lasers, such as the one shown in Figure 1. Lasers only exist because molecular energy is distributed among discrete energy levels.

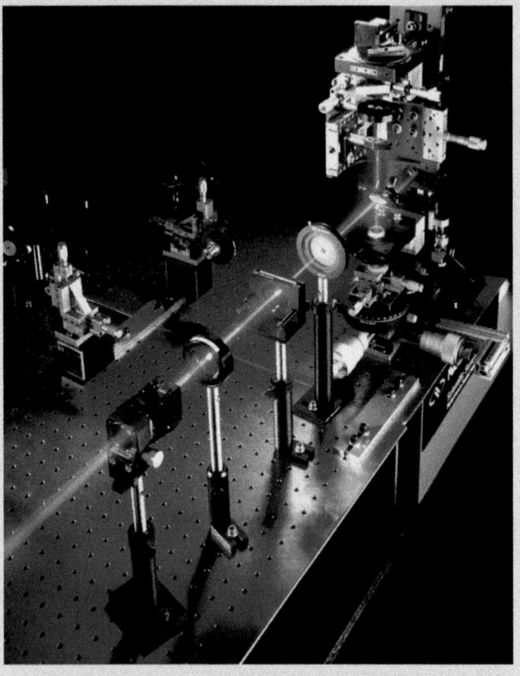

Figure 1 This spectroscopic laser (red beam) is being used to investigate a semiconductor used in solar cells.

Laser stands for light amplification by stimulated emission of radiation. Laser light has some useful properties. It is *monochromatic* (all the photons have the same energy) and *coherent* (all photons are in the same phase). The light is emitted in a narrow beam that does not diverge in all directions. The emission can be very intense and can also be arranged in very short ($<1 \times 10^{-12}$ s) pulses. Applications in chemistry include spectroscopy and photochemistry, which make use of the precise energy and short duration pulses (see Section 9.4, p.389). Other uses are in optical storage devices such as CD or DVD players (see Box 6.3, p.275), in cutting metals and plastics, in bar code readers, and in laser pointers—and in some forms of surgery.

When an atom or molecule absorbs a photon of the correct energy, an electron can be promoted to a higher energy level so that the atom or molecule is in an *excited state*. After a certain time, it will return to the ground state and may emit a photon, a process known as *spontaneous emission*. However, the excited state can be *stimulated* to emit when a photon interacts with the molecule, if the interacting photon has the same energy as the transition to the lower state. This leads to an increase in the number of photons with that energy and so to an *amplification* of the light intensity. Photons from stimulated emission all have the same energy and are all in the same phase. The probability of stimulated emission increases as the number of photons increases.

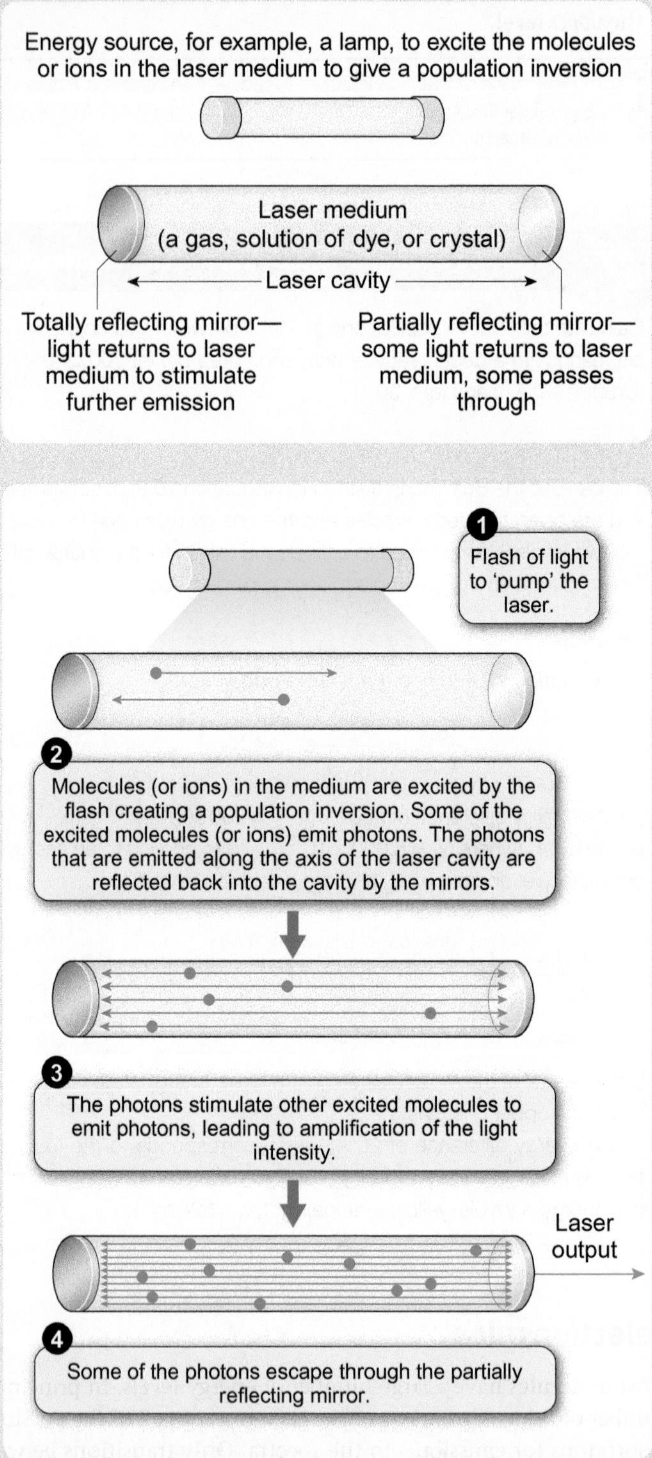

Energy source, for example, a lamp, to excite the molecules or ions in the laser medium to give a population inversion
Laser medium
(a gas, solution of dye, or crystal)
Laser cavity
Totally reflecting mirror—light returns to laser medium to stimulate further emission
Partially reflecting mirror—some light returns to laser medium, some passes through
1
Flash of light to 'pump' the laser.
2
Molecules (or ions) in the medium are excited by the flash creating a population inversion. Some of the excited molecules (or ions) emit photons. The photons that are emitted along the axis of the laser cavity are reflected back into the cavity by the mirrors.
3
The photons stimulate other excited molecules to emit photons, leading to amplification of the light intensity.
Laser output
4
Some of the photons escape through the partially reflecting mirror.

Figure 2 A simple representation of a laser. An energy source, such as a bright flash from a lamp, excites molecules or ions in the laser medium to give a population inversion. Photons are reflected between the mirrors and stimulate emission. Some photons pass through one of the mirrors to give the laser output.

→

→ The substance producing the light (called the *laser medium*) is contained within a *cavity* bounded by mirrors (see Figure 2) so that emitted photons are reflected back through the material where they can stimulate further emission.

To sustain stimulated emission, there must be more molecules in the excited state than in the lower state—this is known as a **population inversion**. Normally only a very small proportion of the molecules are in the excited state as the two states will be in thermal equilibrium (see p.462). To achieve non-equilibrium conditions, an indirect method of populating the excited state is used. This is known as *pumping* and is achieved by using a flash of light or an electrical discharge.

A population inversion can be achieved in atoms and molecules that can exist in more than one excited state. Most modern lasers are *four-level lasers*. Here, an atom (or a molecule) can exist in four energy levels with energies E_1, E_2, E_3, and E_4 as shown in Figure 3.

Initially, the majority of the atoms will be in the ground state. They are promoted (pumped) to level 4, the uppermost level, by a flash from a lamp. Atoms in this upper level rapidly move to the intermediate level 3 in a fast transition that does not involve emission of a photon; it is *radiationless*, with the energy being transferred to vibrational motion (heat). A similar fast transition occurs between level 2 and the ground state. An atom in level 3 may decay by spontaneous emission to level 2, releasing a photon with energy $E_3 - E_2$, called the *laser transition*.

So, atoms are pumped to level 4, decay rapidly to level 3, slowly emit a photon in moving to level 2, and then rapidly move to level 1. This means that the population of level 2 is very small and a population of excited state atoms accumulates in level 3 leading to a population inversion. Laser action is now possible between levels 3 and 2.

A common four-level system uses Nd^{3+} ions in an yttrium aluminium garnet—known as an Nd-YAG laser—which emits at 1064 nm.

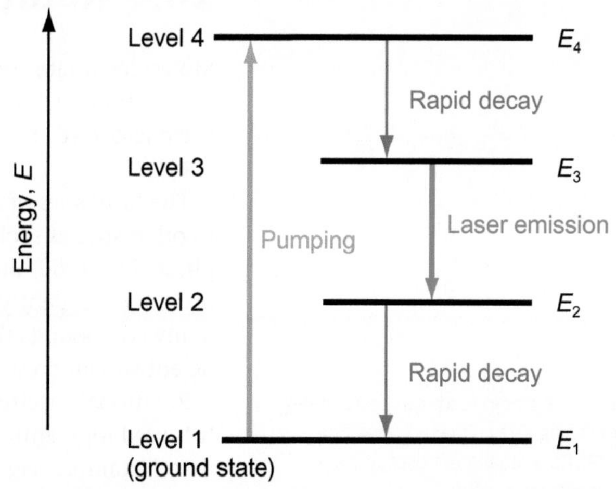

Energy, E
Level 4
E_4
Rapid decay
Level 3
E_3
Pumping
Laser emission
Level 2
E_2
Rapid decay
Level 1
(ground state)
E_1

▲ **Figure 3** Energy levels in a four-level laser, for example, the helium–neon laser, which emits red light.

Not all laser media are solids. Some, as in the helium–neon laser, are gases. Here, helium atoms are excited and transfer energy to neon atoms by collisions. The emission occurs at 632.8 nm. Other, low-power lasers use semiconductors. These are called diode lasers and are used widely in telecommunication systems.

Questions

(a) In what area of the electromagnetic spectrum does the emission from the Nd-YAG laser occur?

(b) For a helium–neon laser, calculate the energy of (i) a single photon and (ii) 1 mol of photons.

Summary

- In absorption spectroscopy, molecules absorb photons with energies that correspond to the difference between energy levels in the molecule. Absorption of radiation promotes the molecule from its ground state to an excited state.
- In emission spectroscopy, radiation is emitted when an excited molecule falls back to the ground state.
- In absorption spectroscopy, a spectrum may be presented as an absorption spectrum or a transmission spectrum.
- An absorption spectrum is recorded by varying the energy of radiation incident on a sample and monitoring how much radiation is absorbed at each energy.
- Radiation can be absorbed when its energy corresponds to the *difference* in energy between energy levels in a molecule.
- The *position* of a peak in a spectrum is related to the *difference* in energy between energy levels, ΔE.
- The *intensity* of the peaks is related to the *number* of transitions that occur when radiation is absorbed. This depends on:
  - the concentration of the solution, c, and the path length, l, according to the Beer–Lambert law: $A = \varepsilon c l$, where A is the absorbance and ε is the molar absorption coefficient;
  - the population of energy levels, according to the Boltzmann distribution;
  - the selection rules for the particular form of spectroscopy, which determine whether a particular transition is allowed.

For practice questions on these topics, see questions 6–12 at the end of this chapter (pp.510–511).

10.4 Rotational spectroscopy

Molecules rotate (see Figure 10.4, p.452) and molecular rotational energy is quantized (see Section 10.2, p.453). The transitions between rotational energy levels correspond to the microwave and far infrared regions of the electromagnetic spectrum (see Figure 10.2, p.451).

The principles of rotational spectroscopy apply not only to neutral molecules but also to other species such as radicals and ions. The measurements are usually made in the gas phase. In a solid, molecules are not free to rotate and, in a liquid or solution, collisions occur too frequently for molecules to rotate freely enough to give rise to spectra. However, many compounds that are liquid or solid at room temperature can be heated to give sufficient vapour pressure for spectra to be recorded.

Bond lengths in solids can be accurately measured using X-ray diffraction (see Box 6.4, p.279). Rotational spectroscopy is a complementary technique.

Rotational spectroscopy is very sensitive so only small quantities of vapour are needed. It is used for identification and to measure molecular properties, especially bond lengths. As an example, Figure 10.11 shows the rotational spectrum of carbon monoxide gas—specifically $^{12}C^{16}O$ since rotational spectra depend critically on the masses of the atoms.

Molecules in solids and liquids absorb microwave radiation but do not give useful spectra. The energy absorbed is distributed through the material as a result of molecular collisions and is converted into vibrations and rotations—so the material heats up. This is the principle behind microwave cooking.

Rotational spectra typically use **wavenumbers**, in units of cm^{-1}, for the *x*-axis. The wavenumber, given the symbol $\tilde{\nu}$, is the number of waves that will fit into unit distance. The SI unit of wavenumber is m^{-1}. However, $\tilde{\nu}$ is usually quoted in cm^{-1} ($1\,cm^{-1} = 100\,m^{-1}$). Wavenumber is the reciprocal of wavelength and is proportional to frequency and energy.

$$\tilde{\nu} = \frac{1}{\lambda} \tag{10.10a}$$

Since $c = \lambda\nu$ (Equation 3.1, p.118):

$$\tilde{\nu} = \frac{\nu}{c} \tag{10.10b}$$

Substituting for ν in $E = h\nu$ (Equation 3.2, p.120) gives

$$E = hc\tilde{\nu} \tag{10.11}$$

Microwave ovens operate by irradiating the contents with microwaves of frequency 2.45 GHz. The energy of these microwaves is absorbed by water and other polar molecules in the food. The microwaves are generated by a magnetron, which is the dark-coloured box at the bottom right of this X-ray photo of a microwave cooker.

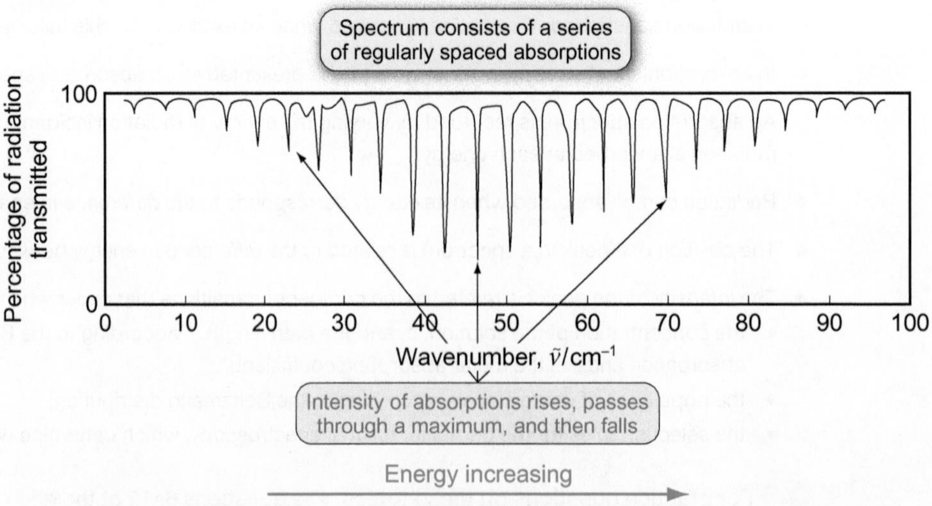
Spectrum consists of a series of regularly spaced absorptions
Percentage of radiation transmitted
100
0
0 10 20 30 40 50 60 70 80 90 100
Wavenumber, $\tilde{\nu}$/cm^{-1}
Intensity of absorptions rises, passes through a maximum, and then falls
Energy increasing

Figure 10.11 The rotational spectrum of CO in the far infrared.

What features of a rotational spectrum, such as the one in Figure 10.11, are significant? Firstly, the *absorptions* occur at wavenumbers up to ~100 cm^{-1}. Secondly, there is *a number of sharp absorptions* (**lines**) that occur at regular, *equally spaced* intervals. Finally, the *line intensities* start off small, get larger as the energy increases, but then pass through a maximum before returning to very small. How can these features be explained?

Rotation of a diatomic molecule

Consider the diatomic molecule in Figure 10.12. To develop a model to explain the spectrum, the molecule is treated as a **rigid rotor**. This means that the bond joining the two atoms has a fixed length, r_0, which does not change during the rotation. This model of the molecule, therefore, resembles a dumbbell. Rotation occurs about the centre of mass, which is the point at which the dumbbell balances.

What properties of the molecule influence the speed of rotation? Imagine giving a rotating dumbbell a small push. If the masses of the atoms, m_1 and m_2, are large or if the distance between them is long, your push will produce only a small increase in speed. These factors are combined in a property of the molecule called the **moment of inertia**, *I*.

For a diatomic molecule, like the one in Figure 10.12, the moment of inertia is given by

$$I = \mu r_0^2 \qquad (10.12)$$

where μ is the reduced mass and is calculated from the masses of the individual atoms

$$\mu = \frac{m_1 m_2}{m_1 + m_2} \qquad (10.13)$$

Worked example 10.3 shows the calculation of the moment of inertia of a diatomic molecule.

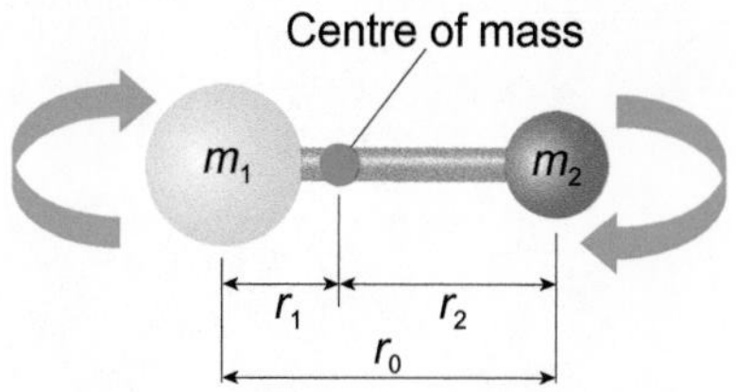
Centre of mass
m_1
m_2
r_1
r_2
r_0

Moment of inertia, $I = \left(\frac{m_1 m_2}{m_1 + m_2}\right) r_0^2$

Figure 10.12 Rotation of a diatomic molecule around its centre of mass is characterized by its moment of inertia, *I*.

The SI units of μ are kg, so the SI units of *I* are kg m^2.

In the classical world, where classical mechanics applies, you can rotate a dumbbell at continuously varying speeds (and so varying energies). Once it is rotating, you could give it a small push and make it rotate faster.

In the quantum world this can't happen. A molecule can only rotate at certain quantized speeds. Energy has to be supplied to increase the speed exactly to correspond to the next energy level. If you try to supply less energy, it won't be absorbed.

Worked example 10.3 Calculating moments of inertia

The bond length in a molecule of $^{12}C^{16}O$, determined by electron diffraction, is 0.113 nm. Calculate the moment of inertia, *I*.

Strategy

Use Equation 10.12 to calculate *I*. The bond length, r_0, is given. Use relative isotopic masses to calculate the reduced mass of the molecule, μ. The SI units for *I* are kg m^2 so, to keep the units consistent, you need to calculate the masses of the atoms in kg.

Solution

Calculate the masses of ^{12}C and ^{16}O.

$$A_r(^{12}C) = 12.00 \quad \text{and} \quad A_r(^{16}O) = 16.00$$

Molar mass of ^{12}C atoms is 12.00 g mol^{-1} = 12.00×10^{-3} kg mol^{-1} (1 g = 1×10^{-3} kg)

The mass of each ^{12}C atom is given by

$$m_{12C} = \frac{12.00 \times 10^{-3}\ \text{kg mol}^{-1}}{6.022 \times 10^{23}\ \text{mol}^{-1}} = 1.993 \times 10^{-26}\ \text{kg}$$

Using the same method, $m_{16O} = 2.657 \times 10^{-26}$ kg

Use Equation 10.13 to calculate the reduced mass, μ.

$$\mu = \frac{m_1 m_2}{m_1 + m_2} \qquad (10.13)$$

$$\mu_{CO} = \frac{(1.993 \times 10^{-26}\ \text{kg}) \times (2.657 \times 10^{-26}\ \text{kg})}{(1.993 \times 10^{-26}\ \text{kg}) + (2.657 \times 10^{-26}\ \text{kg})} = 1.139 \times 10^{-26}\ \text{kg}$$

Use Equation 10.12 to calculate I.

$$I = \mu r_0^2 \qquad (10.12)$$

Convert the bond length from nm to m: r_0 = 0.113 nm = 0.113×10^{-9} m

$$I = (1.139 \times 10^{-26}\ \text{kg}) \times (0.113 \times 10^{-9}\ \text{m})^2 = 1.454 \times 10^{-46}\ \text{kg m}^2$$

Question

The moment of inertia of $^1H^{35}Cl$ is 2.639×10^{-47} kg m^2. Calculate the bond length.

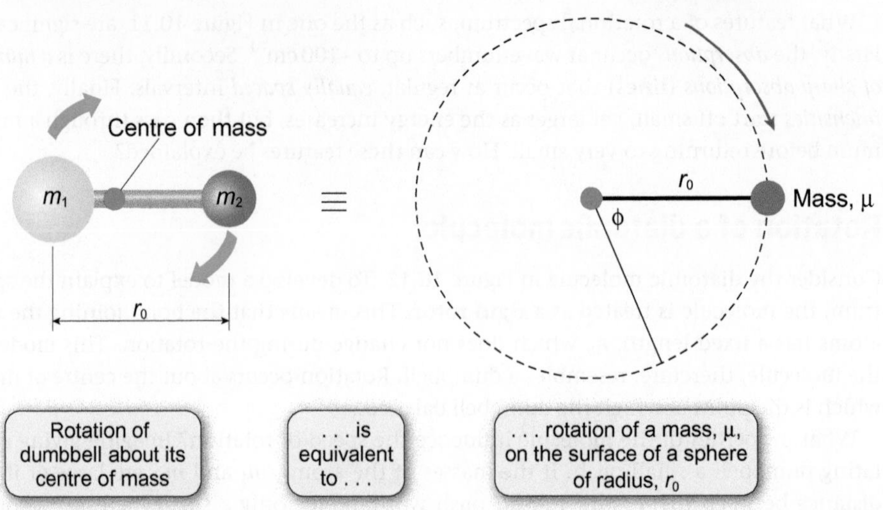
Centre of mass
m1
m2
r0
r0
Mass, μ
ϕ
Rotation of dumbbell about its centre of mass
. . . is equivalent to . . .
. . . rotation of a mass, μ, on the surface of a sphere of radius, r0

Figure 10.13 The rotation of a diatomic molecule can be represented as a mass, μ, moving through an angle ϕ on the surface of a sphere with radius, r_0.

Quantization of rotational energies

A diatomic molecule rotates about its centre of mass, and it can rotate in three dimensions. How can the energy of the different rotational levels be calculated? You can treat the rotational motion of a diatomic molecule with atoms of mass m_1 and m_2 and bond length r_0 as equivalent to the motion of a particle of mass μ moving on the surface of a sphere of radius r_0 with the centre of mass at its centre, as shown in Figure 10.13.

The angular electronic wavefunctions of the hydrogen atom are discussed in Section 3.5 (p.136).

Using this model, the quantum mechanical methods described in Section 10.2 (p.453) can be used to analyse the motion and determine an expression for the wavefunctions. The problem is similar to that encountered in Section 3.5 (p.134) where the angular electronic wavefunctions of the hydrogen atom are discussed. The wavefunctions for the rotating molecule have the same form as those that describe the angular shape of electronic orbitals.

The analysis shows that rotational energy is quantized and the rotational energies are given by

$$E_{\text{rotation}} = E_J = \frac{h^2}{8\pi^2 I} J(J+1) \tag{10.14}$$

where I is the moment of inertia and h is the Planck constant. J is the **rotational quantum number** and can take values 0, 1, . . . ∞. E_J is the energy of the Jth level. The lowest energy level has $J = 0$, and molecules in this level have zero rotational energy, $E_J = 0$. (Contrast this with translational energy (Box 10.1, p.454) where the zero point energy is non-zero.) The next lowest energy level has $J = 1$, and so on.

The wavefunction for $J = 0$ has the same shape as the *s* orbital of the hydrogen atom and so is the same in all directions. There are three wavefunctions for $J = 1$, so the $J = 1$ level is triply degenerate (analogous to the three *p* orbitals of the hydrogen atom). The $J = 2$ level has a degeneracy of 5 (analogous to the 5 *d* orbitals). In general, the degeneracy of level J is $(2J + 1)$.

When constructing an energy level diagram for rotations in the molecule, it is convenient to simplify Equation 10.14 by introducing a **rotational constant,** *B*, which groups together the terms in the first part of Equation 10.14 and is characteristic of the molecule being studied

Table 10.2 Rotational energy levels

J	$J(J+1)$	$E_J = hBJ(J+1)$	Transition	ΔE
0	0	0		
			$J=0 \rightarrow J=1$	2 hB
1	2	2 hB		
			$J=1 \rightarrow J=2$	4 hB
2	6	6 hB		
			$J=2 \rightarrow J=3$	6 hB
3	12	12 hB		
			$J=3 \rightarrow J=4$	8 hB
4	20	20 hB		
			$J=4 \rightarrow J=5$	10 hB
5	30	30 hB		

$$B = \frac{h}{8\pi^2 I} \tag{10.15}$$

The value of B does not change with energy and has units of s^{-1}.

The rotational energies can then be represented in terms of the rotational constant as

$$E_J = hBJ(J+1) \tag{10.16}$$

This equation is used in Table 10.2 to give expressions for the energies of the rotational levels and to show the differences between energy levels. Table 10.2 can be used to construct an energy level diagram as in Figure 10.14. The calculation of energy levels is shown in Worked example 10.4.

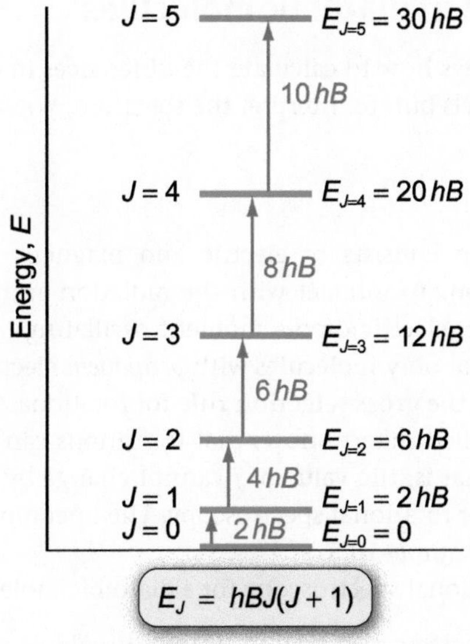
Energy, E
J = 5
E_J=5 = 30 hB
10 hB
J = 4
E_J=4 = 20 hB
8 hB
J = 3
E_J=3 = 12 hB
6 hB
J = 2
E_J=2 = 6 hB
4 hB
J = 1
E_J=1 = 2 hB
J = 0
2 hB
E_J=0 = 0
E_J = hBJ(J + 1)

Figure 10.14 Rotational energy levels. The rigid rotor model gives a series of quantized energy levels.

Units of B

$$B = \frac{h}{8\pi^2 I}$$

So, units of $B = \frac{J \times s}{kgm^2} = \frac{(kg\,m^2\,s^{-2}) \times s}{kg\,m^2} = s^{-1}$

($1\,J = 1\,Nm = 1\,kgm^2s^{-2}$)

Rotational spectra are usually presented in terms of wavenumber, $\tilde{\nu}$, measured in cm^{-1}. Some books refer to the 'energy', E_J, in cm^{-1} since the wavenumber is proportional to energy. In this book, energy is always measured in joules and the rotational constant, B, has units of frequency, s^{-1}. You may see B defined in other units in other books.

Worked example 10.4 Calculating rotational energies

The moment of inertia of $^{12}C^{16}O$ is $I = 1.454 \times 10^{-46}\,\mathrm{kg\,m^2}$.

(a) Calculate the energies of the $J = 0$, $J = 1$, and $J = 2$ energy levels.

(b) At what wavenumber does the $J = 1 \rightarrow J = 2$ transition occur in the spectrum?

(c) To which region of the electromagnetic spectrum does this correspond?

Strategy

(a) Use Equation 10.15 to work out a value for the rotational constant, B, and then use Equation 10.16 to calculate the energies.

(b) The energy of a transition, ΔE, is given by the difference in energies between the two levels. Then, convert ΔE to wavenumber units using Equation 10.11 (p.466).

(c) Use the electromagnetic spectrum (Appendix 11, p.1371) to find in which region the transition occurs.

Solution

(a) *Use Equation 10.15 to calculate the rotational constant,* B.

$$B = \frac{h}{8\pi^2 I} \qquad (h = 6.626\times10^{-34}\,\mathrm{J\,s}) \qquad (10.15)$$

$$B = \frac{6.626\times10^{-34}\,\mathrm{J\,s}}{8 \times (3.142)^2 \times (1.454 \times 10^{-46}\,\mathrm{kg\,m^2})}$$

$$= \frac{6.626 \times 10^{-34}\,\cancel{\mathrm{kg}}\,\cancel{\mathrm{m^2}}\,\mathrm{s^{-2}\,s}}{8 \times (3.142)^2 \times (1.454 \times 10^{-46}\,\cancel{\mathrm{kg}}\,\cancel{\mathrm{m^2}})} \qquad (1\,\mathrm{J} = 1\,\mathrm{kg\,m^2\,s^{-2}})$$

$$= 5.77 \times 10^{10}\,\mathrm{s^{-1}}$$

Use Equation 10.16 to calculate the energies.

$$E_J = hBJ(J+1) \qquad (10.16)$$

For $J = 0$: $E_{J=0} = hB \times 0\,(0+1) = 0$

For $J = 1$: $E_{J=1} = hB \times 1\,(1+1) = 2hB = 7.65 \times 10^{-23}\,\mathrm{J}$

For $J = 2$: $E_{J=2} = hB \times 2\,(2+1) = 6hB = 22.9 \times 10^{-23}\,\mathrm{J}$

(b) *Find the energy of the $J = 1 \rightarrow J = 2$ transition.*

$$\text{Energy difference, } \Delta E = E_{J=2} - E_{J=1}$$
$$= (22.9 \times 10^{-23}\,\mathrm{J}) - (7.65 \times 10^{-23}\,\mathrm{J})$$
$$= 15.3 \times 10^{-23}\,\mathrm{J}$$

Use Equation 10.11 to convert ΔE to wavenumbers.

$$E = hc\tilde{\nu} \quad (c = 2.998 \times 10^8\,\mathrm{m\,s^{-1}}) \qquad (10.11)$$

$$15.3 \times 10^{-23}\,\mathrm{J} = (6.626 \times 10^{-34}\,\mathrm{J\,s}) \times (2.998 \times 10^8\,\mathrm{m\,s^{-1}}) \times \tilde{\nu}$$

$$\tilde{\nu} = \frac{15.3 \times 10^{-23}\,\cancel{\mathrm{J}}}{(6.626 \times 10^{-34}\,\cancel{\mathrm{J}}\,\cancel{\mathrm{s}}) \times (2.998 \times 10^8\,\mathrm{m}\,\cancel{\mathrm{s^{-1}}})}$$

$$= 770\,\mathrm{m^{-1}}$$

$$= 7.7\,\mathrm{cm^{-1}} \qquad (1\,\mathrm{cm^{-1}} = 100\,\mathrm{m^{-1}})$$

(c) *Look at the electromagnetic spectrum.*

From the electromagnetic spectrum in Appendix 11 (p.1371), a wave-number of 7.7 $\mathrm{cm^{-1}}$ corresponds to the microwave region of the spectrum.

Question

Calculate the energies of the $J = 7$ and $J = 8$ levels. At what wave-number does the $J = 7 \rightarrow J = 8$ transition occur?

Rotational spectra of diatomic molecules

The previous section shows how to calculate the differences in energy between the various rotational energy levels but, to interpret the spectrum, you also need to know which transitions can take place.

Selection rules

Gross and specific selection rules are introduced in Section 10.3 (p.458).

Electromagnetic radiation consists of electric and magnetic fields oscillating at the frequency of the radiation. To interact with the radiation and absorb energy from it a molecule must possess an electric dipole moment oscillating (i.e. rotating) at the same frequency. This means that only molecules with *permanent* electric dipole moments have rotational spectra. This is the **gross selection rule** for rotational spectra.

Dipoles and dipole moments in molecules are discussed in Sections 1.8 (p.52), 5.3 (p.235), and 17.3 (p.783).

In addition, more detailed theory shows that transitions can only take place between adjacent energy levels, that is, the value of J cannot change by more than 1. This is the **specific selection rule** for rotational spectroscopy. The operation of these selection rules is illustrated in Worked example 10.5.

Selection rules for rotational spectroscopy for a diatomic molecule:

- the molecule must possess a permanent electric dipole;
- $\Delta J = \pm 1$.

Worked example 10.5 Selection rules in rotational spectroscopy

(a) Which of the following molecules display rotational spectra: H_2; CO; NO; Cl_2?

(b) Which of the following rotational transitions occur in a diatomic molecule:

$J = 0 \rightarrow J = 1$; $J = 6 \rightarrow J = 7$;
$J = 4 \rightarrow J = 3$; $J = 5 \rightarrow J = 3$;
$J = 1 \rightarrow J = 3$?

Strategy

Apply the selection rules to determine: (a) whether the molecule has a permanent electric dipole; (b) whether $\Delta J = \pm 1$. If so, the selection rule is satisfied.

Solution

(a) *Does the molecule have a permanent electric dipole?*

CO and NO have permanent dipoles (gross selection rule). H_2 and Cl_2 are homonuclear diatomics and so have no dipole. Therefore, only CO and NO show rotational spectra.

(b) *Find ΔJ for each transition.*

For $J = 0 \rightarrow J = 1$:	$\Delta J = (1 - 0) = +1$	Allowed
For $J = 4 \rightarrow J = 3$:	$\Delta J = (3 - 4) = -1$	Allowed
For $J = 1 \rightarrow J = 3$:	$\Delta J = (3 - 1) = +2$	Forbidden
For $J = 6 \rightarrow J = 7$:	$\Delta J = (7 - 6) = +1$	Allowed
For $J = 5 \rightarrow J = 3$:	$\Delta J = (3 - 5) = -2$	Forbidden

Since $\Delta J = \pm 1$ for an allowed transition (specific selection rule), $J = 0 \rightarrow J = 1$, $J = 4 \rightarrow J = 3$, and $J = 6 \rightarrow J = 7$ are allowed and will be seen in the spectrum.

Question

Which of the following show rotational spectra: D_2; HCl; ClO; CN?

Populations of rotational energy levels

The energy difference, ΔE, between the $J = 0$ and $J = 1$ rotational levels in CO is 7.65×10^{-23} J (see Worked example 10.4). You can use the Boltzmann distribution (Equation 10.9, p.462) to find the relative populations of these two levels at 298 K.

The degeneracy of rotational levels is equal to $(2J + 1)$ (see p.468), so the degeneracies of the levels are $g_{J=0} = 1$ and $g_{J=1} = 3$. The ratio of the populations is

$$\frac{n_{\text{upper}}}{n_{\text{lower}}} = \frac{g_{\text{upper}}}{n_{\text{lower}}} e^{\left(\frac{-\Delta E}{K_B T}\right)} \qquad (10.9)$$

$$\frac{n_{J=1}}{n_{J=0}} = \frac{3}{1} e^{\left(\frac{-(7.65 \times 10^{-23}\,\text{J})}{1.381 \times 10^{-23}\,\text{JK}^{-1}) \times 298\,\text{K}}\right)}$$

$$= 3 \times 0.982 = 2.95$$

The energy difference, ΔE, is much less than $k_B T$, so the population ratio for these levels is determined largely by the ratio of the degeneracies. Although the degeneracy of the level increases as J increases, the value of ΔE also increases and eventually becomes greater than $k_B T$. Because of these two opposing factors, the ratio of the populations of adjacent levels becomes less than 1 at high values of J (Figure 10.15). Overall, the population of CO molecules is spread throughout many rotational levels.

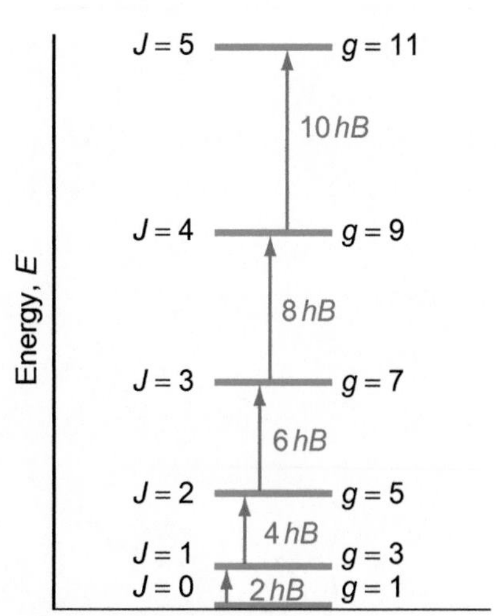

J = 5
g = 11
10 hB
J = 4
g = 9
8 hB
Energy, E
J = 3
g = 7
6 hB
J = 2
g = 5
4 hB
J = 1
g = 3
J = 0
2 hB
g = 1

Figure 10.15 As J increases, the energy difference, ΔE, increases, but so does the degeneracy, g. These two effects work in the opposite direction as far as the ratio of n_{upper} to n_{lower} is concerned.

Transitions between rotational energy levels

You are now in a position to explain the appearance of the rotational spectrum in Figure 10.11 (p.466). The CO molecules are distributed across many rotational energy levels at 298 K. Radiation is absorbed when it corresponds to the gap between *adjacent* energy levels when $\Delta J = +1$ (specific selection rule).

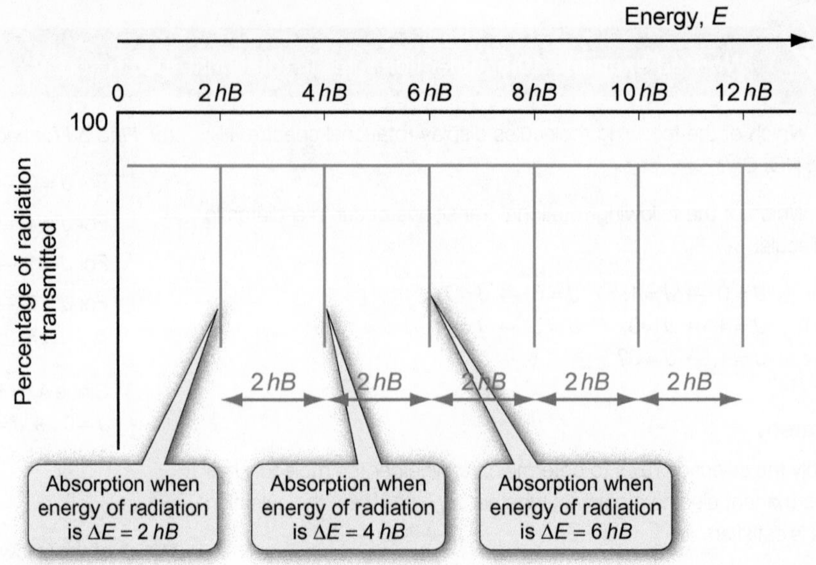
Energy, E
0
2hB
4hB
6hB
8hB
10hB
12hB
100
Percentage of radiation transmitted
2hB
2hB
2hB
2hB
2hB
Absorption when energy of radiation is ΔE = 2 hB
Absorption when energy of radiation is ΔE = 4 hB
Absorption when energy of radiation is ΔE = 6 hB

Figure 10.16 A diagrammatic representation of a rotational spectrum, showing the energy gap between successive lines. Note that in a real spectrum, the lines would not all be of the same height (intensity).

If a molecule in the ground state, $J = 0$, is irradiated with electromagnetic radiation of gradually increasing energy, what happens?

The answer is nothing—until the energy of the photon of radiation exactly corresponds to the energy difference between the $J = 0$ and $J = 1$ states, that is, $\Delta E = 2hB$ (see Table 10.2, p.469). Some of the radiation is then absorbed. If the energy of the radiation is increased further, nothing happens—until it exactly matches the difference between $J = 1$ and $J = 2$, that is, $\Delta E = 4hB$. Increasing the radiation further has no effect until it exactly corresponds to the next $J = 2 \rightarrow J = 3$ transition, that is, $\Delta E = 6hB$.

Energy is only absorbed when it exactly matches the *difference* between energy levels. The result of this is shown in Table 10.2. Absorptions occur at values of $2hB$, $4hB$, $6hB$, $8hB$, and so on. The rotational spectrum, therefore, has a series of regularly spaced lines. *The energy difference between successive lines is always* $2hB$ (measured in energy units, J), as shown in Figure 10.16. However, rotational spectra are often measured in frequency units or wavenumber units. In frequency units, the spacing between the lines is B (units s^{-1}), whereas in wavenumber units, the spacing is $2B/c$ (units m^{-1}), where c is the speed of light in $m\,s^{-1}$.

When an absorption (or emission) peak in a spectrum is sharp, it is often referred to as a *line*.

It is not necessary to have *all* the lines in the rotational spectrum to perform the calculations. The energy gap between consecutive lines is *always* 2 *hB* so that any region of the spectrum will suffice.

Because the population of the energy levels first increases with J, but then decreases as J gets larger, the intensities of the lines in the spectrum show a similar pattern, as seen in Figure 10.11 (p.466).

Finding bond lengths from rotational spectra

Measuring the gap between lines in the spectrum allows the calculation of B, from which the moment of inertia and hence the bond length can be found, as shown in Worked example 10.6.

Worked example 10.6 Calculating bond lengths from rotational spectra

The spectrum of $^{12}C^{16}O$ has absorption lines at:

15.36 cm^{-1}; 19.20 cm^{-1}; 23.04 cm^{-1}; 26.88 cm^{-1}; and 30.72 cm^{-1}.

Calculate the rotational constant, B, and the bond length of $^{12}C^{16}O$.

Strategy

The energy difference between the spectral lines, $\Delta E = 2hB$. Since $\Delta E = h\nu$, the separation of the spectral lines in frequency units is equal to $2B$. Calculate the separation between the spectral lines in wavenumbers and convert this to frequency units. The separation is then equal to $2B$. You can then work out a value for B.

Use Equation 10.15 on p.469 to find the moment of inertia, I, and Equation 10.12 on p.467 to calculate the bond length using the reduced mass from Worked example 10.3. Take care to keep all the units consistent throughout.

(Alternatively, you could convert the separation of the spectral lines into an energy difference using Equation 10.11 (p.466) and hence find a value of B, but this is a longer route.)

Solution

Calculate the separation of the spectral lines in wavenumbers.

Separation between adjacent lines
$$= 3.84\ \text{cm}^{-1}\ (1\ \text{cm}^{-1} = 100\ \text{m}^{-1})$$
$$= 3.84 \times 10^2\ \text{m}^{-1}$$

Rearrange Equation 10.10b to convert this to the equivalent frequency in s^{-1}

$$\tilde{\nu} = \frac{\nu}{c} \quad (c = 2.998 \times 10^8\ \text{m s}^{-1})$$
$$\nu = \tilde{\nu} \times c$$
$$= (3.84 \times 10^2\ \cancel{\text{m}}^{-1}) \times (2.998 \times 10^8\ \cancel{\text{m}}\ \text{s}^{-1})$$
$$= 1.151 \times 10^{11}\ \text{s}^{-1}$$

Calculate B.

The separation between lines is $2B$ in frequency units, so

$$B = \frac{1.151 \times 10^{11}\ \text{s}^{-1}}{2} = 5.756 \times 10^{10}\ \text{s}^{-1}$$

Rearrange Equation 10.15 (p.469) to calculate the moment of inertia of CO.

$$I = \frac{h}{8\pi^2 B} \quad (h = 6.626 \times 10^{-34}\ \text{Js}) \quad \text{(from 10.15)}$$
$$= \frac{6.626 \times 10^{-34}\ \text{Js}}{8 \times (3.142)^2 \times (5.756 \times 10^{10}\ \text{s}^{-1})}$$
$$= 1.46 \times 10^{-46}\ \text{Js}^2$$
$$= 1.46 \times 10^{-46}\ \text{kgm}^2$$

(*Note.* 1 J = 1 N × m = 1 kgms^{-2} × m = 1 kgm^2 s^{-2}. So, the units of I can be expressed as Js^2 or kgm^2 since these are identical.)

Rearrange Equation 10.12 to find the bond length, r_0

$$I = \mu r_0^2 \qquad (10.12)$$

Rearranging

$$r_0 = \left(\frac{I}{\mu}\right)^{1/2}$$

The reduced mass, $\mu_{CO} = 1.139 \times 10^{-26}$ kg (from Worked example 10.3, p.467) so

So, the bond length in $^{12}C^{16}O$ is 0.113 nm (113 pm).

Question

In the rotational spectrum of $^{1}H^{35}Cl$ recorded at 298 K, consecutive lines were observed at the following frequencies (measured in GHz): 621; 1245; 1869; 2493. Calculate the rotational constant, B, and the bond length in $^{1}H^{35}Cl$.

Measurements of the frequency of microwave radiation can be made with very good precision so rotational spectroscopy is a very good method for measuring bond lengths of diatomic molecules in the gas phase.

Polyatomic molecules

Molecules with more than two atoms that have a permanent dipole moment give rise to rotational spectra similar in appearance to those of diatomics. However, the interpretation of the data is more complex.

After diatomics, the next simplest system is a linear triatomic with a dipole, such as O=C=S (see Figure 10.17). The molecule rotates around its centre of mass, which, since S has a greater mass than O, lies between the C and S atoms. Only a single moment of inertia can be calculated but there are two bond lengths, which causes difficulty if you are trying to find these bond lengths. The difficulty can be overcome by taking measurements on

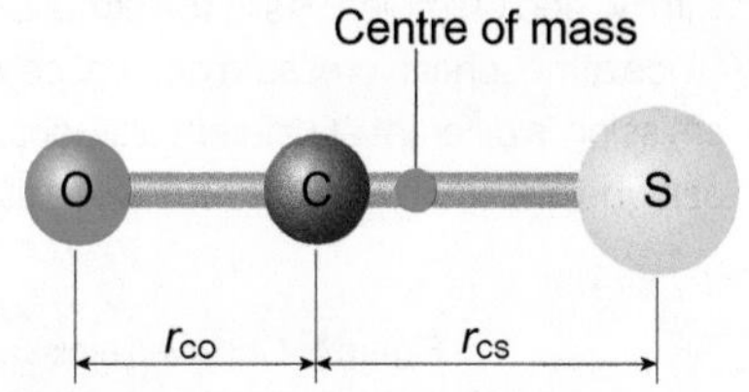

Centre of mass
O
C
S
r_{CO}
r_{CS}

Figure 10.17 The centre of mass of the OCS molecule.

molecules containing different isotopes. For example, for $^{16}O^{12}C^{32}S$, $I = 1.380 \times 10^{-45}\,kg\,m^{-2}$ but, for $^{16}O^{12}C^{34}S$, I changes to $1.414 \times 10^{-45}\,kg\,m^{-2}$. Solving the expressions for I in these cases allows determination of the individual bond lengths ($r_{CO} = 0.117\,nm$ and $r_{OS} = 0.156\,nm$).

For non-linear molecules, the moments of inertia are different with respect to the three coordinate axes so there are three different rotational constants. Analysis of the spectra from these molecules is more complex.

An analogy is to rotate a ball attached to some elastic above your head. As you rotate it faster, the elastic stretches due to the centrifugal force of the ball.

Radio telescopes are used to identify molecules and ions in interstellar space by measuring microwave emission.

Non-rigid rotors

The discussion above has assumed that chemical bonds can be treated as rigid rods with fixed lengths. While this is a good approximation, it is not completely valid. As the energy increases, the molecule rotates faster and the bond lengths get slightly longer. This is known as **centrifugal distortion**.

To account for this, an additional term is added to Equation 10.16 to describe the energy of the rotating diatomic molecule more accurately

$$E_J = hBJ(J+1) - DJ^2(J+1)^2 \qquad (10.17)$$

where D is the *centrifugal distortion constant*. The effect of centrifugal distortion in the spectrum is that the spacing between the lines gets smaller as J increases. High accuracy measurements of bond lengths need to incorporate this effect but, for most purposes, treating molecules as rigid rotors is sufficient.

Applications of rotational spectroscopy

The main application of rotational spectroscopy is in determining accurate bond lengths for molecules in the gas phase. Another use is in identifying the large number of species in interstellar space, detected using radio telescopes such as that at Jodrell Bank in north-west England, as described in the introductory pages to Chapter 4 (p.168). Rotational spectroscopy is also used for the quantitative analysis of atmospheric gases. An example of this is described in Box 10.3.

Box 10.3 Using rotational spectroscopy to monitor $ClO^\bullet$ radicals in the atmosphere

Observing the $ClO^\bullet$ radical in the atmosphere above the Antarctic was key to understanding the role of CFCs in the formation of the Antarctic ozone hole (see Box 27.6, p.1233). Rotational spectroscopy in the microwave region was one of the techniques used to make this observation. Emission was observed at frequencies around 278.6 GHz ($1\,GHz = 1 \times 10^9\,Hz = 1 \times 10^9\,s^{-1}$) corresponding to a rotational transition of $ClO^\bullet$.

Figure 1 shows the observed emission in the microwave region of the spectrum. It is not a sharp line, but has a *width* and a *shape* that depend on both the temperature and the pressure at which emission of radiation occurred. These effects are a result of collisions of $ClO^\bullet$ with N_2 and O_2 molecules in the atmosphere, which *broaden* the line. Since atmospheric pressure decreases with altitude, the shape of the emission is different at different altitudes. The emission line is broader at higher pressures, that is, at lower altitudes.

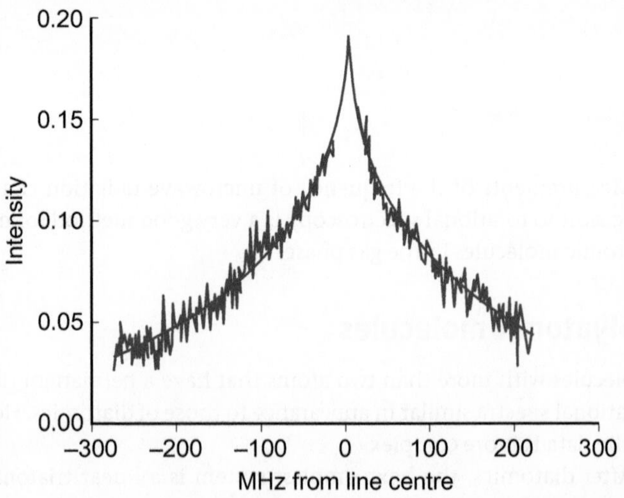
Intensity
0.20
0.15
0.10
0.05
0.00
−300
−200
−100
0
100
200
300
MHz from line centre

Figure 1 Observed emission in the microwave region of the electromagnetic spectrum corresponding to a rotational transition of $ClO^\bullet$. (The smooth curve shows the best fit to the experimental data.)

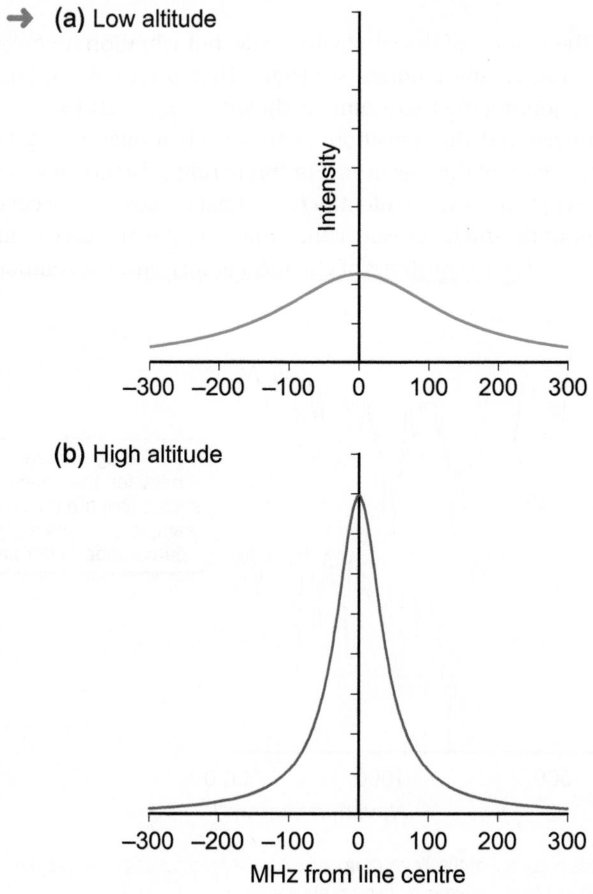

(a) Low altitude
Intensity
−300 −200 −100 0 100 200 300
(b) High altitude
−300 −200 −100 0 100 200 300
MHz from line centre

Figure 2 Intensities of $ClO^{\bullet}$ emission lines at different altitudes, calculated from laboratory data.

In the spectrum in Figure 1, the emission lines at different altitudes are superimposed, but the intensity at larger distances from the centre of the line is mainly due to emission at lower altitudes, while emission at higher altitudes makes a greater contribution to the line centre (see Figure 2).

From knowledge of the dependence of temperature and pressure on altitude, it is possible to analyse the line intensity and its shape and so determine the concentration of $ClO^{\bullet}$ at various altitudes.

Measurements over the period 20–24 September 1987 at the American McMurdo Station in Antarctica showed a sharp increase in the $ClO^{\bullet}$ concentration at an altitude of ~20 km. This altitude corresponds to that of the ozone layer in the stratosphere. The date corresponded to the beginning of spring in the Antarctic and closely followed sunrise after the Antarctic winter night.

Measurements were also made of ozone concentration and the scientists were able to show that the depletion of ozone could be related to the $ClO^{\bullet}$ + $ClO^{\bullet}$ reaction (described in Box 9.4, p.403) to form the ClO dimer.

$$ClO^{\bullet} + ClO^{\bullet} \rightarrow ClOOCl$$

The dimer undergoes a photochemical reaction to produce $Cl^{\bullet}$ radicals, which then react with ozone.

Question

The rotational constant, B, for $^{35}Cl^{16}O$ is 18.69 GHz. Calculate the Cl–O bond length.

Summary

- Rotational spectra are recorded in the gas phase. The transitions correspond to microwave and far infrared radiation.
- Quantized rotational energy is described by $E_J = hBJ(J+1)$ for a diatomic molecule, where B is the rotational constant and J is the rotational quantum number.
- Selection rules:
  - gross (general): the molecules must have a permanent electric dipole to show a spectrum;
  - specific: the rotational quantum number can only change by 1: $\Delta J = \pm 1$.
- At 298 K, the population of molecules is spread through many rotational levels.
- The rotational spectrum of a diatomic molecule consists of a series of evenly spaced lines. The energy difference between successive lines is always $2hB$. (The spacing in frequency units is $2B$ and in wavenumber units is $2B/c$.)
- The rotational constant, $B = \dfrac{h}{8\pi^2 I}$, can be found from the spacing of the lines in the spectrum. (I is the moment of inertia.)
- The bond length, r_0, is found using $I = \mu r_0^2$, where μ is the reduced mass.

For practice questions on these topics, see questions 13–17 at the end of this chapter (p.511).

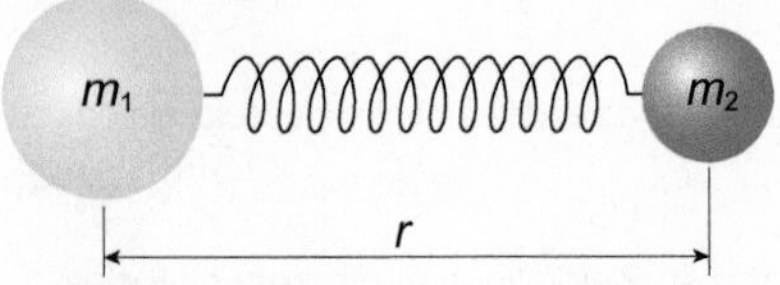

m_1
m_2
r

Figure 10.18 A spring is a good model for the vibration of a chemical bond. A diatomic molecule is like two atoms connected by a spring.

10.5 Vibrational spectroscopy

Molecular rotation involves the motion of the whole molecule, but vibration involves motion of the atoms in individual chemical bonds (see Figure 10.4, p.452). A good model for a chemical bond is a spring joining the two atoms, as shown in Figure 10.18.

Vibrational energy is quantized and the transitions between vibrational energy levels correspond to the infrared (IR) region of the spectrum. For this reason, vibrational spectroscopy is often called IR spectroscopy. It is used to identify functional groups in molecules, to confirm the structure of compounds, and to measure concentrations. It is also used to measure the vibrational frequencies and force constants of chemical bonds within a compound.

Vibrational spectroscopy is often called IR spectroscopy.

Visit the Online Resource Centre to view video 10.2 which gives a simple introduction to IR spectroscopy.

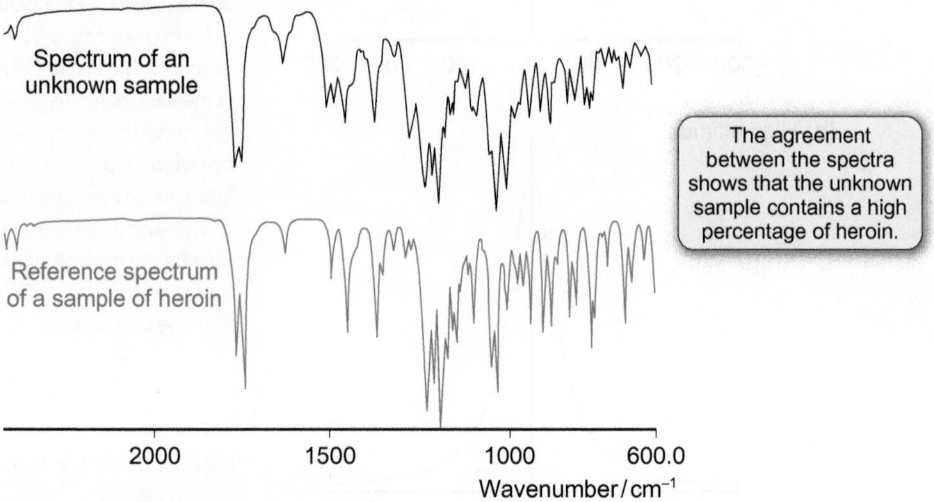

Spectrum of an unknown sample
Reference spectrum of a sample of heroin
The agreement between the spectra shows that the unknown sample contains a high percentage of heroin.
2000
1500
1000
600.0
Wavenumber / cm^{-1}

IR spectroscopy is used to identify the presence of the illegal drug, heroin in a sample. Spectral analysis like this can identify unknown compounds in samples taken from clothing or equipment.

Figure 10.19 shows the IR spectrum of aspirin. A complete analysis of every absorption band is possible but would be complex, particularly in the so-called 'fingerprint' region below 1400 cm^{-1}. However, every sample of pure aspirin gives the same spectrum if recorded in the same way. So, if the spectrum of an unknown sample gives the same spectrum as that of an authentic sample of aspirin, then the unknown compound is confirmed as aspirin. Note that, in an IR spectrum, the wavenumber (and hence energy) increases from right to left.

Many functional groups, for example, the carbonyl group, give rise to absorption bands in easily recognized parts of the spectrum. You can read about the use of IR spectroscopy in identifying functional groups in a molecule in Chapter 12. The present chapter is concerned with the theoretical basis of IR spectroscopy and its use in measuring the vibrational properties of molecules.

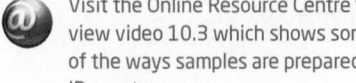

Using IR spectroscopy for the identification of functional groups is described in Section 12.2 (p.570).

Recording IR spectra

Modern infrared spectrometers use the principle of **attenuated total reflection, ATR**. Infrared radiation passes through a crystal at such an angle that it is totally reflected from the face (Figure 10.20). The crystal is made of a material of high refractive index, such as germanium, and the sample is pressed tightly onto the surface of this crystal. Some of the radiation (the *evanescent wave*) passes through the surface of the crystal and interacts with the sample. The radiation penetrates a few micrometres into the sample and some frequencies are absorbed by the vibrations of bonds before the radiation travels through the crystal to the detector. This set-up is used with liquids or with solids that are ground up to a fine powder.

IR spectra of *gases* are recorded using a cell containing the gas which is held in a beam between a source of IR radiation and the detector. The container cannot be made from

Visit the Online Resource Centre to view video 10.3 which shows some of the ways samples are prepared for IR spectroscopy.

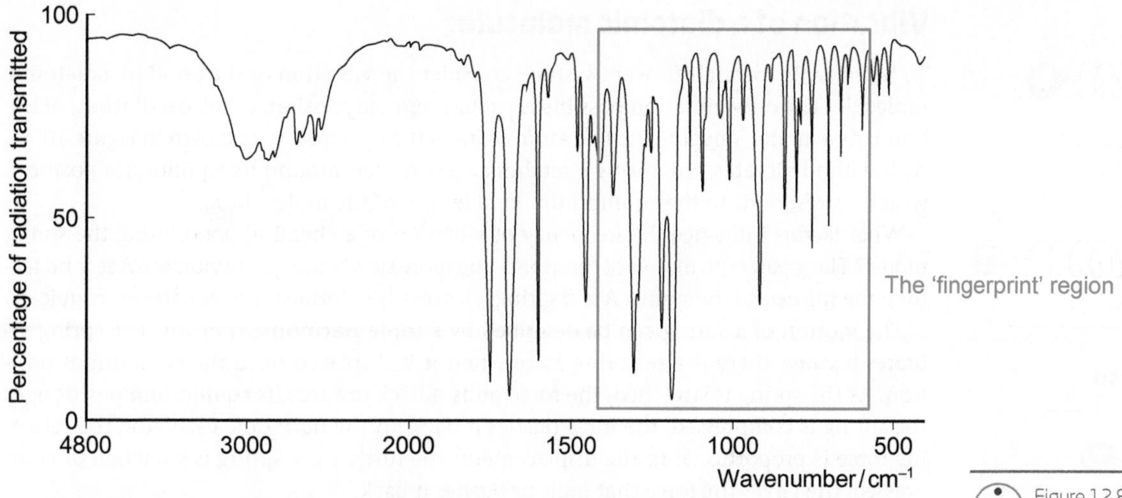
100
50
0
Percentage of radiation transmitted
4800
3000
2000
1500
1000
500
Wavenumber / cm⁻¹
The 'fingerprint' region

Figure 10.19 The 'fingerprint' region of the IR spectrum of aspirin is the region indicated in the box. It would be very difficult to assign all of the peaks to transitions but all samples of pure aspirin give the same pattern so this region of the spectrum can be used for identification.

Figure 12.8 on p.575 shows the infrared spectrum of aspirin taken in a Nujol mull. Compare this spectrum with the one in Figure 10.19.

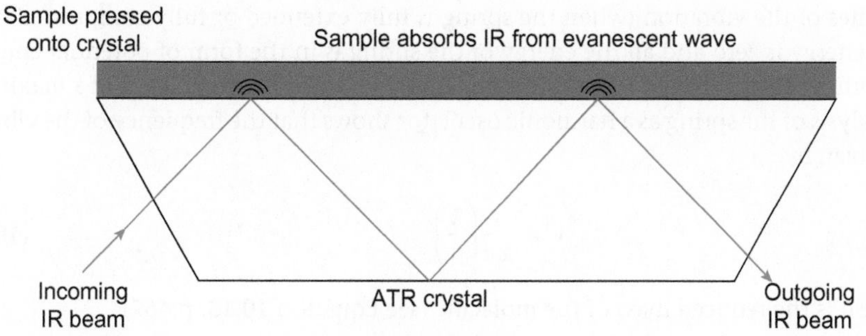
Sample pressed onto crystal
Sample absorbs IR from evanescent wave
Incoming IR beam
ATR crystal
Outgoing IR beam

Figure 10.20 Recording an IR spectrum using attenuated total reflection (ATR).

glass because glass absorbs in the infrared, leading to additional bands in the spectrum. Metal salts such as sodium chloride or potassium bromide absorb outside the range of most organic compounds and can be compressed under high pressure to make plates that are transparent to IR radiation. Gaseous samples are therefore contained in cells fitted with sodium chloride windows (Figure 10.21).

Older IR spectrometers, still in widespread use, use sodium chloride plates to measure the spectra of solid and liquid samples. Liquids are spread onto the plate and solids are ground up to a fine suspension with a hydrocarbon oil called Nujol to form a 'Nujol mull' which can be spread on a NaCl plate. In this case, the spectrum will also contain absorption bands due to the alkanes in Nujol but these additional bands can be subtracted by a computer.

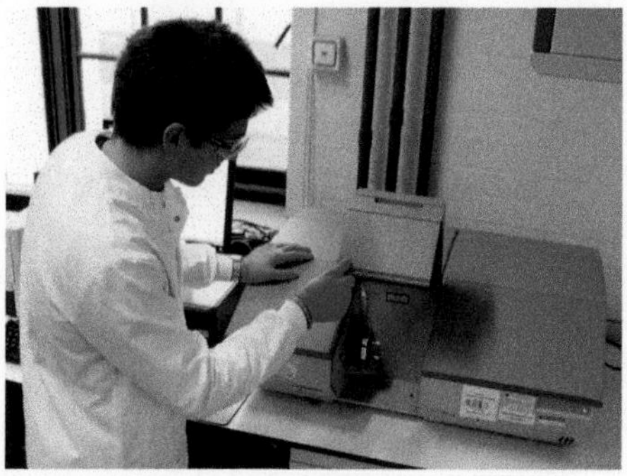

A liquid sample or a Nujol mull is spread between two NaCl plates and placed in the spectrometer beam. Gaseous samples or samples in solution can be placed in cells with NaCl windows.

Figure 10.21 Transparent sodium chloride plates are used to hold the sample in older types of IR spectrometer.

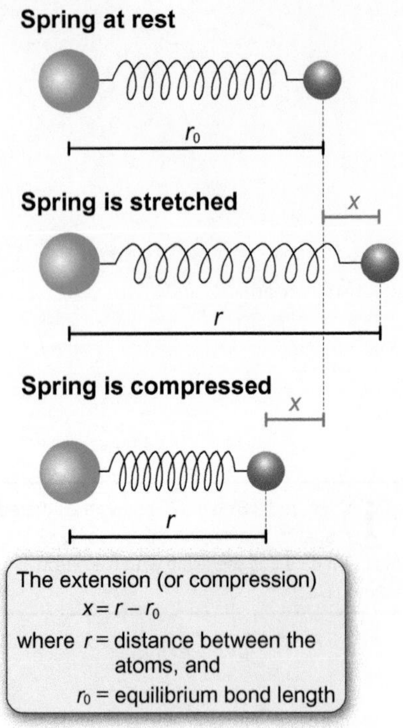

Spring at rest
r_0
Spring is stretched
x
r
Spring is compressed
x
r
The extension (or compression)
$x = r - r_0$
where r = distance between the atoms, and
r_0 = equilibrium bond length

Figure 10.22 The spring model for vibration of a chemical bond in a diatomic molecule. In simple harmonic motion, the restoring force, F, is proportional to the extension, x. $F = -kx$ (Hooke's law).

Vibration of a diatomic molecule

To begin with a straightforward system, consider the vibration of the bond in a diatomic molecule. The only vibration possible is stretching/compression, called **oscillation**, of the bond. As a model, imagine the two atoms connected by a spring, as shown in Figure 10.22. As the bond vibrates, the spring stretches and contracts around its equilibrium position, which corresponds to the equilibrium bond length of the molecule, r_0.

What factors influence the frequency of vibration of a chemical bond, using the spring model? The greater the masses of the atoms, the more slowly the spring vibrates. A second factor is the stiffness of the spring. A stiff spring is less easily deformed, and vibrates more quickly.

The motion of a spring can be described by **simple harmonic motion**. The spring vibrates because there is a restoring force when it is displaced from the equilibrium position. As the spring is stretched, the force pulls it back towards its equilibrium position; as the spring is compressed, the force pushes it back. In the harmonic oscillator, the restoring force is proportional to the displacement: the further the spring is stretched or compressed, the larger the force that pulls or pushes it back.

Using this model, the potential energy curve has the shape of a parabola, the red curve in Figure 10.23. You can see how the equation for the parabola is derived from the behaviour of a simple harmonic oscillator in Box 10.4.

The potential energy of the spring depends on the square of the extension, x. At the extremes of the vibration (when the spring is fully extended or fully compressed) the kinetic energy is zero and all the energy of the spring is in the form of potential energy. As the spring passes through its equilibrium length, r_0, the kinetic energy is at a maximum.

Analysis of the spring as a harmonic oscillator shows that the frequency of the vibration, ν, is given by

$$\nu = \frac{1}{2\pi}\left(\frac{k}{\mu}\right)^{1/2} \qquad (10.18)$$

where μ is the reduced mass of the molecule (see Equation 10.13, p.467)

$$\mu = \frac{m_1 m_2}{m_1 + m_2} \qquad (10.13)$$

The constant, k, is the **force constant** of the bond and is related to its stiffness. As k increases (the spring becomes stiffer), the frequency of vibration increases. As μ increases (the masses of the atoms become larger), the frequency decreases.

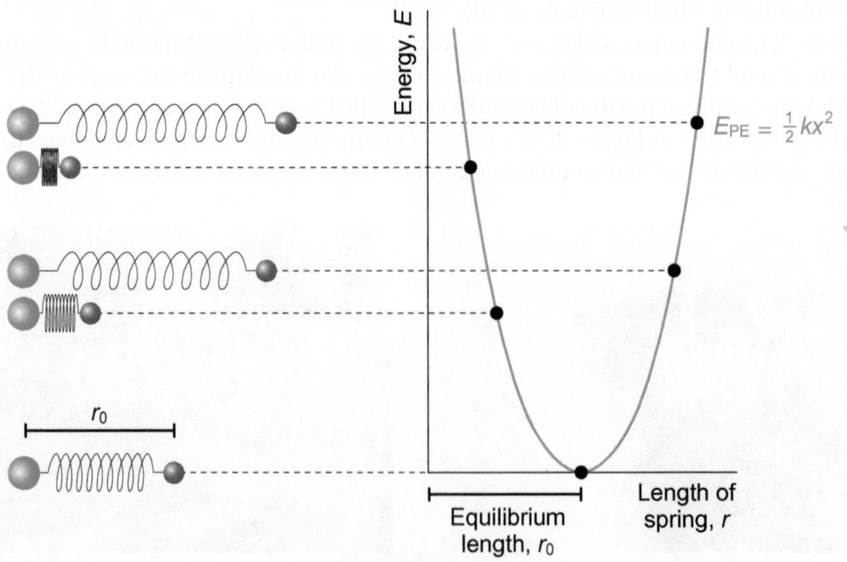

Energy, E
$E_{PE} = \frac{1}{2}kx^2$
r_0
Equilibrium length, r_0
Length of spring, r

Figure 10.23 The energy of a spring increases when it is stretched or compressed. The more the spring is stretched or compressed, the more energy is stored in it.

In molecules with more than two atoms, the term *vibration* describes two general types of movement of atoms within a molecule—the stretching and compressing of individual bonds and the changing of bond angles; see Figure 10.4 on p.452.

In the model for a harmonic oscillator, the potential energy well is a parabola, $E_{PE} = kx^2/2$, where $x = r - r_0$ (see Box 10.4).

The units of k are $kg\,s^{-2} = N\,m^{-1}$. See Worked example 10.7 on p.482.

The reduced mass is used since the frequency depends on the masses of both atoms. In a homonuclear molecule like Cl_2, m_1 and m_2 are equal so that μ simplifies to $\frac{1}{2}m_{Cl}$. In a molecule like HI, where m_I is much greater than m_H, μ is approximately equal to m_H (try substituting in the numbers). During bond vibration, the heavy I atom stays almost still with the light H atom moving much larger distances.

Box 10.4 The simple harmonic oscillator model for a diatomic molecule

For relatively small motions, springs obey *simple harmonic motion.* The restoring force, F, is proportional to the extension, x, as shown in Figure 10.22

$$F = -kx \qquad (10.19)$$

k is the force constant (sometimes called the spring constant). Equation 10.19 is known as Hooke's law. The higher the value of k, the stiffer the spring and the more force is needed to stretch the spring.

In general, the potential energy, k_{PE}, is related to an applied force by

$$E_{PE} = \text{force} \times \text{distance moved}$$

So

$$\text{force} = \frac{E_{PE}}{\text{distance moved}}$$

This equation applies where these quantities have constant values. In simple harmonic motion, the force changes with the distance moved, so that force is related to the potential energy by the differential equation

$$F = -\frac{dE_{PE}}{dx} \quad \text{(negative sign because } F \text{ here is a } \textit{restoring} \text{ force)}$$

(Differentials and differential equations are explained in Maths Toolkit MT6, p.1324.)

The energy in the spring vibration can be found by substituting the expression for F from Equation 10.19 into the equation above and integrating

$$F = -kx = \frac{dE_{PE}}{dx}, \text{ so that } \frac{dE_{PE}}{dx} = kx$$

Rearranging the equation to separate the variables

$$dE_{PE} = kx dx$$

Integrate both sides of the equation from $E_{PE} = 0$ at $x = 0$ to E_{PE} at x

$$\int_0^{E_{PE}} dE_{PE} = k\int_0^x x\,dx$$

To do this you need to use the standard integrals: $\int dE = E$ and $\int x\,dx = \frac{1}{2}x^2$. (Go to Maths Toolkit MT7, p.1327, if you need help with integration.) Then

$$[E_{PE}]_0^{E_{PE}} = \frac{1}{2}k[x^2]_0^x$$

$$E_{PE} = \frac{1}{2}kx^2$$

The potential energy of the spring therefore depends on the square of the extension, x, where $x = r - r_0$. (r is the distance between the atoms and r_0 is the equilibrium bond length.) A plot of E_{PE} against x has the shape of a parabola, as shown below.

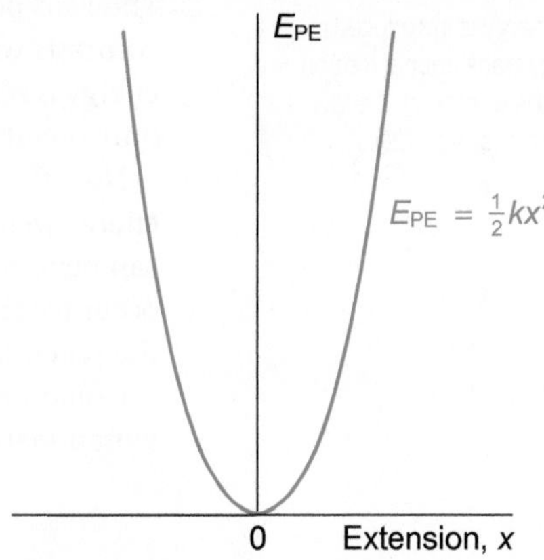
EPE
EPE = ½kx²
0
Extension, x

Quantization of vibrational energies

Vibrational energy is quantized. The wavefunctions and energy levels are found by solving the Schrödinger equation for motion of a particle with energy $E_{PE} = \frac{1}{2}kx^2$. The problem is similar to the one in Box 10.1 (p.454) except that now the potential energy is not constant inside the box but rises smoothly to infinity rather than rising abruptly at the ends of the box. The results are shown in Figure 10.24. The allowed vibrational energies are given by

$$E_{vibration} = E_v = (v + \tfrac{1}{2})\,h\nu \qquad (10.20)$$

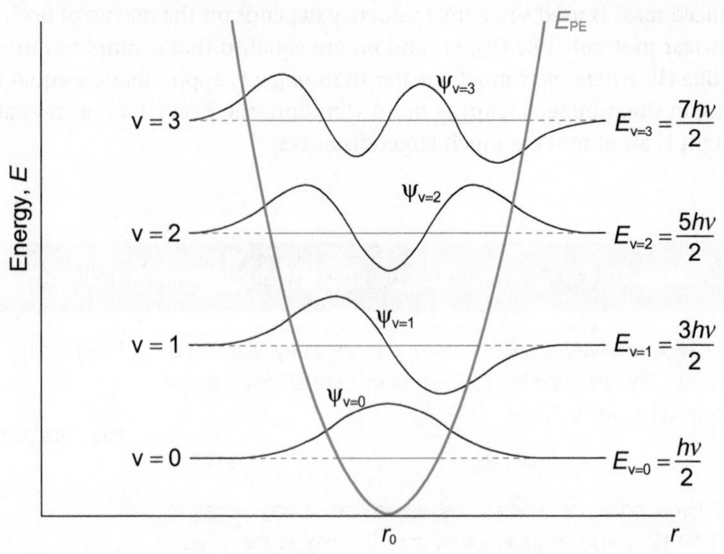

E_{PE}
$\psi_{v=3}$
v = 3
$E_{v=3} = \frac{7hv}{2}$
Energy, E
$\psi_{v=2}$
v = 2
$E_{v=2} = \frac{5hv}{2}$
$\psi_{v=1}$
v = 1
$E_{v=1} = \frac{3hv}{2}$
$\psi_{v=0}$
v = 0
$E_{v=0} = \frac{hv}{2}$
r_0
r

Figure 10.24 Vibrational energy levels and wavefunctions. Solution of the wave equation for a harmonic oscillator gives quantized energy levels.

ⓘ Be careful to distinguish between v ('vee'), the vibrational quantum number, and ν (Greek 'nu'), the frequency of the vibration.

where v is the **vibrational quantum number** and can take the values 0, 1, 2, . . . , ∞ and ν is the vibrational frequency of the bond. E_v is the energy of the vth level.

From Equation 10.20

$$\text{when v} = 0, \quad E_{v=0} = \frac{1}{2}h\nu$$

$$\text{when v} = 1, \quad E_{v=1} = \frac{3}{2}h\nu, \text{ and so on}$$

Zero point energy is introduced in Box 10.1 (p.454) for translational motion. For rotational motion, the energy of the lowest level, with $J = 0$, is zero (see p.469).

The zero point energy, the energy of the lowest level with v = 0, is equal to $\frac{1}{2}h\nu$. This contrasts with the treatment of vibration using classical mechanics where the zero point energy is zero. The vibrational levels for a diatomic molecule are non-degenerate—there is only one state at each allowed energy.

Note that the wavefunctions in Figure 10.24 have non-zero values outside the potential energy well. This is an example of **quantum mechanical tunnelling**, in which a particle can penetrate into a region that is forbidden by classical mechanics. Tunnelling does not occur for the particle in the one-dimensional box considered in Box 10.1, because there the potential energy rises instantly to infinity.

Combining Equation 10.20 with Equation 10.18 (p.478) gives an expression for the vibrational energy, E_v, in terms of the force constant and the reduced mass of the molecule

$$E_v = (\text{v} + \tfrac{1}{2})\frac{h}{2\pi}\left(\frac{k}{\mu}\right)^{1/2} \qquad (10.21)$$

Vibrational spectra of diatomic molecules

The previous section shows how to calculate the energies of the various vibrational energy levels, but to interpret the spectrum you also need to know which transitions can take place.

Selection rules

Gross and specific selection rules are introduced in Section 10.3 (p.458).

Dipoles and dipole moments are discussed in Sections 1.8 (p.52), 5.3 (p.235), and 17.3 (p.783).

The gross selection rule for infrared spectroscopy is that there must be a *change in the dipole moment* of the molecule during the vibration if the vibration is to give rise to a peak in the spectrum. For diatomic molecules this means that the molecule must have a permanent dipole moment; see Figure 10.25. Homonuclear diatomics, such as H_2, do not have dipole moments and do not give IR spectra.

As with rotational energy levels, transitions can only take place between adjacent vibrational levels. The specific selection rule for infrared spectroscopy is therefore $\Delta v = \pm 1$.

Selection rules for infrared spectroscopy:

- the molecular vibration must undergo a change in electric dipole moment;
- $\Delta v = \pm 1$ ($\Delta v = +1$ corresponds to absorption of radiation and $\Delta v = -1$ to emission of radiation).

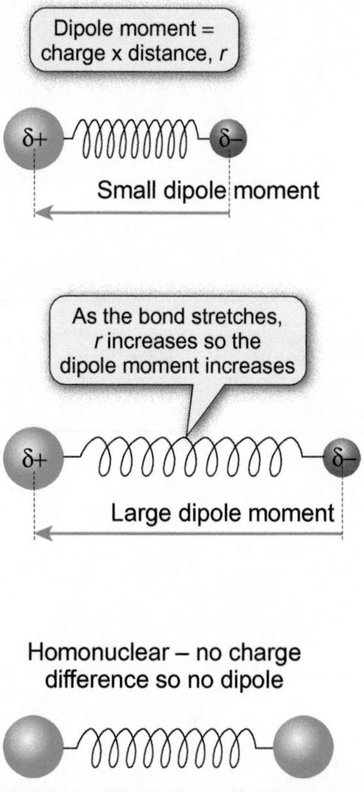
Dipole moment = charge x distance, r
δ+
δ–
Small dipole moment
As the bond stretches, r increases so the dipole moment increases
δ+
δ–
Large dipole moment
Homonuclear – no charge difference so no dipole

Figure 10.25 To be IR active, the molecule must undergo a change of dipole moment during vibration.

Populations of vibrational energy levels

The infrared spectrum for HCl shows that the energy difference, ΔE, between the energy levels v = 0 and v = 1 is 5.94×10^{-20} J. You can use the Boltzmann distribution (Equation 10.9, p.462) to find the relative populations of these two levels at 298 K

$$\frac{n_{\text{upper}}}{n_{\text{lower}}} = \frac{g_{\text{upper}}}{g_{\text{lower}}} e^{\left(\frac{-\Delta E}{K_B T}\right)} \tag{10.9}$$

$$\frac{n_{v=1}}{n_{v=0}} = \frac{1}{1} e^{\left(\frac{-\Delta E}{K_B T}\right)} = e^{\left(\frac{-(5.94 \times 10^{-20}\ \text{J})}{1.381 \times 10^{-23}\ \text{J K}^{-1}) \times 298\ \text{K}}\right)}$$

$$= 5.3 \times 10^{-7}$$

So, at room temperature, virtually all the HCl molecules are in the ground state—only around 1 in every 2 million are in higher states. This means that, in practice, the spectrum is due only to the v = 0 → v = 1 transition. This is generally true for diatomic molecules.

Transitions between vibrational energy levels

Equation 10.20 (p.479) allows an energy level diagram for vibrational energies to be constructed as in Figure 10.26

$$E_{\text{vibration}} = E_v = (v + \tfrac{1}{2})\, h\nu \tag{10.20}$$

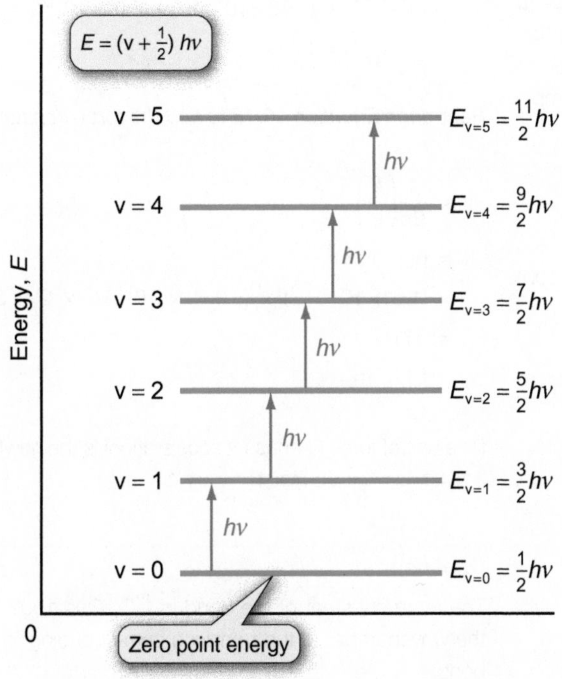
E = (v + 1/2) hv
v = 5
v = 4
v = 3
v = 2
v = 1
v = 0
hv
E_{v=5} = 11/2 hv
E_{v=4} = 9/2 hv
E_{v=3} = 7/2 hv
E_{v=2} = 5/2 hv
E_{v=1} = 3/2 hv
E_{v=0} = 1/2 hv
Energy, E
0
Zero point energy

Figure 10.26 Vibrational energy levels arising from the simple harmonic model for a diatomic molecule.

Vibrational spectra of homonuclear diatomics can be seen in a complementary technique called Raman spectroscopy (Section 10.5, p.488).

The vibrational energy levels for a diatomic molecule are **non-degenerate**—there is only one state at each allowed energy. So, $g = 1$ and the g terms cancel out in the Boltzmann distribution.

Compare this with the population distribution among rotational levels at 298 K (p.471) where the population is spread throughout many levels.

The absorption peaks in an IR spectrum arise from transitions between adjacent energy levels. For a transition v = 0 → v = 1

$$\Delta E_{v=0\to v=1} = E_{vib}(v=1) - E_{vib}(v=0)$$
$$= 1\tfrac{1}{2}h\nu - \tfrac{1}{2}h\nu = h\nu$$

Similarly, for a transition v = 2 → v = 3

$$\Delta E_{v=2\to v=3} = E_{vib}(v=3) - E_{vib}(v=2)$$
$$= 3\tfrac{1}{2}h\nu - 2\tfrac{1}{2}h\nu = h\nu$$

Remember this is for a diatomic molecule. Many more peaks are expected in the infrared spectra of polyatomic molecules.

There is the same energy gap, $h\nu$, between all adjacent vibrational energy levels, so you would expect the same peak in the spectrum irrespective of the level from which the transition originates. For most molecules at room temperature, however, only the v = 0 level is populated, so the observed transition is from v = 0 → v = 1.

Finding force constants from vibrational spectra

One of the major applications of IR spectroscopy is to determine force constants for bonds and this is illustrated in Worked example 10.7. The force constant is a measure of the stiffness of the bond. It is related to the strength of the bond and so depends on the electron density between the atoms. Its measurement can therefore give information about the nature of the bonds between atoms.

Worked example 10.7 Force constants from IR data

$^1H^{35}Cl$ has an absorption at 2990 cm^{-1}. Calculate the force constant, k, of the bond.

Strategy

Equation 10.18 (p.478) gives an expression for the frequency of the absorption in terms of the force constant, k, and the reduced mass, μ, of $^1H^{35}Cl$. You need to convert the wavenumber of the absorption to frequency units using Equation 10.10b on p.466. Calculate the reduced mass using Equation 10.13 (p.467) (see Worked example 10.3, p.467).

Solution

Calculate the frequency of the absorption.

$$\tilde{\nu} = \frac{\nu}{c} \quad (c = 2.998 \times 10^8\,\text{m s}^{-1}) \qquad (10.10b)$$

$$\nu = (2.998 \times 10^8\,\text{m s}^{-1}) \times (2.990 \times 10^5\,\text{m}^{-1}) \quad (1\,\text{cm}^{-1} = 100\,\text{m}^{-1})$$
$$= 8.964 \times 10^{13}\,\text{s}^{-1}$$

Calculate the masses of ^1H and ^{35}Cl.

$$A_r(^1H) = 1.00 \text{ and } A_r(^{35}Cl) = 35.00$$

The mass of each atom is given by

$$m_{^1H} = \frac{1.00 \times 10^{-3}\,\text{kg}\,\cancel{\text{mol}^{-1}}}{6.022 \times 10^{23}\,\cancel{\text{mol}^{-1}}} = 1.66 \times 10^{-27}\,\text{kg}$$

$$m_{^{35}Cl} = \frac{35.00 \times 10^{-3}\,\text{kg}\,\cancel{\text{mol}^{-1}}}{6.022 \times 10^{23}\,\cancel{\text{mol}^{-1}}} = 5.81 \times 10^{-26}\,\text{kg}$$

Use Equation 10.13 (p.467) to calculate the reduced mass.

$$\mu = \frac{m_1 m_2}{m_1 + m_2} \qquad (10.13)$$

$$\mu_{^1H^{35}Cl} = \frac{(1.66 \times 10^{-27}\,\text{kg}) \times (5.81 \times 10^{-26}\,\text{kg})}{(1.66 \times 10^{-27}\,\text{kg}) + (5.81 \times 10^{-26}\,\text{kg})}$$
$$= 1.61 \times 10^{-27}\,\text{kg}$$

Rearrange Equation 10.18 to find the force constant.

$$\nu = \frac{1}{2\pi}\left(\frac{k}{\mu}\right)^{1/2} \qquad (10.18)$$

$$k = \mu \times (\nu 2\pi)^2$$
$$= (1.61 \times 10^{-27}\,\text{kg}) \times (8.964 \times 10^{13}\,\text{s}^{-1} \times 2 \times 3.142)^2$$
$$= 511\,\text{kg s}^{-2}$$
$$= 511\,\text{N m}^{-1}$$

(The unit of force (= mass × acceleration) is the newton. 1 N = 1 kg m s^{-2} so that 1 kg s^{-2} = 1 N m^{-1}.)

Question

The force constant of the bond in $^{12}C^{16}O$ is 1902 N m^{-1}. Calculate the wavenumber of the transition corresponding to vibration of this bond.

Isotopic substitution

Since the energy of a vibration depends on the masses of the atoms, the energy will change if different isotopes are present in the molecule. This gives a powerful application of IR spectroscopy in looking at the structures of compounds.

Changing the mass of one of the atoms in the molecule will of course change the reduced mass, μ, but what about the force constant? The force constant is a property of the chemical bond and so is a function of the electron distribution in the molecule. Adding one or more neutrons to a nucleus has little effect on the distribution of electron density so it is a reasonable assumption that isotopic substitution will not change the force constant.

For example, the bond vibration of HCl gives an absorption band in the IR spectrum at 2990 cm^{-1}. Replacing H by deuterium, D (^2H), changes the reduced mass. Using Equation 10.13 (p.467)

$$\mu = \frac{m_1 m_2}{m_1 + m_2} \tag{10.13}$$

$$m_D = \frac{2.000 \times 10^{-3}\ \text{kg}\ \cancel{\text{mol}^{-1}}}{6.022 \times 10^{23}\ \cancel{\text{mol}^{-1}}} = 3.32 \times 10^{-27}\ \text{kg}$$

$$m_{^{35}Cl} = 5.81 \times 10^{-26}\ \text{kg (see Worked example 10.7)}$$

$$\mu_{D^{35}Cl} = \frac{(3.32 \times 10^{-27}\ \text{kg}) \times (5.81 \times 10^{-26}\ \text{kg})}{(3.32 \times 10^{-27}\ \text{kg}) + (5.81 \times 10^{-26}\ \text{kg})} = 3.14 \times 10^{-27}\ \text{kg}$$

Putting this into Equation 10.18 (and then converting v to $\tilde{v}$) shows that the D–Cl absorption occurs at a lower wavenumber, 2140 cm^{-1}, compared to 2990 cm^{-1} for $^1H^{35}Cl$.

The shift of the peak wavenumber due to isotopic substitution is largest for H and D but is significant for other elements and can be used to confirm the presence or otherwise of elements in a compound.

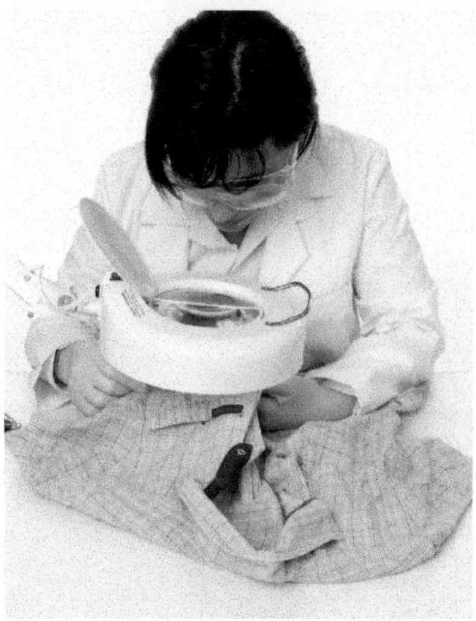

IR spectroscopy is a valuable tool in forensic science. It can be used to identify paints, fibres, explosives, and photocopy toner from a crime scene and match them to evidence collected from a suspect.

The effect of deuteration of a compound on its infrared spectrum is discussed further in Section 25.4 (p.1162).

Anharmonic vibrations

While the assumption that molecules act as harmonic oscillators and undergo simple harmonic motion is a good approximation, it is not exact. The variation of potential energy with bond length in a real molecule has the form shown in Figure 10.27. This is known as a **Morse curve**. It does not correspond exactly to the parabola for the harmonic oscillator, although for low vibrational quantum numbers the agreement is very close.

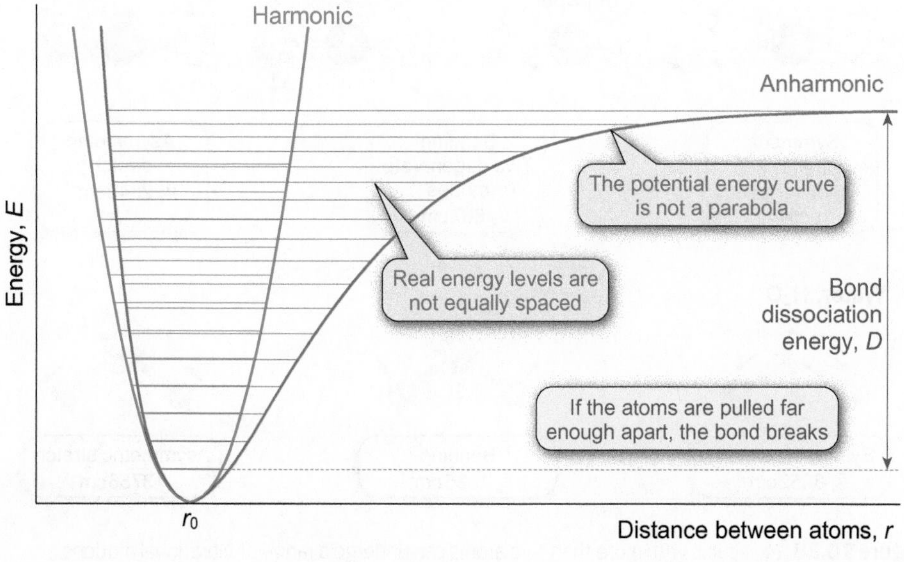
Harmonic
Anharmonic
The potential energy curve is not a parabola
Real energy levels are not equally spaced
Bond dissociation energy, D
If the atoms are pulled far enough apart, the bond breaks
Energy, E
r0
Distance between atoms, r

Figure 10.27 Real chemical bonds do not undergo simple harmonic motion. This means that the potential energy is represented by the curve in green rather than the parabola (shown in red).

At 298 K, the relative populations of the energy levels show that the spectrum mainly features the $v = 0 \rightarrow v = 1$ transition, so the assumption of simple harmonic motion is reasonable. However, it is important to realize that it is an approximation and does not work well for higher states.

As the bond in a real molecule is stretched, the bond becomes weaker until eventually it breaks. To describe this, Equation 10.20 (p.479) is modified by making an **anharmonicity** correction

$$E_v = (v + \tfrac{1}{2})h\nu - x_e(v + \tfrac{1}{2})^2 h\nu \qquad (10.22)$$

where x_e is the anharmonicity constant for a particular molecule. The effect of anharmonicity is that the spacing between energy levels gets slightly smaller as v increases.

The selection rule $\Delta v = \pm 1$ strictly works only for harmonic motion. In a real system, it is possible to have $v = 0 \rightarrow v = 2$ transitions and $v = 0 \rightarrow v = 3$ transitions and these are seen in the spectrum as so-called overtone bands at roughly twice and three times the fundamental frequency (ν in Equation 10.22). However, overtone bands are usually much weaker than the fundamental absorption ($v = 0 \rightarrow v = 1$).

Polyatomic molecules

In a diatomic molecule, the only possible vibration is the bond stretch/compression. As more atoms are introduced, several other types of vibration become possible, as illustrated for CO_2 and H_2O in Figure 10.28. The bond angles can change due to bending, and symmetric and asymmetric stretches are possible. Each of the possible vibrations is known as a **vibrational mode**.

The shapes of molecules are discussed in Section 5.2 (p.223).

Try looking at molecular models of CO_2 and H_2O to see what these vibrational modes look like in three dimensions.

To satisfy the gross selection rule for infrared spectroscopy, the molecular vibration must undergo a change in dipole moment. CO_2 is a linear, non-polar molecule with a zero overall dipole moment since the dipoles of each bond cancel out. The dipoles also cancel during a symmetric stretch ($\tilde{\nu}_1$, in Figure 10.28) so that this vibration is not infrared active. However, the asymmetric stretch ($\tilde{\nu}_3$) and bending modes ($\tilde{\nu}_2$) introduce a dipole that changes during the vibration so they satisfy the selection rule for IR spectroscopy. Figure 10.28 also shows the three possible vibrational modes for water.

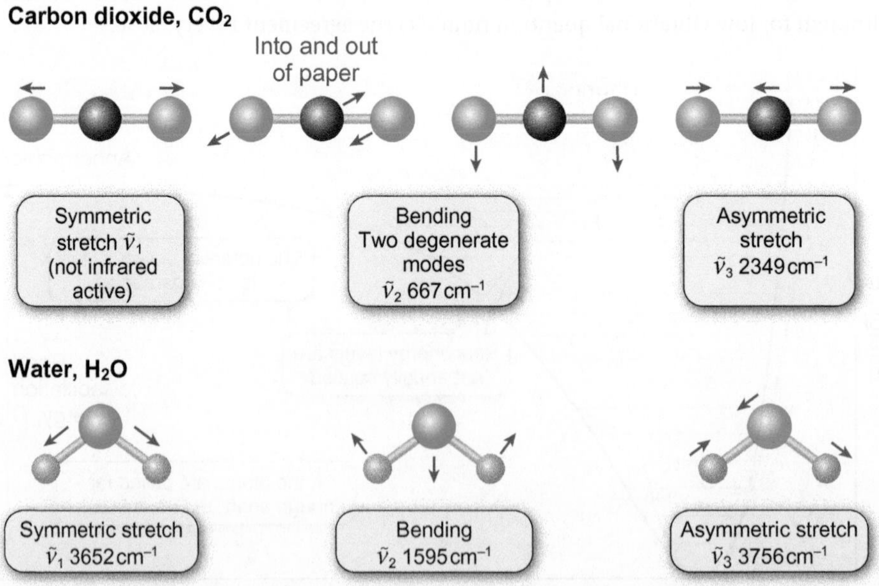
Carbon dioxide, CO₂
Into and out of paper
Symmetric stretch $\tilde{\nu}_1$ (not infrared active)
Bending Two degenerate modes $\tilde{\nu}_2$ 667 cm^{-1}
Asymmetric stretch $\tilde{\nu}_3$ 2349 cm^{-1}
Water, H_2O
Symmetric stretch $\tilde{\nu}_1$ 3652 cm^{-1}
Bending $\tilde{\nu}_2$ 1595 cm^{-1}
Asymmetric stretch $\tilde{\nu}_3$ 3756 cm^{-1}

Figure 10.28 Molecules with more than two atoms can undergo a range of vibrational motions, illustrated here for CO_2 and H_2O. For CO_2, the two bending modes are degenerate (i.e. have the same energy) and give rise to absorption at the same wavenumber.

Box 10.5 describes the use of IR spectroscopy in monitoring the concentration of carbon dioxide in the atmosphere. The absorption band used for the measurement is due to the asymmetric stretching vibration.

Box 10.5 Atmospheric concentrations of carbon dioxide

The rising concentration of carbon dioxide in the atmosphere is causing concern because of its implications for global warming and climate change (see Box 27.2, p.1218). One application of IR spectroscopy is in monitoring the concentration of gases in the atmosphere.

Measurements of CO_2 concentrations have been recorded for almost 50 years at the Mauna Loa Observatory (see Box 1.5, p.34), forming the longest continuous record of precise atmospheric CO_2 concentrations. The graph in Figure 1 has become an icon for the inexorable rise of atmospheric CO_2 and shows a 23.5% increase in the mean annual concentration between 1959 and 2010. In 2015, the CO_2 concentration reached 400 parts per million.

The observatory is located at an altitude of almost 4000 m on volcanic rock in Hawaii and has one of the largest laboratories in the world for atmospheric analysis. Its isolated position makes it ideal for this type of work since there is minimal effect from local vegetation or human activity, so the air contains very little pollution. The values give a good indication of the concentrations of the varying components in the atmosphere and how they are changing.

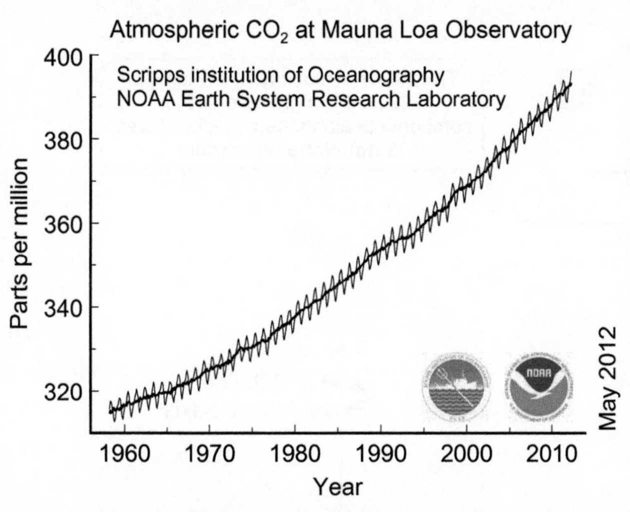
Atmospheric CO_2 at Mauna Loa Observatory
Scripps institution of Oceanography
NOAA Earth System Research Laboratory
Parts per million
400
380
360
340
320
1960
1970
1980
1990
2000
2010
Year
May 2012

▲ **Figure 1** Concentrations of CO_2 in the atmosphere measured using infrared spectroscopy at Mauna Loa in Hawaii.

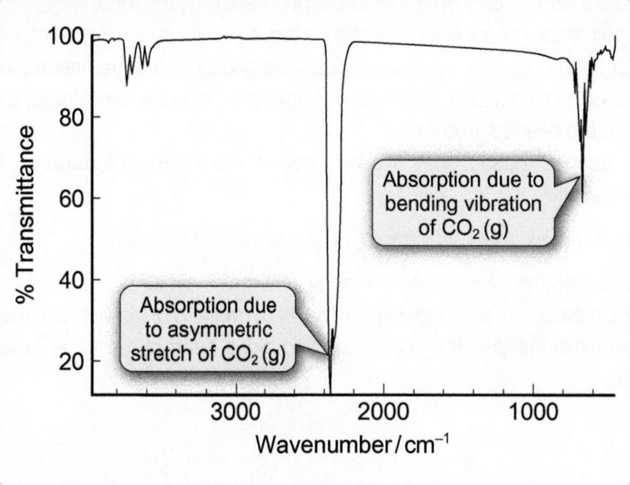
100
80
60
40
20
% Transmittance
Absorption due to bending vibration of CO_2 (g)
Absorption due to asymmetric stretch of CO_2 (g)
3000
2000
1000
Wavenumber / cm^{-1}

▲ **Figure 2** Infrared spectrum of CO_2. The intensity of the absorption is proportional to the concentration of CO_2 in the atmosphere.

Air samples at Mauna Loa are collected each hour and the CO_2 concentrations are measured using an infrared gas analyser. The signals from a stream of air flowing at ~0.5 $dm^3\,min^{-1}$ are compared with a stream of calibrating gas containing CO_2 with known concentration. The size of the main absorption band is proportional to the concentration. The main absorption band in the CO_2 spectrum (Figure 2) occurs at around 2350 cm^{-1}, corresponding to the asymmetric stretching of the CO_2 bonds.

Questions

(a) Suggest explanations for:

(i) the steady rise in CO_2 concentration from 1959 to 2010;

(ii) the zigzag nature of the plot.

(b) Why is the concentration of CO_2 increasing at Mauna Loa which has little traffic or local industry?

As more atoms are introduced into a molecule, more modes become possible. The total number of vibrational modes can be found as outlined in Box 10.6. For a molecule containing N atoms the number of vibrational modes is

- linear molecule: $(3N - 5)$ vibrational modes;
- non-linear molecule: $(3N - 6)$ vibrational modes.

Each vibrational mode has its own potential energy curve and series of vibrational energy levels.

Count the number of atoms in aspirin (look at Figure 10.1, p.450) and calculate how many normal modes of vibration it has. Not all of these will be IR active—but it gives you an idea of how complicated a full analysis of an IR spectrum can be.

Box 10.6 Degrees of freedom and normal modes of vibration

In order to describe the position of a single atom, three coordinates are needed. These may be the *x*, *y*, and *z* distances along each axis. Alternatively, polar coordinates in terms of the distance of the atom from the origin and two angles with axes could be used (see Maths Toolkit MT5, p.1321)—but it still requires three pieces of information for each atom. For a diatomic molecule, six pieces of information (e.g. the *x*, *y*, and *z* coordinates of each atom) are needed to describe its position. For triatomic molecules, nine pieces of information are needed, and so on.

Each of these pieces of information is referred to as a **degree of freedom**.

- A molecule with N atoms has $3N$ degrees of freedom.

The degrees of freedom correlate with movements (modes) within the molecule. For a single atom, the three degrees of freedom are the **translational modes**, that is, motion along each of the three axes as in Figure 1. A diatomic molecule (six degrees of freedom) also has three translational modes that describe the motion of the molecule as a whole through space. In addition, as shown in Figure 1, the molecule can rotate around two axes and so has two **rotational modes**. This accounts for five degrees of freedom and the final degree of freedom is due to the bond vibration. For a diatomic molecule, there is one **vibrational mode**.

- Diatomic molecule: 3 translational + 2 rotational + 1 vibrational = 6 degrees of freedom

All molecules have three translational modes. Molecules with more than two atoms fall into two classes when considering their rotations. Figure 1 shows that a linear molecule has two rotational modes while a non-linear molecule has three. For a triatomic molecule, there are nine degrees of freedom in total: the remainder are taken up by vibrational modes.

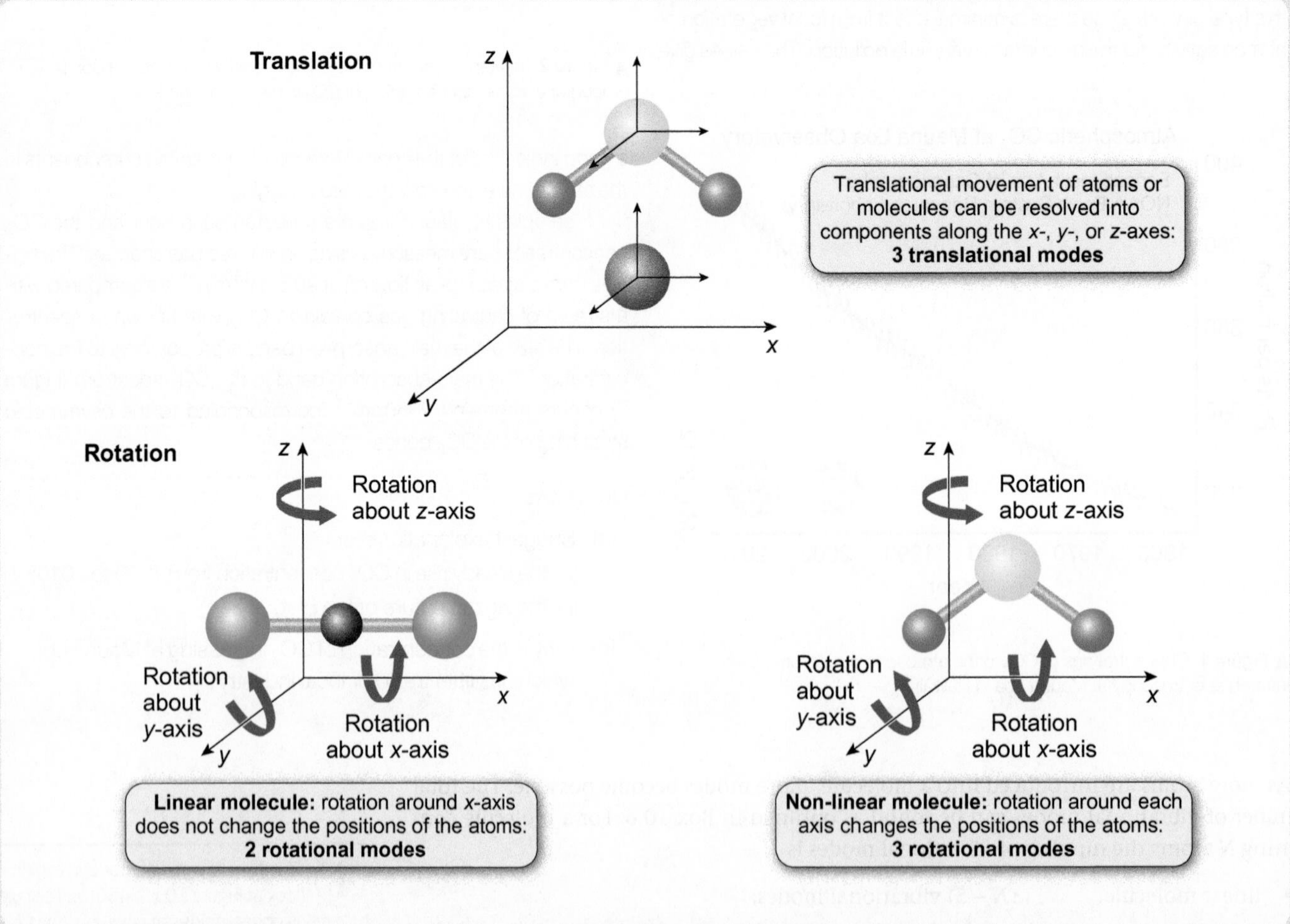
Translation
z
x
y
Translational movement of atoms or molecules can be resolved into components along the x-, y-, or z-axes:
3 translational modes
Rotation
z
Rotation about z-axis
Rotation about y-axis
Rotation about x-axis
x
y
Linear molecule: rotation around x-axis does not change the positions of the atoms:
2 rotational modes
z
Rotation about z-axis
Rotation about y-axis
Rotation about x-axis
x
y
Non-linear molecule: rotation around each axis changes the positions of the atoms:
3 rotational modes

Figure 1 Translational and rotational modes.

→

→

- Linear triatomic molecule:
  3 translational + 2 rotational + 4 vibrational
  = 9 degrees of freedom.
- Non-linear triatomic molecule:
  3 translational + 3 rotational + 3 vibrational
  = 9 degrees of freedom.

The vibrational modes for CO_2 (linear) and H_2O (non-linear) are shown in Figure 2, which summarizes the possibilities for monatomic, diatomic, and triatomic molecules.

In general for a molecule containing N atoms:

- linear molecule: 3 translational + 2 rotational + $(3N - 5)$ vibrational
  $= 3N$ degrees of freedom;
- non-linear molecule:
  3 translational + 3 rotational + $(3N - 6)$ vibrational
  $= 3N$ degrees of freedom.

Question

For a molecule of methane (CH_4) what are: (a) the total number of degrees of freedom; (b) the number of vibrational modes?

Type of molecule	N	Degrees of freedom	Modes
Monatomic, e.g. Ne 	1	3	3 translational 0 rotational 0 vibrational
Diatomic, e.g. HCl 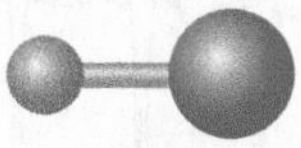	2	6	3 translational 2 rotational 1 vibrational
Triatomic linear, e.g. CO_2 	3	9	3 translational 2 rotational 4 vibrational
Triatomic non-linear, e.g. H_2O 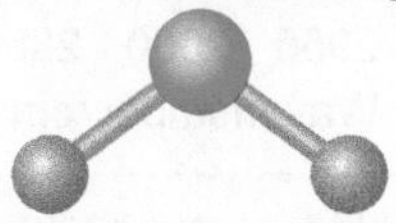	3	9	3 translational 3 rotational 3 vibrational

▲ **Figure 2** The number of degrees of freedom in monatomic, diatomic, and triatomic molecules.

Vibration-rotation spectra

Figure 10.29 shows the IR spectrum of HCl gas recorded at low and high resolution. At high resolution, you can see that the two peaks in the low resolution spectrum are in fact each made up of a series of closely spaced peaks. This is called the **rotational fine structure** of the vibrational spectrum.

Molecules can vibrate and rotate simultaneously and each vibrational energy level has a number of rotational energy levels associated with it. Transitions between rotational levels also occur when the vibrational modes are excited. Each of the equally spaced lines seen in the high-resolution spectrum of HCl arises from a transition between a rotational energy level in the lower vibrational state to one in the upper vibrational state. (Each of the lines is accompanied by a less intense line corresponding to $H^{37}Cl$. ^{35}Cl and ^{37}Cl occur in a ratio of 3 : 1.) Analysis of the spectrum is more complicated but can give the same information—bond lengths, moments of inertia, and force constants—as the two separate spectroscopic methods.

ⓘ As with pure rotational spectra, the rotational transitions are only seen in gas phase spectra (see Section 10.4, p.466).

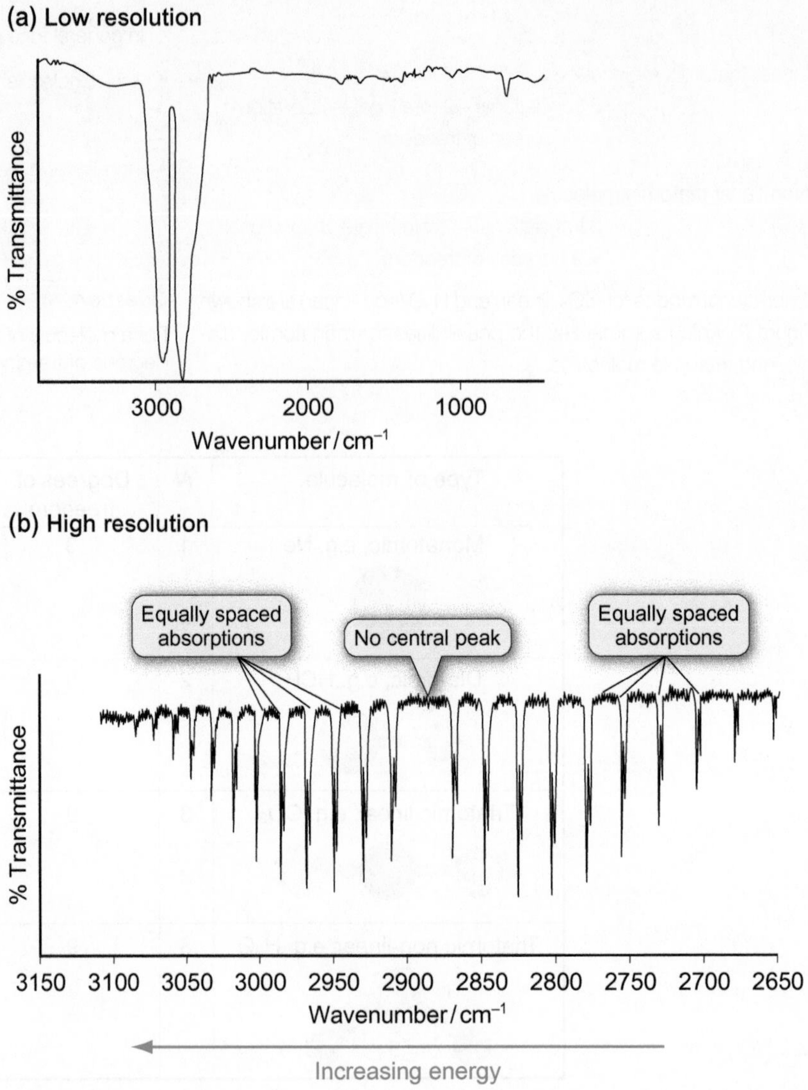
(a) Low resolution
% Transmittance
3000
2000
1000
Wavenumber / cm⁻¹
(b) High resolution
Equally spaced absorptions
No central peak
Equally spaced absorptions
% Transmittance
3150 3100 3050 3000 2950 2900 2850 2800 2750 2700 2650
Wavenumber / cm⁻¹
Increasing energy

Figure 10.29 The IR spectrum of HCl (g) recorded at (a) low resolution and (b) high resolution. The fine structure seen in the high resolution spectrum is due to transitions between rotational levels in the lower and upper vibrational states.

Raman spectroscopy

Raman spectroscopy involves *scattered*, rather than absorbed or emitted, radiation. Scattering occurs when incident radiation interacts with molecules whose size is much smaller than the wavelength of the radiation. Normally this results in *Rayleigh scattering*, in which the scattered radiation has the same frequency as the incident radiation.

> ℹ Rayleigh scattering is the reason the sky is blue: it happens that the blue light in the Sun's spectrum is scattered more than red. When you look up at the sky you see light that has been scattered by air molecules and water vapour, and more of this is at the blue end of the spectrum than the red end.

However, a small part of the scattered radiation has a frequency that is either slightly higher, or slightly lower, than that of the incident radiation (Figure 10.30). The radiation with slightly lower frequency is called **Stokes Raman radiation**, and that with slightly higher frequency is called **anti-Stokes Raman radiation**. These frequency differences are the basis of Raman spectroscopy, which is named after the Indian physicist Sir Chandrasekhara Raman who first discovered the effect.

The reason for the frequency differences is summarized in Figure 10.31. When incident radiation collides with a molecule in the sample, there is an instantaneous interaction involving transfer of energy. You can think of it as the molecule being excited to a 'virtual'

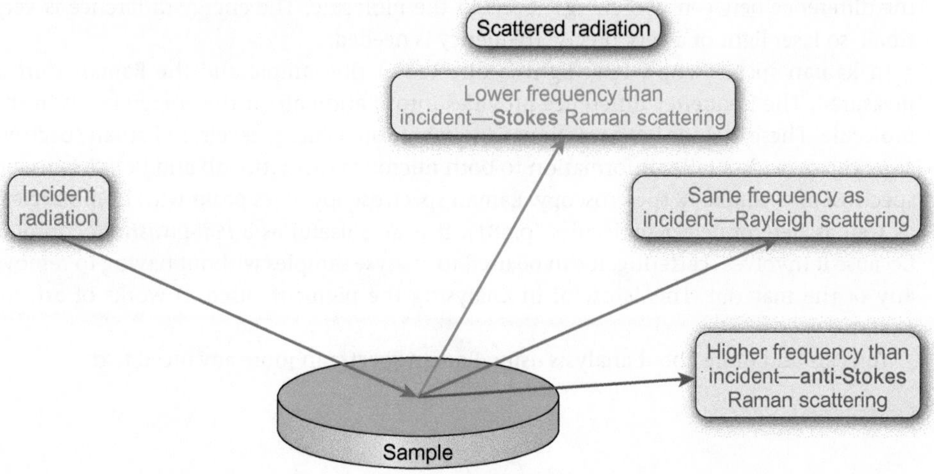
Scattered radiation
Lower frequency than incident—Stokes Raman scattering
Incident radiation
Same frequency as incident—Rayleigh scattering
Higher frequency than incident—anti-Stokes Raman scattering
Sample

Figure 10.30 Rayleigh scattering (no frequency change), Stokes Raman scattering (lower frequency than incident radiation), and anti-Stokes Raman scattering (higher frequency than incident radiation). The scattering takes place in all directions.

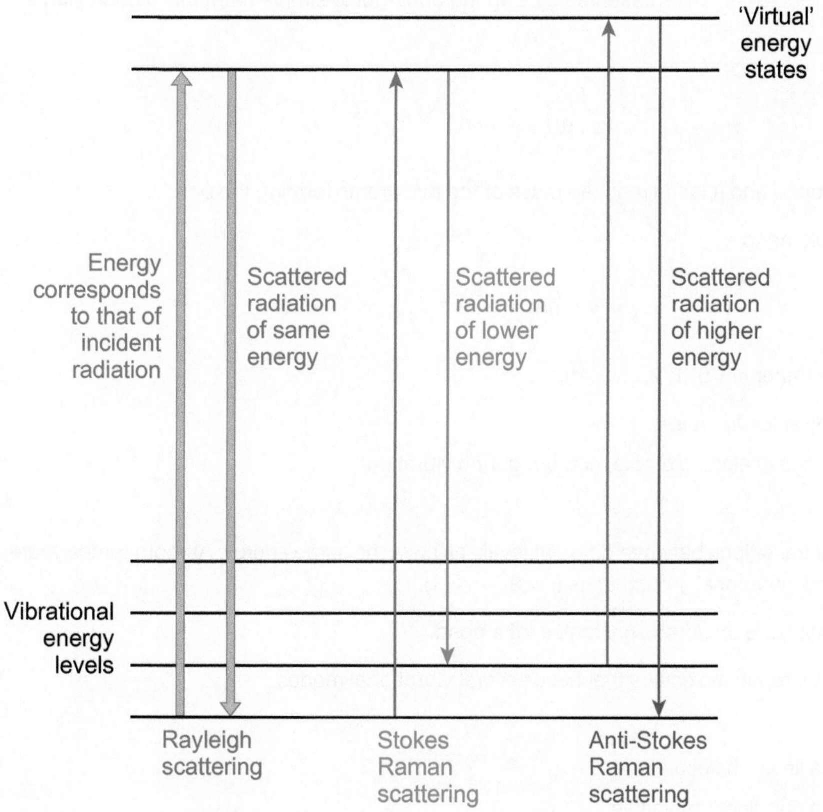
'Virtual' energy states
Energy corresponds to that of incident radiation
Scattered radiation of same energy
Scattered radiation of lower energy
Scattered radiation of higher energy
Vibrational energy levels
Rayleigh scattering
Stokes Raman scattering
Anti-Stokes Raman scattering

Figure 10.31 The energy transitions involved in Raman spectroscopy.

An illustration from the Book of Kells in Trinity College, Dublin, which dates from the year 800. Raman spectroscopy has been used to find out which pigments the illustrator used. Because it only involves scattering of light, Raman spectroscopy does not damage the priceless book.

energy state before returning to a lower energy state. Most molecules return to their original state and the scattered radiation has exactly the same energy as the incident radiation: this is Rayleigh scattering. But a few molecules return to a state that is either of a higher or a lower energy than the one they started from. The energy difference results in a displacement, or Raman shift, of the scattered radiation to either a higher or a lower energy (and therefore frequency) than the incident radiation. This difference in energy corresponds to

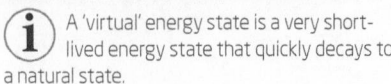

A 'virtual' energy state is a very short-lived energy state that quickly decays to a natural state.

the difference between two energy levels in the molecule. The energy difference is very small, so laser light of a very precise frequency is needed.

In Raman spectroscopy, laser light is directed at the sample and the Raman shift is measured. The frequency difference provides information about the energy levels in the molecule. These include both rotational and vibrational energy levels, so Raman spectroscopy can provide similar information to both microwave (rotational) and IR (vibrational) spectroscopy. Unlike IR spectroscopy, Raman spectroscopy gives peaks with homonuclear as well as heteronuclear molecules (p.481). It is also useful as a *non-invasive* technique: because it involves scattering, it can be used to analyse samples without having to remove any of the material. This is useful in analysing the pigments used in works of art, for example.

You can find more about analysis using Raman spectra in more advanced texts.

Summary

- Infrared radiation is absorbed during transitions between vibrational energy levels.
- The chemical bond in a diatomic molecule can be represented by a spring undergoing simple harmonic motion with a frequency given by

$$\nu = \frac{h}{2\pi}\left(\frac{k}{\mu}\right)^{1/2}$$

  where k is the force constant of the bond and μ is the reduced mass of the two atoms forming the bond.
- Vibrational energy is quantized according to

$$E_v = (v + \tfrac{1}{2})h\nu$$

  where v is the vibrational quantum number ($v = 0, 1, 2, \ldots, \infty$).
- Infrared spectroscopy is governed by selection rules:
  - the molecule must undergo a change in electric dipole moment during vibration;
  - $\Delta v = \pm 1$.
- For a harmonic oscillator, vibrational transitions between different levels all have the same energy. At room temperature, the predominant transition is from the vibrational ground state ($v = 0 \rightarrow v = 1$).
- Vibrational spectra can be used to measure the force constant, k, of a bond.
- Similar models can be applied to polyatomic molecules that have several vibrational modes.
- A molecule containing N atoms has:
  - $(3N - 5)$ vibrational modes if it is a linear molecule;
  - $(3N - 6)$ vibrational modes if it is a non-linear molecule.
- Transitions occur simultaneously between vibrational and rotational states. This results in vibration–rotation spectra.
- Raman spectroscopy uses the Raman effect, in which light scattered by the sample has its frequency shifted to a higher or lower frequency. The energy difference between the incident and scattered radiation corresponds to the difference in energy between vibrational and rotational levels in the molecules.

? For practice questions on these topics, see questions 18–24 at the end of this chapter (p.511).

10.6 Electronic spectroscopy

The energies associated with transitions between electronic states in molecules correspond to the ultraviolet–visible (UV/VIS) region of the electromagnetic spectrum. Absorptions in the visible region (Figure 10.32) are associated with colour in compounds as described in Box 10.7. Spectra are measured using the general methods shown in Figures 10.8 (p.459) and 10.9 (p.461). The detector measures the intensity of a light beam that passes through a sample. The sample solution is usually contained in a plastic or glass cell for work at visible wavelengths. However, these materials are not transparent to UV, so that quartz cells must be used for UV wavelengths below 320 nm.

UV/VIS spectra (see Figures 1 and 2 in Box 10.7) are displayed in terms of wavelength, λ, usually measured in nanometres, nm ($1\ \text{nm} = 1 \times 10^{-9}\ \text{m}$). This is different from rotational and vibrational spectra, which are displayed in terms of *wavenumber*. Wavelength is *inversely* proportional to frequency and energy, so a higher value of λ corresponds to lower v and lower energy. Absorbance is plotted on the *y*-axis (see Section 10.3, p.458).

Colour is the result of electronic transitions in which radiation is absorbed in the visible region ($\lambda = 380\ \text{nm}–750\ \text{nm}$).

In some modern instruments, the sample is irradiated with 'white light'. The radiation emerging from the sample is then split into its component wavelengths and detected by a diode array—several hundred photodiodes, each placed to receive a different wavelength range. These **diode array spectrophotometers** can record a complete spectrum in around 1 second.

Spectrometers that use UV/VIS radiation are often called **spectrophotometers**.

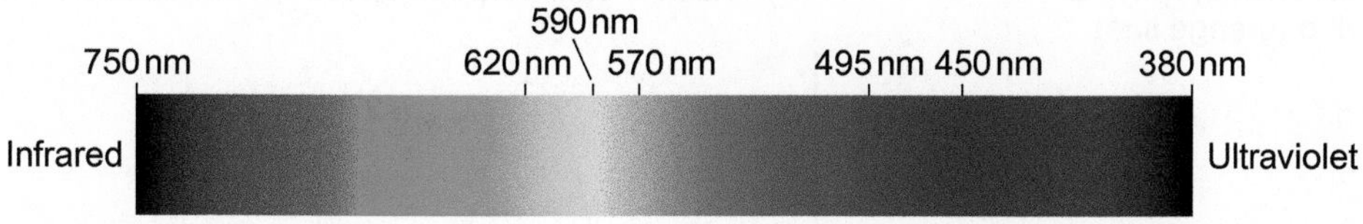
590 nm
750 nm
620 nm
570 nm
495 nm
450 nm
380 nm
Infrared
Ultraviolet

Figure 10.32 The wavelengths of electromagnetic radiation corresponding to visible light ($1\ \text{nm} = 1 \times 10^{-9}\ \text{m}$).

Box 10.7 Electronic spectroscopy and colour

Practically all coloured materials, from paints and dyes to flowers and fruits, owe their colour to electronic transitions that occur due to the absorption of visible light.

Different materials owe their colour to different kinds of electronic transitions. The main types of transitions are:

- between *d* orbitals in transition metal ions, for example, pale blue Cu^{2+} (aq) (see Section 28.6, p.1289);
- between an orbital in a ligand and an orbital in a metal ion, for example, purple $[MnO_4]^-$ (aq) (see Section 28.6, p.1289 and below);
- between π orbitals in conjugated organic compounds, for example, azo dyes (see Chapter 22, p.1002).

In each case, radiation with a specific energy is absorbed, creating an electronically excited state. Absorption of some of the radiation from incident white light makes the reflected or transmitted light appear coloured.

Potassium permanganate ($KMnO_4$) has a strong purple colour due to the presence of $[MnO_4]^-$ (aq) ions. The UV/VIS spectrum is shown in Figure 1. The maximum absorption, λ_{max}, occurs around 520 nm and the absorption band corresponds to green and yellow light (see Figure 10.32). The solution, therefore, absorbs green and yellow light but transmits red/orange and blue/violet. The result is that the solution appears purple (red + blue = purple). Some students say $KMnO_4$ is purple because it *emits* red and blue light. This is not correct; it appears purple because it does not absorb these colours but allows them to pass through.

The strength of absorbance is indicated by the molar absorption coefficient, ε (see Section 10.3, p.458). At 520 nm, $KMnO_4$ in water has a molar absorption coefficient $\varepsilon = 200\ \text{m}^2\,\text{V}\,\text{mol}^{-1}$, indicating →

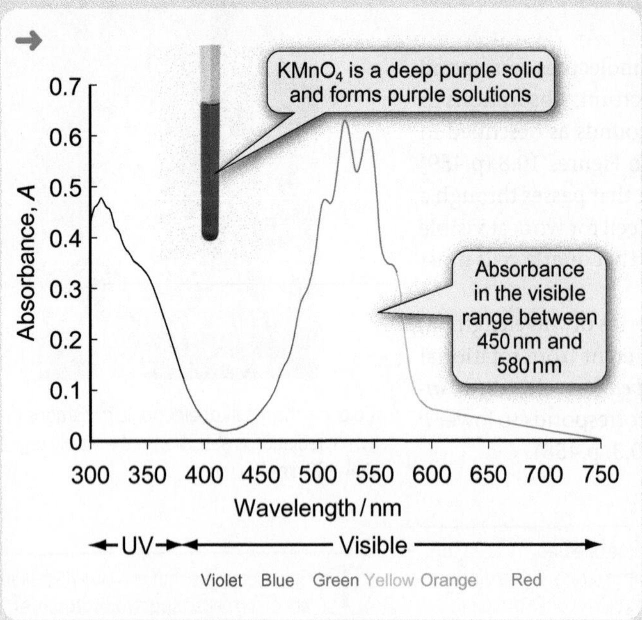

KMnO$_4$ is a deep purple solid and forms purple solutions
Absorbance in the visible range between 450 nm and 580 nm
0.7
0.6
0.5
0.4
0.3
0.2
0.1
0
Absorbance, A
300 350 400 450 500 550 600 650 700 750
Wavelength / nm
UV
Visible
Violet Blue Green Yellow Orange Red

▲ **Figure 1** Potassium permanganate ($KMnO_4$) is purple because it absorbs green and yellow radiation in the visible region.

that it absorbs strongly. So, only a small concentration is needed for a solution of $KMnO_4$ to appear highly coloured.

Questions

Figure 2 shows the visible spectra of two types of chlorophyll and the ultraviolet spectrum of butanone.

(a) Explain why grass is green but butanone is colourless.

(b) Calculate the energies corresponding to the absorption maxima in: (i) $KMnO_4$, λ_{max} = 520 nm; (ii) chlorophyll b, λ_{max} = 456 nm; (iii) butanone, λ_{max} = 273 nm.

(c) Compare the absorption spectrum of chlorophyll with the spectrum of sunlight reaching the surface of the Earth in Box 3.1 (p.120). Can you make any connections?

▼ **Figure 2** The absorption spectra of chlorophyll and butanone.

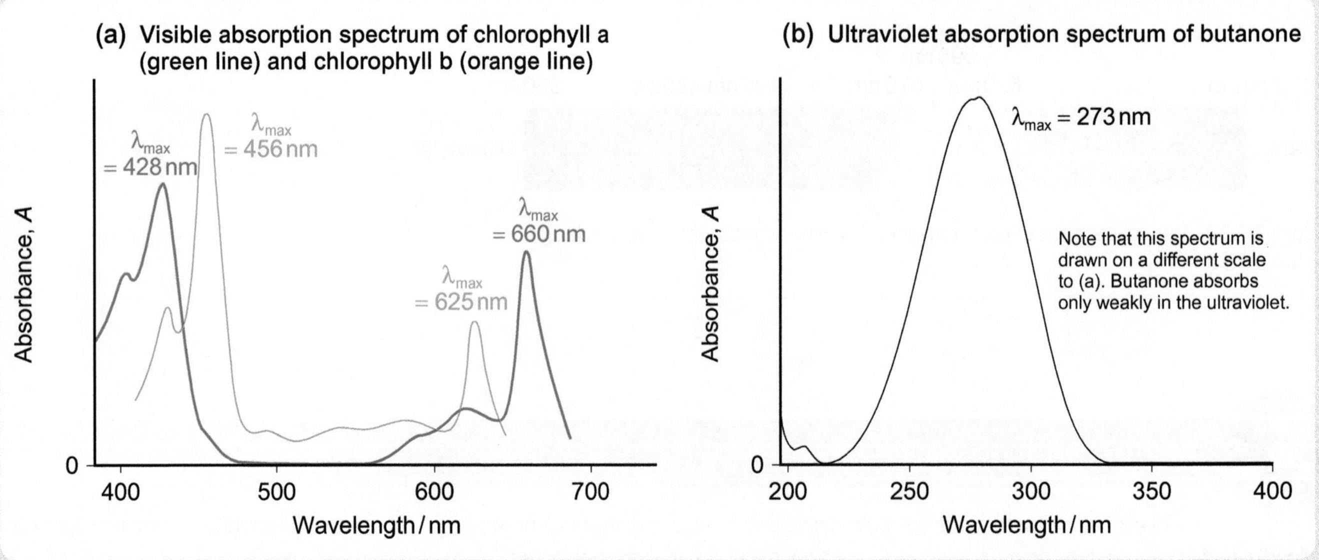

(a) Visible absorption spectrum of chlorophyll a (green line) and chlorophyll b (orange line)
λ_{max} = 428 nm
λ_{max} = 456 nm
λ_{max} = 660 nm
λ_{max} = 625 nm
Absorbance, A
0
400 500 600 700
Wavelength / nm
(b) Ultraviolet absorption spectrum of butanone
λ_{max} = 273 nm
Note that this spectrum is drawn on a different scale to (a). Butanone absorbs only weakly in the ultraviolet.
Absorbance, A
0
200 250 300 350 400
Wavelength / nm

Origin of electronic spectra

The spectra arising from electronic transitions in atoms are described in Sections 3.3 (p.124) and 11.5 (p.547).

Molecular orbitals are discussed in Chapter 4 (Sections 4.6–4.12, from p.184) and Section 5.6 (p.246).

Since both atoms and molecules contain electrons, both show electronic absorption spectra. In molecules, the transitions involve the promotion of an electron from a lower energy occupied molecular orbital to a higher energy unoccupied, or partially occupied, molecular orbital.

Each electronic state has its own set of vibrational energy levels, as shown in Figure 10.33. Promotion of an electron to a higher energy molecular orbital often changes the geometry of a molecule. For example, promotion of an electron in N_2 changes the bond order (see Figure 4.26, p.202) so that, in the excited state, the N–N bond is longer than in the ground state. In Figure 10.33, the Morse curves for both the ground and excited electronic states of a molecule are represented on the same diagram. The curves show how the

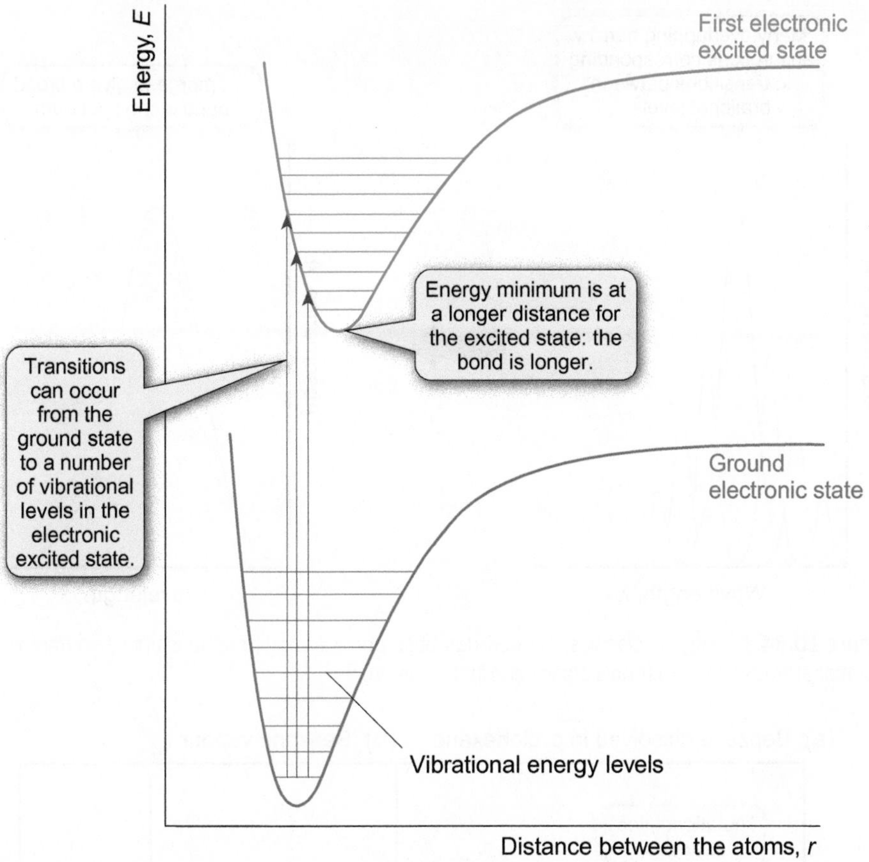
Energy, E
First electronic excited state
Energy minimum is at a longer distance for the excited state: the bond is longer.
Transitions can occur from the ground state to a number of vibrational levels in the electronic excited state.
Ground electronic state
Vibrational energy levels
Distance between the atoms, r

Figure 10.33 Morse curves for the ground and first excited electronic state of a diatomic molecule. The excited state has a longer bond length than the ground state, so absorption of a photon leads to a change in geometry of the molecule.

potential energy in each electronic state changes with the distance between the atoms. The vibrational energy levels in each electronic state are shown as horizontal lines. The spacing of the vibrational levels in the upper electronic state is smaller than that in the ground state because the bond is weaker.

Morse curves are introduced in Section 10.5 (p.476).

Applying the Boltzmann distribution shows that practically all molecules are in their electronic ground states around room temperature (see Worked example 10.2, p.463) and the majority are in their vibrational ground state. As shown in Figure 10.33, transitions can occur into one of a number of vibrational levels of the electronically excited state. This means that there will be a number of absorptions in the spectrum. If these absorptions are closely spaced, they merge to form a broad band in the spectrum, sometimes extending over a wide range of energies as in Figure 10.34 (on the next page).

In benzene (Figure 10.35a), the vibrational levels are sufficiently widely spaced that the overlap shown in Figure 10.34 does not occur. Instead, separate vibrational bands can be seen, representing transitions from the vibrational ground state to a series of vibrational levels in the excited electronic state. Additional bands can also be seen in the gas phase spectrum of benzene (Figure 10.35(b)) if higher resolution spectroscopy is used. Most of these bands arise from transitions from vibrationally excited levels in the ground electronic state and are termed hot bands

Electronic transitions are governed by the Franck–Condon principle which states that:

The excitation of an electron by the absorption of a photon occurs on a much shorter timescale than that for nuclear motions.

This arises because nuclei have much greater masses than electrons and so move more slowly. Electronic excitation happens on a timescale of 10^{-14} s to 10^{-15} s, compared with 10^{-10} s to 10^{-12} s for the movement of nuclei. The photon is therefore absorbed to form the

Bond dissociation energies are discussed in Section 4.1 (p.170).

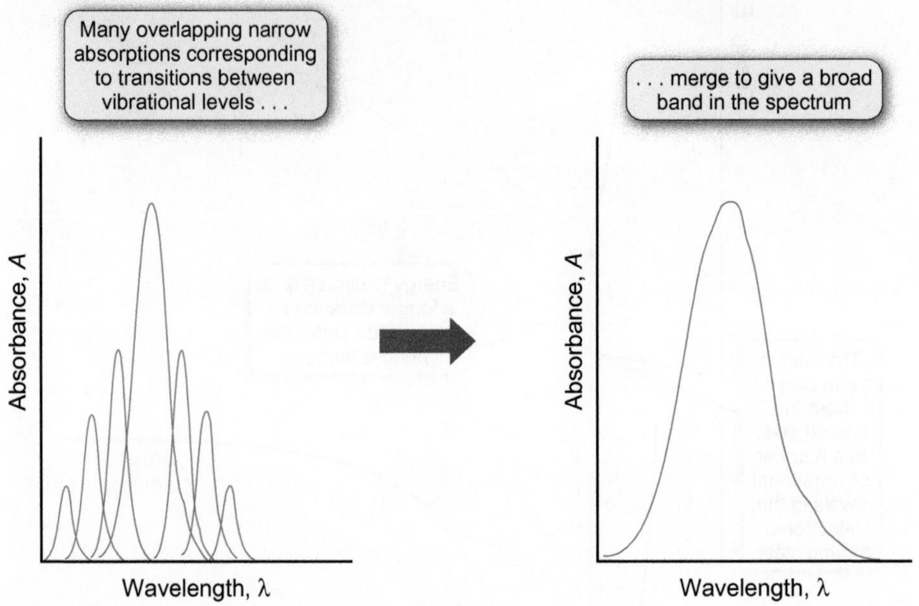
Many overlapping narrow absorptions corresponding to transitions between vibrational levels . . .
. . . merge to give a broad band in the spectrum
Absorbance, A
Wavelength, λ
Absorbance, A
Wavelength, λ

Figure 10.34 For larger molecules, the individual absorptions arising from transitions to different vibrational levels merge to form a broad band in the spectrum.

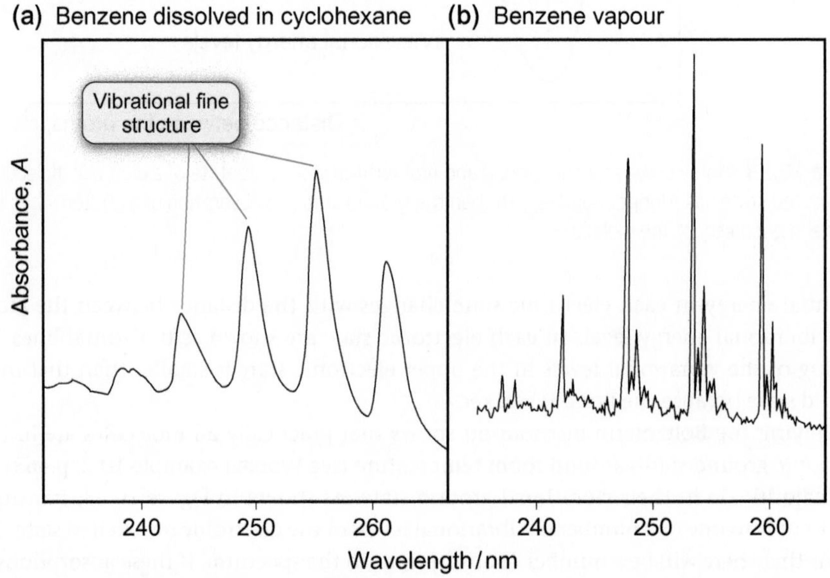
(a) Benzene dissolved in cyclohexane
(b) Benzene vapour
Vibrational fine structure
Absorbance, A
240
250
260
240
250
260
Wavelength / nm

Figure 10.35 UV spectra for benzene. (a) In the solution phase, showing the vibrational fine structure. (b) In the gas phase, where additional vibrational fine structure can be seen if higher resolution spectroscopy is used.

excited state much more quickly than the atoms can move. The absorption first changes the electron distribution around the atoms, which then respond by moving to new equilibrium positions. In some cases, illustrated by Figure 10.36, the absorption of a photon may add enough energy to exceed the bond dissociation energy in the excited state. The bond breaks (called **photodissociation**) and a **photochemical reaction** may result.

A **photochemical reaction** is one in which the first step is the absorption of light. One very important photochemical reaction is the isomerization of retinal, which is a key step in vision. This is described in the introductory pages to Chapter 23 (p.1054).

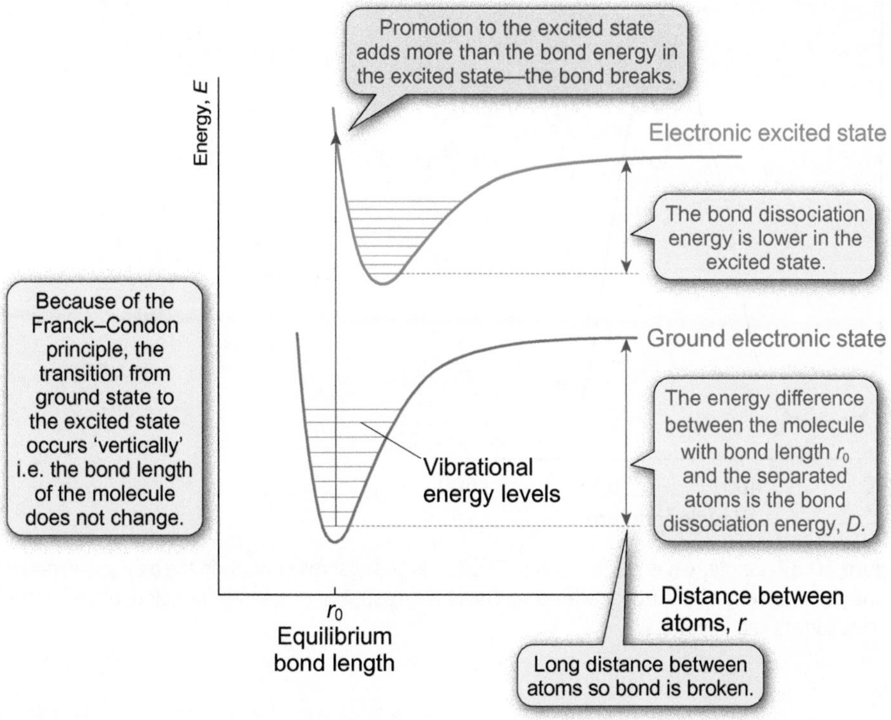
Promotion to the excited state adds more than the bond energy in the excited state—the bond breaks.
Energy, E
Electronic excited state
The bond dissociation energy is lower in the excited state.
Because of the Franck–Condon principle, the transition from ground state to the excited state occurs 'vertically' i.e. the bond length of the molecule does not change.
Ground electronic state
Vibrational energy levels
The energy difference between the molecule with bond length r0 and the separated atoms is the bond dissociation energy, D.
Distance between atoms, r
r0
Equilibrium bond length
Long distance between atoms so bond is broken.

Figure 10.36 Absorption of a photon can add sufficient energy to break a chemical bond—photodissociation.

Chromophores

The part of a molecule that is responsible for the absorption of a photon of UV/VIS radiation is known as a **chromophore**. Various functional groups can act as chromophores but, in organic molecules, the chromophore generally involves a conjugated system in which the electrons are delocalized and the electronic transitions involve π orbitals. For example, a C–C single bond does not absorb even in the far ultraviolet. A carbonyl C=O group absorbs in the UV around 280 nm when it undergoes a transition from a non-bonding orbital to a π* orbital (see the absorption spectrum of butanone in Box 10.7 on p.491). Ethene, containing a C=C double bond, absorbs at 170 nm due to a transition from an occupied π orbital to an unoccupied π* orbital.

Table 10.3 shows the effect of conjugation on absorption, giving the wavelength of the maximum absorption, λ_{max}, and the molar absorption coefficient, ε_{max}. The longer the conjugated system, the longer the wavelength of λ_{max} and the larger the value of ε_{max} (the stronger the absorption). Remember that a higher wavelength corresponds to a lower energy. The β-carotene molecule has 11 conjugated double bonds and absorbs at 451 nm and 497 nm in the visible region (see Figure 10.37).

Conjugation and delocalization of electrons is described in Section 19.1 (p.863).

The molar extinction coefficient, ε, is introduced in Section 10.3 (p.458). The non-SI units in Table 10.3 are commonly used when reporting ε_{max} values in UV/VIS spectra.

In the UV/VIS absorption spectrum of butanone on p.492, ε_{max} is around 20 $dm^3\,mol^{-1}\,cm^{-1}$, showing that butanone absorbs very weakly.

Table 10.3 Effect of conjugation on λ_{max} and ε_{max} in the UV/VIS spectra of ethene and some conjugated polyenes

Compound	Structure	λ_{max}/nm	$\varepsilon_{max}/dm^3\,mol^{-1}\,cm^{-1}$
Ethene		170	10 000
1,3-Butadiene	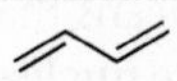	217	21 000
E-1,3,5-Hexatriene	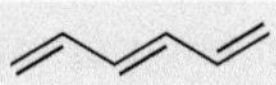	258	35 000
β-Carotene	See Figure 10.37	451	139 500

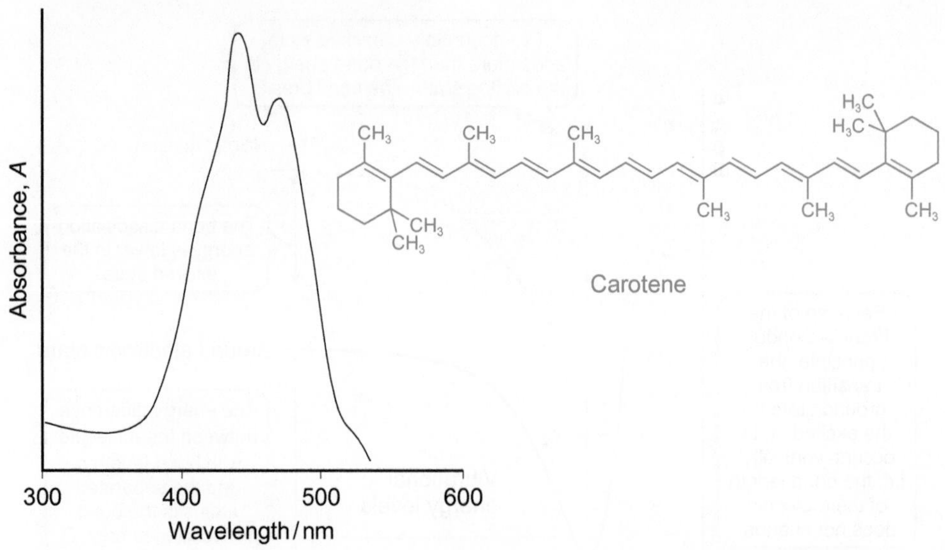
Absorbance, A
300
400
500
600
Wavelength / nm
Carotene

Figure 10.37 Carotene has 11 conjugated C=C bonds and absorbs strongly in the blue-green region of the visible spectrum. Its strong red-orange colour is responsible, in part, for the colour of vegetables such as carrots and tomatoes.

Summary

- Electronic spectroscopy involves the excitation of electrons by absorption of UV or visible radiation.
- The colour of substances is due to the absorption of some wavelengths of visible light by the molecules. The colour observed is due to the light *not* absorbed.
- Each electronic state has several vibrational (and rotational) states associated with it.
- Absorption of photons by electronic states obeys the Frank–Condon principle: the excitation of an electron by the absorption of a photon occurs on a much shorter timescale than that of nuclear motions.
- Electronic excitation may lead to breaking of bonds (photodissociation), causing photochemical reactions.
- A chromophore is the part of a molecule that is responsible for absorption of a photon of UV/VIS radiation.

? For a practice question on these topics, see question 26 at the end of this chapter (p.511).

10.7 Spin resonance spectroscopy

Electron spin is introduced in Section 3.6 (p.143).

Some properties of electrons can be interpreted as their possessing spin, characterized by a spin quantum number, s, where $s = \frac{1}{2}$. Since an electron can spin in one of two directions, there are two possible spin states. The spin magnetic quantum number, m_s, can take values of $+\frac{1}{2}$ or $-\frac{1}{2}$. If all the electrons in an atom are paired, the spins cancel out.

In a classical description, m_s refers to the direction that the electron spins on its axis. In fact, spin is a quantum property and m_s refers simply to one of two possible energy states of the electron.

Neutrons and protons also have spin, which means that atomic nuclei also have a spin quantum number. Nuclear spin is described by the **nuclear spin quantum number,** I (the equivalent of s for electrons). I can take a wider range of values than s, but this section is only concerned with nuclei in which $I = \frac{1}{2}$ (e.g. ^{1}H, ^{13}C, ^{19}F, and ^{31}P). The **nuclear spin magnetic quantum number,** m_I, can then take values of $+\frac{1}{2}$ or $-\frac{1}{2}$.

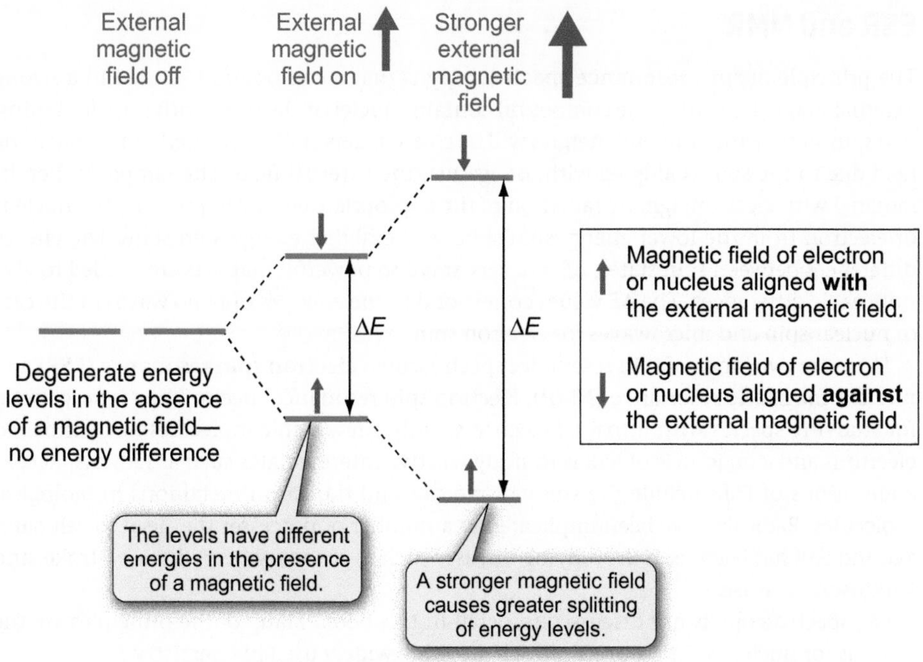
External magnetic field off
External magnetic field on
Stronger external magnetic field
ΔE
ΔE
Degenerate energy levels in the absence of a magnetic field—no energy difference
The levels have different energies in the presence of a magnetic field.
A stronger magnetic field causes greater splitting of energy levels.
Magnetic field of electron or nucleus aligned **with** the external magnetic field.
Magnetic field of electron or nucleus aligned **against** the external magnetic field.

Figure 10.38 Electrons or nuclei with unpaired spins have different energies depending on their alignment in a magnetic field. The energy difference increases as the strength of the magnetic field increases.

Usually, there is no difference in energy between the two spin directions (called **spin states**). However, a spinning charge has an associated magnetic field—it acts as a small magnet. Consider what will happen when a nucleus with spin $I = \frac{1}{2}$ is placed in an external magnetic field. Two states are possible, with different energies depending on whether the magnetic field of the spinning nucleus aligns with (lower energy) or against (higher energy) the external magnetic field. The external magnetic field causes 'splitting' of the energies of the spin states, as shown in Figure 10.38. Figure 10.39 gives a classical analogy involving a bar magnet in a magnetic field.

Do not confuse the nuclear quantum number, I, with moment of inertia, I (see Section 10.4, p.467). The meaning of the symbol, I, is usually clear from the context.

The magnetic fields used in magnetic resonance spectrometers are around 250 000 times higher than the magnetic field of the Earth.

In the promotion of a nucleus or electron to the higher spin state, the spin is often said to 'flip' from the lower to the higher state.

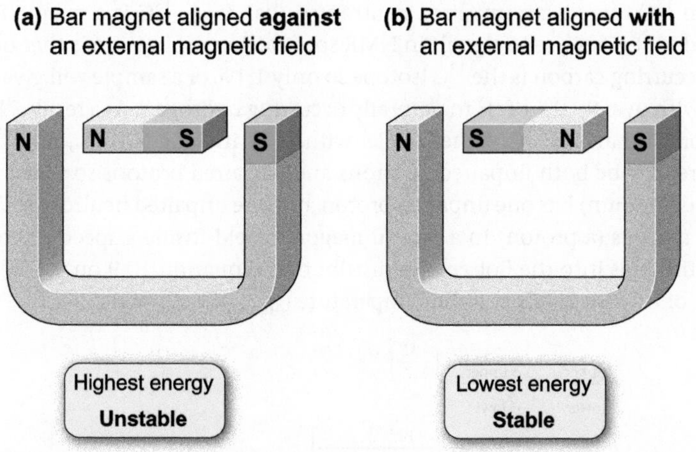
(a) Bar magnet aligned **against** an external magnetic field
(b) Bar magnet aligned **with** an external magnetic field
N
N
S
S
N
S
N
S
Highest energy
Unstable
Lowest energy
Stable

Figure 10.39 A bar magnet can align at *any* angle to an external magnetic field. Each of these arrangements has a different energy. (a) Alignment against the external magnetic field has highest energy. (b) Alignment with the external magnetic field has the lowest energy.

ESR and NMR

The principle of **spin resonance spectroscopy** is that a compound is placed in a strong external magnetic field. If the compound contains nuclei or electrons with unpaired spins, two spin states with different energies will be created, depending on whether the magnetic field due to the spin is aligned with, or against, the external field. The sample is then irradiated with electromagnetic radiation of the appropriate energy to promote the nucleus or electron from the lower energy spin state to the higher energy spin state. The energy differences between spin states, ΔE, are very small so powerful magnets are needed to give noticeable differences. The ΔE values correspond to the energies of radio waves in the case of nuclear spin and microwaves for electron spin.

There are two kinds of spin resonance spectroscopy: **electron spin resonance (ESR)** and **nuclear magnetic resonance (NMR)**. Electron spin resonance spectroscopy is very useful for studying species with unpaired electrons. Only a few stable molecules have unpaired electrons and a major use of ESR is to study reactive intermediates such as radicals. Recent applications of ESR include the study of radicals and transition metal ions in biological molecules. Radicals have been implicated in a number of processes that lead to cell damage and ESR has been used in studying diseases such as coronary heart disease, stroke, and Parkinson's disease.

ESR spectroscopy is not discussed in detail in this book. Many of the principles are the same as for nuclear spin methods, which are more widely used in chemistry.

Nuclear magnetic resonance spectroscopy

NMR spectroscopy is the most useful technique for structure determination in organic chemistry since it is readily applied to the isotopes of hydrogen and carbon, ^{1}H and ^{13}C. NMR spectroscopy also has extensive use in inorganic chemistry with elements such as ^{31}P and ^{19}F. A great deal of information about the structure of molecules—which functional groups are present and how they are arranged—can be derived from NMR spectroscopy. It is widely used to investigate the structure of complex biological molecules, such as proteins, and to aid diagnosis in medicine (see Box 10.9, p.507).

In order to show energy differences in a magnetic field, a nucleus must have a nuclear spin, $I \neq 0$. The simplest example is ^{1}H, which contains only one proton. A proton has $I = \frac{1}{2}$ (the same spin as an electron) and it can adopt two orientations ($m_I = +\frac{1}{2}$ and $m_I = -\frac{1}{2}$), so ^{1}H is NMR active.

Now, consider the isotopes of carbon. ^{12}C has 6 protons (3 pairs) and 6 neutrons (3 pairs). The protons pair up with opposite spins, as do all the neutrons, so the nuclear spin quantum number, $I = 0$. In ^{14}C, there are 8 neutrons, which can pair, so again $I = 0$. However, in ^{13}C there is an unpaired neutron so that $I = \frac{1}{2}$. ^{13}C is a magnetically active nucleus and so it gives rise to signals in NMR spectra. A complication is that only 1.1% of naturally occurring carbon is the ^{13}C isotope so only 1.1% of a sample will give a spectrum compared with over 99.9% of ^{1}H in naturally occurring hydrogen. As a result, NMR is more sensitive for ^{1}H than for ^{13}C. Other nuclei with $I = \frac{1}{2}$ include ^{19}F, ^{29}Si, and ^{31}P. In some nuclei, there may be both unpaired neutrons and unpaired protons so that $I > \frac{1}{2}$. For example, ^{2}H (deuterium) has one unpaired proton and one unpaired neutron, so $I = \frac{1}{2} + \frac{1}{2} = 1$.

For a ^{1}H nucleus (a proton) in a typical magnetic field inside a spectrometer, $\Delta E \approx 6 \times 10^{-26}$ J. Putting this into the Boltzmann distribution (Equation 10.9 on p.462) to find the population of the two levels at room temperature ($g_{\text{upper}} = g_{\text{lower}} = 1$)

$$\frac{n_{\text{upper}}}{n_{\text{lower}}} = \frac{g_{\text{upper}}}{g_{\text{lower}}} e^{\left(\frac{-\Delta E}{k_B T}\right)} \qquad (10.9)$$

$$= e^{\left(\frac{-(6 \times 10^{-26}\,\text{J})}{(1.381 \times 10^{-23}\,\text{J K}^{-1}) \times 298\,\text{K}}\right)} = 0.999985$$

(More significant figures are used in the result here than are justified by the data given to emphasize just how small the energy difference between the two levels is.)

The phenomenon of magnetic resonance was discovered in 1946 in the USA by Felix Bloch and Edward Purcell, who were jointly awarded the Nobel Prize in Physics in 1952. Richard Ernst (Switzerland) was awarded the 1991 Nobel Prize in Chemistry for his contributions to developing chemical applications of NMR.

Protons, neutrons, and electrons all have a spin of $\frac{1}{2}$. They are said to be 'spin $\frac{1}{2}$' particles.

The use of NMR to characterize molecules is described in detail in Section 12.3 (p.578).

Visit the Online Resource Centre to view video 10.4 which gives a simple introduction to NMR.

The pairing process described here is not always so straightforward and it is often difficult to predict how many protons and neutrons will be unpaired.

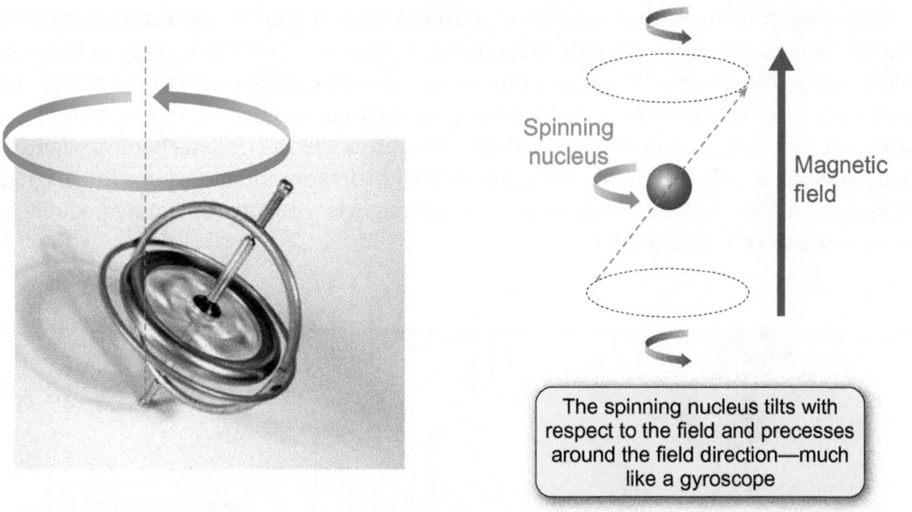
Spinning
nucleus
Magnetic
field
The spinning nucleus tilts with respect to the field and precesses around the field direction—much like a gyroscope

Figure 10.40 A spinning electron or nucleus precesses around the direction of the magnetic field like a gyroscope.

The populations of the two energy levels are almost the same. This means that there is a low probability of observing the transition and very sensitive equipment is needed. The larger the magnetic field, the greater the value of ΔE so stronger magnets give a stronger signal.

In a classical system, a small bar magnet can be held at *any* angle to the external field. When the magnet is aligned exactly *against* the external field, the energy is at its highest; when the alignment is *with* the field, the energy is at its lowest (Figure 10.39). However, analysis using quantum theory shows that nuclear spins can only align at certain angles to the magnetic field. In addition (because of the Heisenberg uncertainty principle) the nuclear spin *precesses* around the direction of the magnetic field, as shown in Figure 10.40. The frequency of this precession is known as the **Larmor frequency**. When the frequency of the applied electromagnetic radiation matches the Larmor frequency, the nucleus comes into **resonance** and energy is absorbed to promote the nucleus to the higher energy spin state. When the radiation is removed, the nucleus **relaxes** back to its original state and releases energy. The energy released mainly goes into the solvent.

The Heisenberg uncertainty principle is discussed in Section 3.4 (p.132).

The term *resonance* is also used to describe bonding in molecules (Sections 5.5, p.243, and 19.1, p.863). Be careful not to confuse the two meanings.

Recording NMR spectra

NMR spectra are recorded by placing the sample inside a tube in a strong magnetic field. Usually, the sample is dissolved in a solvent that does not give an NMR signal. By using specialized equipment, spectra of solids can also be obtained. The sample is irradiated with a brief pulse of radiofrequency electromagnetic radiation which excites some of the nuclei to the higher energy state. The energy that is given out when the molecules return to the ground state is then detected on a sensitive radiofrequency receiver.

There is more detail on running an NMR spectrum in Box 12.3 (p.579).

Visit the Online Resource Centre to view video 10.5 which shows how an NMR sample is prepared and run

The energy difference, ΔE, between the two spin states for an $I = \frac{1}{2}$ nucleus is given by

$$\Delta E = \frac{\gamma B h}{2\pi} \quad (10.23)$$

where h is the Planck constant, B is the strength of the magnetic field, and γ is a constant called the **magnetogyric ratio**.

The value of γ is different for different nuclei, so the energy difference, ΔE, depends on the nucleus involved, for example, whether it is ^{1}H, ^{13}C, or ^{19}F. ΔE is directly proportional to the applied magnetic field—the larger the magnetic field, the larger the energy difference between the spin states and the higher the energy of the radiofrequency radiation required to promote a transition.

The time taken for the excited spin state to relax back to the ground state after the pulse of incident radiation—called the *relaxation time*—depends on the precise environment of the nucleus and gives valuable chemical information.

Box 12.4 (p.580) discusses the relationship between the operating frequency of an NMR spectrometer and the resolution of the spectrum.

The magnet in an NMR spectrometer is usually described by an **operating frequency**, rather than its magnetic strength. The operating frequency is the frequency of the radiation required to bring a ^1H nucleus into resonance when it experiences the full applied magnetic field. For example, a magnet that generates a magnetic field of 18.8 tesla (T) has an operating frequency of 800 MHz. Many of the advances in NMR in chemistry over the past three decades have resulted from the availability of more and more powerful magnets. Today, most NMR instruments have very strong magnets with large operating frequencies, which can be as high as 1000 MHz.

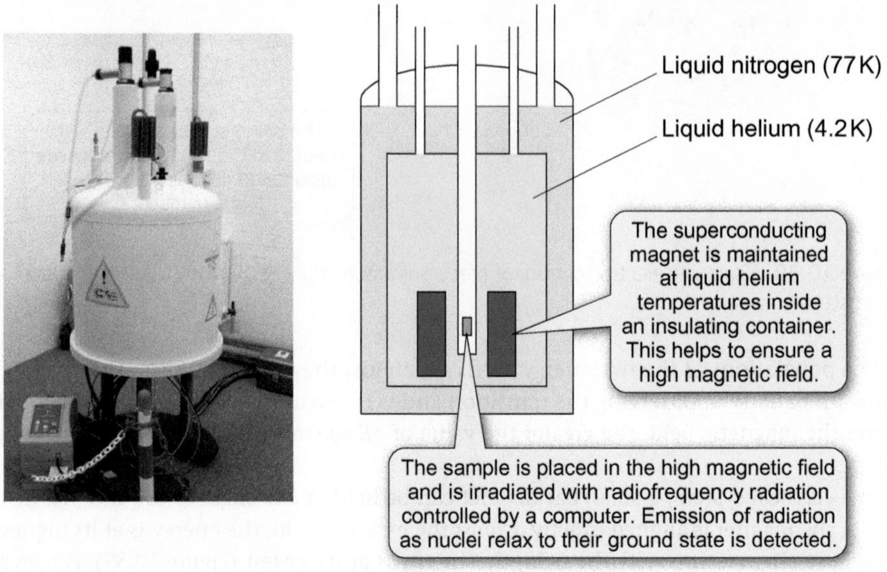
Liquid nitrogen (77 K)
Liquid helium (4.2 K)
The superconducting magnet is maintained at liquid helium temperatures inside an insulating container. This helps to ensure a high magnetic field.
The sample is placed in the high magnetic field and is irradiated with radiofrequency radiation controlled by a computer. Emission of radiation as nuclei relax to their ground state is detected.

An NMR spectrometer.

Chemical shift

Equation 10.23 describes the energy difference between the two spin states, which corresponds to that of the absorbed radiation. From this, the value of γ can be found and used to identify the nucleus involved since different nuclei have different γ values. If this were the only application, NMR would not be very useful. In fact, the exact energy difference between the two spin states of a nucleus depends on the precise local environment of a nucleus.

Think about the different hydrogen atoms in a molecule. If all the ^1H nuclei experience the same applied magnetic field, they will absorb radiofrequency radiation of the *same* energy. However, the different ^1H nuclei in the molecule experience slightly different magnetic fields and so absorb energy at different frequencies. This is because the precise field experienced by any atom in the molecule is not exactly equal to the external applied field.

The applied magnetic field induces motion of the electron clouds in the molecule. This is often described as inducing the electrons 'to circulate', as in Figure 10.41. This, in turn, causes a small local magnetic field that acts against the external field and so slightly reduces the overall field experienced by a nucleus. The nucleus is said to be shielded. The magnetic field experienced by a nucleus is the sum of the applied field and the local field due to electron circulation and depends on the electron density around the nucleus. The precise value of the energy that is absorbed is, therefore, very sensitive to the electron density around the nucleus, which in turn depends on the position of the nucleus in the molecule.

Electron-donating and electron-withdrawing groups are discussed in Section 19.1 (p.870). Their effect on an NMR spectrum is discussed in more detail in Section 12.3 (p.580).

The greater the electron density near a nucleus, the more the nucleus is shielded. An electron-donating group near a ^1H nucleus *increases* the shielding. The nucleus therefore experiences a smaller magnetic field and the frequency (and energy) of the applied radiofrequency radiation needed for resonance is reduced. In contrast, an electron-withdrawing group near a nucleus *decreases* the shielding. The nucleus experiences a larger magnetic

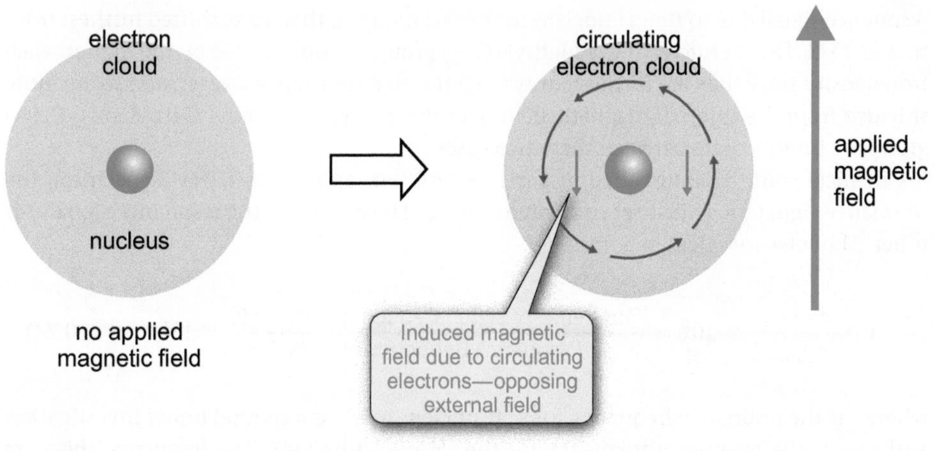
electron cloud
nucleus
no applied magnetic field
circulating electron cloud
applied magnetic field
Induced magnetic field due to circulating electrons—opposing external field

Figure 10.41 An applied magnetic field causes the electrons in an atom to circulate within orbitals. This motion induces a small local magnetic field that opposes the applied magnetic field at the nucleus.

field and the frequency needed for resonance is increased. Thus, nuclei in different electronic environments absorb radiation with slightly different energies, giving rise to an **NMR spectrum.**

In general, the absolute value of the energy of each absorption is unimportant, Because NMR spectra usually display resonance signals in terms of their difference from a reference compound. For ^{1}H and ^{13}C NMR spectra, the usual reference is tetramethylsilane (TMS; Figure 10.42) and spectra are reported in terms of the difference between the position of the resonance signal for the nucleus and that for TMS. This is known as the **chemical shift,** δ.

As an example, Figure 10.43 shows the ^{1}H NMR spectrum of ethanol (CH_3CH_2OH). The three resonance signals in the spectrum show that the ^{1}H nuclei in the CH_3, CH_2, and OH groups experience slightly different local magnetic fields so they are shielded to different extents.

The ^{1}H nucleus attached to the electronegative oxygen atom is the least shielded since the O atom withdraws electrons from it. As a result, it is surrounded by lower electron density than the other ^{1}H nuclei in the molecule. They experience a larger magnetic field and so are brought into resonance at higher frequencies of the applied radiation. The

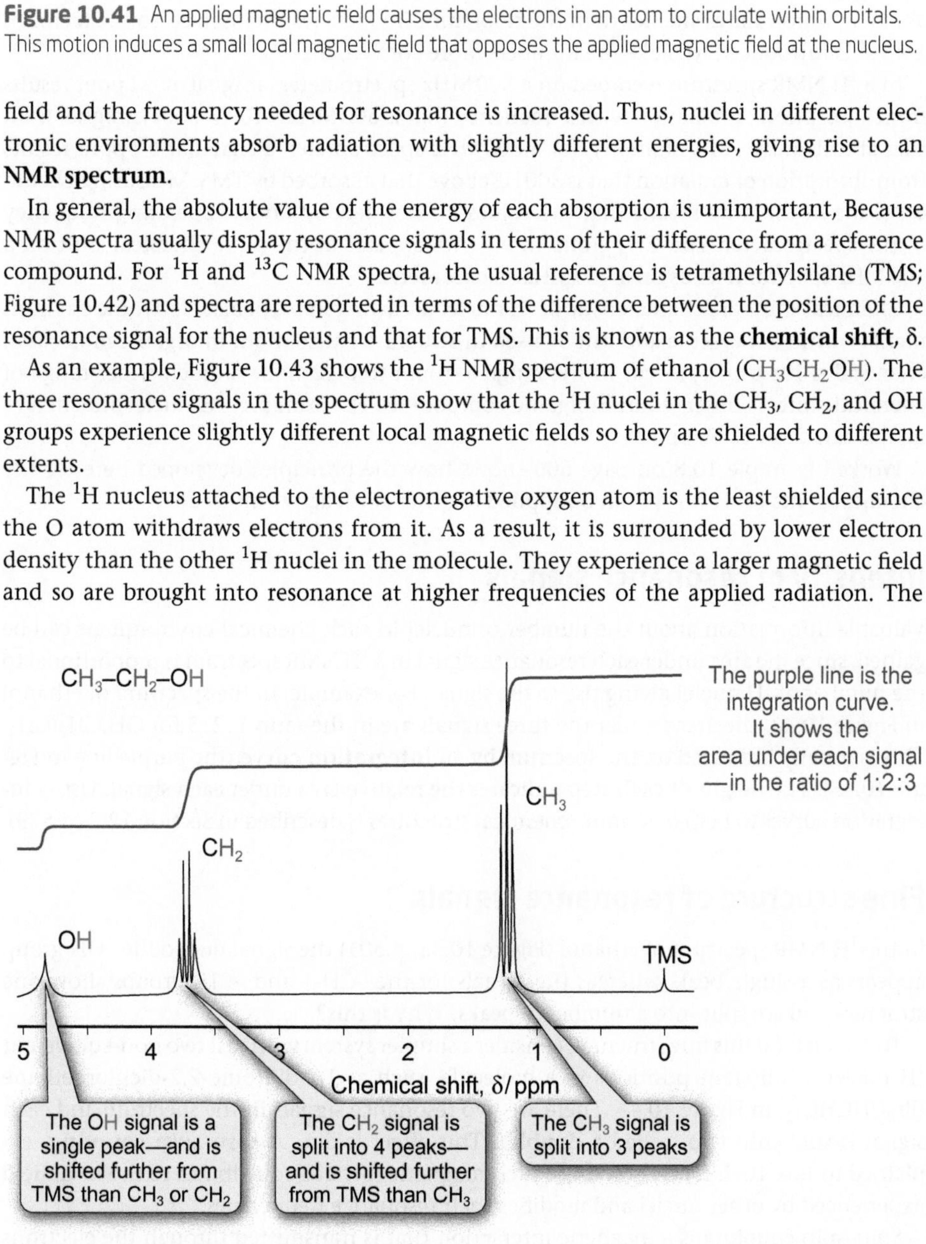
CH3–CH2–OH
The purple line is the integration curve. It shows the area under each signal —in the ratio of 1:2:3
CH3
CH2
OH
TMS
5
4
3
2
1
0
Chemical shift, δ/ppm
The OH signal is a single peak—and is shifted further from TMS than CH3 or CH2
The CH2 signal is split into 4 peaks—and is shifted further from TMS than CH3
The CH3 signal is split into 3 peaks

Figure 10.43 The ^{1}H NMR spectrum of ethanol (CH_3CH_2OH). A small quantity of TMS has been added to the sample.

More information on chemical shifts and how they can be used to elucidate the structures of organic molecules is given in Section 12.3 (p.578).

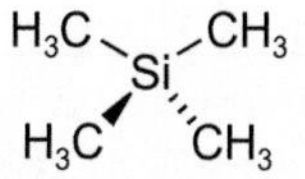
H3C
CH3
Si
H3C
CH3

Figure 10.42 In tetramethylsilane (TMS), the symmetrical arrangement of the methyl groups around Si means that the ^{1}H nuclei are strongly shielded and their signal is outside the range usually seen in most organic compounds. The 12 hydrogen atoms are equivalent to each other, so there is only one ^{1}H peak. The four carbon atoms are also equivalent to each other so there is only one peak in the ^{13}C spectrum. Other reference compounds are used for other NMR active nuclei: $CFCl_3$ is usually used for ^{19}F and H_3PO_4 for ^{31}P.

Compare the NMR spectrum of ethanol in Figure 10.43 with the spectrum of ethanol in Figure 12.16 (p.591). The signal due to the -OH group is 'mobile' and can move around depending on the concentration and on the solvent.

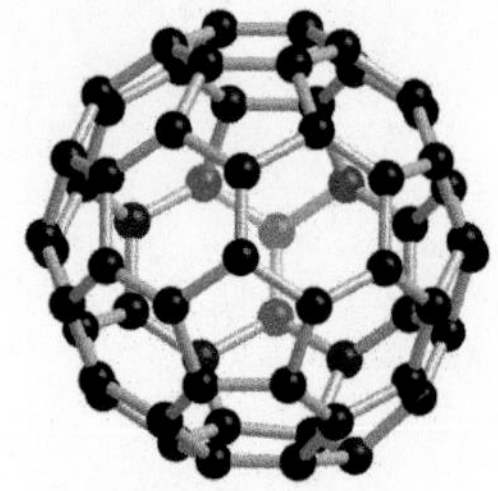

In C_{60}, every C atom is equivalent to every other one. The ^{13}C spectrum of C_{60} therefore shows just a single peak.

resonance signal due to the ^1H nucleus in the OH group is, therefore, shifted furthest from that of TMS. The ^1H nuclei in the methyl (CH_3) group are surrounded by the highest electron density since they are furthest away from the electronegative oxygen and so are more shielded from the external magnetic field. The absorptions due to the methylene ($-CH_2-$) group are intermediate between these two cases.

Chemical shift, δ, is measured in parts per million, ppm. In a ^1H NMR spectrum, the resonance signal for TMS is set to a value of $\delta = 0$. The δ value of the resonance signals for other ^1H nuclei are calculated from

$$\delta = \frac{\nu - \nu_{TMS}}{\nu_{TMS}} \times 10^6 = \frac{\text{Difference of signal frequency from TMS}}{\text{Resonance frequency of TMS}} \times 10^6 \quad (10.24)$$

where ν is the resonance frequency for a ^1H nucleus in the compound under investigation and ν_{TMS} is the resonance frequency for the ^1H nuclei in TMS. The frequency *shifts* are about a million times smaller than the resonance frequencies themselves, so the factor of 1×10^6 is introduced to make the numbers more convenient.

In a ^1H NMR spectrum recorded on a 500 MHz spectrometer, a signal at δ 1 ppm results from absorption of radiation with a frequency that is $500\,\text{MHz} \times 10^{-6} = 500\,\text{Hz}$ higher than that absorbed by TMS. Similarly, for a 300 MHz spectrometer, a signal at δ 1 ppm results from absorption of radiation that is 300 Hz above that absorbed by TMS. Measuring chemical shifts in this way means that the values do not depend on the operating frequency of the spectrometer. The CH_3 group in ethanol appears at the same chemical shift on a 100 MHz spectrometer as on a 900 MHz spectrometer.

A negative chemical shift indicates that the nucleus is *more* shielded than in TMS and occurs to the right of the TMS signal in the spectrum.

Chemical shifts are characteristic of ^1H nuclei occurring in particular electronic environments within a molecule and can be used in structure elucidation. In organic molecules, chemical shifts usually occur in the range δ 0 ppm to δ 12 ppm. A much wider range of chemical shifts is observed in inorganic compounds. For example, ^1H nuclei attached to metals can show δ −50 ppm.

Worked example 10.8 on page 506 shows how the principles developed here for the chemical shifts of ^1H nuclei can be applied to other NMR active molecules.

Intensity of resonance signals

Using integration curves to find the number of nuclei works well for ^1H but does not always work with other nuclei such as ^{13}C.

Valuable information about the number of nuclei in each chemical environment can be gained, since the area under each resonance signal in a ^1H NMR spectrum is proportional to the number of ^1H nuclei giving rise to the signal. For example, in the spectrum of ethanol in Figure 10.43, the areas under the three signals are in the ratio 1:2:3 for $OH:CH_2:CH_3$. This is usually indicated on the spectrum by an **integration curve** (the purple line in Figure 10.43). The height of each step indicates the relative area under each signal. Using integration curves to help determine chemical structures is described in Section 12.3 (p.579).

Fine structure of resonance signals

In the ^1H NMR spectrum of ethanol (Figure 10.43, p.501) the signal due to the −OH group appears as a single peak, whereas the signals for the $-CH_2-$ and $-CH_3$ groups show fine structure and are split into a number of peaks. Why is this?

Bromine and chlorine nuclei are not magnetically active and do not give NMR signals.

To understand this fine structure, consider a simpler system with just two non-equivalent ^1H nuclei in different positions in a molecule, such as 1,1-dibromo-2,2-dichloroethane ($Br_2CHCHCl_2$) in Figure 10.44. There are two resonance signals in the spectrum and each signal is split into two, called a **doublet**. This effect is due to **spin–spin coupling**, explained in Box 10.8. It arises because each magnetic nucleus contributes to the local field experienced by other nuclei and modifies their resonance frequencies.

Spin–spin coupling is a magnetic interaction that is transmitted through the electrons in the chemical bonds joining the nuclei. It occurs when nuclei in different electronic environments are close enough together for their magnetic fields to influence one another.

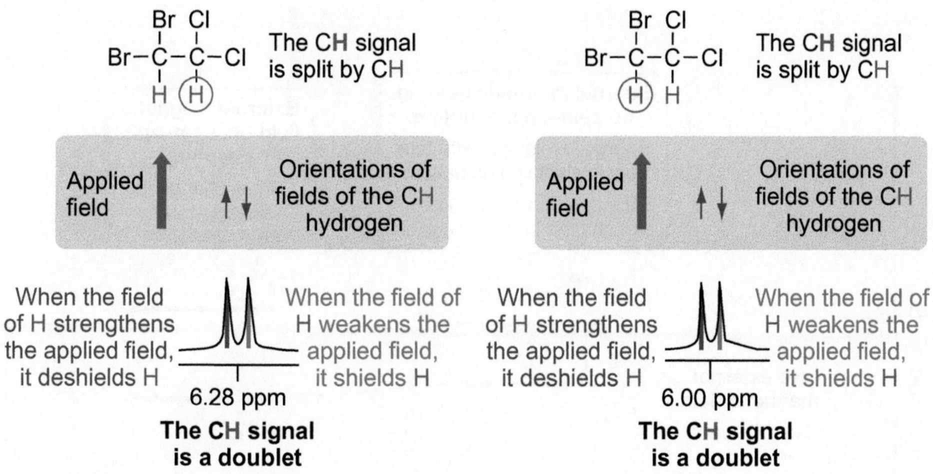
Br Cl
Br–C–C–Cl
H H
The CH signal is split by CH
Applied field
Orientations of fields of the CH hydrogen
When the field of H strengthens the applied field, it deshields H
When the field of H weakens the applied field, it shields H
6.28 ppm
The CH signal is a doublet
Br Cl
Br–C–C–Cl
H H
The CH signal is split by CH
Applied field
Orientations of fields of the CH hydrogen
When the field of H strengthens the applied field, it deshields H
When the field of H weakens the applied field, it shields H
6.00 ppm
The CH signal is a doublet

Figure 10.44 The absorption signals for the hydrogens in 1,1-dibromo-2,2-dichloroethane are both split into doublets.

In organic molecules, this is usually when the hydrogens are on adjacent carbon atoms. As a result, the local field experienced by a nucleus depends on the spins of nuclei on adjacent atoms.

For $Br_2CHCHCl_2$, the resonance signal from H attached to CBr_2 appears as a doublet because of the two spin states of the H attached to CCl_2. The H attached to CCl_2 is also split into a doublet by H because of the two spin states of the H attached to CBr_2. The spectrum appears as a pair of doublets.

> ^{13}C nuclei also show spin–spin coupling. However, since the abundance is only 1.1%, it is unlikely that two ^{13}C atoms will be found adjacent to each other so splitting is not usually observable in the spectrum.

Box 10.8 Spin–spin coupling in NMR spectra

Consider two ^1H atoms with different chemical shifts, labelled H_A and H_X, that are close enough to couple. This is called an AX system (e.g. the H atoms in $Br_2CHCHCl_2$ in Figure 10.44). Each of the ^1H nuclear spins can be aligned with or against the external magnetic field so there are four possible energy levels, as shown in Figure 1 (on the next page). Four transitions are possible since the selection rules allow only one spin to change in any transition.

If there is no interaction between the nuclei, transitions 1 and 4 (involving H_X) have the same energy change. Transitions 2 and 3 (involving H_A) also have the same energy change (although different from that in transitions 1 and 4). As a result, the spectrum would appear as two singlet peaks.

If the spins are now allowed to couple, the energy level diagram changes. In cases where the spins are in the same direction, coupling results in a small increase in energy; when spins are opposite, the energy is slightly decreased. The same four transitions can take place but now transitions 1 and 4 have different energies, as do 2 and 3. Each resonance signal in the spectrum is therefore split into a doublet. Since there is an equal probability of each transition occurring, each signal has the same intensity. The spectrum therefore appears as two doublets.

The magnitude of the coupling is measured by the coupling constant, J, measured in Hz. Coupling constants are independent of the magnetic field strength and so do not vary if measured on spectrometers with different operating frequencies.

Now consider a molecule that has one ^1H atom (H_A) on an atom adjacent to one with two ^1H atoms (H_X). This is called an AX_2 system (Figure 2). Magnetically equivalent nuclei do not cause spin–spin coupling so the two H_X nuclei do not influence each other.

The spectrum of an AX_2 system in Figure 2 is shown along with that of an AX system for comparison. In both spectra, the signal from H_X is split into a doublet by H_A. In the AX_2 spectrum, the H_X nuclei have three possible spin combinations: both aligned with the external field; both aligned against; and one aligned in each direction. The last of these is twice as likely. H_A couples with each spin state. The H_A signal is, therefore, split into three peaks (called a **triplet**) with intensities in the ratio 1 : 2 : 1.

The number of peaks for an NMR signal is called the **multiplicity** of a signal and is calculated using the ***N* + 1 rule**. If a signal is split by N equivalent nuclei, it will be split into $N + 1$ peaks. By considering the possible spin combinations, the relative area of each peak is found from the appropriate line of Pascal's triangle as shown in Table 1. (Each number in the triangle is the sum of the two numbers closest to it in the row above.) Notice that the pattern of peak areas is symmetrical and that the larger peak areas are in the middle of each line. →

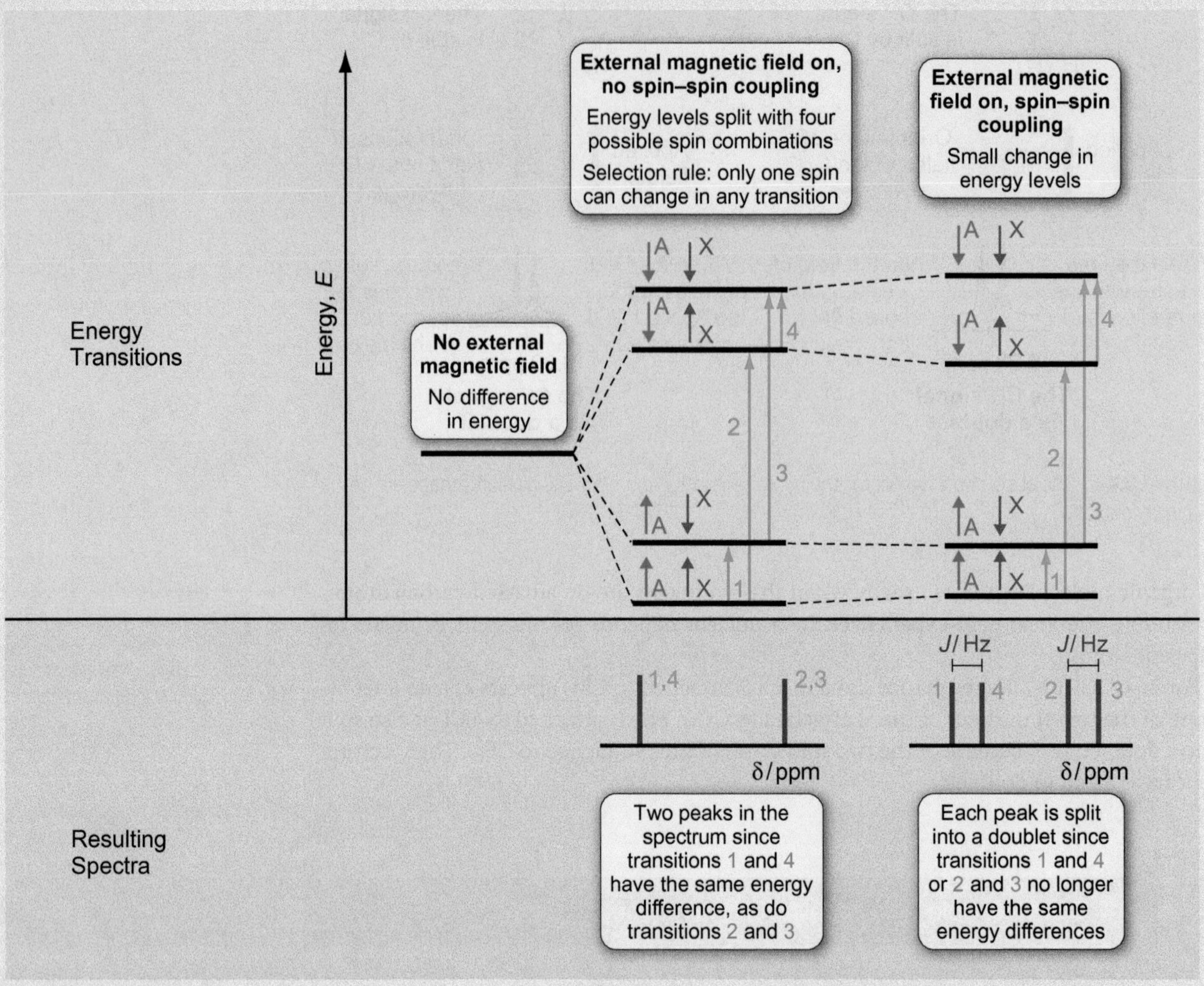
Energy Transitions
Energy, E
No external magnetic field
No difference in energy
External magnetic field on, no spin–spin coupling
Energy levels split with four possible spin combinations
Selection rule: only one spin can change in any transition
External magnetic field on, spin–spin coupling
Small change in energy levels
A X
1 2 3 4
Resulting Spectra
1,4
2,3
J/Hz
δ/ppm
Two peaks in the spectrum since transitions 1 and 4 have the same energy difference, as do transitions 2 and 3
Each peak is split into a doublet since transitions 1 and 4 or 2 and 3 no longer have the same energy differences

Figure 1 Splitting of NMR signals due to spin–spin coupling in an AX system.

Table 1 Splitting of NMR signals due to adjacent groups

Number of equivalent hydrogens causing the splitting (N)	Multiplicity of the signal ($N+1$)	Pattern of peaks (abbreviation in brackets)	Peak area ratios (from Pascal's triangle)
0	1	Singlet (s)	1
1	2	Doublet (d)	1 : 1
2	3	Triplet (t)	1 : 2 : 1
3	4	Quartet (q)	1 : 3 : 3 : 1
4	5	Quintet (quin)	1 : 4 : 6 : 4 : 1
5	6	Sextet (sext)	1 : 5 : 10 : 10 : 5 : 1
6	7	Septet (sep)	1 : 6 : 15 : 20 : 15 : 6 : 1

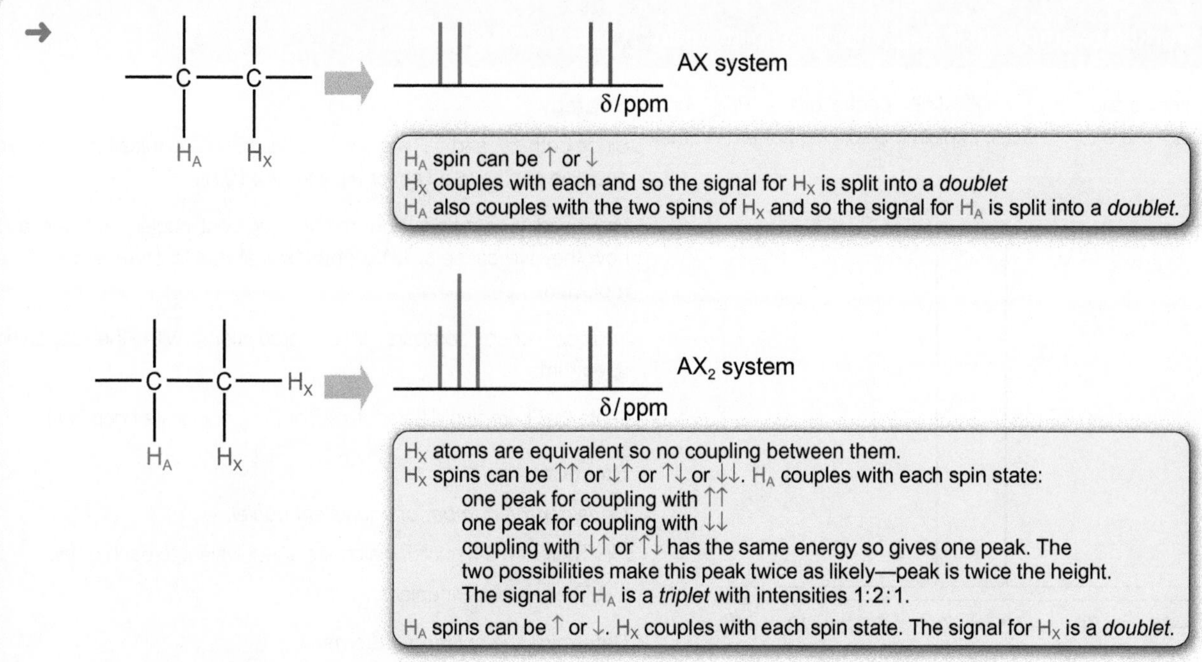
AX system
δ/ppm
H_A H_X
H_A spin can be ↑ or ↓
H_X couples with each and so the signal for H_X is split into a *doublet*
H_A also couples with the two spins of H_X and so the signal for H_A is split into a *doublet.*
H_X
AX$_2$ system
δ/ppm
H_A H_X
H_X atoms are equivalent so no coupling between them.
H_X spins can be ↑↑ or ↓↑ or ↑↓ or ↓↓. H_A couples with each spin state:
one peak for coupling with ↑↑
one peak for coupling with ↓↓
coupling with ↓↑ or ↑↓ has the same energy so gives one peak. The
two possibilities make this peak twice as likely—peak is twice the height.
The signal for H_A is a *triplet* with intensities 1:2:1.
H_A spins can be ↑ or ↓. H_X couples with each spin state. The signal for H_X is a *doublet.*

Figure 2 NMR spectra for an AX system and an AX$_2$ system.

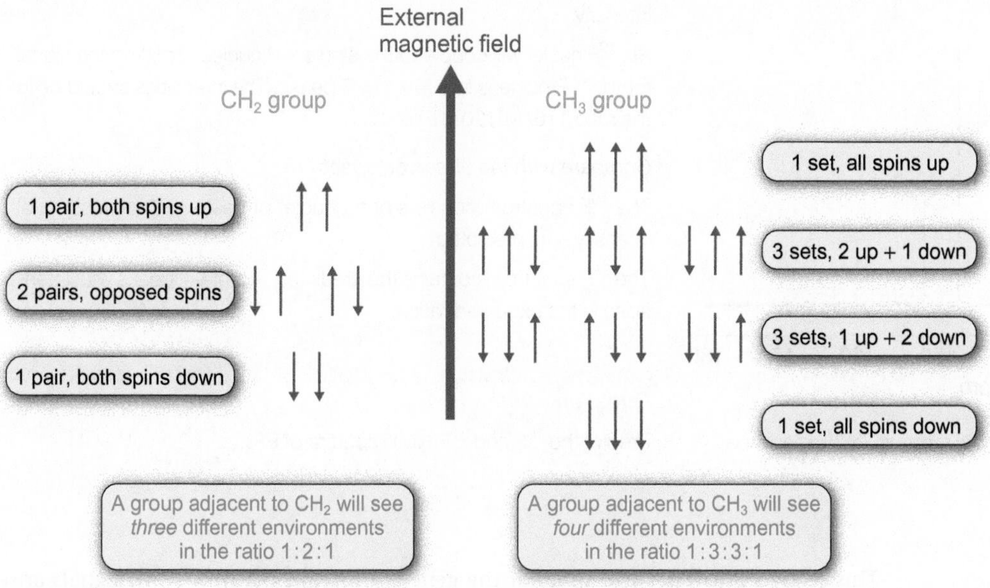
External
magnetic field
CH$_2$ group
CH$_3$ group
1 set, all spins up
1 pair, both spins up
3 sets, 2 up + 1 down
2 pairs, opposed spins
3 sets, 1 up + 2 down
1 pair, both spins down
1 set, all spins down
A group adjacent to CH$_2$ will see *three* different environments in the ratio 1:2:1
A group adjacent to CH$_3$ will see *four* different environments in the ratio 1:3:3:1

Figure 10.45 Spin-spin coupling in the CH_3CH_2 group in ethanol.

The principles in Box 10.8 explain the fine structure of the resonance signals in the spectrum of ethanol (Figure 10.43, p.501). All three hydrogens in the methyl group are equivalent and so do not influence each other. The neighbouring CH_2 group has three possible spin combinations and so splits the methyl signal into three peaks with an area ratio of 1:2:1 (see Table 1, Box 10.8). As shown in Figure 10.45, there are four possibilities for the nuclear spins of the three hydrogens of the methyl group, so the CH_2 signal is split into four peaks—a quartet—in the ratio of 1:3:3:1. The hydrogen in the hydroxyl group does not couple and so appears as a singlet.

The reason why the hydrogen of the OH group does not couple is explained in Section 12.3 (p.590).

Worked example 10.8 Explaining the splittings in ^{19}F and ^{31}P NMR spectra

The figure shows the ^{19}F and ^{31}P NMR spectra of the $[PF_6]^-$ ion. Both ^{19}F and ^{31}P are $I = \frac{1}{2}$ nuclei. Explain the splitting patterns in the spectrum.

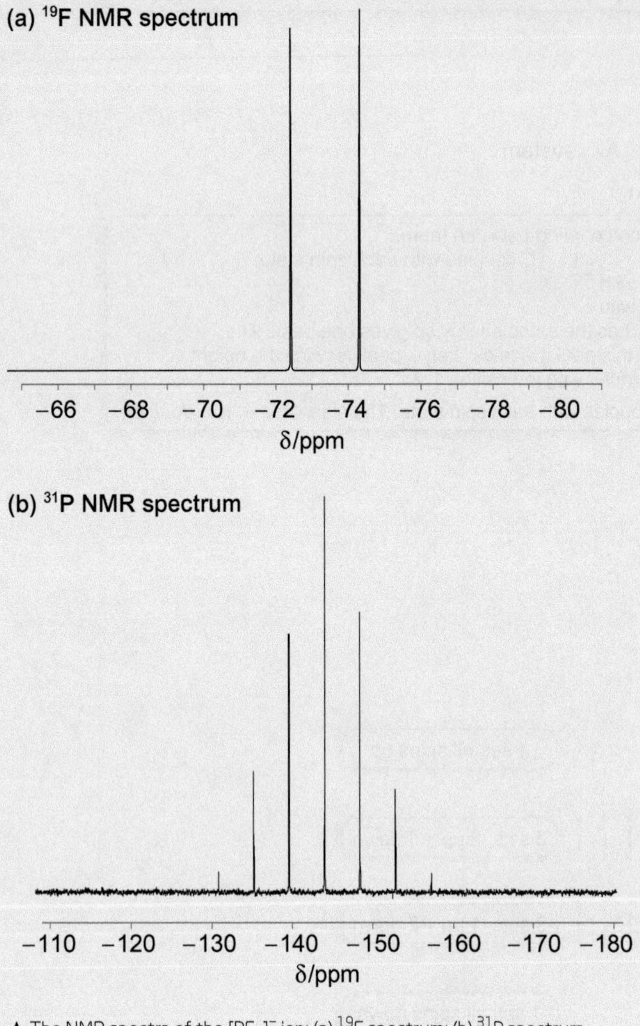
(a) 19F NMR spectrum
−66 −68 −70 −72 −74 −76 −78 −80
δ/ppm
(b) 31P NMR spectrum
−110 −120 −130 −140 −150 −160 −170 −180
δ/ppm

▲ The NMR spectra of the $[PF_6]^-$ ion: (a) ^{19}F spectrum; (b) ^{31}P spectrum.

Strategy

Since both ^{19}F and ^{31}P are $I = \frac{1}{2}$ nuclei, the same rules for spin–spin coupling can be used as for ^{1}H (see Box 10.8).

You need to consider how many equivalent nuclei there are and how they will cause splitting of the signal due to other nuclei in the molecule.

You can then compare these predictions with the observed spectrum.

(Note that here two different types of $I = \frac{1}{2}$ nuclei are coupling.)

Solution

Consider the number of equivalent nuclei.

^{19}F. There are 6 F nuclei, which are all equivalent to each other.

^{31}P. There is one P nucleus.

Determine the splitting patterns.

Since $I = \frac{1}{2}$, N equivalent nuclei give rise to $N + 1$ signals (see Table 1, Box 10.8).

One ^{31}P nucleus will couple to the ^{19}F nuclei, splitting the signal for the ^{19}F nuclei into (1 + 1) = 2 peaks. The peaks should have the same intensity.

Six ^{19}F nuclei will couple to the single ^{31}P nucleus splitting the signal for the ^{31}P nucleus into (6 + 1) = 7 peaks. The intensities should be in the ratio 1 : 6 : 15 : 20 : 15 : 6 : 1.

Compare with the observed spectrum.

The ^{19}F spectrum consists of a doublet of peaks with almost equal intensity—as predicted.

The ^{31}P spectrum contains the predicted septet—7 peaks with intensities in the expected ratios.

Question

Predict the ^{19}F and ^{31}P NMR spectra of PF_3.

The discussion has centred on ^{1}H NMR but the same principles apply to any nucleus with $I = \frac{1}{2}$; examples include ^{13}C, ^{19}F, ^{29}Si, and ^{31}P.

This section illustrates just some of the general principles of how NMR signals arise. Even with this introductory treatment, you can see the power of NMR for looking at the positions of H, C, and other atoms within molecules. NMR, particularly in combination with other spectroscopic methods, is extremely powerful in determining molecular structures. This is discussed in more detail in Section 12.3 (p.578). Box 10.9 describes the use of nuclear magnetic resonance in medical diagnosis.

In 2003, Paul C. Lauterbur of the University of Illinois and Sir Peter Mansfield of the University of Nottingham were awarded the Nobel Prize in Physiology or Medicine for their discoveries concerning magnetic resonance imaging in medical diagnosis. Lauterbur developed the idea of the magnetic field gradient and Mansfield the complex mathematical techniques needed to produce usable images.

Box 10.9 Magnetic resonance imaging (MRI)

The technique of **magnetic resonance imaging (MRI)** extends the uses of NMR spectroscopy to investigating complex assemblies of molecules—to the extent that a whole human body can be imaged. Like NMR, MRI depends on nuclear spin processes. (The term 'nuclear' is not used in the name since it was feared that it would scare some patients!)

MRI was invented to aid medical diagnosis and takes advantage of ^{1}H being a magnetically active nucleus which is present in water throughout the body. The principle, illustrated in Figure 1, is simple—but turning the idea into a practical scanner took many years of effort.

The technique relies on using a magnetic field gradient. The protons in two water molecules in the same environment would resonate at the same frequency if placed in the same magnetic field. Equation 10.23 (p.499) shows that the energy needed to 'flip' the spins, and hence the resonance frequency, depends on the strength of the magnetic field. An MRI scanner, therefore, places the sample (the patient) in a magnetic field that *varies* linearly between the poles of the magnet. Two identical protons will now resonate at different frequencies depending on their position in the magnetic field, that is, on their position in the body. A proton in a region where the magnetic field is higher resonates at a higher frequency than in a region where the field is lower.

A particular position in the body is irradiated with a pulse at a particular radiofrequency and the intensity of the emitted radiation is measured. This makes it possible to measure the number of ^{1}H nuclei in water at that position. This is repeated at the different positions in the scanner with a succession of pulses at differing frequencies to measure the number of ^{1}H nuclei at each position. In this way, a 'map' of the ^{1}H nuclei throughout the body can be built up.

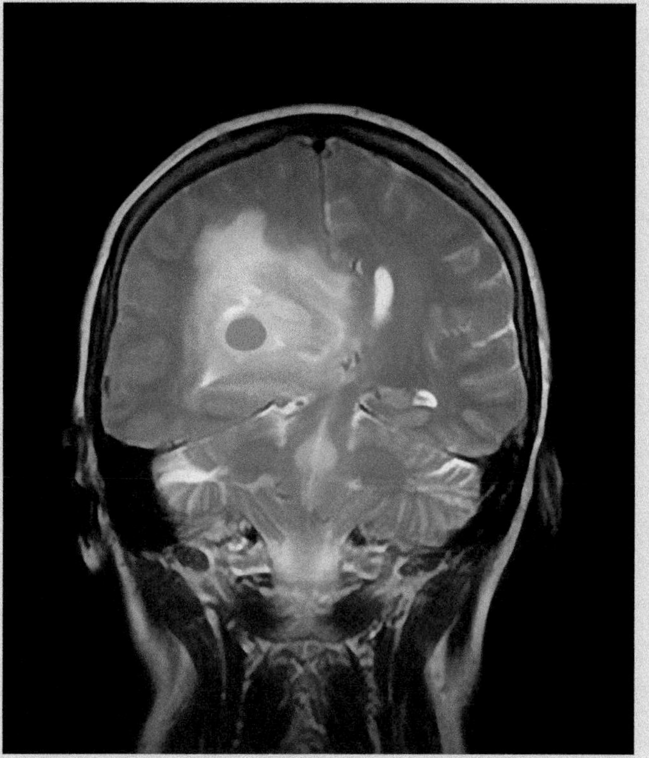

▲ An MRI scan showing a section through the head and neck of a patient with a brain tumour, which shows up red.

The usual arrangement in a medical scanner is to use a system where the magnetic field has gradients in two dimensions to build up an image of a thin 'slice' through the patient. A number of such slices can be combined to give a three-dimensional image. A 'picture' of the body can then be built up with a resolution better than 1 mm. This technique is known as *tomography*.

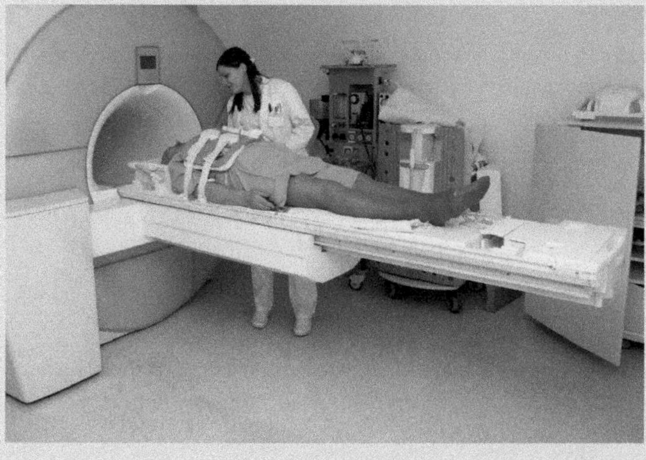

▲ An MRI scanner suitable for scanning a human body. The patient lies in the cavity inside the scanner. The magnetic field surrounds the whole body.

MRI is a non-invasive diagnostic technique that allows cross-sectional images of a human body to be built up without using potentially damaging ionizing radiation such as X-rays. There are differences in water content among tissues and organs and many diseases result in changes of water content. A wide range of diagnoses can be made using MRI including detecting soft-tissue damage and tumours. By looking at how the MRI changes with time, blood flow can be monitored.

In addition to medical applications, MRI has also been applied to a number of chemical problems—for example, in mapping the reactive sites of catalysts and investigating mixing and flow patterns in fluid mixtures.

Question

Why were ^{1}H nuclei, rather than other NMR active nuclei, chosen for monitoring in the MRI technique?

➜

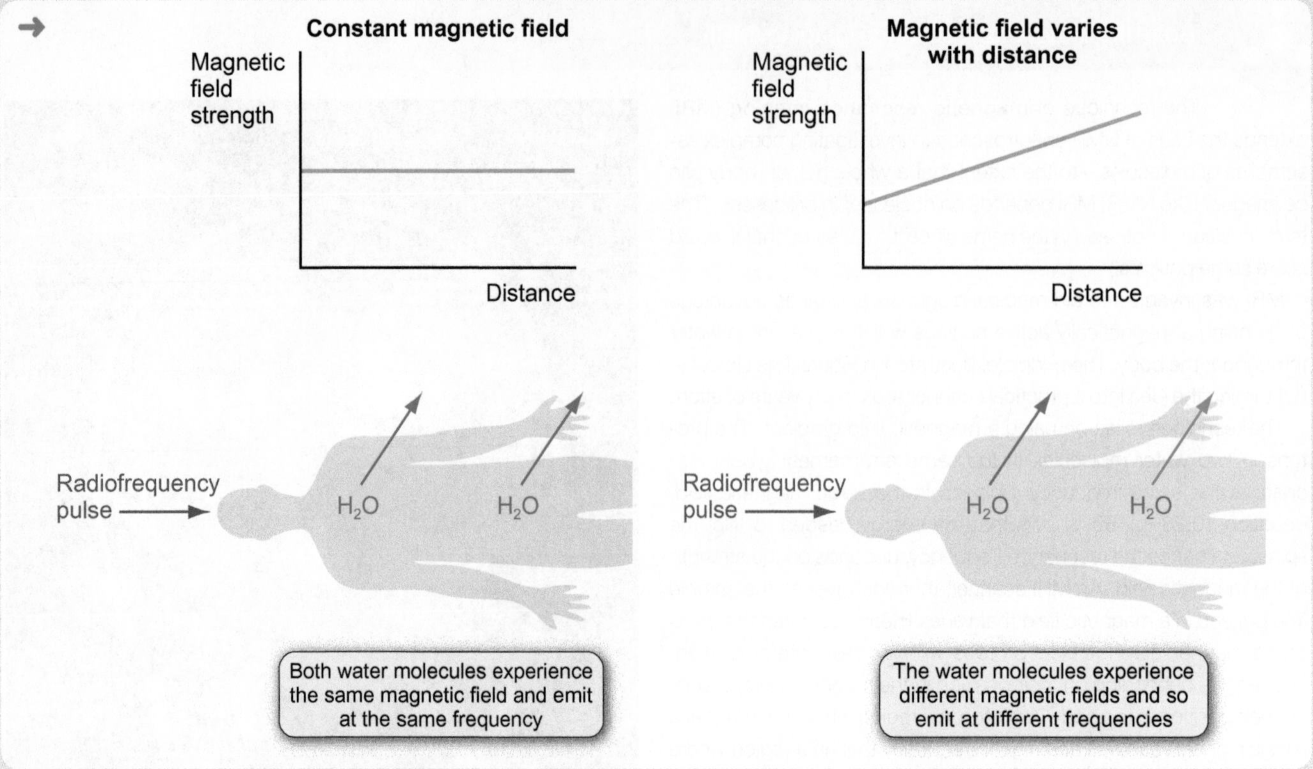
Constant magnetic field
Magnetic field strength
Distance
Magnetic field varies with distance
Magnetic field strength
Distance
Radiofrequency pulse
H₂O
H₂O
Radiofrequency pulse
H₂O
H₂O
Both water molecules experience the same magnetic field and emit at the same frequency
The water molecules experience different magnetic fields and so emit at different frequencies

Figure 1 MRI scanners rely on a magnetic field gradient to give different radio emission from ^{1}H nuclei in different positions in the body.

Summary

- Electrons have spin, which is characterized by a *spin quantum number*, *s*, where $s = \frac{1}{2}$.
- Neutrons and protons also have spin. Atomic nuclei have a *nuclear spin quantum number*, I. $I = \frac{1}{2}$ for ^{1}H, ^{13}C, ^{19}F, and ^{31}P. ($I = 0$ for nuclei where nuclear spins are paired.)
- When placed in an *external magnetic field*, the electron spin states, or the nuclear spin states, have different energies depending on their alignment with or against the magnetic field.
- In spin resonance spectroscopy, the sample is placed in a strong external magnetic field and irradiated with the appropriate energy, ΔE, to promote the electron or nucleus from the lower energy spin state to the higher energy spin state. For electron spin, ΔE corresponds to energy in the microwave region; for nuclear spin, ΔE corresponds to the energy of radio waves.
- *ESR spectroscopy* is used to study species with unpaired electrons.
- *NMR spectra* can be recorded for nuclei with unpaired spins. The most commonly used NMR nuclei have $I = \frac{1}{2}$ but spectra can also be recorded when $I = 1, 1\frac{1}{2}$, etc.
- The nuclear spin comes into resonance when irradiated with electromagnetic radiation with the *Larmor frequency* and the nucleus is promoted into the higher energy spin state.
- *Chemical shift* is characteristic of nuclei (e.g. ^{1}H) in a particular environment within a molecule. It occurs due to shielding effects by electrons around the nucleus and is measured by the difference, δ, of the resonance frequency from that of a standard (e.g. tetramethylsilane for ^{1}H nuclei).
- The relative intensities of the resonance signals in a ^{1}H spectrum depend on the number of ^{1}H nuclei in the molecule giving rise to the signal.
- NMR signals show fine structure due to splitting by adjacent nuclei through *spin–spin* coupling.

For practice questions on these topics, see questions 26–29 at the end of this chapter (p.511).

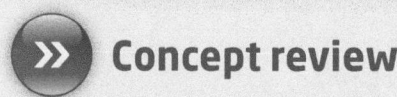

Concept review

By the end of this chapter, you should be able to do the following.

- Understand the basic principles of molecular spectroscopy in terms of the quantization of molecular energy and transitions between molecular energy levels when matter interacts with radiation.
- Use the 'particle in a box' model to account for quantization in a one-dimensional system.
- Calculate the energy of radiation from its wavelength, frequency, or wavenumber and interconvert between these quantities.
- Use the Beer–Lambert law to find the absorbance, A, of the solution and hence the ratio of the intensities of the transmitted and incident radiation, or to find the molar absorption coefficient, depending on what data are available.
- Assign the type of molecular transition associated with radiation of a particular energy in the electromagnetic spectrum.
- Describe in general terms how molecular spectra can be measured.
- Differentiate between absorption and transmission spectra and between absorption and emission spectra.
- Account for the relative populations of energy levels by using the Boltzmann distribution.
- Describe the origin and appearance of rotational spectra.
  - Describe and use the rigid rotor model for a diatomic molecule.
  - State and use the selection rules for rotational spectra.
  - Use spectral data to calculate moments of inertia and bond lengths in diatomic molecules.
- Describe the origin and appearance of infrared spectra.
  - Describe and use the simple harmonic oscillator model for a diatomic molecule.
  - State and use the selection rules for vibrational spectra.
  - Use spectral data to calculate force constants of chemical bonds.
- Describe the types of vibrations that can occur in diatomic and triatomic molecules.
  - Calculate the number of modes of vibration in a molecule.
- Understand what is meant by a vibration–rotation spectrum.
- Understand what is meant by a Raman spectrum.
- Describe the origin and appearance of electronic spectra for molecules.
  - Account for the colours of compounds in terms of the wavelength of radiation absorbed.
  - Explain what is meant by a chromophore.
- Describe the origin and appearance of NMR spectra.
- Determine which nuclei will show NMR spectra.
- Explain what is meant by *chemical shift* and *integration curve* and how these are used to interpret NMR spectra in terms of molecular structure.
- Predict the fine structure in an NMR spectrum of an $I = \frac{1}{2}$ nucleus from the molecular structure.

Table 10.4 Summary of the types of spectra in the chapter

Type of transition	Region of electromagnetic spectrum	Typical units on x-axis of spectrum	Description of spectrum
Electronic	Ultraviolet and visible	Wavelength: λ/nm	Broad absorption bands rising from the baseline
Vibrational	Infrared	Wavenumber: $\tilde{\nu}$/cm^{-1}	Peaks point downwards in a transmission spectrum
Rotational	Far infrared and microwave	Wavenumber: $\tilde{\nu}$/cm^{-1}	Series of evenly spaced lines (pointing downwards in a transmission spectrum or upward in an absorption spectrum)
Nuclear spin	Radiofrequency	Chemical shift: δ/ppm	Resonance signals rising from the baseline. Signals often show fine structure due to spin–spin coupling. Integration curve of ^{1}H spectrum shows area of each signal, which is proportional to the number of ^{1}H nuclei responsible for the signal

Key equations

Born–Oppenheimer approximation	$E_{total} = E_{electronic} + E_{vibration} + E_{rotation} + E_{translation}$	(10.1)
Beer–Lambert law	$A = \varepsilon \times c \times l$	(10.8)
	where $A = \log_{10}\left(\frac{I_0}{I_t}\right)$	(10.7)
Boltzmann distribution of molecules between energy levels	$\frac{n_{upper}}{n_{lower}} = \frac{g_{upper}}{g_{lower}} e^{\left(\frac{-\Delta E}{k_B T}\right)}$	(10.9)
Moment of inertia of a diatomic molecule	$I = \mu r_0^2$	(10.12)
Reduced mass of a diatomic molecule	$\mu = \frac{m_1 m_2}{m_1 + m_2}$	(10.13)
Rotational energy levels of a diatomic molecule	$E_{rotation} = E_J = \frac{h^2}{8\pi^2 I} J(J+1)$	(10.14)
Frequency of vibration of a diatomic molecule acting as a harmonic oscillator	$\nu = \frac{1}{2\pi}\left(\frac{k}{\mu}\right)^{1/2}$	(10.18)
Vibrational energy levels of a diatomic molecule	$E_{vibration} = E_v = \left(v + \frac{1}{2}\right)h\nu$	(10.20)
Energy difference between the two spin states for an	$I = \frac{1}{2}$ nucleus $\Delta E = \frac{\gamma B h}{2\pi}$	(10.23)

Questions

More challenging questions are indicated by an asterisk *.

1 Convert a wavelength of 450 nm to (a) wavenumber, (b) frequency, and (c) energy. (Section 10.2)

2 The yellow colour of sodium emission in a flame is due to emission at 589.0 nm and 589.6 nm. Calculate the difference in energy between the photons emitted at these two wavelengths. (Section 10.2)

3 What wavelength of radiation would provide sufficient energy to break a C–H bond in methane, where the bond dissociation energy is +439 kJ mol^{-1}? (Section 10.2)

$$CH_4 \rightarrow CH_3^{\bullet} + H^{\bullet}$$

4 Calculate the energy difference between the $n = 1$ and $n = 2$ levels for an electron in a one-dimensional box with a length of 4.0×10^{-10} m. At what wavelength would a transition between these levels appear? (Section 10.2)

5* Using the one-dimensional particle in a box model, calculate the first four electronic energy levels for 1,3,5-hexatriene. Assume that a 2*p* electron can move freely along the delocalized π system. At what wavelength would the lowest energy electronic transition appear? Assume the average carbon–carbon distance to be 0.15 nm. (Section 10.2)

6 Calculate the energy differences, ΔE, and the relative populations of the upper and lower energy levels for transitions giving rise to absorption of the following at 298 K (Section 10.3):

(a) an IR photon with wavenumber 2000 cm^{-1};

(b) a microwave photon with frequency 20 GHz;

(c) visible light with wavelength 500 nm;

(d) an X-ray with wavelength 4 nm;

(e) a radio wave with frequency 10 MHz.

(Assume the degeneracy of all the levels is $g = 1$.)

7 What are the transmittance and absorbance of a solution that absorbs: (a) 10%; (b) 90%; (c) 99% of the incident radiation? (Section 10.3)

8 A 5.0×10^{-4} mol dm^{-3} solution of Br_2 in CCl_4 absorbed 64% of the incident light when placed in a 2.0 cm cell at a wavelength where CCl_4 does not absorb. Calculate the molar absorption coefficient of Br_2. (Section 10.3)

9 Naphthalene ($C_{10}H_8$) absorbs around 310 nm in solution. The absorbance of a 1.00×10^{-3} mol dm^{-3} solution is 0.29 in a 1.0 cm cell. Find the molar absorption coefficient. At what concentration will the absorbance be 1.00? (Section 10.3)

10 The percentage transmission of an aqueous solution of fumaric acid is 19.2% for a 0.00050 mol dm^{-3} solution in a 1.0 cm cell. Calculate the absorbance and the molar absorption coefficient. What percentage of light would be transmitted through a solution with double the concentration of this one? (Section 10.3)

11* The molar absorption coefficients for the cobalt and nickel complexes with the ligand, 2,3-quinoxalinedithiol are given below at two different wavelengths. When 2,3-quinoxalinedithiol was added to a solution containing both cobalt and nickel, the solution had an absorbance of 0.446 at 510 nm and 0.326 at 656 nm in a 1.00 cm cell. Calculate the concentrations of both metals in the solution. Assume the absorbances are independent of each other. (Section 10.3)

	$\varepsilon/dm^3\,mol^{-1}\,cm^{-1}$ at 510 nm	$\varepsilon/dm^3\,mol^{-1}\,cm^{-1}$ at 656 nm
Co complex	36400	1240
Ni complex	5520	17500

12 Calculate the relative populations of two non-degenerate energy levels separated at 298 K by (Section 10.3):

(a) an energy corresponding to a frequency 4.3×10^{13} Hz;

(b) an energy corresponding to a wavelength 254 nm;

(c) an energy corresponding to a wavenumber 5 cm^{-1}.

In each case, comment on the value that you calculate.

13 Explain why a rotational spectrum is observed for ICl but not for I_2 or Cl_2. (Section 10.4)

14 The $J = 2 \rightarrow J = 3$ transition occurs in a rotational spectrum at U 30 cm^{-1}. At what wavenumber will the $J = 0 \rightarrow J = 1$ transition be seen? What is the rotational constant for the molecule? (Section 10.4)

15 The $J = 0 \rightarrow J = 1$ transition for $^1H^{79}Br$ occurs at a frequency of 500.7216 GHz. (Section 10.4)

(a) Calculate the bond length for this molecule.

(b) Calculate the relative populations of the $J = 0$ and $J = 1$ levels.

16 The bond length of $^1H^{35}Cl$ is 0.129 nm. Calculate the rotational constant and the separation between peaks (in cm^{-1} and Hz) in the rotational spectrum. (Section 10.4)

17* The lines in the rotational spectrum of HF are separated by 41.90 cm^{-1}. Calculate the bond length. What will be the separation between lines in DF (^2HF)? (Section 10.4)

18 Which of the following molecules is (are) IR active: NO; HBr; O_2? (Section 10.5)

19 Calculate the number of normal modes of vibration of (a) benzene (C_6H_6) and (b) cyclohexane (C_6H_{12}). (Section 10.5)

20 Draw diagrams to represent the normal modes of vibration of CS_2 and OCS. Indicate which modes will show in the IR spectrum. (Section 10.5)

21 HBr shows an IR absorption at 2650 cm^{-1}. The HBr bond length is 0.141 nm. Calculate the force constant of the bond. (Section 10.5)

22 How many normal modes of vibration would you expect for cyclopentane, C_5H_{10}? Why would they not all give rise to peaks in the infrared spectrum? (Section 10.5)

23* Calculate the vibration frequency of a bond between a hydrogen atom and a solid surface if the bond has a force constant of 5 kg s^{-2}. (Section 10.5)

24* The radiation absorbed by $^{12}C^{16}O$ during a vibrational transition occurs at 2168 cm^{-1}. (Section 10.5)

(a) Calculate the ground state vibrational energy in kJ mol^{-1}.

(b) What is the ratio of the number of molecules in the first vibrational excited state (v = 1) compared with the ground state (v = 0) at 298 K?

(c) Calculate the force constant of the bond, assuming $^{12}C^{16}O$ behaves as a simple harmonic oscillator.

(d) Estimate the change in the position of the peak if ^{12}C was replaced by ^{13}C.

25 Phenolphthalein is used as an indicator in acid–base titrations. In solutions at high pH, it is a bright magenta colour with a peak at 553 nm in the absorption spectrum; at low pH phenolphthalein is colourless. (Section 10.6)

(a) In terms of absorption and transmission of light, explain the colour of phenolphthalein at high pH.

(b) At high pH, phenolphthalein is ionized; it is unionized at low pH. Suggest a reason for the colour change.

(c) Does the absorption move to higher energy or lower energy at low pH?

26 Find the nuclear spin quantum number of the following isotopes: ^2H; ^3H; ^{14}N; ^{15}N; ^{16}O. (Section 10.7)

27* Predict the form (approximate chemical shifts, integrals, and splitting patterns) of the ^1H and ^{13}C NMR spectra of pentan-2-one. (Section 10.7)

28 Predict the splitting patterns in the ^1H NMR spectra of the following molecules (Section 10.7):

(a) propanone (CH_3COCH_3);

(b) 1-bromopropane ($CH_3CH_2CH_2Br$);

(c) 1,1-dichloroethene ($CCl_2{=}CH_2$);

(d) *E*-1,2-dichloroethene (CHCl=CHCl); (*Z* and *E* isomers);

(e) nitrobenzene;

(f) 1,2-dinitrobenzene;

(g) 1,3-dinitrobenzene;

(h) 1,4-dinitrobenzene.

29 The magnetogyric ratio for a ^1H nucleus is $26.7519 \times 10^7\,T^{-1}\,s^{-1}$. Calculate the magnetic field strength (in tesla) required to give a Larmor frequency of 400 MHz. (Section 10.7)

Further questions on using IR and NMR spectroscopies in the context of structure elucidation are given at the end of Chapter 12 (p.606).

11

Analytical chemistry

▼ An analytical chemist working with a gas chromatography–mass spectrometry instrument in the Drug Control Centre at Kings College London, one of the accredited testing centres of the World Anti-Doping Agency.

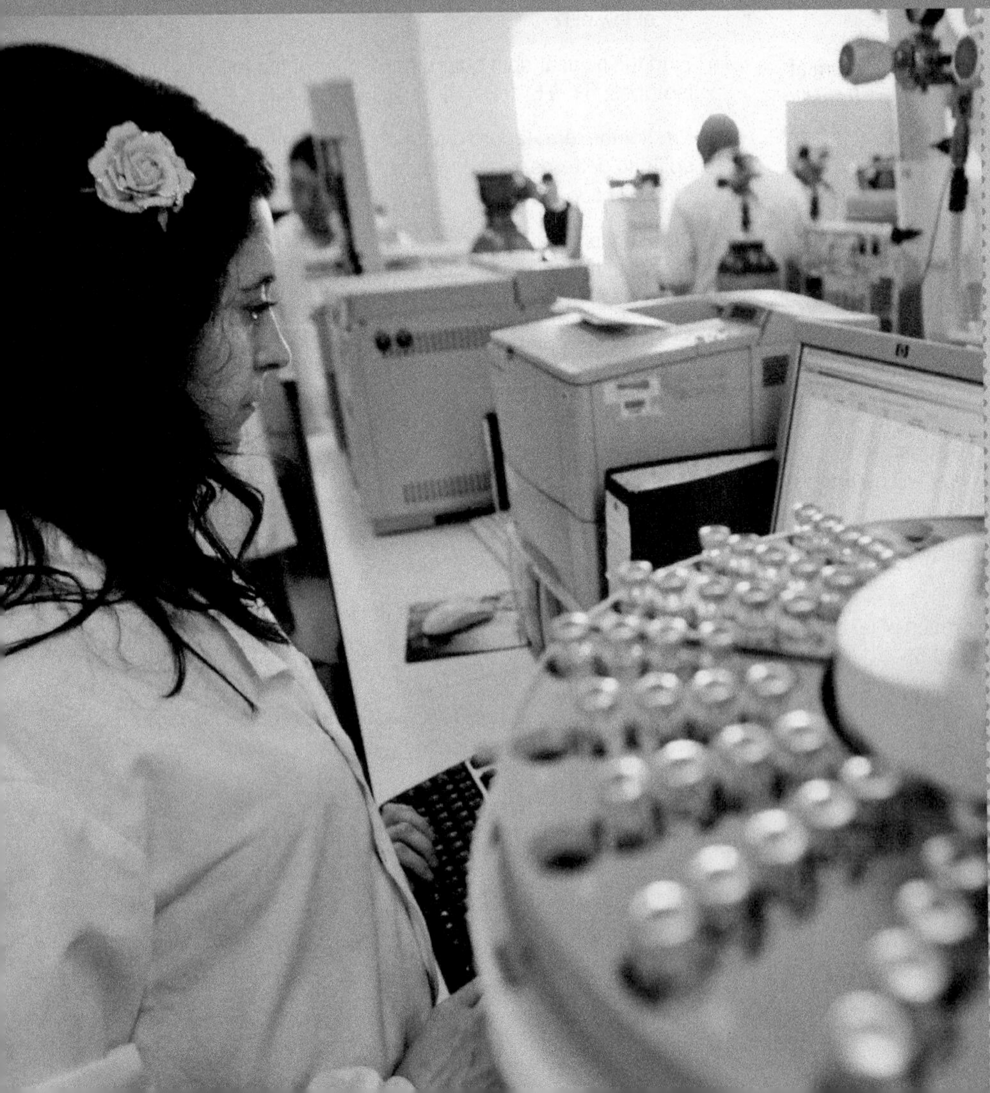

This chapter builds on the following topics:

- Measurement, units, and nomenclature Section 1.2, p.6
- Atoms and the mole Section 1.3, p.12
- Chemical equations Section 1.4, p.21
- Working out how much you have Section 1.5, p.34
- States of matter and phase changes Section 1.7, p.47
- Non-covalent interactions Section 1.8, p.52
- Electromagnetic radiation and quantization Section 3.2, p.117
- Atomic spectra and the Bohr atom Section 3.3, p.124
- The strengths of acids and bases Section 7.2, p.308
- General principles of spectroscopy Section 10.3, p.458
- Vibrational spectroscopy Section 10.5, p.476
- Electronic spectroscopy Section 10.6, p.491
- Electrochemical cells Section 16.3, p.735
- Thermodynamics of electrochemical cells Section 16.4, p.746

Drugs, sport, and analytical chemistry

In elite sport, tiny improvements in performance can make the difference between being good and being the best. For elite athletes, it may be tempting to use drugs to improve performance and so reach the top. Analytical chemists are at the front of keeping athletics, and other sports, drug free.

▼ Testosterone and its stereoisomer epitestosterone

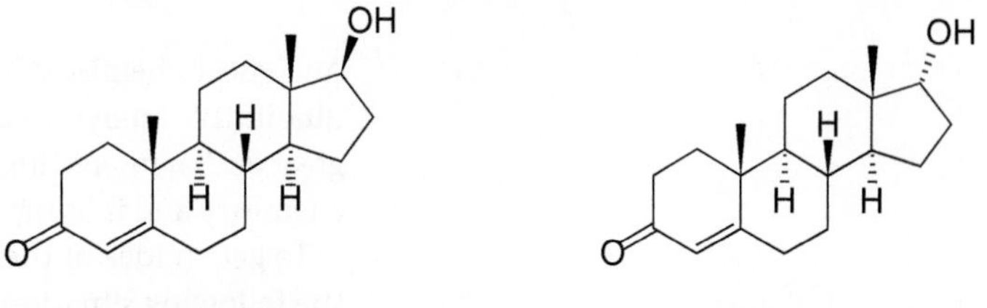

OH
H
H
H
O
OH
H
H
H
O

testosterone   epitestosterone

In preparation for the 2012 London Olympics, the organizers set up what was described as 'the most advanced and up-to-date testing laboratory anywhere in the world', with the capacity to carry out 6250 tests on over 200 banned compounds with results returned within 24 hours. The lab employed more than 150 scientists, with over 1000 additional staff involved in the anti-doping processes such as highly secure sample collection.

But the fight to keep doping out of sport isn't over yet, as new performance-enhancing drugs are produced all the time. The World Anti-Doping Agency (WADA) was set up in 1999 to lead the worldwide mission for drug-free sport, and each year WADA produces a list of banned substances and methodologies. These include blood doping drugs such as erythropoietin (EPO), which increases the blood's ability to carry oxygen, and anabolic steroids to build up athletes' muscle.

Testosterone is a prohibited anabolic steroid used by some male athletes. Testosterone is produced naturally in the testes of adult males and to a lesser extent in the ovaries of females. It makes the male develop active sexual organs and secondary sexual characteristic such as increased muscle and bone mass and body hair. Testosterone can be synthesized and has legitimate medical uses, such as treating osteoporosis. The attraction to athletes is that using extra testosterone—beyond what their body makes—helps them put on extra muscle.

So how can athletes be tested to see if they are taking additional testosterone, when their bodies already contain it naturally? One approach is to define a level of testosterone which would be expected in an average male, and then look for athletes in whom the level is significantly higher. The trouble with this approach is that average testosterone levels vary quite widely among different genetic populations.

The Athlete Biological Passport (ABP) can get round this. An athlete's ABP contains a record of the level in their body of a number of critical substances which may be altered by doping. In the case of testosterone (T), the ABP records the *ratio* between the levels of testosterone and of its stereoisomer epitestosterone (E), which is also produced by the testes but is inactive biologically. The expected ratio of T/E in the body is typically between 0.1:1 and 6:1, and by monitoring the variation in an individual's T/E ratio over time, doping can be identified by checking whether an athlete's T/E ratio is outside their own normal range. The advantage of using this approach is that the ratio remains the same even when the substances pass out of the blood and into the urine, where they are sampled.

The urine sample is tested using a technique called gas chromatography–mass spectrometry, GC–MS (see Box 11.4, p.538). The compounds are carried by a gas along a narrow quartz capillary coated with a polymer, where they are separated. They are detected as they emerge from the capillary, each compound giving a peak on the gas chromatogram. The mass spectrum of each compound is recorded to confirm the identity of each peak. From this spectrum, the levels of T and E can be found, and the ratio can be compared with what would be expected from the athlete's biological passport. GC–MS instruments fit on to a bench top and give results quickly. However, the preparation procedures need much skill and experience, so testing laboratories employ large numbers of highly skilled analytical chemists.

▼ Sample T/E data from a urine sample, analysed using GC–MS. The height of the peaks is used to calculate the T/E ratio in the sample. Significant variations in the T/E ratio over time may suggest doping.

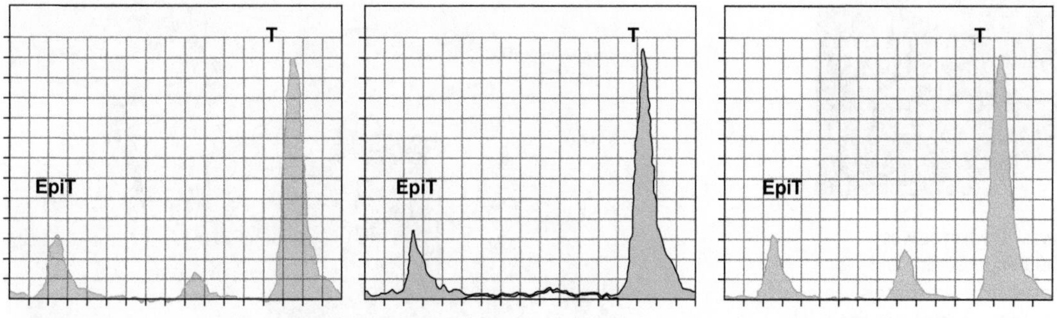

EpiT
T
EpiT
T
EpiT
T

SRM 423 > 196
Peak Height
Testosterone – 11,291
Epi-Test. – 2,727
T/E ratio – 4.14

SRM 423 > 301
Peak Height
Testosterone – 5,445
Epi-Test – 1,364
T/E ratio – 3.99

SRM 423 > 209
Peak Height
Testosterone – 23,736
Epi-Test – 5,687
T/E ratio – 4.17

Analytical chemistry is concerned with determining *which* substances are in a sample—**qualitative analysis**—and *how much* of each substance is present—**quantitative analysis**. A great deal of interesting and exciting scientific discovery has gone into advancing analytical chemistry and it brings together a number of topics from across all branches of the subject.

To get an idea of the range of analytical problems that might be encountered, consider the following situations. Analytical chemists might be analysing samples of:

- *Medicines or the raw materials from which they are manufactured*: to check purity—high purity (99.9+%) must be maintained to avoid the presence of potentially harmful by-products.
- *Medicines, works of art, or gemstones*: to establish their authenticity or otherwise—there is a growing market in counterfeit products.
- *Substances extracted from natural sources as potential active ingredients for new medicines*: only microgram amounts may be available to establish their identity.
- *A batch of steel from a steelworks*: steel is an alloy of iron with small quantities of carbon and up to a dozen other elements. Varying the composition of these elements by just 1% drastically affects the properties of the steel, which may be manufactured in batches of over 20 tonnes. The composition, and consistency from batch to batch, of the product are important for the customer.
- *The effluent from a factory, power station, or industrial plant*: to measure pollutant concentrations and determine whether legal guidelines are being followed.
- *Items of evidence in a forensic investigation*: often, only a tiny sample may be available, such as a paint chip, a single hair, or the DNA extracted from a single fingerprint.
- *Silicon to be used for making electronic devices*: in the semiconductor industry, the silicon used to make chips needs to have a purity of 99.9999999% (see Box 14.7, p.682) so methods that can measure tiny concentrations of impurities are needed.

Clearly, there is huge diversity in the work of analytical chemists. There may be only a few milligrams of sample on which to work—or many tonnes. The result may be needed to 6 decimal places—or a rough estimate may be sufficient. Many different types of sample in many forms must be analysed. The consequences of getting it wrong can be serious. If a manufacturing process goes wrong because the analysis is poor, a company may incur a great deal of waste and expense in product recalls and lost customers. Wrongful convictions or acquittals may result from poor analysis by forensic scientists. In extreme cases, such as failure to detect contamination of drugs or foodstuffs, people may become seriously ill or die.

The aims of this chapter are to describe the chemical principles behind a range of analytical methods and to show how they are used. Table 11.1 lists some of the methods used in analytical chemistry. Some of the methods, such as titrations and gravimetric analysis,

REMEMBER

Stephen Lawrence was 18 when he was murdered in April 1993. Poor handling and analysis of forensic samples meant that his murderers stayed free until new evidence and improved analytical techniques led to their conviction in January 2012.

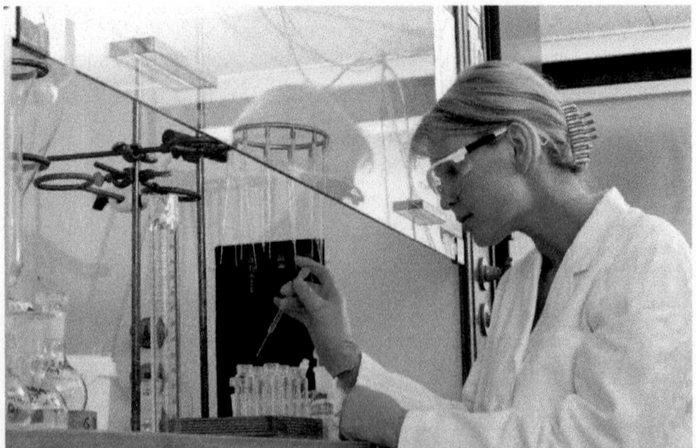

A forensic scientist treats samples to extract tiny amounts of a drug before analysis by mass spectrometry.

Table 11.1 Some methods used in analytical chemistry

Method	Typical applications	See Section
Acid–base titration	Concentration of acids or bases in solution	1.5 (p.34)
Redox titration	Concentration of ions in solution	1.5 (p.34)
Gravimetric analysis	Concentration of ions in solution	1.5 (p.34)
Ion selective electrode	pH, ion concentrations in solution	11.2 (p.523)
Electrochemistry	Concentration of species in solution	11.2 (p.523), 16.4 (p.746)
Atomic spectrometry	Concentrations of elements (usually metals)	11.5 (p.547)
UV/VIS spectroscopy	Concentration of species that absorb in UV or visible region	10.6 (p.491), 11.4 (p.542)
NMR spectroscopy	Identification of functional groups, concentration of species	10.7 (p.496), 12.3 (p.578)
IR spectroscopy	Identification of functional groups, concentration of species	10.5 (p.476), 12.2 (p.570)
Mass spectrometry	Identification of molecules and functional groups	11.3 (p.528), 12.1 (p.558)
Chromatographic methods	Identification, and measurement of concentration, of species in mixtures	11.3 (p.528)

are described in Chapter 1. These 'wet chemical' methods are still used in some analyses. They are absolute methods that give the result directly without the need for calibration. However, they can be time-consuming, must be carried out by skilled analysts, and are usually not very sensitive and so can only be applied when relatively high concentrations are present. This chapter deals with *instrumental* chemical analysis that covers a wide range of concentrations. Before dealing with the types of methods that chemists use, some general features of analytical chemistry are discussed in the next section.

Analytical methods such as titrations that give the amount or concentration of material directly are called **primary methods**. Other methods that involve calibration using solutions with known concentrations are **secondary methods**.

11.1 Carrying out an analysis

A number of factors need to be considered to ensure a successful analysis, in addition to simply selecting the method to use. These are outlined in Figure 11.1.

Choosing an analytical method: what needs to be measured?

In selecting which technique to use for a particular application, there are a number of questions that the analyst must ask. What is the target species? How can I detect it? Can I detect it in the presence of other substances that may be present? What is the concentration that I need to measure and is the technique appropriate for this concentration? What level of accuracy is needed? What will the results be used for? The analyst needs to know what is being measured and what the results will be used for before embarking on the analysis.

The last of these questions is often the major influence in choosing a method. If you are operating an industrial plant and need to check the purity of an intermediate during a reaction sequence, it is unlikely that you will choose a method that takes a week to produce a result. If you are measuring the acidity or alkalinity of the soil in your garden, measuring the pH to ±0.1 is probably sufficient; it is not necessary to know the result to four significant figures.

Obtaining and preparing a sample for analysis

Imagine that an analyst is called in to investigate the suspected contamination of a river with pesticides. The whole river cannot be analysed: a **sample** must be taken. However,

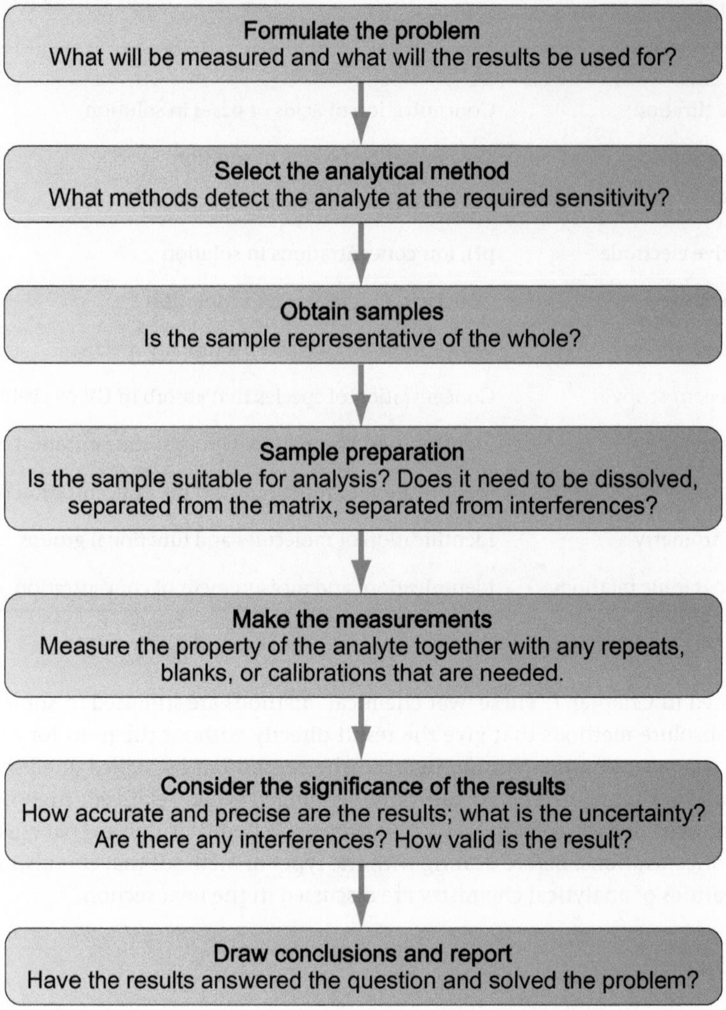
Formulate the problem
What will be measured and what will the results be used for?
Select the analytical method
What methods detect the analyte at the required sensitivity?
Obtain samples
Is the sample representative of the whole?
Sample preparation
Is the sample suitable for analysis? Does it need to be dissolved, separated from the matrix, separated from interferences?
Make the measurements
Measure the property of the analyte together with any repeats, blanks, or calibrations that are needed.
Consider the significance of the results
How accurate and precise are the results; what is the uncertainty? Are there any interferences? How valid is the result?
Draw conclusions and report
Have the results answered the question and solved the problem?

Figure 11.1 An analytical procedure consists of several steps, although not all steps are needed for all analyses.

> There are a number of statistical techniques that can be applied to sampling to ensure that the number and types of samples taken are representative.

the sample taken must be **representative** of the material being analysed. If you simply remove 100 cm^3 from one point in the river, will the result of your analysis be valid? Might the concentration vary with water depth, with the distance from the river bank, with the flow rate, the time of day, or the temperature? All of these factors may affect the concentration of pesticide that you measure. To overcome this, it is usual to take several samples from different places and combine them.

The sample to be analysed contains the target species (molecules, atoms, ions, etc.) in which you are interested—called the **analytes**—usually mixed with a number of other materials, collectively called **the matrix**. The properties of the matrix can have an influence on the analysis since the matrix may contain species that **interfere** with the analysis. Substances in the matrix may absorb an analyte so that it is difficult to obtain for analysis. Other species may also respond to the analytical technique chosen. A technique that responds only to the analyte of interest is a **selective** method. The sample must be stable so that it does not change in the time between it being taken and it being analysed. This is particularly important in biochemical samples. Worked example 11.1 shows some further examples of how samples for analysis are often combined with matrix material.

Once the sample is obtained, it must be converted to a form suitable for use in the chosen method. For solid and liquid analytes, this usually involves dissolving into solution; for example, metals may be treated with acid to form a solution containing the metal ions.

'A man embracing a woman' (1524) by Dosso Dossi, in the National Gallery London. During the restoration of this painting, scientists used gas chromatography–mass spectrometry (Box 11.4, p.538) to analyse the varnish used by the painter.

The resulting solution needs to be in the correct concentration range for the method. For example, pesticides in river water will be present at low concentrations and so must be **pre-concentrated** before analysis. They can, for example, be extracted from the aqueous solution using an organic solvent in which they are more soluble and the solvent is then evaporated to give a solution of the appropriate concentration. Using a solvent that does not dissolve interfering species can be helpful at this stage.

The methods used to prepare samples must not change the nature of the analyte. Samples containing solutions of ions must not be treated in a manner such that the oxidation state of the ion could change. The analyte must not be converted into a volatile species which might be lost. Stirring an acidic solution with a nickel spatula can add Ni^{2+} ions to a solution—stirring should always be done with a glass rod. Each of the processes involved in preparing the sample for analysis should aim for 100% transfer of the analyte—called **quantitative transfer.**

Worked example 11.1 What's in a sample for analysis?

Identify the analyte(s) and the matrix in each of the following systems. In each case, identify potential problems in obtaining a sample for chemical analysis.

(a) A paracetamol pain-relieving tablet.

(b) Greenhouse gases in the atmosphere.

(c) The concentration of alcohol in blood.

Strategy

The analyte is the species that is being measured; the matrix is the remainder of the sample. To highlight potential problems, consider how a representative sample could be obtained and whether components in the matrix might interfere with the analysis.

Solution

(a) A paracetamol pain-relieving tablet.
The analyte is the active ingredient—the paracetamol. The matrix is the other components of the tablet (such as starch, chalk, etc.). Sampling is straightforward—the tablets should all have a uniform composition. The paracetamol can easily be separated from the matrix by dissolving in a suitable solvent that does not dissolve the matrix components.

(b) Greenhouse gases in the atmosphere.
The analytes are, for example, CO_2, CH_4; the matrix is the other components of air. Sampling may be difficult since air composition is affected by seasonal and weather conditions and the local environment. All these factors must be taken into account when drawing conclusions from the measurements.

(c) The concentration of alcohol in blood.
The analyte is ethanol. All the other components of blood constitute the matrix. Only a small quantity of blood will be available—is this representative? Arterial blood differs from venous blood. Alcohol may evaporate; blood changes composition with time so that the sample must be preserved or the analysis carried out quickly. Some of the proteins and cells may interfere with the analysis method.

Question

Identify the analyte(s) and the matrix in each of the following systems:

(a) iron in iron ore from a mine;

(b) pesticide residues in fruit.

Making the measurements

Calibration is carried out to determine the relationship between a measured quantity and the concentration using reference samples of known concentrations. Often, this involves plotting a calibration graph.

A low detection limit indicates that the method is highly sensitive; analytes can be measured at low concentrations.

With the method chosen and the sample prepared, the analysis can now be done. As well as carrying out the analysis on the sample, the chemist also conducts a series of control experiments. For example, the same procedures used with the river water might be applied to a sample of pure water from the laboratory. This **blank** analysis would show if the procedures unintentionally add any of the analytes to the water. The method is checked and **calibrated** by analysing solutions with accurately known concentrations. These may be prepared by the analyst or may be purchased from companies or regulatory authorities who specialize in supplying **reference materials** for calibration. The analysis must be carried out on a number of **replicate** samples so that the **repeatability** (that is, whether the same answer is obtained for each sample) of the analysis can be determined. The method chosen must have a suitable detection limit for the analyte. The **detection limit** is the smallest concentration of substance that can be distinguished from a blank analysis.

Estimating the significance of the results

The analyst can now calculate the concentration of the analyte in the sample. In doing this, it is important to consider how **accurate** and **precise** the results are. In everyday life these two terms are often used as if they have the same meaning, but in scientific use they have particular meanings, as demonstrated in Figure 11.2.

- **Accuracy** refers to how close a result is to the true value.
- **Precision** describes how well a number of measurements of the same quantity under the same conditions agree with each other.

Provided that the correct blank and calibration samples have been analysed, the chemist may have confidence that the analysis is accurate. However, simply repeating the analysis of a sample using the same method and equipment cannot confirm this; a *precise* analysis is not necessarily *accurate*. Obtaining the same result using two different analytical methods will confirm the accuracy. Another way of confirming the accuracy of an analysis is to use the method with a **reference material** whose concentration is accurately known.

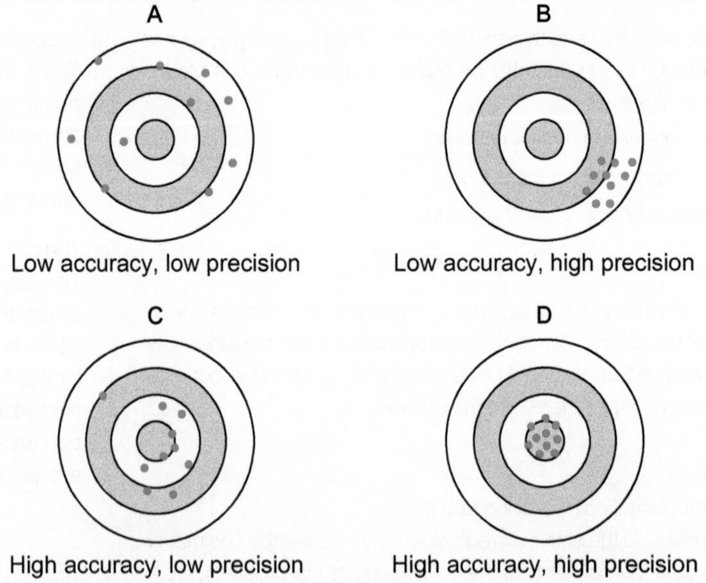
A
B
Low accuracy, low precision
Low accuracy, high precision
C
D
High accuracy, low precision
High accuracy, high precision

Figure 11.2 Four rifle shooters aim at the centre of their target. A is not very good—the shots are not close to the centre (low accuracy) nor close together (low precision). B's shots are close together (high precision) but not close to the aiming point (low accuracy); B is a good shot but the rifle may not be accurate. C's shots are spread out (low precision) but are scattered around the centre of the target (high accuracy); the rifle is good but C is not a very good shot. The shots by D are both precise and accurate.

There are two main types of **error** that can affect the accuracy and precision of a series of measurements.

- **Systematic** errors (also called **determinate** errors) affect all readings in the same way. They are often due to an incorrect instrument setting. For example, if a balance used to weigh compounds when making up standard solutions is always in error by ±0.1 g, all the readings will be affected by this amount. This type of error affects the accuracy of the result and can be eliminated by proper calibration.
- **Random** errors (also called **indeterminate** errors) occur in all experiments and cause results to scatter about the true value. They affect the precision of the result. Even a highly skilled analyst will obtain a spread of results when repeating an experiment under the same conditions. This type of error is dealt with by taking large numbers of measurements, and can be accounted for using statistical methods.

Random errors are often called **uncertainties** to emphasize that they will always occur, even when no mistakes are made.

If all the errors in an experiment are truly random, the measurements follow a **normal distribution**, as shown in Figure 11.3. The values are distributed around the mean value, $\bar{x}$. The width of the distribution is characterized by the **range** of the data (the difference between the highest and lowest values) and by the **standard deviation**, σ. The use of these statistical terms is illustrated in Worked example 11.2.

For a series of n measurements, x_i, the mean, $\bar{x}$, is defined as

$$\bar{x} = \frac{\int_{i=1}^{i=n} x_i}{n} \qquad (11.1)$$

$\Sigma_i x_i$ means the sum of all n values of x_i. So, Equation 11.1 says 'add up all the values and divide by the number of measurements'. $i = 1$ and $i = n$ represent the lowest and highest values to be summed.

The standard deviation, σ, of the results is defined in terms of $(x_i - \bar{x})$, the difference between each measurement and the mean

$$\sigma = \left(\frac{\sum_i (x_i - \bar{x})^2}{n-1} \right)^{1/2} \qquad (11.2)$$

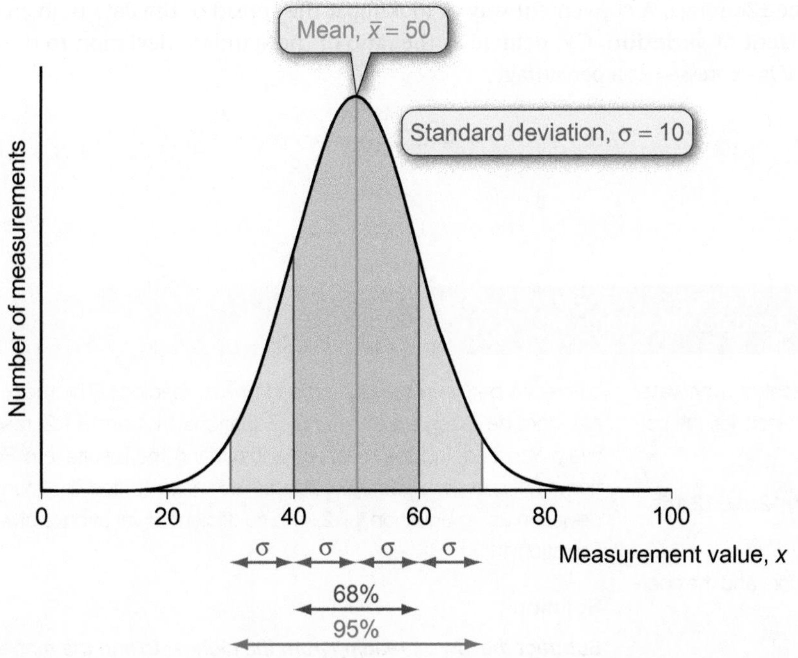

Mean, x̄ = 50
Standard deviation, σ = 10
Number of measurements
0
20
40
60
80
100
Measurement value, x
σ σ σ σ
68%
95%

Figure 11.3 A normal distribution curve. A large number of measurements are made and the number of times a result occurs is plotted against the value of the measurement. For a normal distribution, 68% of the measurements lie within ±1 standard deviation (σ) of the mean, and 95% lie within ±2 standard deviations of the mean.

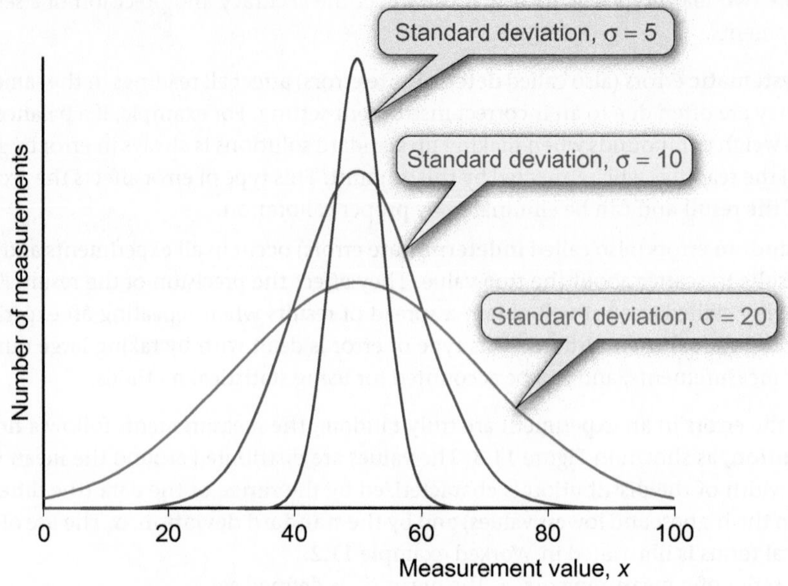
Standard deviation, σ = 5
Standard deviation, σ = 10
Standard deviation, σ = 20
Number of measurements
0
20
40
60
80
100
Measurement value, x

Figure 11.4 A low standard deviation means that all the results lie close together (high precision). A high standard deviation indicates the results are more widely scattered (low precision).

The standard deviation is a measure of how much the measurements differ from the mean value. A large value of σ indicates that there is a large spread of data (low precision) and a small value indicates that all the values lie close to the mean (high precision) as shown in Figure 11.4.

Statistical theory shows that, for a random distribution of a large number of measurements, 68% of the measurements will lie within $\pm\sigma$ of the mean and 95% will lie within $\pm 2\sigma$ of the mean. This means that, in a series of 20 measurements, only one should lie outside the $\pm 2\sigma$ range. A convenient way of looking at the spread of the data is given by the **coefficient of variation**, *CV*, defined as the ratio of the standard deviation to the mean. The *CV* is expressed as a percentage.

> When scientific measurements are reported, there should be some indication of the uncertainty in the values. Results are usually quoted as $\bar{x} \pm 2\sigma$, where σ is the standard deviation and refers to the random errors. (Contributions from any systematic errors will further increase the uncertainty.)

$$CV = \frac{\sigma}{\bar{x}} \times 100 \qquad (11.3)$$

Worked example 11.2 Mean and standard deviation

In a Winkler titration to determine the oxygen content of river water (see Box 1.6, p.38), the following results were obtained for the volume (in cm^3) of sodium thiosulfate required.

12.07, 12.27, 12.15, 12.26, 12.26, 12.21, 12.24, 12.19, 12.16, 12.17, 12.28, 12.19, 12.21

Determine the range, the mean, the standard deviation, and the coefficient of variation of these results.

Strategy

Calculators and computer programs often do these calculations automatically—but you need to know the basic methods. The range is the difference between the highest and lowest readings. The mean and standard deviation are defined by Equations 11.1 and 11.2, respectively. You can find the mean by substituting the results into Equation 11.1 and then use the value of the mean to find the standard deviation using Equation 11.2. The coefficient of variation is given by Equation 11.3.

Solution

Subtract the lowest reading from the highest to find the range.

Highest value = 12.28 cm^3 Lowest value = 12.07 cm^3

Range = (12.28 cm^3 – 12.07 cm^3) = 0.21 cm^3

→

Substitute the results into Equation 11.1 to find the mean, $\bar{x}$.

$$\bar{x} = \frac{\sum_{i=1}^{i=n} x_i}{n} \tag{11.1}$$

$$= \frac{\begin{pmatrix} 12.07+12.27+12.15+12.26+12.21+12.24+12.19 \\ +12.16+12.17+12.28+12.19+12.21 \end{pmatrix} \text{cm}^3}{13}$$

$$= \frac{158.66}{13}\,\text{cm}^3 = 12.20\,\text{cm}^3$$

To use Equation 11.2, first find the difference between each result and the mean and then sum the square of the values.

Result number	Titration volume, x_i/cm^3	$(x_i-\bar{x})$/cm^3	$(x_i-\bar{x})^2$/cm^6
1	12.07	−0.13	0.0169
2	12.27	0.07	0.0049
3	12.15	−0.05	0.0025
4	12.26	0.06	0.0036
5	12.26	0.06	0.0036
6	12.21	0.01	0.0001
7	12.24	0.04	0.0016
8	12.19	−0.01	0.0001
9	12.16	−0.04	0.0016
10	12.17	−0.03	0.0009
11	12.28	0.08	0.0064
12	12.19	−0.01	0.0001
13	12.21	0.01	0.0001
		Sum	**0.0424**

Substitute into Equation 11.2 to find the standard deviation.

$$\sigma = \left(\frac{\sum_i (x_i-\bar{x})^2}{n-1}\right)^{1/2} \tag{11.2}$$

$$= \left(\frac{0.0424\,\text{cm}^6}{13-1}\right)^{1/2}$$

$$= 0.06\,\text{cm}^3$$

Use Equation 11.3 to find the coefficient of variation.

$$CV = \frac{\sigma}{\bar{x}} \times 100 \tag{11.3}$$

$$= \frac{0.06\,\cancel{\text{cm}^3}}{12.10\,\cancel{\text{cm}^3}} \times 100$$

$$= 0.5\%$$

Question

If another 20 titrations were performed under the same conditions, how many results would you expect to lie between 12.14 cm^3 and 12.26 cm^3?

If you look at the titration data in Worked example 11.2 in the form of Figure 11.5, it is clear that one point (12.07 cm^3) is much further from the mean than the others—it is an **outlier**. It may be that a problem occurred in this measurement so including it may lead to an error in the mean. There are a number of statistical tests (such as the Q-test) that can be applied to a series of results to determine whether a particular value is to be expected from an experiment or is an outlier.

An example of how the various aspects of performing an analysis fit into an analytical procedure is given in Box 11.1.

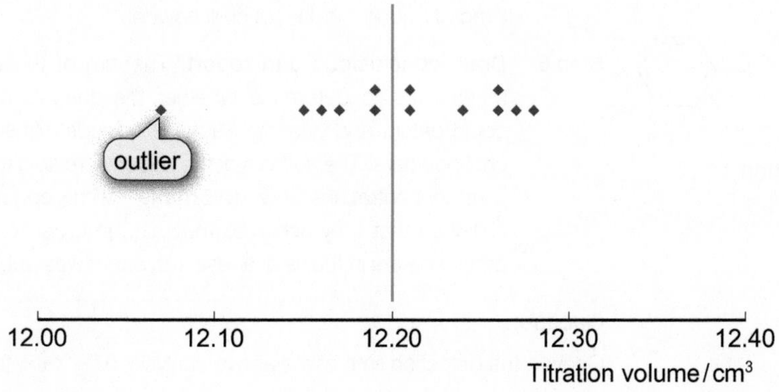

outlier
12.00
12.10
12.20
12.30
12.40
Titration volume/cm3

Figure 11.5 The distribution of the titration results from Worked example 11.2 around the mean. (Since there is a small number of results, they do not form a normal distribution.)

Box 11.1 Analysing food contaminants

In February 2005, the UK Food Standards Agency issued a warning that some food products had been found to contain Sudan I, a dye that is banned from foodstuffs in the EU. The contamination was traced to an imported chilli powder that had been used to make up a sauce, which was then used as an ingredient in over 500 products.

Sudan I is a red azo dye (see Chapter 22, p.1002) that is used to colour waxes, solvents, and polishes. Its use in food is banned since it is suspected of increasing the risk of developing some types of cancer. Laboratory tests in which Sudan I was fed to mice showed that tumours developed in the liver and bladder. At the low concentrations detected in the food products, the risk was said to be very small and there was no immediate danger to human health.

A method for analysing Sudan I in a range of foodstuffs was needed. So, how can the general scheme for an analysis outlined in Figure 11.1 be applied in this case?

Step 1 **Formulate the problem.** Sudan I is banned from food so the method could simply detect its presence (or confirm its absence). However, if it is present, it would be important to know its concentration so that any health risks can be assessed.

Step 2 **Select the analytical method.** Sudan I is coloured so an analyst might first consider using a spectroscopic method (see Section 11.4, p.542). However, the dye was found in chilli powder, which is itself highly coloured, so separation of Sudann I from other components is necessary. The molecule is organic and so unlikely to have simple electrochemistry (see Section 11.2, p.523). Chromatography is a separation technique that also allows measurement of the concentration of each of the components in a mixture. Sudan I is readily soluble in a number of solvents so high performance liquid chromatography (HPLC; see Section 11.3, p.537) could be used. A quick check of the chemistry literature shows that a number of workers have used HPLC to separate Sudan I from a mixture and measure its concentration. The method here is taken from F. Tateo and M. Bononi, *Journal of Agricultural and Food Chemistry*, Vol. 52 (2004).

Step 3 **Obtain samples.** The Sudan I dye may be mixed with dry chilli powder or found as a component of, for example, a tomato sauce. To ensure that the sample is representative, the powder or sauce must be thoroughly mixed and homogenized so that it has a constant composition throughout.

Step 4 **Sample preparation.** The dye must be extracted into solution. This was achieved by treating the chilli powder with ethanol and filtering off the undissolved matrix. The ethanol dissolves Sudan I and some other organic compounds in the sample.

Step 5 **Make the measurements.** The sample in ethanol was injected into a flow of methanol through a chromatography column packed with silica coated with an organic layer. The solution emerging from the column was monitored using a UV detector. Under the conditions used, Sudan I emerged from the column after 4.25 min (see Figure 1). This peak was confirmed as Sudan I by using the same technique on a genuine sample of Sudan I in ethanol. Further confirmation came by recording the mass spectrum of the compound responsible for the peak using HPLC–mass spectrometry. The area under the peak is proportional to the amount of analyte and the concentration of Sudan I was found by calibrating with standard solutions of known concentration. The analysis showed that the detection limit (the lowest detectable concentration) was 5 mg of Sudan I in 1 kg of chilli powder.

Step 6 **Draw conclusions and report.** The aim of the experiments was to determine whether the analysis method could detect and quantify Sudan I in foodstuffs such as chilli powders. The authors of the paper were able to measure concentrations down to 5 $mg\,kg^{-1}$ with good repeatability so that they achieved their aim. In order to inform other workers in the field, a research paper was published.

▲ In February 2005, a batch of imported chilli powder was found to contain the banned dye Sudan I.

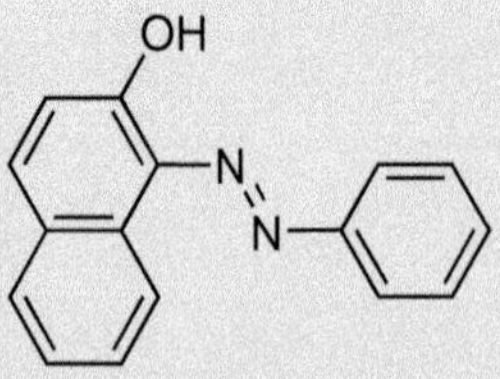
OH
N
N

Sudan I

Question

Convert the detection limit of the above analysis, 5 $mg\,kg^{-1}$, to units of: (a) $mg\,g^{-1}$; (b) $\mu g\,g^{-1}$; and (c) ppm (by mass). (Concentrations in parts per million, ppm, are defined in Box 1.5, p.34.)

➜

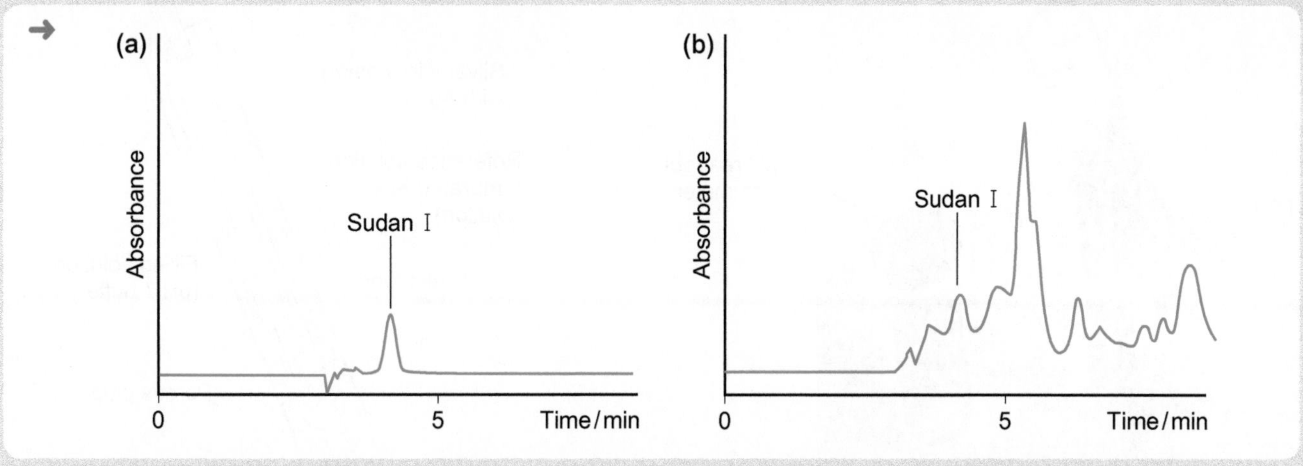
(a)
Absorbance
Sudan I
0
5
Time/min
(b)
Absorbance
Sudan I
0
5
Time/min

Figure 1 HPLC elution profiles (detected at λ = 481 nm). Under the conditions used, the peak for Sudan I appears after 4.25 min. (a) A standard solution of Sudan I in ethanol; (b) an extract from a commercial chilli powder with Sudan I added.

Summary

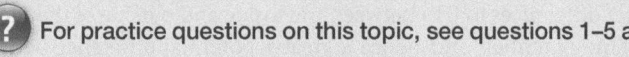

- A sample for analysis contains the analyte (or analytes) dispersed in a matrix. The sample must be representative of the whole amount of substance being analysed.
- Analytical procedures can be checked by comparing results obtained using reference samples or materials which have known concentrations.
- A calibration curve may be constructed to determine the relation between the measured quantity and the concentration, using reference samples of known concentrations.
- Appropriate blank experiments must be performed to ensure that it is the concentration of the analyte of interest that is being measured.
- A good analytical procedure produces results that are accurate and precise.
  - Accuracy refers to how close a result is to the true value.
  - Precision describes how well a number of measurements agree with each other.
- The spread of results is described by the range, the standard deviation, and the coefficient of variation.
- The detection limit is the minimum concentration of an analyte that can be distinguished from a blank experiment.
- Errors (uncertainties) may be systematic or random.

For practice questions on this topic, see questions 1–5 at the end of this chapter (p.553).

11.2 Electrochemical methods of analysis

There are a number of analytical methods based on electrochemistry. These **electroanalytical** methods rely on redox systems. A redox reaction can be separated into two half reactions, which can be arranged to form an electrochemical cell. The size of the cell potential, E_{cell}, depends on the concentrations of the species involved in the reaction. Electroanalytical techniques that measure concentrations from cell potentials are **potentiometric** methods.

Electrochemical cells are described in Sections 16.1 (p.728) and 16.3 (p.735). The cell potential, E_{cell}, is also known as the electromotive force, emf. Redox reactions are discussed in Section 1.4, p.23.

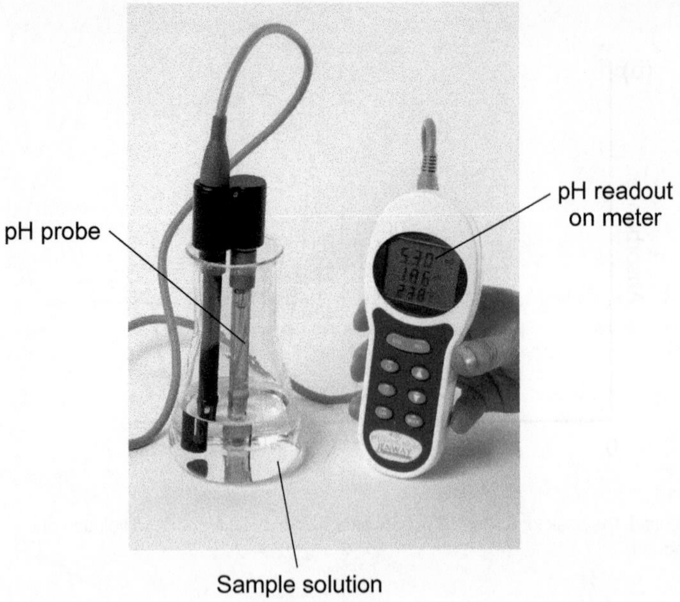
pH probe
pH readout
on meter
Sample solution

Figure 11.6 Using a pH meter. For accuracy pH meters often include, a temperature probe since pH depends on temperature.

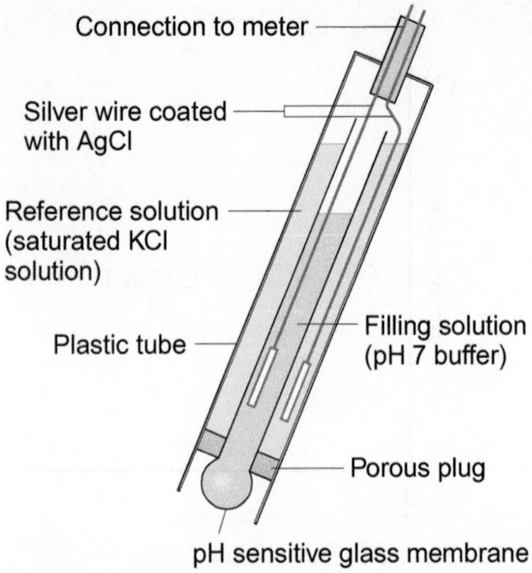
Connection to meter
Silver wire coated
with AgCl
Reference solution
(saturated KCl
solution)
Plastic tube
Filling solution
(pH 7 buffer)
Porous plug
pH sensitive glass membrane

Figure 11.7 The pH probe has a reference Ag/AgCl electrode combined with a glass electrode which uses a glass membrane that is sensitive to pH. The cable contains connections to both electrodes and a meter measures the potential difference between them. The measured potential depends on the pH of the external solution.

Measuring pH

$pH = -\log_{10}[H_3O^+]$. Buffer solutions are described in Section 7.3 (p.319).

Visit the Online Resource Centre to view screencast 11.1 which walks you through the workings of the glass electrode and other ion selective electrodes.

Equation 11.4 has the form of a straight line, $y = mx + c$, see Maths Toolkit MT4 (p.1318).

$$\begin{array}{ccc} y & = & mx \quad + c \\ \text{Measured voltage} & = & 0.0592k\Delta\text{pH} + c \end{array}$$

The concentration dependence of electrochemical cell potentials is described by the Nernst equation, Equation 16.16 (p.750).

The factor 0.0592 in Equation 11.4 arises from terms in the Nernst equation (see Section 16.4, p.746). To keep the units consistent, it has the units of volts.

A familiar example of a potentiometric sensor is a **pH meter** (see Figure 11.6). A pH meter consists of a **probe** that is immersed in a solution and connected to a high resistance voltmeter that measures the voltage produced by the probe. This is converted to a reading of pH on the meter display. The meter is calibrated using two or three buffer solutions with known pH. The pH of a test solution can then be measured.

So, how does a pH meter work? A pH probe consists of a plastic or glass tube containing two electrodes (see Figure 11.7). One electrode is a silver wire coated with solid silver(I) chloride immersed in a saturated solution of Cl^- ions. This produces a constant reference potential. The other electrode is a **glass electrode**. It also has a silver/silver chloride electrode inside, but the important feature is the special glass membrane at the tip of the cell. H^+ ions from the solution interact with the surface of the glass and diffuse into its outermost layers (note that they do not diffuse *through* the glass membrane). This influences the arrangement and movement of Na^+ ions in the glass and so changes the distribution of charge across the membrane. This induces a membrane potential (see Section 16.4, p.753), which changes the potential of the glass electrode relative to the potential of the reference Ag/AgCl electrode. The potential difference between the two electrodes is measured by a high-resistance voltmeter. Greater differences in $[H^+]$ between the inside and outside of the membrane lead to larger membrane potentials and hence a greater potential difference between the two electrodes.

Applying the Nernst equation (Equation 16.16, p.750) shows that, if the pH of the buffer solution inside the glass electrode differs from that in the external solution by 1 unit, the measured voltage changes by 0.0592 V. In a real system, the measured voltage changes according to Equation 11.4.

$$\text{Measured voltage} = 0.0592k\Delta\text{pH} + \text{constant} \tag{11.4}$$

where ΔpH is the difference in pH between the buffer solution in the glass electrode and the external analyte solution. The constant k arises because the two sides of the membrane are not exactly the same. This has a very small effect on the voltage and k is usually very close to 1.0. In practice, the constants are not measured and the calibration of the pH meter with buffer solutions takes them into account. The important conclusion from Equation 11.4 is that the measured voltage depends linearly on the pH of a solution.

Ion selective electrodes

The glass electrode used to measure pH is an example of an **ion selective electrode, ISE.** These are electrodes that respond selectively to a particular ion in solution—for example, the glass electrode responds to H^+(aq) ions. Electrodes are available for the selective analysis of around 35 different ions. The principle of operation is the same as that for the glass electrode. The end of the electrode consists of a barrier that develops a potential that depends on the concentration of the ion in the analyte solution. Several types of barrier can be used.

Like the glass electrode, ion selective electrodes are combined with a reference electrode to form an electrochemical cell. Many ion selective electrodes are used with separate reference electrodes rather than having the reference electrode built in as with the pH probe. Reference electrodes are described in Section 16.3 (p.744).

The fluoride ion selective electrode shown in Figure 11.8(a) is a **solid state electrode.** The ion selective barrier consists of a crystal of lanthanum fluoride (LaF_3) that is doped with a small amount of europium fluoride (EuF_2). A membrane potential (see Section 16.4, p.753) is generated that depends on the difference in F^- ion concentration on each side of the barrier. This potential is measured with reference to a separate electrode. Since the F^- ion concentration inside the electrode is constant, the measured voltage follows a relationship similar to that in Equation 11.4 for the pH probe. The 0.0592 term again arises from the Nernst equation.

$$\text{Measured voltage} = 0.0592\,k\log_{10}[F^-(aq)]_{\text{outside}} + \text{constant} \quad (11.5)$$

The selectivity of the electrode comes from the crystal structure of the doped LaF_3. Other anions, such as Cl^- and Br^-, cannot fit into the lattice structure of the crystal like F^- can,

Visit the Online Resource Centre to view video clip 11.1 which illustrates the use of an ion selective electrode in the laboratory.

Strictly speaking, the measured voltage in Equations 11.4 and 11.5 is a function of the *activity* of the ion in solution, not its molar concentration, although the differences between the two are usually small, particularly at low concentrations. Activities are discussed in Section 14.6 (p.686).

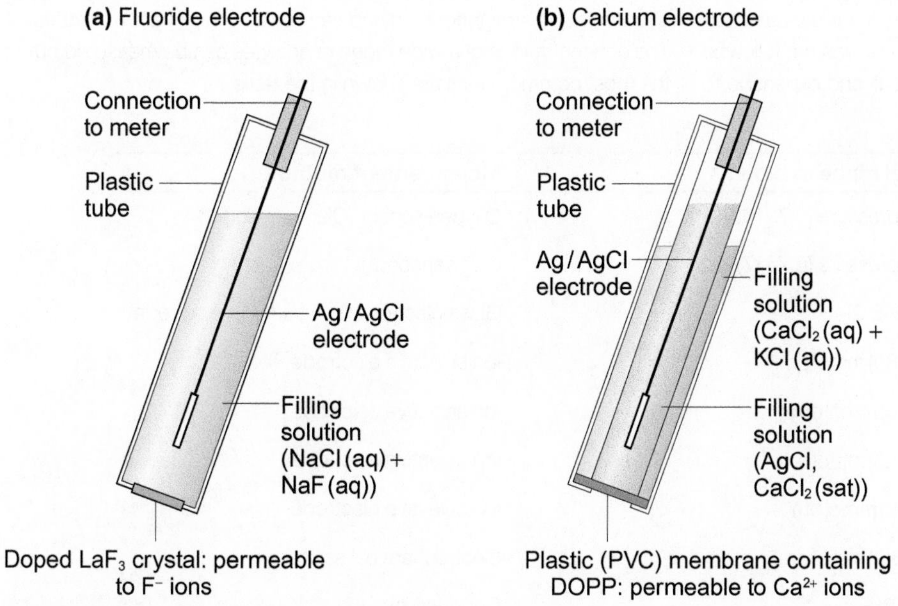
(a) Fluoride electrode
Connection to meter
Plastic tube
Ag/AgCl electrode
Filling solution (NaCl (aq) + NaF (aq))
Doped LaF₃ crystal: permeable to F⁻ ions
(b) Calcium electrode
Connection to meter
Plastic tube
Ag/AgCl electrode
Filling solution (CaCl₂ (aq) + KCl (aq))
Filling solution (AgCl, CaCl₂ (sat))
Plastic (PVC) membrane containing DOPP: permeable to Ca²⁺ ions

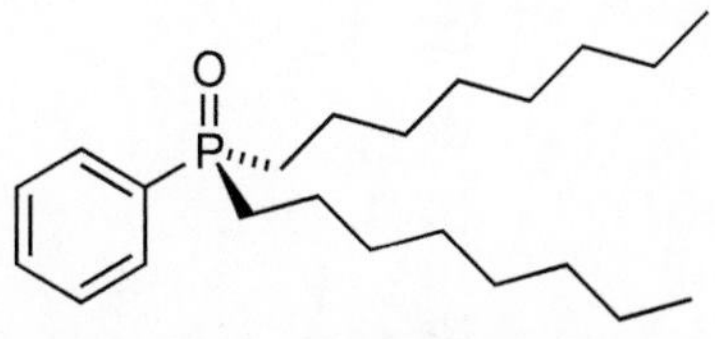

Dioctylphenyl phosphonate, DOPP

Figure 11.8 The most important feature of an ion selective electrode is a barrier that is selectively permeable to the analyte ion. (a) Fluoride electrode. The barrier is a solid. (b) Calcium electrode. The barrier is a liquid membrane.

and so cannot diffuse into it. However, hydroxide ions (OH^-) have a similar size to F^- ions and they can diffuse into the crystal and interfere with the analysis, so the fluoride ion selective electrode is only accurate where $[OH^-(aq)]$ is low.

Other ion selective electrodes use **liquid membranes**. These consist of a polymer film containing a compound that allows the selected ion to be transported into the film to generate a membrane potential. An example is a PVC membrane containing calcium didecyl phosphate and dioctylphenyl phosphonate, as shown in Figure 11.8(b). These compounds bind Ca^{2+} ions from the analyte solution and transport them into the membrane.

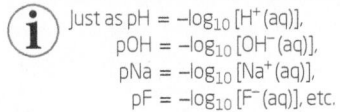

Just as $pH = -\log_{10}[H^+(aq)]$,
$pOH = -\log_{10}[OH^-(aq)]$,
$pNa = -\log_{10}[Na^+(aq)]$,
$pF = -\log_{10}[F^-(aq)]$, etc.

Ion selective electrodes are used in much the same manner as a pH meter. Many are designed to measure the voltage between the electrode and a reference electrode and to display the concentration of the ion directly or to give the 'p' value, which is defined in the same way as pH. For example, for the fluoride electrode

$$pF = -\log_{10}[F^-(aq)] \tag{11.6}$$

Solutions with known concentrations are used to calibrate the ion selective electrode after which unknown concentrations can be measured. Often, this involves plotting a calibration graph of measured voltage versus $\log_{10}$[concentration] and using it to find the concentration of an unknown solution from its measured voltage. Depending on the system, detection limits are usually around 1×10^{-6} mol dm^{-3}. At high concentrations, the diffusion of ions through the membrane is complicated by interactions between the ions so the measured voltage is no longer proportional to $\log_{10}[F^-(aq)]$ (Equation 11.5). Most ion selective electrodes cannot be used to measure accurately concentrations above 1 mol dm^{-3}.

Box 11.2 Rapid blood analysis using electrochemical methods

When a patient is admitted to hospital, a sample of their blood is usually taken. Monitoring blood chemistry is vital to help doctors come to an accurate diagnosis. This used to involve a painstaking procedure to separate the blood cells from the plasma, followed by a series of chemical analyses. This was difficult and expensive to do routinely and also took some time, which is a problem if a patient is seriously ill. Using electrochemical techniques, blood analysis is now economic and straightforward and can be done in just a few minutes. The concentrations of a wide range of analytes can be measured but the most common tests are shown in the table.

Analyte	Normal range in blood	Measurement method
O_2	Partial pressure (0.7–106) kPa	Oxygen sensor (Clark sensor)
CO_2	Partial pressure (0.7–17) kPa	CO_2 sensor
H^+	pH (6.5–8.2)	Glass electrode (ion selective electrode)
Na^+	(100–180) mmol dm^{-3}	Ion selective electrode
K^+	(2.0–9.0) mmol dm^{-3}	Ion selective electrode
Ca^{2+}	(0.25–2.5) mmol dm^{-3}	Ion selective electrode
Cl^-	(65–140) mmol dm^{-3}	Ion selective electrode
Glucose	(1.1–38.9) mmol dm^{-3}	Electrochemical sensor
Haematocrit*	10%–75%	Conductivity

* The haematocrit test gives the percentage by volume of red blood cells in the blood sample.

→

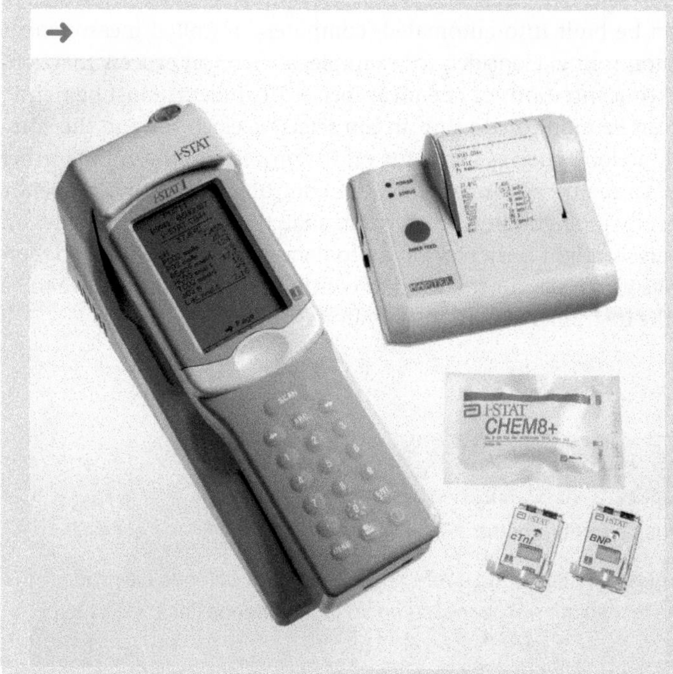
i-STAT
CHEM8+
cTnI
BNP

◄ Analysis of blood gases and electrolyte concentrations can be rapidly carried out on the spot using a handheld analyser. Just a few drops of blood are sufficient to carry out the analysis. The blood is placed on a cartridge that is inserted into the analyser and the results appear in about 2 min.

Around 0.5 cm^3 of blood, treated first with heparin to prevent coagulation, is injected into the analyser that contains the detector probes. The measurements are completed in just a couple of minutes. A calibration solution is injected between samples to keep the electrodes clean and to ensure that the results are accurate. The analysers are compact enough to be located in intensive care wards or in accident and emergency rooms rather than in a central laboratory.

Small handheld versions of the analysers use miniature ion selective electrodes and require only around 0.1 cm^3 of blood. These can be used, for example, in ambulances to provide instant information to guide treatment before the patient reaches hospital.

Question

Explain how the principle of the techniques for measuring concentrations of H^+, Na^+, K^+, and Ca^{2+} ions differs from those for measuring O_2 and CO_2 concentrations in blood.

Using ion selective electrodes

While ion selective electrodes usually only give an approximate measure over a limited range of concentrations, they have several advantages over other methods of analysis. They are relatively cheap and economical to operate. No special training is needed. They give rapid results and are often used as a guide to whether more accurate, but time-consuming, methods might be needed. Most are highly selective and so can be used with little sample pre-treatment. One of their uses is illustrated in Box 11.2.

The Phoenix Lander arrived on Mars in May 2008, on a mission to find out whether conditions on the planet could support life. Phoenix included a robotic chemistry laboratory that used ion selective electrodes to detect ions such as Na^+, Mg^{2+}, Ca^{2+}, and SO_4^{2-}. It also used electrochemical methods to detect O_2 and CO_2.

A major advantage of ion selective electrodes is that they produce an electrical signal. This means that they can be built into automated, computer-controlled measurement systems for continuous monitoring. Consider, for example, a chemical process that generates some effluent. The concentrations of chemical species in effluents must be tightly controlled. The effluent can be monitored using an ion selective electrode and the computer programmed to take actions, such as switching off the process or preventing further discharge of effluent until corrective action is taken. This control mechanism can operate 24 hours a day and detect any problems immediately. An analytical method that relies on sampling and further analysis might only pick up a problem when it is too late for corrective action. Ion selective electrodes for this kind of continuous monitoring have to be much more robust than the instruments used in laboratories.

Summary

- Electrochemical methods of analysis can be potentiometric or amperometric.
- A pH meter measures the potential difference between a glass electrode and a reference silver/silver chloride electrode, usually combined into a single probe that is inserted in the test solution. The measured voltage depends linearly on the pH of the solution.
- Ion selective electrodes respond selectively to a particular ion in solution. They produce an electric potential that is measured with reference to a reference electrode.
- The potential generated by an ion selective electrode depends in a linear fashion on the logarithm of the ionic concentration in solution.
- pH meters and ion selective electrodes are calibrated by using solutions with accurately known concentrations. For pH measurements, buffer solutions are used.
- Amperometric cells measure the current resulting from an electrolysis reaction at a constant applied voltage.

? For practice questions on this topic, see questions 6–9 at the end of this chapter (p.553).

11.3 Chromatography

The first use of chromatography was described by the Russian botanist Mikhail Tsvet (sometimes spelled Tswett) in 1906. He extracted coloured material from vegetables and used a solvent to carry the colours along a length of paper. Using this method he was able to separate α- and β-carotene. A familiar version of these experiments, illustrating the separation of the constituents of some inks, is shown in Figure 11.9. Tsvet coined the name of the method from the Greek for 'colour writing' (*chromos* and *graph*).

Today, the term **chromatography** has developed into a generic description of a number of methods for separating a mixture into its components, which may also involve identifying the components and measuring their concentrations. Such methods range from separating coloured substances on absorbent material to using a sophisticated instrument attached to a mass spectrometer for the analysis of complex mixtures, as in the analysis of a sample of urine for the presence of drugs and drug metabolites described in the opening spread to this chapter, on p.512.

Tsvet defined chromatography as 'a method in which the components of a mixture are separated on an adsorbent column in a flowing system'. In Figure 11.9, the components of the ink are separated as the solvent carries them over the paper. Many modern types of

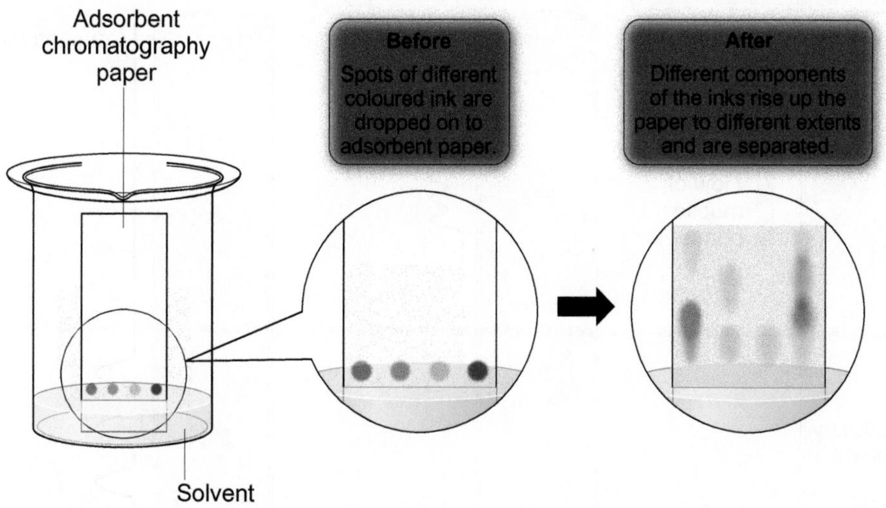
Adsorbent chromatography paper
Before
Spots of different coloured ink are dropped on to adsorbent paper.
After
Different components of the inks rise up the paper to different extents and are separated.
Solvent

Figure 11.9 Chromatography of inks: as the solvent rises up the paper, it carries the ink with it. The different components of the ink separate out since they move to different extents.

chromatography are more complex than this, but all involve the separation of two or more components by distributing them between two phases.

To illustrate the basic principle that underlies all forms of chromatography, consider a solute mixed with two immiscible liquids (see Figure 11.10). The solute dissolves preferentially in one of the liquids so its concentration in that liquid is higher than in the other liquid. The separation does not have to involve two liquids. For example, an adsorbent solid could be added, on to which a compound adsorbs from solution. The solid can then be separated by filtration and the compound recovered by removing it from the adsorbent. The essential principle is that, in the presence of two phases, compounds have a greater affinity for one phase than for the other.

If a system such as that shown in Figure 11.10 is allowed to reach equilibrium, the solute is **partitioned** between the two liquid phases. The concentrations in the two liquids are related by the **partition coefficient, *K***. The higher the value of *K*, the higher the ratio of

Think carefully when you use the words adsorption and absorption. **Adsorption** occurs when a substance sticks to the *surface* of a solid (the adsorbent). **Absorption** is where the substance is distributed throughout the solid (like water in a sponge).

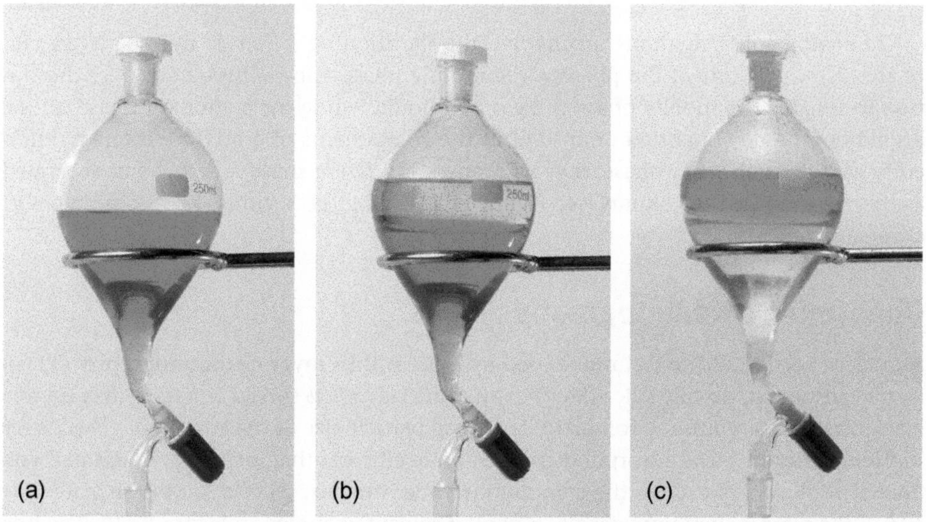
(a)
(b)
(c)

Figure 11.10 Distribution of a solute between two liquid phases. (a) A solution of an orange dye in water. (b) An immiscible organic solvent is added. On shaking, some dye is transferred to the upper organic layer. (c) At equilibrium, the concentration of the dye in the organic layer is much higher than that in the aqueous layer. The organic solvent: water partition coefficient is high.

Distributing a compound between two immiscible liquids forms the basis of solvent extraction, which is used for the purification of many organic compounds (see Box 2.6, p.94).

The thermodynamics of phase equilibrium is discussed in Section 17.2 (p.775).

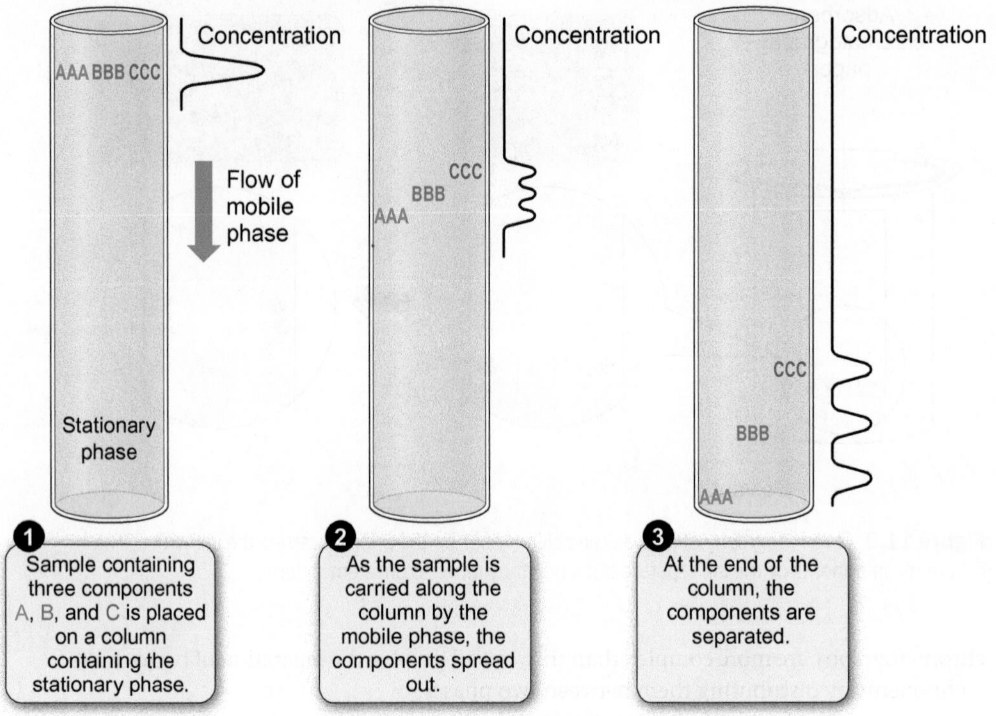
Concentration
AAA BBB CCC
Flow of mobile phase
Stationary phase
Concentration
CCC
BBB
AAA
Concentration
CCC
BBB
AAA
1
Sample containing three components A, B, and C is placed on a column containing the stationary phase.
2
As the sample is carried along the column by the mobile phase, the components spread out.
3
At the end of the column, the components are separated.

Figure 11.11 Components interact with the stationary phase as they are carried along the column by the mobile phase. In this example, C interacts with the stationary phase most strongly. It spends longer on the column and takes longer to be eluted. It comes off the column after A and B. A has the weakest interactions with the stationary phase, and comes off the column before the other components.

the concentrations, and the greater the difference in concentration in the two liquids. If two compounds have different partition coefficients between the two liquids, they can be separated since one can be extracted to a greater extent than the other

$$K = \frac{\text{concentration of solute in phase 1}}{\text{concentration of solute in phase 2}} \qquad (11.8)$$

> **i** The partition coefficient is an equilibrium constant. (See Section 1.9, p.56.) It has a constant value at constant temperature.
>
> Partition coefficients are always written as the ratio of the concentration in the more concentrated phase to that in the less concentrated phase so that the value of *K* is always larger than one.

Chromatographic methods all involve the partitioning of an analyte between two phases. Normally one of the phases—the **mobile** phase—flows through or over the **stationary** phase. The mobile phase is a gas or a liquid, while the stationary phase can be a solid or a liquid supported on a solid. If the components of a mixture interact differently with the stationary phase, they will move at different speeds and become separated as they pass down the column (see Figure 11.11). Table 11.2 summarizes some types of chromatography.

Qualitative chromatography

A modern version of the technique used by Tsvet is **thin layer chromatography (TLC)**. A solid adsorbent, usually silica (SiO_2) or alumina (Al_2O_3), is spread as a thin layer on to a glass, plastic, or aluminium foil plate. The small particle size of the adsorbent gives a very high surface area so the adsorption is much more efficient than using paper. A small volume of a solution of each of the compounds under study is 'spotted' on to the plate. The end of the plate is then immersed in a pool of solvent as shown in Figure 11.12. The line of the spots must be above the surface of the solvent.

> **i** A small quantity of a fluorescent compound is often added to the adsorbent when the TLC plates are made. The plate fluoresces under UV light. The spots quench the fluorescence and appear as dark areas.

As the solvent rises up the plate by capillary action, it carries the compounds with it at rates that depend on how strongly each compound interacts with the solid phase compared with the liquid phase. Each compound then appears in a different region of the

Table 11.2 Different types of chromatography

	Stationary phase	Mobile phase
Paper chromatography	Water absorbed in cellulose fibres	Solvent in which paper dips; solvent rises up the paper
Thin layer chromatography (TLC)	SiO_2 or Al_2O_3 spread on glass, plastic, or aluminium foil plate	Solvent in which plate dips; solvent rises up the plate
Column chromatography	Adsorbent, e.g. SiO_2, packed in a column	Solvent that travels down column
High performance liquid chromatography (HPLC)	Adsorbent, e.g. porous silica, with very high surface area, packed in a column	Solvent that is pumped through column under pressure
Gas–liquid chromatography (GLC)	Liquid coated on a solid support packed in a column housed in an oven	Unreactive gas that flows through column
Gas–solid chromatography (GSC)	Absorbent solid packed in a column housed in an oven	Unreactive gas that flows through column

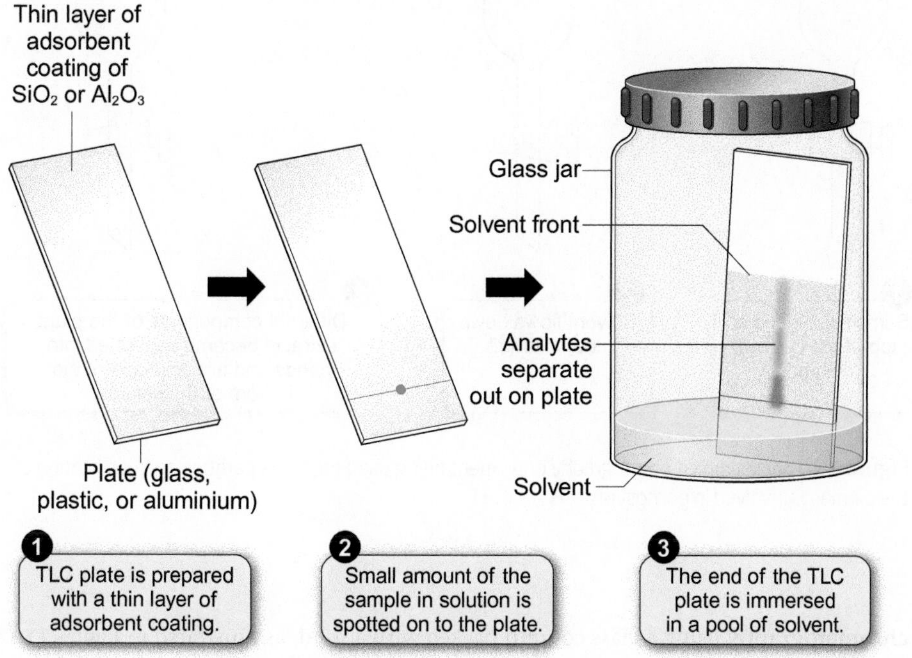

Thin layer of adsorbent coating of SiO_2 or Al_2O_3
Plate (glass, plastic, or aluminium)
Glass jar
Solvent front
Analytes separate out on plate
Solvent
1 TLC plate is prepared with a thin layer of adsorbent coating.
2 Small amount of the sample in solution is spotted on to the plate.
3 The end of the TLC plate is immersed in a pool of solvent.

Figure 11.12 Thin layer chromatography. The TLC plate is coated with silica or alumina. A solution of the analyte is spotted on to the plate and the end of the plate immersed in solvent, so that the spot is just above the surface of the solvent.

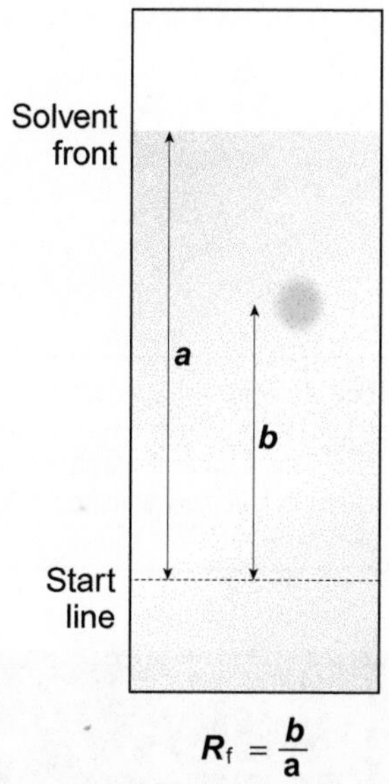

Solvent front
a
b
Start line
$R_f = \frac{b}{a}$

Figure 11.13 The retention factor, R_f, of a compound is the distance the spot moves up the plate relative to the distance moved by the solvent front.

plate, known as a 'spot'. The distance the spot moves up the plate compared with that of the solvent (called the *solvent front*) is known as the **retention factor, R_f,** as shown in Figure 11.13. The R_f values are characteristic of the compound in a particular solvent at a particular temperature.

The compounds do not have to be coloured to be used in TLC. Colourless compounds can be shown up by viewing the plate under UV light or by spraying the plate with a dye or other compound that causes the compounds to become visible.

Thin layer chromatography can be used on a small scale for preparative work by scraping off the spot containing the compound of interest into an appropriate solvent and extracting the compound from the adsorbent. However, this is limited to a few micrograms of material. Preparative chromatography in the laboratory is usually carried out by **column**

TLC is often used by synthetic chemists to follow the course of a reaction. If the products have different R_f values from the starting materials, the chemist can tell how far the reaction has progressed.

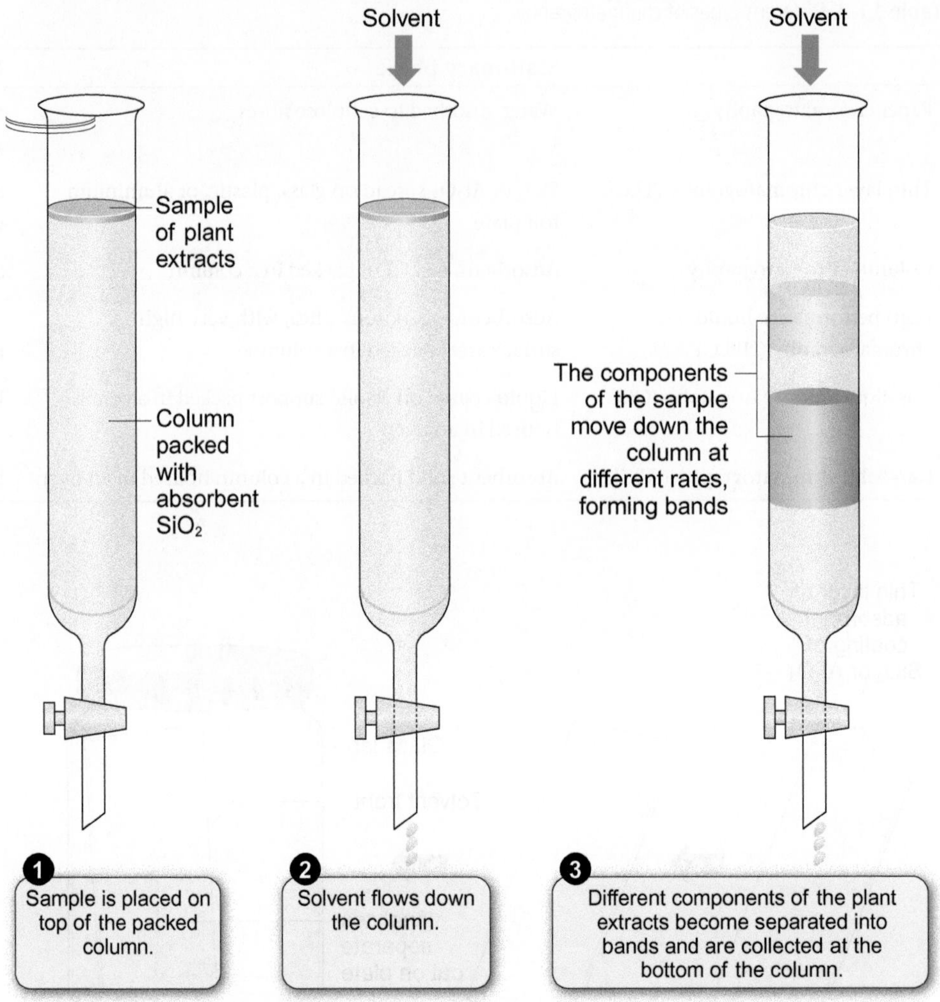
Solvent
Solvent
Sample of plant extracts
Column packed with absorbent SiO_2
The components of the sample move down the column at different rates, forming bands
1
Sample is placed on top of the packed column.
2
Solvent flows down the column.
3
Different components of the plant extracts become separated into bands and are collected at the bottom of the column.

Figure 11.14 Column chromatography equipment being used for the separation and purification of the pigments involved in photosynthesis.

Elution is the process by which an adsorbed material is removed from the adsorbent by movement of a mobile phase. The solution that emerges from a chromatography column is called the **eluate**.

Industrial scale chromatography. Although mainly used in the laboratory, column chromatography can be used on a larger scale. Petrochemical works use columns that are 3 m in diameter and up to 15 m high to separate the components of mixtures.

chromatography using a glass column packed with a solid, as illustrated in Figure 11.14. A mixture is placed on top of the packing and solvent is allowed to flow down the column. The components of the mixture move down the column at different rates. (If the compounds are coloured, they show up as separate bands on the column as in Figure 11.14.) The solvent emerging from the bottom of the column (the *eluate*) is collected in small batches (known as *fractions*). The compounds of interest are then obtained by evaporating the solvent from the relevant fractions.

Quantitative chromatography

Quantitative chromatography involves measuring the quantities of separated components. Using chromatography for quantitative analysis requires more sophisticated apparatus, involving detecting and measuring the concentration of the compounds as they are eluted from the end of the column. The essential components of any quantitative chromatographic instrument are shown in Figure 11.15.

The mobile phase carries the sample over the stationary phase at a constant rate. The detector measures the concentration of the components as they emerge from the column and produces a signal that is proportional to concentration.

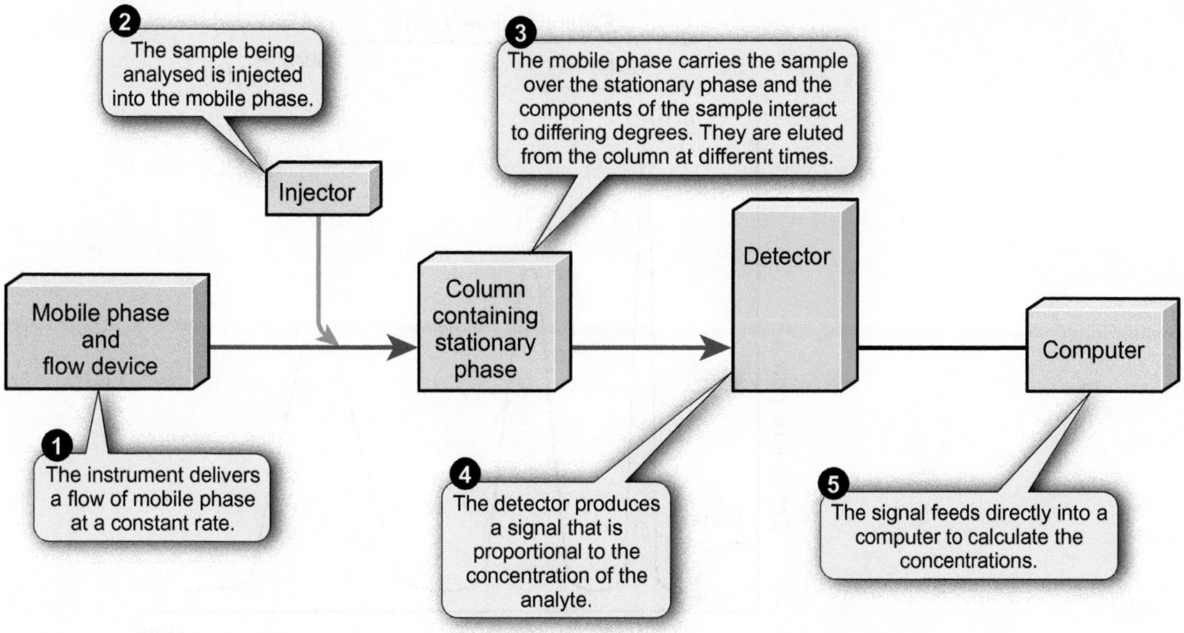

2
The sample being analysed is injected into the mobile phase.
3
The mobile phase carries the sample over the stationary phase and the components of the sample interact to differing degrees. They are eluted from the column at different times.
Injector
Detector
Mobile phase and flow device
Column containing stationary phase
Computer
1
The instrument delivers a flow of mobile phase at a constant rate.
4
The detector produces a signal that is proportional to the concentration of the analyte.
5
The signal feeds directly into a computer to calculate the concentrations.

Figure 11.15 Various different types of quantitative chromatography can be used for chemical analysis—but all have some features in common.

Provided that the temperature and flow of the mobile phase are constant, the time between injection and elution from the column, the elution time, is characteristic of the compound under the conditions used. The elution time is often called the **retention time**, t_r. Comparison of t_r with that of an authentic sample can be used to confirm the identity of an unknown sample.

> **Retention time**, t_r, is the time after injection for the compound to register a maximum signal on the detector (see Figure 11.16).

A plot of detector signal versus time produces a chromatogram of the form shown in Figure 11.16. If the detector signal is directly proportional to concentration, then the amount of each compound eluted is proportional to the area under the peak, corresponding to the shaded areas in Figure 11.16.

The sample is partitioned between the mobile and stationary phases. For a compound A, the partition coefficient, K, is given by Equation 11.8 (p.530)

$$K = \frac{[A]_{stationary}}{[A]_{mobile}}$$

A compound with a large value of K interacts more strongly with the stationary phase, leading to a longer retention time, t_r. The ratio of the retention times of two components A and B is the **relative retention**, α_{AB}, where

$$\alpha_{AB} = \frac{t_r(A)}{t_r(B)} \quad (11.9)$$

> Equation 11.9 assumes that the flow rate of the mobile phase is constant. If the flow rate varies, the retention time will change and this must be taken into account.

By convention, $\alpha > 1$ so A is the compound that is retained longer on the column. The bigger the value of α, the better the separation of the two components.

The detector needs to be sensitive. It must also give a signal where the response is proportional to concentration over a large range of concentrations. However, different compounds may produce a different response in a particular detector even when their concentrations are the same. For example, the detector might measure the absorbance of light. One compound may be more highly coloured than another so that the intensity of light absorbed by solutions of the same concentration may be different, leading to a different signal for each compound. The ratio of the peak areas will then not be the same as the ratio of the amounts of the compounds injected. In calculating concentrations from chromatography, the response of the detector for different compounds must be taken into account.

Absorbance detectors are described later in this section, on p.539.

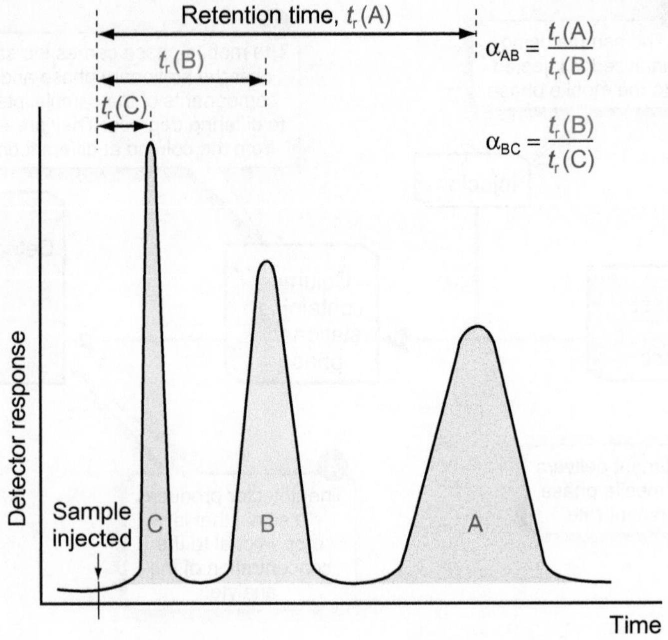

Retention time, t_r(A)
t_r(B)
t_r(C)
$\alpha_{AB} = \frac{t_r(A)}{t_r(B)}$
$\alpha_{BC} = \frac{t_r(B)}{t_r(C)}$
Detector response
Sample injected
C
B
A
Time

Figure 11.16 A schematic chromatogram showing the elution of three analytes, **A**, **B**, and **C**. **A** interacts most strongly with the stationary phase so it is retained and eluted from the column after **B** and **C**. The retention time, t_r, is characteristic of the compound. The area under each peak gives the amount of the analyte eluted.

Gas chromatography (GC)

Much of the work that led to the development of analytical chromatography was carried out in the 1940s by British scientists Archer Martin and Richard Synge who were awarded the Nobel Prize in Chemistry in 1952. They suggested the use of a gaseous mobile phase but the development work took a further 10 years. In fact, Martin took the opportunity of his Nobel lecture to announce the first working gas chromatograph.

As the name suggests, this technique uses a gas as the mobile phase. A small quantity (<1 μg) of the sample is injected into a heated zone where it evaporates into a stream of an inert carrier gas, usually nitrogen or helium (Figure 11.17). The gas then flows over the stationary phase contained in a column housed in an oven. It is often possible to change the temperature during the analysis in a controlled way to improve the separation. Many types of samples can be analysed by gas chromatography as long as they are sufficiently volatile for a small amount to vaporize into the carrier gas.

Detectors

The commonest detector used in gas chromatography is the **flame ionization detector (FID)**. The carrier gas emerging from the column passes into a small flame contained between two electrodes. Combustion causes the formation of ions that are attracted to the electrodes, causing a small current to flow which can be measured. The flame ionization detector is very sensitive for most organic compounds. For example, a sample as small as 2 pg (2×10^{-12} g) of a hydrocarbon can be detected.

Some analyses involve compounds such as water, nitrogen, or carbon dioxide that do not burn, so a flame ionization detector cannot be used. These gases can be detected by monitoring the thermal conductivity of the gas as it emerges from the column since its thermal conductivity will be different from that of the carrier gas. However, the **thermal conductivity detector** is 100–1000 times less sensitive than a flame ionization detector so larger samples are needed. Other specialist detectors have been developed for use with particular categories of compounds (see Box 11.3, p.537).

Visit the Online Resource Centre to watch video clip 11.2 which gives a simple introduction to gas chromatography.

Spectrometers can be used as detectors to measure the concentrations of eluted compounds and can also assist in identifying unknown compounds in samples. A good example is the use of combined **gas chromatography–mass spectrometry, GC–MS**, as described in Box 11.4 (p.538).

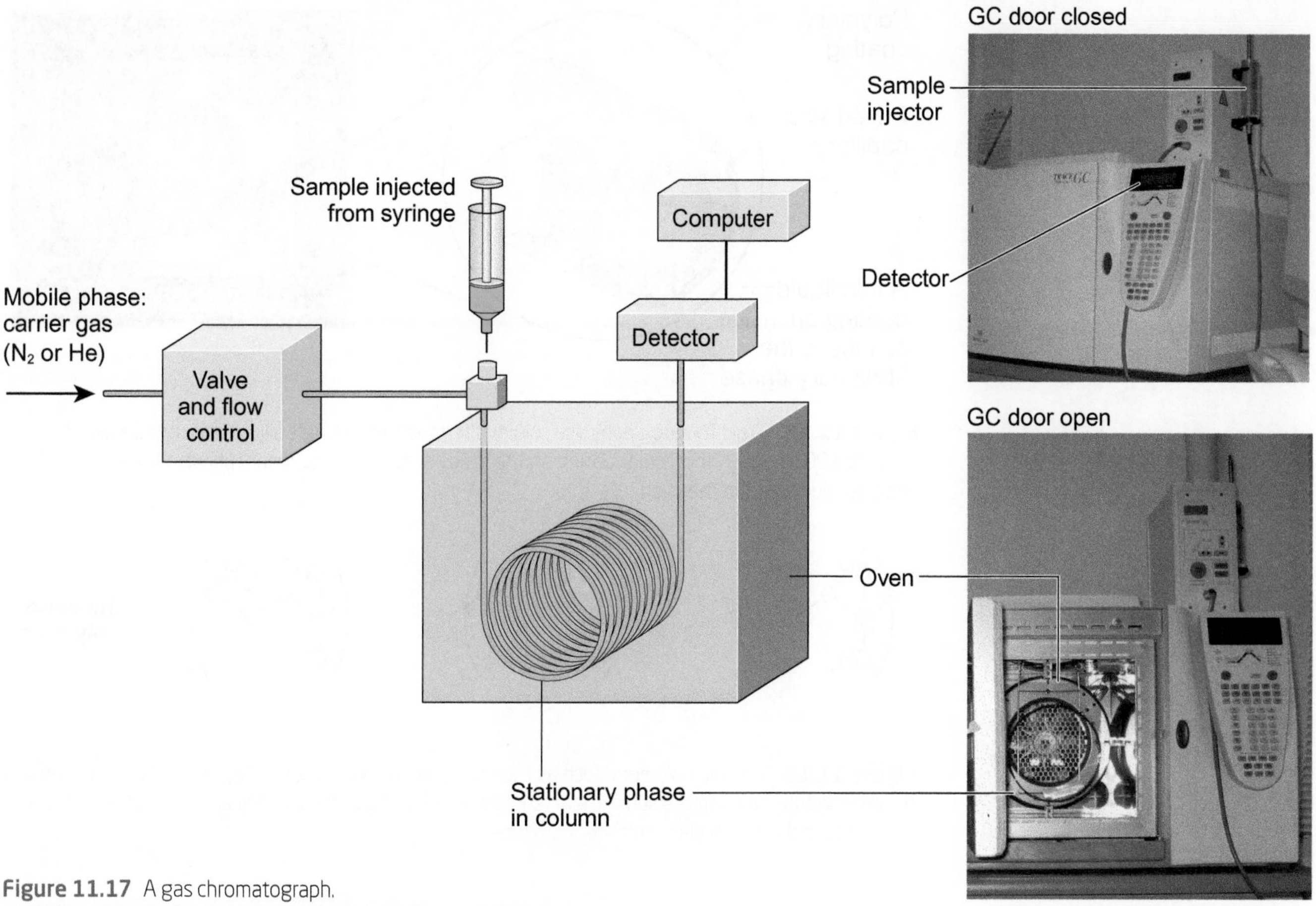
GC door closed
Sample injector
Sample injected from syringe
Computer
Detector
Mobile phase: carrier gas (N_2 or He)
Detector
Valve and flow control
GC door open
Oven
Stationary phase in column

Figure 11.17 A gas chromatograph.

Column packings and stationary phases

Gas chromatography was developed in the 1940s and 1950s using equipment where the gas flowed through a tube packed with finely divided solids such as silica or alumina. This is **gas–solid chromatography, GSC**, which uses a stationary phase consisting of particles of stable, inert solids such as polymers. Other stationary phases consist of a **molecular sieve**—an inorganic compound, usually a mixed metal oxide, that has pores and cavities that can retain and separate small gas molecules such as N_2, CO, and CO_2.

Zeolite molecular sieves are discussed at the start of Chapter 6 (p.254).

A wider range of separations can be achieved by using **gas–liquid chromatography (GLC)**. This uses a non-volatile liquid coated onto an inert solid support and packed in a column, or coated as a thin (~1 μm–5 μm) film on to the *inside* of a **capillary column**. A capillary column is shown in Figure 11.18. It is a silica tube that can be up to 100 m in length and has an inside diameter of 0.2 mm–0.5 mm. The carrier gas flows through the column and separation occurs due to the differing solubilities of the components in the liquid.

The liquids used in the stationary phase have to be stable and non-volatile so that they are not carried off in the carrier gas. Liquid stationary phases are often based on long chain siloxane polymers such as those in Figure 11.19. Siloxanes are polymers in which the chain consists of alternating Si and O atoms. The Si atoms may be substituted with a wide range of functional groups. The polymers are bonded to the wall of the column by chemical reaction with the surface of the silica. Capillary columns can also be coated with porous layers of molecular sieves.

> Siloxane polymers form the basis of silicones such as those used as sealants and waterproof films. Siloxanes are discussed in Box 27.3 (p.1220).

Separations using capillary columns are efficient and rapid. Only small quantities of sample are needed, which can be important, for example, for pharmaceutical intermediates or in forensic investigations. As an example of the power of gas chromatography, the chromatogram in Figure 11.20 shows an analysis of a sample of petrol. In around 2 hours, there is a complete analysis of the identities of over 110 compounds and their concentrations in a very complex mixture.

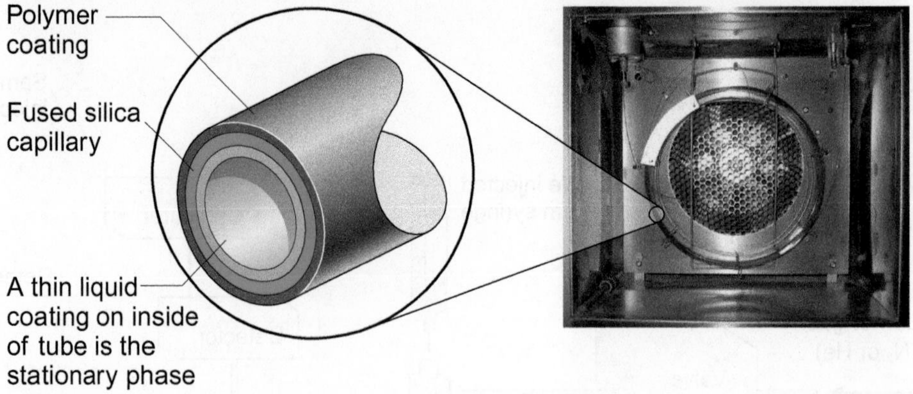

Polymer coating
Fused silica capillary
A thin liquid coating on inside of tube is the stationary phase

Figure 11.18 Gas chromatographs use columns formed from a single silica capillary tube which can be up to 100 m long. A thin liquid coating on the inside acts as the stationary phase. Here you can also see the column within the oven.

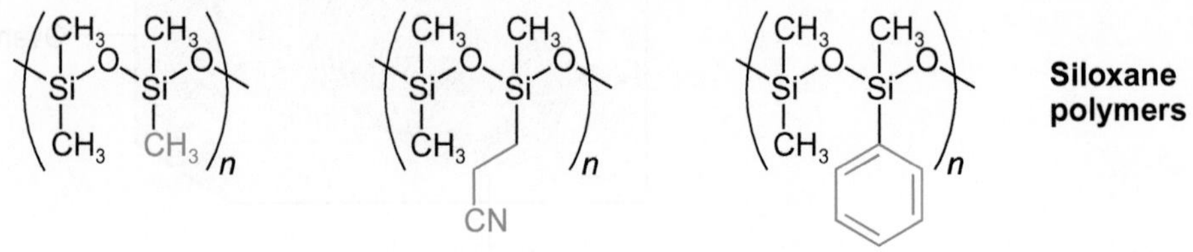

Siloxane polymers

Figure 11.19 Siloxane polymers, such as the ones shown above, are used as stationary phases in gas chromatography. Varying the substituent (shown in red) changes the polarity and interactions of the polymer to make it suitable for different analyses.

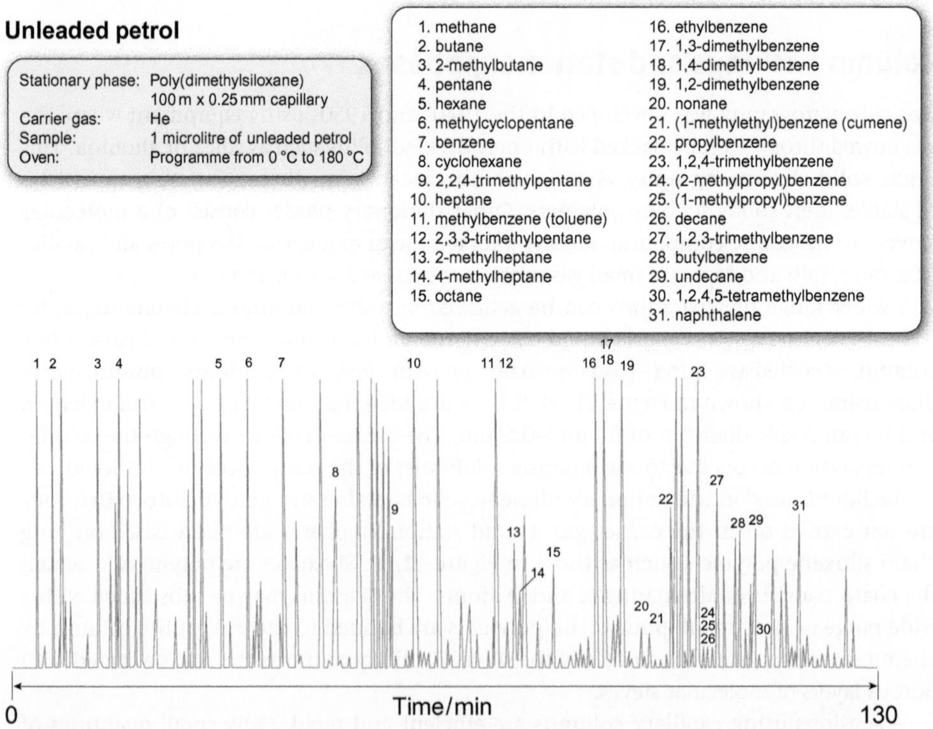

Unleaded petrol
Stationary phase: Poly(dimethylsiloxane) 100 m x 0.25 mm capillary
Carrier gas: He
Sample: 1 microlitre of unleaded petrol
Oven: Programme from 0 °C to 180 °C
1. methane
2. butane
3. 2-methylbutane
4. pentane
5. hexane
6. methylcyclopentane
7. benzene
8. cyclohexane
9. 2,2,4-trimethylpentane
10. heptane
11. methylbenzene (toluene)
12. 2,3,3-trimethylpentane
13. 2-methylheptane
14. 4-methylheptane
15. octane
16. ethylbenzene
17. 1,3-dimethylbenzene
18. 1,4-dimethylbenzene
19. 1,2-dimethylbenzene
20. nonane
21. (1-methylethyl)benzene (cumene)
22. propylbenzene
23. 1,2,4-trimethylbenzene
24. (2-methylpropyl)benzene
25. (1-methylpropyl)benzene
26. decane
27. 1,2,3-trimethylbenzene
28. butylbenzene
29. undecane
30. 1,2,4,5-tetramethylbenzene
31. naphthalene
Time/min
0
130

Figure 11.20 A gas chromatogram of a commercial sample of unleaded petrol. The analysis starts with the column at 0 °C and the oven is heated in stages until it reaches 180 °C. The most volatile compounds are eluted first. There is separation and identification of over 100 compounds (some of which are labelled) in this very complex sample. ($1\ \mu L = 1 \times 10^{-3}\ cm^3$.)

Box 11.3 Monitoring PCBs in the environment

Polychlorinated biphenyls (PCBs) are a group of compounds based on biphenyl with one or more of the hydrogens replaced by chlorine. There are 209 possible compounds containing between one and ten Cl atoms. PCBs were first manufactured in the 1930s and were widely used as electrical insulators and heat transfer fluids in industrial electrical plants and as additives in industrial fluids, oils, and fluorescent light fixtures.

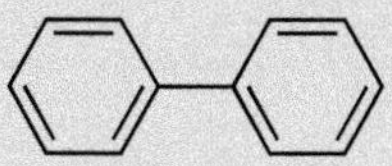

biphenyl

PCBs are not flammable, are highly electrically insulating, and are stable to heat and light. However, by the 1970s, concerns were raised about their toxicity and the harmful effects of long-term exposure. Manufacturing ceased in the 1970s and their use is now banned, but equipment incorporating PCBs is still found and until

Polychlorinated biphenyls (PCBs) have been detected in farmed fish such as salmon. ➤

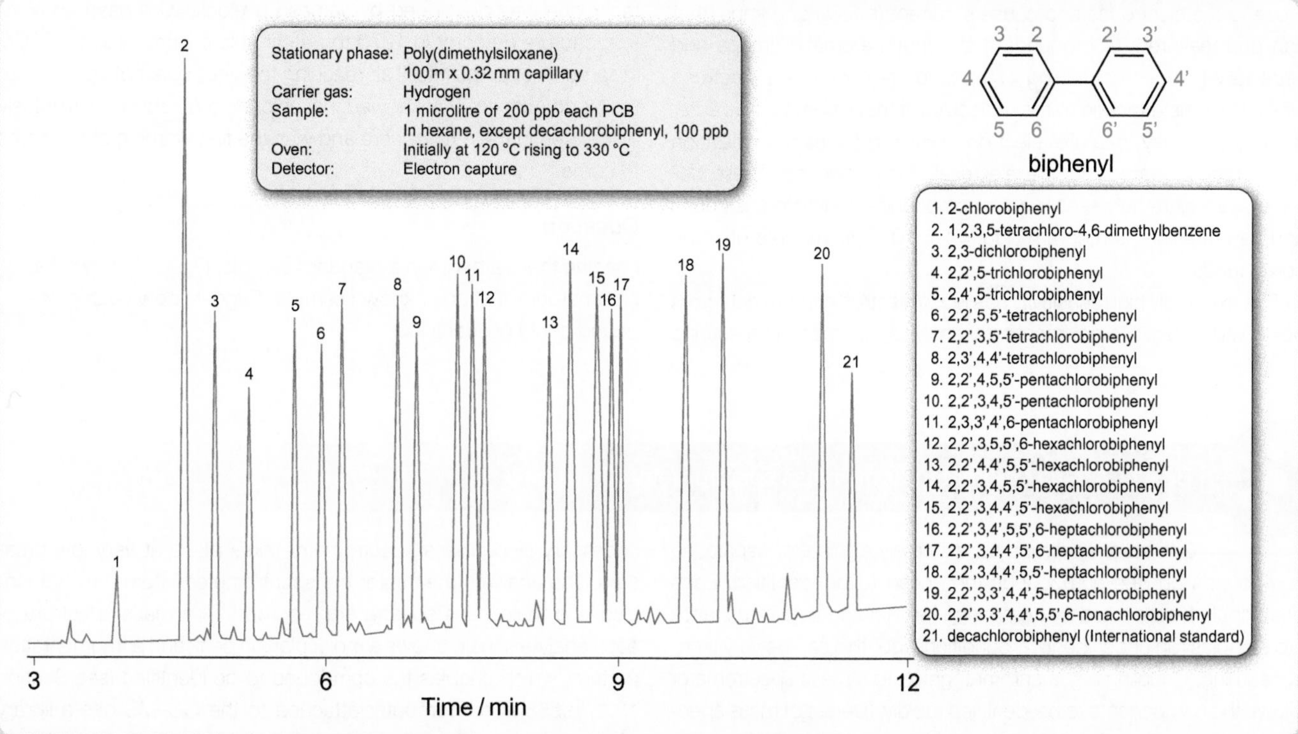
Stationary phase: Poly(dimethylsiloxane)
100 m x 0.32 mm capillary
Carrier gas: Hydrogen
Sample: 1 microlitre of 200 ppb each PCB
In hexane, except decachlorobiphenyl, 100 ppb
Oven: Initially at 120 °C rising to 330 °C
Detector: Electron capture
3 2 2' 3'
4 4'
5 6 6' 5'
biphenyl
1. 2-chlorobiphenyl
2. 1,2,3,5-tetrachloro-4,6-dimethylbenzene
3. 2,3-dichlorobiphenyl
4. 2,2',5-trichlorobiphenyl
5. 2,4',5-trichlorobiphenyl
6. 2,2',5,5'-tetrachlorobiphenyl
7. 2,2',3,5'-tetrachlorobiphenyl
8. 2,3',4,4'-tetrachlorobiphenyl
9. 2,2',4,5,5'-pentachlorobiphenyl
10. 2,2',3,4,5'-pentachlorobiphenyl
11. 2,3,3',4',6-pentachlorobiphenyl
12. 2,2',3,5,5',6-hexachlorobiphenyl
13. 2,2',4,4',5,5'-hexachlorobiphenyl
14. 2,2',3,4,5,5'-hexachlorobiphenyl
15. 2,2',3,4,4',5'-hexachlorobiphenyl
16. 2,2',3,4',5,5',6-heptachlorobiphenyl
17. 2,2',3,4,4',5',6-heptachlorobiphenyl
18. 2,2',3,4,4',5,5'-heptachlorobiphenyl
19. 2,2',3,3',4,4',5-heptachlorobiphenyl
20. 2,2',3,3',4,4',5,5',6-nonachlorobiphenyl
21. decachlorobiphenyl (International standard)
3 6 9 12
Time / min

Figure 1 Gas chromatography allows rapid analysis of a wide range of PCBs. Electron capture detectors are used to measure the tiny quantities of PCBs found.

➜

➔ the 1970s large quantities were released into the environment during manufacture and inappropriate disposal.

PCBs are very stable and break down very slowly over time, so they are persistent organic pollutants. They have low solubility in water but are soluble in organic solvents. As a result, they accumulate in fatty tissues and enter the food chain via small aquatic organisms and fish. Animals that feed on these build up PCBs in their tissue since the compounds are also resistant to biodegradation. PCBs are also found in milk, meat, and eggs.

The Food Standards Agency in the UK has set a 'tolerable daily intake' of 2 pg (2×10^{-12} g) of PCB per kg of bodyweight per day. (The 'tolerable daily intake' is the amount of a contaminant that experts recommend can, on average, be eaten every day over a whole lifetime without causing harm.) A person weighing 60 kg should, therefore, consume no more than an average of 120 pg of PCBs per day. But how are these tiny amounts measured?

Figure 1 shows a gas chromatogram of a mixture of PCBs dissolved in hexane. The concentrations are around 200 parts per billion, ppb (by mass). (1 ppb by mass = 1 μg dm^{-3}.) This is in the range of concentrations that might be found in a concentrated extract from foodstuffs. There is good separation of each of the isomers and a complete run is performed in about 12 min.

But how are these very low concentrations detected? An **electron capture detector** is used. The carrier gas leaves the column and passes over two electrodes that have a high voltage between them (Figure 2). One of the electrodes is coated in a thin foil made from radioactive ^{63}Ni, which emits β-particles (electrons). These electrons cause the carrier gas to ionize. The ions and the electrons move to the electrodes and cause a current (a few nanoamps) that can be measured. Compounds that contain aromatic groups and especially halogen-containing compounds remove these electrons which become attached to the molecules. If these compounds enter the detector they 'capture' electrons, which reduces the ionization of the carrier gas so that the electrical signal changes. These detectors are extremely sensitive for compounds containing halogens and can measure femtogram (1 fg = 1×10^{-15} g) amounts of some compounds.

This extremely high sensitivity has led to electron capture detectors being widely used in environmental analysis, such as in measuring concentrations of pesticides, which often contain halogens. The technique was developed by James Lovelock, who used an electron capture detector in 1970 to detect chlorofluorocarbons (CFCs) in samples of 'unpolluted' air reaching the west coast of Ireland from the North Atlantic. This showed that chlorofluorocarbons were widely distributed in the atmosphere and was the first warning of problems to come.

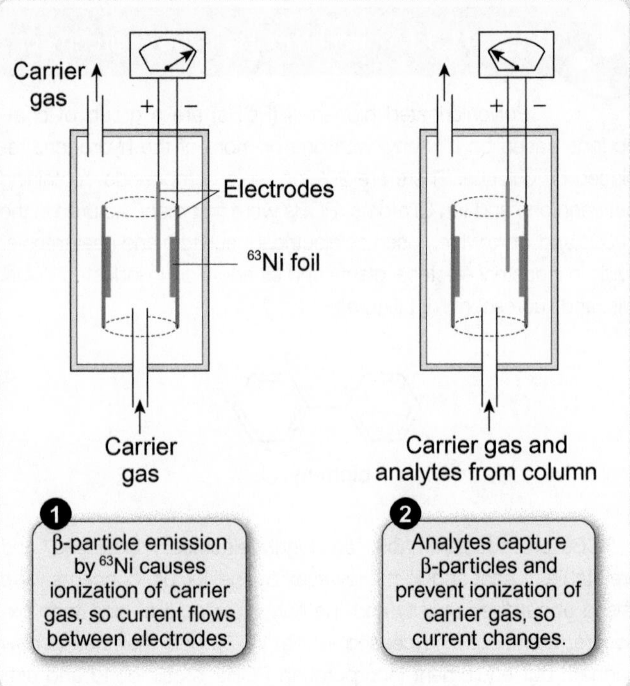

Carrier gas
+
−
Electrodes
^{63}Ni foil
Carrier gas
Carrier gas and analytes from column
1
β-particle emission by ^{63}Ni causes ionization of carrier gas, so current flows between electrodes.
2
Analytes capture β-particles and prevent ionization of carrier gas, so current changes.

▲ **Figure 2** An electron capture detector is extremely sensitive to halogenated compounds. These compounds 'capture' the β-particles (electrons) and prevent ionization of the carrier gas, thus changing the current flowing between the electrodes.

Question

Imagine that, rather than a standard sample, Figure 1 showed a chromatogram of an unknown sample. Suggest how each compound could be identified.

Box 11.4 Gas chromatography–mass spectrometry (GC–MS)

Gas chromatography is extremely efficient at separating compounds but the compounds then need to be identified. Each peak could be identified by comparing its retention time with standard samples run under identical conditions, but this can be very time-consuming. Coupling a gas chromatograph to a mass spectrometer allows the components to be identified rapidly. (Details of mass spectrometry are given in Box 1.2, p.14, and Section 12.1, p.558.)

As the carrier gas leaves the end of the column, it passes into a chamber where the pressure is substantially reduced. This is necessary since mass spectrometry takes place at very low pressure. The analytes then enter the spectrometer where they are ionized and their mass spectra are recorded. The mass spectrum of each analyte usually shows a molecular ion peak and a fragmentation pattern, which enables the compound to be identified (see Section 12.1, p.558). The computer attached to the GC–MS has a library of many thousands of mass spectra that it uses to find a match to identify an unknown sample.

➔

→ As well as identifying each compound, the quantity of each compound is measured, giving a chromatogram similar to those from other detectors. A mass spectrum is measured in a few seconds so that a spectrum is recorded for each peak as it leaves the chromatograph.

GC–MS is widely applied in the pharmaceutical industry and in environmental monitoring. It is particularly suitable where the sample contains a large number of analytes so that identification by comparison of elution times with known samples would be extremely slow. The measurement of drug concentrations in blood or urine described at the start of this chapter (p.512) is another illustration of using GC–MS.

▲ GC–MS is one of the most important methods for the detection of chemical weapons. The Organisation for the Prohibition of Chemical Weapons works to eliminate chemical weapons from the world and won the Nobel Peace Prize in 2013 for its work.

High performance liquid chromatography (HPLC)

This method was developed from column chromatography (p.532). To get good separation, a very large solid surface area is needed so very small particles (around 5–10 μm in diameter) are used. (Compare this with the diameter of a human hair, which is 20–150 μm). Pumping a liquid through a column full of such a solid requires very high pressures—up to around 500 bar—so specialized equipment is needed.

A very large range of samples can be analysed by HPLC—just about anything that can be dissolved in a solvent or mixture of solvents. Using a solvent means that, unlike in gas chromatography, non-volatile samples can be analysed. Since HPLC is usually carried out around room temperature, it can be applied to thermally unstable compounds. Only a small quantity of an analyte is needed and it can be recovered if necessary, which is an advantage when dealing with precious samples. These factors have led to extensive use of HPLC in the pharmaceutical industry. There are also many applications in environmental monitoring.

A typical HPLC apparatus is shown in Figure 11.21. A pump delivers a constant flow of solvent to the column. Due to the high pressures, the sample is injected into the solvent stream via a valve. It then flows through the stationary phase contained in the column and the separated analytes are detected as they emerge.

The most widely used detector is the **UV absorbance detector**. The eluate from the column flows through a small cell which is part of a spectrophotometer (see Section 11.4, p.542). A light beam passes through the cell and its intensity is measured at a wavelength where the solvent does not absorb. As compounds in the sample pass through the detector cell, they absorb UV radiation, giving a signal. Depending on the molar absorption coefficient of the compound, the detection limits are in the region of 1 ng ($1\ \text{ng} = 1 \times 10^{-9}\ \text{g}$) of analyte. Diode array UV detectors can measure the complete UV/VIS spectrum of each of the compounds in <1 s as they are eluted from the column.

As with gas chromatography, the chromatograph in HPLC can be coupled to a mass spectrometer to identify the compound giving rise to a particular peak in the chromatogram. An NMR spectrometer can also be used as the detector. The complete NMR spectrum of each of the compounds can be recorded as they flow through the spectrometer magnet. This is a rather expensive option but it gives a large amount of information very quickly from small amounts of sample and is becoming widely used in the pharmaceutical industry.

ⓘ HPLC is sometimes called high *pressure* liquid chromatography.

@ Visit the Online Resource Centre to view video clip 11.3 which gives a simple introduction to HPLC.

UV/VIS spectra and molar absorption coefficients, ε, are discussed in Section 10.3 (p.458) and in Section 10.6 (p.491). Diode array spectrophotometers are described in an information note on p.491.

The use of NMR spectroscopy in structure determination is discussed in Section 12.3 (p.578).

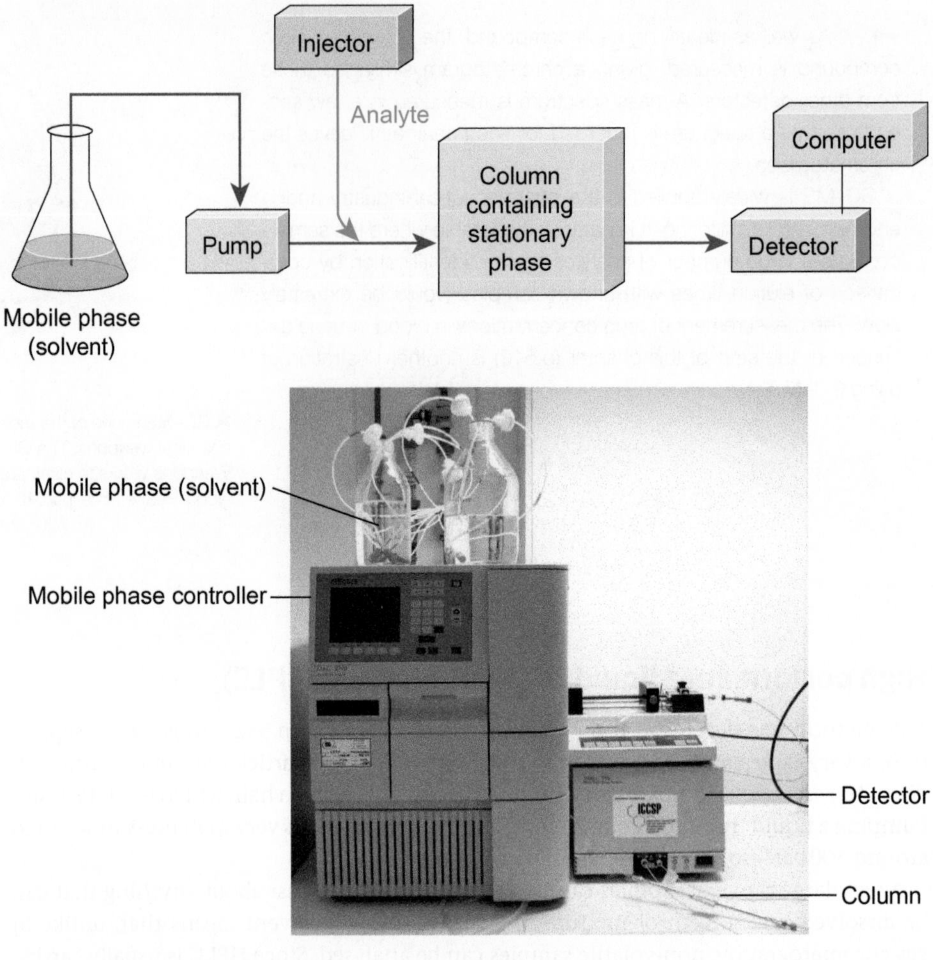
Injector
Analyte
Computer
Column containing stationary phase
Pump
Detector
Mobile phase (solvent)
Mobile phase (solvent)
Mobile phase controller
Detector
Column

Figure 11.21 A typical HPLC system.

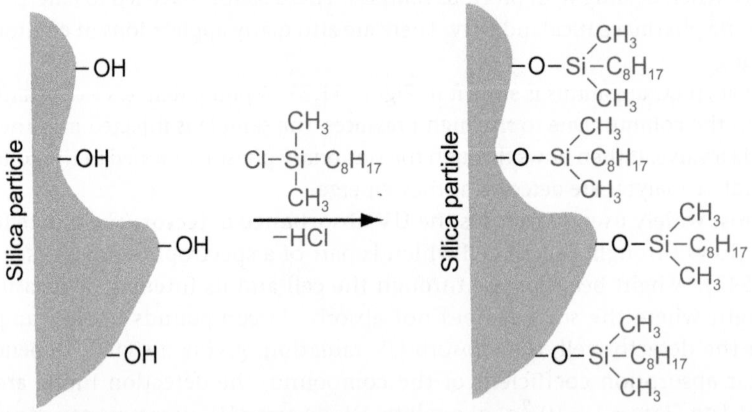
Silica particle
-OH
-OH
-OH
-OH
CH3
Cl-Si-C8H17
CH3
-HCl
Silica particle
CH3
-O-Si-C8H17
CH3

Figure 11.22 In reverse phase HPLC, stationary phases are prepared by reacting small diameter silica particles with organosilicon compounds, which react with OH groups on the silica.

> ℹ A porous solid is one that has pores or channels running through it that are accessible to the solvent and sample. This increases the surface area of the solid. In some cases, the size and shape of the pores can influence which compounds can enter the pores and so assist in the separation.

Separation of the components in a mixture during HPLC occurs as a result of differing interactions with the solid stationary phase. Very small particles are used to maximize the surface area and hence the degree of interaction that can take place. Porous silica (SiO_2) can have a surface area of 500 square metres per gram and is often used as an HPLC stationary phase. The surface has a large number of polar Si–OH groups with which the sample can interact. The sample is usually dissolved in a non-polar solvent such as hexane. This is called **normal phase** HPLC and is best suited to analysing mixtures of non-polar or moderately polar compounds.

Reverse phase HPLC uses a *polar* solvent such as water or CH_3CN to carry the sample over a *non-polar* solid phase. The stationary phase is prepared by bonding organic groups to the surface of the silica by reaction with the Si–OH groups (see Figure 11.22). Various coatings are available that allow the analyst to tailor the interactions to the particular compounds involved. An application of reverse phase HPLC to separate chiral compounds is described in Box 11.5.

Other types of HPLC use different ways to attract the sample to the stationary phase. An **ion exchange column** has a stationary phase that uses ionic interactions to separate dissolved ions in solution. The ions are then detected by their conductivity as they are eluted from the column. Another method is **size exclusion chromatography**, used for analysing large molecules such as polymers and proteins. The separation here depends on the stationary phase having pores with different sizes so that large and small molecules pass through at different rates.

The analysis of food products containing Sudan I described in Box 11.1 (p.522) was carried out using a reverse phase HPLC column packed with silica bonded to octadecyl ($C_{18}H_{37}$) groups and connected to a UV detector or mass spectrometer.

Box 11.5 Chiral HPLC

The pharmaceutical and agrochemical industries increasingly use single enantiomers of chiral compounds (see Section 18.4, p.838) to produce specific pharmacological effects. This means that chemists must be able to separate and analyse the enantiomers. **Chiral HPLC** uses reverse phase HPLC in which chiral groups are attached to the stationary phase.

One type of chiral stationary phase employs a chiral compound attached to the silica packing, as shown in Figure 1. The three functional groups, A, B, and C, complement those on the analyte, A′, B′, and C′, so that there are strong interactions between A and A′, between B and B′, and between C and C′. The stationary phase interacts more strongly where all three interactions can take place (enantiomer A in Figure 1). The interactions with the other enantiomer (enantiomer B) are weaker, so that it is eluted from the column in a shorter time, allowing separation of the two enantiomers. The nature of the three groups, A, B, and C, depends on the target analyte. A, B, and C may, for example, be an aromatic group, a polar substituent, and a hydrogen-bonding group.

Figure 2 shows the result of chiral chromatography of ibuprofen, which is a mixture of two enantiomers.

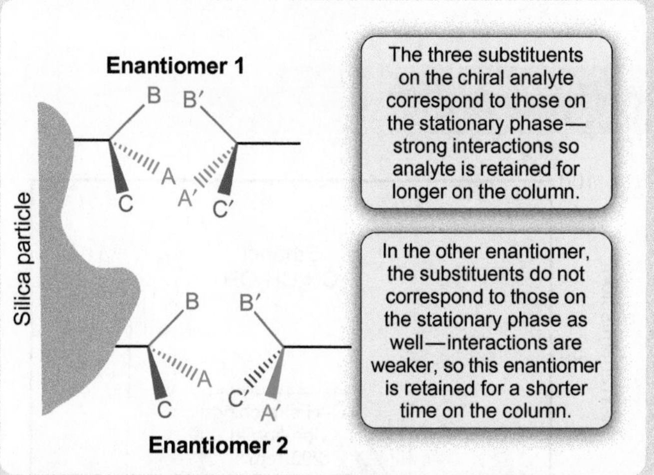
Enantiomer 1
B B′
A A′
C C′
Silica particle
B B′
A C′
C A′
Enantiomer 2
The three substituents on the chiral analyte correspond to those on the stationary phase—strong interactions so analyte is retained for longer on the column.
In the other enantiomer, the substituents do not correspond to those on the stationary phase as well—interactions are weaker, so this enantiomer is retained for a shorter time on the column.

Figure 1 Chiral separations. A, B, and C are functional groups that interact with the corresponding groups A′, B′, and C′ on the chiral analyte. Enantiomer 1 interacts more strongly than enantiomer 2 with the chiral compound on the stationary phase, so that it is retained for longer on the column. The enantiomers are therefore separated.

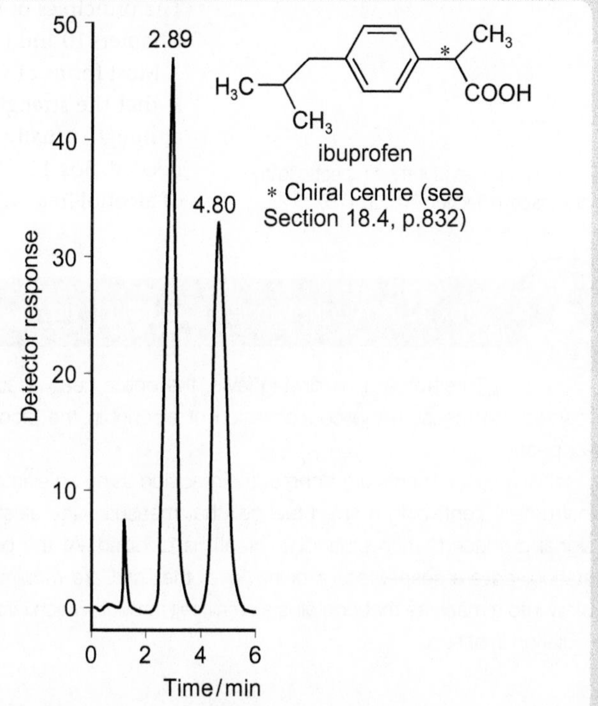
2.89
4.80
Detector response
Time/min
0
10
20
30
40
50
0
2
4
6
H3C
CH3
CH3
COOH
ibuprofen
* Chiral centre (see Section 18.4, p.832)

Figure 2 Many pharmaceutical products consist of enantiomeric mixtures. Ibuprofen is a widely used analgesic (painkiller). The HPLC chromatogram of ibuprofen using a chiral stationary phase shows a separate peak for each enantiomer.

Summary

- Chromatography involves the separation of compounds by distributing them between a mobile phase and a stationary phase.
- The distribution of a compound between two phases is measured by the partition coefficient.
- Analytes interact with the stationary phase to different extents and so move at different speeds through it.
- The degree of separation of two compounds is measured by the relative retention.
- Thin layer chromatography (TLC) is used for qualitative analysis. It involves the separation of compounds by passage of a solution over a plate coated with silica or alumina.
- Column chromatography is used to separate and recover the components of a mixture.
- Gas chromatography (GC) is used to analyse samples that are volatile, using high temperatures if necessary. The detector response is proportional to concentration so the peak area is proportional to the amount of component in the mobile phase.
- Gas chromatography–mass spectrometry (GC–MS) is used to rapidly identify and measure the concentrations of compounds in a mixture.
- Various types of stationary phase can be used in high performance liquid chromatography (HPLC) to analyse a wide range of soluble compounds.

? For practice questions on this topic, see questions 10–13 at the end of this chapter (p.554).

11.4 Spectroscopic methods of analysis

Spectroscopic methods involve the interaction of electromagnetic radiation with matter. The principles of spectroscopy and some examples of *qualitative* analysis are described in Chapters 10 and 12. This section illustrates the use of spectroscopy for *quantitative* analyses.

Most forms of spectroscopy can be used for quantitative analysis. The general principle is that the strength of the signal produced depends on the amount of material present, although usually some form of calibration with samples of known concentration is needed. Box 11.6 illustrates the use of infrared spectroscopy to measure the concentration of alcohol in a sample of blood.

The principles of infrared spectroscopy are discussed in Section 10.5 (p.476).

Box 11.6 Alcohol analysis and drink-driving

To enforce drink-driving laws, the police need accurate methods to measure the concentration of alcohol in the blood of suspects.

Initial measurements are done at the roadside using an electronic instrument containing a small fuel cell that measures the electrical signal produced when ethanol in breath is oxidized. At the police station, drivers suspected of being 'over the limit' are required to blow into a machine that contains a small cell through which infrared radiation is passed.

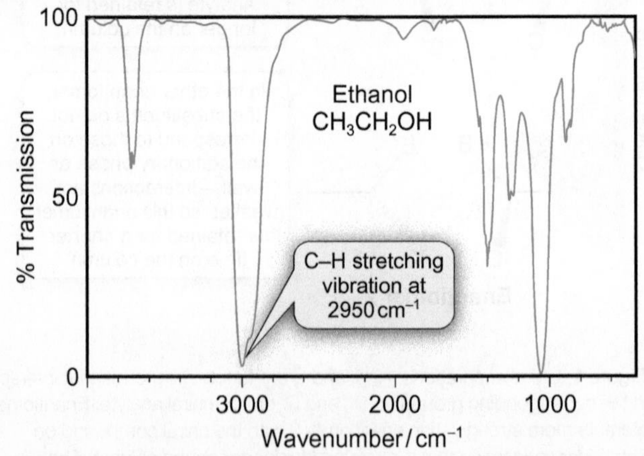
100
50
0
% Transmission
Ethanol
CH3CH2OH
C–H stretching vibration at 2950 cm−1
3000
2000
1000
Wavenumber / cm−1

Figure 1 The IR spectrum of ethanol in the gas phase has a large absorption band at 2950 cm^{-1}. The intensity of this peak is a measure of the alcohol content of the breath.

➜ Aliphatic organic molecules have a characteristic absorption band around 2950 cm^{-1} arising from the C–H bond stretching vibration (see Figure 1). The intensity of the peak in the spectrum is a measure of the alcohol content of the breath. The intensity recorded in the driver's breath is compared with a 'blank' reading when a stream of alcohol-free air is passed through the cell. Other organic compounds in the breath also absorb at 2950 cm^{-1} but the machine takes this into account by taking measurements at several different wavenumbers in the spectrum. Some people have propanone (acetone) present in their breath but this has a characteristic absorption around 1700 cm^{-1} due to the C=O bond. Any contribution to the C–H absorption at 2950 cm^{-1} from propanone can be accounted for and subtracted from the final result. This gives an accurate test for ethanol that produces results in a matter of a few minutes.

The legal limit for driving in the UK is 80 mg of alcohol per 100 cm^3 of blood, which is equivalent to 35 μg of alcohol per 100 cm^3 of breath. A deep breath is required so that air comes from deep in the driver's lungs. Here, ethanol vapour in air in the lungs is in equilibrium with ethanol in the blood so that the concentration in the breath is a direct measure of the concentration in the blood (this is a result of Henry's law, Section 17.4 p.789).

▲ Roadside breath tests are conducted with a handheld meter containing a small fuel cell that measures the oxidation of ethanol.

Question

Suggest a reason why the O–H bond stretching vibration at 3650 cm^{-1} is not used for the analysis of ethanol in breath.

Ultraviolet-visible spectrophotometry

Compounds are coloured because they absorb some wavelengths of visible light and transmit others. For example, $[MnO_4]^-$ ions absorb green and blue light, so solutions of $KMnO_4$ appear a red-purple colour. As shown in Figure 11.23, the intensity of the transmitted light depends on concentration. This is made use of in ultraviolet–visible (UV/VIS) spectrophotometry to measure the concentrations of solutions. The absorbance, rather than the transmission of light, is measured, since the absorbance is proportional to the concentration.

If radiation with intensity I_0 passes through a solution with concentration c and emerges with an intensity I_t, the **absorbance, *A***, of the solution is defined as

$$A = \log_{10}\left(\frac{I_0}{I_t}\right) \quad (10.7)$$

A is proportional to concentration, c, and the path length, l, in the solution through which the radiation passes, as expressed by the **Beer–Lambert law**

$$A = \varepsilon \times c \times l \quad (10.8)$$

The **molar absorption coefficient**, ε, is a measure of how effectively the molecules absorb radiation. A high value of ε indicates that a large amount of radiation is absorbed at a given concentration.

The Beer–Lambert law predicts that, for a fixed path length, the absorbance A is proportional to concentration, as shown in Figure 11.24. The plot is a straight line although at very high concentrations the plot deviates from linearity and the Beer–Lambert law is not obeyed. This region, which is usually when A is over 1.5, is not useful for analysis.

UV/VIS spectrophotometry is a very useful method for measuring the concentrations of coloured species in solution. The usual method is to prepare a calibration graph covering the concentration range of interest. The absorbance of a sample solution can then be measured and its concentration obtained from the graph. Alternatively, the

Coloured compounds and the principles of ultraviolet-visible spectroscopy are discussed in Section 10.6 (p.491). The Beer–Lambert law is introduced in Section 10.3 (p.458).

ℹ Spectrometers that operate in the UV/VIS region are often called **spectrophotometers**.

ℹ The terms **spectrometry** and **spectrophotometry** are usually used to describe experiments where the intensity of light at one wavelength is used to measure the concentrations of different solutions. **Spectroscopy** usually describes experiments where the concentration is fixed and the variation of intensity with different wavelengths is used to record a spectrum.

ℹ The units of ε are often given as $dm^3\,mol^{-1}\,cm^{-1}$ since these are the common units of measurement. The SI units for ε are $m^2\,mol^{-1}$.

5 × 10⁻⁴
1 × 10⁻⁴
5 × 10⁻⁵
2 × 10⁻⁵
1 × 10⁻⁵
5 × 10⁻⁶
Water
Concentration / mol dm⁻³

Figure 11.23 As the solutions of $KMnO_4$ become more dilute, the colour becomes less intense and more light is transmitted through the solution. The absorbance is related to solution concentration by the Beer–Lambert law.

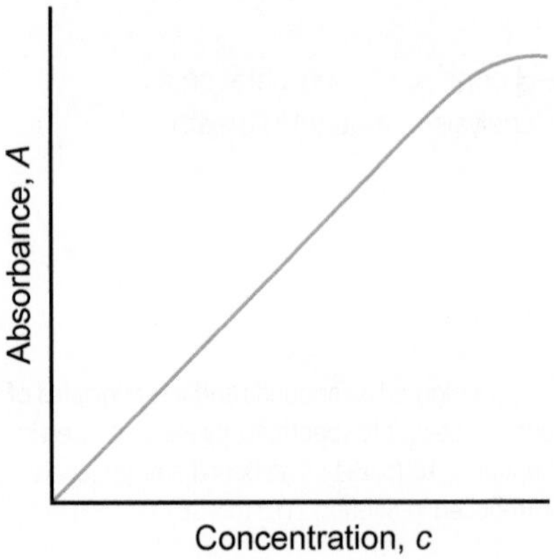
Absorbance, A
Concentration, c

Figure 11.24 The Beer–Lambert law predicts a straight line relation between absorbance and concentration. At high concentrations the line becomes curved.

gradient of an A versus c graph gives the molar absorption coefficient, ε, from which the unknown concentration can be calculated using the Beer–Lambert law (see Worked example 11.3).

Measuring concentrations by spectrophotometry is not limited to single compounds but can also be applied to mixtures (as illustrated in Box 11.7). The spectrophotometer will detect all the species in solution as long as they have distinguishable absorption spectra. The Beer–Lambert law applies separately to each species provided that they do not react with each other.

Worked example 11.3 Spectrophotometric analysis

The absorbances of aqueous solutions of $KMnO_4$ were measured in a 1.0 cm cell using 520 nm radiation. The results are shown in the table below.

Concentration $KMnO_4$(aq)/10^{-4} mol dm^{-3}	0.00	0.10	0.20	0.50	1.00	2.00	3.00	4.00	5.00
Absorbance	0	0.019	0.038	0.096	0.192	0.384	0.576	0.768	0.960

(a) Determine the molar absorption coefficient, ε, of $KMnO_4$ in dm^3 mol^{-1} cm^{-1}.

(b) Convert the value of ε to SI units.

(c) An aqueous solution of $KMnO_4$ has an absorbance of 0.50. What is the concentration of this solution?

Strategy

(a) Use the data to plot a graph of absorbance, A, against concentration, c. From the Beer–Lambert law (Equation 10.8), $A = \varepsilon \times c \times l$, so the graph is a straight line with gradient $\varepsilon \times l$. Since you know the path length of light through the solution, l, you can work out the molar absorption coefficient, ε.

(b) Convert the units of ε from dm^3 mol^{-1} cm^{-1} to m^2 mol^{-1}.

(c) You can find the concentration of the unknown solution either by reading off the appropriate value from the calibration graph or by substituting the values into Equation 10.8.

→

→

Solution

(a) *Plot a graph of the absorbance, A, against concentration, c. Find the gradient of the graph and hence a value for ε.*

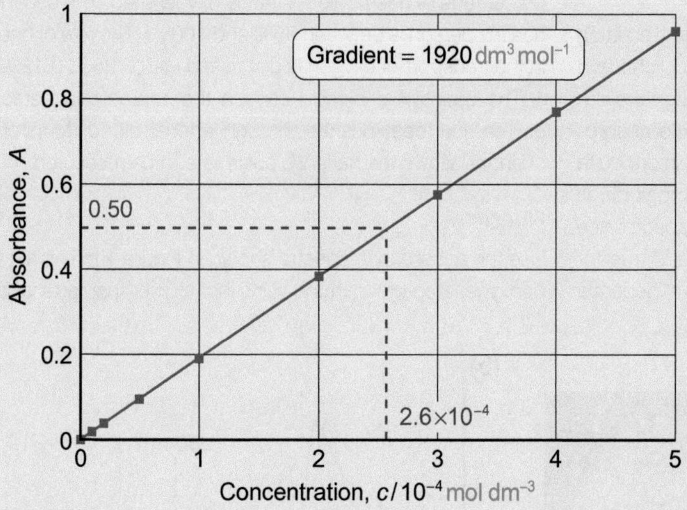

Gradient = 1920 dm³ mol⁻¹
0.50
2.6×10⁻⁴
Absorbance, A
Concentration, c / 10⁻⁴ mol dm⁻³
0
0.2
0.4
0.6
0.8
1
2
3
4
5

Calibration graph.

The plot is a straight line.

Gradient $= \varepsilon \times l = 1920\,\mathrm{dm^3\,mol^{-1}}$

Since the path length is 1.0 cm, $\varepsilon = 1920\,\mathrm{dm^3\,mol^{-1}\,cm^{-1}}$

(b) *Convert the value of ε to SI units.*

Two conversions are needed: $1\,\mathrm{dm^3} = 1 \times 10^{-3}\,\mathrm{m^3}$ and $1\,\mathrm{cm^{-1}} = 100\,\mathrm{m^{-1}}$

$$\varepsilon = 1920\,\mathrm{dm^3\,mol^{-1}\,cm^{-1}} = 1920 \times (1 \times 10^{-3}\,\mathrm{m^3}) \times \mathrm{mol^{-1}} \times (100\,\mathrm{m^{-1}})$$

Rearrange to collect terms together and simplify

$$\varepsilon = 1920 \times (10^{-3} \times 100)\,\mathrm{m^3\,mol^{-1}\,m^{-1}} = 192\,\mathrm{m^2\,mol^{-1}}$$

(c) *Find the concentration when A = 0.50.*

- From the graph, read off the concentration for the solution with $A = 0.50$

  $$c = 2.6 \times 10^{-4}\,\mathrm{mol\,dm^{-3}}$$

- An alternative approach is to substitute values into Equation 10.8

  $$A = \varepsilon \times c \times l$$

  $$0.5 = (1920\,\mathrm{dm^3\,mol^{-1}\,cm^{-1}}) \times c \times (1.0\,\mathrm{cm})$$

  Rearrange to find c

  $$c = \frac{0.50}{(1920\,\mathrm{dm^3\,mol^{-1}\,\cancel{cm}^{-1}}) \times (1.0\,\cancel{cm})} = 2.6 \times 10^{-4}\,\mathrm{mol\,dm^{-3}}$$

Question

0.0277 g of $KMnO_4$ were dissolved in 250 cm³ of water. What would be the absorbance of this solution at 520 nm using the same cell?

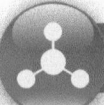

Box 11.7 A pulse oximeter

A **pulse oximeter** is a device that fits on to a patient's fingertip or ear lobe and measures their pulse rate and the concentration of oxygen in their blood.

Oximeters (Figure 1(a)) use a source of light, usually a light-emitting diode (LED) that emits at two wavelengths, 650 nm (red light) and 910 nm (infrared). The intensity of radiation passing through the patient's tissue is measured by a photodetector. The visible absorption spectra of haemoglobin and oxygenated haemoglobin are quite different at these wavelengths (Figure 1(b)). If the haemoglobin is largely oxygenated, the blood looks bright red and the absorbance at 650 nm will be less than that at 910 nm. The reverse is true for deoxygenated haemoglobin. The ratio, A_{910}/A_{650}, therefore, depends on how much of the haemoglobin is oxygenated. The manufacturers calibrate the oximeters to display an 'spO_2' value, which is the percentage of haemoglobin that is oxygenated. The usual range in a healthy person is between 95% and 100%. (The role of haemoglobin in the transport of oxygen is discussed in Box 28.7, p.1298.)

The quantity of blood between the source and detector varies as the heartbeat pushes through arterial blood. Monitoring these changes allows the patient's pulse rate to be measured.

Question

Use the absorbance spectra shown in Figure 1 to explain why deoxygenated blood in the veins looks bluer than arterial blood.

(a)

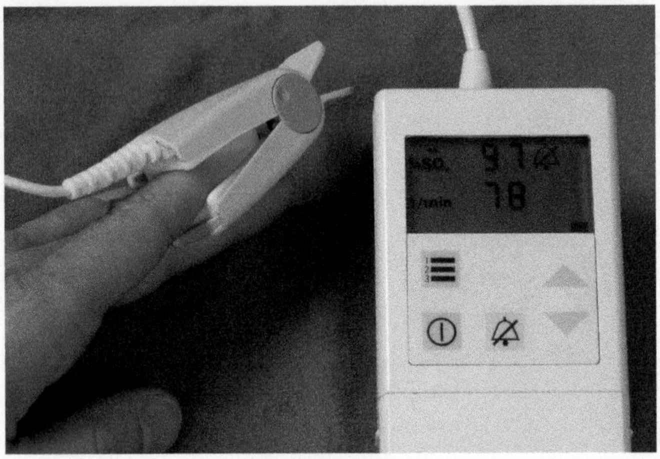
97
78

(b)

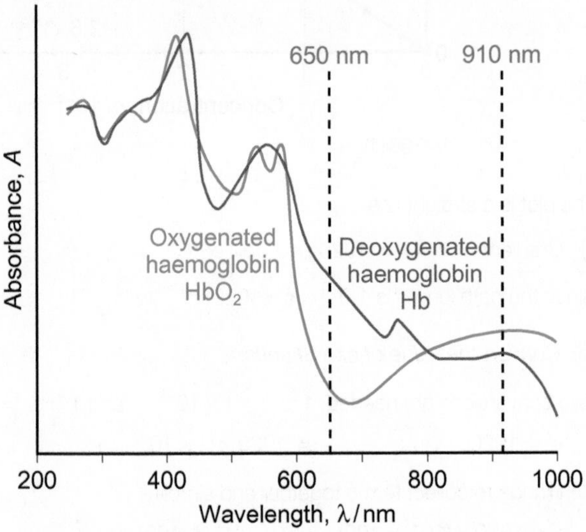
650 nm
910 nm
Absorbance, A
Oxygenated
haemoglobin
HbO2
Deoxygenated
haemoglobin
Hb
200
400
600
800
1000
Wavelength, λ / nm

Figure 1 (a) A pulse oximeter used to monitor oxygen concentrations in blood. (b) Absorbance spectra for oxygenated haemoglobin (HbO_2) and deoxygenated haemoglobin (Hb).

Summary

- Substances appear coloured due to the absorption of some wavelengths of light and transmission of other wavelengths.
- The absorption of radiation is described by the *Beer–Lambert law*: $A = \varepsilon \times c \times l$
- Compounds with high values of ε, the *molar absorption coefficient*, absorb high intensities of light at a given concentration.
- The *absorbance* of a solution is proportional to its concentration, and spectrophotometry can be used to measure the concentration of absorbing species in solution.
- The absorbances of several species in solution are additive so that mixtures of compounds can be analysed provided they have distinguishable absorption spectra.

For practice questions on this topic, see questions 14–19 at the end of this chapter (p.554).

11.5 Atomic spectrometry

The spectroscopic methods described in the previous section involve the absorption of ultraviolet or visible radiation due to transitions between electronic energy levels in *molecules*. Transitions between energy levels in *atoms* form the basis of atomic spectroscopy. Atomic spectroscopy is useful for the analysis of about 75 metals and other elements, which can be detected at very low concentrations.

Electronic energy is quantized. Adding energy to an atom in its ground state can promote electrons into higher orbitals. The atom is said to be in an **excited state**. It is the transfer of electrons between these orbitals that gives rise to the spectra. The absorptions (or emissions) in atomic spectra are usually narrow lines—there are no bond vibrations as in molecular spectra.

Using atomic spectroscopy to measure the concentrations of substances in analytical chemistry does not normally involve recording spectra. Instead, the intensity of radiation that is absorbed or emitted at a particular wavelength is measured and the methods are called **atomic spectrometry**. There are several types of atomic spectrometry but all rely on atomizing the sample, that is, converting it to gaseous atoms by heating. At high temperatures, thermal excitation promotes atoms to excited states and when they return to the ground state, energy is emitted as photons at particular wavelengths. The consequence of this can be seen during flame tests when a small quantity of a metal salt imparts colour to a flame, as illustrated for sodium in Figure 11.25. The salt is atomized in the flame and some atoms are promoted to an excited state, then return to the ground state, emitting photons at a wavelength which, for sodium, corresponds to yellow-orange light.

Flames obtained by burning natural gas, or hydrocarbons such as butane, in air reach temperatures in the region of 1500 K—this is the hottest that a Bunsen flame will reach. Since a larger number of atoms are excited at higher temperatures than this, atomic spectrometry methods use ethyne (acetylene) as the fuel because it typically burns at 2500 K and can reach over 3000 K when burned with oxidants such as N_2O. The importance of a high temperature is discussed in Box 11.8.

Atomic spectroscopy is introduced in Section 3.3 (p.124). Molecular spectroscopy is the subject of Chapter 10 (p.448).

The principles of UV/VIS spectroscopy, in terms of the transition between electronic energy levels in molecules, are discussed in Section 10.6 (p.491).

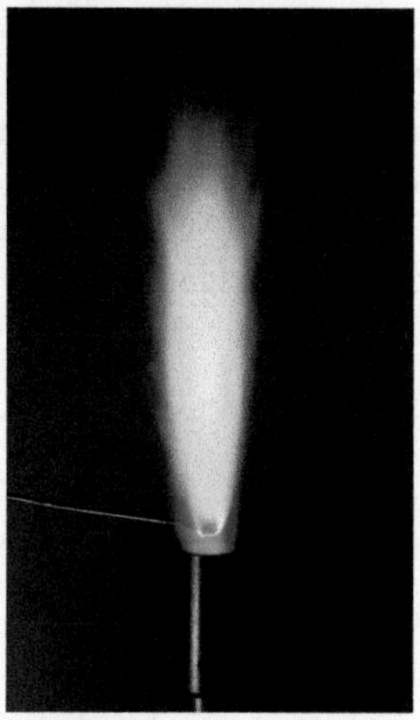

Figure 11.25 A small quantity of a sodium salt in a flame produces a yellow-orange flame due to emission of energy from atoms in excited electronic states as they return to the ground state. The emission from sodium is also responsible for the yellow colour of many street lights.

Atomic emission spectrometry (AES)

The process that gives rise to flame tests forms the basis of atomic emission spectrometry (AES). Part of the energy level diagram for sodium is shown in Figure 11.26. In a ground state sodium atom, the valence electron is in the 3*s* orbital. In atoms that have been thermally excited, this electron can be promoted, for example, to the 3*p* or 4*p* orbitals. When the electron returns to the 3*s* orbital, a photon is emitted with energy corresponding to the difference between the energy levels. The energy difference between the 3*p* and 3*s* orbitals corresponds to visible light with a wavelength of 589 nm, which is in the yellow-orange region in the sodium emission spectrum (see Box 3.2, p.128).

Although thermal excitation can lead to transitions to other orbitals, such as 3*d*, 4*s*, 4*d*, or 5*s* orbitals, only states in which the electrons are in *p* orbitals can emit by returning to the ground state, because of the selection rules for electronic spectroscopy. (Selection rules are introduced in Section 10.3, p.458.)

Box 11.8 Why is the temperature so important in atomic emission spectrometry?

The principle of atomic emission spectrometry is that the intensity of the emitted radiation is proportional to the number of atoms in the excited state.

The number of atoms in a higher energy level relative to those in a lower level is given by the Boltzmann distribution (see Section 10.3, p.458)

$$\frac{n_{\text{upper}}}{n_{\text{lower}}} = \frac{g_{\text{upper}}}{g_{\text{lower}}} e^{\left(\frac{-\Delta E}{k_B T}\right)} \tag{10.9}$$

where *n* is the number of atoms in an energy level and *g* is the degeneracy, that is, the number of states with the same energy level. k_B is the Boltzmann constant, 1.381×10^{-23} J K^{-1}.

Equation 10.9 can be applied to the 589 nm line in the sodium spectrum. Radiation of this wavelength is equivalent to an energy of 3.37×10^{-19} J (using Equations 3.1, p.118, and 3.2, p.120). This is the difference in energy between the 3*p* and 3*s* orbitals. (In a ground state sodium atom, the outer electron is in a 3*s* orbital. This is promoted to a 3*p* orbital in the excited atom, see Figure 11.27.)

→

➜ The 3*p* level has an orbital degeneracy of $g = 3$ (there are three equivalent 3*p* orbitals) while the 3*s* level has an orbital degeneracy of $g = 1$.

Hence, at the temperature of a natural gas flame, 1500 K

$$\frac{n_{3p}}{n_{3s}} = \frac{3}{1} e^{\left(\frac{-(3.37) \times 10^{-19}\,\text{J}}{(1.381 \times 10^{-23}\,\text{J K}^{-1}) \times 1500\,\text{K}}\right)}$$

$$= 2.58 \times 10^{-7}$$

This shows that at the temperature of a natural gas–air flame, only around 1 atom in 4 million is in the upper energy level from which emission takes place. Repeating the calculation at 2500 K gives a ratio of 1.71×10^{-4} or about 1 excited atom in 6000. Thus, a higher temperature results in a significantly stronger signal.

Questions

(a) Calculate the ratio of the numbers of excited state to ground state sodium atoms at 5000 K.

(b) Calculate the temperature needed for this ratio to be 0.1.

(c) Calculate the ratio at 2510 K. Compare your answer with the value given above for 2500 K and comment on the result in terms of the experimental requirements for atomic emission spectrometry.

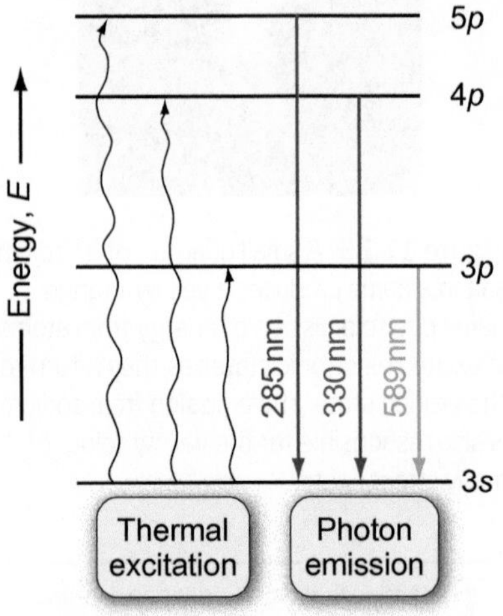
5p
4p
3p
3s
Energy, E
285 nm
330 nm
589 nm
Thermal excitation
Photon emission

Figure 11.27 A schematic energy level diagram for sodium. The 3*s* valence electrons are thermally promoted to higher energy orbitals. When the atom returns to the ground state, the energy is emitted as a photon. The 285 nm and 330 nm transitions are in the ultraviolet; the 589 nm transition gives the characteristic yellow-orange colour.

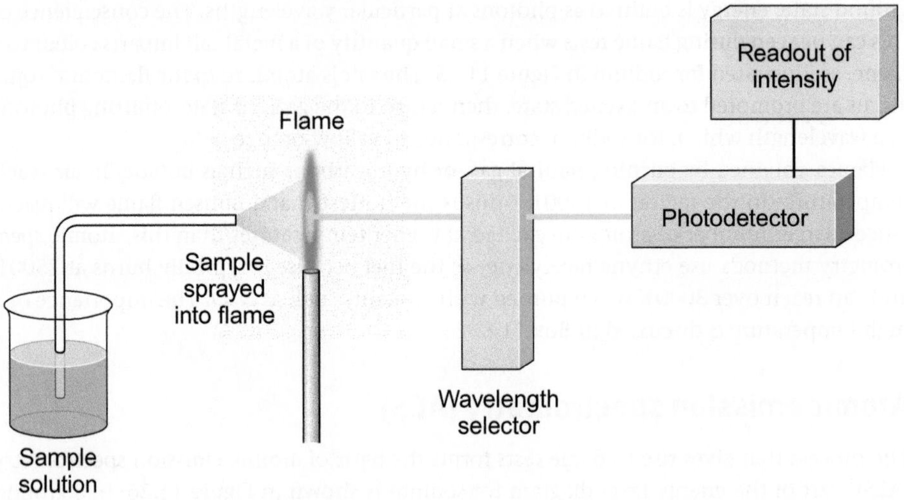
Readout of intensity
Flame
Photodetector
Sample sprayed into flame
Wavelength selector
Sample solution

Figure 11.26 Flame photometer apparatus used for atomic emission spectrometry.

When atomic emission spectrometry uses a flame to excite the atoms, it is also known as **flame photometry**. The sample is dissolved in a solvent and sprayed as small droplets into a flame, as shown in Figure 11.26. The intensity of light emitted is measured at selected wavelengths. Due to the sharpness of the emission lines, elements can be unambiguously identified. Different wavelengths can be selected to analyse for different elements simultaneously—for example, 760 nm for potassium as well as 589 nm for sodium.

Relatively few elements give sufficiently intense emission to be useful at the temperatures achieved in a flame. The range of elements accessible to flame photometry is quite limited and the detection limits are relatively high. Exceptions are sodium and potassium where the detection limits for Na^+ and K^+ are in the region of 50 ppb by mass. One common application of flame photometry is in the measurement of these metals in blood plasma. Normal concentrations in blood are 135 mmol dm^{-3}–145 mmol dm^{-3} for Na^+ and 3.5 mmol dm^{-3}–5.0 mmol dm^{-3} for K^+. Concentrations outside these limits can profoundly affect health and, if they vary widely, can be fatal, so access to a reliable diagnosis method is important.

Inductively coupled plasma (ICP)

Using a flame has disadvantages for very accurate analytical work. Only a small percentage (less than 10%) of the atoms in the flame is excited (see Box 11.8) and careful control of the gas flow rates is needed to keep the temperature of the flame constant. These problems can be largely overcome by replacing the flame by an **inductively coupled plasma (ICP)**. Argon gas flows through a narrow quartz tube around which is wrapped a metal coil. Passing a current through the coil induces a strong magnetic field and a spark starts the ionization of the Ar atoms to form **plasma**—an ionized gas in which ions and electrons undergo energetic collisions. The temperature inside such a plasma can reach 7000 K–10 000 K. The higher temperatures generated in a plasma allow analysis for a wider range of elements. The temperature is so high that atomization of even very stable oxides takes place. Around 60 elements can be analysed in this way, at concentrations down to 1 $\mu g\,dm^{-3}$–5 $\mu g\,dm^{-3}$.

The intensity of emission is proportional to the amount of the species present, so, as with flame photometry, atomic emission spectrometry using the inductively coupled plasma method can be used for measuring the concentration of a sample solution. The spectrophotometer gives a reading of emission intensity and does not give the concentration directly. Solutions with known concentrations of the element under study are prepared and their emission intensity is measured to prepare a calibration graph. The intensity of measured emission from a solution of the analyte allows calculation of its concentration as shown in Worked example 11.4.

1 part per billion (1 ppb) is 1 part in 1×10^9.

For some applications, flame photometry has been replaced by ion selective electrodes for the rapid determination of Na^+ and K^+ concentrations (see Section 11.2, p.525).

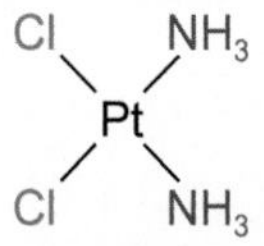
Cl
NH3
Pt
Cl
NH3
cisplatin

Inductively coupled plasma atomic emission spectrometry is used to measure concentrations of the anti-cancer drug, cisplatin (see Box 28.3, p.1275). The high temperature means that samples can be injected directly, without the need to separate the cisplatin from any organic material. Concentrations of Pt as low as 0.1 $mg\,dm^{-3}$ in tissue or body fluid can be measured.

A related technique is **inductively coupled plasma-mass spectrometry**, where the sample is ionized in the plasma and then goes into a mass spectrometer where elements are identified by their mass.

Worked example 11.4 Measuring concentrations using atomic emission spectrometry

The concentration of lithium ions in water is used by geochemists to track the source of underground water.

50.0 cm^3 of water taken from a cave was made up to exactly 1.00 dm^3 with pure water. This solution gave an emission intensity of 497 (arbitrary units) in an inductively coupled plasma spectrometer.

A standard aqueous solution of lithium chloride with a concentration of 5.00 $mg\,dm^{-3}$ was available. Solutions were made by diluting known volumes of this standard solution to 100 cm^3. They were injected onto the spectrometer with the following results.

Volume of standard solution used/cm^3	2.0	4.0	6.0	8.0	10.0
Emission intensity/arbitrary units	142	284	426	568	710

Calculate the concentration of Li^+ ions in the water from the cave.

Strategy

From the volumes of standard solution used, you can calculate the concentration of each diluted solution. Use these values and the emission data to plot a calibration graph. The concentration of the sample solution can then be read off the calibration graph from its emission intensity. Taking the dilution into account, you can now find the concentration of Li^+ ions in the cave water.

Solution

Calculate the concentration of lithium ions in each of the calibration solutions.

For example, 1000 cm^3 of the standard solution contain 5.00 mg of Li^+ ions, so 2.0 cm^3

$$\text{contain } \left(\frac{5.00 \times 2.0}{1000}\right) \text{mg} = 0.010 \text{ mg of } Li^+ \text{ ions}$$

→

→ This was diluted in 100 cm^3 so that the resulting solution had a concentration of 0.10 $mg\ dm^{-3}$. The other calibration data are shown in the following table.

Volume of standard solution used/cm^3	2.0	4.0	6.0	8.0	10.0
Mass of Li^+ used/mg	0.010	0.020	0.030	0.040	0.050
Concentration/$mg\ dm^{-3}$	0.10	0.20	0.30	0.40	0.50
Emission intensity/arbitrary units*	142	284	426	568	710

*The emission intensity depends on the settings on the photometer. This analysis depends on all samples being investigated using the same settings so that the units of the emission intensity do not matter. The units of such measurements are referred to as 'arbitrary units'.

Plot the calibration graph.

The concentration should be on the *x*-axis. The photometer emission intensity is the dependent (*y*-axis) variable. The calibration graph is shown here.

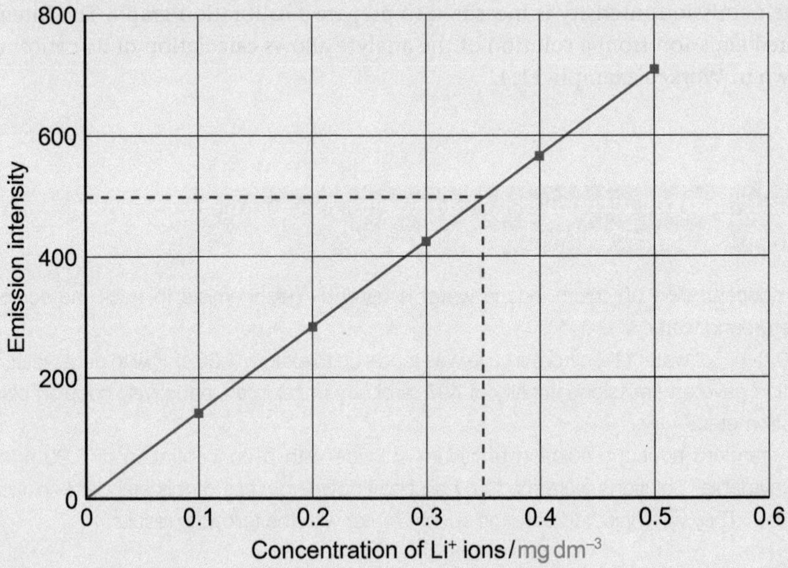
800
600
400
200
0
Emission intensity
0
0.1
0.2
0.3
0.4
0.5
0.6
Concentration of Li⁺ ions / mg dm⁻³

Calibration graph.

Find the concentration of the sample solution.

The sample being analysed gave an emission intensity of 497, which corresponds to a concentration of 0.35 $mg\ dm^{-3}$ on the calibration graph.

Find the concentration of Li^+ ions in the cave water.

First, take into account the dilution of the original sample.

50.0 cm^3 of cave water were made up to 1 dm^3, so 0.35 mg would be present in 50.0 cm^3 of the cave water. So, 1000 cm^3 of the original water would contain

$$\left(\frac{0.35 \times 1000}{50.0}\right) \text{mg} = 7.0 \text{ mg of } Li^+ \text{ ions}$$

Concentration of Li^+ ions in the original water = 7.0 $mg\ dm^{-3}$.

Question

A mineral water has a concentration of Li^+ ions of 0.25 $mg\ dm^{-3}$. What would be the emission intensity from the spectrometer if used under the conditions above?

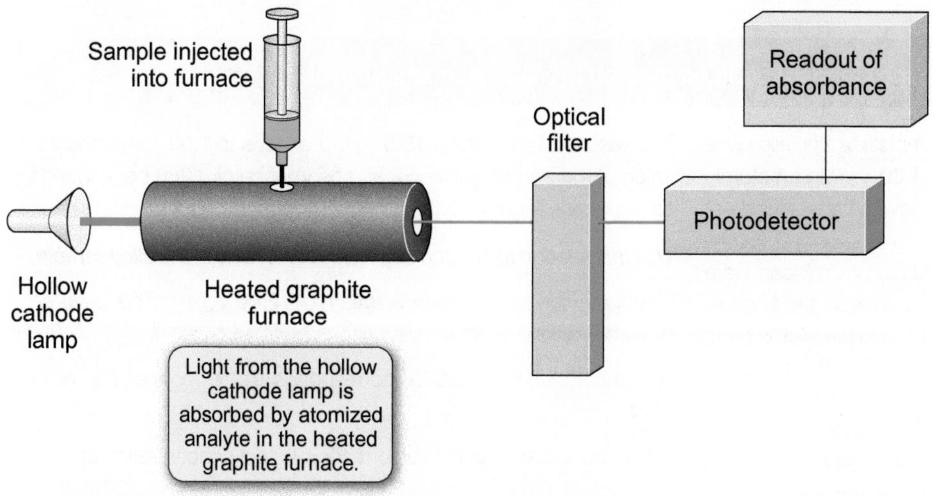
Sample injected into furnace
Readout of absorbance
Optical filter
Photodetector
Hollow cathode lamp
Heated graphite furnace
Light from the hollow cathode lamp is absorbed by atomized analyte in the heated graphite furnace.

Figure 11.28 Schematic diagram for an atomic absorption spectrometer.

Atomic absorption spectrometry (AAS)

In atomic emission spectrometry, only a small fraction of the sample emits radiation which can be analysed. **Atomic *absorption* spectrometry (AAS)** overcomes this drawback by analysing the atoms in the ground state. The atomized sample *absorbs* radiation rather than emitting it.

The apparatus is shown schematically in Figure 11.28. As in atomic emission spectrometry, the sample must be atomized. Most instruments employ a graphite furnace, sometimes known as an **electrothermal analyser**, to heat the sample. A small quantity of solution (5 μL–10 μL) or solid (1 mg–5 mg) is injected into a graphite tube inside the analyser. A large electric current (~100 A) is then passed through the graphite for a few seconds. The temperature rises quickly to ~2800 K and the sample is rapidly and completely atomized. The analyser contains an inert gas such as argon to prevent reaction with the analyte.

> 1 μL (microlitre) = 1×10^{-6} dm^3 = 1×10^{-3} cm^3

This method differs from atomic emission spectrometry in that the atomized sample is irradiated with intense radiation of wavelengths that match the transitions between the energy levels in the elements being analysed. Each element absorbs at a specific wavelength and so a different radiation source, known as a **hollow cathode lamp**, must be used for each element. The wavelength of the lamp is selected so that it corresponds to radiation absorbed by the ground state of the atom so that the analysis is specific for that particular element. When the sample is introduced, radiation is absorbed and the intensity of radiation transmitted is measured. Hollow cathode lamps are available for around 70 elements so that the method can be widely applied. The intensity of radiation transmitted when a sample is introduced is measured, and compared with that due to the pure solvent. These measurements are then used to calculate the absorbance of the solution using Equation 10.7 (see p.461 and Section 10.3, p.458).

The absorbances of a series of standard solutions containing the element of interest are recorded and used to prepare a calibration graph from which the concentrations of unknown solutions can be found (as in Worked example 11.4). The absorption of light follows the Beer–Lambert law (Equation 10.8, p.461) so the absorbance is plotted versus concentration to give a linear calibration graph.

An alternative method for calculating concentrations in atomic absorption spectrometry is known as **standard addition**. The change in absorbance is measured when an exact amount of the element under investigation is added, using a standard solution. The increased absorbance must be due to the additional amount added, so you can work out the relationship between the absorbance and concentration and hence find the concentration of the original solution. This is illustrated in Worked example 11.5.

> The standard addition method helps to compensate for any interferences arising from the matrix that contains the analyte (see Section 11.1, p.515).

Worked example 11.5 Using the standard addition method to measure concentrations

Several metal alloys contain zinc. A 2.00 g sample of a zinc alloy was dissolved in aqueous acid and made up to 1.00 dm^3. In an atomic absorption spectrophotometer, this solution gave an absorbance of 0.179 (arbitrary units).

0.010 cm^3 of a 1.00 $g\,dm^{-3}$ standard solution of Zn^{2+} ions were added to 1.00 dm^3 of the alloy solution. This increased the absorbance to 0.459. Calculate the concentration of zinc in the alloy in parts per million.

Strategy

The absorbance increases when the standard solution is added. The increase must be due to the Zn^{2+} ions added so that the relationship between concentration and absorbance can be found. This can then be applied to the absorbance of the alloy solution to find its concentration.

Solution

Find the mass of zinc added into the standard solution.

1000 cm^3 of the standard solution contain 1.00 g of Zn^{2+} ions.

So, 0.010 cm^3 of the standard solution contain

$$\left(\frac{1.00 \times 0.010}{1000}\right) g = 1.00 \times 10^{-5}\ g$$

$$= 10.0\ \mu g \text{ of } Zn^{2+} \text{ ions}$$

Find the increase in absorbance due to this addition.

Increase of absorbance = (0.459 – 0.179) = 0.280

This resulted from adding 10.0 μg of Zn^{2+} ions to 1.00 dm^3 of the solution of the zinc alloy. Hence, an absorbance of 0.280 corresponds to a concentration of 10.0 $\mu g\,dm^{-3}$.

Use this relation to find the concentration of zinc in the alloy sample.

Concentration of zinc alloy solution = 2.00 $g\,dm^{-3}$. This gave an absorbance of 0.179.

An absorbance of 0.280 corresponds to a concentration of 10.0 $\mu g\,dm^{-3}$.

So, an absorbance of 0.179 corresponds to a concentration of

$$\frac{0.179}{0.280} \times 10.0\ \mu g\,dm^{-3} = 6.39\ \mu g\,dm^{-3}$$

This solution was made by dissolving 2.00 g of alloy in 1.00 dm^3, so

$$\text{concentration of zinc in alloy} = \frac{6.39\ \mu g\,\cancel{dm^{-3}}}{2.00\ \cancel{dm^{-3}}}$$

$$= 3.20\ \mu g\,g^{-1}$$

Since 1 g = 1×10^6 μg, 1 $\mu g\,g^{-1}$ = 1 ppm, concentration of zinc in alloy = 3.20 ppm.

Question

What will be the absorbance of a solution of Zn^{2+} ions with a concentration of 1.0 $\mu mol\,dm^{-3}$, measured under the same conditions as used in the question above?

Summary

- Atomic spectrometry involves atomizing samples at high temperatures.
- Atomic emission spectrometry (AES) measures radiation emitted when thermally excited atoms return to the ground state.
  - A hot flame or, for better results, an inductively coupled plasma, is used to atomize and excite the elements in the sample.
  - Different elements can be analysed by selecting different wavelengths.
  - Concentrations are usually measured by reading from a calibration graph prepared by measuring standard solutions with accurately known concentrations.
- Atomic absorption spectrometry (AAS) measures the absorption of radiation by atoms in their ground state.
  - A flame can be used to vaporize and atomize samples, although better results are obtained with an electrothermal analyser.
  - Samples are irradiated by a hollow cathode lamp that is specific for a particular element.
  - Concentrations can be measured by preparing a calibration graph or by using the standard addition method.

For practice questions on this topic, see questions 20–23 at the end of this chapter (p.555).

Concept review

By the end of this chapter you should be able to do the following.

- Describe the factors to be considered when planning an analysis.
- Define the terms accuracy and precision and discuss the significance of these in analytical measurements.
- Define and calculate the mean, the range, the standard deviation, and the coefficient of variation of a number of measurements.
- Describe the use of pH meters and ion selective electrodes for chemical analysis.
- Account for the general features of chromatography systems.
- Define the retention factor, R_f, and describe how TLC and column chromatography can be used for separation and qualitative analysis.
- Discuss the operation of gas–solid and gas–liquid chromatographs and explain how analytical data can be obtained.
- Describe the use of gas chromatography–mass spectrometry to find the concentrations and identities of components in a mixture.
- Use the Beer–Lambert law to find the concentration of components in solution.
- Describe the principle of operation and basic features of atomic spectrometry.
- Distinguish between atomic emission spectrometry and atomic absorption spectrometry.
- Describe and explain some applications of the analytical methods described.
- Calculate the concentrations of analyte solutions given appropriate calibration data using calibration curves or the standard addition method as appropriate.
- Suggest suitable methods of analysis for a given analyte.

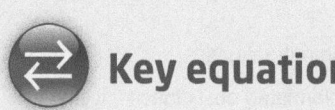

Key equations

Mean of a set of results	$\bar{x} = \dfrac{\sum_{i=1}^{i=n} x_i}{n}$	(11.1)
Standard deviation of a set of results	$\sigma = \left(\dfrac{\sum_i (x_i - \bar{x})^2}{n-1}\right)^{1/2}$	(11.2)
Coefficient of variation of a set of results	$CV = \dfrac{\sigma}{\bar{x}} \times 100$	(11.3)
Partition coefficient	$K = \dfrac{\text{concentration of solute in phase 1}}{\text{concentration of solute in phase 2}}$	(11.8)
Relative retention	$\alpha_{AB} = \dfrac{t_r(A)}{t_r(B)}$	(11.9)

Questions

More challenging questions are indicated by an asterisk *.

1 0.5850 g of NaCl were dissolved in 100.0 cm^3 of water. 10.0 cm^3 of this solution were made up with water to 250.0 cm^3. Calculate the concentration of the resulting solution in: (a) mol dm^{-3}; (b) mol m^{-3}; (c) mg dm^{-3}; (d) ppm (by mass). (Section 11.1)

2 A gravimetric determination (see Section 1.5, p.39) of the chloride ion (Cl^-) content of a mineral water was carried out by precipitating as silver chloride. If the balance available can weigh to ±0.1 mg, what is the detection limit for Cl^-(aq)? (M_r(AgCl) = 143.3.) (Section 11.1)

3 In preparing a standard solution, you forget to zero the balance so that it reads 0.10 g too high. What kind of error is this? What effect would this error have on an analysis if the standard solution was used to prepare a calibration graph by dilution? (Section 11.1)

4 Ten determinations of the sulfate ion content in ppm (by mass) of a water sample were made using two different methods.

Experiment	Method 1	Method 2
1	9.80	10.10
2	10.10	10.11
3	10.54	10.15
4	9.77	10.05
5	10.23	10.03
6	10.69	10.12
7	9.71	10.09
8	9.93	10.11
9	10.01	10.12
10	10.12	10.06

For each method, calculate: (a) the mean value; (b) the standard deviation; (c) the range; (d) the coefficient of variation. Comment on the values you obtain. (Section 11.1)

5 A laboratory operates two methods for determining the concentration of lead in water. To check the method, a Certified Reference Material containing a known Pb concentration of 20.00 μg dm^{-3} is analysed ten times using each of the two methods. The first method gives a mean concentration of 21.9 μg dm^{-3} with a standard deviation of 0.5 μg dm^{-3}. The second method gives a mean and standard deviation of 19.4 μg dm^{-3} and 2.1 μg dm^{-3}, respectively.

Comment on these results in terms of the accuracy and precision of the two analytical methods. (Section 11.1)

6 The electric potential of a glass pH electrode was measured as a function of pH with the following results.

pH	2.0	4.0	5.0	6.0
Potential/V	+0.300	+0.168	+0.081	+0.035
pH	8.0	9.0	10.0	
Potential/V	−0.092	−0.168	−0.235	

Under the same conditions, analysis of a carbonated drink gave a potential of +0.240 V while a sample of water gave a potential of −0.005 V. Find the pH of the drink and of the water. (Section 11.2)

7 An ion selective electrode was used for the analysis of benzoate ion in aqueous solution. The electric potential of the electrode varied with benzoate concentration as shown.

[Benzoate]/mol dm^{-3}	1.0×10^{-3}	2.0×10^{-4}	1.0×10^{-4}
Potential/mV	190	152	136
[Benzoate]/mol dm^{-3}	1.0×10^{-5}	1.0×10^{-6}	
Potential/mV	95	40	

Analysis of a solution of benzoate ions gave a potential of 100 mV under these conditions. Using a calibration curve, determine the concentration of benzoate ions in solution. (Section 11.2)

8 The fluoride ion (F^-(aq)) concentration of a solution was measured using a probe with an ion selective electrode. When immersed in 100 cm^3 of solution, a voltage of 0.505 V was measured. When 2.0 cm^3 of 1.00 mol dm^{-3} NaF were added, the voltage changed to 0.251 V. Estimate the concentration of fluoride ions and the pF in the original solution. (Section 11.2)

9* An ion selective electrode is designed to measure the concentration of perchlorate ions (ClO_4^-). The electrode was immersed in 50.0 cm^3 of an unknown solution of perchlorate ions and registered a potential of 358.7 mV against a standard electrode. 0.50 cm^3 of a solution of NaClO4(aq) of concentration 0.1 mol dm^{-3} was then added and the potential changed to 346.1 mV. What was the concentration of ClO_4^- ions in the unknown solution? (Section 11.2)

10 A solute has a partition coefficient of 28.7 between trichloromethane ($CHCl_3$) and water. 20 cm^3 of $CHCl_3$ were added to 100 cm^3 of an aqueous solution of the solute. At equilibrium, the aqueous layer contained a concentration of 0.005 mol dm^{-3}. What was the concentration of the organic layer? (Assume $k = 1$.) (Section 11.3)

11 You are handed a bottle labelled 'xylene' (dimethylbenzene). A gas chromatogram gives three peaks corresponding to the 1,2-, the 1,3-, and the 1,4-isomers with peak areas in the ratio of 143.1:9.5:6.4, respectively. Assuming that the detector responds equally to each isomer, calculate the composition of the xylene. (Section 11.3)

12 Suggest, with reasons, an appropriate chromatographic technique that could be used for each of the following analyses (Section 11.3):

(a) the concentration of H_2S in natural gas;

(b) trace concentrations of chlorinated pesticides in river water;

(c) the concentrations of each compound in a mixture of several chiral sugars.

13 Suggest suitable forms of chromatography to investigate the following situations. (Section 11.3)

(a) You are an analyst in a dairy. It is claimed that a consignment of milk has been contaminated with a toxic organochlorine pesticide at ppb levels. How would you identify and measure the concentration of the pesticide?

(b) Your colleagues in a pharmaceutical company have developed a new high-yield route to a single enantiomer of a chiral target compound. You are asked to check the purity of the product and to identify any by-product impurities in it and measure their concentrations. How might you do this?

14 Four solutions of a dye were prepared in water. In a 1.00 cm cell, the percentage of light at a particular wavelength transmitted through each solution is given in the following table.

Concentration/mol dm^{-3}	0.004	0.010	0.020	0.040
Percentage light transmitted	79	56	32	10

Confirm that these data obey the Beer–Lambert law and calculate the molar absorption coefficient, ε, at this wavelength. (Section 11.4)

15 The molar absorption coefficient for an aqueous solution of Fe^{2+} ions is ε = 0.3 m^2 mol^{-1} at 325 nm. When the Fe^{2+} is complexed with 1,10-phenanthroline ($C_{12}H_8N_2$), the $[Fe(C_{12}H_8N_2)_3]^{2+}$ ion has a molar absorption coefficient of 1100 m^2 mol^{-1} at 508 nm. If the minimum reading on a spectrophotometer is an absorbance of 0.01, what is the minimum detectable concentration of each of the ions using a 1.0 cm cell? (Section 11.4)

16* As part of an investigation into the harmful effects of petrol engine emissions, a 10 g sample of grass from a roadside verge was analysed for its lead content. The grass was burned to ash in oxygen to remove organic material and the inorganic residue was dissolved in 20 cm^3 of dilute acid. Under certain conditions, this gave an absorbance of 0.72 on an atomic absorption spectrometer.

A standard solution containing 1.0 $\mu g\,cm^{-3}$ of lead was available. Aliquots of this were made up to 50 cm^3 with dilute acid and the absorbances measured as shown in the table.

Volume of standard solution in 50 cm^3/cm^3	5	10	15	20	25
Absorbance	0.26	0.52	0.81	1.04	1.30

Calculate the concentration of lead in the grass in parts per million (by mass). (Section 11.4)

17 A drug with molar mass of 190 g mol^{-1} is formed into a tablet with a matrix that does not absorb visible light. The drug strongly absorbs at 450 nm with a molar absorption coefficient of 1650 $dm^3\,mol^{-1}\,cm^{-1}$. A tablet weighing 0.438 g is dissolved in 500 cm^3 water and gives an absorbance in a 1.0 cm cell of 0.475 at 450 nm. Find the percentage by mass of the drug in the tablet. (Section 11.4)

18 A sample of sodium sulfate is known to be contaminated with copper ions. Analysis of reference solutions containing copper ions at known concentrations gave the absorbances shown in the table. Under the same conditions, a solution made by dissolving 1.00 g of the sodium sulfate sample in 1.00 dm^3 distilled water gave an absorbance of 0.20. What was the purity of the sodium sulfate, assuming that copper ions are the only impurity? (Section 11.4)

Concentration of Cu^{2+}/ppm	2.0	4.0	6.0	8.0	10.0
Absorbance	0.06	0.12	0.18	0.24	0.30

19 A sample of soil is to be analysed for the presence of trace amounts of an organic contaminant. 5.00 g of soil were extracted with 20.0 cm^3 of dichloromethane, a good solvent for the contaminant. An injection of 1.0 microlitre (1.0 μL) of the extract onto a HPLC instrument with a UV–visible detector gave rise to a peak signal with an area of 250 arbitrary units. A separate injection of 1.0 μL of a 0.001 ppm reference solution of the contaminant under the same conditions gave a signal of 310 units. Find the concentration (in ppm) of contaminant in the soil. (Section 11.4)

20 You are provided with 10 kg of soil from a site suspected of uranium contamination. Outline the steps that you would take to obtain an accurate measurement of the uranium concentration in the soil. What blank experiments would you run? (Section 11.5)

21 100 g of a soil sample were treated chemically to extract all the uranium into 1.00 dm^3 of solution. In an atomic absorption spectrometer, this solution gave an absorbance of 0.427. The spectrometer was calibrated using a standard solution with a uranium concentration of 1.00×10^{-6} mol dm^{-3}. Various volumes of the standard solution were diluted with water to prepare 1.00 dm^3 of solution.

Volume of standard solution in 1.00 dm^3/cm^3	5.0	10.0	15.0	20.0	25.0
Absorbance	0.130	0.275	0.412	0.550	0.687

Calculate the concentration of uranium (in ppm by mass) in the soil. (Section 11.5)

22 A water company needed to measure the concentration of iron in a contaminated supply. A determination of Fe by atomic absorption spectroscopy gave an absorbance of the water of 0.323 after diluting it 10 times with pure water.

A standard Fe solution was prepared by dissolving 0.1200 g of iron wire in 1.00 L of acid.

After a further ×100 dilution the solution had an absorbance of 0.813. Find the concentration in ppm of iron in the water supply. How could this estimate be improved?

23 Tributyltin chloride, TBT, was commonly added to marine paints to prevent the growth of barnacles on the hulls of ships although it is now banned due to its toxicity. A common method for analysing TBT in marine sediments is graphite furnace atomic absorption spectroscopy (AAS).

In a particular analysis, 0.500 g of a sediment was dissolved in 10.0 cm^3 of acid and aspirated into a graphite furnace AAS. The sample gave an absorbance of 0.44.

A standard solution of tin with an accurately known concentration of 1.0 $\mu g\,dm^{-3}$ (1.0 ppb) was available. Known volumes of this were added to 1.00 dm^3 volumetric flasks and made up to the mark with pure water. Under identical conditions in the AAS, these solutions gave the following results.

Volume of standard added/cm^3	Absorbance
5	0.13
10	0.26
15	0.39
20	0.52
25	0.65

Find the concentration of TBT in the sediment. (Section 11.5)

24 Suggest suitable analytical procedures to investigate the following situations. More than one technique may be appropriate depending on the circumstances. What are the factors that need to be considered in each case? (Several sections)

(a) A mixture of tablets of painkilling drugs, including aspirin, paracetamol, and morphine.

(b) A mixture of organic dyes in aqueous solution.

(c) A white powder found near the scene of a crime.

(d) The residue from a can of petrol suspected of being used in arson.

(e) The concentration of mercury in a freshwater lake.

12

Molecular characterization

▼ Mass spectrometry is used in flavour research to identify volatile compounds released from drink or food. In this experiment, a mass spectrometer is used to detect the compounds in the nose of a researcher drinking orange juice. This research helps us to understand the ways in which orange juice imparts its taste.

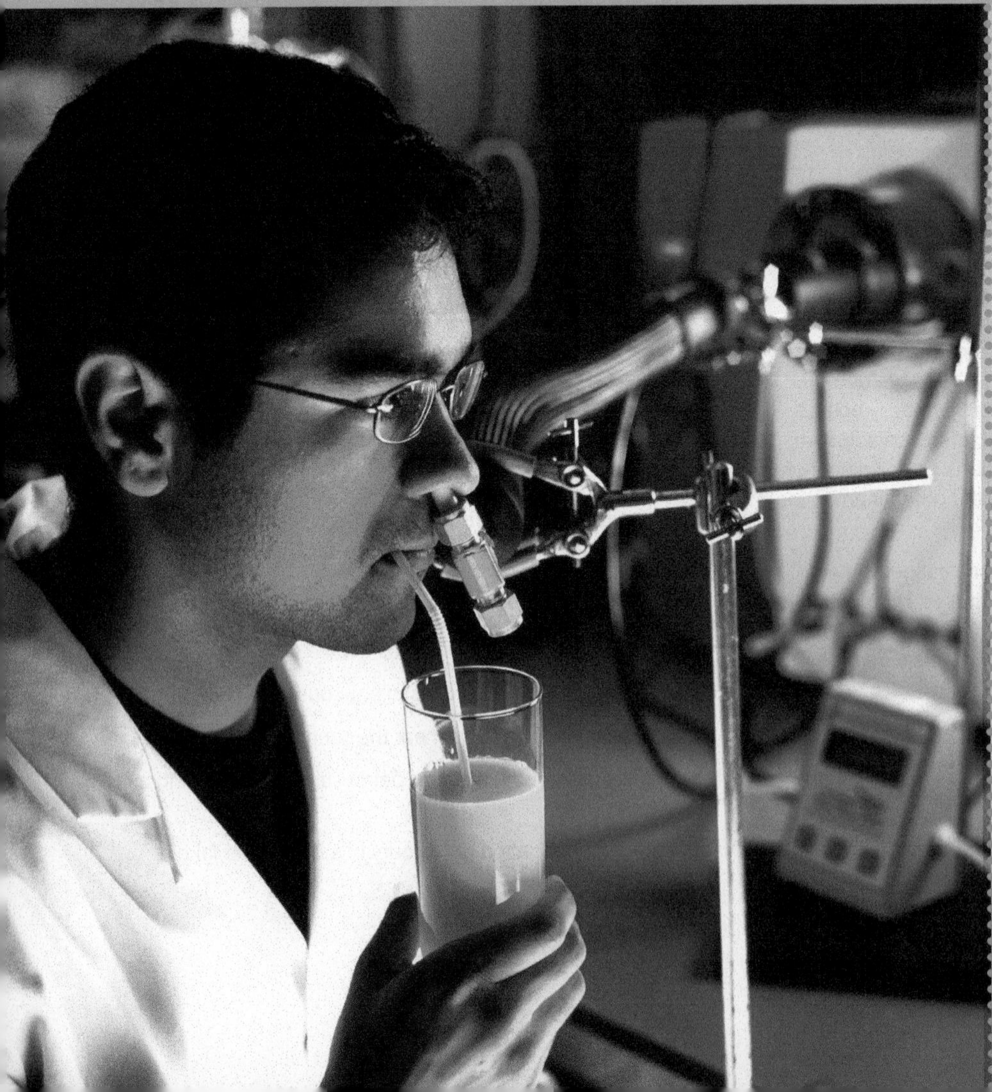

This chapter builds on the following topics:

- **Hydrocarbons** Section 2.5, p.79
- **Functional groups containing one or more heteroatoms** Section 2.6, p.90
- **Functional groups containing carbonyl groups** Section 2.7, p.98
- **Introduction to molecular spectroscopy** Section 10.1, p.450
- **Molecular energies and spectroscopy** Section 10.2, p.453
- **General principles of spectroscopy** Section 10.3, p.458
- **Rotational spectroscopy** Section 10.4, p.466
- **Vibrational spectroscopy** Section 10.5, p.476
- **Electronic spectroscopy** Section 10.6, p.491
- **Spin resonance spectroscopy** Section 10.7, p.496
- **Carrying out an analysis** Section 11.1, p.515
- **Spectroscopic methods of analysis** Section 11.4, p.542

Using isotope ratios to analyse orange juice

A recent survey in the UK showed that around 90% of consumers now look for '100% juice' when buying fruit juices because they believe the product will be healthier and taste better. By far the most popular fruit juice is orange juice. Pure orange juice is relatively expensive and the desire to make increased profits from orange juice sales has led to some illegal practices. There have been several high-profile cases where food manufacturers have claimed that their product is 100% pure orange juice, when it is actually an adulterated and cheaper product.

Reconstituting orange juice from concentrate. To transport orange juice across the globe, fresh orange juice is usually concentrated to remove most of the water. Frozen concentrate is shipped to a customer and then it is reconstituted, by adding water to achieve the desired colour and flavour.

Common methods of adulteration that do not significantly affect the taste of the juice include diluting pure orange juice with water and/or adding inexpensive sugars such as corn sugar, which has a higher fructose content than orange juice. There is a stiff penalty for the guilty—a chief executive of an American juice manufacturer who admitted selling over 40 million gallons of adulterated orange juice over a 12-year period was given a prison sentence. So how do chemists know if a product sold as 100% pure orange juice contains added inexpensive sugars?

$C_6H_{12}O_6$ sugars in natural orange juice (cyclic forms are shown; the wavy line indicates that in some molecules the OH group points up, and in others, it points down). (The stereochemistry of each sugar is indicated by the letter 'D', as described in Box 18.6, p.842.)

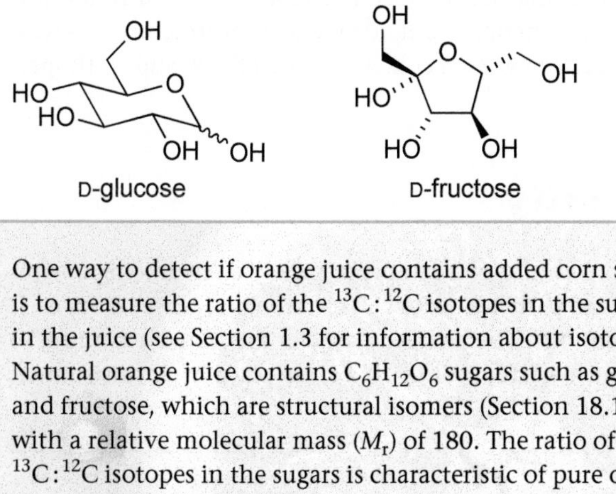
OH
HO
HO
O
OH
OH
OH
O
OH
HO
HO
OH

D-glucose D-fructose

One way to detect if orange juice contains added corn sugar is to measure the ratio of the $^{13}C:^{12}C$ isotopes in the sugars in the juice (see Section 1.3 for information about isotopes). Natural orange juice contains $C_6H_{12}O_6$ sugars such as glucose and fructose, which are structural isomers (Section 18.1, p.816) with a relative molecular mass (M_r) of 180. The ratio of the $^{13}C:^{12}C$ isotopes in the sugars is characteristic of pure orange juice. If, for example, high fructose corn sugar is added, the $^{13}C:^{12}C$ ratio of the sugars in the juice changes very slightly. The change in the ratio is because orange trees synthesize sugars by a different photosynthetic pathway from the one in corn. Both pathways convert $^{12}CO_2$ and $^{13}CO_2$ (from the air) into $C_6H_{12}O_6$ sugars but, while the rate of conversion of $^{12}CO_2$ into ^{12}C-containing sugars is comparable in orange trees and corn, the enzymes in an orange tree convert $^{13}CO_2$ into ^{13}C-containing sugars at a slightly slower rate than the enzymes in corn. Consequently, the sugars produced in the orange tree have a slightly smaller $^{13}C:^{12}C$ ratio than those produced in corn.

The very small differences in $^{13}C:^{12}C$ ratios of sugars from different plants require an extremely accurate method for determining the ratios. One of the most common methods uses mass spectrometry, an analytical technique introduced in Box 1.2 (p.14) for elements and discussed in this chapter for more complex molecules (Section 12.1). Mass spectrometry measures the mass-to-charge ratio (m/z) of ions produced, sometimes by bombarding the substance undergoing analysis with highly energetic electrons. Compounds with different M_r values give different peaks in a mass spectrum. For example, if all six carbon atoms in D-glucose ($C_6H_{12}O_6$) are ^{12}C, a peak is observed at m/z 180, but if one of the carbon atoms is ^{13}C, a new peak is seen at m/z 181. The relative intensities of the two peaks indicate the ratio of $^{13}C:^{12}C$ in the sugar.

Rather than measuring the precise amounts of ^{13}C and ^{12}C in different orange juice samples and comparing them, a more useful measure is to compare the ratio of $^{13}C:^{12}C$ in a juice sample to the $^{13}C:^{12}C$ ratio in a recognized standard. The standard is the CO_2 produced when acid is added to a type of inorganic carbonate called Pee Dee belemnite. The two $^{13}C:^{12}C$ ratios are expressed as a delta ^{13}C value in parts per thousand (ppt). The delta ^{13}C values for sugars in 100% orange juice and those in high fructose corn sugar differ by only around 10 parts per thousand, but mass spectrometry is a sufficiently accurate method to be able to distinguish them.

Sherlock Holmes was the first detective to exploit chemical science as a means of detection. Holmes started, albeit fictionally, linking science with rational thinking—now, an integral part of everyday policing around the globe.

One of the most important advances in modern chemical research has been the development of analytical techniques that allow chemists to accurately and quickly determine the structures of compounds. Research chemists routinely use various analytical techniques to provide different pieces of evidence to help them assign structures to unknown compounds. It is the chemist's job to use the information obtained to solve the chemical puzzle, a bit like Sherlock Holmes using clues to solve a murder mystery. Today, spectroscopic techniques are widely used to determine the structures of a diverse range of compounds, even naturally occurring compounds with complicated structures (Section 2.1, p.70), which are available in, at most, milligram quantities.

The most widely used analytical techniques in organic chemistry are mass spectrometry (MS), infrared (IR) spectroscopy, and, particularly, nuclear magnetic resonance (NMR) spectroscopy. The diagram below gives an overview of the information available from each technique.

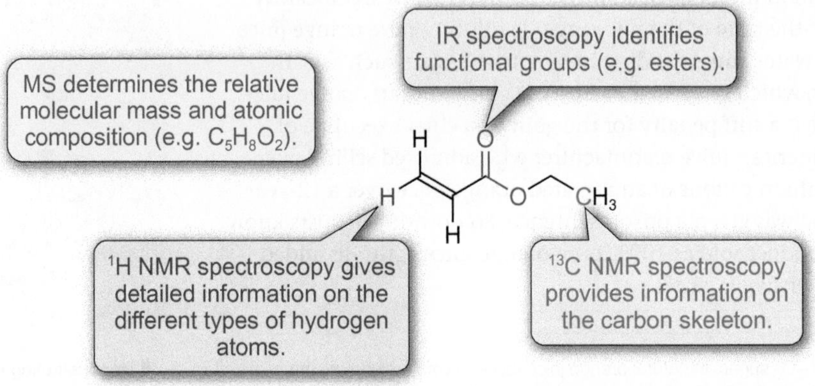
IR spectroscopy identifies functional groups (e.g. esters).
MS determines the relative molecular mass and atomic composition (e.g. $C_5H_8O_2$).
H
O
H
O
CH_3
H
^{1}H NMR spectroscopy gives detailed information on the different types of hydrogen atoms.
^{13}C NMR spectroscopy provides information on the carbon skeleton.

The physical basis of IR and NMR spectra, together with details of the instruments, is discussed in Chapter 10 (p.448). Ultraviolet-visible (UV/VIS) spectrophotometry is a useful spectroscopic method for measuring the concentrations of, for example, coloured species in solution (Section 11.4, p.542).

This chapter discusses each technique in turn, concentrating on how chemists analyse the different types of spectra. The final section (Section 12.4, p.599) illustrates how the techniques can be used together to determine unknown chemical structures. As with many things in life, practice makes perfect when it comes to using MS, IR, and NMR spectra to assign chemical structures.

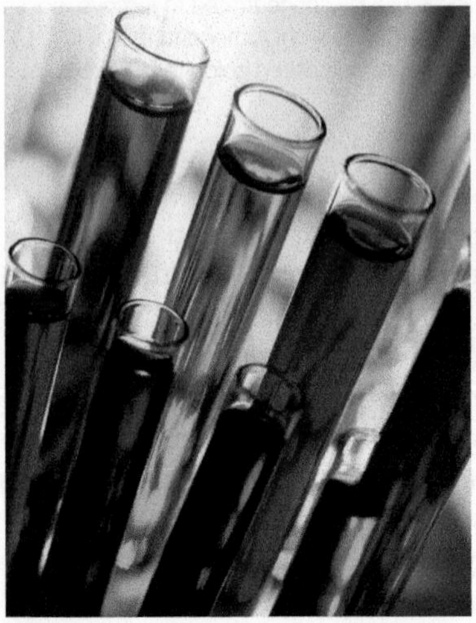

12.1 Mass spectrometry

In mass spectrometry (MS), a small quantity of a sample (typically a few micrograms) is vaporized and ionized to form charged particles—often an electron is removed from each molecule to form positively charged ions. The charged particles are sorted according to their mass-to-charge ratio (*m*/*z*) and detected. All of this takes place in an instrument called a **mass spectrometer**.

The use of a mass spectrometer to determine the isotopic composition of elements is described in Box 1.2 (p.14).

i Mass spectrometry differs from other spectroscopic techniques, because the peaks in the spectrum do not correspond to the absorption or emission of electromagnetic radiation. In mass spectrometry, the mass-to-charge ratio of ions is measured, not the absorption of electromagnetic radiation.

Visit the Online Resource Centre to view video clip 12.1 which illustrates the principles and use of a mass spectrometer.

i A **radical cation** ($M^{+\bullet}$ or $M^{\bullet+}$) has an unpaired electron and a positive charge.

Relative molecular mass, M_r, is discussed in Section 1.3 (p.12).

Ionization is achieved in a number of ways (Box 12.1). In **electron impact (EI)** mass spectrometry, the sample is introduced into a high-vacuum chamber where it is bombarded with highly energetic electrons. The bombarding electrons eject an electron from a molecule (M) of the sample compound to form a radical cation ($M^{+\bullet}$), called the molecular ion. Usually, a relatively high-energy electron is ejected from the sample molecule, such as an electron from a lone pair rather than an electron involved in bonding. The molecular ion is then detected and its mass-to-charge ratio (*m*/*z*) recorded. Because the relative molecular masses of the molecular ion ($M^{+\bullet}$) and the sample (M) are essentially identical (an electron only weighs 9.1×10^{-31} kg), a mass spectrometer determines the M_r of the sample.

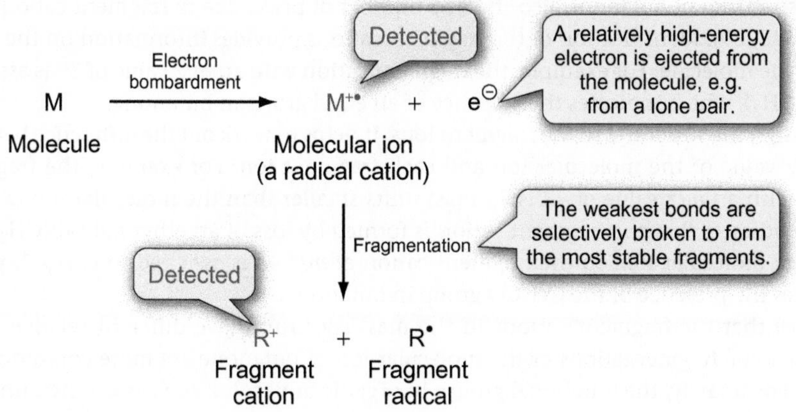
Detected
A relatively high-energy electron is ejected from the molecule, e.g. from a lone pair.
Electron bombardment
M
M+•
+
e⊖
Molecule
Molecular ion (a radical cation)
Fragmentation
The weakest bonds are selectively broken to form the most stable fragments.
Detected
R+
+
R•
Fragment cation
Fragment radical

Normally, a bombarding electron injects more energy into a sample molecule than is required to eject an electron. The excess energy causes the molecular ion to break down, in a process called **fragmentation**, to give smaller **fragment cations** (R^+) and **fragment radicals** ($R^\bullet$). Only charged particles are detected in a mass spectrometer so only the relative molecular masses of the fragment cations are recorded. The information is displayed in a **mass spectrum**, which is a plot of the relative abundance of each positively charged ion against its m/z value. Most of the ions formed in a mass spectrometer have a single charge so m/z is equal to the relative molecular mass of the ion. The most intense peak in the spectrum is called the **base peak**, which is assigned a relative abundance of 100%, and the abundances of the other peaks are shown relative to the base peak.

A mass spectrum usually detects the molecular ion ($M^{+\bullet}$) and cations (such as $CH_3CH_2^+$) formed by fragmentation of the molecular ion.

The base peak can be a peak due to the molecular ion or a fragment cation.

The EI mass spectrum of butanone ($CH_3COCH_2CH_3$, C_4H_8O) is shown in Figure 12.1. The peak with an m/z value of 72 is due to the molecular ion (labelled the M peak). You might expect the molecular ion to have the highest m/z value, but there is a tiny peak with an m/z value of 73. This M + 1 peak is due to one of the four carbon atoms in the molecular ion of butanone having a ^{13}C atom in place of a ^{12}C atom (i.e. $^{13}C^{12}C_3H_8O$). Peaks due to isotopes can be useful in identifying the sample molecule, as discussed on p.561.

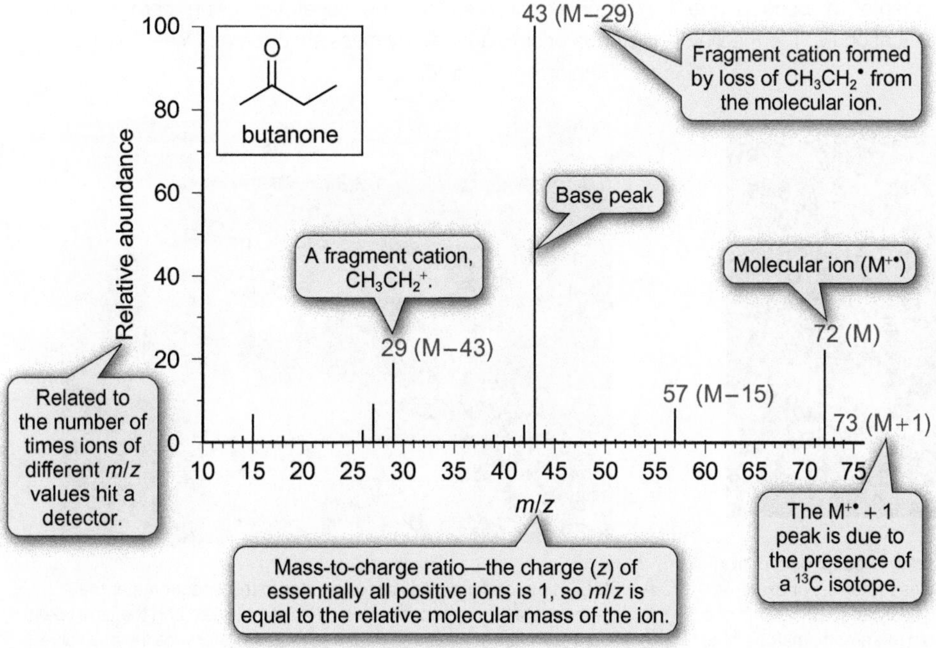
100
80
60
40
20
0
Relative abundance
butanone
43 (M−29)
Fragment cation formed by loss of CH3CH2• from the molecular ion.
Base peak
A fragment cation, CH3CH2+.
29 (M−43)
Molecular ion (M+•)
72 (M)
57 (M−15)
73 (M+1)
Related to the number of times ions of different m/z values hit a detector.
10 15 20 25 30 35 40 45 50 55 60 65 70 75
m/z
The M+• + 1 peak is due to the presence of a 13C isotope.
Mass-to-charge ratio—the charge (z) of essentially all positive ions is 1, so m/z is equal to the relative molecular mass of the ion.

Figure 12.1 EI mass spectrum of butanone.

Several hundred million kilograms of butanone (commonly known as methyl ethyl ketone, or MEK) are manufactured in each year, typically for use in paints and other coatings.

The spectrum of butanone also shows a number of peaks due to fragment cations (e.g. m/z 57, 43, 29). Identification of the fragment cations provides information on the structure of the molecule. For example, the fragment cation with an m/z value of 29 is assigned to $CH_3CH_2^+$, which indicates the presence of an ethyl group in butanone.

To assign the structures to the fragment ions, it helps to work out the difference between the m/z value of the molecular ion and each fragment ion. For example, the fragment cation with an m/z value of 43 is 29 mass units smaller than the molecular ion (M–29), which indicates that the fragment cation is formed by loss of an ethyl radical ($CH_3CH_2^{\bullet}$) from the molecular ion. So the fragment cation at m/z 43 is assigned to $C_2H_3O^+$, which indicates the presence of the CH_3CO group in butanone.

Functional groups (summarized in Figure 2.7, p.574) in organic molecules undergo characteristic fragmentation patterns. A table of common fragmentations of molecular ions is shown on p.569.

Notice that the fragment cations in the mass spectrum have different relative abundances. Some fragmentations of the molecular ion of butanone are more common than others. Fortunately, the functional groups in organic molecules, such as ketones, undergo characteristic fragmentation patterns. So, if you can interpret the fragmentation, you can probably identify the functional group.

Box 12.1 Ionization methods

Electron impact (EI) is not always the most suitable method of ionization. For many compounds, a molecular ion is not observed in the EI mass spectrum, because the molecular ion undergoes fragmentation before it reaches the detector. Without a molecular ion ($M^{+\bullet}$), a mass spectrum is much more difficult to interpret. This problem has been solved by the development of 'softer' methods of ionization that produce less fragmentation of the molecular ion.

One common solution is to use an ionization method called **chemical ionization (CI)**. In this method, a mixture of the sample and a gas, such as ammonia (NH_3), is bombarded with electrons. The ammonia is converted into NH_4^+ that protonates a sample molecule (M), giving an $[M + H]^+$ ion that is detected in the spectrometer. In this way, positive ions are formed that are one mass unit higher than the molecular ion. Importantly, there is only limited fragmentation, because the $[M + H]^+$ ion is more stable than the radical cation ($M^{+\bullet}$) formed by EI ionization.

For molecules to give EI and CI mass spectra they must be relatively volatile because both methods rely on heating the sample to form a vapour, which is bombarded with electrons. But, what if the compound has a relatively high boiling point? For high-boiling compounds, or for compounds that decompose on heating, techniques like **fast atom bombardment (FAB)**, **electrospray ionization (ESI)**, and **matrix-assisted laser desorption/ionization (MALDI)** mass spectrometry have been developed. For example, a MALDI mass spectrometer typically uses a UV laser beam to ionize a sample molecule that is embedded in a chemical matrix. When the matrix material absorbs the UV energy this leads to vaporization and formation of sample molecule ions (by a process that is not fully understood), which are analysed for their mass-to-charge ratio.

Complex mixtures can be separated using chromatography methods prior to analysis by mass spectrometry (see Box 1.2, p.14 and Section 11.3, p.528).

▲ Modern mass spectrometry methods are routinely used for determining the structures of large biological molecules such as proteins. For example, mass spectrometry methods have been used to analyse whether parts of a protein called collagen, from a 0.5-million-year-old mammoth, match amino acid sequences found in collagen of the present day elephant.

REFLEX™ IV

▲ A MALDI time-of-flight mass spectrometer. In this spectrometer the sample molecule ions are accelerated by an electric field and the time taken to reach a detector is measured and then used to determine the m/z value (see Box 1.2, p.14).

Isotope patterns

In an EI mass spectrum, the peak with the highest *m/z* value is normally due to the molecular ion (M). Tiny peaks, one or two mass units higher than the molecular ion (M + 1 or M + 2), arise because many molecules contain heavier isotopes than the common isotopes. For many elements, the percentage abundance of a heavier isotope is below 1% and you can ignore contributions from them. An important exception is ^{13}C.

Natural carbon contains 98.9% of ^{12}C together with 1.1% of ^{13}C. It is the small percentage of molecular ions containing an atom of ^{13}C that is responsible for the tiny M + 1 peak in a mass spectrum. For example, in the EI mass spectrum of methane (CH_4), peaks with *m/z* values of 16 and 17 are due to $^{12}CH_4$ and $^{13}CH_4$, respectively, and the relative abundances of these peaks are approximately 99 : 1.

The relative abundance of the M + 1 peak is important in assigning organic structures, because it indicates the number of carbon atoms in the molecule. The more carbon atoms in a molecule, the greater is the chance of one of them being ^{13}C. For example, for a molecule with 6 carbon atoms, there is a $6 \times 1.1\% = 6.6\%$ chance of one of them being ^{13}C. So, the relative abundance of the M + 1 peak is 6.6% of the value for the M peak. To calculate the number of carbon atoms in an unknown molecule, you need to find the relative abundance of the M + 1 peak expressed as a percentage of the relative abundance of the M peak and divide by 1.1. For example, if the relative abundances of the M + 1 and M peaks are 11% and 100%, respectively, then the compound contains 10 (11/1.1 = 10) carbon atoms (see margin).

The majority of EI mass spectra show a single molecular ion peak with tiny isotope peaks. A notable exception is the case when a compound contains a chlorine or bromine atom.

Naturally occurring chlorine is a mixture of two isotopes in a 3 : 1 ratio—75.8% of ^{35}Cl and 24.2% of ^{37}Cl. So, you can recognize a molecule containing a chlorine atom by a characteristic 3 : 1 ratio for the relative abundances of two molecular ion peaks separated by two mass units. Figure 12.2 shows the EI mass spectrum of 2-chloropropane ($CH_3CHClCH_3$, C_3H_7Cl). There are two molecular ion peaks with *m/z* values of 78 and 80 in an approximately 3 : 1 ratio. A similar ratio is seen for the fragment cations with *m/z* values of 65 and 63, indicating that these cations also contain a chlorine atom (they are formed by loss of a methyl radical, $CH_3^{\bullet}$, from the molecular ion), whereas the fragment ion at *m/z* 43 does not have a partner isotope peak at *m/z* 45, so this cation does not contain a chlorine atom.

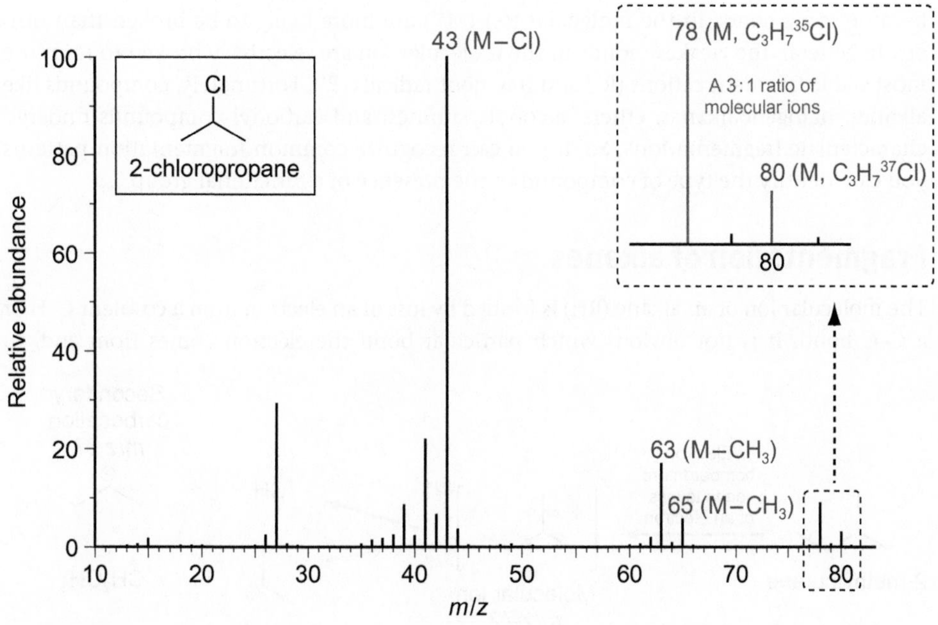

43 (M−Cl)
Cl
2-chloropropane
78 (M, C3H7 35Cl)
A 3 : 1 ratio of molecular ions
80 (M, C3H7 37Cl)
80
63 (M−CH3)
65 (M−CH3)
Relative abundance
100
80
60
40
20
0
10
20
30
40
50
60
70
80
m/z

Figure 12.2 EI mass spectrum of 2-chloropropane.

A list of naturally occurring isotopes and their relative abundances can be found in Table 1.5 (p.18).

Besides ^{13}C, isotopes such as ^{15}N, ^{2}H, and ^{17}O can contribute to the M + 1 peak. However, because the natural abundances of these isotopes are below 0.5%, you can ignore them.

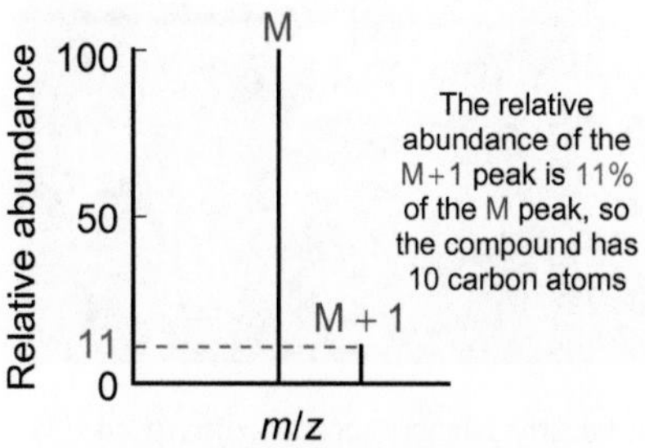

M
100
50
11
0
Relative abundance
M + 1
m/z
The relative abundance of the M + 1 peak is 11% of the M peak, so the compound has 10 carbon atoms

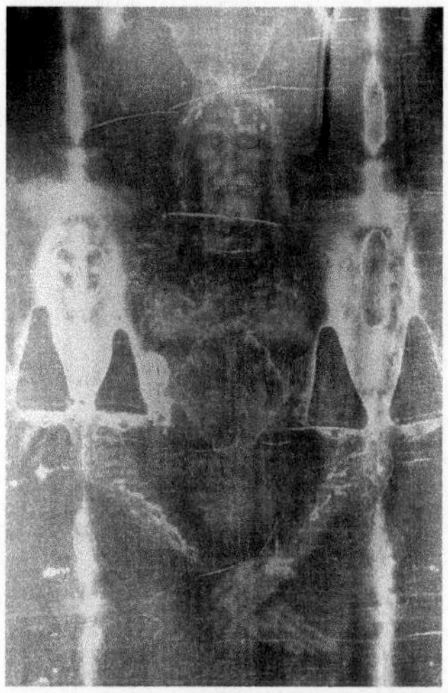

Isotope ratios are important markers of a variety of processes. For example, the ^{12}C:^{14}C isotope ratio can be used to determine the age of a material (as in carbon dating). Mass spectrometry was used to determine the carbon isotope ratio of a swatch taken from a corner of the Shroud of Turin—this established that the shroud came in to existence between AD 1260 and AD 1390, although it has been argued that the swatch was not part of the original cloth.

The presence of a chlorine atom in a molecule is recognized by a 3 : 1 ratio of the relative abundances of the M and M + 2 peaks, respectively.

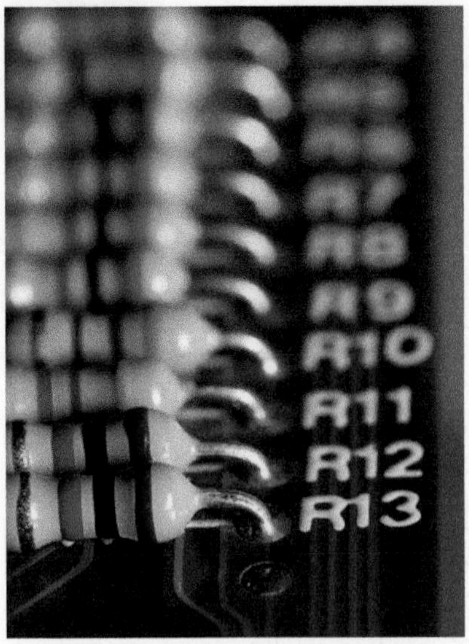

R9
R10
R11
R12
R13

In industry, 1-bromopropane is often used as an alternative to chlorofluorocarbons and 1,1,1-trichloroethane (Cl_3CCH_3), which have the potential to destroy the ozone layer. It has less ozone depleting potential, and also the high volatility (b.p. 71 °C) and non-flammability required for a solvent to clean, for example, circuit boards in the electronics industry.

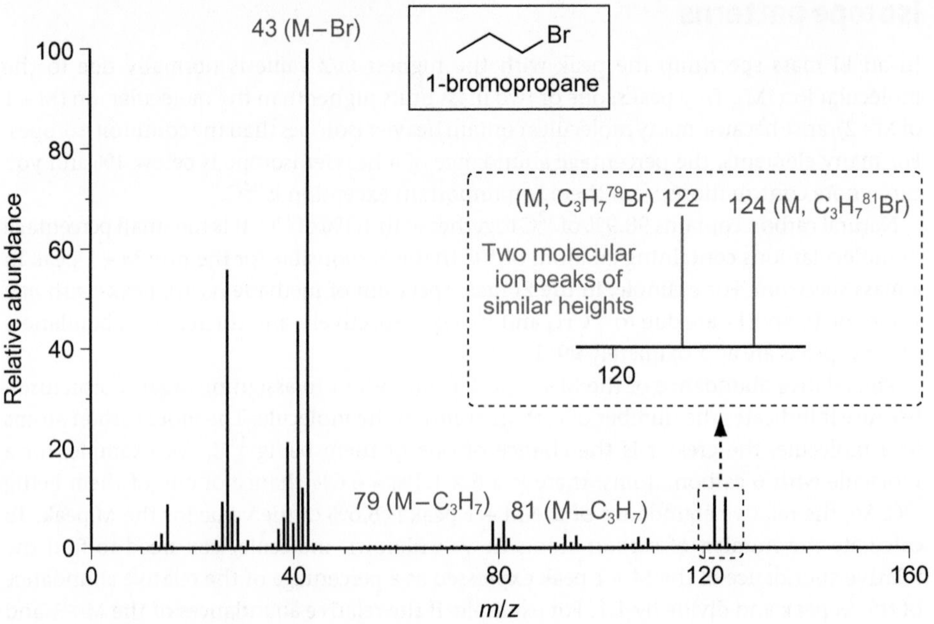

43 (M−Br)
Br
1-bromopropane
100
80
60
40
20
0
Relative abundance
(M, C3H7 79Br) 122
124 (M, C3H7 81Br)
Two molecular ion peaks of similar heights
120
79 (M−C3H7)
81 (M−C3H7)
0
40
80
120
160
m/z

Figure 12.3 EI mass spectrum of 1-bromopropane.

A molecule containing a single bromine atom is recognized by two molecular ion peaks two mass units apart in a characteristic 1 : 1 ratio. The peaks have approximately the same ratio because the natural abundances of ^{79}Br (50.7%) and ^{81}Br (49.3%) are about the same. For example, the mass spectrum of 1-bromopropane ($CH_3CH_2CH_2Br$, C_3H_7Br) has two molecular ion peaks of approximately the same relative abundance at *m/z* values of 122 and 124 (Figure 12.3). There are also peaks for $^{79}Br^+$ and $^{81}Br^+$, which shows that even extremely unstable cations can be formed and detected in an EI mass spectrometer (Br^+ is much less stable than Br^- because Br is electronegative).

The presence of a bromine atom in a molecule is recognized by a 1 : 1 ratio of the relative abundances of the M and M + 2 peaks, respectively.

Molecular ions containing more than one chlorine and/or bromine atom also have characteristic isotope patterns—see a specialist spectroscopy book for details.

Radical cations fragment to form the most stable carbocations and radicals—the stability of carbocations and carbon radicals is discussed in Section 19.1 (p.867).

Fragmentation patterns

In an EI mass spectrum, the relative abundances of the fragment cations are different because some bonds in the molecular ion ($M^{+\bullet}$) are more likely to be broken than others. In general, the weakest bonds in the molecular ion are selectively broken to form the most stable fragment cations (R^+) and fragment radicals ($R^\bullet$). Fortunately, compounds like alkanes, halogenoalkanes, ethers, alcohols, amines, and carbonyl compounds undergo characteristic fragmentations. So, if you can recognize common fragmentation patterns you can identify the type of compound or the presence of a functional group.

Fragmentation of alkanes

The molecular ion of an alkane (RH) is formed by loss of an electron from a covalent C–H or a C–C bond. It is not obvious which particular bond the electron comes from and, to

To help interpret mass spectra, libraries of mass spectra are commercially available. The mass spectrum of an unknown compound can be compared with the spectra of known compounds.

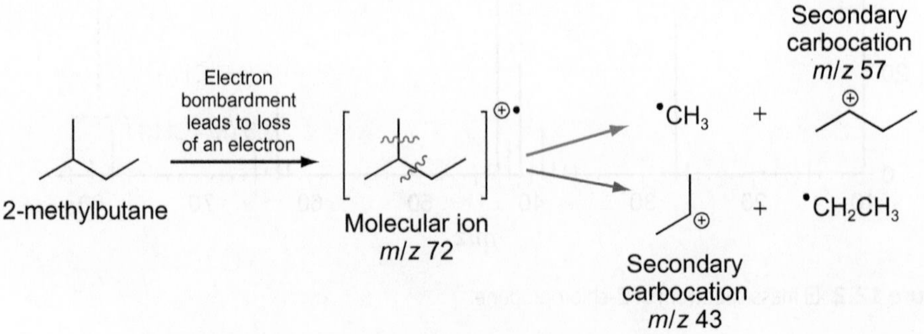

Electron bombardment leads to loss of an electron
2-methylbutane
Molecular ion
m/z 72
•CH3
+
Secondary carbocation
m/z 57
+
•CH2CH3
Secondary carbocation
m/z 43

indicate this, the molecular ion is drawn using square brackets, [], to show that +• is assigned to the whole structure. (For large structures, sometimes the brackets are omitted and the symbol ⏋+• is included at the upper right-hand corner of the structure.)

The molecular ion of an alkane may fragment in several different ways. Normally, it fragments to break a C–C bond in preference to a C–H bond, because a C–H bond is stronger. Where there is a choice of different C–C bonds, the bond leading to the most stable carbocation and carbon radical is selectively broken (see below for the order of carbocation and carbon radical stability). For example, the EI mass spectrum of 2-methylbutane ($(CH_3)_2CHCH_2CH_3$) shows prominent fragment ions at *m*/*z* values of 57 and 43 due to the selective formation of secondary carbocations ($CH_3CH^+CH_2CH_3$ and $(CH_3)_2CH^+$).

Mean bond enthalpies are given in Appendix 10 (p.1369).

In a mass spectrum, you should normally be able to propose fragmentation pathways for two or three of the most abundant fragment ions—you will not be expected to account for all the fragment ions.

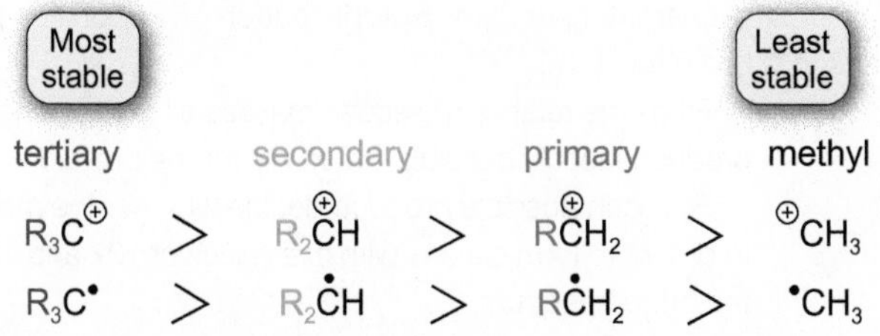
Most stable
Least stable
tertiary
secondary
primary
methyl

Fragmentation of halogenoalkanes

The molecular ion of a halogenoalkane (RX) is formed by loss of an electron from one of the lone pairs on the halogen atom (a lone pair electron is easier to dislodge than an electron in a covalent bond). Fragmentation of the molecular ion ($RX^{+\bullet}$) breaks the relatively weak carbon–halogen bond to form a carbocation (R^+) and a halogen atom or radical ($X^\bullet$). This is illustrated below for the fragmentation of the molecular ions produced from 1-bromopropane ($CH_3CH_2CH_2Br$) and 2-chloropropane ($(CH_3)_2CHCl$). In each case, both electrons in the C–X bond move on to the halogen atom (represented by the curly arrow ⌒) and fragmentation of the molecular ion forms a carbocation with an *m*/*z* value of 43, which is the most abundant fragment cation in the mass spectrum (see Figures 12.2, p.561, and 12.3, p.562).

Notice that a *double-headed* curly arrow (⌒) is used to show the movement of the *two* electrons. A *single-headed* arrow (⌒) is used to show the movement of *one* electron. The use of curly arrows to show the movement of electrons is described in Section 19.1 (p.862).

The molecular ion of a halogenoalkane fragments to break the relatively weak carbon–halogen bond.

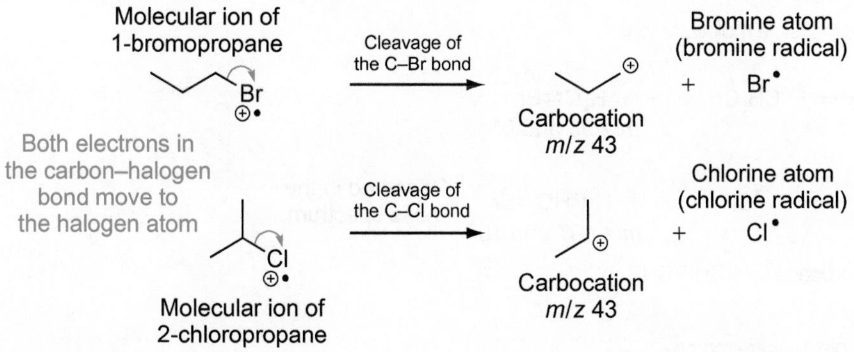
Molecular ion of 1-bromopropane
Cleavage of the C–Br bond
Bromine atom (bromine radical)
Carbocation m/z 43
Both electrons in the carbon–halogen bond move to the halogen atom
Cleavage of the C–Cl bond
Chlorine atom (chlorine radical)
Carbocation m/z 43
Molecular ion of 2-chloropropane

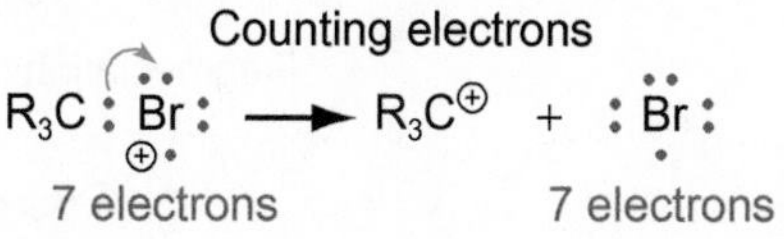
Counting electrons
7 electrons
7 electrons

Homolytic cleavage (homolysis) of a covalent bond occurs when the electrons of a two-electron bond move, one to each of the atoms (Section 19.1, p.866).

The molecular ion of a halogenoalkane may also fragment to give a cation containing a chlorine or bromine atom. For example, the mass spectrum of 2-chloropropane (Figure 12.2, p.561) shows peaks with *m*/*z* values of 63 and 65, in a 3 : 1 ratio. These peaks are assigned to the $[CH_3CHCl]^+$ ion, which is formed by homolytic cleavage of a C–C bond in the molecular ion. Single-headed curly arrows (⌒) are used to show the movement of single electrons. You can practise assigning a structure using a mass spectrum in Worked example 12.1.

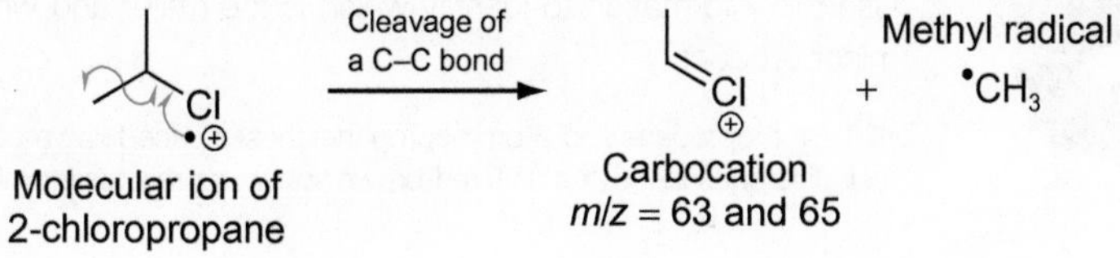
Cleavage of a C–C bond
Methyl radical
Molecular ion of 2-chloropropane
Carbocation m/z = 63 and 65

Worked example 12.1 Identifying a bromoalkane

A mixture of propane and bromine is illuminated with UV radiation to give a mixture of organic products. The products are then separated using gas chromatography and an EI mass spectrum is recorded for each compound. Use the EI mass spectrum shown below to identify one of the products of the reaction. (The mechanism of this reaction is discussed in Section 20.2, p.921.)

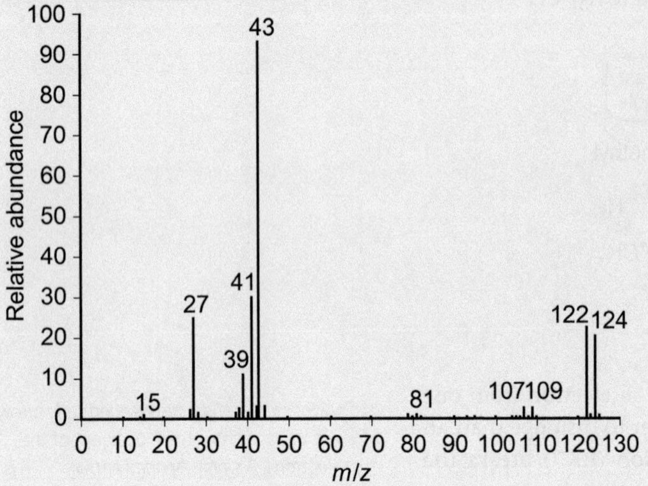

Relative abundance
m/z
43
27
41
39
15
81
107109
122124

Strategy

Draw a reaction scheme, showing the structures of the reactants.

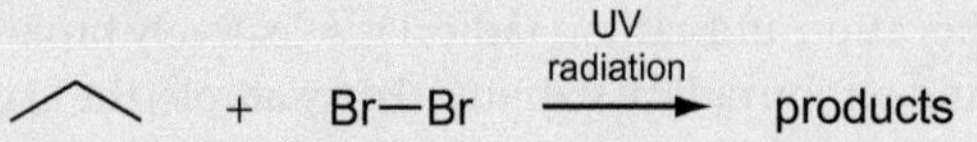

+ Br—Br
UV radiation
products

In the mass spectrum, look at the peak of highest *m/z* value to determine the relative molecular mass of the product. Look for a characteristic 1:1 isotope pattern to spot the presence of a bromine atom.

From the relative molecular masses of the starting materials and product, identify possible structures for the product.

For each possible product, decide whether the molecular ion can fragment to form cations with *m/z* values of 107 and 109 observed in the mass spectrum.

Solution

Possible products of the reaction are $CH_3CH_2CH_2Br$ and $CH_3CHBrCH_3$.

The peaks of highest *m/z* value are at *m/z* 122 and *m/z* 124. The (approximately) 1:1 ratio indicates the presence of a bromine atom. The relative molecular mass of the product with a ^{79}Br atom is 122.

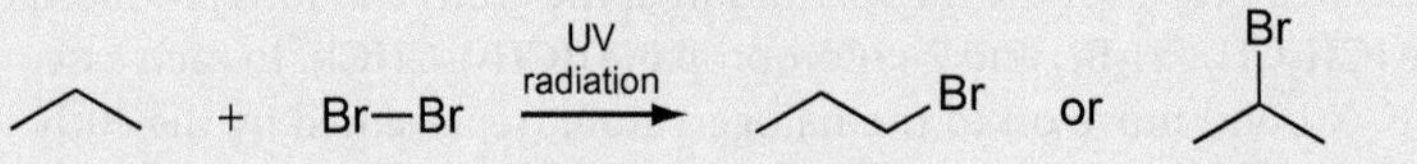

+ Br—Br
UV radiation
Br
or
Br

Molecular formula	C_3H_8	Br_2	C_3H_7Br	C_3H_7Br
Relative molecular mass	44	158 (^{79}Br) 162 (^{81}Br)	122 (^{79}Br) 124 (^{81}Br)	122 (^{79}Br) 124 (^{81}Br)

Fragmentation of the two possible molecular ions:

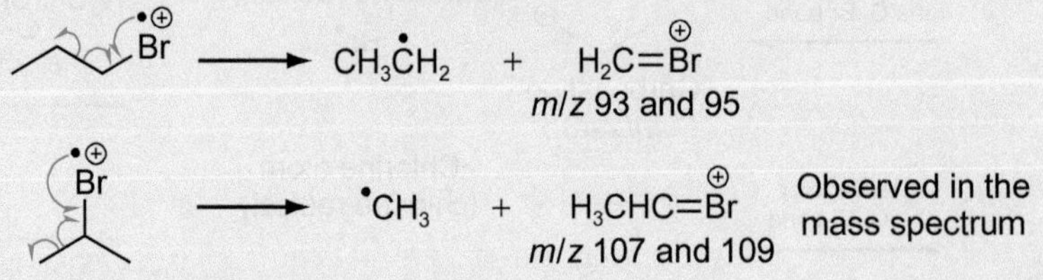

Br
$CH_3\dot{C}H_2$ + $H_2C{=}\overset{\oplus}{Br}$
m/z 93 and 95
Br
$\dot{C}H_3$ + $H_3CHC{=}\overset{\oplus}{Br}$
m/z 107 and 109
Observed in the mass spectrum

The product is 2-bromopropane ($CH_3CHBrCH_3$).

(When using the procedure in this worked example, beware: sometimes a molecular ion is not observed in an EI mass spectrum because it fragments before it reaches the detector.)

Question

When a mixture of ethylbenzene ($PhCH_2CH_3$) and chlorine is illuminated with UV radiation, $PhCHClCH_3$ and $PhCH_2CH_2Cl$ are formed in unequal amounts. After separation, the EI mass spectra of both products were recorded. The mass spectrum of the major product showed peaks at *m/z* values of 125 and 127, in the ratio 3:1. These peaks were not observed in the mass spectrum of the minor product. Use this information to identify which is the major and which is the minor product.

◀ 1-Bromopropane and 2-bromopropane: these space-filling models show their different shapes (the dark red-brown sphere represents bromine).

Fragmentation of ethers, amines, and alcohols

The molecular ion of an ether (ROR), amine (e.g. RNHR), or alcohol (ROH) is formed by loss of an electron from one of the lone pairs on oxygen or nitrogen. Often, the relative abundance of the molecular ion is very small because it readily fragments to form a cation (with a positively charged oxygen or nitrogen atom) and a carbon radical.

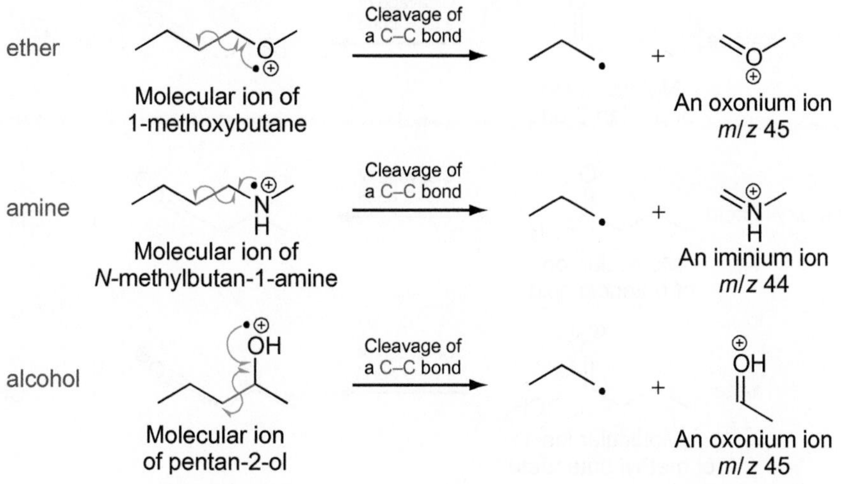
ether
Molecular ion of
1-methoxybutane
Cleavage of
a C–C bond
An oxonium ion
m/z 45
amine
Molecular ion of
N-methylbutan-1-amine
Cleavage of
a C–C bond
An iminium ion
m/z 44
alcohol
Molecular ion
of pentan-2-ol
Cleavage of
a C–C bond
An oxonium ion
m/z 45

> Notice that each molecular ion fragments to cleave a C–C bond adjacent to the oxygen or nitrogen atom. This is sometimes described as alpha (α) cleavage.

> An oxonium ion contains an oxygen atom, which has three covalent bonds and a positive charge (the simplest oxonium ion is H_3O^+). An iminium ion is a protonated or substituted imine. For example, protonation or methylation of the imine $H_2C{=}NMe$, gives iminium ions $H_2C{=}N^+HMe$ and $H_2C{=}N^+Me_2$, respectively.

> The presence of functional groups containing nitrogen, such as amines, can often be inferred using the nitrogen rule. This states that:
>
> - an organic molecule of even-numbered relative molecular mass either contains no nitrogen or contains an even number of nitrogen atoms;
> - an organic molecule of odd-numbered relative molecular mass contains an odd number of nitrogen atoms.
>
> The nitrogen rule arises because, among the common elements, nitrogen is unusual. Nitrogen has a valency of three, which is an odd number, but the most abundant isotope (^{14}N) has an even-numbered relative atomic mass.

Sometimes, alcohols can be identified from a fragment ion that is 18 mass units lower than the molecular ion. In the spectrum of pentan-2-ol ($CH_3CH_2CH_2CH(OH)CH_3$, $C_5H_{12}O$, M_r 88), a small peak at $m/z = 70$ is observed because the molecular ion fragments to lose H_2O. The mechanism of elimination of water starts by abstraction of a hydrogen atom from the aliphatic chain, typically from a carbon atom at the γ-position (via a 5-membered transition state).

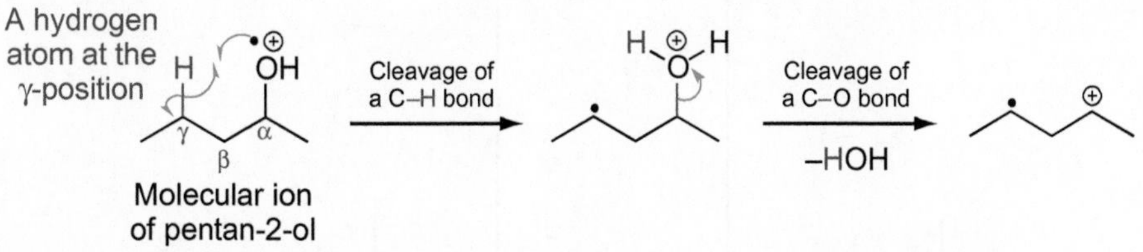
A hydrogen
atom at the
γ-position
Molecular ion
of pentan-2-ol
Cleavage of
a C–H bond
Cleavage of
a C–O bond
–HOH

> Alcohols often show a pronounced fragment peak resulting from loss of H_2O from the molecular ion (i.e. M – 18).

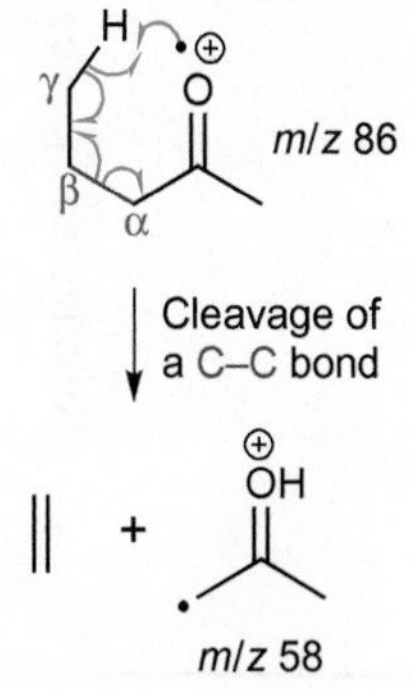
m/z 86
Cleavage of
a C–C bond
m/z 58

Fragmentation of aldehydes, ketones, carboxylic acids, and esters

The molecular ion of a carbonyl compound is formed by loss of an electron from one of the lone pairs on the oxygen atom in the C=O bond. Usually, the molecular ion fragments so as to break a C–H, C–C, or C–O bond adjacent to the C=O group. For an aldehyde (RCHO), it is common to see a peak one mass unit lower than the molecular ion, due to loss of a hydrogen atom. Worked example 12.2 illustrates how you can use a fragmentation pattern to distinguish between an aldehyde (RCHO) and a ketone (RCOR) with the same relative molecular mass.

> Like for the molecular ion of pentan-2-ol, molecular ions of carbonyls, such as ketones (RCOR), can fragment by abstraction of a hydrogen atom at the γ-position (via a 6-membered transition state)—this is called the McLafferty rearrangement, which leads to the cleavage of a $^{\alpha}C{-}C^{\beta}$ bond. For example, in the mass spectrum of pentan-2-one (M_r 86) a peak at $m/z = 58$ is observed due to loss of ethene ($H_2C{=}CH_2$).

Aldehydes often show a pronounced fragment peak resulting from loss of H• from the molecular ion (i.e. M – 1).

The naming of organic carbonyl compounds is discussed in Section 2.7 (p.98).

The cation ($CH_3CH_2CH_2C{\equiv}O^+$) formed from all four fragmentations is an example of an oxonium ion, called an acylium ion. Acylium ions are intermediates in Friedel–Crafts acylation reactions (Chapter 22.2, p.1023).

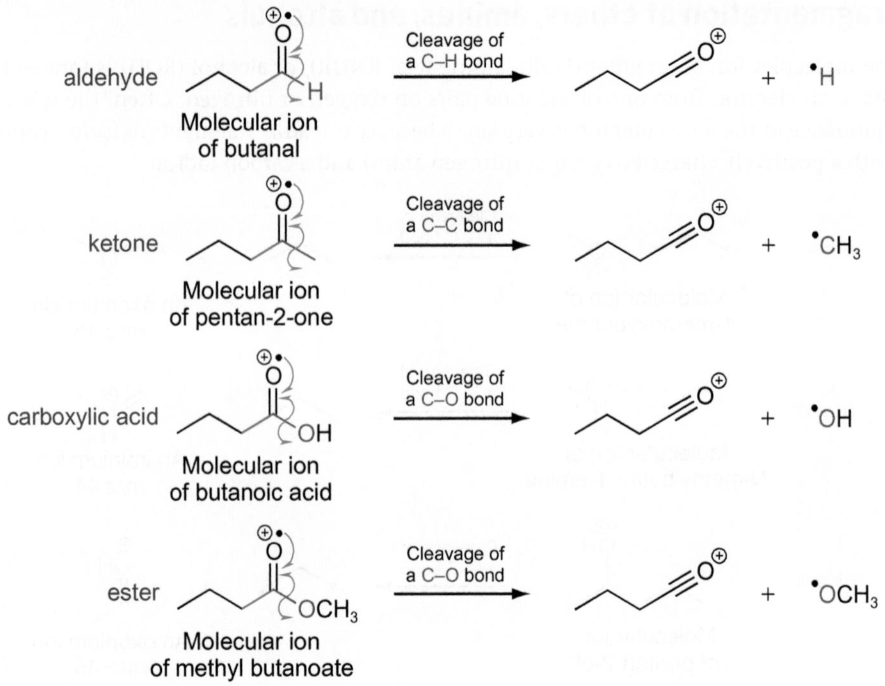
aldehyde
Molecular ion of butanal
Cleavage of a C–H bond
+ •H
ketone
Molecular ion of pentan-2-one
Cleavage of a C–C bond
+ •CH3
carboxylic acid
Molecular ion of butanoic acid
Cleavage of a C–O bond
+ •OH
ester
Molecular ion of methyl butanoate
Cleavage of a C–O bond
+ •OCH3

The molecular ions can also fragment to cleave the C–C bonds

Worked example 12.2 Using a fragmentation pattern to assign a structure

Hydration of phenylethyne ($PhC{\equiv}CH$) can form either acetophenone ($PhCOCH_3$) or 2-phenylethanal ($PhCH_2CHO$) depending on the reaction conditions. When phenylethyne reacts with H^+/H_2O, the major product formed gives the EI mass spectrum shown here. Identify the structure of this product, explaining your reasoning.

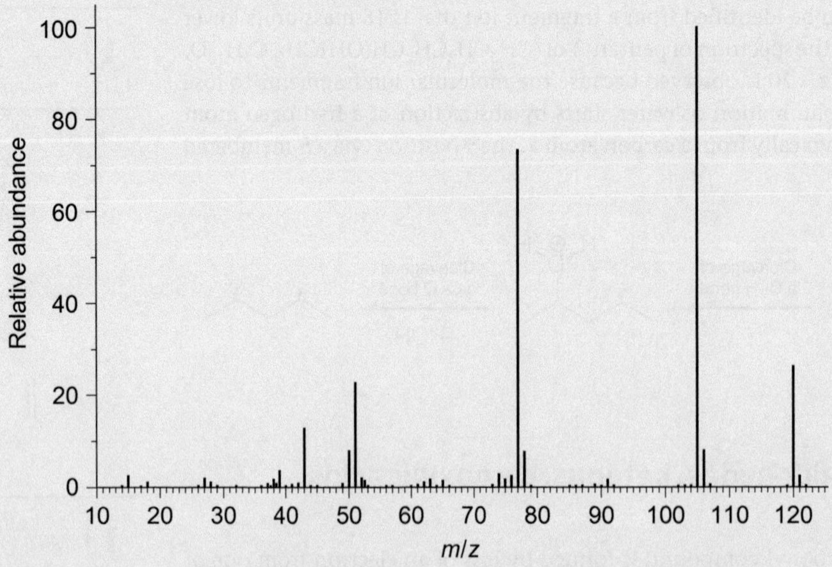
Relative abundance
0
20
40
60
80
100
10 20 30 40 50 60 70 80 90 100 110 120
m/z

Strategy

Draw the structures of the two possible products and calculate their relative molecular masses (M_r). In the mass spectrum, look at the peak of highest m/z value to check that it is equal to the relative molecular mass of the product. Identify the structure of the most abundant fragment cation and decide whether the molecular ions of acetophenone (1-phenylethanone) and 2-phenylethanal can form this fragment cation by breaking a C–C or a C–H bond adjacent to C=O. →

Solution

The two possible products are:

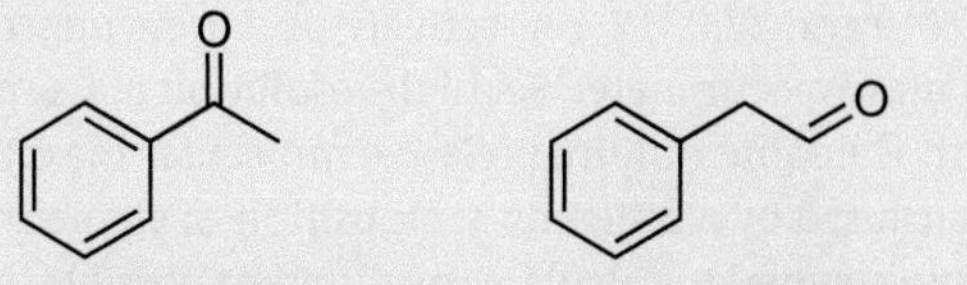

acetophenone (M_r 120) 2-phenylethanal (M_r 120)

The peak of highest *m/z* value in the spectrum occurs at *m/z* 120, so this is the molecular ion peak. The most abundant fragment cation occurs at *m/z* 105.

For acetophenone

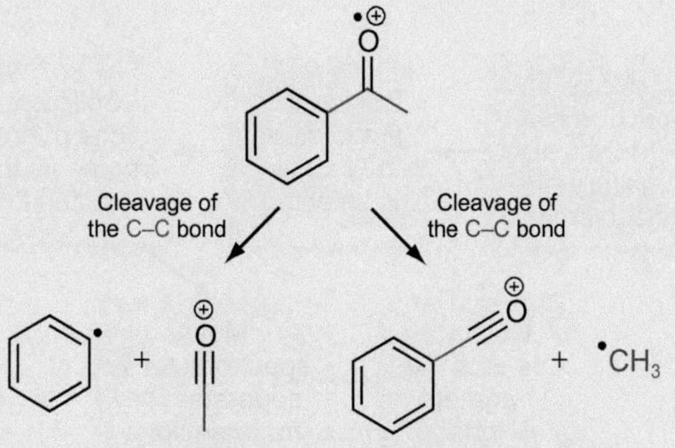
Cleavage of the C–C bond
Cleavage of the C–C bond
$^\bullet CH_3$

m/z 43 (M – 77) *m/z* 105 (M – 15)

For 2-phenylethanal

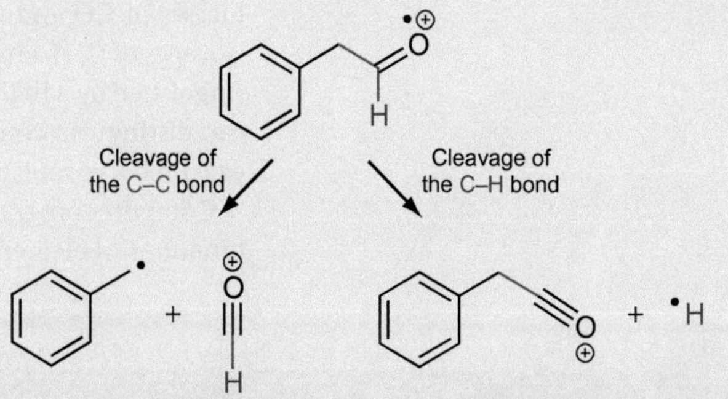
Cleavage of the C–C bond
Cleavage of the C–H bond
$^\bullet H$

m/z 29 (M – 91) *m/z* 119 (M – 1)

The fragmentation pattern observed in the mass spectrum corresponds to that of acetophenone.

Question

The following questions relate to the EI mass spectrum of acetophenone shown at the beginning of Worked example 12.2.

(a) Suggest why the PhC≡O$^+$ fragment cation, but not the $CH_3C{=}O^+$ fragment cation, is the major fragment cation observed in the mass spectrum.

(b) Suggest a structure for the fragment ion that has an *m/z* value of 77.

Determination of molecular formula

The EI mass spectra presented in this section have been recorded on a low-resolution mass spectrometer. Low-resolution spectrometers determine the relative molecular mass of the molecule (and fragment cations) to the nearest whole number. At this level of resolution, it is not possible to determine the molecular formula of an unknown compound. For example, both CO and C_2H_4 have a relative molecular mass (M_r) of 28, to the nearest whole number, and both would give peaks at *m/z* 28 in a low-resolution mass spectrum.

Relative atomic mass (A_r) is introduced in Section 1.3 (p.12). For exact relative isotopic masses see Table 1.5, p.18.

Compound	CO	C_2H_4
Relative molecular mass	C = 12 O = 16 28	2C = (2 × 12) 4H = (4 × 1) 28
Low resolution *m/z*	28	28
Accurate relative molecular mass	^{12}C = 12.0000 ^{16}O = 15.9949 27.9949	2 ^{12}C = (2 × 12.0000) 4 ^{1}H = (4 × 1.0078) 28.0312
High resolution *m/z*	27.9949	28.0312

High-resolution spectrometers can distinguish between compounds that have different compositions of elements so they can be used to determine the molecular formula of an unknown compound. The elemental composition of a compound can also be determined using combustion analysis, in which the compound is burned in a stream of oxygen and the masses of the combustion products found. This gives the empirical formula of the compound (see Worked example 1.5, p.20).

To distinguish CO from C_2H_4 requires a **high-resolution mass spectrometer** that records the accurate mass of a compound to four or five decimal places. The exact relative molecular masses of CO and C_2H_4 (using the exact relative isotopic masses for the most abundant isotopes of C, H, and O) are 27.9949 and 28.0312, respectively, and these masses can be distinguished by a high-resolution mass spectrometer. So, high-resolution mass spectrometers can distinguish compounds with the same nominal relative molecular mass but different exact relative molecular masses (caused by a different composition of elements).

Computer programs are commonly used to calculate the different possible isotopic combinations of elements that produce an observed accurate relative molecular mass.

Box 12.2 Using tandem mass spectrometry in newborn screening

In recent years, by analysis of just a single drop of blood, it has become possible to detect in newborn babies some disorders that lead to disabilities and premature death. One such disorder is phenylketonuria, which leads to abnormally high concentrations of the amino acid (*S*)-phenylalanine in the body.

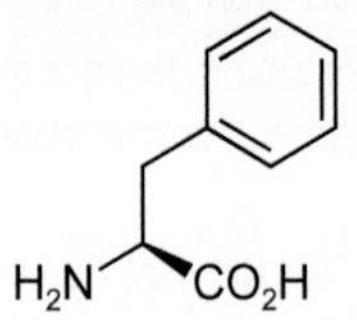
H2N
CO2H

(*S*)-phenylalanine ($C_9H_{11}NO_2$, M_r 165)

High concentrations of (*S*)-phenylalanine in the bloodstream are toxic to the brain causing intellectual disability. A newborn baby diagnosed with this condition is treated by restricting the amount of (*S*)-phenylalanine in the diet. A restricted diet, if started early and maintained, allows for normal development of the baby. But how is it possible to selectively detect a higher than normal concentration of (*S*)-phenylalanine, in the presence of thousands of other compounds, from analysis of just a tiny drop of blood? One technique used is *tandem mass spectrometry*.

In conventional mass spectrometry, the introduction of a mixture of compounds into a mass spectrometer gives a complex mass spectrum that is difficult to interpret. But it is possible to recognize a specific compound in a mixture if two mass spectrometers are joined together via a 'collision chamber' (Figure 1). In tandem mass spectrometry the molecular ions generated in the first mass spectrometer are separated according to their mass-to-charge ratio and the selected ion is passed into the collision chamber where it fragments. The resulting fragment cations are then analysed in a second mass spectrometer—the final spectrum shows the fragmentation pattern of the selected ion. For example, (*S*)-phenylalanine can be identified in a blood sample by selecting the molecular ion with an *m*/*z* value of 165 and then observing if this ion gives the expected fragmentation pattern. By using a known concentration of (*S*)-phenylalanine as an internal standard it is possible to accurately measure the concentration of (*S*)-phenylalanine in blood. Analysis requires just a small blood drop and the mass spectrum is produced in a matter of seconds, so hundreds of samples are quickly screened.

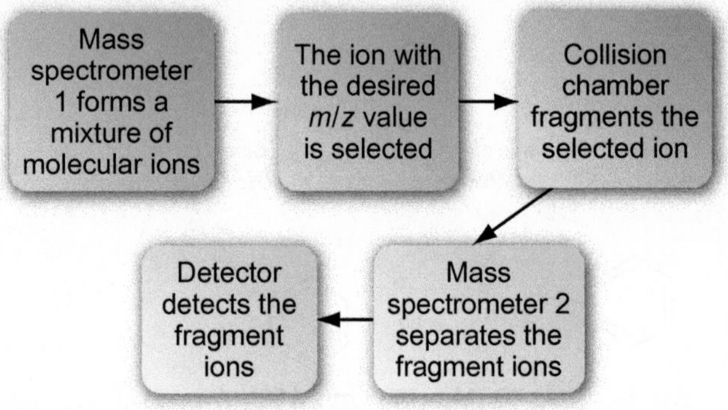
Mass spectrometer 1 forms a mixture of molecular ions
The ion with the desired m/z value is selected
Collision chamber fragments the selected ion
Mass spectrometer 2 separates the fragment ions
Detector detects the fragment ions

▲ **Figure 1** The different stages in tandem mass spectrometry.

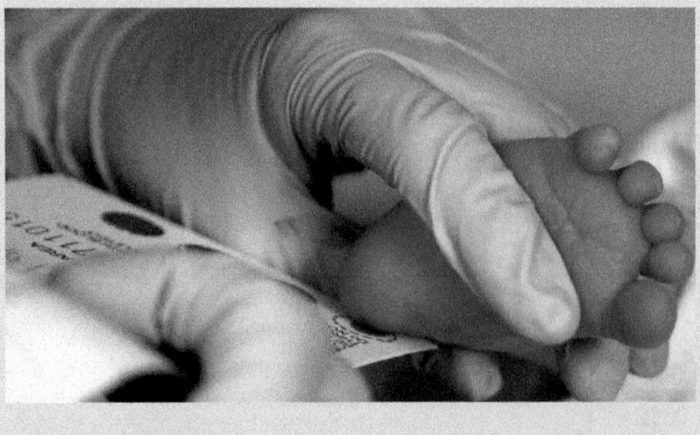

▲ Diagnosis of phenylketonuria requires a small drop of blood, which is taken from the heel of a newborn baby's foot.

An additional benefit of using tandem mass spectrometry is that molecules other than (*S*)-phenylalanine in blood may be detected and measured, so that different disorders can also be identified. Although the instrumentation is expensive and its operation requires specially trained personnel, the use of mass spectrometry is one of the most significant advances in newborn screening.

Question

In the EI mass spectrum of (*S*)-phenylalanine, an intense fragment cation with an *m*/*z* value of 74 is observed. Suggest a structure for this fragment ion and propose a mechanism for its formation.

Summary

- Mass spectrometry (MS) gives information on the relative molecular mass of compounds by measuring the mass-to-charge ratios (*m/z*) of ions produced on ionization.
- In electron impact (EI) mass spectrometry, a sample is ionized by bombardment with high-energy electrons to form a radical cation ($M^{+\bullet}$), called the molecular ion.
- To calculate the number of carbon atoms in the molecule, divide the relative abundance of the M + 1 peak, expressed as a percentage of the molecular ion peak, by 1.1.
- To identify the presence of a chlorine or bromine atom in the compound, look at the relative abundances of the M + 2 and the M peaks.
- If *m/z* for the molecular ion is even, the molecule contains an even number of nitrogen atoms or none at all. If *m/z* for the molecular ion is odd, the molecule contains an odd number of nitrogen atoms.
- The weakest bonds in a molecular ion are often selectively broken to form the most stable fragment cations (which are detected by the spectrometer). Compounds containing functional groups fragment in predictable ways.

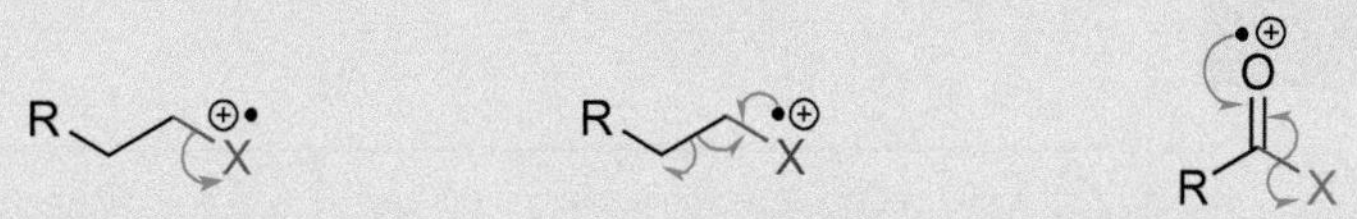

X = Cl, Br, I

X = Cl, Br, I, OH, OR, NHR

X = H, alkyl, OH, OR

- Information on partial structures can be deduced by looking at mass differences between the molecular ion (M) and fragment cations (Table 12.1).

Table 12.1 Common fragmentations of molecular ions

Fragment cation	Typical group that is lost	Possible partial structure
M − 1	H	aldehyde
M − 15	CH_3	methyl group
M − 18	H_2O	alcohol
M − 29	CH_2CH_3	ethyl group
M − 31	CH_3O	methyl ester
	CH_2OH	primary alcohol
M − 35/37	Cl	chloroalkane
M − 43	CH_3CO	methyl ketone
M − 77	C_6H_5	phenyl group
M − 79/81	Br	bromoalkane
M − 91	$C_6H_5CH_2$	benzyl group

- High-resolution spectrometers can distinguish between compounds that have different compositions of elements so they can be used to determine the molecular formula of an unknown compound.

? For practice questions on this topic, see questions 1 and 3–6 at the end of this chapter (pp.606–608).

12.2 Infrared spectroscopy

In infrared (IR) spectroscopy, a chemical sample (a solid, liquid, or gas) is exposed to radiation in the IR region of the electromagnetic spectrum. An **IR spectrometer** measures the radiation that passes through the sample and compares it to the intensity of radiation before it passes through the sample (Figure 10.8, p.459). The result is recorded in an **IR spectrum**, which is a plot of transmission of radiation versus the wavenumber ($\tilde{\nu}$) of the radiation transmitted (typically between 4000 cm^{-1} and 600 cm^{-1}). The wavenumber of the radiation is directly proportional to its energy—the larger the wavenumber, the higher the energy. If the sample absorbs radiation at a particular wavenumber (energy) the transmission decreases and a downward peak called an **absorption band** is observed in the spectrum.

The IR spectrum of an organic molecule is characteristic of the entire molecule—it is often likened to a person's fingerprints. A spectrum typically shows a number of absorption bands, which may be broad or sharp and of differing intensities. For example, the IR spectrum of liquid ethanol (CH_3CH_2OH), shown in Figure 12.4, is characteristic of ethanol and can be used to assign its structure. It is unlikely that any two organic molecules, except enantiomers, will give exactly the same infrared spectrum.

The wavenumber ($\tilde{\nu}$) is usually quoted in cm^{-1} and is proportional to energy (E): $E = hc\tilde{\nu}$ (Equation 10.11, p.466).

The theory of IR spectroscopy, including how to record an IR spectrum, is discussed in Section 10.5 (p.476).

Enantiomers are chiral isomers that are non-superimposable mirror images of one another. Molecules with chiral centres are discussed in Section 18.4 (p.838).

CRIME SCENE DO NOT CROSS

IR spectroscopy is frequently used in forensic analysis. For example, IR spectroscopy can be used to distinguish soil samples of similar appearance—different soil samples have different compositions of organic compounds and this gives rise to dissimilar IR spectra.

IR spectrometers sometimes produce spectra that have a change of scale along the wavenumber axis. Always check the scale carefully to ensure that you have assigned the correct wavenumber to an absorption band.

For example in Figure 12.4, the wavenumber axis is uniform but, in Figure 12.8, the scale changes at 2000 cm^{-1}.

Note that, in an IR spectrum, the wavenumber *increases* from right to left.

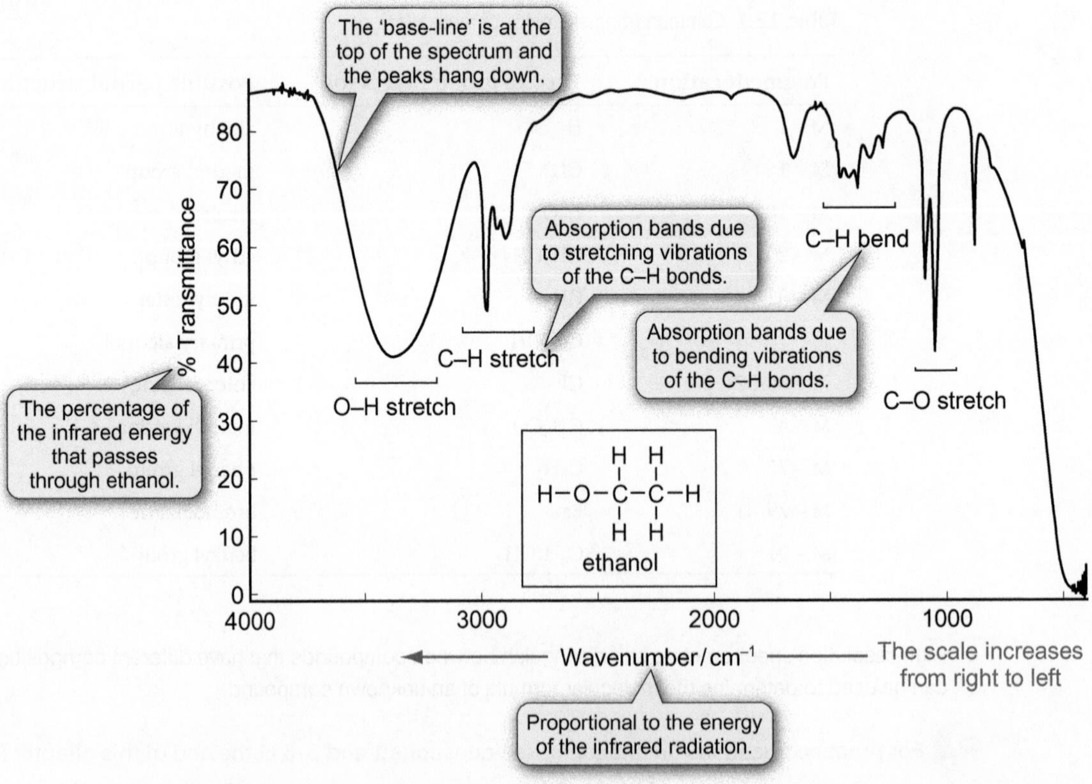
The 'base-line' is at the top of the spectrum and the peaks hang down.
% Transmittance
80
70
60
50
40
30
20
10
0
The percentage of the infrared energy that passes through ethanol.
O–H stretch
C–H stretch
Absorption bands due to stretching vibrations of the C–H bonds.
C–H bend
Absorption bands due to bending vibrations of the C–H bonds.
C–O stretch
ethanol
4000
3000
2000
1000
Wavenumber/cm⁻¹
The scale increases from right to left
Proportional to the energy of the infrared radiation.

Figure 12.4 The IR spectrum of ethanol (liquid film).

Position of absorption bands

The IR spectra of organic molecules show a number of absorption bands because all of the bonds in the molecules can undergo stretching vibrations and bending vibrations. Vibrations that cause a change in the electric dipole moment in the molecule lead to absorption of IR radiation. Each stretching or bending vibration of a bond within a molecule is associated with a specific quantum of energy and, when IR radiation of the same energy interacts with the molecule, the molecule absorbs it and an absorption band is recorded in the IR spectrum. The wavenumber of an absorption band depends on the type of bond vibration, the bond enthalpy, and the masses of the atoms in the bond.

- For a given bond, the stretching vibration gives rise to an absorption band at a higher wavenumber than a bending vibration, because it requires more energy to stretch a bond than to bend it. For example, in the IR spectrum of ethanol (Figure 12.4), the stretching vibration of a C–H bond is at a higher wavenumber than the bending vibration.
- Normally, short strong bonds vibrate at a higher wavenumber than longer weaker bonds because more energy is required to vibrate a short strong bond. For example, the absorption band due to the stretching vibration of a C=C bond is at a higher wavenumber than that for a C–C bond.

2100 C≡C > 1650 C=C > 1000 C–C

2200 C≡N > 1600 C=N > 1100 C–N

1700 C=O > 1150 C–O

Approximate wavenumbers of stretching vibrations / cm^{-1}

- Bonds involving atoms with low mass vibrate at a higher wavenumber than those involving atoms with higher mass (as predicted by Hooke's law). For example, the absorption band due to the stretching vibration of a C–H bond (~3000 cm^{-1}) is at a higher wavenumber than that for a C–D bond (~2200 cm^{-1}).

Atom with lowest mass — Atom with highest mass

3000 C–H > 2200 C–D > 1150 C–O > 700 C–Cl

Approximate wavenumbers of stretching vibrations / cm^{-1}

From the wavenumber of the absorption band, you can identify the type of bond in the molecule that undergoes the stretching or bending vibration. In the IR spectrum of ethanol (Figure 12.4), absorption bands due to the stretching vibrations of O–H and C–H bonds have much higher wavenumbers than for a C–O bond (Figure 12.5). Not only do the O–H and C–H bonds contain a low mass hydrogen atom, but they are also stronger than a C–O bond.

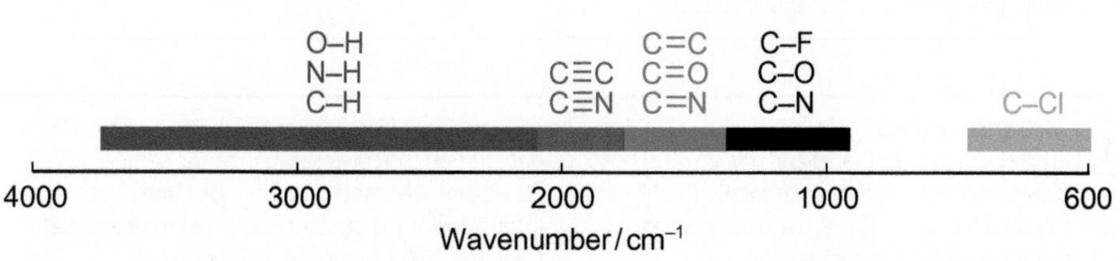
O–H
N–H
C–H
C≡C
C≡N
C=C
C=O
C=N
C–F
C–O
C–N
C–Cl
4000
3000
2000
1000
600
Wavenumber / cm⁻¹

Figure 12.5 Approximate positions of absorption bands due to stretching vibrations.

Molecules absorb those wavenumbers of IR radiation that correspond to the stretching and bending vibrations that lead to a change in the electric dipole moment in the molecule (Section 10.5, p.476).

You can think of a bond as behaving like a spring—the stronger the bond, the tighter the spring.

The simple harmonic oscillator model for a diatomic molecule and Hooke's law are discussed in Box 10.4 (p.479).

Research has shown that fruits flies (*Drosophila melanogaster*) can distinguish acetophenone (1-phenylethanone; $C_6H_5COCH_3$) from deuterated acetophenone (in which all the hydrogen atoms in acetophenone are replaced by deuterium). This observation has supported the controversial theory that the odour of molecules is affected, not only by their molecular shapes, but by the vibrations of their bonds (as C–H and C–D bonds vibrate at different wavenumbers).

Mean bond enthalpies (in kJ mol^{-1}): O–H (+464), C–H (+412), C–O (+358). For a list of mean bond enthalpies see Appendix 10 (p.1369).

Intensity and appearance of absorption bands

The more intense absorption bands arise from vibrations of polar bonds such as O–H and C=O bonds.

The absorption bands in an IR spectrum have different intensities. Usually, the more intense absorption bands are due to the vibrations of polar bonds, which give rise to a relatively large change in dipole moment in the molecule. For example, in the IR spectrum of ethanol (CH_3CH_2OH; Figure 12.4, p.570) the absorption band due to the O–H stretching vibration is more intense than the absorption band due to the stretching vibration of the C–H bonds, because the C–H bonds are much less polar. So, the intensity of an absorption band can be helpful in identifying whether the band arises from vibration of a polar or a non-polar bond. For this reason, in the chemical literature, the intensities of absorption bands are described as strong (s), medium (m), or weak (w).

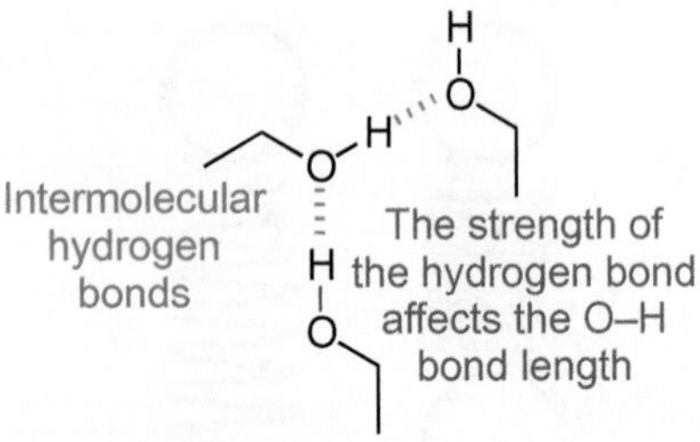
Intermolecular hydrogen bonds
The strength of the hydrogen bond affects the O–H bond length

Notice also that the absorption band due to the stretching vibration of the O–H bond in the IR spectrum of ethanol is particularly broad. The band is broad because of intermolecular hydrogen bonding between the O–H groups (see margin). Hydrogen bonding changes the length of an O–H bond—the stronger the hydrogen bond, the longer the O–H bond, and the easier it is to stretch. In a liquid sample, individual molecules of ethanol hydrogen bond to each other to slightly different extents. Consequently, the O–H bonds in different ethanol molecules have absorption bands at slightly different positions and the end result is a broad absorption band (designated 'br') that is the sum of all the slightly different absorptions, reported as 3650–3000 cm^{-1} (br).

Hydrogen bonding is introduced in Section 1.8 (p.54).

In comparison, the IR spectra of ethanol in the gas phase and of ethanol in a dilute solution of a non-polar solvent (such as hexane) show *sharp* absorption bands for the O–H stretching vibration. A sharp absorption band between 3650 cm^{-1} and 3590 cm^{-1} is observed because there is little or no hydrogen bonding between the O–H groups.

Characteristic group wavenumbers

The precise positions of the absorption bands in an IR spectrum depend on the environment of the bonds in a molecule. For example, in the IR spectrum of ethanol (Figure 12.4, p.570), there are a number of absorption bands between 3000 cm^{-1} and 2800 cm^{-1} because the C–H bonds are in different environments. This explains why an IR spectrum is characteristic of the entire molecule.

An IR spectrum offers a reliable, simple, and quick method for assigning functional groups within a molecule. (Functional groups are introduced in Section 2.3, p.77.)

However, there are certain groups of atoms within a molecule that give rise to absorption bands at the same, or similar, wavenumber, irrespective of the rest of the structure. For example, the broad absorption band around 3400 cm^{-1} in the IR spectrum of ethanol is characteristic of the O–H group of alcohols. The IR spectra of other alcohols also have broad absorption bands close to 3400 cm^{-1}, as shown by the IR spectrum of liquid propan-1-ol ($CH_3CH_2CH_2OH$), in Figure 12.6. Distinctive bands such as these are called **characteristic group wavenumbers**. They are due to stretching vibrations and provide useful structural information—they are normally found between 4000 cm^{-1} and 1400 cm^{-1} in the IR spectrum, which is called the **functional group region**.

The area between 1400 cm^{-1} and 600 cm^{-1} in the IR spectrum is called the **fingerprint region**. This region normally contains a very complicated series of absorption bands,

Infrared spectroscopy provides a relatively cheap and quick way to analyse the ethanol content of beverages, from whisky to wine.

Characteristic group wavenumbers are commonly called characteristic group frequencies, although this is not strictly correct.

Because $\nu = c\tilde{\nu}$, wavenumber ($\tilde{\nu}$) is proportional to frequency (ν). See Equation 10.10b (p.466).

% Transmittance
The IR spectra of ethanol and propan-1-ol have different values for % transmittance for the bands around 3400 cm^{-1}. The actual values for each band are not important in structure determination (although IR bands can be used to measure concentration (Box 10.5, p.485)); the % transmittance depends on the concentration of the sample, which varies between measurements. But, the *relative* intensities of the bands in each spectrum are useful for assigning structures (see Table 12.2).

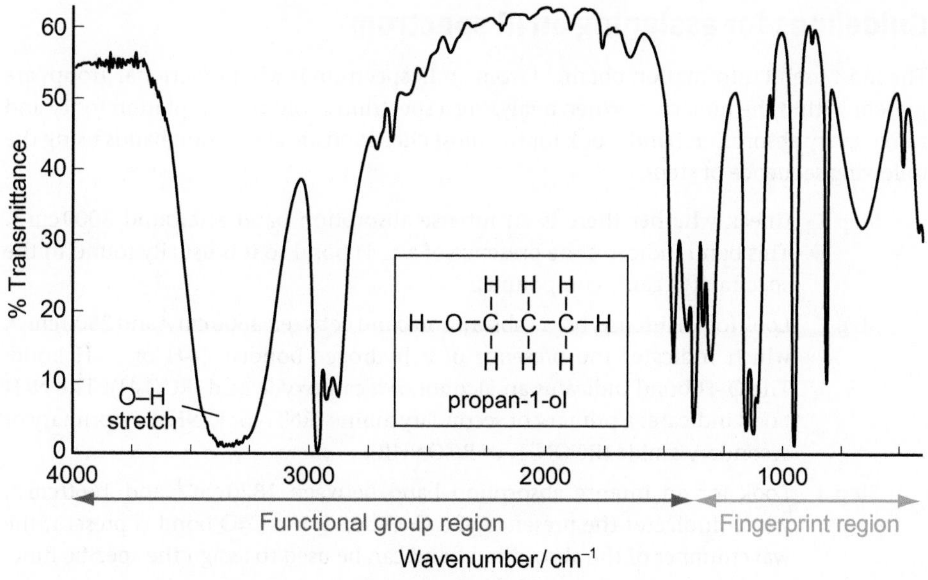

60
50
40
30
20
10
0
% Transmittance
H H H
H–O–C–C–C–H
H H H
propan-1-ol
O–H
stretch
4000
3000
2000
1000
Functional group region
Fingerprint region
Wavenumber / cm⁻¹

Figure 12.6 IR spectrum of propan-1-ol (liquid film).

typically due to bending vibrations. It is much more difficult to identify vibrations of individual bonds in this region than it is for the functional group region. However, the fingerprint region for each compound is unique so it can be useful for identifying compounds. For example, although the IR spectra of ethanol and propan-1-ol (Figures 12.4, p.570, and 12.6, above, respectively) have absorption bands at similar wavenumbers in the functional group region, the patterns in the fingerprint region are completely different, even though both molecules have the same types of bonds. Table 12.2 shows the characteristic group wavenumbers for bonds that are common in organic molecules.

Table 12.2 Characteristic group wavenumbers

Type of bond	Typical wavenumber/cm^{-1}	Typical intensity	
O–H (alcohol)	3600–3200	strong (broad)	↑
N–H	3500–3300	medium (broad)	
O–H (carboxylic acid)	3300–2500	strong (broad)	functional
C–H	3000–2850	medium	group
C≡N	2250–2200	medium	region
C≡C	2200–2100	medium to weak	(stretching
C=O	1820–1650	strong	vibrations)
C=N	1690–1640	medium	
C=C	1680–1620	medium	
C⋯C (benzene)	1600–1500 (2 or 3 bands)	medium to weak	↓
C–O	1250–1050	strong	↑
C–N	1250–1020	medium to weak	fingerprint region (bending vibrations)
C–Cl	800–600	strong	↓

Just like a human fingerprint, the pattern of absorption bands in the fingerprint region is unique to every compound.

ℹ The functional group region is extremely useful in assigning functional groups, while the fingerprint region is helpful in confirming structures.

ℹ In many IR spectra, an absorption band is seen around 2350 cm^{-1}, due to a stretching vibration of CO_2 that is present in the air (see Box 10.5, p.485).

ℹ **Nujol mulls versus ATR**

To record the IR spectrum of a solid sample, the finely ground solid can be mixed with a hydrocarbon oil, called Nujol, to form a Nujol mull (see Section 10.5, p.477). When analysing an IR spectrum of a Nujol mull, it is important to remember that Nujol gives bands around 2950 cm^{-1}–2800 cm^{-1} (C–H stretch) and 1470 cm^{-1}–1350 cm^{-1} (C–H bend). Today, more commonly, a technique called attenuated total reflectance (ATR) is used for solid (and liquid and gas) samples—a solid sample is pressed into direct contact with a crystal (the metal component at the top of the photograph moves down and clamps the solid sample to the crystal surface) as described in Section 10.5, p.477.

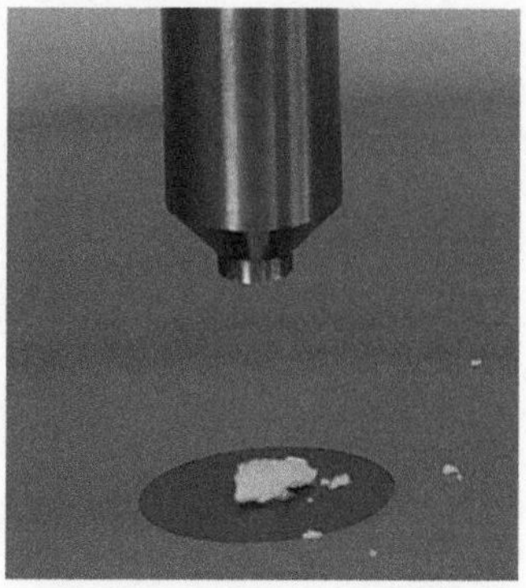

Guidelines for assigning an IR spectrum

The most useful information obtained from an IR spectrum is what functional groups are present within the molecule. When analysing a spectrum avoid the temptation to try and assign every absorption band. Look for the most characteristic absorption bands using the following sequence of steps.

Step 1 Check whether there is an intense absorption band at around 3000 cm^{-1}. This band indicates the presence of a C–H bond so it is usually found in the spectra of organic compounds.

Step 2 Look for an intense, broad absorption band between 3600 cm^{-1} and 2500 cm^{-1}, which indicates the presence of a hydrogen-bonded O–H or N–H bond. The O–H bond indicates an alcohol or a carboxylic acid (RCO_2H). The N–H bond indicates a primary or secondary amine (RNH_2 or R_2NH) or a primary or secondary amide ($RCONH_2$ or RCONHR).

Step 3 Look for an intense absorption band between 1820 cm^{-1} and 1660 cm^{-1}, which indicates the presence of a C=O bond. If a C=O bond is present, the wavenumber of the absorption band can be used to assign the specific functional group (see Figure 12.7 below).

Step 4 Look for medium intensity absorption bands around 2150 cm^{-1} or 1650 cm^{-1}, which indicate the presence of a C≡C or a C=C bond, respectively.

Step 5 Look for 2 or 3 medium to weak intensity absorption bands between 1600 cm^{-1} and 1500 cm^{-1}, which indicate the presence of a benzene ring.

Step 6 The absence of an absorption band can be as useful as the presence of an absorption band, in assigning a structure. If no absorption bands are identified from steps 2–5, then the compound may be an alkane, an ether, or a halogenoalkane.

IR spectroscopy provides information on the presence and absence of functional groups. It cannot be used to determine the precise structure of a compound, unless the compound is already known and its IR spectrum is available for comparison. Today, IR spectra of compounds are routinely assigned by comparison with a database of IR spectra of known compounds.

Absorption of radiation in the ultraviolet-visible (UV/VIS) region of the electromagnetic spectrum can be used to help assign the structure of molecules containing aromatic rings or extended conjugated systems (Section 10.6, p.491). Values of λ_{max} and ε_{max} can provide information about the extent of conjugation.

Cheese sampling. IR spectroscopy can be used to predict the composition and characteristics of foodstuffs. For example, as a cheese matures, the concentrations of naturally occurring carboxylic acids (RCO_2H) and esters (RCO_2R) in the cheese change. The changes can be monitored by looking at the positions and intensities of C=O absorption bands in the 1800 cm^{-1} to 1700 cm^{-1} region of the IR spectrum.

Identifying carbonyl functional groups

The exact position of the absorption band due to the stretching vibration of a C=O bond depends on the substituents joined to the carbon atom. This means that the wavenumber of the C=O stretching vibration is often used to identify a particular carbonyl functional group. So, for example, an acyl chloride (RCOCl) can be distinguished from an ester (RCO_2R) or an amide (such as $RCONH_2$) using the wavenumber of the C=O stretching vibration. Figure 12.7 shows approximate wavenumbers characteristic of C=O stretching vibrations.

The slight differences in wavenumbers of the C=O stretching vibrations are explained by the electronic properties of the substituents attached to C=O. A C=O bond with electron-withdrawing groups normally has an absorption band at a higher wavenumber than a C=O bond with electron-donating substituents. This is because electron-donating substituents with +I and/or +M effects weaken the C=O bond, making it longer and easier

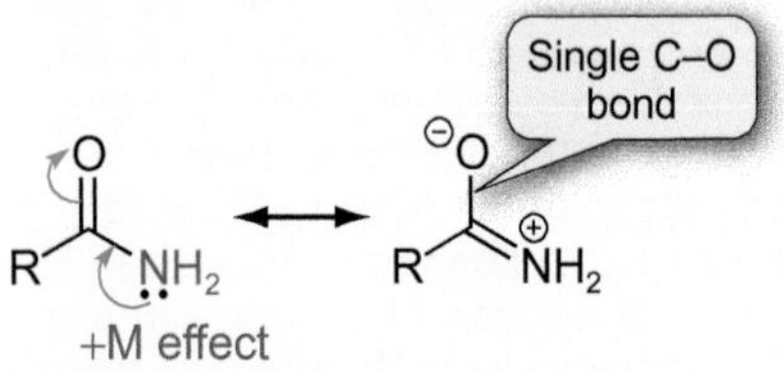
Single C–O bond
+M effect

Resonance forms of an amide

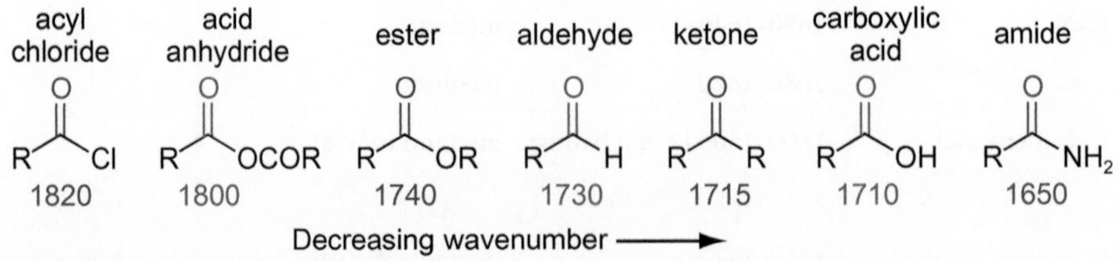
acyl chloride
acid anhydride
ester
aldehyde
ketone
carboxylic acid
amide
1820
1800
1740
1730
1715
1710
1650
Decreasing wavenumber

Figure 12.7 Approximate wavenumbers of absorption bands due to stretching vibrations of C=O bonds/cm^{-1}. (For functional groups with more than one R group, the R groups can be the same or different.)

to vibrate so that the absorption band occurs at a lower wavenumber. The effect is most pronounced for an amide, where the strong +M effect of an NH_2, NHR, or NR_2 group gives a resonance structure with a single C–O bond (see margin). Consequently, the C=O bond of an amide is described as having some single-bond character.

The electron-donating +M effect of a C=C bond and a benzene ring also explains why the C=O stretching vibrations of α,β-unsaturated carbonyls, such as cyclohex-2-enone, and aromatic carbonyls, such as acetophenone (1-phenylethanone; $C_6H_5COCH_3$), are at lower wavenumbers than those for saturated carbonyls, such as propanone (CH_3COCH_3). For both acetophenone and cyclohex-2-enone, electron donation from the benzene ring or C=C bond (+M effects) to oxygen in the C=O bond gives a resonance form that has a C–O bond.

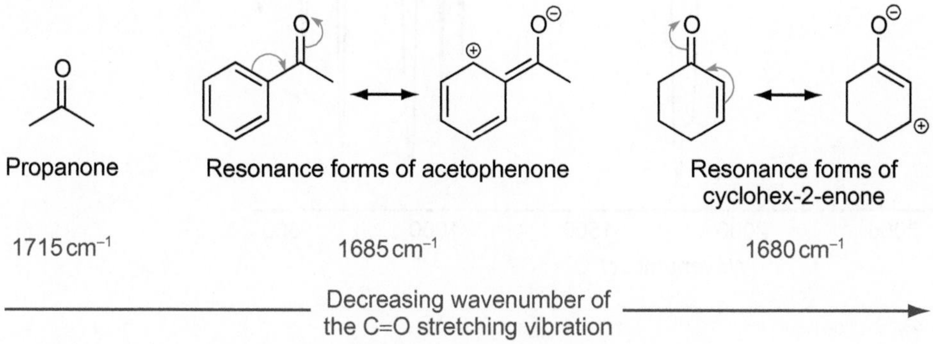
Propanone
1715 cm⁻¹
Resonance forms of acetophenone
1685 cm⁻¹
Resonance forms of cyclohex-2-enone
1680 cm⁻¹
Decreasing wavenumber of the C=O stretching vibration

The presence of characteristic absorption bands due to C=O stretching vibrations is illustrated by the IR spectrum of aspirin (Figure 12.8, shown below), which shows two intense C=O absorption bands—a band at 1750 cm^{-1} due to the C=O bond in the ester group and a band at 1690 cm^{-1} due to the C=O bond in the carboxylic acid group.

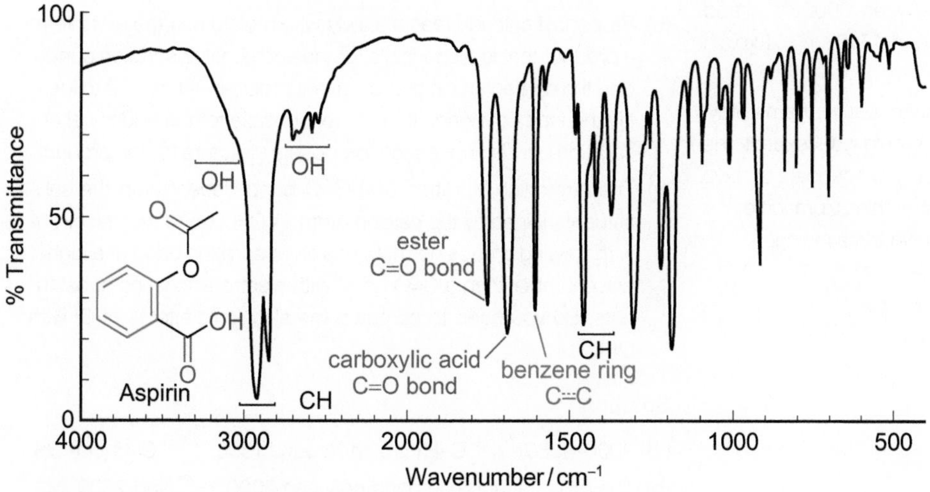
100
50
0
% Transmittance
OH
OH
ester
C=O bond
Aspirin
CH
carboxylic acid
C=O bond
CH
benzene ring
C⩦C
4000
3000
2000
1500
1000
500
Wavenumber / cm⁻¹

Figure 12.8 Assigned IR spectrum of aspirin in a Nujol mull.

The spectrum of aspirin is recorded in a Nujol mull. A **Nujol mull** is a finely divided suspension of a solid sample (such as aspirin) in a hydrocarbon oil, called Nujol. When analysing an IR spectrum of a Nujol mull, remember that Nujol gives absorption bands around 2950–2800 cm^{-1} (C–H stretch) and 1470–1350 cm^{-1} (C–H bend).

Worked example 12.3 illustrates the use of the characteristic group wavenumbers given in Table 12.2 (p.573) to assign structures.

+I and +M effects are discussed in Section 19.1 (p.871). The electronic effects of substituents attached to a C=O bond also explain the different reactivity of carbonyls to nucleophiles (see Box 24.2, p.1108).

Visit the Online Resource Centre to view screencast 12.1 which walks you through how to identify carbonyl functional groups using infrared spectroscopy.

α,β-Unsaturated carbonyls, including enones (RCH=CHCOR) and enals (RCH=CHCHO), are discussed in Section 23.4 (p.1089).

Notice that a double-headed straight arrow, ⟷, is used to indicate resonance forms.

For aspirin, the C=O bond of the ester gives an absorption band *higher* than that for a typical ester (1750 cm^{-1} versus 1740 cm^{-1}), whereas the C=O bond of the carboxylic acid gives an absorption band *lower* than that for a typical carboxylic acid (1690 cm^{-1} versus 1710 cm^{-1}). The C=O bond of the ester has more double bond character (as the OR group has a relatively weak +M effect), whereas the C=O bond of the carboxylic acid has more single bond character (because of the +M effect of the OR group and the benzene ring), as explained by resonance:

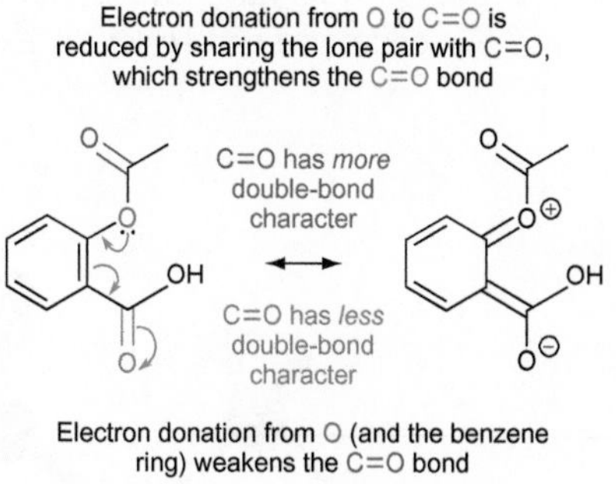
Electron donation from O to C=O is reduced by sharing the lone pair with C=O, which strengthens the C=O bond
C=O has *more* double-bond character
C=O has *less* double-bond character
OH
OH
Electron donation from O (and the benzene ring) weakens the C=O bond

The unassigned IR spectrum of aspirin in KBr disc is shown in Figure 10.1 (p.450). Compare this with the IR spectrum of aspirin in a Nujol mull in Figure 12.8.

Worked example 12.3 Using IR spectroscopy to identify the products of the reactions of propanoyl chloride

The IR spectrum of liquid propanoyl chloride (CH_3CH_2COCl) is shown below.

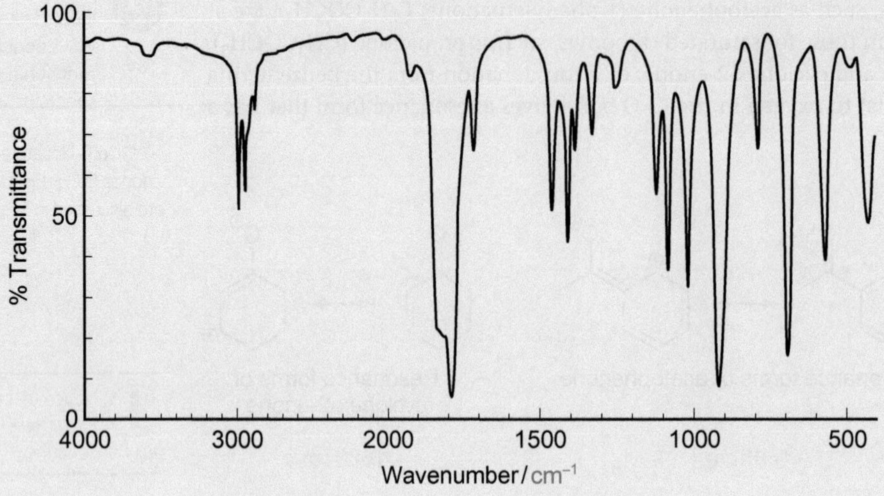
100
50
0
% Transmittance
4000
3000
2000
1500
1000
500
Wavenumber/cm⁻¹

▲ IR spectrum of propanoyl chloride (liquid film).

(a) Identify the characteristic absorption bands in the functional group region of the IR spectrum of propanoyl chloride.

(b) When a sample of propanoyl chloride is exposed to moisture in the air, the IR spectrum shows some additional absorption bands. A broad absorption band appears between 3000 cm^{-1} and 2500 cm^{-1} and also a narrow band around 1710 cm^{-1}. Explain these observations.

(c) Reaction of propanoyl chloride with ethanol (EtOH) forms ethyl propanoate ($CH_3CH_2CO_2Et$) and HCl. Would you expect the stretching vibration of the C=O bond in ethyl propanoate to have a higher or lower wavenumber than the stretching vibration of the C=O bond in propanoyl chloride? Explain your reasoning.

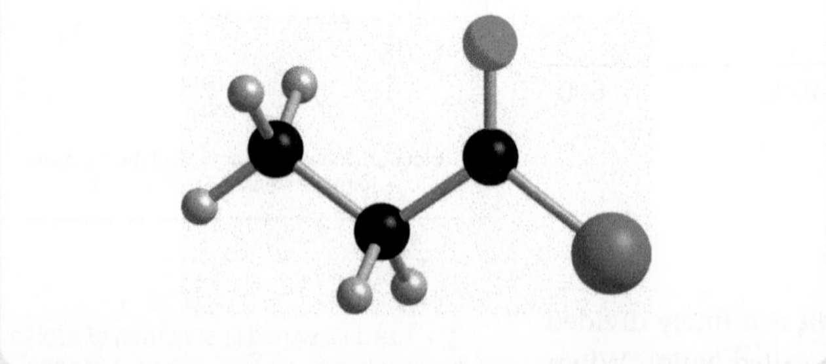

▲ Propanoyl chloride (the red sphere is oxygen and the green one is chlorine). In this computer generated ball and stick model a single stick represents the C=O double bond.

Strategy

(a) Use the characteristic group wavenumbers in Table 12.2 (p.573) to assign the major absorption bands between 4000 cm^{-1} and 1400 cm^{-1} (the functional group region).

(b) Propanoyl chloride reacts slowly with moisture in the air to form a product that is responsible for the additional absorption bands. Use the characteristic group wavenumbers in Table 12.2 to identify the types of bonds that are responsible for the additional absorption bands and, hence, propose a structure for the product.

(c) The stretching vibration of a C=O bond depends on the substituents joined to the carbon atom. A C=O bond with electron-withdrawing groups normally has an absorption band at a higher wavenumber than a C=O bond with electron-donating substituents. So you need to compare the electronic effects of Cl with OEt.

Solution

(a) 3000–2880 cm^{-1}, C–H (stretch); around 1800 cm^{-1}, C=O (stretch).

(b) The broad absorption band between 3000 cm^{-1} and 2500 cm^{-1} is due to an O–H stretching vibration of a carboxylic acid, while the narrow band around 1710 cm^{-1} is due to the C=O stretching vibration of a carboxylic acid. So, propanoyl chloride reacts with moisture in the air to form propanoic acid ($CH_3CH_2CO_2H$). (See Section 24.2, p.1111, for a reaction mechanism.)

(c) The Cl group in propanoyl chloride has a strong −I effect (see Box 24.2, p.1103) and so withdraws electron density from the C=O bond. →

→ For the OEt group in ethyl propanoate, the +M effect is stronger than the −I effect so it donates electron density to the C=O bond.

As a result, the stretching vibration of the C=O bond in ethyl propanoate has a lower wavenumber than the stretching vibration of the C=O bond in propanoyl chloride. (See Figure 12.7, p.574 for typical values.)

Question

Propanoyl chloride (CH_3CH_2COCl) reacts with ammonia to give organic product **A** and HCl. Assign the characteristic absorption bands in the IR spectrum of **A** shown here and use this information to help you propose a structure for **A**.

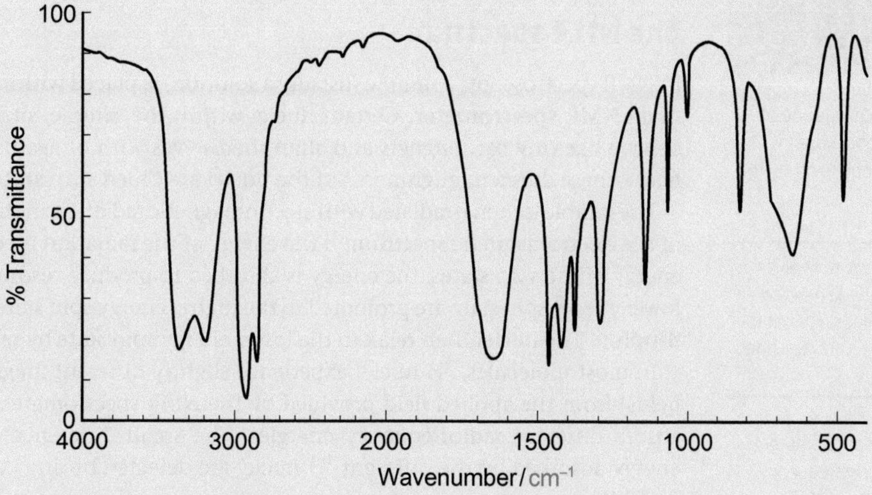
100
50
0
% Transmittance
4000
3000
2000
1500
1000
500
Wavenumber / cm⁻¹

▲ The IR spectrum of compound **A** (Nujol mull).

Summary

- General trends in the position of absorption bands in an IR spectrum.
  1. Absorption bands due to stretching vibrations have higher wavenumbers than the corresponding bending vibrations; for example, for a C–H bond, the stretching vibration has a higher wavenumber than the bending vibration.
  2. Bonds to hydrogen vibrate at a higher wavenumber than those involving heavier atoms; for example, the stretching vibration of a C–H bond has a higher wavenumber than those for C–C, C–O, or C–N bonds.
  3. A triple bond vibrates at a higher wavenumber than a corresponding double bond, which, in turn, vibrates at a higher wavenumber than a single bond (except for bonds to hydrogen).
- In an IR spectrum, the position, relative intensity, and shape of an absorption band are all helpful in structure identification.
- Characteristic group wavenumbers (in cm^{-1}) are due to stretching vibrations of bonds and they provide useful information on functional groups—they are normally found between 4000 cm^{-1} and 1400 cm^{-1}, which is called the functional group region. For example:

3600–2500 cm^{-1} (broad)	O–H	alcohols, carboxylic acids, amines, amides
3000–2850 cm^{-1}	C–H	
1820–1650 cm^{-1}	C=O	carbonyls
1680–1620 cm^{-1}	C=C	alkenes

- Carbonyl functional groups can be identified by the wavenumber of the C=O stretching vibration.
- The area between 1400 cm^{-1} and 600 cm^{-1} in the IR spectrum is called the fingerprint region. Analysis of this region is helpful to confirm structures.

? For practice questions on this topic, see questions 1 and 4–6 at the end of this chapter (pp.606–608).

The Swiss chemist Kurt Wüthrich was one of three chemists awarded the Nobel Prize in 2002 for using NMR spectroscopy to study the structure of proteins.

To use NMR spectroscopy to analyse structures successfully, you need to understand the physical principles on which the method is based. These principles are described in Section 10.7 (p.496).

The magnet in an NMR spectrometer is usually described by an operating frequency, rather than its magnetic field strength. This is discussed in Section 10.7 (p.499) and in Box 12.4 (p.580).

The energy of the radiofrequency radiation that is absorbed is the same as the energy that is emitted when the nuclei relax.

There are other nuclei, besides ^{1}H and ^{13}C, that have a magnetic moment and produce an NMR resonance signal. These include ^{19}F and ^{31}P nuclei. (See Section 10.7, p.498, and Worked example 10.8, p.506.)

Nuclei that experience different local magnetic fields give different resonance signals in an NMR spectrum.

When the magnetic field experienced by a ^{1}H nucleus is a little smaller than the applied magnetic field, the nucleus is said to be **shielded** (see Section 10.7, p.496).

Nuclei in electron-dense environments are the most shielded so they give signals on the right-hand side of the spectrum. This is called the upfield side—the terms upfield and downfield originated when the NMR spectrum was recorded using a constant radiofrequency and the magnetic field was varied.

Nuclei in electron-poor environments are deshielded (the least shielded) and so give signals on the left-hand side of the spectrum (the downfield side).

12.3 Nuclear magnetic resonance spectroscopy

Nuclear magnetic resonance (NMR) spectroscopy is an exceptionally useful analytical technique for determining the structure of a molecule. For example, ^{1}H NMR and ^{13}C NMR spectroscopy provide information about the carbon–hydrogen framework of organic molecules and can be used to determine the entire structure of a molecule.

The NMR spectrum

In NMR spectroscopy, a sample (usually a solution) is placed within a strong magnetic field in an **NMR spectrometer**. Certain nuclei within the sample, such as ^{1}H and ^{13}C nuclei, behave like tiny bar magnets and align themselves with or against the applied magnetic field—these different alignments of the nuclei are called **spin states**.

The sample is then irradiated with electromagnetic radiation in the radiofrequency region of the electromagnetic spectrum. If the energy of the radiation is equal to the difference in energy of the spin states, the energy is absorbed to produce **resonance**—the nuclei in the lower energy spin state are promoted to the higher energy spin state in a process called **spin flipping**. The nuclei then relax to the lower energy spin state by releasing energy.

In most molecules, ^{1}H nuclei experience slightly different magnetic fields (called local fields) from the applied field provided by the NMR spectrometer. So, different ^{1}H nuclei absorb different radiofrequency energies. The small differences in the radiofrequency energy absorbed by the different ^{1}H nuclei are detected by an NMR spectrometer to give an **NMR spectrum.**

Fourier transform NMR spectroscopy

In modern instruments, called **Fourier transform (FT) spectrometers**, the magnetic field is kept constant and a pulse of radiofrequency radiation is applied, which promotes *all* of the nuclei to the higher energy spin state. When the nuclei relax, the energy they release is detected over time and a computer converts the information into a **Fourier transform NMR (FT-NMR) spectrum**. An FT-NMR spectrum is produced in a few seconds using only a few milligrams of compound (see Box 12.3).

Chemical shift

The ^{1}H FT-NMR spectrum of ethyl ethanoate ($CH_3CO_2CH_2CH_3$) is shown in Figure 12.9. The spectrum is a plot of intensity of the resonance signal as the frequency (and energy) of the radiation increases from right to left across the spectrum.

The ^{1}H NMR spectrum of ethyl ethanoate shows **resonance signals**, sometimes called peaks, for the hydrogen atoms. (In this spectrum, the signals are sharp and appear as vertical lines, but this is not always the case.) There are three regions of resonance signals showing that the ^{1}H nuclei in the CH_2, CH_3, and CH_3 groups experience slightly different local magnetic fields (i.e. they are **shielded** to different extents from the applied magnetic field). The two hydrogens in the CH_2 group are the least shielded and sense a larger local magnetic field. So, they are brought into resonance at the higher radiofrequency energy, and the signal appears on the left-hand side of the spectrum—resonance signals at the left-hand side of the spectrum are described as being **deshielded**. Don't be worried at this stage by the different appearance of the signals for the CH_2, CH_3, and CH_3 groups. The reasons for the different patterns are explained on p.502.

The position of a resonance signal on the horizontal axis is called the **chemical shift** (δ), which has units of parts per million (ppm). A chemical shift is a measure of how far a signal is from a reference signal. For hydrogen atoms within molecules dissolved in organic solvents, the reference signal is usually due to tetramethylsilane (TMS, $Si(CH_3)_4$)—the single peak due to the 12 equivalent ^{1}H nuclei in TMS is assigned a chemical shift value of 0 ppm. A small amount of TMS is added to the solvent before the ^{1}H NMR spectrum of the solution

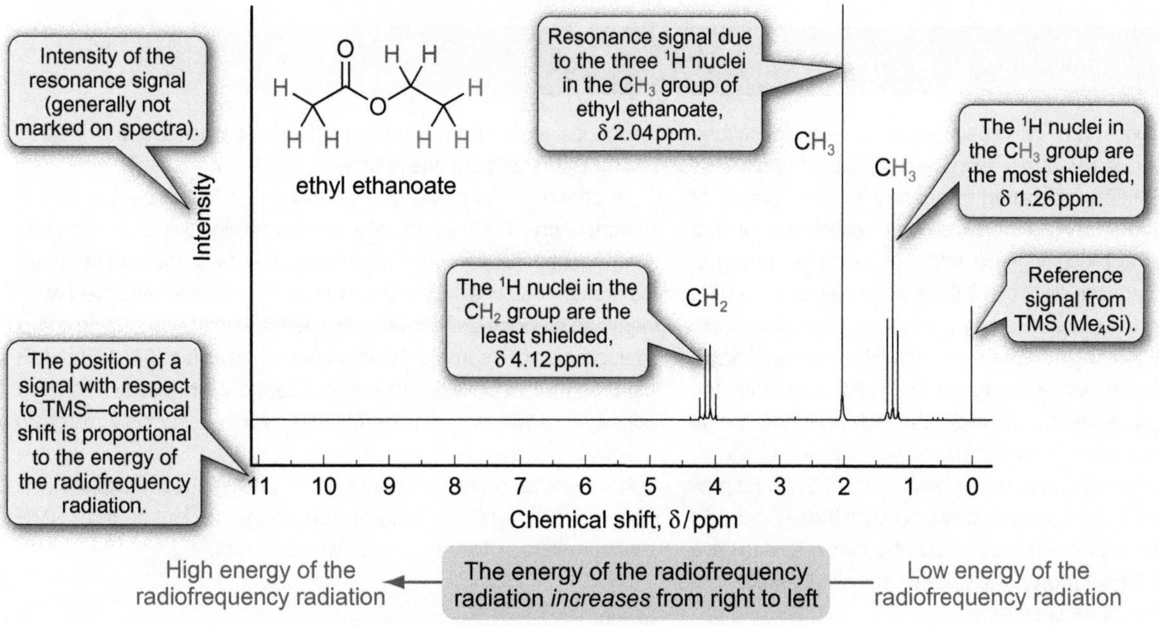
Intensity of the resonance signal (generally not marked on spectra).
Intensity
ethyl ethanoate
Resonance signal due to the three ¹H nuclei in the CH₃ group of ethyl ethanoate, δ 2.04 ppm.
CH₃
CH₃
The ¹H nuclei in the CH₃ group are the most shielded, δ 1.26 ppm.
Reference signal from TMS (Me₄Si).
The ¹H nuclei in the CH₂ group are the least shielded, δ 4.12 ppm.
CH₂
The position of a signal with respect to TMS—chemical shift is proportional to the energy of the radiofrequency radiation.
11 10 9 8 7 6 5 4 3 2 1 0
Chemical shift, δ/ppm
High energy of the radiofrequency radiation
The energy of the radiofrequency radiation increases from right to left
Low energy of the radiofrequency radiation

Figure 12.9 ^{1}H NMR spectrum of ethyl ethanoate using TMS as a reference compound.

Shielding of nuclei and chemical shifts are introduced in Section 10.7 (p.500). Chemical shift is defined in Equation 10.24 (p.502).

is recorded (see Box 12.3, below). Signals at the left-hand side of the ^{1}H NMR spectrum have high chemical shift values, and those at the right-hand side have low chemical shift values. The ^{1}H nuclei in most molecules are less shielded than the ^{1}H nuclei in TMS and so give resonance signals to the left of the TMS signal, typically between 0 ppm and 10 ppm.

To establish whether hydrogen atoms are equivalent you need to compare the electronic environment of each hydrogen atom (Section 12.3, p.583)

Box 12.3 Running an NMR spectrum

NMR spectra can be obtained from just a few milligrams of a solid or liquid, which is usually dissolved in a solvent such as trichlorodeuteriomethane ($CDCl_3$), commonly called 'deuterated chloroform'. One reason for choosing $CDCl_3$ is that it does not contain a hydrogen atom and so does not give a resonance signal in a ^{1}H NMR spectrum. However, the ^{1}H NMR spectrum of a solution of a sample in $CDCl_3$ normally shows a very small peak at δ 7.26 ppm because $CDCl_3$ is never 100% deuterated and a tiny amount of residual trichloromethane (chloroform; $CHCl_3$) is present. Sometimes, the peak at 7.26 ppm is used as a reference signal, in place of the signal from TMS. This explains why some ^{1}H NMR spectra (such as Figure 12.16, p.591) do not have a peak for TMS.

If required, a small amount of tetramethylsilane (TMS) is added to the solution, which is then transferred into a thin-walled cylindrical glass tube called an NMR tube. The NMR tube is inserted into the centre of the magnetic field of the spectrometer and it may be rapidly spun along its long axis using a jet of air—spinning the tube ensures that all the molecules in the sample experience the same magnetic field. To obtain an FT-NMR spectrum, the magnetic field is kept constant while pulses of radiofrequency radiation are broadcast into the sample. Emission of radiofrequency energy is detected and a computer processes the data to produce the spectrum.

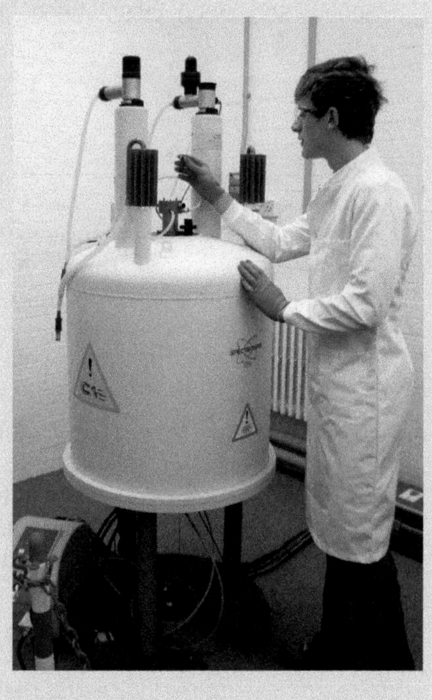

▲ A researcher holding an NMR tube in front of an NMR spectrometer.

Box 12.4 Magnetic field strength and resolution

The magnet in an NMR spectrometer is usually described by an operating frequency, rather than its magnetic strength (see Section 10.7, p.500). The **operating frequency** is the frequency of the radiation required to bring a ^{1}H nucleus into resonance, when it experiences the full applied magnetic field. For example, a magnet that generates a magnetic field of 18.8 tesla (T) has an operating frequency of 800 MHz.

The size of the operating frequency of an NMR instrument affects the separation between resonance signals in an NMR spectrum: the higher the operating frequency, the larger the radiofrequency range and the greater the spacing between resonance signals. As shown below, a spectrometer with an operating frequency of 500 MHz produces spectra with the signals more spread out than those produced by a 300 MHz spectrometer—the 500 MHz instrument is said to have a higher resolution. **Resolution** is the smallest signal separation that can be detected in an NMR spectrum.

As you might expect, high-resolution NMR spectra are easier to interpret than spectra from low-resolution instruments because the resonance signals from different nuclei are sharper and further apart (more dispersed) from one another.

An added bonus of using a high-resolution NMR spectrometer is its **sensitivity**. The intensity of a resonance signal is proportional to the magnetic field strength—the higher the magnetic field strength, the greater the sensitivity of the instrument. A highly sensitive instrument can produce a spectrum with intense signals when using only a small amount of a sample. For example, when using a 500 MHz NMR instrument, it is possible to obtain spectra, with clearly observed signals, in a few seconds or minutes, when using only milligram quantities of a sample.

The benefits of increased resolution and sensitivity explain why research chemists are regularly upgrading to higher field NMR spectrometers as these come on to the market.

Question

For a 900 MHz NMR spectrometer, what is the radiofrequency separation between signals at chemical shifts δ 3 ppm and δ 4 ppm?

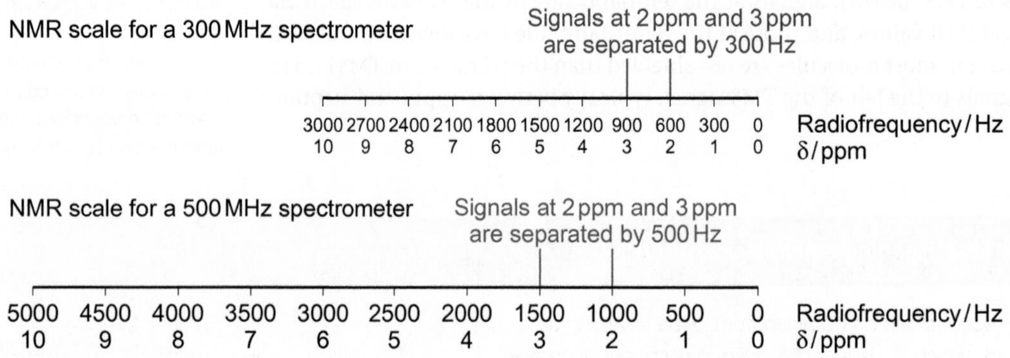
NMR scale for a 300 MHz spectrometer
Signals at 2 ppm and 3 ppm are separated by 300 Hz
3000 2700 2400 2100 1800 1500 1200 900 600 300 0
10 9 8 7 6 5 4 3 2 1 0
Radiofrequency/Hz
δ/ppm
NMR scale for a 500 MHz spectrometer
Signals at 2 ppm and 3 ppm are separated by 500 Hz
5000 4500 4000 3500 3000 2500 2000 1500 1000 500 0
10 9 8 7 6 5 4 3 2 1 0
Radiofrequency/Hz
δ/ppm

In a **proton decoupled** ^{13}C NMR spectrum, a singlet peak is observed for each different type of carbon atom (see p.581).

Notice that the ^{13}C NMR spectrum of ethyl ethanoate looks simpler than the ^{1}H NMR spectrum—in the proton decoupled spectrum each carbon atom gives a single sharp signal, whereas the resonance signals for the hydrogen atoms have clusters of peaks of different patterns (the reasons for the different patterns are discussed in Section 10.7, p.501).

The ^{13}C nuclei in most molecules are less shielded than those in TMS and so give resonances to the left of this signal, typically between 0 ppm and 200 ppm. Figure 12.10 shows the ^{13}C NMR (proton decoupled) spectrum of ethyl ethanoate ($CH_3CO_2CH_2CH_3$). Four resonance signals in the spectrum show that the ^{13}C nuclei in the CO_2, CH_2, CH_3, and CH_3 groups experience slightly different local magnetic fields—the CO_2 carbon is the least shielded (it has the *highest* chemical shift) and the CH_3 carbon is the most shielded (it has the *lowest* chemical shift).

The signals due to ^{1}H and ^{13}C nuclei cannot be observed at the same time so separate spectra are needed. This is because the spin states of these different nuclei have dissimilar energy differences so they absorb radiofrequency radiation of different energy. Both ^{1}H NMR spectra and ^{13}C NMR spectra are used in structure identification. The usual procedure is to record the ^{1}H NMR spectrum first.

^{1}H NMR spectroscopy

In a ^{1}H NMR spectrum, the number, the position, the intensity, and the appearance of resonance signals are all important pieces of information that can be used to assign a structure.

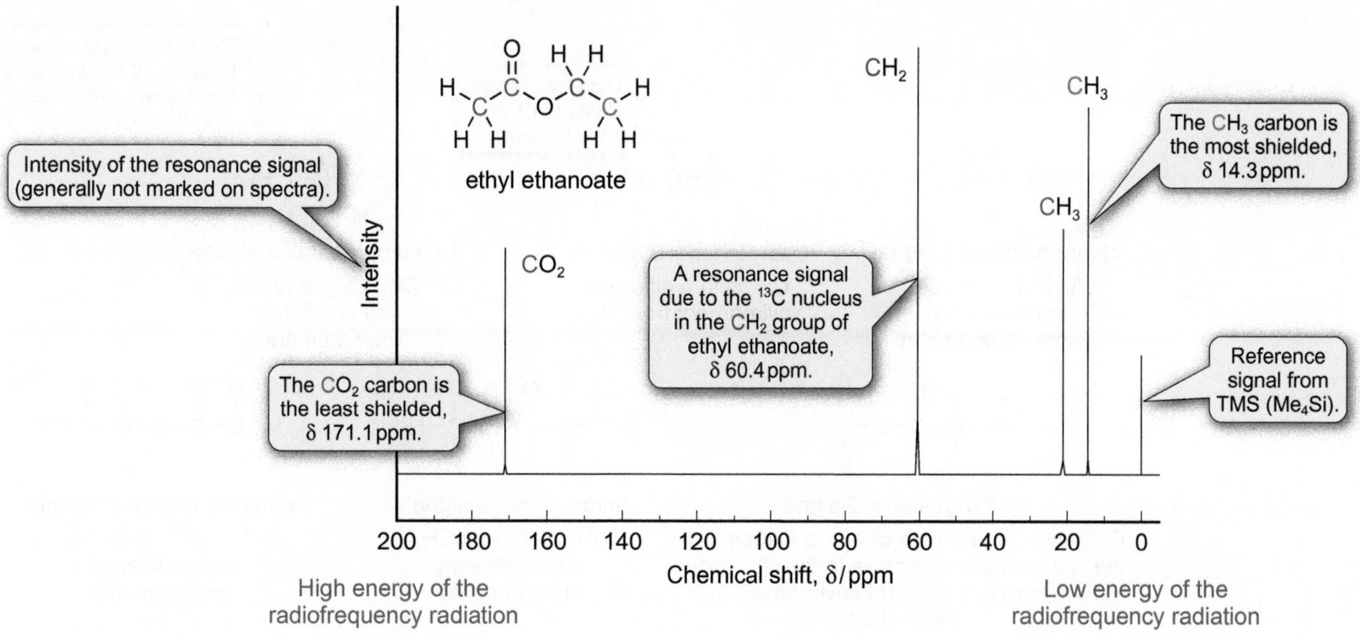
ethyl ethanoate
Intensity of the resonance signal (generally not marked on spectra).
Intensity
CO2
The CO2 carbon is the least shielded, δ 171.1 ppm.
A resonance signal due to the 13C nucleus in the CH2 group of ethyl ethanoate, δ 60.4 ppm.
CH2
CH3
CH3
The CH3 carbon is the most shielded, δ 14.3 ppm.
Reference signal from TMS (Me4Si).
200 180 160 140 120 100 80 60 40 20 0
Chemical shift, δ/ppm
High energy of the radiofrequency radiation
Low energy of the radiofrequency radiation

Figure 12.10 ^{13}C NMR (proton decoupled) spectrum of ethyl ethanoate using TMS as a reference compound.

Number of resonance signals

The number of resonance signals in a ^1H NMR spectrum tells you how many *different* types of hydrogen atoms there are in the molecule.

- Hydrogen atoms in the *same* electronic environment in a molecule produce one resonance signal (it may have a complex pattern) and these hydrogens are called **chemically equivalent hydrogens**.
- Hydrogen atoms in *different* electronic environments in a molecule produce different resonance signals.

To establish whether the hydrogen atoms in a molecule are chemically equivalent, you should draw the structure of the molecule and compare the electronic environment of each hydrogen atom (i.e. identify the groups that each hydrogen atom is next to). For example, in Figure 12.11, the electronic environments of hydrogen atoms depend on their distance from electron-withdrawing chlorine atoms or oxygen atoms.

To identify chemically equivalent hydrogens it also helps to recognize any symmetry in the molecule. If hydrogen atoms in a molecule are in identical positions, relative to a plane of symmetry, then they are chemically equivalent. Some examples are shown in Figure 12.12.

Each group of chemically equivalent hydrogen atoms gives rise to a separate signal in a ^1H NMR spectrum, though the signal may have a complex pattern.

The electronic (I and M) effects of substituents are discussed in Section 19.1 (p.871).

Symmetry in organic molecules is discussed in Section 18.4 (p.838).

Position of resonance signals

The position of a resonance signal in an NMR spectrum is recorded as a chemical shift value (δ in ppm). The chemical shift depends on the electronic environment of the hydrogen atoms that give rise to the signal. Hydrogen atoms in an electron-rich environment are more shielded than those in an electron-poor environment and so have lower chemical shifts.

Chemical shift is introduced in Section 10.7 (p.500).

- Resonance signals with *low* chemical shifts (the right-hand side of the spectrum) are from hydrogen atoms that are close to electron-*donating* groups.
- Resonance signals with *high* chemical shifts (the left-hand side of the spectrum) are from hydrogen atom that are close to electron-*withdrawing* groups.

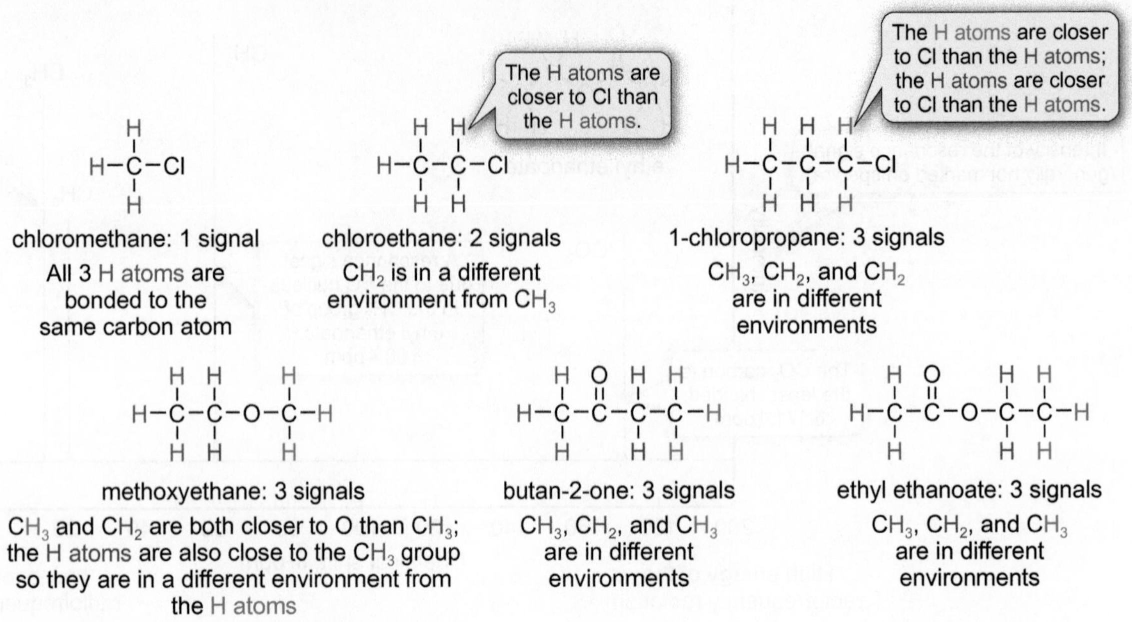
The H atoms are closer to Cl than the H atoms.
The H atoms are closer to Cl than the H atoms; the H atoms are closer to Cl than the H atoms.
chloromethane: 1 signal
All 3 H atoms are bonded to the same carbon atom
chloroethane: 2 signals
CH_2 is in a different environment from CH_3
1-chloropropane: 3 signals
CH_3, CH_2, and CH_2 are in different environments
methoxyethane: 3 signals
CH_3 and CH_2 are both closer to O than CH_3; the H atoms are also close to the CH_3 group so they are in a different environment from the H atoms
butan-2-one: 3 signals
CH_3, CH_2, and CH_3 are in different environments
ethyl ethanoate: 3 signals
CH_3, CH_2, and CH_3 are in different environments

Figure 12.11 Chemically equivalent hydrogens in various organic molecules.

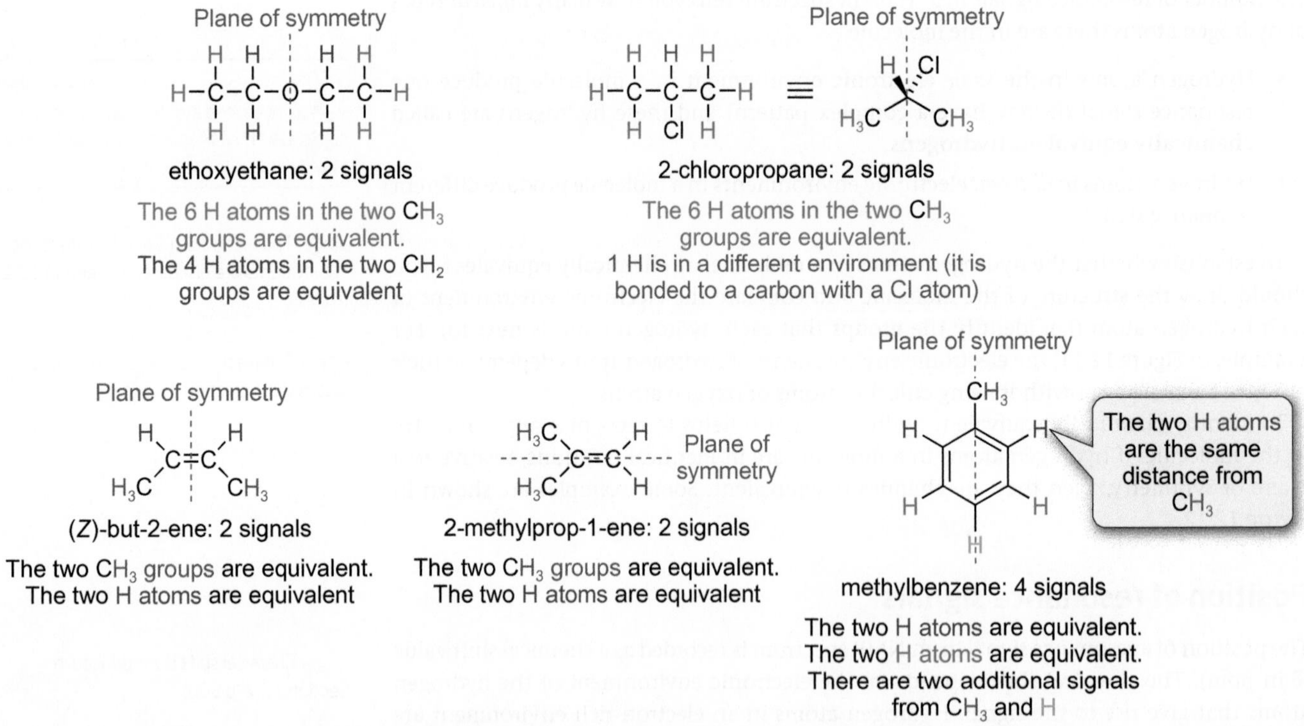
Plane of symmetry
ethoxyethane: 2 signals
The 6 H atoms in the two CH_3 groups are equivalent.
The 4 H atoms in the two CH_2 groups are equivalent
Plane of symmetry
2-chloropropane: 2 signals
The 6 H atoms in the two CH_3 groups are equivalent.
1 H is in a different environment (it is bonded to a carbon with a Cl atom)
Plane of symmetry
(Z)-but-2-ene: 2 signals
The two CH_3 groups are equivalent.
The two H atoms are equivalent
Plane of symmetry
2-methylprop-1-ene: 2 signals
The two CH_3 groups are equivalent.
The two H atoms are equivalent
Plane of symmetry
The two H atoms are the same distance from CH_3
methylbenzene: 4 signals
The two H atoms are equivalent.
The two H atoms are equivalent.
There are two additional signals from CH_3 and H

Figure 12.12 Hydrogen atoms in identical positions relative to a plane of symmetry are chemically equivalent.

The effect of an electron-withdrawing bromine atom is illustrated below by the order of the chemical shifts for the hydrogen atoms in 1-bromopropane ($BrCH_2CH_2CH_3$).

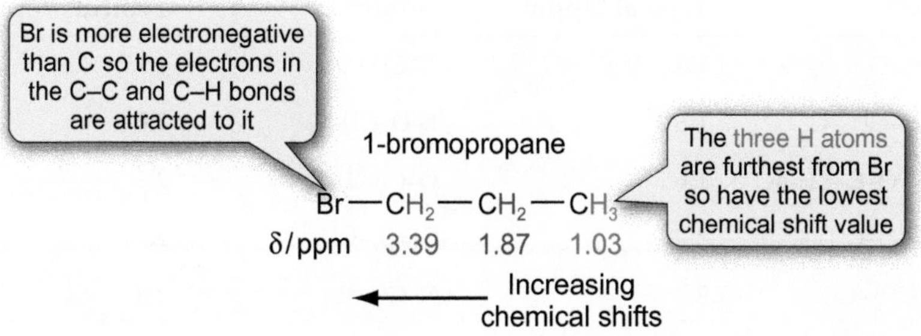
Br is more electronegative than C so the electrons in the C–C and C–H bonds are attracted to it
1-bromopropane
The three H atoms are furthest from Br so have the lowest chemical shift value
Br—CH2—CH2—CH3
δ/ppm 3.39 1.87 1.03
Increasing chemical shifts

As you might expect, hydrogen atoms that are close to a highly electron-withdrawing and electronegative atom (or group), such as fluorine, have a higher chemical shift than those that are close to a weaker electron-withdrawing group, such as chlorine (see margin).

Chemical shift values also depend on whether the hydrogen atom is part of a CH_3 (methyl) group, a CH_2 (methylene) group, or a CH (methine) group. Normally, a CH_3 group gives a signal at a lower chemical shift value than a CH_2 group, and a CH_2 group usually gives a signal at a lower chemical shift than a CH group, when they are in a similar environment. The case when these groups are all bonded to a Br atom is illustrated here.

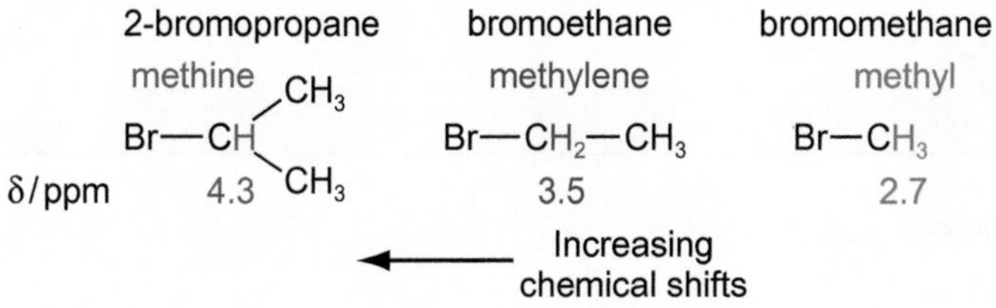
2-bromopropane
bromoethane
bromomethane
methine
methylene
methyl
Br—CH(CH3)CH3
Br—CH2—CH3
Br—CH3
δ/ppm 4.3 3.5 2.7
Increasing chemical shifts

This trend, together with the influence of electron-withdrawing groups, can be used to explain the order of chemical shifts in a variety of organic molecules. For example, the chemical shifts observed for the hydrogen atoms in butanone ($CH_3COCH_2CH_3$) and ethyl ethanoate ($CH_3CO_2CH_2CH_3$) are explained in the following illustration. (The ^1H NMR spectrum of ethyl ethanoate is shown in Figure 12.9, p.579.)

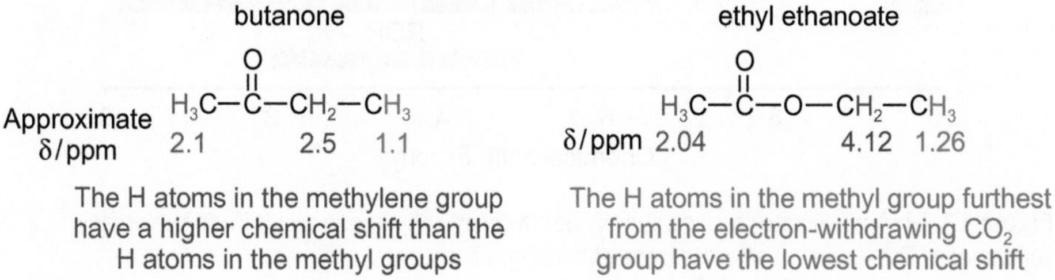
butanone
ethyl ethanoate
H3C—C(=O)—CH2—CH3
H3C—C(=O)—O—CH2—CH3
Approximate δ/ppm 2.1 2.5 1.1
δ/ppm 2.04 4.12 1.26
The H atoms in the methylene group have a higher chemical shift than the H atoms in the methyl groups
The H atoms in the methyl group furthest from the electron-withdrawing CO2 group have the lowest chemical shift

Typical chemical shift values for a variety of methyl, methylene, and methine hydrogens are shown in Table 12.3.

Approximate chemical shift ranges for hydrogen atoms attached to a variety of substituents are also shown in Figure 12.13. The signals for each type of hydrogen appear over a range of chemical shifts because they are affected by the other substituents in a molecule. For example, the chemical shift for a C=C–H hydrogen lies between 4.5 ppm and 6.5 ppm because it is affected by the electronic properties of the other three substituents joined to the C=C bond.

The ^1H NMR spectra for bromoethane (ethyl bromide) and 2-bromopropane (isopropyl bromide) both show *two* resonance signals as the two methyl groups in 2-bromopropane are equivalent. For 2-bromo-2-methylpropane (*tert*-butyl bromide), $BrC(CH_3)_3$, there is only one resonance signal as all three methyl groups are equivalent.

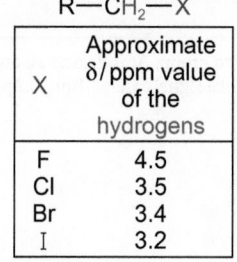

Ethyl ethanoate (ethyl acetate) is a common organic solvent with industrial applications including decaffeinating tea and coffee (Chapter 17 page opener, p.764).

As the electronegativity of the halogen atom decreases, the electron density around the hydrogens increases and the chemical shift decreases.

$R—CH_2—X$

X	Approximate δ/ppm value of the hydrogens
F	4.5
Cl	3.5
Br	3.4
I	3.2

Decreasing electronegativity and decreasing chemical shifts of adjacent CH_2 hydrogen atoms

Computer software is available for predicting chemical shift values. Chemical shift values also depend on the solvent but, typically, for ^1H resonance signals, the predicted values are accurate to within ±0.2 ppm of the actual values.

CH_3 methyl
CH_2 methylene
CH methine

When a hydrogen atom is close to two or more functional groups, the electronic effects of the functional groups are cumulative.

$H_3C—CH_2—OCH_3$
3.4 ppm

$H_3CO—CH_2—OCH_3$
4.6 ppm

Table 12.3 Typical chemical shift values for methyl (CH_3), methylene (CH_2), and methine (CH) hydrogens

Methyl hydrogens		Methylene hydrogens		Methine hydrogens	
Group*	**Typical δ/ppm†**	**Group***	**Typical δ/ppm†**	**Group***	**Typical δ/ppm†**
C–C–CH_3	0.9	C–CH_2–C	1.4	C–CH–C	1.5
R–C(=O)–C–CH_3	1.1	R–O–C–CH_2–C	1.5	R–O–CH–C	2.0
R–O–C–CH_3	1.3	C(=O)–O–C–CH_2–C	1.6	O–C(=O)–CH–C	2.5
C=C–CH_3	1.6	O–C(=O)–CH_2–C	2.2	C(=O)–CH–C	2.7
Br–C–CH_3	1.8	C=C–CH_2–C	2.3	N–CH–C	2.8
O–C(=O)–CH_3	2.0	C(=O)–CH_2–C	2.4	Ar–CH–C	3.0
C–C(=O)–CH_3	2.2	N–CH_2–C	2.5	Ar–C(=O)–CH–C	3.4
Ar–CH_3	2.3	Ar–CH_2–C	2.7	R–O–CH–C	3.7
N–CH_3	2.3	Ar–C(=O)–CH_2–C	2.8	HO–CH–C	3.7
Ar–C(=O)–CH_3	2.5	R–O–CH_2–C	3.4	Br–CH–C	4.3
C(=O)N–CH_3	2.9	Br–CH_2–C	3.4	ArO–CH–C	4.5
R–O–CH_3	3.3	Cl–CH_2–C	3.5	C(=O)O–CH–C	4.8
C(=O)O–CH_3	3.7	HO–CH_2–C	3.6	Ar–C(=O)O–CH–C	5.1
Ar–O–CH_3	3.8	Ar–O–CH_2–C	4.0		
		C(=O)–O–CH_2–C	4.1		

* R indicates an alkyl group; Ar indicates an aromatic group.
† Typically, resonance signals are within ±0.2 ppm of the values quoted.

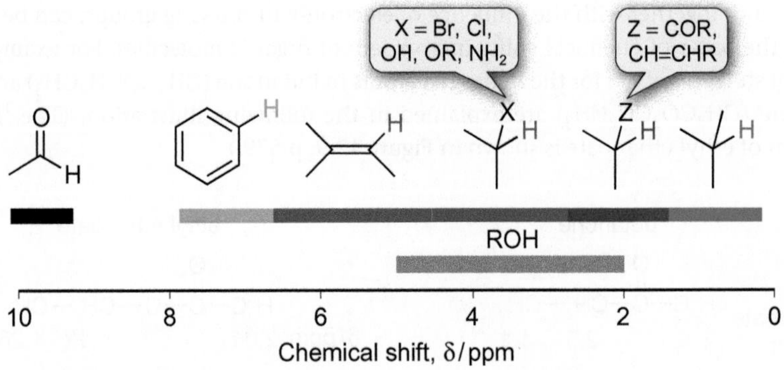
X = Br, Cl, OH, OR, NH$_2$
Z = COR, CH=CHR
ROH
10
8
6
4
2
0
Chemical shift, δ/ppm

Figure 12.13 Approximate chemical shift values (δ / ppm) of hydrogen atoms in various organic compounds. (The values are affected by neighbouring substituents.)

Approximate chemical shifts in ppm (in italic), for the hydrogen atoms in (*E*)-4-(but-1-enyl)benzaldehyde.

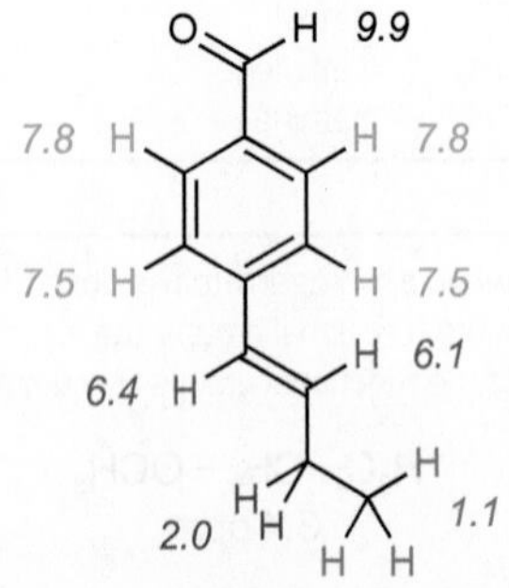
9.9
7.8
7.8
7.5
7.5
6.1
6.4
2.0
1.1

In the margin, notice that the four hydrogen atoms on the benzene ring give two resonance signals—this is because the two substituents either end of the benzene ring are different (two of the hydrogens on the ring are closest to the aldehyde group whereas the remaining two are closest to the alkene). Similarly, the two alkene hydrogen atoms give different resonance signals because the substituents either side of the C=C bond are different.

As predicted from Table 12.3, hydrogen atoms close to electron-withdrawing groups, such as Br, Cl, and OH, give signals at relatively high chemical shifts, up to around 4.3 ppm.

However, you may be surprised by some of the other values. For example, why does a hydrogen atom attached to an electron-rich C=C bond, or one that is part of a benzene ring or an aldehyde group, give a signal at such a high chemical shift? Also, why does the signal for the hydrogen atom in an OH group cover such a wide chemical shift range? These points are addressed in the following subsections.

Benzene, substituted benzenes, alkenes, and aldehydes

The unexpectedly high chemical shift of a hydrogen atom attached to a benzene ring is explained by the circulation of the 6 π electrons within the ring. In benzene, the 6 π electrons are delocalized over the ring and, when placed in a magnetic field, the π electrons circulate to form a **ring current**. The ring current, in turn, produces a small local magnetic field, which acts to increase the magnitude of the applied magnetic field experienced by the six hydrogens that lie outside the ring. As the hydrogens experience a larger magnetic field, radiofrequency of relatively high energy is required to achieve resonance and a single resonance signal appears downfield at δ 7.4 ppm (check you understand why all six hydrogens are equivalent).

Delocalization of the 6 π electrons in benzene is discussed in Section 22.1 (p.1004). The structures of alkenes (such as $R_2C{=}CR_2$) and aldehydes (RCHO) are discussed in Sections 21.1 (p.962) and 23.1 (p.1056), respectively.

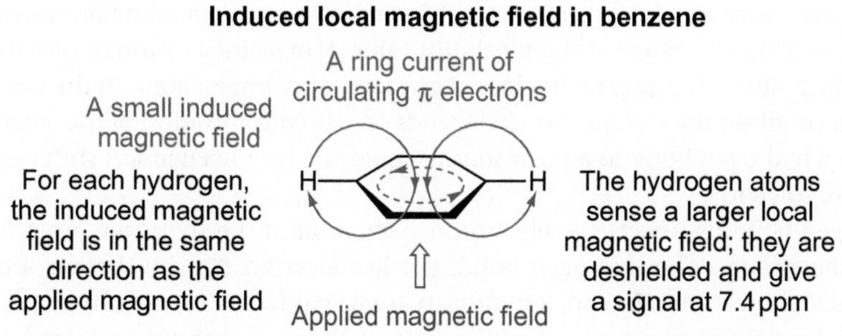
Induced local magnetic field in benzene
A ring current of circulating π electrons
A small induced magnetic field
For each hydrogen, the induced magnetic field is in the same direction as the applied magnetic field
H
H
The hydrogen atoms sense a larger local magnetic field; they are deshielded and give a signal at 7.4 ppm
Applied magnetic field

Induced local magnetic fields are also responsible for the high chemical shifts of a hydrogen atom(s) attached to a C=C or a C=O double bond.

Resonance signals between 6.5 ppm and 8.0 ppm are characteristic of compounds containing a benzene ring. The electronic effects of substituents attached to a benzene ring are discussed in Section 22.3 (p.1029).

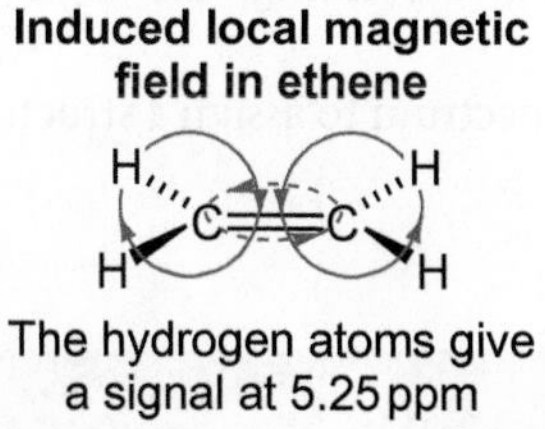
Induced local magnetic field in ethene
H
H
C
C
H
H
The hydrogen atoms give a signal at 5.25 ppm

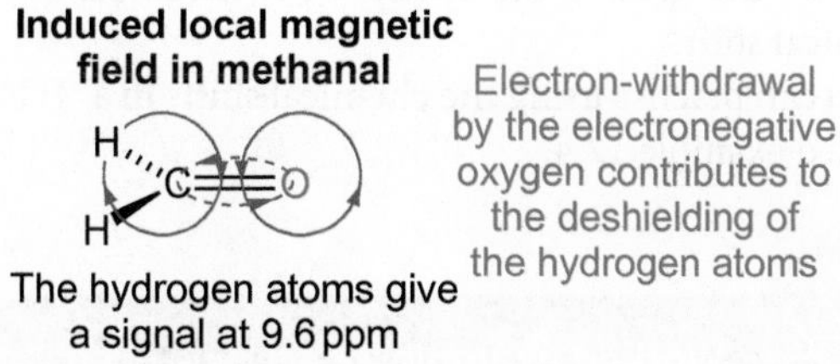
Induced local magnetic field in methanal
H
C
O
H
Electron-withdrawal by the electronegative oxygen contributes to the deshielding of the hydrogen atoms
The hydrogen atoms give a signal at 9.6 ppm

The signals for the ring hydrogens in substituted benzenes like chlorobenzene (C_6H_5Cl) often overlap and appear as a complex broad signal.

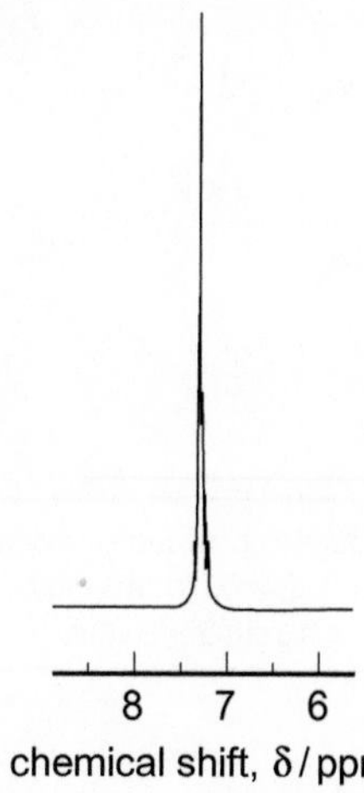
8 7 6
chemical shift, δ / ppm

For an alkene, the induced magnetic field is less than that for benzene, because only two electrons circulate as opposed to six electrons in benzene. So, a hydrogen atom attached to a C=C bond usually gives a signal with a slightly lower chemical shift than that for benzene. For an aldehyde, in addition to the local induced magnetic field caused by the 2 π electrons in the C=O bond, the O=C–H hydrogen is further deshielded by the electron-withdrawing effect of the electronegative oxygen atom. So, the hydrogen atom attached to the C=O bond gives a signal between 9 ppm and 10 ppm.

Some examples of substituted benzenes and alkenes, together with approximate chemical shift values, are shown below. As you may expect, electron-withdrawing Cl and NO_2 groups attached to a benzene ring or a C=C bond lead to deshielding of nearby hydrogen atoms. You will also see that the chemical shift for a hydrogen atom attached directly to either a benzene ring or a C=C bond is much higher than that for a methyl hydrogen (CH_3) attached to a benzene ring or a C=C bond.

Assigning a *Z* configuration to alkenes is discussed in Section 18.3 (p.833).

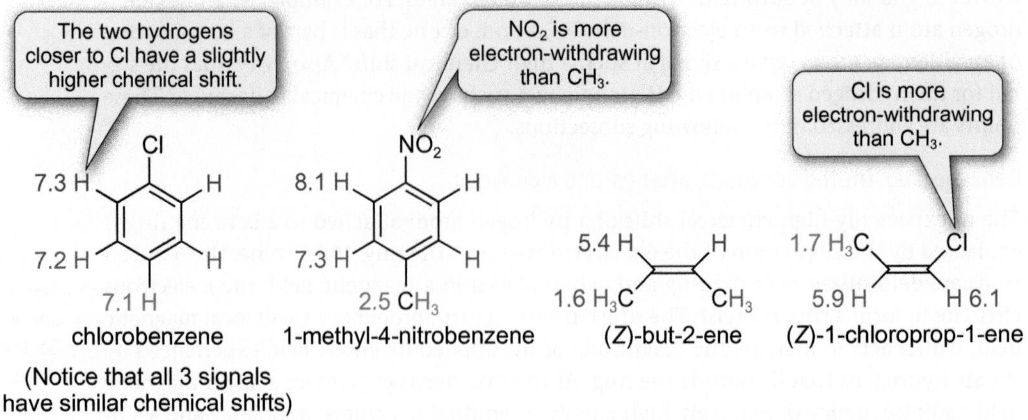
The two hydrogens closer to Cl have a slightly higher chemical shift.
NO2 is more electron-withdrawing than CH3.
Cl is more electron-withdrawing than CH3.
Cl
7.3 H
H
7.2 H
H
7.1 H
NO2
8.1 H
H
7.3 H
H
2.5 CH3
5.4 H
H
1.6 H3C
CH3
1.7 H3C
Cl
5.9 H
H 6.1
chlorobenzene
1-methyl-4-nitrobenzene
(Z)-but-2-ene
(Z)-1-chloroprop-1-ene
(Notice that all 3 signals have similar chemical shifts)

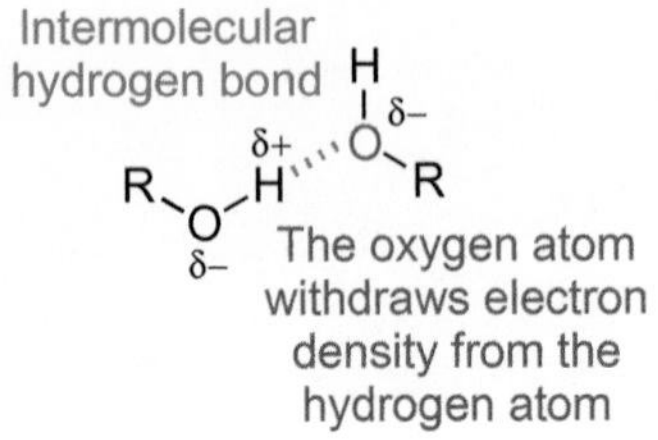
Intermolecular hydrogen bond
δ+
δ−
δ−
R
O
H
R
The oxygen atom withdraws electron density from the hydrogen atom

Hydrogens bonded to oxygen

For an alcohol (ROH), the chemical shift of a resonance signal for the hydrogen atom in the OH group can vary widely, typically from 2.5 ppm to 4 ppm, depending on concentration and solvent. This wide range of chemical shift values is explained by intermolecular hydrogen bonding—the OH group in an alcohol can form a hydrogen bond to the OH group of another alcohol (so the chemical shift depends on the concentration of the alcohol), or it can form a hydrogen bond to a polar solvent molecule (so the chemical shift depends on the type of solvent).

Hydrogen bonding affects the electron density around the hydrogen atom in an OH group—the stronger the hydrogen bond, the less electron density there is around the hydrogen atom (see margin). So, a hydrogen atom that forms a strong hydrogen bond is deshielded and it gives a signal at a relatively high chemical shift (up to 4 ppm).

The extent of intermolecular hydrogen bonding between OH groups is decreased by addition of a non-polar solvent so, under these conditions, an OH group will give a signal at a relatively low chemical shift (down to 2.0 ppm).

For a carboxylic acid (RCO_2H), the hydrogen atom in the CO_2H group gives a resonance signal typically between 10.0 ppm and 13.0 ppm, depending on concentration and solvent—the greater the extent of intermolecular hydrogen bonding, the higher the chemical shift.

You can practise using the chemical shifts in a ^1H NMR spectrum to assign a structure in Worked example 12.4.

> For the ^1H NMR spectrum of ethanol (CH_3CH_2OH) shown in Figure 12.16 (p.591), the signal due to the OH group appears at around 2.5 ppm.

> The influence of hydrogen bonding on the IR spectrum of an alcohol is discussed in Section 12.2 (p.572).

> Structural (constitutional) isomers are compounds in which the atoms are joined together in a different order.

Worked example 12.4 Using chemical shifts to assign structures

An unknown compound with the molecular formula C_4H_9Br gave a ^1H NMR spectrum with resonance signals at δ 4.07, 1.80, 1.68, and 1.00 ppm. Propose a structure for the compound and explain your reasoning.

Strategy

Draw all possible structural isomers that have the molecular formula C_4H_9Br (structural isomers are discussed in Section 18.1, p.816).

For each isomer, decide how many of the hydrogen atoms are chemically non-equivalent, that is, how many sets of hydrogen atoms there are in different electronic environments. (Hydrogens in different electronic environments give separate resonance signals in a ^1H NMR spectrum.)

Use Table 12.3 (p.584) to predict the chemical shift values for the CH_3, CH_2, and CH groups in the isomers. Compare the predicted values with those for the unknown compound. →

➜ Solution

There are four structural isomers, **A–D**, with the molecular formula C_4H_9Br.

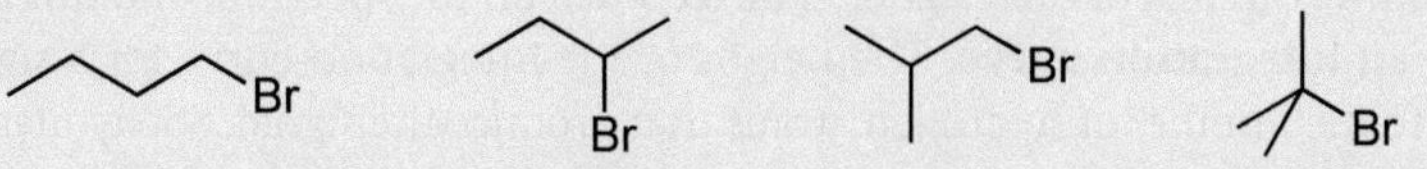

A **B** **C** **D**

In **A** and **B**, there are four sets of hydrogen atoms in different electronic environments. In **C** there are three sets, whereas in **D** all nine hydrogens are chemically equivalent. So, only **A** and **B** give four resonance signals in a ^{1}H NMR spectrum.

Predict 4 signals

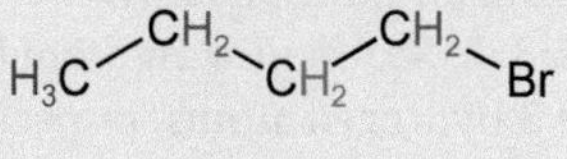

A

Predict 4 signals

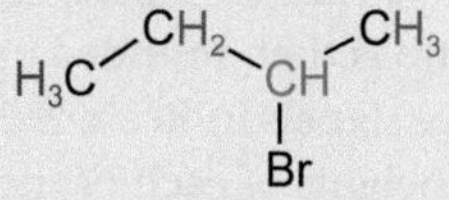

B

Predict 3 signals

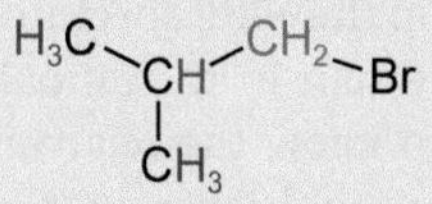

C

Predict 1 signal

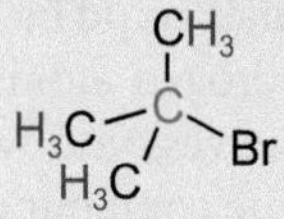

D

From Table 12.3, the predicted chemical shifts (in ppm) for the hydrogen atoms in **A** and **B** are those shown here.

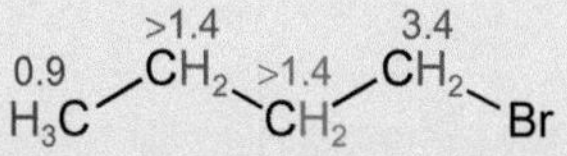

A

The signals for CH_2 and, particularly, CH_2 are expected to be higher than 1.4 ppm, because of the electron-withdrawing Br atom

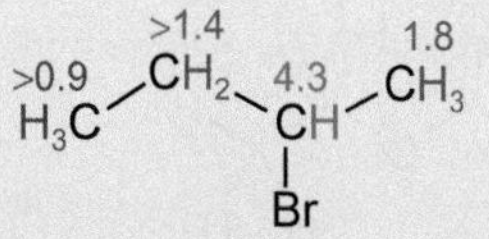

B

The signals for CH_3 and, particularly, CH_2 are expected to be higher than 0.9 ppm and 1.4 ppm, respectively, because of the electron-withdrawing Br atom

The predicted chemical shift values for **B** are a closer match with those for the unknown compound. For example, for **A**, the resonance signal with the highest chemical shift is expected at 3.4 ppm, whereas for **B** it is at 4.3 ppm, which is close to the value of 4.07 ppm quoted for the unknown compound. So, the unknown compound is **B** (2-bromobutane).

Question

An aldehyde or ketone, with the molecular formula C_4H_8O, gave a ^{1}H NMR spectrum with resonance signals at δ 9.57, 2.50, and 1.06 ppm. Propose a structure for the compound and explain your reasoning.

ℹ Organic compounds are commonly represented by skeletal structures, where hydrogen atoms bonded to carbon are not shown (Section 2.2, p.74).

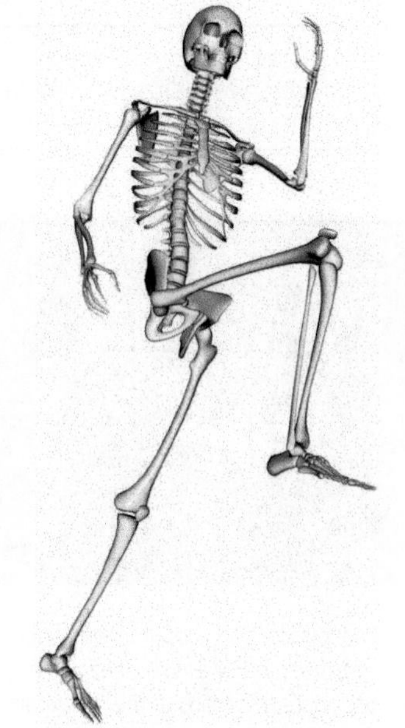

ℹ Compound **A** is called 1-bromobutane (butyl bromide), **B** is 2-bromobutane (*sec*-butyl bromide), **C** is 1-bromo-2-methylpropane (isobutyl bromide), and **D** is 2-bromo-2-methylpropane (*tert*-butyl bromide). Compounds **A** and **C** are primary bromoalkanes, **B** is a secondary (hence *sec*) bromoalkane, and **D** is a tertiary (hence *tert*) bromoalkane.

ℹ Compound **B** is a chiral compound—one of the (tetrahedral) carbon atoms is bonded to four different substituents (Section 18.4, p.838). The two enantiomers are shown below and the chiral centres are indicated by *.

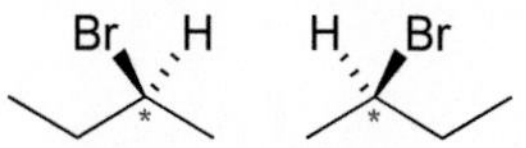

Intensity of resonance signals

Using integration to find the area under a curve is discussed in Maths Toolkit MT7 (p.1327).

The relative numbers of hydrogen atoms that give rise to each signal are determined from the integration curves.

BIODIESEL

^1H NMR spectroscopy is a useful method for analysing a mixture. For a mixture of two or more compounds, the areas of the ^1H signals for each compound are proportional to their concentration. For example, the ^1H NMR spectrum of a biodiesel blend can determine the concentrations of methyl esters in the blend, to ensure that it conforms to regulations of fuel quality.

In an NMR spectrum, the intensity of a resonance signal is determined from the total area of the signal, not from its height. Essentially all NMR instruments are equipped with a device that allows the area of each signal to be recorded on the spectrum and this is normally plotted as an **integration curve**. The height of the integration curve for a signal is proportional to the number of hydrogen atoms that produce the signal. So, by measuring the heights of the integration curves in a spectrum, you can work out the ratio of the hydrogen atoms giving rise to each signal. Modern instruments give a numerical value (in arbitrary units) for each integral curve on the spectrum. Alternatively, you can simply measure the height of the integration curve for each signal using a ruler (in convenient units, such as millimetres or centimetres).

Be aware that the measured heights of the integration curves are only approximate values because of experimental error. For example, in Figure 12.14, the relative heights of the integration curves, on moving from left to right in the spectrum, are approximately 45:67:65 or 1:1.5:1.4. To obtain ratios of hydrogen atoms, all three numbers in the ratio need to be close to whole numbers, so they are all multiplied by two, to give a ratio of 2:3.0:2.8. This ratio is then rounded up to 2:3:3. So, on moving from left to right in the spectrum, the three signals indicate 2, 3, and 3 hydrogen atoms, respectively. (Check this by measuring the heights of each of the signals with a ruler.)

For ethyl ethanoate, the 2:3:3 ratio obtained from the NMR spectrum also equals the actual number of hydrogen atoms in the molecule (i.e. the eight hydrogens in ethyl ethanoate, $CH_3CO_2CH_2CH_3$, give three signals in the ratio 2H:3H:3H). However, the same ratio of resonance signals could be obtained from the NMR spectrum of a larger molecule, with a total of 16 hydrogen atoms, if these are split in the ratio 4H:6H:6H. To decide between these alternatives, you would need to know the relative molecular mass (M_r; Section 1.3, p.14) of the compound. (The relative molecular mass of a compound can be determined using mass spectrometry as discussed in Section 12.1, p.558.)

Appearance of resonance signals

In ^1H NMR spectra, most of the resonance signals appear as distinct clusters of peaks rather than as a single peak. For example, in the ^1H NMR spectrum of ethyl ethanoate (Figure 12.14),

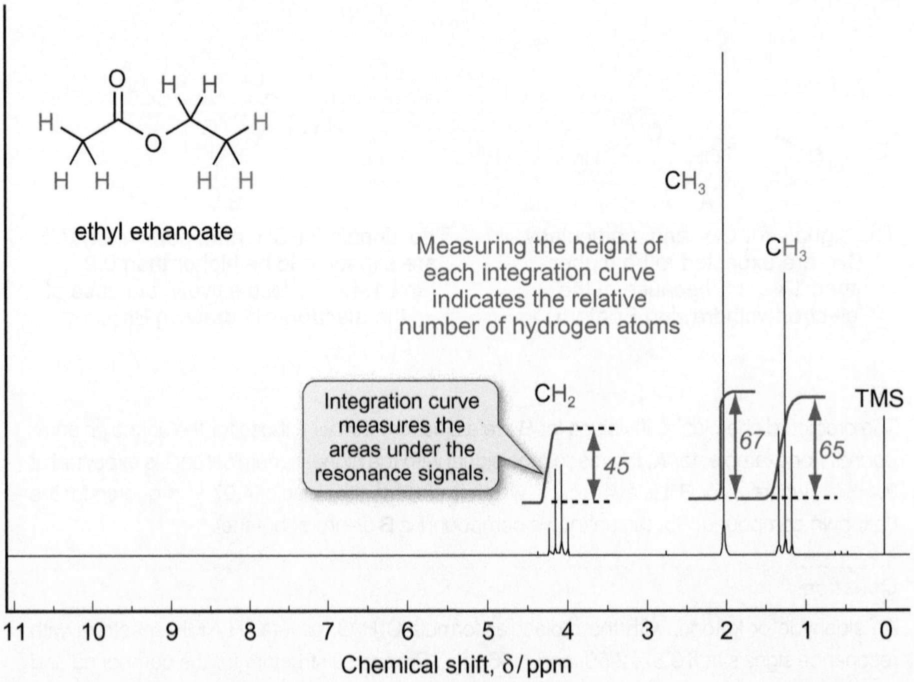
ethyl ethanoate
CH3
CH3
CH2
TMS
Measuring the height of each integration curve indicates the relative number of hydrogen atoms
Integration curve measures the areas under the resonance signals.
45
67
65
11 10 9 8 7 6 5 4 3 2 1 0
Chemical shift, δ/ppm

Figure 12.14 The ^1H NMR spectrum of ethyl ethanoate with superimposed integration curves.

although the signal at 2.04 ppm is a single peak (a **singlet**), the signal at 4.12 ppm appears as a group of four peaks (a **quartet**) and the signal at 1.26 ppm appears as a group of three peaks (a **triplet**). When a resonance signal appears as a group of peaks, the signal is said to be **split**. The splitting pattern arises as a result of **spin–spin coupling**. The splitting of resonance signals provides valuable information on the relative positions of hydrogen atoms within a molecule because it is caused by the interaction of the magnetic fields of neighbouring hydrogens.

For example, as shown in Figure 12.15, the magnetic field of neighbouring hydrogens explains the splitting pattern of signals for the adjacent CH_2 and CH_3 groups in ethyl ethanoate, $CH_3CO_2CH_2CH_3$. The CH_3 signal is affected by the small magnetic fields of the CH_2 hydrogens on the adjacent carbon, which can be aligned in three different ways (Figure 12.15(a)).

1. Both CH_2 magnetic fields can add (↑↑) to the applied magnetic field—the CH_3 hydrogens experience a slightly larger local magnetic field, so radiofrequency radiation of higher energy is required to bring the CH_3 hydrogens into resonance—this results in a slight shift of the signal to a higher chemical shift value.
2. One magnetic field can strengthen the applied magnetic field and the other can weaken it. There are two possible permutations of the magnetic fields (↓↑ and ↑↓) and, in both cases, the effects of the fields cancel each other out so the signal is not shifted.
3. Both CH_2 magnetic fields can subtract (↓↓) from the applied magnetic field—the CH_3 hydrogens experience a smaller local magnetic field, so radiofrequency radiation of lower energy is required to bring the CH_3 hydrogens into resonance—this results in a slight shift of the signal to a lower chemical shift value.

The resonance signal for CH_3 is therefore split into three peaks, in the ratio 1 : 2 : 1, which is called a triplet. (The middle peak is twice as large as the others because it corresponds to two possible permutations of the CH_2 magnetic fields.)

Similarly, the CH_2 signal in the spectrum of ethyl ethanoate is affected by the small magnetic fields of the neighbouring CH_3 hydrogens, as shown in Figure 12.15(b). The magnetic fields for the CH_3 hydrogens can be aligned in four different ways, and this causes the resonance signal for CH_2 to be split into four peaks, with a ratio of peak areas of 1 : 3 : 3 : 1, which is called a quartet. (The middle peaks are three times as large as the outer signals because they each correspond to three possible permutations of the CH_3 hydrogen magnetic fields.)

The origin of spin–spin coupling is discussed in Box 10.8 (p.503).

Coupling constant (*J*)
In a split signal, the distance between two adjacent peaks is called the coupling constant (*J*), which is measured in frequency units, hertz (Hz). The size of a coupling constant depends on the type of bonds that connect two coupled hydrogen atoms, as well as the dihedral angle (Section 18.2, p.822) that separates them.

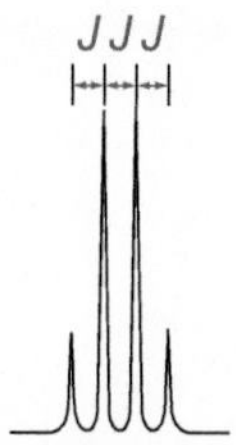
J J J

In most cases, hydrogens on the same carbon are chemically equivalent so they do not split each other.

The splitting pattern here for CH_2CH_3 is the same as that observed in the ^1H NMR spectrum of ethanol (Section 10.7, p.501, and Figure 12.16, p.591).

The number of hydrogens bonded to an adjacent carbon determines the splitting pattern of a resonance signal.

An ethyl group (CH_3CH_2–) is common in structures so learn to recognize an ethyl group from its splitting pattern—a triplet for CH_3 and a quartet for CH_2.

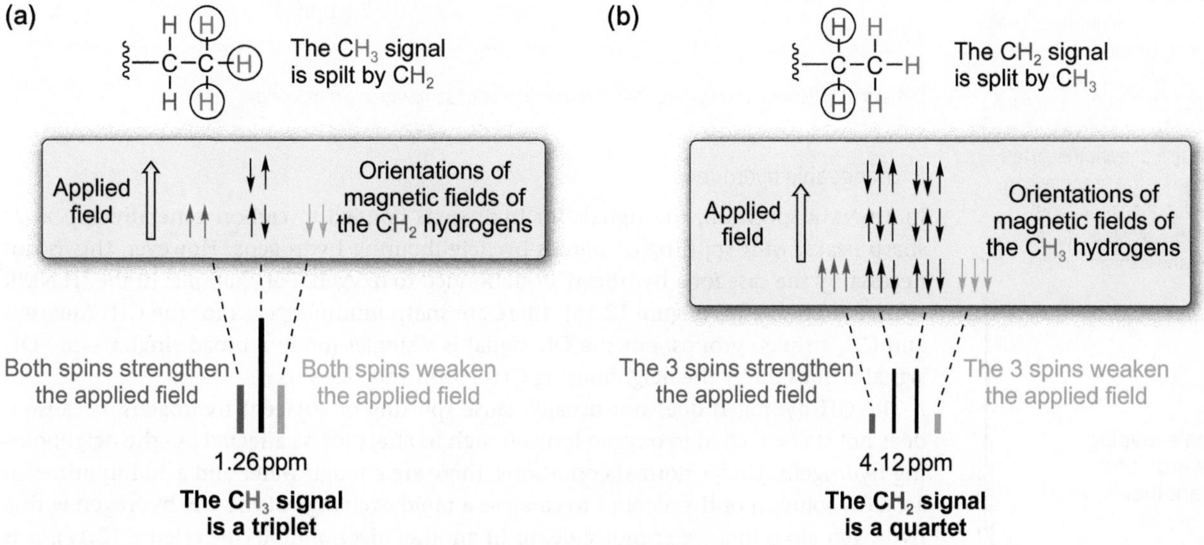
(a)
The CH3 signal is split by CH2
Applied field
Orientations of magnetic fields of the CH2 hydrogens
Both spins strengthen the applied field
Both spins weaken the applied field
1.26 ppm
The CH3 signal is a triplet
(b)
The CH2 signal is split by CH3
Applied field
Orientations of magnetic fields of the CH3 hydrogens
The 3 spins strengthen the applied field
The 3 spins weaken the applied field
4.12 ppm
The CH2 signal is a quartet

Figure 12.15 Splitting of resonance signals in the spectrum of ethyl ethanoate ($CH_3CO_2CH_2CH_3$). (a) Splitting of the signal for the CH_3 hydrogens. (b) Splitting of the signal for the CH_2 hydrogens.

The $N+1$ rule

In general, you can work out the splitting pattern for a resonance signal using the $N+1$ rule, which states that, if a signal is split by N equivalent hydrogens, it is split into $N+1$ peaks. The number of peaks is called the **multiplicity** of the signal. For example, for 2-methoxypropane ($CH_3OCH(CH_3)_2$), there are three sets of chemically equivalent hydrogens, which are predicted to give signals with the following splitting patterns.

The $N+1$ rule for spin-spin coupling is introduced in Box 10.8 (p.503), including using the numbers in Pascal's triangle to determine the relative area of each peak.

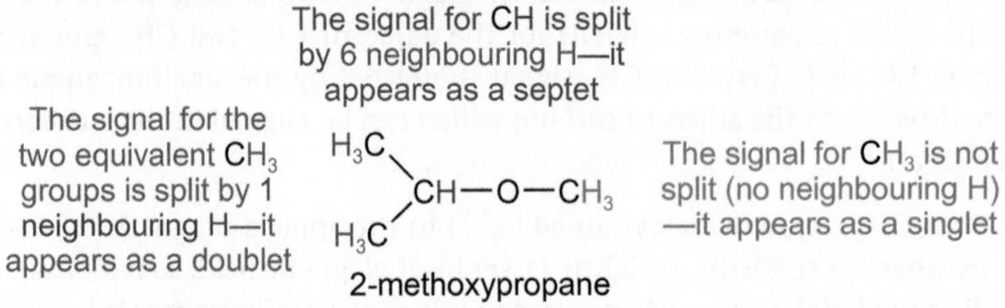
The signal for CH is split by 6 neighbouring H—it appears as a septet
The signal for the two equivalent CH3 groups is split by 1 neighbouring H—it appears as a doublet
The signal for CH3 is not split (no neighbouring H) —it appears as a singlet
2-methoxypropane

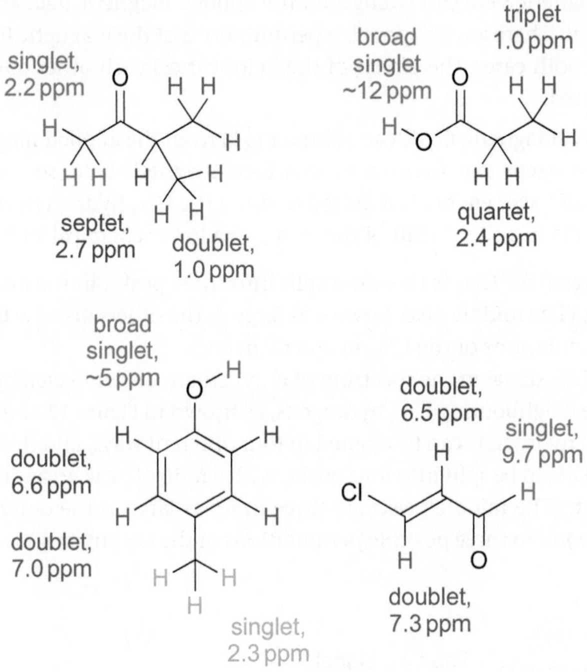
singlet, 2.2 ppm
septet, 2.7 ppm
doublet, 1.0 ppm
broad singlet ~12 ppm
triplet 1.0 ppm
quartet, 2.4 ppm
broad singlet, ~5 ppm
doublet, 6.6 ppm
doublet, 7.0 ppm
singlet, 2.3 ppm
doublet, 6.5 ppm
singlet, 9.7 ppm
doublet, 7.3 ppm

Signal multiplicities and predicted chemical shift values of some organic compounds.

In the spectrum for ethanol in Figure 12.16, notice that the right-hand peak of the quartet and the left-hand peak of the triplet are slightly larger than the corresponding peak on the other side of the multiplet in each case. Ideally, they should be the same height but, in reality, signals that split one another often 'lean' towards one another, in a phenomenon called a **roof effect** (or roofing or tenting). In general, the closer the signals are in the spectrum, the greater the leaning.

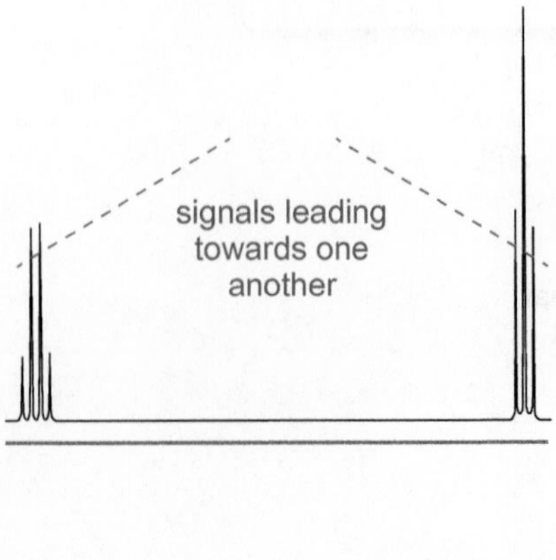
signals leading towards one another

Exchangeable hydrogens

In an NMR spectrum, the signals for hydrogens bonded to carbon generally appear as sharp peaks, with splitting of signals by neighbouring hydrogens. However, this is not necessarily the case for a hydrogen atom bonded to oxygen. For example, in the ^1H NMR spectrum of ethanol (Figure 12.16), there are sharp multiple peaks for the CH_2 (quartet) and CH_3 (triplet) groups, but the OH signal is a singlet (often a broad singlet)—the OH signal is not split by the neighbouring CH_2 group and vice versa.

The OH hydrogen does not usually cause splitting of adjacent hydrogens, because it does not stay attached to oxygen long enough to affect (or be affected by) the neighbouring hydrogens. Under normal conditions, there are enough water and acid impurities in a $CDCl_3$ solution of the alcohol to catalyse a rapid exchange of the OH hydrogen with a hydrogen atom in a water molecule, or in another alcohol molecule (Figure 12.17). It is only when the rate of exchange is very slow that the OH signal is split. If the exchange is moderately slow, then a broad singlet is often observed.

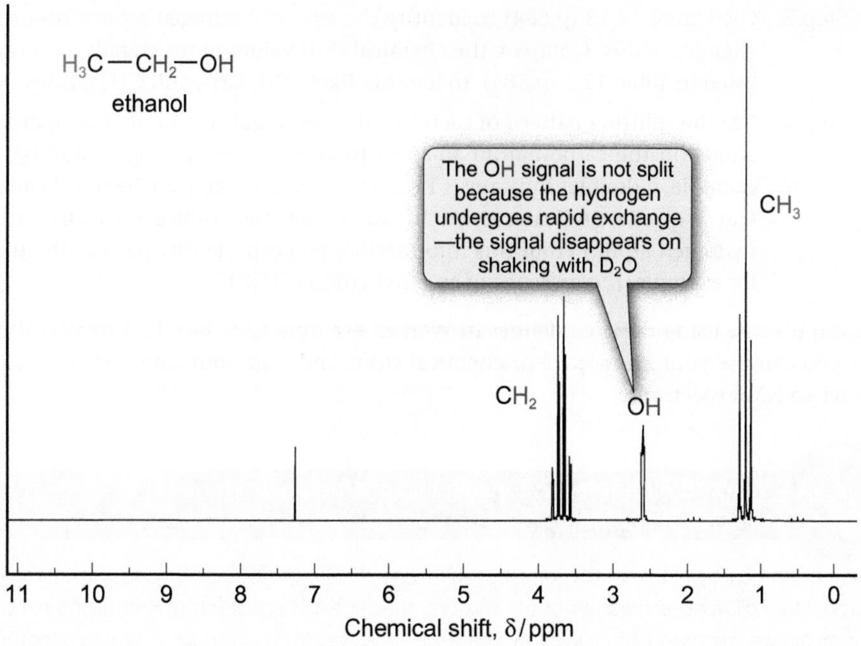
H₃C—CH₂—OH
ethanol
The OH signal is not split because the hydrogen undergoes rapid exchange —the signal disappears on shaking with D₂O
CH₃
CH₂
OH
11 10 9 8 7 6 5 4 3 2 1 0
Chemical shift, δ/ppm

Figure 12.16 ^1H NMR spectrum of a solution of ethanol in trichlorodeuteriomethane ($CDCl_3$). The single peak at 7.26 ppm is due to a tiny amount of residual $CHCl_3$.

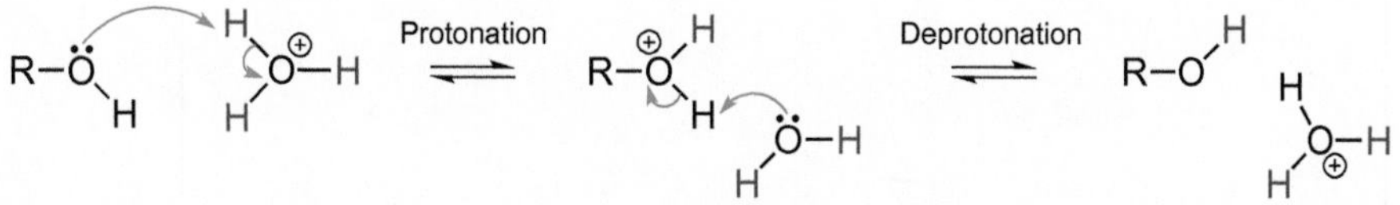
Protonation
Deprotonation

Figure 12.17 Acid-catalysed exchange of an alcohol hydrogen atom.

To confirm that a particular peak in a spectrum of an alcohol (in $CDCl_3$) is due to an OH hydrogen, the solution is usually shaken with excess D_2O and a second spectrum recorded. A deuterium atom from D_2O (deuterium oxide, or 'heavy water') will exchange for the hydrogen atom in an OH group to give an OD group. Deuterium cannot be detected in a ^1H NMR spectrum (see margin) so the signal for the OH hydrogen disappears or becomes less intense. (Note that a signal for HOD appears at around 4.7 ppm.)

Hydrogen exchange also occurs in most carboxylic acids (RCO_2H), amines (RNH_2 or R_2NH), and amides ($RCONH_2$ or RCONHR). So, the OH or NH signals in these compounds can also be detected by looking for the disappearance of certain peaks after the sample has been shaken with D_2O.

Guidelines for assigning a ^1H NMR spectrum

Interpretation of NMR spectra is a technique that requires practice. The following four-step sequence provides you with a systematic method for gathering structural information from a ^1H NMR spectrum.

Step 1 Use the number of resonance signals to determine how many types of hydrogens are in different electronic environments.

Step 2 Use the integration curves to determine the relative number of hydrogen atoms giving rise to each resonance signal. When the molecular formula is known, each integration curve can be assigned to a particular number of hydrogens.

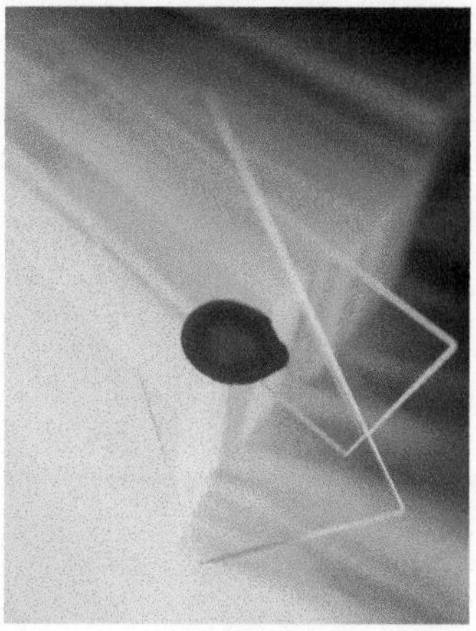

^1H NMR spectroscopy can be used to determine the amount of ethanol in blood, known as the blood alcohol concentration. It offers a rapid and accurate alternative method to the use of gas chromatography in, for example, the analysis of blood samples taken from drivers suspected of being 'over the limit' (Box 11.6, p.542). As this is a non-destructive method, the same sample can subsequently be analysed again, by the same or other analytical techniques, which may be useful in resolving legal disputes.

As an OH hydrogen atom in an alcohol is directly attached to an electronegative oxygen, you might expect it to have a higher chemical shift than 2 ppm to 5 ppm. The relatively low value is possibly due to an induced magnetic field caused by circulation of the lone pairs on oxygen, which shields the OH hydrogen atom. In comparison, in a carboxylic acid, the RCO_2H hydrogen atom gives a (broad) signal at a much higher chemical shift of around 10–13 ppm.

The rate of exchange of an OH hydrogen depends on acidity, solvent, temperature, concentration, and purity.

A deuterium atom is *not* detected in a ^1H NMR spectrum. Although D (^2H) has a nuclear spin, $I = 1$ (see Section 10.7, p.498), the energy of the radiofrequency radiation required for a deuterium atom to achieve resonance (at a given magnetic field strength) is greatly different from that for ^1H.

In structure identification, it is common to use the mass spectrum to identify the relative molecular mass and the IR spectrum to identify functional groups (see Section 12.4, p.599).

Visit the Online Resource Centre to view screencast 12.2 which walks you through how to assign a ^1H NMR spectrum.

Step 3 Use Figure 12.13 (p.584) to identify the *types* of hydrogen atoms from their chemical shifts. Compare the chemical shift values of the signals with those listed in Table 12.3 (p.584), to identify likely CH, CH_2, and CH_3 groups.

Step 4 Use the splitting pattern of each signal to work out the number of hydrogen atoms on the carbon atom adjacent to an observed hydrogen atom(s); for example, a doublet indicates 1 neighbouring hydrogen, a triplet indicates 2, and a quartet indicates 3. (A broad singlet may indicate an OH or NH hydrogen atom.) From this information, you can identify partial structures, for example, the presence of an ethyl group (CH_3CH_2–).

You can practise using these guidelines in Worked example 12.5. Box 12.5 (p.594) shows how you can use your knowledge of chemical shifts and spin–spin coupling patterns to predict an NMR spectrum.

1-Butylbenzene can be prepared from benzene by a Friedel–Crafts acylation reaction to form 1-phenylbutan-1-one ($PhCOCH_2CH_2CH_3$), followed by reduction of the ketone, as described in Section 22.2 (p.1025).

When 1-butylbenzene is oxidized by potassium permanganate ($KMnO_4$), benzoic acid ($PhCO_2H$) is formed (Section 22.4, p.1041). The mechanism of the side-chain oxidation is complex, but probably starts by the removal of a hydrogen atom from the carbon adjacent to the benzene ring (the benzylic position) to form a benzylic radical ($PhCH^{\bullet}CH_2CH_2CH_3$).

Worked example 12.5 Assigning a structure from a ^{1}H NMR spectrum

An enzyme-catalysed oxidation of 1-butylbenzene ($PhCH_2CH_2CH_2CH_3$) gave a mixture of products. Following separation of the mixture, the ^{1}H NMR spectrum of one of the minor products was recorded and it is shown here. Use the spectrum to predict a structure for the product, explaining your reasoning.

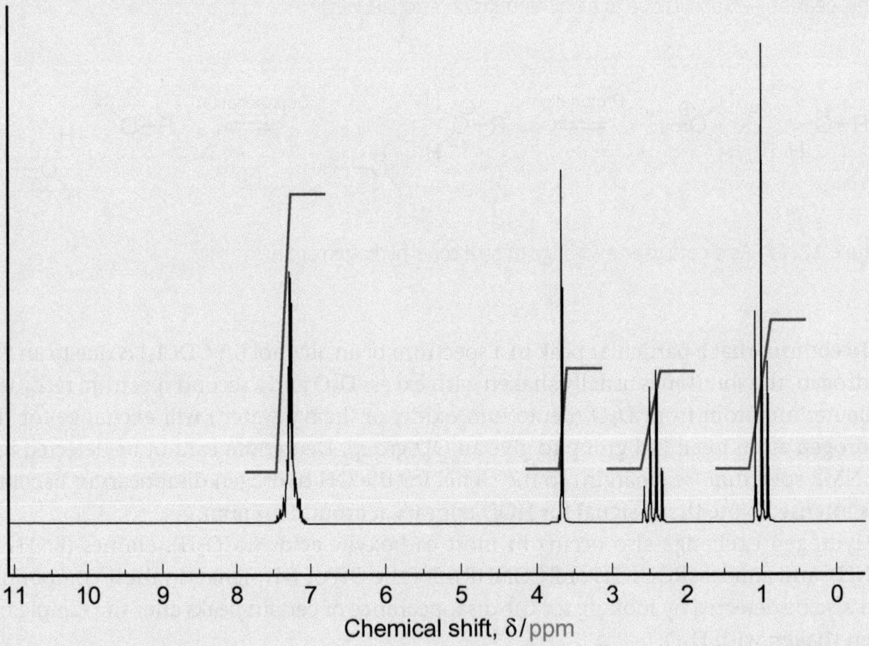
11
10
9
8
7
6
5
4
3
2
1
0
Chemical shift, δ/ppm

Strategy

Use the guidelines for assigning a ^{1}H NMR spectrum on p.591. Combine the partial structures from steps 1–4 to give a structure that corresponds with the spectrum. Check that the structure is consistent with a molecule that could be produced by oxidation of 1-butylbenzene.

Solution

Step 1 There are four resonance signals, so there are four types of hydrogens in different electronic environments. (The broad signal at 7.5–7.1 ppm has a complex pattern, which could be due to hydrogen atoms in slightly different environments.)

Step 2 The heights of the integration curves (from left to right) are 3.7 cm : 1.4 cm : 1.4 cm : 2.0 cm, so the relative number of hydrogen atoms giving rise to each signal, from left to right, is 5 : 2 : 2 : 3, respectively.

→

➜ *Step 3* Using Figure 12.13 (p.584) and Table 12.3 (p.584), the following CH, CH_2, and CH_3 groups can be assigned to the four signals:

7.5–7.1 ppm	5 H	5 × aromatic C**H** (from Figure 12.13)
3.7 ppm	2 H	possibly HO–C**H₂**–C or C(=O)O–C**H₂**–C
2.45 ppm	2 H	possibly C(=O)–C**H₂**–C, N–C**H₂**–C or C=C–C**H₂**–C
1.0 ppm	3 H	C–C**H₃**

Step 4 From the splitting pattern of each signal the following partial structures are identified:

7.5–7.1 ppm	complex signal	C_6H_5–
3.7 ppm	singlet	–CH_2–(no neighbouring hydrogens)
2.45 ppm	quartet	–CH_2–CH_3
1.0 ppm	triplet	–CH_2–CH_3

Combining these partial structures gives the following molecule.

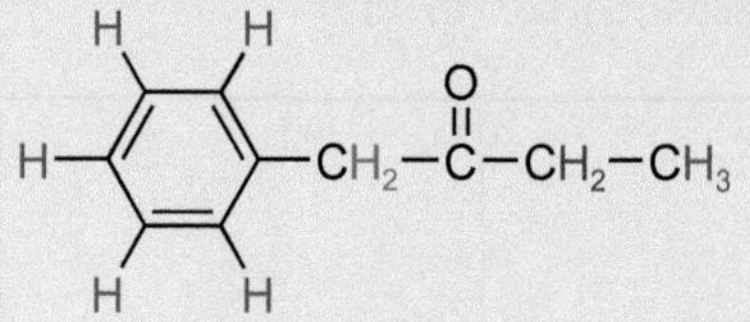

5 × CH complex signal at 7.5–7.1 ppm
CH_2 singlet at 3.7 ppm
CH_2 quartet at 2.45 ppm
CH_3 triplet at 1.0 ppm

From the splitting patterns of each signal, the CH_2 group must be joined to the benzene ring and the C=O bond. In step 3, the possible assignments for the CH_2 group assume that it is adjacent to a single functional group. However, in this molecule, the combined effects of the benzene ring and C=O bond explain why the CH_2 group has a relatively high chemical shift of 3.7 ppm

The minor product is 1-phenylbutan-2-one ($PhCH_2COCH_2CH_3$). This is a possible product of the enzyme-catalysed oxidation of 1-butylbenzene.

Question

The ^{1}H NMR spectrum of the major product from the enzyme-catalysed oxidation of 1-butylbenzene ($PhCH_2CH_2CH_2CH_3$) is shown below. Use the spectrum to predict a structure for the product, explaining your reasoning.

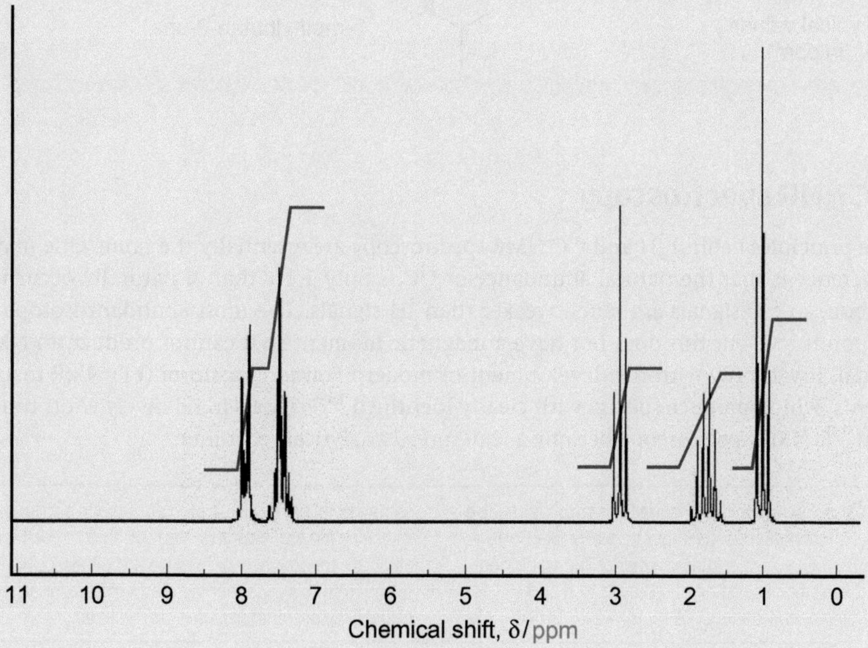
11 10 9 8 7 6 5 4 3 2 1 0
Chemical shift, δ/ppm

Enzymes that catalyse oxidation reactions are called oxidative enzymes. The most common of these are peroxidases, which use hydrogen peroxide as an oxidizing agent, and oxidases which use molecular oxygen. Oxidases are responsible for the browning of fruits, such as apples. The control of enzymatic browning is of great importance to the horticulture industry, because this reaction can affect the colour, taste, flavour, and nutritional value of many fruits and vegetables. It is estimated that more than 50% of fruit market losses are a result of enzymatic browning.

Box 12.5 Drawing a ^{1}H NMR spectrum

Predicting a ^{1}H NMR spectrum of a molecule is a good exercise to help you gain practice in recognizing chemical shifts, areas of peaks, and spin–spin coupling patterns in an actual spectrum. The following stepwise method, using propyl ethanoate ($CH_3CO_2CH_2CH_2CH_3$) as the example, will help you draw a sensible spectrum.

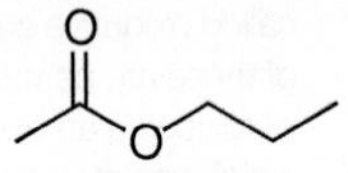

propyl ethanoate

Step 1 From the structure, decide on the number of types of hydrogen atoms that are in different electronic environments.

$H_3C-C(=O)-O-CH_2-CH_2-CH_3$

Predict 4 resonance signals

Step 2 Give the number of hydrogen atoms in each electronic environment.

3 H			2 H	2 H	3 H
H_3C	$-C(=O)-$	O	$-CH_2-$	CH_2-	CH_3

Step 3 Using Table 12.3 (p.584) and Figure 12.13 (p.584), estimate the chemical shift values for each CH_3, CH_2, and CH group.

3 H			2 H	2 H	3 H	
H_3C	$-C(=O)-$	O	$-CH_2-$	CH_2-	CH_3	
2.0			4.1	1.6	0.9	(Typical values in ppm)

Step 4 From the number of hydrogen atoms on an adjacent carbon, use the $N + 1$ rule (see p.590 and Box 10.8, p.503) to calculate the splitting pattern for each resonance signal.

3 H			2 H	2 H	3 H	
H_3C	$-C(=O)-$	O	$-CH_2-$	CH_2-	CH_3	
2.0			4.1	1.6	0.9	(Typical values in ppm)
Singlet ($N = 0$)			Triplet ($N = 2$)	Sextet ($N = 5$)	Triplet ($N = 2$)	

Step 5 Draw the ^{1}H NMR spectrum (remember that the chemical shift scale increases from right to left).

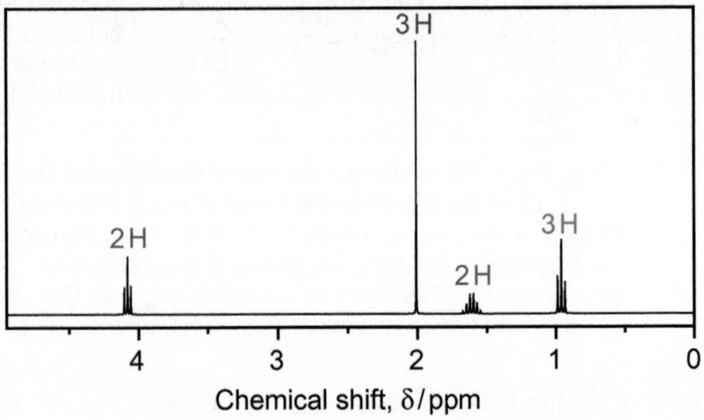
3H
2H
2H
3H
4
3
2
1
0
Chemical shift, δ/ppm

Question

Draw the ^{1}H NMR spectrum for 3-methylbutan-2-one (methyl isopropyl ketone or MIPK; $CH_3COCH(CH_3)_2$).

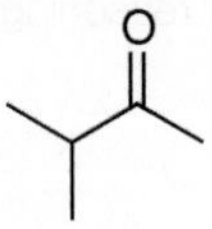

3-methylbutan-2-one

^{13}C NMR spectroscopy

> In FT ^{13}C NMR spectroscopy, many spectra are rapidly recorded and then added together, which enhances the size of the resonance signals relative to the background noise. The greater the size of the desired resonance signals, in comparison to the size of background electronic noise signals, the higher the **signal-to-noise ratio**.

The principles behind ^{1}H and ^{13}C NMR spectroscopy are essentially the same. One major difference is that the natural abundance of ^{13}C is only 1.1% that of naturally occurring carbon, so ^{13}C signals are much weaker than ^{1}H signals. The most abundant isotope of carbon is ^{12}C, but this does not have a magnetic moment so it cannot produce an NMR signal. It was only with the development of modern Fourier transform (FT) NMR instruments, which produce spectra with clearly identified ^{13}C signals in relatively short times, that ^{13}C NMR spectroscopy became a widespread analytical technique.

> For an analogy of Fourier transform and pulse techniques in NMR spectroscopy (Section 10.3, p.460), consider tuning a piano. The traditional approach involves hitting each key in turn and recording each frequency. A faster way of getting the same information is to stretch out an arm and hit all the keys at once. A complicated sound that combines all of the tones is heard. To extract the individual tones from this cacophony, a mathematical analysis called Fourier transformation can be used.

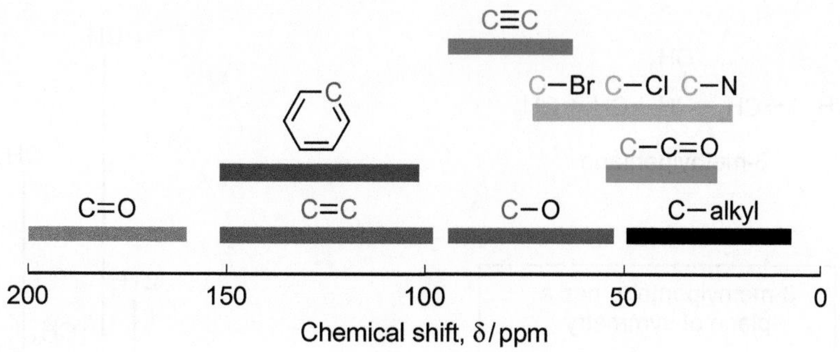
C≡C
C−Br C−Cl C−N
C
C−C=O
C=O
C=C
C−O
C−alkyl
200
150
100
50
0
Chemical shift, δ / ppm

Figure 12.18 Approximate chemical shift values (δ / ppm) of carbon atoms in various organic compounds. (The values are affected by neighbouring substituents.)

If the ^{13}C NMR spectrum is recorded under conditions of proton decoupling, a single line is observed for each different type of carbon atom.

Magnetically active nuclei, such as ^{1}H and ^{13}C, can couple to one another. The number of lines into which a signal from a particular nucleus is split is equal to $2NI + 1$, where N is the number of equivalent nuclei that are coupling with the nucleus, and I is the nuclear spin quantum number of these nuclei. Both ^{1}H and ^{13}C have $I = \frac{1}{2}$. So, for example, the ^{13}C NMR spectrum of $CHCl_3$ shows two lines because the carbon signal is split by the adjacent hydrogen atom, so $N = 1$ and $2NI + 1 = [(2 \times N \times \frac{1}{2}) + 1] = 2$.

Nuclear spin quantum numbers (I) are introduced in Section 10.7 (p.503). For ^{1}H nuclei, $I = \frac{1}{2}$ so a ^{1}H NMR signal is split into $[(2 \times N \times \frac{1}{2}) + 1]$ lines or, more simply, $N + 1$ lines (see the $N + 1$ rule on p.590).

In most cases, the ^{13}C NMR spectrum is produced using a technique called **proton decoupling**. This technique removes the splitting of ^{13}C signals by adjacent ^{1}H nuclei, so that a singlet peak is observed for each *different* type of carbon atom. For example, in Figure 12.10 (p.581), the four carbon atoms in ethyl ethanoate ($CH_3CO_2CH_2CH_3$) give rise to four singlet peaks in the spectrum. Splitting of a signal by a neighbouring ^{13}C atom is not observed because the natural abundance of ^{13}C is so low—in most molecules a ^{13}C atom is bonded to a ^{12}C atom(s), which cannot split a ^{13}C signal.

In a ^{13}C NMR spectrum, the number, position, and, to a lesser degree, intensity of resonance signals provide information on the different types of carbon atoms in an organic molecule.

The ^{13}C NMR (proton decoupled) spectrum of a sample dissolved in $CDCl_3$ usually shows a 1:1:1 pattern of peaks at 77 ppm. Deuterium has a nuclear spin quantum number, $I = 1$, and it splits the carbon signal into three lines, $[(2 \times 1 \times 1) + 1] = 3$. Conditions of proton decoupling do not remove coupling due to deuterium.

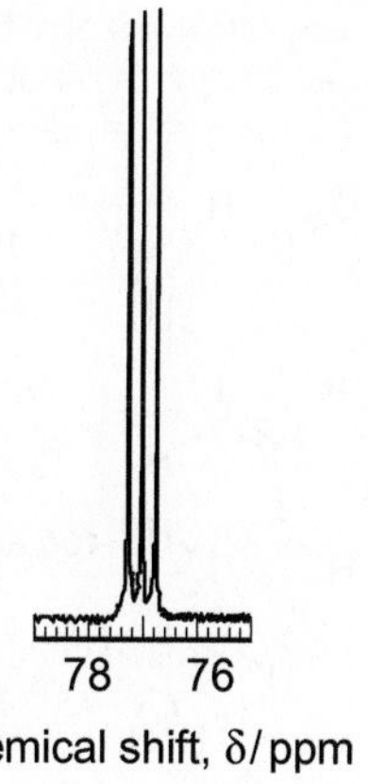
78 76
chemical shift, δ / ppm

Number and position of resonance signals

The number of resonance signals in a ^{13}C NMR spectrum indicates the number of carbon atoms that are in *different* electronic environments. As for ^{1}H NMR spectroscopy, the position of a ^{13}C signal in a spectrum is recorded as a chemical shift value (δ in ppm). The chemical shift values are measured from a reference signal, typically from the ^{13}C signal due to the four equivalent carbon atoms in tetramethylsilane (TMS, $SiMe_4$), which is assigned a chemical shift value of 0 ppm.

Carbon atoms in functional groups have characteristic chemical shifts and the trends in chemical shift values are somewhat similar to those for ^{1}H NMR spectroscopy. So, a carbon atom bonded to an electron-withdrawing group gives a resonance signal with a higher chemical shift than a carbon atom bonded to an electron-donating group (Figure 12.18). As a rule of thumb, the chemical shift of a carbon atom is around 20 times larger than the chemical shift of a hydrogen atom in a similar electronic environment. For example, for an aldehyde functional group, RCHO, the hydrogen atom gives a signal at 9 ppm to 10 ppm in a ^{1}H NMR spectrum, whereas the carbonyl carbon gives a signal around 190 ppm in a ^{13}C NMR spectrum. As the chemical shifts of ^{13}C signals are spread over around 200 ppm, compared to 10 ppm in a ^{1}H NMR spectrum, there is less chance of two ^{13}C signals overlapping, which makes it easier to assign spectra.

Carbon atoms in different electronic environments give different resonance signals in a ^{13}C NMR spectrum.

Chemical shifts for methoxymethane (dimethyl ether)—the chemical shift for the carbons is around 20 times larger than the chemical shift for the hydrogens.

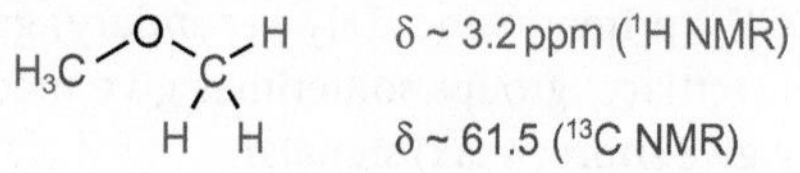
H₃C
O
C
H
H H
δ ~ 3.2 ppm (¹H NMR)
δ ~ 61.5 (¹³C NMR)

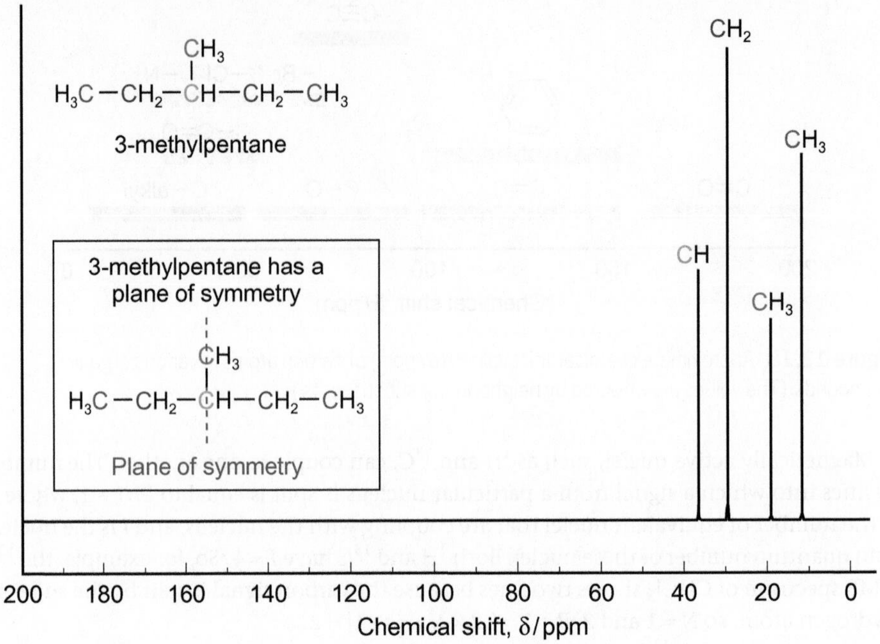
CH3
H3C—CH2—CH—CH2—CH3
3-methylpentane
3-methylpentane has a plane of symmetry
CH3
H3C—CH2—CH—CH2—CH3
Plane of symmetry
CH2
CH3
CH
CH3
200 180 160 140 120 100 80 60 40 20 0
Chemical shift, δ/ppm

Figure 12.19 ^{13}C NMR (proton decoupled) spectrum of 3-methylpentane.

Classification of carbon atoms as primary (1°), secondary (2°), tertiary (3°), and quaternary (4°) is discussed in Section 2.5 (p.83).

Approximate chemical shifts in ppm (in italic), for the carbon atoms in (*E*)-4-(but-1-enyl)benzaldehyde:

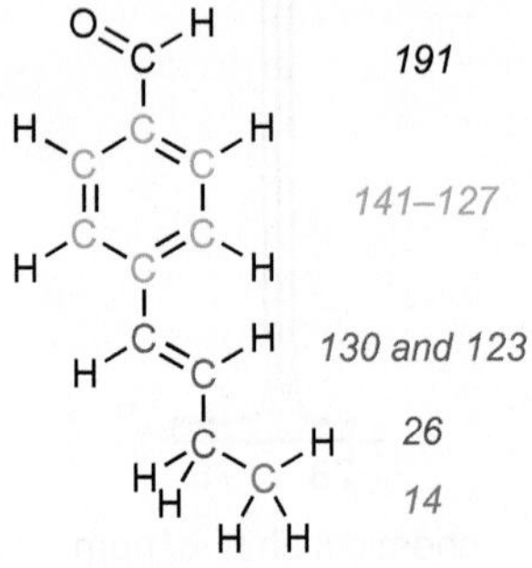
191
141–127
130 and 123
26
14

Hybridization is discussed in Section 5.4 (p.236).

Approximate chemical shifts in ppm (in italic), for the carbon atoms in (*E*)-pent-3-en-2-one:

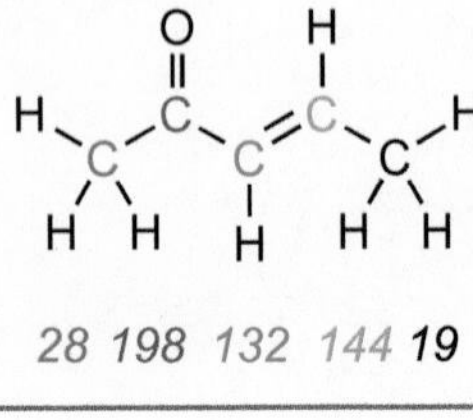
28 198 132 144 19

Chemical shifts are also affected by whether the carbon is primary (CH_3), secondary (CH_2), tertiary (CH), or quaternary (C). Usually, a CH_3 carbon is more shielded than a CH_2 carbon, which, in turn, is more shielded than a CH carbon (see margin). You can see this in the ^{13}C NMR spectrum of 3-methylpentane, $(CH_3CH_2)_2CHCH_3$ (Figure 12.19), where the CH carbon atom has a higher chemical shift than the CH_2, CH_3, or CH_3 carbons. Although 3-methylpentane has 6 carbons, there are only four signals in the spectrum because the molecule has a plane of symmetry—the two CH_2 carbons are identical, as are the two CH_3 carbons.

alkane carbon	CH_3	CH_2	CH	C
Typical chemical shift, δ/ppm	5–20	20–30	30–50	30–45

Hybridization also affects the chemical shift—an *sp*^{3} carbon (typically 5–80 ppm) is usually more shielded than an *sp* carbon (~65–90 ppm), which in turn is more shielded than a *sp*^{2} carbon (~100–200 ppm). Carbonyl (*sp*^{2}) carbon atoms are especially deshielded and so have the highest chemical shift values (usually above 160 ppm).

Intensity of resonance signals

For ^{13}C NMR spectroscopy it is much harder to obtain meaningful integrations of signals than it is for ^{1}H NMR spectroscopy—the area of a ^{13}C signal is not necessarily proportional to the number of carbons giving rise to the peak. For example, in Figure 12.19, the signals for the two CH_2 carbons, and the two CH_3 carbons, are not double the size of the areas of the signals for the other carbons. Under normal conditions, the signal heights can be more indicative of how many hydrogens are attached to each carbon. As a guide, carbons of CH_3 (primary) and CH_2 (secondary) groups often give tall (intense) signals, carbons of CH (tertiary) groups sometimes give medium-sized signals, and quaternary carbons usually give short (weak) signals.

Worked example 12.6 illustrates the use of a ^{13}C NMR spectrum in assigning a structure.

Worked example 12.6 Assigning a structure from a ^{13}C NMR spectrum

Reaction of pentane (C_5H_{12}) with Br_2 in the presence of UV radiation gave a mixture of structural isomers with the molecular formula $C_5H_{11}Br$. Following separation of the mixture, the ^{13}C NMR (proton decoupled) spectrum of one of the products was recorded and is shown here. Use the spectrum to predict a structure for the product, explaining your reasoning. (The mechanism of this reaction is discussed in Section 20.2, p.921.)

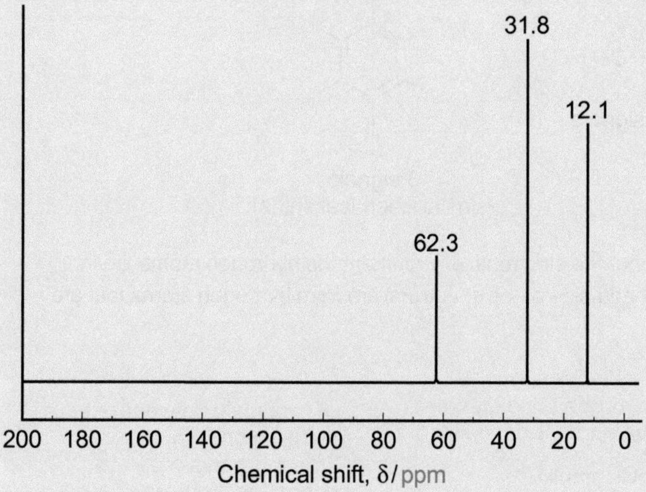
31.8
12.1
62.3
200 180 160 140 120 100 80 60 40 20 0
Chemical shift, δ/ppm

Strategy

Draw all possible structural isomers with the molecular formula $C_5H_{11}Br$, bearing in mind that the starting material is pentane ($CH_3CH_2CH_2CH_2CH_3$).

Decide on the number of types of carbon atoms that are in different electronic environments in each structural isomer.

Use Figure 12.18 (p.595) to check that the chemical shift values in the ^{13}C NMR spectrum are consistent with the structure of the proposed structural isomer.

Solution

There are three possible structural isomers with the molecular formula $C_5H_{11}Br$.

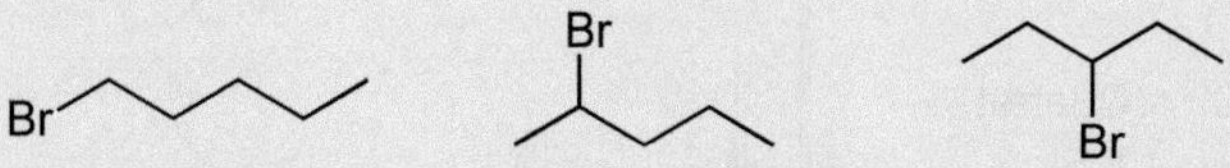
Br
Br
Br

1-bromopentane 2-bromopentane 3-bromopentane

For each structural isomer, the carbon atoms that are in different electronic environments are shown in the following illustration.

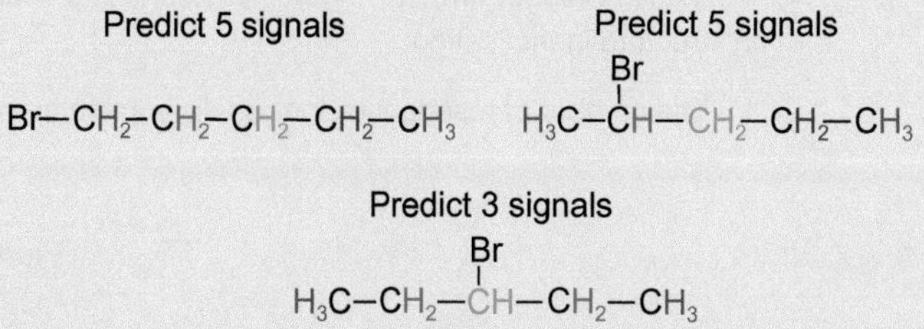
Predict 5 signals
Br—CH2—CH2—CH2—CH2—CH3
Predict 5 signals
Br
H3C—CH—CH2—CH2—CH3
Predict 3 signals
Br
H3C—CH2—CH—CH2—CH3

The product is 3-bromopentane as this is the only structural isomer that gives three signals in the ^{13}C NMR spectrum—1-bromopentane and 2-bromopentane are both expected to give five signals.

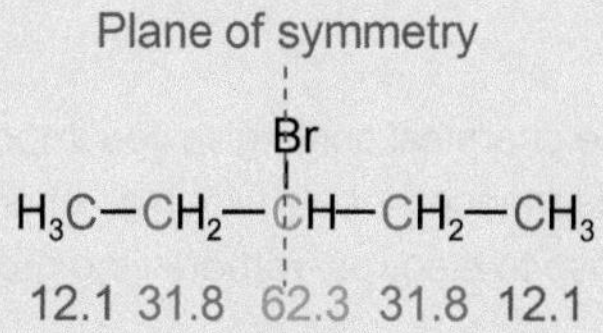
Plane of symmetry
Br
H3C—CH2—CH—CH2—CH3
12.1 31.8 62.3 31.8 12.1

The chemical shift values are consistent with the ranges in Figure 12.18.

Question

Reaction of bromobenzene (C_6H_5Br) with a mixture of Br_2 and $FeBr_3$ gave a mixture of structural isomers with the molecular formula $C_6H_4Br_2$. Following separation of the mixture, the ^{13}C NMR (proton decoupled) spectrum of one of the products was recorded and is shown below. Use the spectrum to predict a structure for the product, explaining your reasoning. (The mechanism of this reaction is discussed in Section 22.2, p.1015.)

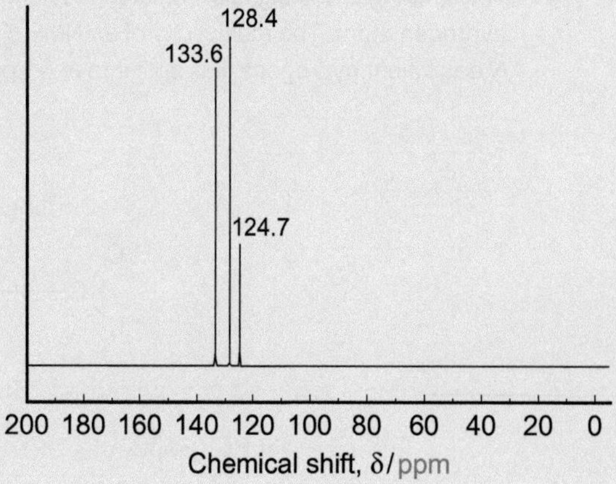
128.4
133.6
124.7
200 180 160 140 120 100 80 60 40 20 0
Chemical shift, δ/ppm

Summary

- An FT-NMR spectrum is a plot of intensity of the resonance signal against chemical shift. The frequency (and energy) of the radiofrequency radiation responsible for the signal increases from right to left across the spectrum.
- In a ^{1}H NMR spectrum, the number, position (chemical shift), area, and appearance of each resonance signal are all helpful in structure identification.

1 The number of resonance signals tells you the number of hydrogen atoms in different electronic environments.

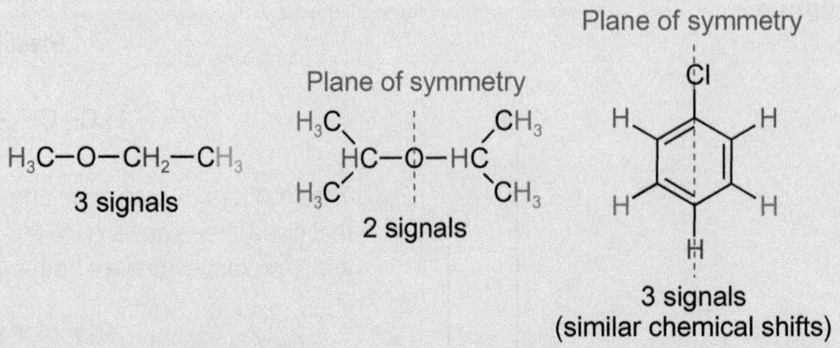

Plane of symmetry
Plane of symmetry
3 signals
2 signals
3 signals
(similar chemical shifts)

2 The chemical shifts give you important information about the electronic environment the hydrogen atoms are in. Resonance signals with high chemical shifts (the left-hand side of the spectrum) are from hydrogen atoms that are close to electron-withdrawing groups.

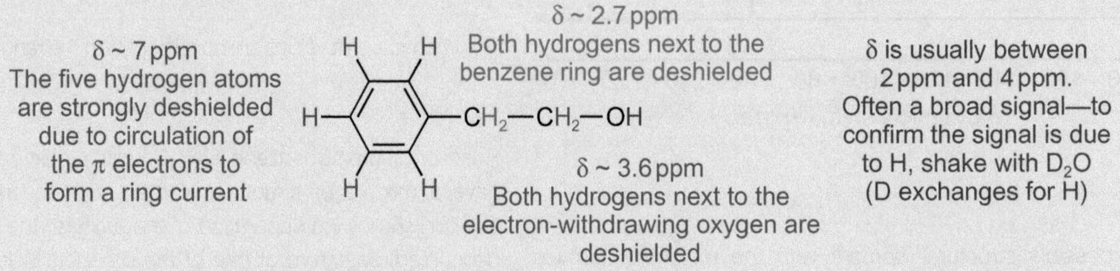

δ ~ 7 ppm
The five hydrogen atoms are strongly deshielded due to circulation of the π electrons to form a ring current
δ ~ 2.7 ppm
Both hydrogens next to the benzene ring are deshielded
δ ~ 3.6 ppm
Both hydrogens next to the electron-withdrawing oxygen are deshielded
δ is usually between 2 ppm and 4 ppm. Often a broad signal—to confirm the signal is due to H, shake with D2O (D exchanges for H)

3 The ratio of the areas under the peaks tells you the ratio of the numbers of hydrogen atoms in each of these environments.

4 The splitting of the signals indicates the number of hydrogen atoms on the carbon atom adjacent to an observed hydrogen atom. The multiplicity of an NMR signal is calculated by the $N + 1$ rule, which states that, if a signal is split by N equivalent hydrogens, it is split into $N + 1$ peaks.

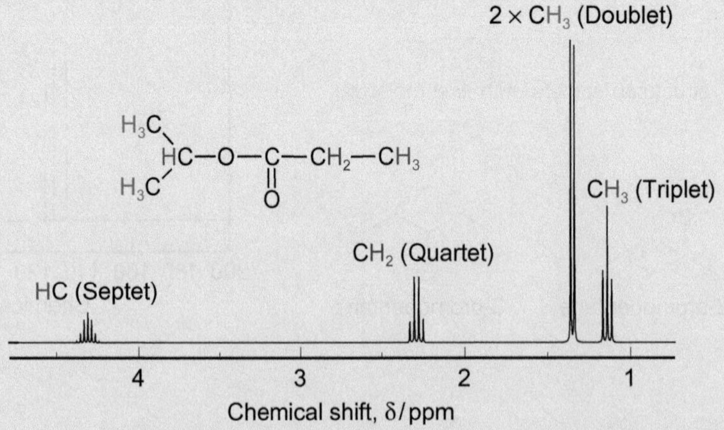

2 × CH3 (Doublet)
CH3 (Triplet)
CH2 (Quartet)
HC (Septet)
4
3
2
1
Chemical shift, δ/ppm

- In a ^{13}C NMR spectrum, the number of different signals indicates the number of carbon atoms in different electronic environments. The chemical shift of a signal indicates the type of carbon atom.

→

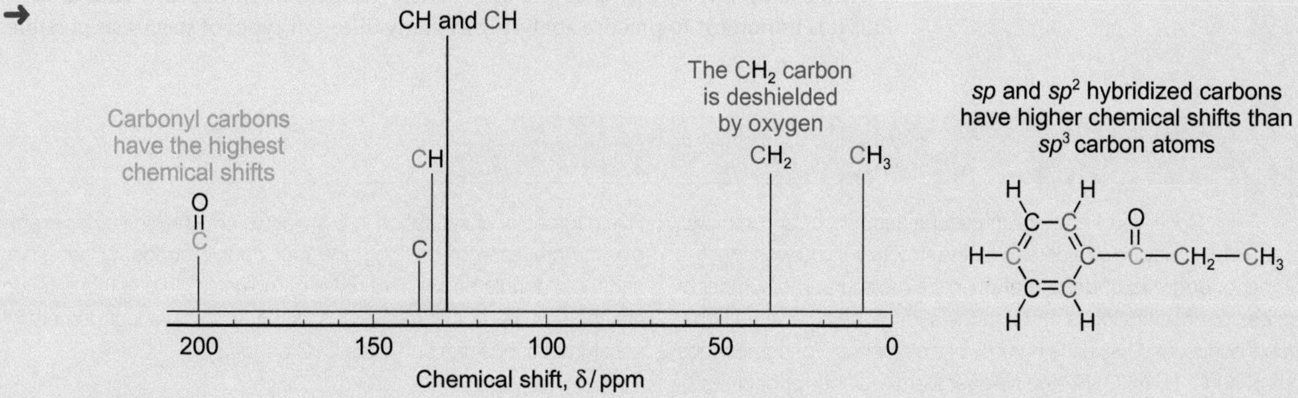
CH and CH
The CH_2 carbon is deshielded by oxygen
Carbonyl carbons have the highest chemical shifts
CH
CH_2
CH_3
C
200
150
100
50
0
Chemical shift, δ/ppm
sp and sp^2 hybridized carbons have higher chemical shifts than sp^3 carbon atoms

For practice questions on these topics, see questions 1–6 at the end of this chapter (pp.606–608).

12.4 Structure determination using a combination of techniques

Chemists routinely use a combination of mass spectra, IR spectra, and NMR spectra to determine the structure of an unknown compound. Each spectrum provides distinctive clues for a structure. To help you develop a method for analysing the different spectra, a six-step approach is given below.

Step 1 Look at the mass spectrum to determine the relative molecular mass of the compound. (Remember that sometimes, a molecular ion is not observed in an EI mass spectrum because it fragments before it reaches the detector.) Look at the isotope pattern of the molecular ion to see if a Cl or Br atom(s) is present (Section 12.1, p.561). Work out the molecular formula from the accurate relative molecular mass (assuming this is available). It may also help to work out the number of degrees of unsaturation, which is discussed in Box 12.6.

Step 2 Look at the IR spectrum to identify any characteristic groups, such as CH, OH, C=C, and C=O groups (Table 12.2, p.573). The exact position of a C=O stretching absorption band can be used to identify the carbonyl functional group, for example, whether an ester (RCO_2R) or an amide (such as $RCONH_2$) is present (Figure 12.7, p.574). The absence of characteristic absorption bands can also be used to rule out the presence of certain functional groups.

Step 3 Look at the ^{1}H NMR spectrum to determine the number of hydrogen atoms in different environments. From the chemical shift values and peak areas identify partial structures, for example, types of CH_3, CH_2, and CH groups (Table 12.3, p.584, and Figure 12.13, p.584). Write down all of the possible partial structures and, from the splitting patterns of the signals, identify how the partial structures can be connected together. Give a list of all possible structures.

Step 4 Look at the ^{13}C NMR spectrum to determine the number of carbon atoms in different environments (Section 12.3, p.594). Check if the chemical shifts of the carbon atoms are consistent with each possible structure from step 3.

Step 5 If more than one structure is still possible, look at the MS spectrum and use the intense fragment ions to identify certain partial groups (e.g. Table 12.1, p.561). To help eliminate structures, decide what differences there would be in the IR, NMR, and MS spectra for each of the possible structures.

Step 6 Double check that your proposed structure is consistent with *all* of the spectra.

Visit the Online Resource Centre to view screencast 12.3 which walks you through structure determination using a combination of spectroscopic techniques.

Research papers in the chemical literature typically include experimental data, such as MS, IR and NMR data. The style in which the spectroscopic data is presented varies depending on the scientific journal, but one common style is shown below, for benzyl 2-methyl-3-oxobutanoate ($CH_3COCH(CH_3)CO_2CH_2Ph$):

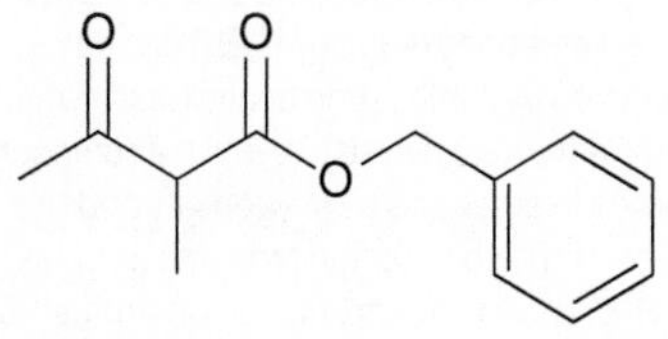

MS data: *m/z* (CI, NH_3) 224 ($M + NH_4^+$, 100%), 207 ($M + H^+$, 50).

IR data: ν_{max} ($CHCl_3$) 3028 (s), 1745 (CO, s), 1723 (CO, s), 1455 (m), and 1229 (s) cm^{-1}.

^{1}H NMR data: δ_H (300 MHz; $CDCl_3$; Me_4Si) 7.37–7.32 (5H, m, Ph), 5.17 (2H, s, CH_2Ph), 3.55 (1H, q, *J* = 7 Hz, $CHCH_3$), 2.19 (3H, s $COCH_3$), and 1.36 (3H, d, *J* = 7 Hz, $CHCH_3$).

^{13}C NMR data: δ_C (75 MHz; $CDCl_3$; Me_4Si) 203.9 (CH_3CO), 170.8 (CO_2), 135.7, 129.0, 128.9, 128.7 (aromatics), 67.5 (CO_2CH_2), 54.0 ($CHCH_3$), 28.9 (CH_3CO), and 13.2 ($CHCH_3$).

(This compound, like other β-keto esters, is subject to keto–enol tautomerism (Section 24.3, p.1126). Only the data for the keto form is given here.)

Worked example 12.7 illustrates the use of these guidelines. To improve your analytical skills it is important to practise analysing as many different types of spectra as possible.

Box 12.6 Determining degrees of unsaturation

Once you know the molecular formula of an unknown compound, it may help to determine the structure if you work out the number of **degrees of unsaturation** in the molecule. The number of degrees of unsaturation is equal to the sum of the number of rings, double bonds, and twice the number of triple bonds. For compounds containing C, H, N, O, S, and halogens, the number of degrees of unsaturation is calculated using Equation 12.1.

$$\text{Degrees of unsaturation} = \text{number of carbons} - \frac{\text{number of hydrogens}}{2} - \frac{\text{number of halogens}}{2} + \frac{\text{number of nitrogens}}{2} + 1 \qquad (12.1)$$

For example, if an unknown compound has the molecular formula $C_7H_5BrO_2$, then the number of degrees of unsaturation is calculated as follows.

$$\text{Degrees of unsaturation} = 7 - \frac{5}{2} - \frac{1}{2} + \frac{0}{2} + 1$$
$$= 5$$

Five degrees of unsaturation in a molecule could indicate, for example, the presence of one ring and four double bonds, or two rings and three double bonds. Examples of molecules that fit this criterion are shown here. (A benzene ring is assumed to have four degrees of unsaturation: one ring and three C=C bonds.)

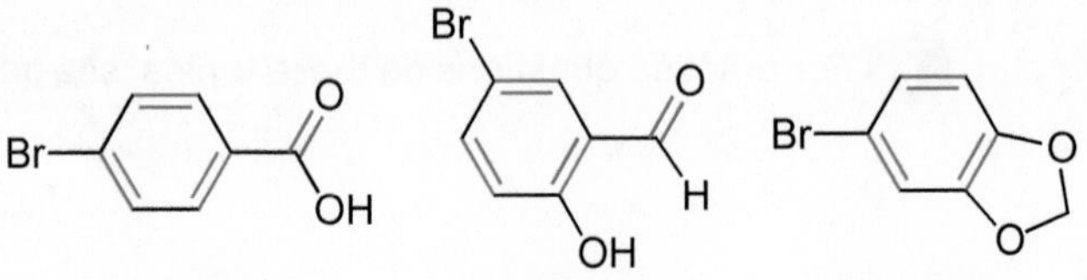
Br
O
OH
Br
O
H
OH
Br
O
O

One ring, four double bonds | One ring, four double bonds | Two rings, three double bonds

▲ Three structures ($C_7H_5BrO_2$) with five degrees of unsaturation.

Question

Calculate the number of degrees of unsaturation for molecules with the following molecular formulae: (a) $C_8H_9NO_2$; (b) C_3H_3ClO.

The empirical formula of a pure organic compound (e.g. $C_{10}H_{12}O_3$) can be determined by combustion analysis (sometimes called elemental analysis), in which a compound is burned in an excess of oxygen gas and the masses of the combustion products (such as CO_2, H_2O, and NO) found. Modern instruments, called combustion or CHN analysers, are sufficiently automated to be able to do these analyses routinely. Only extremely small amounts of samples are required—3 mg of sample is usually sufficient to give an accurate analysis of a sample's carbon, hydrogen, and nitrogen content.

Worked example 12.7 Assigning a structure from a combination of spectra

A sweet-smelling compound with the molecular formula $C_{10}H_{12}O_3$ was isolated from a synthetic perfume. Using the mass spectrum, the IR spectrum, and the ^1H and ^{13}C (proton decoupled) NMR spectra shown here, propose a structure for the unknown compound.

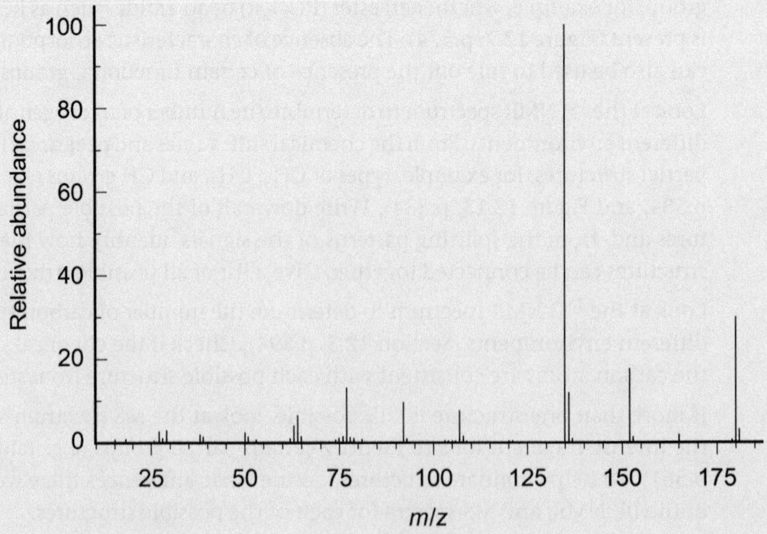
Relative abundance
100
80
60
40
20
0
25
50
75
100
125
150
175
m/z

→

→

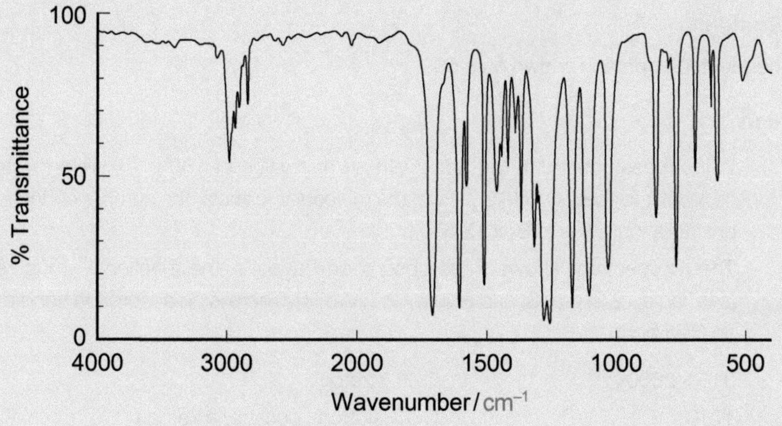
100
50
0
% Transmittance
4000
3000
2000
1500
1000
500
Wavenumber/cm⁻¹

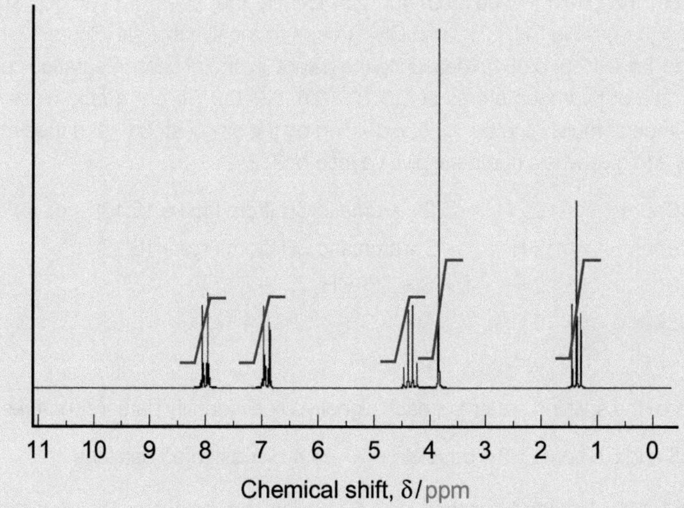
11 10 9 8 7 6 5 4 3 2 1 0
Chemical shift, δ/ppm

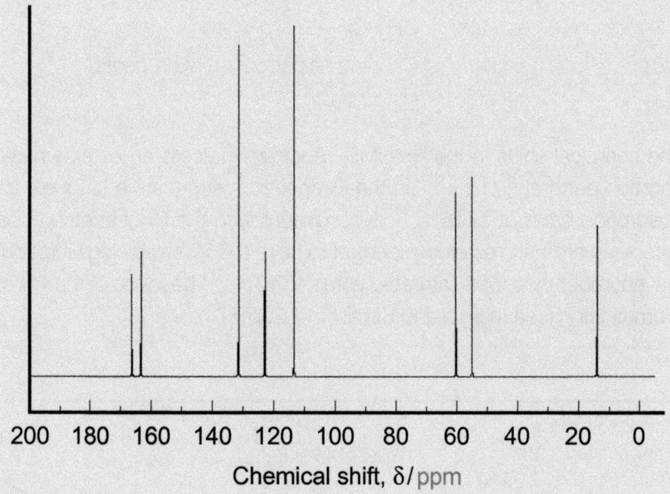
200 180 160 140 120 100 80 60 40 20 0
Chemical shift, δ/ppm

→

→ Strategy

Use the six-step approach shown on p.599.

Solution

Step 1 In the mass spectrum, the peak with an *m/z* value of 180 is consistent with the molecular formula $C_{10}H_{12}O_3$. From the molecular formula, the number of degrees of unsaturation is 5 (see Box 12.6, p.600).

Step 2 The IR spectrum shows three absorption bands in the functional group region (see Table 12.2, p.573, and Figure 12.7, p.574). There is no absorption band due to an O–H bond.

3100–2800 cm^{-1}	C–H (stretch)
1716 cm^{-1}	C=O (stretch), possibly a ketone
1608 cm^{-1}	C=C (stretch), an alkene or a benzene ring

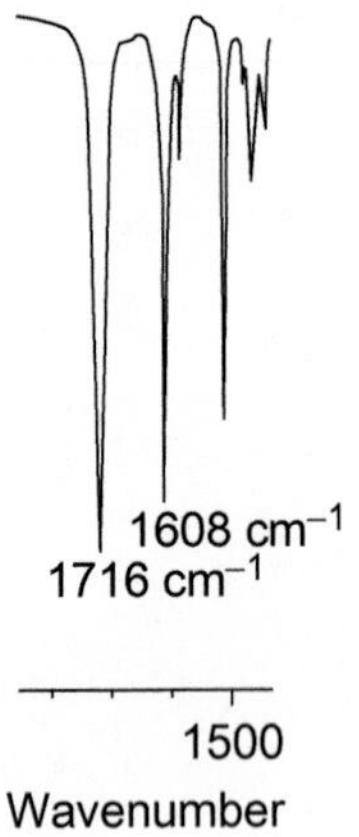
1608 cm−1
1716 cm−1
1500
Wavenumber

Step 3 From the ^1H NMR spectrum you can identify the types of hydrogen atoms and assign possible CH, CH_2, and CH_3 groups to the five signals (Figure 12.13, p.584 and Table 12.3, p.584). Measuring the height (in cm) of each integration curve gives a ratio of hydrogen atoms of 0.6:0.6:0.6:0.9:0.9. To get a ratio close to whole numbers, multiplying by 3.3 (i.e. dividing by the smallest, 0.6, and then multiplying by 2 to give whole numbers) gives a ratio of 2:2:2:3:3.

8.0 ppm	2 H	2 × aromatic CH (from Figure 12.13)
6.9 ppm	2 H	2 × aromatic CH (from Figure 12.13)
4.3 ppm	2 H	C(=O)O–CH_2
3.8 ppm	3 H	Ar–O–CH_3
1.4 ppm	3 H	O–C–CH_3

From the splitting pattern of each signal you can identify partial structures.

8.0 and 6.9 ppm	2 × doublet	1,4-disubstituted benzene

(A 1,4-disubstituted benzene, with two different substituents, has a plane of symmetry, so two sets of doublets are expected—notice that the doublets are leaning towards each other in a roof effect, see margin.)

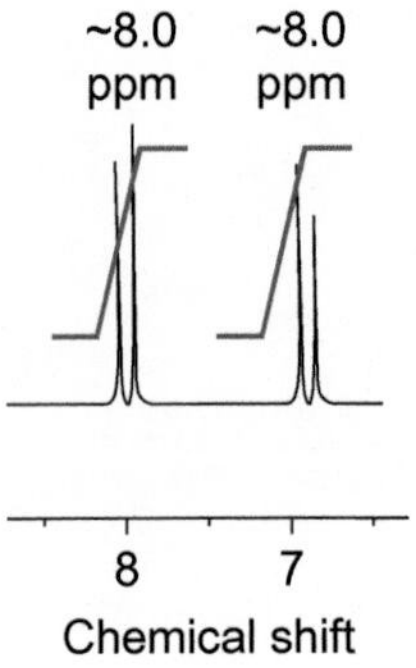
~8.0 ppm
~8.0 ppm
8
7
Chemical shift

4.3 ppm	quartet	CH_3–CH_2–
3.8 ppm	singlet	CH_3–(no neighbouring hydrogens)
1.4 ppm	triplet	CH_3–CH_2–

The chemical shifts in the ^1H NMR spectrum indicate an ester, whereas the absorption band at 1716 cm^{-1} in the IR spectrum suggests a ketone. However, the absorption band at 1716 cm^{-1} is consistent with the C=O stretch of an aromatic ester—an aromatic ester is expected to have a C=O stretching vibration at a lower wavenumber than for a saturated ester (1740 cm^{-1}) because of the +M effect of the aromatic ring (see margin and Section 12.2, p.574). →

ⓘ An aromatic ester has an aromatic ring bonded to the C=O bond of the ester (such as $PhCO_2R$). If an aromatic ring is bonded to the oxygen atom of the ester (such as RCO_2Ph), it is classed as an aliphatic ester.

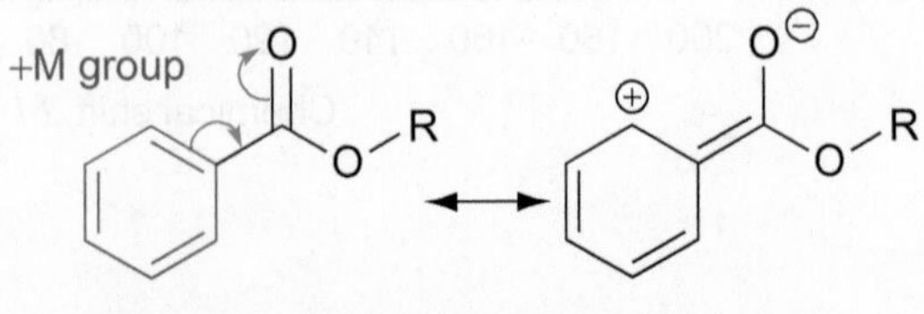
+M group
O
O
R
O
O
R

Resonance forms of an aromatic ester

➜ From the partial structures, the most likely structure is **A** or **B**, which give signals with the same splitting patterns at similar chemical shifts. (Check that both structures have five degrees of unsaturation.)

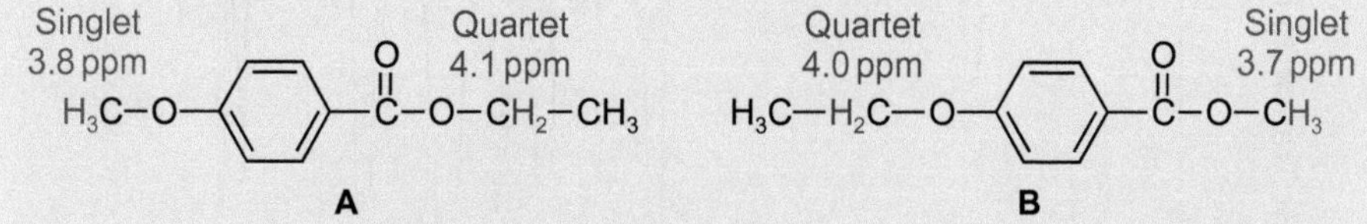
Singlet
3.8 ppm
Quartet
4.1 ppm
$H_3C-O-C_6H_4-C(=O)-O-CH_2-CH_3$
A
Quartet
4.0 ppm
Singlet
3.7 ppm
$H_3C-H_2C-O-C_6H_4-C(=O)-O-CH_3$
B

Step 4 The positions of the eight signals in the ^{13}C NMR spectrum are consistent with structures **A** and **B**. Chemical shifts (in ppm) are indicated below. (Notice that both signals from the benzene ring carbons attached to H atoms are more intense than the signals for the two benzene ring carbons attached to the functional groups.)

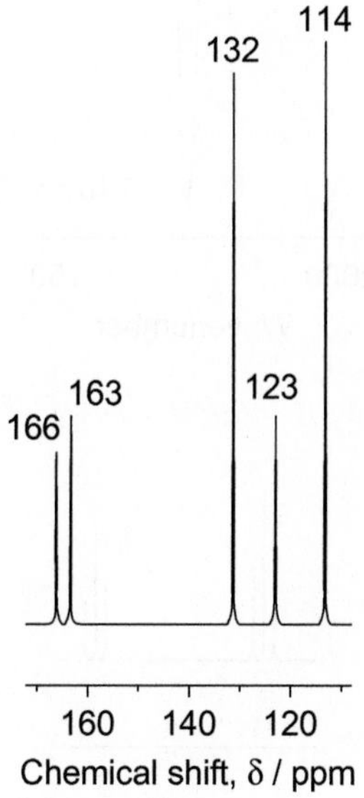
114
132
163
123
166
160
140
120
Chemical shift, δ / ppm

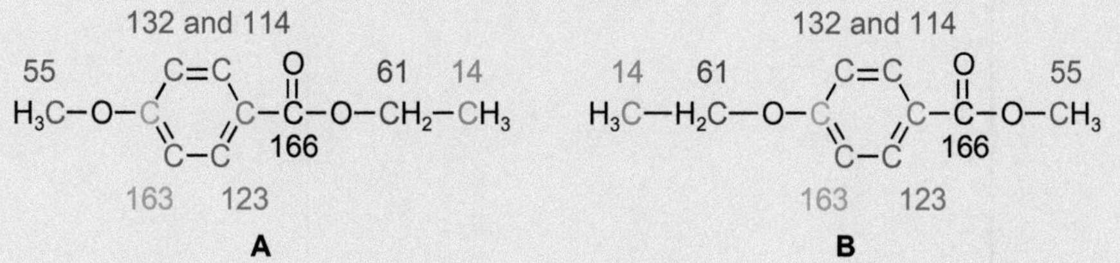
132 and 114
55
61
14
166
163
123
A
132 and 114
14
61
55
166
163
123
B

Steps 5 and 6 To determine if the structure is **A** or **B** look back at the mass spectrum. The major fragment peak has an *m/z* value of 135, which is expected to arise from cleavage of the C–O bond of the ester functional group (Section 12.1, p.565). Cleavage of the C–O bond of **A** gives a fragment peak at *m/z* 135, whereas fragmentation of **B** is expected to give a fragment ion at *m/z* 149.

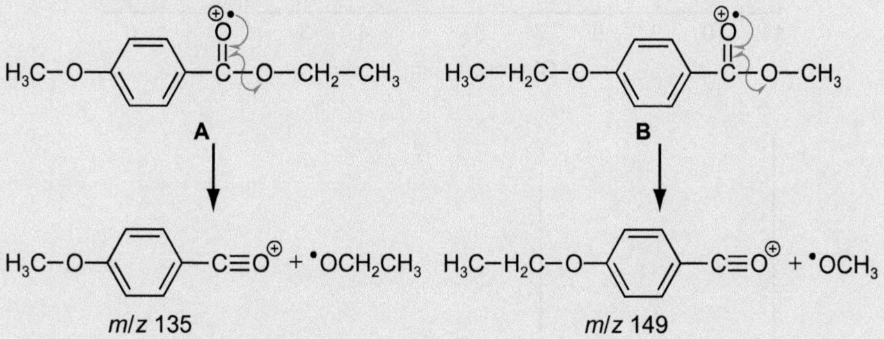
A
B
$+ \cdot OCH_2CH_3$
$+ \cdot OCH_3$
m/z 135
m/z 149

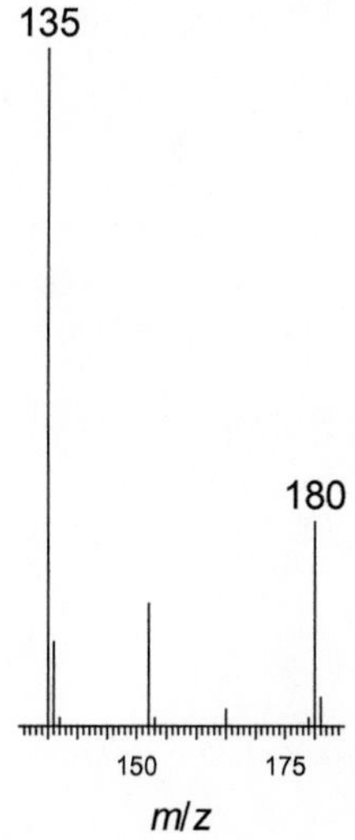
135
180
150
175
m/z

The unknown compound is ethyl 4-methoxybenzoate.

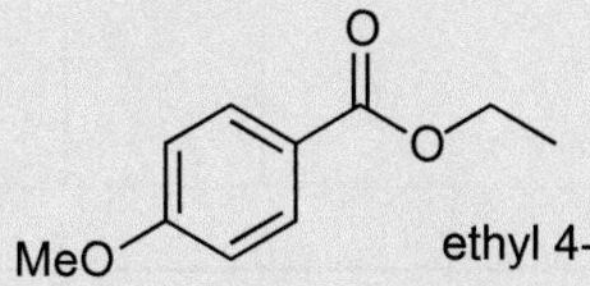
MeO
ethyl 4-methoxybenzoate

Question

A second compound, with the molecular formula $C_{10}H_{12}O_2$, was isolated from the same synthetic perfume. Using the IR spectrum and the ^1H and ^{13}C (proton decoupled) NMR spectra shown here (with expansions in the margin), propose a structure for the unknown compound.

➜

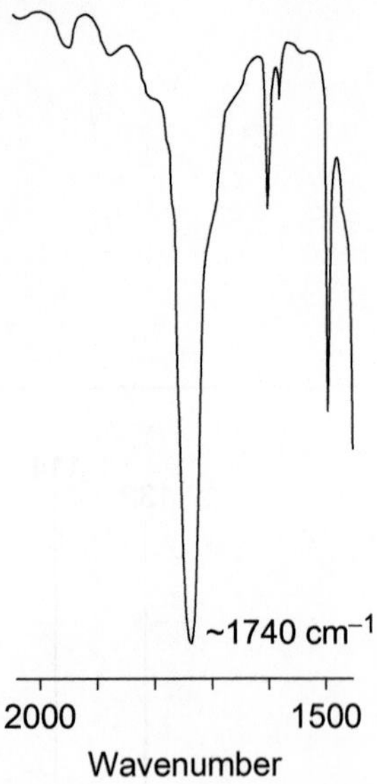

~1740 cm^{-1}
2000
1500
Wavenumber

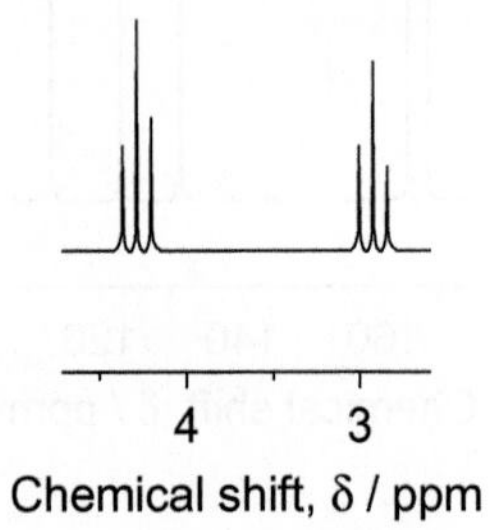

4
3
Chemical shift, δ / ppm

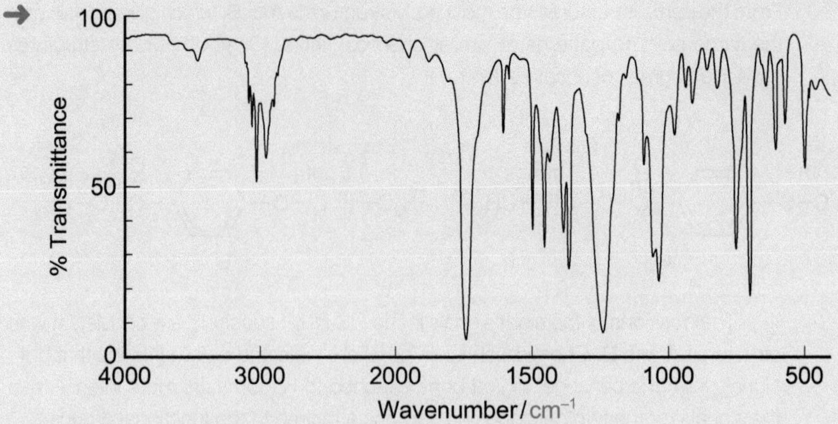

100
50
0
% Transmittance
4000
3000
2000
1500
1000
500
Wavenumber / cm^{-1}

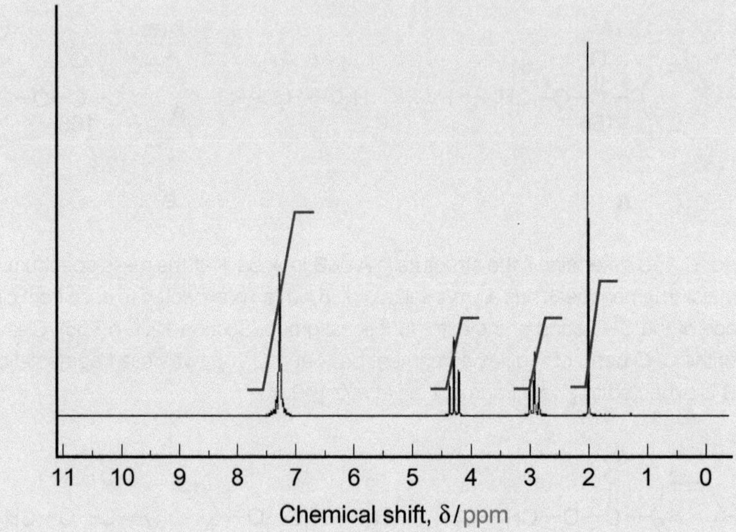

11
10
9
8
7
6
5
4
3
2
1
0
Chemical shift, δ / ppm

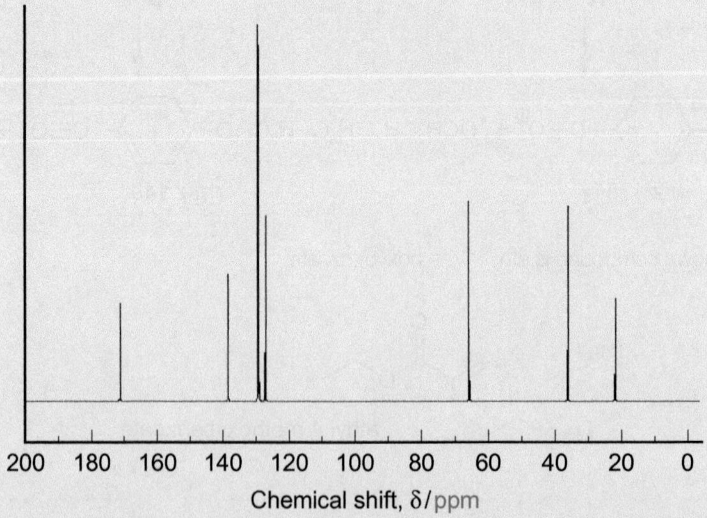

200
180
160
140
120
100
80
60
40
20
0
Chemical shift, δ / ppm

For practice questions on this topic, see questions 4–6 at the end of this chapter (pp.606–608).

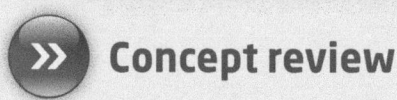

Concept review

By the end of this chapter, you should be able to do the following.

- Describe how an EI mass spectrum is produced.
- Understand that the molecular ion ($M^{+\bullet}$) gives the relative molecular mass of a molecule and that high-resolution mass spectrometers can determine the molecular formula of a molecule.
- Use M + 1 and M + 2 isotope peaks to calculate the number of carbon atoms and/or detect a chlorine or bromine atom(s) in a molecule.
- Identify and predict characteristic fragmentation patterns of molecular ions from compounds containing groups such as alkyl, halogenoalkyl, ether, amine, and carbonyl.
- Suggest partial structures for an unknown molecule, using the differences in mass between the molecular ion and various fragment cations.
- Understand the factors that influence the position of absorption bands in an IR spectrum.
- Use characteristic group wavenumbers to identify functional groups.
- Understand the factors that influence the number of resonance signals in a ^{1}H NMR spectrum and in a ^{13}C NMR spectrum.
- Describe the factors that affect the position of a resonance signal (recorded as a chemical shift) in a ^{1}H NMR spectrum and in a ^{13}C NMR spectrum.
- Use typical ^{1}H or ^{13}C chemical shift values (Figure 12.20) to predict types of CH_3, CH_2, and CH groups in an unknown molecule.
- For a ^{1}H NMR spectrum, use an integration curve to calculate the relative number of hydrogen atoms that give rise to each signal.
- For a ^{1}H NMR spectrum, use the splitting pattern of a signal to determine the number of hydrogen atoms bonded to an adjacent carbon.

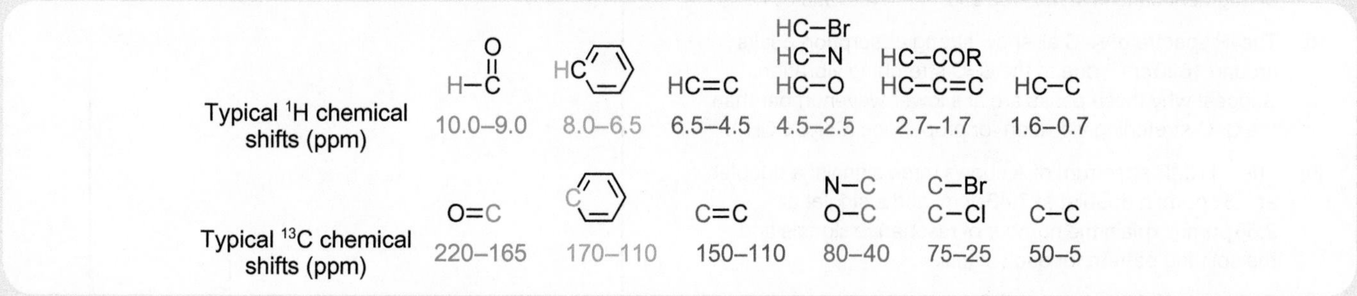
Typical ^{1}H chemical shifts (ppm)
H–C(=O) 10.0–9.0
HC (aromatic) 8.0–6.5
HC=C 6.5–4.5
HC–Br
HC–N
HC–O 4.5–2.5
HC–COR
HC–C=C 2.7–1.7
HC–C 1.6–0.7
Typical ^{13}C chemical shifts (ppm)
O=C 220–165
C (aromatic) 170–110
C=C 150–110
N–C
O–C 80–40
C–Br
C–Cl 75–25
C–C 50–5

Figure 12.20 Summary of typical ^{1}H and ^{13}C chemical shifts.

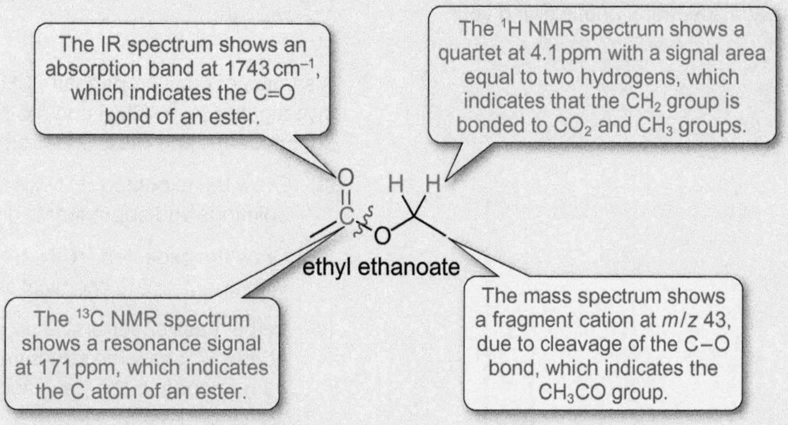
The IR spectrum shows an absorption band at 1743 cm^{-1}, which indicates the C=O bond of an ester.
The ^{1}H NMR spectrum shows a quartet at 4.1 ppm with a signal area equal to two hydrogens, which indicates that the CH_2 group is bonded to CO_2 and CH_3 groups.
ethyl ethanoate
The ^{13}C NMR spectrum shows a resonance signal at 171 ppm, which indicates the C atom of an ester.
The mass spectrum shows a fragment cation at m/z 43, due to cleavage of the C–O bond, which indicates the CH_3CO group.

Questions

More challenging questions are indicated by an asterisk *.

1 The following questions relate to the analysis of the methyl ketones **A**–**C** using different spectroscopic techniques. (Sections 12.1–12.3)

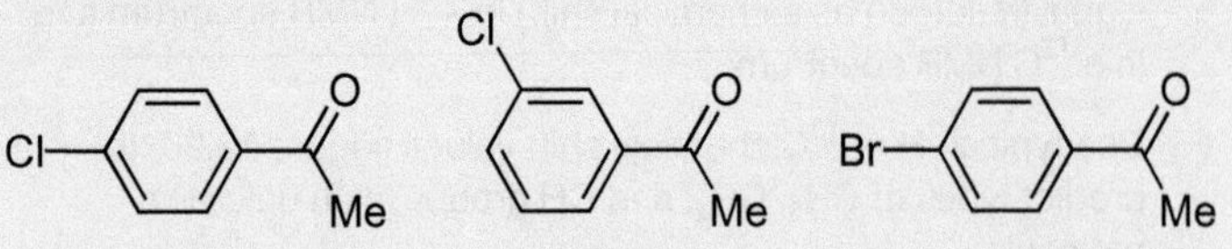
Cl
O
Me
Cl
O
Me
Br
O
Me

A **B** **C**

(a) In the mass spectra of both **A** and **B**, explain the appearance of two molecular ion peaks at *m/z* 154 and 156 in the ratio 3 : 1, respectively.

(b) Would you expect compound **C** to give two molecular ion peaks and, if so, would they also be in a 3 : 1 ratio? Briefly explain your reasoning.

(c) In the mass spectra of both **A** and **B**, explain the appearance of fragment cations at *m/z* 139 and 141 in the ratio 3 : 1.

(d) The IR spectra of **A**–**C** all show strong absorption peaks around 1690 cm^{-1} due to the C=O stretching vibration. Suggest why these peaks are at a lower wavenumber than the C=O stretching vibration for propanone (CH_3COCH_3).

(e) The ^1H NMR spectrum of **A** shows three signals: a doublet at 7.88 ppm; a doublet at 7.43 ppm; and a singlet at 2.58 ppm. Explain the number of resonance signals and the splitting pattern for each signal.

(f) For the ^1H NMR spectrum of **B**, how many resonance signals would you expect? Briefly explain your reasoning.

(g) What major differences would you expect to see in the ^{13}C NMR spectra (proton decoupled) of **A** and **B**?

2 The ^1H NMR spectra (labelled (a), (b), (c)) of three isomers with the molecular formula C_4H_9Cl are shown here. Identify the isomer that produces each spectrum, briefly giving your reasons. (Section 12.3)

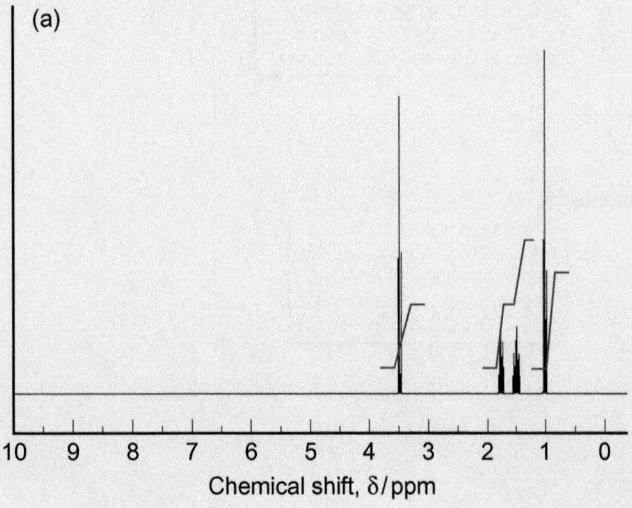
(a)
10 9 8 7 6 5 4 3 2 1 0
Chemical shift, δ/ppm

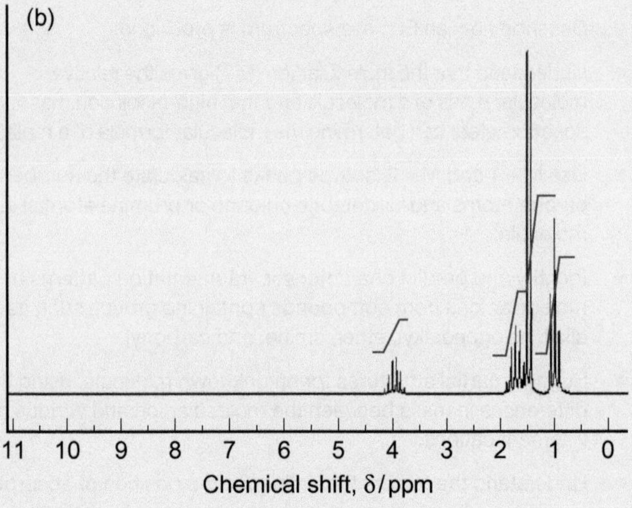
(b)
11 10 9 8 7 6 5 4 3 2 1 0
Chemical shift, δ/ppm

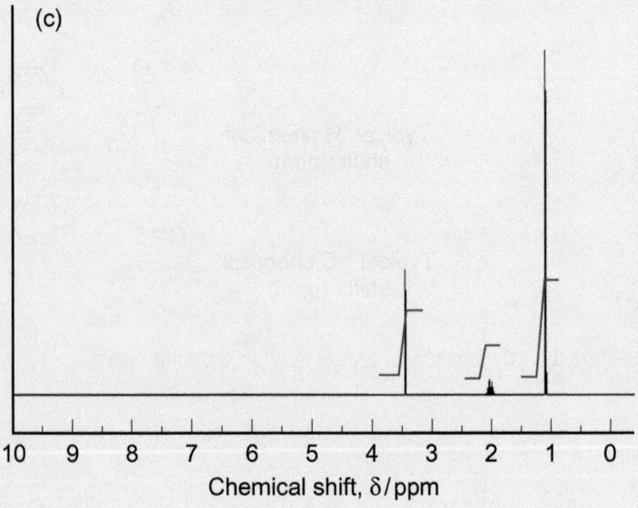
(c)
10 9 8 7 6 5 4 3 2 1 0
Chemical shift, δ/ppm

3 A sample of ethyl propanoate, $CH_3CH_2CO_2CH_2CH_3$, is investigated by ^1H NMR and ^{13}C NMR (proton decoupled) spectroscopy and mass spectrometry. (Sections 12.1–12.3)

(a) Draw the expected ^1H NMR spectrum showing spin–spin splittings and approximate chemical shifts.

(b) Draw the expected ^{13}C NMR spectrum (proton decoupled) with approximate chemical shifts.

(c) In the mass spectrum, an intense peak was observed at *m/z* 57. Draw the structure of the cation that corresponds to this peak and give a mechanism for its formation.

4 A compound was isolated from a synthetic attractant for beetles. Using the mass spectrum, IR spectrum, and ^1H NMR spectrum shown here, propose a structure for this compound. (Sections 12.1–12.4)

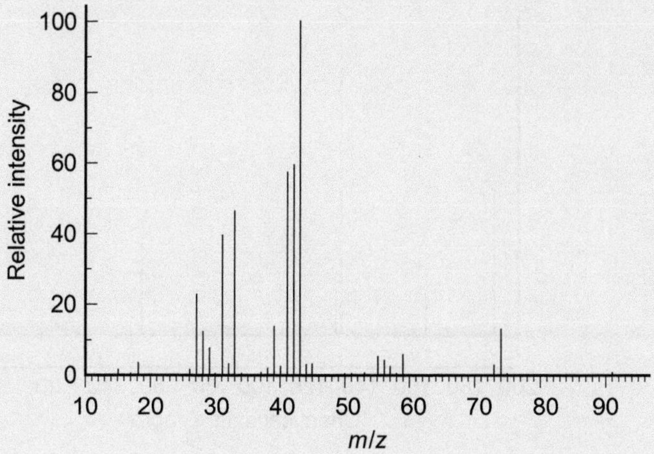
Relative intensity
100
80
60
40
20
0
10 20 30 40 50 60 70 80 90
m/z

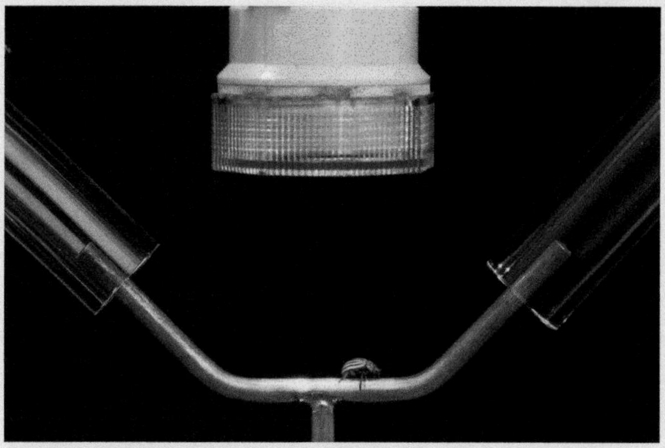

This Colorado potato beetle, a serious pest of potatoes and other crops, is being offered a choice between a chemical attractant and potato leaf aromas and is moving toward the attractant. Research like this helps to develop new methods of crop protection.

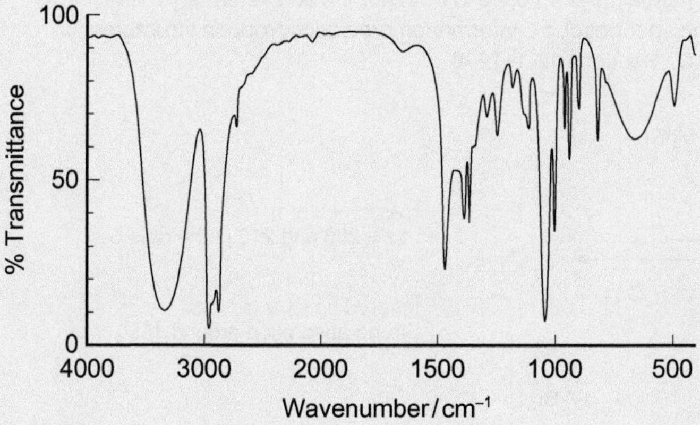
% Transmittance
100
50
0
4000 3000 2000 1500 1000 500
Wavenumber / cm⁻¹

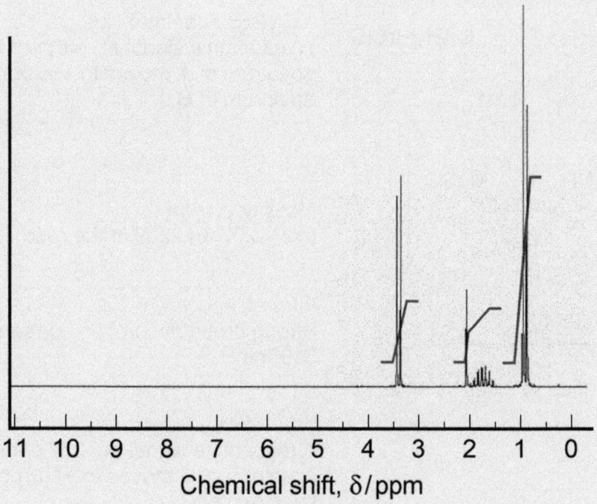
11 10 9 8 7 6 5 4 3 2 1 0
Chemical shift, δ / ppm

5 A strong-smelling compound was isolated from the anise plant. Using the mass spectrum, IR spectrum, ^{1}H NMR spectrum, and ^{13}C NMR spectrum shown here, propose a structure for this compound. (Sections 12.1–12.4)

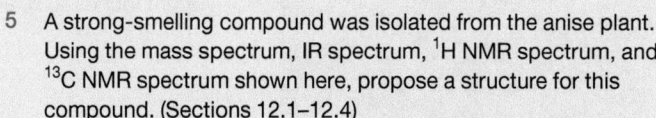

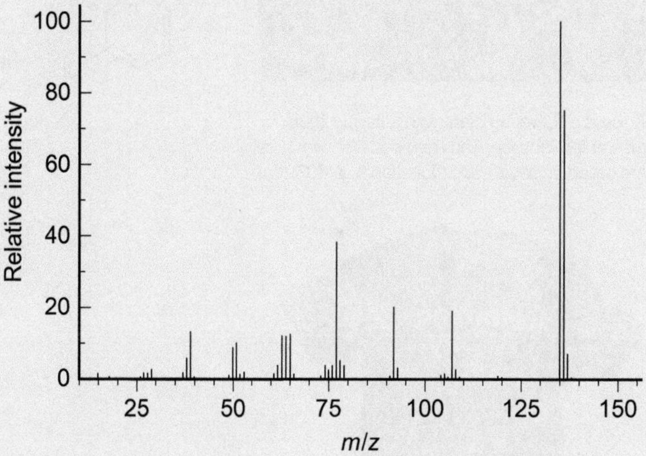
Relative intensity
100
80
60
40
20
0
25 50 75 100 125 150
m/z

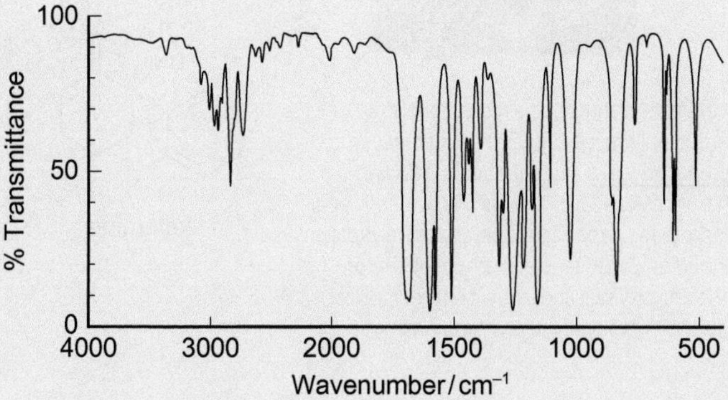
% Transmittance
100
50
0
4000 3000 2000 1500 1000 500
Wavenumber / cm⁻¹

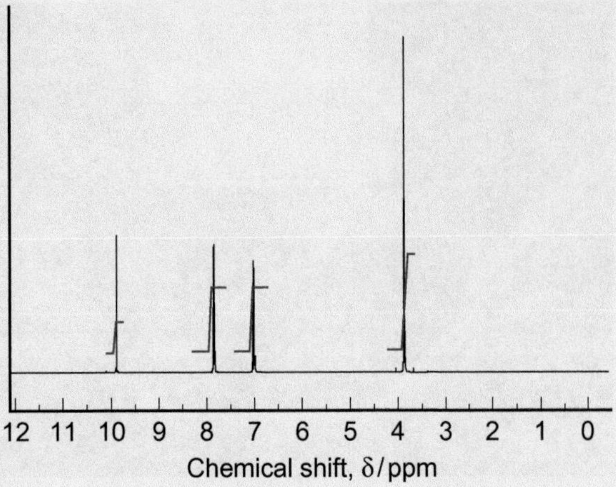
12 11 10 9 8 7 6 5 4 3 2 1 0
Chemical shift, δ/ppm

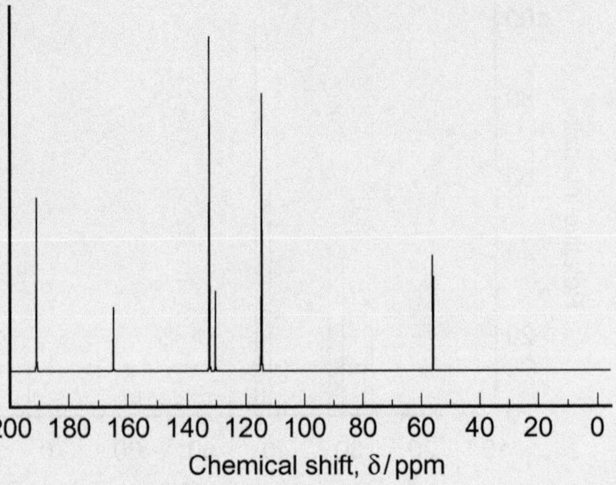
200 180 160 140 120 100 80 60 40 20 0
Chemical shift, δ/ppm

Anise is a flowering plant with a flavour that resembles liquorice. The seed pods (shown here) are called aniseed. (See Question 5, p.607).

Ketamine –
Just Say
Neigh.

Ketamine is abused as a recreational drug (commonly called 'K' or 'Special K') that can produce a wide range of effects including hallucinations. Overdose can cause death, and long-term use can cause psychological problems including dependence on continued use of the drug. Its sale and use, except for medical purposes, is illegal in many countries.

6 *A synthetic route to ketamine, a drug used in both human and veterinary medicine, is shown below. Using the spectroscopic information provided, propose structures for compounds A, B, and C. (Sections 12.1–12.4)

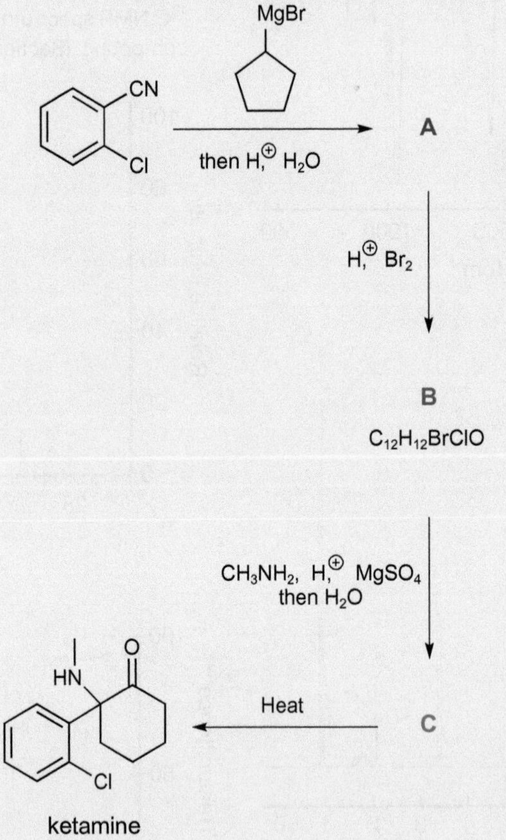
MgBr
CN
Cl
then H⊕, H2O
A
H⊕, Br2
B
C12H12BrClO
CH3NH2, H⊕, MgSO4
then H2O
C
Heat
HN
O
Cl
ketamine

Mass spectrum:
m/z = 208 and 210 in the ratio 3 : 1

Infrared spectrum:
Strong absorption around 1690 cm^{-1}

^{1}H NMR spectrum:
A resonance signal at ~3.0 ppm in the spectrum of A is absent in the spectrum of B

^{13}C NMR spectrum:
A resonance signal at ~46 ppm in the spectrum of A moves to ~59 ppm in the spectrum of B

Mass spectrum:
m/z = 237 and 239 in the ratio 3 : 1

Infrared spectrum:
Strong broad absorption appears around 3400 cm^{-1}

^{13}C NMR spectrum:
A resonance signal at ~207 ppm in the spectrum of B moves to ~165 ppm in the spectrum of C

13

Energy and thermochemistry

▼ Precision docking on the Space Shuttle.

Discovery

This chapter builds on the following topics:

- Measurement, units, and nomenclature Section 1.2, p.6
- Atoms and the mole Section 1.3, p.12
- Chemical equations Section 1.4, p.21
- Concentrations of solutions Section 1.5, p.34
- Energy changes in chemical reactions Section 1.6, p.41
- States of matter and phase changes Section 1.7, p.47
- Non-covalent interactions Section 1.8, p.52
- Using the ideal gas equation Section 8.2, p.349

Launch of the Space Shuttle, a striking example of chemical reactions producing heat and work. The clouds of white smoke are Al_2O_3 and $AlCl_3$ produced by the reaction in the booster rockets. ➤

Launching the Space Shuttle

Launching the Space Shuttle is a spectacular example of using chemical reactions to produce quickly a large quantity of energy. The part of the Shuttle that goes into space—the *orbiter*—weighs around 70 tonnes and can carry an additional 25 tonnes of payload. Getting this into Earth orbit needs a fuel system weighing much more than this.

To launch the shuttle requires a *propellant*—a mixture of a fuel and an oxidizer. The reaction between the fuel and the oxidizer is exothermic (releases heat) and produces a large quantity of gas. The heated gas expands rapidly and is ejected from the engine at high speed, propelling the Shuttle upwards.

The main lifting force needed to launch the Shuttle is provided by two solid rocket boosters, each of which contains over 500 tonnes of propellant. The propellant is a 1 : 1 stoichiometric mixture of powdered aluminium (the fuel) and ammonium perchlorate (NH_4ClO_4; a powerful oxidizer). A small quantity of iron(III) oxide is included as a catalyst. The thermochemical equation (see Section 13.3, p.622) for the reaction is

$$3\,Al(s) + 3\,NH_4ClO_4(s) \rightarrow Al_2O_3(s) + AlCl_3(s) + 6\,H_2O(g) + 3\,NO(g) \qquad \Delta_r H = -2674\,kJ\,mol^{-1}$$

The equation tells you that, when 3 mol of Al react with 3 mol of ammonium perchlorate, 9 mol of gaseous products (water vapour and NO) are produced and 2674 kJ of heat are released. The reaction heats the inside of the solid rocket boosters to over 3000 °C, causing the gases to expand rapidly to lift the Shuttle. The solid products Al_2O_3 and $AlCl_3$ are responsible for the dense white clouds ejected from the rockets on take-off. All the fuel is burned in about 2 minutes, by which time the Shuttle has reached an altitude of 150 000 feet (46 km). The solid rocket boosters are then jettisoned and parachute into the ocean to be recovered and used again.

The main engines on the Shuttle propel it into orbit and rely on a simple chemical reaction

$$2\,H_2(l) + O_2(l) \rightarrow 2\,H_2O(g) \qquad \Delta_r H = -484\,kJ\,mol^{-1}$$

Hydrogen is used since it delivers a larger quantity of energy per kg than most other fuels—it has a higher *energy density* (see Box 13.3, p.628). The huge external fuel tank is attached to the orbiter and contains 102 tonnes of liquid hydrogen (the fuel) and 620 tonnes of liquid oxygen (the oxidizer) in separate compartments. The high temperature produced results in the water vapour being ejected from the engines at around 6000 mph. The fuel is consumed in 8 minutes, by which time the Shuttle is in low orbit. The fuel tank is jettisoned and burns up when it re-enters the atmosphere.

Once the Shuttle is in orbit, it relies on a third propulsion system for manoeuvring. This is a *hypergolic* propellant in which the fuel methylhydrazine (CH_3HNNH_2) and the oxidizer dinitrogen tetroxide (N_2O_4) ignite on contact, without the need for an ignition system.

$$4\,CH_3HNNH_2(l) + 5\,N_2O_4(l) \rightarrow 12\,H_2O(g) + 4\,CO_2(g) + 9\,N_2(g) \qquad \Delta_r H = -4594\,kJ\,mol^{-1}$$

This is a highly reliable and controllable system that can be switched on and off as needed—unlike the system in the solid rocket boosters at the start of the launch that, once ignited, continues to burn until all the fuel is used.

Thermodynamics deals with how and why changes happen; rates of change—how fast or how slowly reactions happen—are discussed in Chapter 9 on reaction kinetics.

Much of chemistry is concerned with substances undergoing change. This may involve changing one substance into another—as with synthesizing new compounds—or it may involve a substance changing phase—such as melting from a solid to a liquid. Understanding how and why changes happen involves a consideration of the *energy* changes that occur. The branch of chemistry known as *thermodynamics* deals with energy in chemistry. This important and powerful topic is based around some simple laws and equations that allow chemists to understand and predict the behaviour of substances and their reactions.

Thermodynamics is concerned with the study of **macroscopic** systems, that is, those consisting of large, measurable amounts of matter. It does not deal with the properties of individual molecules. Indeed, knowing about molecules isn't necessary to discuss thermodynamics, although considering molecular behaviour often helps to understand many of the concepts.

Chemical reactions can absorb or release *heat*. A familiar example is the use of hydrocarbons as fuels. For example, natural gas is largely methane (CH_4), petrol is a mixture of short-chain hydrocarbons such as C_8H_{18}, whereas heating oils contain longer-chain hydrocarbons. Reactions can also produce other changes. While burning fuel in a car engine certainly produces heat, it is the motion imparted to the pistons that drive the vehicle that is important. This is an example of a chemical reaction doing *work*. These concepts—using chemistry to produce heat and work—are linked in thermodynamics. This chapter is concerned with how to calculate and measure the energy changes that occur during reactions.

A hot-air balloon. Burning butane (C_4H_{10}) fuel heats the air in the balloon, which causes the balloon to rise.

In chemical changes, energy can be transferred in a number of ways (see Section 1.6, p.41) but this chapter is mainly concerned with the transfer of energy as heat and work.

Spark plug
Inlet for petrol-air mixture
Outlet for exhaust
Piston

Another example of a chemical reaction causing motion. Here, hydrocarbon fuel burned in the engine cylinder (inset) does work to make the pistons move, which is used to turn the wheels of a car.

13.1 Energy changes in chemistry: heat and work

The Space Shuttle launch described on p.611 is a spectacular example of chemical reactions supplying heat and work. A more straightforward example from a laboratory is discussed in Section 1.6 (see Figure 1.13, p.41) where calcium carbonate is added to a dilute solution of hydrochloric acid. The reaction produces carbon dioxide gas, which pushes up the piston against the pressure of the atmosphere and so does work. The reaction is exothermic and gives out heat so the temperature of the solution rises. So, energy generated by the reactions is transferred as both heat and work.

- **Heat** is energy that is transferred as a result of temperature differences.
- **Work** is done as a result of motion against an opposing force.

Exothermic and endothermic reactions are introduced in Section 1.6 (p.41).

In general, *changes* of energy are more important than the actual energies of products or reactants.

In the earliest days of scientific investigation, heat and work were thought to be separate, unconnected phenomena. The link between them was first demonstrated by some elegant experiments performed in the early nineteenth century by James Joule (1818–1889), after whom the unit of energy is named. As described in Box 13.1, he carried out some very careful measurements and showed that heat and work are alternative ways in which energy can be transferred.

The energy changes that occur when systems do work are discussed further in Section 13.5 (p.636).

Box 13.1 James Joule and the equivalence of heat and work

James Joule conducted experiments in which the water in a beaker was stirred by a paddle that was attached to a falling weight. He observed that the temperature of the water gradually rose as it was stirred mechanically, but that this was not a result of heat transfer from the surroundings. By comparing the work done by the falling weight to the temperature increase, Joule proved that mechanical work could be converted to heat.

It is claimed that Joule was so dedicated to his work that he even used his honeymoon to further his experiments. During a visit to Chamonix in Switzerland, he very carefully measured the temperature of water at the top and bottom of a series of waterfalls. For example, he found that the water was just over 0.5 °C warmer at the bottom of a waterfall that had a drop of 250 m. The values could be exactly accounted for by the difference in potential energy between the top and bottom of the falls and the conversion of potential energy to kinetic energy during the fall. (Potential energy and kinetic energy are discussed in Box 1.7, p.42.)

Question

In a reproduction of the Joule experiment, a stirrer is connected to a mass of 1.0 kg that can fall through a height of 2.0 m. It takes 4200 J to heat 100 g of water by 10 °C. How many times must the mass be dropped to cause this heating?

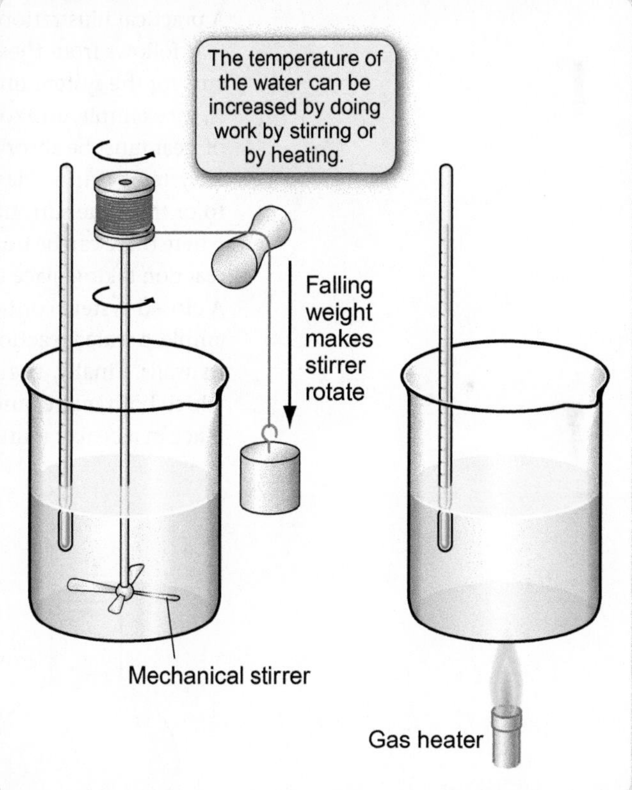
The temperature of the water can be increased by doing work by stirring or by heating.
Falling weight makes stirrer rotate
Mechanical stirrer
Gas heater

Joule showed that heat and work are both methods for transferring energy to a system. Vigorous stirring *or* heating with a gas burner both raise the temperature of the water.➤

Systems and the surroundings

It is common in chemistry to refer to the **system** under study. This is simply the *part of the Universe that is of interest*, that is, the sample of matter or the reaction that is being studied. The rest of the Universe is described as the **surroundings**. This definition allows you to keep track of where energy and matter move during a reaction.

In studying thermodynamics, you need to be careful to make sure that words are used precisely to avoid ambiguity or mistakes. Precise use of symbols and terminology is vital in keeping track of what is going on.

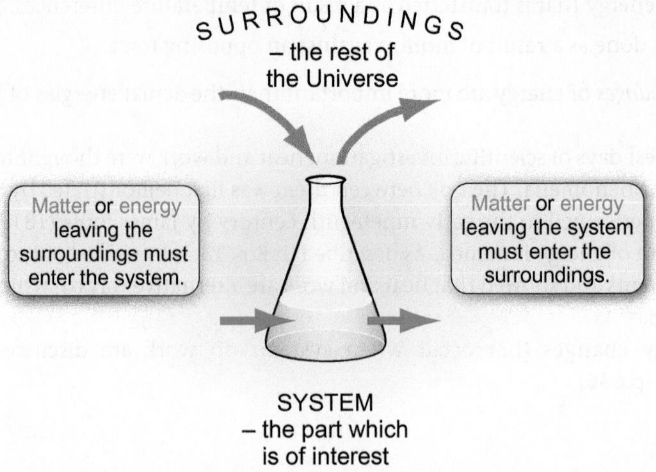
SURROUNDINGS
– the rest of
the Universe
Matter or energy
leaving the
surroundings must
enter the system.
Matter or energy
leaving the system
must enter the
surroundings.
SYSTEM
– the part which
is of interest

Figure 13.1 The system and surroundings. A change for the system causes a change with the same magnitude (size) but with opposite sign in the surroundings.

A practical illustration of different types of system is given in Worked example 13.1.

It follows from these definitions that, as illustrated in Figure 13.1, any change that occurs for the system must result in the opposite change for the surroundings. If the system is, for example, an exothermic reaction that gives out a quantity of heat, the same quantity of heat must be absorbed by the surroundings (assuming that no work is done).

Systems can be classified depending on whether matter or energy can be transferred to or from the surroundings. This is illustrated in Figure 13.2. An **isolated** system is one where there can be no exchange of energy or matter with the surroundings. For example, a reaction taking place in a sealed, well insulated vacuum flask would be an isolated system. A **closed** system contains a fixed amount of matter but allows exchange of energy, for example, a sealed reaction vessel holds all the contents in but heat can be conducted through its walls. Finally, the most common type of system encountered is an **open** system, in which both matter and energy can be exchanged with the surroundings. A reaction taking place in a beaker is an open system.

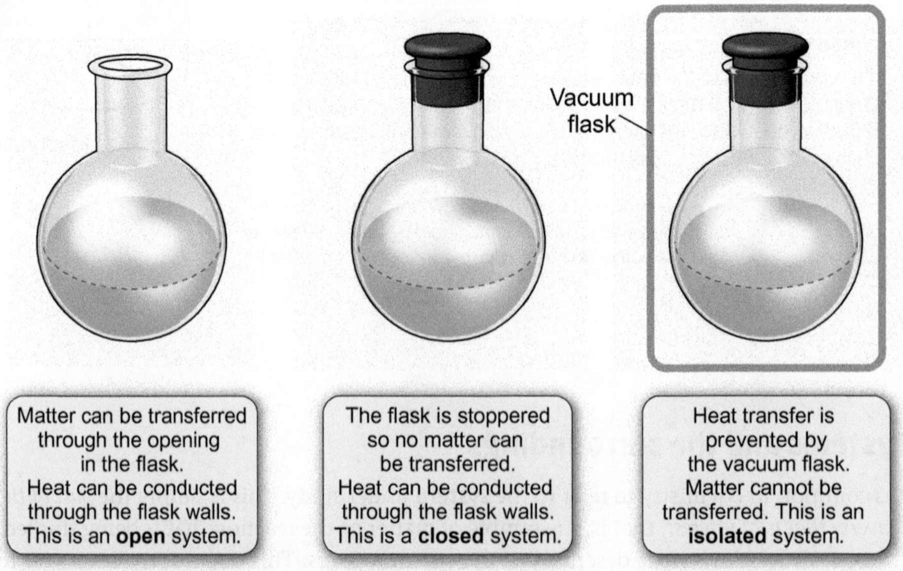
Vacuum
flask
Matter can be transferred through the opening in the flask. Heat can be conducted through the flask walls. This is an open system.
The flask is stoppered so no matter can be transferred. Heat can be conducted through the flask walls. This is a closed system.
Heat transfer is prevented by the vacuum flask. Matter cannot be transferred. This is an isolated system.

Figure 13.2 Types of chemical system.

Worked example 13.1 Types of system: the Breitling Orbiter

In 1999, the Breitling Orbiter 3 became the first piloted balloon to fly non-stop around the world. It was made up of several components, as shown in the figure. The main lift was provided by two sealed helium balloons. These were inside a hot-air balloon that was heated by a gas burner to 'fine-tune' the lift and maintain a controlled altitude.

BREITLING
ORBITER 3

▲ The Breitling Orbiter.

What type of system is represented by each of these components of the Breitling Orbiter: (a) the large helium balloon; (b) the hot-air balloon?

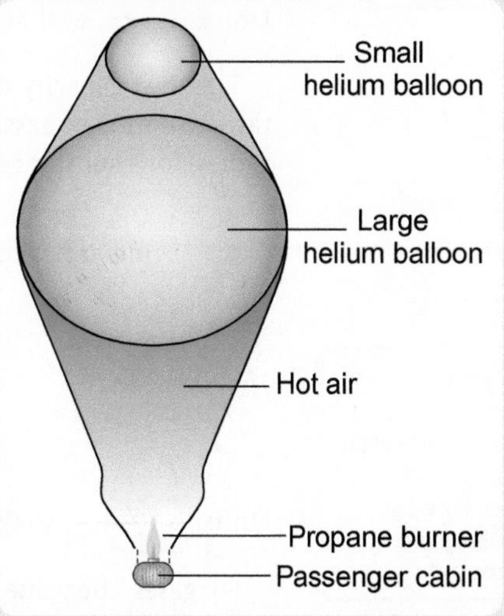
Small
helium balloon
Large
helium balloon
Hot air
Propane burner
Passenger cabin

▲ Two sealed helium balloons are contained within an outer hot-air balloon.

Strategy

Look at the definition for each type of system shown in Figure 13.2 and decide whether matter and heat can be exchanged with the surroundings by each of the components of the balloon.

Solution

(a) The helium balloon is sealed so that no exchange of matter can take place. The balloon material will conduct heat, so it is not an isolated system. It is a *closed* system.

(b) Since there is a gas burner, heat can clearly be transferred to the system. Air can escape from the balloon so that matter transfer can also take place. So, it is an *open* system.

Question

What type of system is the Space Shuttle in orbit around the Earth?

Heat capacities

Heating causes changes of temperature. However, different substances heated with the same amount of energy do not give the same temperature change. In order to quantify energy changes the **heat capacity**, given the **symbol** C, is used. The heat capacity is the heat needed to raise the temperature of a substance by 1 K. The *specific* heat capacity, C_s, refers to *1 gram* of a substance.

If a quantity of energy, q (in joules), is transferred to a mass, m (in grams), of a substance and the temperature rises by ΔT (in kelvin) then

$$\text{specific heat capacity} = \frac{\text{quantity of heat supplied}}{\text{mass of substance} \times \text{temperature rise}}$$

'Specific' values of a quantity refer to unit mass of a substance, usually per gram or per kilogram. Compare this with 'molar' values, which refer to 1 mol of a substance.

$$C_s = \frac{q}{m \times \Delta T} \tag{13.1}$$

Units: $\frac{J}{g \times K}$, so that C_s has units of $J\,K^{-1}\,g^{-1}$.

The heat capacity of a substance can also be defined in terms of the amount in moles rather than the mass. This is the **molar heat capacity**, C_m. Thus, for an amount, n (in moles), of a substance

$$\text{molar heat capacity} = \frac{\text{quantity of heat supplied}}{\text{amount of substance (in mol)} \times \text{temperature rise}}$$

Equations 13.1 and 13.2 are approximations since the heat capacity changes slightly with temperature. For example, C_s for water is 4.180 $J\,K^{-1}\,g^{-1}$ at 25 °C and 4.205 $J\,K^{-1}\,g^{-1}$ at 90 °C. However, since the variation is quite small, it is usually ignored except in the most accurate work.

$$C_m = \frac{q}{n \times \Delta T} \tag{13.2}$$

Units: $\frac{J}{mol \times K}$, so that C_m has units of $J\,K^{-1}\,mol^{-1}$.

For gases, the value of the heat capacity depends on whether the measurement is carried out at **constant pressure** (given the symbol C_p) or at **constant volume** (given the symbol C_V). C_p and C_V refer to molar amounts and more correctly have the symbols $C_{m,p}$ and $C_{m,V}$, respectively, but the 'm' is often omitted for simplicity.

Table 13.1 summarizes the different kinds of heat capacities.

You can find a fuller list of heat capacity values in Appendix 7, p.1352.

Some values of heat capacities are shown in Table 13.2. C_m is similar for all the metals—suggesting that about the same quantity of heat is needed to change their temperature by 1 K. The values are also similar for the diatomic gases N_2, CO, and O_2 but, as the molecules become more complex, C_m increases—compare CO with CO_2 or CH_4, C_2H_6, and C_6H_6. This is because there are more bonds that can absorb the energy in the molecule. Worked example 13.2 shows how values of heat capacities can be used.

Table 13.1 Different kinds of heat capacities

	Symbol	Definition	Units
Heat capacity	C	Heat required to raise the temperature of a substance by 1 K	$J\,K^{-1}$
Specific heat capacity	C_s	Heat required to raise the temperature of 1 gram of a substance by 1 K	$J\,K^{-1}\,g^{-1}$
Molar heat capacity	C_m	Heat required to raise the temperature of 1 mol of a substance by 1 K	$J\,K^{-1}\,mol^{-1}$
For gases			
Molar heat capacity at constant pressure	C_p	Heat required to raise the temperature of 1 mol of a gas by 1 K at constant pressure	$J\,K^{-1}\,mol^{-1}$
Molar heat capacity at constant volume	C_V	Heat required to raise the temperature of 1 mol of a gas by 1 K at constant volume	$J\,K^{-1}\,mol^{-1}$

Table 13.2 Heat capacities of substances at 298 K (25 °C)*

Substance	Heat capacity	
	C_s / J K^{-1} g^{-1}	C_m / J K^{-1} mol^{-1}
Carbon (graphite)	0.67	8.5
Iron (Fe (s))	0.45	25.1
Copper (Cu (s))	0.38	24.4
Gold (Au (s))	0.129	25.4
Hydrogen (H_2 (g))	14.4	28.8
Nitrogen (N_2 (g))	1.04	29.1
Oxygen (O_2 (g))	0.92	29.4
Carbon monoxide (CO (g))	1.04	29.1
Carbon dioxide (CO_2 (g))	0.84	37.1
Methane (CH_4 (g))	2.23	35.7
Ethane (C_2H_6 (g))	1.75	52.5
Benzene (C_6H_6 (g))	1.05	82.4
Water (H_2O (l))	4.18	75.3
Water vapour (H_2O (g))	1.87	33.6

* The values are measured at constant pressure.

Worked example 13.2 Using heat capacities

The specific heat capacity of liquid water is 4.18 J K^{-1} g^{-1}. Calculate the energy required to heat 1.0 mol of water from 298 K (25 °C) to 363 K (90 °C).

Strategy

Note that the question gives you C_s, the *specific* heat capacity. Take the molar mass of water as 18 g mol^{-1} and use Equation 13.1 to find the energy transferred. (You can assume that the heat capacity is constant over this temperature range.)

Solution

Rearrange Equation 13.1 to obtain an expression for the quantity of heat, q.

$$C_m = \frac{q}{m \times \Delta T} \qquad (13.1)$$

$$q = m \times C_s \times \Delta T$$

Substitute in the values directly.

$$q = (1.0\,\cancel{\text{mol}} \times 18\,\cancel{\text{g}}\,\cancel{\text{mol}^{-1}}) \times (4.18\text{ J}\,\cancel{\text{K}^{-1}}\,\cancel{\text{g}^{-1}})$$
$$\times (363\,\cancel{\text{K}} - 298\,\cancel{\text{K}})$$
$$= 4891\text{ J}$$
$$= 4.9\text{ kJ}$$

(Note that the temperatures are in K. Even though it is not essential in this particular problem, since the value of ΔT is the same in both K and °C, it is a good habit *always* to use K as the unit for temperature.)

Question

The specific heat capacity of copper metal is 0.38 J K^{-1} g^{-1}. Calculate the temperature rise of a 100 g bar of copper when 250 J of heat are transferred to it.

Intensive and extensive properties

Specific and molar heat capacities are examples of intensive properties: an **intensive property** of a substance does not depend on the quantity of the substance present. The specific heat capacity of liquid water is 4.18 $J K^{-1} g^{-1}$, whether you have 0.1 g of water or 1.0 kg. Compare this with an **extensive property** like mass or volume that does depend on the quantity of substance present.

Table 13.3 gives more examples of intensive and extensive properties.

Table 13.3 Intensive and extensive properties

Intensive	Extensive
Value does not depend on the quantity of the substance present	Value depends on the quantity of substance present
Examples:	**Examples:**
Specific heat capacity	Mass
Molar heat capacity	Length
Density	Volume
Pressure	
Electrical conductivity	

Summary

- Chemical systems can be classified as:
  - *isolated systems* that have no exchange of energy or matter with the surroundings;
  - *open systems* that have both energy and matter exchange with the surroundings;
  - *closed systems* that contain a fixed amount of matter but allow energy exchange.
- Heat is energy transferred as a result of a temperature difference.
- Work involves energy exchange as a result of motion against an opposing force.
- The heat capacity, C, relates the heat supplied to a substance to the resulting temperature rise.
  - Specific heat capacity, C_s, is the energy required to raise the temperature of 1 g of a substance by 1 K

$$C_s = \frac{q}{m \times \Delta T}$$

  - Molar heat capacity, C_m, is the energy required to raise the temperature of 1 mol of a substance by 1 K

$$C_m = \frac{q}{n \times \Delta T}$$

- For gases, C_p and C_V are the molar heat capacities at constant pressure and constant volume, respectively.
- *Extensive* properties depend on the quantity of substance that is present; *intensive* properties do not.

? For practice questions on these topics, see questions 1–3 at the end of this chapter (p.651).

13.2 Enthalpy and enthalpy changes

Most chemical reactions in the laboratory are carried out in *open systems*. For example, an acid reacting with magnesium in a beaker is an open system (Figure 13.3). The atmospheric pressure does not change during the reaction, so the reaction occurs at *constant pressure*. The reaction is exothermic. The heat released during the reaction is known as the **enthalpy change, ΔH.**

The enthalpy change ΔH is the heat transferred at constant pressure by a chemical reaction or process.

Enthalpy changes for reactions can be measured in a number of ways (see Section 13.6, p.644). Studying heat changes that occur during reactions—a branch of science known as **thermochemistry**—is important in a number of areas. For example, since heat changes during reactions result from making and breaking chemical bonds (see Section 1.6, p.41), the values can give information about reaction mechanisms. If you are operating a chemical plant to carry out a reaction on a large scale, you need to calculate how much heat will be produced in order to be able to design and operate the plant safely.

Enthalpy changes are introduced in Section 1.6 (p.41).

H2
Hydrogen gas escapes and does work against the atmosphere.
Magnesium reacts with acid.

Figure 13.3 Magnesium reacts with acid to produce hydrogen. The hydrogen gas given off in the reaction escapes into the atmosphere. As it escapes, it does work as it expands against the pressure of the atmosphere. Atmospheric pressure does not change, so the reaction has occurred at constant pressure.

This chapter deals mainly with enthalpy changes that occur when compounds react chemically or change their physical state. However, enthalpy changes can be defined for many other processes. Examples include enthalpy changes of ionization and electron gain enthalpy changes described in Section 6.5 (p.290).

Enthalpy as a state function

Enthalpy is an example of a **state function**. Its value depends only on the current state of the system, defined by the pressure and temperature of the system. Another way of expressing this is to say that enthalpy depends on the present state of the system but not on its previous behaviour or on how the system came to be at its present state.

The most important thing to remember about a state function is that the change in value of the function depends only on the final and initial conditions. It is independent of the path between them. Therefore, for any state function, X

$$\Delta X = X(\text{final state}) - X(\text{initial state})$$

Other state functions include the pressure, volume, and temperature of a system

$$\Delta V = V(\text{final}) - V(\text{initial}) \quad \text{change in volume}$$

$$\Delta p = p(\text{final}) - p(\text{initial}) \quad \text{change in pressure}$$

$$\Delta T = T(\text{final}) - T(\text{initial}) \quad \text{change in temperature}$$

For example, suppose you have a beaker of water at 25 °C and you want to raise its temperature to 85 °C. You could simply heat it from 25 °C to 85 °C, or alternatively you could heat it to 95 °C, cool it to 80 °C, heat it to 90 °C, and cool it to 85 °C. Either way, the final temperature is 85 °C and $\Delta T = 60$ °C. An everyday life analogy for a state function is the change in altitude when climbing a mountain (see Figure 13.4).

Change in altitude

Figure 13.4 Altitude as an analogy for a state function. A climber could take two paths up the mountain: straight up (the red, dashed line) or a gentle climb (the blue line). The distance travelled would be very different, although the *change* in altitude would be the same for the two paths.

A **state function** is a quantity whose value depends only on the initial and final state of the system, but not on the route taken to get from the initial to the final state.

Enthalpy changes during changes of phase

Matter can exist as solid, liquid, or gas and you will often have observed the enthalpy changes that accompany changes of phase. Think about putting ice into a drink: the ice melts. In order to melt the ice, the intermolecular attractions between the molecules in ice have to be overcome. This takes energy, which is absorbed from the liquid so the liquid cools down. (Box 1.10, p.47, discusses the phase changes of water and how they affect our everyday lives.)

The transition from solid to liquid, usually known as melting, is more correctly called **fusion**. The **standard enthalpy change of fusion**, $\Delta_{fus}H^{\ominus}$ is the energy required to melt

States of matter and the enthalpy changes during phase changes are introduced in Section 1.7 (p.50). Phase equilibrium is treated in more detail in Section 17.2 (p.775). Intermolecular interactions are introduced in Section 1.8 (p.52) and are discussed in more detail in Section 17.3 (p.783).

The **standard enthalpy change of fusion**, $\Delta_{fus}H^{\ominus}$, is the enthalpy change when 1 mol of a substance melts at its melting point and 1 bar pressure. An earlier name for $\Delta_{fus}H^{\ominus}$, still in common use, is the latent heat of fusion.

The **standard enthalpy change of vaporization**, $\Delta_{vap}H^{\ominus}$, is the enthalpy change when 1 mol of a liquid vaporizes at its boiling point and 1 bar pressure. An earlier name for $\Delta_{vap}H^{\ominus}$, still in common use, is the latent heat of vaporization.

Standard conditions for enthalpy changes, denoted by the superscript symbol $\ominus$, are discussed in Section 1.6 (p.41).

Fusion and vaporization are endothermic processes, so values for the enthalpy changes are *always* positive.

You can find help with carrying out calculations involving moles in Section 1.3 (p.12).

At atmospheric pressure, solid carbon dioxide ('dry ice') turns from solid directly to a gas with no liquid appearing—it sublimes.

one mole of a pure substance at its melting point, T_m, at 1 bar pressure. Some typical values are shown in Table 13.4.

Vaporization is the transformation of a liquid to the gas phase. Similar to $\Delta_{fus}H^{\ominus}$, $\Delta_{vap}H^{\ominus}$ is defined as the energy required to vaporize one mole of a pure liquid at its boiling point, T_b, at 1 bar pressure. The values in Table 13.4 show that $\Delta_{vap}H^{\ominus}$ is considerably larger than $\Delta_{fus}H^{\ominus}$. This is because more energy is needed to overcome the intermolecular attractions holding the liquid together so the molecules are completely separated as a gas, than in melting a solid to a liquid. Vaporizing 1 mol of a liquid overcomes *all* of the forces between the molecules, so $\Delta_{vap}H^{\ominus}$ is used as a measure of the strength of intermolecular forces in liquids. This explains why liquids that form hydrogen bonds, such as water and ethanol, have relatively high values for $\Delta_{vap}H^{\ominus}$ (see Section 1.8, p.52).

The enthalpy changes that occur during changes of phase are illustrated in Worked examples 13.3 and 13.4.

Table 13.4 Standard enthalpy changes of fusion and vaporization, and normal melting points and boiling points, for selected substances

	T_m/K	$\Delta_{fus}H^{\ominus}$ / kJ mol^{-1}	T_b/K	$\Delta_{vap}H^{\ominus}$ / kJ mol^{-1}
Helium (He)	1	+0.02	4.2	+0.1
Argon (Ar)	84	+1.2	87.5	+6.5
Methane (CH_4)	91.1	+0.94	109.1	+8.2
Ammonia (NH_3)	195.3	+5.65	239.7	+23.4
Ethanol (C_2H_5OH)	155.8	+4.60	351.6	+43.5
Water (H_2O)	273.2	+6.01	373.2	+40.7
Mercury (Hg)	234.3	+2.29	629.7	+59.3
Sodium (Na)	371.0	+2.60	1156	+98.0
Silver (Ag)	1235	+11.3	2485	+250.6

Worked example 13.3 Enthalpy change of vaporization

The standard enthalpy change of vaporization of ethanol is +43.5 kJ mol^{-1}. Calculate the enthalpy change when 0.50 g of ethanol vaporizes at its boiling point at 1 bar.

Strategy

The enthalpy change when 1 mol vaporizes is +43.5 kJ, so you need to calculate the amount in moles in 0.50 g and hence the enthalpy change.

Solution

Calculate the amount in moles.

$$M_r(C_2H_5OH) = 46.07$$

$$\text{Amount (in mol)} = \frac{0.50\,\cancel{g}}{46.07\,\cancel{g}\,mol^{-1}} = 0.011\,mol$$

Multiply the amount by the enthalpy change per mole to get ΔH.

$$\Delta H = (0.011\,\cancel{mol}) \times (+43.5\,kJ\,\cancel{mol^{-1}}) = +0.5\,kJ$$

(Note that the amount in moles is quoted to only two significant figures because the mass was given to two significant figures—even though you know M_r to four significant figures.)

Energy is absorbed to evaporate the ethanol so the change is endothermic and the value of ΔH is positive. Alcohols such as ethanol or propanol are often included in the wipes used to sterilize skin before an injection. Heat is absorbed from the skin as the alcohol evaporates—that's why it feels cold.

Question

The standard enthalpy change of vaporization of water is +40.7 kJ mol^{-1}. An electric kettle is rated at 3 kW (1 kW = 1 kJ s^{-1}). How long will it take to evaporate 1.0 kg of water at 100 °C?

Some compounds do not show a liquid phase under certain conditions and the solid vaporizes directly to the gas phase, a process known as **sublimation**. An example is solid carbon dioxide (so-called *dry ice*), which, at atmospheric pressure, sublimes at –78 °C. The enthalpy change for sublimation is the sum of the enthalpy changes for fusion and vaporization.

$$\Delta_{sub}H^{\ominus} = \Delta_{fus}H^{\ominus} + \Delta_{vap}H^{\ominus} \quad (13.3)$$

Equation 13.3 arises as a consequence of Hess's law (Section 13.3, p.622). The overall enthalpy change is the sum of those for the individual steps.

Sublimation is introduced in Section 1.7 (p.47).

Worked example 13.4 The enthalpy change when ice turns into steam

Calculate the total enthalpy change when 1.00 mol of ice at 0 °C is turned into steam at 100 °C. (Take the average molar heat capacity of liquid water to be 75.1 J K^{-1} mol^{-1}. $\Delta_{fus}H^{\ominus}(H_2O) = +6.01$ kJ mol^{-1} and $\Delta_{vap}H^{\ominus}(H_2O) = +40.7$ kJ mol^{-1}.)

Strategy

To turn ice into steam requires three stages. You have to: (i) melt the ice; (ii) heat the water from 0 °C to 100 °C; and then (iii) boil it to form steam. The overall enthalpy change is the sum of the enthalpy changes for the three stages

$$\Delta H_{total} = \Delta H_{(i)} + \Delta H_{(ii)} + \Delta H_{(iii)}$$

You need to find the enthalpy changes for:

(i) 1 mol ice at 0 °C → 1 mol water at 0 °C. This is the enthalpy change of fusion; $\Delta H_{(i)} = \Delta_{fus}H^{\ominus}$.

(ii) 1 mol water at 0 °C → 1 mol water at 100 °C. You can find out how much heat this takes by using the molar specific heat capacity and Equation 13.2.

(iii) 1 mol water at 100 °C → 1 mol steam at 100 °C. This is the enthalpy change of vaporization; $\Delta H_{(iii)} = \Delta_{vap}H^{\ominus}$.

Solution

(i) *The enthalpy change for melting 1.00 mol of ice at 0 °C.*

$$\Delta_{fus}H^{\ominus} = +6.01 \text{ kJ mol}^{-1}$$

(ii) *Rearrange Equation 13.2 to find the enthalpy change for heating 1.00 mol of water from 0 °C to 100 °C.*

$$C_m = \frac{q}{n \times \Delta T}, \quad \text{so} \quad q = C_m \times n \times \Delta T$$

For $n = 1.00$ mol

$$q = 75.1 \text{ J } \cancel{\text{K}^{-1}}\ \cancel{\text{mol}^{-1}} \times 1.00\ \cancel{\text{mol}} \times (373 - 273)\cancel{\text{K}}$$
$$= 7510 \text{ J} = 7.51 \text{ kJ}$$
$$\Delta H = +7.51 \text{ kJ mol}^{-1}$$

(iii) *The enthalpy change for evaporating 1.00 mol of water at 100 °C.*

$$\Delta_{vap}H^{\ominus} = +40.7 \text{ kJ mol}^{-1}$$

Add the three contributions together to get the total enthalpy change.

$$\Delta H = (+6.01 \text{ kJ mol}^{-1}) + (+7.51 \text{ kJ mol}^{-1}) + (+40.7 \text{ kJ mol}^{-1})$$
$$= +54.2 \text{ kJ mol}^{-1}$$

Question

In order to cool a hot room, a blanket is soaked in water and hung up. How much heat is absorbed in evaporating 1.0 kg of water if the room is at 25 °C?

Box 13.2 Which freezes first, hot water or cold?

Water freezes when it is cooled below its freezing point. Freezing is an exothermic process and the heat given out when water freezes is transferred to the surroundings. To make ice cubes, you simply put some water in a freezer and let it cool down. However, when you consider all the energetics and practical consequences of freezing, it is not as straightforward as you might at first think.

When skaters use an ice-rink, the skates scratch the ice and damage its surface. During an ice hockey match, the rink is smoothed out between each period. This is done by a machine that sprays *hot*

An ice rink needs smoothing out between periods of play.

→

water onto the ice, *not* cold. Wouldn't this take longer (and more energy) to freeze?

In fact, cold water does not provide enough energy to melt and smooth out uneven bits of the rink—it freezes *too* quickly. Hot water melts the surface ice to give a smooth layer. Only a thin layer of hot water is used so it does not take very long to freeze. Some of it evaporates (an endothermic process) which speeds up the cooling.

Question

Consider the following experiment. Two dishes containing the same amount of water are placed in a freezer. The water in dish A is at 60 °C and in dish B it is at 100 °C.

(a) The water in dish B freezes first. Why?

(b) Would there be any difference if lids were placed on the dishes?

Summary

- The enthalpy change, ΔH, is the heat transferred at constant pressure by a chemical reaction or process.
- Enthalpy is a state function: $\Delta H = H(\text{final state}) - H(\text{initial state})$.
- Standard enthalpy change of fusion, $\Delta_{fus}H^{\ominus}$, is the enthalpy change when 1 mol of a substance melts at its melting point, T_m, and 1 bar pressure.
- Standard enthalpy change of vaporization, $\Delta_{vap}H^{\ominus}$, is the enthalpy change when 1 mol of a liquid vaporizes at its boiling point, T_b, and 1 bar pressure.
- For sublimation, $\Delta_{sub}H^{\ominus} = \Delta_{fus}H^{\ominus} + \Delta_{vap}H^{\ominus}$.

? For practice questions on these topics, see questions 4–6 at the end of this chapter (p.651).

13.3 Enthalpy changes in chemical reactions

The basic concepts of standard enthalpy changes and thermochemical equations are introduced in Section 1.6 (p.41).

The enthalpy change in a chemical reaction is described in a thermochemical equation. For example, the equation

$$N_2(g) + 3H_2(g) \rightarrow 2NH_3(g) \quad \Delta_r H^{\ominus}_{298} = -92.2\,\text{kJ mol}^{-1}$$

tells you that the standard enthalpy change at 298 K when 1 mol of nitrogen reacts with 3 mol of hydrogen to form 2 mol of ammonia is −92.2 kJ. $\Delta_r H^{\ominus}_{298}$ is a shorthand way of representing the standard enthalpy change of reaction at 298 K. This type of symbol is used for other thermodynamic state functions, as you will see in later sections.

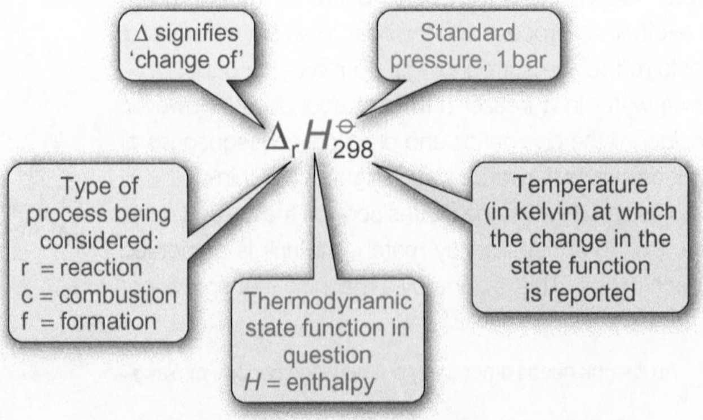
Δ signifies 'change of'
Standard pressure, 1 bar
$\Delta_r H^{\ominus}_{298}$
Type of process being considered:
r = reaction
c = combustion
f = formation
Thermodynamic state function in question
H = enthalpy
Temperature (in kelvin) at which the change in the state function is reported

Using Hess's law to calculate enthalpy change

It is sometimes difficult to measure the enthalpy change of a reaction directly. In such cases, **Hess's law** is very useful.

The total enthalpy change for a chemical reaction is independent of the path by which the reaction occurs, provided the starting and finishing states are the same for each reaction path.

This follows from enthalpy being a state function. The enthalpy change for a reaction is simply the difference between the enthalpy of the final state of the system (i.e. the products) and that of the initial state (i.e. the reactants). Any intermediate stages or reactions do not matter.

To illustrate the use of Hess's law, consider the oxidation of solid carbon (graphite) to carbon dioxide at 1 bar and 298 K (25 °C). The reaction could be conducted directly to measure $\Delta_r H^\ominus$

$$\text{(i)} \quad C(s) + O_2(g) \rightarrow CO_2(g) \qquad \Delta_r H_{(i)}^\ominus = -393.5\,\text{kJ mol}^{-1}$$

Alternatively, graphite could be reacted with oxygen to give carbon monoxide (CO), which could be then oxidized in a separate reaction to form CO_2

$$\text{(ii)} \quad C(s) + \tfrac{1}{2}O_2(g) \rightarrow CO(g) \qquad \Delta_r H_{(ii)}^\ominus = -110.5\,\text{kJ mol}^{-1}$$

$$\text{(iii)} \quad CO(g) + \tfrac{1}{2}O_2(g) \rightarrow CO_2(g) \qquad \Delta_r H_{(iii)}^\ominus = -283.0\,\text{kJ mol}^{-1}$$

Hess's law tells you that these two routes for producing CO_2 from C(s) must give the same overall enthalpy change so that

$$\begin{aligned}\Delta_r H_{(i)}^\ominus &= \Delta_r H_{(ii)}^\ominus + \Delta_r H_{(iii)}^\ominus \\ &= (-110.5\,\text{kJ mol}^{-1}) + (-283.0\,\text{kJ mol}^{-1}) \\ &= -393.5\,\text{kJ mol}^{-1}\end{aligned}$$

This is illustrated in the energy cycle in Figure 13.5.

It follows directly from Hess's law that chemical equations can be treated in the same way as algebraic equations. They can be added, subtracted, or multiplied by a constant value. Adding together reactions (ii) and (iii) for the CO_2 system gives

$$[C(s) + \tfrac{1}{2}O_2(g) \rightarrow CO(g)] + [C(s) + \tfrac{1}{2}O_2(g) \rightarrow CO_2(g)] \qquad \Delta_r H_{(ii)}^\ominus + \Delta_r H_{(iii)}^\ominus = \Delta_r H^\ominus$$

Grouping the reactants together and grouping the products together gives

$$C(s) + \tfrac{1}{2}O_2(g) + \cancel{CO}(g) + \tfrac{1}{2}O_2(g) \rightarrow \cancel{CO}(g) + CO_2(g) \quad \Delta_r H^\ominus = (-110.5 + (-283.0))\,\text{kJ mol}^{-1}$$

Since CO (g) appears in both the reactants and the products, it may be cancelled. The result of the addition is

$$C(s) + O_2(g) \rightarrow CO_2(g) \qquad \Delta_r H^\ominus = -393.5\,\text{kJ mol}^{-1}$$

which is the same as reaction (i). You can try using this approach in Worked example 13.5.

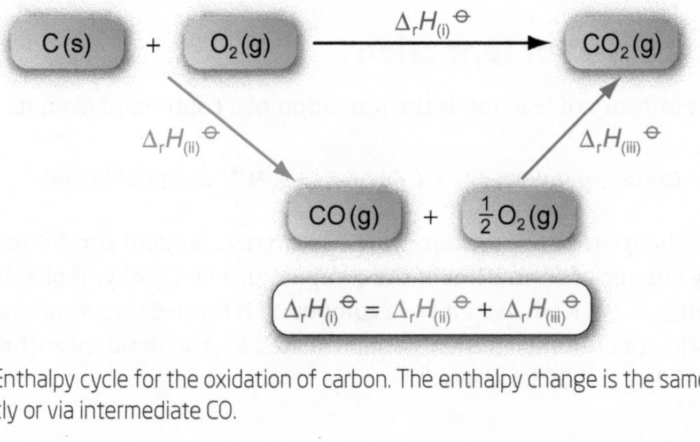
C(s)
+
O₂(g)
$\Delta_r H_{(i)}^\ominus$
CO₂(g)
$\Delta_r H_{(ii)}^\ominus$
$\Delta_r H_{(iii)}^\ominus$
CO(g)
+
½O₂(g)
$\Delta_r H_{(i)}^\ominus = \Delta_r H_{(ii)}^\ominus + \Delta_r H_{(iii)}^\ominus$

Figure 13.5 Enthalpy cycle for the oxidation of carbon. The enthalpy change is the same whether CO_2 is formed directly or via intermediate CO.

The measurement of enthalpy changes is described in Section 13.6 (p.644).

When carbon burns in air, it may be oxidized directly to CO_2—this is what is happening where it glows red. But where the air supply is limited, it is oxidized first to CO, which then burns with a pale flame to form CO_2.

This method is particularly useful for finding $\Delta_r H^\ominus$ values using enthalpy changes of formation (see p.624).

At first, this 'algebraic' approach may seem more difficult than using enthalpy cycles. However, in some cases, particularly when reactions occur under a range of conditions or in sequences of more than two reactions, it can be more straightforward to apply and saves you from making mistakes. After doing a number of examples, many students find this approach easier.

One of the major uses of calculations involving Hess's law is to calculate the enthalpy changes for reactions that would be difficult (or even impossible) to carry out in practice. It is often used for reactions in biochemical systems or when looking at the properties of reactive intermediates that cannot be isolated.

Worked example 13.5 Using Hess's law

The gaseous hydrocarbons ethene (C_2H_4) and ethane (C_2H_6) are products from the cracking of oil. The standard enthalpy changes at 298 K for the reactions of graphite and hydrogen gas to form 1 mol of each of these compounds are +52.5 kJ mol^{-1} and −83.8 kJ mol^{-1}, respectively.

Calculate the standard enthalpy change for the hydrogenation of ethene to ethane, at 298 K. (Hydrogenation is the addition of hydrogen to a C=C (or C≡C) bond; see Section 21.2, p.967.)

Strategy

The first step in solving this type of problem is to write down the thermochemical equation for the reaction of interest. Then write the equations for all the known reactions, including any thermochemical information given.

Manipulate (add, subtract, multiply, etc.) these equations to give the chemical reaction you want. Remember that any manipulation applied to the chemical equation must also be done to the values for the enthalpy changes. If you reverse the equation, you must reverse the sign of the enthalpy change.

Solution

Write the equation for the reaction of interest, the hydrogenation of ethene.

$$C_2H_4(g) + H_2(g) \rightarrow C_2H_6(g) \qquad \Delta H = ?$$

Write the thermochemical equations for the known reactions.

$$\text{Ethene: } 2C(\text{graphite}) + 2H_2(g) \rightarrow C_2H_4(g)$$
$$\Delta_r H^{\ominus} = +52.5 \text{ kJ mol}^{-1} \qquad \textbf{(A)}$$
$$\text{Ethane: } 2C(\text{graphite}) + 3H_2(g) \rightarrow C_2H_6(g)$$
$$\Delta_r H^{\ominus} = -83.8 \text{ kJ mol}^{-1} \qquad \textbf{(B)}$$

Manipulate equations (A) and (B) to give the reaction of interest.

You want to end up with $C_2H_6(g)$ on the right-hand side and $C_2H_4(g)$ on the left-hand side of your final equation. So, subtract (**A**) from (**B**)

$$[2C(\text{graphite}) + 3H_2(g) \rightarrow C_2H_6(g)] - [2C(\text{graphite}) + 2H_2(g) \rightarrow C_2H_4(g)]$$

Rearrange to group terms and cancel.

$$[2C(\text{graphite}) + 3H_2(g)] - [2C(\text{graphite}) + 2H_2(g)] \rightarrow [C_2H_6(g)] - [C_2H_4(g)]$$
$$[2\not{C}(\text{graphite}) - 2\not{C}(\text{graphite})] + [3H_2(g) - 2H_2(g)] \rightarrow [C_2H_6(g) - C_2H_4(g)]$$
$$H_2(g) \rightarrow C_2H_6(g) - C_2H_4(g)$$

Rearrange to form the reaction you need.

$$C_2H_4(g) + H_2(g) \rightarrow C_2H_6(g)$$

Now subtract $\Delta_r H^{\ominus}$(A) from $\Delta_r H^{\ominus}$(B) to calculate the required enthalpy change.

$$\Delta_r H^{\ominus} = \Delta_r H^{\ominus}(\textbf{B}) - \Delta_r H^{\ominus}(\textbf{A})$$
$$\Delta_r H^{\ominus} = -83.8 \text{ kJ mol}^{-1} - (+52.5 \text{ kJ mol}^{-1})$$
$$= -136.3 \text{ kJ mol}^{-1}$$

The standard enthalpy change for the hydrogenation of ethene at 298 K is −136.3 kJ mol^{-1}.

Question

The reaction of 1.00 mol of phosphorus with chlorine gas to form $PCl_3(g)$ releases 287.0 kJ. The reaction of 1 mol of phosphorus with chlorine gas to form PCl_5 (g) releases 374.9 kJ. Calculate the enthalpy change for the reaction $PCl_3(g) + Cl_2(g) \rightarrow PCl_5(g)$ under these conditions.

Enthalpy changes of formation

An important category of reaction is the formation of a compound from its elements. For example

$$C(\text{graphite}) + O_2(g) \rightarrow CO_2(g) \qquad \Delta_r H^{\ominus} = -393.5 \text{ kJ mol}^{-1}$$

The enthalpy change for this reaction carried out under standard conditions is the **standard enthalpy change of formation** of the compound, given the symbol $\Delta_f H^{\ominus}$. This is the enthalpy change at 298 K when 1 mol of a compound is formed under standard conditions from its constituent elements in their standard states. So, the standard enthalpy change of formation of $CO_2(g)$ is −393.5 kJ mol^{-1}.

The **standard enthalpy change of formation**, $\Delta_f H^{\ominus}_{298}$, is the enthalpy change at 298 K when 1 mol of a compound is formed under standard conditions from its constituent elements in their standard states.

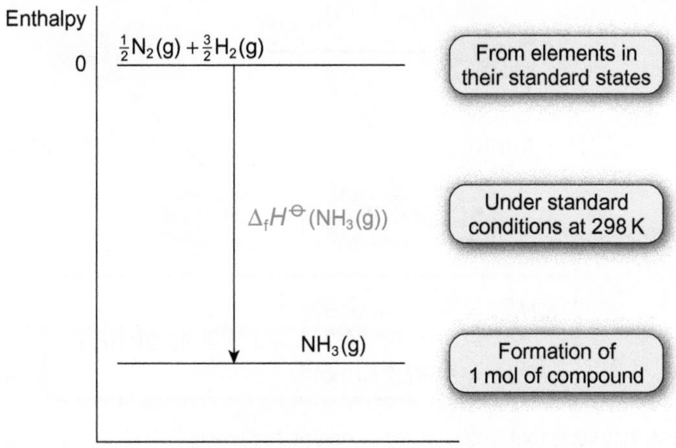

Enthalpy
0
$\frac{1}{2}N_2(g) + \frac{3}{2}H_2(g)$
From elements in their standard states
$\Delta_f H^\ominus(NH_3(g))$
Under standard conditions at 298 K
$NH_3(g)$
Formation of 1 mol of compound

Figure 13.6 The enthalpy change of formation of ammonia. It is important to use the precise definition.

It is not possible to define an absolute measure of enthalpy (you can measure enthalpy *changes* but not enthalpy) so a reference state is needed. By convention, the enthalpy of formation, $\Delta_f H^\ominus$, for an element in its **standard state** is zero. The standard state is defined as the most stable form of the element at 298 K and 1 bar. This provides a reference point for enthalpy changes.

For example, $\Delta_f H^\ominus$ for ammonia is the standard enthalpy change for the reaction

$$\tfrac{1}{2}N_2(g) + \tfrac{3}{2}H_2(g) \rightarrow NH_3(g) \qquad \Delta_r H^\ominus_{298} = \Delta_f H^\ominus_{298}(NH_3(g)) = -46.1\,\text{kJ mol}^{-1}$$

Note that the equation is written so as to form 1 mol of ammonia (Figure 13.6). $\Delta_f H^\ominus_{298}$ for $N_2(g)$ and $H_2(g)$ are zero since they are elements in their standard states.

As usual in thermodynamics, the definition of $\Delta_f H^\ominus$ is a very precise statement and there are some traps that need to be considered. The key phrase is *standard state*. Where several forms (allotropes) of an element exist, the one with lowest energy is used as the standard state. A good example is carbon where solid graphite is taken as standard in preference to diamond or fullerene. Take care with elements, such as bromine and iodine, that change state around 298 K and 1 bar. Although bromine and iodine often react as vapours, they exist under standard conditions as liquid and solid, respectively. Look at the following examples.

$$H_2(g) + \tfrac{1}{2}O_2(g) \rightarrow H_2O(l) \qquad \Delta_r H^\ominus_{298} = \Delta_f H^\ominus_{298}(H_2O(l)) = -285.8\,\text{kJ mol}^{-1}$$
$$C(\text{graphite}) + O_2(g) \rightarrow CO_2(g) \qquad \Delta_r H^\ominus_{298} = \Delta_f H^\ominus_{298}(CO_2(g)) = -393.5\,\text{kJ mol}^{-1}$$
$$H_2(g) + Br_2(l) \rightarrow 2HBr(g) \qquad \Delta_r H^\ominus_{298} = 2 \times \Delta_f H^\ominus_{298}(HBr(g)) = -72.8\,\text{kJ mol}^{-1}$$

In each of the three reactions, the enthalpy change of formation for the reactants (i.e. for the elements in their standard states) is zero so that the enthalpy change for the reaction represents the enthalpy change of formation of the product.

Using values of $\Delta_f H^\ominus$

$\Delta_f H^\ominus$ is an important property of a compound and is very useful in considering energy changes during chemical reactions.

To illustrate the use of $\Delta_f H^\ominus$ values, consider the gas phase reaction of ethyne (C_2H_2) to form benzene (C_6H_6)

$$3C_2H_2(g) \rightarrow C_6H_6(g) \qquad (13.4)$$

The reaction can be represented by the enthalpy cycle in Figure 13.7 where it proceeds via the intermediate elements. The principle here is Hess's law so that the overall enthalpy change for the reaction is the same whether it occurs directly or via the intermediates.

$$\Delta_r H^\ominus_{298} = \Delta H^\ominus_{(i)} + \Delta H^\ominus_{(ii)}$$

Tables of $\Delta_f H^\ominus_{298}$ are available and a selection of values is listed in Appendix 7 (p.1352).

The standard enthalpy change of formation of an *element* in its standard state is zero.

Although not stated at each step, all the enthalpy changes considered in this section are standard enthalpy changes under standard conditions at 298 K. You can see how to treat reactions taking place at other than 298 K in Section 13.4 (p.633).

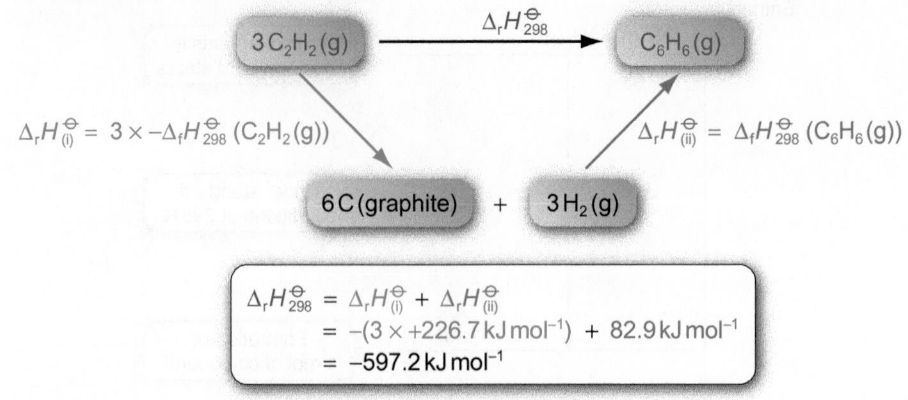

$\Delta_r H^\ominus_{298}$
$3C_2H_2(g)$
$C_6H_6(g)$
$\Delta_r H^\ominus_{(i)} = 3 \times -\Delta_f H^\ominus_{298}(C_2H_2(g))$
$\Delta_r H^\ominus_{(ii)} = \Delta_f H^\ominus_{298}(C_6H_6(g))$
$6C(graphite) + 3H_2(g)$
$\Delta_r H^\ominus_{298} = \Delta_r H^\ominus_{(i)} + \Delta_r H^\ominus_{(ii)}$
$= -(3 \times +226.7\,kJ\,mol^{-1}) + 82.9\,kJ\,mol^{-1}$
$= -597.2\,kJ\,mol^{-1}$

Figure 13.7 Enthalpy cycle for the conversion of ethyne to benzene.

In the reaction of ethyne to benzene via the elements, the second reaction step involves the formation of benzene from its elements. So, $\Delta H_{(ii)}^\ominus$ is the enthalpy change of formation of benzene vapour

$$6C(graphite) + 3H_2(g) \rightarrow C_6H_6(g) \qquad \Delta H_{(ii)}^\ominus = \Delta_f H^\ominus_{298}(C_6H_6(g))$$

ⓘ $\Delta_f H^\ominus_{298}(C_6H_6(g)) = +82.9\,kJ\,mol^{-1}$

The first reaction step is the conversion of ethyne into its constituent elements. $\Delta H_{(i)}^\ominus$ is given by the reverse of the enthalpy change of formation of ethyne.

ⓘ $\Delta_f H^\ominus_{298}(C_2H_2(g)) = +226.7\,kJ\,mol^{-1}$

$$C_2H_2(g) \rightarrow 2C(graphite) + H_2(g) \qquad -\Delta_f H^\ominus(C_2H_2(g))$$
$$3C_2H_2(g) \rightarrow 6C(graphite) + 3H_2(g) \qquad \Delta H_{(i)}^\ominus = 3 \times -\Delta_f H^\ominus_{298}(C_2H_2(g))$$

The overall enthalpy change of reaction for forming 1 mol of benzene from 3 mol of ethyne (as in Equation 13.4) is

$$\begin{aligned}\Delta_r H^\ominus_{298} &= \Delta H_{(i)}^\ominus + \Delta H_{(ii)}^\ominus \\ &= [3 \times -\Delta_f H^\ominus_{298}(C_2H_2(g))] + \Delta_f H^\ominus_{298}(C_6H_6(g)) \\ &= \Delta_f H^\ominus_{298}(C_6H_6(g)) - [3 \times \Delta_f H^\ominus_{298}(C_2H_2(g))]\end{aligned}$$

Substituting values

$$\begin{aligned}\Delta_f H^\ominus_{298} &= +82.9\,kJ\,mol^{-1} - [3 \times (+226.7)\,kJ\,mol^{-1}] \\ &= -597.2\,kJ\,mol^{-1}\end{aligned}$$

The enthalpy change for the reaction has been calculated from the enthalpy changes of formation of the compounds involved.

This approach to thermochemistry can be expressed in a more general way that is extremely useful in performing calculations. In Equation 13.5, the enthalpy changes of formation for the reactants are subtracted from those of the products. This gives a general way of expressing enthalpy changes of reaction, as shown in Equation 13.6.

The measurement of enthalpy changes of formation is described in Section 13.6 (p.644).

$$\Delta_r H^\ominus_{298} = \sum \nu_i \Delta_f H^\ominus_{298}\,(\text{products}) - \sum \nu_i \Delta_f H^\ominus_{298}\,(\text{reactants}) \qquad (13.6)$$

The sum of all the standard enthalpy changes of formation of the *products*

The sum of all the standard enthalpy changes of formation of the *reactants*

The symbol Σ is shorthand for the summation of the terms that follow. For example,

$\sum_1^5 (a_i) = a_1 + a_2 + a_3 + a_4 + a_5$

(see Maths Toolkit MT1, p.1312).

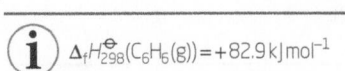

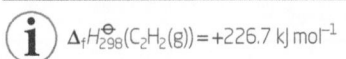

The symbol 'ν_i' in Equation 13.6 is the **stoichiometric coefficient**, the number of moles of each reactant or product involved in the balanced thermochemical equation. In the example above, $\Delta_f H^{\ominus}_{298}(C_2H_2(g))$ was multiplied by three, its stoichiometric coefficient in Equation 13.4. The stoichiometry must be taken into account in this way since enthalpy is an extensive quantity and the heat transferred depends on how much of the compounds you are using. For a general reaction

$$\alpha A + \beta B \rightarrow \gamma C + \delta D$$

$$\Delta_f H^{\ominus}_{298} = [\gamma \Delta_f H^{\ominus}_{298}(C) + \delta \Delta_f H^{\ominus}_{298}(D)] - [\alpha \Delta_f H^{\ominus}_{298}(A) + \beta \Delta_f H^{\ominus}_{298}(B)]$$

> Note that values of $\Delta_f H^{\ominus}$ may be positive or negative.

To calculate $\Delta_f H^{\ominus}_{298}$ for any reaction, you need to list all the $\Delta_f H^{\ominus}_{298}$ values for the reactants and products and multiply each by the stoichiometric coefficient in the thermochemical equation. Sum up the resulting values for the reactants and subtract from the sum of those for the products.

Equation 13.6 can be used to calculate the enthalpy change for *any* reaction where $\Delta_f H^{\ominus}_{298}$ data is available for the components. You can see an example of its use in Worked example 13.6.

> While the enthalpy cycle approach illustrated in Figure 13.7 works well in straightforward reactions, it is easy to make errors in more complicated cases. Equation 13.6 is more readily applied to all cases.

Worked example 13.6 Using standard enthalpy changes of formation

Calculate the standard enthalpy change of reaction at 298 K for the reaction

$$3Fe_2O_3(s) + 2NH_3(g) \rightarrow 6FeO(s) + 3H_2O(l) + N_2(g)$$

Strategy

Use Equation 13.6 directly after looking up the $\Delta_f H^{\ominus}_{298}$ data in Appendix 7 (p.1352).

Solution

From Appendix 7, the values of $\Delta_f H^{\ominus}_{298}$ for reactants and products are

$$\Delta_f H^{\ominus}_{298}/\text{kJ mol}^{-1}: \quad Fe_2O_3(s) = -824.2;$$
$$NH_3 = -46.1;$$
$$FeO(s) = -266.3;$$
$$H_2O(l) = -285.8;$$
$$N_2(g) = 0$$

Use Equation 13.6 to calculate the enthalpy change of reaction.

$$\Delta_r H^{\ominus}_{298} = \sum \nu_i \Delta_f H^{\ominus}_{298}(\text{product}) - \sum \nu_i \Delta_f H^{\ominus}_{298}(\text{reactants}) \quad (13.6)$$

$$= [6 \times \Delta_f H^{\ominus}_{298}(FeO(s)) + 3 \times \Delta_f H^{\ominus}_{298}(H_2O)(l) + \Delta_f H^{\ominus}_{298}(N_2(g))] - [3 \times \Delta_f H^{\ominus}_{298}(Fe_2O_3)(s) + 2 \times \Delta_f H^{\ominus}_{298}(NH_3(g))]$$

$$= [6 \times (-266.3\,\text{kJ mol}^{-1}) + 3 \times (-285.8\,\text{kJ mol}^{-1}) + 0] - [3 \times (-824.2\,\text{kJ mol}^{-1}) + 2 \times (-46.1\,\text{kJ mol}^{-1})]$$

$$= [-2455.2\,\text{kJ mol}^{-1}] - [-2564.8\,\text{kJ mol}^{-1}]$$

$$= +109.6\,\text{kJ mol}^{-1}$$

Question

Use data in Appendix 7 to calculate the enthalpy change when 1.00 mol of solid ammonium nitrate (NH_4NO_3) decomposes at 298 K to N_2O gas and water.

$$NH_4NO_3(s) \rightarrow N_2O(g) + 2H_2O(l)$$

Enthalpy changes of combustion

The standard enthalpy change of combustion, given the symbol $\Delta_c H^{\ominus}$, gives a measure of the heat released when a substance burns at constant pressure. This is often referred to simply as the *heat of combustion*.

> The **standard enthalpy change of combustion**, $\Delta_c H^{\ominus}$, is the enthalpy change when 1 mol of a substance reacts completely with oxygen gas at 1 bar pressure. Values are usually quoted at 298 K. The temperature quoted indicates that the reactants start at 298 K and the products finish at this temperature. During the combustion, of course, the temperature goes up, but what matters are the starting and finishing temperatures.

The combustion of 1 mol of methane gas at 298 K gives out 890.3 kJ. This can be summarized by the thermochemical equation

$$CH_4(g) + 2O_2(g) \rightarrow CO_2(g) + 2H_2O(l) \qquad \Delta_c H^{\ominus}_{298} = -890.3\,\text{kJ mol}^{-1}$$

> Enthalpy changes of combustion are *always* negative values since combustion always gives out heat. You should always include the negative sign so that the arithmetic works out correctly in calculations.

The production of hydrogen and its use as a fuel are described in Section 25.1 (p.1140).

An oxyacetylene torch burns a mixture of ethyne (acetylene) and oxygen. The highly negative enthalpy change of combustion gives a flame temperature of 3500 °C, which easily melts steel.

A variation of $\Delta_c H^{\ominus}$ that is in common use is the enthalpy change per unit mass of a compound. This is often called the **energy density** and has units of $kJ\,g^{-1}$ or $MJ\,kg^{-1}$. It is the enthalpy change on burning 1 g or 1 kg of a compound and is used as a measure of the efficiency of fuels, which are often purchased by mass. You can compare the energy densities of various fuels in Box 13.3. For example, the energy density of carbon (in the form of graphite) is $32.8\,kJ\,g^{-1}$. This means the enthalpy change when 1 g of graphite burns is −32.8 kJ. The energy density of saturated alkanes such as octane (a major component of petroleum) is around $50\,kJ\,g^{-1}$. One of the highest values is that of hydrogen, $143\,kJ\,g^{-1}$, which explains its use as a rocket fuel (see p.611). In mass terms, hydrogen is one of the best sources of energy, although other factors, primarily production costs, storage, and safety considerations, mean that currently it is not widely used.

Enthalpy changes of combustion are easy to measure (see Section 13.6). As well as providing information about the properties of fuels, $\Delta_c H^{\ominus}$ values are used to calculate enthalpy changes of formation for compounds. The method is illustrated in Worked example 13.7.

Worked example 13.7 Calculating $\Delta_f H^{\ominus}$ from values of $\Delta_c H^{\ominus}$

The main form of sugar in your diet is probably sucrose ($C_{12}H_{22}O_{11}$). Its combustion at 1 bar and 298 K releases 5647 $kJ\,mol^{-1}$ of heat. Calculate the standard enthalpy change of formation of sucrose.

$$(\Delta_f H^{\ominus}_{298}(CO_2(g)) = -393.5\,kJ\,mol^{-1} \text{ and } \Delta_f H^{\ominus}_{298}(H_2O(l)) = -285.8\,kJ\,mol^{-1})$$

Strategy

The first step is to write the thermochemical equation for the reaction, in this case the combustion of $C_{12}H_{22}O_{11}$.

Then use Equation 13.6 (p.626). Remember that $\Delta_f H^{\ominus}_{298}$ for oxygen gas is zero, since it is an element in its standard state.

Solution

Write the thermochemical equation for the combustion of $C_{12}H_{22}O_{11}$.

$$C_{12}H_{22}O_{11}(s) + 12\,O_2(g) \rightarrow 12\,CO_2(g) + 11\,H_2O(l)$$

$$\Delta_c H^{\ominus}_{298} = -5647\,kJ\,mol^{-1}$$

Use Equation 13.6 directly with the data provided.

$$\Delta_r H^{\ominus}_{298} = \sum \nu_i\, \Delta_f H^{\ominus}_{298}(\text{products}) - \sum \nu_i\, \Delta_f H^{\ominus}_{298}(\text{reactants}) \quad (13.6)$$

$$-5647\,kJ\,mol^{-1} = [12 \times \Delta_f H^{\ominus}_{298}(CO_2(g)) + 11 \times \Delta_f H^{\ominus}_{298}(H_2O(l))] - [\Delta_f H^{\ominus}_{298}(C_{12}H_{22}O_{11}(s)) + 12 \times \Delta_f H^{\ominus}_{298}(O_2(g))]$$

$$-5647\,kJ\,mol^{-1} = 12 \times (-393.5\,kJ\,mol^{-1}) + 11 \times (-285.8\,kJ\,mol^{-1}) - [\Delta_f H^{\ominus}_{298}(C_{12}H_{22}O_{11}(s)) - (12 \times 0)]$$

$$-5647\,kJ\,mol^{-1} = (-4722\,kJ\,mol^{-1}) + (-3144\,kJ\,mol^{-1}) - \Delta_f H^{\ominus}_{298}(C_{12}H_{22}O_{11}(s))$$

$$-5647\,kJ\,mol^{-1} = -7866\,kJ\,mol^{-1} - \Delta_f H^{\ominus}_{298}(C_{12}H_{22}O_{11}(s))$$

Rearranging

$$\Delta_f H^{\ominus}_{298}((C_{12}H_{22}O_{11})(s)) = -7866\,kJ\,mol^{-1} + 5647\,kJ\,mol^{-1} = -2219\,kJ\,mol^{-1}$$

Question

$\Delta_c H^{\ominus}_{298}$ for methane (CH_4) is −890.3 $kJ\,mol^{-1}$. Calculate $\Delta_f H^{\ominus}_{298}$ or methane.

Box 13.3 Comparing fuels

Burning fuels produces most of our energy needs in heating, transport, and electrical power. Although alternative fuel sources are under active study, most of these needs are met at present by burning fossil fuels. The main factor when considering substances as fuels is the quantity of energy they can provide.

However, a highly exothermic combustion reaction is not the only factor. A vehicle has to carry the fuel it uses and so some of the energy will be consumed in transporting the fuel as well as the vehicle. This energy consumption depends on the mass of fuel carried. For this reason, a good fuel needs to provide high energy per unit mass, that is, it must have a *high energy density* (see above). Values of the enthalpy change of combustion for some fuels are shown in the table.

Up to 15% of MTBE can be added to petrol to improve its ignition properties and reduce ‘knocking’. It is known as an *oxygenate* because the oxygen atom in MTBE promotes complete oxidation

→

	Typical composition	M_r	Δ_cH/kJ mol^{-1}
Petrol	C_8H_{18}*	114	−5470
Diesel fuel	$C_{20}H_{42}$*	282	−8090
Natural gas	CH_4	16	−890
Hydrogen	H_2	2	−286
Methanol	CH_3OH	32	−726
Ethanol	C_2H_5OH	46	−1367
MTBE†	$CH_3OC_4H_9$	88	−3369

*Petrol and diesel are complex mixtures of hydrocarbons. On average, their properties resemble those of C_8H_{18} and $C_{20}H_{42}$, respectively.
†Methyl *tert*-butyl ether (2-methoxy-2-methylpropane).

and lowers the CO present in the exhaust gases. However, MTBE is being phased out, and has been banned in some countries, since it is water soluble (around 40 g dm^{-3} at 20 °C) and can be washed from spills and leak into groundwater. The most promising replacement seems to be ethanol, which is available from renewable resources.

Questions

(a) For each compound, calculate its energy density. Suggest why the Space Shuttle uses hydrogen as its primary fuel source.

(b) Formula 1 racing cars use petrol as their fuel. In the USA, 'Indy cars' use methanol. Assuming the engines perform equally well, which are more fuel efficient: Formula 1 cars or Indy cars?

(c) Compare the energy density of petrol containing 10% MTBE (by mass) with the energy density of petrol. Comment on your answer.

Enthalpy changes of solution

Many chemical reactions occur in solution. Several different types of enthalpy change can be defined for reactions associated with solutions. An additional complication is that the values often depend on the concentration of solutions involved.

The enthalpy change when 1 mol of a substance dissolves in a large excess of *pure solvent* at 1 bar pressure is called the **standard enthalpy change of solution**, $\Delta_{sol}H^{\ominus}$. For example, when hydrogen chloride gas dissolves in water at 298 K

$$HCl(g) + (aq) \rightarrow HCl(aq) \quad \Delta_{sol}H^{\ominus}_{298} = -72.4\,kJ\,mol^{-1}$$

Here, the symbol (aq) indicates a large amount of water. Under these conditions, the enthalpy change is independent of the amount of water present. However, when hydrogen chloride gas reacts with a smaller quantity of water, a different enthalpy change is measured. For example

$$HCl(g) + H_2O(l) \rightarrow HCl(aq,\,0.5\,mol\,dm^{-3}) \quad \Delta_{sol}H^{\ominus} = -49.0\,kJ\,mol^{-1}$$

The value of $\Delta_{sol}H$ depends on concentration because the interaction between ions in solution depends on how far the ions are apart—which changes with the concentration. The measurement of enthalpy changes of solution is described in Section 13.6 (p.644).

i The **standard enthalpy change of solution**, $\Delta_{sol}H^{\ominus}$, is the enthalpy change when 1 mol of a substance dissolves in a large excess of pure solvent at 1 bar pressure. This is more correctly called the **integral enthalpy change of solution**, which signifies a large excess of solvent.

i Note that values of $\Delta_{sol}H^{\ominus}$ may be positive or negative.

Box 13.4 Thermochemistry for faster food

One application of thermochemistry is in the provision of hot food in situations where cooking is difficult or impossible. Much of the development work in this area was done for military uses. Fast food is also useful in emergency situations where electricity is not available, such as disaster relief, and for people involved in outdoor pursuits.

'Flameless heaters' rely on an exothermic reaction to provide hot food. In the *Heatermeal*™ products, the food is sealed inside a metal foil pouch and this is placed inside a plastic bag containing a heating pad made up of a mixture of magnesium and iron. To start the heating process, salt water (~30 cm^3) is poured into the bag onto the heating pad. (The addition of NaCl in the water activates the finely divided magnesium by reacting with the surface layer of MgO to form MgClOH, which is water-soluble.) An exothermic reaction starts and the foil pouch containing the food gets hot. About 300 g of food or water can be heated to 80 °C from room temperature in around 10 minutes.

Endothermic reactions can also be useful, an example being cooling packs used to treat sporting injuries. In one type of cooling pack, a small sachet of water is placed inside a bag filled with ammonium nitrate. To use the pack, the water sachet is broken and the bag shaken to mix the contents thoroughly. The enthalpy change of solution of ammonium nitrate is +25.7 kJ mol^{-1}. The pack stays cold for 20–30 min, enough to reduce swelling injuries. Similar packs can be used to chill food. →

HEATER MEALS
Self Heating
Ready to Eat
Long Shelf Life
Lamb Casserole
Delicious! 300g
THIS MEAL HEATS ITSELF!
NO REFRIGERATION OR MICROWAVE NEEDED
PACK INCLUDES - MEAL, SPOON & SELF HEATER
Best Before End
FOR INGREDIENTS SEE INSIDE
For more information please visit www.heatermeals.co.uk

▲ Self-heating meals. Salt water is poured into the bag containing a heating pad with a mixture of magnesium and iron, and the exothermic reaction heats the food.

KOOL PAK
SPORT
Instant Ice Pack for Sport
Designed to treat sports injuries
IMPORTANT: DO NOT OPEN OR PUNCTURE PACK

▲ Cooling packs are used for the treatment of minor sporting injuries.

Questions

(a) Write the equation for the reaction of magnesium metal with water. Using enthalpy changes of formation from Appendix 7, calculate the standard enthalpy change for the reaction.

(b) Assume that there is 300 g of food in the bag and that the food has the same heat capacity as water. How much energy is needed to heat the food from 20 °C to 80 °C? How much magnesium is needed to provide this heat?

Bond enthalpy

Chemical reactions involve breaking and making chemical bonds. The enthalpy change in a reaction is due to the difference in energy between bond breaking in the reactants and bond formation in the products (see Section 1.6, p.41). The enthalpy change when a particular chemical bond is broken is the bond dissociation enthalpy, $\Delta_{diss}H$, often given the symbol D.

The **bond dissociation enthalpy**, ***D*(A–B)**, is the enthalpy change per mole when a particular chemical bond, A–B, is broken under standard conditions in the gas phase.

> The **bond dissociation enthalpy**, *D*(A–B), is the enthalpy change per mole when a particular chemical bond, A–B is broken under standard conditions in the gas phase. Values are *always* positive—bond-breaking is an endothermic process.

Consider one mole of methane molecules (CH_4). One of the C–H bonds could be broken to form one mole each of hydrogen atoms and methyl radicals

$$CH_4(g) \rightarrow CH_3^{\bullet}(g) + H^{\bullet}(g) \qquad \Delta_{diss}H = D(\text{C–H}) = +439\,\text{kJ mol}^{-1}$$

The bond dissociation enthalpy for this particular C–H bond is therefore $+439\,\text{kJ mol}^{-1}$. Successively breaking the other C–H bonds in methane requires different quantities of energy, so each successive bond has a different bond dissociation enthalpy

$$CH_4(g) \rightarrow CH_3(g) + H(g) \qquad H_3C\text{–}H \qquad D_1(\text{C–H}) = +439\,\text{kJ mol}^{-1}$$

$$CH_3(g) \rightarrow CH_2(g) + H(g) \qquad H_2C\text{–}H \qquad D_2(\text{C–H}) = +462\,\text{kJ mol}^{-1}$$

$$CH_2(g) \rightarrow CH(g) + H(g) \qquad HC\text{–}H \qquad D_3(\text{C–H}) = +424\,\text{kJ mol}^{-1}$$

$$CH(g) \rightarrow C(g) + H(g) \qquad C\text{–}H \qquad D_4(\text{C–H}) = +338\,\text{kJ mol}^{-1}$$

(The radical dots are omitted in the above equations for clarity.)

The variation arises since the electron density in a particular C–H bond, and hence the energy needed to break it, depends on the arrangement of electrons around the carbon atom and this will change as the C–H bonds in methane are successively broken.

The total enthalpy change ($D_1 + D_2 + D_3 + D_4$) when *all* the C–H bonds in methane are broken is +1663 kJ mol^{-1}. So, the average value of D_1 to D_4 is +416 kJ mol^{-1}. This is called the **mean bond enthalpy**, $\bar{D}$(C–H) for methane.

Mean bond enthalpy

For a series of related compounds, the difference in the values of the bond dissociation enthalpies of the C–H bonds are small and it is possible to use an average value. For saturated hydrocarbons, the mean bond enthalpy of C–H is +412 kJ mol^{-1} (see Appendix 10, p.1368). This is slightly different from the mean bond enthalpy for the C–H bonds in methane, because the value is averaged over a range of compounds.

Table 13.5 gives selected values for mean bond enthalpies. Always remember, though, that the values shown in tables such as Table 13.5 are approximate values. They are the average calculated from measurements on a large number of compounds. Mean bond enthalpies can be used to estimate enthalpy changes during reactions, as shown in Worked example 13.8.

The **mean bond enthalpy**, $\bar{D}$(A–B), is the mean value of the bond dissociation enthalpy for the A–B bond, averaged across a range of related compounds.

Mean bond enthalpies are used extensively in inorganic and organic chemistry. For example, see Sections 20.1 (p.929) and 27.1 (p.1203).

Table 13.5 Selected values of mean bond enthalpies at 298 K

	$\bar{D}$ / kJ mol^{-1}		$\bar{D}$ / kJ mol^{-1}		$\bar{D}$ / kJ mol^{-1}
C–H	+412	C–Cl	+346	O–H	+464
C–C	+347	C–N	+286	O–O	+144
C=C	+612	C–O	+358	Si–H	+318
C≡C	+838	C=O	+742	Si–O	+466
C–F	+467	N–N	+158	Si–Si	+226

Worked example 13.8 Using mean bond enthalpies

Use values of mean bond enthalpies to estimate the standard enthalpy change for the gas phase isomerization of methoxymethane to form ethanol

$$CH_3\text{–}O\text{–}CH_3(g) \rightarrow CH_3\text{–}CH_2\text{–}OH(g)$$

Strategy

Assume the reaction can take place by breaking all the chemical bonds in the reactant to form gaseous atoms. The atoms then rearrange and form new bonds in the product.

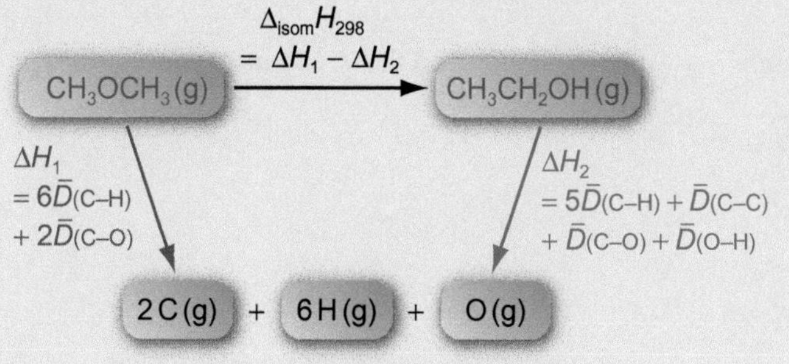

$\Delta_{isom}H_{298}$
$= \Delta H_1 - \Delta H_2$
CH_3OCH_3(g)
CH_3CH_2OH(g)
ΔH_1
$= 6\bar{D}$(C–H)
$+ 2\bar{D}$(C–O)
ΔH_2
$= 5\bar{D}$(C–H) $+ \bar{D}$(C–C)
$+ \bar{D}$(C–O) $+ \bar{D}$(O–H)
2C(g) + 6H(g) + O(g)

- ΔH_1 is the sum of the mean bond enthalpies for the bonds in CH_3–O–CH_3. This is the enthalpy change when all the bonds are broken in the reactant.
- ΔH_2 is the sum of the mean bond enthalpies for the bonds in CH_3–CH_2–OH. This is the enthalpy change when all the bonds are broken in the product. So, the enthalpy change to *make* all the bonds in the product from the gaseous atoms is $-\Delta H_2$.

$$\Delta_{isom}H = \Delta H_1 - \Delta H_2$$

Solution

Add the mean bond enthalpies for CH_3–O–CH_3 to find the enthalpy change ΔH_1.

CH_3–O–CH_3 consists of six C–H bonds and two C–O bonds.

$$\Delta H_1 = 6 \times \bar{D}(\text{C–H}) + 2 \times \bar{D}(\text{C–O})$$
$$= 6 \times (+412\ \text{kJ mol}^{-1}) + 2 \times (+358\ \text{kJ mol}^{-1})$$
$$= +3188\ \text{kJ mol}^{-1}$$

→

→ *Add the mean bond enthalpies for CH_3–CH_2–OH to find the enthalpy change ΔH_2.*

CH_3–CH_2–OH consists of five C–H bonds, one C–C, one C–O, and one O–H.

$$\Delta H_2 = 5\times\bar{D}(\text{C–H}) + \bar{D}(\text{C–C}) + \bar{D}(\text{C–O}) + \bar{D}(\text{O–H})$$
$$= 5\times(+412\ \text{kJ mol}^{-1}) + (+347\ \text{kJ mol}^{-1}) + (+358\ \text{kJ mol}^{-1}) + (+464\ \text{kJ mol}^{-1})$$
$$= +3229\ \text{kJ mol}^{-1}$$

Find the enthalpy change for the isomerization, $\Delta_{\text{isom}}H$.

$$\Delta_{\text{isom}}H = \Delta H_1 - \Delta H_2$$
$$= +3188\ \text{kJ mol}^{-1} - (+3229)\ \text{kJ mol}^{-1}$$
$$= -41\ \text{kJ mol}^{-1}$$

You can also arrive at a value for $\Delta_{\text{isom}}H$ using $\Delta_f H^\ominus$ values. $\Delta_f H^\ominus_{298}$ for dimethyl ether and ethanol are −184.0 kJ mol^{-1} and −235.1 kJ mol^{-1}, respectively. Using Equation 13.6 (p.626), you can work out that ΔH for the isomerization is −51.1 kJ mol^{-1}. The values obtained by the two methods are close but the agreement is not perfect. Using mean bond enthalpies typically gives a result with around 10–20% error from the true value. It provides a useful method for obtaining an estimate but not for determining accurate values.

Question

Use mean bond enthalpies to estimate $\Delta_r H$ for the following reaction of methanol

$$CH_3OH(g) + 1.5O_2(g) \rightarrow CO_2(g) + 2H_2O(g)$$

» Summary

- The standard enthalpy change of reaction, $\Delta_r H^\ominus$, is defined as the enthalpy change at 1 bar for a reaction with all components in their standard states. The value refers to the molar amounts given in the accompanying thermochemical equation.
- Values for standard enthalpy changes are often reported at 298 K (25 °C).
- The standard enthalpy change of formation, $\Delta_f H^\ominus$, is the enthalpy change when 1 mol of a substance is formed under standard conditions from its constituent elements in their standard states.
- For elements in their standard states, $\Delta_f H^\ominus = 0$.
- Hess's law states that the total enthalpy change for a chemical reaction is independent of the path by which the reaction occurs, provided the starting and finishing states are the same for each reaction path.
- Enthalpy changes for reactions can be calculated from $\Delta_f H^\ominus$ values for the components

$$\Delta_r H^\ominus_{298} = \sum \nu_i \Delta_f H^\ominus_{298}\,(\text{products}) - \sum \nu_i \Delta_f H^\ominus_{298}\,(\text{reactants})$$

- The standard enthalpy change of combustion, $\Delta_c H^\ominus$, is the enthalpy change when 1 mol of a substance reacts completely with excess oxygen gas at 1 bar pressure.
- The bond dissociation enthalpy, D(A–B), is the enthalpy change per mole when a *particular* chemical bond, A–B, is broken under standard conditions in the gas phase.
- The mean bond enthalpy, $\bar{D}$(A–B), is the mean value of the bond dissociation enthalpy for the A–B bond, averaged across a range of related compounds.

? **For practice questions on these topics, see questions 7–10, 12, 13, 20, 22, and 29–31, 34 at the end of this chapter (pp.651–653).**

13.4 Variation of enthalpy with temperature

All of the enthalpy changes discussed so far have been associated with reactions occurring under standard conditions and at 298 K (25 °C). Of course, not all chemical reactions are carried out at 25 °C, so how can enthalpy changes at other temperatures be calculated?

The enthalpy change is also different if a reaction is conducted at pressures other than 1 bar. However, the variation is small. Treatment of the pressure variation of enthalpy change can be found in more advanced texts.

How does the enthalpy of a substance change with temperature?

The enthalpy of a substance increases when the temperature increases (Figure 13.8). By definition, the enthalpy change is the heat transferred at constant pressure. Therefore, if q_p is the energy used to heat a substance in an open container from an initial temperature T_1 to a final temperature T_2, the enthalpy must change by

$$\Delta H_{T_1 \rightarrow T_2} = H_{T_2} - H_{T_1} = q_p \qquad (13.7)$$

where $\Delta H_{T_1 \rightarrow T_2}$ is the difference in enthalpy of the substance between the final and initial temperatures. (The subscript 'p' in q_p is included as a reminder that, for an enthalpy change, the heat transfer takes place at constant pressure.) In this case, $\Delta H_{T_1 \rightarrow T_2}$ is not an enthalpy change of *reaction* of the type discussed earlier in this chapter. It is the enthalpy change of the *substance* when its temperature is raised.

You can use Equation 13.2 (p.616) to find q_p from the molar heat capacity at constant pressure, C_p, of the substance and the temperature change. For n mol

$$q_p = n \times C_p \times \Delta T = n \times C_p \times (T_2 - T_1) \qquad (13.8)$$

ΔT has units of K, C_p has units of J K^{-1} mol^{-1}, so q_p determined from Equation 13.8 has units of J (*not* kJ).

Heat capacities are introduced in Section 13.1 (p.612).

Figure 13.9 shows the increase in enthalpy when 1 mol of water is heated from 25 °C (298 K) to 50 °C (323 K). The heat capacity of water, C_p = 75.3 J K^{-1} mol^{-1}. From Equations 13.7 and 13.8, the heat transferred, q_p, is

$$q_p = \Delta H_{298\text{ K} \rightarrow 323\text{ K}} = n \times C_p \times \Delta T$$
$$= 1 \cancel{\text{mol}} \times (75.3 \text{ J } \cancel{\text{K}^{-1}} \cancel{\text{mol}^{-1}}) \times (323 - 298) \cancel{\text{K}}$$
$$= +1883 \text{ J} = +1.88 \text{ kJ}$$

This means the enthalpy of the water increases by 1.88 kJ.

In a general case, if the enthalpy of a substance is H_{T_1} at temperature T_1 and increases to H_{T_2} when it is heated to T_2, then for 1 mol

$$\Delta H_{T_1 \rightarrow T_2} = H_{T_2} - H_{T_1} = C_p \Delta T$$

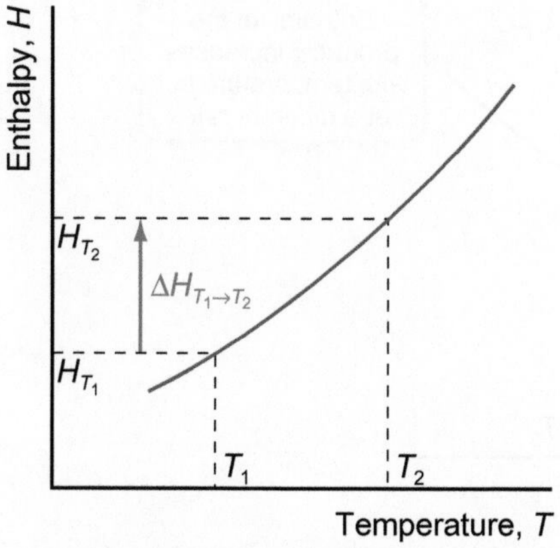
Enthalpy, H
H_T2
ΔH_T1→T2
H_T1
T1
T2
Temperature, T

Figure 13.8 Raising the temperature from T_1 to T_2 raises the enthalpy by $\Delta H_{T_1 \rightarrow T_2}$.

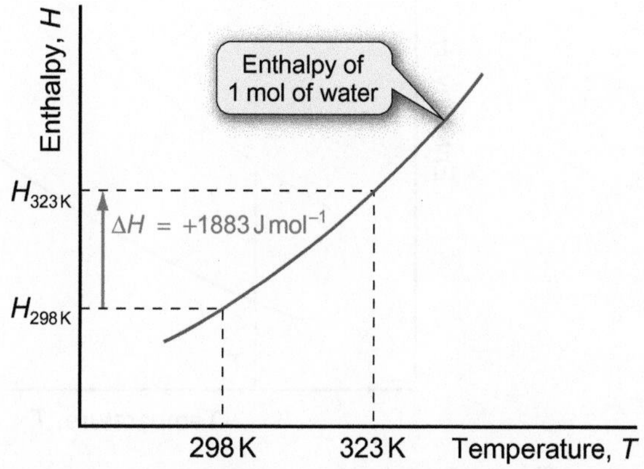
Enthalpy, H
Enthalpy of 1 mol of water
H323 K
ΔH = +1883 J mol−1
H298 K
298 K
323 K
Temperature, T

Figure 13.9 The enthalpy of 1 mol of water increases by 1883 J when it is heated from 25 °C (298 K) to 50 °C (323 K).

Visit the Online Resource Centre to watch screencast 13.1 which explains the variation of enthalpy with temperature as in Equation 13.9.

This can be written as

$$H_{T_2} = H_{T_1} + C_p\Delta T \qquad (\text{where } \Delta T = T_2 - T_1) \qquad (13.9)$$

In Equation 13.9, the enthalpies and heat capacity refer to 1 mol so $n = 1$.

Notice that the graphs in Figures 13.8 and 13.9 are *curves* rather than straight lines. This is because C_p also depends on temperature. However, for small changes in temperature, the change in C_p is small. The value of C_p used in Equation 13.9 represents the *mean* heat capacity of the compound over the temperature range T_1 to T_2.

Assuming an average value of the heat capacity usually gives sufficient accuracy. For more exact work, you would need to take into account the way that C_p depends on temperature.

How does $\Delta_r H$ change with temperature?

Equation 13.9 is also very useful for finding the enthalpy change for a *reaction* at temperatures other than 298 K. Figure 13.10 illustrates how the enthalpy change of a reaction changes with temperature.

If the enthalpy change for the reaction at a temperature T_1, $\Delta_r H_{T_1}$, is known, then the value at a different temperature, $\Delta_r H_{T_2}$, can be found using Equation 13.10.

$$\Delta_r H_{T_2} = \Delta_r H_{T_1} + \Delta C_p \Delta T$$

You can see how this equation is derived in Box 13.5. The equation can be written as

Standard enthalpy change of reaction at temperature, T_2

Standard enthalpy change of reaction at T_1

Temperature change (T in kelvin)

$$\Delta_r H^{\ominus}_{T_2} = \Delta_r H^{\ominus}_{T_1} + \Delta C_p (T_2 - T_1) \qquad (13.10)$$

The difference in heat capacity of reactants and products in the reaction

Equation 13.10 is known as the **Kirchhoff equation**.

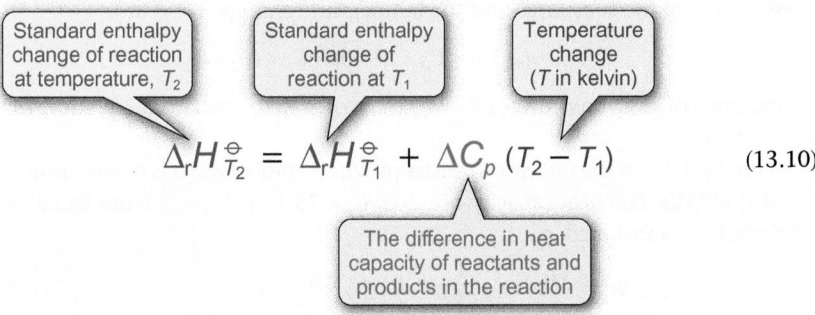

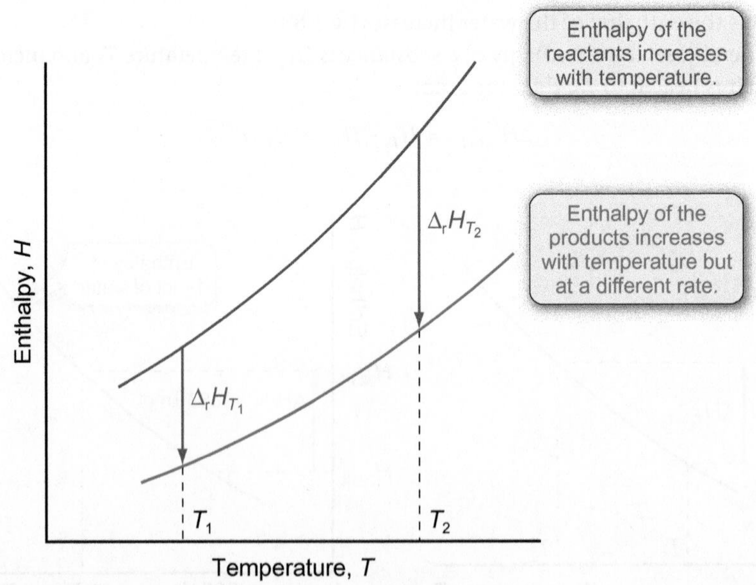
Enthalpy of the reactants increases with temperature.
Enthalpy of the products increases with temperature but at a different rate.
$\Delta_r H_{T_2}$
$\Delta_r H_{T_1}$
Enthalpy, H
T_1
T_2
Temperature, T

Figure 13.10 In this reaction, the enthalpy of the reactants increases with temperature at a faster rate than that of the products. It is an exothermic reaction, which becomes more exothermic as the temperature is raised.

In Equation 13.10, ΔC_p is the difference between the heat capacity of the products and the heat capacity of the reactants, given by

$$\Delta C_p = \sum \nu_i C_p\,(\text{products}) - \sum \nu_i C_p\,(\text{reactants}) \qquad (13.11)$$

Visit the Online Resource Centre to watch screencast 13.2 which walks you through Worked example 13.9, showing how to use the Kirchhoff equation.

Equation 13.10 is known as the **Kirchhoff equation**. It allows you to calculate the enthalpy change for a reaction at any temperature, providing you know the value of the enthalpy change at one temperature, and the heat capacities of the reactants and products. Its application is illustrated in Worked example 13.9.

As in Equation 13.6 (p.626), ν is the stoichiometric coefficient and represents the number of moles of reactant or product appearing in the thermochemical equation.

Box 13.5 Deriving the Kirchhoff equation

Consider a simple reaction where 1 mol of a reactant, R, is converted into 1 mol of a product, P. An example might be the isomerization of methoxymethane to ethanol in Worked example 13.8.

$$\text{R} \rightarrow \text{P}$$

The enthalpy change of the reaction is given by $\Delta_r H = H(\text{P}) - H(\text{R})$, where $H(\text{P})$ and $H(\text{R})$ are the enthalpies of product and reactant, respectively.

Consider the product, P, first. From Equation 13.9, the enthalpy of P at two temperatures T_2 and T_1 is related by

$$H(\text{P})_{T_2} = H(\text{P})_{T_1} + C_p(\text{P})\,(T_2 - T_1) \qquad (13.12)$$

where $C_p(\text{P})$ is the molar heat capacity at constant pressure for P.

The same relations hold for the reactant, R

$$H(\text{R})_{T_2} = H(\text{R})_{T_1} + C_p(\text{R})\,(T_2 - T_1) \qquad (13.13)$$

The enthalpy change of *reaction* at T_2 is given by

$$\Delta_r H_{T_2} = H(\text{P})_{T_2} - H(\text{R})_{T_2}$$

Substituting into this equation the expressions for $H(\text{P})_{T_2}$ and $H(\text{R})_{T_2}$ (Equations 13.12 and 13.13) gives

$$\Delta_r H_{T_2} = [H(\text{P})_{T_1} + C_p(\text{P})\,(T_2 - T_1)] - [H(\text{R})_{T_1} + C_p(\text{R})\,(T_2 - T_1)]$$

Grouping together the H terms and the C_p terms

$$\Delta_r H_{T_2} = (H(\text{P})_{T_1} - H(\text{R})_{T_1}) + (C_p(\text{P}) - C_p(\text{R})) \times (T_2 - T_1)$$

$$= \Delta_r H_{T_1} + (C_p(\text{P}) - C_p(\text{R})) \times \Delta T$$

$$\text{So, } \Delta_r H_{T_2} = \Delta_r H_{T_1} + \Delta C_p \Delta T \qquad (13.10)$$

where ΔC_p is the difference between the heat capacity of the products and the heat capacity of the reactants, given by Equation 13.11.

This derivation is for a simple reaction involving only 1 mol each of reactant and product. The same method can be applied to more complex reactions and leads to the same equation as long as the stoichiometry is taken into account.

Worked example 13.9 The temperature dependence of enthalpy changes of reaction

The enthalpy change for the complete combustion of ethane at 1 bar and 298 K is −1558.8 kJ mol^{-1}. Calculate $\Delta_c H^\ominus$ at 373 K.

(Mean C_p/J K^{-1} mol^{-1}: C_2H_6(g) 52.5; O_2(g) 29.4; CO_2(g) 37.1; H_2O(l) 75.1.)

Strategy

You are given the enthalpy change for the reaction at 298 K. Use the Kirchhoff equation (Equation 13.10) to calculate $\Delta_c H^\ominus_{373}$.

Write a thermochemical equation for the combustion reaction and use the stoichiometric coefficients to work out ΔC_p from Equation 13.11 above. (Equation 13.11 assumes that C_p values remain constant over the temperature change.)

Remember to include a sign when writing a value for $\Delta_c H^\ominus$.

Solution

Write the thermochemical equation for the reaction.

$$C_2H_6\,(g) + 3\tfrac{1}{2}O_2\,(g) \rightarrow 2CO_2\,(g) + 3H_2O\,(l)$$

$$\Delta_c H^\ominus_{298} = -1558.8\,\text{kJ mol}^{-1}$$

Use Equation 13.11 to calculate ΔC_p.

$$\Delta C_p = \sum \nu_i C_p(\text{products}) - \sum \nu_i C_p(\text{reactants})$$

$$= [2 \times C_p(CO_2) + 3 \times C_p(H_2O)] - [C_p(C_2H_6) + 3\tfrac{1}{2} \times C_p(O_2)]$$

$$= (2 \times 37.1\,\text{J K}^{-1}\text{mol}^{-1}) + (3 \times 75.1\,\text{J K}^{-1}\text{mol}^{-1})$$

$$-52.5\,\text{J K}^{-1}\text{mol}^{-1} - (3\tfrac{1}{2} \times 29.4\,\text{J K}^{-1}\text{mol}^{-1})$$

$$= 144.1\,\text{J K}^{-1}\text{mol}^{-1}$$

→

→ *Use the Kirchhoff equation, Equation 13.10, to calculate $\Delta_c H^{\ominus}_{373}$.*

$$\Delta_c H^{\ominus}_{373} = \Delta_c H^{\ominus}_{298} + \Delta C_p \Delta T \quad (13.10)$$
$$= -1558.8 \times 10^3 \text{ J mol}^{-1} + (144.1 \text{ J } \cancel{\text{K}^{-1}} \text{ mol}^{-1} \times 75 \cancel{\text{K}})$$
$$= -1558.8 \times 10^3 \text{ J mol}^{-1} + 10810 \text{ J mol}^{-1}$$
$$= -1548 \text{ kJ mol}^{-1}$$

The most common error in performing this type of calculation is to forget to use the same energy units for $\Delta_c H$ and C_p. In this example, the value of $\Delta_c H^{\ominus}_{298}$ is changed from kJ to J to use in Equation 13.10.

Notice that the difference in ΔH at the two temperatures is quite small: a change of 10.8 kJ mol^{-1}, less than 1%. In general, $\Delta_r H$ values change relatively little for small temperature changes.

Question

Calculate the enthalpy change for the gas phase oxidation of sulfur dioxide at 798 K.

$\Delta_f H^{\ominus}_{298}$ / kJ mol^{-1}	SO_2	−296.8	SO_3	−395.7		
Mean C_p/J K^{-1} mol^{-1}	SO_2	39.9	SO_3	50.7	O_2	29.4

Summary

- The change of enthalpy of a *substance* with temperature is given by

$$H_{T_2} = H_{T_1} + C_p(T_2 - T_1)$$

- The difference in heat capacity between products and reactants is

$$\Delta C_p = \sum v_i C_p\text{(products)} - \sum v_i C_p\text{(reactants)}$$

where v_i is the stoichiometric coefficient in the balanced equation.

- Enthalpy change of reaction at temperatures other than 298 K can be found from the Kirchhoff equation

$$\Delta_r H_{T_2} = \Delta_r H_{T_1} + \Delta C_p(T_2 - T_1)$$

? For practice questions on these topics, see questions 14, 23, 24, and 37 at the end of this chapter (pp.651–653).

13.5 Internal energy and the First Law of thermodynamics

Work is introduced in Section 1.6 (p.41).

An enthalpy change is the energy transferred as heat by a chemical change at constant pressure. Energy may also be transferred as a result of **work**.

Energy changes due to work

A common type of energy change in chemistry arises from a volume change—*expansion*. The chemical reaction in Figure 13.11 produces a gas that pushes the piston out against the external atmospheric pressure. Pushing out the piston requires energy, which comes from the reaction. Energy is transferred to the surroundings as a result of **expansion work**.

In chemical reactions involving *gases*, volume changes can be large so that significant amounts of work can be done. In general, volume changes in systems containing only liquids or solids are much smaller and, as a first approximation, can usually be ignored.

The discussion here involves only expansion work done by a system. Another type of work done by chemical reactions is *electrical* work, done when electrons move around a circuit. The thermodynamics of electrochemical systems is discussed in Section 16.4 (p.746).

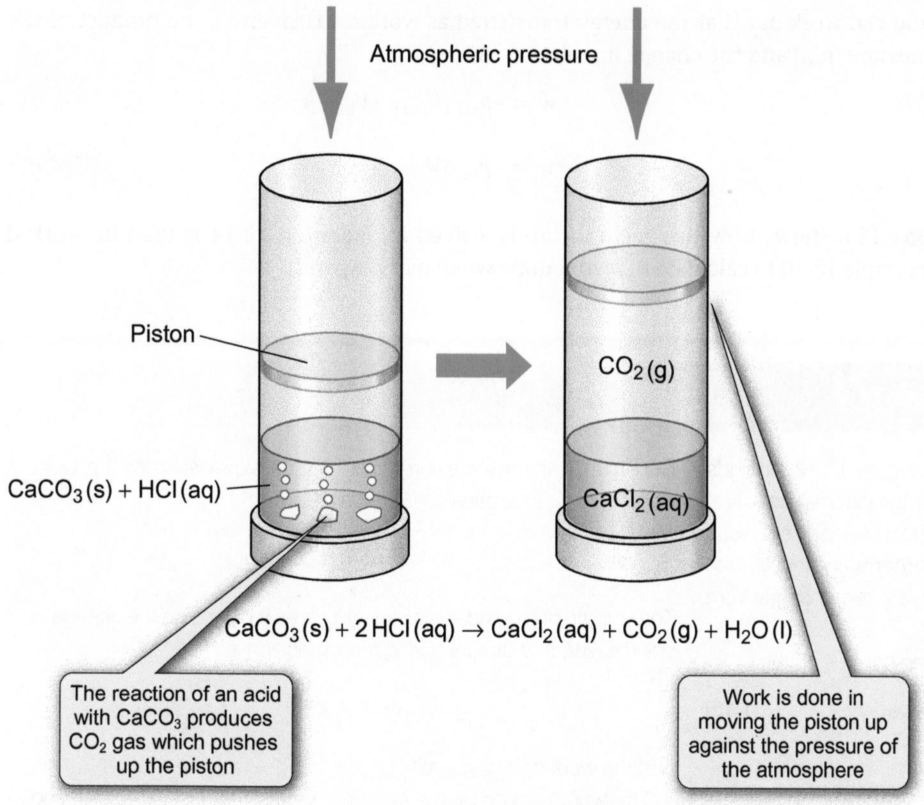
Atmospheric pressure
Piston
CO2 (g)
CaCO3 (s) + HCl (aq)
CaCl2 (aq)
CaCO3 (s) + 2 HCl (aq) → CaCl2 (aq) + CO2 (g) + H2O (l)
The reaction of an acid with CaCO3 produces CO2 gas which pushes up the piston
Work is done in moving the piston up against the pressure of the atmosphere

Figure 13.11 An example of a chemical reaction that does work in pushing back the atmosphere.

To derive an expression for expansion work, consider a quantity of gas trapped in a cylinder by a piston, as shown in Figure 13.12. The gas expands so that the piston moves upwards to a new position against the external pressure, p_{ext}, changing the volume from $V_{initial}$ to V_{final}. Using the definition of work given in Box 1.7 (p.47)

$$\text{work} = \text{force} \times \text{distance moved} \qquad (1.15)$$

An ideal gas is one that obeys the ideal gas equation; see Section 8.1 (p.348).

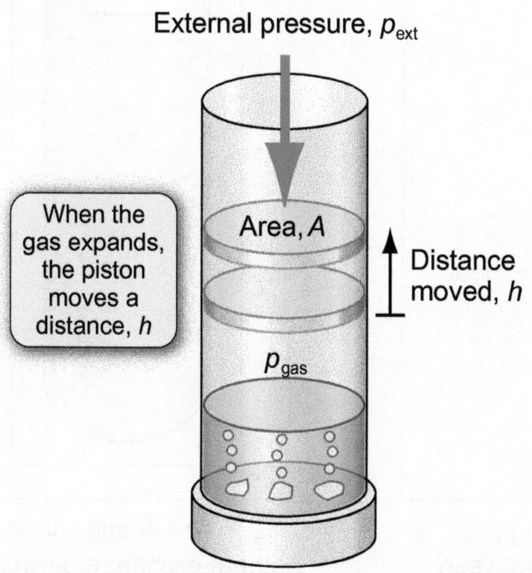
External pressure, p_ext
When the gas expands, the piston moves a distance, h
Area, A
Distance moved, h
p_gas

Figure 13.12 The expansion of an ideal gas against a constant external pressure. The system does work when the piston moves upwards a distance *h*. The system loses energy as work. The surroundings gain energy.

you can work out that the energy transferred as work, w, is given by the product of the pressure (p_{ext}) and the change in volume, ΔV

$$w = -p_{ext}\,(V_{final} - V_{initial})$$

$$w = -p_{ext}\Delta V \qquad (13.14)$$

Box 13.6 shows how this relationship is arrived at. Equation 13.14 is used in Worked example 13.10 to calculate the work done when a gas expands.

Box 13.6 Isothermal expansion of an ideal gas

The gas in the cylinder shown in Figure 13.12 expands so that the piston moves a distance h against the external pressure, p_{ext}. Assume that there is no friction and no heat change as a result of the motion so that the expansion occurs **isothermally**, that is, with no change of gas temperature. The volume of the gas changes from $V_{initial}$ to V_{final}.

The work done during the expansion is given by

$$\text{work} = \text{force} \times \text{distance moved} \qquad (1.15)$$
$$= F \times h$$

The force, F, acting on the piston is calculated from the pressure and the area of the piston, A

$$\text{pressure} = \frac{\text{force}}{\text{area}} \quad \text{so } p_{ext} = \frac{F}{A} \quad \text{and } F = p_{ext}A$$

Substituting this expression for F into the expression for the work done during the expansion (Equation 1.15)

$$\text{work done} = p_{ext}A \times h$$

The volume of a cylinder is given by its area × height, so the change in volume, ΔV, that occurs during the expansion is

$$\Delta V = A \times h,$$

and the work done = $p_{ext}\Delta V$.

The work is done by the system so the system *loses* energy and the value of the energy transferred as work, w, has a negative sign

$$w = -p_{ext}\Delta V \text{ (where } \Delta V = V_{final} - V_{initial}) \qquad (13.14)$$

An **isothermal** change takes place at constant temperature—the temperature of the system stays the same as that of the surroundings (Figure 13.13). This may require exchange of heat with the surroundings. In an **adiabatic** process, there is no exchange of heat with the surroundings. If a gas expands adiabatically, its temperature will fall.

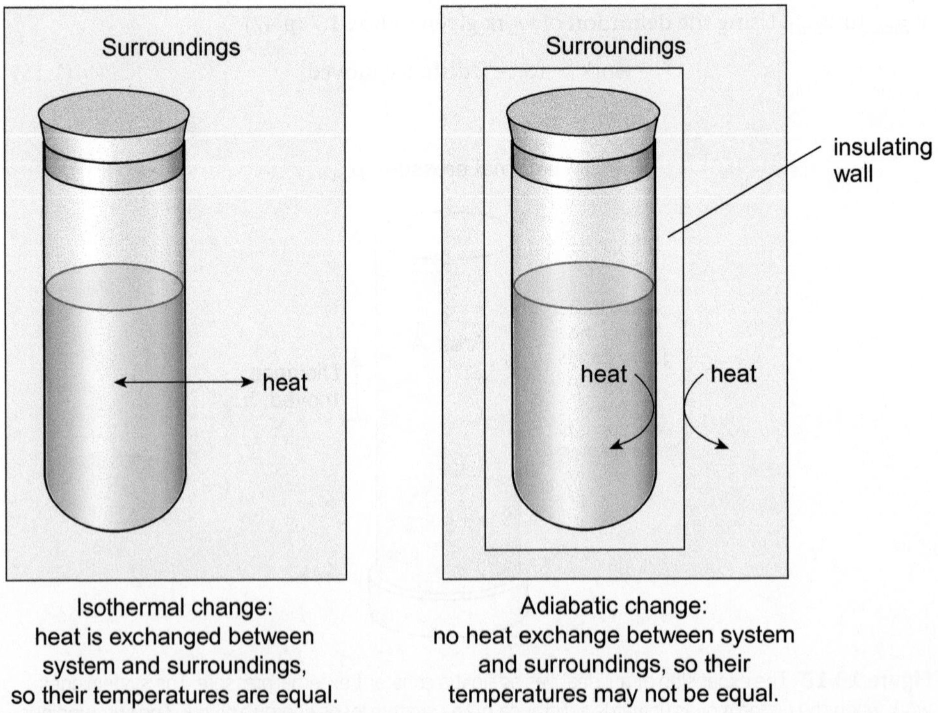
Surroundings
heat
Isothermal change:
heat is exchanged between
system and surroundings,
so their temperatures are equal.
Surroundings
insulating
wall
heat
heat
Adiabatic change:
no heat exchange between system
and surroundings, so their
temperatures may not be equal.

Figure 13.13 Isothermal and adiabatic changes

Worked example 13.10 Work of expansion

Calculate the energy transferred as work, w, when 1 mol of a gas expands from a volume of 5 dm^3 to 10 dm^3 against a constant pressure of 1 atm.

Strategy

The external pressure is constant so you can use Equation 13.14 directly. To obtain the energy in joules, the pressure and volumes must be converted to SI units (see Section 1.2, p.6).

Solution

Convert the values given for pressure and volumes to SI units.

$$1\,\text{atm} = 101\,325\,\text{N m}^{-2} \quad \text{and} \quad 1\,\text{dm}^3 = 1\times10^{-3}\,\text{m}^3$$

Use Equation 13.14 to calculate the energy transferred as work.

$$w = -p_{ext}\Delta V \qquad (13.14)$$
$$= -(101\,325\,\text{N m}^{-2}) \times [(10-5)\times10^{-3}\,\text{m}^3]$$
$$= -506.6\,\text{N m} = -506.6\,\text{J} \qquad (1\,\text{J} = 1\,\text{N m})$$
$$= -500\,\text{J}$$

Note that the value of w is negative since work is done by the gas in opposition to the external pressure and the gas (the system) loses energy.

Question

Calculate w for a reaction that produces 1.00 dm^3 of CO_2 at 298 K and 1 bar.

An accounting convention for energy changes

When a system does work or changes temperature, you need to look at the energy changes from the point of view of the *system* rather than the *surroundings*, so that the value of:

- energy *gained* (heat *absorbed* by or work done *on*) a system is *positive*;
- energy *lost* (heat *released* from or work done *by*) a system is *negative*.

For example, in an exothermic reaction, heat is released and the system loses energy so that the numerical value of the heat change has a negative sign. Similarly, the numerical value of work done by a chemical reaction has a negative sign since the system again loses energy. This is shown schematically in Figure 13.14.

This convention for accounting for energy changes is rather like the one you use to keep track of your personal finances. Your bank account is like the chemical system. Money given to you is added to your account and values are positive since your balance increases. Values of money you spend are negative since your balance goes down.

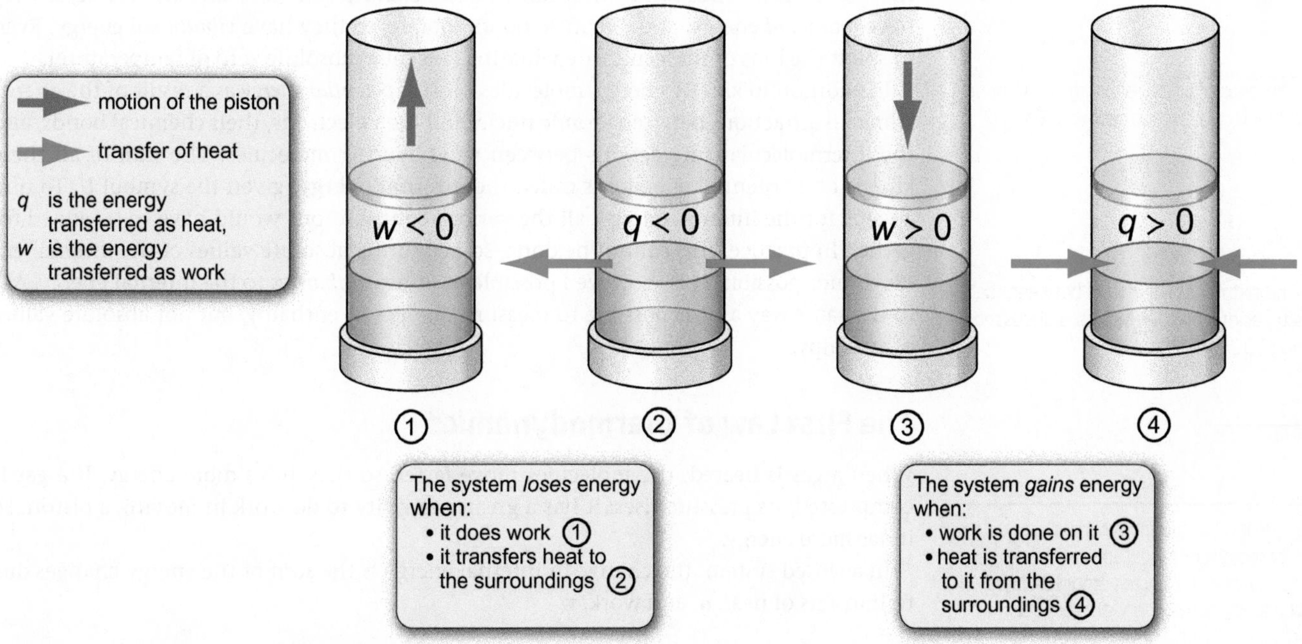

motion of the piston
transfer of heat
q is the energy transferred as heat,
w is the energy transferred as work
w < 0
q < 0
w > 0
q > 0
1
2
3
4
The system loses energy when:
• it does work 1
• it transfers heat to the surroundings 2
The system gains energy when:
• work is done on it 3
• heat is transferred to it from the surroundings 4

Figure 13.14 Work done on a system or heat gained by a system has a positive value. Work done by or heat lost from a system has a negative value.

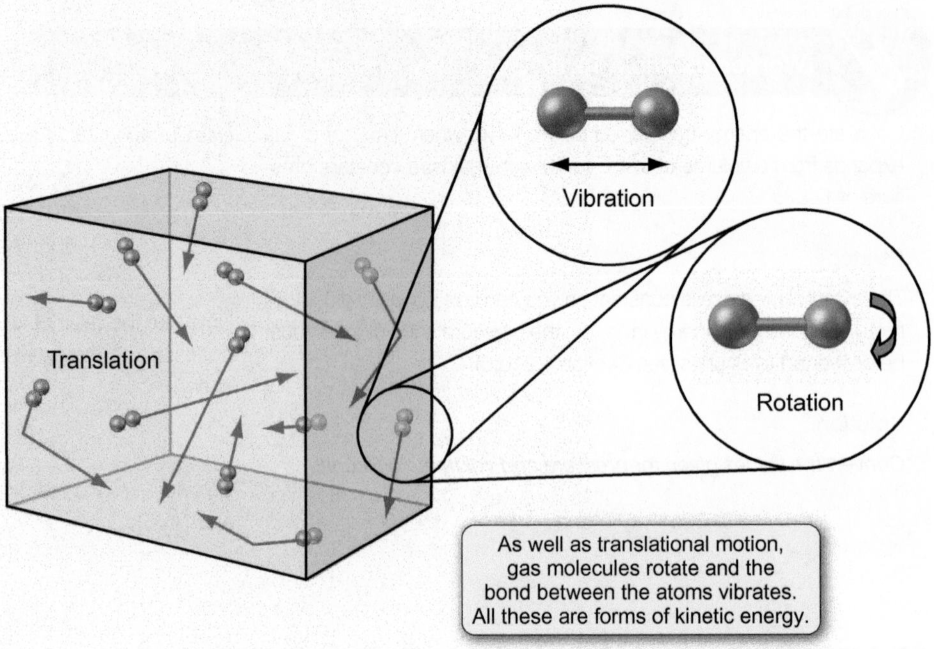
Vibration
Rotation
Translation
As well as translational motion, gas molecules rotate and the bond between the atoms vibrates. All these are forms of kinetic energy.

Figure 13.15 Molecules in a gas have kinetic energy from three different forms of motion: translation; rotation; and bond vibrations.

In order to keep track of all the energy changes due to transfer of heat and work that can take place in a chemical system, a new state function, the *internal energy*, is needed.

Internal energy

All molecules have energy, which takes various forms. In all states of matter, atoms and molecules are moving and so have *kinetic energy*, as illustrated in Figure 13.15. In fluids, the molecules move around—they have *translational* energy. Molecules also rotate, so they have *rotational* energy, and chemical bonds vibrate, so they have *vibrational* energy. Even in solids the ions or molecules are vibrating (except at absolute zero of temperature).

The ideas of potential energy and kinetic energy in molecules are introduced in Box 1.7 (p.42).

In addition to kinetic energy, molecules possess *potential energy* as a result of the energy stored in attractions between atomic nuclei and their electrons, their chemical bonds, and the intermolecular interactions between neighbouring molecules. The sum of all these kinetic and potential energies is called the **internal energy**, given the symbol ***U***. To find a value for the internal energy, all the various contributions would have to be added together. In practice, this cannot be done, so measuring absolute values of the internal energy is not possible; it is, however, possible to measure *changes* to the internal energy, ΔU, in the same way as it is possible to measure changes in enthalpy, but not absolute values of enthalpy.

The distribution of energy between the various types of molecular motion is discussed in Section 10.2 (p.453).

The First Law of thermodynamics

When a gas is heated, the molecules move faster, so they have more energy. If a gas is compressed, its pressure rises: it has a greater capacity to do work in moving a piston, so it has more energy.

In a closed system, the change in internal energy is the sum of the energy changes due to transfers of heat, q, and work, w

$$\Delta U = q + w \tag{13.15}$$

The laws of thermodynamics are not based on any model and cannot be derived from theory. They are **empirical laws** that summarize many observations of the behaviour of systems. The Second and Third Laws of thermodynamics are discussed in Sections 14.2 (p.660) and 14.3 (p.664).

Worked example 13.11 uses this expression to calculate the change in the internal energy of a system.

Equation 13.15 is one way of expressing the **First Law of thermodynamics**. There are a number of other ways of expressing this law in words. Perhaps the most common statement is the *principle of conservation of energy*, which states that 'energy can neither be created nor destroyed, only interconverted between forms' or 'the total quantity of energy in the Universe is constant'. If a system is heated then its energy increases; if it does work then its energy decreases. It is not possible to obtain heat or work from a system without it losing energy to the surroundings.

Visit the Online Resource Centre to watch screencast 13.3 which explains the significance of internal energy change and Equation 13.15.

An isolated system is one where no exchange of matter or energy can occur with the surroundings, so $q = w = 0$. A consequence of the First Law is that the internal energy of such a system is constant.

An isolated system is illustrated in Figure 13.2 (p.614).

$$\text{For an } \textit{isolated} \text{ system: } \Delta U = 0 \qquad (13.16)$$

The First Law as expressed here discounts any nuclear changes, where mass is converted to energy according to Einstein's equation $E = mc^2$. It also assumes that no other energy is lost, for example, emission of photons in a photochemical reaction.

To summarize, the **First Law of thermodynamics** can be stated in a number of ways.

- ***Energy can neither be created nor destroyed, only interconverted between forms.***
- ***The total quantity of energy in the Universe is constant.***
- ***The internal energy of an* isolated *system is constant, $\Delta U = 0$.***
- **$\Delta U = q + w$.**

Worked example 13.11 Calculating an internal energy change

The temperature of 1 mol of a liquid is raised by heating it with 750 J of energy. It expands and does 200 J of work. Calculate the change in internal energy of the liquid.

Strategy

You are given the heat transferred, q, and the work transferred, w, so you can use the First Law directly in the form of Equation 13.15. Remember the sign convention: if the system gains heat the value of q is a positive; if it does work, the value of w is negative.

Solution

Decide the signs of the heat and work transfers.

- The temperature increases so $q = +750$ J (the system gains energy).
- The system expands and so does work, $w = -200$ J (the system loses energy).

Use Equation 13.15 to calculate ΔU.

$$\Delta U = q + w \qquad (13.15)$$
$$= (+750\,\text{J}) + (-200\,\text{J}) = +550\,\text{J}$$

Question

A gas is compressed by doing 480 J of work on it and it absorbs 289 J of heat from the surroundings. What is the change in internal energy of the gas?

Chemical reactions at constant volume

Imagine the reaction shown in Figure 13.11 (p.637) being conducted so that the piston cannot move. The gas could not expand so it could not do any work. The reaction takes place **at constant volume**, so energy changes can only occur by the transfer of heat (Figure 13.16). Combining Equations 13.15 and 13.14 gives

$$\Delta U = q + w = q - p_{\text{ext}}\Delta V$$

But $\Delta V = 0$, so

$$\Delta U = q_V \qquad (13.17)$$

In practice, few reactions are carried out under conditions of constant volume so internal energy changes are not considered as often as enthalpy changes. Examples include reactions in geological systems deep in the Earth's crust where the tremendous pressures mean that volumes do not change.

where q_V indicates that the heat is transferred at constant volume. ΔU can therefore be thought of as the heat change for a process carried out at constant volume. Compare

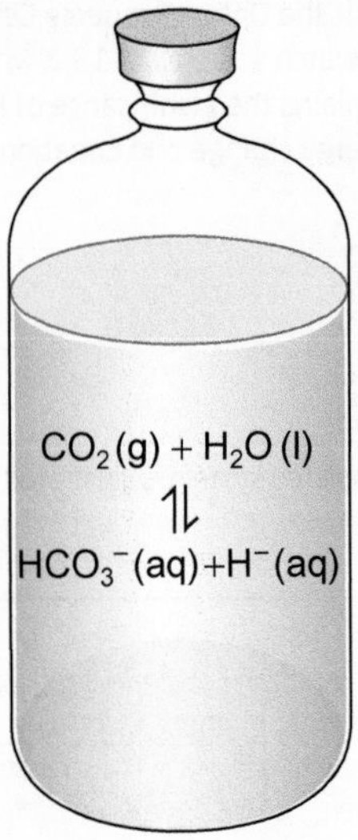
CO₂ (g) + H₂O (l)
HCO₃⁻ (aq)+H⁻ (aq)

Figure 13.16 In a sealed bottle of fizzy water, reactions occur at constant volume. The position of the equilibrium changes when the temperature changes, but there is no change in volume.

this with ΔH, which is the heat change at constant pressure. Carrying out reactions in sealed chambers and measuring the heat changes gives a method for measuring ΔU (see Section 13.6, p.644).

Enthalpy changes and internal energy

The relation between ΔH and ΔU can be found by looking at the energy changes that occur during a reaction such as that shown in Figure 13.11 (p.637). The reaction does work as the piston moves out against the atmospheric pressure. The pressure of the gas stays constant. Energy is transferred between the system (the reaction) and surroundings as heat and work. From Equation 13.15

$$\Delta U = q + w \qquad (13.15)$$

Remember that the enthalpy change, ΔH, is the heat change at constant pressure. Since this reaction is conducted at constant pressure, $q = \Delta H$ so

$$\Delta U = \Delta H + w$$

w is the work involved in the expansion, given by $w = -p_{ext}\Delta V$ (Equation 13.14, p.638), so

$$\Delta U = \Delta H + (-p\Delta V)$$

Rearranging to give an expression for ΔH gives Equation 13.18.

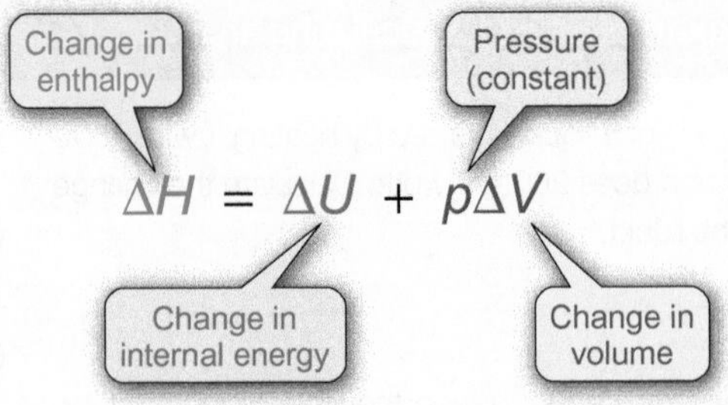
Change in enthalpy
Pressure (constant)
ΔH = ΔU + pΔV
Change in internal energy
Change in volume

One way of thinking about ΔH is that, for a reaction at constant pressure, it is the proportion of the internal energy change that is transferred as heat; the $p\Delta V$ term is the proportion transferred to the surroundings as work.

It is important to appreciate the difference in values between ΔU and ΔH. As shown in Equation 13.18, the difference between ΔU and ΔH arises from the $p\Delta V$ term. In general, the volume changes involved in reactions where all the components are liquids, solids, or in solution are very small and can be neglected with little error. Thus, in systems containing no gases, $\Delta H \approx \Delta U$. However, when gases are involved, the volume changes can be large.

If the gas behaves ideally (i.e. it obeys the ideal gas equation) at temperature, T, then

$$pV = nRT \qquad (8.5)$$

and

$$\Delta pV = \Delta(nRT)$$

At constant temperature, T, and pressure, p

$$p\Delta V = \Delta n_{gas}RT$$

Substituting for $p\Delta V$ in Equation 13.18

$$\Delta H = \Delta U + \Delta n_{gas}RT \qquad (13.19)$$

Visit the Online Resource Centre to watch screencast 13.4 which explains the relationship between ΔH and ΔU in Equations 13.18 and 13.19.

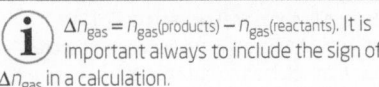

$\Delta n_{gas} = n_{gas}(\text{products}) - n_{gas}(\text{reactants})$. It is important always to include the sign of Δn_{gas} in a calculation.

where Δn_{gas} is the change in number of moles of gas between reactants and products. At 298 K (25 °C) and 1 atm, $(\Delta H - \Delta U) = +2.48$ kJ for each mole of gas that changes between reactants and products; at 773 K (500 °C), $(\Delta H - \Delta U) = +6.43$ kJ mol^{-1}. For chemical reactions involving a change in the number of moles of gas, the differences between ΔH and ΔU are significant although often small compared with the magnitude of ΔH itself (see Worked example 13.12).

In Box 13.7, some of the ideas in this section are applied to the explosion of nitroglycerine.

Only gases are considered in Equation 13.19 since the molar volumes of solids and liquids are so much smaller than gases. 1 mol of an ideal gas at 25 °C and 1 atm occupies ~24 500 cm^3. Compare this with the molar volumes for water, ~18 cm^3 mol^{-1}, and for benzene, 89 cm^3 mol^{-1}. In these cases, the $p\Delta V$ term is negligible when used to convert ΔU to ΔH.

Worked example 13.12 Comparing enthalpy changes and internal energy changes

Calculate the difference between ΔH and ΔU for the reaction of zinc metal with nitric acid at 298 K.

Strategy

You can find the difference from Equation 13.19 so you need to find Δn_{gas}. Write the balanced equation for the reaction and compare the number of moles of gas in the products with the reactants. When using Equation 13.19, the temperature must be in K.

Solution

Write the balanced equation for the reaction.

$$Zn(s) + 2HNO_3(aq) \rightarrow H_2(g) + Zn(NO_3)_2(aq)$$

Calculate Δn_{gas}

$$\begin{aligned}\Delta n_{gas} &= n_{gas}(\text{products}) - n_{gas}(\text{reactants})\\ &= 1\text{ mol} - 0\text{ mol}\\ &= +1\text{ mol (Always give a sign for the value of } \Delta n_{gas}.)\end{aligned}$$

Rearrange Equation 13.19 to find the difference between ΔH and ΔU.

$$\Delta H = \Delta U + \Delta n_{gas}RT \qquad (13.19)$$

$$\begin{aligned}\Delta H - \Delta U &= \Delta n_{gas}RT\\ &= +1\,\cancel{\text{mol}} \times 8.314\text{ J}\,\cancel{\text{K}^{-1}}\,\cancel{\text{mol}^{-1}} \times 298\,\cancel{\text{K}}\\ &\qquad (R = 8.314\text{ J K}^{-1}\text{ mol}^{-1})\\ &= +2.48\text{ kJ (for the molar amounts in the equation)}\end{aligned}$$

Question

$\Delta_c H^{\ominus}_{298}$ for methane (CH_4) is −890.3 kJ mol^{-1}. Calculate $\Delta_c H^{\ominus}_{298}$ for methane.

Box 13.7 Nitroglycerine: the chemistry of an explosive

Nitroglycerine, a trinitrate of glycerol (propane-1,2,3-triol), is an explosive liquid. It is made by treating glycerol with a mixture of concentrated sulfuric acid and nitric acid.

Nitroglycerine was discovered in Italy in 1846 but was too dangerous in its pure form to be used in practice since it is extremely sensitive to shock. Among the students working on nitroglycerine was Alfred Nobel, a young Swedish chemist and engineer. When he returned home to Sweden, he carried on his research and set up a family business making explosives.

Handling nitroglycerine was dangerous and a number of workers, including Alfred's younger brother Emil, were killed. Nobel's research led him to discover that nitroglycerine could be adsorbed onto solids such as kieselguhr, a soft chalk-like rock found in the hills near one of his factories in Germany. Mixing nitroglycerine with kieselguhr, or other solids such as sawdust or with sand or clay, made a useful explosive, *dynamite*. This is safer and more easily handled since it does not explode until detonated.

Dynamite is still one of the most commonly used explosives in the mining and construction industries. The invention made Nobel a very rich man and he founded the Nobel Institute, which to this day awards the prizes that are named in his honour.

The overall equation for the decomposition of nitroglycerine can be written as

$$4C_3H_5N_3O_9(l) \rightarrow 6N_2(g) + O_2(g) + 12CO_2(g) + 10H_2O(g)$$

→

▲ Nitroglycerine is easily detonated, decomposing to form 10 000 times its own volume of gas causing a powerful explosion. In 1941 in Marshburg, Pennsylvania, a truck carrying 200 litres of nitroglycerine was blown to bits when the nitroglycerine exploded. The driver of this car was unlucky enough to be overtaking the truck at the time.

In addition to being an explosive, nitroglycerine is also used as a therapeutic treatment for angina and heart disease. It acts as a vasodilator and widens the blood vessels. Indeed, Nobel himself was treated with nitroglycerine late in his life.

Questions

(a) Given that $\Delta_f H^{\ominus}_{298}$ for nitroglycerine is $-372.4\ \text{kJ mol}^{-1}$, use data from Appendix 7 to calculate the enthalpy change at 298 K for:
   (i) the reaction represented by the above equation;
   (ii) the reaction of 1 mol of nitroglycerine.

(b) Assume that all the gases given off behave as ideal gases and the reaction takes place at 298 K and at constant pressure. Calculate:
   (i) the energy transferred as work, w, when 1 mol of nitroglycerine reacts;
   (ii) the change in internal energy when 1 mol of nitroglycerine reacts.

Summary

- The sign convention used for values of energy changes is:
  - energy *gained* by a system is *positive* (heat absorbed by or work done on a system);
  - energy *lost* by a system is *negative* (heat released from or work done by a system).
- The sum of the kinetic and potential energies of molecules in a sample is the internal energy, U.
- First Law of thermodynamics: $\Delta U = q + w$ where q is the energy transferred as heat, and w is the energy transferred as work. This is also called the Law of Conservation of Energy.
- For an isothermal expansion of an ideal gas: $\Delta U = 0$; $q = -w$.
- $\Delta U = q_V$ (the heat change for a process at constant volume).
- Relation between enthalpy change and change of internal energy:

$$\Delta H = \Delta U + p\Delta V, \quad \text{or}$$
$$\Delta H = \Delta U + \Delta n_{\text{gas}}RT$$

For practice questions on these topics, see questions 15–19, 33 and 36 at the end of this chapter (p.651–653).

13.6 Measuring energy changes

A device for measuring heat changes during chemical reactions is known as a **calorimeter**. There are various types of calorimeter depending on the type of reaction under study. The measurement of heat changes using a calorimeter is called **calorimetry**.

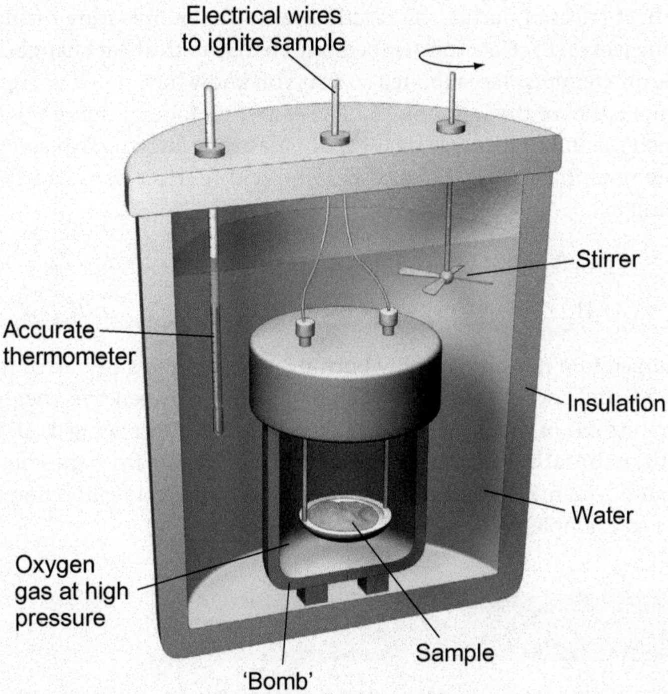
Electrical wires
to ignite sample
Stirrer
Accurate
thermometer
Insulation
Water
Oxygen
gas at high
pressure
Sample
'Bomb'

Figure 13.17 A bomb calorimeter. The liquid or solid sample is placed in a tightly sealed 'bomb'. After pressurizing the bomb with oxygen, an electric current is passed through a piece of wire in contact with the substance to start the reaction. The bomb is surrounded by a known volume of water, which absorbs any heat given out by the reaction. The temperature of the water is measured precisely to ±0.001 °C with an accurate thermometer. The water container, in turn, is surrounded by insulation to minimize heat losses.

Visit the Online Resource Centre to watch video clip 13.1 which illustrates the use of one type of bomb calorimeter.

The bomb calorimeter

A type of apparatus suitable for studying combustion or other solid–gas reactions is the **bomb calorimeter**, shown in Figure 13.17. The tightly sealed reaction chamber or (totally misnamed!) *bomb* is constructed from thick (~1 cm) stainless steel, which means that the volume of the system does not change. From Equation 13.17 (p.641), the heat change during a reaction conducted at constant volume is the internal energy change. Thus, *a bomb calorimeter measures* ΔU.

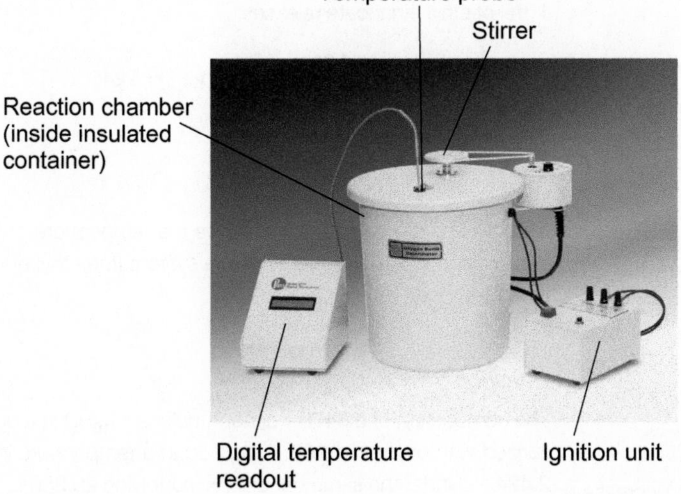
Temperature probe
Stirrer
Reaction chamber
(inside insulated
container)
Digital temperature
readout
Ignition unit

A commercial bomb calorimeter.

benzoic acid

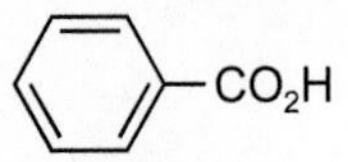
CO_2H

naphthalene

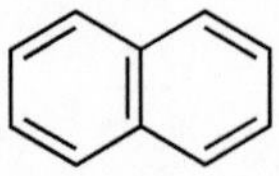

You might think that, if a known mass of water were used in the calorimeter, the specific heat capacity of water could be used in an analogous way to Worked example 13.2 (p.617). However, this would not account for any heat absorbed by the metal bomb nor for any heat losses from the casing, which cannot be completely eliminated.

The heat released during the reaction raises the temperature of the water in the surrounding jacket. To relate this temperature change to the heat change during the reaction, the calorimeter must be calibrated so that you know how much energy it takes to change the temperature of the water by 1 K. The easiest method of doing this is to use a reaction for which the internal energy change, ΔU, is already known. Two compounds commonly used are naphthalene ($C_{10}H_8$) and benzoic acid ($C_6H_5CO_2H$), which react with oxygen according to

$$C_{10}H_8\,(s) + 12\,O_2\,(g) \rightarrow 10\,CO_2\,(g) + 4\,H_2O\,(l) \qquad \Delta_c U^{\ominus}_{298} = -5142\,\text{kJ mol}^{-1}$$

$$C_6H_5CO_2H\,(s) + 7\tfrac{1}{2}O_2\,(g) \rightarrow 7\,CO_2\,(g) + 3\,H_2O\,(l) \qquad \Delta_c U^{\ominus}_{298} = -3251\,\text{kJ mol}^{-1}$$

The temperature rise produced by burning a known mass of one of these compounds is measured, and used to calculate the relationship between the energy released and the temperature rise of the water for the particular calorimeter being used. This relationship is called the **calibration factor**. To find the energy released from the combustion of another compound, you measure the temperature rise and use this calibration factor as explained in Worked example 13.13.

Worked example 13.13 Using a bomb calorimeter

The combustion of 0.8750 g of benzoic acid (M_r = 122.1) in a bomb calorimeter at 298 K caused a temperature rise of 2.279 K, while combustion of 0.783 g ethyl ethanoate (M_r = 88.1) caused a temperature rise of 1.951 K. Calculate the internal energy change of combustion, $\Delta_c U$, for ethyl ethanoate. ($\Delta_c U$ (benzoic acid) = −3251 kJ mol^{-1}.)

Strategy

First, you need to find the calibration factor for the calorimeter. Calculate the heat released by burning the benzoic acid and use this together with the temperature rise to calculate the calibration factor.

Knowing this, you can then calculate the heat released by the combustion of ethyl ethanoate. Knowing the heat released for the mass used, you can use M_r to find $\Delta_c U$ per mole.

Solution

Calculate the amount (in mol) of benzoic acid used.

$$\text{Amount (in moles)} = \frac{0.8750\,\cancel{g}}{122.1\,\cancel{g}\,\text{mol}^{-1}} = 7.17 \times 10^{-3}\,\text{mol}$$

Calculate the heat released by the combustion of benzoic acid.

$$\text{Heat released} = \text{number of moles} \times \text{heat released per mole}$$
$$= 7.17 \times 10^{-3}\,\cancel{\text{mol}} \times 3251\,\text{kJ}\,\cancel{\text{mol}^{-1}}$$
$$= 23.31\,\text{kJ}$$

Calculate the calibration factor for the calorimeter.

The release of 23.31 kJ caused a temperature rise in the water of 2.279 K. So the calibration factor is given by

$$\text{calibration factor} = \frac{\text{heat supplied to the water}}{\text{temperature change}}$$

$$= \frac{23.31\,\text{kJ}}{2.279\,\text{K}} = 10.23\,\text{kJ K}^{-1}$$

Thus, it takes a heat release of 10.23 kJ from the reaction to increase the temperature of the water in the calorimeter by 1 K.

Use the calibration factor to find the heat released by the combustion of ethyl ethanoate.

The temperature rise was 1.951 K, so

$$\text{heat released} = 10.23\,\text{kJ}\,\cancel{\text{K}}^{-1} \times 1.951\,\cancel{\text{K}} = 19.96\,\text{kJ}$$

Calculate the heat released by the combustion of 1 mol of ethyl ethanoate.

19.96 kJ was released by the combustion of 0.783 g of ethyl ethanoate

$$\text{Amount (in mol)} = \frac{0.783\,\cancel{g}}{88.1\,\cancel{g}\,\text{mol}^{-1}}$$
$$= 8.89 \times 10^{-3}\,\text{mol of ethyl ethanoate}$$

So

8.89×10^{-3} mol of ethyl ethanoate released 19.96 kJ

1 mol of ethyl ethanoate releases

$$\frac{19.96\,\text{kJ}}{8.89 \times 10^{-3}\,\cancel{\text{mol}}} \times 1\,\cancel{\text{mol}} = 2245\,\text{kJ}$$

And therefore

$$\Delta_c U = -2245\,\text{kJ mol}^{-1}$$

Note that heat is *released* and *raises* the temperature of the water. The change in internal energy of the system (the ethyl ethanoate) is therefore *negative*.

Question

The combustion of 0.6475 g of naphthalene ($C_{10}H_8$) in a bomb pressurized with oxygen at 298 K resulted in a temperature increase of 2.424 K. Under the same conditions, supplying 20.250 kJ of energy raised the temperature by 1.890 K. Calculate $\Delta_c U$ for naphthalene.

Using bomb calorimetry measurements to determine $\Delta_c H^{\ominus}$

Bomb calorimetry can also be used to find values of $\Delta_c H^{\ominus}$ The value of $\Delta_c H^{\ominus}$ found from bomb calorimetry is converted to $\Delta_c H^{\ominus}$ using the relationship between the changes in enthalpy and internal energy given in Equation 13.19. Applying this to a combustion reaction gives Equation 13.20

$$\Delta_c H^{\ominus} = \Delta_c U^{\ominus} + \Delta n_{gas} RT \qquad (13.20)$$

In using this expression, you are assuming that the gas obeys the ideal gas equation. The assumption of ideal gas behaviour is reasonable for pressures up to several hundred bar. The procedure is illustrated in Worked example 13.14.

Visit the Online Resource Centre to watch screencast 13.5 which walks you through Worked Example 13.14.

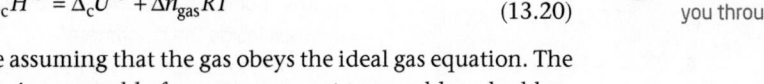

Worked example 13.14 Using measurements from a bomb calorimeter to determine $\Delta_c H^{\ominus}$

From bomb calorimetry, $\Delta_c U^{\ominus}_{298}$ for naphthalene was found to be -5142 kJ mol^{-1}. Calculate $\Delta_c H^{\ominus}$ for naphthalene at 298 K.

Strategy

The first step is to write the thermochemical equation for the combustion reaction and use this to calculate the change in number of moles of gas during the reaction.

Then use Equation 13.20 to find $\Delta_c H^{\ominus}$. Be careful to keep all values in the same energy units and not to mix J and kJ in the same equation. (This is the most common source of error in these calculations.)

Solution

Write the thermochemical equation for the reaction.

$$C_{10}H_8(s) + 12O_2(g) \rightarrow 10CO_2(g) + 4H_2O(l)$$

$$\Delta_c U^{\ominus}_{298} = -5142 \text{ kJ mol}^{-1}$$

Calculate Δn_{gas}

$$\Delta n_{gas} = n_{gas}(\text{products}) - n_{gas}(\text{reactants}) = 10 - 12 = -2 \text{ mol}$$

Use Equation 13.20 to find $\Delta_c H^{\ominus}$.

$$\Delta_c H^{\ominus} = \Delta_c U^{\ominus} + \Delta n_{gas} RT \qquad (13.20)$$

$$\Delta_c U^{\ominus} \text{ (for 1 mol of naphthalene)} = -5142 \text{ kJ} = -5\,142\,000 \text{ J}$$

$$\Delta_c H^{\ominus} = -5\,142\,000 \text{ J} + (-2 \text{ mol} \times 8.314 \text{ J K}^{-1} \text{ mol}^{-1} \times 298 \text{ K}) = -5\,147\,000 \text{ J}$$

This is for the combustion of 1 mol of naphthalene so that

$$\Delta_c H^{\ominus} = -5147 \text{ kJ mol}^{-1}$$

The difference between the internal energy change and the enthalpy change is relatively small in this case. This is usually true at low temperatures.

Question

$\Delta_c U^{\ominus}_{298} = -2245$ kJ mol^{-1} for ethyl ethanoate ($CH_3CO_2CH_2CH_3$). Calculate $\Delta_c H^{\ominus}$ at 298 K.

Biochemists use $\Delta_c H^{\ominus}$ values to indirectly provide information on reactions which are difficult to study directly. Many metabolic processes essentially involve the reaction of food components with oxygen to form compounds such as water, carbon dioxide, and urea. These processes occur at constant pressure inside a living cell rather than in a bomb calorimeter, so $\Delta_c H$ applies rather than $\Delta_c U$.

Even at rest, a human requires about 6000 kJ per day, which is provided by food intake, and this is clearly much higher if you do anything other than lie in bed all day! As a guide, proteins and carbohydrates produce around 15 kJ g^{-1}–20 kJ g^{-1}, while the metabolism of fats yields 35 kJ g^{-1}– 40 kJ g^{-1}. Consumption of alcohol gives around 30 kJ g^{-1}. Box 13.8 illustrates a more unusual form of calorimetry.

Although a bomb calorimeter is mainly used for studying combustion reactions, other reactions of solids and liquids with gases can also be studied in this way. If enthalpy changes are to be calculated, the chemical equation for the reaction must be known in order to calculate Δn_{gas}. With hydrocarbons, this is straightforward since the products are CO_2 and water. However, if other elements such as nitrogen or phosphorus are present, a number of reaction products are possible. For example, nitrogen may form N_2, NO, or NO_2. In these cases, careful analysis to determine the products of the reaction is needed.

The energy content of foods is often given in non-SI units. 1 Calorie = 4.184 kJ. By definition, this is the amount of heat needed to raise the temperature of 1 kg of water by 1 °C.

Box 13.8 Calorimetry, food, and metabolism

The energy content of foods can be determined from experiments using a bomb calorimeter. For example, for sugar (sucrose, $C_{12}H_{22}O_{11}$)

$$C_{12}H_{22}O_{11}\,(s) + 12\,O_2\,(g) \rightarrow 12\,CO_2\,(g) + 11\,H_2O\,(l)$$

$$\Delta_c H^{\ominus} = -5647.5\ \text{kJ mol}^{-1}$$

During respiration, our bodies carry out essentially the same reaction although the oxidation pathway is very different. Hess's law means that the energy production will be the same by both pathways.

While the bomb calorimeter gives a good measure of the energy content of foods, it does not, of course, take into account how food is processed in the body. Different people have different rates of consuming energy and turning food into energy—their *metabolisms* differ.

Detailed study of energy consumption is possible using a **whole body calorimeter**. This is a sealed chamber, or even a room in sophisticated versions, that is thermally insulated and in which the temperature can be accurately measured. The air supply is tightly controlled and the composition of gases in the atmosphere is carefully monitored. Volunteers can spend anything from 12 hours to a week inside the calorimeter.

The controlled environment means that the energy usage can be determined while performing a range of tasks, from rest to vigorous exercise. It also means that the food intake of the volunteers can be monitored.

A major area of study is the investigation of factors that lead to weight problems and obesity. The production of energy from foodstuffs can be followed in detail and correlated with other factors in patients' lifestyles. The effects of potentially therapeutic drugs can also be monitored under controlled conditions. Other studies have shown that some diseases, including many cancers, cause significant changes to metabolism.

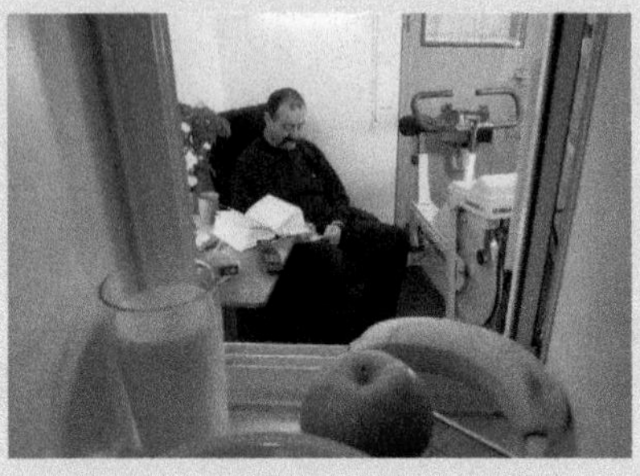

▲ A whole body calorimeter. These calorimeters monitor gaseous exchange to continually measure energy exchange and consumption for up to seven consecutive days and nights. The small room in this picture is actually a calorimeter and the volunteer lives in it for that length of time. Food is passed into the calorimeter through a two-door hatchway.

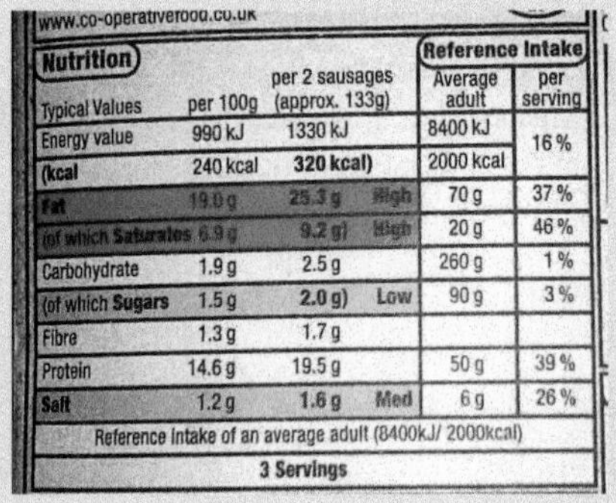
www.co-operativefood.co.uk
Nutrition
Reference Intake
Typical Values | per 100g | per 2 sausages (approx. 133g) | Average adult | per serving
Energy value | 990 kJ | 1330 kJ | 8400 kJ | 16 %
(kcal | 240 kcal | 320 kcal) | 2000 kcal
Fat | 19.0 g | 25.3 g | High | 70 g | 37 %
(of which Saturates | 6.9 g | 9.2 g) | High | 20 g | 46 %
Carbohydrate | 1.9 g | 2.5 g | 260 g | 1 %
(of which Sugars | 1.5 g | 2.0 g) | Low | 90 g | 3 %
Fibre | 1.3 g | 1.7 g
Protein | 14.6 g | 19.5 g | 50 g | 39 %
Salt | 1.2 g | 1.6 g | Med | 6 g | 26 %
Reference intake of an average adult (8400kJ/ 2000kcal)
3 Servings

▲ The energy content of food is now displayed in kJ and calories on packaging labels.

Solution calorimetry

Reactions occurring in aqueous solution are particularly important, especially in biochemical systems. Enthalpy changes for reactions in solution are measured in a **solution calorimeter**. The principle can be illustrated in a simple apparatus used in elementary studies, commonly called a 'coffee cup calorimeter', as shown in Figure 13.18.

The reaction takes place in a container that is thermally insulating—hence the use of a polystyrene coffee cup—and the temperature change is measured during the reaction. Since the calorimeter is not sealed, reactions are performed at *constant pressure*.

As with the bomb calorimeter described earlier, the relationship between the temperature change and the heat output from the reaction must be established by calibration, often using an acid–base neutralization. The enthalpy change for this type of reaction is $-56.9\ \text{kJ mol}^{-1}$, irrespective of the acid and base used as long as they are strong acids and bases.

For accurate work, a coffee cup or even a vacuum flask is not sufficient. A commercial apparatus for performing accurate enthalpy change measurements in solution is shown in Figure 13.19. It allows two (or more) reactants to be placed in an insulated, thermostatically controlled chamber until the temperature is constant. The reactants can then be rapidly mixed to start the reaction, either by tilting the reaction vessel or, in other versions, using a magnet to break a seal between the reactant chambers.

Enthalpy changes in ionic reactions can also be measured by monitoring electrochemical properties. This is discussed further in Section 16.4 (p.746).

For the reaction of a strong acid with a strong base, the ionic equation is $H^+(aq) + OH^-(aq) \rightarrow H_2O\,(l)$; see Section 7.1 (p.304).

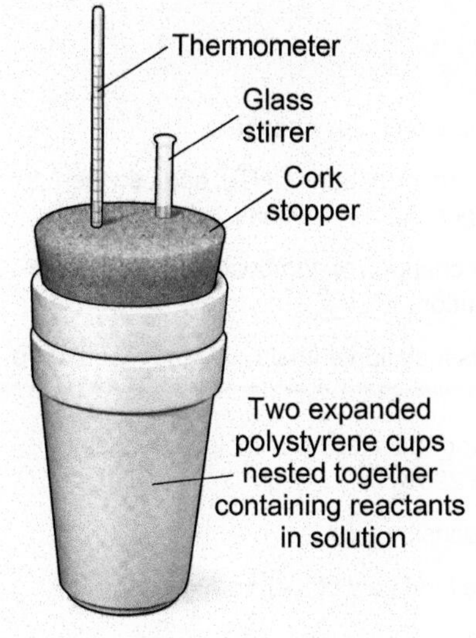
Thermometer
Glass stirrer
Cork stopper
Two expanded polystyrene cups nested together containing reactants in solution

Figure 13.18 A simple 'coffee cup' calorimeter. The polystyrene foam acts as insulation and stops heat from the reaction being lost to the surroundings. An improved version uses a vacuum flask as the reaction vessel to minimize heat losses.

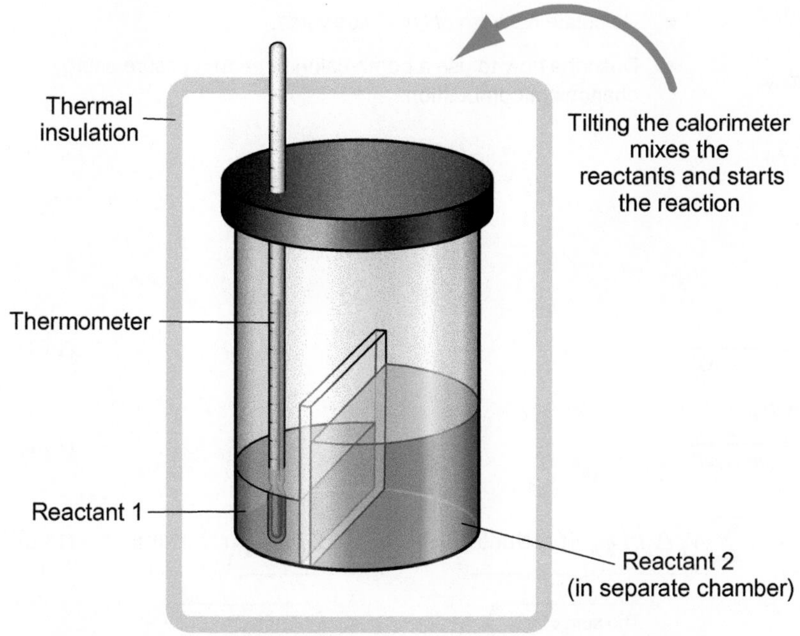
Thermal insulation
Tilting the calorimeter mixes the reactants and starts the reaction
Thermometer
Reactant 1
Reactant 2 (in separate chamber)

Figure 13.19 A calorimeter used to measure enthalpy changes for reactions in solution. The vessel is tilted to mix the reactants and start the reaction.

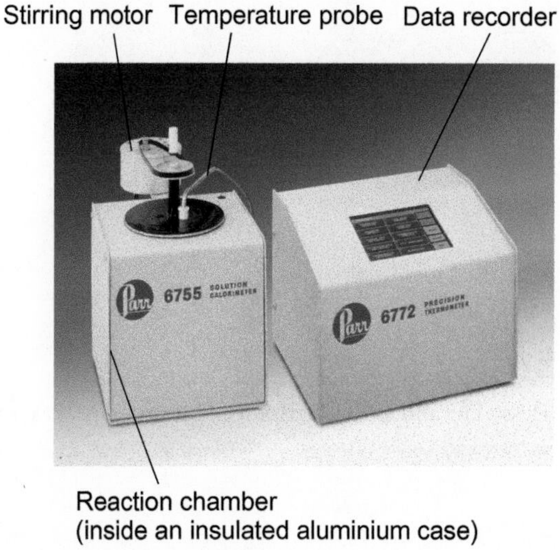
Stirring motor
Temperature probe
Data recorder
6755
6772
Reaction chamber (inside an insulated aluminium case)

A commercial solution calorimeter.

Summary

- $\Delta_c U$ can be measured using a bomb calorimeter.
- Calorimeters can be calibrated by measuring the energy change for compounds with known $\Delta_c U^{\ominus}$.
- $\Delta_c H^{\ominus}$ can be calculated from measurements of $\Delta_c U^{\ominus}$, using

$$\Delta_c H^{\ominus} = \Delta_c U^{\ominus} + \Delta n_{gas} RT$$

- Solution calorimeters are used to study reactions that take place in solution.
- Solution calorimeters can be calibrated by the reaction of a strong acid and a strong base which has $\Delta_r H = -56.9\,\text{kJ mol}^{-1}$.

For practice questions on these topics, see questions 11, 21, 25–28, 32, and 35 at the end of this chapter (pp.651–653).

Concept review

By the end of this chapter you should be able to do the following.

- Give examples of the different ways that energy can be transferred in chemical systems.
- Understand what is meant by the terms *system* and *surroundings*.
- Explain the difference between *isolated*, *open*, and *closed* systems.
- Define, and perform calculations involving, heat and work.
- Understand the difference between *heat capacity*, *specific heat capacity*, and *molar heat capacity*, and carry out calculations using these properties.
- Explain the difference between *extensive* and *intensive* properties.
- Describe the characteristics of a state function.
- Define enthalpy change and calculate changes of enthalpy during phase changes and chemical reactions.
- Use Hess's law to calculate enthalpy changes.
- Define enthalpy changes of formation, $\Delta_f H^{\ominus}_{298}$, and use them to calculate enthalpy changes during reactions.
- Define and use enthalpy changes of combustion, $\Delta_c H^{\ominus}_{298}$, and enthalpy changes of solution, $\Delta_{sol} H^{\ominus}_{298}$.
- Explain and use bond dissociation enthalpy, D, and mean bond enthalpy, $\overline{D}$.
- Use the Kirchhoff equation to calculate enthalpy changes at temperatures other than 298 K.
- Define and calculate expansion work.
- Describe and explain the First Law of thermodynamics.
- Define internal energy, U.
- Calculate ΔU from ΔH and vice versa.
- Describe how to use a bomb calorimeter to measure energy changes of combustion.

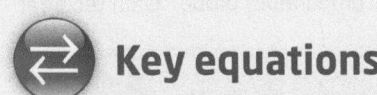

Key equations

Specific heat capacity	$C_s = \dfrac{q}{m \times \Delta T}$	(13.1)
Molar heat capacity	$C_s = \dfrac{q}{m \times \Delta T}$	(13.2)
Hess's law and enthalpy change of formation	$\Delta_r H^{\ominus}_{298} = \sum \nu_i \Delta_f H^{\ominus}_{298}\text{(products)} - \sum \nu_i \Delta_f H^{\ominus}_{298}\text{(reactants)}$ The sum of all the standard enthalpy changes of formation of the *products* The sum of all the standard enthalpy changes of formation of the *reactants*	(13.6)
Kirchhoff equation	$\Delta_r H^{\ominus}_{T_2} = \Delta_r H^{\ominus}_{T_1} + \Delta_r C_p (T_2 - T_1)$	(13.10)
Expansion work	$w = -p_{ext}\Delta V$	(13.14)
The First Law of thermodynamics	$\Delta U = q + w$	(13.15)
Enthalpy change and internal energy change	$\Delta H = \Delta U + p\Delta V$	(13.18)
	$\Delta H = \Delta U + \Delta n_{gas} RT$	(13.19)

? Questions

More challenging questions are indicated by an asterisk *.

Some problems require the use of $\Delta_f H^{\ominus}$ and/or heat capacity data from Appendix 7 (p.1352).

1. Classify the following properties as intensive or extensive: density; amount in moles; pressure; length; temperature. (Section 13.1)

2. The heat capacity of air at room temperature (20 °C) is approximately 21 J K^{-1} mol^{-1}. (Section 13.1)

   (a) How much heat is required to raise the temperature of a 5 m × 5 m × 3 m room by 10 °C?

   (b) How long will it take a 1 kW heater to achieve this? (Assume that the volume of 1 mol of air is 24 dm^3 at 20 °C.)

3. A 50.0 g block of copper at 90.0 °C is placed in a beaker containing 200.0 g of water at 20.0 °C. Calculate the final temperature of the copper and water. (The specific heat capacities of copper and water are 0.38 J K^{-1} g^{-1} and 4.18 J K^{-1} g^{-1}, respectively.) (Section 13.1)

4. Calculate the enthalpy change when 100 g of water freezes at 0 °C. (The standard enthalpy change of fusion of water is +6.01 kJ mol^{-1}.) (Section 13.2)

5. Calculate the energy needed to melt 750 kg of sodium metal at 371 K. (The standard enthalpy change of fusion of sodium is +2.60 kJ mol^{-1}.) (Section 13.2)

6. Calculate how much heat is required to convert 10 g of ice at 0 °C to steam at 100 °C. (The enthalpy change of fusion of ice is +6.01 kJ mol^{-1} and the enthalpy change of vaporization of water is +40.7 kJ mol^{-1}.) (Section 13.2)

7. The decomposition at constant volume of 1 mol of gaseous krypton difluoride (KrF_2) to its elements at 298 K gives out 59.4 kJ of heat. Calculate the standard enthalpy change of formation, $\Delta_f H^{\ominus}_{298}$, of solid KrF_2. (The enthalpy change of sublimation of solid KrF_2 is +41 kJ mol^{-1}.) (Section 13.3)

8. Use the following thermochemical equations to calculate the standard enthalpy change of formation of HNO_3 (l).

$$H_2O_2\,(l) + 2NO_2\,(g) \rightarrow 2HNO_3\,(l) \qquad \Delta_r H^{\ominus} = -226.8\ \text{kJ mol}^{-1}$$
$$N_2\,(g) + 2O_2\,(g) \rightarrow 2NO_2\,(g) \qquad \Delta_r H^{\ominus} = +66.4\ \text{kJ mol}^{-1}$$
$$H_2\,(g) + O_2\,(g) \rightarrow H_2O_2\,(l) \qquad \Delta_r H^{\ominus} = -187.8\ \text{kJ mol}^{-1}$$

   (Section 13.1)

9. The reaction

$$2Al\,(s) + Fe_2O_3(s) \rightarrow 2Fe\,(s) + Al_2O_3\,(s)$$

   is known as the *thermite* reaction and is a spectacular demonstration of an exothermic chemical reaction. Use data from Appendix 7 to calculate the enthalpy change when 54.0 g of aluminium react at 298 K (25 °C). (Section 13.3)

10. At 298 K, the standard enthalpy change of combustion of hydrogen is −286 kJ mol^{-1}. The corresponding values for graphite and methanol are −394 kJ mol^{-1} and −727 kJ mol^{-1}, respectively. Calculate the standard enthalpy change of formation of methanol at 298 K. (Section 13.3)

11. When 1.54 g of solid biphenyl ($(C_6H_5)_2$) were burned in excess oxygen at 298 K in a bomb calorimeter to form CO_2 (g) and H_2O (l), 64.23 kJ of heat were released. Calculate $\Delta_f H^{\ominus}_{298}$ for biphenyl. (Section 13.5)

12. Calcite and aragonite are two forms of calcium carbonate. Calculate the enthalpy change for the transition from calcite to aragonite. ($\Delta_f H$(calcite) = −1206.9 kJ mol^{-1} and $\Delta_f H$(aragonite) = −1207.1 kJ mol^{-1}.) (Section 13.3)

13. Calculate the enthalpy change when gaseous benzene (C_6H_6) dissociates into gaseous atoms at 298 K. Carry out the calculation by two different methods using the data in (a) and (b) below. Comment on the difference in the values you obtain by the two methods. (Section 13.3)

    (a) Assume benzene molecules contain three single and three double carbon–carbon bonds, and use mean bond enthalpy data. (Mean bond enthalpies/kJ mol^{-1}: C–C, 347; C=C, 612; C–H, 412.)

    (b) The enthalpy change of combustion of liquid benzene at 298 K is −3267.4 kJ mol^{-1}. The enthalpy change of vaporization of benzene at 298 K is +33.9 kJ mol^{-1}.

    ($\Delta_f H^{\ominus}_{298}$ / kJ mol^{-1}: CO_2 (g), −393.5; H_2O (l), −285.8; C (g), 716.7; H (g), 218.)

14. The thermochemical equation for the following reaction at 300 K is

$$N_2\,(g) + 3H_2\,(g) \rightarrow 2NH_3\,(g) \qquad \Delta_r H = -92.4\ \text{kJ mol}^{-1}$$

    Estimate the enthalpy change for the reaction at 800 K. (Section 13.4)

15. Calculate the energy transferred as work, w, when a gas is compressed from 250 cm^3 to 125 cm^3 by an external pressure of 10 kPa. (Section 13.5)

16. For a reaction at constant pressure, the enthalpy change is +30 kJ. During the reaction, the system expands and does 25 kJ of work. What is the change in internal energy for the reaction? (Section 13.5)

17. Calculate the energy transferred as work, w, for a system that releases 450 J of heat in a process for which the internal energy of the system decreases by 135 J. (Section 13.5)

18. A 100 W electric heater is used to heat gas in a cylinder for 10 min. The gas expands from 500 cm^3 to 21.4 dm^3 against a pressure of 1.10 atm. What is the change in internal energy, ΔU, of the gas? (Section 13.5)

19. 0.10 mol of calcium carbonate react with excess dilute hydrochloric acid at 298 K. Calculate the energy transferred as work. (Section 13.5)

20. The molar heat capacity of water at constant pressure is $75.3\,J\,K^{-1}\,mol^{-1}$ at room temperature. Calculate the mass of methanol that must be burned to heat $1\,dm^3$ of water from 20 °C to 50 °C. (Assume the density of water is $1.0\,g\,cm^{-3}$.) (Section 13.3)

21. A slice of banana weighing 2.7 g was burned in oxygen in a bomb calorimeter and produced a temperature rise of 3.05 K. In the same calorimeter, the combustion of 0.316 g of benzoic acid produced a temperature rise of 3.24 K. $\Delta_c U$ for benzoic acid is $-3251\,kJ\,mol^{-1}$. If the average mass of a banana is 125 g, how much energy in (a) kJ and (b) calories (kcal) can be obtained on average from a banana? (1 calorie = 4.18 kJ.) (Section 13.5)

22. Car safety airbags inflate when the car undergoes a sudden deceleration, setting off a reaction which produces a large amount of gas. One of the reactions used is

$$NaN_3(s) \rightarrow Na(s) + 1.5N_2(g) \qquad \Delta_r H^{\circ}_{298} = -21.7\,kJ\,mol^{-1}$$

(Section 13.3)

(a) What is the value of $\Delta_f H^{\ominus}_{298}$ for NaN_3?

(b) An airbag system is inflated by 2.4 mol (around $60\,dm^3$) of $N_2(g)$. Calculate the enthalpy change when this amount of nitrogen is produced at 298 K.

(c) Assuming this energy all goes into heating the $N_2(g)$, what will be the final temperature of the gas? (C_p for $N_2(g) = 29.1\,J\,K^{-1}\,mol^{-1}$.)

23. The standard enthalpy change of vaporization, $\Delta_{vap}H^{\ominus}$, of ethanol is $42.30\,kJ\,mol^{-1}$ at 298 K. Calculate the value of $\Delta_{vap}H^{\ominus}$ at 340 K. (Section 13.4)

24.* Using values of $\Delta_f H^{\ominus}_{298}$ from Appendix 7, calculate the enthalpy change for the oxidation of liquid ethanol to ethanoic acid by molecular oxygen at the human body temperature of 37 °C. (Section 13.4)

25. $50.0\,cm^3$ of HCl(aq) of concentration $1.0\,mol\,dm^{-3}$ were mixed with $50.0\,cm^3$ of NaOH(aq) of concentration $1.0\,mol\,dm^{-3}$ at 25 °C in a constant pressure solution calorimeter. After the reactants were mixed the temperature increased to 31.8 °C. Assuming that the solutions have the same density and specific heat capacity as water, calculate the enthalpy change for the neutralization reaction. (Section 13.5)

26. A flame calorimeter is used to measure enthalpy changes in gaseous reactions. In order to calibrate a particular flame calorimeter, a quantity of methane was burned in oxygen to release 12.54 kJ of energy which resulted in a temperature rise of 1.000 K.

Using the same flame calorimeter, the combustion of 0.02 mol of butane, C_4H_{10}, raised the temperature by 5.017 K while combustion of the same amount of methylpropane caused a temperature rise of 4.575 K.

Calculate the enthalpy change of isomerization for the two butane isomers. (Section 13.5)

27.* A chemist is trying to measure the enthalpy change for the hydration reaction:

$$Na_2CO_3(s) \rightarrow Na_2CO_3{\cdot}10H_2O(s)$$

Calibration of a solution calorimeter shows that 357.9 J of heat raises the temperature by 1.00 K. When 2.500 g of anhydrous sodium carbonate, Na_2CO_3, was dissolved in $100.00\,cm^3$ of water, the temperature increased by 1.550 K. On dissolving 3.500 g of sodium carbonate decahydrate, $Na_2CO_3{\cdot}10H_2O$ in $100.00\,cm^3$ of water, the temperature decreased by 2.310 K.

Calculate enthalpy change of hydration of sodium carbonate. (Section 13.5)

28.* A bomb calorimeter was calibrated by igniting a 0.7807 g sample of benzoic acid in excess oxygen. The temperature of the calorimeter rose by 1.940 K from 298 K. The standard internal energy change of combustion of benzoic acid is $-3251\,kJ\,mol^{-1}$. Combustion of 0.9008 g of α-D-glucose in the same calorimeter gave a temperature rise of 1.311 K. Use this information to find:

(a) the enthalpy change of formation of α-D-glucose at 298 K;

(b) the enthalpy change of combustion of α-D-glucose at the human body temperature of 37 °C.

Use relevant data from Appendix 7. C_p for $C_6H_{12}O_6(s) = 330.0\,J\,K^{-1}\,mol^{-1}$. (Section 13.5)

29.* A mixture of sulfur dioxide and air containing 10% (by volume) SO_2 is passed over a catalyst at 700 K at a rate of $100\,mol\,min^{-1}$. If, at constant pressure, all of the SO_2 is converted to SO_3, what is the rate of evolution of heat? Use relevant data from Appendix 7. (Section 13.3)

30. Using data from Appendix 7, calculate the standard enthalpy change at 25 °C for the reaction

$$CH_4(g) + 2H_2S(g) \rightarrow CS_2(l) + 4H_2(g)$$ (Section 13.3)

31. Cinnabar is an ore from which mercury metal can be obtained by reaction with oxygen. It mainly comprises mercury sulfide, HgS. If $\Delta_f H^{\ominus}$ for HgS(s) is $-58.2\,kJ\,mol^{-1}$, calculate the standard enthalpy change at 25 °C for

$$HgS(s) + O_2(g) \rightarrow SO_2(g) + Hg(l)$$ (Section 13.3)

32. A student used a calorimeter containing 100 g of deionized water which required an energy change of 818 J to cause a temperature change of 1 K.

An unknown mass of sodium hydroxide, NaOH, was dissolved in the water and the temperature rose from 25.00 °C to 31.00 °C. Given that the molar enthalpy of solution of NaOH is $-44.51\,kJ\,mol^{-1}$, what mass of sodium hydroxide was dissolved? (Section 13.6)

33. Water has a standard enthalpy of vaporization of $+40.7\,kJ\,mol^{-1}$. Calculate q, w, ΔH, and ΔU when 5 mol are vaporized at 373 K and 1 bar. (Section 13.5)

34. Calculate the standard enthalpy of formation of $N_2O_5(g)$ from the following data

$$N_2(g) + O_2(g) \rightarrow 2NO(g) \qquad \Delta_r H^\ominus = -180.5\,kJ\,mol^{-1}$$

$$2NO(g) + O_2(g) \rightarrow 2NO_2(g) \qquad \Delta_r H^\ominus = -114.4\,kJ\,mol^{-1}$$

$$2N_2O_5(g) \rightarrow 4NO_2(g) + O_2(g) \qquad \Delta_r H^\ominus = -110.2\,kJ\,mol^{-1}$$

(Section 13.3)

35. Calculate $\Delta_f H^\ominus$ for benzene ($C_6H_6(l)$) at 298 K given that the standard molar enthalpy of combustion of benzene is $-3268\,kJ\,mol^{-1}$. (Section 13.6)

36. Calculate the change in internal energy at 298 K when one mol of benzene vapour reacts with excess oxygen to form carbon dioxide gas and liquid water. (Section 13.5)

37. The water gas shift reaction is used to remove CO from industrial gases. It is commonly operated over a catalyst at 623 K. Using data in Appendix 7, calculate the standard enthalpy change for the reaction at this temperature.

$$CO(g) + H_2O(g) \rightarrow CO_2(g) + H_2(g)$$ (Section 13.4)

14

Entropy and Gibbs energy

▼ Haemoglobin is a globular protein. It is found in red blood cells, which transport oxygen around the body.

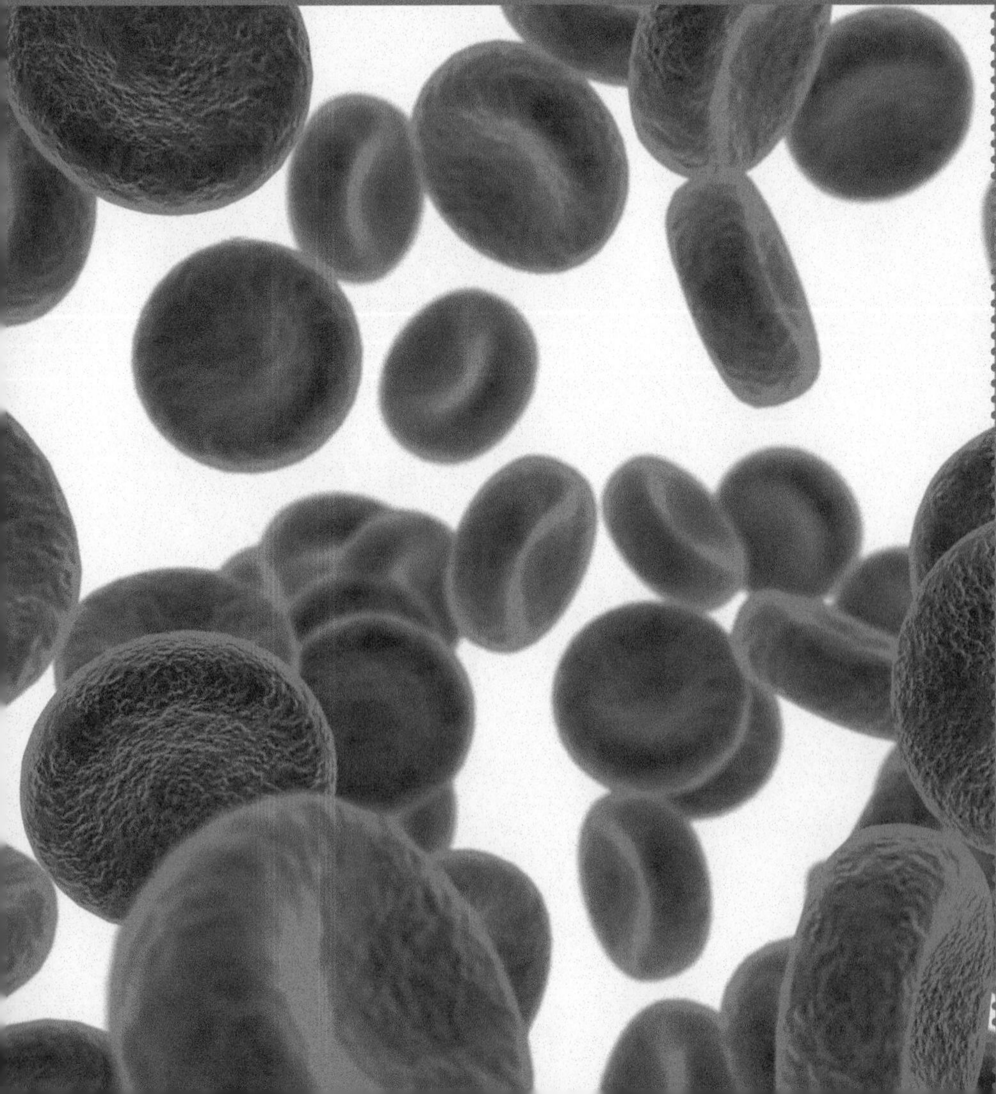

This chapter builds on the following topics:

- **Measurement, units, and nomenclature** Section 1.2, p.6
- **Atoms and the mole** Section 1.3, p.12
- **Chemical equations** Section 1.4, p.21
- **Concentrations of solutions** Section 1.5, p.34
- **Energy changes in chemical reactions** Section 1.6, p.41
- **States of matter and phase changes** Section 1.7, p.47
- **Non-covalent interactions** Section 1.8, p.52
- **Using the ideal gas equation** Section 8.2, p.349
- **Energy changes in chemistry: heat and work** Section 13.1, p.612
- **Enthalpy and enthalpy changes** Section 13.2, p.619
- **Enthalpy changes in chemical reactions** Section 13.3, p.622
- **Variation of enthalpy with temperature** Section 13.4, p.633
- **Internal energy and the First Law of thermodynamics** Section 13.5, p.636

Protein folding

Proteins are made up of amino acid units joined together to form long chains (see Box 2.9 (p.104) for the structure of amino acids). The secondary amide bonds that link the amino acid units together are called **peptide bonds** and so the chains are known as **polypeptide chains**. The order of amino acids in the polypeptide chain is the **primary structure** of the protein (Figure 1).

Figure 1 A section of a polypeptide chain. The R groups are different in different amino acids.

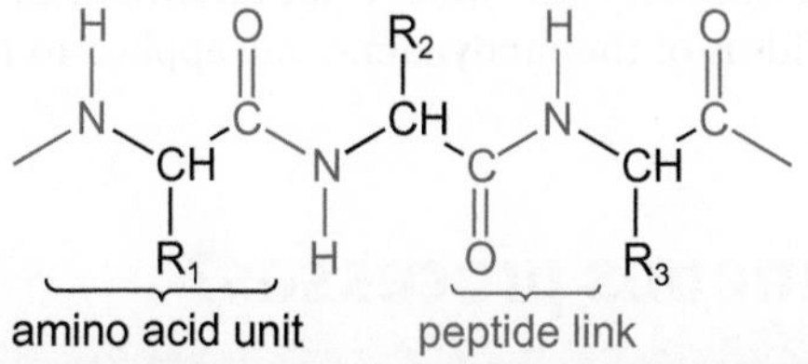
amino acid unit
peptide link

In cells, formation of the primary structure takes place on *ribosomes*, where amino acids are assembled like beads on a necklace. The formation of the polypeptide chain is **non-spontaneous** and requires an input of energy. ATP molecules supply energy when they are hydrolysed to ADP (see Box 14.8, p.683). However, once the chain is formed, it **spontaneously**, and almost instantaneously, self-assembles (folds) to give the characteristic shape of the protein. Spontaneous processes are introduced in Section 14.1 and are the major theme throughout the chapter.

You can think of the folding process as taking place in stages as shown in Figure 2.

1. Hydrogen bonding causes the polypeptide chains to fold or twist to give the **secondary structure**, which is either an *α-helix* or a *β-pleated sheet*. A protein may have areas of both α-helix and β-pleated sheet in its overall structure, as well as more random areas.

2. The α-helical and β-pleated sheet structures then fold further to form a **tertiary structure**.

3. Sometimes the functional protein contains several of these tertiary structures grouped together. This is called the quaternary structure of the protein. Figure 2 shows the tertiary and **quaternary structures** of haemoglobin, the protein in blood that carries oxygen around the body.

The folded form of a protein is called the *native state* and the resulting shape gives the protein its function. For example, an enzyme folds to form an active site to which just one specific reactant can bind. Any changes to the shape or folding of a protein make it inactive. This process is called *denaturation* and renders the protein unable to perform its specific function.

It is important to understand how protein folding occurs in order to understand many biological processes, including the causes of diseases such as Alzheimer's disease, cystic fibrosis, and even some forms of cancer. Box 14.4 (p.673) describes the misfolded proteins in prions responsible for BSE.

The driving force for protein folding comes from the balance between the **energy changes** when folding occurs and the **entropy changes** as the protein structure becomes more ordered. In this chapter, you will find out about this crucial balance, which is the key to understanding all chemical processes.

Figure 2 The folding of a polypeptide chain to form a globular protein, such as haemoglobin, takes place in three stages.

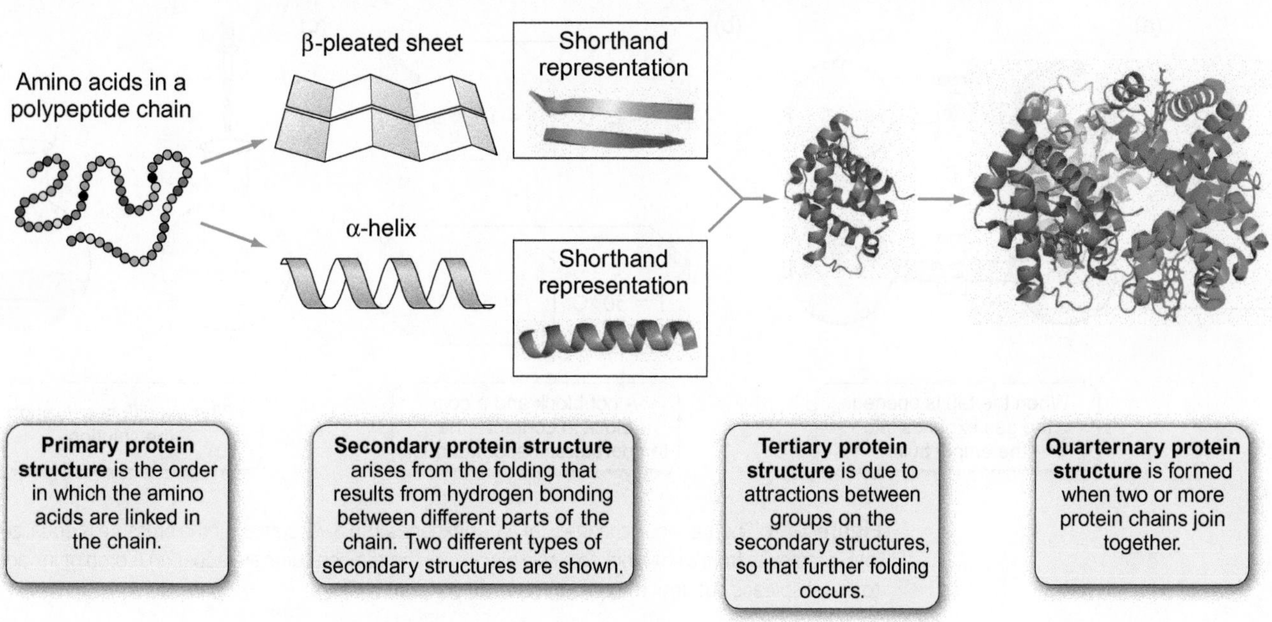
β-pleated sheet
Shorthand representation
Amino acids in a polypeptide chain
α-helix
Shorthand representation
Primary protein structure is the order in which the amino acids are linked in the chain.
Secondary protein structure arises from the folding that results from hydrogen bonding between different parts of the chain. Two different types of secondary structures are shown.
Tertiary protein structure is due to attractions between groups on the secondary structures, so that further folding occurs.
Quarternary protein structure is formed when two or more protein chains join together.

You will have encountered many types of chemical reaction in your studies. Some reactions occur very readily and go effectively to completion to form 100% products. Examples are the decomposition of sodium azide used in a car airbag (see Box 8.1, p.353)

$$2\,NaN_3\,(s) \rightarrow 2\,Na\,(s) + 3\,N_2\,(g)$$

and the combustion of hydrocarbon fuels

$$CH_4\,(g) + 2\,O_2\,(g) \rightarrow CO_2\,(g) + 2\,H_2O\,(l)$$

Some substances such as diamond or gold seem to be totally unreactive. Some reactions proceed at high temperatures but not at low. Some are exothermic; some are endothermic (see Section 1.6, p.41). Many reactions occur readily but do not go to completion: they reach ***equilibrium*** with a final mixture of reactants and products (see Section 1.9, p.56). In this chapter you will see how thermodynamics can be used to predict these different types of behaviour—whether or not reactions will occur and under what conditions their yields can be maximized. In Chapter 15, the ideas of thermodynamics are applied to reactions that come to equilibrium.

14.1 What are spontaneous processes?

Some things happen naturally and other things don't. Think about the situations depicted in Figure 14.1. When the tap in Figure 14.1(a) is opened, gas flows into the evacuated vessel until the pressure is equal on both sides. In the future, it is very unlikely that the reverse process will occur—that all the gas will diffuse into one of the bulbs and leave a vacuum in the other. If a block of hot metal is placed against a cold one, as in Figure 14.1(b), it is very unlikely that it will get hotter while the cold one cools down. Heat will be transferred until both blocks are the same temperature. In Figure 14.1(c), a drop of ink gradually spreads throughout a flask of water until its concentration is uniform. It is unlikely that all the ink will appear in one spot in the water at some later time.

One way of splitting water into hydrogen and oxygen is by electrolysis; see Section 16.5 (p.775).

Consider the reaction of hydrogen with oxygen to form water (Figure 14.2). A mixture of the two gases will, if given a small push in the form of a spark or shock or the addition of a catalyst, readily form water (very quickly and spectacularly!) Around room temperature,

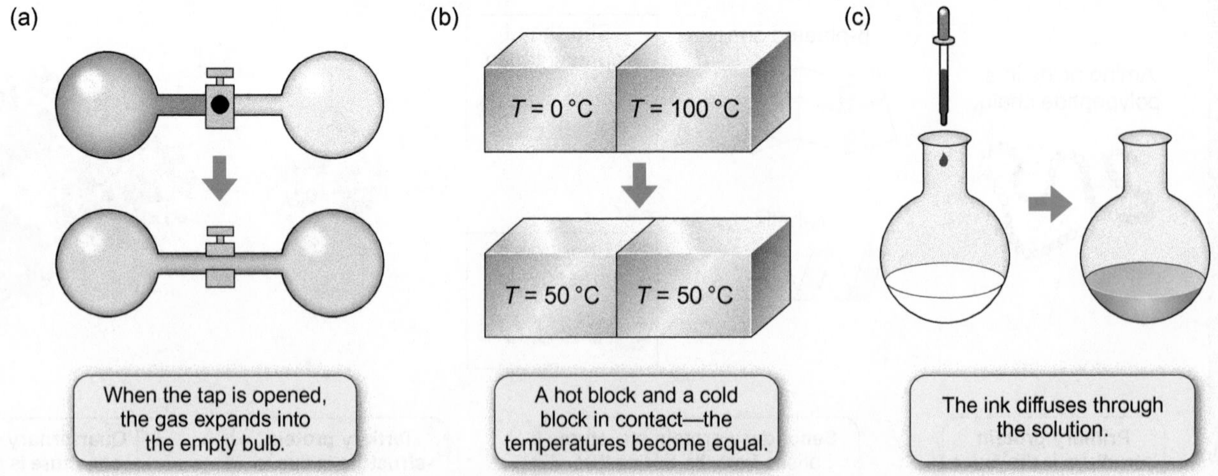
(a)
(b)
(c)
T = 0 °C
T = 100 °C
T = 50 °C
T = 50 °C
When the tap is opened, the gas expands into the empty bulb.
A hot block and a cold block in contact—the temperatures become equal.
The ink diffuses through the solution.

Figure 14.1 Spontaneous changes. (a) A gas expands into a vacuum until the pressure is equalized. (b) Heat moves from a hot block to a cold block until the temperatures are equal. (c) A drop of ink added to water spreads out until the concentration is uniform.

the reverse reaction does not take place easily. Water *can* be split into its constituent elements—but it takes a great deal of energy to do so.

The reaction of hydrogen and oxygen and the processes in Figure 14.1 are said to be **spontaneous**. Spontaneous reactions are those that, once started, will continue without any outside intervention. Another definition is that *spontaneous processes move towards their equilibrium state* without being driven by an external influence. Can we account for the spontaneity of processes in rather more scientific terms than just common sense?

$H_2(g) + \frac{1}{2}O_2(g) \longrightarrow H_2O(l)$

Figure 14.2 The reaction of hydrogen with oxygen is a spontaneous process.

In the early days of physical chemistry it was thought that systems reacted or changed so as to minimize their energy. In the nineteenth century, Marcellin Berthelot suggested that all exothermic reactions are spontaneous. However, things are not quite that simple since many endothermic processes do take place spontaneously. For example, some salts, such as ammonium nitrate, dissolve endothermically in water

$$NH_4NO_3\,(s) + aq \rightarrow NH_4NO_3\,(aq) \qquad \Delta_{sol}H^{\ominus} = +26\,kJ\,mol^{-1}$$

The system gains energy by absorbing heat from the surroundings but the process occurs readily.

Think also of the process in Figure 14.1(a). An ideal gas expands spontaneously into a vacuum at constant temperature (isothermal expansion). There is no exchange of heat with the surroundings and there is no change in internal energy. The change is neither exothermic nor endothermic—but it happens anyway. Whether a process is spontaneous also depends on the conditions. Ice forms liquid water spontaneously at temperatures above 0 °C but the reverse is spontaneous at temperatures below 0 °C.

Spontaneous does not necessarily mean fast. Hydrogen and oxygen gases will combine to form water even without a spark or a catalyst—but the reaction is so slow that it cannot be observed. Rusting of iron is spontaneous—but slow. The thermodynamic spontaneity of a reaction tells nothing about *how fast* it occurs—that's the subject of chemical kinetics as discussed in Chapter 9.

Entropy and disorder

The endothermic change when ammonium nitrate dissolves in water is used in cooling packs; see Box 13.4 (p.629).

An isothermal change takes place at constant temperature; see Box 13.6 (p.638). An ideal gas is one that obeys the ideal gas equation; see Section 8.2 (p.349).

Clearly, there is some factor other than the energy change that plays a part in determining whether changes happen. This factor is called **entropy** and is given the symbol *S*. There are a number of ways of looking at entropy but the way most relevant to chemistry is to consider what is happening to the organization of the molecules in the system and the way in which energy is shared between them.

In each of the situations illustrated in Figure 14.1 the changes lead to a *more random, less ordered* distribution of matter or energy. In Figure 14.1(a), the gas molecules are distributed over a greater volume after opening the tap, so they are *less ordered* than before. In Figure 14.1(b), the thermal energy is shared over a larger number of atoms than before, so the distribution of energy is *less ordered*. In Figure 14.1(c), the ink molecules are mixed with a large amount of water and so are spread out in a *less ordered* way than before. Entropy is identified with the amount of randomness or disorder. In each case, the system becomes more random so that entropy increases and the entropy change, ΔS, has a positive value. Further illustrations are given in Worked example 14.1.

It is more precise to define entropy as a measure of the occupation of the various energy levels in molecules. However, associating it with disorder gives a convenient way of thinking about systems.

Like enthalpy, *H*, and internal energy, *U*, entropy is a *state function*. The change of entropy is given by the difference between the final and initial states of a system

State functions are introduced in Section 13.2 (p.619).

$$\Delta S = S(\text{final}) - S(\text{initial}) \qquad (14.1)$$

As you will see in Section 14.3, it is possible to measure the entropy of pure substances (compare this with enthalpy, where only enthalpy *differences* can be measured). Some values of standard entropy, $S^{\ominus}_{298}$, are given in Table 14.1 on page 665. For example, the standard entropies of the three states of water are:

$$H_2O(s) \quad S^{\ominus}_{298} = 45.0\,J\,K^{-1}\,mol^{-1}$$
$$H_2O(l) \quad S^{\ominus}_{298} = 69.9\,J\,K^{-1}\,mol^{-1}$$
$$H_2O(g) \quad S^{\ominus}_{298} = 188.8\,J\,K^{-1}\,mol^{-1}$$

The formation of frost and ice from gaseous or liquid water is spontaneous below 0 °C but not above 0 °C.

Notice how the standard entropy increases as you go from the more ordered solid state to the less ordered liquid state and to the disordered gaseous state.

Worked example 14.1 Predicting the sign of an entropy change

Predict the sign of the value of ΔS in the system for the following changes:

(a) diffusion of a perfume through a room;

(b) sublimation of 'dry ice', $CO_2(s) \rightarrow CO_2(g)$;

(c) $N_2O_4(g) \rightarrow 2NO_2(g)$;

(d) $CaCO_3(s) \rightarrow CaO(s) + CO_2(g)$;

(e) $Ag^+(aq) + Cl^-(aq) \rightarrow AgCl(s)$.

Strategy

Look at each process and decide whether there is greater disorder in the system after the change than before. If so, the entropy increases and the value of ΔS is positive. If the final situation is more ordered, then the value of ΔS is negative.

Solution

(a) Diffusion of a perfume through a room. Perfume molecules spread through the room mixing with the molecules in the air; there is an increase in disorder compared with all the perfume molecules being in the bottle. ΔS *is positive*.

(b) Sublimation of 'dry ice', $CO_2(s) \rightarrow CO_2(g)$. A gas is more disordered than a solid because the molecules move around to a much greater extent than in a solid. The molecules in a solid are arranged in an ordered way and only vibrate about fixed positions. ΔS *is positive*.

(c) $N_2O_4(g) \rightarrow 2NO_2(g)$. There are more molecules after the change and there are many more ways of arranging these additional molecules, so there is an increase in disorder. ΔS *is positive*.

(d) $CaCO_3(s) \rightarrow CaO(s) + CO_2(g)$. A gas is produced that is more disordered than the solid reactant. ΔS *is positive*.

(e) $Ag^+(aq) + Cl^-(aq) \rightarrow AgCl(s)$. Two types of ions in solution combine to form one solid compound. The ions are free to move in solution but become bound to other ions in the solid. There is an increase in order. ΔS *is negative*.

Question

What will be the sign of the value of ΔS in the system for:

(a) crystallization of salt from a solution;

(b) condensation of a vapour to a liquid;

(c) dissolving sugar in water?

Entropy and probability

Look again at the isothermal expansion of an ideal gas in Figure 14.1(a) (p.656). To start with, consider only *two* molecules. Figure 14.3 shows the possible arrangements of the molecules after the tap is opened to allow the expansion. Assuming that each molecule is equally likely to be in either bulb, there are four possible arrangements of the molecules. Two of these involve one molecule in each bulb so this arrangement is twice as likely as both molecules being in a particular bulb.

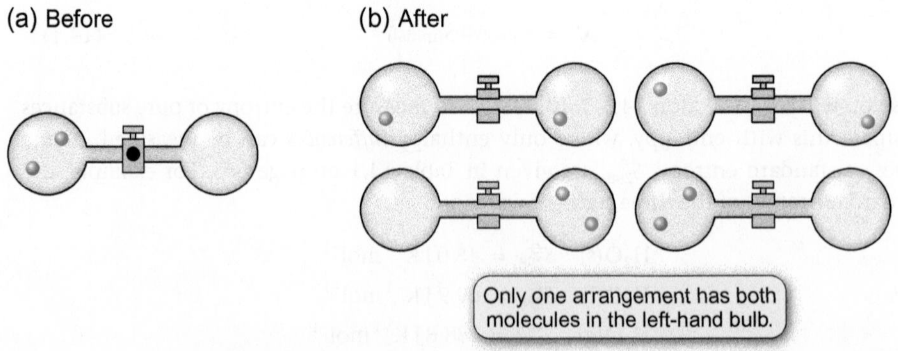
(a) Before
(b) After
Only one arrangement has both molecules in the left-hand bulb.

Figure 14.3 (a) One bulb is empty; the other contains two molecules of gas. (b) The tap is opened. There are four possible arrangements of the molecules.

The chance of any one molecule being in the left bulb is 1 in 2, so its probability is 0.5. The probability of *both* molecules being in the left bulb is $0.5 \times 0.5 = (0.5)^2 = 0.25$. An equivalent way of saying this is that it is one arrangement out of a possible four so the probability is $1 \div 4 = 0.25$.

So, even with only two molecules, the arrangement that spreads the molecules out is more probable. With three molecules, the probability of all being in the left bulb is $0.5 \times 0.5 \times 0.5 = (0.5)^3 = 0.125$. With four molecules the probability is 0.0625 and it decreases rapidly as the number of molecules increases. With 20 molecules it is $\sim 1 \times 10^{-6}$. When you consider any chemically reasonable number of molecules, the value becomes infinitesimally small. For one mole of gas containing 6.022×10^{23} molecules, the chance of finding all the molecules in one bulb is $(0.5)^{6.022\times10^{23}}$, which is less than 1×10^{-100}. In practical terms, there is no realistic chance of ever observing all the gas in one side of the apparatus. Arrangements where the molecules are mixed up and distributed between the two bulbs are much more likely.

If you follow this kind of analysis for all the possible arrangements of molecules, the one with equal numbers of molecules in each bulb is always the most probable.

LVDWIG
BOLTZMANN

Boltzmann's formula, $S = k_B \ln W$, became his epitaph—it is carved into his gravestone in Vienna. His ideas attracted much criticism and were slow to be accepted, but now form the basis of many branches of science. Shortly before experiments confirmed many of his suggestions, Boltzmann committed suicide in 1906 aged 62, depressed and in ill health.

The Boltzmann formula

Entropy is a measure of the randomness or disorder in a system, reflected in the number of ways that the molecules and energy in a system can be arranged. The Austrian physicist, Ludwig Boltzmann, suggested that a quantitative measure of entropy could be defined by the **Boltzmann formula**.

$$S = k_B \ln W \qquad (14.2)$$

Number of ways of arranging molecules and their energies

Entropy

Boltzmann constant $= 1.381 \times 10^{-23}\,\mathrm{J\,K^{-1}}$

W is the number of ways of arranging the molecules and their energy in the system and k_B is a constant, known as the **Boltzmann constant**. The larger the number of arrangements, that is, the less ordered the system, the larger the entropy. The value of k_B is $1.381 \times 10^{-23}\,\mathrm{J\,K^{-1}}$. Note that W is a measure of the ways of arranging molecules *and their energies*. Energy comes in discrete packages called quanta, so it is possible to count the arrangements of quanta in a similar way to counting the arrangements of the molecules themselves.

The concept of energy existing in discrete quanta is discussed in Sections 3.2 (p.117) and 10.2 (p.453).

The Boltzmann constant is related to two other fundamental constants. It is the gas constant expressed per molecule rather than per mole, that is, R divided by the Avogadro constant, N_A. The Boltzmann constant, $k_B = R / N_A$

so $k_B = 8.314\,\mathrm{J\,K^{-1}\,mol^{-1}} / 6.022\times10^{23}\,\mathrm{mol^{-1}}$

$= 1.381\times10^{-23}\,\mathrm{J\,K^{-1}}$.

Summary

- Spontaneous reactions, once started, will continue towards equilibrium without any outside intervention.
- Entropy is a measure of the randomness or disorder in a system.
- Entropy is related to the number of arrangements that a system can adopt by the Boltzmann formula

$$S = k_B \ln W \qquad k_B = 1.381 \times 10^{-23}\,\mathrm{J\,K^{-1}}$$

For practice questions on these topics, see questions 1 and 2 at the end of this chapter (p.692).

14.2 Entropy and the Second Law of thermodynamics

The main use of entropy in chemistry is in predicting the direction of chemical (or any other) change. This is summarized in the **Second Law of thermodynamics**. The Second Law can be stated in a number of ways, but the most straightforward is:

Spontaneous processes are those that increase the total entropy of the Universe.

In thermodynamics, the Universe is the system under consideration together with the surroundings. The system and surroundings are defined in Section 13.1 (p.612).

This may seem a rather grand definition and, in fact, it is not necessary to consider the whole Universe every time you think about a change. You only need to think about the *system* that is changing and its *surroundings*, which are equivalent to the rest of the Universe. If the total entropy change for a process can be determined, the direction of the change can be predicted. Since entropy is related to disorder, the Second Law shows that spontaneous processes are those that lead to an *increase in disorder* of the Universe.

The *total* entropy change for a process is made up of the changes for the system and the surroundings. The Second Law can therefore be expressed as

$$\Delta S(\text{total}) = \Delta S(\text{system}) + \Delta S(\text{surroundings}) \quad (14.3)$$

When a process is at equilibrium, there is no tendency for any net change to occur.

- $\Delta S(\text{total}) = \Delta S(\text{system}) + \Delta S(\text{surroundings}) > 0$ spontaneous process
- $\Delta S(\text{total}) = \Delta S(\text{system}) + \Delta S(\text{surroundings}) < 0$ non-spontaneous process
- $\Delta S(\text{total}) = \Delta S(\text{system}) + \Delta S(\text{surroundings}) = 0$ process is at equilibrium

A quantitative measure of entropy change

In the nineteenth century, quantitative models of the entropy of systems were developed by considering heat–work cycles and the efficiency of heat engines. Later it became clear that entropy has a much wider application to physical and chemical changes.

If a quantity of heat, q_{rev}, is added *reversibly* to a system at temperature, T, then the *entropy change for the system*, ΔS(system), is given by

The population of energy levels is described by the Boltzmann distribution, which is discussed in Section 10.3 (p.462).

$$\Delta S = \frac{q_{rev}}{T} \quad (14.4)$$

Quantity of heat added reversibly (J)
temperature (K)
Entropy change (J K^{-1})

Adding the heat 'reversibly' means that heat is added very slowly so that at any stage the temperature difference between the system and its surroundings is infinitesimally small, and so is always close to thermal equilibrium.

Why does Equation 14.4 have this form? Think about what happens when a system is heated by having a quantity of heat q_{rev} added. Molecules move around faster, chemical bonds vibrate more, and higher energy levels are populated. This leads to the system having greater disorder and hence higher entropy. But why is q_{rev} divided by T? Adding the same amount of heat has a greater effect at a lower temperature than at a higher temperature. Adding 10 J to a system at a high temperature will not make as much difference to its entropy as 10 J added to a system at a low temperature.

Units of entropy

In Equation 14.4, the units of q are J and the units of T are K, so the units of S and ΔS are J K^{-1}. If the entropy change is measured for 1 mol of substance, the units are J K^{-1} mol^{-1}.

Entropy changes during changes of phase

Entropy changes during changes of phase—fusion (melting), vaporization, sublimation—provide straightforward applications of Equation 14.4. A change of phase is reversible since it takes place at equilibrium. At the normal temperature of the phase change, the reversible heat change at constant pressure is the enthalpy change so that $q_{rev} = \Delta H$.

For vaporization and fusion

$$\Delta_{vap}S = S(\text{vapour}) - S(\text{liquid}) \quad \text{and} \quad q_{rev} = \Delta_{vap}H$$

$$\Delta_{fus}S = S(\text{liquid}) - S(\text{solid}) \quad \text{and} \quad q_{rev} = \Delta_{fus}H$$

So, from Equation 14.4

$$\Delta_{vap}S^{\ominus} = \frac{\Delta_{vap}H^{\ominus}}{T_b} \qquad (14.5)$$

The superscript $^{\ominus}$ in the symbol $\Delta_{vap}S^{\ominus}$ indicates that it is the entropy change measured at standard pressure, 1 bar.

$$\Delta_{fus}S^{\ominus} = \frac{\Delta_{fus}H^{\ominus}}{T_m} \qquad (14.6)$$

where $\Delta_{vap}H^{\ominus}$ and $\Delta_{fus}H^{\ominus}$ are the standard enthalpy changes of vaporization and fusion at the boiling point and melting point, T_b and T_m, respectively. The entropy changes for the reverse processes—condensation and freezing—have the opposite sign but the same magnitude as those for vaporization and fusion.

Note that the expression $\Delta S = \frac{q}{T}$ only applies when heat is transferred *reversibly*. For example, at 1 bar, vapour and liquid are in equilibrium at T_b, so the change is reversible.

$\Delta_{vap}H^{\ominus}$ and $\Delta_{fus}H^{\ominus}$ have positive values since vaporization and fusion are endothermic processes and an input of energy is required. So, from Equations 14.5 and 14.6, the values of ΔS must be positive for these phase changes. (T_b and T_m are measured in kelvin and so must be positive numbers.) This is consistent with the idea of entropy as a measure of disorder. A vapour is more disordered—and so has higher entropy—than a liquid as illustrated in Figure 14.4. The value of $\Delta_{vap}S$ is therefore positive. The same argument holds for melting and $\Delta_{fus}S$. You can use Equation 14.5 to calculate a value for $\Delta_{vap}S^{\ominus}$ in Worked example 14.2.

For water

$$\Delta_{vap}S^{\ominus} = +109\,\text{J K}^{-1}\,\text{mol}^{-1}$$

$$\Delta_{fuss}S^{\ominus} = +22\,\text{J K}^{-1}\,\text{mol}^{-1}$$

This shows that there is a bigger change in disorder going from liquid water to gaseous water than going from solid water to liquid water.

The enthalpy changes accompanying phase changes are discussed in Sections 1.7 (p.47) and 13.2 (p.619). Further discussion of these phase changes appears in Sections 17.1 (p.766) and 17.2 (p.775).

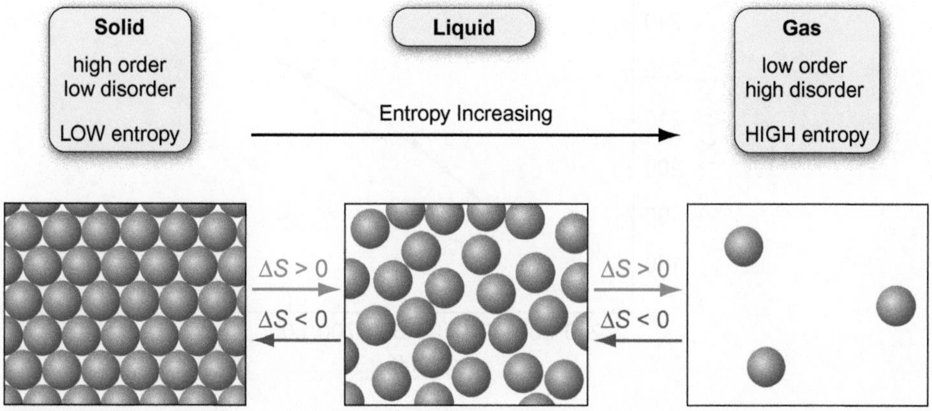
Solid
high order
low disorder
LOW entropy
Liquid
Entropy Increasing
Gas
low order
high disorder
HIGH entropy
ΔS > 0
ΔS < 0
ΔS > 0
ΔS < 0

Figure 14.4 The entropy of the system increases as solids melt and liquids vaporize. The entropy of the system decreases for freezing and condensation.

Worked example 14.2 Entropy change of vaporization

Calculate the entropy change when 1.00 mol of water at 100 °C vaporizes to steam at 100 °C. ($\Delta_{vap}H^{\ominus}$ for water is +40.7 kJ mol^{-1}.)

Strategy

You can use Equation 14.5 directly since all the information is given. Remember to convert the temperature to kelvin and $\Delta_{vap}H^{\ominus}$ to J.

Solution

Convert the temperature to kelvin: 100 °C = 373 K.

Now use Equation 14.5,

$$\Delta_{vap}S^{\ominus} = \frac{\Delta_{vap}H^{\ominus}}{T_b} = \frac{+40700\ \text{J mol}^{-1}}{373\ \text{K}} = 109\ \text{J K}^{-1}\ \text{mol}^{-1} \qquad (14.5)$$

Question

Calculate the entropy change when 1.00 mol of water at 0 °C freezes to form ice. (For water, $\Delta_{fus}H^{\ominus}$ = +6.02 kJ mol^{-1}.)

How the entropy of a system changes with temperature

As the temperature increases, matter generally becomes more disordered. The entropy of a system, therefore, increases with temperature—but by how much? Equation 14.4 shows that the change in entropy of a substance is related to the heat supplied, q_{rev}, by

$$\Delta S(\text{system}) = \frac{q_{rev}}{T} \qquad (14.4)$$

Heat capacities are defined in Section 13.1 (p.616). The variation of enthalpy of a system with temperature is discussed in Section 13.4 (p.633).

From Equation 13.8 (p.633), q for n moles of a substance is related to its molar heat capacity at constant pressure, C_p, and to the temperature change, ΔT, by the expression

$$q = n \times C_p \times \Delta T \qquad (13.8)$$

For 1 mol of substance

$$q = C_p \times \Delta T$$

The difference between the entropy of 1 mol of a substance (its molar entropy) at two temperatures is given by Equation 14.7. S_{T_i} is the molar entropy of the substance at an initial temperature T_i and S_{T_f} is the molar entropy at the final temperature T_f

$$S_{T_f} = S_{T_i} + C_p \ln\frac{T_f}{T_i} \qquad (14.7)$$

The **molar entropy** is the entropy of 1 mol of a substance. This is discussed further in Section 14.3 (p.664).

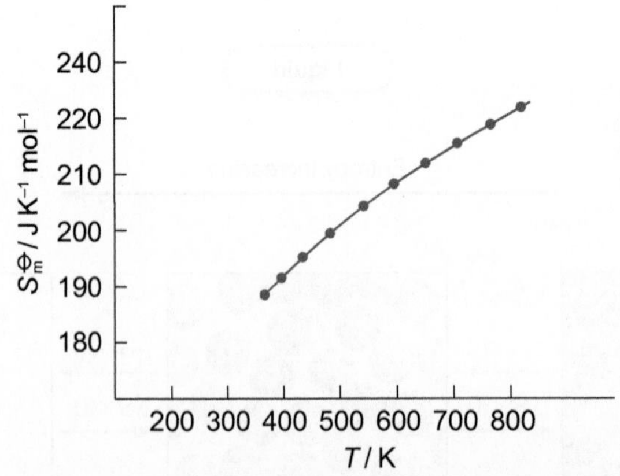
240
220
210
200
190
180
200 300 400 500 600 700 800
T / K

Figure 14.5 The variation of the standard molar entropy of H_2O (g) between 300 K and 800 K. You can see that entropy increases with temperature, but the *rate of increase* becomes lower as the temperature increases (the gradient of the plot gets smaller as the temperature increases).

Figure 14.5 uses Equation 14.7 to plot the variation of the molar entropy of water vapour, H_2O (g) over a range of temperatures.

Box 14.1 shows how this expression is derived. If T_f is larger than T_i, the logarithm term is positive, and S_{T_f} is larger than S_{T_i}; heating a substance increases its entropy, as expected. An application of Equation 14.7 is illustrated in Worked example 14.3.

Visit the Online Resource Centre to watch screencast 14.1 which explains the variation of entropy with temperature.

Box 14.1 Variation of the entropy of a substance with temperature

To derive an expression for the difference in entropy of a substance at two temperatures, you need first to express Equations 14.4 and 13.8 in terms of differentials (see Maths Toolkit MT6, p.1324).

From Equation 13.8, the heat change at constant pressure when 1 mol of a substance is heated through a small temperature change dT is given by

$$dq = C_p dT$$

where C_p is the molar heat capacity at constant pressure.

From Equation 14.4, the corresponding entropy change is given by

$$dS = \frac{dq_{rev}}{T}$$

Combining these two equations, the entropy change caused by a temperature change, dT, is given by

$$dS = \frac{dq_{rev}}{T} = \frac{C_p dT}{T} \qquad (14.8)$$

To find ΔS, Equation 14.8 is integrated between the initial and final temperatures, T_i and T_f

$$\int_{S_{T_i}}^{S_{T_f}} dS = \int_{T_i}^{T_f} \frac{C_p}{T} dT$$

Integration is equivalent to summing a number of small changes and is described in Maths Toolkit MT7 (p.1327). You need to use the following standard integrals

$$\int_{x_1}^{x_2} dx = [x]_{x_1}^{x_2} = x_2 - x_1 \quad \text{and} \quad \int_{x_1}^{x_2} \frac{dx}{x} = [\ln x]_{x_1}^{x_2} = \ln x_2 - \ln x_1$$

Assuming that C_p does not vary with temperature, it can be treated as a constant and taken outside the integration, leading to

$$\int_{S_{T_i}}^{S_{T_f}} dS = C_p \int_{T_i}^{T_f} \frac{1}{T} dT$$

Carrying out the integration of each side of the equation separately

$$\int_{S_{T_i}}^{S_{T_f}} dS = [S]_{T_i}^{T_f} = S_{T_f} - S_{T_i}$$

$$C_p \int_{T_i}^{T_f} \frac{1}{T} dT = C_p [\ln T]_{T_i}^{T_f} = C_p (\ln T_f - \ln T_i) = C_p \ln \frac{T_f}{T_i}$$

Putting the two expressions together leads to

$$S_{T_f} - S_{T_i} = C_p \ln \frac{T_f}{T_i}$$

$$S_{T_f} = S_{T_i} + C_p \ln \frac{T_f}{T_i} \qquad (14.7)$$

The derivation of Equation 14.7 assumes that no phase change occurs within the temperature range so the compound remains in the same state (solid, liquid, or gas) between T_i and T_f.

Equation 14.7 gives sufficient accuracy over small temperature ranges, where C_p is the mean value over the range. For greater accuracy, the temperature variation of C_p must be taken into account.

Worked example 14.3 Calculating entropy changes on heating

Calculate the change of entropy when 1.00 mol of water is heated at constant pressure from: (a) 0 °C to 25 °C and (b) 50 °C to 75 °C. (The molar heat capacity, C_p, of water is 75.2 J K^{-1} mol^{-1}.)

Strategy

The change in entropy is the difference between the values for the molar entropy of water at the two temperatures, $\Delta S = S_{T_f} - S_{T_i}$, which can be calculated from Equation 14.7. Since the temperature range is small, you can assume that C_p is constant.

Solution

Use Equation 14.7, remembering that temperatures must be in kelvin: 0 °C = 273 K.

$$S_{T_f} = S_{T_i} + C_p \ln \frac{T_f}{T_i} \qquad (14.7)$$

Rearranging the equation to give an expression for ΔS.

$$\Delta S = S_{T_f} - S_{T_i} = C_p \ln \frac{T_f}{T_i}$$

→

→

(a) $\Delta S = S_{298\,K} - S_{273\,K} = (75.2\,J\,K^{-1}\,mol^{-1}) \times \ln\left(\frac{298\,K}{273\,K}\right)$

$= (75.2\,J\,K^{-1}\,mol^{-1}) \times \ln(1.09)$

$= +6.59\,J\,K^{-1}\,mol^{-1}$

(b) $\Delta S = S_{348\,K} - S_{323\,K} = (75.2\,J\,K^{-1}\,mol^{-1}) \times \ln\left(\frac{348\,K}{323\,K}\right)$

$= +5.61\,J\,K^{-1}\,mol^{-1}$

As expected, both the entropy changes are positive. The water molecules have more energy at the higher temperature and so are less ordered. The entropy change is smaller for the same quantity of heat added at higher temperature (see p.660).

Question

Calculate the entropy change when 3.00 mol of nitrogen gas are heated from 25 °C to 50 °C at constant pressure. (C_p for N_2(g) is 29.1 J K^{-1} mol^{-1}.)

Summary

- The Second Law of thermodynamics: spontaneous processes increase the total entropy of the Universe

$$\Delta S(\text{total}) = \Delta S(\text{system}) + \Delta S(\text{surroundings})$$

- The entropy change of a system is related to the heat transferred and the temperature by

$$\Delta S(\text{system}) = \frac{q_{rev}}{T}$$

- A change of state at T_b or T_m is a reversible process at 1 bar since it occurs at equilibrium

$$\Delta_{vap}S^{\ominus} = \frac{\Delta_{vap}H^{\ominus}}{T_b} \quad \text{and} \quad \Delta_{fus}S^{\ominus} = \frac{\Delta_{fus}H^{\ominus}}{T_m}$$

- The variation of the molar entropy of a substance with temperature is given by the equation

$$S_{T_f} = S_{T_i} + C_p \ln\frac{T_f}{T_i}$$

? For practice questions on these topics, see questions 3, 4, 6, and 12 at the end of this chapter (p.692).

14.3 The Third Law and absolute entropies

To define the entropy of a compound in absolute terms, a reference value is needed. In discussing *enthalpy* changes, the reference used is that $\Delta_f H^{\ominus}_{298} = 0$ for elements in their standard states. Because it is not possible to define a zero point for enthalpy, enthalpy *changes* rather than absolute enthalpies are always discussed. However, for entropy it *is* possible to define a zero point.

Entropy is related to disorder, so a perfectly ordered system must, by definition, have zero entropy. Such a system would be a perfect solid crystal at absolute zero of temperature so that there is no disorder or motion of any type. A perfect crystal is one with all the atoms, ions, or molecules aligned perfectly and with no defects. At 0 K, all molecules will be in their ground state. There is only one way of arranging the molecules, so that $W = 1$ in the Boltzmann formula (Equation 14.2, p.659) and hence $S = 0$. These ideas form the basis of the **Third Law of thermodynamics**, which states that:

The entropy of a perfect crystal at zero kelvin is zero.

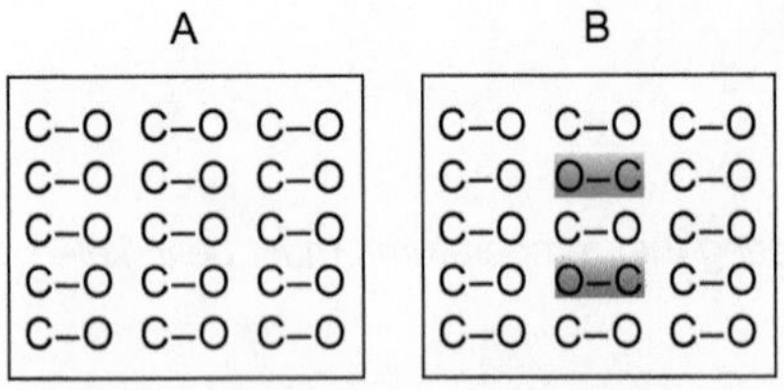
A
B
C–O C–O C–O
C–O O–C C–O

Figure 14.6 The dipole moment in CO is small. The energy difference between the two different orientations of CO in A and B is small. The presence of different orientations means that a perfect crystal does not form even at 0 K.

In the real world, a perfect crystal is impossible to achieve. All crystals contain defects. For example, Figure 14.6 shows an arrangement of molecules in solid carbon monoxide. The dipole in CO is very small so that there is only a very small difference in energy between the orientation of molecules in the crystal, and the presence of different orientations means that there is some disorder even at 0 K. Even when all the molecules in a crystal are aligned perfectly, other effects can result in a **residual entropy** at 0 K. For example, molecules containing different isotopes (e.g. $H^{35}Cl$ and $H^{37}Cl$) may be present, and this creates disorder. However, for many substances the residual entropy is close to zero and the Third Law provides a useful reference state.

Standard entropy

Equations 14.5, 14.6, and 14.7 on pages 661 and 662 can be used to calculate the *changes* in entropy that occur when a substance changes phase or when it is heated. Think about a substance that is a liquid at room temperature. From the Third Law, the entropy of the substance as a perfect crystal at 0 K is zero. If the changes in entropy on heating from 0 K to room temperature are calculated, the sum of these gives an *absolute* value of the entropy at room temperature. In calculating entropy changes, the same set of standard conditions is used for ΔS as is used for ΔH and ΔU in Chapter 13.

> ***The standard entropy, $S^{\ominus}_{298}$, of a substance is the entropy of 1 mol at 298 K and 1 bar pressure. The units are J K^{-1} mol^{-1}.***

Standard entropies, $S^{\ominus}$, are those measured at a pressure of 1 bar. They are most often quoted at 298 K. The values are relative to the reference state where $S = 0$ at 0 K.

To determine $S^{\ominus}_{298}$, you measure the entropy changes on heating 1 mol of the substance from 0 K to 298 K taking into account any phase changes that occur (melting for liquids, melting and vaporization for gases) as shown in Box 14.2. Worked example 14.4 shows how the method is applied in practice.

$S^{\ominus}_{298}$ values are sometimes called **absolute** or **Third Law** entropies and tables of values are published. Some examples of $S^{\ominus}_{298}$ values are given in Table 14.1. A more comprehensive selection is listed in Appendix 7 (p.1362). $S^{\ominus}_{298}$ values provide a convenient way to calculate entropy changes during chemical reactions (see Section 14.4, p.669).

Table 14.1 Some values of standard entropies $S^{\ominus}_{298}$ / J K^{-1} mol^{-1}

Solids		Liquids		Gases	
C (s) (diamond)	2.4	**Water (H_2O (l))**	**69.9**	H_2 (g)	130.7
C (s) (graphite)	5.7	Hg (l)	76.0	**Water vapour (H_2O (g))**	**188.8**
Cu (s)	33.1	Br_2 (l)	152.2	CO (g)	197.7
Ice (H_2O (s))	**45.0**	Ethanol (C_2H_5OH (l))	159.9	CO_2 (g)	213.7
NaCl (s)	72.1	Benzene (C_6H_6 (l))	173.3	Br_2 (g)	245.5
D-Glucose ($C_6H_{12}O_6$ (s))	212.0	Octane (C_8H_{18} (l))	361.2	Benzene (C_6H_6 (g))	269.3
Sucrose ($C_{12}H_{22}O_{11}$ (s))	360.2			Octane (C_8H_{18} (g))	467.1

Box 14.2 Determination of standard entropy, $S^{\ominus}_{298}$

The entropy of a substance is zero at 0 K and increases as it is heated and also when it melts and vaporizes. Consider a substance that is a gas at room temperature.

To find its *standard entropy*, $S^{\ominus}_{298}$, you need to calculate each of the entropy changes (at 1 bar) in taking 1 mol of the substance through the stages shown in Figure 1. The expressions for the entropy changes for each of these stages are derived using Equations 14.5, 14.6, and 14.7.

From the Third Law, the entropy is zero at 0 K, so $S^{\ominus}_{298}$ is the sum of the five entropy changes

$$S^{\ominus}_{298} = \Delta S_1^{\ominus} + \Delta S_2^{\ominus} + \Delta S_3^{\ominus} + \Delta S_4^{\ominus} + \Delta S_5^{\ominus} \quad (14.10)$$

→

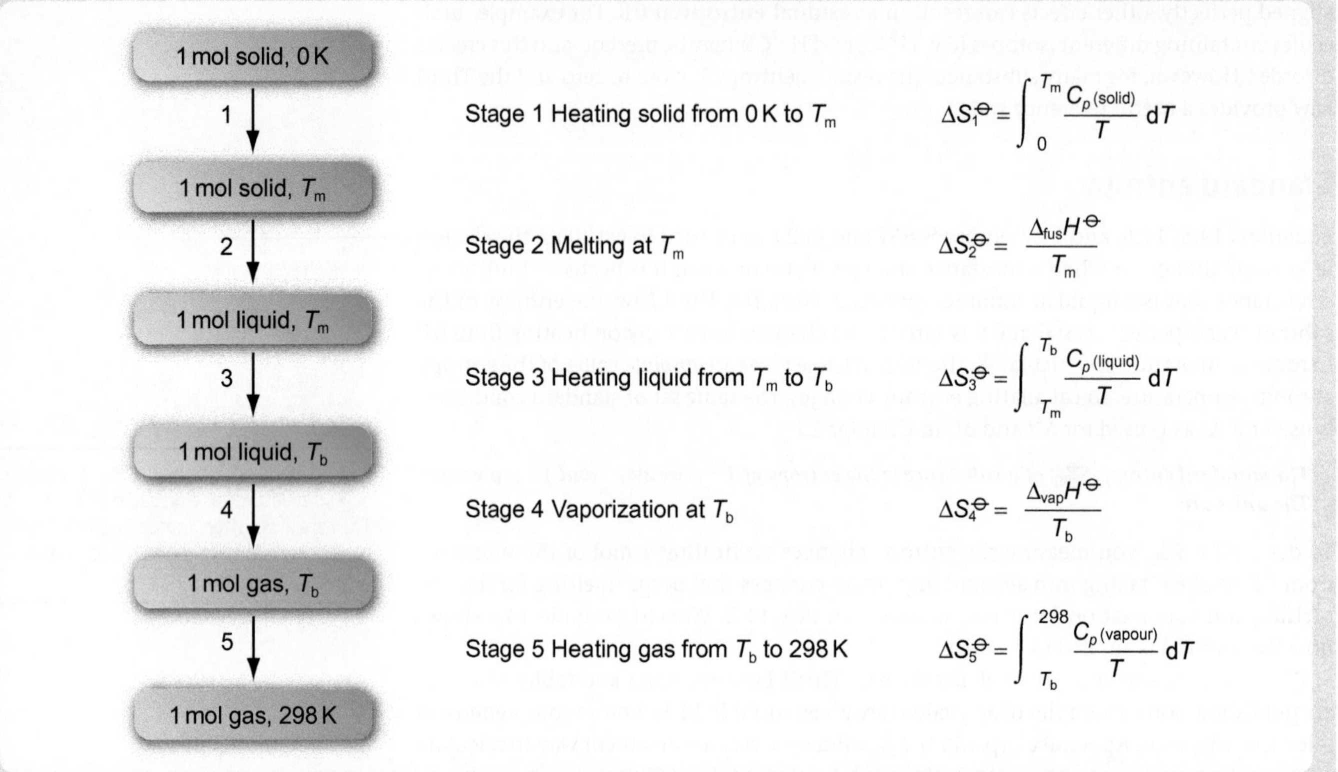
1 mol solid, 0 K
1
Stage 1 Heating solid from 0 K to T_m
$\Delta S_1^\ominus = \int_0^{T_m} \frac{C_p\text{(solid)}}{T}\,dT$
1 mol solid, T_m
2
Stage 2 Melting at T_m
$\Delta S_2^\ominus = \frac{\Delta_{fus}H^\ominus}{T_m}$
1 mol liquid, T_m
3
Stage 3 Heating liquid from T_m to T_b
$\Delta S_3^\ominus = \int_{T_m}^{T_b} \frac{C_p\text{(liquid)}}{T}\,dT$
1 mol liquid, T_b
4
Stage 4 Vaporization at T_b
$\Delta S_4^\ominus = \frac{\Delta_{vap}H^\ominus}{T_b}$
1 mol gas, T_b
5
Stage 5 Heating gas from T_b to 298 K
$\Delta S_5^\ominus = \int_{T_b}^{298} \frac{C_p\text{(vapour)}}{T}\,dT$
1 mol gas, 298 K

▲ **Figure 1** The entropy changes on heating a substance from 0 K to a gas at 298 K.

For compounds that are liquid at 298 K and 1 bar, stages 4 and 5 are omitted and T_b is replaced by 298 K in the integral for $\Delta S_3^\ominus$.

The entropy increases with temperature as in Figure 2. A value for $S_{298}^\ominus$ can also be found using a graphical method with experimental data for C_p and T. A plot of C_p/T against T is constructed as in Figure 3 and extrapolated to 0 K. Finding the area under the curves is equivalent to performing the integrals. (See Maths Toolkit MT7, p.1327, for an explanation of using integrals to find areas under graphs.)

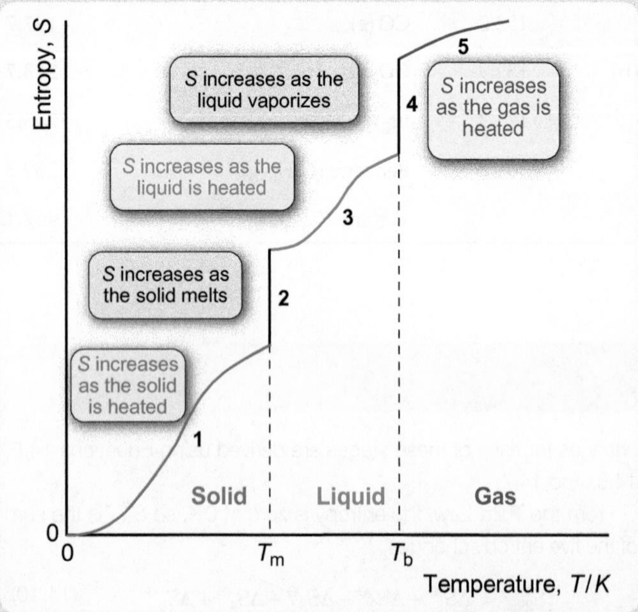
Entropy, S
S increases as the liquid vaporizes
S increases as the gas is heated
S increases as the liquid is heated
S increases as the solid melts
S increases as the solid is heated
1
2
3
4
5
Solid
Liquid
Gas
0
0
T_m
T_b
Temperature, T/K

▲ **Figure 2** The entropy of a substance increases on heating. There are sharp increases in entropy on melting, at T_m, and vaporization, at T_b.

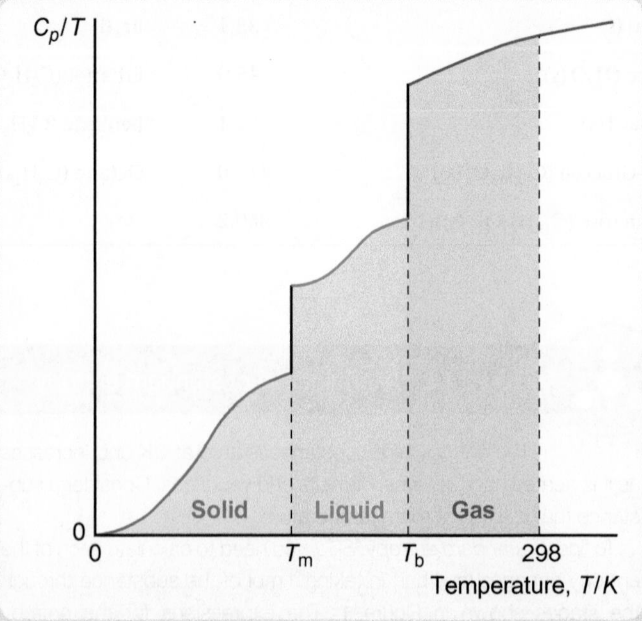
C_p/T
Solid
Liquid
Gas
0
0
T_m
T_b
298
Temperature, T/K

▲ **Figure 3** When C_p/T is plotted against T, the entropy changes due to heating the solid, liquid, and gas are given by the shaded areas.

Worked example 14.4 Finding a standard entropy

Find the standard entropy at 298 K, $S^{\ominus}_{298}$, for chlorine gas, Cl_2(g), given the following data.

Melting point of Cl_2(s)	172 K
Boiling point of Cl_2(l)	239 K
Entropy change when Cl_2(s) is heated from 0 K to 172 K without melting (found graphically)	+70.7 J K^{-1} mol^{-1}
Entropy change when Cl_2(l) is heated from 172 K to 239 K without vaporizing (found graphically)	+21.9 J K^{-1} mol^{-1}
Entropy change when Cl_2(g) is heated from 239 K to 298 K (found graphically)	+7.90 J K^{-1} mol^{-1}
Standard enthalpy change of fusion, $\Delta_{fus}H^{\ominus}$, of Cl_2(s)	+6.40 kJ mol^{-1}
Standard enthalpy change of vaporization, $\Delta_{vap}H^{\ominus}$, of Cl_2(s)	+20.4 kJ mol^{-1}

Strategy

From the Third Law, chlorine has zero entropy at 0 K. Identify the stages that occur as Cl_2(s) at 0 K is heated to become Cl_2(g) at 298 K.

Use the data to calculate the entropy change for each stage: ΔS_1 for stage 1, ΔS_2, for stage 2, etc. Use Equations 14.5 and 14.6 to calculate the entropy changes when Cl_2(l) vaporizes and when Cl_2(s) melts.

Then add together the entropy changes for all the stages to find $S^{\ominus}_{298}$.

Solution

Chlorine has zero entropy at 0 K. Identify the stages that occur as Cl_2(s) at 0 K is heated to become Cl_2(g) at 298 K.

See Figure 1.

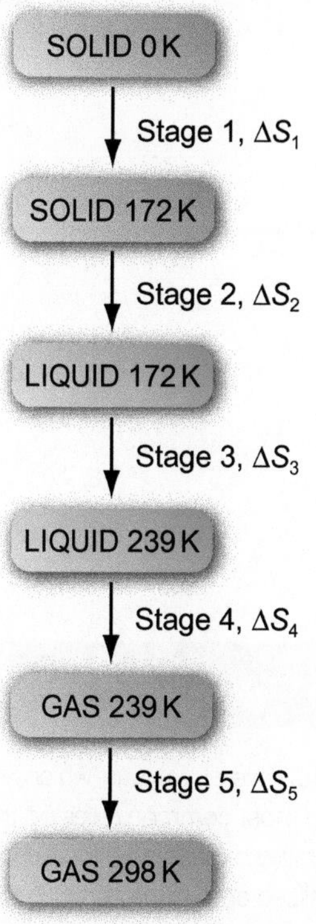

SOLID 0 K
Stage 1, ΔS1
SOLID 172 K
Stage 2, ΔS2
LIQUID 172 K
Stage 3, ΔS3
LIQUID 239 K
Stage 4, ΔS4
GAS 239 K
Stage 5, ΔS5
GAS 298 K

Figure 1 Stages involved in heating Cl_2 from 0 K to 298 K.

Use the data to calculate the entropy change for each stage.

Stage 1: $\Delta S_1 = +70.7$ J K^{-1} mol^{-1} (given)

$$\text{Stage 2:} \quad \Delta S_2 = \Delta_{fus}S(Cl_2(s)) = \frac{\Delta_{fus}H^{\ominus}}{T_m} = +6400\ \text{J mol}^{-1}/172\ \text{K}$$

$$= +37.2\ \text{J K}^{-1}\ \text{mol}^{-1} \qquad (14.6)$$

Note that in $\Delta_{fus}H^{\ominus}$, kJ have been converted to J to get consistent units.

Stage 3: $\Delta S_3 = +21.9$ J K^{-1} mol^{-1} (given)

$$\text{Stage 4:} \quad \Delta S_4 = \Delta_{vap}S(Cl_2(l)) = \frac{\Delta_{vap}H^{\ominus}}{T_b} = +20400\ \text{J mol}^{-1}/239\ \text{K}$$

$$= +85.4\ \text{J K}^{-1}\ \text{mol}^{-1} \qquad (14.5)$$

Stage 5: $\Delta S_5 = +7.90$ J K^{-1} mol^{-1} (given)

Then add together the entropy changes for all the stages to find $S^{\ominus}_{298}$.

$$S^{\ominus}_{298}(Cl_2) = \Delta S_1 + \Delta S_2 + \Delta S_3 + \Delta S_4 + \Delta S_5$$

$$= +(70.7 + 37.2 + 21.9 + 85.4 + 7.90)\ \text{J K}^{-1}\ \text{mol}^{-1}$$

$$= +223\ \text{J K}^{-1}\ \text{mol}^{-1}$$

Question

Draw a diagram, similar to Figure 1 in this example, to identify the stages involved in finding $S^{\ominus}_{298}$ (H_2O).

Look at the values of $S^{\ominus}_{298}$ in Table 14.1 on p.665. Are they consistent with the idea of entropy as a measure of disorder? The three-dimensional diamond lattice is more ordered than the layer structure of graphite with its delocalized electrons, so it has a lower value of $S^{\ominus}_{298}$. Solids with more complex structures, particularly those with large numbers of atoms in the molecules such as glucose and sucrose, have higher values of $S^{\ominus}_{298}$, since larger molecules have more ways of distributing energy within the vibrations of their

Visit the Online Resource Centre to watch screencast 14.2 which walks you through Worked example 14.4, the calculation of the standard entropy of chlorine gas at 298 K.

many bonds. You can think of $S^{\ominus}_{298}$ as increasing with the molecular complexity of the compound.

As a general rule, gases have higher $S^{\ominus}_{298}$ values than liquids, which in turn have higher values than solids (for molecules containing about the same number of atoms). This is best seen by comparing the values for ice, water, and water vapour in Table 14.1.

The standard entropy at temperatures other than 298 K can be found from $S^{\ominus}_{298}$ values using Equation 14.7 on p.662. For example, the standard entropy at any temperature T, $S^{\ominus}_{T}$, is related to that at 298 K by

$$S^{\ominus}_{T} = S^{\ominus}_{298} + C_p \ln \frac{T}{298\,\mathrm{K}} \qquad (14.9)$$

assuming that the substance remains in the same phase (solid, liquid or gas) between 298 K and T and that C_p remains constant over the temperature range.

Box 14.3 What is the lowest temperature that can be reached?

There are several ways in which very low temperatures can be reached. The most common relies on the cooling effect when a gas undergoes a rapid expansion—the principle used in domestic refrigerators. Specialized apparatus can give temperatures as low as 2 K. Liquid helium, which boils at 4.2 K, is produced by this method.

To achieve even lower temperatures, a different principle must be used. Some compounds with unpaired electrons are extremely sensitive to magnetic fields. The electron spins behave as small magnets and line up when a magnetic field is applied. (Magnetic properties of compounds are discussed in Section 28.7, p.1292.) Gadolinium sulfate ($Gd_2(SO_4)_3 \cdot 8H_2O$) is an example of such a compound.

A sample of this salt is cooled down to ~2 K using liquid helium. A strong magnetic field is then applied, causing the electron spins in the Gd^{3+} ions to align. The entropy of the system decreases since the aligned spins are more ordered. The change takes place at constant temperature and, in the plot of entropy against temperature shown in the figure, the system moves from **A** to **B**.

The coolant is removed and the sample is thermally isolated by a high vacuum. The magnetic field is then switched off. The electron spins no longer align in the magnetic field and so adopt random orientations. Since the system is isolated, the energy needed to convert to the higher energy arrangement can only come from within the system so the temperature falls. The system moves from **B** to **C** at constant entropy—the increase in entropy from the more random arrangement is compensated for by the decrease in entropy due to the fall in temperature.

Using this **adiabatic demagnetization** principle, temperatures as low as 0.005 K (5 mK) can be achieved. (In an adiabatic process, there is no exchange of heat with the surroundings; see Section 13.5, p.638.) Still lower temperatures (0.1 mK–1 μK) can be achieved by exploiting the magnetic properties of atomic nuclei.

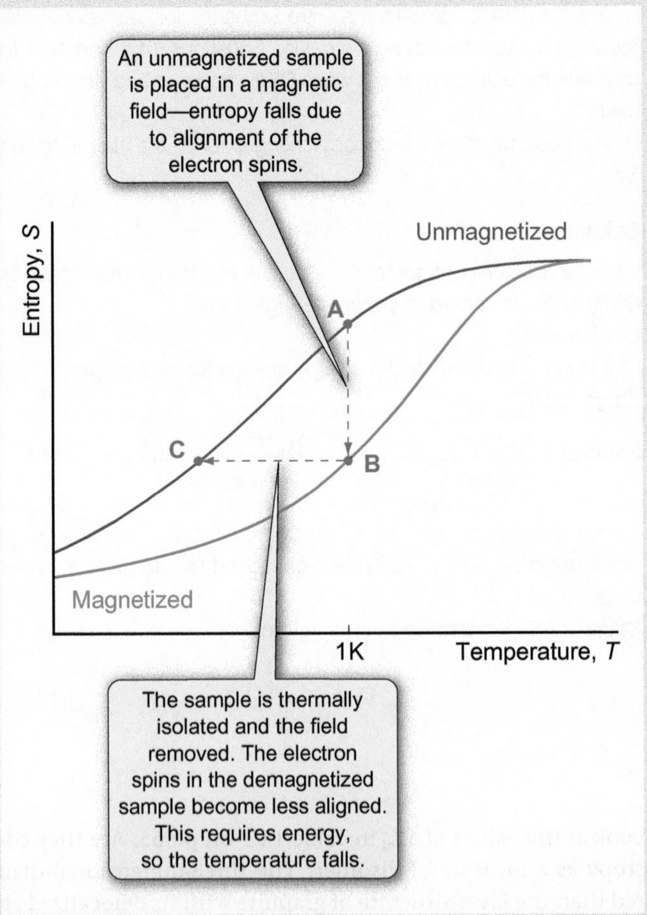
An unmagnetized sample is placed in a magnetic field—entropy falls due to alignment of the electron spins.
Entropy, S
Unmagnetized
A
C
B
Magnetized
1K
Temperature, T
The sample is thermally isolated and the field removed. The electron spins in the demagnetized sample become less aligned. This requires energy, so the temperature falls.

▲ Adiabatic demagnetization allows scientists to reach extremely low temperatures.

Summary

- The Third Law of thermodynamics: the entropy of a perfect crystal at zero kelvin is zero.
- The standard entropy, $S^{\ominus}_{298}$, of a substance is the entropy of 1 mol at 298 K and 1 bar pressure. This refers to a reference state where $S^{\ominus} = 0$ at 0 K.
- Values of $S^{\ominus}_{298}$ are determined by adding the entropy changes involved in heating 1 mol of the substance from 0 K to 298 K, taking into account any phase changes that occur.
- The standard entropy of a substance at temperatures other than 298 K can be found using

$$S^{\ominus}_{T} = S^{\ominus}_{298} + C_P \ln\frac{T}{298\,\text{K}}$$

assuming the substance remains in the same phase between 298 K and T.

For practice questions on these topics, see questions 7, 8, and 9 at the end of this chapter (p.692).

14.4 Entropy changes in chemical reactions

Standard entropy change of reaction

The standard entropy change of reaction, $\Delta_r S^{\ominus}$, is defined as the difference in standard entropy between the products and reactants of a reaction system. For values at 298 K

$$\Delta_r S^{\ominus}_{298} = \sum \nu_i S^{\ominus}_{298}\,(\text{products}) - \sum \nu_i S^{\ominus}_{298}\,(\text{reactants}) \qquad (14.11)$$

The sum of the standard entropies of all of the *products*

The sum of the standard entropies of all of the *reactants*

Since $S^{\ominus}_{298}$ refers to one mole of a compound, it must be multiplied by the stoichiometric coefficient, ν_i, in the balanced equation for the reaction. The use of this equation is illustrated in Worked example 14.5.

The definition of the standard entropy change of reaction, $\Delta_r S^{\ominus}$, is analogous to the definitions of the standard enthalpy change of reaction, $\Delta_r H^{\ominus}$ (Section 13.3, p.622), and the standard internal energy change of reaction, $\Delta_r U^{\ominus}$ (Section 13.5, p.636).

The symbol Σ is shorthand for summing the terms that follow. For example,

$$\sum_1^5 (a_i) = a_1 + a_2 + a_3 + a_4 + a_5$$

See Maths Toolkit MT1 (p.1304).

Worked example 14.5 Entropy changes for reactions at 298 K

Using data in Appendix 7 (p.1352), calculate the standard entropy change of reaction at 298 K for each of the following reactions:

(a) $C(\text{graphite}) + O_2(g) \rightarrow CO_2(g)$

(b) $N_2(g) + 3H_2(g) \rightarrow 2NH_3(g)$

(c) $C_2H_5OH(l) + CH_3CO_2H(l) \rightarrow CH_3CO_2C_2H_5(l) + H_2O(l)$

Strategy

Substitute values for the standard entropies at 298 K for reactants and products into Equation 14.11 to calculate $\Delta_r S^{\ominus}_{298}$.

→

→

Solution

Remember to multiply each standard entropy by its stoichiometric coefficient in the balanced equation for the reaction.

$$\Delta_r S^{\ominus}_{298} = \sum \nu_i S^{\ominus}_{298}(\text{products}) - \sum \nu_i S^{\ominus}_{298}(\text{reactants}) \qquad (14.11)$$

(a) $\Delta_r S^{\ominus}_{298} = [S^{\ominus}_{298}(CO_2\,(g))] - [S^{\ominus}_{298}(C(\text{graphite})) + S^{\ominus}_{298}(O_2\,(g))]$

$= [213.7\,J\,K^{-1}\,mol^{-1}] - [5.7\,J\,K^{-1}\,mol^{-1} + 205.1\,J\,K^{-1}\,mol^{-1}]$

$= +2.9\,J\,K^{-1}\,mol^{-1}$

(b) $\Delta_r S^{\ominus}_{298} = [2\times S^{\ominus}_{298}(NH_3\,(g))] - [S^{\ominus}_{298}(N_2\,(g)) + 3\times S^{\ominus}_{298}(H_2\,(g))]$

$= [2\times 192.5\,J\,K^{-1}\,mol^{-1}] - [191.6\,J\,K^{-1}\,mol^{-1} + (3\times 130.7\,J\,K^{-1}\,mol^{-1})]$

$= -198.7\,J\,K^{-1}\,mol^{-1}$

(c) $\Delta_r S^{\ominus}_{298} = [S^{\ominus}_{298}(H_2O\,(l)) + S^{\ominus}_{298}(CH_3CO_2C_2H_5\,(l))] - [S^{\ominus}_{298}(C_2H_5OH\,(l)) + S^{\ominus}_{298}(CH_3CO_2H\,(l))]$

$= [69.9\,J\,K^{-1}\,mol^{-1} + 259.4\,J\,K^{-1}\,mol^{-1}] - [159.9\,J\,K^{-1}\,mol^{-1} + 158.0\,J\,K^{-1}\,mol^{-1}]$

$= +11.4\,J\,K^{-1}\,mol^{-1}$

The entropy change in (a) is relatively small since there are the same number of moles of gas in the reactants and the products. In (b), the number of moles of gas is halved during the reaction, accounting for the large negative value of ΔS. In reaction (c), all the components are in the liquid phase so that the change in entropy is relatively small.

Question

Calculate the standard entropy changes of reaction at 298 K for the following reactions:

(a) $CaCO_3\,(s) \rightarrow CaO\,(s) + CO_2\,(g)$

(b) $N_2O_4\,(g) \rightarrow 2NO_2\,(g)$

Reactions at other temperatures

Equation 14.11 gives the entropy change for a reaction when it is conducted at a standard pressure of 1 bar and at a temperature of 298 K. The entropy change at other temperatures can be calculated by using Equation 14.9 (p.669) to find the standard entropies of the reactants and products at other temperatures. The standard entropy change of reaction at temperature T, $\Delta_r S^{\ominus}_T$, is given by

Standard entropy change of reaction at temperature, T

Temperature of reaction

$$\Delta_r S^{\ominus}_T = \Delta_r S^{\ominus}_{298} + \Delta C_p \ln \frac{T}{298\,K} \qquad (14.12)$$

Standard entropy change of reaction at 298 K

The molar heat capacity change for the reaction (the difference between products and reactants)

Equation 14.12 works on the assumption that the heat capacities of the components in the reaction can be treated as constant over the temperature range involved.

ΔC_p is the difference in molar heat capacities of the products and reactants given by Equation 13.11 (p.635).

$$\Delta C_p = \sum v_i C_p(\text{products}) - \sum v_i C_p(\text{reactants}) \quad (13.11)$$

Equation 14.12 enables the calculation of entropy changes for reactions at any temperature as long as heat capacity data are available. It is analogous to the Kirchhoff equation (Equation 13.10, p.634) that is used to calculate enthalpy changes of reactions at different temperatures. You can practise using Equation 14.12 in Worked example 14.6.

Note that $S^{\ominus}_{298}$, or indeed the entropy of a compound at any other temperature, S_T, *must* be positive since it must have greater disorder than a perfect crystal at 0 K. However, $\Delta_r S^{\ominus}_{298}$ or $\Delta_r S^{\ominus}_T$ can be positive or negative since they refer to *differences* in entropy between products and reactants. It is important to keep this distinction in mind.

Worked example 14.6 Entropy changes for reactions at other temperatures

The entropy changes at 298 K for the following reactions are as shown. Calculate $\Delta_r S^{\ominus}$ for each reaction at 1023 K.

(a) $C_{(graphite)} + O_2(g) \rightarrow CO_2(g)$ $\quad \Delta_r S^{\ominus}_{298} = +2.96\ \text{J K}^{-1}\ \text{mol}^{-1}$

(b) $N_2(g) + 3H_2(g) \rightarrow 2NH_3(g)$ $\quad \Delta_r S^{\ominus}_{298} = -198.7\ \text{J K}^{-1}\ \text{mol}^{-1}$

Strategy

Use Appendix 7 (p.1352) to find the value of C_p for each component. You can then calculate ΔC_p from Equation 13.11 and substitute this into Equation 14.12 to find $\Delta_r S^{\ominus}_{1023}$. Assume that C_p is constant over the temperature range.

Solution

(a) *Use Equation 13.11 to calculate* ΔC_p.

$$\begin{aligned}\Delta C_p &= \sum v_i C_p(\text{products}) - \sum v_i C_p(\text{reactants}) \quad (13.11)\\ &= [C_p(CO_2(g))] - [C_p(C(\text{graphite})) + C_p(O_2(g))]\\ &= [37.1\ \text{J K}^{-1}\ \text{mol}^{-1}] - [8.5\ \text{J K}^{-1}\ \text{mol}^{-1} + 29.4\ \text{J K}^{-1}\ \text{mol}^{-1}]\\ &= -0.8\ \text{J K}^{-1}\ \text{mol}^{-1}\end{aligned}$$

Use the value of $\Delta_r S^{\ominus}_{298}$ *(from Worked example 14.5, p.669) and Equation 14.12 to find* $\Delta_r S^{\ominus}_T$.

From Worked example 14.5

$$\Delta_r S^{\ominus}_{298} = +2.9\ \text{J K}^{-1}\ \text{mol}^{-1}$$

$$\Delta_r S^{\ominus}_T = \Delta_r S^{\ominus}_{298} + \Delta C_p \ln\left(\frac{T}{298\ \text{K}}\right)$$

$$\begin{aligned}\Delta_r S^{\ominus}_{1023} &= +2.9\ \text{J K}^{-1}\ \text{mol}^{-1} + \left[-0.8\ \text{J K}^{-1}\ \text{mol}^{-1} \times \ln\left(\frac{1023\ \text{K}}{298\ \text{K}}\right)\right]\\ &= -1.9\ \text{J K}^{-1}\ \text{mol}^{-1}\end{aligned}$$

(b) *Use the same reasoning as in (a).*

From Worked example 14.5

$$\begin{aligned}\Delta_r S^{\ominus}_{298} &= -198.7\ \text{J K}^{-1}\ \text{mol}^{-1}\\ \Delta C_p &= [2 \times C_p(NH_3(g))] - [C_p(N_2(g)) + 3 \times C_p(H_2(g))]\\ &= -45.3\ \text{J K}^{-1}\ \text{mol}^{-1}\end{aligned}$$

$$\begin{aligned}\Delta_r S^{\ominus}_{1023} &= -198.7\ \text{J K}^{-1}\ \text{mol}^{-1} + \left[-45.3\ \text{J K}^{-1}\ \text{mol}^{-1} \times \ln\left(\frac{1023\ \text{K}}{298\ \text{K}}\right)\right]\\ &= -246\ \text{J K}^{-1}\ \text{mol}^{-1}\end{aligned}$$

Question

Calculate the standard entropy changes of reaction at 100 °C (373 K) for the following reactions:

(a) $CaCO_3(s) \rightarrow CaO(s) + CO_2(g)$ $\quad \Delta_r S^{\ominus}_{298} = +160.6\ \text{J K}^{-1}\ \text{mol}^{-1}$

(b) $N_2O_4(g) \rightarrow 2NO_2(g)$ $\quad \Delta_r S^{\ominus}_{298} = +175.9\ \text{J K}^{-1}\ \text{mol}^{-1}$

Entropy as a predictor of the spontaneity of chemical change

The Second Law shows that reactions that increase the entropy of the Universe are spontaneous (see p.660). Consider a familiar chemical reaction

$$H_2(g) + \tfrac{1}{2}O_2(g) \rightarrow H_2O(l)$$

Using values of $S^{\ominus}_{298}$ from Appendix 7 (p.1352) you can calculate $\Delta_r S^{\ominus}$ at 298 K using Equation 14.11 (p.669)

$$\begin{aligned}\Delta_r S^\ominus_{298} &= [S^\ominus_{298}(H_2O(l))] - [S^\ominus_{298}(H_2(g)) + 0.5\ S^\ominus_{298}(O_2(g))] \\ &= [69.9\ \mathrm{J\,K^{-1}\,mol^{-1}}] - [130.7\ \mathrm{J\,K^{-1}\,mol^{-1}} + (0.5 \times 205.1\ \mathrm{J\,K^{-1}\,mol^{-1}})] \\ &= -163.4\ \mathrm{J\,K^{-1}\,mol^{-1}}\end{aligned}$$

The standard entropy change of reaction at 298 K, for the molar amounts in the equation, is $-163.4\ \mathrm{J\,K^{-1}\,mol^{-1}}$. There is a large decrease in the entropy of the *system* on forming water from hydrogen and oxygen, as you would expect for a reaction in which 1.5 mol of disordered, high entropy gas forms 1 mol of relatively ordered and hence lower entropy liquid.

But doesn't the Second Law state that the entropy *increases* in a spontaneous reaction? This apparent difference arises since $\Delta_r S^\ominus_{298}$ calculated here refers to the $H_2/O_2/H_2O$ *system*. The Second Law refers to the *total entropy of the Universe* increasing as a requirement for spontaneity. Remember that the Universe consists of the system *and* the surroundings.

Entropy change in the surroundings

The system and surroundings are defined in Section 13.1 (p.614).

The total entropy change of the Universe is made up of two parts: the entropy change in the system, ΔS(system), plus the entropy change in the rest of the Universe surrounding the system, ΔS(surroundings)

$$\Delta_r S(\text{total}) = \Delta_r S(\text{system}) + \Delta_r S(\text{surroundings}) \qquad (14.13)$$

Fortunately, there is a straightforward way to calculate ΔS(surroundings) using Equation 14.4 on p.660

$$\Delta_r S(\text{surroundings}) = \frac{q_{rev}(\text{surroundings})}{T}$$

When the system reacts at constant pressure, the enthalpy change is $\Delta_r H$. A quantity of heat equal in magnitude to $\Delta_r H$ passes from the system to the surroundings, or vice versa. The exchange of heat is effectively reversible, because the surroundings are so large that they can easily absorb the heat change without the temperature or pressure changing. The heat exchanged with the surroundings is $-\Delta_r H$ (if the system loses heat, the surroundings gain heat). So

$$\begin{aligned}q_{rev}(\text{surroundings}) &= -q_{rev}(\text{system}) \\ &= -\Delta_r H \qquad \text{(at constant pressure)}\end{aligned}$$

and so

$$\Delta_r S(\text{surroundings}) = \frac{-\Delta_r H}{T} \qquad (14.14)$$

Equation 14.14 is a very useful and important result that enables you to look at changes in the *surroundings* from the point of view of the *system*.

Applying this to the reaction of hydrogen and oxygen to form water, $\Delta_f H_{298}$ for water is $-285.8\ \mathrm{kJ\,mol^{-1}}$ so for the reaction

$$H_2(g) + \tfrac{1}{2}O_2(g) \rightarrow H_2O\ (l) \quad \Delta_r H^\ominus = -285.8\ \mathrm{kJ}$$

The heat given out by the system must be gained by the surroundings (see Figure 14.7).

$$\Delta_r H^\ominus_{298}(\text{surroundings}) = -\Delta_r H^\ominus_{298}(\text{system}) = +285.8\ \mathrm{kJ}$$

From Equation 14.14

$$\begin{aligned}\Delta_r S^\ominus_{298}(\text{surroundings}) &= \frac{-\Delta_r H^\ominus_{298}(\text{system})}{T} \\ &= \frac{-(-285.8 \times 10^3\ \mathrm{J})}{298\ \mathrm{K}} \\ &= +959.1\ \mathrm{J\,K^{-1}}\end{aligned}$$

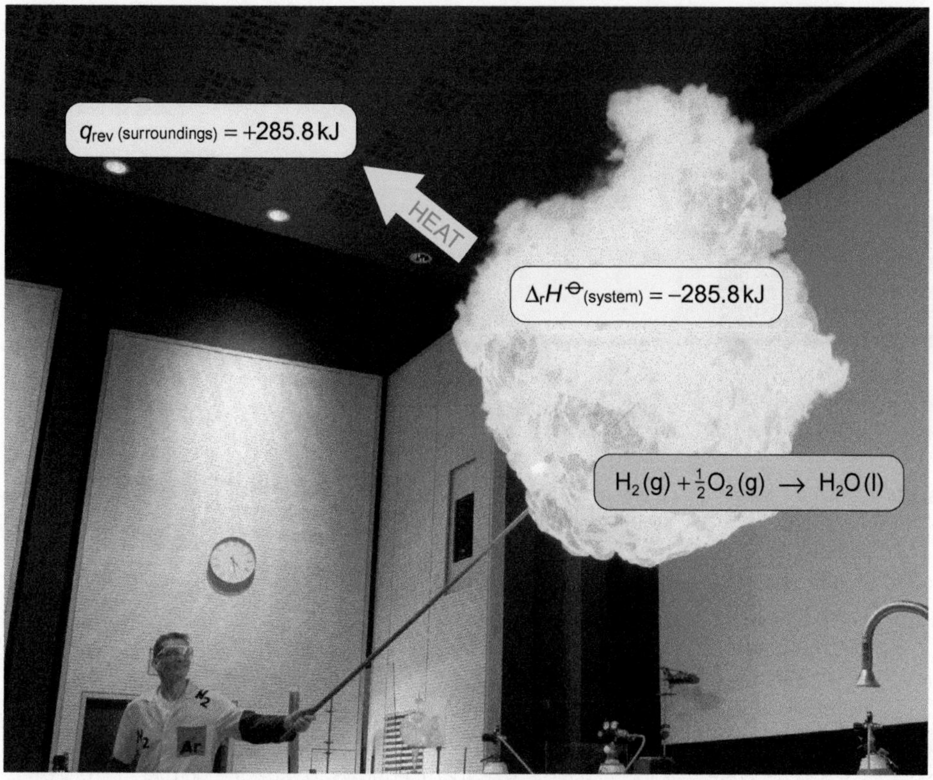
q_{rev} (surroundings) = +285.8 kJ
HEAT
$\Delta_r H^{\ominus}$(system) = −285.8 kJ
$H_2(g) + \frac{1}{2}O_2(g) \rightarrow H_2O(l)$

Figure 14.7 The heat given out by the system is gained by the surroundings.

From Equation 14.13, the total entropy change for the Universe is given by

$$\begin{aligned}\Delta_r S^{\ominus}_{298}(\text{total}) &= \Delta_r S^{\ominus}_{298}(\text{system}) + \Delta_r S^{\ominus}_{298}(\text{surroundings})\\ &= -163.4\,\text{J K}^{-1} + 959.1\,\text{J K}^{-1}\\ &= +795.7\,\text{J K}^{-1}\end{aligned}$$

So, when the surroundings are taken into account, there is an overall *increase* in the total entropy of the Universe, in accordance with the Second Law.

So, when $\Delta_r S_{total}$ is calculated by taking both system and surroundings into account, the direction of a spontaneous reaction can be predicted. However, it is not very convenient to have to consider both the system and the surroundings for every reaction. It would be useful to have some property of the system that would allow you to predict the spontaneity or otherwise of a reaction. This is provided by the Gibbs energy.

Visit the Online Resource Centre to watch screencast 14.3 which explains entropy change in the surroundings and introduces Gibbs Energy change.

Box 14.4 What happens when proteins don't fold correctly?

For water soluble proteins in living cells, one of the main driving forces for folding (see p.655) is the stability gained by placing amino acids with non-polar R groups, such as alanine ($R = CH_3$), inside the folded structure. This positions them away from the surrounding water molecules and so reduces unfavourable interactions. Polar amino acids with groups that can form hydrogen bonds to water are on the outer surface of the protein. The main driving force for the folding of proteins in membranes is hydrogen bond formation between N–H and C=O in peptide bonds. Non-polar R groups are buried in the hydrophobic lipid membrane. This folded *native state* is generally the most stable arrangement for the protein. But what happens when a protein does not fold correctly?

Mad cow disease, BSE (bovine spongiform encephalopathy), and its human equivalent Creutzfeldt–Jakob disease, are believed to result from protein misfolding. The diseases are transmitted by *prions*. These proteins can cause disease in a susceptible animal or human when introduced in tiny quantities.

Prions are misfolded forms of a protein called PrP (prion-related protein). Normal PrP occurs naturally and is tethered to the cell membranes of nerve tissue. It has a large percentage of α-helix →

▲ This BSE-infected cow shows some hallmark symptoms of the disease: abnormal posture; difficulty standing; and weight loss.

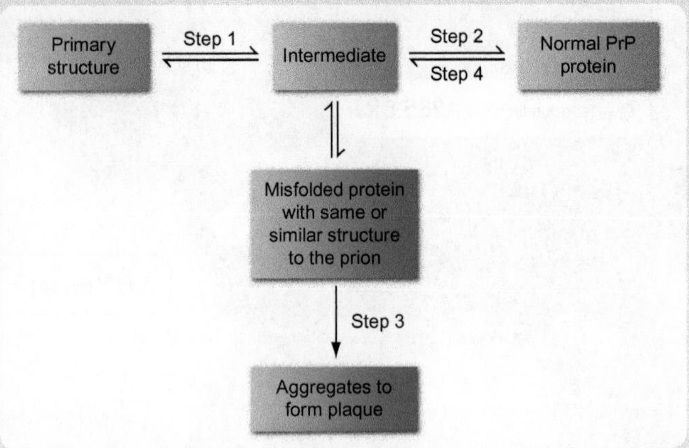

Primary structure
Step 1
Intermediate
Step 2
Step 4
Normal PrP protein
Misfolded protein with same or similar structure to the prion
Step 3
Aggregates to form plaque

▲ **Figure 2** The misfolded protein that causes BSE is a conformer of an intermediate in the formation of a normal protein.

in its structure, whereas the misfolded form is a rogue conformer with a large percentage of β-pleated sheet (see Figure 1). (You can read about the conformation of molecules in Section 18.2, p.819.) The areas of β-pleated sheet on adjacent protein molecules interact causing the molecules to aggregate to form insoluble deposits called *plaques*. The plaques are often found in brain tissue—a condition also seen in Alzheimer's disease. As the nerve tissues degenerate, progressive mental deterioration, dementia, and finally death occur.

Normal PrP protein is constantly being produced in the body. Usually, it folds properly, remains functional and tethered to the membrane. When released from the membrane, it becomes soluble and is disposed of without causing any problems. However, there is an intermediate, misfolded form of the protein that is also stable and this can cause problems. When this misfolded form of the protein (a prion) is introduced into the body, it changes the folding process so that the protein produced in the cells folds to form the prion even though it has the normal primary structure. A very small amount of the prion initially causes the body to keep on producing more and more of the misfolded form. In effect, the prion is 'replicating' itself.

How this happens is not fully understood. Figure 2 shows one route being investigated. Step 1 is the folding of the primary structure into an intermediate that then folds to form the normal PrP protein in step 2. Under some conditions or when the protein sequence is changed by mutation, the intermediate misfolds to form a conformer with an identical (or similar) structure to the prion. This misfolded conformer self-associates and is stabilized by plaque formation in an essentially irreversible reaction (step 3).Under these conditions the folding reaction to the normal PrP protein (step 2) can be reversed (step 4) resulting in more prion formation. Research is still being carried out to establish the exact role of prions in causing the disease.

The discovery of prions and their link to infectious diseases won Stanley Prusiner the Nobel Prize in Physiology or Medicine in 1997.

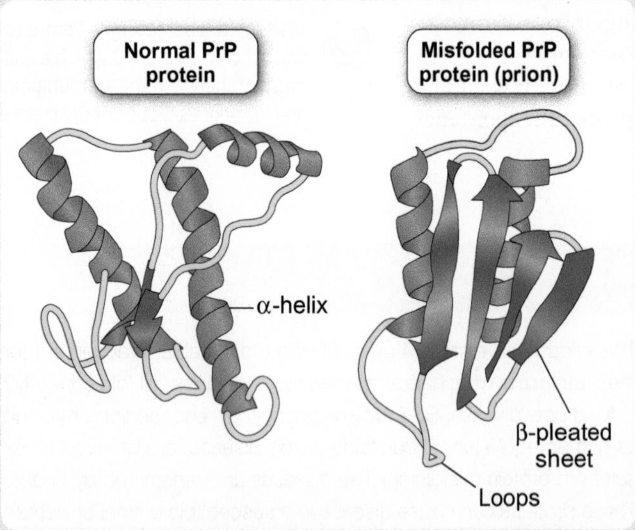

Normal PrP protein
Misfolded PrP protein (prion)
α-helix
β-pleated sheet
Loops

▲ **Figure 1** Normal PrP protein and a likely structure of the misfolded PrP protein.

Questions

The thermodynamics of protein folding can be considered using the relationship in Equation 14.13

$$\Delta S(\text{total}) = \Delta S(\text{system}) + \Delta S(\text{surroundings})$$

(a) The folding of a protein leads to a highly ordered structure. What is the sign of the value of ΔS(system) for the process?

(b) Folding is a spontaneous process under physiological conditions. What does this tell you about the value of ΔS(total)? What implications does this have for the value of ΔS(surroundings)? Explain how this happens.

(c) Suggest why the formation of the misfolded form of PrP may not take place spontaneously from the normal form of the protein.

Acknowledgement

Prof Sheena Radford, Astbury Centre for Structural Molecular Biology, University of Leeds

Summary

- The standard entropy change for a reaction at 298 K is given by

$$\Delta_r S^{\ominus}_{298}\text{(reaction)} = \sum v_i S^{\ominus}_{298}\text{(products)} - \sum v_i S^{\ominus}_{298}\text{(reactants)}$$

- The standard entropy change for a reaction at another temperature, T, is given by

$$\Delta_r S^{\ominus}_{T} = \Delta_r S^{\ominus}_{298} + \Delta C_p \ln\left(\frac{T}{298\,\text{K}}\right)$$

$$\text{where } \Delta C_p = \sum v_i C_p\text{(products)} - \sum v_i C_p\text{(reactants)}$$

- For a chemical change to be spontaneous, the total entropy change of the Universe must have a positive value.

$$\Delta_r S\text{(total)} = \Delta_r S\text{(system)} + \Delta_r S\text{(surroundings)}$$

$$\Delta_r S\text{(surroundings)} = \frac{-\Delta_r H}{T} \quad \text{(at constant pressure)}$$

? For practice questions on these topics, see questions 5, 10, 11, and 13 at the end of this chapter (p.692).

14.5 Gibbs energy

Total entropy change and Gibbs energy change

The sign of the total entropy change for a reaction indicates whether or not it will be spontaneous. The Gibbs energy change, ΔG, combines changes in enthalpy and entropy into a single state function that describes the spontaneity of a process at constant temperature and pressure and uses only properties of the system.

From the Second Law (p.660), it follows that for a spontaneous process at constant temperature and pressure

$$\Delta S\text{(total)} > 0$$

and

$$\Delta S\text{(system)} + \Delta S\text{(surroundings)} > 0$$

Using Equation 14.14 on p.672 to substitute an expression for ΔS(surroundings) into Equation 14.13

$$\Delta S\text{(total)} = \Delta_r S\text{(system)} + \left(\frac{-\Delta_r H}{T}\right)$$

The right-hand side of this equation now only contains terms referring to the system under study. Multiplying each side by $-T$ and rearranging the equation gives

$$-T\Delta S\text{(total)} = \Delta_r H - T\Delta_r S\text{(system)} \tag{14.15}$$

The term $(\Delta H - T\Delta S)$ referring to a system has a special significance in thermodynamics. It is called the change in Gibbs energy, and is given the symbol ΔG

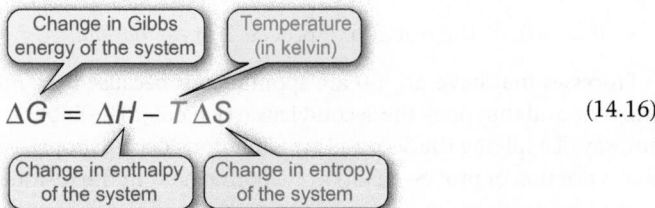

$$\Delta G = \Delta H - T\Delta S \tag{14.16}$$

Change in Gibbs energy of the system
Temperature (in kelvin)
Change in enthalpy of the system
Change in entropy of the system

Gibbs energy is named after the American chemist Josiah Willard Gibbs who developed a great deal of the thermodynamic theory that is used today.

It is generally understood that, if a symbol is not labelled as referring to the system or the surroundings, it refers to the system.

Sometimes ΔG is known as Gibbs free energy. The reason for the term 'free energy' will be apparent later in this section (see p.685).

Since enthalpy and entropy are state functions, Gibbs energy must also be a state function.

The requirement for a change to be spontaneous is $\Delta S(\text{total}) > 0$. Since $\Delta G = -T\Delta S(\text{total})$, the requirement for a spontaneous change can also be written as

The symbol $(\Delta G)_{p,T}$ is shorthand for 'the change in Gibbs energy at constant temperature and pressure'. It is often shortened simply to ΔG.

$$\text{for a spontaneous change} \qquad (\Delta G)_{p,T} < 0 \qquad (14.17)$$

so that a spontaneous process at constant temperature and pressure has a negative value for the change in Gibbs energy. This is illustrated in Worked example 14.7.

Worked example 14.7 Gibbs energy change and spontaneity

Calculate the change in Gibbs energy when 1.00 mol of ice melts at: (a) 0 °C; (b) 10 °C; and (c) −10 °C. Comment on the results.

($\Delta_{\text{fus}}H^{\ominus}(H_2O) = +6.01\ \text{kJ mol}^{-1}$ and $\Delta_{\text{fus}}S^{\ominus}(H_2O) = +22.0\ \text{J K}^{-1}\ \text{mol}^{-1}$).

Strategy

Use Equation 14.16, to find the change in Gibbs energy

$$\Delta G = \Delta H - T\Delta S \qquad (14.16)$$

Take care to use the correct units at each stage; the most common error here is to mix J and kJ in the same calculation.

Solution

(a) *Convert 0 °C to kelvin:* 0 °C = 273 K.
Use Equation 14.16 and substitute values for $\Delta_{\text{fus}}H^{\ominus}_{298}$ *and* $\Delta_{\text{fus}}S^{\ominus}_{298}$.

$$\begin{aligned}\Delta_{\text{fus}}G^{\ominus} &= \Delta_{\text{fus}}H^{\ominus} - T\Delta_{\text{fus}}S^{\ominus} \qquad \text{(from 14.16)}\\ &= +6.01\ \text{kJ mol}^{-1} - [273\ \cancel{\text{K}} \times (+22.0\times10^{-3}\ \text{kJ}\ \cancel{\text{K}}^{-1}\ \text{mol}^{-1})]\\ &= 0\ \text{kJ mol}^{-1}\end{aligned}$$

The entropy change is multiplied by 1×10^{-3} to keep the units consistent since the entropy change is given in J K^{-1} mol^{-1} whereas the enthalpy change is given in kJ mol^{-1}.

Apply the same method to (b) and (c).

(b) At 10 °C (283 K)

$$\begin{aligned}\Delta_{\text{fus}}G^{\ominus} &= \Delta_{\text{fus}}H^{\ominus} - T\Delta_{\text{fus}}S^{\ominus}\\ &= +6.01\ \text{kJ mol}^{-1} - [283\ \cancel{\text{K}} \times (+22.0\times10^{-3}\ \text{kJ}\ \cancel{\text{K}}^{-1}\ \text{mol}^{-1})]\\ &= -0.22\ \text{kJ mol}^{-1}\end{aligned}$$

(c) At −10 °C (263 K)

$$\begin{aligned}\Delta_{\text{fus}}G^{\ominus} &= \Delta_{\text{fus}}H^{\ominus} - T\Delta_{\text{fus}}S^{\ominus}\\ &= +6.01\ \text{kJ mol}^{-1} - [263\ \cancel{\text{K}} \times (+22.0\times10^{-3}\ \text{kJ}\ \cancel{\text{K}}^{-1}\ \text{mol}^{-1})]\\ &= +0.22\ \text{kJ mol}^{-1}\end{aligned}$$

At the melting point, T_m, $\Delta_{\text{fus}}G^{\ominus} = 0$. Above T_m, melting is spontaneous—as reflected in the negative value of $\Delta_{\text{fus}}G^{\ominus}$. Below T_m ice does not spontaneously melt—hence the positive value of $\Delta_{\text{fus}}G^{\ominus}$.

Question

For the melting of sodium chloride (NaCl) $\Delta_{\text{fus}}H^{\ominus} = +30.2\ \text{kJ mol}^{-1}$ and $\Delta_{\text{fus}}S^{\ominus} = +28.1\ \text{J K}^{-1}\ \text{mol}^{-1}$. Estimate the melting point of NaCl.

Thermodynamics is applied to chemical equilibrium in Chapter 15 and to phase changes in Chapter 17.

The relationship in Equation 14.17 is perhaps the most important relationship in chemical thermodynamics. It is the fundamental basis for the explanation of chemical spontaneity and the direction of chemical change, and also for the explanation of chemical equilibrium and the phase behaviour of compounds. The key relationships can be summarized as:

- If $\Delta G < 0$ the reaction or process is *spontaneous*.
- If $\Delta G > 0$ the reaction or process is *non-spontaneous*.
- If $\Delta G = 0$ the reaction or process is at *equilibrium*.

Processes that have $\Delta G < 0$ are spontaneous because they increase the entropy of the Universe and thus obey the Second Law (see p.660). The Gibbs energy just gives a convenient way of applying the Second Law while considering properties only of the system. Note that a reaction or process that is non-spontaneous in the forward direction is spontaneous in the reverse direction.

Box 14.5 Gibbs energy: the balance between enthalpy change and entropy change

Look again in detail at the equation defining Gibbs energy change (Figure 1).

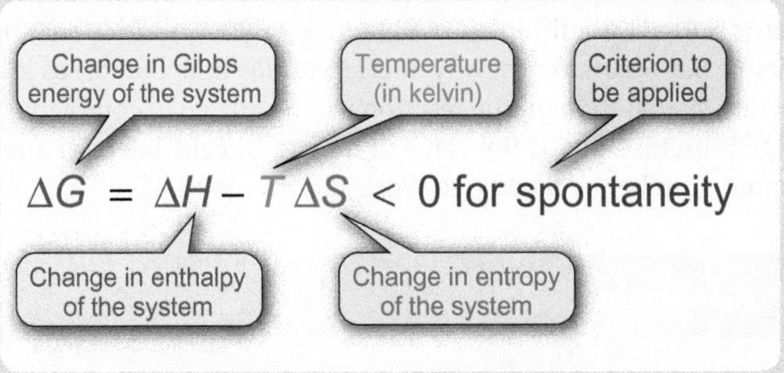
Change in Gibbs energy of the system
Temperature (in kelvin)
Criterion to be applied
ΔG = ΔH – TΔS < 0 for spontaneity
Change in enthalpy of the system
Change in entropy of the system

Figure 1 The definition of ΔG.

The Gibbs energy change must be negative for a process to be spontaneous at constant T and p. What about if the contributions from the enthalpy change and entropy change?

- $\Delta H < 0$ (*negative*, exothermic change) makes a *favourable* contribution to spontaneity.
- $\Delta H > 0$ (*positive*, endothermic change) makes an *unfavourable* contribution to spontaneity.
- $\Delta S < 0$ (*negative*, entropy decreases) makes an *unfavourable* contribution to spontaneity.
- $\Delta S > 0$ (*positive*, entropy increases) makes a *favourable* contribution to spontaneity.

The contribution of ΔS works in the opposite way to that of ΔH because of the minus sign in the ($-T\Delta S$) term. The temperature, T, determines the relative importance of these two contributions.

An exothermic reaction that also has an increase in entropy will be spontaneous at all temperatures, since both contributions lead to a negative value for ΔG. An endothermic reaction that also has a decrease of entropy will be non-spontaneous at all temperatures, since both contributions lead to a positive value for ΔG.

Where ΔH and ΔS have opposite signs, the situation is more complicated. At *low temperatures*, the $T\Delta S$ term is relatively small, and the sign of ΔG is mainly determined by that of ΔH. (Since temperatures are in kelvin, the value of T is always positive.) Think about an exothermic reaction, $\Delta H < 0$. Even if the $T\Delta S$ term is unfavourable, at low temperatures it will be outweighed by the negative value of ΔH so the value of ΔG will also be negative. Thus, at low temperatures, exothermic reactions are usually spontaneous.

As the temperature increases, the $T\Delta S$ term becomes more important. At sufficiently *high temperatures*, it becomes dominant.

The contributions to the Gibbs energy change are summarized in Figure 2 and in Table 14.2.

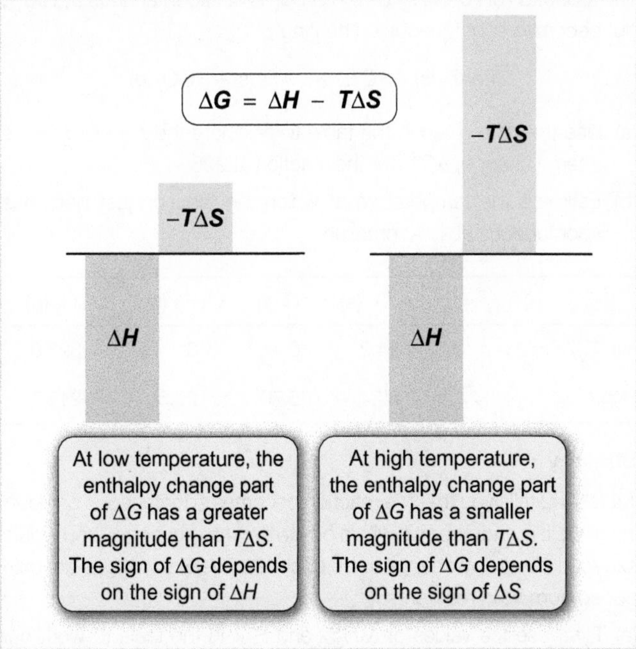
ΔG = ΔH – TΔS
–TΔS
ΔH
–TΔS
ΔH
At low temperature, the enthalpy change part of ΔG has a greater magnitude than TΔS. The sign of ΔG depends on the sign of ΔH
At high temperature, the enthalpy change part of ΔG has a smaller magnitude than TΔS. The sign of ΔG depends on the sign of ΔS

Figure 2 The effect of temperature on the sign of ΔG. This example is for a reaction in which ΔH and ΔS are both negative, so that ΔH has a negative value and $-T\Delta S$ has a positive value.

Table 14.2 Contributions to Gibbs energy change

Reaction	Value of ΔH	Value of ΔS	Value of ΔG ($= \Delta H - T\Delta S$)	Spontaneity
Endothermic	Positive, > 0	Negative, < 0	Positive for all T	Never spontaneous
Endothermic	Positive, > 0	Positive, > 0	Positive at low T; negative at high T	Becomes spontaneous on heating
Exothermic	Negative, < 0	Positive, > 0	Negative for all T	Always spontaneous
Exothermic	Negative, < 0	Negative, < 0	Positive at high T; negative at low T	Becomes spontaneous on cooling

$\Delta G^{\ominus}$ means the Gibbs energy change measured under standard conditions of 1 bar, which are the same as those used for $\Delta H^{\ominus}$ and $\Delta S^{\ominus}$.

The relationship in Equation 14.17 shows that if a process or reaction results in a lowering of the Gibbs energy, it is spontaneous—it is favourable and *can* happen. It does not mean that it *will* happen. For example, as has been seen, the reaction of hydrogen gas and oxygen gas to form water at 1 bar and 298 K has a highly negative value for the change in Gibbs energy

$$H_2(g) + \tfrac{1}{2}O_2(g) \rightarrow H_2O(l) \quad \Delta G^{\ominus} = -237.1\,\text{kJ mol}^{-1}$$

The rates of chemical reactions and activation energies are discussed in Sections 9.2 (p.384) and 9.7 (p.425). The formation of thermodynamic products and kinetic products is described in Section 9.8 (p.436).

However, a mixture of hydrogen and oxygen in a container will exist without reacting for a long time until the reaction is started by ignition or by adding a catalyst. The reaction is *thermodynamically* very favourable but occurs very slowly. It is *kinetically* unfavourable as it has a high activation energy. Once started though, the reaction will readily go to completion.

The Gibbs energy change is useful since, if the value of ΔG can be calculated for a reaction or process, its spontaneity can be predicted (see Worked example 14.8).

Worked example 14.8 Calculating the Gibbs energy change of reaction at 298 K

The reduction of iron(III) oxide by carbon is not spontaneous at 298 K, but becomes spontaneous on heating.

$$2\,Fe_2O_3(s) + 3\,C(s) \rightarrow 4\,Fe(s) + 3\,CO_2(g)$$

(a) Use the data given in the table to calculate the standard Gibbs energy change, $\Delta G^{\ominus}$, for the reaction at 298 K.

(b) Estimate the temperature at which the reaction just becomes spontaneous at 1 bar pressure.

	Fe_2O_3 (s)	C (s)	Fe (s)	CO_2 (g)
$\Delta_f H^{\ominus}_{298}$/kJ mol^{-1}	−824.2	0	0	−393.5
$H^{\ominus}_{298}$/J K^{-1} mol^{-1}	87.4	5.70	27.3	213.7

Strategy

Table 14.2 tells you that the reaction becomes spontaneous on heating since it is endothermic with a positive entropy change. (You would expect the entropy change to be positive since 3 mol of gas are produced from solid reactants.)

(a) First, use the values of $\Delta_f H^{\ominus}_{298}$ and $S^{\ominus}_{298}$ in the table to work out the changes in enthalpy and entropy for the reaction at 298 K. The Gibbs energy change can then be calculated using Equation 14.16

$$\Delta_r G^{\ominus} = \Delta_r H^{\ominus} - T\Delta_r S^{\ominus} \qquad \text{(from 14.16)}$$

(b) As the temperature increases, the $(-T\Delta_r S^{\ominus})$ term becomes more important. The reaction is endothermic so it has a positive value of $\Delta_r H^{\ominus}$. It becomes spontaneous at temperatures above that at which $\Delta_r G^{\ominus} = 0$. So, you need to calculate the temperature at which $\Delta_r G = 0$. Assume that $\Delta_r H^{\ominus}$ and $T\Delta_r S^{\ominus}$ do not change very much with temperature and so use the values calculated at 298 K.

Solution

(a) *Use Equation 13.6 (p.626) to calculate $\Delta_r H^{\ominus}_{298}$.*

$$\Delta_r H^{\ominus}_{298} = \sum \nu_i \Delta_f H^{\ominus}_{298}(\text{products}) - \sum \nu_i \Delta_f H^{\ominus}_{298}(\text{reactants}) \qquad (13.6)$$

$$= [0 + 3 \times (-393.5\,\text{kJ mol}^{-1})] - [2 \times (-824.2\,\text{kJ mol}^{-1}) + 0]$$

$$= +467.9\,\text{kJ mol}^{-1}$$

Use Equation 14.11 (p.669) to calculate. $\Delta_r S^{\ominus}_{298}$

$$\Delta_r S^{\ominus}_{298} = \sum \nu_i S^{\ominus}_{298}(\text{products}) - \sum \nu_i S^{\ominus}_{298}(\text{reactants})$$

$$= [4 \times (27.3\,\text{J K}^{-1}\text{mol}^{-1}) + 3 \times (213.7\,\text{J K}^{-1}\text{mol}^{-1})]$$

$$- [2 \times (87.4\,\text{J K}^{-1}\text{mol}^{-1}) + 3 \times (5.70\,\text{J K}^{-1}\text{mol}^{-1})]$$

$$= +558.4\,\text{J K}^{-1}\text{mol}^{-1}$$

Use Equation 14.16 (p.675) to calculate. $\Delta_r G^{\ominus}_{298}$

$$\Delta_r G^{\ominus}_{298} = \Delta_r H^{\ominus}_{298} - T\Delta_r S^{\ominus}_{298}$$

$$= +467.9\,\text{kJ mol}^{-1} - [298\,\cancel{\text{K}} \times (+558.4 \times 10^{-3}\,\text{kJ}\,\cancel{\text{K}}^{-1}\,\text{mol}^{-1})]$$

$$= +302\,\text{kJ mol}^{-1}$$

The entropy change is multiplied by 1×10^{-3} to convert the value from J K^{-1} mol^{-1} to kJ K^{-1} mol^{-1} to be consistent with the enthalpy change in kJ mol^{-1}.

(b) The Gibbs energy change is positive so the reaction is non-spontaneous at 298 K. It becomes spontaneous at temperatures above that at which

$$\Delta_r G^{\ominus} = \Delta_r H^{\ominus} - T\Delta_r S^{\ominus} = 0$$

Substitute the values of $\Delta_r H^{\ominus}_{298}$ and $\Delta_r S^{\ominus}_{298}$ to find T when $\Delta_r G^{\ominus} = 0$.

When $\Delta_r G^{\ominus} = 0$, $T\Delta_r S^{\ominus} = \Delta_r H^{\ominus}$ so that

$$T = \frac{\Delta_r H^{\ominus}}{\Delta_r S^{\ominus}} = \frac{+467.9\,\cancel{\text{kJ}}\,\cancel{\text{mol}^{-1}}}{+558.4 \times 10^{-3}\,\cancel{\text{kJ}}\,\text{K}^{-1}\,\cancel{\text{mol}^{-1}}}$$

$$= 838\,\text{K}$$

Question

Calculate the standard Gibbs energy change at 298 K for the reaction

$$4\,HCl(g) + O_2(g) \rightarrow 2\,Cl_2(g) + 2\,H_2O(l)$$

(Use Appendix 7 (p.1352), to find values of $\Delta_f H^{\ominus}_{298}$ and standard entropies for reactants and products.)

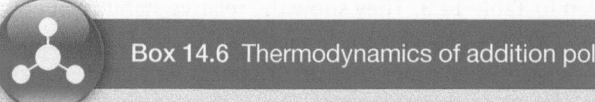

Box 14.6 Thermodynamics of addition polymerization

Many of the plastic articles that you see around you are made using **addition polymerization**, in which a number of small molecules **(monomers)** containing a double bond are linked together in a **chain polymer**. There can be up to 10 000 monomers in a single chain.

The figure shows the synthesis of poly(methyl methacrylate), commonly known as Perspex or Lucite. The reaction is often started by adding an initiator, which, on heating, forms radicals to start the chains growing.

A related polymer, poly(methyl cyanoacrylate) (in which the CH_3 group attached to the double bond is replaced by a CN group) is the polymer that forms 'superglue'. (You can see the mechanism for the formation of a poly(cyanoacrylate) in Box 19.8, p.904.)

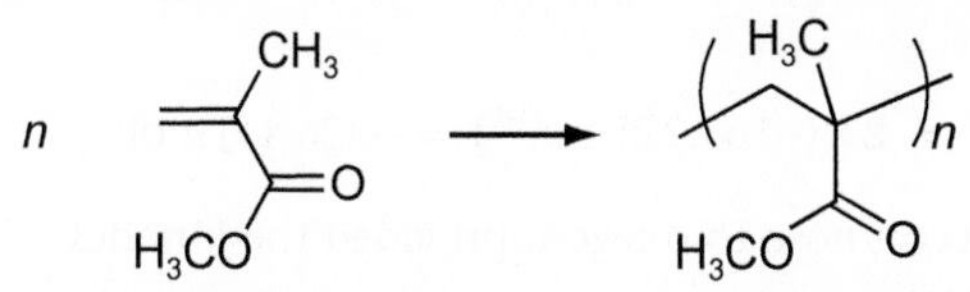
n
CH3
O
H3CO
H3C
n
H3CO
O

methyl methacrylate, MMA — poly(methyl methacrylate), PMMA

▲ Poly(methyl methacrylate) is a polymer that consists of a large number, *n*, of methyl methacrylate monomers joined in a chain.

The entropy changes for addition polymerization reactions are negative and usually in the region of $-100\,J\,K^{-1}\,mol^{-1}$ to $-120\,J\,K^{-1}\,mol^{-1}$. A large number of small molecules come together to form a single chain. The chains are much more ordered than the individual monomers, so that the entropy of the system decreases by a large amount.

For a spontaneous reaction, $(\Delta H - T\Delta S) < 0$, so values of ΔH must be negative and quite large—polymerization reactions are highly exothermic. As the temperature increases, the entropy term becomes more important. Eventually, the magnitude of the $(-T\Delta S)$ term will be bigger than that of ΔH, so ΔG becomes positive and the reaction is not spontaneous. The polymerization will not take place. The temperature at which the Gibbs energy of polymerization is exactly zero is known as the *ceiling temperature*.

Questions

At 298 K, $\Delta_{pol}H$ (the enthalpy change of polymerization) for styrene is $-68.5\,kJ\,mol^{-1}$. The value of $\Delta_{pol}H$ for α-methyl styrene is $-35.1\,kJ\,mol^{-1}$. $\Delta_{pol}S_{298} = -105\,J\,K^{-1}\,mol^{-1}$ for both reactions.

H
H
H
H
H
CH3

styrene — α-methylstyrene

▲ Styrene monomers.

(a) Calculate the Gibbs energy change of polymerization at 298 K for both monomers. Estimate the ceiling temperature at which $\Delta_{pol}G = 0$ in each case.

(b) What are the implications of the ceiling temperatures for these monomers?

▲ This aquarium tank has windows made from poly(methyl methacrylate). This polymer is more transparent than glass so the windows can be made thicker. The polymer is also lighter and can be made into very large sheets up to 35 cm thick so it is an ideal material for large water tanks where the windows must be thick in order to withstand the high pressure from millions of gallons of water. This would not be possible using glass.

Gibbs energy changes of formation and reaction

Standard Gibbs energy changes of formation at 298 K, $\Delta_f G^{\ominus}_{298}$, are defined in a similar way to standard enthalpy changes of formation, $\Delta_f H^{\ominus}_{298}$ (see Section 13.3, p.622).

> ***$\Delta_f G^{\ominus}_{298}$ is the change in Gibbs energy when 1 mol of a compound is formed at 1 bar and 298 K from its elements in their standard states.***

As with $\Delta_f H^{\ominus}_{298}$, an absolute value of $\Delta_f G^{\ominus}_{298}$ cannot be defined so that the convention is to use a reference state where $\Delta_f G^{\ominus}_{298}$ is zero for a pure element in its standard state.

Some values of $\Delta_f G^{\ominus}_{298}$ are given in Table 14.3. They show the relative stability of compounds. For example, $\Delta_f G^{\ominus}_{298}$ for CO_2 is −394.4 kJ mol^{-1}; that for CO is −137.2 kJ mol^{-1}. Thus, both compounds have lower Gibbs energies than their constituent elements and are, therefore, more stable with respect to their elements—but CO_2 is more stable with respect to CO.

> The standard Gibbs energy change of formation, $\Delta_f G^{\ominus}_{298}$, is the change in Gibbs energy when 1 mol of a compound is formed at 1 bar and 298 K from its constituent elements in their standard states.

> More extensive lists of values of $\Delta_f G^{\ominus}_{298}$ are given in Appendix 7 (p.1352).

Table 14.3 Standard Gibbs energy change of formation, $\Delta_f G^{\ominus}_{298}$, for selected compounds

	$\Delta_f G^{\ominus}_{298}$/kJ mol^{-1}
C (graphite)	0
C (diamond)	+2.9
CO (g)	−137.2
CO_2 (g)	−394.4
HF (g)	−273.2
H_2O (l)	−237.1
H_2O (g)	−228.6
NH_3 (g)	−16.5
O_2 (g)	0
O_3 (g)	+163.2

The value of $\Delta_f G^{\ominus}_{298}$ is not always negative, as demonstrated by ozone, O_3 (g), where $\Delta_f G^{\ominus}_{298}$ = +163.2 kJ mol^{-1} This means that O_3 (g) is unstable with respect to the standard state of diatomic oxygen gas, O_2 (g).

$$1\tfrac{1}{2}O_2(g) \rightarrow O_3(g) \qquad \Delta_f G^{\ominus} = +163.2\ \text{kJ mol}^{-1}$$

so

$$2O_3(g) \rightarrow 3O_2(g) \quad \Delta_r G^{\ominus} = 2\times(-163.2\ \text{kJ mol}^{-1}) = -326.4\ \text{kJ mol}^{-1}$$

and ozone should spontaneously decompose to oxygen, provided the kinetics allow the reaction to occur at an observable rate.

Diamond is a less thermodynamically favourable form of carbon than graphite under standard conditions

$$\text{C(diamond)} \rightarrow \text{C(graphite)} \quad \Delta_r G^{\ominus} = -2.9\ \text{kJ mol}^{-1}$$

However, the conversion reaction is infinitely slow at room temperature, a fact that will be a welcome relief to buyers of expensive jewellery!

Gibbs energy changes of formation can be used to calculate the change in Gibbs energy during a reaction in the same manner as $\Delta_f H^{\ominus}_{298}$ is used to calculate the enthalpy change of a reaction (see Section 13.3, p.622). Some examples are shown in Worked example 14.9. Since G is a state function, changes during a reaction can be calculated using Equation 14.18

> Note that values of $\Delta_f G^{\ominus}_{298}$ may be positive or negative.

$$\Delta_r G^{\ominus}_{298} = \sum \nu_i \Delta_f G^{\ominus}_{298}(\text{products}) - \sum \nu_i \Delta_f G^{\ominus}_{298}(\text{reactants}) \qquad (14.18)$$

> Equations similar to Equation 14.18 are used to calculate $\Delta_r H^{\ominus}_{298}$ (Equation 13.6, p.626) and $\Delta_r S^{\ominus}_{298}$ (Equation 14.11, p.669).

$\Delta_f G^{\ominus}_{298}$ values give the Gibbs energy change under only one particular set of conditions—at 298 K and 1 bar pressure. Nonetheless, they give a useful indication of the tendency of a reaction to occur, as illustrated in Worked examples 14.9 and 14.10.

Barratt
Sherbet
fountain
OPEN
Sherbet with
a liquorice dip

Sherbet contains a mixture of solid edible acid, such as citric acid, together with sodium hydrogen carbonate. In the aqueous medium in your mouth, a reaction takes place producing carbon dioxide. The reaction is endothermic, so your mouth feels cool. The negative entropy change in the surroundings (your mouth) is more than compensated for by the positive entropy change in the system when the solid reactants are converted to gas and aqueous ions, so the Gibbs energy change is negative and the reaction is spontaneous.

Worked example 14.9 Using Gibbs energy changes of formation to calculate Gibbs energy changes of reaction

Using data from Appendix 7 (p.1352), calculate the standard Gibbs energy change, $\Delta_r G^{\ominus}_{298}$, for each of the following reactions and comment on whether the reactions are likely to occur.

(a) $N_2O_4(g) \rightarrow 2NO_2(g)$

(b) $3Fe_2O_3(s) + 2NH_3(g) \rightarrow 6FeO(s) + 3H_2O(l) + N_2(g)$

Strategy

Use Equation 14.18 directly to calculate $\Delta_r G^{\ominus}_{298}$. The reaction is spontaneous if $\Delta G < 0$.

Solution

Use Equation 14.18 and substitute in the values of $\Delta_r G^{\ominus}_{298}$ for the reactants and products.

$$\Delta_r G^{\ominus}_{298} = \sum \nu_i \Delta_f G^{\ominus}_{298}(\text{products}) - \sum \nu_i \Delta_f G^{\ominus}_{298}(\text{reactants}) \qquad (14.18)$$

(a)
$$\begin{aligned}\Delta_r G^{\ominus}_{298} &= [2 \times \Delta_f G^{\ominus}_{298}(NO_2(g))] - [\Delta_f G^{\ominus}_{298}(N_2O_4(g))] \\ &= [2 \times 51.3\,\text{kJ mol}^{-1}] - [+97.9\,\text{kJ mol}^{-1}] \\ &= +4.70\,\text{kJ mol}^{-1}\end{aligned}$$

(b)
$$\begin{aligned}\Delta_r G^{\ominus}_{298} &= [6\,\Delta_f G^{\ominus}_{298}(FeO(s)) + 3\,\Delta_f G^{\ominus}_{298}(H_2O(l)) \\ &\quad + \Delta_f G^{\ominus}_{298}(N_2(g))] - [3\,\Delta_f G^{\ominus}_{298}(Fe_2O_3(s)) \\ &\quad + 2\,\Delta_f G^{\ominus}_{298}(NH_3(g))] \\ &= [6 \times (-244.1\,\text{kJ mol}^{-1}) + 3 \times (-237.1\,\text{kJ mol}^{-1}) + 0] \\ &\quad - [3 \times (-742.2\,\text{kJ mol}^{-1}) + 2 \times (-16.5\,\text{kJ mol}^{-1})] \\ &= [-2175.9\,\text{kJ mol}^{-1}] - [-2259.6\,\text{kJ mol}^{-1}] \\ &= +83.7\,\text{kJ mol}^{-1}\end{aligned}$$

In each case $\Delta_r G^{\ominus}_{298} > 0$, so neither of these reactions is spontaneous under these conditions.

Question

Calculate the standard Gibbs energy change for the oxidation of ammonia at 298 K.

$$5O_2(g) + 4NH_3(g) \rightarrow 6H_2O(l) + 4NO(g)$$

Worked example 14.10 Calculating the Gibbs energy change of reaction at other temperatures

Use the data in the following table to:

(a) calculate the standard Gibbs energy change for the following reaction at 298 K;

(b) estimate the value at 1298 K.

$$MgO(s) + CO(g) \rightarrow Mg(s) + CO_2(g)$$

	MgO (s)	CO (g)	Mg (s)	CO_2 (g)
$\Delta_f H^{\ominus}_{298}$/kJ mol^{-1}	–601.7	–110.5	0	–393.5
$S^{\ominus}_{298}$/J K^{-1} mol^{-1}	26.9	197.7	32.7	213.7
C_p/J K^{-1} mol^{-1}	37.1	29.1	24.9	37.1

Strategy

(a) Use the enthalpy changes of formation to calculate $\Delta_r H^{\ominus}_{298}$ and the standard entropies to calculate. $\Delta_r S^{\ominus}_{298}$ Then, use Equation 14.16 (p.675) to find $\Delta_r G^{\ominus}_{298}$.

(b) To change these to the higher temperature, you first need to calculate ΔC_p using Equation 13.11 (p.635). Then, apply the Kirchhoff equation (Equation 13.10, p.634) to find $\Delta_r H^{\ominus}_{1298}$ and Equation 14.12 (p.670) to find $\Delta_r S^{\ominus}_{1298}$. Finally, use Equation 14.16 to calculate the Gibbs energy change at the higher temperature.

Solution

(a) *Use Equation 13.6 (p.626) to calculate $\Delta_r H^{\ominus}_{298}$.*

$$\begin{aligned}\Delta_r H^{\ominus}_{298} &= \sum \nu_i \Delta_f H^{\ominus}_{298}(\text{products}) - \sum \nu_i \Delta_f H^{\ominus}_{298}(\text{reactants}) \qquad (13.6) \\ &= [0 + (-393.5\,\text{kJ mol}^{-1})] - [(-601.7\,\text{kJ mol}^{-1}) \\ &\quad + (-110.5\,\text{kJ mol}^{-1})] \\ &= +318.7\,\text{kJ mol}^{-1}\end{aligned}$$

Use Equation 14.11 (p.669) to calculate $\Delta_r S^{\ominus}_{298}$.

$$\begin{aligned}\Delta_r S^{\ominus}_{298} &= \sum \nu_i S^{\ominus}_{298}(\text{products}) - \sum \nu_i S^{\ominus}_{298}(\text{reactants}) \qquad (14.11) \\ &= [32.7\,\text{J K}^{-1}\text{mol}^{-1} + 213.7\,\text{J K}^{-1}\text{mol}^{-1}] \\ &\quad - [26.9\,\text{J K}^{-1}\text{mol}^{-1} + 197.7\,\text{J K}^{-1}\text{mol}^{-1}] \\ &= +21.8\,\text{J K}^{-1}\text{mol}^{-1}\end{aligned}$$

Use Equation 14.16 (p.675) to calculate $\Delta_r G^{\ominus}_{298}$.

$$\begin{aligned}\Delta_r G^{\ominus}_{298} &= \Delta_r H^{\ominus}_{298} - T\Delta_r H^{\ominus}_{298} \qquad (\text{from } 14.16) \\ &= +318.7\,\text{kJ mol}^{-1} - [298\,\cancel{\text{K}} \times (+21.8 \times 10^{-3}\,\text{kJ}\,\cancel{\text{K}}^{-1}\,\text{mol}^{-1})] \\ &= +312.2\,\text{kJ mol}^{-1}\end{aligned}$$

Note the conversion of the entropy change from J K^{-1} mol^{-1} to kJ K^{-1} mol^{-1} to keep the units consistent.

➜

(b) *Use Equation 13.11 (p.635) to calculate ΔC_p so you can convert $\Delta_r H^{\ominus}_{298}$ and $\Delta_r S^{\ominus}_{298}$ to the higher temperature.*

$$\Delta C_p = \sum v_i C_p\text{(products)} - \sum v_i C_p\text{(reactants)} \qquad (13.11)$$
$$= [24.9\,\text{J K}^{-1}\,\text{mol}^{-1} + 37.1\,\text{J K}^{-1}\,\text{mol}^{-1}]$$
$$-[37.1\,\text{J K}^{-1}\,\text{mol}^{-1} + 29.1\,\text{J K}^{-1}\,\text{mol}^{-1}]$$
$$= -4.2\,\text{J K}^{-1}\,\text{mol}^{-1}$$

Convert $\Delta_r H^{\ominus}_{298}$ to the higher temperature using the Kirchhoff equation (Equation 13.10, p.634).

$$\Delta_r H_{T_2} = \Delta_r H_{T_1} + \Delta C_p \Delta T \qquad (13.10)$$
$$\Delta_r H^{\ominus}_{1298} = \Delta_r H^{\ominus}_{298} + \Delta C_p (1298\,\text{K} - 298\,\text{K})$$
$$= +318.7\,\text{kJ mol}^{-1} + [(-4.2\times10^{-3}\,\text{kJ}\,\cancel{\text{K}}^{-1}\,\text{mol}^{-1})\times 1000\,\cancel{\text{K}}]$$
$$= +314.5\,\text{kJ mol}^{-1}$$

Convert $\Delta_r S^{\ominus}_{298}$ to the higher temperature using Equation 14.12 (p.670).

$$\Delta_r S^{\ominus}_T = \Delta_r S^{\ominus}_{298} + \Delta C_p \ln\left(\frac{T}{298\,\text{K}}\right) \qquad (14.12)$$
$$\Delta_r S^{\ominus}_{1298} = \Delta_r S^{\ominus}_{298} + \Delta C_p \ln\left(\frac{1298\,\cancel{\text{K}}}{298\,\cancel{\text{K}}}\right)$$
$$= +21.8\,\text{J K}^{-1}\,\text{mol}^{-1} + [-4.2\,\text{J K}^{-1}\,\text{mol}^{-1} \times 1.471]$$
$$= +15.62\,\text{J K}^{-1}\,\text{mol}^{-1}$$

Use Equation 14.16 (p.675) to calculate $\Delta_r G^{\ominus}$ at the higher temperature

$$\Delta_r G^{\ominus}_{1298} = \Delta_r H^{\ominus}_{1298} - T\Delta_r S^{\ominus}_{1298} \qquad \text{(from 14.16)}$$
$$= +314.5\,\text{kJ mol}^{-1} - [1298\,\cancel{\text{K}} \times (+15.62\times10^{-3}\,\text{kJ}\,\cancel{\text{K}}^{-1}\,\text{mol}^{-1})]$$
$$= +294\,\text{kJ mol}^{-1}$$

Reduction of a metal oxide with CO is sometimes used to obtain a metal from its ore (see Box 14.10, p.689). For magnesium oxide this method is not feasible since the Gibbs energy change for the reduction has a high positive value at both temperatures. Magnesium metal is obtained by electrolysis of molten $MgCl_2$, see Section 16.5 (p.775) and Section 26.6 (p.1183).

Question

Calculate the standard Gibbs energy change for the oxidation of ammonia at 650 K. Compare your answer with the value calculated for 298 K in the question at the end of Worked example 14.9.

$$5\,O_2(g) + 4\,NH_3(g) \rightarrow 6\,H_2O(l) + 4\,NO(g)$$

Box 14.7 Obtaining silicon for use in silicon chips

The integrated circuits that power microprocessors are manufactured on thin wafers sliced from a large single crystal of pure silicon. The crystals used must be of extremely high quality as the electrical properties of silicon deteriorate if the crystal is slightly imperfect or if small amounts of impurities are present. For this purpose the silicon needs to be 99.9999999% pure (i.e. only 1 non-silicon atom in 10 billion is allowed).

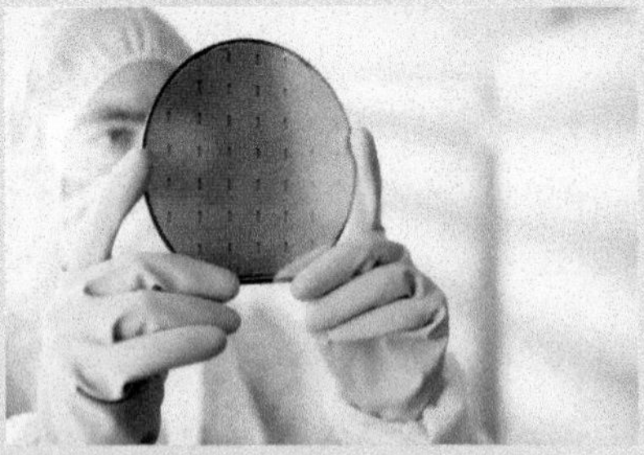

▲ A silicon wafer for making silicon chips. The silicon must be extremely pure, so ultra-clean manufacturing conditions are essential.

Silicon comprises 28% by mass of the Earth's crust but elemental silicon does not occur naturally. It has to be obtained from rocks where it is mainly in the form of SiO_2 in quartz and sand (see Section 27.3, p.1216).

Questions

Use the thermodynamic data in the table to answer the following questions.

	Si	SiO_2	O_2	C	CO	CO_2
$\Delta_f H^{\ominus}_{298}$/kJ mol^{-1}	0	–910.9	0	0	–110.5	–393.5
$S^{\ominus}_{298}$/J K^{-1} mol^{-1}	18.8	41.8	205.1	5.7	197.7	213.7

(a) Explain, in terms of entropy, why it is so difficult to prepare ultra-high purity silicon crystals.

Silicon could be prepared using the reaction

$$SiO_2(s) \rightarrow Si(s) + O_2(g) \qquad \text{(Reaction 1)}$$

(b) What are the standard enthalpy change and the standard entropy change at 298 K for this reaction? Comment on the value of $\Delta_r S^{\ominus}_{298}$ that you obtain.

(c) Calculate the value of $\Delta_r G^{\ominus}_{298}$ for the reaction and use it to explain why elemental silicon does not exist in the Earth's crust.

→

(d) Estimate the temperature at which $\Delta G^{\ominus} = 0$. What is the significance of this for preparing silicon using this reaction?

In practice, the reduction of SiO_2 is combined with a second reaction that helps to drive it forward. For example, combining it with $2C(s) + O_2(g) \rightarrow 2CO(g)$ gives

$$SiO_2(s) + 2C(s) \rightarrow Si(s) + 2CO(g) \qquad \text{(Reaction 2)}$$

(e) Calculate the value of $\Delta_r G^{\ominus}_{298}$ or this reaction and comment on the value you obtain.

(f) Estimate the temperature at which $\Delta G^{\ominus} = 0$ for this reaction. What are the implications of your answer for making silicon?

This is an example of a 'coupled' reaction. The reaction is spontaneous at high temperatures because the reduction of SiO_2 is coupled to the reaction of carbon to form CO. The process is carried out in an electric arc furnace at about 2500 K.

Even at this temperature, the reaction does not go to completion and the silicon produced is contaminated with sand and carbon and must be purified before it can be used. The impure silicon is treated with chlorine to produce $SiCl_4$, which is volatile and can be distilled to separate it from the reactants and impurities. The $SiCl_4$ is then converted back to silicon.

A possible alternative reaction for the reduction of SiO_2 by carbon might be

$$SiO_2(s) + C(s) \rightarrow Si(s) + CO_2(g) \qquad \text{(Reaction 3)}$$

(g) Repeat the calculations of $\Delta_r G^{\ominus}_{298}$ and the temperature at which $\Delta G^{\ominus} = 0$ for this reaction. Comment on the values you obtain.

(h) How does the value of the entropy change influence whether Reaction 2 or Reaction 3 actually occurs in practice?

Coupled reactions

The Gibbs energy change for a reaction can be used to predict whether or not a reaction will be spontaneous. However, even if a reaction has a positive value of $\Delta_r G^{\ominus}$ that does not necessarily mean that it cannot be used. The reaction in isolation will not be spontaneous but it can be made to occur if it is **coupled** to another reaction with a larger, negative $\Delta_r G^{\ominus}$. Box 14.7 describes how the preparation of silicon from SiO_2 is driven by coupling it to the oxidation of carbon. Such reactions are sometimes called **tandem** reactions (see Figure 14.8).

Many biochemical processes rely on coupled reactions. One of the most important examples is the series of reactions involved with the storage and use of energy in living cells. This is described in Box 14.8.

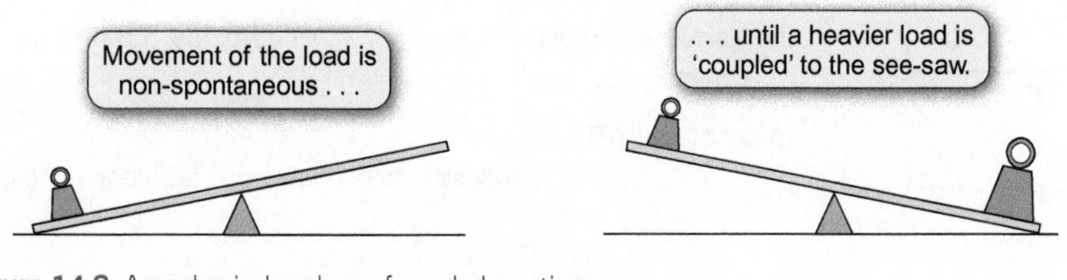
Movement of the load is non-spontaneous . . .
. . . until a heavier load is 'coupled' to the see-saw.

Figure 14.8 A mechanical analogy of coupled reactions.

Box 14.8 Energetics of biochemical reactions

All living organisms require energy and this comes from the oxidation of glucose in a complex sequence of reactions. In plants, glucose is a product of photosynthesis; in animals it comes from the breakdown of food.

The first step in the metabolism of glucose is its reaction with hydrogen phosphate ions to form glucose-6-phosphate

$$\underset{\text{glucose}}{C_6H_{12}O_6(aq)} + \underset{\text{Pi}}{HPO_4^{2-}(aq)} \rightarrow \underset{\text{glucose-6-phosphate}}{C_6H_{11}O_6PO_3^{2-}(aq)} + \underset{\text{water}}{H_2O(l)}$$

(Reaction 1)

where P_i indicates the inorganic hydrogen phosphate ion, HPO_4^{2-}. The Gibbs energy change for this reaction, $\Delta G^{\ominus\prime}$, is +13.4 kJ mol^{-1} at 37 °C, so the reaction is not spontaneous and it does not occur of its own accord.

Note that $\Delta G^{\ominus\prime}$ here refers to the biochemical standard state rather than the standard state used in most chemical systems. The conventional standard state uses a concentration of 1 mol dm^{-3}. In a reaction involving H^+ ions, this would be equivalent to pH = 0 so it is not appropriate in living systems where most cells have a

→

→

physiological pH close to neutral, pH = 7. This leads to the definition of the **biochemical standard state** where the standard concentration for $[H^+]$ is taken as 1×10^{-7} mol dm^{-3}. Standard Gibbs energy changes occurring under these conditions are designated $\Delta G^{\ominus\prime}$.

Inside a cell Reaction 1 does occur because it is coupled to a second reaction which drives it forward. This 'driving force' comes from the conversion of adenosine triphosphate (ATP; see Figure 1) to adenosine diphosphate (ADP). The terminal phosphate group in ATP is readily hydrolysed to form adenosine diphosphate ADP

$$\text{ATP(aq)} + \text{H}_2\text{O(l)} \rightarrow \text{ADP(aq)} + \text{H}^+\text{(aq)} + \text{P}_\text{i}\text{(aq)} \qquad \text{(Reaction 2)}$$

For this reaction, $\Delta G^{\ominus\prime} = -30.5$ kJ mol^{-1} at 37 °C. This is more negative than the value of $\Delta G^{\ominus\prime}$ or Reaction 1 so the overall coupled process is spontaneous.

Simply mixing the reactants will not ensure a coupled reaction. The impressive trick that living cells play is to use enzymes to catalyse the coupling. In this way, the phosphate ions react specifically with glucose rather than other species in solution.

The function of ATP is to store the energy made available when glucose is metabolized, and then supply it on demand for a wide variety of biological processes, such as the contraction of muscles, the transport of materials across cell membranes, and protein synthesis (Figure 2). ATP has been described as 'the single most important substance in biochemistry'. It can act as a store of chemical energy in biological systems because the reverse reaction can also be performed, in which ADP is converted to ATP, providing chemical energy to drive other reactions. When glucose is oxidized to carbon dioxide and water, 38 molecules of ATP are produced from ADP for each molecule of glucose metabolized.

The overall reaction for the complete oxidation of glucose has a large negative Gibbs energy change.

$$\text{C}_6\text{H}_{12}\text{O}_6\text{(aq)} + 6\text{O}_2\text{(g)} \rightarrow 6\text{CO}_2\text{(g)} + 6\text{H}_2\text{O(l)} \qquad \text{(Reaction 3)}$$

$$\Delta_r G^{\ominus\prime} = -2872 \text{ kJ mol}^{-1} \text{ at } 37\text{ °C}$$

Part of this Gibbs energy is used to produce ATP from ADP. Under conditions where the cell has a plentiful supply of oxygen, this conversion occurs via a complex sequence of reactions that can be summarized as

$$\text{C}_6\text{H}_{12}\text{O}_6\text{(aq)} + 38\text{H}^+\text{(aq)} + 38\text{ADP(aq)} + 38\text{P}_\text{i}\text{(aq)} \rightarrow 38\text{ATP(aq)} + 6\text{CO}_2\text{(g)} + 44\text{H}_2\text{O(l)} \qquad \text{(Reaction 4)}$$

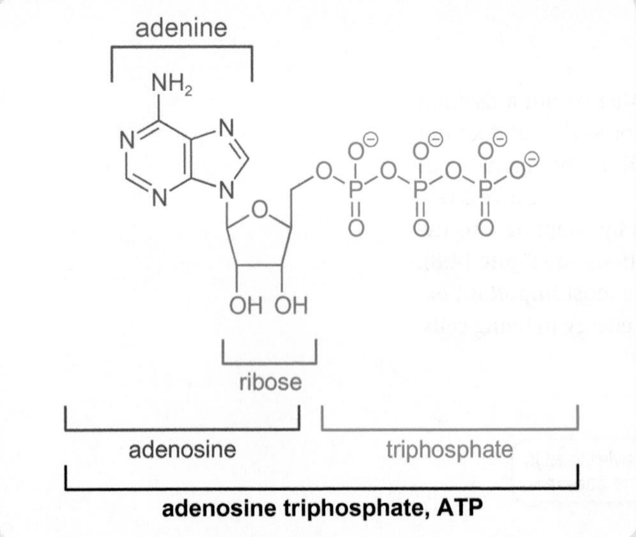
adenine
NH2
N
N
N
N
O
O
P
O
P
O
P
O
OH OH
ribose
adenosine
triphosphate
adenosine triphosphate, ATP

Figure 1 The structure of ATP (adenosine triphosphate). ADP (adenosine diphosphate) has one less phosphate group.

Questions

(a) The enthalpy change for the hydrolysis of ATP under biochemical standard conditions (*Reaction 2*) at 37 °C is $\Delta_r H^{\ominus\prime} = -20$ kJ mol^{-1}. Calculate the entropy change under these conditions and comment on the value.

(b) Calculate the overall Gibbs energy change for the following reaction at 37 °C

$$\text{glucose(aq)} + \text{ATP(aq)} \rightarrow \text{glucose-6-phosphate(aq)} + \text{ADP(aq)} + \text{H}^+\text{(aq)}$$

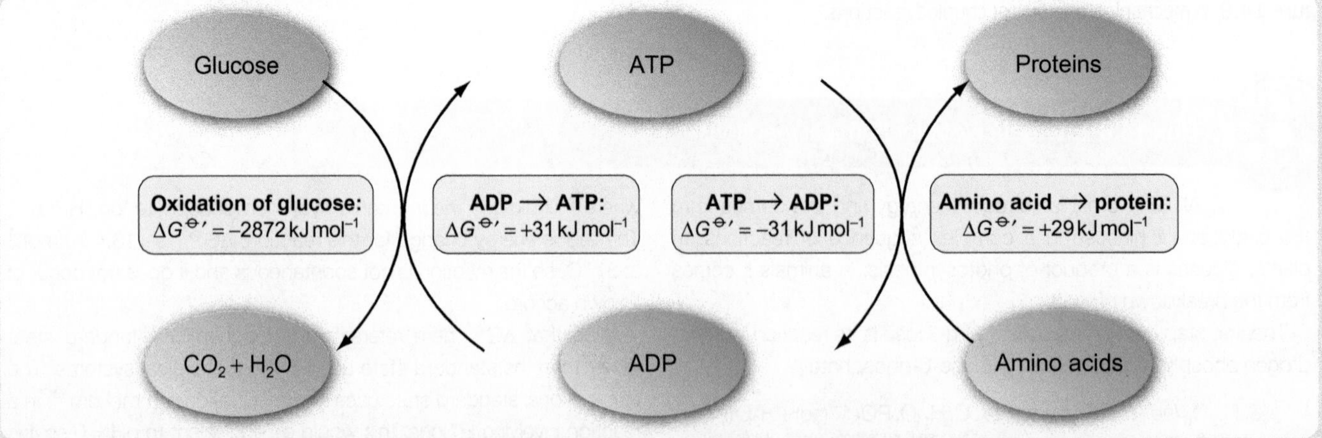
Glucose
ATP
Proteins
Oxidation of glucose:
ΔG⊖′ = −2872 kJ mol−1
ADP → ATP:
ΔG⊖′ = +31 kJ mol−1
ATP → ADP:
ΔG⊖′ = −31 kJ mol−1
Amino acid → protein:
ΔG⊖′ = +29 kJ mol−1
CO2 + H2O
ADP
Amino acids

Figure 2 The oxidation of glucose to CO_2 and H_2O drives the conversion of ADP to ATP. The reverse reaction ATP → ADP then drives other non-spontaneous reactions such as the synthesis of proteins.

Gibbs energy and maximum work

The Gibbs energy change shows whether a process is spontaneous or not, but it gives other useful information about a reaction as well. Its value is equivalent to the **maximum non-expansion work** that can be obtained from a change. This is the maximum amount of work that the change can do after any work due to change in volume has been accounted for.

For example, for the formation of water from hydrogen and oxygen gases, $\Delta_r G^{\ominus}_{298} = -237.1\,\text{kJ mol}^{-1}$. This means that, if the reaction takes place in circumstances that allow the work to be used, such as in a fuel cell, up to $237.1\,\text{kJ mol}^{-1}$ of work can be obtained.

Using Gibbs energy changes to determine the available work is important when considering:

- the electrical work that can be obtained from chemical reactions in batteries.
- the mechanical work that can be obtained from a chemical reaction, for example, when using fuels to drive vehicles.
- biochemical reactions, for example metabolizing glucose to power a muscle contraction, as described in Box 14.9.

The proof that ΔG represents the maximum non-expansion work that can be obtained from a reaction can be found in more advanced textbooks. The link to maximum work is why ΔG is sometimes called the 'free energy'. It is the energy that is 'free' to do useful work after expansion work is accounted for.

Electrochemical work in chemical systems is discussed in Section 16.4 (p.746).

Box 14.9 How much work can you get from glucose?

Glucose is oxidized to carbon dioxide and water in muscle cells—a process known as cellular respiration. The energy produced by cellular respiration is converted to mechanical energy when the muscles contract. For the oxidation of glucose at body temperature (37 °C) and under biochemical standard conditions (see Box 14.8)

$$C_6H_{12}O_6\,(aq) + 6O_2\,(g) \rightarrow 6CO_2\,(g) + 6H_2O\,(l)$$

$$\Delta_r H^{\ominus'} = -2816\,\text{kJ mol}^{-1}$$

$$\Delta_r S^{\ominus'} = +182\,\text{J K}^{-1}\text{mol}^{-1}$$

In order to find the maximum amount of work that this reaction can do, the change in Gibbs energy for the reaction is calculated.

Using Equation 14.16 (p.675)

$$\Delta G = \Delta H - T\Delta S \qquad (14.16)$$

$$\Delta_r G^{\ominus'}_{310} = -2816\,\text{kJ mol}^{-1} - [310\,\text{K} \times (+182 \times 10^{-3}\,\text{kJ K}^{-1}\,\text{mol}^{-1})]$$

$$= -2872\,\text{kJ mol}^{-1}$$

This means that the oxidation of 1 mol of glucose can be used to produce up to 2872 kJ of work, equivalent to 16 kJ for each gram of glucose.

The oxidation of glucose has a highly negative Gibbs energy change and is a spontaneous change. However, you can go out and buy glucose-containing sweets safe in the knowledge that they will not burst into flames. This is another example of a reaction being thermodynamically (very) favourable but kinetically extremely slow. In cellular respiration, the reaction occurs very rapidly in the presence of specific enzymes.

Note that the maximum work that can be done is greater than the energy released as heat. This is because the reaction is accompanied by a positive entropy change that arises because large glucose molecules are converted to several smaller molecules including a gas. The system therefore 'draws' energy ($T\Delta S$) from the surroundings (where the entropy decreases) and this energy is available for doing work.

Question

The Shard in London has an observation platform at a height of 244 m. Rather than taking the lift, a person weighing 65 kg climbs the stairs to the platform. How much glucose would the person need to consume to replace the energy used in climbing the stairs?

▲ At 309.6 m (1016 ft) to the top, the Shard, opened in 2012, is one of the tallest buildings in Europe.

Summary

- The Gibbs energy, G, is a state function.
- The change in Gibbs energy, ΔG, is defined by

$$\Delta G = \Delta H - T\Delta S$$

- The requirement for a change to be spontaneous is

$$(\Delta G)_{p,T} < 0$$

- The Gibbs energy change for a reaction, $\Delta_r G$, is given by

$$\Delta_r G = \Delta_r H - T\Delta_r S$$

- The Gibbs energy change for a reaction can be used to predict whether or not a reaction is spontaneous under a given set of conditions:
  - if $\Delta G < 0$, the reaction or process is spontaneous;
  - $\Delta G > 0$, the reaction or process is non-spontaneous;
  - if $\Delta G = 0$, the reaction or process is at equilibrium.
- The standard Gibbs energy change of formation at 298 K, $\Delta_f G^{\ominus}_{298}$ is the change in Gibbs energy when 1 mol of a compound is formed at 1 bar and 298 K from its elements in their standard states.
- Gibbs energy changes of formation can be used to calculate the change in Gibbs energy during a reaction using

$$\Delta_r G^{\ominus}_{298} = \sum v_i \Delta_f G^{\ominus}_{298}\,(\text{products}) - \sum v_i \Delta_f G^{\ominus}_{298}\,(\text{reactants})$$

- The value of $\Delta_r G^{\ominus}$ is negative for a spontaneous reaction at 1 bar pressure.
- A reaction may be spontaneous but may occur so slowly that it does not appear to take place.
- A non-spontaneous reaction may be made to take place if it is coupled to a reaction that has a larger, negative value of $\Delta_r G$.
- The value of the Gibbs energy change of reaction is equivalent to the maximum non-expansion work that can be obtained from a system.

? For practice questions on these topics, see questions 14 and 16–21 at the end of this chapter (pp.692–693).

14.6 Variation of Gibbs energy with conditions

The standard Gibbs energy change of reaction at 298 K is defined in terms of the pure reactants and products under standard conditions (1 bar) for the molar amounts stated in the chemical equation. For example, for the reaction:

$$N_2O_4\,(g) \rightarrow 2\,NO_2\,(g)$$

$\Delta_r G^{\ominus}_{298}$ is the change in Gibbs energy when 1.0 mol of N_2O_4 (g) completely reacts at 1 bar to form 2.0 mol of NO_2 (g) at 298 K.

Of course, most chemical reactions occur under non-standard conditions, at temperatures other than 298 K, and do not necessarily go to completion. To fully understand the role that Gibbs energy plays in chemistry, you need to know how G varies when conditions such as concentration, pressure, and temperature are changed. To describe the effects of concentration and pressure a term called the *thermodynamic activity* is used.

Thermodynamic activity

Look again at the reaction of N_2O_4 above. The Gibbs energies of N_2O_4 and NO_2 will change throughout the reaction because the concentrations of N_2O_4 and NO_2 change as the reaction proceeds.

It is convenient to have one equation to describe how the Gibbs energy depends on concentration whether the reaction occurs in solution or in the gas phase. To do this, a property called the **thermodynamic activity, *a*,** is used. The activity is sometimes called the 'effective concentration' of the component and is the ratio of the pressure or concentration to a standard value.

- For ideal gases, activities are defined in terms of pressures using a standard pressure of 1 bar.
- For substances in solution, activities are defined in terms of their molar concentrations using a standard concentration of $1\,\text{mol dm}^{-3}$.

Ideal gases extstyle: $$a = \frac{p}{p^{\ominus}} \qquad p^{\ominus} = 1\,\text{bar} \qquad (14.19a)$$

Ideal solutions: $$a = \frac{[A]}{[A]^{\ominus}} \qquad [A]^{\ominus} = 1\,\text{mol dm}^{-3} \qquad (14.19b)$$

Pure liquid or solid: $$a(\text{pure solid}) = a(\text{pure liquid}) = 1 \qquad (14.19c)$$

Equation 14.19c arises because pure liquids and solids have fixed composition and their activity is defined as 1.

For an ideal gas, *a* is given by the ratio of the gas pressure measured in bar to the standard pressure, 1 bar. It is, therefore, numerically equal to the pressure in bar. Similarly, for an ideal solution, *a* is numerically equal to the concentration in mol dm^{-3}. Note that, because the activity is a ratio of concentrations or partial pressures, it does not have units.

Using the concept of activity allows a single equation to be used to describe how Gibbs energy depends on pressure or concentration. The equation is

$$G = G^{\ominus} + RT\ln a \qquad (14.20)$$

where R is the gas constant ($R = 8.314\,\text{J K}^{-1}\,\text{mol}^{-1}$).

Ideal gases obey the ideal gas equation, $pV = nRT$; see Section 8.2 (p.349).

Ideal solutions are solutions that obey Raoult's law (Section 17.4, p.789). The more dilute a solution, the more ideally it behaves.

The use of thermodynamic activity simplifies the treatment of chemical equilibrium (Section 15.1, p.697).

The origin and derivation of Equation 14.20 can be found in more advanced textbooks.

Variation of *G* with pressure for an ideal gas

Substituting the expression for the activity of an ideal gas (Equation 14.19a) into Equation 14.20 gives an expression for G_p, the molar Gibbs energy at pressure p

$$G_p = G^{\ominus} + RT\ln\frac{p}{p^{\ominus}} \qquad (14.21)$$

where $p^{\ominus} = 1\,\text{bar}$. This is often written as $G_p = G^{\ominus} + RT\ln p$, but this is only the case if the pressure is measured in units of bar. Note that, if $p = 1\,\text{bar}$, $G_p = G^{\ominus}$.

Variation of *G* with concentration

Substituting the expression for the activity of a substance in solution (Equation 14.19b) into Equation 14.20 gives an expression for G_A, the molar Gibbs energy at a concentration, [A]

$$G_A = G_A^{\ominus} + RT\ln\frac{[A]}{[A]^{\ominus}} \qquad (14.22)$$

where $[A]^{\circ} = 1\,\text{mol dm}^{-3}$.

This is often written as

$$G_A = G_A^{\ominus} + RT\ln[A] \qquad (14.23)$$

where G_A is the Gibbs energy of the substance in solution at a concentration of [A] (in mol dm^{-3}). $G_A^{\ominus}$ is the value at a concentration of $1.0\,\text{mol dm}^{-3}$.

Variation of *G* with temperature

Knowing how G and $\Delta_r G^\ominus$ vary with temperature is important in predicting the effect of temperature on the spontaneity of chemical reactions—for example, in obtaining metals from ores (see Box 14.10).

The molar Gibbs energy, G_m, of a substance is defined as

$$G_m = H_m - TS_m \tag{14.24}$$

The subscripts in Equation 14.24 indicate that molar quantities are being considered.

Assuming that H_m and S_m do not vary with temperature, Equation 14.24 has the form of a straight line relationship so that a plot of G_m against T is a straight line with a gradient of $-S_m$, as shown in Figure 14.9. This can be expressed as a differential by differentiating Equation 14.24 with respect to temperature. This gives

$y = mx + c$

$G_m = -S_m T + H_m$

The equation of a straight line is discussed in Maths Toolkit MT4 (p.1318).

$$\left(\frac{dG_m}{dT}\right)_p = -S_m \tag{14.25}$$

Differentiating an expression is equivalent to finding the instantaneous gradient of a graph and is discussed in Maths Toolkit MT6 (p.1324).

Since in real systems H_m and S_m do depend slightly on temperature, the plots in Figure 14.9 will not be perfectly linear except over small temperature intervals.

Substances with a large molar entropy, S_m, such as gases, show the greatest variation of G_m with temperature (the plot in Figure 14.9 for gases has the steepest slope). For solids and liquids G_m does not vary greatly with temperature, since their entropy is low.

Plots like the one in Figure 14.9 are important in determining which phase (solid, liquid, or gas) will be stable under a particular set of conditions (Section 17.2, p.775).

Equation 14.25 refers to the Gibbs energy of a single substance but a similar relation holds for the Gibbs energy change in chemical reactions.

$$\left(\frac{d\Delta_r G}{dT}\right)_p = -\Delta_r S \tag{14.26}$$

This section assumes that the temperature is changing while the pressure remains constant: hence the subscript 'p' on the differential in Equations 14.25 and 14.26.

When considering reactions at temperatures other than 298 K, it is usually more straightforward to calculate $\Delta_r H^\ominus$ and $\Delta_r S^\ominus$ at the appropriate temperature and then to use Equation 14.16 to calculate $\Delta_r G^\ominus$ (see, for example, Worked example 14.10, p.681).

Looking at $\Delta_r S^\ominus$ as the slope of a plot of $\Delta_r G^\ominus$ versus temperature helps to interpret the Ellingham diagram in Figure 1 in Box 14.10.

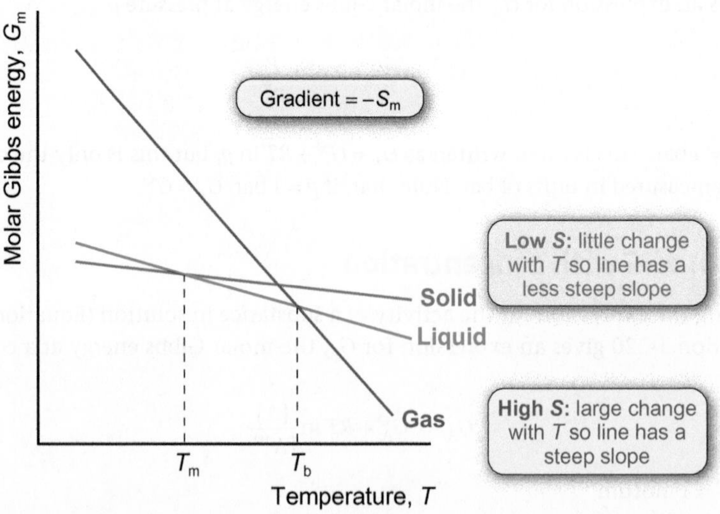
Molar Gibbs energy, G_m
Gradient = $-S_m$
Low S: little change with T so line has a less steep slope
Solid
Liquid
Gas
High S: large change with T so line has a steep slope
T_m
T_b
Temperature, T

Figure 14.9 Variation of G_m with temperature at constant pressure (1 bar). The gradient of each line is equal to $-S_m$. These graphs also show the phase that has the lowest Gibbs energy at a particular temperature. This determines which phase is stable at that temperature (see Chapter 17). The intersections of the solid and liquid lines, and of the liquid and gas lines, give the positions of the normal melting point, T_m, and normal boiling point, T_b.

Visit the Online Resource Centre to watch screencast 14.4 which explains the variation of Gibbs energy change with temperature.

Box 14.10 Obtaining metals from ores

Many metals are produced by reducing their oxides with carbon. Historically, the production of iron was the first large-scale industrial process of this type and the availability of large quantities of iron was a major factor in the industrial revolution of the eighteenth century. Deposits of iron ores largely exist as a mixture of iron oxides such as FeO and Fe_2O_3. These can be reduced by reaction with carbon—readily available as coal or charcoal obtained from controlled burning of wood.

▲ The early production of iron was spectacular—but wasteful of energy as shown by this painting of 'Coalbrookdale by night' by Philip de Loutherbourg. Modern production is rather more efficient.

For the production of a metal, M, there are two possible reduction reactions

$$\text{MO(s)} + \text{C(s)} \rightarrow \text{M(s)} + \text{CO(g)}$$

$$2\text{MO(s)} + \text{C(s)} \rightarrow 2\text{M(s)} + \text{CO}_2\text{(g)}$$

The metal will be produced if either of these reactions has a negative Gibbs energy change. (Many metals produce oxides with formulae other than MO. You can treat the reductions of these oxides in a similar way, although the stoichiometry is different.)

From Equation 14.18, the Gibbs energy change for a reaction is given by

$$\Delta_r G^{\ominus} = \sum \nu_i \Delta_f G^{\ominus}\text{(products)} - \sum \nu_i \Delta_f G^{\ominus}\text{(reactants)} \qquad (14.18)$$

$\Delta_f G^{\ominus}$ for the elements C (s) and M (s) is zero. Consider $\Delta_f G^{\ominus}$ for the compounds MO (s), CO (g), and CO_2 (g)

$$\text{M(s)} + \tfrac{1}{2}\text{O}_2\text{(g)} \rightarrow \text{MO(s)} \qquad \Delta_r G^{\ominus} = \Delta_f G^{\ominus}(\text{MO(s)}) \qquad \text{(Reaction 1)}$$

$$\text{C(s)} + \tfrac{1}{2}\text{O}_2\text{(g)} \rightarrow \text{CO(g)} \qquad \Delta_r G^{\ominus} = \Delta_f G^{\ominus}(\text{CO(g)}) \qquad \text{(Reaction 2)}$$

$$\text{C(s)} + \text{O}_2\text{(g)} \rightarrow \text{CO}_2\text{(g)} \qquad \Delta_r G^{\ominus} = \Delta_f G^{\ominus}(\text{CO}_2\text{(g)}) \qquad \text{(Reaction 3)}$$

If $\Delta_f G^{\ominus}$ for CO (g) or CO_2 (g) is more negative than that for MO (s), $\Delta_r G^{\ominus}$ for the reaction of MO (s) with C (s) will be negative. The reaction will be spontaneous and the carbon will remove the oxygen from the metal oxide.

The Gibbs energy change for each reaction is given by

$$\Delta_r G^{\ominus} = \Delta_r H^{\ominus} - T\Delta_r S^{\ominus} \qquad (14.16)$$

so you can predict the effect of temperature on the values of $\Delta_r G^{\ominus}$ by considering the entropy changes for Reactions 1–3. For each reaction, the value of $\Delta_r H^{\ominus}$ is negative. This means that, if the value of $\Delta_r S^{\ominus}$ is positive, then $\Delta_r G^{\ominus}$ will become more negative as T increases. Conversely, if the value of $\Delta_r S^{\ominus}$ is negative then $\Delta_r G^{\ominus}$ becomes less negative as T increases. Reactions with a small entropy change will show only a small dependence of $\Delta_r G^{\ominus}$ on temperature.

Plots of $\Delta_f G^{\ominus}$ for MO (s), CO (g), and CO_2 (g) against temperature, known as an **Ellingham diagram**, provide a useful way of deciding whether the reduction of a metal oxide will be a spontaneous reaction. A simplified example is shown in Figure 1.

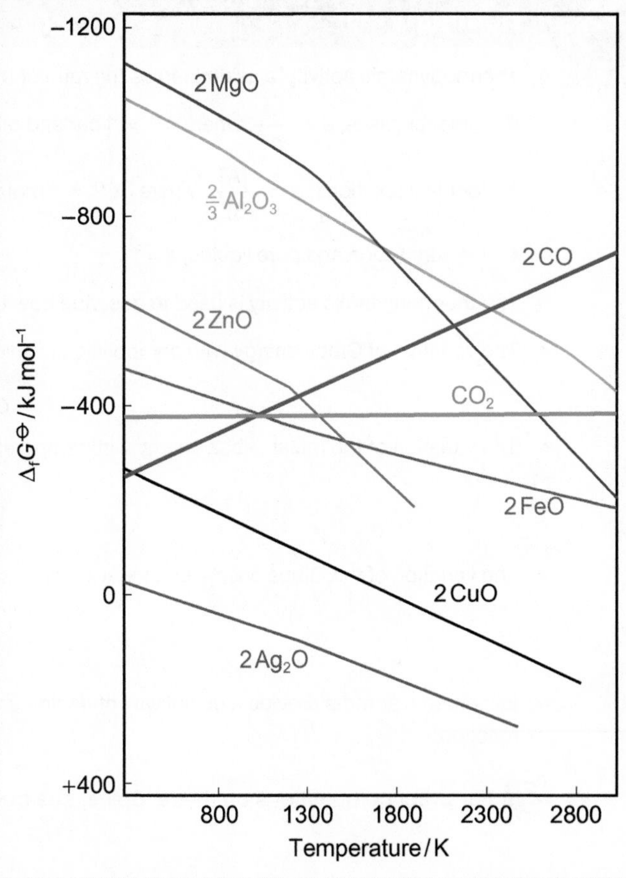

−1200
−800
−400
0
+400
2 MgO
2/3 Al2O3
2 CO
2 ZnO
CO2
2 FeO
2 CuO
2 Ag2O
800
1300
1800
2300
2800
Temperature / K
ΔfG⊖ / kJ mol−1

▲ **Figure 1** An Ellingham diagram. The Gibbs energy changes of formation of metal oxides and carbon oxides are plotted as a function of temperature. Note that $\Delta_f G^{\ominus}$ becomes *more* negative as you go up the vertical axis.

→

In an Ellingham diagram, $\Delta_f G^\ominus$ is plotted such that it becomes *more negative* as you go *up* the *y*-axis. Reactions at the top are more likely to be spontaneous than those lower in the diagram. Thus, any metal oxide for which the line is *below* the lines for CO(g) or CO_2(g) will be reduced by carbon.

For example, the line for CuO lies below that for CO_2 at all temperatures so that CuO can be reduced by carbon across the whole temperature range. Conversely, the line for MgO lies above CO(g) and CO_2(g) up to 2000 K, so MgO can only be reduced by carbon at temperatures above 2000 K—too high to be used in practice. FeO cannot be reduced at low temperatures but the reaction becomes feasible above around 1000 K. (see Worked Example 14.10).

Questions

(a) Predict the signs of the entropy changes for Reactions 1–3.

(b) Explain why the line for CO slopes upwards, that for CO_2 is almost horizontal, while those for the metal oxides slope downward.

(c) Use the Ellingham diagram to estimate the lowest temperature at which zinc oxide can be reduced to zinc metal by carbon.

(d) Suggest why aluminium is produced using electrolysis (see Section 16.5, p.755 and Section 27.2, p.1209) rather than by reduction of the oxide.

The production of metals from metal ores provides good examples of coupled reactions. $\Delta_r G^\ominus$ for the oxidation of carbon has a large negative value and this reaction is coupled to the unfavourable reduction of the metal oxide to give an overall reaction that is thermodynamically favourable. (This is an analogous situation to the reduction of SiO_2 in Box 14.7, p.682.)

Summary

- Thermodynamic activity, *a*, is defined as the ratio of the concentration or pressure to a standard value:
  - for ideal gases, $a = \frac{p}{p^\ominus}$, where $p^\ominus$ = 1 bar and *p* is the partial pressure in bar;
  - for ideal solutions, $a = \frac{[A]}{[A]^\ominus}$, where $[A]^\ominus$ = 1 mol dm^{-3} and [A] is the concentration in mol dm^{-3};
  - for pure solids and pure liquids, $a = 1$.
- The thermodynamic activity is used to describe how the Gibbs energy depends on the composition of a mixture.
- The variation of Gibbs energy with composition is given by

$$G = G^\ominus + RT\ln a.$$

- The variation of the molar Gibbs energy with temperature is given by

$$\left(\frac{dG_m}{dT}\right)_p = -S_m$$

- The variation of the Gibbs energy change with temperature for a reaction is given by

$$\left(\frac{d\Delta_r G}{dT}\right)_p = -\Delta_r S$$

- Ellingham diagrams provide a useful way of deciding whether the reduction of a metal oxide will be a spontaneous reaction.

? For practice questions on these topics, see questions 15, 22, and 23 at the end of this chapter (pp.692–693).

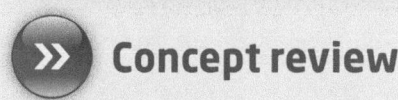

Concept review

By the end of this chapter, you should be able to do the following.

- Describe, and give examples of, spontaneous changes.
- Use changes in entropy and Gibbs energy to assess the spontaneity of processes or reactions.
- Calculate the temperature dependence of entropy using heat capacities.
- Account for the sign of entropy changes in terms of the disorder of a system.
- Calculate absolute entropies of compounds and describe the origin of residual entropies.
- Calculate entropy changes of reaction from absolute entropies.
- Calculate Gibbs energy changes from enthalpy and entropy changes.
- Describe how coupling of reactions allows non-spontaneous reactions to take place.
- Describe how Gibbs energy changes with temperature, pressure, and concentration.
- Define the thermodynamic activity of a substance.

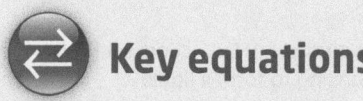

Key equations

Entropy and number of ways of organizing a system	$S = k_B \ln W$	(14.2)
Total entropy change: the Second Law of thermodynamics	$\Delta S(\text{total}) = \Delta S(\text{system}) + \Delta S(\text{surroundings})$	(14.3)
Entropy change in the system	$\Delta S(\text{system}) = \dfrac{q_{rev}}{T}$	(14.4)
Change of entropy with temperature	$\Delta S_{T_f} = S_{T_i} + C_p \ln \dfrac{T_f}{T_i}$	(14.7)
Entropy change of reaction	$\Delta_r S^{\ominus}_{298} = \sum \nu_i S^{\ominus}_{298}(\text{products}) - \sum \nu_i S^{\ominus}_{298}(\text{reactants})$ $\sum \nu_i S^{\ominus}_{298}(\text{products})$: The sum of the standard entropies of all of the *products* $\sum \nu_i S^{\ominus}_{298}(\text{reactants})$: The sum of the standard entropies of all of the *reactants*	(14.11)
Standard entropy change at temperature T	$\Delta_r S^{\ominus}_T = \Delta_r S^{\ominus}_{298} + \Delta C_p \ln \dfrac{T}{298\,\text{K}}$ where $\Delta C_p = \sum \nu_i C_p(\text{products}) - \sum \nu_i C_p(\text{reactants})$	(14.12)
Entropy change in the surroundings	$\Delta_r S(\text{surroundings}) = \dfrac{-\Delta_r H}{T}$	(14.14)
Gibbs energy change	$\Delta G = \Delta H - T\Delta S$	(14.16)
Gibbs energy and spontaneity	For a spontaneous change $(\Delta G)_{p,T} < 0$	(14.17)
Gibbs energy change of reaction	$\Delta_r G^{\ominus}_{298} = \sum \nu_i \Delta_f G^{\ominus}_{298}(\text{products}) - \sum \nu_i \Delta_f G^{\ominus}_{298}(\text{reactants})$	(14.18)
Thermodynamic activity	Ideal gases: $a = \dfrac{p}{p^{\ominus}}$ $\quad p^{\ominus} = 1\,\text{bar}$	(14.19a)
	Ideal solutions: $a = \dfrac{[A]}{[A]^{\ominus}}$ $\quad [A]^{\ominus} = 1\,\text{mol dm}^{-3}$	(14.19b)
	Pure liquid or solid: $a(\text{pure solid}) = a(\text{pure liquid}) = 1$	(14.19c)
Variation of G with activity	$G = G^{\ominus} + RT \ln a$	(14.20)
Variation of G with temperature	$\left(\dfrac{dG_m}{dT}\right)_p = -S_m$	(14.25)

? Questions

More challenging questions are indicated by an asterisk *.
Some problems require the use of data from Appendix 7 (p.1352).

1 (a) Explain the meaning of the term 'spontaneous' as applied to a chemical or physical change.

(b) Explain why some reactions which appear to be spontaneous from thermodynamic data do not in fact take place. Illustrate your answer by referring to the following changes:

(i) $H_2O\,(s) \rightarrow H_2O\,(l)$

(ii) Oxidation of aluminium in air. (Section 14.1)

2 Predict the sign of the change in entropy *in the system* for the following processes (Section 14.1)

(a) Steam condensing on a cold window

(b) A cloud forming in the atmosphere

(c) Inflating a bicycle tyre with air

(d) Dissolving sugar in hot coffee

(e) $PCl_3\,(g) + Cl_2\,(g) \rightarrow PCl_5\,(g)$

(f) $H_2O\,(g) + CaSO_4\,(s) \rightarrow CaSO_4{\cdot}H_2O\,(s)$

(g) $SO_3\,(g) + H_2O\,(l) \rightarrow H_2SO_4\,(aq)$

(h) $2\,KCl\,(s) + H_2SO_4\,(l) \rightarrow K_2SO_4\,(s) + 2\,HCl\,(g)$

(i) $C_2H_4\,(g) + H_2O\,(g) \rightarrow C_2H_5OH\,(l)$

3 An apparatus consists of two bulbs of the same volume connected by a tap. Initially, the tap is closed with one bulb containing nitrogen gas and the other oxygen gas. Both bulbs are at the same temperature and pressure. (Section 14.2)

(a) What happens when the tap is opened? What will be the equilibrium state of the system?

(b) What are the signs of ΔH, ΔS, and ΔG for the process in (a)?

(c) Is this consistent with the Second Law of thermodynamics?

4 Estimate the change in entropy when 1.00 mol of argon is heated from 300 K to 1200 K. What assumptions have you made and how could you make your estimate more accurate? (Section 14.2)

(The heat capacity, C_p, of argon gas is $20.8\,J\,K^{-1}\,mol^{-1}$.)

5 For each of the following reactions, suggest whether the entropy change in the system would be: (i) near zero; (ii) positive; or (iii) negative. Explain your answers. (Section 14.4)

(a) $N_2\,(g) + 3\,H_2\,(g) \rightarrow 2\,NH_3\,(g)$

(b) $Zn\,(s) + Cu^{2+}\,(aq) \rightarrow Zn^{2+}\,(aq) + Cu\,(s)$

(c) $3\,Mg\,(s) + 2\,Fe^{3+}\,(aq) \rightarrow 3\,Mg^{2+}\,(aq) + 2\,Fe\,(s)$

(d) $CH_4\,(g) + 2\,O_2\,(g) \rightarrow CO_2\,(g) + 2\,H_2O\,(g)$

(e) C (diamond) → C (graphite)

6 Dissolving solid potassium iodide in water results in a lowering of the temperature. Explain why this endothermic process can be spontaneous. (Section 14.2)

7 Calculate the entropy change when 1 mol of water is heated from 250 K to 300 K.

The enthalpy change of fusion for water is at 0 °C is $+6.01\,kJ\,mol^{-1}$ and the heat capacities are $C_p = 75.3\,J\,K^{-1}\,mol^{-1}$ and $37.2\,J\,K^{-1}\,mol^{-1}$ for water and ice, respectively. (Section 14.3)

8 100.0 g of water at 30 °C were placed in a refrigerator at 4 °C. When the water cools down, what is the entropy change of (a) the water and (b) the refrigerator? What is the overall entropy change? Comment on the results. (Section 14.3)

9 Given that the normal melting and boiling points of CO are 74 K and 82 K, respectively, sketch a plot to show how the entropy of CO varies with temperature between 0 K and 273 K. (Section 14.3)

10 What are the signs of the values of ΔH, ΔS, and ΔG for the process of a vapour condensing to form a liquid? (Section 14.4)

11 An ice cube of mass 18 g is added to a large glass of water just above 0 °C. Calculate the change of entropy for the ice and for the water (without the ice). (Section 14.4) (The enthalpy change of fusion for water is $+\,6.01\,kJ\,mol^{-1}$.)

12 Calculate the change in entropy when 100 g of water at 90 °C are added to an insulated flask containing 100 g of water at 10 °C. (Section 14.2)

13 Calculate the standard entropy change at 298 K for the reactions (Section 14.4):

(a) $N_2\,(g) + 3\,H_2\,(g) \rightarrow 2\,NH_3\,(g)$

(b) $Hg\,(l) + Cl_2\,(g) \rightarrow HgCl_2\,(s)$

(c) $C_6H_{12}O_6\,(s) + 6\,O_2\,(g) \rightarrow 6\,CO_2\,(g) + 6\,H_2O\,(g)$

14 Calculate the entropy changes for the system and surroundings when 1.00 mol of NaCl melts at 1100 K. Calculate $\Delta_{fus}G$ and estimate the melting point of NaCl. (Section 14.5)

(For the melting of sodium chloride (NaCl), $\Delta_{fus}H = +30.2\,kJ\,mol^{-1}$ and $\Delta_{fus}S = +28.1\,J\,K^{-1}\,mol^{-1}$.)

15 Calcium carbonate ($CaCO_3$) decomposes to form CaO and CO_2 with $\Delta_r H_{298} = +178\,kJ\,mol^{-1}$ and $\Delta_r S_{298} = +161\,J\,K^{-1}\,mol^{-1}$. Estimate the temperature at which the decomposition becomes spontaneous. (Section 14.6)

16 You expend about 100 kJ a day keeping your heart beating. What is the minimum mass of glucose you must oxidize per day in order to produce this much energy? (Section 14.5)

$$C_6H_{12}O_6\,(s) + 6O_2\,(g) \rightarrow 6CO_2\,(g) + 6H_2O\,(l)$$
$$\Delta G^{\ominus} = -2872\,kJ\,mol^{-1}$$

17 For the reaction

$$CH_4\,(g) + 2O_2\,(g) \rightarrow CO_2\,(g) + 2H_2O\,(l)$$
$$\Delta_c H^{\ominus}_{298} = -890\,kJ\,mol^{-1}.$$

(a) Calculate $\Delta_c G^{\ominus}_{298}$ for the combustion of methane.

(b) How much of the heat produced by burning 1.00 mol of methane cannot be used to do work?

(c) A heat engine uses methane as a fuel. What height could a 1.00 kg mass be raised to by burning 1.00 dm^3 of methane? (Section 14.5)

18 Ethanoic acid can be produced by a number of methods, including:

(a) the reaction of methanol with carbon monoxide

$$CH_3OH\,(g) + CO\,(g) \rightarrow CH_3CO_2H\,(l)$$

(b) the oxidation of ethanol with oxygen gas

$$CH_3CH_2OH\,(l) + O_2\,(g) \rightarrow CH_3CO_2H\,(l) + H_2O\,(l)$$

(c) the reaction of carbon dioxide with methane

$$CO_2\,(g) + CH_4\,(g) \rightarrow CH_3CO_2H\,(l)$$

Calculate the standard Gibbs energy change for each of these reactions at (i) 298 K and (ii) 773 K. (Section 14.5)

19 The reaction of methanol (CH_3OH (l)) with oxygen can be used in a fuel cell. Calculate the enthalpy change for the reaction at 298 K and the maximum work that can be produced by the oxidation of 1.00 mol of methanol. (Section 14.5)

20 Calculate the normal boiling point of ethanol given that $\Delta_{vap}H = +42.6\,kJ\,mol^{-1}$ and $\Delta_{vap}S = +122.0\,J\,K^{-1}\,mol^{-1}$. (Section 14.5)

21* 'Synthesis gas' is a mixture of hydrogen and carbon monoxide, prepared by reacting water or steam with a source of carbon such as coal. The reaction can be represented as:

$$C\,(s) + H_2O\,(g) \rightarrow CO\,(g) + H_2\,(g)$$

Estimate the temperature at which the reaction becomes thermodynamically spontaneous. (Section 14.5)

22* Use the mean heat capacities and standard entropies listed in Appendix 7 to estimate the Gibbs energy change for the decomposition of water vapour into hydrogen and oxygen at 2000 °C and 1 bar pressure. (Section 14.6)

(The standard enthalpy change of formation of water vapour is $-241.8\,kJ\,mol^{-1}$.)

23* The standard Gibbs energy for a reaction is $-332.9\,kJ\,mol^{-1}$ at 298 K and $-339.5\,kJ\,mol^{-1}$ at 500 K. Estimate the standard entropy change for the reaction. (Section 14.6)

24* Find the changes in enthalpy, entropy, and Gibbs energy for the freezing of water at −10 °C at 1 bar pressure. (The enthalpy change of fusion for water at 0 °C is $+6.01\,kJ\,mol^{-1}$ and the heat capacities are $C_p = 75.3\,J\,K^{-1}\,mol^{-1}$ and $37.2\,J\,K^{-1}\,mol^{-1}$ for water and ice, respectively.) (Several sections)

25 Use data from Appendix 7 to consider the following reaction: $2\,NO\,(g) + O_2\,(g) \rightarrow 2\,NO_2\,(g)$

(a) Calculate the changes in enthalpy, entropy and Gibbs energy at 298 K.

(b) Is the reaction spontaneous at 25 °C and 1 bar? Explain the signs of $\Delta_r H$ and $\Delta_r S$.

26 Impure nickel metal is purified using the Mond process where it is first reacted at 80 °C with carbon monoxide to form $Ni(CO)_4$ (g)

$$Ni\,(s) + 4\,CO\,(g) \rightarrow Ni(CO)_4\,(g)$$

followed by heating to 200 °C when the reverse reaction occurs. Use the following thermodynamic data to show that this approach is thermodynamically feasible.

	Ni (s)	$Ni(CO)_4$ (g)	CO (g)
$\Delta_f G^{\ominus}_{298}/kJ\,mol^{-1}$:	0	−601.6	−110.5
$S^{\ominus}_{298}/J\,K^{-1}\,mol^{-1}$:	29.9	415.5	197.6

27 Determine the standard Gibbs energy change for the acid–base neutralization reaction

$$NaOH\,(aq) + HCl\,(aq) \rightarrow NaCl\,(aq) + H_2O\,(l)$$

$\Delta_f G^{\ominus}_{298}/kJ\,mol^{-1}$: OH^- (aq) −157.3; H^+ (aq) 0; H_2O (l) −237.1

28 Two chemists were discussing the possibility of using the waste methane from an oil field to produce useful chemicals. The first suggested it could be reacted with atmospheric CO_2 (g) to produce ethanoic acid.

$$CH_4\,(g) + CO_2\,(g) \rightarrow CH_3COOH\,(l)$$

The other chemist suggested that the methane should be converted to ethanol by reacting with water.

$$2\,CH_4\,(g) + H_2O\,(l) \rightarrow C_2H_5OH\,(l) + 2\,H_2\,(g)$$

Use Gibbs energy of formation data in Appendix 7 to examine the thermodynamic feasibility of these reactions around room temperature.

29 Nitrogen and oxygen have very similar electronegativities. Estimate the entropy of NO at 0 K (the 'zero-point entropy').

15

Chemical equilibrium

▼ Corals are under major threat from rising acidity in the oceans. Their structures largely consist of aragonite, a more soluble form of $CaCO_3$(s), so they may be particularly hard hit. It has been suggested that their rate of growth may be reduced by up to a half.

This chapter builds on the following topics:

- Measurement, units, and nomenclature Section 1.2, p.6
- Atoms and the mole Section 1.3, p.12
- Chemical equations Section 1.4, p.21
- Concentrations of solutions Section 1.5, p.34
- Energy changes in chemical reactions Section 1.6, p.41
- States of matter and phase changes Section 1.7, p.47
- Chemical equilibrium: how far has a reaction gone? Section 1.9, p.56
- The strengths of acids and bases Section 7.2, p.308
- Buffer solutions Section 7.3, p.319
- Enthalpy and enthalpy changes Section 13.2, p.619
- Enthalpy changes in chemical reactions Section 13.3, p.622
- Entropy changes in chemical reactions Section 14.4, p.669
- Gibbs energy Section 14.5, p.675
- Variation of Gibbs energy with conditions Section 14.6, p.686

Equilibria in the oceans

Pure water is neither acidic nor basic—it has a neutral pH = 7. However, if you take a glass of pure water and leave it open to the atmosphere, the pH of the water gradually falls to around 5.7, where it remains. The change in pH is mainly due to the water absorbing carbon dioxide (CO_2) from the atmosphere. No further change in pH occurs when the system comes to equilibrium.

$$CO_2(g) \rightleftharpoons CO_2(aq) \quad (1)$$

CO_2 is a weak acid and reacts with water to form carbonic acid ($H_2CO_3(aq)$), which reacts further with water to form the hydrogencarbonate (HCO_3^-) and carbonate (CO_3^{2-}) anions.

$$CO_2(aq) + H_2O(l) \rightleftharpoons H_2CO_3(aq) \quad (2)$$

$$H_2CO_3(aq) + H_2O(l) \rightleftharpoons HCO_3^-(aq) + H_3O^+(aq) \quad (3)$$

$$HCO_3^-(aq) + H_2O(l) \rightleftharpoons CO_3^{2-}(aq) + H_3O^+(aq) \quad (4)$$

At a fixed temperature, each of these reactions is governed by an **equilibrium constant** (Sections 7.2, p.308, and 15.1, p.697) so that, when the system comes to equilibrium, the concentration of each of the species remains constant. If the pH changes, the changing concentration of H_3O^+ ions affects the three equilibria, so that the relative amounts of H_2CO_3, HCO_3^-, and CO_3^{2-} in the solution change, as shown in Figure 1.

This series of equilibria has potentially serious implications for anything living in the sea. The key factor in regulating atmospheric CO_2 concentration is the amount of CO_2 that dissolves in the oceans: it is estimated that more than fifty times as much CO_2 is dissolved in the sea than is present in the atmosphere. The present atmospheric concentration of CO_2 would be much higher than the current value of ~400 ppm if most of the CO_2 from human sources had not been absorbed into the sea. This additional absorption has produced a lowering in the pH of sea water by about 0.1 from its pre-industrial value of ~8.2. This might seem like a small change but pH is a logarithmic scale so this indicates an approximately 30% increase in H_3O^+ concentration. Does this 'acidification of the oceans' matter?

Over many millions of years, an equilibrium has built up in sea water between atmospheric CO_2, dissolved HCO_3^- and CO_3^{2-} anions, and the solid calcium carbonate ($CaCO_3$) found in rocks, such as chalk or limestone, and in the shells and skeletons of many marine animals. $CaCO_3$ is only sparingly soluble in water (see Box 15.1) so that only low concentrations of $CO_3^{2-}(aq)$ are present

$$CaCO_3(s) \rightleftharpoons Ca^{2+}(aq) + CO_3^{2-}(aq) \quad (5)$$

Figure 1 When CO_2 dissolves in water, it forms H_2CO_3 and both HCO_3^- and CO_3^{2-} ions. The mole fraction of each species depends on the pH. At low pH, most of the CO_2 exists as H_2CO_3; at high pH most exists as CO_3^{2-} anions.

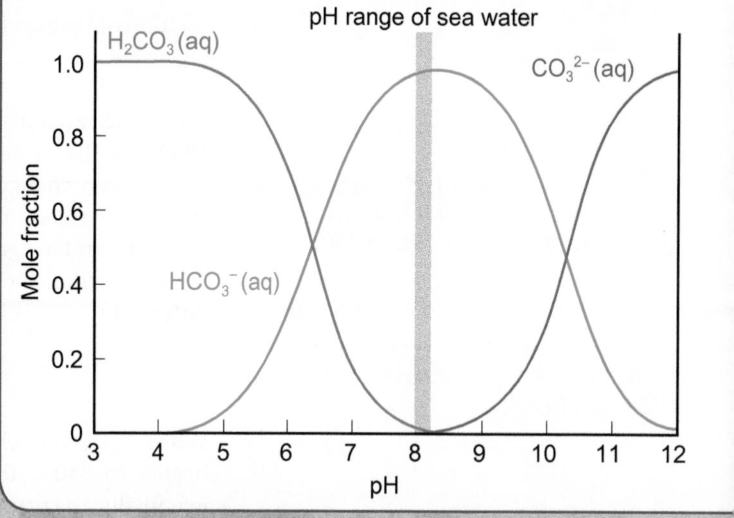

pH range of sea water
H₂CO₃(aq)
CO₃²⁻(aq)
HCO₃⁻(aq)
Mole fraction
1.0
0.8
0.6
0.4
0.2
0
3
4
5
6
7
8
9
10
11
12
pH

Figure 2 shows the equilibria in sea water. These equilibria serve to buffer the ocean against large changes in acidity. Increased absorption of CO_2 into the oceans drives reactions (2), (3), and (4) forward. However, the increased acidity causes $CaCO_3(s)$ to dissolve, adding more $CO_3^{2-}(aq)$ to reverse the change and hence maintain the equilibrium. (The action of buffers is discussed in Section 7.3, p.319.)

It was long assumed that this natural buffering capacity of the ocean would prevent significant changes in acidity—even if CO_2 concentrations increased massively. However, recent work has shown that this would be the case only if the increase happens over hundreds of thousands of years. The rate at which atmospheric CO_2 is currently rising is simply too fast for the natural system to respond. The most immediate effects are likely to be noticed in marine creatures. The fall in $CO_3^{2-}(aq)$ concentration as the pH falls (see Figure 1) means that organisms will not be able to build shells—and the shells of existing creatures will start to dissolve.

Figure 2 The calcium carbonate/carbon dioxide equilibria in the oceans act as a buffer against changing H_3O^+ (aq) concentration.

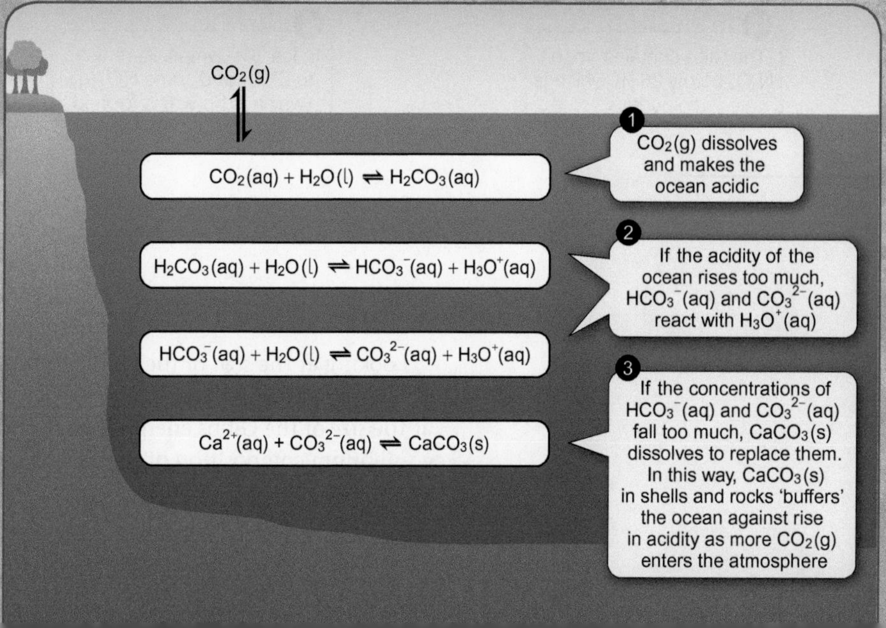

CO2(g)
CO2(aq) + H2O(l) ⇌ H2CO3(aq)
H2CO3(aq) + H2O(l) ⇌ HCO3⁻(aq) + H3O⁺(aq)
HCO3⁻(aq) + H2O(l) ⇌ CO3²⁻(aq) + H3O⁺(aq)
Ca²⁺(aq) + CO3²⁻(aq) ⇌ CaCO3(s)
1
CO2(g) dissolves and makes the ocean acidic
2
If the acidity of the ocean rises too much, HCO3⁻(aq) and CO3²⁻(aq) react with H3O⁺(aq)
3
If the concentrations of HCO3⁻(aq) and CO3²⁻(aq) fall too much, CaCO3(s) dissolves to replace them. In this way, CaCO3(s) in shells and rocks 'buffers' the ocean against rise in acidity as more CO2(g) enters the atmosphere

No chemical reaction proceeds only in one direction. One that very nearly does is the reaction of hydrogen and oxygen gases to form water at 298 K

$$2\,H_2(g) + O_2(g) \rightarrow 2\,H_2O(l)$$

When the reaction stops the amount of hydrogen remaining is immeasurably small: at 298 K, less than 1 molecule in about 1×10^{80} remains unconverted to water. The standard Gibbs energy change, $\Delta_r G^{\ominus}$, for this reaction is $-237.1\,\text{kJ mol}^{-1}$. The reaction is spontaneous and goes virtually to completion. At 298 K, the reverse reaction in which water forms hydrogen and oxygen does not occur to any measurable extent.

$\Delta_r G^{\ominus}$ is introduced in Section 14.5 (p.675), and the basic ideas of chemical equilibrium are described in Section 1.9 (p.56).

The reaction of dinitrogen tetroxide (N_2O_4) to form nitrogen dioxide (NO_2) behaves differently

$$\underset{\text{colourless}}{N_2O_4(g)} \rightleftharpoons \underset{\text{brown}}{2\,NO_2(g)}$$

The effect of changes of temperature on the $N_2O_4(g)/NO_2(g)$ equilibrium is discussed in Section 1.9 (p.56).

When $N_2O_4(s)$ is placed in a sealed tube at low temperature and the tube is gradually heated to 350 K, the reaction mixture gradually turns brown as $NO_2(g)$ is formed. Eventually, the reaction comes to equilibrium with both $N_2O_4(g)$ and $NO_2(g)$ present. No further change in the concentrations of the two gases occurs however long the mixture is left at this temperature. $\Delta_r G^{\ominus}$ for this reaction is $-4.5\,\text{kJ mol}^{-1}$ at 350 K. The reaction is spontaneous at this temperature but it does not go to completion.

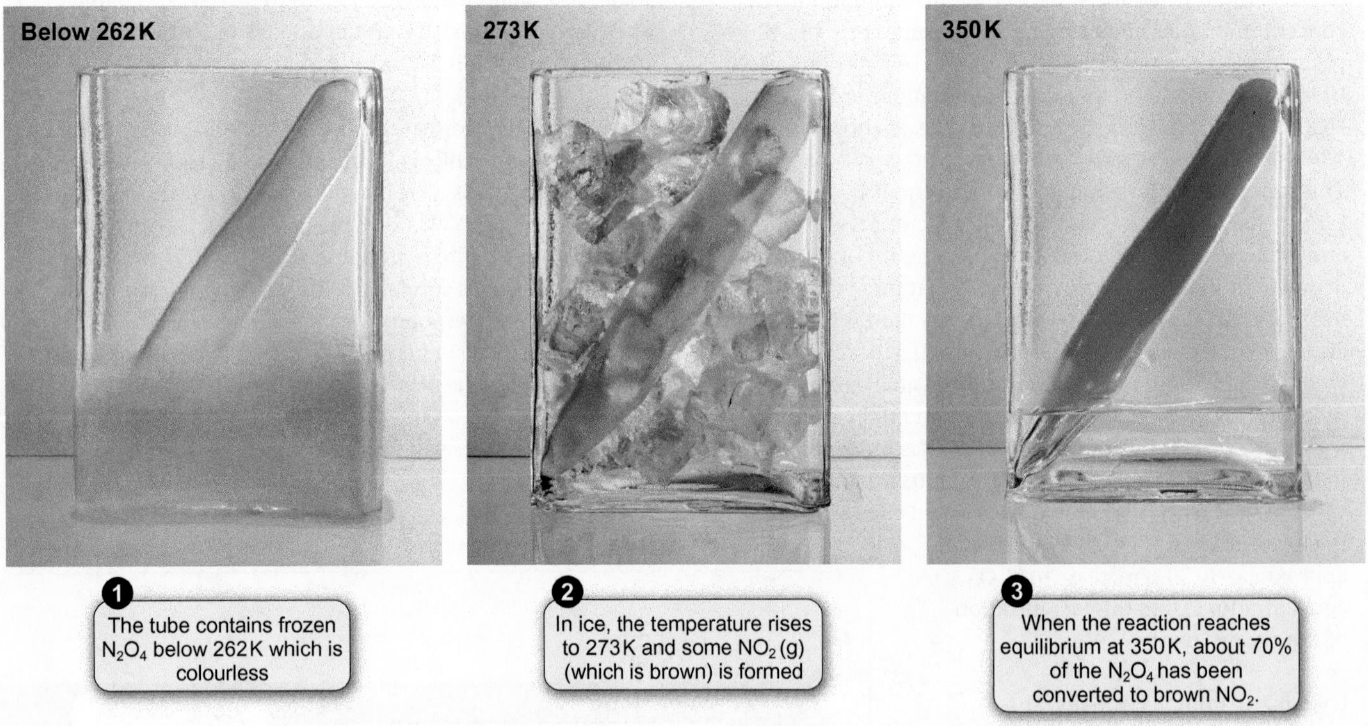
Below 262 K
273 K
350 K
1
The tube contains frozen N₂O₄ below 262 K which is colourless
2
In ice, the temperature rises to 273 K and some NO₂ (g) (which is brown) is formed
3
When the reaction reaches equilibrium at 350 K, about 70% of the N₂O₄ has been converted to brown NO₂.

N_2O_4 is colourless. As it reacts to form NO_2, the brown colour gets stronger until the reaction reaches equilibrium after which no further change in colour is observed.

Looking at the *sign* of the change in Gibbs energy that occurs during a reaction allows you to predict whether or not the reaction is spontaneous (Section 14.5, p.675). Looking at the *size* of the Gibbs energy changes helps you make *quantitative* predictions about the equilibrium composition of the reaction mixture. This is the subject of this chapter.

15.1 Gibbs energy and equilibrium

In a spontaneous reaction at constant temperature and pressure, $\Delta_r G < 0$ so the system moves to a lower Gibbs energy. Think about a straightforward model reaction where 1 mol of reactant, R, forms 1 mol of product, P

$$R \rightarrow P$$

$\Delta_r G$ is the difference in molar Gibbs energy between the product and the reactant. The Gibbs energy of the system changes as the reaction proceeds since the amounts of R and P change.

Figure 15.1 shows how G changes for a reaction that goes virtually to completion. At any stage during the reaction, the Gibbs energy of the system decreases as the reaction proceeds. The lowest Gibbs energy is achieved when the reactants are completely converted to products. This type of behaviour would be shown during the formation of water from hydrogen and oxygen.

Contrast this with a reaction that comes to equilibrium with observable amounts of reactants and products present, such as the $N_2O_4(g)/NO_2(g)$ reaction. In this case, the Gibbs energy plot looks like Figure 15.2. Starting with the reactants, the system can move to lower G by forming products and proceeding in the forward direction. However, once the system reaches point E, G has reached a minimum. If the reaction continued, the Gibbs energy would increase and this process would be non-spontaneous. So, the composition of the reaction mixture does not change beyond point E, which represents the equilibrium composition of the reaction mixture.

This approach demonstrates one of the fundamental statements of chemistry:

A system comes to equilibrium when it reaches its minimum Gibbs energy.

Figure 15.2 also shows that, if you start the reaction with 100% products, the system can lower its Gibbs energy by the reverse reaction taking place until point E is reached. The same equilibrium composition is obtained whatever the direction from which equilibrium is approached.

Reactions do not stop when equilibrium is reached. Although the overall proportion of reactants and products does not change, individual molecules may still react—but, for every molecule that undergoes the forward reaction, another molecule will undergo the reverse reaction. There is a **dynamic** equilibrium.

The criterion that a system comes to equilibrium when it reaches its minimum Gibbs energy applies to changes taking place at constant temperature and pressure.

Dynamic equilibrium is discussed in Section 1.9 (p.56). Another definition of equilibrium is where the forward and reverse reactions proceed at the same rate; see Section 9.6 (p.416).

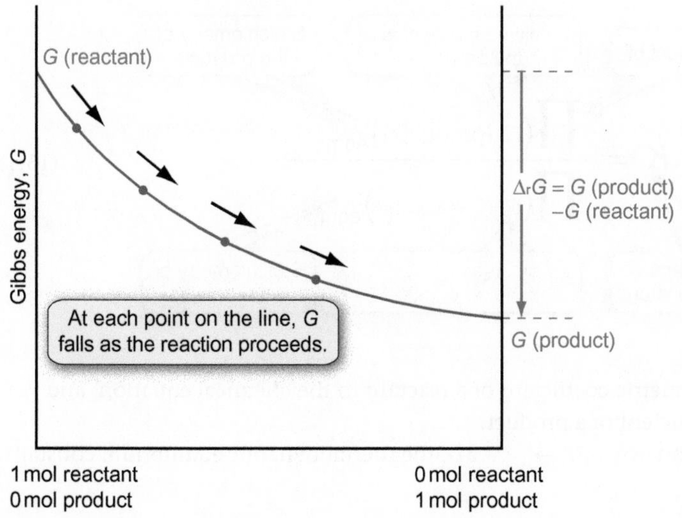
G (reactant)
$\Delta_r G = G$ (product) $-G$ (reactant)
G (product)
Gibbs energy, G
At each point on the line, G falls as the reaction proceeds.
1 mol reactant
0 mol product
0 mol reactant
1 mol product

Figure 15.1 A reaction that goes virtually to completion. At any point in the reaction, G falls as the reactants form products.

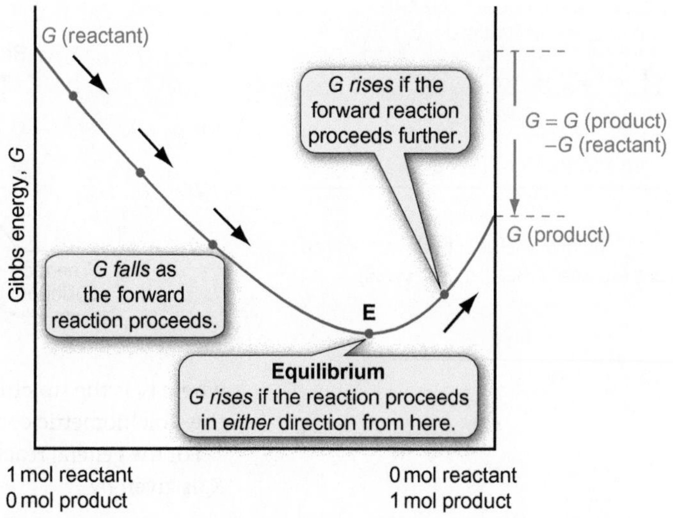
G (reactant)
G rises if the forward reaction proceeds further.
G = G (product) −G (reactant)
G (product)
Gibbs energy, G
G falls as the forward reaction proceeds.
E
Equilibrium
G rises if the reaction proceeds in either direction from here.
1 mol reactant
0 mol product
0 mol reactant
1 mol product

Figure 15.2 A reaction that comes to equilibrium with observable amounts of reactants and products present. As the reaction proceeds, G falls as the reactants form products until point E. Further reaction would lead to an increase in G. E is the position of equilibrium in the reaction.

Thermodynamic equilibrium constant, *K*

When an ant bites its victim, it sprays methanoic acid into the wound made by its jaws. The $H_3O^+(aq)$ formed by ionization of the acid causes an intense stinging pain. Methanoic acid is also known as formic acid after the Latin *formica*, meaning ant.

The composition of a reaction mixture when it comes to equilibrium is described by an **equilibrium constant, *K***. Different types of equilibrium constant can be defined. For gas phase reactions, an equilibrium constant, K_p, calculated in terms of the equilibrium partial pressures, p_{eqm}, of the components is usually used.

For example, for the reaction $N_2O_4(g) \rightleftharpoons 2NO_2(g)$

$$K_p = \frac{(p_{NO_2})^2_{eqm}}{(p_{N_2O_4})_{eqm}}$$

The units of K_p depend on the stoichiometry of the reaction. For this reaction, K_p has the same units as used for the pressure measurement.

For reactions in solution, K_c, calculated in terms of the molar concentrations is more useful. For example, methanoic acid (HCO_2H) reacts with water to form methanoate (HCO_2^-) ions and H_3O^+ ions. The reaction does not go to completion and comes to equilibrium

$$HCO_2H(aq) + H_2O(l) \rightleftharpoons HCO_2^-(aq) + H_3O^+(aq)$$

$$K_c = \frac{[H_3O^+(aq)]_{eqm}\,[HCO_2^-(aq)]_{eqm}}{[HCO_2H(aq)]_{eqm}\,[H_2O(l)]_{eqm}}$$

The equilibrium constants K_c and K_p are discussed in Section 1.9 (p.56). The use of acidity constants, K_a, is described in Section 7.2 (p.308).

In dilute solutions, the term $[H_2O(l)]$ is essentially constant and a new equilibrium constant, K_a, is defined that incorporates $[H_2O(l)]$.

$$K_a = \frac{[H_3O^+(aq)]_{eqm}\,[HCO_2^-(aq)]_{eqm}}{[HCO_2H(aq)]_{eqm}}$$

Remember that [A] represents the concentration of A measured in mol dm^{-3}.

K_a is called the acidity constant and is a measure of the strength of methanoic acid.

It is usually clear which type of equilibrium constant is appropriate for a reaction. However, these two types of equilibrium constant can be unified by defining a **thermodynamic equilibrium constant**, given the symbol ***K***.

In some books, thermodynamic equilibrium constants are referred to as **standard equilibrium constants**. An alternative symbol is $K^{\ominus}$.

K is defined as the ratio of the thermodynamic activities of the products at equilibrium to those of the reactants (taking into account the stoichiometry), as shown in Equation 15.1.

The symbol $\prod$ is shorthand for 'the product of' (in the same way as $\sum$ represents 'the sum of'). So,

$\prod_1^5(a_1) = a_1 \times a_2 \times a_3 \times a_4 \times a_5$ and

$\prod(2,3,4) = 2 \times 3 \times 4 = 24.$

See Maths Toolkit MT1 (p.1304).

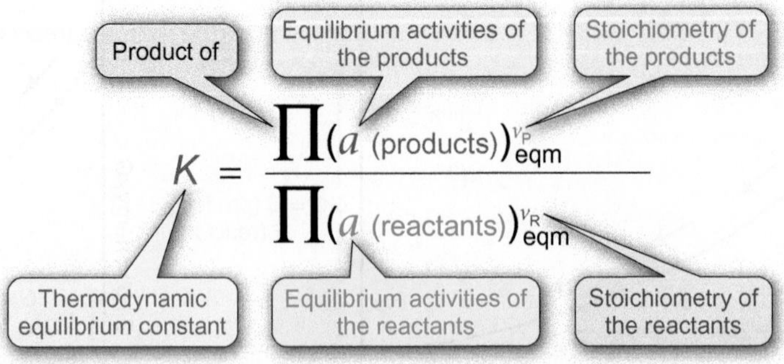
Product of
Equilibrium activities of the products
Stoichiometry of the products
$K = \dfrac{\prod(a\ (\text{products}))^{\nu_P}_{eqm}}{\prod(a\ (\text{reactants}))^{\nu_R}_{eqm}}$
Thermodynamic equilibrium constant
Equilibrium activities of the reactants
Stoichiometry of the reactants

(15.1)

Thermodynamic activities are defined and explained in Section 14.6 (p.686).

where ν_R is the stoichiometric coefficient of a reactant in the chemical equation, and ν_P is the stoichiometric coefficient of a product.

For the general reaction, $\alpha A + \beta B \rightarrow \gamma C + \delta D$, the thermodynamic equilibrium constant, *K*, is given by

$$K = \frac{(a_C)^{\gamma}_{eqm} \times (a_D)^{\delta}_{eqm}}{(a_A)^{\alpha}_{eqm} \times (a_B)^{\beta}_{eqm}}$$

where the subscript 'eqm' indicates that the activities are measured when the reaction has come to equilibrium. Examples of writing thermodynamic equilibrium constants are given in Worked example 15.1.

The activity of a gas is the ratio of its partial pressure measured in bar to the standard state pressure: $p/p^{\ominus} = p/1\,\text{bar}$. In solution, the activity of a species is given by the ratio of its molar concentration to the standard: $[A]/[A]^{\ominus} = [A]/1\,\text{mol dm}^{-3}$. This means that K has the same *numerical value* as K_p if partial pressures are measured in bar, or as K_c if concentrations are measured in mol dm^{-3}. However, it is important to remember that activities have no units, so that K also *has no units and is dimensionless*.

For example, in the reaction $N_2O_4(g) \rightleftharpoons 2NO_2(g)$

$$K = \frac{(a_{NO_2})^2_{eqm}}{(a_{N_2O_4})_{eqm}} = \frac{\left(\dfrac{p_{NO_2}}{p^{\ominus}}\right)^2_{eqm}}{\left(\dfrac{p_{N_2O_4}}{p^{\ominus}}\right)_{eqm}}$$

$$= 15.5 \text{ (no units) at } 373\,\text{K}$$

The expression 'p_{NO_2}/p' means 'the partial pressure of NO_2 measured in bar, divided by 1 bar'. This gives a quantity that is numerically equal to p_{NO_2} in bar but has no units.

So, K is numerically the same as K_p, providing the pressures are expressed in bars, though the two differ in terms of units.

For some types of reaction, such as heterogeneous equilibria involving gases and pure solids or liquids, using K, rather than K_p or K_c, makes the discussion of the equilibrium much more straightforward. For example, for the decomposition of silver carbonate

A **heterogeneous reaction** is one where the component reactants are in different phases, for example, a reaction mixture containing solids and gases.

$$Ag_2CO_3(s) \rightleftharpoons Ag_2O(s) + CO_2(g)$$

$$K = \frac{(a_{CO_2})_{eqm} \times (a_{Ag_2O})_{eqm}}{(a_{Ag_2CO_3})_{eqm}}$$

The activity of pure solids is 1 (see Section 14.6, p.686), so

$$K = \frac{(a_{CO_2})_{eqm} \times (1)}{(1)} = (a_{CO_2})_{eqm}$$

$$= \left(\frac{p_{CO_2}}{p^{\ominus}}\right)_{eqm} = \frac{(p_{CO_2})_{eqm}}{1\,\text{bar}}$$

The three types of equilibrium constant are summarized in Table 15.1. Note that care is needed in stating and using equilibrium constants. For example, the statement 'The equilibrium constant for the reaction of $N_2O_4(g)$ to form $NO_2(g)$ at 373 K is 15.5' is at best ambiguous and often meaningless! For the reaction

$$N_2O_4(g) \rightleftharpoons 2NO_2(g)$$

Table 15.1 Types of equilibrium constant

K_c	K_p	K
Equilibrium constant in terms of concentrations	Equilibrium constant in terms of pressures	Equilibrium constant in terms of activities
Units in terms of concentrations, normally mol dm^{-3}	Units in terms of pressure units, e.g. Pa or bar	No units (activities are dimensionless)
Used for reactions in solution (and sometimes for reactions in the gas phase)	Used for reactions in the gas phase	Used for *all* reactions

Worked example 15.1 Writing expressions for thermodynamic equilibrium constants

Write expressions for the thermodynamic equilibrium constants for the following reactions.

$$N_2(g) + 3H_2(g) \rightleftharpoons 2NH_3(g)$$

$$CaCO_3(s) \rightleftharpoons CaO(s) + CO_2(g)$$

$$CH_3CO_2H(aq) + H_2O(l) \rightleftharpoons CH_3CO_2^-(aq) + H_3O^+(aq)$$

Strategy

Use Equation 15.1 to define K in terms of the activities of the reactants and products, paying attention to the stoichiometry. Then express the activity in terms of pressure or concentration as appropriate.

Solution

(a) *Decide how the activity of each component is defined.*

The system contains only gases so, assuming ideal gas behaviour, the activity, a, is given by

$$a = p/p^{\ominus}$$

Express K in terms of the activities, including the stoichiometric coefficients.

$$K = \frac{(a_{NH_3})^2_{eqm}}{(a_{H_2})^3_{eqm} \times (a_{N_2})_{eqm}}$$

Substitute $a = p/p^{\ominus}$

$$K = \frac{(p_{NH_3}/p^{\ominus})^2_{eqm}}{(p_{H_2}/p^{\ominus})^3_{eqm} \times (p_{N_2}/p^{\ominus})_{eqm}} \quad \text{where } p^{\ominus} = 1\text{ bar}$$

(b) The system contains both gases and solids. Remember that the activity of a pure solid is 1.

$$K = \frac{(a_{CO_2})_{eqm} \times (a_{CaO})}{(a_{CaCO_3})}$$

$$= \frac{(p_{CO_2}/p^{\ominus})_{eqm} \times (1)}{(1)}$$

$$= (p_{CO_2}/p^{\ominus})_{eqm} \quad \text{where } p^{\ominus} = 1\text{ bar}$$

(c) The system contains substances in solution, so activity, a, is given by $[A]/[A]^{\ominus}$

$$K = \frac{(a_{CH_3CO_2^-})_{eqm} \times (a_{H^+})_{eqm}}{(a_{CH_3CO_2H})_{eqm} \times (a_{H_2O})_{eqm}}$$

Since $(a_{H_2O})_{eqm} = 1$ and $[A]^{\ominus} = 1\text{ mol dm}^{-3}$

$$K = \frac{([CH_3CO_2^-]/1\text{ mol dm}^{-3})_{eqm} \times ([H^+]/1\text{ mol dm}^{-3})_{eqm}}{([CH_3CO_2H]/1\text{ mol dm}^{-3})_{eqm}}$$

Question

Write expressions for the thermodynamic equilibrium constants for the following reactions.

(a) $4NH_3(g) + 5O_2(g) \rightleftharpoons 4NO(g) + 6H_2O(l)$

(b) $Ca(OH)_2(s) \rightleftharpoons Ca^{2+}(aq) + 2OH^-(aq)$

(c) $Ni(CO)_4(g) \rightleftharpoons Ni(s) + 4CO(g)$

In practice, using K is not very different from using K_p and K_c. K has the same numerical value as K_p when the partial pressures are in bar and as K_c when molar concentrations are used.

K is indeed 15.5. However, if the equation for the reaction is written as

$$\tfrac{1}{2}N_2O_4(g) \rightleftharpoons NO_2(g)$$

then $K = 3.94$. For the reverse reaction, $2NO_2(g) \rightleftharpoons N_2O_4(g)$, $K = 0.065$. An equilibrium constant *must* be associated with the equation for the reaction, and the temperature, being considered. Worked example 15.2 shows you how to calculate a value for a thermodynamic equilibrium constant and Box 15.1 shows how a thermodynamic equilibrium constant is used to describe the heterogeneous equilibrium established when a sparingly soluble salt dissolves in water.

Worked example 15.2 Calculating a thermodynamic equilibrium constant, K

In an ammonia synthesis (see Box 15.6, p.720), the reaction comes to equilibrium with partial pressures $p_{H_2} = 0.810$ bar, $p_{N_2} = 1.20$ bar, and $p_{NH_3} = 0.150$ bar. Calculate the thermodynamic equilibrium constant, K, under these conditions.

$$N_2(g) + 3H_2(g) \rightleftharpoons 2NH_3(g)$$

Strategy

You are given the partial pressures of each component. Use Equation 15.1 (p.698) to obtain an expression for K and then substitute in the values of the partial pressures.

→

➜ Solution

Use Equation 15.1 to write the expression for the thermodynamic equilibrium constant, K.

$$K = \frac{(p_{NH_3}/p^{\ominus})^2_{eqm}}{(p_{H_2}/p^{\ominus})^3_{eqm} \times (p_{N_2}/p^{\ominus})_{eqm}} \quad \text{where } p^{\ominus} = 1\text{ bar}$$

Substitute in the values of the partial pressures and solve to find K.

$$K = \frac{(0.150\text{ bar}/1\text{ bar})^2}{(0.810\text{ bar}/1\text{ bar})^3 \times (1.20\text{ bar}/1\text{ bar})} = \frac{(0.150)^2}{(0.810)^3 \times (1.20)}$$

$$= 0.035$$

Question

At 298 K, the thermodynamic equilibrium constant, K, for the reaction

$$N_2O_4(g) \rightleftharpoons 2NO_2(g)$$

is 0.150. An equilibrium mixture contains a partial pressure of 0.05 bar of $NO_2(g)$. Calculate the equilibrium partial pressure of $N_2O_4(g)$.

Box 15.1 Solubility equilibria

Calcium carbonate ($CaCO_3$) is the major constituent of chalk and limestone rocks. It dissolves only to a small degree in water—it is **sparingly soluble**—and the aqueous ions formed are in equilibrium with the undissolved solid

$$CaCO_3(s) \rightleftharpoons Ca^{2+}(aq) + CO_3^{2-}(aq)$$

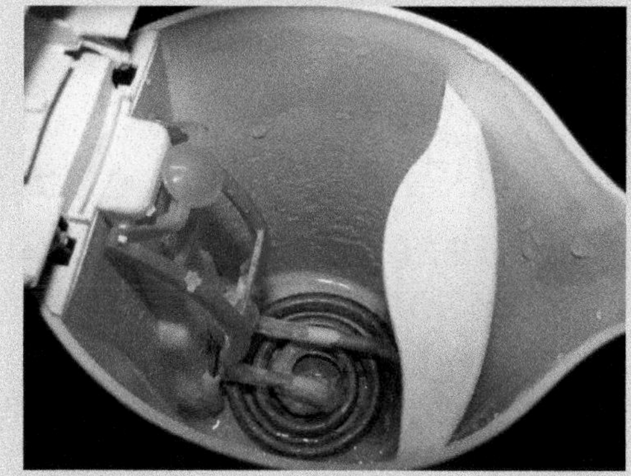

▲ Hard water contains relatively high concentrations of $Ca^{2+}(aq)$ and $HCO_3^-(aq)$ ions. When the water is heated, $HCO_3^-(aq)$ decomposes to form $CO_3^{2-}(aq)$ and the concentrations of $Ca^{2+}(aq)$ and $CO_3^{2-}(aq)$ become high enough to exceed the solubility product. Solid $CaCO_3$ precipitates as 'scale' in a pipe or a kettle.

The equilibrium constant for this process is given by

$$K = \frac{(a_{Ca^{2+}})_{eqm} \times (a_{CO_3^{2-}})_{eqm}}{(a_{CaCO_3})_{eqm}}$$

Since the activity of a pure solid is 1, this can be written as:

$$K = (a_{Ca^{2+}})_{eqm} \times (a_{CO_3^{2-}})_{eqm}$$
$$= ([Ca^{2+}(aq)]/1\text{ mol dm}^{-3})_{eqm} \times ([CO_3^{2-}(aq)]/1\text{ mol dm}^{-3})_{eqm}$$

This type of equilibrium constant is often called a **solubility product** or **solubility constant**. It is a thermodynamic equilibrium constant, though it is sometimes defined in terms of concentrations rather than activities, in which case it has units and is given the symbol $\boldsymbol{K_{sp}}$.

Solubility products are useful for calculating the solubility of sparingly soluble compounds. The molar solubility, s, measured in mol dm^{-3}, gives the number of moles of solute dissolved in 1 dm^3 of a saturated solution. Each 1 mol of $CaCO_3$ that dissolves produces 1 mol of $Ca^{2+}(aq)$ and 1 mol of $CO_3^{2-}(aq)$. So for 1 : 1 compounds, such as $CaCO_3$

$$s = [Ca^{2+}(aq)] = [CO_3^{2-}(aq)]$$

The expression for K can now be written in terms of the solubility, s

$$K = ([Ca^{2+}(aq)]/1\text{ mol dm}^{-3})_{eqm} \times ([CO_3^{2-}(aq)]/1\text{ mol dm}^{-3})_{eqm}$$
$$= (s/1\text{ mol dm}^{-3}) \times (s/1\text{ mol dm}^{-3})$$
$$= (s/1\text{ mol dm}^{-3})^2$$

K for $CaCO_3$ is 8.7×10^{-9} so the solubility, s, is 9.3×10^{-5} mol dm^{-3}.

K is an equilibrium constant and does not change at a fixed temperature. Therefore, addition of one of the ions will reduce the concentration of the other. For example, when solid $CaCO_3(s)$ is added to a 0.01 mol dm^{-3} solution of calcium nitrate (which completely dissolves in water to form $Ca^{2+}(aq)$ ions and $NO_3^-(aq)$ ions), the concentration of $Ca^{2+}(aq)$ already in solution is 0.01 mol dm^{-3}.

$$K = (a_{Ca^{2+}})_{eqm} \times (a_{CO_3^{2-}})_{eqm}$$
$$8.7\times10^{-9} = (0.01\text{ mol dm}^{-3}/1\text{ mol dm}^{-3}) \times (s/1\text{ mol dm}^{-3})$$
$$s = \frac{8.7\times10^{-9}}{0.01} \times (1\text{ mol dm}^{-3}) = 8.7\times10^{-7}\text{ mol dm}^{-3}$$

so the solubility of $CaCO_3$ is reduced to 8.7×10^{-7} mol dm^{-3}. (This is a slight approximation since it ignores the small additional ➜

→ concentration of Ca^{2+} that comes from the $CaCO_3$.) The reduction in solubility of a salt due to the presence in solution of one of the ions from another source is known as **the common ion effect**.

Questions

(a) Write expressions for the solubility products for MgF_2, Ag_2CO_3, $Al(OH)_3$, and Bi_2S_3 in terms of activities.

(b) Write the solubility products for MgF_2, Ag_2CO_3, $Al(OH)_3$, and Bi_2S_3 in terms of their molar solubilities, *s*.

(c) *K* for MgF_2 is 6.4×10^{-9} and *K* for $Al(OH)_3$ is 1.0×10^{-33}. Calculate the molar solubility of each compound in water.

(d) Calculate the molar solubility of: (i) MgF_2 in 0.1 mol dm^{-3} sodium fluoride solution; (ii) $Al(OH)_3$ in 0.02 mol dm^{-3} sodium hydroxide solution.

Summary

- Chemical systems come to equilibrium when they reach a state of minimum Gibbs energy.
- Chemical equilibrium is a dynamic equilibrium.
- The thermodynamic equilibrium constant, *K*, is defined in terms of activities by

$$K = \frac{\prod (a(\text{products}))_{\text{eqm}}^{\nu_P}}{\prod (a(\text{reactants}))_{\text{eqm}}^{\nu_R}}$$

- *K* is dimensionless, so has no units.
- Equilibrium constants can also be written in terms of partial pressures (K_p) or concentrations (K_c).

? For practice questions on these topics, see questions 1–3 at the end of this chapter (p.721).

15.2 The direction of a reaction: the reaction quotient

If a reaction is not at equilibrium, the amounts of products and reactants will change until they reach the equilibrium composition. The Gibbs energy of the system changes as the reaction proceeds since the proportions of reactant and product change (see Figure 15.2, p.697). To investigate how *G* changes, you need a measure of the composition of the reaction mixture as it changes during the reaction.

An expression similar to that in Equation 15.1 (p.698) for equilibrium constants is used—except the activities are not those at equilibrium. The expression is called the **reaction quotient**, *Q*, and is defined as

$$Q = \frac{\prod (a(\text{products}))^{\nu_P}}{\prod (a(\text{reactants}))^{\nu_R}} \qquad (15.2)$$

So, for the $N_2O_4(g)/NO_2(g)$ reaction, *Q* is given by

$$Q = \frac{(a_{NO_2})^2}{(a_{N_2O_4})} \qquad (15.3)$$

where a represents the thermodynamic activity. When the system reaches equilibrium, the value of Q is the same as the value of the equilibrium constant, K.

The value of Q can be calculated at *any* stage during the reaction and changes as the reaction proceeds towards equilibrium. If Q does not equal K, the reaction moves so as to bring it to equilibrium, so that $Q = K$.

To illustrate this, look at the following example.

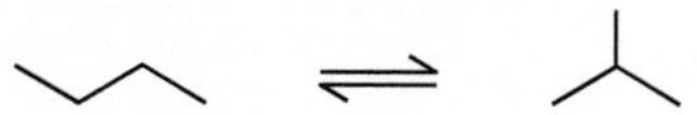

butane methylpropane

$$K = \frac{(p_{\text{(methylpropane)}}/p^{\ominus})_{\text{eqm}}}{(p_{\text{(butane)}}/p^{\ominus})_{\text{eqm}}} = 2.5 \qquad \text{where } p^{\ominus} = 1\,\text{bar}$$

For the gas phase isomerization of butane at 298 K, the equilibrium constant is 2.5. This means that if the equilibrium partial pressure of butane, $p_{\text{(butane)}}$ is 1 bar then the equilibrium value of $p_{\text{(methylpropane)}}$ *must* be 2.5 bar. Any other value would mean that the reaction is not at equilibrium.

Say a reaction mixture has $p_{\text{(methylpropane)}} = 0.5$ bar and $p_{\text{(butane)}} = 1$ bar. $Q = 0.5$ so that $Q < K$ and the system is not at equilibrium. To move the reaction towards equilibrium, more butane must react to increase the pressure of methylpropane, that is, the *forward* reaction occurs until the system comes to equilibrium. Conversely, if a reaction mixture has $p_{\text{(methylpropane)}} = 5$ bar and $p_{\text{(butane)}} = 1$ bar, the value of Q is 5. Since $Q > K$, the pressure of methylpropane must go down while that of butane must increase to reach equilibrium. The *reverse* reaction takes place.

The reaction quotient provides an additional way to predict the direction of a reaction. This is illustrated in Figure 15.3.

- If $\Delta_r G < 0$ or $Q < K$, the *forward* reaction proceeds.
- If $\Delta_r G > 0$ or $Q > K$, the *reverse* reaction proceeds.
- If $\Delta_r G = 0$ or $Q = K$, the system is at *equilibrium*.

Worked example 15.3 provides another example of using Q to predict the direction of a reaction. An everyday example is described in Box 15.2.

The *expression* to calculate Q is always the same as that for K but the *value* of Q is not the same, except when the reaction is at equilibrium.

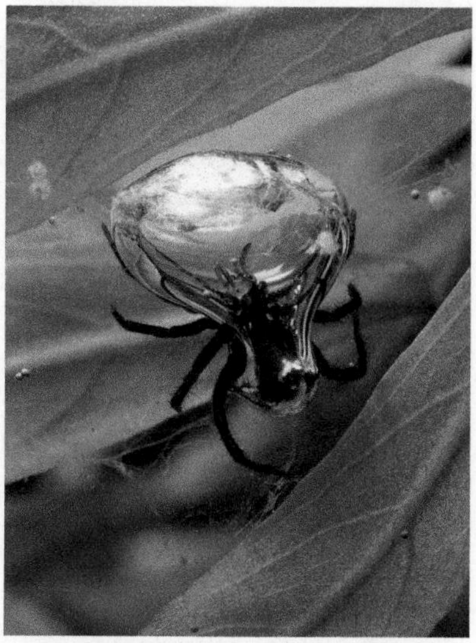

The diving bell spider spends its entire life underwater, relying on the equilibrium that exists between O_2(aq) and O_2(g). It spins a web to trap an air bubble to breathe. As oxygen is removed from the air, it is replaced by fresh oxygen diffusing into the bubble from the aqueous phase.

The conditions for spontaneous reactions involve $\Delta_r G$ at constant temperature and pressure.

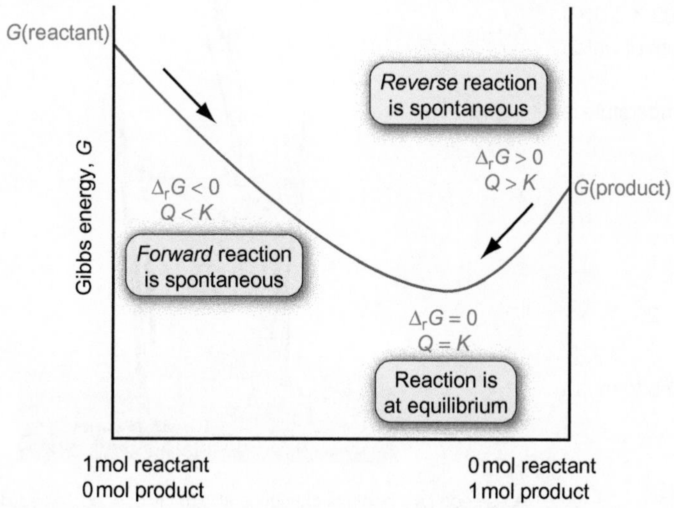
G(reactant)
G(product)
Gibbs energy, G
Reverse reaction is spontaneous
ΔrG > 0
Q > K
ΔrG < 0
Q < K
Forward reaction is spontaneous
ΔrG = 0
Q = K
Reaction is at equilibrium
1 mol reactant
0 mol product
0 mol reactant
1 mol product

Figure 15.3 A reaction comes to equilibrium when it reaches its minimum Gibbs energy and $Q = K$.

Worked example 15.3 Using the reaction quotient, Q

For the reaction of hydrogen and iodine below, $K = 46$ at 510 °C.

$$H_2(g) + I_2(g) \rightleftharpoons 2HI(g)$$

If hydrogen gas, iodine vapour, and hydrogen iodide, each at a partial pressure of 0.55 bar, are mixed in a vessel, predict what will happen.

Strategy

You are given the partial pressures of each of the components and so can calculate the reaction quotient, *Q*. By comparing *Q* with *K* you can decide in which direction the reaction will proceed.

Solution

Use Equation 15.2 on p.702 to write the expression for Q. Remember that the activity of a gas is given by p/p⊖.

$$Q = \frac{\prod (a(\text{products}))^{\nu_P}}{\prod (a(\text{reactants}))^{\nu_R}} \qquad (15.2)$$

$$= \frac{(p_{HI}/p^{\ominus})^2}{(p_{H_2}/p^{\ominus}) \times (p_{I_2}/p^{\ominus})} \qquad \text{where } p^{\ominus} = 1 \text{ bar}$$

Substitute in the values of the partial pressures.

$$Q = \frac{(0.55/1\text{ bar})^2}{(0.55/1\text{ bar}) \times (0.55/1\text{ bar})}$$

$$= 1.00$$

Decide if Q is greater than or less than K and in which direction the reaction will proceed.

Since $Q < K$, the forward reaction will take place so that more HI will form.

Question

What will happen in the above reaction if the starting partial pressures of H_2(g), I_2(g), and HI (g) are 0.2 bar, 0.5 bar, and 1.4 bar, respectively?

Box 15.2 Equilibrium in fizzy water

A bottle of sparkling mineral water can be used to illustrate the use of *Q* and *K*. A few mineral waters are naturally carbonated and contain dissolved carbon dioxide (CO_2; see Box 15.5, p.713), but most mineral waters have CO_2 forced into them under pressure at the bottling plant. The solubility of CO_2 at 293 K (20 °C) is around 0.036 mol dm^{-3} in water (1.7 mg of CO_2 per gram of water) when the CO_2 pressure above the water is 1 bar.

A closed bottle of sparkling mineral water at a fixed temperature is at equilibrium. For the CO_2, the significant equilibrium is

$$CO_2(aq) \rightleftharpoons CO_2(g)$$

The equilibrium constant (at 293 K) is given by

$$K = \frac{(a_{CO_2(g)})_{eqm}}{(a_{CO_2(aq)})_{eqm}} = \frac{(p_{CO_2(g)}/p^{\ominus})_{eqm}}{([CO_2(aq)]/1\text{ mol dm}^{-3})_{eqm}} = 28$$

The expression for the reaction quotient, *Q*, has the same form but does not refer to equilibrium conditions

$$Q = \frac{(a_{CO_2(g)})}{(a_{CO_2(aq)})} = \frac{(p_{CO_2(g)}/p^{\ominus})}{([CO_2(aq)]/1\text{ mol dm}^{-3})}$$

▲ As soon as a bottle of fizzy mineral water is opened, the equilibrium $CO_2(aq) \rightleftharpoons CO_2(g)$ is disturbed.

→

→ When you open the bottle, gas escapes since the pressure of the gas inside is above atmospheric pressure ($p_{CO_2(g)} > 1$ atm, 1 atm = 1.013 bar), and so $p_{CO_2(g)}$ over the water decreases. The value of Q also decreases, and the system is no longer at equilibrium—even if you put the cap back on. At this point, the value of Q is less than K.

The expression for the equilibrium constant shows that $[CO_2\,(aq)]$, the concentration of dissolved CO_2 (its *solubility*), is proportional to the partial pressure. (This is discussed further in Section 17.4, p.789.) So, when the pressure is reduced, gas leaves the solution. CO_2 escapes from solution by forming bubbles in the liquid and rising to the surface.

If you put the top back on the bottle, some CO_2 continues to escape from the water into the space at the top of the bottle, reducing $[CO_2\,(aq)]$, and increasing $p_{CO_2(g)}$.

The value of Q increases until $Q = K$, when the system has reached a new equilibrium with the mineral water containing less CO_2. Next time you open the bottle, the mineral water will have lost some of its 'fizz'.

Questions

(a) A bottle of fizzy water is opened at room temperature and the water is poured into a glass. The glass is then left to stand for several hours. What happens to: (i) K; (ii) Q, over this period of time?

(b) What will be the solubility of CO_2 in mol dm^{-3} at 293 K (20 °C) if the gas is injected into pure water at 2 bar pressure?

Summary

- The thermodynamic equilibrium constant, K, is defined in terms of the activities of components in a reaction at equilibrium by

$$K = \frac{\prod (a(\text{products}))^{\nu_P}_{\text{eqm}}}{\prod (a(\text{reactants}))^{\nu_R}_{\text{eqm}}}$$

- The reaction quotient, Q, is defined in terms of the activities by

$$Q = \frac{\prod (a(\text{products}))^{\nu_P}}{\prod (a(\text{reactants}))^{\nu_R}}$$

- If $Q < K$, the forward reaction proceeds. If $Q > K$, the reverse reaction proceeds. If $Q = K$, the system is at equilibrium.

? For practice questions on these topics, see questions 4, 5, and 10 at the end of this chapter (p.721).

15.3 Gibbs energy and equilibrium constants

A reaction reaches equilibrium when it reaches its minimum Gibbs energy at constant temperature and pressure (see Figure 15.3, p.703). The concentrations (or pressures) of reactant and product at equilibrium are related by the equilibrium constant, K. There must, therefore, be a quantitative relationship between Gibbs energy and K.

In fact, the relationship is quite straightforward and is given by Equation 15.4. You can see how Equation 15.4 is derived in Box 15.3.

Standard Gibbs energy change | Gas constant = 8.314 J K^{-1} mol^{-1} | Natural logarithm

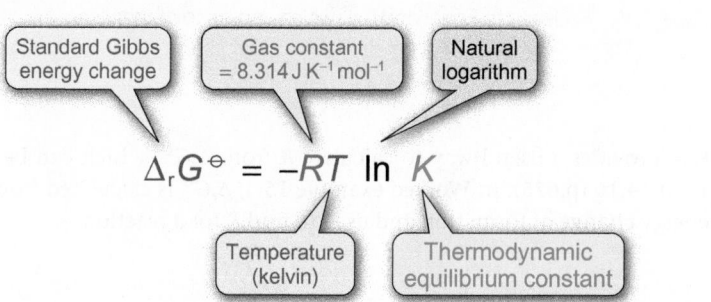

$$\Delta_r G^\ominus = -RT \ln K \qquad (15.4)$$

Temperature (kelvin) | Thermodynamic equilibrium constant

@ Visit the Online Resource Centre to watch screencast 15.1 which walks you through the derivation of Equation 15.4.

i It is important to remember that the *standard* Gibbs energy change, $\Delta_r G^\ominus$, must be used in Equation 15.4. $\Delta_r G^\ominus$ is defined in Section 14.5 (p.675).

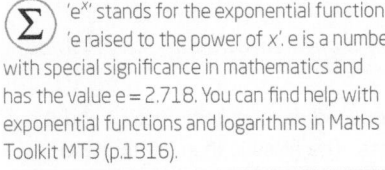

'e^x' stands for the exponential function 'e raised to the power of x'. e is a number with special significance in mathematics and has the value e = 2.718. You can find help with exponential functions and logarithms in Maths Toolkit MT3 (p.1316).

Equation 15.4 allows the equilibrium constant to be calculated from the Gibbs energy change for the reaction. Rearranging Equation 15.4 to give an expression for K

$$\ln K = -\Delta_r G^\ominus/RT$$

and

$$K = e^{\left(\frac{-\Delta_r G^\ominus}{RT}\right)} \quad (15.5)$$

Equations 15.4 and 15.5 are perhaps the most important relationships in considering chemical reactions at equilibrium.

Box 15.3 Deriving the relationship between $\Delta_r G^\ominus$ and K

Consider a general reaction of 1 mol reactant, R, forming 1 mol product, P, at constant temperature

$$R \rightleftharpoons P$$

The standard Gibbs energy change for the reaction is the difference between the standard Gibbs energies of 1 mol of product and 1 mol of reactant (see Figure 15.2). It represents the change in Gibbs energy when 1 mol of P is formed from 1 mol of R.

$$\Delta_r G^\ominus = G^\ominus(\text{product}) - G^\ominus(\text{reactant}) \quad (15.6)$$

$G^\ominus$ is the molar Gibbs energy of a species under standard conditions.

As the reaction proceeds, the activities of the components change—the activity of R falls and that of P increases. The Gibbs energy of each component, G, is related to its activity by Equation 14.20, p.687.

$$G(\text{product}) = G^\ominus(\text{product}) + RT\ln(a(\text{product}))$$

$$G(\text{reactant}) = G^\ominus(\text{reactant}) + RT\ln(a(\text{reactant}))$$

The change in Gibbs energy at any stage during the reaction, $\Delta_r G$, is the difference in the Gibbs energies of the components at that composition

$$\Delta_r G = G(\text{product}) - G(\text{reactant})$$

Substituting for the product and reactant

$$\Delta_r G = [G^\ominus(\text{product}) + RT\ln(a(\text{product}))] - [G^\ominus(\text{reactant}) + RT\ln(a(\text{reactant}))]$$

Grouping together similar terms

$$\Delta_r G = [G^\ominus(\text{product}) - G^\ominus(\text{reactant})] + [RT\ln(a(\text{product})) - RT\ln(a(\text{reactant}))]$$

$$= [G^\ominus(\text{product}) - G^\ominus(\text{reactant})] + RT[\ln(a(\text{product})) - \ln a(\text{reactant}))]$$

From Equation 15.6, $[G^\ominus(\text{product}) - G^\ominus(\text{reactant})] = \Delta_r G^\ominus$.

The terms in the second bracket can also be rearranged, leading to

$$\Delta_r G = \Delta_r G^\ominus + RT\ln\left(\frac{a(\text{product})}{a(\text{reactant})}\right) \quad (15.7)$$

The final term arises from the rules for manipulating logarithms:

$$\ln x + \ln y = \ln(x \times y) \quad \text{and} \quad \ln x - \ln y = \ln(x/y)$$

(See Maths Toolkit MT3, p.1316.)

The final term in Equation 15.7 can be simplified further since

$$Q = \frac{a(\text{products})}{a(\text{reactants})} \quad (15.3)$$

Hence

$$\Delta_r G = \Delta_r G^\ominus + RT\ln Q \quad (15.8)$$

What happens when the reaction reaches equilibrium? The value of Q becomes equal to the equilibrium constant, K. Also, the reaction has reached its minimum Gibbs energy and there is no tendency to react further. This means that there is no further change and $\Delta_r G = 0$. Substituting these factors into Equation 15.8

$$0 = \Delta_r G^\ominus + RT\ln Q_{eqm}$$

$$= \Delta_r G^\ominus + RT\ln K$$

so that

$$\Delta_r G^\ominus = -RT\ln K \quad (15.4)$$

Although this derivation applies to a reaction with 1 : 1 stoichiometry, the same method applies to *any* reaction with *any* stoichiometry and leads to Equation 15.4, which applies to *all* reactions.

Equation 14.16 is the relationship: $\Delta_r G^\ominus = \Delta_r H^\ominus - T\Delta_r S^\ominus$. Section 14.5 (p.675) describes how to calculate $\Delta_r G$ using tables of thermodynamic data, such as enthalpy changes of formation, standard entropies, and heat capacities.

Equation 15.4 provides a useful way to calculate K from $\Delta_r G^\ominus$, which can be calculated using Equation 14.16 (p.675). In Worked example 15.4, $\Delta_r G^\ominus$ is calculated from values of the Gibbs energy change of formation and used to find K for a reaction.

Worked example 15.4 Using $\Delta_r G^{\ominus}_{298}$ to calculate a value for K

Industrially, methanol is synthesized using the reaction

$$CO(g) + 2H_2(g) \rightleftharpoons CH_3OH(g)$$

Calculate the equilibrium constant for the reaction at 298 K.

($\Delta_f G^{\ominus}_{298}$/kJ mol^{-1}: CO(g), −137.2; H_2(g), 0; CH_3OH(g), −166.3)

Strategy

You can use Equation 15.4 to calculate K at 298 K if you know $\Delta_r G^{\ominus}_{298}$. This can be calculated from the $\Delta_f G^{\ominus}_{298}$ data in the question, using Equation 14.18 (p.680).

Solution

Use Equation 14.18 to calculate $\Delta_r G^{\ominus}_{298}$.

$$\begin{aligned}\Delta_r G^{\ominus}_{298} &= \sum \nu_i \Delta_f G^{\ominus}_{298}(\text{products}) - \sum \nu_i \Delta_f G^{\ominus}_{298}(\text{reatants}) \\ &= [\Delta_f G^{\ominus}_{298}(CH_3OH(g))] - [\Delta_f G^{\ominus}_{298}(CO(g)) \\ &\quad + 2 \times \Delta_f G^{\ominus}_{298}(H_2(g))] \qquad (14.18) \\ &= [-166.3\ \text{kJ mol}^{-1}] - [-137.2\ \text{kJ mol}^{-1} + 0] \\ &= -29.1\ \text{kJ mol}^{-1}\end{aligned}$$

($\Delta_f G^{\ominus}_{298}(H_2(g)) = 0$ since H_2 is an element in its standard state.)

Substitute values into Equation 15.4, taking care to make the units consistent (i.e. convert the units of $\Delta_r G^{\ominus}_{298}$ to J mol^{-1} to be consistent with the units for R).

$$\Delta_r G^{\ominus} = -RT \ln K \qquad (15.4)$$

$$K = e^{\left(\frac{-\Delta_r G^{\ominus}}{RT}\right)} = e^{\left(\frac{-(-29.1 \times 10^3\ \text{J mol}^{-1})}{(8.314\ \text{J K}^{-1}\text{mol}^{-1}) \times (298\ \text{K})}\right)}$$

$$K = e^{(11.7)}$$

$$= 1.2 \times 10^5$$

The high value of K shows that the equilibrium lies very much in favour of methanol under these conditions.

Question

The thermodynamic equilibrium constant, K, is 15.51 at 100 °C for the following reaction

$$N_2O_4(g) \rightleftharpoons 2NO_2(g)$$

Calculate the standard Gibbs energy change for the reaction.

Box 15.4 illustrates the use of K at different temperatures in some current research and shows how thermodynamic relationships can be used to calculate quantities that are not directly measurable.

Box 15.4 Contrails from jet aircraft

The exhaust gases from jet aircraft contain a wide range of species including CO_2, water vapour, NO_x, SO_2, SO_3, H_2SO_4, and soot particles. Water vapour condenses as droplets, containing H_2SO_4 and soot particles, and freezes to form ice crystals. It has been estimated that 0.1% of the sky is covered by jet contrails at any given time. They warm the atmosphere in a similar way to high thin clouds and are believed to have an effect on climate. As air traffic continues to increase, there is concern about their effect on global warming.

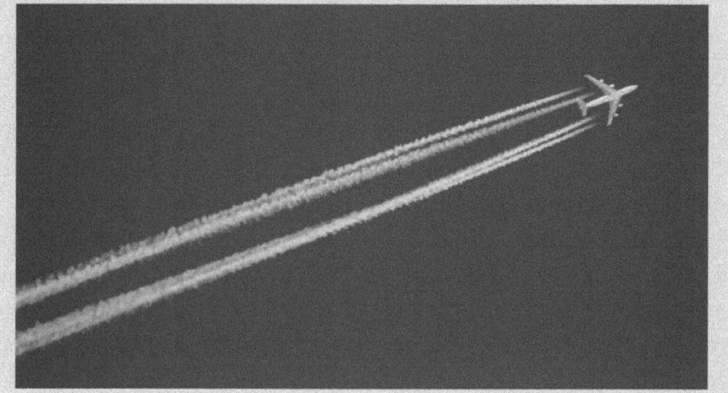

Aviation fuel (kerosene) contains 0.4 g of sulfur per kg of fuel and forms sulfur dioxide (SO_2) during combustion. The temperature of the gases leaving a jet aircraft is about 600 K. As the gases cool the SO_2 reacts with hydroxyl radicals ($^{\bullet}OH$) formed during combustion and starts a sequence of reactions ending in the formation of sulfuric acid (H_2SO_4)

$$SO_2(g) + {}^{\bullet}OH(g) \rightleftharpoons {}^{\bullet}HOSO_2(g) \qquad \text{(Reaction 1)}$$

$$^{\bullet}HOSO_2(g) + O_2(g) \rightleftharpoons {}^{\bullet}HO_2(g) + SO_3(g) \qquad \text{(Reaction 2)}$$

$$SO_3(g) + H_2O(g) \rightleftharpoons H_2SO_4(g) \qquad \text{(Reaction 3)}$$

◀ Contrails are the visible clouds that form in a line behind aircraft flying at high altitude in cold air.

→

→ An important factor in the formation of the droplets is the ratio of SO_2 to the fully oxidized SO_3 and H_2SO_4 in the exit gases. The higher the proportion of SO_3 and H_2SO_4 the greater is the tendency to form contrails.

Reactions 1–3 also occur in the atmosphere outside the jet contrails—but much more slowly since the concentrations of the reactants, particularly ˙OH radicals, are very much smaller. At high temperature, reaction 1 occurs much faster than reaction 2 and soon reaches equilibrium. It is the key reaction in the sequence and its equilibrium constant is an important quantity.

Chemists at the University of Leeds studied reaction 1 under controlled conditions in the laboratory. They measured the rates of the forward and back reactions at different temperatures using flash photolysis (see Section 9.4, p.402). At equilibrium, the rates of the forward and back reactions are equal so that the equilibrium constant, K, could be found (see Section 9.6, p.418).

From K, they used thermodynamic relationships to calculate values for $\Delta_r G^\ominus$, $\Delta_r S^\ominus$, and $\Delta_r H^\ominus$ for the reaction. Once these are known, it is possible to work out the equilibrium constant under any set of conditions.

The key to the formation of the fully oxidized compounds, SO_3 and H_2SO_4, is the competition between the back reaction in reaction 1 and the forward reactions 2 and 3. High equilibrium concentrations of ˙$HOSO_2$ lead to a high ratio of SO_3 and H_2SO_4 compounds in the exit gases and so increase the formation of contrails.

Questions

The Leeds group obtained values of $\Delta_r H^\ominus_{298} = -113.3$ kJ mol^{-1} and $\Delta_r S^\ominus_{298} = -142$ J K^{-1} mol^{-1} for the forward reaction in reaction 1.

(a) Write an expression for the equilibrium constant, K, in reaction 1.

(b) Calculate $\Delta_r G^\ominus$ and the equilibrium constant K at 298 K for reaction 1.

(c) Assuming $\Delta_r S^\ominus$ and $\Delta_r H^\ominus$ are independent of temperature, calculate K at 500 K and 1100 K.

(d) Suggest what effect the temperature of the exhaust gases has on the ratio of SO_2 to $SO_3 + H_2SO_4$ in the jet exhaust.

Gibbs energy change and equilibrium composition

So far, the value of the Gibbs energy change has been used to indicate whether a reaction will be spontaneous or not. The relationship between $\Delta_r G^\ominus$ and K in Equations 15.4 and 15.5 allows you to find out much more about a reaction.

The value of K gives some information about the composition of the reaction mixture at equilibrium. If K is large ($K \gg 1$) then the equilibrium mixture will contain mainly products. If K is small ($K \ll 1$) then the reaction does not proceed vary far and little product forms. If $K \approx 1$, then the equilibrium mixture will contain substantial amounts of both product and reactant.

Consider the reaction

$$\mathrm{R} \rightleftharpoons \mathrm{P}$$

Around room temperature, say at 298 K, if the value of $\Delta_r G^\ominus$ for the reaction is negative, the reaction is spontaneous. From Equation 15.4 (p.705), $\Delta_r G^\ominus = -RT\ln K$, so, if $\Delta_r G^\ominus$ is negative, K must be >1, and, at equilibrium, the products are present in larger concentrations than the reactants. A more negative value of $\Delta_r G^\ominus$ will give a larger value of K and greater proportion of products at equilibrium.

Say a complete reaction is one that gives a yield of 99.99% products. The corresponding equilibrium constant is

$$K = \frac{\text{Concentration of product}}{\text{Concentration of reactant}} = \frac{[\mathrm{P}]}{[\mathrm{R}]} = \frac{99.99\%}{0.01\%} = 9999 \approx 1\times10^4$$

Using $\Delta_r G^\ominus = -RT\ln K$ shows that, when $K = 9999$ at 298 K, $\Delta_r G^\ominus = -22.8$ kJ mol^{-1}. This means that any reaction with a value of $\Delta_r G^\ominus$ more negative than -22.8 kJ mol^{-1} will effectively go to completion. This value of $\Delta_r G^\ominus$ is sometimes called the *threshold* value. The same argument shows that any reaction with $\Delta_r G^\ominus$ more positive than $+22.8$ kJ mol^{-1} will not occur to any noticeable extent—at most 0.01% of the products will form. Between these extremes, the behaviour of a reaction varies as summarized in Figure 15.4.

The threshold value of $\Delta_r G^\ominus$ required for a reaction that 'goes to completion' depends on temperature. In Worked example 15.5, you can calculate the threshold value of $\Delta_r G^\ominus$ for a reaction to go to completion at 750 K.

	R reactants → P products				
	'No reaction'	'Unfavourable reaction'	'50–50 reaction'	'Favourable reaction'	'Complete reaction'
	< 0.01% P > 99.99% R		50% P 50% R		> 99.99% P < 0.01% R
K	$< 1 \times 10^{-4}$		1		$> 1 \times 10^{+4}$
$\Delta_r G^{\ominus}$/kJ mol^{-1}	> +22.8		0		< −22.8

Figure 15.4 The Gibbs energy change not only tells you whether the reaction R ⇌ P is spontaneous but also indicates the equilibrium composition of the reaction.

Worked example 15.5 $\Delta_r G^{\ominus}$ for a reaction that goes to completion at 750 K

What is the Gibbs energy change at 750 K for a reaction, R ⇌ P, that produces 99.99% product at equilibrium? Compare your answer with the value for the reaction at 298 K.

Strategy

From the concentrations of reactant and product at equilibrium, you can calculate the equilibrium constant at 750 K. Then use Equation 15.4 to calculate the corresponding $\Delta_r G^{\ominus}$ at 750 K.

Solution

Calculate the value of K.

$$K = \frac{[P]}{[R]} = \frac{99.99\%}{0.01\%} = 9999$$

Use Equation 15.4 to find the Gibbs energy change at 750 K.

$$\begin{aligned} \Delta_r G^{\ominus}_{750} &= -RT \ln K \quad \text{(from 15.4)} \\ &= -(8.314\ \text{J K}^{-1}\ \text{mol}^{-1}) \times (750\ \text{K}) \times \ln(9999) \\ &= -57400\ \text{J mol}^{-1} \\ &= -57.4\ \text{kJ mol}^{-1} \end{aligned}$$

So, a reaction with a value of $\Delta_r G^{\ominus}$ more negative than −57.4 kJ mol^{-1} at 750 K will effectively go to completion.

The threshold value of $\Delta_r G^{\ominus}$ for a reaction to go to completion becomes more negative as the temperature increases.

Question

Calculate the $\Delta_r G^{\ominus}$ for a reaction, R ⇌ P, that comes to equilibrium with equal concentrations of product and reactant at 473 K. Comment on your answer.

Summary

- The thermodynamic equilibrium constant, K, for a reaction is linked to the standard Gibbs energy change by

$$\Delta_r G^{\ominus} = -RT \ln K$$

- If a reaction, R ⇌ P, has a Gibbs energy change more negative than −22.8 kJ mol^{-1} at 298 K, then the reaction will effectively go to completion.

For practice questions on these topics, see questions 6–8 and 19–21 at the end of this chapter (pp.722–723).

15.4 Calculating the composition of a reaction at equilibrium

The amounts of each of the components in a reaction at equilibrium can be found from the equilibrium constant. If the concentration or pressure of one or more components is known, all the others can be found. An example is given in Worked example 15.6.

Worked example 15.6 Calculating the partial pressure of a product at equilibrium

The Gibbs energy change at 1000 K for the reaction

$$\tfrac{1}{2}N_2(g) + \tfrac{1}{2}O_2(g) \rightleftharpoons NO(g)$$

is $\Delta_r G^{\ominus} = +77.8\,kJ\,mol^{-1}$. Estimate the equilibrium partial pressure of NO in air at 1000 K and 1 bar pressure. Assume that air is a mixture of 80% N_2 and 20% O_2 (by volume).

Strategy

You know the partial pressures of N_2 and O_2 from the composition of air. To calculate the partial pressure of NO, you therefore need a value for K. You can use Equation 15.4 to find this from $\Delta_r G^{\ominus}$ at 1000 K.

(This calculation ignores the fact that some N_2 and O_2 molecules react, so their partial pressures will be very slightly less than 0.80 bar and 0.20 bar. However, the high positive value of $\Delta_r G^{\ominus}$ indicates a low value of K, so this assumption is justified.)

Solution

Use Equation 15.1 (p.698) to write an expression for K.

$$K = \frac{(p_{NO}/p^{\ominus})_{eqm}}{(p_{N_2}/p^{\ominus})^{1/2}_{eqm} \times (p_{O_2}/p^{\ominus})^{1/2}_{eqm}} \quad \text{where } p^{\ominus} = 1\,bar$$

Use Equation 15.4 (p.705) to find the value of K. (Remember that the units for $\Delta_r G^{\ominus}$ and R must be consistent.)

$$\Delta_r G^{\ominus} = -RT \ln K \qquad (15.4)$$

$$+77.8 \times 10^3\,J\,mol^{-1} = -(8.314\,J\,K^{-1}\,mol^{-1}) \times (1000\,K) \times \ln K$$

$$-\ln K = \frac{+77.8 \times 10^3\,J\,mol^{-1}}{8.314\,J\,K^{-1}\,mol^{-1} \times 1000\,K} = 9.36$$

$$K = e^{(-9.36)}$$

$$= 8.6 \times 10^{-5} \quad \text{(at 298 K)}$$

Find the partial pressures of the components.

The total pressure is 1 bar. Since 80% of the air is N_2, $p_{N_2} = 0.80\,bar$. Similarly, $p_{O_2} = 0.20\,bar$.

Substitute into the expression for K.

$$K = \frac{(p_{NO}/p^{\ominus})_{eqm}}{(0.80)^{1/2} \times (0.20)^{1/2}} = 8.6 \times 10^{-5}$$

$$p_{NO} = (8.6 \times 10^{-5}) \times (0.80)^{1/2} \times (0.20)^{1/2} \times 1\,bar$$

$$= 3.4 \times 10^{-5}\,bar$$

Even at this high temperature, only a very small concentration of NO exists at equilibrium.

Question

The thermodynamic equilibrium constant, K, is 15.5 at 100 °C for the reaction

$$N_2O_4(g) \rightleftharpoons 2\,NO_2(g)$$

In an equilibrium mixture at 100 °C, the partial pressure of $N_2O_4(g)$ is 0.14 bar. What is the partial pressure of $NO_2(g)$ at equilibrium?

Calculating equilibrium compositions from starting quantities

In Worked example 15.6, the partial pressures of all the components except one at *equilibrium* are known, so that substituting into the expression for K gives the unknown quantity. A rather more challenging problem arises if the *starting* quantities and the value of K are given. Can the partial pressures of products and reactants present at equilibrium be found?

For many reactions the answer is yes if you use the balanced equation for the reaction and consider the amounts of each component that react.

The most straightforward case can be illustrated by a gas phase isomerization reaction, such as the isomerization of butane to methylpropane (see p.703)

Visit the Online Resource Centre to watch screencast 15.2 which walks you through the butane/methylpropane calculation.

$$\text{butane} \rightleftharpoons \text{methylpropane}$$

At 298 K

$$K = \frac{a_{\text{(methylpropane)}}}{a_{\text{(butane)}}} = \frac{p_{\text{(methylpropane)}}/p^{\ominus}}{p_{\text{(butane)}}/p^{\ominus}} = 2.5$$

Suppose the reaction starts with 1 mol of butane in a sealed container. When the reaction reaches equilibrium at 298 K, the amount of butane will be less than 1 mol because some has reacted. Call the fraction that reacts α.

α is defined as the mole fraction of the starting material that reacts. It is sometimes called by its older name, the degree of dissociation.

When the reaction has come to equilibrium, there will be $(1 - \alpha)$ mol of butane remaining and α mol of methylpropane will have been formed. Because of the stoichiometry, the total number of moles of gas present is always 1. A convenient way of keeping track of these changes is to list them as follows

	butane	⇌ methylpropane	
At start:	1 mol	0 mol	Total number of moles = 1 mol
At equilibrium:	$(1 - \alpha)$ mol	α mol	Total number of moles = $(1 - \alpha) + \alpha = 1$ mol

At constant volume and temperature, $p \propto n$ (see Section 8.1, p.345).

Remember, $p^{\ominus} = 1$ bar.

Since, for ideal gases, pressure is proportional to the number of moles, n

$$K = 2.5 = \frac{p_{\text{(methylpropane)}}/p^{\ominus}}{p_{\text{(butane)}}/p^{\ominus}} = \frac{n_{\text{(methylpropane)}}}{n_{\text{(butane)}}}$$

So,
$$2.5 = \frac{\alpha}{(1-\alpha)}$$

A value for α can be found by multiplying each side by $(1 - \alpha)$ and rearranging

$$2.5(1 - \alpha) = \alpha$$
$$2.5 - 2.5\alpha = \alpha$$
$$2.5 = 3.5\alpha$$
$$\alpha = 0.71$$

So, at equilibrium, the flask will contain 0.71 mol of methylpropane and 0.29 mol of butane.

This example shows how equilibrium constants can be used to predict the yields of reactions. Unfortunately, reactions with more complex stoichiometry can lead to complicated equations to find the value of α. In many cases, quadratic or cubic equations are obtained. Some of these can be difficult to solve without a computer.

The usefulness of this method is that, if K can be calculated from thermodynamic data the value of α can be calculated. From this, the equilibrium composition of the mixture and the expected yield of product can be found. Another example is given in Worked example 15.7.

The subscript 'eqm' can make expressions for K look very cluttered. Since the reactions in this section, and for the remainder of the chapter, are all at equilibrium, the subscript 'eqm' has been omitted for simplicity—but always be aware of whether or not a reaction mixture you are dealing with is at equilibrium.

Worked example 15.7 Calculating equilibrium yields from starting quantities

For the gas phase decomposition of phosphorus pentachloride

$$PCl_5(g) \rightleftharpoons PCl_3(g) + Cl_2(g)$$

the equilibrium constant is 1.7 at 500 K. 1 mol of PCl_5 is placed in a sealed container and allowed to come to equilibrium at 500 K. At equilibrium the total pressure in the container is 0.2 bar. Calculate the amount of Cl_2 present at equilibrium.

Strategy

First, write the expression for the equilibrium constant in terms of the partial pressures of each component. Then, use the stoichiometry to determine how many moles of each component will be present when the reaction comes to equilibrium.

Use these expressions to find the partial pressure of each component. Substitute these into the expression for K and solve to find the amount of Cl_2 present at equilibrium.

Solution

Write the expression for K in terms of the partial pressures.

$$K = \frac{(a_{PCl_3}) \times (a_{Cl_2})}{(a_{PCl_5})} = \frac{(p_{PCl_3}/p^{\ominus}) \times (p_{Cl_2}/p^{\ominus})}{(p_{PCl_5}/p^{\ominus})} \quad \text{where } p^{\ominus} = 1\text{ bar}$$

Determine how many moles of each component will be present at equilibrium.

At the start of the reaction, there is 1 mol of PCl_5. Let α be the fraction of PCl_5 that reacts.

	$PCl_5(g)$	$\rightleftharpoons$	$PCl_3(g)$ +	$Cl_2(g)$
At start of reaction:	1 mol		0	0
At equilibrium:	$(1-\alpha)$ mol		α mol	α mol

Total number of moles at equilibrium = $((1-\alpha) + \alpha + \alpha)$ mol = $(1+\alpha)$ mol

Find the partial pressure of each component.

The partial pressure of a gas in a mixture is given by Equation 8.10 (p.356)

$$\text{partial pressure of a gas A} = p_A = x_A \times p_{total} \qquad (8.10)$$

where x_A is the mole fraction of A and p_{total} is the total pressure. So

$$p_{PCl_5} = x_{PCl_5} \times p_{total}$$

where x_{PCl_5} is the mole fraction of PCl_5 at equilibrium.

From Equation 8.9 (p.352)

$$x_{PCl_5} = \frac{\text{number of moles } PCl_5}{\text{total number of moles}} = \frac{1-\alpha}{1+\alpha}$$

$$\text{so} \quad p_{PCl_5} = \frac{1-\alpha}{1+\alpha} \times p_{total}$$

Similarly

$$p_{Cl_2} = p_{PCl_3} = \frac{\alpha}{(1+\alpha)} \times p_{total}$$

Substitute the partial pressures into the expression for K.

$$K = \frac{(p_{PCl_3}/p^{\ominus}) \times (p_{Cl_2}/p^{\ominus})}{(p_{PCl_5}/p^{\ominus})}$$

$$= \frac{(x_{PCl_3} \times (p_{total}/p^{\ominus})) \times (x_{Cl_2} \times (\cancel{p_{total}}/\cancel{p^{\ominus}}))}{(x_{PCl_5} \times (\cancel{p_{total}}/\cancel{p^{\ominus}}))}$$

$$= \frac{\left(\frac{\alpha}{(1+\alpha)}\right) \times \left(\frac{\alpha}{\cancel{(1+\alpha)}}\right)}{\left(\frac{1-\alpha}{\cancel{(1+\alpha)}}\right)} \times (p_{total}/p^{\ominus})$$

$$= \frac{\alpha^2}{(1+\alpha)(1-\alpha)} \times (p_{total}/p^{\ominus})$$

$$= \frac{\alpha^2}{(1-\alpha^2)} \times (p_{total}/p^{\ominus})$$

Substitute values for K and p_{total} and solve to find α.

At 500 K, $p_{total} = 0.2$ bar and $K = 1.7$. So

$$1.7 = \frac{\alpha^2}{(1-\alpha^2)} \times \frac{0.2\ \cancel{\text{bar}}}{1\ \cancel{\text{bar}}} \qquad p^{\ominus} = 1\text{ bar}$$

Multiply both sides by $(1-\alpha^2)$	$1.7 \times (1-\alpha^2) = \alpha^2 \times 0.2$
Multiply out the bracket	$1.7 - 1.7\alpha^2 = 0.2\alpha^2$
Rearrange	$1.7 = (1.7 + 0.2)\alpha^2$
Solve for α^2	$\frac{1.7}{1.9} = \alpha^2$

$$\alpha^2 = 0.895 \quad \text{so} \quad \alpha = 0.95$$

α is the fraction of moles of PCl_5 that has reacted. Since you started with 1 mol of PCl_5, 0.95 mol will react forming 0.95 mol of chlorine at equilibrium.

Question

The thermodynamic equilibrium constant, K, is 15.5 at 100 °C for the reaction

$$N_2O_4(g) \rightleftharpoons 2NO_2(g)$$

1 mol of N_2O_4 (g) is introduced into a sealed container at 100 °C. The final pressure in the container is 1.5 bar when the system comes to equilibrium. Find the composition of the mixture at equilibrium.

Box 15.5 Chalk, lime, and mineral water: an example of heterogeneous equilibrium

Limestone rocks are mainly composed of calcium carbonate ($CaCO_3$). On heating to very high temperatures, limestone decomposes to calcium oxide (CaO) releasing carbon dioxide gas.

$$CaCO_3(s) \rightleftharpoons CaO(s) + CO_2(g)$$

At 298 K, $\Delta_r H^\ominus = +178.3\,kJ\,mol^{-1}$ and $\Delta_r S^\ominus = +160.6\,J\,K^{-1}\,mol^{-1}$.

Calcium oxide readily absorbs water and is used as a desiccant.

The decomposition reaction also occurs naturally if the limestone rock gets hot enough. An example is the vicinity of Vergèze in southern France, which is the source of the Perrier brand of carbonated mineral water. Limestone rocks beneath the Perrier spring are in contact with hot volcanic rocks at a temperature of over 1000 °C. Limestone decomposes and carbon dioxide gas rises through the cracks in the limestone and dissolves in the underground water. The increased pressure below ground causes the carbon dioxide to be more soluble than at the surface.

In the original Perrier spring, the water emerged naturally sparkling. Today the water and gas are piped separately to the surface and mixed together before bottling.

▲ Much of the bedrock in the UK is limestone, often in the form of chalk. A spectacular example of this white stone is the White Horse carved into a hillside at Westbury, Wiltshire, which measures 54.9 m from head to tail.

Questions

(a) Write an expression for the equilibrium constant, *K*, for the decomposition of $CaCO_3(s)$.

(b) Suppose you heat some $CaCO_3(s)$ in a closed oven. You then add some more $CaCO_3(s)$ (without changing the temperature). What happens to the pressure of $CO_2(g)$?

(c) In contrast, you heat some $CaCO_3(s)$ in an open crucible. What happens?

(d) Calculate $\Delta_r G^\ominus$ and *K* for the decomposition of $CaCO_3(s)$ at 298 K and estimate their values at 1273 K. Comment on the results of your calculations.

(e) Estimate the temperature at which the decomposition of $CaCO_3(s)$ just becomes spontaneous at 1 bar.

(f) What is the partial pressure of $CO_2(g)$ in equilibrium with $CaCO_3(s)$ at 1273 K?

Summary

- If the equilibrium concentrations of all except one component are known, using the equilibrium constant allows calculation of the unknown concentration.
- The equilibrium composition of a reaction can be found from the equilibrium constant and starting quantities by considering the fraction that reacts, α.

? For practice questions on these topics, see questions 9 and 13–16 at the end of this chapter (p.722).

15.5 Effect of conditions on reaction yields and *K*

Synthetic chemists are often interested in obtaining the maximum yield of a desired product. For reactions that come to equilibrium, it is important to understand the effect of changing conditions on the composition of the equilibrium mixture, so that conditions can be chosen to maximize the yield.

Le Chatelier's principle

Studies of what happened to reaction equilibria when, for example, the temperature or pressure was changed led Henri Le Chatelier in 1885 to suggest **Le Chatelier's principle**, which, in modern language, can be stated as:

When a change is made to a system in dynamic equilibrium, the system responds to minimize the effect of the change.

Le Chatelier's principle is introduced in Section 1.9 (p.56).

Chapter 1 gives some examples of how Le Chatelier's principle can be used.

The reason that pressure and temperature have an effect on reactions is because of the way they influence the Gibbs energy change of reaction and the equilibrium constant, and hence the position of equilibrium. So, how can thermodynamics explain Le Chatelier's principle?

Changing the total pressure

The equilibrium constant K depends on $\Delta_r G^{\ominus}$, the standard Gibbs energy change for a reaction at 1 bar pressure

$$\Delta_r G^{\ominus} = -RT \ln K \tag{15.4}$$

$\Delta_r G^{\ominus}$ is the value at 1 bar and so, by definition, does not change with changing pressure. Therefore, K is a thermodynamic constant at constant temperature so that:

The equilibrium constant does not vary with pressure at constant temperature.

However, this does *not* mean that pressure has no effect on the *position* of equilibrium and the composition of the equilibrium mixture.

This is most easily shown by looking at an example. For $N_2O_4(g) \rightleftharpoons 2NO_2(g)$ the equilibrium constant is

$$K = \frac{(a_{NO_2})^2}{(a_{N_2O_4})} = \frac{(p_{NO_2}/p^{\ominus})^2}{(p_{N_2O_4}/p^{\ominus})} \quad \text{where } p^{\ominus} = 1\text{ bar} \tag{15.9}$$

Partial pressures of gases are discussed in Section 8.3 (p.355).

The partial pressure, p, of a component A in a mixture is given by

$$p_A = x_A \times p_{total} \tag{8.10}$$

where x_A is the mole fraction of component A and p_{total} is the total pressure of the system. So, Equation 15.9 can be written as

$$K = \frac{(x_{NO_2} \times (p_{total}/p^{\ominus}))^2}{(x_{N_2O_4} \times (p_{total}/p^{\ominus}))} = \frac{(x_{NO_2})^2}{x_{N_2O_4}} \times (p_{total}/p^{\ominus})$$

For the reaction $N_2O_4(g) \rightleftharpoons 2NO_2(g)$, Le Chatelier's principle says increasing the pressure will move the position of equilibrium to the left since it is the side with the lower number of moles of gas. This reduces the pressure and minimizes the effect of the imposed change.

K is constant at constant temperature so, if p_{total} changes, the mole fractions must change to compensate. If p_{total} increases then $\frac{(x_{NO_2})^2}{(x_{N_2O_4})}$ must go down to keep K constant, so x_{NO_2} decreases and $x_{N_2O_4}$ increases. Thus, less N_2O_4 will react and the yield of NO_2 will go down. This is what you would predict using Le Chatelier's principle.

Changing the pressure does not always affect the position of equilibrium. For example, in the reactions

$$\text{butane} \rightleftharpoons \text{methylpropane}$$

and

$$CO(g) + H_2O(g) \rightleftharpoons CO_2(g) + H_2(g)$$

A change in pressure does not affect the equilibrium if there is the same number of moles of gas on both sides of the chemical equation.

there are the same number of moles of gas on each side of the equation. The p_{total} term will therefore cancel out in the expression for K. An example of how pressure affects equilibrium is given in Worked example 15.8.

Worked example 15.8 Effect of pressure on an equilibrium

The reaction

$$PCl_5(g) \rightleftharpoons PCl_3(g) + Cl_2(g)$$

comes to equilibrium at a pressure of 1 bar. By formulating the equilibrium constant, predict what will happen if the pressure is increased at the same temperature. Compare your prediction with that using Le Chatelier's principle.

Strategy

Write the expression for the equilibrium constant in terms of the mole fractions of each component and the total pressure. You know that K must remain constant since the temperature is fixed, so you can see what happens to the mole fractions of each compound if p_{total} is increased.

Solution

Write the expression for the equilibrium constant in terms of mole fractions and p_{total}.

From Worked example 15.7

$$K = \frac{(x_{PCl_3} \times (p_{total}/p^{\ominus})) \times (x_{Cl_2} \times (\cancel{p_{total}}/\cancel{p^{\ominus}}))}{(x_{PCl_5} \times (\cancel{p_{total}}/\cancel{p^{\ominus}}))}$$

Cancelling terms

$$K = \frac{x_{PCl_3} \times x_{Cl_2}}{x_{PCl_5}} \times (p_{total}/p^{\ominus})$$

Determine the effect of changing p_{total} on the mole fractions.

At fixed temperature, K is constant. $p^{\ominus} = 1$ bar so, if p_{total} increases, the ratio of the mole fractions

$$\frac{x_{PCl_3} \times x_{Cl_2}}{x_{PCl_5}}$$

must go down to compensate. Hence, x_{PCl_3} and x_{Cl_2} must go down while x_{PCl_5} increases. This means that the reverse reaction is promoted and more PCl_5 is present in the equilibrium mixture than at the lower pressure.

The same result is predicted by Le Chatelier's principle; increasing the pressure will drive the reaction towards the side containing the lower number of moles of gas.

Question

Without using Le Chatelier's principle, predict the effect of changing the pressure in a vessel containing the following equilibrium reaction at constant temperature.

$$C(s) + CO_2(g) \rightleftharpoons 2\,CO(g)$$

Changing the amount of a component

Adding or removing one of the components from a reaction will affect the concentration (or pressure) of the others in the equilibrium mixture. This can be seen by examining the expression for the equilibrium constant. As an example

$$N_2(g) + 3H_2(g) \rightleftharpoons 2NH_3(g)$$

$$K = \frac{(p_{NH_3}/p^{\ominus})^2}{(p_{N_2}/p^{\ominus}) \times (p_{H_2}/p^{\ominus})^3} \quad \text{where } p^{\ominus} = 1\,\text{bar}$$

$$= \frac{p_{NH_3}^{\,2}}{p_{N_2} \times p_{H_2}^{\,3}} \times (p^{\ominus})^2$$

If some NH_3 is added to an equilibrium mixture at the same total pressure, the partial pressure of NH_3 will increase and the reaction is no longer at equilibrium. Since K must remain constant at fixed temperature, the partial pressures of N_2 and H_2 must also increase. Some of the added NH_3 reacts to form N_2 and H_2 until a new equilibrium is established with the same value of K.

If N_2 is added to an equilibrium mixture, it follows that, to re-establish the equilibrium, some of it must react to form more NH_3. This in turn reduces the partial pressure of H_2. These changes are illustrated in Figure 15.5.

In a sealed bottle of fizzy drink, CO_2(g) and CO_2(aq) are in equilibrium. When the cap is unscrewed, pressure is reduced and CO_2(g) comes out of solution. This normally happens slowly, unless there are nucleation sites around which CO_2(g) bubbles can grow. For example, dropping mint sweets into cola provides so many nucleation sites that the CO_2(g) and the contents erupt from the bottle.

The direction that the reaction will take after changing the amounts of a component can also be assessed by calculating the reaction quotient, Q, and comparing with K.

Visit the Online Resource Centre to view screencast 15.3 which explains the graph in Figure 15.5 by walking you through the changes that happen when reactant or product is added to an equilibrium mixture

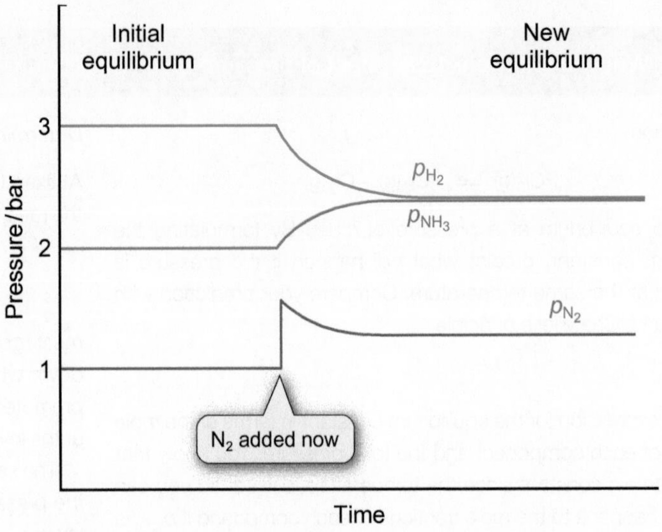

Initial equilibrium
New equilibrium
Pressure / bar
3
2
1
p_{H_2}
p_{NH_3}
p_{N_2}
N_2 added now
Time

Figure 15.5 If one of the components is added to an equilibrium mixture in the gas phase, the partial pressures change to keep the value of K constant. For the reaction $N_2(g) + 3H_2(g) \rightleftharpoons 2NH_3(g)$, if N_2 is added, the partial pressure of H_2 falls and that of NH_3 rises.

Changing the temperature

Le Chatelier's principle states that the equilibrium position for an exothermic reaction will move to favour the reactants if the temperature is increased. For an endothermic reaction, an increased temperature favours the products.

This qualitative observation can be explained quantitatively by combining Equations 15.4 (p.705) and 14.16 (p.675) for the standard Gibbs energy change for the reaction

$$\Delta_r G^{\ominus} = -RT \ln K \qquad (15.4)$$

$$\Delta_r G^{\ominus} = \Delta_r H^{\ominus} - T\Delta_r S^{\ominus} \qquad (14.16)$$

Equating these

$$-RT \ln K = \Delta_r H^{\ominus} - T\Delta_r S^{\ominus}$$

Dividing through by $-RT$ gives

$$\ln K = \frac{-\Delta_r H^{\ominus}}{RT} + \frac{\cancel{T}\Delta_r S^{\ominus}}{R\cancel{T}}$$

which simplifies to

$$\ln K = \frac{\Delta_r S^{\ominus}}{R} - \frac{\Delta_r H^{\ominus}}{RT} \qquad (15.10)$$

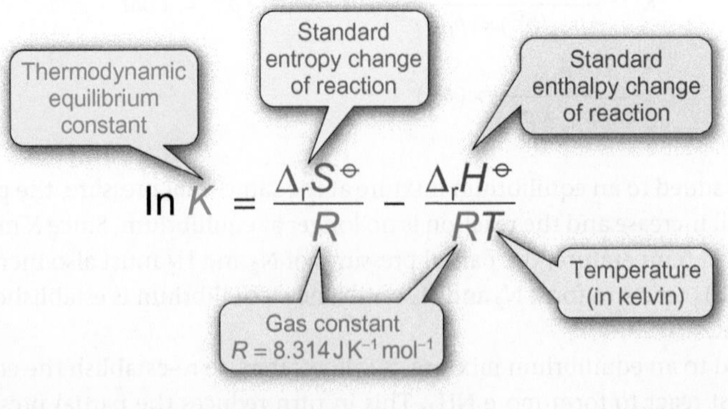

Thermodynamic equilibrium constant
Standard entropy change of reaction
Standard enthalpy change of reaction
Temperature (in kelvin)
Gas constant $R = 8.314\ J\ K^{-1}\ mol^{-1}$

The Dutch chemist, Jacobus van 't Hoff, was the winner of the first Nobel Prize in Chemistry (awarded in 1901) for his studies in solution chemistry.

This is known as the **van 't Hoff equation.**

A common way of using the van 't Hoff equation is in the form of a graph. If $\Delta_r H^\ominus$ and $\Delta_r S^\ominus$ are independent of temperature, as is usually the case over small temperature intervals, Equation 15.10 can be written as

$$\ln K = \text{constant} - \frac{\Delta_r H^\ominus}{R}\left(\frac{1}{T}\right) \qquad (15.11)$$

where the constant is given by $(\Delta_r S^\ominus/R)$. A graph of $\ln K$ against $1/T$ is a straight line with a gradient of $-\Delta_r H^\ominus/R$, as shown in Figure 15.6. (The intercept with the y-axis (when $1/T = 0$) is equal to $\Delta_r S^\ominus/R$.)

How does this equation relate to Le Chatelier's principle? Remember that $\Delta_r H^\ominus$ and $\Delta_r S^\ominus$ do not vary much with temperature. For an *endothermic* reaction, the $(-\Delta_r H^\ominus/RT)$ term will be *negative*. As temperature increases, the magnitude of this term gets *smaller*. Subtracting a smaller value means that $\ln K$, and hence K, will be *bigger* than at a lower temperature. If K is bigger, a larger proportion of products will form. This agrees with Le Chatelier's principle, which states that the position of equilibrium moves in the direction of the endothermic change when the temperature is raised, because this lowers the temperature and minimizes the effect of the change.

The effect of temperature on a reaction equilibrium is best illustrated by looking at examples such those as in Worked example 15.9 and Box 15.6 (p.720). The effects are summarized in Table 15.2.

$$\begin{array}{cccc} y = & mx & + & c \\ \ln K = & -\frac{\Delta_r H^\ominus}{R}\left(\frac{1}{T}\right) & + & \frac{\Delta_r S^\ominus}{R} \end{array}$$

The equation for a straight line is discussed in Maths Toolkit MT4 (p.1318).

Visit the Online Resource Centre to view screencast 15.4 which walks you through the derivation of Equation 15.10 and shows how it is used to find $\Delta_r H$ and $\Delta_r S$.

Measuring the variation of an equilibrium constant with temperature gives a convenient method for obtaining the standard enthalpy change for a reaction using Equation 15.10. The concentrations of components can often be accurately measured so that K can be found precisely. This method is often more convenient and accurate than using the calorimetric methods described in Section 13.6 (p.644).

Table 15.2 Effect of temperature on reaction equilibria

	Temperature change	K	Equilibrium composition
Exothermic	Increase	Smaller	Favours reactants
	Decrease	Bigger	Favours products
Endothermic	Increase	Bigger	Favours products
	Decrease	Smaller	Favours reactants

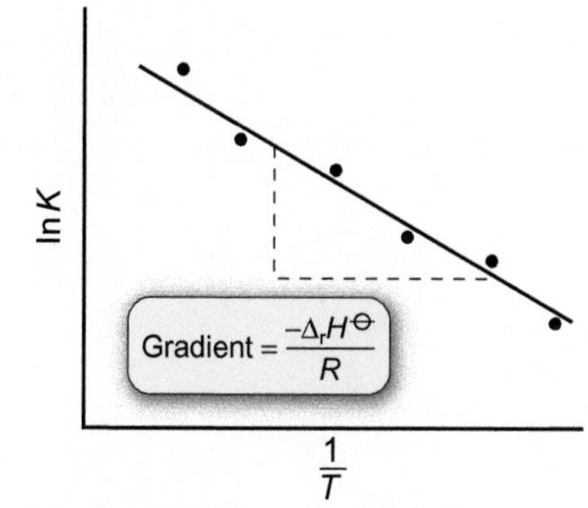
ln K
Gradient = $-\Delta_r H^\ominus / R$
1/T

Figure 15.6 A plot of $\ln K$ against $1/T$ is a straight line. The standard enthalpy change for the reaction can be found from the gradient.

A similar graphical approach is used in the treatments of the Arrhenius equation in Section 9.7 (p.425), and the Clausius–Clapeyron equation in Section 17.2 (p.775).

Worked example 15.9 Temperature dependence of K

The equilibrium constant for the reaction

$$2NO_2(g) \rightleftharpoons N_2O_4(g)$$

was measured over a range of temperatures. The results are shown in the table. Calculate $\Delta_r H^\ominus$ and $\Delta_r S^\ominus$ for the reaction.

Temperature/K	282.2	293.2	298.2	306.2	325.2	333.2	343.2
Equilibrium constant, K	33.2	13.1	8.80	4.80	1.30	0.75	0.41

Strategy

Use the van 't Hoff equation 15.10.

$$\ln K = \frac{\Delta_r S^\ominus}{R} - \frac{\Delta_r H^\ominus}{RT} \qquad (15.10)$$

Plot a graph of $\ln K$ against $1/T$. The gradient of the line is $-\Delta_r H^\ominus/R$. The intercept of the graph, that is, the value of $\ln K$ when $1/T = 0$, is equal to $\Delta_r S^\ominus/R$. →

➜ Solution

Calculate and tabulate the data needed to plot the graph of ln*K against* 1/*T. Take care to use the correct units.*

Temperature / K	282.2	293.2	298.2	306.2	325.2	333.2	343.2
$(1/T)/10^{-3}\ K^{-1}$	3.54	3.41	3.35	3.27	3.08	3.00	2.91
Equilibrium constant, K	33.2	13.1	8.80	4.80	1.30	0.75	0.41
$\ln K$	3.50	2.57	2.17	1.57	0.26	−0.29	−0.89

Plot the graph and calculate the gradient and intercept of the line.

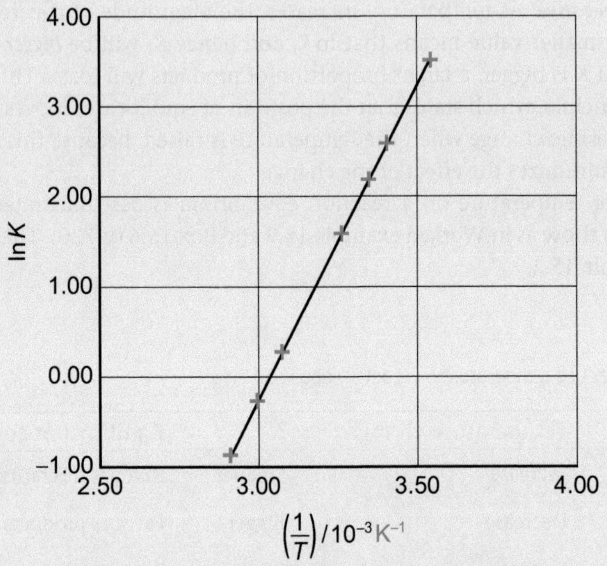

4.00
3.00
2.00
1.00
0.00
−1.00
ln K
2.50
3.00
3.50
4.00
(1/T)/10⁻³K⁻¹

Using graph-plotting software, the equation of the straight line is given as

$$\ln(K) = 6970(1/T) - 21.2$$

so, gradient = 6970 K and intercept = −21.2

(In this example, if you plot the graph manually, you will find it difficult to obtain a value for the intercept because the extrapolation to $1/T = 0$ is so long.)

Calculate $\Delta_r H^{\ominus}$ *and* $\Delta_r S^{\ominus}$.

$$\text{Gradient} = -\frac{\Delta_r H^{\ominus}}{R} = 6970\ \text{K} \qquad R = 8.314\ \text{J K}^{-1}\ \text{mol}^{-1}$$

$$\Delta_r H^{\ominus} = -6970\ \text{K} \times 8.314\ \text{J K}^{-1}\ \text{mol}^{-1}$$

$$= -57.9\ \text{kJ mol}^{-1}$$

$$\text{Intercept} = \frac{\Delta_r S^{\ominus}}{R} = -21.2$$

$$\Delta_r S^{\ominus} = -21.2 \times 8.314\ \text{J K}^{-1}\ \text{mol}^{-1}$$

$$= -176\ \text{J K}^{-1}\ \text{mol}^{-1}$$

Question

The equilibrium constant for a reaction was measured and found to fit the straight-line relationship

$$\ln(K) = 7.55 - 4844/T$$

Calculate: (a) $\Delta_r H^{\ominus}$ and $\Delta_r S^{\ominus}$; (b) $\Delta_r G^{\ominus}$ for the reaction at 500 K.

Adding a catalyst

A catalyst is a substance that increases the rate of a reaction without being used up. However, the rates of *both* the forward and reverse reactions are increased, so that a catalyst has *no effect* on the position of equilibrium. This makes sense in thermodynamic terms since K depends on $\Delta_r G^{\ominus}$, which is a state function. $\Delta_r G^{\ominus}$ depends only on the Gibbs energies of the products and reactants and not on the path by which a reaction occurs. Catalysts work by providing a different reaction path with a lower activation energy, but do not change the initial and final states.

The effect of catalysts on the rates of reactions is discussed in Section 9.9 (p.437).

Summary

- The effect of changing conditions on chemical reactions can be predicted qualitatively using Le Chatelier's principle: when a change is made to a system in dynamic equilibrium, the system responds to minimize the effect of the change.
- The equilibrium constant does not vary with pressure at constant temperature but the position of equilibrium can depend on the total pressure.
- The effect of temperature on the equilibrium constant is described by the van't Hoff equation

$$\ln K = \frac{\Delta_r S^{\ominus}}{R} - \left(\frac{\Delta_r H^{\ominus}}{R}\right)\frac{1}{T}$$

- Addition of a catalyst accelerates both forward and reverse reactions and so does not influence the position of equilibrium.

? For practice questions on these topics, see questions 11, 12, 17, 18, and 22–25 at the end of this chapter (pp.722–723).

15.6 Applying the thermodynamics in Chapters 13, 14, and 15

This chapter builds on the concepts introduced in Chapters 13 and 14, and applies these ideas to chemical equilibrium. To complete this chapter, one further example is given in Box 15.6. It shows how a few thermodynamic properties readily available from tables such as those in Appendix 7 (p.1352) can be used to estimate the product yields from an important chemical reaction: the industrial synthesis of ammonia.

While $\Delta_r G^{\ominus}$ and K can be used to predict the equilibrium position of a chemical reaction, from which some idea of the reaction yields can be gained, it does not always tell the whole story. Maximizing the equilibrium yield isn't always the best approach to operating an industrial process. For example, think about an exothermic reaction. Le Chatelier's principle states that the yield of products will be maximized by reducing the temperature. However, this will slow down the reaction. In order to optimize a reaction, a process chemist must balance the yield of products with the length of time for which the reaction must be carried out and its energy demand. A 70% yield today may be better than a 100% yield next week!

Box 15.6 A case study in reaction thermodynamics: the Haber process

Nitrogen-containing compounds are very useful chemicals, two examples being their use as agricultural fertilizers and as explosives. You might think there is little to link these applications—but in fact there is a common factor that provides an interesting example of how chemical equilibria are considered in industrial processes.

Until the early part of the twentieth century, the main sources of nitrogen compounds for use as fertilizers were sodium nitrate and potassium nitrate, which were mined from deposits in the ground. In the early twentieth century, the food requirements of an increasing population in Europe meant that mining could no longer provide sufficient nitrates for agricultural needs. In 1912, the German chemist Fritz Haber developed a process to 'fix' atmospheric nitrogen to manufacture ammonia (NH_3)

$$N_2(g) + 3\,H_2(g) \rightleftharpoons 2\,NH_3(g)$$

Haber realized that, in order to maximize the production of ammonia, the reaction should be operated at high pressure. Also, the removal of ammonia as it is formed favours the forward reaction. The forward reaction is exothermic and so, in principle, should be more efficient at low temperatures. However, at low temperatures the reaction is too slow to be used commercially and an important discovery for the production of large amounts of ammonia was an iron-based catalyst that allows the reaction to proceed at usable rates. Haber used a kinetic and thermodynamic analysis of the reactions to determine that the optimum conditions were 450 °C and 200 atm pressure. The process is sometimes known as the 'Haber–Bosch' process to recognize the contribution of Karl Bosch, the engineer who designed and built the plant to operate under these (in those days) challenging conditions. The ammonia produced is converted to compounds such as ammonium sulfate or ammonium phosphate, which are used as fertilizers. Today, most reactors operate between 25 atm and 150 atm.

▲ High-pressure reaction vessels for the industrial synthesis of ammonia.

Shortly after its discovery the Haber process became even more vital to Germany. At the outbreak of the First World War in 1914, Germany relied on imports of nitrates from South America. These were cut off by a naval blockade so the Haber process became the source of all nitrogen compounds for fertilizers and also for the explosives in weapons used by the military. Haber was awarded the 1918 Nobel Prize in Chemistry for his discovery of the catalysts and optimization of the nitrogen fixation reaction. This was controversial, however, since he also was instrumental in developing the manufacture of poisonous gases such as chlorine and phosgene, used as weapons during the war.

The Haber process remains the major source of ammonia to this day and is of tremendous commercial significance, with a worldwide production in 2011 estimated as 1.5×10^{11} kg.

Questions

(a) Use data from Appendix 7 (p.1352) on enthalpy change of formation and heat capacity to calculate $\Delta_r H^{\ominus}$ for the synthesis of ammonia at: (i) 298 K; (ii) 773 K; (iii) 1000 K.

(b) Calculate the standard Gibbs energy change of the reaction at: (i) 298 K; (ii) 773 K; (iii) 1000 K.

(c) Comment on the significance of the results from (b).

(d) Write the expression for the equilibrium constant and use it, together with the van 't Hoff equation (Equation 15.10, p.716), to explain the effect on the yield of ammonia of: (i) increasing the pressure; (ii) increasing temperature; (iii) adding hydrogen gas to the reaction mixture; (iv) changing the amount of catalyst in the reaction chamber.

(e) Calculate and comment on the equilibrium constant for the reaction at: (i) 298 K; (ii) 773 K; (iii) 1000 K.

(f) A reaction mixture at 298 K has the following partial pressures: 1 bar N_2, 3 bar H_2, and 0.5 bar NH_3. In which direction will the reaction move towards equilibrium?

Concept review

By the end of this chapter, you should be able to do the following.

- Define and use the three types of equilibrium constant: K_c, K_p, and the thermodynamic equilibrium constant, K.
- Define and use the reaction quotient, Q.
- Calculate Q given the composition of a reaction mixture.
- Calculate K from $\Delta_r G^{\ominus}$ and vice versa.
- Calculate the composition of an equilibrium mixture from K, given appropriate data.
- Describe and explain how the equilibrium constant and composition change when experimental conditions such as pressure, concentration, and temperature are varied.
- Use the van 't Hoff equation to describe the effect of changing temperature on chemical equilibrium and to obtain values for $\Delta_r H^{\ominus}$ and $\Delta_r S^{\ominus}$ from a plot of $\ln K$ against $1/T$.
- Apply equilibrium thermodynamics to real chemical situations.

Key equations

Thermodynamic equilibrium constant	$K = \dfrac{\prod (a(\text{products}))^{\nu_P}_{\text{eqm}}}{\prod (a(\text{reactants}))^{\nu_R}_{\text{eqm}}}$	(15.1)
The reaction quotient	$Q = \dfrac{\prod (a(\text{products}))^{\nu_P}}{\prod (a(\text{reactants}))^{\nu_R}}$	(15.2)
Relationship between $\Delta_r G^{\ominus}$ and K	$\Delta_r G^{\ominus} = -RT\ln K$	(15.4)
	$K = e^{\left(\frac{-\Delta_r G^{\ominus}}{RT}\right)}$	(15.5)
van 't Hoff equation showing variation of K with temperature	$\ln K = \dfrac{\Delta_r S^{\ominus}}{R} - \dfrac{\Delta_r H^{\ominus}}{RT}$	(15.10)

Questions

More challenging questions are indicated by an asterisk *.

Some problems require the use of data from Appendix 7 (p.1352).

1 Write the expressions for the thermodynamic equilibrium constant K for the following reactions. (Section 15.1)

(a) $4\,NH_3(g) + 7\,O_2(g) \rightleftharpoons 4\,NO_2(g) + 6\,H_2O(l)$

(b) $HCN(aq) + H_2O(l) \rightleftharpoons CN^-(aq) + H_3O^+(aq)$

(c) $PCl_5(g) \rightleftharpoons PCl_3(g) + Cl_2(g)$

(d) $3O_2(g) \rightleftharpoons 2O_3(g)$

(e) $2\,H_2O(l) \rightleftharpoons H_3O^+(aq) + OH^-(aq)$

(f) $3\,Zn(s) + 2\,Fe^{3+}(aq) \rightleftharpoons 2\,Fe(s) + 3\,Zn^{2+}(aq)$

2 The solubility of silver chloride in water at 25 °C is 1.27×10^{-5} mol dm^{-3}. Calculate

(a) the solubility product of AgCl

(b) the solubility of AgCl in 0.01 mol dm^{-3} aqueous sodium chloride solution. (Section 15.1)

3 The equilibrium constants for two gas phase reactions at 1000 °C are shown.

$$CO_2(g) \rightleftharpoons CO(g) + \tfrac{1}{2}O_2(g) \quad K_1 = 9.1\times10^{-12}$$
$$H_2O(g) \rightleftharpoons H_2(g) + \tfrac{1}{2}O_2(g) \quad K_2 = 7.1\times10^{-12}$$

Use these data to find the equilibrium constant at the same temperature for the reaction:

$$CO_2(g) + H_2(g) \rightleftharpoons CO(g) + H_2O(g)$$

(Section 15.1)

4 The equilibrium constant for the following reaction is $K = 1.5 \times 10^4$.

$$CO(g) + Cl_2(g) \rightleftharpoons COCl_2(g)$$

In a reaction vessel, the partial pressures of the reaction mixture are:

$COCl_2$ 0.050 bar, CO 0.0010 bar, and Cl_2 0.0001 bar.

(a) Calculate the value for the reaction quotient, Q, for this mixture.

(b) What will happen to the composition of the reaction mixture as it moves to equilibrium? (Section 15.2)

5 An important reaction in the formation of smog is:

$$O_3(g) + NO(g) \rightleftharpoons O_2(g) + NO_2(g)$$

Under certain conditions, the equilibrium constant for this reaction is $K = 6.0 \times 10^{34}$. If the partial pressures of each gas in the air over your home town were 1.0×10^{-6} bar O_3, 1.0×10^{-5} bar NO, 2.5×10^{-4} bar NO_2, and 8.2×10^{-3} bar O_2, what could you say about the course of the reaction as it moves to equilibrium? (Section 15.2)

6 Calculate the equilibrium constant, K, at 298 K for the reaction

$$H_2O(l) \rightleftharpoons H_2(g) + \tfrac{1}{2}O_2(g)$$

The standard Gibbs energy change of formation of H_2O (l) at 298 K is $-237.1\ kJ\ mol^{-1}$. (Section 15.3)

7 Use the data below to calculate the standard Gibbs energy change and the equilibrium constant, K, at 298 K for the reaction

$$CO(g) + H_2O(g) \rightleftharpoons CO_2(g) + H_2(g)$$

(Section 15.3)

	$\Delta_f H^\ominus_{298}/kJ\ mol^{-1}$	$S^\ominus_{298}/J\ K^{-1}\ mol^{-1}$
CO (g)	−110.5	197.7
H_2O (g)	−241.8	188.8
CO_2 (g)	−393.5	213.7
H_2 (g)	0	130.7

8 A vessel initially contains graphite and partial pressures of O_2 (g) and CO_2 (g) of 0.02 bar and 0.001 bar, respectively, at 298 K. The reaction that occurs is

$$C(s) + O_2(g) \rightleftharpoons CO_2(g)$$

In what direction will the reaction proceed? (The standard Gibbs energy change of formation of CO_2 (g) is $-394.4\ kJ\ mol^{-1}$ at 298 K.) (Section 15.3)

9 2.0 mol of carbon disulfide and 4.0 mol of chlorine react at constant temperature according to this equation

$$CS_2(g) + 3Cl_2(g) \rightleftharpoons S_2Cl_2(g) + CCl_4(g)$$

At equilibrium, 0.30 mol of tetrachloromethane are formed. How much of each of the other components is present in this equilibrium mixture? (Section 15.4)

10 Nitrosyl chloride (NOCl) decomposes to nitric oxide and chlorine when heated

$$2\,NOCl(g) \rightleftharpoons 2\,NO(g) + Cl_2(g)$$

In a mixture of all three gases at 600 K, the partial pressure of NOCl is 0.88 bar, that of NO is 0.06 bar, and the partial pressure of chlorine is 0.03 bar. At 600 K, the equilibrium constant, K, is 0.060. (Section 15.2 and several others)

(a) What is the value of the reaction quotient for this mixture? Is the mixture at equilibrium?

(b) In which direction will the system move to reach equilibrium?

(c) What will happen if an additional amount of NOCl (g) is injected into the reaction?

11 The following reaction is exothermic

$$Ti(s) + 2Cl_2(g) \rightleftharpoons TiCl_4(g)$$

How could the yield of $TiCl_4$ be increased? (Section 15.5)

12 The following gas phase reaction is exothermic

$$CO(g) + \tfrac{1}{2}O_2(g) \rightleftharpoons CO_2(g)$$

What will be the effect of (a) increasing the pressure, (b) increasing the temperature, and (c) adding a catalyst on (i) the equilibrium constant, K, and (ii) the yield of CO_2? (Section 15.5)

13 For the gas phase reaction

$$COCl_2(g) \rightleftharpoons CO(g) + Cl_2(g)$$

at 100 °C and 2 bar pressure, the fraction, α, of phosgene ($COCl_2$) that reacts is 6.3×10^{-5}. Calculate the equilibrium constant, K, for the reaction. (Section 15.4)

14 Bromine and chlorine react to produce bromine monochloride according to the equation

$$Br_2(g) + Cl_2(g) \rightleftharpoons 2\,BrCl(g)$$

0.2 mol of bromine gas and 0.2 mol of chlorine gas are introduced into a sealed flask with a volume of $5.0\ dm^3$. Under the conditions of the experiment, $K = 36.0$. How much BrCl will be present at equilibrium? (Section 15.4)

15 When ammonia dissolves in water, the following equilibrium is established

$$NH_3(aq) + H_2O(l) \rightleftharpoons NH_4^+(aq) + OH^-(aq)$$

Calculate the hydroxide ion concentration in the solution formed when 0.10 mol of ammonia are dissolved in sufficient water to make $500\ cm^3$ of solution. (At the temperature of the reaction, $K = 1.8 \times 10^{-5}$.) (Section 15.4)

16 The standard Gibbs energy change for the gas phase isomerization of *cis*-2-pentene to *trans*-2-pentene is $-3.67\ kJ\ mol^{-1}$. (Section 15.4)

(a) Calculate the equilibrium constant for the reaction.

(b) What are the equilibrium mole fractions of the *cis* and *trans* isomers at 298 K?

17* What is the standard enthalpy change for a reaction in which the equilibrium constant doubles when the temperature increases from 298 K to 308 K? (Section 15.5)

18* CO_2 decomposes into CO and O_2 over a platinum catalyst

$$2\,CO_2(g) \rightleftharpoons 2\,CO(g) + O_2(g)$$

At 1 bar pressure, the fraction, α, of CO_2 that reacts is 0.014 at 1395 K, 0.025 at 1443 K, and 0.047 at 1498 K. (Section 15.5)

(a) Calculate the equilibrium constant at 1443 K.

(b) Calculate $\Delta_r H^\ominus$ and $\Delta_r S^\ominus$ for the reaction.

19* For a general reaction

$$\alpha A + \beta B \rightleftharpoons \gamma C + \delta D$$

derive the relationship between the Gibbs energy change of the reaction and the reaction quotient. Use the relationship to show that $\Delta_r G^\ominus = -RT \ln K$. (Section 15.3)

20 Silver carbonate decomposes on heating. Calculate the equilibrium constant at 383 K for the reaction. Would it be appropriate to dry a sample of silver carbonate in an oven at 383 K? (Section 15.3)

	Ag_2CO_3 (s)	Ag_2O (s)	CO_2 (g)
$\Delta_f H^\ominus_{298}$/kJ mol^{-1}	−501.4	−31.1	−393.5
$S^\ominus_{298}$/J K^{-1} mol^{-1}	167.3	121.3	213.7
C_p/J K^{-1} mol^{-1}	109.6	65.9	37.1

21* For the following esterification reaction

$$C_2H_5OH(l) + CH_3CO_2H(l) \rightleftharpoons CH_3CO_2C_2H_5(l) + H_2O(l)$$

the equilibrium constant at 298 K is 3.8. A mixture containing 0.5 mol dm^{-3} each of ethanol and ethanoic acid was reacted in a sealed flask at 298 K. After a certain time, the concentrations of each had changed to 0.39 mol dm^{-3}. (Section 15.3)

(a) Had the reaction reached equilibrium?

(b) If not, what would the concentration of $CH_3CO_2C_2H_5$ (aq) be at equilibrium?

(c) In practice, the reaction is carried out so as to remove the water as it forms. Explain why this is done.

22 The equilibrium constant for the reaction

$$SO_3(g) \rightleftharpoons SO_2(g) + \tfrac{1}{2}O_2(g)$$

has been measured as 0.157 at 900 K and 0.513 at 1000 K. Assuming the values of ΔH and ΔS are constant over this temperature range, calculate the standard enthalpy change and the standard entropy change for the reaction. (Section 15.5)

23 An equilibrium constant, K, is five times larger when a reaction is performed at 200 K than at 150 K. Assuming the enthalpy change is constant over this temperature range, calculate $\Delta_r H^\ominus$ for the reaction. (Section 15.5)

24 If the following reaction was at equilibrium in a closed vessel at a controlled temperature, what would be the effect of adding more H_2 to the reaction vessel and permitting the reaction to approach equilibrium again? (Section 15.5)

$$CO(g) + H_2O(g) \rightleftharpoons CO_2(g) + H_2(g)$$

25* The folding and unfolding of proteins have important biological effects (see Chapter 14, p.673). Studies have been carried out on protein G which shows a reversible transition from a folded to an unfolded form as the temperature is raised.

$$\text{protein G (folded)} \rightleftharpoons \text{protein G (unfolded)}$$

For the unfolding of protein G at the normal cell temperature of 37 °C (310 K), $\Delta H^\ominus = +210.9$ kJ mol^{-1} and $\Delta S^\ominus = +616.7$ J K^{-1} mol^{-1}. (Section 15.5)

(a) Write an expression for K for this equilibrium.

(b) What is the value of K at the temperature when the protein is half-folded and half-unfolded?

(c) What is the value of $\Delta G^\ominus$ at this temperature?

(d) Calculate $\Delta G^\ominus$ for the unfolding process at 310 K.

(e) Calculate K at 310 K and comment on the value you obtain.

(f) Calculate a value for K at 69 °C (342 K) and comment on the value you obtain.

26* For the reaction

$$CO(g) + H_2O(g) \rightleftharpoons CO_2(g) + H_2(g)$$

	$\Delta_f H^\ominus_{298}$/kJ mol^{-1}	$S^\ominus_{298}$/J K^{-1} mol^{-1}	C_p/J K^{-1} mol^{-1}
CO (g)	−110.5	197.7	29.1
H_2O (g)	−241.8	188.8	33.6
CO_2 (g)	−393.5	213.7	37.1
H_2 (g)	0	130.7	28.8

(Section 15.6)

(a) Use the data to calculate the Gibbs energy change and the equilibrium constant at 798 K.

(b) 1 mol of CO (g) and 1 mol of H_2O (g) are introduced into a container at 798 K and allowed to come to equilibrium at a total pressure of 0.1 bar. Write an expression to relate the equilibrium constant, K, to the number of moles of each component in the equilibrium mixture and calculate the number of moles of H_2 gas present at equilibrium.

27 Calculate the maximum quantity (in mol) of KIO_3 that can be added to 250 cm^3 of a solution containing 1.00×10^{-3} mol dm^{-3} of Cu^{2+} (aq) without precipitating Cu(IO3)$_2$ (s). $K_{sp} = 1.4 \times 10^{-7}$ for $Cu(IO_3)_2$ (s).

28 For the reaction, $H_2(g) + I_2(s) \rightleftharpoons 2HI(g)$, $\Delta_r G^\ominus = 3.40$ kJ mol^{-1} at 298.15 K.

(a) Calculate the equilibrium constant.

(b) Does the reaction favour the products or reactants?

(c) If additional H_2 (g) was added to the equilibrium mixture at the same temperature, predict what would happen to the position of equilibrium.

29 Use data in Appendix 7 to answer the following for the reaction.

$$2\,NO(g) + O_2(g) \rightleftharpoons 2\,NO_2(g)$$

(a) Calculate the enthalpy change and the entropy change for the reaction at 25 °C. Is the reaction spontaneous at this temperature?

(b) Find the equilibrium constant under these conditions?

30 A student was investigating the following equilibrium reaction which has an equilibrium constant of 0.220 at 800 °C

$$CaCO_3(s) \rightleftharpoons CaO(s) + CO_2(g)$$

and did four experiments.

(a) 0.2 g of $CaCO_3$ (s) was heated to 800 °C in a 1.0 dm^3 container

(b) 2.0 g of $CaCO_3$ (s) was heated to 800 °C in a 1.0 dm^3 container

(c) 0.2 g of $CaCO_3$ (s) was heated to 800 °C in a 500 cm^3 container

(d) 2.0 g of $CaCO_3$ (s) was heated to 800 °C in a 500 cm^3 container

The pressure of CO_2 (g) measured in each case was (i) 0.18 bar, (ii) 0.22 bar, (iii) 0.22 bar, (iv) 0.22 bar. Explain these observations.

31 The equilibrium constants for the gas-phase dissociation of molecular iodine,

$$I_2(g) \rightleftharpoons 2I(g)$$

have been measured at the following temperatures:

T/K	872	973	1073	1173
K_p	1.8×10^{-4}	1.8×10^{-3}	1.08×10^{-2}	4.8×10^{-2}

(a) Use a graphical method to determine $\Delta_r H^{\ominus}$ and $\Delta_r S^{\ominus}$ for this reaction.

(b) Determine $\Delta_r G^{\ominus}$ at 1000 K.

(c) What would be the effect on the position of equilibrium of increasing the temperature?

16

When you are far away from mains electricity supply, portable power is vital. ➤

Electrochemistry

This chapter builds on the following topics:

- Chemical equations Section 1.4, p.21
- Energy changes in chemical reactions Section 1.6, p.41
- The strengths of acids and bases Section 7.2, p.308
- Gibbs energy Section 14.5, p.765
- Variation of Gibbs energy with conditions Section 14.6, p.686
- Gibbs energy and equilibrium constants Section 15.3, p.705

◄ A portable fuel cell can be used to recharge batteries when there's no mains electricity.

Electrical energy on the move

To use an electrical device on the move—like a smart phone or laptop computer—you need a portable electricity source. Portable fuel cells have arrived and portable solar cells are on the way, but for the time being batteries, or *electrochemical cells* as they are more accurately described, are essential.

All electrochemical cells work on the same basic chemical principle (Section 16.3, p.735). There are two half cells (commonly called electrodes): the anode, where oxidation occurs, and the cathode, where reduction occurs. These two redox reactions are separated so that electrons have to flow down a wire to get from the anode to the cathode, providing an electric current.

The most widely sold batteries are alkaline cells. These involve the following two redox reactions.

At the zinc anode:

$$Zn\,(s) + 2\,OH^-\,(aq) \rightarrow ZnO\,(s) + H_2O\,(l) + 2\,e^- \quad \text{oxidation}$$

At the graphite cathode:

$$2\,MnO_2\,(s) + H_2O\,(l) + 2\,e^- \rightarrow Mn_2O_3\,(s) + 2\,OH^-\,(aq) \quad \text{reduction}$$

Adding these two half reactions together gives the overall cell reaction:

$$Zn\,(s) + 2\,MnO_2\,(s) \rightarrow ZnO\,(s) + Mn_2O_3\,(s)$$

This type of cell produces a higher voltage and is much more stable over time than older types of batteries. It can also deliver higher power, making it useful for a wide range of applications. While rather more expensive, alkaline cells last much longer when used under the same conditions as older batteries (see Box 16.2, p.737).

Lithium cells use lithium instead of zinc as the anode in a variety of cells. This makes the cell lighter and gives it a higher voltage, because lithium is a more reactive metal than zinc.

Alkaline batteries and lithium batteries cannot be recharged. The redox reactions they employ are not reversible—once the reaction has reached equilibrium, the battery is 'dead'. Other batteries can be recharged by passing an electric current through them and so reversing the redox reaction. The lead–acid battery in a car is rechargeable—but it is far too heavy to be any use in a mobile phone.

Battery technology is developing all the time, producing lighter and more powerful rechargeable cells. Most mobile phones and laptops use the lightweight rechargeable lithium-ion battery (different from the lithium batteries above), which takes advantage of the low atomic mass of lithium.

Reversing a current-producing reaction to recharge a cell involves electrolysis: using electricity to drive a chemical reaction in the non-spontaneous direction (Section 16.5, p.755). Some examples of rechargeable batteries are described in Box 16.6 (p.759).

Modern life would be impossible without electrochemistry. Much of the energy that powers our mobile existence comes from batteries (see introductory section). A car relies on a rechargeable battery—chemical reactions store energy and release it when needed to start the engine. Electrochemical reactions are used on a much larger scale in the industrial manufacture of some metals (e.g. aluminium) and other widely used substances (e.g. chlorine) by electrolysis. The splitting of water by electrolysis into hydrogen and oxygen could be the basis of hydrogen-based clean power sources in the future.

The 'hydrogen economy' is discussed in Box 25.1 (p.1141).

In studying electrochemistry in this chapter, you will find that, as well as explaining how batteries work, electrochemistry explains the rusting of iron and lies behind some important biochemical processes, such as the transmission of nerve impulses. Section 16.4 shows how electrochemistry is linked to the fundamental laws of thermodynamics covered in Chapters 14 and 15.

16.1 What is electrochemistry?

Electrochemistry deals with the interconversion of electrical energy and chemical energy. Electrical effects occur as a result of the movement of electrical charge, either as ions in solution or electrons in a conductor. This chapter is concerned with the factors that govern the transfer of ions and electrons in chemical reactions.

Some terms used in electrochemistry

The SI unit of electric charge is the coulomb, C. An electron or a singly charged negative ion carries a charge of -1.6×10^{-19} C; a proton or a singly charged positive ion carries a charge of $+1.6 \times 10^{-19}$ C. The rate at which charge passes a point is called the electric current, given the symbol I. Current is measured in amperes (amps), A. 1 A is the current when 1 C of charge flows per second

> The units used in electricity are named after some of the pioneers of the subject: the amp is named after the French physicist André-Marie Ampère (1775–1836), the ohm after the German physicist Georg Simon Ohm (1789–1854), and the volt after the Italian physicist Count Alessandro Volta (1745–1827), the inventor of the first battery.

$$1\,\mathrm{A} = 1\,\mathrm{C\,s^{-1}} \tag{16.1}$$

An electric current flows if there is a **potential difference (voltage)** between two points in a circuit. The potential difference represents the 'driving force' that pushes the electrons or ions around the circuit. Potential difference is measured in volts, V. The current, I, and potential difference, V, are related by **Ohm's law**

$$V = I \times R \tag{16.2}$$

R is the resistance of the material through which the current is flowing and is measured in ohms, Ω. If a potential difference of 1 V results in a current of 1 A, then the material has a resistance of 1 Ω. For a constant potential difference, the lower the resistance in a circuit, the higher the current.

If an electrochemical cell is used to produce a high current, the voltage falls as the reactants are used up. However, if the resistance of the circuit is high, the current is low and the voltage does not fall so quickly. When the resistance is extremely high, an infinitesimally small current flows and the voltage is at its maximum. This maximum voltage is called the **zero-current cell potential**, also known as the **electromotive force (emf)** of the cell, given the symbol E_{cell}.

Some fundamentals of electrochemistry

Consider this example of a simple electrochemical system. If a piece of zinc metal is placed in a solution of a copper salt, say, copper(II) sulfate, the zinc rapidly becomes coated with a brown film of copper metal, as shown in Figure 16.1. Over a period of time, the quantity of solid copper deposited increases while the zinc dissolves. The solution gradually changes from blue to colourless. So, what's going on?

Zinc strip
Copper(II) sulfate solution
Zinc strip becomes covered in copper metal
Zinc(II) sulfate solution with a little copper(II) sulfate remaining

Figure 16.1 A piece of zinc immersed in copper(II) sulfate solution gradually dissolves while solid copper is deposited on the zinc. The blue colour of the solution disappears as colourless Zn^{2+}(aq) replaces blue Cu^{2+}(aq).

The overall reaction that takes place is

$$Zn\,(s) + Cu^{2+}(aq) \rightarrow Zn^{2+}(aq) + Cu\,(s) \tag{16.3}$$

This reaction can be thought of as consisting of two **half reactions.** The zinc metal forms zinc ions and dissolves; as a consequence it loses two electrons. These electrons are transferred to copper ions in solution, which form copper metal. These two half reactions added together give the overall reaction shown in Equation 16.3

Half reactions and redox reactions are discussed in Section 1.4 (p.23).

$$Zn\,(s) \rightarrow Zn^{2+} + 2\,e^- \qquad \text{oxidation} \tag{16.4}$$

$$Cu^{2+}(aq) + 2\,e^- \rightarrow Cu\,(s) \qquad \text{reduction} \tag{16.5}$$

Reactions in which electrons are gained are called **reductions. Oxidations** are those reactions in which electrons are lost. Thus, zinc metal forming zinc ions is an *oxidation*; Cu^{2+} ions are *reduced* to copper metal. Reactions where oxidation and reduction occur simultaneously are called **redox** reactions.

If the reaction is performed as shown in Figure 16.1, all the electron transfers take place at the surface of the metal, as illustrated in Figure 16.2. To be useful and generate electrical energy, the reaction has to be arranged so that the electron transfer can be harnessed by making the electrons flow along an external circuit. This forms the basis of the electrochemical cells described in Section 16.3. Figure 16.2 illustrates some of the key ideas that you need to understand: movement of ions in solution; electron transfer reactions

In this chapter, only systems involving water as the solvent are considered. However, electrochemistry can be carried out in any solvent in which ions can form and electron transfer reactions can take place.

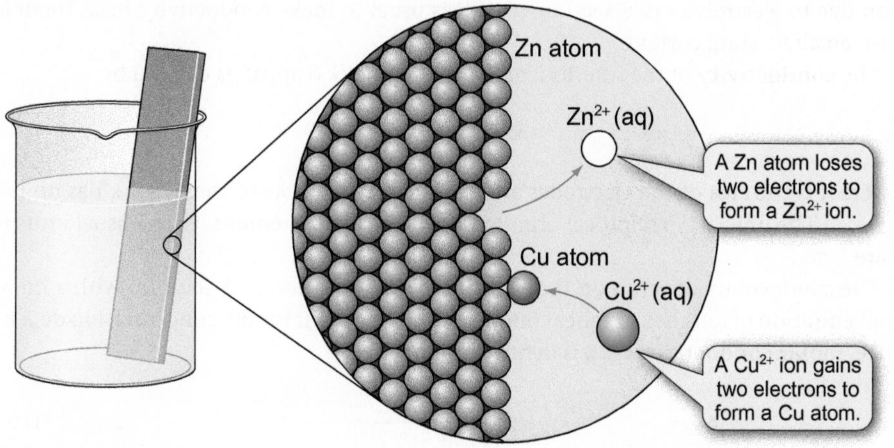
Zn atom
Zn2+(aq)
A Zn atom loses two electrons to form a Zn2+ ion.
Cu atom
Cu2+(aq)
A Cu2+ ion gains two electrons to form a Cu atom.

Figure 16.2 The electron transfer reactions take place at the surface of the zinc. Zinc gradually dissolves to form Zn^{2+}(aq) ions while Cu^{2+}(aq) ions form Cu metal, which is deposited on the surface.

The British scientist, Michael Faraday (1791–1867), did much of the pioneering work in electrochemistry in the nineteenth century.

at electrode surfaces; and the relation between the chemical change and the energy produced. This chapter describes each of these ideas in turn, starting with the properties of ions in solution.

Summary

- Electrochemistry involves the interconversion of electrical energy and chemical energy.
- Charge can be moved by electron transfer through conductors and by ion transport in solutions.
- The maximum voltage produced by an electrochemical cell is called the zero-current cell potential, E_{cell}. It is also known as the electromotive force, emf.
- Redox reactions involve simultaneous reduction and oxidation half reactions.
  - Reductions involve gain of electrons.
  - Oxidations involve loss of electrons.

16.2 Ions in solution

The term 'electrolyte' is sometimes used to describe the conducting solution as well as the substance dissolved in it.

Ions in solution do not exist as isolated species, but are solvated (see Section 1.8, p.54). In aqueous solution the ions are hydrated and, as they move, they carry with them a surrounding layer of water molecules.

An electrolyte is a substance that, when dissolved, gives a solution that conducts electricity. Most electrolytes are ionic substances, such as sodium chloride (NaCl), potassium chloride (KCl), and sodium sulfate (Na_2SO_4), although some molecular substances, such as hydrogen chloride (HCl) and ammonia (NH_3), form ions when they dissolve in water. It is useful to be able to measure the conductivity of solutions, because this provides information about the ions that carry the current.

Conductivity due to ions in solution

In an electrolyte solution, the charge is carried between the electrodes by dissolved ions. The properties of electrolyte solutions can be studied in a conductivity cell such as that shown in Figure 16.3. It consists of two platinum electrodes held apart at a fixed distance in a solution. Platinum is used since it is inert, that is, very unreactive. A voltage is applied between the electrodes and the resistance, R, is measured. The resistance is a measure of how difficult the motion of ions is between the electrodes. To prevent any chemical reaction due to electrolysis (see Section 16.5) it is usual to make conductivity measurements with an alternating current.

The **conductivity** of the solution, given the symbol κ (kappa), is defined by

$$\kappa = \frac{d}{A} \times \frac{1}{R} \tag{16.6}$$

where A is the area of the electrodes and d the distance between them. So, κ has units of $\Omega^{-1}\,m^{-1}$. The unit Ω^{-1} ('reciprocal ohm') is given the name **siemens**, S. The usual units for κ are $S\,m^{-1}$.

The conductivity depends on the number of ions in solution. A solution with a higher concentration of ions has a higher conductivity. To account for the concentration dependence, **molar conductivity**, Λ_m, is defined as

$$\Lambda_m = \frac{\kappa}{c} \qquad \left(\text{Units: } \frac{S\,m^{-1}}{mol\,m^{-3}} = S\,m^2\,mol^{-1}\right) \tag{16.7}$$

Concentrations are often reported in $mol\,dm^{-3}$. Here, SI units ($mol\,m^{-3}$) are being used, so you need to take care to be consistent when doing calculations.

where c is the concentration of the solution in $mol\,m^{-3}$. Molar conductivities are often reported in units of $mS\,m^2\,mol^{-1}$ (where $1\,mS\,m^2\,mol^{-1} = 1 \times 10^{-3}\,S\,m^2\,mol^{-1}$).

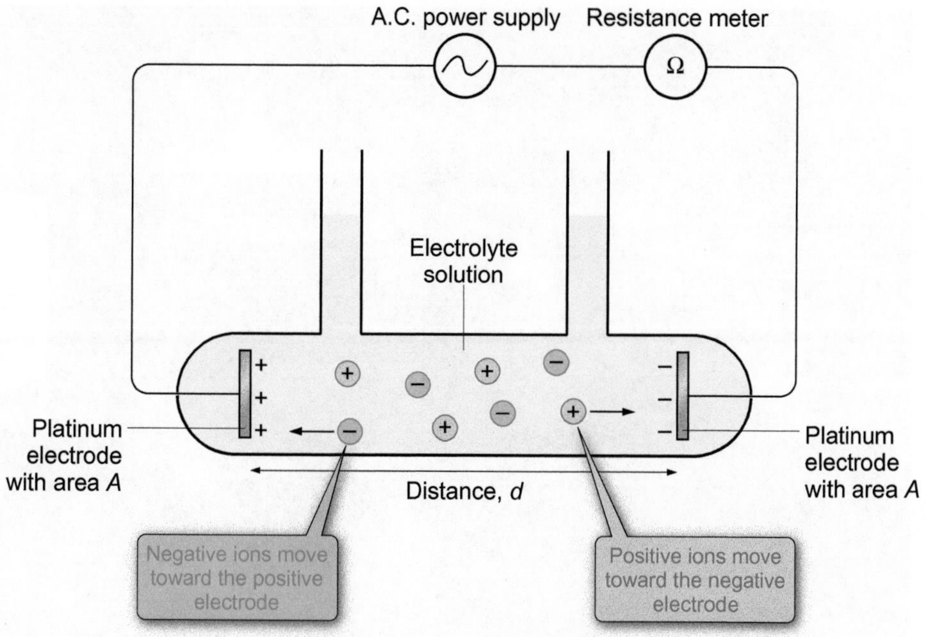
A.C. power supply
Resistance meter
Ω
Electrolyte solution
Platinum electrode with area A
Distance, d
Platinum electrode with area A
Negative ions move toward the positive electrode
Positive ions move toward the negative electrode

Figure 16.3 A conductivity cell. The resistance of the cell is measured in such a way that current is not drawn from the cell so that no chemical change takes place.

Ionic substances like KCl or Na_2SO_4 are fully ionized in solution. They are called **strong electrolytes**. For a strong electrolyte, measurements using a conductivity cell show that the molar conductivity *increases* as the solution becomes more dilute. The interactions between ions in a solution affect the movement of ions through the solution. These interactions are less important in dilute solutions, so Λ_m increases with dilution. Eventually, the molar conductivity ceases to increase with dilution and approaches a limiting value, called the **limiting molar conductivity**, Λ_m°.

When the solution is very dilute, the ions are far enough apart that they do not interact with each other and so behave independently of one another. Each ion makes its own independent contribution to the molar conductivity. A K^+ ion in dilute solution always has the same value of **ionic conductivity**, λ, whether it is in a solution of KCl, KBr, KNO_3, K_2SO_4, or any other potassium salt. The same is true for Cl^-; it has the same value of ionic conductivity in KCl, NaCl, HCl, NH_4Cl, etc. Some values of λ are shown in Table 16.1. These are known as limiting values since they are only valid in very dilute solutions—at **infinite dilution.**

Greek symbols

Λ is a Greek upper-case (capital) lambda

λ is lower-case lambda

For any electrolyte, the limiting molar conductivity is given by

$$\Lambda_m^\circ = \nu_+\lambda_+ + \nu_-\lambda_- \qquad (16.8)$$

where λ_+ and λ_- are the ionic conductivities of the cation and anion, respectively, while ν_+ and ν_- are, respectively, the numbers of cations and anions formed when the formula unit of the electrolyte dissolves. For example, $\nu_+ = \nu_- = 1$ for KCl and $CuSO_4$; $\nu_+ = 1$, $\nu_- = 2$ for $MgCl_2$; $\nu_+ = 2$, $\nu_- = 1$ for Na_2SO_4. Equation 16.8 is known as the **law of independent migration of ions.**

Look at the values shown in Table 16.1. Most of the singly charged anions or cations have similar values of λ. This makes sense since each ion carries the same amount of charge. Divalent ions carry more charge so they have higher values of λ. However, the values for the hydrogen ion and the hydroxide ion are clearly out of line with the other species.

The original explanation for the abnormally high value for H^+ was that the small size of an H^+ ion (a proton) meant that it would move rapidly through solution. However, free

The conductivity of fresh water gives a good indication of how pure the water is, because ionic contaminants increase water's conductivity.

In a Greek bus, the entrance is labelled '**ΑΝΟΔΟΣ**' (ANODOS) and the exit is labelled '**ΚΑΤΟΔΟΣ**' (KATODOS). The names 'anode' and 'cathode' are derived from ancient Greek words meaning 'way up' and 'way down'.

Anions are negative ions.
Cations are positive ions.

Table 16.1 Limiting ionic conductivities, λ, in water / 10^{-3} S m^2 mol^{-1}

Cations		Anions	
H^+ (H_3O^+)	35.0	OH^-	19.9
Li^+	3.9	F^-	5.5
Na^+	5.0	Cl^-	7.6
K^+	7.4	Br^-	7.8
Cs^+	7.7	I^-	7.7
Ag^+	6.2	HCO_2^-	5.5
NH_4^+	7.4	NO_3^-	7.2
Mg^{2+}	10.6	SO_4^{2-}	13.9
Ca^{2+}	11.9	CO_3^{2-}	17.0
Sr^{2+}	11.9	$CH_3CO_2^-$	4.1

protons do not occur in aqueous solution—they are always associated with water molecules and the size of H_3O^+(aq) is comparable to that for Na^+(aq). The explanation lies in a proton hopping, chain mechanism known as the **Grotthus mechanism** illustrated in Figure 16.4.

Although the ions are shown in Table 16.1 simply as M^{n+} or A^{n-}, they generally adopt more complex structures in solution, where they are *solvated*. The solvent molecules move along with the ion under the influence of an electric field. Highly solvated ions with lots of associated solvent molecules move more slowly.

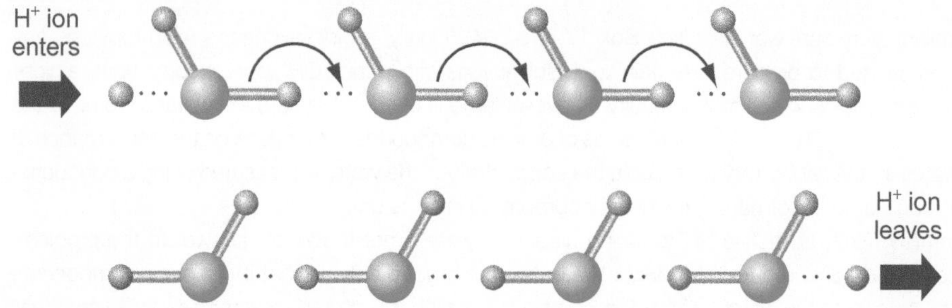
H^+ ion enters
H^+ ion leaves

Figure 16.4 The Grotthus mechanism. In solutions of H^+ ions, charge is not conducted by a single ion moving through solution. Hydrogen ion transfer takes place between adjacent, hydrogen bonded, water molecules so that the conduction is faster than expected. (The blue curved arrows indicate movement of H^+—*not* movement of electron pairs.)

The high conductivity is due to the swapping of H^+ from one molecule to another. The result is that the charge moves a large distance even though each individual H^+ ion moves only a small distance—it is the electrons in the bonds that move. The Grotthus mechanism also explains the unusually high conductivity of the OH^- ion.

Some compounds form ions when they dissolve in water but do not completely ionize. These are **weak electrolytes**. An example is ethanoic acid (CH_3CO_2H). When it dissolves in water, it forms H_3O^+ ions and ethanoate ($CH_3CO_2^-$) ions but most molecules remain as unionized acid. Weak electrolytes, such as acids, are characterized in terms of their acidity constants, K_a. For ethanoic acid

$$CH_3CO_2H\,(aq) + H_2O\,(l) \rightleftharpoons CH_3CO_2^-\,(aq) + H_3O^+\,(aq)$$

$$K_a = \frac{[H_3O^+\,(aq)][CH_3CO_2^-\,(aq)]}{[CH_3CO_2H\,(aq)]}$$

$$= 1.8 \times 10^{-5}\ \text{mol dm}^{-3}$$

An acid with a higher value of K_a ionizes to a greater extent to form H_3O^+ ions (it is a stronger acid). Conversely, an acid with low K_a exists mainly in the unionized form in solution. The conductivity of a solution of a weak electrolyte has a more complex dependence on concentration than that of a strong electrolyte and is largely influenced by the degree of ionization.

Acidity constants, K_a, and the strengths of acids are discussed in Section 7.2 (p.308).

Box 16.1 describes the production of 'ultrapure' water and asks you to calculate values of Λ_m° and κ for the water produced.

Box 16.1 Ultrapure water and conductivity

The quality of the water we drink is vital to our quality of life. Tap water is filtered and disinfected to remove biological contaminants—but it is far from pure in a chemical sense. Chemically pure water would be H_2O and nothing else. Water is a very good solvent and it contains dissolved substances from the ground it passes over (or through) before collection. For example, water from an area where the ground rock, such as chalk, is predominantly calcium carbonate will contain appreciable concentrations of calcium ions and hydrogencarbonate ions and will have a pH of around 8.0–8.5 (see Box 26.4, p.1191). Other substances, such as carbon dioxide, sulfur dioxide, or oxides of nitrogen, are absorbed from the atmosphere. These tend to decrease the pH of the water.

Many of these dissolved substances are ionic and change the conductivity of water. Measuring the conductivity does not tell you which compounds are present, or their amounts, but it does give a good overall measure of the water purity. In particular, the conductivity is a good indication of the total concentration of inorganic contaminants.

In the UK, tap water has conductivity in the range of 200 mS cm^{-1} –1000 mS cm^{-1} depending on its source. (By contrast, sea water has conductivity greater than 40 000 mS cm^{-1}.) While the impurities in tap water may impart a slight taste, they are not usually a problem for health. In contrast, water used in the microelectronics industry must be very pure so as to avoid contamination, and a conductivity →

→ of less than $0.5\,\mathrm{mS\,cm^{-1}}$ is required. Similarly, very pure water, with a conductivity of less than $1.5\,\mathrm{mS\,cm^{-1}}$, is needed to prepare medicines and other pharmaceutical products. So, how is water of this purity produced?

A typical unit for producing 'ultrapure' water in the laboratory is shown in Figure 1. Tap water is passed through a series of cartridges in which the water is successively treated. First, an active charcoal adsorbent removes most of the dissolved organic matter; then a reverse osmosis module uses high pressure to force the water through a membrane that does not allow dissolved ions to pass (see Box 17.5, p.804). Finally, an ion-exchange resin replaces any remaining dissolved ions with H^+ and OH^- ions. In some units, a high-intensity ultraviolet lamp is used to irradiate the water to remove the final traces of organic compounds. As a check on the performance of the unit, the conductivity of the water is measured using a conductivity meter incorporated into the unit.

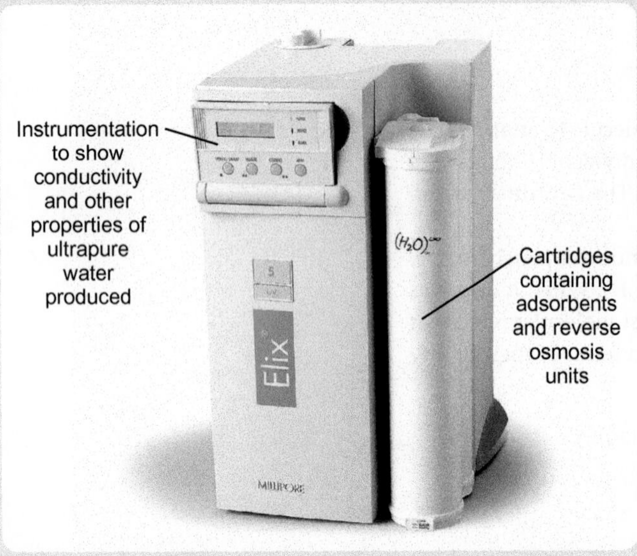
Instrumentation to show conductivity and other properties of ultrapure water produced
Cartridges containing adsorbents and reverse osmosis units
Elix

▲ **Figure 1** A water purification unit.

Producing ultrapure water is not the whole answer to the problem, though. Water is such a good solvent that it dissolves components from the container in which it is stored—even glass to a very small extent. More importantly, water rapidly absorbs carbon dioxide from the air. This forms hydrogencarbonate ions (HCO_3^-) and oxonium ions (H_3O^+) so that the pH of water rapidly approaches 5.7 at room temperature.

Questions

(a) Pure water is only ionized to a small extent. It has a self-ionization constant, $K_w = [H^+(aq)][OH^-(aq)] = 1 \times 10^{-14}\,\mathrm{mol^2\,dm^{-6}}$ around room temperature (see Section 7.2, p.308). Calculate the concentrations of $H^+(aq)$ and $OH^-(aq)$ in pure water in $\mathrm{mol\,m^{-3}}$.

(b) From data in Table 16.1, predict the molar conductivity, Λ_m°, of pure water.

(c) Calculate the conductivity, κ, of pure water in $\mu\mathrm{S\,cm^{-1}}$. (In order to use convenient numbers, the conductivity of water is often reported in units of $\mu\mathrm{S\,cm^{-1}}$, where micro, μ, is the SI prefix for 1×10^{-6}; $1\,\mathrm{m} = 100\,\mathrm{cm}$.)

(d) Assuming that the conductivity of a sample of water is due only to dissolved sodium chloride, calculate the concentration of NaCl needed to give a solution with $\kappa = 100\,\mu\mathrm{S\,cm^{-1}}$.

Summary

- Electrolytes are substances that dissolve to give conducting solutions, usually due to the movement of ions.
- The conductivity, κ, measures how easily charge is transported through a solution.
- The molar conductivity of a compound, Λ_m, is given by κ/c, where c is the concentration of the solution in $\mathrm{mol\,m^{-3}}$.
- For a strong electrolyte, Λ_m increases with dilution, up to a value at infinite dilution called the limiting molar conductivity, Λ_m°.
- For a weak electrolyte, the conductivity is largely influenced by the degree of ionization.
- The ionic conductivity, λ, is characteristic of an individual ion and independent of the other ions present.
- The conductivities of H^+ ions and OH^- ions are high due to the Grotthus mechanism.
- For any electrolyte, $\Lambda_m^\circ = \nu_+\lambda_+ + \nu_-\lambda_-$, where ν_+ and ν_- are the numbers of cations and anions formed when one formula unit of the electrolyte dissolves.

? **For practice questions on these topics, see questions 1 and 2 at the end of this chapter (p.761).**

16.3 Electrochemical cells

Look again at the redox reaction between zinc and copper(II) ions shown in Figure 16.1. An alternative arrangement for performing the same reaction is shown in Figure 16.5.

$$Zn\,(s) + Cu^{2+}\,(aq) \rightarrow Zn^{2+}\,(aq) + Cu\,(s) \qquad (16.3)$$

A piece of zinc metal is placed in a solution of zinc(II) sulfate and, in a separate beaker, a piece of copper is in contact with a solution of copper(II) sulfate. The oxidation and reduction reactions are now physically separate and the electron transfer takes place through an electrically conducting wire between the metals. The flow of electrons provides the current. To complete the circuit, the two solutions are connected by a **salt bridge**, where the charge is conducted by ions in solution. In Figure 16.5, the salt bridge is a strip of absorbent material soaked in a solution of sodium sulfate, although in more sophisticated versions the salt bridge is provided by a glass tube filled with a conducting gel. The potential difference between the metals (the **electrodes**) is measured with a voltmeter. If the resistance of the voltmeter is very high, an infinitesimally small current flows and the voltmeter records the emf of the cell. When both solutions have a concentration of 1.00 mol dm^{-3}, the measured emf is 1.10 V.

This arrangement of separate redox reactions is an example of an **electrochemical cell**. Figure 16.6 shows what happens when current flows in the cell. The electrons move through the wire giving an electric current to light the lamp.

Electrochemical cells are sometimes called **galvanic** or **voltaic** cells, named after the Italian scientists Luigi Galvani and Alessandro Volta, both of whom did early pioneering work in electrochemistry. The cell shown in Figures 16.5 and 16.6 involving Cu and Zn is known as a **Daniell cell**, named after John Daniell who first used it in 1836. A modern 'dry cell' used in torch batteries is described in Box 16.2.

In the Daniell cell, the copper and zinc metals that form the electrodes also take part in the cell reactions. These are known as **metal–ion electrodes**. Many other types of reaction that produce or consume electrons can be used as components of electrochemical cells. For example, many metal ions can exist in more than one oxidation state. Iron can switch in solution between the Fe^{2+} and Fe^{3+} states and this redox couple can provide a half reaction for an electrode, called a **redox electrode**.

$$Fe^{2+}\,(aq) \rightarrow Fe^{3+}\,(aq) + e^-$$

In order to collect the electrons from a redox couple of this kind, an electrode made from inert platinum metal is used, as shown in Figure 16.7.

Electrochemistry is not limited to metals and ions in solution; half reactions involving gases can also be used. A half reaction using gaseous reagents, the **hydrogen electrode**, is shown in Figure 16.8.

$$2\,H^+\,(aq) + 2\,e^- \rightarrow H_2\,(g)$$

The general features of an electrochemical cell are summarized in Figure 16.9. Every electrochemical cell is made up of two **half cells**. These are commonly called **electrodes**, but note that the term 'electrode' can be used in two ways. It can refer to the half cell that is producing or consuming electrons, but it is also used to mean the piece of metal that conducts electrons in and out of the solution and on which the transfer of electrons takes place (e.g. the inert platinum wire in the redox half cell in Figure 16.7).

When performing an experiment such as that in Figure 16.5 it is important to use a voltmeter with a very high resistance to allow only minimal electron flow. Under these conditions, the measured voltage is equal to the emf. If electron flow takes place, the measured voltage will decrease because the chemical reaction proceeds and the concentrations of the ions changes (see Section 16.4, p.749).

The sulfate salts of zinc and copper are used in the Daniell cells in Figures 16.5 and 16.6, but the nature of the anion is unimportant because it does not take part in the reaction. Any soluble zinc(II) and copper(II) salts can be used. Ions that do not take part in a reaction are sometimes called *spectator ions* (Section 1.4, p.26).

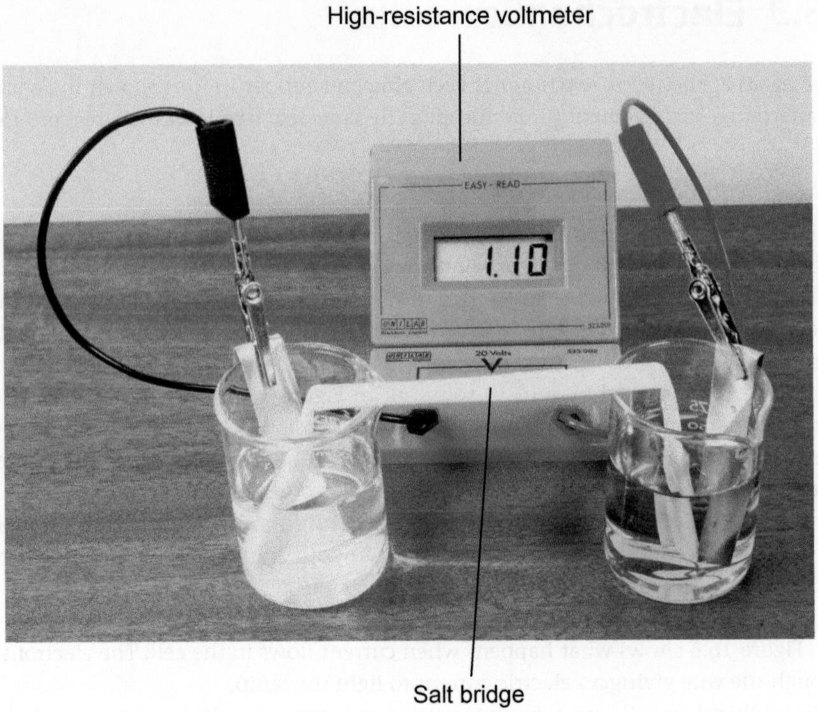

High-resistance voltmeter
1.10
Salt bridge

Figure 16.5 A Daniell cell. Zinc metal (in the left-hand beaker) is immersed in 1.0 mol dm^{-3} $ZnSO_4$(aq). Copper metal (in the right-hand beaker) is immersed in 1.0 mol dm^{-3} $CuSO_4$(aq). The cell produces an emf of 1.10 V.

Visit the Online Resource Centre to view screencast 16.1 which walks you through the working of the Daniell cell shown in Figures 16.5 and 16.6

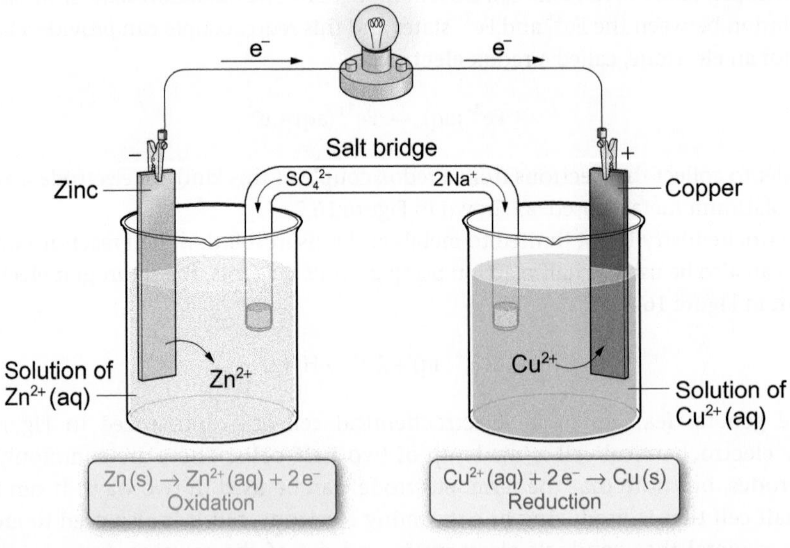

e⁻
e⁻
Salt bridge
−
+
Zinc
SO₄²⁻
2 Na⁺
Copper
Solution of Zn²⁺(aq)
Zn²⁺
Cu²⁺
Solution of Cu²⁺(aq)
Zn(s) → Zn²⁺(aq) + 2e⁻
Oxidation
Cu²⁺(aq) + 2e⁻ → Cu(s)
Reduction

Figure 16.6 The processes in a Daniell cell. The oxidation and reduction reactions occur in separate beakers joined by a salt bridge. Electrons flow through the external circuit. In the solutions and in the salt bridge, the current is carried by ions.

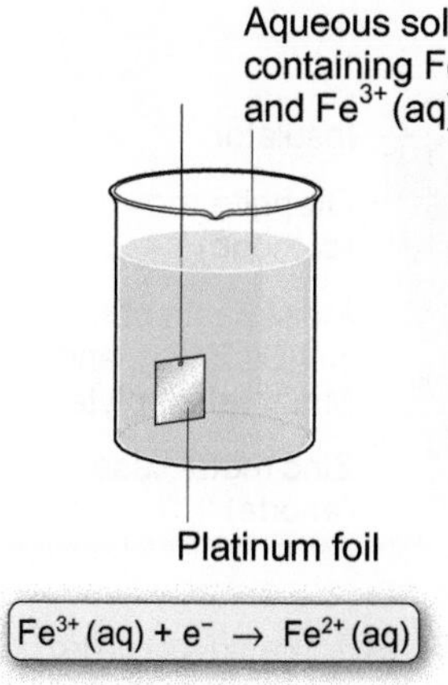
Aqueous solution
containing Fe^{2+}(aq)
and Fe^{3+}(aq)
Platinum foil
$Fe^{3+}(aq) + e^- \rightarrow Fe^{2+}(aq)$

Figure 16.7 In a redox electrode, an inert platinum foil collects the electrons.

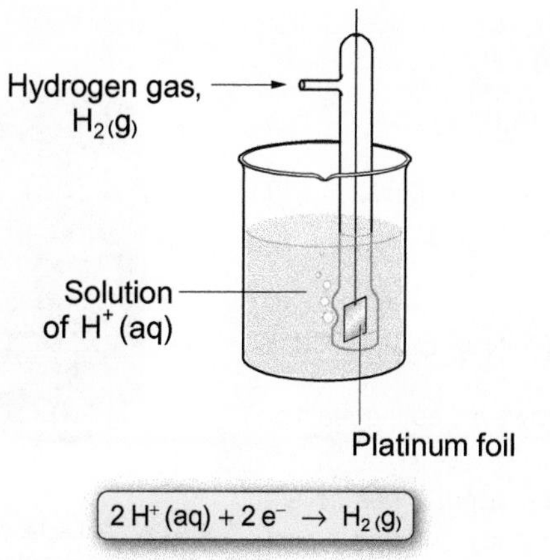
Hydrogen gas,
$H_2(g)$
Solution
of H^+(aq)
Platinum foil
$2H^+(aq) + 2e^- \rightarrow H_2(g)$

Figure 16.8 In the hydrogen electrode, hydrogen gas is bubbled over a platinum foil immersed in a solution containing H^+(aq) ions. The electron transfer reaction takes place at the metal surface.

Visit the Online Resource Centre to view screencast 16.2 which walks you through the working of the redox electrodes shown in Figures 16.7 and 16.8.

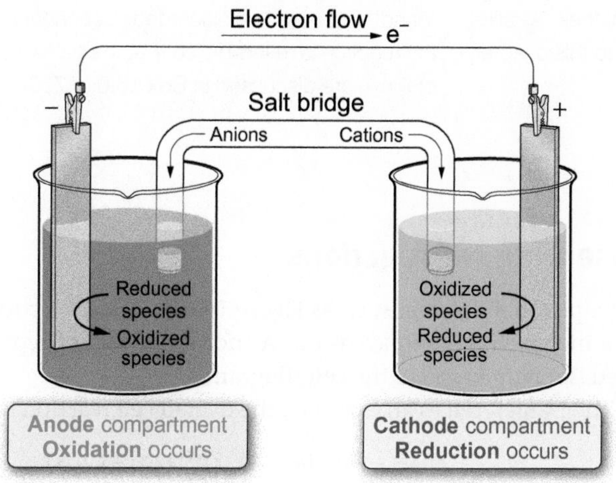
Electron flow
e^-
Salt bridge
Anions
Cations
−
+
Reduced
species
Oxidized
species
Oxidized
species
Reduced
species
Anode compartment
Oxidation occurs
Cathode compartment
Reduction occurs

Figure 16.9 A generalized electrochemical cell.

Box 16.2 Electrochemical cells as portable energy sources

One of the most widespread applications of redox chemistry is in batteries (see opening spread to this chapter, p.726). Originally, the term *battery* referred to several electrochemical cells connected together to increase the available voltage from that of a single cell, although now it is commonly used to describe any portable electrochemical power source.

The Daniell cell (Figures 16.5 and 16.6) was used as a power source in the early days of electrical communication by wire telegraph. However, since it uses solutions of a copper salt and a zinc salt, it is not very convenient as a portable battery for modern applications.

The first battery to be widely used was the **dry cell**, so-called since it eliminates the use of solutions. It is also known as a **Leclanché cell** after its inventor Georges Leclanché.

The zinc metal that makes up the case of the battery acts as the anode. The reaction that takes place there is:

Anode: $Zn(s) \rightarrow Zn^{2+}(aq) + 2e^-$ oxidation

A complex sequence of reactions takes place at the surface of the graphite cathode. The first step is the reduction of ammonium ions

Cathode: $2NH_4^+(aq) + 2e^- \rightarrow 2NH_3(g) + H_2(g)$ reduction

Both of these products are gases and must be removed; otherwise pressure would build up and the battery would explode. The hydrogen reacts with MnO_2 and the ammonia is complexed by the Zn^{2+} ions formed at the anode

➜

$$2MnO_2(s) + H_2(g) \rightarrow Mn_2O_3(s) + H_2O(l)$$

$$Zn^{2+}(aq) + 2NH_3(aq) \rightarrow [Zn(NH_3)_2]^{2+}(aq)$$

$[Zn(NH_3)_2]^{2+}$ ions quickly react with Cl^- ions to form solid $[Zn(NH_3)_2Cl_2]$. The overall cell reaction can be represented as

$$2MnO_2(s) + 2NH_4Cl(s) + Zn(s) \rightarrow Mn_2O_3(s) + [Zn(NH_3)_2Cl_2](s) + H_2O(l)$$

This gives the battery a voltage of +1.5 V. However, as the reaction proceeds, the voltage falls to around +0.8 V.

While this type of battery has been widely used, it has several disadvantages. The zinc and ammonium chloride undergo a slow reaction separate from that shown, so that the battery cannot be stored for long periods of time because the zinc case corrodes and the contents can leak. Also, if a large current is drawn, the gases produced cannot react quickly enough so that the pressure builds up and the components can leak out. The voltage also falls off quickly under these conditions. This makes these batteries unsuitable for many modern applications. However, they are the cheapest batteries on the market and they are commonly used in torches. Alkaline batteries, which are discussed in the opening spread to this chapter, p.726, have replaced Leclanché cells for many uses.

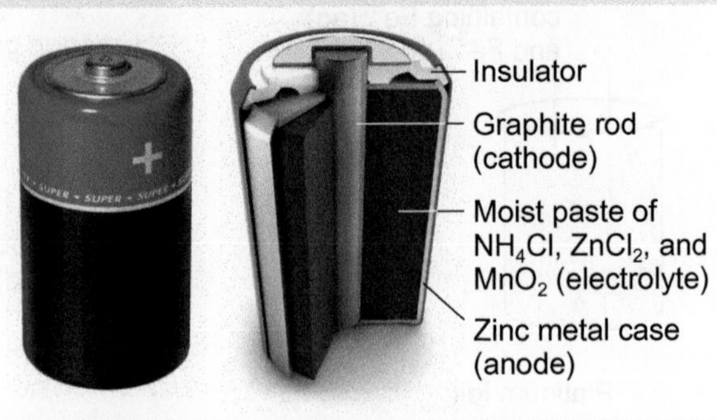
Insulator
Graphite rod (cathode)
Moist paste of NH_4Cl, $ZnCl_2$, and MnO_2 (electrolyte)
Zinc metal case (anode)

▲ A dry cell Leclanché battery. The cathode is a graphite rod surrounded by a moist paste containing ammonium chloride (NH_4Cl), zinc chloride ($ZnCl_2$), and manganese(IV) oxide (MnO_2). This is contained in a zinc case that also acts as the anode. The paste is moist to give a medium through which ions can move. The cell is protected by an outer steel case.

Most batteries are primary cells: the reactions involved are not reversible. When all the reactants are used up and the reaction comes to equilibrium, no further energy can be produced. The battery is 'dead' and must be discarded. Secondary cells are those in which the reaction is reversible so that the battery can be recharged. Some of these are discussed in Box 16.6 (p.759).

Representing cell reactions

Drawing a pictorial diagram such as Figure 16.6 for every electrochemical cell you study would be inconvenient and awkward. A shorthand way of representing a cell has been developed, sometimes called the **cell diagram**.

Using the Daniell cell as an example, the overall cell reaction is

$$Zn(s) + Cu^{2+}(aq) \rightarrow Zn^{2+}(aq) + Cu(s) \qquad (16.3)$$

The zinc half cell is written as $Zn(s)|Zn^{2+}(aq)$, indicating that solid zinc is in contact with aqueous zinc ions. The vertical line indicates that there is a boundary between phases, in this case between solid and solution. The copper half cell is represented as $Cu^{2+}(aq)|Cu(s)$. The conducting salt bridge that connects the two half cells is represented by two vertical lines ||. The Daniell cell is then represented as

$$Zn(s)|Zn^{2+}(aq)||Cu^{2+}(aq)|Cu(s)$$

The cell components are listed, as far as possible, in the order that indicates the overall reaction proceeding from left to right. In this case, the cell reaction occurs in the direction shown in Equation 16.3, so the cell diagram has Zn (s) on the left and Cu (s) on the right. Note that:

- the metal electrodes are positioned outermost;
- the anode, at which the oxidation reaction occurs, is placed on the left;
- the cathode, at which the reduction reaction occurs, is placed on the right.

In a redox half cell, both components are in solution and the ions are shown in the cell diagram separated by a comma, with the more reduced form nearest to the platinum. Thus, the Fe^{3+}/Fe^{2+} electrode in Figure 16.7 is represented as

$$Pt(s)|Fe^{2+}(aq), Fe^{3+}(aq)||$$

ⓘ In an electrochemical cell, electrons flow around the *external* circuit from the anode to the cathode. Oxidation occurs at the anode and reduction occurs at the cathode.

ⓘ Fuel cells are electrochemical cells which carry out the oxidation of a fuel such as hydrogen on an electrode, creating a flow of electrons rather than an exothermic reaction.

ⓘ Later in this section, you will see that the potential of a cell depends on the concentrations in the half cells, so a full description of the half cell should include the solution concentration, for example, $Zn(s)|Zn^{2+}(aq, 1.0\,mol\,dm^{-3})$.

The notations for half cells involving reactions with a gas or containing solid components are complicated by the presence of the additional phase. The hydrogen electrode shown in Figure 16.8 is represented as

$$Pt\,(s)|H_2\,(g)|H^+\,(aq)||$$

A general representation of the cell diagram for an electrochemical cell is given in Figure 16.10. You can practise drawing and interpreting cell diagrams in Worked examples 16.1 and 16.2.

LITHIUM BATTERY
CR2032
3V
HL

A lithium cell. The anode is made of Li metal which is oxidized to Li^+ when current is drawn from the cell. Unlike *lithium-ion* cells, lithium cells are not rechargeable.

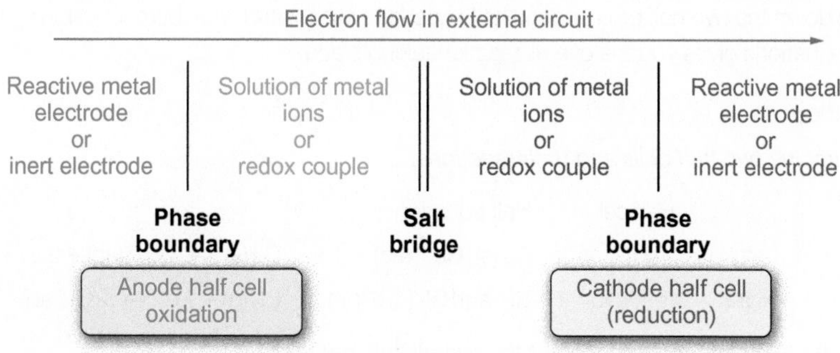
Electron flow in external circuit
Reactive metal electrode or inert electrode
Solution of metal ions or redox couple
Solution of metal ions or redox couple
Reactive metal electrode or inert electrode
Phase boundary
Salt bridge
Phase boundary
Anode half cell oxidation
Cathode half cell (reduction)

Figure 16.10 A generalized cell diagram to represent an electrochemical cell.

Worked example 16.1 Drawing a cell diagram

Draw a cell diagram for an electrochemical cell that involves the following overall reaction

$$Fe\,(s) + Sn^{2+}\,(aq) \rightarrow Sn\,(s) + Fe^{2+}\,(aq)$$

Strategy

First, separate the overall reaction into two separate half reactions. Then, decide where the oxidation and reduction reactions occur so that you can assign the anode and cathode to the left and right sides of the diagram, respectively. Finally, look at where the phase boundaries between the components occur and draw the cell diagram in the form of that in Figure 16.10.

Solution

Write the two half cell reactions and identify the oxidation and reduction reactions.

$Fe\,(s) \rightarrow Fe^{2+}\,(aq) + 2\,e^-$	Fe (s) loses electrons:	oxidation
$Sn^{2+}\,(aq) + 2\,e^- \rightarrow Sn\,(s)$	Sn^{2+} (aq) gains electrons:	reduction

Write the two sides of the cell diagram.

Left:	oxidation	$Fe\,(s)\,|\,Fe^{2+}\,(aq)$
Right:	reduction	$Sn^{2+}\,(aq)\,|\,Sn\,(s)$

Put the two halves together, joined by a salt bridge.

$$Fe\,(s)\,|\,Fe^{2+}\,(aq)\,||\,Sn^{2+}\,(aq)\,|\,Sn\,(s)$$

Question

Draw a cell diagram for an electrochemical cell with the overall reaction

$$H_2\,(g) + Cl_2\,(g) \rightarrow 2\,HCl\,(aq)$$

Worked example 16.2 Interpreting a cell diagram

Write equations for the half reactions and the overall reaction that occurs in the following electrochemical cell

$$Cu(s)\,|\,Cu^{2+}(aq)\,||\,Cl^{-}(aq)\,|\,Cl_2(g)\,|\,Pt(s)$$

Strategy

You essentially reverse the procedure used in Worked example 16.1. From the cell diagram, write down the two half cells and the half equations associated with them. Combining the two half equations gives you the overall reaction taking place.

Solution

Identify the two half cells and their reactions.

	Half cell	Half equation	
Left:	oxidation	$Cu(s)\,|\,Cu^{2+}(aq)$	$Cu(s) \rightarrow Cu^{2+}(aq) + 2e^{-}$
Right:	reduction	$Cl^{-}(aq)\,|\,Cl_2(g)\,|\,Pt(s)$	$Cl_2(g) + 2e^{-} \rightarrow 2Cl^{-}(aq)$

Add the two half equations to get the overall cell reaction.

Always check when adding half equations that the number of electrons transferred is the same in both equations.

$$Cu(s) + Cl_2(g) \rightarrow 2Cl^{-}(aq) + Cu^{2+}(aq)$$

Note that the right-hand half cell is a redox half cell. The Pt electrode is inert and plays no part in the reaction. It simply acts as a surface where the electron transfer between Cl_2 and Cl^{-} can take place.

Question

What is the overall reaction occurring in the following cell?

$$Pt(s)\,|\,V^{2+}(aq), V^{3+}(aq)\,||\,Fe^{3+}(aq), Fe^{2+}(aq)\,|\,Pt(s)$$

Standard reduction potentials, $E^{\ominus}$

When the concentrations of Cu^{2+}(aq) and Zn^{2+}(aq) in a Daniell cell are both 1.0 mol dm^{-3}, the cell produces an emf of 1.10 V. How much does each half cell contribute to this total emf? The high-resistance voltmeter in Figure 16.5 measures the potential difference between the two half cells, but it is not possible to measure the potential of a single half cell on its own. However, by fixing a standard value against which all other half cells can be compared, it is possible to draw up a table of half cell potentials relative to the standard value.

The reaction chosen to be the standard is the **standard hydrogen half cell**, commonly called the **standard hydrogen electrode, SHE**

$$2H^{+}(aq) + 2e^{-} \rightarrow H_2(g) \tag{16.9}$$

Strictly speaking, standard conditions should refer to unit *activity* rather than unit concentration (see Section 14.6, p.686).

Under standard conditions (a hydrogen gas pressure of 1.00 bar, and $[H^{+}(aq)]$ = 1.00 mol dm^{-3}) at 298 K, the potential of this half cell is set to 0.00 V. Figure 16.11 demonstrates the use of a standard hydrogen electrode to find the relative potentials of the two half cells of the Daniell cell.

In Figure 16.11(a), the $Cu^{2+}(aq)\,|\,Cu(s)$ half cell is set up as the cathode and the standard hydrogen electrode as the anode. By convention, the potential for the standard hydrogen

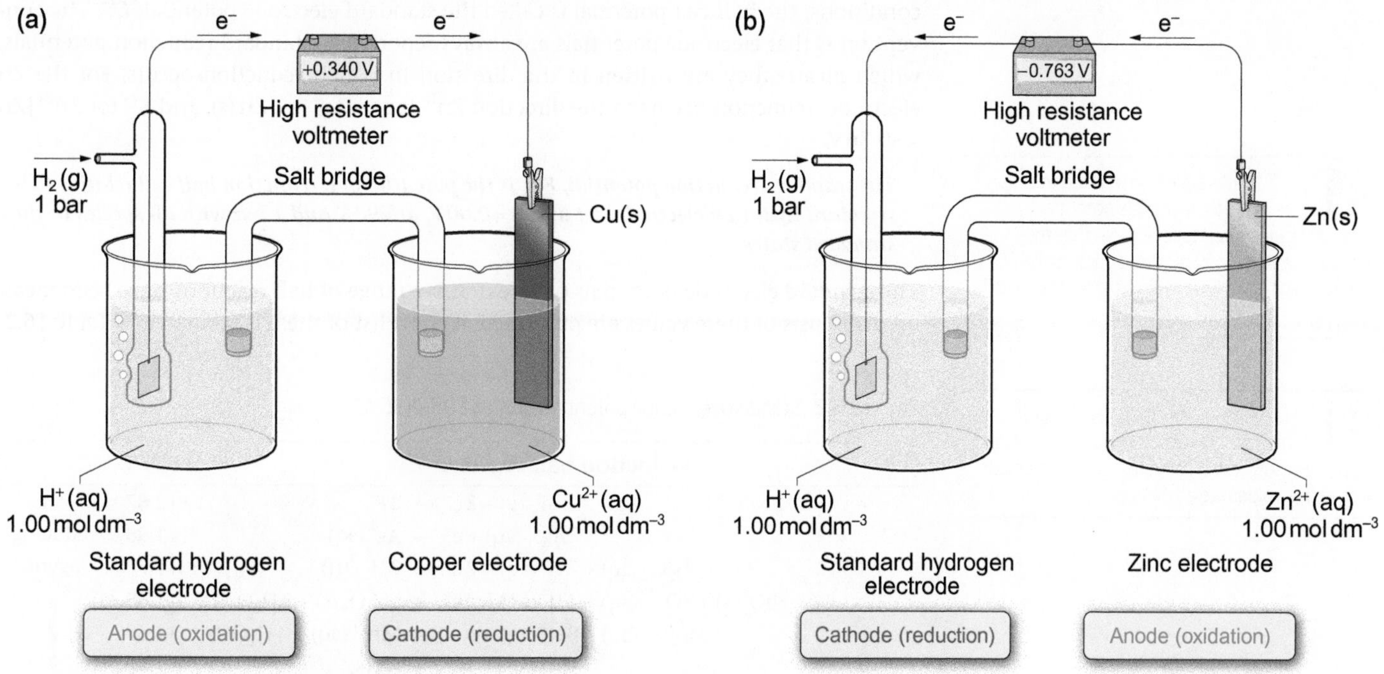
(a)
e⁻
e⁻
+0.340 V
High resolution voltmeter
Salt bridge
H₂ (g)
1 bar
Cu(s)
H⁺(aq)
1.00 mol dm⁻³
Cu²⁺(aq)
1.00 mol dm⁻³
Standard hydrogen electrode
Copper electrode
Anode (oxidation)
Cathode (reduction)
(b)
e⁻
e⁻
−0.763 V
High resistance voltmeter
Salt bridge
H₂ (g)
1 bar
Zn(s)
H⁺(aq)
1.00 mol dm⁻³
Zn²⁺(aq)
1.00 mol dm⁻³
Standard hydrogen electrode
Zinc electrode
Cathode (reduction)
Anode (oxidation)

Figure 16.11 Measuring electrode potentials against the standard hydrogen electrode. In (a) the standard hydrogen electrode is the anode so Cu^{2+} is reduced. In (b) the standard hydrogen electrode is the cathode so Zn is oxidized. The arrows show the direction in which electrons tend to flow in the external circuit but, if the resistance of the voltmeter is very high, no electron flow actually takes place.

electrode is set at zero and the value of the overall emf, the zero-current cell potential, E_{cell} is assigned to the copper half cell. For this electrochemical cell, E_{cell} is measured as +0.340 V, so the **electrode potential** of the copper half cell, $E(Cu^{2+}(aq)/Cu(s))$, is +0.340 V. The half reactions are

$$Cu^{2+}(aq) + 2e^- \rightarrow Cu(s)$$
$$H_2(g) \rightarrow 2H^+(aq) + 2e^-$$

so the overall cell reaction is

$$Cu^{2+}(aq) + H_2(g) \rightarrow Cu(s) + 2H^+(aq)$$

The copper half cell gains electrons from the hydrogen half cell, so it must be more positive than the hydrogen half cell, in order to attract electrons from it. Therefore, its potential is *positive* relative to the standard hydrogen half cell.

When the zinc half cell is set up as in Figure 16.11(b), E_{cell} is measured as −0.763 V. Again, the potential of the standard hydrogen electrode is set at zero, so the electrode potential of the zinc half cell, $E(Zn^{2+}(aq)/Zn(s))$, is −0.763 V. The half reactions are

$$Zn(s) \rightarrow Zn^{2+}(aq) + 2e^-$$
$$2H^+(aq) + 2e^- \rightarrow H_2(g)$$

and the overall cell reaction is

$$Zn(s) + 2H^+(aq) \rightarrow Zn^{2+}(aq) + H_2(g)$$

The zinc half cell gives electrons to the hydrogen half cell and its potential is *negative* relative to the standard hydrogen half cell. In this electrochemical cell, the anode is on the right and the electrons flow in the external circuit in the opposite direction to that in Figure 16.11(a).

This procedure can be followed in principle for *any* half cell so that its electrode potential relative to the standard hydrogen electrode can be determined. Under standard

The Nissan Leaf all-electric car uses a pack of lithium-ion batteries (containing a total of 192 cells) to drive its electric motor. Unlike *lithium* cells, lithium-ion cells are rechargeable. They are also lightweight, because lithium has low density unlike other metals used in car batteries, such as lead.

> $E^{\ominus}$ values are often referred to simply as standard electrode potentials. However, you need to remember that they always refer to *reduction* reactions so for emphasis they will be referred to as standard reduction potentials in this book.

> Take care when using tabulated values of standard reduction potentials from different sources. In some textbooks, the most negative values are at the top.

conditions, the half cell potential is called the standard electrode potential, $E^{\ominus}$. The convention is that electrode potentials are always reported as standard reduction potentials, which means they are written in the direction in which reduction occurs. For the Zn electrode, reduction occurs in the direction $Zn^{2+}(aq) + 2e^- \rightarrow Zn(s)$, and $E^{\ominus}$ for $Zn^{2+}|Zn = -0.76$ V.

The standard reduction potential, $E^{\ominus}$, is the potential of a reduction half cell relative to the standard hydrogen electrode, set at $E^{\ominus} = 0.00$ V, at 298 K and 1 bar with all species in their standard states.

The standard electrode potentials of an extensive range of half reactions have been measured and lists of these values are published. A short list of these is tabulated in Table 16.2.

Table 16.2 Standard reduction potentials, $E^{\ominus}$, at 298 K (25 °C)

	Reduction half reaction	$E^{\ominus}$/V	
Stronger oxidizing agent	$F_2(g) + 2e^- \rightarrow 2F^-$	+2.87	Weaker reducing agent
	$Ag^{2+}(aq) + e^- \rightarrow Ag^+(aq)$	+1.98	
	$H_2O_2(aq) + 2H^+(aq) + 2e^- \rightarrow 2H_2O(l)$	+1.77	
	$PbO_2(s) + SO_4^{2-}(aq) + 4H^+(aq) + 2e^- \rightarrow PbSO_4(s) + 2H_2O(l)$	+1.68	
	$MnO_4^-(aq) + 8H^+(aq) + 5e^- \rightarrow Mn^{2+}(aq) + 4H_2O(l)$	+1.52	
	$Au^{3+}(aq) + 3e^- \rightarrow Au(s)$	+1.50	
	$Cl_2(g) + 2e^- \rightarrow 2Cl^-(aq)$	+1.36	
	$Cr_2O_7^{2-}(aq) + 14H^+(aq) + 6e^- \rightarrow 2Cr^{3+}(aq) + 7H_2O(l)$	+1.33	
	$O_2(g) + 4H^+(aq) + 4e^- \rightarrow 2H_2O(l)$	+1.23	
	$Br_2(l) + 2e^- \rightarrow 2Br^-(aq)$	+1.08	
	$NO_3^-(aq) + 4H^+ + 3e^- \rightarrow NO(g) + 2H_2O(l)$	+0.96	
	$OCl^-(aq) + H_2O(l) + 2e^- \rightarrow Cl^-(aq) + 2OH^-(aq)$	+0.89	
	$Hg^{2+}(aq) + 2e^- \rightarrow Hg(l)$	+0.86	
	$Ag^+(aq) + e^- \rightarrow Ag(s)$	+0.80	
	$Hg_2^{2+}(aq) + 2e- \rightarrow 2Hg(l)$	+0.79	
	$Fe^{3+}(aq) + e^- \rightarrow Fe^{2+}(aq)$	+0.77	
	$I_2(s) + 2e^- \rightarrow 2I^-(aq)$	+0.54	
	$O_2(g) + 2H_2O(l) + 4e^- \rightarrow 4OH^-(aq)$	+0.40	
	$Cu^{2+}(aq) + 2e- \rightarrow Cu(s)$	+0.34	
	$Hg_2Cl_2(s) + 2e^- \rightarrow 2Hg(l) + 2Cl^-(aq)$	+0.27	
	$AgCl(s) + e^- \rightarrow Ag(s) + Cl^-(aq)$	+0.22	
	$Sn^{4+}(aq) + 2e^- \rightarrow Sn^{2+}(aq)$	+0.15	
	$AgBr(s) + e^- \rightarrow Ag(s) + Br^-(aq)$	+0.07	
	$2H^+(aq) + 2e^- \rightarrow H_2(g)$	**0.00**	
	$Sn^{2+}(aq) + 2e^- \rightarrow Sn(s)$	−0.14	
	$Ni^{2+}(aq) + 2e^- \rightarrow Ni(s)$	−0.25	
	$V^{3+}(aq) + e^- \rightarrow V^{2+}(aq)$	−0.26	
	$Co^{2+}(aq) + 2e^- \rightarrow Co(s)$	−0.28	
	$PbSO_4(s) + 2e^- \rightarrow Pb(s) + SO_4^{2-}(aq)$	−0.36	
	$Cd^{2+}(aq) + 2e^- \rightarrow Cd(s)$	−0.40	
	$Fe^{2+}(aq) + 2e^- \rightarrow Fe(s)$	−0.44	
	$Zn^{2+}(aq) + 2e^- \rightarrow Zn(s)$	−0.76	
	$2H_2O(l) + 2e^- \rightarrow H_2(g) + 2OH^-(aq)$	−0.83	
	$Cr^{2+}(aq) + 2e^- \rightarrow Cr(s)$	−0.91	
	$Al^{3+}(aq) + 3e^- \rightarrow Al(s)$	−1.66	
	$Mg^{2+}(aq) + 2e^- \rightarrow Mg(s)$	−2.37	
	$Na^+(aq) + e^- \rightarrow Na(s)$	−2.71	
	$K^+(aq) + e^- \rightarrow K(s)$	−2.93	
Weaker oxidizing agent	$Li^+(aq) + e^- \rightarrow Li(s)$	−3.05	Stronger reducing agent

Using $E^{\ominus}$ values to decide whether a redox reaction is spontaneous

In Table 16.2, the half reactions are written as reductions and are listed in order of decreasing $E^{\ominus}$. The list is in order of the tendency to accept electrons, relative to the hydrogen half reaction. This allows you to assess the likelihood of a redox reaction occurring. The value of $E^{\ominus}$ gives a measure of how strongly the reduction is driven. Half reactions with highly positive $E^{\ominus}$ values accept electrons readily so are good oxidizing agents. Half reactions with highly negative $E^{\ominus}$ values donate electrons readily and are good reducing agents.

A half reaction with a high positive $E^{\ominus}$ value has a stronger tendency to occur in the forward direction than one with a less positive value, that is, it is a stronger oxidizing agent. The more negative the value, the stronger the tendency for the half reaction to proceed in the reverse direction and to act as a reducing agent. Fluorine ($E^{\ominus}(F_2/F^-)$ = +2.87 V) is, therefore, the strongest oxidizing agent in Table 16.2, since it has the greatest tendency to be reduced, and lithium *ions* ($E^{\ominus}(Li^+/Li)$ = −3.05 V) are the weakest oxidizing agent. The lithium half reaction has a strong tendency to occur in the reverse direction, so lithium *metal* is a strong reducing agent. A list of half reactions ordered by $E^{\ominus}$ gives the order of oxidizing power and is known as the electrochemical series. In general

The standard reduction potential, $E^{\ominus}$, is the potential of a reduction half cell relative to the standard hydrogen electrode, set at $E^{\ominus}$ = 0.00 V, at 298 K and 1 bar with all species in their standard states.

Worked example 16.3 illustrates how you can use $E^{\ominus}$ values and the electrochemical series to predict whether a redox reaction will occur spontaneously.

A species that is *easily reduced* gains electrons easily, causing another species to be oxidized. It is thus a good *oxidizing agent*. Similarly, a good *reducing agent* is one that loses electrons easily and is *easily oxidized*, causing another species to be reduced.

Spontaneous reactions are discussed in Section 14.4 (p.671) and Section 14.5 (p.676).

Visit the Online Resource Centre to watch screencast 16.3 which walks you through Worked example 16.3 and explains the use of standard reduction potentials to predict the direction of redox reactions.

Values of $E^{\ominus}$ reflect the thermodynamic tendency for a reaction to happen under standard conditions. In some cases, however, kinetic factors mean that thermodynamically spontaneous reactions do not occur under standard conditions.

Worked example 16.3 Predicting the direction of redox reactions

What reactions would you expect to take place when bromine (Br_2) is added to an aqueous solution containing both NaCl and NaI, if each is present at a concentration of 1.0 mol dm^{-3}?

Strategy

Write equations for the half reactions that might occur for the species involved. Then consider their $E^{\ominus}$ values and decide which overall reactions should be spontaneous under standard conditions.

Solution

Write equations for the possible half reactions and their $E^{\ominus}$ values. Write all of these as reduction reactions, with the most positive $E^{\ominus}$ values at the top.

$Cl_2(g) + 2e^- \rightarrow 2Cl^-(aq)$ $\quad E^{\ominus} = +1.36$ V

$Br_2(l) + 2e^- \rightarrow 2Br^-(aq)$ $\quad E^{\ominus} = +1.07$ V

$I_2(s) + 2e^- \rightarrow 2I^-(aq)$ $\quad E^{\ominus} = +0.53$ V

Decide which overall reactions can take place on the basis of the $E^{\ominus}$ values.

The value of $E^{\ominus}$ for Br_2 is less positive than that for Cl_2 so Br_2 cannot oxidize Cl^-. However, it is more positive than the value for I_2 so that Br_2 will oxidize I^-. The half reaction for I_2 written above will proceed in the reverse direction, and the reactions will be

$$Br_2(l) + 2e^- \rightarrow 2Br^-(aq)$$
$$2I^-(aq) \rightarrow I_2(s) + 2e^-$$

Overall: $$Br_2(l) + 2I^-(aq) \rightarrow 2Br^-(aq) + I_2(s)$$

The Cl^- ions in the mixture will not react. The Na^+ ions do not play a part in the reaction—they are spectator ions.

Question

Predict whether acidified aqueous potassium permanganate ($KMnO_4$) could be used to oxidize Fe^{2+} to Fe^{3+} under standard conditions.

Remember that the cathode is the electrode at which reduction (electron gain) is taking place and the anode is where oxidation (electron loss) is taking place.

Right and left in Equation 16.11 refer to the side of the *cell diagram* on which the half reaction appears; it has, of course, nothing to do with the order in which apparatus might be assembled on the bench!

Using $E^{\ominus}$ values to calculate $E^{\ominus}_{cell}$

Using $E^{\ominus}$ values makes the calculation of cell potentials straightforward. $E^{\ominus}$ for the overall cell is given by the difference in the $E^{\ominus}$ values for the two half cells.

$$E^{\ominus}_{cell} = E^{\ominus}_{cathode} - E^{\ominus}_{anode} \quad (16.10)$$

By convention, in a cell diagram the anode is always written on the left. Therefore, another way of expressing Equation 16.10 is

$$E^{\ominus}_{cell} = E^{\ominus}_{R} - E^{\ominus}_{L} \quad (16.11)$$

where $E^{\ominus}_{R}$ is the reduction potential of the right-hand half cell in the cell diagram and $E^{\ominus}_{L}$ is that of the left-hand half cell in the diagram. **The sign of $E^{\ominus}_{cell}$ will always be positive if the cell reaction occurs spontaneously**. If your calculation of $E^{\ominus}_{cell}$ gives a negative answer, it means the cell reaction is non-spontaneous—it will occur spontaneously in the opposite direction to that in the cell diagram.

Continuing with the Daniell cell as an example, the cell diagram is

$$Zn\,(s)|Zn^{2+}\,(aq)||Cu^{2+}\,(aq)|Cu\,(s)$$

So, from Equations 16.10 and 16.11, the cell potential is given by

$$\begin{aligned} E^{\ominus}_{cell} &= E^{\ominus}_{cathode} - E^{\ominus}_{anode} \\ &= E^{\ominus}_{R} - E^{\ominus}_{L} \\ &= E^{\ominus}(Cu^{2+}/Cu) - E^{\ominus}(Zn^{2+}/Zn) \\ &= +0.34\,V - (-0.76\,V) \\ &= +1.10\,V \end{aligned}$$

The use of Equations 16.10 and 16.11 is further illustrated in Worked example 16.4 and in Box 16.3, which describes how cell potentials are often measured in practice.

Worked example 16.4 Calculating cell potentials

The half reactions that occur in a lead–acid battery of the type used to start a car engine (see Box 16.6, p.759) are

$$PbSO_4\,(s) + 2e^- \rightarrow Pb\,(s) + SO_4^{2-}\,(aq) \qquad E^{\ominus} = -0.36\,V$$

$$PbO_2\,(s) + 4H^+\,(aq) + SO_4^{2-}\,(aq) + 2e^- \rightarrow PbSO_4\,(s) + 2H_2O\,(l) \qquad E^{\ominus} = +1.69\,V$$

Calculate the overall potential for the cell, $E^{\ominus}_{cell}$.

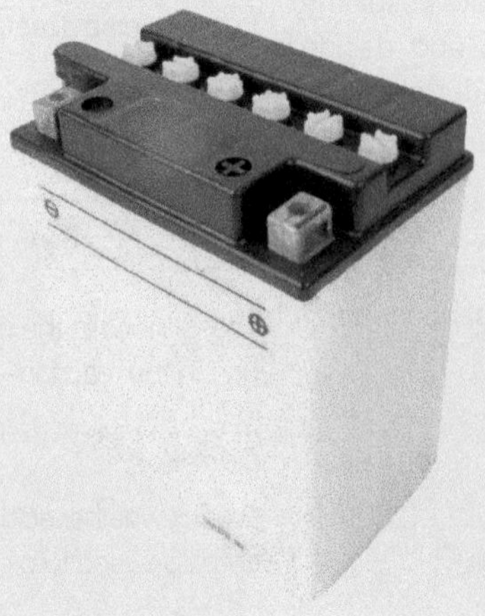

Most cars have a 12 volt electrical system. A car battery has 6 cells arranged in series, each with an emf of about 2 V, giving a total of 12 V.

→

→ Strategy

Use the standard reduction potentials for the two half reactions to decide the direction of the overall cell reaction. You can then decide which half reaction is the anode (oxidation) and which is the cathode (reduction) and use Equation 16.10 to calculate $E^{\ominus}_{cell}$.

Solution

Use the $E^{\ominus}$ values to decide the direction of the overall cell reaction.

The $E^{\ominus}$ value for the reduction of PbO_2 is more positive than that for the reduction of $PbSO_4$, so the PbO_2 half reaction will oxidize Pb to $PbSO_4$.

Decide which is the anode and which is the cathode.

Remember that oxidation occurs at the anode and reduction occurs at the cathode.

Anode $\quad Pb(s) + SO_4^{2-}(aq) \rightarrow PbSO_4(s) + 2e^-$

Cathode $\quad PbO_2(s) + 4H^+(aq) + SO_4^{2-}(aq) + 2e^- \rightarrow PbSO_4(s) + 2H_2O(l)$

Use Equation 16.10 to calculate $E^{\ominus}_{cell}$.

$$E^{\ominus}_{cell} = E^{\ominus}_{cathode} - E^{\ominus}_{anode} \qquad (16.10)$$

$$= (+1.69\,V) - (-0.36\,V)$$

$$= +2.05\,V$$

Question

Use values of standard reduction potentials from Table 16.2 to calculate $E^{\ominus}_{cell}$ for the following electrochemical cell

$$Fe(s)|Fe^{2+}(aq)||Fe^{3+}(aq), Fe^{2+}(aq)|Pt(s)$$

Box 16.3 Practical measurement of $E^{\ominus}$ values

The standard hydrogen electrode is the reference half cell for measuring other electrodes against, but it is not very convenient to use in the laboratory. The need for hydrogen gas means that it is difficult to use and also causes safety problems. In practice, other reference half cells are usually used. A common one is the silver–silver chloride electrode, which is illustrated in Figure 1.

The **silver–silver chloride electrode** consists of a silver wire coated with solid silver chloride dipping into a saturated solution of KCl. The half reaction for the reduction of silver chloride under these conditions is

$$AgCl(s) + e^- \rightarrow Ag(s) + Cl^-(aq)$$

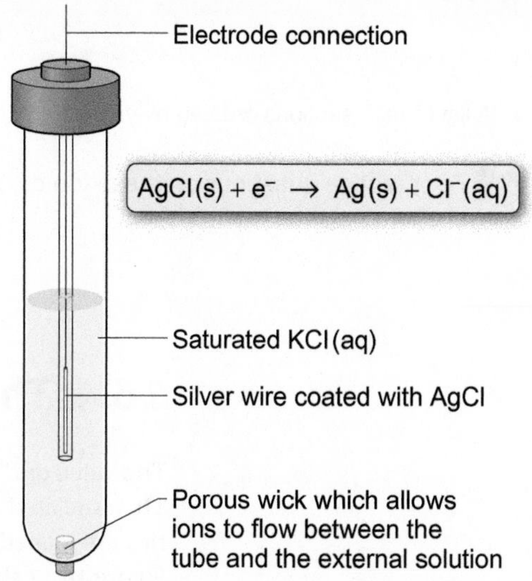
Electrode connection
AgCl(s) + e⁻ → Ag(s) + Cl⁻(aq)
Saturated KCl(aq)
Silver wire coated with AgCl
Porous wick which allows ions to flow between the tube and the external solution

The Ag/AgCl electrode is a more convenient standard to use than the standard hydrogen electrode. The electrode is placed in a solution of the half cell for which the potential is being measured. ➤

→

➜ and the half cell diagram is Ag (s) | AgCl (s) | Cl^- (aq). This gives a half cell with very reproducible properties for which $E^\ominus = +0.22$ V at 25 °C relative to the standard hydrogen electrode.

The potential of this half cell has been very accurately determined against the standard hydrogen electrode. Since the half reaction involves solids, liquids, and saturated solutions, its potential is constant at a particular temperature since the concentrations of the components do not vary (see Section 16.4, p.749).

When using a reference half cell other than the standard hydrogen electrode, the *relative* positions of all of the potentials for other half cells are unchanged and may be simply converted into values relative to the standard hydrogen electrode.

Question

Calculate the cell potentials, $E^\ominus_{cell}$, at 25 °C for the electrochemical cells

reference half cell || Zn (s) | Zn^{2+} (aq)

and

reference half cell || Cu^{2+} (aq) | Cu (s)

where the reference half cell is the silver–silver chloride electrode, AgCl (s) | Ag (s), Cl^- (aq).

Use your answer to find the value of $E^\ominus_{cell}$ at 25 °C for the electrochemical cell

Zn (s) | Zn^{2+} (aq) || Cu^{2+} (aq) | Cu (s)

Compare this value with the value found using the standard hydrogen electrode on pp.741 and 744.

Summary

- Electrochemical cells involve reduction and oxidation reactions arranged so that electrons can flow through an external circuit.
- A cell consists of two half reactions. Reduction occurs at the cathode; oxidation occurs at the anode.
- A cell diagram is a convenient way of representing an electrochemical cell on paper (see Figure 16.10, p.739).
- The standard hydrogen electrode is a half cell in which hydrogen gas is bubbled over a platinum foil immersed in a solution of H^+(aq) (see Figure 16.8, p.737). It is assigned a potential of $E^\ominus = 0.00$ V under standard conditions of 298 K (25 °C), hydrogen gas pressure of 1.00 bar, and H^+ (aq) concentration = 1 mol dm^{-3}.
- The standard reduction potential, $E^\ominus$, is the potential of a reduction half cell relative to the standard hydrogen electrode set to $E^\ominus = 0.00$ V with all species in their standard states.
- Under standard conditions, a half reaction with a high (positive) $E^\ominus$ will oxidize a half reaction with lower (less positive) $E^\ominus$.
- The overall cell potential (emf) is given by

$$E^\ominus_{cell} = E^\ominus_{cathode} - E^\ominus_{anode}$$

- A list of half reactions ordered by value of $E^\ominus$ and oxidizing power is known as the electrochemical series.

? For practice questions on these topics, see questions 3–11 at the end of this chapter (pp.761–762).

16.4 Thermodynamics of electrochemical cells

The value of $E^\ominus_{cell}$ gives a measure of the tendency for the overall cell reaction to happen. This is related to the Gibbs energy change for the reaction. To see how these two quantities are related, consider the work involved in moving the electrons around the circuit. Figure 16.12 shows one example of an electrochemical cell doing work.

$E^{\ominus}_{cell}$ and Gibbs energy change

If a reaction produces z mol of electrons, the total charge, Q, that is produced is given by

$$Q = zF$$

where F is the charge on 1 mol of electrons. This quantity is called the **Faraday constant** and has a value of 96 485 coulombs.

$$F = 96\,485\ \mathrm{C\,mol^{-1}} \qquad (16.12)$$

The movement of the charge around a circuit against the resistance requires energy and work is done by the cell. The energy required to transfer this charge represents the electrical work that the cell can perform.

When a charge of 1 coulomb passes through a potential difference of 1 volt, 1 joule of work is done: 1 J = 1 C × 1 V.

$$\text{Electrical work} = -(\text{charge} \times \text{potential difference})$$

$$w_{elec} = -Q \times V$$

The value of w_{elec} is a negative quantity because the reaction system loses energy as it does work.

If the charge is zF and the potential difference is $E^{\ominus}_{cell}$, the electrical energy transferred as work is

$$w_{elec} = -zF \times E^{\ominus}_{cell} \qquad (16.13)$$

In reality, the cell voltage is $E^{\ominus}_{cell}$ only when the resistance in the circuit is very large and zero current flows. In order for the cell to do work, the resistance has to be lowered so that a current flows. This, in turn, lowers the cell potential, so Equation 16.13 shows the *maximum* electrical work that can be transferred.

This work is equal to the change in Gibbs energy for the cell reaction, $\Delta_r G^{\ominus}$. From Section 14.5 (p.685), $\Delta_r G^{\ominus}$ this is equal to the maximum work that the cell reaction can perform

$$\Delta_r G^{\ominus} = w_{elec}$$

$$\Delta_r G^{\ominus} = -z F \times E^{\ominus}_{cell} \qquad (16.14)$$

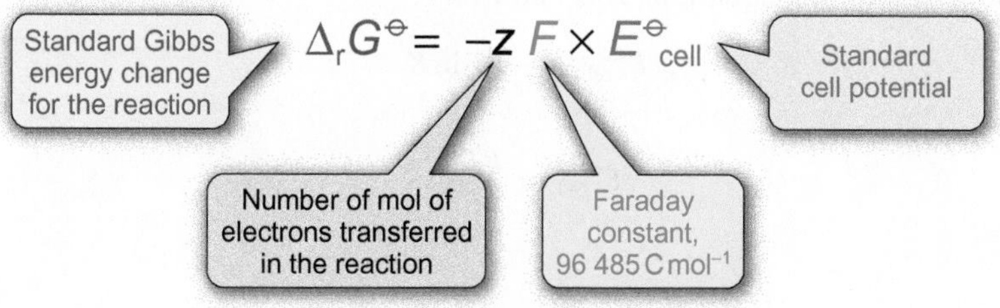
Standard Gibbs energy change for the reaction
Standard cell potential
Number of mol of electrons transferred in the reaction
Faraday constant, 96 485 C mol⁻¹

Equation 16.14 shows that a reaction with a positive cell potential has a negative value for Gibbs energy change and is, therefore, spontaneous under standard conditions. The use of Equation 16.14 to calculate $\Delta_r G^{\ominus}$ for a reaction is illustrated in Worked example 16.5.

Equation 16.14 is written in terms of $\Delta_r G^{\ominus}$ and $E^{\ominus}$ and refers to reactions at 298 K (25 °C) and conditions of standard pressure and unit activities. However, the same form of equation can be used under other conditions, $\Delta_r G^{\ominus} = -zFE$. (The absence of the superscript $^{\ominus}$ shows that the reaction is not occurring under standard conditions.)

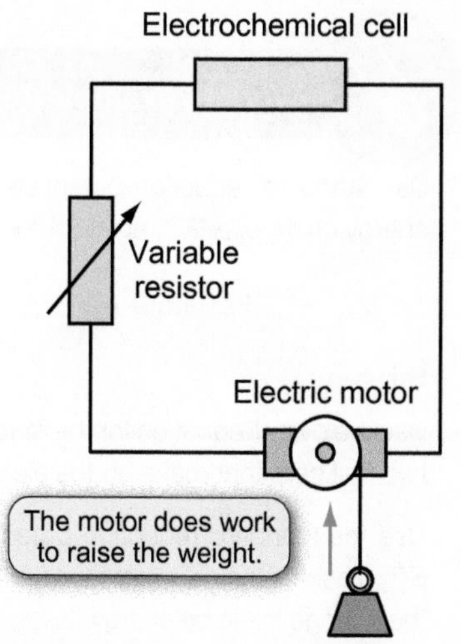
Electrochemical cell
Variable resistor
Electric motor
The motor does work to raise the weight.

Figure 16.12 The electrochemical cell drives the electric motor, raising the weight and doing work.

The Gibbs energy change for a reaction, $\Delta_r G$, is introduced in Section 14.5 (p.675).

The Faraday constant, F, is the charge on 1 mol of electrons. It is equal to the Avogadro constant, N_A, multiplied by the charge, e, on an electron:

$$F = N_A e.$$

Maximum work is done under reversible conditions. For a cell, this corresponds to a very high resistance which gives a very small current. Under these conditions, the measured voltage corresponds to $E^{\ominus}_{cell}$

Note that z is the *number* of moles of electrons transferred. It has no units.

Worked example 16.5 Gibbs energy changes from electrochemistry

Use standard reduction potentials to calculate the standard Gibbs energy change, $\Delta_r G^{\ominus}$, at 298 K for the reaction

$$Fe^{2+}(aq) + Ag^{+}(aq) \rightarrow Fe^{3+}(aq) + Ag(s)$$

Strategy

Use the overall equation for the reaction to write half equations for the two half cells that make up the electrochemical cell.

Use the standard reduction potentials of the half cells to calculate $E^{\ominus}_{cell}$ as in Worked example 16.4. Then use Equation 16.14 to find the change in Gibbs energy.

Solution

Write the two half equations and identify the oxidation and reduction.

$Fe^{2+}(aq) \rightarrow Fe^{3+}(aq) + e^{-}$ Fe^{2+}(aq) loses electrons: oxidation

$Ag^{+}(aq) + e^{-} \rightarrow Ag(s)$ Ag^{+}(aq) gains electrons: reduction

Using data from Table 16.2 (p.742) and Equation 16.10 (p.744), calculate $E^{\ominus}_{cell}$. (Remember that reduction occurs at the cathode and oxidation at the anode.)

$$\begin{aligned} E^{\ominus}_{cell} &= E^{\ominus}_{cathode} - E^{\ominus}_{anode} \qquad (16.10)\\ &= E^{\ominus}_{(Ag^{+}/Ag)} - E^{\ominus}_{(Fe^{3+}/Fe^{2+})}\\ &= (+0.80\,V) - (+0.77\,V)\\ &= +0.03\,V\\ &= +1.10\,V \end{aligned}$$

Use Equation 16.14 to calculate the standard Gibbs energy change of reaction.

The reaction involves the transfer of 1 mol of electrons, so $z = 1$

$$\begin{aligned} \Delta_r G^{\ominus} &= -zF \times E^{\ominus}_{cell} \qquad (16.14)\\ &= -(1) \times (96\,485\,C\,mol^{-1}) \times (+0.03\,V)\\ &= -2900\,C\,V\,mol^{-1}\\ &= -2900\,J\,mol^{-1} \quad (1\,C \times 1\,V = 1\,J)\\ &= -2.90\,kJ\,mol^{-1} \quad (1\,kJ = 1 \times 10^{3}\,J) \end{aligned}$$

Question

Calculate the standard Gibbs energy change at 298 K for the reaction in the Daniell cell

$$Zn(s)|Zn^{2+}(aq)||Cu^{2+}(aq)|Cu(s)$$

The relationship between the standard Gibbs energy change for a reaction and the thermodynamic equilibrium constant, *K*, is derived in Box 15.3 (p.706).

The equilibrium constant in Equations 15.4 and 16.15 is the thermodynamic equilibrium constant, *K*. This is defined in terms of activities (p.698) and is dimensionless (has no units).

Supports
battery recycling

When a battery has gone 'flat', the cell reaction has reached equilibrium and the battery is no longer useful, unless it is rechargeable. Proper disposal at a recycling point is needed, because batteries often contain metals which are harmful to the environment.

$E^{\ominus}_{cell}$ and the thermodynamic equilibrium constant

The standard Gibbs energy change for a reaction and the thermodynamic equilibrium constant are related by Equation 15.4 (p.705)

$$\Delta_r G^{\ominus} = -RT \ln K \qquad (15.4)$$

where T is the temperature in kelvin and R is the gas constant, $8.314\,J\,K^{-1}\,mol^{-1}$.

This means that $E^{\ominus}_{cell}$ can be used to find the thermodynamic equilibrium constant, K, for a reaction. Combining Equations 16.14 and 15.4

$$-zFE^{\ominus}_{cell} = -RT \ln K$$

so

$$E^{\ominus}_{cell} = \frac{RT}{zF} \ln K$$

and

$$\ln K = \frac{zF}{RT} E^{\ominus}_{cell} \qquad (16.15)$$

What does it mean when a cell reaction reaches equilibrium? $E^{\ominus}_{cell}$ is measured when the reactants are present under standard conditions and no current flows. If current is drawn from the cell, reactions occur in the half cells and the concentrations of species change. Eventually, the concentrations stop changing and no current flows. The cell reaction has reached equilibrium—the cell has 'run out' or 'gone flat'.

Worked example 16.6 calculates the thermodynamic equilibrium constant for the reaction in the Daniell cell. This shows that for a cell potential of +1.1 V, an extremely large value of K is obtained and the reaction proceeds effectively to completion in the forward direction. Measurement of cell potentials is a very convenient way of finding equilibrium constants for a wide range of reactions, particularly where K is very large or very small, in which case accurate measurement of the concentrations of the products or reactants at equilibrium would be difficult.

Worked example 16.6 Calculating a thermodynamic equilibrium constant

The cell potential, $E^{\ominus}_{cell}$, for the Daniell cell (see Section 16.3, p.736) is +1.10 V at 298 K under standard conditions. Calculate the thermodynamic equilibrium constant, K, for the reaction

$$Zn(s) + Cu^{2+}(aq) \rightarrow Zn^{2+}(aq) + Cu(s)$$

Strategy

Use Equation 16.15 directly to find a value for the thermodynamic equilibrium constant, K. You need to take care with units. Remember, K is defined in terms of activities (see Section 14.6, p.698) and is dimensionless (has no units).

Solution

Use Equation 16.15 and substitute in the values. 2 mol of electrons are transferred in the reaction, so z = 2.

$$\ln K = \frac{zF}{RT} E^{\ominus}_{cell} \qquad (16.15)$$

$$= \frac{2 \times 96\,485\,\text{C}\,\cancel{\text{mol}^{-1}}}{(8.314\,\text{J}\,\cancel{\text{K}^{-1}}\,\cancel{\text{mol}^{-1}}) \times (298\,\cancel{\text{K}})} \times 1.10\,\text{V} = 85.7\,\text{C}\,\text{V}\,\text{J}^{-1}$$

Check the units and calculate K.

Since 1 J = 1 C × 1 V, the units cancel so $\ln K$ is dimensionless as required

$$\ln K = 85.7$$

$$K = 1.6 \times 10^{37}$$

This very large value means the reaction effectively goes to completion with complete conversion of Cu^{2+} to Cu metal.

Question

Calculate the thermodynamic equilibrium constant at 298 K for the reaction

$$Cu(s) + 2Ag^{+}(aq) \rightarrow Cu^{2+}(aq) + 2Ag(s)$$

Table 16.3 Relationships between the values of $E^{\ominus}_{cell}$, $\Delta_r G^{\ominus}$, and K

$E^{\ominus}_{cell}$	$\Delta_r G^{\ominus}$	K	Reaction
Positive, >0	Negative, <0	>1	Spontaneous in the forward direction
Zero	Zero	1	No tendency to occur—at equilibrium
Negative, <0	Positive, >0	<1	Spontaneous in the reverse direction

The interrelationships between the cell potential, $E^{\ominus}_{cell}$, the standard Gibbs energy change for the reaction, $\Delta_r G^{\ominus}$, and the thermodynamic equilibrium constant, K, are summarized in Table 16.3.

The relationship between $E^{\ominus}_{cell}$ and $\Delta_r G^{\ominus}$ in Equation 16.14 (p.747) allows other thermodynamic data to be determined from electrochemical measurements. The variation of $E^{\ominus}_{cell}$ with temperature can be measured easily in many cases. From this, the temperature variations of $\Delta_r G^{\ominus}$ can be found, which, in turn, lead to values for the entropy change, $\Delta_r S^{\ominus}$, and enthalpy change, $\Delta_r H^{\ominus}$, for the reaction. This method is very useful for studying reactions for which calorimetry would be difficult or impossible. Electrochemical methods are particularly useful in biological studies where small electrodes can be implanted to monitor biochemical processes, for example, in single-celled organisms.

The temperature variation of $\Delta_r G^{\ominus}$ (leading to a value of $\Delta_r S^{\ominus}$) is discussed in Section 14.6 (p.688). $\Delta_r H^{\ominus}$ can be found by substituting values for $\Delta_r G^{\ominus}$ and $\Delta_r S^{\ominus}$ into Equation 14.16 (p.675).

Concentration dependence of E_{cell}

All of the cell reactions discussed so far in this chapter have taken place under standard conditions with activities of all components = 1. What happens if the concentrations or gas pressures are changed?

This problem can be tackled by considering how the Gibbs energy change, $\Delta_r G$, for the reaction varies with the activities of the reactants and products. This is given by Equation 15.8, which is derived in Box 15.3 (p.706)

The reaction quotient, Q, is introduced in Section 15.2 (p.702). The variation $\Delta_r G$ of with the activities of reactants and products is discussed in Section 14.6 (p.686).

$$\Delta_r G = \Delta_r G^{\ominus} + RT \ln Q \qquad (15.8)$$

Q is the reaction quotient, given by

$$Q = \frac{\prod (a(\text{products}))^{\nu_P}}{\prod (a(\text{reactants}))^{\nu_R}} \qquad (15.2)$$

The symbol Π is shorthand for 'the product of' (in the same way as Σ represents 'the sum of'). So $\Pi(2,3,4) = 2 \times 3 \times 4 = 24$.
See Maths Toolkit MT1 (p.1304).

where ν_R and ν_P are the stoichiometric coefficients of the reactants and products, respectively, in the chemical equation.

For example, for the reaction

$$H_2(g) + 2AgCl(s) \rightarrow 2H^+(aq) + 2Cl^-(aq) + 2Ag(s)$$

$$Q = \frac{(a_{H^+})^2 \times (a_{Cl^-})^2}{a_{H_2}}$$

The activity of a gas is given by $a = p/p^{\ominus}$, where p is the pressure of a gas expressed in bar and $p^{\ominus} = 1$ bar. For substances in solution, $a = [A]/[A]^{\ominus}$, where [A] is the concentration in mol dm^{-3} and $[A]^{\ominus} = 1$ mol dm^{-3}. a(pure solid) $= a$(pure liquid) $= 1$ (see Section 14.6, p.686).

The expressions for a_{AgCl} and a_{Ag} are omitted because they are solids and so $a = 1$. Q has the same general form as the equilibrium constant K, but Q refers to a reaction at *any* stage, not only at equilibrium.

In Equation 15.8, $\Delta_r G^{\ominus}$ refers to the reaction taking place under standard conditions while $\Delta_r G$ is the Gibbs energy change under different conditions. Q is the reaction quotient for the cell reaction. Equation 16.14 (p.747) can be used to substitute expressions involving the cell potential in place of the Gibbs energy changes in Equation 15.8

$$-zFE_{cell} = -zFE^{\ominus}_{cell} + RT \ln Q$$

where E_{cell} is the cell potential recorded at conditions other than standard. Rearranging and dividing each term by zF gives

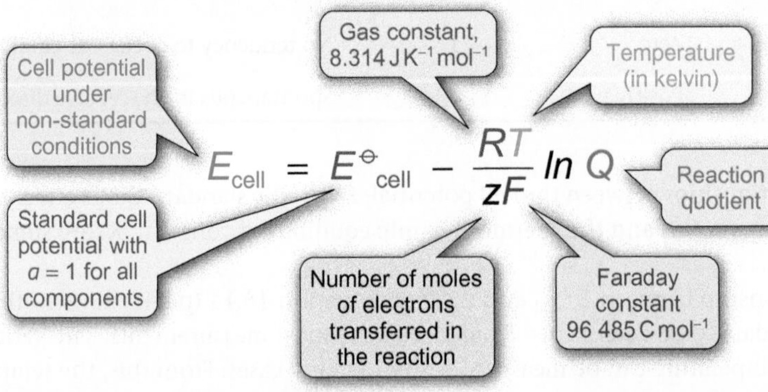
Cell potential under non-standard conditions
Gas constant, 8.314 J K^{-1} mol^{-1}
Temperature (in kelvin)
Reaction quotient
Standard cell potential with a = 1 for all components
Number of moles of electrons transferred in the reaction
Faraday constant 96 485 C mol^{-1}

$$E_{cell} = E^{\ominus}_{cell} - \frac{RT}{zF} \ln Q \qquad (16.16)$$

Visit the Online Resource Centre to view screencast 16.4 which explains the use of the Nernst equation and Worked example 16.7.

Equation 16.16 is known as the **Nernst equation**.

Using the Nernst equation, you can calculate the cell potential under any set of conditions, as shown in Worked example 16.7. The Nernst equation can also be used to calculate the concentration of a species involved in an electrochemical cell, and is the basis of a number of methods in analytical chemistry.

Some uses of electrochemistry in analytical chemistry are discussed in Section 11.2 (p.523).

The Nernst equation can be applied to individual half cells as well as to complete cells. So

$$E_{half\ cell} = E^{\ominus}_{half\ cell} - \frac{RT}{zF} \ln Q \qquad (16.17)$$

where $E_{\text{half cell}}$ is the reduction potential for the half cell relative to the standard hydrogen electrode (see Table 16.2, p.742). For example, for the $Cu^{2+}|Cu$ half cell, the reduction reaction is

$$Cu^{2+}(aq) + 2e^- \rightarrow Cu(s)$$

So

$$\begin{aligned} E(Cu^{2+}/Cu) &= E^{\ominus}(Cu^{2+}/Cu) - \frac{RT}{zF}\ln\left(\frac{a_{Cu}}{a_{Cu^{2+}}}\right) \\ &= E^{\ominus}(Cu^{2+}/Cu) - \frac{RT}{zF}\ln\left(\frac{1}{a_{Cu^{2+}}}\right) \end{aligned}$$

Box 16.4 shows how the mechanism of the rusting of iron can be understood in terms of the electrochemical processes taking place.

Worked example 16.7 Using the Nernst equation

What is the cell potential, E_{cell}, of a Daniell cell (see Section 16.3, p.736) with a copper(II) sulfate concentration of 0.0050 mol dm^{-3} and a zinc sulfate concentration of 0.10 mol dm^{-3}? ($E^{\ominus}_{\text{cell}} = 1.10$ V)

Strategy

Use the Nernst equation (Equation 16.16) to calculate E_{cell} under these conditions using the known value of $E^{\ominus}_{\text{cell}}$. In order to do this, you need to write an equation for the overall cell reaction and work out a value for Q under the given conditions.

Solution

Write an equation for the overall cell reaction.

$$Zn(s) + Cu^{2+}(aq) \rightarrow Cu(s) + Zn^{2+}(aq)$$

Work out a value for the reaction quotient, Q, using Equation 15.2.

Remember that the activity of a pure solid is 1 and that of a species in solution is given by $[A]/[A]^{\ominus}$, where $[A]$ is its concentration in mol dm^{-3} and $[A]^{\ominus} = 1$ mol dm^{-3}.

$$Q = \frac{\prod (a(\text{products}))^{\nu_R}}{\prod (a(\text{reactants}))^{\nu_R}} = \frac{a_{Cu} \times a_{Zn^{2+}}}{a_{Cu^{2+}} \times a_{Zn}} \qquad (15.2)$$

$$= \frac{1 \times 0.10}{0.0050 \times 1} = 20$$

Use the Nernst equation (Equation 16.16) to calculate E_{cell}.

$$E_{\text{cell}} = E^{\ominus}_{\text{cell}} - \frac{RT}{zF}\ln Q \qquad (16.16)$$

$$= 1.10\,\text{V} - \frac{(8.314\,\text{J}\,\cancel{\text{K}^{-1}}\,\cancel{\text{mol}^{-1}}) \times (298\,\cancel{\text{K}})}{2 \times 96\,485\,\text{C}\,\cancel{\text{mol}^{-1}}} \times \ln 20$$

$$= 1.10\,\text{V} - 0.038\,\text{J}\,\text{C}^{-1} \qquad (1\,\text{J} = 1\,\text{C} \times 1\,\text{V})$$

$$= 1.06\,\text{V}$$

Questions

(a) Write an equation for the overall reaction that occurs in the following cell

$$Pt(s)|H_2(g)|H^+(aq, 1\,\text{mol dm}^{-3})||Cu^{2+}(aq, 1\,\text{mol dm}^{-3})|Cu(s)$$

(b) Calculate the value of $E^{\ominus}_{\text{cell}}$.

(c) The potential, E_{cell}, for the cell below is measured and found to be +0.25 V.

$$Pt(s)|H_2(g)|H^+(aq, 1\,\text{mol dm}^{-3})||Cu^{2+}(aq, c)|Cu(s)$$

What is the concentration, c, of the $Cu^{2+}(aq)$?

Box 16.4 Corrosion as an electrochemical process

While the electrochemical cells represented in Figures 16.5 and 16.6 (p.736) are a convenient way of studying redox reactions in the laboratory, redox reactions occur in a variety of other situations. A familiar example is the rusting of iron.

A piece of bare iron left outside where it is exposed to moisture will quickly become coated with rust, which is a mixture of hydrated oxides of iron. This form of corrosion has serious economic consequences. It has been estimated that 20–25% of all new steel is produced to replace corroded metal objects and structures.

The mechanism of rusting

Although rusting is a complex process, the overall reaction can be represented as

$$4Fe(s) + 3O_2(g) + H_2O(l) \rightarrow 2Fe_2O_3 \cdot xH_2O(s)$$

→

▲ Rusting is accelerated in the presence of salt water.

where $Fe_2O_3 \cdot xH_2O$ represents a mixture of hydrated oxides of iron whose exact composition is not known. The mechanism involves several stages. Iron metal is first oxidized to Fe^{2+} ions.

$$Fe(s) \rightarrow Fe^{2+}(aq) + 2e^-$$

For the reverse process $Fe^{2+}(aq) + 2e^- \rightarrow Fe(s) \quad E^\ominus = -0.44\,V$

In principle, any half reaction with $E^\ominus$ more positive than −0.44 V will promote the oxidation of Fe(s). However, water is needed for rusting to occur and it is significant that there is little reaction of iron with water unless the water contains an appreciable concentration of dissolved oxygen. For example, some iron water supply pipes that are buried deep in the ground, away from oxygen, last for many years.

In the presence of water and oxygen, rusting occurs readily. The half equation for the reduction of oxygen in aqueous solution depends on the conditions, especially pH. Two possible half reactions are

Reaction 1: $O_2(g) + 4H^+(aq) + 4e^- \rightarrow 2H^2O(l) \quad E^\ominus = +1.23\,V$

Reaction 2: $O_2(g) + 2H_2O(l) + 4e^- \rightarrow 4OH^-(aq) \quad E^\ominus = +1.23\,V$

Combining these half equations with the one above for the oxidation of iron, the overall equations for the initial stage in rusting are then

$$2Fe(s) + O_2(g) + 4H^+(aq) \rightarrow 2Fe^{2+}(aq) + 2H_2O(l)$$

$$2Fe(s) + O_2(g) + 2H_2O(l) \rightarrow 2Fe^{2+}(aq) + 4OH^-(aq)$$

The reactions take place in droplets of water on the surface of the iron, as shown in Figure 1. Once $Fe^{2+}(aq)$ ions have formed, they are oxidized by dissolved oxygen to $Fe^{3+}(aq)$ and form the insoluble hydrated oxide known as rust.

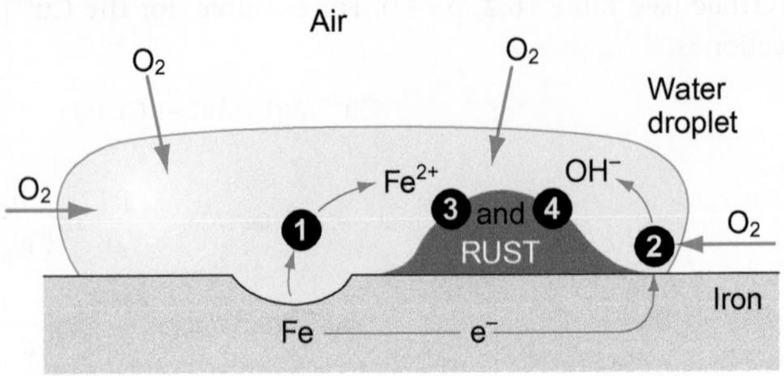

Air
O2
O2
Water droplet
O2
Fe2+
OH−
3 and 4
RUST
1
2
O2
Iron
Fe
e−

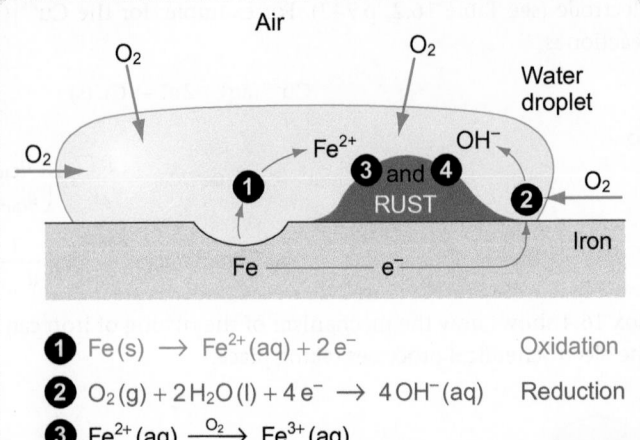

1 $Fe(s) \rightarrow Fe^{2+}(aq) + 2e^-$ Oxidation

2 $O_2(g) + 2H_2O(l) + 4e^- \rightarrow 4OH^-(aq)$ Reduction

3 $Fe^{2+}(aq) \xrightarrow{O_2} Fe^{3+}(aq)$

4 $Fe^{3+}(aq) \xrightarrow{OH^-(aq),\ H_2O} Fe_2O_3 \cdot xH_2O\ (rust)$

▲ **Figure 1** A water droplet on the surface of iron forms an electrochemical cell, which leads to rusting. Atmospheric oxygen plays a major part in the reaction.

Preventing rusting

There are a number of ways of preventing rusting. The simplest way is to paint or otherwise coat the metal so that water and oxygen cannot reach the iron. This is only effective until the coating is scratched or broken and the iron is exposed to water and oxygen. Rather more effective is the process of *galvanization*. This involves coating iron with a thin film of zinc. Zinc forms a protective layer of oxide over itself and so corrodes only slowly. Even if the zinc coating breaks and iron is exposed, the zinc reduces the $Fe^{2+}(aq)$ ions back to Fe(s) since Zn has a more negative reduction potential than Fe: $E^\ominus(Zn^{2+}/Zn) = -0.76\,V$; $E^\ominus(Fe^{2+}/Fe) = -0.44\,V$.

The same principle is used in *cathodic protection*. Here, an article made from iron that needs protection—a pipe, water tank, or even a ship—is attached by a wire to a block of a different, more easily oxidized metal such as magnesium (Figure 2). This is known as a *sacrificial anode* since it reacts in preference to the iron. It must be replaced periodically as it is used up but this is much cheaper and more convenient than replacing the article that is being protected.

$$Mg^{2+}(aq) + 2e^- \rightarrow Mg(s) \quad E^\ominus = -2.37\,V$$

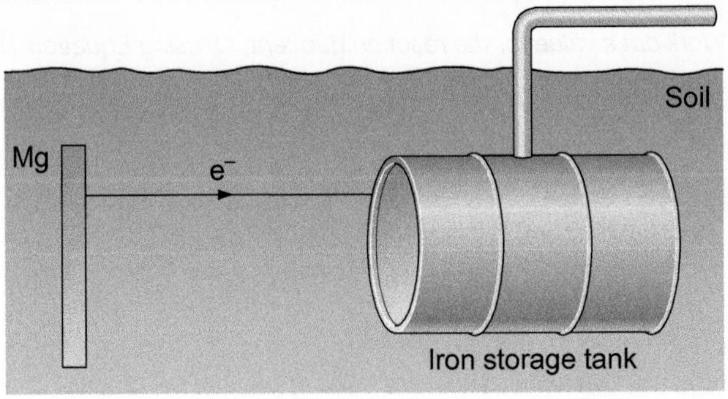

Soil
Mg
e−
Iron storage tank

$Mg(s) \rightarrow Mg^{2+}(aq) + 2e^-$ Oxidation

$O_2(g) + 2H_2O(l) + 4e^- \rightarrow 4OH^-(aq)$ Reduction

▲ **Figure 2** Magnesium is more easily oxidized than iron and so it will react in preference and prevent the buried iron tank from corroding.

Questions

(a) Suggest why rusting occurs much more readily in solutions containing dissolved electrolytes, for example, in sea water or in cars during the winter months when salt is spread onto roads.

(b) Which of the following, sodium, copper, tin, zinc, or graphite, could be used as sacrificial anodes for an iron structure?

Concentration cells

Because reduction potentials change with concentration, it is possible to construct an electrochemical cell using two half cells with the same chemical reaction but with different concentrations. Such cells are called **concentration cells.**

Think about the cell

$$\mathrm{Ag(s)\,|\,Ag^+(aq,\ 0.1\,mol\,dm^{-3})\,||\,Ag^+(aq,\ 0.5\,mol\,dm^{-3})\,|\,Ag(s)}$$

Both half cells are silver metal–silver ion electrodes but the concentration of silver ions is different in each. The two half reactions are

Left:	$\mathrm{Ag(s)} \rightarrow \mathrm{Ag^+}(\mathrm{aq},\ a = 0.1) + \mathrm{e^-}$
Right:	$\mathrm{Ag^+}(\mathrm{aq},\ a = 0.5) + \mathrm{e^-} \rightarrow \mathrm{Ag(s)}$
Overall:	$\mathrm{Ag^+}(\mathrm{aq},\ a = 0.5) \rightarrow \mathrm{Ag^+}(\mathrm{aq},\ a = 0.1)$

Using the Nernst equation (Equation 16.6) to calculate the cell potential, E_{cell}

$$E_{\text{cell}} = E^{\ominus}_{\text{cell}} - \frac{RT}{zF}\ln\left(\frac{a_{\mathrm{Ag^+(right)}}}{a_{\mathrm{Ag^+(left)}}}\right)$$

$E^{\ominus}_{\text{cell}}$ is zero because, under standard conditions, $[\mathrm{Ag^+}] = 1.0\,\mathrm{mol\,dm^{-3}}$ and $a = 1.0$ in both half cells, so

$$E_{\text{cell}} = 0\,\mathrm{V} - \frac{(8.314\,\mathrm{J\,\cancel{K^{-1}}\,\cancel{mol^{-1}}})\times(298\,\cancel{\mathrm{K}})}{1\times(96485\,\mathrm{C\,\cancel{mol^{-1}}})} \times \ln\left(\frac{0.1}{0.5}\right)$$

$$= +0.04\,\mathrm{V} \qquad (1\,\mathrm{J} = 1\,\mathrm{C}\times 1\,\mathrm{V})$$

The value of E_{cell} for concentration cells is usually quite small. When current is drawn from the cell, the cell potential decreases as the reaction proceeds. The concentrations in the two half cells tend to equalize; when the concentrations are the same in both half cells, then the reaction comes to equilibrium and $E_{\text{cell}} = 0$.

The principle underlying concentration cells is particularly important where solutions of electrolytes with different concentrations are separated by a porous membrane. If one of the ions—say, the positive ion as shown in Figure 16.13—can pass through the membrane but others cannot, then passage of ions will occur, tending to equalize the concentrations. This leads to an excess of the positive ions on one side of the membrane.

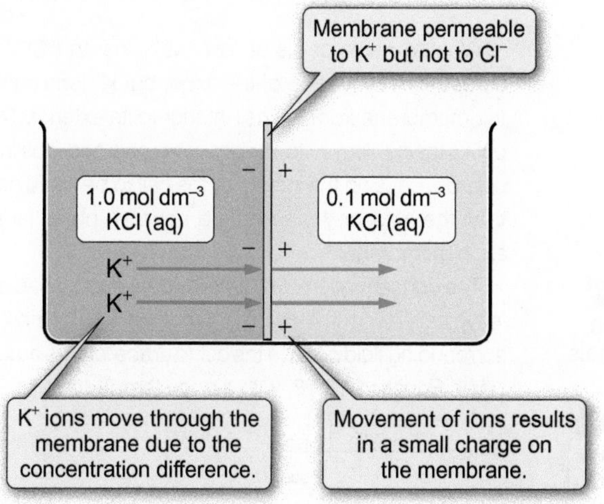
Membrane permeable to K+ but not to Cl−
1.0 mol dm−3 KCl (aq)
0.1 mol dm−3 KCl (aq)
K+
K+
K+ ions move through the membrane due to the concentration difference.
Movement of ions results in a small charge on the membrane.

Figure 16.13 Development of a membrane potential. K^+ ions move through the membrane due to the difference in concentration, but Cl^- ions cannot get through. This puts a small positive charge on the right-hand side of the membrane. Equilibrium is established when the 'push' on K^+ is balanced by the repulsion of K^+ by the positive charge. Only a very small fraction of the K^+ ions take part in this process so the concentrations do not noticeably change.

The imbalance of ions on each side leads to a potential difference across the membrane, which slows down and stops the movement of ions. This is known as a **membrane potential**, sometimes called a **Nernst potential**. Membrane potentials are often found in biological cells (see Box 16.5) where the concentrations of K^+, Na^+, Cl^-, and other ions are different inside the cell compared with outside. The situation can be analysed in the same way as concentration cells to show that, for any ion, M^+, the membrane potential, ΔE_{M^+}, is given by

$$\Delta E_{M^+} = \frac{RT}{zF}\ln\frac{[M^+]_{outside}}{[M^+]_{inside}} \quad (16.18)$$

The symbol ΔE for the membrane potential emphasizes that it is a *difference* in potentials due to a concentration gradient sometimes, the symbol φ (phi is used).

Box 16.5 Bioelectrochemistry: nerve cells and ion channels

Biological cells rely on a complex interplay of membrane potentials and ionic concentrations to exchange material between their interiors and their surroundings. A particularly good example is a nerve cell, shown in Figure 1.

A human nerve cell consists of the main body of the cell, surrounded by *dendrites* that branch off from the cell body and detect information from adjacent cells, and a long extension called an *axon*. The purpose of the axon is to transmit information (in the form of an electrical pulse) along the nerve. This is done by controlling the concentrations of ions, particularly Na^+, K^+, and Cl^- ions inside and outside the cell.

The membrane of a cell consists mainly of a lipid bilayer (Figure 2). However, within the membrane are embedded many proteins, including those known as ion channels. Ion channels allow the passage of specific ions across the cell membrane.

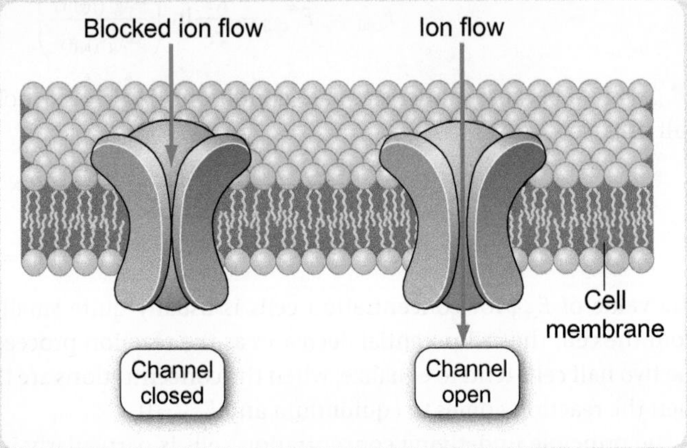

Blocked ion flow
Ion flow
Cell membrane
Channel closed
Channel open

▲ **Figure 2** A cell membrane showing the channels through which ion transport takes place. There are separate channels for sodium, potassium, and chloride ions. At rest, the sodium channels are closed and prevent ion movement. They open when the cell is triggered. After the cell fires, the resting concentrations are restored by 'ion pump' proteins that move ions against the concentration gradient. This requires energy, which comes from the hydrolysis of ATP.

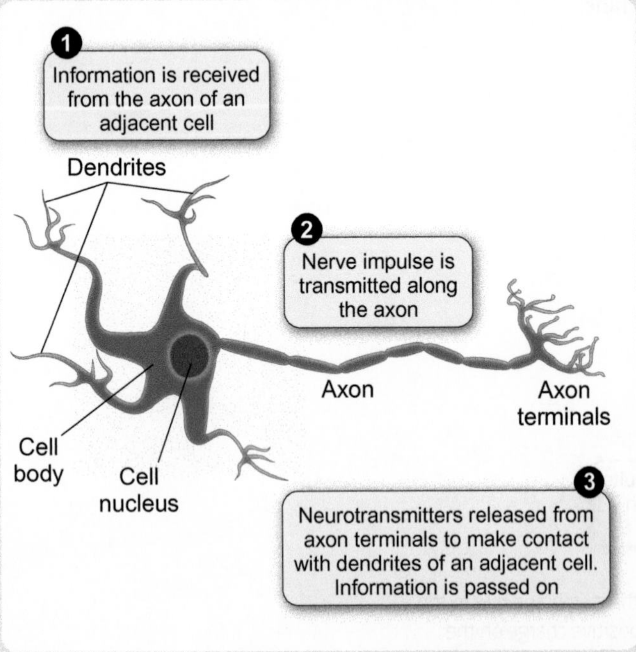

1
Information is received from the axon of an adjacent cell
Dendrites
2
Nerve impulse is transmitted along the axon
Axon
Axon terminals
Cell body
Cell nucleus
3
Neurotransmitters released from axon terminals to make contact with dendrites of an adjacent cell. Information is passed on

▲ **Figure 1** A human nerve cell.

When a nerve cell is at rest, Na^+ ions and Cl^- ions cannot pass across the membrane of the axon, but K^+ ions can—flowing across the membrane from the cell interior to its exterior. The flow of K^+ ions out of the cell leaves fewer positively charged ions inside the cell than outside, causing the interior of the cell to be more negatively charged than the exterior. As a result, a negative potential exists across the axon membrane.

The concentration of K^+ ions *inside* a human nerve cell (*intracellular*) is typically 150 mmol dm^{-3} (1 mmol = 1 × 10^{-3} mol), while that in the surrounding fluid *outside* the cell (*extracellular*) is around 5 mmol dm^{-3}. Using Equation 16.18, with a body temperature of 37 °C (310 K)

$$\Delta E_{K^+} = \frac{RT}{zF}\ln\frac{[K^+]_{outside}}{[K^+]_{inside}} \quad (16.18)$$

$$= \frac{(8.314\,\text{J}\,\cancel{\text{mol}^{-1}}\,\cancel{\text{K}^{-1}})\times(310\,\cancel{\text{K}})}{1\times(96485\,\text{C}\,\cancel{\text{mol}^{-1}})}\times\ln\left(\frac{5\times10^{-3}\,\cancel{\text{mol dm}^{-3}}}{150\times10^{-3}\,\cancel{\text{mol dm}^{-3}}}\right)$$

$$= 0.0267\,\text{J C}^{-1}\times(-3.40)$$

$$= -0.09\,\text{V} = -90\,\text{mV} \quad (1\,\text{J} = 1\,\text{C}\times1\,\text{V})$$

→

→ Thus, the potential across the membrane due to the difference in K^+ concentration is around −90 mV. For Na^+ ions, the intracellular and extracellular concentrations are 15 mmol dm^{-3} and 150 mmol dm^{-3}, respectively. A similar calculation shows that the potential across the membrane due to the difference in Na^+ concentrations is about +62 mV.

All the various ions that exist in and around nerve cells contribute to the overall membrane potential, which, when the cell is at rest and not transmitting information, is around −60 mV to −75 mV. The exact ionic concentrations depend on the situation of the cell and the ambient conditions. The values given here are typical of those found in the human body.

When the nerve cell is stimulated, sodium channels at the start of the axon open. This allows Na^+ ions to enter the cell, making the potential inside the cell less negative. As more Na^+ ions enter the cell, the potential increases to around +20 mV to +40 mV. At this point, the sodium channels close and separate potassium channels open so that K^+ ions leave the cell, changing the potential back towards the resting value. When the resting value is reached, the potassium channels close. The normal balance of ions is restored by ion pumps, proteins that can transport particular ions against the concentration gradient. This requires energy which is supplied by the hydrolysis of ATP inside the cell (see Box 14.8, p.683).

This change in potential—known as an action potential—triggers similar changes in neighbouring ion channels (and a similar movement of ions across the axon membrane) so that the action potential is transmitted along the axon. This is a very rapid process. All the various changes take place in a few milliseconds so that the impulses travel along the axon at speeds approaching 30 m s^{-1}. When the impulse reaches the end of the axon, it triggers the release of chemicals known as neurotransmitters, which transmit the nerve impulse—chemically rather than electrically—to adjacent nerve cells. Acetylcholine (see Boxes 4.11, p.212, and 18.1, p.822) is one example of a neurotransmitter.

Questions

(a) The intracellular concentration of Cl^- ions is 7 mmol dm^{-3}. If the membrane potential due to Cl^- is +76 mV, calculate the extracellular concentration of Cl^-.

(b) Calculate the Gibbs energy changes for the transport of K^+ ions and Na^+ ions against the membrane potentials. Compare these with the energy available from the hydrolysis of ATP described in Box 14.8.

Summary

- The Faraday constant F is the charge on 1 mol of electrons and has a value of 96 485 coulombs.
- $E^{\ominus}_{cell}$ is related to the standard Gibbs energy change for the reaction by

$$\Delta_r G^{\ominus} = -zFE^{\ominus}_{cell}$$

- $E^{\ominus}_{cell}$ is related to the thermodynamic equilibrium constant by

$$E^{\ominus}_{cell} = \frac{RT}{zF}\ln K$$

- A positive value of E_{cell} indicates that the cell reaction will be spontaneous in the forward direction.
- The variation in cell potential with changing concentration of the cell contents is given by the Nernst equation

$$E_{cell} = E^{\ominus}_{cell} - \frac{RT}{zF}\ln Q$$

- A concentration cell comprises two half cells that are identical apart from the concentration of the ion.

? For practice questions on these topics, see questions 12–16 at the end of this chapter (p.762).

16.5 Electrolysis

In the cells described in Section 16.4, chemical reactions are used to produce energy. The reverse process is also possible. Electrolysis involves driving a chemical reaction by supplying it with electrical energy. The input of energy means that non-spontaneous reactions can be made to occur and this principle forms the basis of a number of industrially very important reactions. The same principle is used in the recharging of batteries as described in Box 16.6 (p.759).

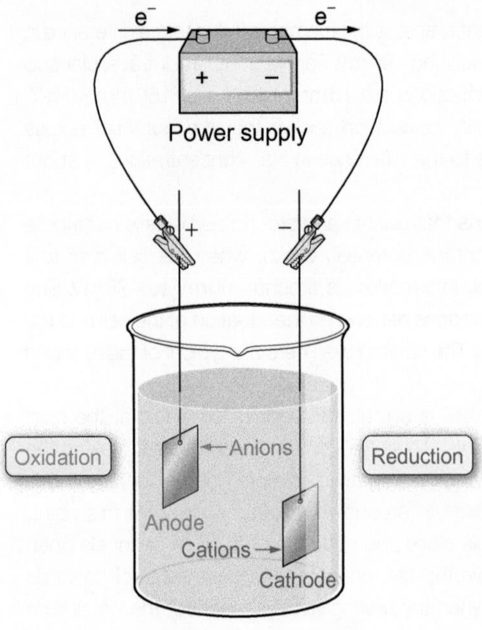
e⁻
e⁻
+
−
Power supply
+
−
Oxidation
Anions
Reduction
Anode
Cations
Cathode

Figure 16.14 A schematic electrolysis cell. The voltage applied is larger than that of the cell potential, so the reaction is driven in the reverse direction.

Copper metal is purified using electrolysis. Impure copper is made the anode and a thin sheet of pure copper is the cathode. During electrolysis, copper dissolves from the anode as Cu^{2+}(aq) and redeposits on the cathode as pure copper, leaving the impurities behind.

An **electrolysis cell** looks very similar to an electrochemical cell except that the electrodes are connected to an electrical power source such as a battery, as shown in Figure 16.14. (Compare this to Figure 16.9, p.737, to see the differences between electrochemical and electrolytic cells.)

Electrolysis of water

Water is stable with respect to hydrogen and oxygen since the decomposition of water has a large, positive change in Gibbs energy

$$2H_2O(l) \rightarrow 2H_2(g) + O_2(g) \quad \Delta_r G^\ominus = +474.4\ \text{kJ mol}^{-1}$$

The production of hydrogen and oxygen from water is, therefore, non-spontaneous at 25 °C and 1 bar. In order to make the reaction proceed, energy needs to be supplied.

If two electrodes (made from a non-reactive material such as platinum) are placed in water and a battery connected between them, very little happens even at high voltages. This is because the concentration of ions in pure water ($[H^+] \approx 1 \times 10^{-7}\ \text{mol dm}^{-3}$) is very low, so the conductivity of the water is also low and little current flows. However, if a small quantity of a strong electrolyte, such as sulfuric acid, is added, bubbles of gases appear at the electrodes.

The process at the anode is the oxidation of water

$$2H_2O(l) \rightarrow 4H^+(aq) + O_2(g) + 4e^- \qquad (16.19)$$

whereas the process at the cathode is the reduction of water

$$2H_2O(l) + 2e^- \rightarrow H_2(g) + 2OH^-(aq) \qquad (16.20)$$

To find the overall equation for the reaction, you first need to multiply Equation 16.20 by two, so that the same number of electrons is transferred in each half reaction, and then add the resulting half equation to Equation 16.19. The H^+(aq) and OH^-(aq) ions formed combine to form water, so the overall reaction is

$$2H_2O(l) \rightarrow 2H_2(g) + O_2(g)$$

In electrolysis, the cathode is more negative than the anode. In an electrochemical cell, it is the other way round. What you need to remember is that, in all cases, electrons flow from the anode to the cathode; oxidation occurs at the anode and reduction at the cathode.

At 25 °C and pH = 7, E_{cell} for this reaction is −1.23 V. The negative value indicates that the reaction is not spontaneous under these conditions. If a voltage of greater than +1.23 V is applied from a power source, then the reaction is driven in the non-spontaneous direction and water is decomposed into hydrogen and oxygen.

If the electrolysis is carried out in a cell such as that in Figure 16.15, the two gases can be separated, collected, and stored. On a much larger scale, this reaction is becoming very important for producing hydrogen as an alternative to fossil fuels, though the overall process is not carbon-neutral unless solar power or other renewable sources are used to generate the required electricity.

In practice, an applied voltage significantly larger than +1.23 V is needed to electrolyse water. It is often found that a voltage in excess of the E_{cell} value—an overvoltage—must be applied for electrolysis to occur at an observable rate. This overcomes kinetic limitations on the reaction such as the rate of electron transfer processes at the electrodes. Reactions involving O_2 gas at electrodes are often kinetically very slow.

The Hydrosol II pilot-scale plant, launched in Almeria, Spain, in 2008, uses solar power to generate hydrogen from water. The hydrogen is obtained in a thermal decomposition reaction that takes place around 1000 °C on the surface of a ceramic honeycomb structure coated with metal oxides. The team claims that this is a more efficient process than using electrolysis to split water.

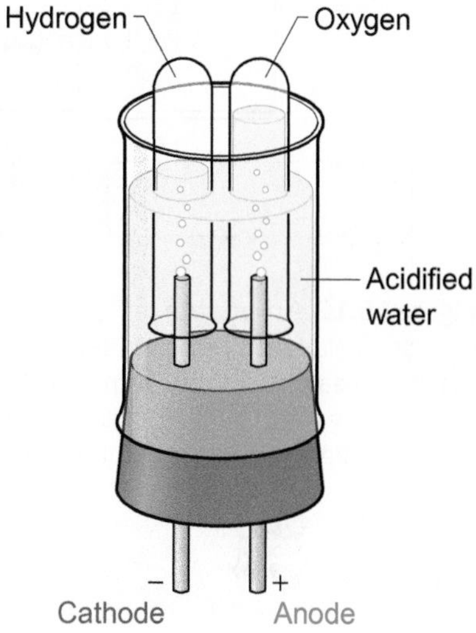
Hydrogen
Oxygen
Acidified water
−
+
Cathode
Anode

Figure 16.15 Electrolysis of water. Water can be split into its elements by applying a voltage to a solution of acidified water.

Electrolysis in the chemical industry

Large-scale electrochemical processes are used to produce some important industrial chemicals. Two particularly important processes are the electrolysis of sodium chloride solution and the production of some metals.

The industrial electrolysis of sodium chloride solution

An aqueous solution of sodium chloride conducts electricity due to the presence of Na^+ ions and Cl^- ions. If the solution is reasonably concentrated, the reaction at the anode is the oxidation of chloride ions (rather than the oxidation of water as in Equation 16.19)

$$2\,Cl^-(aq) \rightarrow Cl_2(g) + 2\,e^-$$

The standard reduction potential for $Cl_2(g) + 2e^- \rightarrow 2Cl^-(aq)$ is $E^\ominus = +1.36$ V (see Table 16.2, p.742) so the potential for the *oxidation* of chloride ions is $E^\ominus_{oxid} = -1.36$ V. Although the oxidation of water ($E^\ominus_{oxid} = -1.23$ V) would be expected to occur at a slightly lower applied voltage, experimentally, it is found that chlorine gas is evolved at the anode. The overvoltage for the production of O_2 is very high so that the oxidation of chloride is kinetically more favourable.

Given the very highly negative reduction potential for $Na^+(aq)$ ($E^\ominus = -2.71$ V), the reaction at the cathode is the reduction of water (Equation 16.20) rather than the reduction of $Na^+(aq)$

$$2\,H_2O(l) + 2\,e^- \rightarrow H_2(g) + 2\,OH^-(aq) \qquad E^\ominus = -0.83\text{ V}$$

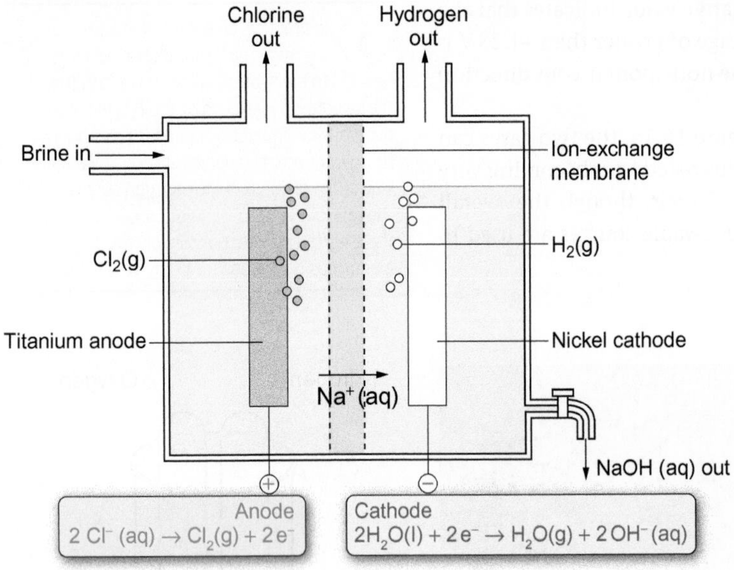
Chlorine out
Hydrogen out
Brine in
Ion-exchange membrane
$Cl_2(g)$
$H_2(g)$
Titanium anode
Nickel cathode
Na^+ (aq)
NaOH (aq) out
Anode
$2\,Cl^-(aq) \rightarrow Cl_2(g) + 2e^-$
Cathode
$2H_2O(l) + 2e^- \rightarrow H_2O(g) + 2\,OH^-(aq)$

Figure 16.16 A membrane cell for the industrial electrolysis of sodium chloride solution. The two half cells are in separate compartments linked by a membrane that is permeable to $Na^+(aq)$ but not to $OH^-(aq)$ or $Cl^-(aq)$. This prevents chlorine from reacting with the NaOH product. At the anode, $Cl_2(g)$ is collected. At the cathode, water is reduced to $OH^-(aq)$, forming NaOH (aq).

Industrially, electrolysis of NaCl solution is carried out on a massive scale. Each one of the square tanks contains over 30 anodes and cathodes. The chlorine produced is used in water treatment; it is also used in making a wide range of chemicals such as pharmaceuticals, bleaches, plastics (e.g. PVC), and textiles.

So the overall electrolysis reaction that occurs in the cell is

$$2\,H_2O\,(l) + 2\,Cl^-(aq) \rightarrow Cl_2(g) + H_2(g) + 2\,OH^-(aq) \qquad (16.21)$$

The overall cell potential for this reaction is then

$$E^{\ominus}_{cell} = (-0.83\,V) - (+1.36\,V) = -2.19\,V$$

The negative value of $E^{\ominus}_{cell}$ shows the reaction to be non-spontaneous. For the reaction to occur under standard conditions, a potential higher than +2.19 V has to be applied. The reaction products are chlorine gas at the anode and hydrogen gas at the cathode. This leaves behind a solution containing undischarged Na^+ and OH^- ions, in other words a solution of sodium hydroxide, NaOH (aq). The production of NaOH and Cl_2 from the electrolysis of brine (naturally occurring salt solution) is the basis of the commercially very important **chloralkali industry** (Figure 16.16). Both chlorine and sodium hydroxide are in the top 10 chemicals in terms of the quantity produced, with the global annual production of chlorine being nearly 70 million tonnes.

Aluminium metal is obtained from its ore by electrolysis (Section 27.2, p.1210). The production of sodium by electrolysis is discussed in Section 26.1 (p.1171).

Box 14.10 (p.689) shows how Ellingham diagrams use Gibbs energy changes to understand the methods needed to extract metals from their ores.

The industrial production of metals by electrolysis

Reactive metals such as sodium, magnesium, and aluminium are produced by electrolysis because their ores cannot be reduced with carbon. Sodium is produced by the electrolysis of *molten* sodium chloride rather than the aqueous solution.

Calculating reacting quantities in electrolysis

The Faraday constant gives the charge on one mole of electrons

$$F = 96\,485\,C\,mol^{-1} \qquad (16.12)$$

You can use the Faraday constant to calculate the reacting quantities involved in electrolysis, as described in Worked example 16.8.

Worked example 16.8 Reacting quantities in electrolysis

Calculate the mass of H_2(g) that will be collected at the cathode when an aqueous solution of Na_2SO_4 is electrolysed for 120.0 min with a current of 10.0 A.

Strategy

From the current and the time for which it was passed, you can find the quantity of charge that passed through the solution. You can then use the Faraday constant to calculate the number of moles of electrons that this charge is equivalent to. Finally, use the balanced equation to find the number of moles of hydrogen gas that are formed and hence the mass.

Solution

From the current and the time for which it was passed, find the amount of charge that passed through the solution.

$1\,\text{A} = 1\,\text{C}\,\text{s}^{-1}$. Hence:

$$\text{Amount of charge} = 10.0\,\text{C}\,\cancel{\text{s}^{-1}} \times 120.0\,\cancel{\text{min}} \times 60\,\cancel{\text{s}}\,\cancel{\text{min}^{-1}} = 72000\,\text{C}$$

Use the Faraday constant to calculate the number of moles of electrons that this charge is equal to.

1 mol of electrons carries 96 485 C

$$7200\,\text{C is therefore equivalent to } \frac{72000\,\cancel{\text{C}}}{96485\,\cancel{\text{C}}\,\text{mol}^{-1}}$$

$$= 0.746\,\text{mol of electrons}$$

Use the balanced equation to find the number of moles of hydrogen gas that are formed and hence the mass.

The balanced equation for the reaction at the cathode is

$$2\,H_2O(l) + 2\,e^- \rightarrow H_2(g) + 2\,OH^-(aq)$$

So 1 mol of H_2(g) arises from the passage of 2 mol of electrons.

0.746 mol of electrons therefore gives rise to 0.373 mol of H_2(g).

0.373 mol H_2 (g) have a mass of $0.373\,\cancel{\text{mol}} \times 2.02\,\text{g}\,\cancel{\text{mol}^{-1}} = 0.753\,\text{g}$.

Question

In the electrolysis of Na_2SO_4, O_2(g) is produced at the anode. For how long would a current of 10 A need to pass in order to produce 0.50 g of O_2(g) at the anode?

Box 16.6 Electrolysis and rechargeable batteries

The batteries described in Box 16.2 are primary cells—the cell reaction is not reversible. Secondary cells are those in which the cell reactions are reversible so that the battery can be recharged.

For a long time, the most common secondary cell was the lead–acid battery, which is still used in motor vehicles. Invented by Gaston Planté in 1859, this battery produces enough energy to start a car engine and is recharged when the engine is running. The battery operates on the overall cell reaction:

$$PbO_2(s) + Pb(s) + 4\,H^+(aq) + 2\,SO_4^{2-}(aq)$$

Discharging (electricity produced) ⇅ Charging (electricity used)

$$2\,PbSO_4(s) + 2\,H_2O(l)$$

When operating as an electrochemical cell to produce energy, the lead electrode is oxidized to Pb^{2+}(aq), which reacts with SO_4^{2-}(aq) to form solid lead sulfate.

Anode: $Pb(s) + SO_4^{2-}(aq) \rightarrow PbSO_4(s) + 2\,e^-$ $\quad E^{\ominus} = -0.36\,\text{V}$

The cathode reaction involves the reduction of PbO_2(s) (oxidation state: +4) to Pb^{2+}(aq) (oxidation state: +2)

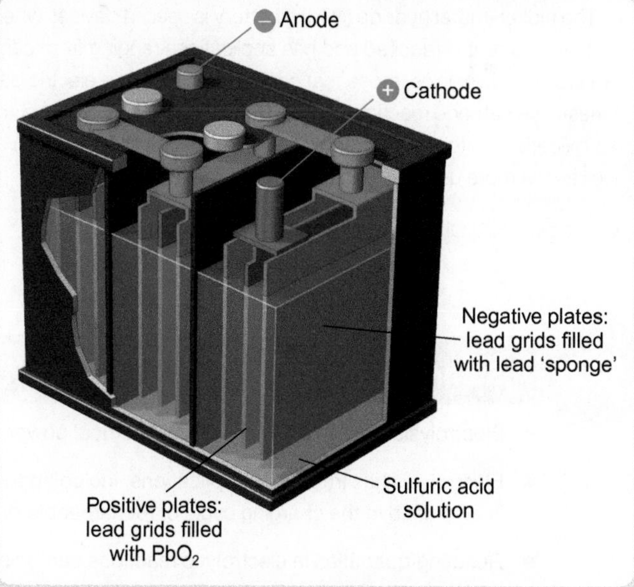

Anode
Cathode
Negative plates: lead grids filled with lead 'sponge'
Sulfuric acid solution
Positive plates: lead grids filled with PbO_2

▲ A lead–acid car battery contains six cells connected in series. The electrodes consist of grids of lead, filled alternately with solid lead(IV) oxide (PbO_2) and lead metal. They are immersed in a conducting solution of sulfuric acid. Each cell produces about 2 V.

→

→ Cathode: $PbO_2(s) + 4H^+(aq) + SO_4^{2-}(aq) + 2e^- \rightarrow PbSO_4(s) + 2H_2O(l) \quad E^\ominus = +1.68\,V$

Using Equation 16.10 (p.744)

$$\begin{aligned} E^\ominus_{cell} &= E_{cathode} - E_{anode} \\ &= +1.68\,V - (-0.36\,V) \\ &= +2.04\,V \end{aligned}$$

The two half reactions leave both electrodes covered with lead sulfate. The crucial thing is that the $PbSO_4$ remains attached to the electrodes so that, when a power source delivering greater than 2.04 V is connected to the battery, the reverse reactions are driven and the original reactants are restored. When all the sulfuric acid has been converted to lead sulfate, the battery is fully discharged or 'flat'. It is important not to leave the battery discharged for long because the lead sulfate undergoes a slow recrystallization into a different form that is difficult to reverse. This reduces the capacity of the battery to hold charge.

Most car batteries have six of these cells arranged in series in a single casing to give an overall voltage of 6 × 2.04 V, or just over 12 V.

This type of battery has been in use for a long time and its major advantage is that it can deliver the very large currents needed to turn the starter motor in a car. However, this battery is very heavy due to the quantity of lead involved and the 'energy density' (available energy per unit mass) is relatively low. Although lead–acid batteries generally last a long time, their eventual disposal has environmental consequences due to the harmful effects of lead.

A lead–acid battery would be difficult to use for powering laptop computers, phones, or other portable appliances and the development of these devices has led to major improvements in battery technology.

The **nickel–metal hydride (NiMH) battery** is used in devices where a steady current is required and has applications ranging from cameras to hybrid vehicles (those that can run on petrol and electric batteries). The cathode reaction exploits the ability of nickel to adopt the +3 oxidation state under strongly basic conditions in NiO(OH)(s) as well as the more usual +2 oxidation state in $Ni(OH)_2(s)$.

Cathode: $NiO(OH)(s) + H_2O(l) + e^- \rightarrow Ni(OH)_2(s) + OH^-(aq)$

The anode consists of a metal alloy, often containing Ni, Co, and Mn. The metals in the alloy form hydrides but can easily give up the hydrogen in a reversible reaction. The anode reaction involves the transfer of hydrogen from the metal hydride to a hydroxide ion

Anode: $OH^-(aq) + MH \rightarrow H_2O(l) + M(s) + e^-$

The overall reaction produces 1.2 V per cell. The reaction is easily reversed on recharging and the batteries can be recharged up to 1000 times with little loss of power.

Energizer
RECHARGEABLE
15 min CHARGER
4 x 2200mAh
Energizer
LASTS UP TO
4X LONGER
IN DIGITAL CAMERAS*

▲ Nickel–metal hydride (NiMH) rechargeable batteries.

Questions

(a) Draw cell diagrams for: (i) the lead–acid battery; (ii) the nickel–metal hydride battery.

(b) What is the balanced equation for the overall cell reaction for the nickel–metal hydride battery?

» Summary

- Electrolysis reactions involve using electrical power to drive non-spontaneous reactions.
- Electrolysis has important applications, including the manufacture of metals, chlorine, sodium hydroxide, and hydrogen. It is also used in the charging cycle of rechargeable batteries.
- Reacting quantities in electrolytic reactions can be calculated using the Faraday constant.
- The cell reaction in a primary cell is not reversible, so batteries using primary cells cannot be recharged.
- Secondary cells employ reversible reactions and can be recharged.

? **For practice questions on these topics, see questions 17–19 at the end of this chapter (p.762).**

Concept review

By the end of this chapter you should be able to do the following.

- Describe, and give examples of, the uses of electrochemical reactions.
- Define conductivity and molar conductivity for a solution of an electrolyte and explain how the molar conductivity of a strong electrolyte depends on concentration.
- Calculate the limiting molar conductivity of a strong electrolyte from the individual ionic conductivities of the cations and anions present.
- Understand that an electrochemical cell is made up of two half cells; reduction occurs at the cathode and oxidation at the anode.
- Construct and use cell diagrams.
- Define the standard reduction potential, $E^{\ominus}$, and use it to calculate $E^{\ominus}_{cell}$.
- Use $E^{\ominus}_{cell}$ to predict the spontaneity of a redox reaction.
- Use $E^{\ominus}_{cell}$ data to calculate $\Delta_r G^{\ominus}$ and K, and vice versa.
- Use the Nernst equation to calculate the cell potential, E_{cell}, under non-standard conditions.
- Explain what is meant by a concentration cell.
- Describe the basic principles of electrolysis and how it is used in chemical processes.
- Use the Faraday constant to calculate reacting quantities in electrolytic reactions.
- Give some examples of primary and secondary cells and their underlying electrochemistry.

Key equations

Law of independent migration of ions	$\Lambda^{\circ}_{m} = \nu_+\lambda_+ + \nu_-\lambda_-$	(16.8)
Calculation of $E^{\ominus}$ for a cell	$E^{\ominus}_{cell} = E^{\ominus}_{cathode} - E^{\ominus}_{anode}$	(16.10)
The Faraday constant	F = charge on 1 mol of electrons = 96 485 C mol^{-1}	(16.12)
Gibbs energy change and $E^{\ominus}$ for a cell	$\Delta_r G^{\ominus} = -zF \times E^{\ominus}_{cell}$	(16.14)
Equilibrium constant and $E^{\ominus}$ for a cell	$\ln k = \dfrac{zF}{RT} E^{\ominus}_{cell}$	(16.15)
The Nernst equation	$E^{\ominus}_{cell} = E^{\ominus}_{cell} - \dfrac{RT}{zF}\ln Q$	(16.16)
	$E^{\ominus}_{half\ cell} = E^{\ominus}_{half\ cell} - \dfrac{RT}{zF}\ln Q$	(16.17)
Membrane potential	$\Delta E_{M^+} = \dfrac{RT}{zF}\ln\dfrac{[M^+]_{outside}}{[M^+]_{inside}}$	(16.18)

Questions

More challenging questions are indicated by an asterisk *.

You will need to use standard reduction potentials, $E^{\ominus}$, from Table 16.2 (p.742). (Section 16.2)

1 (a) Use the data in Table 16.1 to calculate the limiting molar conductivities of magnesium sulfate and sodium carbonate in water.

(b) Estimate the conductivity of a solution of magnesium sulfate with a concentration of 1×10^{-5} mol dm^{-3}.

2* The conductivity of a saturated solution of silver chloride, AgCl, at 25 °C is 1.89×10^{-4} S m^{-1}. Given that the molar conductivities at infinite dilution of KCl(aq), KNO_3(aq), and $AgNO_3$(aq) are 15.0 mS m^2 mol^{-1}, 14.6 mS m^2 mol^{-1}, and 13.4 mS m^2 mol^{-1}, calculate the solubility product of AgCl in water at 25 °C. (Section 16.2)

3 Consider the cell

$$Al(s)\,|\,Al^{3+}(aq)\,||\,Au^{3+}(aq)\,|\,Au(s)$$

(Section 16.3)

(a) At which electrode does reduction occur?

(b) Which electrode is the anode?

(c) Which electrode will lose mass in the cell reaction?

(d) To which electrode will cations migrate?

(e) Which substance is acting as a reducing agent?

(f) Write the half reaction for the gold half cell.

(g) Calculate a value for $E^\ominus_{cell}$.

4 Draw cell diagrams for electrochemical cells that use the following reactions (Section 16.3):

(a) $Cd(s) + Sn^{2+}(aq) \rightarrow Sn(s) + Cd^{2+}(aq)$

(b) $H_2(g) + O_2(g) \rightarrow H_2O_2(aq)$

(c) $Br_2(aq) + Sn^{2+}(aq) \rightarrow Sn^{4+}(aq) + 2Br^-(aq)$

(d) $Cu^{2+}(aq) + 2Ag(s) + 2Br^-(aq) \rightarrow Cu(s) + 2AgBr(s)$

(e) $MnO_4^-(aq) + 8H^+(aq) + 5Fe^{2+}(aq) \rightarrow Mn^{2+}(aq) + 4H_2O(l) + 5Fe^{3+}(aq)$

5 For each of the following electrochemical cells: (i) write the half cell reactions; (ii) calculate $E^\ominus_{cell}$. (Section 16.3)

(a) $Fe(s)|Fe^{2+}(aq)||Zn^{2+}(aq)|Zn(s)$

(b) $Pt(s)|H_2(g), H^+(aq)||Cl^-(aq), AgCl(s)|Ag(s)$

(c) $Hg(l)|Hg_2Cl_2(s), Cl^-(aq)||Cl^-(aq), AgCl(s)|Ag(s)$

(d) $Pt(s)|Fe^{2+}(aq), Fe^{3+}(aq)||Sn^{4+}(aq), Sn^{2+}(aq)|Pt(s)$

6 Which of the following is the strongest oxidizing agent? (Section 16.3)

(a) H_2O_2 in acid solution

(b) H_2O_2 in basic solution

(c) MnO_4^- in acid solution

(d) MnO_4^- in basic solution

(e) CrO_4^{2-} in acid solution.

7* Using standard reduction potentials, explain why copper metal does not dissolve in 1 mol dm^{-3} hydrochloric acid (HCl (aq)) but does dissolve in 1 mol dm^{-3} nitric acid (HNO_3 (aq)). (Section 16.3)

8 Use values of standard reduction potentials, $E^\ominus$, in Table 16.2 to decide whether the following reactions occur spontaneously. (Section 16.3)

(a) $Cr^{2+}(aq) + Ni(s) \rightarrow Cr(s) + Ni^{2+}(aq)$

(b) $Br_2(l) + 2I^-(aq) \rightarrow 2Br^-(aq) + I_2(s)$

(c) $Cl_2(s) + Sn^{2+}(aq) \rightarrow 2Cl^-(aq) + Sn^{4+}(aq)$

(d) $Al(s) + Au^{3+}(aq) \rightarrow Al^{3+}(aq) + Au(s)$

9 Silver articles sometimes become tarnished with a black coating of Ag_2S. The tarnish can be removed by placing the silverware in an aluminium pan and covering the articles with a solution of an inert electrolyte, such as NaCl. Explain the electrochemical basis for this procedure. (Section 16.3)

10 Use data from Table 16.2 to calculate the standard Gibbs energy change, $\Delta_r G^\ominus$, and the thermodynamic equilibrium constant, K, at 298 K for the reactions (Section 16.4):

(a) $Mg(s) + Zn^{2+}(aq) \rightarrow Zn(s) + Mg^{2+}(aq)$

(b) $Cl_2(g) + 2I^-(aq) \rightarrow I_2(s) + 2Cl^-(aq)$

(c) $Cr_2O_7^{2-}(aq) + 3Fe(s) + 14H_3O^+(aq) \rightarrow 2Cr^{3+}(aq) + 3Fe^{2+}(aq) + 21H_2O(l)$

(d) $2Br^-(aq) + Cl_2(g) \rightarrow Br_2(l) + 2Cl^-(aq)$

11 Calculate $E^\ominus_{cell}$ for the cell $Co(s)|Co^{2+}(aq)||Ni^{2+}(aq)|Ni(s)$ when (Section 16.4):

(a) $[Ni^{2+}] = 1.0$ mol dm^{-3}, $[Co^{2+}] = 0.10$ mol dm^{-3};

(b) $[Ni^{2+}] = 0.01$ mol dm^{-3}, $[Co^{2+}] = 1.0$ mol dm^{-3}.

12* Given the following half cell reactions, calculate the solubility product (see Box 15.1, p.701) of silver bromide at 298 K. (Section 16.4)

$$AgBr(s) + e^- \rightarrow Ag(s) + Br^-(aq) \quad E^\ominus = +0.07\,V$$
$$Ag^+(aq) + e^- \rightarrow Ag(s) \quad E^\ominus = +0.80\,V$$

13 Calculate E_{cell} and the Gibbs energy change for the following cells (Section 16.4):

(a) $Ag(s)|Ag^+(aq, 1.0\,mol\,dm^{-3})||Cu^{2+}(aq, 1.0\,mol\,dm^{-3})|Cu(s)$

(b) $Ag(s)|Ag^+(aq, 0.1\,mol\,dm^{-3})||Cu^{2+}(aq, 0.1\,mol\,dm^{-3})|Cu(s)$

(c) $Ag(s)|Ag^+(aq, 1.0\,mol\,dm^{-3})||Cu^{2+}(aq, 0.1\,mol\,dm^{-3})|Cu(s)$.

14 Use the following standard reduction potentials to calculate the equilibrium constant for the formation of the $Zn(NH_3)_4^{2+}$ ion. (Section 16.4)

$$Zn(NH_3)_4^{2+}(aq) + 2e^- \rightleftharpoons Zn(s) + 4NH_3(aq) \quad E^\ominus = -1.04\,V$$
$$Zn^{2+}(aq) + 2e^- \rightarrow Zn(s) \quad E^\ominus = -0.76\,V$$

15 Calculate the standard electrode potential for a cell in which the reaction forms 99.99% products at equilibrium under standard conditions at 298 K. Assume $z = 1$ for the reaction. (Section 16.4)

16* A student measured the emf of the following electrochemical cell at 25 °C.

$$Cu(s)|CuSO_4(aq, 0.050\,mol\,dm^{-3})||CuSO_4(aq, 0.500\,mol\,dm^{-3})|Cu(s)$$

The student connected a piece of copper wire between the electrodes and left the experiment to go for lunch. Sometime later the student removed the wire and repeated the emf measurement, recording a value of +0.027 V. (Section 16.4)

(a) Write the reactions which take place at each electrode.

(b) Calculate the emf recorded during the first measurement.

(c) Describe briefly what happened in the cell over lunch.

(d) Calculate the concentration of copper sulfate solution in each cell compartment after lunch.

17 Electrolysis of a molten chromium salt for 1.5 h with a 5.00 A current deposited 4.835 g of chromium metal. Find the charge on the chromium ion in the salt. (Section 16.5)

18 A Leclanché cell (see Section 16.5) produces a current of 0.002 A. If the battery contains 2.5 g of MnO_2, how long will the battery last under these conditions? (Section 16.5)

19 Aluminium is manufactured by the electrolysis of molten aluminium oxide, Al_2O_3 (l). If an aluminium plant produces 1000 tonnes of aluminium every 24 hours, what total electric current does it need? (1 tonne = 1000 kg) (Section 16.5)

17

Phase equilibrium and solutions

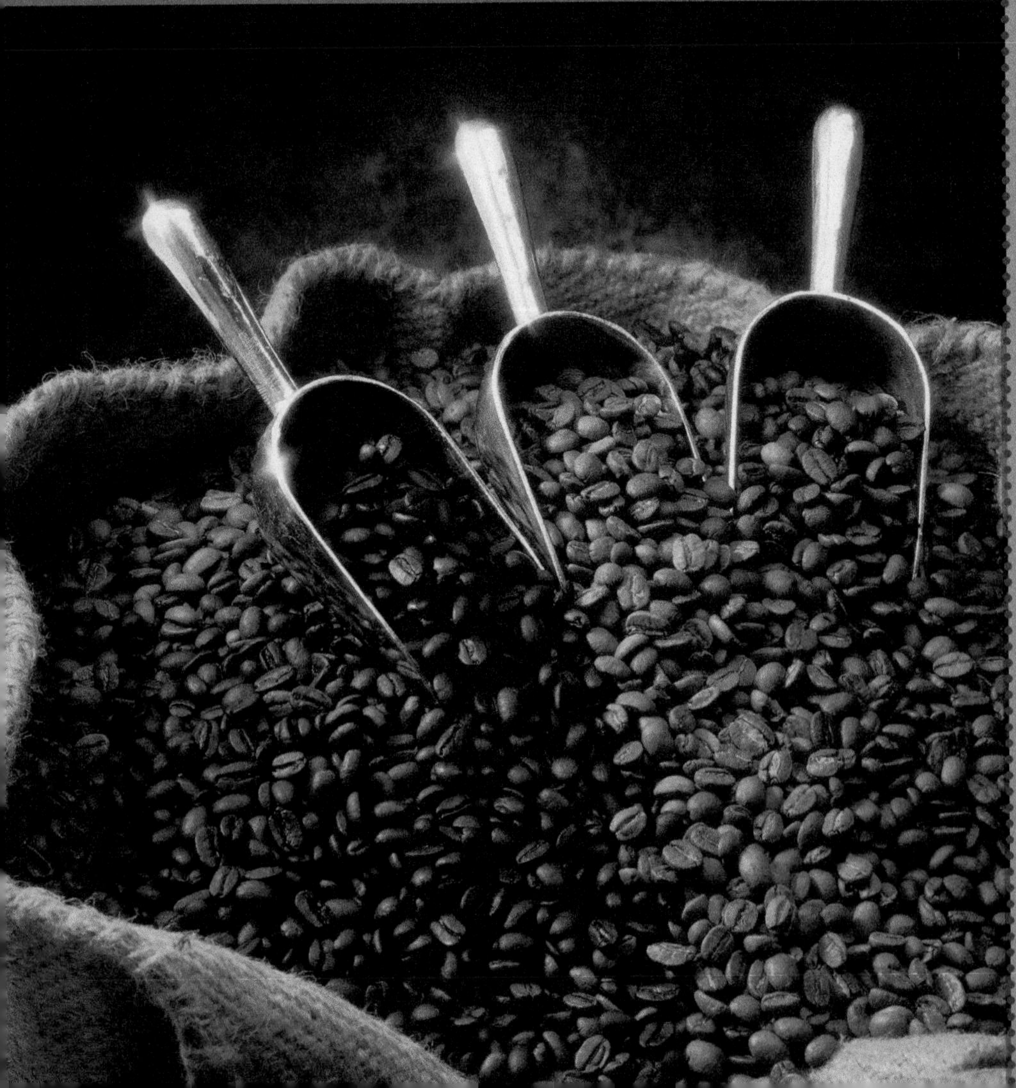

This chapter builds on the following topics:

- **States of matter and phase changes** Section 1.7, p.47
- **Non-covalent interactions** Section 1.8, p.52
- **Chemical equilibrium: how far has a reaction gone?** Section 1.9, p.56
- **Using the ideal gas equation** Section 8.2, p.349
- **Mixtures of gases** Section 8.3, p.355
- **Kinetic molecular theory and the gas laws** Section 8.4, p.358
- **Gibbs energy** Section 14.5, p.675
- **Variation of Gibbs energy with conditions** Section 14.6, p.686
- **Gibbs energy and equilibrium** Section 15.1, p.697

◄ Supercritical carbon dioxide can be used to decaffeinate coffee beans.

Supercritical fluids

If a liquid is heated in an open container, it gradually evaporates away as it turns to vapour. If the heating is carried out in a sealed container, liquid evaporates, increasing the pressure and density of the vapour. Eventually, as the temperature increases, a point is reached where the vapour and liquid have the same density—and the surface between the two phases disappears as in Figure 1. This is known as the critical point. For water, it occurs at 647 K and 220 bar. Above this temperature and pressure, the distinction between the gas and liquid phases does not apply and the substance can only be described as a fluid. Fluids above their critical temperature and pressure are known as supercritical fluids.

Supercritical fluids combine some of the properties of both liquids and gases. For example, they are effective solvents for a range of substances (a property normally associated with liquids) but have relatively low viscosities and high diffusion rates (properties normally associated with gases). The physical properties vary widely with density, which can be controlled by small variations in the temperature and pressure. Over recent years, supercritical fluids have found a range of uses in chemistry and are now used on an industrial scale.

Supercritical water has very different properties from normal liquid water. The H_2O molecules are closely packed together so it acts as a good solvating agent, but it also behaves as a relatively non-polar solvent and dissolves organic species such as hydrocarbons. This is because, although individual water molecules are still polar, they are moving around rapidly and do not have time to line up with molecules as they do at room temperature. As a result, dipolar interactions and hydrogen bonding are much less effective. Supercritical H_2O is a strong oxidizing agent and can be used for destroying pollutants and for decontamination of polluted materials. However, it is highly corrosive and this, along with the high temperature and pressure needed for its generation, means that highly specialized (and expensive) equipment is needed whenever supercritical H_2O is used.

A more commonly used supercritical fluid is carbon dioxide. Liquid CO_2 does not exist at atmospheric pressure but the gas can be liquefied at pressures above 5 bar. It becomes supercritical above 304 K and 73.8 bar: a much lower temperature and pressure than those needed for supercritical H_2O. One of the first applications of supercritical CO_2 was in extraction of flavour components of food. Many people like the taste of coffee but prefer it without caffeine. Decaffeinated coffee was originally made by extracting the caffeine with hot water, steam, or organic solvents such as dichloromethane; however, these solvents also remove some of the compounds that give coffee its distinctive flavour. Using supercritical carbon dioxide as the solvent can counter this problem.

▼ Caffeine

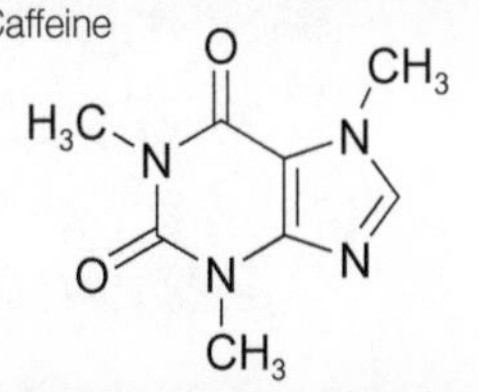

O
CH_3
H_3C
N
N
N
O
N
CH_3

The use of supercritical CO_2 allows the removal of 97–99% of the caffeine from the beans, while leaving most of the flavour components behind. The supercritical CO_2 is circulated through the beans, then into a second vessel at a lower pressure which contains water. The caffeine dissolves in the water, and the CO_2 is re-pressurized and recycled. This means there are no problems with contamination of the beans with solvent, and there is no waste to dispose of. The extracted caffeine is sold to manufacturers of soft drinks and pharmaceuticals.

Supercritical CO_2 can also be used as a solvent for chemical reactions. One advantage of using supercritical CO_2 is that, when the reaction is complete, the pressure is released and the CO_2 evaporates (and is collected for reuse) leaving the products behind free from solvents. No further purification is needed, making the overall process very efficient.

▼ **Figure 1** Formation of a supercritical fluid in a sealed container

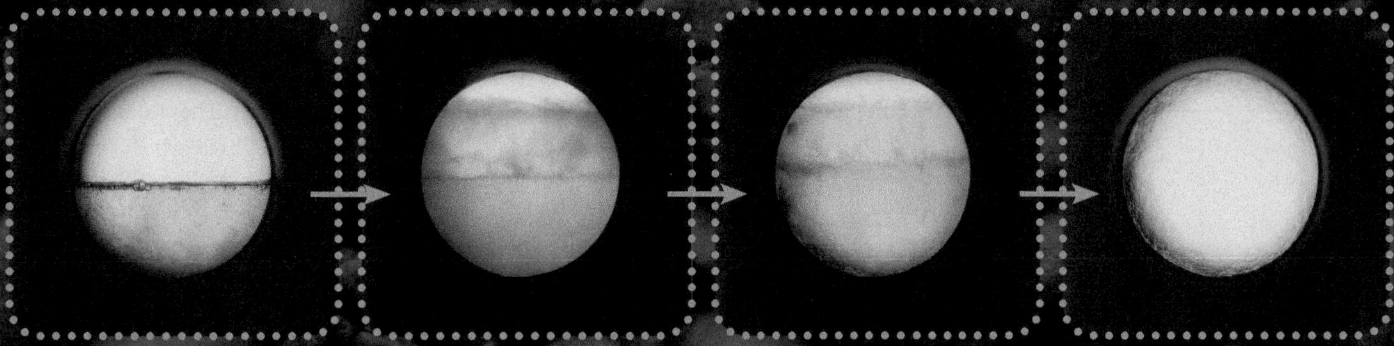

This photo shows distinct liquid and gaseous phases. The meniscus between the liquid and gas is clearly visible.

As the temperature increases the meniscus begins to disappear.

As the temperature increases further, the densities of the gaseous and liquid phases become nearly equal and the meniscus is indistinct.

Once the critical temperature and pressure have been reached, the two separate phases of liquid and gas are no longer visible. One single phase is present that behaves like both a liquid and gas: the supercritical fluid.

Gibbs energy and equilibrium are discussed in Section 15.1 (p.697).

Chapter 15 introduces the idea that a chemical system comes to equilibrium when the Gibbs energy is at a minimum. This can be used to define the position of chemical equilibrium in reacting systems. In this chapter, the same idea is applied to the equilibrium between different phases of substances.

A phase can be defined as follows.

A phase is a part of a system that is homogeneous in chemical and physical state throughout and is separated from other phases by a definite boundary.

In gases and vapours, there is plenty of space for the molecules to mix so that there is only ever one phase, irrespective of how many components are involved. For liquids, the number of phases depends on the nature of the components. For a single component (i.e. a single element or compound) there can be only one liquid phase. When more than one component is involved, they can form one phase if they mix—such as ethanol and water—or two phases if they don't—such as oil and water. Whether two liquids mix may change depending on the conditions.

Two immiscible liquids. They form two separate phases with a phase boundary in between.

Helium provides an exception to the rule that single-component liquids only exist in one phase. Pure ^{4}He forms two liquid phases: one behaves as a normal liquid, whereas the other, existing below 2 K, is a superfluid with very unusual properties, such as being able to creep out of its container.

Mixtures of liquids that form a single phase are said to be **miscible**; those that form more than one phase are said to be **immiscible**. Mixtures that form one phase under some conditions and more than one under other conditions are **partially miscible**.

Two allotropes of sulfur are discussed in Section 1.7 (p.48). The structures of graphite, diamond, and buckminsterfullerene (C_{60}) are described in Section 6.1 (p.258).

Although the term 'molecule' has been used here, solids and liquids can also be made up of atoms or ions. A solid may also have a covalent network structure.

Transitions between phases and the energy changes that take place during phase changes are discussed in Sections 1.7 (p.47) and 13.2 (p.619).

Phase behaviour is more complex in solids, where even a single component may exhibit several phases. Familiar examples are carbon, which can exist as graphite, diamond, or buckminsterfullerene, and sulfur, which can exist as one of several crystal allotropes.

Chemists need to understand how and why this phase behaviour occurs. Much of chemistry and virtually all of biochemistry occur in solution, so this phase is particularly important.

This chapter introduces some of the ideas and concepts that govern phase behaviour. It looks at the factors that determine why compounds exist in a particular phase under a particular set of conditions and what governs the transitions between phases, and discusses some properties of mixtures and solutions.

17.1 Phase behaviour of single components

Matter exists in three phases: solid; liquid; or gas (vapour). The solid shown in Figure 17.1 has an ordered array of molecules held close together by intermolecular interactions. Melting of the solid occurs when the molecules gain sufficient energy to move around, but not enough to completely overcome the attractive interactions. In the liquid phase, the molecules remain close together and collide with each other frequently. The distance between them is only a little larger than in the solid. In the vapour phase, molecules have sufficient energy to overcome the intermolecular interactions and move around independently, occupying a much larger volume than the same amount of liquid or solid.

Vapour pressure

At 1 atm pressure, ice melts to water at 0 °C and water boils to form steam at 100 °C. However, it is important to realize that some water enters the vapour phase even at temperatures well below the boiling point. Some molecules near the surface of the liquid will have

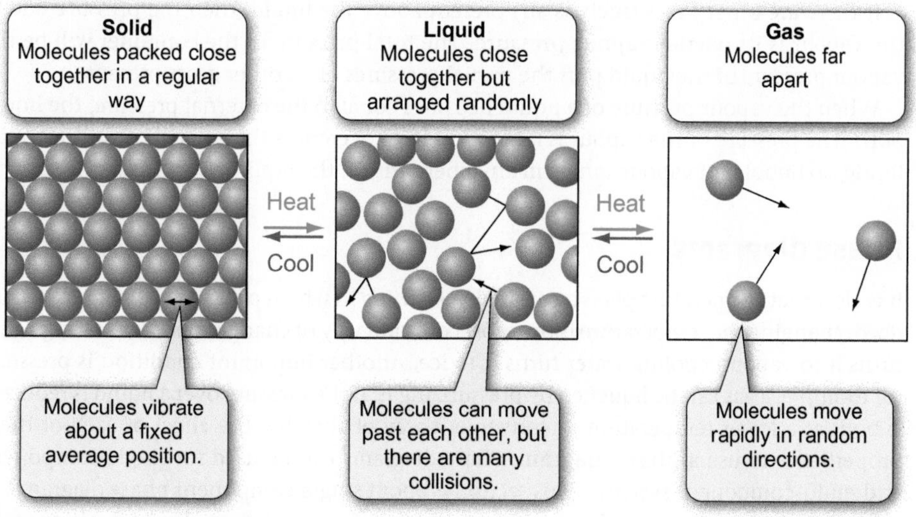

Solid
Molecules packed close together in a regular way
Liquid
Molecules close together but arranged randomly
Gas
Molecules far apart
Heat
Cool
Heat
Cool
Molecules vibrate about a fixed average position.
Molecules can move past each other, but there are many collisions.
Molecules move rapidly in random directions.

Figure 17.1 Solids, liquids, and gases.

sufficient energy to overcome the attractive interactions with neighbouring molecules and escape into the vapour phase. If some liquid is placed in a container, the vaporized material will exert a pressure—the **vapour pressure**. If the container is now closed and the temperature is constant, the vapour and liquid come to equilibrium and the equilibrium vapour pressure is then called the **saturated vapour pressure, $p°$**. This is shown schematically in Figure 17.2. At higher temperatures, liquids have a higher vapour pressure since more molecules have sufficient energy to escape the liquid. Solids also exert a vapour pressure, although the stronger interactions between molecules mean that it is usually lower than for liquids.

ℹ Solids and liquids that have a high vapour pressure are said to be **volatile**.

ℹ The term **gas** is usually used to describe substances that are gases at room temperature, while **vapour** describes the gas phase of substances that are liquids or solids at room temperature—such as water. (A more precise definition is given in Section 8.6, p.374.) However, in a chemical context, the two terms essentially mean the same.

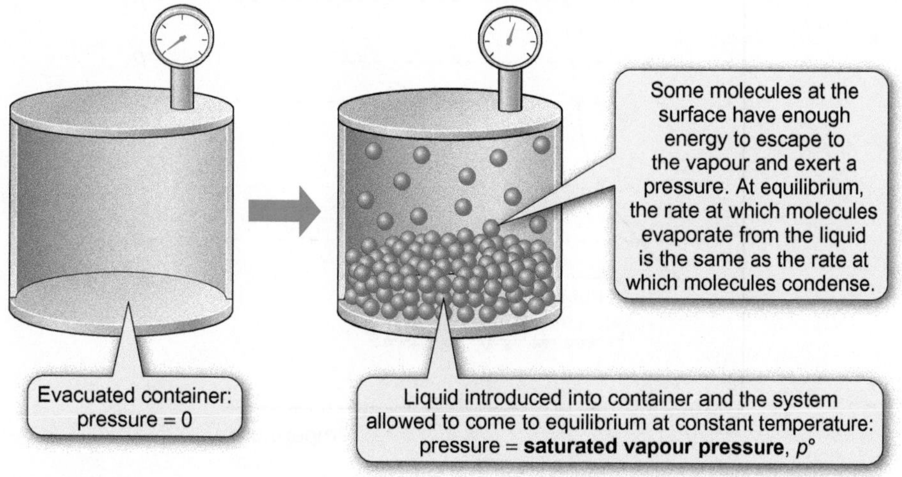

Some molecules at the surface have enough energy to escape to the vapour and exert a pressure. At equilibrium, the rate at which molecules evaporate from the liquid is the same as the rate at which molecules condense.
Evacuated container: pressure = 0
Liquid introduced into container and the system allowed to come to equilibrium at constant temperature: pressure = saturated vapour pressure, p°

Figure 17.2 The equilibrium vapour pressure over a pure liquid in a closed system is known as the saturated vapour pressure.

ℹ The situation in the sealed container illustrated in Figure 17.2 is an example of dynamic equilibrium (Section 1.9, p.56). In this closed system, molecules evaporate from the liquid but also condense back. At equilibrium, the rates of vaporization and condensation are equal. In an open container, molecules vaporize faster than they condense, and the liquid is lost through evaporation.

Partial pressures are introduced in Section 8.3 (p.355).

If there are other gases (such as air) present above the liquid, then the pressure due to the vapour is its **partial vapour pressure**. The total pressure in the container will be the vapour pressure of the liquid plus the partial pressures of all other gases present.

When the vapour pressure of a liquid becomes equal to the external pressure, the liquid boils. The pressure of the vapour is large enough to overcome the external pressure on the liquid, so bubbles of vapour can form anywhere within the liquid, as well as at the surface.

Phase diagrams

The lowering of boiling point at reduced pressure is used in **vacuum distillation**. The distillation apparatus is attached to a vacuum pump and the pressure is reduced in the apparatus. This makes it possible to purify compounds that would decompose if heated to their normal boiling points.

Laboratory equipment for vacuum distillation.

It is clear that temperature plays a part in determining in which phase a substance exists. Indeed, changing the temperature is the most common way of changing phase: heating water turns it to vapour; cooling water turns it to ice. Another important condition is pressure: for example, a gas can be liquefied by pressurizing it. If the pressure over a liquid is reduced, it boils at a lower temperature. A convenient way of showing the effect of each of these properties is to use a **phase diagram**. Phase diagrams can be used for single-component and multi-component systems. This section is about single-component phase diagrams.

A phase diagram is a plot of the pressures and temperatures where phase transitions take place. The diagram is constructed by measuring the melting, boiling, and sublimation temperatures at a series of pressures and plotting them as shown in Figure 17.3. The resulting phase diagram is shown in Figure 17.4.

The shaded regions in Figure 17.4(a) indicate which phase is stable under a particular set of conditions. At lower temperatures and higher pressures, matter exists in the solid phase. At higher temperatures and lower pressures, the vapour phase is most stable.

The lines on the diagram represent the conditions where phase transitions take place and two phases are in equilibrium. The lines are sometimes known as the **phase boundaries**. For example, in Figure 17.4(a), line T–A shows how the melting temperature changes with pressure. At the melting temperature, solid and liquid exist in equilibrium. Line T–C shows how the boiling temperature changes with pressure.

The solid–vapour curve (B–T) and the liquid–vapour curve (T–C) also show how the equilibrium vapour pressure for solid and liquid, respectively, depends on temperature.

Because very large pressure ranges are involved, phase diagrams are usually not drawn to scale. They are usually drawn so as to emphasize the points of interest for the compound concerned.

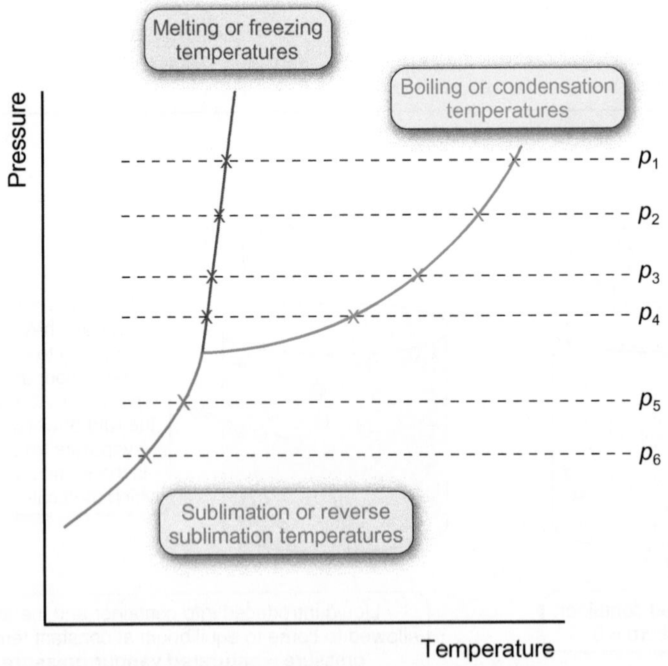

Melting or freezing temperatures
Boiling or condensation temperatures
Pressure
p_1
p_2
p_3
p_4
p_5
p_6
Sublimation or reverse sublimation temperatures
Temperature

Figure 17.3 Constructing a phase diagram. Start with a solid under pressure p_1. This is heated and the melting and boiling temperatures are plotted on the graph. This is repeated at pressures p_2, p_3, and p_4. Joining the temperatures measured gives the lines on the phase diagram.

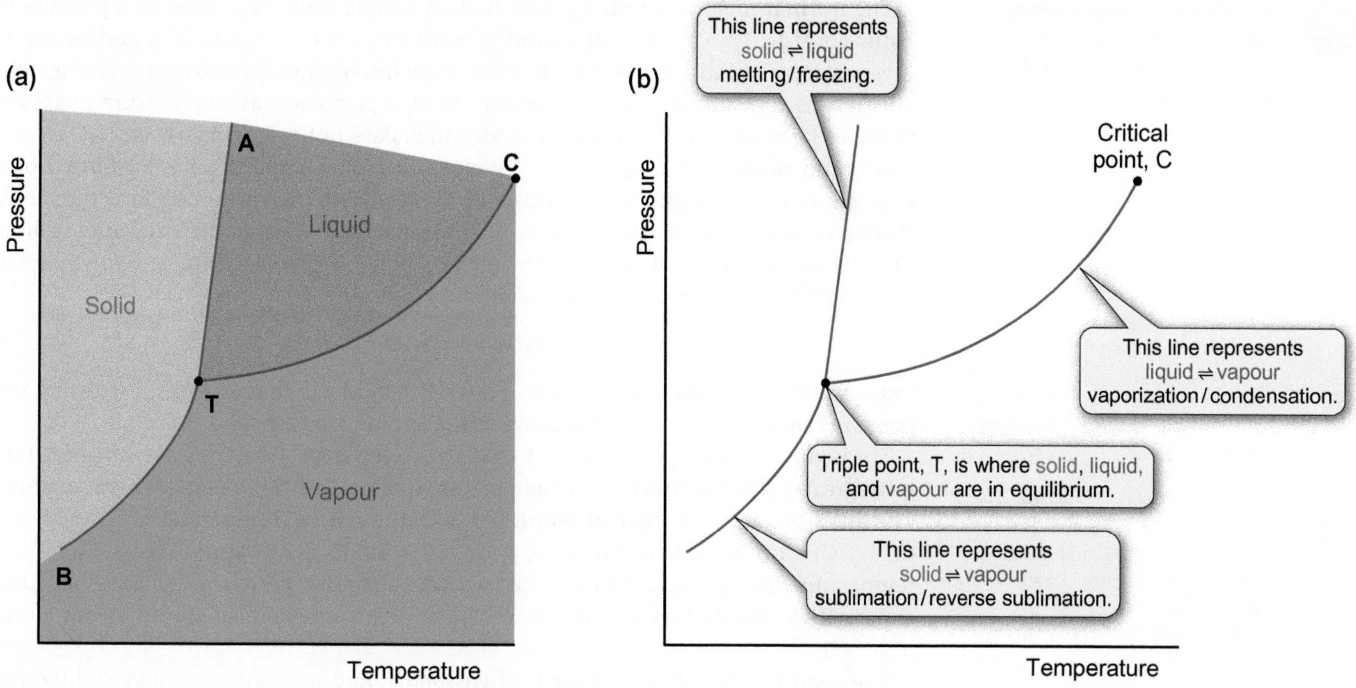
(a)
Pressure
A
C
Liquid
Solid
T
Vapour
B
Temperature
(b)
This line represents
solid ⇌ liquid
melting/freezing.
Pressure
Critical
point, C
This line represents
liquid ⇌ vapour
vaporization/condensation.
Triple point, T, is where solid, liquid,
and vapour are in equilibrium.
This line represents
solid ⇌ vapour
sublimation/reverse sublimation.
Temperature

Figure 17.4 A typical single-component phase diagram. It shows how pressure influences the melting, boiling, and sublimation temperatures: (a) shows which phase is stable at a given temperature and pressure; (b) indicates the location of the triple point, T, critical point, C, and the equilibria represented by each line.

Phase diagram for H_2O

To illustrate what phase diagrams can tell you, Figure 17.5 shows the phase diagram for H_2O.

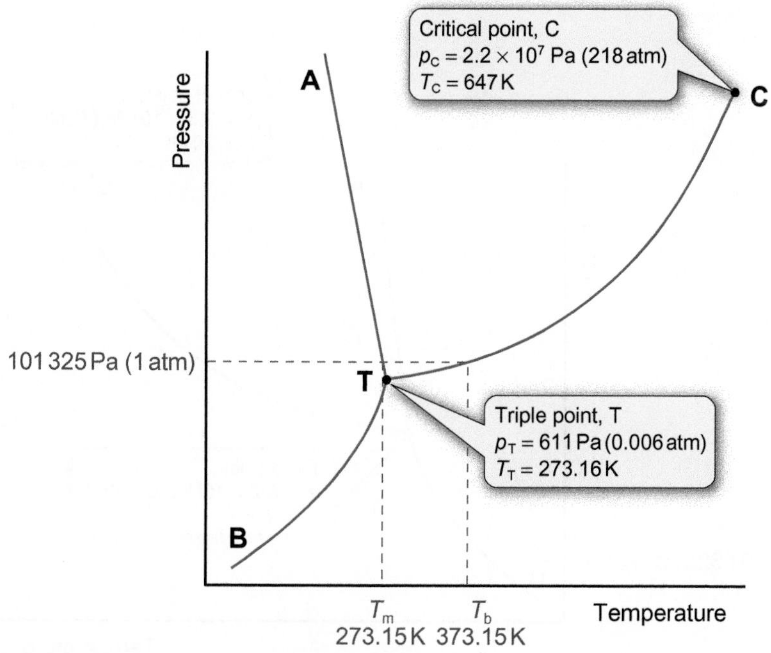
Critical point, C
$p_C = 2.2 \times 10^7$ Pa (218 atm)
$T_C = 647$ K
Pressure
A
C
101 325 Pa (1 atm)
T
Triple point, T
$p_T = 611$ Pa (0.006 atm)
$T_T = 273.16$ K
B
T_m
273.15 K
T_b
373.15 K
Temperature

Figure 17.5 The phase diagram for H_2O. (Phase diagrams are not usually drawn to scale.) The horizontal, dashed purple line at 1 atm crosses the solid–liquid line A–T at the normal melting point of H_2O, and crosses the liquid–vapour line C–T at the normal boiling point of H_2O.

Visit the Online Resource Centre to view video clip 17.1 which demonstrates the formation of supercritical carbon dioxide.

The **normal** melting point, T_m, and normal boiling point, T_b, relate to a pressure of 1 atm. For H_2O, $T_m = 273.15\,K$ (0 °C) and $T_b = 373.15\,K$ (100 °C). The solid–liquid line (A–T) is very steep and has a *negative* slope. (The slope has been exaggerated in this diagram.) This means that the melting temperature of H_2O *goes down* as the pressure increases, though a large increase in pressure is needed to make a noticeable difference. H_2O is unusual in this respect: for most other substances, the solid–liquid line has a *positive* slope, as in Figure 17.4. The unusual behaviour of H_2O is due to the very 'open' structure of ice (H_2O (s)) caused by hydrogen bonding. This means that H_2O (s) occupies a larger volume than the same amount of H_2O (l) at the melting point. So, by Le Chatelier's principle, increasing the pressure makes the equilibrium

$$H_2O(s) \rightleftharpoons H_2O(l)$$

move to the right, towards the liquid phase. For most other substances, increasing the pressure causes the equilibrium to move to the left towards the solid phase.

The structure of ice is illustrated in Box 1.10 (p.50). The volume change on melting ice is discussed further in Section 17.2 (p.777).

The liquid–vapour line for H_2O (T–C) has a less steep slope. Changing pressure therefore has a greater effect on the boiling temperature than it does on the melting temperature. The line stops at C, the **critical point**. For H_2O, the **critical temperature**, T_C, is 647.6 K (374.5 °C) and the **critical pressure**, p_C, is 2.2×10^7 Pa (217.7 atm). Above the critical temperature and pressure, liquids and gases do not exist as separate phases, but form **supercritical fluids** (see p.764). Above T_C, applying pressure to the fluid can never liquefy it.

The triple point of H_2O is used as a fixed reference point on the international temperature scale and is used to calibrate accurate thermometers.

The point T on the phase diagram is where solid, liquid, and vapour exist in equilibrium. This is called the **triple point** and is the only set of conditions where the three phases can be in equilibrium. The conditions for H_2O are $T_T = 273.16\,K$ (0.01 °C) and $p_T = 611\,Pa$ (0.006 atm). The triple point also represents the lowest temperature at which the substance can exist in the liquid state.

The normal melting and boiling points, T_m and T_b, respectively, are those at 1 atm pressure. At pressures other than 1 atm, the terms **melting temperature** and **boiling temperature** are used (see Section 1.7, p.47).

Phase diagram for carbon dioxide

The phase diagram for carbon dioxide (CO_2) is shown in Figure 17.6. While it looks similar to the diagram for H_2O, there are some important differences. The triple point for carbon dioxide is at $T_T = 216\,K$ (−57 °C) and $p_T = 5.2 \times 10^5$ Pa (5.1 atm). At atmospheric pressure,

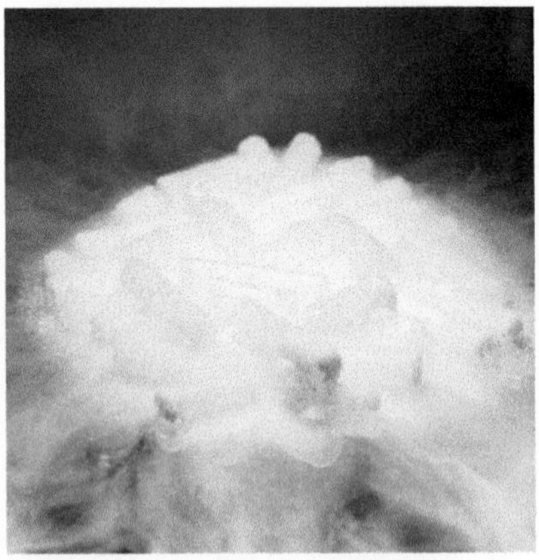

Solid CO_2, known as 'dry ice', sublimes at atmospheric pressure so that it turns directly to a gas with no liquid phase being observed. In this photo, the clouds are due to water vapour in the atmosphere condensing at the low temperature.

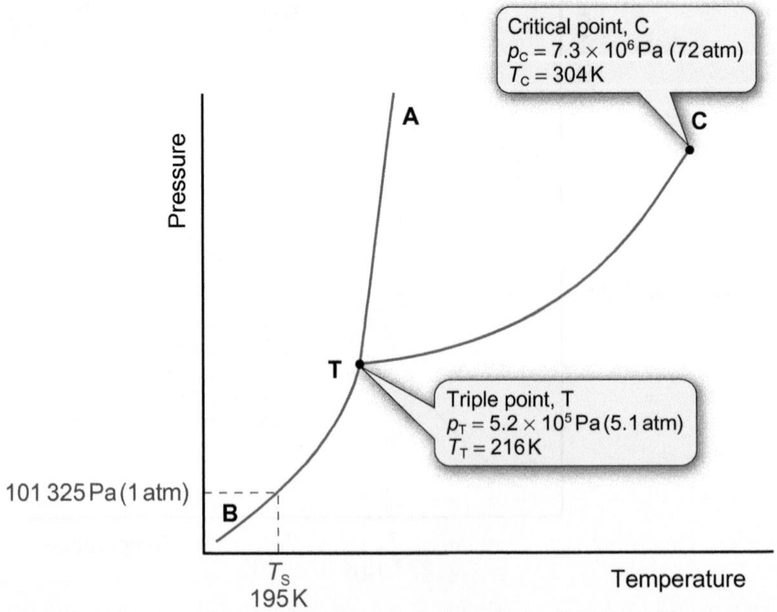
Critical point, C
$p_C = 7.3 \times 10^6$ Pa (72 atm)
$T_C = 304$ K
A
C
Pressure
T
Triple point, T
$p_T = 5.2 \times 10^5$ Pa (5.1 atm)
$T_T = 216$ K
101 325 Pa (1 atm)
B
T_S
195 K
Temperature

Figure 17.6 Phase diagram for carbon dioxide. The dashed purple line crosses the solid–vapour line B–T at the normal sublimation temperature of carbon dioxide, T_S.

therefore, liquid CO_2 cannot exist. This explains why solid CO_2 sublimes under normal laboratory conditions. To make liquid CO_2 requires a pressure of at least 5.1 atm. Carbon dioxide fire extinguishers usually contain the liquid. At room temperature, the equilibrium vapour pressure of liquid carbon dioxide is around 67 atm, so this is the pressure of CO_2(g) inside a fire extinguisher.

The solid–liquid equilibrium line for CO_2 (T–A) has a positive slope so that the melting temperature increases as the pressure rises. This is the opposite behaviour to that of H_2O but is much more common.

Phase diagram for sulfur

Sulfur has two stable solid forms—allotropes—with different crystalline structures. This is shown in the phase diagram in Figure 17.7.

At low temperatures, the rhombic crystalline form is more stable. If this is heated at 1 atm, it slowly changes into the monoclinic crystalline form at 368 K (95 °C). This form remains stable until it melts to form liquid sulfur at 388 K (115 °C). The line T_1–A on the diagram between the solid phases represents the conditions under which the rhombic–monoclinic transition takes place under equilibrium conditions. The change from one solid phase to another can take place very slowly. If rhombic sulfur is heated quickly, it melts without passing through the monoclinic phase.

Some compounds have a number of solid phases. Figure 17.8 shows the phase diagram for H_2O up to much higher pressures than in Figure 17.5. The green shaded areas in Figure 17.8 each represent ice with a different crystal structure. Under normal conditions, H_2O exists as Ice I. The other phases only exist at extreme pressures, such as those at the bottom of a glacier.

A further illustration of the interpretation of phase diagrams is shown in Worked example 17.1. Box 17.1 describes the special types of phase changes that take place in liquid crystals.

Crystals of monoclinic sulfur in a cave in the Wai-o-Tapu volcanic region of New Zealand. The crystals are formed by reverse sublimation from sulfur vapour coming out of the rocks: referring to Figure 17.7(b), what phase transition has taken place?

The allotropes of sulfur are described in Sections 1.7 (p.48) and 27.5 (p.1234).

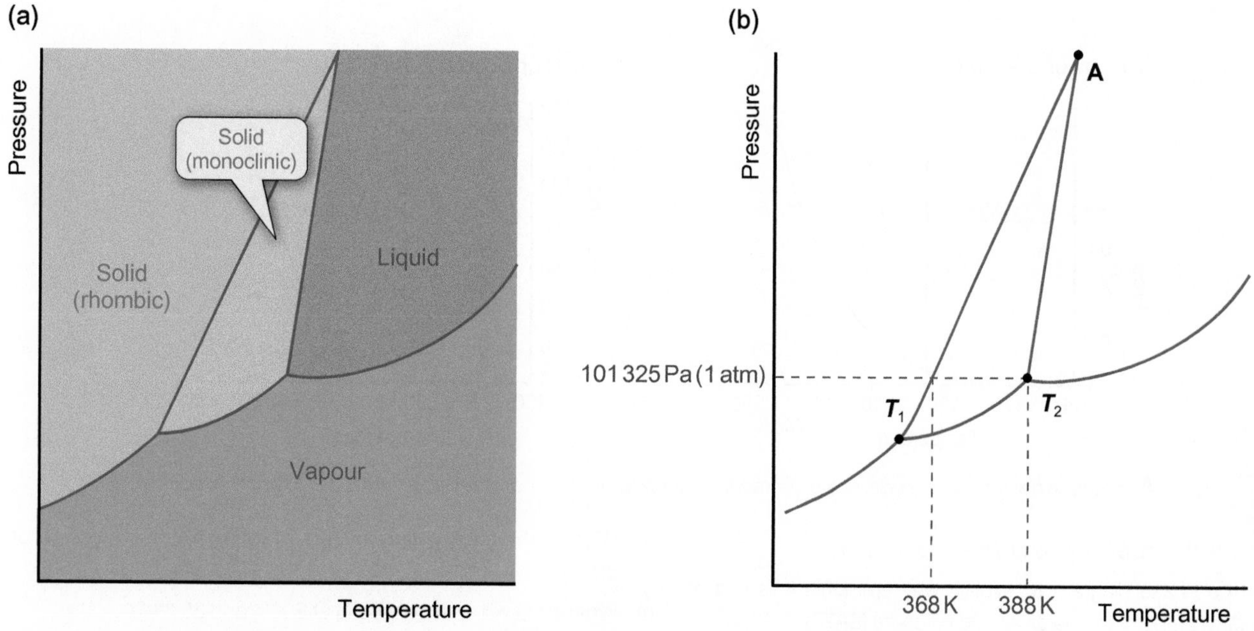
(a)
Pressure
Solid (monoclinic)
Solid (rhombic)
Liquid
Vapour
Temperature
(b)
Pressure
A
101325 Pa (1 atm)
T_1
T_2
368 K
388 K
Temperature

Figure 17.7 The phase diagram for sulfur. Sulfur has two stable solid forms, rhombic and monoclinic. The phase diagram shows: (a) which form is stable under particular conditions; (b) the detailed conditions where transitions between the various phases occur. Note that the diagram is not to scale, and the slopes of the lines are exaggerated.

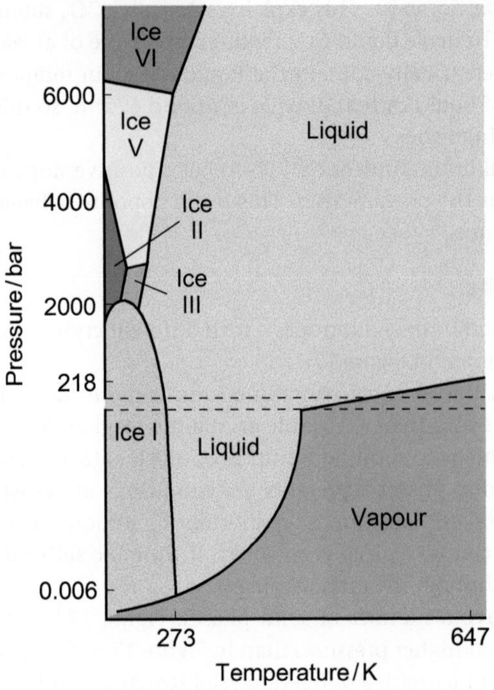
Ice
VI
6000
Ice
V
Liquid
4000
Ice
II
Ice
III
2000
218
Pressure/bar
Ice I
Liquid
Vapour
0.006
273
647
Temperature/K

Figure 17.8 At high pressures, ice has several different crystalline forms. The lines on the phase diagram show where the transitions between the phases occur. The two horizontal dashed lines indicate a change of scale on the vertical axis.

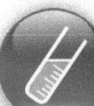

Worked example 17.1 Using a phase diagram

The phase diagram for ammonia (NH_3) is shown in the figure.

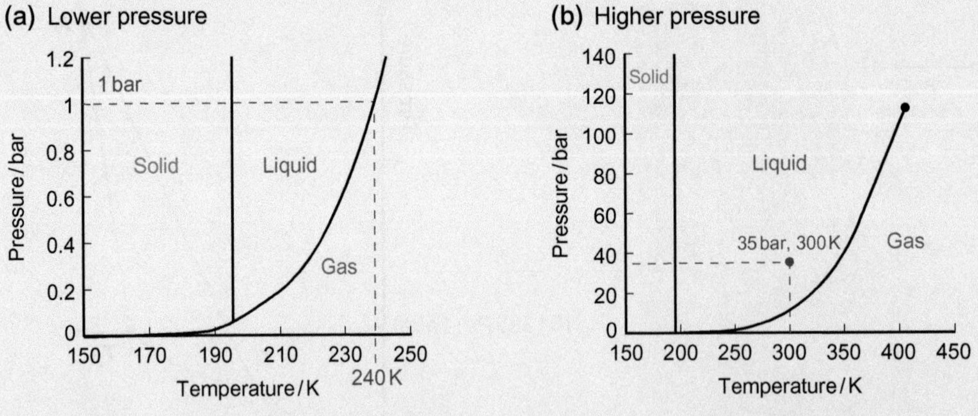
(a) Lower pressure
1 bar
Solid
Liquid
Gas
Pressure/bar
Temperature/K
240 K
(b) Higher pressure
Solid
Liquid
35 bar, 300 K
Gas
Pressure/bar
Temperature/K

▲ The phase diagram for ammonia: (a) the diagram for lower pressures; (b) the same diagram extended to higher pressures.

(a) Estimate the triple point conditions for ammonia.

(b) Use the phase diagram to explain why ammonia is a gas at standard ambient temperature and pressure (SATP).

(c) What are the normal melting and boiling points, T_m and T_b, for ammonia?

(d) Estimate the critical pressure and critical temperature of ammonia.

(e) What is the melting temperature of ammonia at a pressure of 80 bar?

(f) What state will ammonia be in at 300 K and 35 bar?

(g) Describe the phase changes that occur when ammonia is heated from 150 K to 250 K at a pressure of (i) 0.01 bar, (ii) 0.4 bar.

→

→ Strategy

In each case, you will need to decide whether to use the low pressure or high pressure phase diagram to read off the values.

(a) The triple point lies where the solid, liquid, and vapour phases coexist (corresponding to point T in Figure 17.4).

(b) SATP is 298 K (25 °C) and 1 bar (see p.349).

(c) The normal melting point and boiling point are the values at 1 atm (= 1.013 bar) pressure.

(d) The critical point lies at the end of the liquid–vapour line (corresponding to point C in Figure 17.4).

(e) The melting temperature at a pressure of 80 bar can be read from the solid–liquid boundary line.

(f) You need to find in which area of the phase diagram these conditions lie.

(g) Draw a horizontal line at the appropriate pressure and follow it as the temperature increases. Phase changes occur where this line intersects the lines of the phase diagram.

Solution

(a) From diagram (a), the triple point is at 195 K and 0.06 bar.

(b) From diagram (a), at 1 bar pressure, the boiling temperature is around 240 K. This is less than 298 K so ammonia is a gas at SATP.

(c) From diagram (a), $T_m = 195$ K and $T_b = 240$ K.

(d) From diagram (b), the critical point is at 405 K and 113 bar.

(e) From diagram (b), the melting temperature at 113 bar is 196 K.

(f) Liquid (diagram (b)).

(g) Using diagram (a): (i) solid → gas; (ii) solid → liquid → gas.

Question

The normal melting and boiling points, T_m and T_b, for methanol (CH_3OH) are 175.7 K and 337.8 K, respectively. The critical temperature and pressure are 512.8 K and 81 bar. The triple point occurs at 175.5 K. Sketch (not to scale) the phase diagram for methanol.

Box 17.1 Liquid crystals

Liquid crystals combine some of the properties of a liquid, for example, the ability to flow, with some properties of a solid such as directional order of the molecules.

Thermotropic liquid crystals change their properties with temperature. They normally show liquid crystal **mesophase** behaviour just above the melting point: the compound will flow but is cloudy rather than forming a clear liquid. The cloudiness arises because the molecules retain some degree of order and alignment and so scatter light. At higher temperatures, the molecules lose this order and a conventional, clear liquid state is formed. The phenomenon was first reported in 1888 by the Austrian botanist Friedrich Reinitzer who was studying cholesteryl benzoate, an ester of cholesterol.

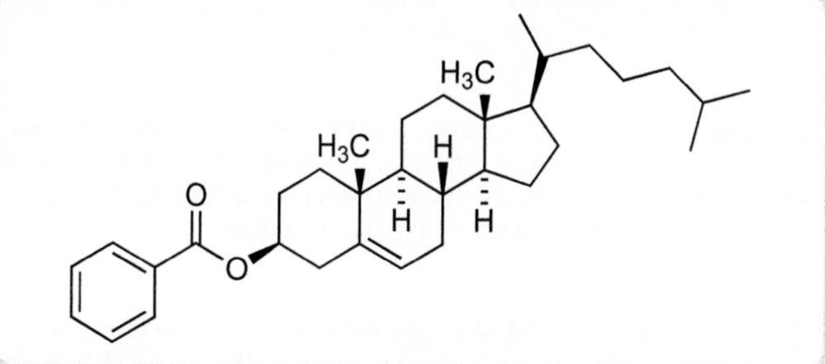
H3C
H3C
H
H
H
O
O

▲ Cholesteryl benzoate was one of the first compounds reported to show liquid crystalline behaviour.

The properties of liquid crystal mesophases are generally *anisotropic*, which means that they vary according to direction. This behaviour is unlike that of *isotropic* materials, such as conventional liquids, whose properties are the same in every direction. Compounds that form mesophases often have molecules that are longer in one direction than the others, widely used examples being the alkyl cyanobiphenyls, which form liquid crystals phases at around room temperature. Various types of mesophase can form. In an isotropic liquid, the molecules are oriented in random directions (see Figure 1), but in a **nematic** mesophase, they preferentially point about one direction called the **director**. Some compounds form **smectic** mesophases in which, as well as pointing in a particular direction, the molecules form layered structures.

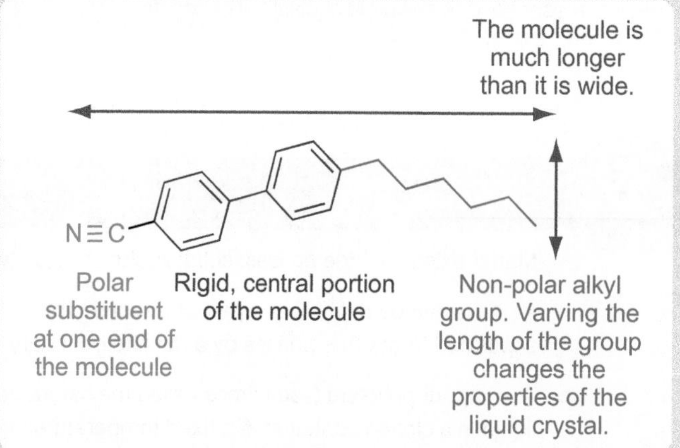
The molecule is much longer than it is wide.
N≡C
Polar substituent at one end of the molecule
Rigid, central portion of the molecule
Non-polar alkyl group. Varying the length of the group changes the properties of the liquid crystal.

▲ An alkyl cyanobiphenyl, a compound that forms liquid crystals.

If the molecule is also chiral (see Section 18.4, p.838), **chiral nematic** mesophases can form. The molecules align along the →

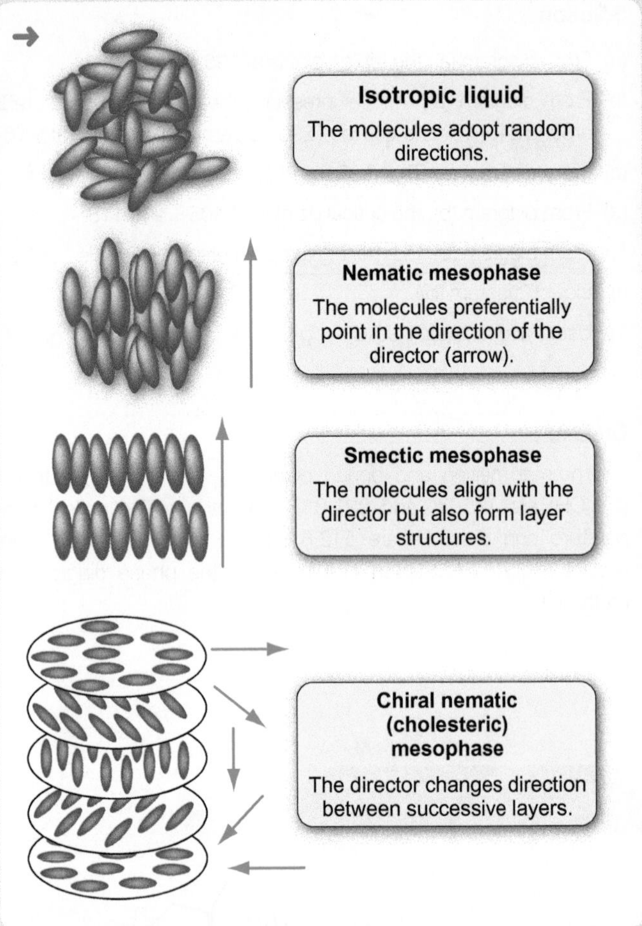
Isotropic liquid
The molecules adopt random directions.
Nematic mesophase
The molecules preferentially point in the direction of the director (arrow).
Smectic mesophase
The molecules align with the director but also form layer structures.
Chiral nematic (cholesteric) mesophase
The director changes direction between successive layers.

▲ **Figure 1** Liquid crystalline compounds form mesophases in which some directional order of molecules is retained in the liquid. The diagram shows the different types of mesophase with differing arrangements of the molecules. Cholesteryl benzoate forms a chiral nematic mesophase.

director, but the director changes between successive layers. The change of direction of the director traces out a helix with its axis perpendicular to the director, as illustrated in Figure 1. Chiral nematic mesophases are sometimes said to be **cholesteric** since this is the form of mesophase observed in cholesteryl benzoate.

▲ Liquid crystals give beautiful optical textures when viewed through a polarizing microscope. (Photo courtesy of Professor Duncan Bruce, University of York.)

Since their commercialization in the 1970s, many uses for liquid crystals have been developed. Perhaps the best known is in liquid crystal displays (LCD). These rely on the polar nature of liquid crystals, such as cyanobiphenyls, which means that the direction of alignment is influenced by an electric field. The passage of light through a thin liquid crystal layer can be controlled by switching the orientation of the molecules in the layer.

Liquid crystal thermometers rely on the fact that the *pitch* (the distance over which the helical twist in a chiral nematic liquid crystal repeats) is about the same as the wavelength of visible light and so reflection from thin films of these compounds is coloured. However, the pitch changes with temperature so the colour also changes—the compounds are **thermochromic**. By varying the chemical structure of the liquid crystal or the composition of a mixture of liquid crystals, the precise temperature at which colours become visible can be controlled.

Question

Suggest some reasons why cyanobiphenyls form mesophases.

Summary

- Matter exists in three phases: solid; liquid; and gas (vapour).
- A phase can be defined as a part of a system that is homogeneous in chemical and physical state throughout and is separated from other phases by a definite boundary.
- The vapour pressure (sometimes called the saturated vapour pressure) is the equilibrium partial pressure exerted by a liquid in a closed container at a fixed temperature.
- A phase diagram is a plot of the pressures and temperatures where phase transitions take place.

- Phase diagrams show the conditions of temperature and pressure under which the various phases are stable (the areas on the diagram) and the conditions where phase transitions take place (the lines on the diagram).
- The triple point of a compound is the temperature and pressure at which solid, liquid, and vapour phases are in equilibrium.
- The critical point is the temperature and pressure where the distinction between liquid and vapour properties disappears.
- Supercritical fluids exist above the critical temperature and critical pressure of a compound. Above the critical temperature, a gas cannot be liquefied by the application of pressure alone.

For practice questions on these topics, see questions 1–4 at the end of this chapter (p.810).

17.2 Quantitative treatment of phase transitions

A chemical system is at equilibrium when its Gibbs energy is at a minimum (see Section 15.1, p.697). The most stable phase at a particular temperature and constant pressure is the phase with the lowest Gibbs energy at that temperature.

For a single, pure component at constant pressure, the dependence of its molar Gibbs energy, G_m, on temperature is given by Equation 14.25 (p.688)

$$\left(\frac{dG_m}{dT}\right)_p = -S_m \qquad (14.25)$$

where S_m is the molar entropy of the compound.

The value of S_m is always positive so that G_m decreases as the temperature rises, as shown in Figure 17.9. The gradient of a plot of G_m against T is negative and equal to $-S_m$. The molar entropy of the solid, $S_m(s)$, is *smaller* than that of the liquid, which is in turn smaller than that of the vapour. Therefore, the line for the solid in Figure 17.9 has the smallest

When considering the Gibbs energy of a single compound, it is important to use the symbol G_m, which indicates that the value applies to 1 mol. (For a chemical reaction $\Delta_r G$ applies to the molar amounts, as shown in the equation.)

Equation 14.25 assumes that S_m does not change significantly with temperature.

Molar entropies are introduced in Section 14.2 (p.662) and discussed further in Section 14.6 (p.688).

The differential term $\left(\frac{dG_m}{dT}\right)_p$ is the change of molar Gibbs energy with temperature, at constant pressure. This is equivalent to the gradient of a plot of G_m versus T at constant pressure. Help with differential terms is in Maths Toolkit MT6 (p.1324).

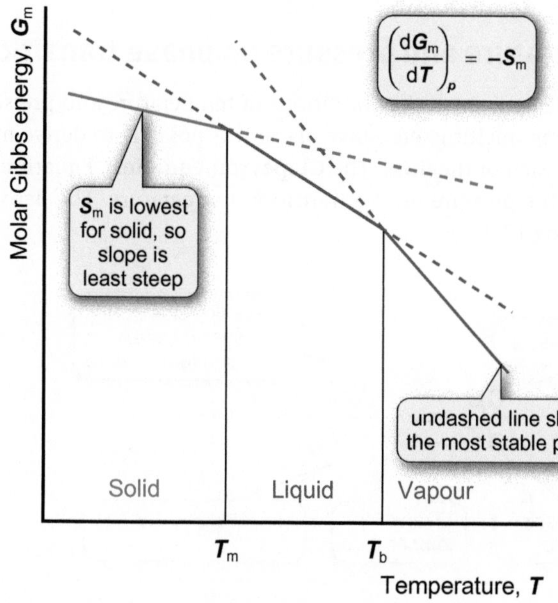
Molar Gibbs energy, G_m
$\left(\frac{dG_m}{dT}\right)_p = -S_m$
S_m is lowest for solid, so slope is least steep
undashed line shows the most stable phase
Solid
Liquid
Vapour
T_m
T_b
Temperature, T

Figure 17.9 The gradient of a graph of G_m against T for a single substance is equal to $-S_m$. At low temperatures, solids have the lowest G_m and so form the stable phase. At higher temperatures, gases are the most stable phase. Liquids are stable at intermediate temperatures. (*Note.* If G_m for the liquid is higher than for the gas at all temperatures, the solid sublimes instead of melting.)

slope, and the line for the vapour has the steepest slope. At low temperatures, the solid phase has the smallest G_m, so this is the most stable phase. Between the melting and boiling temperatures, G_m is smallest for the liquid phase. The gas phase is more stable above the boiling temperature since it has the smallest G_m at these temperatures.

The phase transitions, melting or freezing at T_m and boiling or condensation at T_b, occur where the lines on Figure 17.9 intersect. At these temperatures and pressures, G_m values for the two phases are equal.

Consider the melting transition (also known as **fusion**). The change in Gibbs energy, $\Delta_{fus}G$, is the difference between G_m for the liquid and for the solid

$$\Delta_{fus}G = G_m(\text{liquid}) - G_m(\text{solid})$$

Since $G_m(\text{liquid}) = G_m(\text{solid})$ at T_m

$$\Delta_{fus}G = 0 \quad (\text{at } T_m) \qquad (17.1)$$

Gibbs energy is introduced in Section 14.5 (p.675). Enthalpy changes of state are introduced in Section 1.7 (p.47) and discussed further in Section 13.2 (p.619). Entropy changes of state are discussed in Section 14.2 (p.660).

From the definition of Gibbs energy change (Equation 14.16, p.675), it follows that

$$\Delta_{fus}G = \Delta_{fus}H - T_m\Delta_{fus}S \qquad (17.2)$$

where $\Delta_{fus}H$ and $\Delta_{fus}S$ are the changes in enthalpy and entropy that occur on melting. Combining Equations 17.1 and 17.2 and rearranging the resulting equation

$$\Delta_{fus}G = 0 = \Delta_{fus}H - T_m\Delta_{fus}S$$
$$\Delta_{fus}H = T_m\Delta_{fus}S$$

$$\Delta_{fus}S = \frac{\Delta_{fus}H}{T_m} \qquad (17.3a)$$

Equations 17.3a and b are also used in Section 14.2 (p.661). There, the enthalpy change, $\Delta H^{\ominus}$, refers to a pressure of 1 bar, but the equation can also be used at other pressures, although the melting temperature and boiling temperature will be different.

You can apply the same argument to the boiling transition at T_b to give

$$\Delta_{vap}S = \frac{\Delta_{vap}H}{T_b} \qquad (17.3b)$$

Effects of temperature and pressure on phase transitions

The lines on a phase diagram show the conditions of temperature and pressure at which phase transitions (boiling, melting, etc.) take place. It is possible to derive mathematical expressions to describe each of the lines. The **Clapeyron equation**, Equation 17.4, shows the effect of changing the pressure on the transition temperature. The derivation of the equation is shown in Box 17.2.

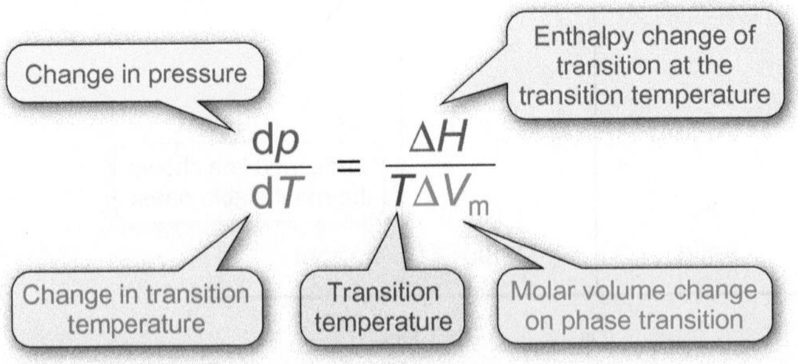
Change in pressure
Enthalpy change of transition at the transition temperature
Change in transition temperature
Transition temperature
Molar volume change on phase transition

$$\frac{dp}{dT} = \frac{\Delta H}{T\Delta V_m} \qquad (17.4)$$

Visit the Online Resource Centre to view screencast 17.1 which walks you through the derivation and use of the Clapeyron equation.

The Clapeyron equation gives the gradient of the lines on a phase diagram in terms of the molar entropy change and molar volume change at the transition.

T is the transition temperature and ΔV_m is the change in volume of 1 mol of the substance during the phase transition.

The value of $\frac{dp}{dT}$ is the gradient of a plot of pressure versus temperature and so gives the gradient of the lines in the phase diagram. When a solid melts, there is only a small change in volume, so ΔV_m is small and therefore $\frac{dp}{dT}$ is large. This means that a large pressure change is needed to have a significant effect on the melting temperature. The volume of 1mol of gas is much larger than that of the liquid or solid it comes from, so, for vaporization or sublimation, ΔV_m is large and $\frac{dp}{dT}$ is smaller than for the melting transition. Compare the relative slopes of the different lines in the phase diagrams in Figures 17.4 and 17.5. The solid–liquid transition lines have their slope exaggerated—they are really almost vertical.

The contraction of ice on melting means that ice has a *lower* density than water at 0 °C. This is true up to around 4 °C, above which water expands like other liquids. The lower density of ice explains why icebergs float on cold water.

Melting and freezing transitions

Applying the Clapeyron equation to melting gives

$$\frac{dp}{dT} = \frac{\Delta_{fus}H}{T\Delta_{fus}V_m} \tag{17.5}$$

Enthalpy changes of fusion always have positive values; energy has to be supplied to the system to overcome the attractions holding the solid together. Most compounds expand slightly on melting so that, at the melting temperature, the liquid occupies a larger volume than the solid. So $\Delta_{fus}V$ is positive and $\frac{dp}{dT}$ has a positive value. The solid–liquid line in the phase diagram for CO_2, for example, has a positive gradient (see Figure 17.6, p.770) and the melting temperature increases as the pressure rises. Since $\Delta_{fus}V$ does not change very much with pressure and $\Delta_{fus}S$ does not change very much with temperature, $\frac{dp}{dT}$ has a reasonably constant value and so the solid–liquid line in the phase diagram is almost straight.

Russian Alfa class submarines were powered by nuclear reactors cooled by liquid lead. However, they had problems because the lead tended to solidify at the wrong time. It is important to understand the effect of pressure on freezing temperature.

Water is the best known example of a small number of liquids which have a *negative* value for $\Delta_{fus}V$. At 0°C, an amount of water occupies a smaller volume than the same amount of ice due to changes in the hydrogen bonded structure (see Box 1.10, p.50). This means that $\frac{dp}{dT}$ for water has a negative value and the solid–liquid line in the phase diagram has a negative gradient and slopes the opposite way to that for most substances (see Figure 17.5). The melting temperature of water *decreases* as the pressure rises.

The application of the Clapeyron equation to melting is illustrated in Worked example 17.2.

Box 17.2 Deriving the Clapeyron equation

To describe the conditions for phase equilibrium, you need to know the dependence of molar Gibbs energy on both temperature and pressure. The variation with temperature is connected to the molar entropy, described by Equation 14.25

$$\left(\frac{dG_m}{dT}\right)_p = -S_m \tag{14.25}$$

This applies at fixed pressure. Rearranging Equation 14.25 gives $dG_m = -S_m dT$.

A similar equation can be derived to describe the effect of changing the pressure at a fixed temperature

$$\left(\frac{dG_m}{dp}\right)_T = V_m$$

so, $dG_m = V_m dp$ at fixed temperature. V_m is the molar volume (i.e. the volume of 1 mol) of the substance.

To see how the molar Gibbs energy changes when both pressure and temperature vary, the two equations are combined to give

$$dG_m = V_m dp - S_m dT$$

dG_m is the change in molar Gibbs energy when the pressure changes by a very small amount dp and the temperature changes by a very small amount dT.

If a pure substance exists in equilibrium between two phases, phase 1 and phase 2, then the molar Gibbs energy of each phase must be the same, so that

$$G_m(1) = G_m(2)$$

→

→ If the temperature and pressure are changed by dp and dT, then the changes in molar Gibbs energies are

$$dG_m(1) = V_m(1)dp - S_m(1)dT$$

and

$$dG_m(2) = V_m(2)dp - S_m(2)dT$$

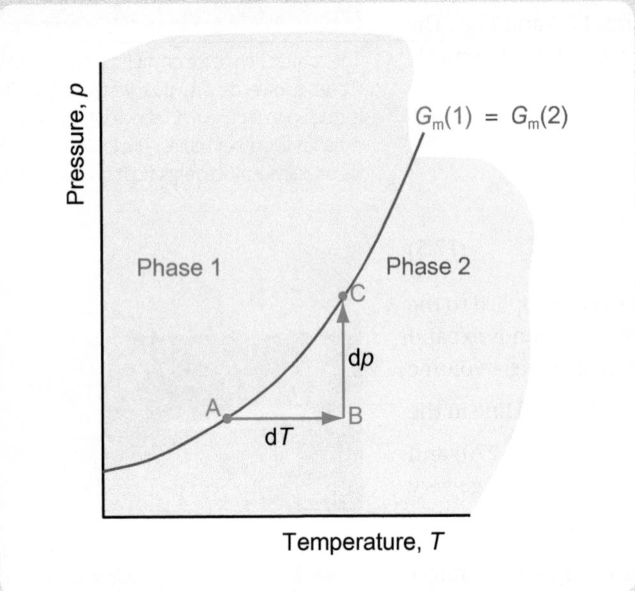
Pressure, p
$G_m(1) = G_m(2)$
Phase 1
Phase 2
C
dp
A
B
dT
Temperature, T

▲ The blue line shows the temperatures and pressures at which phase 1 and phase 2 are in equilibrium and $G_m(1) = G_m(2)$. Say the system is in equilibrium at point A. If the temperature changes by dT (i.e. it moves to B) then it is no longer in equilibrium. To return to equilibrium (point C), the pressure must increase by dp.

If the phases remain in equilibrium, the changes in molar Gibbs energies must be equal (see figure) so

$$dG_m(1) = dG_m(2)$$

and therefore

$$V_m(1)dp - S_m(1)dT = V_m(2)dp - S_m(2)dT$$

Rearranging

$$V_m(2)dp - V_m(1)dp = S_m(2)dT - S_m(1)dT$$

$$dp[V_m(2) - V_m(1)] = dT[S_m(2) - S_m(1)]$$

$V_m(2) - V_m(1)$ is the change of volume, ΔV_m, and $S_m(2) - S_m(1)$ is the change in entropy, ΔS_m, when phase 1 changes to phase 2 so that

$$dp\Delta V_m = dT\Delta S_m \quad \text{and} \quad \frac{dp}{dT} = \frac{\Delta S_m}{\Delta V_m}$$

From Equation 17.3, $\Delta S = \frac{\Delta H}{T}$ at the temperature of a phase transition, so

$$\frac{dp}{dT} = \frac{\Delta S_m}{\Delta V_m} = \frac{\Delta H}{T\Delta V_m} \qquad (17.4)$$

where ΔH is the enthalpy change for the phase transition that occurs at temperature T. Equation 17.4 is the Clapeyron equation. It gives the equation of any phase boundary line on a phase diagram in terms of the molar entropy change and molar volume change at the transition.

Worked example 17.2 Effect of pressure on melting temperature

The melting point of sodium is 97.8 °C at 1.00 atm pressure. The densities at this temperature of solid and liquid sodium are 0.952 g cm^{-3} and 0.929 g cm^{-3}, respectively. The enthalpy change of fusion is +3.00 kJ mol^{-1}. Calculate the melting temperature of sodium at a pressure of 120 atm. (A_r(Na) = 23.0.)

Strategy

The change of melting temperature with pressure is described by the Clapeyron equation, Equation 17.5. $\Delta_{fus}H$ is given and you can find the change in molar volume, $\Delta_{fus}V$, from the densities of the liquid and solid. Substitute values for $\Delta_{fus}H$ and $\Delta_{fus}V$ into Equation 17.5 to find how much the melting temperature changes with pressure and hence calculate the melting temperature at 120 atm. Remember to convert to SI units: volumes in m^3; temperature in K; pressure in Pa; energy in J.

Solution

Use the densities of liquid and solid to find $\Delta_{fus}V$.

$$\Delta_{fus}V = V_{m(liquid)} - V_{m(solid)}$$

For 1 mol, $\text{density} = \frac{\text{molar mass}}{\text{molar volume}}$ so $V_m = \frac{\text{molar mass}}{\text{density}}$

$$V_{m\,(liquid)} = \frac{23.0\,\text{g mol}^{-1}}{0.929\,\text{g mol}^{-3}} = 24.76\,\text{cm}^3\,\text{mol}^{-1}$$

$$V_{m\,(solid)} = \frac{23.0\,\text{g mol}^{-1}}{0.929\,\text{g cm}^{-3}} = 24.16\,\text{cm}^3\,\text{mol}^{-1}$$

$$\begin{aligned}\Delta_{fus}V &= 24.76\,\text{cm}^3\,\text{mol}^{-1} - 24.16\,\text{cm}^3\,\text{mol}^{-1}\\ &= +0.60\,\text{cm}^3\,\text{mol}^{-1}\\ &= +0.60\times10^{-6}\,\text{m}^3\,\text{mol}^{-1} \quad (1\,\text{cm}^3 = 1\times10^{-6}\,\text{m}^3)\end{aligned}$$

→

→ *Use the Clapeyron equation to find how melting temperature varies with pressure.*

$$\frac{dp}{dT} = \frac{\Delta_{fus}H}{T\Delta_{fus}V} \qquad (17.5)$$

The melting point at 1 atm = T_m = 97.8 °C = 371.0 K

$$\frac{dp}{dT} = \frac{+3.00 \times 10^3\ \text{J}\ \cancel{\text{mol}^{-1}}}{(371.0\ \text{K}) \times (+0.60 \times 10^{-6}\ \text{m}^3\ \cancel{\text{mol}^{-1}})} = 1.35 \times 10^7\ \text{J m}^{-3}\ \text{K}^{-1}$$

$$= 1.35 \times 10^7\ \text{Pa K}^{-1}\ (1\ \text{J m}^{-3} = 1 (\text{N m})\ \text{m}^{-3} = 1\ \text{N m}^{-2} = 1\ \text{Pa})$$

Therefore, a change of pressure of 1.35×10^7 Pa changes the melting temperature by 1 K. (Note that this assumes a linear relationship between p and T. This is an approximation but holds very well for most substances.)

Use the given data to find the change in pressure and hence the change in melting temperature.

The melting temperature at 1.00 atm is given and needs to be found at 120 atm. The change in pressure is (120 atm − 1.00 atm) = 119 atm. Since 1 atm = 1.01×10^5 Pa

$$119\ \text{atm} = (119 \times 1.01 \times 10^5)\ \text{Pa} = 1.20 \times 10^7\ \text{Pa}$$

A pressure difference of 1.35×10^7 Pa changes the melting temperature by 1.0 K, so a pressure difference of 1.20×10^7 Pa will cause the melting temperature to change by

$$\Delta T = \frac{1.20 \times 10^7\ \cancel{\text{Pa}}}{1.35 \times 10^7\ \cancel{\text{Pa}}} \times 1.0\ \text{K} = 0.89\ \text{K}$$

Find the new melting temperature.

At 1 atm, sodium melts at 371.0 K. At 120 atm, this changes by 0.89 K. So

$$\begin{aligned}\text{melting temperature at 120 atm} &= 371.0\ \text{K} + 0.9\ \text{K} \\ &= 371.9\ \text{K} \\ &= 372\ \text{K}\end{aligned}$$

Question

Calculate the pressure at which the melting temperature of sodium is 373 K.

Sublimation and vaporization

For phase changes involving gases or vapours, the Clapeyron equation can be modified as shown in Box 17.3 to give the **Clausius–Clapeyron equation**

$$\ln p = \text{constant} - \frac{\Delta H}{R}\left(\frac{1}{T}\right) \qquad (17.6)$$

where p is the vapour pressure and ΔH is either $\Delta_{vap}H$ or $\Delta_{sub}H$.

If the vapour pressures, p_1 and p_2, are known at two temperatures, T_1 and T_2, then an alternative form of Equation 17.6 is more useful.

Enthalpy change of transition

$$\ln\frac{p_2}{p_1} = \frac{\Delta H}{R}\left(\frac{1}{T_1} - \frac{1}{T_2}\right) \qquad (17.7)$$

Pressures

Transition temperatures

The Clausius–Clapeyron equation relates pressure and temperature for phase transitions of vaporization or sublimation, so ΔH is $\Delta_{vap}H$ or $\Delta_{sub}H$.

Water boils at a higher temperature under higher pressure. Conversely, water boils at a lower temperature under lower pressure. This has to be taken into account when cooking at high altitudes where the atmospheric pressure is low.

ⓘ A similar approach is used in the treatments of the temperature variation of rate constants in the Arrhenius equation in Section 9.7 (p.428) and the temperature variation of equilibrium constants in the van 't Hoff equation in Section 15.5 (p.716).

Equation 17.7 is particularly useful if the normal boiling point, T_b, of a substance is known. At this temperature, the vapour pressure is 1 atm (1.013×10^5 Pa). The boiling temperature, T, at a different pressure, p, can be found from

$$\ln\left(\frac{1.013 \times 10^5\,\text{Pa}}{p}\right) = \frac{\Delta_{\text{vap}}H}{R}\left(\frac{1}{T} - \frac{1}{T_b}\right) \quad (17.8)$$

The use of Equation 17.8 is illustrated in Worked example 17.3.

The **normal boiling point**, T_b, is defined as the temperature at which the vapour pressure of a substance is 1 atm (1.013×10^5 Pa; see Section 1.7, p.47).

IUPAC now recommends the use of **standard boiling point**, $T_{b(s)}$, defined as the temperature at which the vapour pressure becomes 1 bar. $T_{b(s)}$ is slightly smaller than T_b. For water, T_b is 99.97 °C at 1 atm, whereas $T_{b(s)}$ is 99.61 °C at 1 bar.

Box 17.3 Deriving the Clausius–Clapeyron equation

The Clapeyron equation, Equation 17.4, derived in Box 17.2

$$\frac{dp}{dT} = \frac{\Delta S_m}{\Delta V_m} = \frac{\Delta H}{T\Delta V_m} \quad (17.4)$$

applies to all phase transitions. For solid–liquid transitions, the use of the equation can be simplified by assuming that the molar entropy, S_m, does not change very much with temperature and that the molar volume, V_m, does not change significantly with pressure. With gases and vapours, the latter is true only over a very small range of pressures. To use the equation with gases, it needs to be modified. Three approximations can be introduced.

- The molar volume, V_m, of a solid or liquid is insignificant compared with that of a gas (see Section 8.2, p.354), so ΔV_m is approximately equal to $V_{m(gas)}$.
- The gas or vapour acts as an ideal gas so that, for 1 mol, $V_{m(gas)} = \frac{RT}{p}$ (see Section 8.1, p.349).
- $\Delta_{vap}H$ and $\Delta_{sub}H$ do not vary significantly over the temperature range being considered.

Introducing these approximations into Equation 17.4 leads to

$$\frac{dp}{dT} = \frac{\Delta H}{TV_{m(gas)}} = \frac{\Delta H}{T\left(\frac{RT}{p}\right)} = \frac{p\Delta H}{RT^2}$$

Rearranging to separate the variables so that p is on one side of the equation and T is on the other

$$\frac{dp}{p} = \frac{\Delta H dT}{RT^2} = \left(\frac{\Delta H}{R}\right)\frac{dT}{T^2} \quad (17.9)$$

To solve this differential equation and obtain a relationship between p and T, integrate both sides of the equation.

(*Note.* Go to Maths Toolkit MT7 (p.1327) if you need help with carrying out the integration.)

Integrating the two sides of the equation separately

$$\int \frac{dp}{p} = \ln p + \text{constant}$$

and $\int \frac{dT}{T^2} = \int T^{-2}\,dT = -T^{-1} + \text{constant} = -\left(\frac{1}{T}\right) + \text{constant}$

$$\text{So } \frac{\Delta H}{R}\int \frac{dT}{T^2} = \text{constant} - \frac{\Delta H}{R}\left(\frac{1}{T}\right)$$

The result of integrating Equation 17.9 is then

$$\ln p = \text{constant} - \frac{\Delta H}{R}\left(\frac{1}{T}\right) \quad (17.6)$$

If the vapour pressures are p_1 at temperature T_1 and p_2 at T_2, then integrating between these limits

$$\int_{p_1}^{p_2} \frac{dp}{p} = \frac{\Delta H}{R}\int_{T_1}^{T_2} \frac{dT}{T^2}$$

gives

$$[\ln p]_{p_1}^{p_2} = \frac{\Delta H}{R}\left[\frac{1}{T}\right]_{T_1}^{T_2}$$

So

$$(\ln p_2 - \ln p_1) = \frac{\Delta H}{R}\left(\frac{1}{T_1} - \frac{1}{T_2}\right)$$

Using the expression $\ln\frac{a}{b} = \ln a - \ln b$ (Maths Toolkit MT3, p.1316), the equation can be rewritten as

$$\ln\frac{p_2}{p_1} = \frac{\Delta H}{R}\left(\frac{1}{T_1} - \frac{1}{T_2}\right) \quad (17.7)$$

Equations 17.6 and 17.7 are versions of the **Clausius–Clapeyron equation**. This equation is used for vaporization and sublimation phase changes, so ΔH is $\Delta_{vap}H$ or $\Delta_{sub}H$.

Worked example 17.3 Using the Clausius–Clapeyron equation

The normal boiling point, T_b, of benzene is 353.3 K (80.1 °C) and its enthalpy change of vaporization is +30.8 kJ mol^{-1}. Calculate the boiling temperature of benzene at a pressure of 5.00 kPa.

Strategy

The vapour pressure is 1 atm (101.3 kPa) at the normal boiling point, T_b = 353.3 K, so you can use Equation 17.8 directly to find T since $\Delta_{vap}H$ is given.

Solution

Substitute the given data into Equation 17.8.

$$\ln\left(\frac{1.013\times10^5\,\text{Pa}}{p}\right)=\frac{\Delta_{vap}H}{R}\left(\frac{1}{T_1}-\frac{1}{T_b}\right)\qquad(17.8)$$

where $R = 8.314\,\text{J K}^{-1}\,\text{mol}^{-1}$.

$$\ln\left(\frac{101.3\,\cancel{\text{kPa}}}{5.00\,\cancel{\text{kPa}}}\right)=\frac{+30.8\times10^3\,\cancel{\text{J}}\,\cancel{\text{mol}^{-1}}}{8.314\,\cancel{\text{J}}\,\text{K}^{-1}\,\cancel{\text{mol}^{-1}}}\left(\frac{1}{T}-\frac{1}{353.3\,\text{K}}\right)$$

$$\ln\left(\frac{101.3}{5.00}\right)=\frac{+30.8\times10^3}{8.314\,\text{K}^{-1}}\left(\frac{1}{T}-\frac{1}{353.3\,\text{K}}\right)$$

Solve to find the new boiling temperature.

Dividing the terms

$$3.01 = 3705\,\text{K}\times\left(\frac{1}{T}-0.002833\,\text{K}^{-1}\right)$$

Multiplying out the brackets

$$3.01=\frac{3705\,\text{K}}{T}-10.50$$

Adding 10.50 to both sides

$$13.51=\frac{3705\,\text{K}}{T}$$

Rearranging

$$T=\frac{3705\,\text{K}}{13.51}=274\,\text{K}$$

Question

Find the vapour pressure of benzene at 325 K.

Enthalpy change of vaporization and entropy change of vaporization

Values for the enthalpy change and entropy change of vaporization for a number of substances are shown in Table 17.1. The enthalpy change of vaporization is related to the

Enthalpy change of vaporization is discussed in Section 13.2 (p.619).

Table 17.1 Normal boiling points and standard enthalpy changes and standard entropy changes of vaporization

	b.p./°C	$\Delta_{vap}H^{\ominus}$/kJ mol^{-1}	$\Delta_{vap}S^{\ominus}$/J K^{-1} mol^{-1}
Benzene (C_6H_6)	80.1	+30.8	+87.2
Carbon disulfide (CS_2)	46.3	+26.7	+83.7
Hydrogen sulfide (H_2S)	−60.4	+18.7	+87.9
Tetrachloromethane (CCl_4)	76.7	+30.0	+85.8
Cyclohexane (C_6H_{12})	80.7	+30.1	+85.1
Decane ($C_{10}H_{22}$)	174	+38.8	+86.7
Methoxymethane ($(CH_3)_2O$)	−23.0	+21.5	+86.0
Methanol (CH_3OH)	65.0	+35.2	+104.1
Ethanol (C_2H_5OH)	78.3	+38.6	+110.0
Water (H_2O)	100.0	+40.7	+109.1
Butane (C_4H_{10})	−0.50	+22.4	+82.1
Methylbenzene ($C_6H_5CH_3$)	110.7	+33.5	+87.2
Ethanoic acid (CH_3COOH)	118.3	+24.4	+61.9

energy needed to overcome the intermolecular interactions for 1 mol of liquid. It is, therefore, a good measure of the strength of these interactions.

Many of the liquids in Table 17.1 have values of $\Delta_{vap}S^{\ominus}$ between $+80\,J\,K^{-1}\,mol^{-1}$ and $+90\,J\,K^{-1}\,mol^{-1}$. This is summarized in an empirical statement known as **Trouton's rule**

An empirical rule is based simply on experimental observations rather than theoretical predictions.

$$\text{Trouton's rule:}\quad \Delta_{vap}S^{\ominus} = +85\,J\,K^{-1}\,mol^{-1}$$

Even liquids with very different boiling points and chemical natures have similar values of $\Delta_{vap}S$—compare, for example, the values for benzene, methoxymethane, and hydrogen sulfide. Why should this be?

Intermolecular interactions are discussed in Section 17.3 (p.783).

During vaporization, a liquid turns into a vapour. Provided that there are no very strong interactions between the molecules, such as hydrogen bonding, most liquids have similar *arrangements* of molecules. The vapour phase is highly random and disordered, no matter what substance is involved. Therefore, the same degree of change in molecular order, and hence in entropy, is to be expected as a result of vaporization, whatever the substance, provided there are no strong interactions between the molecules.

Deviations from Trouton's rule result from unusually strong interactions between the molecules in the liquid or vapour. Water has a relatively ordered liquid structure due to hydrogen bonding, which is not the case in water vapour. Therefore, there is a greater than usual increase in disorder on vaporization of water so $\Delta_{vap}S$ ($+109.1\,J\,K^{-1}\,mol^{-1}$) is higher than predicted by Trouton's rule. In contrast, ethanoic acid shows a smaller $\Delta_{vap}S$ ($+61.9\,J\,K^{-1}\,mol^{-1}$) than would be predicted. Pure ethanoic acid is a molecular liquid that does not ionize to a great extent. The vapour contains a high proportion of dimers consisting of two molecules held together by hydrogen bonds (Figure 17.10). The gas phase of ethanoic acid is therefore more ordered than for most gases so that there is a lower change of entropy on vaporization. These differences are summarized in Figure 17.11.

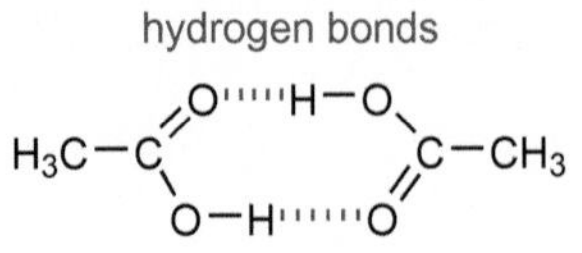
hydrogen bonds
H3C
CH3

Figure 17.10 In the vapour phase, ethanoic acid forms dimers held together by hydrogen bonds.

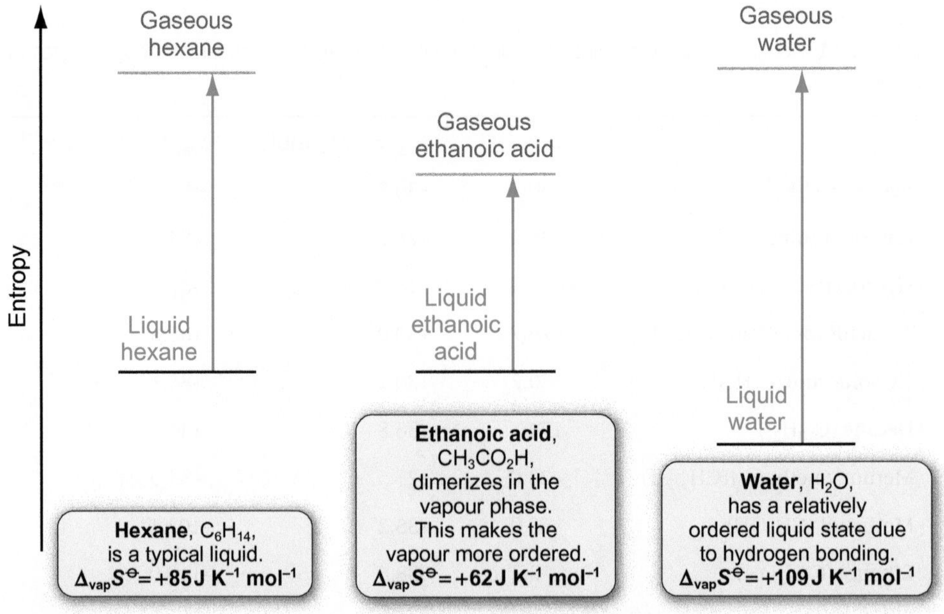
Entropy
Gaseous hexane
Liquid hexane
Gaseous ethanoic acid
Liquid ethanoic acid
Gaseous water
Liquid water
Hexane, C6H14, is a typical liquid. ΔvapS⊖ = +85 J K−1 mol−1
Ethanoic acid, CH3CO2H, dimerizes in the vapour phase. This makes the vapour more ordered. ΔvapS⊖ = +62 J K−1 mol−1
Water, H2O, has a relatively ordered liquid state due to hydrogen bonding. ΔvapS⊖ = +109 J K−1 mol−1

Figure 17.11 The size of the standard entropy change on vaporization depends on the relative degrees of order in the liquid and vapour. Most liquids act like hexane. Some substances, such as ethanoic acid, can dimerize in the vapour phase so that the vapour is more ordered than an ideal gas and there is a smaller change in entropy on vaporization. Conversely, some liquids (such as water) have a more ordered structure so there is a larger change in entropy on vaporization.

Summary

- The phase with the lowest Gibbs energy at a particular temperature and pressure is the most stable.
- At a phase transition temperature, $\Delta S = \dfrac{\Delta H}{T}$.
- The gradients of the lines on the phase diagrams are given by the Clapeyron equation

$$\frac{dp}{dT} = \frac{\Delta S_m}{\Delta V_m} = \frac{\Delta H}{T\Delta V_m}$$

- The Clapeyron equation applies to *any* phase transition. When applied to the melting/freezing transition, it is written as

$$\frac{dp}{dT} = \frac{\Delta_{fus}H}{T\Delta_{fus}V}$$

  where T is the melting temperature.

- The Clapeyron equation can be modified for transitions involving gases and vapours. The variation of vapour pressure with temperature is then described by the Clausius–Clapeyron equation

$$\ln p = \text{constant} - \frac{\Delta H}{T}\left(\frac{1}{T}\right)$$

  where ΔH is $\Delta_{vap}H$ or $\Delta_{sub}H$.

- A useful form of the Clausius–Clapeyron equation is

$$\ln\frac{p_2}{p_1} = \frac{\Delta H}{T}\left(\frac{1}{T_1} - \frac{1}{T_2}\right)$$

- The normal boiling point, T_b, is defined as the temperature at which the vapour pressure of a substance is 1 atm (1.013×10^5 Pa).
- The standard boiling point, $T_{b(s)}$, is defined as the temperature at which the vapour pressure of a substance is 1 bar (1.0×10^5 Pa).
- Trouton's rule says that many substances have an entropy change of vaporization $\Delta_{vap}S^{\ominus} \approx +85\ \mathrm{J\,K^{-1}\,mol^{-1}}$. Exceptions to this rule occur when molecular interactions are unusually strong in the liquid or the vapour.

? For practice questions on these topics, see questions 5–14 at the end of this chapter (pp.810–811).

17.3 Intermolecular interactions

The phase in which a substance exists at a particular temperature and pressure is largely governed by the interactions between its molecules. (Here the word 'molecule' is used to include molecules, atoms, and ions.)

In gases, interactions between molecules are small except at high pressures. Solids and liquids (*condensed phases*) are held together by attractive interactions between the molecules. There are also repulsive interactions due to the overlapping of electron clouds, and the balance between attractive and repulsive interactions keeps the molecules at equilibrium separation.

An introduction to the origin of intermolecular interactions is in Section 1.8 (p.52).

In describing interactions between molecules, the term *intermolecular forces* is often used. However, chemists are also interested in the *energy* of the interactions. The force, F,

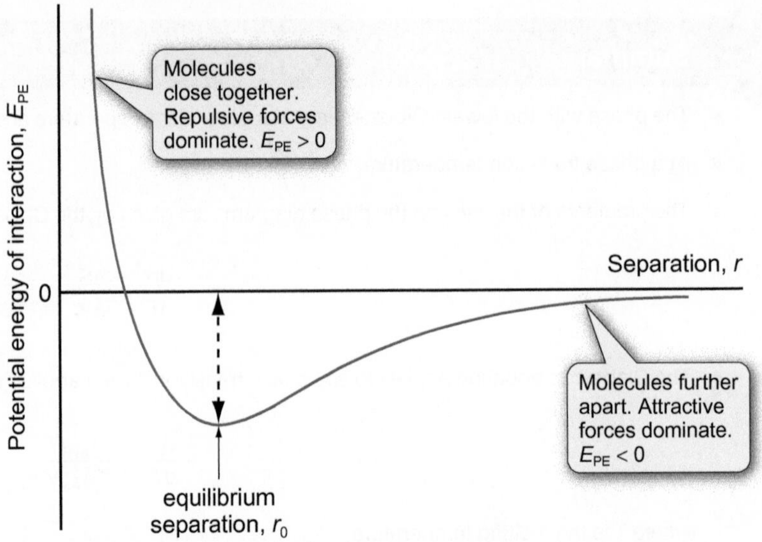
Molecules close together. Repulsive forces dominate. $E_{PE} > 0$
Potential energy of interaction, E_{PE}
0
Separation, r
Molecules further apart. Attractive forces dominate. $E_{PE} < 0$
equilibrium separation, r_0

Figure 17.12 When two molecules are far apart, E_{PE} is zero. As they approach one another, attractive interactions lower the potential energy and E_{PE} becomes increasingly negative. At the equilibrium separation, r_0, the energy of interaction is at its minimum.

Potential energy, E_{PE}, is discussed in Section 1.6 (p.42).

	Value of F	Value of E_{PE}
Attraction:	Positive	Negative
Repulsion:	Negative	Positive

The shape of Figure 17.12 is explained in Figure 17.15 (p.787).

between the two molecules and the potential energy, E_{PE}, of the interaction both vary with the distance between the molecules, r. This is described by

$$F = -\frac{dE_{PE}}{dr} \tag{17.10}$$

When there are attractive interactions (positive forces between the molecules), the potential energy becomes more negative as the distance between the molecules decreases. When there are repulsive interactions (negative forces between the molecules), the potential energy becomes more positive as the molecules move closer together. This is illustrated in Figure 17.12.

Ionic interactions

The interaction between ions is used to calculate the lattice energy of an ionic solid in Section 6.6. Compare Equation 17.11 with Equation 6.4 on p.291.

The strongest interactions occur in ionic systems. The energy of interaction, E_{PE}, between two ions with charges q_1 and q_2 separated by a distance r is

$$E_{PE}\,(\text{ionic}) = \frac{q_1 q_2}{4\pi\varepsilon r} \tag{17.11}$$

where ε is the **permittivity** of the medium surrounding the ions.

The energy of interaction differs according to whether the two ions are in a vacuum, in air, in water, or in a hydrocarbon solvent. The permittivity accounts for this difference. It is usually expressed as

$$\varepsilon = \varepsilon_0 \varepsilon_r$$

An older name for the relative permittivity, ε_r, is the dielectric constant.

where ε_0 is the **permittivity of a vacuum** ($\varepsilon_0 = 8.85 \times 10^{-12}\ \text{C}^2\ \text{J}^{-1}\ \text{m}^{-1}$) and ε_r is the **relative permittivity**. ε_r for water at 25 °C is 78 so the energy of interaction between two ions in water is 78 times less than if the two ions were in a vacuum.

If the ions have opposite charges, q_1 and q_2 have opposite signs, and the potential energy, E_{PE}, is negative, meaning that there is *attraction* between the ions; E_{PE} becomes more negative as the ions come closer together. The energy becomes less negative as the distance between the ions increases, but the $(1/r)$ dependence shows that ions interact even at quite long distances compared with the size of the ions.

Consider two ions, one with +1 charge and one with −1 charge, at a distance of 0.2 nm apart in a vacuum. The charge on an electron, $e = 1.602 \times 10^{-19}$ C, so $q_1 = +1.062 \times 10^{-19}$ C, and $q_2 = -1.062 \times 10^{-19}$ C. From Equation 17.11

$$E_{PE}(\text{ionic}) = \frac{(+1.602 \times 10^{-19}\,\text{C}) \times (-1.602 \times 10^{-19}\,\text{C})}{4 \times 3.142 \times (8.85 \times 10^{-12}\,\text{C}^2\,\text{J}^{-1}\,\text{m}^{-1}) \times (2 \times 10^{-10}\,\text{m})}$$
$$= -1.15 \times 10^{-18}\,\text{J}$$

Multiplying this value by the Avogadro constant shows that this energy is equivalent to $-695\,\text{kJ mol}^{-1}$. Remember that this calculation is for just two ions in a vacuum. In reality, the energy would be reduced by the relative permittivity of the medium. For example, in water the energy would be reduced by a factor of 78, to $-8.9\,\text{kJ mol}^{-1}$. Furthermore, each ion would also interact with all others surrounding it. Nonetheless, the value is useful in comparing with interaction energies in non-ionic systems.

Non-covalent interactions are introduced in Section 1.8 (p.52). The different types of non-covalent interactions are summarized in Table 1.10 (p.53).

Non-covalent interactions

Interactions that occur between molecules that have no overall charge are known as **non-covalent interactions**, to distinguish them from covalent bonds *within* molecules. They are also sometimes known as van der Waals interactions, but this term has a variety of meanings and it will not be used in this book.

Many neutral molecules contain dipoles arising from the uneven distribution of electronic charge in the molecule. The strength of the dipole is measured by the **dipole moment**, given the symbol μ and defined as in Equation 17.12 (see Figure 17.13)

$$\mu = q \times r \qquad (17.12)$$

Because the charges and distances in Equation 17.12 are small, the SI unit for dipole moments (C m) gives rather small values. For example, μ for HCl is $3.60 \times 10^{-30}\,\text{C m}$. Dipole moments are often reported using the debye, D, where

$$1\,\text{D} = 3.3356 \times 10^{-30}\,\text{C m}$$

Using this unit, HCl has a dipole moment of 1.08 D.

The Dutch physicist Johannes van der Waals developed an equation of state to explain the non-ideal behaviour of gases in terms of intermolecular attractions and the finite size of molecules (see Equation 8.25, p.376). In this context van der Waals interactions include *all* types of non-covalent interactions. Sometimes, however, you will see the term applied only to London dispersion interactions.

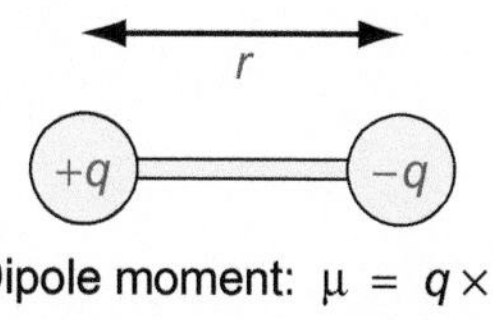
r
+q
−q
Dipole moment: μ = q × r

Figure 17.13 If a dipole has a charge of $+q$ at one end and $-q$ at the other, separated by a distance r, the dipole moment is defined as $\mu = q \times r$.

Dipole-dipole interactions

Two molecules that have dipoles can interact (see Figure 1.25, p.53). The larger the dipole moment, the stronger the interaction. The potential energy of interaction varies depending on the relative orientation of the molecules but the average interaction energy for two molecules with dipole moments μ_1 and μ_2 at temperature T is given by

$$E_{PE}(\text{dipole–dipole}) = -\frac{0.66(\mu_1\mu_2)^2}{(4\pi\varepsilon_0)^2 k_B T} \times \frac{1}{r^6} \qquad (17.13)$$

where k_B is the Boltzmann constant. The ε_0 term, the permittivity of a vacuum, is used because the molecules are considered to be interacting in a vacuum.

For example, the HCl molecule has a dipole moment, $\mu = 1.08\,\text{D}$. For two HCl molecules separated by 0.3 nm, Equation 17.13 gives an energy of interaction of $-3.0 \times 10^{-21}\,\text{J}$ at 25 °C. (Remember that a *negative* sign for the value of the interaction energy means an *attractive* interaction.) This energy corresponds to $-1.8\,\text{kJ mol}^{-1}$, which is much smaller in magnitude than the energy of covalent bonds or of ionic interactions.

The $1/r^6$ dependence in Equation 17.13 means that the potential energy approaches zero very quickly as the molecules get further apart, so that the interactions are only significant when the molecules are very close together (within a few molecular diameters). The $1/T$ dependence means that the *magnitude* of the interaction energy is smaller at higher temperatures. This is because, at higher temperatures, the molecules have more thermal energy which leads to a more random orientation of the dipoles.

Dipole moments are introduced in Section 1.8 (p.52). Bond polarity and polar molecules are discussed in Section 5.3 (p.235).

The Boltzmann constant, k_B, is introduced in Section 10.3 (p.462).

Dipole-induced dipole interactions

When a molecule with a dipole approaches a molecule without one, it *induces* a movement of the electrons causing a temporary dipole in the non-polar molecule, so there is an

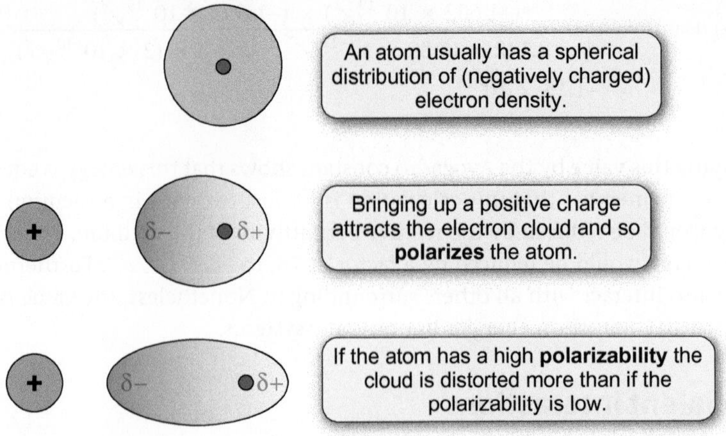
An atom usually has a spherical distribution of (negatively charged) electron density.
+
δ−
δ+
Bringing up a positive charge attracts the electron cloud and so **polarizes** the atom.
+
δ−
δ+
If the atom has a high **polarizability** the cloud is distorted more than if the polarizability is low.

Figure 17.14 Polarizability is a measure of how easily the electron cloud in an atom or molecule can be influenced by a nearby charge.

attractive interaction (see Section 1.8, p.52). The strength of the induced dipole depends on how easily the electron cloud of the non-polar molecule can be moved around. This is measured by the **polarizability** of the molecule, given the symbol, α. The higher the value of α, the more easily the electrons can be distorted by a nearby dipole or charge (see Figure 17.14).

The energy of interaction depends on the dipole moment of the polar molecule, μ_1, and the polarizability of the second molecule, α_2, and is given by

$$E_{PE}\,(\text{dipole–induced dipole}) = -\frac{\mu_1{}^2\alpha_2}{(4\pi\varepsilon_0)^2} \times \frac{1}{r^6} \qquad (17.14)$$

Taking HCl as the polar molecule and using a value of polarizability typical for molecules such as benzene or tetrachloromethane, Equation 17.14 gives a value of $E_{PE} = -1.4 \times 10^{-24}$ J when $r = 0.3$ nm. This energy corresponds to –0.8 kJ mol^{-1} and is smaller in magnitude than that for dipole–dipole interaction, as you would expect.

Induced dipole–induced dipole interactions

Non-polar molecules have attractive interactions due to the formation of 'instantaneous' induced dipoles (see Figure 1.24, p.52). These are known as **dispersion** or **London** interactions. A theoretical model for dispersion forces was worked out by the German physicist Fritz London, who showed that the strength of interaction depends on how easily the electrons in the molecules can be distorted and hence on their polarizability, α, as well as their ionization energy, I

> ℹ The ionization energy of a molecule, I, is the energy required to remove an electron, producing a positive ion. Ionization energies of atoms are discussed in Section 3.7 (p.154).

$$E_{PE}\,(\text{dispersion}) = -\frac{1.5\alpha_1\alpha_2}{(4\pi\varepsilon_0)^2} \times \left(\frac{I_1 I_2}{I_1 + I_2}\right) \times \frac{1}{r^6} \qquad (17.15)$$

Using the 0.3 nm separation between molecules as in previous examples, and with methane (CH_4) molecules as an example of a non-polar molecule, Equation 17.15 gives $E_{PE} = -8.30 \times 10^{-21}$ J, which corresponds to –5 kJ mol^{-1}.

While the magnitudes of dispersion energies are much smaller than those of chemical bonds, they are comparable with, and usually slightly larger than, dipole–dipole interactions. They are usually the major contribution to intermolecular interactions. Table 17.2 gives values for an HCl molecule.

Table 17.2 Comparing the energies of non-covalent interactions for two HCl molecules at a separation of 0.3 nm

Interaction	Energy / kJ mol^{-1}
Dipole–dipole	–1.8
Dipole–induced dipole	–0.3
Dispersion (London)	–9.9

Overall attractive interaction

The overall attractive interaction is the sum of those given by Equations 17.13, 17.14, and 17.15. For interactions between molecules of a single compound with dipole moment μ, polarizability α, ionization energy I, and intermolecular separation r, the terms can be collected together, giving

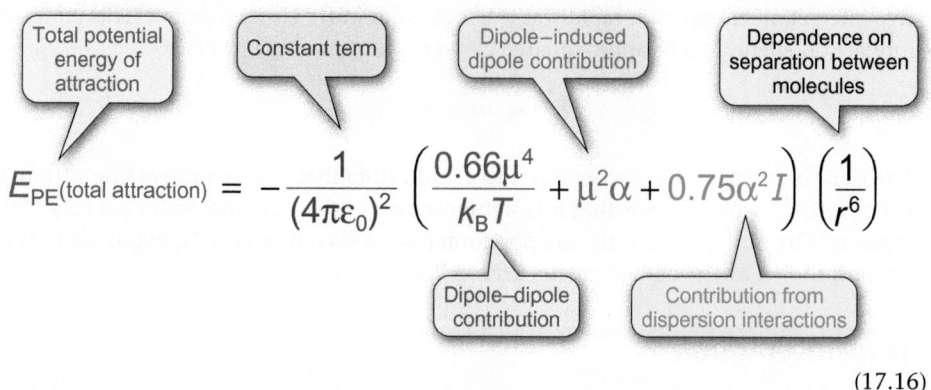

$$E_{PE}(\text{total attraction}) = -\frac{1}{(4\pi\varepsilon_0)^2}\left(\frac{0.66\mu^4}{k_B T} + \mu^2\alpha + 0.75\alpha^2 I\right)\left(\frac{1}{r^6}\right) \qquad (17.16)$$

Total potential energy of attraction — Constant term — Dipole–induced dipole contribution — Dependence on separation between molecules — Dipole–dipole contribution — Contribution from dispersion interactions

The $(0.75\alpha^2 I)$ term in Equation 17.16 arises if you put $\alpha_1 = \alpha_2$ and $I_1 = I_2$ in Equation 17.15, as in the case where the two molecules are identical.

All of the contributions to the attractive interactions depend on separation as $(1/r^6)$. This means that they are *short range* and their magnitudes rapidly decrease as the separation increases. They become very small once the separation is more than a few molecular diameters.

Using Equation 17.16 is straightforward since, at constant temperature, all the terms except r^6 are constant for a particular compound so that it can be simplified to

$$E_{PE}\,(\text{attraction}) = -\frac{A}{r^6} \qquad (17.17)$$

where A is a constant and represents all the attractive terms. The value of A is different for different compounds. A sketch of how E_{PE}(attraction) varies with intermolecular separation, r, is shown in Figure 17.15 (the dashed red curve).

One other category of attractive intermolecular interaction is the hydrogen bond between a hydrogen atom and an electronegative atom on a neighbouring molecule. These usually have energies in the range $-10\,\text{kJ mol}^{-1}$ to $-40\,\text{kJ mol}^{-1}$ and so are intermediate between ionic and non-covalent interactions.

The hydrogen bond is discussed in more detail in Sections 1.8 (p.54) and 25.3 (p.1156).

Repulsive interactions

The electron clouds of two adjacent molecules repel each other. The repulsive force is small until the molecules approach close to each other. At very small separations, the repul-

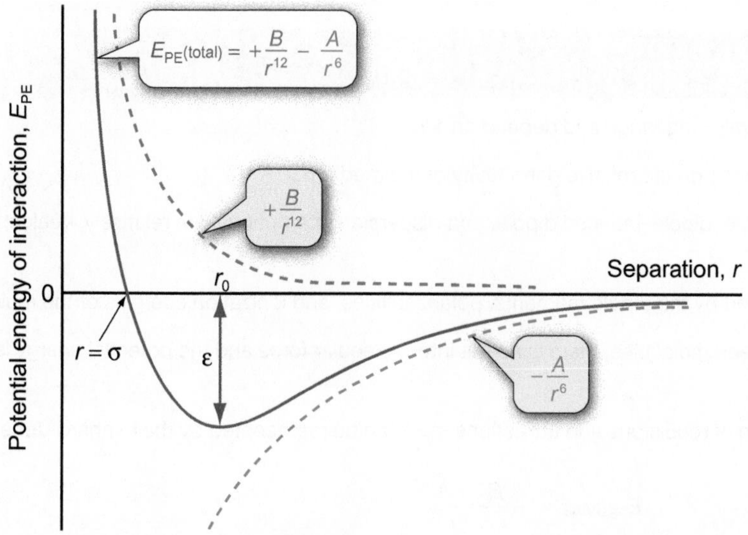

$E_{PE}(\text{total}) = +\frac{B}{r^{12}} - \frac{A}{r^6}$
$+\frac{B}{r^{12}}$
$-\frac{A}{r^6}$
Separation, r
r_0
0
$r = \sigma$
ε
Potential energy of interaction, E_{PE}

Figure 17.15 The overall interaction energy between two molecules (blue line) is called the Lennard-Jones potential. It is a combination of the attractive (dashed red line) and repulsive (dashed green) interactions.

sive interaction energy rises rapidly as the molecules move closer together. This repulsive interaction energy can be approximated by an expression of the form

$$E_{PE}\,(\text{repulsion}) = +\frac{B}{r^{12}}$$

The value of the potential energy is positive, indicating that it is a repulsive interaction. It varies as $1/r^{12}$, which means that it is significant only when the molecules are very close together. The $1/r^{12}$ term is only an approximation. A sketch of how E_{PE}(repulsion) varies with r is shown by the dashed green curve in Figure 17.15.

Total interaction

For two molecules, the overall interaction energy is the sum of the attractive and repulsive contributions

$$E_{PE}\,(\text{total}) = E_{PE}\,(\text{repulsion}) + E_{PE}\,(\text{attraction}) = +\frac{B}{r^{12}} - \frac{A}{r^6} \qquad (17.18)$$

Visit the Online Resource Centre to view screencast 17.2 which walks you through the idea of total interaction and the Lennard-Jones potential.

The Lennard-Jones potential is named after John Lennard-Jones who proposed this model in 1931. It is sometimes called the 'LJ-6,12 potential'.

Note that here the symbol ε represents an energy, not the permittivity as above.

The sum of the repulsive and attractive energy given by Equation 17.18 is called the **Lennard-Jones potential**. It is shown by the blue curve in Figure 17.15.

At large separations, there is no interaction between the molecules. As they approach, they begin to attract each other and the potential energy falls. When the molecules come very close, the repulsive interactions become dominant and the potential energy rises again and eventually becomes positive.

The minimum potential energy corresponds to the equilibrium separation between the molecules, r_0. At this minimum, E_{PE}(total) is equal to $-\varepsilon$, where ε is the energy needed to overcome the attractive interactions and completely separate the molecules. The larger the value of ε, the stronger the interaction between the molecules. The distance r_0 is the equilibrium separation where the net force between the molecules is zero, that is, $\frac{dE_{PE}}{dr} = 0$. At $r = \sigma$ (see Figure 17.15), the attractive energy is equal to the repulsive energy and $E_{PE} = 0$.

Using ε and σ, the Lennard-Jones potential becomes

$$E_{PE}\,(\text{total}) = 4\varepsilon\left\{\left(\frac{\sigma}{r}\right)^{12} - \left(\frac{\sigma}{r}\right)^{6}\right\} \qquad (17.19)$$

Summary

- Ionic interactions are strong and relatively long range and depend on $1/r$.
- The strength of ionic interactions depends on the relative permittivity of the medium.
- Non-covalent interactions (dipole–dipole, dipole–induced dipole, and dispersion interactions) are relatively weak and depend on $1/r^6$.
- Non-covalent interactions are influenced by the dipole moments, polarizabilities, and ionization energies of the molecules.
- At the equilibrium separation, r_0, between molecules, there is no net intermolecular force and the potential energy is at a minimum.
- Intermolecular interactions are the sum of repulsions and attractions, and can be represented by the Lennard-Jones potential

$$E_{PE}\,(\text{total}) = +\frac{B}{r^{12}} - \frac{A}{r^6}$$

For practice questions on these topics, see questions 15–20 at the end of this chapter (p.811).

17.4 Phase behaviour in two-component systems

How do systems with two components differ from single-component systems? There can be only one vapour phase no matter how many gases are present since gases mix in all proportions. Mixtures of two solids can, like single components, display any number of solid phases. The major difference is that, whereas pure substances only have one liquid phase, mixtures can have more than one. This is an everyday occurrence. For example, water and ethanol mix readily in all proportions: they are **miscible**; water and petrol hardly mix at all—in effect, they are **immiscible** in all proportions.

The Phase Rule

The Phase Rule, developed by Willard Gibbs, is a simple rule that helps you interpret phase diagrams. For a system at equilibrium

$$F = C - P + 2 \tag{17.20}$$

where C is the number of components, P is the number of phases, and F is the number of degrees of freedom. **Degrees of freedom** means the number of intensive variables (such as temperature and pressure) that can be varied independently without changing the number of phases in equilibrium.

The degrees of freedom of a system in phase equilibrium means the number of intensive variables that can be independently varied without changing the number of phases.

Here's an example. Look at the phase diagram for H_2O in Figure 17.5, page 769. H_2O is a pure, single substance, so it has only one component, $C = 1$. Now consider a point on the line AT. AT represents the equilibrium between solid and liquid, so on this line $P = 2$. Applying the Phase Rule (Equation 17.20) gives $F = C - P + 2 = 1 - 2 + 2 = 1$

So on the line AT there is only *one* degree of freedom. For example, if you fix the temperature, you have used up your one degree of freedom, so you cannot change the pressure without changing the number of phases. But in the region bounded by points A, T, and C there is only one phase: liquid. So $P = 1$, and $F = C - P + 2 = 1 - 1 + 2 = 2$

There are now *two* degrees of freedom—you can vary both temperature and pressure without changing the number of phases.

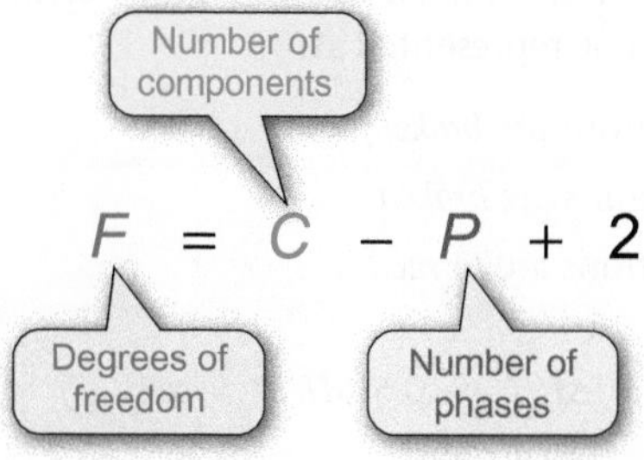
Number of components
F = C − P + 2
Degrees of freedom
Number of phases
(17.20)

The Phase Rule

Liquid-liquid phase behaviour

What determines whether two liquids will mix? Mixing is spontaneous if it results in a lowering of the Gibbs energy of the system. When two liquids A and B mix to form a solution, the change in Gibbs energy, $\Delta_{mix}G$, is the difference between the molar Gibbs energy of the solution and that of the separate components.

In this section, discussion is confined to solutions of non-electrolytes, which contain no ions. The behaviour of electrolyte solutions containing ions is described in Section 16.2 (p.730).

$$\begin{aligned}\Delta_{mix}G &= G_m(\text{solution}) - G_m(\text{pure components})\\ &= G_m(\text{solution}) - [x_A G_m(A) + x_B G_m(B)]\end{aligned} \tag{17.21}$$

The composition of the mixture is represented by the mole fraction of each component, x_A and x_B.

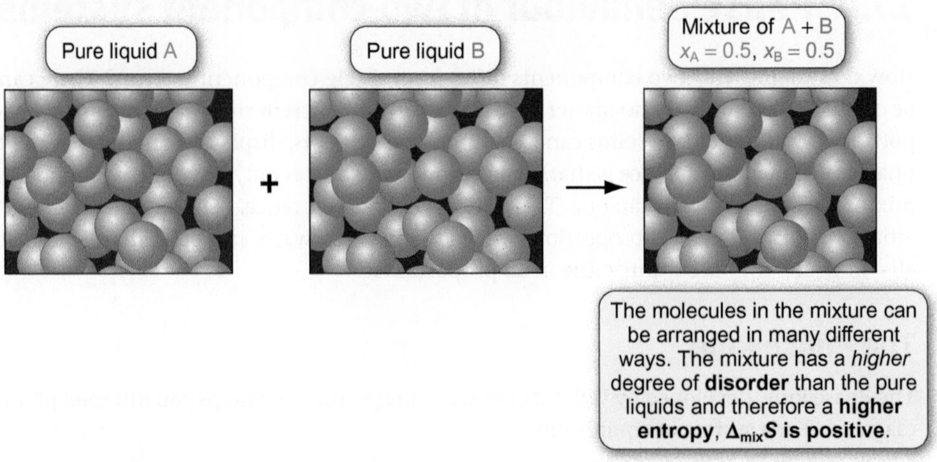
Pure liquid A
Pure liquid B
Mixture of A + B
$x_A = 0.5$, $x_B = 0.5$
The molecules in the mixture can be arranged in many different ways. The mixture has a *higher* degree of **disorder** than the pure liquids and therefore a **higher entropy, $\Delta_{mix}S$ is positive.**

Figure 17.16 A solution usually has higher entropy than its constituent liquids so that the value of the entropy change of mixing is positive.

The mole fraction of a component is defined as

$$x_A = \frac{\text{number of moles of A}}{\text{total number of moles present}} = \frac{n_A}{n_{total}}$$

(see Section 8.3, p.356).

The value of $\Delta_{mix}S$ is positive for mixtures of most compounds. However, it can be negative if the two compounds have unusually strong interactions, such as hydrogen bonding, that result in the mixture having a higher degree of order than the pure components. The entropy contribution is then unfavourable to mixing but this is almost always overcome by a large negative value for $\Delta_{mix}H$ associated with the strong interactions. This commonly occurs with solutions in water.

From the definition of Gibbs energy (Equation 14.16, p.675), $\Delta_{mix}G$ can be written as

$$\Delta_{mix}G = \Delta_{mix}H - T\Delta_{mix}S \tag{17.22}$$

The value of $\Delta_{mix}S$ is usually positive since the mixture will be more disordered and will have a higher entropy than the separate components (see Figure 17.16). The ($-T\Delta_{mix}S$) term in Equation 17.22 will, therefore, usually be negative and so make a negative contribution to the change in Gibbs energy. The *entropy* change favours mixing: whether two liquids actually mix depends on the *enthalpy* change of mixing.
The enthalpy change on mixing, $\Delta_{mix}H$, arises from differences in the intermolecular interactions between molecules in each of the pure liquids and between the components in solution.

If 0.5 mol each of A and B are mixed to form 1 mol of liquid mixture, the enthalpy changes that take place can be represented as:

- 0.5 mol: A–A interactions are *broken*;
- 0.5 mol: B–B interactions are *broken*;
- 1 mol: A–B interactions are *formed*.

$$\Delta_{mix}H = \Delta H_{(A-B)} - 0.5(\Delta H_{(A-A)} + \Delta H_{(B-B)}) \tag{17.23}$$

This is illustrated in Figure 17.17.

Equation 17.23 shows that if the A–B interactions in the mixture are stronger than the A–A and B–B interactions, then the value of $\Delta_{mix}H$ will be negative. If the interactions are weaker in the mixture than in the pure components, then $\Delta_{mix}H$ will be positive. The former usually occurs where there is an interaction, such as hydrogen bonding, between the components in the mixture that does not occur in the pure components.

For example, propanone and trichloromethane form hydrogen bonds when mixed (Figure 17.18), although there is no hydrogen bonding in the pure components.

What effect does this have on the phase behaviour? If the value of $\Delta_{mix}H$ is negative, it reinforces the negative $-T\Delta_{mix}S$ contribution to $\Delta_{mix}G$, so the mixing of the two liquids is spontaneous.

If the value of $\Delta_{mix}H$ is positive, then it makes an unfavourable contribution to $\Delta_{mix}G$. If $\Delta_{mix}H$ is large and positive (arising from much weaker interactions in the mixture) then its

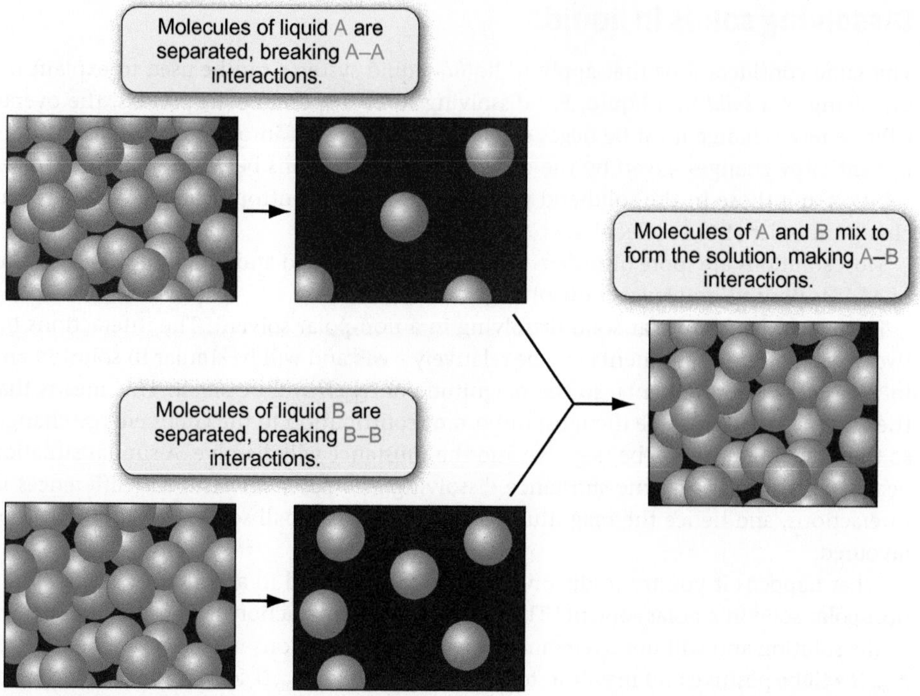
Molecules of liquid A are separated, breaking A–A interactions.
Molecules of A and B mix to form the solution, making A–B interactions.
Molecules of liquid B are separated, breaking B–B interactions.

Figure 17.17 Forming a solution can be thought of as separating the molecules in the pure liquids followed by mixing them to form a solution.

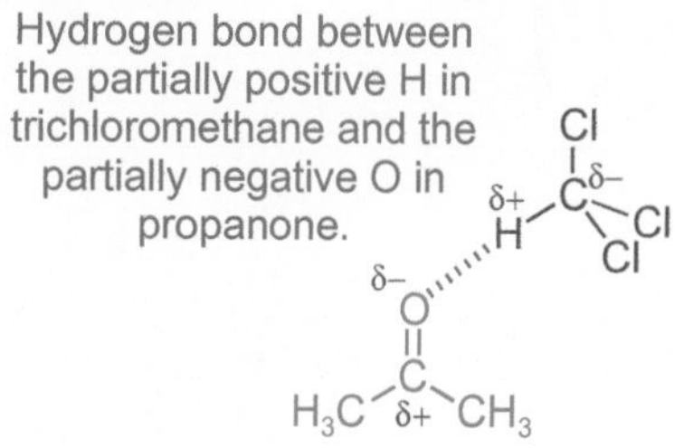
Hydrogen bond between the partially positive H in trichloromethane and the partially negative O in propanone.
Cl
δ+
δ–
H
C
Cl
Cl
δ–
O
C
H3C
δ+
CH3

Figure 17.18 Hydrogen bonding in mixtures of propanone and trichloromethane. The three chlorines in trichloromethane withdraw electrons resulting in polarization of the C–H bond with hydrogen slightly positive. The carbonyl group in propanone is polarized so that the oxygen atom is the negative end of a dipole.

magnitude may be greater than that of $-T\Delta_{mix}S$. In this case, $\Delta_{mix}G$ will be positive so the separate components will be thermodynamically more stable than the mixture—they do not mix spontaneously. However, if the value of $\Delta_{mix}H$ is small and positive, mixing may occur if the magnitude of $\Delta_{mix}H$ is less than that of $-T\Delta_{mix}S$.

The value of $\Delta_{mix}G$ changes with temperature since the $T\Delta_{mix}S$ term becomes more significant at high temperatures. Two liquids that do not mix at low temperatures may mix on heating. The thermodynamic contributions to miscibility are summarized in Table 17.3.

Phase behaviour in solutions can get more complicated than this. Since the enthalpy change depends on the relative amounts of components involved, it can be small for some compositions but high for others. Hence, some compounds mix in some compositions but not others—they are said to be **partially miscible**. Also, the enthalpy change can be temperature dependent. For example, an equimolar mixture of nicotine (Figure 17.19) and water forms a solution at room temperature. On heating, it separates into two phases around 60 °C but mixes again on further heating.

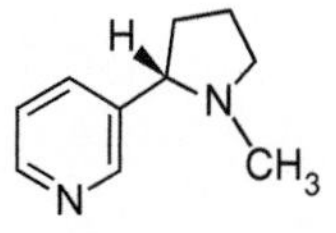
H
N
CH3
N

Figure 17.19 The structure of nicotine.

Table 17.3 Contributions to liquid–liquid miscibility

$-T\Delta_{mix}S$	Interactions	$\Delta_{mix}H$	$\Delta_{mix}G$	
Negative	A–B stronger than A–A, B–B	Negative	Negative, large magnitude	Components mix
Negative	A–B much weaker than A–A, B–B	Positive, large	Positive, large	Components do not mix (but may do so at high temperatures)
Negative	A–B weaker than A–A, B–B	Positive, small	Negative, small magnitude	Components may mix, depending on temperature and composition

Dissolving solids in liquids

The same considerations that apply to liquid–liquid systems can be used to explain the dissolving of a *solid* in a liquid. For dissolving to be a spontaneous process, the overall Gibbs energy change must be negative and this can be split into the contributions from the enthalpy change—given by the difference in interactions between molecules in the solution and those in the solid and pure liquid—and the entropy change related to the change in order of the molecules.

Most solutions are more disordered than the separate solid and liquid so that the value of $\Delta_{mix}S$ is positive and favours dissolving.

Think about a non-polar solid dissolving in a non-polar solvent. The interactions between each of the components will be relatively weak and will be similar in solution and in the separate components, so the magnitude of $\Delta_{mix}H$ will be small. This means that the entropy change will be the most important contribution to the Gibbs energy change, so the value of $\Delta_{mix}G$ will be negative and the substance will dissolve. A similar situation occurs with a polar or ionic substance dissolving in a polar solvent—the differences in interactions, and hence the magnitude of $\Delta_{mix}H$, will be small so that dissolving will be favoured.

The so-called 'like dissolves like' principle says that non-polar substances dissolve in non-polar solvents and polar substances dissolve in polar solvents. But there are many exceptions to this principle.

What happens if you try to dissolve a polar or ionic solid in a non-polar solvent, or a non-polar solid in a polar solvent? The solute–solvent interactions will be relatively weak in the solution and will not overcome the stronger interactions in the solid. The value of $\Delta_{mix}H$ will be positive and may lead to an overall positive $\Delta_{mix}G$ so that the solid does not dissolve.

However, there are many exceptions to the 'like dissolves like' rule. For example, calcium carbonate ($CaCO_3$) is an ionic solid but is only sparingly soluble in water (see Box 15.1, p.701). In this case, the value of $\Delta_{mix}S$ is negative. The clustering of water molecules due to their interaction with the highly charged Ca^{2+} and CO_3^{2-} ions in solution leads to a significant increase in order in the solution.

The thermodynamics of dissolving Group 1 salts in water is discussed in Section 26.3 (p.1180). The entropy change, $\Delta_{mix}S$, is important when considering the pK_a values of weak acids (see margin note in Section 19.2, p.889).

Vapour-liquid equilibrium in mixtures of liquids

The term 'solution' is used to describe a mixture of liquids as well as a mixture of a solid in a liquid.

The vapour pressure of a liquid is related to the intermolecular forces in the liquid (see Section 17.3, p.783). If the molecules strongly attract each other, few molecules escape the liquid to enter the vapour phase so the liquid exerts a low vapour pressure. How can this model be related to mixtures of liquids?

Working toward the end of the nineteenth century in France, Francois Raoult measured the vapour pressures of a large number of compounds and their solutions. He found that, for many solutions, the vapour pressure behaved as shown in Figure 17.20(a). The vapour pressure of each component, p, was proportional to its mole fraction, x. For a solution with two components A and B

Vapour pressure of A in mixture

Vapour pressure of pure A

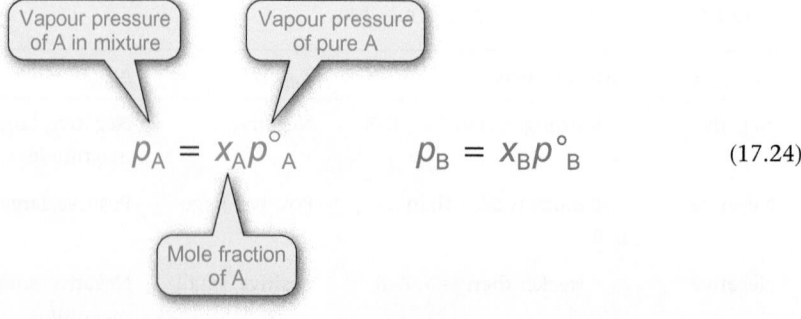

$$p_A = x_A p^{\circ}_A \qquad p_B = x_B p^{\circ}_B \tag{17.24}$$

Mole fraction of A

Raoult's law

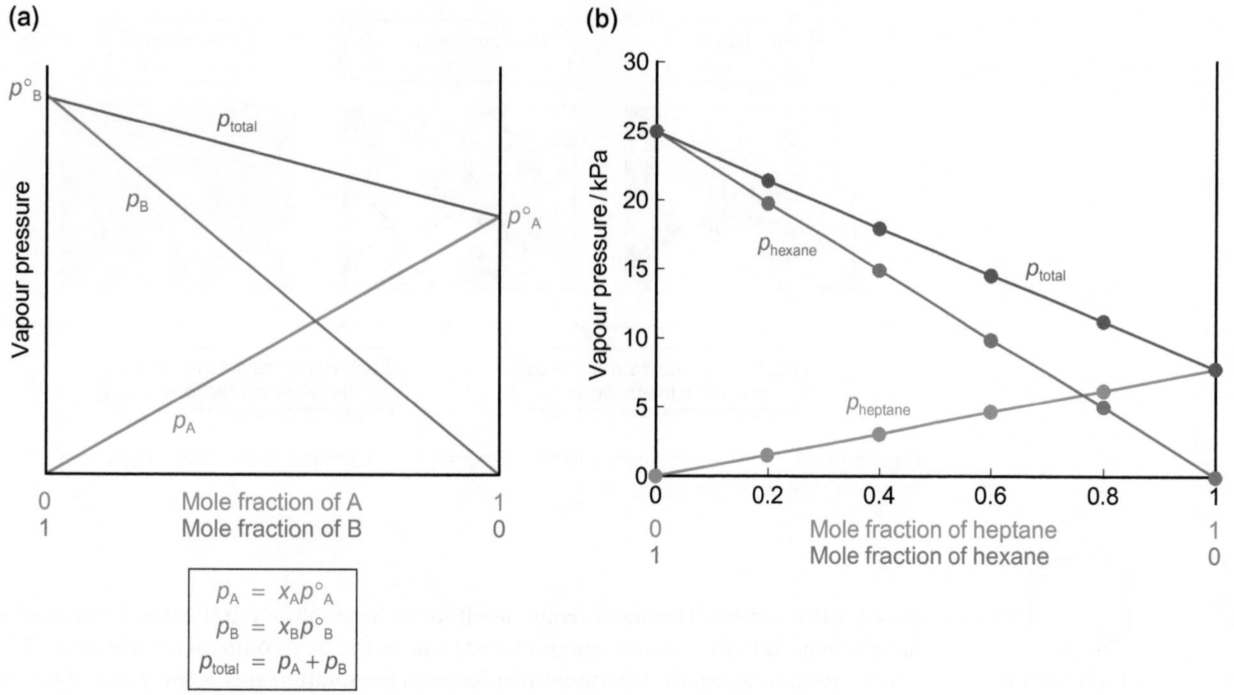
(a)
p°B
ptotal
pB
p°A
pA
Vapour pressure
0
1
Mole fraction of A
Mole fraction of B
1
0
pA = xAp°A
pB = xBp°B
ptotal = pA + pB
(b)
30
25
20
15
10
5
0
Vapour pressure / kPa
phexane
ptotal
pheptane
0
0.2
0.4
0.6
0.8
1
0
1
Mole fraction of heptane
Mole fraction of hexane
1
0

Figure 17.20 (a) In an ideal solution, the vapour pressure of each component is a linear function of its mole fraction. (b) For example, mixtures of hexane and heptane form ideal solutions.

and

$$p_{\text{total}} = p_A + p_B = x_A p^\circ_A + x_B p^\circ_B \qquad (17.25)$$

where p° is the vapour pressure of the pure component (i.e. $x = 1$) at the temperature of the solution.

The relationships in Equation 17.24 are known as **Raoult's law**. In practice, few solutions obey Raoult's law precisely and it can only be regarded as an approximation. However, it works quite well for mixtures of similar compounds, such as hexane and heptane (see Figure 17.18(b)) or benzene and methylbenzene. Mixtures that obey Raoult's law are known as **ideal solutions**. An illustration of Raoult's law is given in Worked example 17.4.

Visit the Online Resource Centre to view screencast 17.3 which walks you through Raoult's law.

Consider a mixture of hexane and heptane with a mole fraction $x_{\text{hexane}} = 0.8$. Equation 17.24 tells you that the vapour pressure of hexane above the solution will be 80% of that of pure hexane. Now consider a mixture of hexane and decane with $x_{\text{hexane}} = 0.8$. The vapour pressure of hexane over this solution will also be 80% of that of pure hexane, because the nature of the other component does not change the likelihood of a hexane molecule moving from the liquid to the vapour. This gives a clue to what makes a solution ideal.

As Figure 17.21 shows, a hexane molecule in the solution experiences the same interactions that it would in pure hexane, and the molecules mix randomly. So, there are two requirements for a solution to behave ideally. The solution should consist of molecules that:

- are similar in chemical nature so that they have similar intermolecular interactions;
- are similar in size and shape.

The first of these conditions is usually more important. Differences in the size and shape of the components are only significant if the difference is large, such as in a solution of a

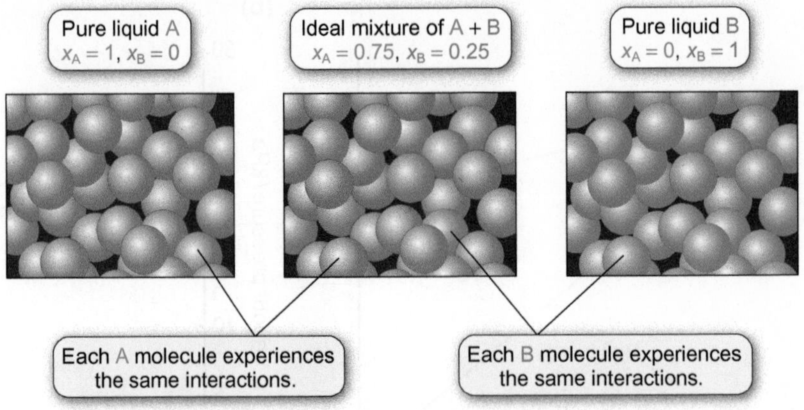
Pure liquid A
$x_A = 1, x_B = 0$
Ideal mixture of A + B
$x_A = 0.75, x_B = 0.25$
Pure liquid B
$x_A = 0, x_B = 1$
Each A molecule experiences the same interactions.
Each B molecule experiences the same interactions.

Figure 17.21 In an ideal solution, molecules experience the same intermolecular interactions in solution as they do in the pure liquids.

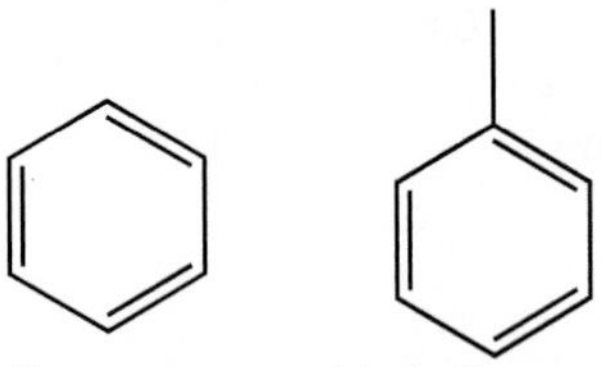
Benzene
Methylbenzene

Figure 17.22 Benzene and methylbenzene are similar in chemical nature, so they have similar intermolecular interactions and form a near-ideal solution.

Ideal gases are discussed in Section 8.2 (p.349).

polymer in a solvent. The major requirement for an ideal solution is that the intermolecular interactions in the solution are similar to those in the pure components (Figure 17.22).

In an ideal solution, the intermolecular forces in the solution are the same as in the pure components, so the enthalpy change of mixing is zero ($\Delta_{mix}H = 0$) and the value of $\Delta_{mix}G$ is negative (see Table 17.4).

Like an ideal gas, an ideal solution is a useful reference state. However, while the ideal gas model describes the behaviour of most gases over a wide range of conditions, few solutions display ideal behaviour. In fact, it is in looking at *deviations* from ideal solution behaviour where the model is useful, because these deviations provide information about the interactions that occur between molecules in the solution.

Compare Table 17.4 with Table 17.3 on p.791 for non-ideal solutions.

Table 17.4 Contributions to liquid–liquid miscibility for an ideal solution

$-T\Delta_{mix}S$	Interactions	$\Delta_{mix}H$	$\Delta_{mix}G$	
Negative	A–A, B–B, and A–B equal	Zero	Negative	Components mix

Worked example 17.4 Using Raoult's law

The saturated vapour pressures at 298 K for benzene and methylbenzene are 12.85 kPa and 3.85 kPa, respectively. What is the total vapour pressure over a solution containing 2.00 mol of benzene and 3.00 mol of methylbenzene?

Strategy

Since they are similar compounds, you can assume that a mixture of benzene and methylbenzene acts as an ideal solution. The vapour pressure of each component in the mixture is given by Raoult's law, Equation 17.24. You can calculate p for each component and add them to get the total pressure. To use Equation 17.24, you first need to calculate the mole fraction of each component in the solution.

Solution

Calculate the mole fraction of each component: x_b for benzene, x_m for methylbenzene.

$$x_b = \frac{\text{number of moles of benzene}}{\text{total number of moles}} = \frac{n_b}{n_b + n_m}$$

$$= \frac{2.00\,\cancel{\text{mol}}}{(2.00 + 3.00)\,\cancel{\text{mol}}} = 0.400$$

$$x_m = (1 - 0.400) = 0.600$$

→

➜ *Use Equation 17.24 to find the vapour pressure of each component.*

$$p_A = x_A p_A \quad (17.23)$$

$$p_b = x_b p_b = 0.400 \times 12.85\,\text{kPa} = 5.14\,\text{kPa}$$

$$p_m = x_m p_m = 0.600 \times 3.85\,\text{kPa} = 2.31\,\text{kPa}$$

Add the vapour pressures of each component to give the total vapour pressure.

$$p_{total} = p_b + p_m$$
$$= 5.14\,\text{kPa} + 2.31\,\text{kPa}$$
$$= 7.45\,\text{kPa}$$

Question

What is the percentage composition of a solution of benzene and methylbenzene with a total vapour pressure of 9.50 kPa?

Non-ideal solutions

Raoult's law states that the vapour pressure of a component, A, in an ideal solution is given by

$$p_{A(\text{ideal})} = x_A p^{\circ}_A \quad (17.24)$$

Few solutions behave ideally and the experimentally determined vapour pressure, $p_{A(\text{real})}$, usually differs from $p_{A(\text{ideal})}$. The behaviour of real solutions can be described by modifying Raoult's law and introducing an additional parameter, the **activity coefficient**, γ

$$p_{A(\text{real})} = \gamma_A x_A p^{\circ}_A \quad (17.26)$$

It follows from Equations 17.23 and 17.25 that

$$p_{A(\text{real})} = \gamma_A p_{A(\text{ideal})}$$

so that

$$\gamma_A = \frac{p_{A(\text{real})}}{p_{A(\text{ideal})}} \quad (17.27)$$

For an ideal solution, $\gamma = 1.0$. If $\gamma > 1$, the vapour pressure of the component over the solution is higher than it would be if the solution behaved ideally; if $\gamma < 1$, the vapour pressure is lower than for an ideal solution.

What *chemical* information can the value of γ give about the solution?

The ideal model describes solution behaviour when the intermolecular interactions between the components are identical. In most solutions, the interactions in solution are *weaker* than those in the pure liquids. In this case, molecules are 'held less tightly' into the solution and can more readily escape into the vapour (see Figure 17.23). As a result, the vapour pressure is *greater* than would be the case if the solution conformed to ideal behaviour. The vapour pressure curves have the form shown in Figure 17.24(a), an example being the propanone–carbon disulfide system shown in Figure 17.24(b). Such systems are said to display **positive deviations** from Raoult's law, since $p > p(\text{ideal})$ so that $\gamma > 1$. The higher the value of γ, the greater the difference in intermolecular interactions between the solution and pure liquid (see Worked example 17.5).

Different behaviour arises if the components show *stronger* intermolecular interactions in solution than in the pure liquids. A common example is where hydrogen bonding occurs between the two components, which cannot occur in the pure liquids. Here, the interactions in solution are *stronger* than those in the pure liquids so that *fewer* molecules

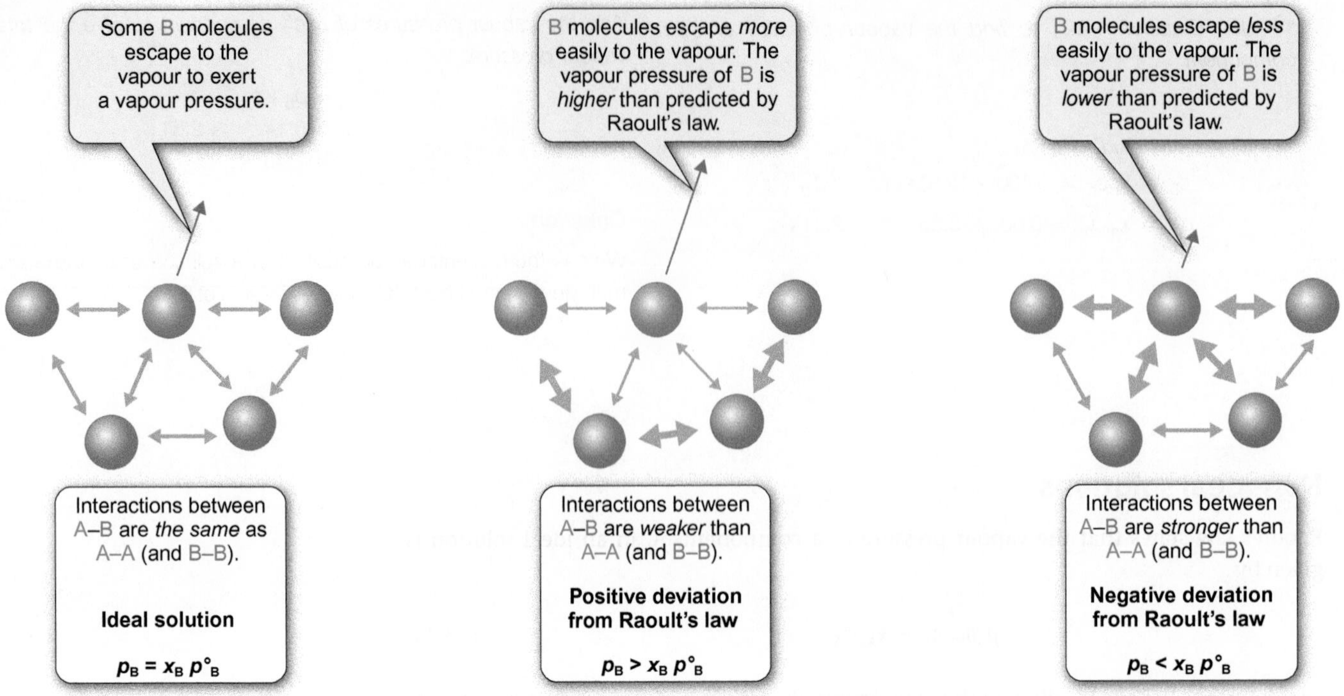
Some B molecules escape to the vapour to exert a vapour pressure.
Interactions between A–B are the same as A–A (and B–B).
Ideal solution
$p_B = x_B p°_B$
B molecules escape more easily to the vapour. The vapour pressure of B is higher than predicted by Raoult's law.
Interactions between A–B are weaker than A–A (and B–B).
Positive deviation from Raoult's law
$p_B > x_B p°_B$
B molecules escape less easily to the vapour. The vapour pressure of B is lower than predicted by Raoult's law.
Interactions between A–B are stronger than A–A (and B–B).
Negative deviation from Raoult's law
$p_B < x_B p°_B$

Figure 17.23 Molecular interpretation of non-ideal behaviour. The vapour pressure of a liquid depends on how easily its molecules can leave the surface. The three diagrams compare how easily molecules of B leave a mixture of A and B, with a low concentration of B, for different strengths of intermolecular interactions.

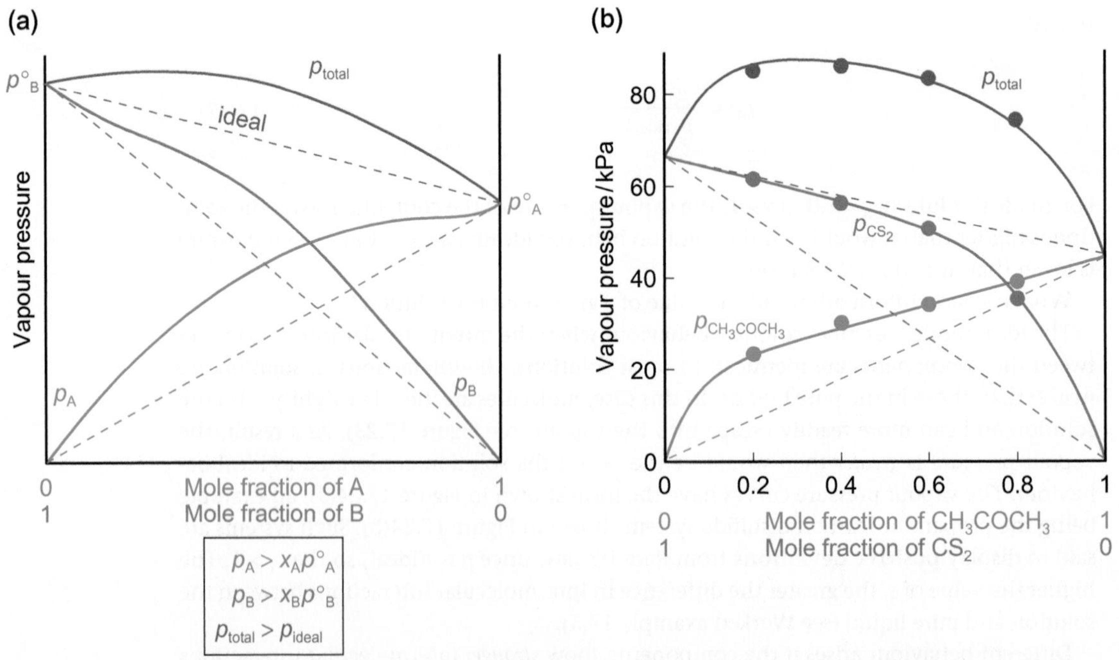
(a)
$p°_B$
p_{total}
ideal
$p°_A$
Vapour pressure
p_A
p_B
0
1
Mole fraction of A
Mole fraction of B
1
0
$p_A > x_A p°_A$
$p_B > x_B p°_B$
$p_{total} > p_{ideal}$
(b)
80
60
40
20
0
Vapour pressure / kPa
p_{total}
p_{CS_2}
$p_{CH_3COCH_3}$
0
0.2
0.4
0.6
0.8
1
0
1
Mole fraction of CH_3COCH_3
Mole fraction of CS_2
1
0

Figure 17.24 Positive deviations from Raoult's law. (a) In most real solutions, the vapour pressure over the solution is greater than in an ideal solution. (b) An example is a mixture of propanone (CH_3COCH_3) and carbon disulfide (CS_2). The dashed lines show ideal behaviour. These data are at 308 K.

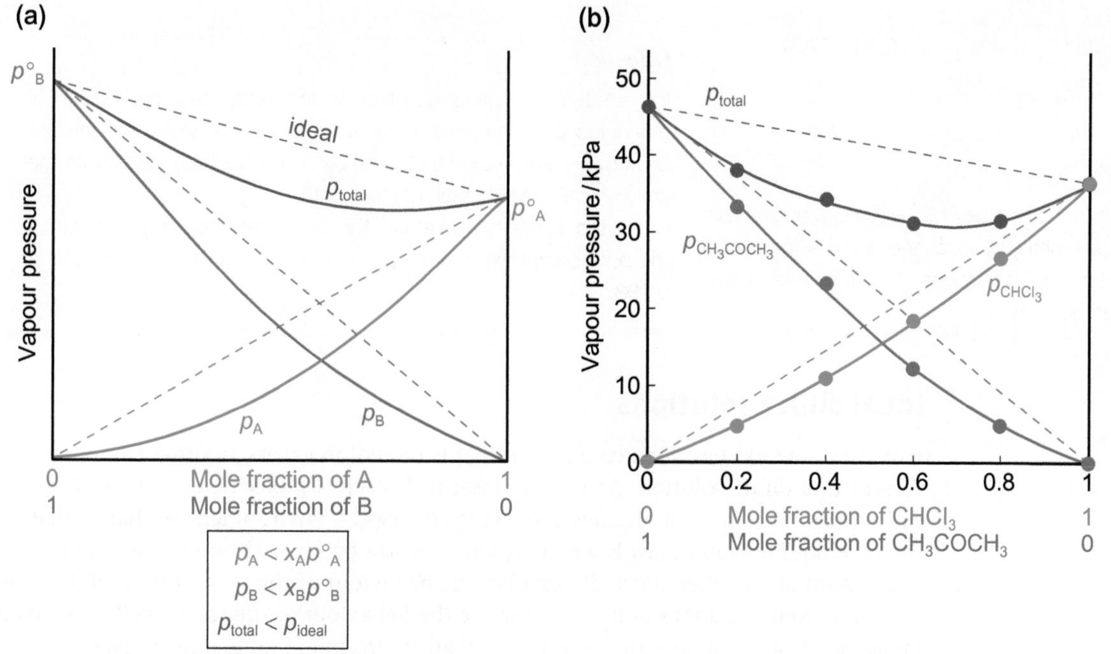
(a)
(b)
$p°_B$
ideal
p_{total}
$p°_A$
Vapour pressure
p_A
p_B
0
1
Mole fraction of A
Mole fraction of B
1
0
$p_A < x_A p°_A$
$p_B < x_B p°_B$
$p_{total} < p_{ideal}$
50
40
30
20
10
0
p_{total}
$p_{CH_3COCH_3}$
p_{CHCl_3}
Vapour pressure/kPa
0.2
0.4
0.6
0.8
Mole fraction of $CHCl_3$
Mole fraction of CH_3COCH_3

Figure 17.25 Negative deviations from Raoult's law. (a) In some solutions, the vapour pressure over a solution is lower than in an ideal solution. (b) An example is a mixture of propanone (CH_3COCH_3) and trichloromethane ($CHCl_3$). The dashed lines show ideal behaviour. These data are at 308 K.

escape to the vapour. As a result the vapour pressure is *smaller* than would be the case if the solution conformed to ideal behaviour, so $p < p(\text{ideal})$ and $\gamma < 1$. The vapour pressure curve takes the form shown in Figure 17.25. Such systems are said to display **negative deviations** from Raoult's law. A good example of this behaviour is in mixtures of trichloromethane and propanone. In the mixture, hydrogen bonding is possible between the dipole of the C–H in $CHCl_3$ and the C=O in the carbonyl group in propanone (see Figure 17.18) and molecules escape less readily into the vapour than from an ideal solution.

Visit the Online Resource Centre to view screencast 17.4 which explains deviations from Raoult's law.

Worked example 17.5 Deviations from Raoult's law

A solution of 0.18 mol ethanol (C_2H_5OH) and 0.32 mol of methylcyclohexane (C_7H_{14}) was placed in a sealed container at 55 °C. At equilibrium, the vapour pressures of ethanol and methylcyclohexane were 26.2 kPa and 24.4 kPa, respectively.

The saturated vapour pressures at 55 °C are: for ethanol 37.3 kPa, and for methylcyclohexane 22.4 kPa.

Calculate the activity coefficient for each component and comment on its value.

Strategy

The mole fraction of each component in the solution can be calculated from the numbers of moles. The vapour pressure of each component over the solution is given so the activity coefficient can be calculated from Equation 17.26. Since $\gamma = 1$ for an ideal solution where the intermolecular interactions are identical to those in the pure liquids, the values of γ allow you to comment on the types of interaction in the solutions.

Solution

Calculate the mole fractions of the liquids in solution.

$$x_{C_2H_5OH} = \frac{n_{C_2H_5OH}}{n_{C_2H_5OH} + n_{C_7H_{14}}} = \frac{0.18\,\cancel{\text{mol}}}{(0.18 + 0.32)\,\cancel{\text{mol}}} = 0.36$$

Since $x_{C_2H_5OH} + x_{C_7H_{14}} = 1$

$$x_{C_7H_{14}} = (1 - 0.36) = 0.64$$

Use Equation 17.26 to calculate the activity coefficients.

$$p_{A(\text{real})} = \gamma_A x_A p°_A \qquad (17.26)$$

→

→ For ethanol: $26.2\,\text{kPa} = \gamma_{C_2H_5OH} \times 0.36 \times 37.3\,\text{kPa}$

$\gamma_{C_2H_5OH} = 1.95$

For methylcyclohexane: $24.4\,\text{kPa} = \gamma_{C_2H_{14}} \times 0.64 \times 22.4\,\text{kPa}$

$\gamma_{C_7H_{14}} = 1.70$

Both activity coefficients are >1, indicating positive deviations from Raoult's law. The non-polar methylcyclohexane disrupts the hydrogen bonding in ethanol so that interactions in solution are weaker than those in the pure components.

Question

In a mixture of ethanol and trichloromethane, the mole fraction of ethanol is 0.6. The vapour pressure of ethanol above the mixture is 0.087 bar and that of trichloromethane is 0.256 bar. Calculate the activity coefficient of each component.

(At the same temperature, the saturated vapour pressures of the pure components are: ethanol 0.137 bar and trichloromethane 0.393 bar.)

Ideal dilute solutions

Raoult's law works best for a component that is present in excess, in other words, for the solvent in a dilute solution. As x_A gets closer to 1, each A molecule is more likely to be surrounded by other A molecules and so experiences an environment similar to that in the pure liquid. Raoult's law is a *limiting law*. It works better as the concentration of one component approaches 100%. It describes the behaviour of the solvent in a dilute solution quite well but does not usually describe the behaviour of the solute well. The *solute* molecules, B, are in a minority, so they are in an environment quite unlike that of pure B. This means that Raoult's law does not describe the vapour pressure of B at all well. This is particularly the case for gases dissolved in liquids, such as a dilute solution of oxygen in water. The water molecules are in an environment similar to that of pure water, whereas the oxygen molecules are in a very different environment from that of pure oxygen.

Experimentally, it has been found that the vapour pressure of a solute, B, in dilute solution is proportional to the mole fraction x_B *but it does not obey Raoult's law*. Instead, the vapour pressure obeys a different law. This is Henry's law and the proportionality constant K_B is known as the **Henry's law constant**

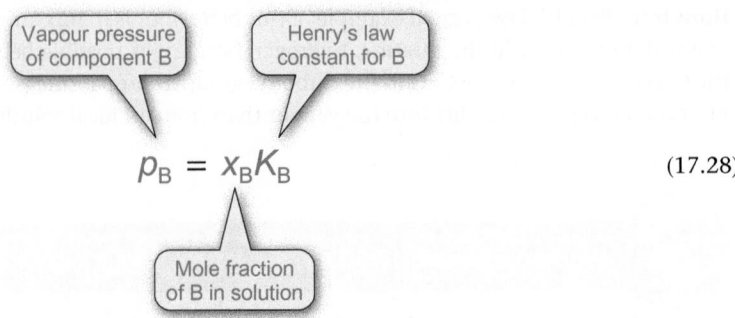

Vapour pressure of component B — Henry's law constant for B

$$p_B = x_B K_B \qquad (17.28)$$

Mole fraction of B in solution

Henry's law

ⓘ Henry's law is named after the English chemist William Henry who found in the nineteenth century that the mass of gas that dissolves in a liquid is proportional to its pressure over the liquid.

Solutions that behave according to Henry's law are called **ideal dilute solutions**. Henry's law is also a *limiting law* and solutions obey it better and better as the solution becomes more dilute.

To summarize:

- For solvent A, Raoult's law: as $x_A \to 1$, $p_A = x_A p^\circ_A$
- For solute B, Henry's law: as $x_B \to 0$, $p_B = x_B K_B$

Figure 17.26 summarizes the conditions under which a single substance obeys these two laws. Table 17.5 gives Henry's law constants, K_B, for some common gases. The *more soluble* the gas, the *smaller* the value of K_B.

One consequence of Henry's law is that, for a given mass of solvent, the amount of gas dissolved is proportional to the partial pressure of the gas over the solution. This is explored in Worked example 17.6. The solubility of the gas rises as the pressure increases. When the pressure is released, the solubility falls and the gas rapidly comes out of solution.

Table 17.5 Henry's law constants, K_B, for gases in water at 298 K

Gas	K_B/bar
O_2	4.3×10^4
N_2	9.0×10^4
CO_2	0.2×10^4
H_2	7.2×10^4
He	15.2×10^4
CO	5.9×10^4

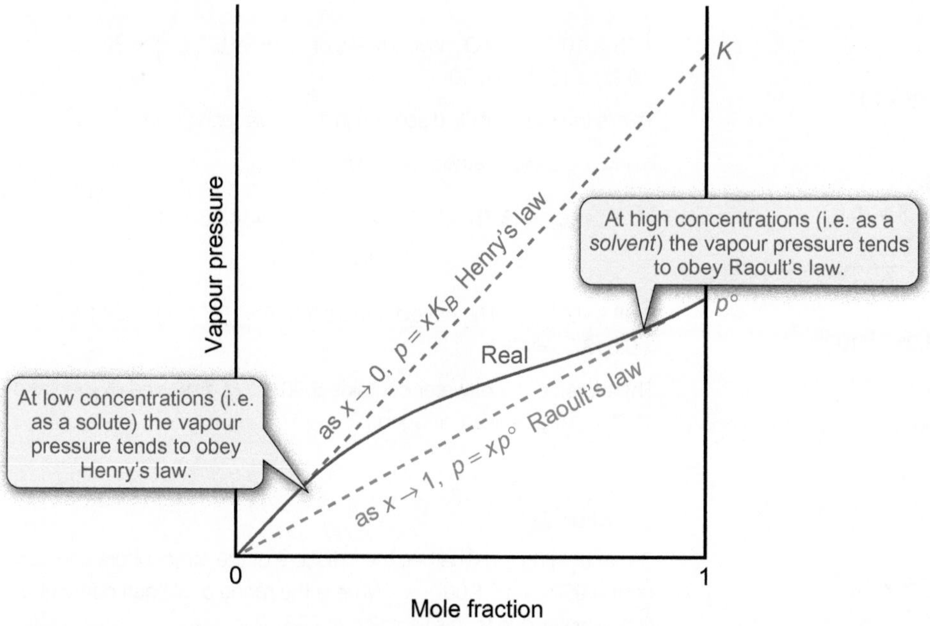
K
Vapour pressure
as x → 0, p = xK_B Henry's law
At high concentrations (i.e. as a solvent) the vapour pressure tends to obey Raoult's law.
p°
Real
At low concentrations (i.e. as a solute) the vapour pressure tends to obey Henry's law.
as x → 1, p = xp° Raoult's law
0
1
Mole fraction

Figure 17.26 At low concentrations, a component in a mixture can be thought of as the solute and tends to obey Henry's law. At high concentrations, it can be thought of as the solvent and tends to obey Raoult's law.

When carbonated drinks are opened, the pressure above the liquid is reduced so that the CO_2 becomes less soluble and forms bubbles. In champagne, the CO_2 is a result of natural fermentation rather than being added artificially as in soft drinks.

When you open a bottle of a fizzy drink, there is a sudden release of pressure over the solution and gas bubbles form as the gas comes out of solution (see Box 15.2, p.704).

The same phenomenon leads to a problem for deep sea divers. As they descend, the pressure increases so more nitrogen from the compressed air that they breathe dissolves in their blood and tissue. If the diver comes up too quickly, the gas comes out of solution and forms bubbles in the blood stream causing decompression sickness, known as 'the bends'. This is very painful and, in severe cases, can be fatal. For diving to extreme depths, divers use a mixture of helium and oxygen since helium is much less soluble in the blood than nitrogen. The application of Henry's law in artificial blood is described in Box 17.4.

The problems of breathing underwater are discussed in more detail in Chapter 8, p.342.

Worked example 17.6 Henry's law and gas solubilities

The Henry's law constants, K_B, at 298 K for N_2(g) and O_2(g) in water are 9.0×10^4 bar and 4.3×10^4 bar, respectively. Estimate the concentration of each gas in water, in mg dm^{-3}, under 1 atm of dry air (assume 78.0% N_2 and 21.0% O_2).

Strategy

You first need to calculate the partial pressure of each gas in the air above the solution. From this and the Henry's law constant, you can use Equation 17.28 to calculate the mole fraction of each gas in solution and hence its concentration.

Solution

Calculate the partial pressure of each gas over the solution.

Since the Henry's law constants are given in bar, it makes sense to use this unit for the gas pressures, 1 atm = 1.013 bar. For an ideal gas, the molar percentage is the same as the percentage by volume.

Using Equation 8.10 (p.356)

$$p_A = x_A p_{total} \quad (8.10)$$

$$p_{O_2} = 0.210 \times 1.013\,\text{bar} = 0.213\,\text{bar}$$

$$p_{N_2} = 0.780 \times 1.013\,\text{bar} = 0.790\,\text{bar}$$

Use Equation 17.28 (Henry's law) to calculate the mole fractions of each gas in solution.

$$p_B = x_B K_B \quad (17.28)$$

For O_2(g)

$$p_{O_2} = 0.213\,\text{bar} = x_{O_2} \times 4.3 \times 10^4\,\text{bar}$$

Rearranging

$$x_{O_2} = \frac{0.213\,\cancel{\text{bar}}}{4.3 \times 10^4\,\cancel{\text{bar}}} = 4.95 \times 10^{-6}$$

(Remember, mole fraction has no units.)

→

→ Similarly, for N_2(g)

$$x_{N_2} = \frac{0.790\,\cancel{\text{bar}}}{9.0\times10^4\,\cancel{\text{bar}}} = 8.78\times10^{-6}$$

Calculate the concentrations of the gases.

1 dm^3 of water has a mass of 1000 g. $M_r(H_2O) = 18.0$

$$\text{Amount (in moles) of water in 1000 g} = \frac{1000\,\cancel{\text{g}}}{18.0\,\cancel{\text{g}}\,\text{mol}^{-1}} = 55.6\,\text{mol}$$

$$x_{O_2} = \frac{n_{O_2}}{n_{O_2} + n_{N_2} + n_{H_2O}} = 4.95 \times 10^{-6}$$

n_{O_2} and n_{N_2} are very small compared with n_{H_2O}, so to a good approximation

$$x_{O_2} \approx \frac{n_{O_2}}{n_{H_2O}}$$

so that $n_{O_2} \approx (4.95 \times 10^{-6}) \times 55.6\,\text{mol} = 2.75 \times 10^{-4}\,\text{mol}$

$$M_r(O_2) = 32.0$$

2.75×10^{-4} mol of O_2 have a mass of $(2.75\times10^{-4}\,\text{mol}) \times 32.0\,\text{g mol}^{-1} = 8.80\times10^{-3}\,\text{g} = 8.80\,\text{mg}$

This is the mass of O_2 dissolved in 1 dm^3 water.

Using the same method for N_2(g)

$$n_{N_2} \approx (8.78 \times 10^{-6}) \times 55.6\,\text{mol} = 4.88 \times 10^{-4}\,\text{mol}$$

$$M_r(N_2) = 28.0$$

4.88×10^{-4} mol of N_2 have a mass of $(4.88\times10^{-4}\,\text{mol}) \times 28.0\,\text{g mol}^{-1} = 0.0137\,\text{g} = 13.7\,\text{mg}$

Therefore, 1 dm^3 of water contains 8.80 mg of dissolved oxygen and 13.7 mg of dissolved nitrogen.

Question

Depending on the weather, the pressure of the atmosphere can vary from 0.960 bar to 1.067 bar. What is the range of oxygen concentration in water due to this variation at 298 K?

Box 17.4 Perfluorocarbons and blood substitutes

Blood transfusions are a routine procedure during surgery or in the treatment of accident victims. Blood is collected from volunteers and can be stored for future use but remains usable for only about 5 to 6 weeks. It must be screened and treated to remove potentially dangerous micro-organisms.

Even minor surgery can require several litres of blood and major heart surgery or serious accident victims can require 10 litres or more.

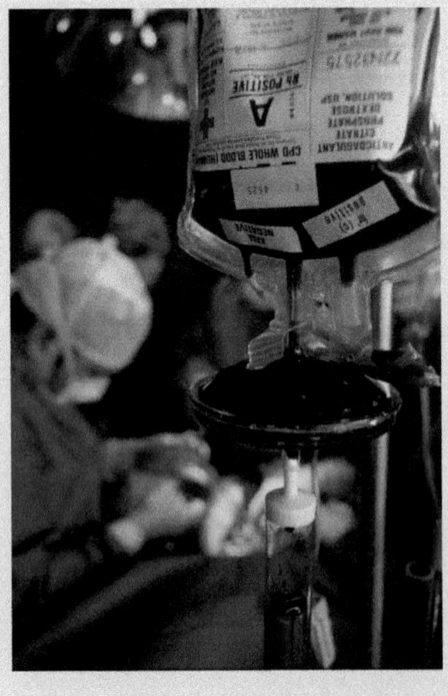

Since the usual volunteer donation is around 0.5 litres, the demand for blood products is often higher than the supply from blood donations. In some cultures it is not permitted to donate—or to receive—blood. A safe, synthetic blood substitute is needed.

One of the main functions of blood is to transport oxygen to the tissues throughout the body and to remove waste carbon dioxide to the lungs so that it can be exhaled. Transport of O_2 takes place by formation of a complex with haemoglobin (see Box 28.7, p.1298) contained within the red blood cells. One approach to artificial blood is to use haemoglobin-based products, but these are not without problems.

A different approach relies on a class of synthetic chemicals called **perfluorocarbons**. These are hydrocarbons in which all the hydrogen atoms are replaced by fluorine (see Figure 1). These compounds are very stable and unreactive due to the high strength of the C–F bond and so can be safely injected into the body. The key property of perfluorocarbons for this application is that they can dissolve large amounts of gases, such as oxygen and CO_2. For example, at body temperature and atmospheric pressure, 100 cm^3 of a perfluorocarbon can dissolve 40 cm^3–50 cm^3 of O_2 and as much as 250 cm^3 of CO_2. Compare this with water, which dissolves around 2.5 cm^3 of O_2 and 80 cm^3 of CO_2 under the same conditions. In contrast to blood, there is no reaction between the dissolved gas and the perfluorocarbon solvent; the gas is simply physically dissolved.

◀ A blood transfusion is a routine procedure—but supplies of real blood cannot keep up with demands so substitutes are needed.

→

➜ In 1966 Leland Clark showed that mice and rats could survive when completely immersed in a perfluorocarbon liquid. Sufficient oxygen transport occurred to keep the animals alive, albeit only for periods up to a few days. Today, perfluorocarbons, such as perfluorodecalin, can be synthesized in purer form with no toxic by-products and so are safer. They are insoluble in water and so are injected as an emulsion in water, containing droplets of perfluorocarbon around 0.1–0.2 µm in diameter. Rather than substitutes for blood, they are used as 'extenders' to add to the blood of a patient who has lost blood.

A comparison of the oxygen content of blood with a perfluorocarbon emulsion is shown in Figure 2. Blood, which becomes saturated with oxygen even at low O_2 pressures, can carry around 200 cm^3 of O_2 per dm^3 in contrast to 30 cm^3 of O_2 per dm^3 in plasma (blood from which the cells have been removed). Emulsions containing perfluorocarbons are intermediate but carry much less O_2 than an equivalent amount of blood. However, the volume of O_2 that is dissolved does not tell the whole story. Since the oxygen is simply dissolved in the perfluorocarbon and not chemically bound as it is in blood, the exchange of gases between the perfluorocarbon and tissues occurs faster than with blood and this, in part, makes up for the lower amount dissolved.

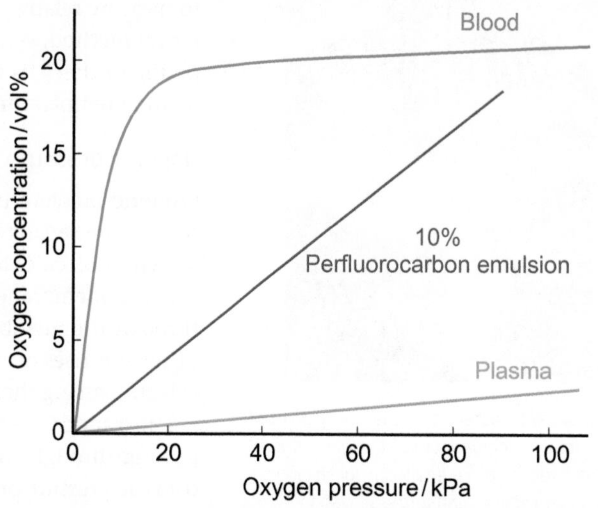
Blood
10%
Perfluorocarbon emulsion
Plasma
Oxygen concentration / vol%
Oxygen pressure / kPa
0
5
10
15
20
0
20
40
60
80
100

▲ **Figure 2** The solubility of oxygen gas in blood, blood plasma, and perfluorocarbon increases with pressure. At low pressures, whole blood contains higher concentrations of oxygen than the plasma. At higher pressures, the solubility in a perfluorocarbon emulsion can approach that in blood.

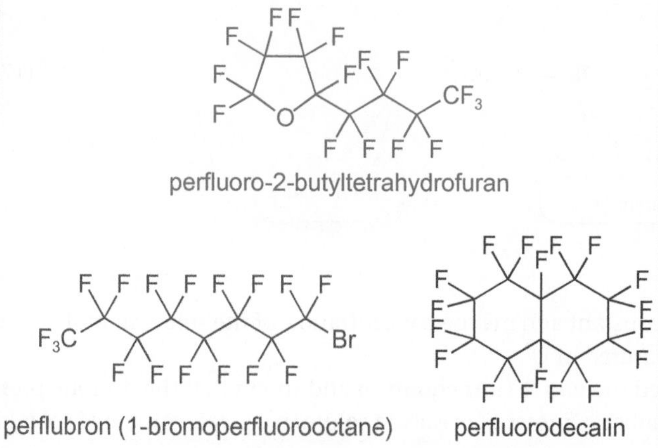
perfluoro-2-butyltetrahydrofuran
perflubron (1-bromoperfluorooctane)
perfluorodecalin

▲ **Figure 1** Perfluorocarbons such as these are used as synthetic blood substitutes and extenders.

Questions

(a) Suggest explanations for the different shapes of the plots in Figure 2. (*Hint.* Look at Box 28.7 (p.1298) on the way O_2 binds to haemoglobin.)

(b) What is the partial pressure of oxygen in air? Estimate the percentage of perfluorocarbon emulsion that would be needed to carry the same amount of oxygen as blood at this partial pressure of oxygen.

Colligative properties

Raoult's law shows that the presence of a second component (a solute) will reduce the vapour pressure of the solvent over a solution. In an ideal solution, the lowering of the vapour pressure does not depend on the nature of the solute but only on its mole fraction in solution. Properties that depend only on the number of dissolved molecules are called **colligative properties.**

Two related colligative properties are the elevation of the boiling temperature of a liquid when a solute is added and the lowering of the freezing temperature. A familiar application is the use of salt to lower the freezing point of water on roads in wintry weather. The colligative effect of reducing the freezing point of a solvent is the reason that sea water, with its high salt content, freezes at a lower temperature than fresh water.

Adding sugar to water raises the boiling point of the water. The temperature of the water–sugar mixture needs to be carefully controlled when making jam, caramel, and toffee. Elevation of boiling point is an example of a colligative property.

Before the advent of mass spectrometry, the lowering of freezing temperature was used to measure relative molecular masses of compounds. Although no longer used as an analytical method, synthetic chemists routinely measure the melting temperature of a compound to check its purity. The compound is repeatedly purified until no further change of melting temperature occurs, indicating that the compound is pure.

Osmotic pressure

Osmotic pressure is a colligative property. Osmosis occurs when a solution and a pure solvent are separated by a **semi-permeable membrane**. This is a membrane that allows solvent molecules to pass through but not solute molecules, as shown in Figure 17.27.

If the membrane is used in the arrangement shown in Figure 17.28, solvent passes through the membrane, tending to equalize the concentration on both sides. The level of solvent rises on the left-hand side, exerting a pressure that eventually stops any more solvent passing through.

The osmotic pressure, π, is the external pressure that must be applied to prevent solvent passing through the membrane. If the solution acts ideally, then at temperature T, the osmotic pressure of a solution is proportional to its concentration.

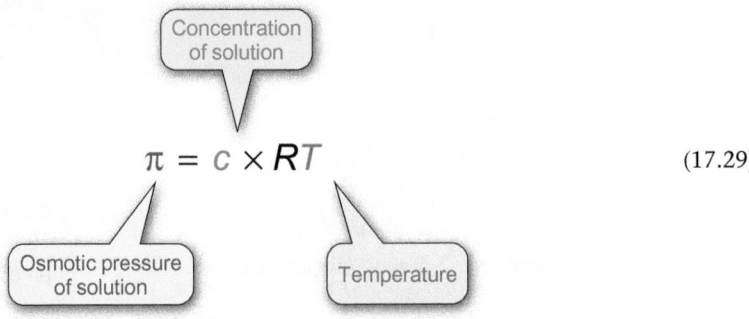

Concentration of solution

$$\pi = c \times RT \qquad (17.29)$$

Osmotic pressure of solution

Temperature

The Dutch physical chemist Jacobus van 't Hoff was awarded the first Nobel Prize for Chemistry in 1901, partly for his work on osmosis. Note that as well as Equation 17.29, Equation 15.10 (p.716) is also called the van 't Hoff equation.

where R is the ideal gas constant and c is the concentration of the solute in mol m^{-3}. π is the osmotic pressure measured in Pa.

Equation 17.29 is called the **van 't Hoff equation** and shows that the osmotic pressure depends only on the number of moles of dissolved solute in a given volume of solvent. At

Osmosis also occurs when two solutions of differing concentration are separated by a semi-permeable membrane, that is, where there is a *concentration gradient*.

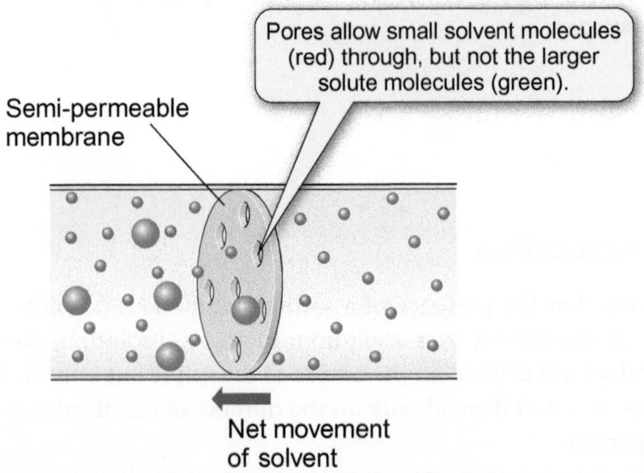
Pores allow small solvent molecules (red) through, but not the larger solute molecules (green).
Semi-permeable membrane
Net movement of solvent

Figure 17.27 A semi-permeable membrane allows solvent molecules to pass through but not solute molecules. In this case, the effect is due to the size of the pores, although other effects can occur, for example, the passage of ions can be prevented by incorporating charged groups into the membrane.

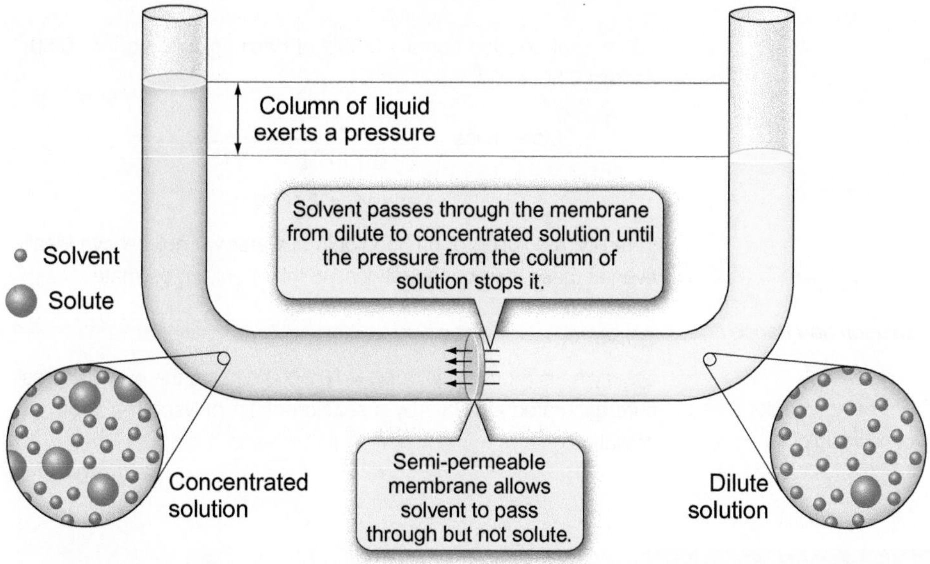
Column of liquid exerts a pressure
Solvent passes through the membrane from dilute to concentrated solution until the pressure from the column of solution stops it.
Solvent
Solute
Concentrated solution
Semi-permeable membrane allows solvent to pass through but not solute.
Dilute solution

Figure 17.28 Solvent passes through a semi-permeable membrane until the pressure is sufficient to prevent passage. The system then comes to equilibrium.

Even when the system reaches equilibrium, solvent molecules pass through the membrane in both directions. Strictly speaking, π is the pressure that is needed to equalize the rates at which solvent passes through the membrane.

Since the molar concentration, c, is the number of moles, n, of solute divided by the volume, V, of solvent, Equation 17.29 can also be written as $\pi V = nRT$. This has a similar form to the ideal gas equation and so is easy to remember.

the same molar concentration, all solutes give the same osmotic pressure. This phenomenon can be used to measure the relative molecular mass of the solute. It is particularly useful for substances with high M_r, such as proteins and polymers, because a measurable osmotic pressure can be achieved even with low molar concentrations (see Worked example 17.7). However, very few solutes, particularly in solvents such as water, give ideal solutions. The mathematical treatment of real solutions is rather more complicated than Equation 17.29, which only gives an estimate of M_r. The equation is obeyed best when the solution is very dilute.

Colligative properties depend on the *total number of particles in solution* whether they are molecules or ions. Therefore, a 1 mol dm^{-3} solution of sodium chloride in water generates double the osmotic pressure of a sugar solution of the same concentration since 1 mol of NaCl in water is made up of 1 mol of Na^+ and 1 mol Cl^- ions, giving a total ionic concentration of 2 mol dm^{-3}.

Osmosis is of great biological importance. Many biological cell membranes are semi-permeable. They allow water, gases such as O_2, and ions to pass through but keep larger molecules and structures inside the cell.

Box 17.5 describes how the application of pressure can force a solvent to flow in the opposite direction (reverse osmosis). This is used in water purification.

Red blood cells in a solution that is more concentrated than that inside the cell. Osmosis causes water to flow out of the cells, which have started to change their normal round shape and pucker.

Determination of relative molecular masses by mass spectrometry (Section 12.1, p.558) has largely replaced osmotic pressure measurements, though they are still used to characterize macromolecules.

Worked example 17.7 Using osmosis to find relative molecular mass, M_r

A solution of 0.50 g of haemoglobin in 100 cm^3 of water exerts an osmotic pressure of 193 Pa. Assuming that the solution is dilute and acts ideally, estimate the relative molecular mass of haemoglobin.

Strategy

Since you are told to assume ideal behaviour, Equation 17.29 can be used to calculate the molar concentration. From this, you can find the number of moles in solution and, since the mass is known, you can then calculate M_r.

Solution

Use Equation 17.29 to find the molar concentration.

$$\pi = c \times RT \qquad (17.29)$$

→

➜ *Rearrange to give an expression for c.*

$$c = \frac{\pi}{RT} = \frac{193\,\text{Pa}}{(8.314\,\text{J K}^{-1}\,\text{mol}^{-1}) \times (298\,\text{K})}$$

$$= \frac{193\,\text{J m}^{-3}}{(8.314\,\text{J K}^{-1}\,\text{mol}^{-1}) \times (298\,\text{K})} \qquad (1\,\text{Pa} = 1\,\text{J m}^{-3})$$

$$= 7.80 \times 10^{-2}\,\text{mol m}^{-3}$$

$$= 7.80 \times 10^{-5}\,\text{mol dm}^{-3} \qquad (1\,\text{dm}^{-3} = 1 \times 10^{-3}\,\text{m}^{-3})$$

Calculate the number of moles in 100 cm^3 solution and hence find M_r for haemoglobin.

1000 cm^3 of solution contain 7.80×10^{-5} mol of haemoglobin, so 100 cm^3 of solution contain 7.80×10^{-6} mol of haemoglobin.

100 cm^3 of solution contain 0.50 g of haemoglobin, so that 0.50 g contain 7.80×10^{-6} mol.

$$\text{Molar mass} = \frac{0.50\,\text{g}}{7.80 \times 10^{-6}\,\text{mol}} = 64\,100\,\text{g mol}^{-1}$$

$$\text{and } M_r = 64\,000$$

In reality, a solution of haemoglobin in water will not behave ideally even at these low concentrations, so this is only an estimate.

Question

Lysozyme is a protein with M_r = 16 500. What is the osmotic pressure generated at 298 K by a solution of 1 g of lysozyme in 50 cm^3 of water?

Box 17.5 Reverse osmosis and water purification

Osmosis is the process by which a solvent passes spontaneously through a semi-permeable membrane and into a solution. It occurs when there is a concentration gradient across the membrane—the movement of solvent tends to equalize the concentrations. Osmosis can be prevented by applying a pressure to the solution. In fact, if the pressure applied is *greater* than the osmotic pressure, the solvent can be made to flow from the solution into the pure solvent. This is known as **reverse osmosis**, as illustrated in the figure.

Reverse osmosis is used on a large scale for water purification. It can be used in small units in the laboratory for producing very pure water (see Box 16.1, p.733) but is also used on a large scale to produce drinking water. Many regions of the world lack an adequate supply of fresh water but have supplies of water from boreholes or even sea water. Plants using reverse osmosis have been built to supply whole towns or regions with fresh, drinkable water by removal of salt from sea water, **desalination**.

No heating is required and, even taking into account the high pressure needed, reverse osmosis desalination has lower energy consumption than, for example, distillation. Reverse osmosis membranes do not rely on the size of pores in the same way as those in Figure 17.26 but have chemically modified surfaces that allow water to diffuse through but that will not allow ions such as Na^+ or Cl^- to pass through. An added factor in purifying water by reverse osmosis is that most bacteria and viruses do not pass through the membrane so that some disinfection also takes place. High pressures are needed to force the water through the membranes so that the membranes must be carefully designed and constructed to withstand these pressures.

Question

1 dm^3 of sea water contains about 35 g of dissolved substances. Assuming that this is all sodium chloride, estimate the minimum pressure needed for reverse osmosis to take place at 25 °C.

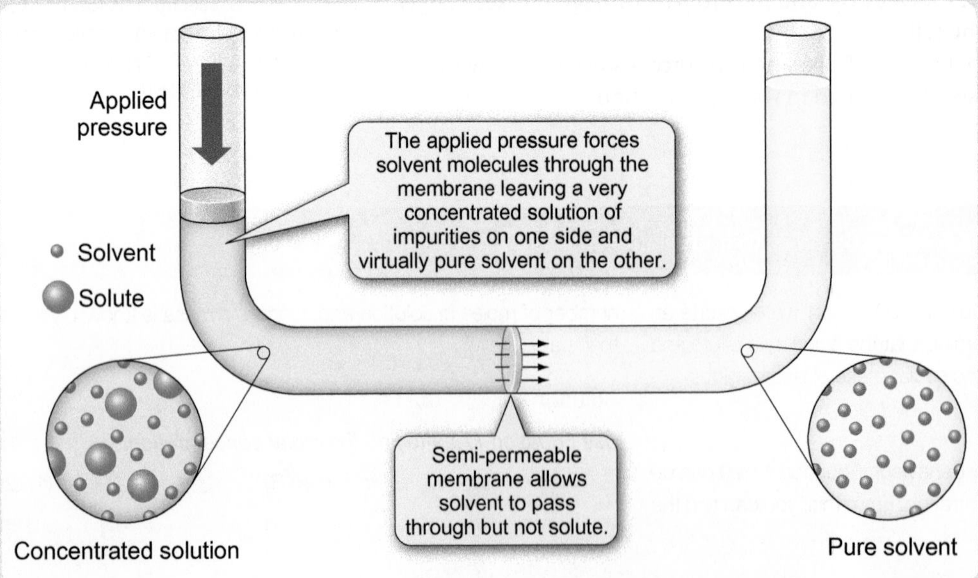
Applied pressure
The applied pressure forces solvent molecules through the membrane leaving a very concentrated solution of impurities on one side and virtually pure solvent on the other.
Solvent
Solute
Semi-permeable membrane allows solvent to pass through but not solute.
Concentrated solution
Pure solvent

◀ Reverse osmosis. Applying pressure can overcome the osmotic pressure and force solvent through the membrane, leaving behind a concentrated solution and producing pure solvent.

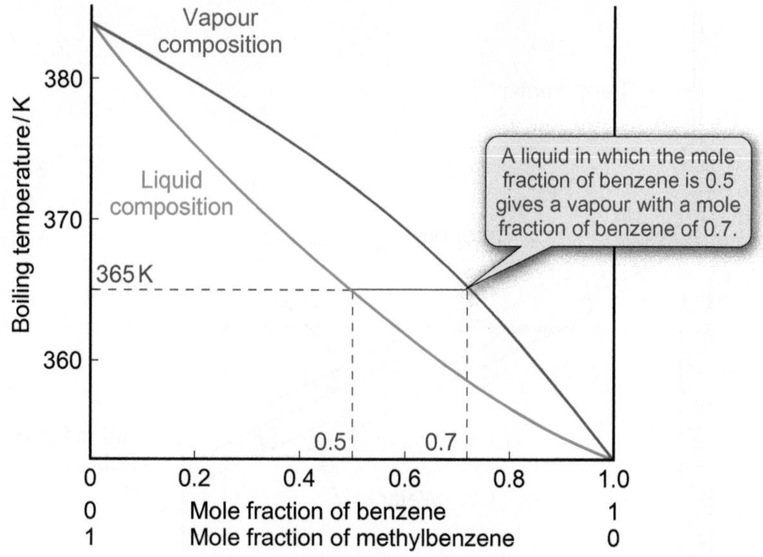
Vapour composition
Liquid composition
A liquid in which the mole fraction of benzene is 0.5 gives a vapour with a mole fraction of benzene of 0.7.
380
370
365 K
360
Boiling temperature / K
0.5
0.7
0
0.2
0.4
0.6
0.8
1.0
0
Mole fraction of benzene
1
1
Mole fraction of methylbenzene
0

Figure 17.29 Boiling an equimolar mixture of benzene and methylbenzene gives a vapour richer in benzene. Notice that the mole fraction scale runs in opposite directions for the two components, so when the mole fraction of benzene is 1, that of methylbenzene is 0, and vice versa.

Distillation of liquid mixtures

A mixture of benzene (T_b = 353.3 K) and methylbenzene (T_b = 383.8 K) forms an ideal solution. If an equimolar mixture (i.e. $x_{benzene} = x_{methylbenzene} = 0.5$) is heated, the vapour in equilibrium with the liquid mixture will be richer in benzene since benzene is more volatile than methylbenzene. This principle forms the basis of distillation, which is the main separation and purification method used for liquids.

The process can be represented on a **temperature–composition diagram** such as that shown in Figure 17.29. This is a phase diagram that plots the boiling temperature as a function of mole fraction of the solution. Also plotted is the composition of the vapour that is in equilibrium with the liquid, usually at 1 atm.

Figure 17.29 shows that an equimolar mixture of benzene and methylbenzene boils at 365 K. The vapour above this solution has a benzene mole fraction of 0.7. If this vapour is then condensed, it produces a liquid with this composition. Therefore, a single distillation gives a liquid product with $x_{benzene} = 0.7$. This new mixture could be redistilled to increase the concentration of benzene further, and the process of repeated distillation could continue until the two liquids were completely separated. However, this batch distillation is not very efficient and normally a process of **fractional distillation** is used.

Fractional distillation uses a column containing a large surface area so that the vapour is continually condensing and re-evaporating as it passes up the column. In the laboratory a glass column containing glass beads as a packing is often used (see Figure 17.30). The temperature varies up the column with the top being cooler than the liquid in the boiler. When the apparatus is used to distil a mixture of benzene and methylbenzene, the mixture undergoes many cycles of evaporation and condensation as it passes up the column and the distillate emerges from the top of the column as almost pure benzene.

The separation is represented by the example in Figure 17.31. A solution with mole fraction of benzene, $x_{benzene} = 0.2$, boils initially at 375 K. As it passes up the column, the vapour becomes progressively more concentrated until the **distillate** is pure benzene, $x_{benzene} = 1$. The **residue** in the distillation flask is pure methylbenzene.

Scotch whisky is made by distilling a mixture of ethanol and water, produced by fermenting barley. The liquid mixture is heated from below, and the vapour condenses in the copper pipes above, giving a liquid that is richer in ethanol. This is an example of batch distillation.

Visit the Online Resource Centre to view screencast 17.5 which explains distillation in terms of phase equilibria.

Azeotropic mixtures

Most solutions are not ideal, so relatively few have temperature–composition diagrams like Figure 17.31. Solutions showing large *positive* deviations from Raoult's law may have a

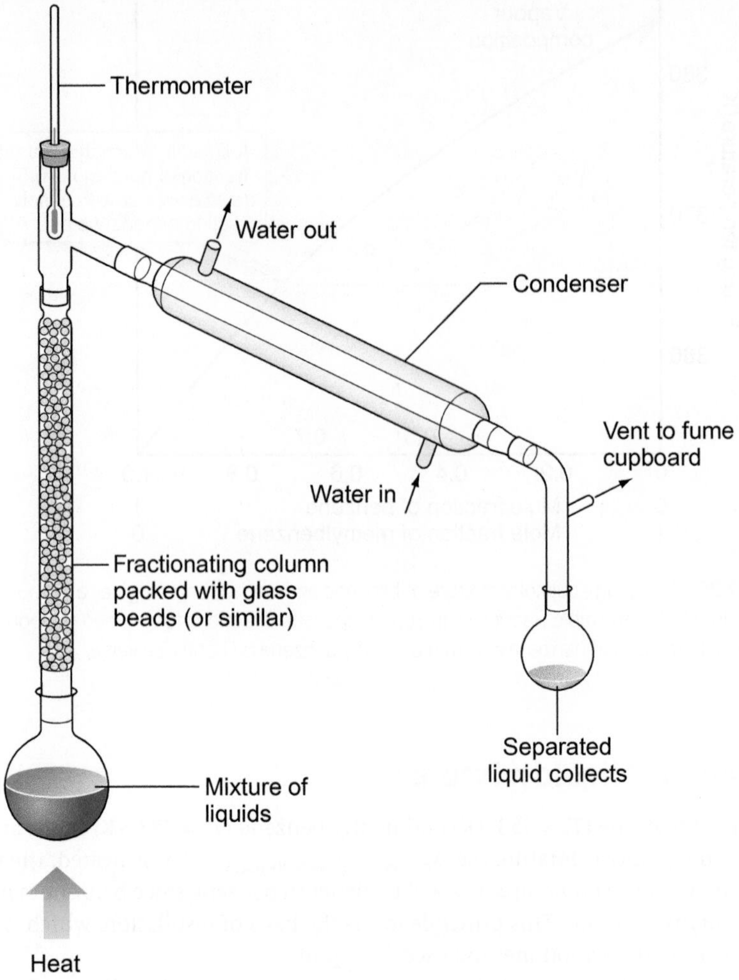
Thermometer
Water out
Condenser
Vent to fume cupboard
Water in
Fractionating column packed with glass beads (or similar)
Separated liquid collects
Mixture of liquids
Heat

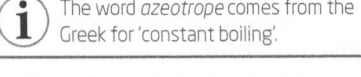

See Section 22.4, p.1038 for a photo of the apparatus for a laboratory fractional distillation.

Figure 17.30 Apparatus for fractional distillation. Industrially, a large metal column containing a number of plates or trays may be used.

minimum boiling temperature that is lower than either of the pure components. Ethanol and water form such a mixture, and their temperature–composition diagram has the form shown in Figure 17.32.

If a solution with an ethanol mole fraction of 0.2 is fractionally distilled, the vapour becomes progressively richer in ethanol. However, when it reaches the minimum in the curve, corresponding to 95.6% ethanol, the composition of the liquid and the vapour does not change further. The mixture forms an **azeotrope**. When this composition has been reached, no further separation is possible since the distillate has the same composition as the liquid. This means that distillation of ethanol–water mixtures can only ever give 95.6% pure ethanol. This behaviour occurs because ethanol–water mixtures show *positive deviations* from Raoult's law.

The word *azeotrope* comes from the Greek for 'constant boiling'.

If ethanol with a purity >95% is needed, the water must be removed by another method such as adsorption by a drying agent. Another method is to add a small amount of benzene to the distillation. Mixtures of ethanol, water, and benzene form a *ternary* azeotrope

If deviations from Raoult's law are positive ($\gamma_A > 1$), molecules escape to the vapour phase more easily. This is the case with the ethanol-water system, because each pure liquid disrupts the other's hydrogen bonding so that molecules are held in the solution less strongly. The vapour pressure is higher than in an ideal solution so that the mixture boils at a lower temperature. The reverse is true for solutions showing negative deviations ($\gamma_A < 1$).

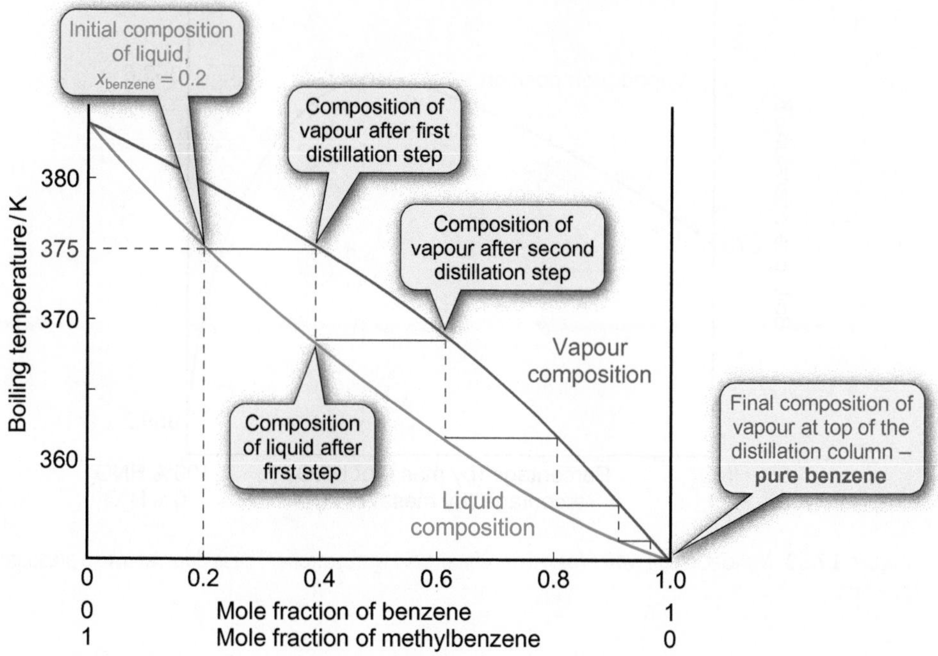
Initial composition of liquid, $x_{benzene} = 0.2$
Composition of vapour after first distillation step
Composition of vapour after second distillation step
Vapour composition
Composition of liquid after first step
Liquid composition
Final composition of vapour at top of the distillation column – pure benzene
Boiling temperature / K
380
375
370
360
0
0.2
0.4
0.6
0.8
1.0
0
1
Mole fraction of benzene
Mole fraction of methylbenzene
1
0

Figure 17.31 Fractional distillation can completely separate benzene and methylbenzene. In effect, a large number of successive distillations occurs as the vapour passes up the column.

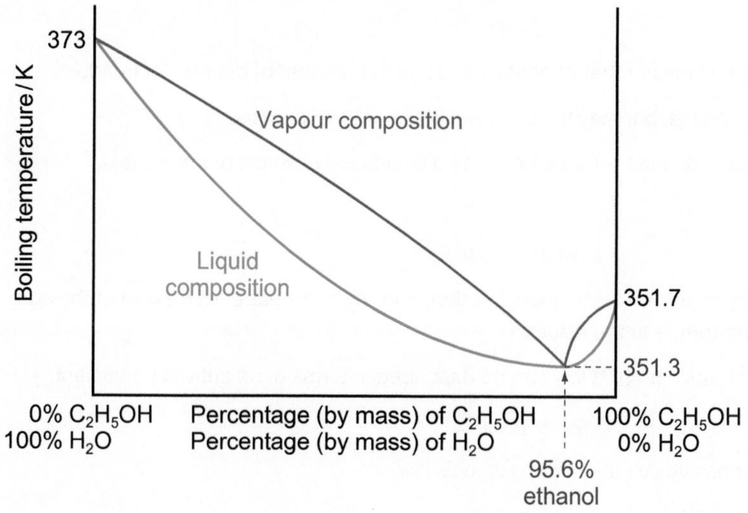
373
Boiling temperature / K
Vapour composition
Liquid composition
351.7
351.3
0% C_2H_5OH
100% H_2O
Percentage (by mass) of C_2H_5OH
Percentage (by mass) of H_2O
100% C_2H_5OH
0% H_2O
95.6% ethanol

Figure 17.32 Ethanol and water form a minimum boiling azeotrope. These data refer to a pressure of 1 atm.

with a boiling point of 64.9 °C. This can be fractionally distilled, removing essentially all of the water in the process and leaving anhydrous ethanol with just a few parts per million of residual benzene. This 'absolute alcohol' is almost pure, but the small content of toxic benzene makes it dangerous to drink.

Mixtures with strong interactions between the components show large *negative* deviations from Raoult's law. They may have a maximum boiling temperature as shown in Figure 17.33. Again, azeotropes are formed so that the two liquids cannot be separated by distillation. Nitric acid and water form this kind of mixture, with a maximum boiling point mixture containing 68% nitric acid. This means that distillation of nitric acid–water mixtures can only ever give 68% nitric acid (by mass).

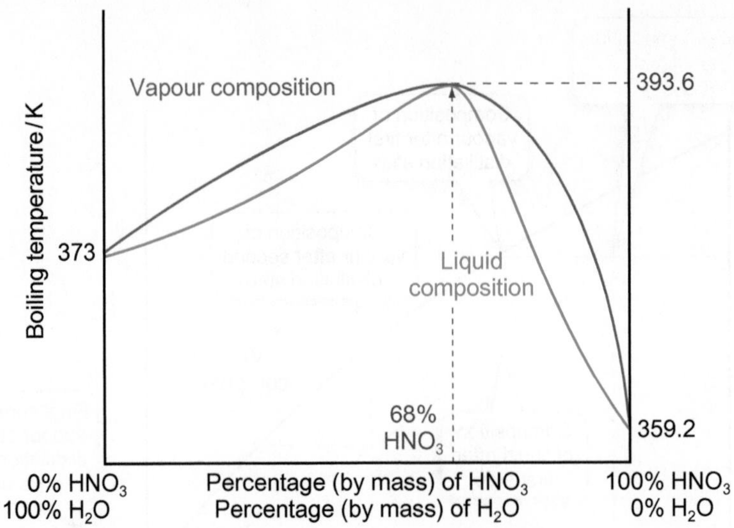
Vapour composition
393.6
Boiling temperature / K
373
Liquid
composition
68%
HNO3
359.2
0% HNO3
100% H2O
Percentage (by mass) of HNO3
Percentage (by mass) of H2O
100% HNO3
0% H2O

Figure 17.33 Nitric acid and water form a maximum boiling azeotrope. These data refer to a pressure of 1 atm.

Summary

- The Phase Rule

$$F = C - P + 2$$

where C is the number of components, P is the number of phases and F is the number of degrees of freedom.

- Two-component mixtures have one gas phase, but may have several liquid or solid phases.
- The enthalpy change of mixing, $\Delta_{mix}H$, usually determines whether two liquids are miscible or immiscible.
- Ideal solutions obey Raoult's law

$$p_{A(ideal)} = x_A p°_A$$

where p_A is the vapour pressure of component A, x_A is its mole fraction, and $p°_A$ is the vapour pressure of the pure component (i.e. when $x_A = 1$) at the temperature of the solution.

- Most solutions are non-ideal. Deviations from Raoult's law can be described in terms of an activity coefficient, γ

$$p_A = \gamma_A x_A p°_A$$

  - When $\gamma_A > 1$, the system displays a positive deviation from Raoult's law.
  - When $\gamma_A > 1$, the system displays a negative deviation from Raoult's law.

- The solute in dilute solutions obeys Henry's law for ideal dilute solutions

$$p_B = x_B K_B$$

where K_B is the Henry's law constant.

- Colligative properties depend only on the number of molecules (or ions) of the dissolved species, not on their nature.
- Osmosis occurs when a solution and pure solvent are separated by a semi-permeable membrane. Solvent passes out of the solution through the membrane.
- Distillation can be used to separate the components of liquid mixtures.
- Non-ideal solutions often form azeotropes, so the components cannot be completely separated by distillation.

? For practice questions on these topics, see questions 21–32 at the end of this chapter (pp.811–812).

Concept review

By the end of this chapter, you should be able to do the following.

- Draw and interpret single-component phase diagrams and account for their main features.
- Use phase diagrams to interpret the phase behaviour of pure elements and compounds.
- State and use the Clapeyron equation to describe the effect of pressure on melting temperatures.
- State and use the Clausius–Clapeyron equation to describe the effect of pressure on vaporization and sublimation temperatures.
- Describe the molecular properties that influence intermolecular interactions.
- Describe the origin and uses of the Lennard-Jones potential.
- Account for miscibility behaviour of mixtures in terms of the enthalpy change, $\Delta_{mix}H$, and entropy change, $\Delta_{mix}S$, of mixing.
- State and use Raoult's law to explain the properties of ideal solutions.
- Calculate activity coefficients and relate their values to the intermolecular interactions in solution.
- State and use Henry's law to explain the properties of ideal dilute solutions.
- Explain what is meant by colligative properties and apply the idea to osmotic pressure.
- Describe the separation of liquids by distillation and account for the formation of azeotropes.

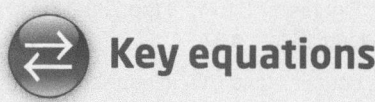

Key equations

The Clapeyron equation	$\frac{dp}{dT} = \frac{\Delta S_m}{\Delta V_m} = \frac{\Delta H}{T\Delta V_m}$	(17.4)
The Clausius–Clapeyron equation	$\ln p = \text{constant} - \frac{\Delta H}{R}\left(\frac{1}{T}\right)$	(17.6)
	$\ln\frac{p_2}{p_1} = \frac{\Delta H}{R}\left(\frac{1}{T_1} - \frac{1}{T_2}\right)$	(17.7)
Energy of interaction between two ions	$E_{PE\,(\text{ionic})} = \frac{q_1 q_2}{4\pi\varepsilon r}$	(17.11)
Lennard-Jones potential	$E_{PE\,(\text{total})} = +\frac{B}{r^{12}} - \frac{A}{r^6}$	(17.18)
The Phase Rule	$F = C - P + 2$	(17.20)
Raoult's law	$p_A = x_A p^\circ_A;\ p_B = x_B p^\circ_B$	(17.24)
Activity coefficient	$\gamma_A = \frac{p_{A(\text{real})}}{p_{A(\text{ideal})}}$	(17.27)
Henry's law	$p_B = x_B K_B$	(17.28)
van 't Hoff equation for osmotic pressure	$\pi = c \times RT$	(17.29)

Questions

More challenging questions are indicated by an asterisk *.

1 The figure shows a generalized phase diagram. (Section 17.1)

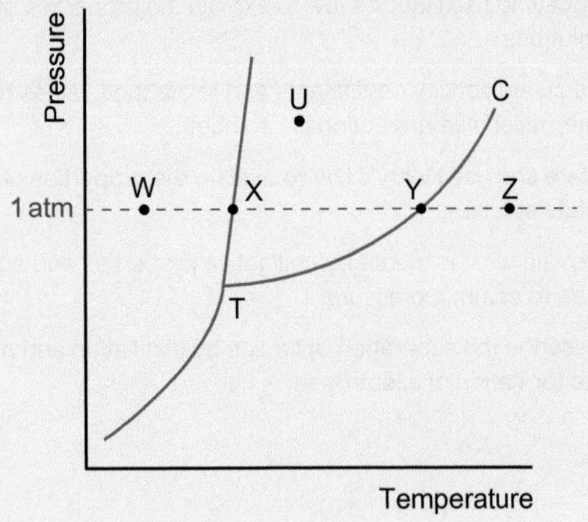
Pressure
U
C
W
X
Y
Z
1 atm
T
Temperature

(a) What phase(s) is/are present at W, X, Y, and Z?

(b) What is the significance of points X, Y, T, and C?

(c) What phase change would occur if you started at point U and:

(i) increased the temperature with the pressure constant;

(ii) increased the pressure with the temperature constant?

2 For the phase diagram below (Section 17.1):

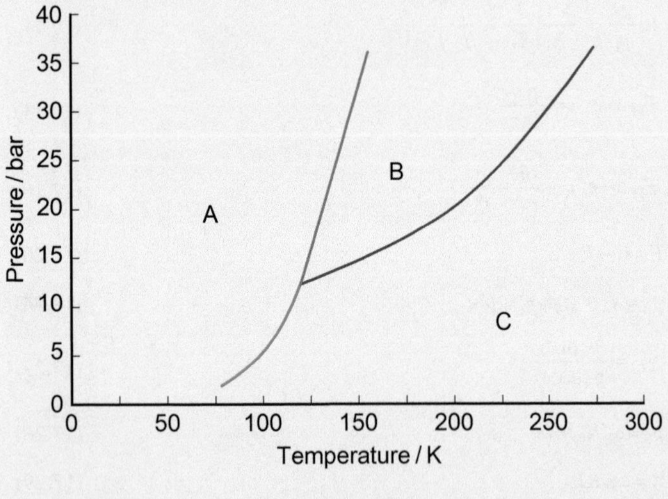
Pressure / bar
40
35
30
25
20
15
10
5
0
0
50
100
150
200
250
300
Temperature / K
A
B
C

(a) Which area of the diagram (A, B, or C) represents the liquid phase?

(b) Which phase(s) can exist at SATP?

(c) If a sample of this substance is cooled from 250 K to 50 K at a pressure of 20 bar, what phase changes will occur?

(d) What name is given to the change from phase A to phase C?

(e) What sign do the values of enthalpy change and entropy change have for the change from phase B to phase A?

(f) Estimate the conditions of the triple point of this substance.

(g) What are the melting and boiling points of this substance at 25 bar?

(h) What phase(s) can exist at 300 K and 40 bar?

(i) At a pressure of 25 bar, does the liquid or the solid have the higher density at the melting point?

(j) What is the lowest temperature at which the liquid phase can exist?

(k) Place the enthalpy changes for the following phase changes in increasing order of energy: A → B, A → C, B → C.

(l) What changes would happen when the pressure over a quantity of the substance is changed from 35 bar to 5 bar at a constant temperature of 135 K?

3 Use the phase diagrams in Figures 17.5 to 17.8 (pp.769–772) to answer the following questions. (Section 17.1)

(a) CO_2 fire extinguishers contain liquid CO_2 in equilibrium with gas. At 298 K, estimate the pressure of CO_2 (g) above the liquid.

(b) Which form of sulfur will crystallize from liquid sulfur at normal atmospheric pressure?

(c) Which has the lower density: Ice I or Ice II?

(d) What will happen when water is:

(i) heated from 250 K to 300 K at a pressure of 0.005 bar; then

(ii) compressed to 10 bar with the temperature held at 300 K; then

(iii) cooled to 250 K with the pressure held at 10 bar?

4 The conditions at the triple point of methanol are –98.15 °C and 0.5 atm while those at its critical point are 240 °C and 78.5 atm.

Use this information to *sketch* (not to scale) and fully label a phase diagram for methanol. (Section 17.1)

5 Mercury has a melting point of 234.3 K. At this temperature, the density of Hg (l) is 13.690 $g\,cm^{-3}$ and that of Hg (s) is 14.193 $g\,cm^{-3}$. The enthalpy change of fusion is 9.75 $J\,g^{-1}$. Find the pressure needed to change the melting temperature by 1 K. (Section 17.2)

6 Sulfur melts at a temperature of 115.2 °C under a pressure of 1 atm. In one process for its mining, sulfur is melted and pumped from underground deposits as a liquid under a pressure of 6 bar. Use the Clapeyron equation to estimate the lowest temperature at which sulfur is a liquid at this pressure. (Section 17.2)

(The enthalpy change of fusion at 115.2 °C is 53.67 $J\,g^{-1}$ and the densities of solid and liquid sulfur are 2.15 $g\,cm^{-3}$ and 1.811 $g\,cm^{-3}$, respectively.)

7 Explain the observation that, if a liquid evaporates from an open container, the temperature remains about constant but, if it evaporates from an insulated flask, the temperature falls. (Section 17.2)

8 The vapour pressure of tetrachloromethane (CCl_4) at 0 °C is 44.0 mbar and at 50 °C it is 422.0 mbar. Use these data to estimate the enthalpy change of vaporization of CCl_4 and to estimate its standard boiling point. (Section 17.2)

9* Mount Everest is the highest mountain on Earth with a height of 8850 m. At this altitude, the atmospheric pressure is about one-third that at sea-level. Find the melting temperature and boiling temperature of water at this pressure. (Section 17.2) (The densities of ice and water at 0 °C are 0.92 g cm^{-3} and 1.00 g cm^{-3}, respectively, and the enthalpy change of fusion of ice is +6.01 kJ mol^{-1}. The enthalpy change of vaporization of H_2O(l) at the normal boiling point is +40.7 kJ mol^{-1}.)

10* Determine the melting point of ice under a pressure of 500 bar. The density of ice at 0 °C is 0.9168 g cm^{-3} and that of water is 0.9987 g cm^{-3}. The enthalpy of fusion of ice is 6.008 kJ mol^{-1}. (Section 17.2)

11* The vapour pressure exerted by carbon disulfide (CS_2) was measured at different temperatures.

Temperature/°C	0	10	20	30	40
Pressure/bar	0.168	0.262	0.394	0.576	0.820

Determine the enthalpy and entropy changes of vaporization of CS_2. (Section 17.2)

12* The vapour pressures of sulfur dioxide in the solid and liquid states are given by:

$$\text{Solid: } \ln(p/\text{Pa}) = 29.28 - \left(\frac{4308}{T/K}\right)$$

$$\text{Liquid: } \ln(p/\text{Pa}) = 24.055 - \left(\frac{3284}{T/K}\right)$$

Calculate (i) the temperature and pressure of the triple point of sulfur dioxide and (ii) the enthalpy and entropy changes of fusion at the triple point. (Section 17.2)

13* The vapour pressure of a liquid is 2.026 kPa at 20.0 °C and 44.209 kPa at 80 °C.

(a) Find the enthalpy change of vaporization of the liquid.

(b) Estimate the normal boiling point of the liquid.

(c) Trouton's rule states that $\Delta_{vap}S$ for liquids ≈ 88 J K^{-1} mol^{-1}. Comment on the liquid under investigation here. (Section 17.2)

14 At 1 bar, two solid forms of tin (grey tin and white tin) are in equilibrium at 291 K. The densities of the two forms are 5750 kg m^{-3} and 7280 kg m^{-3} respectively. The entropy change for the transition is 8.8 J K^{-1} mol^{-1}. At what temperature will the two forms be in equilibrium at a pressure of 200 bar? (Section 17.2)

15 Calculate the energy of attraction between a Mg^{2+} ion and a Cl^- ion separated by 0.2 nm in a vacuum. (Section 17.3)

16 Compare the energy of interaction between two singly charged ions separated by a distance of 0.1 nm in (a) vacuum, (b) water (relative permittivity = 78), (c) a hydrocarbon oil (relative permittivity = 2). (Section 17.3)

17* Show that for a molecule following the Lennard-Jones potential, the equilibrium separation, r_0, is given by $r_0 = 2^{1/6}\sigma$.

18 For each of the following pairs, which compound would you expect to have the higher enthalpy change of vaporization? Give the reason for your answer in each case. (Section 17.3)

(a) Ethanol and methoxymethane

(b) Propane and methoxymethane

(c) Butane and propanone

(d) (*E*)-1,2-dichloroethene and (*Z*)-1,2-dichloroethene.

19 Arrange the following in order of increasing boiling point, giving detailed reasons for your answer. (Section 17.3)

He, H_2O, Ar, N_2, HCl, NaCl

20 The figure below shows the Lennard-Jones potential for three noble gases. Explain the variation in shapes of the three curves. (Section 17.3)

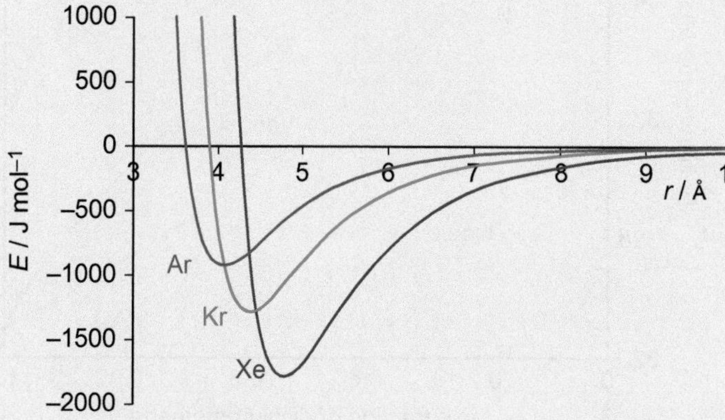
1000
500
0
−500
−1000
−1500
−2000
E / J mol−1
3
4
5
6
7
8
9
10
r / Å
Ar
Kr
Xe

21 A diver descends to a depth where the pressure is 5 atm. The diver's body contains around 5 dm^3 of blood. The Henry's law constant for N_2(g) in water at 310 K is 1.07 × 10^5 atm. (Section 17.4)

(a) Calculate the amount of nitrogen gas absorbed from the air in the diver's blood at a pressure of 1 atm and at 5 atm. (Assume that the solubility of nitrogen in the blood is the same as in water.)

(b) If all the gas dissolved at 5 atm was suddenly released, what volume would it occupy at 1 atm and 298 K?

22 The vapour pressure of pure $CHCl_3$ at 318 K is 58 kPa. What would be the partial vapour pressure of $CHCl_3$ above a mixture of 1 mol of $CHCl_3$ with 1 mol of ethanol, assuming the mixture behaves as an ideal solution? Comment on the fact that the measured value of the partial vapour pressure of $CHCl_3$ above the mixture is 42 kPa. (Section 17.4)

23 Ethanol and methanol form nearly ideal solutions. At 20 °C, the vapour pressure of pure ethanol is 5930 Pa and of methanol is 11 830 Pa. (Section 17.4) Calculate

(a) the vapour pressure of each compound and

(b) the total vapour pressure over a solution formed by mixing 50 g of each.

($M_r(CH_3OH)$ = 32.04; $M_r(C_2H_5OH)$ = 46.07.)

24* In a solution of ethanol ($p°$ = 0.174 bar) and 2-methylhexane ($p°$ = 0.059 bar), the mole fraction of ethanol, $x_{ethanol}$, is 0.90. The vapour in equilibrium with the solution has a total vapour pressure of 0.248 bar. The mole fraction of ethanol in the vapour is 0.67. Calculate the activity coefficients of each component in the solution. (Section 17.4)

25 A protein has a molar mass of 69 000 g mol^{-1}. Assuming ideal behaviour, calculate the osmotic pressure of a solution containing 20 g dm^{-3} of the protein at 298 K. (Section 17.4)

26 The figure shows a sketch of the vapour–liquid phase diagram for mixtures of butanone and dichloromethane. (Section 17.4)

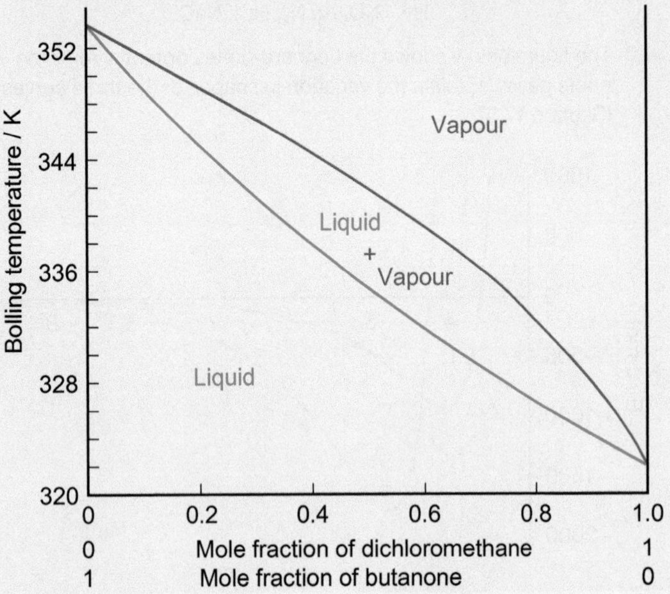

352
344
336
328
320
Boiling temperature / K
Vapour
Liquid
+
Vapour
Liquid
0
0.2
0.4
0.6
0.8
1.0
0
Mole fraction of dichloromethane
1
1
Mole fraction of butanone
0

For a solution with composition $x_{butanone}$ = 0.4 and $x_{dichloromethane}$ = 0.6:

(a) Estimate the boiling temperature of this mixture.

(b) Estimate the composition of the vapour that boils from it.

(c) If this vapour was condensed, what would be the boiling temperature of the resulting liquid?

(d) If the boiling and condensing cycles continued, what would be the composition of (i) the distillate and (ii) the residue?

27 A solution of iodoethane, I, and propanone, P, with a mole fraction of I, x_I = 0.55, had a partial vapour pressure of I of 28.44 kPa and a partial vapour pressure of P of 19.21 kPa at 50 °C. At this temperature, the saturated vapour pressure of I is 47.12 kPa and of P is 37.38 kPa. Calculate the activity coefficients of both components in the solution. (Section 17.4)

28 When a patient suffers severe dehydration, why will medical treatment involve giving 'saline' solution rather than pure water? (Section 17.4)

29 Use the Phase Rule to calculate the number of degrees of freedom at the triple point of water. Comment on your answer. (Section 17.4)

30 The vapour pressure over a mixture of 2 mol of hexane and 1 mol of octane is 9.6 kPa at 40 °C. An equimolar solution has a vapour pressure of 8.2 kPa at the same temperature. What are the vapour pressures of the two pure liquids? (Section 17.4)

31 The vapour pressure of pure toluene is 0.0285 bar at 20 °C and that of benzene is 0.0974 bar. One mole of each compound were mixed and formed an ideal solution. Calculate the mole fraction of each compound in the vapour over the solution. (Section 17.4)

32 The transport of water up a tree occurs partially through osmosis; the concentration of sugar in the tree sap is higher than the water around the tree roots. The sap in a certain species of tree can be represented as a 30 g dm^{-3} solution of sucrose in water. Find the osmotic pressure generated by this solution. (Section 17.4)

18

Isomerism and stereochemistry

This chapter builds on the following topics:

- Chemical formulae Section 1.3, p.19
- Non-covalent interactions Section 1.8, p.52
- Drawing organic compounds Section 2.2, p.73
- Carbon frameworks and functional groups Section 2.3, p.77
- Naming organic compounds Section 2.4, p.79
- Hydrocarbons Section 2.5, p.79
- Valence shell electron pair repulsion theory Section 5.2, p.223
- Bond polarity and polar molecules Section 5.3, p.235
- Valence bond theory for polyatomic molecules Section 5.4, p.236

◀ Wort is boiled with hops in a vessel called a copper for at least one hour.

Bitter isomers in beer

There are a large number of beers with different tastes and strengths. In the UK alone, over 2000 ales, keg beers, and lagers are produced. A distinctive feature of many popular beers is the bitter flavour and aroma, which are due to the use of hops in the brewing process.

Hops are the dried ripe flowers of a twining vine plant (*Humulus lupulus*). Of the hundreds of compounds present in hops, brewers have a particular interest in compounds called humulones because these compounds produce the bitter taste. In particular, (–)-humulone, though not bitter itself, rearranges on heating to form bitter compounds. (–)-Humulone contains a carbon atom (labelled * in the figure) with four different substituents. Such a carbon is called a **chiral centre**. There are two ways of arranging the four different groups around the chiral centre, as shown by the structures of (–)-humulone and (+)-humulone, but only (–)-humulone occurs in hops. (Two compounds related in this way are special types of chiral isomers and are called **enantiomers**, as discussed in Section 18.4, p.838.)

During the brewing process, another kind of isomerism comes into play. Hops are heated with a liquid called wort, which is produced from malted barley (barley that is allowed to sprout and is then dried). On heating, (–)-humulone from the hops undergoes a **rearrangement reaction** to form *cis*- and *trans*-isohumulone and it is these compounds that provide most of the bitterness in beer. In both compounds the 5-membered ring is planar, but in *cis*-isohumulone the HO– and $Me_2C{=}CHCH_2$– groups are on the same side of the ring, whereas *trans*-isohumulone has these groups on opposite sides of the ring.

▲ During the 16-day Oktoberfest in Munich in 2014, around seven million litres of beer were drunk by over six million people.

Cis- and *trans*-isohumulone are both **structural isomers** of (–)-humulone as they have the same molecular formula as (–)-humulone ($C_{21}H_{30}O_5$) but the atoms are joined together in a different way. The rearrangement of atoms in (–)-humulone to form the bitter tasting compounds in beer is called an **isomerization reaction**. Without this isomerization, the traditional pint of bitter would not live up to its name.

▼ During the beer-brewing process, (–)-humulone forms *cis*- and *trans*-isohumulone.

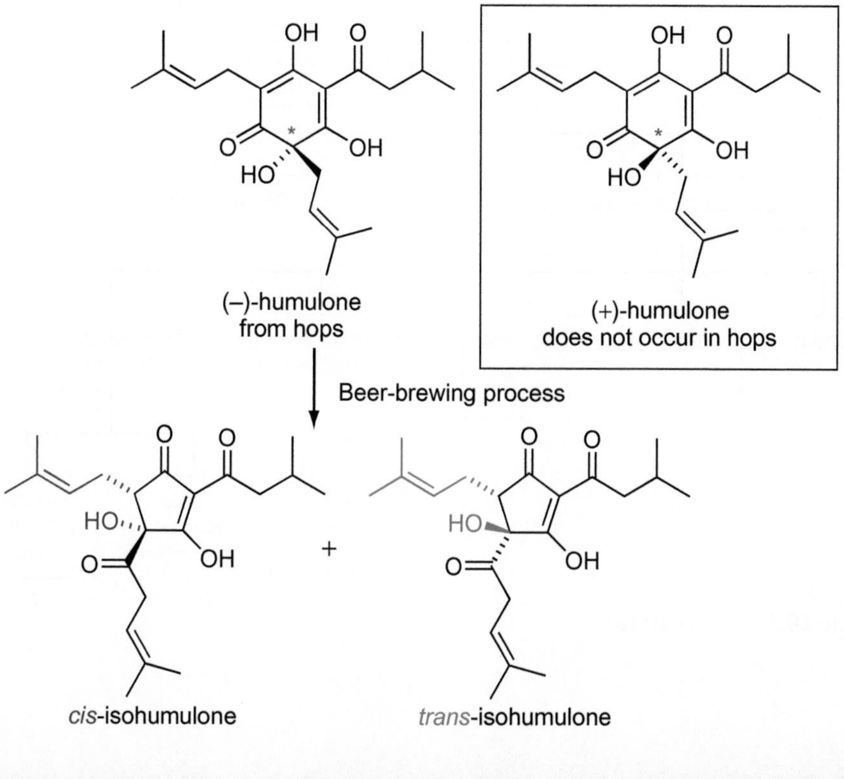
OH
O
OH
HO
*
(–)-humulone
from hops
(+)-humulone
does not occur in hops
Beer-brewing process
HO
OH
+
cis-isohumulone
trans-isohumulone

▲ Light-struck beer: beer decomposes on exposure to sunlight to form a 'skunky' flavour, due to the formation of a thiol, RSH (Section 2.6, p.95). Hence, beer is stored in opaque cans, or green or brown bottles.

Functional groups are introduced in Section 2.3 (p.77).

For simple hydrocarbons, such as methane (CH_4), ethane (C_2H_6), and ethene (C_2H_4), the molecular formula provides enough information to work out the structure of the compound. However, for larger, more complex molecules, this is not the case because the atoms can be arranged in different ways. Molecules that have the same molecular formula but differ in the way in which their atoms are arranged are called **isomers.** Isomers may have different functional groups, different carbon skeletons, or just different shapes. It is important to understand the shape of a molecule because it influences physical and chemical properties and the way the molecule acts in a biological system. The presence of different isomers is called **isomerism**.

Molecular models are invaluable when learning about the stereochemistry of organic molecules. Having a physical model that you can construct and play with in your hands is an important way to learn about the shapes of molecules.

The different types of isomerism in organic compounds are introduced in Section 18.1. An important aim of this chapter is to show how the spatial arrangements of atoms in molecules determine their **stereochemistry**, and hence their shape, and this is discussed in the following sections. An understanding of stereochemistry will prove invaluable when you study the reactions of functional groups—and it is crucial to the molecular machinery of life itself.

18.1 Isomerism

Isomerism in metal-containing compounds is discussed in Section 28.3 (p.1272).

Isomerism is the term used to describe the existence of different isomers. The different types of isomers are shown in Figure 18.1. There are two main classes of isomerism depending on the different ways in which the atoms are arranged in the isomers.

- **Structural (constitutional) isomers** are compounds in which the atoms are joined together in a different order.
- **Stereoisomers** are compounds in which the atoms are joined in the same order but the positions of the atoms in space are different.

The word isomer comes from the Greek words, *isos* meaning equal and *meros*, meaning part, or to share.

Structural isomers

Structural isomers are further subdivided into three types:

- chain isomers;
- position isomers;
- functional group isomers.

To help visualize different structural isomers, as a very approximate analogy, consider two bead bracelets containing the same number of blue and silver coloured beads. If the order of the blue and silver beads on the bracelets is different, then this can be likened to having different structural isomers.

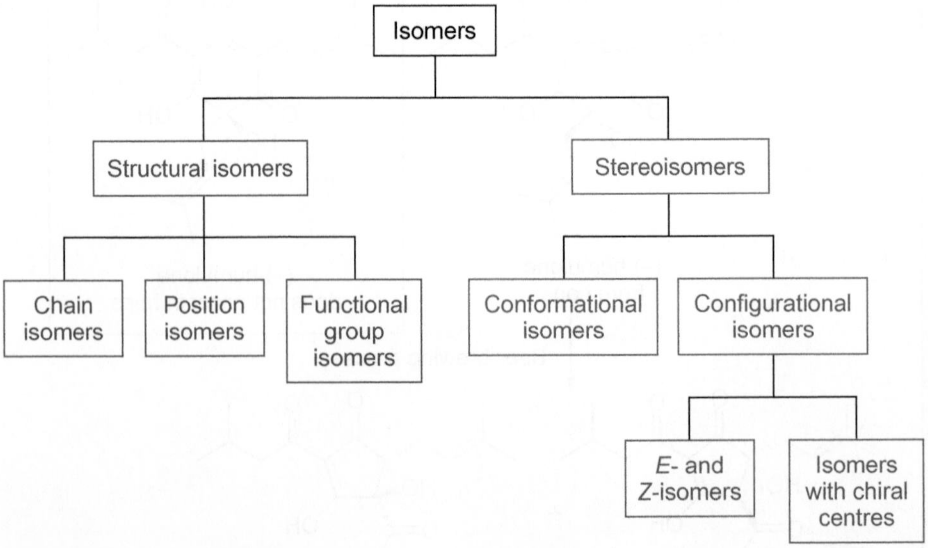
Isomers
Structural isomers
Stereoisomers
Chain isomers
Position isomers
Functional group isomers
Conformational isomers
Configurational isomers
E- and Z-isomers
Isomers with chiral centres

Figure 18.1 Types of isomers.

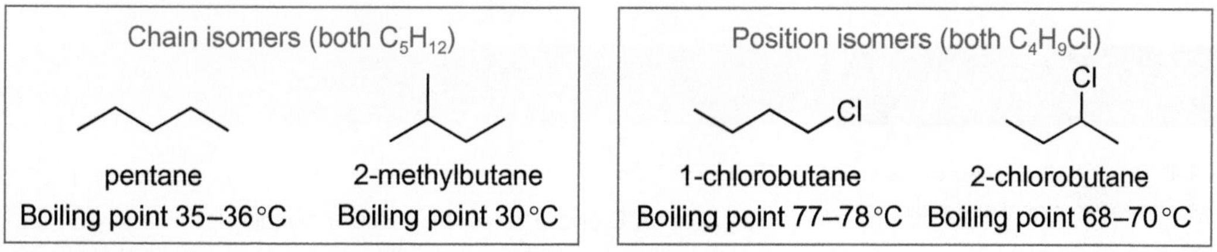
Chain isomers (both C_5H_{12})
pentane
Boiling point 35–36 °C
2-methylbutane
Boiling point 30 °C
Position isomers (both C_4H_9Cl)
Cl
Cl
1-chlorobutane
Boiling point 77–78 °C
2-chlorobutane
Boiling point 68–70 °C

Figure 18.2 Examples of chain isomers and position isomers.

Chain isomers are molecules in which the carbon chain is connected in different ways. Pentane and 2-methylbutane are examples of chain isomers (Figure 18.2). They both have the molecular formula C_5H_{12} but, whereas pentane ($CH_3CH_2CH_2CH_2CH_3$) is a straight-chain alkane, 2-methylbutane ($(CH_3)_2CHCH_2CH_3$) has a branched chain (Section 2.5, p.81). The different structures of these alkanes lead to differing properties. For example, 2-methylbutane has a lower boiling point than pentane.

Position isomers are molecules in which one (or more) functional group(s) is at a different position on the carbon chain. For example, there are two possible straight-chain chloroalkanes with the molecular formula C_4H_9Cl (Figure 18.2). These molecules arise because the –Cl functional group can be situated on one of two carbon atoms. A further important example relates to benzene derivatives containing two or more substituents. For dichlorobenzene, for example, there are three position isomers because there are three ways of arranging the two –Cl functional groups on the benzene ring.

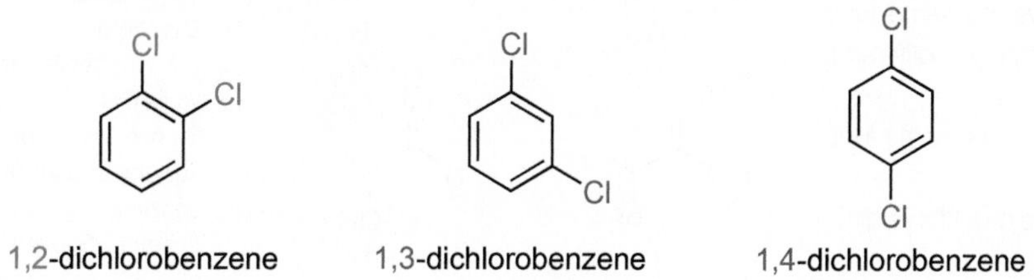
Cl
Cl
Cl
Cl
Cl
Cl
1,2-dichlorobenzene
1,3-dichlorobenzene
1,4-dichlorobenzene

Functional group isomers are molecules with the same molecular formula but different functional groups. An example is propan-1-ol ($CH_3CH_2CH_2OH$) and methoxyethane ($CH_3CH_2OCH_3$), which both have the molecular formula C_3H_8O. In addition to having different physical properties (such as boiling points), they have quite different chemical properties because of their dissimilar functional groups.

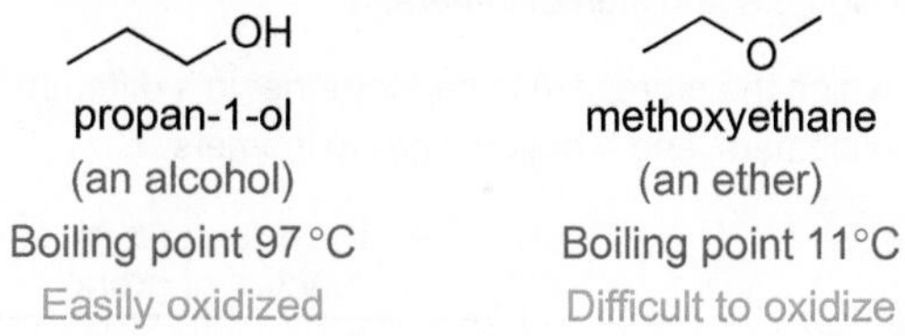
OH
O
propan-1-ol
(an alcohol)
Boiling point 97 °C
Easily oxidized
methoxyethane
(an ether)
Boiling point 11°C
Difficult to oxidize

Worked Example 18.1 will help you to practice identifying structural isomers.

Stereoisomers

The atoms in stereoisomers are connected in the same way, but are arranged differently in space. There are two kinds of stereoisomers, called conformational isomers and configurational isomers. **Conformational isomers** usually interconvert rapidly at room temperature by rotation about carbon–carbon bonds. This is discussed in the following section. In contrast, **configurational isomers** do not usually interconvert at room temperature since this normally involves breaking and reforming bonds. There are two types of configurational isomers, ***E*- and *Z*-isomers** and **isomers with chiral centres**, and these are discussed in Sections 18.3 and 18.4, respectively.

Position isomers are often called **regioisomers** (see Section 19.3, p.909).

The names *ortho-*, *meta-*, and *para-* can also be used to designate the positions of the groups in disubstituted benzenes (Section 2.5, p.88).

1,4-Dichlorobenzene (or *para*-dichlorobenzene) has been used as a disinfectant in, for example, deodorizer blocks found in urinals. It is less commonly used today because of concerns over safety—it has been identified as a possible carcinogen.

Alcohols (ROH) have higher boiling points than ethers (ROR) because of hydrogen bonding (Section 1.8, p.54). In alcohols, hydrogen bonds form between molecules. It requires energy to break these bonds so alcohols have comparatively high boiling points. Ethers cannot form hydrogen bonds so they have much lower boiling points.

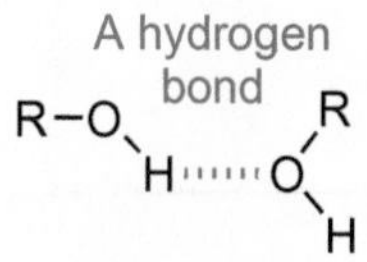
A hydrogen bond
R–O
H
O
R
H

Conformational isomers are sometimes called **conformers** or **rotational isomers** or **rotamers**.

Isomers with chiral centres are sometimes called **isomers with stereogenic centres**.

Worked example 18.1 Structural isomers

For the following six compounds, **A–F**, identify which are chain isomers, which are position isomers, and which are functional group isomers.

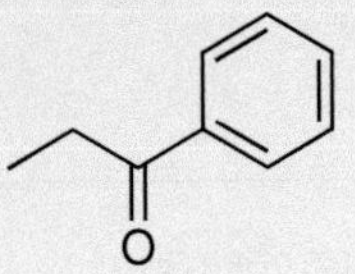
A

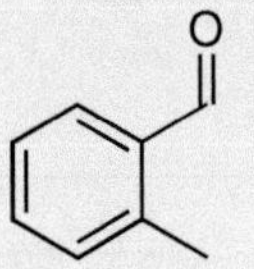
B

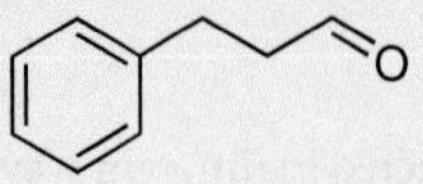
C

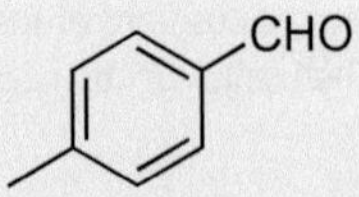
D

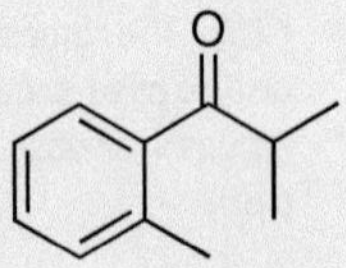
E

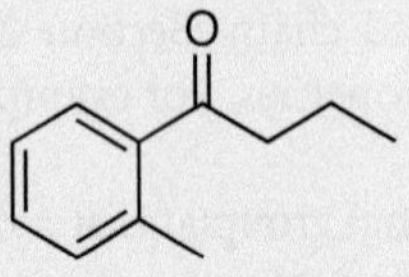
F

Strategy

Determine the molecular formula of each compound. Those with the same molecular formula, and different structures, are isomers.

- To be *chain* isomers, the molecules must have the same functional groups but the carbon chain must be arranged differently, for example, straight or branched.
- To be *position* isomers, the molecules must have the same functional groups but at different positions.
- To be *functional group* isomers, the compounds must have different functional groups.

Solution

Compounds **E** and **F** have the same molecular formula ($C_{11}H_{14}O$) and the same functional groups. The carbon chain of the ketone is straight in **F** but branched in **E**. They are chain isomers.

Compounds **B** and **D** have the same molecular formula (C_8H_8O) and the same functional groups. The methyl groups on the benzene rings are at different positions in **B** and **D**, so these compounds are position isomers.

Compounds **A** and **C** have the molecular formula $C_9H_{10}O$. As compound **A** is a ketone and **C** is an aldehyde, they are functional group isomers.

Question

For the following four compounds, **G–J**, identify which are chain isomers, which are position isomers, and which are functional group isomers. (Ph stands for phenyl, C_6H_5–.)

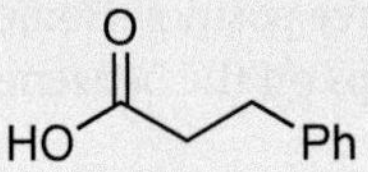
G

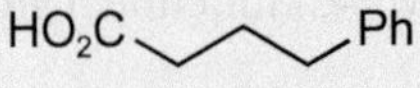
H

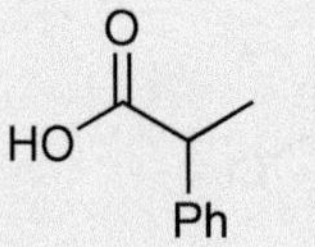
I

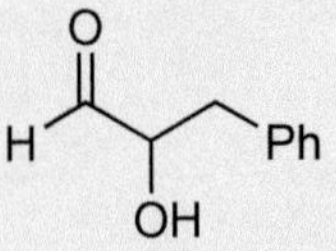
J

Compound **B** is called 2-methylbenzaldehyde, while **G** is 3-phenylpropanoic acid. To name substituted benzene rings, see Chapter 2.5 (p.88).

Summary

- Isomers are non-identical compounds with the same molecular formula.
- The two main classes of isomers are called structural isomers and stereoisomers.
- Structural (constitutional) isomers are compounds in which the atoms are joined together in a different order. The three types of structural isomers are chain isomers, position isomers, and functional group isomers.

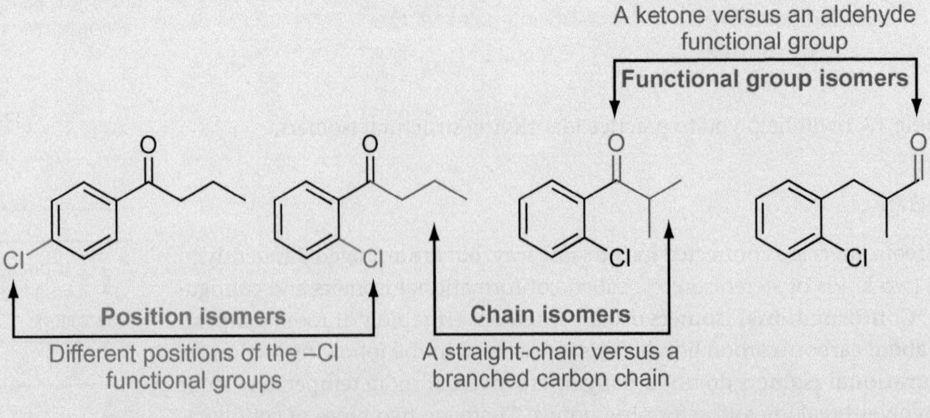
A ketone versus an aldehyde functional group
Functional group isomers
Position isomers
Different positions of the –Cl functional groups
Chain isomers
A straight-chain versus a branched carbon chain

- Stereoisomers are compounds in which the atoms are joined in the same order but the positions of the atoms in space are different. The two types of stereoisomers are *E*- and *Z*-isomers and isomers with chiral centres.

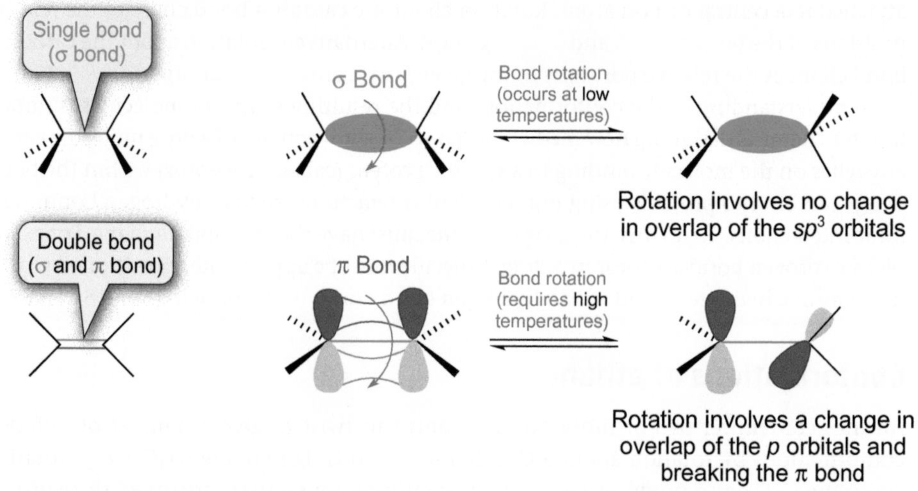
Single bond
(σ bond)
Double bond
(σ and π bond)
σ Bond
Bond rotation
(occurs at low
temperatures)
Rotation involves no change
in overlap of the sp3 orbitals
π Bond
Bond rotation
(requires high
temperatures)
Rotation involves a change in
overlap of the p orbitals and
breaking the π bond

Figure 18.3 Rotation about single and double bonds.

The hashed-wedged line notation shows the three-dimensional arrangement of atoms (Section 2.2, p.75).

sp^3 hybridization is discussed in Section 5.4 (p.237).

Conformers cannot be separated from one another because they rapidly interconvert.

Analysis of the physical and chemical properties of molecules as a function of their molecular shape is called conformational analysis. **Conformational analysis** was developed by Derek H.R. Barton and Odd Hassel who, in 1969, shared a Nobel Prize for their work.

18.2 Conformational isomers

The different three-dimensional arrangements of atoms that result from rotation about a single bond are called **conformations.** A specific conformation of a molecule that is relatively stable is called a **conformer** (**conformational isomer**). Conformers are different spatial arrangements of the same compound that are readily interconverted by rotation about a single bond. So, at normal temperatures, there is no way you can tell the difference between conformers from their bulk properties because they interconvert so readily. This is a major difference from configurational isomers, which have observable differences in their bulk properties (Sections 18.3, p.833, and 18.4, p.838).

For single bonds, such as C–C bonds, there is generally no restriction to rotation because rotation does not change the orbital overlap (see Figure 18.3). This is not the case for double bonds. For example, rotation about the C=C bond of alkenes is severely restricted because this requires breaking the π bond (so that high temperatures are needed).

This section discusses how rotation about a C–C bond changes the three-dimensional shape of acyclic (non-cyclic) alkanes, such as ethane (CH_3CH_3) and butane ($CH_3CH_2CH_2CH_3$), and cyclic alkanes, such as cyclohexane (C_6H_{12}) and its derivatives. Rotation about a C–C bond changes the relative positions of groups in space and to understand this you need to be able to visualize bond rotation in molecules (the best way to do this is to build molecular models and investigate what happens when the single bonds are rotated). Figure 18.4 shows a molecule with four different groups (represented by different coloured circles)

A **drug** is a substance that, when taken into an organism's body, affects the functions of that organism. Chemists commonly use the term drug to describe a medicine. More correctly, a **medicine** is the product taken by the patient (the tablet, suspension, etc.), which contains the drug and other components.

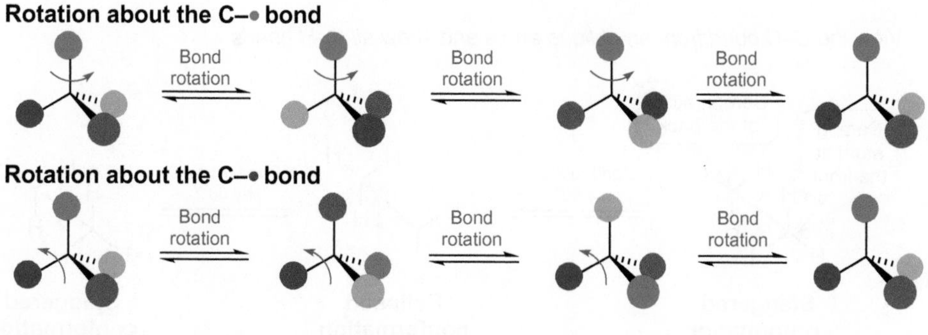
Rotation about the C–• bond
Bond
rotation
Bond
rotation
Bond
rotation
Rotation about the C–• bond
Bond
rotation
Bond
rotation
Bond
rotation

Figure 18.4 Rotation about single bonds in a molecule with four different groups attached to a central carbon atom.

In modern research, computerized molecular modelling is a standard method for explaining and predicting the structure and reactions of compounds, including drug molecules.

attached to a central carbon atom. Rotation about the carbon–• bond changes the relative positions of the purple, pink, and orange groups. Alternatively, rotation about the carbon–• bond changes the relative positions of the green, pink, and orange groups.

An understanding of the conformation, and the resulting shape, of molecules is important for chemists designing new medicines. The biological activity of a drug molecule generally relies on the molecule binding to a specific protein (called a **receptor**) within the body. Drugs can bind to proteins using non-covalent interactions, such as hydrogen bonds, and ionic interactions. To do this, the drug molecule must have the appropriate shape. For example, a hydrogen bond acceptor in a drug molecule must be appropriately positioned so that it can form a hydrogen bond with a hydrogen bond donor in the protein (Box 18.1, p.822).

Non-covalent interactions are introduced in Section 1.8 (p.52), and are used to explain the biological action of captopril (p.69).

Conformations of ethane

Alkanes containing two or more carbon atoms can twist to give a number of different conformations by rotation about a C–C bond or bonds. For ethane ($H_3C–CH_3$), rotation about the C–C bond produces two extreme conformations—these are called the **eclipsed conformation** and the **staggered conformation** (shown in Figure 18.5). In the eclipsed conformation, the three hydrogen atoms, on each of the carbon atoms, are as *close* as possible to one another. In the staggered conformation, the three hydrogen atoms, on each of the carbon atoms, are as *far apart* as possible from one another.

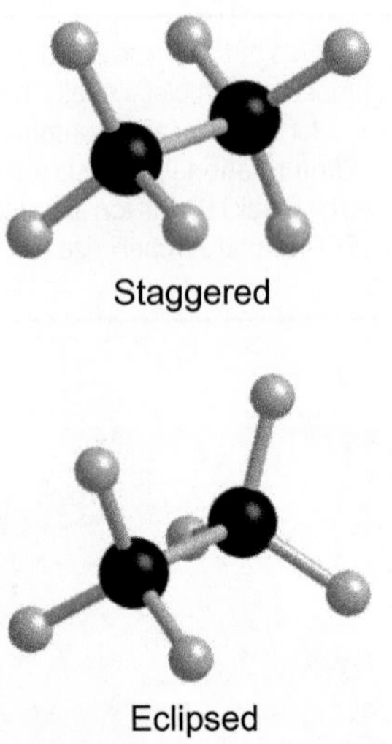
Staggered
Eclipsed

Models of the staggered and eclipsed conformations of ethane ($H_3C–CH_3$).

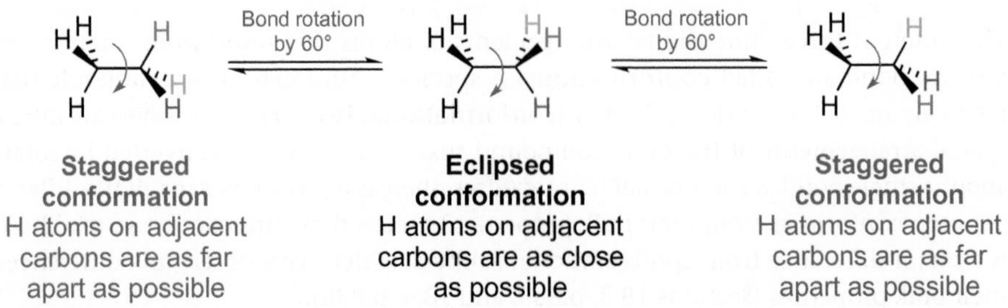
Bond rotation by 60°
Bond rotation by 60°
Staggered conformation
H atoms on adjacent carbons are as far apart as possible
Eclipsed conformation
H atoms on adjacent carbons are as close as possible
Staggered conformation
H atoms on adjacent carbons are as far apart as possible

Figure 18.5 Staggered and eclipsed conformations of ethane.

On rotation about the C–C bond in ethane, the positions of the H, H, and H atoms change with respect to the neighbouring three H atoms. In the staggered conformation, the H, H, and H atoms are the greatest distance from the three H atoms. However, on rotation of the C–C bond by 60°, the shape of ethane changes to the eclipsed conformation, in which the H, H, and H atoms are the shortest distance from the three H atoms.

The three-dimensional arrangement of the atoms in the different conformations of ethane can also be represented using **sawhorse projections** or **Newman projections**. In a sawhorse projection, the C–C bond is drawn at an oblique angle. The carbon atom nearer to you is on the left and is drawn below the carbon atom that is further from you (Figure 18.6).

Sawhorse projections are named after a primitive wooden structure used for sawing logs.

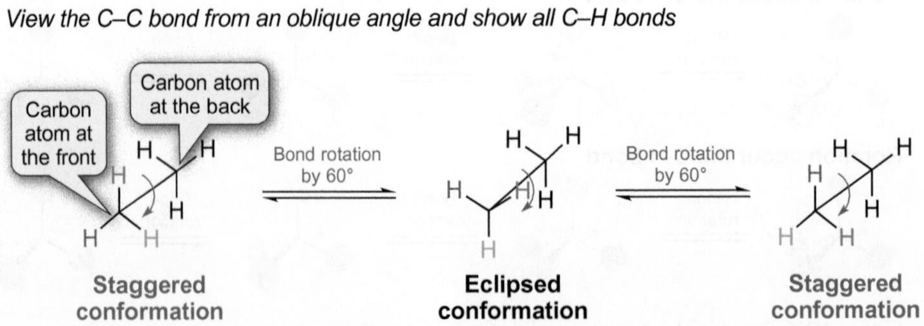
View the C–C bond from an oblique angle and show all C–H bonds
Carbon atom at the front
Carbon atom at the back
Bond rotation by 60°
Bond rotation by 60°
Staggered conformation
Eclipsed conformation
Staggered conformation

Figure 18.6 Sawhorse projections of the staggered and eclipsed conformations of ethane.

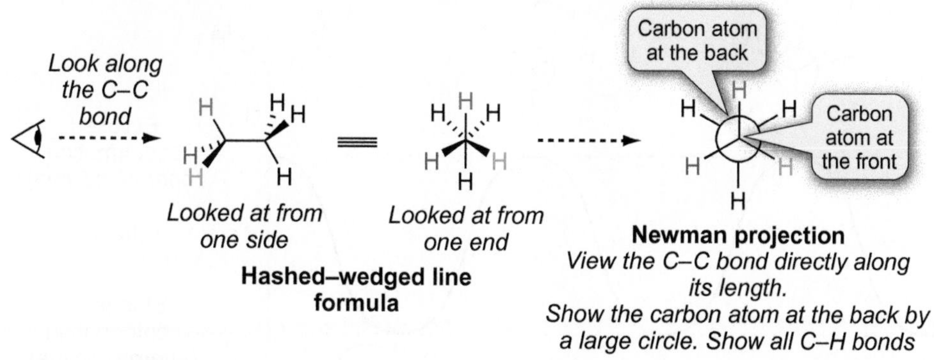

Look along the C–C bond
Carbon atom at the back
Carbon atom at the front
Looked at from one side
Looked at from one end
Hashed–wedged line formula
Newman projection
View the C–C bond directly along its length.
Show the carbon atom at the back by a large circle. Show all C–H bonds

Figure 18.7 Converting a hashed–wedged line formula into a Newman projection.

For a Newman projection, you need to imagine you are looking directly along the C–C bond (Figure 18.7). The carbon atom nearer to you is represented by the point at which the three other bonds to carbon intersect. The carbon atom further from you is represented by a circle and the three bonds are drawn from the edge of the circle.

Newman projections are most useful for showing the positions of substituents on adjacent carbon atoms, as shown in Figure 18.8 for ethane. The angle between the C–H bonds on the front and back carbon atoms is called the **dihedral (torsional) angle**. In the staggered conformation, the smallest dihedral angle is 60° (the dihedral angles between H and the three H atoms are 60°, 180°, and 240°). In the eclipsed conformation, the C–H bonds on each carbon atom are superimposed and so the smallest dihedral angle is 0° (the dihedral angles between H and the three H atoms are 0°, 120°, and 240°).

The size of the dihedral angle determines the different energies of the different conformations of ethane. A plot of the relative energy of the conformation versus the dihedral angle is shown in Figure 18.9. As the dihedral angle changes from 0° to 360°, ethane adopts three identical eclipsed and three identical staggered conformations. The graph shows that the eclipsed conformation is at a higher energy than the staggered conformation so the staggered conformation is more stable than the eclipsed conformation. You can explain the difference in energies by considering the electrons in the C–H bonds. Pairs of electrons in different C–H bonds repel one another and, when they get very close to one another, as in the eclipsed conformation, the repulsion is particularly great and this raises the energy of the conformation. The energy difference between the staggered and eclipsed conformations, due to the repulsion of bonding electrons, is called **torsional strain (torsional energy)**. In ethane, the torsional strain is approximately 12 kJ mol^{-1}.

Since the eclipsed and staggered conformations in ethane have different energies, rotation about the C–C bond is not completely free. To convert one staggered conformation into another, by rotation about the C–C bond, ethane requires an additional 12 kJ mol^{-1} of energy so as to overcome the torsional strain—this additional energy is also called the

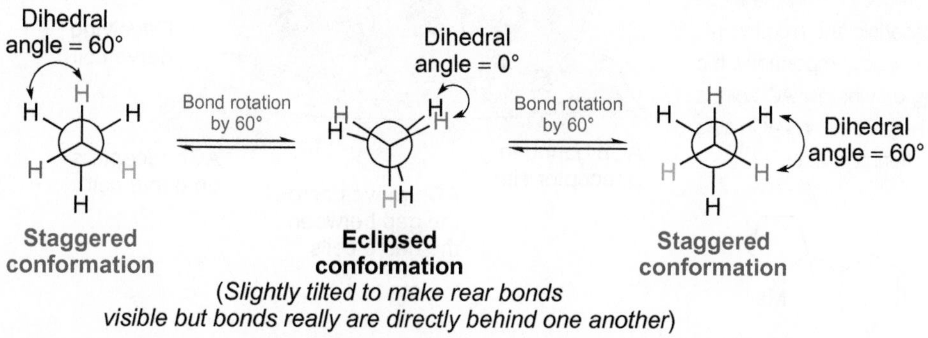

Dihedral angle = 60°
Dihedral angle = 0°
Dihedral angle = 60°
Bond rotation by 60°
Bond rotation by 60°
Staggered conformation
Eclipsed conformation
Staggered conformation
(Slightly tilted to make rear bonds visible but bonds really are directly behind one another)

Figure 18.8 Newman projections for the staggered and eclipsed conformations of ethane.

The Newman projection is named after American chemist Melvin Spencer Newman who proposed the representation in 1952.

In the eclipsed conformation, although the hydrogen atoms are close to one another, there is little interaction because hydrogen atoms are so small. With larger atoms, the interaction can be much greater.

For the eclipsed conformation, the Newman projection clearly shows the hydrogen atoms on the front carbon block our view of the hydrogen atoms on the rear carbon–like in a solar eclipse, when the Moon blocks our view of the Sun.

Different conformations of molecules are detected (usually at low temperatures) by spectroscopic techniques including NMR spectroscopy (Sections 10.7, p.496, and 12.3, p.578).

In general, the higher the temperature, the more probable is rotation about the C–C bond. This is because molecules have greater kinetic energy at higher temperature. As the temperature is lowered, the amount of energy available to cause rotation about the C–C bond decreases so interconversion is less probable.

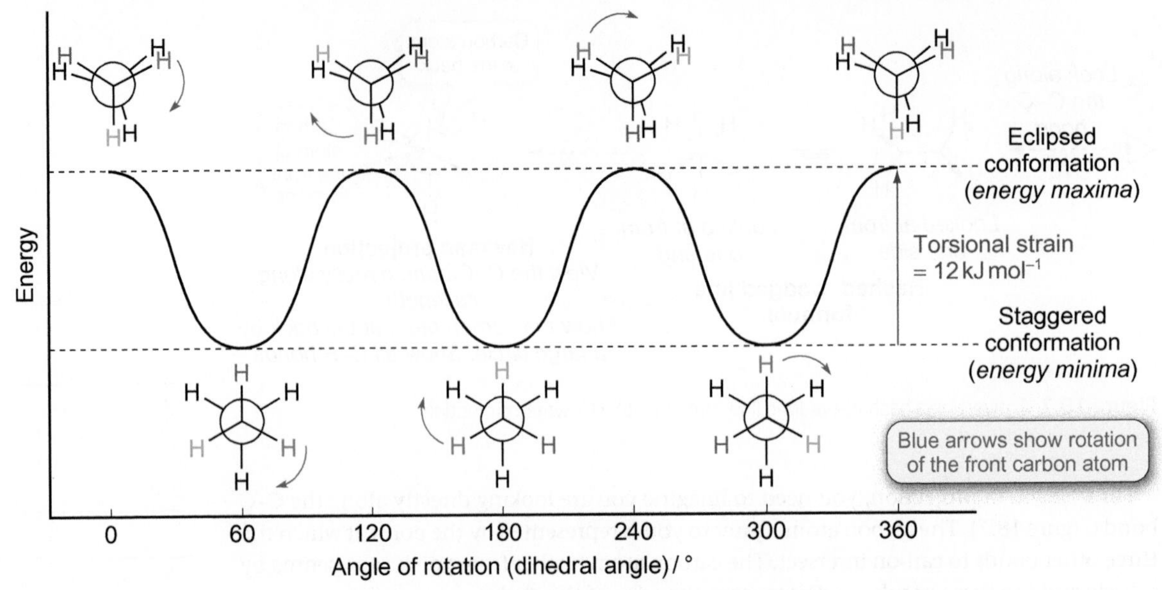
Eclipsed conformation (energy maxima)
Torsional strain = 12 kJ mol−1
Staggered conformation (energy minima)
Blue arrows show rotation of the front carbon atom
Energy
0
60
120
180
240
300
360
Angle of rotation (dihedral angle) / °

Figure 18.9 The relative energies of the different conformations of ethane.

> The plot shown in Figure 18.9 gives a cosine wave—a curve representing periodic oscillations of constant amplitude (like alternating current in an electrical circuit).

energy barrier to rotation. However, this is such a small energy barrier that, even at room temperature, an ethane molecule has enough energy to undergo C–C bond rotation. At 25 °C, one staggered conformation is converted into another at a rate of about 1×10^{11} times per second. (Even though the bond is rotating so quickly, ethane spends more time in the staggered conformation.)

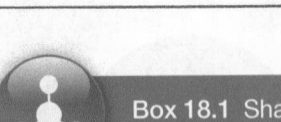

Box 18.1 Shaping up for maximum activity

Acetylcholine (abbreviated ACh) is found in the central nervous system of the human body and it plays an important role in transmitting nerve impulses (Box 4.11, p.212 and Box 16.5, p.754).

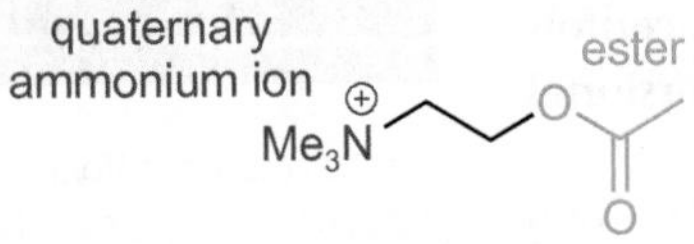
quaternary ammonium ion
ester
Me3N

acetylcholine

ACh triggers nerve impulses by binding to two different types of receptors. These are called the *muscarinic receptor* and the *nicotinic receptor*. When bound to ACh, the muscarinic and nicotinic receptors exert different biological effects in the body. Importantly, the heartbeat increases or decreases depending on whether ACh binds to nicotinic or muscarinic receptors, respectively. So how is ACh able to bind to two different receptors that have differing shapes?

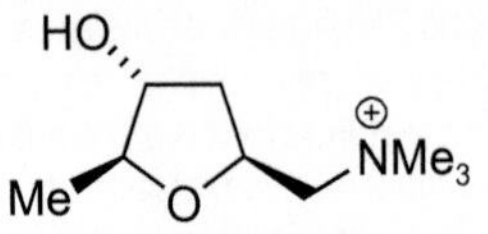

muscarine

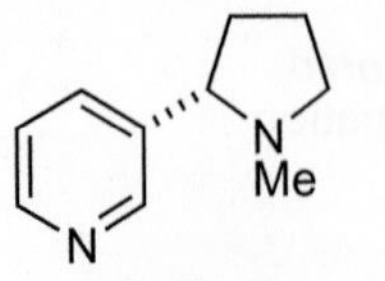

nicotine

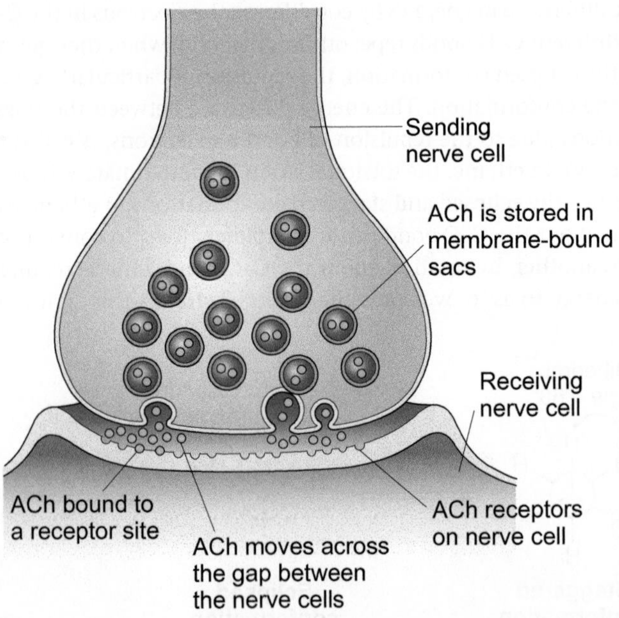
Sending nerve cell
ACh is stored in membrane-bound sacs
Receiving nerve cell
ACh bound to a receptor site
ACh moves across the gap between the nerve cells
ACh receptors on nerve cell

▲ ACh moves from one nerve cell to receptors in another nerve cell. When ACh is bound to the receptors, the nerve impulse is continued.

➜

▲ The muscarinic receptor is named after a compound called muscarine, which selectively binds to this type of receptor. Muscarine is a natural product found in the poisonous mushroom *Amanita muscaria* (fly agaric).

▲ Nicotine, found in the tobacco plant (*Nicotiana tabacum*), selectively binds to the nicotinic receptor, which induces the relaxing or stimulating effect of smoking a cigarette.

The answer lies in the ability of ACh to adopt different conformations by rotation about the central C–C bond (as shown using sawhorse projections in Figure 1). The relative positions of the quaternary ammonium ion and ester groups change on C–C bond rotation and this is important because these groups form noncovalent bonds with the nicotinic and muscarinic receptors. For binding to the muscarinic receptor, ACh adopts a conformation that has the ester and quaternary ammonium ion groups in the appropriate position to match the shape of this receptor. For binding to the nicotinic receptor, ACh adopts a different conformation so as to match the shape of this receptor.

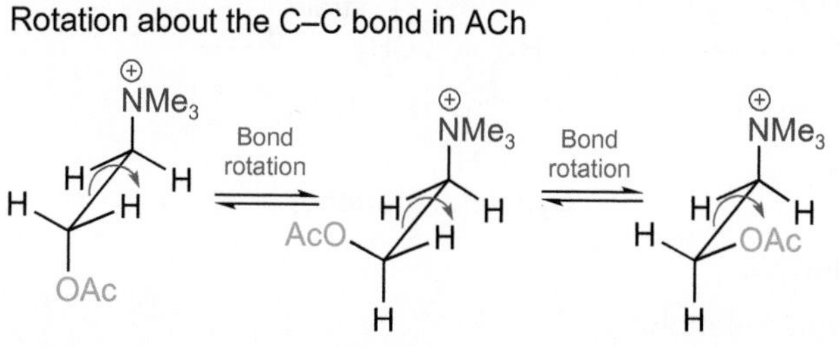
Rotation about the C–C bond in ACh
NMe3
Bond rotation
NMe3
Bond rotation
NMe3
H
H
H
H
OAc
AcO
H
H
H
H
H
H
H
OAc
H

▲ **Figure 1** Sawhorse projections showing rotation about the C–C bond in ACh. (Ac is an abbreviation for the CH_3CO– group.)

But which conformation of ACh binds to the muscarinic receptor and which to the nicotinic receptor? To answer this question medicinal chemists have prepared *conformationally rigid* analogues of ACh in the laboratory. These molecules have the same (or similar) functional groups as ACh but can only adopt one conformation because the central C–C bond is part of a ring. (This is a general method used by chemists to lock a molecule into a particular conformation.)

For example, in the *cis*- and *trans*-cyclopropane derivatives shown in Figure 2, a CH_2 group links to the C–C bond in ACh and this prevents bond rotation. The relative positions of the quaternary ammonium ion and ester groups are fixed and these molecules have different activities at the muscarinic receptor—although less potent than ACh, the *trans*-cyclopropane produces the same type of response as ACh, while the *cis*-cyclopropane is essentially inactive. So, when ACh binds to the muscarinic receptor, the ester and quaternary ammonium ion groups point in opposite directions.

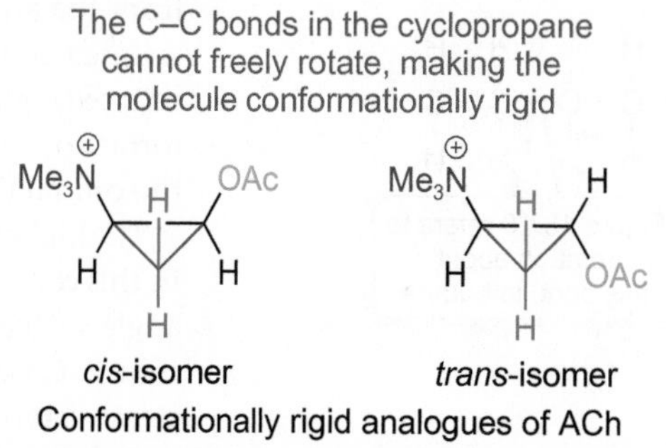
The C–C bonds in the cyclopropane cannot freely rotate, making the molecule conformationally rigid
Me3N
OAc
H
H
H
H
Me3N
H
H
H
OAc
H
cis-isomer
trans-isomer
Conformationally rigid analogues of ACh

▲ **Figure 2** *cis*- and *trans*-Cyclopropane derivatives.

Question

(a) Which of the cyclopropane derivatives shown in Figure 2 would you expect to be a more effective medicine for tachycardia (an abnormally fast heart rate)? Explain your reasoning.

(b) The amino acid L-glutamic acid is another important neurotransmitter. Like ACh, L-glutamic acid is a flexible molecule, which binds to different receptors by adopting different conformations. Explain why all three of the following molecules, **A–C**, are conformationally rigid analogues of L-glutamic acid.

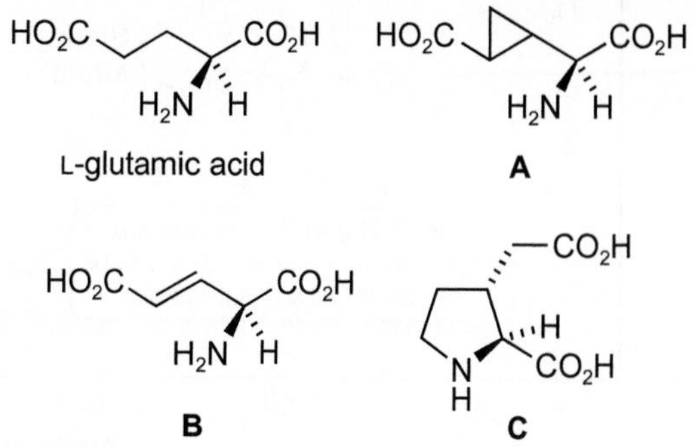
HO2C
CO2H
H2N H
L-glutamic acid
HO2C
CO2H
H2N H
A
HO2C
CO2H
H2N H
B
CO2H
H
N
H
CO2H
C

Conformations of butane

In alkanes, the preferred C–C–C bond angle is around 112°, not 109.5° (the tetrahedral angle).

Butane ($CH_3CH_2CH_2CH_3$) can adopt a large number of different conformations by rotation about each of the three C–C bonds. For example, rotation about one of the C–C bonds at the end of the chain produces the staggered and eclipsed conformations shown below. As with ethane (CH_3CH_3), the staggered conformation is more stable than the eclipsed conformation.

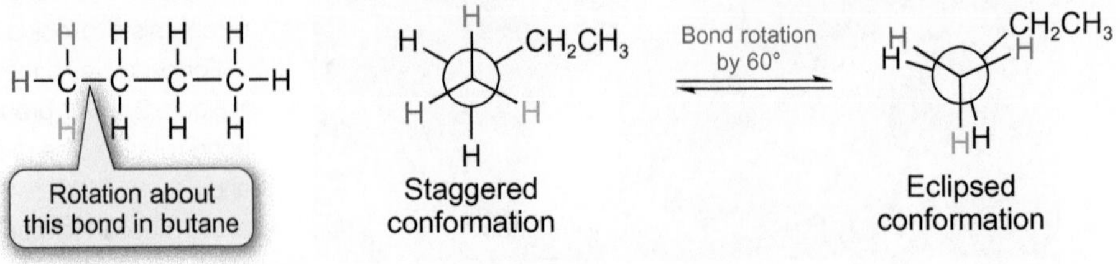

Rotation about this bond in butane
Bond rotation by 60°
Staggered conformation
Eclipsed conformation

In Greek, 'anti' is opposite, 'syn' is together, and 'clinal' is slant.

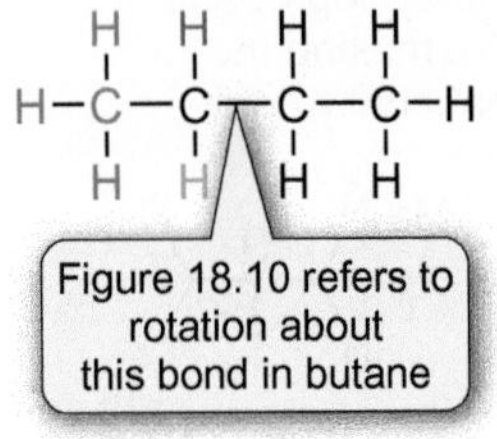

Figure 18.10 refers to rotation about this bond in butane

Rotation about the central C–C bond in butane (in 60° intervals) leads to two types of **eclipsed conformations** (called **syn-periplanar** and **anticlinal** conformations) and two types of **staggered conformations** (called **anti-periplanar** and **synclinal (gauche)** conformations) depending on the relative positions of the CH_3 groups. As you would expect from the analysis of ethane, the staggered conformations are of lower energy than the eclipsed conformations because of lower torsional strain. In addition, the two types of eclipsed conformations have different energies. Similarly, the two types of staggered conformations have different energies. The diagram in Figure 18.10 refers to rotation about the central C–C bond in butane (shown in margin).

The highest energy eclipsed conformation of butane is the syn-periplanar conformation. In this conformation, the two CH_3 groups are as close as possible to one another (dihedral angle = 0°) and, in addition to torsional strain (due to repulsion of electrons in the adjacent C–C bonds), this close arrangement of atoms in the two groups gives rise to **steric strain (steric hindrance)**. Steric strain is the repulsive interaction between two atoms

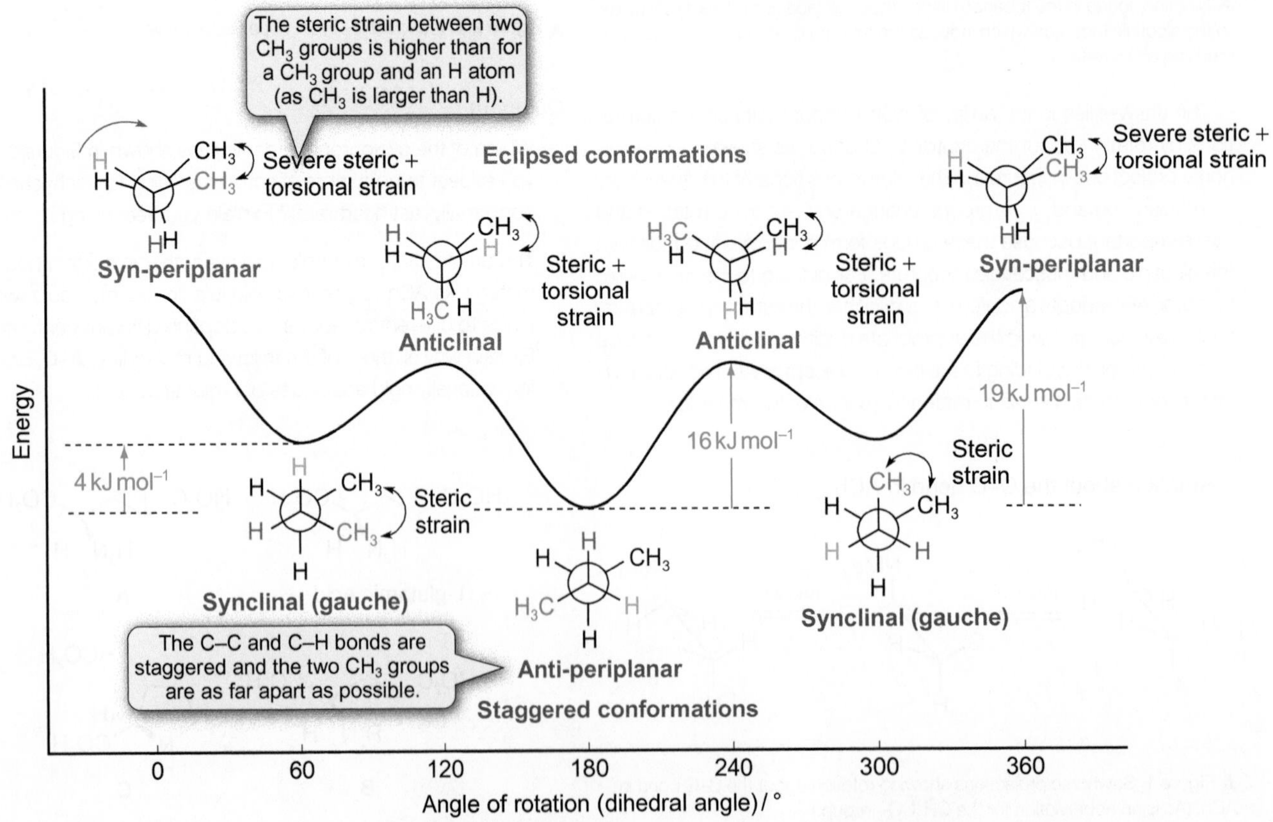

The steric strain between two CH3 groups is higher than for a CH3 group and an H atom (as CH3 is larger than H).
Severe steric + torsional strain
Eclipsed conformations
Syn-periplanar
Anticlinal
Steric + torsional strain
Anticlinal
Steric + torsional strain
Severe steric + torsional strain
Syn-periplanar
19 kJ mol−1
16 kJ mol−1
4 kJ mol−1
Steric strain
Steric strain
Synclinal (gauche)
Synclinal (gauche)
The C–C and C–H bonds are staggered and the two CH3 groups are as far apart as possible.
Anti-periplanar
Staggered conformations
Energy
0
60
120
180
240
300
360
Angle of rotation (dihedral angle)/ °

Figure 18.10 The relative energies of the different conformations of butane.

(or groups of atoms) that are closer to one another than their atomic radii allow. In the syn-periplanar conformation of butane, the two relatively large-sized CH_3 groups (in comparison to smaller-sized H atoms) are forced abnormally close to one another. As the two CH_3 groups cannot occupy the same space, they repel one another and this increases the energy of the conformation. (The steric strain in propane is discussed in Worked Example 18.2.)

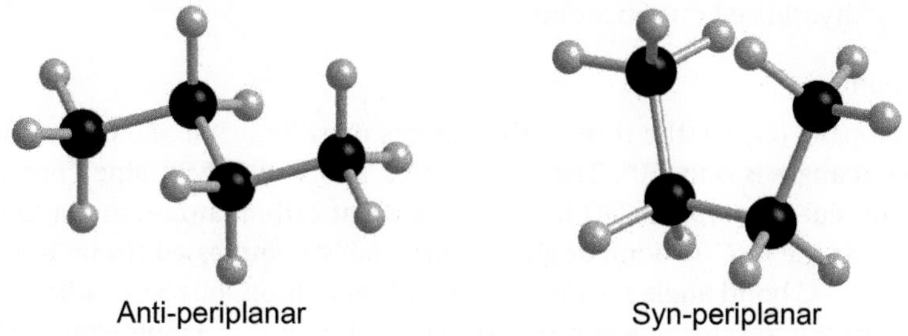
Anti-periplanar
Syn-periplanar

Models of the anti-periplanar and syn-periplanar conformations of butane.

As the anti-periplanar conformation of butane has the lowest energy, this is the most stable conformation and butane molecules will adopt this shape for most of the time. At room temperature, at any time there are around twice as many molecules of butane in the anti-periplanar conformation than in the synclinal (gauche) conformation, which has the second lowest energy. These conformations rapidly interconvert at room temperature, via the anticlinal conformation, but, as the eclipsed conformations are so unstable, they are not present to any significant extent.

The preference for a staggered conformation means that, in the most stable conformation, the carbon chain in butane adopts a zigzag shape. This zigzag shape also is observed for alkanes with longer carbon chains, as shown below for octane.

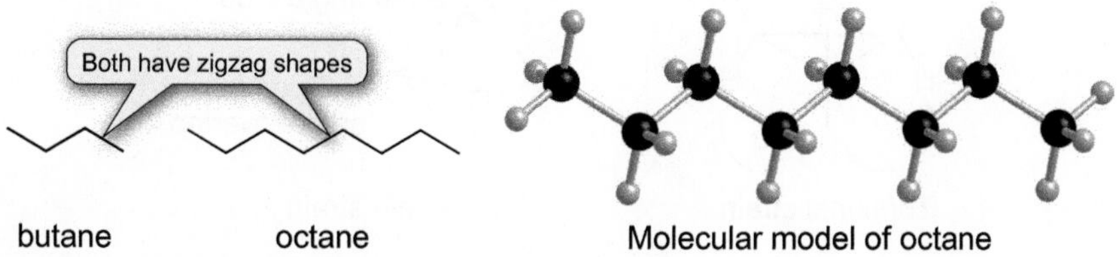
Both have zigzag shapes
butane
octane
Molecular model of octane

In general, the larger the size of groups in eclipsed conformations, the higher the steric strain in the molecule.

Steric strain is sometimes called **van der Waals strain.** van der Waals strain is considered a form of strain where the interacting atoms are at least four bonds away from each other.

Sometimes, the names syn-coplanar and anti-coplanar are used in place of syn-periplanar and anti-periplanar.

Worked example 18.2 Using Newman projections

Use Newman projections to draw the eclipsed and staggered conformations of propane. Why is the steric strain in propane higher than the steric strain in ethane?

Strategy

There are two equivalent C–C bonds in propane ($H_3C–CH_2–CH_3$) so rotation about either of these bonds will lead to identical conformations. It is easiest to consider propane as a substituted ethane and replace one H on ethane with a CH_3 group.

Solution

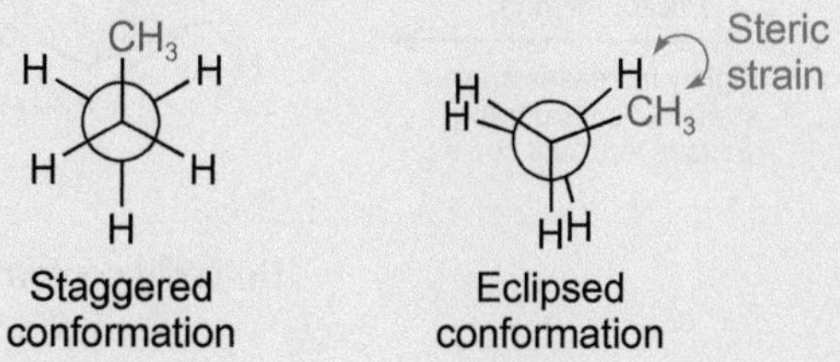
CH3
Steric strain
Staggered conformation
Eclipsed conformation

The higher steric strain in propane, compared to ethane, is due to a CH_3 group being larger than an H atom. In propane, there is an unfavourable steric interaction between the CH_3 substituent and an eclipsing H atom. In ethane, a steric interaction between eclipsing H atoms is not observed because the two H atoms are too small.

Question

Use Newman projections to draw the eclipsed and staggered conformations of chloroethane. Explain which conformation you would expect to have the higher energy.

◀ Chloroethane is a mild anaesthetic that has been used to numb the skin prior to ear piercing.

Conformations of cycloalkanes

The structure and naming of cycloalkanes is discussed in Section 2.5 (p.84).

In a similar way to ethane and butane, the conformations of cycloalkanes are determined by torsional strain and steric strain. However, cycloalkanes also suffer from an additional strain, called **angle strain (Baeyer strain)**. Angle strain arises when the C–C–C bond angle in cycloalkanes is different from the normal tetrahedral bond angle (109.5°), which is preferred for sp^3 hybridized carbon atoms.

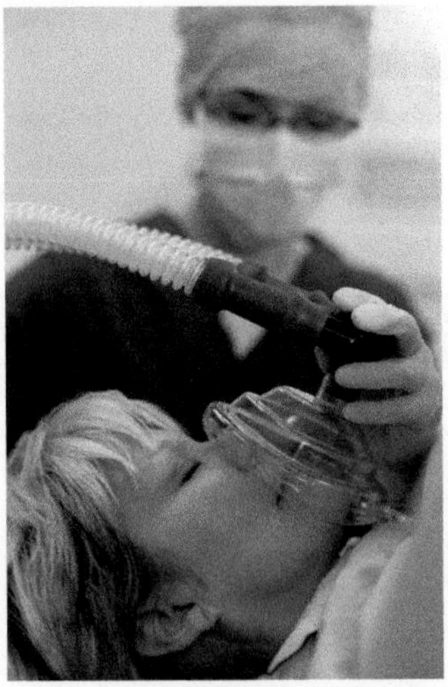

Cyclopropane is a colourless, flammable gas, once used in medicine as a general anaesthetic. Its use became discontinued when electrical equipment started being used in operating theatres (cyclopropane forms explosive mixtures with air).

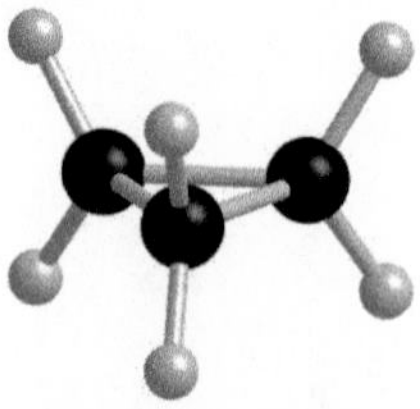

Molecular model of cyclopropane

Cyclopropane

For cyclopropane (C_3H_6), the three carbon atoms must lie in the same plane so that the C–C–C bond angle is only 60°. The 3-membered ring is highly strained because of torsional strain, due to eclipsed C–H bonds on adjacent carbon atoms, and also because of angle strain as the C–C–C bond angle is considerably compressed (from 109.5° to 60°). The small C–C–C bond angle means that the sp^3 orbitals on adjacent carbon atoms cannot point directly towards one another so the C–C bonds in cyclopropane are bent. This means that the overlap between the two sp^3 orbitals in cyclopropane is less than that for a normal C–C σ bond in an alkane. The C–C bond is therefore much weaker than a normal C–C bond (270 kJ mol^{-1} compared to 360–375 kJ mol^{-1}) so cyclopropane is more reactive than other alkanes, such as propane.

> **i** The bent shape of a C–C bond in cyclopropane has been likened to a banana and so these bonds are sometimes called **banana bonds**. (There is another way of describing the bonding in cyclopropane, called the Walsh model. In this model, the three C–C bonds are treated like π bonds, but this lies outside the scope of this book.)

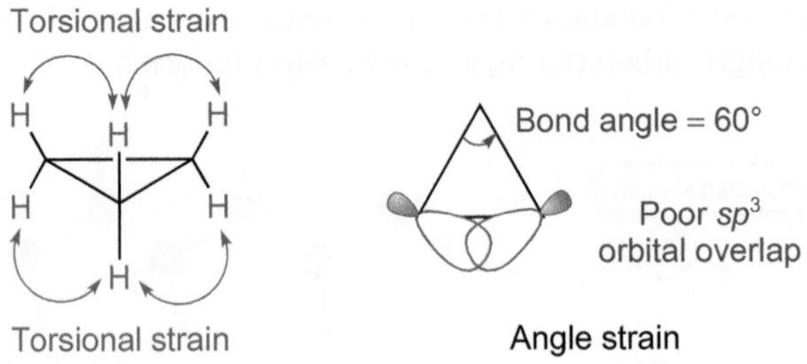
Torsional strain
H
H
H
H
H
H
Torsional strain
Bond angle = 60°
Poor sp^3 orbital overlap
Angle strain

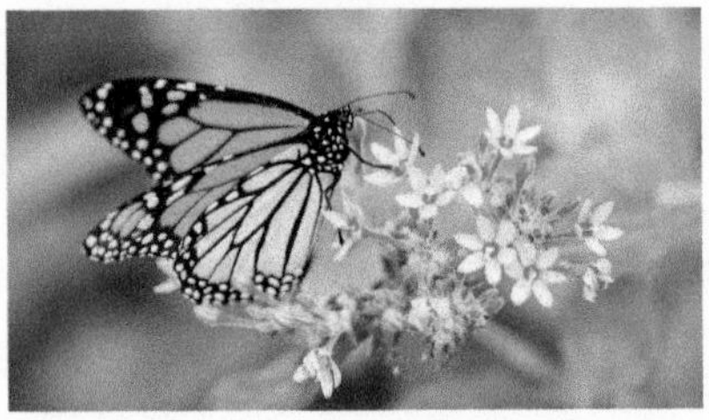

The conformation of cyclobutane resembles a butterfly.

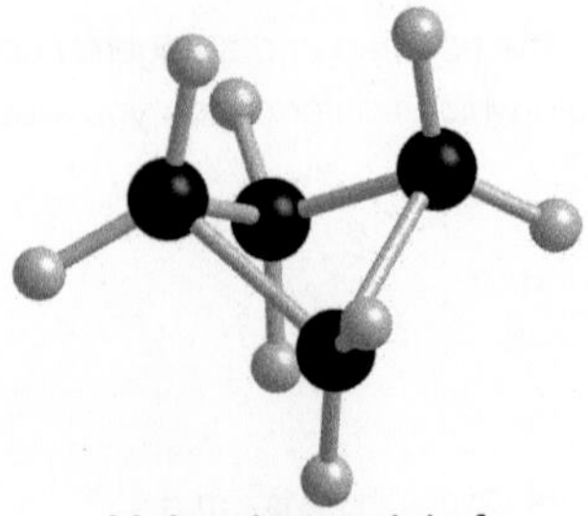

Molecular model of cyclobutane

Cyclobutane

If all four carbon atoms of cyclobutane (C_4H_8) were in the same plane, the C–C–C bond angle would be 90°. This planar conformation is highly strained because of considerable torsional strain due to eight pairs of eclipsing C–H bonds from the four carbons. To reduce torsional strain, one of the four carbon atoms moves out of the plane and the ring puckers to form a shape like the wings of a butterfly. On changing to the **butterfly conformation**, the adjacent C–H bonds become more staggered and this reduces the torsional strain. Puckering the planar 4-membered ring does increase angle strain because the C–C–C bond angle changes from 90° to 88°. However, the relief in torsional strain more than compensates for the increase in angle strain.

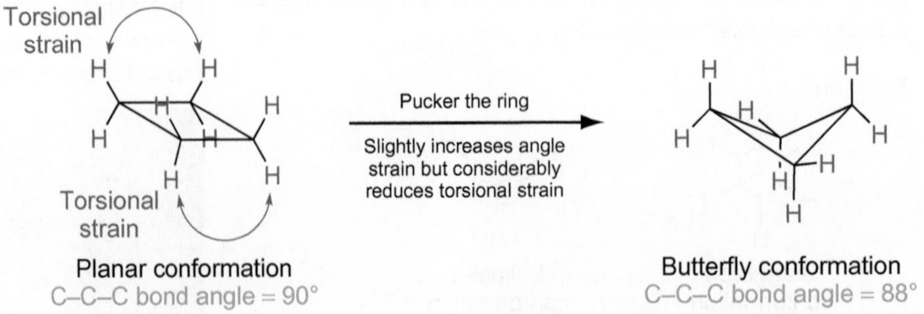
Torsional strain
Torsional strain
Pucker the ring
Slightly increases angle strain but considerably reduces torsional strain
Planar conformation
C–C–C bond angle = 90°
Butterfly conformation
C–C–C bond angle = 88°

Cyclopentane

For cyclopentane (C_5H_{10}), a planar conformation has C–C–C bond angles of 108°, which is almost the same as the normal tetrahedral bond angle (109.5°). Although the planar conformation of cyclopentane has very little angle strain, the torsional strain is considerable because of ten pairs of eclipsing C–H bonds from the five carbons. To minimize torsional strain, the ring puckers and one of the carbon atoms moves out of the plane to form a shape like that of an open envelope. On changing to the **envelope conformation**, the adjacent C–H bonds become nearly staggered and this reduces the torsional strain. Puckering the planar 5-membered ring slightly increases the angle strain but the relief in torsional strain compensates for this.

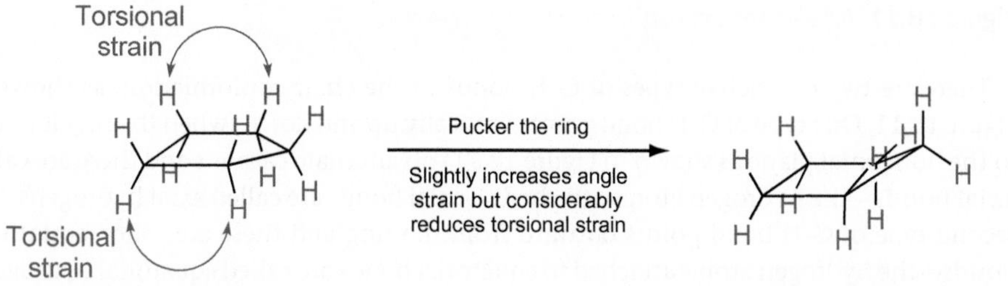
Torsional strain
Torsional strain
Pucker the ring
Slightly increases angle strain but considerably reduces torsional strain

Planar conformation
C–C–C bond angle = 108°

Envelope conformation
C–C–C bond angle = 105°

Cyclohexane

For cyclohexane (C_6H_{12}), a planar conformation has C–C–C bond angles of 120°, which is well above the normal tetrahedral bond angle (109.5°). As a result, this conformation suffers from angle strain. In addition, the energy of this conformation is raised by torsional strain due to the twelve pairs of eclipsing C–H bonds from the six carbons. To reduce both angle strain and torsional strain, the planar ring puckers. One carbon moves above the plane and the carbon on the opposite side of the ring moves below the plane. This forms a shape a bit like a reclining garden chair. On changing to the **chair conformation**, the angle strain is reduced because the C–C–C bond angle (111°) becomes closer to the tetrahedral angle, and the torsional strain is reduced because the adjacent C–H bonds become staggered.

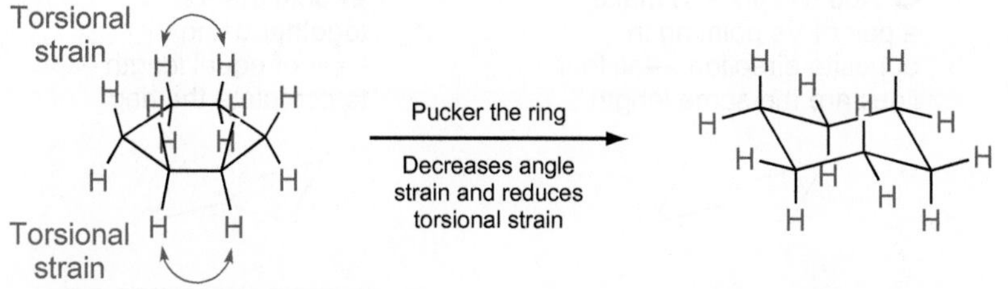
Torsional strain
Torsional strain
Pucker the ring
Decreases angle strain and reduces torsional strain

Planar conformation
C–C–C bond angle = 120°

Chair conformation
C–C–C bond angle = 111°

The staggered arrangement of the C–H bonds in the chair conformation is most easily seen using a Newman projection, as shown below.

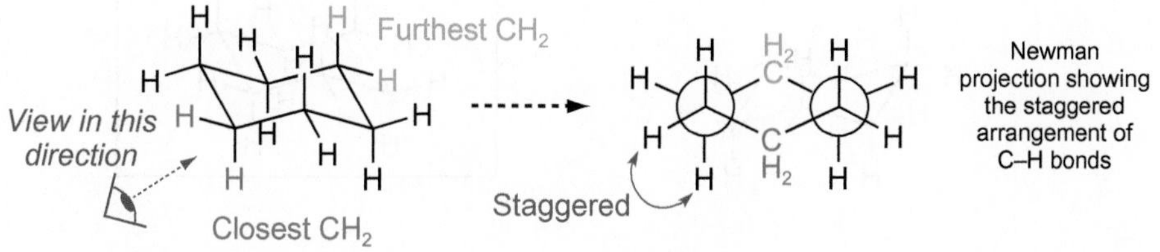
Furthest CH2
View in this direction
Closest CH2
Staggered
Newman projection showing the staggered arrangement of C–H bonds

The angle strain in cyclopropane and cyclobutane was first recognized by Johann Friedrich Wilhelm Adolf von Baeyer (1835–1917). Baeyer was also the first to prepare the naturally occurring dye, indigo, in the laboratory. Synthetic indigo is used to colour over one billion pairs of blue jeans around the world annually. For his work on organic dyes and other chemical achievements, Baeyer was awarded the Nobel Prize in Chemistry in 1905.

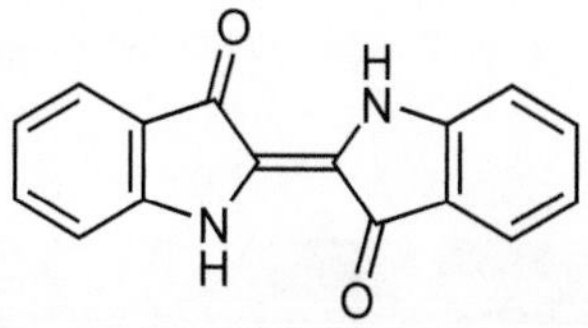

indigo

The conformation of cyclopentane resembles an open envelope.

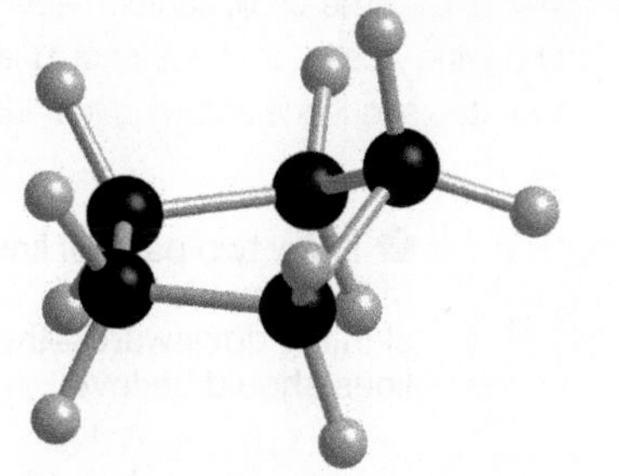

Molecular model of cyclopentane

The cyclohexane ring is present in many medicines and natural products (see Box 18.3, p.830).

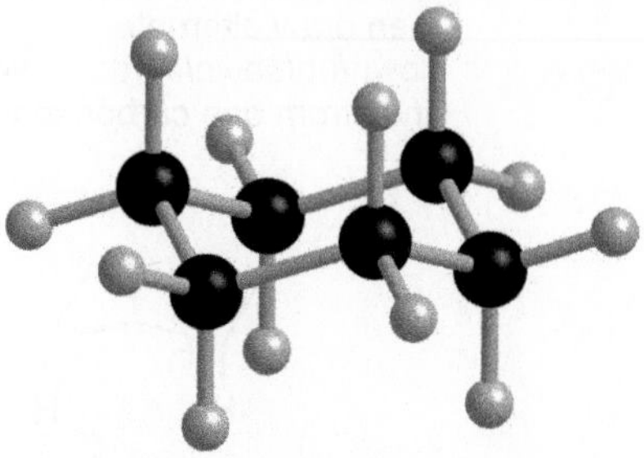

Molecular model of cyclohexane

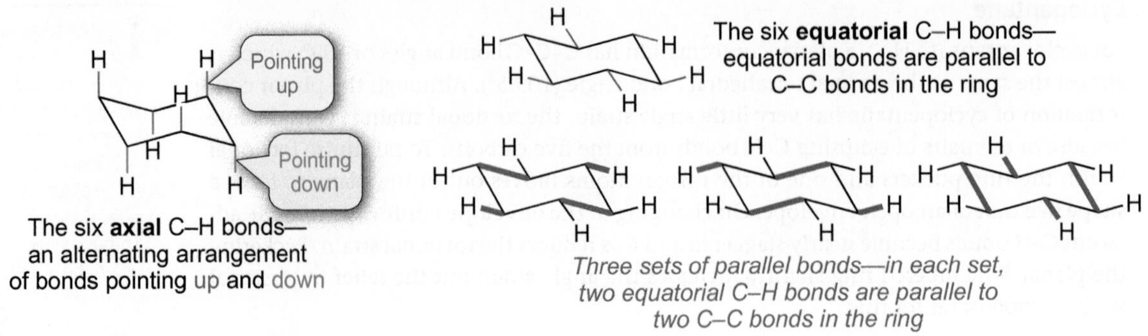
Pointing up
Pointing down
The six **axial** C–H bonds—an alternating arrangement of bonds pointing up and down
The six **equatorial** C–H bonds—equatorial bonds are parallel to C–C bonds in the ring
Three sets of parallel bonds—in each set, two equatorial C–H bonds are parallel to two C–C bonds in the ring

Figure 18.11 Axial and equatorial C-H bonds in cyclohexane.

The conformation of cyclohexane resembles a reclining garden chair.

There are two distinctive types of C–H bonds in the chair conformation, as shown in Figure 18.11. One type of C–H bond points vertically up and down (when the ring is drawn in the horizontal plane as shown in Figure 18.11) on alternate carbons and these are called **axial bonds**—the hydrogen atoms attached to axial bonds are called axial hydrogens. The second type of C–H bond points outward from the ring and these are called **equatorial bonds**—the hydrogen atoms attached to equatorial bonds are called equatorial hydrogens. Every carbon atom in the chair conformation has one axial bond and one equatorial bond. (For a simple way to accurately draw the chair conformation of cyclohexane, see Box 18.2.)

Since the six C–C bonds in cyclohexane are interconnected, independent rotation about a single C–C bond in cyclohexane is impossible. However, the carbon atoms within the 6-membered ring can move together and, when this occurs, it leads to conversion of one chair conformation into a second chair conformation. This process of interconverting chair conformations is called a **ring-flip (ring inversion)**.

Box 18.2 Drawing cyclohexane rings

The chair conformation of cyclohexane is so important, and widely used, that you should learn how to draw it following the five-step sequence shown in the illustration. A common mistake is to draw the axial and equatorial bonds at the wrong angles, so it is wise to practise.

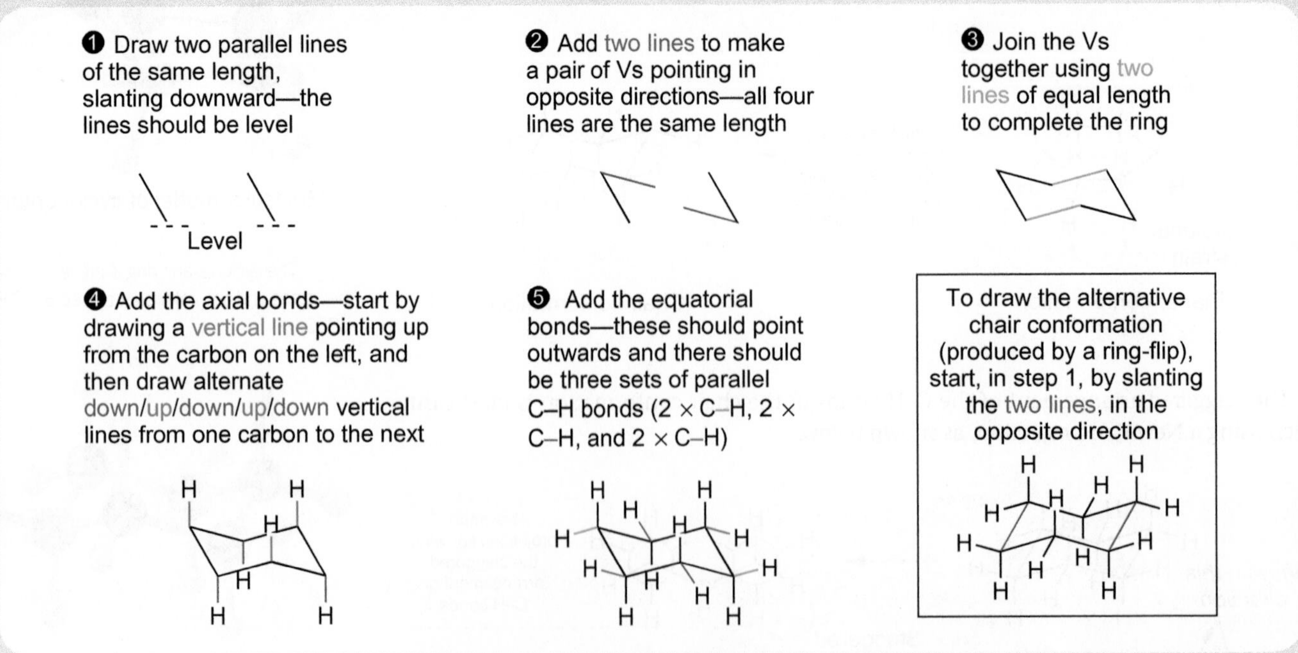
❶ Draw two parallel lines of the same length, slanting downward—the lines should be level
Level
❷ Add two lines to make a pair of Vs pointing in opposite directions—all four lines are the same length
❸ Join the Vs together using two lines of equal length to complete the ring
❹ Add the axial bonds—start by drawing a vertical line pointing up from the carbon on the left, and then draw alternate down/up/down/up/down vertical lines from one carbon to the next
❺ Add the equatorial bonds—these should point outwards and there should be three sets of parallel C–H bonds (2 × C–H, 2 × C–H, and 2 × C–H)
To draw the alternative chair conformation (produced by a ring-flip), start, in step 1, by slanting the two lines, in the opposite direction

When the chair conformation of cyclohexane undergoes a ring-flip, three of the carbon atoms in the ring move up and the remaining three carbon atoms move down. The change in position of the carbon atoms results in all axial hydrogens becoming equatorial hydrogens and all equatorial hydrogens becoming axial hydrogens. The energy barrier for the ring-flip is only 45 kJ mol^{-1} so the chair conformation rapidly interconverts—at room temperature, the ring flips at a rate of around 1×10^5 times per second. This rapid change in the positions of the hydrogen atoms means that, when averaged over time, the axial and equatorial hydrogen atoms are equivalent. It is not possible to distinguish an axial hydrogen atom from an equatorial hydrogen atom in cyclohexane.

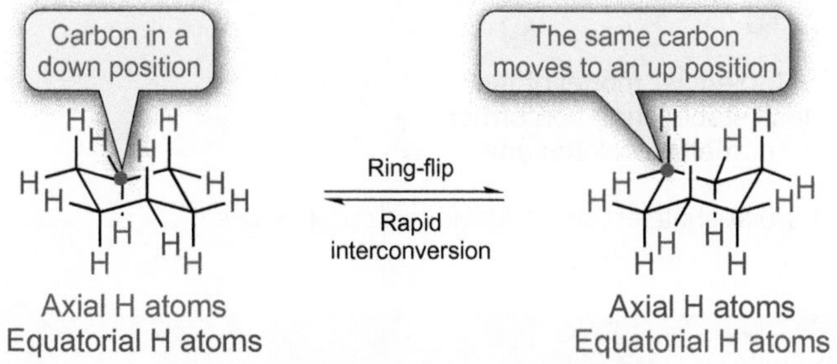
Carbon in a down position
The same carbon moves to an up position
Ring-flip
Rapid interconversion
Axial H atoms
Equatorial H atoms
Axial H atoms
Equatorial H atoms

The terms axial and equatorial are also used to indicate the positions of atoms or ligands in structures based on the trigonal bipyramid and octahedra (Section 5.2, p.227). For example, $[Cu(H_2O)_6]^{2+}$ has two axial ligands and four equatorial ligands (Section 28.5, p.1282).

Conversion of one chair conformation of cyclohexane into another chair conformation (by a ring-flip) is a complicated process that involves several less-stable conformations of cyclohexane–the most important of these are called the **half-chair**, **twist-boat** (skew-boat), and **boat** conformations. For the half-chair conformation, four adjacent carbons lie in the same plane with the fifth carbon pointing above the plane and the sixth one pointing below the plane—cyclohexene prefers to adopt the half-chair conformation (Box 18.3, p.830 and Section 21.3, p.976).

Conformations of substituted cyclohexanes

Cyclohexane has two equivalent chair conformations but, if a group other than hydrogen is bonded to the ring, then the two chair conformations are no longer equivalent. The two chair conformations of a substituted cyclohexane can interconvert by a ring-flip, but the proportions of the two chair conformations differ because they have different energies. In general, the chair conformation with the substituent in an axial position has the higher energy and so will be less stable. Conversely, the chair conformation with the substituent in an equatorial position has the lower energy and so will be more stable. This is illustrated by the alkyl-substituted cyclohexanes shown in Figure 18.12, which favour the alkyl group in the equatorial position.

The axial position is less favoured for the substituent because this conformation has more steric strain and more torsional strain. When in an axial position, the substituent is close to the axial hydrogens at positions 3 and 5, which are parallel to the substituent and on the same side of the ring. This leads to steric strain, which increases the energy of the conformation. Three carbon atoms separate the axial substituent from the axial hydrogens, so these two steric interactions are called **1,3-diaxial interactions** (see Figure 18.13).

The size of the substituent on the ring has an effect on the size of 1,3-diaxial interactions. If the substituent is large, it is close to the 1,3-diaxial hydrogens, resulting in strong diaxial interactions so the substituent favours the equatorial position. For example, a *tert*-butyl group ($C(CH_3)_3$) is larger than a methyl group (CH_3), so the *tert*-butyl group in *tert*-butylcyclohexane sits in the equatorial position to a greater extent than the methyl group in methylcyclohexane (see Figure 18.12).

Different conformers of substituted cyclohexanes give different peaks in a ^{1}H NMR spectrum—the spectrum is recorded at low temperature where the ring-flip is not so fast. The ratio of conformers is determined by integration of the absorption signals for each conformer (Sections 10.7, p.500, and 12.3, p.578).

The dissimilar reactivities of different halocyclohexane conformers in E2 elimination reactions are discussed in Section 20.4 (p.943).

Methylcyclohexane (C_7H_{14}) is a component of jet fuel, which is designed for use by aircraft powered by gas-turbine engines.

Abbreviations for alkyl groups, such as ^{t}Bu for the *tert*-butyl group, are introduced in Box 2.3 (p.81).

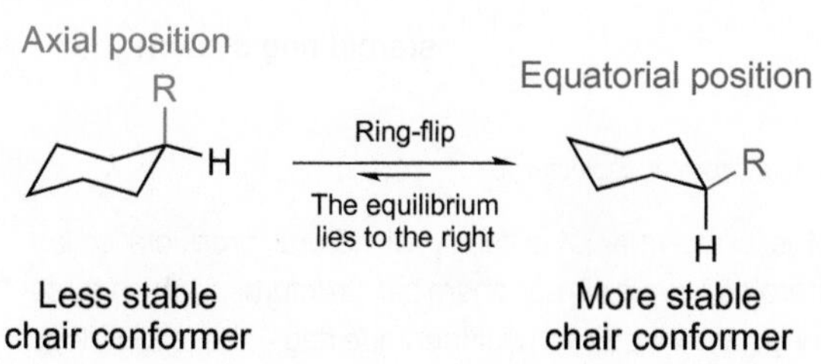
Axial position
R
H
Ring-flip
The equilibrium lies to the right
Equatorial position
R
H
Less stable chair conformer
More stable chair conformer

R	% of the cyclohexane with R in the equatorial position at room temperature
H	50
Me	95
^{i}Pr	98
^{t}Bu	>99

Figure 18.12 Less stable and more stable chair conformers of substituted cyclohexanes.

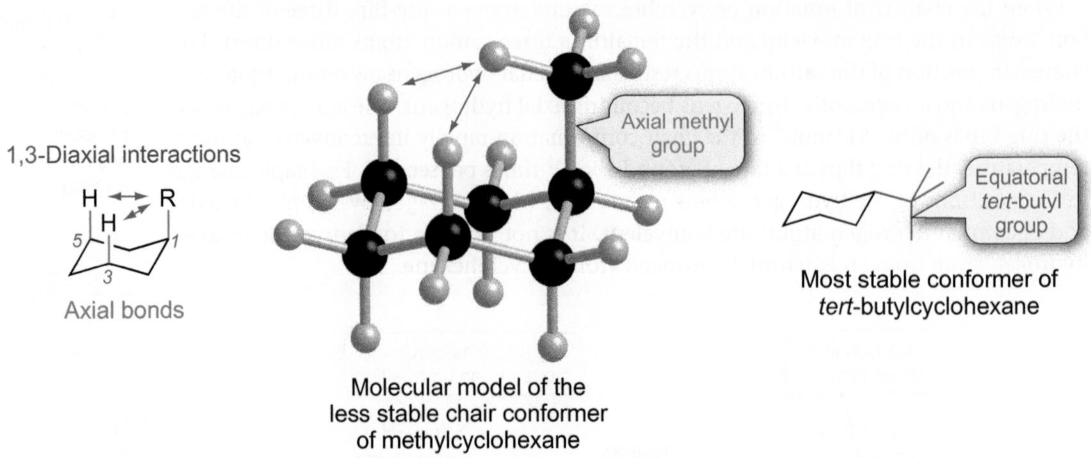
1,3-Diaxial interactions
H
R
H
5
1
3
Axial bonds
Axial methyl group
Molecular model of the less stable chair conformer of methylcyclohexane
Equatorial *tert*-butyl group
Most stable conformer of *tert*-butylcyclohexane

Figure 18.13 1,3-Diaxial interactions of substituted cyclohexanes.

Box 18.3 The shape of cholesterol and its role in the body

Cholesterol is essential to life. It is vital as a major constituent of cell membranes. It is also the starting material for the production of bile salts (which help to digest fatty food), vitamin D, and the sex hormones, such as oestrogen and testosterone. The adult human body contains around 150 g of cholesterol, which is present in the bloodstream and in every cell in the body. About 80% is produced in the liver, with our food supplying the remaining 20%. The concentration of cholesterol in the bloodstream is important as too much can increase the risk of coronary heart disease and disease of arteries elsewhere in the body (Box 2.4, p.86).

▲ Cholesterol is present in animal products such as meat, fish, milk, and egg yolk. It is not found in plants.

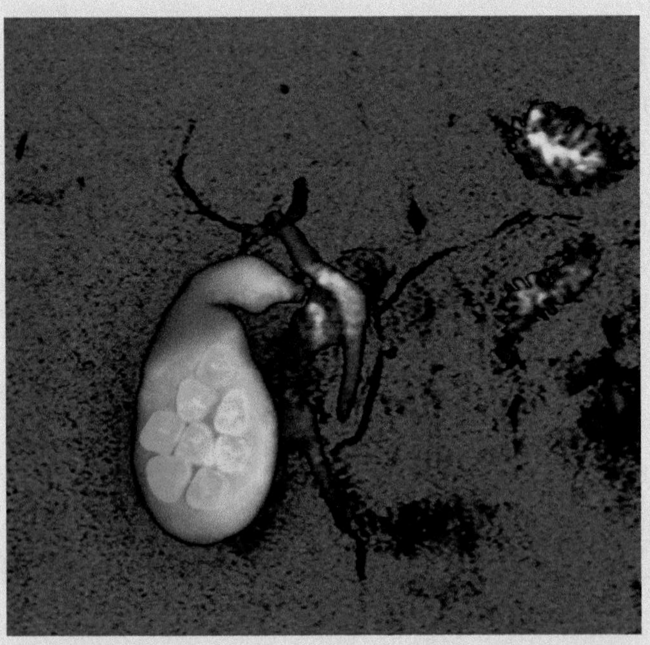

▲ Cholesterol is found in gall stones, which are rock-like materials formed in the gall bladder of the body. Gall stones vary in size from a few millimetres to a few centimetres.

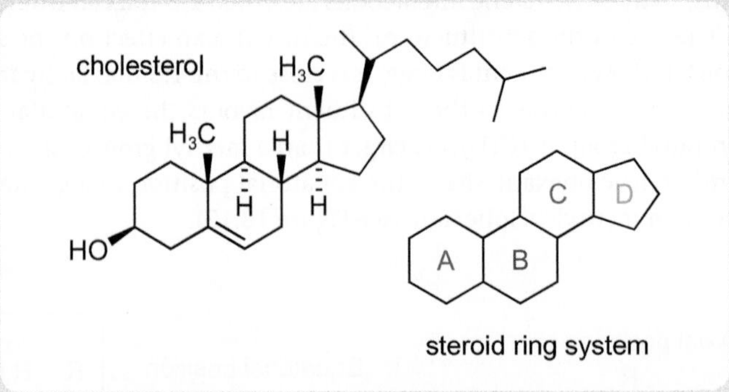
cholesterol
H_3C
H_3C
H
H
H
HO
A
B
C
D
steroid ring system

▲ The steroid ring system of cholesterol.

Cholesterol is a member of a family of natural products called steroids. All steroids have a similar chemical structure, in which three cyclohexane rings are linked to a cyclopentane ring—these four rings are labelled A–D, as shown in the illustration. →

➜ The cyclohexane rings A and C in cholesterol adopt chair conformations and the cyclopentane ring (ring D) adopts an open-envelope conformation to minimize strain. Ring B is a cyclohexene and the presence of the C=C bond flattens the structure of half of the ring so this is called a **half-chair conformation**.

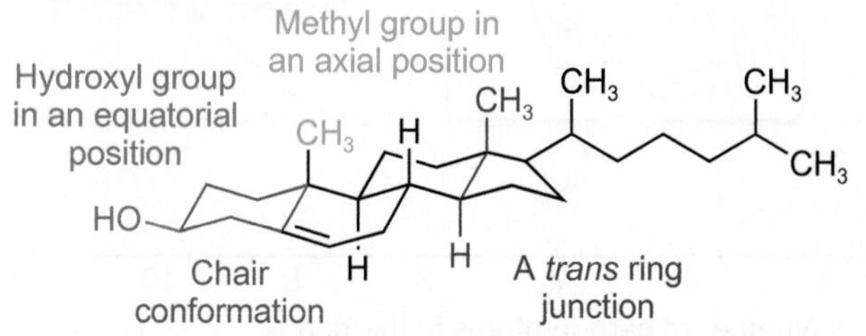
Methyl group in
an axial position
Hydroxyl group
in an equatorial
position
CH3
CH3
CH3
CH3
H
CH3
HO
Chair
conformation
H
H
A *trans* ring
junction

At the ring junctions (where the rings meet) of cholesterol, the small-sized methyl groups and hydrogen atoms are in axial positions. To minimize steric strain, the two groups at a ring junction are on opposite sides of the molecule, so the ring junction is said to have *trans* stereochemistry. (This is opposed to the *cis* stereochemistry when the two groups would be on the same side of the molecule.)

The C–C bonds in the ring system of cholesterol cannot freely rotate so the ring system provides a rigid carbon framework that separates the polar hydroxyl group (HO–) from the non-polar alkyl chain.

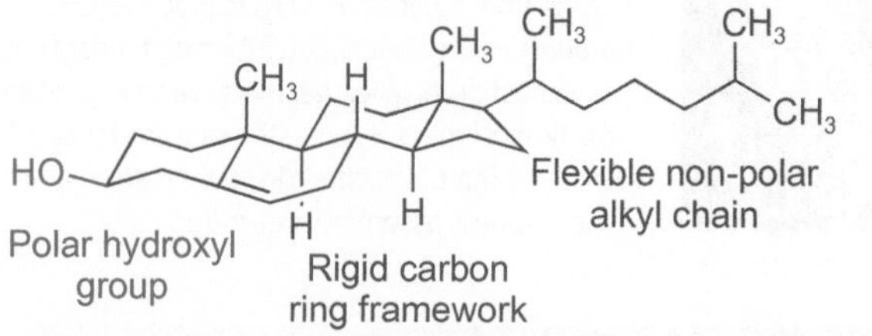
CH3
CH3
CH3
CH3
H
CH3
HO
Polar hydroxyl
group
H
H
Rigid carbon
ring framework
Flexible non-polar
alkyl chain

In nature, when a cholesterol molecule inserts into a cell membrane, the surface of the membrane becomes more rigid. This is because the polar –OH group in cholesterol aligns with the polar groups on the surface of the membrane. This positions the rigid ring system of cholesterol near the surface of the membrane. The amount of cholesterol in a cell membrane can thus alter the strength of the membrane, which in turn determines its biological function.

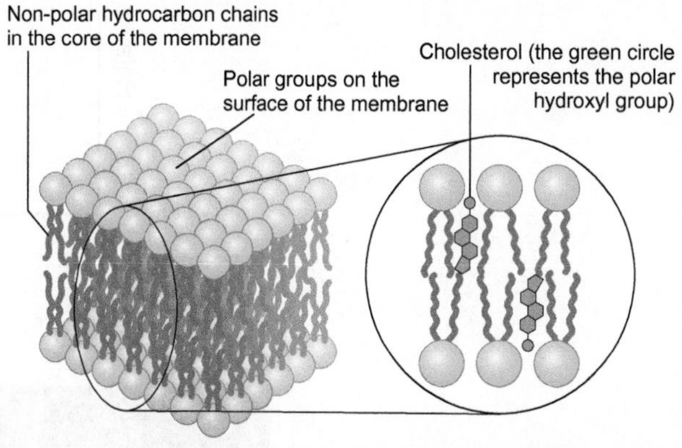
Non-polar hydrocarbon chains
in the core of the membrane
Polar groups on the
surface of the membrane
Cholesterol (the green circle
represents the polar
hydroxyl group)

▲ Cholesterol molecules within a cell membrane.

Question

Anabolic–androgenic steroids are synthetic compounds that do not occur in nature. They have similar structures to male sex hormones. One example is nandrolone, which has caused controversy because of its use by athletes to enhance performance (see Chapter 11, p.513).

Draw the most stable conformation of nandrolone.

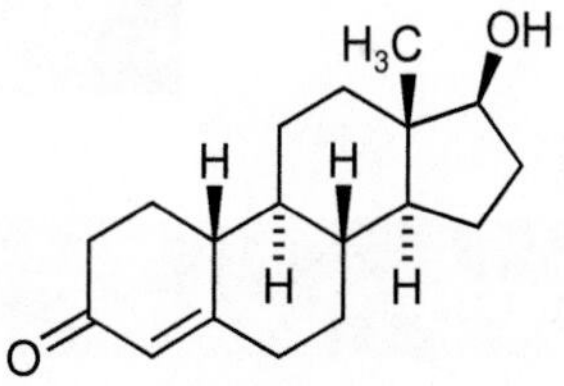
H3C
OH
H
H
H
H
O

nandrolone

Relative stability of cycloalkanes

The stability of cycloalkanes is determined by torsional strain, steric strain, and angle strain. As shown in Figure 18.14, cyclopropane and cyclobutane are more highly strained than other cycloalkanes. This is predominantly due to angle strain in the 3- and 4-membered rings. As the size of the cycloalkane increases, so the angle strain reduces and cyclopentane has low strain and cyclohexane has no strain. This explains why 5- and 6-membered cycloalkane rings are the most stable and the easiest to form. Cycloalkanes containing seven carbons or more are slightly less stable than cyclohexane (due to increased torsional strain and also steric strain caused by the interaction of hydrogen atoms across the ring), but after nine carbons the rings become more stable. Large-ring cycloalkanes have almost strain-free conformations because the long chains can adopt a structure that is very similar to that of straight-chain alkanes.

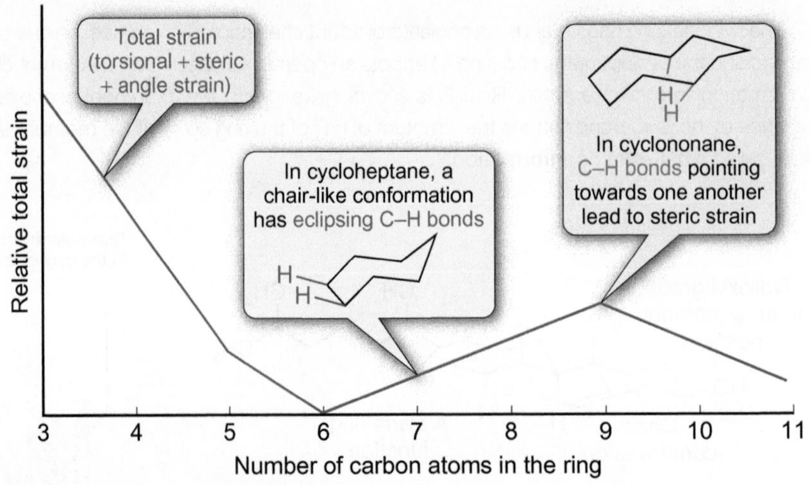
Total strain
(torsional + steric
+ angle strain)
In cycloheptane, a
chair-like conformation
has eclipsing C–H bonds
In cyclononane,
C–H bonds pointing
towards one another
lead to steric strain
Relative total strain
3
4
5
6
7
8
9
10
11
Number of carbon atoms in the ring

Figure 18.14 Total strain of cycloalkanes relative to cyclohexane.

Angle strain can be experimentally determined by measuring the standard enthalpy of combustion (Section 13.3, p.627)–strained cycloalkane rings, which have more energy than unstrained ones, should give off more heat per CH_2 unit when they combust to form carbon dioxide and water. The standard enthalpy of combustion at 298 K of cyclopropane is $-2091\ kJ\,mol^{-1}$ (or $-697\ kJ\,mol^{-1}$ per CH_2) while for cyclohexane it is $-3952\ kJ\,mol^{-1}$ (or $-659\ kJ\,mol^{-1}$ per CH_2). As cyclohexane has the lower enthalpy of combustion per CH_2 unit, it is more stable than cyclopropane.

Also, for combustion of straight-chain alkanes, on moving from ethane (CH_3CH_3) to propane ($CH_3CH_2CH_3$) to butane ($CH_3CH_2CH_2CH_3$) etc., for the standard enthalpies of combustion, on average, there is a change of around $-660\ kJ\,mol^{-1}$ per CH_2 unit—the same as for cyclohexane. Assuming there is no strain in straight-chain alkanes, then cyclohexane is essentially strain free.

Summary

- Three-dimensional arrangements of atoms that result from rotation about single bonds are called conformations (or conformers).
- Different conformations can have different amounts of strain and so have different energies.
- There are three types of strain: torsional strain; steric strain; and angle strain.
  - Torsional strain is due to electron–electron repulsions in adjacent bonds.
  - Steric strain is due to crowding of (particularly large-sized) groups in a certain conformation.
  - Angle strain (Baeyer strain) arises when the C–C–C bond angle (in non-aromatic cyclic compounds) is different from the normal tetrahedral bond angle (109.5°).

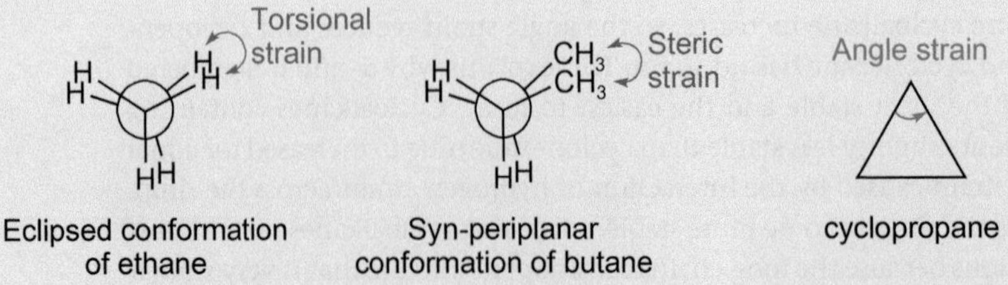
Torsional
strain
Steric
strain
Angle strain
Eclipsed conformation
of ethane
Syn-periplanar
conformation of butane
cyclopropane

→

- For ethane, the staggered conformation has a lower energy than the eclipsed conformation.
- For butane, the anti-periplanar conformation has the lowest energy and the syn-periplanar conformation has the highest energy.
- Cyclopropane and cyclobutane are the least stable of the cycloalkanes because of high angle strain.
- Cyclopentane and cyclohexane are the most stable cycloalkanes (and therefore the easiest to make) because of low angle strain.
- The lowest energy conformations of cyclohexane are two chair conformations that can be interconverted by a ring-flip.
- For substituted cyclohexanes, the substituents favour equatorial positions, rather than axial positions, in order to minimize 1,3-diaxial interactions.

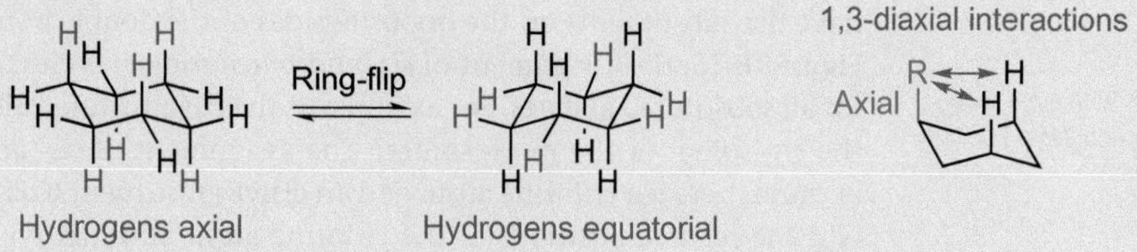

Ring-flip
Hydrogens axial
Hydrogens equatorial
1,3-diaxial interactions
Axial

For practice questions on these topics, see questions 1, 2, 4, and 6 at the end of this chapter (pp.858–859).

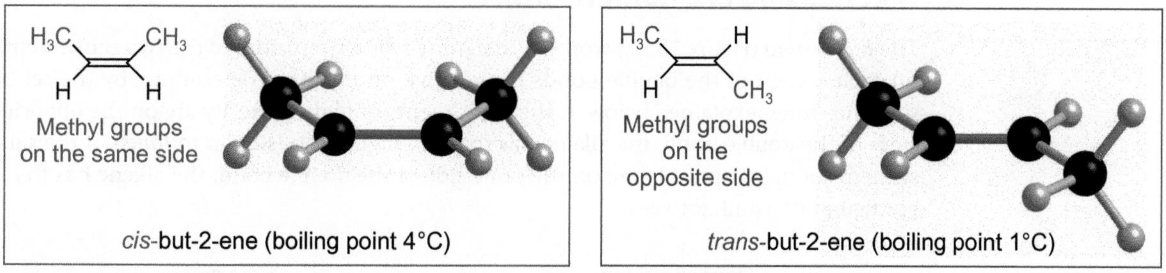

Methyl groups on the same side
cis-but-2-ene (boiling point 4°C)
Methyl groups on the opposite side
trans-but-2-ene (boiling point 1°C)

Figure 18.15 Two configurational isomers: *cis*- and *trans*-isomers of but-2-ene. (In these computer generated ball and stick models, a single stick represents the C=C double bond.)

18.3 Configurational isomers: *E*- and *Z*-isomers

Configurational isomers are compounds with the same molecular formula and the same types of bonds (Section 18.1, p.817). Unlike conformational isomers, configurational isomers normally cannot be interconverted without breaking a bond. Configurational isomers occur in disubstituted alkenes (RCH=CHR) because there is no rotation about the C=C bond at normal temperatures. This means that two substituents can be on the same side of the double bond or on the opposite side of the double bond, as shown in Figure 18.15. The two isomers can be separated from one another and they have different physical properties (such as different boiling points and different melting points).

Some configurational isomers of alkenes also have differing chemical properties. For example, maleic acid forms an anhydride more easily than fumaric acid.

Conformational isomers of a molecule are interconverted by rotation about single bonds. **Configurational** isomers usually can only be interconverted by breaking bonds.

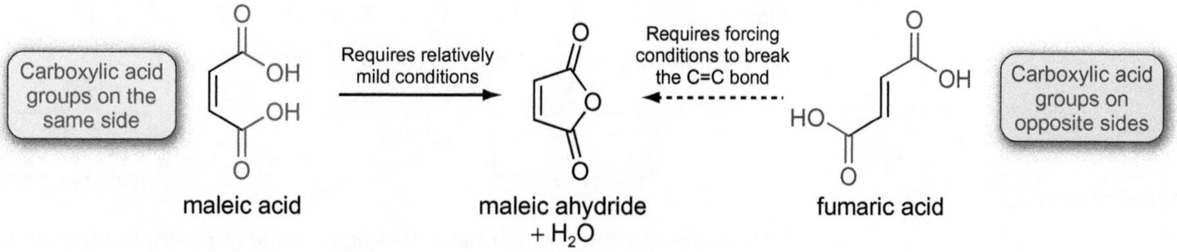

Carboxylic acid groups on the same side
Requires relatively mild conditions
Requires forcing conditions to break the C=C bond
Carboxylic acid groups on opposite sides
maleic acid
maleic ahydride + H_2O
fumaric acid

From the Latin, *cis* means 'on this side of' and *trans* means 'across' (as in *trans*atlantic).

The terms *cis* and *trans* are useful for naming disubstituted alkenes, isomers in coordination chemistry (Section 28.3, p.1274), and also for rings that are joined together (Box 18.3, p.830). For tri- and tetra-substituted alkenes (RCH=CR_2 and R_2C=CR_2), it is better to use *E* and *Z*.

For naming compounds with more than one functional group, see Section 2.8, p.106.

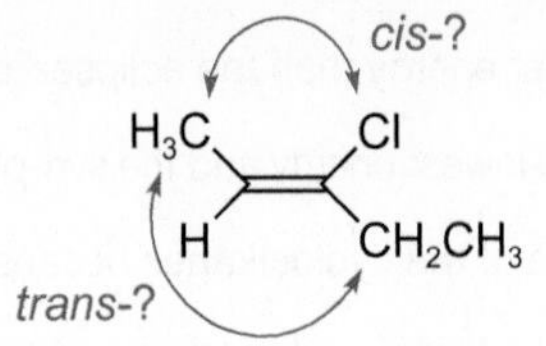
cis-?
H3C
Cl
H
CH2CH3
trans-?

Is this the *cis*- or *trans*-isomer of 3-chloropent-2-ene?

Figure 18.16 Isomer of 3-chloropent-2-ene.

In Section 9.5, configurational isomers of alkenes are called *cis*- and *trans*-isomers. *Cis*-isomers have the substituents on the same side of the double bond, while *trans*-isomers have the substituents on the opposite sides of the double bond. For but-2-ene, shown in Figure 18.15, the assignment of *cis*- and *trans*-isomers is clear cut, but this is not the case for all substituted alkenes. For example, is the trisubstituted alkene shown in Figure 18.16 the *cis*-isomer or the *trans*-isomer? The assignment of *cis*- and *trans*-isomers is difficult because there is a chlorine atom and an ethyl substituent (CH_3CH_2–) on the same carbon. Is it the relative positions of the chlorine atom and methyl group (CH_3–), or the ethyl group and methyl group, that determine whether the alkene is the *cis*- or *trans*-isomer? To overcome this problem, there is a systematic approach for assigning the stereochemistry of variously substituted alkenes. This is called the ***E/Z* system of nomenclature**.

Alkenes and *E/Z* nomenclature

The choice of *E* and *Z* letters comes from the German words for opposite (*entgegen*) and together (*zusammen*). So, for disubstituted alkenes, *E* corresponds to *trans* and *Z* corresponds to *cis*.

The alphabetical order of '*E*' and '*Z*' in the term *E/Z* isomers, and of '*cis*' and '*trans*' in the term *cis/trans* isomers, is unfortunate, because *E*-isomers correspond to *trans*-isomers and *Z*-isomers correspond to *cis*-isomers.

The *E*/*Z* system (Figure 18.17) works by assigning relative priorities to the two substituents on each carbon of the double bond. The relative priorities are determined by the set of sequence rules, explained below. If the substituents of higher priority are on the opposite sides of the double bond, the alkene has the *E* configuration (similar to *trans*). If the substituents of higher priority are on the same side of the double bond, the alkene has the *Z* configuration (similar to *cis*).

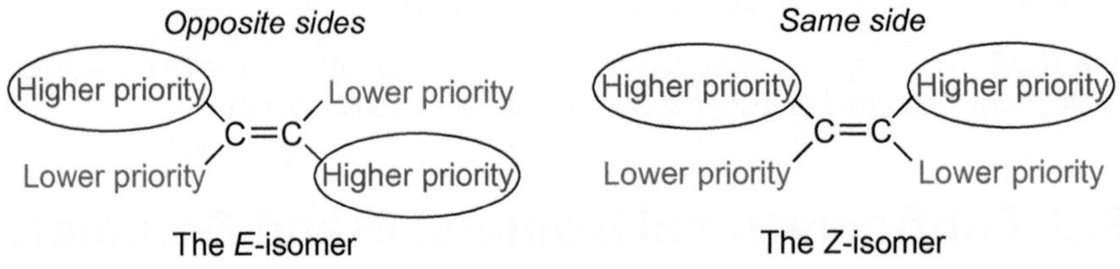
Opposite sides
Higher priority
Lower priority
C=C
Lower priority
Higher priority
The E-isomer
Same side
Higher priority
Higher priority
C=C
Lower priority
Lower priority
The Z-isomer

Figure 18.17 *E*- and *Z*-isomers.

Determining the priorities of substituents on alkene double bonds

This unambiguous method for naming alkenes is part of a general system for naming configurational isomers, called the Cahn–Ingold–Prelog (CIP) nomenclature (Section 18.4, p.842).

Step 1 Rank the atoms *directly attached to each carbon atom in the double bond* in the order of decreasing atomic number so that the atom of highest atomic number has the highest priority.

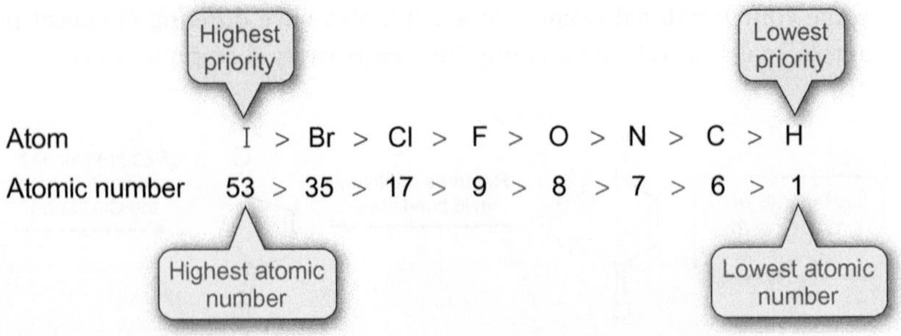
Highest priority
Lowest priority
Atom I > Br > Cl > F > O > N > C > H
Atomic number 53 > 35 > 17 > 9 > 8 > 7 > 6 > 1
Highest atomic number
Lowest atomic number

Isotopes are introduced in Section 1.3, p.14.

For isotopes of atoms (which have the same atomic number), use the atomic mass number to prioritize the group. The isotope of highest atomic mass number is ranked first and this isotope has the highest priority (e.g. deuterium has a higher priority than hydrogen).

Step 2 If the two atoms directly attached to the same carbon are the same, then rank the second, third, fourth, and so on atoms (working away from the C=C bond) one at a time until a difference is found. This is often called the **first point of difference rule**.

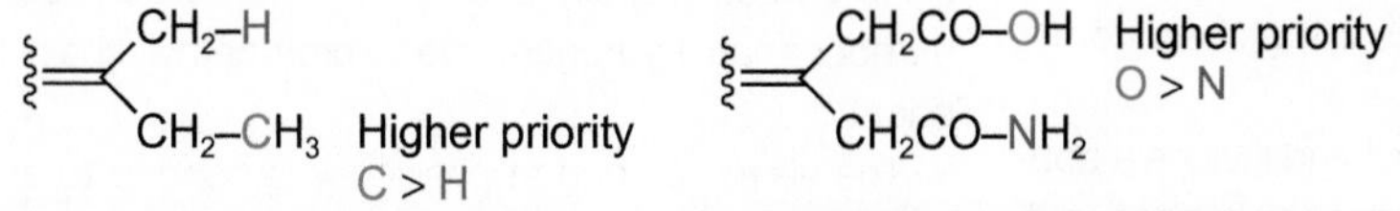

CH_2–H
CH_2–CH_3
Higher priority
C > H
CH_2CO–OH
CH_2CO–NH_2
Higher priority
O > N

Step 3 If substituents contain double and triple bonds, treat the atoms joined by the double or triple bonds as if they were linked to two or three single-bonded atoms.

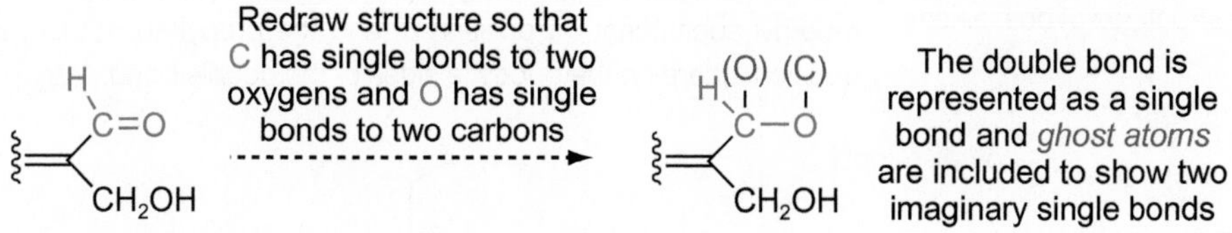

H
C=O
CH_2OH
Redraw structure so that C has single bonds to two oxygens and O has single bonds to two carbons
H
(O) (C)
C–O
CH_2OH
The double bond is represented as a single bond and *ghost atoms* are included to show two imaginary single bonds

For example,

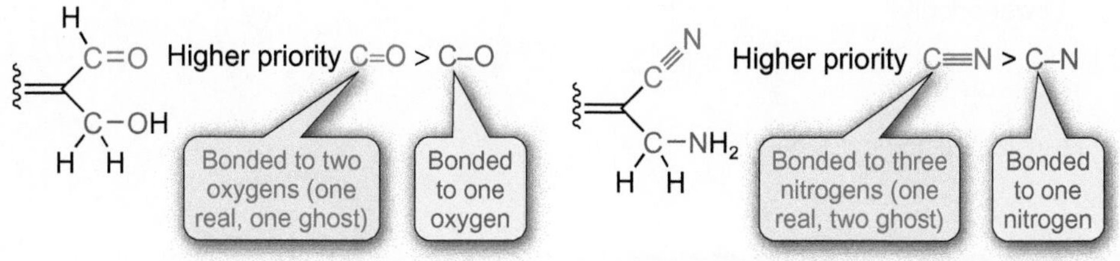

H
C=O
C–OH
H H
Higher priority C=O > C–O
Bonded to two oxygens (one real, one ghost)
Bonded to one oxygen
C≡N
C–NH_2
H H
Higher priority C≡N > C–N
Bonded to three nitrogens (one real, two ghost)
Bonded to one nitrogen

The examples in Figure 18.18 show how the sequence rules are used to assign *E*/*Z* configurations to some substituted alkenes and Worked Example 18.3 provides practice in translating the name of an alkene into its skeletal structure.

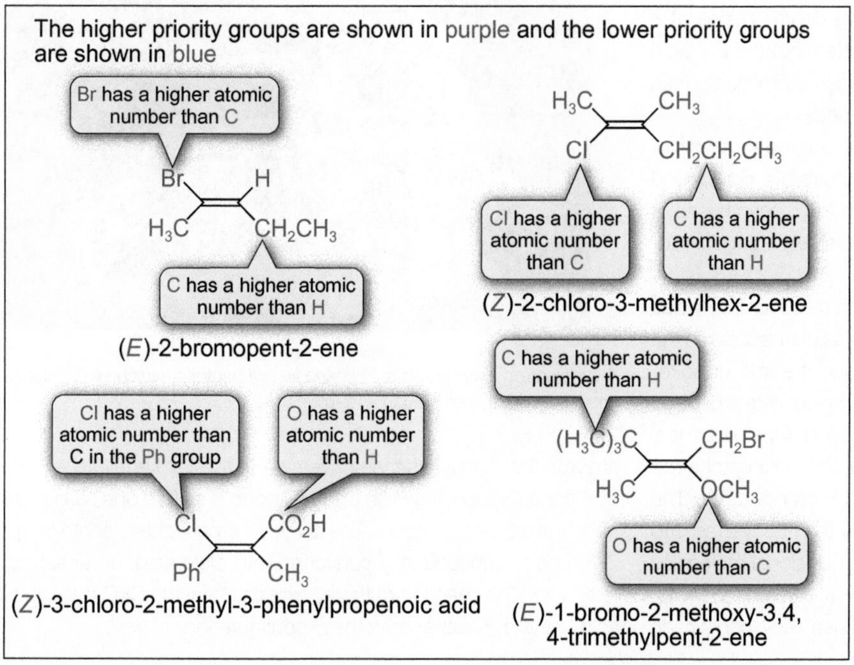

The higher priority groups are shown in purple and the lower priority groups are shown in blue
Br has a higher atomic number than C
Br
H
H_3C
CH_2CH_3
C has a higher atomic number than H
(*E*)-2-bromopent-2-ene
H_3C
CH_3
Cl
$CH_2CH_2CH_3$
Cl has a higher atomic number than C
C has a higher atomic number than H
(*Z*)-2-chloro-3-methylhex-2-ene
Cl has a higher atomic number than C in the Ph group
O has a higher atomic number than H
Cl
CO_2H
Ph
CH_3
(*Z*)-3-chloro-2-methyl-3-phenylpropenoic acid
C has a higher atomic number than H
$(H_3C)_3C$
CH_2Br
H_3C
OCH_3
O has a higher atomic number than C
(*E*)-1-bromo-2-methoxy-3,4,4-trimethylpent-2-ene

Figure 18.18 Applying the sequence rules to assign *E*/*Z* configurations to some substituted alkenes.

Remember that for an alkene to have *E*/*Z* isomers, each carbon of the C=C bond must be bonded to *two* different substituents (e.g. MeCH=CHMe can have *E*/*Z* isomers, but Me_2C=CH_2 cannot).

The order of priority of functional groups containing a C=O bond is: CO_2Me > CO_2H > $CONH_2$ > COMe > CHO

A double or triple bond is converted into a hypothetical 'saturated' equivalent system using ghost atoms (that are shown in brackets). Notice that a ghost atom is only attached to the real atom—it does not have any additional substituents.

Conversion of a C=C bond from *Z* configuration into *E* configuration is important in vision (Chapter 23, p.1055).

When an alkoxy group, RO–, is directly attached to a C=C bond then the functional group is called an enol ether (Section 2.6, p.95).

To test your knowledge of naming *E*- and *Z*-isomers, see the drag and drop activity on the Online Resource Centre.

Worked example 18.3 *E/Z* nomenclature

Draw a skeletal structure for (*E*)-1-bromo-4-ethylhept-3-ene.

Strategy

Draw a zigzag structure of heptane and number the chain from *1* to *7*.

Draw a Br atom attached to carbon number *1* and include a double bond at position *3*. The ethyl group then needs to be at position *4*.

Use the sequence rules to prioritize the four substituents on the double bond. The *E*-isomer has the two groups of higher priority on opposite sides of the double bond. (To help with the assignment of configuration, it generally helps to draw the structures of both alkene isomers before prioritizing the substituents.)

Solution

The alkene carbon at position *3* is bonded to a hydrogen atom and a 1-bromoethyl chain ($BrCH_2CH_2$–). As carbon has a higher atomic number than hydrogen, the 1-bromoethyl chain has the higher priority.

The alkene carbon at position *4* is bonded to an ethyl substituent (–CH_2CH_3) and a propyl substituent (–$CH_2CH_2CH_3$). The propyl substituent has a higher priority than the ethyl substituent because of the additional carbon atom in the chain.

For the *E*-isomer, the two groups with higher priority, the 1-bromoethyl substituent (at position *3*) and the propyl substituent (at position *4*), are on the opposite sides of the double bond.

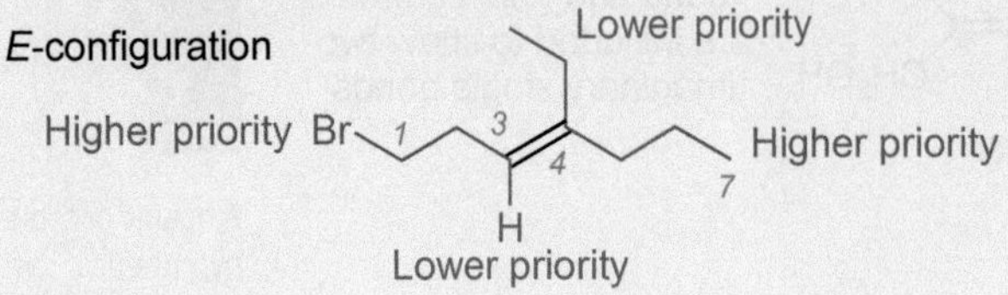

E-configuration
Lower priority
Higher priority Br
1
3
4
7
Higher priority
H
Lower priority

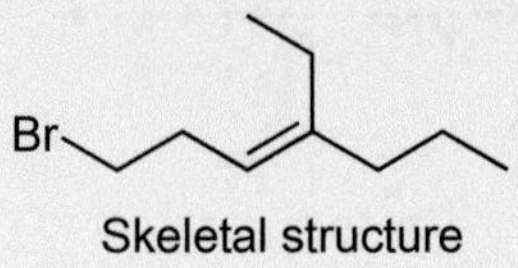

Br
Skeletal structure

Question

Draw a skeletal structure for (*Z*)-3-ethoxybut-2-en-1-ol.

Box 18.4 A natural selection: biosynthesis of fatty acids

The stereochemistry of alkene double bonds in naturally occurring compounds is important as *E*- and *Z*-isomers of the same alkene can have different biological activities. For example, naturally occurring unsaturated fatty acids generally have double bonds of *Z* (*cis*) configuration. If the double bond is converted into the *E* (*trans*) configuration, the shape of the molecule changes dramatically (see Box 2.4, p.86) and this affects its activity in the body.

Alkene stereochemistry plays an interesting role in the synthesis of saturated and unsaturated fatty acids. The synthesis of a compound by a living organism is called **biosynthesis** and the biosyntheses of saturated and unsaturated fatty acids in bacteria share a similar reaction pathway, as shown in the illustration.

Ethanoic acid (CH_3CO_2H) is used as a building block for fatty acids such as stearic acid and oleic acid. The ethanoic acid molecules (in the form of acetyl-coenzyme A) are joined together, one at a time, using reactions that are catalysed by different enzymes. The enzyme is bonded to the growing carbon chain and is removed at the end of the sequence.

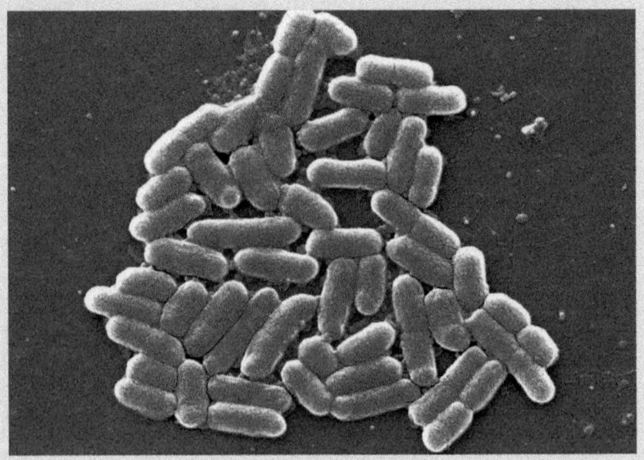

▲ An electron micrograph of *Escherichia coli* bacteria, which synthesize unsaturated fatty acids such as *Z*-oleic acid.

In bacteria, the formation of stearic acid starts by linking together six ethanoic acid molecules to form an alkene of *E*-configuration. The twelve-carbon chain containing the *E*-alkene is then converted into stearic acid (a saturated fatty acid) using three additional molecules of ethanoic acid. However, in some bacteria, a different pathway is possible. These bacteria are unusual because they contain an enzyme that catalyses the conversion of the *E*-alkene into an alkene of *Z*-configuration, in which the C=C bond is shifted one carbon away from the carbonyl group. The twelve-carbon chain containing the *Z*-alkene is subsequently converted into oleic acid, an unsaturated fatty acid. This reaction pathway explains how the C=C bond in oleic acid has the *Z*- rather than the *E*-configuration.

→

→

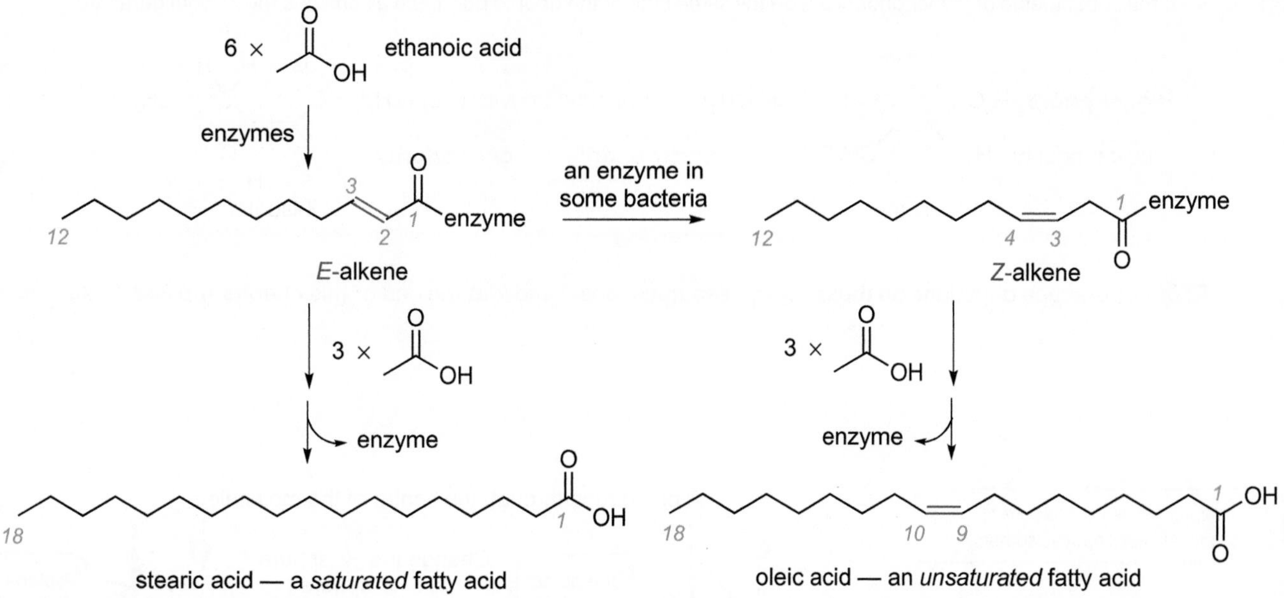
6 ×
O
OH
ethanoic acid
enzymes
an enzyme in
some bacteria
enzyme
E-alkene
Z-alkene
3 ×
enzyme
stearic acid — a *saturated* fatty acid
oleic acid — an *unsaturated* fatty acid

Question

Leukotrienes are a family of naturally occurring molecules that are biologically synthesized from a polyunsaturated fatty acid called arachidonic acid. In the body, leukotrienes play a role in allergic reactions and can give rise to the symptoms of asthma. Molecules that prevent the formation of leukotrienes from arachidonic acid are often used to treat asthma.

Assign the stereochemistry of each of the C=C bonds in arachidonic acid and leukotriene B_4 using *E*/*Z* nomenclature.

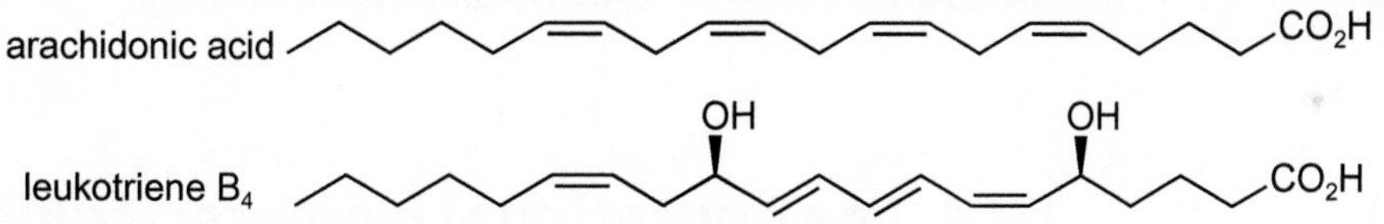
arachidonic acid
CO_2H
OH
OH
leukotriene B_4
CO_2H

Summary

- The *E*/*Z* system of nomenclature is a general system for naming the configurational isomers of substituted alkenes.
- *trans*-Alkenes have *E*-configuration, whereas *cis*-alkenes have *Z*-configuration. The *E*/*Z* system for naming alkenes is preferable to using *cis*-/*trans*- because *cis*-/*trans*- is limited to naming simple alkenes (typically 1,2-disubstituted alkenes).
- The two substituents at each end of the double bond are ranked in priority using a sequence of steps.

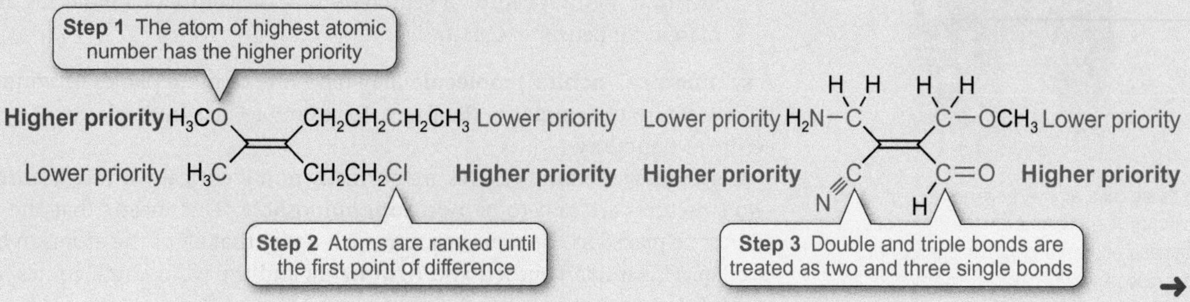
Step 1 The atom of highest atomic number has the higher priority
Higher priority H_3CO
$CH_2CH_2CH_2CH_3$ Lower priority
Lower priority H_3C
CH_2CH_2Cl
Higher priority
Step 2 Atoms are ranked until the first point of difference
Lower priority H_2N
OCH_3 Lower priority
Higher priority
Higher priority
Step 3 Double and triple bonds are treated as two and three single bonds

→

- If the substituents of higher priority are on the *opposite* sides of the double bond, the alkene has the *E*- configuration.
- If the substituents of higher priority are on the *same* side of the double bond, the alkene has the *Z*- configuration.

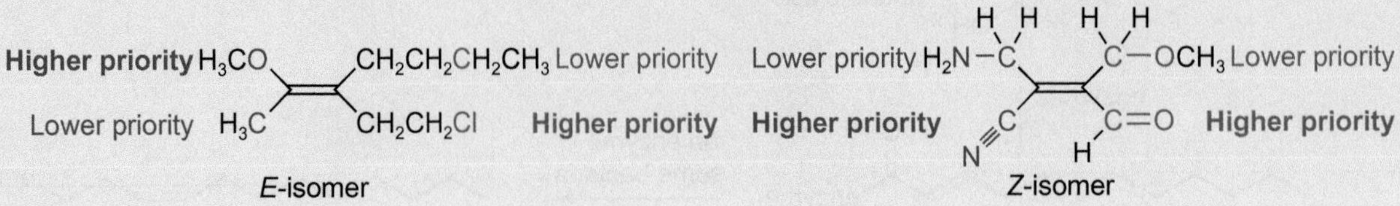
Higher priority H_3CO
$CH_2CH_2CH_2CH_3$ Lower priority
Lower priority H_3C
CH_2CH_2Cl Higher priority
E-isomer
Lower priority H_2N-C
$C-OCH_3$ Lower priority
Higher priority
$C=O$ Higher priority
Z-isomer

For practice questions on these topics, see questions 3 and 7 at the end of this chapter (pp.858–859).

Isomers with chiral centres are sometimes called **optical isomers.**

Not all molecules containing more than one chiral centre are asymmetric. A small number are symmetrical (see *meso* compounds later in this section on p.851).

Isomers with chiral centres are configurational isomers (p.817). Most configurational isomers cannot be interconverted without breaking a chemical bond.

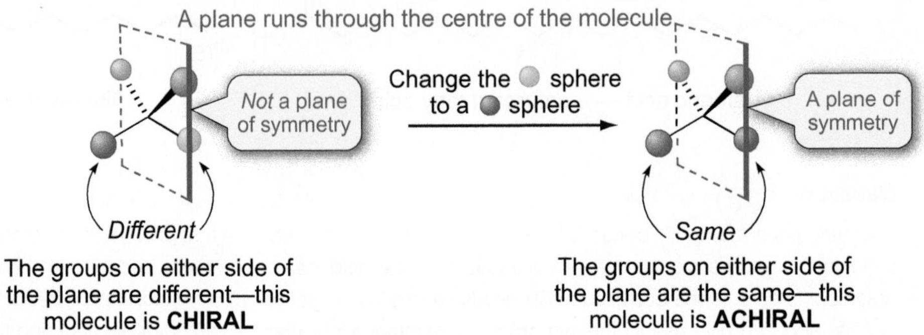
A plane runs through the centre of the molecule
Not a plane of symmetry
Change the sphere to a sphere
A plane of symmetry
Different
Same
The groups on either side of the plane are different—this molecule is **CHIRAL**
The groups on either side of the plane are the same—this molecule is **ACHIRAL**

Figure 18.19 Chiral molecules do not have a plane of symmetry.

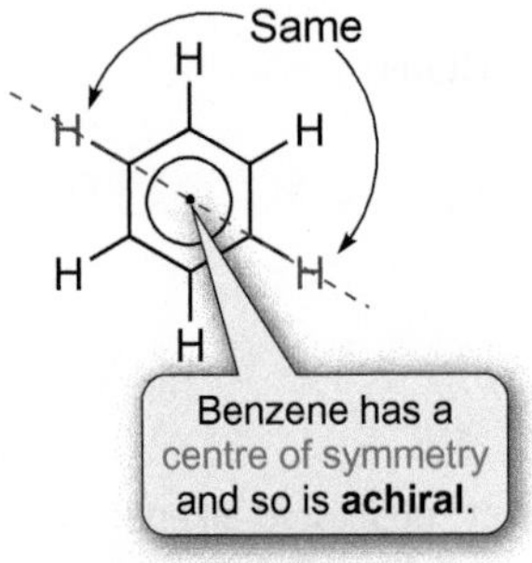
Same
Benzene has a centre of symmetry and so is **achiral**.

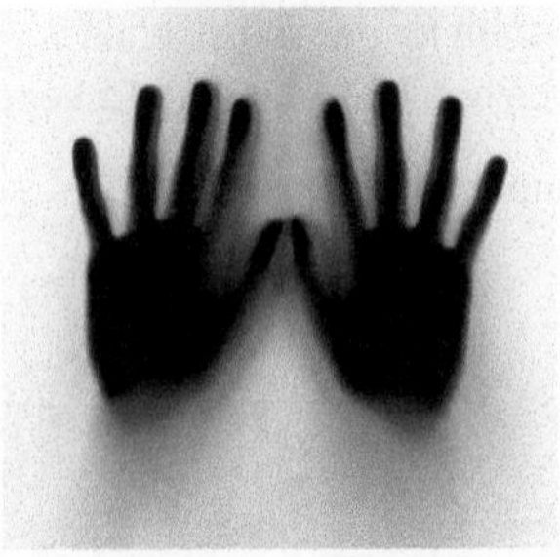

Hands are chiral as they have no plane of symmetry. A left hand is a mirror image of a right hand, and left and right hands are not superimposable. When placed side by side the thumbs point in opposite directions.

18.4 Configurational isomers: isomers with chiral centres

Isomers with chiral centres are usually **asymmetric molecules** that do *not* have a plane or centre of symmetry.

- As shown in Figure 18.19, a **plane of symmetry** is a plane drawn through the centre of a molecule, such that the two halves of the molecule are identical in size and shape, but the halves are mirror images of each other.
- An object has a **centre of symmetry** if you can draw a straight line from any position, through the centre of the object to an equal distance the other side, and arrive at an identical point. A centre of symmetry is at the centre of a molecule as shown in the margin for benzene (C_6H_6).

A **symmetrical (achiral) molecule** may have one or more planes of symmetry or it may have a centre of symmetry. (Benzene, for example, has both planes of symmetry and a centre of symmetry.)

Molecules with chiral centres are normally non-identical with their mirror image—the two structures are said to be **non-superimposable**. This means that the two structures cannot be placed in the same space in such a way that all of the atoms in both structures overlap. This is like your left and right hand, and the word chiral comes from the Greek word *cheir*, which means hand.

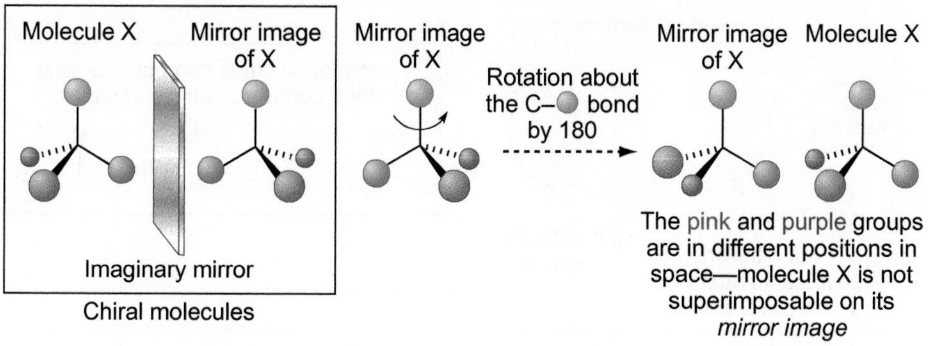
Molecule X
Mirror image of X
Imaginary mirror
Chiral molecules
Mirror image of X
Rotation about the C–● bond by 180
Mirror image of X
Molecule X
The pink and purple groups are in different positions in space—molecule X is not superimposable on its *mirror image*

Many objects we use in our everyday world have non-identical mirror image forms and so are chiral. This includes golf clubs, certain types of scissors, and shoes.

> ℹ A molecule is chiral if it *cannot* be superimposed upon its mirror image. A molecule is achiral if it *can* be superimposed upon its mirror image. If in doubt, use molecular models to help you work out if a molecule is chiral.

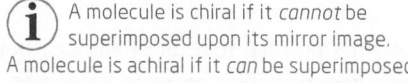

Box 18.5 Some landmarks in chirality

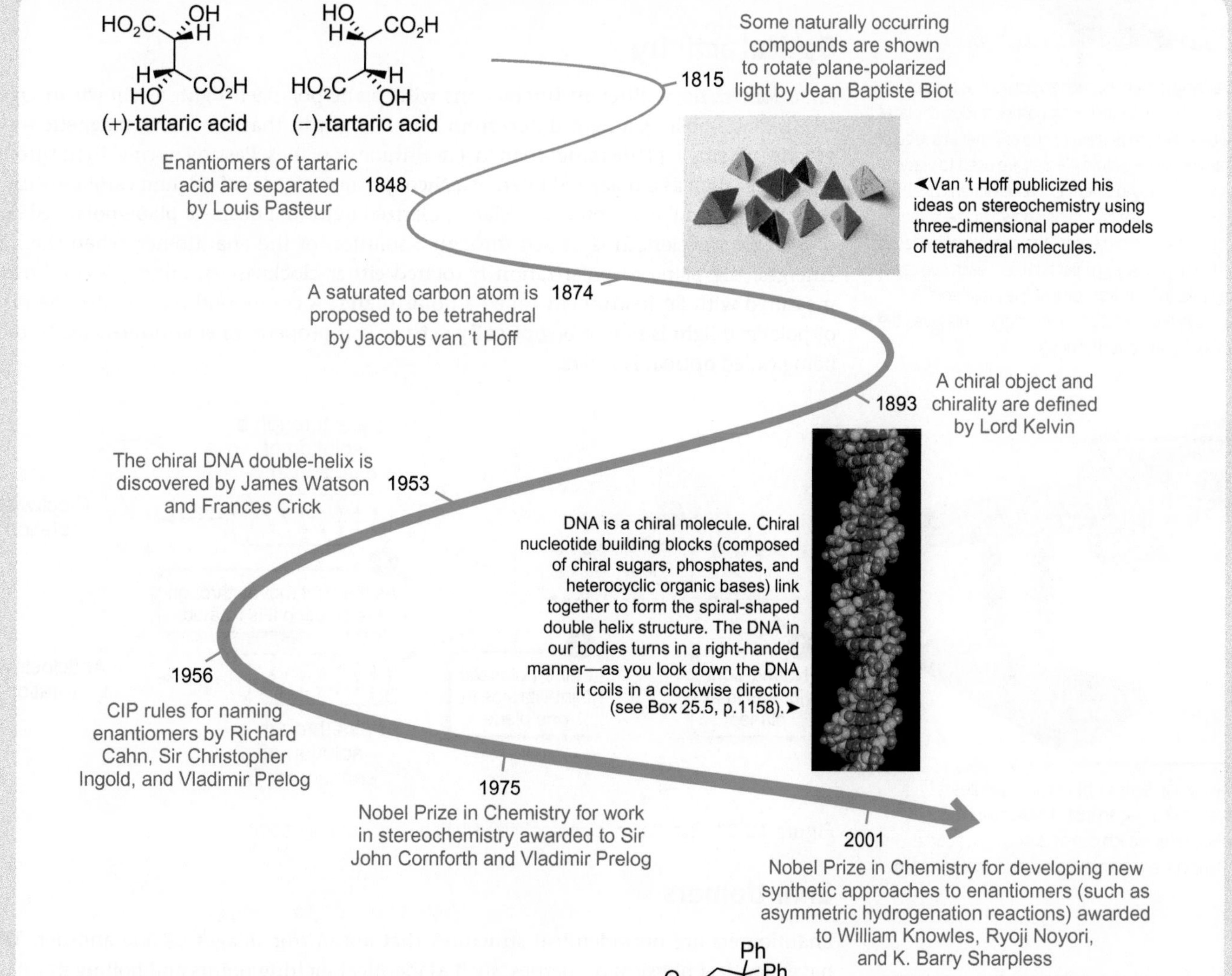
(+)-tartaric acid
(−)-tartaric acid
1815 Some naturally occurring compounds are shown to rotate plane-polarized light by Jean Baptiste Biot
1848 Enantiomers of tartaric acid are separated by Louis Pasteur
◄Van 't Hoff publicized his ideas on stereochemistry using three-dimensional paper models of tetrahedral molecules.
1874 A saturated carbon atom is proposed to be tetrahedral by Jacobus van 't Hoff
1893 A chiral object and chirality are defined by Lord Kelvin
1953 The chiral DNA double-helix is discovered by James Watson and Frances Crick
DNA is a chiral molecule. Chiral nucleotide building blocks (composed of chiral sugars, phosphates, and heterocyclic organic bases) link together to form the spiral-shaped double helix structure. The DNA in our bodies turns in a right-handed manner—as you look down the DNA it coils in a clockwise direction (see Box 25.5, p.1158).►
1956 CIP rules for naming enantiomers by Richard Cahn, Sir Christopher Ingold, and Vladimir Prelog
1975 Nobel Prize in Chemistry for work in stereochemistry awarded to Sir John Cornforth and Vladimir Prelog
2001 Nobel Prize in Chemistry for developing new synthetic approaches to enantiomers (such as asymmetric hydrogenation reactions) awarded to William Knowles, Ryoji Noyori, and K. Barry Sharpless

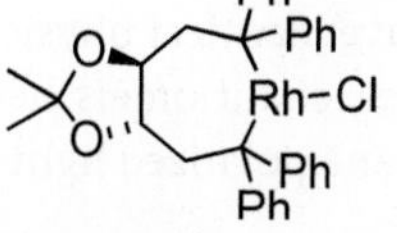

◄A chiral catalyst for asymmetric hydrogenation reactions

The terms **chirality centre** and **stereogenic centre** are sometimes used in place of chiral centre.

The three-dimensional arrangement of atoms and substituents attached to a chiral centre is called the **configuration**.

REAL

Polarizing materials, such as polaroid (a type of plastic sheet), contain long particles, rods, or plates that are aligned parallel to one another in a regular arrangement. Polaroid glasses are used to view some 3-D films in cinemas. Three-dimensional films are actually two films, shown at the same time, using two projectors that display polarized light. When wearing polaroid filters each eye can only take in light from one of the projectors so each eye receives a different image. This gives the viewer a perception of depth.

The optical activity of chiral molecules is measured using an instrument called a polarimeter, which consists of a light source, polarizing lens, sample tube, and analysing lens.

Enantiomers of coordination complexes, including octahedral complexes, are discussed in Section 28.3, p.1276.

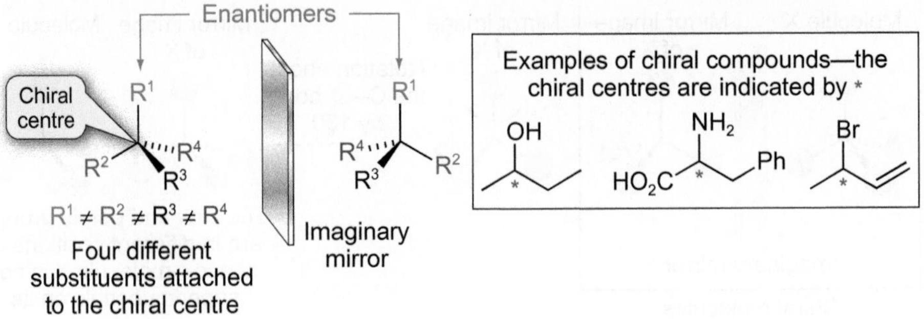
Enantiomers
Chiral centre
R¹ ≠ R² ≠ R³ ≠ R⁴
Four different substituents attached to the chiral centre
Imaginary mirror
Examples of chiral compounds—the chiral centres are indicated by *
OH
NH2
HO2C
Ph
Br

Figure 18.20 Organic compounds with four different substituents (R^1–R^4) bonded to a tetrahedral carbon.

Organic compounds with four different substituents bonded to a tetrahedral carbon (Figure 18.20) are chiral compounds and the tetrahedral carbon atom is known as a **chiral (asymmetric) centre**. There are two ways of arranging the four different groups around a tetrahedral carbon atom and these give mirror-image structures. The three-dimensional arrangement of the substituents is indicated using hashed–wedged-line notation (Section 2.2, p.75). A chiral molecule and its non-superimposable mirror image are special types of configurational isomers called **enantiomers**.

Optical activity

Enantiomers have different interactions with plane-polarized light, as shown in Figure 18.21. Plane-polarized light differs from normal light in that the electromagnetic waves vibrate in a single plane rather than in a multitude of planes. Passing normal light through a polarizer such as a polarized filter, or a Nicol prism (a crystal of calcium carbonate called calcite or Iceland Spar), produces plane-polarized light. If a beam of plane-polarized light of a single wavelength is passed through a solution of the enantiomer, when the light emerges, the plane of polarization is rotated either clockwise or anticlockwise—this is measured with an instrument called a **polarimeter**. A compound that rotates the plane of polarized light is said to be **optically active**. This property of enantiomers led to them being called **optical isomers**.

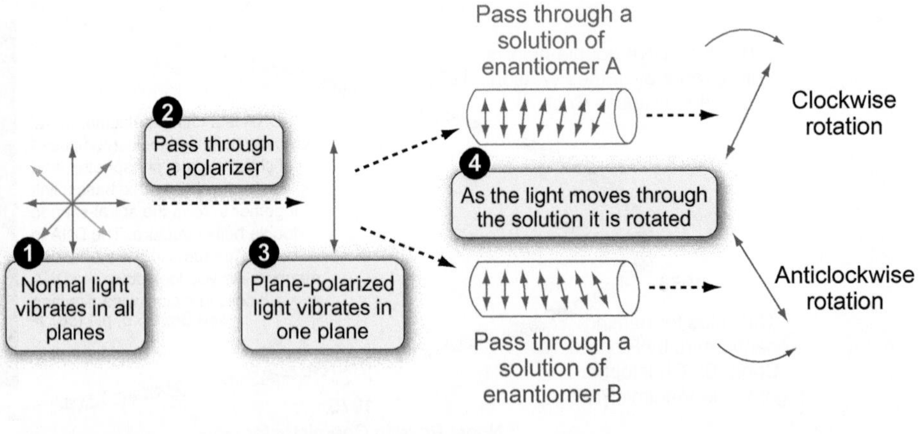
Pass through a solution of enantiomer A
Clockwise rotation
2 Pass through a polarizer
4 As the light moves through the solution it is rotated
1 Normal light vibrates in all planes
3 Plane-polarized light vibrates in one plane
Anticlockwise rotation
Pass through a solution of enantiomer B

Figure 18.21 Rotation of plane-polarized light by a pair of enantiomers.

Enantiomers

Enantiomers are non-identical structures that are mirror images of one another. They have identical physical properties, such as identical melting points and boiling points. In fact, enantiomers have identical chemical properties until they come into contact with plane-polarized light—or something else that is chiral.

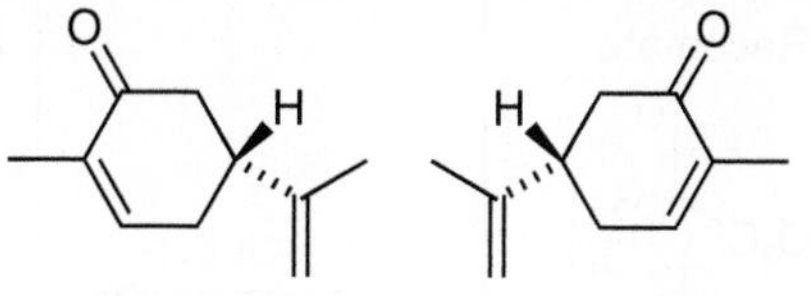
O
H
H
O

	L-(−)-carvone	D-(+)-carvone	Enantiomers of carvone
Boiling point	227–230 °C	227–230 °C	*Same* physical property
Optical rotation of polarized light	−61°	+61°	*Opposite* optical rotation
Biological activity (smell/taste)	spearmint	caraway	*Different* biological activity (because taste receptors in the body are chiral)

Figure 18.22 Properties of enantiomers.

The enantiomers of carvone shown in Figure 18.22 have different effects on plane-polarized light. One enantiomer rotates plane-polarized light in a clockwise direction and is called the (+)-**enantiomer** or the **dextrorotatory** enantiomer. The other enantiomer rotates plane-polarized light in an anticlockwise direction and is called the (−)-**enantiomer** or the **laevorotatory** enantiomer. (Latin: *dextra*, right hand; *laevus*, left hand.) The angle through which the plane-polarized light is rotated is called the **optical rotation** and is given the symbol α.

The optical rotation of a compound is the quantitative measure of its optical activity. It is proportional to the number of optically active molecules present in the beam of polarized light so it depends on the concentration (c) of the optically active compound and the path length (l). The path length is how far the light has to travel through the sample of the optically active compound. When the effects of concentration and path length are corrected for, you have a standard for comparison of all molecules, which is called the **specific rotation**, $[\alpha]$. In Equation 18.1, a subscript indicates the wavelength of plane-polarized light used (normally $[\alpha]_D$ because the D line of a sodium lamp (589 nm) is used) and a superscript indicates the temperature in degrees Celsius.

Temperature at which the optical rotation is measured in °C

$$[\alpha]_D^{20} = \frac{\text{observed rotation (degrees)} \times 100}{\text{path length (dm)} \times \text{concentration (g per 100 cm}^3\text{)}} = \frac{\alpha \times 100}{l \times c} \quad (18.1)$$

Polarized light from a sodium lamp of wavelength 589 nm

By convention, the units of $[\alpha]$ (10^{-1} deg cm^2 g^{-1}) are generally not quoted

Visit the Online Resource Centre to view screencast 18.1 which defines enantiomers and walks you through how to assign the configuration of an enantiomer as *R* or *S* (p.845).

Caraway seeds are used in cooking. The oil from the seeds contains D-(+)-carvone.

If only one enantiomer is present, a sample is considered to be **optically pure**. When a sample consists of a mixture of enantiomers, the effect of each enantiomer cancels out, molecule for molecule. Therefore, a 50:50 mixture of two enantiomers will not rotate plane-polarized light and is **optically inactive**. This is because, for every molecule that rotates plane-polarized light in one direction, there is a mirror-image molecule that rotates plane-polarized light in the opposite direction. An equal mixture of two enantiomers is called a **racemate** (**racemic mixture**) and is designated as (±). A racemate may have different physical properties from the enantiomers as illustrated by the different melting points of lactic acid (2-hydroxypropanoic acid, $CH_3CH(OH)CO_2H$) in Figure 18.23.

A racemate may be separated into its component enantiomers by a process called **resolution** (Box 18.8, p.850). Conversely, a pure enantiomer may be transformed into a racemate. This process is called **racemization**. Racemization requires a change of configuration and so the covalent bonds of the chiral carbon atom need to be broken and reformed. The

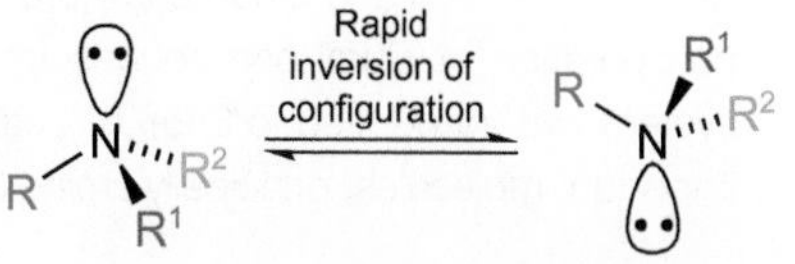
Rapid
inversion of
configuration
R
N
R2
R1
R
N
R2
R1

The nitrogen is tetrahedral if the lone pair of electrons is considered to be a substituent.

Usually, configurational isomers do not interconvert at room temperature since this normally involves breaking and reforming bonds. However, there are some exceptions. For example, chiral tertiary amines (with four different groups on nitrogen) undergo inversion so rapidly that it is not possible to obtain a single enantiomer.

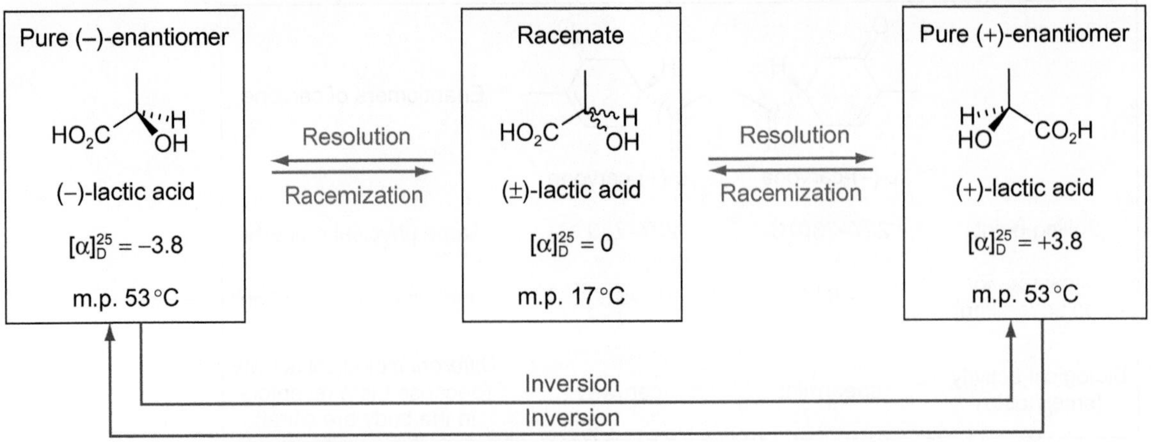
Pure (−)-enantiomer
HO2C
H
OH
(−)-lactic acid
[α]D25 = −3.8
m.p. 53 °C
Resolution
Racemization
Racemate
HO2C
H
OH
(±)-lactic acid
[α]D25 = 0
m.p. 17 °C
Resolution
Racemization
Pure (+)-enantiomer
H
HO
CO2H
(+)-lactic acid
[α]D25 = +3.8
m.p. 53 °C
Inversion
Inversion

Figure 18.23 Properties of the two enantiomers of lactic acid (2-hydroxypropanoic acid) compared with those of the racemate.

Changing the configuration of a chiral carbon atom requires bonds to be broken and reformed. This occurs in a nucleophilic substitution reaction, called an S_N2 reaction (Section 20.3, p.930).

change of configuration leading to the formation of one enantiomer from another enantiomer is called **inversion.**

The **enantiomeric purity** (sometimes called **optical purity**) of a compound is measured by a quantity called the **enantiomeric excess (ee)**. For a racemate, the enantiomeric excess is 0% because there is 50% of one enantiomer and 50% of the other enantiomer. A 40% ee corresponds to a mixture of 70% of one enantiomer and 30% of the other enantiomer (i.e. 70% – 30% = 40%)

The two wavy lines 〰〰 in the structure of (±)-lactic acid show that both isomers are present (Figure 2.5, Section 2.2, p.75).

$$\text{Enantiomeric excess, ee (\%)} = \text{\% of the major enantiomer} - \text{\% of the other enantiomer} = \frac{\text{observed } [\alpha]_D}{[\alpha]_D \text{ of pure enantiomer}} \times 100 \qquad (18.2)$$

(with the same sign as the observed $[\alpha]_D$)

Note that it is *not* possible to deduce if an enantiomer has a (+) or (–) optical rotation from its molecular structure—this requires a polarimeter. The (+) or (–) sign of the optical rotation is an experimental measurement and *cannot* tell us the **configuration** of the chiral molecule (i.e. the spatial arrangement of the four groups about the chiral centre). The **Cahn–Ingold–Prelog (CIP)** system of nomenclature unambiguously describes the configuration of groups at a chiral centre in a molecule (see p.845).

Box 18.6 The D/L nomenclature

The D and L nomenclature is a system for assigning the configuration of chiral compounds that was used before the CIP system was introduced (p.845). It is still useful for some biologically important molecules, principally amino acids and sugars. In this system, glyceraldehyde (2,3-dihydroxypropanal, $HOCH_2CH(OH)CHO$), a simple chiral compound, is used as the standard against which the configurations of other chiral centres are compared. The (+)-enantiomer of glyceraldehyde is given the label D (from dextrorotatory) and the (–)-enantiomer of glyceraldehyde is given the label L (from laevorotatory). Any single enantiomer of a compound that is prepared from D-glyceraldehyde is given the label D. Any single enantiomer of a compound that can be converted into D-glyceraldehyde is also given

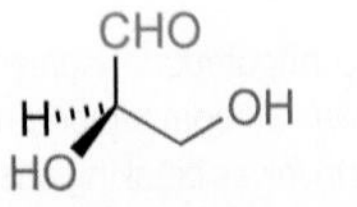
CHO
H
OH
HO

D-(+)-glyceraldehyde

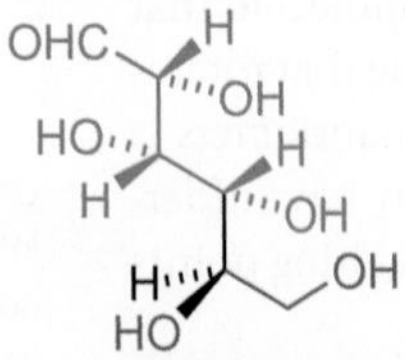
OHC
H
OH
HO
H
H
OH
H
OH
HO

D-(+)-glucose (open-chain form)

A naturally occurring sugar

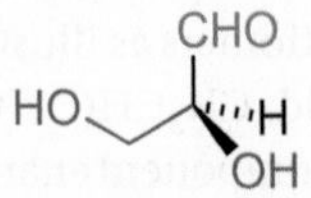
CHO
HO
H
OH

L-(–)-glyceraldehyde

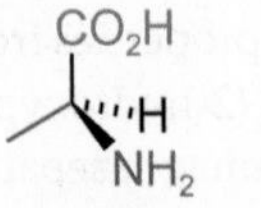
CO2H
H
NH2

L-(+)-alanine

A naturally occurring amino acid

→

→ the label D. For example, naturally occurring glucose is D-glucose (sometimes called dextrose). Similarly, single enantiomers related to L-glyceraldehyde in these ways are given the label L.

The structural relationship of glyceraldehyde with sugars is relatively easy to see, so the arrangements of atoms at the chiral centres of glyceraldehyde and the sugar are easily compared. This is why the use of D and L is so useful for these compounds. In a similar way, the structures of amino acids can easily be related to the structure of glyceraldehydes.

Most naturally occurring amino acids are L-amino acids and almost all naturally occurring sugars are D-sugars. Note that the assignment of D or L to a structure is not related to the (+) or (–) sign of the optical rotation. For example, a D-amino acid could be a (+)-enantiomer or a (–)-enantiomer.

Question

Draw the two enantiomers of phenylalanine using hashed–wedged line notation and indicate which of these is more likely to be found in nature.

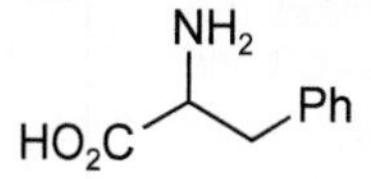
NH2
HO2C
Ph
(±)-phenylalanine

Biological properties of enantiomers

In addition to having different interactions with plane-polarized light, enantiomers usually have different biological properties. For example, one enantiomer of a molecule may function as a therapeutic medicine in the human body, while the other enantiomer could be toxic (see Box 18.7, p.848, which is about thalidomide). The enantiomers have different biological properties because the human body is made up of enantiomerically pure compounds, such as proteins, which are made by joining together single enantiomers of amino acids. When in the body, the enantiomer is in a chiral environment and different enantiomers of a molecule (with different three-dimensional shapes) react in different ways with chiral proteins—just like a left glove will fit a left hand but not a right hand. Active sites in enzymes and protein receptor sites in membranes are all chiral.

A good analogy for the interaction of two chiral molecules comes from the world of golf. As most people have a particular handedness they require different golf clubs, as both golf clubs and our hands are chiral. A right-handed person requires a right-handed golf club, rather than a left-handed golf club, to be able to hit the ball.

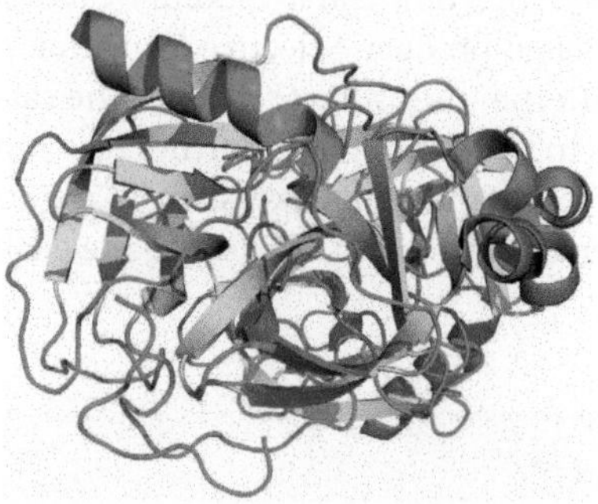

Chymotrypsin is a protein found in our digestive systems (Box 19.6, p.897). A molecule of chymotrypsin contains 251 chiral centres. With this number of chiral centres, 2^{251} stereoisomers are possible, but in nature, only one of these stereoisomers is found. (For molecules with more than one chiral centre, see p.849.) The complex protein structure is represented here using a ribbon diagram (see Chapter 2 opening page p.69 and Chapter 14, p.655).

The different interactions of enantiomers with the same protein (or receptor) can be understood using a model called **three-point attachment**. In this model, only one of the enantiomers has the substituents (about the chiral centre) in the appropriate spatial positions to make three intermolecular interactions with the protein. The other enantiomer, which has a different three-dimensional shape, can only form two intermolecular interactions with the protein.

This model explains why the (–)-enantiomer of epinephrine (adrenaline) binds more strongly than the (+)-enantiomer to a protein in the body called the α-adrenergic receptor. When (–)-epinephrine binds to this receptor it causes multiple effects, one of which is to increase blood pressure.

Non-covalent (intermolecular) attractions are discussed in Section 1.8, p.52.

The three-point attachment model is also used in developing chromatography stationary phases for analysing chiral compounds (Section 11.3, p.541).

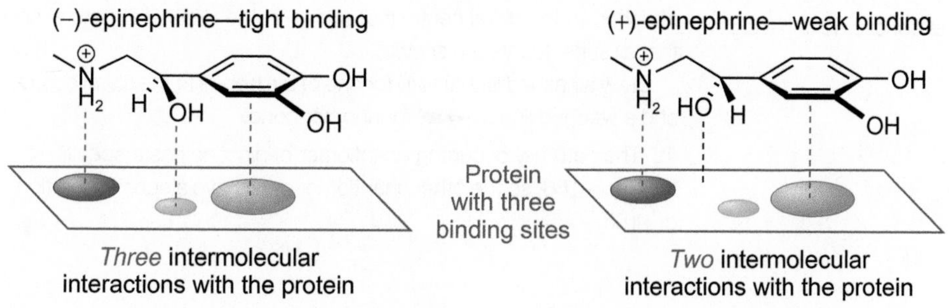
(–)-epinephrine—tight binding
(+)-epinephrine—weak binding
N
H2
H
OH
OH
OH
HO
H
Protein with three binding sites
Three intermolecular interactions with the protein
Two intermolecular interactions with the protein

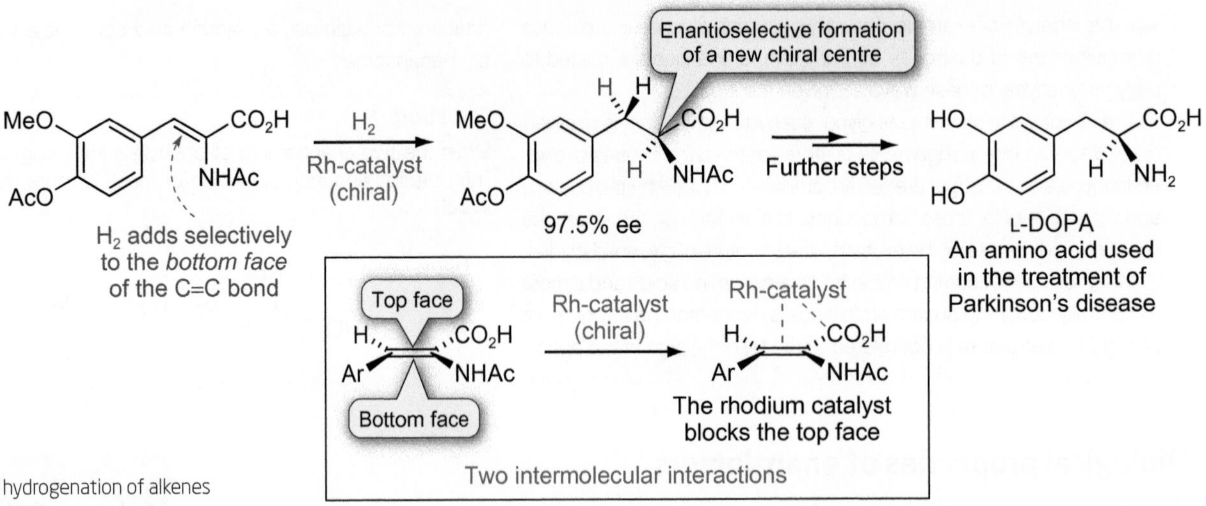
Enantioselective formation of a new chiral centre
MeO
AcO
CO2H
NHAc
H2
Rh-catalyst (chiral)
97.5% ee
Further steps
HO
NH2
L-DOPA
An amino acid used in the treatment of Parkinson's disease
H2 adds selectively to the bottom face of the C=C bond
Top face
Bottom face
Ar
The rhodium catalyst blocks the top face
Two intermolecular interactions

For catalytic hydrogenation of alkenes see Section 21.2 (p.968).

Figure 18.24 An enantioselective reaction in the synthesis of L-DOPA.

The use of chiral building blocks to form the shell of the edible snail (*Helix pomatia*) results in 99.995% of these snails having a shell that spirals in the same direction (right-handed shells). The shells of the remaining 0.005% of the snails spiral in the opposite direction (left-handed shells).

Since enantiomers can have different biological activities, both enantiomers of any chiral drug molecule must be tested for biological activity including side-effects. The preparation of enantiomerically pure drugs, and the resolution of racemate drugs, are, therefore, major areas of research today. Of particular interest is the development of new **enantioselective (asymmetric) reactions**, which produce an excess of one enantiomer of the product.

An example of an enantioselective reaction used in the pharmaceutical industry is shown in Figure 18.24. It produces L-DOPA, a medicine used to treat Parkinson's disease. Addition of H_2 to a trisubstituted alkene, in the presence of a chiral rhodium catalyst, leads to essentially one enantiomer of the hydrogenated product. Selective formation of one enantiomer is due to the catalyst interacting with the trisubstituted alkene. The catalyst blocks the top face of the C=C bond, so H_2 adds to the bottom face. Further reactions of the hydrogenated product give L-DOPA in 97.5% enantiomeric excess (ee). Worked example 18.4 provides further practice in drawing enantiomers and calculating the enantiomeric excess of a mixture of enantiomers.

Worked example 18.4 Enantiomers and enantiomeric excess

◀ About 1 mg of nicotine is absorbed into the body on smoking an average cigarette.

Nicotine is found as a single enantiomer in the leaves of the tobacco plant (*Nicotiana tabacum*). People have smoked or chewed tobacco leaves for over 8000 years to experience the effect of the nicotine. (Nicotine induces the relaxing or stimulating effect of smoking a cigarette (Box 18.1, p.822).)

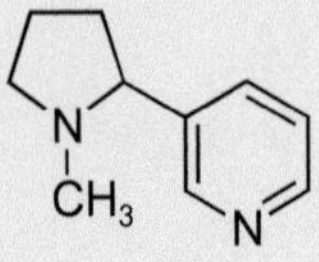
N
CH3
N

nicotine

(a) Draw both enantiomers of nicotine using hashed–wedged line notation.

(b) The naturally occurring enantiomer of nicotine has a specific rotation, $[\alpha]_D$, of −166. If the observed specific rotation of a mixture of the two enantiomers of nicotine is +60, what is the enantiomeric excess (ee) of the mixture?

Strategy

(a) Identify the chiral centre.

Although not strictly necessary, it helps to draw the hydrogen atom attached to the chiral centre (not shown in the skeletal structure) so that all substituents are shown.

Draw a hashed line (IIIIIIIII) for one of the bonds to the chiral centre and a wedged line (▬) for another bond.

(b) The naturally occurring enantiomer of nicotine has a specific rotation of −166, so the other enantiomer must have a specific rotation of +166.

➜

→ As the observed specific rotation is +60, there is an excess of the enantiomer with a specific rotation of +166 (because the observed specific rotation is closer to +166 than −166).

To determine the ee of the (+)-enantiomer, use Equation 18.2 (p.841).

Solution

(a) Enantiomers of nicotine.

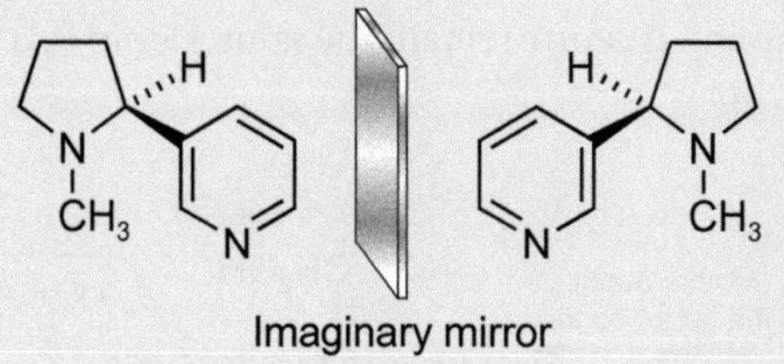
H
N
CH3
N
H
N
CH3
N
Imaginary mirror

(b) Using Equation 18.2 on p.841.

$$\text{Enantiomeric excess of the mixture, ee (\%)} = \frac{\text{observed } [\alpha]_D}{[\alpha]_D \text{ of pure enantiomer}} \times 100 \quad (18.2)$$

$$= \frac{+60}{+166} \times 100$$

$$= 36\%$$

Question

What is the observed specific optical rotation, $[\alpha]_D$, of a mixture containing 25% (−)-nicotine and 75% (+)-nicotine?

Cahn-Ingold-Prelog nomenclature

To be able to name individual enantiomers, a system of nomenclature is required that indicates the three-dimensional arrangement (the configuration) of all atoms or substituents attached to a chiral centre. To do this, chemists use the letters *R* and *S* in a nomenclature system devised by Robert Cahn, Sir Christopher Ingold, and Vladimir Prelog. (This is sometimes abbreviated to **CIP nomenclature**.) For any pair of enantiomers containing a single chiral centre, one enantiomer will have the *R* configuration and the other will have the *S* configuration. The system relies on a five-step approach, which begins by ranking the atoms or groups about a chiral centre using the same rules as for the *E*/*Z* nomenclature for configurational isomers of alkenes (Section 18.3, p.834).

ⓘ CIP nomenclature does not rely on reference to a standard such as D-glyceraldehyde in the D/L system.

ⓘ To name a particular enantiomer, *R* or *S* is added as a prefix, often in brackets (e.g. (*R*)-2-bromobutane).

ⓘ The atomic number is the number of protons in the nucleus of an atom and the values for all elements are given on the inside front cover.

ⓘ For isotopes (which have the same atomic number), the atomic mass number is used to rank the atom or group (deuterium has a higher priority than hydrogen).

Working out the CIP nomenclature for a chiral compound

Step 1 Rank the four atoms *directly attached to a carbon chiral centre* in order of decreasing atomic number, so that the atom of highest atomic number is ranked first and the atom of lowest atomic number is ranked fourth. The group ranked number *4* has the lowest priority.

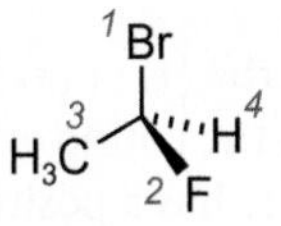
1 Br
3 H3C
4 H
2 F

1 = Br (highest)
2 = F
3 = C
4 = H (lowest)

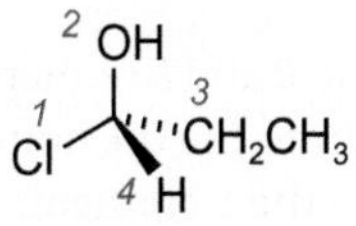
2 OH
1 Cl
3 CH2CH3
4 H

1 = Cl (highest)
2 = O
3 = C
4 = H (lowest)

Start by ranking the four groups attached to the chiral carbon.

Step 2 If the atoms directly attached to the chiral centre are the same, then rank the second, third, fourth, and so on atoms (working away from the chiral centre) one at a time until a difference is found.

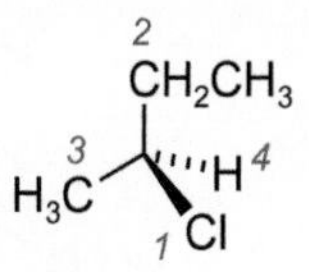
2 CH2CH3
3 H3C
H 4
1 Cl

1 = Cl (highest)
2 = C–C
3 = C–H
4 = H (lowest)

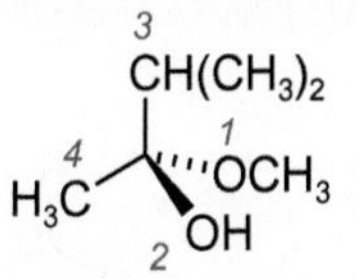
3 CH(CH3)2
4 H3C
1 OCH3
2 OH

1 = O–C (highest)
2 = O–H
3 = C–C
4 = C–H (lowest)

ⓘ Keep in mind that the ranking is determined by the first point of difference along the chain of two similar substituents. After the first point of difference, the rest of the chain is irrelevant. For example, the order of priority of the following groups is: $-CHBrCH_3 > -CHClCH_3 > -CH(CH_3)_2 > -CH_2CH_2CH_3$

Step 3 If substituents contain double and triple bonds, treat the atoms joined by the double or triple bonds as if they were linked to two or three single-bonded atoms (see p.835).

The simplest approach to drawing a particular stereoisomer of *R* or *S* configuration is to begin by drawing the group or atom of lowest priority pointing away from you.

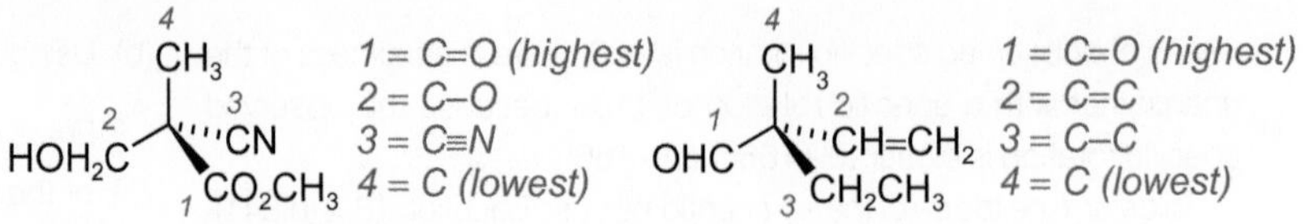

1 = C=O (highest)
2 = C–O
3 = C≡N
4 = C (lowest)
1 = C=O (highest)
2 = C=C
3 = C–C
4 = C (lowest)

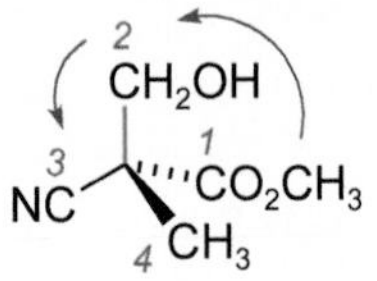

The arrow indicates the configuration is *S*, but, the lowest priority group is pointing towards you, so the configuration is *R*

If the group of lowest priority is pointing towards you, then the configuration is the *opposite* of that indicated by the curved arrow. This approach saves redrawing the structure with the lowest priority group pointing away from you.

Step 4 Orientate the molecule so that the atom of lowest (fourth) priority is pointing away from you. (After rotation, take care to ensure that the relative positions of the groups about the chiral centre remain the same.) From this point, ignore the lowest priority group.

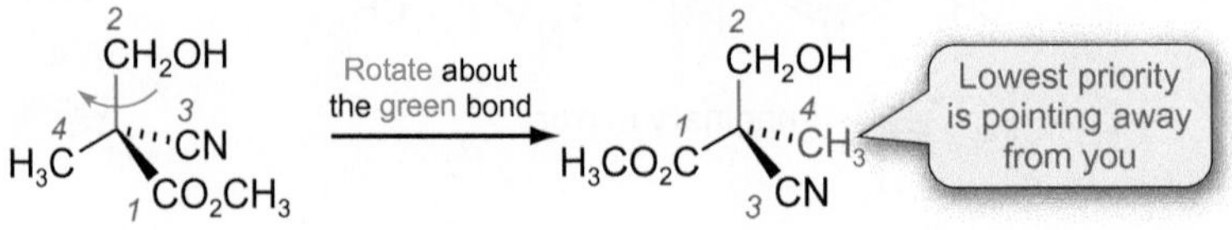

Rotate about the green bond
Lowest priority is pointing away from you

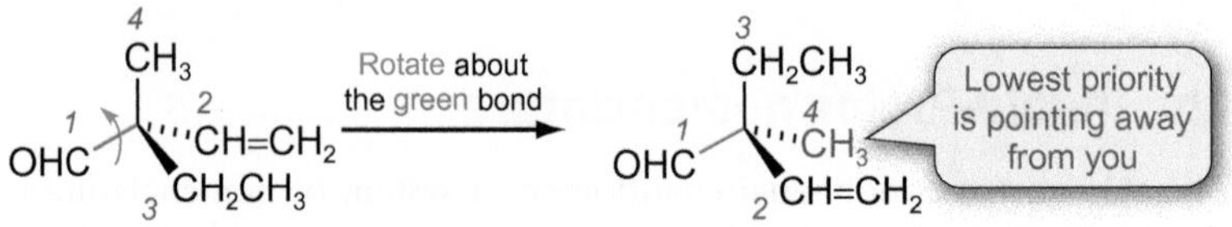

Rotate about the green bond
Lowest priority is pointing away from you

The choice of the letters *R* and *S* comes from the Latin words for proper or right (*rectus*) and left-hand side (*sinister*). You can remember the assignment of *R* by analogy to a car steering wheel making a right turn.

Step 5 Draw a curved arrow from the group of highest priority to the group of next highest priority (i.e. 1 → 2 → 3).

If the arrow points in a clockwise direction (↻), then the chiral centre has the *R* configuration. If the arrow points in an anticlockwise direction (↺), then the chiral centre has the *S* configuration. Worked example 18.5 provides practice in naming enantiomers as *R* or *S*.

Clockwise = *R* *R* Anticlockwise = *S* *S*

If a structure has the *R* configuration, you can change it to *S* by swapping the positions of *two* of the groups on the chiral centre.

Remember that the *R* and *S* nomenclature is not related to the (+) or (–) sign of the optical rotation value nor to the D/L nomenclature. For example, the amino acids L-serine and L-tyrosine both have the *S*-configuration but, whereas L-serine has a positive optical rotation value, L-tyrosine has a negative value.

Worked example 18.5 *R/S* nomenclature

There are more than 20 common naturally occurring amino acids, and many of these have the *S* configuration at the α-position. All are L-amino acids.

(a) Draw the *S* enantiomer of the amino acid serine.

(b) Does L-cysteine have the *R* or *S* configuration?

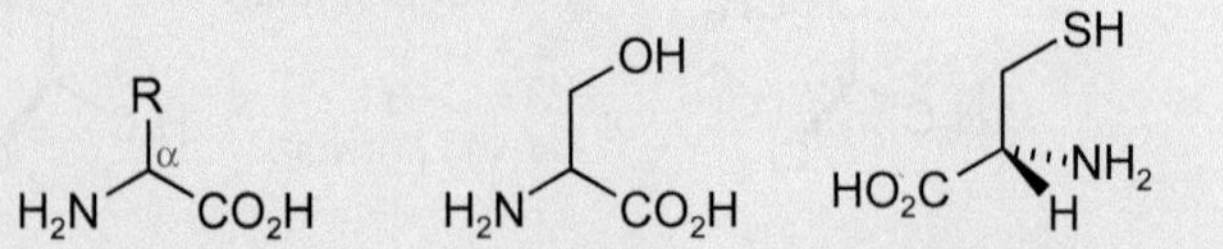

General formula for an α-amino acid showing the α-position — serine — L-cysteine (naturally occurring cysteine)

Strategy for part (a)

Using the CIP sequence rules, rank the groups (*1–4*) directly attached to the chiral centre.

Draw serine in hashed–wedged line notation—show a hashed bond to the hydrogen atom attached to the α-position and a wedged bond to one of the other groups (e.g. NH_2). The hydrogen atom has the lowest priority of substituents on the chiral centre so, if you draw this bond pointing away from you, it makes the assignment of *R* and *S* easier.

Draw a curved arrow from the group of highest priority to the group of next highest priority (i.e. *1* → *2* → *3*).

If the arrows point in a clockwise direction, the chiral centre has the *R* configuration. If the arrows point in an anticlockwise direction, the chiral centre has the *S* configuration.

Solution to part (a)

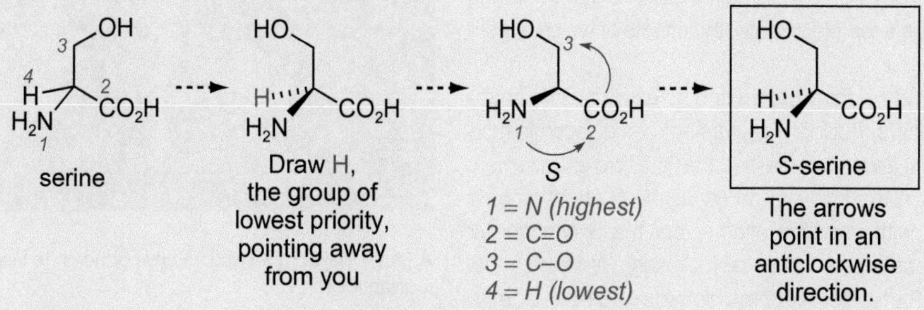
serine
Draw H,
the group of
lowest priority,
pointing away
from you
S
1 = N (highest)
2 = C=O
3 = C–O
4 = H (lowest)
S-serine
The arrows
point in an
anticlockwise
direction.

Strategy for part (b)

Rank the groups (*1–4*) directly attached to the chiral centre and then follow approach 1 or 2.

Approach 1: change the orientation

Orientate the molecule so that the atom of lowest (fourth) priority is drawn so that it is pointing away from you.

Draw a curved arrow from the group of highest priority to the group of next highest priority (i.e. *1* → *2* → *3*).

Use the direction of the arrows to determine whether the chiral centre has the *R* or *S* configuration.

Approach 2: stick with the orientation

With the group of lowest priority pointing towards you as in the structure in the question, draw a curved arrow from the group of highest priority to the group of next highest priority (i.e. *1* → *2* → *3*). Remember the configuration is the *opposite* of that indicated by the curved arrows.

Solution to part (b)

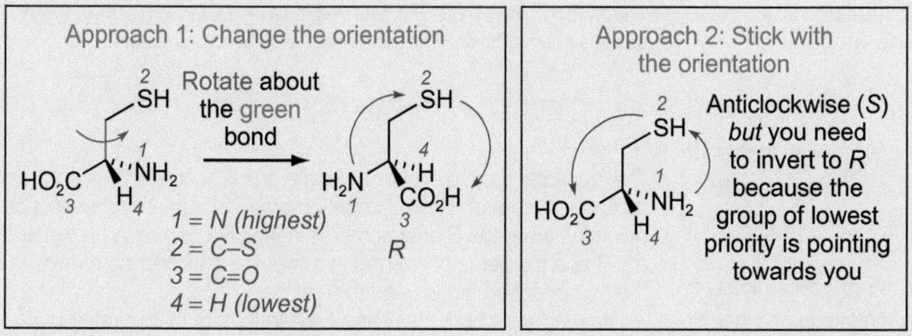
Approach 1: Change the orientation
Rotate about
the green
bond
1 = N (highest)
2 = C–S
3 = C=O
4 = H (lowest)
R
Approach 2: Stick with
the orientation
Anticlockwise (S)
but you need
to invert to R
because the
group of lowest
priority is pointing
towards you

So, L-cysteine has the *R*-configuration.

▲ Sheep require L-cysteine to produce wool. It is an essential amino acid taken in by sheep through eating grass.

Question

Draw a hashed–wedged line structure of (*R*)-butan-2-ol.

To test your knowledge of naming *R*- and *S*-enantiomers, see the drag and drop activity on the Online Resource Centre.

Box 18.7 Thalidomide—from villain to hero?

Thalidomide was first introduced in the late 1950s as a sedative and, because it was thought to be so safe, it was prescribed for morning sickness and insomnia in pregnant women. It was very effective and was soon being prescribed to thousands of women across the world. Unfortunately, thalidomide was subsequently found to be the cause of severe birth defects in children whose mothers had taken it in the early stages of pregnancy. It is estimated that over 10 000 children were born with major malformations, many missing arms and legs. This was a terrible tragedy, the effects of which are still being felt today.

Thalidomide has a single chiral centre and so exists as two enantiomers. In the late 1950s, thalidomide was sold as a racemic mixture of enantiomers but it was not realized that the two enantiomers have different effects in the body. In humans, the *R* enantiomer is an effective sedative with few side effects, but the *S* enantiomer is teratogenic (it causes malformation of foetuses). We now know that, even if the pure *R* enantiomer of thalidomide had been administered, the disaster was unlikely to have been prevented because the *R* enantiomer can be converted into the *S* enantiomer in the body.

▲ Many thalidomide victims have gone on to lead successful and fulfilling lives.

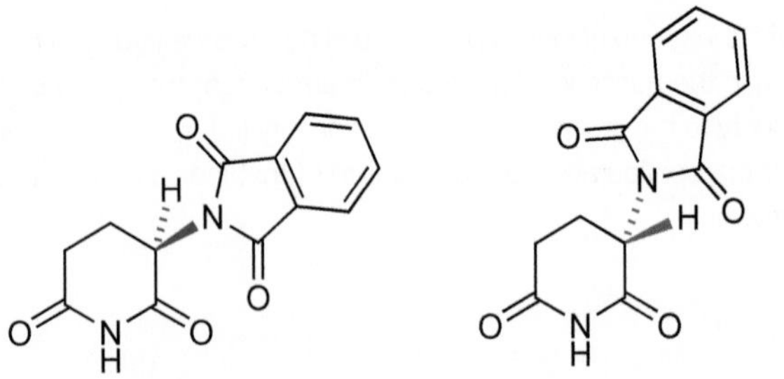

R-thalidomide (sedative) *S*-thalidomide (teratogenic)

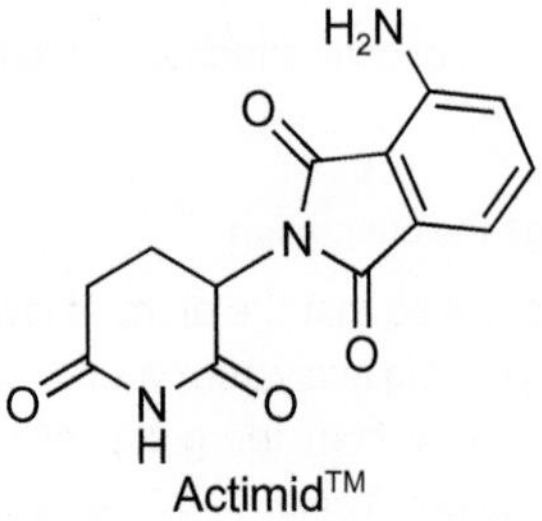

Actimid™

▲ In 1998, the pharmaceutical company Celgene received approval from the US Food and Drug Administration (FDA) to market thalidomide (as Thalomid™). The company is also investigating analogues of thalidomide, such as Actimid™, which is thought to have fewer side effects.

A positive outcome of this disaster is that it led to new legislation on the testing of medicines. Now, all chiral forms of a drug have to be tested rigorously for side effects and chiral stability *in vivo* (in a living organism) before approval.

Today, racemic thalidomide is used, with strict precautions, to treat Hansen's disease (leprosy), which is an infectious and horrific disease that attacks the skin, nerves, and mucous membranes (particularly the eyes). Studies are also being carried out to determine the effectiveness of thalidomide in treating symptoms associated with AIDS and some cancers. The remarkable positive effects of thalidomide in the body mean that it is hard to deprive those who could benefit from its use, but it is a difficult decision to risk using this drug because of the teratogenic properties.

An important step forward is to learn how thalidomide works in the body and to produce medicines that only have the positive effects of thalidomide.

Question

The painkiller ibuprofen has a single chiral centre and is sold as a racemic mixture of *R* and *S* enantiomers. The *S* enantiomer acts as a painkiller, while the *R* enantiomer is ineffective. However, in the body, the *R* enantiomer is converted into the *S* enantiomer so a mixture of the two enantiomers is used to treat pain.

Draw structures of the *R* and *S* enantiomers of ibuprofen.

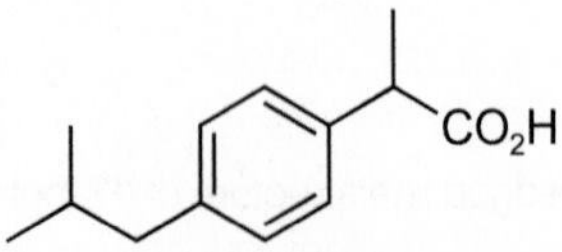

ibuprofen

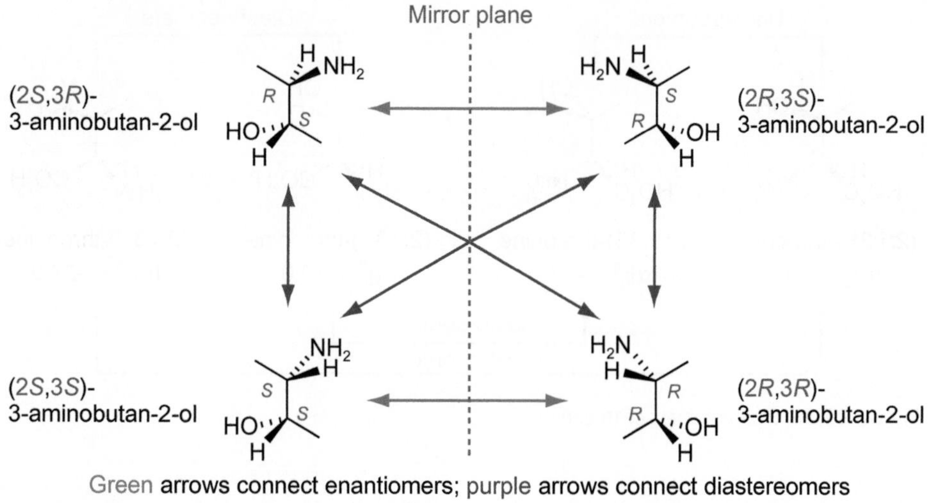
Mirror plane
(2S,3R)-
3-aminobutan-2-ol
(2R,3S)-
3-aminobutan-2-ol
(2S,3S)-
3-aminobutan-2-ol
(2R,3R)-
3-aminobutan-2-ol
Green arrows connect enantiomers; purple arrows connect diastereomers

Figure 18.25 The stereoisomers of 3-aminobutan-2-ol.

Not all molecules containing more than one chiral centre are chiral. A small number are achiral (see *meso* compounds on p.851).

Diastereomers

Many organic compounds contain more than one chiral centre. As the number of chiral centres increases in a molecule, so does the number of possible stereoisomers. You can calculate the maximum number of stereoisomers from the number of chiral centres. Each chiral centre can have two configurations so that, for a molecule containing *n* chiral centres, the maximum number of stereoisomers is 2^n. For example, the sugar D-glucose (Box 23.5, p.1073) has four chiral centres so it is one member of a set of 16 ($2^4 = 16$) stereoisomers.

Diastereomers are stereoisomers with more than one chiral centre that are not mirror images.

Whereas the NMR spectra of enantiomers are identical, the NMR spectra of diastereomers are different. The spectra of diastereomers may be similar in terms of the chemical shift values (and coupling constant values of the resonance signals), but overall, the spectra will be different and so the compounds will be distinguishable (Section 12.3, p.578). For example, for the ^{1}H NMR spectra of the (2*R*,3*R*) and (2*R*,3*S*) diastereomers of threonine (Figure 18.26), the CH hydrogens adjacent to the NH_2 group give resonance signals that differ by around 0.2 ppm (with coupling constants that differ by 1.0 Hz).

In IUPAC nomenclature, the configuration of each chiral centre is shown at the start of the name, typically in brackets. For example, for (2*S*,3*R*)-3-aminobutan-2-ol, the chiral centre at position 2 has the *S* configuration and the chiral centre at position 3 has the *R* configuration. For convenience, this notation is shortened in the text to *SR*.

A molecule with two chiral centres can have four stereoisomers ($2^2 = 4$). The configuration of each chiral centre can be *R* or *S* and so the four stereoisomers will have the *SS*, *SR*, *RS*, and *RR* configurations. This is illustrated in Figure 18.25 by the four possible stereoisomers of 3-aminobutan-2-ol ($CH_3CH(NH_2)CH(CH_3)OH$).

Look carefully at the four structures of 3-aminobutan-2-ol in Figure 18.25. The *RS*/*SR* and the *SS*/*RR* pairs of stereoisomers are mirror images of each other and so are enantiomers. Each pair of enantiomers has different configurations at *both* chiral centres (i.e. *SS* and *RR* or *RS* and *SR*).

In contrast, the *RS*/*SS*, *RS*/*RR*, *SR*/*SS*, and *SR*/*RR* pairs of stereoisomers are not mirror images of one another. These stereoisomers are called **diastereomers** (also spelled **diastereoisomers**). Diastereomers have a different configuration at *one* of the two chiral centres (e.g. *RS* and *SS*).

Unlike enantiomers, diastereomers have different physical properties, such as solubility, melting point, and boiling point. This means that you can separate diastereomers using, for example, distillation or crystallization. Diastereomers also have different chemical properties and so will react with the same reagent at different rates. The different chemical and physical properties of diastereomers are due to the different distances between the groups and the different dihedral angles in the molecules, which give rise to different energies and different shapes. For enantiomers, the distances between groups and the dihedral angles are the same so they have the same energies. Also, whereas the optical rotation values of enantiomers are always equal and of opposite sign, there is no such relationship between the optical rotation values of diastereomers. This is illustrated in Figure 18.26 for the amino acid threonine ($CH_3CH(OH)CH(NH_2)CO_2H$).

Liquids with significantly different boiling points can be separated by distillation (Section 17.4, p.805).

The dihedral angle is the angle between two bonds on neighbouring atoms (Section 18.2, p.824).

In terms of the D/L nomenclature (Box 18.6, p.842), the 2*S*,3*R* isomer is L-threonine and the 2*R*,3*S* isomer is D-threonine. The 2*S*,3*S* isomer is called L-*allo*-threonine and the 2*R*,3*R* isomer is D-*allo*-threonine. Amino acids with two chiral centres were named by assigning a name to the first diastereomer to be discovered. The second diastereomer, when found or synthesized, was then assigned the same name but with the prefix *allo*-.

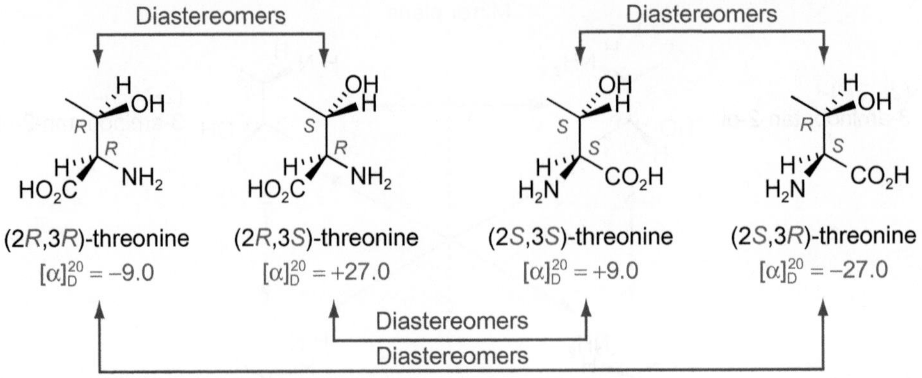
Diastereomers
Diastereomers
(2R,3R)-threonine
[α]D20 = −9.0
(2R,3S)-threonine
[α]D20 = +27.0
(2S,3S)-threonine
[α]D20 = +9.0
(2S,3R)-threonine
[α]D20 = −27.0
Diastereomers
Diastereomers

Figure 18.26 Diastereomers of threonine.

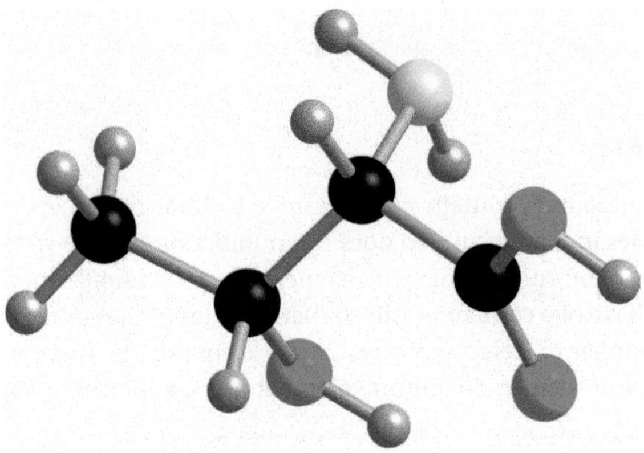

A molecular model of the (2*S*,3*R*)-isomer of threonine (in this computer generated ball and stick model, a single stick represents the C=O double bond; the red spheres represent oxygen and the blue sphere nitrogen).

Box 18.8 Using diastereomers to separate enantiomers: resolution

Enantiomers have identical physical properties so it is not possible to separate them using standard methods of purification. However, diastereomers have different physical properties, such as different solubilities, so they can be separated by crystallization. You can use these different physical properties of diastereomers to separate a racemic mixture of enantiomers in a process called **resolution**.

Resolution involves reacting the racemic mixture of enantiomers with a *single enantiomer* of a chiral reagent called a **resolving agent.** This reaction produces an equal mixture of two diastereomers. Because they are diastereomers, not enantiomers, they have different properties, so they can be separated by, for example, crystallization. The resolving agent is then removed to produce the pure enantiomers.

Naturally occurring chiral amines, such as the alkaloid quinine (Box 2.2, p.72), are commonly used as resolving agents for carboxylic acids (RCO_2H). Quinine exists in nature as the single isomer (–)-quinine. On reaction of (–)-quinine with a racemic mixture of a chiral carboxylic acid, two diastereomeric salts are formed. In this acid–base reaction, there is no change in the configuration of the carboxylic acid or the amine. The two diastereomeric salts, with differing solubilities, are then separated by crystallization—one of the two diastereomeric salts generally crystallizes more easily than the other. This process is illustrated here for the separation of the two enantiomers of (±)-2-phenylpropanoic acid ($CH_3CH(Ph)CO_2H$).

On treatment of each of the separated diastereomeric salts (salts 1 and 2) with aqueous acid, the chiral carboxylic acids are reformed. (You should check that the (–)-isomer of 2-phenylpropanoic acid has the *R* configuration and the (+)-isomer has the *S* configuration.) The chiral carboxylic acid is uncharged and so separates as crystals from the aqueous layer, whereas the positively charged (–)-quinine remains dissolved in the aqueous acid. The enantiomerically pure carboxylic acid is then removed by filtration. (The resolving agent can be recovered (and reused) by deprotonation of the amine functional group with a base such as sodium hydroxide.)

A more modern approach to separating enantiomers uses a technique called chiral chromatography, which is described in Box 11.5 (p.541).

→

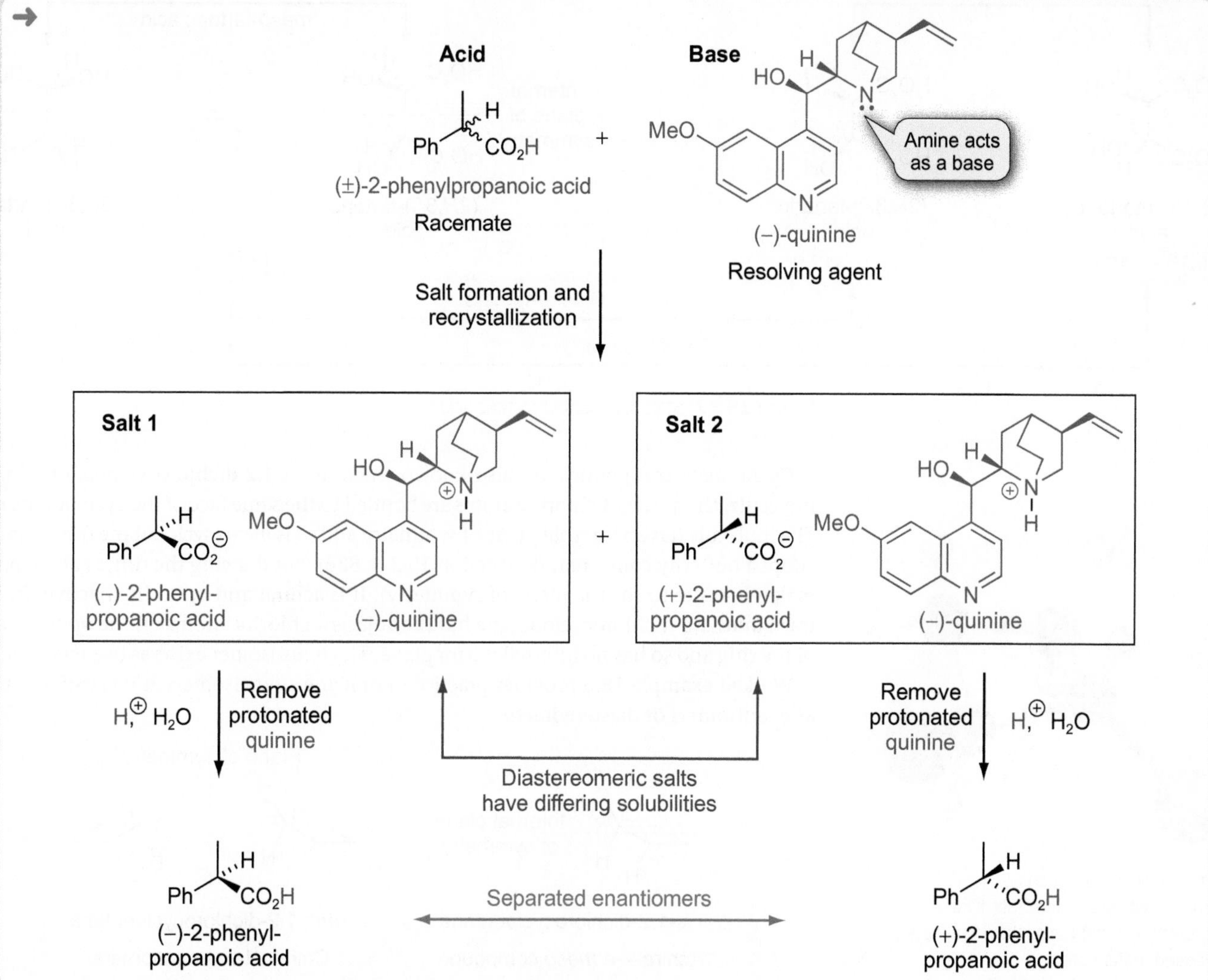
Acid
Base
(±)-2-phenylpropanoic acid
Racemate
(–)-quinine
Resolving agent
Amine acts as a base
Salt formation and recrystallization
Salt 1
Salt 2
(–)-2-phenyl-propanoic acid
(+)-2-phenyl-propanoic acid
Remove protonated quinine
Diastereomeric salts have differing solubilities
Separated enantiomers

Question

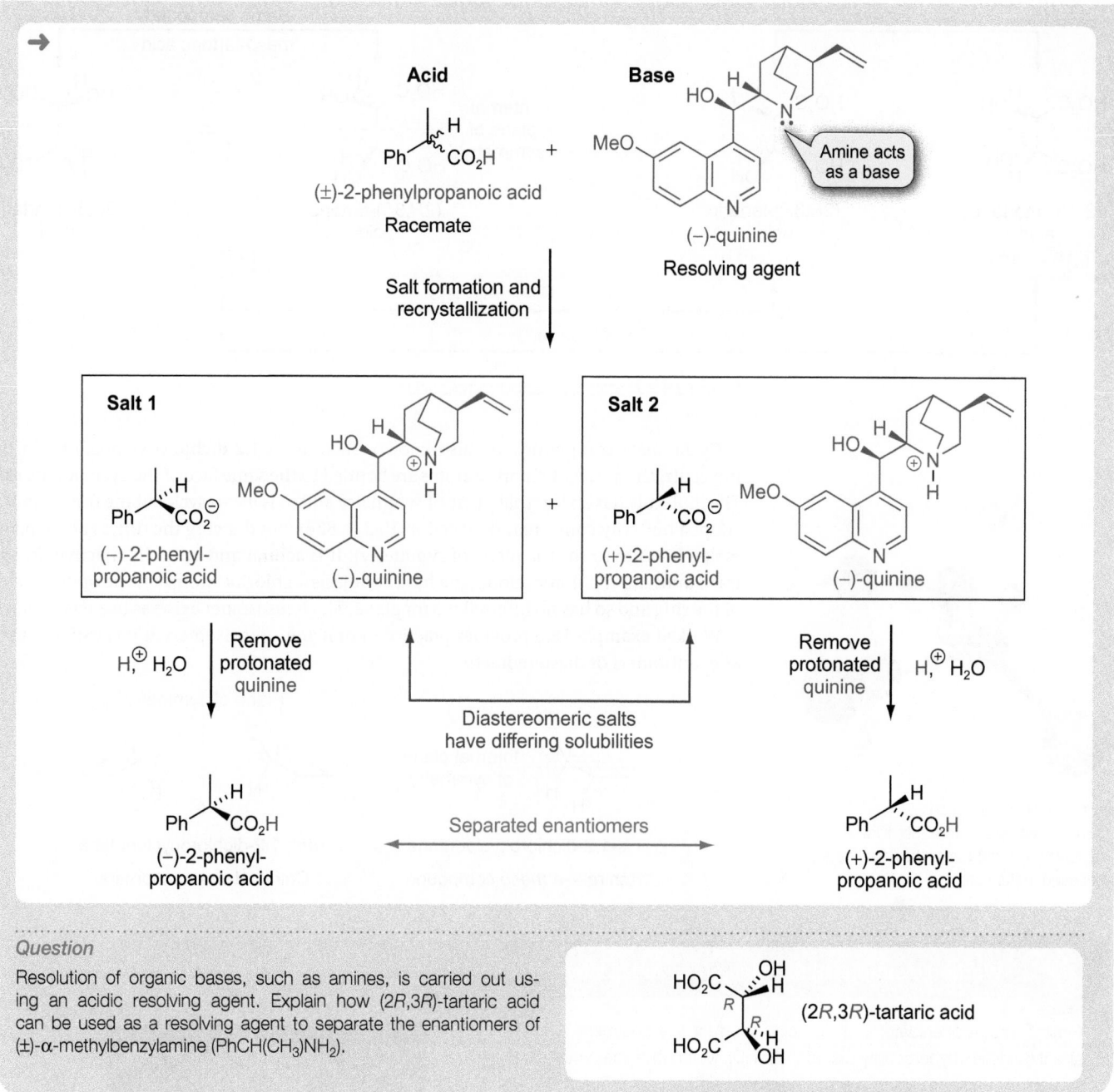

Resolution of organic bases, such as amines, is carried out using an acidic resolving agent. Explain how (2*R*,3*R*)-tartaric acid can be used as a resolving agent to separate the enantiomers of (±)-α-methylbenzylamine ($PhCH(CH_3)NH_2$).

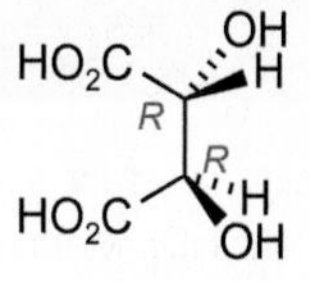

(2*R*,3*R*)-tartaric acid

Meso compounds

Diastereomers are not necessarily chiral. In some cases, a diastereomer can be achiral because it contains an internal plane of symmetry. The plane of symmetry divides the molecule into halves that contribute equally but in opposite directions to the rotation of plane-polarized light (i.e. the halves cancel each other out and the molecule is optically inactive). A compound with two chiral centres, with the same four groups bonded to each chiral centre, has three, rather than four, stereoisomers. For example, tartaric acid ($HO_2CCH(OH)CH(OH)CO_2H$) has two enantiomers plus a third stereoisomer with an internal plane of symmetry (Figure 18.27). This symmetrical stereoisomer is called a ***meso* compound.**

Enantiomers are always chiral, but diastereomers are not necessarily chiral. For example, in tartaric acid, one of the three diastereomers is not chiral (Figure 18.27).

A compound that contains more than one chiral centre but is achiral because it has an internal plane of symmetry is called a *meso* compound.

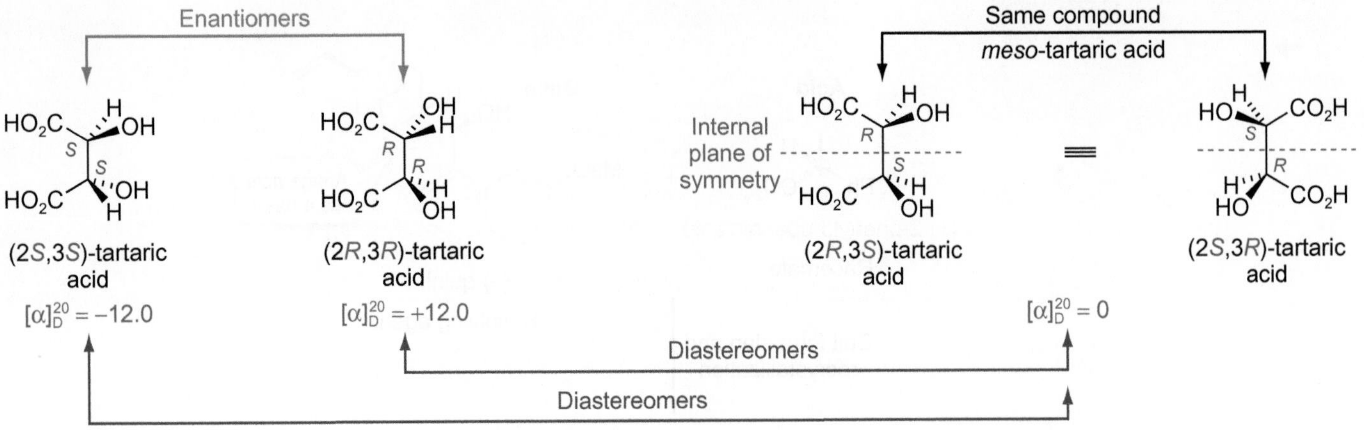
Enantiomers
Same compound
meso-tartaric acid
(2*S*,3*S*)-tartaric acid
[α]$_D^{20}$ = −12.0
(2*R*,3*R*)-tartaric acid
[α]$_D^{20}$ = +12.0
Internal plane of symmetry
(2*R*,3*S*)-tartaric acid
(2*S*,3*R*)-tartaric acid
[α]$_D^{20}$ = 0
Diastereomers
Diastereomers

Figure 18.27 Stereoisomers of tartaric acid.

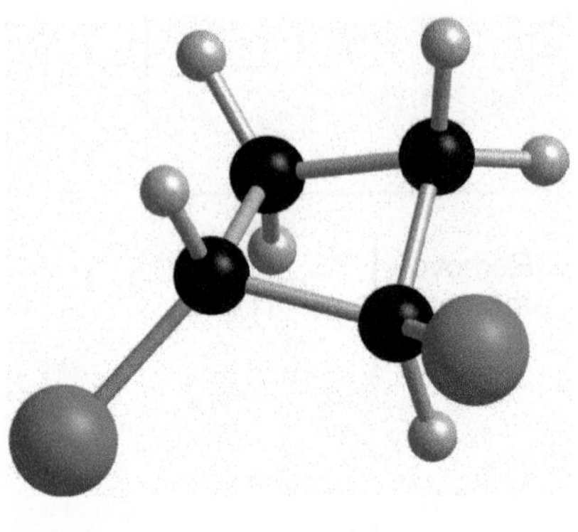

A molecular model of *trans*-1,2-dichlorocyclobutane (the green spheres represent chlorine); notice the ring adopts a puckered conformation.

Cyclic *meso* compounds are also known, such as *cis*-1,2-dichlorocyclobutane. In this molecule, the adjacent chlorine atoms are bonded to the same face of the cyclobutane ring. The molecule has an internal plane of symmetry and so is *meso* (cyclobutane rings typically adopt a butterfly conformation (Section 18.2, p.826), but drawing the ring as being planar makes it easier to see the plane of symmetry). It is achiral and optically inactive. In contrast, *trans*-1,2-dichlorocyclobutane has the adjacent chlorine atoms on the opposite sides of the ring and so has no internal mirror plane. The *trans*-isomer exists as two enantiomers.

Worked example 18.6 provides practice in drawing stereoisomers and classifying them as enantiomers or diastereomers.

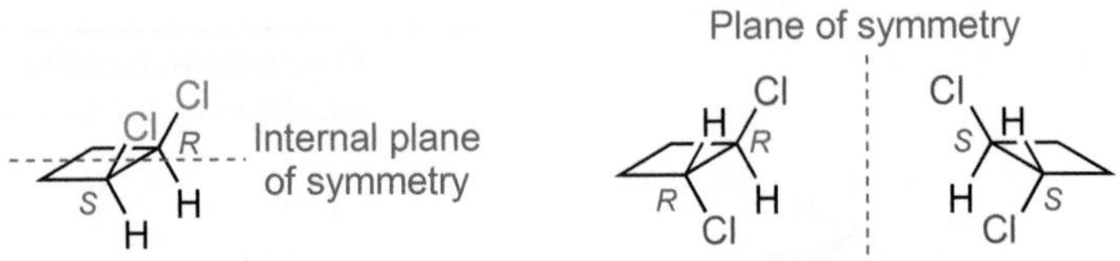
Plane of symmetry
Internal plane of symmetry

cis-1,2-dichlorocyclobutane
Achiral—a *meso* compound

trans-1,2-dichlorocyclobutane
Chiral—two enantiomers

Worked example 18.6 Enantiomers and diastereomers

Ritalin® (methylphenidate) is a medicine used for the treatment of attention deficit hyperactivity disorder (ADHD). Methylphenidate has two chiral centres so there are four possible stereoisomers. Ritalin® is as an equal mixture of two of the four possible stereoisomers, namely, the *RR* and *SS* isomers. The *RR* isomer is used to treat attention deficit disorder while the *SS* isomer is an antidepressant.

Draw the *RR* and *SS* isomers of Ritalin®. Are these pairs of isomers enantiomers or diastereomers?

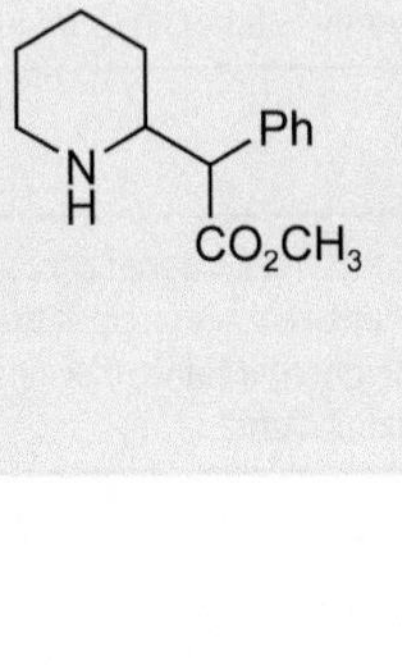

methylphenidate
(A mixture of *RR* and *SS* stereoisomers is marketed as Ritalin®)

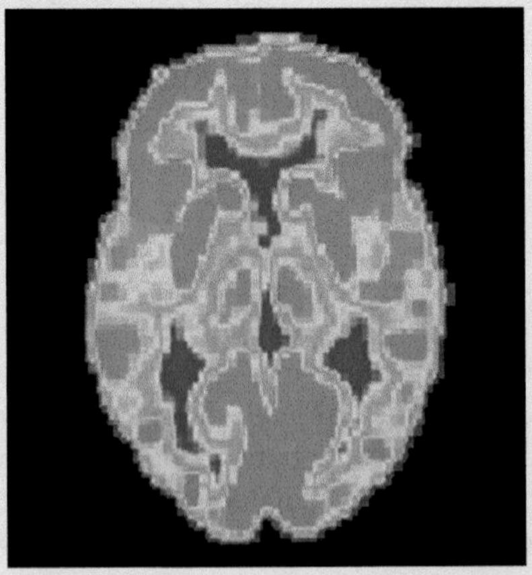

▲ Ritalin® increases the levels of dopamine (4-(2-aminoethyl)benzene-1,2-diol) in the brain. The levels can be recorded by injecting the patient with a radiotracer and scanning the brain using positron emission tomography (PET).

→

Strategy

Identify the two chiral centres in the molecule.

Draw the hydrogen atoms that are attached to each of the chiral carbons, and assume these are both pointing away from you (i.e. use two hashed lines). Draw a wedged line to any one of the other substituents on each of the chiral centres.

Rank the groups (*1–4*) directly attached to each chiral centre.

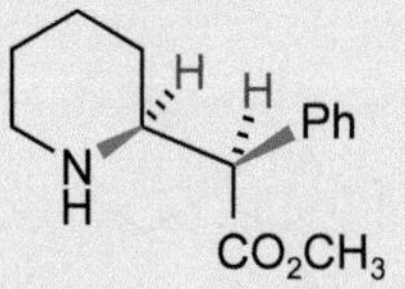

H
H
Ph
N
H
CO_2CH_3

On each chiral centre, the hydrogen atom is the lowest priority atom and this is pointing away from you. Draw a curved arrow from the highest (*1*) to next highest (*2*) priority group on each chiral centre.

If the arrow points in a clockwise direction, the chiral centre has the *R* configuration. If the arrow points in an anticlockwise direction, the chiral centre has the *S* configuration.

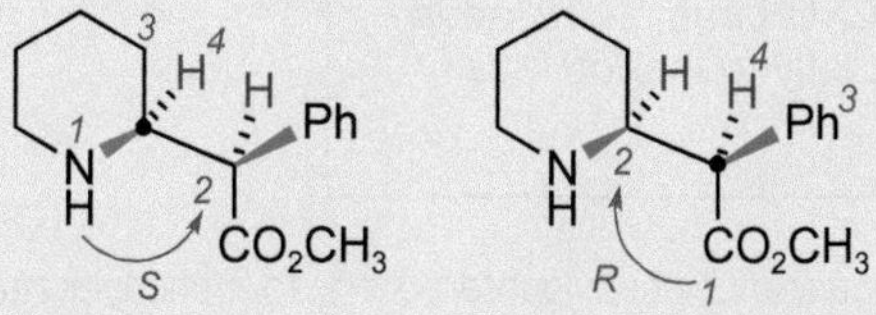

H
H
Ph
N
H
CO_2CH_3
S
R
Ph

1 = N (highest)
2 = C–O
3 = C–C
4 = H (lowest)

1 = C–O (highest)
2 = C–N
3 = C=C, Ph
4 = H (lowest)

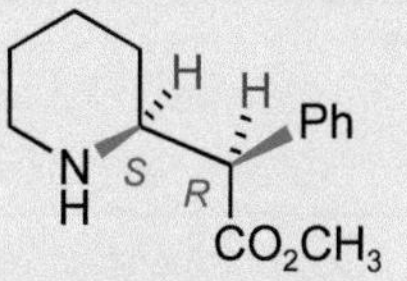

H
H
Ph
N
H
S
R
CO_2CH_3

The structure drawn above is the *SR* isomer

Change the configuration of the *S* chiral centre by exchanging the hashed bond for a wedged bond, and the wedged bond for a hashed bond (this inverts the stereochemistry from *S* to *R*). The same approach can be used to change the configuration of the *R* chiral centre (to *S*).

Solution

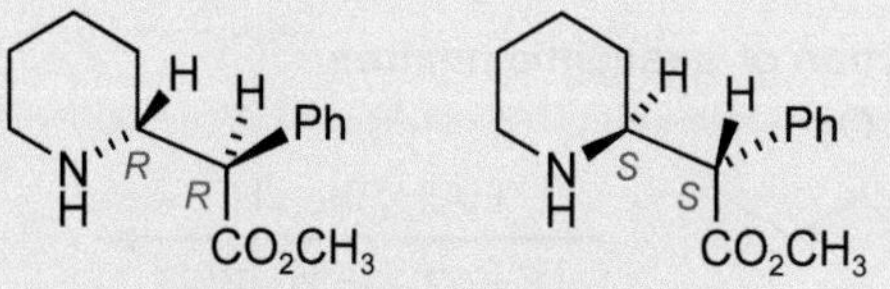

H
H
Ph
N
H
R
R
CO_2CH_3
S
S

These isomers have a different configuration at both chiral centres, so they are enantiomers.

Question

Ephedrine is a naturally occurring compound that is used to treat asthma.

(a) Assign a configuration to each of the chiral centres in ephedrine.

(b) Draw the structure of a diastereomer of ephedrine.

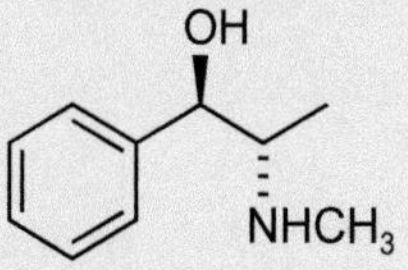

OH
$NHCH_3$

ephedrine

For a discussion of medicines that relieve the symptoms of asthma (triggered by allergens such as pollen) see Box 22.8, p.1049.

Diastereoselective reactions

Reactions that lead to the preferential formation of one diastereomer of the product over the other diastereomer are known as **diastereoselective reactions**. An example of a diastereoselective reaction is shown in Figure 18.28. Reduction of the C=O bond in 2-phenylpentan-3-one ($CH_3CH(Ph)COCH_2CH_3$) leads to the selective formation of one diastereomer of 2-phenylpentan-3-ol ($CH_3CH(Ph)CH(OH)CH_2CH_3$). One diastereomer is selectively formed because the large phenyl (Ph) group hinders the approach of the reducing agent, lithium aluminium hydride ($LiAlH_4$), to one face of the C=O bond.

If the *S*-enantiomer of the ketone is used in the reduction, the two diastereomers of the alcohol formed will be *single* enantiomers with the *SR* and *SS* configurations—the structures shown in Figure 18.28(a) represent the **absolute configuration** (**absolute stereochemistry**) of the alcohols. Alternatively, if a racemic mixture of the ketone is used in the reduction, two diastereomers of the alcohol will also be formed (in the same 3 : 1 ratio) but each diastereomer will be a racemate. The major diastereomer is an *equal mixture*

Visit the Online Resource Centre to view screencast 18.2 which walks you through some examples of diastereoselective reactions, including nucleophilic addition reactions.

The mechanism of reduction of a ketone (RCOR) using $LiAlH_4$ is given in Section 23.2 (p.1062).

The **absolute configuration** of a diastereomer is the configuration of a single enantiomer. The **relative configuration** of a diastereomer is the configuration of a diastereomer relative to another diastereomer. Two diastereoisomers have different relative stereochemistry.

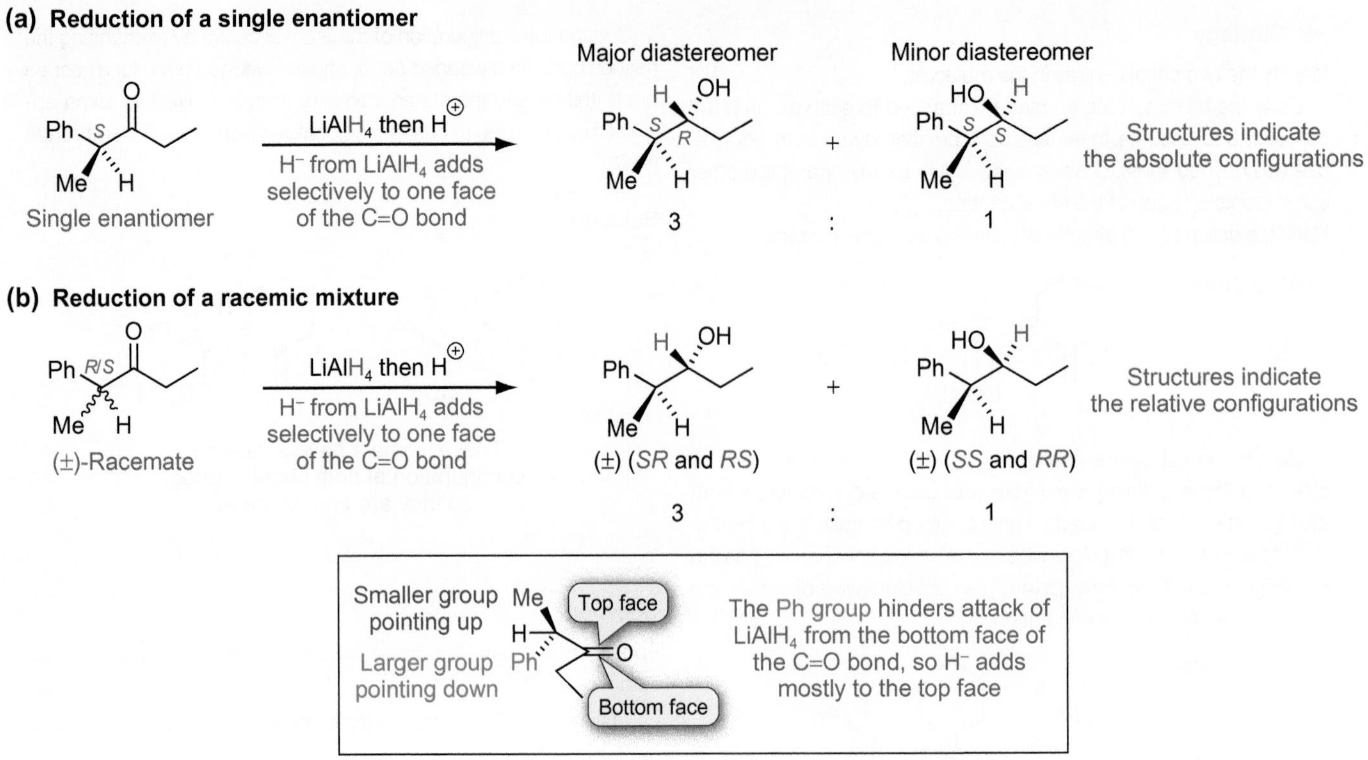
(a) Reduction of a single enantiomer
Single enantiomer
LiAlH4 then H⊕
H⁻ from LiAlH4 adds selectively to one face of the C=O bond
Major diastereomer
Minor diastereomer
3 : 1
Structures indicate the absolute configurations
(b) Reduction of a racemic mixture
(±)-Racemate
(±) (SR and RS)
(±) (SS and RR)
3 : 1
Structures indicate the relative configurations
Smaller group pointing up
Larger group pointing down
Top face
Bottom face
The Ph group hinders attack of LiAlH4 from the bottom face of the C=O bond, so H⁻ adds mostly to the top face

Figure 18.28 Diastereoselective reduction of 2-phenylpentan-3-one to 2-phenylpentan-3-ol.

of enantiomers of configurations *SR* and *RS*. In this case, the structures shown for the racemic alcohol products in Figure 18.28(b) indicate the **relative configuration** (**relative stereochemistry**).

ⓘ A sugar is a carbohydrate that is crystalline, very soluble in water, and usually has a sweet taste.

ⓘ Remember that, when naming substituents on carbon chains, the lowest possible numbers are used (e.g. 2-phenylpentan-3-one not 4-phenylpentan-3-one). The use of wavy lines (∿∿∿) indicates that, for half of the molecules, the Me group points behind the plane of the paper, and for the other half, the Me group points in front of the plane of the paper.

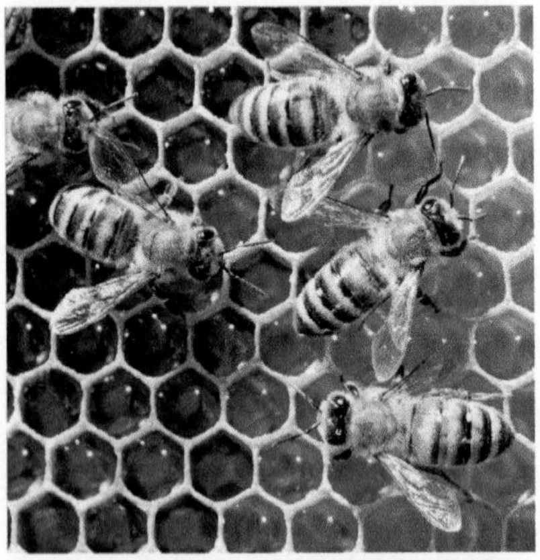

ⓘ Fischer projections are named after the German chemist Emil Fischer (1852–1919). In 1902, he was awarded the Nobel Prize in Chemistry for his work on sugars. Part of this work involved establishing the configuration of naturally occurring sugars such as glucose, fructose (both found in honey), and mannose.

Fischer projections

A **Fischer projection** is a way of representing tetrahedral carbon atoms and their substituents on paper. Fischer projections are particularly useful for showing a simple and concise representation of molecules, such as sugars, that have a number of chiral centres. For these molecules the use of hashed–wedged line notation can look cluttered.

In a Fischer projection, the molecule is drawn in the form of a cross and the tetrahedral carbon atom is located at the point of intersection of the lines. The vertical lines represent the bonds pointing away from you, and the horizontal lines represent the bonds pointing towards you. Figure 18.29 shows you how to convert a hashed–wedged line structure into a Fischer projection. Remember, however, that Fischer projections, unlike hashed–wedged line notation, do not provide an accurate representation of the shape of the molecule. In effect, a Fischer projection is a flattened-out picture of the molecule.

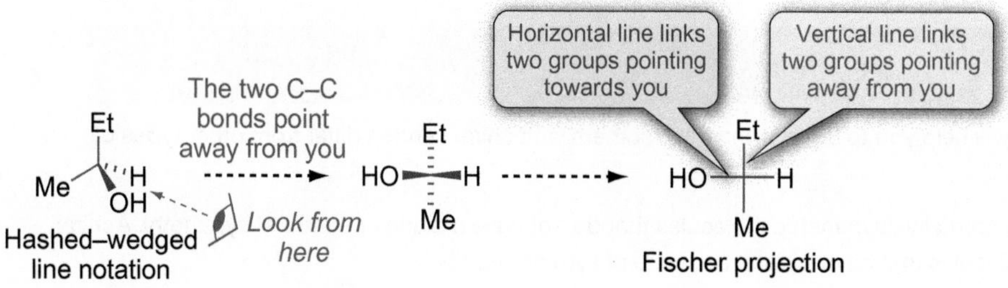
Horizontal line links two groups pointing towards you
Vertical line links two groups pointing away from you
The two C–C bonds point away from you
Look from here
Hashed–wedged line notation
Fischer projection

Figure 18.29 Converting a hashed–wedged line structure into a Fischer projection.

> The compound in Figure 18.29 is (*S*)-butan-2-ol. Check that you understand why this compound has *S* configuration.

The Fischer projection of a compound is usually drawn so that the carbon framework of the molecule is vertical and the highest priority functional group (assigned using the CIP sequence rules) is positioned at the top of the vertical line. To draw Fischer projections of molecules containing two or more chiral centres, you need to arrange the carbon skeleton in an eclipsed conformation rather than the more stable staggered conformation (Section 18.2, p.824).

The CIP nomenclature is described earlier in this Section, on page 845.

Fischer projections allow the absolute configuration of each chiral centre in different molecules to be easily compared. This is particularly useful in the assignment of the D/L nomenclature to sugars (Box 18.6, p.842). For example, in the Fischer projection of D-glyceraldehyde the OH group attached to the chiral centre is pointing to the right.

> Glyceraldehyde (2,3-dihydroxypropanal, $HOCH_2CH(OH)CHO$) is the simplest example of an aldotriose—this is a monosaccharide (a simple sugar) with only one aldehyde group (an aldose) that has three carbon atoms. D-Glyceraldehyde has *R* configuration and L-glyceraldehyde has *S* configuration.

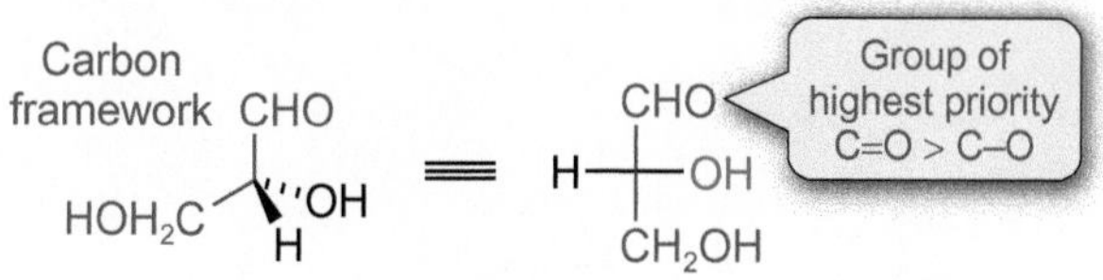
Carbon framework
Group of highest priority
C=O > C–O

D-glyceraldehyde

If the OH group on the carbon adjacent to the CH_2OH group in a sugar (such as erythrose below) is also pointing to the right then this molecule is assigned as a D sugar. For an L sugar, the OH group on the carbon adjacent to the CH_2OH group will point to the left (as for L-glyceraldehyde).

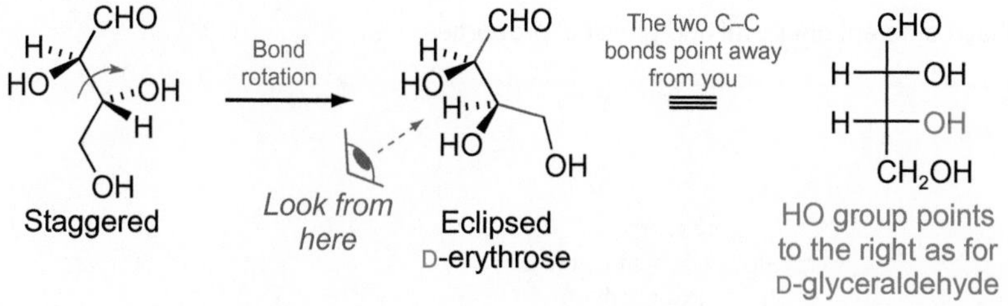
Bond rotation
The two C–C bonds point away from you
Look from here
Staggered
Eclipsed D-erythrose
HO group points to the right as for D-glyceraldehyde

> The IUPAC name for D-erythrose, an aldotetrose (one aldehyde and four carbons), is (2*R*,3*R*)-2,3,4-trihydroxybutanal.

The use of Fischer projections also helps to identify *meso* compounds and spot pairs of enantiomers and diastereomers.

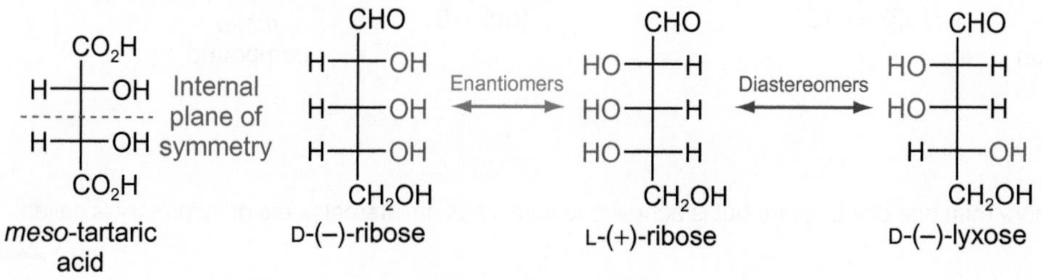
Internal plane of symmetry
Enantiomers
Diastereomers
meso-tartaric acid
D-(–)-ribose
L-(+)-ribose
D-(–)-lyxose

> The open-chain forms of sugars, shown here, are in equilibrium with 5- and 6-membered cyclic hemiacetals (see Box 23.5, p.1073). D-Ribose exists predominantly as a 5-membered ring; it is found in nature and is one of the building blocks used to make ribonucleic acid (RNA).

Summary

- The flow chart in Figure 18.30 will help you to understand how isomers with chiral centres differ from other types of isomers.
- Isomers with chiral centres are normally asymmetric molecules that do not have a plane or centre of symmetry. Achiral molecules are symmetrical molecules that have a plane or centre of symmetry.
- Enantiomers are non-identical molecules that are mirror images of one another.
- Enantiomers rotate plane-polarized light in opposite directions and often have different biological properties.
- The three-dimensional arrangement of atoms or groups attached to a chiral centre is called the configuration.
- The configuration of chiral centres is assigned as *R* or *S* using the Cahn–Ingold–Prelog (CIP) sequence rules.

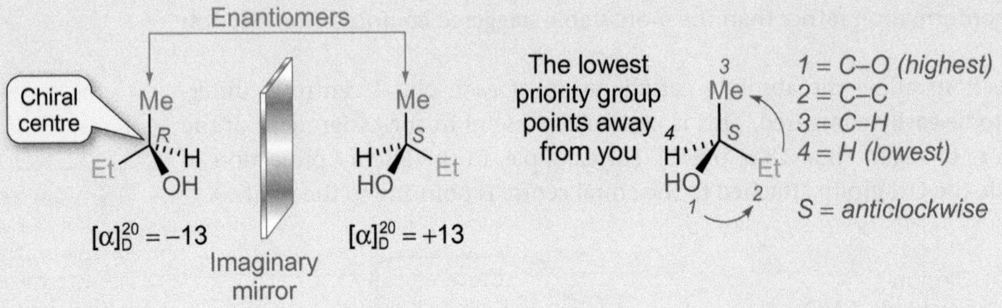
Enantiomers
Chiral centre
Me
R
Et
H
OH
$[\alpha]_D^{20} = -13$
Imaginary mirror
Me
S
H
Et
HO
$[\alpha]_D^{20} = +13$
The lowest priority group points away from you
3
Me
4
S
H
Et
HO
1
2
1 = C–O (highest)
2 = C–C
3 = C–H
4 = H (lowest)
S = anticlockwise

The Cahn–Ingold–Prelog (CIP) sequence rules.

- The D/L nomenclature is a system for assigning the configurations of chiral amino acids and sugars, using glyceraldehyde as the standard.
- Fischer projections show the spatial arrangement of groups attached to a chiral centre—horizontal lines represent bonds that point toward you and vertical lines represent bonds that point away from you.
- Diastereomers are stereoisomers with more than one chiral centre that are not mirror images.
- Unlike enantiomers, diastereomers have different chemical and physical properties.

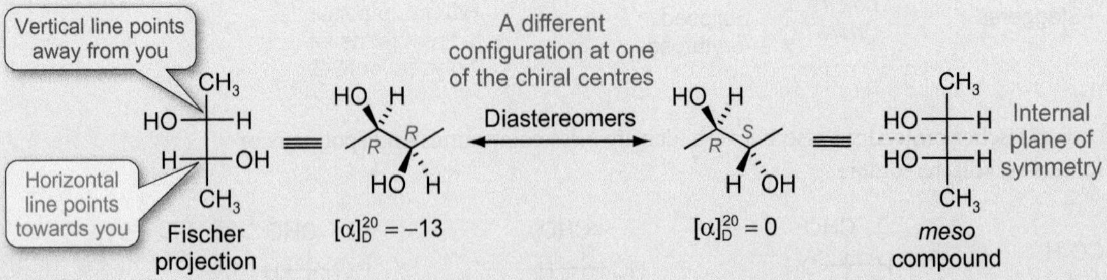
Vertical line points away from you
Horizontal line points towards you
CH_3
HO
H
H
OH
CH_3
Fischer projection
HO H
R
R
HO H
$[\alpha]_D^{20} = -13$
A different configuration at one of the chiral centres
Diastereomers
HO H
S
R
H OH
$[\alpha]_D^{20} = 0$
CH_3
HO
H
HO
H
CH_3
Internal plane of symmetry
meso compound

- A compound that contains more than one chiral centre but is achiral because it has an internal plane of symmetry is called a *meso* compound.

→

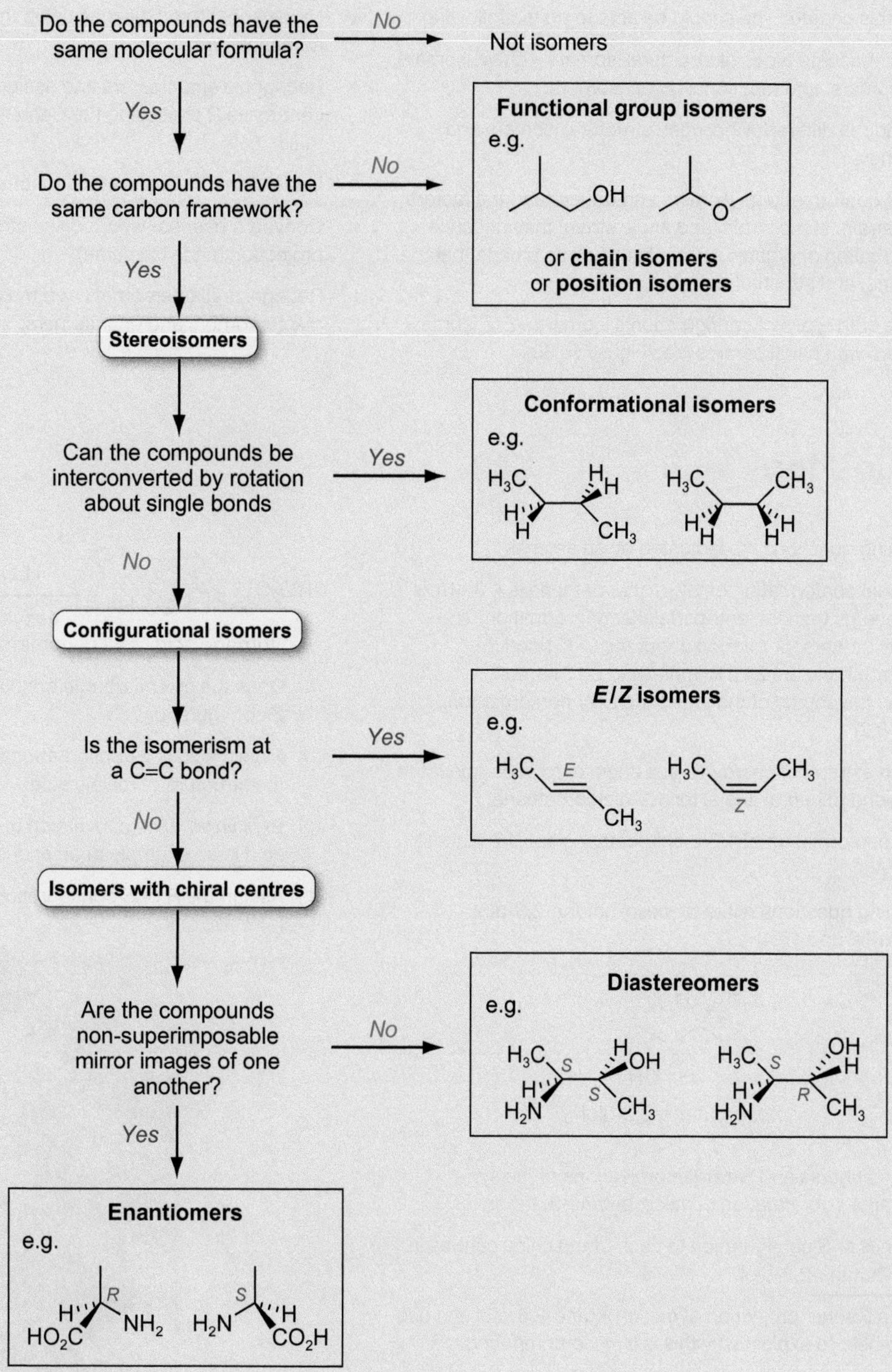
Do the compounds have the same molecular formula?
No
Not isomers
Yes
Do the compounds have the same carbon framework?
No
Functional group isomers
e.g.
OH
O
or chain isomers
or position isomers
Yes
Stereoisomers
Can the compounds be interconverted by rotation about single bonds
Yes
Conformational isomers
e.g.
No
Configurational isomers
Is the isomerism at a C=C bond?
Yes
E/Z isomers
e.g.
E
Z
No
Isomers with chiral centres
Are the compounds non-superimposable mirror images of one another?
No
Diastereomers
e.g.
Yes
Enantiomers
e.g.

Figure 18.30 Recognizing different isomers.

For practice questions on these topics, see questions 2, 3, and 5–7 at the end of this chapter (pp.858–859).

Concept review

By the end of this chapter, you should be able to do the following.

- Recognize the three types of structural isomers—chain isomers, position isomers, and functional group isomers.
- Understand the difference between structural isomers and stereoisomers.
- Recognize conformational isomers and understand the factors (torsional strain, steric strain, and angle strain) that influence the conformation of organic molecules, such as ethane, butane, cyclohexane, and substituted cyclohexanes.
- Recognize both types of configurational isomers—*E*/*Z* isomers and isomers with chiral centres (see Figure 18.30).
- Name substituted alkenes using the *E*/*Z* system of nomenclature.
- Recognize enantiomers and assign the configuration of chiral centres as *R* or *S* using the Cahn–Ingold–Prelog sequence rules.
- Understand the D/L system of nomenclature.
- Convert a hashed–wedged line structure into a Fischer projection (and vice versa).
- Recognize diastereomers and the difference between diastereomers and enantiomers.

Questions

More challenging questions are indicated by an asterisk *.

1 The synclinal conformation of 1,2-dichloroethane is 4.8 kJ mol^{-1} higher in energy than the anti-periplanar conformation. The two energy barriers for rotation about the C–C bond in 1,2-dichloroethane are 21.5 kJ mol^{-1} and 38.9 kJ mol^{-1} higher than the energy of the anti-periplanar conformation. (Section 18.2)

(a) Sketch a graph of energy versus angle of rotation about the C–C bond (dihedral angle) for 1,2-dichloroethane.

(b) What conformation of 1,2-dichloroethane has the highest energy?

2 The following questions relate to *meso*-butane-2,3-diol. (Sections 18.2 and 18.4)

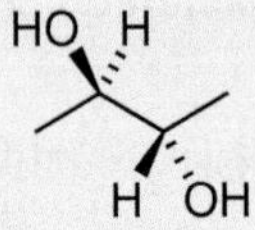
HO H
H OH

meso-butane-2,3-diol

(a) Draw sawhorse and Newman projections of the syn-periplanar conformation of *meso*-butane-2,3-diol.

(b) Assign *R* or *S* configuration to each of the chiral centres in *meso*-butane-2,3-diol.

(c) Draw a Fischer projection of *meso*-butane-2,3-diol and use this to help to explain why this is a *meso* compound.

(d) Draw a Fischer projection of a diastereomer of *meso*-butane-2,3-diol.

3 The citric acid cycle (Krebs cycle) is a series of reactions involved in the oxidation of fats, proteins, and carbohydrates in the body to form carbon dioxide and water (p.960). One step in the citric acid cycle is the addition of water to fumaric acid to make malic acid. (Sections 18.3 and 18.4)

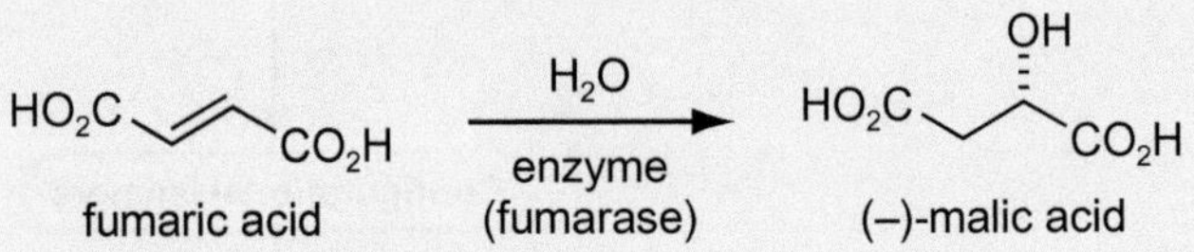
HO2C
CO2H
H2O
enzyme
(fumarase)
OH
HO2C
CO2H
fumaric acid
(–)-malic acid

(a) Does the alkene double bond in fumaric acid have the *E* or *Z* configuration?

(b) Maleic acid is a configurational isomer of fumaric acid. Draw the structure of maleic acid.

(c) Explain why the conversion of fumaric acid into (–)-malic acid is an example of an enantioselective reaction.

(d) Assign the *R* or *S* configuration to the chiral centre in (–)-malic acid.

Rhubarb leafstalks are rich in (–)-malic acid and citric acid, which produces the tart taste—dipping raw stalks in sugar can remove some of the tartness.

4 Tetrodotoxin is a potent poison that is found in the puffer fish. It is estimated to be more than 10 000 times deadlier than cyanide ions. The structure of tetrodotoxin contains a group of interconnected 6-membered rings. Which of the bonds on the cyclohexane ring of tetrodotoxin are in an axial position and which are in an equatorial position? (Section 18.2)

Many in Japan consider the puffer fish (fugu) a delicacy. Because of the risk of poisoning, the Japanese government licenses sushi chefs who wish to prepare fugu.

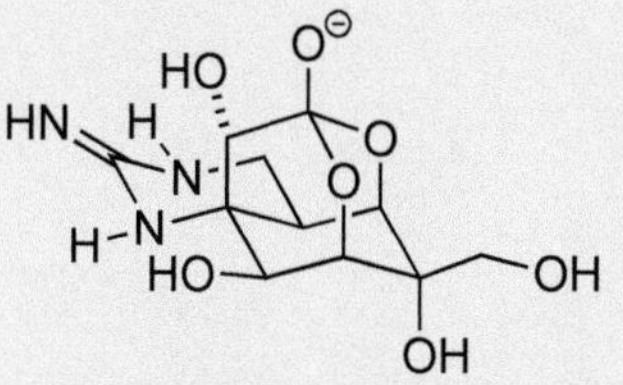

tetrodotoxin

5 Lamivudine (Epivir) is used for the treatment of HIV (AIDS) and is an example of a class of drugs called *nucleoside reverse transcriptase inhibitors*. These inhibitors stop HIV from infecting cells in the body. Only the (–)-enantiomer of lamivudine is registered for the treatment of HIV because it is more active and less toxic than either the (+)-enantiomer or the racemate. In the (–)-enantiomer, the chiral centre at position *2* has the *R* configuration and the chiral centre at position *5* has the *S* configuration. Draw the structure of (–)-lamivudine using hashed–wedged line notation. (Section 18.4)

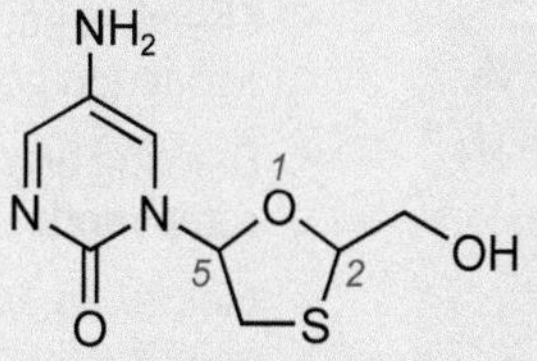

(±)-lamivudine

6 The following questions relate to (–)-menthol, a naturally occurring compound isolated from peppermint oil. (Sections 18.2 and 18.4)

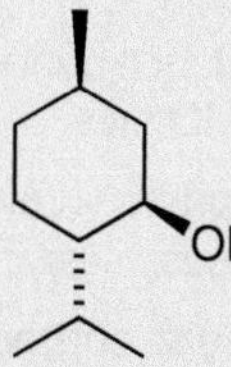

(–)-menthol

(a) Draw the preferred chair conformation of (–)-menthol.

(b) Assign *R* or *S* configuration to each of the chiral centres in (–)-menthol.

(c) Draw the enantiomer of (–)-menthol.

(d) Draw a diastereomer of (–)-menthol.

Some cigarette brands use menthol as a smoking tobacco additive. People who smoke menthol cigarettes can be less likely to stop smoking; recent research suggests this could be due to menthol making the brain more sensitive to nicotine.

7* Anatoxin-*a*, otherwise known as Very Fast Death Factor, is an alkaloid isolated from blue-green algae. It is a potent neurotoxin, so affects the functioning of the nervous system, often causing death due to paralysis of the respiratory muscles. The following questions relate to part of a synthesis of the (–)-enantiomer of anatoxin-*a*. (Sections 18.3 and 18.4)

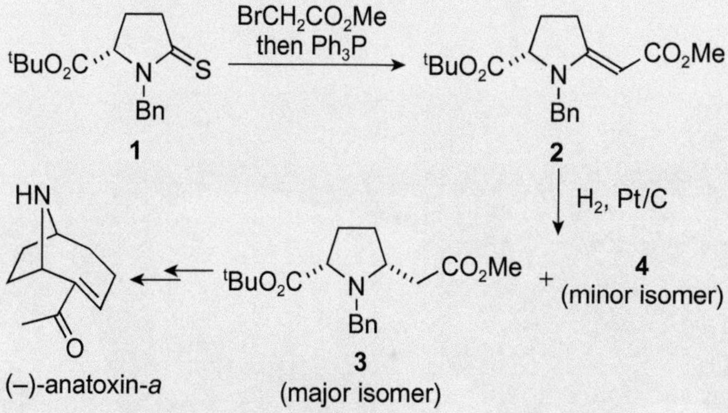
BrCH2CO2Me
then Ph3P
1
2
H2, Pt/C
3
(major isomer)
4
(minor isomer)
(–)-anatoxin-a

(a) Assign the *R* or *S* configuration to the chiral centre in compound **1**.

(b) Draw the enantiomer of compound **1**.

(c) Does the C=C bond in compound **2** have the *E* or *Z* configuration?

(d) Draw the structure of compound **4** and assign the *R* or *S* configuration to both chiral centres.

(e) Are compounds **3** and **4** enantiomers or diastereomers?

(f) Draw the structure of (+)-anatoxin-*a* and identify the chiral centres.

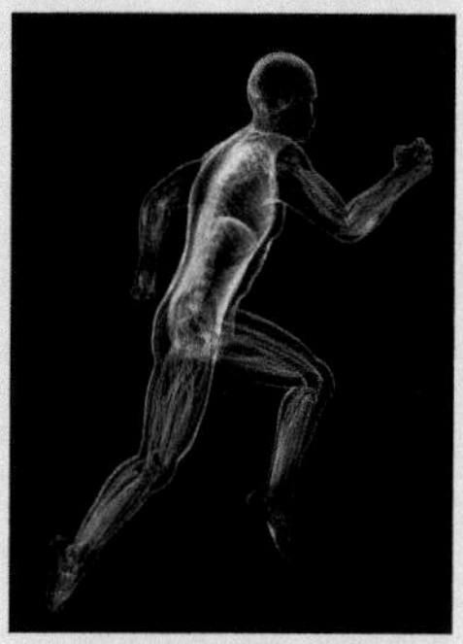

◀ Anatoxin-*a* causes muscles to become over-stimulated, fatigued and then paralysed. When respiratory muscles are affected, convulsions occur due to a lack of oxygen supply to the brain. This is followed by suffocation and death, only a few minutes after ingesting the toxin.

19

Organic reaction mechanisms

This chapter builds on the following topics:

- **Redox in organic reactions: oxidation levels** Section 1.4, p.21
- **Enthalpy profiles for exothermic and endothermic processes** Section 1.6, p.41
- **Chemical equilibrium: how far has a reaction gone?** Section 1.9, p.56
- **Drawing organic compounds** Section 2.2, p.73
- **Electronegativity and bond polarity** Section 4.3, p.177
- **Molecular orbitals** Sections 4.7 to 4.9, pp.184–196
- **Bond polarity and polar molecules** Section 5.3, p.235
- **Valence bond theory for polyatomic molecules** Section 5.4, p.236
- **Brønsted–Lowry acids and bases** Section 7.1, p.304
- **The strengths of acids and bases** Section 7.2, p.308

◀ Japanese police carry out a terror attack drill at a subway in Tokyo. In March 1995, 12 people were killed and over 5000 were injured when sarin was released in the Tokyo subway system.

Antidotes for nerve agents

Nerve agents are quick-acting poisons that enter the body through the skin or the lungs and attack the nervous system. They were developed as chemical warfare agents during the Second World War (1939–45). Some of the most potent nerve agents are organophosphorus compounds (organic compounds that contain phosphorus) such as sarin.

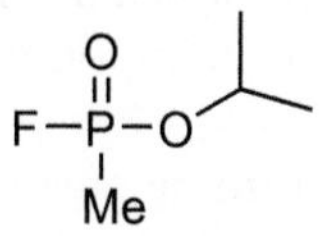

sarin

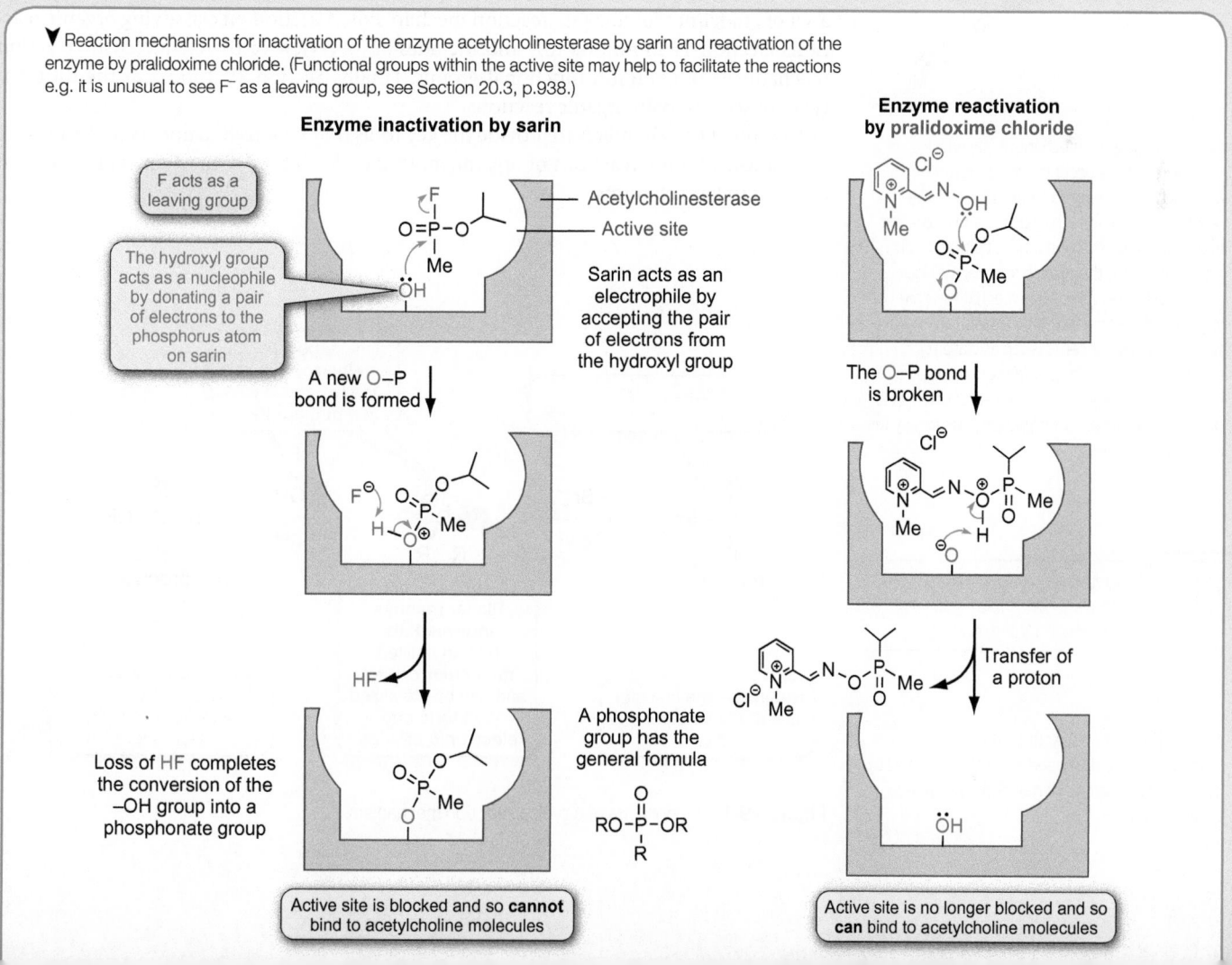

▲ Notice that sarin is a chiral molecule—the phosphorus atom is bonded to four different groups (Section 18.4, p.838)

Sarin ($MeP(O)(F)OCHMe_2$) interferes with nerve communication in the body by inhibiting an enzyme called *acetylcholinesterase*. This enzyme is responsible for breaking down acetylcholine, which acts as a neurotransmitter and carries nerve signals across gaps between nerve cells (see Box 18.1, p.822). To send a nerve signal, a nerve cell releases acetylcholine ($MeCO_2CH_2CH_2N^+Me_3$), which then moves across a gap between nerve cells. The nerve signal is transmitted when acetylcholine interacts with receptors on a receiving nerve cell. Once acetylcholine has transmitted a nerve signal it is usually destroyed by acetylcholinesterase. However, when acetylcholinesterase is inhibited it cannot destroy the acetylcholine and so the amount of acetylcholine rises in the nervous system. This leads to constant nerve transmission and contraction of muscles. Eventually, the muscles tire leading to violent muscle spasms.

In order to develop antidotes for sarin poisoning, chemists have investigated how sarin inhibits acetylcholinesterase in the body. Research has shown that sarin binds to the enzyme blocking its active site. In this process, a hydroxyl group (–OH) in the active site is converted into a phosphonate group (–OP(O)(Me)$OCHMe_2$). The phosphonate group cannot react with acetylcholine and so the action of the enzyme acetylcholinesterase is inhibited. The step-by-step description of how sarin binds to the enzyme is given by the reaction mechanism on the left of the illustration shown here. In reaction mechanisms, curly arrows (⌒) are used to show the movement of electrons and terms such as leaving group, nucleophiles, and electrophiles are used to explain the reactivity of different parts of the reacting molecules. In this example, double-headed curly arrows are used to show the movement of pairs of electrons.

Reaction mechanisms are the subject of this chapter and an understanding of them is extremely important. In this example, an understanding of how sarin reacts with acetylcholinesterase led to the development of antidotes for nerve agent poisoning. One antidote is pralidoxime chloride, which works by converting the phosphonate group on the inactivated acetylcholinesterase back into a hydroxyl group. The sequence of events that lead to reactivation of the enzyme is shown by the reaction mechanism on the right of the illustration. Pralidoxime chloride can be administered using an injection device and in many countries these antidote kits are available to the armed forces.

▼ Reaction mechanisms for inactivation of the enzyme acetylcholinesterase by sarin and reactivation of the enzyme by pralidoxime chloride. (Functional groups within the active site may help to facilitate the reactions e.g. it is unusual to see F⁻ as a leaving group, see Section 20.3, p.938.)

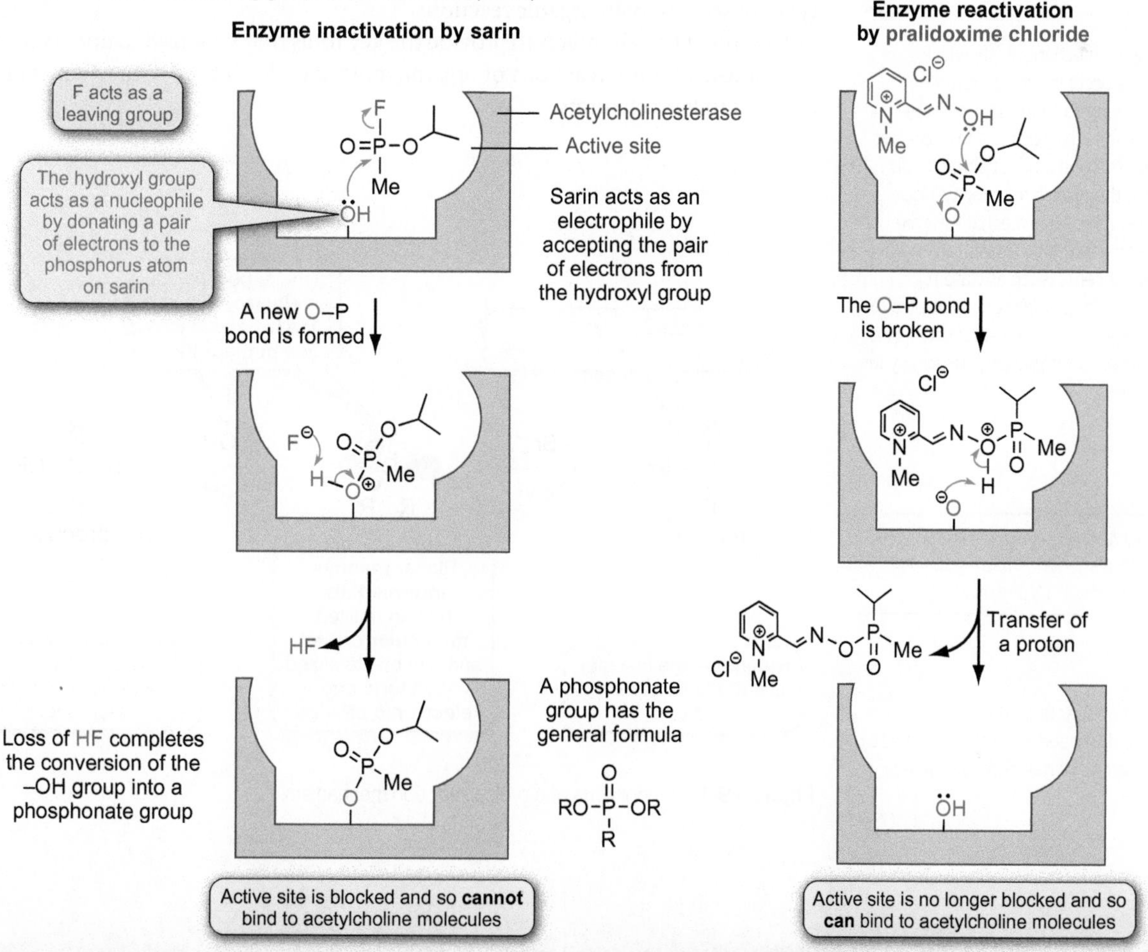

Enzyme inactivation by sarin
Enzyme reactivation by pralidoxime chloride
F acts as a leaving group
The hydroxyl group acts as a nucleophile by donating a pair of electrons to the phosphorus atom on sarin
Acetylcholinesterase
Active site
Sarin acts as an electrophile by accepting the pair of electrons from the hydroxyl group
A new O–P bond is formed
The O–P bond is broken
Transfer of a proton
Loss of HF completes the conversion of the –OH group into a phosphonate group
A phosphonate group has the general formula
Active site is blocked and so cannot bind to acetylcholine molecules
Active site is no longer blocked and so can bind to acetylcholine molecules

Reaction mechanisms give step-by-step pictures of how reactants are converted into products. Typically, a reaction mechanism shows the position of attack by a reagent on a reactant, together with an indication of which bonds break, and which bonds form, and in what order.

For some reactions, the reactants are converted into products in one step. The transition from reactants into products takes place smoothly—as old bonds break in the reactants, new bonds are formed in the products. Reactions with mechanisms of this type are called **concerted reactions**. More commonly, reactions take place in two (or more) steps. For these **stepwise reactions**, an unstable **reactive intermediate** is formed in the first step, which subsequently reacts to form products in the second step (and so on for each step).

An important class of concerted reactions are pericyclic reactions, which proceed through cyclic transition states (Section 19.2, p.908).

Important spectroscopic methods for identifying, or detecting, intermediates are UV/VIS spectroscopy, IR spectroscopy, NMR spectroscopy and ESR spectroscopy (Chapters 10 and 12).

Understanding reaction mechanisms is the most useful approach to learning organic chemistry. Once you understand the fundamental principles of organic mechanisms you can apply them to any organic reaction. As well as explaining the results of known reactions, an understanding of the principles of mechanisms can allow you to predict the products from new reactions as well as forecast the effect of changing the reaction conditions.

To understand and be confident in writing organic reaction mechanisms you need a good understanding of a small number of concepts. These concepts include the use of curly arrows and molecular orbitals, an understanding of electronic and steric effects, electrophiles, nucleophiles, and intermediates, together with an appreciation of reaction kinetics and thermodynamics. Figure 19.1 shows how these concepts can be used to explain in detail how a tertiary bromoalkane (R_3CBr) is converted into a tertiary alcohol (R_3COH) in the presence of hydroxide ion. The mechanism involves two steps and proceeds via a reactive intermediate containing a positively charged carbon atom (R_3C^+) called a **carbocation**.

Section 19.1 discusses each of these important concepts separately and concludes with a set of guidelines for drawing reaction mechanisms. A section on classifying organic reactions (Section 19.2) follows this and includes an overview of the characteristic reaction mechanisms of common functional groups. Finally, Section 19.3 looks at the different types of selectivity in organic reactions.

The aim of this chapter is to provide the key tools that you need to understand and write mechanisms for the reactions of organic molecules. Each type of mechanism is discussed in detail in later chapters.

Understanding the mechanisms by which organic reactions occur is a much better approach than trying to memorize all of the reactions. There is such a wide range of organic reactants, reagents, and products, that if you try to memorize them, each reaction will look new and different. Reaction mechanisms are logical and best mastered by understanding the underlying concepts (such as classifying reactions) and then applying these principles to new examples, again and again—practice makes perfect and you will soon find you are using your eraser less and less!

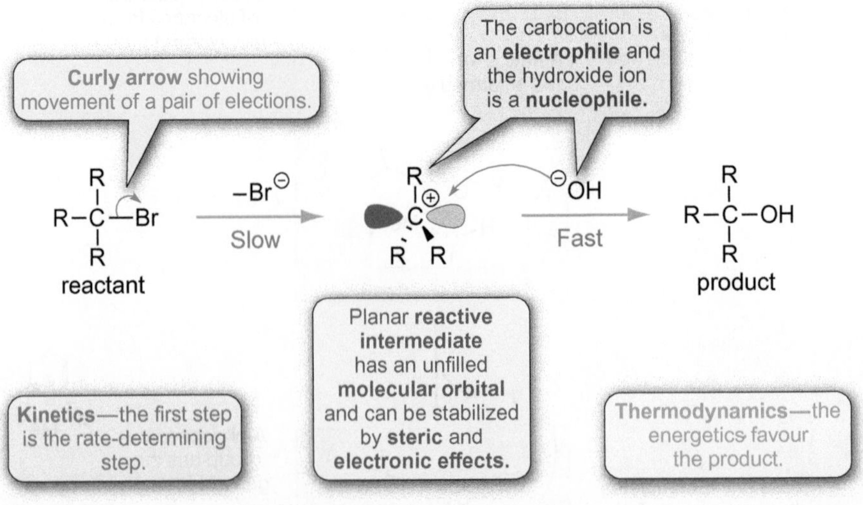
Curly arrow showing movement of a pair of elections.
The carbocation is an electrophile and the hydroxide ion is a nucleophile.
R-C(R)(R)-Br
reactant
$-Br^{\ominus}$
Slow
R R R
$^{\ominus}OH$
Fast
R-C(R)(R)-OH
product
Planar reactive intermediate has an unfilled molecular orbital and can be stabilized by steric and electronic effects.
Kinetics—the first step is the rate-determining step.
Thermodynamics—the energetics favour the product.

Figure 19.1 Components of a typical reaction mechanism.

A carbocation is a molecule that has three covalent bonds to a positively charged carbon (Section 19.1, p.867)

The kinetic aspects of reaction mechanisms are discussed in Section 9.6 (p.416). The thermodynamics of reactions is covered in Chapters 13–15.

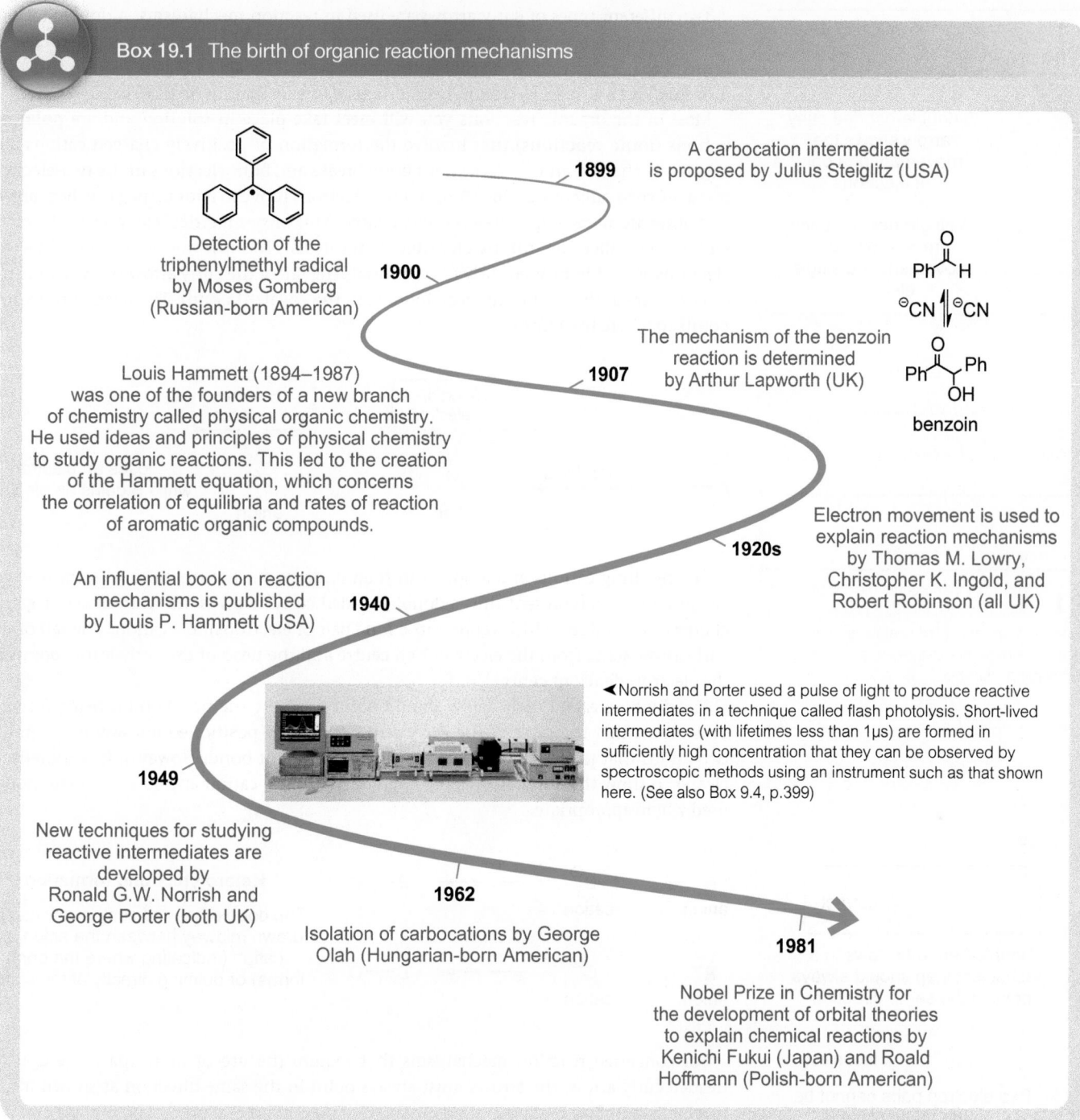

Box 19.1 The birth of organic reaction mechanisms

1899 A carbocation intermediate is proposed by Julius Steiglitz (USA)

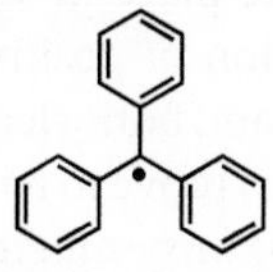

1900 Detection of the triphenylmethyl radical by Moses Gomberg (Russian-born American)

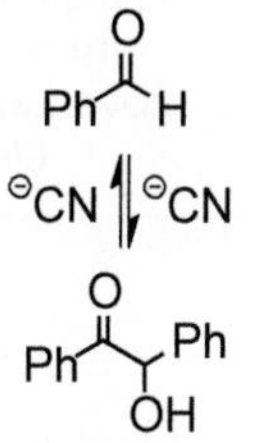

benzoin

1907 The mechanism of the benzoin reaction is determined by Arthur Lapworth (UK)

Louis Hammett (1894–1987) was one of the founders of a new branch of chemistry called physical organic chemistry. He used ideas and principles of physical chemistry to study organic reactions. This led to the creation of the Hammett equation, which concerns the correlation of equilibria and rates of reaction of aromatic organic compounds.

1920s Electron movement is used to explain reaction mechanisms by Thomas M. Lowry, Christopher K. Ingold, and Robert Robinson (all UK)

1940 An influential book on reaction mechanisms is published by Louis P. Hammett (USA)

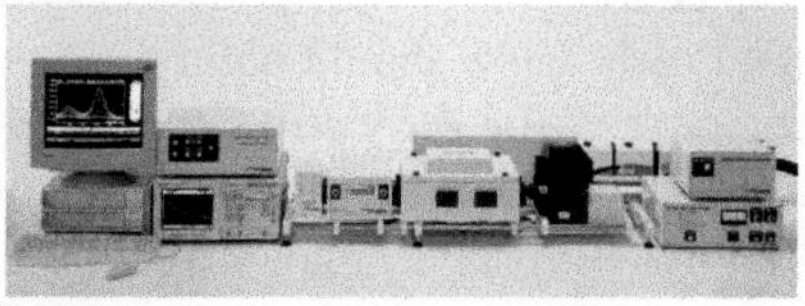

◀ Norrish and Porter used a pulse of light to produce reactive intermediates in a technique called flash photolysis. Short-lived intermediates (with lifetimes less than 1μs) are formed in sufficiently high concentration that they can be observed by spectroscopic methods using an instrument such as that shown here. (See also Box 9.4, p.399)

1949 New techniques for studying reactive intermediates are developed by Ronald G.W. Norrish and George Porter (both UK)

1962 Isolation of carbocations by George Olah (Hungarian-born American)

1981 Nobel Prize in Chemistry for the development of orbital theories to explain chemical reactions by Kenichi Fukui (Japan) and Roald Hoffmann (Polish-born American)

19.1 Fundamental concepts of organic reaction mechanisms

Curly arrow notation

The conversion of reactants into products involves breaking and making covalent bonds. When covalent bonds break and form, the pairs of electrons in the bonds move to different atoms. To show the movement of electrons in reactions, chemists use **curly arrows**.

Straight arrows, →, are used to show the conversion of reactants into products. Double-headed straight arrows, ↔, are used to connect resonance forms (p.874).

The tail of a curly arrow shows where the electrons start from; the arrowhead shows where they finish.

A double-headed curly arrow shows the movement of a pair of electrons

A single-headed curly arrow shows the movement of a single electron

Two different types of curly arrows are used in reaction mechanisms. A double-headed curly arrow (↷) is used to show the movement of a pair of electrons in reaction mechanisms that involve ions. A single-headed (fishhook) curly arrow (↷) is used to show the movement of a single electron in reaction mechanisms that involve radicals.

Most of the organic reactions you will meet take place in solution and are **polar reactions (ionic reactions)** that involve the formation of positively charged cations and negatively charged anions. A covalent bond breaks and both electrons in the two-electron bond move to only one atom. The atom the electrons move to ends up negatively charged. The other atom ends up with a positive charge. This unsymmetrical cleavage of the covalent bond is called **heterolytic cleavage (heterolysis)**. To show the movement of the two electrons, a double-headed curly arrow is used. The tail of the curly arrow shows where the electrons are at the start of the reaction (i.e. in the covalent bond). The head of the arrow points to where they finish.

The circles drawn around positive (⊕) and negative (⊖) charges help them to stand out.

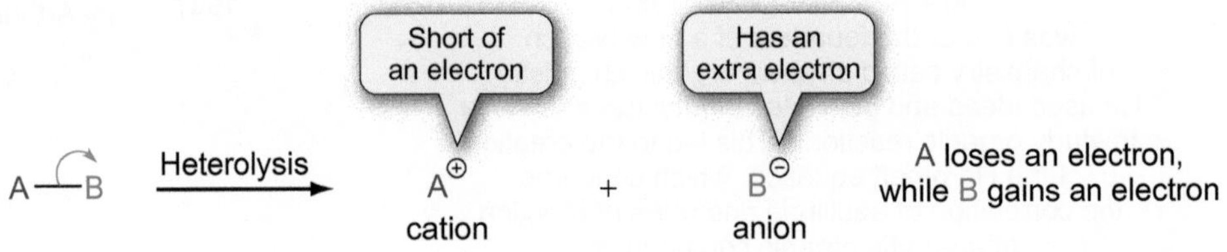
Short of
an electron
Has an
extra electron
A—B
Heterolysis
A⊕
+
B⊖
cation
anion
A loses an electron,
while B gains an electron

The atom that accepts the pair of electrons from the bond now has a lone pair of electrons. It is sometimes drawn with both a negative charge and two dots to represent the lone pair.

The resulting cations and anions can then undergo further reactions to form new covalent bonds. To represent this, a double-headed curly arrow is drawn from the negative charge of the anion (which represents a lone pair of electrons) to a cation. The tail of the curly arrow starts from the electron-rich centre and the head of the curly arrow points to the electron-deficient centre.

The double-headed curly arrow should point to where the new bond is being formed. This means that the head of the curly arrow should be positioned midway between the two atoms that join together to make the new covalent bond. However, it is sometimes clearer to show the curly arrow pointing directly to the cation and so this convention is used where appropriate.

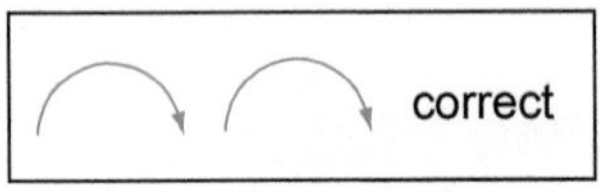
correct

Double-headed arrows in a mechanism step should always point in the same direction

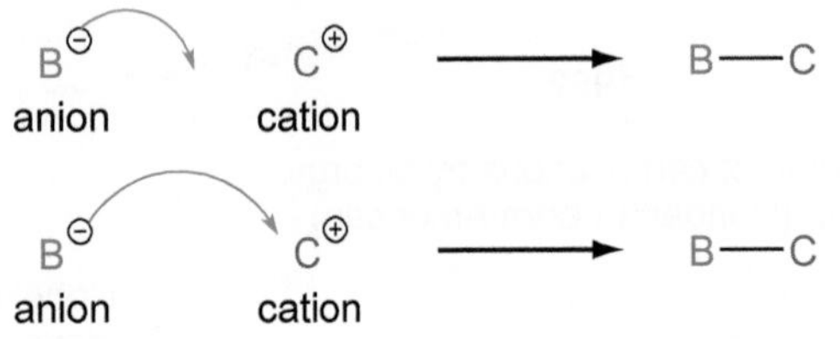
B⊖
C⊕
B—C
anion
cation
B⊖
C⊕
B—C
anion
cation

Heterolytic bond formation

The double-headed curly arrow can be drawn midway between the anion and cation (indicating where the bond forms) or pointing directly at the cation

incorrect

Two electron pairs cannot be in the same place

For concerted reaction mechanisms that require the use of more than one double-headed curly arrow, the arrows must always point in the same direction as shown in the mechanism in Box 19.2.

Box 19.2 Natural cascades

In nature, steroids are prepared by a remarkable series of cyclizations of a polyene (a compound with many C=C bonds) called squalene oxide. Opening of the epoxide ring (see Figure 1) of squalene oxide occurs on addition of acid. When this occurs in the active site of an enzyme, the ring opening triggers the series of four sequential cyclizations shown. Follow the curly arrows in the mechanism to see how the four rings form.

➜

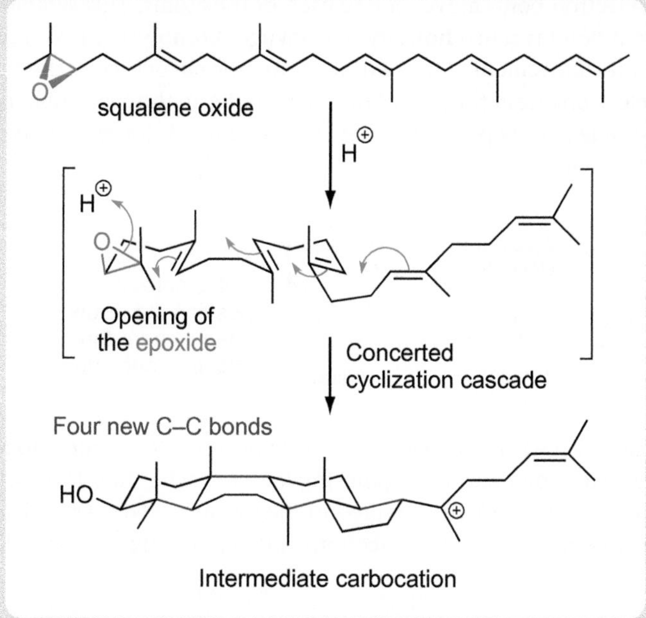
squalene oxide
H⊕
H⊕
O
Opening of
the epoxide
Concerted
cyclization cascade
Four new C–C bonds
HO
Intermediate carbocation

Figure 1

These cyclizations are shown using five red double-headed curly arrows, which all point in the same direction (i.e. from right to left). The cyclizations produce a carbocation. (Check that you agree that the curly arrows lead to this carbocation.) In total, the successive cyclizations of squalene oxide produce four new C–C bonds in a single sequence. These types of successive reactions are called **cascade reactions**.

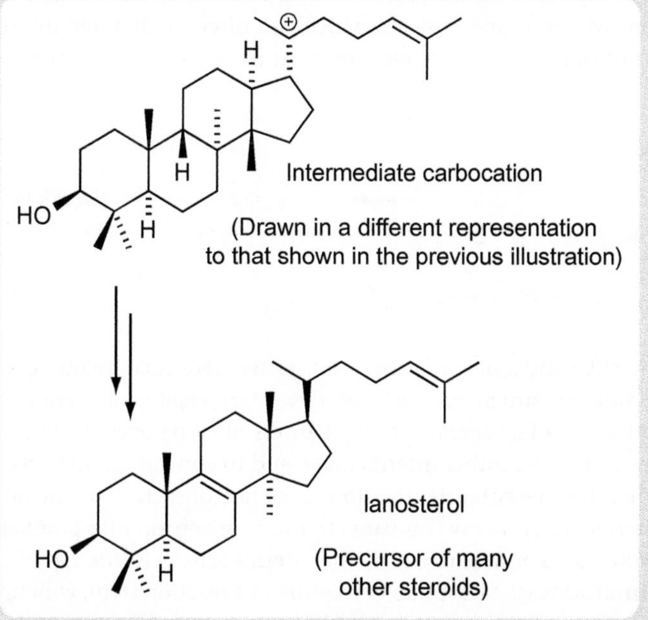
H
H
HO
H
Intermediate carbocation
(Drawn in a different representation
to that shown in the previous illustration)
lanosterol
HO
H
(Precursor of many
other steroids)

Figure 2

Naming successive organic reactions as cascade reactions comes from the use of 'cascade' to describe a series of small waterfalls.

The carbocation is subsequently converted into a steroid called lanosterol (Figure 2). Lanosterol is the precursor of many other steroids including cholesterol. (The structure and shape of cholesterol (and other steroids) are discussed in Box 18.3, p.830.)

Question

The occurrence of a cascade reaction (to form steroids) in nature has inspired researchers to mimic this type of reaction in the laboratory. This is called **biomimetic synthesis**. An example of a biomimetic cascade reaction, to form a similar ring system to that found in steroids, is shown here.

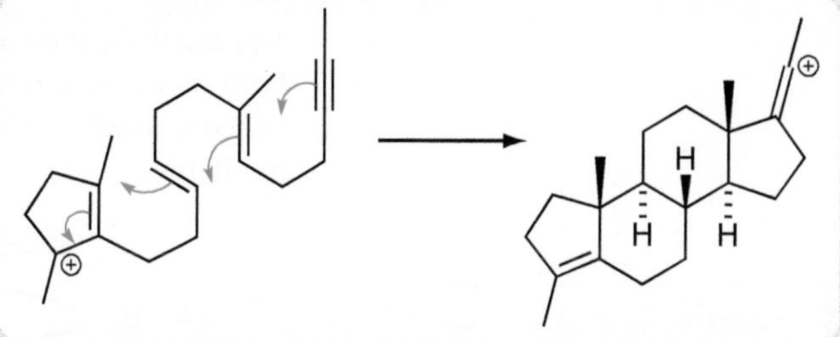
H
H
H

(a) Identify the new bonds that are formed in the cascade reaction.

(b) Identify the chiral centres in the product. Draw the enantiomer of the product.

Lanosterol is a precursor of sex hormones including testosterone. Testosterone helps in sexual maturation and deepening of the voice as well as the growth of testes, muscles, bones, and body hair. ➤

Radical reactions are common in the gas phase (e.g. in reactions in the atmosphere, as illustrated in the opening page of Chapter 9, p.382 and Box 27.6, p.1233).

The name homolysis comes from the Greek words *homoios* meaning 'equal' and *lysis* meaning 'loosening'.

Organic molecules also undergo **radical reactions**, in which radicals are formed. A **radical** is an atom or molecule containing an unpaired electron. As the covalent bond breaks, the electrons in the two-electron bond move, one to each of the atoms. This symmetrical cleavage of the covalent bond is called **homolytic cleavage** (**homolysis**). Two single-headed curly arrows show the movement of the two electrons. The tail of each of the curly arrows shows where the electrons are at the start of the reaction (i.e. in the covalent bond). The head of each single-headed arrow points to where they end up. A dot represents the unpaired electron on each atom.

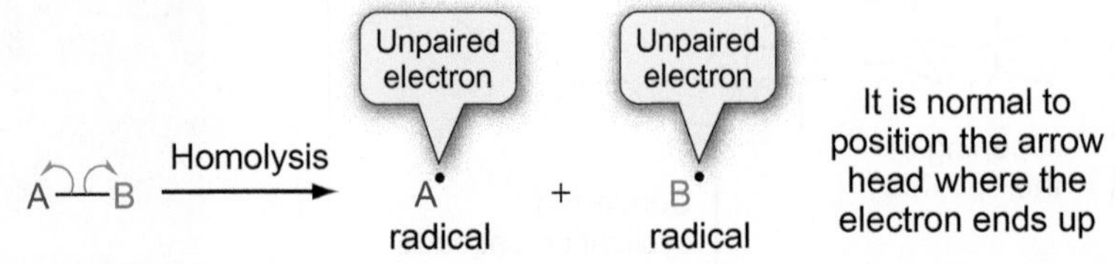
Unpaired electron
Unpaired electron
A—B
Homolysis
A•
+
B•
radical
radical
It is normal to position the arrow head where the electron ends up

Radicals are often described as free radicals because they exist independently, free of any support from other species. This contrasts with cations and anions, which exist together and can be stabilized in polar solvents by solvation (see p.880).

The resulting radicals can react with other radicals to form new covalent bonds. To represent this, two single-headed arrows are drawn pointing *towards* each other. The tails of the curly arrows start from the two dots (which represent the two unpaired electrons) and the heads of the curly arrows point to where the electrons end up, forming a new covalent bond.

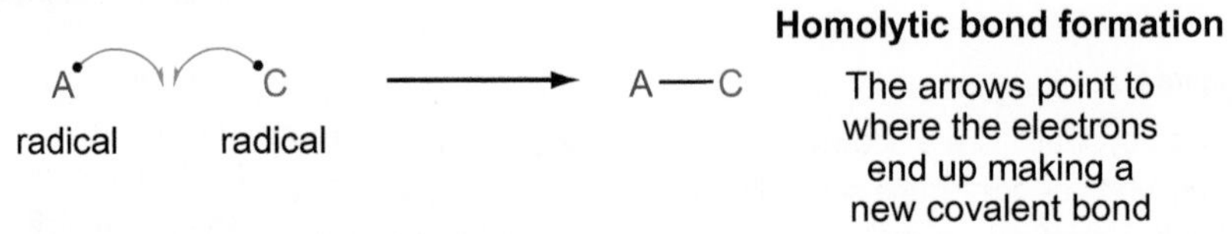
Homolytic bond formation
A•
•C
radical
radical
A—C
The arrows point to where the electrons end up making a new covalent bond

Reactive intermediates

The reactions shown above involve the formation of ions or radicals that then react to form products. The short-lived ions and radicals, which can often be detected (by spectroscopic methods) but not isolated from the reaction mixture, are called **reactive intermediates**.

Reactions with activation energies (E_a) above +20 kJ mol^{-1} generally require heating so that reactants can be converted into products. On heating, a greater proportion of colliding molecules have sufficient energy to overcome the barrier. Most organic reactions have activation energies between +40 kJ mol^{-1} and +150 kJ mol^{-1}, so they require heating.

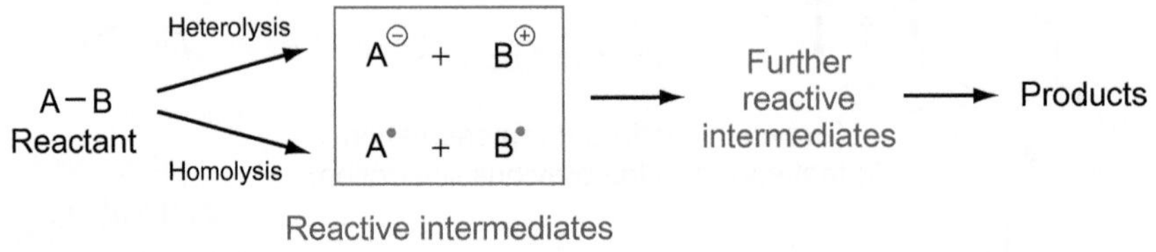
A–B
Reactant
Heterolysis
Homolysis
A⊖ + B⊕
A• + B•
Further reactive intermediates
Products
Reactive intermediates

Organic reactions generally consist of a number of successive steps each involving reactive intermediates with different structures. Each reactive intermediate corresponds to a shallow dip in the **reaction profile**. Separate energy barriers must be overcome to form an intermediate, to convert it into another intermediate, and to convert an intermediate into a product—the higher the energy barrier the slower the rate of each step. The rate of formation of reactive intermediates is slow compared to their conversion into product. (If this were not the case, you would be able to isolate the intermediates.) At the top of each energy barrier is a **transition state** (the least stable structure in a reaction step), which, unlike a reactive intermediate, cannot be detected. A typical reaction profile for a three-step organic reaction is shown in Figure 19.2.

A reaction profile is a plot of Gibbs energy versus the progress of reaction and is sometimes called a **Gibbs energy profile**. Gibbs energy profiles are introduced in Section 9.8 (p.435). The kinetics of complex reactions involving a series of steps is discussed in Section 9.6 (p.419).

Gibbs energy change of reaction and chemical equilibrium are discussed in Section 15.1 (p.697).

In step 1 of the reaction in Figure 19.2, reactants are converted into the first reactive intermediate. This step has the highest energy transition state (TS‡1) and so this is the **rate-determining step** (i.e. the slowest step of the reaction). The energy difference between the reactants and TS‡1 is the **Gibbs energy of activation** and the higher this activation energy

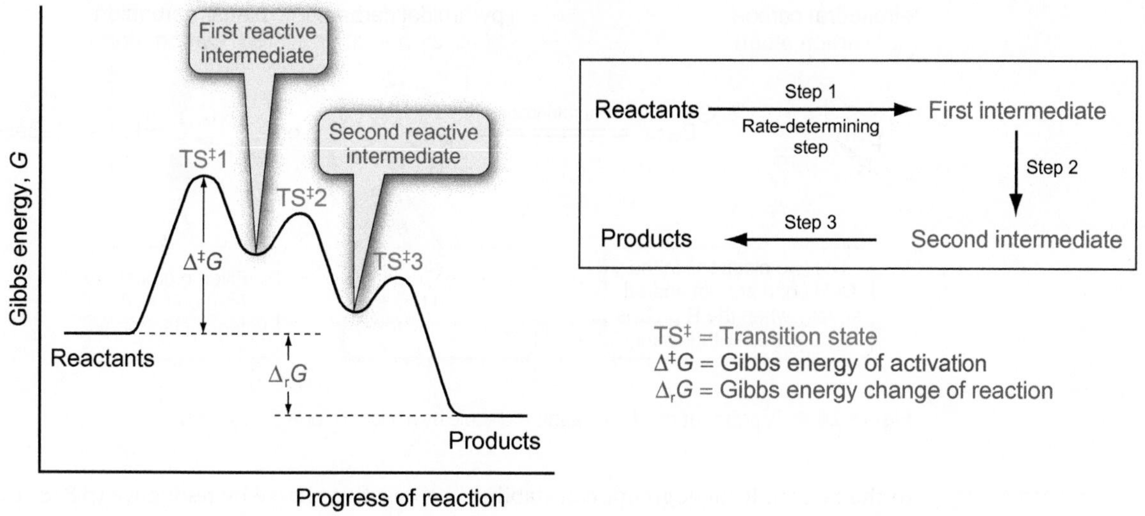
First reactive intermediate
Second reactive intermediate
TS‡1
TS‡2
TS‡3
Δ‡G
ΔrG
Reactants
Products
Gibbs energy, G
Progress of reaction
Reactants
Step 1
Rate-determining step
First intermediate
Step 2
Second intermediate
Step 3
Products
TS‡ = Transition state
Δ‡G = Gibbs energy of activation
ΔrG = Gibbs energy change of reaction

Figure 19.2 A typical reaction profile for a three-step organic reaction.

the slower the reaction (Section 9.8, p.434). In the remaining two steps, the first reactive intermediate is converted into the second reactive intermediate and this is subsequently converted into products. Overall, the value of the Gibbs energy change of this particular reaction is negative (as the energy of the products is less than that of the reactants so that energy is lost from the reacting system), so the products will be favoured over reactants at equilibrium.

The relationship between the Gibbs energy of activation ($\Delta^{‡}G$) and the activation energy (E_a) is discussed in Section 9.8 (p.434).

Reactive intermediates: carbocations and carbanions

Charged reactive intermediates, called carbocations and carbanions, are formed on heterolysis of covalent bonds to carbon. A **carbocation** has three covalent bonds to a positively charged carbon (e.g. R_3C^+). A **carbanion** has three covalent bonds to a negatively charged carbon (e.g. R_3C^-). For example, heterolysis of a C–Br bond in a tertiary bromoalkane (R_3CBr) produces a carbocation and a bromide ion, as shown in Figure 19.3. The bromide ion is called the **leaving group** because it departs from the organic molecule. The two electrons in the polar $^{\delta+}$C–Br$^{\delta-}$ bond move on to the bromine atom (not the carbon atom) because bromine is more electronegative than carbon. Carbocations are generally planar and contain an empty *p* orbital on an sp^2 hybridized carbon atom.

A carbanion is formed when a base removes a proton (H^+) from an organic molecule. On deprotonation, the pair of electrons that was once in the C–H bond becomes a non-bonding lone pair on carbon, as shown in Figure 19.4. The resulting carbanion is pyramidal or planar (or something in between) depending on the nature of the R groups attached

Formally, the name carbocation refers to both carbenium and carbonium ions. Carbenium ions have three bonds to the positively charged carbon (e.g. Me_3C^+), whereas carbonium ions have five bonds to the positively charged carbon (e.g. H_5C^+). However, the term carbocation is commonly used in place of carbenium ion and it is used to describe R_3C^+ in this book.

Leaving groups, including Br^-, are discussed in more detail in Section 20.2 (p.925).

Electronegativity is the tendency of each atom in a stable molecule to attract electrons. In the Pauling electronegativity scale, Br has a value of 2.96 and C has a value of 2.55 (see Section 4.3, p.177).

sp^2 and sp^3 hybridization of carbon are discussed in Section 5.4 (p.237).

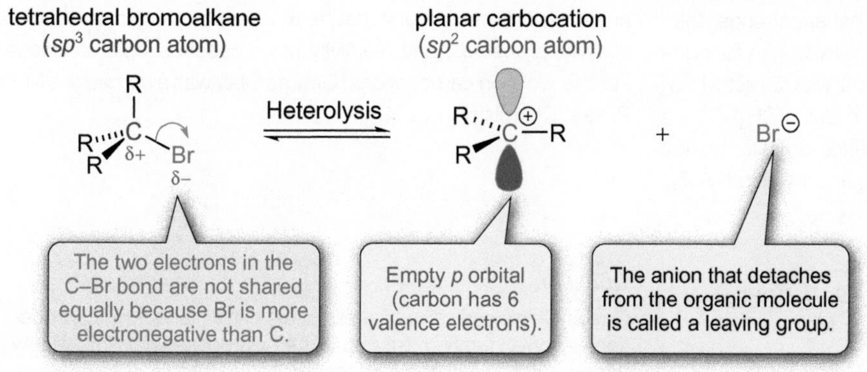
tetrahedral bromoalkane
(sp3 carbon atom)
planar carbocation
(sp2 carbon atom)
Heterolysis
δ+
δ–
Br
+
Br⊖
The two electrons in the C–Br bond are not shared equally because Br is more electronegative than C.
Empty p orbital (carbon has 6 valence electrons).
The anion that detaches from the organic molecule is called a leaving group.

Figure 19.3 Heterolysis of a C–Br bond to form a carbocation and a bromide ion.

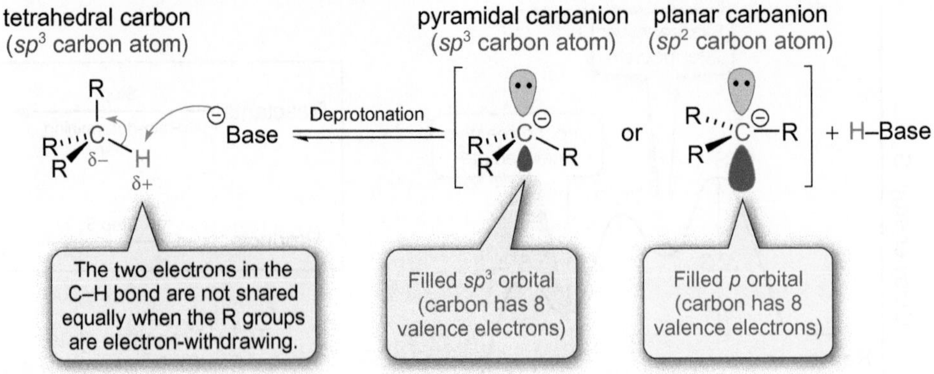
tetrahedral carbon
(sp³ carbon atom)
pyramidal carbanion
(sp³ carbon atom)
planar carbanion
(sp² carbon atom)
Base
Deprotonation
or
+ H–Base
δ−
δ+
The two electrons in the C–H bond are not shared equally when the R groups are electron-withdrawing.
Filled sp³ orbital (carbon has 8 valence electrons)
Filled p orbital (carbon has 8 valence electrons)

Figure 19.4 Deprotonation of an organic molecule by a base to form a carbanion.

to the carbon. If the R groups can stabilize the negative charge by resonance (p.873) the carbanion is planar; otherwise, it is pyramidal.

The formations of both carbocations and carbanions are equilibrium processes. The identity of the R groups in the carbanion and carbocation can influence the equilibrium constants of the heterolysis and deprotonation reactions. Changing the R substituents alters the stability of the carbanions and carbocations, which affects the equilibria of the reactions. *In general, the more stable the carbanion or carbocation, the higher the equilibrium constant (K) and the greater the concentration of the product carbanion or carbocation at equilibrium.* Chemists explain how the R groups affect the stability of carbocations and carbanions in terms of electronic and steric factors, which are discussed in the following sections.

Alkyl anions, such as the ethyl anion ($CH_3CH_2^-$), are pyramidal (sp^3 orbitals form the three σ bonds). The benzyl anion ($C_6H_5CH_2^-$), which is stabilized by resonance, is planar (sp^2 orbitals form the three σ bonds).

Acid–base reactions of organic molecules are discussed in Section 19.2 (p.888).

The stability of carbocations (R_3C^+) and carbanions (R_3C^-) depends on the nature of the R groups attached to the carbon atom. Relatively stable carbocations and carbanions have relatively low energies.

All organic reactions are reversible, and reactants and products interconvert to varying degrees. For most organic reactions in this book, the products are strongly favoured at equilibrium ($K \gg 1$) and a single reaction arrow (→) shows this. Reversible equilibrium reaction arrows (⇌) are used for those reactions in which the product is not strongly favoured at equilibrium ($K \approx 1$).

Box 19.3 George Olah and carbocations

It was not until 1962 that carbocations (e.g. R_3C^+) were proved to be intermediates in many organic reactions. The Hungarian-born chemist George Olah was able to prepare and examine carbocations using reagents called **superacids** at very low temperatures (Box 7.7, p.337). Superacids are acids that are stronger than pure sulfuric acid. An example of a superacid is hydrogen fluoride–antimony pentafluoride (HF^-SbF_5), which reacts with 2-methylpropane ($(CH_3)_3CH$) to generate the *tert*-butyl cation ($(CH_3)_3C^+$)—a relatively stable carbocation. 2-Methylpropane loses H^- (a hydride ion), which is protonated by $HF–SbF_5$ to form H_2 and the anion $^-SbF_6$.

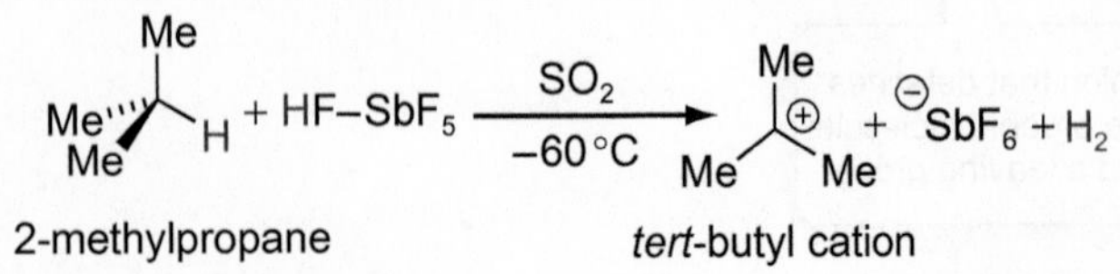
Me
Me
Me
H
+ HF–SbF5
SO2
−60 °C
+ SbF6 + H2
2-methylpropane
tert-butyl cation

Olah found that when he carried out this reaction at low temperature in a solvent such as SO_2, which reacts only very slowly with carbocations, carbocations were formed in high concentrations. Some carbocations had such long lifetimes (at low temperatures, some are stable for many hours) that he was able to study their structure, stability, properties, and reactivity using spectroscopic techniques. For this work on carbocations, George Olah was awarded the Nobel Prize in Chemistry in 1994.

Question

One of Olah's early and successful experiments to form carbocations involved treating 2-fluoro-2-methylpropane with SbF_5 at low temperature. Suggest a balanced equation for this reaction.

Reactive intermediates: radicals

Carbon radicals are formed on homolysis of covalent bonds to carbon. For example, in Figure 19.5 homolysis of a C–Br bond in a bromoalkane (R_3C–Br) produces a carbon radical ($R_3C^\bullet$) and a bromine radical ($Br^\bullet$). There are three bonds attached to the carbon carrying the unpaired electron. This carbon atom is sp^2 hybridized and the arrangement around it is planar.

Homolysis of the C–Cl bond in benzyl chloride ($PhCH_2Cl$) is compared with heterolysis in Worked example 19.1.

Like carbocations and carbanions, the stability of carbon radicals depends on the electronic and steric effects of the R groups attached to the carbon as discussed in the following sections.

A bromine radical ($Br^\bullet$) is a bromine atom; it has seven valence electrons.

Radicals can be detected, and their structures determined, using ESR spectroscopy (Section 10.7, p.498). For example, ESR spectroscopy has shown that most carbon-centred radicals are planar, or nearly planar.

In a bromoalkane, the C–Br bond is weaker than the C–H and C–C bonds. For example, in bromoethane (CH_3CH_2Br), here are approximate values for the bond enthalpies: +421 kJ mol^{-1} for C–H, +370 kJ mol^{-1} for C–C, and +293 kJ mol^{-1} for C–Br.

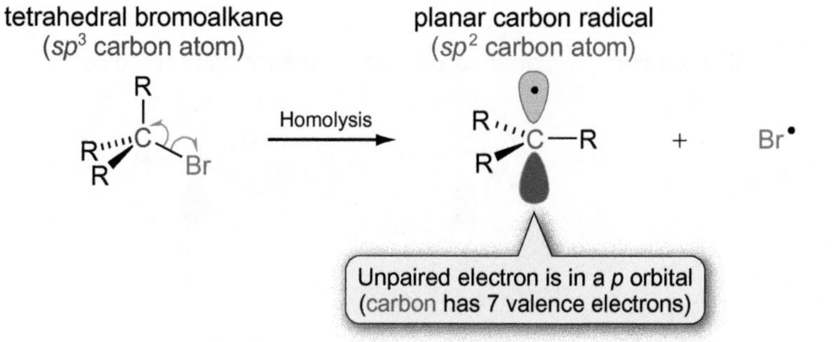
tetrahedral bromoalkane
(sp3 carbon atom)
planar carbon radical
(sp2 carbon atom)
Homolysis
+
Br•
Unpaired electron is in a p orbital
(carbon has 7 valence electrons)

Figure 19.5 Homolysis of a C–Br bond to form a carbon radical and a bromine radical.

Worked example 19.1 Breaking bonds to form ions or radicals

Use curly arrows to show what products are formed on (a) heterolytic cleavage and (b) homolytic cleavage of the C–Cl bond in benzyl chloride ($PhCH_2Cl$).

▲ A molecular model of benzyl chloride (the green sphere represents chlorine).

Strategy

(a) On heterolytic cleavage the C–Cl bond breaks to form ions. The two electrons in the C–Cl bond are not shared equally because Cl is more electronegative than C. To show the polarization of electrons towards Cl, draw a partial negative charge (δ–) on Cl and a partial positive charge (δ+) on C. Then, use a double-headed arrow (↷) to show the movement of the two electrons in the polar C–Cl bond towards the chlorine atom.

(b) On homolytic cleavage the C–Cl bond breaks to form radicals. Use two single-headed arrows (⇀) to show the movement of the two electrons in the C–Cl bond. One electron in the C–Cl bond moves on to carbon and the other electron moves on to chlorine.

Solution

(a)

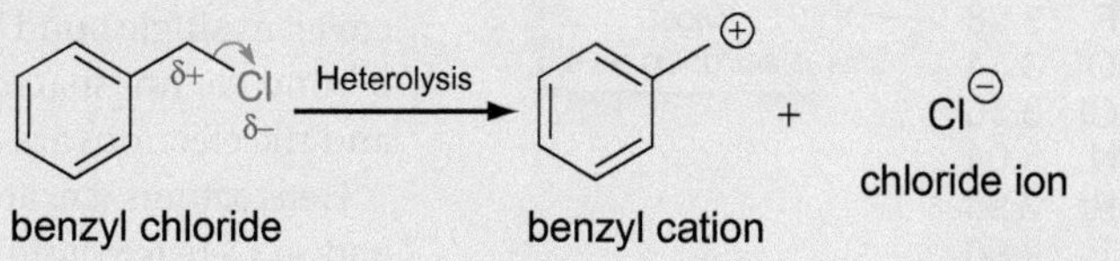
δ+
Cl
δ–
Heterolysis
+
Cl⊖
chloride ion
benzyl chloride
benzyl cation

(b)

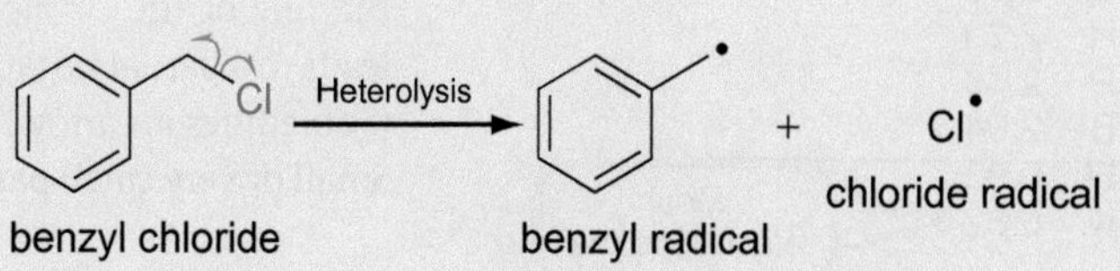
Cl
Heterolysis
+
Cl•
chloride radical
benzyl chloride
benzyl radical

Question

Use curly arrows to show the products formed on (a) heterolysis and (b) homolysis of the C–Br bond in Me_3C–Br.

Summary

- Double-headed curly arrows (⌒) are used to represent the movement of electrons in polar reactions.
- Single-headed curly arrows (⌒) are used to represent the movement of electrons in radical reactions.
- Heterolytic cleavage (heterolysis) is the unsymmetrical cleavage of a covalent bond to form a cation and an anion.
- Homolytic cleavage (homolysis) is the symmetrical cleavage of a covalent bond to form two radicals.
- Short-lived carbocations, carbanions, and carbon radicals, which can often be detected (but not isolated) in polar and radical reactions, are called reactive intermediates.
- The shape of the reactive intermediate depends on the hybridization of the carbon atom carrying the charge or unpaired electron.

planar carbocation pyramidal carbanion planar carbanion planar carbon radical

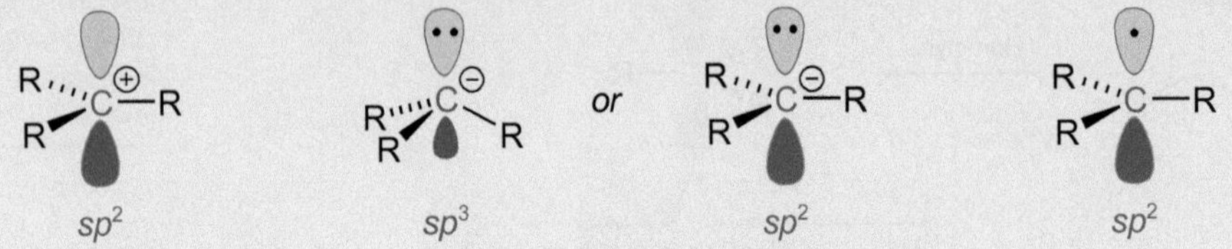
sp2
sp3
or
sp2
sp2

Electronic effects

All single covalent bonds in organic molecules are σ bonds, while double or triple bonds are composed of one σ bond and one or two π bonds, respectively (Section 5.4, p.239).

The Pauling electronegativity scale is introduced in Chapter 4; see Figure 4.6 (p.178).

Recognizing the electronic properties of atoms, or groups of atoms, helps you understand how organic compounds react. If you know whether an atom or a group of atoms is electron-withdrawing or electron-donating, it can help to determine the position of curly arrows in reaction mechanisms and also to assess the relative stability of reactive intermediates and leaving groups.

The three factors that influence the electronic properties of an atom or group are **inductive effects, hyperconjugation**, and **mesomeric effects**. Inductive effects and hyperconjugation refer to the polarization of electrons in σ bonds, while mesomeric effects refer to the delocalization of electron density through π bonds.

Inductive effects

F	3.98	Most electronegative
O	3.44	
Cl	3.16	
N	3.04	
Br	2.96	
I	2.66	
S	2.58	
C	2.55	
H	2.20	
P	2.19	
B	2.04	
Li	0.98	
K	0.82	Most electropositive

Inductive effects are due to the electronegativity differences that exist in a molecule. In a covalent single bond between two different types of atoms, the two electrons forming the σ bond are not shared equally. Different types of atoms have different electronegativities and the electrons are attracted towards the more electronegative atom.

Heteroatoms (i.e. atoms other than C or H) are usually more electronegative than carbon and so C–heteroatom bonds are generally polar bonds. For example, in the C–Cl bond the electrons are attracted from carbon towards chlorine. An arrow (+→), drawn above or below the line of the covalent bond represents polarization of the electrons in the bond, which leads to a partial negative charge (δ–) on chlorine and a partial positive charge (δ+) on carbon. (Sometimes an arrowhead is drawn in the middle of the covalent bond, e.g. R_3C —→— Cl.) Small or very small partial charges are sometimes shown by δδ and δδδ, respectively.

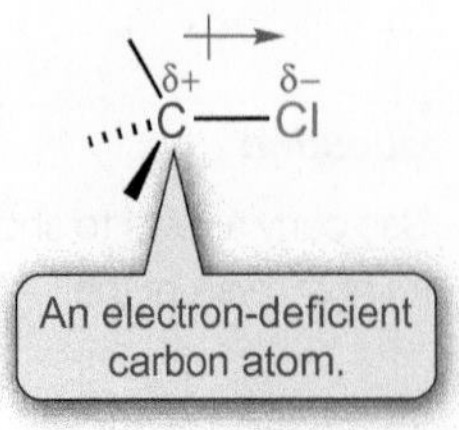
Polarization
δ+
δ–
C
Cl
An electron-deficient carbon atom.

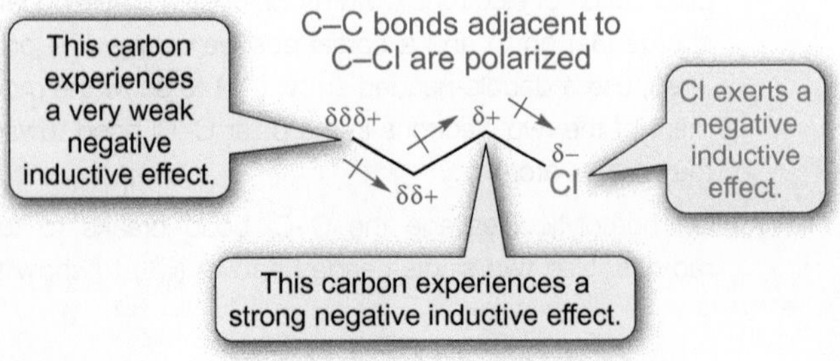
This carbon experiences a very weak negative inductive effect.
C–C bonds adjacent to C–Cl are polarized
δδδ+
δ+
δδ+
δ–
Cl
Cl exerts a negative inductive effect.
This carbon experiences a strong negative inductive effect.

C–H bonds are considered to be non-polar bonds because C and H have similar electronegativities.

If the carbon atom in the C–Cl bond is attached to further carbon atoms, the effect can be transmitted further so that adjacent C–C bonds are also polarized. The electrons in an adjacent σ bond are attracted to the electron-deficient carbon atom and this causes the next carbon atom to be slightly electron-deficient—the nearer the carbon atom is to chlorine, the more electron-deficient it is. Transmission quickly dies away in a saturated carbon chain and is not really noticeable after the second carbon. The polarization of the C–C bonds caused by the C–Cl bond is called a **negative inductive effect** and the chlorine atom is referred to as a **–I group**. It is called a –I group because it draws electrons away from the carbon chain.

Other –I groups are shown in Figure 19.6. HO and CO_2H groups, which have electronegative oxygen atoms that can attract electrons from adjacent σ bonds, are both –I groups. For the CO_2H group, electrons are attracted to both oxygen atoms and this causes the carbon atom of the CO_2H group to be electron-deficient, as shown in Figure 19.6. Notice that the inductive effect of the CO_2H group is transmitted through the C–C bond to the α-carbon, which is also electron-deficient.

An **inductive effect** is the polarization of electrons through σ bonds. An electron-withdrawing group exerts a **–I effect**, whereas an electron-donating group exerts a **+I effect**.

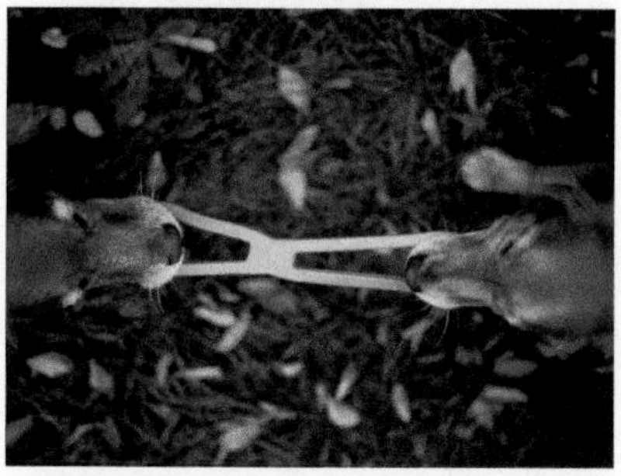

▲ The negative inductive effect can be likened to a tug of war. The atom, or group of atoms, that pulls the electrons more strongly will win out.

A carbon atom next to a C=O bond is called an α-carbon (Section 2.7, p.100).

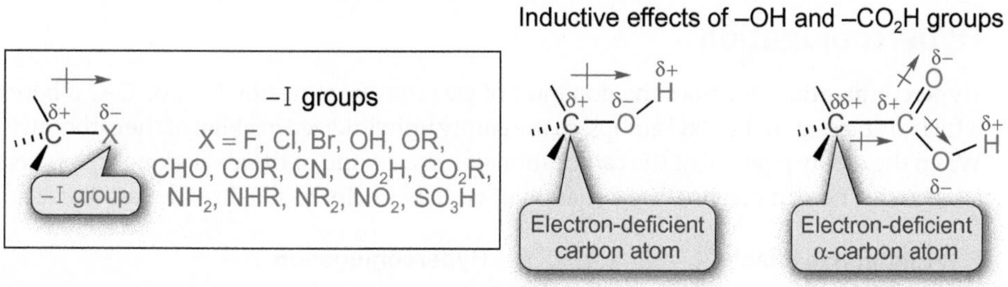

Inductive effects of –OH and –CO2H groups
–I groups
X = F, Cl, Br, OH, OR, CHO, COR, CN, CO2H, CO2R, NH2, NHR, NR2, NO2, SO3H
–I group
δ+ δ–
C X
C O H
δδ+ δ+ δ–
Electron-deficient carbon atom
Electron-deficient α-carbon atom

Figure 19.6 Some groups with a negative inductive effect.

In general, the greater the number of heteroatoms, and the more electronegative the heteroatom, the stronger the inductive effect of the –I group. For example, carbon atoms bonded to the strongly electron-withdrawing NO_2 (nitro) group are particularly electron-deficient.

Examples of electron-withdrawing groups in decreasing order of their –I effect: $NO_2 > SO_3H > CO_2H > Cl > Br > OH > NH_2$

In comparison to –I groups, there are comparatively few electron-donating +I groups. Some examples are shown in Figure 19.7. Metals such as lithium and magnesium are +I groups because these atoms are more electropositive than carbon. The Li–C and Mg–C bonds are polarized and the electrons are attracted to carbon.

Organometallic compounds, such as organolithiums (RLi) and Grignard reagents (RMgX), have the general structure $^{\delta-}R–M^{\delta+}$, where R is an organic group and M is a metal (Section 26.10, p.1195). The electron-rich carbon atom can donate an electron pair to H^+ and act as a base, or donate an electron pair to a δ^+ carbon atom and act as a nucleophile.

The most important +I groups are alkyl groups (R). The fact that alkyl groups are electron-donating explains the order of stability of carbocations. Tertiary carbocations (R_3C^+) are more stable than secondary carbocations (R_2CH^+) because tertiary carbocations contain a greater number of electron-donating alkyl groups to stabilize the positive charge. All three alkyl groups in the tertiary carbocation act as +I groups because the electrons in the three $C–C^+$ σ bonds are attracted towards the positively charged carbon (i.e. a positively charged carbon is more electropositive than a neutral carbon). Similarly, a secondary carbocation (R_2CH^+) is more stable than a primary carbocation (RCH_2^+), which is more stable than the methyl carbocation, CH_3^+ (Figure 19.8).

The order of carbocation stability suggests that alkyl groups must be stronger +I groups than hydrogen atoms. On the basis of the electronegativities of hydrogen (2.20) and

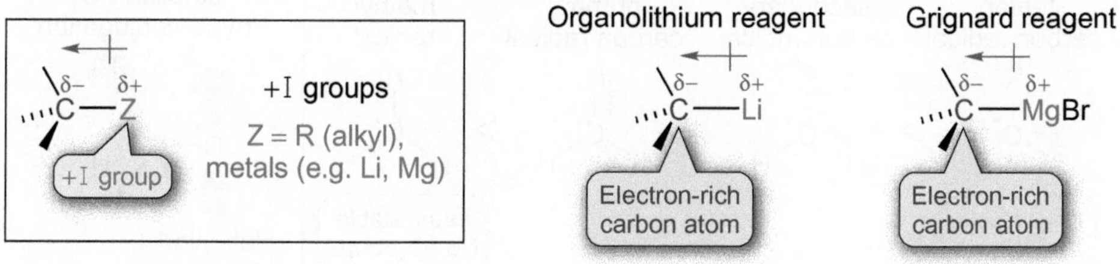

Organolithium reagent
Grignard reagent
+I groups
Z = R (alkyl), metals (e.g. Li, Mg)
+I group
δ– δ+
C Z
C Li
C MgBr
Electron-rich carbon atom
Electron-rich carbon atom

Figure 19.7 Some groups with a positive inductive effect.

WRONG WAY
RIGHT WAY

▲The blue arrow, ⟶ (crossed), is used in the organic chapters to represent the polarization of electrons. The head of the arrow points towards the more *negative* part of the distribution. (Don't confuse this with the symbol for a dipole moment, ⟵, which, by convention, points towards the *positive* end of the dipole, see Section 5.3, p.253.)

ℹ The greater the number of alkyl groups (R) bonded to the positive carbon in a carbocation the more stable it is. Primary carbocations, RCH_2^+, are so unstable that they are rarely observed in solution (in organic reactions), whereas the methyl carbocation, CH_3^+, is never observed in solution.

ℹ Examples of electron-donating groups in decreasing order of their +I effect: $C(CH_3)_3 > CH(CH_3)_2 > CH_2CH_3 > CH_3$

ℹ The alignment allows for side-on overlap (p.885) between the unhybridized *p* orbital and a molecular orbital formed from overlap of a carbon sp^3 orbital and a hydrogen 1*s* orbital.

ℹ In general, the more spread out, or **delocalized**, a positive charge is, the more stable the carbocation is. The delocalization of electrons by overlap of a σ orbital with an empty *p* orbital is called **hyperconjugation**.

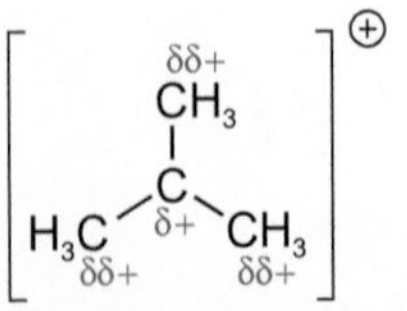

Delocalization of the positive charge in the *tert*-butyl cation

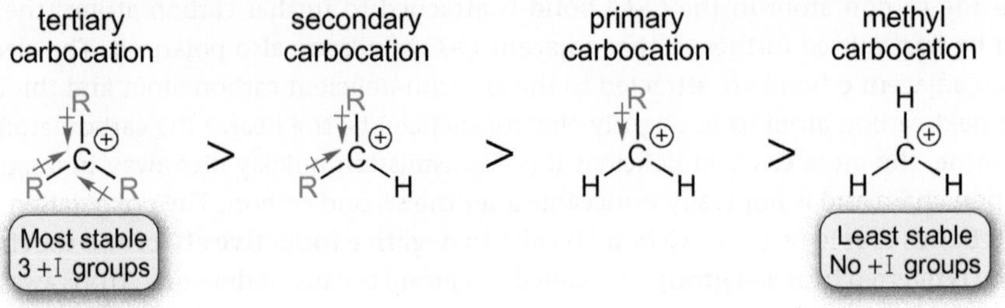
tertiary carbocation
secondary carbocation
primary carbocation
methyl carbocation
Most stable
3 +I groups
Least stable
No +I groups

Figure 19.8 Relative stabilities of carbocations.

carbon (2.55), it is not obvious why the donation of electrons to the carbocation in a C–C$^+$ bond is stronger than that in an H–C$^+$ bond. To explain this you need to take into account the effect the σ bonds of the alkyl group(s) have on the carbocation. The σ bonds of the alkyl groups stabilize carbocations by a process called hyperconjugation.

Hyperconjugation

Hyperconjugation describes the donation of electrons from nearby C–H or C–C σ bonds, which are present in the alkyl groups, to the empty unhybridized *p* orbital of the carbocation. When the empty *p* orbital of the carbocation and the σ bond of the alkyl group are properly aligned (see note in margin), the donation of electrons helps to stabilize the carbocation.

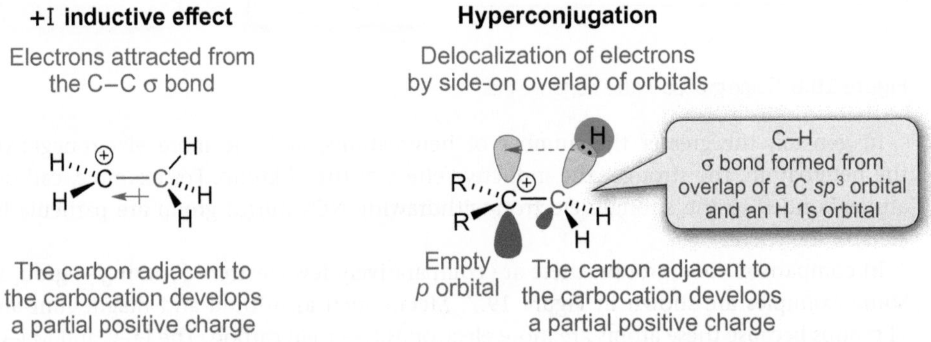
+I inductive effect
Electrons attracted from the C–C σ bond
The carbon adjacent to the carbocation develops a partial positive charge
Hyperconjugation
Delocalization of electrons by side-on overlap of orbitals
C–H σ bond formed from overlap of a C sp3 orbital and an H 1s orbital
Empty p orbital
The carbon adjacent to the carbocation develops a partial positive charge

Electron donation from an alkyl group by both inductive effects and hyperconjugation causes partial positive charges to develop on carbon atoms in the alkyl groups. The positive charge in the carbocation is no longer localized on one carbon atom: it is spread out over a larger area of the molecule. This spreading out of charge over more than one atom is called **delocalization.**

The order of stability of carbon radicals in Figure 19.9 is the same as that for carbocations (see Figure 19.8). This is because carbon-centred radicals, like carbocations, are electron-deficient intermediates. Whereas a positively charged carbon of a carbocation lacks two valence electrons, the neutral carbon atom of a carbon-centred radical is short of one

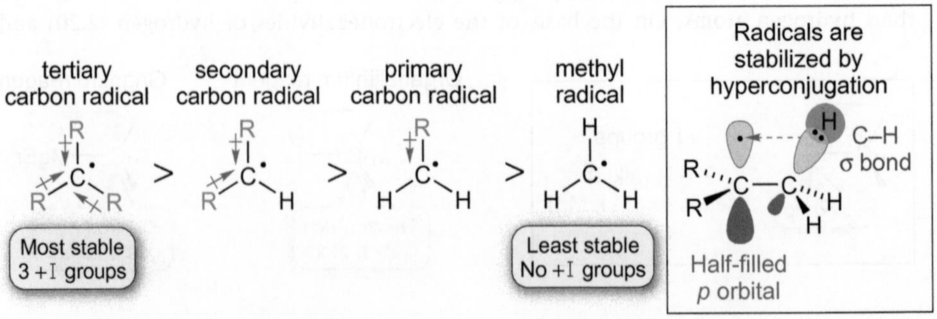
tertiary carbon radical
secondary carbon radical
primary carbon radical
methyl radical
Most stable
3 +I groups
Least stable
No +I groups
Radicals are stabilized by hyperconjugation
C–H σ bond
Half-filled p orbital

Figure 19.9 Relative stabilities of carbon radicals.

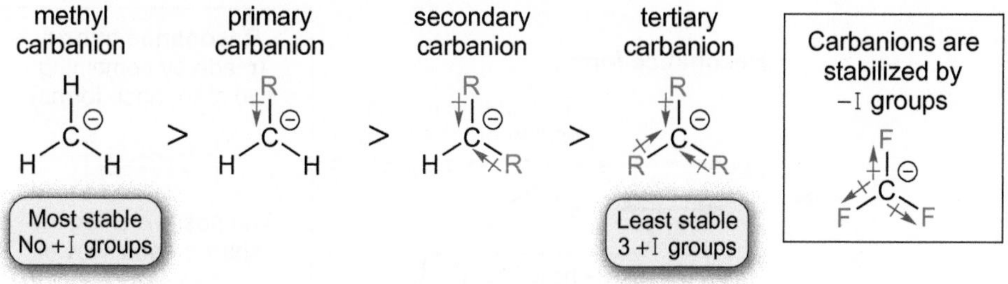
methyl carbanion
primary carbanion
secondary carbanion
tertiary carbanion
Most stable
No +I groups
Least stable
3 +I groups
Carbanions are stabilized by −I groups

Figure 19.10 Relative stabilities of carbanions.

The relative stabilities of carbanions can be determined by comparing the pK_a values of the anions' conjugate acids. For example, methane (CH_4) has a very high pK_a value of 48, whereas trifluoromethane (fluoroform, CHF_3) has a lower pK_a value of around 31. As CHF_3 has the lower pK_a value, it is a stronger acid than CH_4 (i.e. CHF_3 is more easily deprotonated), which means that $^-CF_3$ is more stable than $^-CH_3$ (Section 19.2, p.888).

electron. Tertiary carbon-centred radicals ($R_3C^•$) are more stable than secondary ($R_2CH^•$), primary ($RCH_2^•$), and methyl ($CH_3^•$) radicals, as electron-donating alkyl groups stabilize the radical by inductive effects and hyperconjugation.

In support of this explanation, the order of stability of carbanions in Figure 19.10 is exactly the reverse of that for carbocations and carbon-centred radicals. A tertiary carbanion (R_3C^-) is less stable than secondary (R_2CH^-), primary (RCH_2^-), and methyl (CH_3^-) carbanions because the electron-donating (+I) alkyl groups increase the negative charge and so destabilize the carbanion.

However, electron-withdrawing groups stabilize carbanions by inductive effects. An electron-withdrawing –I group, such as an F atom, attracts electrons from the negatively charged carbon causing a partial negative charge to develop on the fluorine atom. The negative charge in the carbanion is no longer localized on one carbon, so the carbanion is stabilized by delocalization.

In general, the more spread out (delocalized) the negative charge, the more stable the carbanion is. For example, $^-CH_2NO_2$ is more stable than $^-CH_2CH_3$ because the negative charge is spread over the adjacent NO_2 group (Section 19.2, p.889).

Mesomeric effects

Whereas inductive effects push and pull electrons in the σ *bonds* of organic molecules, **mesomeric effects** involve the delocalization of electron density through π *bonds*. Atoms, or groups of atoms, are classified as having a **negative mesomeric effect (–M)** or a **positive mesomeric effect** (+M). Like inductive effects, mesomeric effects can lead to stabilization of carbocations, carbanions, and radicals (and also neutral non-radical molecules) by delocalization.

–M groups accept electrons and +M groups donate electrons.

Representative examples of important –M and +M groups are shown in Figure 19.11. Groups that contain double bonds or triple bonds, often incorporating an electronegative atom, are –M groups because they can *accept* a pair of electrons. Groups with single bonds to atoms that contain a lone pair of electrons, or a C=C π bond, are classed as +M groups because they can *donate* a pair of electrons. This means that alkenyl (R_2C=C(R)–) and phenyl (Ph, C_6H_5–) groups can be either –M or +M groups depending on what groups they are bonded to.

Electron-donating +M groups, such as a C=C bond, stabilize adjacent carbocations by donation of electrons from the π bond (if the *p* orbital of the carbocation and the π bond are properly aligned (see p.885)). The delocalization of a pair of electrons gives a different **resonance form** of the same cation. Curly arrows (⌒) are used to draw resonance forms, which contribute to the overall delocalized structure. As a result, mesomeric effects are

Stabilization of neutral molecules or ions by delocalization of bonding or non-bonding electrons is called **resonance** (Section 5.5, p.243). Strictly, hyperconjugation is a form of resonance, but resonance normally refers to mesomeric effects (sometimes called **resonance effects**) and the delocalization of electrons in π bonds.

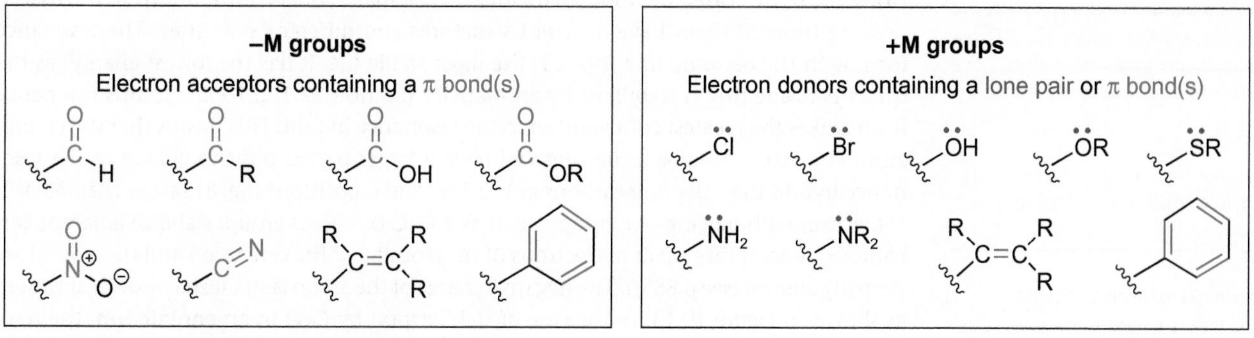
–M groups
Electron acceptors containing a π bond(s)
+M groups
Electron donors containing a lone pair or π bond(s)

Figure 19.11 Some –M and +M groups.

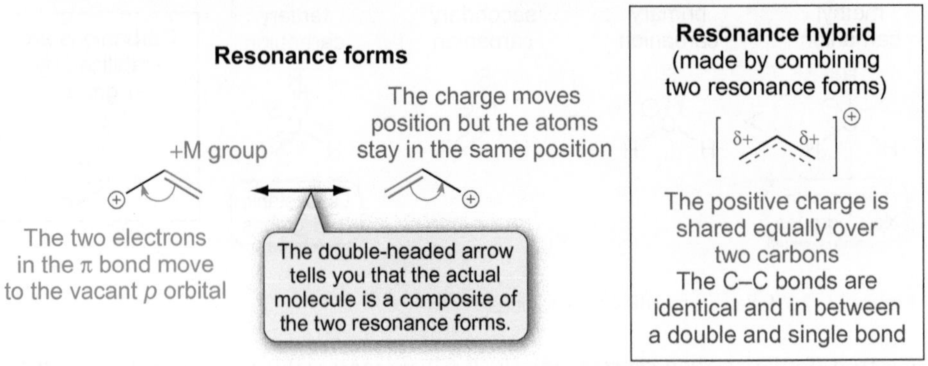
Resonance forms
+M group
The charge moves position but the atoms stay in the same position
The two electrons in the π bond move to the vacant p orbital
The double-headed arrow tells you that the actual molecule is a composite of the two resonance forms.
Resonance hybrid (made by combining two resonance forms)
δ+
δ+
The positive charge is shared equally over two carbons
The C–C bonds are identical and in between a double and single bond

Figure 19.12 Resonance stabilization of the allyl cation.

Visit the online resource centre to view screencast 19.1 which walks you through some examples of resonance-stabilization of neutral molecules and ions. It also shows how resonance can explain the reactivity of enol ethers (e.g. $ROCH{=}CH_2$) and enones (e.g. $RCOCH{=}CH_2$), and the ^1H NMR chemical shifts of substituted benzenes.

Resonance forms show all the possible distributions of electrons in a molecule or ion (Section 5.5, p.243). All resonance forms must have the same overall charge and obey the normal rules of valency (i.e. resonance forms with five bonds to carbon are not possible).

Resonance forms are sometimes called **resonance structures** or **canonical** forms.

A **resonance hybrid** is a molecule, ion, or radical that is a combination of a number of contributing resonance forms.

Benzyl and allyl cations are formed in S_N1 reactions of benzyl halides, $PhCH_2X$, and allyl halides, $H_2C{=}CHCH_2X$, respectively (where X = Cl, Br, or I). See Section 20.3, p.935.

The more stable a resonance form is, the greater its contribution to the resonance hybrid. In general, the most stable resonance forms have:

- an aromatic ring;
- the greatest number of covalent bonds; and/or
- atoms with a complete valence shell of electrons.

The formation of enolate ions (e.g. $RCOCH_2^-$) and their reactions with electrophiles are discussed in Section 23.3 (p.1082).

sometimes called resonance effects and the molecule is said to be **resonance-stabilized**. This is shown in Figure 19.12 for the allyl cation, $H_2C{=}CH{-}CH_2^+$.

In both resonance forms of the allyl cation in Figure 19.12 the position of the atoms is the same. The only difference is the position of the electrons and, consequently, the positive charge. Note that the resonance forms do *not* rapidly interconvert. The two-headed straight arrow, ⟷, (*not* an equilibrium arrow ⇌) connecting the resonance forms tells you this. Each resonance form shows the limit of the electron distribution in the cation and the actual structure of the cation lies in between the two resonance forms. The actual structure, called a **resonance hybrid**, shows that the cation is stabilized by delocalization of the charge over two carbon atoms.

You can use the same approach to show that the benzyl cation ($PhCH_2^+$) is stabilized by a +M effect. This carbocation is stabilized by delocalization of the positive charge around the 2, 4, and 6 positions of the benzene ring. In total, you can draw four resonance forms, two of which are equivalent. *As a rule of thumb, the more resonance forms you can draw for an ion or neutral molecule, the more stable it is.* Therefore, you would expect the benzyl cation to be more stable than the allyl cation because it has two additional resonance forms. As a result, the positive charge in the benzyl cation is more delocalized.

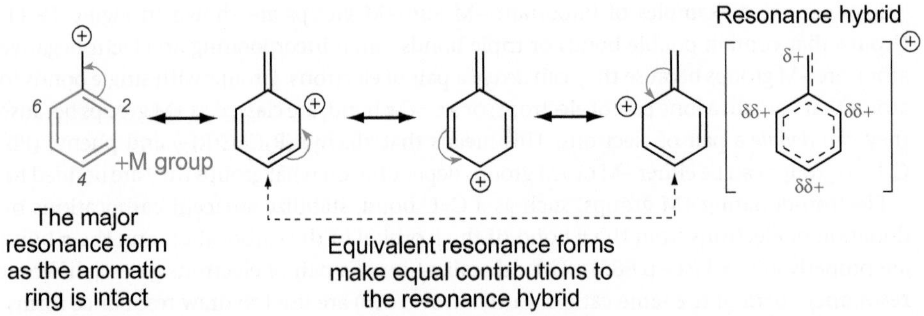
Resonance hybrid
6
2
4
+M group
δ+
δδ+
δδ+
δδ+
The major resonance form as the aromatic ring is intact
Equivalent resonance forms make equal contributions to the resonance hybrid

The actual structure of the benzyl cation is a hybrid of the four resonance forms. However, in this case, the resonance forms do not make equal contributions to the hybrid because three of them have different structures and different stabilities. The resonance form with the benzene ring intact is the most stable (i.e. it has the lowest energy) as the intact benzene ring is stabilized by aromaticity (Section 22.1, p.1005). So this resonance form makes the greatest contribution to the resonance hybrid. This means that the carbon atom attached to the benzene ring will have a larger partial positive charge in the resonance hybrid than the ring carbons at the 2, 4, and 6 positions (i.e. δ+ rather than δδ+).

Electron-withdrawing –M groups, such as a C=C or a C=O group, stabilize adjacent carbanions by accepting a pair of electrons (if the *p* orbital of the carbanion and the π bond are properly aligned (see p.885)). The negative charge of the anion is stabilized by delocalization, as shown in Figure 19.13. In the case of stabilization by C=O in an **enolate ion**, the resonance form with the negative charge on the electronegative oxygen atom is the more stable.

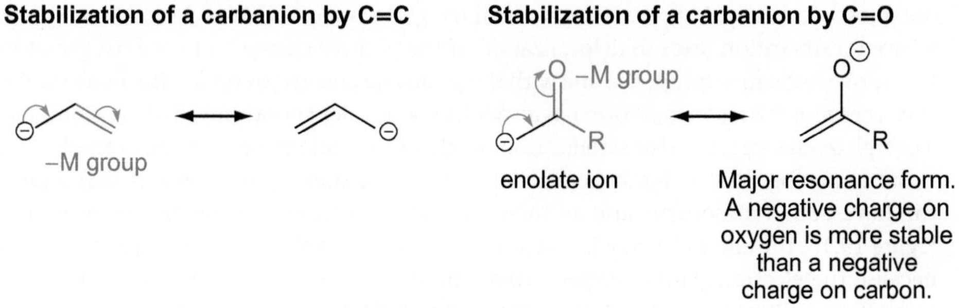
Stabilization of a carbanion by C=C
−M group
Stabilization of a carbanion by C=O
−M group
R
R
enolate ion
Major resonance form.
A negative charge on
oxygen is more stable
than a negative
charge on carbon.

Figure 19.13 Resonance stabilization of carbanions.

Reactions of amides ($RCONH_2$, RCONHR, and $RCONR_2$) are discussed in Section 24.2 (p.1123).

Resonance explains why amides are protonated on oxygen, rather than on nitrogen (Section 24.2, p.1123), and why enol ethers are protonated on carbon, rather than on oxygen.

Neutral molecules also have their stability and reactivity influenced by mesomeric effects. If neutral non-radical molecules contain both a +M and a −M group and these groups are appropriately positioned so that the −M group can accept the electrons donated by the +M group, then the molecule is stabilized by resonance. For example, the resonance forms of an amide ($RCONR_2$) and an enol ether ($ROCH{=}CH_2$) are shown in Figure 19.14. In both cases, the resonance forms without charges make a larger contribution to the resonance hybrids. This is because separating opposite charges requires energy.

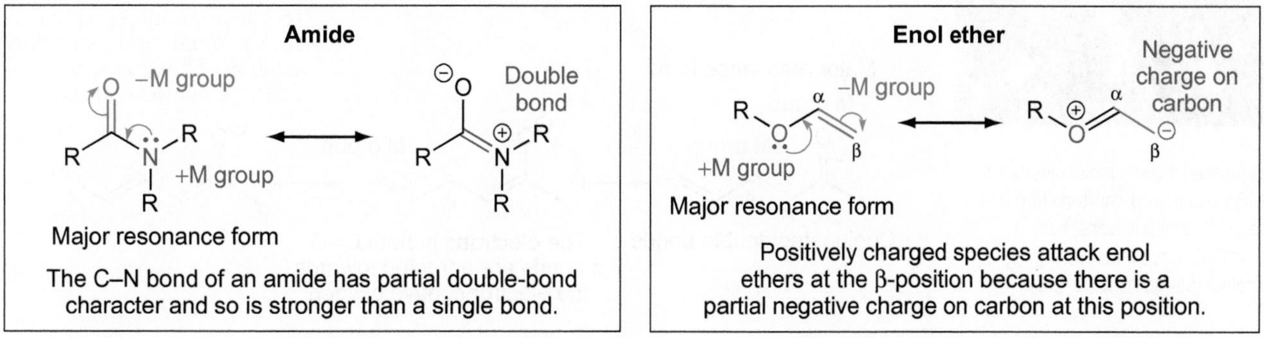
Amide
−M group
+M group
R
R
R
Double bond
R
R
R
Major resonance form
The C–N bond of an amide has partial double-bond character and so is stronger than a single bond.
Enol ether
−M group
+M group
α
β
R
Negative charge on carbon
α
β
R
Major resonance form
Positively charged species attack enol ethers at the β-position because there is a partial negative charge on carbon at this position.

Figure 19.14 Examples of neutral molecules stabilized by mesomeric effects.

Competing mesomeric and inductive effects

Sometimes mesomeric and inductive effects work in the same direction but more often they oppose one another. Atoms carrying a lone pair of electrons are +M groups and these groups can stabilize adjacent carbocations by donating the lone pair. For example, in Figure 19.15

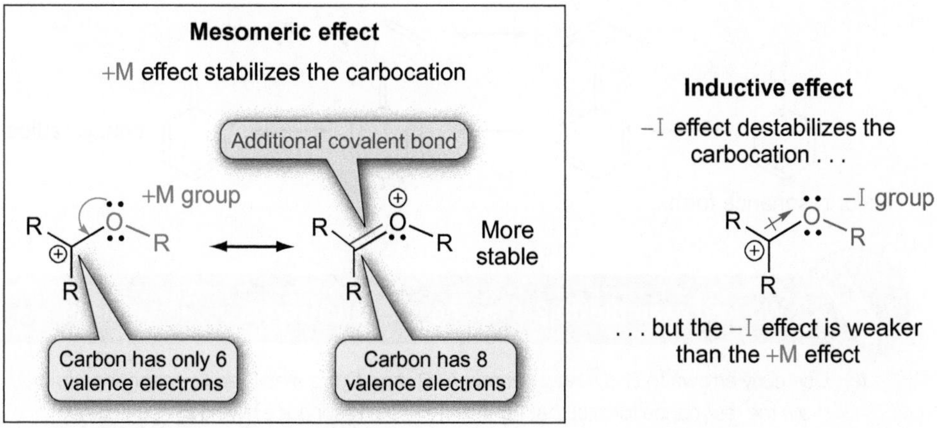
Mesomeric effect
+M effect stabilizes the carbocation
Additional covalent bond
+M group
R
R
R
R
R
R
More stable
Carbon has only 6 valence electrons
Carbon has 8 valence electrons
Inductive effect
−I effect destabilizes the carbocation . . .
−I group
R
R
R
. . . but the −I effect is weaker than the +M effect

Figure 19.15 The +M effect of the alkoxyl group stabilizes the carbocation, while the −I effect destabilizes it. The +M effect is stronger.

The alkoxymethyl cation ($ROCH_2^+$) is formed in an S_N1 reaction of an alkoxymethyl halide, $ROCH_2X$ (where X = Cl, Br or I). See Section 20.3, p.935.

Notice that in $ROCR_2^+$, the two R groups bonded to the positively charged carbon stabilize the positive charge by +I effects.

donation of a lone pair of electrons from an oxygen atom of an alkoxyl group (–OR) to an adjacent carbocation leads to delocalization of the positive charge on to the oxygen atom. Of the two resonance forms, the one with the positively charged oxygen is the more stable—this is because this resonance form has an additional covalent bond and all of the atoms have a complete valence shell. (For stabilization of other carbocations, see Worked example 19.2.)

But remember that inductive effects also affect the stability of ions and that a group can have both a mesomeric and an inductive effect at the same time. In the example in Figure 19.15, the alkoxyl group is a –I group and will destabilize the adjacent carbocation because the electronegative oxygen atom will attract the electrons in the C–O σ bond. Overall, the alkoxyl group stabilizes the carbocation because electron donation by the mesomeric effect is stronger than electron withdrawal by the inductive effect (+M > –I). This is a general point and, usually, the +M or –M effect of an atom or group of atoms has a stronger electronic effect than the +I or –I effect of the same group.

ⓘ The stability of ions depends on both the mesomeric and inductive effects of the substituents. Usually, mesomeric effects are stronger than inductive effects so a +M group is likely to stabilize a cation more effectively than a +I group.

Mesomeric effects in conjugated molecules

Mesomeric effects can be effective over much longer distances than inductive effects, provided that the molecules are conjugated. The alternate single and double bonds in conjugated molecules are appropriately positioned for the movement of pairs of electrons, as shown in Figure 19.16 (for (*E*)-hexa-3,5-dien-2-one).

▲ Highly conjugated organic molecules can be coloured. For example, β-carotene (present in vegetables, including pumpkins) has 11 conjugated C=C bonds and is a strong red-orange colour (Section 10.6, p.495).

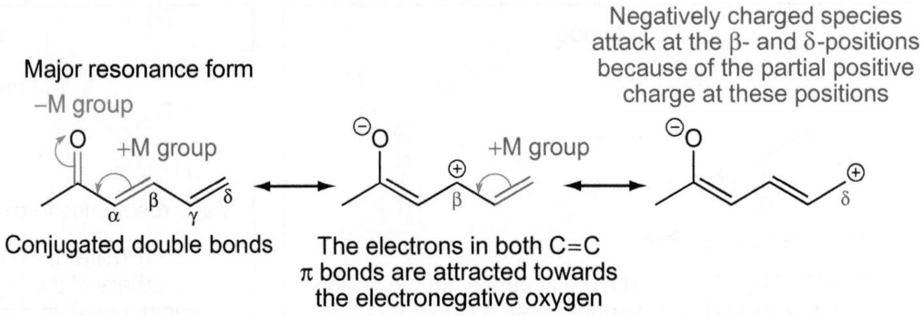

Negatively charged species attack at the β- and δ-positions because of the partial positive charge at these positions
Major resonance form
–M group
+M group
α
β
γ
δ
+M group
β
δ
Conjugated double bonds
The electrons in both C=C π bonds are attracted towards the electronegative oxygen

Figure 19.16 Movement of pairs of electrons in conjugated molecules.

ⓘ **Conjugated molecules** contain alternating single and double bonds. The term conjugation describes the interaction of a π bond with another π bond or a *p* orbital. For buta-1,3-diene ($H_2C=CH-CH=CH_2$) the two π bonds interact, whereas in the allyl cation ($CH_2=CH-CH_2^+$) the π bond interacts with an empty *p* orbital. Therefore, both buta-1,3-diene and the allyl cation can be described as conjugated.

The unpaired electron of a carbon-centred radical can also be stabilized by delocalization over an adjacent π bond if the *p* orbital of the carbon radical and the π bond are properly aligned (see p.885). Allyl and benzyl radicals ($H_2C=CH-CH_2^{\bullet}$ and $PhCH_2^{\bullet}$, respectively), are important examples. However, the unsaturated groups (attached to the carbon-centred radical) are not described as +M or –M groups because the delocalizations involve the movement of a single electron and not a pair of electrons.

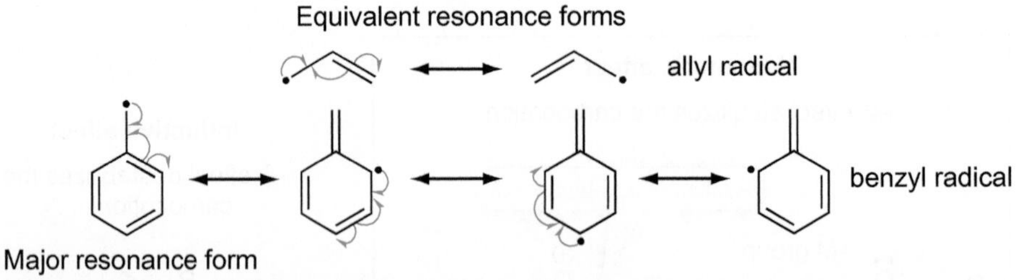

Equivalent resonance forms
allyl radical
benzyl radical
Major resonance form

ⓘ As allyl and benzyl radicals are so stable, they are often easier to form than other carbon-centred radicals. This explains why cyclohexene forms 3-bromocyclohex-1-ene on reaction with Br_2 and UV radiation (Worked example 20.1, p.922).

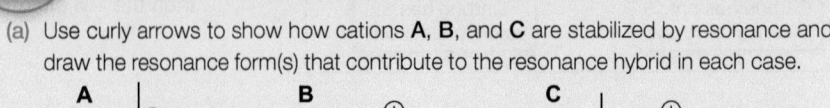

Worked example 19.2 Drawing resonance forms of carbocations

(a) Use curly arrows to show how cations **A**, **B**, and **C** are stabilized by resonance and draw the resonance form(s) that contribute to the resonance hybrid in each case.

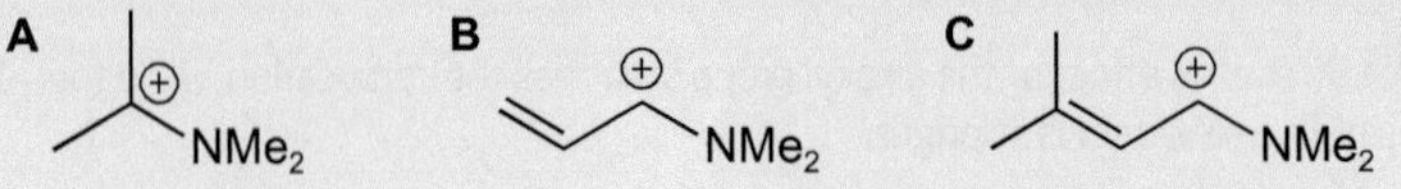

A
NMe2
B
NMe2
C
NMe2

→

→

(b) Considering both mesomeric and inductive effects, would you expect **A**, **B**, or **C** to be the more stable? Briefly explain your reasoning.

Strategy

(a) Classify the groups attached to the positively charged carbons as +M or −M groups. +M groups are able to delocalize an adjacent positive charge.

The nitrogen atom in the amine group (NMe_2) has a lone pair of electrons and it acts as a +M group.

The C=C bond also acts as a +M group by donating two electrons from the π bond.

Draw curly arrows from the lone pair on nitrogen or from the C=C bond to stabilize the charge. Check that you have drawn all the possible resonance forms for each cation.

(b) Remember these rules.

- The inductive and mesomeric effects of all substituents attached to the positively charged carbon need to be considered.
- Generally, mesomeric effects are more effective in stabilizing a carbocation than inductive effects.
- As a rule of thumb, the more resonance forms a cation has, the more stable it is (because the positive charge is more delocalized).

Solution

(a) +M groups are shown in green

A

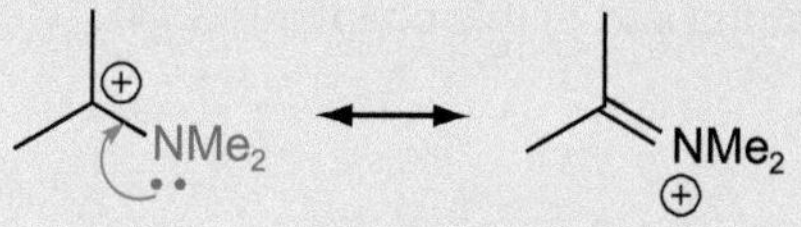

B

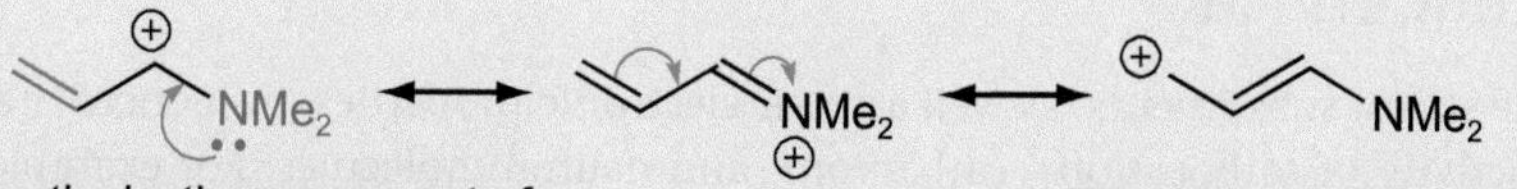

Alternatively, the movement of electrons in the C=C bond could be drawn first (as shown for C)

(A mixture of *E*– and *Z*– isomers is possible)

C

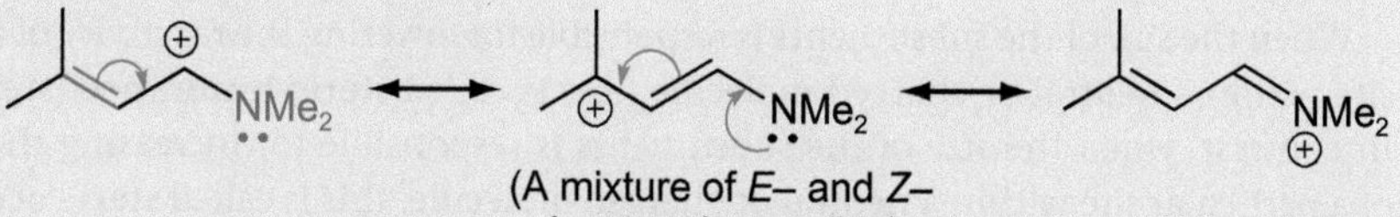

(A mixture of *E*– and *Z*– isomers is possible)

(b) Carbocation **C** is expected to be the most stable. **A** has the +M effect of the amine and the +I effects of the two methyl groups, but **C** has two +M groups and so an additional resonance form can be drawn for **C**. In comparison to **B**, **C** has two additional methyl groups that stabilize the carbocation by +I effects as shown here.

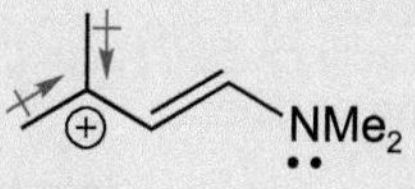

Question

By considering mesomeric effects, explain why $PhCH{=}CHCH_2^+$ is a more stable carbocation than $CH_3CH{=}CHCH_2^+$.

Summary

- Inductive effects and hyperconjugation refer to the polarization of electrons in σ bonds, while mesomeric effects refer to the delocalization of lone pairs of electrons and electrons in π bonds.
- +I/+M groups are electron-donating groups that stabilize carbocations and −I/−M groups are electron-withdrawing groups that stabilize carbanions.

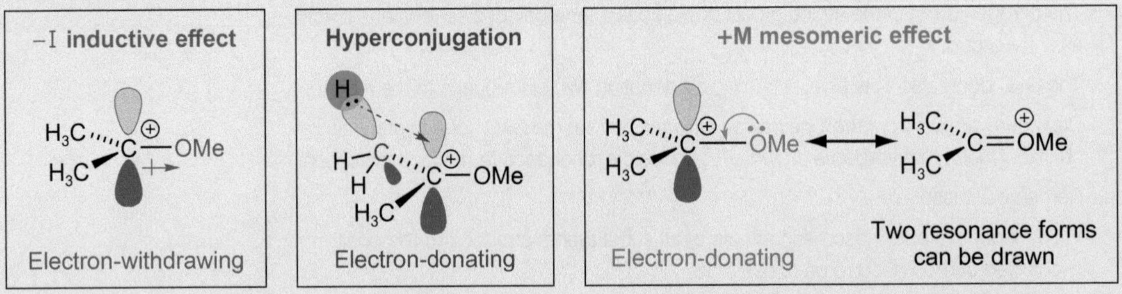
−I inductive effect
H_3C
H_3C
C
OMe
Electron-withdrawing
Hyperconjugation
H
C
OMe
H_3C
Electron-donating
+M mesomeric effect
H_3C
H_3C
C
OMe
C=OMe
Electron-donating
Two resonance forms can be drawn

- A resonance hybrid is a molecule, ion, or radical that is a combination of a number of contributing resonance forms.
- Sometimes mesomeric and inductive effects work in the same direction, but more often they oppose one another.
- Mesomeric effects are generally stronger, and can be more effective over longer distances, than inductive effects and hyperconjugation.
- Conjugated molecules contain alternating single and double bonds.

For practice questions on these topics, see questions 1–4 at the end of this chapter (p.914).

Steric effects

Steric hindrance (steric strain) is also discussed in Section 18.2 (p.819).

Steric hindrance explains why tertiary halogenoalkanes (R_3CX) do not undergo S_N2 reactions (Section 20.3, p.930). There is steric acceleration in the S_N1 reaction (Section 20.3, p.933).

The size of substituents, as well as their electronic properties, influences the stability and reactivity of carbocations, carbanions, and neutral molecules. For example, when large bulky groups surround a carbon-centred radical, this reduces the susceptibility of the radical to attack by other groups. This is because the bulky groups hinder the approach of other groups by **steric effects**.

When the size of the substituents is responsible for lowering the reactivity of a particular site within a neutral or charged molecule, this is called **steric hindrance (steric strain)**. In contrast, when the size of the substituents is responsible for increasing the reactivity of a particular site within a neutral or charged molecule, this is called **steric acceleration**. Examples of steric hindrance and steric acceleration are shown in Figure 19.17.

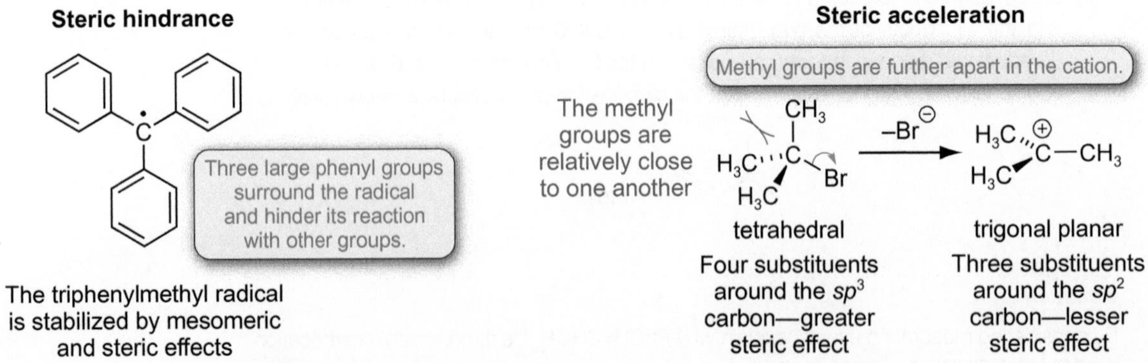
Steric hindrance
C
Three large phenyl groups surround the radical and hinder its reaction with other groups.
The triphenylmethyl radical is stabilized by mesomeric and steric effects
Steric acceleration
Methyl groups are further apart in the cation.
The methyl groups are relatively close to one another
CH_3
H_3C
C
Br
H_3C
−Br
H_3C
C−CH_3
H_3C
tetrahedral
Four substituents around the sp^3 carbon—greater steric effect
trigonal planar
Three substituents around the sp^2 carbon—lesser steric effect

Figure 19.17 Steric effects: steric hindrance and steric acceleration.

Nucleophiles and electrophiles

In some polar reaction mechanisms, negatively charged ions react with positively charged ions to form covalent bonds. The negatively charged ion acts as the electron donor and the positively charged ion acts as the electron acceptor. The charges in the reactants may be full-blown charges, as shown here, or partial charges (δ+ or δ–) caused by polarization of electrons within a molecule. Rather than using the terms 'electron donor' and 'electron acceptor', organic chemists classify reactants as nucleophiles or electrophiles, respectively.

The name nucleophile comes from the Greek words for nucleus (*nucleo*) and loving (*philos*). Electrophile comes from the Greek words for electron (*electros*) and loving (*philos*).

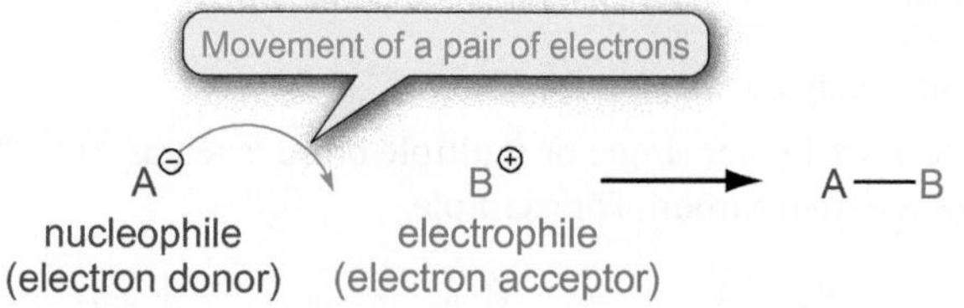
Movement of a pair of electrons
A
B
A—B
nucleophile
(electron donor)
electrophile
(electron acceptor)

- Nucleophiles are negatively charged ions or neutral molecules that *donate* a pair of electrons (to an electron acceptor) to form a covalent bond (e.g. Br^-, R_3C^-).
- Electrophiles are positively charged ions or neutral molecules that *accept* a pair of electrons (from an electron donor) to form a covalent bond (e.g. R_3C^+).

Nucleophiles are Lewis bases and electrophiles are Lewis acids (see Section 7.8, p.336).

Spotting which reactant is the nucleophile and which is the electrophile is fundamental to drawing curly arrow mechanisms. The double-headed curly arrow is drawn from the nucleophile towards the electrophile.

Notice that the curly arrows point to where the new bonds are being formed. For some mechanisms it is clearer to show the curly arrow pointing directly to the cation (p.864).

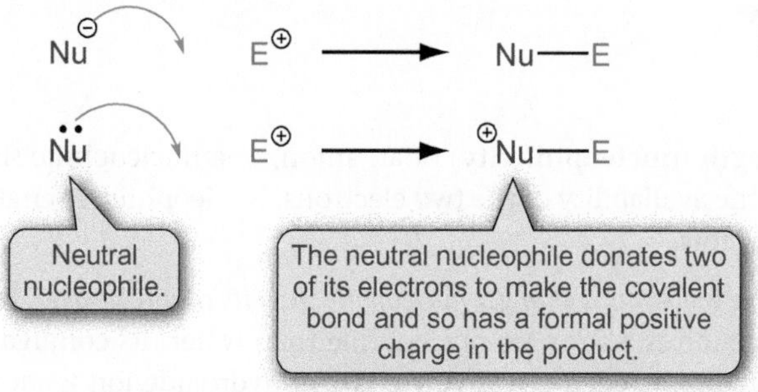
Nu
E
Nu—E
Nu
E
Nu—E
Neutral nucleophile.
The neutral nucleophile donates two of its electrons to make the covalent bond and so has a formal positive charge in the product.

Nu represents a nucleophile and E represents an electrophile

The overall charge of the reactants must be the same as that of the products

In terms of frontier orbitals, the (filled) highest occupied molecular orbital (HOMO) of the nucleophile interacts with the (empty) lowest unoccupied molecular orbital (LUMO) of the electrophile (Section 4.10, p.196). The best nucleophiles have high-energy HOMOs and the best electrophiles have low-energy LUMOs (see p.880).

For reactions of ions, it is relatively easy to spot the nucleophile and electrophile but this is not always the case for reactions involving neutral molecules. The most difficult situation is when a neutral nucleophile donates a pair of electrons to a neutral electrophile to form a covalent bond.

For neutral molecules, different sites within the molecules are classified as being nucleophilic (electron-rich sites) or electrophilic (electron-deficient sites).

Nucleophilic sites may have:

- a heteroatom, such as nitrogen or oxygen, with a lone pair of electrons;
- a multiple carbon–carbon bond, such as C=C in alkenes or C≡C in alkynes, which are of high electron density (single C–C bonds are not nucleophilic sites);
- a carbon atom in a polar single bond, such as $^{\delta-}C–Z^{\delta+}$, where carbon is more electronegative than Z.

Other heteroatoms that are nucleophilic sites include the phosphorus atom in triphenylphosphine (Ph_3P:) and the sulfur atom in a thiol (RSH)—the sulfur atom has two lone pairs.

$R_3C–Li$ is called an organolithium compound. The extremely polar $^{\delta-}C–Li^{\delta+}$ bond makes organolithium compounds strong nucleophiles and strong bases (Section 23.2, p.1068).

For example,

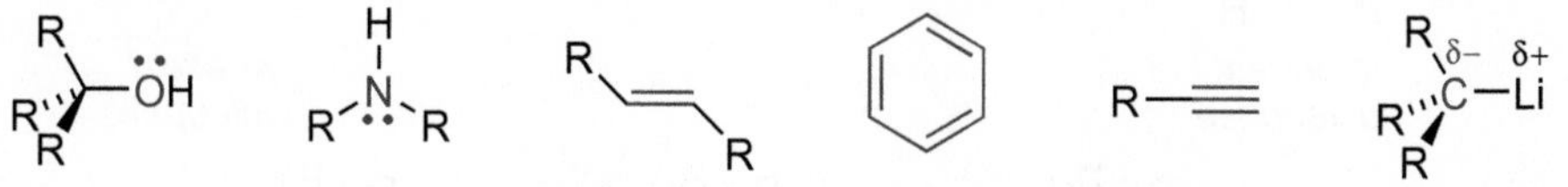
R
OH
R
R
H
N
R
R
R
R
R—
R
δ–
δ+
C—Li
R
R

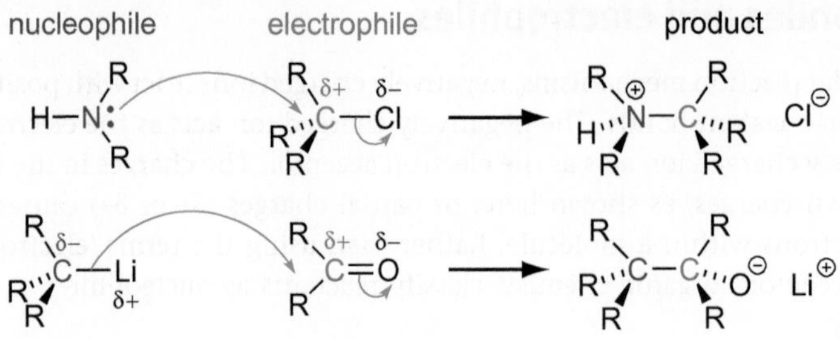

nucleophile
electrophile
product

Figure 19.18 Examples of reactions of neutral nucleophiles with neutral electrophiles.

Electrophilic sites may have:

- a carbon atom in a polar single or multiple bond, such as $^{\delta+}C\text{–}X^{\delta-}$, where X is more electronegative than carbon. For example,

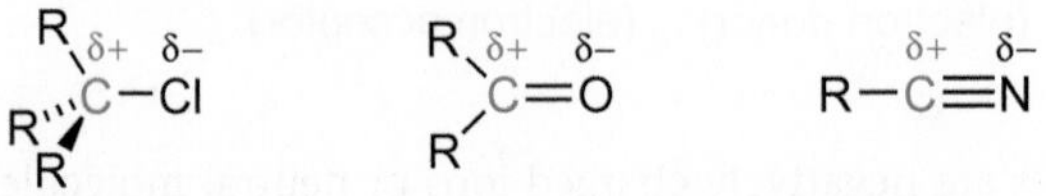

Examples of reactions of neutral nucleophiles with neutral electrophiles, to form new C–N and C–C bonds, are shown in Figure 19.18. For practice in identifying nucleophiles and electrophiles see Worked example 19.3 (p.883)

For reactions involving polyfunctional molecules, there can be more than one nucleophile/nucleophilic site and more than one electrophile/electrophilic site. To determine which particular nucleophile donates a pair of electrons and which electrophile accepts a pair of electrons, you need to understand the factors that influence the relative strengths of nucleophiles and electrophiles.

The reaction of R_2NH with R_3CCl is an example of a nucleophilic *substitution* reaction; the reaction of R_3CLi with R_2CO is an example of a nucleophilic *addition* reaction (see Section 19.2, p.903). For the reaction involving R_3CLi notice that the curly arrow starts from the centre of the C–Li bond.

Nucleophilic strength

The relative **nucleophilic strength** (**nucleophilicity**) of an anion, or a nucleophilic site within a molecule, depends on the availability of the two electrons. Nucleophilic strength is influenced by the following factors.

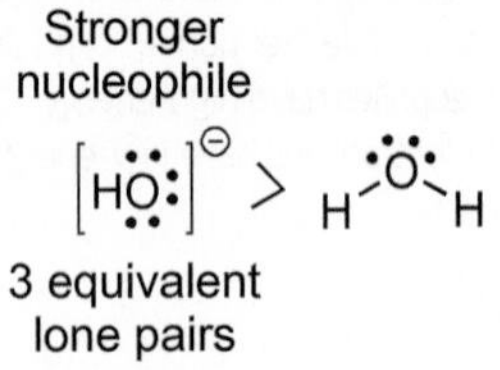

Stronger nucleophile
3 equivalent lone pairs

1. **Charge.** *A negatively charged ion is a stronger nucleophile than its conjugate acid.* For example, the hydroxide ion is a stronger nucleophile than water (its conjugate acid). As you may have predicted, a negatively charged hydroxide ion is more strongly attracted to an electrophile than a neutral water molecule.

For an introduction to conjugate acids see Section 7.1 (p.304).

$$B^{\ominus} + H^{\oplus} \rightleftharpoons BH$$

base — conjugate acid

2. **Electronegativity and electronic effects.** *Highly electronegative atoms are weak nucleophiles, whereas less electronegative atoms (particularly those bonded to +I groups) are stronger nucleophiles.* The nucleophilic strengths of anions within the same row of the Periodic Table follow the same order as electronegativity. For example, the fluoride ion is a much weaker nucleophile than the methyl anion (Figure 19.19). This is because the electrons on fluorine are held much tighter to the nucleus than the electrons on carbon.
   - The lower electronegativity of nitrogen compared to oxygen also explains why amines (e.g. $R\text{–}NH_2$) are stronger nucleophiles than alcohols (R–OH).

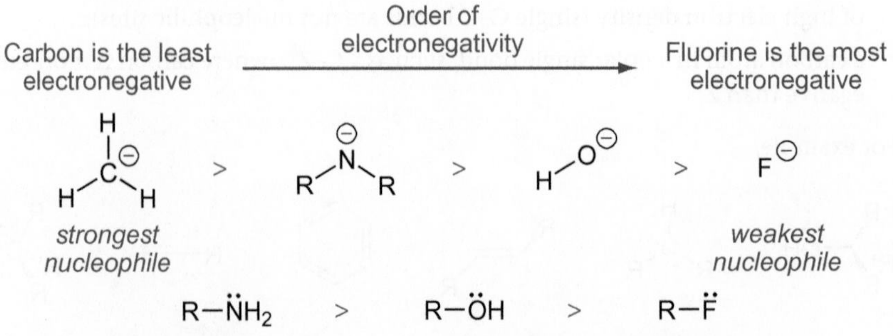

Carbon is the least electronegative
Order of electronegativity
Fluorine is the most electronegative
strongest nucleophile
weakest nucleophile

Figure 19.19 Nucleophilic strength decreases with increasing electronegativity.

- In general, electron-donating +I groups increase the nucleophilicity of neighbouring groups (although this depends on the size of the +I group, see point 3 below). Conversely, electron-withdrawing –I groups decrease the nucleophilicity of neighbouring groups.

ⓘ H_2N^- is called an amide ion and R_2N^- is a dialkylamide ion. The use of 'amide' for these ions is not to be confused with the same name for the carboxylic acid derivatives $RCONH_2$, RCONHR, and $RCONR_2$.

ⓘ Anions can act as nucleophiles and/or bases. When the anion donates a pair of electrons to an atom other than hydrogen it acts as a **nucleophile**. When the anion donates a pair of electrons to hydrogen (or H^+) it acts as a Brønsted–Lowry **base** (see Section 7.1, p.304). The basicity of anions within the same row of the Periodic Table follows the same order as nucleophilic strength.

For example, H_2N^- is a stronger base and a stronger nucleophile than RO^-.

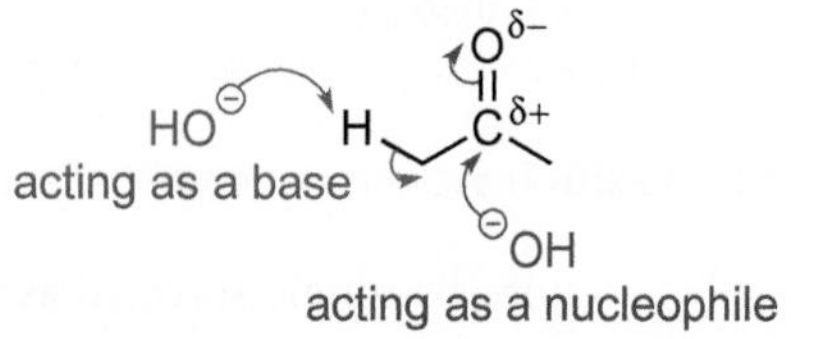
HO
acting as a base
OH
acting as a nucleophile

3 **Polarizability.** *Large nucleophilic sites are often strong nucleophiles.* The nucleophilic strength of neutral nucleophilic sites increases on going down a group of the Periodic Table. For example, an alcohol (ROH) is a much weaker nucleophile than a thiol (RSH). This is because the pairs of electrons on oxygen are held tighter to the nucleus than the electrons in the larger sulfur atom. As the electrons in a larger sulfur atom are not held as strongly, they can have a stronger interaction with the electrophile. For charged nucleophiles, of different sizes, the order of nucleophilicity can depend on the solvent; for example, the larger iodide ion is a stronger nucleophile than a fluoride ion in the solvent methanol. Relatively large atoms, such as sulfur and iodine, are said to have a higher polarizability than small atoms, such as oxygen and fluorine.

- For molecules, such as thiols (RSH), alcohols (ROH), amines (RNH_2, R_2NH, R_3N), and alkoxide ions (RO^-), the size of the groups bonded to the attacking S, O, or N atom affects the nucleophilicity. For example, the methoxide ion (CH_3O^-) is a stronger nucleophile than the *tert*-butoxide ion (Me_3CO^-). In Me_3CO^-, the large Me_3C group hinders the oxygen atom from approaching an electrophile (see Section 20.5, p.954). So, molecules with large groups close to the nucleophilic site are usually weak nucleophiles.

ⓘ **Polarizability** is a measure of how the electron cloud of an uncharged molecule distorts when placed in an electric field. In an electric field, the electrons tend to be drawn away from the nucleus and larger atoms are more polarizable than small atoms because their electron clouds are more easily distorted (see Sections 1.8, p.13, and 17.3, p.785).

ⓘ Highly polarizable nucleophiles, with a negative charge on a large atom (e.g. I^-, RS^-), are sometimes called **soft nucleophiles**. Nucleophiles with a negative charge on a small atom (e.g. F^-, HO^-) are sometimes called **hard nucleophiles**. Hard and soft metals and ligands are discussed in Section 28.3 (p.1277).

4 **Solvation.** *Anions are generally more nucleophilic in polar aprotic solvents than in polar protic solvents.* You might expect anions to be stronger nucleophiles in non-polar solvents than in polar solvents. In a non-polar solvent, the anion and the non-polar solvent do not interact, whereas in a polar solvent an interaction takes place, which reduces the reactivity of the anion. But, ions are insoluble in most non-polar solvents so reactions of ions are usually carried out in polar solvents. The extent to which an anion interacts with a polar solvent depends on whether the solvent is a polar protic solvent or a polar aprotic solvent.

- **Polar protic solvents** (such as methanol, MeOH, and ethanoic acid, $MeCO_2H$) contain polar O–H (or N–H) groups. When ions dissolve in a protic solvent, the O–H groups of the solvent molecules form hydrogen bonds to the anions and cations. A shell of protic solvent molecules surrounds the anion and this is called **solvation**. Solvation reduces the reactivity of the anion as the solvent molecules hinder attack at an electrophile (for a reaction, a certain amount of energy is required to remove the solvent shell around the anion). For halide ions, the F^- ion (with a negative charge concentrated in the smallest volume) is strongly solvated while the much larger I^- ion is weakly solvated. So, less energy is required to remove the solvent shell around I^- and it is a stronger nucleophile. Therefore, the order of nucleophilicity of halide ions is $I^- > Br^- > Cl^- > F^-$ (Figure 19.20).
- **Polar aprotic solvents**, such as dimethyl sulfoxide (Me_2SO), contain polar groups (e.g. a $^{\delta+}S{=}O^{\delta-}$ bond) but no O–H (or N–H) groups. When ions dissolve in an aprotic solvent, the polar groups in the aprotic solvent surround the ions and solvate them. For example, in Me_2SO, the δ– oxygen atom is attracted to a cation and the δ+ sulfur atom is attracted to an anion. However, the ions are

Solvation is introduced in Box 7.1 (p.305). Solvation of ions by polar water molecules is called **hydration** (see Section 1.8, p.54).

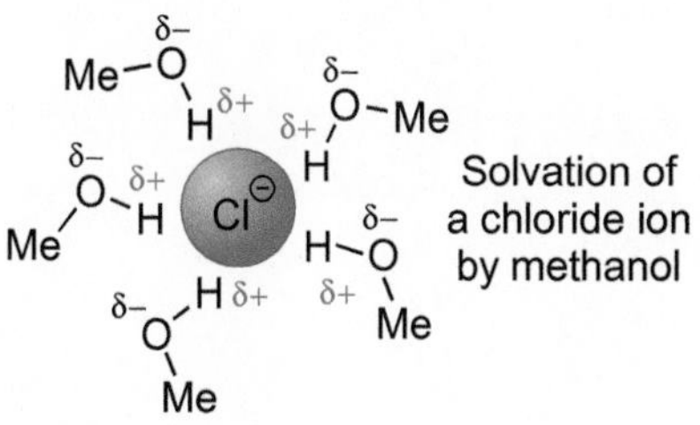
Me
Cl
Solvation of a chloride ion by methanol

ⓘ Polar protic solvents contain polar O–H or N–H groups, whereas polar aprotic solvents contain polar groups (e.g. an $^{\delta+}S{=}O^{\delta-}$ bond) but no O–H or N–H groups.

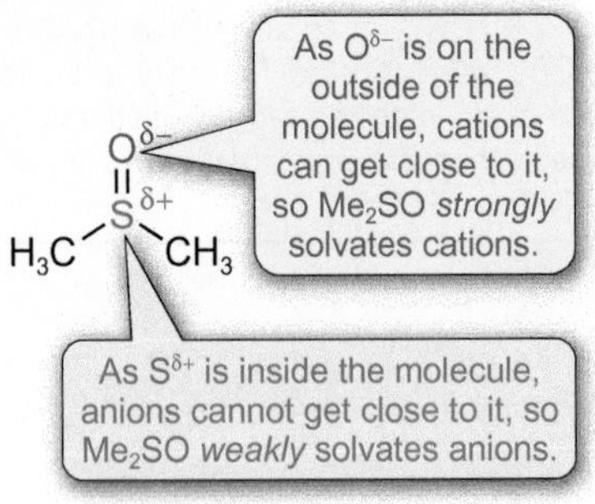
As $O^{\delta-}$ is on the outside of the molecule, cations can get close to it, so Me_2SO *strongly* solvates cations.
As $S^{\delta+}$ is inside the molecule, anions cannot get close to it, so Me_2SO *weakly* solvates anions.

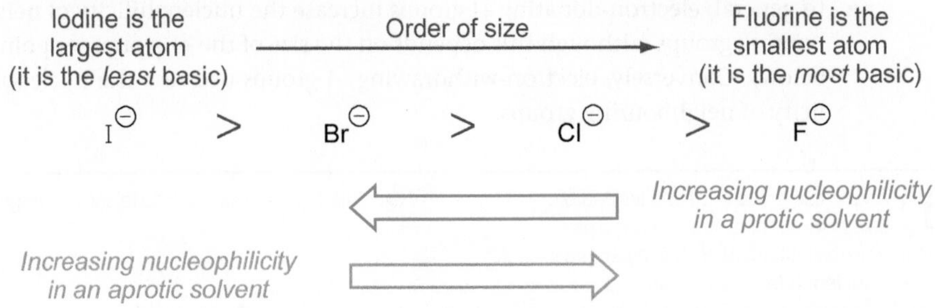
Iodine is the largest atom (it is the *least* basic)
Order of size
Fluorine is the smallest atom (it is the *most* basic)
Increasing nucleophilicity in a protic solvent
Increasing nucleophilicity in an aprotic solvent

Figure 19.20 The relative nucleophilic strengths of halide ions in polar protic and polar aprotic solvents.

not as strongly solvated as they are in a polar protic solvent. In Me_2SO, the anions are very weakly solvated because the $\delta+$ sulfur atom is in the middle of the Me_2SO molecule (it is surrounded by the two methyl groups and an oxygen atom), which means that it is difficult for an anion to get close to it. So, in Me_2SO the anion is 'naked' and it is a stronger nucleophile than when it is surrounded by a shell of protic solvent molecules. All halide ions are stronger nucleophiles in an aprotic solvent, but because the small F^- ion has the negative charge concentrated in the smallest volume, it is the least stable and so the strongest nucleophile (even though fluorine is highly electronegative). Therefore, the order of nucleophilicity of halide ions is $F^- > Cl^- > Br^- > I^-$ (which is the same order as the basicity of these ions; Figure 19.20).

The fact that F^- is less stable than I^- explains why it binds more strongly to H^+ (i.e. why F^- is a stronger base than I^-, see Section 25.2, p.1153).

Electrophilic strength

The relative **electrophilic strength** (**electrophilicity**) of a cation, or an electrophilic site within a molecule, depends on the stability of the cation or the size of the partial positive charge, which are influenced by the following factors.

An acid catalyst is often added to increase the rate of addition of a neutral nucleophile to a C=O bond (Section 23.2, p.1062).

1 **Charge.** *A positively charged ion is a stronger electrophile than its conjugate base.* For example, a protonated ketone ($R_2C{=}OH^+$, formed in the presence of acids) is a stronger electrophile than a ketone ($R_2C{=}O$, its conjugate base) because a nucleophile is more strongly attracted to the positive charge (Figure 19.21).

For an introduction to conjugate bases see Section 7.1 (p.304).

acid
conjugate base

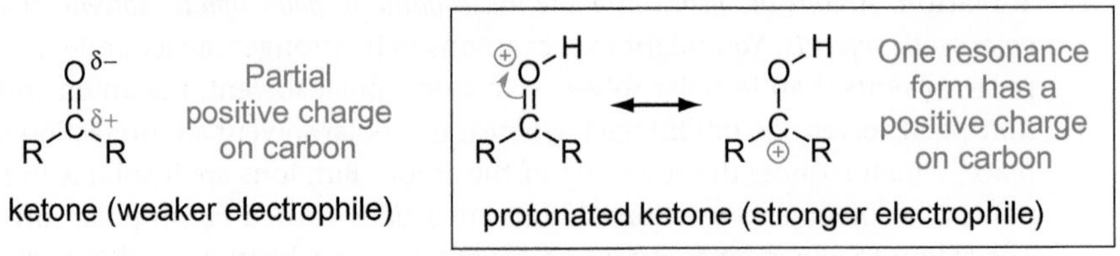
Partial positive charge on carbon
ketone (weaker electrophile)
One resonance form has a positive charge on carbon
protonated ketone (stronger electrophile)

Figure 19.21 A protonated ketone is a stronger electrophile than a ketone (its conjugate base).

2 Electronegativity and electronic effects. *Carbon atoms bonded to highly electronegative atoms are stronger electrophiles.* The more electronegative the atom bonded to carbon, the greater the size of the partial positive charge ($\delta+$) on carbon. For example, the carbon atom in the C=O bond of an acyl chloride (RCOCl) is more electrophilic than the carbonyl carbon in an acyl bromide (RCOBr) or ketone (RCOR), because the carbon is bonded to a strongly electronegative chlorine atom (Figure 19.22).

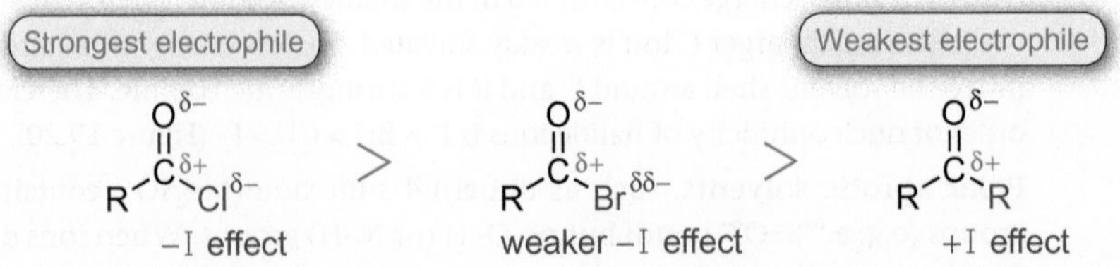
Strongest electrophile
Weakest electrophile
−I effect
weaker −I effect
+I effect

Figure 19.22 The carbon atom in the C=O bond of an acyl chloride is more electrophilic than the carbonyl carbon in an acyl bromide or ketone, because the carbon is bonded to a strongly electronegative chlorine atom.

- The electronegative chlorine atom exerts a strong –I effect (and a weak +M effect—see Section 22.3, p.1030) and, in general, electron-withdrawing –I and/or –M groups increase the electrophilicity of adjacent carbon atoms. Conversely, electron-donating +I and/or +M groups decrease the electrophilicity of adjacent carbon atoms.

The influence of inductive (+I and –I) and mesomeric (+M and –M) effects on carbonyl reactivity is discussed in more detail in Box 24.2 (p.1103).

Worked example 19.3 Spotting nucleophiles and electrophiles

(a) In the following reaction, which reactant is the nucleophile and which is the electrophile?

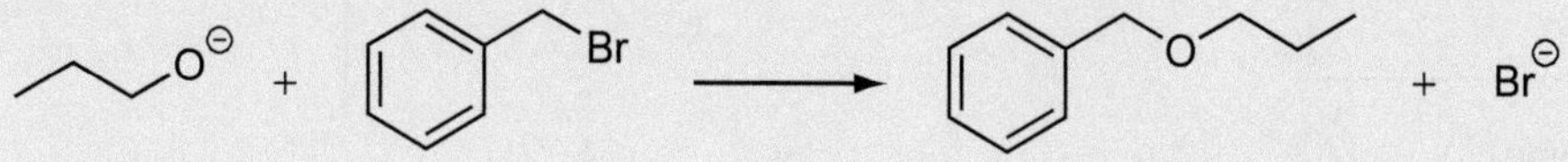

(b) Use curly arrows to show how the nucleophile reacts with the electrophile to form the products.

Strategy

(a) To help spot the nucleophile, remember that lone pairs of electrons are more nucleophilic than bonding electrons and negatively charged molecules are more nucleophilic than neutral molecules.

To help spot the electrophile, draw partial (δ+ and δ–) charges on the polar bond in the neutral molecule.

(b) Use double-headed curly arrows to show the mechanism—start an arrow from the nucleophilic site and move to the electrophilic site.

Solution

(a) and (b)

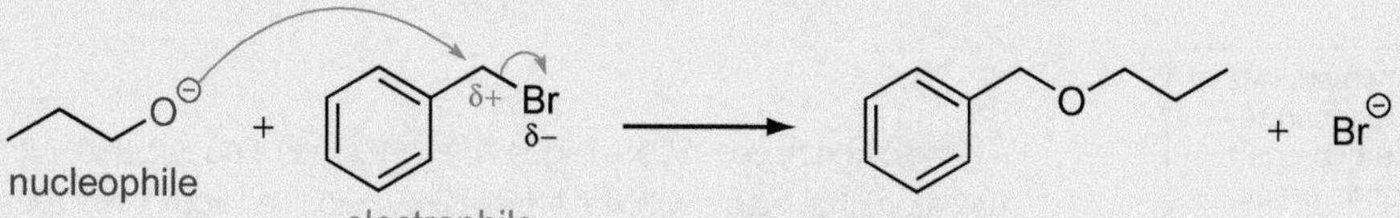
nucleophile
electrophile

Question

(a) In the following reaction, which reactant is the nucleophile and which is the electrophile?

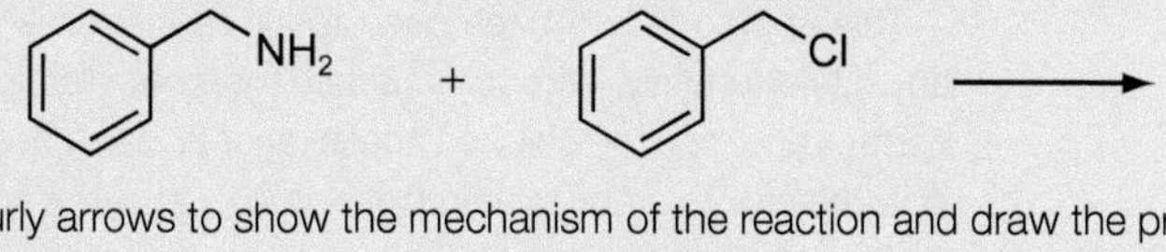

(b) Use curly arrows to show the mechanism of the reaction and draw the product that is formed.

▲ This one-step nucleophilic substitution is an example of an S_N2 reaction (Section 20.3, p.930). Reaction of alkoxide ions (RO^-) with halogenoalkanes (RX) is commonly used to make ethers (ROR). For example, reaction of the phenoxide ion (PhO^-) with bromomethane (CH_3Br) forms anisole (methoxybenzene, $PhOCH_3$). Aromatic ethers (ArOR), like anisole, have pleasant odours and they are used in perfumes and as flavourings in foods.

Box 19.4 Anti-cancer agents

Nitrogen mustards, such as mechlorethamine ($MeN(CH_2CH_2Cl)_2$), are a class of poisonous compounds originally developed for military use.

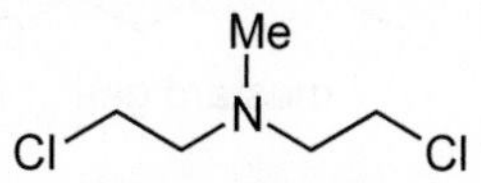

mechlorethamine

In the body, mechlorethamine undergoes an intramolecular (internal) reaction to form an aziridinium ion, containing a positively charged nitrogen atom in a 3-membered ring. The aziridinium ion is a powerful electrophile, which reacts with nucleophilic sites in DNA. DNA consists of two strands of linked nucleotides and each nucleotide unit contains a phosphate, a deoxyribose sugar, and a heterocyclic organic base. For more information on DNA, see Figure 1.1 (p.5) and Box 25.5 (p.1158).

→

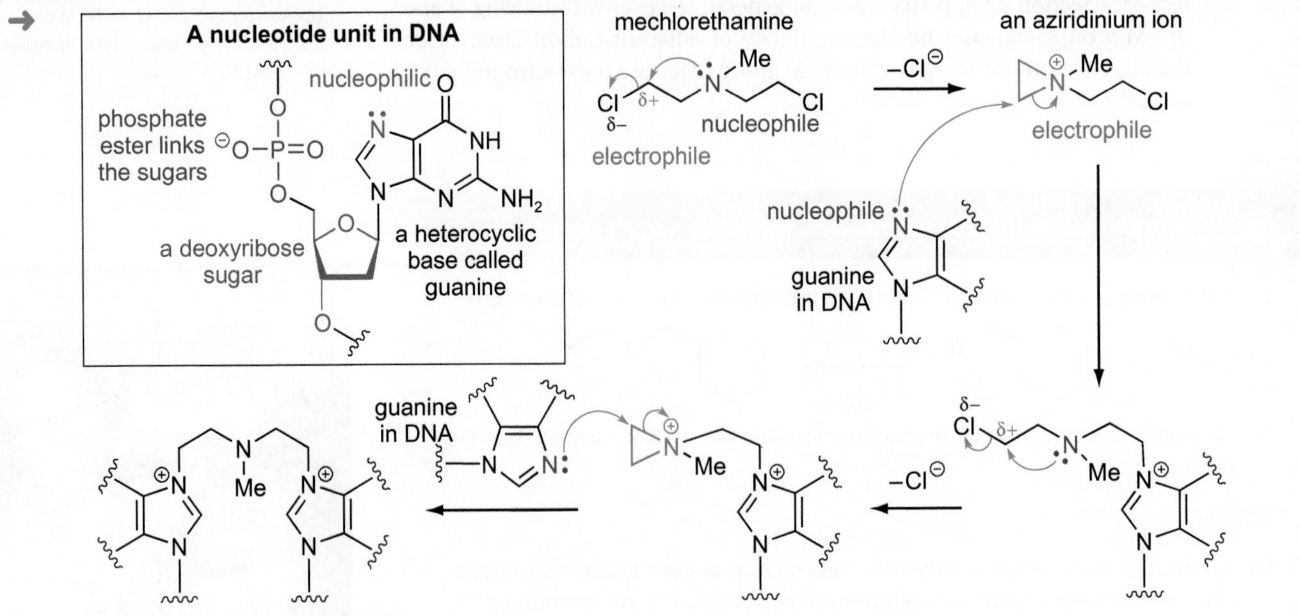
A nucleotide unit in DNA
nucleophilic
phosphate ester links the sugars
a deoxyribose sugar
a heterocyclic base called guanine
mechlorethamine
nucleophile
electrophile
an aziridinium ion
$-Cl^{\ominus}$
electrophile
nucleophile
guanine in DNA
$-Cl^{\ominus}$
guanine in DNA

Cross-linking of two DNA strands

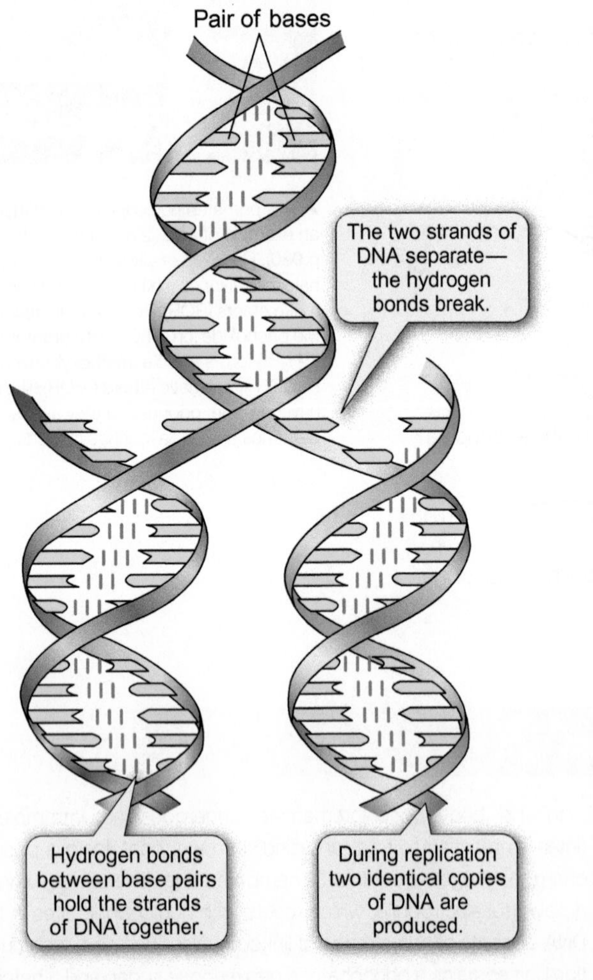
Pair of bases
The two strands of DNA separate—the hydrogen bonds break.
Hydrogen bonds between base pairs hold the strands of DNA together.
During replication two identical copies of DNA are produced.

◀ Before a cell can divide, it must replicate (duplicate) its DNA. To do this, the two strands in DNA must be separated so that copies of each strand can be made. This is not possible if the strands are cross-linked by covalent bonds. (In normal DNA, the strands are held together by hydrogen bonding between pairs of bases.)

One of the bases in DNA is guanine, which has a nucleophilic nitrogen atom that reacts with the aziridinium ion formed from the nitrogen mustard. This is followed by the formation of a second aziridinium ion and reaction with a further molecule of guanine from a different strand of DNA. When two DNA strands are linked together, cell division is effectively stopped and the cell dies.

This approach has been used to kill cancer cells in chemotherapy. Unfortunately, nitrogen mustards cannot distinguish between normal and cancer cells so these agents can cause unpleasant side-effects. Current research is looking at ways to improve the selectivity of these types of reagents so that they only kill cancer cells (see Box 28.3, p.1275).

Question

Mustard gas ($S(CH_2CH_2Cl)_2$) is a chemical warfare agent, first used in the First World War (1914–18), which stops cell division by cross-linking DNA. Propose a mechanism to show how mustard gas is able to cross-link DNA.

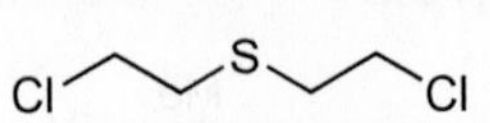

mustard gas

Summary

- Nucleophiles are negatively charged ions or neutral molecules that *donate* a pair of electrons.
- Electrophiles are positively charged ions or neutral molecules that *accept* a pair of electrons.
- Strong nucleophiles are anions (often of large atoms that are not highly electronegative) or neutral molecules containing a lone pair of electrons (on an atom that is not highly electronegative).

Strong nucleophiles	Moderate nucleophiles	Weak nucleophiles
I^-, NC^-, HO^-, RO^-, RS^-, R_3N	Br^-, Cl^-, H_3N, R_2S	RCO_2^-, F^-, H_2O, ROH

Approximate order in polar protic solvents—order depends on the solvent

- Strong electrophiles are carbocations or neutral molecules with carbon bonded to highly electronegative atoms.

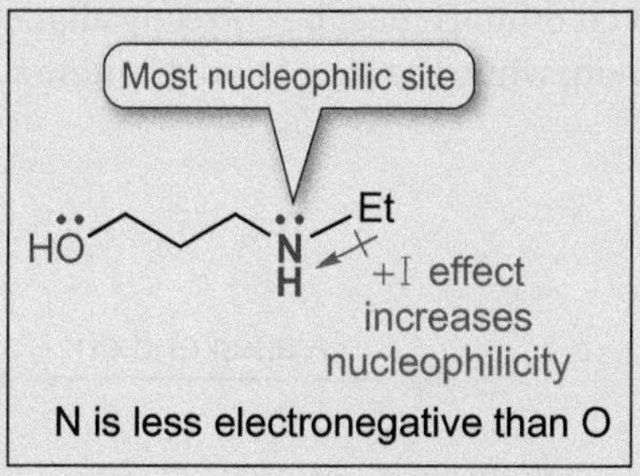
Most nucleophilic site
HO
Et
N
H
+I effect increases nucleophilicity
N is less electronegative than O

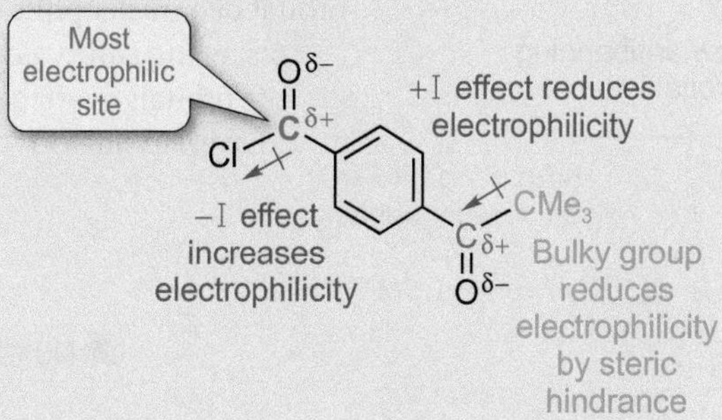
Most electrophilic site
O δ−
C δ+
Cl
+I effect reduces electrophilicity
−I effect increases electrophilicity
CMe3
C δ+
O δ−
Bulky group reduces electrophilicity by steric hindrance

For a practice question on these topics, see question 5 at the end of this chapter (p.914).

Orbital overlap in resonance and reaction mechanisms

When you use curly arrows to show mesomeric effects or to show how nucleophiles react with electrophiles, it helps to consider what atomic and/or molecular orbitals are involved. For example, the allyl cation ($H_2C{=}CH{-}CH_2^+$) shown in Figure 19.12 (p.874) is stabilized by resonance, which involves the donation of electrons from the adjacent π bond to the vacant unhybridized *p* orbital of the carbocation. For the π bond to interact with the *p* orbital, the orbitals must have the correct symmetry (the two lobes of a *p* orbital have opposite phases; see Section 5.7, p.249) and be able to overlap efficiently. For maximum orbital overlap, the *p* orbital and π bond need to be parallel to one another, which occurs when the allyl cation is planar.

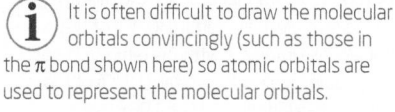

It is often difficult to draw the molecular orbitals convincingly (such as those in the π bond shown here) so atomic orbitals are used to represent the molecular orbitals.

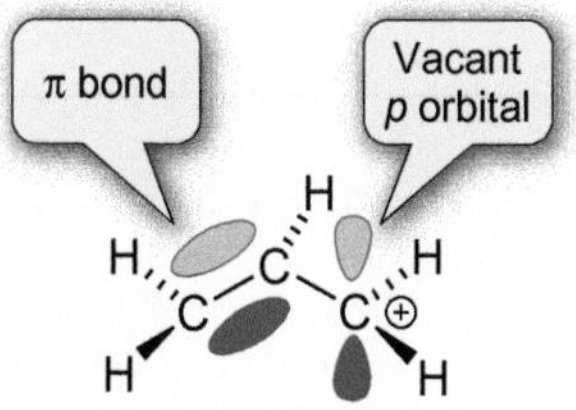
π bond
Vacant p orbital
H
H
C
C
C
H
H
H

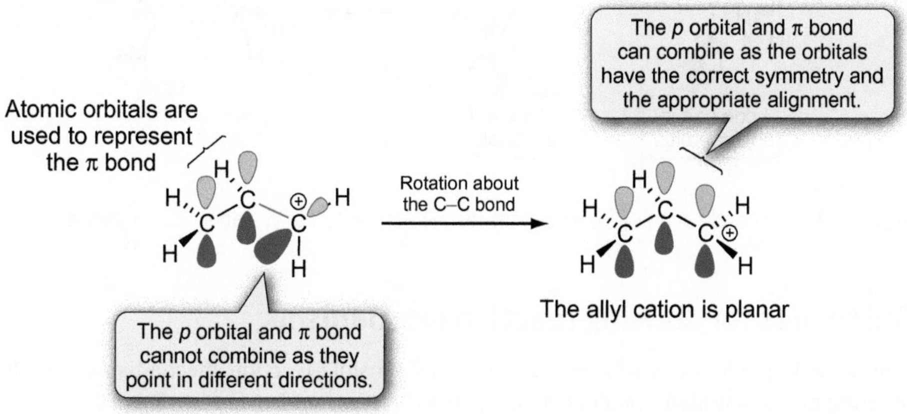
Atomic orbitals are used to represent the π bond
The p orbital and π bond cannot combine as they point in different directions.
Rotation about the C–C bond
The p orbital and π bond can combine as the orbitals have the correct symmetry and the appropriate alignment.
The allyl cation is planar

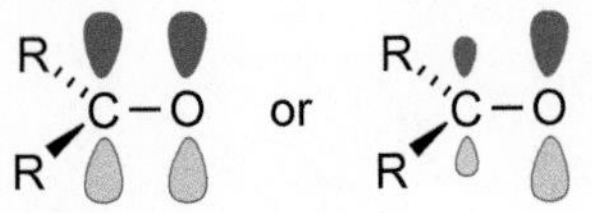
or

▲ In a C=O π bond, as oxygen is more electronegative than carbon, it has a larger share of the electron population. To indicate this, the lobes of the atomic orbitals on oxygen are sometimes drawn larger than those on carbon.

Visit the Online Resource Centre to view screencast 19.2 which walks you through orbital overlap in reaction mechanisms, including how orbital overlap can explain the direction of attack of nucleophiles on electrophilic sites.

The benzyl cation ($PhCH_2^+$) and benzyl radical ($PhCH_2^{\bullet}$) are also planar because in this conformation there is greatest orbital overlap between the vacant or half-filled *p* orbital and the π electrons in the benzene ring. Orbital overlap also explains why the amide group ($RCONR_2$) is planar—in this conformation there is greatest overlap between the lone pair of electrons in the *p* orbital on nitrogen and the π bond of the C=O group.

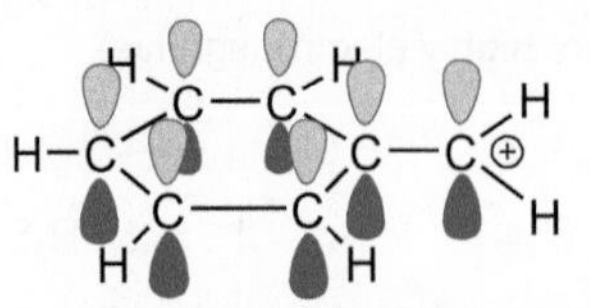

The benzyl cation is planar

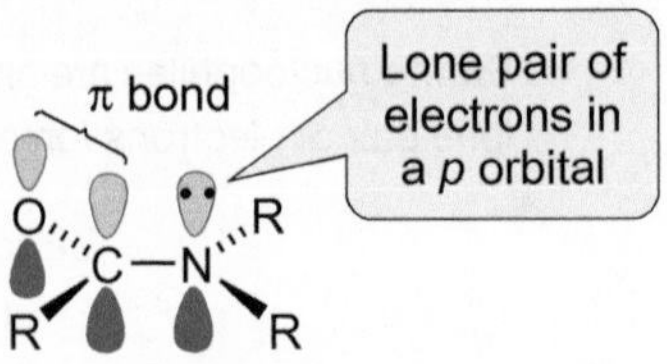
π bond
Lone pair of electrons in a *p* orbital

The amide group is planar

When a nucleophile reacts with an electrophile, two electrons are donated from a filled orbital of a nucleophile into an empty orbital of the electrophile. For the reaction to take place, the empty and filled orbitals must be correctly aligned. To make a new σ bond, the orbitals overlap end-on, while the formation of a new π bond requires the orbitals to overlap side-on.

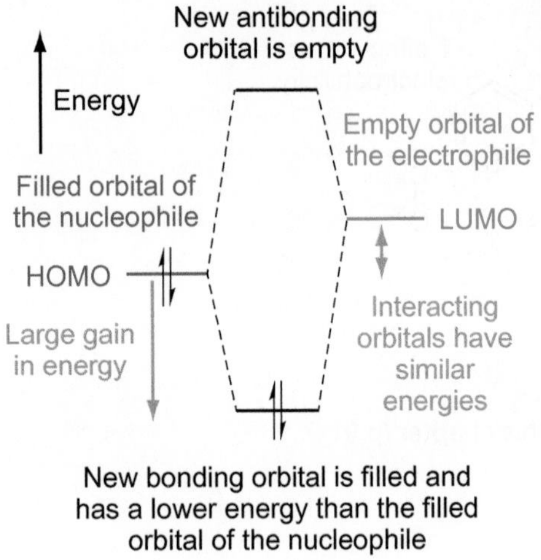
New antibonding orbital is empty
Energy
Empty orbital of the electrophile
Filled orbital of the nucleophile
LUMO
HOMO
Large gain in energy
Interacting orbitals have similar energies
New bonding orbital is filled and has a lower energy than the filled orbital of the nucleophile

▲ In terms of frontier orbitals, the (filled) highest occupied molecular orbital (HOMO) of the nucleophile interacts with the (empty) lowest unoccupied molecular orbital (LUMO) of the electrophile (Section 4.10, p.196).

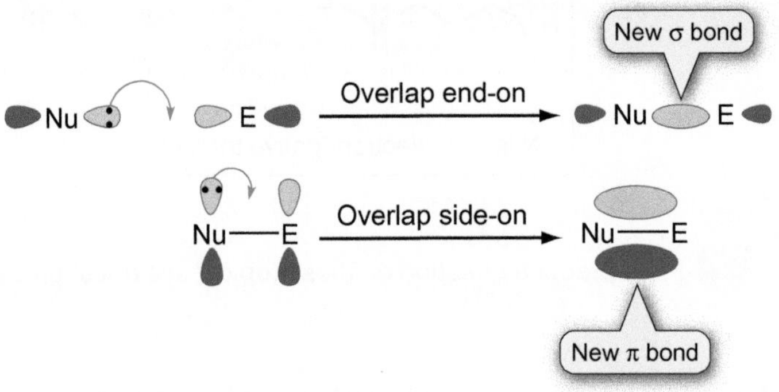
New σ bond
Overlap end-on
Overlap side-on
New π bond

The importance of orbital overlap is shown by the reaction of the *tert*-butyl cation (Me_3C^+) with ammonia (NH_3) to form a new C–N σ bond in Figure 19.23.

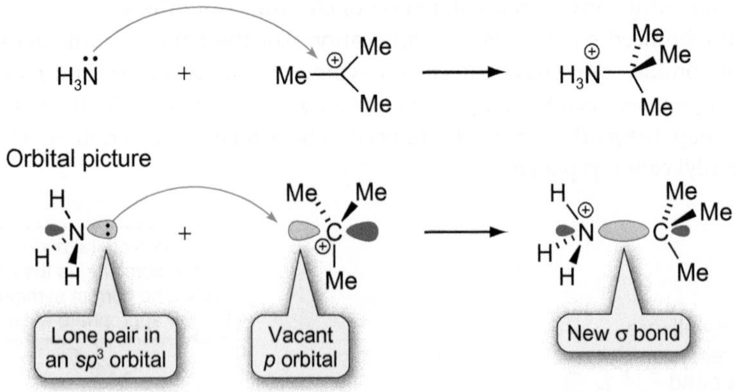
Orbital picture
Lone pair in an sp^3 orbital
Vacant *p* orbital
New σ bond

Figure 19.23 The role of orbital overlap in the reaction of the *tert*-butyl cation with ammonia.

For a reaction to take place, to form a new bond, the HOMO and LUMO orbitals must be correctly aligned and also have similar energies. When the interacting HOMO and LUMO orbitals are of similar energy, when the bond forms, this results in the most gain in energy. (This occurs when a strong nucleophile, which has a high-energy HOMO, interacts with a strong electrophile, which has a low-energy LUMO).

A lone pair can be in a *p*, *sp*, sp^2, or sp^3 orbital. For example, the nitrogen atom in an amide ($RCONR_2$) is sp^2 hybridized while the nitrogen atom in ammonia (NH_3) is sp^3 hybridized.

Guidelines for drawing reaction mechanisms

The following guidelines will help you to draw sensible reaction mechanisms—as with learning many new skills, practice makes perfect!

Step 1 Draw skeletal structures of the reactants and include lone pairs on any O or N heteroatoms (lone pairs on electronegative halogen atoms are held so tightly that they are rarely involved in the formation of bonds and so they do not need to be included). Show the polarities of polar bonds using δ+ and δ–.

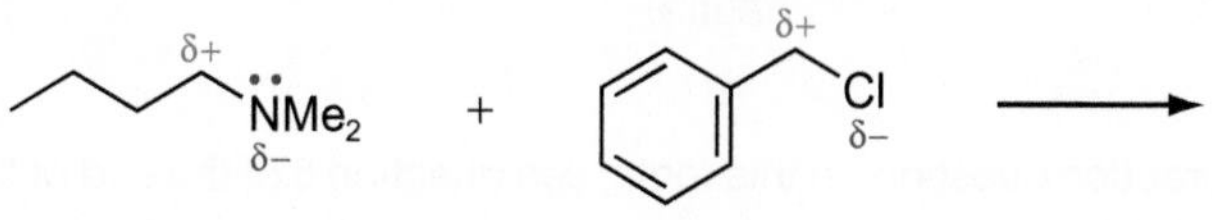
δ+
NMe2
δ–
+
δ+
Cl
δ–

Step 2 Identify any reactants or bases that are acids or bases and then assess if the first step of the mechanism involves the transfer of H^+ (to make a more reactive nucleophile or a more reactive electrophile).

Step 3 Identify which reactant is the nucleophile and which is the electrophile, and which part of each reactant is responsible.

- For nucleophilic sites, lone pairs of electrons are stronger nucleophiles than π electrons in multiple carbon–carbon bonds.
- For electrophilic sites on carbon atoms, carbons that are bonded to the more electronegative atoms, or groups of atoms, are the stronger electrophiles.

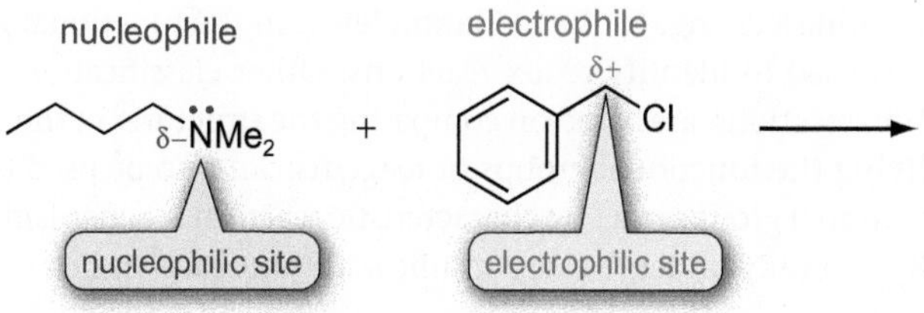
nucleophile
electrophile
δ–NMe2
+
δ+
Cl
nucleophilic site
electrophilic site

Step 4 Draw a double-headed curly arrow from the nucleophilic site (which can have a negative charge, lone pair, or a multiple bond) in the nucleophile to the electrophilic site (which has a positive charge or a partial positive charge) in the electrophile. By comparing the structures of the starting materials and products check that the curly arrow indicates the formation of a sensible bond.

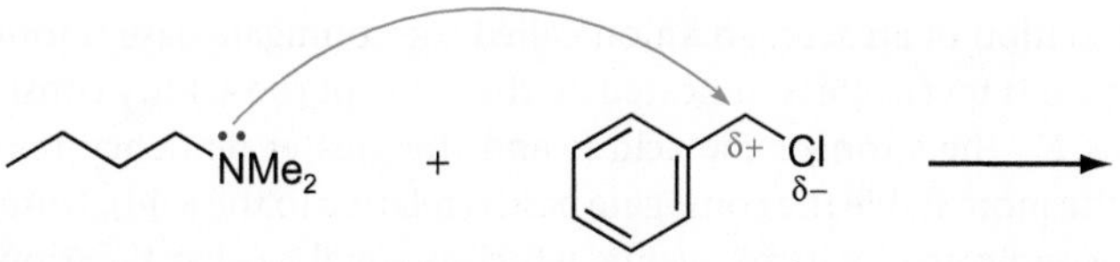
NMe2
+
δ+
Cl
δ–

Step 5 If the nucleophilic centre is a negatively charged atom, this atom will be uncharged in the product. If the nucleophilic centre is an uncharged atom, this atom will have a positive charge in the product.

Step 6 If the electrophilic centre is a positively charged atom, this atom will be neutral in the product. If a new bond is made to a neutral atom in the electrophile, break one of the existing bonds and draw a second curly arrow (pointing in the same direction as the first arrow). Draw the second curly arrow so that the most electronegative atom accepts the electrons from the bond. By comparing the structures of the starting materials and products check that the curly arrow indicates the cleavage of a sensible bond.

Step 7 The overall charge of the reactants is the same as the overall charge of the products.

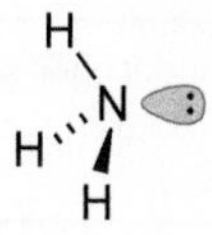
H
N
H
H

▲ For the nitrogen atom in ammonia, lobes of unequal size show the lone pair is in an sp^3 orbital; sometimes just one lobe is drawn.

Filled (electron-donating) orbitals of **nucleophiles** include a lone pair in an sp^3 orbital in ammonia (as in Figure 19.23), the π orbital of the C=C bond in an alkene (Section 21.3, p.970) and the σ orbital of the C–M bond in an organometallic reagent (Section 23.2, p.1069).

Empty (electron-accepting) orbitals of **electrophiles** include the *p* orbital of a carbocation (as in Figure 19.23 and Section 20.3, p.930), the π* orbital of the C=O bond in a carbonyl compound (Section 23.1, p.1057) and the σ* orbital of the C–X bond in a halogenoalkane (Section 20.3, p.931).

▲ Learning how to draw organic reaction mechanisms is like learning to play the piano. In order to play at all, you need to practice every day—it cannot be learned the night before an examination.

Visit the Online Resource Centre to view screencast 19.3 which walks you through guidelines for drawing reaction mechanisms. Examples include the addition of HBr to an unsymmetrical alkene.

There are a few important reactions where σ bonds are nucleophilic sites. For example, a B–H bond is a nucleophilic site when using the $[BH_4^-]$ ion to reduce a carbonyl compound (Section 23.2, p.1063). Also, a covalent C–metal bond, of an organometallic reagent, can act as a nucleophile (Section 23.2, p.1069).

This is a nucleophilic substitution reaction, called an S_N2 reaction (see Section 20.3, p.930).

The product of this reaction is called a quaternary ammonium salt (Section 2.6, p.95).

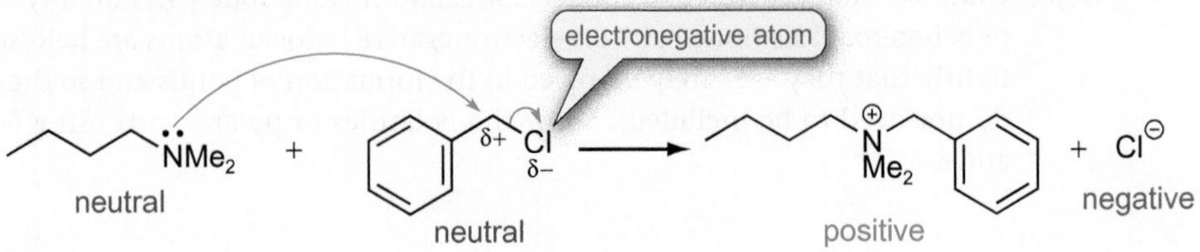
electronegative atom
NMe2
δ+
Cl
δ−
N
Me2
Cl
neutral
neutral
positive
negative

To practise drawing curly arrow mechanisms for organic reactions, follow the link on the Online Resource Centre.

For a practice question on this topic, see question 5 at the end of this chapter (p.914).

19.2 Classification of organic reaction mechanisms

In terms of frontier orbitals, the nitrogen lone pair of the tertiary amine (the HOMO) overlaps with the empty σ* orbital of the C–Cl bond (the LUMO). The weakest bond to carbon, usually to the most electronegative atom, will typically have the LUMO associated with it.

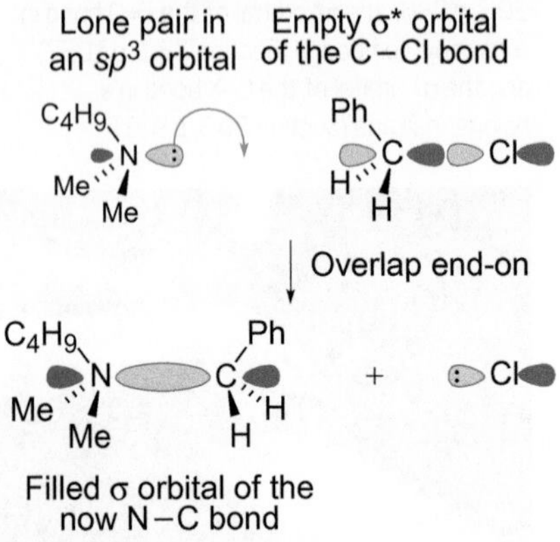
Lone pair in an sp3 orbital
Empty σ* orbital of the C–Cl bond
C4H9
N
Me
Me
Ph
C
Cl
H
H
Overlap end-on
C4H9
N
Me
Me
Ph
C
H
H
Cl
Filled σ orbital of the now N–C bond

Notice that the σ* orbital results from an out-of-phase combination of two *p* orbitals (sometimes the two inner lobes, with opposite phases, are drawn smaller than the outer lobes to show a reduced electron density between the nuclei), whereas the new σ orbital results from an in-phase combination of two *p* orbitals (with the same phases). See Section 4.9, p.193.

A common difficulty that students find when studying organic chemistry is that there are simply too many reactions to learn. Learning all of the reactions covered in an undergraduate course can seem a daunting prospect. But you can make this process much simpler by *classifying* reactions of organic compounds. Once the reaction has been classified, this can help enormously when writing a reaction mechanism.

Reactions are classified in different ways (see Section 1.4, p.23). Two obvious classes are acid–base reactions and redox reactions. The transfer of a proton identifies acid–base reactions, while a change in the oxidation levels in the functional groups of reactants and products is used to identify redox reactions. Other classifications, such as addition and elimination reactions, are based on comparing the structures of the reactants and products.

Identifying the functional groups of reagents can also be used to classify reactions because functional groups react by characteristic reaction mechanisms. For example, alkenes ($R_2C{=}CR_2$) typically undergo electrophilic addition reactions.

▲ Organic reactions can be classified as acid–base reactions, redox reactions, polar reactions (additions, eliminations, and substitutions), radical reactions, and pericyclic reactions.

For a summary of the different classes of functional groups see Figure 2.7 (p.106).

Acid–base reactions

To understand acid–base reactions in organic chemistry, you need to be familiar with the types of organic compounds that commonly act as acids or bases and the factors that influence their acidity or basicity.

Organic acids

On deprotonation of an acid, an anion called the conjugate base is formed. Whether the deprotonation is reversible is indicated by the value of the acidity constant, K_a. The larger the value of K_a, the stronger the acid is and the greater tendency for deprotonation to occur (i.e. the more stable the conjugate base is relative to the acid). Normally, the strength of an acid is indicated by its pK_a value, which is equal to –log K_a. Strong acids have relatively stable conjugate bases and low pK_a values.

$$\underset{\text{acid}}{HA} + H_2O \rightleftharpoons \underset{\text{conjugate base}}{A^{\ominus}} + H_3O^{\oplus}$$

$$K_a = \frac{[H_3O^{\oplus}]\,[A^{\ominus}]}{[HA]}$$

$$pK_a = -\log_{10} K_a$$

For example, water (H_2O, pK_a 16) is a much stronger acid than methane (CH_4, pK_a 48) for the following reasons. Firstly, the electronegative oxygen atom has a –I effect so it pulls electron density away from the hydrogen atom in the O–H bond. This weakens the bond, meaning the proton is more easily removed. Secondly, the electronegative oxygen atom stabilizes the negative charge on the conjugate base more effectively than a carbon atom (i.e. HO^- is more stable than H_3C^-).

$$H_2O + H_2O \rightleftharpoons HO^{\ominus} + H_3O^{\oplus}$$
pK_a 16

$$CH_4 + H_2O \rightleftharpoons {}^{\ominus}CH_3 + H_3O^{\oplus}$$
pK_a 48

The equilibrium lies heavily to the left

Acidic organic molecules, such as alcohols (ROH, $pK_a = 16$), contain H atoms bonded to electronegative atoms. Stronger organic acids, such as carboxylic acids (RCO_2H) and phenols (e.g. PhOH), also have –I and –M groups that help to stabilize the negative charge in the conjugate base as discussed later in this section.

The carboxylate ion (RCO_2^-) formed when a carboxylic acid reacts with a base is stabilized by resonance.

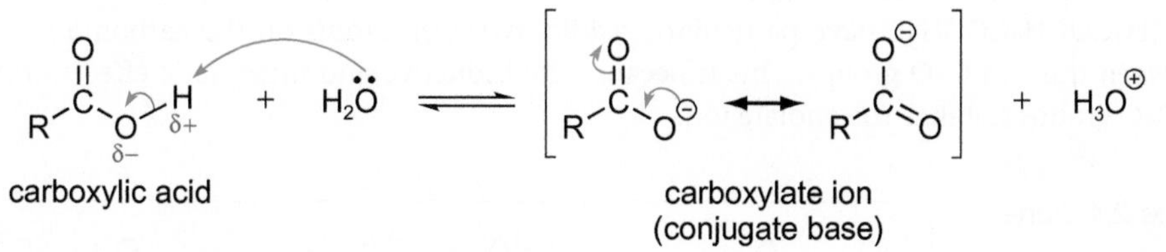
carboxylic acid
carboxylate ion
(conjugate base)

Electron-withdrawing (–I) groups at the α-position (the 2-position) on the carboxylic acid increase the acidity of the acid while electron-donating (+I) groups reduce the acidity. For example, 2-fluoroethanoic acid (FCH_2CO_2H) is a stronger acid than ethanoic acid (CH_3CO_2H). This is because the electron-withdrawing fluorine atom stabilizes the negative charge on the conjugate base.

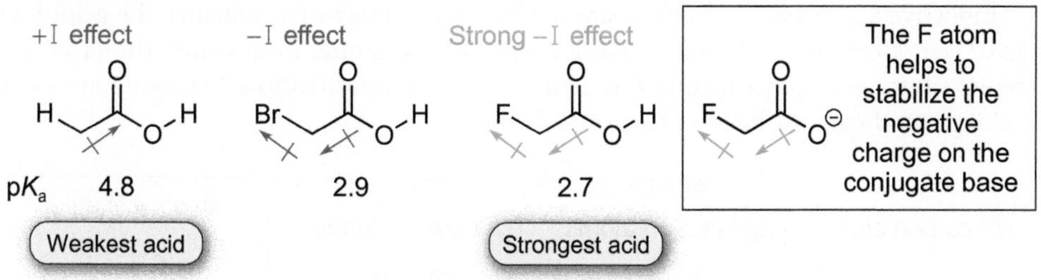
+I effect
−I effect
Strong −I effect
The F atom helps to stabilize the negative charge on the conjugate base
pKa 4.8
2.9
2.7
Weakest acid
Strongest acid

On deprotonation of phenol (PhOH, pK_a 9.9), a phenoxide ion is formed that is stabilized by resonance—the negative charge is delocalized on to the 2, 4, and 6 positions of the ring (compare this with the resonance stabilization of the benzyl cation, $PhCH_2^+$, in Section 19.1, p.874).

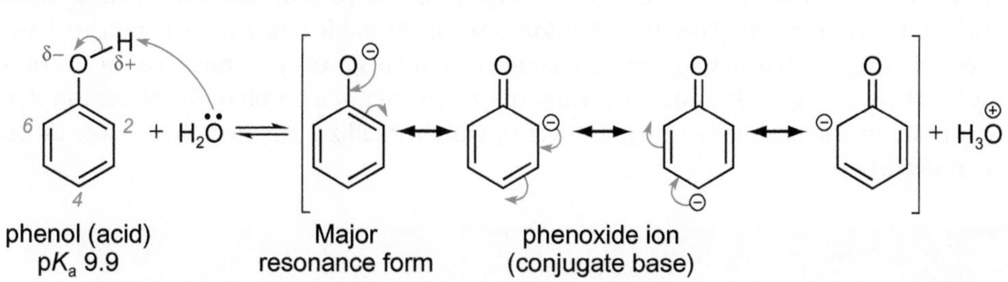
phenol (acid)
pKa 9.9
Major
resonance form
phenoxide ion
(conjugate base)

Substituted phenols with –I groups and, particularly, –M groups at the 2, 4, and/or 6 positions of the ring are more acidic than phenol. The conjugate bases of these compounds are stabilized by delocalization of the negative charge on to the –I or –M group. For example, 2-nitrophenol is a stronger acid than phenol because the strongly electron-withdrawing NO_2 group stabilizes the conjugate base by inductive and mesomeric effects.

This section discusses reactions of Brønsted–Lowry acids and bases, rather than Lewis acids and bases (Sections 7.1, p.304, and 7.8, p.336).

For an introduction to acids and bases, including pK_a, see Sections 7.1 (p.304) and 7.2 (p.308). You can find a table of pK_a values for various organic and inorganic compounds in Appendix 9 (p.1366).

The pK_a value is equal to the pH of the acid when it is half ionized (see Section 7.4, p.322). Approximate pK_a values measured in water are quoted in this section.

The position of the equilibrium in an acid–base reaction depends on the relative strengths of the acid and base (see Section 7.2, p.308). Equilibrium can be modelled as two people playing on a see-saw—when the weight of the two people is evenly balanced, the see-saw is at equilibrium.

The role of entropy in solvation is discussed in Section 17.4 (p.792). The effect of entropy changes on chemical reactions is introduced in Section 14.4 (p.669). Chemical equilibrium and the direction of change are discussed in Chapter 15.

The pK_a values of the acids are affected by entropy. In water, deprotonation of RCO_2H to form RCO_2^- and H_3O^+ leads to a decrease in entropy, because the ions are solvated by a shell of 'ordered' water molecules. In comparison to $CH_3CO_2^-$, the more delocalized negative charge on $FCH_2CO_2^-$ is not as strongly solvated, so the water molecules are less ordered. There is a smaller decrease in entropy, so the formation of the conjugate base is more favoured.

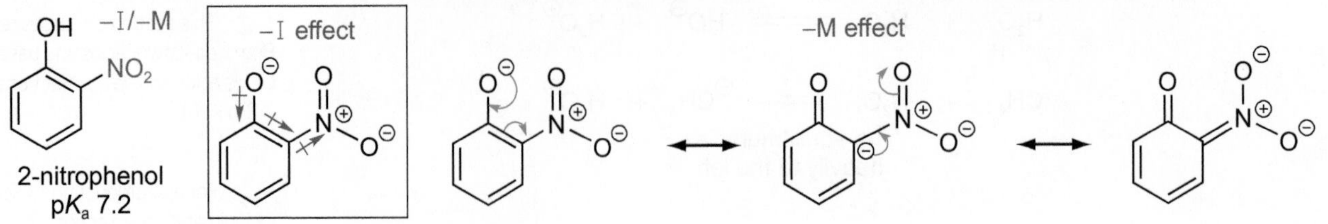

OH
−I/−M
NO_2
2-nitrophenol
pK_a 7.2
−I effect
−M effect

For reactions of enolate ions see Section 23.3 (p.1062).

Hybridization is discussed in Section 4.5 (p.180) and Section 5.4, p.236.

Some organic molecules that contain –I and –M groups have hydrogen atoms bonded to carbon that behave in an acidic way. For example, hydrogen atoms at the α-position of a carbonyl compound are much more acidic than the hydrogen atoms of an alkane. The C=O group stabilizes the conjugate base (e.g. $CH_3COCH_2^-$), called an enolate ion, by inductive effects and mesomeric effects. Dicarbonyl compounds, such as pentane-2,4-dione ($CH_3COCH_2COCH_3$), have particularly acidic hydrogen atoms on the carbon atom between the two C=O groups. This is because the inductive and mesomeric effects of *both* C=O groups stabilize the enolate ion.

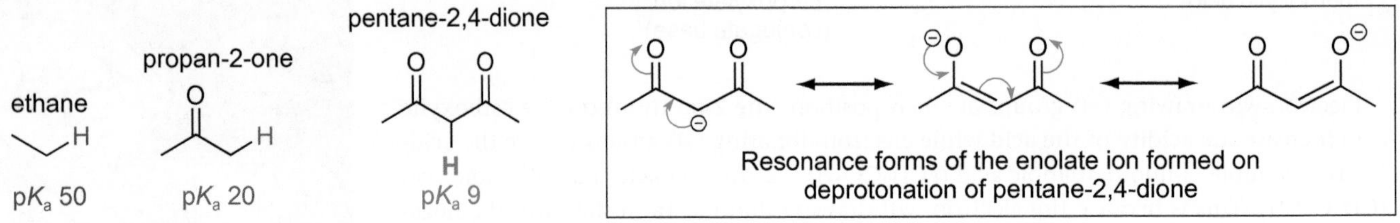

ethane
pK_a 50
propan-2-one
pK_a 20
pentane-2,4-dione
pK_a 9
Resonance forms of the enolate ion formed on deprotonation of pentane-2,4-dione

For a –M group to stabilize a conjugate base by mesomeric effects, the negative charge must be on a carbon atom adjacent to the –M group in one of the resonance forms of the conjugate base.

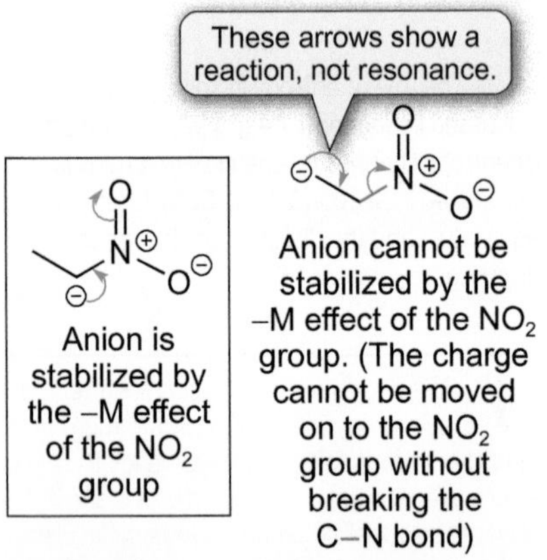

These arrows show a reaction, not resonance.
Anion is stabilized by the –M effect of the NO_2 group
Anion cannot be stabilized by the –M effect of the NO_2 group. (The charge cannot be moved on to the NO_2 group without breaking the C–N bond)

Inductive and mesomeric effects are not the only factors that influence the acidity of hydrogen atoms in C–H bonds. Hybridization of the orbital from which the proton is removed, for example, explains why hydrogen atoms attached to a C≡C bond are more acidic than those attached to a C=C or C–C bond.

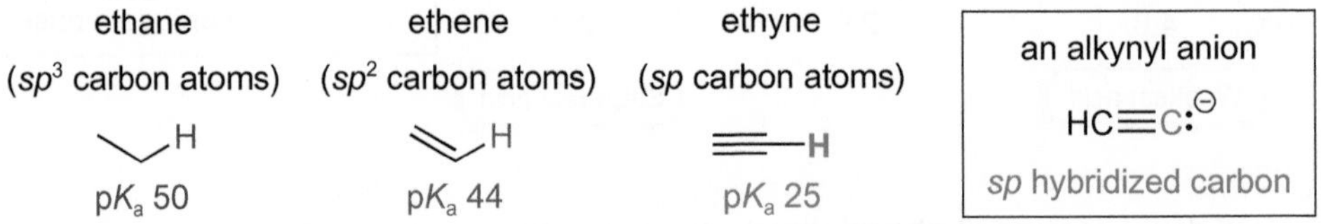

ethane
(sp^3 carbon atoms)
pK_a 50
ethene
(sp^2 carbon atoms)
pK_a 44
ethyne
(*sp* carbon atoms)
pK_a 25
an alkynyl anion
HC≡C:
sp hybridized carbon

When ethyne (HC≡CH) loses a proton, the conjugate base formed is an alkynyl anion ($HC{\equiv}C^-$). The lone pair is in an *sp* orbital. In comparison, the conjugate bases formed from ethene ($H_2C{=}CH_2$) or ethane ($H_3C{-}CH_3$) have lone pairs of electrons in an sp^2 or an sp^3 orbital, respectively. The alkynyl anion is the most stable of the three conjugate bases because an *sp* orbital has a greater *s* character than an sp^2 or sp^3 orbital (i.e. $sp = 50\%\ s$, $sp^2 = 33\%\ s$, and $sp^3 = 25\%\ s$). The greater the *s* character of an orbital, the closer the electrons are to the positively charged nucleus, which stabilizes the negative charge of the conjugate base.

Summary

- Acidic compounds have low pK_a values and are good proton donors.
- The more stable the conjugate base, the stronger the acid.
- Organic acids generally contain O–H and N–H bonds; relatively strong acids also have –I and –M groups that stabilize the conjugate base.
- Alkynes are more acidic than alkenes or alkanes because the lone pair of electrons in the conjugate base is in an *sp* orbital.

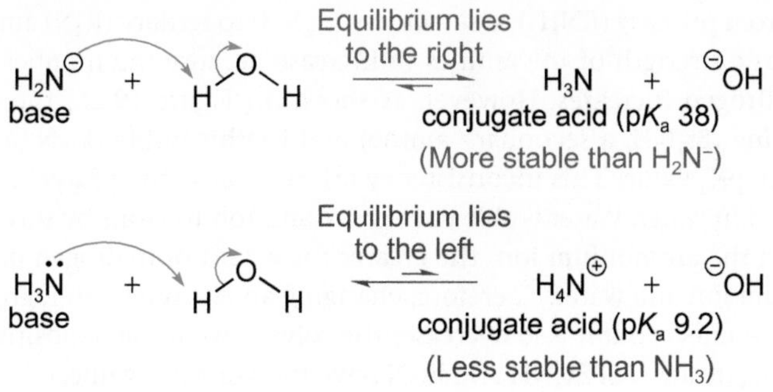
Equilibrium lies
to the right
$H_2N^{\ominus}$
base
+
H O H
H_3N + $^{\ominus}OH$
conjugate acid (pK_a 38)
(More stable than H_2N^-)
Equilibrium lies
to the left
H_3N
base
+
H O H
$H_4N^{\oplus}$ + $^{\ominus}OH$
conjugate acid (pK_a 9.2)
(Less stable than NH_3)

Figure 19.24 Protonation of a base gives a conjugate acid. The more stable the conjugate acid (and the higher its pK_a), the stronger the base.

Alkynyl anions ($HC\equiv C^-$ and $RC\equiv C^-$) are strong nucleophiles that are commonly used in organic synthesis to form C–C bonds. For example, they react with halogenoalkanes (Section 21.2, p.969) and carbonyl compounds (Section 23.2, p.1068).

Organic bases

Basic compounds are good proton acceptors. Bases can be negatively charged ions or neutral molecules containing lone pairs of electrons. Protonation of a negatively charged base gives a neutral conjugate acid, while protonation of a neutral base forms a positively charged conjugate acid. To determine if a molecule is a strong base, you need to look at the pK_a value of the conjugate acid (see Figure 19.24). A strong base has a relatively stable conjugate acid and the conjugate acid has a high pK_a value (i.e. it is a weak acid with a low tendency to donate a proton to a base). In general, organic molecules with a formal negative charge, such as alkoxides (RO^-; the pK_a of the conjugate acid is 16), are relatively strong bases. Neutral molecules with lone pairs of electrons on nitrogen or oxygen, such as amines (e.g. RNH_2) and carbonyl compounds (e.g. RCOR), are weaker bases.

H_2N^- is called an amide ion, not to be confused with the same name for $RCONR_2$ compounds. Sodium amide (sodamide, $NaNH_2$) is a strong base used to deprotonate acidic organic molecules (e.g. alkynes and alcohols) and to promote elimination reactions of halogenoalkanes, including 1,2-dibromoalkanes (Section 21.2, p.969).

Protonation of the C=O bond of carbonyl compounds is commonly used to convert carbonyl compounds into stronger electrophiles, that react more rapidly with nucleophiles (see, for example, Section 23.2, p.1072).

For convenience, when talking about the strength of a base, organic chemists often use a shorthand terminology in which they assign a pK_a value to the base. For example, NH_3 is sometimes quoted as being a weak base because it has a relatively low pK_a value of 9.2. It is important to remember that 9.2 is the pK_a value of *the conjugate acid of ammonia* (i.e. the ammonium ion, $^+NH_4$), not of ammonia itself.

For example, $^-NH_2$ is a strong base because protonation forms NH_3, an uncharged conjugate acid with a pK_a value of 38, which is more stable than $^-NH_2$. In comparison, NH_3 is a weak base because protonation forms $^+NH_4$, a positively charged conjugate acid with a pK_a value of 9.2, which is less stable than NH_3.

The basic strength (basicity) of neutral organic molecules is increased if +I and/or +M groups are able to stabilize the positive charge in the conjugate acid. For example, tertiary amines (NR_3) are stronger bases than NH_3 because the three alkyl groups bonded to nitrogen stabilize the conjugate acid ($^+HNR_3$, called a trialkylammonium ion) by inductive effects (Figure 19.25).

When comparing the strengths of different bases it is normal to compare the pK_a values of their conjugate acids (Section 7.2, p.308). The conjugate acid of a strong base has a high pK_a value, i.e. it is a weak acid.

Amines (e.g. RNH_2) are generally stronger bases than alcohols (ROH) because nitrogen is less electronegative than oxygen so nitrogen is a better electron donor.

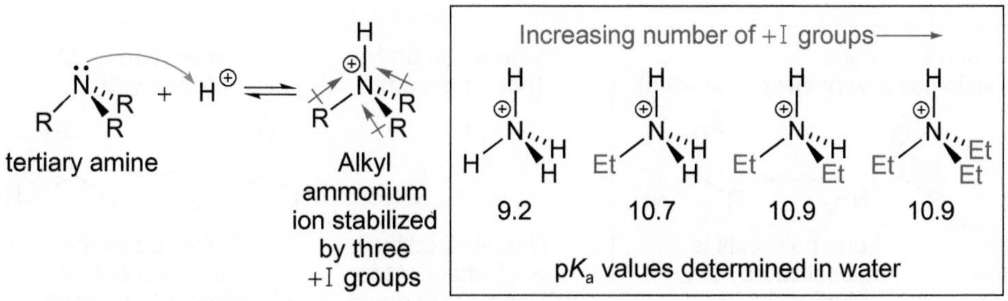
tertiary amine
Alkyl ammonium ion stabilized by three +I groups
Increasing number of +I groups
9.2
10.7
10.9
10.9
pK_a values determined in water

Figure 19.25 The conjugate acid (a trialkylammonium ion) of a tertiary amine is stabilized by three +I groups.

Whereas basicity relates to the reaction of an electron donor with a hydrogen atom (or H⁺), nucleophilicity relates to the reaction of an electron donor with an atom other than hydrogen, typically carbon. A fundamental difference between basicity and nucleophilicity relates to how they are measured.
Acid–base reactions tend to be *reversible* and so we can measure the basicity from the position of equilibrium—as we are measuring the relative stabilities of the reactants and products, this is a **thermodynamic property**.
Reactions of nucleophiles are typically *irreversible* and so we can measure the nucleophilicity from the rate of the reaction—measuring reaction rates is a **kinetic property**.
Most strong bases are also strong nucleophiles (and vice versa).

On going from primary (RNH_2) to secondary (R_2NH) to tertiary (R_3N) amines, you might expect the basic strength of the amines to increase because the number of alkyl groups bonded to nitrogen increases. However, as shown in Figure 19.25, the conjugate acids of diethylamine (Et_2NH, a secondary amine) and triethylamine (Et_3N, a tertiary amine) have the same pK_a value. This inconsistency arises because the pK_a values of the amines are determined in water. Water is able to stabilize ammonium ions by solvation. The more N–H bonds in the ammonium ion, the greater the extent of hydrogen bonding between the ammonium ion and water. Therefore, changing an H atom to an R group on nitrogen increases the +I effects but it also decreases the solvation—these opposing effects explain why the conjugate acids of Et_2NH and Et_3N have the same pK_a value.

The influence of +M effects on increasing the basic strength of nitrogen is illustrated by guanidine ($HN{=}C(NH_2)_2$). This is one of the strongest neutral nitrogen bases because the conjugate acid of guanidine ($^+H_2N{=}C(NH_2)_2$) is stabilized by extensive delocalization. Following protonation of the sp^2 hybridized nitrogen atom, a cation is formed that is stabilized by the +M effects of the two NH_2 groups.

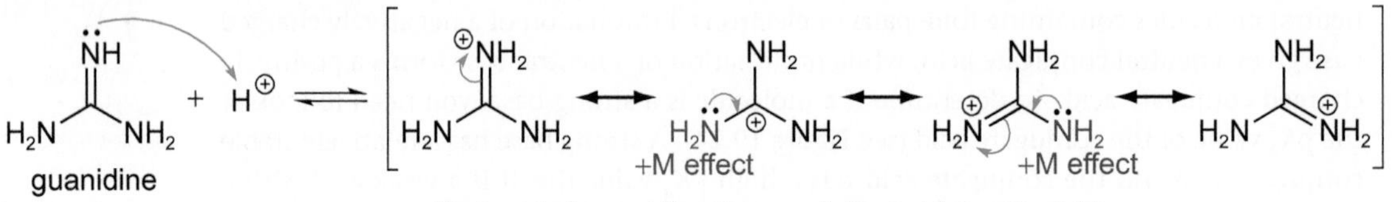
guanidine
+M effect
+M effect

Protonated guanidine has a relatively high pK_a of 13

In water, pK_a values are influenced by solvation. Solvation is introduced in Box 7.1, p.305.

A lone pair in an sp^2 orbital is held more tightly to the nucleus and is usually harder to protonate than a lone pair in an sp^3 orbital. However, if one of the sp^3 nitrogen atoms of guanidine is protonated, the conjugate acid cannot be stabilized by delocalization.

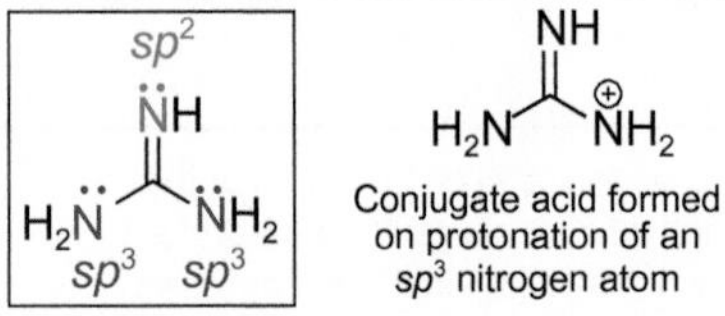
sp2
sp3 sp3
Conjugate acid formed on protonation of an sp3 nitrogen atom

A +M group is likely to stabilize a cation more effectively than a +I group (Section 19.1, p.875).

Whereas electron-donating (+I/+M) groups increase the basic strength of nitrogen, electron-withdrawing (–I/–M) groups reduce the basic strength. This is because the strength of a neutral base is determined by the availability of a lone pair of electrons. For example, the nitrogen atom in an amide (e.g. $RCONH_2$) is much less basic than a nitrogen atom in an amine because of the –I and –M effects of the adjacent C=O group (Figure 19.26).

An amide can only be protonated in the presence of a strong acid and even then the H⁺ adds to the oxygen atom of the amide, not the nitrogen atom. This is because protonation on oxygen gives a conjugate acid that is stabilized by resonance, whereas protonation on nitrogen gives a much less stable conjugate acid. This is shown in Figure 19.26.

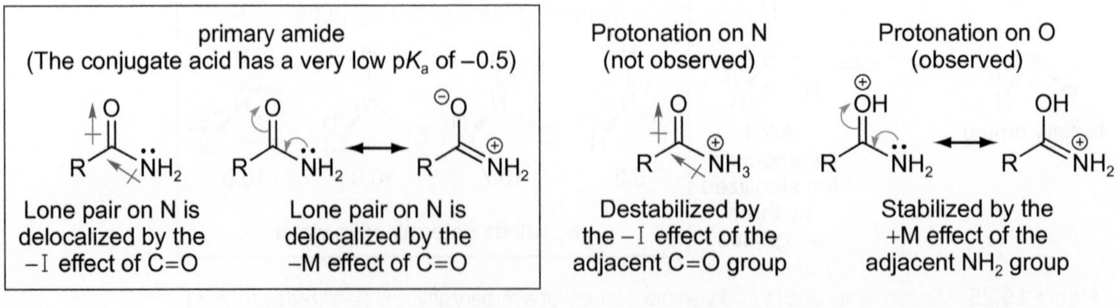
primary amide
(The conjugate acid has a very low pKa of –0.5)
Lone pair on N is delocalized by the –I effect of C=O
Lone pair on N is delocalized by the –M effect of C=O
Protonation on N
(not observed)
Destabilized by the –I effect of the adjacent C=O group
Protonation on O
(observed)
Stabilized by the +M effect of the adjacent NH2 group

Figure 19.26 Amides are less basic than amines due to the –I and –M effects of the C=O group.

The aromatic amine phenylamine (aniline, $PhNH_2$) is less basic than aliphatic amines because the lone pair of electrons on nitrogen is delocalized on to the 2, 4, and 6 positions of the benzene ring.

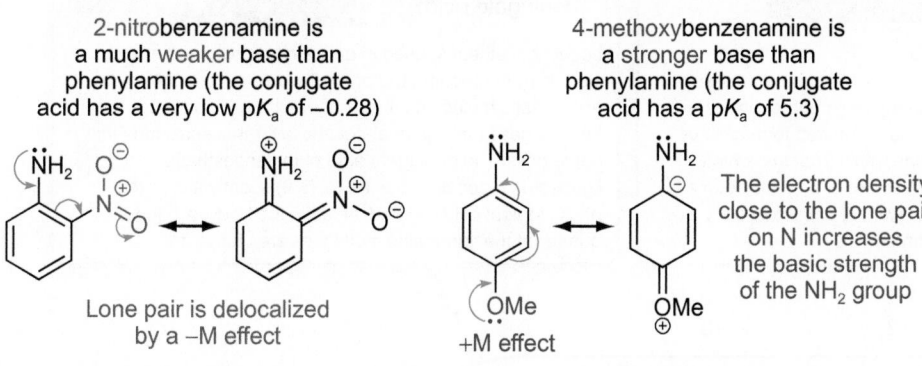

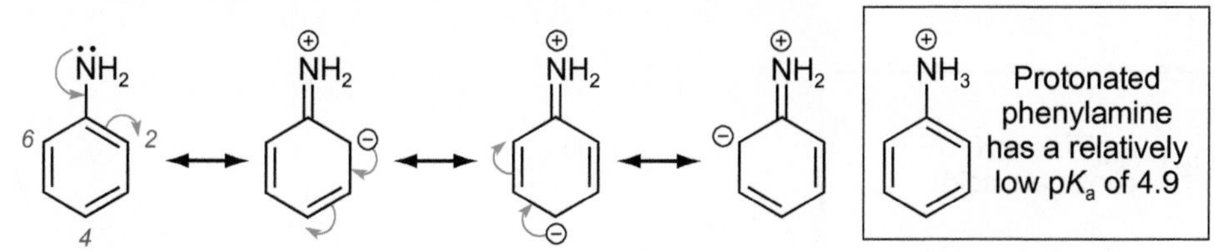
Protonated phenylamine has a relatively low pK_a of 4.9

phenylamine

Guanidine nitrate, $^{+}C(NH_2)_3\ ^{-}NO_3$, is the salt formed from reaction of guanidine with nitric acid. This salt is a high energy fuel (when it burns, it produces a large amount of expanding hot gas), which has been used to propel model aeroplanes.

Replacing hydrogen atoms on the benzene ring of phenylamine by electron-withdrawing or electron-donating substituents influences the basicity of the nitrogen atom.

- Electron-withdrawing substituents (e.g. NO_2) reduce the basic strength of the nitrogen atom by inductive effects, particularly if a –I group is close to the lone pair in the 2-position, and by mesomeric effects if a –M group is at the 2, 4, or 6 position of the ring.

2-nitrobenzenamine is a much weaker base than phenylamine (the conjugate acid has a very low pK_a of –0.28)

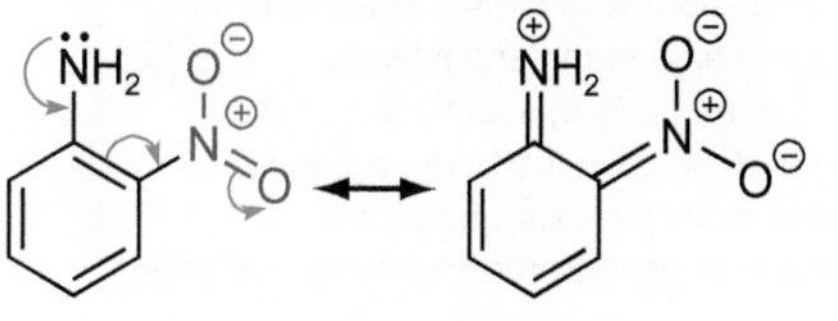

Lone pair is delocalized by a –M effect

4-methoxybenzenamine is a stronger base than phenylamine (the conjugate acid has a pK_a of 5.3)

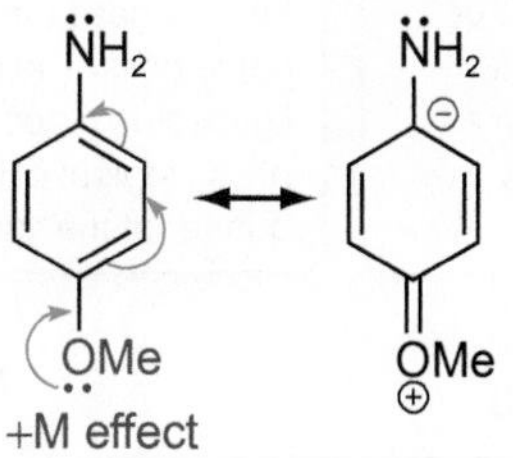

+M effect

The electron density close to the lone pair on N increases the basic strength of the NH_2 group

Carboxylic acids (RCO_2H), like amides (e.g. $RCONH_2$), are protonated on the oxygen atom of the C=O group because this produces the most stable conjugate acid.

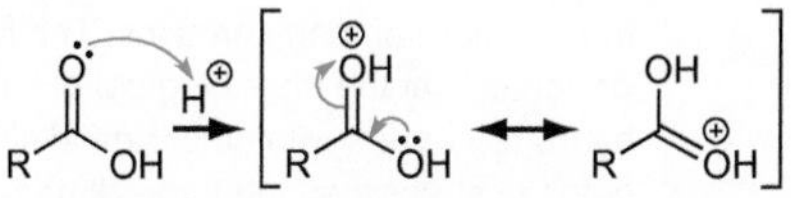

In 4-methoxybenzenamine (4-methoxyaniline or *p*-anisidine), the –I effect of the OMe group is much weaker than the +M effect. This is because the OMe group is well apart from the NH_2 group.

- Conversely, electron-donating substituents *increase* the basic strength of the nitrogen atom by inductive effects, particularly if a +I group is close to the lone pair in the 2-position, and by mesomeric effects if a +M group (e.g. OMe) is at the 2, 4, or 6 position of the ring.

Some nitrogen heterocycles, such as pyridine (C_5H_5N), also act as bases. The lone pair of electrons on nitrogen in pyridine is in an sp^2 orbital and this orbital is not involved with the delocalized electrons of the aromatic ring. On protonation of pyridine, the pyridinium ion ($C_5H_5NH^+$) formed is also aromatic because the proton reacts with the lone pair on nitrogen and this does not disrupt the 6 π electrons in the aromatic ring. However, pyridine is a weaker base than aliphatic amines, such as Et_3N, because the lone pair is in an sp^2 rather than an sp^3 orbital and so it is held closer to the nucleus. Pyridine is, however, a much stronger base than pyrrole (C_4H_4NH). For pyrrole, the lone pair of electrons on nitrogen contributes to the delocalized 6 π electrons and so the lone pair is much less available for protonation.

Worked example 19.4 (p.898) provides practice in comparing the strengths of bases.

For the use of pyridine as a base, see, for example, Section 20.2 (p.927).

Aromatic nitrogen heterocycles are discussed in Section 22.1 (p.1009).

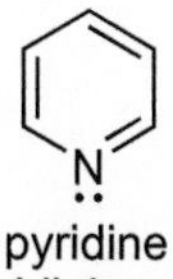

pyridine (pyridinium ion has a pK_a of 5)

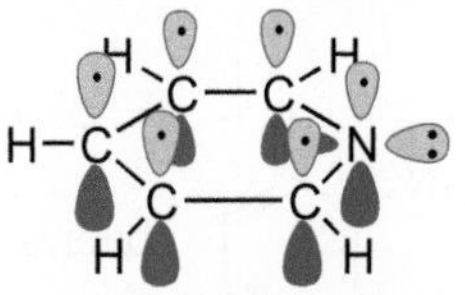

Lone pair does not contribute to the 6 π electrons in the aromatic ring

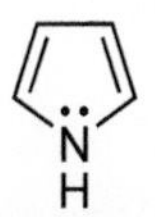

pyrrole (conjugate acid has a low pK_a of –4)

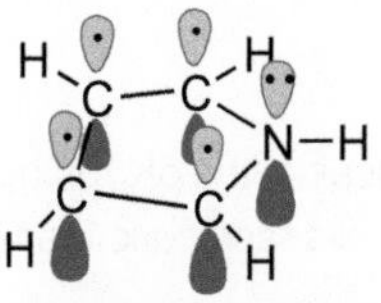

Lone pair does contribute to the 6 π electrons in the aromatic ring

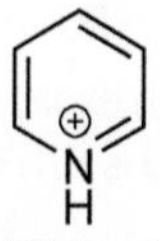

pyridinium ion (aromatic)

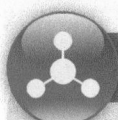

Box 19.5 Natural acids and bases

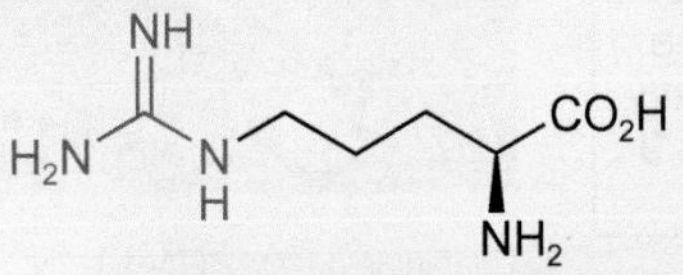
NH
H_2N
N
H
CO_2H
NH_2

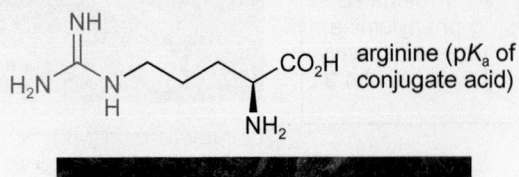

arginine (pK_a of conjugate acid)

Arginine is a common natural amino acid that plays an important role in the structure of many proteins. The basic guanidino group of arginine is protonated at the physiological pH of 7.4. This group can form ionic or hydrogen bonds with other amino side chains within proteins, which helps to strengthen the three-dimensional structure of proteins such as collagen. The skin of a rhinoceros contains layers of crisscrossed collagen, which forms a thick, protective armour.

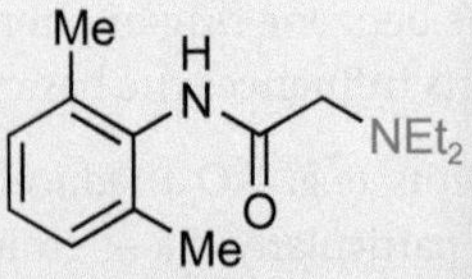
Me
H
N
NEt_2
O
Me

lignocaine (pK_a of conjugate acid)

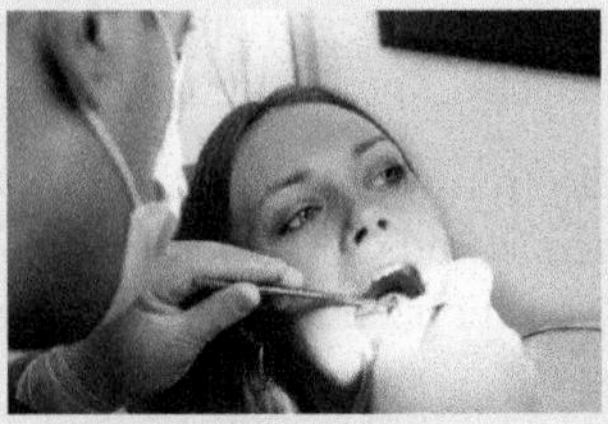

Local anaesthetics, used in dentistry, stop some nerve signals from reaching the brain. To be effective, the anaesthetic molecules must not be charged, because they have to pass through a fat-soluble layer that surrounds the nerve cells. The conjugate acid of the anaesthetic lignocaine (lidocaine) has a pK_a (7.9) slightly above the physiological pH (7.4) and, when in the body, around a quarter of the lignocaine molecules are uncharged.

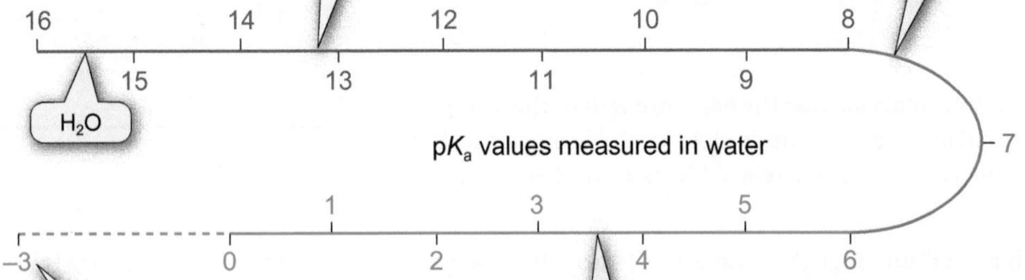
16
14
12
10
8
15
13
11
9
H_2O
7
pK_a values measured in water
1
3
5
−3
0
2
4
6

O
O
S
HO
OH

sulfuric acid

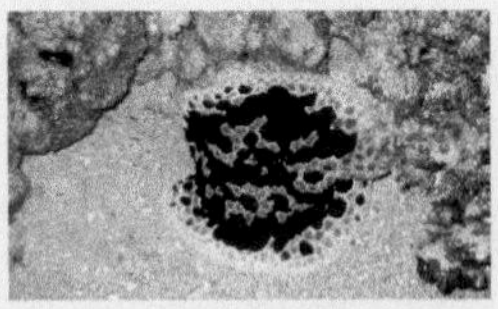

Some sea slugs secrete sulfuric acid as a mechanism of self-defence against predators.

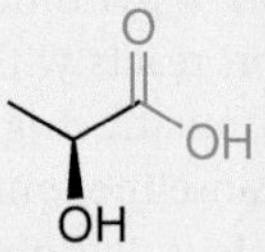
O
OH
OH

lactic acid

When the body makes lactic acid (the *S* isomer is formed), it reacts with water to form the conjugate base (called lactate) and H_3O^+. The production of H_3O^+ lowers the pH in muscles and this causes the ache that is observed during intense activity.

Question

(a) Suggest an explanation for why bupivacaine (the pK_a of the conjugate acid is 8.1) is a slower-acting local anaesthetic than lignocaine.

(b) Draw the structure of the conjugate acid of bupivacaine.

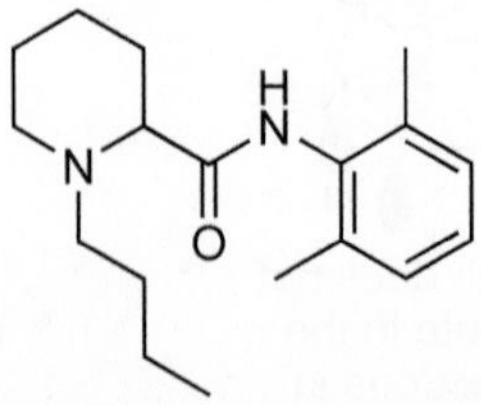
N
H
N
O

bupivacaine

Summary

- Basic compounds are good proton acceptors.
- A strong base has a relatively stable conjugate acid, with a high pK_a value.
- Organic bases generally have a formal negative charge or a lone pair of electrons on nitrogen or oxygen.
- Relatively strong organic bases have +I and +M groups that stabilize the conjugate acid.
- Relatively weak neutral organic bases have –I and –M groups.
- Pyridine is a much stronger base than pyrrole.

Characteristics of acid–base reactions

All acid–base reactions are equilibrium reactions, and the position of the equilibrium depends on the strength of the acid (HA) and the strength of the base (B^-). The strengths of the acids HA and BH (the conjugate acid of the base) are shown by their pK_a values, so you can use pK_a values of the reactants to predict if an acid–base reaction can take place. For an acid to donate a proton to a base, it must be a stronger acid than the conjugate acid of the base, so the conjugate acid of the base must have the higher pK_a value. If this is the case, the products will be more stable than the reactants so the position of equilibrium will lie over to the product side.

$$\underset{\text{acid}}{HA} + \underset{\text{base}}{B^{\ominus}} \rightleftharpoons \underset{\text{conjugate base}}{A^{\ominus}} + \underset{\text{conjugate acid}}{BH}$$

To help you identify and design acid–base reactions, the approximate pK_a values of a selection of acids measured in water are listed in Table 19.1. In Table 19.1, the pK_a values of the compounds gradually increase on going down the table. Strong acids are at the top of the table, while weak acids are at the bottom. Thus, any acid in the table will react with the conjugate base of any acid that is below it in the table. For example, an alcohol (ROH), such as ethanol (EtOH), is deprotonated by hydride ions (H^-) from sodium hydride (Na^+ H^-) to form an alkoxide ion (RO^-) and hydrogen.

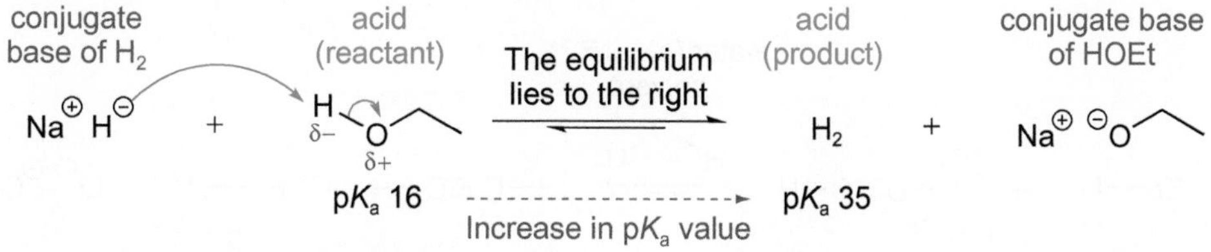

conjugate base of H_2
acid (reactant)
The equilibrium lies to the right
acid (product)
conjugate base of HOEt
Na⊕ H⊖
+
δ–
δ+
H_2
+
Na⊕ ⊖O
pK_a 16
pK_a 35
Increase in pK_a value

For acid–base reactions, you also need to consider the effect of the solvent in which the reaction is taking place. For example, ethyllithium (EtLi) deprotonates a terminal alkyne in a solvent such as hexane (C_6H_{14}). But if the solvent is changed to water, ethyllithium prefers to deprotonate water, as water is a stronger acid than an alkyne (see Figure 19.27). It is not possible to deprotonate a compound with a pK_a value well above 16 in water, because the water will act as the acid and the hydroxide ion will be formed. This is called the **levelling effect** of the solvent.

Acids and bases are useful catalysts in many organic reactions. Addition of small amounts of an acid or base can alter the reactivity of organic molecules and this can increase the rates of reactions. For example, in Figure 19.28 (diagram A) protonation of a ketone (R_2CO) makes the ketone more electrophilic and so it reacts more rapidly with nucleophiles. In Figure 19.28 (diagram B), deprotonation of an alcohol (ROH) forms an alkoxide ion (RO^-) that is more nucleophilic than the alcohol so it reacts more rapidly with electrophiles.

The difference in pK_a values between an acid and the conjugate acid of a base is equal to the log of the equilibrium constant for the acid–base reaction—the larger the difference in pK_a values, the larger the equilibrium constant (see Equation 7.14, p.317).

$$pK_c = pK_a(\text{HA}) - pK_a(\text{BH})$$

So that

$$K_c = 10^{(pKa(BH) - pKa(AH))}$$

Sodium hydride (NaH) can ignite in air and so it is typically sold as a dispersion in mineral oil (a mixture of alkanes), which is safer to handle.

Many organic reactions involve the transfer of a proton from one electronegative atom to another (see, for example, Box 19.6, p.897)—to draw mechanisms it is important to draw out all of the hydrogen atoms near the areas of the molecules that change.

The equilibrium constant (K_c) for the reaction of sodium hydride with ethanol is equal to $10^{pK_a\text{ acid product} - pK_a\text{ acid reactant}} = 10^{(35-16)} = 10^{19}$. To put this into context, 10 quintillion (10^{19}) is the estimated number of individual insects alive today. If K_c is greater than 10^8, for all practical purposes the proton transfer can be considered irreversible.

The levelling effect is introduced in Section 7.2 (p.309).

Water will donate a proton to any base stronger than the hydroxide ion.

On deprotonation of ethyne using ethyllithium, the curly arrows drawn actually lead to formation of a salt containing Li^+ and ethynyl ions ($^-C{\equiv}CH$). In fact, organometallic compounds can have covalent carbon–metal bonds and it is more accurate to draw 1-ethynyllithium with a polar C–Li bond (see Section 23.2, p.1068).

Table 19.1 pK_a values of some acids at 298 K (measured in water)

	Acid	Approximate pK_a	Conjugate base	
Strongest acid	$R_2C{=}OH^+$	−7.2	$R_2C{=}O$	Weakest base
	HCl	−7	Cl^-	
	R_2OH^+	−3.5	R_2O	
	H_2SO_4	−3	HSO_4^-	
	ROH_2^+	−2.4	ROH	
	H_3O^+	−1.7	H_2O	
	RCO_2H	4–5	RCO_2^-	
	Pyridinium ion	5	Pyridine	
	$CH_3COCH_2COCH_3$	9	$CH_3COCH^-COCH_3$	
	PhOH	9.9	PhO^-	
	R_3NH^+	10–11	R_3N	
	H_2O	15.7	HO^-	
	ROH	16–17	RO^-	
	CH_3COCH_3	20	$CH_3COCH_2^-$	
	$HC{\equiv}CH$	25	$HC{\equiv}C^-$	
	H_2	35	H^-	
	NH_3	38	H_2N^-	
Weakest acid	$H_2C{=}CH_2$	44	$H_2C{=}CH^-$	Strongest base
	CH_3CH_3	50	$CH_3CH_2^-$	

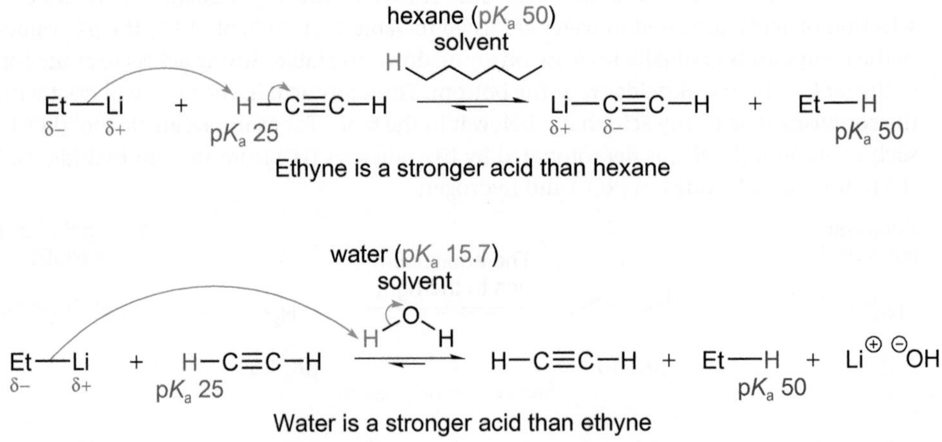

hexane (pK_a 50) solvent
Et—Li + H—C≡C—H ⇌ Li—C≡C—H + Et—H
δ− δ+ pK_a 25 δ+ δ− pK_a 50
Ethyne is a stronger acid than hexane
water (pK_a 15.7) solvent
Et—Li + H—C≡C—H ⇌ H—C≡C—H + Et—H + Li⊕ ⊖OH
δ− δ+ pK_a 25 pK_a 50
Water is a stronger acid than ethyne

Figure 19.27 The effect of the solvent on an acid–base reaction.

Catalysts increase the rate of a reaction by allowing the reaction to proceed by a different pathway (see Section 9.9, p.437). For an example of an acid-catalysed esterification reaction see Section 24.2 (p.1108).

Amino acids, including α-amino acids, $H_2NCH(R)CO_2H$, can be considered to undergo an intramolecular acid–base reaction. As they contain both a basic amine group (H_2N) and an acidic carboxylic acid group (CO_2H), there is an internal transfer of H^+ from the CO_2H group to the NH_2 group. The resulting ion, $^+H_3NCH(R)CO_2^-$, contains both a positive charge and a negative charge and is called a **zwitterion**. α-Amino acids exist in the zwitterion form in both the solid state and in solution, although they are often drawn as $H_2NCH(R)CO_2H$.

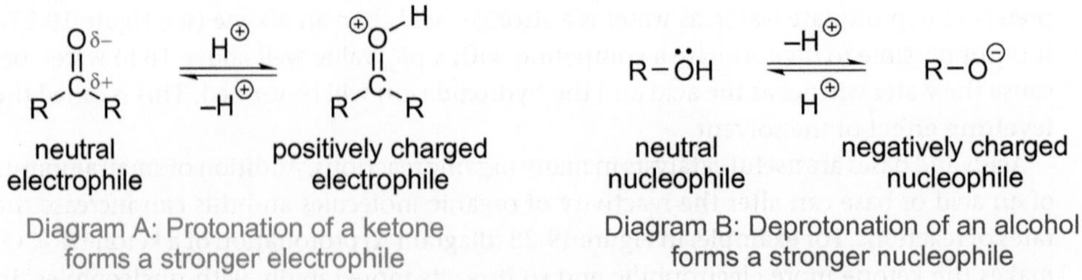

neutral electrophile
positively charged electrophile
Diagram A: Protonation of a ketone forms a stronger electrophile
neutral nucleophile
negatively charged nucleophile
Diagram B: Deprotonation of an alcohol forms a stronger nucleophile

Figure 19.28 Examples of the use of acid or base to increase the reactivity of a ketone or an alcohol, respectively.

Box 19.6 Natural acid–base catalysis

Chymotrypsin is a digestive enzyme that is secreted into the small intestine of the body, where it acts as a catalyst for the hydrolysis of amide bonds (e.g. RCONHR) in proteins to form amines (RNH_2) and carboxylic acids (RCO_2H).

The first step in the likely reaction mechanism involves nucleophilic attack by an OH group in the active site of the enzyme on to the C=O bond of the amide group in a protein chain. Normally, alcohols (ROH) react extremely slowly with amides because alcohols are relatively weak nucleophiles and amides are weak electrophiles. However, in the enzyme, the nucleophilicity of the OH group is increased by acid–base catalysis. This involves an adjacent imidazole group that is ideally positioned to interact and remove a proton from the OH group.

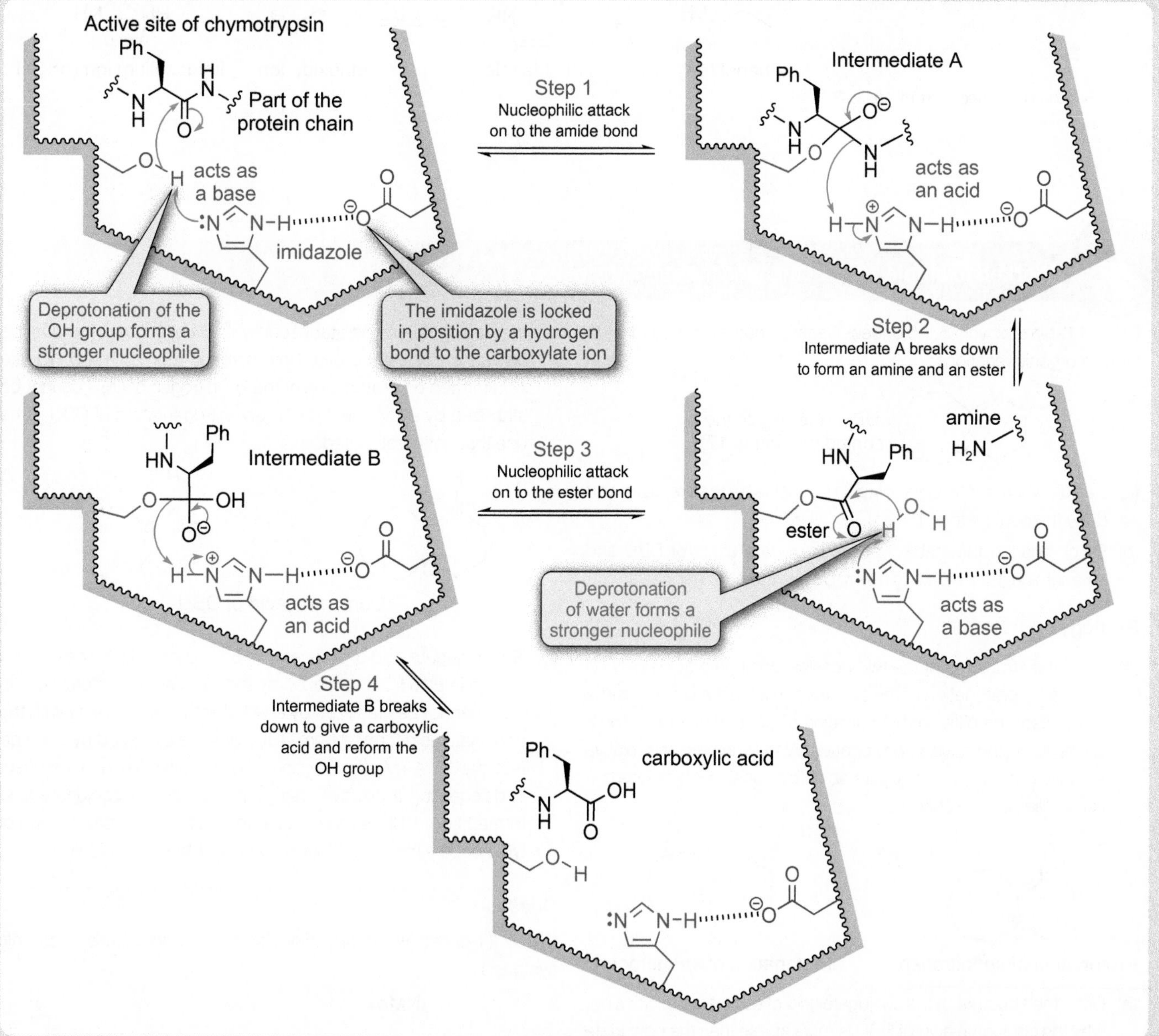

Active site of chymotrypsin
Part of the protein chain
acts as a base
imidazole
Deprotonation of the OH group forms a stronger nucleophile
The imidazole is locked in position by a hydrogen bond to the carboxylate ion
Step 1
Nucleophilic attack on to the amide bond
Intermediate A
acts as an acid
Step 2
Intermediate A breaks down to form an amine and an ester
amine
ester
acts as a base
Deprotonation of water forms a stronger nucleophile
Step 3
Nucleophilic attack on to the ester bond
Intermediate B
acts as an acid
Step 4
Intermediate B breaks down to give a carboxylic acid and reform the OH group
carboxylic acid

Follow the curly arrows in step 1, where the imidazole group removes a proton from the OH group. The resulting O⁻ group is a better nucleophile than OH and attacks the C=O bond of the amide to form an intermediate (labelled intermediate A) with a tetrahedral carbon atom at the point of attack. In step 2 the proton is transferred back to the protein chain and the C–N bond in intermediate A breaks to give an amine and an ester (RCO_2R).

In step 3, the amine moves out of the active site of the enzyme and is replaced by water, which reacts with the ester. Water usually reacts only slowly with esters (as water is a weak nucleophile), but →

in the active site of the enzyme the imidazole group acts as a base and water is converted into the hydroxide ion. The hydroxide ion is a stronger nucleophile than water and attacks the C=O bond of the ester to form a second intermediate (labelled B). Finally, in step 4, the proton transfer is reversed again and intermediate B breaks down to form a carboxylic acid and regenerate the OH group in the enzyme.

Question

In step 1, in the above mechanism, chymotrypsin produces an alkoxide ion by deprotonation of an alcohol using an imidazole group as the base. In the absence of the enzyme, where would you expect the equilibrium of the reaction of ethanol with imidazole (shown below) to lie?

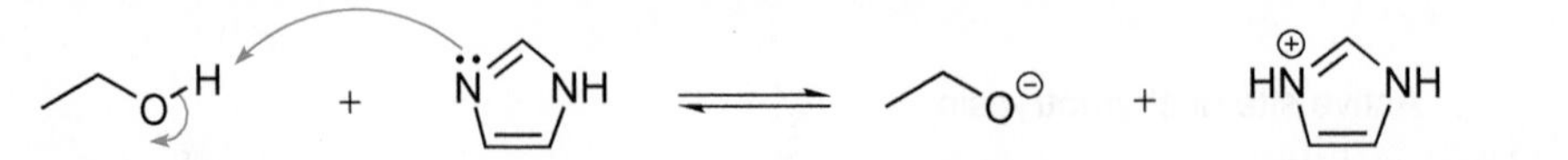

ethanol (pK_a 16) imidazole ethoxide ion imidazolium ion (pK_a 7)

Reaction of ethanol with imidazole ➤

Worked example 19.4 Comparing the strengths of bases

DBU (1,8-diazabicyclo[5.4.0]undec-7-ene) is commonly used as a base in organic reactions.

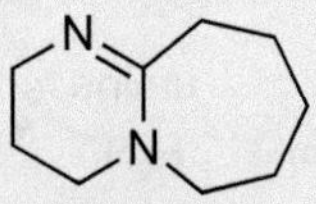

DBU (the pK_a of the conjugate acid is 12)

(a) Decide where protonation of DBU is most likely to take place and draw the conjugate acid.

(b) By comparing the stability of the conjugate acids of DBU and ethanamine ($EtNH_2$), explain why DBU is the stronger base.

Strategy

(a) Draw the structures of the two possible conjugate acids of DBU formed by protonation of the sp^2 or sp^3 nitrogen atom. Consider what effect the different hybridization of the two nitrogen atoms will have on the ease of protonation. Also, compare the relative stabilities of the two conjugate acids by considering delocalization of the positive charge.

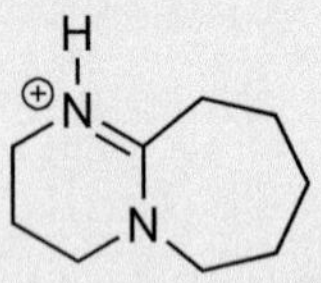

Protonation of sp^2 nitrogen

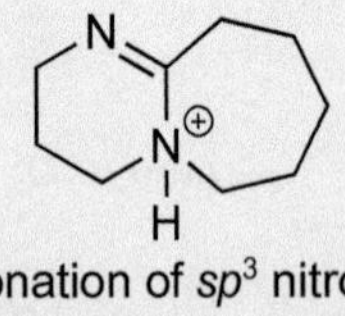

Protonation of sp^3 nitrogen

(b) Draw the structure of the conjugate acid of ethanamine. Consider why the conjugate acid of DBU is more stable than the conjugate acid of ethanamine.

Solution

(a) A lone pair in an sp^2 orbital is held more tightly to the nucleus and is usually harder to protonate than a lone pair in an sp^3 orbital (p.890). However, if the sp^2 nitrogen atom of DBU is protonated, the conjugate acid is stabilized by a +M effect and the positive charge is delocalized over both nitrogen atoms. The conjugate acid formed on protonation of the sp^3 nitrogen of DBU cannot be stabilized by a +M effect so the sp^2 nitrogen atom of DBU is the one that gets protonated.

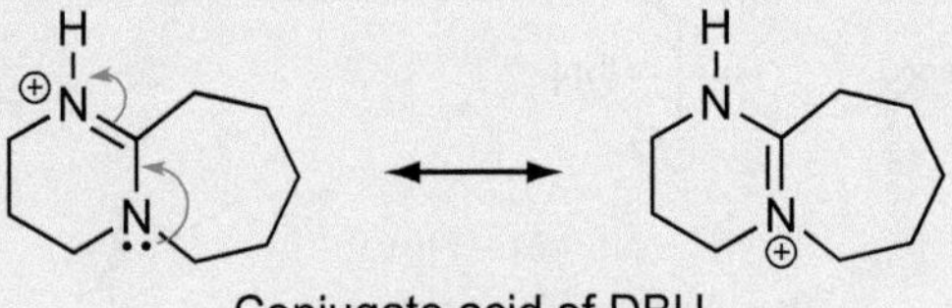

Conjugate acid of DBU

(b) The conjugate acid of ethanamine ($EtNH_2$) is $EtNH_3^+$. The positive charge in $EtNH_3^+$ is stabilized by the +I effect of the ethyl group but it cannot be stabilized by a +M effect. This contrasts with the conjugate acid formed from DBU, which is stabilized by a +M effect (part a). A +M effect is generally more effective than a +I effect in delocalizing a positive charge (p.875), so the conjugate acid formed from DBU is more stable than $EtNH_3^+$. As protonation of DBU occurs more readily, it is a stronger base than $EtNH_2$.

Question

DMAP ((4-dimethylamino)pyridine) is another useful base in organic synthesis.

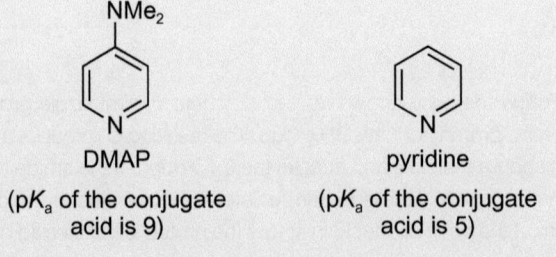

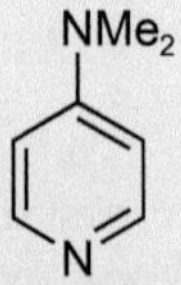

DMAP

(pK_a of the conjugate acid is 9)

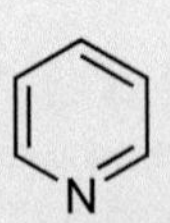

pyridine

(pK_a of the conjugate acid is 5)

➜

→

(a) Draw the structure of the more stable conjugate acid of DMAP.

(b) By comparing the stability of the conjugate acids of DMAP and pyridine, explain why DMAP is the stronger base.

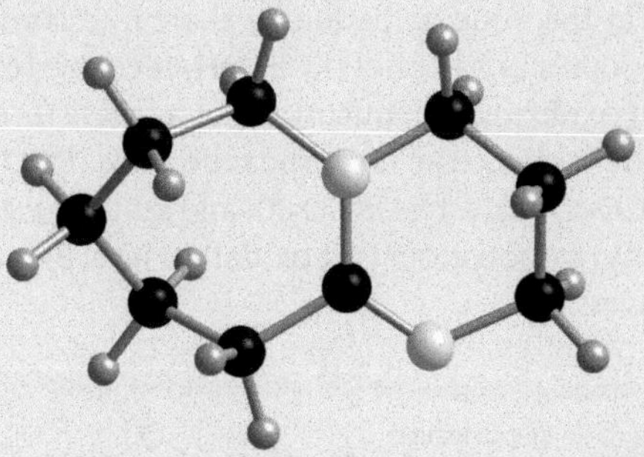

A molecular model of DBU (a common base for reactions of organic molecules).

Summary

- Acid–base reactions are equilibrium reactions, and the position of the equilibrium depends on the pK_a values of the acid and the conjugate acid of the base.
- For an acid to donate a proton to a base, the acid must have a lower pK_a value than the conjugate acid of the base.
- Acids and bases are useful catalysts. They can alter the reactivity of some organic molecules and increase the rate of reaction.

For a practice question on this topic, see question 5 at the end of this chapter (p.914).

Redox reactions

Another way of classifying organic reactions is by considering whether oxidation or reduction has taken place. To do this, you need to determine the **oxidation levels** of the carbon atoms that are part of, or attached to, the functional groups in the reactants and products using the following guidelines.

- Every bond the carbon atom makes to a heteroatom (such as oxygen or nitrogen) increases the oxidation level of the carbon atom by +1.
- The greater the degree of unsaturation, the higher the oxidation level. Alkenes are assigned an oxidation level of 1 while alkynes have an oxidation level of 2.

Table 19.2 shows the oxidation levels of carbon atoms associated with some common functional groups, which have oxidation levels from 0 to 4. For example, the oxidation

Oxidation levels are different from oxidation states. Carbon has nine oxidation states ranging from –4 (in CH_4) to +4 (in CO_2 or CCl_4). Oxidation states can be used to assess if an organic reaction involves redox, but oxidation levels are more useful because they concentrate on the functional groups. Oxidation states and oxidation levels are introduced in Section 1.4 (p.28).

Table 19.2 Oxidation levels (0–4) of carbon atoms associated with some functional groups (X = Cl, Br, I)

0	1	2	3	4
alkane (R_4C)	alkene (RC=CR)	alkyne (RC≡CR)		
	monohalogenoalkane (RCH_2X)	dihalogenoalkane ($RCHX_2$)	trihalogenoalkane (RCX_3)	tetrahalogenoalkane (CX_4)
	ether (RCH_2OR)	acetal ($RCH(OR)_2$)		
	primary/secondary alcohol (RCH_2OH/R_2CHOH)	aldehyde/ketone ($RCHO/R_2CO$)	carboxylic acid (RCO_2H)	carbon dioxide (CO_2)
	primary amine (RCH_2NH_2)	imine (RCH=NH)	nitrile (RC≡N)	

A reaction that produces an ether (ROR) is called etherification (Section 20.3, p.931), while a reaction that forms an ester (RCO_2R) is called esterification (Section 24.2, p.1108).

level of the carbon in the primary alcohol CH_3OH is 1 because there is one bond to oxygen. This classification helps you to identify the type of reaction required to interconvert functional groups. Carbon atoms at the same oxidation level are interconverted without oxidation or reduction. Therefore, a primary alcohol (RCH_2OH) is converted into an ether (RCH_2OR) without the need for an oxidizing or reducing agent as these functional groups are at the same oxidation level. The same is true for carboxylic acids (RCO_2H), acyl chlorides (RCOCl), acid anhydrides (RCO_2COR), esters (RCO_2R), and amides (e.g. $RCONH_2$). In all of these compounds, the carbonyl carbon atom is at oxidation level 3 so redox reagents are not required for their interconversion.

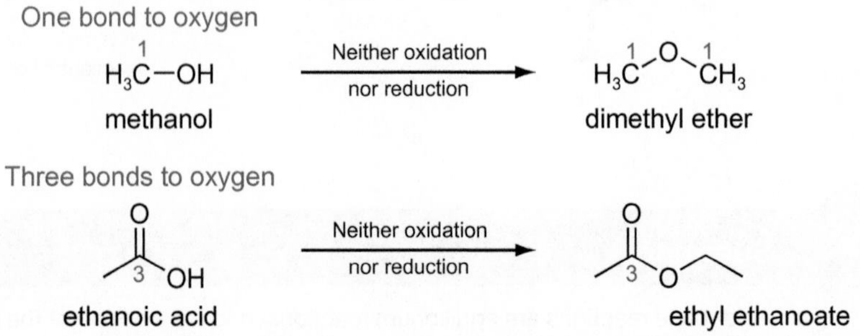
One bond to oxygen
1
H₃C—OH
methanol
Neither oxidation
nor reduction
1 O 1
H₃C CH₃
dimethyl ether
Three bonds to oxygen
O
3 OH
ethanoic acid
Neither oxidation
nor reduction
O
3 O
ethyl ethanoate

Alcohols are oxidized to carbonyl compounds using oxidizing agents such as potassium permanganate ($KMnO_4$), sodium dichromate ($Na_2Cr_2O_7$), or chromium trioxide (CrO_3) (Box 23.2, p.1064).

Carbonyl compounds are reduced to alcohols using reducing agents such as sodium borohydride ($NaBH_4$) or lithium aluminium hydride ($LiAlH_4$) (Section 23.2, p.1063).

Alkynes and alkenes are reduced to alkanes by reaction with H_2 and a Pd/C catalyst (Section 21.2, p.968).

Carbon atoms at different oxidation levels are interconverted by oxidation or reduction. If the carbon atom of the product is at a higher oxidation level than the corresponding carbon atom in the reactant, then an oxidizing agent is required. For example, to convert a primary alcohol (RCH_2OH) into a ketone (RCOR) requires an oxidation. Conversely, a reducing agent is required if the carbon atom in the product is at a lower oxidation level than the corresponding carbon atom in the starting material.

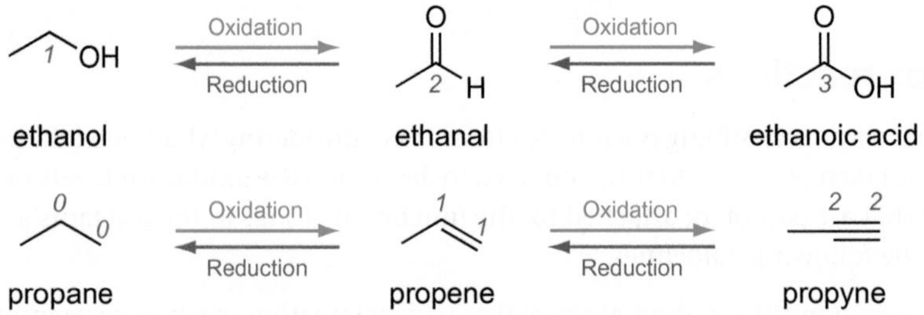
1 OH
ethanol
Oxidation
Reduction
O
2 H
ethanal
Oxidation
Reduction
O
3 OH
ethanoic acid
0
0
propane
Oxidation
Reduction
1
1
propene
Oxidation
Reduction
2 2
propyne

For oxidation or reduction of alkenes and alkynes, the oxidation level of *both* carbon atoms in the double or triple bond must either increase or decrease. To convert an alkyne (RC≡CR) into an alkane (RCH_2–CH_2R) requires a reduction as the oxidation levels of both carbons in the triple bond are decreased.

Worked example 19.5 provides practice in assigning oxidation levels.

Box 19.7 Aspirin

Aspirin (acetylsalicylic acid) has been widely used for over 100 years to alleviate the swelling, redness, and pain of inflammation, to reduce a fever, and to relieve a headache. Bayer and Co. first put aspirin on the market in 1899 as a safe and convenient way of supplying the body with the active substance salicylic acid. Taking salicylic acid itself can cause severe stomach and mouth irritation. It is believed that salicylic acid inhibits an enzyme that produces a family of biologically active compounds called prostaglandins (such as $PGF_{2\alpha}$, p.71) that cause inflammation in the body.

Although aspirin has been around for a long time, its biological activity is still being investigated. Recent studies have focused on the potential of aspirin to prevent such diverse ailments as heart attack and colon cancer. For example, clinical studies performed over many years have shown that regular low-dose aspirin use can

▲ The bark of the willow tree (genus *Salix*) contains salicin from which aspirin is derived. Extracts from willow trees have been used medicinally for thousands of years.

reduce the recurrence of heart attack or stroke by up to 50%. Today, the world production of aspirin has been estimated at 36 000 tonnes a year. In developed countries, this averages to a consumption of around 70 tablets per person per year. Aspirin was even included in the first-aid kit on USA's Apollo 11 mission to the moon.

Question

Salicin, isolated from the willow tree, reacts with water (in a hydrolysis reaction) to form salicyl alcohol and glucose. The conversion of salicyl alcohol into aspirin involves the three steps in the reaction scheme shown below. Salicylic acid is converted into aspirin using ethanoyl chloride (CH_3COCl).

Assign oxidation levels to the carbon atoms in ethanoyl chloride and to each of the four molecules in the reaction scheme. Use your values to identify which steps involve redox reactions.

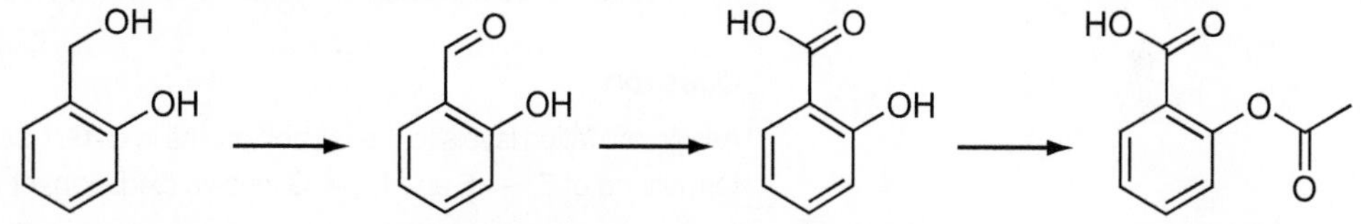
OH
OH
O
OH
HO
O
OH
HO
O
O
O

salicyl alcohol salicaldehyde salicylic acid aspirin

Conversion of salicyl alcohol ► into aspirin

Worked example 19.5 Assigning oxidation levels

Assign oxidation levels to the carbon atoms in **A–D**. Use these values to determine if the transformations of **A** → **B**, **A** → **C**, and **A** → **D** involve oxidation, or reduction, or neither.

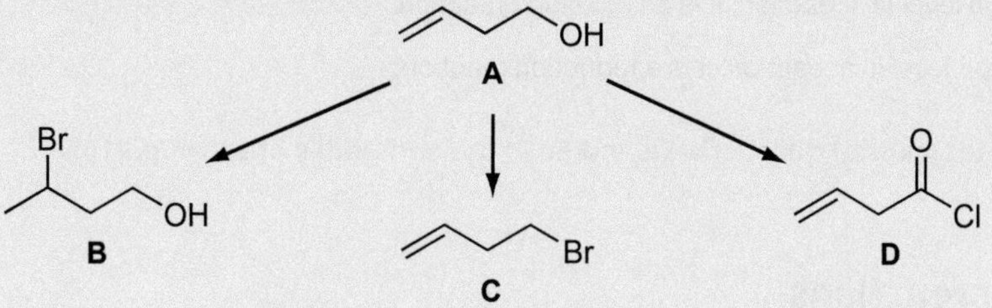
OH
A
Br
OH
B
Br
C
O
Cl
D

Strategy

To determine the oxidation levels, count the number of bonds to oxygen, bromine, and chlorine for each carbon atom in all four molecules. Carbon atoms in alkenes are at oxidation level 1.

If the oxidation level increases in the reaction, this is an oxidation reaction.

If the oxidation level decreases in the reaction, this is a reduction reaction.

If there is no change in oxidation level, the reaction does not involve redox.

→

Solution

The blue numbers indicate oxidation levels in the diagram shown here.

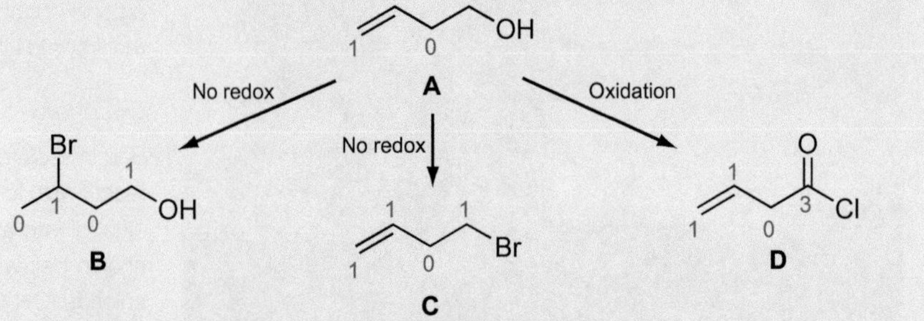

1
1
OH
1
0
A
No redox
No redox
Oxidation
Br
1
1
0
0
OH
B
1
1
Br
1
0
C
O
1
3
Cl
1
0
D

The conversion of A (but-3-en-1-ol) into B (3-bromobutan-1-ol) requires addition of HBr to the C=C bond (Section 21.3, p.970). The conversion of A into C (4-bromobut-1-ene) requires a brominating agent such as phosphorus tribromide, PBr_3 (Section 20.2, p.926). The conversion of A into D (but-3-enoyl chloride) involves oxidation of the primary alcohol to form a carboxylic acid (Box 23.2, p.1064) followed by reaction with a chlorinating agent such as thionyl chloride, $SOCl_2$ (Box 24.4, p.1110).

The conversion of **A** → **B** is not a redox reaction because the oxidation level of *both* alkene carbons in **A** does not change on going to **B** (p.900).

The conversion of **A** → **C** is not a redox reaction because the C–OH and C–Br carbon atoms are at the same oxidation level.

The conversion of **A** → **D** is an oxidation reaction because the C–OH carbon (level 1) is transformed into the COCl carbon of the acyl chloride (level 3).

Question

Assign oxidation levels to the carbon atoms in **E–G**. Use these values to determine if the transformations of **E** → **F** and **F** → **G** involve oxidation, or reduction, or neither.

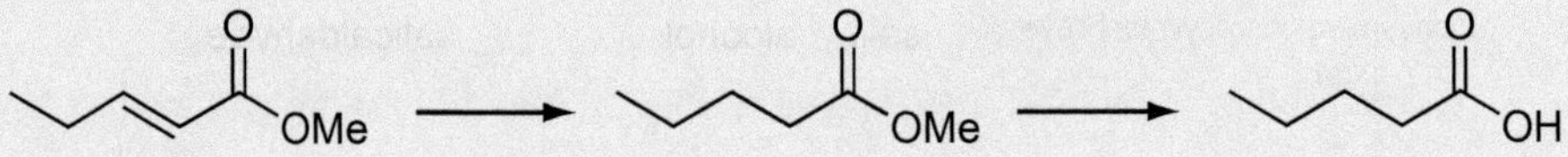

O
OMe
O
OMe
O
OH

E **F** **G**

Summary

- Carbon atoms at the *same* oxidation level are interconverted without oxidation or reduction.
- Carbon atoms at *different* oxidation levels are interconverted by oxidation or reduction.
- When carbon atoms increase oxidation level in a reaction, it is an oxidation reaction.
- When carbon atoms decrease oxidation level in a reaction, it is a reduction reaction.

For practice questions on these topics, see questions 7a and 7c at the end of this chapter (p.915).

A polar molecule has a partial positive charge (δ+) at one end of the molecule and a complementary partial negative charge (δ–) at the other end. Polar bonds include the $^{\delta+}C–X^{\delta-}$ bond in a halogenoalkane and the $^{\delta+}C=O^{\delta-}$ bond in a carbonyl compound.

Nucleophiles *donate* a pair of electrons to form a covalent bond. Electrophiles *accept* a pair of electrons to form a covalent bond.

Polar reactions

Polar reactions are the most common reactions of organic molecules. These reactions take place between two polar molecules or between a polar molecule and a charged molecule. The three main classes of polar reactions are addition, elimination, and substitution reactions.

- **Addition reactions** involve the addition of two reactants to form a single product. Unsaturated compounds with double or triple bonds, such as alkenes ($R_2C=CR_2$), alkynes (RC≡CR), aldehydes (RC(=O)H), and ketones (RC(=O)R), mainly undergo addition reactions. Examples of two types of addition reaction are shown in Figure 19.29.

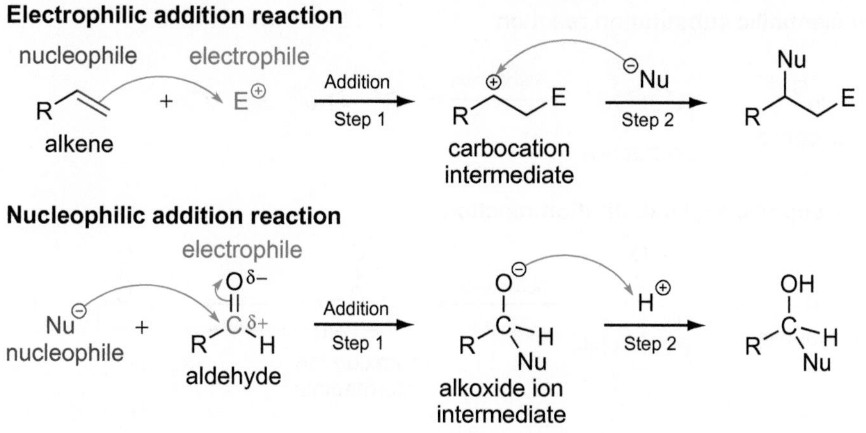
Electrophilic addition reaction
nucleophile
electrophile
R
alkene
+
E
Addition
Step 1
R
E
carbocation
intermediate
Nu
Step 2
Nu
R
E
Nucleophilic addition reaction
electrophile
O δ−
C δ+
R
H
aldehyde
Nu
nucleophile
+
Addition
Step 1
O
C
R
H
Nu
alkoxide ion
intermediate
H
Step 2
OH
C
R
H
Nu

Figure 19.29 Addition reactions can involve attack by an electrophile or by a nucleophile.

For addition reactions of alkenes and alkynes, the C=C and C≡C bonds act as the nucleophile and the reactant acts as the electrophile. As an electrophile adds to the alkene or alkyne in the first step of the reaction, these are called **electrophilic addition reactions.**

The mechanism of electrophilic addition is discussed in detail in Section 21.3 (p.970).

For addition reactions of aldehydes (RCHO) and ketones (RCOR) the positively polarized carbon atom in the C=O bond is the electrophile and the reagent is the nucleophile. As the C=O bond is attacked by a nucleophile in the first step of the reaction, these are called **nucleophilic addition reactions.**

The mechanism of nucleophilic addition is discussed in detail in Section 23.2 (p.1061).

- **Elimination reactions** are the opposite of addition reactions, as one reactant molecule is converted into two products. Saturated compounds are converted into unsaturated compounds in elimination reactions. For example, in one type of mechanism halogenoalkanes (RX) lose HX to form alkenes in elimination reactions. This occurs, for example, when a halogenoalkane reacts with a base, as shown in Figure 19.30. The base removes a proton from the halogenoalkane. This initiates the formation of a new C=C bond and loss of the halide ion (X^-), in a concerted (one-step) elimination reaction.

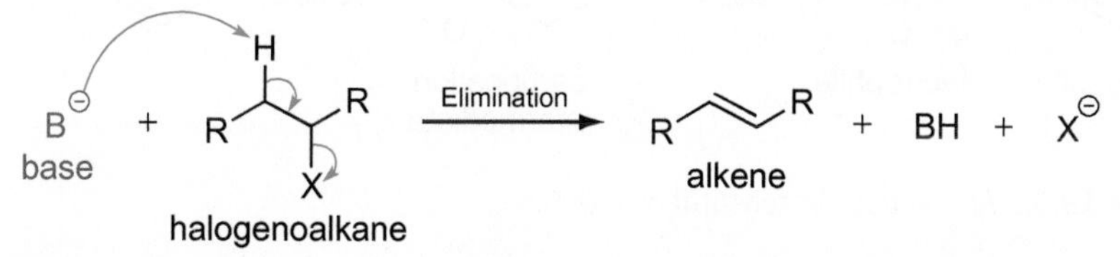
B
base
+
H
R
R
X
halogenoalkane
Elimination
R
R
alkene
+ BH + X

Figure 19.30 An example of an elimination reaction.

Elimination reactions that take place by a concerted mechanism are called E2 reactions. Some elimination reactions take place in two steps, via intermediate carbocations (R^+). The mechanisms of elimination are discussed in Section 20.4 (p.943).

- In a **substitution reaction**, a functional group in a particular compound is replaced by another group. Various functional groups undergo substitution reactions including halogenoalkanes (RX), carboxylic acid derivatives such as esters (RCO_2R) and acyl chlorides (RCOCl), and compounds containing a benzene ring. Two examples of nucleophilic substitution are shown in Figure 19.31.

Halogenoalkanes (e.g. RCH_2X) undergo substitution reactions that result in the halogen atom (X) being replaced by another atom or group of atoms. In one type of mechanism, a nucleophile attacks the positively polarized carbon attached to the halogen atom. This initiates the formation of a new C–Nu bond and loss of the halide ion (X^-) in a concerted substitution reaction. Since a nucleophile attacks the C–X bond in the halogenoalkane and replaces the halide atom, this is called a **nucleophilic substitution reaction.**

Nucleophilic substitution reactions that take place by a concerted mechanism are called S_N2 reactions. Some nucleophilic substitution reactions take place in two steps, via intermediate carbocations (R^+). The mechanisms of nucleophilic substitution are discussed in Section 20.3 (p.930).

Carboxylic acid derivatives, such as acyl chlorides (RCOCl), also undergo nucleophilic substitution reactions. The nucleophile attacks the positively polarized carbon atom of the C=O bond of an acyl chloride to produce an intermediate alkoxide ion,

The mechanism of nucleophilic acyl substitution is discussed in Section 24.2 (p.1107).

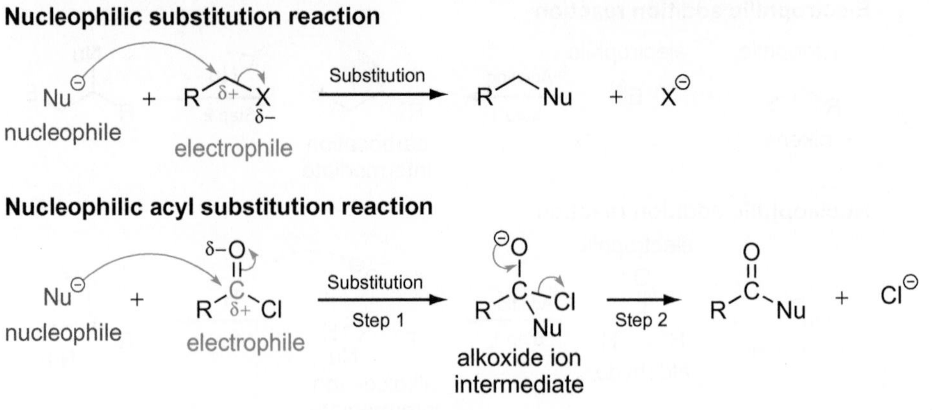

Nucleophilic substitution reaction
Nu
nucleophile
+
R
δ+
X
δ−
electrophile
Substitution
R
Nu
+
X
Nucleophilic acyl substitution reaction
δ−O
C
R
δ+
Cl
Nu
+
nucleophile
electrophile
Substitution
Step 1
O
C
Cl
Nu
R
alkoxide ion
intermediate
Step 2
O
C
R
Nu
+
Cl

Figure 19.31 Two examples of nucleophilic substitution.

> The use of 'acyl' in nucleophilic acyl substitution shows that substitution takes place at a C=O bond. Sometimes these reactions are called nucleophilic addition–elimination reactions.

with a negative charge on oxygen. In a second step, the alkoxide ion loses a chloride ion (Cl^-) to form a new carboxylic acid derivative, RCONu. Overall, the Cl group of the acyl chloride is substituted by the nucleophile in a **nucleophilic acyl substitution reaction**.

Electrophilic substitution reactions are characteristic of benzene (and derivatives). In these reactions, another atom or group of atoms replaces one or more hydrogen atoms on the benzene ring. The benzene ring contains 6 π delocalized electrons, which attract electrophiles. For example, benzene (C_6H_6) attacks reactive electrophiles such as a nitronium ion ($^+NO_2$; formed from a mixture of concentrated nitric acid and sulfuric acid) to produce an intermediate carbocation (see Figure 19.32). In a second step, the carbocation loses a proton to reform the aromatic ring (to give nitrobenzene, $C_6H_5NO_2$). Overall, an electrophilic reagent reacts with the benzene ring and replaces a hydrogen atom, so this is called an **electrophilic substitution reaction.**

Electrophilic substitution reactions of benzene are discussed in Section 22.2 (p.1013).

Worked example 19.6 (p.906) provides practice in classifying reactions.

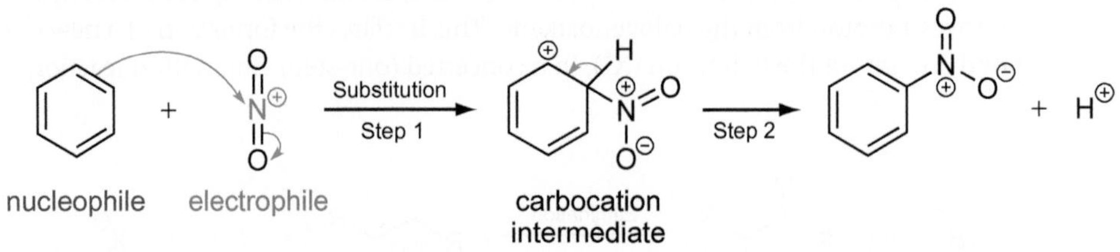

O
N
O
+
Substitution
Step 1
H
N
O
O
Step 2
O
N
O
+
H
nucleophile
electrophile
carbocation
intermediate

Figure 19.32 An electrophilic substitution reaction.

Box 19.8 Superglue and fingerprints

To identify criminals from fingerprints left at the scenes of crimes you need a method of making the fingerprints visible. Most fingerprints are deposits of sweat, grease, and various amino acids and sugars. One way to show up a fingerprint is to spray it with ethyl 2-cyanoacrylate ($H_2C=C(CN)CO_2Et$), the monomer in superglue. Hydroxide ions produced from water in the air start the polymerization of the cyanoacrylate. A white polymer forms along the ridges of

Robots are able to collect fingerprint evidence from packages that might be too dangerous for a person to approach. One device consists of a heating element, a cartridge of superglue, and a short pipe. Using remote controls, the superglue is heated and the fumes directed towards the package. The fumes react with any fingerprints to form white polymers, which are photographed (using the robot's high-definition camera) before safe disposal of the package. ➤

➜

→

the fingerprint to give an image of the fingerprint that can be photographed and identified.

The mechanism of polymerization of ethyl 2-cyanoacrylate in the presence of hydroxide ions is shown below. In the first step, the hydroxide ion attacks the C=C bond at the β-position of the acrylate. This is unusual because C=C bonds usually react with electrophiles, not with nucleophiles. However, the nitrile (CN) and ester (CO_2Et) substituents on the C=C bond withdraw electron density (by mesomeric effects) from the β-position of the acrylate and this promotes the nucleophilic attack of HO^-. A carbanion is produced at the α-position, which is stabilized by the nitrile and ester groups. The carbanion acts as a nucleophile and attacks another molecule of cyanoacrylate, to form a second carbanion, and so on. The additions continue to form a long polymer chain of repeating cyanoacrylate units (*n* denotes that a large number of units are linked together). Finally, the growth of the chain is stopped by protonation of one of the intermediate carbanions. The mechanism of the polymerization involves a series of addition reactions, so this is called **addition polymerization**.

Resonance forms of ethyl 2-cyanoacrylate explaining why the β position is susceptible to nucleophilic attack

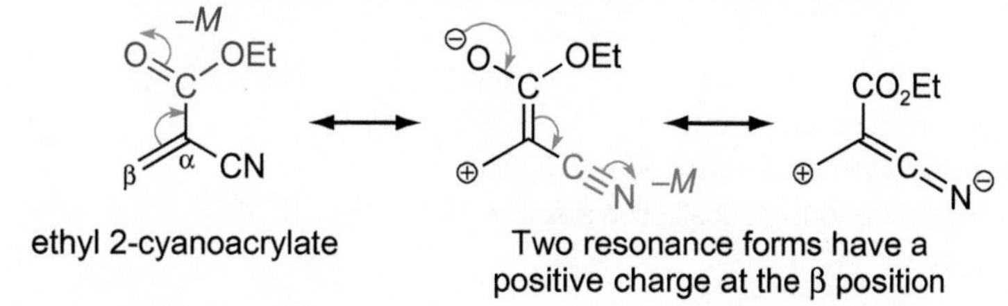
–M
O
OEt
C
β
α
CN
OEt
CO2Et
C≡N –M
C=N
ethyl 2-cyanoacrylate
Two resonance forms have a positive charge at the β position

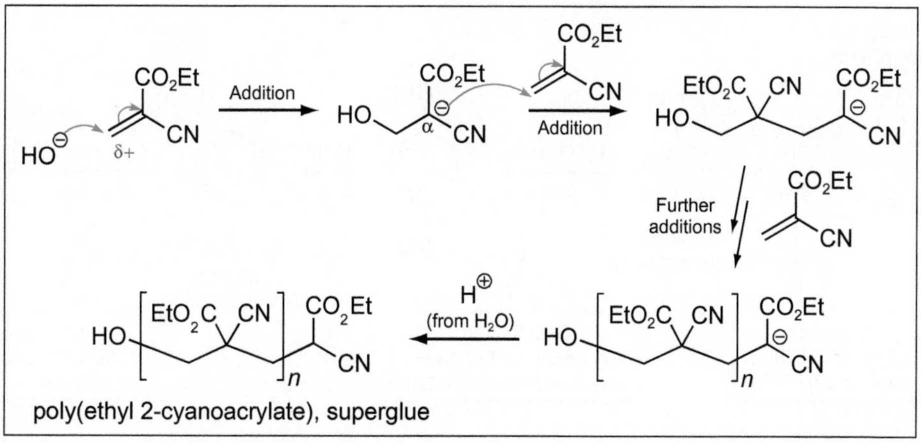
CO2Et
Addition
CN
HO
δ+
HO
α
CN
Addition
EtO2C
CN
Further additions
H⊕
(from H2O)
n
poly(ethyl 2-cyanoacrylate), superglue

Question

The polymerization of ethyl 2-cyanoacrylate by hydroxide ion is an example of an **anionic polymerization** (because the first step involves an anion). Alkenes also undergo polymerization by reaction mechanisms that involve radicals or carbocations. For example, the **cationic polymerization** of 2-methylpropene (isobutylene; $Me_2C{=}CH_2$) can be promoted by the addition of a small amount of an acid as shown here.

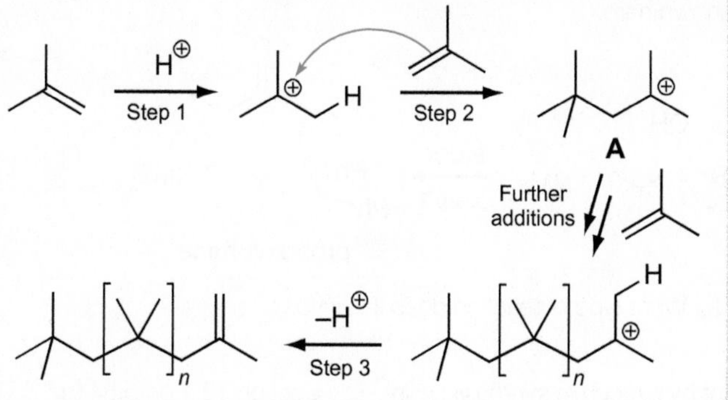
H⊕
Step 1
H
Step 2
A
Further additions
H
–H⊕
Step 3
n

(a) Use a curly arrow to show the mechanism of step 1.

(b) Explain why carbocation **A** is formed in step 2, rather than carbocation **B** (formed by the mechanism shown below).

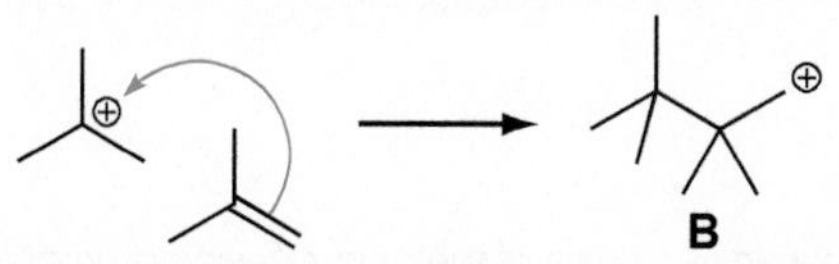
B

(c) Use a curly arrow to show the mechanism of step 3.

Worked example 19.6 Classifying reactions

A four-step synthesis of the pain-relieving drug phenacetin is illustrated here. Identify the reactions (electrophilic or nucleophilic addition, elimination, nucleophilic or electrophilic substitution) taking place in steps 1, 2, and 4.

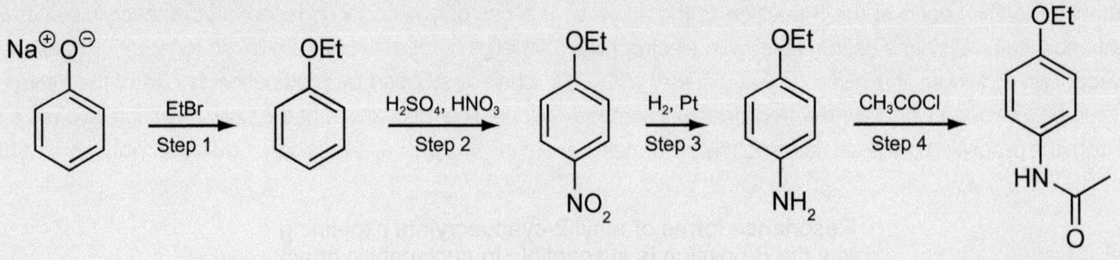
EtBr
Step 1
H_2SO_4, HNO_3
Step 2
H_2, Pt
Step 3
CH_3COCl
Step 4

phenacetin

Strategy

Identify the functional groups that react in steps 1, 2, and 4.

Classify the functional groups as being either nucleophilic or electrophilic.

Classify the reaction as an addition, elimination, or substitution reaction.

Solution

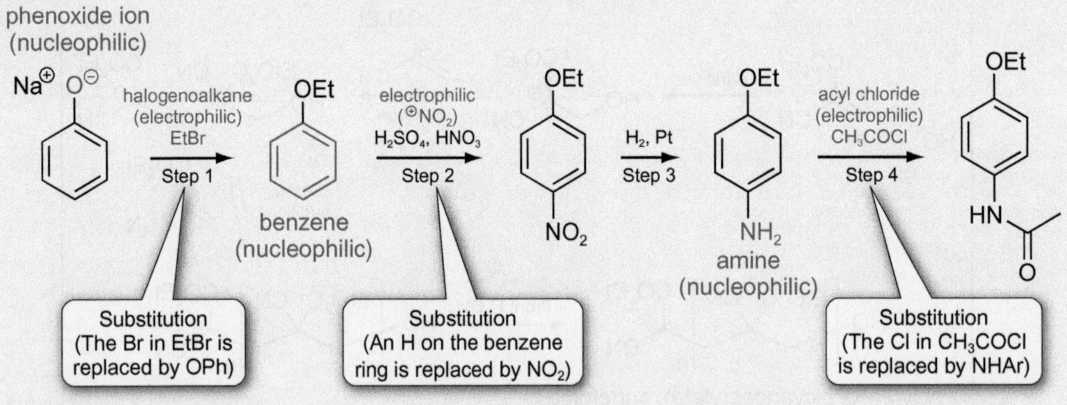
phenoxide ion (nucleophilic)
halogenoalkane (electrophilic) EtBr
Step 1
benzene (nucleophilic)
electrophilic ($^{\oplus}NO_2$) H_2SO_4, HNO_3
Step 2
H_2, Pt
Step 3
amine (nucleophilic)
acyl chloride (electrophilic) CH_3COCl
Step 4
Substitution (The Br in EtBr is replaced by OPh)
Substitution (An H on the benzene ring is replaced by NO_2)
Substitution (The Cl in CH_3COCl is replaced by NHAr)

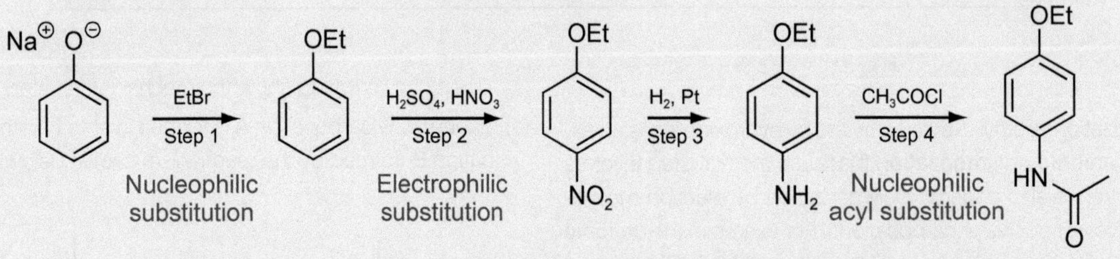
EtBr
Step 1
Nucleophilic substitution
H_2SO_4, HNO_3
Step 2
Electrophilic substitution
H_2, Pt
Step 3
CH_3COCl
Step 4
Nucleophilic acyl substitution

Question

Part of a racemic synthesis of the pain-relieving drug propoxyphene is shown here.

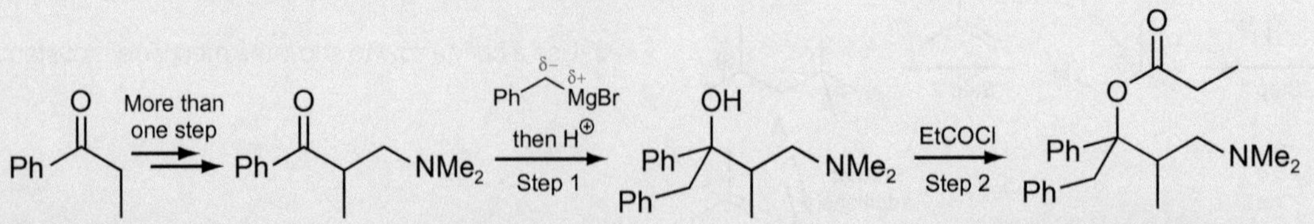
More than one step
Step 1
then $H^{\oplus}$
EtCOCl
Step 2

A propoxyphene

(a) Compound **A** is prepared by reacting benzene with EtCOCl and $AlCl_3$. What type of reaction does this involve?

(b) What types of reactions are taking place in steps 1 and 2?

(c) Identify the chiral centres in propoxyphene and explain what is meant by a *racemic synthesis*. (Hint. See Section 18.4 on p.841.)

Summary

- An addition reaction involves the addition of two reactants to form a single product.
- An elimination reaction is the opposite of an addition reaction, as one reactant molecule is converted into two products.
- A substitution reaction involves a functional group in a particular compound being replaced by another group.
- Functional groups react by characteristic reaction mechanisms:

electrophilic addition	alkene
nucleophilic addition	aldehyde, ketone
elimination	halogenoalkane
nucleophilic substitution	halogenoalkane
nucleophilic acyl substitution	carboxylic acid derivatives
electrophilic substitution	benzene (and derivatives)

For practice questions on these topics, see questions 6a and 6b at the end of this chapter (p.914).

Radical reactions

Many reactions involving radicals are **chain reactions**. These reactions involve a series of steps in which each step generates a radical product.

Heating or shining UV radiation on to a reactant starts a chain reaction by forming radicals in an **initiation step**. Reactants, such as peroxides (ROOR), that start radical chain reactions are called **initiators**. These have a particularly weak bond that is selectively broken on heating or irradiation. The bond breaks homolytically to generate two radicals.

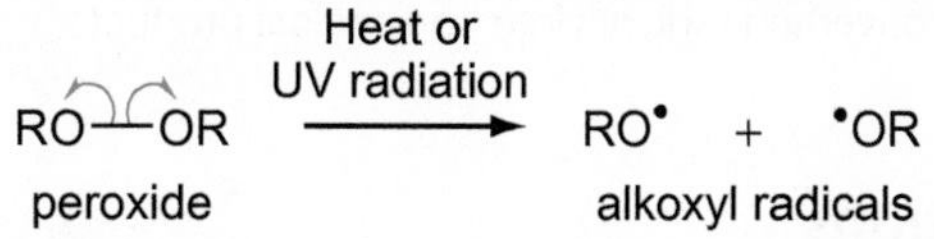
RO—OR
Heat or
UV radiation
RO•
+
•OR
peroxide
alkoxyl radicals

In radical chain reactions, radicals undergo a series of **propagation steps** in which a radical reactant forms a radical product (see Figure 19.33). Propagation steps include abstraction and addition reactions as shown below. In step 1, the RO• radical removes a hydrogen atom from a non-radical reactant. This is called a **radical abstraction reaction**. In step 2, the $R^{1\bullet}$ radical adds to a C=C bond in a **radical addition reaction**. Finally, in step 3, the $R^1CH_2CH^{\bullet}R^2$ radical removes a hydrogen atom from H–R^1 in another radical abstraction reaction. Overall, these three sequential steps lead to the addition of R^1H to $H_2C{=}CHR^2$ to produce $R^1CH_2CH_2R^2$.

In a chain reaction, a reactant radical must be regenerated in at least one of the propagation steps. For example, in the reaction in Figure 19.33, radical $R^{1\bullet}$ is first formed in step 1; it undergoes an addition reaction in step 2, and then is regenerated in step 3.

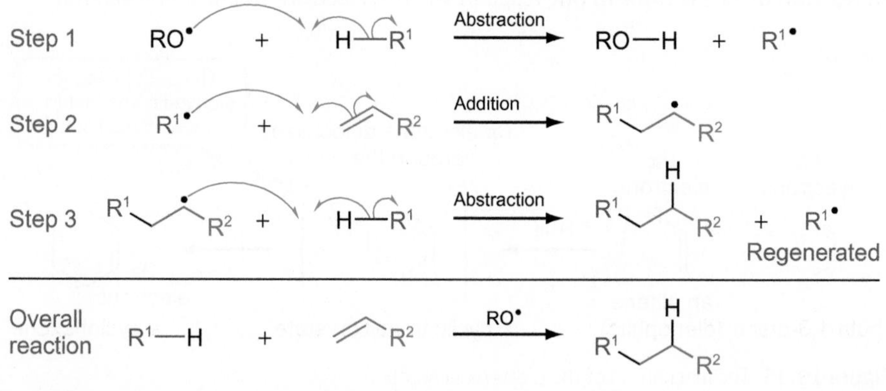
Step 1
RO•
+
H—R1
Abstraction
RO—H
+
R1•
Step 2
R1•
+
R2
Addition
R1
R2
Step 3
R1
R2
+
H—R1
Abstraction
H
R1
R2
+
R1•
Regenerated
Overall
reaction
R1—H
+
R2
RO•
H
R1
R2

Figure 19.33 Propagation steps in a radical chain reaction.

Single-headed curly arrows (⌒) are used to show the mechanisms of radical reactions.

The energy required to cleave a bond homolytically, to produce radicals, is called the bond enthalpy. (For a specific bond this is sometimes called the bond dissociation enthalpy; see Section 13.3, p.630 and Appendix 10, p.1368.) The mean bond enthalpy of the O–O bond of peroxides is approximately +150 kJ mol^{-1}. For comparison, the mean bond enthalpies of the C–H and C–C bonds in alkanes are around +412 kJ mol^{-1} and +347 kJ mol^{-1}, respectively.

$R^{1\bullet}$ adds selectively to the least hindered end of the C=C bond and the addition is described as regioselective (see Section 19.3, p.909). For the regioselective addition of HBr to an alkene by a radical mechanism, see Box 21.2 (p.973).

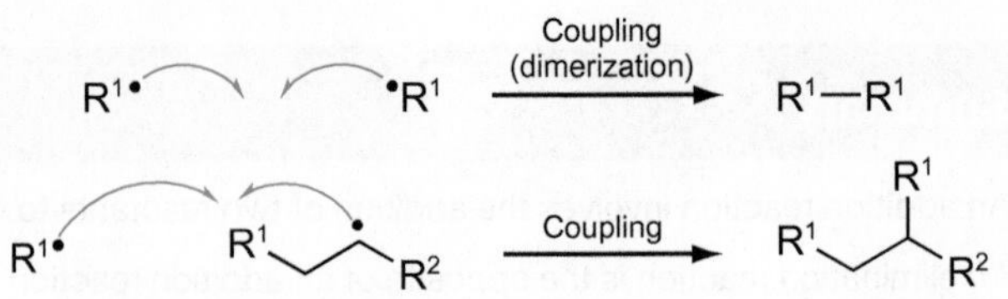
Coupling
(dimerization)
R¹•
•R¹
R¹—R¹
Coupling
R¹
R²

Figure 19.34 The coupling of radicals stops chain reactions by converting radicals into non-radical products.

In theory, only one molecule of $R^{1\bullet}$ needs to be formed in step 1 for the complete conversion of $H_2C{=}CHR^2$ into $R^1CH_2CH_2R^2$. This is because, for every molecule of $R^{1\bullet}$ that reacts in step 2, another molecule of $R^{1\bullet}$ is formed in step 3. In practice, however, more than one molecule of $R^{1\bullet}$ needs to be formed in step 1 because of competing **termination steps**.

The kinetics of chain reactions are discussed in Section 9.6 (p.423), for the reaction of H_2 and Br_2.

In a termination step, such as the **coupling reaction** in Figure 19.34, two radicals react to form only non-radical products. Coupling reactions can involve the reaction of two identical radicals or two different radicals. The coupling of radicals stops chain reactions so, for the reaction in Figure 19.34, further molecules of $R^{1\bullet}$ need to be generated (in step 1) to restart the chain.

Summary

- An initiation step produces radicals that can react in chain reactions.
- Radical chain reactions are composed of a number of propagation steps. In a propagation step, a radical reactant forms a radical product.
- A termination step stops a radical chain reaction by converting radicals into non-radical products.

Pericyclic reactions

Pericyclic reactions include cycloadditions, chelotropic reactions, sigmatropic reactions, and electrocyclic reactions. Only cycloaddition reactions lie within the scope of this book.

Pericyclic reactions are concerted (single-step) reactions that involve a change in the position of bonding electrons via cyclic transition states. A common class of pericyclic reactions is **cycloaddition reactions**. In cycloaddition reactions, two reactants form a cyclic product. For example, reaction of a 1,3-diene (e.g. $H_2C{=}CH{-}CH{=}CH_2$) with an alkene (called a dienophile) forms a 6-membered ring in the **Diels–Alder reaction** (after Otto Diels (1876–1952) and Kurt Alder (1902–1958) who discovered the reaction in 1928).

Notice that no ionic (R^+ or R^-) or radical ($R^\bullet$) intermediates are formed in a Diels–Alder reaction. The electrons move in a circle: the electrons can be drawn rotating anticlockwise or clockwise—both lead to the same result.

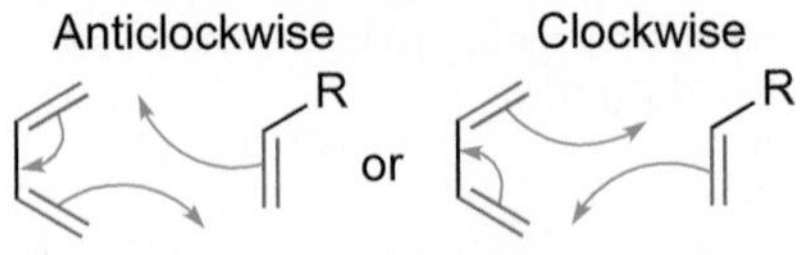
Anticlockwise
Clockwise
R
or
R

The mechanism of the Diels–Alder reaction is shown in Figure 19.35. Three double-headed curly arrows show simultaneous movement of three electron pairs to form a 6-membered ring. Overall, the three C=C bonds become C–C bonds and two new C–C bonds and one C=C bond are formed in the cyclic product. The Diels–Alder reaction belongs to a class of reactions called **4 + 2 cycloadditions**, which are characterized by the formation of a ring on reaction of 4π electrons in one reactant with 2π electrons in another reactant.

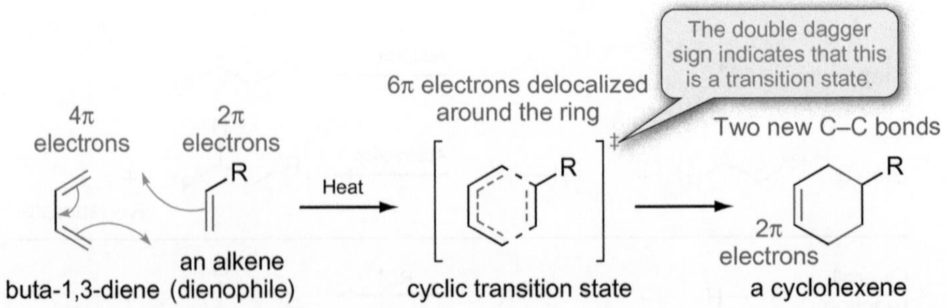
The double dagger sign indicates that this is a transition state.
6π electrons delocalized around the ring
4π electrons
2π electrons
R
Heat
R
‡
Two new C–C bonds
R
2π electrons
an alkene
buta-1,3-diene (dienophile)
cyclic transition state
a cyclohexene

Figure 19.35 The mechanism of the Diels–Alder reaction.

19.3 Reaction selectivity

Organic compounds containing one or more functional groups may react to give many different products. In order to form one product in preference to others, chemists design and use *selective* organic reactions. There are different types of selectivity depending on: (1) which functional group reacts; (2) in what position the functional group reacts; and (3) the stereochemistry of the product.

Visit the Online Resource Centre to view screencast 19.4 which walks you through examples of chemoselective, regioselective, and stereoselective reactions.

Chemoselectivity

A **chemoselective reaction** is a reaction in which one functional group reacts in preference to another functional group or groups. For example, hept-6-en-2-one ($CH_3COCH_2CH_2CH_2CH{=}CH_2$) is chemoselectively reduced using either sodium borohydride ($NaBH_4$) followed by acid, or by catalytic hydrogenation (H_2, Pd/C) (see Figure 19.36). These reactions are chemoselective because the C=O and C=C bonds have such different reactivities that, under one set of conditions, only the C=O bond is reduced, whereas, under a different set of conditions, only the C=C bond is reduced.

For naming compounds with more than one functional group, such as hept-6-en-2-one, see Section 2.8 (p.106).

Reductions of ketones and alkenes are discussed in Sections 23.2 (p.1062) and 21.2 (p.968), respectively.

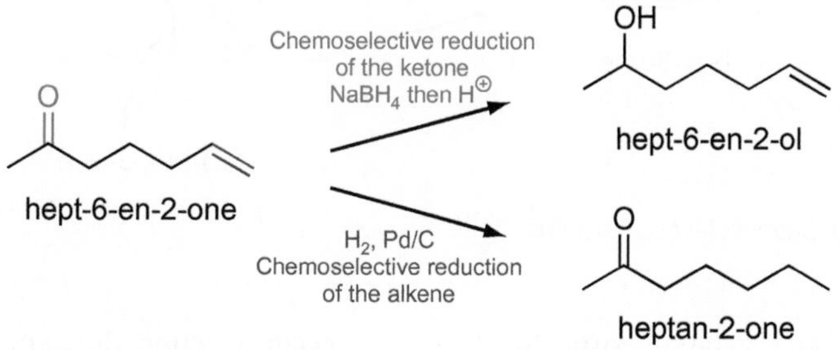
Chemoselective reduction
of the ketone
NaBH4 then H⊕
OH
hept-6-en-2-ol
O
hept-6-en-2-one
H2, Pd/C
Chemoselective reduction
of the alkene
O
heptan-2-one

Figure 19.36 Chemoselective reactions.

Blue cheeses have a distinctive aroma caused by certain methyl ketones (alkan-2-ones), including heptan-2-one. Heptan-2-one is the most abundant ketone in Blue Stilton.

Regioselectivity

A **regioselective reaction** is a reaction in which one position of a functional group, within an unsymmetrical molecule, reacts selectively over another. This leads to the selective formation of one structural isomer. A reaction can be moderately regioselective, highly regioselective, or completely regioselective, according to the relative amounts of the structural isomers formed in the reaction. For example, a 3 : 1 mixture of structural isomers may be described as moderately regioselective, whereas a 25 : 1 mixture may be described as highly regioselective.

For example, addition of HCl to an unsymmetrical alkene, such as 2-methylpropene ($(H_3C)_2C{=}CH_2$), is a highly regioselective addition reaction (see Figure 19.37). Two structural isomers are possible from the addition reaction and these are formed in different amounts. The hydrogen atom from HCl adds selectively to the CH_2 carbon atom of the C=C bond and the chlorine atom adds to the carbon atom that is bonded to two methyl groups. The regioselectivity arises because the two carbon atoms in the C=C bond are not equivalent. The substituents on the two carbon atoms are different and so the carbon atoms have different reactivity.

Structural (constitutional) isomers (Section 18.1, p.817) are sometimes called **regioisomers**.

Regioselective addition of HCl to alkenes is discussed in detail in Section 21.3 (p.970). Regioselective electrophilic substitution reactions are discussed in Section 22.3 (p.1027).

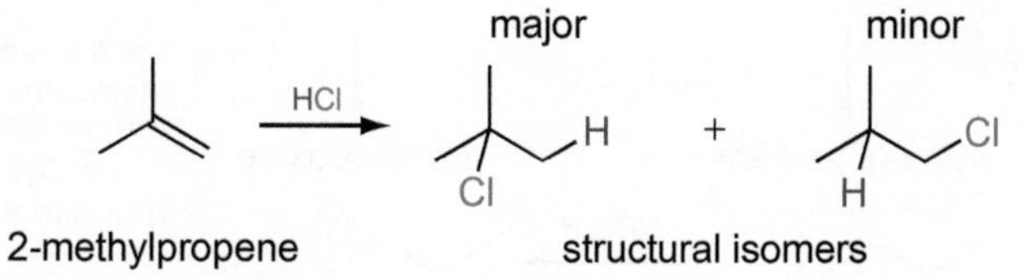
major
minor
HCl
H
+
Cl
Cl
H
2-methylpropene
structural isomers

Figure 19.37 A regioselective reaction.

Stereoselectivity

The *E*/*Z* system of nomenclature is discussed in Section 18.3 (p.833).

The Lindlar catalyst (which is a mixture of Pd, $CaCO_3$, and PbO) is a less active catalyst than palladium for hydrogenation reactions and permits an alkyne to be reduced to an alkene without further reduction to an alkane (Section 21.2, p.968).

Reduction of an alkyne using sodium in liquid ammonia involves the formation of a radical anion, which has both a negative charge and an unpaired electron (Section 21.2, p.967).

A **stereoselective reaction** is a reaction in which one stereoisomer is formed selectively over others. A reaction can be moderately stereoselective, highly stereoselective, or completely stereoselective according to the relative amounts of the stereoisomers formed in the reaction.

For example, the stereoselective reduction of an alkyne to an alkene occurs on hydrogenation in the presence of the Lindlar catalyst or by reaction with sodium in liquid ammonia, as shown in Figure 19.38. Reaction of the alkyne with H_2/Lindlar catalyst produces mainly, or even exclusively, the *Z*-isomer of the alkene. Reaction with Na/NH_3 produces mainly, or even exclusively, the *E*-isomer of the alkene. The high or even complete stereoselectivity of these reductions is due to the reagents selectively adding the two hydrogen atoms to the same or the opposite sides of the C≡C bond.

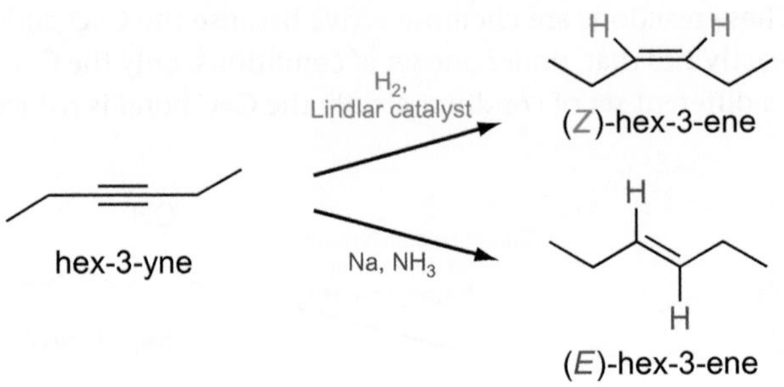
H
H
H_2,
Lindlar catalyst
(*Z*)-hex-3-ene
hex-3-yne
Na, NH_3
H
H
(*E*)-hex-3-ene

Figure 19.38 Stereoselective reactions.

For the mechanism of the epoxidation reaction see Section 21.3 (p.986).

When an achiral peroxycarboxylic acid (RCO_3H) reacts with a C=C bond, a racemic mixture of epoxides is formed. Either enantiomer of the epoxide is drawn to show the *relative* configuration of the product (Section 18.4, p.853). Although the epoxidation reactions shown are not enantioselective (as both enantiomers are formed), they are diastereoselective (as one diastereomer is favoured over another).

Epoxides are important compounds in synthesis because they undergo a variety of reactions (including nucleophilic substitutions) to make useful compounds (Section 21.3, p.986).

Reactions can also be **stereospecific**. In a stereospecific reaction, different stereoisomers react differently. For example, Figure 19.39 shows different stereoisomers of an alkene reacting with a peroxycarboxylic acid (RCO_3H) to give different stereoisomers of the epoxide product (in an **epoxidation reaction**).

Both of the stereospecific epoxidation reactions shown in Figure 19.39 are also stereoselective. This is because the *cis*-alkene selectively forms the *cis*-epoxide and the *trans*-alkene selectively forms the *trans*-epoxide. Indeed, all stereospecific reactions are also stereoselective.

However, not all stereoselective reactions are stereospecific. For example, the stereoselective reduction of the alkyne hex-3-yne to give an *E*-hex-3-ene (shown in Figure 19.38) is not stereospecific. This is because the alkyne reactant cannot exist as stereoisomers. For a stereospecific reaction, the reactant must have an *E* or Z double bond, or a chiral centre, so that the reactant can exist as stereoisomers.

Worked example 19.7 provides practice in identifying selectivity in organic reactions.

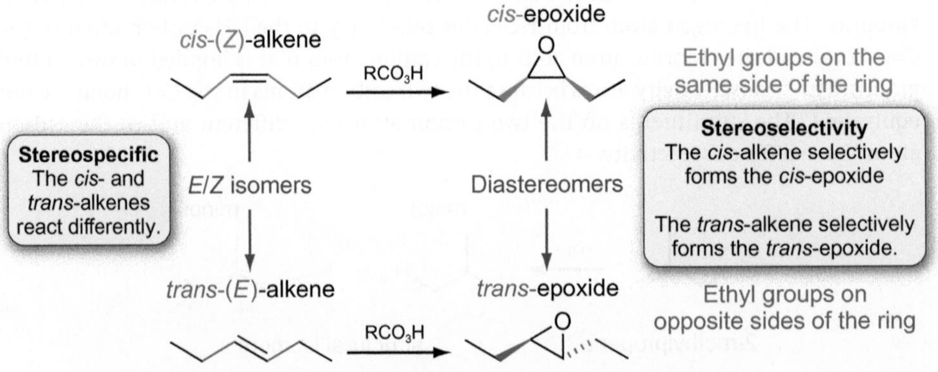
cis-(*Z*)-alkene
cis-epoxide
O
RCO_3H
Ethyl groups on the same side of the ring
Stereospecific
The *cis*- and *trans*-alkenes react differently.
E/*Z* isomers
Diastereomers
Stereoselectivity
The *cis*-alkene selectively forms the *cis*-epoxide
The *trans*-alkene selectively forms the *trans*-epoxide.
trans-(*E*)-alkene
trans-epoxide
O
RCO_3H
Ethyl groups on opposite sides of the ring

Figure 19.39 In stereospecific reactions, different stereoisomers react differently.

Worked example 19.7 Selective reactions

A synthesis of methyl 9-bromo-8-hydroxy-non-5(*Z*)-enoate is shown here.

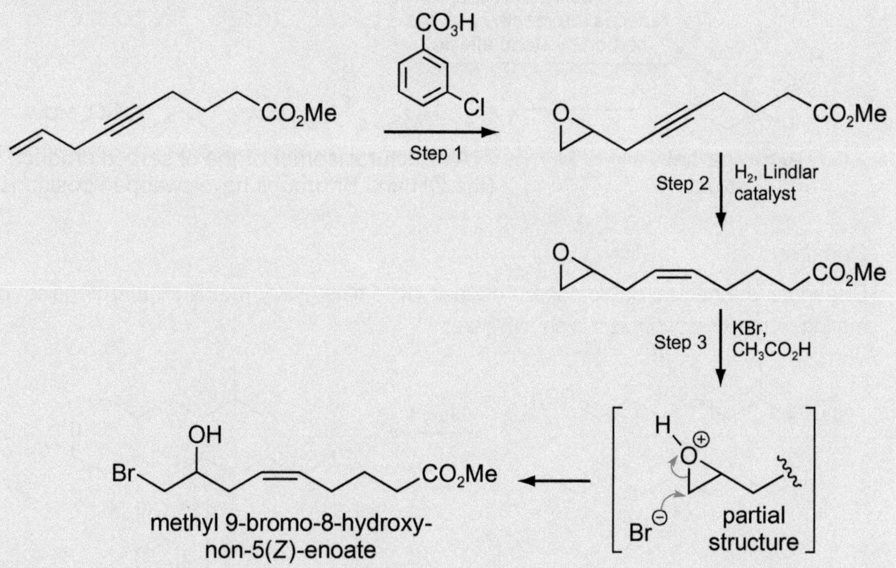

CO_3H
Cl
CO_2Me
Step 1
O
CO_2Me
Step 2
H_2, Lindlar catalyst
O
CO_2Me
Step 3
KBr, CH_3CO_2H
OH
Br
CO_2Me
methyl 9-bromo-8-hydroxy-non-5(Z)-enoate
H
O
Br
partial structure

(a) Decide which of the terms *chemoselective*, *regioselective*, and *stereoselective* apply to steps 1 and 2.

(b) A key stage in the mechanism of step 3 involves attack of the bromide ion on to the protonated epoxide as shown in the square brackets. Use this to help explain why step 3 is regioselective and draw the structure of the other possible structural isomer.

> The reagent used in step 1 is called *meta*-chloroperbenzoic acid, which has the abbreviation *m*CPBA (Section 21.3, p.986). A peroxycarboxylic acid (RCO_3H) has an extra oxygen atom between the C=O and OH groups of a carboxylic acid (RCO_2H).

Strategy

(a) To identify a chemoselective reaction, look at the functional groups in the reactants and products and decide whether one functional group has reacted preferentially.

To identify a regioselective reaction, decide whether alternative structural isomers could have been formed.

To identify a stereoselective reaction, decide whether an alternative enantiomer, diastereomer, or double bond isomer could have been formed.

(b) For step 3 to be regioselective, the bromide ion must be able to react with the protonated epoxide to give an alternative structural isomer.

Solution

(a) Steps 1 and 2 are chemoselective.

In step 1, the C=C bond, rather than the C≡C bond or CO_2Me group, reacts. In step 2, the C≡C bond, rather than the epoxide ring or CO_2Me group, reacts.

Steps 1 and 2 are not regioselective. It is not possible to form structural isomers from either of these reactions.

In step 1, formation of the epoxide ring introduces a chiral centre in the product. However, the reaction is not stereoselective because the product is racemic. Step 2 is stereoselective because a *Z*-alkene, rather than an *E*-alkene, is formed.

> In step 1, if the C≡C bond reacted in the same way as the C=C bond, it would form an oxirene. A 3-membered ring containing a C=C bond and an oxygen atom is extremely strained (try making a molecular model), and hence it is very difficult to form.

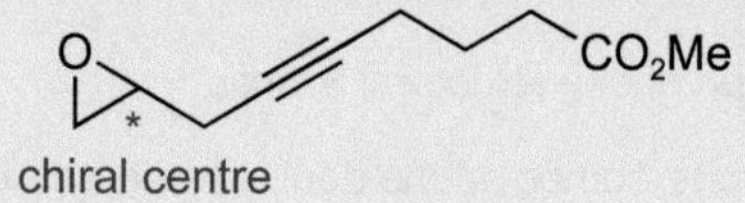

O
*
CO_2Me

chiral centre

The absolute stereochemistry is not shown and so you can assume that a 1:1 mixture of enantiomers is formed.

→

→ (b) Step 3 is regioselective because the bromide ion could attack either of the two carbon atoms in the epoxide ring to form two structural isomers. Attack at the more sterically hindered carbon atom of the epoxide would form the product shown here.

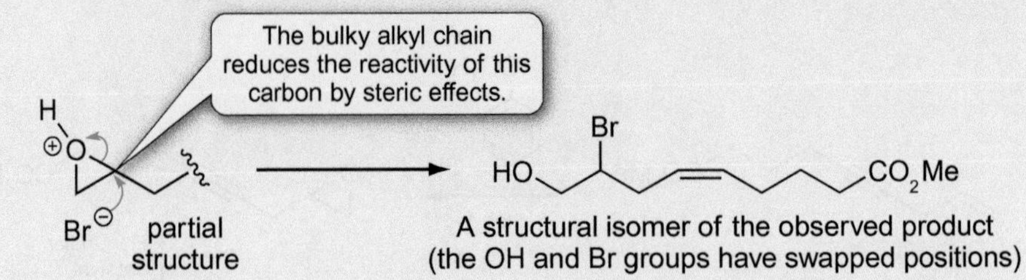
The bulky alkyl chain reduces the reactivity of this carbon by steric effects.
H
O
Br
partial structure
Br
HO
CO_2Me
A structural isomer of the observed product (the OH and Br groups have swapped positions)

Question

> The IUPAC name for racemic disparlure is (7*RS*,8*SR*)-*cis*-7,8-epoxy-2-methyloctadecane. In step 1, the aldehyde (RCHO) reacts with a phosphorane ($RCH{=}PPh_3$), to form the C=C bond, in a Wittig reaction (Section 23.2, p.968).

Disparlure is a natural pheromone produced by female gypsy moths to attract males for mating. A racemic synthesis is shown here.

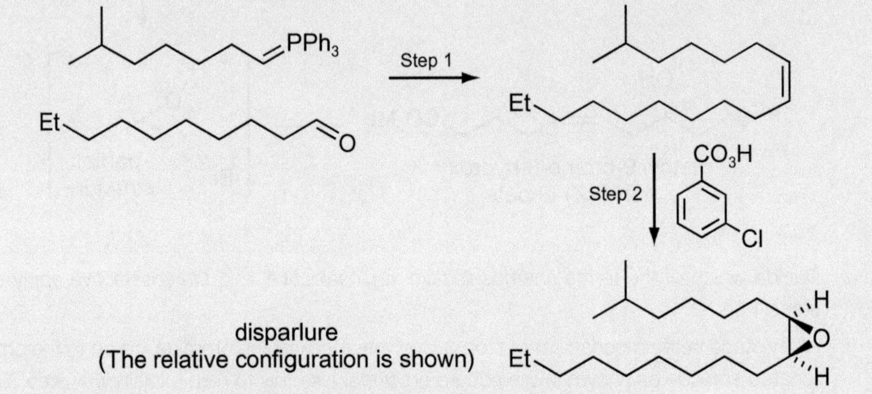
PPh_3
Et
O
Step 1
Et
Step 2
CO_3H
Cl
H
O
H
Et
disparlure
(The relative configuration is shown)

(a) Explain why steps 1 and 2 are both classed as stereoselective reactions.

(b) Why is step 2 classed as a stereospecific reaction?

(c) Is step 2 an enantioselective or diastereoselective reaction?

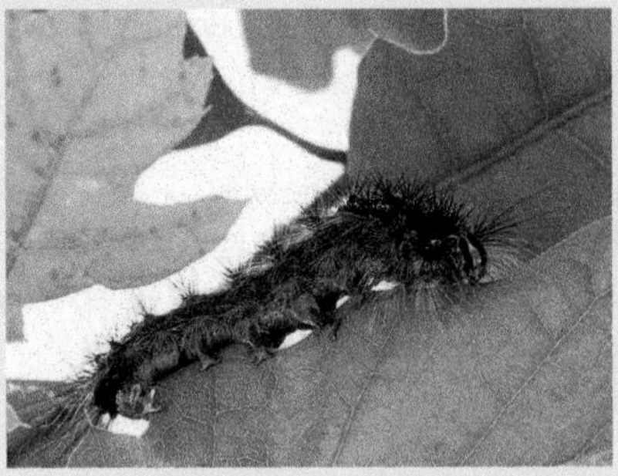

▲ Caterpillar of the gypsy moth (*Lymantria dispar*).

» Summary

- A chemoselective reaction is a reaction in which one functional group reacts in preference to another functional group or groups.
- A regioselective reaction is a reaction that leads to the selective formation of one structural isomer.
- A stereoselective reaction is a reaction in which one enantiomer, one diastereomer, or one double bond isomer is formed selectively over others.

→

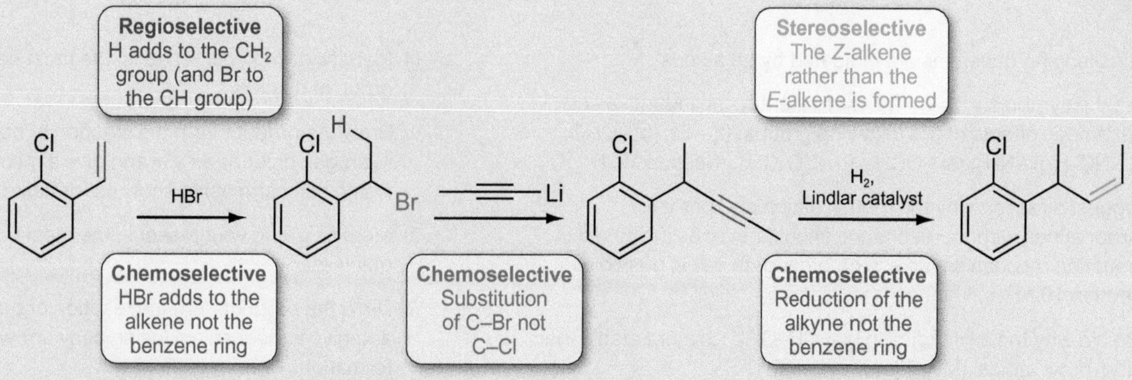
Regioselective
H adds to the CH2 group (and Br to the CH group)
Stereoselective
The Z-alkene rather than the E-alkene is formed
Cl
HBr
H
Br
Li
H2, Lindlar catalyst
Chemoselective
HBr adds to the alkene not the benzene ring
Chemoselective
Substitution of C–Br not C–Cl
Chemoselective
Reduction of the alkyne not the benzene ring

- In a stereospecific reaction, different stereoisomers react differently.

For practice questions on these topics, see questions 6c and 6d at the end of this chapter (p.914).

Concept review

By the end of this chapter, you should be able to do the following.

- Use double-headed and single-headed arrows to represent the movement of electrons in polar and radical reactions, respectively.
- Recognize carbocations, carbanions, and carbon-centred radicals.
- Understand how hyperconjugation, inductive effects, and mesomeric effects affect the stability of ions and neutral organic molecules. Table 19.3 shows the inductive and mesomeric effects of some functional groups.

Table 19.3 Inductive and mesomeric effects of some functional groups. (Alkenes and aromatic rings can be +M or –M.)

–I, –M groups*	–I, +M groups†	+I groups‡
$-NO_2$	–OH	–R (alkyl)
$-CO_2H$	–OR	–Li
$-CO_2R$	$-NH_2$	–MgBr
–COR	$-NR_2$	
–CHO	–SR	
–CN	–Br	
	–Cl	

* –I, –M groups contain an electronegative atom(s) and a double or triple bond.
† –I, +M groups contain an electronegative atom with a lone pair.
‡ +I groups contain an electropositive atom or hydrocarbon chain.

- Understand how steric effects can influence organic reactions.
- Recognize, and compare the relative strengths of, nucleophiles and electrophiles.
- Recognize the importance of orbital overlap in resonance and organic reactions.
- Draw sensible curly arrow mechanisms for reactions of nucleophiles with electrophiles.
- Recognize organic acids and the factors that influence acidic strength.
- Recognize organic bases and the factors that influence basic strength.
- Recognize acid–base reactions involving organic compounds.
- Classify functional groups by their oxidation level and use oxidation levels to recognize redox reactions.
- Classify polar reactions as electrophilic additions, nucleophilic additions, eliminations, nucleophilic substitutions, nucleophilic acyl substitutions, or electrophilic substitutions.
- Recognize radical initiation, propagation, and termination reactions.
- Recognize the Diels–Alder reaction as an example of a pericyclic reaction.
- Recognize chemoselective, regioselective, stereoselective, and stereospecific reactions.

Questions

More challenging questions are indicated by an asterisk *.

1 Use the symbols +I, −I, +M, −M to identify the inductive and mesomeric effects of the following groups: (a) –Et; (b) –CHO; (c) $-NO_2$; (d) $-NH_2$; (e) $-OCH_3$; (f) $-CO_2CH_3$. (Section 19.1)

2 Suggest a reason why the methyl cation and primary carbocations without resonance should never be proposed in a reaction mechanism unless no other pathway is possible. (Section 19.1)

3 Explain why the allyl radical ($H_2C{=}CH{-}CH_2^{\bullet}$) is more stable than a *tert*-butyl radical ($Me_3C^{\bullet}$). (Section 19.1)

4 Rank the following carbanions (**A–C**) in order of increasing stability. Explain your reasoning. (Section 19.1)

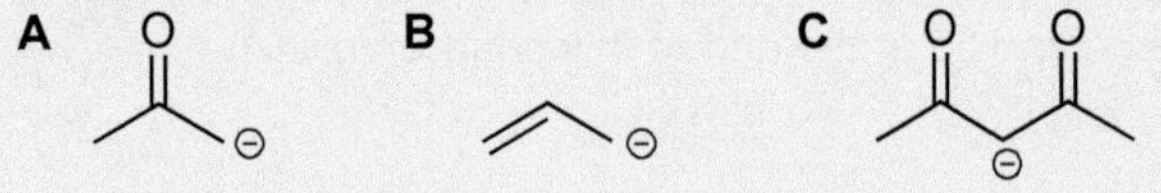
A
B
C

5 The following questions relate to quinine, a naturally occurring antimalarial agent. (Sections 19.1 and 19.2)

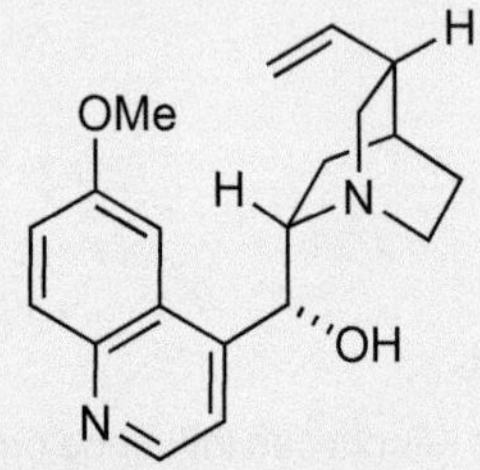
OMe
H
N
H
OH
N

quinine

(a) (i) Identify, giving your reasons, the most acidic hydrogen atom in quinine.

(ii) Draw the product from the reaction of quinine with sodium hydride ($Na^+ H^-$) and give a mechanism (using curly arrows) to show its formation.

(b) (i) Identify, giving your reasons, the most basic functional group in quinine.

(ii) Draw the product from the reaction of quinine with hydrogen chloride (H–Cl) and give a mechanism (using curly arrows) to show its formation.

(c) (i) Identify, giving your reasons, the most nucleophilic site in quinine.

(ii) Draw the product from the reaction of quinine with CH_3Br and give a mechanism (using curly arrows) to show its formation.

Tonic water will fluoresce (it emits blue-green visible light) under ultraviolet radiation owing to the presence of quinine. The sensitivity of quinine to ultraviolet radiation is such that it will appear visibly fluorescent in direct sunlight.

6 The following questions relate to the sequence of reactions shown below. (Sections 19.2 and 19.3)

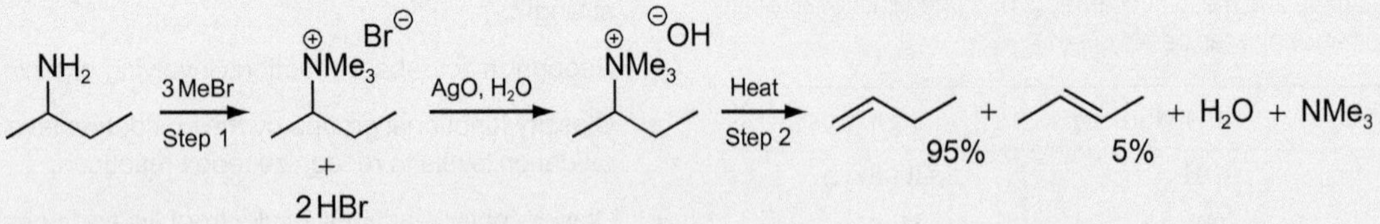
NH2
3 MeBr
Step 1
NMe3 Br
+
2 HBr
AgO, H2O
NMe3 OH
Heat
Step 2
+
95%
5%
+ H2O + NMe3

(a) What type of polar reaction is taking place in step 1?

(b) What type of polar reaction is taking place in step 2?

(c) Explain why the reaction in step 2 is a regioselective reaction.

(d) Explain why the reaction in step 2 is a stereoselective reaction.

7 (*S*)-(+)-Clopidogrel bisulfate (Plavix) is an orally active medicine that helps to prevent harmful blood clots and it is used as an antithrombotic agent. It is prepared by treating the *S* isomer of clopidogrel with sulfuric acid in an acid–base reaction. A racemic synthesis of clopidogrel is shown below. (Sections 19.2 and 19.3)

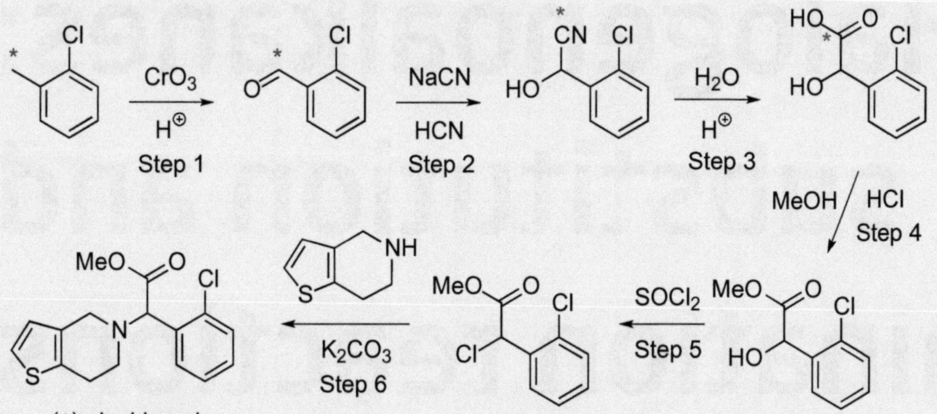
CrO3
H⊕
Step 1
NaCN
HCN
Step 2
H2O
H⊕
Step 3
MeOH HCl
Step 4
SOCl2
Step 5
K2CO3
Step 6

(±)-clopidogrel

(a) In step 1, assign oxidation levels to the two carbon atoms identified by the purple asterisk (*). Use these values to determine if the transformation in step 1 involves oxidation, reduction, or neither.

(b) What type of polar reaction is taking place in step 2?

(c) In step 3, assign oxidation levels to the two carbon atoms identified by the dark red asterisk (*). Use these values to determine if the transformation in step 3 involves oxidation, reduction, or neither.

(d) What type of polar reaction is taking place in step 4?

(e) Decide which of the terms *chemoselective*, *regioselective*, and *stereoselective* apply to step 4.

(f) What type of polar reaction is taking place in step 5?

(g) What type of polar reaction is taking place in step 6?

(h) Giving your reasons, draw the *S* isomer of clopidogrel (Hint: See Section 18.4, p. 845).

(i) Draw the structure of (*S*)-(+)-clopidogrel bisulfate (Plavix).

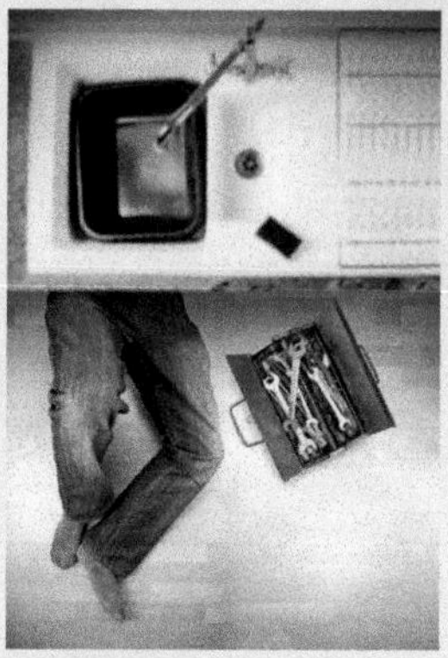

In question 6, the major alkene product from step 2 is called but-1-ene. In industry, but-1-ene is converted into a polymer called polybutylene (or poly(1-butene), $[-CH_2CH(CH_2CH_3)-]_n$). This polymer has a range of uses, from flexible packaging to shoe soles and piping in home plumbing systems. Between 1978 and 1994 around 6–10 million homes in the United States had pipes made from polybutylene. However, production stopped in 1994 after numerous claims that the pipes were rupturing and causing property damage. It was subsequently shown that oxidants, including certain disinfectants, react with polybutylene causing the tubing to become brittle and flake apart.

When a blood vessel is injured, the body uses platelets and a protein called fibrin to form a blood clot, which prevents loss of blood.

20

Halogenoalkanes: substitution and elimination reactions

This chapter builds on the following topics:

- **Functional groups containing one or more heteroatoms** Section 2.6, p.90
- **Conformational isomers** Section 18.2, p.819
- **Configurational isomers: *E*- and *Z*-isomers** Section 18.3, p.833
- **Configurational isomers: isomers with chiral centres** Section 18.4, p.838
- **Fundamental concepts of organic reaction mechanisms** Section 19.1, p.863
- **Polar reactions** Section 19.2, p.888
- **Regioselectivity** Section 19.3, p.909
- **Stereoselectivity** Section 19.3, p.910

◀ In 1997, more than 13 655 tons of sweet cherries were shipped from America to Japan and all of these cherries were fumigated with bromomethane in order to kill the codling moth. The caterpillar of the codling moth bores into fruit, causing considerable damage to the crop. As of 2009, sweet cherries have been allowed into Japan without bromomethane fumigation providing certain strict conditions are met. For example, the cherries must be inspected three times after harvest and the designated orchards are audited periodically by Japanese government inspectors.

Alternative pesticides to bromomethane

Bromomethane (CH_3Br) has been widely used as a pesticide. It is effective against a wide range of pests in a variety of situations (hence it is called a *broad-spectrum* pesticide). For example, soils prepared for planting, edible fruits and vegetables, and cargoes of timber were once routinely sprayed with bromomethane. Unfortunately, bromomethane is toxic to other living organisms, not just to pests. Its effect on humans is minimized by carefully controlling the concentrations used and limiting the length of exposure. Care is needed when spraying and precautions must be taken to protect the people involved.

◀ Bromomethane gas is applied to a field—the soil is then covered with a plastic sheet to minimize loss of bromomethane from the soil.

Halogenoalkanes (RX) are reactive compounds that undergo a wide range of substitution and elimination reactions—it is the reactivity of bromomethane that is responsible for the problems associated with its use as a pesticide. Proteins and DNA contain nucleophilic groups (such as OH, NH_2, and SH) that can donate an electron pair to a carbon atom. These groups react with bromomethane in **nucleophilic substitution** reactions. These reactions are characteristic of halogenoalkanes and their mechanisms are discussed in this chapter.

Recently, farmers have started to move away from using bromomethane because of the introduction of legislation to control its use. In addition to problems associated with its toxicity, bromomethane has been shown to be an important ozone-depleting substance in the stratosphere. The C–Br bond in bromomethane is broken by high-energy ultraviolet radiation forming $Br^\bullet$ and $CH_3^\bullet$ radicals. These radicals react with O_3 in a catalytic cycle and cause thinning of the ozone layer (see Boxes 9.4, p.403, and 27.8, p.1241). The destructive effects on the ozone layer were identified in the Montreal Protocol of 1987. This agreement (and subsequent amendments) urged developed countries to find replacements to bromomethane and to ban its use by 2005.

▼ Bromomethane is a colourless, odourless gas. It is a naturally occurring **halogenoalkane** that is released into the atmosphere by marine organisms. Numerous plants and vegetables, such as broccoli and cabbage, also produce it.

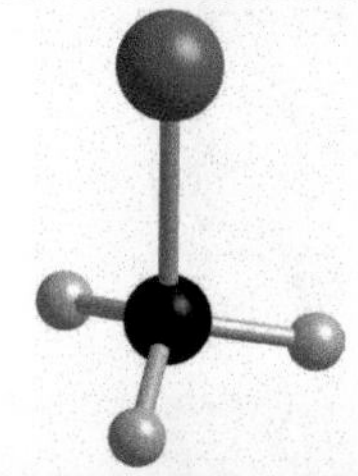

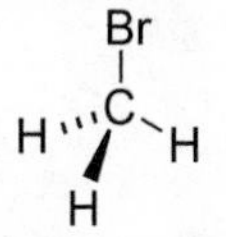

bromomethane

▼ Reaction of DNA with bromomethane in a nucleophilic substitution reaction.

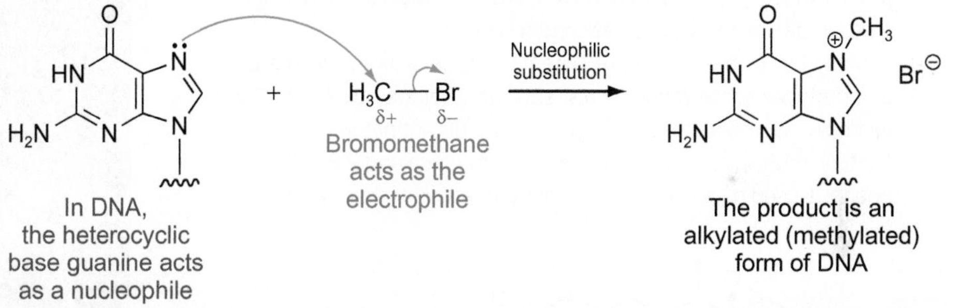
Nucleophilic substitution
In DNA, the heterocyclic base guanine acts as a nucleophile
Bromomethane acts as the electrophile
The product is an alkylated (methylated) form of DNA

In nucleophilic substitution reactions involving bromomethane, the nucleophilic group donates a pair of electrons to the carbon atom of bromomethane. For example, reaction with a nucleophilic OH group in an alcohol forms a new O–CH_3 bond at *the same time* as the C–Br bond in bromomethane breaks (a **concerted reaction**).

▼ A nucleophilic substitution reaction of ROH with bromomethane.

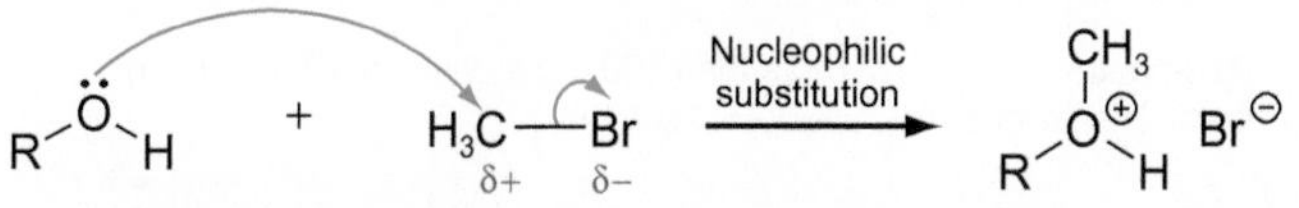
Nucleophilic substitution

In living cells, bromomethane reacts first with proteins and then, on prolonged exposure, with DNA. The reaction above right shows how bromomethane reacts with DNA and this may explain its pesticide action and also its toxicity in humans.

Currently, no low-cost, easy-to-use chemicals that have the same range of pesticidal activity as bromomethane are available and research into the development of alternative pesticides is actively being pursued. The compounds to the right are possible alternatives that do not affect the ozone layer.

The key to finding safe replacement pesticides lies in understanding and controlling their reactivity to minimize environmental effects and in ensuring that their biological effect is much more specific, for example, by targeting them at specific enzymes.

chloropicrin (trichloronitromethane)

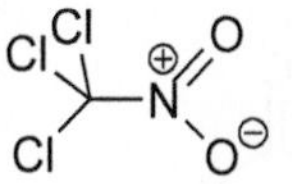

Used against nematodes, some bacteria, and fungi

telone (1,3-dichloropropene)

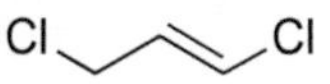

Used against nematodes, some fungi, and insects

vapam (sodium *N*-methyldithiocarbamate)

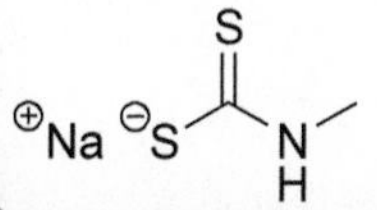

Used against some weeds, nematodes, and fungi

Halogenoalkane

$\overset{\delta+}{R}$—$\overset{\delta-}{X}$

alkyl group — halogen atom (F, Cl, Br, or I)

Nucleophilic substitution and elimination reactions are introduced in Section 19.2 (p.902).

Many natural products (see Box 20.1), as well as synthetic pesticides, disinfectants, and the polymer PVC, that are used in our everyday lives contain carbon–halogen bonds. **Halogenoalkanes** (RX) are a large class of compounds that contain an alkyl group (R) bonded to a halogen atom (X = F, Cl, Br, or I).

Halogenoalkanes are particularly useful intermediate compounds in organic synthesis. This is because they react with nucleophiles and bases to form valuable products. Carbon–halogen bonds are generally polar bonds and nucleophiles attack the slightly positive carbon atom in the $^{\delta+}C–X^{\delta-}$ bond in **nucleophilic substitution** reactions, which lead to the nucleophile replacing X in the halogenoalkane. These reactions are discussed in Section 20.3 (p.930) and are used to prepare molecules with various functional groups. Halogenoalkanes react with bases in **elimination** reactions. As discussed in Section 20.4 (p.943), this is an important way of making alkenes ($R_2C{=}CR_2$). Whether the halogenoalkane undergoes nucleophilic substitution or elimination depends on the structure of the halogenoalkane, the reaction conditions, and the nature of the nucleophile or base (Section 20.5, p.953).

This chapter begins with an overview of the structure and reactivity of halogenoalkanes (Section 20.1) and this is followed by a discussion of the most important methods of preparing these compounds in Section 20.2.

Box 20.1 Natural organohalogens

Over 4000 naturally occurring compounds are known to contain one or more halogen atoms. The structures of these organohalogens are diverse and range from small molecules, such as bromomethane (CH_3Br, p.917) and 2,2-dibromoethanal (Br_2CHCHO), to much larger molecules such as cryptophycin A.

Many organohalogens have potent biological activities with roles as hormones, pheromones, repellents, and natural pesticides. For example, fluoroethanoate ($FCH_2CO_2^-$) is a potent natural pesticide found in many plants in Australia and South Africa and 2,6-dichlorophenol is used as a sexual attractant by female American dog ticks.

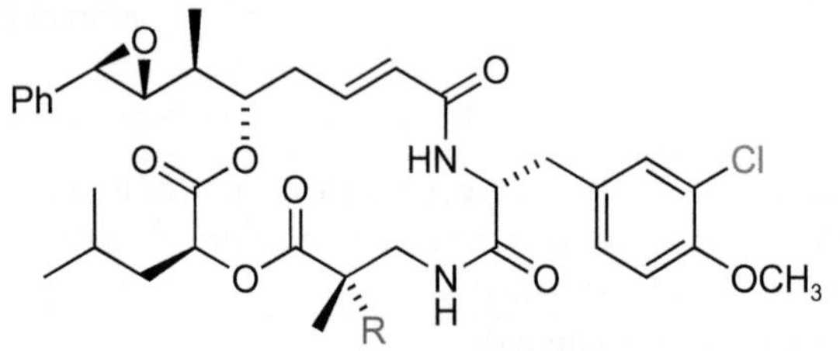

Ph
O
HN
Cl
OCH3
N
H
R

R = H; cryptophycin A
R = Me; cryptophycin 52

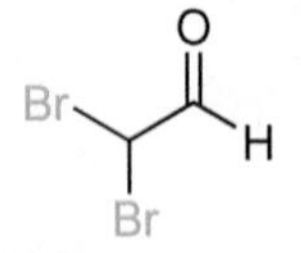

Br
O
H
Br

2,2-dibromoethanal

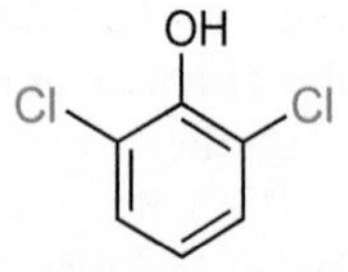

OH
Cl
Cl

2,6-dichlorophenol

▲ 2,2-Dibromoethanal is one of more than 100 organohalogens found in an edible Hawaiian seaweed (*Asparagopsis taxiformis*).

→

Online Support. The icon in the margin indicates that accompanying interactive resources are provided online to help your understanding: just type www.chemtube3d.com/burrows/123 into your browser, replacing 123 with the number of the page where you see the icon. For pages linking to more than one resource, type 123-1, 123-2, etc. (replacing 123 with the page number) for access to successive links.

◀ The Australian gidgee tree (*Acacia georginae*) uses fluoride ions obtained from the soil, water, or air to make fluoroethanoate.

Cryptophycin A, isolated from blue-green algae in 1990, was tested and shown to have potent antitumour activity. It was used as the starting point for the development of a drug. Over 450 analogues of cryptophycin A have been prepared in the laboratory including cryptophycin 52. This compound, which has a methyl group (CH_3) in place of a hydrogen atom in cryptophycin A, entered clinical trials for the treatment of solid tumours in humans.

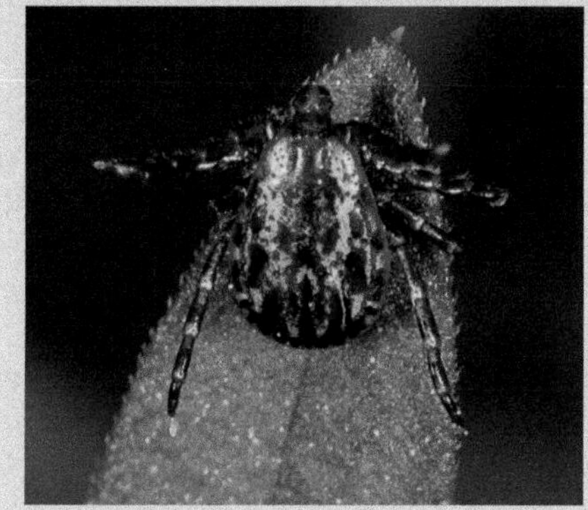

▲ The American dog tick (*Dermacentor variabilis*) is a carrier of a disease called Rocky Mountain spotted fever. This disease, which can be treated with antibiotics, affects about 800 people in America each year.

Question

Benzastatin C is a member of a family of structurally similar compounds that are produced by a bacterium (*Streptomyces nitrosporeus*).

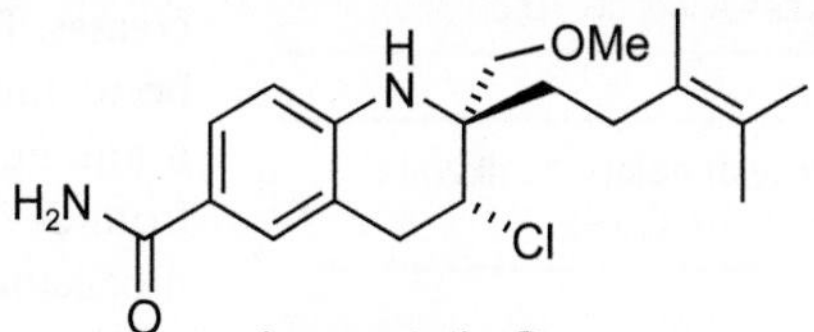
H
N
OMe
H_2N
O
Cl

benzastatin C

(a) Is the chlorine atom in benzastatin C bonded to a primary, secondary, or tertiary carbon atom?

(b) What are the configurations (*R* or *S*) of the two chiral centres in benzastatin C? (*Hint*. See Section 18.4 on p.845.)

20.1 Structure and reactivity of halogenoalkanes

In halogenoalkanes (RX), an sp^3 carbon atom in an alkyl group is joined to a fluorine, chlorine, bromine, or iodine atom by a single covalent bond. Halogenoalkanes are divided into **aliphatic compounds**, where the carbon atoms are in a chain, and **alicyclic compounds**, where the carbon atoms form a ring.

The naming of halogenoalkanes is described in Section 2.6 (p.90).

sp^3 and sp^2 hybridization is introduced in Section 5.4 (p.236).

Halogen atoms in halogenobenzenes (e.g. chlorobenzene, C_6H_5Cl) and halogenoalkenes (e.g. chloroethene, or vinyl chloride, $H_2C{=}CHCl$) are bonded to sp^2 carbon atoms. These compounds do not undergo the same types of reactions as halogenoalkanes. The chemistry of halogenobenzenes is discussed in Section 22.3 (p.1027).

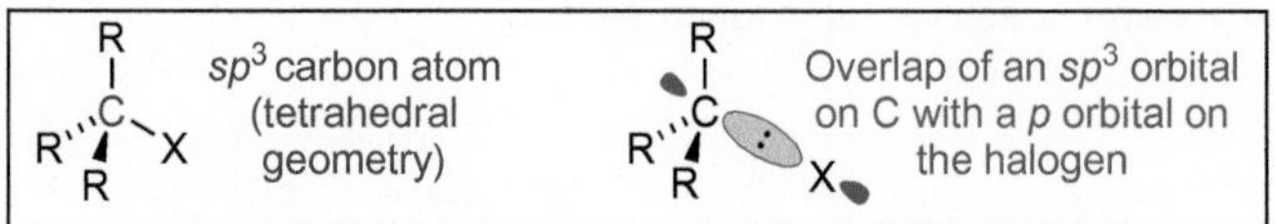
R
R
X
R
*sp*³ carbon atom (tetrahedral geometry)
Overlap of an *sp*³ orbital on C with a *p* orbital on the halogen

Structure of a halogenoalkane (X = F, Cl, Br, or I)

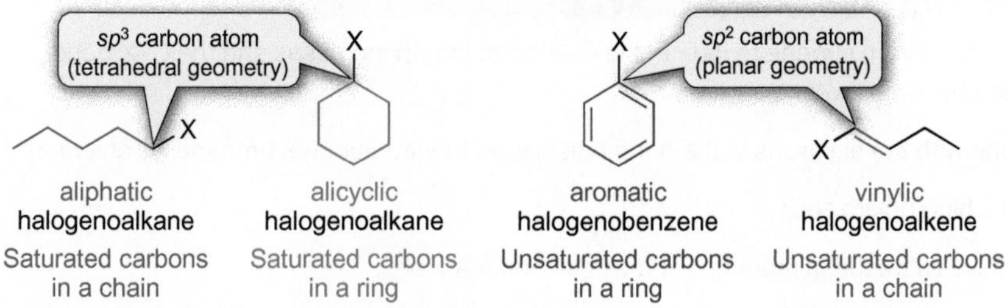
*sp*³ carbon atom (tetrahedral geometry)
*sp*² carbon atom (planar geometry)
X
aliphatic halogenoalkane
Saturated carbons in a chain
alicyclic halogenoalkane
Saturated carbons in a ring
aromatic halogenobenzene
Unsaturated carbons in a ring
vinylic halogenoalkene
Unsaturated carbons in a chain

Table 20.1 Mean bond lengths and mean bond enthalpies of carbon–halogen bonds together with electronegativities of halogens

Bond	Mean bond lengths / pm	Mean bond enthalpies / kJ mol^{-1}	Electronegativity (χ^P) of halogen
C–F	139 (shortest bond)	+467 (strongest bond)	F 3.98 (most electronegative)
C–Cl	179	+346	Cl 3.16
C–Br	194	+290	Br 2.96
C–I	213 (longest bond)	+228 (weakest bond)	I 2.66 (least electronegative)

The bond lengths and bond enthalpies quoted are average values that are calculated from the bond lengths and bond enthalpies of a large number of halogenoalkanes. (For other bond lengths and bond enthalpies see Appendix 10, p.1368.)

Reactivity of halogenoalkanes
R–I > R–Br > R–Cl >> R–F

Insertion of magnesium into carbon–halogen bonds, to form Grignard reagents (RMgX), is discussed in Section 23.2 (p.1068). Whereas halogenoalkanes are electrophiles, Grignard reagents are strong nucleophiles.

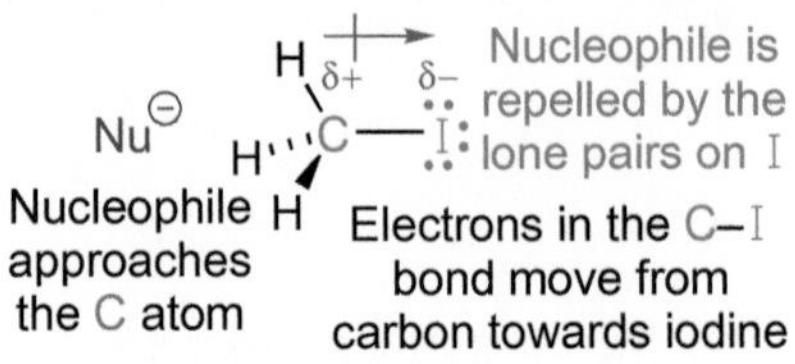
Nucleophile is repelled by the lone pairs on I
Nucleophile approaches the C atom
Electrons in the C–I bond move from carbon towards iodine

The length and strength of the carbon–halogen bond in halogenoalkanes change as the size of the halogen atom changes (Table 20.1). Halogen atoms increase in size as you go down the Periodic Table, from fluorine to iodine, and, consequently, so does the mean carbon–halogen bond length.

As the length of the carbon–halogen bond increases, so the strength of the bond decreases. For example, the C–F bond is both the shortest and strongest carbon–halogen bond. Indeed, the C–F bond is so strong that a large amount of energy is required to break it and fluoroalkanes (RF) are virtually inert. Carbon–halogen bonds get weaker on going from C–Cl to C–Br to C–I so that chloroalkanes (RCl) usually react more slowly than bromoalkanes (RBr), which in turn react more slowly than iodoalkanes (RI).

With the exception of iodine, all of the other halogens are more electronegative than carbon, so C–F, C–Cl, and C–Br bonds are polar. The electrons in the bond are attracted towards the halogen atom, leading to a partial negative charge (δ–) on F, Cl, or Br and a partial positive charge (δ+) on carbon.

The polarization of the carbon–halogen bond explains why halogenoalkanes react with nucleophiles in nucleophilic substitution reactions. Figure 20.1 shows the general reaction. A nucleophile attacks the slightly positively charged carbon atom of the halogenoalkane and this leads to the substitution of the halogen atom (X) by the nucleophile. The halide ion (X^-) formed is called the **leaving group**.

Even though the C–I bond is not polar, iodoalkanes do undergo nucleophilic substitution reactions. This is because, when a nucleophile approaches the C–I bond, the bond becomes polarized—the electrons in the C–I bond move from carbon to iodine so the carbon becomes partially positive. The C–I bond is said to have a high **polarizability** (see p.939).

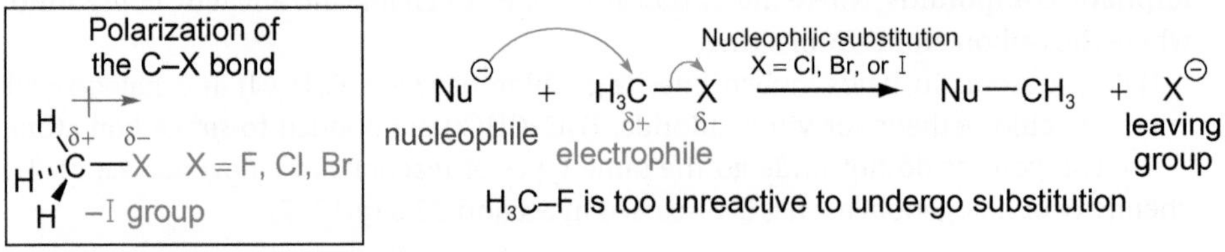
Polarization of the C–X bond
X = F, Cl, Br
–I group
Nucleophilic substitution
X = Cl, Br, or I
nucleophile
electrophile
leaving group
H₃C–F is too unreactive to undergo substitution

Figure 20.1 A general nucleophilic substitution reaction.

Summary

- As the halogen atom (X) increases in size, from fluorine to iodine, the C–X bond length increases and the C–X bond enthalpy decreases, that is, the bond becomes weaker.
- C–X bonds, apart from C–I, are polar with the electrons in the bonds displaced toward fluorine, bromine, or chlorine.
- The C–I bond is non-polar but has a high polarizability.
- Halogenoalkanes undergo nucleophilic substitution reactions and elimination reactions.

20.2 Preparation of halogenoalkanes

Halogenation of alkanes

Alkanes (RH) react with molecular chlorine (Cl_2) or bromine (Br_2), on heating or on irradiation with ultraviolet (UV) radiation, to form chloroalkanes (RCl) or bromoalkanes (RBr), respectively. The mechanism of these reactions involves a radical chain reaction.

In the initiation step, homolysis of the Cl–Cl or Br–Br bond forms chlorine or bromine radicals, respectively. Homolysis of a covalent bond is an endothermic process and the energy required to break the relatively weak Cl–Cl (+242 kJ mol^{-1}) or Br–Br (+194 kJ mol^{-1}) bond is supplied by heating or by absorption of UV radiation.

Radical chain reactions of organic compounds are introduced in Section 19.2 (p.907). The kinetics of chain reactions is discussed in Section 9.6 (p.423).

Remember that single-headed curly arrows are used to represent the movement of single electrons in radical reactions. This is discussed in Section 19.1 (p.866).

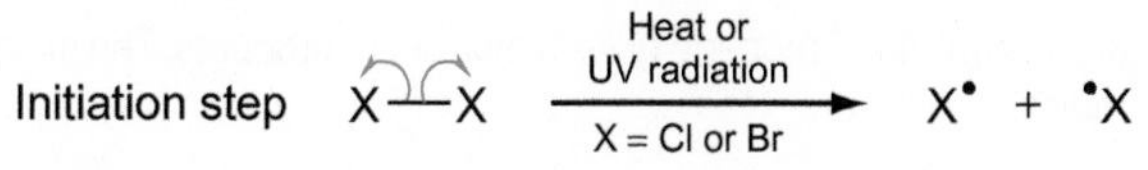
Initiation step
X—X
Heat or
UV radiation
X = Cl or Br
X• + •X

Bond enthalpies (Section 13.3, p.630) are a good guide to predicting products from radical reactions. In general, radical reactions involve homolysis of the weakest bonds in the reactants to form radicals that react to form stronger bonds in the products.

In the **propagation** stage, a chlorine or bromine radical abstracts a hydrogen atom from an alkane molecule to form HCl or HBr and generate an alkyl radical (R$^\bullet$). For example, reaction of Cl$^\bullet$ with ethane (CH_3CH_3) forms the ethyl radical ($CH_3CH_2^\bullet$) together with HCl (see the reaction scheme below). This reaction is exothermic because a stronger H–Cl bond (+431 kJ mol^{-1}) is formed at the expense of a weaker C–H bond in ethane (+423 kJ mol^{-1}).

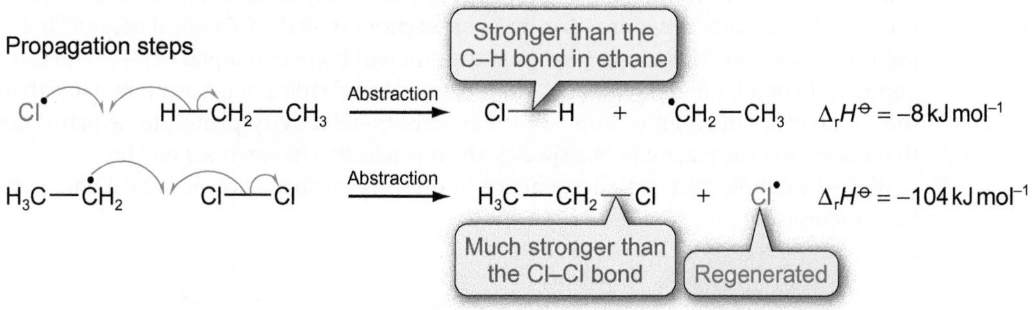
Propagation steps
Stronger than the C–H bond in ethane
Abstraction
Cl• H—CH2—CH3 → Cl—H + •CH2—CH3 ΔrH⊖ = −8 kJ mol−1
Abstraction
H3C—•CH2 Cl—Cl → H3C—CH2—Cl + Cl• ΔrH⊖ = −104 kJ mol−1
Much stronger than the Cl–Cl bond
Regenerated

The ethyl radical ($CH_3CH_2^\bullet$) then abstracts a chlorine atom from Cl_2 in a second propagation step to form chloroethane (CH_3CH_2Cl) and Cl$^\bullet$. This is a highly exothermic reaction as the C–Cl bond in chloroethane (+346 kJ mol^{-1}) is much stronger than the Cl–Cl (+242 kJ mol^{-1}) bond. The Cl$^\bullet$ radical then reacts with another molecule of ethane to continue the chain reaction.

Overall, the chain reaction leads to the formation of chloroethane (CH_3CH_2Cl) from ethane (CH_3CH_3). It is an example of a **radical substitution** reaction (i.e. an H atom in ethane is substituted by a Cl atom). These reactions are used on a large scale in industry because they convert unreactive alkanes into reactive and useful halogenoalkane products.

Chain reactions are stopped when two radicals react in a coupling reaction, called a **termination** step (see Section 19.2, p.908). For example, $CH_3CH_2^\bullet + {}^\bullet CH_2CH_3 \rightarrow CH_3CH_2CH_2CH_3$. However, the concentration of radicals in the reaction mixture is low, so the chances of two radicals reacting are small and chain termination is rare.

Radical chlorinations of alkanes give a mixture of products. This is because the chloroalkane products (e.g. RCH_2Cl) can undergo further chlorinations to give dichloro- and trichloroalkanes (e.g. $RCHCl_2$ and $RCCl_3$), although this can be minimized by using an excess of the alkane.

The selective formation of one chloroalkane product can be particularly difficult when the reacting alkane has non-equivalent hydrogen atoms. Propane ($CH_3CH_2CH_3$), for example, contains six hydrogen atoms attached to primary carbon atoms and two hydrogen atoms attached to a secondary carbon atom. If each hydrogen atom on propane reacts equally in the chlorination, then you would expect a 3 : 1 mixture of 1-chloropropane : 2-chloropropane (or $CH_3CH_2CH_2Cl : CH_3CHClCH_3$). However, the observed product ratio is actually 1 : 1.3 in favour of 2-chloropropane! (See Figure 20.2.) You can explain this result by considering the different C–H bond dissociation enthalpies in propane and the different stabilities of the intermediate carbon-centred radicals.

The order of C–H bond enthalpies (approximate values are given) is: methane C–H (+440 kJ mol^{-1}) > primary C–H (+420 kJ mol^{-1}) > secondary C–H (+410 kJ mol^{-1}) > tertiary C–H (+400 kJ mol^{-1}).

The six primary C–H bonds at the ends of the propane molecule are stronger than the two secondary C–H bonds in the middle. This is because the magnitude of a bond enthalpy

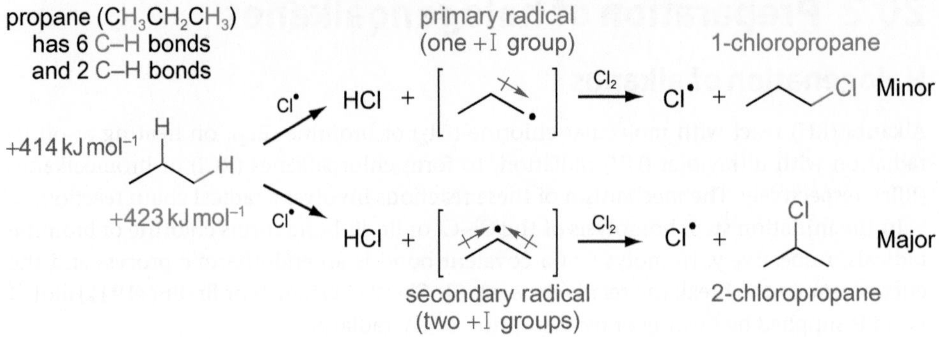

propane ($CH_3CH_2CH_3$)
has 6 C–H bonds
and 2 C–H bonds
+414 kJ mol^{-1}
+423 kJ mol^{-1}
primary radical
(one +I group)
secondary radical
(two +I groups)
HCl +
Cl_2
Cl• +
1-chloropropane
2-chloropropane
Minor
Major

Figure 20.2 Radical chlorination of propane gives a mixture of products. The major product is formed from the secondary radical.

depends on the stability of the radicals that are formed on homolysis of the bond—the more stable the radicals, the lower the bond enthalpy. Abstraction of a hydrogen atom from the secondary carbon atom forms an intermediate secondary carbon-centred radical ($CH_3CH^{\bullet}CH_3$). This radical is more stable than the primary carbon-centred radical ($CH_3CH_2CH_2^{\bullet}$) produced on abstraction of a hydrogen atom from one of the primary carbon atoms.

Interestingly, the bromination of propane is considerably more selective than the chlorination. The Br• radical is more selective in its reactions than the Cl• radical because Br• is a less reactive radical. This can be seen from the different bond enthalpies of C–Br and C–Cl bonds. C–Br bonds are weaker than C–Cl bonds so the Br• radical must be more stable than the Cl• radical. This is an example of the **reactivity–selectivity principle**, which states that the greater the reactivity of a species, the less selective its reactions will be.

Worked example 20.1 provides practice in drawing mechanisms for radical halogenation reactions.

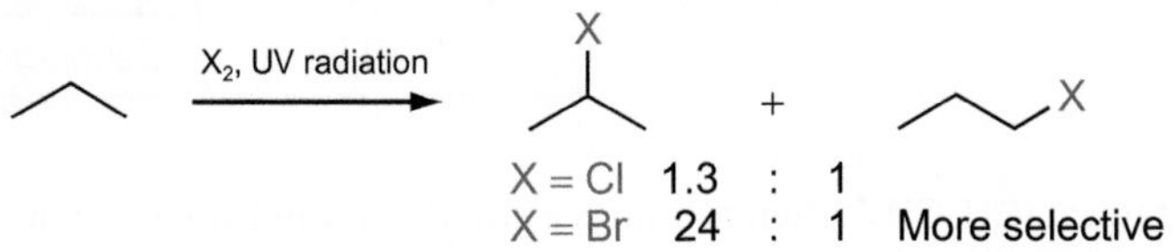

X_2, UV radiation
X = Cl 1.3 : 1
X = Br 24 : 1 More selective

Carbon-centred radicals, such as $(H_3C)_3C^{\bullet}$, are stabilized by +I groups and hyperconjugation, as described in Section 19.1 (p.869).

Radical bromination is more regioselective than radical chlorination. For an introduction to regioselectivity see Section 19.3, p.909.

Only radical bromination and chlorination reactions are of synthetic use. Fluorine reacts so violently with alkanes that fluorinations are difficult to control, while radical iodinations do not take place. The order of reactivity of halogen radicals is F• > Cl• > Br• > I•.

In place of using hazardous liquid bromine in this reaction (which fumes and is tricky to handle), *N*-bromosuccinimide (abbreviated NBS) can be used—NBS is an easily handled solid reagent. Trace amounts of HBr are usually present in NBS that reacts with NBS to form succinimide and Br_2. This provides a constant, low concentration of Br_2, which is beneficial for allylic bromination. A high concentration of Br_2 can lead to a competing reaction, namely electrophilic addition of Br_2 to the C=C bond.

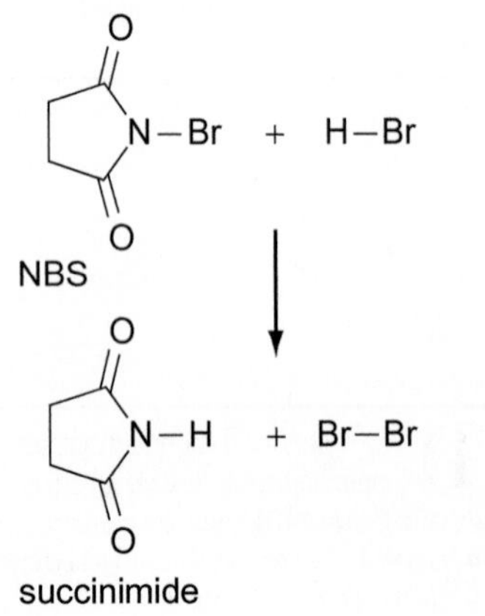

N–Br + H–Br
NBS
N–H + Br–Br
succinimide

Worked example 20.1 Allylic halogenation

A saturated carbon atom next to a C=C bond is called the **allylic** position—the hydrogen atoms attached to this carbon atom are said to be **allylic hydrogen atoms**.

When cyclohexene (C_6H_{10}) is treated with bromine in the presence of UV radiation, an allylic hydrogen atom is substituted by a bromine atom, to give 3-bromocyclohex-1-ene (C_6H_9Br).

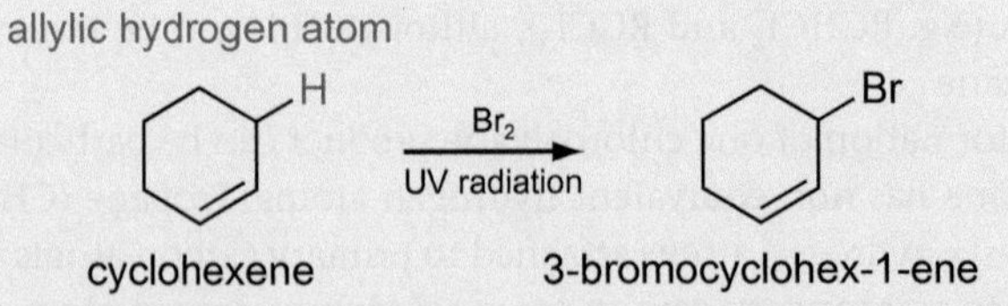

allylic hydrogen atom
Br_2
UV radiation
cyclohexene
3-bromocyclohex-1-ene

Suggest a mechanism to explain this reaction.

Strategy

Show the initiation and propagation steps (using single-headed curly arrows to represent the movement of electrons) that lead to the substitution of an allylic hydrogen atom by a bromine atom. →

→

Solution

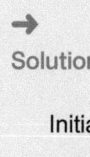

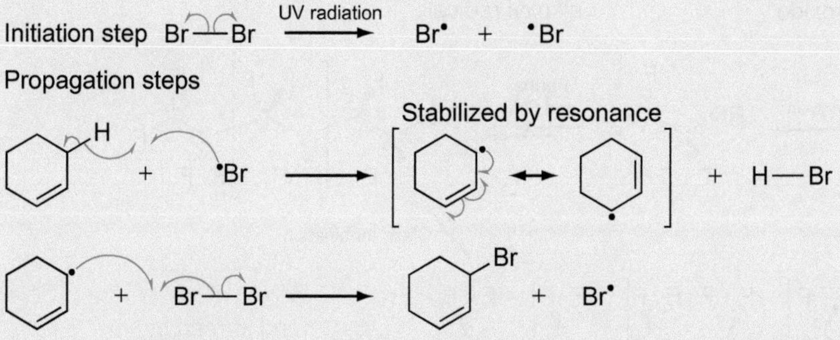
Initiation step Br—Br
UV radiation
Br• + •Br
Propagation steps
Stabilized by resonance
H
+ •Br
+ H—Br
Br
+ Br—Br
+ Br•

Reaction of either resonance form of the allylic radical with Br_2 gives the same product.

Note that if bromine is mixed with cyclohexene in the absence of UV radiation, then bromine radicals are not formed, and the C=C bond of cyclohexene reacts with Br_2 in an electrophilic addition reaction (see Section 21.3, p.975).

Question

1-(Chloromethyl)benzene (benzyl chloride, $PhCH_2Cl$) is made on a large scale in industry by reacting methylbenzene (toluene, $PhCH_3$) with chlorine in the presence of UV radiation. Suggest a mechanism to explain the formation of 1-(chloromethyl)benzene from methylbenzene.

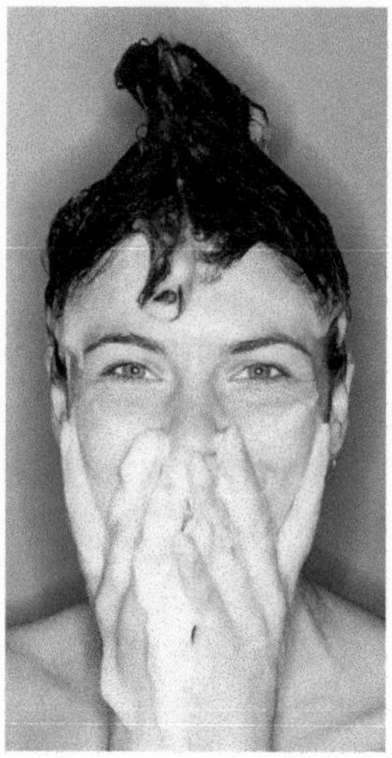

Industrially, benzyl chloride is used to make a range of useful products including quaternary ammonium salts ($R_4N^+ X^-$). For example, reaction of benzyl chloride with a long chain alkyldimethylamine [$CH_3(CH_2)_nN(CH_3)_2$] forms benzalkonium chloride, [$CH_3(CH_2)_nN(CH_3)_2CH_2Ph]^+$ Cl^- (where $n = 7, 9, 11$, etc.). This salt is used in many consumer products from ear drops, to wet wipes, disinfectants, and shampoos.

Box 20.2 Teflon and radical polymerization

Teflon® is the trademark name for a family of fluorinated polymers that includes poly(tetrafluoroethene), PTFE ([–CF_2–CF_2–]$_n$). PTFE was first made by accident by the American chemist Roy Plunkett when he was working on the development of new non-toxic refrigerants for the chemical company, DuPont. Plunkett used pressurized cylinders of gaseous tetrafluoroethene ($F_2C{=}CF_2$) in his research and on one occasion, on the morning of 6 April 1938, he found that there was no pressure in the cylinder but it weighed the same as when filled with gas. On opening the cylinder, he found a waxy white solid had formed.

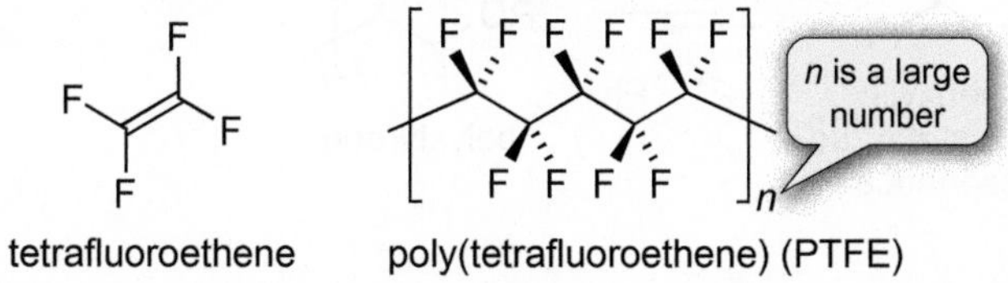
F
F
F
F
F F F F F F
n is a large number
F F F F
n

tetrafluoroethene   poly(tetrafluoroethene) (PTFE)

To his surprise, Plunkett found that the tetrafluoroethene had polymerized in the bottle to form PTFE. PTFE has some remarkable properties. It is inert to virtually all chemicals as it contains only strong C–F and C–C bonds, and it is considered the most slippery material in existence because the surface is covered with fluorine atoms that do not interact with atoms in other compounds. These properties have made PTFE one of the most useful polymers ever invented. PTFE tubes hold cables and wires in aircraft and cars, in medicine it is used in reconstructive and cosmetic facial surgery, and it has become a household name through its use as a coating on non-stick cookware.

▲ The Apollo 11 space suits worn by Neil Armstrong and crew during the first human exploration of the moon were made of layers of Teflon. An estimated 500 million people watched Armstrong walk on the moon and scoop up some lunar soil into a Teflon bag attached to his spacesuit. →

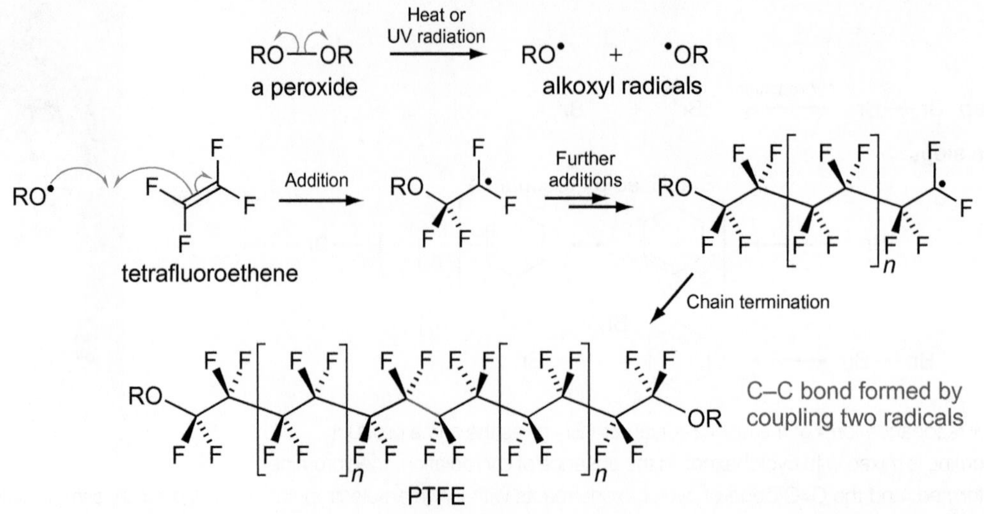
Heat or
UV radiation
RO—OR
a peroxide
RO• + •OR
alkoxyl radicals
Addition
tetrafluoroethene
Further
additions
Chain termination
C–C bond formed by
coupling two radicals
PTFE

The mechanism of polymerization of tetrafluoroethene involves radical intermediates. To start the polymerization an initiator is required and this is often an organic peroxide (ROOR). On heating or on irradiation with UV radiation, the weak O–O bond (+125 kJ mol^{-1} to +155 kJ mol^{-1}) of the peroxide breaks to form two alkoxyl radicals (RO•) in an initiation step. An alkoxyl radical then adds to the C=C bond of tetrafluoroethene to form a carbon radical. This is an exothermic reaction as the O–C bond in the product (~+350 kJ mol^{-1}) is stronger than the C=C π bond (~+275 kJ mol^{-1}). In the presence of further molecules of tetrafluoroethene, the radical addition process continues many times—the additions are exothermic as stronger C–C σ bonds are formed at the expense of weaker C=C π bonds. Eventually, coupling of the polymer radical with another radical (either another polymer radical or RO•) terminates the growing polymer chain to form PTFE.

In the chemical industry, the radical polymerization of alkenes is an important method for making addition polymers, such as polyethene (polythene), poly(vinyl chloride) (PVC), and polystyrene.

When RO• adds to an unsymmetrical alkene, such as styrene (PhCH=CH$_2$), the radical adds regioselectively (p.921) to the less substituted carbon atom of the C=C bond. This is explained by steric effects and also by the stability of the radical that is formed. Addition of the RO• radical to the less substituted carbon atom forms a carbon radical (RCH•Ph), called a benzylic radical, which is stabilized by resonance. The benzylic radical is more stable than the radical formed from addition of RO• to the more substituted carbon atom of styrene (check this for yourself).

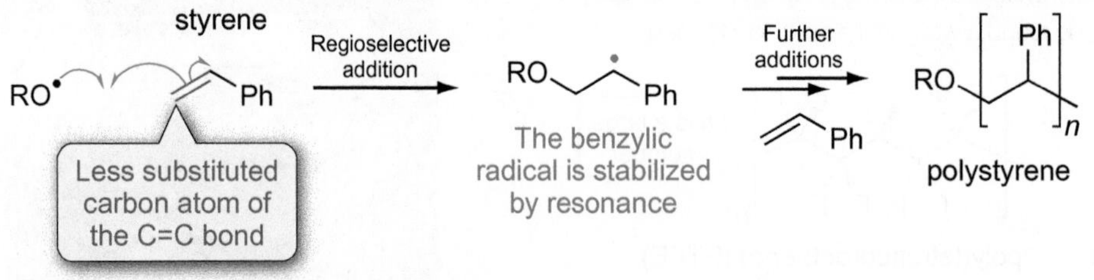
styrene
RO•
Ph
Less substituted
carbon atom of
the C=C bond
Regioselective
addition
The benzylic
radical is stabilized
by resonance
Further
additions
polystyrene

Question

The polymer PVC is prepared by heating chloroethene (vinyl chloride) with an organic peroxide (ROOR). A key step in the mechanism is shown here.

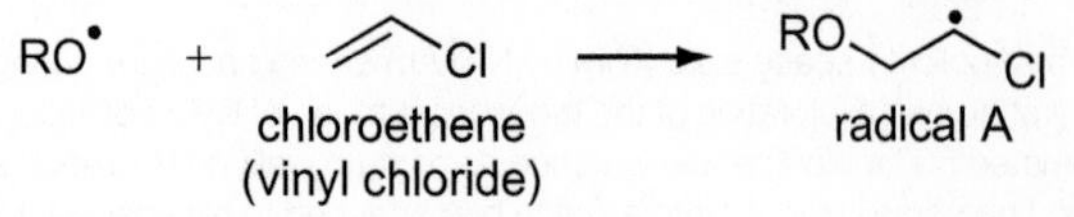
RO• + Cl
chloroethene
(vinyl chloride)
radical A

(a) Suggest an explanation for the regioselective addition of RO• to the C=C bond (i.e. why does the RO• radical add to the less substituted carbon atom in the C=C bond?)

(b) Suggest an explanation for the chemoselectivity of the reaction (i.e. why does the RO• radical add to the C=C bond and not abstract a chlorine atom?)

(c) Suggest a structure for the product formed from reaction of radical A with one molecule of chloroethene.

Halogenation of alcohols

There are a variety of methods for converting alcohols (ROH) into halogenoalkanes (RX). In these nucleophilic substitution reactions, the –OH group of the alcohol is substituted by a halogen atom.

All substitution reactions of alcohols require activation of the –OH group to convert it into a good leaving group (Figure 20.3). Good leaving groups are neutral molecules or stable anions. As a rough guide, you can assess whether a group is a good or a poor leaving group from its basic strength—*the weaker the basic strength of a group, the better leaving group it is*. A good leaving group readily accepts a pair of electrons to give a neutral molecule or anion that has a low pK_a value. A low pK_a value indicates that the molecule or anion is relatively stable (Section 19.2, p.888). So, because the pK_a of H_2O is relatively high (15.74), the hydroxide ion is a strong base and a poor leaving group.

pK_a and leaving groups. pK_as are only a *guide* for assessing if a group is a good leaving group. To assess leaving group ability, you really need to compare the stabilities of the groups relative to their precursors. For example, N_2 is a better leaving group than HO^-, because N_2 is more stable relative to RN_2^+ than HO^- is relative to ROH.

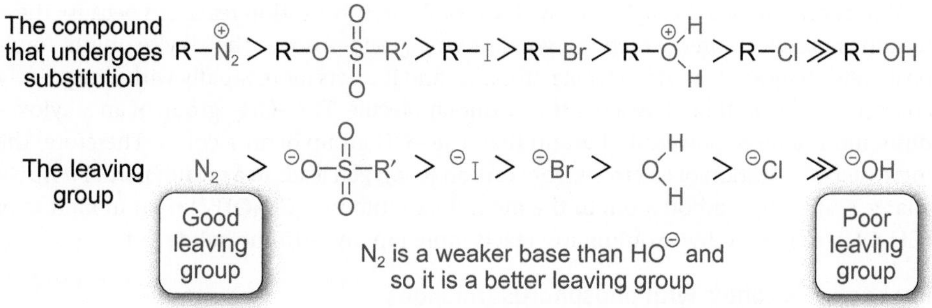
The compound that undergoes substitution
The leaving group
Good leaving group
Poor leaving group
N₂ is a weaker base than HO⁻ and so it is a better leaving group

Figure 20.3 Approximate order of leaving groups attached to a saturated carbon atom.

Halide ion leaving groups are discussed in Section 20.3 (p.938).

To determine if an anion or neutral molecule is a good leaving group, look at the pK_a value of the conjugate acid. As a rule of thumb, if the conjugate acid of the leaving group has a pK_a value above 13 then it is a poor leaving group. Whereas Br^- is an excellent leaving group (the pK_a of HBr is −7), H_2O is a good leaving group (the pK_a of H_2OH^+ is −1.7), $CH_3CO_2^-$ is a moderately good leaving group (the pK_a of CH_3CO_2H is 4.76) and CH_3O^- is a poor leaving group (the pK_a of CH_3OH is 15.5). If the conjugate acid of a group has a pK_a value above 30, then it cannot act as a leaving group.

There are different ways of activating the –OH group of an alcohol to produce compounds that will react with halide ions in nucleophilic substitution reactions. These reactions convert alcohols into halogenoalkanes. The mechanisms of the nucleophilic substitution reactions depend on whether the alcohol is primary (RCH_2OH), secondary (R_2CHOH), or tertiary (R_3COH). Some substitutions are concerted (one-step) reactions, whereas others involve the formation of intermediates, as discussed in Section 20.3 (p.930). The following sections concentrate on the mechanisms of the activation steps. Methanol (CH_3OH) is used as a representative alcohol because the substitution reactions are all concerted. The general process of activation is summarized in Figure 20.4.

Primary, secondary, and tertiary alcohols are introduced in Section 2.6 (p.92).

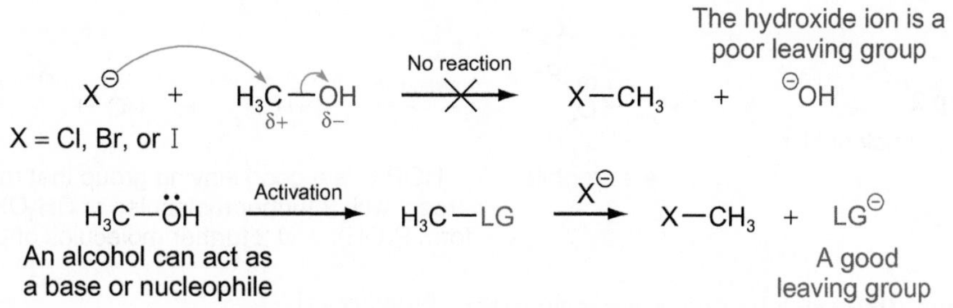
The hydroxide ion is a poor leaving group
No reaction
X = Cl, Br, or I
Activation
An alcohol can act as a base or nucleophile
A good leaving group

Figure 20.4 An alcohol group is activated by converting it into a compound that gives a better leaving group.

Reaction of alcohols with hydrogen halides

Reaction of alcohols (ROH) with acids, such as HCl or HBr, produces halogenoalkanes. In the first step of the reaction, the –OH group of the alcohol is protonated to form an alkyloxonium ion (ROH_2^+). Protonation of the –OH group activates the alcohol towards nucleophilic substitution by chloride or bromide ions, which takes place in the second step of the reaction to give a halogenoalkane, RCl or RBr (see Figure 20.5).

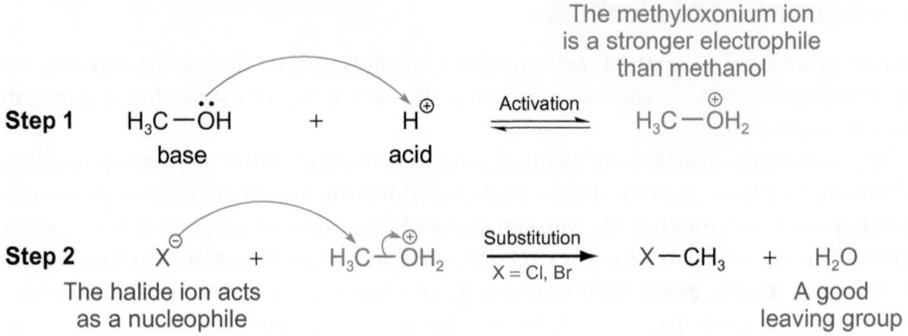
The methyloxonium ion
is a stronger electrophile
than methanol
Step 1
base
acid
Activation
Step 2
Substitution
X = Cl, Br
The halide ion acts
as a nucleophile
A good
leaving group

Figure 20.5 Protonation of the –OH group activates the alcohol to nucleophilic substitution.

The negative inductive effect and –I groups are introduced in Section 19.1, p.870.

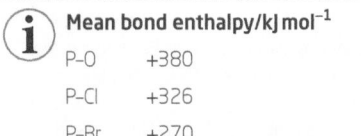

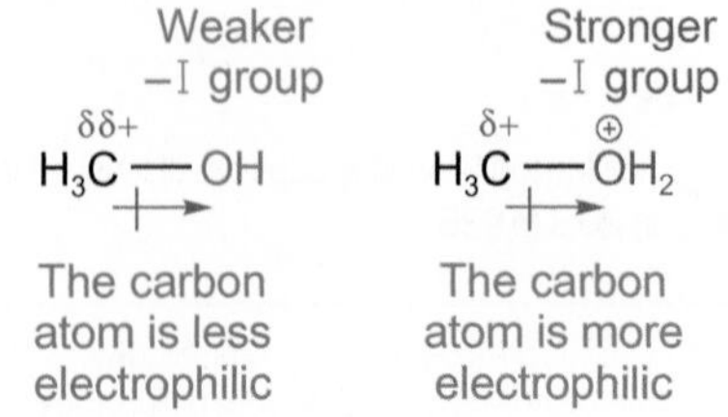
Weaker
–I group
Stronger
–I group
The carbon
atom is less
electrophilic
The carbon
atom is more
electrophilic

Alkyloxonium ions (ROH_2^+) react with Cl^- or Br^- in substitution reactions because these reactions produce water, which is a good leaving group. Also, as the alkyloxonium ion is positively charged, it is a stronger electrophile, and it reacts more rapidly with a negatively charged halide ion than does a neutral alcohol molecule. The $-OH_2^+$ group of an alkyloxonium ion is a more powerful –I group than the –OH group of an alcohol. Therefore, the increased polarization of electrons from carbon to oxygen leads to a greater partial positive charge (δ+) on the carbon atom in the methyloxonium ion ($CH_3OH_2^+$) than in methanol (CH_3OH), so the methyloxonium ion reacts more rapidly with the halide ion.

Reaction of alcohols with phosphorus trihalides

Phosphorus trihalides, such as phosphorus trichloride (PCl_3) or phosphorus tribromide (PBr_3), react with alcohols to form halogenoalkanes. In the first step of the reaction, the alcohol reacts with the phosphorus trihalide to form an intermediate, which then reacts with a chloride or bromide ion to give the halogenoalkane, as shown in Figure 20.6. The phosphorus trihalide can react with 1, 2, or 3 molecules of methanol. (Practise your curly arrow pushing by showing how $HOPX_2$ reacts with two molecules of methanol.)

Mean bond enthalpy/kJ mol^{-1}

P–O	+380
P–Cl	+326
P–Br	+270

Notice that the halide ion (X^-) does not deprotonate the positively charged intermediate formed in step 1. This is because halide ions are extremely weak bases–the conjugate acids of weak bases have low pK_a values and HBr and HCl have exceptionally low values of –9 and –7, respectively.

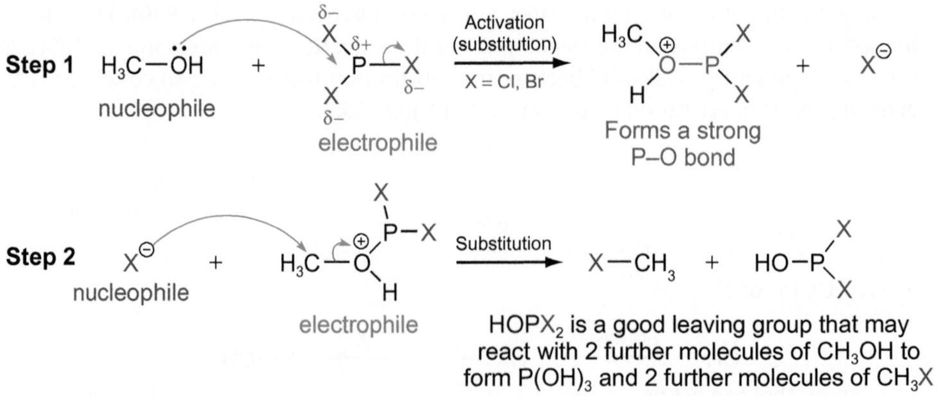
Step 1
nucleophile
electrophile
Activation
(substitution)
X = Cl, Br
Forms a strong
P–O bond
Step 2
nucleophile
electrophile
Substitution
HOPX₂ is a good leaving group that may
react with 2 further molecules of CH₃OH to
form P(OH)₃ and 2 further molecules of CH₃X

Figure 20.6 The –OH group is activated by reaction with PX_3 (X = Cl or Br).

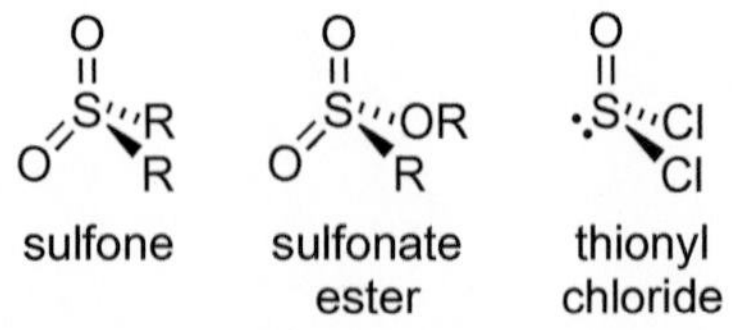
sulfone
sulfonate
ester
thionyl
chloride

Reaction of alcohols with thionyl chloride

Alcohols are converted into chloroalkanes by reaction with thionyl chloride ($SOCl_2$) and a base such as pyridine. The reaction takes place in three steps as shown in Figure 20.7.

- In **step 1**, the alcohol reacts with thionyl chloride in a nucleophilic substitution reaction. The alcohol replaces a chlorine atom in $SOCl_2$ and expels Cl^- as the leaving group.
- In **step 2**, the activated alcohol reacts with pyridine in an acid–base reaction in which pyridine removes H^+. Over the two steps, the –OH group of the alcohol is converted into an –OSOCl group.

Functional groups that contain a sulfur atom attached to four groups have a tetrahedral shape. Thionyl chloride has a shape based on a tetrahedron because there is a lone pair on sulfur, which is one of the four groups. The shapes of these molecules can be predicted using VSEPR theory, as discussed in Section 5.2 (p.223).

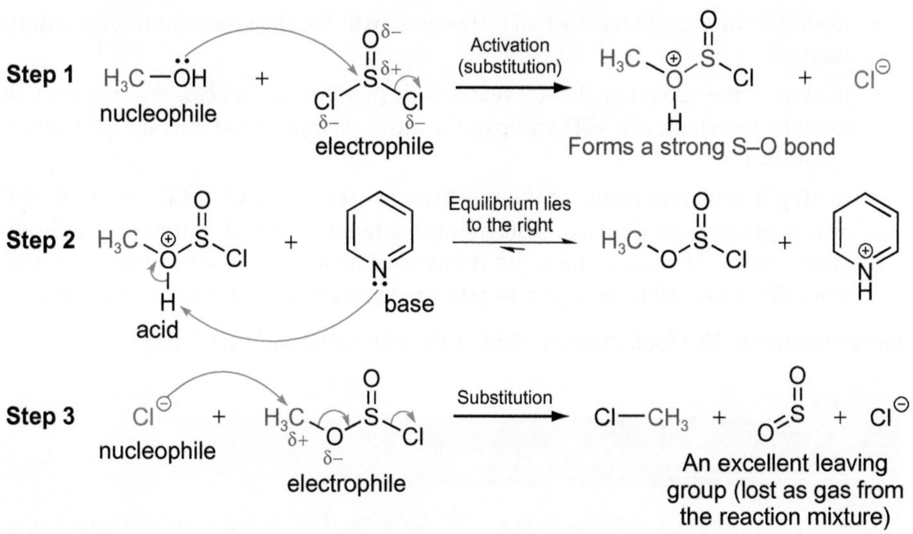
Step 1
$H_3C-\ddot{O}H$
nucleophile
+
electrophile
Activation (substitution)
Forms a strong S–O bond
Step 2
acid
base
Equilibrium lies to the right
Step 3
nucleophile
electrophile
Substitution
An excellent leaving group (lost as gas from the reaction mixture)

Figure 20.7 The –OH group is activated by reaction with $SOCl_2$ and then pyridine.

- This is followed by a second nucleophilic substitution reaction in **step 3**. This involves nucleophilic attack by the chloride ion (formed in step 1) to give the chloroalkane. The –OSOCl group leads to a much better leaving group than ^{-}OH because the $^{-}OSOCl$ anion breaks down to form Cl^- and SO_2 gas. The formation of SO_2 gas is favourable because a strong S=O bond is formed, and also because this results in a large increase in entropy (as gases are more disordered than liquids) and because the substitution in step 3 gives three molecules of products from just two molecules of reactants (Section 14.4, p.669).

Reaction of alcohols with 4-toluenesulfonyl chloride

Alcohols can be converted into a range of halogenoalkanes by activation of the –OH group using 4-toluenesulfonyl chloride ($4\text{-}CH_3C_6H_4SO_2Cl$, TsCl), sometimes called tosyl chloride. Reaction of an alcohol with TsCl, in the presence of a base such as pyridine, produces a 4-toluenesulfonate ester ($4\text{-}CH_3C_6H_4SO_2OR$), which is often called a tosylate (TsOR). The reaction takes place in three steps, as shown in Figure 20.8.

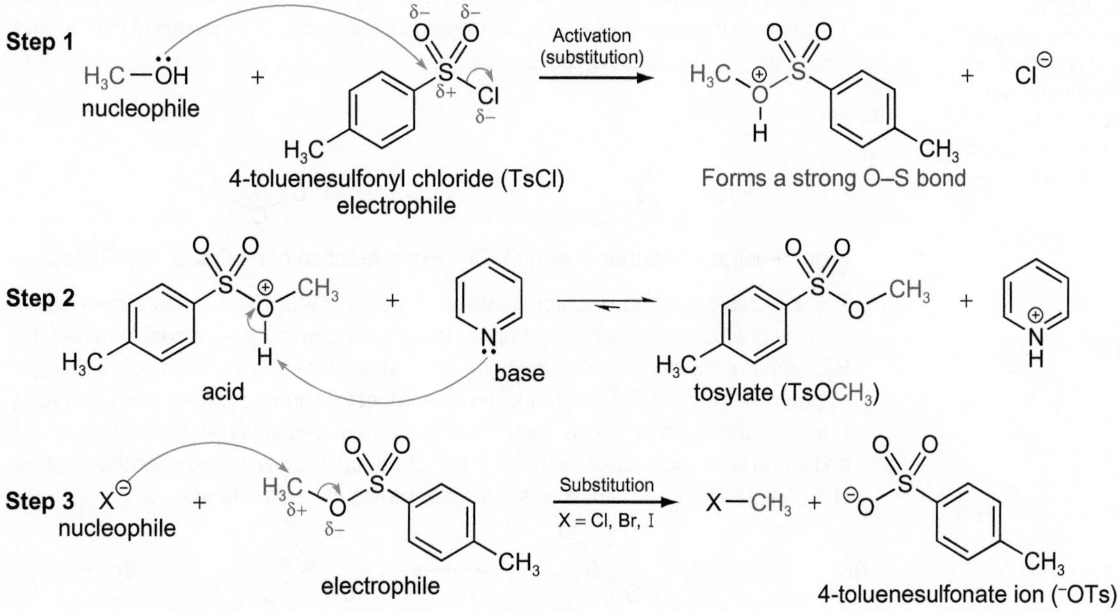
Step 1
nucleophile
Activation (substitution)
4-toluenesulfonyl chloride (TsCl)
electrophile
Forms a strong O–S bond
Step 2
acid
base
tosylate ($TsOCH_3$)
Step 3
nucleophile
electrophile
Substitution
X = Cl, Br, I
4-toluenesulfonate ion (^{-}OTs)

Figure 20.8 The –OH group is activated by reaction with TsCl and then pyridine.

Pyridine (the pK_a of the conjugate acid is 5.23) is not a strong enough base to deprotonate an alcohol (pK_a = 16–17), but it can deprotonate a protonated alcohol (pK_a = –2.4). Organic acids and bases are discussed in Section 19.2 (p.888).

Pyridine has a conjugated system of 6 π electrons—the lone pair of electrons on nitrogen is not part of the π ring system (Section 22.1, p.1009). It is widely used as a weak base in organic synthesis and also as a ligand in coordination chemistry, along with its derivatives 2,2′-bipyridine and 2,2′:6,2′-terpyridine (Section 28.3, p.1266).

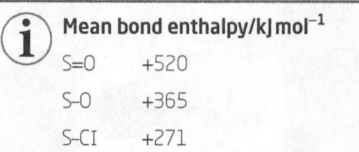

Mean bond enthalpy/kJ mol^{-1}

S=O	+520
S–O	+365
S–Cl	+271

- In **step 1**, the alcohol is activated by reaction with TsCl in a nucleophilic substitution reaction.
- In **step 2**, the activated alcohol reacts with pyridine in an acid–base reaction and, over the two steps, the –OH group of the alcohol is converted into an –OTs group of the tosylate.
- In **step 3**, a second nucleophilic substitution takes place with Cl^-, Br^-, or I^- as the attacking nucleophile. The 4-toluenesulfonate ion (^-OTs) is a much better leaving group than ^-OH because the anion is stabilized by resonance—the negative charge is delocalized over all three electronegative oxygen atoms (but not the benzene ring).

Resonance forms of the 4-toluenesulfonate ion

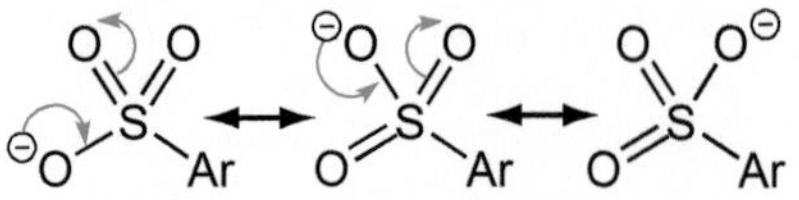
Ar
Ar
Ar

(Ar stands for the aromatic ring)

Worked example 20.2 looks at activation of alcohols using sulfonyl chlorides.

Visit the Online Resource Centre to view screencast 20.1 which walks you through activation of alcohols using various sulfonyl chlorides.

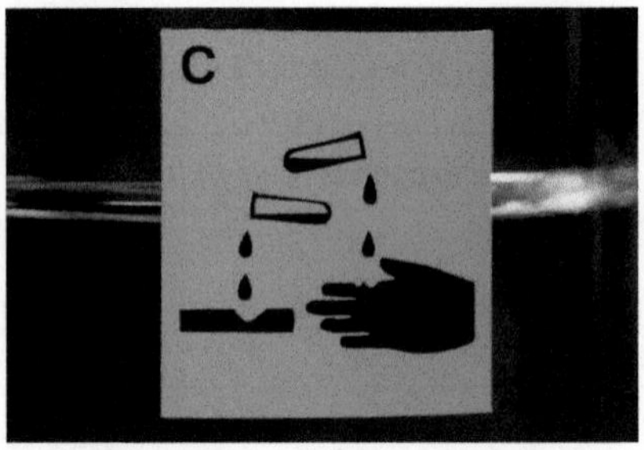
C

The stability of sulfonate ions (RSO_3^-) is reflected in the low pK_a values of the conjugate acids. Sulfonic acids (RSO_3H) are very strong and corrosive acids. For example, methanesulfonic acid (CH_3SO_3H) has a pK_a value of around −2 and trifluoromethanesulfonic acid (or triflic acid, CF_3SO_3H) has a pK_a value of around −15.

Worked example 20.2 Activation of alcohols

Other sulfonyl halides, besides 4-toluenesulfonyl chloride (TsCl), are used to activate alcohols for nucleophilic substitution reactions. The two sulfonyl halides shown here are often used.

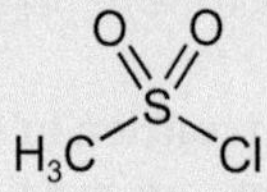

methanesulfonyl chloride (MsCl for short)

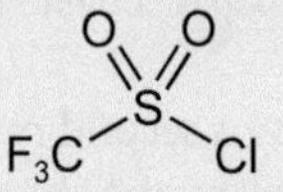

trifluoromethanesulfonyl chloride (TfMsCl for short)

(a) Draw the structures of the products formed from reaction of MsCl and TfMsCl with methanol in the presence of pyridine.

(b) Which of the sulfonates from part (a) would you expect to react more rapidly with bromide ions? Explain your reasoning.

Strategy

(a) There are two steps to consider. First, methanol reacts with MsCl or TfMsCl in a nucleophilic substitution reaction—methanol is the nucleophile and the sulfonyl chloride is the electrophile. Then, an acid–base reaction with pyridine removes H^+. (See the reaction of methanol with TsCl/pyridine in Figure 20.8 on p.927.)

(b) Consider whether the $-OCH_3$ carbon atom in the sulfonate derived from MsCl or from TfMsCl is the stronger electrophile and also, following attack by Br^-, which of the two leaving groups is the more stable. To do this, compare the electronic properties of the CF_3 and CH_3 groups in the products from part (a).

Solution

(a)

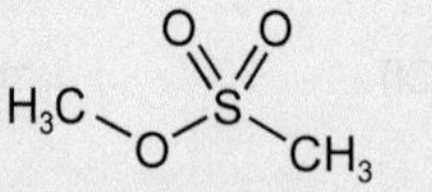

From reaction of methanol with MsCl

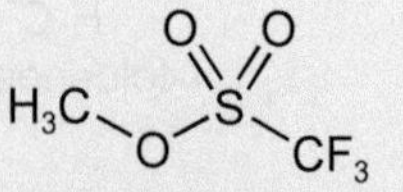

From reaction of methanol with TfMsCl

(b) The activated alcohol, formed from reaction of methanol with TfMsCl, reacts more rapidly with the bromide ion. The F_3C– group is electron-withdrawing (a −I group), whereas the H_3C– group is electron-donating (a +I group). As a result, the CF_3SO_2O– group is a stronger −I group than the CH_3SO_2O– group. This means that the carbon atom of the H_3CO– group in the activated alcohol formed using TfMsCl will have a greater partial positive charge (δ+) and so will react more rapidly with Br^-. The F_3C– group also helps to stabilize the negative charge in the $CF_3SO_2O^-$ ion, so this is a better leaving group than $CH_3SO_2O^-$.

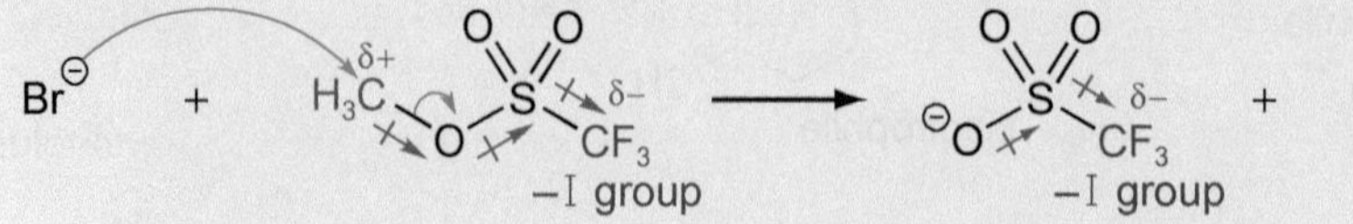
−I group
−I group

Question

The 4-toluenesulfonate ion ($^{-}$OTs) is such a good leaving group that tosylates react with a wide variety of nucleophiles (in addition to halide ions). For the substitution reaction illustrated here, suggest structures for products **A** and **B** and propose a mechanism for their formation.

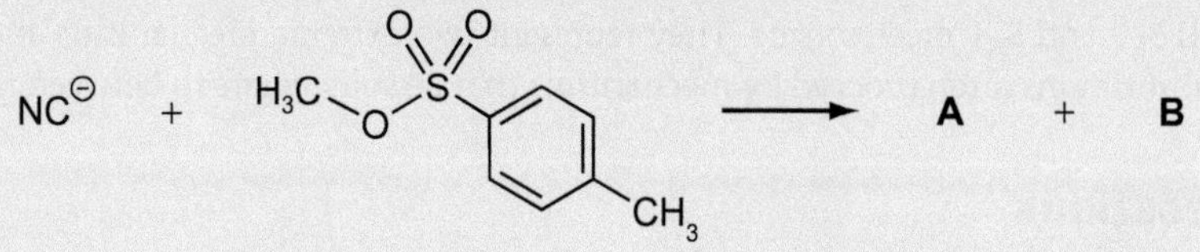
NC⊖
+
H3C
O
S
O
O
CH3
A
+
B

Addition of hydrogen halides to alkenes

Halogenoalkanes can be prepared by addition of hydrogen halides or bromine to the C=C bond of alkenes. Chloro-, bromo-, and iodoalkanes are formed by addition of HCl, HBr, and HI to the C=C bond, respectively, while addition of Br_2 forms dibromoalkanes that have bromine atoms on adjacent carbons.

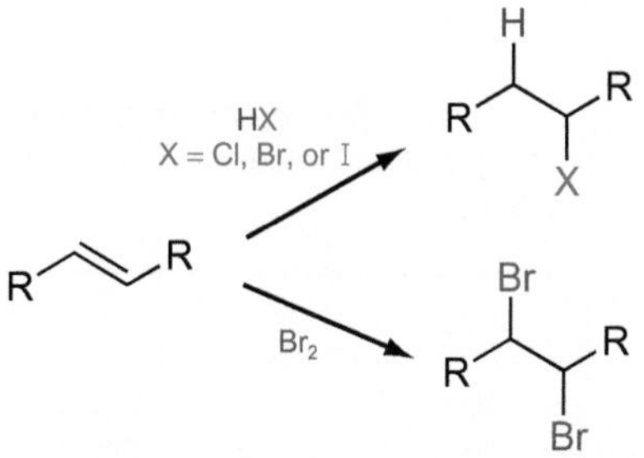
HX
X = Cl, Br, or I
Br2
R
R
H
X
Br
Br

In these reactions, the alkene is the nucleophile and the hydrogen halide or Br_2 is the electrophile. The mechanisms of these **electrophilic addition** reactions are discussed in Section 21.3 (p.975). HBr also adds to a C=C bond in the presence of a peroxide (ROOR), which acts as a radical initiator. This is an example of a radical addition reaction, the mechanism of which is discussed in Box 21.2 (p.973).

Summary

- Radical substitution reactions of alkanes (using Cl_2 or Br_2) produce chloro- or bromoalkanes.
- Radical bromination of alkanes is more selective than chlorination.
- The relative order of reactivity of C–H bonds toward radical halogenation is tertiary C–H > secondary C–H > primary C–H > methyl C–H.
- Halogenoalkanes are formed from alcohols using reagents that convert the –OH group into a good leaving group. Good leaving groups are neutral molecules or stable anions that are weak bases. A halogen atom then displaces the leaving group in a nucleophilic substitution reaction.

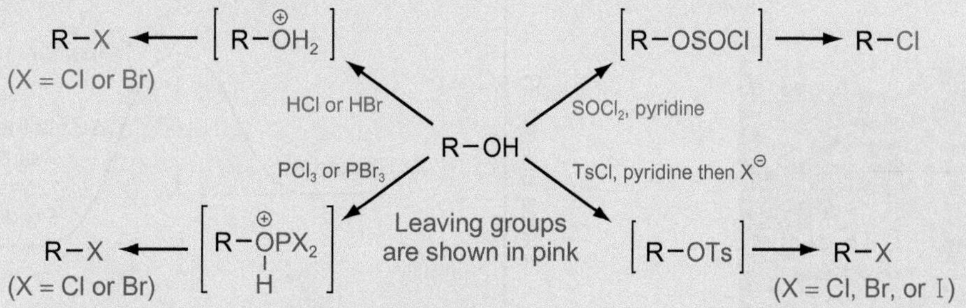
R–X
(X = Cl or Br)
[R–OH2 ⊕]
HCl or HBr
[R–OSOCl]
R–Cl
SOCl2, pyridine
R–OH
PCl3 or PBr3
TsCl, pyridine then X⊖
R–X
(X = Cl or Br)
[R–OPX2 ⊕ H]
Leaving groups
are shown in pink
[R–OTs]
R–X
(X = Cl, Br, or I)

- Halogenoalkanes are formed on addition of HCl, HBr, HI, or Br_2 to alkenes.

For practice questions on this topic, see questions 1 and 7 at the end of this chapter (pp.958–959).

20.3 The mechanisms of nucleophilic substitution reactions

A nucleophilic substitution reaction of a halogenoalkane (RX) leads to the replacement of the halogen atom by another substituent. The halide ion (X^-) produced is called a **leaving group**.

There are two distinct mechanisms for nucleophilic substitution of halogenoalkanes. These are called S_N2 and S_N1 mechanisms. They represent two extreme mechanisms and nucleophilic substitutions often proceed by mechanisms that lie somewhere in between the two.

S_N2 reactions

Bimolecular elementary reactions are discussed in Sections 9.6, (p.416) and 9.8 (p.432). Rate-determining steps and the use of kinetic experimental evidence to support mechanisms are covered in Section 9.6 (p.416).

Substitution reactions that take place by a concerted (single-step) mechanism are called **S_N2 reactions**—where S stands for *substitution*, N stands for *nucleophilic*, and 2 stands for *bimolecular*. The mechanism was deduced from kinetic studies. The rate of the substitution depends on the concentration of both the halogenoalkane and the nucleophile, so the reaction is second order with the following rate equation:

$$\text{rate of an } S_N2 \text{ reaction} = k \times [\text{halogenoalkane}] \times [\text{nucleophile}]$$

Reaction profiles are sometimes called Gibbs energy profiles. Their use is explained in Section 9.8 (p.435).

where k is the rate constant. The kinetics suggest that both the halogenoalkane and the nucleophile are involved in the transition state of the rate-determining step, which is therefore bimolecular.

Transition states are often shown inside square brackets marked with a double dagger symbol (‡). The δ– charges on Nu and X indicate around half a minus charge (they must add up to one).

In an S_N2 reaction, the nucleophile attacks the carbon atom in the C–X bond, as shown in Figure 20.9. The nucleophile forms a new bond to the carbon atom *at the same time* as the C–X bond is broken. In the transition state, the carbon atom is bonded to three R groups (which are now in the same plane) and partially bonded to both the nucleophile and the halogen atom (which are 180° apart). The reaction profile for an S_N2 reaction is shown in Figure 20.10.

The transition state has five groups around the central carbon atom.

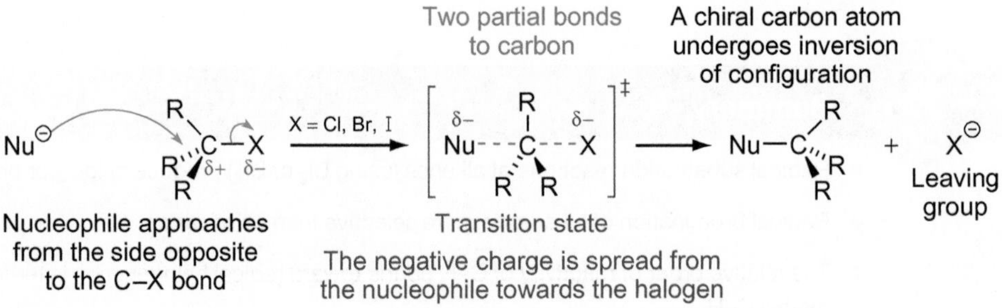
Two partial bonds to carbon
A chiral carbon atom undergoes inversion of configuration
X = Cl, Br, I
Leaving group
Nucleophile approaches from the side opposite to the C–X bond
Transition state
The negative charge is spread from the nucleophile towards the halogen

Figure 20.9 The mechanism of an S_N2 reaction.

Interactive 3D animation of an S_N2 reaction.

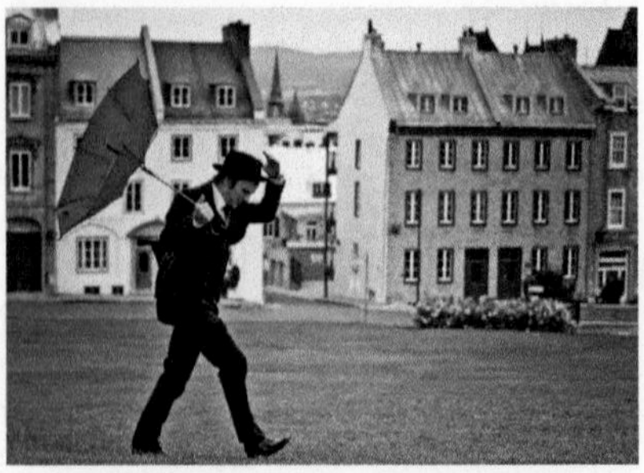

In an S_N2 mechanism the molecule inverts, like an umbrella being blown inside out when caught in the wind.

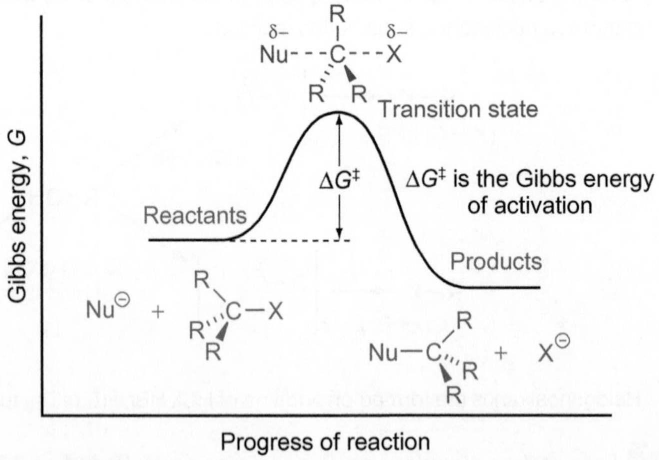
Gibbs energy, G
Transition state
ΔG‡
ΔG‡ is the Gibbs energy of activation
Reactants
Products
Progress of reaction

Figure 20.10 The reaction profile for an S_N2 reaction.

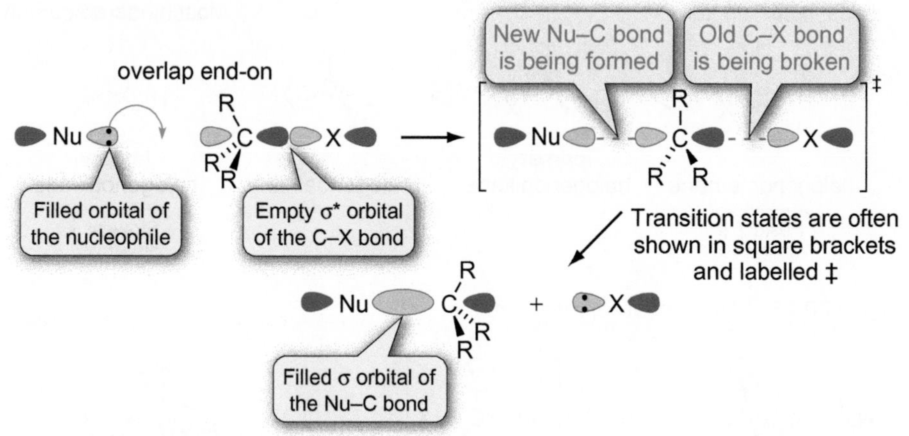
overlap end-on
Filled orbital of the nucleophile
Empty σ* orbital of the C–X bond
New Nu–C bond is being formed
Old C–X bond is being broken
Transition states are often shown in square brackets and labelled ‡
Filled σ orbital of the Nu–C bond

Figure 20.11 An S_N2 reaction showing orbital overlap.

The σ* orbital is the antibonding orbital. Orbital labels are introduced in Section 4.7 (p.186).

The molecular orbital energy level diagram below (see Sections 4.8, p.188, and 5.6, p.246), shows that the interaction between a nucleophile lone pair (the HOMO) and a σ* orbital (the LUMO) leads to a bonding interaction. For a σ* orbital, sometimes the two inner lobes (with opposite phases) are drawn smaller than the outer lobes to show a reduced electron density between the nuclei.

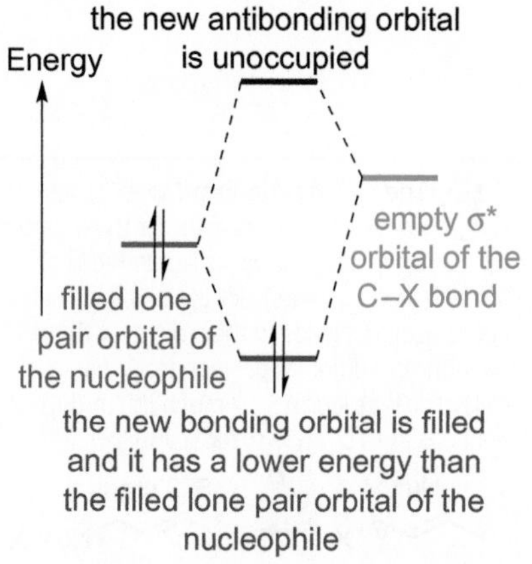
the new antibonding orbital is unoccupied
Energy
empty σ* orbital of the C–X bond
filled lone pair orbital of the nucleophile
the new bonding orbital is filled and it has a lower energy than the filled lone pair orbital of the nucleophile

Note that the nucleophile approaches the carbon atom from the opposite side to the C–X bond (this is sometimes called 'attack from the back'). The nucleophile approaches the C–X bond at an angle of 180° because this maximizes the overlap of the orbital of the nucleophile containing the electron pair with the empty σ* orbital of the C–X bond (see Figure 20.11). It is the σ* orbital of the C–X bond that overlaps because this orbital has the lowest energy of the empty orbitals of the halogenoalkane so it is the best 'electron-accepting' orbital.

Since the nucleophile attacks on the opposite side to the C–X bond, the groups on the carbon atom 'turn inside out'—like an umbrella blowing inside out. If the carbon atom in the C–X bond is a chiral centre, the S_N2 reaction leads to an **inversion of configuration**—this is also called a **Walden inversion** (after the chemist Paul Walden (1863–1957) who discovered this). For example, (*S*)-2-bromooctane is converted into (*R*)-2-ethoxyoctane on reaction with ethoxide ion (EtO^-) in an S_N2 reaction.

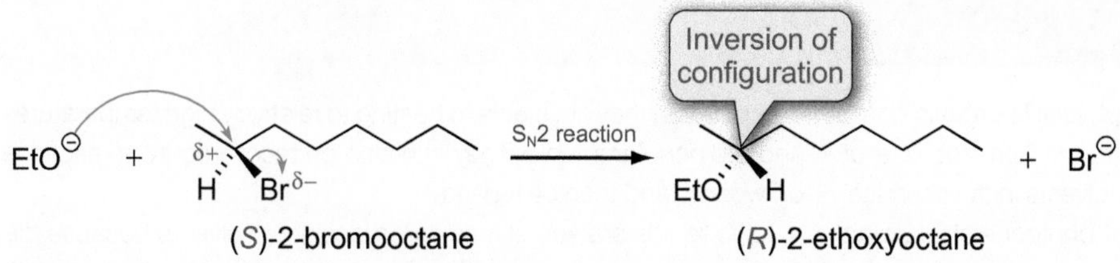
Inversion of configuration
EtO
δ+
H
Br δ−
SN2 reaction
EtO
H
+ Br
(S)-2-bromooctane
(R)-2-ethoxyoctane

Formation of an ether (RCH_2OR) by reaction of an alkoxide ion (^-OR) with a primary halogenoalkane (RCH_2X), via an S_N2 mechanism, is called the **Williamson ether synthesis**.

Similarly, (*R*)-2-bromooctane is converted into (*S*)-2-ethoxyoctane on reaction with ethoxide ion in an S_N2 reaction. Different enantiomers of the reactant produce different enantiomers of the product, so S_N2 reactions are stereospecific.

For assigning *R* and *S* configurations to chiral centres see Section 18.4 (p.845). Stereospecific reactions are introduced in Section 19.3 (p.910).

Effect of R groups on the rate of an S_N2 reaction

The size of the substituents bonded to the carbon atom in the C–X bond influences the rate of an S_N2 reaction—the larger the substituents, the slower the S_N2 reaction. This is because approach of the nucleophile is sterically hindered when large substituents are bonded to the carbon atom in the C–X bond. This helps explain why halogenomethanes (CH_3X) and primary halogenoalkanes (RCH_2X) readily undergo S_N2 reactions, whereas S_N2 reactions of secondary halogenoalkanes (R_2CHX) are slower and tertiary halogenoalkanes (R_3CX) do not undergo S_N2 reactions (although these compounds undergo substitution by an alternative S_N1 reaction (see p.934)). The relative rates are shown in Figure 20.12.

Note that the energy of the transition state of an S_N2 reaction is also affected by the size of the substituents bonded to the carbon atom in the C–X bond. The transition state is crowded because there are five groups surrounding the central atom. The larger the size of the groups, the greater the steric interactions, and the higher the energy of the transition state (Figure 20.13).

Relative rates of S_N2 reactions of bromoalkanes with the hydroxide ion

Bromoalkane	Class	Relative rate
CH_3Br	Methyl	1500
CH_3CH_2Br	Primary	50
$(CH_3)_2CHBr$	Secondary	1
$(CH_3)_3CBr$	Tertiary	negligible

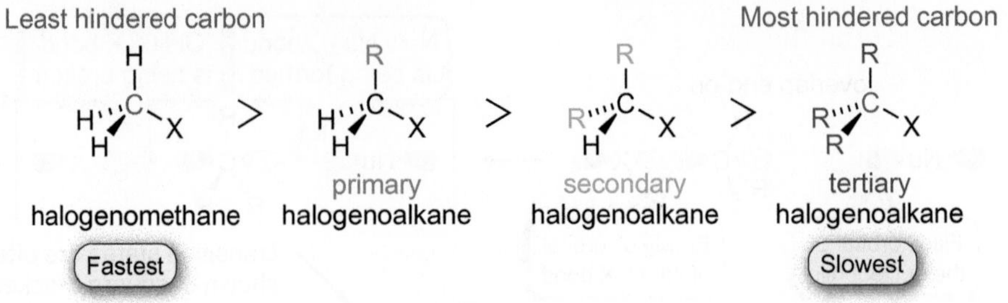
Least hindered carbon
Most hindered carbon
halogenomethane
primary halogenoalkane
secondary halogenoalkane
tertiary halogenoalkane
Fastest
Slowest

Figure 20.12 Relative rates of S_N2 reactions.

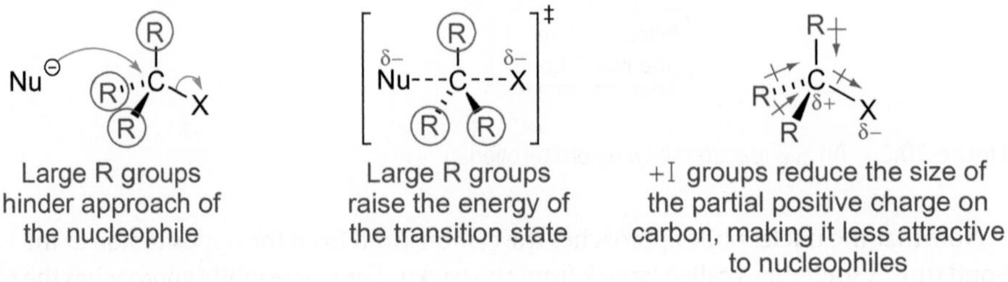
Large R groups hinder approach of the nucleophile
Large R groups raise the energy of the transition state
+I groups reduce the size of the partial positive charge on carbon, making it less attractive to nucleophiles

Figure 20.13 Effects of R groups on the rate of an S_N2 reaction.

The electronic effects of the alkyl groups (R) bonded to the carbon atom in the C–X bond also influence the rate of an S_N2 substitution. In an S_N2 reaction, the nucleophile attacks the partially positive carbon atom bonded to the electronegative halogen atom. Alkyl groups bonded to the carbon atom exert a positive inductive effect (+I), which reduces the partial positive charge (δ+) on the carbon.

The greater the number of alkyl groups on the carbon atom, the weaker the partial positive charge on the carbon atom, and the slower the rate of attack by the nucleophile.

The S_N2' mechanism. Nucleophiles can attack allyl halides at either the α-carbon atom or the γ-carbon atom as illustrated here. The two products are structural isomers. The mechanism that leads to the product with the C=C bond in a different position from that in the allyl halide is called an **S_N2' mechanism**. (S_N1' reactions are also known; see p.933.)

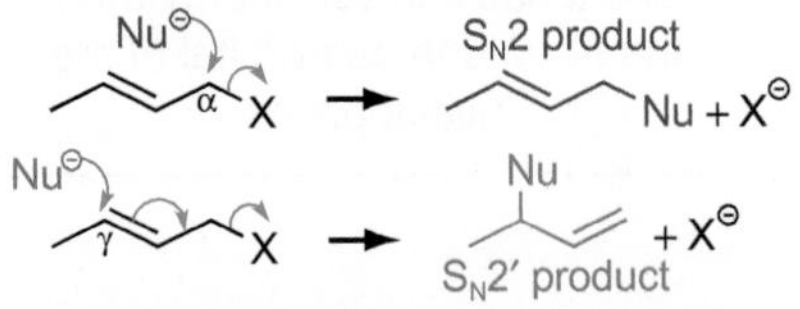
S_N2 product
S_N2' product

Box 20.3 Designer solvents

There is increasing concern about volatile organic compounds (VOCs) in the atmosphere. These come from the use of petrol in motor vehicles and from the use of solvents in the chemical industry. They contribute to the formation of photochemical smog (see Chapter 27, p.1201).

Traditionally, organic reactions are carried out in volatile organic solvents such as diethyl ether ($CH_3CH_2OCH_2CH_3$) and dichloromethane (CH_2Cl_2). Many organic compounds are soluble in these solvents and the solvents are easily removed from products by evaporation. Recently, however, some chemists have moved away from using VOCs as solvents and are using ionic liquids instead.

Ionic liquids are salts that are liquids at room temperature and their use offers several advantages over traditional organic solvents. For example, ionic liquids are not volatile and so do not escape into the atmosphere; they dissolve a wide range of polar organic compounds; they are stable to heating to relatively high temperatures and are non-flammable. They can also be recovered from reactions on workup and then be reused.

Ionic liquids are sometimes called designer solvents because different combinations of cations and anions produce ionic liquids with different physical and chemical properties. This means that an ionic liquid solvent can be designed for a specific purpose.

An important class of ionic liquids are bmim ionic liquids, which contain a 1-butyl-3-methylimidazolium cation and an anion such as PF_6^-.

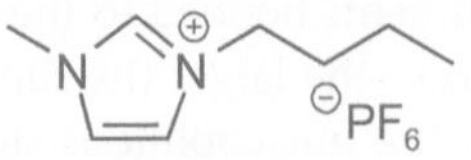

[bmim][PF_6]

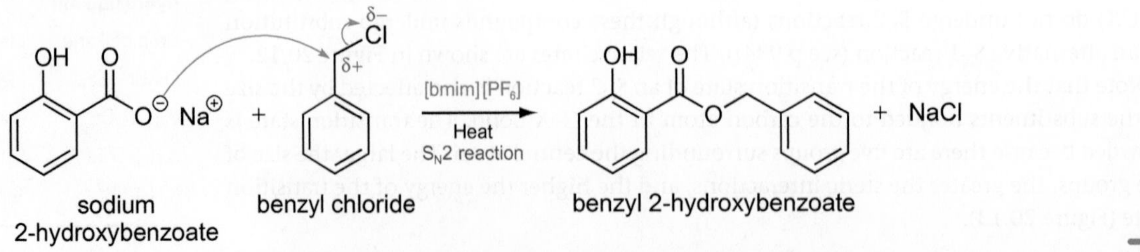
[bmim][PF_6]
Heat
S_N2 reaction
+ NaCl
sodium 2-hydroxybenzoate
benzyl chloride
benzyl 2-hydroxybenzoate

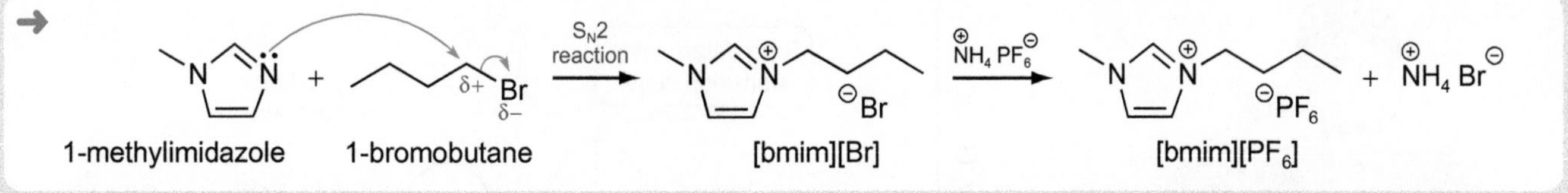

S_N2 reaction
$NH_4\ PF_6$
$NH_4\ Br$
1-methylimidazole
1-bromobutane
[bmim][Br]
[bmim][PF6]

A range of organic reactions, including nucleophilic substitution reactions, can be carried out in 1-butyl-3-methylimidazolium hexafluorophosphate ([bmim][PF_6]). For example, heating sodium 2-hydroxybenzoate and benzyl chloride in [bmim][PF_6] produces benzyl 2-hydroxybenzoate (an ester) in an S_N2 reaction. At the end of the reaction, the ionic liquid can be recovered on workup of the reaction (Box 2.6, p.94), then recycled and reused.

[bmim][PF_6] is itself easily prepared from 1-methylimidazole and 1-bromobutane using a nucleophilic substitution reaction in the first step, as shown in the reaction scheme above.

Question

Suggest a two-step synthesis of the ionic liquid 1-ethyl-3-methylimidazolium tetrafluoroborate, [emim][BF_4], starting from 1-methylimidazole.

S_N1 reactions

Substitution reactions that take place by a two-step mechanism are called **S_N1 reactions**—where S stands for *substitution*, N stands for *nucleophilic*, and 1 stands for *unimolecular*. The mechanism was deduced from kinetic studies. The rate of the substitution depends on the concentration of only the halogenoalkane (and not the nucleophile), so the reaction is first order with the following rate equation:

$$\text{rate of an } S_N1 \text{ reaction} = k \times [\text{halogenoalkane}]$$

where k is the rate constant. The kinetics suggest that only the halogenoalkane is involved in the transition state of the rate-determining step, which is therefore unimolecular.

The mechanism of an S_N1 reaction is shown in Figure 20.14. In the first step, the C–X bond breaks to form a carbocation (R_3C^+) and a halide ion (X^-). In the second step, the nucleophile attacks the carbocation to form a new bond.

Step 1 is the slower step so it is rate-determining. This explains why the rate of an S_N1 reaction depends only on the concentration of the halogenoalkane and not that of the nucleophile. In the second step the nucleophile rapidly attacks the carbocation and, as the carbocation is planar, the nucleophile can equally approach the planar carbocation from above or below. This means that, if a single enantiomer of a halogenoalkane undergoes an S_N1 reaction, the product will be racemic (e.g. a halogenoalkane of *S* configuration will react to form a 1 : 1 mixture of *R* and *S* enantiomers of the product).

Unimolecular elementary reactions are discussed in Section 9.6 (p.416).

Carbocations are generally planar and contain an empty *p* orbital on an sp^2 hybridized carbon atom (Section 19.1, p.867).

Assigning the configuration of chiral centres as *R* or *S* (CIP nomenclature) is introduced in Section 18.4 (p.845).

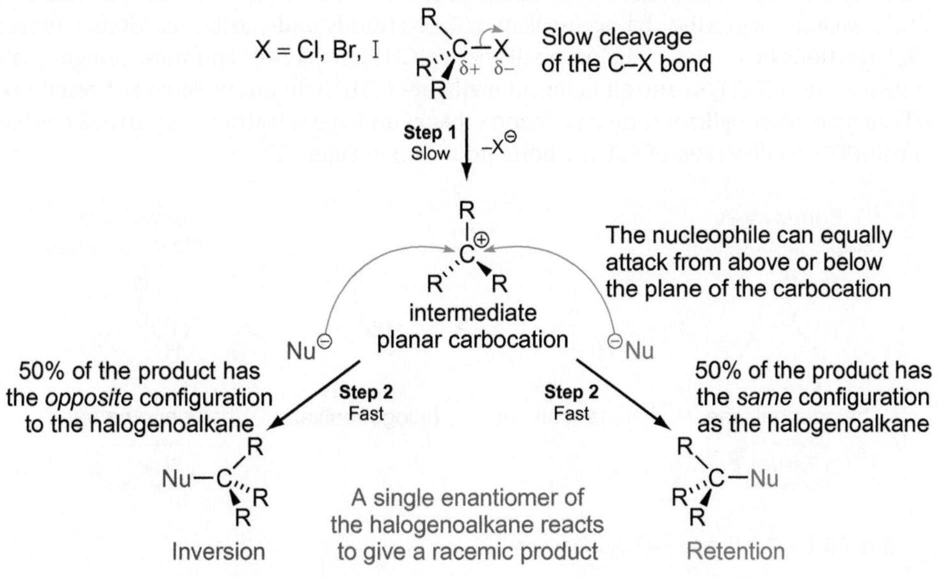

X = Cl, Br, I
Slow cleavage of the C–X bond
Step 1 Slow
$-X^{\ominus}$
The nucleophile can equally attack from above or below the plane of the carbocation
intermediate planar carbocation
50% of the product has the *opposite* configuration to the halogenoalkane
Step 2 Fast
Step 2 Fast
50% of the product has the *same* configuration as the halogenoalkane
A single enantiomer of the halogenoalkane reacts to give a racemic product
Inversion
Retention

Figure 20.14 The mechanism of an S_N1 reaction.

The nucleophile can approach the carbocation from either side of the plane, typically in a 1 : 1 ratio.

Interactive 3D animation of S_N1 reaction.

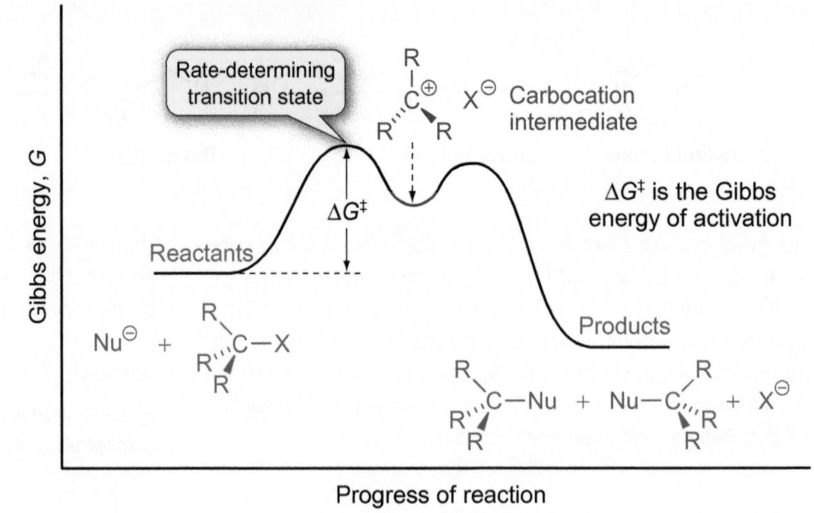

Rate-determining transition state
Carbocation intermediate
Gibbs energy, G
ΔG‡
Reactants
ΔG‡ is the Gibbs energy of activation
Products
Progress of reaction

Figure 20.15 Reaction profile for an S_N1 reaction.

The reaction profile for an S_N1 reaction is shown in Figure 20.15, while the orbital overlap in an S_N1 reaction is shown in the illustration below.

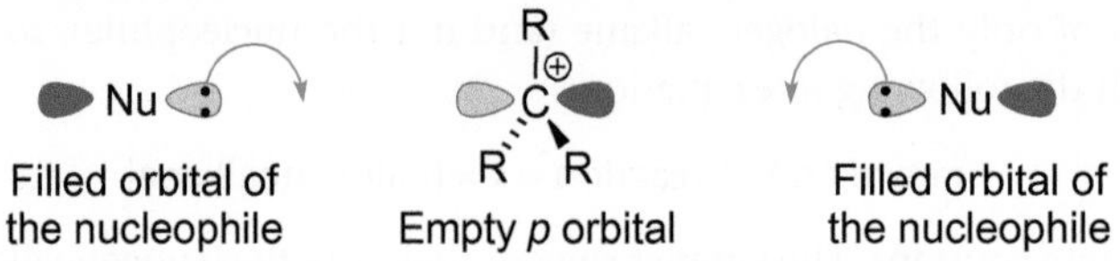

Filled orbital of the nucleophile
Empty p orbital
Filled orbital of the nucleophile

Many S_N1 reactions do not lead to complete racemization and the product derived from inversion of configuration is formed in excess. When the C–X bond breaks, the carbocation (R_3C^+) and X^- are attracted to one another and do not separate immediately—they form a **tight (intimate) ion pair.** Because X^- blocks one face of the carbocation, the nucleophile enters mostly from the opposite side to the departing X^- (see margin).

Tight ion pairs and S_Ni reactions

Tight ion pair—the carbocation and halide ion are close together and surrounded by solvent molecules.

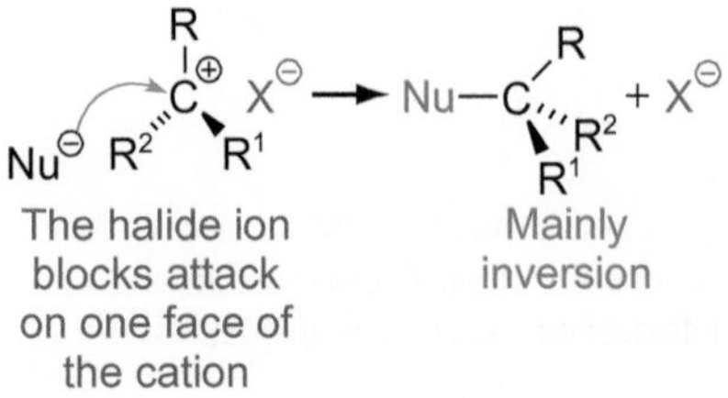

The halide ion blocks attack on one face of the cation
Mainly inversion

This type of S_N1 reaction is called an **S_Ni reaction**—where S stands for *substitution*, N stands for *nucleophilic*, and i stands for *internal*.

Effect of R groups on the rate of an S_N1 reaction

The number of alkyl groups (R) bonded to the carbon atom in the C–X bond influences the rate of an S_N1 reaction—the greater the number of alkyl substituents, the faster the S_N1 reaction. Alkyl groups are electron-donating (+I) groups so they stabilize the intermediate carbocation. This affects the rate of the reaction because, the more stable the carbocation is, the faster it is formed by cleavage of the C–X bond (in step 1 of the reaction). This explains why tertiary halogenoalkanes (R_3CX) readily undergo S_N1 reactions, whereas S_N1 reactions of secondary halogenoalkanes (R_2CHX) are slower and most primary halogenoalkanes (RCH_2X) and all halogenomethanes (CH_3X) do not undergo S_N1 reactions. (Primary halogenoalkanes and halogenomethanes undergo substitution by an S_N2 mechanism.) The relative rates of S_N1 reactions are shown in Figure 20.16.

For an S_N1 reaction, it is assumed that lowering the energy of the carbocation intermediate lowers the energy of the transition state and this reduces the Gibbs energy of activation.

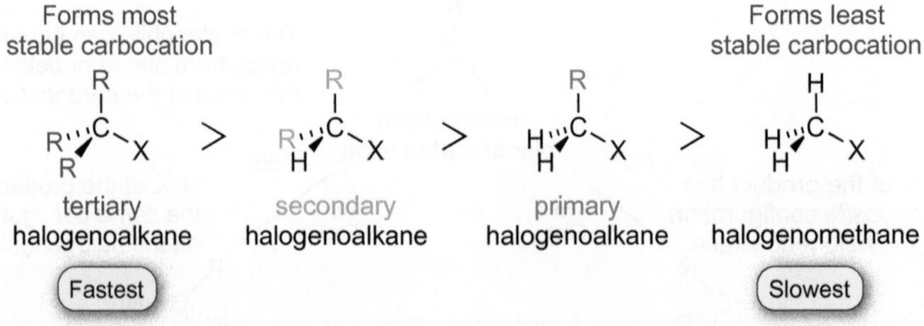

Forms most stable carbocation
Forms least stable carbocation
tertiary halogenoalkane
secondary halogenoalkane
primary halogenoalkane
halogenomethane
Fastest
Slowest

Figure 20.16 Relative rates of S_N1 reactions.

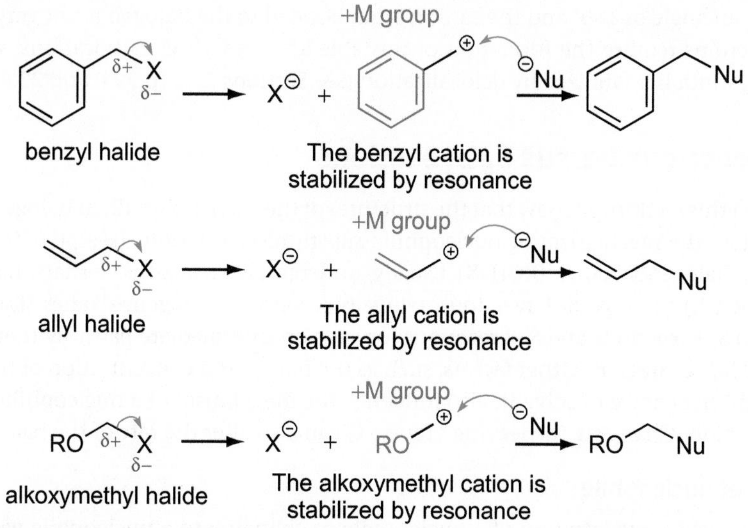
+M group
benzyl halide
The benzyl cation is stabilized by resonance
+M group
allyl halide
The allyl cation is stabilized by resonance
+M group
alkoxymethyl halide
The alkoxymethyl cation is stabilized by resonance

Figure 20.17 Examples of primary halogenoalkanes that undergo S_N1 reactions.

Relative reaction rates of S_N1 reactions of bromoalkanes with water

Bromoalkane	Class	Relative rate
CH_3Br	Methyl	negligible
CH_3CH_2Br	Primary	negligible
$(CH_3)_2CHBr$	Secondary	1
$(CH_3)_3CBr$	Tertiary	104 000

A molecular model of the benzyl cation ($PhCH_2^+$).

Although most primary halogenoalkanes do not undergo S_N1 reactions, there are some important exceptions, shown in Figure 20.17. A substituent that can stabilize a carbocation by mesomeric effects will promote the S_N1 mechanism. For example, benzyl halides ($PhCH_2X$), allyl halides (e.g. $H_2C{=}CH{-}CH_2X$), and alkoxymethyl halides ($ROCH_2X$) often react by an S_N1 mechanism, even though the carbon atom bonded to the halogen atom is primary. This is because the intermediate benzyl, allyl, and alkoxymethyl carbocations are stabilized by the positive mesomeric effects (+M) of the benzene ring, the C=C bond, and the OR group, respectively. (Check this by drawing resonance forms for each of the three carbocations.)

The size, as well as the electronic effects, of the alkyl groups influences the rates of S_N1 reactions (Figure 20.18). For example, increasing the size of the alkyl groups in tertiary halogenoalkanes (R_3CX) increases the rate of S_N1 reactions. This is explained by steric acceleration. When a halogenoalkane is converted into a carbocation there is a release of steric strain because the alkyl groups are further apart in the carbocation than in the halogenoalkane. Therefore, the larger the alkyl groups in the tertiary halogenoalkane, the greater the release of steric strain on forming the carbocation and the faster the rate of heterolytic cleavage of the C–X bond.

The stability of carbocations is discussed in Section 19.1 (p.867). Sometimes the carbocations formed in the first step are able to rearrange to form more stable carbocations (see Section 21.3, p.971).

The S_N1' mechanism
Nucleophiles can react with either resonance form of the carbocation formed from an allyl halide as illustrated here. The mechanism that forms the product with the C=C bond in a different position from that in the allyl halide is called an **S_N1' mechanism**. (S_N2' reactions are also known; see p.932.)

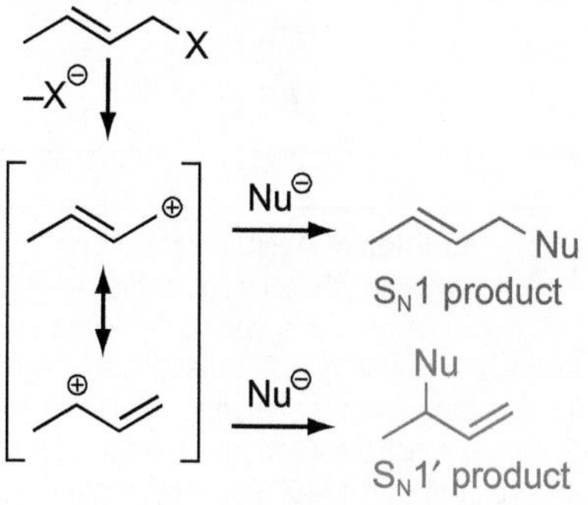
S_N1 product
S_N1' product

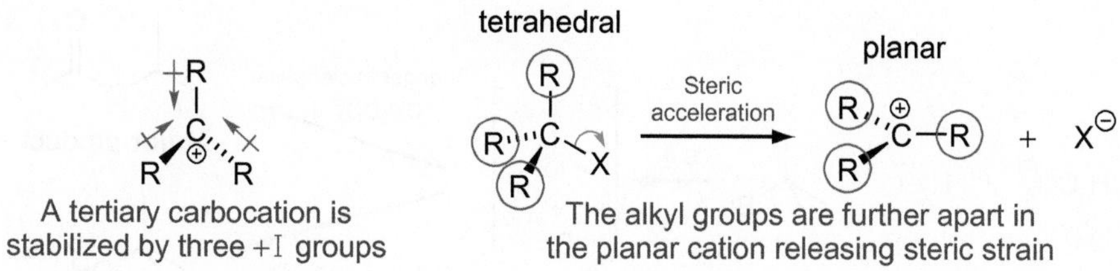
A tertiary carbocation is stabilized by three +I groups
tetrahedral
Steric acceleration
planar
The alkyl groups are further apart in the planar cation releasing steric strain

Figure 20.18 Effects of R groups on the rate of an S_N1 reaction.

Steric acceleration is introduced in Section 19.1 (p.878).

Vinyl halides (e.g. RCH=CHX) and aryl halides (ArX) do not undergo S_N1 or S_N2 reactions, which means they are unusually unreactive for halogen-containing compounds. S_N2 reactions are not possible because the nucleophile cannot approach an sp^2 carbon

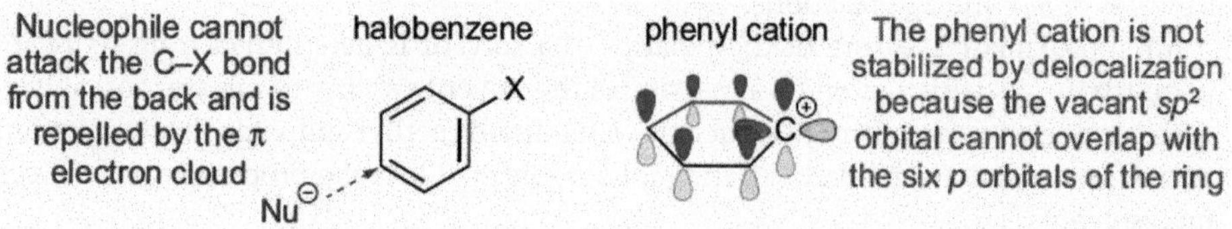
Nucleophile cannot attack the C–X bond from the back and is repelled by the π electron cloud
halobenzene
phenyl cation
The phenyl cation is not stabilized by delocalization because the vacant sp^2 orbital cannot overlap with the six p orbitals of the ring

Aryl cations, such as the phenyl cation, are formed from reactions of diazonium ions (ArN_2^+) because N_2 is an exceptionally good leaving group (Section 22.4, p.1041).

atom at an angle of 180° and the carbon atom bonded to the halogen is not very positive. S_N1 reactions require the formation of unstable aryl and vinyl carbocations, where the charge cannot be stabilized by delocalization (see Sections 21.5, p.994, and 22.4, p.1042).

S_N2 reactions versus S_N1 reactions

For a discussion of the factors that influence nucleophilic strength see Section 19.1 (p.880).

Earlier in this section, you saw that the structures of the alkyl groups (R) in halogenoalkanes (RX) affect the mechanism of nucleophilic substitution reactions (see pp.930 and 933). Primary halogenoalkanes (RCH_2X) usually undergo S_N2 reactions, tertiary halogenoalkanes (R_3CX) undergo S_N1 reactions, whereas secondary halogenoalkanes (R_2CHX) can undergo S_N1 reactions and S_N2 reactions (or even an intermediate pathway that has both S_N1 and S_N2 character). Other factors, such as the nature and concentration of the nucleophile and the choice of solvent, also influence the mechanism of a nucleophilic substitution reaction. Changing the leaving group (X) can also alter the rate of the reaction.

Choice of nucleophile

S_N2 reaction: rate = k[RX][Nu]
S_N1 reaction: rate = k[RX]

Increasing the concentration of a nucleophile or changing to a nucleophile with greater nucleophilic strength (nucleophilicity) increases the rate of an S_N2, but not an S_N1, reaction. This is because the nucleophile takes part in the rate-determining step of an S_N2 reaction, but not an S_N1 reaction. Therefore, a relatively high concentration of a strong nucleophile (such as HO^-) reacts more rapidly with a halogenomethane (CH_3X) in an S_N2 reaction than a relatively low concentration of a weak nucleophile (such as H_2O). Also, a relatively high concentration of a strong nucleophile is likely to react with a secondary halogenoalkane (R_2CHX) by an S_N2 mechanism rather than by an S_N1 mechanism.

Relative reaction rates of S_N2 reactions of bromomethane (CH_3Br) with different nucleophiles (in ethanol).

Nucleophile	Relative rate
I^-	120 000
$CH_3CH_2O^-$	60 000
HO^-	12 000
Br^-	5000
PhO^-	2000
$CH_3CO_2^-$	900
H_2O	1

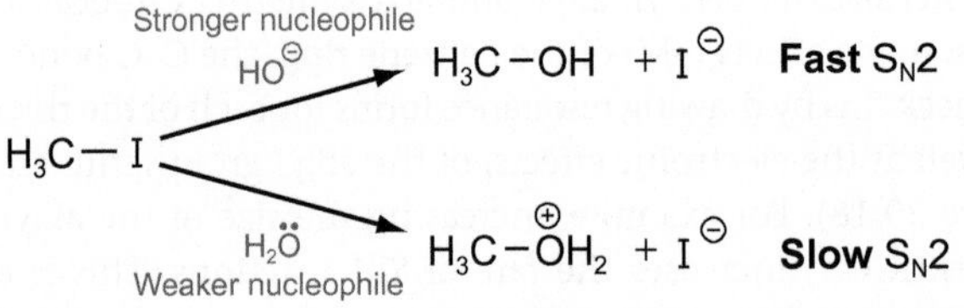
Stronger nucleophile
HO⊖
H₃C−OH + I⊖
Fast S_N2
H₃C−I
H₂Ö
Weaker nucleophile
H₃C−ÖH₂ + I⊖
Slow S_N2

For an S_N1 reaction, the strength and concentration of the nucleophile do not affect the rate of the reaction, but they can affect what products are formed. If two or more nucleophiles compete for reaction with the same carbocation, then the strengths and concentrations of the nucleophiles affect the products that are formed.

Addition of $AgNO_3$. A solution of $AgNO_3$ in ethanol is sometimes added to halogenoalkanes to speed up the formation of the carbocation in an S_N1 reaction. The Ag^+ ion associates with the halogen, weakening the C–X bond, which breaks to form the carbocation and a precipitate of AgX. Since AgX is no longer in solution, the equilibrium shifts to form more of the carbocation.

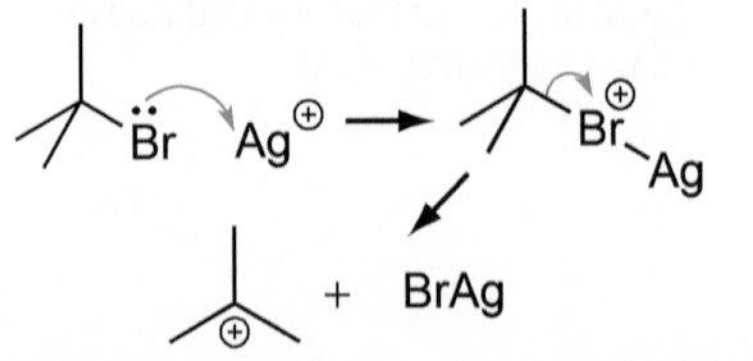
Br
Ag⊕
Br
Ag
+ BrAg

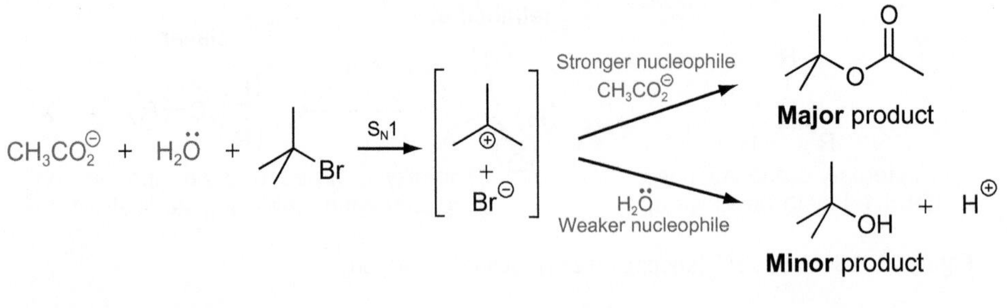
CH₃CO₂⊖ + H₂Ö +
Br
S_N1
Br⊖
Stronger nucleophile
CH₃CO₂⊖
Major product
H₂Ö
Weaker nucleophile
OH
+ H⊕
Minor product

Choice of solvent

Changing the polarity of the solvent and also changing from a polar protic solvent (such as methanol, MeOH) to a polar aprotic solvent (such as dimethyl sulfoxide, Me_2SO) alters the rates of S_N1 and S_N2 reactions.

An approximate measure of the polarity of a solvent is determined from its relative permittivity (ε_r), which provides a measure of the solvent's ability to solvate ions. Solvents with high relative permittivities, such as water and dimethyl sulfoxide, have polar functional groups and are good at solvating ions. Table 20.2 lists the properties of some common solvents.

Polar protic solvents contain polar O–H or N–H groups, whereas polar aprotic solvents contain polar groups (e.g. an $^{\delta+}S{=}O^{\delta-}$ bond) but no O–H or N–H groups (Box 7.1, p.305).

Table 20.2 Relative permittivities (at room temperature) and boiling points of common solvents. Protic solvents are shown in red and aprotic solvents are shown in black*

Solvent	Relative permittivity	Boiling point/°C	
Water (H_2O)	79	100	**Most polar**
Dimethyl sulfoxide, DMSO (Me_2SO)	47	189	↑
Acetonitrile (MeCN)	38	81–82	
N,N-Dimethylformamide, DMF (Me_2NCHO)	37	153	
Methanol (MeOH)	33	65	
Propanone (MeCOMe)	21	56	
Ethanoic acid ($MeCO_2H$)	6	117–118	
Diethyl ether (EtOEt)	4	35	
Hexane ($CH_3CH_2CH_2CH_2CH_2CH_3$)	2	69	**Least polar**

*Note that the choice of solvent also determines the maximum temperature at which a reaction can be carried out, since the temperature of a reaction cannot exceed the boiling point of a solvent at atmospheric pressure.

Relative permittivity is discussed in Box 7.1 (p.305) and Section 17.3 (p.784).

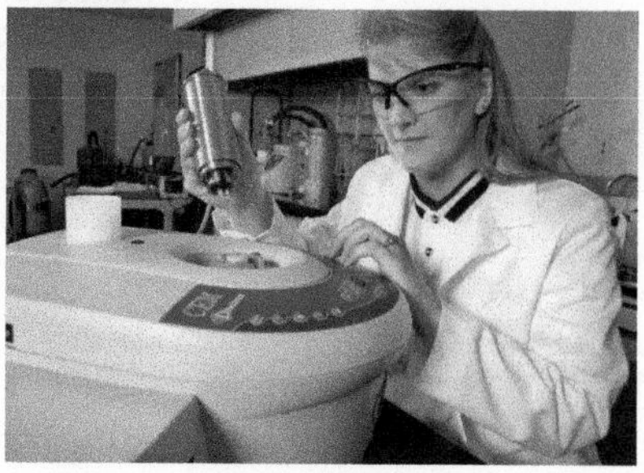

A commercial microwave reactor used in chemical laboratories. Microwaves are now routinely used in research laboratories to heat reaction mixtures rapidly. Reactions can be completed more quickly in a microwave reactor than by using conventional heating methods. Only polar molecules (with high relative permittivities) will absorb microwaves and heat up, so water, dimethyl sulfoxide, and acetonitrile are commonly used as solvents for reactions in a microwave reactor.

Effect of the solvent on S_N2 reactions

For the reaction of a negatively charged nucleophile with a halogenoalkane, increasing the polarity of the solvent results in a *slight decrease* in the rate of reaction by an S_N2 mechanism. (Note that relatively polar solvents are required for these reactions to dissolve the ionic reactants.) This is because a polar solvent solvates (and stabilizes) the small negatively charged nucleophile much better than it solvates the larger transition state, where the negative charge is spread over two atoms (see Figure 20.9, p.930). The more polar the solvent, the more strongly the negatively charged nucleophile is solvated, and the slower the rate of the S_N2 reaction (see Figure 20.19).

In contrast, increasing the polarity of the solvent results in a *slight increase* in the rate of an S_N2 reaction of an uncharged nucleophile with a halogenoalkane. This is because polar solvents solvate (and stabilize) the dipolar transition state more strongly than the uncharged nucleophile.

Changing from a polar protic to a polar aprotic solvent has a dramatic effect on the rate of an S_N2 reaction. The rates of S_N2 reactions are fastest in polar aprotic solvents such as dimethyl sulfoxide (Me_2SO). This is because polar aprotic solvents cannot solvate negatively charged nucleophiles by forming hydrogen bonds to *anions* (although they can solvate *cations* by donating lone-pair electrons from heteroatoms). These 'naked' (unsolvated) anions are highly reactive and are free to approach and react with the halogenoalkane. In contrast, polar protic solvents (such as methanol, MeOH) readily solvate negatively charged anions by forming hydrogen bonds. Solvation reduces the reactivity of the nucleophile since the solvent molecules hinder approach to the halogenoalkane. This is illustrated in the diagram on the next page.

Dimethyl sulfoxide (DMSO, Me_2SO) penetrates the skin very readily, so suitable protective gloves should be worn when handling it. This property of dimethyl sulfoxide has been exploited in medicine, where it can be used to carry drugs through the skin and into the body.

Many nucleophiles used in substitution reactions are not soluble in non-polar solvents so polar solvents are generally used. For example, reaction of chloroalkanes (RCl) with I^- is usually carried out in propanone (MeCOMe).

Solvation of nucleophiles by protic and aprotic solvents is also discussed in Section 19.1 (p.881).

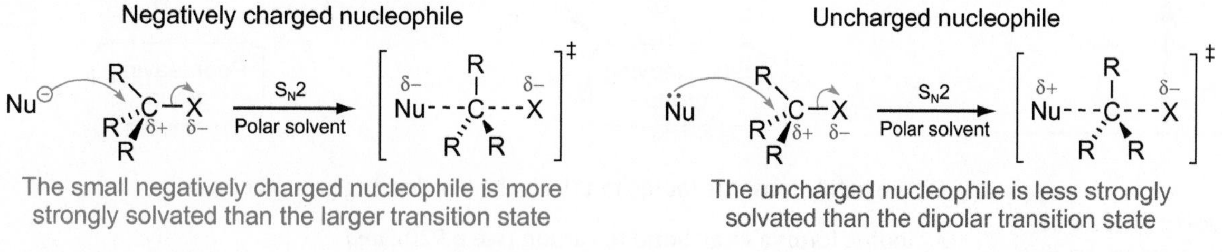
Negatively charged nucleophile
Uncharged nucleophile
SN2
Polar solvent
The small negatively charged nucleophile is more strongly solvated than the larger transition state
The uncharged nucleophile is less strongly solvated than the dipolar transition state

Figure 20.19 Effect of solvent polarity on the rate of an S_N2 reaction.

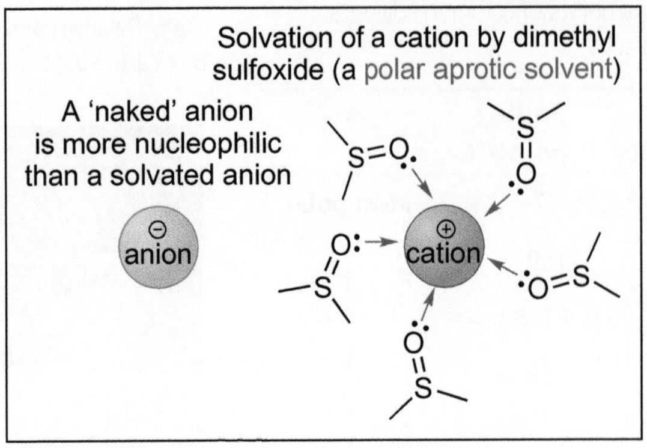
Solvation of a cation by dimethyl sulfoxide (a polar aprotic solvent)
A 'naked' anion is more nucleophilic than a solvated anion
anion
cation

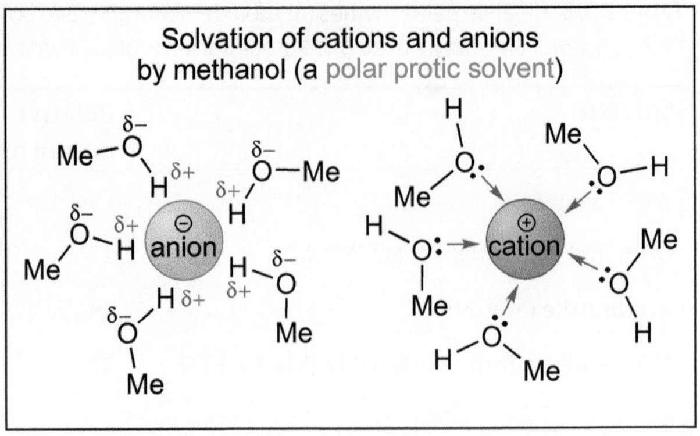
Solvation of cations and anions by methanol (a polar protic solvent)
anion
cation

Effect of the solvent on S_N1 reactions

Increasing the polarity of the solvent results in a *significant increase* in the rate of an S_N1 reaction of both negatively charged and uncharged nucleophiles with halogenoalkanes. Polar solvents solvate (and stabilize) the carbocations formed in the rate-determining step of the S_N1 reaction. Solvation provides much of the energy necessary for heterolytic cleavage of the C–X bond in the halogenoalkane, which explains why S_N1 reactions do not occur in the gas phase.

Relative reaction rates of S_N2 reactions of azide ion (N_3^-) with 1-bromobutane (a primary bromoalkane, BuBr) in different solvents.

Solvent (type)	Relative rate
MeCN (aprotic)	5000
Me_2SO (aprotic)	1300
H_2O (protic)	7
MeOH (protic)	1

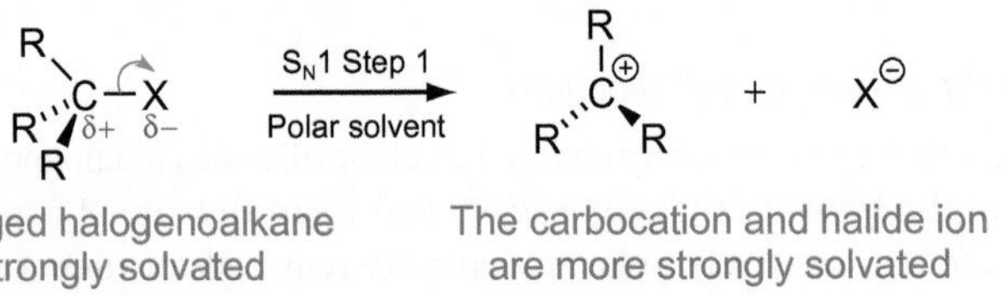
S_N1 Step 1
Polar solvent
The uncharged halogenoalkane is less strongly solvated
The carbocation and halide ion are more strongly solvated

Changing from a polar protic to a polar aprotic solvent *decreases* the rate of an S_N1 reaction. A polar protic solvent solvates (and stabilizes) both the carbocation (R^+) and halide ion (X^-) formed on heterolytic cleavage of the C–X bond, whereas polar aprotic solvents solvate only the carbocation. As a result, solvents such as methanol, or mixtures of methanol and water, are commonly used as the solvent in S_N1 reactions.

Note that, even though a nucleophile is less reactive in a polar protic solvent (because of solvation), the rate of an S_N1 reaction is not reduced because the rate-determining step of an S_N1 reaction does not involve the nucleophile.

Even though the sulfur atom in $Me_2S^{\delta+}{=}O^{\delta-}$ is partially positive it does not interact with an anion, whereas the partially positive $^{\delta-}O{-}H^{\delta+}$ hydrogen in MeOH does. One reason for this is that the sulfur atom in Me_2SO is more sterically hindered than the OH hydrogen in methanol.

Solvents such as water and methanol are also nucleophiles. A reaction in which the solvent reacts with a halogenoalkane in a nucleophilic substitution reaction is called **solvolysis**.

Choice of leaving group

Changing from a poor leaving group to a good leaving group in a halogenoalkane increases the rates of *both* S_N2 and S_N1 reactions with the same nucleophile. The rate-determining steps in both mechanisms involve heterolytic cleavage of the C–X bond and, the better the leaving group, the faster the cleavage of the C–X bond. Good leaving groups are neutral molecules or stable anions that are weakly basic (Section 20.2, p.921).

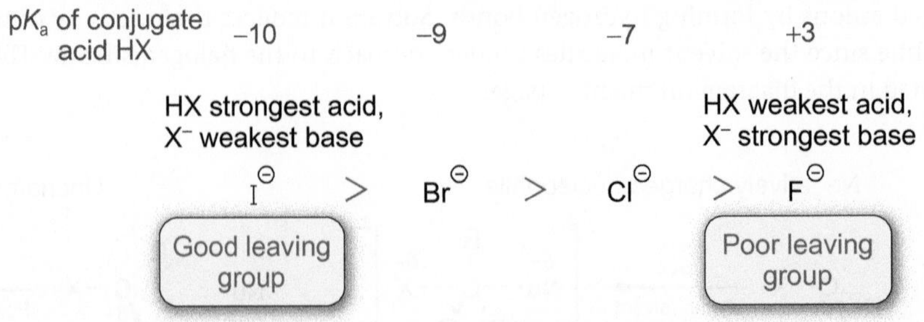
pK_a of conjugate acid HX
−10
−9
−7
+3
HX strongest acid, X^- weakest base
HX weakest acid, X^- strongest base
I^- > Br^- > Cl^- > F^-
Good leaving group
Poor leaving group

Relative reaction rates of S_N1 reactions of 2-bromo-2-methylpropane (a tertiary bromoalkane, Me_3CBr) in solvents of different polarity

Solvent	Relative rate	
100% H_2O	1200	↑ increasing polarity
50% H_2O: 50% EtOH	60	
100% EtOH	1	

Of the halide ions, the iodide ion (I^-) is the best leaving group because:

- iodine forms a weak bond to carbon (see p.920); *and*
- the I^- ion is a weak base.

pK_a values are a guide for assessing if a group is a good leaving group. See Section 20.2 (p.925).

Iodoalkanes (RI), therefore, react more rapidly than other halogenoalkanes in S_N1 and S_N2 reactions.

The iodide ion is unusual because, not only is it a good leaving group, it is also a good nucleophile because the iodine atom is large and polarizable (see p.920). Hence, the iodide ion is used as a catalyst to increase the rates of slow S_N2 reactions. In these **nucleophilic catalysis** reactions, two fast S_N2 reactions (using I^- as first a nucleophile and then a leaving group) replace one slow S_N2 reaction. For example, the S_N2 reaction of a primary chloroalkane (RCH_2Cl) with ethanoate ions ($MeCO_2^-$) is slow, but is speeded up by the addition of iodide ions.

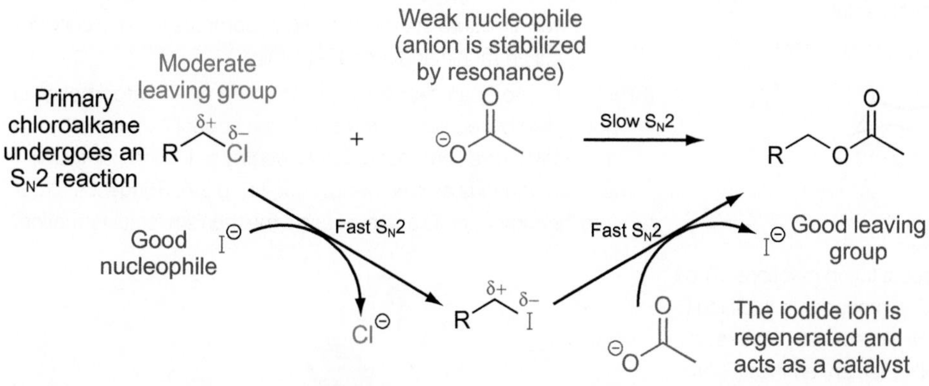
Primary chloroalkane undergoes an S_N2 reaction
Moderate leaving group
Weak nucleophile (anion is stabilized by resonance)
Slow S_N2
Good nucleophile
Fast S_N2
Fast S_N2
Good leaving group
The iodide ion is regenerated and acts as a catalyst

Worked example 20.3 provides practice in predicting whether a substitution reaction involves an S_N1 or S_N2 pathway.

Relative reaction rates of S_N2 reactions of primary halogenoalkanes, RCH_2X, with the hydroxide ion

Leaving group (X)	Relative rate
I^-	30 000
Br^-	10 000
Cl^-	200
F^-	1

Notice that the carboxylate ion (RCO_2^-) is converted into an ester (RCO_2R). As this involves an S_N2 reaction(s), the method works well for primary and secondary halogenoalkanes, to give esters of type RCO_2CH_2R and RCO_2CHR_2, respectively. An alternative and popular method of making esters, including those of type RCO_2CR_3, is to use a Fischer esterification reaction (Section 24.2, p.1108).

Worked example 20.3 Predicting substitution pathways

Predict, giving your reasons, whether the following reactions will take place by S_N1 or S_N2 mechanisms. Give the structures of the organic products, showing the stereochemistry where appropriate.

(a) 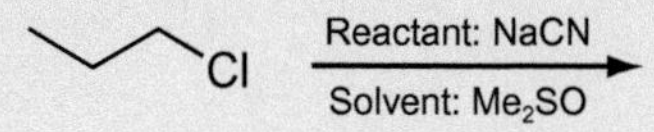
Cl
Reactant: NaCN
Solvent: Me_2SO

(b) 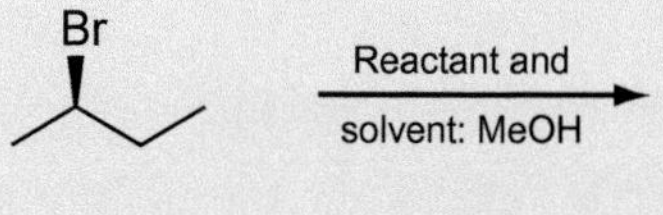
Br
Reactant and
solvent: MeOH

(c) 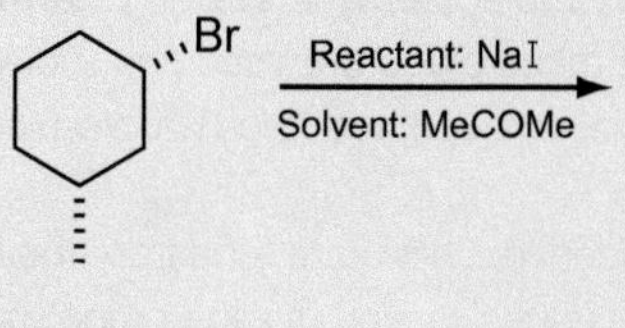
Br
Reactant: NaI
Solvent: MeCOMe

Strategy

In each case identify the nucleophile and the leaving group. Identify the halogenoalkane as primary, secondary, or tertiary.

Decide whether the reaction takes place by an S_N1 or S_N2 mechanism by considering the structure of the halogenoalkane, the strength (nucleophilicity) of the nucleophile, and the nature of the solvent.

Remember that an S_N2 reaction of a single enantiomer of a halogenoalkane gives a product with inverted configuration.

Remember that an S_N1 reaction of a single enantiomer of a halogenoalkane gives a racemic product.

Solution

(a) The cyanide ion (^-CN) is the nucleophile and Cl^- is the leaving group.

1-Chloropropane is a primary halogenoalkane and so reacts by an S_N2 mechanism. Also, ^-CN is a strong nucleophile (p.880) and, because Me_2SO is a polar aprotic solvent, it does not solvate the ^-CN ion (p.936). Therefore, an S_N2 mechanism is favoured to form butanenitrile.

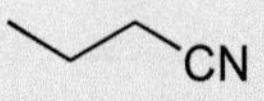
CN

(b) Methanol is the nucleophile and Br^- is the leaving group.

A secondary bromoalkane can react by an S_N2 or S_N1 mechanism. An S_N2 mechanism would be favoured if a strong nucleophile and a polar aprotic solvent were used. However, methanol is a poor nucleophile (p.880) and a polar protic solvent (p.936), so an S_N1 mechanism is favoured. In the S_N1 mechanism, the intermediate secondary carbocation ($CH_3CH^+CH_2CH_3$) reacts with methanol to give a racemic mixture of two methyl ethers.

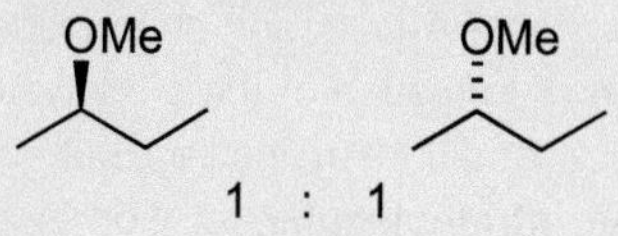
OMe
OMe
1 : 1

→

(c) The iodide ion (I^-) is the nucleophile and Br^- is the leaving group.

A secondary bromoalkane can react by an S_N2 or S_N1 mechanism. An S_N1 mechanism would be favoured if a poor nucleophile and a polar protic solvent were used. However, the iodide ion is a strong nucleophile (p.880) and propanone is a polar aprotic solvent (p.936), so an S_N2 mechanism is favoured. This produces a product with an inverted configuration at the C-1 chiral centre.

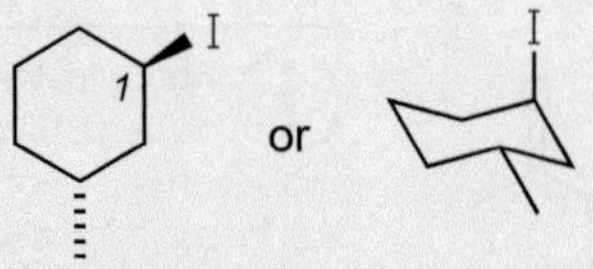
I
1
or
I

Question

Ammonia reacts with halogenoalkanes in substitution reactions. The reactions usually produce a mixture of amine products. It is difficult to obtain good yields of primary amines (RNH_2) because these react further to produce secondary amines (R_2NH), tertiary amines (R_3N), and quaternary ammonium salts ($R_4N^+X^-$).

The reaction scheme that follows shows the preparation of propan-1-amine from 1-bromopropane by two routes.

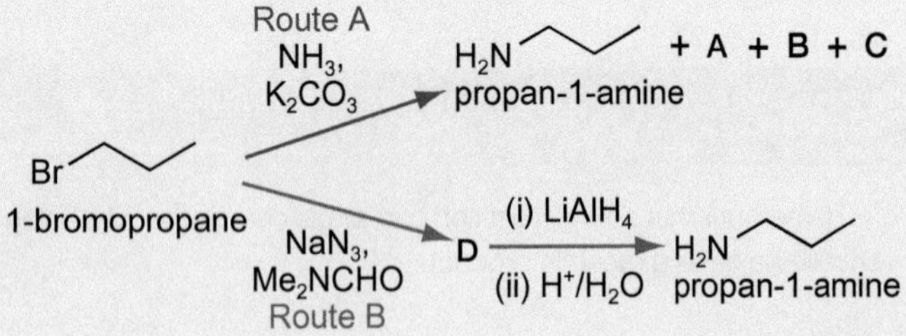
Route A
NH3,
K2CO3
H2N
+ A + B + C
propan-1-amine
Br
1-bromopropane
NaN3,
Me2NCHO
Route B
D
(i) LiAlH4
(ii) H+/H2O
H2N
propan-1-amine

(a) In route A, 1-bromopropane reacts with ammonia, in the presence of a base such as potassium carbonate, to give propan-1-amine in low yield and three additional organic compounds (**A**, **B**, and **C**).

(i) Propose a mechanism to explain the formation of propan-1-amine from 1-bromopropane and ammonia.

(ii) Suggest structures for the three compounds **A**, **B**, and **C**.

(iii) Suggest why the reaction of 1-bromopropane with ammonia gives a mixture of products. (*Hint*. Compare the nucleophilicity of ammonia and propan-1-amine.)

(b) In route B, propan-1-amine is prepared from 1-bromopropane in good yield by reaction with NaN_3 (in the solvent *N,N*-dimethylformamide) to give compound **D**, followed by reduction using $LiAlH_4$ (and an aqueous acid work-up, Box 2.6, p.94). Suggest a structure for compound **D** and propose a mechanism for its formation.

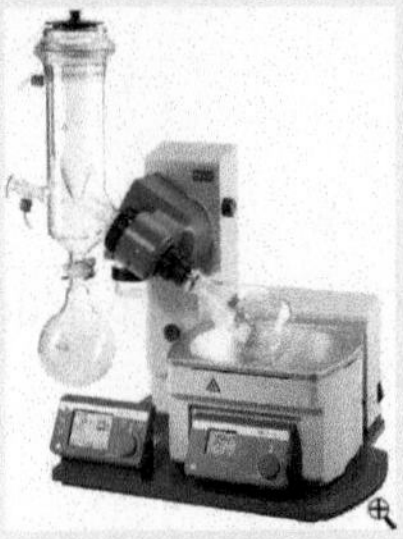

▲ The impure products from these substitution reactions are isolated by evaporation of the solvent. Commonly, this is achieved using a piece of equipment called a rotary evaporator. Using this technique, the reaction mixture is gently heated in a rotating glass flask, which is held under reduced pressure. Using a reduced pressure lowers the boiling point of the solvent and this ensures that the solvent can be removed efficiently using only gentle heating.

Box 20.4 Epoxy resins

Epoxy resin adhesives, such as Araldite®, are extremely strong polymers that can stick to almost all surfaces. They are resistant to solvents and many can withstand extremes of high temperatures. Some have been formulated to set under water so they can be used for repairing and joining underwater pipes and sticking tiles in swimming pools.

A typical epoxy resin adhesive comes in two parts—a prepolymer and a hardener. The prepolymer is formed by reacting bisphenol A with epichlorohydrin and hydroxide ion to make a relatively short-chain polymer called a *prepolymer*—this contains about 8–10 bisphenol A units and an epoxide ring at each end. The mechanism of formation of the prepolymer involves a series of intramolecular and intermolecular S_N2 reactions as shown in Figure 1. (An **intramolecular** reaction is a reaction between different parts of the *same* molecule, whereas an **intermolecular** reaction is a reaction that takes place between *two* molecules.) In epichlorohydrin, notice that the epoxide ring is more electrophilic than the C–Cl bond.

The prepolymer is provided in a separate tube from the hardener. When you want to use the adhesive, you mix the prepolymer with a triamine (the hardener) to form the epoxy resin. Each amine group in the hardener can react with an epoxide ring on the prepolymer in an intermolecular S_N2 reaction (see the insert box in Figure 2 for the mechanism). The three amine groups (two –NH_2 and one –NH–) can each react with different molecules of the prepolymer, forming a densely cross-linked structure that is extremely strong. (Cross-links are covalent bonds that join polymer chains together.)

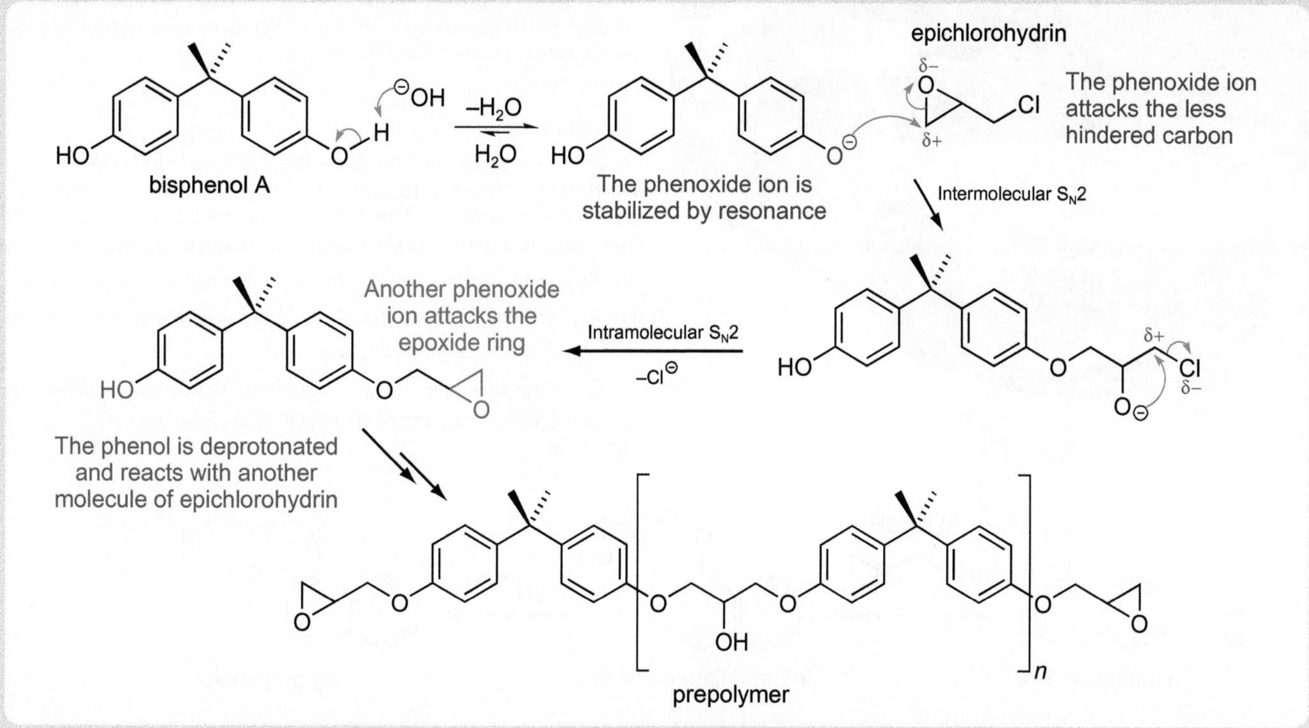
epichlorohydrin
The phenoxide ion attacks the less hindered carbon
bisphenol A
The phenoxide ion is stabilized by resonance
Intermolecular S_N2
Another phenoxide ion attacks the epoxide ring
Intramolecular S_N2
$-Cl^{\ominus}$
The phenol is deprotonated and reacts with another molecule of epichlorohydrin
prepolymer

▲ **Figure 1** The mechanism of formation of the prepolymer (*n* denotes the number of polymerized bisphenol A units; it can be as high as 25, but is usually lower than this).

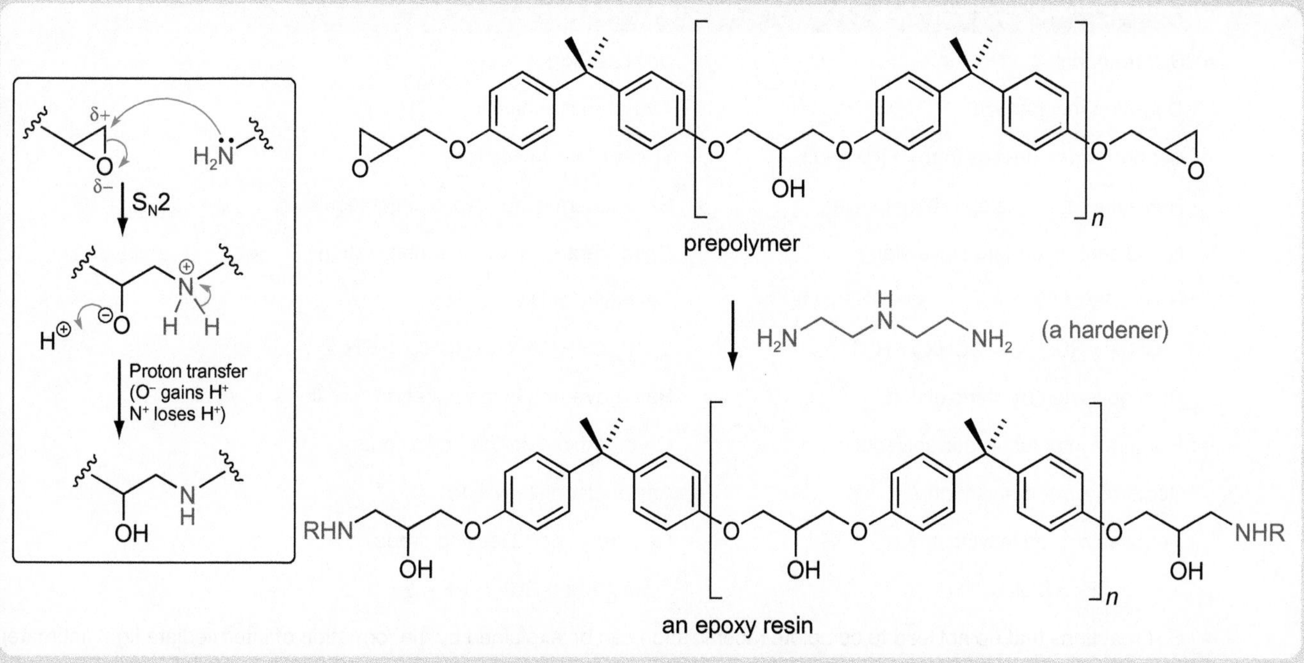
S_N2
Proton transfer
(O^- gains H^+
N^+ loses H^+)
prepolymer
(a hardener)
an epoxy resin

▲ **Figure 2** The prepolymer is mixed with a triamine (the hardener) to form the epoxy resin.

◀ As Araldite® is able to withstand extreme fatigue and environmental conditions it has a range of industrial uses. For example, it has been used to bond components in an Audi R8 sports car.

Question

Propanolol (Inderal®) is a medicine that reduces blood pressure. It binds to and blocks β_1-receptors that control muscles in the heart, so it is called a beta-blocker. Propanolol is prepared in racemic form from naphthalen-1-ol, as illustrated in the reaction scheme below.

(a) Suggest reaction mechanisms for both steps in the synthesis.

(b) Explain why the reaction of the intermediate epoxide with propan-2-amine is regioselective.

(c) Draw the structure of the product from reaction of the (*S*)-isomer of the intermediate epoxide with propan-2-amine.

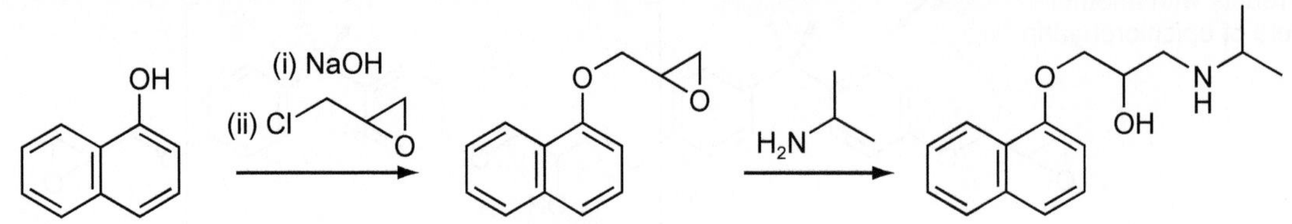
OH
(i) NaOH
(ii) Cl
O
O
O
H_2N
O
OH
N
H

naphthalen-1-ol intermediate epoxide (±)-propanolol

▲ The synthesis of propanolol (in racemic form) from naphthalen-1-ol.

Summary

• S_N2 reaction	S_N1 reaction
One-step mechanism	Two-step mechanism
Second-order kinetics (rate = *k*[RX][Nu])	First-order kinetics (rate = *k*[RX])
Rate-determining step is bimolecular	Rate-determining step is unimolecular
Chiral centres undergo inversion	Chiral centres undergo racemization
Reactivity of RX:	Reactivity of RX:
$CH_3X > RCH_2X > R_2CHX >> R_3CX$	$R_3CX > R_2CHX >> RCH_2X > CH_3X$
Rate governed by steric effects	Rate governed by carbocation stability
Favoured in polar aprotic solvents	Favoured in polar protic solvents
Requires a good nucleophile	Any nucleophile will do
Requires a good leaving group:	Requires a good leaving group:
RI > RBr > RCl >> RF	RI > RBr > RCl >> RF

• S_N1 reactions that do not lead to complete racemization can be explained by the formation of intermediate tight (intimate) ion pairs. These are S_Ni reactions.

? For practice questions on these topics, see questions 2–6 at the end of this chapter (pp.957–958).

20.4 The mechanisms of elimination reactions

Halogenoalkanes react with bases to produce alkenes (e.g. $R_2C{=}CR_2$). A molecule of HX is lost so the reaction is an elimination reaction (Section 19.2, p.903).

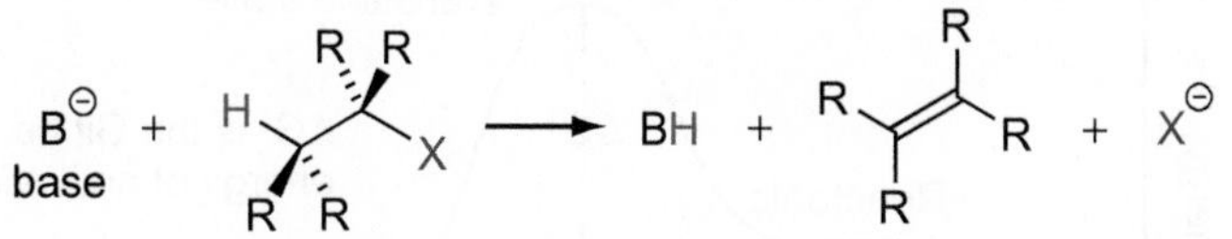
base

The two main mechanisms of elimination of HX from halogenoalkanes are called E2 and E1. These represent two extreme mechanisms and, often, eliminations proceed by mechanisms that lie somewhere in between the two (i.e. there is a continuum of mechanisms that range from E2 to E1).

Bases such as HO^- and RO^- are used in elimination reactions. These anions can also act as nucleophiles, so elimination reactions are in competition with substitution reactions. This is discussed in Section 20.5 (p.953).

E2 reactions

Elimination reactions that take place by a concerted (single-step) mechanism are called **E2 reactions**—where E stands for *elimination* and 2 stands for *bimolecular*. The mechanism was deduced from kinetic studies. The rate of the elimination depends on the concentration of both the halogenoalkane and the base, so the reaction is second order with the following rate equation:

Bimolecular elementary reactions are discussed in Sections 9.6 (p.416) and 9.8 (p.432). Rate-determining steps and the use of kinetic experimental evidence to support mechanisms are covered in Section 9.5 (p.405).

$$\text{rate of an E2 reaction} = k \times [\text{halogenoalkane}] \times [\text{base}]$$

where k is the rate constant. The kinetics suggest that both the halogenoalkane and the base are involved in the transition state of the rate-determining step, which is therefore bimolecular.

The mechanism of an E2 reaction is shown in Figure 20.20. The base (B^-) attacks a hydrogen atom on the carbon atom adjacent to one carrying the halogen atom X (the β-carbon atom)—even though the hydrogen atom is not acidic, loss of H^+ can occur because X^- is a good leaving group. The base forms a new bond to the hydrogen atom *at the same time* as the C–X bond is broken and the C=C bond is formed. The reaction profile for an E2 reaction is shown in Figure 20.21.

Note that the halogenoalkane adopts an anti-periplanar conformation in which the H and X groups are in the same plane and on opposite sides of the molecule (see the Newman projections in the margin).

The σ orbital of the C–H bond is parallel to the σ* orbital of the C–X bond. When these orbitals are parallel, the orbitals can overlap (side-on) as the H and X groups leave and rehybridize to form the overlapping *p* orbitals of the new C=C π bond, as shown in Figure 20.22.

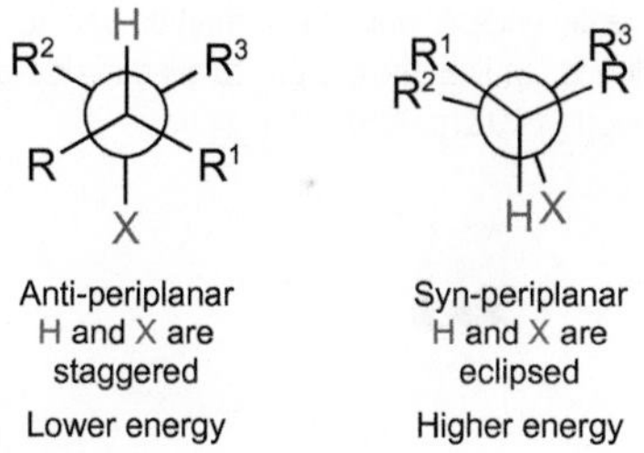
Anti-periplanar
H and X are staggered
Lower energy
Syn-periplanar
H and X are eclipsed
Higher energy

Newman projections (see Section 18.2, p.821).

Stereoselectivity in E2 reactions

When halogenoalkanes can adopt two anti-periplanar conformations, one leading to the *E*-isomer of the alkene and the other leading to the *Z*-isomer, in general, the *E*-isomer is formed in higher yield. An example is shown in Figure 20.23. The stereoselectivity arises because the conformation leading to the *E*-isomer is more stable than the conformation leading to the *Z*-isomer as there is less steric strain.

Interactive 3D animation of an E2 raction

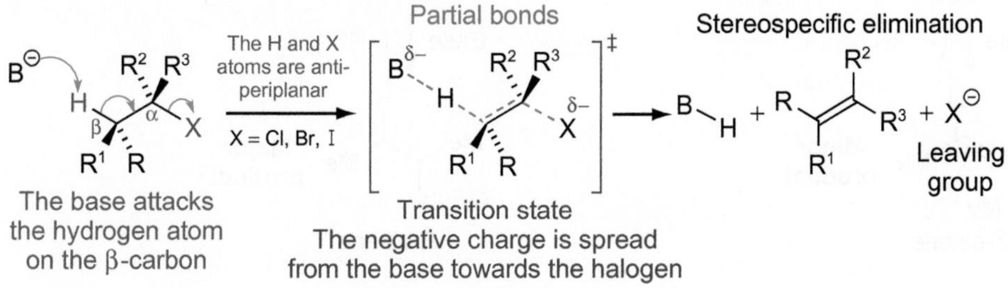
Partial bonds
Stereospecific elimination
The H and X atoms are anti-periplanar
X = Cl, Br, I
Leaving group
The base attacks the hydrogen atom on the β-carbon
Transition state
The negative charge is spread from the base towards the halogen

Figure 20.20 The mechanism of an E2 reaction.

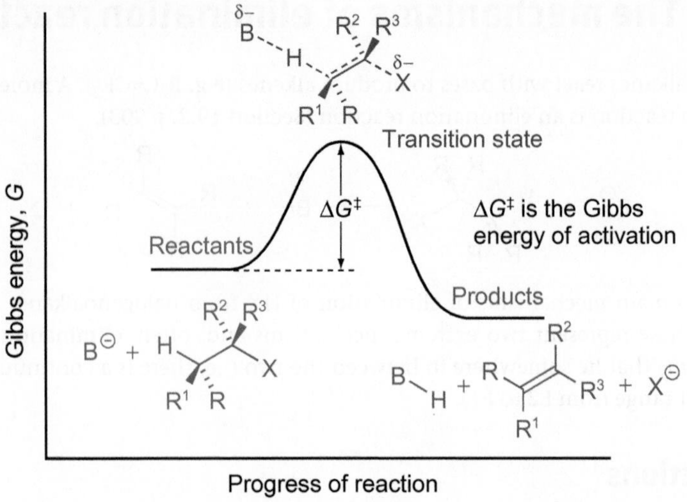

Gibbs energy, G
Transition state
ΔG‡
ΔG‡ is the Gibbs energy of activation
Reactants
Products
Progress of reaction

Figure 20.21 The reaction profile for an E2 reaction.

The anti-periplanar conformation is preferred over other conformations (see Section 18.2, p.824) because the groups on neighbouring carbon atoms are staggered, so this conformation has the lowest energy.

To make a new σ bond, the orbitals overlap end-on, while the formation of a new π bond requires the orbitals to overlap side-on (Section 19.1, p.885).

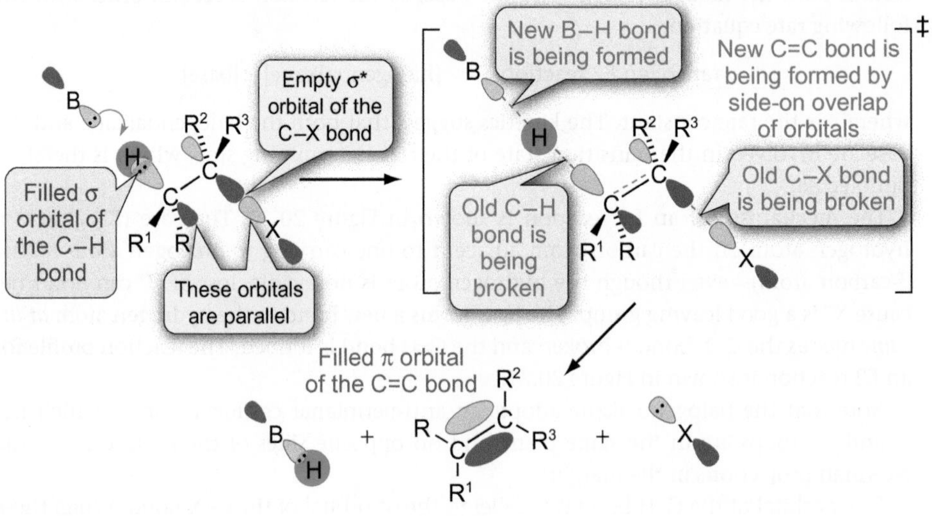

Filled σ orbital of the C–H bond
Empty σ* orbital of the C–X bond
These orbitals are parallel
New B–H bond is being formed
New C=C bond is being formed by side-on overlap of orbitals
Old C–X bond is being broken
Old C–H bond is being broken
Filled π orbital of the C=C bond

Figure 20.22 An E2 reaction showing orbital overlap.

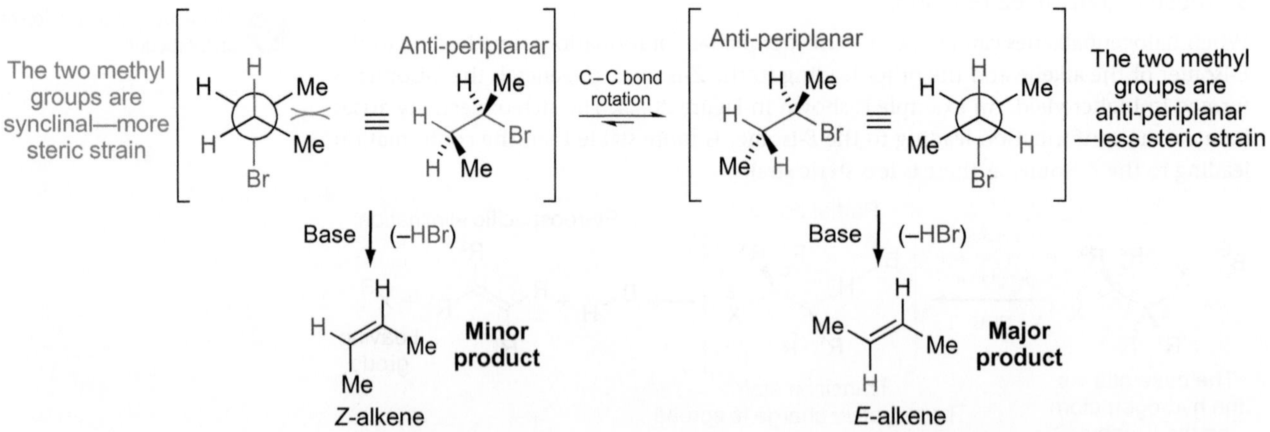

The two methyl groups are synclinal—more steric strain
Anti-periplanar
C–C bond rotation
Anti-periplanar
The two methyl groups are anti-periplanar —less steric strain
Base (–HBr)
Minor product
Z-alkene
Base (–HBr)
Major product
E-alkene

Figure 20.23 E2 elimination products from 2-bromobutane ($CH_3CHBrCH_2CH_3$), which can adopt two conformations where H is anti-periplanar to Br.

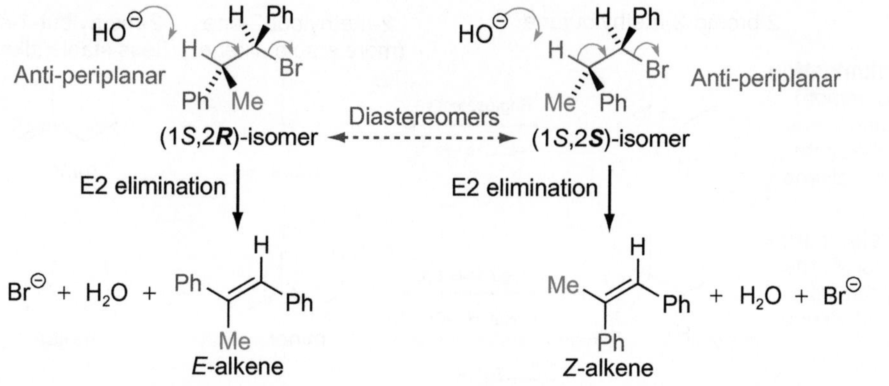
Anti-periplanar
(1*S*,2**R**)-isomer
Diastereomers
(1*S*,2**S**)-isomer
Anti-periplanar
E2 elimination
E2 elimination
E-alkene
Z-alkene

Figure 20.24 Stereospecific E2 elimination of two diastereomers of 1-bromo-1,2-diphenylpropane.

Section 18.3 (p.883) shows you how to assign *E* and *Z* configurations to alkenes. Note that the *E*-isomer of an alkene is more stable than the *Z*-isomer (see Section 21.1, p.963).

For halogenoalkanes that can only adopt one anti-periplanar conformation, the E2 reaction is stereospecific (i.e. different stereoisomers of the halogenoalkane form different stereoisomers of the alkene). The reaction scheme in Figure 20.24 shows the stereospecific E2 elimination of two diastereomers of 1-bromo-1,2-diphenylpropane (PhCH(Me)CH(Br)Ph).

For an introduction to stereospecific reactions, see Section 19.3 (p.910).

To determine if an alkene has *E* or *Z* configuration, spectroscopic methods such as ^1H NMR spectroscopy can be used—see a specialist spectroscopy book for details.

Regioselectivity in E2 reactions

The need for an anti-periplanar conformation in the E2 mechanism also explains why some substituted halocyclohexanes undergo regioselective E2 reactions (i.e. elimination forms one structural isomer of an alkene in preference to another). For example, the E2 reaction of *trans*-1-bromo-2-methylcyclohexane with *tert*-butoxide ion ($^tBuO^-$ or Me_3CO^-) forms only 3-methylcyclohex-1-ene (*trans* means that the Br and Me groups are on opposite sides of the ring). This is surprising because elimination of HBr could also form 1-methylcyclohex-1-ene, which is a more stable alkene than 3-methylcyclohex-1-ene, because the C=C bond is more substituted (see p.963).

You can explain the preferential formation of 3-methylcyclohex-1-ene by the different reactivities of the two chair conformations of *trans*-1-bromo-2-methylcyclohexane, as shown in Figure 20.25. The conformer with both Me and Br groups in axial positions is only present

The greater the number of alkyl groups (R) attached to a C=C bond, the more stable the alkene is (e.g. $R_2C=CHR$ is more stable than $RCH=CH_2$), as discussed in Section 21.1 (p.963).

The *tert*-butoxide ion (Me_3CO^-) is a strong base (the pK_a of the conjugate acid is around 17)—the large size of the Me_3C group hinders the anion from participating in nucleophilic substitution reactions (see p.954).

For chair conformations of substituted cyclohexanes, see Section 18.2 (p.827).

For halocyclohexanes to undergo E2 reactions, both the C–H and C–X bonds must be in axial positions. The greater the proportion of a chair conformer(s) with C–H and C–X bonds in axial positions the faster the rate of E2 elimination.

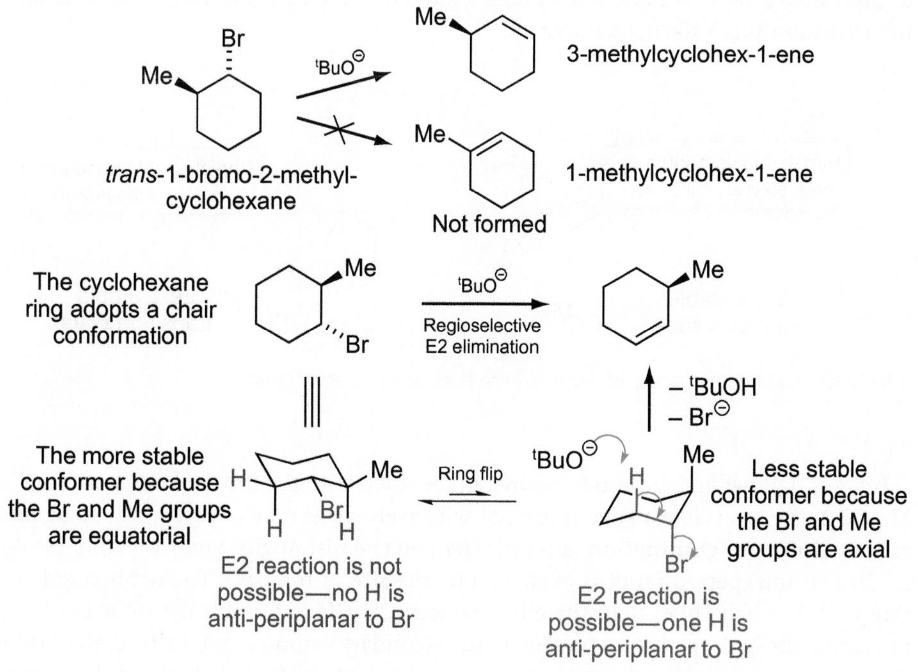
3-methylcyclohex-1-ene
trans-1-bromo-2-methyl-cyclohexane
1-methylcyclohex-1-ene
Not formed
The cyclohexane ring adopts a chair conformation
Regioselective E2 elimination
The more stable conformer because the Br and Me groups are equatorial
Ring flip
Less stable conformer because the Br and Me groups are axial
E2 reaction is not possible—no H is anti-periplanar to Br
E2 reaction is possible—one H is anti-periplanar to Br

Figure 20.25 Regioselective E2 elimination of *trans*-1-bromo-2-methylcyclohexane.

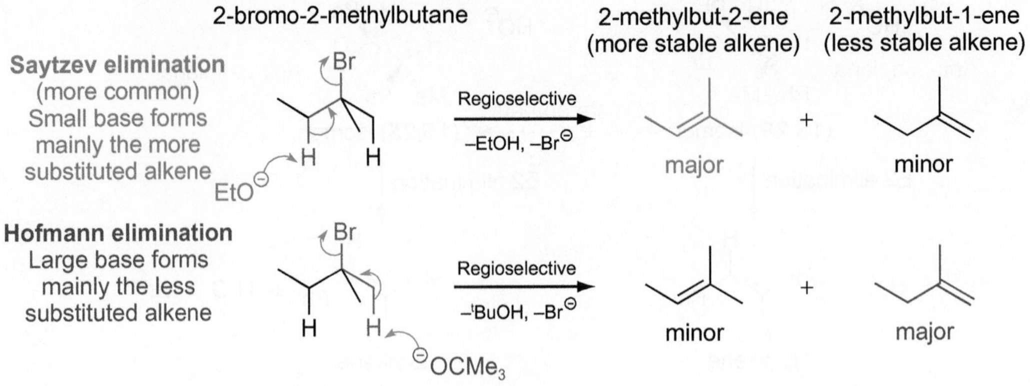
2-bromo-2-methylbutane
2-methylbut-2-ene (more stable alkene)
2-methylbut-1-ene (less stable alkene)
Saytzev elimination
(more common)
Small base forms mainly the more substituted alkene
Br
H H
EtO⊖
Regioselective
–EtOH, –Br⊖
major
minor
Hofmann elimination
Large base forms mainly the less substituted alkene
Br
H H
⊖OCMe3
Regioselective
–tBuOH, –Br⊖
minor
major

Figure 20.26 Saytzev versus Hofmann elimination.

as a small fraction of the equilibrium mixture (and hence the rate of elimination is relatively slow), but it is the only chair conformer from which elimination can occur.

Saytzev versus Hofmann elimination

For aliphatic halogenoalkanes the carbon atoms are in a chain, whereas for alicyclic halogenoalkanes the carbon atoms form a ring (Section 20.1, p.919).

E2 reactions of aliphatic halogenoalkanes that can rotate freely about C–C bonds normally produce the most substituted alkene. This type of reaction is called a **Saytzev** (sometimes spelt **Zaitsev**) **elimination** (after the Russian chemist Alexander Saytzev, 1841–1910). In the transition state of the E2 reaction, the C=C bond is partially formed so the structure of the transition state resembles the alkene product—the more stable the alkene, the more stable the transition state. Consequently, an E2 reaction gives the more substituted alkene (with the more stable C=C bond) because the transition state leading to the more substituted alkene will be the more stable (i.e. lower in energy).

Notice that neither 2-methylbut-2-ene nor 2-methylbut-1-ene have *E*/*Z* isomers.

For example, in Figure 20.26, the reaction of 2-bromo-2-methylbutane ($CH_3CH_2C(CH_3)BrCH_3$) with ethoxide ion (EtO^- or $CH_3CH_2O^-$) produces mainly 2-methylbut-2-ene, $CH_3CH{=}C(CH_3)_2$ (a trisubstituted alkene). The transition state to 2-methylbut-2-ene is lower in energy than the transition state leading to 2-methylbut-1-ene ($CH_3CH_2C(CH_3){=}CH_2$, a disubstituted alkene) because it resembles a more stable alkene (i.e. a trisubstituted rather than a disubstituted alkene; see above).

August von Hofmann (1818–1892) also discovered the Hofmann rearrangement reaction, which involves the conversion of a primary amide ($RCONH_2$) into an amine (RNH_2), with one less carbon atom, using bromine and aqueous sodium hydroxide. Bromine and sodium hydroxide generate sodium hypobromite, NaOBr (see Section 27.6, p.1243), which reacts with the primary amide.

Remember, for an E2 reaction, the H atom and leaving group must be in the same plane and pointing in opposite directions.

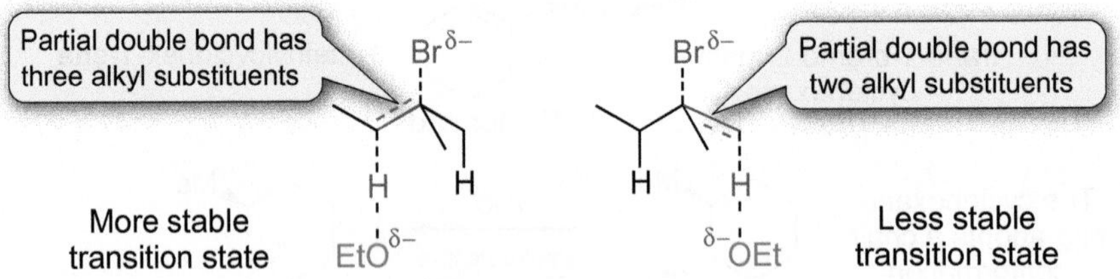
Partial double bond has three alkyl substituents
Brδ–
H H
EtOδ–
More stable transition state
Brδ–
Partial double bond has two alkyl substituents
H H
δ–OEt
Less stable transition state

Transition states for the reaction of 2-bromo-2-methylbutane with ethoxide ion.

Interestingly, when 2-bromo-2-methylbutane reacts with the *tert*-butoxide ion ($^tBuO^-$ or Me_3CO^-), the less stable alkene is formed selectively—this type of reaction is sometimes called a **Hofmann elimination** (after the German chemist August von Hofmann; see Box 20.5). This unexpected result is explained by the size of the base. The *tert*-butoxide ion (Me_3CO^-) is much larger than the ethoxide ion ($CH_3CH_2O^-$) and it has difficulty in approaching the hydrogen atom bonded to the secondary carbon atom. Instead, the *tert*-butoxide ion selectively removes one of the more accessible hydrogen atoms on the primary carbon atom at the end of the chain.

The name 'Hofmann elimination' is often associated with the E2 elimination of a quaternary ammonium ion (e.g. the reaction of RCH_2–CH_2–NMe_3^+ with HO^-, to give $RCH{=}CH_2$, H_2O and NMe_3)—see the question in Box 20.5.

Box 20.5 Muscle relaxants

Muscle relaxants are administered to patients before surgery so that muscles can be touched or cut without them contracting. This makes it easier for the surgeon to operate and the operation is safer.

The starting point for the development of modern muscle relaxants was curare (an extract from tropical vines), which was used as a poison on the tips of arrows by some South American peoples. When introduced into the bloodstream, curare paralyses the victim by interrupting the flow of nerve impulses in muscles. In 1935, D-tubocurarine was identified as the major active component of curare and it has been used in surgery since 1942.

More recently, synthetic drugs with similar actions to those of D-tubocurarine have been developed. To be biologically active the compounds require two quaternary ammonium groups ($^+NR_4$). An important example is atracurium, a particularly short-acting muscle relaxant, which was developed at the University of Strathclyde in Scotland. Atracurium was designed to break down in the body after a short time to form laudanosine, which does not act as a muscle relaxant. In the slightly alkaline conditions (pH 7.4) of the body, atracurium is converted into laudanosine by two Hofmann elimination reactions (see p.946). Today, atracurium is used in more than half of surgical operations worldwide, with sales of around £100 million per year.

▲ South American natives hunt small animals using curare-tipped arrows fired from a blow pipe.

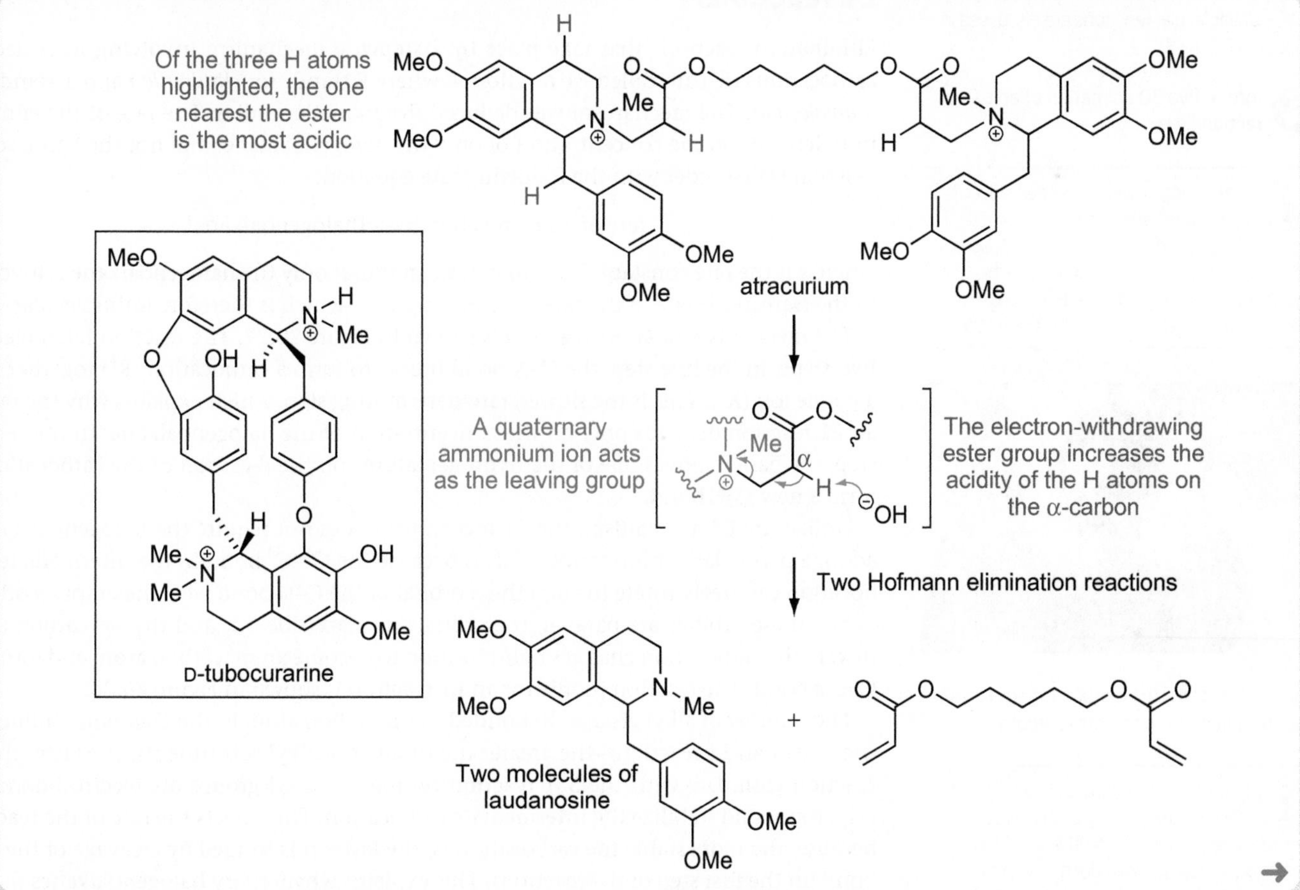
Of the three H atoms highlighted, the one nearest the ester is the most acidic
atracurium
D-tubocurarine
A quaternary ammonium ion acts as the leaving group
The electron-withdrawing ester group increases the acidity of the H atoms on the α-carbon
Two Hofmann elimination reactions
Two molecules of laudanosine

→

Question

An example of a Hofmann elimination is shown here.

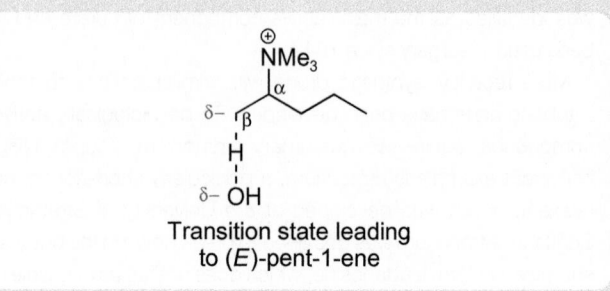

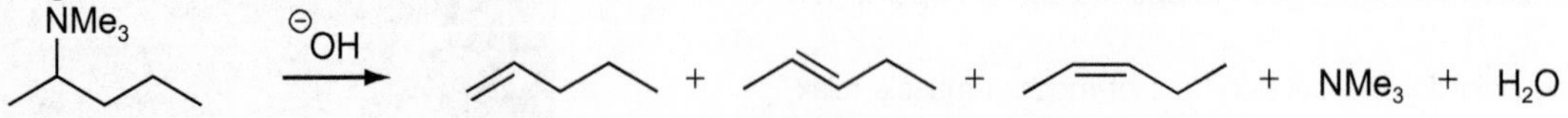
NMe3
OH
NMe3
H2O

pent-1-ene Major product — (*E*)-pent-2-ene Minor product — (*Z*)-pent-2-ene Minor product

(a) Suggest reaction mechanisms for the formation of pent-1-ene and (*E*)-pent-2-ene.

(b) The regioselective formation of the less substituted alkene is explained by the mechanism of the elimination. The C–H bond on the β-carbon starts to break before the C–N bond, so that the transition state has a carbanion-like structure. The transition state leading to pent-1-ene is shown here.

Draw the structure of the transition state that leads to (*E*)-pent-2-ene and suggest why the transition state that leads to pent-1-ene is lower in energy. (*Hint*. Consider the relative stabilities of carbanions.)

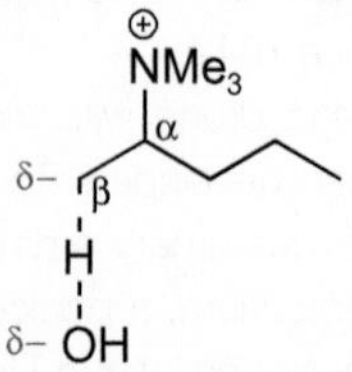
NMe3
α
δ−
β
H
δ−
OH

Transition state leading to (*E*)-pent-1-ene

E1 reactions

Unimolecular reactions are discussed in Section 9.6 (p.416).

Interactive 3D animation of an E1 raction

Weak bases, even uncharged solvent molecules, can remove a proton (H^+) from the carbocation. Sometimes the base is not even shown in an E1 mechanism—the H^+ is shown just 'dropping-off'. In reality, a base (often the solvent) is always needed to remove the H^+.

For an E1 reaction, the C–H bond needs to align with the empty *p* orbital of the carbocation.

For an E1 reaction, it is assumed that lowering the energy of the carbocation intermediate (R^+) lowers the energy of the transition state and this reduces the activation energy.

Elimination reactions that take place by a stepwise mechanism involving intermediate carbocations (R^+) are called E1 reactions—where E stands for *elimination* and 1 stands for *unimolecular*. The mechanism was deduced from kinetic studies. The rate of the elimination depends on the concentration of only the halogenoalkane (and not the base), so the reaction is first order with the following rate equation:

$$\text{rate of an E1 reaction} = k \times [\text{halogenoalkane}]$$

where k is the rate constant. The kinetics suggest that only the halogenoalkane is involved in the transition state of the rate-determining step, which is therefore unimolecular.

The mechanism of an E1 reaction is shown in Figure 20.27. The reaction takes place in two steps. In the first step, the C–X bond breaks to form a carbocation (R^+) together with a halide ion (X^-). This is the slower, rate-determining step, which explains why the rate of an E1 reaction depends only on the concentration of the halogenoalkane. In the second step, the base removes one of the hydrogen atoms on the β-carbon of the carbocation to form a new C=C bond.

Unlike the E2 mechanism, the E1 mechanism does not require the halogenoalkane to adopt a particular conformation. This is because the C^α–C^β bond in the intermediate carbocation can freely rotate to align the σ orbital of the C–H bond with the empty *p* orbital. When these orbitals are parallel, the orbitals overlap (side-on) and the sp^3 carbon atom next to the carbocation changes hybridization to become an sp^2 carbon atom and form the C=C π bond. The reaction profile for an E1 reaction is shown in Figure 20.28.

The number of alkyl groups (R) bonded to the carbon atom in the C–X bond influences the rate of an E1 reaction—the greater the number of alkyl substituents, the faster the E1 reaction (compare with the S_N1 reaction on p.911). Alkyl groups are electron-donating (+I) groups and stabilize the intermediate carbocation. This affects the rate of the reaction because, the more stable the carbocation is, the faster it is formed by cleavage of the C–X bond (in the first step of the reaction). This explains why tertiary halogenoalkanes (R_3CX) readily undergo E1 reactions, whereas E1 reactions of secondary halogenoalkanes (R_2CHX)

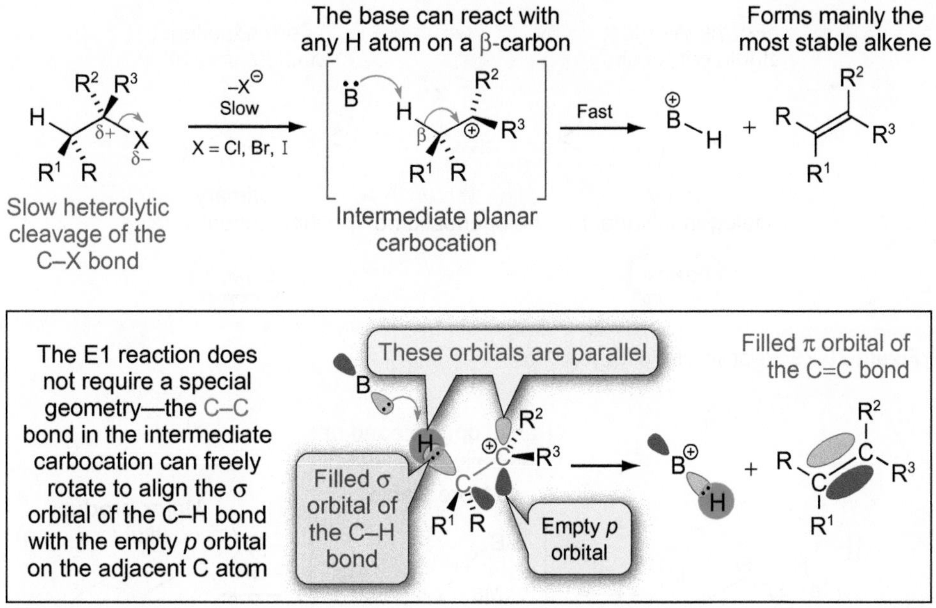

The base can react with any H atom on a β-carbon
Forms mainly the most stable alkene
Slow
X = Cl, Br, I
Fast
Slow heterolytic cleavage of the C–X bond
Intermediate planar carbocation
The E1 reaction does not require a special geometry—the C–C bond in the intermediate carbocation can freely rotate to align the σ orbital of the C–H bond with the empty p orbital on the adjacent C atom
These orbitals are parallel
Filled π orbital of the C=C bond
Filled σ orbital of the C–H bond
Empty p orbital

Figure 20.27 The mechanism of an E1 reaction, showing orbital overlap.

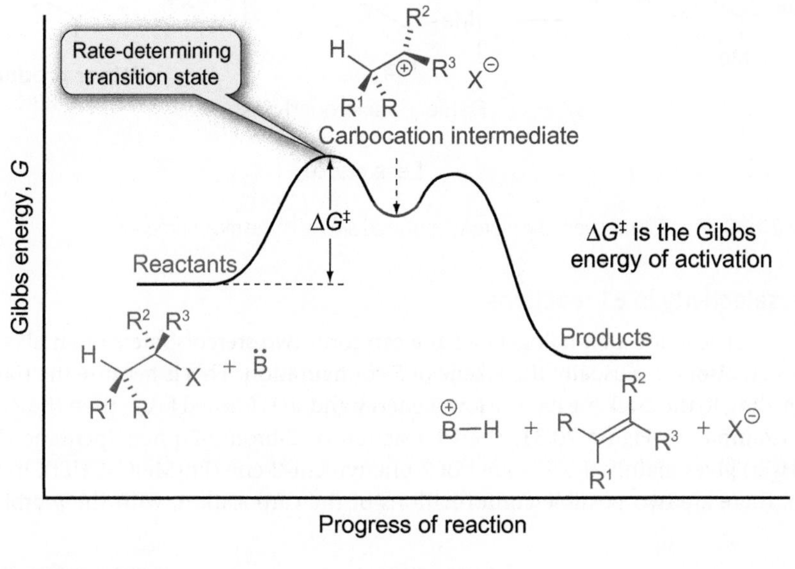

Rate-determining transition state
Carbocation intermediate
ΔG‡
ΔG‡ is the Gibbs energy of activation
Reactants
Products
Gibbs energy, G
Progress of reaction

Figure 20.28 The reaction profile for an E1 reaction.

are slower and primary halogenoalkanes (RCH_2X) do not undergo E1 reactions. The relative rates of E1 reactions are shown in Figure 20.29.

Regioselectivity in E1 reactions

An E1 reaction of a halogenoalkane can often form two structural isomers of an alkene. Where this happens, the major product is usually the most substituted structural isomer of the alkene. This is because the transition state leading to the more substituted alkene has the lower energy and so the more substituted alkene is formed faster. For example, in the E1 reaction of 2-bromo-3-methylbutane ($Me_2CHCHBrCH_3$, a secondary bromoalkane) shown in Figure 20.30, the major product is a trisubstituted alkene rather than a monosubstituted alkene.

The stability of carbocations (R^+) is discussed in Section 19.1 (p.867). Remember that the order of carbocation stability is tertiary (R_3C^+) > secondary (R_2CH^+) > primary (RCH_2^+), and conjugation with multiple bonds and/or lone pairs of electrons stabilizes carbocations.

The greater the number of alkyl groups (R) attached to a C=C bond, the more stable the alkene is (e.g. $R_2C{=}CHR$ is more stable than $RCH{=}CH_2$), as discussed in Section 21.1 (p.963).

2-Methylbut-2-ene ($Me_2C{=}CHMe$) can be distinguished from 3-methylbut-1-ene ($Me_2CHCH{=}CH_2$) using spectroscopic techniques, such as ^1H NMR spectroscopy. Whereas 2-methylbut-2-ene has one resonance signal between δ 4.5–6.5 ppm (characteristic of an alkene hydrogen, HC=C), 3-methylbut-1-ene has three resonance signals (Section 12.3, p.580).

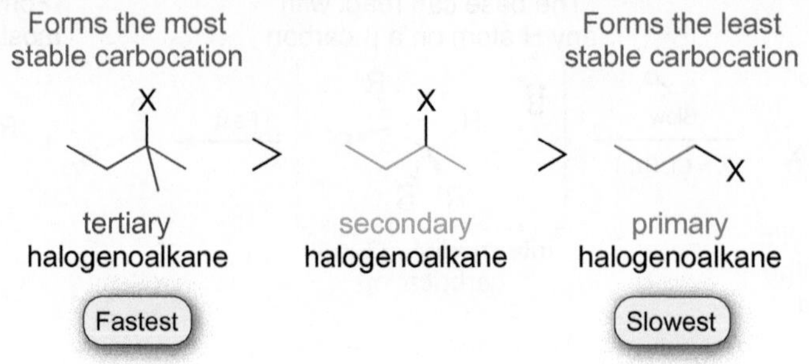
Forms the most stable carbocation
Forms the least stable carbocation
X
X
X
>
>
tertiary halogenoalkane
secondary halogenoalkane
primary halogenoalkane
Fastest
Slowest

Figure 20.29 Relative rates of E1 reactions.

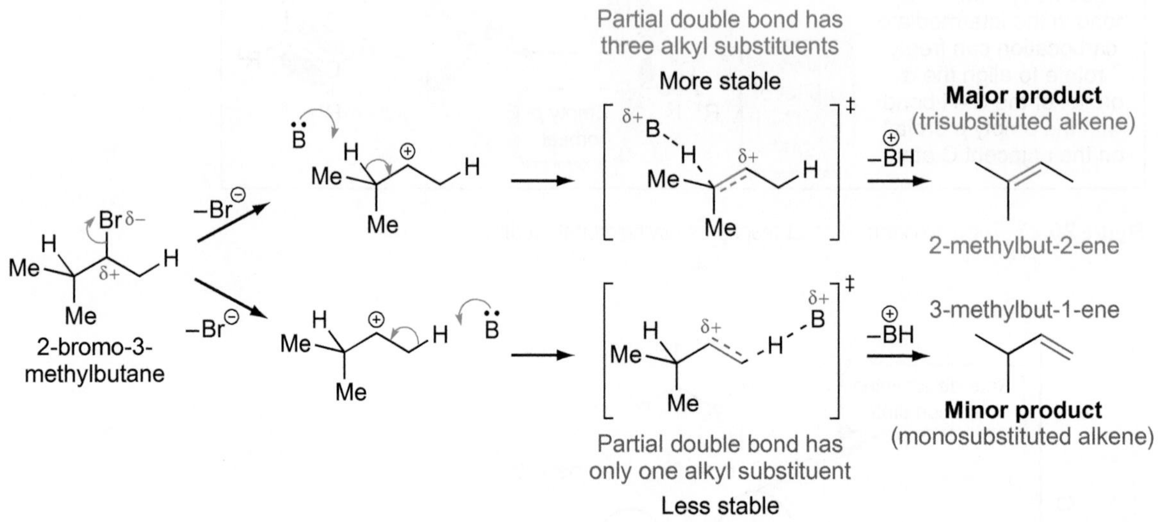
Partial double bond has three alkyl substituents
More stable
Major product
(trisubstituted alkene)
2-methylbut-2-ene
3-methylbut-1-ene
Minor product
(monosubstituted alkene)
Partial double bond has only one alkyl substituent
Less stable
2-bromo-3-methylbutane
−Br⁻
−BH
Me
H
B

Figure 20.30 In an E1 reaction, the more substituted alkene is the major product.

Section 18.3 (p.833) shows you how to assign *E* and *Z* configurations to alkenes. Note that the *E*-isomer of an alkene is more stable than the *Z*-isomer (see p.963).

To determine if an alkene has *E* or *Z* configuration, spectroscopic methods such as ^{1}H NMR spectroscopy can be used—see a specialist spectroscopy book for details.

Stereoselectivity in E1 reactions

When an E1 reaction of a halogenoalkane can form two stereoisomers of an alkene, the major stereoisomer is usually the alkene of *E* configuration. This is because the transition state leading to the *E*-alkene has the lower energy and so is formed faster than the *Z*-alkene.

For example, in Figure 20.31, the E1 reaction of 2-bromo-2-phenylpentane (Ph(Me)CBrCH$_2$Et) gives mainly the *E*-isomer of 2-phenylpent-2-ene (Ph(Me)C=CHEt). In this reaction, there are two possible conformations of the carbocation, with the *p* orbital and

2-Phenylpent-2-ene (both *E* and *Z* isomers) is an example of a conjugated alkene (Section 19.1, p.873). Resonance forms can be drawn showing interaction of the π bond of the alkene with the π electrons of the benzene ring, which makes the C=C bond more stable (for practice, draw the resonance forms).

Notice that 2-bromo-2-phenylpentane contains a chiral centre and that the *R* enantiomer is drawn here (for assigning *R* and *S* configurations to chiral centres see Section 18.4, p.845). E1 reactions of the *S* and *R* enantiomers gives the same products, as cleavage of the C–Br bond in each enantiomer forms the same carbocation intermediate.

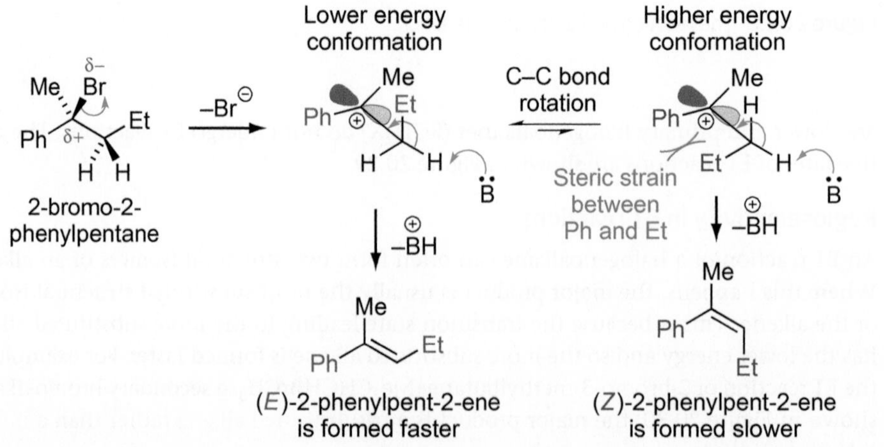
Lower energy conformation
Higher energy conformation
C–C bond rotation
2-bromo-2-phenylpentane
−Br⁻
Steric strain between Ph and Et
−BH
(E)-2-phenylpent-2-ene is formed faster
(Z)-2-phenylpent-2-ene is formed slower

Figure 20.31 The E1 reaction of 2-bromo-2-phenylpentane gives mainly the *E*-isomer of the alkene.

$^{\beta}$C–H bond aligned parallel, that can react with base to form the C=C bond. The conformation with the Ph and Et groups furthest apart is more stable because there is less steric strain. This is also true for the transition states that lead to both alkenes and so the transition state with the Ph and Et groups farthest apart will be the most stable and the *E*-alkene will be formed faster.

E2 reactions versus E1 reactions

The factors affecting the rates of E1 and E2 reactions are similar to those affecting S_N1 and S_N2 reactions (see p.930). The structures of the alkyl groups (R) in halogenoalkanes (RX) affect their reactivity in elimination reactions. Primary halogenoalkanes (RCH_2X) only undergo slow E2 reactions, whereas secondary (R_2CHX) and particularly tertiary (R_3CX) halogenoalkanes undergo faster E2 reactions and also E1 reactions. Other factors, such as the nature and concentration of the base and the choice of solvent, influence the mechanism of elimination reactions. Changing the leaving group also alters the rates.

Worked example 20.4 provides practice in predicting whether an elimination reaction involves an E1 or E2 pathway.

Choice of base

Increasing the concentration of the base or changing to a base with greater basic strength increases the rate of an E2 reaction, but not that of an E1 reaction. This is because the base takes part in the rate-determining step of an E2 reaction, but not that of an E1 reaction. Therefore, a relatively high concentration of a strong base (e.g. HO^- or RO^-) reacts more rapidly with a primary halogenoalkane in an E2 reaction than a relatively low concentration of a weaker base (e.g. R_3N), as shown below. Also, a relatively high concentration of a strong base is likely to react with a secondary halogenoalkane (such as $RCHXCH_2R$) in an E2 reaction rather than in an E1 reaction.

(i) A strong negatively charged base is used in an E2 reaction, whereas any base can be used in an E1 reaction.

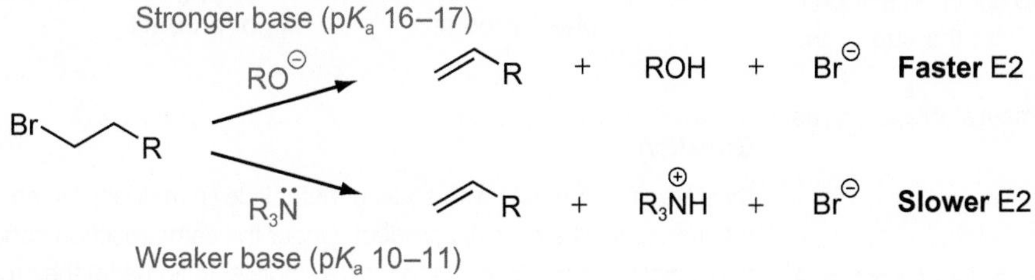
Stronger base (pK_a 16–17)
RO⊖
R + ROH + Br⊖ Faster E2
Br
R
R₃N
R + R₃NH⊕ + Br⊖ Slower E2
Weaker base (pK_a 10–11)

For a discussion of the factors that influence the strength of bases see Section 19.2 (p.888).

Changing the solvent

Changing the polarity of the solvent and also changing from a polar protic solvent (such as methanol, MeOH) to a polar aprotic solvent (such as dimethyl sulfoxide, Me_2SO) alter the rates of both E1 and E2 reactions (Table 20.3). The solvent effects are the same as for S_N1 and S_N2 reactions (p.937).

Polar protic and aprotic solvents are introduced in Box 7.1 (p.305) and discussed further in Section 19.1 (p.881).

An approximate measure of the polarity of a solvent is determined from its relative permittivity (see Section 20.3, p.937).

Choice of leaving group

Changing from a poor leaving group to a good leaving group in a halogenoalkane increases the rates of *both* E2 and E1 reactions. This is because the rate-determining steps in both elimination reactions involve heterolytic cleavage of the C–X bond and, the better the leaving group, the faster the cleavage of the C–X bond.

Classifying a group as a good or poor leaving group is discussed in Section 20.2 (p.925).

Table 20.3 Effect of changing the solvent on E1 and E2 reactions

	E2 reaction	E1 reaction
Increasing the solvent polarity	Slight decrease in rate (polar solvents solvate the base)	Significant increase in rate (polar solvents solvate the carbocation and halide ion)
Changing from a polar protic to a polar aprotic solvent	Increase in rate (aprotic solvents do not solvate the base)	Decrease in rate (aprotic solvents solvate the carbocation but not the halide ion)

Worked example 20.4 Predicting elimination pathways

Predict, giving your reasons, whether the reactions (a)–(c) shown here will take place by E1 or E2 mechanisms. Give the structures of the major organic products, showing the stereochemistry where appropriate.

(a)

Br
Reactant and solvent:
H_2O

(b)

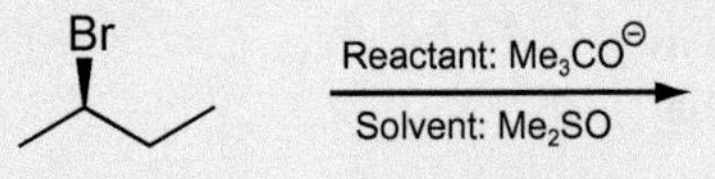
Br
Reactant: $Me_3CO^{\ominus}$
Solvent: Me_2SO

(c)

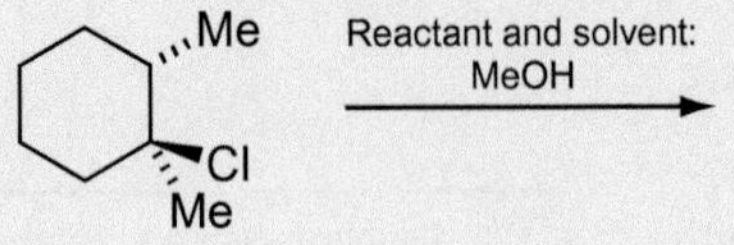
Me
Cl
Me
Reactant and solvent:
MeOH

Strategy

In each case, identify the base and the leaving group. Identify the halogenoalkane as primary, secondary, or tertiary.

Decide whether the reaction takes place by an E1 or E2 mechanism by considering the structure of the halogenoalkane, the strength of the base, and the nature of the solvent.

Remember that the halogenoalkane needs to adopt an anti-periplanar conformation in an E2 mechanism and that the size of the base affects the regioselectivity of the elimination.

Remember that an E1 mechanism gives the most stable alkene as the major product.

Solution

(a) Water is the base (and the solvent) and Br^- is the leaving group. A tertiary halogenoalkane can react by an E2 or E1 mechanism. An E2 mechanism is favoured using a strong base and a polar aprotic solvent. However, water is a weak base (p.896) and a polar protic solvent (p.936), so an E1 mechanism is favoured. The product is 2-methylpropene.

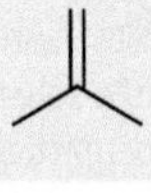

(b) The *tert*-butoxide ion (Me_3CO^-) is the base and Br^- is the leaving group. A secondary halogenoalkane can react by an E2 or E1 mechanism. Me_3CO^- is a strong base (p.896) and Me_2SO is a polar aprotic solvent (p.936), so an E2 mechanism is favoured. The E2 elimination could produce a monosubstituted and/or a disubstituted alkene. However, Me_3CO^- is a large anion and it selectively removes one of the hydrogen atoms on the β-carbon atom at the end of the chain to form mainly a monosubstituted alkene (a Hofmann elimination).

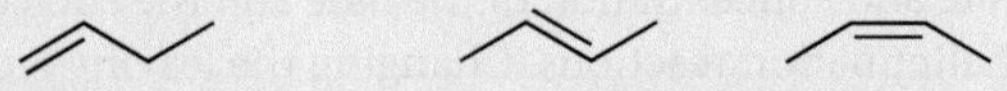

Major product — Minor products

(c) Methanol is the base (and solvent) and Cl^- is the leaving group. A tertiary halogenoalkane can react by an E2 or E1 mechanism. An E1 mechanism is favoured in methanol because this is a weak base (p.896) and a polar protic solvent (p.936). The E1 elimination could lead to a trisubstituted and/or a tetrasubstituted alkene. However, the E1 reaction is expected to give mainly the tetrasubstituted alkene because this alkene is the more stable.

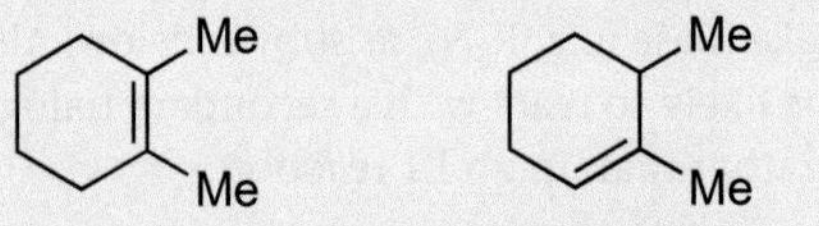
Me
Me
Me
Me

Major product — Minor product

Question

Diastereomer **A** reacts with sodium methoxide (in methanol) in an E2 reaction to form a racemic product. Under the same reaction conditions, diastereomer **B** reacts to form a single enantiomer of the product. Suggest an explanation for why **A** and **B** react differently.

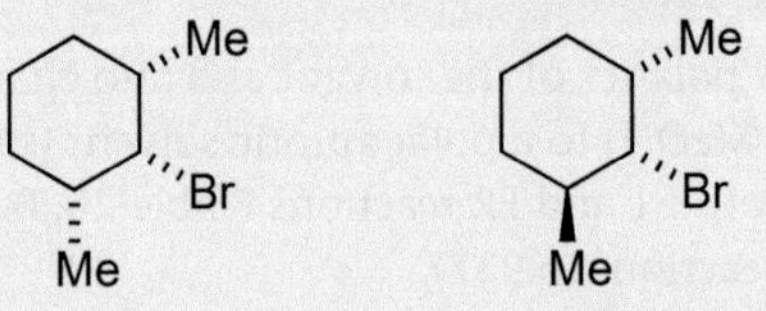
Me
Br
Me
Me
Br
Me

A — B

Summary

E2 reaction	E1 reaction
One-step mechanism	Two-step mechanism
Second-order kinetics (rate = k[RX][B])	First-order kinetics (rate = k[RX])
Rate-determining step is bimolecular	Rate-determining step is unimolecular
Anti-periplanar conformation required	No specific conformation of RX required

→

Reactivity of RX: $R_3CX > R_2CHX > RCH_2X$	Reactivity of RX: $R_3CX > R_2CHX >> RCH_2X$
Saytzev or Hofmann elimination	Most stable alkene formed
Favoured by polar aprotic solvents	Favoured by polar protic solvents
Requires a strong base (e.g. RO^-)	Any base will do (usually weak, e.g. ROH)
Requires a good leaving group: RI > RBr > RCl >> RF	Requires a good leaving group: RI > RBr > RCl >> RF

For practice questions on these topics, see questions 4, 6, and 7 at the end of this chapter (pp.957–959).

20.5 Substitution versus elimination reactions

Sections 20.3 and 20.4 describe how halogenoalkanes react with nucleophiles and bases in substitution and elimination reactions and discuss the competition between S_N2 and S_N1 mechanisms, and between E1 and E2 mechanisms.

Most nucleophiles can also act as bases, so there is also competition between substitution and elimination reactions. The major product of a reaction will depend on:

- the structure of the halogenoalkane;
- the nature of the nucleophile or base;
- the solvent;
- the reaction temperature.

Generally, S_N2 reactions are in competition with E2 reactions and S_N1 reactions are in competition with E1 reactions, because the conditions that favour S_N2 reactions also favour E2 reactions, and the conditions that favour S_N1 reactions also favour E1 reactions.

Visit the Online Resource Centre to view screencast 20.2 which walks you through examples of substitution and elimination reactions.

Substitution and elimination reactions are almost always in competition with each other—sometimes one mechanism wins out, but sometimes there is no clear winner and a number of products are formed. The aim is to identify which is/are the major product(s).

S_N2 versus E2 reactions

Tertiary halogenoalkanes (R_3CX) rapidly form elimination products by E2 mechanisms, but do not form substitution products by S_N2 reactions. Therefore, when tertiary halogenoalkanes react in conditions that favour $E2/S_N2$ reactions (e.g. use of a polar aprotic solvent), only elimination products are formed.

Primary (RCH_2X) and secondary (R_2CHX) halogenoalkanes can react by both S_N2 and E2 mechanisms. The relative amounts of the substitution and elimination products formed from $S_N2/E2$ reactions of primary and secondary halogenoalkanes depend on the strength of the nucleophile/base and the reaction temperature.

The factors that affect the strengths of nucleophiles and bases are discussed in Sections 19.1 (p.880) and 19.2 (p.888), respectively.

S_N2 reaction

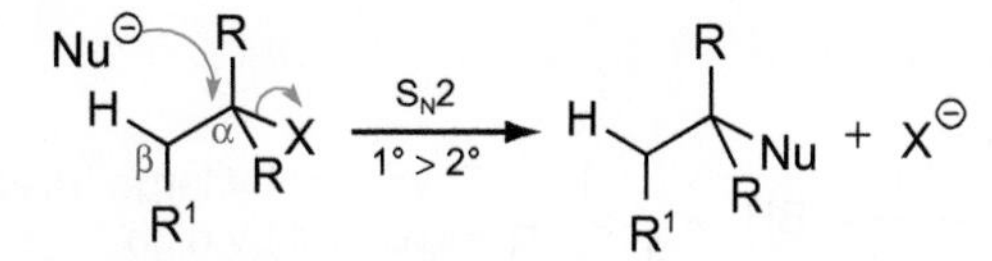
Nu⊖
R
H
α
X
β
R
R¹
S_N2
1° > 2°
H
Nu
R
R¹
+ X⊖

Favoured by small strong nucleophiles

1°, 2°, and 3° stand for primary, secondary, and tertiary, respectively (Section 2.6, p.81).

E2 reaction

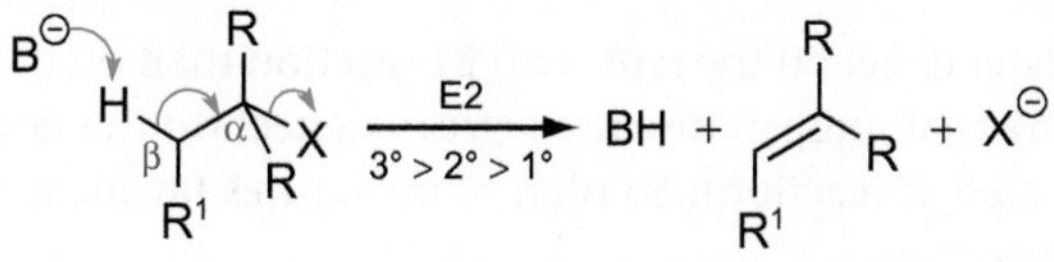
B⊖
H
R
α
X
β
R
R¹
E2
3° > 2° > 1°
BH +
R
R
R¹
+ X⊖

Favoured for tertiary halogenoalkanes
Favoured by large strong bases
Favoured by high temperatures

> **i** Note that the term 'amide' is used to name both R_2N^- ions and $RCONR_2$ compounds. For example, $NaNH_2$ is called sodamide (see Section 26.5, p.1182) and $LiN(CHMe_2)_2$ is called lithium diisopropylamide (often abbreviated as LDA). LDA is usually formed by deprotonation of diisopropylamine ($HN(CHMe_2)_2$) with butyllithium (BuLi).

> **i Ratio of substitution:elimination for reaction of $CH_3(CH_2)_{15}CH_2CH_2Br$ with different alkoxides**

S_N2 : E2	Alkoxide (RO^-)
99 : 1	MeO^-
15 : 85	Me_3CO^-

The product from substitution is $CH_3(CH_2)_{15}CH_2CH_2OR$ (R = Me or CMe_3) and the product from elimination is $CH_3(CH_2)_{15}CH{=}CH_2$.

> **i** Not all strong bases are large anions. A notable exception is the hydride ion, H^-. H_2 has a high pK_a value of 35 and so H^- is a strong base, which deprotonates various organic compounds (including alcohols and phenols) to form a strong H–H bond (436 kJ mol^{-1}).

> **i** Raising the temperature usually increases the rate of elimination reactions proportionally more than substitution reactions.

> **i Ratio of substitution:elimination for solvolysis of Me_3CCl in 80% aqueous ethanol**

S_N1 : E1	Temperature (°C)
5.00 : 1	25
1.75 : 1	55
1.25 : 1	75

The product from substitution is Me_3COH and the product from elimination is $Me_2C{=}CH_2$.

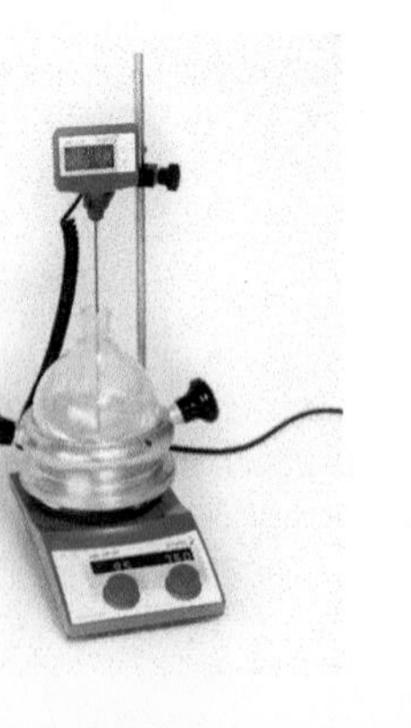

A popular experimental method for heating an organic reaction mixture (that contains a flammable solvent) uses a hotplate stirrer and an oil bath or an aluminium heating block (shown here).

Reaction of primary or secondary halogenoalkanes with strong nucleophiles favours the formation of substituted products by S_N2 reactions. Strong nucleophiles are anions that are relatively small molecules (e.g. HO^- or MeO^-), which have little difficulty in approaching the α-carbon atom ($^{\alpha}C$–X) from the opposite side to the leaving group (which is required for an S_N2 reaction).

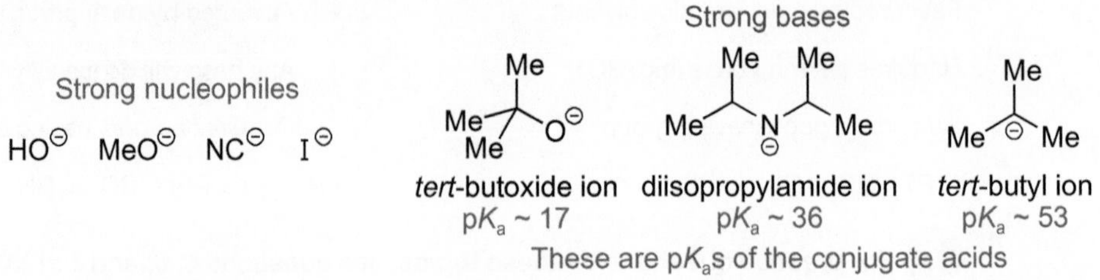

Strong nucleophiles
HO⊖ MeO⊖ NC⊖ I⊖
Strong bases
tert-butoxide ion
pKa ~ 17
diisopropylamide ion
pKa ~ 36
tert-butyl ion
pKa ~ 53
These are pKas of the conjugate acids

Reaction of primary or secondary halogenoalkanes with strong bases favours the formation of elimination products by E2 reactions. Strong bases are anions with high pK_a values that are relatively large molecules (e.g. Me_3CO^-), which have difficulty in attacking the α-carbon atom (from the opposite side to the leaving group) in an S_N2 mechanism. It is much easier for the large anion to attack a hydrogen atom on the β-carbon (H–$^{\beta}C$–$^{\alpha}C$–X) in an E2 mechanism, because the β-hydrogen atom is less hindered than the α-carbon. As a result, large anions are better bases than nucleophiles.

Increasing the temperature at which the reaction is carried out increases the rates of both S_N2 and E2 reactions. However, on increasing the temperature, the rate of an E2 reaction increases more than the rate of an S_N2 reaction so high temperatures favour the formation of elimination products. An E2 reaction has a higher activation energy than an S_N2 reaction because *two* bonds need to be broken in an E2 reaction compared to only *one* bond in an S_N2 reaction. Higher temperatures are, therefore, required to form elimination products in E2 reactions.

S_N1 versus E1 reactions

In both S_N1 and E1 reactions, tertiary halogenoalkanes react more rapidly than secondary halogenoalkanes, and primary halogenoalkanes do not react at all. When tertiary and secondary halogenoalkanes react in conditions that favour E1/S_N1 reactions (e.g. use of a polar protic solvent), the relative amounts of the substitution and elimination products depend on the size of the nucleophile/base and the reaction temperature.

Reaction of secondary or tertiary halogenoalkanes with small size anions favours the formation of substituted products by S_N1 reactions as the anion can easily attack the α-carbon of the carbocation (e.g. $R_3{}^{\alpha}C^+$). Larger size bases favour the formation of elimination products, as it is easier for large anions to attack a hydrogen atom on the β-carbon rather than the α-carbon (H–$^{\beta}C$–$^{\alpha}C$–X).

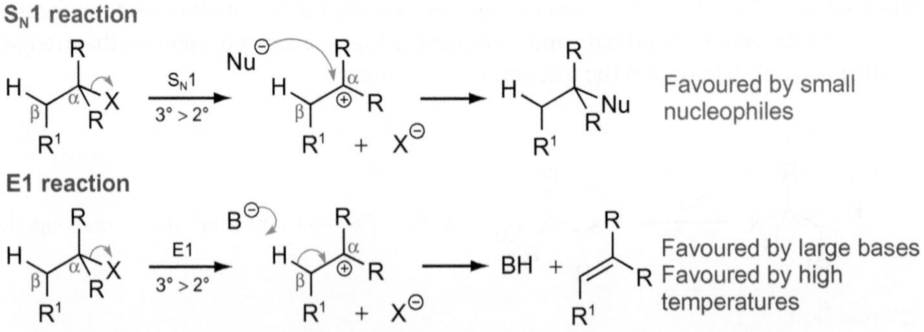

SN1 reaction
SN1
3° > 2°
Favoured by small nucleophiles
E1 reaction
E1
3° > 2°
Favoured by large bases
Favoured by high temperatures

Increasing the temperature has a greater effect on the rate of an E1 reaction than on the rate of an S_N1 reaction (in the same way that temperature has a greater effect on the rate of an E2 reaction than on the rate of an S_N2 reaction). So high temperatures favour the formation of elimination products.

S_N2/E2 versus S_N1/E1 reactions

Secondary halogenoalkanes (R_2CHX) react with anions to form substitution products in S_N1 or S_N2 reactions, or elimination products in E1 or E2 reactions. Whether a secondary halogenoalkane reacts by S_N2/E2 or S_N1/E1 pathways depends on the solvent and the strength and concentration of the nucleophile/base.

- S_N1/E1 reactions are favoured over S_N2/E2 reactions in polar protic solvents (such as MeOH) rather than polar aprotic solvents (such as Me_2SO).
- S_N2/E2 reactions are favoured over S_N1/E1 reactions when high concentrations of strong nucleophiles (such as HO^-) or strong bases (such as Me_3CO^-) are used.

Worked example 20.5 provides practice in assigning mechanisms to substitution and/or elimination reactions.

Worked example 20.5 Assigning mechanisms to the formation of products

Predict, giving your reasons, whether the following products are formed by S_N1, S_N2, E1, or E2 mechanisms.

(a) 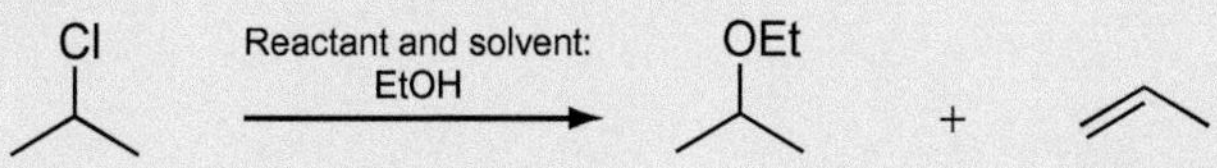
Cl
Reactant and solvent:
EtOH
OEt
+

(b) 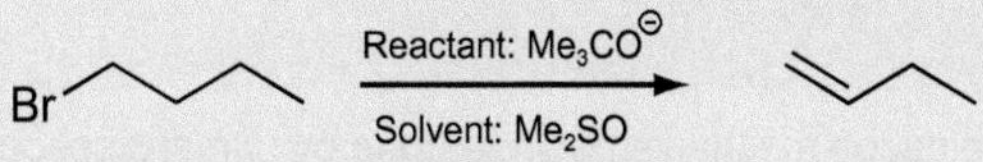
Br
Reactant: Me₃CO⊖
Solvent: Me₂SO

(c) 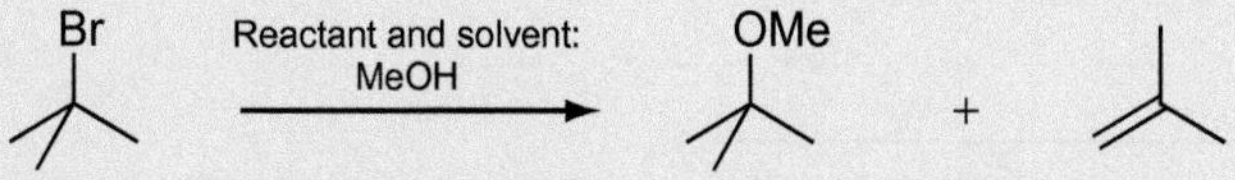
Br
Reactant and solvent:
MeOH
OMe
+

Strategy

In each case, decide whether the product is formed by a substitution or elimination reaction. Identify the halogenoalkane as primary, secondary, or tertiary.

Decide whether the halogenoalkane reacts by S_N2/E2 or S_N1/E1 pathways by considering the structure of the halogenoalkane, the type of solvent, and the strength and concentration of the nucleophile/base.

Solution

(a) 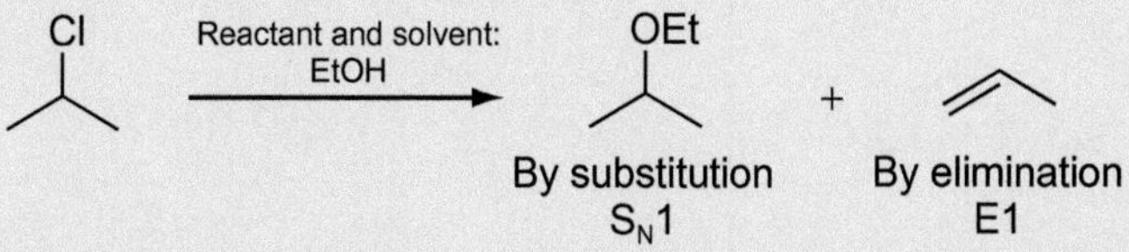
Cl
Reactant and solvent:
EtOH
OEt
+
By substitution
SN1
By elimination
E1

A secondary halogenoalkane can undergo S_N2/E2 or S_N1/E1 reactions. However, as ethanol is a polar protic solvent and a weak nucleophile, S_N1 and E1 reactions are favoured.

(b) The alkene is formed by an elimination reaction. A primary halogenoalkane undergoes elimination by an E2 mechanism rather than by an E1 mechanism. An E2 reaction is favoured because the *tert*-butoxide ion is a strong base and a polar aprotic solvent (dimethyl sulfoxide) is used. (The terminal alkene is formed because the *tert*-butoxide ion removes one of the hydrogen atoms on the β-carbon, Br–$^{\alpha}$C–$^{\beta}$C–H.)

→

The ether product from reaction (c) is 2-methoxy-2-methylpropane (or **m**ethyl ***t**ert*-**b**utyl **e**ther), which is commonly abbreviated as MTBE. MTBE can be added to petrol to improve its ignition properties (Box 13.3, p.628)

(c)

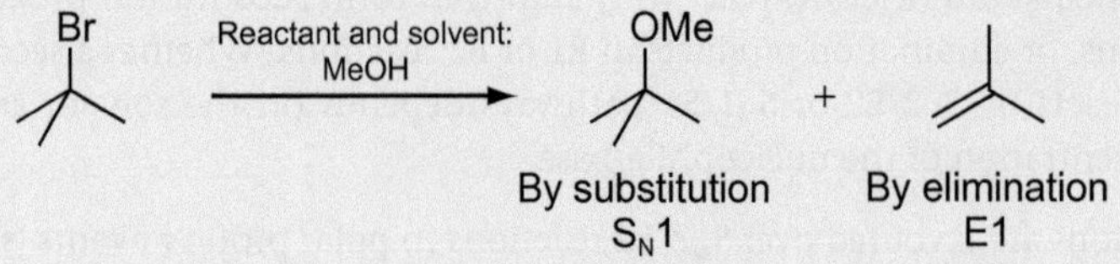
Br
Reactant and solvent:
MeOH
OMe
+
By substitution
S_N1
By elimination
E1

A tertiary halogenoalkane can undergo S_N1/E1 or E2 reactions. As methanol is a polar protic solvent and a weak base, S_N1 and E1 reactions are favoured.

Question

Suggest structures for the major substitution product and the major elimination product from reaction of (*R*)-2-bromobutane with ethoxide ion in acetonitrile.

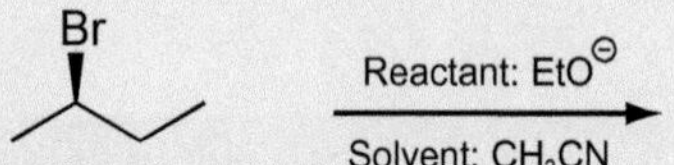
Br
Reactant: $EtO^{\ominus}$
Solvent: CH_3CN

(*R*)-2-bromobutane

Summary

- The following table summarizes the favoured substitution and elimination mechanisms undergone by primary, secondary, and tertiary halogenoalkanes.

Halogenoalkane	Favoured substitution and elimination mechanisms
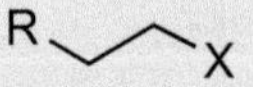 primary (1°)	S_N2 E2
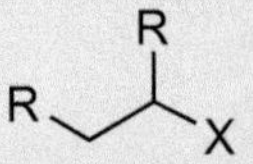 secondary (2°)	S_N1 S_N2 E1 E2
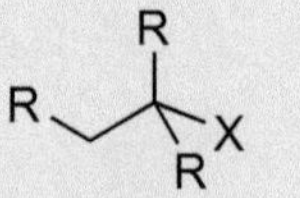 tertiary (3°)	S_N1 E1 E2

Favoured when using a strong nucleophile

Favoured when using a strong base

For practice questions on these topics, see questions 4, 6, and 7 at the end of this chapter (pp.957–959).

Concept review

By the end of this chapter you should be able to do the following.

- Understand why different carbon–halogen bonds in halogenoalkanes react differently from one another.
- Describe how halogenoalkanes can be prepared from alkanes, alcohols, and alkenes.
- Understand how halogenoalkanes react in S_N2 reactions.
- Understand how halogenoalkanes react in S_N1 reactions.
- Recognize the importance of tight ion pairs in some S_N1 reactions.
- Understand how halogenoalkanes react in E2 reactions.
- Understand how halogenoalkanes react in E1 reactions.
- Understand the factors that affect S_N2 versus S_N1 reactions.
- Understand the factors that affect E2 versus E1 reactions.
- Understand the factors that influence substitution versus elimination reactions.

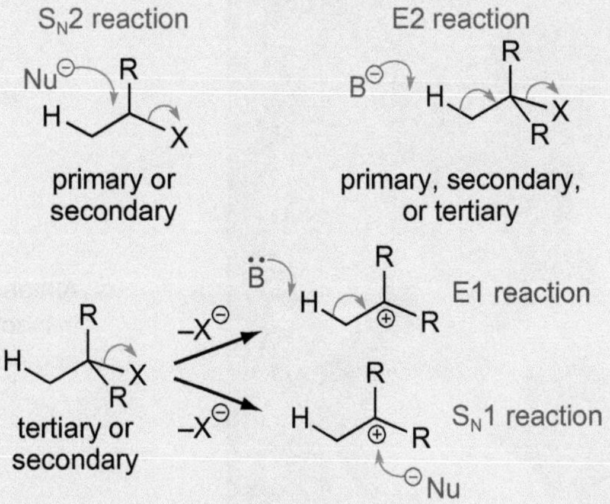
S$_N$2 reaction
E2 reaction
primary or secondary
primary, secondary, or tertiary
E1 reaction
S$_N$1 reaction
tertiary or secondary

Questions

More challenging questions are indicated by an asterisk *.

1 Suggest a mechanism for this reaction:

$$MeOH + PCl_5 \rightarrow MeCl + POCl_3 + HCl$$

(Section 20.2)

2 The relative rates of solvolysis of 2-bromo-2-methylpropane in three solvents are shown below.

Solvent:	CH_3CO_2H	CH_3OH	H_2O
Relative rate of solvolysis:	1	4	150000

Suggest an explanation for these observations. (Section 20.3)

3 Explain why the relative rate of reaction of 1-bromobutane with azide ion (N_3^-) increases from 1 to 2800 on changing the solvent from methanol to dimethylformamide. (Section 20.3)

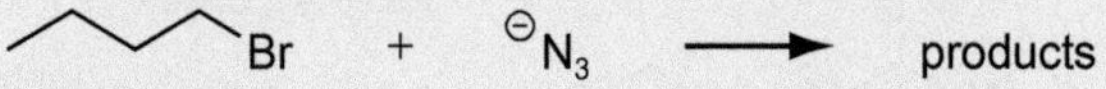
Br
+
N$_3$
products

1-bromobutane

4 Suggest structures for the major products formed in the following reactions and state whether these are formed by S_N1, S_N2, E1, or E2 mechanisms. Explain your reasoning.

(Sections 20.3 and 20.5)

(a)

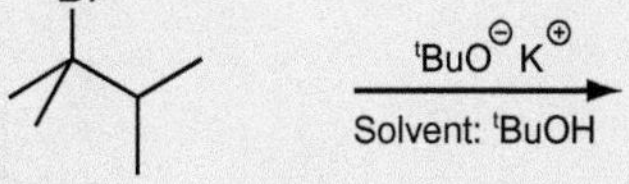
Br
$^tBuO^{\ominus}K^{\oplus}$
Solvent: ^tBuOH

(b)

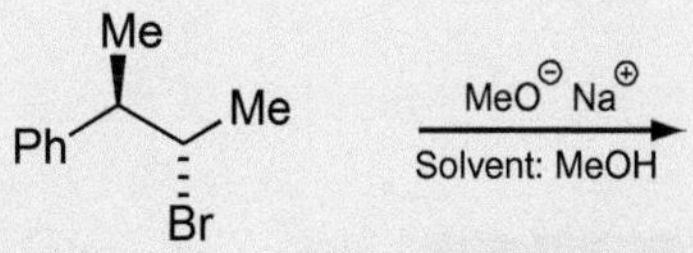
Me
Ph
Me
Br
$MeO^{\ominus}Na^{\oplus}$
Solvent: MeOH

(c)

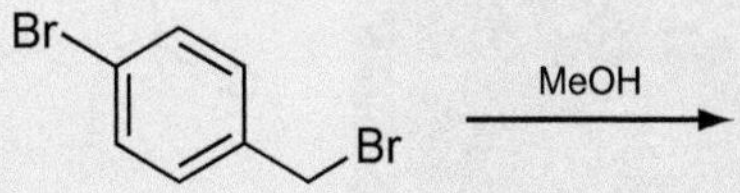
Br
Br
MeOH

5 Compound SB-207266 is a medicine developed for the treatment of irritable bowel syndrome. An outline synthesis is shown below. (Sections 19.2 and 20.3)

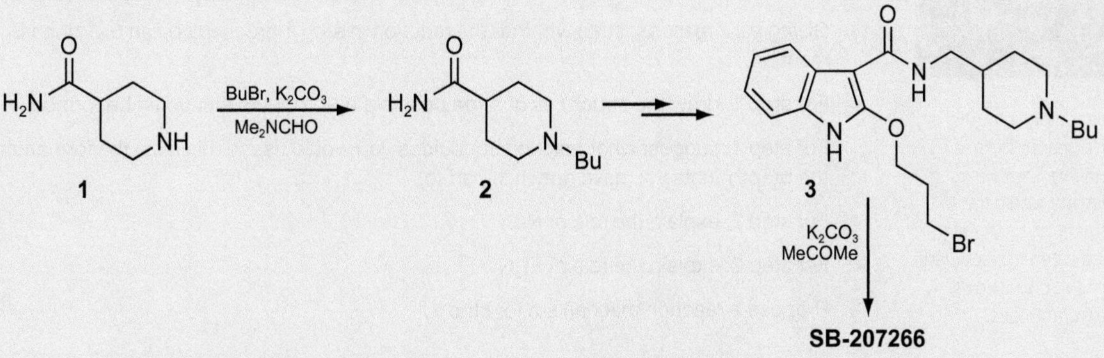
BuBr, K$_2$CO$_3$
Me$_2$NCHO
1
2
3
K$_2$CO$_3$
MeCOMe
SB-207266

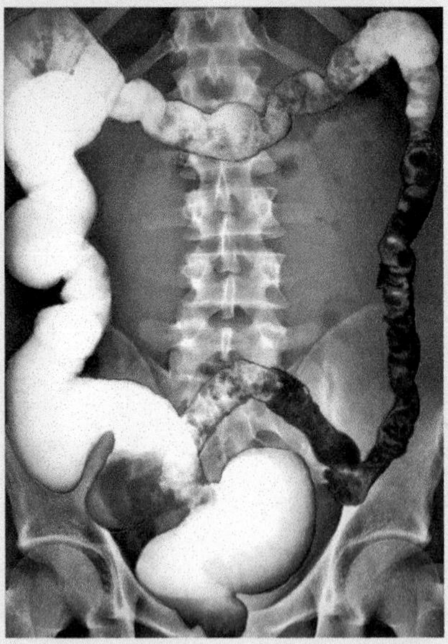

Irritable bowel syndrome is a condition in which the usual rhythmic contraction of muscles in the colon wall is disrupted. One diagnostic procedure uses X-rays to assess the size and shape of the colon—first, barium sulfate is used to coat the lining of the colon so that it shows up more clearly.

(a) Using the order of acid strength shown below, propose a mechanism for the formation of compound **2** from compound **1**.

Order of acid strength in DMF: $R_3\overset{\oplus}{N}H$ > $HCO_3^{\ominus}$ > RNHCOR > R_2NH

Strongest acid ... Weakest acid

(b) Although the CO_3^{2-} ion is not a sufficiently strong base to deprotonate the NH group in indole itself (see below), it is able to deprotonate the NH group in the indole ring of compound **3**. Suggest an explanation for the different acidities of these two indole rings.

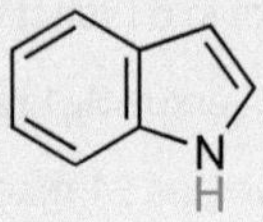

indole

(c) The formation of SB-207266 from compound **3** involves deprotonation of the indole ring to form an anion, which acts as a nucleophile in a nucleophilic substitution reaction. Suggest a structure for SB-207266 and give a mechanism for its formation. Does the substitution involve an S_N1 or an S_N2 mechanism?

6 Aripiprazole is an antipsychotic and antidepressant medicine used to treat schizophrenia, bipolar disorder, and clinical depression. It works by acting on various receptors in the brain. A synthesis of aripiprazole is shown below. (Sections 20.3 and 20.5)

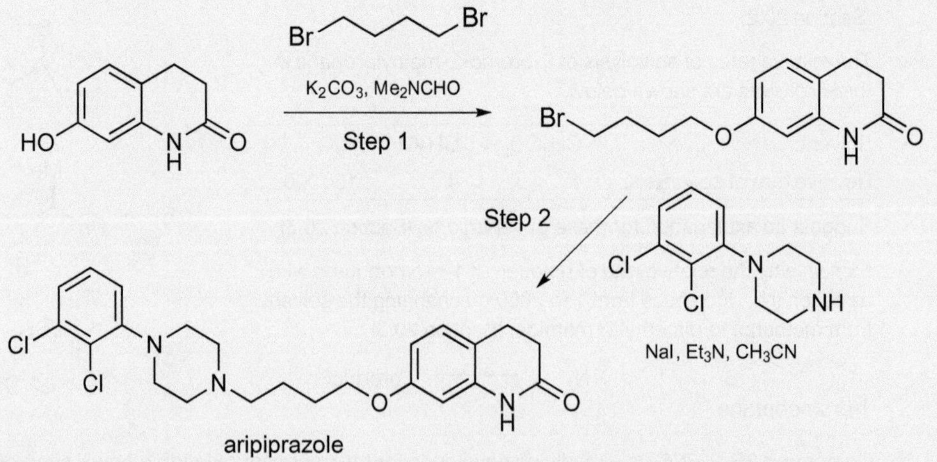
Br
Br
K₂CO₃, Me₂NCHO
Step 1
HO
N
H
O
Br
O
Step 2
Cl
Cl
N
NH
NaI, Et₃N, CH₃CN
aripiprazole

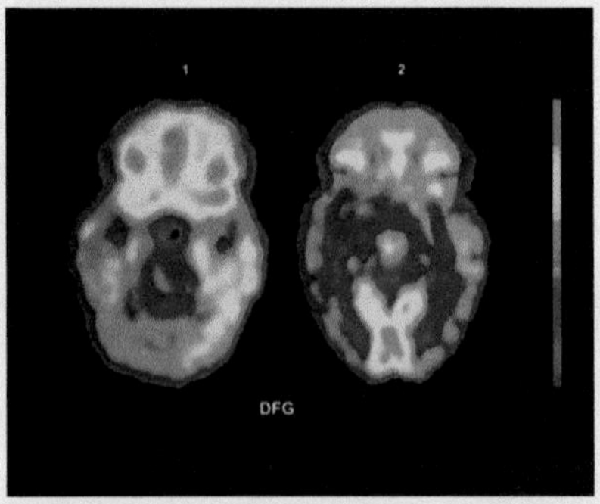
1
2
DFG

Coloured positron emission tomography (PET) scans of sections through a healthy brain (left) and a schizophrenic brain (right). The colours show different levels of activity within the brain during an attention test. Red shows high activity, through yellow and green to black (very low activity). The schizophrenic brain shows much lower activity in the frontal lobes.

(a) Giving your reasons, state whether the reaction in step 1 proceeds by an S_N1 or an S_N2 reaction.

(b) For step 1, draw the structures of *three* possible by-products that could be formed.

(c) For step 1, suggest what reaction conditions you would use to minimize the formation of the by-products you have drawn in part (b).

(d) For step 2, explain the role of NaI.

(e) For step 2, explain the role of Et_3N.

(f) Propose a reaction mechanism for step 2.

7* Pyrethroids are a family of synthetic compounds with insecticidal properties that function by damaging the nervous system of insects. They have found widespread use as they have low toxicity in mammals and are broken down naturally in the environment.

A synthetic route to pyrethroid **6** is shown below. (Sections 19.2, 20.2, and 20.5)

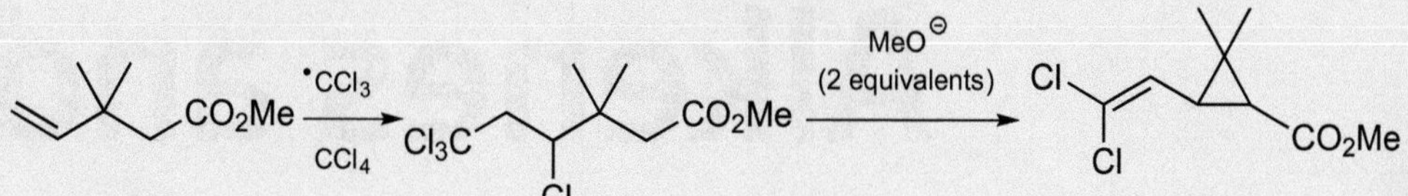
CO2Me
•CCl3
CCl4
Cl3C
CO2Me
Cl
MeO⊖
(2 equivalents)
Cl
Cl
CO2Me
4
5
6

(a) Propose a reaction mechanism to explain the formation of compound **5** from compound **4**.

(b) Suggest an explanation for the regioselective formation of compound **5**.

(c) Why is compound **5** formed as an equal mixture of enantiomers?

(d) Using the pK_a values given below, propose *three* possible mechanisms for the formation of pyrethroid **6** from compound **5**.

Acid	pK_a
$CH_3CO_2CH_3$	25.0
CH_3OH	15.5

(e) Giving your reasons, draw the major diastereomer of **6** that is formed from compound **5**.

Pyrethroids are analogues of a pair of naturally occurring compounds called pyrethrins. Pyrethrins are found in the seed cases of *Chrysanthemum cinerariaefolium* and the dried flower heads were used for hundreds of years to ward off insects from crops.

21

Alkenes and alkynes: electrophilic addition and pericyclic reactions

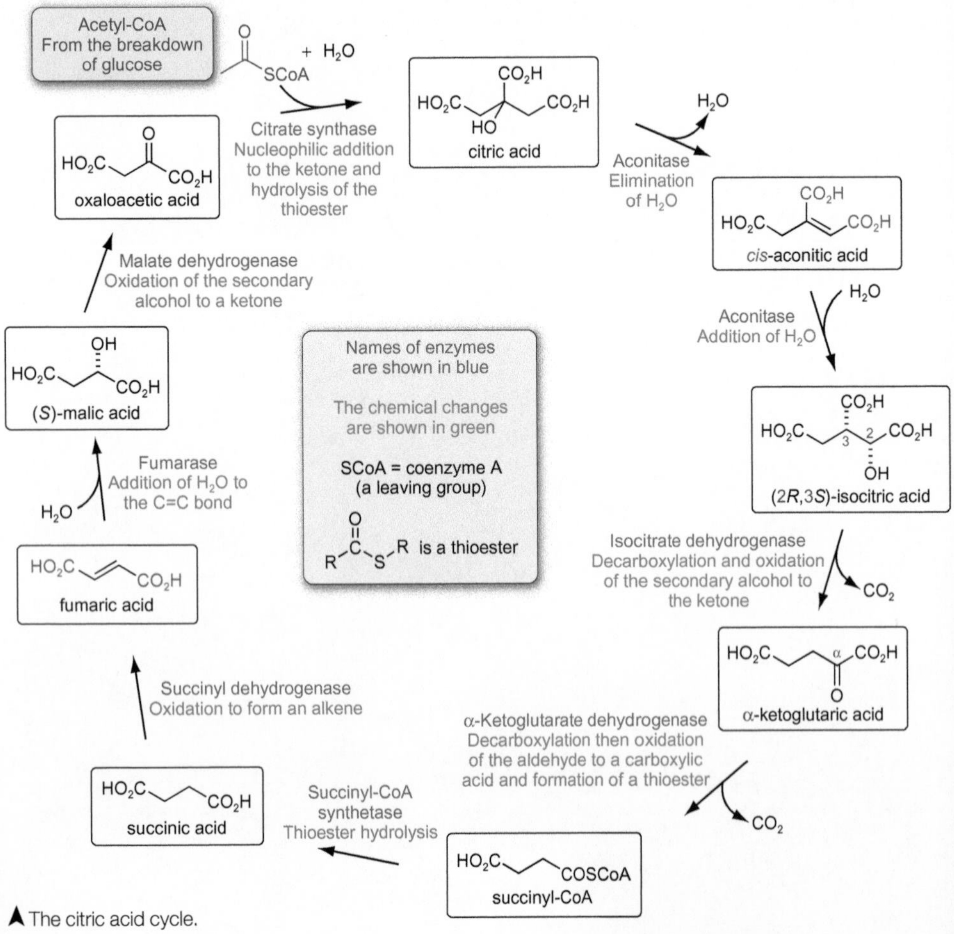

Acetyl-CoA
From the breakdown of glucose
SCoA
+ H2O
Citrate synthase
Nucleophilic addition to the ketone and hydrolysis of the thioester
citric acid
H2O
Aconitase
Elimination of H2O
cis-aconitic acid
Aconitase
Addition of H2O
(2R,3S)-isocitric acid
Isocitrate dehydrogenase
Decarboxylation and oxidation of the secondary alcohol to the ketone
CO2
α-ketoglutaric acid
α-Ketoglutarate dehydrogenase
Decarboxylation then oxidation of the aldehyde to a carboxylic acid and formation of a thioester
succinyl-CoA
Succinyl-CoA synthetase
Thioester hydrolysis
succinic acid
Succinyl dehydrogenase
Oxidation to form an alkene
fumaric acid
Fumarase
Addition of H2O to the C=C bond
(S)-malic acid
Malate dehydrogenase
Oxidation of the secondary alcohol to a ketone
oxaloacetic acid
Names of enzymes are shown in blue
The chemical changes are shown in green
SCoA = coenzyme A (a leaving group)
is a thioester

▲ The citric acid cycle.

This chapter builds on the following topics:

- Hydrocarbons Section 2.5, p.79
- Conformational isomers Section 18.2, p.819
- Configurational isomers: *E*- and *Z*-isomers Section 18.3, p.833
- Configurational isomers: isomers with chiral centres Section 18.4 p.838
- Fundamental concepts of organic reaction mechanisms Section 19.1 p.863
- Redox reactions Section 19.2, p.899
- Polar reactions Section 19.2, p.902
- Radical reactions Section 19.2, p.907
- Pericyclic reactions Section 19.2 p.908
- Regioselectivity Section 19.3, p.909
- Stereoselectivity Section 19.3, p.910
- The mechanisms of nucleophilic substitution reactions Section 20.3, p.930
- The mechanisms of elimination reactions Section 20.4, p.943

The citric acid cycle

The reactions that living organisms use to generate the energy they need, and to prepare the compounds they require, are called metabolism. There are two types of metabolism, depending on whether energy is generated or used. In anabolism, enzyme-catalysed reactions are used to make large complex molecules. These processes require an input of energy. In catabolism, nutrients are broken down in enzyme-catalysed reactions to provide simple starting materials for synthesis and also to supply energy. The energy released is used to convert adenosine diphosphate (ADP) into adenosine triphosphate (ATP), which acts as a mobile source of energy in living systems (see Box 14.8, p.683).

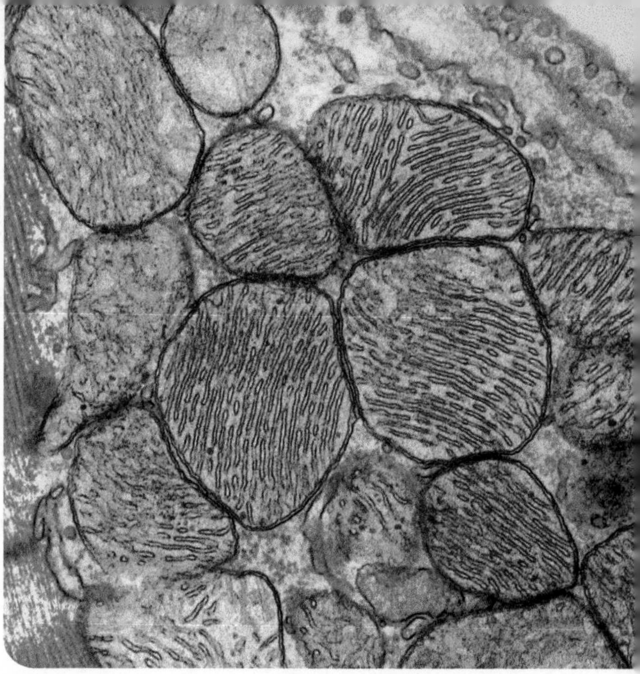

▲ The citric acid cycle occurs inside the mitochondria of cells. Mitochondria are usually rod-shaped structures and ATP is generated on the surface of the folded inner membrane.

An important stage in catabolism is the **citric acid cycle** (sometimes called the Krebs cycle or tricarboxylic acid cycle), which is the series of nine enzyme-catalysed reactions illustrated here. In the first step of the cycle (shown at the top left of the figure), acetyl-CoA ($CH_3COSCoA$, formed from glucose in the body) reacts with oxaloacetic acid to form citric acid (the IUPAC name is 2-hydroxypropane-1,2,3-tricarboxylic acid). Citric acid then undergoes a series of eight reactions in a cycle that reforms oxaloacetic acid and produces two molecules of CO_2. In the process, 11 molecules of ADP are converted into ATP. The cycle continues by reaction of oxaloacetic acid with another molecule of acetyl-CoA.

Of particular interest to this chapter are the reactions of *cis*-aconitic acid and the *trans* acid, fumaric acid, with water. In each of these enzyme-catalysed reactions, water adds to the C=C bond to form a secondary alcohol as a single enantiomer. The addition of water to a C=C bond is called a **hydration** reaction and this is a characteristic reaction of alkenes described in Section 21.3.

The conversion of citric acid into (2*R*,3*S*)-isocitric acid, via *cis*-aconitic acid, takes place in the active site of the enzyme aconitase and produces a structural isomer of citric acid by exchanging the positions of one H atom and the OH group. In the first step, an elimination reaction forms the C=C bond of *cis*-aconitic acid and, in the second step, H_2O adds regioselectively (see Section 19.3, p.909) to the C=C bond.

◄ Sir Hans Krebs (1900–1981) left Germany in 1933 to carry out research in biochemistry at the University of Cambridge, subsequently moving to the University of Sheffield and then to Oxford. He was part of a team that discovered the citric acid cycle and to mark this achievement he was awarded the Nobel Prize in Medicine in 1953.

Citric acid is responsible for the tart taste of many fruits including lemons, limes, and gooseberries. It is widely used as a flavouring and preservative in food and beverages, particularly soft drinks, and it has the E number E330.

R• and RO• radicals also add to C=C bonds and this type of reaction is commonly used to make polymers, as illustrated in Box 20.2 (p.923).

The naming of alkenes and alkynes is described in Section 2.5 (p.85).

Overlap of two sp^2 orbitals to give a σ bond, and two *p* orbitals to give a π bond, is shown in Section 5.4 (p.239).

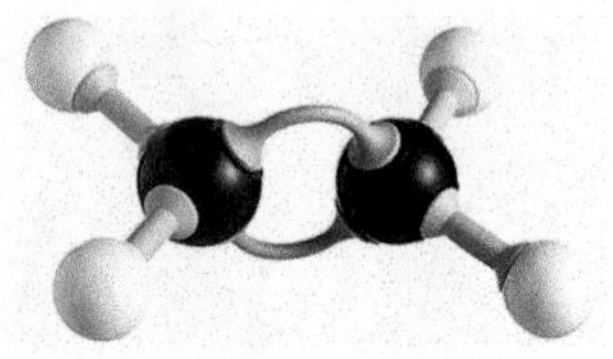

Often, molecular models of alkenes show two curved bonds in-between the two carbons (as shown here for ethene, $H_2C{=}CH_2$). Remember that this is not correct because the C=C bond has one σ bond and one π bond.

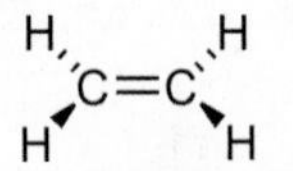

ethene

Both carbon atoms and all four hydrogen atoms are in the same plane.

H−C≡C−H

ethyne

A linear molecule

A hydrogen atom attached to a C≡C bond is more acidic than a hydrogen atom attached to a C=C bond (Section 19.2, p.890). Deprotonation of alkynes (RC≡CH) forms alkynyl ions ($RC{\equiv}C^-$), which react with electrophiles such as the δ+ carbon atom in a C–X bond (Section 21.2, p.969) and that in a C=O bond (Section 23.2, p.1069). Both ethynylsodium and sodium acetylide are acceptable names for $HC{\equiv}C^-\ Na^+$.

Alkenes are an important class of organic compounds that contain a C=C bond. Many natural products contain C=C bonds, such as unsaturated fatty acids (see Box 2.4, p.86). Alkenes are especially useful compounds in organic synthesis, because the C=C bond reacts with numerous electrophiles in electrophilic addition reactions. These reactions, which are discussed in Section 21.3, can be used to prepare molecules containing a range of functional groups. Alkenes also react in pericyclic reactions (Section 19.2, p.908) and some examples are described in Section 21.4.

Alkynes are a class of organic compounds that contain a C≡C bond. Like the C=C bond in alkenes, the C≡C bond in alkynes reacts with electrophiles in electrophilic addition reactions (Section 21.5). Alkynes are less reactive than alkenes with electrophiles so there are fewer synthetically useful electrophilic addition reactions.

This chapter begins with an overview of the structure and reactivity of alkenes and alkynes (Section 21.1) and this is followed by a discussion of the most important methods of preparing these compounds (Section 21.2).

21.1 Structure and reactivity of alkenes and alkynes

The C=C bond of an alkene contains one strong σ bond and one weaker π bond. The double bond is formed by overlap of the orbitals on the two sp^2 hybridized carbons. A σ bond is formed by end-on overlap of an sp^2 orbital of one carbon with an sp^2 orbital of another carbon. The π bond is formed by side-on overlap of the unhybridized *p* orbitals. Each carbon atom in the C=C bond forms two other σ bonds (to atoms in the R groups, where R is an alkyl group). The three σ bonds formed by each carbon atom are in the same plane and are separated by angles of 120°.

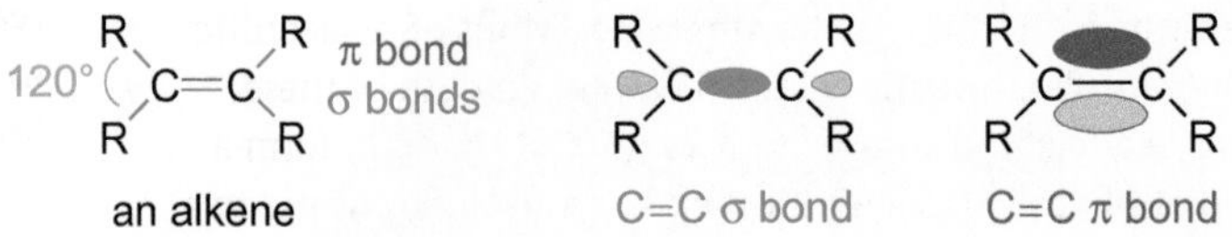
120°
π bond
σ bonds
an alkene
C=C σ bond
C=C π bond

The C≡C bond of an alkyne contains one strong σ bond and two weaker π bonds. The triple bond is formed by overlap of the orbitals on the two *sp* hybridized carbons. A σ bond is formed by end-on overlap of an *sp* orbital of one carbon with an *sp* orbital of another carbon. The π bonds are formed by side-on overlap of the two unhybridized *p* orbitals on each carbon atom. Each carbon atom in the C≡C bond forms one other σ bond (to atoms in the R groups). The two σ bonds formed by each carbon atom are separated by an angle of 180°.

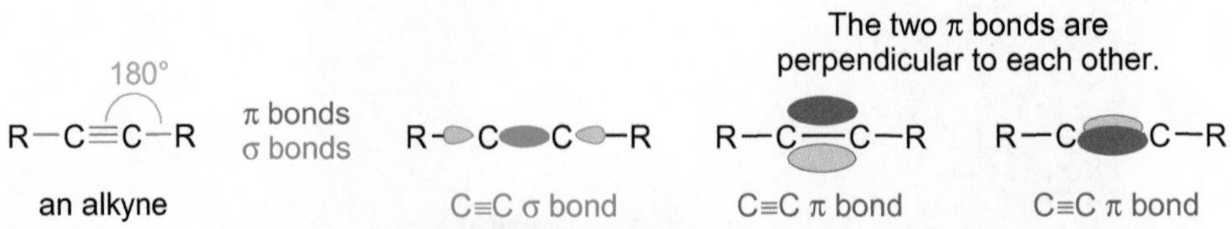
The two π bonds are perpendicular to each other.
180°
π bonds
σ bonds
an alkyne
C≡C σ bond
C≡C π bond
C≡C π bond

C=C, C≡C, and C–C bonds have different bond lengths and bond enthalpies (Table 21.1). As the number of covalent bonds that hold the two carbon atoms together increases, so the bond length decreases and the bond enthalpy increases (the bond becomes shorter and stronger). You can work out from the data in Table 21.1 that a C–C σ bond (+347 kJ mol^{-1}) is

Online Support. The icon in the margin indicates that accompanying interactive resources are provided online to help your understanding: just type www.chemtube3d.com/burrows/123 into your browser, replacing 123 with the number of the page where you see the icon. For pages linking to more than one resource, type 123-1, 123-2, etc. (replacing 123 with the page number) for access to successive links.

Table 21.1 Mean bond lengths and mean bond enthalpies

Bond	Mean bond lengths / pm	Mean bond enthalpies / kJ mol^{-1}
C–C	153 (longest bond)	+347 (weakest bond)
C=C	134	+612
C≡C	120 (shortest bond)	+838 (strongest bond)

Notice that *mean* values of bond lengths and bond enthalpies are shown in Table 21.1. This is because the substituents that are attached to the bonds affect the bond lengths and strengths. The effect of substituents on the stability and reactivity of C=C bonds is particularly important.

Tables of bond lengths and bond enthalpies are given in Appendix 10 (p.1368).

around 82 kJ mol^{-1} stronger than a π bond in C=C (+265 kJ mol^{-1}), and around 101.5 kJ mol^{-1} stronger than a π bond in C≡C (+245.5 kJ mol^{-1}). This explains why a π bond, and not a σ bond, is broken in reactions of C=C and C≡C bonds.

Factors affecting the stability and reactivity of C=C bonds

Whereas a C–C bond can freely rotate (i.e. one carbon atom can rotate relative to the other), rotation about a C=C bond is severely restricted (and requires high temperatures) because this requires the π bond to break (Section 18.2, p.819). As a result, a C=C bond in the middle of a carbon chain can have different configurational isomers, which are called *E*- and *Z*-isomers. The relative stabilities of the *E*- and *Z*-isomers of an alkene depend on the sizes of the substituents attached to the C=C bond. In general, the more stable configurational isomer has the two largest substituents on the opposite sides of the C=C bond, because this minimizes steric strain. It is often easier to make the *E*- (or *trans*-) isomer of a disubstituted alkene, rather than the *Z*- (or *cis*-) isomer, because the *E*- (or *trans*-) isomer is more stable.

Steric strain is also present in disubstituted terminal alkenes. For example, in 2-methylpropene ($H_2C{=}C(CH_3)_2$), there is steric strain due to interaction of the two methyl groups.

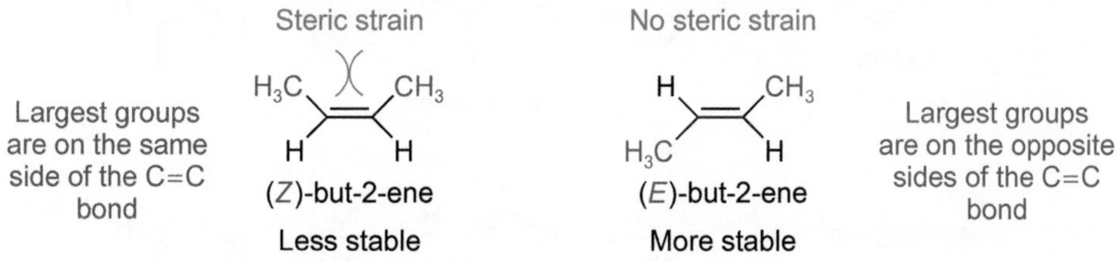
Steric strain
No steric strain
Largest groups are on the same side of the C=C bond
(Z)-but-2-ene
Less stable
(E)-but-2-ene
More stable
Largest groups are on the opposite sides of the C=C bond

The stability of alkenes is also influenced by the electronic effects of the substituents attached to the C=C bonds (see Figure 21.1). It is generally found that the greater the number of alkyl groups attached to the C=C bond, the more stable the alkene is. The relative stabilities of different C=C bonds explain why some alkenes react faster than others.

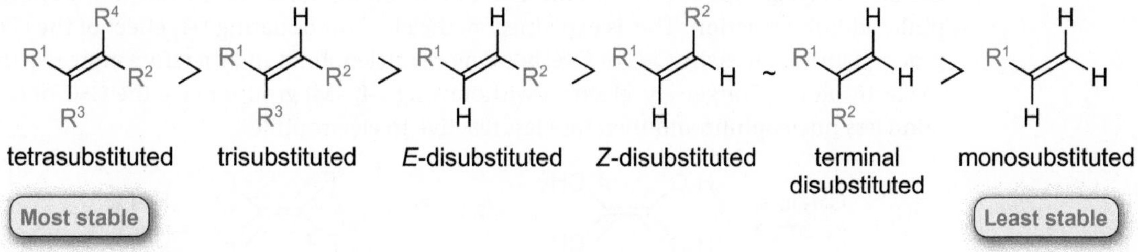
tetrasubstituted
trisubstituted
E-disubstituted
Z-disubstituted
terminal disubstituted
monosubstituted
Most stable
Least stable

Figure 21.1 Relative stabilities of alkyl-substituted alkenes.

Assigning the stereochemistry of alkenes using *E/Z* nomenclature is described in Section 18.3 (p.833). For tri- and tetrasubstituted alkenes, the *E*-isomer is not necessarily the most stable, because the priorities used to assign the *E/Z* configuration are *not* based on the size of the substituents.

Selective formation of the *E*-isomer of an alkene, in an E2 elimination reaction, is discussed in Section 20.4 (p.943).

The *E*- and *Z*-isomers of an alkene can have different biological activities (Box 2.4, p.86).

To determine if an alkene has *E* or *Z* configuration, spectroscopic methods such as ^{1}H NMR spectroscopy can be used—see a specialist spectroscopy book for details.

The relative stability of a C=C bond is determined from the enthalpy change ($\Delta H^{\ominus}$) for the reaction of the C=C bond with H_2 to form a saturated alkane (see p.967–968). In a series of alkenes, the less negative the value of $\Delta H^{\ominus}$, the more stable the alkene. The procedure is explained for benzene in Section 22.1 (p.1005).

Conjugated alkenes, which have alternating single and double bonds (e.g. $H_2C{=}CH{-}CH{=}O$), are more stable than isolated alkenes (e.g. $H_2C{=}CH{-}CH_2{-}CH_2{-}CH{=}O$). The greater stability of conjugated alkenes is explained by interaction of the C=C π bond with another π bond, as discussed in Section 19.1 (p.873).

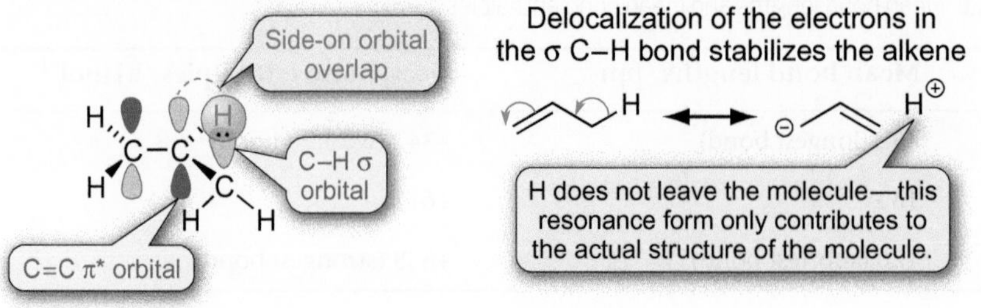
Side-on orbital overlap
C–H σ orbital
C=C π* orbital
Delocalization of the electrons in the σ C–H bond stabilizes the alkene
H does not leave the molecule—this resonance form only contributes to the actual structure of the molecule.

Figure 21.2 Hyperconjugation.

Hyperconjugation

Hyperconjugation is introduced in Section 19.1 (p.872).

For hyperconjugation, a C–H bond needs to be adjacent to the C=C bond.

The reasons why alkyl groups stabilize C=C bonds are not well understood. One explanation uses the concept of hyperconjugation. In an alkene, **hyperconjugation** involves the interaction of a pair of electrons in a filled C–H σ orbital with the empty π* orbital of the C=C bond. This interaction leads to the two electrons in the σ bond being spread over the molecule, and the delocalization of these electrons stabilizes the C=C bond (see Figure 21.2). As the number of alkyl groups on a C=C bond increases, so does the number of C–H bonds that can interact with, and thereby stabilize, the C=C bond.

Electrophilic addition

Generally, increasing the number of alkyl groups attached to a C=C bond increases the stability of the alkene, but it also makes the C=C bond more reactive to electrophilic addition.

The carbon atoms in C=C and C≡C bonds are held together by four and six electrons, respectively, so these bonds are electron-rich and act as nucleophiles in reactions. The reactions of nucleophilic C=C and C≡C bonds with electrophiles, E^+, are called **electrophilic additions**—this name shows that the first step in these reactions involves addition of the electrophile to the multiple bond. The carbocation formed then reacts with a nucleophile, Nu^- (Figure 21.3). Overall, two new σ bonds replace the π bond. Figure 21.3 shows general addition reactions for an alkene and an alkyne.

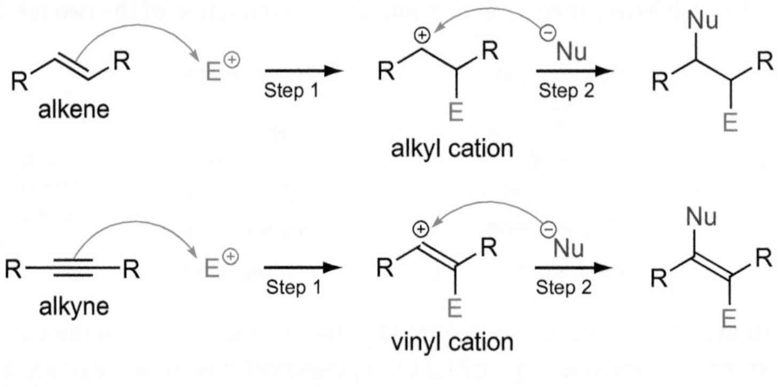
alkene
Step 1
alkyl cation
Step 2
alkyne
Step 1
vinyl cation
Step 2

Figure 21.3 Electrophilic addition to alkenes and alkynes.

In some cases, C=C bonds that have electron-withdrawing substituents can react with anions (see Box 19.8, p.904).

The reactivity of C=C and C≡C bonds towards electrophiles depends on the steric and electronic effects of the substituents that are attached to the multiple bonds. Generally, the more alkyl groups attached to the C=C or C≡C bond, the faster the rate of the electrophilic addition reaction. This is explained by the electron-donating (+I) effect of the alkyl groups, which makes the C=C or C≡C bond more nucleophilic and therefore more reactive to electrophiles. Conversely, electron-withdrawing (–I, –M) groups make the C=C or C≡C bond less nucleophilic and therefore less reactive to electrophiles.

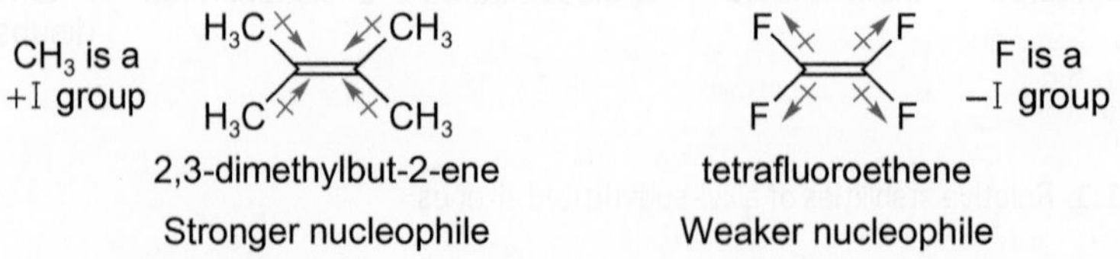
CH3 is a +I group
2,3-dimethylbut-2-ene
Stronger nucleophile
F is a –I group
tetrafluoroethene
Weaker nucleophile

Although tetrafluoroethene does not react easily with electrophiles, it does react with radicals—the radical polymerization of tetrafluoroethene forms poly(tetrafluoroethene), or PTFE, which is used as a non-stick coating for cookware (Box 20.2, p.923).

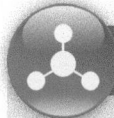

Box 21.1 Ethene production in plants

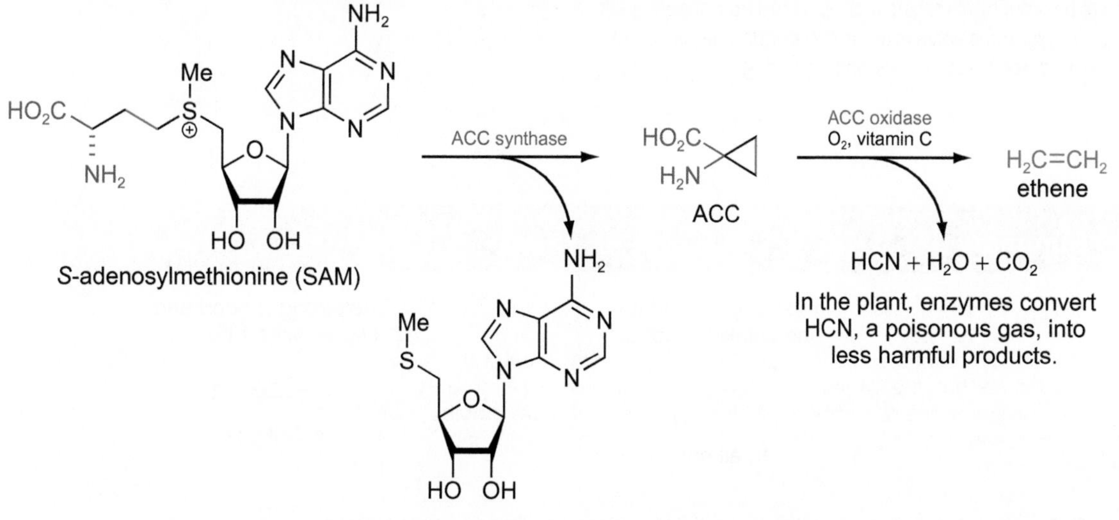
S-adenosylmethionine (SAM)
ACC synthase
ACC
ACC oxidase
O2, vitamin C
H2C=CH2
ethene
HCN + H2O + CO2
In the plant, enzymes convert HCN, a poisonous gas, into less harmful products.

Ethene ($H_2C{=}CH_2$) is a plant growth substance (hormone) that promotes the ripening of fruits such as apples, bananas, and tomatoes (Box 5.5, p.240). Ethene is produced in plants from *S*-adenosylmethionine (SAM). In the first step, SAM is converted into the amino acid 1-aminocyclopropane-1-carboxylic acid (ACC) by the enzyme ACC synthase. ACC is then converted into ethene in a second step, using the enzyme ACC oxidase in the presence of vitamin C and oxygen.

→

▲The genetic engineering of *Flavr Savr* tomatoes allows them to avoid spoilage during storage and transportation for much longer periods than normal tomatoes. This tomato was the first genetically engineered whole food to be brought to market. It disappeared from supermarket shelves in 1997, just three years after its introduction, partly due to supply problems, mixed taste reviews, and marketing controversies.

Several varieties of tomatoes have been genetically engineered to ripen more slowly. One way of doing this is to alter the gene sequence of the tomato so that production of the enzyme ACC synthase is decreased, which, in turn, decreases the production of ethene. This makes the fruit last longer in shipment and, on arrival, the genetically engineered fruit is treated with ethene to induce ripening.

Question

In living cells, *S*-adenosylmethionine (SAM) acts as an efficient methylating agent. Methylation reactions in the body regulate the biological activities of various hormones and neurotransmitters. For example, SAM reacts with norepinephrine in the presence of an enzyme to form epinephrine, which triggers a rise in blood pressure and heart rate in the body.

Suggest a mechanism (using curly arrows) for the reaction below and explain why SAM is such a reactive methylating agent. (*Hint.* Use your knowledge of nucleophilic substitution reactions (Section 20.3, p.930) to help you propose a mechanism.)

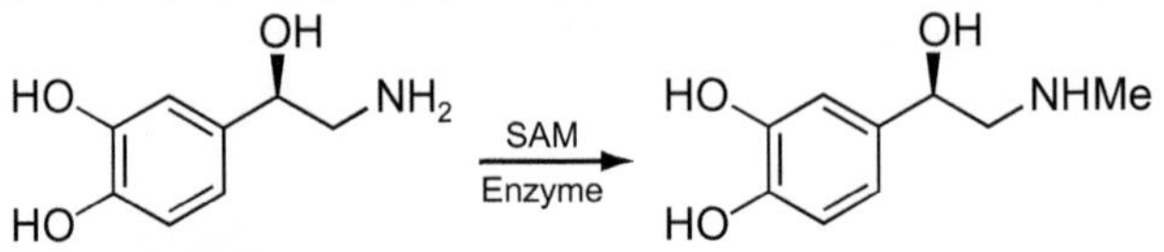
OH
HO
NH$_2$
HO
SAM
Enzyme
OH
HO
NHMe
HO

norepinephrine (noradrenaline) — epinephrine (adrenaline)

Summary

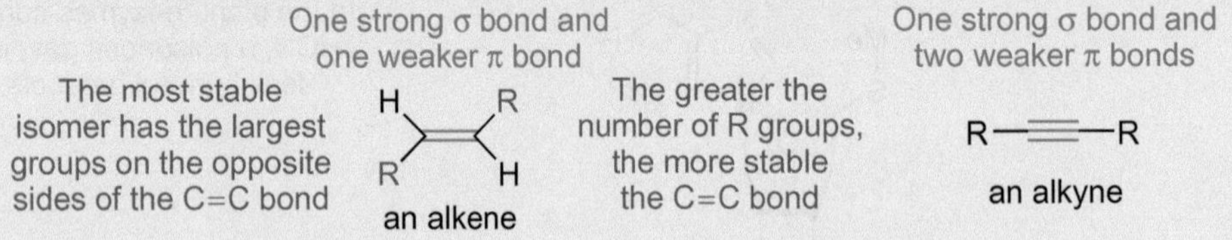
One strong σ bond and one weaker π bond
The most stable isomer has the largest groups on the opposite sides of the C=C bond
H R R H
an alkene
The greater the number of R groups, the more stable the C=C bond
One strong σ bond and two weaker π bonds
R—≡—R
an alkyne

- C=C and C≡C bonds react with electrophiles in electrophilic addition reactions.

R R E⊕
R—≡—R E⊕

21.2 Preparation of alkenes and alkynes

Preparation of alkenes

From halogenoalkanes and alcohols

Elimination reactions are commonly used to make alkenes. Halogenoalkanes (RX) react with bases in E1 and E2 reactions to form a range of alkenes (see Section 20.4, p.943).

Alcohols (ROH) undergo elimination reactions when using reagents that convert the OH group of the alcohol into a compound that provides a good leaving group (Section 20.2, p.925). For example, heating propan-2-ol (Me_2CHOH) with an acid (e.g. H_2SO_4) forms propene ($H_2C{=}CHCH_3$) in an E1 reaction, as shown in Figure 21.4. This type of reaction, which leads to the loss of water from a molecule, is called a **dehydration** reaction. The acid acts as a catalyst and all of the steps in this E1 reaction are reversible (the reverse reaction, namely addition of H_2O to an alkene, in a hydration reaction, is discussed in Section 21.3 on p.980). You can force the equilibrium to favour the alkene by distilling the reaction mixture—the alkene has a lower boiling point than the alcohol so it is selectively removed from the reaction mixture, displacing the position of equilibrium.

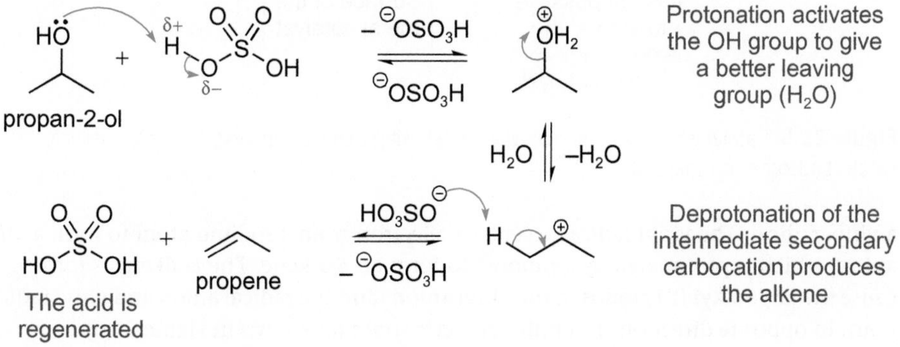
propan-2-ol
$-^{\ominus}OSO_3H$
$^{\ominus}OSO_3H$
Protonation activates the OH group to give a better leaving group (H_2O)
H_2O
$-H_2O$
$HO_3SO^{\ominus}$
$-^{\ominus}OSO_3H$
Deprotonation of the intermediate secondary carbocation produces the alkene
The acid is regenerated
propene

Figure 21.4 Formation of an alkene from an alcohol in an E1 reaction.

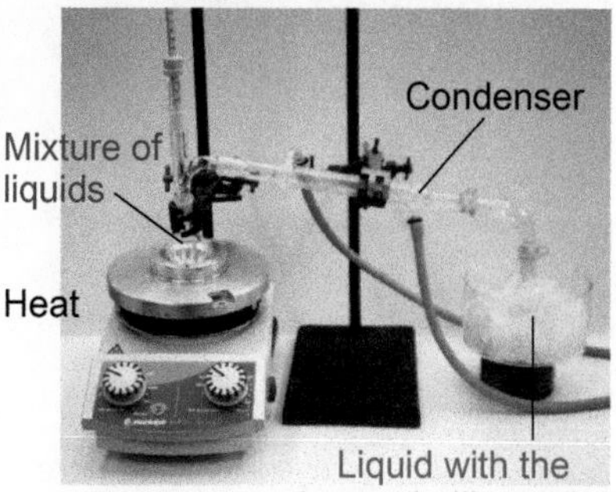
Thermometer
Condenser
Mixture of liquids
Heat
Liquid with the lowest boiling point

Liquids with significantly different boiling points can be separated by distillation. The mixture is heated so that the liquid with the lowest boiling point is converted into a gas. The gas separates from the liquid mixture, is cooled, and condenses back into liquid form (see Section 17.4, p.805).

Alcohols (ROH) have much higher boiling points than alkenes (of comparable molecular mass) because alcohols form intermolecular hydrogen bonds. Hydrogen bonds hold the alcohol molecules together and so extra energy is required to convert the liquid into a gas (Section 1.8, p.54).

By partial reduction of alkynes using a Lindlar catalyst

Alkenes are formed by partial reduction of the C≡C bond of alkynes using hydrogen in the presence of a Lindlar catalyst ($Pd/CaCO_3/PbO$). In the Lindlar catalyst, $CaCO_3$ and PbO are used to reduce the catalytic activity of Pd (it is 'poisoned') and further reduction of the alkene bond is slow. If a platinum (or palladium) on carbon catalyst is used, addition of hydrogen to both C≡C and C=C bonds is fast and the alkyne is converted into an alkane. Addition of hydrogen to an alkene or alkyne is called **hydrogenation**. The catalytic hydrogenation of alkynes is summarized in Figure 21.5.

Notice that the hydrogen atoms add to the same face of the C≡C or C=C bond which is described as a **syn addition** reaction (see Section 21.3, p.975). The hydrogen molecules and the C≡C or C=C bond come together on the surface of the metal catalyst and this leads to the transfer of hydrogen atoms to the same face of the C≡C or C=C bond. In the absence of the metal catalyst nothing happens (at room temperature and pressure) because of the high activation energy of the hydrogenation—the catalyst provides an alternative pathway with a lower activation energy by breaking the strong H–H bond (+436 kJ mol^{-1}).

The Lindlar catalyst, introduced in 1952, is named after Dr Herbert H. M. Lindlar who worked at the chemical company Hoffman La Roche in Switzerland.

Visit the Online Resource Centre to view screencast 21.1 which walks you through some selective syntheses, which involve the partial reduction of alkynes.

By partial reduction of alkynes using sodium in liquid ammonia

Whereas alkynes react with hydrogen and Lindlar catalyst to form *Z*-disubstituted alkenes, alkynes react with Na/NH_3 to form *E*-disubstituted alkenes. When sodium metal dissolves in liquid ammonia at low temperature, the sodium atoms lose electrons to form Na^+ ions (Section 26.5, p.1182). The electrons add to the C≡C bonds to form radical anions. As the name suggests, a radical anion has both a negative charge and an unpaired electron.

The mechanism of the reduction is shown in Figure 21.6. Addition of an electron to a disubstituted alkyne forms a radical anion, which is then protonated by ammonia to form

Addition of H_2 in the presence of a metal catalyst is called **catalytic hydrogenation**—this is an example of a reduction reaction (Appendix 4, p.1338).

Catalytic hydrogenation of unsaturated fatty acids in vegetable oils produces solid fats (Box 2.4, p.86).

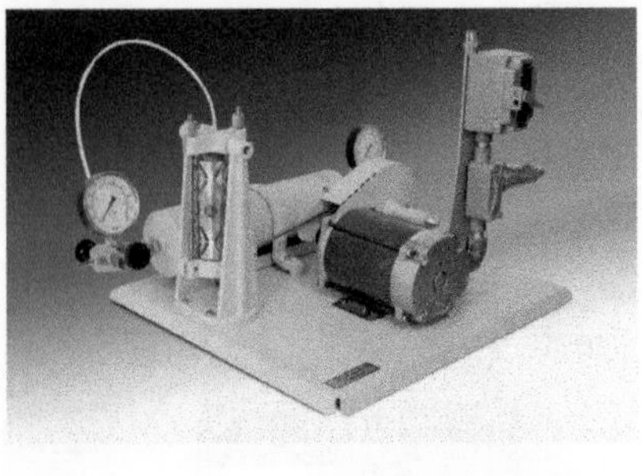

Some catalytic hydrogenations only take place at a reasonable rate using a high pressure of hydrogen gas. These reactions are usually carried out in a sealed reaction bottle in a Parr hydrogenator (above).

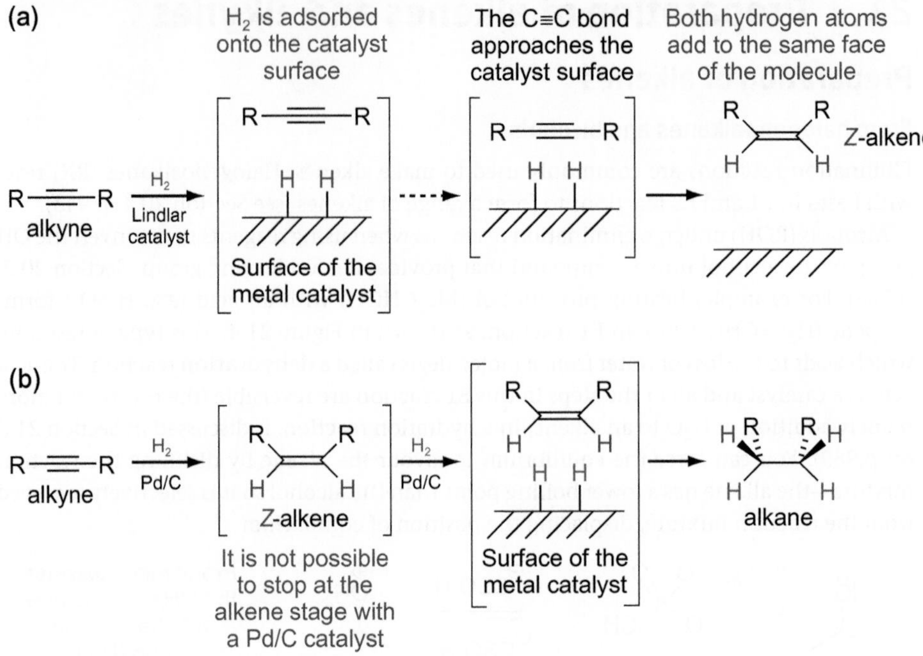
(a)
H_2 is adsorbed onto the catalyst surface
The C≡C bond approaches the catalyst surface
Both hydrogen atoms add to the same face of the molecule
R≡R
alkyne
H_2
Lindlar catalyst
Surface of the metal catalyst
Z-alkene
(b)
Pd/C
Z-alkene
It is not possible to stop at the alkene stage with a Pd/C catalyst
Surface of the metal catalyst
alkane

Figure 21.5 Catalytic hydrogenation of alkynes: (a) using a Lindlar catalyst; (b) using a palladium catalyst adsorbed on charcoal.

a vinyl radical. The vinyl radical accepts an electron from a sodium atom to form a vinyl anion, which is protonated by ammonia to form an *E*-alkene. The *E*-alkene is formed because the large alkyl (R) groups in the vinyl anion (and the radical anion and vinyl radical) point in opposite directions to minimize steric strain (as shown in Figure 21.6).

Sodium in liquid ammonia is a powerful reducing agent, which gives electrons to organic compounds in reactions called **dissolving metal reductions.**

Ammonia acts as an acid in this reaction because the radical anion and the vinyl anion are exceptionally strong bases (pK_a (NH_3) ~38 and pK_a ($H_2C{=}CH_2$) ~44; Section 19.2, p.896).

Adding one electron to a neutral organic molecule (RH) forms a radical anion ($RH^{\bullet-}$). Removing one electron from a neutral organic molecule forms a radical cation ($RH^{\bullet+}$). Radical cations are formed in a mass spectrometer (Section 12.1, p.559).

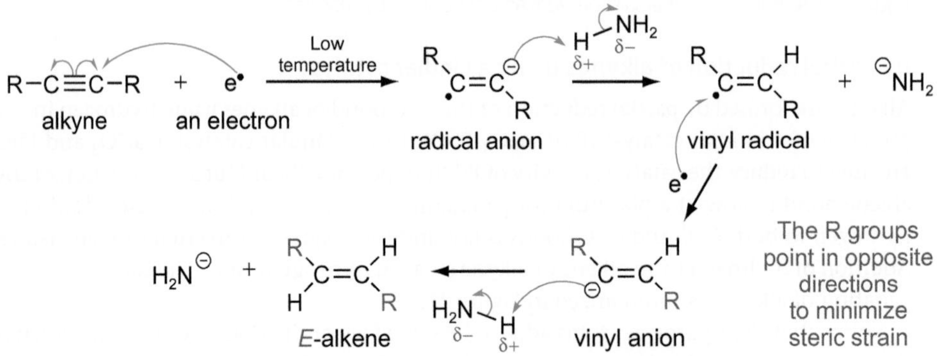
alkyne
an electron
Low temperature
radical anion
vinyl radical
vinyl anion
E-alkene
The R groups point in opposite directions to minimize steric strain

Figure 21.6 Reduction of an alkyne using sodium in liquid ammonia.

By a Wittig reaction

Alkenes are also prepared from the reaction of a phosphorane (phosphonium ylide; $Ph_3P{=}CR_2$, $Ph_3P{=}CHR$, or $Ph_3P{=}CH_2$) with an aldehyde (RCHO) or ketone (RCOR) in a **Wittig reaction**, as shown in Figure 21.7. The overall effect of the Wittig reaction is to replace the oxygen atom in the aldehyde or ketone with CR_2, CHR, or CH_2—it is perhaps the most general and versatile method for making C=C bonds. Wittig reactions are discussed in Section 23.2 (p.1071).

The synthesis of the alkene (in Figure 21.7, the alkene is methylenecyclohexane) is driven forward by the formation of the P=O bond in triphenylphosphine oxide, which is extremely strong (around +575 kJ mol^{-1}).

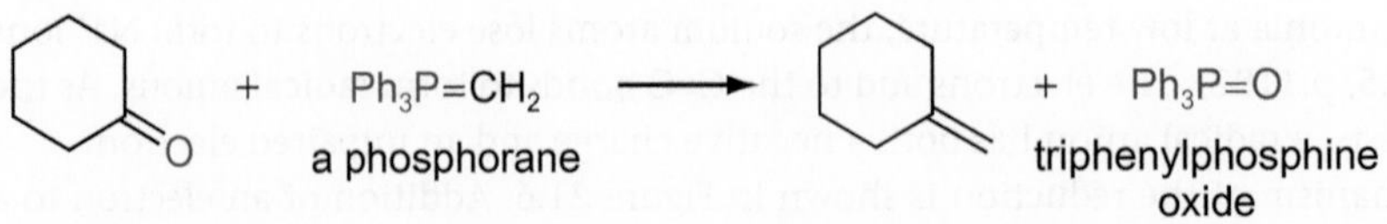
$Ph_3P{=}CH_2$
a phosphorane
$Ph_3P{=}O$
triphenylphosphine oxide

Figure 21.7 A Wittig reaction.

Preparation of alkynes

Alkynes are usually prepared by treating 1,2-dibromoalkanes (**vicinal** dibromides, RCH(Br)–CH_2Br) with two equivalents of a strong base, such as sodium amide (sodamide, $NaNH_2$). The mechanisms of these reactions can involve two consecutive E2 reactions that lead to the elimination of two molecules of HBr from the 1,2-dibromoalkane.

Sodium amide (sodamide, $NaNH_2$) can be prepared by reacting sodium with liquid ammonia, as described in Section 26.5 (p.1182). The $^-NH_2$ ion is a strong base—the conjugate acid, NH_3 (ammonia), has a high pK_a value of 38 (Table 19.1, p.896).

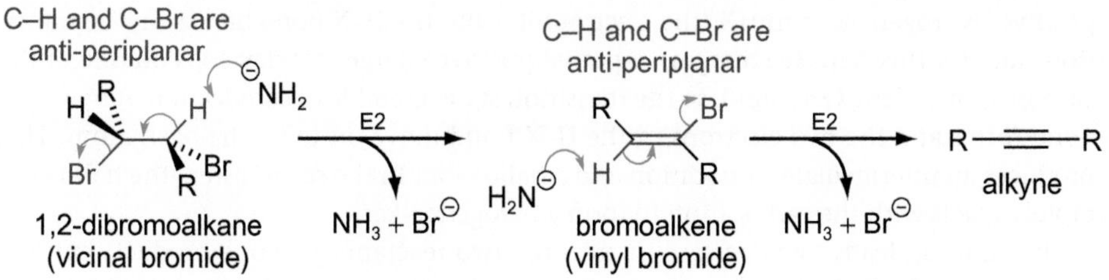
C–H and C–Br are anti-periplanar
NH2
E2
NH3 + Br
1,2-dibromoalkane (vicinal bromide)
C–H and C–Br are anti-periplanar
H2N
bromoalkene (vinyl bromide)
E2
NH3 + Br
R—≡—R
alkyne

Two **equivalents** is a short way of saying 'two stoichiometric equivalents', that is, twice as many moles of base as there are of the 1,2-dibromoalkane.

1,2-Dibromoalkanes are formed by addition of Br_2 to the C=C bond of an alkene (Section 21.3, p.975), so bromination followed by consecutive elimination of two molecules of HBr is a simple way to convert alkenes into alkynes.

Elimination of HBr from a bromoalkene of type RCH=C(Br)–CH_2R could form an allene (Section 2.5, p.87), RCH=C=CHR. However, an alkyne (RC≡C–CH_2R) is normally formed, as this is more stable than the allene.

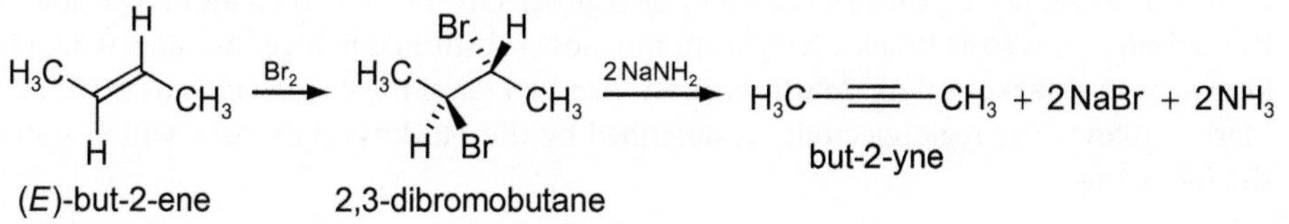
Br2
2 NaNH2
H3C—≡—CH3 + 2 NaBr + 2 NH3
(E)-but-2-ene
2,3-dibromobutane
but-2-yne

Notice that the elimination of two molecules of HBr from 2,3-dibromobutane forms an alkyne, rather than a 1,3-diene (H_2C=CH–CH=CH_2).

Substituted alkynes are also formed from the reaction of alkynyl anions with halogenoalkanes (RX) in nucleophilic substitution reactions. Alkynyl anions are strong nucleophiles that react with various primary halogenoalkanes (RCH_2X) in S_N2 reactions to form a new carbon–carbon bond.

Deprotonation of terminal alkynes (RC≡CH) to form alkynyl anions (RC≡C⁻) is introduced in Section 19.2 (p.890).

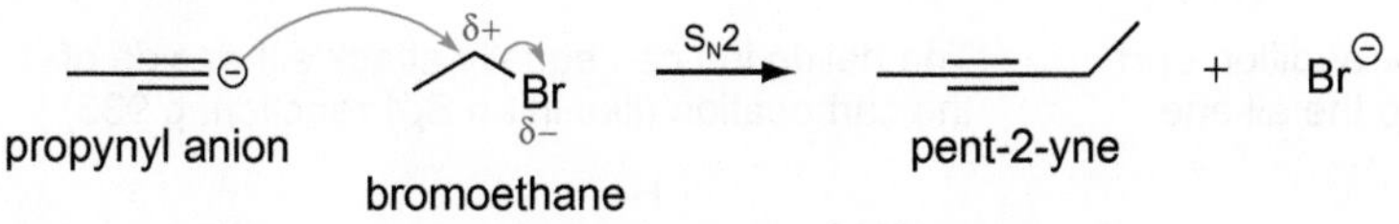
propynyl anion
δ+
Br
δ−
bromoethane
SN2
pent-2-yne
+ Br

S_N2 reactions are introduced in Section 20.3 (p.930).

Summary

- Alkenes are prepared from halogenoalkanes, alcohols, alkynes, and aldehydes/ketones.

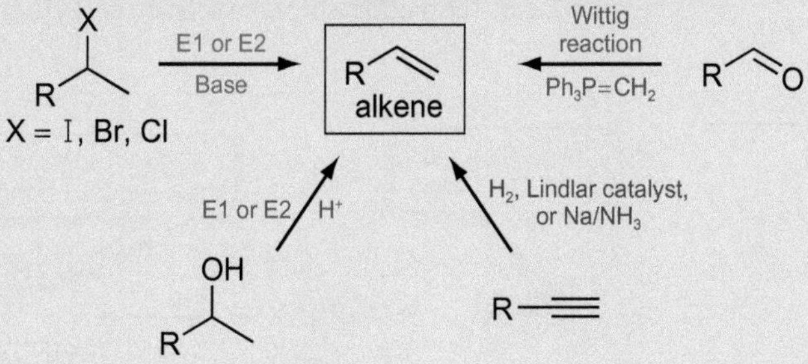
X = I, Br, Cl
E1 or E2
Base
alkene
Wittig reaction
Ph3P=CH2
E1 or E2
H+
OH
H2, Lindlar catalyst, or Na/NH3

- Alkynes are prepared from 1,2-dibromoalkanes.
- Substituted alkynes are prepared by alkylation of alkynyl anions with halogenoalkanes.

21.3 Electrophilic addition reactions of alkenes

Addition of hydrogen halides

Alkenes react with HX (where X = Cl, Br, or I) in a two-step reaction to produce halogenoalkanes. The mechanism is shown in Figure 21.8. In the first and slower step of the reaction, the two electrons in the C=C π bond (the nucleophile) are attracted towards the partially positive hydrogen atom in HX (the electrophile) and the H–X bond breaks. In the transition state for this step (see margin), a partial positive charge (δ+) develops on one of the carbon atoms of the C=C bond, so the transition state resembles a carbocation. A new C–H bond forms and the two electrons in the H–X bond move on to the halogen atom. This produces an intermediate carbocation and a halide ion. In the second step, the halide ion rapidly reacts with the carbocation to form a halogenoalkane.

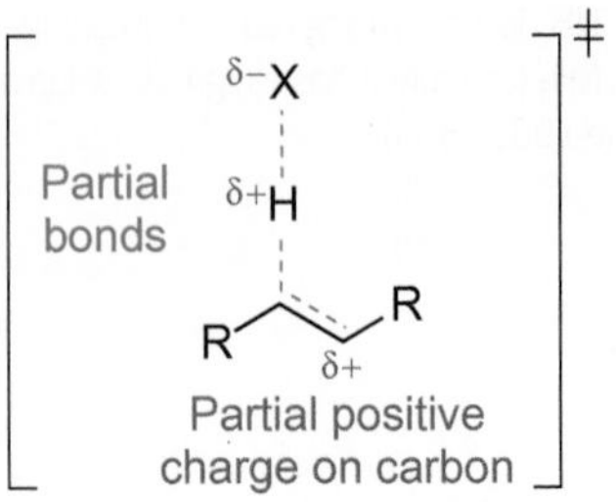
δ−X
δ+H
Partial bonds
R
R
δ+
Partial positive charge on carbon
‡

The transition state for addition of HX to an alkene resembles a carbocation.

The addition leads to a decrease in entropy (two reactants give one product) but it is thermodynamically favoured because of the enthalpy change, which has a large and negative value. Additions are exothermic because the C–H and C–X bonds that are formed are much stronger than the H–X σ bond and C=C π bond that are broken. Figure 21.9 shows the mechanism of the reaction in terms of orbital overlap.

For a spontaneous reaction, the Gibbs energy change for the reaction, ΔG, is < 0, where $\Delta G = \Delta H - T\Delta S$; see Section 14.5 (p.675).

When HX adds to **unsymmetrical alkenes** (that have different groups on each carbon atom in the C=C bond), the reactions may be regioselective. The H and X atoms can add to the carbon atoms so as to selectively form the more substituted halogenoalkane, which is known as the **Markovnikov (Markovnikoff) product** (after the Russian chemist Vladimir Markovnikov). The regioselectivity is described by the **Markovnikov rule**, which states the following.

On addition of HX to an alkene, H attaches to the carbon with fewest alkyl groups and X attaches to the carbon with most alkyl groups.

The selective formation of the most substituted halogenoalkane is explained by the stabilities of the two intermediate carbocations. In the example shown in Figure 21.10,

A **regioselective reaction** is a reaction that leads to the selective formation of one structural isomer (i.e., the reaction may give two or more structural isomers as products, but one is the major product); see Section 19.3 (p.909).

The order of reactivity of addition of hydrogen halides to C=C bonds is usually: HI > HBr > HCl > HF.

The greater the number of (electron-donating) alkyl groups bonded to a positively charged carbon, the more stable the carbocation is (Section 19.1, p.871).

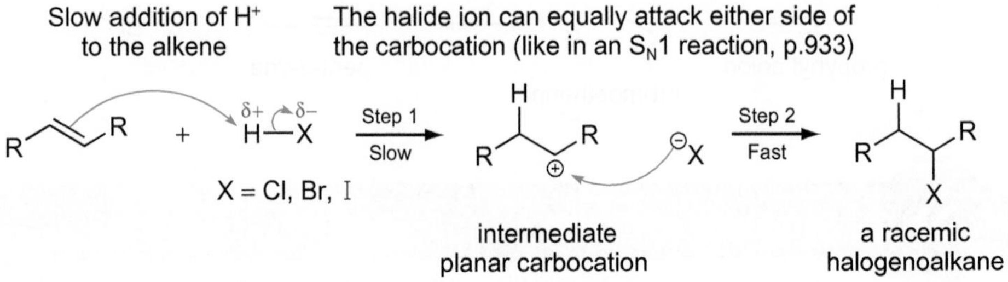
Slow addition of H⁺ to the alkene
The halide ion can equally attack either side of the carbocation (like in an S_N1 reaction, p.933)
R
R
+
δ+ δ−
H–X
X = Cl, Br, I
Step 1
Slow
H
R
R
X
Step 2
Fast
H
R
R
X
intermediate planar carbocation
a racemic halogenoalkane

Figure 21.8 The mechanism of the addition of HX to an alkene.

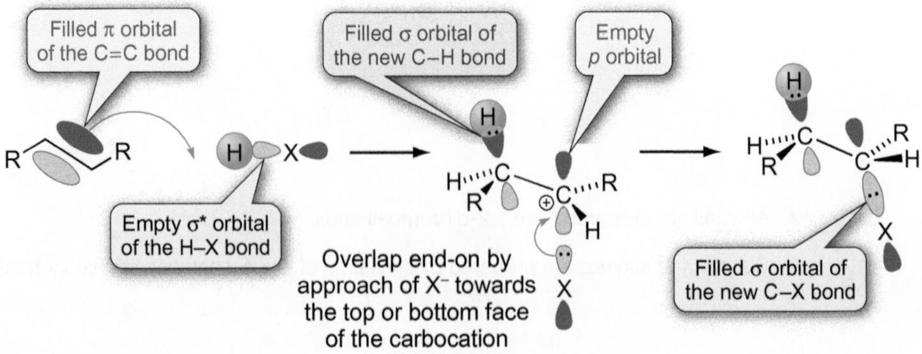
Filled π orbital of the C=C bond
Empty σ* orbital of the H–X bond
Filled σ orbital of the new C–H bond
Empty p orbital
Overlap end-on by approach of X⁻ towards the top or bottom face of the carbocation
Filled σ orbital of the new C–X bond

Figure 21.9 Addition of HX to an alkene showing orbital overlap.

Interaction between the π orbital of the alkene (the HOMO) and the σ* orbital of HX (the LUMO) leads to a bonding interaction. Orbital overlap in reaction mechanisms is introduced in Section 19.1 (p.885).

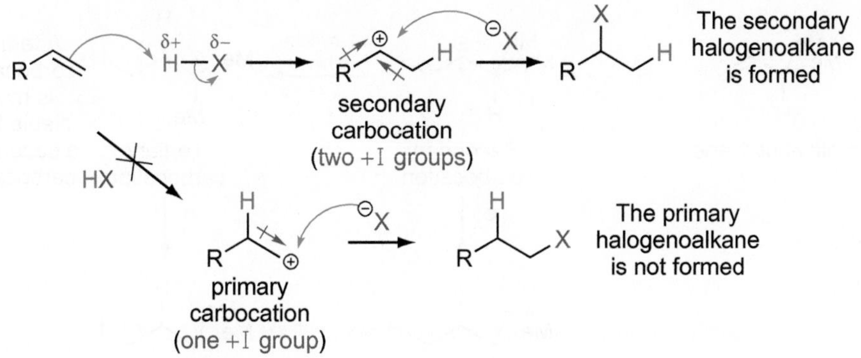
δ+ δ−
H—X
HX
secondary
carbocation
(two +I groups)
The secondary
halogenoalkane
is formed
primary
carbocation
(one +I group)
The primary
halogenoalkane
is not formed

Figure 21.10 Regioselective addition of HX to a terminal alkene.

Alternatively, for the first step of the mechanism, you could show the C=C bond reacting with H^+ (rather than HX).

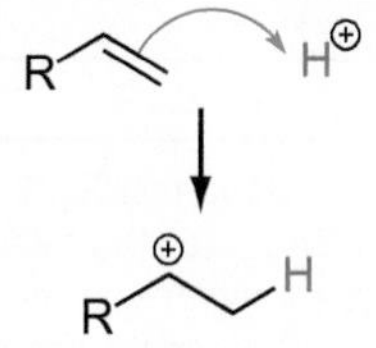

addition of H^+ to a terminal alkene ($RCH=CH_2$) can form either a secondary carbocation ($RCH^+–CH_2H$) or a primary carbocation ($RCHH–CH_2^+$). The secondary carbocation is selectively formed because this is more stable than the primary carbocation—two electron-donating (+I) alkyl groups stabilize a secondary carbocation, whereas only one alkyl group stabilizes a primary carbocation. Primary carbocations are so unstable that they are rarely formed so addition of HX to a terminal alkene only produces the secondary halogenoalkane.

Reaction profiles for the addition of HX to the terminal alkene are shown in Figure 21.11. The transition states resemble carbocations, so the reaction to form the more stable carbocation has the lower Gibbs energy of activation.

Reaction profiles are sometimes called Gibbs energy profiles. Their use is explained in Section 9.8 (p.435).

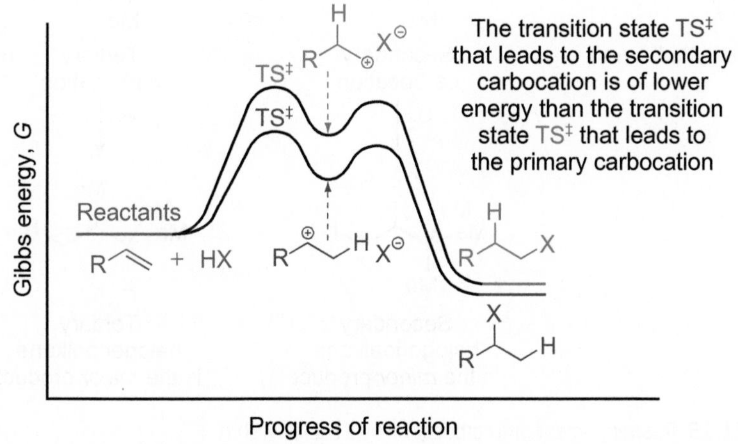
Gibbs energy, G
Reactants
+ HX
TS‡
TS‡
The transition state TS‡
that leads to the secondary
carbocation is of lower
energy than the transition
state TS‡ that leads to
the primary carbocation
Progress of reaction

Figure 21.11 Reaction profiles for the addition of HX to a terminal alkene to form a primary halogenoalkane and a secondary halogenoalkane.

Rearrangement of intermediate carbocations

Sometimes, the intermediate carbocations formed in the addition reactions undergo rearrangements to form more stable carbocations. The rearrangements involve the movement of a hydrogen atom or an alkyl group.

When a hydrogen atom moves from a carbon atom next to the carbocation to the positively charged carbon, this is called a **1,2-hydride shift** (see Figure 21.12). The term hydride is used because hydrogen moves with a pair of electrons (it is H^- that moves) and 1,2- is used to show that H^- moves to an adjacent carbon atom.

Interactive 3D animation of electrophilic addition of HBr to an alkene.

A **rearrangement** changes the way that the atoms are connected.

The use of 1,2- indicates that the hydride ion moves to an adjacent carbon. It is not related to the numbering of the carbon chain.

A tertiary carbocation (R_3C^+) is more stable than a secondary carbocation (R_2CH^+) because a tertiary carbocation contains a greater number of electron-donating alkyl (R) groups to stabilize the positive charge (Section 19.1, p.871).

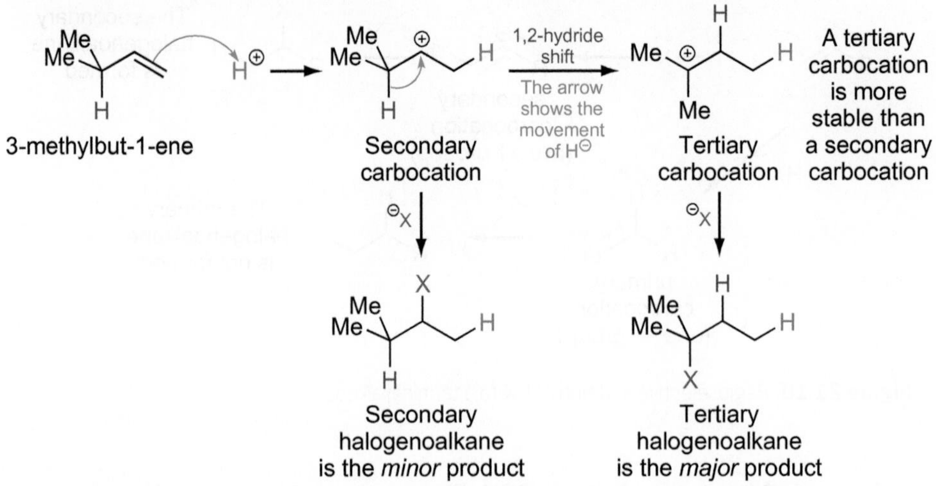
3-methylbut-1-ene
Secondary carbocation
1,2-hydride shift
The arrow shows the movement of H⊖
Tertiary carbocation
A tertiary carbocation is more stable than a secondary carbocation
Secondary halogenoalkane is the *minor* product
Tertiary halogenoalkane is the *major* product

Figure 21.12 Rearrangement of a carbocation by a 1,2-hydride shift.

If a reaction involves an intermediate carbocation, always check to see if the carbocation can rearrange to form a more stable carbocation.

Worked example 21.1 examines how vinylcyclohexane reacts with HBr to give two addition products, one being derived from a 1,2-hydride shift.

When an alkyl group moves from a carbon atom next to the carbocation to the positively charged carbon, this is called a **1,2-alkyl shift (Wagner–Meerwein rearrangement)** (see Figure 21.13). The alkyl group moves with a pair of electrons (it is R^- that moves, e.g. Me^-) to form a more stable carbocation.

The Wagner–Meerwein rearrangement is named after the Russian chemist Egor Vagner (1849–1903), also known as Georg Wagner, and the German chemist Hans Meerwein (1879–1965).

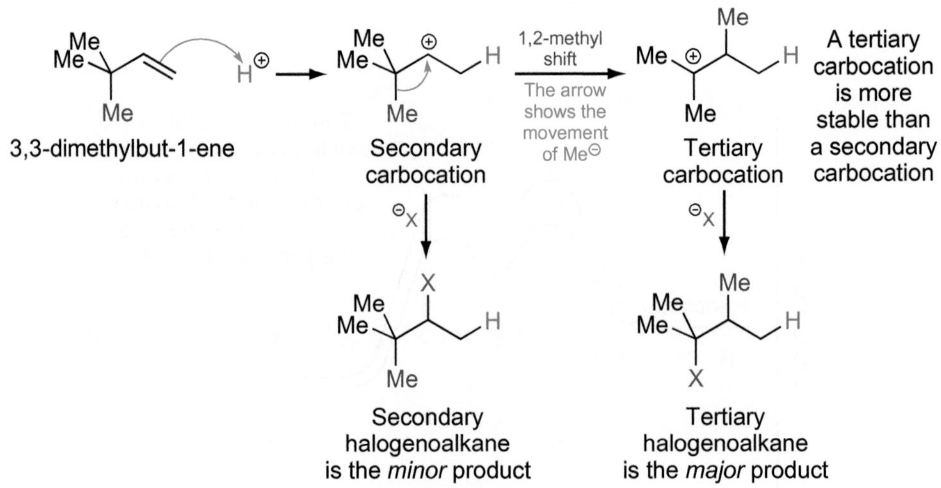
3,3-dimethylbut-1-ene
Secondary carbocation
1,2-methyl shift
The arrow shows the movement of Me⊖
Tertiary carbocation
A tertiary carbocation is more stable than a secondary carbocation
Secondary halogenoalkane is the *minor* product
Tertiary halogenoalkane is the *major* product

Figure 21.13 Rearrangement of a carbocation by a 1,2-alkyl shift.

Visit the Online Resource Centre to view screencast 21.2 which walks you through examples of carbocation rearrangements, including those involved in the biosynthesis of terpenes. (Terpenes are a large class of natural compounds, found in plants, which are derived from a precursor containing five carbon atoms, called isoprene, or 2-methylbuta-1,3-diene. Linking together isoprene units forms terpenes with 10, 15, 20, etc. carbon atoms.)

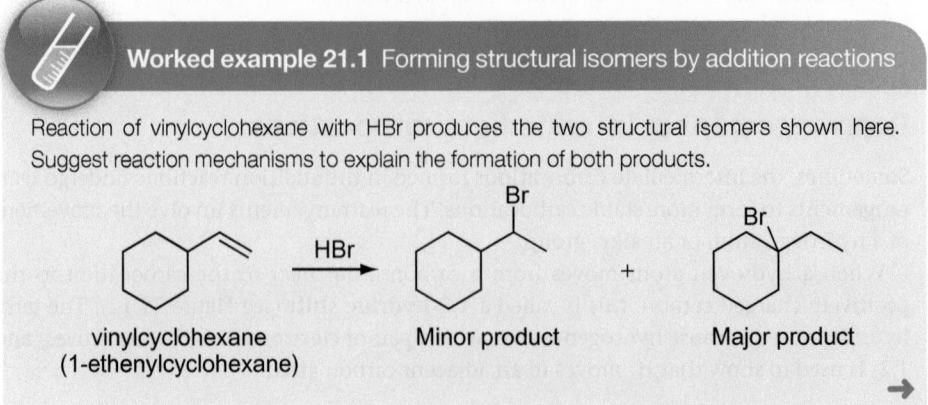

Worked example 21.1 Forming structural isomers by addition reactions

Reaction of vinylcyclohexane with HBr produces the two structural isomers shown here. Suggest reaction mechanisms to explain the formation of both products.

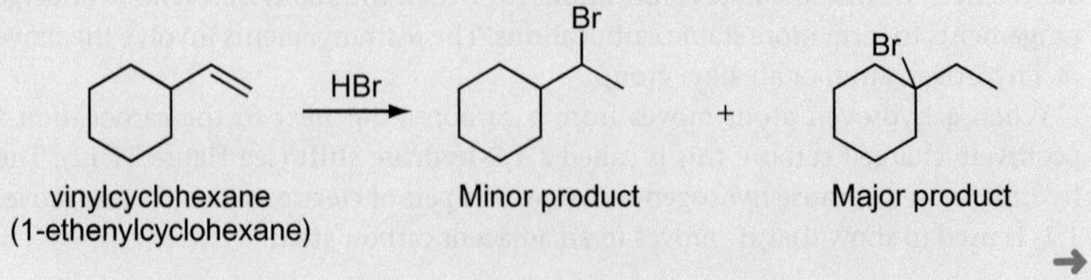
HBr
Br
Br
+
vinylcyclohexane
(1-ethenylcyclohexane)
Minor product
Major product

→

→ Strategy

Identify the nucleophilic site, the electrophilic site, and the leaving group.

Consider the stability of the two possible carbocations formed by addition of H^+ to the terminal C=C bond.

Check if the carbocations can rearrange to form more stable carbocations.

Solution

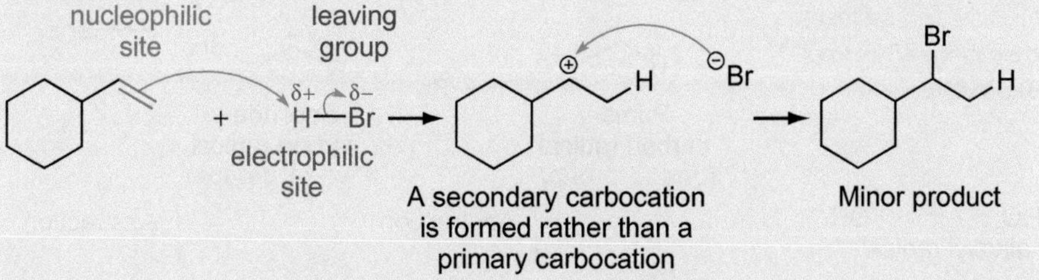
nucleophilic site
leaving group
δ+ δ−
H—Br
electrophilic site
Br
H
A secondary carbocation is formed rather than a primary carbocation
Minor product

The minor product ((1-bromoethyl)cyclohexane) is formed by Markovnikov addition.

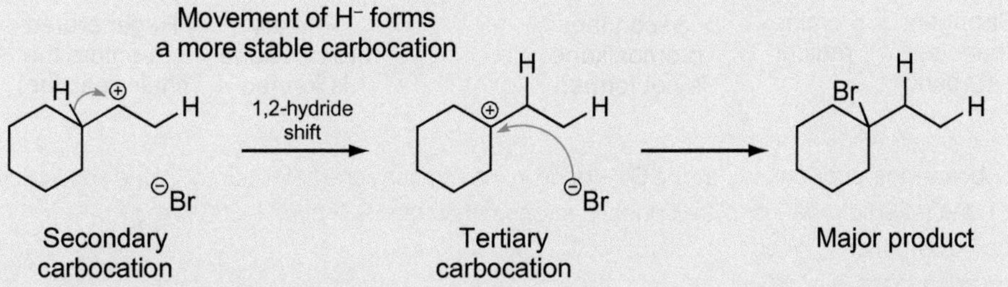
Movement of H⁻ forms a more stable carbocation
1,2-hydride shift
Br
H
Secondary carbocation
Tertiary carbocation
Major product

The major product (1-bromo-1-ethylcyclohexane) is formed by carbocation rearrangement.

Question

Addition of HCl to the C=C bond of ethylidenecyclohexane forms products **A** and **B** in unequal amounts. Neither product arises from a rearrangement of a carbocation intermediate. Suggest structures for **A** and **B** and explain why **A** is formed in higher yield than **B**.

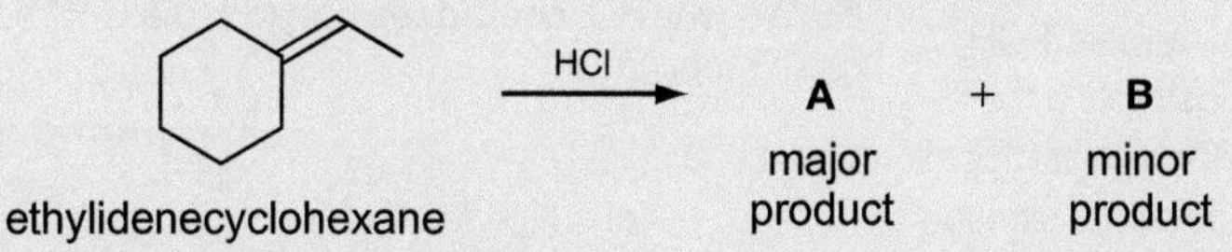
HCl
A + B
major product
minor product
ethylidenecyclohexane

(Ethylidene is the name for the $CH_3CH=$ group)

The concentrations of the carbocations in these reactions are extremely low and so their existence cannot be detected by common spectroscopic techniques. However, in certain conditions, a secondary (R_2CH^+) or tertiary (R_3C^+) carbocation can be stabilized and identified experimentally. For example, in antimony pentafluoride ($SbCl_5$), the C–Cl bond in 2-chloropropane ($CH_3CHClCH_3$) undergoes heterolysis, to form the salt $(CH_3)_2CH^+\ SbClF_5^-$ (Box 19.3, p.868). As $SbClF_5^-$ is such a weak base and a weak nucleophile, the concentration of $(CH_3)_2CH^+$ can be increased to a point where the carbocation can be detected by ^1H NMR spectroscopy. The ^1H NMR spectrum shows resonance signals at 13.5 ppm (for the CH group) and 5.1 ppm (for the two CH_3 groups)—these high chemical shift values are due to the strong deshielding effect of the positive charge (Section 12.3, p.581).

X
X

Remember that carbocations are generally planar and so the halide ion can equally attack either side.

Box 21.2 Adding HBr to alkenes in the presence of peroxides

When HBr adds to C=C bonds by a polar mechanism, the Markovnikov product is selectively formed. However, when a mixture of HBr, an alkene, and a peroxide (ROOR) is heated or irradiated with UV radiation (conditions that favour the formation of radicals), the regioselectivity of the addition reaction changes and the less substituted bromoalkane is selectively formed—this is called the **anti-Markovnikov product.** The change in regioselectivity, in the presence of a peroxide, is called the **peroxide effect.**

The following reaction scheme shows the addition of HBr to but-1-ene ($CH_3CH_2CH=CH_2$) by a polar mechanism and by a radical mechanism.

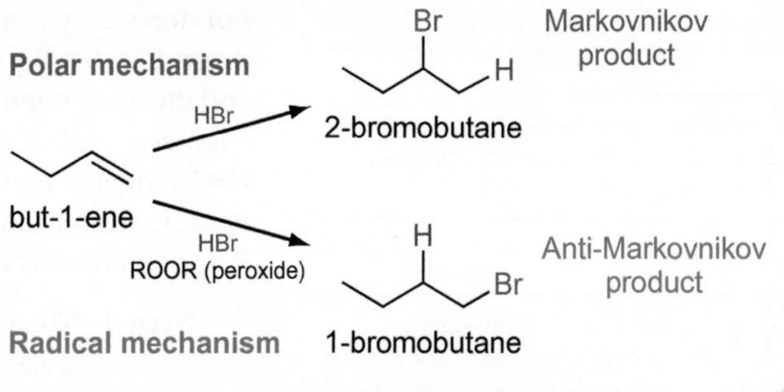
Polar mechanism
Br
H
Markovnikov product
HBr
2-bromobutane
but-1-ene
HBr
ROOR (peroxide)
H
Br
Anti-Markovnikov product
Radical mechanism
1-bromobutane

→

➜ In the presence of a peroxide, the regioselectivity of the addition changes because the reaction mechanism involves radical intermediates, not charged intermediates, and a rapid chain reaction (Section 20.2, p.921). On heating or irradiating the reaction mixture with UV radiation the weak O–O bond of the peroxide (RO–OR) selectively breaks to form two alkoxyl radicals (RO•) in an initiation step. An alkoxyl radical then abstracts a hydrogen atom from HBr to form an alcohol (ROH) and a bromine radical (atom)—this is an exothermic reaction because the O–H bond formed in the alcohol (~464 kJ mol^{-1}) is stronger than the H–Br bond broken (~366 kJ mol^{-1}).

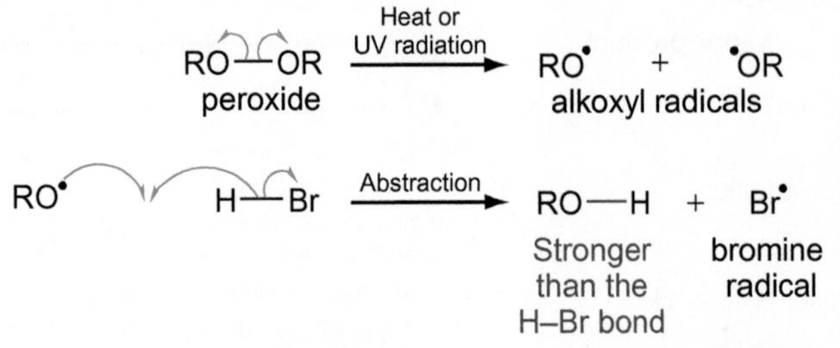

Heat or UV radiation
RO—OR peroxide
RO• + •OR alkoxyl radicals
RO• H—Br
Abstraction
RO—H + Br•
Stronger than the H–Br bond
bromine radical

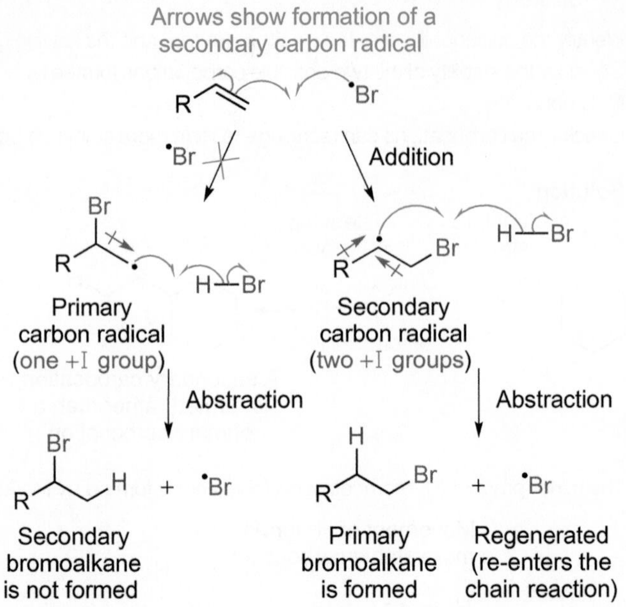

Arrows show formation of a secondary carbon radical
Addition
Primary carbon radical (one +I group)
Secondary carbon radical (two +I groups)
Abstraction
Abstraction
Secondary bromoalkane is not formed
Primary bromoalkane is formed
Regenerated (re-enters the chain reaction)

The bromine radical then adds to a C=C bond—the addition is exothermic because a slightly stronger C–Br bond (~290 kJ mol^{-1}) is formed at the expense of a weaker C=C π bond (~265 kJ mol^{-1}). When Br• reacts with a C=C bond, the Br• selectively adds to the least hindered end of the C=C bond so the addition is regioselective. This is explained by steric effects and also by the stability of the radical that is formed—Br• adds to the less hindered end of the C=C bond to form the more stable carbon-centred radical. (Br• is larger than H$^+$ so, when Br• adds to the C=C bond, the steric interactions are greater than when H$^+$ adds.) Carbon-centred radicals are stabilized by +I groups (Section 19.1, p.869) so, for example, secondary carbon-centred radicals ($R_2CH^•$) are more stable than primary carbon-centred radicals ($RCH_2^•$).

Finally, the intermediate carbon-centred radical abstracts a hydrogen atom from HBr to form the bromoalkane and Br•, which can react with another molecule of the alkene. This is an exothermic reaction as the C–H bond in the bromoalkane (~410 kJ mol^{-1} for a secondary C–H bond) is stronger than the H–Br bond (~366 kJ mol^{-1}).

Question

For hydrogen halides to add to a C=C bond in a radical reaction, both addition and abstraction steps must be exothermic. If either step is endothermic, then the propagation step is too slow for the chain reaction to proceed.

Reaction of HCl or HI with a terminal alkene, in the presence of a peroxide, does not produce the anti-Markovnikov addition products. Using the approximate bond enthalpies given below, explain why HCl or HI does not add to a C=C bond in a radical chain reaction.

H–Cl	C–Cl	H–I	C–I
431 kJ mol^{-1}	346 kJ mol^{-1}	298 kJ mol^{-1}	228 kJ mol^{-1}

ℹ Interhalogen compounds with polar covalent bonds, such as $^{\delta+}$Br–Cl$^{\delta-}$ and $^{\delta+}$I–Cl$^{\delta-}$, also add to C=C bonds. (Interhalogens are compounds formed between halogen elements, Section 27.6, p.1246.)

ℹ 1,2-Dibromoalkanes are also called **vicinal** dibromides.

ℹ Addition of Br_2 to an alkene is **stereospecific**: different stereoisomers of the alkene produce different stereoisomers of the product.

Addition of bromine

Bromine adds to the C=C bond of alkenes to form 1,2-dibromoalkanes in electrophilic addition reactions. (The use of 1,2- indicates that the Br atoms are on adjacent atoms, it is not necessarily related to the numbering of the carbon chain.) At first sight, it may seem surprising that Br_2 can act as an electrophile—the Br_2 molecule is not positively charged and the Br–Br bond is not polarized. However, the Br_2 molecule becomes polarized when it approaches a C=C bond. The electrons in the Br–Br bond are repelled by the electron-rich C=C bond and move away from the Br atom that is nearer the C=C bond. This makes the Br atom nearer the C=C bond electrophilic (δ+). The mechanism of addition of Br_2 to a C=C bond involves two steps, as shown in Figure 21.14.

Step 1 The C=C bond acts as a nucleophile and donates a pair of electrons to the polarized $^{\delta+}$Br–Br$^{\delta-}$ molecule to form a C–Br bond and Br$^-$. At the same time, the partially positive bromine atom donates a lone pair of electrons to one of the carbon atoms to form a second C–Br bond. The bromine atom

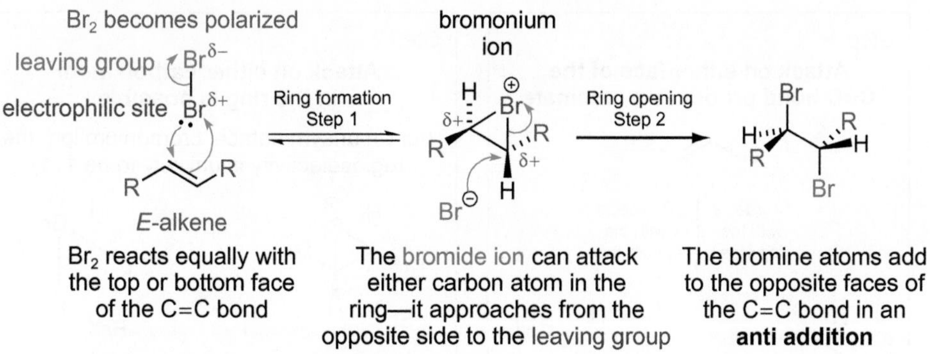

Br2 becomes polarized
leaving group
electrophilic site
E-alkene
Ring formation
Step 1
bromonium ion
Ring opening
Step 2
Br2 reacts equally with the top or bottom face of the C=C bond
The bromide ion can attack either carbon atom in the ring—it approaches from the opposite side to the leaving group
The bromine atoms add to the opposite faces of the C=C bond in an anti addition

Figure 21.14 Addition of Br_2 to an alkene.

simultaneously bonds to both carbons to make an intermediate 3-membered ring called a **bromonium ion**. Note that, in the bromonium ion, the positively charged bromine is a strong –I group so the electrons in the two C–Br bonds are attracted away from both carbon atoms. This makes the carbon atoms in the ring electrophilic (δ+). The formation of the bromonium ion is a reversible process.

ⓘ Although bromonium ions are extremely reactive, some have been detected at low temperature and the bromonium ion shown here has been isolated. The two extremely bulky adamantyl groups, attached to the 3-membered ring, hinder the approach of Br^- and this reduces the rate of ring opening.

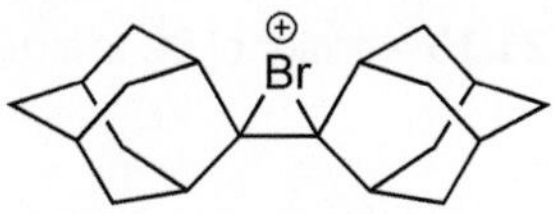

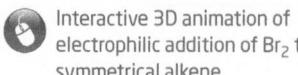

Step 2 Br^- acts as a nucleophile and rapidly reacts with the bromonium ion in an S_N2 reaction. The Br^- approaches the bromonium ion from the opposite side to the positively charged bromine. It attacks either carbon atom in the ring to form a new C–Br bond at the same time as one C–Br^+ bond is broken.

An S_N2 reaction is a concerted bimolecular nucleophilic substitution reaction, which is discussed in Section 20.3 (p.930).

Overall, two Br atoms add to the opposite faces of the C=C bond and this is described as an **anti addition** (see below).

ⓘ In an **anti addition**, two atoms or groups of atoms add to the *opposite* faces of the reactant. In a **syn addition**, two atoms or groups of atoms add to the *same* face of the reactant.

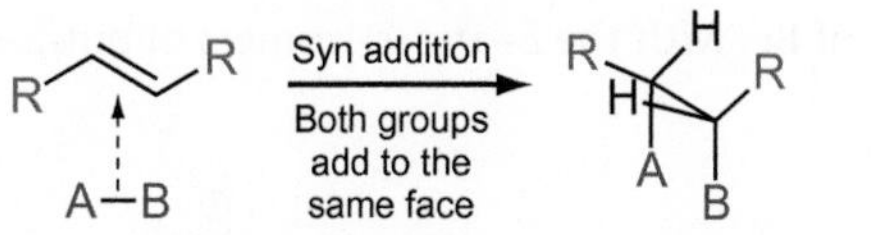

Syn addition
Both groups add to the same face

In a hydroboration reaction, H and BH_2 groups add to a C=C bond in a syn addition (p.982).

In the reaction scheme in Figure 21.14, the Br_2 molecule attacks only the top face of the C=C bond and the Br^- attacks only one carbon atom in the bromonium ion. However, the Br_2 molecule can react equally well with the bottom face of the C=C bond and Br^- can react with either carbon atom in the bromonium ion. Because of this, the bromonium ion is formed as a racemate (a 1 : 1 mixture of enantiomers), which means that the dibromide product is also a racemate. So, it is the relative configuration, *not* the absolute configuration, that is shown for both the bromonium ion and the dibromide in Figure 21.14. This is illustrated in Figure 21.15 for the addition of bromine to the unsymmetrical alkene, (*E*)-pent-2-ene, to form a racemic mixture of the enantiomers of the dibromide product.

Interactive 3D animation of electrophilic addition of Br_2 to a symmetrical alkene.

Reaction of Br^- with the bromonium ion of *R,R* configuration gives the same enantiomers, in the same ratio, as reaction of Br^- with the bromonium ion of *S,S* configuration. (Use molecular models to check this for yourself.)

The **relative** configuration shows the configuration of a racemate, but the **absolute** configuration shows the configuration of a single enantiomer (Section 18.4, p.854).

Interactive 3D animation of electrophilic addition of Br_2 to an unsymmetrical alkene.

Envelope and chair conformations of cycloalkanes are discussed in Section 18.2 (p.826). Whereas cyclohexane adopts a chair conformation, cyclohexene adopts a half-chair conformation. In the half-chair conformation, four of the six carbons are in the same plane and this arrangement of atoms accommodates the planar C=C bond.

Interactive 3D animation of electrophilic addition of Br_2 to cyclohexene.

From your previous studies, you may have seen a mechanism for bromination of a C=C bond that involves an intermediate carbocation and not a bromonium ion. This mechanism does not tell the complete story because Br^- could attack either face of the carbocation and this would form a mixture of anti and syn addition products.

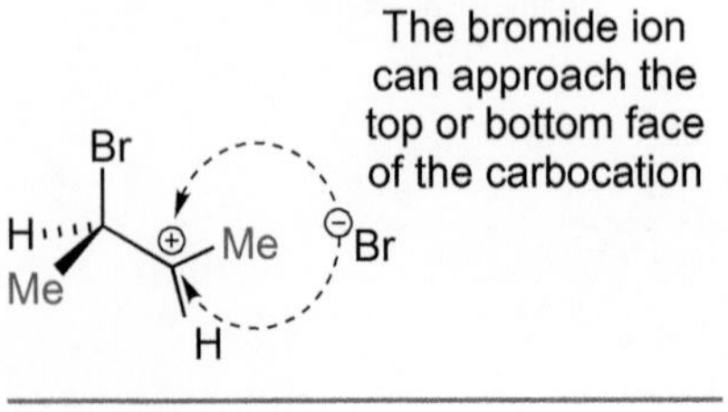
The bromide ion can approach the top or bottom face of the carbocation
Br
H
Me
Me
Br
H

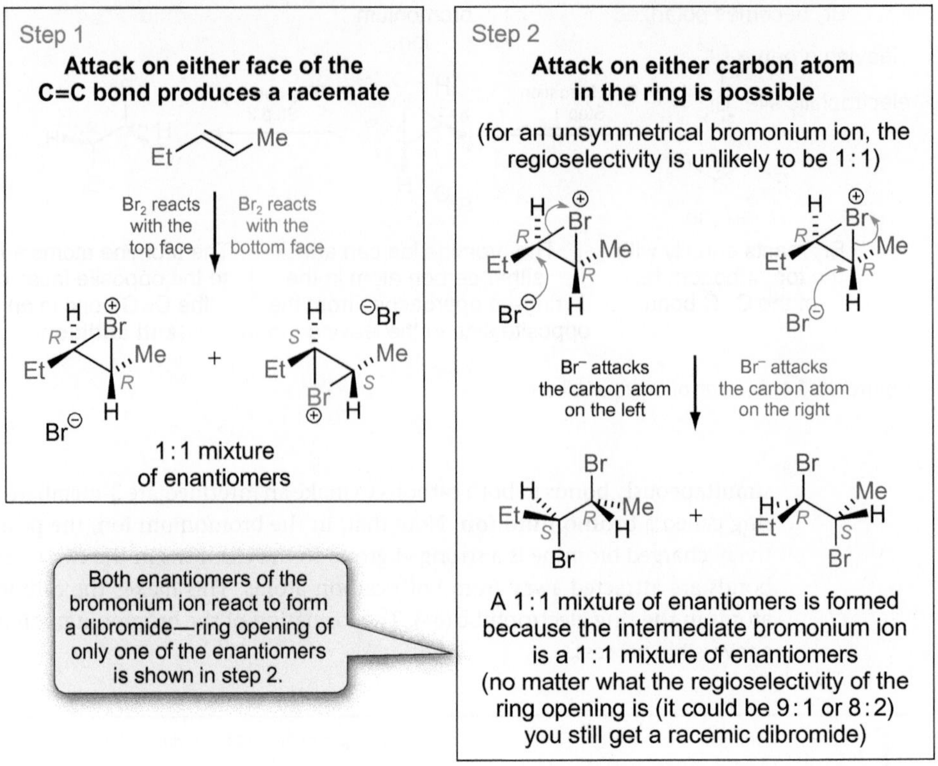
Step 1
Attack on either face of the C=C bond produces a racemate
Et
Me
Br2 reacts with the top face
Br2 reacts with the bottom face
1:1 mixture of enantiomers
Step 2
Attack on either carbon atom in the ring is possible
(for an unsymmetrical bromonium ion, the regioselectivity is unlikely to be 1:1)
Br⁻ attacks the carbon atom on the left
Br⁻ attacks the carbon atom on the right
Both enantiomers of the bromonium ion react to form a dibromide—ring opening of only one of the enantiomers is shown in step 2.
A 1:1 mixture of enantiomers is formed because the intermediate bromonium ion is a 1:1 mixture of enantiomers (no matter what the regioselectivity of the ring opening is (it could be 9:1 or 8:2) you still get a racemic dibromide)

Figure 21.15 Addition of Br_2 to an unsymmetrical alkene to form a racemic mixture.

Evidence for the mechanism for addition of bromine to a C=C bond

Strong evidence for the formation of a bromonium ion and the overall anti addition of Br_2 to a C=C bond is provided by the reaction of cyclic alkenes with bromine to produce cyclic 1,2-dibromides, as shown in Figure 21.16. Rotation about the BrC–CBr bond is not possible in these compounds, so the relative positions of the two bromine atoms in the products are fixed. For example, addition of Br_2 to cyclopentene or cyclohexene produces a *trans*-1, 2-dibromide (as a racemate), which has the Br atoms on opposite sides of the rings.

The formation of an intermediate bromonium ion explains why addition of Br_2 to a substituted alkene is stereospecific. In a stereospecific reaction, different stereoisomers of the reactant produce different stereoisomers of the product. This is shown by the reaction of Br_2 with the *E*- and *Z*-isomers of but-2-ene ($CH_3CH{=}CHCH_3$) in Figure 21.17.

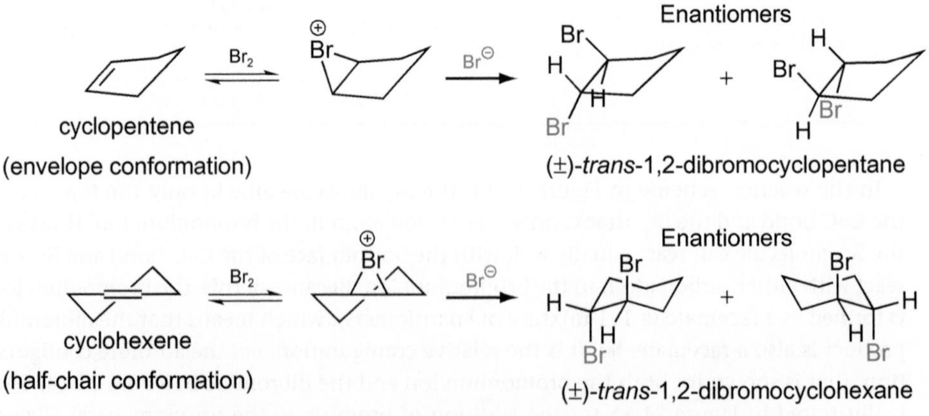
Enantiomers
Br2
Br⁻
cyclopentene
(envelope conformation)
(±)-trans-1,2-dibromocyclopentane
Enantiomers
cyclohexene
(half-chair conformation)
(±)-trans-1,2-dibromocyclohexane

Figure 21.16 Addition of Br_2 to cyclic alkenes provides evidence for the intermediate bromonium ions.

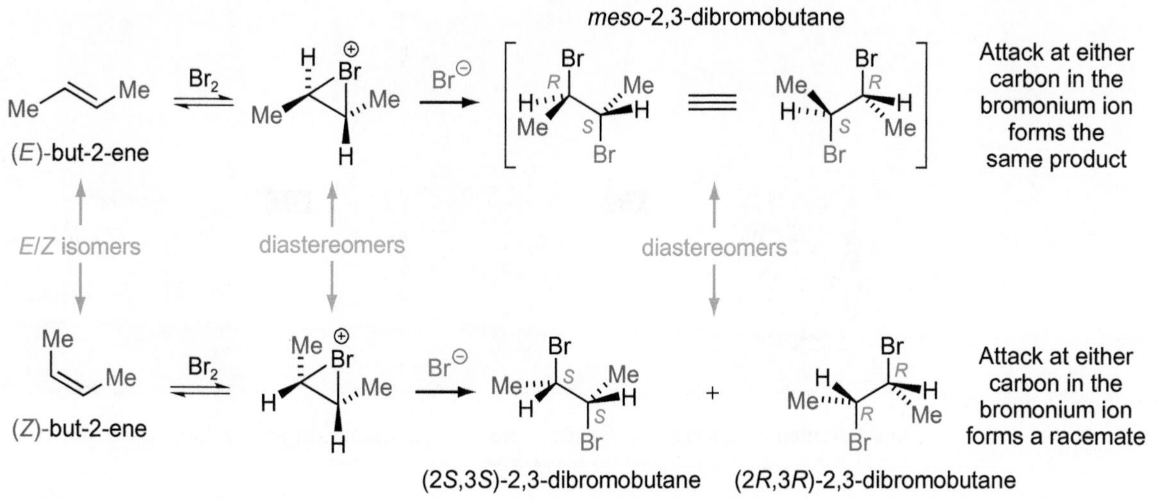
meso-2,3-dibromobutane
Attack at either carbon in the bromonium ion forms the same product
(E)-but-2-ene
E/Z isomers
diastereomers
diastereomers
(Z)-but-2-ene
Attack at either carbon in the bromonium ion forms a racemate
(2S,3S)-2,3-dibromobutane
(2R,3R)-2,3-dibromobutane

Figure 21.17 Stereospecific addition of Br_2 to an alkene.

Addition of Br_2 to the *E*-isomer forms a *meso* compound—this compound has a plane of symmetry (see margin) and so is achiral (see Section 18.4, p.851). In contrast, addition of Br_2 to the *Z*-isomer forms a racemate. Compare the structures of the products from both reactions—molecular models will help you do this. The meso compound and the racemate are diastereomers (diastereomers have a different configuration at *one* of the two chiral centres). The *E*- and *Z*-isomers of but-2-ene react to form different diastereomers of 2,3-dibromobutane ($CH_3CHBr–CHBrCH_3$), so the addition of Br_2 is stereospecific.

Bromination of an alkene is stereospecific because the bromine atoms add to the opposite faces of the C=C bond in an anti addition.

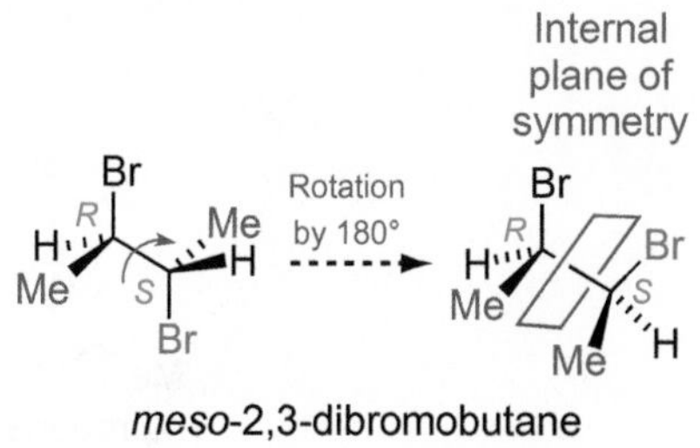
Internal plane of symmetry
Rotation by 180°
meso-2,3-dibromobutane

Use molecular models to help you understand this.

Addition of bromine in the presence of water

When water is present in the reaction of Br_2 with a C=C bond, a 1,2-bromoalcohol (RCHBr–CH(OH)R; sometimes called a **bromohydrin**), rather than a 1,2-dibromoalkane, is formed, as shown in Figure 21.18. The reaction is an electrophilic addition reaction, and the mechanism is similar to that for the addition of bromine shown in Figure 21.14, (p.974), and in Figure 21.17 above.

The first step in the reaction involves the formation of a bromonium ion (as described in the previous section). In the second step, water acts as a nucleophile (rather than Br^-) and rapidly reacts with the bromonium ion in an S_N2 reaction. The H_2O approaches the bromonium ion from the opposite side to the positively charged bromine and attacks either carbon atom in the ring to form a new C–O bond at the same time as one C–Br^+ bond is broken. Following deprotonation (in the third step), a 1,2-bromoalcohol is formed as a racemate. Note that the Br and OH groups add to the opposite sides of the C=C bond in an anti addition.

Both H_2O and Br^- are nucleophiles and if both reacted with the bromonium ion this would produce a mixture of a 1,2-bromoalcohol (e.g. RCHBr–CH(OH)R) and a 1,2-dibromide (RCHBr–CHBrR). Br^- is a stronger nucleophile than H_2O (as Br^- is negatively charged), so you might expect the 1,2-dibromide to be the major product in these reactions. However, the rate of ring opening of the bromonium ion depends on the concentration of the

The use of 1,2- indicates that the Br and OH groups are on adjacent atoms; it is not necessarily related to the numbering of the carbon chain.

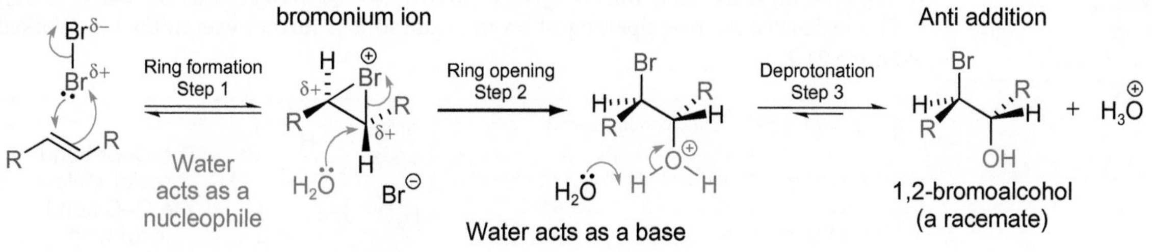
bromonium ion
Anti addition
Ring formation
Step 1
Water acts as a nucleophile
Ring opening
Step 2
Water acts as a base
Deprotonation
Step 3
1,2-bromoalcohol
(a racemate)

Figure 21.18 Addition of Br_2/H_2O to an alkene.

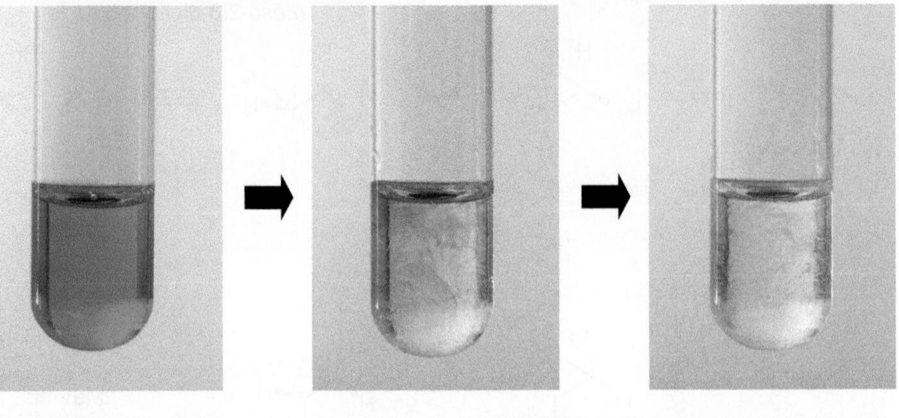

Decolorization of an orange-brown solution of bromine water is often used as a test for the presence of a C=C bond. When Br_2 adds to a C=C bond, the orange-brown colour disappears.

Br^- is a very weak base—the conjugate acid, HBr, has a very low pK_a value of −9 (Appendix 9, p.1366).

nucleophile—the higher the concentration of the nucleophile, the faster the rate of ring opening. To form the 1,2-bromoalcohol selectively, you need to ensure that the H_2O molecules far outnumber the Br^- ions so water is used in excess. Also, as there are many more H_2O molecules than Br^- ions in the reaction mixture, H_2O, rather than Br^-, is shown as the base in step 3 in Figure 21.18.

When Br_2/H_2O reacts with unsymmetrical alkenes, the reactions are regioselective. The Br and OH groups add to selectively form the 1,2-bromoalcohol with the Br atom on the *less* substituted carbon as illustrated by the reaction in Figure 21.19.

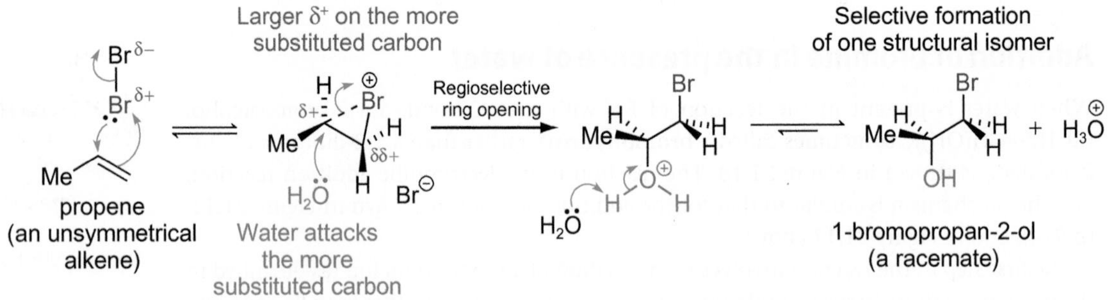
Larger δ+ on the more substituted carbon
Selective formation of one structural isomer
Regioselective ring opening
propene (an unsymmetrical alkene)
Water attacks the more substituted carbon
1-bromopropan-2-ol (a racemate)

Figure 21.19 Addition of Br_2/H_2O to an unsymmetrical alkene is regioselective.

Notice that the product, 1-bromopropan-2-ol ($CH_3CH(OH)–CH_2Br$), has one chiral centre. An equal mixture of *R* and *S* enantiomers is formed because Br_2 reacts equally with the top and bottom faces of propene to form the bromonium ion intermediate as a racemate.

The regioselectivity is explained by the structure of the bromonium ion. In the bromonium ion in Figure 21.19, the two C–Br bonds are not the same length. The more substituted carbon atom forms a longer and weaker bond to bromine because this carbon atom is better at stabilizing a partial positive charge (each carbon atom has a partial positive charge because the electrons in the C–Br^+ bond are attracted towards the positively charged bromine atom). The methyl group is an electron-donating +I group so it helps to stabilize the partial positive charge on this carbon atom. Because the more substituted carbon has the greater partial positive charge, water selectively attacks this carbon, even though it is more sterically hindered (as the methyl group, CH_3, is larger than an H atom).

The regioselective ring opening of bromonium ions is further exemplified in Worked example 21.2.

Steric hindrance occurs when the size of substituents is responsible for lowering the reactivity of a particular site within a neutral or charged molecule (Section 19.1, p.878).

In the transition state for ring opening of the bromonium ion, breaking of the C–Br bond occurs to a greater extent than formation of the new C–O bond—this is described as a **loose S_N2 transition state**.

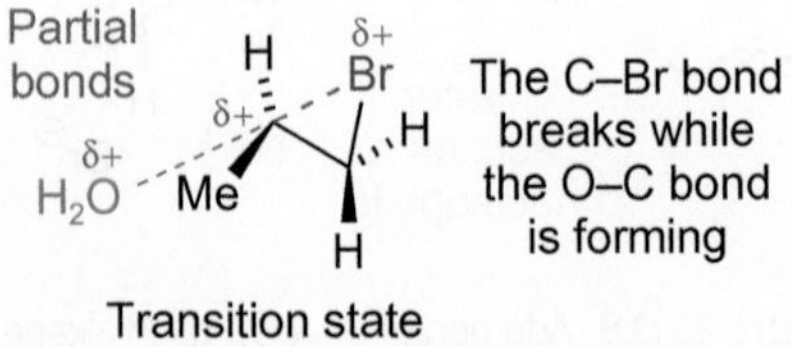
Partial bonds
The C–Br bond breaks while the O–C bond is forming
Transition state

Worked example 21.2 Opening bromonium ions

Suggest a mechanism for the reaction of 1-methylcyclohex-1-ene with bromine in water. Give the structure of the major product and explain its stereochemistry.

Strategy

Draw the structures of the reactants. The conformation of 1-methylcyclohex-1-ene does not have to be shown (although you are always encouraged to draw a structure that shows the three-dimensional shape of the molecule).

Identify the nucleophilic site, the electrophilic site, and the leaving group.

Consider the structure of the bromonium ion formed by addition of Br^+ to the C=C bond and decide whether the reaction with water is regioselective.

Draw the structure of the product, paying particular attention to stereochemistry.

Solution

Using drawings that do not show the conformation of molecules

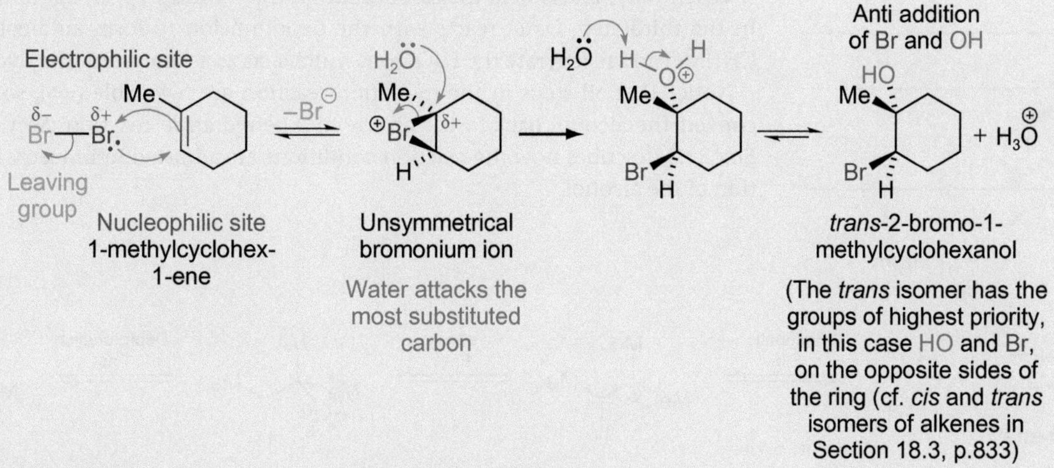
Electrophilic site
Leaving group
Nucleophilic site 1-methylcyclohex-1-ene
Unsymmetrical bromonium ion
Water attacks the most substituted carbon
Anti addition of Br and OH
trans-2-bromo-1-methylcyclohexanol
(The trans isomer has the groups of highest priority, in this case HO and Br, on the opposite sides of the ring (cf. cis and trans isomers of alkenes in Section 18.3, p.833)

Using drawings that show the conformation of molecules

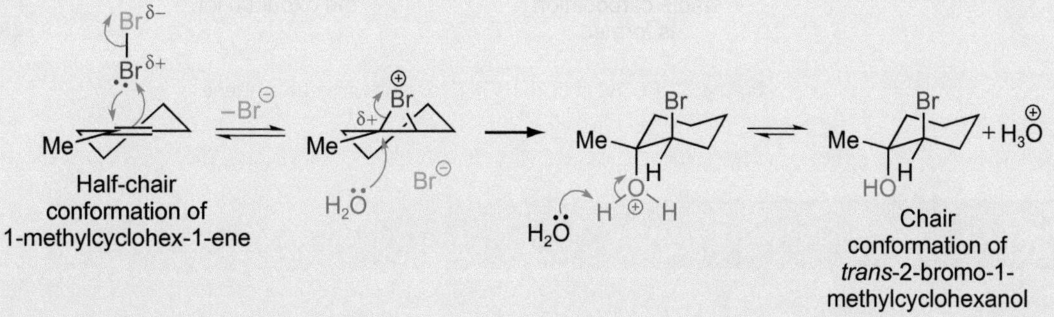
Half-chair conformation of 1-methylcyclohex-1-ene
Chair conformation of trans-2-bromo-1-methylcyclohexanol

The product is formed as a racemate because Br_2 can react equally well with either face of the C=C bond.

The two possible chair conformations of cyclohexanes interconvert by a ring-flip (Section 18.2, p.827). For a substituted cyclohexane, the most stable chair conformation usually has more of the substituents in equatorial positions.

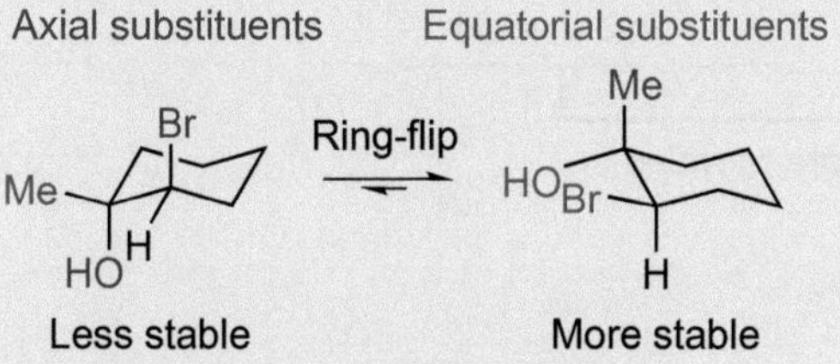
Axial substituents
Equatorial substituents
Ring-flip
Less stable
More stable

Question

Cl_2 reacts with C=C bonds to form intermediate chloronium ions that are converted into 1,2-dichlorides. Assuming that, in the presence of water, both Cl_2 and Br_2 react with C=C bonds by the same type of mechanism, suggest a mechanism for the reaction of (*Z*)-but-2-ene with Cl_2 and water, and give the structure of the major product.

Addition of water in the presence of acid

Strong acids are introduced in Section 7.2 (p.308).

Hydration of an alkene forms an alcohol. **Dehydration** of an alcohol forms an alkene.

Hydration of alkenes is one of a small number of reactions that has a theoretical atom efficiency of 100% (Box 1.4, p.25). All of the atoms in water and the alkene make their way into the alcohol. In theory, there is no chemical waste from the reaction but, in practice, side-reactions (including acid-catalysed polymerization) can occur (Box 19.8, p.904).

An **oxonium ion** is a compound containing a positively charged oxygen (e.g. R_3O^+, R_2O^+H, $R_2C{=}O^+H$).

Alkenes react with H_2O in the presence of a strong acid to form alcohols (ROH). These addition reactions are called **hydration reactions**, because water adds to the C=C bond.

Reaction of a C=C bond with water and a strong acid (e.g. aqueous H_2SO_4) starts by protonation of the C=C bond to form a carbocation. For unsymmetrical alkenes, the addition of H^+ is regioselective (this is the same as for addition of HBr to alkenes, p.970). For example, in Figure 21.20, addition of H^+ to 2-methylbut-2-ene ($Me_2C{=}CHMe$) forms a tertiary carbocation ($Me_2C^+{-}CHHMe$) rather than a less stable secondary carbocation ($Me_2CH{-}^+CHMe$).

In the second step, a nucleophile reacts with the carbocation. The nucleophile can be water or the conjugate base of the acid (e.g. HSO_4^- for sulfuric acid). The nature of the nucleophile depends on the concentration of the acid. In dilute sulfuric acid, water reacts with the carbocation because the concentration of water is much greater than the concentration of HSO_4^-. However, in concentrated sulfuric acid, HSO_4^- reacts with the carbocation to form an alkyl hydrogen sulfate, $ROSO_3H$ (see Box 21.3).

When water reacts with the carbocation in the second step, an oxonium ion is formed. In the third step, water reacts with the oxonium ion to form an alcohol ($Me_2C(OH){-}CHHMe$) and regenerate the H_3O^+ ion, which acts as a catalyst for the hydration reaction.

Notice that all steps in the hydration reaction are reversible (⇌), so it is possible to convert the alcohol back to the alkene in a dehydration reaction (Section 21.2, p.967). Box 21.3 describes how the reaction conditions are adjusted in industry to favour formation of the alcohol.

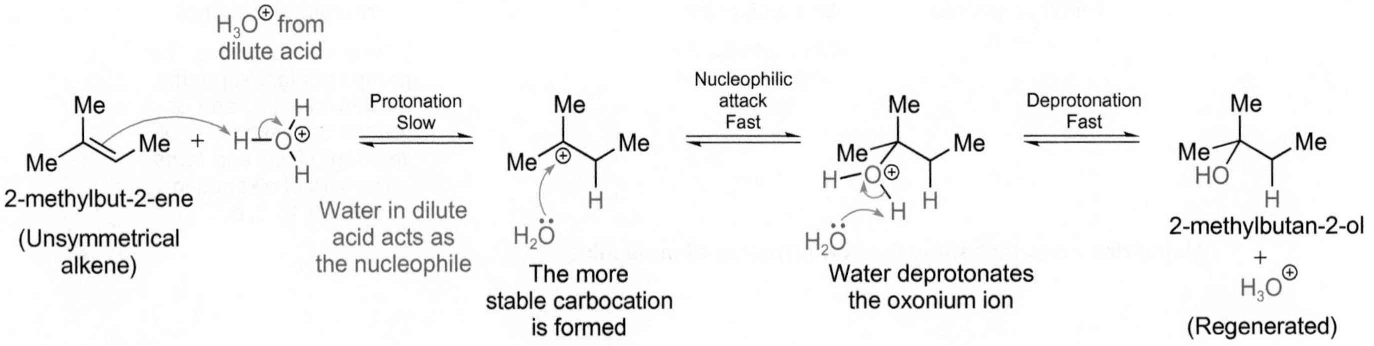
H3O⊕ from dilute acid
2-methylbut-2-ene
(Unsymmetrical alkene)
Protonation Slow
Water in dilute acid acts as the nucleophile
The more stable carbocation is formed
Nucleophilic attack Fast
Water deprotonates the oxonium ion
Deprotonation Fast
2-methylbutan-2-ol
+ H3O⊕
(Regenerated)

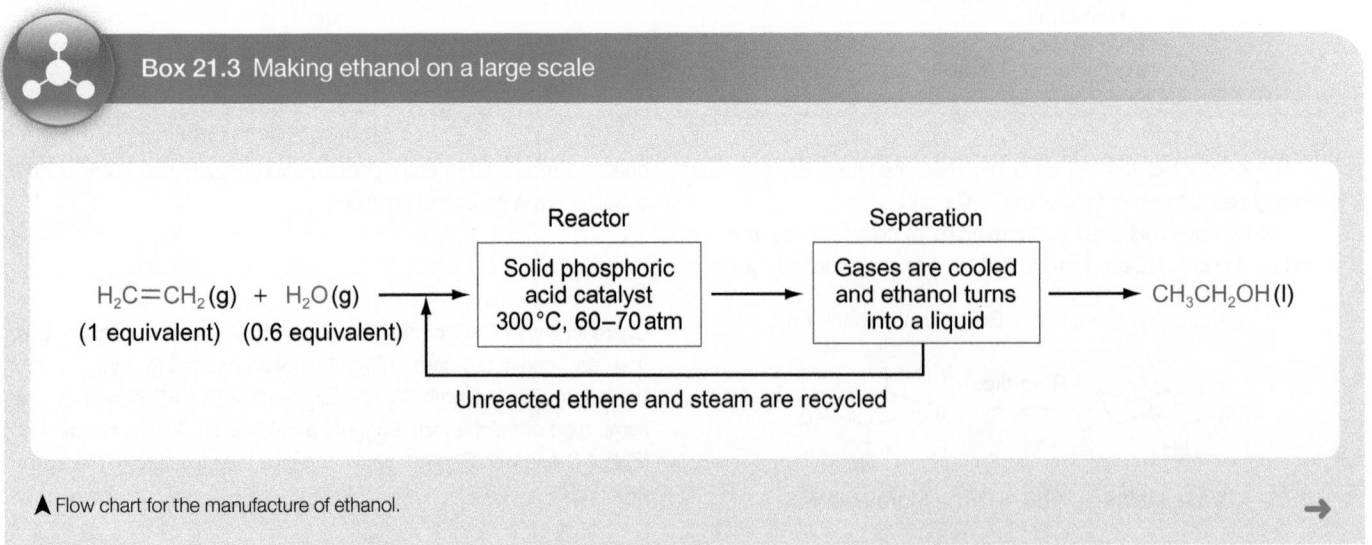

Figure 21.20 Addition of H_2O/H^+ to an unsymmetrical alkene.

Box 21.3 Making ethanol on a large scale

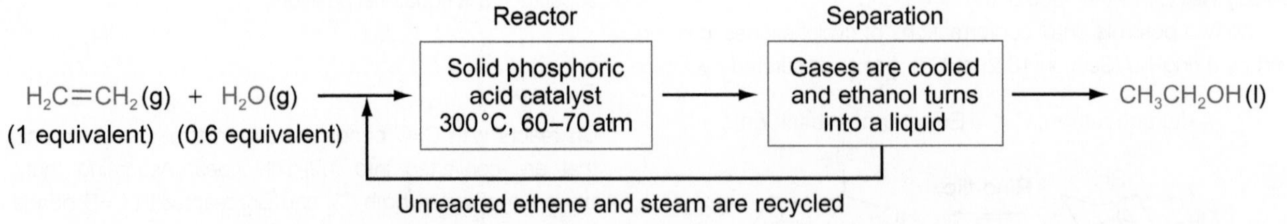
H2C=CH2(g) + H2O(g)
(1 equivalent) (0.6 equivalent)
Reactor
Solid phosphoric acid catalyst 300°C, 60–70 atm
Separation
Gases are cooled and ethanol turns into a liquid
CH3CH2OH(l)
Unreacted ethene and steam are recycled

▲ Flow chart for the manufacture of ethanol.

→

The worldwide use of ethanol (CH_3CH_2OH) has steadily grown in recent years, particularly after research showed that ethanol-containing petrols produce fewer pollutants than petrol itself.

Currently, the most popular method for making ethanol in industry is by hydration of ethene ($H_2C{=}CH_2$) in the presence of an acid catalyst.

$$H_2C{=}CH_2(g) + H_2O(g) \xrightleftharpoons{H^+} H_3C{-}CH_2OH(g) \quad \Delta H^{\ominus} = -46\,kJ\,mol^{-1}$$

Addition of H_2O to ethene is reversible, so reaction conditions need to be chosen so as to force the equilibrium to favour the formation of ethanol. The formation of ethanol from ethene is an exothermic process, so, by Le Chatelier's principle (Section 1.9, p.57), the reaction is favoured by using a low temperature. It is also favoured by a high pressure since it takes place in the gas phase and two reactant molecules are converted into a single product molecule. The rate at which ethanol is formed depends on the concentration of the reactants, so you might expect a high concentration of steam to be used. In practice, however, these conditions do not give the optimum yield of ethanol. A temperature of 300 °C and a ratio of ethene : steam of 1 : 0.6 are maintained in the reactor. The pressure is 60–70 atmospheres.

Steam and ethene are passed over a solid acid catalyst in the reactor (for an introduction to catalysis, see Section 9.9, p.437). The rate of the reaction also depends on the temperature. The high temperature (300 °C) ensures the gases reach equilibrium within the short time during which they come into contact with the acid catalyst. Only 0.6 equivalent (i.e. 0.6 of the stoichiometric amount) of steam is used because increasing the amount of steam is found to reduce the efficiency of the acid catalyst. Unfortunately, using a high temperature and a relatively low concentration of steam means that only around 5% of the ethene is converted into ethanol per pass through the catalyst. However, the process is made efficient, with an overall yield of 95%, by continually recycling the unreacted ethene and steam through the acid catalyst.

Although aqueous acid catalysts such as sulfuric acid (H_2SO_4) and phosphoric acid (H_3PO_4) can be used to prepare ethanol, solid acid catalysts are used in industry because these are easier and safer to handle than corrosive liquids. For this reason, phosphoric acid groups are coated on to the surface of zeolites (Chapter 6, p.255) to make a solid acid catalyst. Their structures have large pores and cages, and ethene and steam can pass into the interior of the zeolite where the surface is covered with phosphoric acid groups.

This ethanol manufacturing plant in West Burlington, Iowa (USA), produces ethanol from corn. Sugars derived from corn are fermented to give ethanol, which is purified by distillation and stored in large tanks.

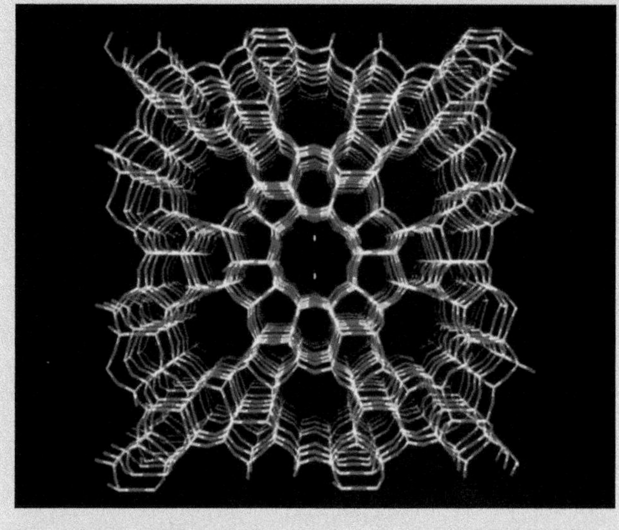

Computer graphic of the molecular structure of a zeolite—notice the various pores and channels.

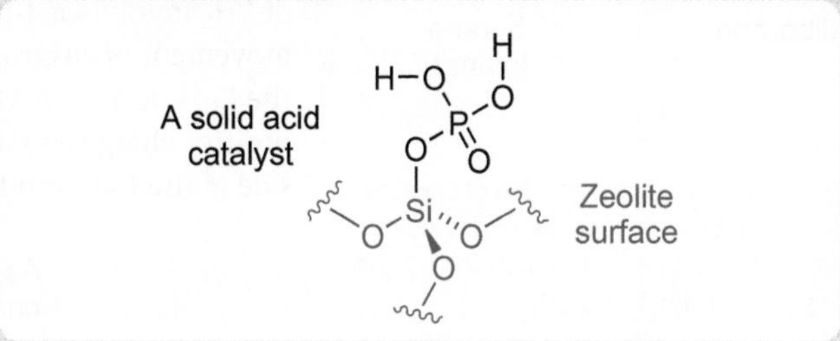
A solid acid catalyst
Zeolite surface

Question

Concentrated sulfuric acid reacts with ethene to form ethyl hydrogen sulfate (sulfovinic acid), which can be converted into ethanol by reaction with water, as shown below. Suggest a mechanism for the formation of ethyl hydrogen sulfate in the first stage.

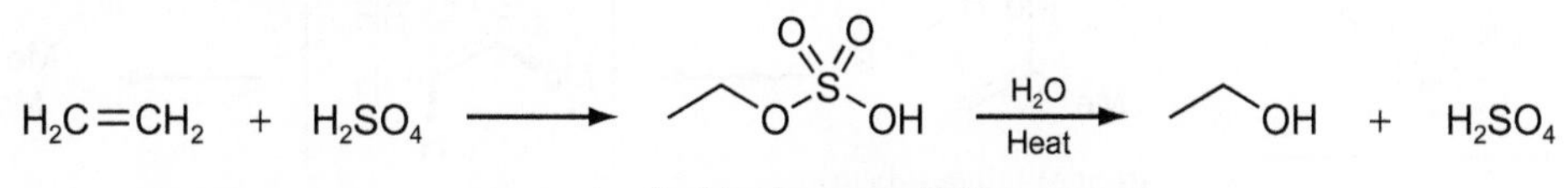
H₂C=CH₂ + H₂SO₄
H₂O
Heat
+ H₂SO₄
ethyl hydrogensulfate

Conversion of ethene into ethanol using an aqueous acid catalyst.

The hydration starts by protonation of the C=C bond to form the more stable carbocation. As the carbocation is planar, water can equally approach the planar carbocation from above or below (like in an S_N1 reaction, Section 20.3, p.933). This means that, if the OH group is bonded to a chiral carbon, then the alcohol will be racemic.

Hydration of unsymmetrical alkenes

Hydration of unsymmetrical C=C bonds using H^+/H_2O results in alcohols (ROH) that have the OH group attached to the more substituted carbon. By analogy to the addition of HX to a C=C bond (see p.970), these products can be described as the Markovnikov products and the hydrations considered to follow the Markovnikov rule.

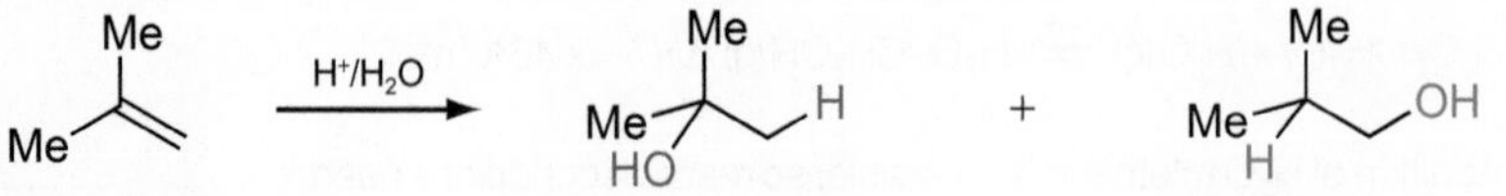
Me
Me
H^+/H_2O
Me
Me
H
HO
+
Me
Me
OH
H

2-methylprop-1-ene (Unsymmetrical alkene) | 2-methylpropan-2-ol **Selectively formed** Markovnikov product | 2-methylpropan-1-ol **Not formed** Anti-Markovnikov product

2-Methylpropan-2-ol (Me_3COH, *tert*-butanol) is commonly used as a solvent for reactions above ~25 °C—below this temperature it is a solid.

Hydrating unsymmetrical C=C bonds to form alcohols that have the OH group attached to the less substituted carbon (the anti-Markovnikov product) requires a different approach, that starts by addition of borane (BH_3) to the C=C bond (see next section).

Addition of borane followed by oxidation

Alcohols (ROH) are formed from alkenes by reaction of the C=C bond with diborane (B_2H_6) followed by oxidation using a solution of hydrogen peroxide (H_2O_2) and aqueous sodium hydroxide. The mechanism for this transformation involves the following four stages. Look carefully at these stages first to get an overview of the process.

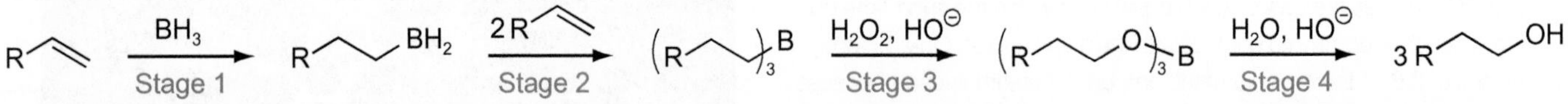
R
BH_3
Stage 1
R
BH_2
2 R
Stage 2
(R)$_3$B
H_2O_2, $HO^⊖$
Stage 3
(R O)$_3$B
H_2O, $HO^⊖$
Stage 4
3 R
OH

Syn addition of H and BH_2 | Two further hydroborations | Oxidation: B–C bonds converted to B–O–C bonds | Hydrolysis to give alcohols

Stage 1 Addition of BH_3 to C=C

Diborane (B_2H_6) is in equilibrium with a small amount of borane (BH_3) and it is the electron-deficient boron atom (see margin) of BH_3 that acts as the electrophile in the electrophilic addition to the C=C bond.

Empty *p* orbital (the boron atom can accept a pair of electrons)

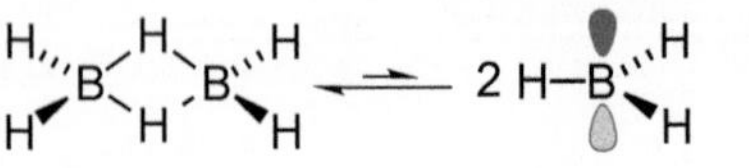
H
H
B
B
H
H
H
H
2 H–B
H
H

diborane | borane (planar)

BH_3 adds to the C=C bond to form an alkylborane in a **hydroboration** reaction. The mechanism is shown in Figure 21.21. The C=C bond attacks the boron atom to form a new C–B bond. As the C–B bond is forming, a new C–H bond starts to form so the addition of BH_3 involves a 4-membered transition state. Two curly arrows are used to indicate the movement of electrons but you should remember that the C–B bond starts to form before the C–H bond—this explains why one of the carbon atoms from the alkene has a partial positive charge (δ+) in the transition state. Overall, the H and BH_2 groups add to the same side of the C=C bond in a syn addition reaction.

The structure and properties of diborane (B_2H_6) and other boron hydrides are described in Section 25.2 (p.1147). The bonding in B_2H_6 is discussed in Section 5.7 (p.250).

Diborane must be handled with extreme caution because it is a highly flammable gas that ignites spontaneously in moist air at room temperature.

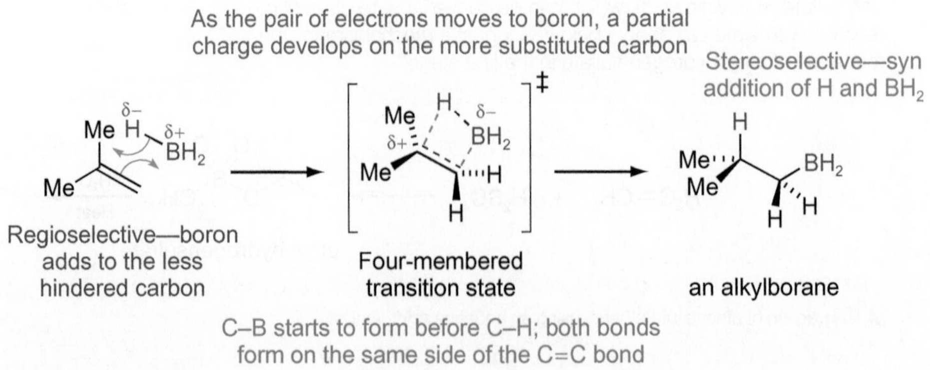
As the pair of electrons moves to boron, a partial charge develops on the more substituted carbon
Stereoselective—syn addition of H and BH_2
δ–
Me H δ+
BH_2
Me
Me
H δ–
δ+
BH_2
Me
H
H
‡
H
Me
BH_2
Me
H
H
Regioselective—boron adds to the less hindered carbon
Four-membered transition state
an alkylborane
C–B starts to form before C–H; both bonds form on the same side of the C=C bond

Figure 21.21 Mechanism of the hydroboration reaction in stage 1.

Addition of BH_3 to the C=C bond is regioselective. The C=C bond attacks the boron atom to give a partial positive charge on the more substituted carbon atom in the 4-membered transition state—the more substituted the carbon atom, the more stable the partial positive charge. This explains why the BH_2 group selectively adds to the less hindered carbon atom in the C=C bond. This is also favoured for steric reasons because the BH_2 group is larger than an H atom.

The hydroboration reaction was discovered by Herbert C. Brown from Purdue University, who was awarded a Nobel Prize in Chemistry in 1979 for his work on organoboranes. Organoboranes contain hydrogen, carbon, and boron atoms for which the chemical symbols are, by coincidence, the initials of Herbert C. Brown.

Stage 2 Further hydroboration to give a trialkylborane

Like BH_3, alkylboranes add to C=C bonds in hydroboration reactions. The two B–H bonds in the alkylborane (RBHH) are replaced successively by two B–C bonds to give first a dialkylborane (R_2BH) and then a trialkylborane (R_3B). These hydroboration reactions are regioselective and stereoselective, because they follow the same reaction mechanism as for the formation of the alkylborane.

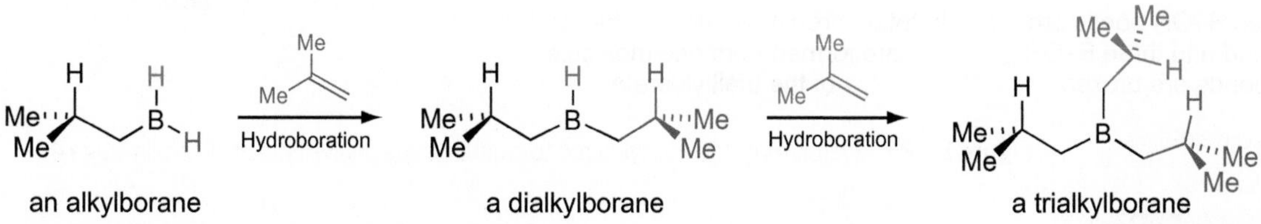
Hydroboration
Hydroboration
an alkylborane
a dialkylborane
a trialkylborane

Stage 3 Oxidation of the trialkylborane to produce a trialkylborate

To make an alcohol (ROH), the trialkylborane is reacted with hydrogen peroxide and aqueous sodium hydroxide. The first part of this process is an oxidation reaction that converts the trialkylborane (R_3B) into a trialkylborate ($(RO_3)B$). The three C–B bonds in the trialkylborane are replaced by three stronger B–O bonds and three C–O bonds in the trialkylborate (see margin).

When a trialkylborane is converted into a trialkylborate, 3 B–C (3 × +395 kJ mol^{-1}) and 3 O–O (3 × +150 kJ mol^{-1}) bonds are broken and 3 B–O (3 × +515 kJ mol^{-1}) and 3 C–O (3 × ~+360 kJ mol^{-1}) bonds are formed. The enthalpy change ($\Delta H^{\ominus}$) is equal to 1635 – 2625 = –990 kJ mol^{-1}.

The mechanism of the oxidation, shown in Figure 21.22, starts by deprotonation of hydrogen peroxide (H_2O_2) by the hydroxide ion to form a hydroperoxide ion (HOO^-). The HOO^- ion donates a pair of electrons (the HOMO) to the boron atom of the trialkylborane (the empty *p* orbital is the LUMO)—the hydroperoxide ion acts as a nucleophile and the trialkylborane acts as an electrophile. Note that the planar boron atom in the

Hydrogen peroxide (pK_a 11.62) is a stronger acid than water (pK_a 15.7)—the hydroperoxide ion (HOO^-) is more stable than the hydroxide ion (HO^-) as the adjacent electronegative oxygen atom helps to stabilize the negatively charged oxygen.

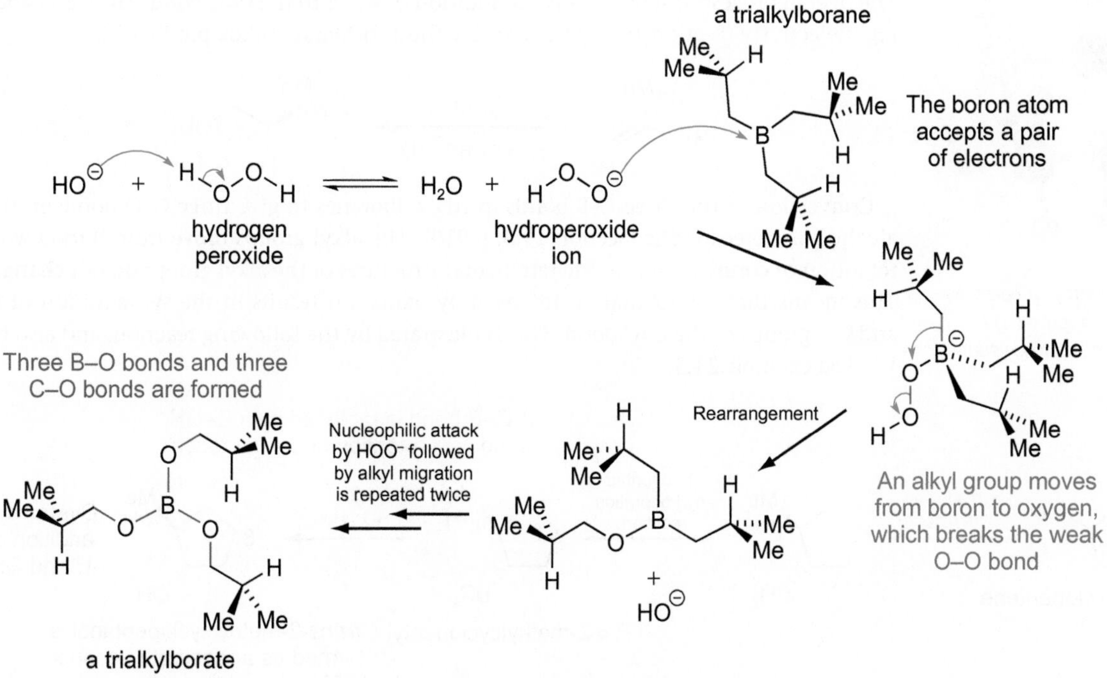
a trialkylborane
The boron atom accepts a pair of electrons
hydrogen peroxide
hydroperoxide ion
Three B–O bonds and three C–O bonds are formed
Nucleophilic attack by HOO⁻ followed by alkyl migration is repeated twice
Rearrangement
An alkyl group moves from boron to oxygen, which breaks the weak O–O bond
a trialkylborate

Figure 21.22 Oxidation of a trialkylborane to produce a trialkylborate in stage 3.

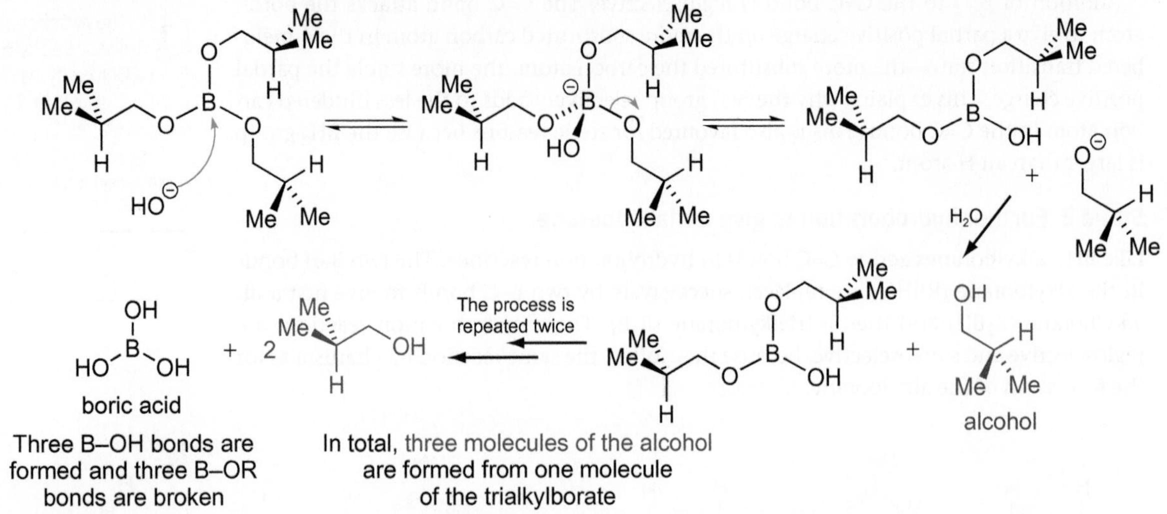
The process is repeated twice
boric acid
alcohol
Three B–OH bonds are formed and three B–OR bonds are broken
In total, three molecules of the alcohol are formed from one molecule of the trialkylborate

Figure 21.23 Hydrolysis of the trialkylborate to produce three equivalents of alcohol in stage 4.

The chemistry of boric acid ($B(OH)_3$) is discussed in Section 27.2 (p.1212).

2-Methylpropan-1-ol (isobutanol), Me_2CHCH_2OH, is a common solvent and is being investigated as an alternative to petrol, to fuel combustion engines.

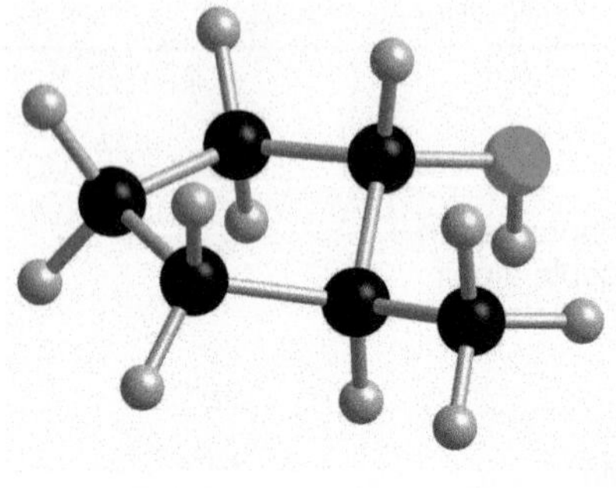

A molecular model of *trans*-2-methylcyclopentanol (the black spheres represent carbon; the grey spheres hydrogen, and the red sphere oxygen).

trialkylborane becomes tetrahedral as the B–O bond forms. One of the alkyl groups then moves from boron to oxygen and the hydroxide ion acts as a leaving group. Even though HO^- is a poor leaving group, migration of the alkyl group occurs because a C–O bond is formed in place of a much weaker O–O bond. The boron atom becomes planar once more. Attack by HOO^- followed by alkyl group migration is repeated until all three B–C bonds in the trialkylborane are converted into three B–O bonds in a trialkylborate.

Stage 4 Hydrolysis of the trialkylborate to produce three equivalents of alcohol

In the final stage, the trialkylborate ($(RO)_3B$) reacts with hydroxide ion and water in a series of three nucleophilic substitution reactions to form three molecules of the alcohol and boric acid, ($(HO)_3B$) (see Figure 21.23).

Regiochemistry and stereochemistry of the overall process

The overall process in stages 1–4 is the addition of water to the C=C bond. The water adds regioselectively (Section 19.3, p.909) to give the anti-Markovnikov product.

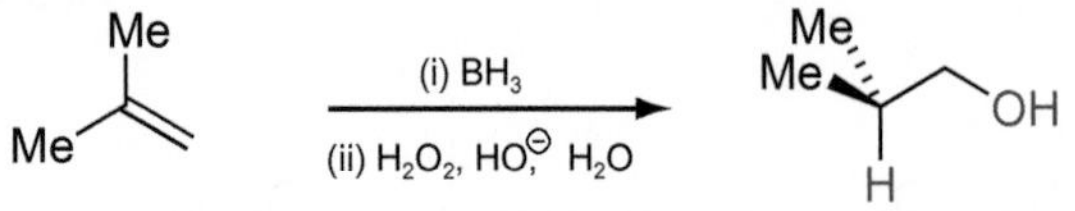
(i) BH_3
(ii) H_2O_2, HO^-, H_2O

Conversion of the three C–B bonds in trialkylboranes to give three C–O bonds in the alcohols is stereospecific (Section 19.3, p.910). The alkyl groups move from B to O with retention of configuration so the tetrahedral structures of the alkyl groups do not change. This means that hydroboration followed by oxidation results in the syn addition of H and OH groups to the C=C bond. This is illustrated by the following reaction, and also by Worked example 21.3.

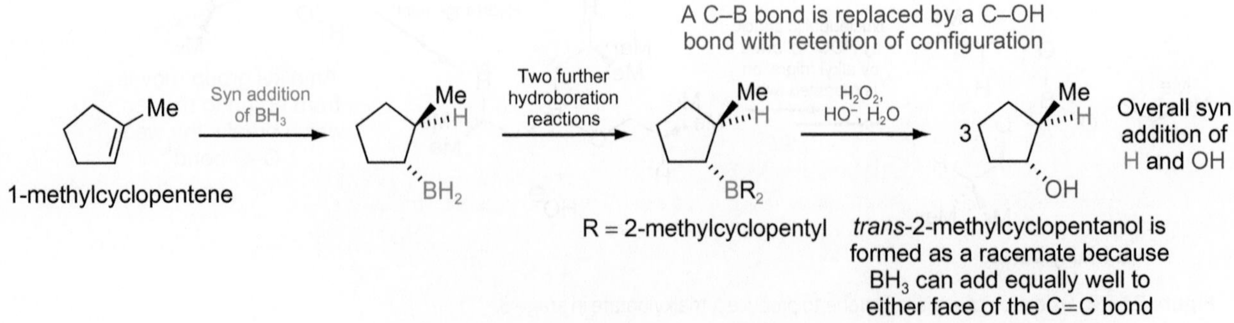
A C–B bond is replaced by a C–OH bond with retention of configuration
Syn addition of BH_3
Two further hydroboration reactions
H_2O_2, HO^-, H_2O
Overall syn addition of H and OH
1-methylcyclopentene
R = 2-methylcyclopentyl
trans-2-methylcyclopentanol is formed as a racemate because BH_3 can add equally well to either face of the C=C bond

Worked example 21.3 Selectivities in hydroboration–oxidation reactions

1-Methylcyclohex-1-ene is treated with diborane and then with a solution of hydrogen peroxide in aqueous sodium hydroxide.

(a) Suggest structures for the intermediates and the major product of the reaction.

(b) Does the process result in the selective formation of a single enantiomer (enantioselective) or of a single diastereomer (diastereoselective)?

Strategy

Draw the structures of the reactants. The conformation of 1-methylcyclohex-1-ene does not have to be shown (although you are always encouraged to draw a structure that shows the three-dimensional shape of the molecule).

Remember that the reactant is BH_3 not B_2H_6.

For addition of BH_3 to the C=C bond, consider the regioselectivity and stereoselectivity of the addition.

For oxidation of the intermediate trialkylborane (using H_2O_2 and HO^-) consider how the stereochemistry changes when the B–C bonds are converted into O–C bonds.

Draw the structure of the product, paying particular attention to stereochemistry.

Consider whether the transformation results in the selective formation of a single enantiomer or a single diastereomer.

Solution

(a)

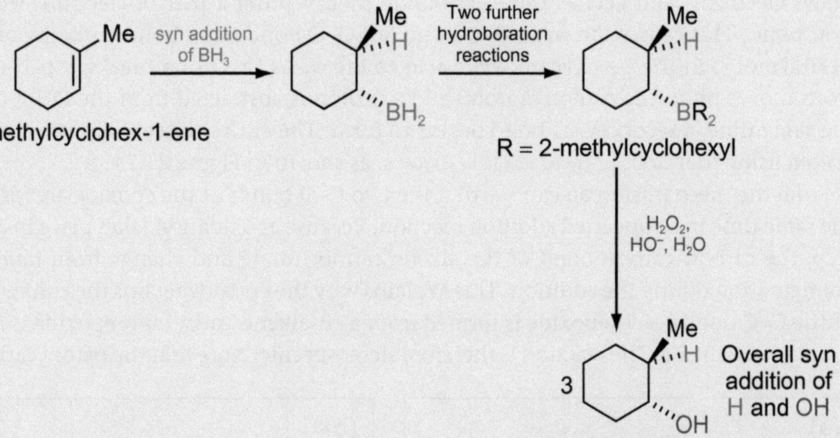

Me
syn addition of BH_3
1-methylcyclohex-1-ene
Me
H
BH_2
Two further hydroboration reactions
Me
H
BR_2
R = 2-methylcyclohexyl
H_2O_2, HO^-, H_2O
Me
H
OH
3
Overall syn addition of H and OH

trans-2-methylcyclohexanol is formed as a racemate because BH_3 can add equally well to either face of the C=C bond

(b) This is an example of a diastereoselective transformation as *trans*- rather than *cis*-2-methylcyclohexanol is selectively formed. (The *trans* isomer has the groups of highest priority, in this case HO and Me groups, on opposite sides of the ring.) The product is formed as a racemate because BH_3 can add equally well to either side of the C=C bond in 1-methylcyclohex-1-ene.

→

ℹ As dialkylboranes (R_2BH) are larger than BH_3 they add to alkenes with higher levels of regioselectivity–the larger the dialkylborane the higher the regioselectivity. For example, reaction of styrene ($PhCH{=}CH_2$) with BH_3 followed by H_2O_2/HO^- gives a 4:1 mixture of $PhCH_2CH_2OH$:$PhCH(OH)CH_3$. When the bulkier dialkylborane 9-BBN is used (followed by H_2O_2/HO^-), a 99:1 mixture of the same alcohols is formed.

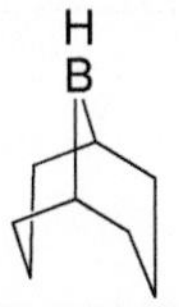

H
B

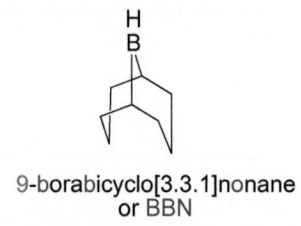

9-borabicyclo[3.3.1]nonane or BBN

ℹ To determine that the *trans*- isomer of 2-methylcyclohexanol is formed, and not the *cis*- isomer, spectroscopic methods such as ^1H NMR spectroscopy can be used—see a specialist spectroscopy book for details.

Remember that substituted cyclohexanes adopt chair conformations and that the most stable chair conformation usually has more of the substituents in equatorial positions (Section 18.2, p.827).

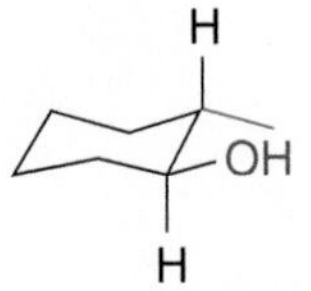
H
OH
H

The OH and Me groups are in equatorial positions

Most stable conformer of the *trans*- isomer

An equal mixture of enantiomers is formed

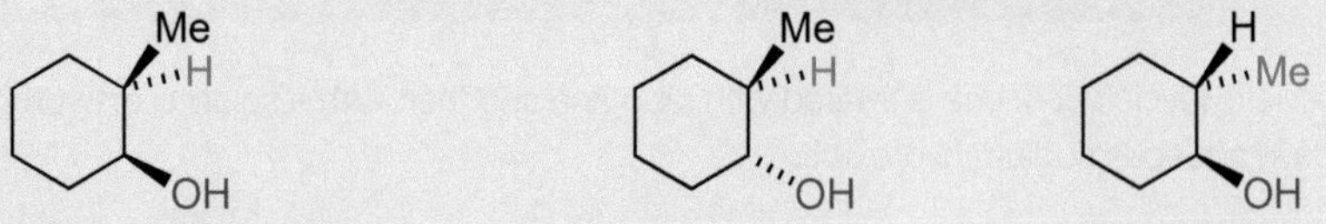
Me
H
OH
Me
H
OH
H
Me
OH

cis-2-methylcyclohexanol is not formed—this requires anti addition of H and OH

From addition of BH_3 to the bottom face of the C=C bond

From addition of BH_3 to the top face of the C=C bond

Question

2-Methylhex-2-ene is treated with diborane and then with a solution of hydrogen peroxide in aqueous sodium hydroxide.

(a) Suggest a structure for the major product of the reaction.

(b) Explain why the process is regioselective.

Reaction with peroxycarboxylic acids

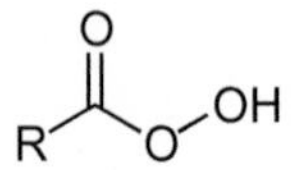
O
R
O
OH

A peroxycarboxylic acid (peracid)

O
H
H
R
R

An epoxide

Peroxycarboxylic acids (peracids, RCO_3H) have an oxygen atom between the C=O and OH groups of a carboxylic acid (RCO_2H). They react with alkenes by donating an oxygen atom to a C=C bond to form an epoxide, in a reaction called an **epoxidation**. The other product is a carboxylic acid.

This reaction is an electrophilic addition. The oxygen atom in the OH group in the peroxycarboxylic acid is partially positive (because the RCO_2 group is a –I group and withdraws electrons) and acts as the electrophile by accepting a pair of electrons from the C=C bond. This causes the weak O–O bond (an O–O bond has a bond enthalpy of about +150 kJ mol^{-1}) in the peroxycarboxylic acid to break. As this bond breaks, a pair of electrons moves on to the carbonyl group and a proton is abstracted from the OH group. At the same time, a second C–O bond begins to form. The entire concerted process is represented using four double-headed curly arrows, as shown in Figure 21.24.

From the mechanism you can see that the two C–O bonds of the epoxide are formed at the same time in a concerted addition reaction. Because epoxidation takes place in a single step, the carbon–carbon bond of the alkene cannot rotate and change from *trans* to *cis* configuration during the addition. This explains why the epoxide retains the configuration of the C=C bond—a *cis*-epoxide is formed from a *cis*-alkene and a *trans*-epoxide is formed from a *trans*-alkene. The reaction is therefore stereospecific. Note that the peroxycarboxylic

Compare the reaction mechanisms leading to epoxides and bromonium ions (p.975) to note the similarities.

meta-Chloroperbenzoic acid (abbreviated *m*CPBA) is a commonly used peroxycarboxylic acid. Despite being a reactive oxidizing agent, *m*CPBA is easy to handle, as it is a solid and it can be stored in a refrigerator.

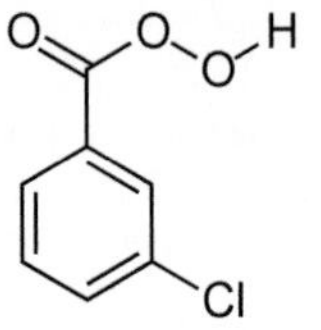
O
O
O
H
Cl

The C–C–O bond angle in an epoxide is 60° rather than the normal tetrahedral bond angle of 109.5°. Epoxides therefore have appreciable angle strain (Section 18.2, p.826) and they react to break open the ring and form products with tetrahedral bond angles.

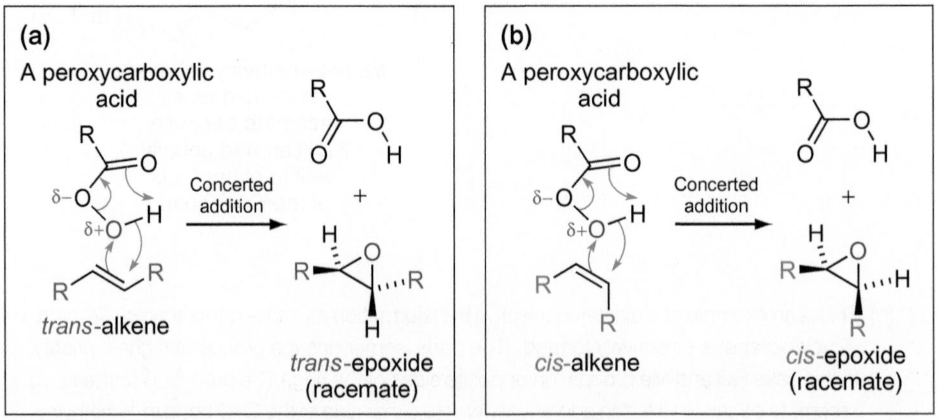
(a)
A peroxycarboxylic acid
Concerted addition
trans-alkene
trans-epoxide (racemate)
(b)
A peroxycarboxylic acid
Concerted addition
cis-alkene
cis-epoxide (racemate)

Figure 21.24 Epoxidation is stereospecific: (a) a *trans*-alkene produces a *trans*-epoxide; (b) a *cis*-alkene produces a *cis*-epoxide.

Interactive 3D animation of epoxidation of an alkene.

acid can approach either side of the C=C bond equally well and so the epoxide is formed as a racemate.

Reactions of epoxides

Epoxides are important compounds in synthesis because they undergo a variety of reactions to make useful compounds. The 3-membered epoxide ring is highly strained and can be opened by reaction with a range of nucleophiles. For example, water reacts with epoxides in the presence of an acid or base catalyst to form 1,2-diols (or glycols; e.g. RCH(OH)CH(OH)R).

Acid-catalysed ring opening

Water is such a weak nucleophile that it does not react with an epoxide but it will react with a protonated epoxide, as shown in Figure 21.25. Protonation of the epoxide produces a much stronger electrophile that reacts with water in a nucleophilic substitution reaction. This is an S_N2 type reaction (Section 20.3, p.930) as the water molecule approaches the epoxide from the opposite side to one of the C–O bonds. Substitution produces a new C–O bond at the same time as one of the C–O bonds in the epoxide is broken. Finally, deprotonation of the oxonium ion forms a 1,2-diol that has the two OH groups in an anti arrangement. Notice that the acid is regenerated in the last step, so H_3O^+ acts as a catalyst for the reaction.

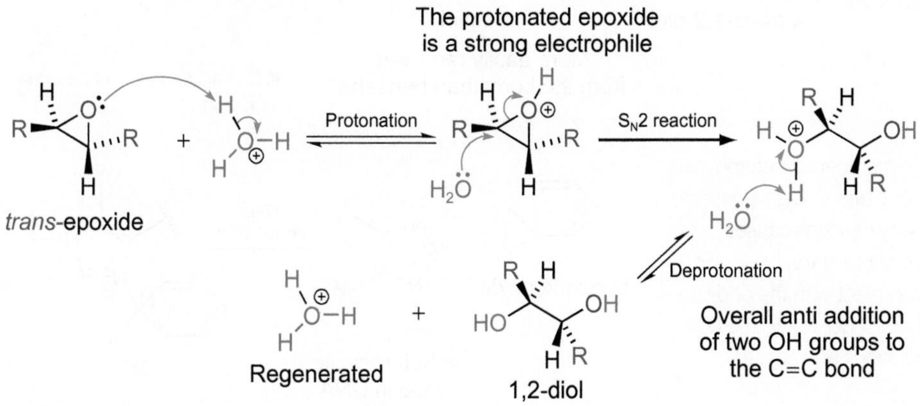
The protonated epoxide is a strong electrophile
Protonation
S_N2 reaction
trans-epoxide
Deprotonation
Regenerated
1,2-diol
Overall anti addition of two OH groups to the C=C bond

Figure 21.25 Acid-catalysed ring opening of an epoxide.

Alkali-catalysed ring opening

Water also reacts with epoxides in the presence of a small amount of the hydroxide ion, as shown in Figure 21.26. The hydroxide ion is a stronger nucleophile than water and reacts with the epoxide in a nucleophilic substitution reaction. This is an S_N2 type reaction (Section 20.3, p.930) as the hydroxide ion approaches the epoxide from the opposite side to one of the C–O bonds. Finally, protonation of the alkoxide ion forms a 1,2-diol that has the two OH groups in an anti arrangement. Notice that the hydroxide ion is regenerated in the last step so it acts as a catalyst for the reaction.

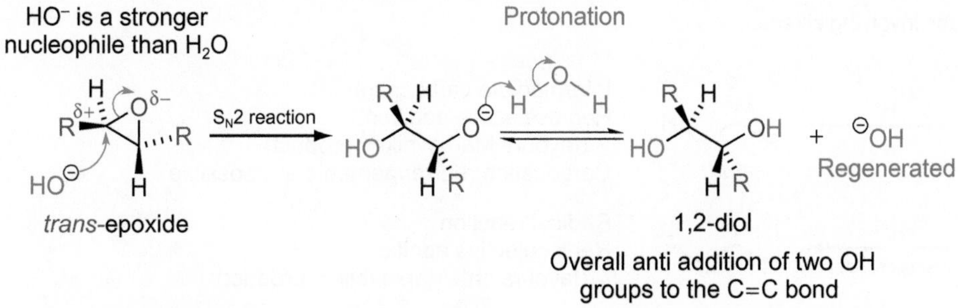
HO^- is a stronger nucleophile than H_2O
S_N2 reaction
Protonation
trans-epoxide
Regenerated
1,2-diol
Overall anti addition of two OH groups to the C=C bond

Figure 21.26 Base-catalysed ring opening of an epoxide.

1,2-Diols, such as ethane-1,2-diol ($HOCH_2CH_2OH$), are commonly used in the protection of aldehydes and ketones (Section 23.2, p.1075) and also to form polyesters including polyethylene terephthalate (PET) (Chapter 24 opener, p.1099).

Water reacts with protonated unsymmetrical epoxides to selectively attack the more substituted carbon. As in bromonium ions (p.975), the more substituted carbon atom in the 3-membered ring has the greater partial positive charge (δ+).

Reaction of a C=C bond with RCO_3H followed by H_2O/H^+ introduces two OH groups to the opposite sides of the C=C bond in an anti addition. Reactions that lead to the introduction of two OH groups are called **dihydroxylations**.

Remember that all acid–base reactions are equilibrium reactions. The position of the equilibrium depends on the difference in pK_a values between reactants and products, as discussed in Section 7.2 (p.309).

The ether group (R–O–R) is a weak base. Protonation of R_2O gives the conjugate acid (R_2OH^+), which has a very low pK_a value of −3.5 (Table 19.1, p.896). Thus a 1.0 mol dm^{-3} solution of HCl (pH 0) will only protonate around 1 in $10^{3.5}$ molecules of the ether. Even at these low concentrations, the protonated ethers increase the rate of nucleophilic substitution because they are extremely reactive electrophiles.

The hydroxide ion often acts a base, rather than a nucleophile. On reaction with an epoxide it acts as a nucleophile because the hydrogen atoms in an epoxide are not acidic.

Visit the Online Resource Centre to view screencast 21.3 which walks you through the use of epoxides in synthesis.

Box 21.4 Why is benzene carcinogenic?

Benzene (C_6H_6) is a useful component of petrol because it has a high octane number and helps to reduce the tendency of the petrol–air mixture to pre-ignite in the engine (called 'knocking'). However, benzene is known to be carcinogenic, so it is limited to a maximum of 1% by volume in petrol.

Your body is constantly exposed to foreign organic compounds, such as benzene. These organic compounds are soluble in fat and they tend to build up in the membranes of cells. To remove these unwanted compounds, your body uses enzymes in the liver to catalyse a number of chemical reactions. These reactions, called **biotransformations**, convert fat-soluble compounds into water-soluble compounds that can be excreted from the body.

An important biotransformation of aromatic organic compounds is epoxidation, which is catalysed by an enzyme called cytochrome P450. In the body, benzene reacts with cytochrome P450 and O_2 to form benzene oxide (C_6H_6O), which contains an epoxide ring. Because of the strain in the epoxide ring, benzene oxide reacts with various nucleophiles including water. Reaction with water opens the epoxide ring to produce catechol (1,2-$C_6H_4(OH)_2$), which is a more polar and water-soluble molecule than benzene.

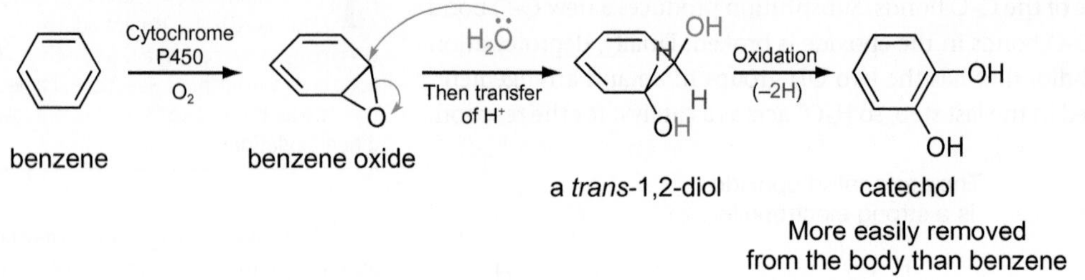
Cytochrome
P450
O_2
benzene
benzene oxide
H_2O
Then transfer
of H+
OH
H
H
OH
a *trans*-1,2-diol
Oxidation
(–2H)
OH
OH
catechol
More easily removed
from the body than benzene

◀ Biotransformation of benzene in the body.

Benzene oxide is in equilibrium with a 7-membered heterocycle containing three C=C bonds called oxepine (or oxepin). The high reactivity of oxepine and benzene oxide explains why benzene causes cancer (carcinogenic). For example, other nucleophiles in the body, such as the heterocyclic bases present in DNA, can react with the epoxide ring of benzene oxide to form products that no longer fit into the DNA double helix (see p.864). This can lead to mutations and cancer.

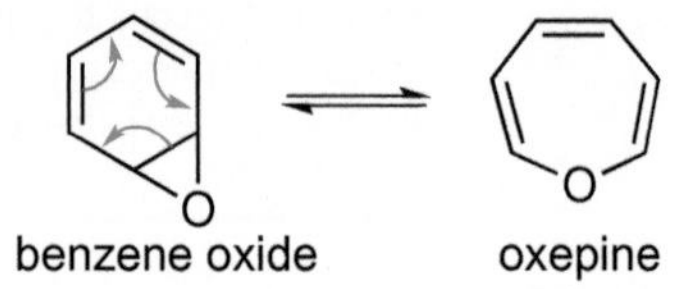
benzene oxide
oxepine

▲ Benzene oxide is in equilibrium with oxepine.

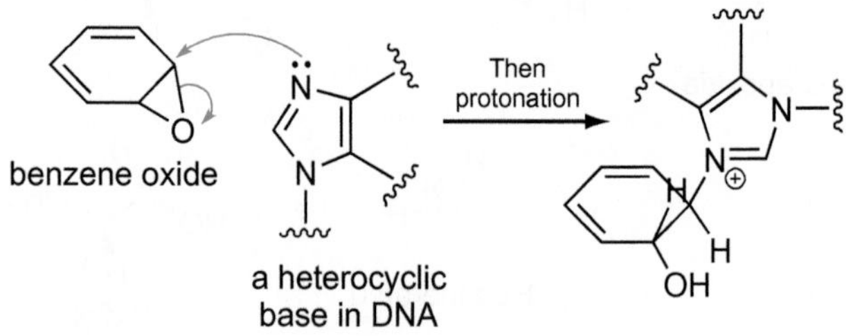
O
benzene oxide
N
N
Then
protonation
a heterocyclic
base in DNA
N
N
H
H
OH

▲ Reaction of benzene oxide with DNA.

Question

Suggest a mechanism for the reaction of 1,2-epoxy-1-methylcyclohexane (or 1-methylcyclohexene oxide) with aqueous acid and give the structure of the product.

Summary

Seven important addition reactions involving alkenes.

1 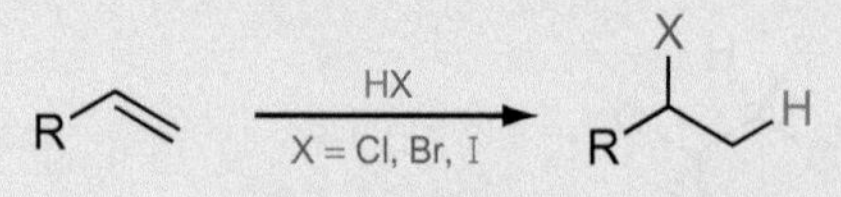
R
HX
X = Cl, Br, I
X
H
R

Intermediate carbocation
Regioselective addition
(favours Markovnikov product)
Carbocation rearrangements are possible

2 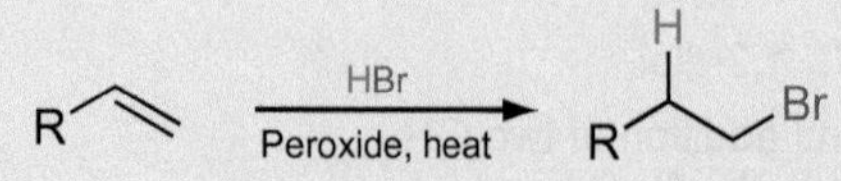
R
HBr
Peroxide, heat
H
Br
R

Radical reaction
Regioselective addition
(favours anti-Markovnikov product)

→

3 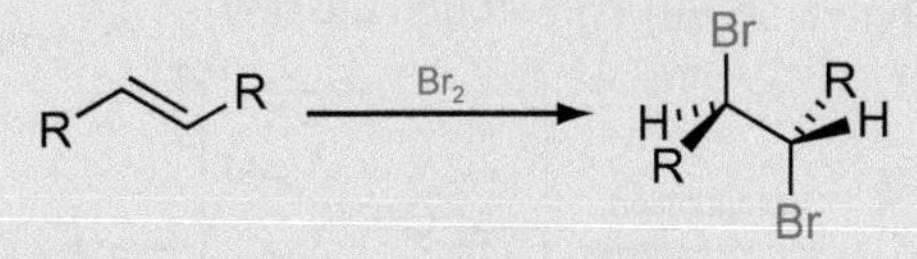
Br₂

Intermediate bromonium ion
Stereospecific addition (anti addition)

4 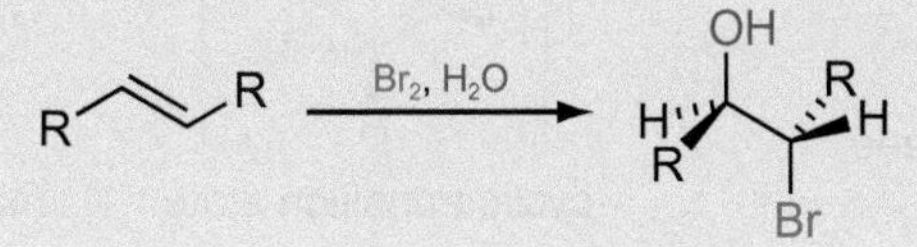
Br₂, H₂O

Intermediate bromonium ion
Stereospecific addition (anti addition)
Regioselective addition

5 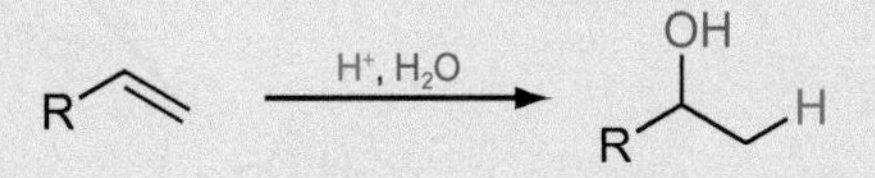
H⁺, H₂O

Hydration via an intermediate carbocation
Regioselective addition
(favours Markovnikov product)

6 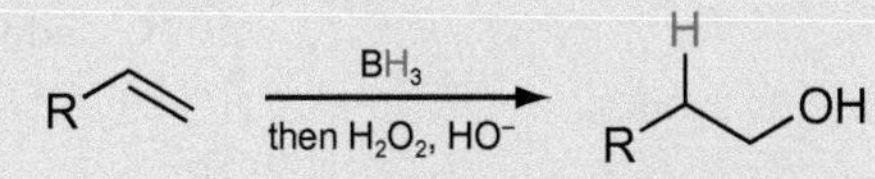
BH₃
then H₂O₂, HO⁻

Hydration by hydroboration then oxidation
Regioselective addition
(favours anti-Markovnikov product)
Stereospecific addition
(overall syn addition)

7 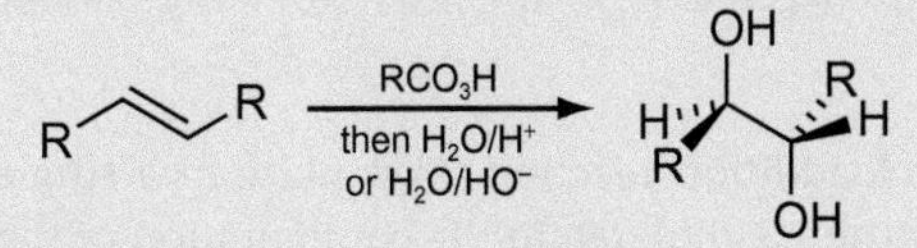
RCO₃H
then H₂O/H⁺
or H₂O/HO⁻

1,2-Dihydroxylation by epoxidation then hydrolysis
Stereospecific addition
(overall anti addition)

For practice questions on these topics, see questions 1, 2, and 6 at the end of this chapter (pp.1000-1001).

21.4 Pericyclic reactions of alkenes

Pericyclic reactions are concerted (single-step) reactions that involve a change in the position of bonding electrons via cyclic transition states. The most common pericyclic reactions involving C=C bonds are **cycloaddition reactions**. In these reactions the C=C bond of an alkene reacts with another molecule to form a cyclic product.

A pericyclic reaction of an alkene with a 1,3-diene to give a 6-membered ring is called the Diels–Alder reaction (Section 19.2, p.908).

Reaction with potassium permanganate or osmium tetroxide

Potassium permanganate ($KMnO_4$) or osmium tetroxide (OsO_4) are used to convert an alkene into a 1,2-diol (e.g. RCH(OH)CH(OH)R). Reaction of a C=C bond to form a 1,2-diol is called a **dihydroxylation reaction**.

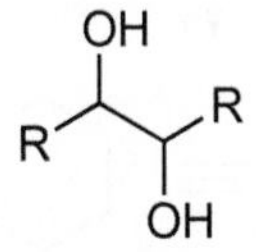

A 1,2-diol

Reaction with potassium permanganate followed by hydrolysis

A cold, dilute solution of potassium permanganate ($KMnO_4$) reacts with a C=C bond in a pericyclic reaction to produce, after reaction with aqueous sodium hydroxide, a 1,2-diol. The mechanism for this reaction, shown in Figure 21.27, involves a concerted cycloaddition reaction to form a manganate ester.

The C=C bond acts as a nucleophile and attacks an oxygen atom on MnO_4^- to form a C–O bond. As this happens, a pair of electrons moves on to manganese and a pair of electrons in an adjacent Mn=O bond moves to form a second C–O bond. Both C–O bonds are formed on the same face of the alkene in a syn addition.

The reaction of $KMnO_4$ with a C=C bond is usually carried out in the presence of aqueous hydroxide so that the manganate ester is immediately hydrolysed to form a 1,2-diol (together with MnO_2).

Reaction of a C=C bond with $KMnO_4$/HO^-/H_2O introduces two OH groups on the *same* side of the C=C bond in a syn addition.

Decolorization of a purple aqueous solution of $KMnO_4$ is often used as a test for the presence of a C=C bond in a molecule. $KMnO_4$ is purple because it absorbs green and yellow radiation in the visible region (Box 10.7, p.491).

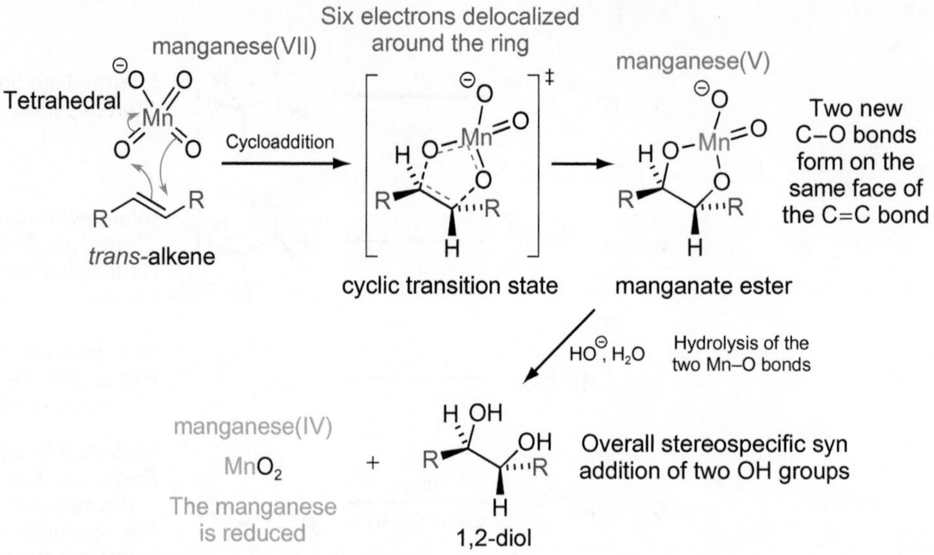
Six electrons delocalized around the ring
manganese(VII)
Tetrahedral
Cycloaddition
trans-alkene
manganese(V)
cyclic transition state
manganese ester
Two new C–O bonds form on the same face of the C=C bond
HO⁻, H₂O
Hydrolysis of the two Mn–O bonds
manganese(IV)
MnO₂
The manganese is reduced
1,2-diol
Overall stereospecific syn addition of two OH groups

Figure 21.27 Reaction of an alkene with MnO_4^- followed by alkaline hydrolysis gives a 1,2-diol.

In the process of oxidizing a substance, $KMnO_4$ produces the deep brown MnO_2, forming a stain that will discolour practically anything organic. This discolouration has been put to good effect in the film and TV business where $KMnO_4$ is used to artificially age props and set dressings.

Because the cycloaddition reaction takes place in a single step, the manganate ester, and subsequently the 1,2-diol, retain the configuration of the C=C bond—a *cis*-1,2-diol is formed from a *cis*-alkene and a *trans*-1,2-diol is formed from a *trans*-alkene (see margin). The reaction is therefore stereospecific. Note that $KMnO_4$ can react equally well with the top or bottom face of the C=C bond so the 1,2-diol is formed as a racemate.

An equal mixture of two enantiomers is called a racemate (Section 18.4, p.841). In a molecule with two chiral centres, a pair of enantiomers has different configurations at both chiral centres (e.g. *SS* and *RR*, or *RS* and *SR*).

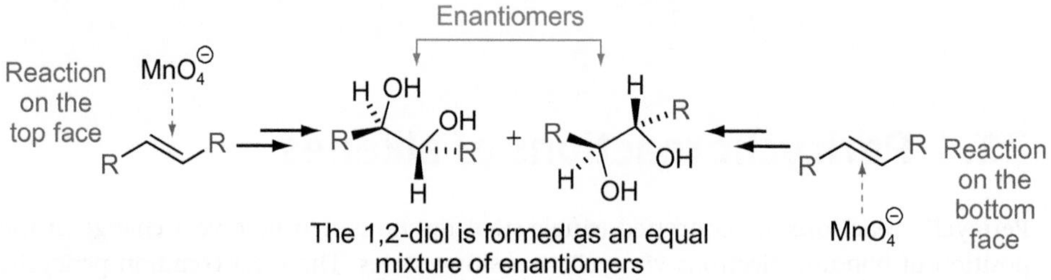
Enantiomers
Reaction on the top face
Reaction on the bottom face
The 1,2-diol is formed as an equal mixture of enantiomers

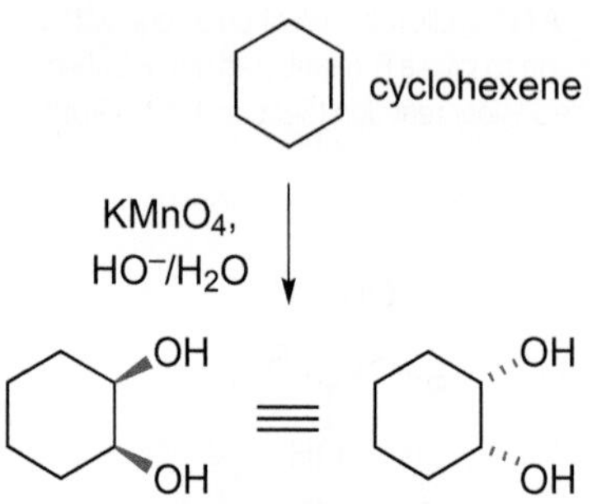
cyclohexene
KMnO₄, HO⁻/H₂O

cis-cyclohexane-1,2-diol (a *meso* compound, p.851)

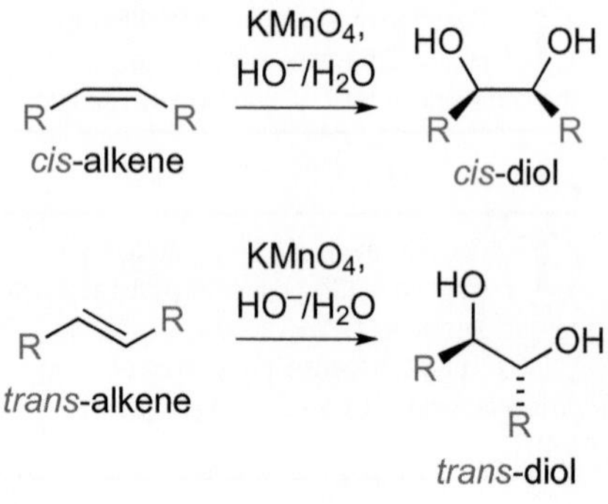
KMnO₄, HO⁻/H₂O
cis-alkene
cis-diol
KMnO₄, HO⁻/H₂O
trans-alkene
trans-diol

To obtain a 1,2-diol from the reaction of an alkene with a dilute solution of $KMnO_4$, the reaction mixture must be kept cold. $KMnO_4$ is such a powerful oxidizing agent that, at room temperature or above, it oxidizes 1,2-diols to form ketones (RCOR) and carboxylic acids (RCO_2H). This is an example of **oxidative cleavage**, which is discussed later in this section.

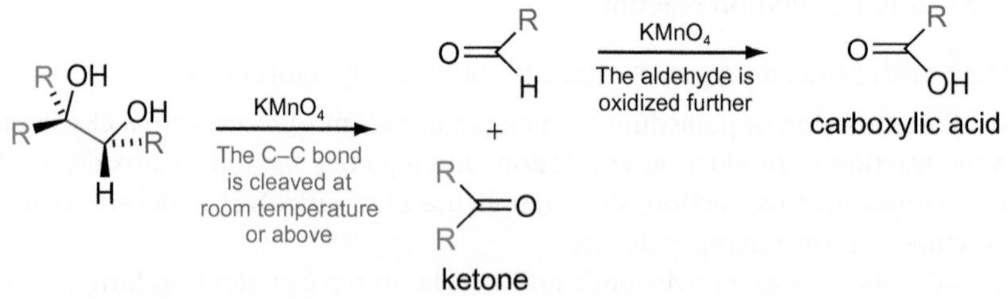
KMnO₄
The C–C bond is cleaved at room temperature or above
KMnO₄
The aldehyde is oxidized further
carboxylic acid
ketone

Convincing evidence for the overall syn addition of two OH groups to a C=C bond is provided by the reaction of cyclic alkenes with $KMnO_4$ to produce *cis*-cycloalkane-1,2-diols (see margin). ($KMnO_4$ reacts equally well with the top or bottom face of the C=C bond.)

ℹ $KMnO_4$ in aqueous hydroxide is used as a stain in thin layer chromatography (TLC) to show up the positions of colourless compounds on the TLC plate (Section 11.3, p.530). Compounds containing C=C bonds react with the $KMnO_4$ to give a yellow spot on a purple background. A pencil is used to mark the plate, to show where the compounds are applied, and to indicate the position reached by the solvent.

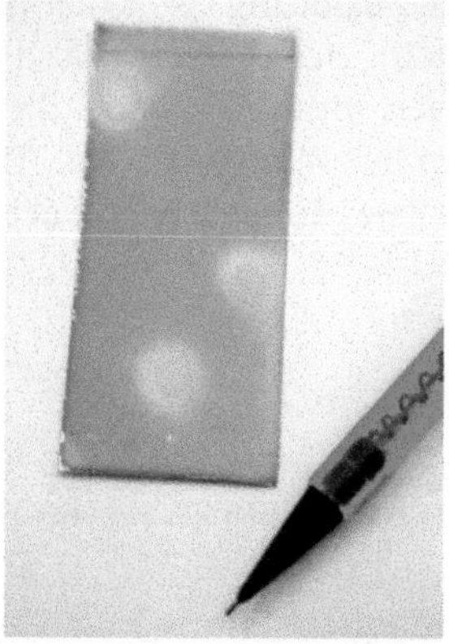

ℹ Reaction of a C=C bond with OsO_4/H_2O introduces two OH groups on the *same* side of the C=C bond in a syn addition.

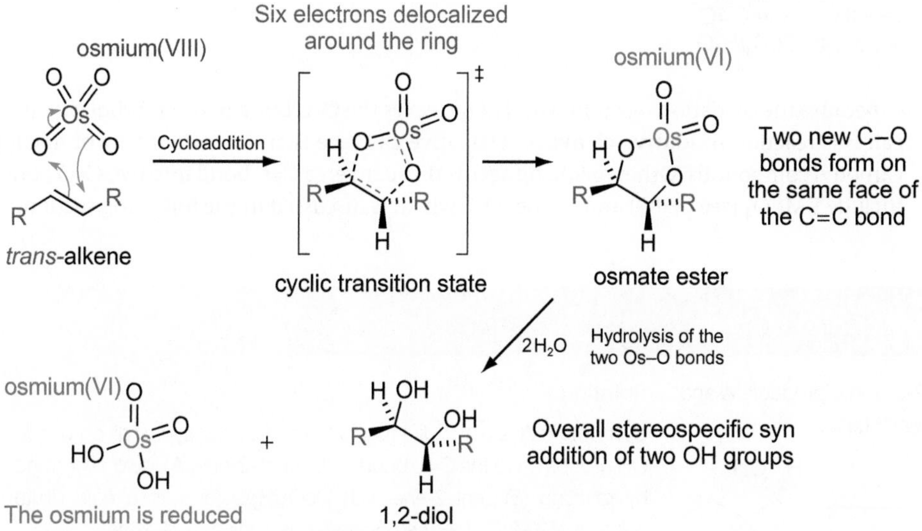
Six electrons delocalized around the ring
osmium(VIII)
Cycloaddition
osmium(VI)
Two new C–O bonds form on the same face of the C=C bond
trans-alkene
cyclic transition state
osmate ester
2H2O
Hydrolysis of the two Os–O bonds
osmium(VI)
Overall stereospecific syn addition of two OH groups
The osmium is reduced
1,2-diol

Figure 21.28 Reaction of an alkene with osmium tetroxide followed by hydrolysis to give a 1,2-diol.

Reaction with osmium tetroxide followed by hydrolysis

Like $KMnO_4$, osmium tetroxide (OsO_4) reacts with a C=C bond in a pericyclic reaction to produce, after reaction with water, a 1,2-diol. The mechanism for this reaction is still under debate but the simplest possible mechanism involves a concerted cycloaddition reaction to form an osmate ester (see Figure 21.28). The C=C bond acts as a nucleophile and attacks an oxygen atom on OsO_4 to form a C–O bond. As this happens, a pair of electrons moves on to osmium (which is in the +8 oxidation state) and a pair of electrons in an adjacent Os=O bond moves to form a second C–O bond. Both C–O bonds are formed on the same face of the alkene in a syn addition.

The reaction of OsO_4 with a C=C bond is usually carried out in the presence of water so that the osmate ester is immediately hydrolysed to form a 1,2-diol together with $Os(OH)_2O_2$. Note that the dihydroxylation reaction is stereospecific and, as OsO_4 can react equally well with the top or bottom face of the C=C bond, the osmate ester is formed as a racemate.

Worked example 21.4 provides practice in identifying suitable oxidizing agents for preparing 1,2-diols.

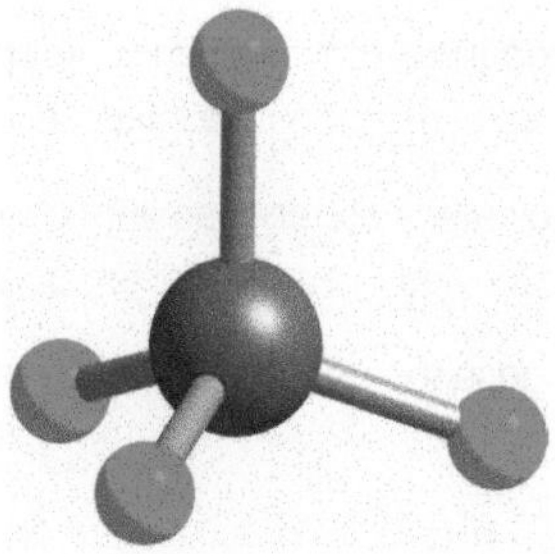

Osmium tetroxide (OsO_4) is tetrahedral. The colourless solid is usually contaminated by a small amount of osmium dioxide (OsO_2), which is yellow-brown in colour. In 2004, a plot to detonate a bomb, believed to contain osmium tetroxide, was foiled by British intelligence sources.

Reaction with osmium tetroxide in the presence of an oxidizing agent

Unfortunately, osmium tetroxide is expensive and toxic so, to reduce the amount of OsO_4 needed in dihydroxylation reactions, an oxidizing agent is usually added to the reaction mixture. As soon as OsO_4 is converted into $Os(OH)_2O_2$, the oxidizing agent selectively oxidizes $Os(OH)_2O_2$ back to OsO_4 so only a small amount of OsO_4 is required. A common oxidizing agent that is used is potassium hexacyanoferrate(III) (potassium ferricyanide, $K_3Fe(CN)_6$).

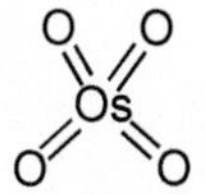

OsO_4
osmium tetroxide
(osmium(VIII) oxide)

HIO_4 and HIO_3 are called hypervalent compounds because the iodine atoms in these compounds have more than eight electrons in the valence shell. Hypervalency is discussed in Section 5.5 (p.245).

Another oxidizing agent that can be used to convert $Os(OH)_2O_2$ into OsO_4 is periodic acid (HIO_4). However, HIO_4 also oxidizes 1,2-diols to form carbonyls. So, when OsO_4/HIO_4 reacts with a C=C bond, the product is an aldehyde (RCHO) and/or ketone (RCOR) as shown below, not a 1,2-diol. The mechanism of oxidation of the 1,2-diol involves an intermediate periodate ester that breaks down to form two C=O bonds together with iodic acid (HIO_3).

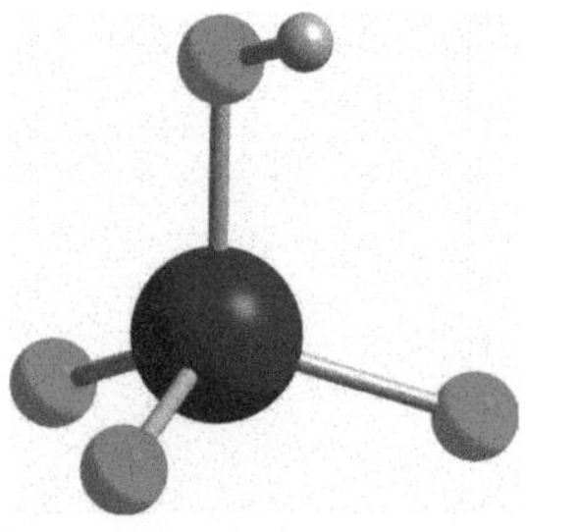

Periodic acid (HIO_4). For HIO_4, sometimes you will see the name metaperiodic acid used, to distinguish it from the related iodine(VII) compound, H_5IO_6 (O=I(OH)$_5$), which is called orthoperiodic acid. (The prefix meta means less water, whereas ortho means more water.)

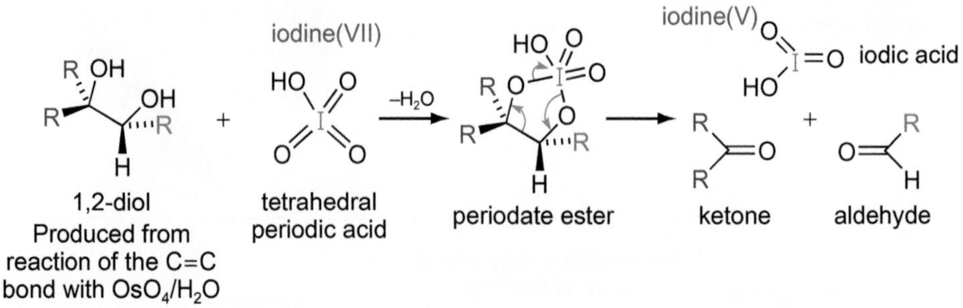
iodine(VII)
iodine(V)
iodic acid
$-H_2O$
1,2-diol
Produced from reaction of the C=C bond with OsO_4/H_2O
tetrahedral periodic acid
periodate ester
ketone
aldehyde

Because the oxidation reaction with HIO_4 cleaves the C–C bond in the 1,2-diol, chemists call this reaction **oxidative cleavage**. Oxidative cleavage is an important way of making carbonyl compounds. Other oxidizing agents that convert a C=C bond into two C=O bonds include $KMnO_4$ (see p.989) and ozone (O_3), which is discussed in the following section.

Worked example 21.4 Oxidation of C=C bonds

Suggest reagents for converting (*E*)-pent-2-ene into products **A** and **B**. More than one step may be required in each case.

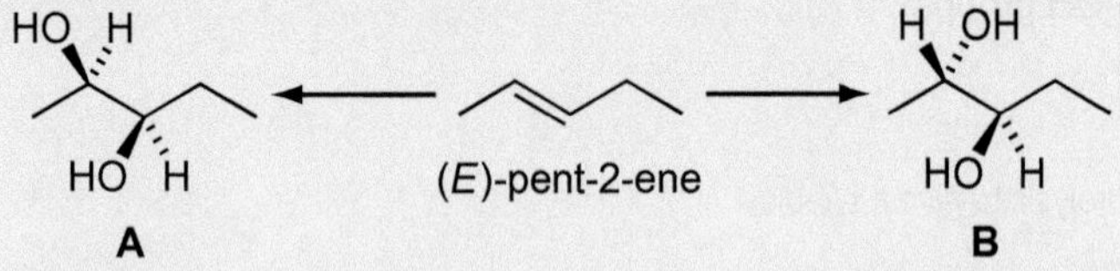
(*E*)-pent-2-ene
A
B

Strategy

Recognize the functional groups in the products to identify what types of reactions are involved.

Look at the stereochemistry of products **A** and **B** to determine if they are formed in a syn addition or an anti addition reaction.

Choose the appropriate reagents.

Solution

Compound **A** is a 1,2-diol that is formed by the syn addition of two OH groups on to the C=C bond of (*E*)-pent-2-ene. **A** could be formed by reacting (*E*)-pent-2-ene with OsO_4/H_2O (or with a cold dilute solution of $KMnO_4$ followed by aqueous sodium hydroxide).

Compound **B** is a 1,2-diol that is formed by the anti addition of two OH groups on to the C=C bond of (*E*)-pent-2-ene. **B** is formed by reacting (*E*)-pent-2-ene with RCO_3H (to make an epoxide) followed by H^+/H_2O (or HO^-/H_2O).

Question

Give the structures of the products from reaction of (*E*)-pent-2-ene with OsO_4/H_2O followed by HIO_4.

Reaction with ozone

The preparation and properties of ozone are described in Section 27.5 (p.1233).

Ozone (O_3) is used to convert a C=C bond into two C=O bonds.

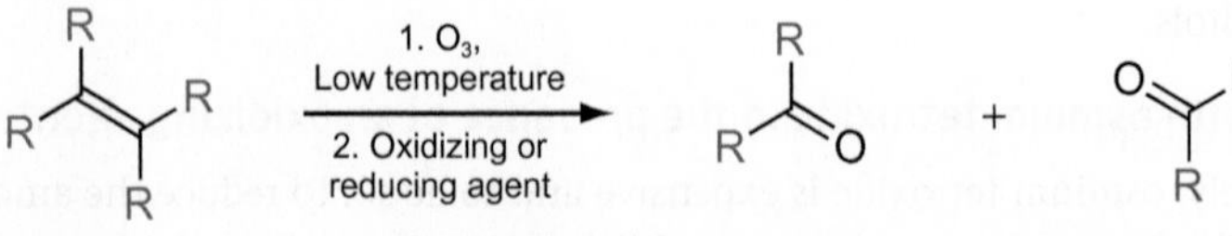
1. O_3, Low temperature
2. Oxidizing or reducing agent

Aldehydes, (RCHO), ketones (RCOR), and/or carboxylic acids (RCO_2H) are formed, depending on the structure of the alkene and whether, after the reaction with ozone, an oxidizing or reducing agent is used.

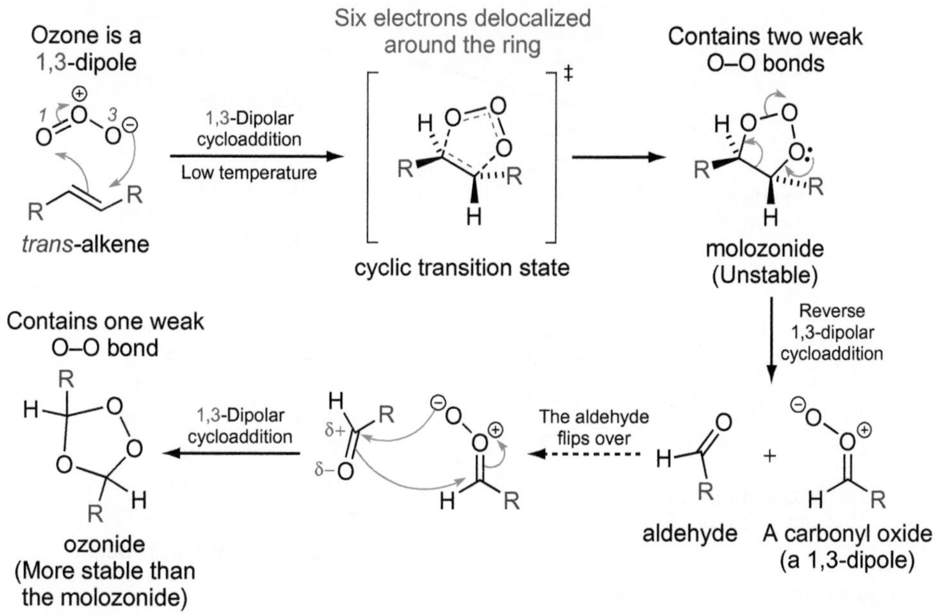
Ozone is a 1,3-dipole
trans-alkene
1,3-Dipolar cycloaddition
Low temperature
Six electrons delocalized around the ring
cyclic transition state
Contains two weak O–O bonds
molozonide (Unstable)
Reverse 1,3-dipolar cycloaddition
aldehyde
A carbonyl oxide (a 1,3-dipole)
The aldehyde flips over
1,3-Dipolar cycloaddition
Contains one weak O–O bond
ozonide (More stable than the molozonide)

Figure 21.29 Reaction of an alkene with ozone to form an ozonide.

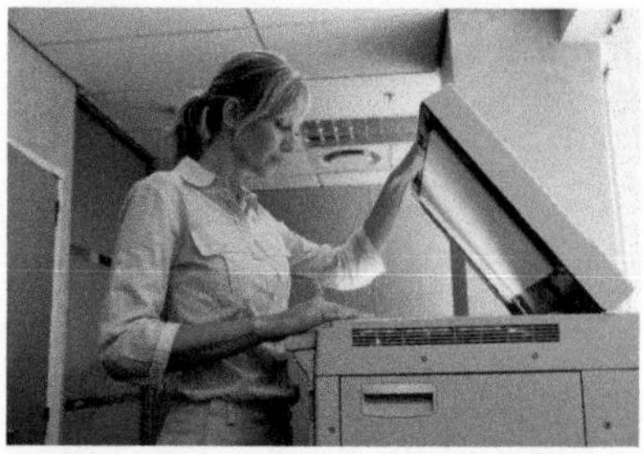

Ozone is produced by some electrical devices, including laser printers and photocopiers.

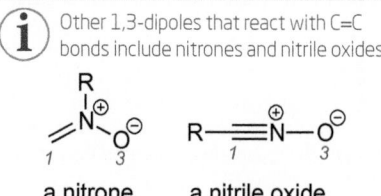

Other 1,3-dipoles that react with C=C bonds include nitrones and nitrile oxides.

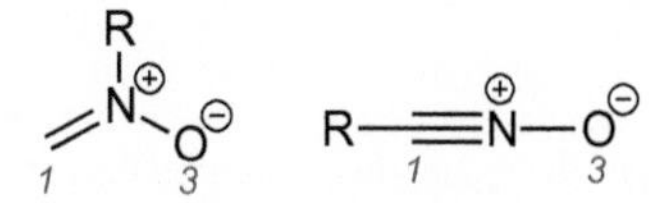

a nitrone a nitrile oxide

The carbonyl oxide ($RCH=O^+-O^-$ or $R_2C=O^+-O^-$) is sometimes called a Criegee intermediate after the German chemist Rudolf Criegee (1902–1975) who elucidated the reaction mechanism for ozonolysis.

Addition of nucleophiles to C=O bonds is discussed in Chapters 23 and 24.

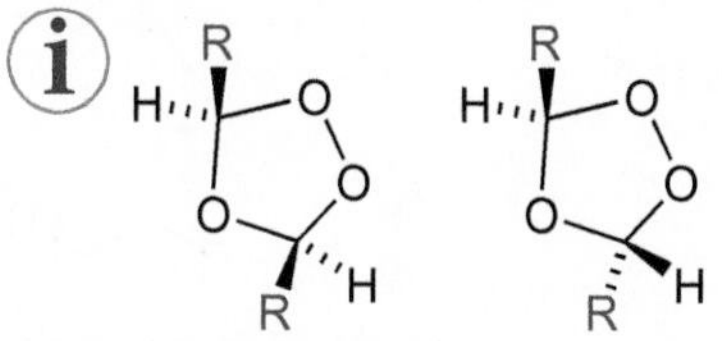

Diastereomers of an ozonide.

Ozone (O_3) is a symmetrical bent molecule. The two oxygen atoms at the ends of the molecule share a negative charge and the central oxygen atom is positively charged. Because O_3 has both a positive charge and a negative charge it is called a **dipolar reagent**.

The major resonance forms of ozone

O_3 reacts with a C=C bond (usually at low temperature) in a concerted cycloaddition reaction to form a molozonide (primary ozonide or 1,2,3-trioxolane), as shown in Figure 21.29. The C=C bond donates two electrons to the positively charged oxygen atom in O_3 to form a C–O bond. As this happens, one of the partially negatively charged oxygen atoms in O_3 donates two electrons to one of carbon atoms of the C=C bond to form a second C–O bond. In O_3, the two oxygen atoms that react to form the C–O bonds have a 1,3-relationship. Because of this, O_3 is called a **1,3-dipole** and the reaction of O_3 with a C=C bond is called a **1,3-dipolar cycloaddition**.

The molozonide is unstable because it contains two weak O–O bonds (an O–O bond has a bond enthalpy of around +150 kJ mol^{-1}) and it immediately undergoes a reaction that is the reverse of a 1,3-dipolar cycloaddition. The electrons move in the molozonide to break an O–O and a C–C bond to produce a C=O bond and a carbonyl oxide, which is a 1,3-dipole. Finally, the carbonyl oxide reacts with the C=O bond in a second 1,3-dipolar cycloaddition to form an ozonide (or a 1,2,4-trioxolane). This requires the aldehyde to flip over so that the negatively charged oxygen atom in the 1,3-dipole can attack the partially positive (δ+) carbon atom in the C=O bond. Note that the ozonide is formed as a mixture of diastereomers (see margin) because the 5-membered ring is formed with the R groups on the same or the opposite sides of the ring.

Although ozonides are more stable than molozonides, ozonides are extremely reactive (they can explode) and are not usually isolated from reactions. Instead, the ozonide is converted into more stable carbonyl compounds by reaction with a reducing or oxidizing agent, as shown in Figure 21.30. Reaction of an ozonide with a reducing agent produces aldehydes (RCHO) and/or ketones (RCOR). Reaction with an oxidizing agent produces carboxylic acids (RCOOH) and/or ketones (RCOR). Transformation of a C=C bond into two carbonyl compounds by reaction with O_3, followed by an oxidizing or reducing agent, is called **ozonolysis**.

Ozonizers are used in industry and in households to make ozone. Ozone is used to purify water or air, as it rapidly oxidizes and destroys any bacteria and viruses that are present. The picture shows an ozonizer used for purifying water supplies.

Dimethyl sulfide (Me_2S) has a distinctive smell often described as cabbage-like. It is a component of the smell produced from cooking certain vegetables, including cabbage and asparagus.

PPh_3 is called triphenylphosphine. It reacts with an ozonide to form triphenylphosphine oxide, $Ph_3P{=}O$, which has a strong P=O double bond (see also the Wittig reaction, Section 23.2, p.1071).

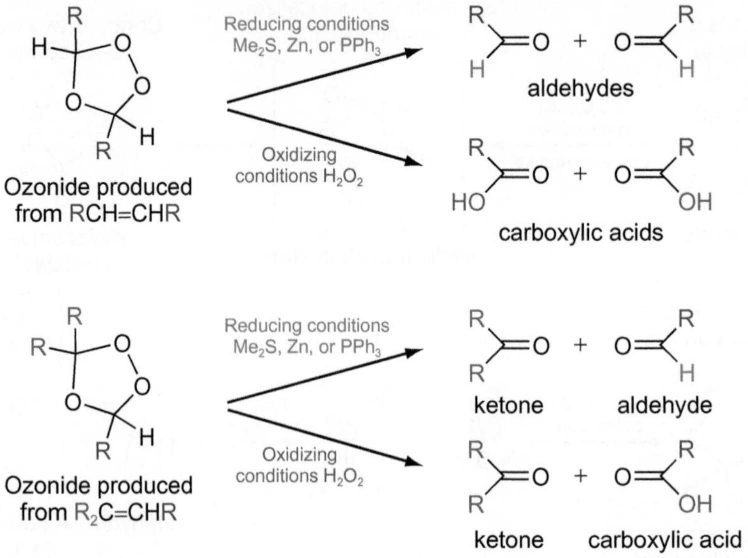
Ozonide produced from RCH=CHR
Reducing conditions Me2S, Zn, or PPh3
aldehydes
Oxidizing conditions H2O2
carboxylic acids
Ozonide produced from R2C=CHR
Reducing conditions Me2S, Zn, or PPh3
ketone
aldehyde
Oxidizing conditions H2O2
ketone
carboxylic acid

Figure 21.30 Reactions of ozonides.

Summary

Important pericyclic reactions involving alkenes:

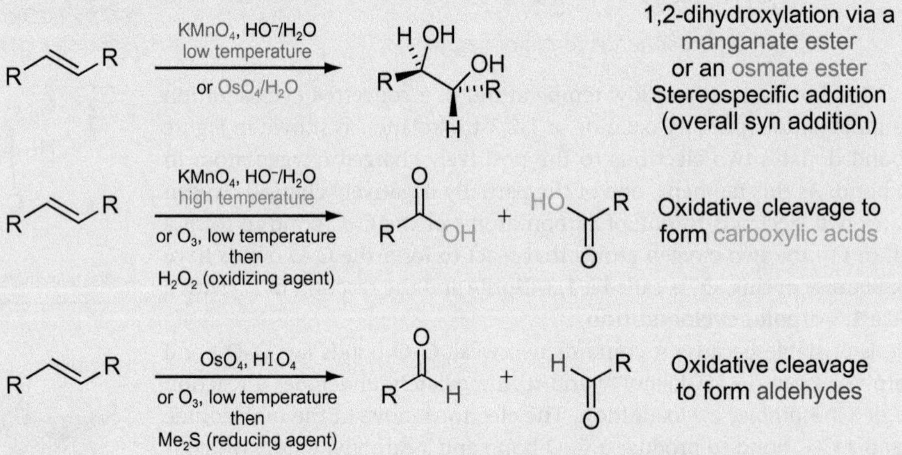
KMnO4, HO−/H2O low temperature
or OsO4/H2O
1,2-dihydroxylation via a manganate ester or an osmate ester
Stereospecific addition (overall syn addition)
KMnO4, HO−/H2O high temperature
or O3, low temperature then H2O2 (oxidizing agent)
Oxidative cleavage to form carboxylic acids
OsO4, HIO4
or O3, low temperature then Me2S (reducing agent)
Oxidative cleavage to form aldehydes

For practice questions on these topics, see questions 2 and 3 at the end of this chapter (pp.999-1000).

HC≡CH and RC≡CH react with strong bases to form the alkynyl ions HC≡C⁻ and RC≡C⁻, respectively (Section 19.2, p.890). Alkynylsodium salts (e.g. RC≡C⁻ Na⁺) are strong nucleophiles that react with various electrophiles (Section 23.2, p.1068).

21.5 Electrophilic addition reactions of alkynes

Alkynes undergo similar electrophilic addition reactions to those of alkenes. The electron-rich C≡C bond acts as a nucleophile in these reactions. When comparing the relative reactivity of C≡C and C=C bonds, the relative nucleophilicity of the bonds needs to be considered as does the stability of any intermediates, such as carbocations.

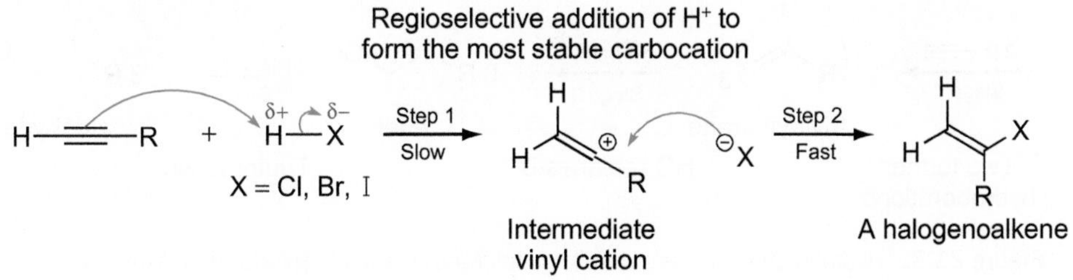
Regioselective addition of H+ to form the most stable carbocation
δ+ δ−
X = Cl, Br, I
Step 1
Slow
Step 2
Fast
Intermediate vinyl cation
A halogenoalkene

Figure 21.31 Alkynes react with HX (where X = Cl, Br, or I) to produce halogenoalkenes.

Addition of hydrogen halides to alkynes

Alkynes react with HX (where X = Cl, Br, or I) in a two-step reaction (Figure 21.31) to produce halogenoalkenes (or vinyl halides, e.g. $RC(X)=CH_2$). The mechanism of addition of HX to a C≡C bond is analogous to the addition of HX to a C=C bond (p.970).

HX adds selectively to unsymmetrical alkynes to form the more substituted halogenoalkene as predicted by Markovnikov's rule. This is because addition of H^+ to the C≡C bond selectively forms the more stable vinyl cation. In Figure 21.31, a secondary vinyl carbocation ($RC^+=CH_2$) is selectively formed because this cation is more stable than a primary vinyl carbocation ($RCH=CH^+$) (see margin).

Because a C≡C bond is more electron-rich than a C=C bond, you might expect the C≡C bond to react faster with HX. However, addition of HX to a C≡C bond is *slower* than addition to a C=C bond. The reason for this is the different stabilities of vinyl and alkyl cations.

Addition of H^+ to a C≡C bond forms a vinyl cation (e.g. $RC^+=CH_2$), whereas addition of H^+ to a C=C bond forms an alkyl cation (e.g. RCH^+CH_3). A vinyl cation is less stable than a similarly substituted alkyl cation, because of the electronic effects of an alkylidene group (e.g. $=CH_2$) versus an alkyl substituent (e.g. $-CH_3$). The greater the *s* character of a hybrid orbital on carbon, the closer the electrons are held to the positively charged nucleus, and the more electron-withdrawing it is. So, the sp^2 carbon (33% *s* character) of the alkylidene group is more electron-withdrawing than an sp^3 carbon of an alkyl substituent (25% *s* character), and it cannot stabilize the adjacent positive charge as well as an alkyl group. Because the vinyl cation is less stable than a similarly substituted alkyl cation, the Gibbs energy of activation for its formation will be higher and it will be formed more slowly.

Addition of HX to a C≡C bond can often be stopped after addition of one equivalent of HX. If two equivalents of HX react with a C≡C bond, a second molecule of HX adds to the halogenoalkene to form a geminal dihalogenoalkane (**geminal** indicates that both halogen atoms are on the same carbon atom).

Organosodium compounds are normally ionic

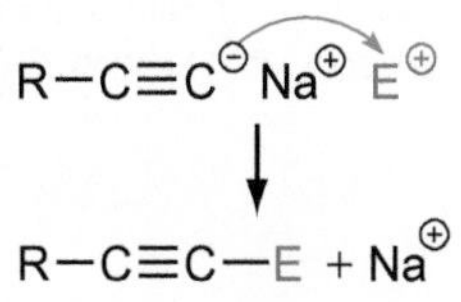

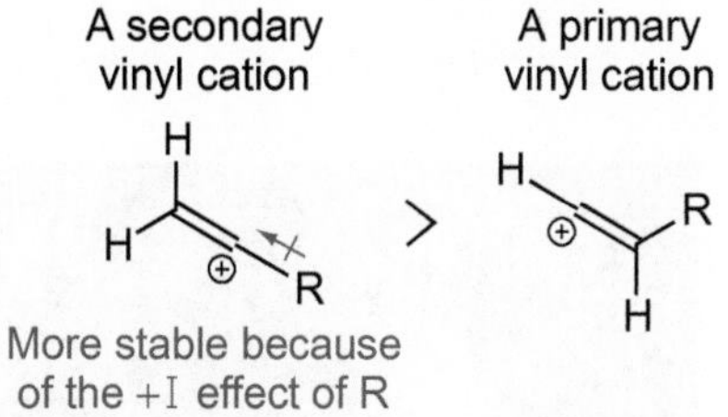
A secondary vinyl cation
A primary vinyl cation
More stable because of the +I effect of R

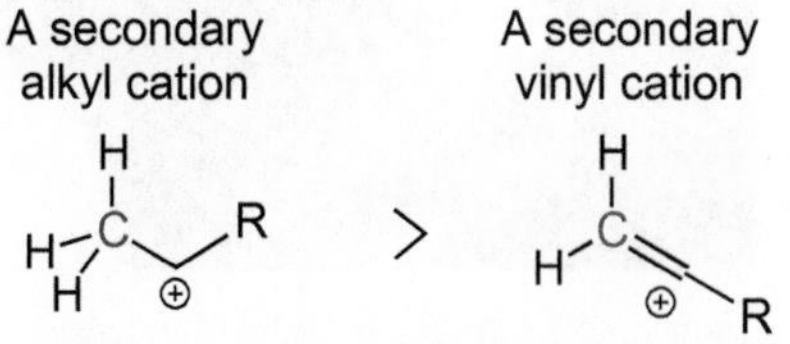
A secondary alkyl cation
A secondary vinyl cation

The sp^2 carbon is more electron-withdrawing than the sp^3 carbon, so the $H_2C=$ group cannot stabilize the cation so well as the H_3C- group

The Gibbs energy of activation is introduced in Section 9.8 (p.433).

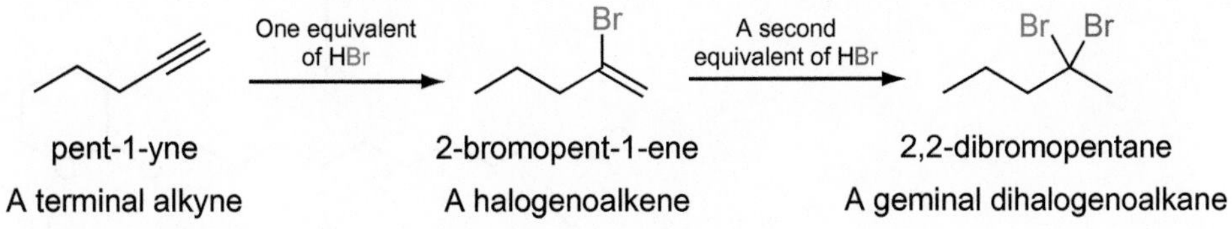
One equivalent of HBr
A second equivalent of HBr
pent-1-yne
A terminal alkyne
2-bromopent-1-ene
A halogenoalkene
2,2-dibromopentane
A geminal dihalogenoalkane

Addition of borane to alkynes followed by oxidation

Aldehydes (RCHO) or ketones (RCOR) are formed from alkynes by reaction of the C≡C bond with diborane (B_2H_6) followed by oxidation using a solution of hydrogen peroxide (H_2O_2) and aqueous sodium hydroxide. A terminal alkyne (RC≡CH) reacts to form an aldehyde (RCH_2CHO), while an internal alkyne (RC≡CR) reacts to form a ketone ($RCOCH_2R$).

The aldehyde (RCH_2CHO) or ketone ($RCOCH_2R$) is formed by overall addition of H_2O to the C≡C bond so this is an example of a hydration reaction, as illustrated in Figure 21.32. The mechanism for this transformation involves four stages and is very similar to

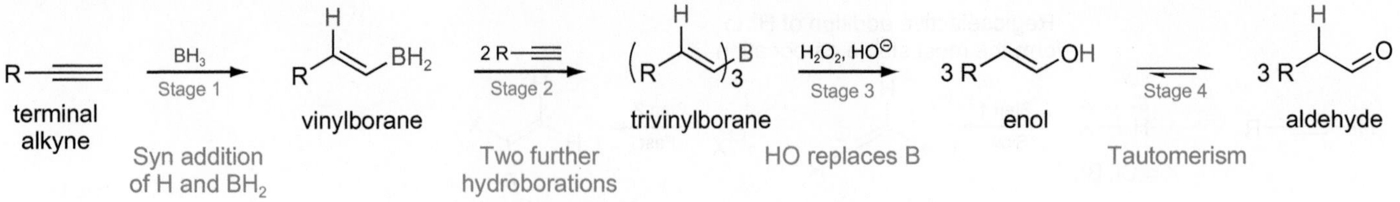

BH3
Stage 1
terminal alkyne
Syn addition of H and BH2
vinylborane
2 R
Stage 2
Two further hydroborations
trivinylborane
H2O2, HO⊖
Stage 3
HO replaces B
3 R
enol
Stage 4
Tautomerism
aldehyde

Figure 21.32 Hydroboration of a terminal alkyne followed by oxidation produces an aldehyde.

the reaction of an alkene with B_2H_6 (which reacts as BH_3) then H_2O_2/HO^- (Section 21.3, p.983). Notice that stage 3 produces an enol (RCH=CHOH) that rearranges to form an aldehyde (RCHHCHO)—an H atom moves from oxygen to carbon and the position of the double bond changes. The equilibrium between an aldehyde (or ketone) and an enol is called **tautomerism**.

Enols have an OH group attached to a C=C bond; see Section 2.6 (p.95). Tautomerism is discussed in Section 23.3 (p.1082).

Worked example 21.5 discusses the synthesis of a pheromone, which includes the hydration of a terminal alkyne.

Worked example 21.5 Synthesis of a pheromone

Pheromones are volatile chemicals that are emitted by an individual and trigger a response in an individual of the same species. For example, the female pea moth (*Cydia nigricana*) attracts males for mating by producing a sex pheromone that contains (*E*)-dodec-10-enyl ethanoate. Farmers use this compound as a lure to trap pea moths and prevent destruction of valuable pea crops.

An outline of a synthesis of (*E*)-dodec-10-enyl ethanoate is shown below. The synthesis uses a tetrahydropyranyl (THP) group, which is a protecting group for alcohols. A **protecting group** converts a reactive functional group into an unreactive form such that, after the desired transformation(s), the original functional group can be regenerated.

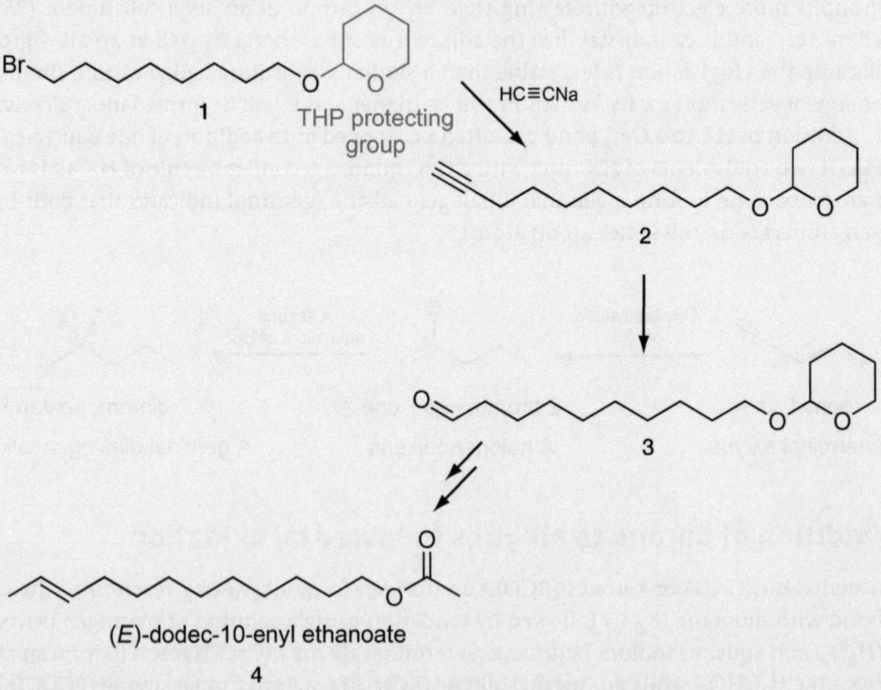

Br
1
THP protecting group
HC≡CNa
2
3
(E)-dodec-10-enyl ethanoate
4

(a) Provide a mechanism to explain how compound **1** is converted into alkyne **2**.

(b) Suggest reagents for converting alkyne **2** into aldehyde **3**.

Pea moths attack peas, sweet peas, and vetch in the UK. The moths lay eggs on pea flowers and the resulting caterpillars tunnel into the pod and feed on the peas.

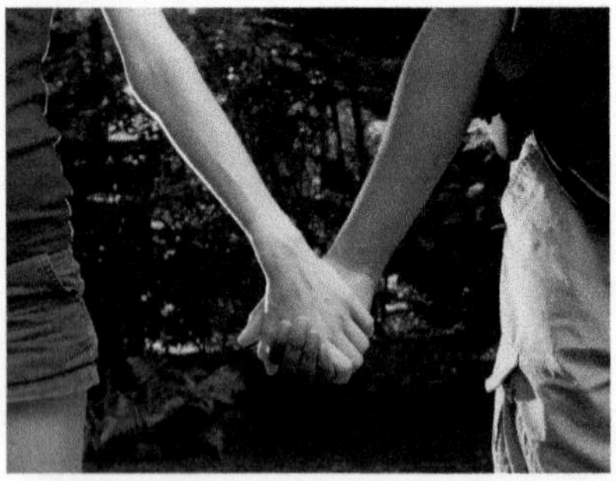

Despite more than 50 years of research, there is no direct evidence for the existence of a human pheromone. Scientists have had some success in demonstrating that exposure to body odour can elicit responses in other humans—for example, research has showed that the scent of ovulating women could cause testosterone levels to increase in men—but no human pheromones have been identified.

➜ Strategy

(a) Consider the type of reaction involved and identify the nucleophile and electrophile.

(b) Look at the structures of the reactant and product to determine what reaction is required. Consider how a RC≡CH bond can be regioselectively hydrated to form an aldehyde, RCH_2CHO.

Solution

(a) This is a nucleophilic substitution reaction. The partially positive carbon adjacent to the Br atom in compound **1** is the electrophile and $Na^+ \, ^-C{\equiv}CH$ is the nucleophile.

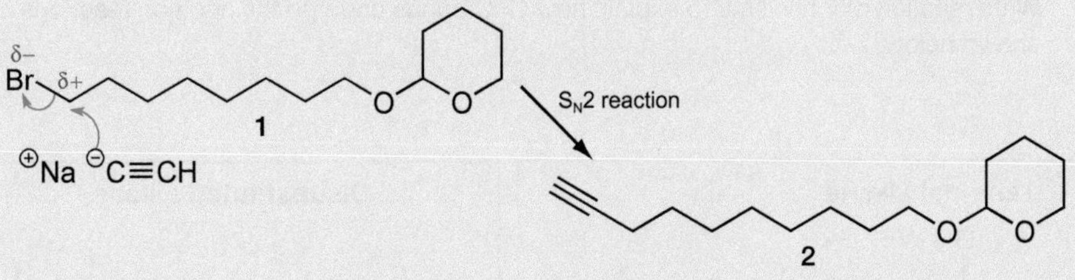
δ−
Br
δ+
1
⊕Na ⊖C≡CH
O
O
S$_N$2 reaction

2

+

NaBr

(b) Addition of B_2H_6 (which acts as BH_3) followed by H_2O_2/HO^-.

Question

Suggest reagents for converting alkyne **2** into compound **5**.

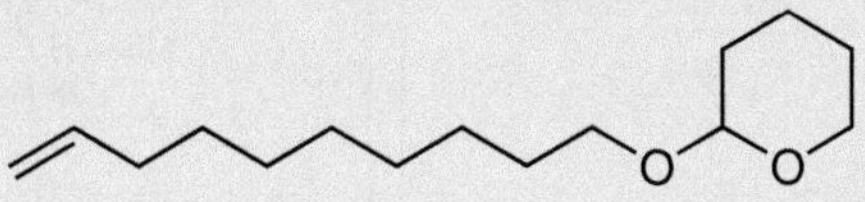

5

» Summary

Two important electrophilic addition reactions of alkynes:

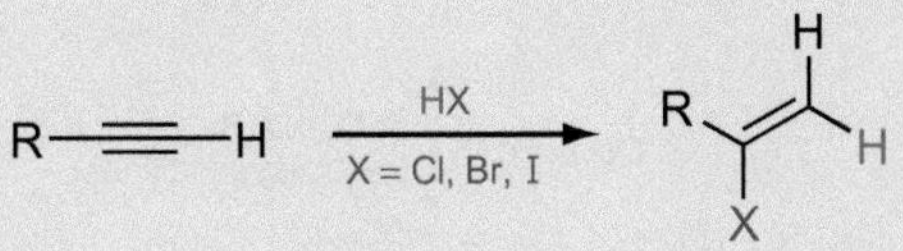
R—≡—H
HX
X = Cl, Br, I
R
H
H
X

Via an intermediate vinyl carbocation
Regioselective addition (favours Markovnikov product)
Slower than addition of HX to a C=C bond

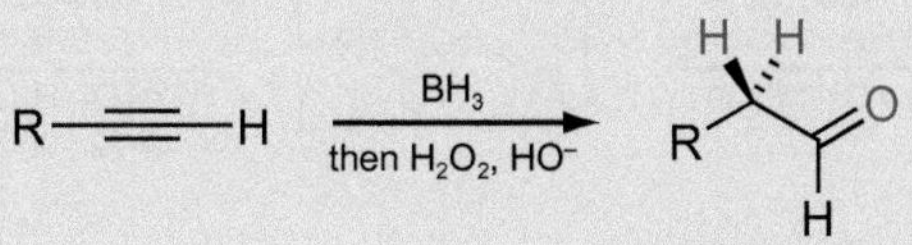
R—≡—H
BH3
then H2O2, HO−
H H
R
O
H

Hydration via an intermediate enol
Regioselective addition to give an aldehyde

- HC≡CH and RC≡CH react with strong bases to form alkynylmetal reagents (e.g. $HC{\equiv}C^- \, ^+Na$), which can act as nucleophiles.

? For a practice question on this topic, see question 4 at the end of this chapter (p.1000).

Concept review

By the end of this chapter, you should be able to do the following.

- Describe how alkenes are prepared from halogenoalkanes, alcohols, alkynes, and aldehydes/ketones.
- Describe how alkynes are prepared from 1,2-dibromoalkanes and how substituted alkynes are prepared by alkylation of alkynyl anions ($RC{\equiv}C^-$) with halogenoalkanes.
- Understand how C=C and C≡C bonds react in electrophilic addition reactions.
- Write reaction mechanisms to explain how C=C bonds undergo the addition reactions shown below.

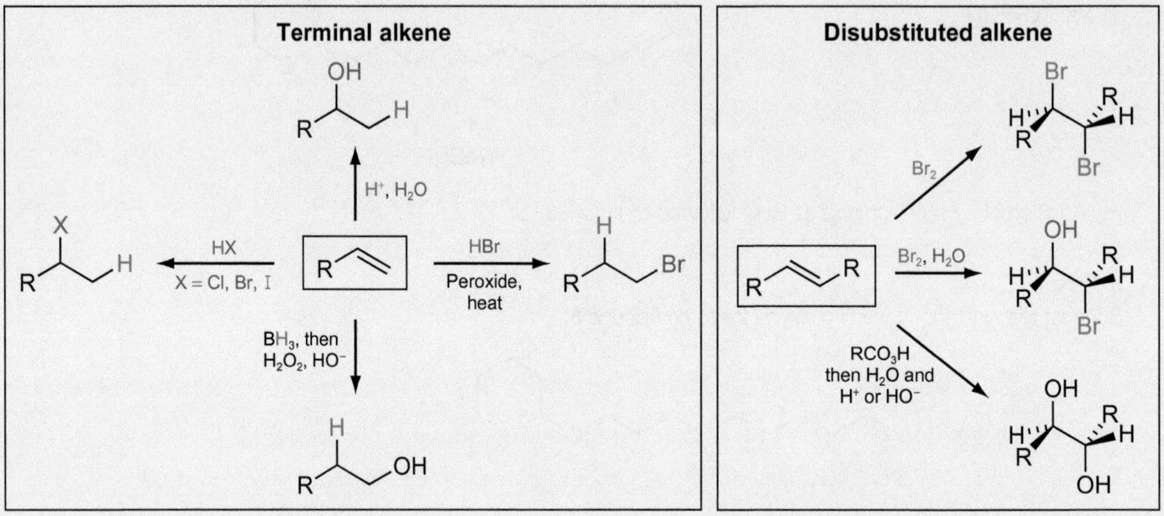

Terminal alkene
OH
H
R
H⁺, H₂O
X
H
R
HX
X = Cl, Br, I
R
HBr
Peroxide, heat
H
R
Br
BH₃, then H₂O₂, HO⁻
H
R
OH
Disubstituted alkene
R
R
Br₂
Br
H
R
R
H
Br
Br₂, H₂O
OH
H
R
R
H
Br
RCO₃H then H₂O and H⁺ or HO⁻
OH
H
R
R
H
OH

- Write reaction mechanisms to explain how C=C bonds undergo the pericyclic reactions shown below.

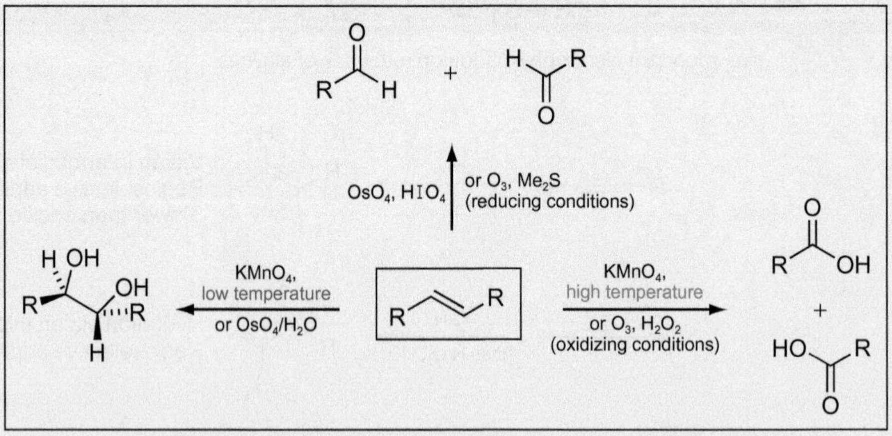

O
R
H
+
H
R
O
OsO₄, HIO₄
or O₃, Me₂S (reducing conditions)
H OH
R
OH
R
H
KMnO₄, low temperature
or OsO₄/H₂O
R
R
KMnO₄, high temperature
or O₃, H₂O₂ (oxidizing conditions)
O
R
OH
+
HO
R
O

- Write reaction mechanisms to explain how C≡C bonds undergo the addition reactions shown below.

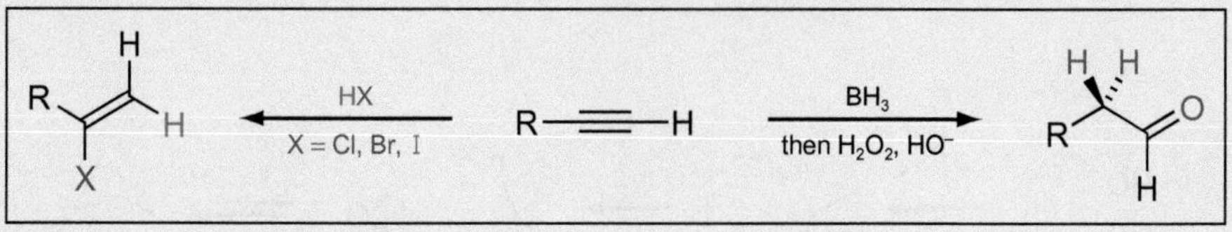

Questions

More challenging questions are indicated by an asterisk *.

1. Suggest mechanisms for the following electrophilic addition reactions. (Section 21.3)

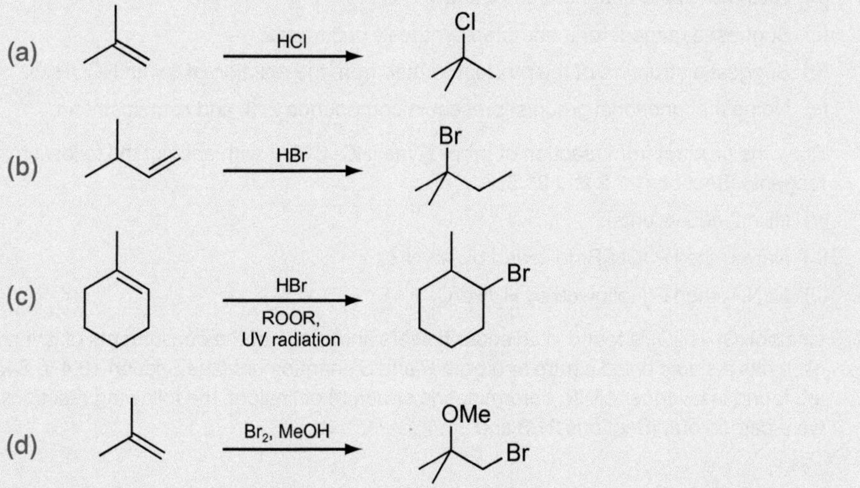

2. Suggest reagents (a–h) for the following transformations. More than one step may be required for each transformation. (The relative stereochemistry of the products is shown, not the absolute stereochemistry.) (Sections 21.3 and 21.4)

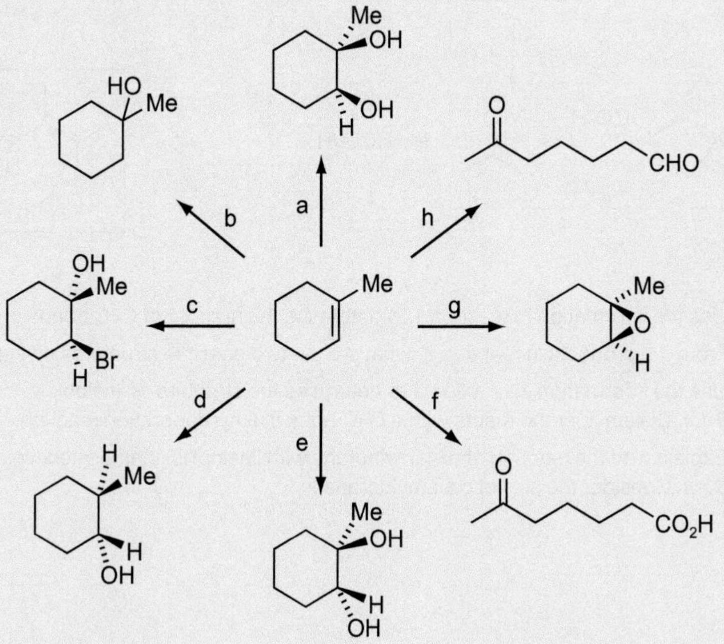

The European elm bark beetle (*Scolytus multistriatus*) carries Dutch elm disease fungus from tree to tree. The female beetle burrows into the bark creating long narrow chambers.

3. An outline of a racemic synthesis of α-multistriatin, a pheromone of the elm bark beetle, is shown below. (Sections 21.3 and 21.4)

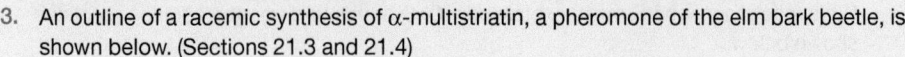

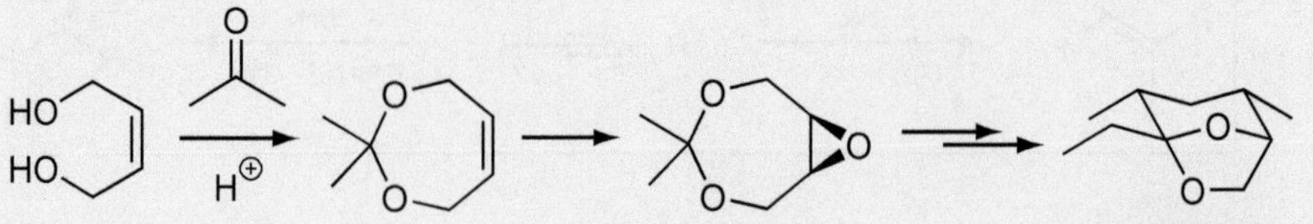
HO
HO
O
H⊕
O
O
O
O
O
O
O

1 2 3 (The relative stereochemistry is shown) α-multistriatin (A 1:1 mixture of enantiomers)

(a) Suggest reagents for a one-step synthesis of **1** from but-2-yne-1,4-diol.

(b) Give the IUPAC name of compound **1**.

(c) Suggest a reagent for a one-step synthesis of **3** from **2**.

(d) Suggest a structure of the product formed from the reaction of **3** with HO^-/H_2O.

(e) Name the functional group(s) present in compounds **2**, **3**, and α-multistriatin.

4. Draw the product from reaction of prop-1-yne ($HC{\equiv}CCH_3$) with each of the following reagents (Sections 21.2 and 21.5):

(a) HBr (2 equivalents);

(b) $NaNH_2$ then $PhCH_2Br$ followed by Na/NH_3;

(c) $NaNH_2$ then EtI followed by $H_2/Pd/C$.

5. Linalool ($C_{10}H_{18}O$) is found in lavender flowers and is one of the constituents of lavender oil. It has a single chiral centre and both *R* and *S* enantiomers (see Section 18.4, p.845) are found in lavender oil. To determine the structure of linalool, the following reactions were carried out. (Sections 21.3 and 21.4)

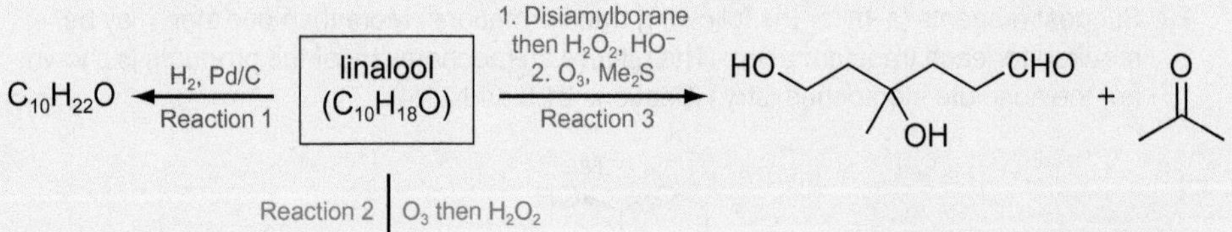
C10H22O
H2, Pd/C
Reaction 1
linalool
(C10H18O)
1. Disiamylborane
then H2O2, HO−
2. O3, Me2S
Reaction 3
HO
CHO
OH
+
O
Reaction 2
O3 then H2O2

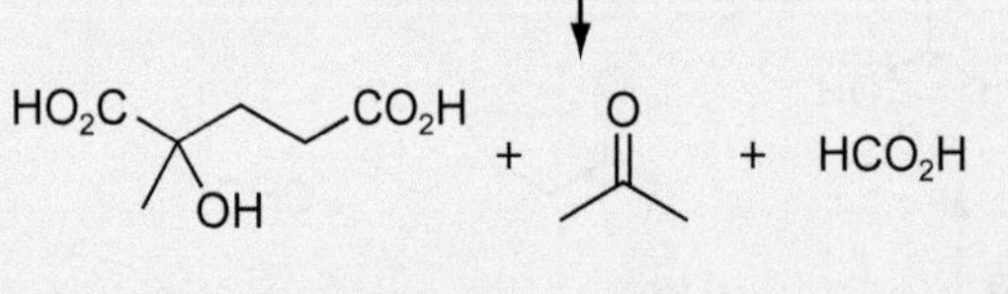
HO2C
CO2H
OH
+
O
+ HCO2H

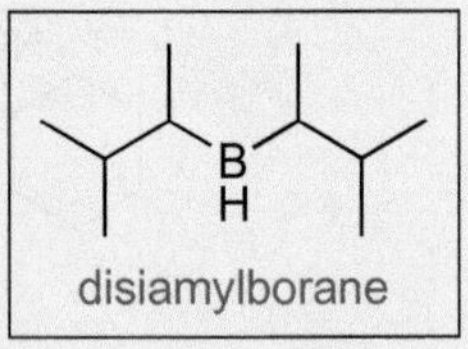
B
H
disiamylborane

Lavender oil from the lavender plant is widely used in perfumery.

(a) Use the information in reaction 1 to determine the number of C=C bonds in linalool.

(b) From the products in reaction 2, what are the two possible structures of linalool?

(c) Use the information in reaction 3 to determine the structure of linalool. (*Hint.* Disiamylborane reacts with a C=C bond in a hydroboration reaction.)

(d) Explain why the reaction of disiamylborane with linalool is chemoselective. (*Hint.* Consider the size of disiamylborane.)

6.* Reserpine is a natural product that has been used in medicine to control high blood pressure and to relieve psychotic symptoms. Part of a 21-step synthesis of the natural product reserpine, published by the American chemist R. B. Woodward, is shown below. (Section 21.3)

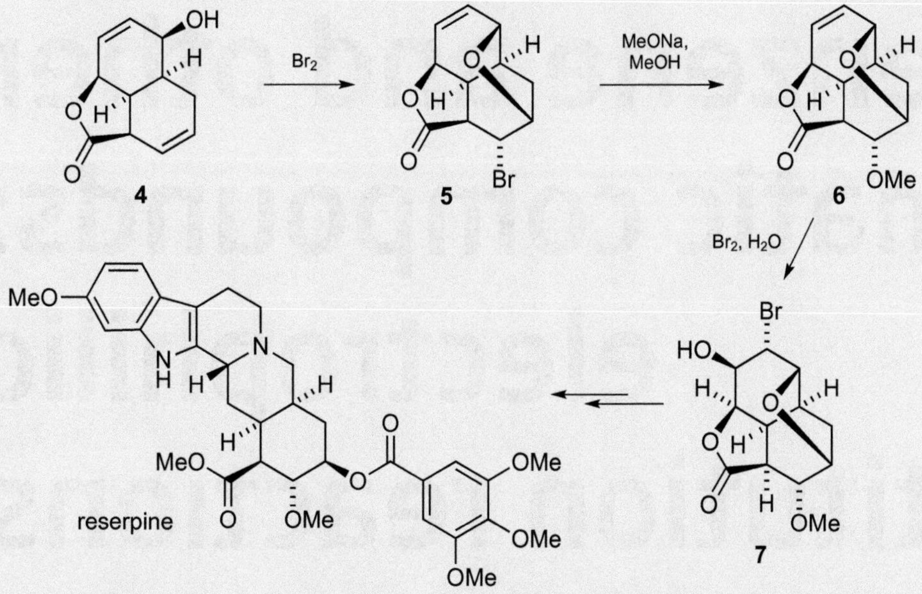
OH
H
H
O
O
4
Br_2
H
O
H
O
O
Br
5
MeONa,
MeOH
H
O
H
O
O
OMe
6
Br_2, H_2O
Br
HO
H
H
O
H
O
O
H
OMe
7
MeO
N
NH
H
H
H
H
MeO
O
OMe
O
O
OMe
OMe
OMe
reserpine

(a) Give a reaction mechanism that explains how **4** is converted into **5**. (*Hint:* the reaction involves an intermediate bromonium ion).

(b) The first step in the conversion of **5** into **6** involves deprotonation of **5** and elimination of bromide ion to form an alkene. Suggest a structure of the intermediate alkene.

(c) Give a reaction mechanism that explains how **6** is converted into **7**. (*Hint:* the reaction involves an intermediate bromonium ion).

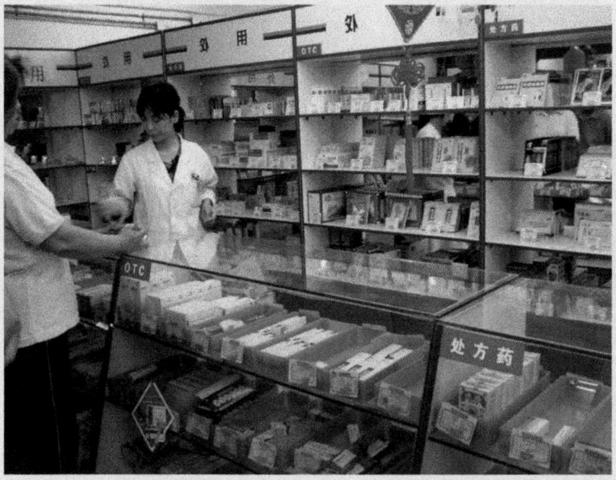
OTC
处方药

More than 10 million patients use reserpine tablets in China, with between 8 and 9 billion pills taken annually.

Painters use pigments in oil paints, acrylics, and watercolour paints. Organic and inorganic pigments are mixed with a binder that prevents the paint from running as it is applied. ➤

22

Benzene and other aromatic compounds: electrophilic substitution reactions

This chapter builds on the following topics:

- **Benzene and arenes** Section 2.5, p.88
- **Fundamental concepts of organic reaction mechanisms** Section 19.1, p.863
- **Polar reactions** Section 19.2, p.902
- **Regioselectivity** Section 19.3, p.909
- **The mechanisms of nucleophilic substitution reactions** Section 20.3, p.930
- **Electrophilic addition reactions of alkenes** Section 21.3, p.970

◀ Today, there are several types of hair dyes on the market. Temporary hair dyes contain large molecules that coat the surface of hair. Semi-permanent dyes contain smaller molecules that can temporarily pass into hair, whereas permanent dyes are small enough to pass into hair and, once there, they react to form larger molecules that become trapped.

Azo dyes: the start of the rainbow

Dyes are generally organic compounds that colour materials by forming covalent or non-covalent bonds with the molecules in the materials, such as hair and cloth.

Naturally occurring dyes are extracted from animals, insects, and vegetables, and evidence for their use dates back to 2600 BC. However, natural dyes are not very colour-fast and the range of colours is limited. Until about 150 years ago, good dyes were expensive and were only for the wealthy. Most people wore drab clothes dyed with cheap vegetable dyes.

It was not until the middle of the nineteenth century that chemists started to prepare synthetic dyes and pigments (Box 22.3, p.1011). The breakthrough came in 1856 when William Perkin accidentally discovered the first synthetic dye by oxidizing phenylamine (aniline; $C_6H_5NH_2$) to form a complex organic salt called Perkin's mauve (after the French word for the mallow plant), mauveine, or aniline purple. This discovery encouraged other chemists to investigate the preparation of new dyes. In 1862, Charles Méne prepared a yellow-brown product that 'dyes cotton and silk perfectly'. Marketed as Aniline Yellow, this discovery saw the birth of the most important class of synthetic dyes called azo dyes.

▲ This silk dress was dyed mauve in 1862.

Azo compounds contain the azo functional group (–N=N–) and in azo dyes this group is usually linked to two aromatic rings (Ar–N=N–Ar; Section 2.6, p.98). Azo compounds are prepared by joining together (**coupling**) a diazonium salt ($R–N_2^+ X^-$) with an aromatic compound called a **coupling agent**. For example, Aniline Yellow (4-benzeneazoaniline) is prepared by coupling benzenediazonium chloride ($C_6H_5–N_2^+ Cl^-$) with phenylamine. In the first step, phenylamine acts as a nucleophile and reacts with benzenediazonium chloride (the electrophile) to form a new C–N bond. The product then loses H^+ to reform an aromatic ring and this gives the azo dye (Cl^- or a solvent molecule can act as the base). Because aniline reacts with an electrophile to substitute the H atom on the ring by a PhN=N– group, this is called an **electrophilic substitution reaction**. This is the most important type of reaction of aromatic compounds and these reactions are the main topic of this chapter.

The part of the dye molecule that is mostly responsible for the colour is called a **chromophore**. In azo compounds, the chromophore is the Ar–N=N–Ar group, where Ar are aromatic rings. By changing the structures of the aromatic rings, or by changing the substituents on the rings, it is possible to produce new colours. Sudan I, for example, is a red azo dye. Many azo dyes are now available to give a rainbow of colours, primarily yellow, orange, or red, with a small number of blues and greens.

▼ Aniline Yellow (4-benzeneazoaniline) is prepared by coupling benzenediazonium chloride with phenylamine.

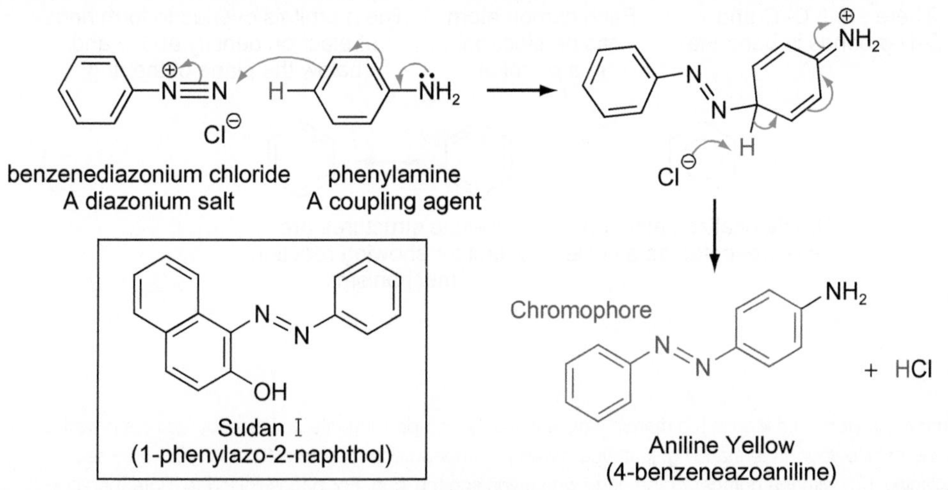
benzenediazonium chloride
A diazonium salt
phenylamine
A coupling agent
Sudan I
(1-phenylazo-2-naphthol)
Chromophore
+ HCl
Aniline Yellow
(4-benzeneazoaniline)

In recent years, the use of some azo dyes has been curtailed because of safety concerns and their potential to cause cancer (carcinogenic). For example, Sudan I causes cancer and so is not authorized as a food additive in the European Union (Box 11.1, p.522). Since July 2003, cargoes of red chilli powder coming into any EU Member State must be tested and found to be free of Sudan I. Even so, prepared foods contaminated with Sudan I were found in 2005 and were traced back to an imported batch of chilli powder. Affected products had to be removed from shelves in shops. Many azo dyes, however, are non-carcinogenic and will continue to be used well into the twenty-first century.

Benzene and arenes are introduced in Section 2.5 (p.88).

Sweet smelling aromatic spices have been used in Asian cuisine for centuries. For example, spices including turmeric, coriander, and cumin are used in curries.

Functional groups are introduced in Section 2.3 (p.77).

Benzene (C_6H_6) is one of the most fascinating organic molecules and it has an interesting history (Box 22.1, p.1006). The six carbon atoms in benzene are linked in a planar hexagon and, as each carbon atom is bonded to only one hydrogen atom, benzene is an unsaturated hydrocarbon. For an unsaturated hydrocarbon, benzene is unusually unreactive. Its stability is attributed to the cyclic arrangement in which the 6 π electrons are delocalized over the ring. Benzene is described as **aromatic**. The term aromatic was originally used to describe compounds with pleasant fragrances but is now used to define a class of organic compounds called arenes, with related chemical structures (Section 22.1, p.1008).

Benzene reacts with strong electrophiles in electrophilic substitution reactions. These reactions form substituted benzenes that retain the aromatic ring (Section 22.2, p.1013). Many important functional groups can be attached to the benzene ring to give substituted benzenes that can act as medicines, dyes, insecticides, and pesticides (Sections 22.3, p.1027, and 22.4, p.1036).

i Alkenes (e.g. $R_2C{=}CR_2$) and benzene are both unsaturated hydrocarbons that react with electrophiles. Alkenes react with electrophiles in electrophilic *addition* reactions (Section 21.3, p.970). Benzene reacts with electrophiles in electrophilic *substitution* reactions (Section 22.2, p.1013).

22.1 The structure of benzene and other aromatic compounds

i Sometimes, bond lengths are quoted in angstroms (Å). 1 Å = 10^{-10} m so 1.54 Å = 154 pm.

The bonding in benzene is discussed in Section 5.5 (p.224).

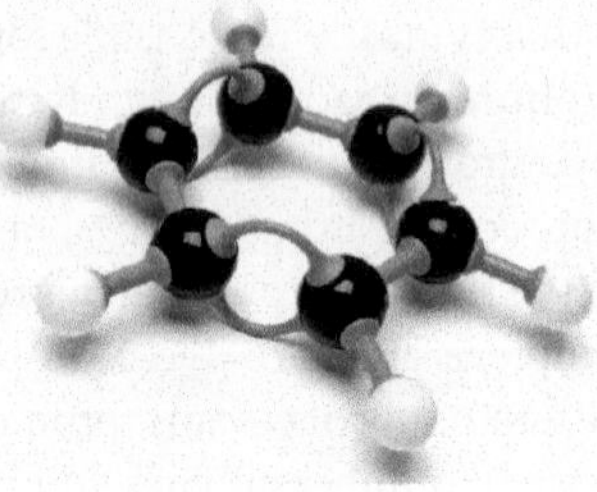

A molecular model of benzene.

Benzene (C_6H_6) has six sp^2 hybridized carbon atoms that link together to form a planar ring. The ring is hexagonal, with all six C–C–C bond angles equal to 120°. The six carbon–carbon bonds are all 139 pm long, which is in between the normal values for a C–C (154 pm) single bond and a C=C (134 pm) double bond.

Each of the six sp^2 hybridized carbon atoms in benzene has an unhybridized *p* orbital, at right angles to the plane of the ring, containing a single electron. The *p* orbitals are parallel and are close enough to overlap side-on with *p* orbitals on adjacent carbon atoms. Because the *p* orbitals combine side-on, the electrons in the *p* orbitals are called π electrons. The overlapping *p* orbitals produce two continuous rings of π electrons above, and below, the plane of the benzene ring. These electrons are not bound to any particular carbon atom and are said to be **delocalized** (spread out) over the ring. Drawing a circle in the middle of the hexagon represents the delocalization of electrons.

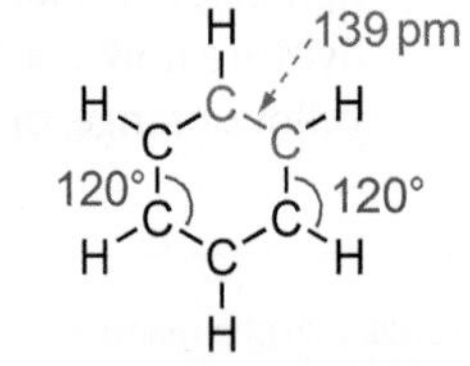
139 pm
120°
120°

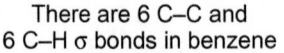

There are 6 C–C and 6 C–H σ bonds in benzene

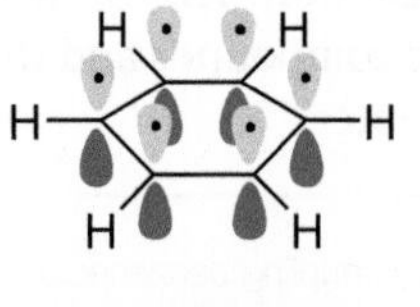

Each carbon atom has an electron in a *p* orbital

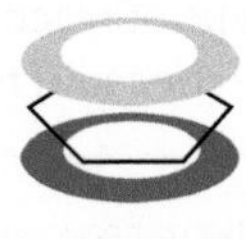

The *p* orbitals overlap to form rings of electron density above and below the plane of the ring

The delocalized electrons are represented as a circle

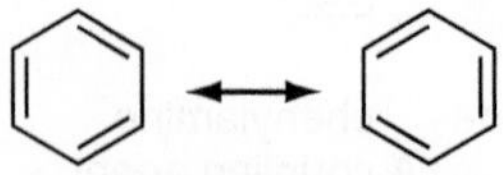

Kekulé structures are useful for showing reaction mechanisms

Each continuous ring of π electrons is sometimes described as being doughnut-shaped.

Online Support. The icon in the margin indicates that accompanying interactive resources are provided online to help your understanding: just type www.chemtube3d.com/burrows/123 into your browser, replacing 123 with the number of the page where you see the icon. For pages linking to more than one resource, type 123-1, 123-2, etc. (replacing 123 with the page number) for access to successive links.

In this book, a Kekulé structure of benzene with alternating C–C and C=C bonds is used (see Box 22.1, p.1006), since this allows you to keep track of the movement of electron pairs in reaction mechanisms. Remember, however, that a Kekulé structure of benzene corresponds to just one resonance structure. Chemists know that benzene does not have alternating single and double bonds for the following reasons.

- In the Kekulé structure, the three C=C bonds would be shorter than the three C–C bonds but all six carbon–carbon bonds in benzene are the same length.
- In the Kekulé structure, the three C=C bonds would be expected to react with electrophiles in electrophilic *addition* reactions (like the C=C bonds in alkenes). However, benzene undergoes electrophilic *substitution* reactions.
- Benzene is much more stable than the Kekulé structure suggests, because the *p* electrons are delocalized around the ring, rather than being localized in three π bonds.

Molecules containing alternating single and double bonds, such as (2*E*,4*E*)-hexa-2,4-dienal, are called **conjugated molecules** (Section 19.1, p.876).

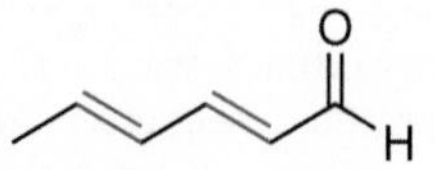
O
H

(2*E*,4*E*)-hexa-2,4-dienal

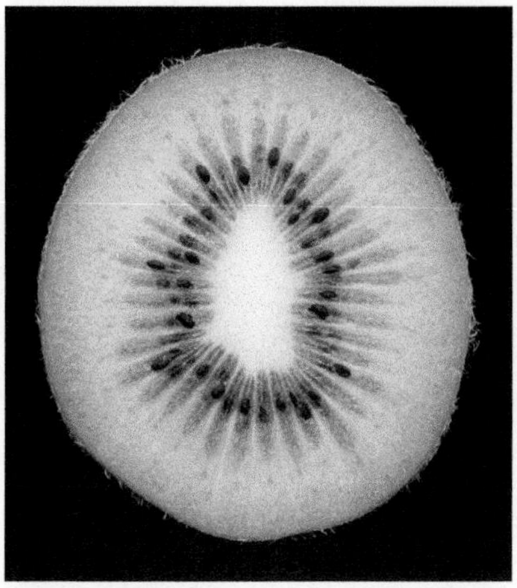

(2*E*,4*E*)-Hexa-2,4-dienal has been detected in numerous foods ranging from olives to roasted peanuts and kiwi fruit.

Stability of benzene

The unusual stability of benzene can be determined experimentally by measuring the enthalpy changes that occur on hydrogenation of cyclohexene (C_6H_{10}), cyclohexa-1,3-diene (C_6H_8), and benzene (C_6H_6). The results are shown in Figure 22.1. Hydrogenation of cyclohexene (using H_2 and a platinum catalyst) produces cyclohexane and the standard enthalpy change for this reaction (at 298 K) is $-120\ \text{kJ mol}^{-1}$. The hydrogenation is exothermic because two C–H bonds are formed in cyclohexane and these bonds are stronger than the C=C and H–H bonds that are broken (i.e. more energy is released when two C–H bonds are formed than is absorbed when one H–H bond is broken and a C=C bond is converted into C–C). Cyclohexa-1,3-diene has two C=C bonds and you would expect the enthalpy change for hydrogenation of both C=C bonds to be twice that of cyclohexene, that is, $-240\ \text{kJ mol}^{-1}$. In fact, the actual enthalpy change is very close to this ($-232\ \text{kJ mol}^{-1}$).

If the value of the enthalpy change is negative, heat is released in an exothermic reaction (Section 1.6, p.41).

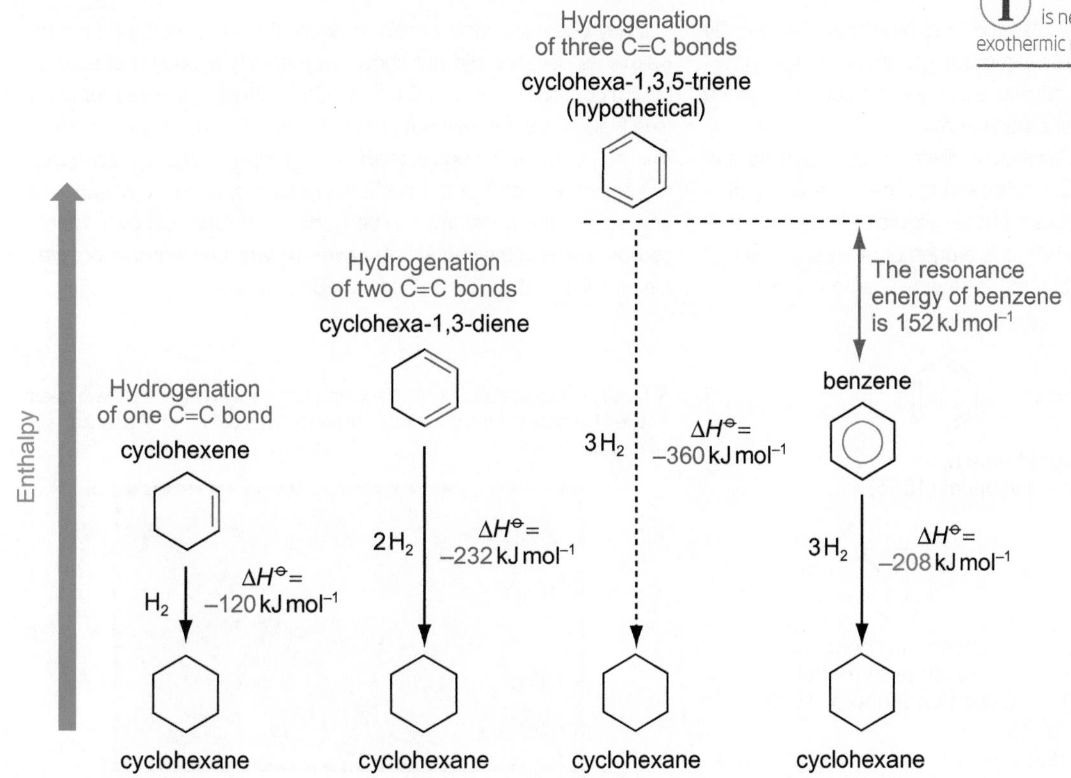
Hydrogenation of three C=C bonds
cyclohexa-1,3,5-triene (hypothetical)
Hydrogenation of two C=C bonds
cyclohexa-1,3-diene
The resonance energy of benzene is 152 kJ mol⁻¹
benzene
Hydrogenation of one C=C bond
cyclohexene
Enthalpy
3 H2
ΔH⦵ = −360 kJ mol⁻¹
2 H2
ΔH⦵ = −232 kJ mol⁻¹
3 H2
ΔH⦵ = −208 kJ mol⁻¹
H2
ΔH⦵ = −120 kJ mol⁻¹
cyclohexane
cyclohexane
cyclohexane
cyclohexane

Figure 22.1 Catalytic hydrogenation of benzene and related alkenes.

Applying the same argument to hydrogenation of a hypothetical single Kekulé structure (which could be called cyclohexa-1,3,5-triene), you would expect the enthalpy change for hydrogenation of the three C=C bonds to be $-360\,kJ\,mol^{-1}$ ($3 \times -120\,kJ\,mol^{-1}$). In fact, the measured enthalpy change for hydrogenation of benzene ($-208\,kJ\,mol^{-1}$) is considerably different. You can see from Figure 22.1 that benzene is more stable than the hypothetical Kekulé structure. The difference between the enthalpy change observed for benzene and that calculated for the Kekulé structure is $152\,kJ\,mol^{-1}$. This is called the **resonance energy** (or sometimes the **delocalization energy**) of benzene.

> The resonance energy is a measure of how much more stable a compound is because π electrons are delocalized rather than localized.

The stability of benzene is due to the delocalization of the electrons in the ring and this is an example of resonance (see Sections 4.5, 5.5, and 19.1). The two Kekulé structures are the resonance forms of benzene and you can show the delocalization of electrons using three double-headed curly arrows.

Resonance forms of benzene   Resonance hybrid

> A two-headed straight arrow ↔ is used to link resonance forms. (Resonance forms of carbocations, R^+, and carbanions, R^-, are discussed in Section 19.1, p.873.)

> Remember that benzene has a hydrogen atom attached to each carbon atom, although this is not shown in the skeletal structure (Section 2.2, p.74).

Box 22.1 A short history of benzene

Benzene was first isolated in 1825 by Michael Faraday. By heating crude oil, he extracted a flammable gas that he called 'bicarburet of hydrogen'. Distilling coal was subsequently found to give the same product, which was later called benzene.

The structure of benzene fascinated chemists and, in 1865, the German chemist August Kekulé proposed that benzene was a mixture of two compounds in rapid equilibrium (Figure 1). They both contained a hexagonal structure with a hydrogen atom at each corner, with alternate single and double bonds between carbon atoms.

Many chemists soon began to doubt the structure proposed by Kekulé as benzene did not show the reactivity expected of a compound containing three C=C bonds. Other structures were proposed (see Figure 1) but it was not until the work of Linus Pauling in 1931 that the key breakthrough came. Pauling proposed that benzene was a resonance hybrid of the two structures proposed by Kekulé. He argued that the π electrons in benzene are delocalized over the six carbon atoms in the ring and this explains why benzene is more stable than if it contained three isolated C=C bonds.

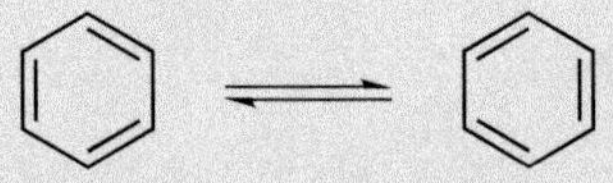

Kekulé thought these two structures were in equilibrium (1865)

'Dewar benzene' model (James Dewar, 1867)

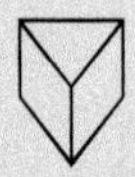

'Ladenburg benzene' (prismane) model (Albert Ladenburg, 1869)

▲ **Figure 1** Early attempts at proposing a structure for benzene. All are now known to be incorrect.

▼ Friedrich August Kekulé von Stradonitz (1829–1896) claimed to have been inspired to deduce the ring structure of benzene by dreaming of a snake seizing its own tail.

DDR
1829
1896
FRIEDR. AUGUST KEKULÉ
25
1979

Box 22.2 Making phenol from benzene

Phenol (C_6H_5OH) is an important starting material for the production of pharmaceuticals, weedkillers, and synthetic resins such as Bakelite. In industry, phenol is made from benzene using the *cumene hydroperoxide* process. This process involves the conversion of two relatively inexpensive reactants, benzene (C_6H_6) and propene ($CH_3CH{=}CH_2$), into two more valuable products, phenol and propanone (acetone, CH_3COCH_3; see Box 1.4, p.25). In step 1 of the process, benzene reacts with propene and H^+ to form 1-methylethylbenzene (cumene, $C_6H_5CH(CH_3)_2$). In step 2, 1-methylethylbenzene is converted into a hydroperoxide by reaction with a radical initiator and O_2. In step 3, the hydroperoxide reacts with H^+ to form phenol and propanone.

Currently, the demand for phenol is much greater than that for propanone so researchers are devising alternative synthetic routes to phenol that avoid the co-production of propanone. Alternatively, ways of recycling propanone are being explored, which involve converting propanone into propene, which is a reactant in step 1.

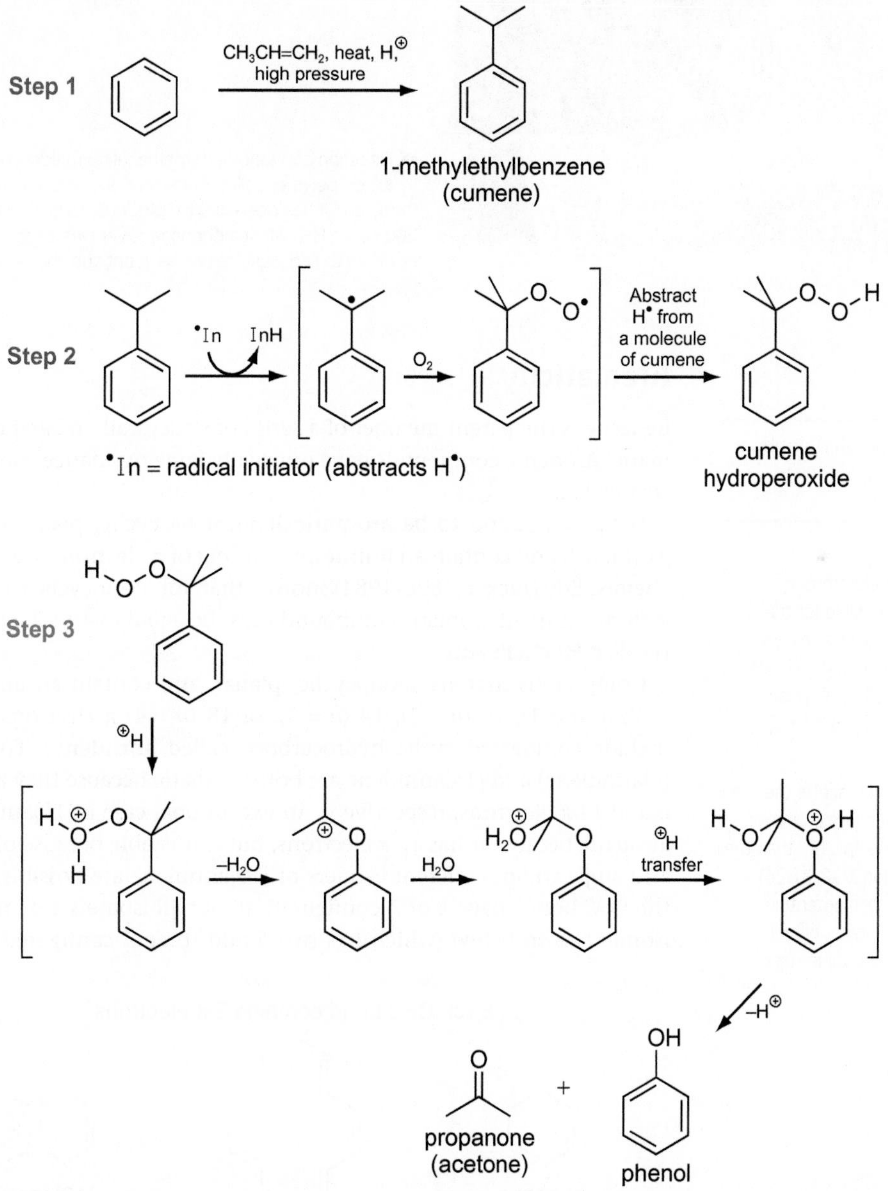

Step 1
CH3CH=CH2, heat, H⊕
high pressure
1-methylethylbenzene
(cumene)
Step 2
•In
InH
O2
Abstract
H• from
a molecule
of cumene
cumene
hydroperoxide
•In = radical initiator (abstracts H•)
Step 3
⊕H
–H2O
H2O
⊕H
transfer
–H⊕
OH
propanone
(acetone)
phenol

▲ The cumene hydroperoxide process.

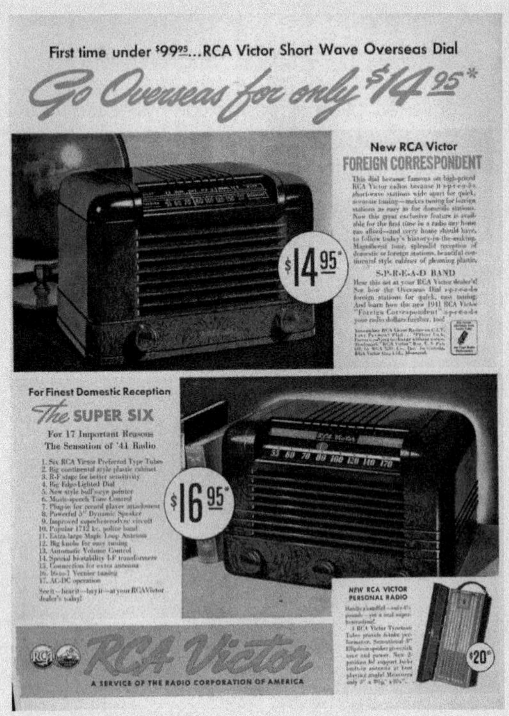
First time under $99.95...RCA Victor Short Wave Overseas Dial
Go Overseas for only $14.95*
New RCA Victor
FOREIGN CORRESPONDENT
$14.95*
For Finest Domestic Reception
The SUPER SIX
For 17 Important Reasons
The Sensation of '41 Radio
$16.95*
NEW RCA VICTOR
PERSONAL RADIO
$20*
RCA Victor
A SERVICE OF THE RADIO CORPORATION OF AMERICA

Reaction of phenol with methanal (formaldehyde; HCHO) forms a tough synthetic polymer called Bakelite. Bakelite can be moulded into thousands of forms and it has been used in products ranging from radios to jewellery. The 200 series Bakelite telephone receiver was designed in 1929, and today you are likely to find Bakelite pieces in antique shops.

Questions

(a) In step 1, an intermediate carbocation is formed by regioselective addition of H^+ to propene. Draw the structure of the carbocation and explain why the addition is regioselective (*Hint*. See Section 21.3, p.970.)

(b) Explain why the radical initiator ($In^•$) selectively abstracts a hydrogen atom from the tertiary carbon atom of 1-methylethylbenzene in step 2.

(c) Suggest a reaction mechanism, using curly arrows, to explain how cumene hydroperoxide is converted into phenol and propanone in step 3. (Note that this mechanism involves ionic intermediates, not radical intermediates as in step 2.)

Aromaticity

Monocyclic compounds that are planar and have an uninterrupted ring of $(4n+2)$ π electrons are aromatic.

The delocalization of electrons in aromatic compounds is responsible for the hydrogens bonded to aromatic rings having characteristic chemical shifts in the ^1H NMR spectra (typically around δ 7.4 ppm). This is discussed in detail in Section 12.3 (p.585).

Steric strain (steric hindrance) is the repulsive interaction between two atoms (or groups) that are closer to one another than their atomic radii allow (see Section 18.2, p.824).

Angle strain is the increase in potential energy of a molecule due to one or more bond angles deviating from the ideal values (see Section 18.2, p.826)

Benzene is the parent member of a series of structurally related compounds that are aromatic. Aromatic compounds have unusually large resonance energies so they are particularly stable.

For a compound to be **aromatic** it must be cyclic, planar (to allow overlap of the *p* orbitals), and contain an uninterrupted ring of π electrons. Calculations by the German chemist Eric Hückel (1896–1984) showed that, for monocyclic compounds, the number of π electrons in an aromatic compound must be equal to $4n + 2$, where $n = 1, 2, 3$, etc. This is called **Hückel's rule**.

Compounds that are monocyclic, planar, and contain an uninterrupted ring of 2 ($n = 0$), 6 ($n = 1$), 10 ($n = 2$), 14 ($n = 3$), or 18 ($n = 4$) π electrons are aromatic. Examples include conjugated cyclic hydrocarbons called **annulenes**. For example, benzene (or [6]annulene) and [14]annulene are both aromatic because they are planar and they have 6 and 14 π electrons, respectively. An exceptional case is [10]annulene, which should be aromatic because it has 10 π electrons, but is unstable because of a combination of steric and angle strain. Different isomers of [10]annulene are possible, depending on whether the C=C bonds have *E* or *Z* configuration, but all isomers are unstable. For example, the isomer shown below (which has two *E* and three Z configured C=C bonds) is unstable

Each C=C bond contains 2 π electrons

H H

There is steric strain between the two hydrogens

benzene or [6]annulene (6 π electrons)

[14]annulene (14 π electrons)

[10]annulene (10 π electrons) should be aromatic but it is unstable

because of a steric interaction between the two internal hydrogen atoms, which prevents the ring from being planar.

Aromatic compounds can be neutral or charged. For example, in the cyclopentadienyl anion ($C_5H_5^-$, sometimes abbreviated Cp^-), the two electrons on the negatively charged carbon are in a *p* orbital that is part of the π ring system (see margin).

Each C=C bond contains 2 π electrons

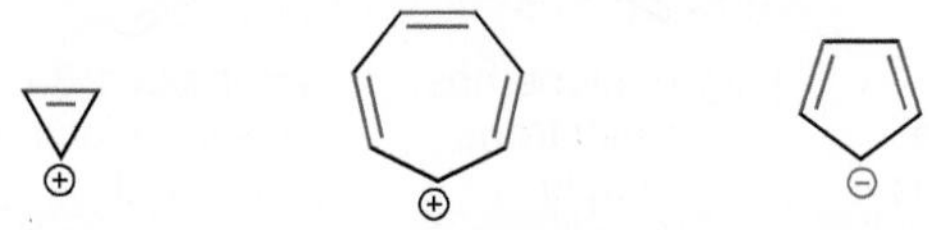

Structure			
Name	cyclopropenyl cation	cycloheptatriene cation	cyclopentadienyl anion
Number of π electrons	2	6	6

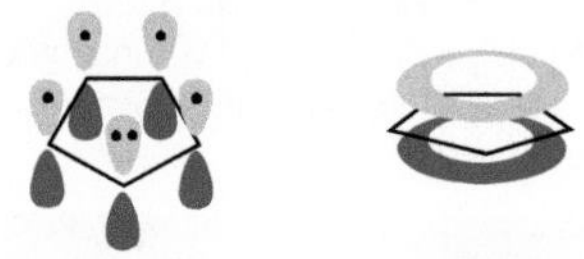

Delocalization of 6 π electrons in the cyclopentadienyl anion

Heteroatoms can also be part of the aromatic ring. In some compounds, a lone pair of electrons on the heteroatom is part of the ring of π electrons. For example, in pyrrole (C_4H_4NH), the lone pair of electrons on nitrogen is in a *p* orbital that is part of the π ring system (see margin). This contrasts with pyridine (C_5H_5N), where the lone pair of electrons on nitrogen is not part of the π ring system.

i Pyridine is a much stronger base than pyrrole, because the lone pair of electrons on nitrogen is not part of the π ring system (see Section 19.2, p.893).

Each C=C bond contains 2 π electrons
•• represents a lone pair of electrons that is part of the ring of π electrons
•• represents a lone pair of electrons that is not part of the ring of π electrons

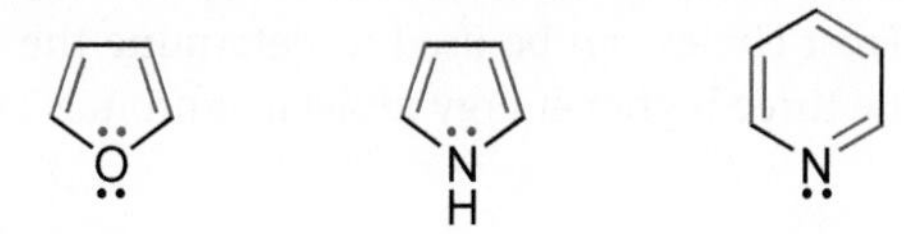

Structure			
Name	furan	pyrrole	pyridine
Number of π electrons	6	6	6

Delocalization of 6 π electrons in pyrrole

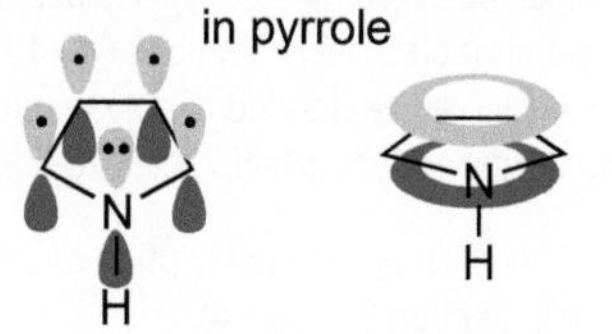

Delocalization of 6 π electrons in pyridine

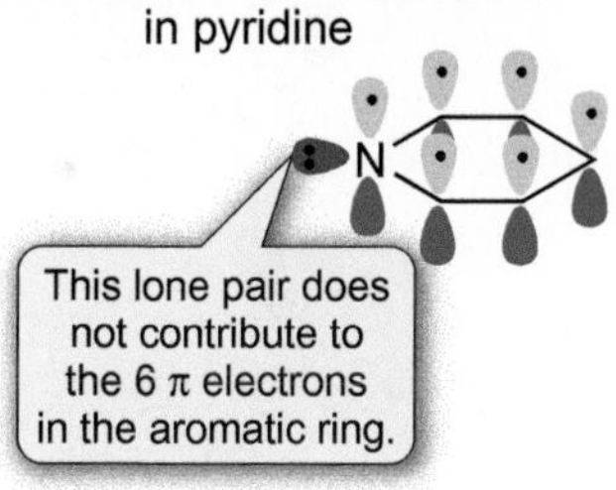
This lone pair does not contribute to the 6 π electrons in the aromatic ring.

The unusual stability of aromatic molecules explains their chemical reactivity. It also explains why these compounds are relatively easy to prepare. For example, the aromatic cyclopentadienyl anion ($C_5H_5^-$) is formed by deprotonation of cyclopenta-1,3-diene (C_5H_6) using the *tert*-butoxide ion (Me_3CO^-). Cyclopenta-1,3-diene is unusually acidic ($pK_a = 16$) for a hydrocarbon and this is because deprotonation forms such a stable aromatic anion—the negative charge is equally spread around all five carbon atoms in the ring.

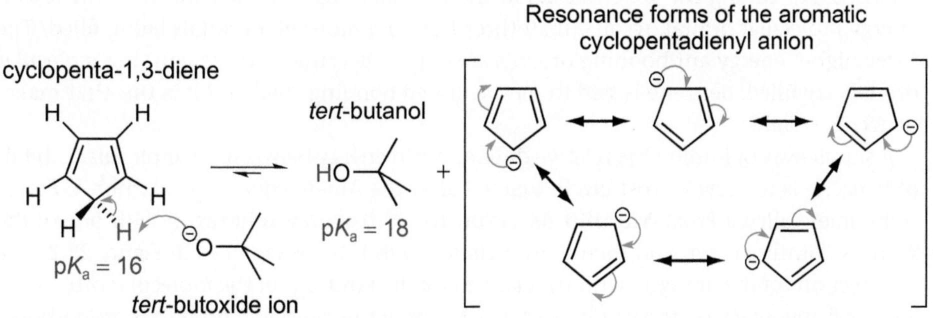
Resonance forms of the aromatic cyclopentadienyl anion
cyclopenta-1,3-diene
pKa = 16
tert-butoxide ion
tert-butanol
pKa = 18

The strengths of organic acids are discussed in Section 19.2 (p.888). Hydrocarbons are normally extremely weak acids. For example, ethene ($H_2C{=}CH_2$) has an exceptionally high pK_a value of 44.

Some molecules that are cyclic, planar, and contain an uninterrupted ring of π electrons are *less* stable than a similar compound with localized electrons. These molecules contain $4n$ π electrons and are called **antiaromatic** compounds. You can explain the instability and high reactivity of antiaromatic compounds by considering the molecular orbitals. This is discussed in the next section.

i The pK_a values of cyclopenta-1,3-diene and *tert*-butanol are 16 and 18, respectively, so the equilibrium constant for deprotonation is equal to 10^2. Therefore, the position of the equilibrium lies over towards the cyclopentadienyl anion ($C_5H_5^-$).

The [12]annulene (1,3,5,7,9,11-cyclododecahexaene; $C_{12}H_{12}$) shown here has a series of alternating *E* and *Z* double bonds. It is a very unstable compound that can only be stored at low temperature.

[8]Annulene (C_8H_8), also known as 1,3,5,7-cyclooctatetraene or COT, is a stable compound that has a shape like a tub (it bends to avoid being antiaromatic). COT undergoes electrophilic addition reactions typical of alkenes.

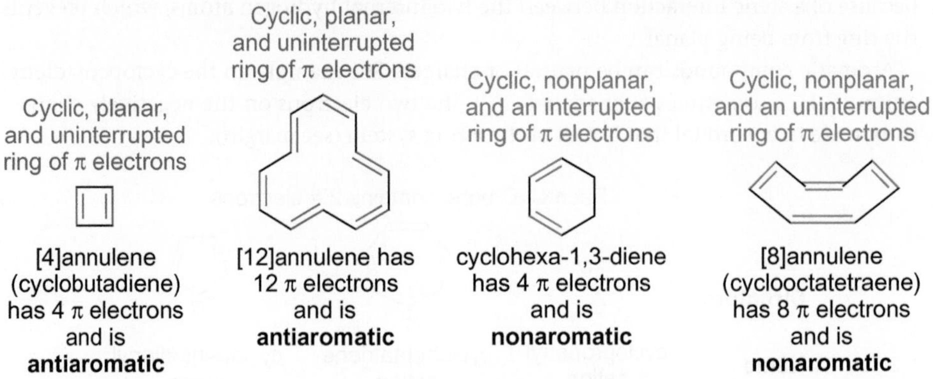
Cyclic, planar, and uninterrupted ring of π electrons
[4]annulene (cyclobutadiene) has 4 π electrons and is antiaromatic
Cyclic, planar, and uninterrupted ring of π electrons
[12]annulene has 12 π electrons and is antiaromatic
Cyclic, nonplanar, and an interrupted ring of π electrons
cyclohexa-1,3-diene has 4 π electrons and is nonaromatic
Cyclic, nonplanar, and an uninterrupted ring of π electrons
[8]annulene (cyclooctatetraene) has 8 π electrons and is nonaromatic

Other cyclic molecules that are nonplanar and have an interrupted or uninterrupted ring of π electrons are called **nonaromatic** compounds. These molecules have similar stabilities to related compounds that are not cyclic, such as buta-1,3-diene ($H_2C{=}CH{-}CH{=}CH_2$).

To gain practice in assigning a compound as aromatic, antiaromatic, or nonaromatic, see Worked example 22.1.

Molecular orbital diagrams

Molecular orbital theory is introduced for diatomic molecules in Sections 4.6–4.12 (pp.184–212) and extended to polyatomic molecules in Sections 5.6–5.7 (pp.246–252).

The valence bond description of benzene is given in Section 5.5 (p.244).

For benzene, molecular orbital theory states that the six *p* atomic orbitals combine to form six molecular orbitals as shown below. The three lower energy molecular orbitals, which have lower energies than the six *p* atomic orbitals, are the bonding molecular orbitals. Notice that two of the bonding molecular orbitals have the same energy (later in this section you will see how Frost circles can be used to determine the relative energies of the molecular orbitals). The three higher energy molecular orbitals are the antibonding orbitals.

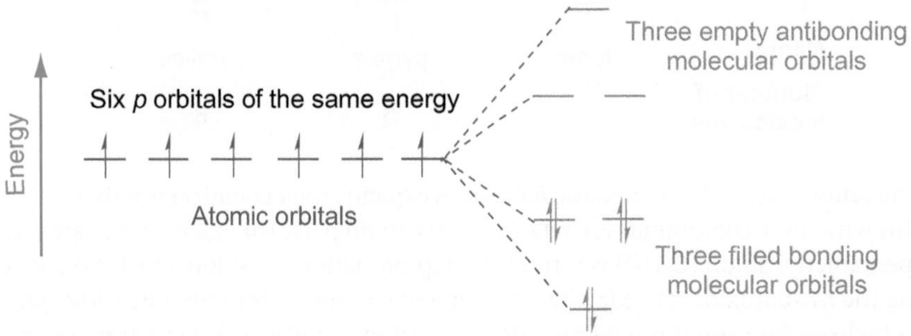
Energy
Six p orbitals of the same energy
Atomic orbitals
Three empty antibonding molecular orbitals
Three filled bonding molecular orbitals

Notice that the upper bonding pair of orbitals are the HOMOs—as these two orbitals are of equal energy they are called degenerate orbitals (Section 3.6, p.143). There are also two degenerate (but unfilled) orbitals that are the LUMOs.

Each molecular orbital can accommodate two electrons if the spins of the electrons are opposed. So, adding the six electrons to the molecular orbitals, starting with the lowest energy molecular orbital, results in all three bonding molecular orbitals being filled. The three higher energy antibonding orbitals are empty. Because all three bonding molecular orbitals are filled, benzene is said to have a closed bonding shell and it is this that makes benzene so stable.

A simple way of finding the relative differences in energy between the molecular orbitals of benzene is to draw a **Frost circle** (named after the American scientist Arthur A. Frost), sometimes called a **Frost–Musulin diagram**. To do this, draw a hexagon with one of the vertices pointing down and then draw a circle around the hexagon as in Figure 22.2. The intersections of the hexagon with the circle mark the positions of the molecular orbitals on a vertical energy scale—the vertical distance between the molecular orbitals is proportional to the relative energy difference between the actual molecular orbitals. The molecular orbitals below the midpoint of the hexagon are the bonding molecular orbitals. The molecular orbitals above the midpoint of the hexagon are the antibonding molecular orbitals.

Frost circles can be used to determine the number and relative energies of molecular orbitals of other planar, cyclic compounds. For cyclobutadiene (a square shape), C_4H_4, drawing a circle around a square marks the position of the four molecular orbitals. The two

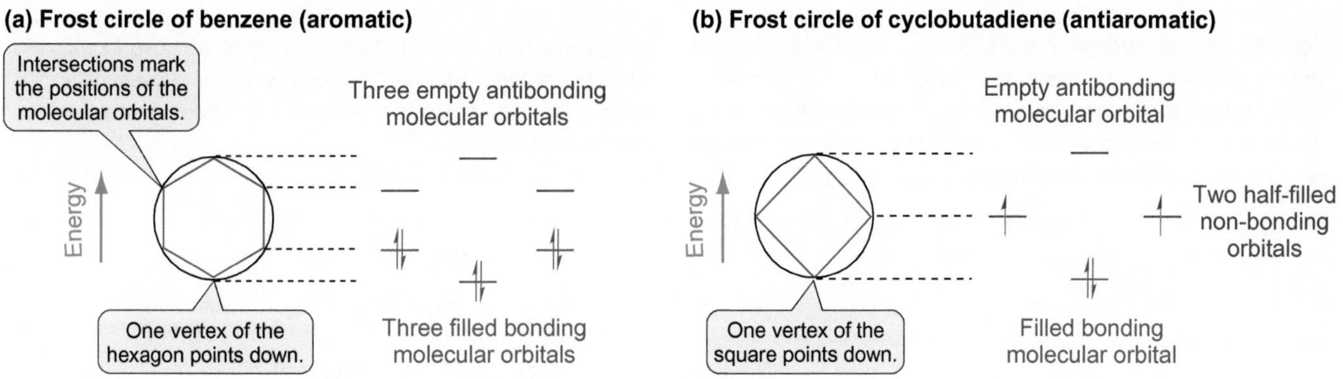
(a) Frost circle of benzene (aromatic)
Intersections mark the positions of the molecular orbitals.
Three empty antibonding molecular orbitals
Energy
One vertex of the hexagon points down.
Three filled bonding molecular orbitals
(b) Frost circle of cyclobutadiene (antiaromatic)
Empty antibonding molecular orbital
Two half-filled non-bonding orbitals
Energy
One vertex of the square points down.
Filled bonding molecular orbital

Figure 22.2 Frost circles for (a) benzene and (b) cyclobutadiene.

molecular orbitals at the midpoint of the circle are non-bonding orbitals. Adding the four electrons to the molecular orbitals results in a filled bonding molecular orbital and two half-filled non-bonding orbitals. Because cyclobutadiene contains two unpaired electrons in the non-bonding orbitals, it is unstable and readily reacts to pair up the electrons. It is called an antiaromatic compound.

Worked example 22.1 Aromatic and antiaromatic compounds

Predict whether the following two compounds are aromatic, antiaromatic, or nonaromatic. Explain your reasoning.

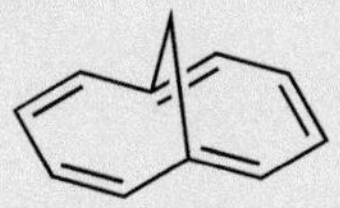

1,6-methano[10]annulene

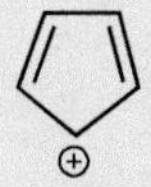

cyclopentadienyl cation

Strategy

Cyclic, planar compounds that contain an uninterrupted ring of π electrons can be aromatic or antiaromatic.

If the number of π electrons is equal to $4n + 2$ (where $n = 1, 2, 3$, etc.), the compound is aromatic.

If the number of π electrons is equal to $4n$ (where $n = 1, 2, 3$, etc.), the compound is antiaromatic.

If the compound cannot be planar, it will be nonaromatic.

Solution

1,6-Methano[10]annulene is cyclic, approximately planar, and contains an uninterrupted ring of π electrons. It has 10 π electrons (two for each C=C bond) so it obeys Hückel's rule ($n = 2$; $4 \times 2 + 2 = 10$) and is aromatic.

The cyclopentadienyl cation is cyclic, planar, and contains an uninterrupted ring of π electrons. The cyclopentadienyl cation has 4 π electrons (two for each C=C bond) so it is antiaromatic ($n = 1$; $4 \times 1 = 4$).

Question

5-Bromocyclopenta-1,3-diene is insoluble in water, whereas adding water to 7-bromocyclohepta-1,3,5-triene rapidly produces a water-soluble salt. Suggest an explanation for the different behaviour.

H
Br

5-bromocyclopenta-1,3-diene

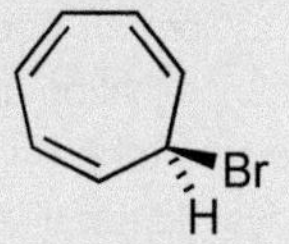
Br
H

7-bromocyclohepta-1,3,5-triene

Box 22.3 From coal tar dyes to pharmaceuticals

When William Perkin (1838–1907) accidentally prepared the purple dye, mauve, in 1856 (see p.1003), he could hardly have imagined where his discovery would lead. Mauve was the first synthetic dye. Its discovery prompted the development of many other synthetic dyes and the world became much more colourful—but it also heralded the birth of the chemical industry, and led to developments in immunology and chemotherapy.

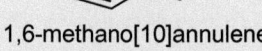

Perkin made mauve by oxidizing aniline (phenylamine, $C_6H_5NH_2$) derived from coal tar. At this time, gas lamps lighted big cities like London. The gas was made in the local gas works by heating →

➜ coal or coke strongly in the absence of air. The process left behind a black sticky residue called coal tar, which was originally a waste product. It was soon realized, however, that the black tar was a rich source of organic chemicals and coal tar became the raw material for the emerging chemical industry.

▲ William Perkin was just 19 when, in 1857, he opened a factory at Greenford Green, near London to manufacture mauve on a large scale. He became the first chemist to become rich from chemistry when he subsequently sold his company, which eventually became part of ICI. The fastest growth in dye production came in Germany and chemical companies including BASF, Bayer, and AGFA have their roots in early dye production.

Soon after Perkin's discovery, the German chemist, Peter Griess, while working at a brewery in Burton-on-Trent, discovered the diazotization reaction and the coupling of diazonium ions ($Ar–N_2^+$) to produce coloured azo compounds (Ar–N=N–Ar; see Section 22.4, p.1043). This provided a general method for making a whole range of azo dyes—all based on coal tar. Dye works sprang up in different parts of the UK and Europe (particularly in Germany) to meet the demand for the new dyes.

In 1891, the story of synthetic dyes took an unexpected turn. Paul Ehrlich, a German doctor, who was using dyes to stain body tissues for examination under the microscope, found that certain cells or organisms took up some dyes selectively. He reasoned that, if such a dye did not affect other cells, he should be able to inject it in sufficiently large doses to kill pathogenic microorganisms in the body. The search was now on for suitable dyes, or other chemicals, to do this—the so-called 'magic bullets'. By this time, Perkin had sold his factory and the UK dye industry was in decline, so most of this work was carried out in Germany. There were some early successes. Ehrlich used the dye Methylene Blue to target the *Plasmodium* parasite responsible for malaria, and a compound, marketed as Salvarsan (an organoarsenic compound), was successful against syphilis—the first time a chemical was used to selectively kill bacteria in the body.

The real breakthrough came in 1935 at the German syndicate, I G Farbenindustrie, where chemists and microbiologists were working in close collaboration. The biochemist Gerhard Domagk was responsible for screening large numbers of compounds for possible biological action, when he discovered a red azo dye (compound 1932) that showed potential. The compound was subsequently marketed as Prontosil and was effective against streptococcal bacterial infections in humans.

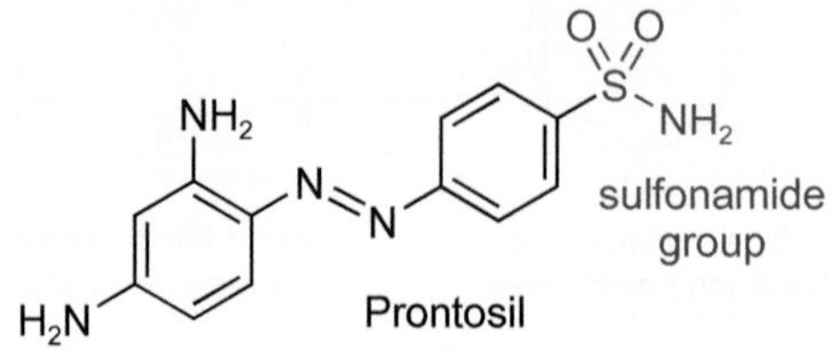
NH2
H2N
N
N
O O
S
NH2
sulfonamide group

Prontosil

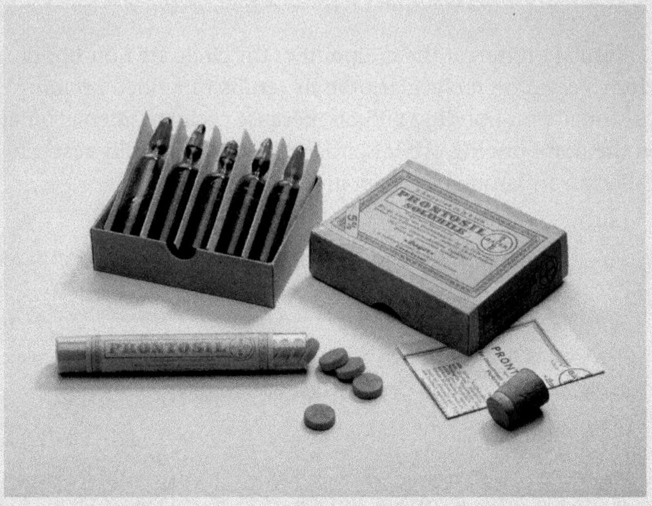
PRONTOSIL

▲ The red azo dye, Prontosil, was found to be effective against streptococcal bacterial infections in humans.

It was subsequently found that the azo part of the dye molecule was not necessary for the antibacterial action of Prontosil—in fact, all that was needed was 4-aminobenzenesulfonamide ($4\text{-}H_2NC_6H_4SO_2NH_2$), known then as sulfanilamide. A whole range of sulfonamide drugs then followed. Perhaps the most famous of these was sulfapyridine (better known as M&B 693), which quickly cured Winston Churchill from a bout of pneumonia during the Second World War (1939–45).

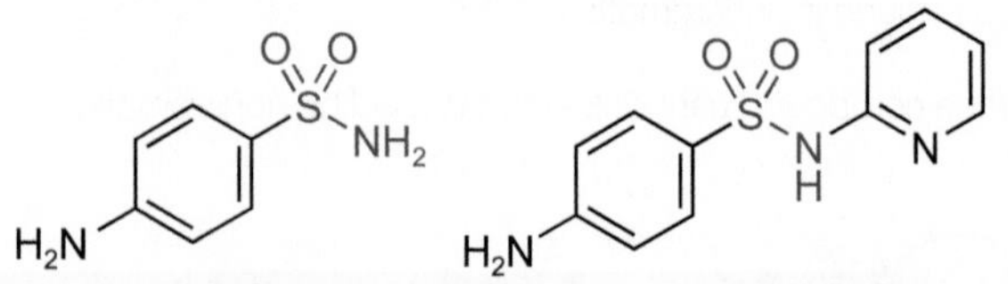
O O
S
NH2
H2N
O O
S
N
H
N
H2N

sulfanilamide (4-aminobenzenesulfonamide)

sulfapyridine (M&B 693)

The modern era of antibiotics dates from 1928 when Alexander Fleming discovered the action of the mould *Penicillium notatum*, though the compound penicillin was not isolated and characterized until the work of Howard Florey and Ernst Chain in Oxford some ten years later. Manufacture of penicillin began in 1943. ➜

➜ It transformed the treatment of bacterial diseases and led to a whole range of related penicillins and other antibiotics.

During the twentieth century, the source of organic chemicals for the chemical industry changed from coal tar to crude oil, and the range of chemicals manufactured expanded to include other products such as polymers, perfumes, and pesticides. However, the close link between the dye and pharmaceutical industries remains. Many of the companies that originally manufactured dyes also took up the synthesis of pharmaceuticals. Large pharmaceutical companies, such as AstraZeneca, BASF, Bayer, and Novartis (a merger of Ciba-Geigy and Sandoz), all have their roots in early dye production.

Question

Suggest a diazonium salt and a coupling agent that could be used to prepare the azo dye, Prontosil.

Summary

Planar hexagon with bond lengths in between C–C and C=C

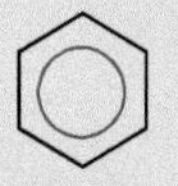

benzene

Cyclic delocalization of 6 π electrons explains the unusual stability of benzene

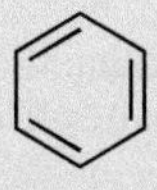

Kekulé structure of benzene

Benzene has $(4n + 2)$ π electrons (where $n = 1$) and it is aromatic

- Aromatic compounds are cyclic, planar, and have an uninterrupted ring of π electrons.
- According to Hückel's rule, the number of π electrons in an aromatic monocyclic compound must be equal to $4n + 2$, where $n = 1, 2, 3$, etc.
- Aromatic compounds can be neutral or charged, and heteroatoms can be part of the ring.
- Antiaromatic compounds are cyclic, planar, and have an uninterrupted ring of $4n$ π electrons.
- Frost circles provide a convenient way of determining the number of molecular orbitals, and their relative energies, in aromatic or antiaromatic compounds.

? For a practice question on this topic, see question 1 at the end of this chapter (p.1052).

22.2 Electrophilic substitution reactions of benzene

Benzene is said to be aromatic (Section 22.1). It is an unusually stable compound and does not undergo the electrophilic addition reactions typical of compounds containing C=C bonds. For example, benzene does not decolorize bromine water. However, the electron-rich benzene ring does react with strong electrophiles in *substitution*, rather than addition, reactions. In an **electrophilic substitution reaction**, an atom or group of atoms replaces a hydrogen atom on the benzene ring so that the product retains the stable aromatic ring, as shown in Figure 22.3. If benzene reacted with electrophiles in electrophilic *addition* reactions, the aromatic ring would be destroyed and the products would be much less stable than benzene.

Alkenes (e.g. $R_2C{=}CR_2$) react with electrophiles in addition reactions and they decolorize bromine water; see Section 21.3 (p.978).

The mechanism of the reaction of benzene with an electrophile (E^+) is shown in Figure 22.4. It involves two steps. In step 1, which is the slower step of the reaction, two electrons from the aromatic ring of benzene (the nucleophile) are attracted towards the electrophile. The aromaticity is broken as a new C–E bond forms and this produces a non-aromatic carbocation (called an arenium ion). Although the carbocation is not aromatic it is stabilized by resonance. Three resonance forms contribute to the resonance hybrid and show the delocalization of electrons in the carbocation.

Compare the mechanism of electrophilic substitution of benzene with the mechanism of electrophilic addition of an alkene (Section 21.1, p.964).

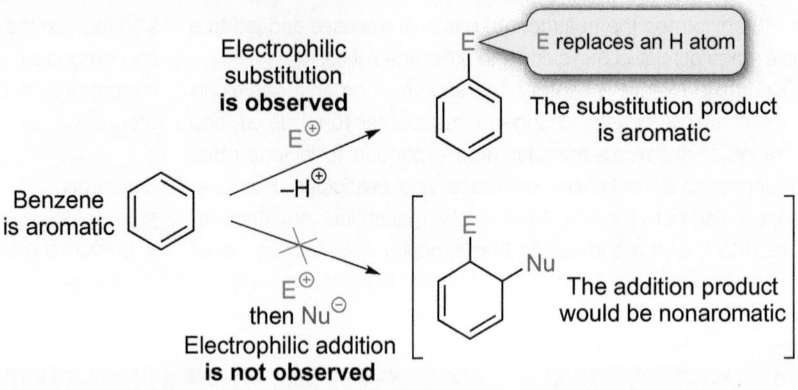

Electrophilic
substitution
is observed
E replaces an H atom
The substitution product
is aromatic
Benzene
is aromatic
–H
Nu
The addition product
would be nonaromatic
then Nu
Electrophilic addition
is not observed

Figure 22.3 Reaction of benzene with an electrophile, E^+.

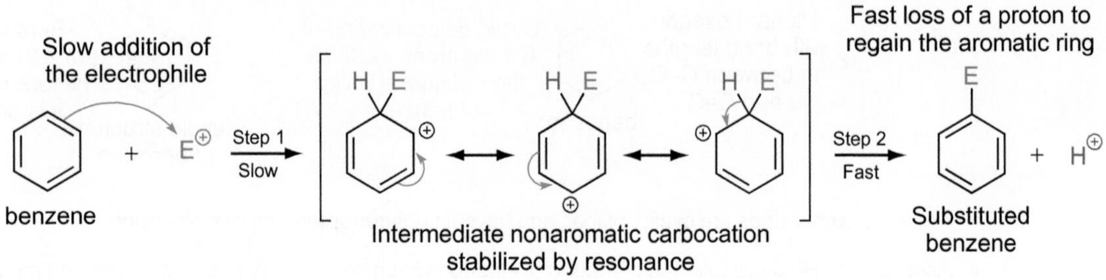

Slow addition of
the electrophile
Fast loss of a proton to
regain the aromatic ring
Step 1
Slow
Step 2
Fast
benzene
Intermediate nonaromatic carbocation
stabilized by resonance
Substituted
benzene

Figure 22.4 Mechanism of the electrophilic substitution of benzene.

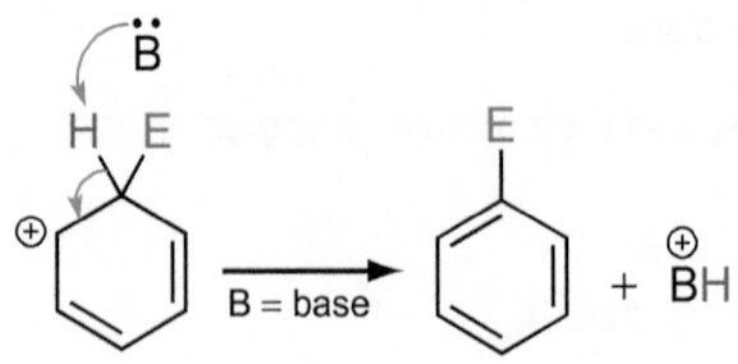

B = base

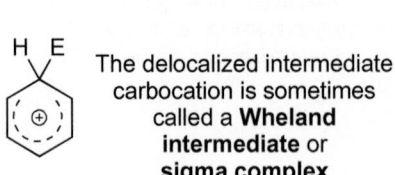

In step 2, the carbocation loses a proton (H^+), usually by reacting with a base (as shown in the margin) and the two electrons in the C–H bond move into the ring to restore aromaticity. The nonaromatic carbocation always loses the H^+ from the carbon atom that forms a bond to the electrophile. Overall, the electrophilic substitution of benzene produces a substituted benzene (C_6H_5E).

Figure 22.5 shows the reaction profile for a typical electrophilic substitution reaction of benzene. From this diagram you can draw the following conclusions.

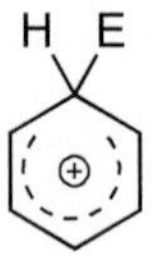

The delocalized intermediate carbocation is sometimes called a **Wheland intermediate** or **sigma complex**

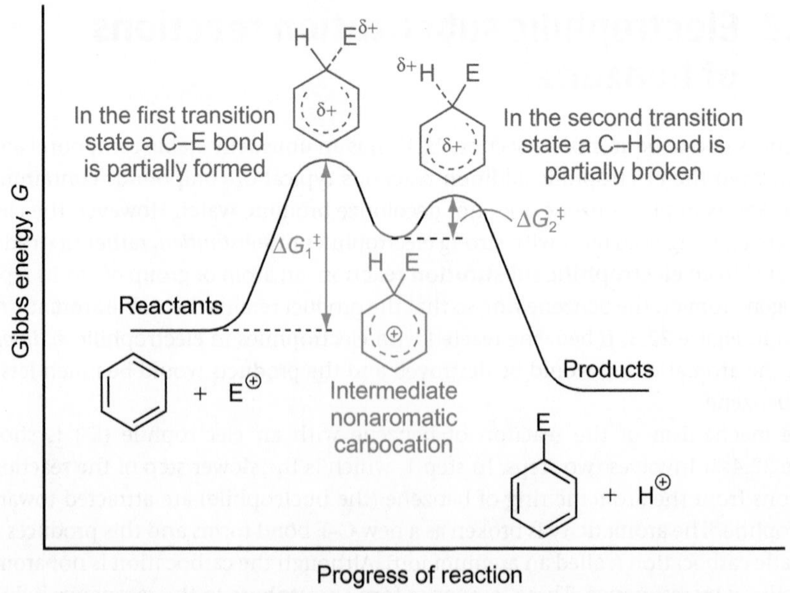

In the first transition
state a C–E bond
is partially formed
In the second transition
state a C–H bond is
partially broken
Gibbs energy, G
ΔG1‡
ΔG2‡
Reactants
Products
Intermediate
nonaromatic
carbocation
Progress of reaction

Figure 22.5 Reaction profile for a typical electrophilic substitution reaction of benzene.

- The first step of the reaction is the slower step because the Gibbs energy of activation for formation of the intermediate carbocation ($\Delta G_1^{\ddagger}$) is much greater than the Gibbs energy of activation ($\Delta G_2^{\ddagger}$) for deprotonation of the carbocation.
- The intermediate carbocation has a much higher Gibbs energy than the reactants or products—this is because the carbocation is nonaromatic whereas benzene and the substituted benzene are aromatic.
- Overall, the value of the Gibbs energy change for this reaction is negative (as the energy of the products is less than that of the reactants) so the products will be favoured over reactants at equilibrium.

Gibbs energy reaction profiles are explained in Section 9.8 (p.435). Gibbs energy change of reaction and the direction of change are discussed in Sections 14.5 (p.675), and 15.2 (p.702), respectively.

A Lewis acid is a compound that accepts a lone pair of electrons (see Section 7.8, p.336). A coordination complex (the adduct) is the product of a Lewis acid–Lewis base reaction.

The five most common electrophilic substitution reactions of benzene are halogenation, nitration, sulfonation, alkylation, and acylation. These are discussed in the following sections.

Halogenation of benzene

Benzene reacts with bromine or chlorine in the presence of a Lewis acid catalyst (typically $FeBr_3$, $FeCl_3$, or $AlCl_3$) to form bromobenzene (C_6H_5Br) or chlorobenzene (C_6H_5Cl), respectively. The mechanism of halogenation follows the general mechanism for electrophilic substitution reactions of benzene in Figure 22.4, but includes an extra first step in which the halogen reacts with the Lewis acid to form a powerful electrophile. This is illustrated in Figure 22.6 for the reaction of benzene with Br_2 and $FeBr_3$.

- In step 1, a bromine atom in Br_2 donates a pair of electrons to the Lewis acid to form a coordination complex. The halogen atom that donates the electrons becomes positively charged and the Br–Br bond becomes polarized—the pair of electrons in the Br–Br bond is attracted to the positive charge and this produces a partial positive charge (δ+) on the singly bonded bromine atom. By forming a complex with the Lewis acid, the Br_2 is converted into a stronger electrophile. Without the Lewis acid, Br_2 does not react with benzene.
- In step 2, benzene acts as a nucleophile and attacks the partially positive bromine atom in the coordination complex. This produces a nonaromatic carbocation together with $[FeBr_4]^-$. To simplify the reaction mechanism, only one of the three resonance forms of the carbocation is drawn (compare this with Figure 22.4).

Alkenes and alkynes are more reactive to electrophiles than benzene. It is easier to form a carbocation from reaction of an electrophile with a C=C or C≡C bond because an aromatic ring does not have to be broken. This explains why C=C and C≡C bonds react with Br_2 in the *absence* of a Lewis acid (Sections 21.3, p.975, and 21.5, p.994).

A catalyst increases the rate of a chemical reaction without being consumed in the reaction (see Section 9.9, p.437).

You may also see a variation of this mechanism in which the Br–Br bond in $Br–Br^+–^-FeBr_3$ breaks to form Br^+ and $^-FeBr_4$. The Br^+ ion is then the electrophile and reacts with benzene to form bromobenzene.

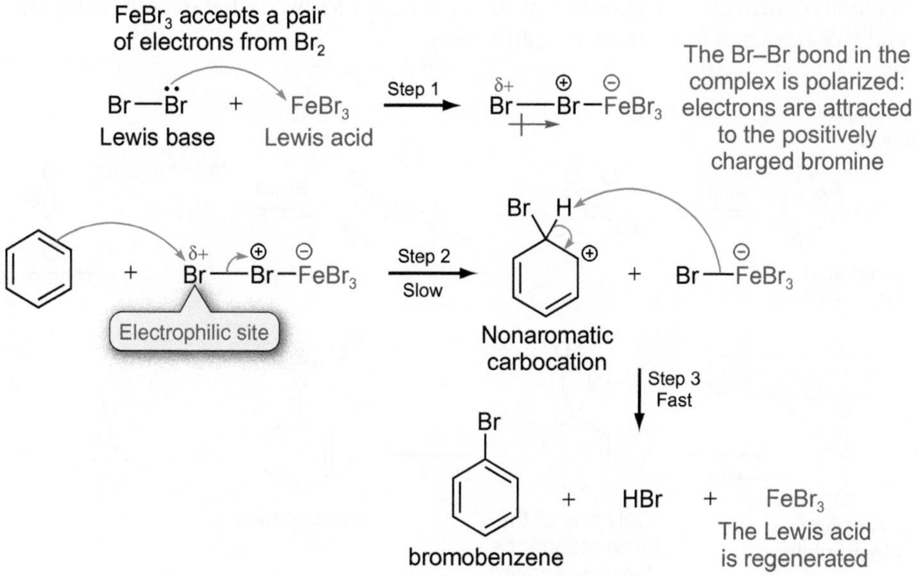
FeBr3 accepts a pair of electrons from Br2
Br—Br + FeBr3
Lewis base
Lewis acid
Step 1
δ+ Br—Br—FeBr3
The Br–Br bond in the complex is polarized: electrons are attracted to the positively charged bromine
Electrophilic site
Step 2
Slow
Nonaromatic carbocation
Br—FeBr3
Step 3
Fast
bromobenzene
HBr
FeBr3
The Lewis acid is regenerated

Figure 22.6 Bromination of benzene.

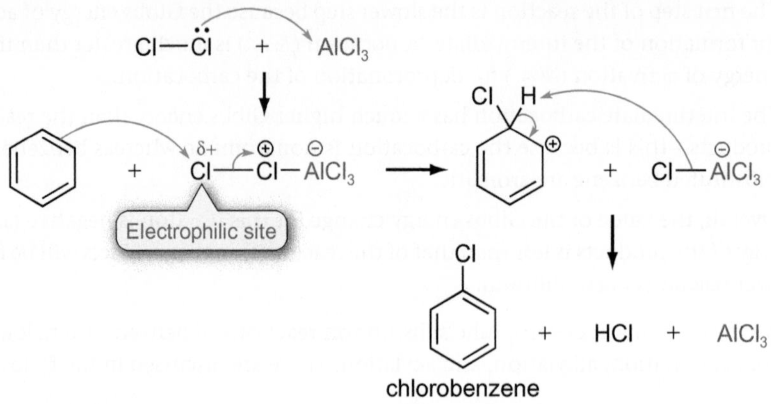
Cl—Cl + AlCl3
δ+
Cl—Cl—AlCl3
Electrophilic site
Cl H
Cl—AlCl3
Cl
HCl + AlCl3
chlorobenzene

Figure 22.7 Chlorination of benzene.

ⓘ Bromobenzene (C_6H_5Br) and chlorobenzene (C_6H_5Cl) are useful starting materials for organic synthesis. For example, reaction of bromobenzene with magnesium forms a Grignard reagent, C_6H_5MgBr, that is a more reactive nucleophile than benzene (the carbon atom bonded to Mg has a partial negative charge, $^{\delta-}C\text{–}Mg^{\delta+}$). C_6H_5MgBr reacts with a wide range of electrophiles, including aldehydes and ketones (Section 23.2, p.1068), to form substituted benzenes.

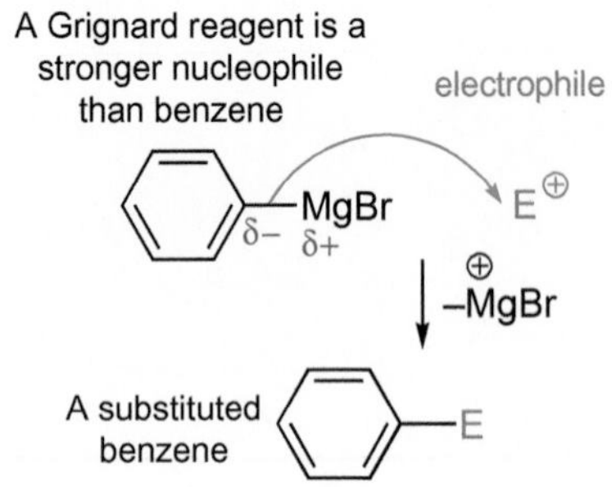
A Grignard reagent is a stronger nucleophile than benzene
electrophile
MgBr
δ– δ+
E
–MgBr
A substituted benzene
E

$$I_2 \longrightarrow 2\,I^{\oplus} + 2e^{\ominus}$$

The electrophile I^+ is formed in the presence of an oxidizing agent, which removes two electrons from I_2.

ⓘ Using IUPAC nomenclature, the nitronium ion (NO_2^+) is called a nitryl ion.

- In step 3, $[FeBr_4]^-$ removes a proton from the nonaromatic carbocation. The aromatic ring is reformed to give bromobenzene together with HBr and $FeBr_3$. Since the $FeBr_3$ is regenerated, it acts as a catalyst for bromination of benzene.

Similar reaction mechanisms can be drawn for chlorination of benzene using $Cl_2/FeCl_3$ or $Cl_2/AlCl_3$ (see Figure 22.7).

Bromination and chlorination of benzene are useful synthetic reactions but there are problems with fluorination of benzene. Fluorine reacts extremely rapidly with benzene and it is difficult to control the number of fluorine atoms that are introduced on the benzene ring. A number of fluorinated products are formed, so the reaction is not used in synthesis. In contrast to fluorine, iodine does not react with benzene. To form iodobenzene (C_6H_5I) from benzene and I_2 requires an oxidizing agent, such as nitric acid. The oxidizing agent probably removes two electrons from I_2, to form two I^+ ions ($2H^+ + 2HNO_3 + I_2 \rightarrow 2I^+ + 2NO_2 + 2H_2O$), which are strong electrophiles and react with benzene in an electrophilic substitution (see margin).

Nitration of benzene

Benzene is converted into nitrobenzene ($C_6H_5NO_2$) by reaction with a mixture of concentrated nitric acid (HNO_3) and concentrated sulfuric acid (H_2SO_4). The mechanism of nitration, shown in Figure 22.8, follows the general mechanism for electrophilic substitution reactions of benzene in Figure 22.4 (p.1014), but includes two initial steps in which HNO_3 and H_2SO_4 react to form the electrophile $^+NO_2$.

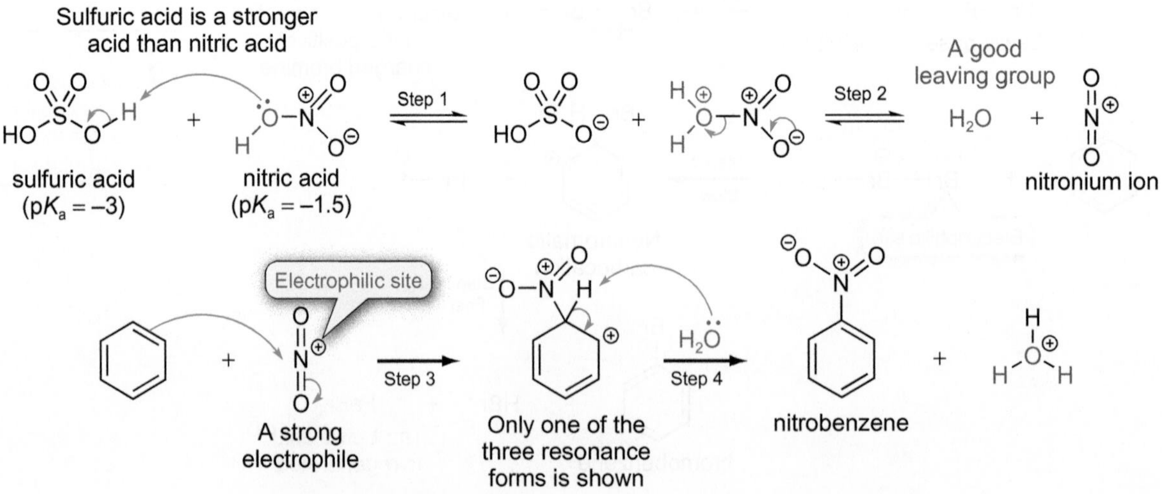
Sulfuric acid is a stronger acid than nitric acid
Step 1
Step 2
A good leaving group
H2O
sulfuric acid (pKa = –3)
nitric acid (pKa = –1.5)
nitronium ion
Electrophilic site
Step 3
Step 4
H2O
A strong electrophile
Only one of the three resonance forms is shown
nitrobenzene

Figure 22.8 Nitration of benzene.

- In step 1, sulfuric acid donates a proton to nitric acid in an acid–base reaction.
- The protonated OH group of nitric acid provides a good leaving group and, in step 2, loss of H_2O produces a $^+NO_2$ ion (a nitronium ion), which is a strong electrophile.
- In step 3, benzene donates a pair of electrons to the $^+NO_2$ ion to form a nonaromatic carbocation intermediate. Two electrons move from the benzene ring to form a new C–N bond and, at the same time, a pair of π electrons in one of the N=O bonds in the nitronium ion moves onto one of the O atoms (otherwise nitrogen would have to make five bonds).
- Finally, in step 4, water deprotonates the carbocation to form nitrobenzene together with H_3O^+.

Sulfonation of benzene

Benzene is converted into benzenesulfonic acid ($C_6H_5SO_3H$) by reaction with fuming sulfuric acid (oleum). Fuming sulfuric acid is a mixture of sulfuric acid (H_2SO_4) and sulfur trioxide (SO_3). The mechanism of sulfonation, shown in Figure 22.9, follows the general mechanism for electrophilic substitution reactions of benzene in Figure 22.4 (p.1014), but includes an initial step in which H_2SO_4 and SO_3 react to form the electrophile HSO_3^+.

- In step 1 of Figure 22.9, sulfuric acid reacts with sulfur trioxide in an acid–base reaction. The protonated sulfur trioxide molecule is a strong electrophile because the electrons in the S=$^+$OH bond are attracted to the positive charge and this produces a partial positive charge (δ+) on the sulfur atom.
- In step 2, the HSO_3^+ ion reacts with benzene to form a nonaromatic carbocation. Two electrons move from the benzene ring to form a new C–S bond and, at the same time, a pair of electrons in the S=$^+$OH bond moves to the positively charged oxygen.
- Finally, in step 3, the $^-OSO_3H$ ion deprotonates the carbocation to form benzenesulfonic acid and regenerate sulfuric acid.

Notice that all three steps in the sulfonation of benzene are reversible, unlike other electrophilic aromatic substitution reactions of benzene. When benzenesulfonic acid is heated with *dilute* sulfuric acid, the reverse reaction occurs and benzenesulfonic acid is converted into benzene in a **desulfonation reaction** (see Figure 22.10).

The mechanism of desulfonation starts by protonation of the aromatic ring of benzenesulfonic acid to give a nonaromatic carbocation (in step 1). In step 2, the carbocation loses the HSO_3^+ ion to form benzene. The HSO_3^+ ion could then react with benzene to re-form benzenesulfonic acid but, in dilute sulfuric acid, it is more likely to react with water (in step 3) to give sulfuric acid (in step 4). This is because dilute sulfuric acid contains a relatively high concentration of water and water is a stronger nucleophile than benzene.

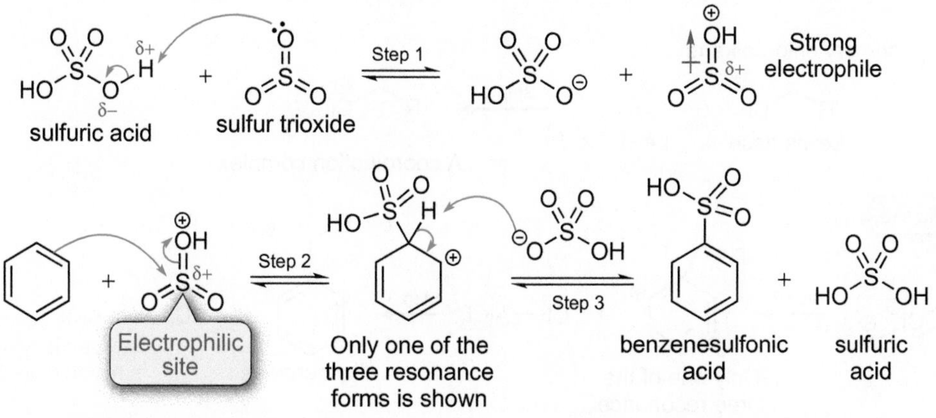
Step 1
sulfuric acid
sulfur trioxide
Strong electrophile
Step 2
Electrophilic site
Only one of the three resonance forms is shown
Step 3
benzenesulfonic acid
sulfuric acid

Figure 22.9 Sulfonation of benzene using fuming sulfuric acid.

Although H_2O is shown as the base in step 4, $HOSO_3^-$ could also act as the base.

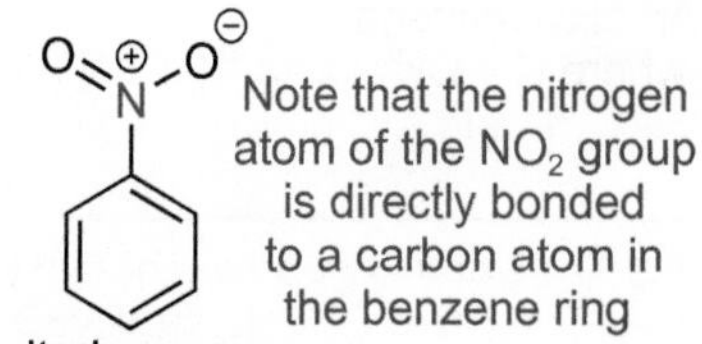
Note that the nitrogen atom of the NO_2 group is directly bonded to a carbon atom in the benzene ring

nitrobenzene

(Only one resonance form is shown—both N–O bonds are equivalent.)

Interactive 3D animation of nitration of benzene.

Notice that the sulfur atom of the SO_3H group is directly bonded to a carbon atom in the benzene ring.

Even without protonation, the sulfur atom in SO_3 is electrophilic because oxygen is more electronegative than sulfur (electrophilic strength is discussed in Section 19.1, p.882). However, protonation of SO_3 increases the magnitude of the partial positive charge on sulfur and this makes HSO_3^+ a much stronger electrophile than SO_3.

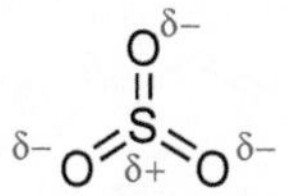

Because the SO_3H group can be introduced and removed from benzene rings, it is used as a temporary directing group in the synthesis of substituted benzenes. Temporary substituents are discussed in Section 22.4 (p.1043).

Lone pairs of electrons are more easily donated than bonding electrons, hence water is a stronger nucleophile than benzene.

Sulfonic acids (and related compounds) are introduced in Box 2.8 (p.101).

Interactive 3D animation of sulfonation of benzene.

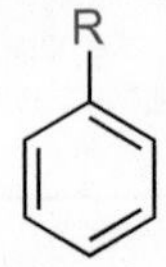
R

Friedel–Crafts alkylation introduces an alkyl group, R, onto the ring

An alkylbenzene where R = CH_3, C_2H_5, etc.

Alkylation is the transfer of an alkyl group (CH_3, C_2H_5, etc.; see Box 2.3, p.81) from one molecule to another.

Primary, secondary, and tertiary halogenoalkanes are introduced in Section 2.6 (p.91).

Charles Friedel (1832–1899) and James Mason Crafts (1839–1917) discovered the alkylation of benzene by chance. They expected the reaction of a chloroalkane with iodine and aluminium chloride to form an iodoalkane in a nucleophilic substitution reaction, but they were surprised to find that, in the presence of an aromatic compound, the aromatic ring replaced the chlorine atom in the chloroalkane.

Vinyl halides (e.g. RCH=CHX) and aryl halides (ArX) do not react with benzene in Friedel–Crafts reactions. Coordination complexes from vinyl and aryl halides do not form vinyl or aryl carbocations because these cations are so unstable.

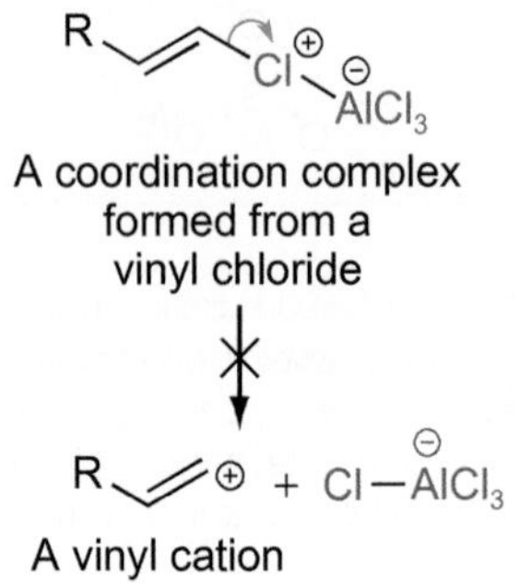
R
Cl
AlCl3
A coordination complex formed from a vinyl chloride
R
+ Cl—AlCl3
A vinyl cation

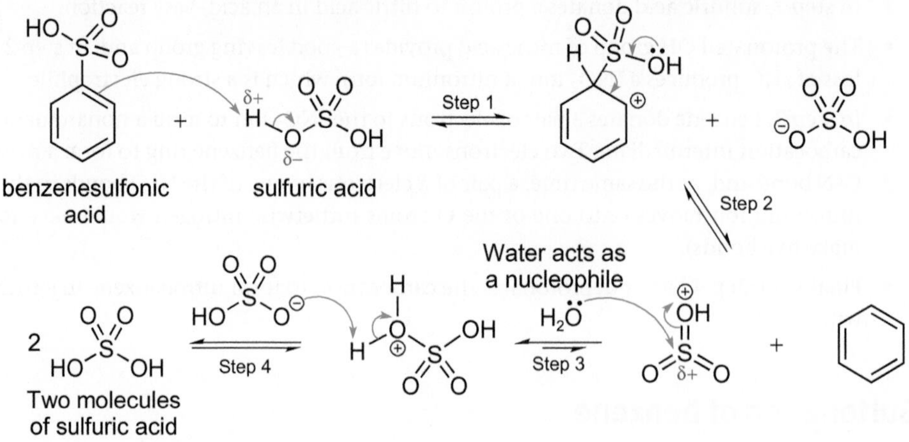
HO
S
O
O
δ+
H
OH
δ−
Step 1
H
OH
+
benzenesulfonic acid
sulfuric acid
Step 2
Water acts as a nucleophile
H2O
Step 3
δ+
Step 4
2
HO
OH
Two molecules of sulfuric acid

Figure 22.10 Desulfonation of benzenesulfonic acid using dilute sulfuric acid.

Alkylation of benzene

Reaction of benzene with chloro- or bromoalkanes (RCl or RBr) and a Lewis acid catalyst (typically $AlCl_3$, $FeCl_3$, or $FeBr_3$) produces alkylbenzenes (C_6H_5R, where R = CH_3, C_2H_5, etc.) in **Friedel–Crafts alkylation** reactions. The mechanism of alkylation follows the general mechanism for electrophilic substitution reactions of benzene in Figure 22.4 (p.1014), but includes an initial step in which the chloro- or bromoalkane reacts with the Lewis acid to form a strong electrophile.

With a primary chloroalkane

The mechanism of reaction of benzene with a primary chloroalkane (RCH_2Cl) and $AlCl_3$ is shown in Figure 22.11.

- In step 1, the primary chloroalkane reacts with $AlCl_3$ to form a coordination complex. By forming a coordination complex with the Lewis acid, the primary chloroalkane is converted into a stronger electrophile because the electrons in the C–Cl bond of the coordination complex are strongly attracted to the positively charged chlorine.
- In step 2, benzene attacks the partially positive carbon atom in the complex to produce a nonaromatic carbocation together with $[AlCl_4]^-$.
- In step 3, $[AlCl_4]^-$ removes a proton from the nonaromatic carbocation to give an alkylbenzene together with HCl and $AlCl_3$. Since $AlCl_3$ is regenerated, it acts as a catalyst for alkylation of benzene.

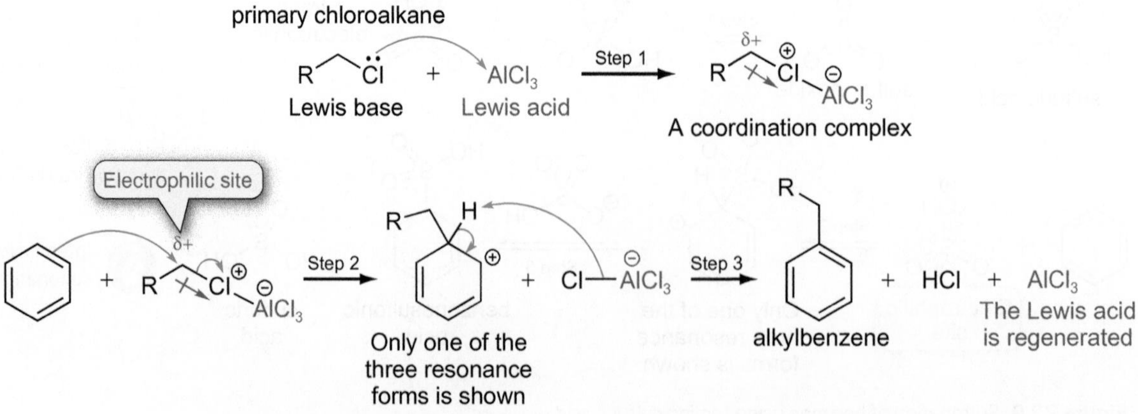
primary chloroalkane
R
Cl
+
AlCl3
Lewis base
Lewis acid
Step 1
δ+
A coordination complex
Electrophilic site
Step 2
H
Only one of the three resonance forms is shown
Cl—AlCl3
Step 3
alkylbenzene
+ HCl + AlCl3
The Lewis acid is regenerated

Figure 22.11 A Friedel–Crafts alkylation using a primary chloroalkane.

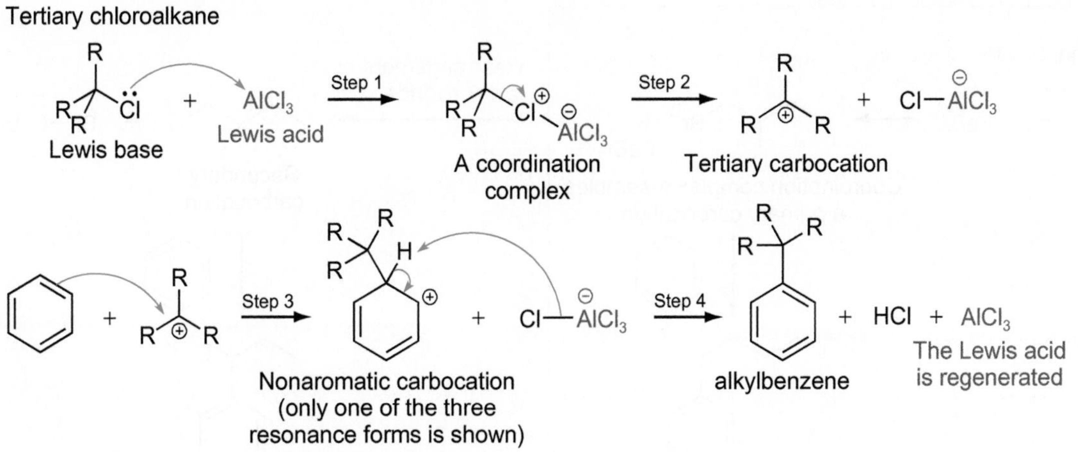
Tertiary chloroalkane
Lewis base
AlCl3
Lewis acid
Step 1
A coordination complex
Step 2
Tertiary carbocation
Cl—AlCl3
Step 3
Nonaromatic carbocation (only one of the three resonance forms is shown)
Step 4
alkylbenzene
+ HCl + AlCl3
The Lewis acid is regenerated

Figure 22.12 A Friedel–Crafts alkylation using a tertiary chloroalkane.

With secondary and tertiary chloroalkanes

The reaction of benzene with $AlCl_3$ and *secondary and tertiary* chloroalkanes (R_2CHCl and R_3CCl, respectively) also produces alkylbenzenes, although the reaction mechanism is usually slightly different from that in Figure 22.11 for a *primary* chloroalkane. The mechanism for a tertiary chloroalkane is shown in Figure 22.12.

The coordination complex formed by reaction of a secondary or a tertiary chloroalkane with $AlCl_3$ (in step 1) usually breaks down to form a carbocation and $[AlCl_4]^-$ (in step 2). The rate of formation of the carbocation from the coordination complex depends on the stability of the carbocation. As tertiary carbocations are more stable than secondary, and particularly primary, carbocations, tertiary carbocations are formed most rapidly from coordination complexes. The secondary or tertiary carbocation is a strong electrophile and, in step 3, two electrons move from the benzene ring to the carbocation. The mechanism is then the same as that for the reaction with a primary chloroalkane in Figure 22.11.

Worked Example 22.2 provides practice in drawing a reaction mechanism for an alkylation involving a tertiary carbocation.

Tertiary carbocations (R_3C^+) are more stable than secondary carbocations (R_2CH^+), which are more stable than primary carbocations (RCH_2^+), as discussed in Section 19.1 (p.872).

Interactive 3D animation of Friedel–Crafts alkylation.

Rearrangements in Friedel–Crafts alkylations and polyalkylation

A Friedel–Crafts alkylation reaction generally gives more than one product, particularly when using primary chloro- or bromoalkanes (RCH_2Cl or RCH_2Br), and this restricts the use of this reaction in synthesis. Take, for example, the reaction of benzene with 1-bromobutane ($CH_3CH_2CH_2CH_2Br$) and $FeBr_3$ (Figure 22.13). Apart from 1-butylbenzene, which is the expected product of the Friedel–Crafts alkylation, 1-*sec*-butylbenzene (or (1-methylpropyl)benzene), 1,4-dibutylbenzene, and other substituted benzenes are formed. So why do Friedel–Crafts alkylations of benzene give so many products? There are two reasons.

One reason is that the intermediate coordination complex rearranges. In the reaction in Figure 22.13, the coordination complex formed from 1-bromobutane and $FeBr_3$ can

Because it is so difficult to obtain a good yield of a single alkylbenzene from benzene using a Friedel–Crafts *alkylation*, chemists generally use an alternative two-step synthesis that starts with a Friedel–Crafts *acylation* reaction (see p.1023). This reaction introduces an acyl group (RCO) that is then reduced to RCH_2.

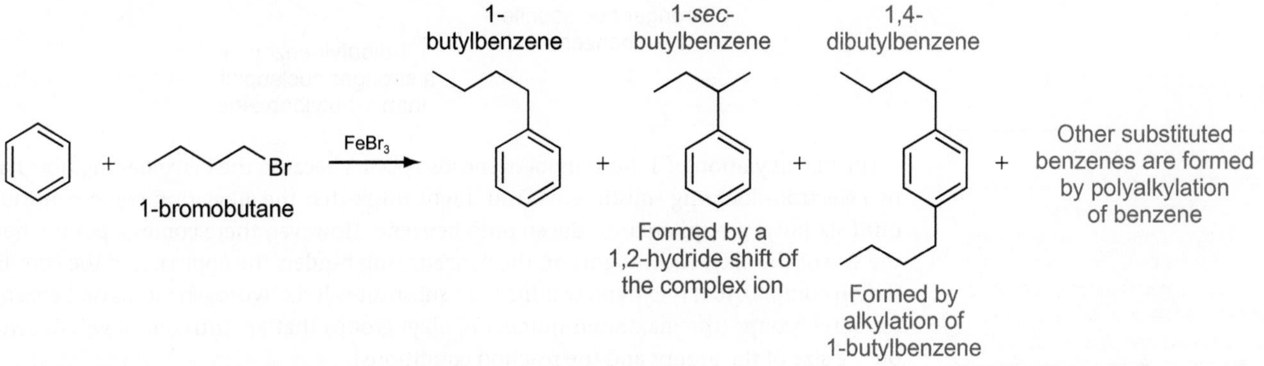
1-butylbenzene
1-sec-butylbenzene
1,4-dibutylbenzene
Br
1-bromobutane
FeBr3
Other substituted benzenes are formed by polyalkylation of benzene
Formed by a 1,2-hydride shift of the complex ion
Formed by alkylation of 1-butylbenzene

Figure 22.13 A Friedel–Crafts alkylation reaction leading to several products.

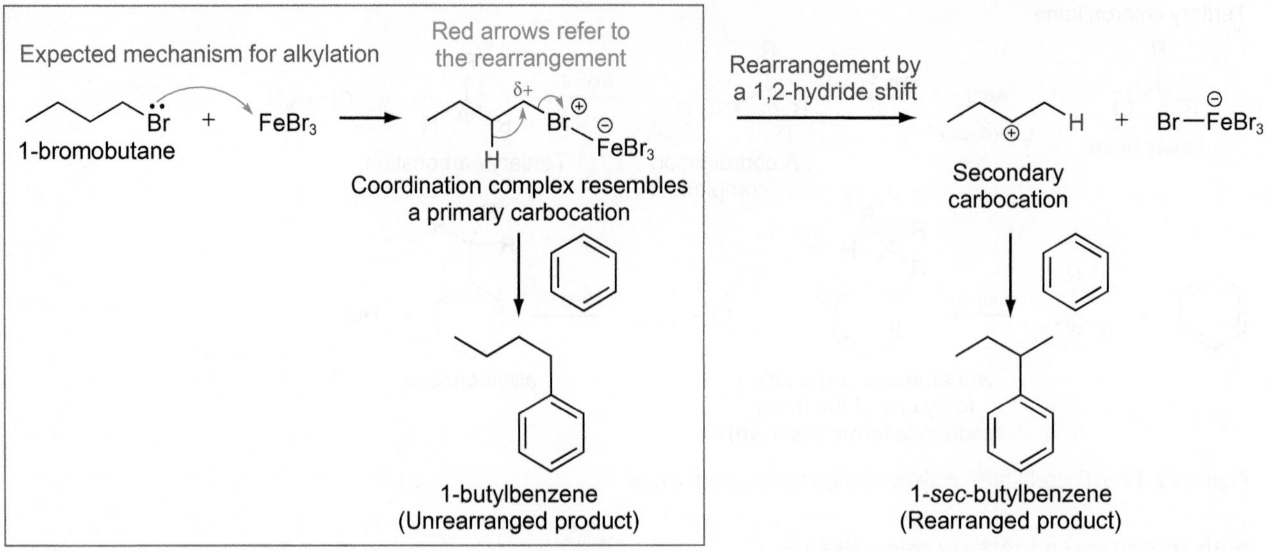
Expected mechanism for alkylation
Red arrows refer to the rearrangement
Br
FeBr3
1-bromobutane
δ+
Br
FeBr3
H
Coordination complex resembles a primary carbocation
Rearrangement by a 1,2-hydride shift
H
Br—FeBr3
Secondary carbocation
1-butylbenzene
(Unrearranged product)
1-sec-butylbenzene
(Rearranged product)

Figure 22.14 Rearrangement of the coordination complex formed from 1-bromobutane and $FeBr_3$.

Similar carbocation rearrangements take place when HX adds to some C=C bonds (see Section 21.3, p.971). The mechanism of 1,2-hydride shifts is discussed in more detail here.

undergo a 1,2-hydride shift to form the secondary carbocation $CH_3CH_2CH^+CH_3$ (see Figure 22.14). It is this secondary carbocation that reacts with benzene to form 1-*sec*-butylbenzene. The driving force for the 1,2-hydride shift is the stability of the secondary carbocation. Because the coordination complex has a partial positive charge (δ+) on a primary carbon atom, the coordination complex resembles a primary carbocation and the coordination complex undergoes a 1,2-hydride shift to produce a more stable secondary carbocation (see Section 21.3, p.971). Conversion of the coordination complex to the secondary carbocation is described as a **rearrangement** and the substituted benzene formed from the secondary carbocation is called the **rearranged product.**

You can see why the second butyl group substitutes in the 4-position in Section 22.3 (p.1030).

Alkyl groups are electron-donating groups that exert a positive inductive effect, +I (see Section 19.1, p.870).

The second reason why Friedel–Crafts reactions of benzene give many products is that polyalkylation (multiple alkylation) takes place. It is difficult to introduce just one alkyl substituent on to a benzene ring because replacing an H atom by an electron-donating (+I) alkyl substituent makes the ring more nucleophilic and therefore more reactive to further electrophilic substitution. This explains why 1-butylbenzene reacts with 1-bromobutane and $FeBr_3$ to form 1,4-dibutylbenzene.

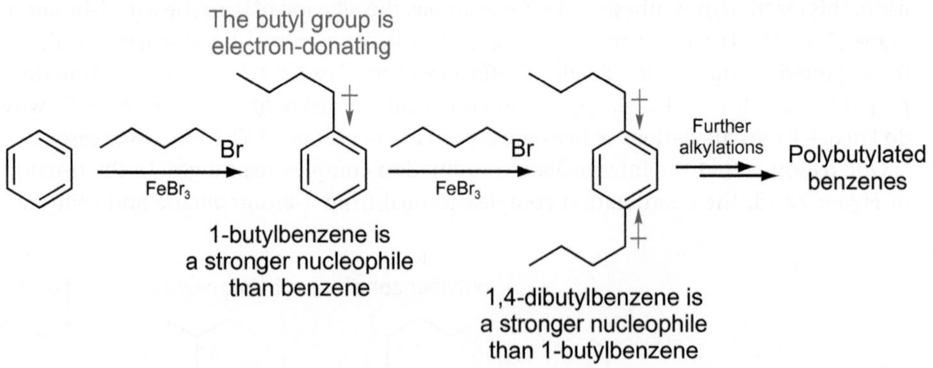
The butyl group is electron-donating
Br
FeBr3
Br
FeBr3
Further alkylations
Polybutylated benzenes
1-butylbenzene is a stronger nucleophile than benzene
1,4-dibutylbenzene is a stronger nucleophile than 1-butylbenzene

Halogenation, nitration, and sulfonation of benzene all stop after the introduction of one Br, Cl, NO_2, or SO_3H substituent. These substituents are electron-withdrawing so bromobenzene, chlorobenzene, nitrobenzene, and benzenesulfonic acid are all *less* reactive than benzene in electrophilic substitution reactions (see Section 22.3, p.1029).

Further alkylation of 1,4-dibutylbenzene also occurs because the benzene ring now has two electron-donating substituents. You might think that the alkylation could continue until six butyl groups are introduced onto benzene. However, there comes a point when the size of the alkyl substituents on the benzene ring hinders the approach of the coordination complex so it is extremely difficult to substitute all six hydrogen atoms on benzene for butyl groups (the maximum number of alkyl groups that are introduced will depend on the size of the groups and the reaction conditions).

Box 22.4 Substituted benzenes in sport

The sport and fitness industry owes much to developments in chemistry. A number of synthetic compounds, many of which contain substituted benzenes, play a key role in the performance of athletes in training, in competition, and in recovering from an injury. Some examples follow.

Spandex

Spandex (Lycra®) is an elastic polymeric material (polymers are composed of repeating structural units). Athletes wear spandex to support muscles, which reduces injury. The properties of spandex are due to the presence of both flexible segments and rigid segments in the polymer chain. The rigid segments contain substituted benzenes and urea linkages (–NHCONH–). The urea linkages within different chains form hydrogen bonds to one another (C=O⋯H–N) and this aligns the rigid segments in different chains in the fibres. Because flexible segments link the rigid segments in each chain, the fibre can stretch, and this explains why spandex is so elastic.

Lexan®

The outer shell of helmets used in American football is made from a lightweight polycarbonate called Lexan®. This polymer has high impact strength and does not easily shatter. Lexan® contains substituted benzene rings that are linked by carbonate ($-OCO_2-$) groups.

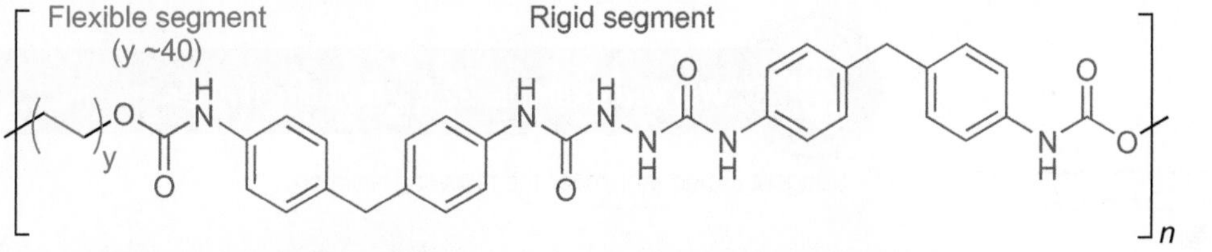
Flexible segment
(y ~40)
Rigid segment
y
n

Spandex (*n* and y indicate repeating units)

▲ The track cycling team of Great Britain competed in the 2012 Olympic Games in London wearing spandex. The tight-fitting spandex bodysuits help smooth the flow of air over their bodies.

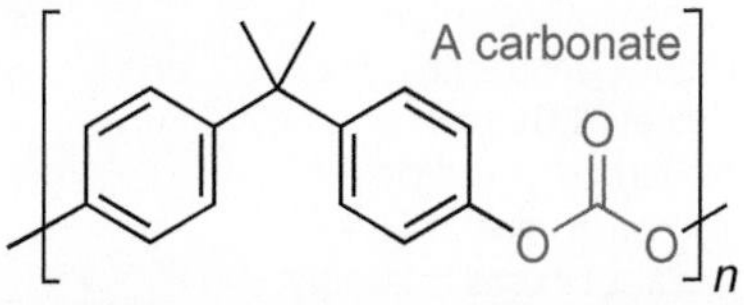
A carbonate
n

Lexan® (*n* is the number of repeat units)

▲ American football helmets contain the polycarbonate Lexan®. Kevlar® is another polymer that is lightweight and exceptionally strong (Box 1.11, p.55). It is used to make golf club shafts and lightweight bicycles.

Lignocaine (lidocaine)

Sprays that contain local anaesthetics such as lignocaine are widely used in contact sports, such as football, for fast pain relief from minor injuries (see Box 19.5, p.894). Lignocaine contains a substituted benzene ring that is linked to a tertiary amine group by an amide bond (–CONH–).

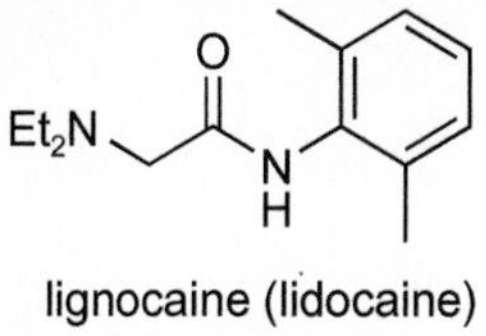
Et_2N

lignocaine (lidocaine)

→

→ Question

Ibuprofen is an anti-inflammatory drug used, for example, by athletes to relieve minor aches and pains and to help reduce minor swelling. Ibuprofen is sold as a racemate (see Box 18.7, p.848) and is prepared on an industrial scale from isobutylbenzene, $C_6H_5CH_2CH(CH_3)_2$.

Isobutylbenzene is prepared in industry using a two-step sequence (Friedel–Crafts acylation followed by reduction of the C=O bond (see p.1025)). This is because attempts to prepare isobutylbenzene in a single step, by reacting benzene with 1-chloro-2-methylpropane ($(CH_3)_2CHCH_2Cl$) and $AlCl_3$, gave *tert*-butylbenzene, $C_6H_5C(CH_3)_3$, as the major product. Suggest a reaction mechanism to explain the formation of *tert*-butylbenzene, rather than isobutylbenzene, from the reaction below.

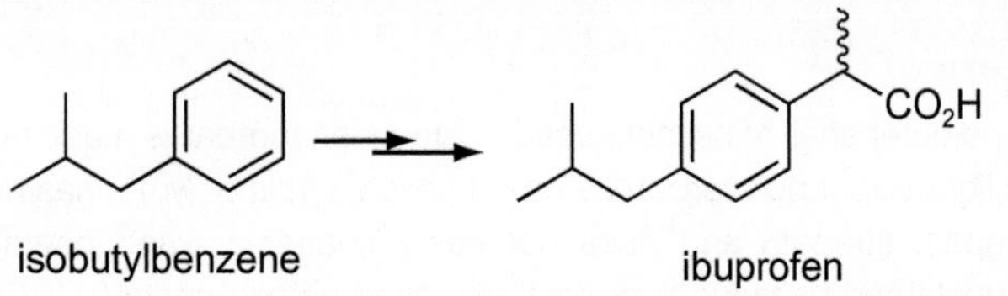
CO2H
isobutylbenzene
ibuprofen

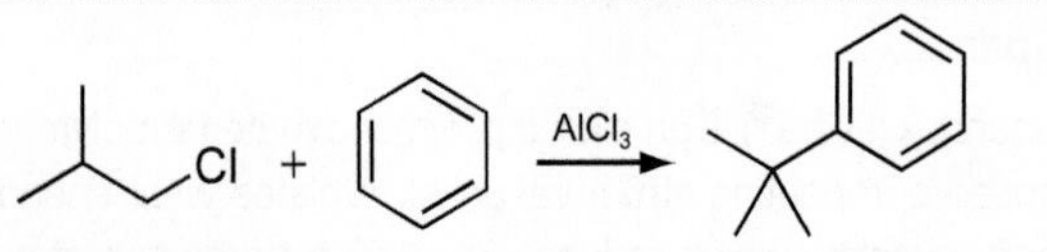
Cl
+
AlCl3
1-chloro-2-methylpropane
tert-butylbenzene

Worked example 22.2 Alkylating benzene

Suggest a mechanism for the following reaction.

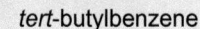

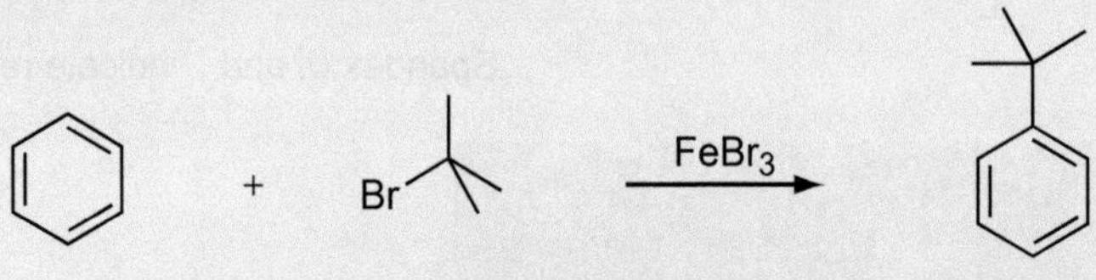
+
Br
FeBr3
2-bromo-2-methylpropane
tert-butylbenzene

▲ Molecular model of *tert*-butylbenzene.

Strategy

Draw the structure of the coordination complex formed from 2-bromo-2-methylpropane and $FeBr_3$.

Decide whether the coordination complex is likely to break down to form a carbocation by considering the stability of the carbocation.

Consider if rearrangement of the coordination complex or the carbocation is likely.

React the coordination complex or the carbocation with benzene to form a nonaromatic carbocation.

Remove a proton from the nonaromatic carbocation to form a substituted benzene.

→

The conversion of 2-bromo-2-methylpropane into *tert*-butylbenzene can be monitored by ^1H NMR spectroscopy (Section 12.3, p.580). In 2-bromo-2-methylpropane, the 9 chemically equivalent hydrogens in the *tert*-butyl group appear as a singlet resonance signal at ~1.8 ppm, whereas the same hydrogens in *tert*-butylbenzene appear as a singlet resonance signal at ~1.4 ppm. Mass spectrometry can also be used to identify *tert*-butylbenzene: the EI mass spectrum shows a molecular ion peak with an *m/z* value of 134. A fragment ion with an *m/z* value of 119 is also observed, due to the formation of a benzylic cation (PhC^+Me_2) on loss of $CH_3^\bullet$ (Section 12.1, p.558).

Iron(III) bromide (ferric bromide) is a red-brown solid that can be prepared from iron metal and bromine ($2\,Fe + 3\,Br_2 \rightarrow 2\,FeBr_3$).

→ Solution

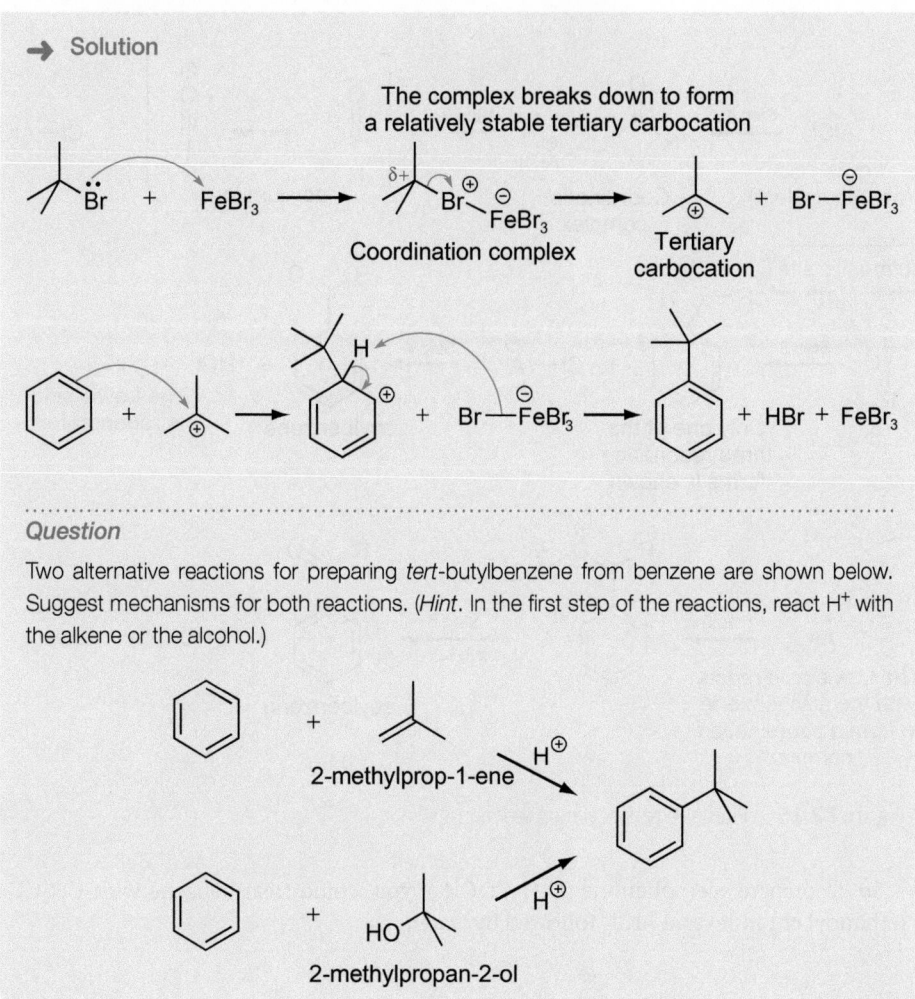

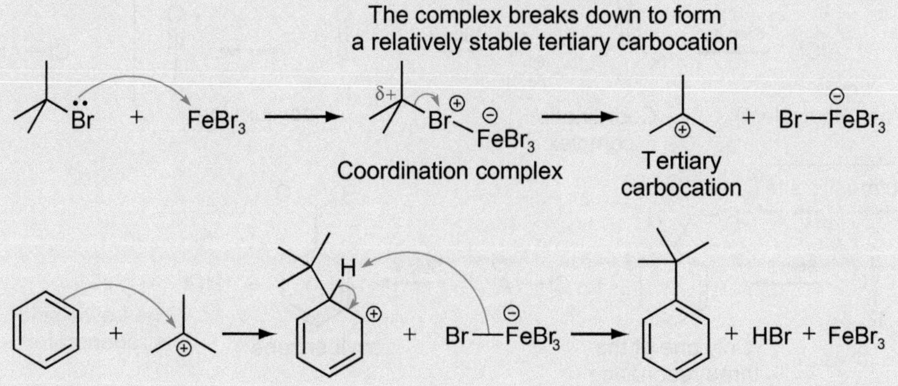
The complex breaks down to form
a relatively stable tertiary carbocation
Br
FeBr3
δ+
Br
FeBr3
Coordination complex
Tertiary
carbocation
Br—FeBr3
H
Br—FeBr3
+ HBr + FeBr3

Question

Two alternative reactions for preparing *tert*-butylbenzene from benzene are shown below. Suggest mechanisms for both reactions. (*Hint*. In the first step of the reactions, react H^+ with the alkene or the alcohol.)

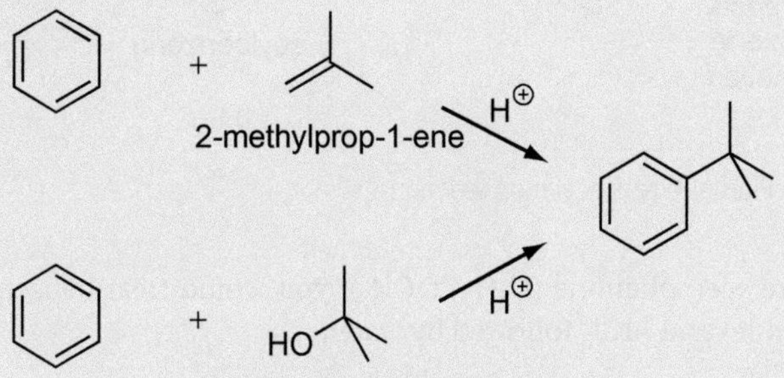
2-methylprop-1-ene
HO
2-methylpropan-2-ol

2-Methylprop-1-ene is also called isobutylene and the common name for 2-methylpropan-2-ol is *tert*-butanol or *tert*-butyl alcohol.

Acylation of benzene

Reaction of benzene with acyl chlorides (acid chlorides, RCOCl) and a Lewis acid (typically $FeCl_3$ or $AlCl_3$) produces acylbenzenes in **Friedel–Crafts acylation reactions**. The mechanism of acylation, shown in Figure 22.15, follows the general mechanism for electrophilic substitution reactions of benzene in Figure 22.4 (p.1014), but includes two initial steps in which the acyl chloride reacts with the Lewis acid (to form a strong electrophile) and two extra steps at the end.

- In step 1, the acyl chloride reacts with $AlCl_3$ to form a coordination complex, which breaks down (in step 2) to form an **acylium ion** (R–C$^+$=O) and $[AlCl_4]^-$.
- The positively charged acylium ion is stabilized by resonance. It is a stronger electrophile than the acyl chloride so the acylium ion reacts with benzene in step 3 to produce a nonaromatic carbocation.
- In step 4, $[AlCl_4]^-$ removes a proton from the nonaromatic carbocation to give an acylbenzene (C_6H_5COR) together with HCl and $AlCl_3$.
- In step 5 the regenerated $AlCl_3$ forms a coordination complex with the acylbenzene.
- When the reaction is completed, water is added to the reaction mixture (in step 6), $AlCl_3$ is hydrolysed to $Al(OH)_3$, and the acylbenzene is released from the complex.

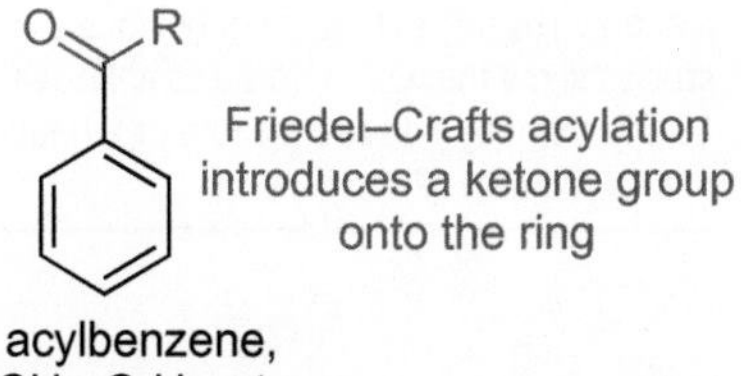
O
R
Friedel–Crafts acylation
introduces a ketone group
onto the ring

An acylbenzene, R = CH_3, C_2H_5, etc.

In a Friedel–Crafts *acylation*, the Lewis acid forms a coordination complex with the product of the reaction, the acylbenzene (C_6H_5COR), so one equivalent of the Lewis acid is required. In the last stage of the reaction, the Lewis acid is hydrolysed. In a Friedel–Crafts *alkylation*, the Lewis acid is regenerated and behaves as a catalyst.

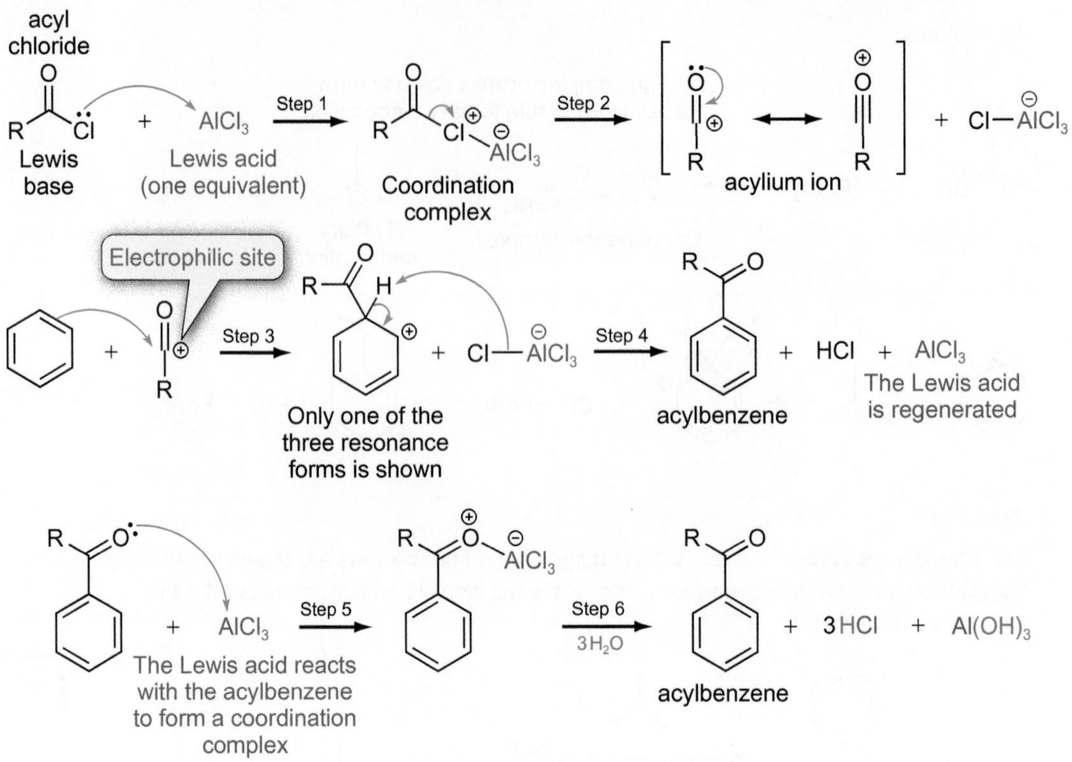
acyl chloride
Lewis base
Lewis acid (one equivalent)
Step 1
Coordination complex
Step 2
acylium ion
Electrophilic site
Step 3
Only one of the three resonance forms is shown
Step 4
acylbenzene
The Lewis acid is regenerated
Step 5
The Lewis acid reacts with the acylbenzene to form a coordination complex
Step 6
acylbenzene

Figure 22.15 A Friedel-Crafts acylation reaction.

Aluminium chloride is a white-coloured solid that hydrolyses rapidly (Box 22.5) to form aluminium hydroxide and hydrogen chloride:

$AlCl_3(s) + 3H_2O(l) \rightarrow Al(OH)_3(s) + 3HCl(g)$.

Interactive 3D animation of Friedel-Crafts acylation.

A mixture of $AlCl_3$ and CuCl is a common catalyst in the Gattermann-Koch reaction. The role of CuCl is not obvious, but studies show that CuCl forms a complex with CO and this may increase the rate of formation of HC^+=O from HCl and CO.

Benzaldehyde (PhCHO) has a strong almond odour and it is the main component of bitter almond oil. This oil is widely used in the food industry, for example, in making marzipan.

So, to prepare acetophenone ($C_6H_5COCH_3$), you would treat benzene with CH_3COCl (ethanoyl chloride) and $AlCl_3$ followed by water.

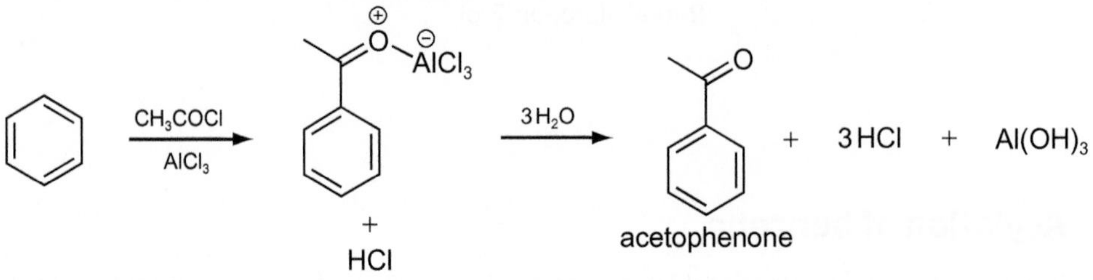
acetophenone

To convert benzene into benzaldehyde (PhCHO) requires reaction with the acylium ion HC^+=O. However, HC^+=O cannot be formed from HCOCl (methanoyl chloride) and $AlCl_3$ because HCOCl is not readily available as it is highly unstable. To overcome this problem, a mixture of benzene, CO, HCl, and $AlCl_3$ is reacted under high pressure in the **Gattermann–Koch reaction**. At high pressure, CO, HCl, and $AlCl_3$ produce the HC^+=O ion, which then reacts with benzene to form benzaldehyde, as shown in Figure 22.16.

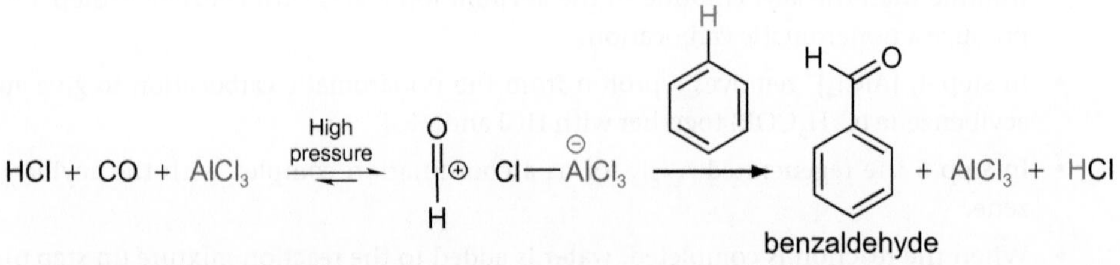
High pressure
benzaldehyde

Figure 22.16 The Gattermann-Koch reaction.

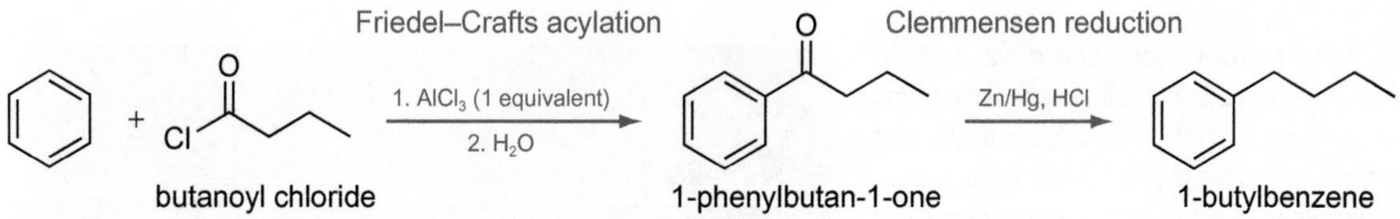
Friedel–Crafts acylation
Clemmensen reduction
1. AlCl3 (1 equivalent)
2. H2O
Zn/Hg, HCl
Cl
O
butanoyl chloride
1-phenylbutan-1-one
1-butylbenzene

Figure 22.17 Using Friedel-Crafts acylation followed by reduction to prepare an alkylbenzene.

Unlike Friedel–Crafts alkylations, where many alkylated benzenes are formed by rearrangement of coordination complexes/carbocations and by polyalkylation, the reaction of benzene in a Friedel–Crafts acylation forms only one acylbenzene product. This is because acylium ions do not rearrange (unlike coordination complexes from halogenoalkanes or secondary carbocations). Also, an acylbenzene is *less* reactive to electrophilic substitution than benzene (because the acyl group (RC(=O)–) is electron-withdrawing), so the only product from acylation of benzene is the unrearranged monoacylated benzene.

Inductive and mesomeric effects are introduced in Section 19.1 (p.870).

Alternative methods to reduce RC(=O)R to RCH_2R are discussed in Section 23.2 (pp.1077 and 1080).

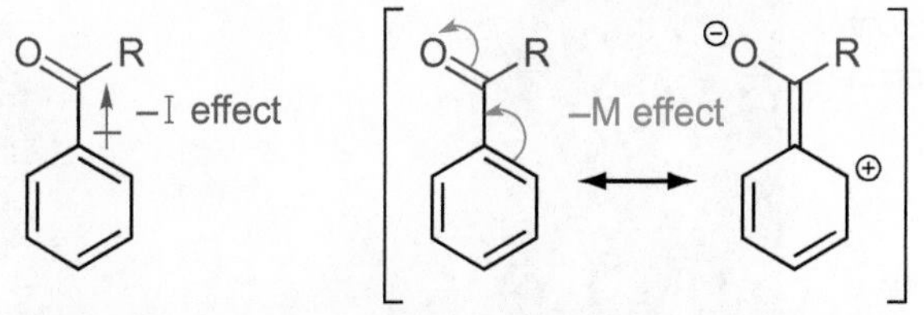
O
R
–I effect
–M effect

An acylbenzene is a weaker nucleophile than benzene because the acyl group is electron-withdrawing (–I, –M)

For this reason, Friedel–Crafts acylation of benzene is a more useful synthetic reaction than alkylation, which generally gives mixtures of alkylated benzenes. Indeed, acylation of benzene is normally the first stage of the preferred route to alkylbenzenes, as shown in Figure 22.17. The acylbenzene is then reacted with a reducing agent that converts the C=O bond into a CH_2 group. A common reducing agent for this transformation is an acidic solution of zinc dissolved in mercury (zinc amalgam). The reduction is called the **Clemmensen reduction** (Figure 22.17).

Erik Christian Clemmensen (1876–1941) was born in Denmark and emigrated to America, where he worked as a research chemist. The University of Copenhagen awarded him a doctorate for his development of the Clemmensen reduction. The reduction takes place at the surface of the zinc, which acts as the reducing agent:

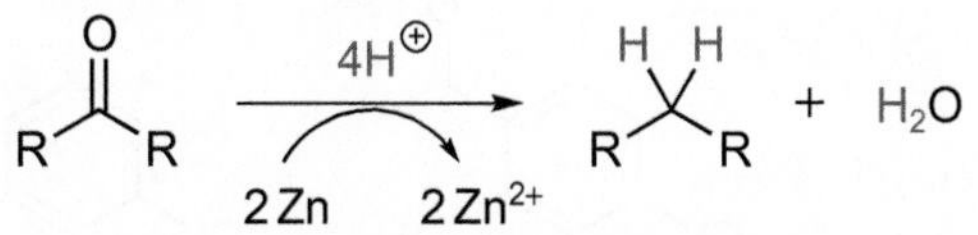
O
R
R
4H⊕
H H
+ H2O
2 Zn
2 Zn2+

Box 22.5 Cleaning up the Friedel-Crafts acylation

The growing pressure from legislation and environmental bodies has led to the development of an important area of modern research called *clean (green) chemistry*. Clean chemistry involves the development of efficient and eco-friendly synthetic methods that reduce or remove the use or generation of hazardous substances in the production of valuable products. Many useful reactions generate large amounts of waste products and chemists are currently looking at ways to treat or clean up the waste products—or, even better, how not to form the waste products in the first place.

One industrially important reaction that produces lots of waste products is the Friedel–Crafts acylation reaction. Conventional Friedel–Crafts acylation reactions use at least a stoichiometric →

➜ amount of $AlCl_3$, which is a corrosive solid. Unfortunately, $AlCl_3$ is not recovered at the end of the acylation reaction because it forms a complex with the ketone product, which is hydrolysed at the end of the reaction to release the ketone (see p.1024). Hydrolysis of $AlCl_3$ produces a large rise in temperature and creates large volumes of corrosive and toxic aluminium-containing waste as well as HCl gas.

To avoid producing large quantities of aluminium-containing waste, chemists have investigated alternative Lewis acids. One exciting development is the use of a clay called montmorillonite. Montmorillonite is found in the deposits of weathered volcanic ash and its structure contains layers of aluminosilicate. Negatively charged layers in the clay are held together by positively charged Al^{3+} ions. Research has shown that the Al^{3+} ions can be exchanged for other positively charged ions such as Fe^{3+}, Ni^{2+}, Zr^{4+}, Ce^{3+}, Cu^{2+}, Zn^{2+}, or even H_3O^+. The metal ions or H_3O^+ in the 'cation-exchanged' montmorillonites act as Lewis acids in Friedel–Crafts acylations.

There are many advantages of using cation-exchanged montmorillonites in place of $AlCl_3$ in Friedel–Crafts acylations. Cation-exchanged montmorillonites:

- are more easily handled than corrosive $AlCl_3$;
- do not form complexes with ketones so they are regenerated and act as catalysts;
- have large surface areas so that small amounts of the catalysts can be used in large-scale reactions;
- are solids and are easily removed from liquid products by filtration and reused.

▲ The violent reaction of H_2O with $AlCl_3$.

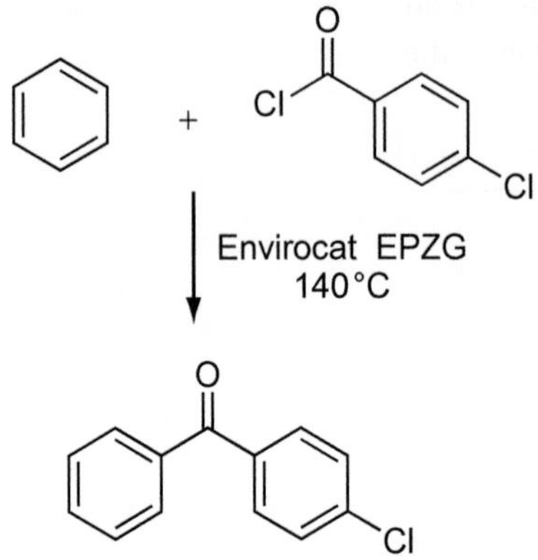
O
Cl
Cl
Envirocat EPZG
140 °C
O
Cl

▲ An industrial Friedel–Crafts acylation using an Envirocats™ catalyst.

Today, cation-exchanged montmorillonite catalysts are used in the chemical industry in place of $AlCl_3$. Some of these catalysts are sold under the name Envirocats™, as shown in the reaction scheme above.

Question

Ketones **1–3** are prepared industrially by Friedel–Crafts acylations of chlorobenzene using an Envirocat and a suitable acyl chloride. Suggest structures for the acyl chlorides that are used to produce **1–3**.

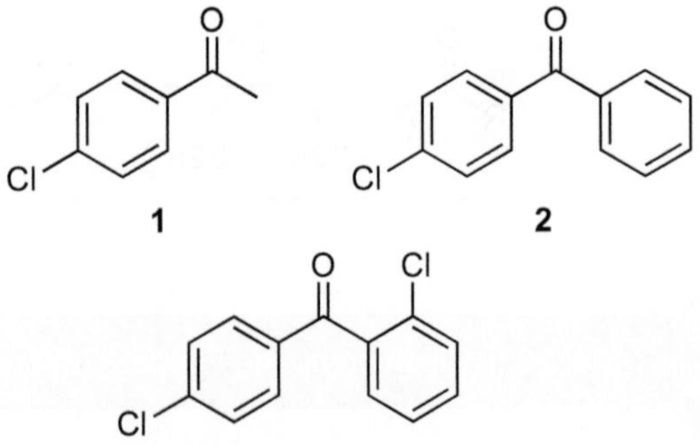
O
Cl
O
Cl
O
Cl
Cl

1 **2** **3**

Summary

- Benzene reacts with electrophiles in **electrophilic substitution reactions**. Another atom, or group of atoms, replaces a hydrogen atom on the benzene ring, and the product retains the stable aromatic ring.
- There are five important electrophilic substitution reactions of benzene.

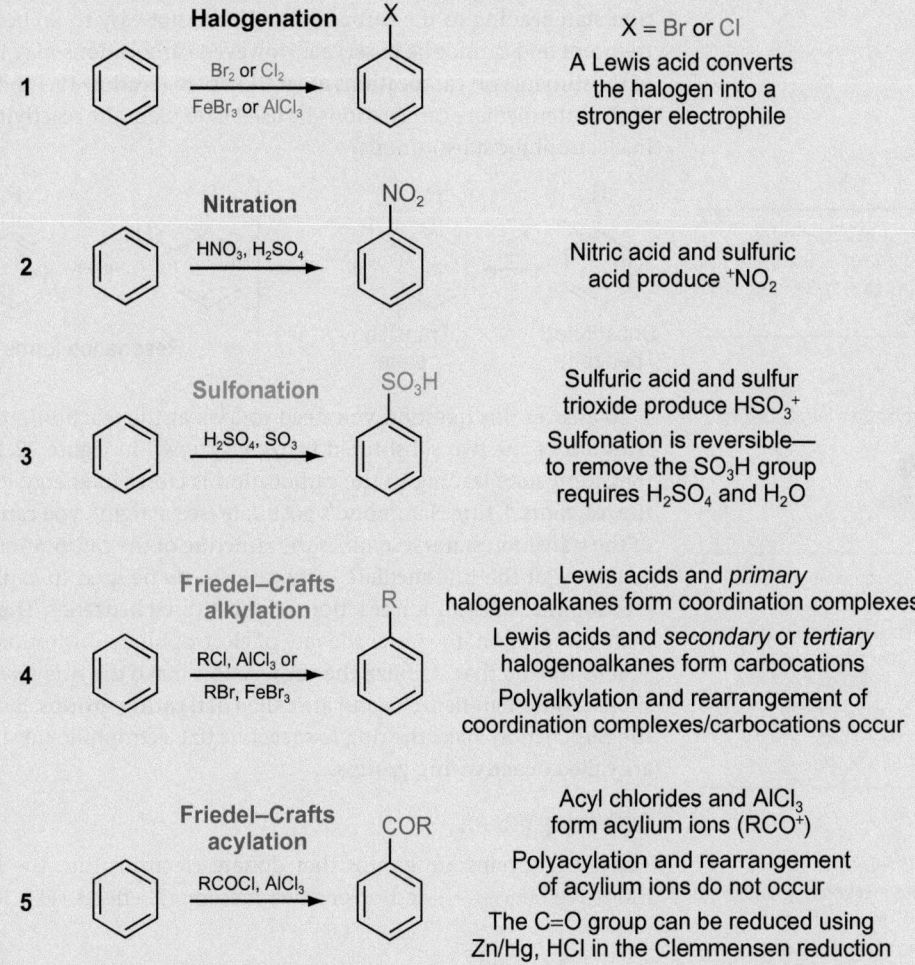

1 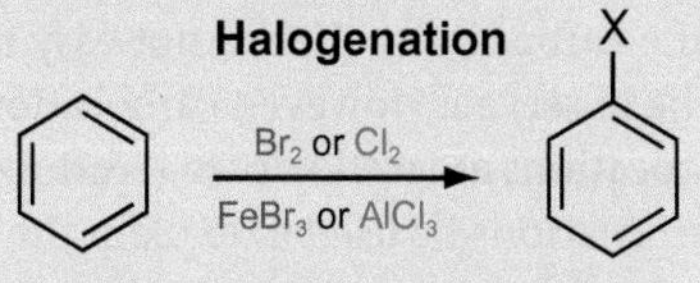
Halogenation
X
Br2 or Cl2
FeBr3 or AlCl3

X = Br or Cl
A Lewis acid converts the halogen into a stronger electrophile

2 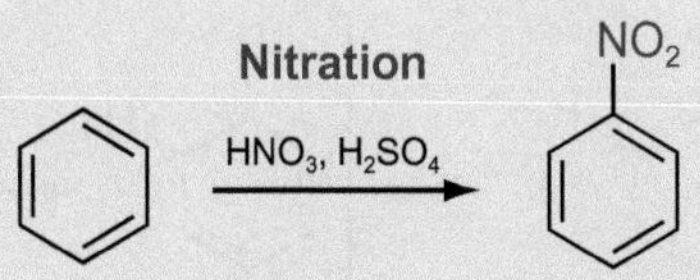
Nitration
NO2
HNO3, H2SO4

Nitric acid and sulfuric acid produce $^{+}NO_2$

3 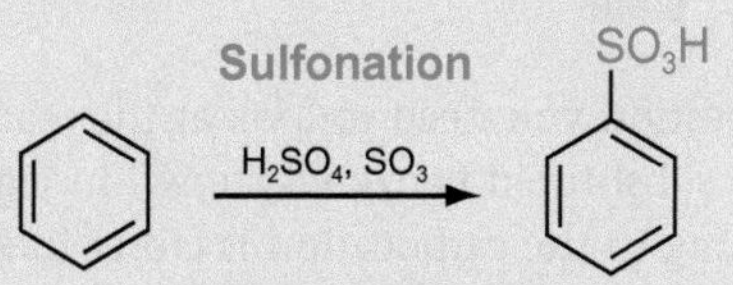
Sulfonation
SO3H
H2SO4, SO3

Sulfuric acid and sulfur trioxide produce HSO_3^{+}
Sulfonation is reversible—to remove the SO_3H group requires H_2SO_4 and H_2O

4 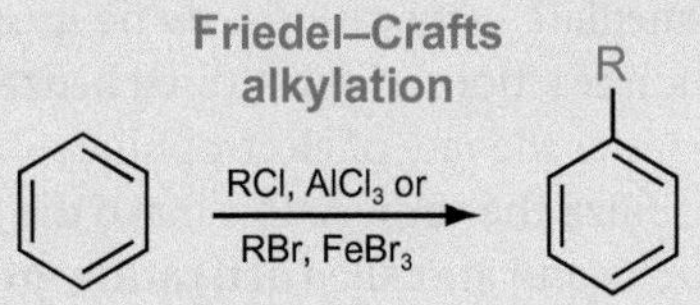
Friedel–Crafts alkylation
R
RCl, AlCl3 or
RBr, FeBr3

Lewis acids and *primary* halogenoalkanes form coordination complexes
Lewis acids and *secondary* or *tertiary* halogenoalkanes form carbocations
Polyalkylation and rearrangement of coordination complexes/carbocations occur

5 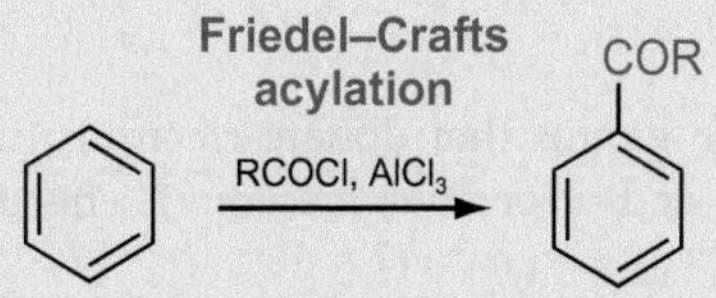
Friedel–Crafts acylation
COR
RCOCl, AlCl3

Acyl chlorides and $AlCl_3$ form acylium ions (RCO^{+})
Polyacylation and rearrangement of acylium ions do not occur
The C=O group can be reduced using Zn/Hg, HCl in the Clemmensen reduction

- These reactions are useful in synthesis as a way of introducing reactive groups onto a benzene ring.

? For a practice question on this topic, see question 2 at the end of this chapter (p.1095).

22.3 Reactivity of substituted benzenes in electrophilic substitutions

In the previous section, electrophilic substitution reactions of benzene were shown to introduce substituents on the ring. But what happens when substituted benzenes (e.g. C_6H_5R) undergo electrophilic substitution reactions? There are two issues to consider:

- the effect of the existing substituent on the rate of further electrophilic substitution;
- the position of further electrophilic substitution.

Does R make the ring more or less reactive than benzene?

At which position does the electrophile substitute in the benzene ring?

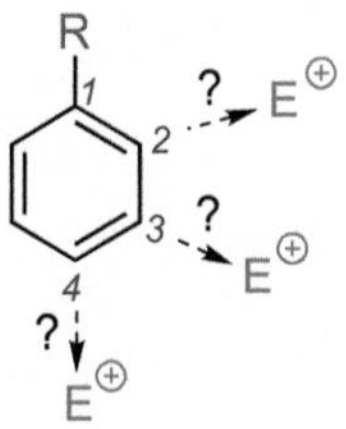
R
1
2
3
4
?
E⊕

Relative reactivity of substituted benzenes

Transition state theory is introduced in Section 9.8 (p.434).

In the mechanism of electrophilic substitution in Figure 22.4 (p.1014) the slower step is the formation of an intermediate carbocation by reaction of the benzene ring with an electrophile. The lower the energy of the transition state leading to the carbocation, the lower the activation energy and the faster the rate of electrophilic substitution.

To predict the effect of an existing substituent, R, on the rate of further electrophilic substitution of C_6H_5R you need to consider whether R stabilizes or destabilizes the transition state leading to the carbocation. This is not easy to do because transition states are transient and cannot be observed. However, carbocations may be isolated and the effects of substituents on carbocations are well known (Section 19.1, p.870). So can the stabilities of the intermediate carbocations be used to explain the reactivity of substituted benzenes in electrophilic substitutions?

Hammond's postulate states that the structure of the transition state resembles the structure of the stable species that is nearest in energy.

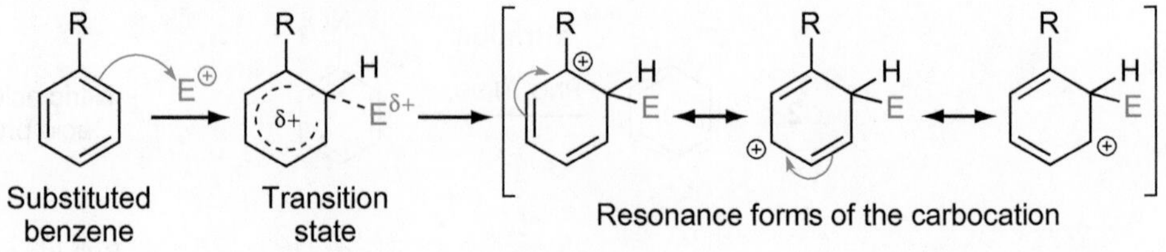
Substituted benzene
Transition state
Resonance forms of the carbocation

Gibbs energy profiles are introduced in Section 9.8 (p.435).

To answer this question you need to look at the reaction profile for electrophilic substitution of the two substituted benzenes shown in Figure 22.18. For each reaction, the transition state leading to the carbocation is closer in energy to the carbocation than to the reactants. Using Hammond's postulate (see margin), you can assume that the structure of the transition states resembles the structure of the carbocation in each reaction. So, the stabilities of the intermediate carbocations *can* be used to explain the different rates of electrophilic substitution reactions of substituted benzenes. The more stable the intermediate carbocation, the faster the rate of electrophilic substitution.

The intermediate carbocations, called arenium ions, can be observed. For example, the simplest such ion, the benzenium ion ($C_6H_7^+$), forms when benzene is dissolved in a superacid (such as a mixture of HSO_3F, SbF_5, SO_2ClF, and SO_2F_2); at very low temperature (about −140 °C) the benzenium ion can be detected by ^1H and ^{13}C NMR spectroscopy.

Substituents that stabilize the carbocation make the ring *more* reactive to electrophilic substitution than benzene and are called **activating groups**. Substituents that destabilize the carbocation make the ring *less* reactive to electrophilic substitution than benzene and are called **deactivating groups**.

Activating groups

Inductive effects (+I/−I) refer to polarization of electrons in σ bonds, while mesomeric effects (+M/−M) refer to the movement of lone pairs of electrons and electrons in π bonds (see Section 19.1, p.870).

Activating groups are groups that donate electrons into the benzene ring by positive inductive effects (+I) and/or positive mesomeric effects (+M). Electron-donating groups

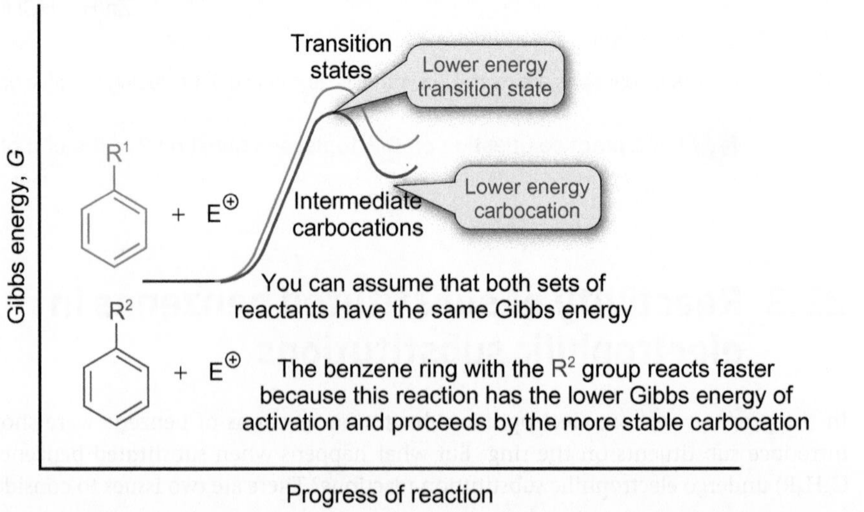
Transition states
Lower energy transition state
Intermediate carbocations
Lower energy carbocation
You can assume that both sets of reactants have the same Gibbs energy
The benzene ring with the R2 group reacts faster because this reaction has the lower Gibbs energy of activation and proceeds by the more stable carbocation
Gibbs energy, G
Progress of reaction

Figure 22.18 Reaction profiles for the formation of the intermediate carbocations in the electrophilic substitution of two substituted benzenes. (R^1 and R^2 are groups with different electronic effects.)

Table 22.1 Activating groups on a benzene ring and their electronic effects

Activating groups†	Electronic effects*	
$-NH_2$, –NHR, $-NR_2$	+M > –I	↑
–OH, –OR, $-O^-$	+M > –I	
–NHCOR, –OCOR	+M > –I	Increasing reactivity of the benzene ring with electrophiles
–Ph, $-CH=CH_2$	+M, +I	
–R (alkyl)	+I	

* Usually mesomeric effects are stronger than inductive effects (Section 19.1, p.875).
† Ph stands for phenyl (C_6H_5–).

> The relative rates of nitration of phenol (C_6H_5OH), methylbenzene ($C_6H_5CH_3$), and benzene (C_6H_6) are 1000 : 25 : 1.

increase the reactivity of the ring to reaction with electrophiles because the carbocation is stabilized by electron-donating groups. Some common activating groups are shown in Table 22.1.

> An **activating group** on a benzene ring makes the ring *more* reactive to electrophilic substitution than benzene.

+M effects are stronger than +I effects so electron-donating groups with +M effects stabilize carbocations more effectively than electron-donating groups with +I effects. For example, the $-NH_2$ group has a +M effect and is a strong electron donor, whereas the –Me group has a +I effect and is a weaker electron donor. This explains why aniline ($C_6H_5NH_2$) reacts faster with electrophiles than methylbenzene (toluene; $C_6H_5CH_3$).

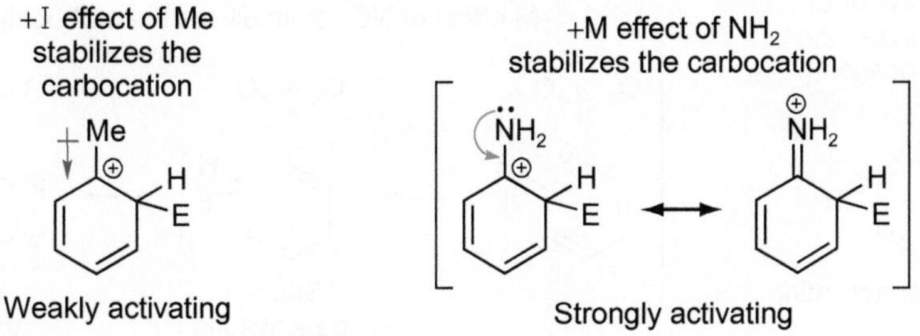
+I effect of Me stabilizes the carbocation
Me
H
E
Weakly activating
+M effect of NH2 stabilizes the carbocation
NH2
H
E
Strongly activating

> The reactivity of phenylamine (aniline; $C_6H_5NH_2$) and phenol (C_6H_5OH) towards electrophiles depends on the reaction conditions. For aniline, the $-NH_2$ group can act as a base and on protonation the resulting $-NH_3^+$ group is electron-withdrawing; this *decreases* the reactivity of the benzene ring to reaction with electrophiles. In contrast, phenol is a weak acid and on deprotonation the resulting $-O^-$ group is a stronger electron donor than the OH group; this *increases* the reactivity of the benzene ring to reaction with electrophiles.

Electron-donating groups with strong +M effects stabilize carbocations more effectively than electron-donating groups with weaker +M effects.

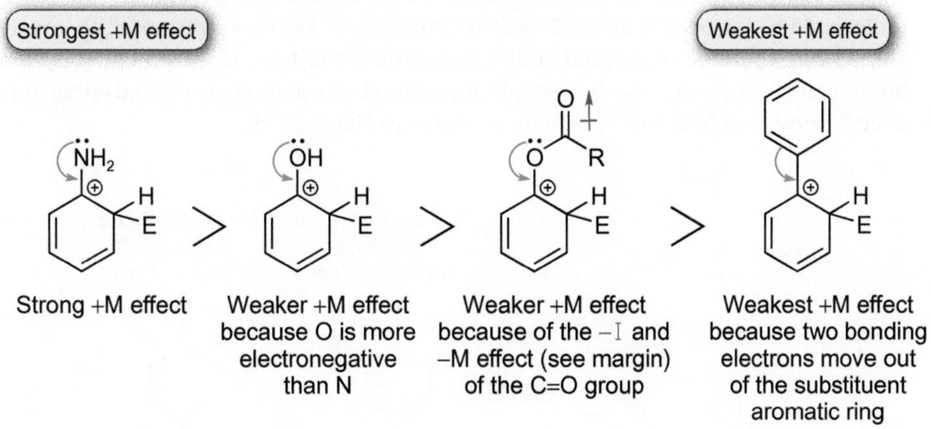
Strongest +M effect
Weakest +M effect
NH2
OH
O
R
H
E
Strong +M effect
Weaker +M effect because O is more electronegative than N
Weaker +M effect because of the –I and –M effect (see margin) of the C=O group
Weakest +M effect because two bonding electrons move out of the substituent aromatic ring

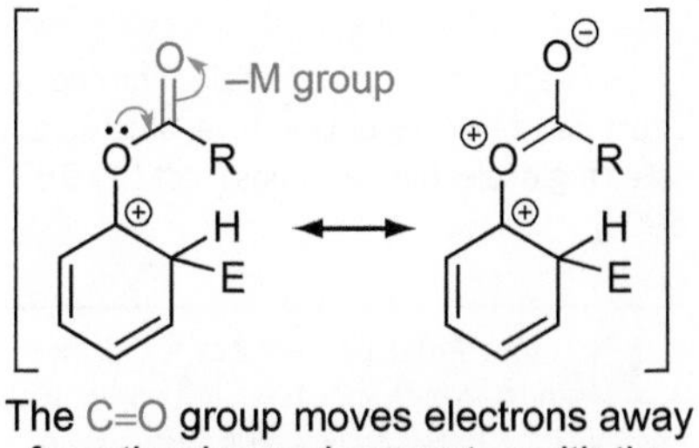
–M group
O
R
H
E

The C=O group moves electrons away from the ring and competes with the +M effect of the O atom, directly attached to the ring

Deactivating groups

Deactivating groups are groups that withdraw electrons from the benzene ring by negative inductive effects (–I) and/or negative mesomeric effects (–M). Electron-withdrawing groups decrease the reactivity of the ring to reaction with electrophiles because the carbocation is destabilized by electron-withdrawing groups. Some common deactivating groups are shown in Table 22.2.

> A **deactivating group** on a benzene ring makes the ring *less* reactive to electrophilic substitution than benzene.

Table 22.2 Deactivating groups on a benzene ring and their electronic effects

Deactivating groups	Electronic effects*
–Cl, –Br, –I	–I > +M
–CHO, –COR	–M, –I
$-CO_2H$, $-CO_2R$	–M, –I
$-SO_3H$	–M, –I
$-NO_2$, $-NR_3^+$	–M, –I

Decreasing reactivity of the benzene ring with electrophiles

* –M and –I effects reinforce one another.

The relative rates of nitration of benzene (C_6H_6), chlorobenzene (C_6H_5Cl), and nitrobenzene ($C_6H_5NO_2$) are $1 : 3 \times 10^{-2} : 6 \times 10^{-8}$.

Halogen atoms are unusual in that their –I effect is greater than their +M effect. The reasons for this are discussed on p.1033.

The characteristic (almond-like) aroma of traditional shoe and floor polishes is due to nitrobenzene, added to mask unpleasant odours.

Electron-withdrawing groups with –M effects destabilize carbocations more effectively than electron-withdrawing groups with –I effects because –M effects are stronger than –I effects. For example, the NO_2 group has a –M effect and is a strong electron acceptor, whereas the Cl group has a –I effect and is a weaker electron acceptor. This explains why chlorobenzene (C_6H_5Cl) reacts faster with electrophiles than nitrobenzene ($C_6H_5NO_2$).

The strongest electron-withdrawing groups include SO_3H (sulfonic acid) and NO_2 (nitro) groups. These groups contain three or four highly electronegative atoms that exert a strong –I effect and have at least one polar double bond that accepts a pair of electrons (and so exerts a strong –M effect), that is, strong –I and strong –M effects reinforce one another.

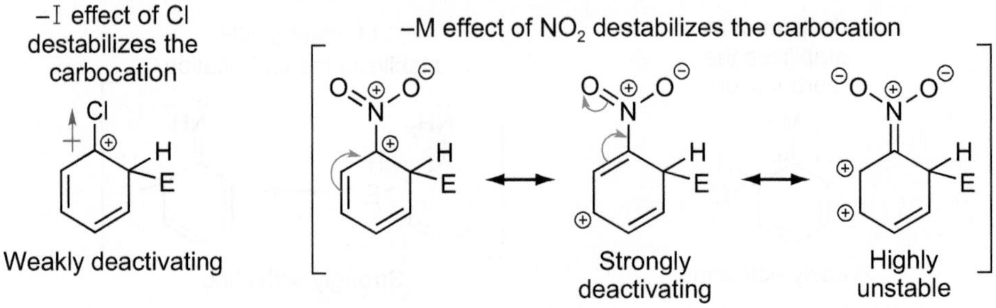
–I effect of Cl destabilizes the carbocation
Cl
H
E
Weakly deactivating
–M effect of NO_2 destabilizes the carbocation
O
N
O
H
E
Strongly deactivating
Highly unstable

The position of substitution

When a substituted benzene undergoes an electrophilic substitution reaction, the electrophile could be introduced at the 2 (*ortho*-), 3 (*meta*-), or 4 (*para*-) position. In theory, the electrophile could be introduced at all three positions to form three structural isomers, but in practice there are only two possibilities. The electrophile is introduced either at the 2- and 4-positions or at the 3-position, as shown in Figure 22.19.

An electrophile is introduced either at the 2- and 4-positions or at the 3-position of a substituted benzene.

Reactions that preferentially form one structural isomer (regioisomer) over others are called **regioselective reactions** (Section 19.3, p.909).

A substituted benzene has two *ortho*- and two *meta*-positions, but only one *para*-position. Based on probability, you would expect the reaction of a substituted benzene (C_6H_5R) with an electrophile to give a 2 : 2 : 1 ratio of *ortho*- : *meta*- : *para*-isomers.

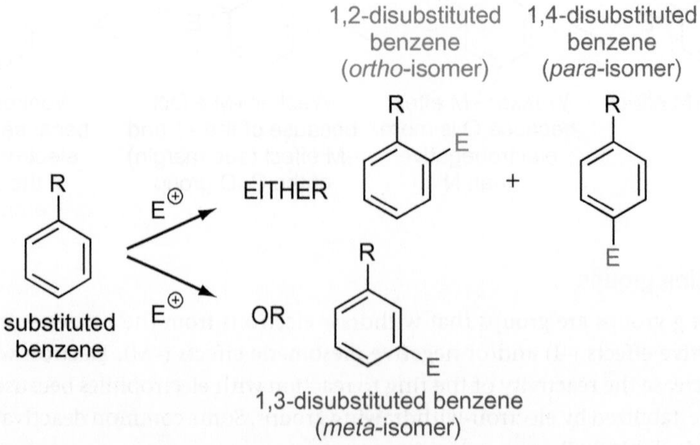
1,2-disubstituted benzene (*ortho*-isomer)
1,4-disubstituted benzene (*para*-isomer)
R
E
EITHER
+
OR
substituted benzene
1,3-disubstituted benzene (*meta*-isomer)

Figure 22.19 Position of electrophilic substitution in a substituted benzene.

Table 22.3 Classification of substituents as activators or deactivators

Class	Substituent†	Strength
2,4-directing activators	$-NH_2$, $-NHR$, $-NR_2$	very strongly activating
	$-OH$, $-OR$, $-O^-$	very strongly activating
	$-NHCOR$, $-OCOR$	strongly activating
	$-Ph$, $-CH{=}CH_2$, $-CH_3$	weakly activating
2,4-directing deactivators	$-Cl$, $-Br$, $-I$	weakly deactivating
3-directing deactivators	$-CHO$, $-COR$	strongly deactivating
	$-CO_2H$, $-CO_2R$, $-CN$, $-CX_3$	strongly deactivating
	$-SO_3H$	strongly deactivating
	$-NO_2$	very strongly deactivating
	$-NR_3^+$	very strongly deactivating

† Ph stands for phenyl (C_6H_5-); X = halogen

The electronic effects of the R group in the substituted benzene determine which of the two possibilities occurs. Electron-donating (activating) groups direct electrophiles to the 2- and 4-positions whereas electron-withdrawing (deactivating) groups direct electrophiles to the 3-position. The halogens are an exception. When R = Cl, Br, or I, electrophiles are directed to the 2- and 4-positions, even though these groups are deactivating (see p.1030). Substituents are, therefore, classified as either 2,4-directing activators, 2,4-directing deactivators, or 3-directing deactivators (see Table 22.3).

Activating groups direct electrophiles to the 2- and 4-positions. Deactivating groups direct electrophiles to the 3-position, except for the halogens, which direct electrophiles to the 2- and 4-positions.

The rates of electrophilic substitution at the 2-, 3-, and 4-positions of substituted benzenes have been determined experimentally and are reported as partial rate factors.

Partial rate factors compare the rate of attack at one position in a substituted benzene with the rate of attack at one position in benzene.

Partial rate factors for nitration of methylbenzene at 0 °C

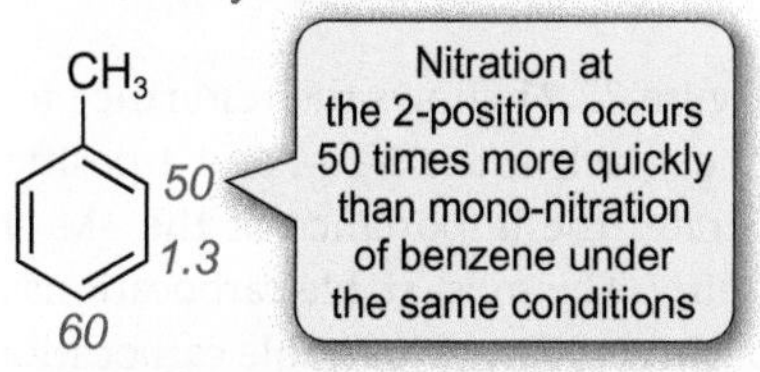
CH3
50
1.3
60
Nitration at the 2-position occurs 50 times more quickly than mono-nitration of benzene under the same conditions

All three positions are more reactive to $^+NO_2$ than benzene (as the partial rate factors are above 1), but the 2- and 4-positions are the most reactive

Worked example 22.3 discusses the sulfonation of substituted benzenes and looks at how the existing substituent influences the rate and regioselectivity of the electrophilic substitution.

2,4-Directing activators

All activating substituents with positive inductive (+I) and positive mesomeric (+M) effects direct the incoming electrophile to the 2- and 4-positions of the ring. This is because the intermediate carbocation is stabilized by the +I and/or +M effect of the activating substituent. In contrast, introducing the electrophile at the 3-position produces an intermediate carbocation that cannot be effectively stabilized by the +I and/or +M effect of the activating substituent. Because the carbocations formed from attack at the 2- and 4-positions are the more stable, electrophilic substitutions at these positions will have lower Gibbs energies of activation.

For nitration of methylbenzene (at 25 °C), the product composition is: 58% of the 2-nitro isomer, 37% of the 4-nitro isomer, and just 5% of the 3-nitro isomer (see p.1038).

Figure 22.20 shows the resonance forms of possible carbocations produced by electrophilic attack at the 2-, 3-, and 4-positions of methylbenzene (toluene; $C_6H_5CH_3$) and demonstrates the importance of the +I effect of an activating substituent on carbocation stability. The most stable carbocations are produced by electrophilic attack at the 2- and 4-positions. The most stable resonance forms have a positively charged carbon atom next to the CH_3 group. They are the most stable because the CH_3 group can stabilize the carbocation by exerting a strong +I effect.

A +I activator group is 2,4-directing. The +I effect of a substituent or group diminishes as the chain length increases (Section 19.1, p.870).

A +M (weak −I) activator group is 2,4-directing.

As a rule of thumb, the more resonance forms a cation has, the more stable it is because the positive charge is more delocalized (Section 19.1, p.874).

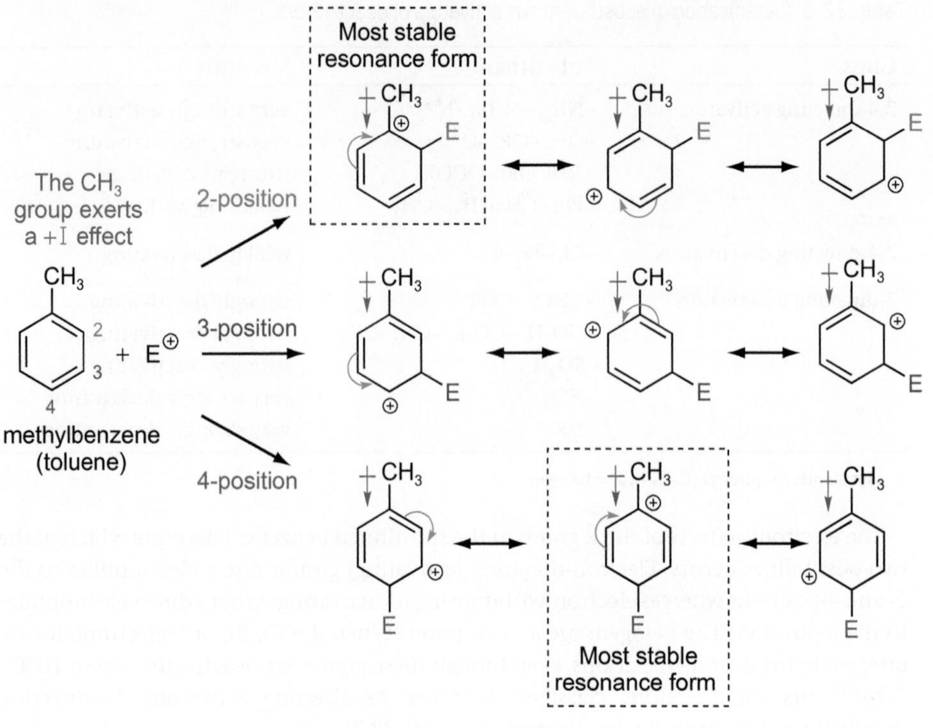
Most stable resonance form
The CH_3 group exerts a +I effect
2-position
3-position
4-position
methylbenzene (toluene)
Most stable resonance form

Figure 22.20 Reaction of methylbenzene ($C_6H_5CH_3$) with an electrophile, E^+, showing the resonance stabilization of the intermediate carbocations.

Mechanisms of electrophilic substitutions of substituted benzenes with +M groups usually show the lone pair of electrons being pushed round the ring to react with the electrophile.

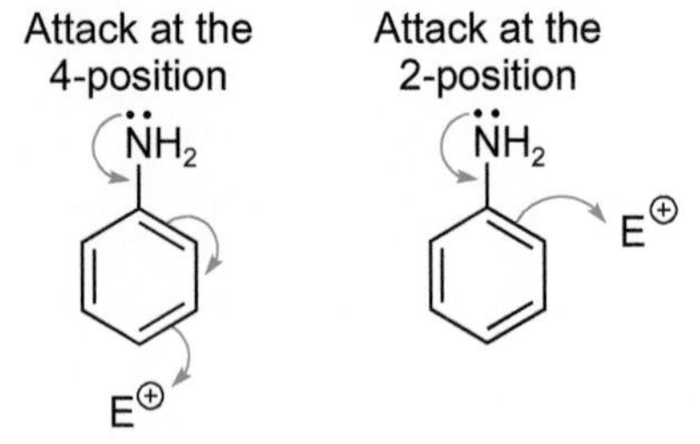
Attack at the 4-position
Attack at the 2-position

Figure 22.21 shows the resonance forms of possible carbocations produced by electrophilic attack at the 2-, 3-, and 4-positions of phenylamine (aniline; $C_6H_5NH_2$) and demonstrates the importance of the +M effect of an activating substituent on carbocation stability. The most stable carbocations are produced by electrophilic attack at the 2- and 4-positions. The most stable carbocations have an additional resonance form because the lone pair on the NH_2 group is delocalized over the ring.

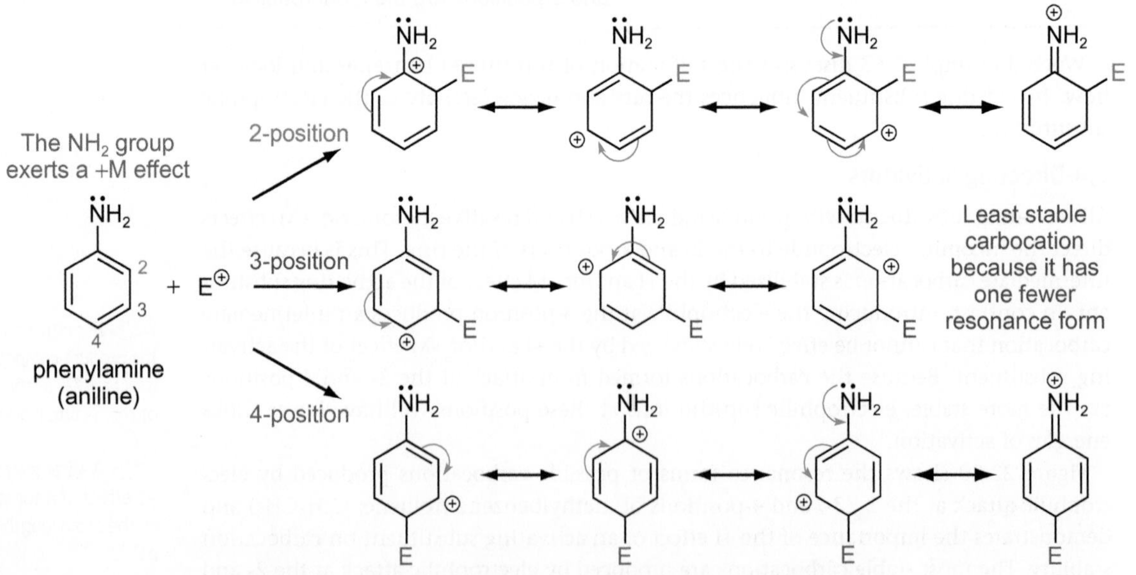
The NH_2 group exerts a +M effect
2-position
3-position
4-position
phenylamine (aniline)
Least stable carbocation because it has one fewer resonance form

Figure 22.21 Reaction of phenylamine ($C_6H_5NH_2$) with an electrophile, E^+, showing the resonance stabilization of the intermediate carbocations.

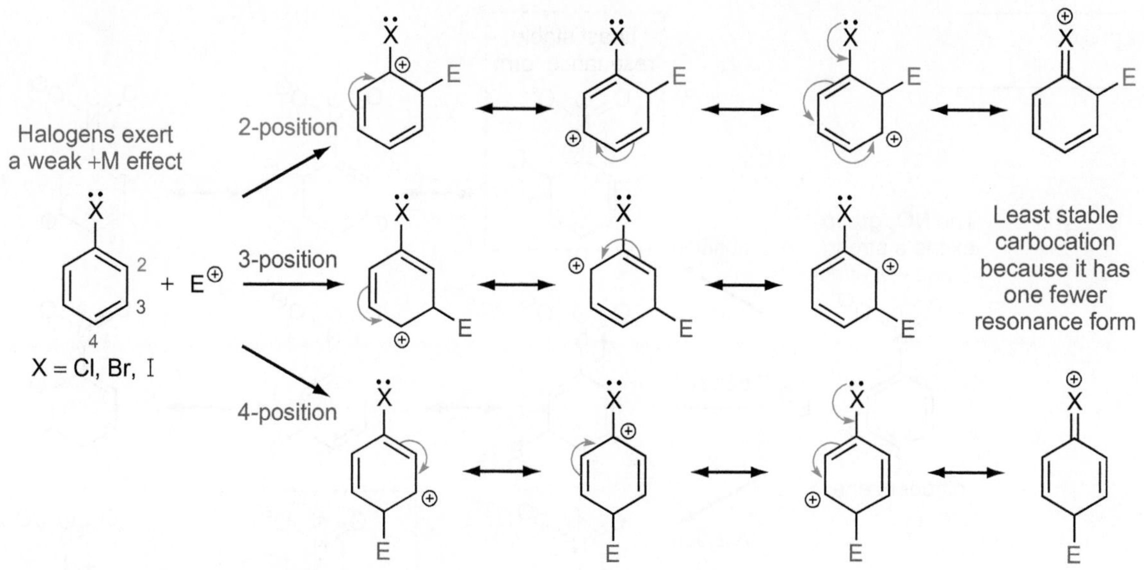

Halogens exert a weak +M effect
2-position
3-position
4-position
+ E⊕
X = Cl, Br, I
Least stable carbocation because it has one fewer resonance form

Figure 22.22 Reaction of a halogenobenzene (C_6H_5X) with an electrophile, E^+, showing the resonance stabilization of the intermediate carbocations.

2,4-Directing deactivators

Cl, Br, and I substituents are unique in that, although deactivating, they direct the incoming electrophile to the 2- and 4-positions of the ring (Figure 22.22). The halogens are deactivating because they are so highly electronegative that the –I effect of these atoms is stronger than the +M effect. However, the weak +M effect of the halogens does explain why these substituents direct electrophiles to the 2- and 4-positions. As shown in Figure 22.22, attack at the 2- and 4-positions produces the most stable carbocations, which have an additional resonance form because the lone pair on the halogen atom is delocalized over the ring.

For nitration of chlorobenzene (at 25 °C), the product composition is: 64% of the 4-nitro isomer, 35% of the 2-nitro isomer, and only 1% of the 3-nitro isomer.

A –I (weak +M) deactivator group is 2,4-directing.

The weak +M effect of Cl, Br, and I is also explained by the larger size of these atoms compared to carbon. The orbitals that contain the lone pairs of electrons on the halogens (e.g. 3*p* for chlorine) are larger and do not overlap well with the 2*p* orbital of carbon.

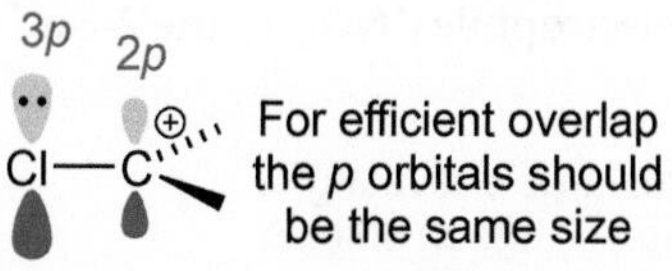

3p
2p
Cl—C
For efficient overlap the p orbitals should be the same size

3-Directing deactivators

All deactivating substituents with negative inductive (–I) and negative mesomeric (–M) effects direct the incoming electrophile to the 3-position of the ring.

Figure 22.23 shows the resonance forms of possible carbocations produced by electrophilic attack at the 2-, 3-, and 4-positions of nitrobenzene ($C_6H_5NO_2$). The most stable resonance forms are formed by attack at the 3-position. The most unstable resonance forms have a positively charged carbon atom next to the positively charged nitrogen of the NO_2 group. These are the most unstable because the NO_2 group destabilizes the carbocation by exerting strong –I and –M effects.

For nitration of nitrobenzene (at 100 °C), the product composition is: 93% of the 1,3-dinitro isomer, 6% of the 1,2-dinitro isomer, and only 1% of the 1,4-dinitro isomer.

The three dinitrobenzene isomers have different boiling points and so the isomers can be separated from one another, and their quantities determined, using, for example, distillation (see p.805) or gas chromatography (Section 11.3, p.534).

Steric effects are important

From the previous sections it is clear that the electronic effects of an existing substituent on a benzene ring determine which structural isomers of the disubstituted benzene are

A –M, –I deactivator group is 3-directing.

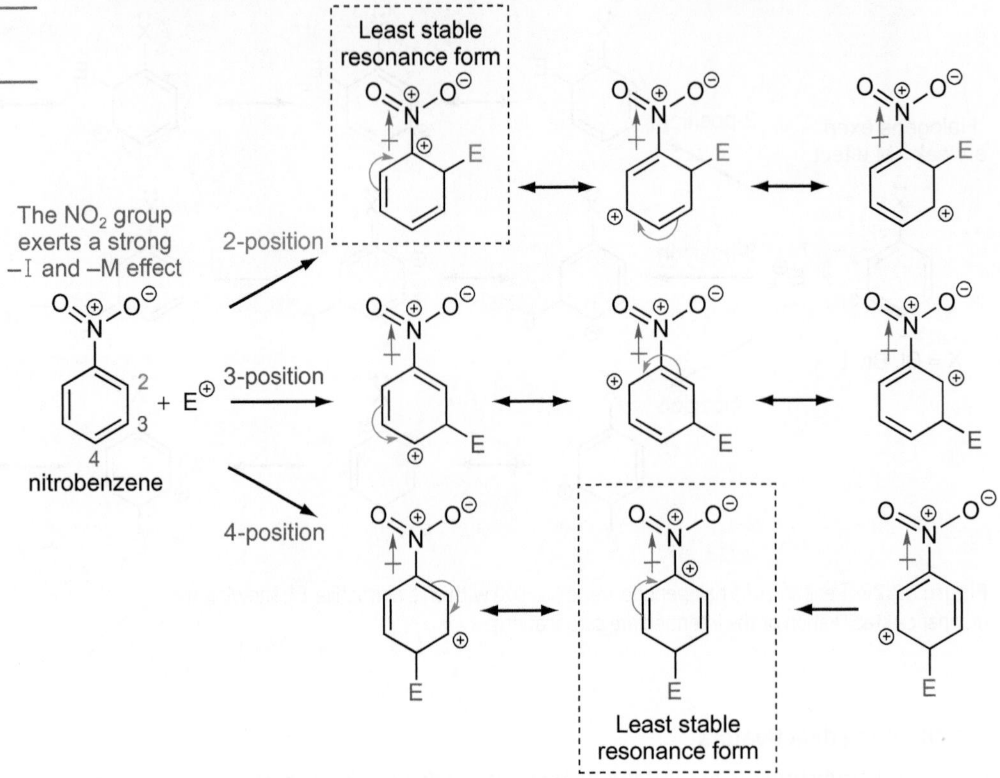

Least stable resonance form
The NO_2 group exerts a strong –I and –M effect
2-position
3-position
4-position
nitrobenzene
Least stable resonance form

Figure 22.23 Reaction of nitrobenzene ($C_6H_5NO_2$) with an electrophile, E^+, showing the resonance stabilization of the intermediate carbocations.

formed preferentially. But, electronic effects are not the complete story, particularly when the substituent is large. This is illustrated below for the nitration of *tert*-butylbenzene ($C_6H_5C(CH_3)_3$). The *tert*-butyl group activates the ring to nitration (+I effect) and directs the electrophile ($^+NO_2$) to the 2- and 4-positions.

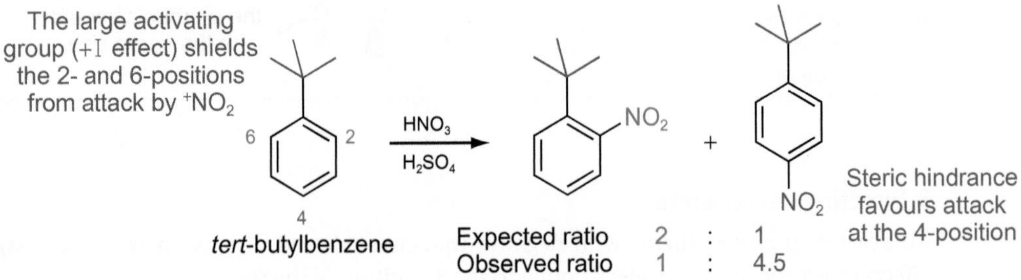

The large activating group (+I effect) shields the 2- and 6-positions from attack by $^+NO_2$
HNO_3
H_2SO_4
tert-butylbenzene
Expected ratio 2 : 1
Observed ratio 1 : 4.5
Steric hindrance favours attack at the 4-position

Steric effects and steric hindrance are discussed in Section 19.1 (p.878).

On nitration of ethylbenzene ($C_6H_5CH_2CH_3$), the ratio of 2-nitro : 4-nitro isomers is 1 : 1.1. As the ethyl group (CH_3CH_2–) is smaller than the *tert*-butyl group the 2- and 6-positions in ethylbenzene are not as hindered as those in *tert*-butylbenzene.

Based on probability, the reaction should give a 2 : 1 ratio of 2-nitro : 4-nitro isomers because the 2- and 6-positions in *tert*-butylbenzene are identical. However, by far the major product is the 4-nitro isomer, because the size of the $C(CH_3)_3$ group influences the site of attack by $^+NO_2$. The 2- and 6-positions of the ring are sterically hindered so the electrophile is directed predominantly to the 4-position. As you may expect, the larger the activating group on the benzene ring, the greater the proportion of attack at the 4-position.

Worked example 22.3 The effect of existing substituents on electrophilic substitutions

The following questions are based on the sulfonation of compounds **1–3** using H_2SO_4 and SO_3 to give monosulfonated products.

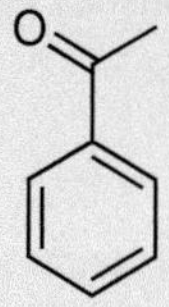
O

1

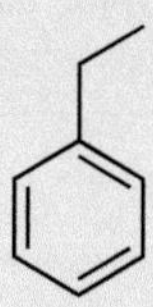

2

Br

3

(a) Do each of the compounds **1–3** react faster or slower than benzene with H_2SO_4 and SO_3?

(b) Give the structure(s) of the major product(s) from reaction of each of **1–3** with H_2SO_4 and SO_3.

Strategy

(a) Assign inductive (+I or −I) and mesomeric (+M or −M) effects to the substituent on the benzene ring.

Use the inductive and mesomeric effects to determine whether the substituent is activating (electron-donating) or deactivating (electron-withdrawing).

Activating groups increase the reactivity of the ring to reaction with electrophiles.

Deactivating groups decrease the reactivity of the ring to reaction with electrophiles.

(b) Activating groups direct electrophiles to the 2- and 4-positions. Deactivating groups direct electrophiles to the 3-position, except for the halogens, which direct electrophiles to the 2- and 4-positions.

Solution

(a)

−M and −I
Deactivating

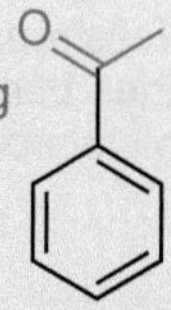
O

1 is *less* reactive than benzene and so reacts more slowly

+I
Activating

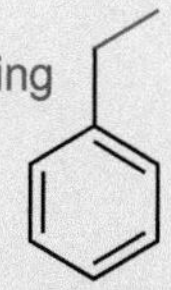

2 is *more* reactive than benzene and so reacts faster

−I > +M
Deactivating

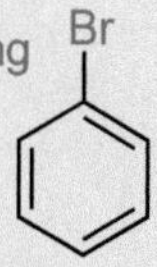
Br

3 is *less* reactive than benzene and so reacts more slowly

(b)

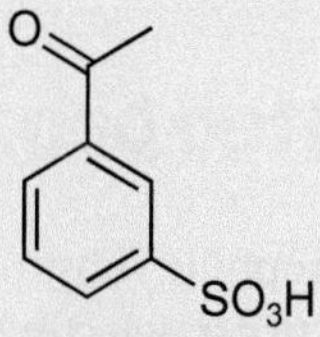
O
SO3H

from 1

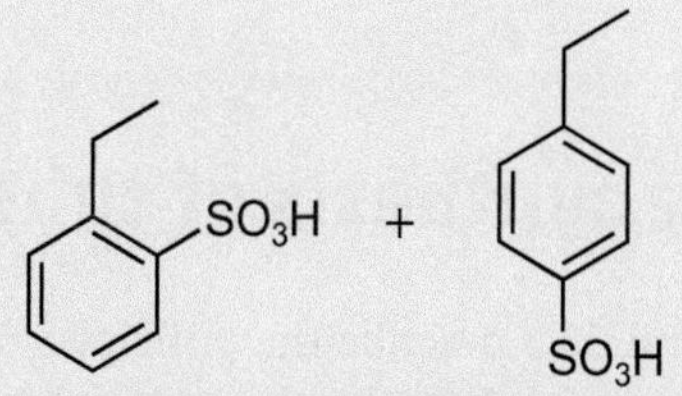
SO3H
+
SO3H

from 2

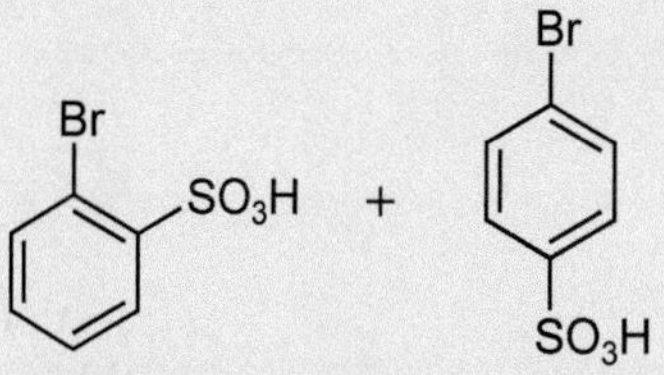
Br
SO3H
+
Br
SO3H

from 3

(expect more of the 1,4-disubstituted products based on steric effects)

Question

The following questions are based on the bromination of compounds **4–6** using Br_2 and $FeBr_3$ to give monobrominated products.

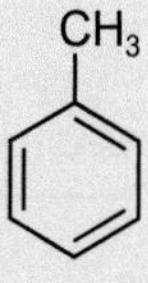
CH3

4

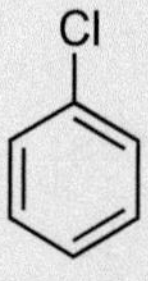
Cl

5

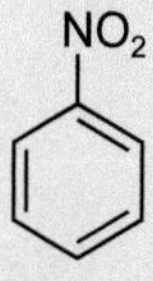
NO2

6

(a) Do each of the compounds **4–6** react faster or slower than benzene with Br_2 and $FeBr_3$?

(b) Give the structure(s) of the major product(s) from reaction of each of **4–6** with Br_2 and $FeBr_3$.

Summary

- Activating groups direct electrophiles to the 2- and 4-positions. The larger the activating group on the ring, the greater the proportion of attack at the 4-position.
- Deactivating groups direct electrophiles to the 3-position, except for the halogens, which direct electrophiles to the 2- and 4-positions.

	Substituents	Electronic effects	
2,4-directing activators	$-NH_2$, $-NHR$, $-NR_2$	+M > −I	Increasingly more reactive than benzene to electrophiles
	−OH, −OR	+M > −I	
	−NHCOR, −OCOR	+M > −I	
	−Ph, $-CH{=}CH_2$	+M, +I	
	−R	+I	
	−H		benzene
2,4-directing deactivators	−Cl, −Br, −I	−I > +M	Increasingly less reactive than benzene to electrophiles
3-directing deactivators	−CHO, −COR	−M, −I	
	$-CO_2H$, $-CO_2R$	−M, −I	
	$-SO_3H$	−M, −I	
	$-NO_2$	−M, −I	

For a practice question on this topic, see question 2 at the end of this chapter (p.1052).

22.4 The synthesis of substituted benzenes

Sections 22.2 and 22.3 describe the principles of electrophilic substitution reactions. In this section, you can see how this theory is put into practice in selecting the best synthetic routes to substituted benzenes.

In most syntheses of substituted benzenes, only one structural isomer (or regioisomer) is required. This can be difficult to achieve by electrophilic substitution because, for example, activating groups direct electrophiles to both the 2- and 4-positions of a benzene ring. To maximize the yield of the desired substituted benzene, a synthesis needs to take into account the following questions.

There is usually more than one way of making a substituted benzene—to decide on the most effective route there are various factors to consider, including regioselectivity and reaction conditions.

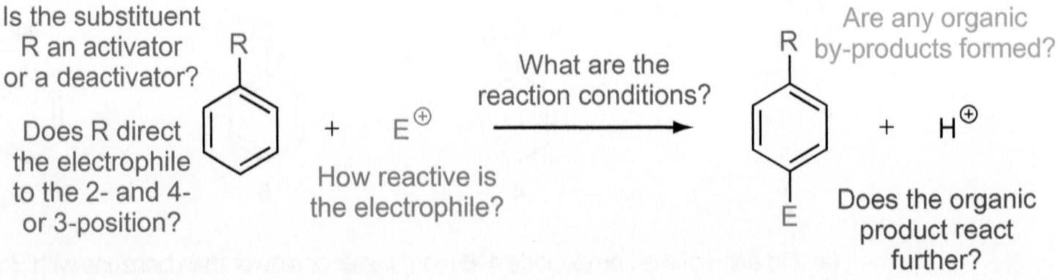
Is the substituent R an activator or a deactivator?
Does R direct the electrophile to the 2- and 4- or 3-position?
R
+ E⊕
How reactive is the electrophile?
What are the reaction conditions?
R
E
+ H⊕
Are any organic by-products formed?
Does the organic product react further?

The points raised by these questions are illustrated using specific reactions in the following sections.

Reactivity of the electrophile and choice of reaction conditions

The reactivity of substituted benzenes to electrophilic substitution depends on the existing substituent(s) on the benzene ring. Benzene rings containing activating (electron-donating) substituents are more reactive to electrophiles than benzene rings containing deactivating (electron-withdrawing) substituents. This explains why activated benzene rings react with relatively weak electrophiles whereas deactivated benzene rings only react with strong electrophiles. The different reagents used for the bromination of nitrobenzene ($C_6H_5NO_2$) and phenol (C_6H_5OH) illustrate this (Figure 22.24).

The mechanism of bromination of benzene using Br_2 and $FeBr_3$ is discussed in Section 22.2 (p.1015).

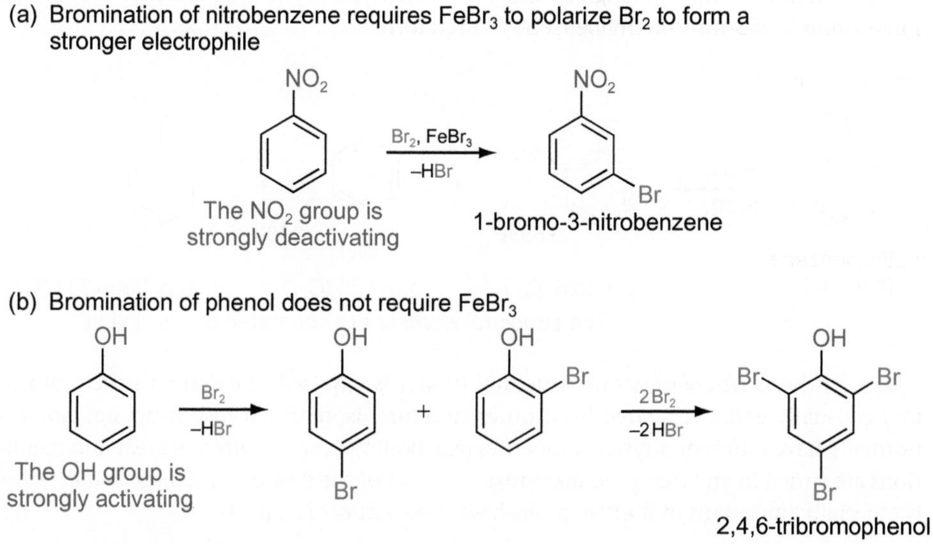
(a) Bromination of nitrobenzene requires FeBr3 to polarize Br2 to form a stronger electrophile
NO2
Br2, FeBr3
–HBr
NO2
Br
The NO2 group is strongly deactivating
1-bromo-3-nitrobenzene
(b) Bromination of phenol does not require FeBr3
OH
Br2
–HBr
OH
Br
+
OH
Br
2 Br2
–2 HBr
OH
Br
Br
Br
The OH group is strongly activating
2,4,6-tribromophenol

Figure 22.24 Bromination of (a) nitrobenzene ($C_6H_5NO_2$) and (b) phenol (C_6H_5OH).

2,4,6-Tribromophenol is a common fungicide. It made the headline news in 2010 when a number of medicines were recalled because of a musty smell. Patients experienced nausea, vomiting, stomach pains, and diarrhoea. The culprit was found to be 2,4,6-tribromoanisole, derived from 2,4,6-tribromophenol, which was used to protect wooden pallets on which packaging materials were stored and transported. After the outbreak, it was concluded that a species of fungus converted the OH group of 2,4,6-tribromophenol into an OCH_3 group.

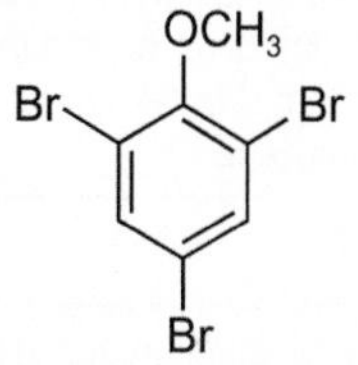
OCH3
Br
Br
Br

2,4,6-tribromoanisole

To brominate nitrobenzene a strong electrophile is needed because the NO_2 group in nitrobenzene is so strongly deactivating. In contrast, bromination of phenol requires only Br_2, a weak electrophile (that becomes polarized, $^{\delta+}Br–Br^{\delta-}$, when it approaches the electron-rich benzene ring), because the OH group in phenol is strongly activating. Even 2- and 4-bromophenol react with a further two equivalents of Br_2 to form 2,4,6-tribromophenol (abbreviated TBP). This is impressive because a Br substituent is a deactivator, so the greater the number of Br atoms on the ring, the less reactive it is to Br_2.

The importance of matching the electrophile to the substituted benzene is also illustrated by treating nitrobenzene ($C_6H_5NO_2$) or phenylamine (aniline; $C_6H_5NH_2$) with RCl and $AlCl_3$, or RCOCl and $AlCl_3$. Neither aniline, nor nitrobenzene, undergo Friedel–Crafts reactions, though for different reasons.

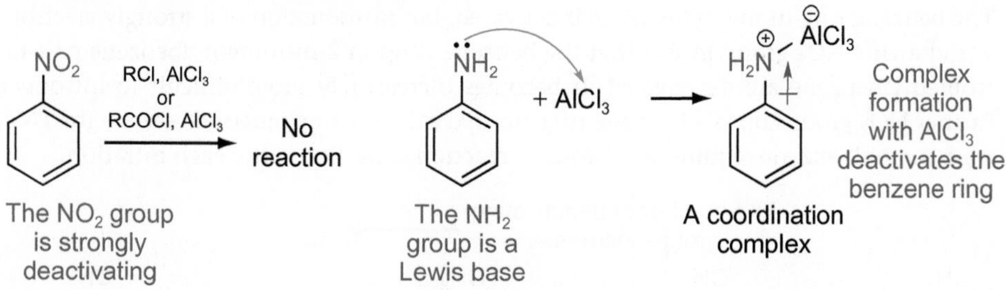
NO2
RCl, AlCl3
or
RCOCl, AlCl3
No reaction
The NO2 group is strongly deactivating
NH2
+ AlCl3
The NH2 group is a Lewis base
H2N
AlCl3
A coordination complex
Complex formation with AlCl3 deactivates the benzene ring

Friedel–Crafts alkylations and acylations are discussed in Section 22.2 (p.1018 and p.1023).

Another complication is that primary amines (RNH_2) react with acyl chlorides (RCOCl) to form secondary amides (RNHCOR) and HCl (see Section 24.2, p.1111).

No reaction is observed with nitrobenzene because the ring is too unreactive to react with the coordination complex or carbocation. Indeed, *no* substituted benzene with a deactivating group undergoes Friedel–Crafts alkylation or acylation reactions.

You would not expect this to be the case for phenylamine since the $–NH_2$ substituent is activating. So why doesn't phenylamine react with RCl and $AlCl_3$, or RCOCl and $AlCl_3$? The problem is that $AlCl_3$ reacts with the $–NH_2$ group of phenylamine in preference to RCl or RCOCl. The $–NH_2$ group donates a lone pair of electrons to $AlCl_3$ and this produces a coordination complex, which has a positively charged nitrogen atom. The positively

Unlike phenylamine, phenol (C_6H_5OH) *does* undergo Friedel–Crafts alkylation and acylation reactions. Phenol does not form a coordination complex with $AlCl_3$ because the OH group on phenol is a weaker base than the NH_2 group on phenylamine (a lone pair on oxygen is less readily donated than the lone pair on nitrogen).

The mechanism of nitration of benzene is discussed in Section 22.2 (p.1016).

charged nitrogen atom exerts a strong –I effect so the ring becomes deactivated and does not react in Friedel–Crafts alkylation or acylation reactions.

To separate a mixture containing several miscible liquids, each with different boiling points, it is common to use fractional distillation and a fractionating column (see Section 17.4, p.805).

Formation of by-products and further substitution

Electrophilic substitution reactions of substituted benzenes rarely form one structural isomer (or regioisomer) of a product. Usually a mixture of structural isomers is formed. For example, in the nitration of methylbenzene (toluene; $C_6H_5CH_3$) using HNO_3 and H_2SO_4, the CH_3 group of methylbenzene activates the ring and directs the $^+NO_2$ ion to the 2- and 4-positions of the ring. Although 2- and 4-nitromethylbenzene are the major products, a small amount of 3-nitromethylbenzene is often formed.

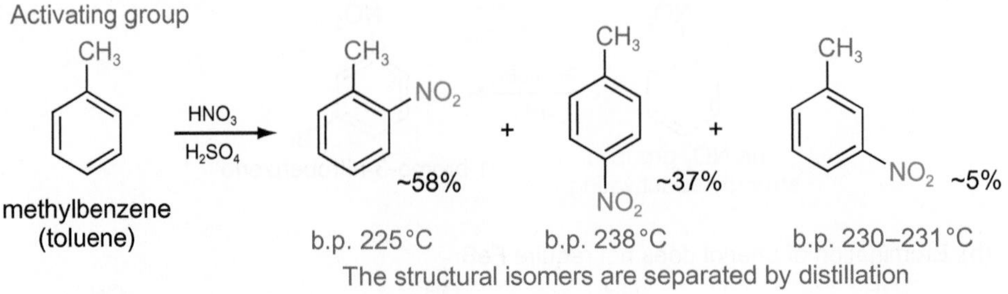
Activating group
CH3
HNO3
H2SO4
NO2
~58%
~37%
~5%
methylbenzene
(toluene)
b.p. 225 °C
b.p. 238 °C
b.p. 230–231 °C
The structural isomers are separated by distillation

In a synthesis where only one structural isomer is required, the desired structural isomer is usually easily separated from other structural isomers because structural isomers normally have different physical properties (e.g. boiling points). Often the reaction conditions are varied in an attempt to maximize the yield of the desired structural isomer—this is especially important in multistep syntheses (see Section 1.4, p.24).

To emphasize the positions in which the nitro groups are introduced on the methylbenzene ring, names such as 2-nitromethylbenzene are used here. However, the IUPAC name for 2-nitromethylbenzene is 1-methyl-2-nitrobenzene, and for 2,4,6-trinitromethylbenzene it is 2-methyl-1,3,5-trinitrobenzene.

For an electrophilic substitution reaction of an activated benzene you should always show that both 1,2- and 1,4-disubstituted benzenes are formed—if you use this type of reaction to make a 1,2-disubstituted benzene, you should indicate that the 1,4-disubstituted benzene is also formed, and note that the 1,2-disubstituted benzene needs to be separated from the 1,4-disubstituted benzene.

Nitration, sulfonation, and Friedel–Crafts acylation reactions can all be stopped after the introduction of a single substituent.

Other potential by-products in electrophilic substitutions are formed by polysubstitution of the benzene ring. The likelihood of polysubstitution depends on the reactivity of the first formed products and reactants to electrophiles. For example, reaction of methylbenzene (toluene; $C_6H_5CH_3$) with one (stoichiometric) equivalent of $^+NO_2$ forms 2- or 4-nitromethylbenzene in preference to 2,4-dinitromethylbenzene or 2,4,6-trinitromethylbenzene (commonly called 2,4,6-trinitrotoluene, or TNT) because methylbenzene is more reactive to $^+NO_2$ than either 2-nitromethylbenzene or 4-nitromethylbenzene. The benzene ring in methylbenzene is activated, but introduction of a strongly electron-withdrawing NO_2 group means that the benzene rings in 2-nitromethylbenzene or 4-nitromethylbenzene are deactivated. It becomes increasingly more difficult to introduce further NO_2 groups onto a benzene ring. To make 2,4,6-trinitromethylbenzene (or TNT) from methylbenzene requires more forcing reaction conditions after each nitration.

TNT (2,4,6-trinitrotoluene) is a powerful explosive prepared by nitration of toluene ($C_6H_5CH_3$). When TNT burns, it produces a large amount of expanding hot gas that can be used to propel a projectile.

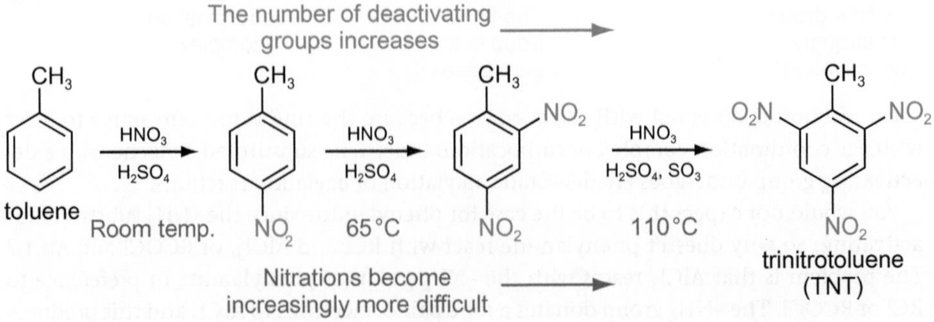
The number of deactivating groups increases
CH3
toluene
HNO3
H2SO4
Room temp.
NO2
65 °C
H2SO4, SO3
110 °C
O2N
trinitrotoluene
(TNT)
Nitrations become increasingly more difficult

Box 22.6 The biological action and synthesis of Viagra

On 27 March 1998, a drug for the treatment of male erectile dysfunction was approved for sale in the US. Male erectile dysfunction affects around 10% of all men and the drug created history by being the first oral anti-impotence drug. This created tremendous media interest and within a few months the drug, marketed as Viagra (Viagra is the brand name; sildenafil is the generic name), became a household name.

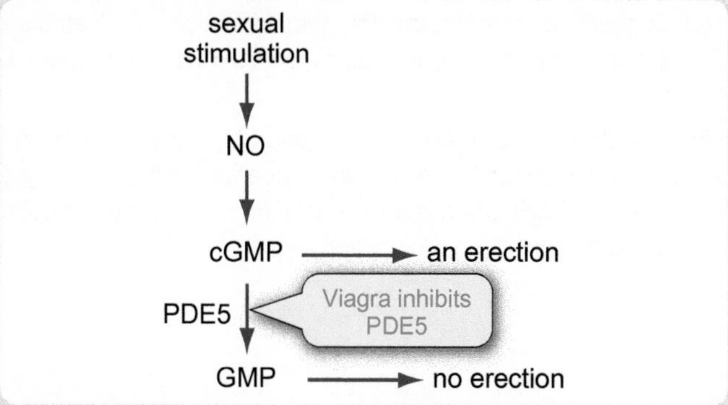
sexual stimulation
NO
cGMP
an erection
PDE5
Viagra inhibits PDE5
GMP
no erection

Figure 1 How Viagra works.

Viagra works as an anti-impotence drug by affecting the concentration of cyclic guanosine monophosphate (cGMP) in the body. Normally, on sexual stimulation, nitric oxide (NO) is released in the body (see Box 4.11, p.212) and this leads to the production of cGMP. cGMP triggers relaxation of smooth muscle and an increase in blood flow to the penis, which causes an erection (Figure 1). The effect is temporary because an enzyme called cGMP phosphodiesterase type 5 (PDE5) destroys cGMP and this causes loss of the erection. Viagra works by inhibiting PDE5, thereby maintaining the concentration of cGMP in the body and the duration of an erection. The similarity in the structures of Viagra and cGMP explains why both molecules fit into the active site of PDE5 (Figure 2).

VIAGRA
VIAGRA
6
TAURUS

Ford racing car sponsored by Pfizer.

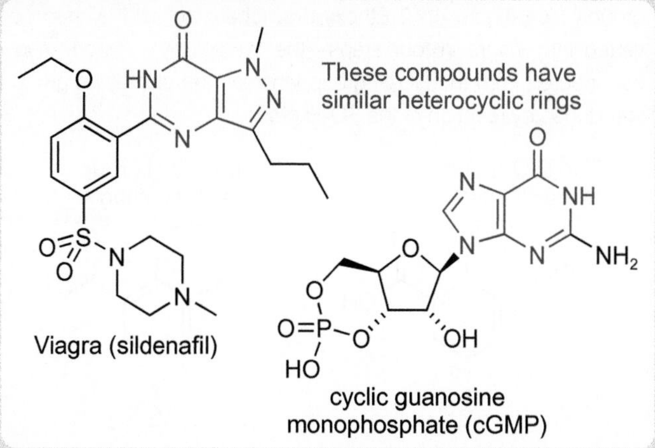
These compounds have similar heterocyclic rings
Viagra (sildenafil)
cyclic guanosine monophosphate (cGMP)

Figure 2

To keep up with the increasing demand, the chemical company Pfizer currently produces around 45 tonnes of Viagra each year. Preparing Viagra on such a large scale is a challenge because Viagra contains a 1,2,5-trisubstituted benzene ring with three different substituents (Figure 3).

To introduce the three substituents at the correct positions of the ring requires strategic use of electrophilic aromatic substitution reactions. For example, one approach to Viagra involves reacting 2-ethoxybenzoic acid with chlorosulfonic acid ($ClSO_3H$). ($ClSO_3H$ is sometimes used in place of H_2SO_4/SO_3 because it can lead to fewer by-products). In this electrophilic substitution reaction an

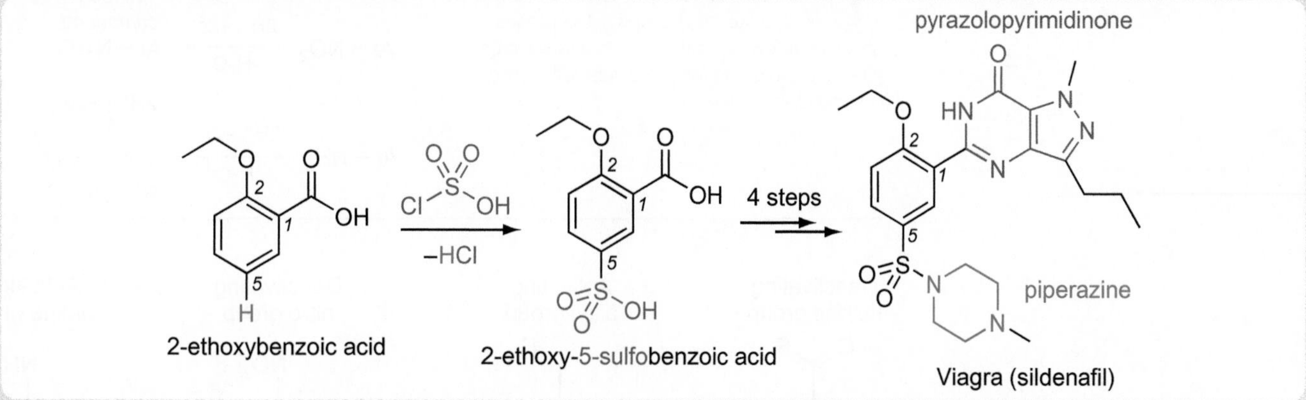
pyrazolopyrimidinone
2-ethoxybenzoic acid
–HCl
2-ethoxy-5-sulfobenzoic acid
4 steps
piperazine
Viagra (sildenafil)

Figure 3 A synthetic route to Viagra

→

→ H atom in 2-ethoxybenzoic acid is replaced by an SO_3H group to give 2-ethoxy-5-sulfobenzoic acid. The SO_3H group is directed to the 5-position of 2-ethoxybenzoic acid by both the EtO and CO_2H groups (see Figure 4). 2-Ethoxy-5-sulfobenzoic acid is then converted into Viagra in four steps—the pyrazolopyrimidinone ring is constructed from the CO_2H group whereas the piperazine group is introduced by reaction of the SO_3H group.

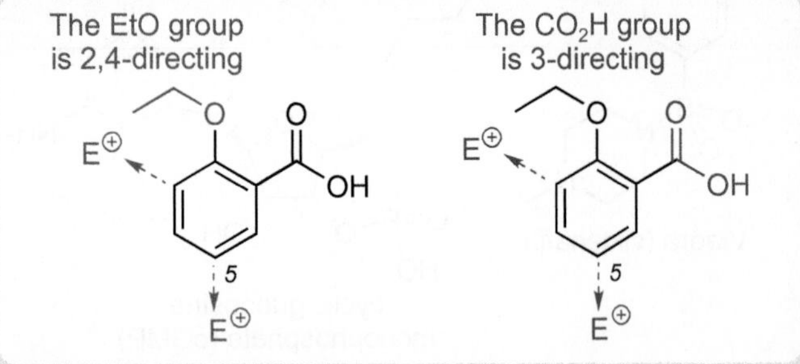
The EtO group
is 2,4-directing
The CO_2H group
is 3-directing
OH
E⊕
5

▲ **Figure 4** The EtO and CO_2H groups both direct the SO_3H group into the 5-position.

The actual synthetic route used by Pfizer to make Viagra on a large scale has not been made public, but it is known to be extremely efficient. Currently, only 1.5 kilograms of reagents and starting materials are needed to make 1.0 kilogram of Viagra (enough for around 20000 tablets) and just 10 litres of chemical waste are produced. This process compares favourably with the production of other medicines, which generally produce 12–100 litres of chemical waste per kilogram of medicine.

Question

(a) Suggest a mechanism for the formation of 2-ethoxy-5-sulfobenzoic acid from 2-ethoxybenzoic acid and chlorosulfonic acid.

(b) In 2-ethoxybenzoic acid, the EtO and CO_2H groups direct an incoming electrophile to the 3- and 5-positions of the ring. Suggest why the SO_3H group is selectively introduced at the 5-position, rather than the 3-position, of the ring.

Transforming substituents

The Clemmensen reduction is introduced in Section 22.2 (p.1025).

Reduction of a ketone (C=O) or nitro group (NO_2) in the presence of a benzene ring are examples of chemoselective reactions (Section 19.3, p.909).

Only a limited range of electrophiles reacts with benzene in electrophilic substitution reactions. Using these electrophiles, it is possible to introduce halogen atoms, NO_2, SO_3H, and alkyl or acyl substituents on to a benzene ring (Section 22.2), but what about the introduction of other groups? There are various methods in the synthetic chemist's toolkit for transforming existing substituents into other functional groups. Some of these methods are considered now.

Redox reactions

Oxidation and reduction of substituents on benzene rings is very useful in synthesis. For example, in the **Clemmensen reduction**, zinc amalgam (Zn/Hg) and HCl are used to reduce a ketone group (C=O) into a CH_2 group. Similarly, a nitro substituent (NO_2) is reduced to an amine substituent (NH_2) using tin (Sn) and HCl. Notice that both reductions convert a deactivating substituent into an activating substituent.

ⓘ The mechanism of reduction of aromatic nitro compounds using Sn/HCl is not well understood. Electron transfer from tin to the nitro group is presumed to form an amine via a nitroso compound and hydroxylamine:

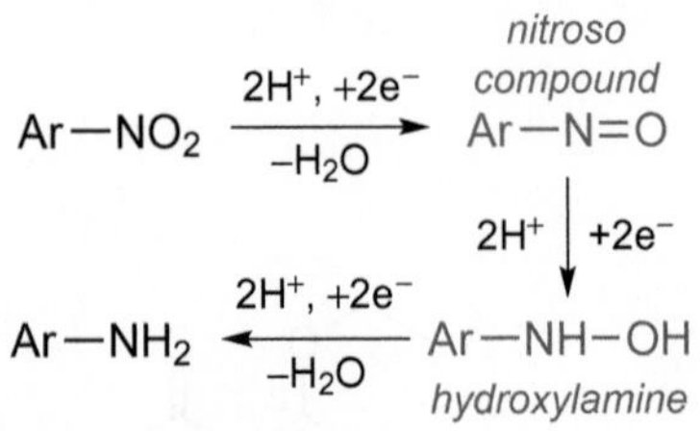
nitroso
compound
Ar—NO2
2H+, +2e−
−H2O
Ar—N=O
2H+
+2e−
Ar—NH2
2H+, +2e−
−H2O
Ar—NH—OH
hydroxylamine

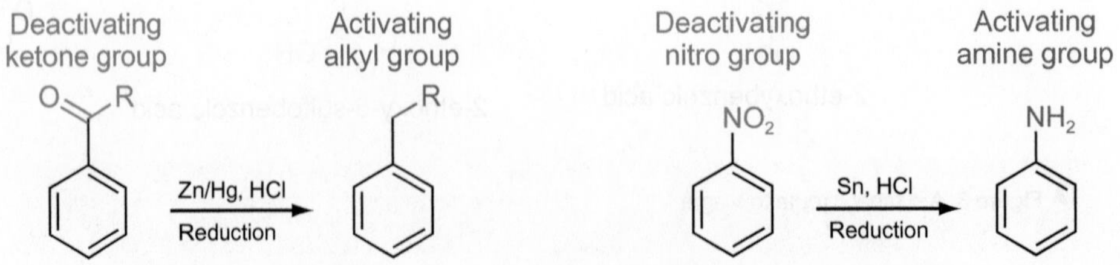
Deactivating
ketone group
Activating
alkyl group
Deactivating
nitro group
Activating
amine group
O
R
Zn/Hg, HCl
Reduction
NO2
Sn, HCl
NH2

Another useful redox reaction is the oxidation of a CH_3 substituent to a carboxylic acid (CO_2H) using potassium permanganate ($KMnO_4$). This oxidation converts an activating substituent into a deactivating substituent.

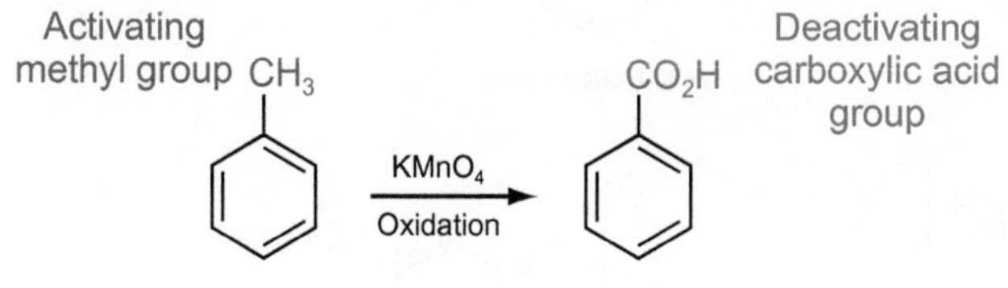
Activating methyl group CH_3
$KMnO_4$
Oxidation
CO_2H
Deactivating carboxylic acid group

Diazonium ions

One of the most versatile methods for introducing different substituents onto a benzene ring involves reactions of aryl diazonium ions ($ArN^+{\equiv}N$, where Ar is an aromatic group). Aryl diazonium ions are prepared by reacting aromatic amines ($ArNH_2$) with the NO^+ ion (a nitrosonium ion) between 0 °C and 5 °C.

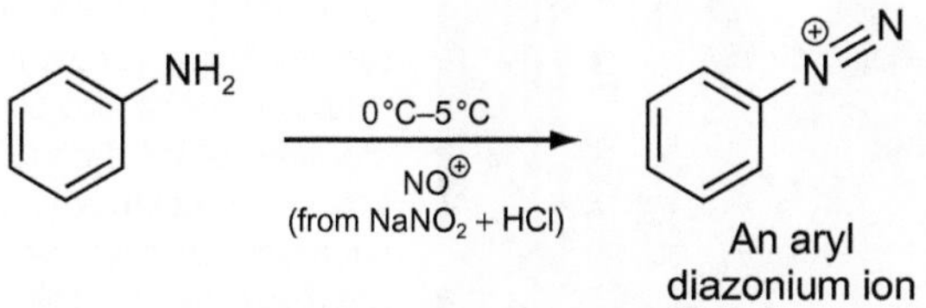
NH_2
0 °C–5 °C
$NO^{\oplus}$
(from $NaNO_2$ + HCl)
An aryl diazonium ion

The NO^+ ion is generated from HNO_2 (nitrous acid). Because nitrous acid is unstable, it is formed *in situ* by reacting $NaNO_2$ (sodium nitrite) with HCl at 0 °C. The mechanism is shown in Figure 22.25.

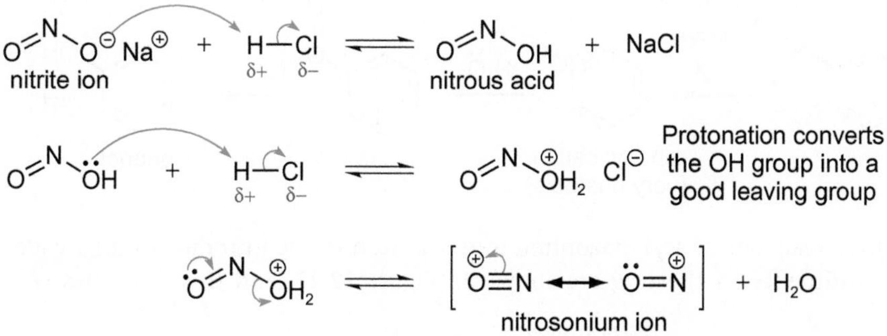
nitrite ion
nitrous acid
+ NaCl
Protonation converts the OH group into a good leaving group
nitrosonium ion
+ H_2O

Figure 22.25 Formation of the nitrosonium ion (NO^+).

The NO^+ ion is a strong electrophile and it reacts with a nucleophilic amine at 0 °C to 5 °C to form a compound with a new N–N bond. After deprotonation, a nitrosamine is formed and this reacts by the series of protonations and deprotonations shown in Figure 22.26 to form the aryl diazonium ion and water—this process is called **diazotization**. The diazonium ion is formed with an accompanying anion and, if HCl is used in the synthesis, an arenediazonium chloride salt ($ArN_2^+\ Cl^-$) is formed.

Aryl diazonium ions are unstable and, on warming to room temperature or above, they lose N_2 (which is an excellent leaving group). This forms a very unstable aryl cation (Ar^+) that reacts with nucleophiles in an S_N1 reaction. Reaction with water introduces an OH substituent on the ring and this is an important way of making phenols (they cannot be prepared directly by electrophilic substitution reactions because HO^+ cannot be generated).

Oxidation levels are used to classify organic oxidations and reductions, as discussed in Section 19.2 (p.899).

It is known that oxidizing metal ions can accept an electron from methylbenzene (toluene; $C_6H_5CH_3$) to form a radical cation. The radical cation can be converted into benzyl alcohol (phenylmethanol) following deprotonation, oxidation, and reaction with water:

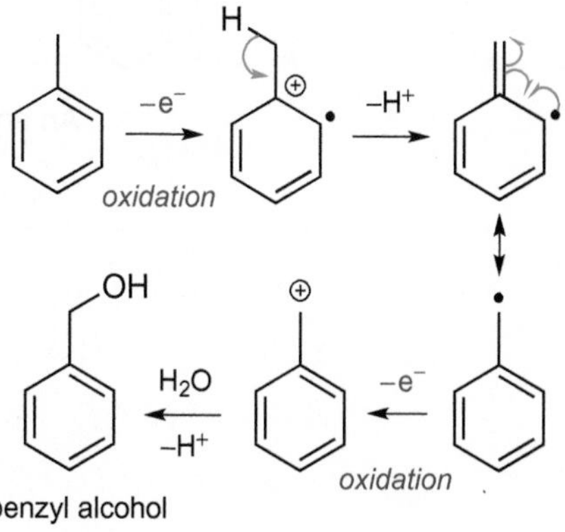
–e⁻
oxidation
–H⁺
H_2O
–H⁺
–e⁻
oxidation
benzyl alcohol

Benzyl alcohol ($C_6H_5CH_2OH$) can then be oxidized further to form benzoic acid ($C_6H_5CO_2H$).

Using IUPAC nomenclature, the nitrosonium ion NO^+ is called a nitrosyl ion.

The $N{\equiv}O^+$ resonance form is more stable than $O{=}N^+$ because it has more bonds (relative stabilities of resonance forms are discussed in Section 19.1, p.876).

Problems associated with nitrite ions (NO_2^-) in drinking water are discussed in Box 5.4 (p.232).

HNO_2 decomposes rapidly at room temperature so, as soon as it is made, it is used in a subsequent reaction (in the same reaction vessel).

Many reaction mechanisms involve successive protonation and deprotonation steps. When an H^+ ion changes position in a molecule this is called a **proton transfer**.

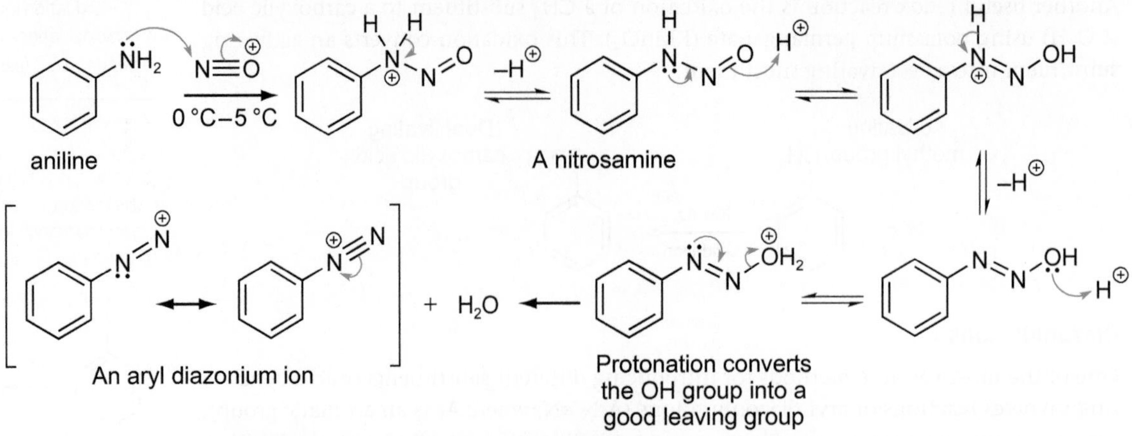
0 °C–5 °C
aniline
A nitrosamine
An aryl diazonium ion
Protonation converts the OH group into a good leaving group

Figure 22.26 Mechanism for the formation of an aryl diazonium ion (ArN_2^+).

Nitrosamines (R_2N-N=O) are formed in beer and some foods, such as cured meats, by reaction of naturally occurring secondary amines (R_2NH) with the sodium nitrite added as a food preservative. There is a cause for concern because nitrosamines are potent carcinogens, although there is no evidence that very small concentrations of nitrosamines cause cancer in humans. Changes in the malting process mean that beer now contains only 2% of the concentration of Me_2N-N=O that was present over 20 years ago.

S_N1 reactions are discussed in detail in Section 20.3 (p.933).

Aryl cations, such as the phenyl cation (Ph^+ or $C_6H_5^+$), are unstable because these cations are not stabilized by resonance—the vacant sp^2 orbital cannot overlap with the *p* orbitals of the ring (Section 20.3, p.935). Even though aryl cations are so unstable, they are formed from diazonium ions because N_2 is an excellent leaving group (Section 20.2, p.925).

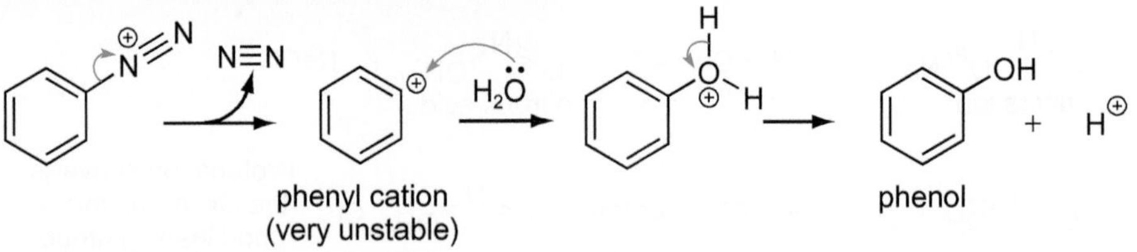
phenyl cation
(very unstable)
phenol

Other reactions of aryl diazonium ions that lead to the introduction of a wide array of substituents on the ring are shown in Figure 22.27, and this illustrates why aryl

Formation of an intermediate phenyl radical ($Ph^{\bullet}$ or $C_6H_5^{\bullet}$) in a Sandmeyer reaction (notice that the Cu(I) ion is regenerated and so only a small quantity is required):

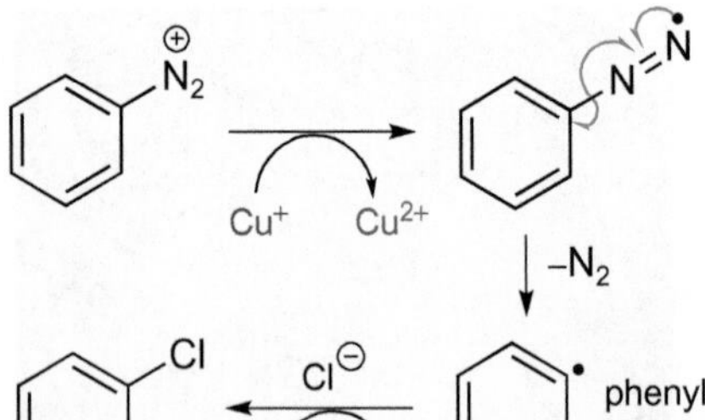
phenyl
radical

Aryl diazonium ions ($ArN^+\equiv N$) are relatively weak electrophiles because the positive charge is stabilized by resonance (by the aromatic ring). Only relatively strong nucleophiles react with aryl diazonium ions, which explains why coupling reactions require strongly activated substituted benzenes, such as phenol.

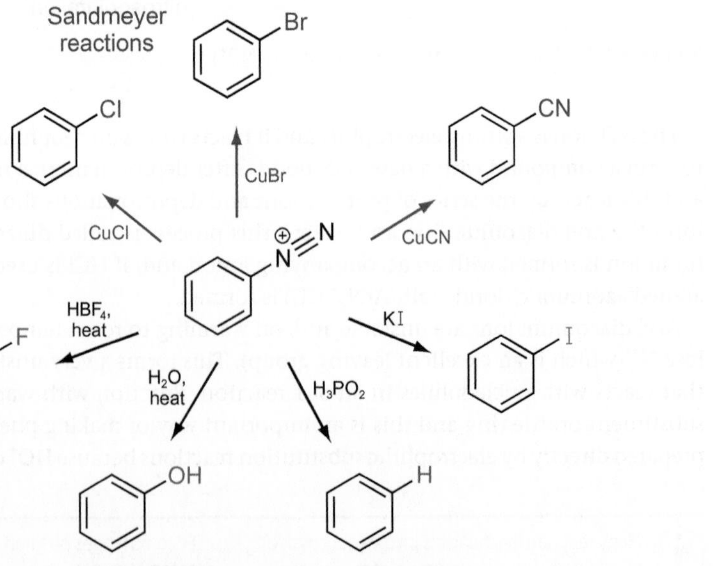
Sandmeyer
reactions
CuBr
CuCl
CuCN
HBF4, heat
KI
H2O, heat
H3PO2

Figure 22.27 Reactions of an aryl diazonium ion (ArN_2^+).

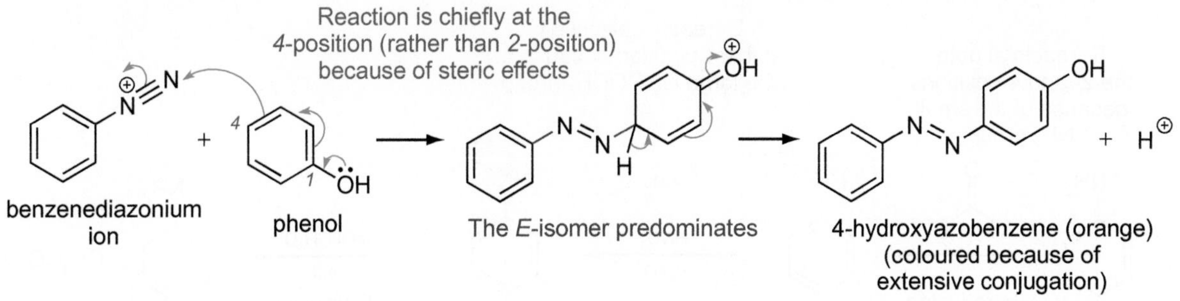
Reaction is chiefly at the
4-position (rather than 2-position)
because of steric effects
benzenediazonium
ion
phenol
The E-isomer predominates
4-hydroxyazobenzene (orange)
(coloured because of
extensive conjugation)

Figure 22.28 A coupling reaction to produce an azo compound (Ar–N=N–Ar).

diazonium ions are commonly used in synthesis. The three reactions using copper(I) salts are called **Sandmeyer reactions** (after the Swiss chemist Traugott Sandmeyer). It is likely that these reactions involve aryl radical intermediates rather than aryl cations (see margin note on p.1042).

Aryl diazonium ions react with phenols (ArOH) or anilines (phenylamines, $ArNH_2$) in electrophilic aromatic substitution reactions (see p.1032). Because aryl diazonium ions are positively charged they act as electrophiles and react with strongly activated substituted benzenes to form highly coloured azo compounds in **coupling reactions**. Figure 22.28 shows a coupling reaction between the benzenediazonium ion ($PhN^+{\equiv}N$) and phenol (C_6H_5OH).

The lone pair of electrons on oxygen in phenol attacks through the benzene ring—if the OH group attacked the diazonium ion directly, the product would be a less stable product with a weak O–N bond. Also, notice that phenol attacks the terminal nitrogen atom in the diazonium ion—attack at the more hindered, positively charged nitrogen atom, would require nitrogen to be able to form five bonds.

Highly conjugated compounds are coloured because they absorb light in the visible region of the electromagnetic spectrum (see Section 10.6, p.495).

The more stable *E*-isomer of the azo compound is formed as the major product.

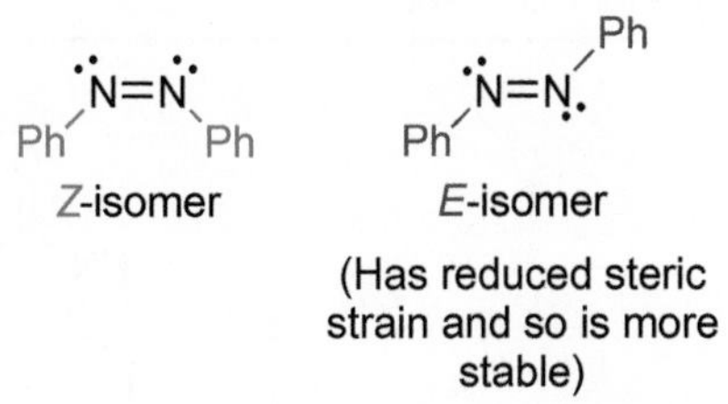
Z-isomer
E-isomer
(Has reduced steric
strain and so is more
stable)

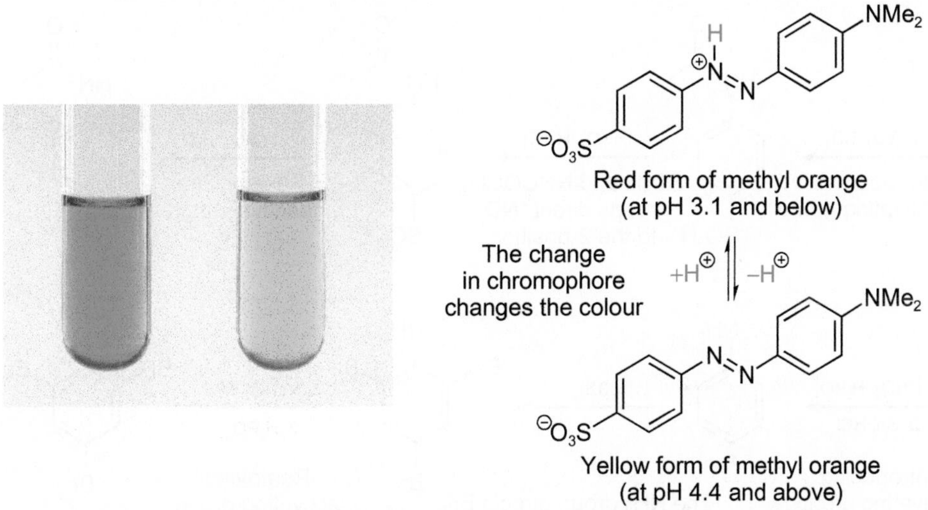
Red form of methyl orange
(at pH 3.1 and below)
The change
in chromophore
changes the colour
Yellow form of methyl orange
(at pH 4.4 and above)

Methyl orange (Section 7.5, p.328) is an azo dye that changes from red to orange to yellow when the pH is increased from 3.1 to 4.4. It is used as an indicator in the titration of weak bases with strong acids.

Temporary substituents

An activating group on a benzene ring (such as OH or NH_2) directs the electrophile to the 2- and 4-positions of the ring. The ratio of structural isomers formed depends on the size of the activating group—the larger the activating group, the greater the proportion of electrophilic attack at the 4-position (Section 22.3, p.1034). With small-sized activating groups on benzene (e.g. NH_2), a relatively large proportion of 1,2-disubstituted benzenes are formed, which is a problem when only 1,4-disubstituted benzenes are required.

To increase the selectivity for attack of an electrophile E^+ at the 4-position of an activated benzene, the size of the activating group is temporarily increased. For example, in Figure 22.29, the NH_2 substituent is converted into a larger $NHCOCH_3$ (amide) substituent

Formation and reactions of amides (RNHCOR), including hydrolysis of amides to form amines, are discussed in Section 24.2 (p.1123).

The NHCOR substituent is not so strongly activating as the NH_2 substituent (Section 22.3, p.1031) so multiple substitutions are less common.

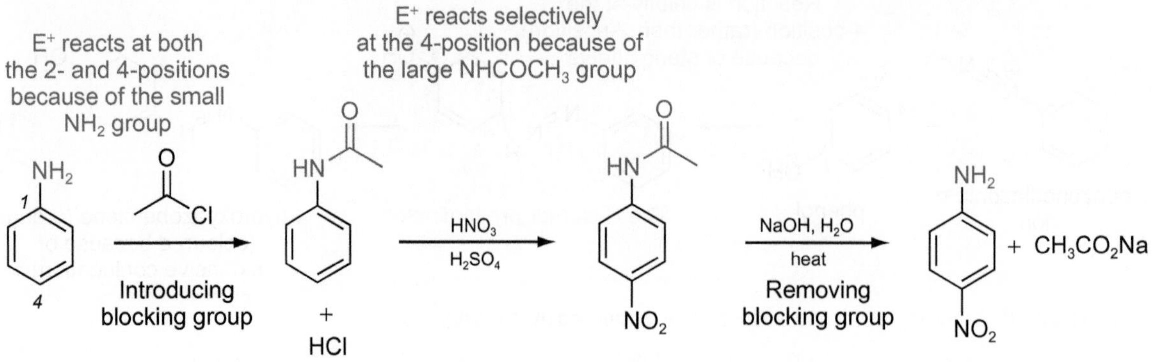
E+ reacts at both the 2- and 4-positions because of the small NH2 group
E+ reacts selectively at the 4-position because of the large NHCOCH3 group
NH2
Cl
Introducing blocking group
HN
+ HCl
HNO3
H2SO4
NO2
NaOH, H2O
heat
Removing blocking group
+ CH3CO2Na

Figure 22.29 Use of a blocking group to direct electrophilic attack to the 4-position.

by reaction with ethanoyl chloride (CH_3COCl). The $NHCOCH_3$ substituent is activating and directs the electrophile selectively to the 4-position of the ring. The size of the $NHCOCH_3$ group minimizes attack of the electrophile at the 2- and 6-positions, so it is called a **blocking group**. After the electrophilic substitution reaction, the $NHCOCH_3$ substituent is converted back to an NH_2 substituent by heating with aqueous NaOH. For a blocking group to be useful, it needs to be easily introduced and removed.

For some reactions, introducing a temporary substituent, such as SO_3H or NH_2, onto a benzene ring is used to direct the position of electrophilic attack to form the desired structural isomer. Afterwards, they are removed from the ring and replaced by H. Two examples of using temporary substituents in this way are shown in Figure 22.30.

i 1,3,5-Tribromobenzene cannot be formed by bromination of benzene because Br directs electrophiles to the 2- and 4-positions of the ring. Hence, bromination of bromobenzene (using Br_2, $FeBr_3$) forms a mixture of 1,2-dibromobenzene and 1,4-dibromobenzene.

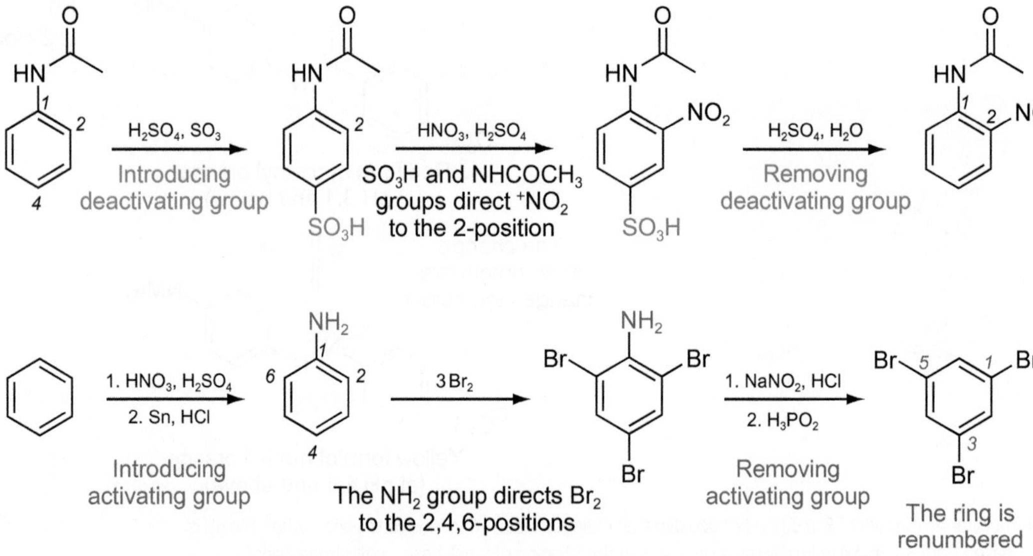
HN
H2SO4, SO3
Introducing deactivating group
SO3H
HNO3, H2SO4
SO3H and NHCOCH3 groups direct +NO2 to the 2-position
NO2
H2SO4, H2O
Removing deactivating group
NH2
1. HNO3, H2SO4
2. Sn, HCl
Introducing activating group
3Br2
The NH2 group directs Br2 to the 2,4,6-positions
Br
1. NaNO2, HCl
2. H3PO2
Removing activating group
The ring is renumbered

Figure 22.30 Introducing temporary substituents to direct the position of electrophilic attack.

The mechanisms of sulfonation of benzene and desulfonation of benzenesulfonic acid ($C_6H_5SO_3H$) are discussed in Section 22.2 (p.1018).

i H_3PO_2 (phosphinic acid or hypophosphorous acid, HPA) reacts with an aryl diazonium ion to form an intermediate aryl radical (like in the Sandmeyer reaction), which abstracts a hydrogen atom from H_3PO_2.

Strategies for making polysubstituted benzenes

When planning syntheses of polysubstituted benzenes, you need to consider the directing effects of each substituent on the ring. For example, 1-bromo-3-nitrobenzene is prepared from benzene by nitration followed by bromination (see Figure 22.31). The different directing effects of the NO_2 and Br substituents explain the order of these reactions. The NO_2 substituent is first introduced onto the ring because this directs Br^+ to the 3-position. If the Br substituent is introduced onto the ring first, then nitration produces a mixture of 2- and 4-bromonitrobenzene.

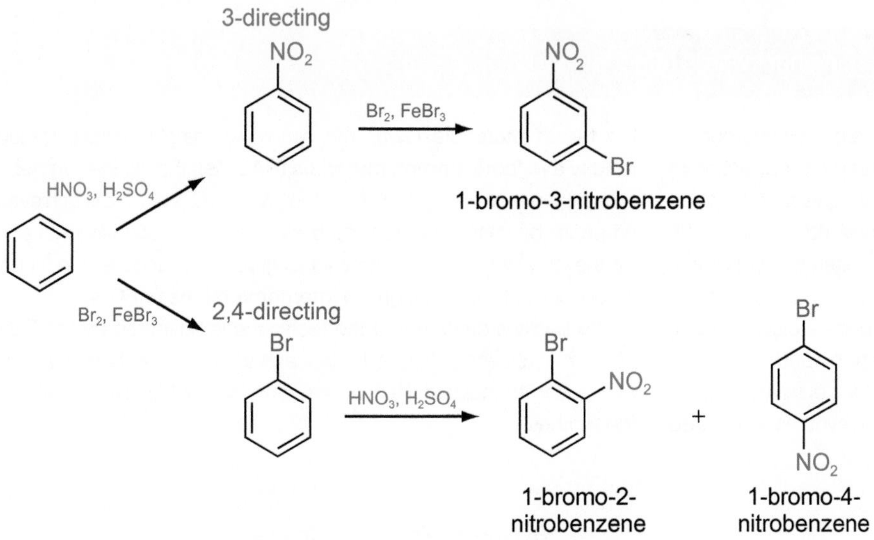
3-directing
NO_2
HNO_3, H_2SO_4
Br_2, $FeBr_3$
Br
1-bromo-3-nitrobenzene
Br_2, $FeBr_3$
2,4-directing
HNO_3, H_2SO_4
+
1-bromo-2-nitrobenzene
1-bromo-4-nitrobenzene

Figure 22.31 Deciding on the order of electrophilic substitution reactions.

The directing effects of substituents determine the order of electrophilic substitutions. To make a *1,3-disubstituted benzene*, the substituent introduced first should direct the electrophile in the second substitution to the 3-position. To make a *1,4-disubstituted benzene*, the substituent introduced first should direct the electrophile in the second substitution to the 4-position.

As the NO_2 group in nitrobenzene is a strong deactivating substituent, bromination of nitrobenzene (using Br_2, $FeBr_3$) requires a relatively high reaction temperature of around 140 °C.

When preparing trisubstituted benzenes from disubstituted benzenes, the position of electrophilic attack depends on the electronic and steric effects of both substituents. But what if the two substituents direct the electrophile to different positions on the ring? Are all possible structural isomers formed in equal amounts or is one structural isomer favoured over others? These questions are answered by the following guidelines.

The structures of the isomers can be determined using spectroscopic techniques such as ^1H NMR spectroscopy (Section 12.3, p.580). For example, 1-bromo-2-nitrobenzene gives four signals in the ^1H NMR spectrum whereas 1-bromo-4-nitrobenzene gives only two signals.

Guidelines for making polysubstituted benzenes

1. The more powerful activating group on a benzene ring controls the position of attack by an electrophile.

The more powerful activating group is shown in blue

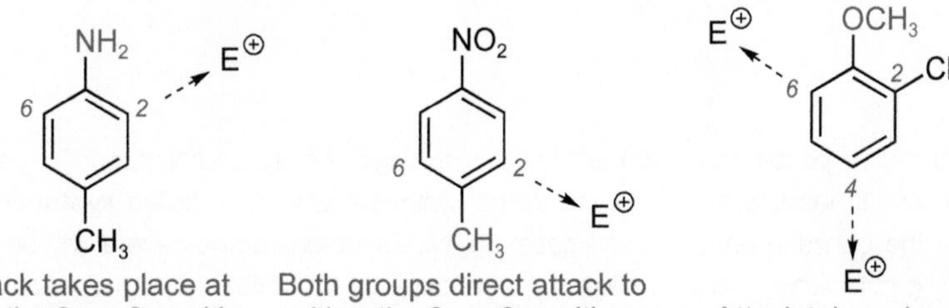
NH_2
$E^{\oplus}$
6
2
CH_3
NO_2
CH_3
OCH_3
Cl
4

Attack takes place at either the 2- or 6-position (these are equivalent)

Both groups direct attack to either the 2- or 6-position (these are equivalent)

Attack takes place at the 4- and 6-positions

2. For 1,3-disubstituted benzenes, the electrophile rarely attacks in between the two substituents because of steric hindrance.

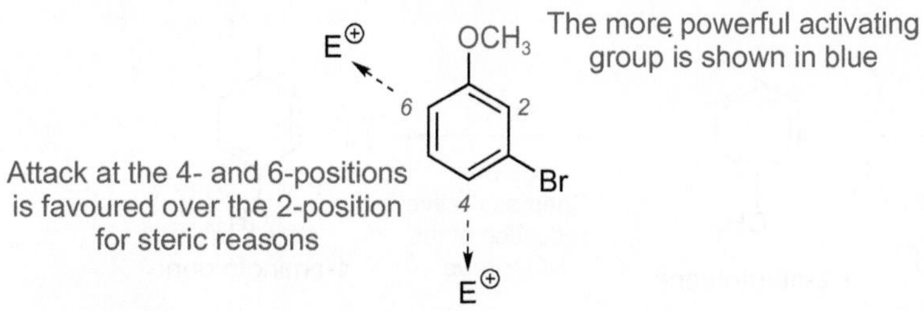
$E^{\oplus}$
OCH_3
The more powerful activating group is shown in blue
6
2
Br
4
Attack at the 4- and 6-positions is favoured over the 2-position for steric reasons

Worked example 22.4 provides practice in designing efficient syntheses of substituted benzenes.

1-Bromo-2-nitrobenzene can be separated from 1-bromo-4-nitrobenzene by recrystallization. Recrystallization of a mixture of these compounds, using ethanol as the solvent, gives crystals of 1-bromo-4-nitrobenzene (this isomer is less soluble in ethanol) while 1-bromo-2-nitrobenzene remains in solution (as this isomer is more soluble in ethanol).

Visit the Online Resource Centre to view screencast 22.1 which walks you through the synthesis of polysubstituted benzenes.

Box 22.7 Retrosynthesis—working backwards to go forwards

Prior to the 1950s, the synthesis of organic compounds, such as substituted benzenes, relied on a trial and error selection of commercially available starting materials on the basis that they had a similar structure to the molecules to be synthesized. However, the situation changed in the 1960s when a more systematic approach, called **retrosynthetic analysis**, was introduced by the American chemist E. J. Corey. It proved to be such an important discovery that Corey was awarded the Nobel Prize in Chemistry in 1990.

Retrosynthetic analysis (RSA) is a technique for planning organic syntheses by breaking down the molecule to be synthesized, called the **target molecule (TM)**, into simpler starting materials, called **readily available starting materials (RASMs)**, that is, the synthesis is designed by working back from the TM to RASMs. This is achieved on paper by imaginary breaking bonds (called disconnections) and by the conversion of one functional group into another functional group (called **functional group interconversions** or **FGIs**).

The backwards nature of the technique is illustrated by the RSA of 4-aminotoluene in Figure 1. Notice that a particular type of arrow, ⇒, is used to indicate that you are working backwards in planning the synthesis.

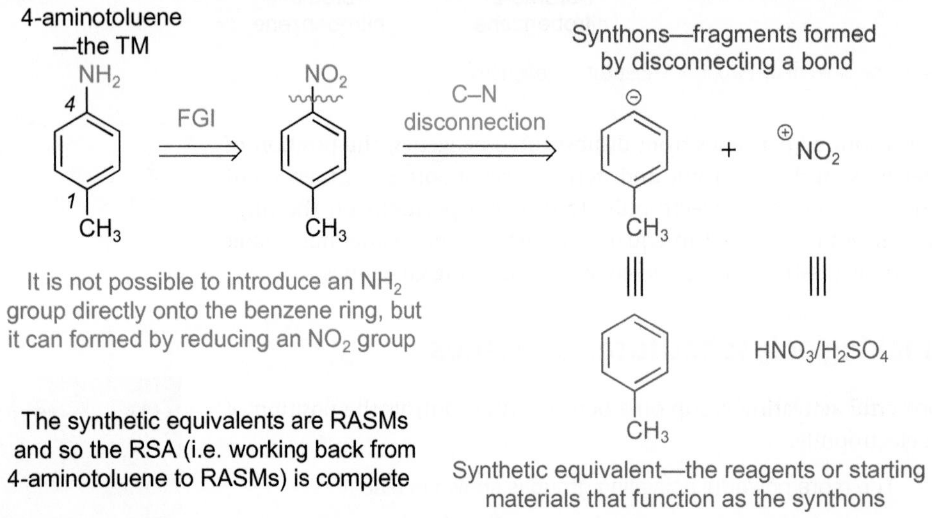

4-aminotoluene
—the TM
NH_2
4
1
CH_3
FGI
NO_2
CH_3
C–N
disconnection
Synthons—fragments formed
by disconnecting a bond
⊖
+
⊕
NO_2
CH_3
It is not possible to introduce an NH_2
group directly onto the benzene ring, but
it can formed by reducing an NO_2 group
HNO_3/H_2SO_4
CH_3
The synthetic equivalents are RASMs
and so the RSA (i.e. working back from
4-aminotoluene to RASMs) is complete
Synthetic equivalent—the reagents or starting
materials that function as the synthons

▲ **Figure 1** An RSA of 4-aminotoluene.

The RSA starts by an FGI, which transforms the NH_2 group back to an NO_2 group. This FGI is used for two reasons: (i) because an NH_2 group cannot be introduced directly onto the benzene ring; (ii) the NH_2 group can be formed from an NO_2 group, which can be introduced directly onto the benzene ring. The C–NO_2 bond of 4-nitrotoluene is then disconnected (shown by a red wavy line, ∿∿∿). Disconnection of a bond produces charged fragments, called **synthons**. Often synthons do not actually exist but they help to identify the actual nucleophilic and electrophilic reagents that are required in the synthesis, which are called **synthetic equivalents**. In this case the $^{+}NO_2$ synthon can be generated from HNO_3/H_2SO_4 while the synthetic equivalent for the aryl anion is toluene ($C_6H_5CH_3$)—the benzene ring in toluene acts as a nucleophile and the CH_3 group directs an electrophile predominantly to the 4-position.

From the RSA in Figure 1, a two-step synthesis of 4-aminotoluene can be proposed, which is shown in Figure 2.

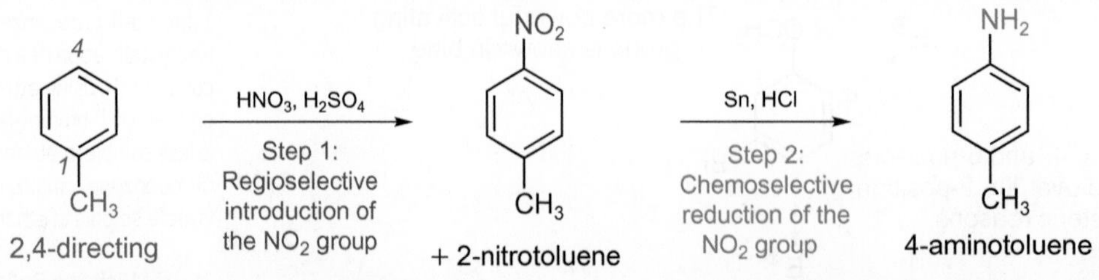

4
1
CH_3
2,4-directing
HNO_3, H_2SO_4
Step 1:
Regioselective
introduction of
the NO_2 group
NO_2
CH_3
+ 2-nitrotoluene
Sn, HCl
Step 2:
Chemoselective
reduction of the
NO_2 group
NH_2
CH_3
4-aminotoluene

▲ **Figure 2** Synthesis of 4-aminotoluene based on the RSA in Figure 1.

→

Retrosynthesis is an excellent technique for discovering different synthetic routes and comparing them in a logical and straightforward fashion. Indeed, the systematic approach has allowed the development of computer programs, also pioneered by Corey, that can rapidly evaluate different retrosynthetic pathways.

E. J. Corey formalized the concept of RSA in his book '*The Logic of Chemical Synthesis*'.

Question

Propose a synthesis of 1-(4-bromophenyl)ethanone based on the RSA shown below.

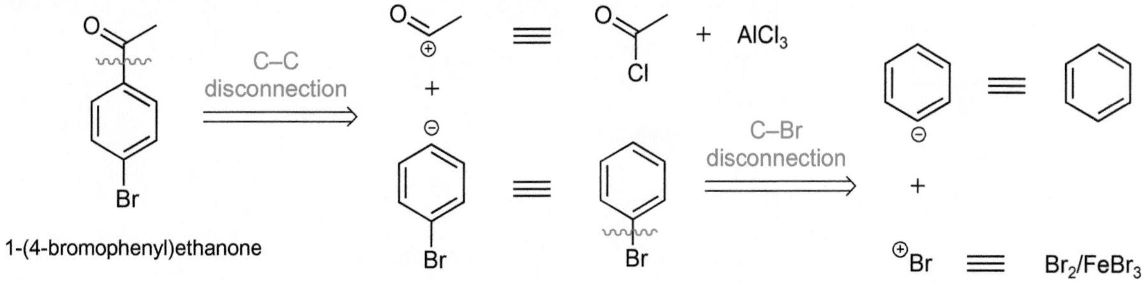

C–C disconnection
1-(4-bromophenyl)ethanone
Br
+ AlCl3
Cl
C–Br disconnection
Br2/FeBr3

Worked example 22.4 Preparing substituted benzenes

Propose efficient syntheses of the following compounds from the starting material indicated in each case. Each synthesis requires more than one step.

(a)

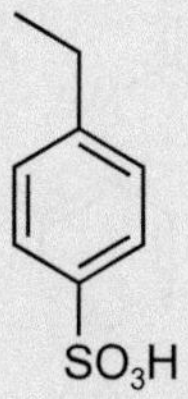

1
(from benzene)

(b)

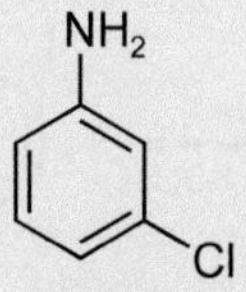

2
(from benzene)

(c)

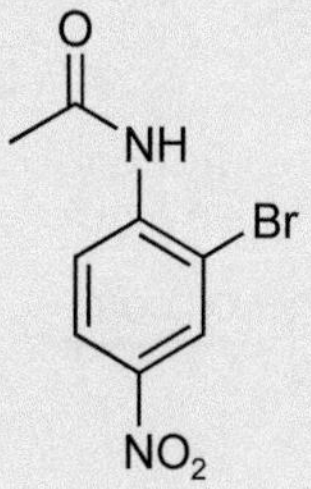

3
(from aniline)

Visit the Online Resource Centre to view screencast 22.2 which introduces you to the concepts of retrosynthetic analysis and its use in planning efficient syntheses of substituted benzenes.

Compound **1** is called 4-ethylbenzenesulfonic acid, compound **2** is 3-chloroaniline (3-chlorophenylamine), and compound **3** is *N*-(2-bromo-4-nitrophenyl) ethanamide.

3-Chloroaniline (**2**) is used in the dye industry, where it is commonly called Fast Orange GC Base. The NH_2 is converted into a diazonium ion, which this is then coupled to phenols or anilines to form azo dyes.

Ethylbenzene ($C_6H_5CH_2CH_3$) is made on a large scale in industry, mainly by reacting benzene with ethene ($H_2C{=}CH_2$) in the presence of an acid catalyst. Protonation of ethene gives the ethyl cation ($CH_3CH_2^+$), which reacts with benzene in an electrophilic substitution reaction. This approach uses inexpensive starting materials and it is a shorter synthesis than using acylation followed by reduction. However, depending on the catalyst and reaction conditions, varying amounts of di-, tri-, and tetraethylbenzenes are formed, along with ethylbenzene. To improve the efficiency of the process, reaction conditions have been developed for a transalkylation reaction; this is where an ethyl group is transferred from di-, tri-, and/or tetraethylbenzenes to benzene to form more ethylbenzene.

→ Strategy

Assign directing effects to each substituent on the ring.

Consider how each substituent is introduced onto the ring. For substituents other than halogen atoms, NO_2, SO_3H, alkyl (R), or acyl (RCO) substituents, more than one step is required.

Use the directing effects to determine the order the substituents are introduced on to the ring.

If introducing an activating group, consider the possibility that more than one of these groups will be introduced onto the ring.

When one substituent is converted into another substituent on the ring, use the directing effects of both substituents to determine the order of electrophilic substitution reactions.

When particularly large substituents are on the ring, consider how steric effects could influence the regioselectivity of electrophilic substitution reactions.

Solution

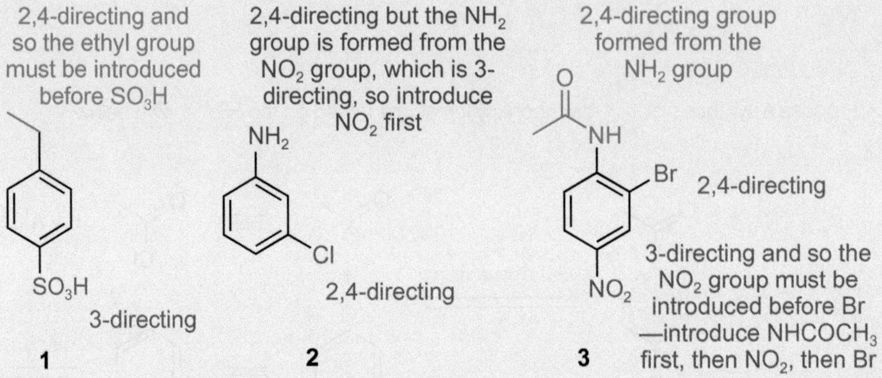
2,4-directing and so the ethyl group must be introduced before SO3H
SO3H
3-directing
1
2,4-directing but the NH2 group is formed from the NO2 group, which is 3-directing, so introduce NO2 first
NH2
Cl
2,4-directing
2
2,4-directing group formed from the NH2 group
O
NH
Br
2,4-directing
NO2
3-directing and so the NO2 group must be introduced before Br —introduce NHCOCH3 first, then NO2, then Br
3

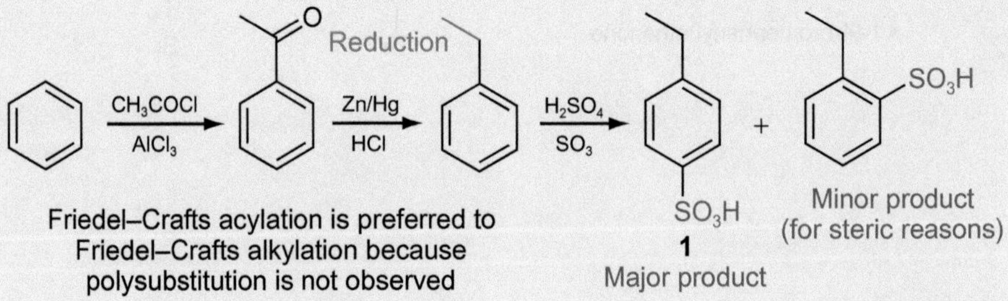
O
CH3COCl
AlCl3
Reduction
Zn/Hg
HCl
H2SO4
SO3
+
SO3H
SO3H
1
Major product
Minor product (for steric reasons)
Friedel–Crafts acylation is preferred to Friedel–Crafts alkylation because polysubstitution is not observed

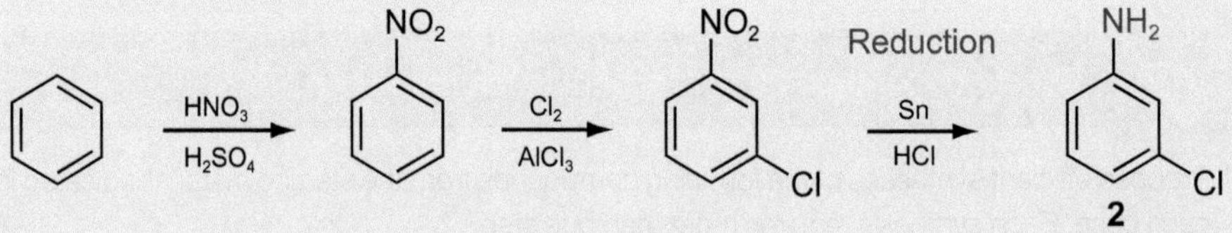
HNO3
H2SO4
NO2
Cl2
AlCl3
NO2
Cl
Reduction
Sn
HCl
NH2
Cl
2

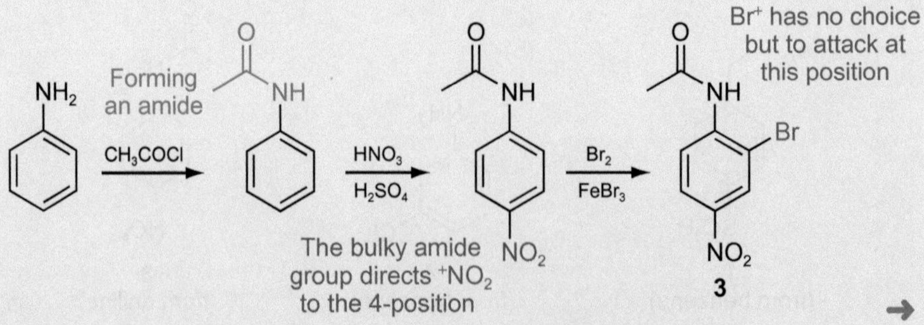
NH2
Forming an amide
CH3COCl
O
NH
HNO3
H2SO4
O
NH
NO2
The bulky amide group directs +NO2 to the 4-position
Br2
FeBr3
O
NH
Br
NO2
3
Br+ has no choice but to attack at this position

→

➜ *Question*

Propose efficient syntheses of the following compounds starting from the starting material indicated in each case. Each synthesis requires more than one step.

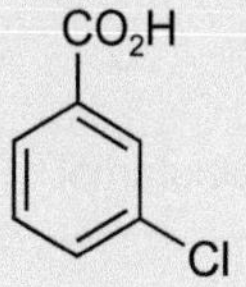

4

(from methylbenzene)

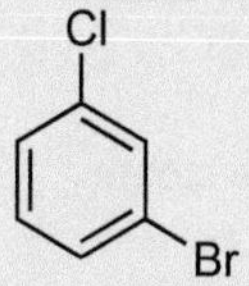

5

(from nitrobenzene)

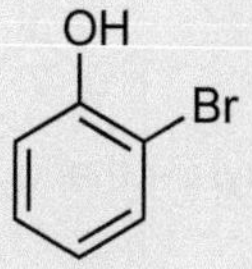

6

(from phenol)

Hint: consider using a temporary substituent

Compound **4** is called 3-chlorobenzoic acid, compound **5** is 1-bromo-3-chlorobenzene, and compound **6** is 2-bromophenol.

Box 22.8 The synthesis and biological action of salbutamol

Around 150 million people worldwide suffer from asthma. In the UK, the disease affects about 5% of the population. The usual symptoms of an asthma attack are wheezing and breathlessness, which may be caused by allergens such as pollen. Although there is currently no cure for asthma, sufferers get quick relief from the debilitating breathlessness by using Ventolin inhalers, which contain salbutamol. Salbutamol belongs to a group of medicines called bronchodilators. These medicines relieve the symptoms of asthma by opening up air passages in the body so that air can flow into the lungs more easily, making breathing easier.

◀ An asthma sufferer using a Ventolin inhaler.

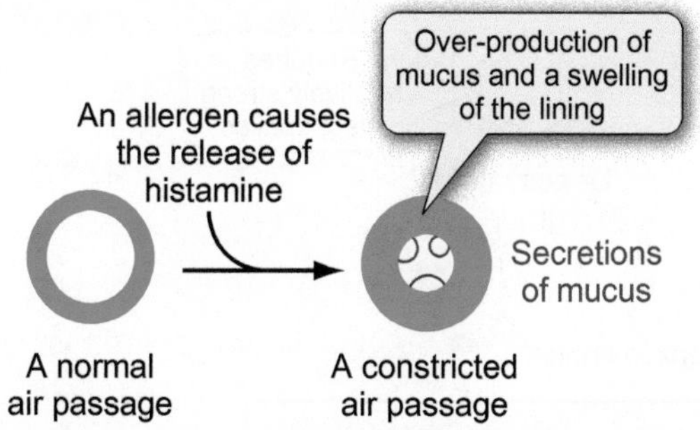
Over-production of mucus and a swelling of the lining
An allergen causes the release of histamine
Secretions of mucus
A normal air passage
A constricted air passage

▲ How an allergen can trigger asthma.

Salbutamol works as a bronchodilator by mimicking adrenaline (epinephrine, p.843). Your body produces adrenaline when you are frightened or excited or feel the need to act quickly. A burst of adrenaline causes excessive stimulation of the brain, a faster-beating heart, an increase in blood pressure, and, importantly if you suffer from asthma, relaxation of smooth muscles that keep your airways open. The adverse biological effects of adrenaline mean it cannot be used to treat asthma. What is needed is a compound with a structure related to adrenaline that will react more selectively with the receptors and produce just one of the biological effects of adrenaline—the bronchodilating effect. Salbutamol was developed to do this.

To prepare salbutamol, three different groups need to be introduced at the correct positions of the benzene ring by electrophilic aromatic substitution reactions. One approach to salbutamol involves reacting methyl 2-hydroxybenzoate with CH_3COCl and $AlCl_3$ in a Friedel–Crafts acylation reaction. The $COCH_3$ group is directed to the 5-position of the ring and methyl 5-ethanoyl-2-hydroxybenzoate is formed. This is then converted into salbutamol in four steps—the amino-alcohol side chain is produced from the $COCH_3$ group,

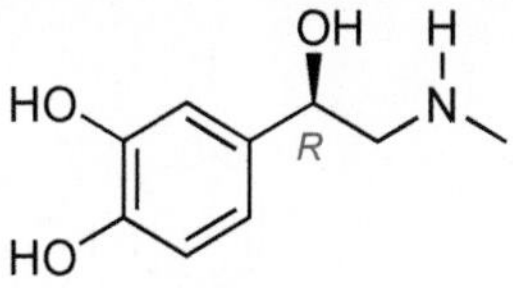

adrenaline

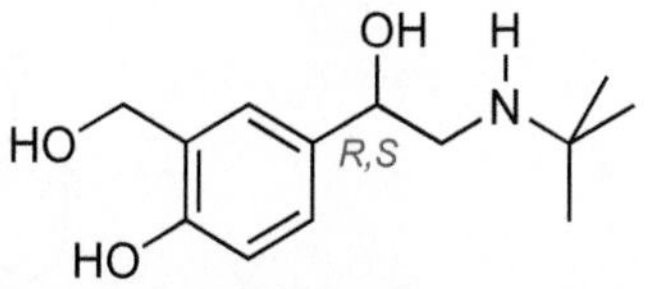

salbutamol
Marketed as a racemate or more recently as the *R*-enantiomer

➜

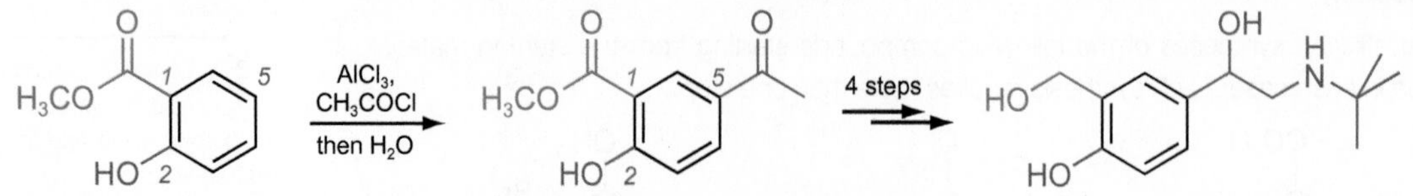

H3CO
HO
AlCl3,
CH3COCl
then H2O
4 steps
OH
H
N
HO

methyl 2-hydroxybenzoate | methyl 5-ethanoyl-2-hydroxybenzoate | salbutamol

whereas the primary alcohol is formed from the CO_2CH_3 group. Whether this approach is used to manufacture salbutamol on a large scale is not clear. Several companies currently manufacture salbutamol and the process is commercially sensitive.

Until recently, only the racemic form of salbutamol was available. However, research has showed that the *S*-enantiomer of salbutamol may worsen the effects of asthma so companies are now using enantioselective approaches to produce the *R*-enantiomer. (The two enantiomers of salbutamol can be separated using chiral chromatography; see Box 11.5, p.541.)

Question

(a) Suggest a mechanism for the formation of methyl 5-ethanoyl-2-hydroxybenzoate from methyl 2-hydroxybenzoate, $AlCl_3$, and CH_3COCl.

(b) Suggest why the $COCH_3$ group is selectively introduced at the 5-position of the ring.

(c) Would you expect methyl 5-ethanoyl-2-hydroxybenzoate or methyl 2-hydroxybenzoate to react more quickly with electrophiles in electrophilic substitution reactions?

Summary

- You need to consider the following factors when planning syntheses of substituted benzenes by electrophilic substitution reactions.

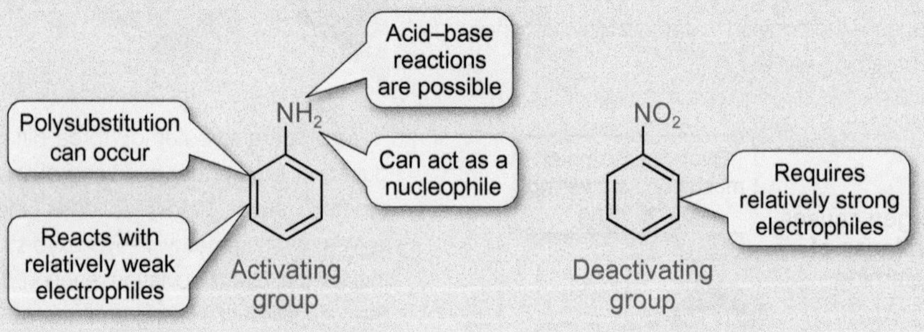

Acid–base reactions are possible
NH2
Polysubstitution can occur
Can act as a nucleophile
Reacts with relatively weak electrophiles
Activating group
NO2
Requires relatively strong electrophiles
Deactivating group

- The following reactions are used to transform substituents on benzene rings.

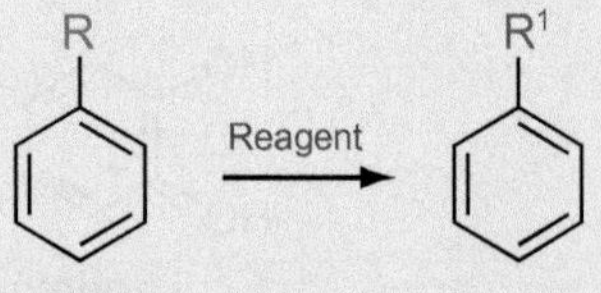

R
Reagent
R1

R	R^1	Reagent
$COCH_3$	CH_2CH_3	Zn/Hg, HCl
NO_2	NH_2	Sn, HCl
CH_3	CO_2H	$KMnO_4$

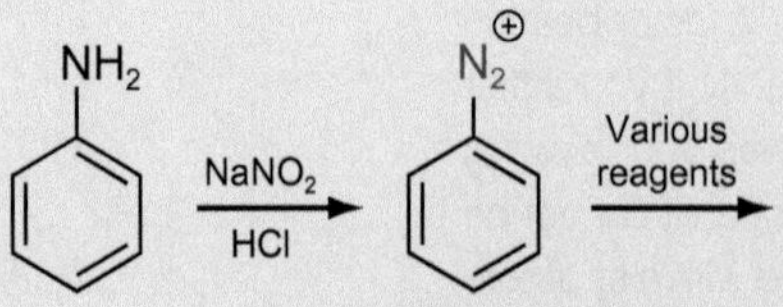

NH2
NaNO2
HCl
N2
Various reagents
R

aryl diazonium ion (nucleophile)

R = F, Cl, Br, I, CN, OH, H, N=NAr

➜ • Regioselective formation of di- and trisubstituted benzenes is possible.

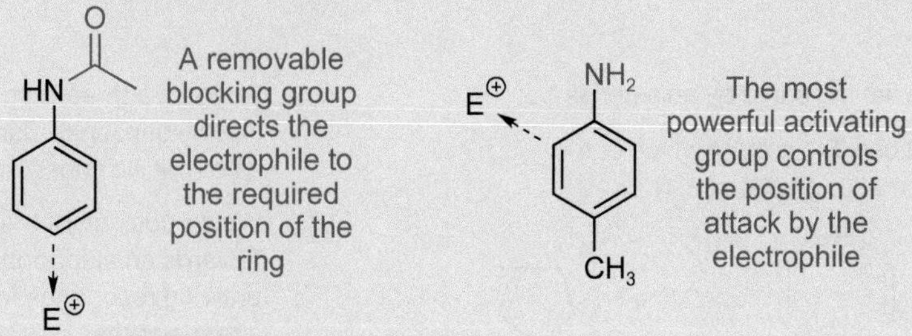
HN
O
A removable blocking group directs the electrophile to the required position of the ring
E⊕
E⊕
NH₂
CH₃
The most powerful activating group controls the position of attack by the electrophile

For practice questions on this topic, see questions 3–5 at the end of this chapter (pp.1052–1053).

Concept review

By the end of this chapter, you should be able to do the following.

- Know what aromatic, antiaromatic, and nonaromatic compounds are and give examples of each.
- Understand why benzene reacts in electrophilic substitution reactions.
- Give reagents and write reaction mechanisms to explain how benzene undergoes halogenation, nitration, sulfonation, Friedel–Crafts alkylation, and Friedel–Crafts acylation.
- Classify substituents on benzene rings as 2,4-directing activators, 2,4-directing deactivators, or 3-directing deactivators.
- Understand how the electronic and steric effects of substituents on benzene rings influence the rates and regioselectivities of electrophilic substitution reactions.
- Describe how substituents on benzene rings can be converted into other substituents by redox reactions or by forming diazonium ions.
- Design efficient syntheses of polysubstituted benzenes from benzene.

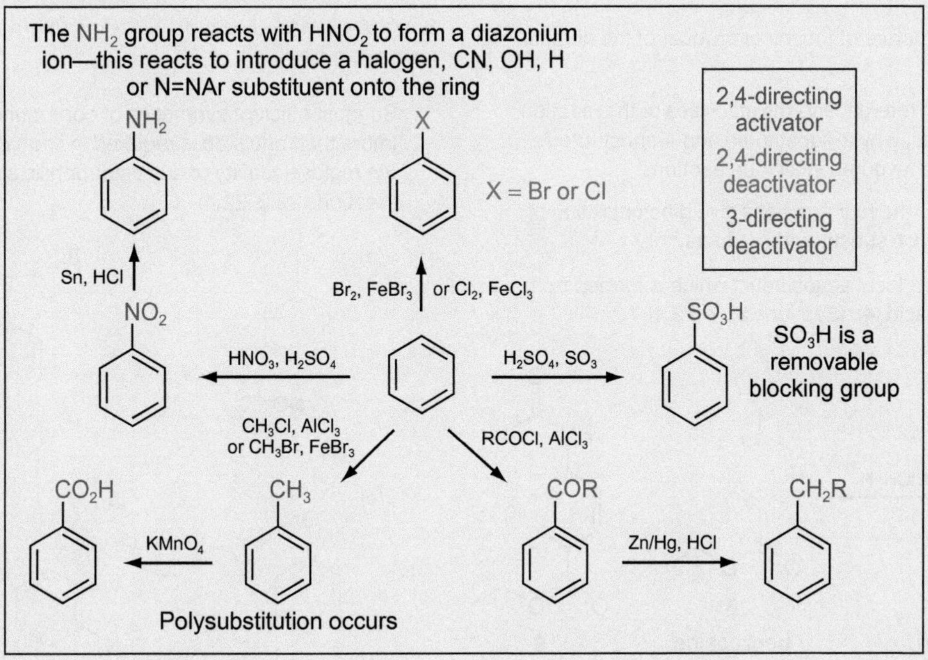
The NH₂ group reacts with HNO₂ to form a diazonium ion—this reacts to introduce a halogen, CN, OH, H or N=NAr substituent onto the ring
NH₂
X
X = Br or Cl
2,4-directing activator
2,4-directing deactivator
3-directing deactivator
Sn, HCl
Br₂, FeBr₃
or Cl₂, FeCl₃
NO₂
SO₃H
HNO₃, H₂SO₄
H₂SO₄, SO₃
SO₃H is a removable blocking group
CH₃Cl, AlCl₃
or CH₃Br, FeBr₃
RCOCl, AlCl₃
CO₂H
CH₃
COR
CH₂R
KMnO₄
Zn/Hg, HCl
Polysubstitution occurs

Questions

More challenging questions are indicated by an asterisk *.

1 State Hückel's rule, and use it to predict which of the compounds **1**–**3** are aromatic. (Section 22.1)

1

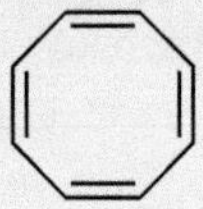

2

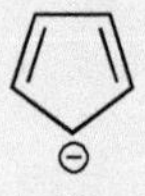

3

2 The following table gives the results of four electrophilic substitution reactions of chlorobenzene. (Sections 22.2–22.3)

Reaction	% 2-substitution	% 3-substitution	% 4-substitution
Chlorination	39	6	55
Nitration	30	0	70
Bromination	11	2	87
Sulfonation	1	0	99

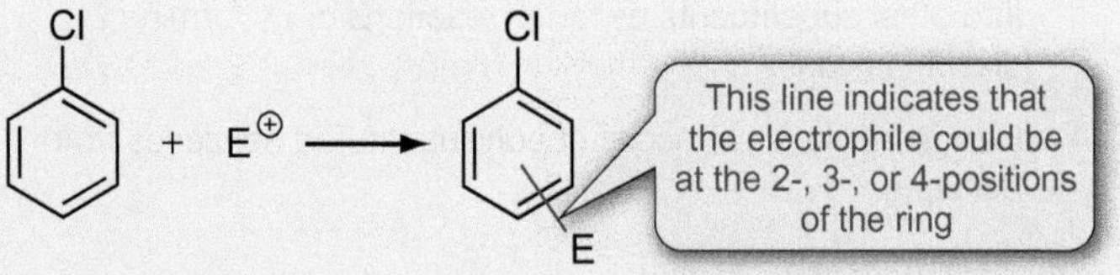
Cl
+ E⊕
Cl
E
This line indicates that the electrophile could be at the 2-, 3-, or 4-positions of the ring

(a) Draw the structure of the major product of the sulfonation reaction.

(b) Draw the structure of the minor product of the nitration reaction.

(c) Explain, with reference to the stabilities of the reaction intermediates, why 2-substituted and 4-substituted products are favoured in all four reactions.

(d) Suggest why the four reactions give different ratios of 2-substituted:4-substituted products.

3 Benzocaine (**5**) is a local anaesthetic, which is formed from 4-aminobenzoic acid (**4**). (Sections 22.2–22.4)

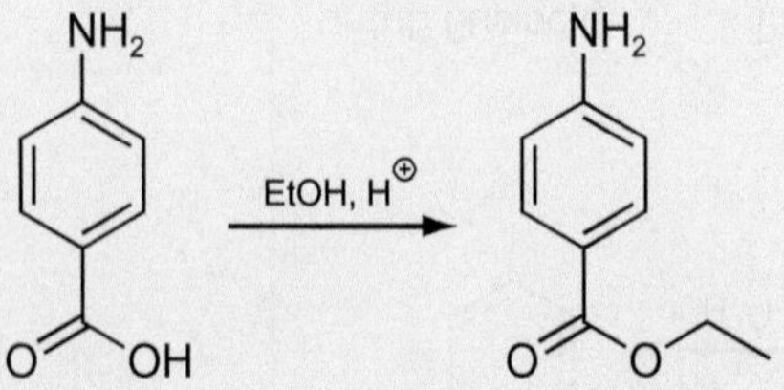
NH2
EtOH, H⊕
NH2

4
4-aminobenzoic acid

5
benzocaine

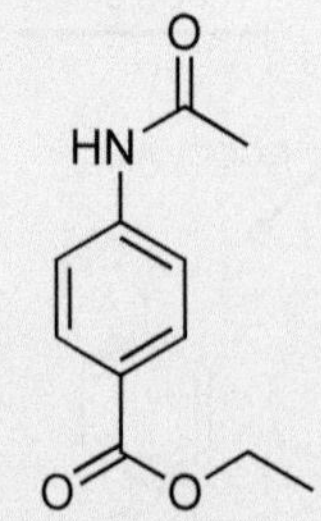

6

(a) Suggest a three-step synthesis of **4** starting from toluene (methylbenzene). Comment on the regioselectivity of any electrophilic substitution reactions.

(b) Which position of the ring in benzocaine (**5**) is most reactive towards an electrophile (E^+)? Explain your reasoning by drawing resonance forms of the carbocation reaction intermediates.

(c) Would you expect benzocaine (**5**) or compound **6** to react more rapidly with an electrophile? Explain your reasoning.

Benzocaine is widely used in first aid creams, sore throat sprays and lozenges, and sunburn remedies.

4 Suggest efficient syntheses of compounds **7**–**10** from benzene (more than one step is required in each case). Comment on the regioselectivity of any electrophilic substitution reactions. (Sections 22.2–22.4)

(a)

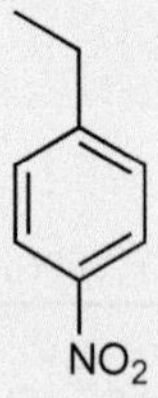

7

(b)

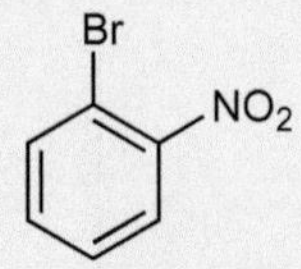

8

(c)

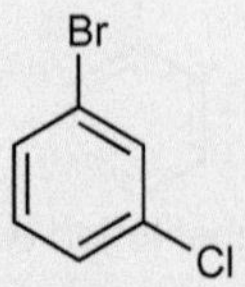

9

(d)

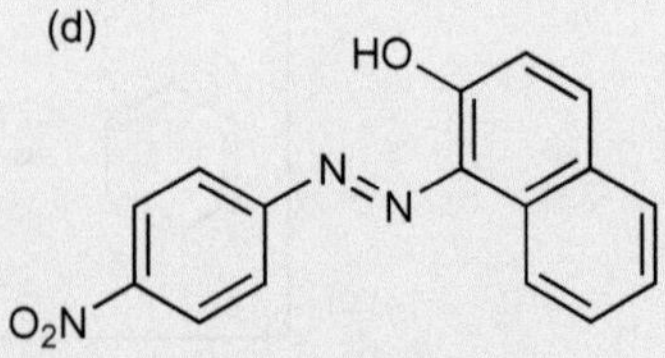

10

5* The following questions are based on the RSA of paracetamol shown below. (Section 22.4)

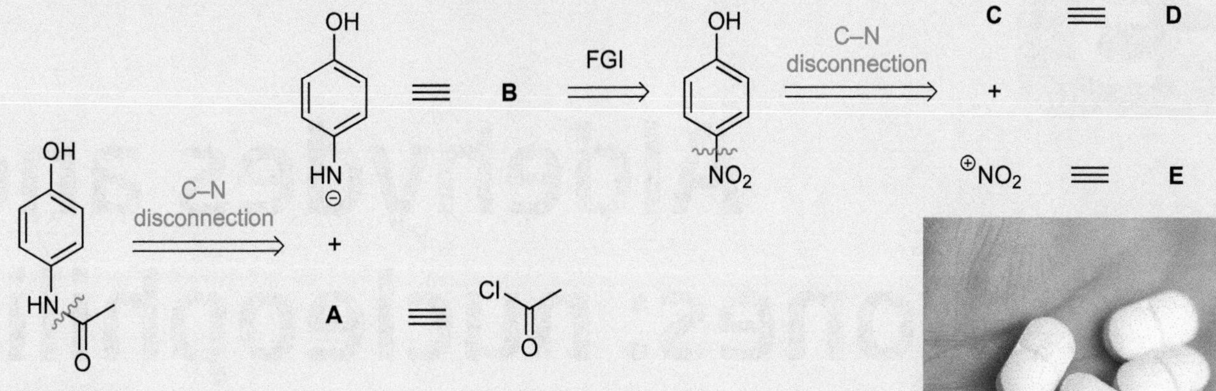
OH
HN
paracetamol
C–N
disconnection
OH
HN
+
A
Cl
O
B
FGI
OH
NO2
C–N
disconnection
C
D
+
NO2
E

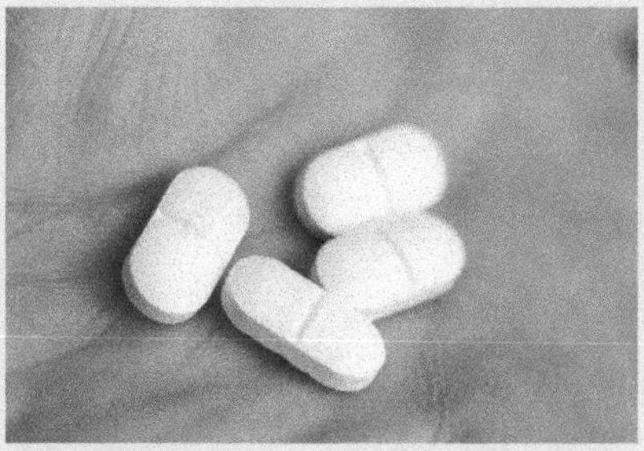

Paracetamol is one of the most common painkillers available worldwide. More than 960 million paracetamol tablets are prescribed by GPs each year in the UK alone. The groove down the middle of each tablet is used to help split the tablet in half if a lower dose is required.

(a) Draw the missing structures, **A**–**E**.

(b) Use the RSA to propose a synthesis of paracetamol.

(c) Give the structures of two possible by-products from your proposed synthesis.

23

Aldehydes and ketones: nucleophilic addition and α-substitution reactions

EURO 500 EYPΩ
500

This chapter builds on the following topics:

- Functional groups containing carbonyl groups Section 2.7, p.98
- Identifying carbonyl functional groups Section 12.2, p.574
- Fundamental concepts of organic reaction mechanisms Section 19.1, p.863
- Acid-base reactions Section 19.2, p.888
- Redox reactions Section 19.2, p.899
- Polar reactions Section 19.2, p.902

◀ A 500 euro note has a holographic watermark to protect against frauds and counterfeiters. Retinal is being investigated for use in holographic data storage techniques. These techniques offer the possibility of storing digital or analogue data throughout the entire thickness of a material. Currently, CDs and DVDs store data only on the surface of a disk so less than 0.01% of the total volume of the disk is used to store the information. To improve the efficiency of data storage, holography aims to record information in a volume that is greater than 90% of the total volume of the material.

Rhodopsin and vision

The human eye is remarkable. Apart from the brain, the eye is the most complex organ in the body. Its job is to respond rapidly to changing conditions of light and to focus light rays originating from different distances. The eye works like a sophisticated camera, with each individual component playing a vital role in providing clear vision. When all components function properly, light is converted into electrical impulses that are sent to the brain. The brain decodes the impulses to produce the image that you see. If you have an average life span, then your eyes will bring you almost 24 million images of the world around you!

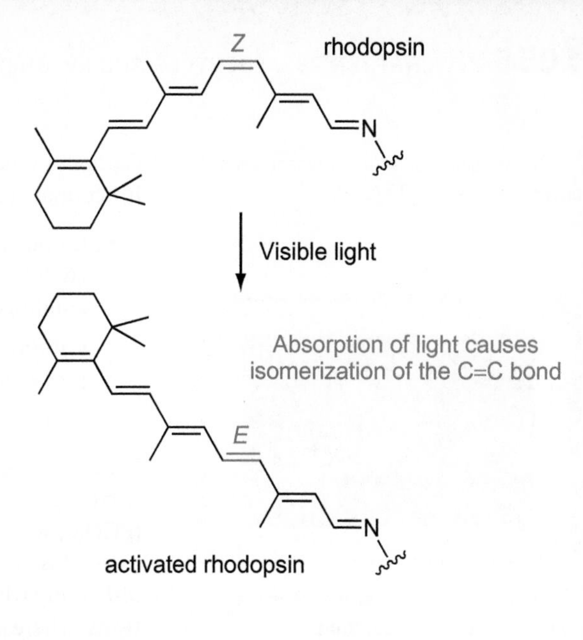
Z
rhodopsin
N
Visible light
Absorption of light causes isomerization of the C=C bond
E
N
activated rhodopsin

▲ When light strikes a rhodopsin molecule, the C=C bond with the *Z*-configuration is converted into the *E*-configuration.

At the back of the human eye is the retina, which is made up of light-sensitive cells called cones and rods. These have different jobs—the cone cells are responsible for colour vision and for vision in bright light, while the rod cells are responsible for vision in dim light. Rod cells are light-sensitive because they contain a compound called *rhodopsin*.

Rhodopsin is a conjugated imine. It is formed from reaction of the aldehyde, (*Z*)-retinal, with a protein called opsin. In the first step of the reaction, an NH_2 group in opsin acts as a nucleophile and donates a pair of electrons to the partially positively charged carbon atom of the C=O bond in (*Z*)-retinal. In step 2, a proton is transferred from nitrogen to oxygen. Overall, steps 1 and 2 constitute a **nucleophilic addition reaction**, which is the most common type of reaction of aldehydes (and ketones). Finally, in step 3, reaction with an acid catalyst leads to elimination of water (a **condensation** reaction) and formation of the imine C=N bond in rhodopsin. In the body, the mechanism is likely to be a little different from that shown below as other functional groups in opsin's active site may be involved in stabilizing the intermediates.

Rhodopsin is light-sensitive because, when visible light strikes it, the C=C bond with the *Z*-configuration is rapidly converted into the *E*-configuration. This gives a product called *activated rhodopsin*. Because rhodopsin and activated rhodopsin are *E*- and *Z*-isomers, they have different shapes and it is the change in shape (from 'curved' to 'straight') that triggers a cascade process that leads, virtually instantaneously, to a nerve impulse being sent to the brain. This is a remarkably sensitive mechanism, as the eye can register even a single photon hitting the retina. Activated rhodopsin then isomerizes back to rhodopsin, so that it can register any further photons.

The *Z*- to *E*-isomerization of retinal is one of the most remarkable systems created by nature. It has inspired chemists to mimic the *Z*- to *E*-isomerization of retinal and to search for possible applications of retinal derivatives for storage of data. Optical data storage relies on altering the refractive index of an organic or inorganic material by using light. Information from a data-encoded laser beam is stored in the material through a series of changes in the refractive index. Retinal derivatives are of interest in data storage, because the light-induced *Z*- to *E*-isomerization leads to a change in the refractive index of the substance.

▼ (*Z*)-Retinal, an aldehyde, undergoes a nucleophilic addition reaction and then a condensation reaction to form rhodopsin, which contains an imine C=N bond.

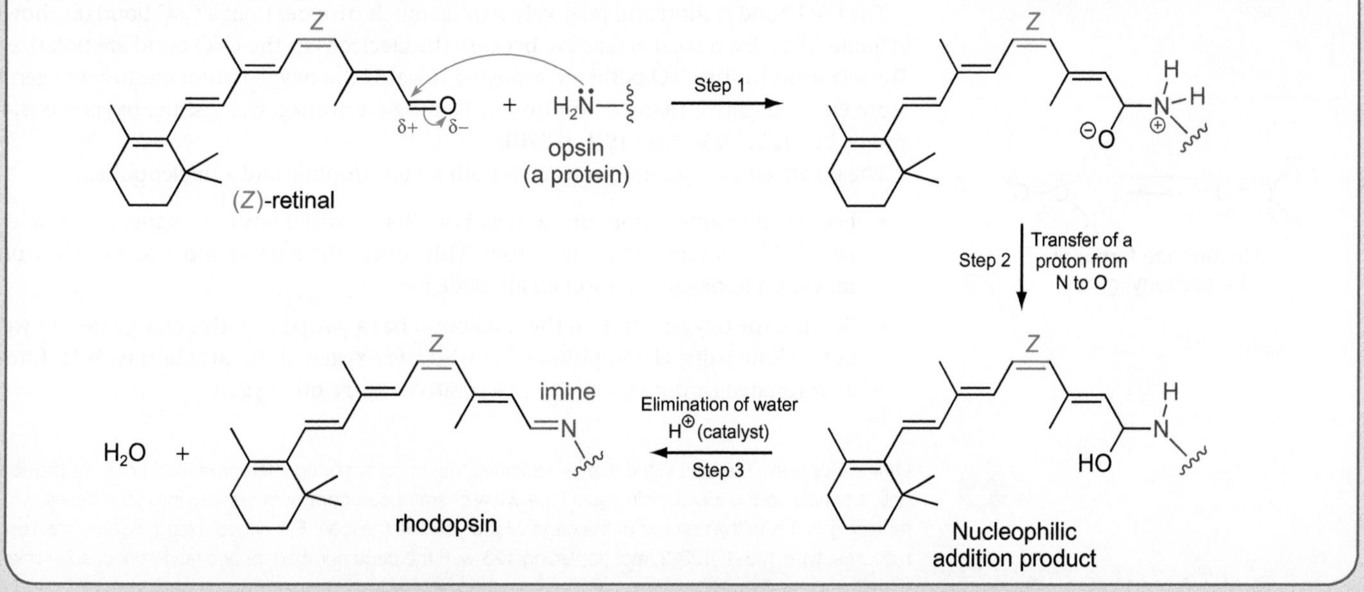
Z
O
δ+
δ−
+
H2N
opsin
(a protein)
(Z)-retinal
Step 1
Z
H
N
H
O
Step 2
Transfer of a proton from N to O
Z
H
N
HO
Nucleophilic addition product
Elimination of water
H⊕ (catalyst)
Step 3
Z
imine
N
H2O
+
rhodopsin

The naming of carbonyl compounds is discussed in Section 2.7 (p.98).

Carbonyl compounds all contain the C=O group. This is probably the most important functional group in organic chemistry for the following reasons.

- Carbonyl compounds make up an extremely large class of compounds, which includes aldehydes, ketones, carboxylic acids, and their derivatives (such as esters, amides, etc.).
- Carbonyl compounds are widespread in nature and many have important biological properties (Box 23.1, p.1060).
- Carbonyls are amongst the most useful compounds in organic synthesis.

2+2=4

The most common reaction of aldehydes and ketones is nucleophilic **addition**.

Because there are so many types of carbonyl compounds, their chemistry is divided into two chapters. This chapter concentrates on the chemistry of aldehydes (RCHO) and ketones (RCOR), whereas Chapter 24 discusses the chemistry of carboxylic acids and their derivatives. This division is made because aldehydes/ketones and carboxylic acid derivatives react differently with nucleophiles. Aldehydes and ketones react in **nucleophilic *addition* reactions**, whereas carboxylic acid derivatives react **in nucleophilic *substitution* reactions**.

23.1 The structure and reactions of aldehydes and ketones

H
C=O
R
R
C=O
R

An aldehyde (names end in -al)   A ketone (names end in -one)

An aldehyde contains a C=O group bonded to at least one hydrogen atom (RCHO or HCHO), whereas a ketone has a carbonyl group bonded to two alkyl or aryl groups (RCOR).

The C=O bond contains one strong σ bond and one weaker π bond. The double bond is formed by overlap of the orbitals of an sp^2 hybridized carbon atom with an sp^2 hybridized oxygen atom (hybridization is introduced in Section 5.4, p.236). A σ bond is formed by end-on overlap of an sp^2 orbital of carbon with an sp^2 orbital of oxygen. The π bond is formed by side-on overlap of an unhybridized p orbital on carbon with an unhybridized p orbital on oxygen.

The two remaining sp^2 orbitals of the oxygen atom each contain a lone pair of electrons, whereas the two remaining sp^2 orbitals of the carbon atom form two other σ bonds. All three of the σ bonds to carbon are in the same plane and are separated by angles of 120°.

Often, molecular models of carbonyls show two curved bonds between C and O (as shown here for ethanal). Remember that this is not correct because the C=O bond has one σ bond and one π bond.

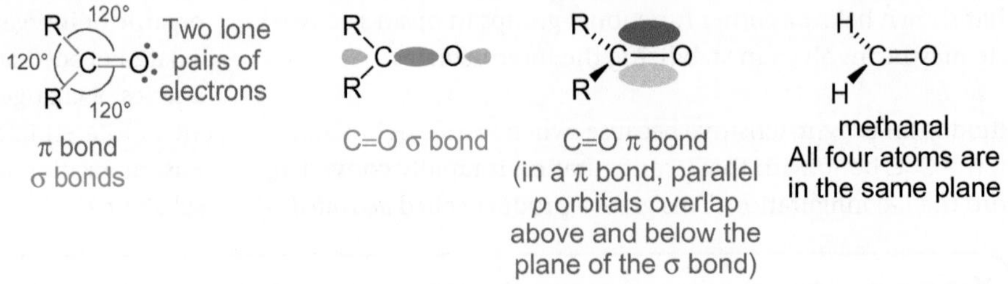
120°
R
120°
C=O
R
120°
Two lone pairs of electrons
π bond
σ bonds
R
C
O
R
C=O σ bond
R
C—O
R
C=O π bond
(in a π bond, parallel p orbitals overlap above and below the plane of the σ bond)
H
C=O
H
methanal
All four atoms are in the same plane

The C=O bond is short and relatively strong, much stronger than a C=C bond (as shown in Table 23.1). Even so, it is reactive because the electrons in the C=O bond are polarized. The electrons in the C=O bond are attracted towards the oxygen atom because oxygen is more electronegative than carbon (in the Pauling electronegativity scale, oxygen is 3.44 and carbon is 2.55; Section 19.1, p.870).

The polarized C=O group can react as both an electrophile and a nucleophile.

R
C=O
R
δ+ δ−
R
C—O
R

Resonance forms of the carbonyl group

- Because the carbon atom in the C=O bond has a partial positive charge (δ+), nucleophiles (Nu^-) attack the carbon atom. This breaks the π bond and a pair of electrons moves on to oxygen to form an alkoxide ion.
- Because the oxygen atom in the C=O bond has a partial negative charge (δ–), as well as two lone pairs, electrophiles react with the oxygen atom. Reaction with H^+ forms a protonated carbonyl, which has a positive charge on oxygen.

Online Support. The icon in the margin indicates that accompanying interactive resources are provided online to help your understanding: just type www.chemtube3d.com/burrows/123 into your browser, replacing 123 with the number of the page where you see the icon. For pages linking to more than one resource, type 123-1, 123-2, etc. (replacing 123 with the page number) for access to successive links.

Table 23.1 Mean bond lengths and mean bond enthalpies of C–C, C=C, and C=O

Bond	Mean bond lengths / pm	Mean bond enthalpies / kJ mol^{-1}
C–C	153 (longest bond)	+347 (weakest bond)
C=C	134	+612
C=O	121 (shortest bond)	+742 (strongest bond)

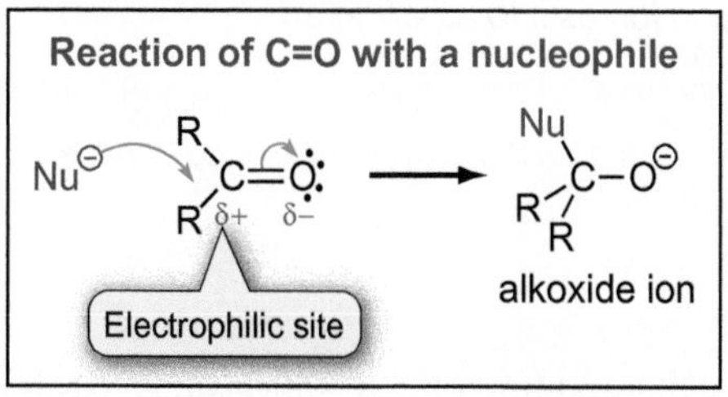
Reaction of C=O with a nucleophile
Nu
R
C=O:
R δ+ δ−
Nu
C–O
R
R
alkoxide ion
Electrophilic site

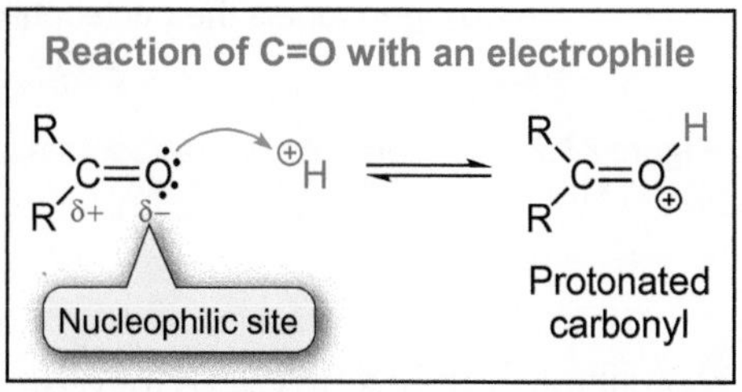
Reaction of C=O with an electrophile
R
C=O:
R δ+ δ−
H
R
H
C=O
R
Protonated carbonyl
Nucleophilic site

There are three general mechanisms by which aldehydes and ketones react. These are:

1 nucleophilic addition reactions;
2 α-substitution reactions;
3 carbonyl–carbonyl condensation reactions.

A brief overview of each general mechanism is presented in the remainder of this section. Each type of reaction is then considered in detail in the following sections.

Overview of nucleophilic addition reactions

The most common reaction of an aldehyde or ketone involves reaction with a negatively charged or neutral nucleophile in a nucleophilic addition reaction.

Reaction of the C=O bond with a negatively charged nucleophile

Reaction of the C=O bond with a negatively charged nucleophile (Nu$^-$), such as H$^-$ (hydride ion), R$^-$ (carbanions), or $^-$CN (cyanide ion), produces an intermediate alkoxide ion (see Figures 23.1 and 23.2).

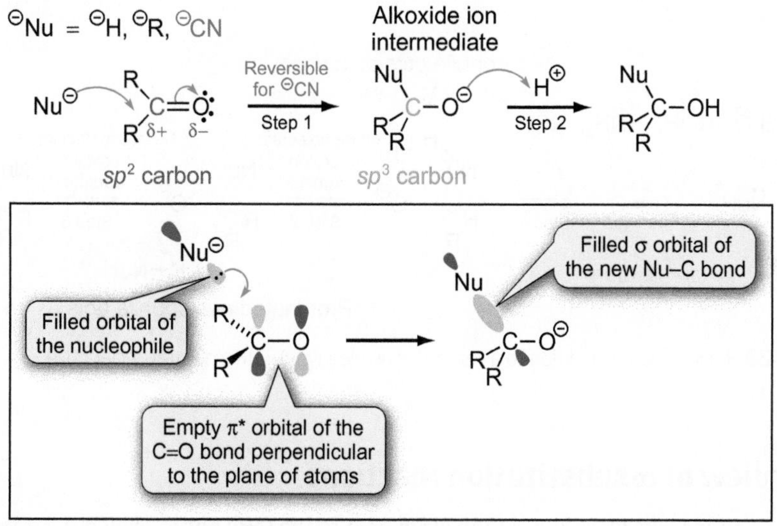
Nu = H, R, CN
Alkoxide ion intermediate
Nu
R
C=O:
R δ+ δ−
Reversible for CN
Step 1
Nu
C–O
R
R
H
Step 2
Nu
C–OH
R
R
sp² carbon
sp³ carbon
Nu
Filled orbital of the nucleophile
R
C–O
R
Empty π* orbital of the C=O bond perpendicular to the plane of atoms
Filled σ orbital of the new Nu–C bond
Nu
C–O
R
R

Figure 23.1 Nucleophilic addition to a ketone (or an aldehyde) with a negatively charged nucleophile containing a lone pair in a *p* orbital.

There are three general mechanisms by which aldehydes and ketones react.

Classification of organic reaction mechanisms is introduced in Section 19.2 (p.888).

Nucleophilic addition reactions are discussed in detail in Section 23.2 (p.1061).

Orbital overlap in organic reaction mechanisms is introduced in Section 19.1 (p.885). For the nucleophilic addition shown in Figure 23.1, the filled *p* orbital of the nucleophile is the HOMO and this interacts with the empty π* orbital of the carbonyl, which is the LUMO. Filling the π* orbital causes the C=O π bond to break.

The HOMO of the nucleophile depends on the identity of the nucleophile. For the cyanide ion the HOMO is an *sp* orbital on carbon, for ammonia the HOMO is an sp^3 orbital on nitrogen and for an organometallic reagent the HOMO is a σ orbital of the carbon–metal bond.

Crystallographers Hans-Beat Bürgi and Jack D. Dunitz examined the crystal structures of a large number of compounds that contained both a carbonyl group and a nucleophile. They showed the geometric relationship between the carbonyl group and nucleophile was not random—its value depends on the exact system, but it is generally taken to be 107°. The results from the experimental data have been supported by computational studies.

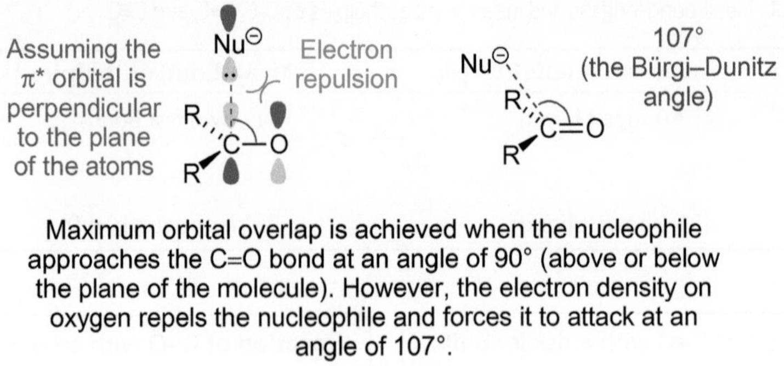
Assuming the
π* orbital is
perpendicular
to the plane
of the atoms
Nu⊖
Electron
repulsion
107°
(the Bürgi–Dunitz
angle)
Maximum orbital overlap is achieved when the nucleophile approaches the C=O bond at an angle of 90° (above or below the plane of the molecule). However, the electron density on oxygen repels the nucleophile and forces it to attack at an angle of 107°.

Figure 23.2 The nucleophile (Nu^-) attacks the carbon of the C=O bond at an angle of 107° to the plane of the molecule.

Addition of H^- or R^- to C=O is irreversible, while addition of ^-CN is reversible. ^-CN is a much better leaving group than H^- or R^- (see margin) and the C=O bond can be reformed from the intermediate alkoxide ion. In all cases, the alkoxide ion is protonated in a second step (usually, H^+ is added in a separate stage) to form a product containing an OH group.

The fact that ^-CN is a much better leaving group than H^- or R^- is indicated by the lower pK_a value of HCN (9.21) compared to H_2 (35) or an alkane, RH (50). Using pK_a values to assess leaving groups is discussed in Section 20.2 (p.925).

Addition of the nucleophile is energetically favourable because a strong C–Nu σ bond is formed at the expense of the weaker π bond in C=O (e.g. the mean bond enthalpy for a C–H σ bond is +412 kJ mol^{-1}, whereas a C=O π bond has a lower mean bond enthalpy of +384 kJ mol^{-1}). The net result is that Nu^- and H^+ add across the C=O bond.

For a series of related reactions with similar entropy changes, the numerical value and sign of the enthalpy change are important in helping to decide whether or not a reaction is feasible.

Reaction of the C=O bond with a neutral nucleophile

Reaction of the C=O bond with a neutral nucleophile (N̈uH), with at least one hydrogen atom bonded to a heteroatom, such as H_2O, ROH (an alcohol), RNH_2 (a primary amine), or RSH (a thiol), produces an intermediate containing both a negative and a positive charge (this is an example of a **zwitterion** (Section 19.2, p.896)).

The mechanism is shown in Figure 23.3. Transfer of a proton from Nu to O, in two subsequent steps, is facilitated by the nucleophile (or the solvent; see margin), to form a product containing an OH group. All three steps are reversible. An acid catalyst is often added to increase the rate of addition of a neutral nucleophile to the C=O bond (this is because protonation of the C=O bond forms a more reactive electrophile; see Section 23.2, p.1072). The net result is addition of NuH across the C=O bond.

Proton transfer steps Any base present in a reaction mixture can remove a proton, in step 2. Any acid present in a reaction mixture can donate a proton to the alkoxide ion in step 3. Sometimes, a solvent can facilitate proton transfer, for example, in an aqueous solvent, H_2O can act as a base and H_3O^+ can act as an acid.

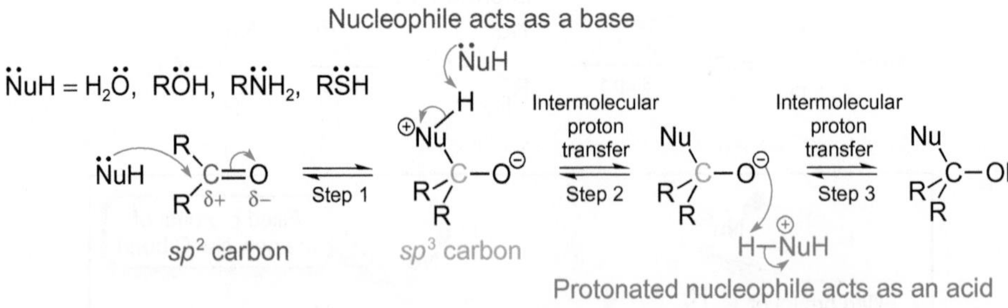
Nucleophile acts as a base
N̈uH = H₂Ö, RÖH, RN̈H₂, RS̈H
Intermolecular proton transfer
Intermolecular proton transfer
Step 1
Step 2
Step 3
δ+ δ−
sp² carbon
sp³ carbon
Protonated nucleophile acts as an acid

Figure 23.3 Nucleophilic addition to a ketone (or an aldehyde) with a neutral nucleophile.

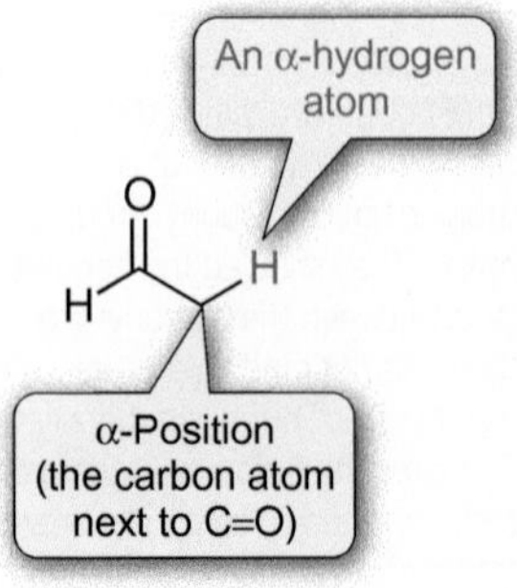
An α-hydrogen atom
α-Position (the carbon atom next to C=O)

Overview of α-substitution reactions

These reactions take place at a carbon atom next to the C=O bond—called the α-position. A hydrogen atom attached to the α-position is replaced by another group in a two-step

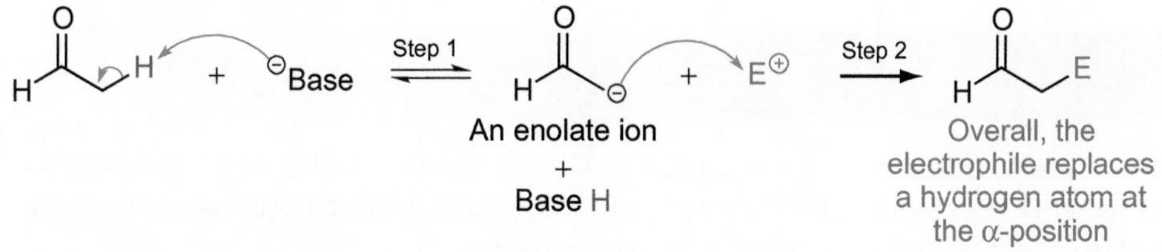
Step 1
Base
Step 2
An enolate ion
+
Base H
Overall, the electrophile replaces a hydrogen atom at the α-position

Figure 23.4 An α-substitution reaction of an aldehyde.

α-Substitution reactions are discussed in detail in Section 23.3 (p.1082).

reaction that occurs under acidic or, more usually, basic conditions. For example, on reaction with a base, an aldehyde or ketone is deprotonated at the α-position to form an enolate ion (see Figure 23.4). Typically, an α-hydrogen atom in an aldehyde or ketone has a pK_a value between 17 and 20. Enolate ions are strong nucleophiles that react with various electrophiles to introduce a new substituent at the α-position.

In an enolate ion, the negative charge is stabilized by delocalization, so α-hydrogen atoms in aldehydes and ketones are relatively acidic (see Section 19.1, p.876).

Overview of carbonyl-carbonyl condensation reactions

As the name suggests, these reactions involve two carbonyl compounds. The two carbonyl compounds react with one another so that one carbonyl compound undergoes an α-substitution reaction and the other undergoes a nucleophilic addition reaction.

This type of reaction is illustrated by the **aldol condensation reaction** of ethanal (CH_3CHO) shown in Figure 23.5. In the first step of the reaction, a base removes a proton from the α-position of ethanal to form an enolate ion ($^-CH_2CHO$). In step 2, the enolate ion reacts as a nucleophile and attacks the partially positive carbon atom of the $^{\delta+}C{=}O^{\delta-}$ group in a molecule of ethanal that has not been deprotonated. A new carbon–carbon bond is formed and the product is an alkoxide ion. In step 3, protonation of the alkoxide ion forms a product containing both a CHO (ald) and an OH (ol) functional group and the common name of this type of product is **aldol** (hence the reaction name). Notice that the base is regenerated in step 3 so it acts as a catalyst in the reaction.

Often, the aldol addition product is heated with acid or base (in step 4) to form a conjugated enal (the C=C (en) and C=O (al) bonds are conjugated) by elimination of water. The overall conversion of ethanal into the conjugated enal is an example of a **condensation reaction**.

Carbonyl-carbonyl condensation reactions are discussed in detail in Section 23.4 (p.1089).

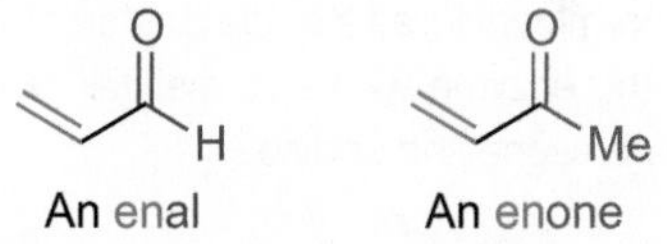
H
Me
An enal
An enone

Reactions like the aldol condensation reaction, that form new carbon–carbon bonds, are very useful in synthesis as they can convert reactants with smaller carbon chains into products with larger ones.

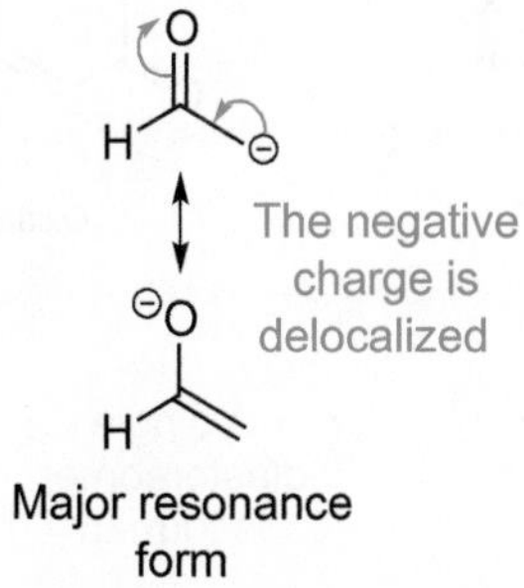
The negative charge is delocalized
Major resonance form

Enolate ions are stabilized by resonance—the resonance form with the negative charge on the electronegative oxygen is the more stable (Section 23.3, p.1083).

In a **condensation reaction**, two reactants combine to form a product with loss of a small molecule, typically water.

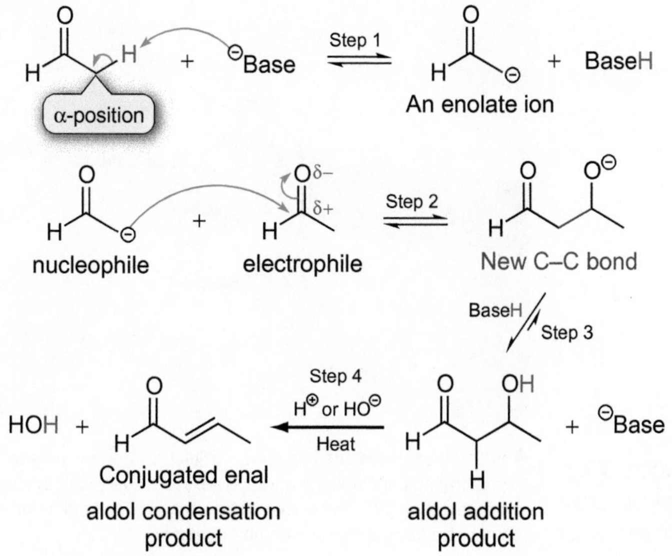
α-position
Base
Step 1
BaseH
An enolate ion
nucleophile
electrophile
Step 2
New C–C bond
BaseH
Step 3
Step 4
H⊕ or HO⊖
Heat
HOH +
Conjugated enal
aldol condensation product
aldol addition product
Base

Figure 23.5 An aldol condensation reaction.

Box 23.1 Cortisone and cortisol

Cortisone contains three C=O bonds. It belongs to a family of compounds called steroids that are produced in the body by the adrenal cortex, near the kidneys. Steroids contain one cyclopentane ring and three cyclohexane rings joined together (cholesterol is also a steroid; see Boxes 18.3, p.830, and 19.2, p.864). The production of cortisone helps to control the concentrations of fats, proteins, carbohydrates, and calcium ions in the body.

Many synthetic routes to cortisone and related compounds have been developed because of the important biological properties of these compounds. Cortisone injections, using synthetically produced cortisone, are used to reduce inflammation in a variety of ailments such as rheumatoid arthritis, tuberculosis, and severe asthma. Cortisone is also used in organ transplants to reduce the defence reaction of the body to foreign proteins in the implanted organ.

Within the body, cortisone is converted into cortisol by the enzyme 11-beta-hydroxysteroid dehydrogenase. This involves stereoselective reduction of the C=O bond at position 11 in cortisone to form cortisol as a single diastereomer. The reduction is reversible and the enzyme works to maintain a balance in the concentrations of cortisone and cortisol.

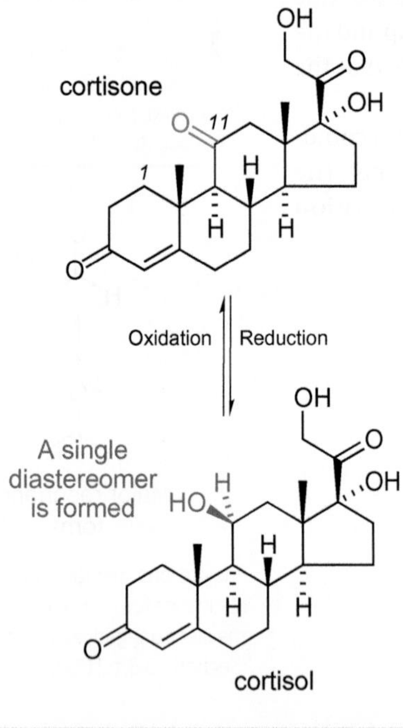
cortisone
OH
O
OH
O
11
1
H
H
H
O
Oxidation
Reduction
OH
O
A single
diastereomer
is formed
H
HO
OH
H
H
H
O
cortisol

Cortisol has many functions within the body, one of which is regulation of blood pressure. Your body also produces cortisol (from cortisone) in response to stressful situations, such as when you sit examinations. As the concentration of cortisol rises, the concentrations of sugars and fatty acids in the blood also rise and these are metabolized to produce energy.

▲ Cortisone injections are used by athletes such as tennis players to give short-term relief and reduce the swelling from inflammation of a joint or tendon. These injections are controversial because, although they can speed up recovery from an injury in the short term, they can leave lasting damage to joints and tendons.

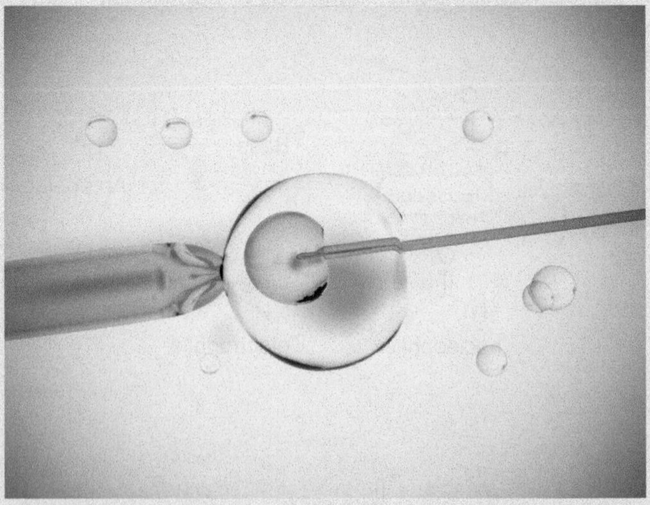

▲ **IVF babies**. Research has shown a link between the relative concentrations of cortisone and cortisol in the ovary with the success of *in vitro* fertilization (IVF). Patients with high ratios of cortisol : cortisone were found to be more likely to become pregnant by IVF.

➜

➜ Question

Triamcinolone acetonide is a synthetic steroid, used to help treat inflammation in patients suffering from asthma and arthritis.

(a) Which of the two C=O bonds in triamcinolone acetonide is not conjugated?

(b) For the C=O bond that is not conjugated, indicate the α-positions and determine the number of α-hydrogen atoms.

(c) Draw the structure of the product formed on oxidation of the secondary alcohol in triamcinolone acetonide.

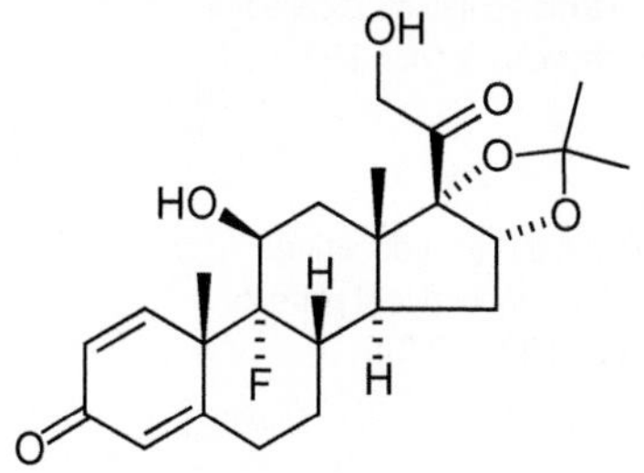
OH
O
O
O
HO
H
F
H
O

triamcinolone acetonide

Summary

- All aldehydes (RCHO) and ketones (RCOR) contain a C=O bond.
- The C=O bond is polar and the electrons are attracted to the oxygen atom ($^{\delta+}C{=}O^{\delta-}$)—the oxygen atom reacts with electrophiles and the carbon atom reacts with nucleophiles.

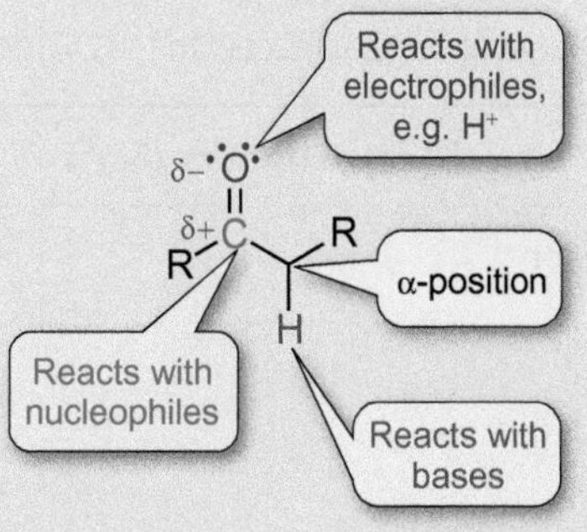
Reacts with electrophiles, e.g. H+
δ–
O
δ+
C
R
R
H
α-position
Reacts with nucleophiles
Reacts with bases

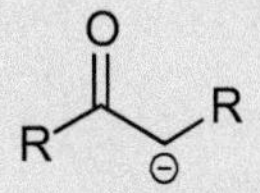
O
R
R

An enolate ion

- The position of a carbon atom next to C=O is called the α-position.
- Deprotonation of an aldehyde or ketone at the α-position forms an enolate ion.
- Aldehydes and ketones undergo nucleophilic addition reactions, α-substitution reactions, and carbonyl–carbonyl condensation reactions.

23.2 Nucleophilic addition reactions of aldehydes and ketones

Relative reactivity and types of nucleophiles

The C=O bonds in aldehydes (RCHO) and ketones (RCOR) react at different rates with nucleophiles in nucleophilic addition reactions. Usually, aldehydes react faster than ketones and this can be explained by steric and electronic factors. In ketones, the two alkyl or aryl groups (R) hinder the approach of the nucleophile and also reduce the size of partial positive charge on the carbon atom in the C=O bond (by exerting a +I effect), which makes the carbon atom less electrophilic.

When bulky groups lower the reactivity of a particular site within a molecule, this is called steric hindrance (steric strain) (Section 19.1, p.878).

Alkyl groups (R) are electron-donating and they exert a +I effect—inductive (I) effects are introduced in Section 19.1 (p.870).

Negatively charged nucleophiles usually react more rapidly with electrophiles than neutral nucleophiles (Section 19.1, p.880).

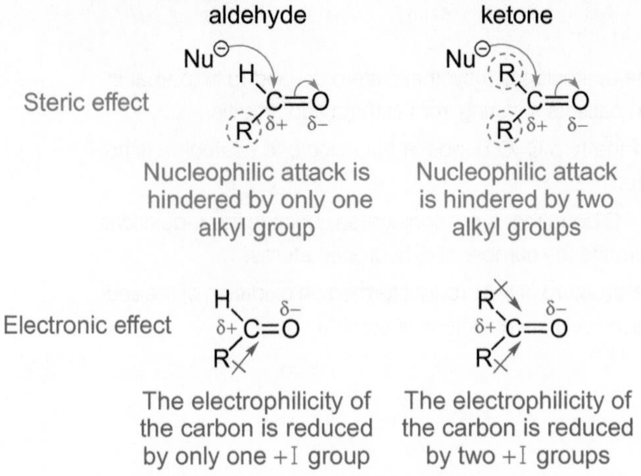
aldehyde
ketone
Steric effect
Nucleophilic attack is hindered by only one alkyl group
Nucleophilic attack is hindered by two alkyl groups
Electronic effect
The electrophilicity of the carbon is reduced by only one +I group
The electrophilicity of the carbon is reduced by two +I groups

Compounds containing the cyanide ion, such as sodium cyanide (NaCN), are fast-acting poisons that can be lethal. Apricot kernels, and seeds of other fruits, contain a compound called amygdalin. When consumed, digestive enzymes hydrolyse amygdalin, producing benzaldehyde (PhCHO), two equivalents of D-glucose and poisonous hydrogen cyanide (HCN). Fortunately, a significant number of fruit seeds are required to reach a lethal dose of HCN.

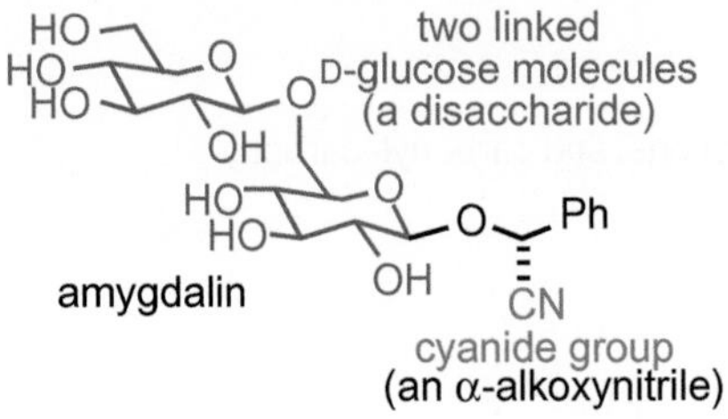
two linked D-glucose molecules (a disaccharide)
amygdalin
cyanide group (an α-alkoxynitrile)

Some common negatively charged (Nu^-) and neutral ($\ddot{N}uH$) nucleophiles that react with aldehydes and ketones are shown in Table 23.2. Reactions of aldehydes and ketones with each of these nucleophiles are discussed in this section, starting with negatively charged nucleophiles that add rapidly to the C=O bond. Later in this section, you will see that relatively weak nucleophiles, such as water and alcohols, react only slowly with the C=O bond and, in these cases, an acid catalyst is used to increase the rate of addition.

Table 23.2 Nucleophiles that react with the C=O bond in aldehydes and ketones

Negatively charged nucleophiles ($Nu^{\ominus}$)	Neutral nucleophiles ($\ddot{N}uH$)
$H^{\ominus}$, hydride ion	$H_2\ddot{O}$, water
$R^{\ominus}$, carbanion	$R\ddot{O}H$, alcohol
$NC^{\ominus}$, cyanide ion	$R\ddot{S}H$, thiol
$HO^{\ominus}$, hydroxide ion	$R\ddot{N}H_2$, primary amine
$RO^{\ominus}$, alkoxide ion	$R_2\ddot{N}H$, secondary amine

Reaction with hydrides

Aldehydes and ketones react with hydride ions (H^-), followed by H^+ (in a separate stage), to form alcohols (e.g. RCH_2OH). In these reactions the C=O bond is reduced.

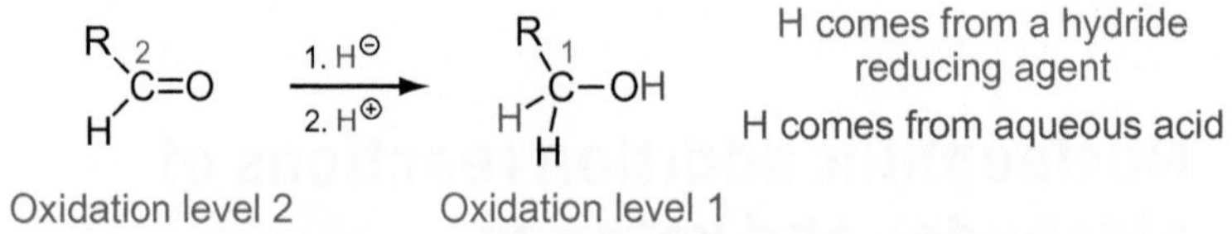
1. H⊖
2. H⊕
H comes from a hydride reducing agent
H comes from aqueous acid
Oxidation level 2
Oxidation level 1

Oxidation levels are used to assess if an organic reaction involves redox (Section 19.2, p.899). Organic oxidation and reduction reactions are summarized in Appendix 4 (p.1338).

The hydride ions can be derived from various reducing agents, as described in the following subsections.

Using complex metal hydrides

The most popular method of reducing aldehydes and ketones is to react them with a complex metal hydride, such as sodium borohydride ($NaBH_4$) or lithium aluminium hydride ($LiAlH_4$) followed by treatment with acid in a separate stage (called an 'acidic work-up', see Box 2.6, p.94). Reduction of aldehydes (RCHO) forms *primary* alcohols (RCH_2OH) while reduction of ketones (RCOR) produces *secondary* alcohols (R_2CHOH).

In the second step, aqueous acid is used so H_3O^+, rather than a free H^+ ion, is the likely proton donor. However, for simplicity, H^+ is often written in equations.

Sodium borohydride ($NaBH_4$) and lithium aluminium hydride ($LiAlH_4$) are also called sodium tetrahydridoborate(III) and lithium tetrahydridoaluminate(III), respectively. In $[BH_4]^-$ and $[AlH_4]^-$ there are four bonding electron pairs and the anions are tetrahedral in shape (Section 5.2, p.223).

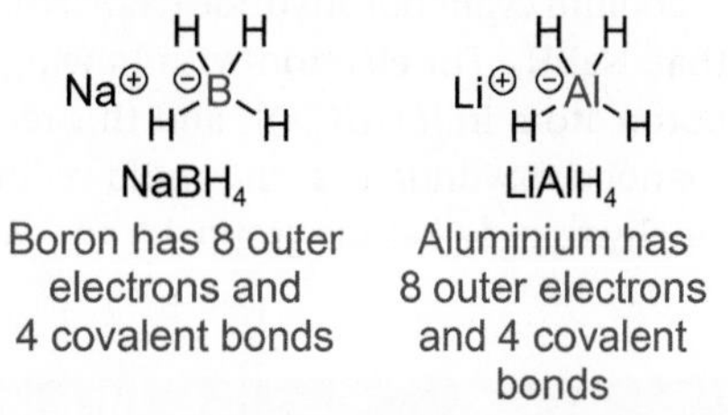
NaBH4
Boron has 8 outer electrons and 4 covalent bonds
LiAlH4
Aluminium has 8 outer electrons and 4 covalent bonds

In the Natrium car (made by DaimlerChrysler), hydrogen for the fuel cell is generated from $NaBH_4$. An aqueous solution of $NaBH_4$ is pumped through a chamber containing a catalyst to produce hydrogen at a rate equal to the demand of the fuel cell (see Box 25.3, p.1147).

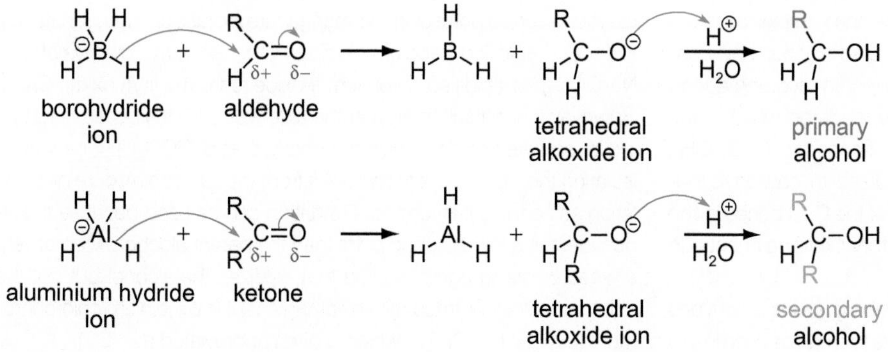
borohydride ion
aldehyde
tetrahedral alkoxide ion
primary alcohol
aluminium hydride ion
ketone
tetrahedral alkoxide ion
secondary alcohol

Figure 23.6 Reduction of aldehydes and ketones using $[BH_4]^-$ or $[AlH_4]^-$.

Sodium borohydride ($NaBH_4$) reduces aldehydes and ketones, but sodium hydride (NaH) does not. NaH acts as a base (Sections 19.2, p.895, and 25.2, p.1145), not a reducing agent. In NaH, the H^- ion is 'free' and as hydrogen does not effectively stabilize the negative charge it is a strong base. In comparison, in $NaBH_4$, the H atoms form covalent bonds to boron and this increases the nucleophilicity of the hydrogen atoms. The $^-BH_4$ ion is a stronger nucleophile than H^- because boron is larger than hydrogen and it is more polarizable (larger, more polarizable anions are stronger nucleophiles, see Section 19.1, p.881).

Interactive 3D animation of borohydride ion reduction of an aldehyde.

An electropositive element has a low electronegativity (Section 4.3, p.177).

The reaction mechanisms, shown in Figure 23.6, involve breaking a B–H or Al–H bond in the $[BH_4]^-$ or $[AlH_4]^-$ ion to donate a hydride ion (H^-) to the carbon atom of the C=O bond to form a tetrahedral alkoxide ion. (The negatively charged boron or aluminium atom does not attack the C=O bond—neither atom has a lone pair of electrons, so the electrons must come from a covalent B–H or Al–H bond.) Protonation of the alkoxide ion forms an alcohol. For clarity, the mechanisms shown in Figure 23.6 are simplified because they assume that BH_3 (borane) and AlH_3 (aluminium hydride) are products of the reactions, whereas these hydrides actually undergo further reactions. Also, for $NaBH_4$, the reaction mechanism can depend on the solvent (as discussed below).

$LiAlH_4$ is a much stronger reducing agent than $NaBH_4$ and it reacts more rapidly with aldehydes and ketones. The Al–H bonds in $LiAlH_4$ are more reactive (and weaker) than the B–H bonds in $NaBH_4$ because aluminium is more electropositive than boron. The electropositive aluminium atom in $[AlH_4]^-$ rapidly donates a hydride ion to the C=O bond and so loses its negative charge.

Because of the greater reactivity, $LiAlH_4$ must be handled more carefully than $NaBH_4$. Reactions using $LiAlH_4$ are carried out in dry aprotic solvents (such as dry diethyl ether, EtOEt) because it reacts violently with protic solvents, such as water and alcohols, to produce hydrogen gas. At the end of the reaction, when aqueous acid is added to protonate the intermediate alkoxide ion, the reaction mixture is cooled to 0 °C and the aqueous acid is added dropwise. In marked contrast, because $NaBH_4$ is significantly less reactive, reductions using $NaBH_4$ can use a protic solvent, such as an alcohol or even water!

Worked example 23.1 provides practice in drawing the mechanism of a reduction reaction using sodium borohydride.

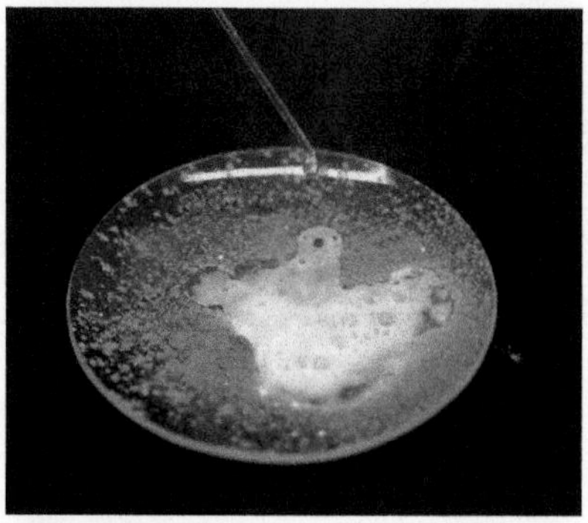

$LiAlH_4$ reacts violently with water, releasing hydrogen gas ($LiAlH_4 + 4\,H_2O \rightarrow LiOH + Al(OH)_3 + 4\,H_2$). The δ– hydrogen atoms in $LiAlH_4$ react with the δ+ hydrogen atoms in water, in an acid–base reaction.

When $NaBH_4$ reduces an aldehyde or ketone in a protic solvent, such as MeOH, the solvent may participate in the reduction. Under these conditions, the alcohol (rather than the alkoxide ion) is formed in the reaction mixture, so addition of H^+ (in a second stage) is not required.

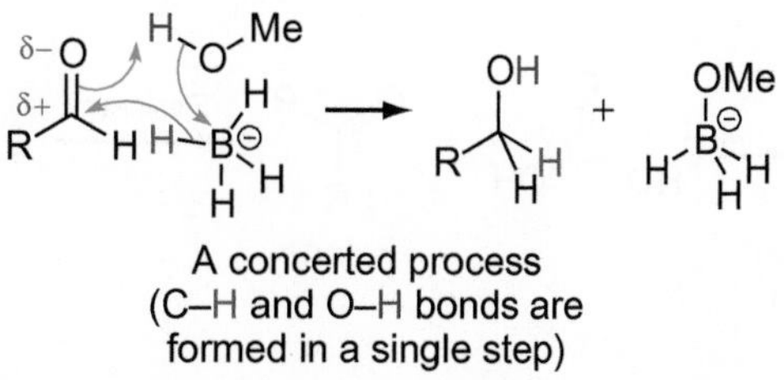
A concerted process
(C–H and O–H bonds are formed in a single step)

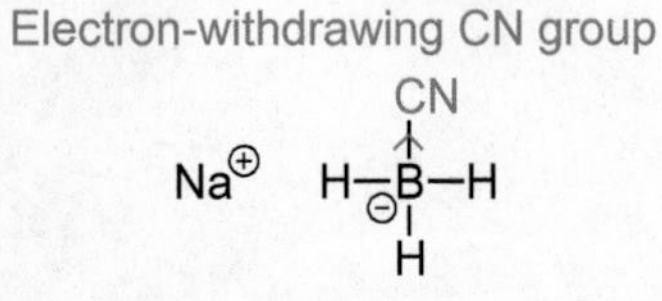
Electron-withdrawing CN group
CN
Na⊕ H–B–H
H
sodium cyanoborohydride

Sodium cyanoborohydride ($Na(CN)BH_3$) is another useful hydride, which is less reactive than $NaBH_4$. The electron-withdrawing CN group (–I) stabilizes the negative charge on the boron atom in $[BH_3(CN)]^-$ and this reduces the rate of transfer of a hydride ion. Sodium cyanoborohydride is such a mild reducing agent that, in molecules containing both an aldehyde and a ketone group, it can selectively reduce the (more reactive) aldehyde group.

Box 23.2 Oxidizing alcohols to carbonyls

Reduction of aldehydes and ketones using complex metal hydrides is not reversible. To do the reverse reaction and convert alcohols into aldehydes or ketones requires an oxidizing agent. Whether an aldehyde or ketone is formed depends on whether the alcohol being oxidized is primary (RCH_2OH) or secondary (R_2CHOH). Tertiary alcohols (R_3COH) are not easily oxidized to carbonyls because formation of a C=O bond requires one of the C–C bonds in the tertiary alcohol to be broken. Standard laboratory oxidizing agents do not efficiently do this.

Oxidation of *primary* alcohols (RCH_2OH) initially forms aldehydes (Figure 1). There are a number of oxidizing agents that can be used in this reaction such as potassium permanganate (potassium manganate(VII), $KMnO_4$), acidified sodium dichromate (sodium dichromate(VI), $Na_2Cr_2O_7$), or acidified chromium trioxide (chromium(VI) oxide, CrO_3). However, it is difficult to stop at the aldehyde (RCHO) because further oxidation takes place to give a carboxylic acid (RCO_2H). One way of isolating the aldehyde is to remove it from the reaction mixture by distillation as soon as it is formed. Distillation can be used because the aldehyde has a lower boiling point than the parent alcohol. Alternatively, a weak oxidizing agent is used that oxidizes the alcohol but not the aldehyde. One common mild oxidizing agent is pyridinium chlorochromate ($C_5H_5NH^+$ $ClCrO_3^-$, which is often abbreviated as PCC).

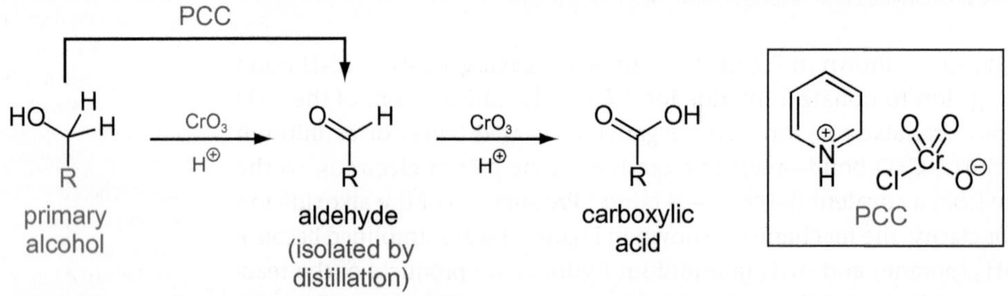
PCC
HO H H
R
primary alcohol
CrO_3
H⊕
O H
R
aldehyde (isolated by distillation)
CrO_3
H⊕
O OH
R
carboxylic acid
N H
Cl Cr O O O⊖
PCC

▲ **Figure 1** Oxidation of a primary alcohol.

Oxidation of *secondary* alcohols (R_2CHOH) using potassium permanganate ($KMnO_4$), acidified sodium dichromate ($Na_2Cr_2O_7$), or acidified chromium trioxide (CrO_3) forms ketones. Reactions using CrO_3 and H^+ are called Jones oxidations. The colour of these reactions changes from orange to green as the chromium is reduced from Cr(VI) to Cr(III). The mechanism of the reaction, shown in Figure 2, involves the formation of an intermediate chromate ester ($R_2CHOCrO_2OH$), which undergoes an elimination reaction to form the ketone and a Cr(IV) compound. During the reaction the Cr(IV) compound is subsequently reduced to green Cr(III). (See Appendix 4 (p.1338), for a table of common oxidizing and reducing agents used in organic reactions.)

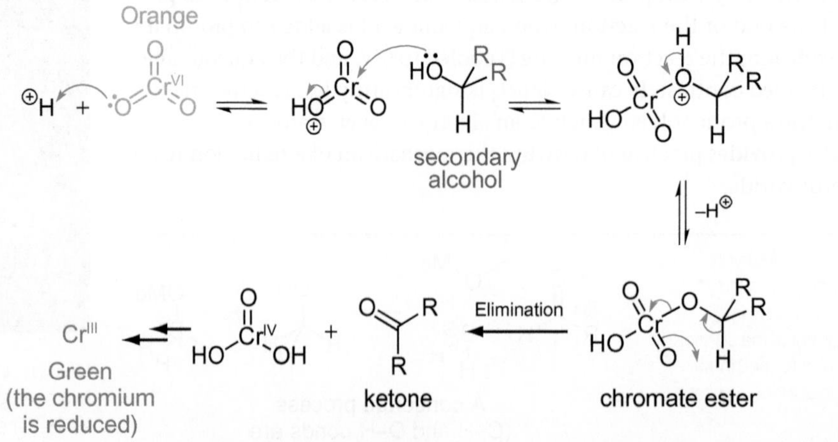
Orange
secondary alcohol
–H⊕
Elimination
Cr^{III}
Green (the chromium is reduced)
ketone
chromate ester

▲ **Figure 2** Oxidation of a secondary alcohol to a ketone using CrO_3 and H^+.

Interactive 3D animation of oxidation of a primary alcohol to an aldehyde using CrO_3 and H^+.

→ *Question*

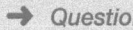

The following questions relate to the mechanism of oxidation of the secondary alcohol shown in Figure 2.

(a) What is the role of H^+ in the reaction?

(b) Formation of the ketone from the chromate ester may take place by a different elimination mechanism. If the chromate ester is protonated, it can react with water in an E2-type elimination to form the ketone (see Figure 3 below). Suggest a mechanism for this reaction. (E2 reactions are discussed in Section 20.4, p.943.)

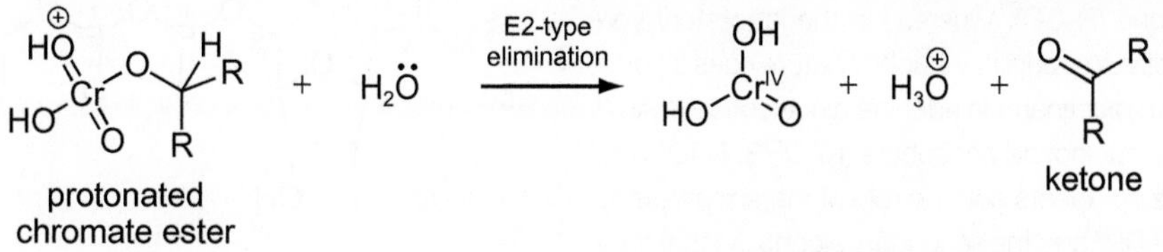

E2-type elimination
protonated chromate ester
ketone

▲ **Figure 3** Formation of a ketone from the protonated chromate ester.

Cannizzaro reaction

In the Cannizzaro reaction, the hydride ion (H^-) is generated from the aldehyde itself by reaction with hydroxide ions. The aldehyde must not contain any α-hydrogen atoms because the hydroxide ion must react with the C=O bond of the aldehyde in the first step. (Normally, the hydroxide ion is a better base than a nucleophile, and so it deprotonates α-hydrogen atoms from aldehydes to form enolate ions (p.1059).) The reaction scheme in Figure 23.7 shows the mechanism of the Cannizzaro reaction.

Methanal (HCHO) has no α-hydrogen atoms so, in step 1, the hydroxide ion attacks the carbon atom of the C=O bond to form a tetrahedral anion, $HOCH_2O^-$. The tetrahedral anion can either lose the hydroxide ion, to re-form starting materials, or transfer a hydride ion to reduce the C=O bond of another molecule of methanal. Transfer of the hydride ion forms the methoxide ion (HCH_2O^-) together with methanoic acid, HCO_2H (step 2). In the final step (step 3), a proton transfer takes place to give the methanoate ion (HCO_2^-) and methanol (HCH_2OH).

In this redox reaction, for every molecule of methanal that is reduced to methanol, a molecule of methanal is oxidized to the methanoate ion. The Cannizzaro reaction is an example of a **disproportionation** reaction.

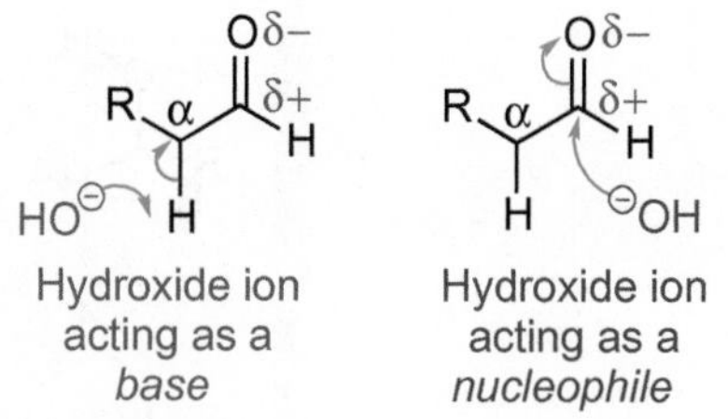

Hydroxide ion acting as a *base*
Hydroxide ion acting as a *nucleophile*

The Cannizzaro reaction is named after the Italian chemist Stanislao Cannizzaro who discovered the reaction in 1853.

A **disproportionation** reaction is a reaction where the starting material is simultaneously oxidized and reduced to form two different products (Section 26.2, p.1173).

Methanoic acid (formic acid, HCO_2H) has a low pK_a value of 3.75 and it is a much stronger acid than methanol, CH_3OH (pK_a 15.5), hence the proton transfer in Step 3.

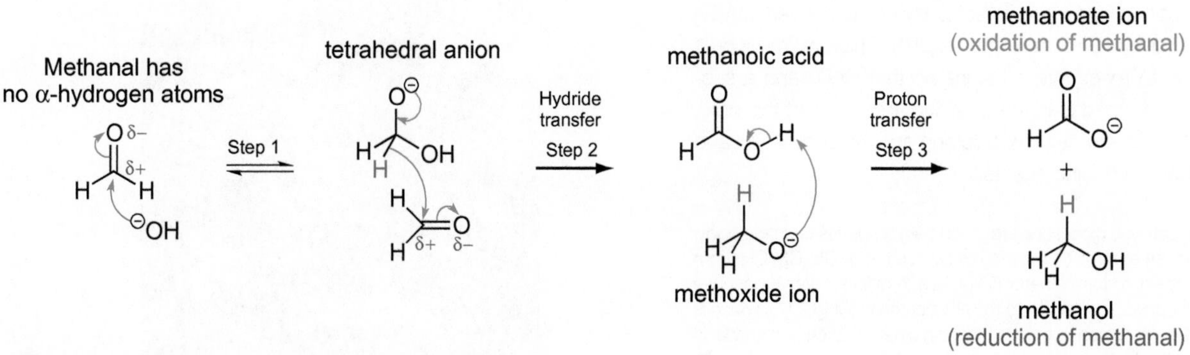

Methanal has no α-hydrogen atoms
tetrahedral anion
methanoic acid
methanoate ion (oxidation of methanal)
Step 1
Hydride transfer
Step 2
Proton transfer
Step 3
methoxide ion
methanol (reduction of methanal)

Figure 23.7 Mechanism of the Cannizzaro reaction.

Box 23.3 Hydride transfer in nature

Redox reactions are important in nature, as well as in the laboratory. For example, in the citric acid cycle (see the introduction to Chapter 21, p.961), which is involved in the metabolism of glucose, two of the nine steps involve oxidation of a secondary alcohol (R_2CHOH) to form a ketone (R_2CO). Whereas in the laboratory you might use CrO_3/H^+ to oxidize a secondary alcohol, nature uses a combination of an enzyme with nicotinamide adenine dinucleotide (NAD^+) or nicotinamide adenine dinucleotide phosphate ($NADP^+$). NAD^+ and $NADP^+$ are nature's oxidizing agents and the role of the enzyme is to bring together NAD^+ or $NADP^+$ and the secondary alcohol so that they react rapidly at room temperature. The enzyme-catalysed oxidation of an alcohol is much faster and usually takes place under milder conditions than with the conventional oxidants used in the laboratory.

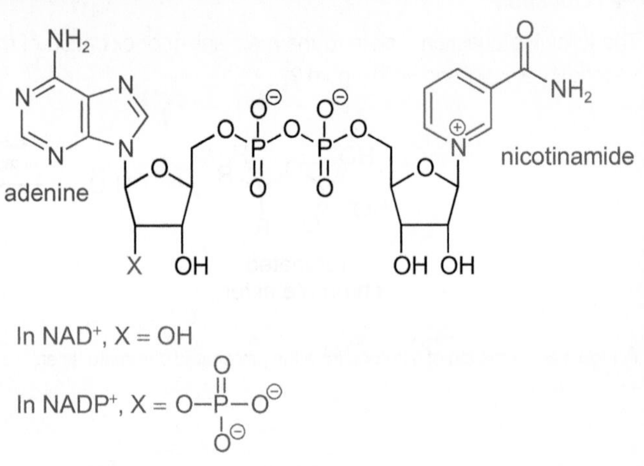
adenine
nicotinamide
In NAD+, X = OH
In NADP+, X = O–P(O)(O⁻)–O⁻

Figure 1 The important role of the enzyme in the oxidation of a secondary alcohol to give a ketone.

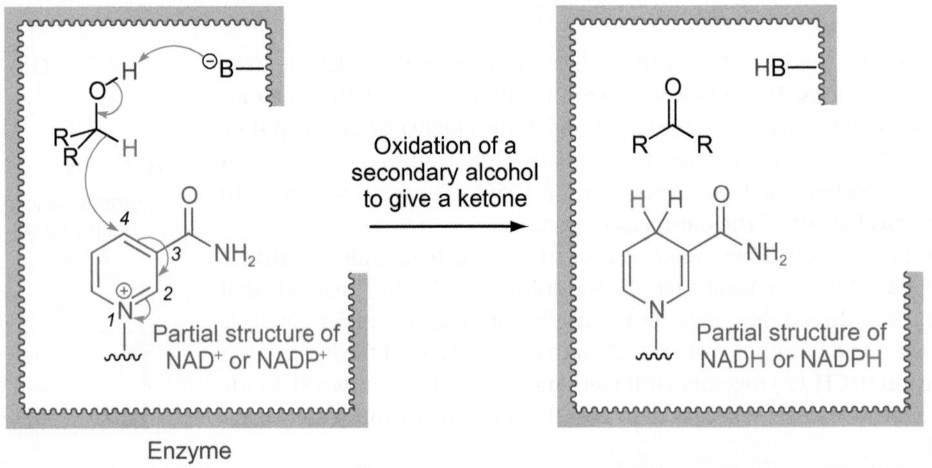
Oxidation of a secondary alcohol to give a ketone
Partial structure of NAD+ or NADP+
Partial structure of NADH or NADPH
Enzyme

The important role of the enzyme in the oxidation of an alcohol is shown in Figure 1. A basic group (B^-) within the active site of the enzyme deprotonates the OH group of the alcohol and this leads to the transfer of a hydride ion from the alcohol to the 4-position of the nicotinamide ring (compare this to oxidation using CrO_3 and H^+ on p.1064). This converts the secondary alcohol into a ketone. Addition of the hydride ion to NAD^+ (or $NADP^+$) neutralizes the positively charged nitrogen atom in the ring to form NADH (or NADPH), for example, $NAD^+ + H^- \rightarrow NADH$.

Interestingly, nature uses a different enzyme to convert NADH back into NAD^+ (or NADPH back into $NADP^+$). This enzyme works by forcing NADH to react with a ketone so that NAD^+ and a secondary alcohol are formed (see the Question at the end of this box). So, NAD^+ and $NADP^+$ are nature's oxidizing agents, and NADH and NADPH are nature's reducing agents.

Enzymes that catalyse oxidations are called *dehydrogenases*. In the body, a dehydrogenase enzyme (in the liver) oxidizes ethanol (CH_3CH_2OH) from alcoholic drinks to form ethanal (CH_3CHO). A different dehydrogenase enzyme then converts ethanal into the ethanoate ion ($CH_3CO_2^-$), which is further metabolized. Research suggests that high concentrations of ethanal in the body is one of the factors responsible for the symptoms of a hangover. ➤

➜

→ *Question*

In the reaction shown here, use curly arrows to show where the electrons move when NADH or NADPH reduces the ketone.

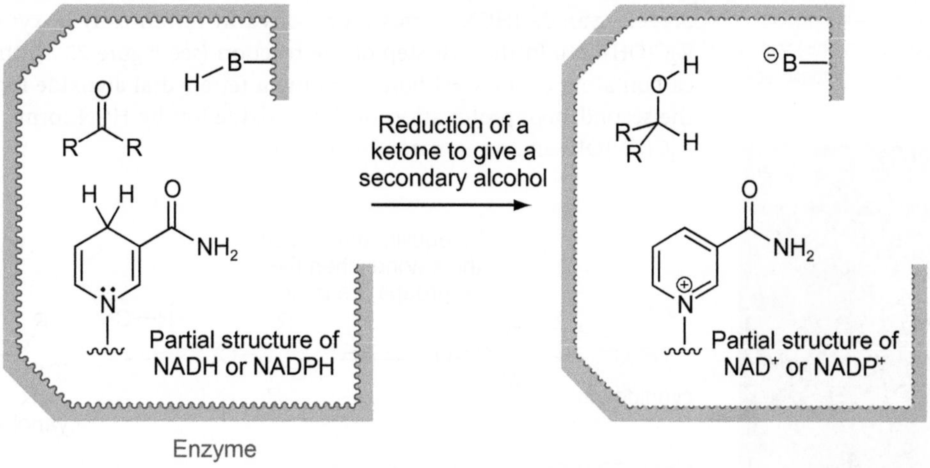
Reduction of a ketone to give a secondary alcohol
Partial structure of NADH or NADPH
Partial structure of NAD⁺ or NADP⁺
Enzyme

Worked example 23.1 Reduction of an aldehyde

The organic product of the reaction of butanal with $NaBH_4$ followed by aqueous acid is butan-1-ol.

(a) Give a mechanism for this reaction.

(b) Suggest a method for converting butan-1-ol back into butanal.

Strategy

(a) Draw the skeletal structures of butanal and butan-1-ol.

For the C=O bond in butanal, show δ+ on carbon and δ– on oxygen.

Recognize that $[BH_4]^-$ donates an H^- ion to the carbon atom of the C=O bond to form an alkoxide ion. Protonation of the alkoxide ion gives the alcohol. The overall process is a nucleophilic addition to the C=O bond.

(b) Recognize that converting butan-1-ol into butanal requires an oxidizing agent (use oxidation levels to work this out, if this is not obvious (p.899)).

Recognize that 'over-oxidation' of the aldehyde to the carboxylic acid is possible.

Solution

(a)

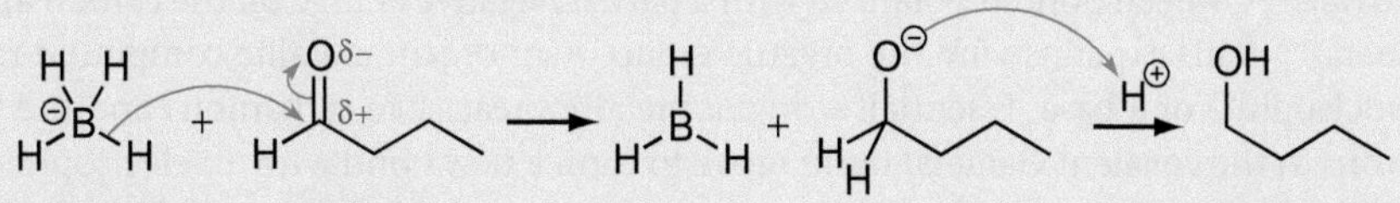
borohydride ion
butanal
butan-1-ol

(b) React butan-1-ol with CrO_3 and H^+ (Jones oxidation) and remove the butanal from the reaction mixture by distillation as soon as it is formed. (Alternatively, react butan-1-ol with a milder oxidizing agent such as pyridinium chlorochromate (PCC).)

Question

The organic product of the reaction of butanone (or butan-2-one) with $LiAlH_4$ followed by aqueous acid is butan-2-ol. Give a mechanism for this reaction.

Butan-1-ol ($CH_3CH_2CH_2CH_2OH$) has been proposed as a 'green' transportation fuel to substitute for diesel and petrol. Small quantities of butan-1-ol are produced in most bacterial fermentations and current research is genetically altering bacteria so that butan-1-ol can be produced from glucose on an industrial scale.

Hydrogen cyanide (HCN) is a toxic gas. It is usually generated as needed by reacting HCl with NaCN.

Steric strain is the repulsive interaction between two atoms (or groups of atoms) that are closer to one another than their atomic radii allow (Section 18.2, p.824).

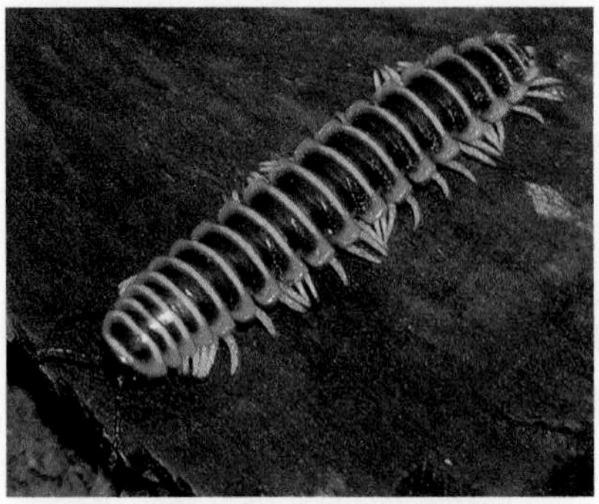

A cyanohydrin (called mandelonitrile, PhCH(OH)-CN) protects the millipede *Apheloria corrugata*. When attacked by ants the millipede secretes the cyanohydrin, together with an enzyme, through pores in its body. The enzyme converts the cyanohydrin into an aldehyde (PhCHO) and HCN, and it is the poisonous HCN that deters the attackers.

Interactive 3D animation showing formation of a cyanohydrin.

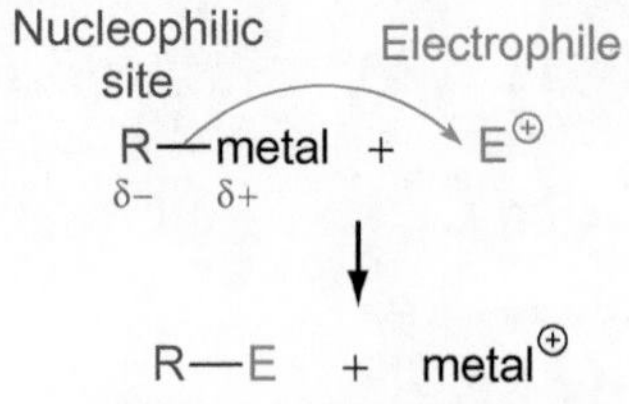
Nucleophilic site
Electrophile
R—metal + E⊕
δ– δ+
R—E + metal⊕

The synthesis and structures of organolithium compounds (RLi) and organomagnesium compounds (R_2Mg and RMgX) are discussed in Section 26.10 (p.1195).

Usually, organolithium (RLi) and organomagnesium (e.g. RMgX) compounds have very polar covalent $^{\delta-}C{-}Li^{\delta+}$ and $^{\delta-}C{-}Mg^{\delta+}$ bonds, whereas organosodium (RNa) compounds are essentially ionic.

Alkynyl anions, $RC{\equiv}C^-$, are also called alkynide anions, so, for example, $HC{\equiv}C^-$ may be called the ethynyl ion or the ethynide ion. (An older name is the acetylide ion.) (Section 26.2, p.1177).

Organolithium compounds (RLi) are particularly strong bases that are commonly used in synthesis (for an introduction to organic bases see Section 19.2, p.891). For example, *tert*-butyllithium ($LiC(CH_3)_3$, *t*-BuLi) has an extremely high pK_a value of 53.

Reaction with carbon nucleophiles

Reaction with cyanide to form cyanohydrins

Reaction of an aldehyde or ketone with a mixture of sodium cyanide (NaCN) and hydrogen cyanide (HCN) forms a cyanohydrin (or an α-hydroxynitrile, RCH(OH)CN or $R_2C(OH)CN$). In the first step of the reaction (see Figure 23.8), the ^-CN ion attacks the carbon atom of the C=O bond forming a tetrahedral alkoxide ion (e.g. $R_2C(CN)O^-$). In the second step, protonation of the alkoxide ion by HCN forms the cyanohydrin (e.g. $R_2C(CN)OH$) and regenerates the ^-CN ion.

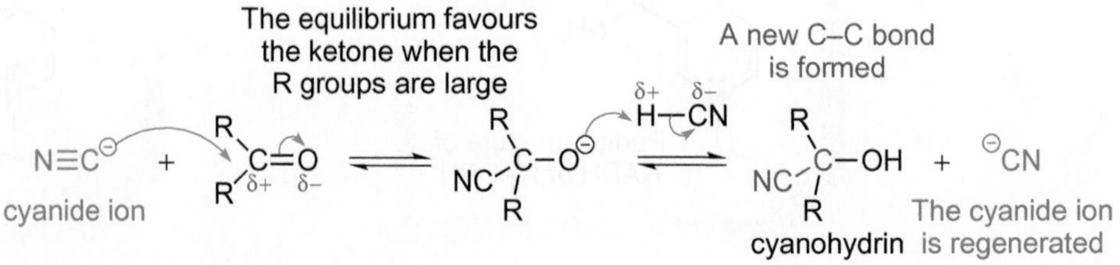
The equilibrium favours the ketone when the R groups are large
A new C–C bond is formed
N≡C⊖ + R2C=O ⇌ NC–CR2–O⊖ ⇌ NC–CR2–OH + ⊖CN
δ+ δ–
cyanide ion
cyanohydrin
The cyanide ion is regenerated

Figure 23.8 Formation of a cyanohydrin.

Formation of the cyanohydrin is reversible and the position of equilibrium depends on the size of the R groups attached to the C=O bond.

- For *ketones* (RCOR) with large R groups, the equilibrium favours the ketone rather than the cyanohydrin ($R_2C(OH)CN$). In the ketone, the two R groups are separated by a bond angle of 120°, but this angle is reduced to 109° in the cyanohydrin and the R groups are pushed closer together. If the R groups are large enough, the resulting steric strain causes the equilibrium to favour the ketone.
- For *aldehydes* (RCHO), the equilibrium generally favours the cyanohydrin. The steric strain in a cyanohydrin (RCH(OH)CN) formed from an aldehyde is less than that in a cyanohydrin formed from a ketone because a very small hydrogen atom replaces one of the R substituents.

Cyanohydrins are useful intermediates in synthesis because a new C–C bond is formed and the CN group can then be converted into a CO_2H group (Box 24.7, p.1125).

Reaction with organometallics to form alcohols

Organometallics are a large class of compounds that contain an organic group bonded to a metal (R–metal, where R is an alkyl, vinyl, or aryl group). Whether the C–metal bond is ionic or covalent depends on the electropositivity of the metal (Section 26.10, p.1195) and the structure of the organic group. For example, carbon forms covalent bonds to lithium and magnesium, but an ionic bond to sodium. Metals are more electropositive than carbon, so a covalent C–metal bond is polarized with a partial negative charge on the carbon atom ($^{\delta-}$C–metal$^{\delta+}$). This explains why the organic group of an organometallic compound reacts as a nucleophile or a base. Essentially, organometallics react like carbanions and the two electrons in the covalent C–metal bond move to form a new bond with an electrophile or H^+.

Common organometallic reagents used in synthesis include organolithium compounds (R–Li), Grignard reagents (R–MgX, where X = Cl, Br, or I), and alkynylsodium salts ($RC{\equiv}C^-$ Na^+). These compounds are prepared using the reactions shown in Figure 23.9.

These organometallics are extremely reactive compounds and they all react with water in acid–base reactions. The organometallic reagents are strong bases and deprotonate even a very weak acid like water (which has a pK_a of 15.7) to form alkanes.

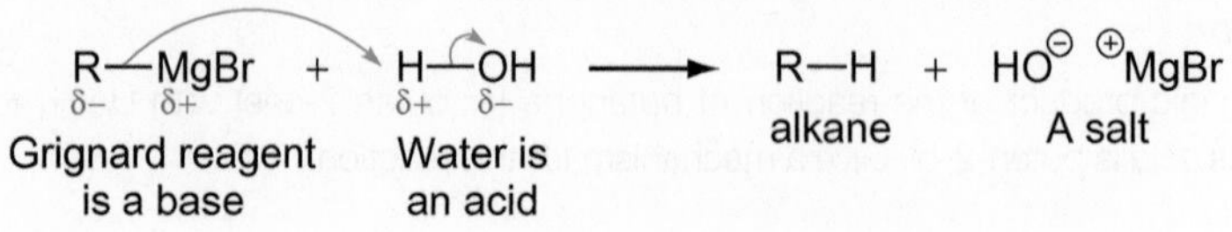
R—MgBr + H—OH ⟶ R—H + HO⊖ ⊕MgBr
δ– δ+ δ+ δ–
Grignard reagent is a base
Water is an acid
alkane
A salt

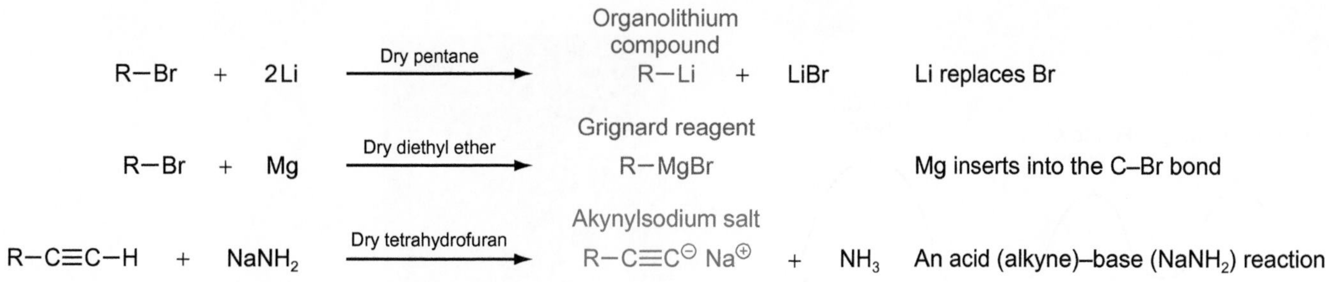
Organolithium compound
R—Br + 2Li
Dry pentane
R—Li + LiBr
Li replaces Br
Grignard reagent
R—Br + Mg
Dry diethyl ether
R—MgBr
Mg inserts into the C–Br bond
Akynylsodium salt
R—C≡C—H + NaNH2
Dry tetrahydrofuran
R—C≡C⊖ Na⊕ + NH3
An acid (alkyne)–base (NaNH2) reaction

Figure 23.9 Preparation of some organometallic reagents used in synthesis.

If alkanes are not required, the organometallics must be prepared and undergo reactions in a nitrogen atmosphere (to exclude water vapour) using 'dry' solvents (solvents that contain little, if any, water).

Organometallic reagents react with aldehydes and ketones to form alcohols. The organic group acts as a nucleophile and attacks the carbon atom of the C=O bond to form an alkoxide ion. The alkoxide ion then reacts with H^+ in a separate second stage. These reactions are very useful in synthesis because a new C–C bond is formed and the –OH group can be further modified. The reaction schemes in Figure 23.10 show a Grignard reagent (RMgBr) reacting with methanal, an aldehyde, and a ketone to give a primary alcohol (RCH_2OH), a secondary alcohol (RR^1CHOH), and a tertiary alcohol (RR^1R^2COH), respectively—notice that the mechanisms of these reactions are identical. In all three reactions, the $[MgBr]^+$ ion reacts with water to produce HOMgBr. However, HOMgBr may react further, depending on the concentration of the aqueous acid, to form water and hydrated Mg^{2+} and Br^- ions.

Worked example 23.2 discusses the mechanism of the reaction of an organolithium compound with an aldehyde.

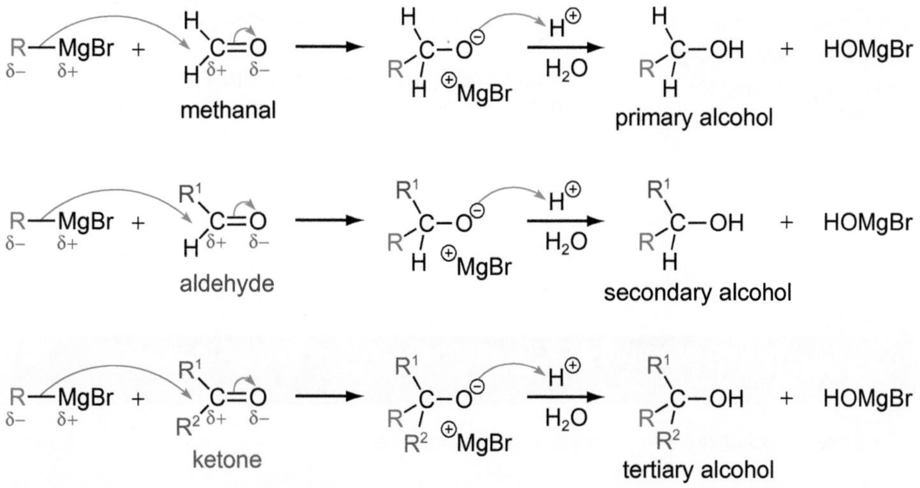
methanal
primary alcohol
aldehyde
secondary alcohol
ketone
tertiary alcohol
HOMgBr

Figure 23.10 Reactions of a Grignard reagent with methanal, an aldehyde, and a ketone.

Grignard reagents are named after the French chemist François Auguste Victor Grignard. He was awarded the Nobel Prize in Chemistry in 1912 for the discovery of these reagents. Although Grignard reagents are normally represented by the formula RMgX, the actual structures are more complex (Section 26.10, p.1195).

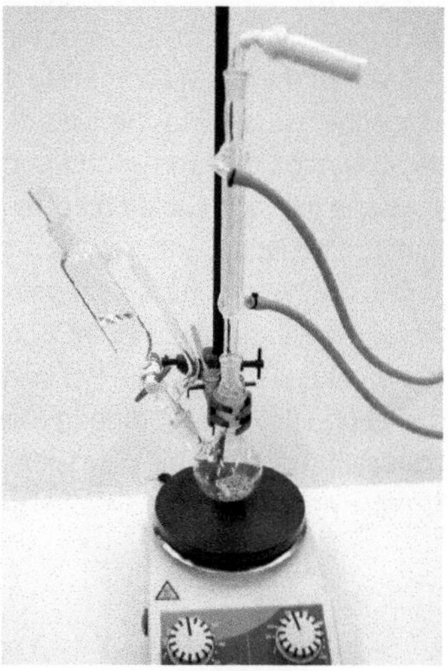

An experimental set-up for preparing a Grignard reagent (the Grignard reagent is prepared and then used immediately). You will see a drying tube (containing calcium chloride as the desiccant) is inserted at the top of the condenser to help keep atmospheric moisture out of the apparatus.

Interactive 3D animation showing nucleophilic addition of an organolithium and a Grignard reagent to an aldehyde.

Box 23.4 A short history of organometallics

Like many discoveries in science, the birth of organometallics started by accident. The story starts in a French pharmacy in 1760, when Louis Claude Cadet de Gassicourt accidentally prepared a vile smelling liquid, known as 'Cadet's fuming liquid'. Later, Robert Bunsen (better known for his invention of the Bunsen burner) showed that Cadet had prepared a mixture of tetramethyldiarsine ($Me_2As–AsMe_2$) and the oxide $Me_2AsOAsMe_2$, the first synthetic organometallic compounds. However, the true founder of organometallic chemistry was a student of Bunsen's called Edward Frankland. In 1849, while trying to make ethyl radicals, Frankland reacted iodoethane (Et–I) with zinc, but instead of making ethyl radicals he prepared diethylzinc (Et_2Zn). Starting from this accidental discovery, Frankland prepared other dialkylzincs and began investigating their use in synthesis. →

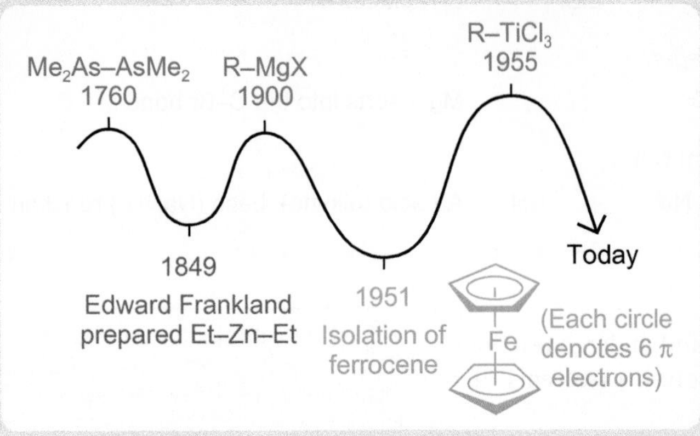

$Me_2As–AsMe_2$
1760
R–MgX
1900
$R–TiCl_3$
1955
1849
Edward Frankland prepared Et–Zn–Et
1951
Isolation of ferrocene
Fe
(Each circle denotes 6 π electrons)
Today

◄ Developed in the 1950s, Ziegler–Natta catalysts are used to make the poly(propene) in plastic containers, carpeting, and ropes. The world production of poly(propene) is in excess of 52 million tonnes per year.

The prominence of dialkylzincs (R_2Zn) as reagents in synthesis lasted until 1900, when Victor Grignard introduced Grignard reagents (RMgX). This discovery led to the preparation of other organometallics containing a σ bond between a main group metal and carbon.

The next major development took place in 1951, with the isolation of ferrocene ($Fe(C_5H_5)_2$), a kind of molecular sandwich. Ferrocene demonstrated that iron can form bonds to π electrons in aromatic rings and this kick-started a tremendous interest in organotransition metal compounds, which continues today.

One of the important achievements of this work has been the production of industrially useful catalysts. For example, in Ziegler–Natta catalysts, titanium(IV) chloride ($TiCl_4$) and a trialkylaluminium compound (R_3Al) react to form alkyltitanium compounds (with Ti–C bonds) that catalyse the polymerization of propene ($CH_3CH{=}CH_2$) to form poly(propene) (or polypropylene, $[–CH(CH_3)–CH_2–]_n$).

Question

In poly(propene), the methyl groups on the carbon chain can have different relative orientations. In *isotactic* poly(propene), all of the methyl groups are on the same side of the carbon chain; in *syndiotactic* poly(propene), the positions of the methyl groups alternate; whereas, in *atactic* poly(propene), the positions of the methyl groups are random. Given the partial structure of atactic poly(propene) shown here, draw partial structures of both isotactic and syndiotactic poly(propene).

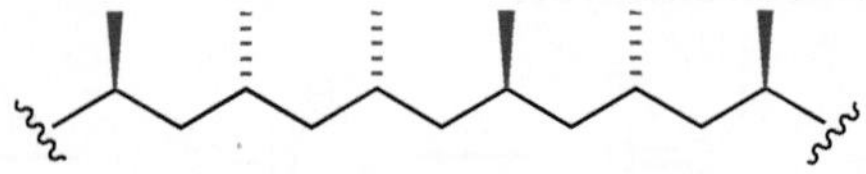

Atactic poly(propene)—the methyl groups are on random sides of the carbon backbone

Worked example 23.2 Reaction of carbonyl compounds with organometallics

Visit the Online Resource Centre to view screencast 23.1 which walks you through the reaction of various aldehydes and ketones with different organometallics, to form alcohols.

Propose a mechanism to explain the reaction shown here.

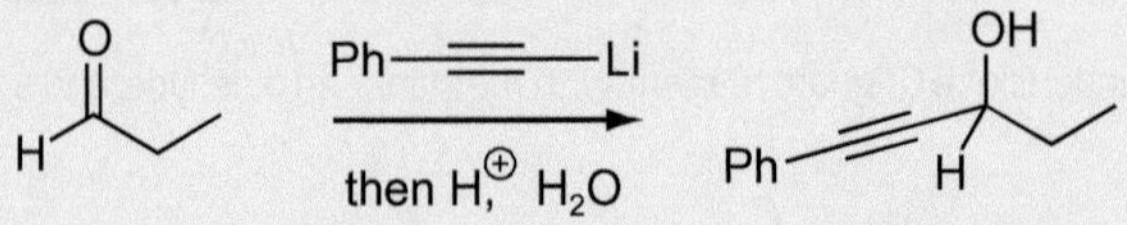

O
H
Ph—≡—Li
then $H^{\oplus}$, H_2O
OH
Ph
H

Strategy

For the C=O bond in propanal, show δ+ on carbon and δ– on oxygen.

For the C–Li bond in the alkynyllithium, show δ+ on lithium and δ– on carbon.

Recognize that the alkynyllithium reacts with the C=O bond in a nucleophilic addition reaction.

→ Solution

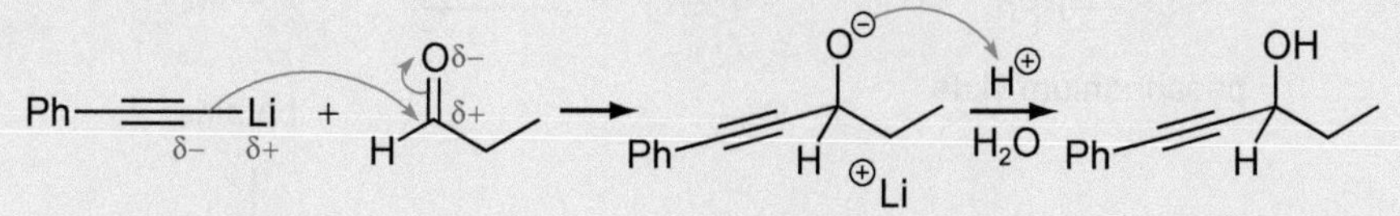

Question

Propose a mechanism to explain the following reaction.

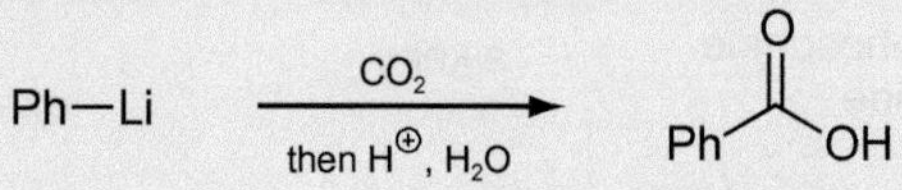

Organolithium reagents (RLi) are corrosive, flammable, and in some cases pyrophoric (they can ignite spontaneously on contact with air). Solutions of organolithiums in dry solvents (such as anhydrous diethyl ether) can be transferred from storage bottles to reaction flasks, while constantly under an inert atmosphere, using a syringe or a hollow flexible tube called a cannula.

The Wittig reaction

Aldehydes and ketones react with **phosphonium ylides** (ylids) to form alkenes in the **Wittig reaction** (after the German chemist Georg Wittig). Phosphonium ylides have a positively charged phosphorus atom and an adjacent negatively charged carbon atom (e.g. $R_3P^+–^-CR_2$, $R_3P^+–^-CHR$, or $R_3P^+–^-CH_2$). A resonance form of the ylide has a C=P bond and this is called a phosphorane (see margin; notice that in this resonance form, phosphorus has 10 valence electrons).

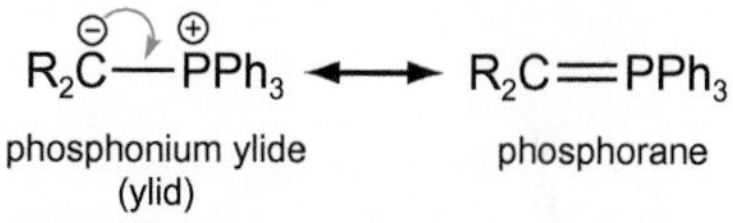
phosphonium ylide (ylid)
phosphorane

Phosphonium ylides are prepared in two steps from primary or secondary halogenoalkanes (RCH_2X or R_2CHX, respectively). In the first step, reaction of the halogenoalkane with a trialkyl- or triaryl phosphine (PR_3), typically triphenylphosphine (PPh_3), which is a strong nucleophile, produces a **phosphonium** salt in an S_N2 reaction (Section 20.3, p.930). In the phosphonium salt (e.g. $Ph_3P^+–CHR_2\ X^-$), the positively charged phosphorus atom is electron-withdrawing and this makes the hydrogen atom on the adjacent carbon relatively acidic. Treatment with a strong base, such as *n*-butyllithium (*n*-BuLi or $CH_3CH_2CH_2CH_2Li$; butane has a very high pK_a value of around 50), removes H^+ to form the phosphonium ylide. Some phosphonium ylides can be isolated, although they are generally formed and used immediately in a reaction.

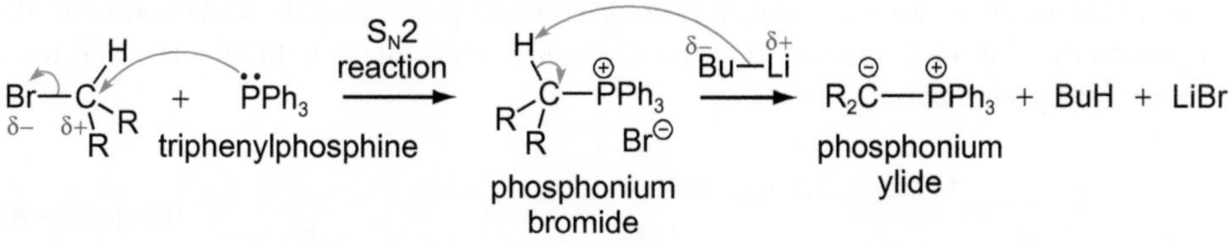
S_N2 reaction
triphenylphosphine
phosphonium bromide
phosphonium ylide

Phosphorus, like nitrogen, has five electrons in its valence shell and so trialkyl- and triaryl phosphines have a lone pair on phosphorus (R_3P:). Triphenylphosphine (Ph_3P:) is a stronger nucleophile than triphenylamine (Ph_3N:) as the phosphorus lone pair is stabilized less effectively by resonance with the benzene rings than the nitrogen lone pair. This can be explained by less efficient overlap between the lone pair orbital on the large phosphorus atom with the adjacent *p* orbital on a smaller carbon atom.

The acidity of the hydrogen atom in $Ph_3P^+–CHR_2$ is affected by the nature of the R groups. When the R groups are electron-donating (+I), the pK_a of the hydrogen atom is relatively high (20–25) and so a strong base is required. However, if one or both R groups are electron-withdrawing, then the pK_a value of the hydrogen atom drops and a weaker base can be used (R groups that are −I/−M can stabilize the negative charge in the phosphonium ylide). For example, sodium ethoxide (NaOEt; ethanol has a pK_a value of around 16) can be used to deprotonate $Ph_3P^+–CH_2CO_2Et$.

The mechanism of the Wittig reaction, shown in Figure 23.11, involves reaction of the negatively charged carbon atom in the ylide (e.g. $R_2C^-–^+PPh_3$) with the carbon atom of the C=O bond (e.g. R^1CHO) to form a **betaine**. The betaine then reacts to form an unstable 4-membered ring called an **oxaphosphetane**. In the final step, the oxaphosphetane ring breaks open to form an alkene ($R_2C=CHR^1$) and triphenylphosphine oxide (Ph_3PO). The P=O bond in triphenylphosphine oxide is extremely strong (around +575 kJ mol^{-1}) and the formation of this bond drives the reaction forward. The overall effect of the Wittig reaction is to replace the oxygen atom in the carbonyl compound by the CR_2 group in the phosphonium ylide.

The Wittig reaction is a popular method for preparing alkenes. Usually, mono-, di-, and trisubstituted alkenes ($RCH=CH_2$, $R_2C=CH_2$, $RCH=CHR$, and $R_2C=CHR$) can all be prepared in good yield (Section 21.2, p.968).

A **betaine** is an organic intermediate containing both a positively charged functional group, which does not contain a hydrogen atom(s) (e.g. a phosphonium ion, R_4P^+), and a negatively charged functional group (e.g. an alkoxide ion, RO^-).

Interactive 3D animation showing the mechanism of the Wittig reaction.

Recent research suggests that the oxaphosphetane is actually formed in a concerted reaction (the C–C and P–O bonds form at the same time), although the exact mechanism is still unclear.

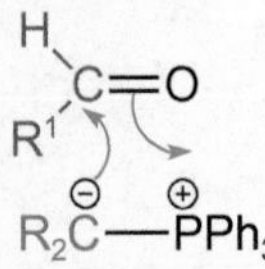

Concerted addition of the ylide to the C=O bond

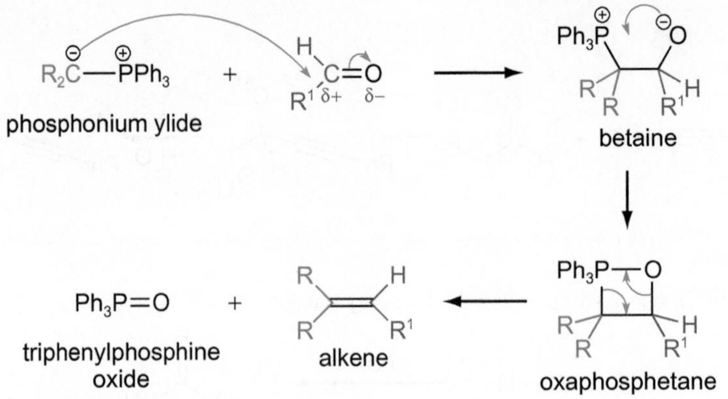
phosphonium ylide
betaine
triphenylphosphine oxide
alkene
oxaphosphetane

Figure 23.11 Mechanism of the Wittig reaction.

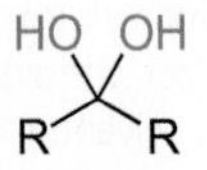

A hydrate

A hydrate is a 1,1-diol sometimes called a *gem*-diol

Reaction with oxygen nucleophiles

Addition of water to form hydrates

Water adds reversibly to the C=O bond of aldehydes and ketones to form **hydrates** ($RCH(OH)_2$ or $R_2C(OH)_2$). A lone pair of electrons on the oxygen atom in water attacks the carbon atom in the C=O bond to form an intermediate containing both a positive and a negative charge (e.g. $RCH(O^-)O^+H_2$). Deprotonation of the positively charged oxygen atom (using H_2O as the base), followed by protonation of the negatively charged oxygen atom (using H_3O^+ as the acid), forms the hydrate.

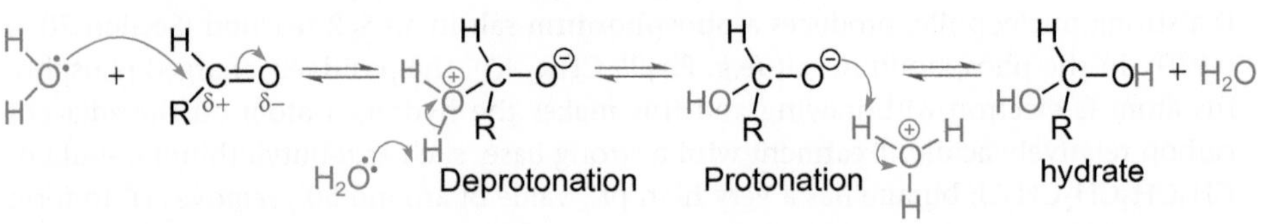
All steps are reversible
Deprotonation
Protonation
hydrate

Proton transfer steps

You will see that formation of the hydrate involves proton transfers (in two of the three steps, an H^+ ion changes position). Proton transfers between oxygen atoms are fast and reversible. Sometimes, in step 2 of the reaction mechanism, an intramolecular proton transfer is drawn, in which the negatively charged oxygen acts as the base, as shown below.

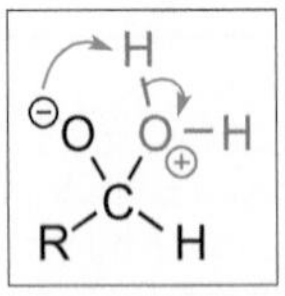

However, this might not occur because the negatively charged oxygen may not be close enough to react with one of the protons on the positively charged oxygen. Alternatively, as the solvent (water) surrounds the reactive intermediates it can assist in the transfer of H^+. So, usually, two intermolecular proton transfers are drawn, where the water acts as a proton shuttle.

Water is a weak nucleophile and adds only slowly to the C=O bond. To increase the rate of the reaction, an acid catalyst is used. The acid protonates the C=O bond and this converts the carbonyl compound into a stronger electrophile (e.g. $RCH{=}OH^+$) that reacts more rapidly with water.

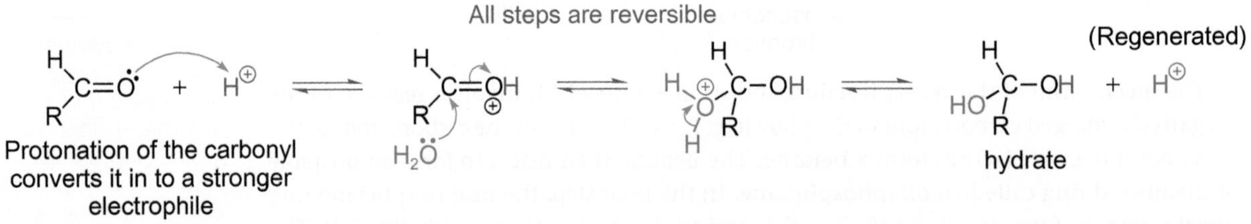
All steps are reversible
(Regenerated)
Protonation of the carbonyl converts it in to a stronger electrophile
hydrate

Although using an acid catalyst affects the rate of addition, it does *not* affect the position of the equilibrium. The position of the equilibrium is determined by the stability of the carbonyl compound relative to the hydrate, which is determined by the steric and electronic effects of the groups in both compounds (Figure 23.12).

- The *steric* effects of the R groups explain why aldehydes are usually hydrated to a greater extent than ketones—there is more steric strain in a hydrate produced from a ketone so it is less stable relative to the ketone (this is the same as in cyanohydrin formation; see p.1068).
- The *electronic* effects of the R groups explain why carbonyls with strongly electron-withdrawing groups, such as trichloroethanal (or chloral, Cl_3CCHO), are usually

The carbonyl group is a weak base. Protonation of $R_2C{=}O$ gives the conjugate acid ($R_2C{=}OH^+$), which has a very low pK_a value of –7. Thus a 1.0 mol dm^{-3} solution of HCl (pH 0) will only protonate around 1 in 10^7 molecules of the carbonyl. Even at these low concentrations, the protonated carbonyls increase the rate of nucleophilic additions because they are extremely reactive electrophiles.

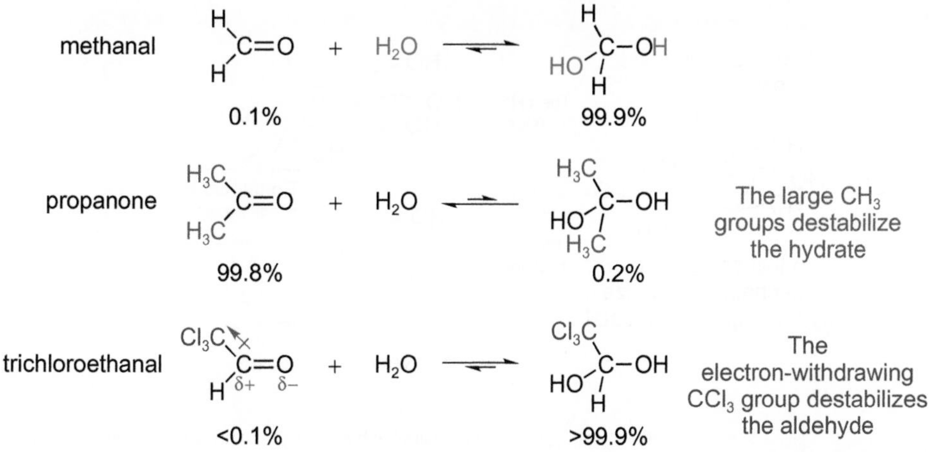
methanal
0.1%
99.9%
propanone
99.8%
0.2%
The large CH3 groups destabilize the hydrate
trichloroethanal
<0.1%
>99.9%
The electron-withdrawing CCl3 group destabilizes the aldehyde

Figure 23.12 Position of equilibrium in the reaction of aldehydes and ketones with water.

Living on the edge, it's the only way he lives.
TIMOTHY DALTON
JAMES BOND 007
THE LIVING DAYLIGHTS

Chloral hydrate ($Cl_3CCH(OH)_2$) induces sleep and it is the 'knock out drops' or 'Mickey Finns' used in crime stories and thrillers such as 'The Living Daylights' (starring Timothy Dalton as James Bond).

hydrated in solution. In trichloroethanal, the CCl_3 group withdraws electron density from the partially positive carbon atom and this destabilizes the C=O bond. The electron-withdrawing effect of the CCl_3 group does not affect the stability of the hydrate to the same extent so the hydrate of trichloroethanal (chloral hydrate, $Cl_3CCH(OH)_2$) is more stable than trichloroethanal itself.

Addition of alcohols to form hemiacetals and acetals

Alcohols (ROH) add reversibly to the C=O bond of aldehydes and ketones to form **hemiacetals** (RCH(OR)OH or $R_2C(OR)OH$). The mechanism of the reaction is the same as that for addition of water—H_2O is simply replaced by ROH. Like water, alcohols are relatively weak nucleophiles so an acid catalyst is used to increase the rate of formation of the hemiacetal. The position of equilibrium, as in the formation of hydrates, depends on the electronic and steric effects of the groups in the carbonyl compound.

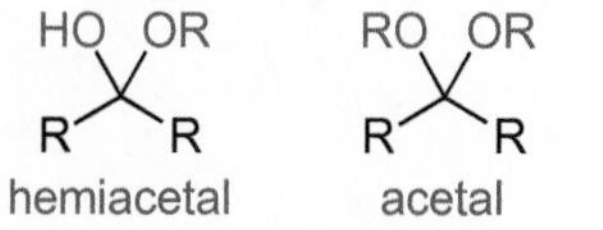
hemiacetal
acetal

Hemi- is a prefix meaning half, so a hemiacetal is half an acetal.

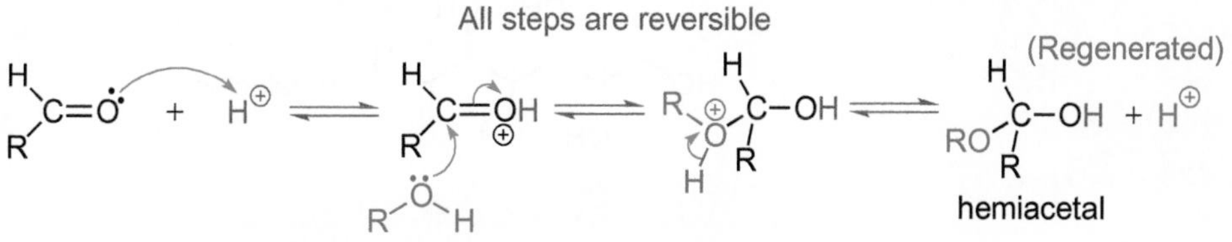
All steps are reversible
(Regenerated)
hemiacetal

For simplicity, in a reaction mechanism, the H^+ is shown just 'dropping off'. In reality, a base (often the solvent) is always needed to remove the H^+.

Interactive 3D animation showing the mechanism of formation of a hemiacetal.

Box 23.5 The structure of glucose

Glucose is one of the most important sources of energy for animals and plants (Box 14.9, p.685). This sugar is produced in plants (by photosynthesis) as a single enantiomer, which has the D-configuration (Box 18.6, p.842).

$$C_6H_{12}O_6 + 6O_2 \underset{\text{Photosynthesis}}{\overset{\text{Respiration}}{\rightleftharpoons}} 6CO_2 + 6H_2O + \text{energy}$$

D-glucose

D-Glucose is a polyhydroxy aldehyde with the formula $C_6H_{12}O_6$. Drawing the structure of D-glucose is not straightforward because there are a number of structures to choose from. There is an open-chain form, with a CHO group and five OH groups, and also two major cyclic forms that are hemiacetals.

The hemiacetals are formed by cyclization of the open-chain form. One of the OH groups in the chain attacks the carbon atom in the C=O bond to form a 6-membered ring. There are two possible 6-membered rings, both of which are hemiacetals, and they have chair-like conformations (p.827). The cyclization produces a new →

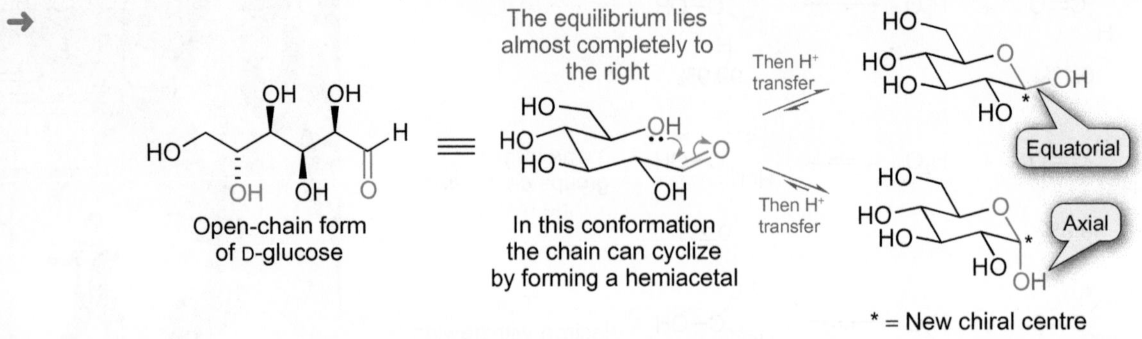

Open-chain form of D-glucose
The equilibrium lies almost completely to the right
In this conformation the chain can cyclize by forming a hemiacetal
Then H+ transfer
Then H+ transfer
Equatorial
Axial
* = New chiral centre

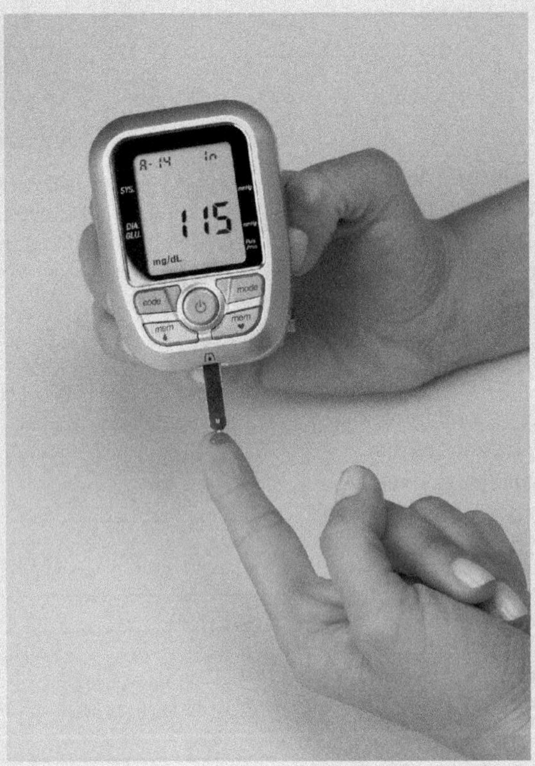

115
mg/dL
mode

Patients suffering from diabetes have a higher than normal concentration of glucose in their blood. Glucose concentrations can be monitored using a device that measures the change in colour of a dye when it reacts with glucose (in a drop of blood), oxygen, and water.

chiral centre and the hemiacetals are diastereomers because the OH group of the hemiacetal can adopt an axial or equatorial position (p.828). Notice that all of the other groups on the rings are in equatorial positions.

In aqueous solution, the three forms of D-glucose are in equilibrium with one another. At equilibrium, more than 99% of D-glucose exists in the hemiacetal forms. The name pyranose is used to describe a 6-membered ring structure of a sugar so the hemiacetals of glucose are called glucopyranoses.

Question

The structure of the open-chain form of the sugar D-galactose is shown here. Draw the structures of the two galactopyranoses.

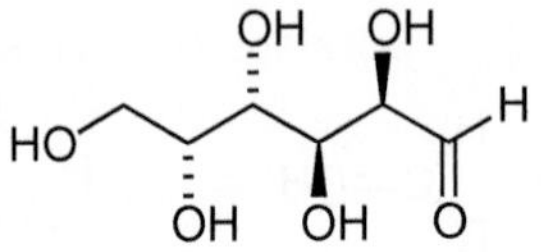

D-galactose

The name acetal used to be restricted to the product from an aldehyde ($RCH(OR)_2$) and the name ketal was used for a product from a ketone ($R_2C(OR)_2$), but chemists now use acetal to describe both types of products.

In the hemiacetal (RCH(OR)OH), the two oxygen atoms are both basic so either can be protonated. Protonation of either the OH group or OR group is reversible.

Interactive 3D animation showing the mechanism of formation of an acetal.

Hemiacetals react with a second equivalent of an alcohol, in the presence of an acid catalyst, to form an **acetal** ($RCH(OR)_2$ or $R_2C(OR)_2$). There are four steps in the reaction mechanism shown in Figure 23.13. First, protonation of the hemiacetal converts the –OH group into a good leaving group, $-^{+}OH_2$. In the second step, water is eliminated—a lone pair on the oxygen atom of the OR group helps to 'kick' the water out—and a π bond is formed in the resulting oxonium ion ($RCH{=}OR^+$). The oxonium ion is a strong electrophile and, in the third step, the alcohol attacks the oxonium ion to break the π bond. (Notice how similar this step is to the reaction of $RCH{=}OH^+$ with an alcohol to give a hemiacetal, p.1073). Finally, deprotonation of the product ($RCH(OR)O^+HR$) forms an acetal, $RCH(OR)_2$.

Every step in the mechanism of formation of an acetal is reversible. To ensure that the position of equilibrium lies on the side of the acetal, an excess of the alcohol is used and a dehydrating agent (such as anhydrous magnesium sulfate or molecular sieves) is added. The dehydrating agent removes water from the reaction mixture and this forces the reaction to produce more of the acetal (Le Chatelier's principle). For the same reason, an anhydrous acid (such as dry HCl gas) is often used to supply H^+, because an aqueous acid would move the position of the equilibrium to the left.

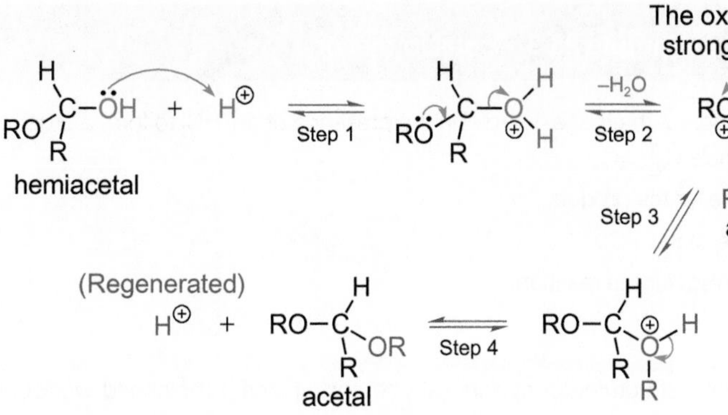
The oxonium ion is a strong electrophile
hemiacetal
Step 1
$-H_2O$
Step 2
Step 3
alcohol
(Regenerated)
Step 4
acetal

Figure 23.13 Formation of an acetal from a hemiacetal.

> Sometimes, in step 2 of the mechanism, loss of a molecule of water (from the protonated hemiacetal) is shown to give a carbocation (like in an S_N1 reaction):
>
> 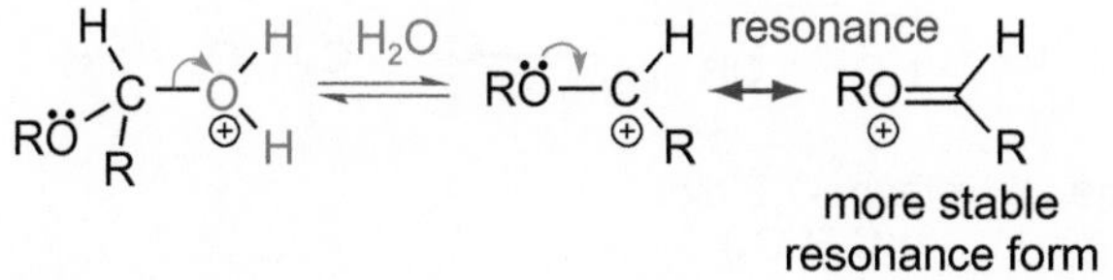
> H_2O
resonance
more stable resonance form
>
> The carbocation is stabilized by resonance—donation of a lone pair from the adjacent OR group gives the oxonium ion, which is the more stable resonance form (Section 19.1, p.876). Either way, it is important to recognize that the neighbouring OR group facilitates the loss of water from the protonated hemiacetal. This explains why the protonated hemiacetal loses water before it reacts with the alcohol (i.e. why the protonated hemiacetal does not react with the alcohol in an S_N2 reaction).

Acetals as protecting groups

Acetals do not react with bases, redox reagents, or nucleophiles, and this, together with their reversible formation, explains why acetals are important protecting groups for aldehydes and ketones. (A **protecting group** converts a reactive functional group into an unreactive form such that, after the desired transformation(s), the original functional group can be regenerated.) The acetal protecting group is introduced by reacting an aldehyde or ketone with two equivalents of an alcohol, or by reaction with a single equivalent of a diol (a compound containing two OH groups). For example, reaction of a C=O bond with ethane-1,2-diol (ethylene glycol, $HOCH_2CH_2OH$) forms a cyclic acetal (called a 1,3-dioxolane) in a *protection* step. To convert the acetal back into the C=O bond requires hydrolysis by heating with aqueous acid—this is called a *deprotection* step. Figure 23.14 shows an example of using ethane-1,2-diol to protect an aldehyde group in a molecule.

The formation of a cyclic acetal is discussed in Worked example 23.3.

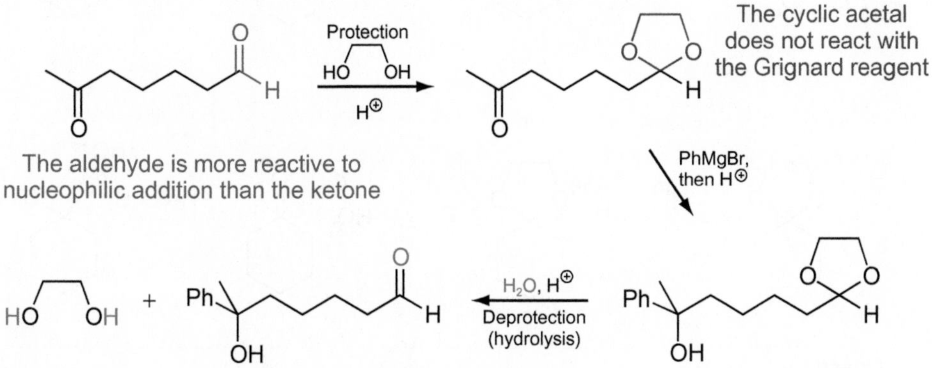
Protection
The cyclic acetal does not react with the Grignard reagent
The aldehyde is more reactive to nucleophilic addition than the ketone
PhMgBr, then H⊕
H_2O, H⊕
Deprotection (hydrolysis)

Figure 23.14 Using acetal formation to protect an aldehyde group while carrying out a Grignard reaction on a ketone group in the same molecule.

Molecular sieves are a type of zeolite clay that has cavities into which water molecules enter and bind tightly (Chapter 6, p.255). Larger organic molecules will not fit into the cavities. Molecular sieves are added in the formation of an acetal from a hemiacetal to remove water so the equilibrium moves towards acetal formation.

> A **protecting group** is used to prevent a functional group from undergoing an unwanted reaction(s).

> A **hydrolysis reaction** is a reaction involving water.

Visit the Online Resource Centre to view screencast 23.2 which walks you through the use of acetals as protecting groups, including why aldehydes and ketones need protecting.

Notice that following addition of PhMgBr to the ketone group, the acidic work-up (at room temperature) selectively protonates the alkoxide ion to give the tertiary alcohol, and the cyclic acetal is not hydrolysed. Combining protonation of the alkoxide ion with hydrolysis of the cyclic acetal, in a single step, requires heating with aqueous acid.

Ethane-1,2-diol ($HOCH_2CH_2OH$) is a cheap commercially available diol, which is commonly used as an antifreeze in cars. A mixture of 60% ethane-1,2-diol and 40% water does not freeze until temperatures fall below −45 °C.

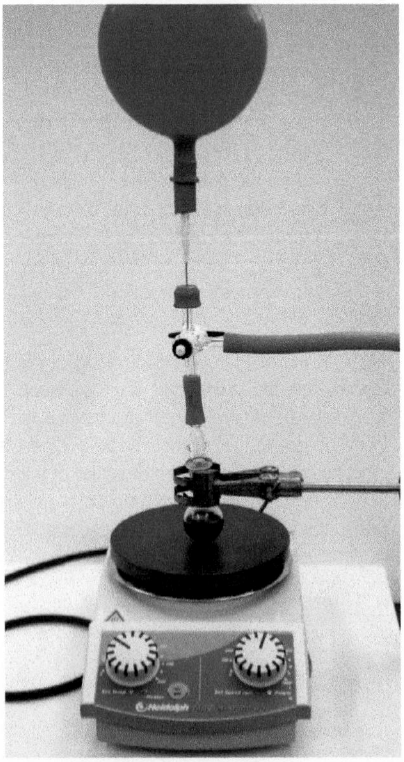

Raney nickel is made up of extremely finely divided grains of nickel. It is commonly used as a catalyst for hydrogenation reactions. For example, see the Mozingo reduction on the next page.

Worked example 23.3 Acetal formation

Cyclohexanone reacts with ethane-1,2-diol in the presence of dry HCl to form a product with the molecular formula $C_8H_{14}O_2$.

(a) Give a structure for the product.

(b) Explain the role of the HCl.

(c) Give a mechanism for the reaction.

Strategy

(a) Draw the skeletal structures of cyclohexanone and ethane-1,2-diol and work out their molecular formulae.

Compare the molecular formulae of the reactants with the product to show that H_2O must also be a product of the reaction.

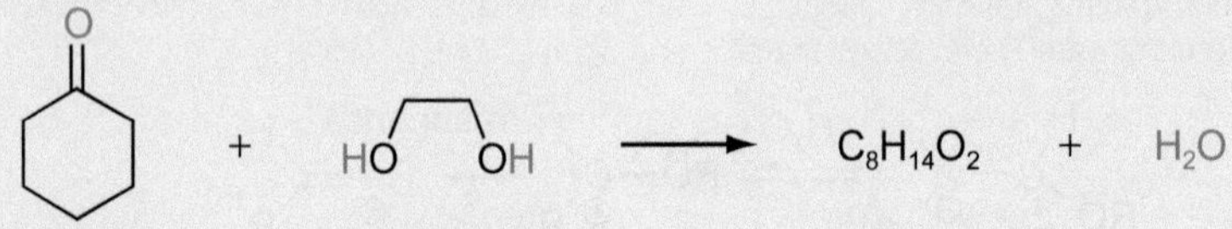
O
+
HO OH
$C_8H_{14}O_2$ + H_2O

cyclohexanone ($C_6H_{10}O$) ethane-1,2-diol ($C_2H_6O_2$)

Recognize that a diol reacts with a ketone to form an acetal and water.

(b) Consider the effect of protonating the oxygen atom in the C=O bond of cyclohexanone by HCl. (*Dry* HCl is used to move the position of the equilibrium to the right.)

(c) Outline the mechanism of acetal formation by nucleophilic addition of the two ROH groups to the C=O bond.

Solution

(a)

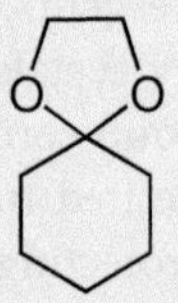
O
O

(b) Protonation of the C=O bond makes the ketone a stronger electrophile so it reacts more rapidly with ethane-1,2-diol.

(c)

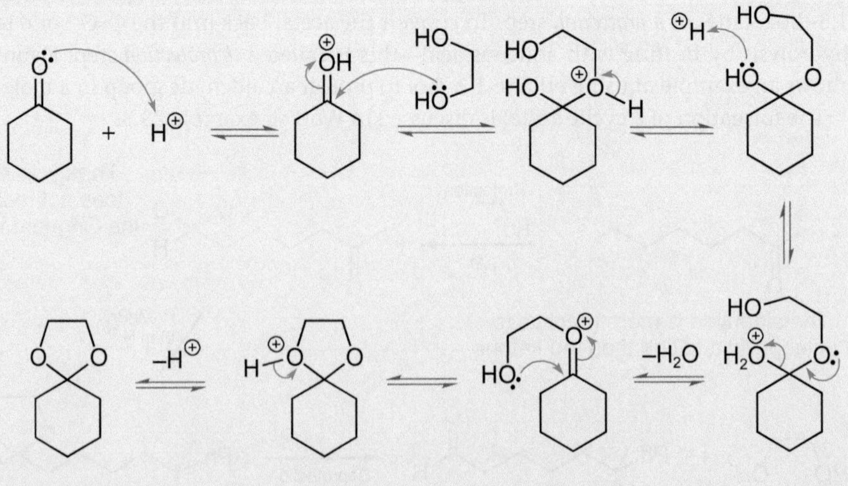
O
+ H⊕
ÖH
HO
HÖ
HO
HO
O
H
⊕H
HO
HÖ
O
−H⊕
H
O
O
HO
−H_2O
H_2O

Question

Give a mechanism to show how the acetal is converted back into cyclohexanone and ethane-1,2-diol using H^+/H_2O.

Reaction with sulfur nucleophiles

Addition of thiols to form thioacetals

A thiol (RSH) adds reversibly to the C=O bond of aldehydes and ketones to form a **thiohemiacetal** (RCH(OH)SR or $R_2C(OH)SR$) and then a **thioacetal** ($RCH(SR)_2$ or $R_2C(SR)_2$). The mechanism of the reaction is the same as that for the formation of acetals (see p.1075)—ROH is simply replaced by RSH. Like alcohols, thiols are relatively weak nucleophiles so an acid catalyst is used to increase the rate of formation of the thioacetal.

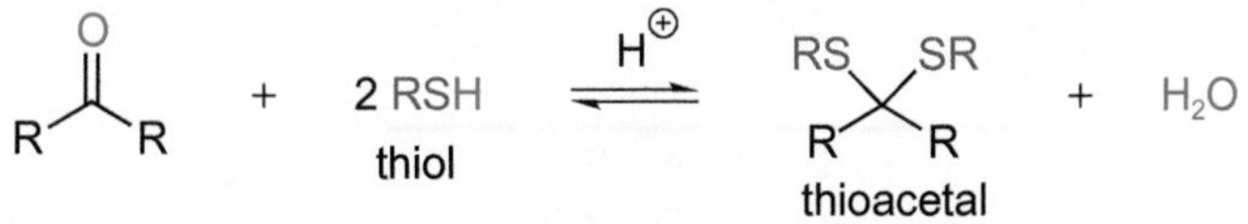
2 RSH
thiol
thioacetal
H_2O

Thioacetals are useful compounds in synthesis because they are reduced using Raney nickel and hydrogen in the **Mozingo reduction** (see Figure 23.15). Both C–S bonds are converted into C–H bonds so this method is used to convert aldehydes and ketones into alkanes. (In contrast, acetals do not react with Raney Ni and hydrogen, which is one of the reasons why acetals are such good protecting groups.)

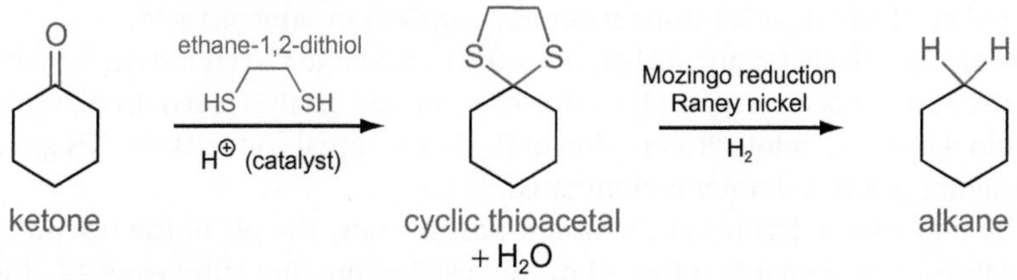
ethane-1,2-dithiol
H⊕ (catalyst)
Mozingo reduction
Raney nickel
H_2
ketone
cyclic thioacetal
+ H_2O
alkane

Figure 23.15 Using thioacetal formation to convert a ketone into an alkane.

Reaction with nitrogen nucleophiles

Reaction with primary amines to form imines

A primary amine (RNH_2) adds reversibly to the C=O bond of an aldehyde or ketone to form an **imine** (RN=CHR, sometimes called an aldimine, or $RN=CR_2$, sometimes called a ketimine). You can think of the mechanism as taking place in two parts. In the first part, the amine adds to the C=O bond and, after transfer of a proton from nitrogen to oxygen, a hemiaminal ($RNH–C(OH)R_2$) is formed. In the second part, the hemiaminal is protonated and elimination of H_2O (facilitated by the RNH– group), followed by deprotonation, forms the imine ($RN=CR_2$). To describe both parts of the mechanism, the name **addition–elimination reaction** is commonly used. The mechanism is shown in Figure 23.16.

Part 1: Addition of RNH_2 to the C=O bond to form a hemiaminal

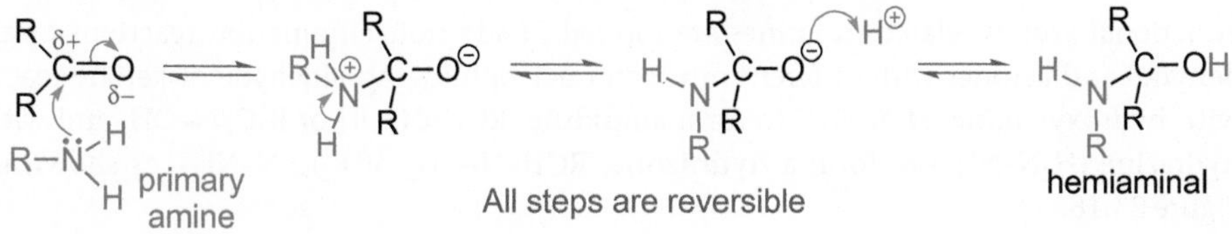
primary amine
All steps are reversible
hemiaminal

Part 2: Elimination of water from the hemiaminal

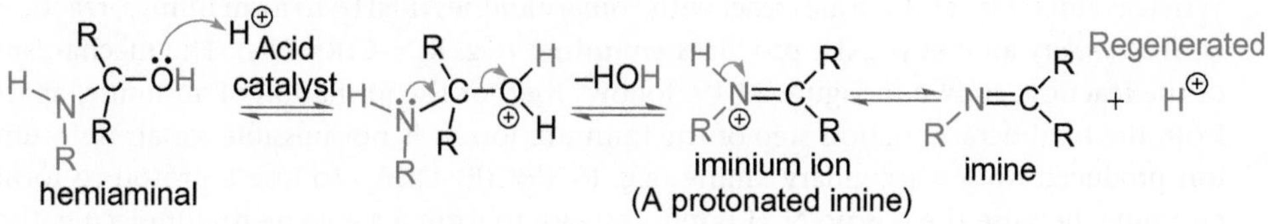
Acid catalyst
–HOH
Regenerated
hemiaminal
iminium ion
(A protonated imine)
imine

Figure 23.16 Reaction of a primary amine with an aldehyde or ketone in an addition–elimination reaction to form an imine.

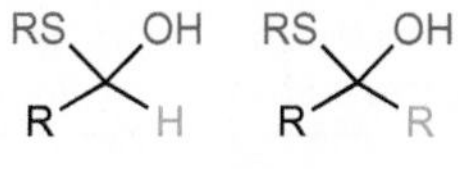
thiohemiacetals

Aldehydes and ketones are also converted into alkanes using the Clemmensen reduction, Zn/Hg/HCl (Section 22.4, p.1040) or the Wolff–Kishner reaction of a hydrazone, with HO^-/heat (Box 23.6, p.1079).

The Mozingo reduction is believed to involve absorption of the thioacetal on the nickel surface. Research indicates that absorption is followed by homolysis of the C–S bonds to form carbon-centred radicals (R•), which, in the presence of hydrogen, are reduced to form alkanes (RH).

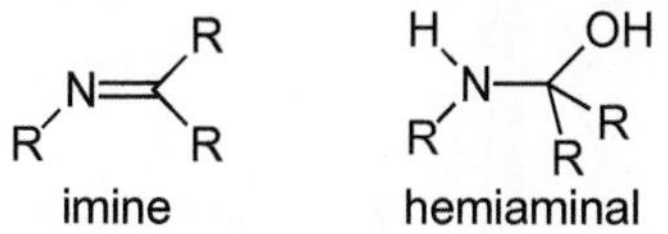
imine
hemiaminal

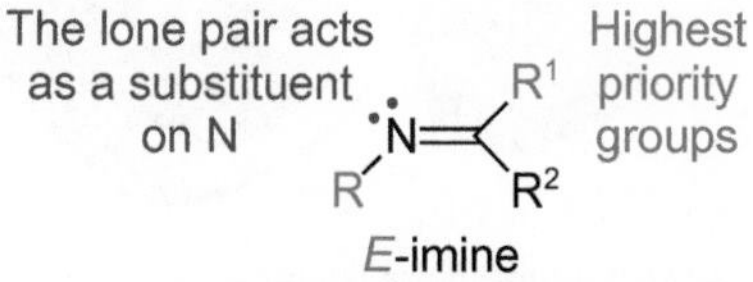
The lone pair acts as a substituent on N
Highest priority groups
E-imine

An imine with different R^1 and R^2 groups has *E*- and *Z*-isomers, just like an alkene (Section 18.3, p.833)

Overall, the formation of an imine from a primary amine and an aldehyde or ketone is an example of a condensation reaction. In a **condensation reaction**, two reactant molecules combine to form a product with loss of a small molecule, typically water.

The conversion of (*Z*)-retinal into rhodopsin (p.1055) is an example of an addition–elimination reaction.

Protonation of a hemiaminal ($R_2C(OH)$-NHR) leads to loss of water and formation of an imine ($R_2C=NR$). In contrast, protonation of a hemiacetal ($R_2C(OH)OR$), in the presence of ROH, forms an acetal ($R_2C(OR)_2$). Both reactions start with protonation of the –OH group followed by loss of water. But, whereas the iminium ion ($R_2C=N^+HR$) can form a stable neutral product by losing a proton, this is not possible for the corresponding oxonium ion ($R_2C=O^+R$), which is a strong electrophile and so reacts with a second equivalent of the alcohol.

Replacing the OH group in a hemiaminal ($R_2C(OH)NR_2$) with NR_2 gives an aminal, $R_2C(NR_2)_2$. Aminals are often unstable and this has limited their usefulness in synthesis.

A hemiaminal ($R_2C(OH)NHR$) can be protonated on nitrogen or oxygen. An equilibrium will be set up and although protonation on nitrogen is preferred to oxygen (as nitrogen is less electronegative than oxygen it is a better electron donor), protonation on oxygen leads to loss of water. Water is a better leaving group than a primary amine (H_3O^+ has a much lower pK_a value than H_3N^+R, see Section 20.2, p.925), and the position of the equilibrium can be forced to the imine by using a drying agent.

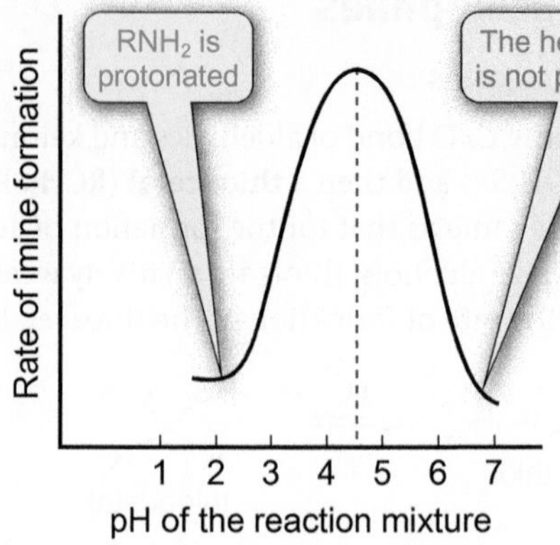
RNH2 is protonated
The hemiaminal is not protonated
Rate of imine formation
1 2 3 4 5 6 7
pH of the reaction mixture

Figure 23.17 Effect of pH on the rate of formation of the imine.

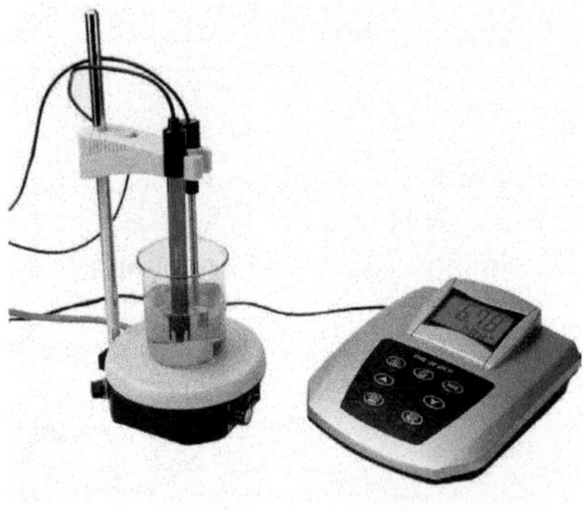

The pH of the reaction mixture can be determined using a pH meter probe (see Section 11.2, p.524).

Interactive 3D animation showing the mechanism of formation of an imine.

All of the steps in the reaction are reversible and, to force the position of the equilibrium to the right to form the imine, a drying agent is added to the reaction mixture to remove water (as in acetal formation, p.1075). Conversely, to convert an imine back into a primary amine and an aldehyde or a ketone requires hydrolysis by aqueous acid.

Amines are relatively strong nucleophiles and they add to C=O bonds to form hemiaminals in the absence of an acid catalyst. However, an acid catalyst is required to convert the hemiaminal into the imine. Protonation of the hemiaminal converts the OH group into a better leaving group and water is eliminated.

For the imine to be formed at the fastest possible rate, the pH of the reaction mixture is maintained at around 4.5. If the pH of the reaction mixture falls below 4.5, the rate of imine formation is slower (see Figure 23.17) because, at lower pH values, the primary amine becomes protonated. The protonated amine (RNH_3^+) cannot act as a nucleophile and the rate of imine formation decreases. If the pH rises above 4.5, the rate of imine formation is slower because there is not enough acid to protonate the OH group of the hemiaminal.

Although an acid catalyst is not required to form a hemiaminal, when the pH of the reaction mixture is 4.5, it is likely that the hemiaminal is formed by addition of the amine to a protonated C=O bond.

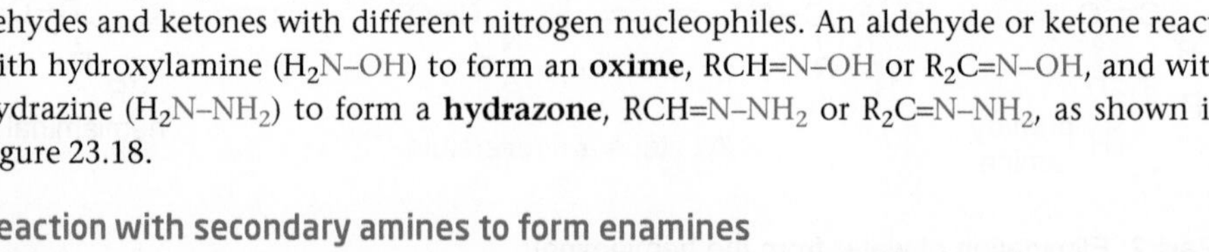

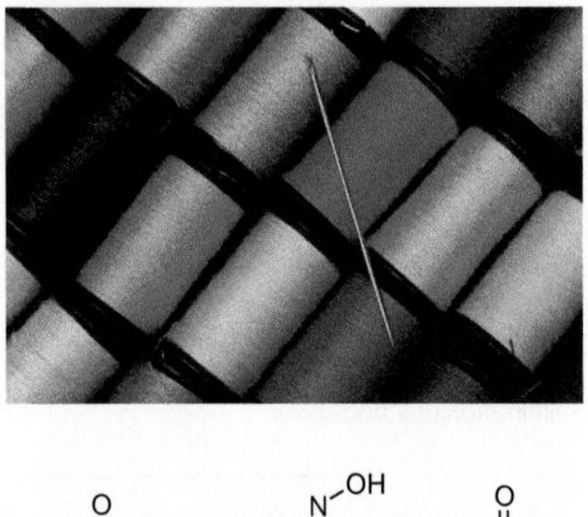

Reaction with other nitrogen nucleophiles

Functional groups related to imines are formed in addition–elimination reactions of aldehydes and ketones with different nitrogen nucleophiles. An aldehyde or ketone reacts with hydroxylamine ($H_2N–OH$) to form an **oxime**, RCH=N–OH or $R_2C=N–OH$, and with hydrazine ($H_2N–NH_2$) to form a **hydrazone**, $RCH=N–NH_2$ or $R_2C=N–NH_2$, as shown in Figure 23.18.

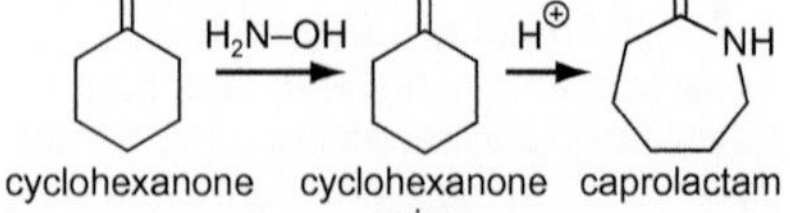

cyclohexanone cyclohexanone oxime caprolactam

Nylon 6 is a tough polymer that is used to make threads and ropes. Nylon 6 is prepared by polymerization of caprolactam, which is formed from cyclohexanone oxime. (There are six carbons in caprolactam hence the name Nylon 6.)

Reaction with secondary amines to form enamines

Whereas aldehydes and ketones react with *primary* amines (RNH_2) to form imines, reaction with *secondary* amines (R_2NH) produces **enamines** (e.g. $R_2N–C(R)=CR_2$). The mechanism of the reaction, shown in Figure 23.19, follows that for the formation of an imine, apart from the final deprotonation step of the iminium ion. It is not possible for an iminium ion produced from a secondary amine (e.g. $R_2N^+=C(R)–CHR_2$) to lose a proton to form an imine, because there is no N–H bond. Instead, to form a stable neutral molecule, the iminium ion loses a proton from the α-carbon to form a new C=C bond with an NR_2 substituent, which is called an enamine.

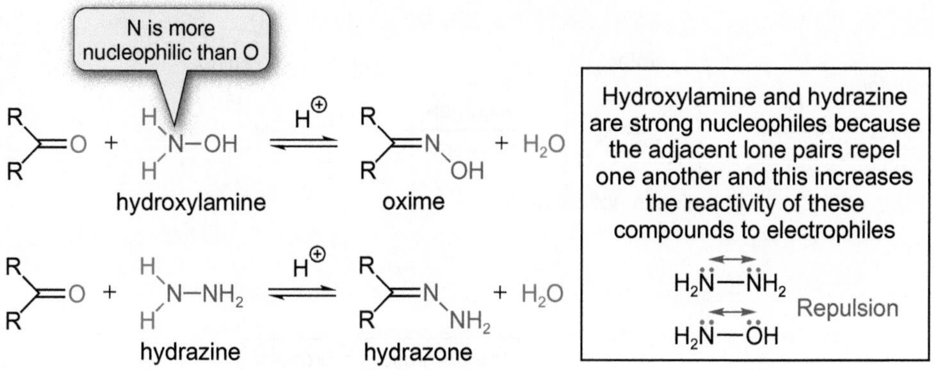
N is more nucleophilic than O
hydroxylamine
oxime
hydrazine
hydrazone
Hydroxylamine and hydrazine are strong nucleophiles because the adjacent lone pairs repel one another and this increases the reactivity of these compounds to electrophiles
Repulsion

Figure 23.18 Formation of oximes and hydrazones.

Like imines, oximes or hydrazones with different groups attached to the C=N bond have *E*- and *Z*-isomers.

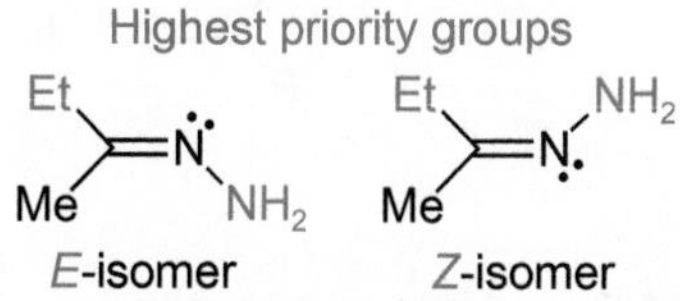
Highest priority groups
E-isomer
Z-isomer

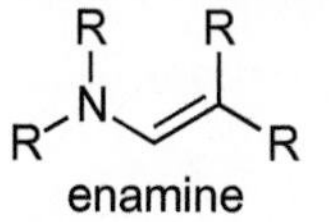
enamine

Interactive 3D animation showing the mechanism of formation of an enamine.

Enamines ($RC(NR_2)=CR_2$) and enols ($RC(OH)=CR_2$) have similar structures and reactivities. For example, they can both react with an electrophile to introduce a new substituent at the α-position (see Section 23.3, p.1082, for reactions of enols).

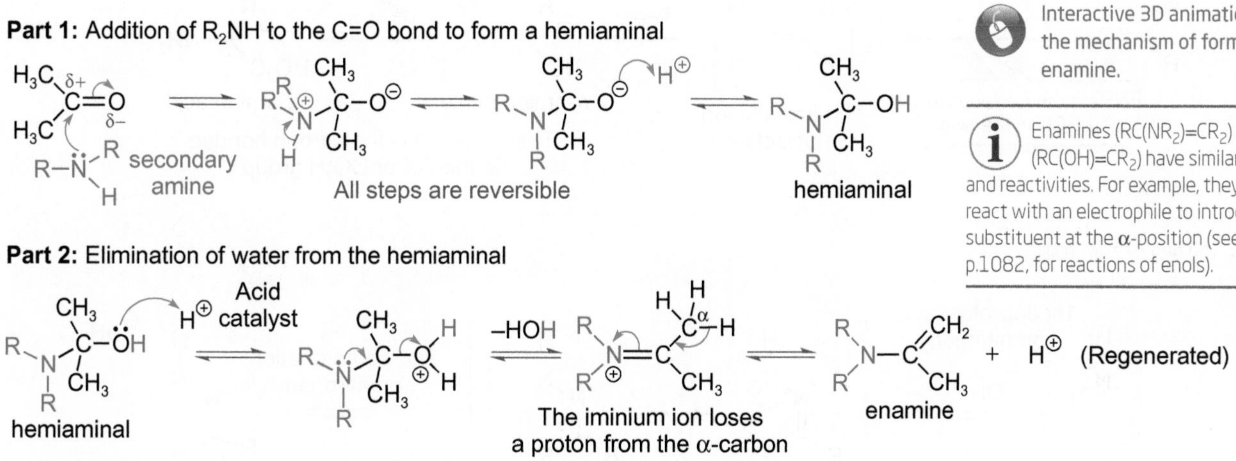
Part 1: Addition of R2NH to the C=O bond to form a hemiaminal
secondary amine
All steps are reversible
hemiaminal
Part 2: Elimination of water from the hemiaminal
Acid catalyst
hemiaminal
–HOH
The iminium ion loses a proton from the α-carbon
enamine
(Regenerated)

Figure 23.19 Reaction of a secondary amine (R_2NH) with an aldehyde or ketone in an addition–elimination reaction to form an enamine.

Box 23.6 Imines and hydrazones in organic synthesis

Imines and hydrazones are useful intermediates in synthesis because they are easily converted into other functional groups by reaction with nucleophiles. Nitrogen is more electronegative than carbon, so the C=N bond is polarized (like the C=O bond) and it should come as no surprise that imines react with nucleophiles in nucleophilic addition reactions. The nucleophile attacks the carbon atom in the $^{\delta+}C=N^{\delta-}$ bond to form a tetrahedral intermediate with a negative charge on nitrogen. In a second step, protonation of the nitrogen anion forms the organic product.

Imines react with nucleophiles such as the hydride ion (H^-) and the cyanide ion (^-CN). An imine prepared from an aldehyde or ketone and a primary amine (RNH_2) reacts with H^- followed by H^+ to form a secondary amine (R_2NH). It is often possible to make the imine and immediately reduce it in the same reaction vessel—sodium cyanoborohydride ($Na(CN)BH_3$; see p.1064) is generally used as the source of hydride ions because it reduces imines much more quickly than aldehydes or ketones. The overall effect of converting an aldehyde or ketone into an imine and then an amine is called **reductive amination** (Figure 1). It is used in synthesis to convert ammonia or primary amines efficiently into primary and secondary amines, respectively.

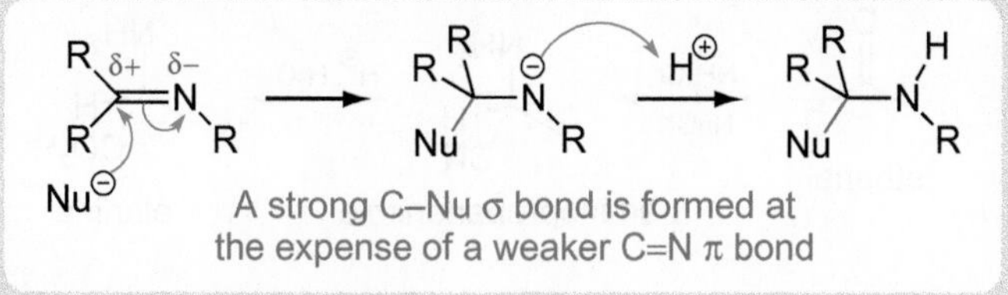
A strong C–Nu σ bond is formed at the expense of a weaker C=N π bond

The reaction of imines with the ^-CN ion is used to make α-amino acids in the **Strecker synthesis** (Figure 2). The German chemist Adolph Strecker (1822–1871) found that mixing an aldehyde or ketone with NH_4Cl and NaCN produced an α-amino nitrile ($RCH(CN)NH_2$ or $R_2C(CN)NH_2$). In aqueous solution, NH_4Cl and NaCN are in equilibrium with NH_3, HCN, and NaCl, and it is the NH_3 and HCN that react with the C=O bond to form an iminium ion (a protonated imine, for example, $R_2C=NH_2^+$). As soon as the iminium ion is formed, the ^-CN ion attacks the carbon atom of the C=N bond to form an α-amino nitrile. The α-amino nitrile is converted into the α-amino acid ($RCH(CO_2H)NH_2$ or $R_2C(CO_2H)NH_2$) in a second reaction, by heating with acid and water (the mechanism of nitrile hydrolysis is covered in Box 24.7, p.1125). ➜

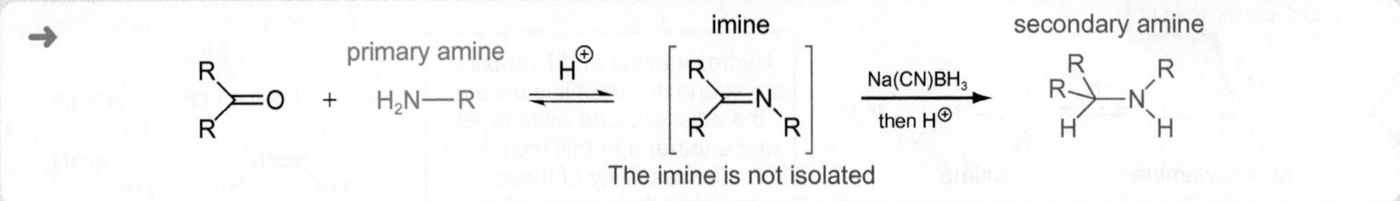

primary amine
imine
secondary amine
Na(CN)BH3
then H⊕
The imine is not isolated

▲ **Figure 1** Reductive amination.

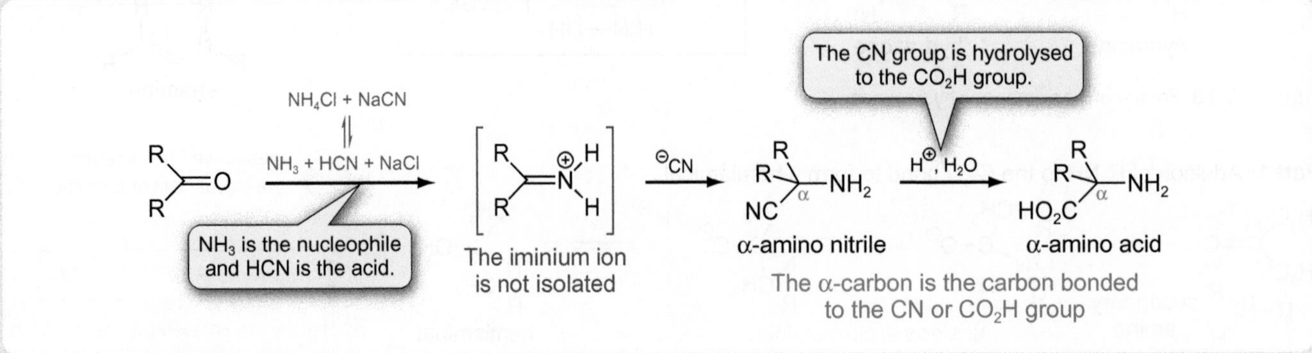

NH4Cl + NaCN
NH3 + HCN + NaCl
NH3 is the nucleophile and HCN is the acid.
The iminium ion is not isolated
The CN group is hydrolysed to the CO2H group.
α-amino nitrile
α-amino acid
The α-carbon is the carbon bonded to the CN or CO2H group

▲ **Figure 2** The Strecker synthesis.

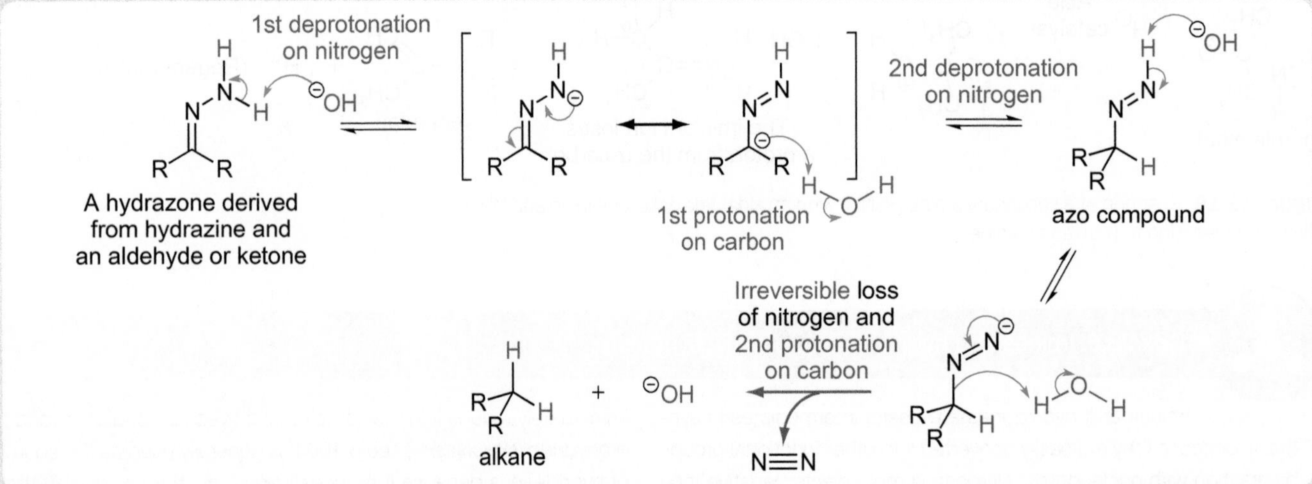

1st deprotonation on nitrogen
A hydrazone derived from hydrazine and an aldehyde or ketone
1st protonation on carbon
2nd deprotonation on nitrogen
azo compound
Irreversible loss of nitrogen and 2nd protonation on carbon
alkane

▲ **Figure 3** The Wolff–Kishner reaction.

Since 1850, the Strecker synthesis has been used to prepare various amino acids efficiently in the laboratory. However, the synthesis may have much earlier origins, as some scientists believe that this was how the first amino acids on Earth were made (see also Box 4.7, p.198).

Hydrazones (prepared from an aldehyde or ketone and hydrazine) also react with nucleophiles in nucleophilic addition reactions. However, reaction of a hydrazone with a base, such as hydroxide ions, is especially useful in synthesis. When a hydrazone (RCH=N–NH_2 or R_2C=N–NH_2) is heated with a concentrated solution of hydroxide ions, it forms an alkane (RCH_3 or R_2CH_2) and nitrogen gas in the **Wolff–Kishner reaction**. The mechanism of the reaction involves two cycles of deprotonation on nitrogen followed by protonation on carbon, as shown in Figure 3. The reaction is driven to completion in the second cycle, where the irreversible loss of nitrogen gas (an excellent leaving group) occurs simultaneously with the second protonation on carbon.

Question

The amino acid alanine is prepared from ethanal by the Strecker synthesis shown here. Suggest a mechanism to explain how ethanal is converted into 2-aminopropanenitrile.

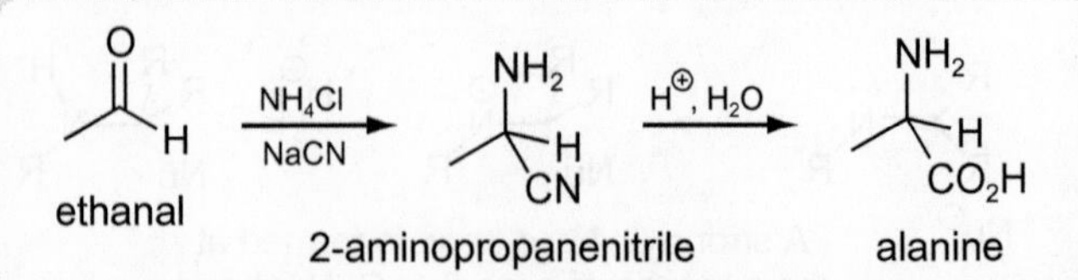

ethanal
NH4Cl
NaCN
2-aminopropanenitrile
H⊕, H2O
alanine

Summary

- The C=O bond of aldehydes and ketones reacts with nucleophiles (such as H^-, an organometallic reagent, or ^-CN) in nucleophilic addition reactions.
- Nucleophiles add more rapidly to aldehydes (RCHO) than ketones (R_2CO) because of steric and electronic effects.
- Reaction of a phosphonium ylide (ylid) with an aldehyde or ketone forms an alkene in the Wittig reaction.

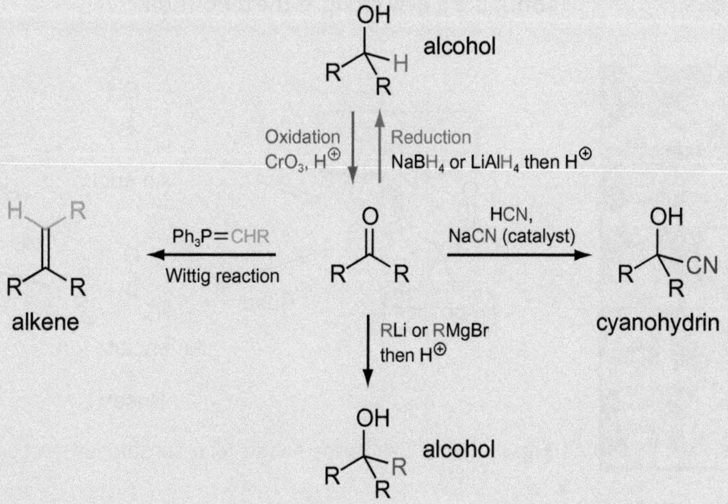

alcohol
Oxidation
CrO3, H⊕
Reduction
NaBH4 or LiAlH4 then H⊕
Ph3P=CHR
Wittig reaction
alkene
HCN,
NaCN (catalyst)
cyanohydrin
RLi or RMgBr
then H⊕
alcohol

- The reactions of nucleophilic groups containing oxygen, sulfur, or nitrogen with the C=O bond are reversible and are catalysed by an acid.

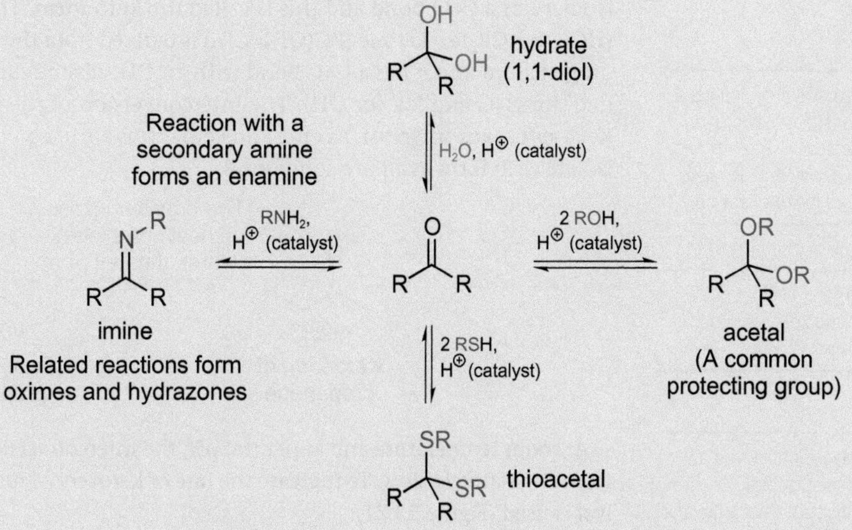

hydrate
(1,1-diol)
H2O, H⊕ (catalyst)
Reaction with a
secondary amine
forms an enamine
RNH2,
H⊕ (catalyst)
imine
Related reactions form
oximes and hydrazones
2 ROH,
H⊕ (catalyst)
acetal
(A common
protecting group)
2 RSH,
H⊕ (catalyst)
thioacetal

For practice questions on these topics, see questions 1–4 at the end of this chapter (pp.1095–1096).

23.3 α-Substitution reactions of aldehydes and ketones

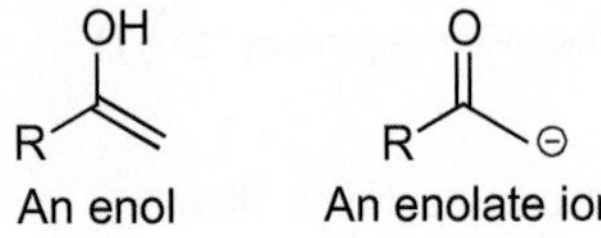
OH
R
O
R
An enol
An enolate ion

Enols are described in Section 2.6 (p.95) and enolate ions in Section 19.2 (p.890).

α-Substitution reactions of aldehydes and ketones involve replacement of a hydrogen atom by another group at the carbon atom next to the C=O bond (this is called the **α-position**; Section 23.1, p.1058). The mechanisms of α-substitution reactions involve intermediate **enols** ($RC(OH){=}CH_2$) or **enolate ions** ($RC({=}O)CH_2^-$). A general mechanism is shown in Figure 23.20. Enols and enolate ions are nucleophiles and they react with electrophiles to introduce a new group at the α-position.

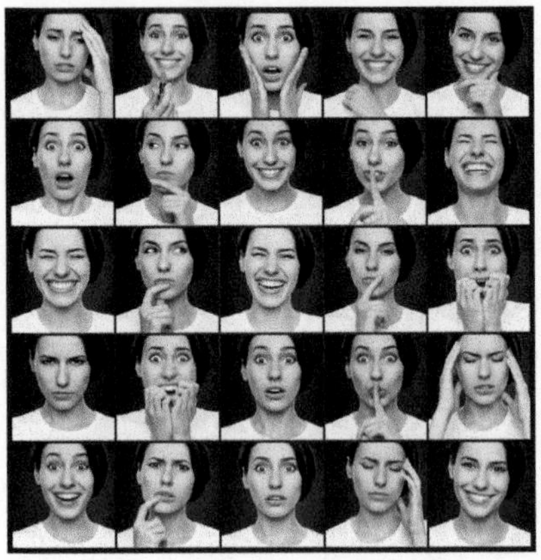

Note that whereas the keto form is *electrophilic* and reacts with nucleophiles at the carbonyl carbon, the enol form is *nucleophilic* and reacts with electrophiles at the α-position. So, the same aldehyde or ketone can show contrasting reactivities, depending on the situation, like us having mood swings.

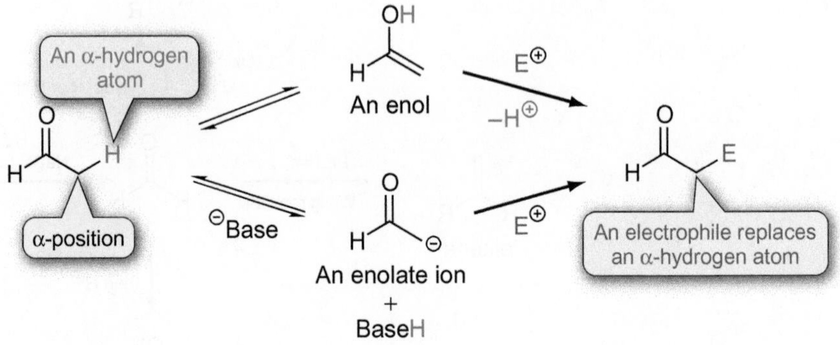
An α-hydrogen atom
O
H
H
α-position
Base
OH
H
An enol
O
H
An enolate ion
+
BaseH
E
−H
E
O
H
E
An electrophile replaces an α-hydrogen atom

Figure 23.20 General mechanism for α-substitution reactions of aldehydes and ketones.

Keto-enol tautomerism

Aldehydes and ketones with hydrogen atoms at the α-position are in equilibrium with structural isomers called enols. For most aldehydes and ketones the most stable structural isomer has a C=O bond and this is called the **keto form**. The **enol form** of the aldehyde ($HC(OH){=}CR_2$) or ketone ($RC(OH){=}CR_2$) is derived from the keto form by transfer of a hydrogen atom and it has a C=C bond with an OH substituent—hence this is called an enol ('en' for C=C and 'ol' for OH). The interconversion of the keto and enol forms is called **keto–enol tautomerism**. As enol and keto forms are in equilibrium it is usually difficult to isolate each form as a pure compound.

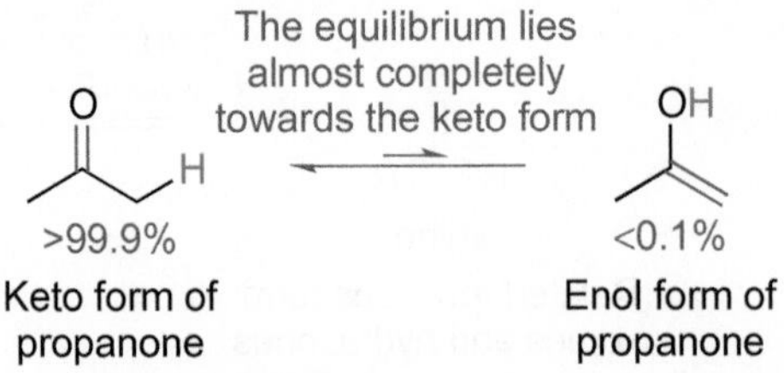
The equilibrium lies almost completely towards the keto form
O
H
OH
>99.9%
Keto form of propanone
<0.1%
Enol form of propanone

i **Tautomerism** is a general term for the intermolecular transfer of a hydrogen atom (or proton), accompanied by the shift of a double bond. The process of converting a keto form into the enol form is called **enolization**.

i **Tautomers** are organic compounds that are interconvertible by a chemical reaction called tautomerism.

At room temperature and at neutral pH, the interconversion between the keto and enol forms is relatively slow. To increase the rate of keto–enol tautomerism an acid or base catalyst is used (Figure 23.21).

- In acid-catalysed tautomerism, the C=O bond is first protonated on oxygen and this is followed by loss of H^+ from the α-position (e.g. in aqueous solvents, water can act as a base).

i The ratio of keto:enol forms can be determined using spectroscopic techniques such as ^1H NMR spectroscopy (Section 12.3, p.580). Keto and enol forms give resonance signals at different chemical shifts. For example, for $RC(OH){=}CH_2$ and $RCOCH_3$, the integration curves for the signals (at ~5 ppm and 2.1 ppm) can determine the relative proportion of each form.

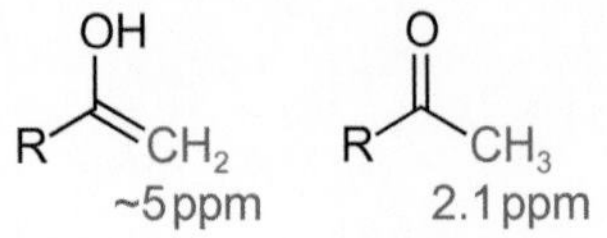
OH
R
CH2
~5 ppm
O
R
CH3
2.1 ppm

i **Acid-catalysed tautomerism** The pK_a values of H_3O^+ and $R_2C{=}OH^+$ are −1.7 and −7.2, respectively. The equilibrium constant for protonation is equal to $10^{-5.5}$ ($10^{-pK_a\,\text{acid reactant}}/10^{-pK_a\,\text{acid product}} = 10^{1.7}/10^{7.2} = 10^{-5.5}$) so the position of equilibrium lies over towards the ketone (Section 19.2, p.895).

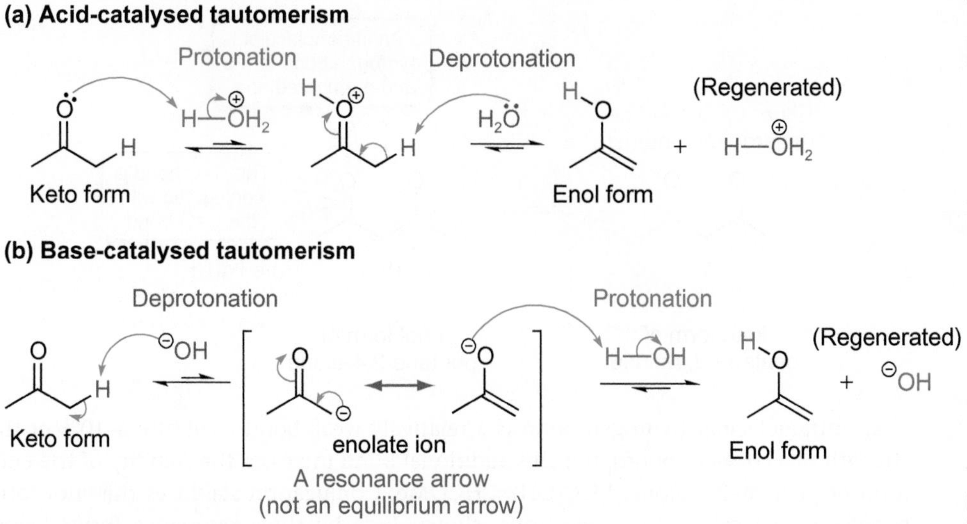
(a) Acid-catalysed tautomerism
Protonation
Deprotonation
(Regenerated)
Keto form
Enol form
(b) Base-catalysed tautomerism
Deprotonation
Protonation
(Regenerated)
Keto form
enolate ion
Enol form
A resonance arrow
(not an equilibrium arrow)

Figure 23.21 Mechanisms of keto–enol tautomerism.

WRONG WAY
RIGHT WAY

The mechanisms for converting the enol form into the keto form are the reverse of those shown in Figure 23.21. For practice, draw mechanisms to show how the enol form ($MeC(OH){=}CH_2$) is converted into the keto form (MeCOMe) in the presence of an acid or a base catalyst.

- In base-catalysed tautomerism, a base such as hydroxide ions deprotonates the aldehyde or ketone at the α-position to first form an enolate ion (e.g. $RCOCH_2^-$). The enolate ion is stabilized by delocalization of the negative charge onto the electronegative oxygen atom, which is then protonated in a second step.

Base-catalysed tautomerism The pK_a values of a methyl ketone ($RCOCH_3$) and H_2O are ~20 and 16, respectively. The equilibrium constant for deprotonation is equal to 10^{-4} ($10^{-pK_a \text{ acid reactant}}/10^{-pK_a \text{ acid product}} = 10^{-20}/10^{-16} = 10^{-4}$) so the position of equilibrium lies over towards the methyl ketone.

Both steps in the acid- or base-catalysed tautomerism are reversible. The acid or base catalyst increases the rate of interconversion but it does *not* alter the position of the equilibrium.

For most aldehydes and ketones, the position of the equilibrium lies heavily towards the keto form because the keto form is usually more stable than the enol form. You can use the mean bond enthalpies for C=O, C–H, O–H, and C=C bonds (the bonds that are broken or formed) to estimate the enthalpy change for the tautomerism.

You can find out about using mean bond enthalpies to estimate an enthalpy change for a reaction in Section 13.3 (p.631). Note that a C=O bond is much stronger than a C=C bond, and a O–H bond is stronger than a C–H bond.

For propanone (acetone), the combined bond enthalpy of the C=O and C–H bonds (in the keto form) is higher than the combined bond enthalpy for the H–O and C=C bonds (in the enol form), so the equilibrium lies towards the keto form (CH_3COCH_3).

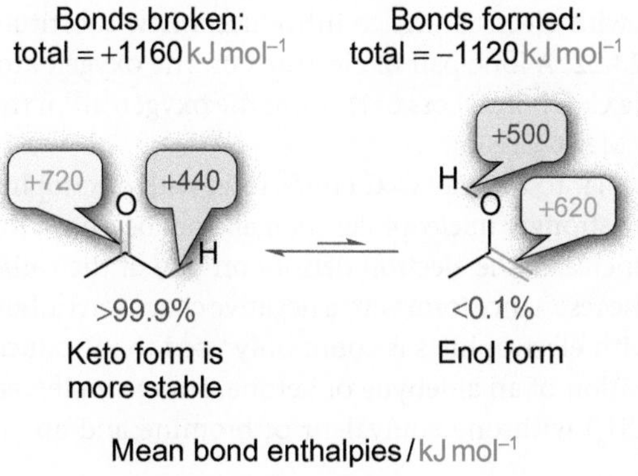
Bonds broken:
total = +1160 kJ mol⁻¹
Bonds formed:
total = −1120 kJ mol⁻¹
+720
+440
+500
+620
>99.9%
Keto form is
more stable
<0.1%
Enol form
Mean bond enthalpies / kJ mol⁻¹

For conversion of the keto form into the enol form:

$$\Delta_r H = (+720 + 440)\,\text{kJ mol}^{-1} - (500 + 620)\,\text{kJ mol}^{-1}$$
$$= +1160\,\text{kJ mol}^{-1} - 1120\,\text{kJ mol}^{-1}$$
$$= +40\,\text{kJ mol}^{-1}$$

However, the enol forms of 1,3-dicarbonyl compounds are unusually stable so the equilibrium lies more towards the enol form. For pure pentane-2,4-dione liquid ($MeCOCH_2COMe$), around 15% of an enol form ($MeC(OH){=}CHCOMe$) is present at equilibrium. An enol form of pentane-2,4-dione is stabilized by an intramolecular hydrogen bond (represented by dotted, dashed, or hashed lines, e.g. ııııııı) and by conjugation of the C=O bond with the C=C bond. The conjugated enol is formed preferentially (a less stable enol form is shown in the margin on p.1084.

The ratio of the keto:enol forms of pentane-2,4-dione can be determined from the integration curves of the resonance signals in the ¹H NMR spectrum (Section 12.3, p.580).

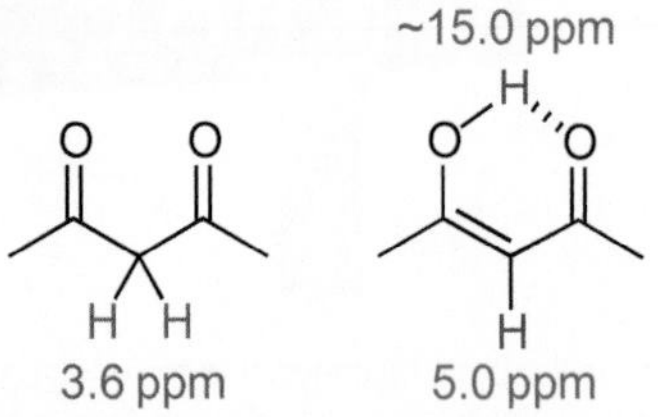
~15.0 ppm
H H
3.6 ppm
H
5.0 ppm

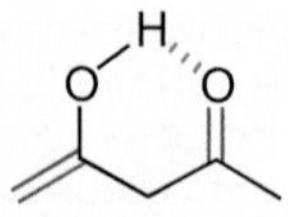

This enol form is much less stable because the C=O and C=C bonds are not conjugated

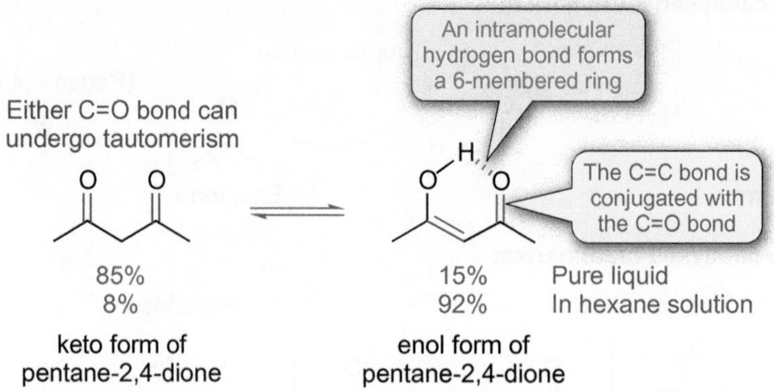
An intramolecular hydrogen bond forms a 6-membered ring
Either C=O bond can undergo tautomerism
The C=C bond is conjugated with the C=O bond
85%
8%
15%
92%
Pure liquid
In hexane solution
keto form of pentane-2,4-dione
enol form of pentane-2,4-dione

An intramolecular hydrogen bond is a relatively weak bond (typically 5–10% of the strength of a covalent bond), but this additional bond increases the stability of the enol form of pentane-2,4-dione, MeC(OH)=CHCOMe. Conjugation stabilizes the enol form because of delocalization of electrons, shown here by three resonance forms below (Section 19.1, p.876).

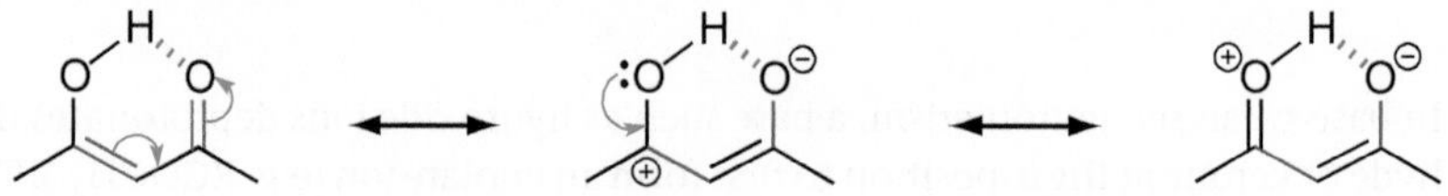

Delocalization of electrons stabilizes the enol

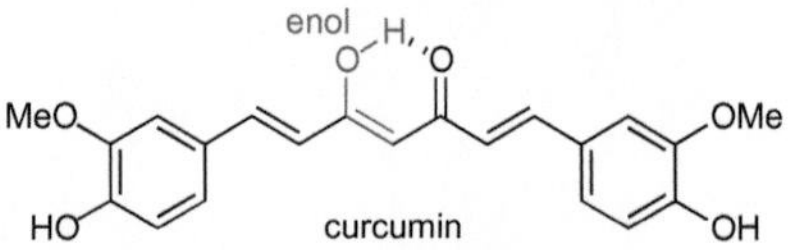
enol
MeO
OMe
HO
curcumin
OH

Curcumin, found in the yellow Indian spice turmeric, exists predominantly in the enol form in a variety of organic solvents.

Be careful when quoting the amounts of keto and enol forms at equilibrium because the presence of a solvent affects the ratio. When pentane-2,4-dione is dissolved in hexane (C_6H_{14}), around 92% of the enol form is present at equilibrium. The enol form is more stable in a non-polar solvent such as hexane because it is less polar than the keto form (in the keto form there are two very polar C=O bonds). However, the equilibrium is pushed back towards the keto form by adding a solvent that forms intermolecular hydrogen bonds to the two C=O bonds (i.e. enol stabilization by intramolecular hydrogen bonding will be less pronounced when intermolecular hydrogen bonding with the solvent competes with it).

Reactions of enols with electrophiles

Enols are nucleophiles and they react with electrophiles to introduce a new substituent at the α-position, as shown in Figure 23.22. A lone pair of electrons on the oxygen atom attacks the electrophile (E^+) through the C=C bond. Loss of H^+ from the oxygen atom then leads to the α-substituted product (e.g. $MeCOCH_2E$).

The first step in this mechanism is similar to the way C=C bonds react with electrophiles (Section 21.1, p.964). However, enols are stronger nucleophiles than alkenes because a lone pair of electrons on the oxygen atom increases the electron density on one of the carbon atoms in the C=C bond (see margin for the resonance form with a negative charge on carbon).

In synthesis, the reaction of enols with electrophiles is commonly used to introduce a halogen atom (Cl, Br, or I) at the α-position of an aldehyde or ketone. For example, reaction of propanone (acetone; CH_3COCH_3) with one equivalent of bromine and an acid

Solvent polarity and its influence on nucleophilic substitution reactions are discussed in Section 20.3, p.936.

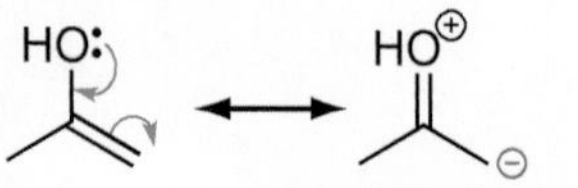

A nucleophilic carbon atom

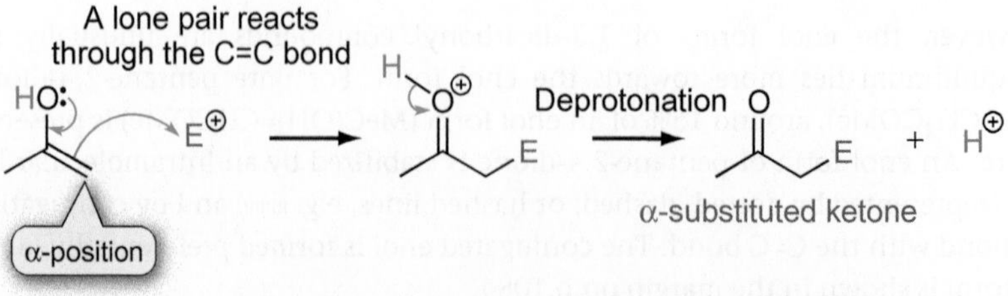
A lone pair reacts through the C=C bond
Deprotonation
α-position
α-substituted ketone

Figure 23.22 Reaction of an enol with an electrophile to form an α-substituted ketone.

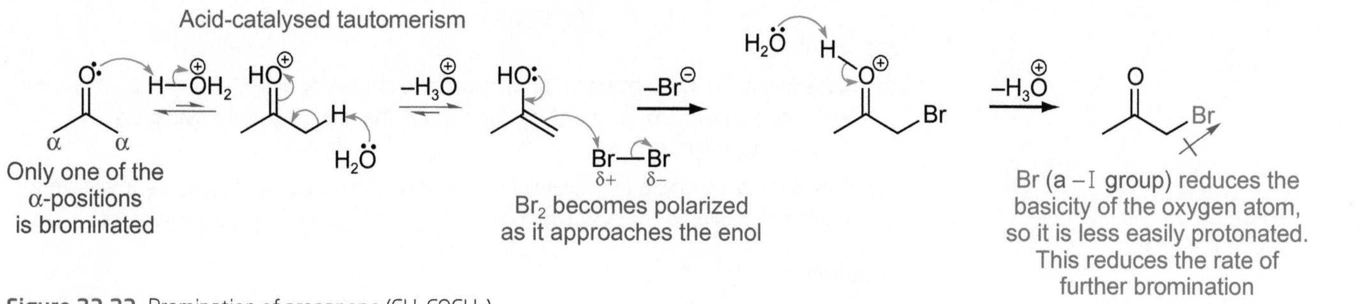
Acid-catalysed tautomerism
α
α
Only one of the α-positions is brominated
$H_2\ddot{O}$
$-H_3O^+$
$-Br^-$
Br—Br
δ+ δ−
Br_2 becomes polarized as it approaches the enol
Br
$-H_3O^+$
Br (a −I group) reduces the basicity of the oxygen atom, so it is less easily protonated. This reduces the rate of further bromination

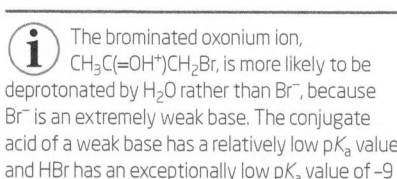

Figure 23.23 Bromination of propanone (CH_3COCH_3).

catalyst (in a solvent such as water) forms bromopropanone, CH_3COCH_2Br (see Figure 23.23). Bromination occurs because the keto form of propanone (CH_3COCH_3) is in equilibrium with a small amount of the enol form ($CH_3C(OH){=}CH_2$). The enol form reacts rapidly with Br_2 to introduce a bromine atom at the α-position. Then, deprotonation of the oxonium ion ($CH_3C({=}OH^+)CH_2Br$) forms bromopropanone. Acid-catalysed tautomerism converts more of the keto form of propanone into the enol form, to restore the equilibrium, and the process continues.

One of the reasons synthetic chemists use this reaction is because it is very selective. Propanone has six α-hydrogen atoms and you might expect to obtain a mixture of brominated products. However, this is not the case and, with one equivalent of Br_2, only one of the α-hydrogen atoms is substituted by a bromine atom. The bromination is selective because bromopropanone is less reactive than propanone. The –I effect of the bromine atom reduces the basicity of the oxygen atom in bromopropanone, so it is more difficult to protonate the C=O bond in bromopropanone than in propanone, and the rate of formation of the enol form is lower than for propanone.

Selective bromination of an unsymmetrical ketone, which can form two different enols, is illustrated in Worked example 23.4.

The products of these reactions, α-halo aldehydes (RCHXCHO) or ketones (RCHXCOR), are useful in synthesis (see margin for the Reformatsky reagent). On reaction with a base they can undergo elimination (to lose HX), to give α,β-unsaturated carbonyls (e.g. $RCH_2CHXCOR$ can be converted into RCH=CHCOR). For an alternative synthesis of α,β-unsaturated carbonyls, from aldehydes and ketones, see Section 23.4.

The brominated oxonium ion, $CH_3C({=}OH^+)CH_2Br$, is more likely to be deprotonated by H_2O rather than Br^-, because Br^- is an extremely weak base. The conjugate acid of a weak base has a relatively low pK_a value and HBr has an exceptionally low pK_a value of −9 (Section 7.2, p.310).

Compare the mechanism of bromination of an enol with that of an alkene (Section 21.3, p.974).

Interactive 3D animation of acid-catalysed bromination of propanone (CH_3COCH_3).

The Reformatsky reagent

α-Halo carbonyls ($RC({=}O)CH_2X$, where X = Br, Cl, I) are useful starting materials in synthesis. For example, the α-bromo ester $BrCH_2CO_2Et$ reacts with Zn to form a carbon nucleophile called the Reformatsky (Reformatskii) reagent. This reagent is usually drawn with a Zn–C bond, but it may contain a Zn–O bond instead.

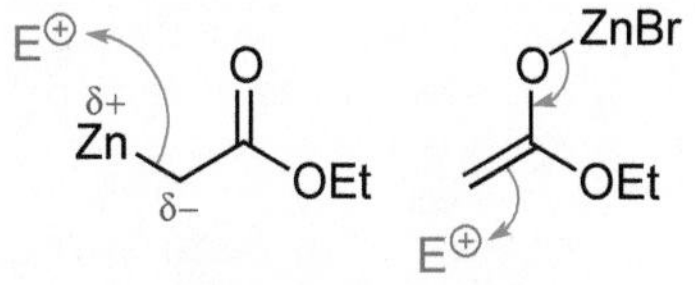
E⊕
δ+
Zn
δ−
O
OEt
ZnBr
O
OEt
E⊕

The Reformatsky reagent acts as a nucleophile

Worked example 23.4 Bromination of an unsymmetrical ketone

Unsymmetrical ketones can have more than one enol form (see, for example, the two enol forms of pentane-2,4-dione on p.1084). At equilibrium, the percentage of each enol form depends on their relative stabilities. Usually, the more substituted the C=C bond in the enol form, the more stable it is. Also, the more substituted the C=C bond, the more reactive it is to electrophiles. This is similar to the trends in stabilities and reactivities of C=C bonds in alkenes (Section 21.1, p.963).

(a) Draw the different enol forms of 2-methylcyclohexanone and predict which of these is the more stable.

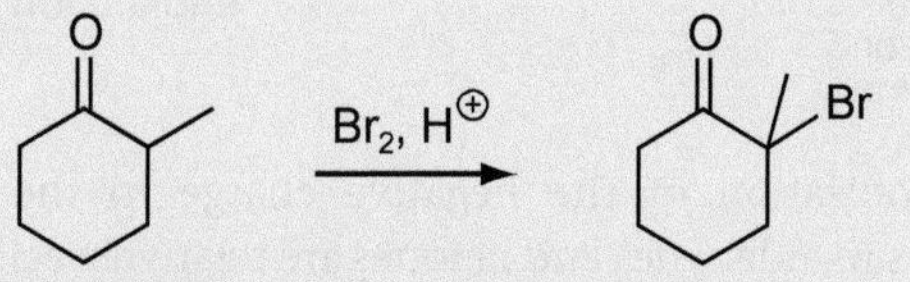
O
Br_2, H^+
O
Br

2-methylcyclohexanone

(Note that the most stable chair conformation of 2-methylcyclohexanone has the methyl group in an equatorial position, see margin (Section 18.2, p.828).)

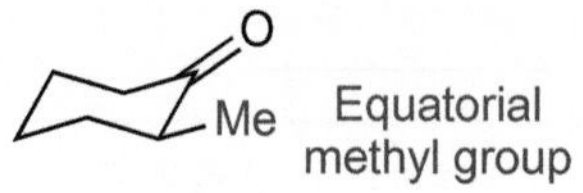
O
Me
Equatorial methyl group

(b) Suggest a mechanism for the acid-catalysed bromination of 2-methylcyclohexanone. ➜

➜ Strategy

(a) Replace the CH–C=O group in 2-methylcyclohexanone by a C=C–OH group. There are two possible positions for the C=C bond within the ring. The more stable enol has the more substituted C=C bond.

(b) Outline the mechanism of α-bromination, which involves acid-catalysed tautomerism to give the more stable enol, followed by reaction with Br_2 and, finally, deprotonation.

Solution

(a)

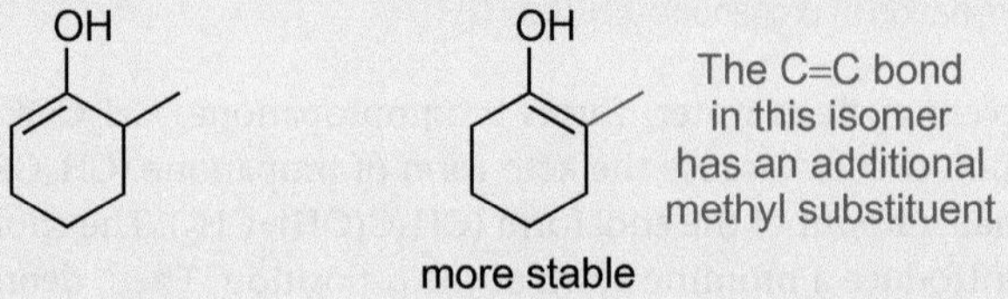
OH
OH
The C=C bond in this isomer has an additional methyl substituent
more stable

(b)

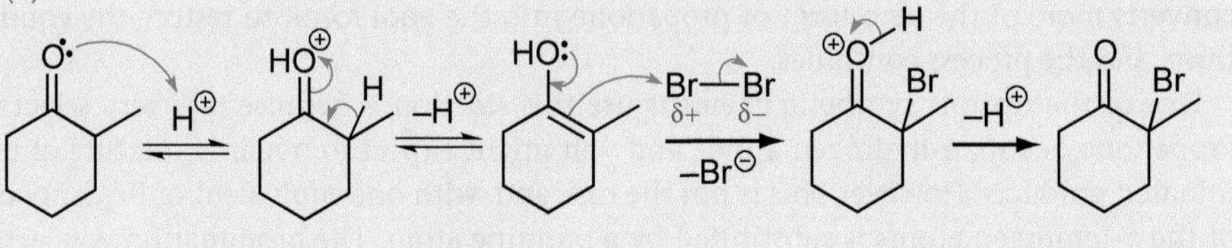
HO
H
H⊕
–H⊕
HO:
Br—Br
δ+ δ–
–Br⊖
⊕O–H
Br
–H⊕
O
Br

Question

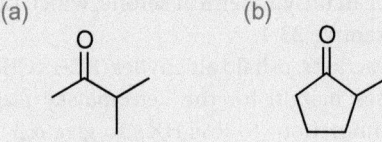

Draw the different enol forms of the following unsymmetrical ketones and, for each compound, predict which is the more stable enol form.

(a)

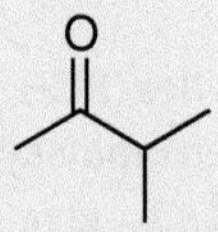

(b)

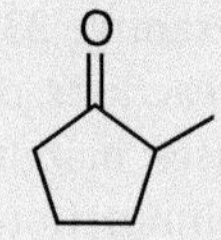

> ℹ Notice that both 2-methylcyclohexanone and the brominated product (2-bromo-2-methylcyclohexanone) contain a chiral carbon atom. Both the *R* and *S* enantiomers of 2-methylcyclohexanone give the same brominated product, because they form the same enol. Also, as the intermediate enol is planar, Br_2 can react equally well with the top and bottom faces of the enol C=C bond. Consequently, the brominated product is formed as a racemate (this is similar to the bromination of an alkene, Section 21.3, p.975).

> ℹ As drawn here, often, for simplicity, the H^+ is shown just 'dropping off'. In reality, a base (typically the solvent) is always needed to remove the H^+.

Reaction of enolate ions with electrophiles

An **enolate ion** (sometimes simply called an enolate; e.g. $RCOCH_2^-$) is formed by deprotonation of an aldehyde or ketone at the α-position in the presence of base. The negatively charged enolate ion is stabilized by delocalization and the major resonance form has the negative charge on the electronegative oxygen atom.

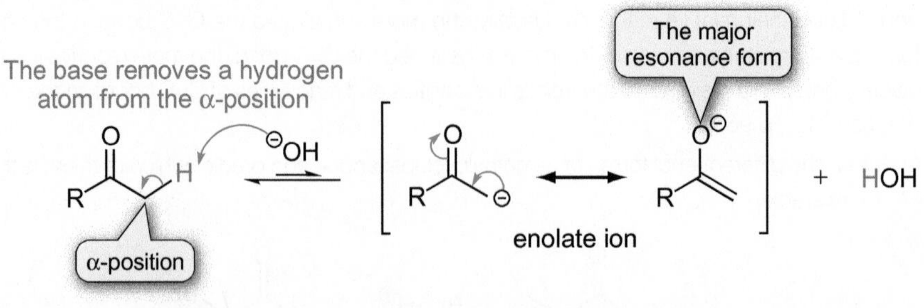
The base removes a hydrogen atom from the α-position
O
R
H
⊖OH
α-position
O
R
O⊖
R
The major resonance form
enolate ion
+ HOH

> ℹ The pK_a values of a methyl ketone ($RCOCH_3$) and H_2O are ~20 and 16, respectively. The equilibrium constant for deprotonation is equal to 10^{-4} ($10^{-pK_a\,\text{acid reactant}}/10^{-pK_a\,\text{acid product}} = 10^{-20}/10^{-16} = 10^{-4}$) so the position of equilibrium lies over towards the methyl ketone (Section 19.2, p.895). There is a list of pK_a values in Appendix 9 (p.1366).

Resonance stabilization of the negative charge in the enolate ion explains why α-hydrogen atoms in aldehydes and ketones are relatively acidic. Typically, aldehydes and ketones with α-hydrogen atoms have pK_a values around 16–21 (in comparison, alkanes have extremely high pK_a values of around 50).

Enolate ions are strong nucleophiles and, because they are negatively charged, they react more rapidly with electrophiles than enols do. Enolate ions have two electron-rich sites, so electrophiles can react at either the oxygen or carbon site, as shown in Figure 23.24. In

> ℹ Nucleophiles such as enolate ions that react with electrophiles at two different sites, to give two different products, are called **ambident nucleophiles**. Another ambident nucleophile is the cyanide ion, $^-$CN, which has lone pairs of electrons on both carbon and nitrogen, $^-$:C≡N:. Reactions normally take place on carbon as illustrated by the formation of cyanohydrins (Section 23.2, p.1068).

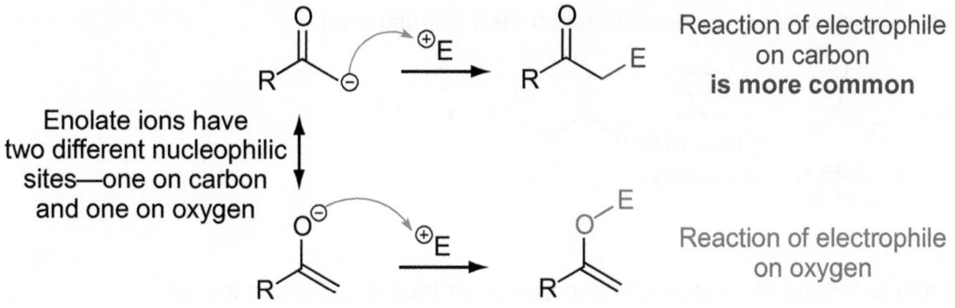
Reaction of electrophile
on carbon
is more common
Enolate ions have
two different nucleophilic
sites—one on carbon
and one on oxygen
Reaction of electrophile
on oxygen

Figure 23.24 Enolate ions can react at carbon or at oxygen.

> The fact that most electrophiles react with enolate ions on carbon can be explained by the 'hard-soft-acid-base' (HSAB) concept, which lies outside the scope of this book.

practice, most electrophiles react at the carbon site to introduce a new substituent at the alpha position of the carbonyl (though protonation usually takes place on oxygen).

For example, deprotonation of cyclopentanone ($O{=}C(CH_2)_4$) with the base sodium hydride (NaH) forms an enolate ion that reacts with bromoethane (in an S_N2 reaction; see Section 20.3, p.930), to form 2-ethylcyclopentanone. This is a useful reaction in synthesis because two smaller molecules are joined together by a new C–C bond.

> The pK_a values of cyclopentanone and H_2 are ~20 and 35, respectively. The equilibrium constant for deprotonation is equal to 10^{15} ($10^{-pK_a\,acid}/10^{-pK_a\,conjugate\,base} = 10^{-20}/10^{-35} = 10^{15}$) so the position of equilibrium lies over towards the enolate ion. Notice that as cyclopentanone is a symmetrical ketone, deprotonation at either of the alpha positions will give the same enolate ion.

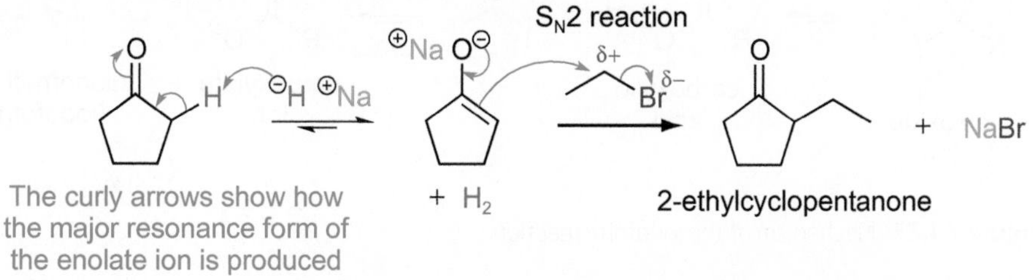
S_N2 reaction
+ NaBr
The curly arrows show how
the major resonance form of
the enolate ion is produced
+ H_2
2-ethylcyclopentanone

Triiodomethane (iodoform, CHI_3) crystals formed from reaction of ethanal (CH_3CHO) with HO^-/I_2.

Enolate ions, like enols, also react with halogens (Cl_2, Br_2, or I_2) to introduce a halogen atom at the α-position of aldehydes and ketones. Before the introduction of spectroscopic methods, this type of reaction was commonly used to identify the presence of a methyl ketone group ($RCOCH_3$) in an unknown compound. The unknown compound was treated with excess hydroxide ion and excess iodine. The formation of a yellow crystalline solid showed the presence of a methyl ketone group. The yellow crystalline solid is CHI_3 (triiodomethane, or iodoform) and the reaction is called the **iodoform reaction**.

The mechanism of the iodoform reaction is divided into three parts in Figure 23.25 and, at first sight, looks rather long. On closer inspection, you will see that the first two parts of the reaction are relatively straightforward because they involve three identical cycles of deprotonation by the hydroxide ion and subsequent reaction with I_2. These cycles lead to substitution of all three α-hydrogen atoms by iodine atoms to form $RCOCI_3$. In part 3, the hydroxide ion then acts as a nucleophile and attacks the carbon atom of the C=O bond in $RCOCI_3$. The alkoxide ion then reacts to reform the C=O bond by expelling the $^-CI_3$ ion, which is a better leaving group than ^-OH (the pK_a of CHI_3 is 14, while the pK_a of H_2O is 16). Loss of the $^-CI_3$ ion produces a carboxylic acid. The overall conversion of $RCOCI_3$ into RCO_2H is called a **nucleophilic acyl substitution reaction** (Section 24.2, p.1107)—the CI_3 group bonded to C=O is replaced by an OH group. In the last step of the reaction, the $^-CI_3$ ion deprotonates the carboxylic acid (RCO_2H) to give triiodomethane (CHI_3) and a carboxylate ion (RCO_2^-). The position of equilibrium lies towards triiodomethane and the carboxylate ion because the carboxylate ion is stabilized by resonance, so it is more stable than $^-CI_3$.

> The pK_a values of a methyl ketone ($RCOCH_3$) and H_2O are ~20 and ~16, respectively, so the equilibrium constant for deprotonation is 10^{-4}. It helps to understand the mechanism if the equilibrium arrows show that the position of equilibrium lies towards the methyl ketone. (However, many reversible reactions are drawn with two opposing arrows of equal length, ⇌, even if the equilibrium lies to one side.)

A similar mechanism explains how a methyl ketone ($RCOCH_3$) reacts with excess hydroxide ion and excess bromine to form RCO_2^- and $CHBr_3$ (tribromomethane, or bromoform). It is interesting to compare the reaction of a methyl ketone with one equivalent of bromine under acidic and basic conditions. Under acidic conditions, an enol ($RC(OH){=}CH_2$) is formed and this reacts to form a product containing one α-bromine atom, $RCOCH_2Br$ (p.1085). In contrast, under basic conditions, an enolate ion ($RCOCH_2^-$) is formed and, even when using just one equivalent of Br_2/HO^-, products containing more than one bromine atom are formed. So why is it not possible to introduce just one bromine atom under basic conditions?

> In a **nucleophilic acyl substitution reaction**, a group (or atom) bonded to C=O is replaced by another group (or atom). These reactions are discussed in more detail in Section 24.2 (p.1107).

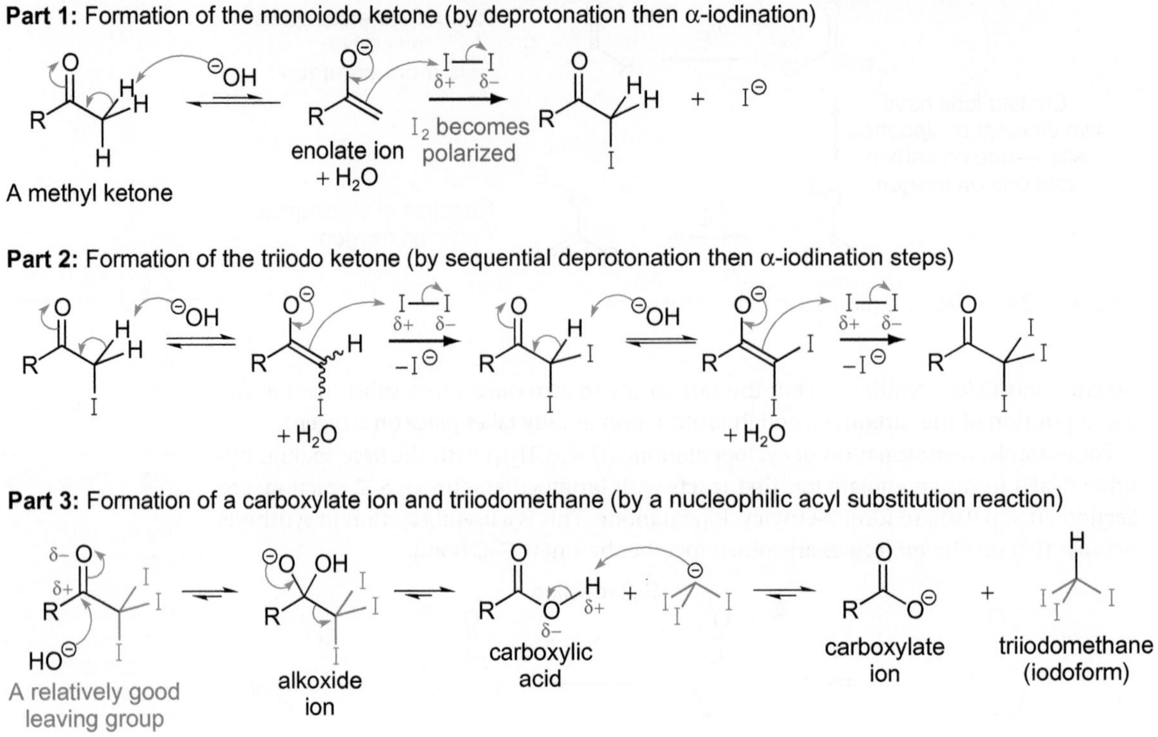
Part 1: Formation of the monoiodo ketone (by deprotonation then α-iodination)
A methyl ketone
enolate ion
+ H2O
I2 becomes polarized
Part 2: Formation of the triiodo ketone (by sequential deprotonation then α-iodination steps)
+ H2O
+ H2O
Part 3: Formation of a carboxylate ion and triiodomethane (by a nucleophilic acyl substitution reaction)
A relatively good leaving group
alkoxide ion
carboxylic acid
carboxylate ion
triiodomethane (iodoform)

Figure 23.25 Mechanism of the iodoform reaction.

In part 2 of the mechanism, the two wavy lines (∿∿∿) attached to the C=C bond in the enolate ion, RC(O⁻)=CHI, show that the configuration of the double bond could be *E*, or *Z*, or a mixture of the two. As the *E*- and *Z*-isomers react to form the same product ($RCOCHI_2$), the configuration of the enolate ion is not important.

Interactive 3D animation of base-catalysed bromination of propanone (CH_3COCH_3).

Products with more than one bromine atom are formed because introducing a bromine atom at the α-position of a methyl ketone makes it more reactive to Br_2/HO^- (i.e. the order of relative reaction rates to Br_2/HO^- is: $RCOCHBr_2 > RCOCH_2Br > RCOCH_3$). Bromine is electron-withdrawing (a –I group) and it makes the remaining α-hydrogen atoms more acidic—the greater the number of α-bromine atoms, the more acidic the remaining α-hydrogen atoms. Acidity affects the rate of bromination because, the more acidic the α-hydrogen atoms, the more rapidly they react with HO^- to form enolate ions, which subsequently react with Br_2 to introduce α-bromine atoms.

Summary

- α-Substitution reactions involve substitution of a hydrogen atom by another group, at the carbon atom next to the C=O bond.
- α-Substitution reactions involve intermediate enols ($RC(OH)=CH_2$) or enolate ions ($RC(=O)CH_2^-$).
- The interconversion of the keto and enol forms of an aldehyde or ketone is called tautomerism.
- Formation of an enol and reaction with an electrophile (E^+):

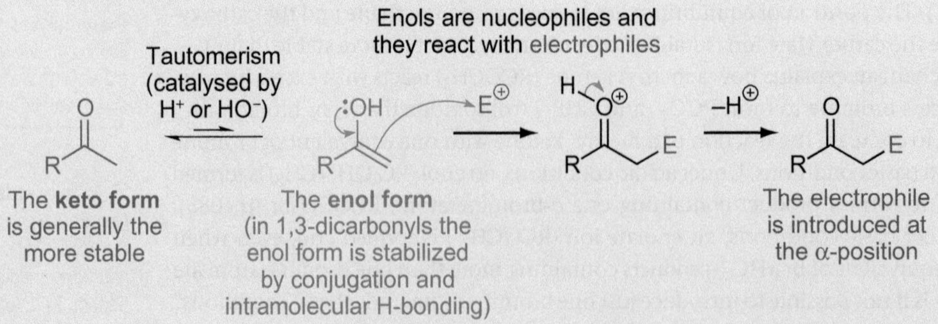
Enols are nucleophiles and they react with electrophiles
Tautomerism (catalysed by H⁺ or HO⁻)
The keto form is generally the more stable
The enol form (in 1,3-dicarbonyls the enol form is stabilized by conjugation and intramolecular H-bonding)
–H⁺
The electrophile is introduced at the α-position

➜

- Formation of an enolate ion and reaction with an electrophile (E^+):

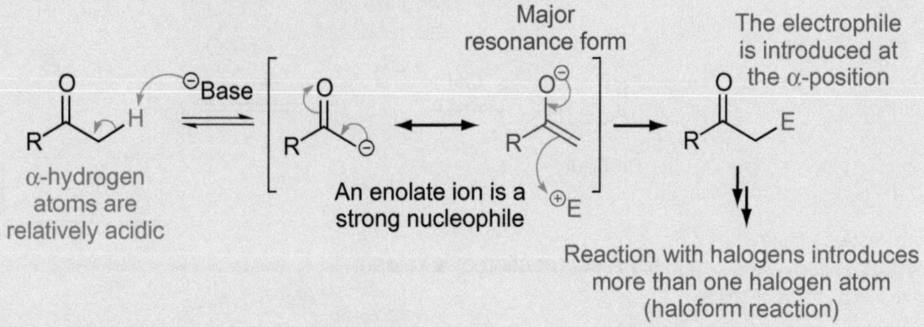

Major resonance form
The electrophile is introduced at the α-position
Base
α-hydrogen atoms are relatively acidic
An enolate ion is a strong nucleophile
Reaction with halogens introduces more than one halogen atom (haloform reaction)

For a practice question on these topics, see question 5 at the end of this chapter (p.1097).

23.4 Carbonyl-carbonyl condensation reactions

In carbonyl–carbonyl condensation reactions, two carbonyl molecules react together to form a single organic product together with a small molecule, typically water. For aldehydes and ketones, the mechanisms of these reactions involve two parts. In the first part, two aldehyde or ketone molecules react so that one of the molecules undergoes an α-substitution reaction (Section 23.3) and the other molecule undergoes a nucleophilic addition reaction (Section 23.2). In the second part of the reaction, the organic product loses water, so, overall, the reaction is a condensation.

There are different classes of carbonyl–carbonyl condensation reactions. Reactions of two molecules of the *same* aldehyde or ketone are called **aldol condensations**. Reactions of two *different* aldehyde or ketone molecules are called **crossed (mixed) aldol condensations**. Finally, when one molecule containing two aldehyde groups (a dial; e.g. hexanedial, $OHC(CH_2)_4CHO$) or two ketone groups (a dione; e.g. hexane-2,5-dione, $CH_3CO(CH_2)_2COCH_3$) reacts with itself, this is called an **intramolecular aldol condensation**. These reactions are some of the most useful in synthesis because they make larger molecules by joining together small molecules by forming a new C–C bond. They are also used to convert linear molecules into rings.

CHANDOS HISTORICAL
CHANDOS
BORODIN
String Quartets Nos 1 & 2
BORODIN QUARTET

The Russian composer Alexander P. Borodin (1833–1887) earned a living as a chemist. He discovered the aldol reaction in 1872.

The nucleophilic addition step in an aldol reaction requires one molecule of the aldehyde or ketone to act as a nucleophile and another molecule to act as an electrophile.

Aldol condensations

The mechanism of an aldol condensation is shown in Figures 23.26 and 23.27. In the first stage, two molecules of an aldehyde or a ketone react (in the presence of a base at room temperature or below) in an aldol reaction (Figure 23.26). To start the reaction, a base (e.g. HO^-) removes an α-hydrogen atom from one molecule of the aldehyde or ketone to form an enolate ion (such as $O{=}CH{-}CH_2^- \leftrightarrow {}^-O{-}CH{=}CH_2$). The enolate ion is a nucleophile and it attacks a second molecule of the aldehyde or ketone to form an alkoxide ion (such as $OHCCH_2{-}CH(O^-)CH_3$)—a new bond forms between an α-carbon atom of one molecule and the carbon atom in the C=O bond of a second molecule. In the last step, protonation of the alkoxide ion forms a β-hydroxy-aldehyde (an aldol, such as $OHCCH_2{-}CH(OH)CH_3$) or a β-hydroxy-ketone (such as $CH_3C({=}O)CH_2{-}C(OH)(CH_3)_2$). The name aldol comes from the name of the product formed from an aldehyde, which contains both an *ald*ehyde and an alcoh*ol* group.

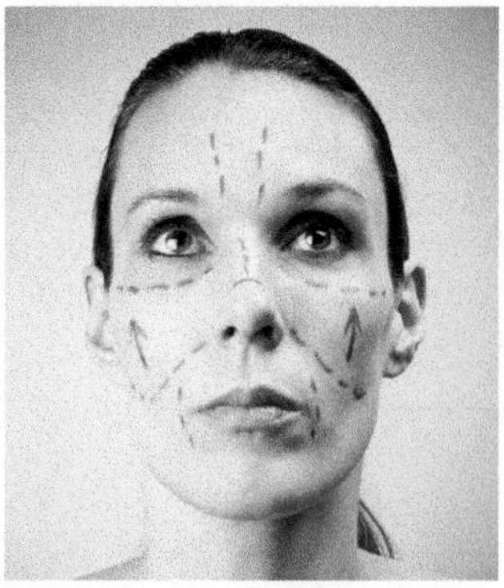

Collagen injections are widely used in plastic surgery, for example, to increase the fullness of lips. In our bodies, collagen molecules are cross-linked using aldol condensation reactions. Collagen is a fibrous protein that holds the cells together in your body and the cross-linking increases its rigidity.

Notice that all three steps in an aldol reaction are reversible (the reverse of an aldol reaction is called a **retro-aldol reaction**). The position of the equilibrium depends on the nature of the carbonyl compound. With an aldehyde (such as ethanal, CH_3CHO), the position of equilibrium usually lies more to the aldol product. For a ketone substrate (such as propanone, CH_3COCH_3), the position of equilibrium usually lies more to the ketone starting material. Conversion of an aldehyde into an aldol product is faster than the corresponding reaction of a ketone because the C=O bond in an aldehyde is more reactive

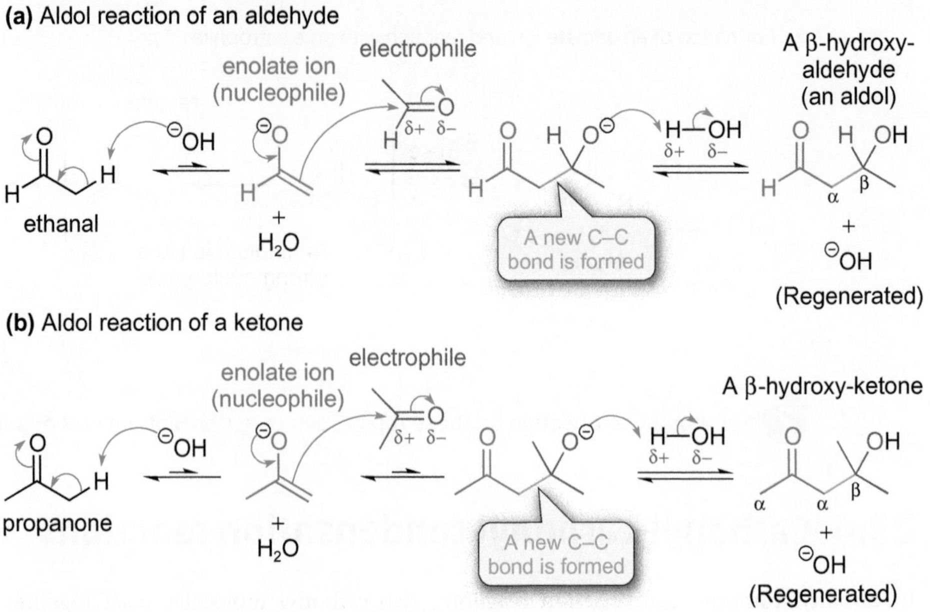
(a) Aldol reaction of an aldehyde
enolate ion
(nucleophile)
electrophile
A β-hydroxy-
aldehyde
(an aldol)
ethanal
+
H_2O
A new C–C
bond is formed
(Regenerated)
(b) Aldol reaction of a ketone
enolate ion
(nucleophile)
electrophile
A β-hydroxy-ketone
propanone
+
H_2O
A new C–C
bond is formed
(Regenerated)

Visit the Online Resource Centre to view screencast 23.3 which walks you through the mechanisms of some aldol condensations.

In parts (a) and (b), notice that for the deprotonation steps, the curly arrows are drawn to show the major resonance forms of the enolate ions (which have the negative charge on oxygen).

Figure 23.26 Mechanism of the aldol reaction for (a) an aldehyde and (b) a ketone.

Reaction of two molecules of an aldehyde or ketone to form a β-hydroxy-aldehyde or β-hydroxy-ketone, respectively, is called an **aldol reaction**. Formation of the β-hydroxy-aldehyde or β-hydroxy-ketone followed by elimination of water is an **aldol condensation**. Sometimes the name aldol reaction is used to describe an aldol condensation reaction.

to nucleophilic addition by an enolate ion than the C=O bond in a ketone (for steric and electronic reasons, Section 23.2, p.1061).

Acids catalyse the elimination of water from a β-hydroxy-aldehyde or β-hydroxy-ketone. The elimination proceeds via an enol, which expels H_2O (a good leaving group).

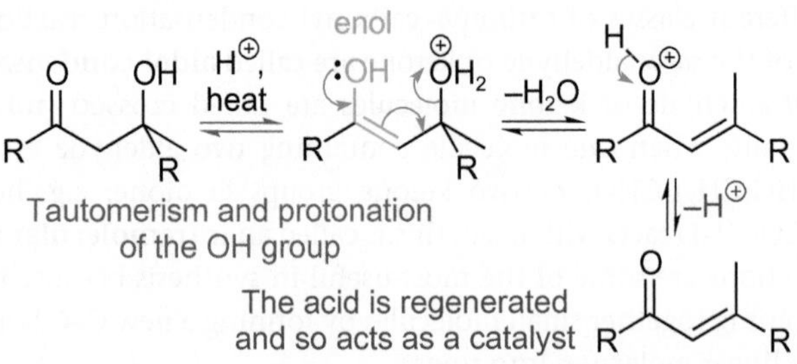
heat
enol
$-H_2O$
Tautomerism and protonation
of the OH group
The acid is regenerated
and so acts as a catalyst

In the second stage of an aldol condensation, the β-hydroxy-aldehyde or β-hydroxy-ketone is heated with an acid catalyst (see the note above) or a base catalyst, such as HO^-, to eliminate water. When using a base, the mechanism of the elimination (see Figure 23.27) involves removing an α-hydrogen atom from the β-hydroxy-aldehyde or β-hydroxy-ketone to form an enolate ion (even though the α-hydrogen atoms are less acidic than the H atom in the OH group, a small amount of the enolate ion is formed on heating). The enolate ion then expels HO^- to form a new C=C bond. Usually, HO^- is a poor leaving group but, in this elimination, the position of equilibrium lies to the right because the product is so stable. Either an enal (such as HC(=O)–CH=$CHCH_3$; but-2-enal) or an enone (such as CH_3C(=O)–CH=$C(CH_3)_2$; 4-methylpent-3-en-2-one) is formed, both of which have conjugated C=O and C=C bonds. Because of the position of the C=C

The stability of C=C bonds is discussed in Section 21.1 (p.962). A conjugated enal or enone is more stable than a non-conjugated enal or enone (such as $H_3C(=O)CH_2CH=CH_2$) because of the interaction of the C=C π bond with the C=O π bond. Overlap of orbitals on adjacent double bonds allows electrons to be delocalized over the molecule, lowering the energy of the molecule and making it more stable.

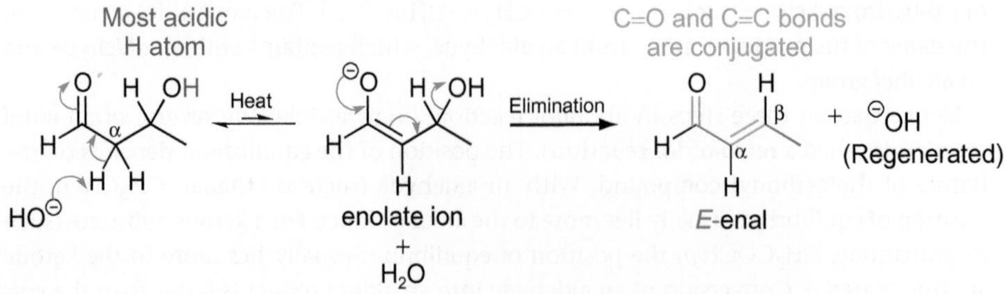
Most acidic
H atom
Heat
C=O and C=C bonds
are conjugated
Elimination
enolate ion
+
H_2O
E-enal
(Regenerated)

Interactive 3D animation showing the mechanism of an acid-catalysed aldol reaction.

Figure 23.27 Base-catalysed elimination of water in an aldol condensation.

bond, relative to the C=O bond, these compounds are collectively called **α,β-unsaturated carbonyls**. Note that the *E*-isomer of the C=C bond is usually formed, because this isomer is more stable than the *Z*-isomer (in the *E*-isomer the largest groups are on the opposite sides of the C=C bond).

The aldol condensation reaction of cyclohexanone, $(CH_2)_5C{=}O$, is discussed in Worked example 23.5.

Worked example 23.5 An aldol condensation reaction

Reaction of cyclohexanone with hydroxide ions forms product **A** from an aldol reaction. On heating with an acid catalyst, **A** is converted into the aldol condensation product **B** and water.

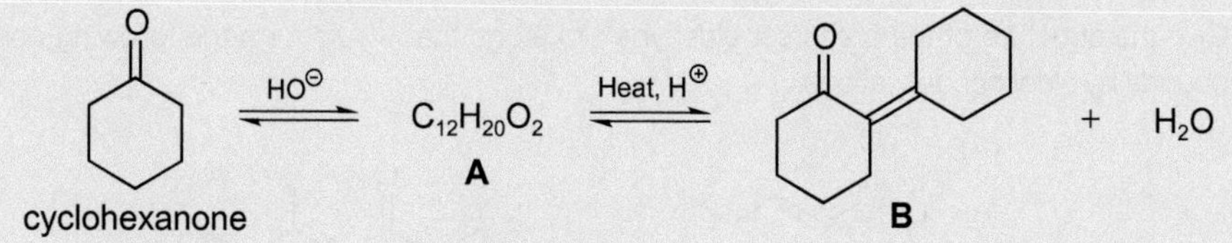
HO⊖
$C_{12}H_{20}O_2$
A
Heat, H⊕
+ H_2O
cyclohexanone
B

(a) Give a structure for **A**.

(b) Give a mechanism for the formation of **A**.

(c) Give a mechanism to explain how **A** is converted into **B** via an enol.

Strategy

(a) The molecular formula for **A** shows that it is formed from two molecules of cyclohexanone. In an aldol reaction, two molecules of a ketone react together to form a β-hydroxy-ketone.

(b) The nucleophilic addition step in an aldol reaction requires one molecule of cyclohexanone to act as a nucleophile and another molecule to act as an electrophile.

In the presence of hydroxide ions, the reaction proceeds via an enolate ion, which reacts as the nucleophile.

(c) The acid catalyses the elimination of H_2O from **A** to form the C=C bond in **B**.

Acid-catalysed tautomerism of the β-hydroxy-ketone gives an enol form, which reacts as a nucleophile.

Protonation of **A** converts the OH group into a good leaving group.

Solution

(a)

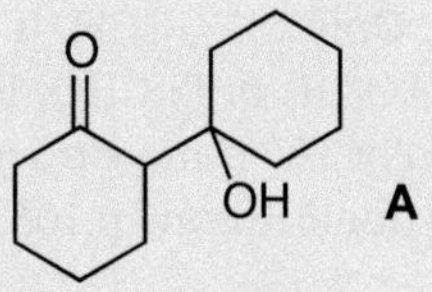
O
OH
A

(b)

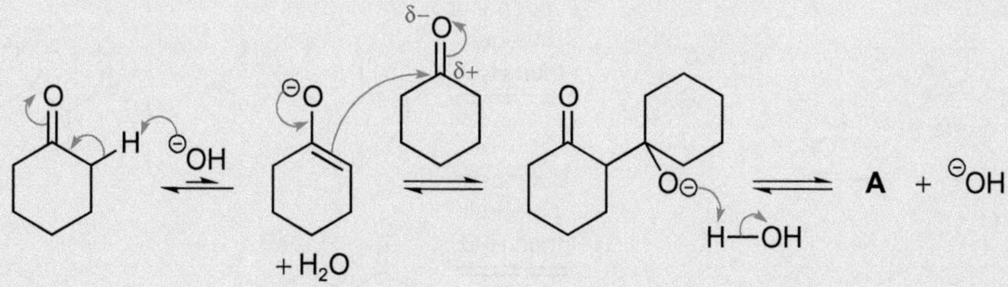
H
⊖OH
⊖O
+ H_2O
δ−
δ+
O⊖
H—OH
A + ⊖OH

→

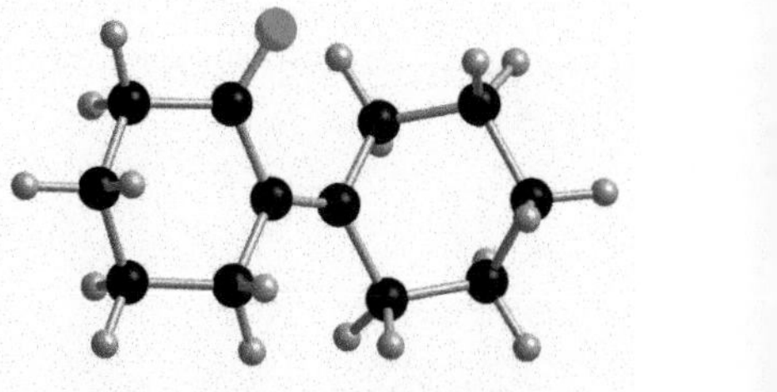

A molecular model of the aldol condensation product **B** (2-cyclohexylidenecyclohexanone). In this computer generated ball and stick model a single stick represents the C=O and C=C double bonds. Remember that 6-membered saturated carbon rings, such as cyclohexanone, typically adopt a chair conformation (Section 18.2, p.827).

Notice that cyclohexanone, like propanone, is a symmetric ketone and so deprotonation at either of the alpha positions will give the same enolate ion. In contrast, deprotonation of an asymmetric ketone with hydrogen atoms at both α-carbons (such as $H_3CC({=}O)CH_2CH_3$; butan-2-one) can lead to two different enolate ions, which can each react with a molecule of the ketone to form a mixture of aldol condensation products. To maximize the yield of the desired aldol condensation product, synthetic chemists have developed methods to selectively form just one of the enolate ions (such as $^-H_2C({=}O)CH_2CH_3$ rather than $H_3CC({=}O)CH^-CH_3$). See a specialist organic synthesis book for further details.

Notice that compound **A** (2-(1′-hydroxycyclohexyl) cyclohexanone) contains a chiral carbon atom. As the intermediate cyclohexanone enolate ion is planar, cyclohexanone can react equally well with the top and bottom faces of the enolate ion. Consequently, compound **A** is formed as a racemate.

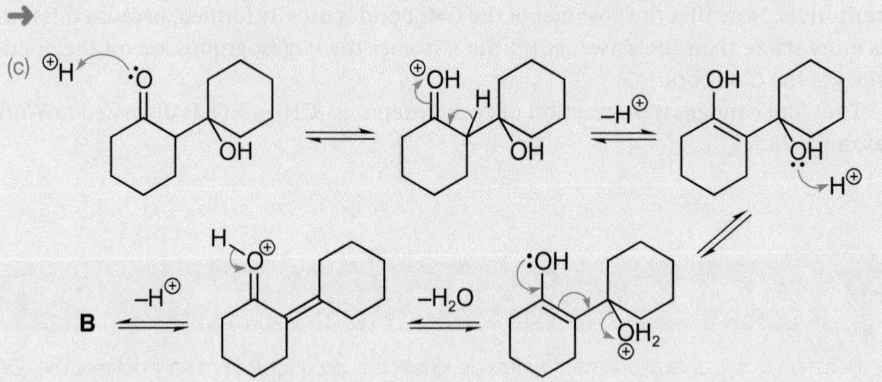
(c)
-H⊕
-H₂O
B

Question

Give the structure of the precursor aldehyde or ketone that would form the following compounds by aldol condensations.

(a)

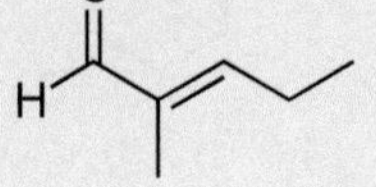

(b)

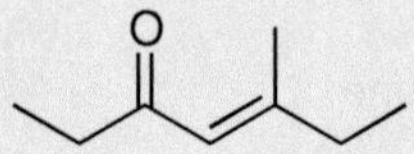

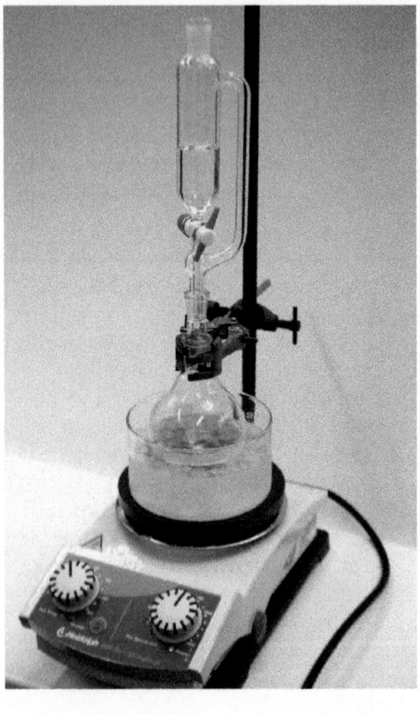

A typical experimental set-up for slowly adding a solution of a reactant to a stirred solution of another reactant.

Crossed aldol condensations

Crossed (mixed) aldol reactions involve the reaction of two *different* aldehydes or ketones. If the two aldehydes and ketones can form enolate ions equally well, and can both act as electrophiles in nucleophilic additions, then reaction with a base (at room temperature) produces a mixture of all four possible aldol products, as shown in Figure 23.28 for the reaction between propanal (CH_3CH_2CHO) and butanal ($CH_3CH_2CH_2CHO$). Because the maximum yield of any one product is low, this type of reaction is rarely used in synthesis.

A crossed aldol reaction does give a single product in high yield under the following conditions:

- when only *one* of the carbonyls is able to form an enolate ion;
- when the carbonyl compound unable to form an enolate ion is more reactive to nucleophilic addition than the carbonyl compound that can form an enolate ion;
- when the carbonyl compound able to form an enolate ion is added slowly to the reaction mixture.

For example, PhCHO reacts with CH_3COCH_3 in a crossed aldol reaction to selectively form one product, as shown in Figure 23.29. Only CH_3COCH_3 is able to form an enolate ion so $CH_3COCH_2^-$ can react with either CH_3COCH_3 or PhCHO. But the reaction of $CH_3COCH_2^-$ with PhCHO is faster because aldehyde PhCHO is more reactive to nucleophilic addition than ketone CH_3COCH_3 (for steric and electronic reasons, p.1061). Also, the CH_3COCH_3 is added drop-

The wavy bonds (〰) in the enolate ions show that *E*- and *Z*-isomers are possible.

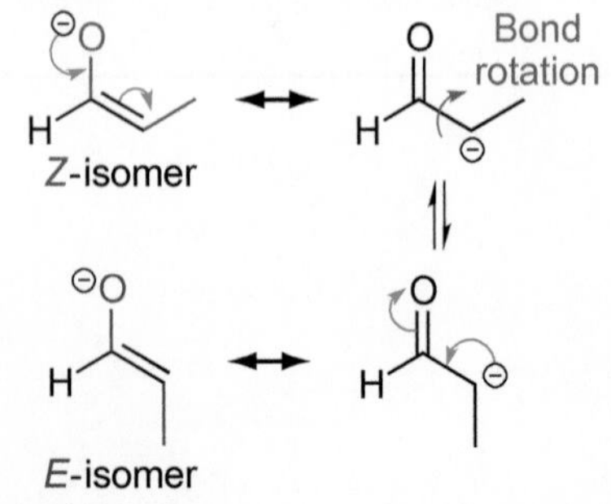
Bond rotation
Z-isomer
E-isomer

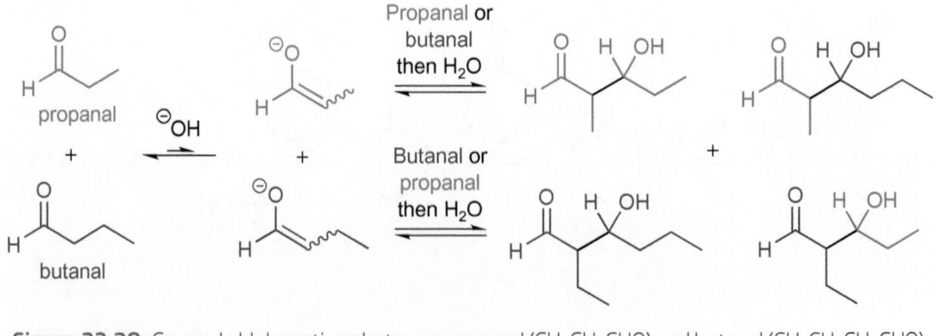
propanal
butanal
Propanal or butanal then H₂O
Butanal or propanal then H₂O

Figure 23.28 Crossed aldol reactions between propanal (CH_3CH_2CHO) and butanal ($CH_3CH_2CH_2CHO$) to give four aldol products.

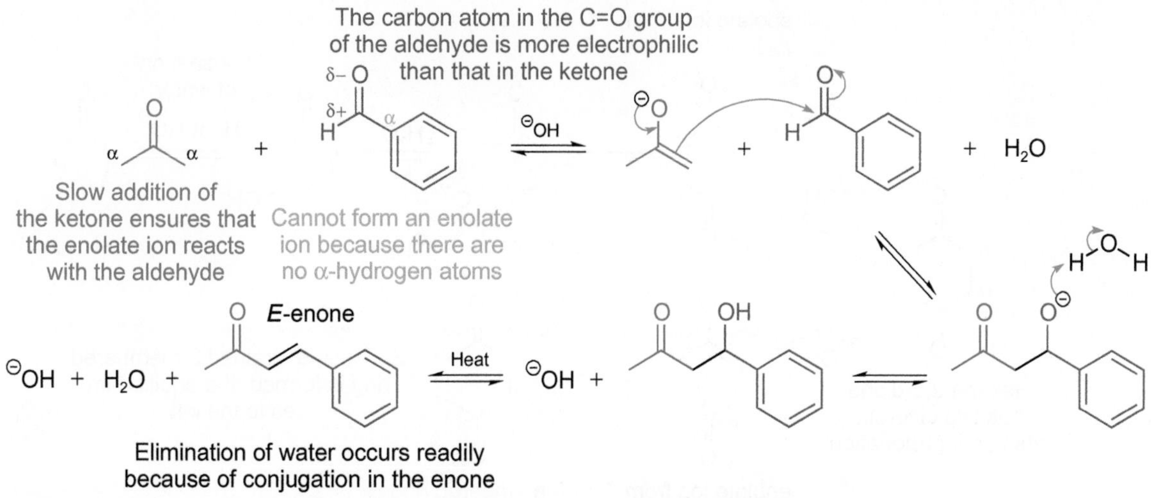
The carbon atom in the C=O group
of the aldehyde is more electrophilic
than that in the ketone
Slow addition of
the ketone ensures that
the enolate ion reacts
with the aldehyde
Cannot form an enolate
ion because there are
no α-hydrogen atoms
E-enone
Heat
Elimination of water occurs readily
because of conjugation in the enone

Figure 23.29 Crossed aldol condensation between propanone (CH_3COCH_3) and benzaldehyde (PhCHO) to give a single α,β-unsaturated ketone.

wise to a mixture of PhCHO/HO^-, so that, as soon as $CH_3COCH_2^-$ forms, it reacts with PhCHO (because there is very little CH_3COCH_3 in the reaction mixture). Heating the β-hydroxy-ketone ($CH_3COCH_2CH(OH)Ph$; 4-hydroxy-4-phenylbutan-2-one) with hydroxide ions leads to the elimination of water, to give a single α,β-unsaturated ketone ($CH_3COCH{=}CHPh$; 4-phenylbut-3-en-2-one) derived from an aldol condensation reaction. Note that the *E*-isomer of the C=C bond is usually formed, because this isomer is more stable than the *Z*-isomer.

The IUPAC name for the *E*-isomer of $CH_3COCH{=}CHPh$ is (*E*)-4-phenylbut-3-en-2-one (Section 18.3, p.833). Notice that the newly formed C=C bond is stabilized by conjugation with the C=O bond *and* the benzene ring.

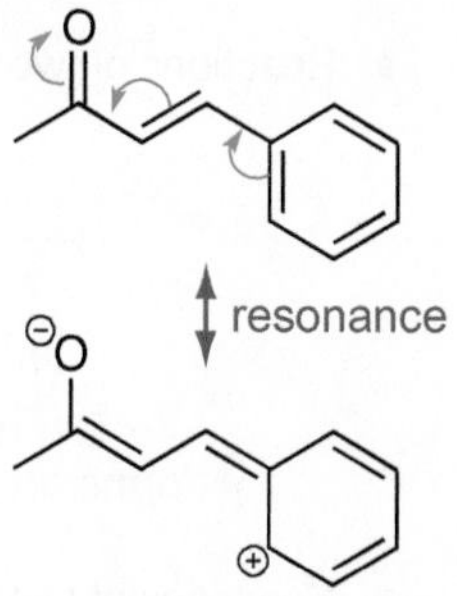
resonance

Intramolecular aldol condensations

In an **intramolecular aldol condensation reaction**, a starting material with two C=O bonds (a dicarbonyl) is heated with a base and this forms a cyclic product. For example, the intramolecular aldol condensation reaction of hexanedial ($OHC(CH_2)_4CHO$) shown in Figure 23.30 gives a product with a 5-membered ring (called cyclopent-1-ene-1-carbaldehyde, C_6H_8O). The ring is formed by nucleophilic addition of the intermediate enolate ion onto a C=O bond within the same molecule. This type of intramolecular reaction to form a ring is called a **cyclization** reaction. When drawing mechanisms of cyclizations, it helps to draw the precursor in a conformation that resembles the ring that is being formed, so the structure of hexanedial is drawn to resemble a 5-membered ring.

In theory, some dicarbonyl compounds could react to form mixtures of various sized rings but, in practice, a 5- or 6-membered ring is formed because these rings are the most stable and the easiest to make. Angle strain explains why the intramolecular aldol condensation reaction of hexane-2,5-dione ($CH_3COCH_2CH_2COCH_3$) shown in Figure 23.31 gives a product with a 5-membered ring (called 3-methylcyclopent-2-en-1-one, C_6H_8O) rather than a 3-membered ring.

An **intramolecular** reaction is a reaction between different parts of the same molecule and, when a ring is formed, the reaction is called a **cyclization**.

Rings containing more than 6 atoms are difficult to make because, as the chain length increases, the likelihood of the two reacting centres coming together to form the ring decreases. 3- and 4-Membered rings are also difficult to make because they have high strain energy due to the bond angles of the small ring (Section 18.2, p.826).

Cyclic alkenes with 3–7 carbons in the ring are fixed in the *cis*- configuration as the *trans*- configuration has too much ring strain. So, because the C=C bond in cyclopentene must be *cis*-, it is not shown in the name (Section 2.5, p.85).

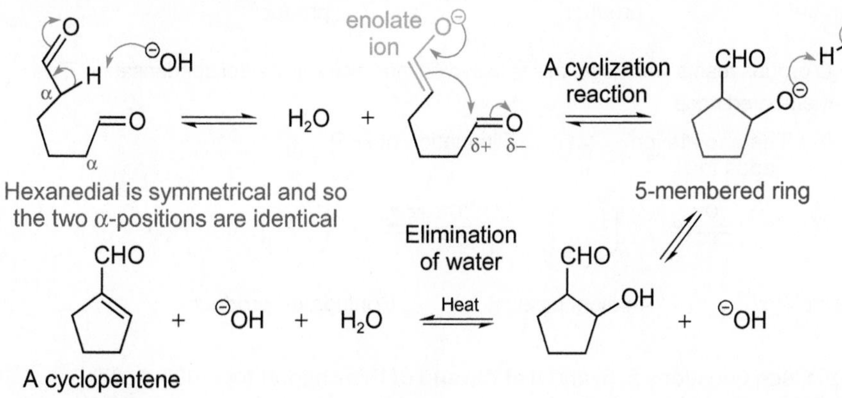
enolate
ion
A cyclization
reaction
Hexanedial is symmetrical and so
the two α-positions are identical
5-membered ring
Elimination
of water
Heat
A cyclopentene

Figure 23.30 Intramolecular aldol condensation reaction of hexanedial ($OHC(CH_2)_4CHO$) to give a 5-membered ring.

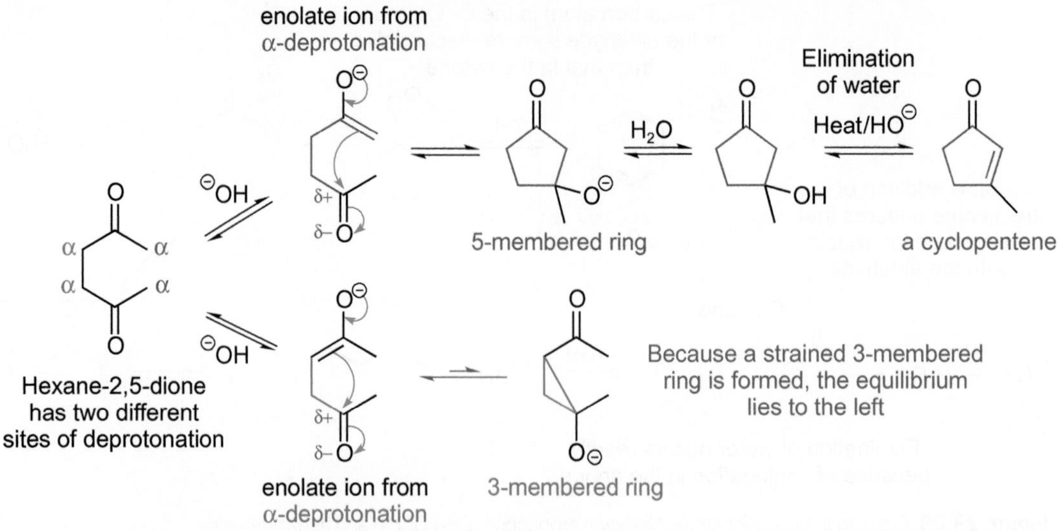
enolate ion from
α-deprotonation
Elimination of water
H_2O
Heat/HO⊖
⊖OH
δ+
δ–
α α α α
5-membered ring
a cyclopentene
⊖OH
Hexane-2,5-dione
has two different
sites of deprotonation
Because a strained 3-membered
ring is formed, the equilibrium
lies to the left
enolate ion from
α-deprotonation
3-membered ring

Figure 23.31 Intramolecular aldol condensation of hexane-2,5-dione ($CH_3COCH_2CH_2COCH_3$) gives a 5-membered ring product rather than a 3-membered ring.

Summary

- In carbonyl–carbonyl condensation reactions, two carbonyl molecules react to form a single organic product together with a molecule of water.
- Reactions of two molecules of the *same* aldehyde or ketone are called aldol condensations.

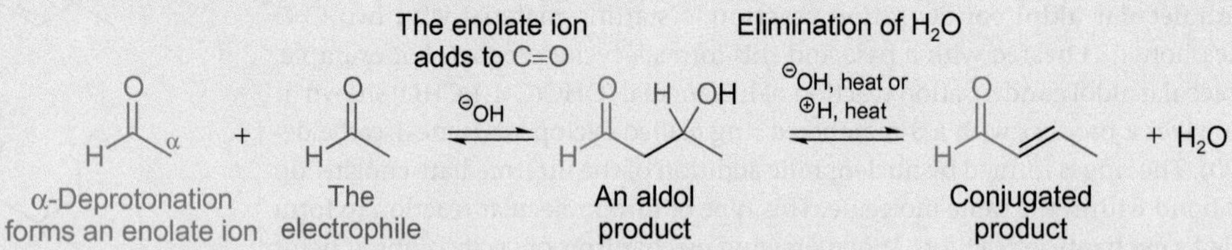
The enolate ion
adds to C=O
Elimination of H_2O
⊖OH
⊖OH, heat or
⊕H, heat
+ H_2O
α-Deprotonation
forms an enolate ion
The
electrophile
An aldol
product
Conjugated
product

- Reactions of two molecules of different aldehydes or ketones are called crossed (mixed) aldol condensations—a mixture of products is formed, except under certain conditions.

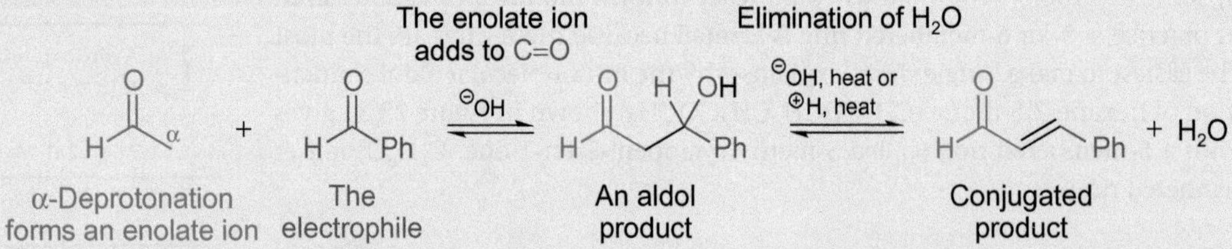
The enolate ion
adds to C=O
Elimination of H_2O
⊖OH
⊖OH, heat or
⊕H, heat
+ H_2O
α-Deprotonation
forms an enolate ion
The
electrophile
An aldol
product
Conjugated
product

- When one molecule containing two C=O groups reacts with itself, this is called an intramolecular aldol condensation. This is a good method for making 5- and 6-membered rings.

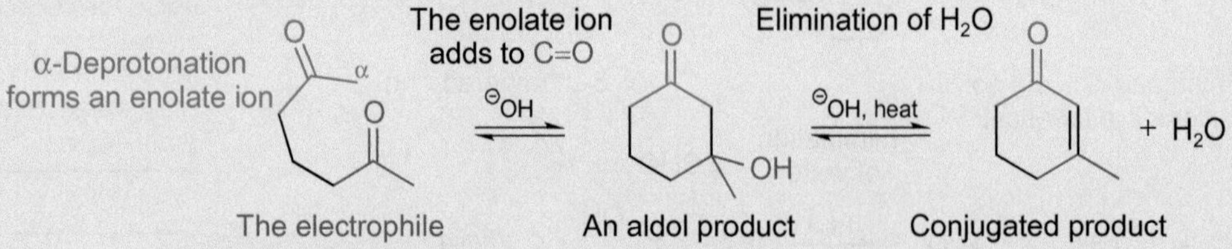
α-Deprotonation
forms an enolate ion
The enolate ion
adds to C=O
Elimination of H_2O
⊖OH
⊖OH, heat
+ H_2O
The electrophile
An aldol product
Conjugated product

For practice questions on this topic, see questions 2, 5, and 6 at the end of this chapter (pp.1095–1097).

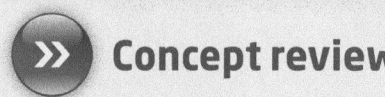

Concept review

By the end of this chapter, you should be able to do the following.

- For an aldehyde or ketone, describe how the C=O bond reacts with H^+.
- For an aldehyde or ketone, describe how the C=O bond reacts with a nucleophile (in a nucleophilic addition) and how enolate ions are formed by deprotonation of α-hydrogen atoms.

Nucleophilic addition versus α-deprotonation

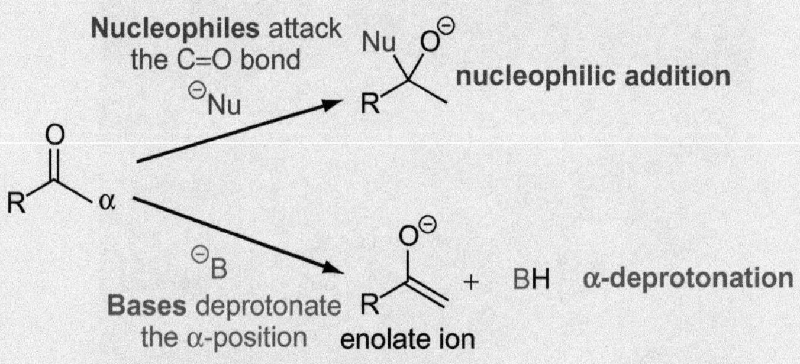
Nucleophiles attack the C=O bond
Nu
nucleophilic addition
R
O
α
B
Bases deprotonate the α-position
+ BH
α-deprotonation
enolate ion

- Draw general mechanisms for a nucleophilic addition reaction, an α-substitution reaction, and a carbonyl–carbonyl condensation reaction.
- Understand why aldehydes are more reactive to nucleophiles than ketones.
- Draw mechanisms for nucleophilic addition reactions of aldehydes and ketones using the following reagents:

$NaBH_4$ or $LiAlH_4$ then H^+	HCN, NaCN (catalyst)
RLi or RMgBr then H^+	$Ph_3P{=}CHR$
H_2O, H^+ (catalyst)	ROH, H^+ (catalyst)
RSH, H^+ (catalyst)	RNH_2, H^+ (catalyst)

- Understand how aldehydes and ketones are prepared by oxidation of alcohols.
- Recognize that addition of oxygen, sulfur, or nitrogen nucleophiles to the C=O bond is reversible and is catalysed by an acid.
- Discuss keto–enol tautomerism and understand the factors that influence the stability of keto and enol forms.
- Recognize that α-substitution reactions involve intermediate enols or enolate ions.
- Draw mechanisms for α-substitution reactions of aldehydes and ketones using the following reagents: H^+, Br_2; base, RX (where X is a good leaving group); HO^-, I_2 (both in excess).
- Draw mechanisms for aldol condensations, crossed (mixed) aldol condensations, and intramolecular aldol condensations.
- Predict the structure of a product derived from a nucleophilic addition, an α-substitution, or a carbonyl–carbonyl condensation reaction of an aldehyde or ketone.
- Propose reagents for converting an aldehyde or ketone into a product derived from a nucleophilic addition reaction, an α-substitution reaction, or a carbonyl–carbonyl condensation reaction.

Questions

More challenging questions are marked with an asterisk *.

1 The following questions are based on the reactions of ethanal (**1**) shown here. (Section 23.2)

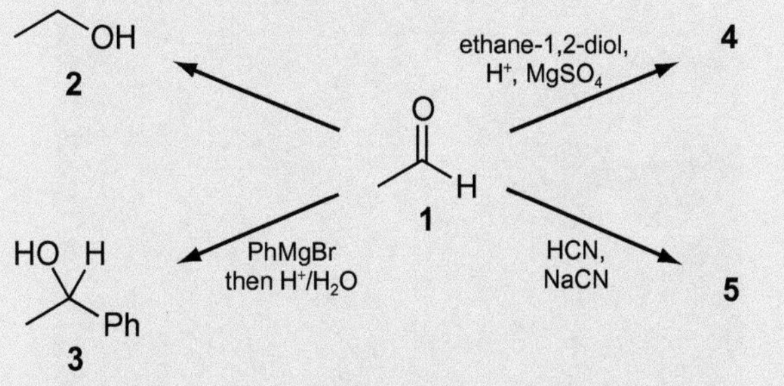
OH
2
ethane-1,2-diol, H⁺, MgSO₄
4
O
H
1
HO H
Ph
3
PhMgBr then H⁺/H₂O
HCN, NaCN
5

(a) Give appropriate reagents for converting **1** into **2**. Is this an example of an oxidation or a reduction reaction?

(b) Draw a reaction mechanism to show how **1** is converted into **3**.

(c) Suggest a method for preparing PhMgBr.

(d) Give structures for organic compounds **4** and **5**.

2 The following questions are based on the reactions of benzaldehyde (**6**) shown here. (Sections 23.2 and 23.4)

(a) Draw the structures of organic compounds **7–9**.

(b) Suggest reagents that will convert **6** into **10**, and **10** into **11**.

(c) Give a mechanism that explains how **6** is converted into **12**.

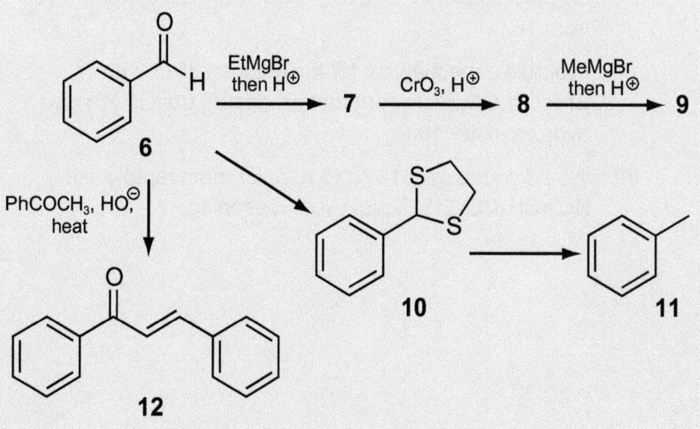
O
H
6
EtMgBr then H⊕
7
CrO₃, H⊕
8
MeMgBr then H⊕
9
PhCOCH₃, HO⊖, heat
S
S
10
11
O
12

3 In nature, pyridoxal phosphate reacts with an enzyme (abbreviated as H_2N–Enzyme) to form a coenzyme that catalyses the conversion of α-amino acids into α-keto acids. (Section 23.2)

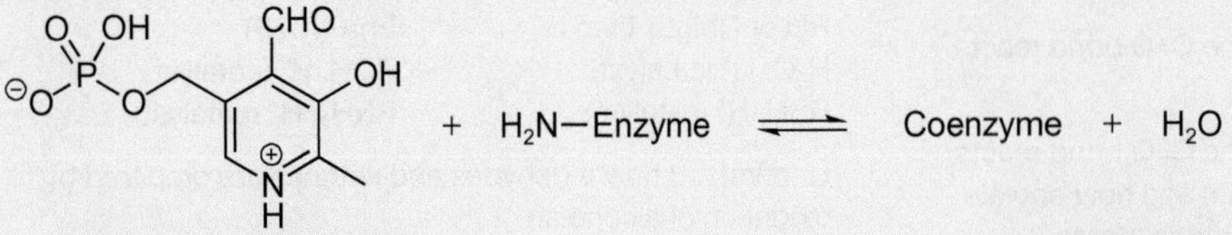
CHO
OH
O OH
P
O
N
H
+ H₂N—Enzyme
Coenzyme + H₂O

pyridoxal phosphate

(a) Draw the structure of the coenzyme and name the functional group that is formed.

(b) Is the coenzyme formed by an addition reaction, a substitution reaction, or an addition–elimination reaction? Explain your reasoning.

(c) An acid within the active site of the enzyme plays an important part in the mechanism of the reaction to produce the coenzyme. Explain the role of an acid in this type of transformation.

(d) Give the general structure of an α-amino acid, with one α-hydrogen atom.

(e) Suggest a method for converting aldehydes and ketones into α-amino acids in the laboratory.

Meats such as turkey contain high concentrations of vitamin B_6, which is a source of pyridoxal phosphate.

4 A synthetic route to the naturally occurring terpene geraniol (**13**) is shown here. (Section 23.2)

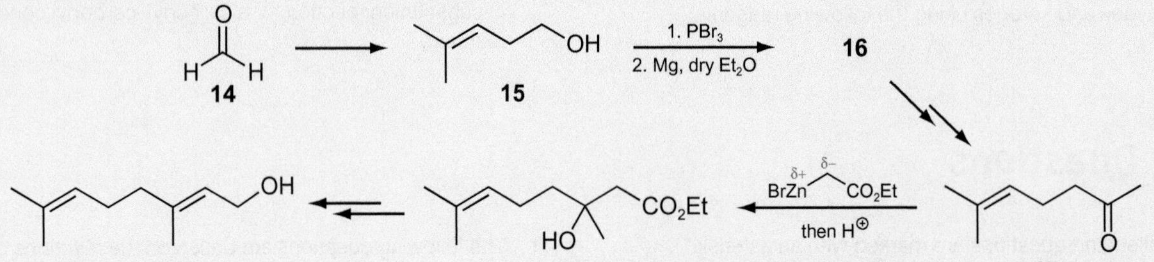
O
H H
14
OH
15
1. PBr₃
2. Mg, dry Et₂O
16
δ+ δ−
BrZn CO₂Et
then H⊕
O
17
CO₂Et
HO
18
OH
13

(a) Suggest reagents that will convert **14** into **15**.

(b) Give the structure of organic compound **16**. Why is it important to use *dry* Et_2O?

(c) Suggest reagents for a *two-step* synthesis of ketone **17** from **16**.

(d) Propose a mechanism for the reaction of **17** with $BrZnCH_2CO_2Et$ (Reformatsky reagent), then H^+, to give hydroxy-ester **18**.

(e) Would you expect **14** or **17** to react more rapidly with $BrZnCH_2CO_2Et$? Explain your reasoning.

Geraniol is an effective insecticide against insects such as fire ants, which feed on young plants and seeds.

5 The following questions are based on the reactions of acetophenone (**19**) shown here. (Sections 23.3 and 23.4)

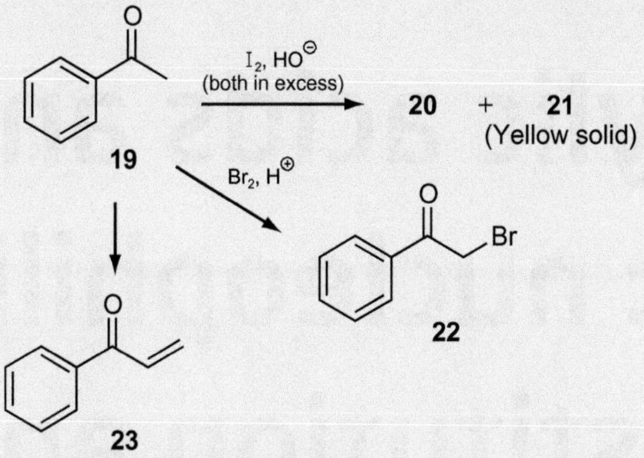
O
I_2, $HO^{\ominus}$
(both in excess)
20 + 21
(Yellow solid)
19
Br_2, $H^{\oplus}$
O
Br
22
O
23

(a) Draw the structures of organic compounds **20** and **21**.

(b) Give a mechanism that explains how **19** is converted into **22**.

(c) Suggest reagents that will convert **19** into **23**.

6* Isophorone is prepared on a large scale in industry for use as a solvent in, for example, some printing inks. A synthetic route to isophorone is shown below. (Section 23.5)

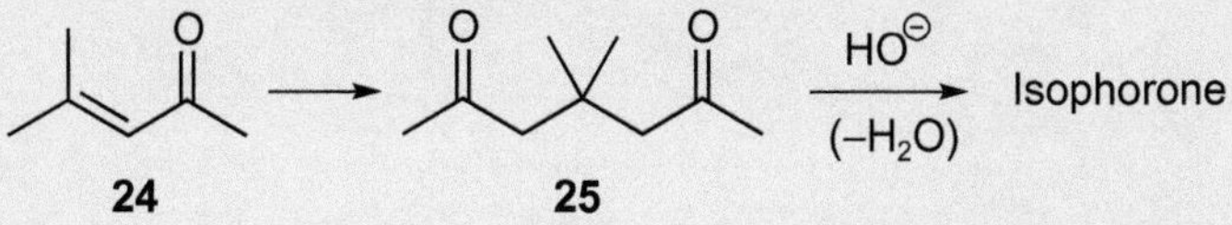
O
O
O
$HO^{\ominus}$
($-H_2O$)
Isophorone
24
25

(a) Give a reaction scheme to show how **24** could be prepared from propanone.

(b) Draw compound **25** and indicate all of the α-hydrogen atoms.

(c) Given that isophorone is formed from **25** in an intramolecular aldol condensation reaction, propose a structure for isophorone.

(d) Give a reaction mechanism to explain the formation of isophorone from **25**.

Isophorone has a peppermint-like odour and is found naturally in cranberries.

24

Carboxylic acids and derivatives: nucleophilic acyl substitution and α-substitution reactions

This chapter builds on the following topics:

- Functional groups containing carbonyl groups Section 2.7, p.98
- Infrared spectroscopy: identifying carbonyl functional groups Section 12.2, p.574
- Fundamental concepts of organic reaction mechanisms Section 19.1, p.863
- Acid-base reactions Section 19.2, p.888
- Redox reactions Section 19.2, p.899
- Polar reactions Section 19.2, p.902
- The mechanisms of nucleophilic substitution reactions Section 20.3, p.930
- Nucleophilic addition reactions of aldehydes and ketones Section 23.2, p.1061
- α-Substitution reactions of aldehydes and ketones Section 23.3, p.1082
- Carbonyl-carbonyl condensation reactions (of aldehydes and ketones) Section 23.4, p.1089

◀ Recycled PET is used in a range of products from car bumpers to surfboards.

PET plastics

Polyester fabrics, the filling in many duvets, and most fizzy drinks bottles are all made from the polyester, poly(ethylene terephthalate) (PET). Polyesters are synthetic polymers with repeating ester groups as part of the main polymer chain. Globally, around 22 million tonnes of PET are made in industry each year. Over 60% of the PET produced is melted and spun into fibres, which are marketed under trade names such as Terylene® or Dacron®. The fibres are used in furnishing materials and clothing, often blended with other fibres such as cotton. About 7% of PET is made into polyester film, some of which is used to make overhead projection transparencies, while the remaining ~30% is used in food packaging. When used for bottles, the PET granules are heated to 240 °C (a process called *curing*) and the polymer is then stretched and moulded. The plastic becomes impermeable to gases and is widely used for bottling carbonated drinks.

PET is made by heating ethane-1,2-diol ($HOCH_2CH_2OH$) with dimethyl terephthalate (1,4-$C_6H_4(CO_2Me)_2$). The steps in the mechanism are shown below. In the first step, a lone pair of electrons on an oxygen atom in the diol reacts with one of the C=O bonds in the dimethyl ester. An unstable tetrahedral intermediate is formed, which, after transfer of a proton from one oxygen atom to another, collapses by loss of methanol (CH_3OH) as a leaving group. Overall, the $-CO_2CH_3$ group in the starting material is converted into a $-COOCH_2CH_2OH$ group in the product hydroxyethyl ester.

This type of reaction is called a **nucleophilic acyl substitution** and you will meet many other examples in Section 24.2. The hydroxyethyl ester contains both an alcohol and a **methyl ester** so further nucleophilic acyl substitutions can take place at each of these positions, and it is the repeating substitution reactions that form PET.

▲ PET is an ideal material for fire hoses because it is strong, resistant to oil, and does not absorb water.

Notice that each step in the nucleophilic acyl substitution reaction is reversible (⇌). To shift the position of the equilibrium to the hydroxyethyl ester and methanol, the reaction mixture is heated so that the methanol is removed from the reaction mixture as soon as it is formed (methanol has a low boiling point of 65 °C).

PET is an example of a thermoplastic polymer, which means that it softens when heated and hardens again when cooled. Because PET can go through many heating/freezing cycles, with no significant chemical change, it can be recycled. PET bottles are collected in bottle banks, sorted according to colour, and then processed to reuse the PET. The amount of PET that is recycled is on the increase (over 66 billion PET bottles were recycled in Europe in 2014) and, controversially, the recycled PET is normally used to make other products such as carpeting, rather than bottles. Some people believe that more PET should be recycled back into bottles because the demand for plastic drinks bottles is so great and the PET in bottles (but not carpets) can be recycled more than once. Others argue that, because recycled PET for bottles needs to be of a higher quality, the process requires more energy and is less cost-effective than recycling PET for use in alternative products.

▼ Formation of PET.

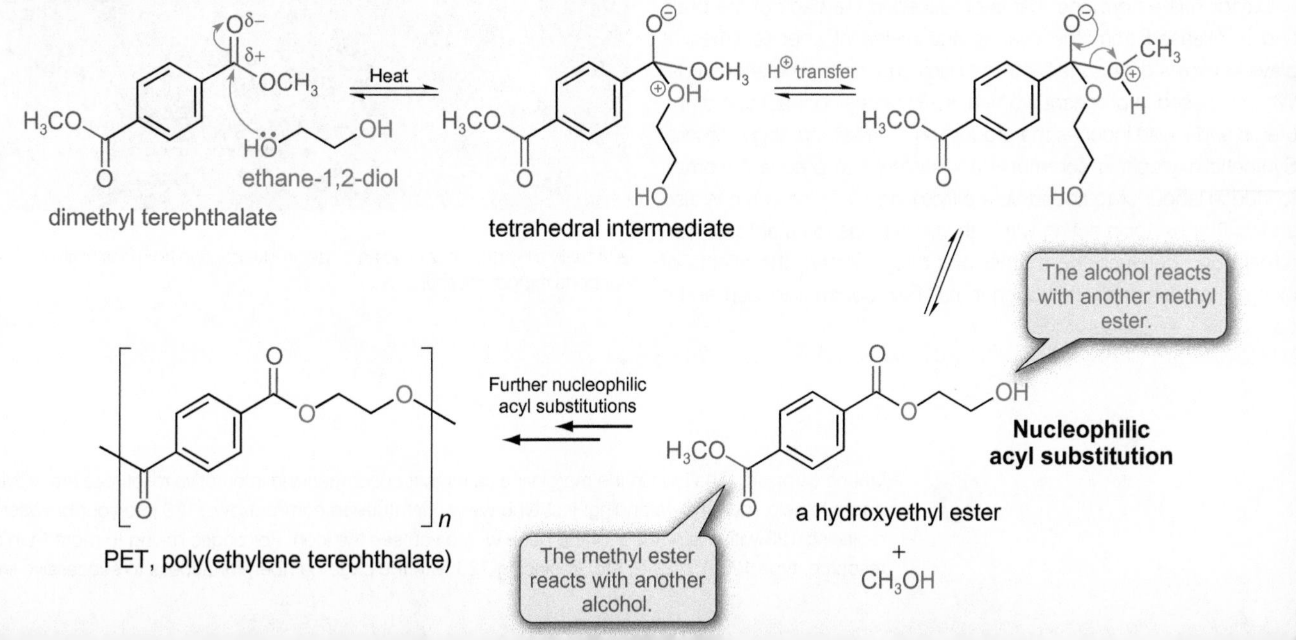

δ−
δ+
OCH3
H3CO
HO
OH
ethane-1,2-diol
dimethyl terephthalate
Heat
OCH3
OH
H⊕ transfer
tetrahedral intermediate
CH3
H
HO
The alcohol reacts with another methyl ester.
Further nucleophilic acyl substitutions
Nucleophilic acyl substitution
a hydroxyethyl ester
+
CH3OH
PET, poly(ethylene terephthalate)
n
The methyl ester reacts with another alcohol.

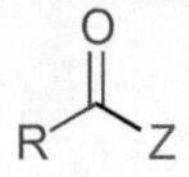
R
Z
O

acyl group

Z = –OH
–OR
–OCOR
$–NH_2$, –NHR, $–NR_2$
–Cl

Carboxylic acids and derivatives.

Carboxylic acids and derivatives are introduced in Section 2.7 (p.100).

Carbonyl functional groups are easily identified using infrared spectroscopy, as discussed in Section 12.2 (p.574).

For esters, acid anhydrides, and amides with two or more R groups, the R groups can be the same or different. In a symmetrical acid anhydride both R groups are the same, whereas in a mixed anhydride they are different.

24.1 Structure and reactions of carboxylic acids and derivatives

Carboxylic acids and derivatives are a large family of compounds that have similar structures —they all contain a C=O bond joined to an electronegative atom, typically oxygen, nitrogen, or a halogen atom. Esters (RCO_2R), acid anhydrides (RCO_2COR), amides ($RCONH_2$, RCONHR, and $RCONR_2$), and acyl chlorides (RCOCl) are called carboxylic acid derivatives because they can be prepared from carboxylic acids (RCO_2H). This family of compounds is important for a number of reasons. Many naturally occurring compounds are carboxylic acids, esters, or amides, as are many synthetic compounds (such as aspirin, paracetamol, poly(ethylene terephthalate) (PET, see p.1099), and nylon). Acid anhydrides and acyl chlorides, on the other hand, are particularly reactive compounds and are widely used in synthesis.

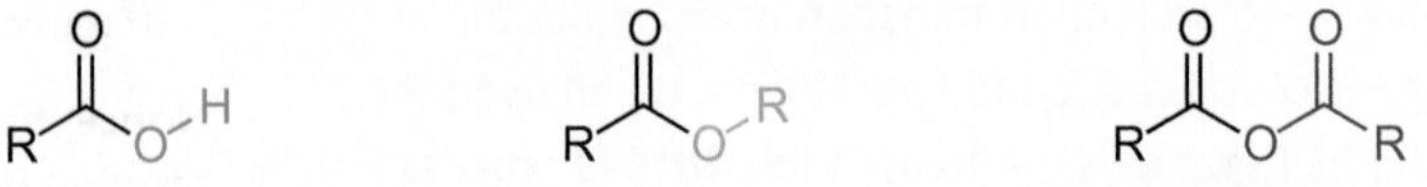

carboxylic acid, RCO_2H — ester, RCO_2R — acid anhydride, RCO_2COR (acyl anhydride)

amide, RCONHR (also $RCONH_2$ and $RCONR_2$) — acyl chloride, RCOCl (acid chloride)

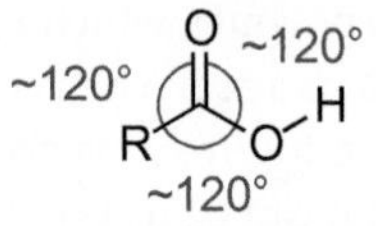
~120°
~120°
~120°
O
R
O
H

Planar arrangement about the carbon atom in the C=O bond.

In all of these compounds, the carbon atom in the C=O bond is sp^2 hybridized, as it is in aldehydes (RCHO) and ketones (RCOR) (Section 23.1, p.1056). All three σ bonds to carbon are in the same plane and are separated by angles of approximately 120°.

Box 24.1 Oxytocin, the hormone of love

Oxytocin is a member of a family of naturally occurring compounds called peptides (Box 2.9, p.104). Peptides are made from amino acids ($H_2NCHRCO_2H$, which exist as zwitterions, $^+H_3NCHRCO_2^-$, see Section 19.2, p.896), which are joined together by secondary amide bonds, –NH–CO– (sometimes called **peptide bonds**). For oxytocin, nine amino acids are linked together to form a complicated structure that contains a 20-membered ring.

Oxytocin is a hormone that is produced in the back of the brain and is released from the ovaries and testes at specific times. It plays key roles during the birth of a baby and in the time afterwards. When released in pregnant women, it stimulates contractions of the uterus and it also induces the production of milk in nursing mothers. Synthetic oxytocin is commonly administered to pregnant women to induce labour or to speed up a difficult labour. Research has also shown that oxytocin acting within the brain helps to establish a maternal bond between the mother and baby. In men, the effects of oxytocin are less clear, although it may help sperm transport and it could affect sexual behaviour.

▲Virtually all vertebrates have an oxytocin-like peptide hormone that supports reproductive functions.

→

Online Support. The icon in the margin indicates that accompanying interactive resources are provided online to help your understanding: just type www.chemtube3d.com/burrows/123 into your browser, replacing 123 with the number of the page where you see the icon. For pages linking to more than one resource, type 123-1, 123-2, etc. (replacing 123 with the page number) for access to successive links.

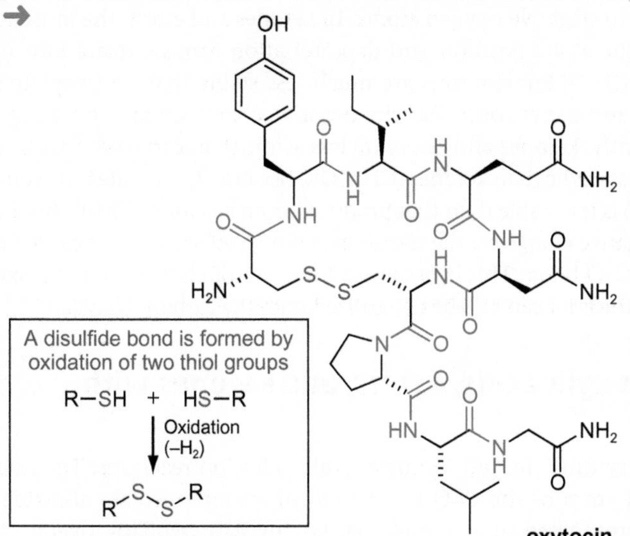
A disulfide bond is formed by oxidation of two thiol groups
R–SH + HS–R
Oxidation
(–H2)

The amide bonds that link together amino acids are called **peptide bonds**

oxytocin

Recent research into the effects of oxytocin has produced some interesting results. People who claim to be falling in love have higher than average concentrations of oxytocin in their blood so oxytocin may help bonding between adults. For this reason, oxytocin is sometimes called the 'hormone of love'.

Oxytocin also appears to make us more trusting. In an investment game that involved risk, players who were given oxytocin were judged to have a higher level of trust than those who were not. Some people have suggested that sufferers of social anxieties and also individuals with autism may benefit from the trust-inducing effect of oxytocin, but others have pointed out the possible abuse of oxytocin by confidence tricksters.

Question

Amino acids link together to form peptides (see Box 2.9, p.104). Draw the structures of the eight amino acids that, when linked together with $H_2NCH_2CONH_2$, form oxytocin. (Assume the S–S bond is not formed at this stage.)

Although carboxylic acids and derivatives have similar structures to aldehydes and ketones, their reactions are very different. The electronegative atom attached to the C=O bond affects the reactivity of the C=O bond. The effect depends on the nature of the electronegative atom or the group containing the electronegative atom. These effects are illustrated by the differences in acidity between carboxylic acids, esters, and ketones, and also their contrasting reactions with nucleophiles.

Acidity of carboxylic acids, esters, and ketones

Carboxylic acids (RCO_2H, pK_a 4–5) are much stronger acids than ketones ($RCOCH_3$, pK_a ~20), which are stronger acids than esters ($ROCOCH_3$, pK_a ~25). In carboxylic acids, the hydrogen atom on the oxygen is lost most easily to form a carboxylate ion, RCO_2^- (see Figure 24.1).

The strengths of acids are covered in Section 7.2 (p.309)—note that carboxylic acids (RCO_2H) are generally classed as weak acids. Factors affecting the pK_a values of organic acids are discussed in Section 19.2 (p.888). For a table of pK_a values, see Appendix 9 (p.1366).

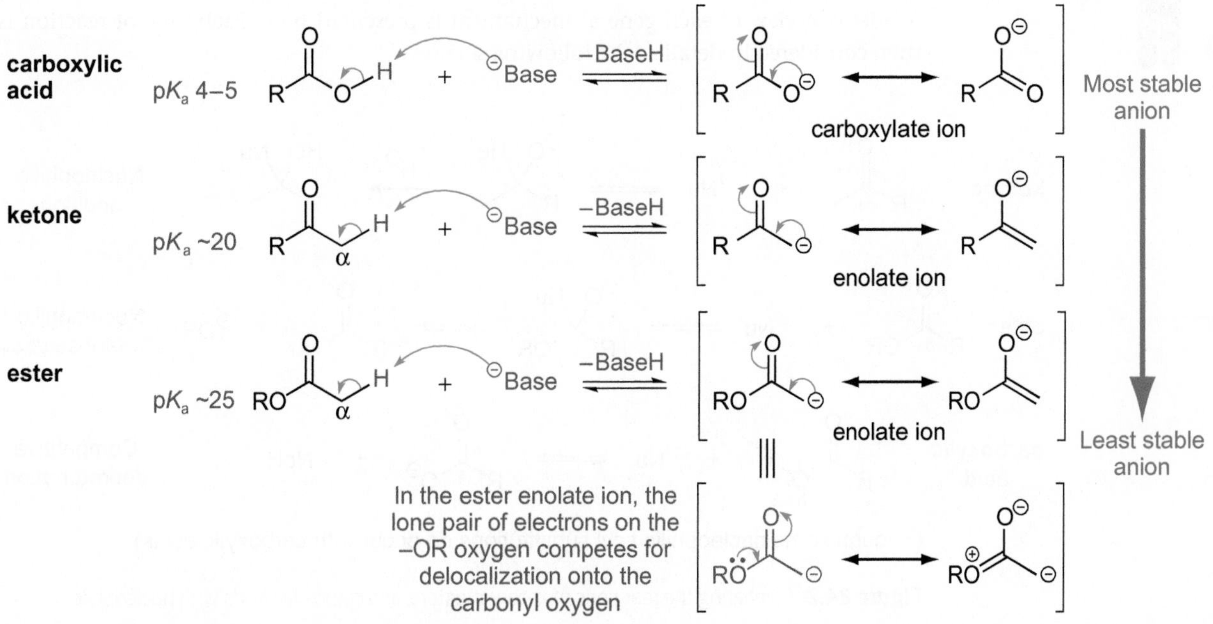
carboxylic acid
pKa 4–5
Base
–BaseH
carboxylate ion
Most stable anion
ketone
pKa ~20
enolate ion
ester
pKa ~25
enolate ion
Least stable anion
In the ester enolate ion, the lone pair of electrons on the –OR oxygen competes for delocalization onto the carbonyl oxygen

Figure 24.1 Comparing the acidity of carboxylic acids, ketones, and esters.

Carboxylate ions (RCO_2^-) are relatively stable anions because the negative charge is delocalized over two electronegative oxygen atoms. In ketones and esters, the most acidic hydrogen atoms are at the alpha position and deprotonation forms enolate ions (such as $RCOCH_2^-$ and $ROCOCH_2^-$). Enolate ions are much less stable than carboxylate ions because only one of the resonance forms has the negative charge on an electronegative oxygen atom. Consequently, ketones and esters are less acidic than carboxylic acids.

Esters (CH_3CO_2R) are less acidic than ketones (CH_3COR) because the enolate ion produced from an ester ($^-CH_2CO_2R$) is less stable than that produced from a ketone ($^-CH_2COR$). In the ester enolate ion, the negative charge on the α-carbon is not so effectively delocalized onto the oxygen atom in the C=O bond. This is because a lone pair of electrons on the oxygen atom in the OR group of the ester can also be delocalized onto the carbonyl oxygen.

The formation of enolate ions from aldehydes (such as $^-CH_2CHO$) and ketones (such as $^-CH_2COCH_3$) is discussed in Section 23.3 (p.1082).

In primary and secondary amides, the hydrogen atoms bonded to nitrogen are weakly acidic.

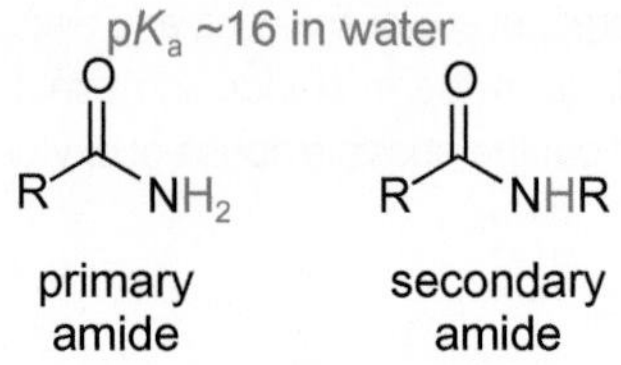
pKa ~16 in water
primary amide
secondary amide

Reactions of carboxylic acids, esters, and ketones with nucleophiles

Nucleophiles react with ketones (RCOR) in nucleophilic addition reactions. The nucleophile attacks the carbon atom of the C=O bond to form an intermediate alkoxide ion ($R_2C(O^-)Nu$), which is protonated in a second step (giving $R_2C(OH)Nu$)—overall, NuH adds to the C=O bond (Figure 24.2).

In contrast, esters (RCO_2R, and other carboxylic acid derivatives) react with nucleophiles in **nucleophilic acyl substitution reactions.** The first step of the mechanism is the same as for nucleophilic addition—the nucleophile attacks the carbon atom of the C=O bond of the ester to form an alkoxide ion ($RC(O^-)(Nu)OR$), as shown in Figure 24.2. However, the second step is different because the alkoxide ion produced from the ester is unstable and it loses the RO^- group, to re-form the C=O bond (giving RCONu). In this substitution reaction, RO^- acts as a leaving group and the nucleophile takes the place of the RO group in the ester.

Carboxylic acids also react with some nucleophiles in nucleophilic acyl substitution reactions, in which the nucleophile replaces the OH group of the carboxylic acid. However, because carboxylic acids are so acidic, some nucleophiles act as bases and deprotonate the carboxylic acid (to form a carboxylate ion, RCO_2^-), rather than react with the C=O bond.

There are three general mechanisms by which esters and other carboxylic acid derivatives react. These are:

1. nucleophilic acyl substitution reactions;
2. α-substitution reactions;
3. carbonyl–carbonyl condensation reactions.

A brief overview of each general mechanism is presented here. Each type of reaction is then considered in detail in the following sections.

Nucleophilic addition reactions of aldehydes and ketones are described in Section 23.2 (p.1061).

The term 'acyl' in nucleophilic acyl substitution, indicates that substitution takes place at the C=O bond of an R–C=O (acyl) group.

Classification of organic reaction mechanisms is introduced in Section 19.2 (p.888). In a substitution reaction, a functional group in a particular compound is replaced by another group (like a player replacing a teammate in a football match).

11 | 23

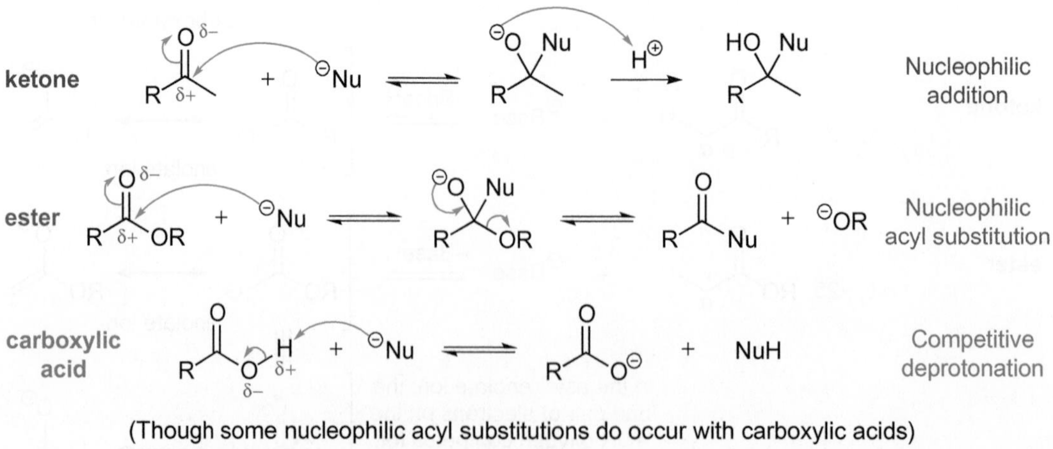
ketone
Nucleophilic addition
ester
Nucleophilic acyl substitution
carboxylic acid
Competitive deprotonation
(Though some nucleophilic acyl substitutions do occur with carboxylic acids)

Figure 24.2 Comparing the reactions of ketones, esters, and carboxylic acids with nucleophiles.

Overview of nucleophilic acyl substitution reactions

The most common reaction of carboxylic acid derivatives involves reaction with a negatively charged or neutral nucleophile in a nucleophilic acyl substitution reaction.

With a negatively charged nucleophile

Reaction of a negatively charged nucleophile (e.g. H^- or R^-) with the C=O bond of acyl chlorides, acid anhydrides, or esters forms unstable tetrahedral alkoxide ions. The alkoxide ions are unstable because $^-$Cl, $^-$OCOR, or $^-$OR ions act as leaving groups, which are rapidly expelled from the alkoxide ion to re-form the (strong) C=O bond. For amides, under certain conditions, $^-NH_2$, $^-$NHR, or $^-NR_2$ can act as leaving groups. For the equilibrium to lie towards the substituted product (RCONu in Figure 24.3), the leaving group ($^-$X) must be more stable than the attacking nucleophile ($^-$Nu), otherwise, the reverse reaction takes place.

For the reaction of the ester, the instability of the alkoxide ion (RC(O$^-$)(Nu)OR) can be explained by the mean bond enthalpies of the bonds that are formed or broken when the alkoxide ion is converted into RCONu. A very strong C=O bond (+742 kJ mol^{-1}) is formed at the expense of two weaker C–O bonds (2 × +358 kJ mol^{-1} = +716 kJ mol^{-1}), so RCONu will be favoured over RC(O$^-$)(Nu)OR at equilibrium.

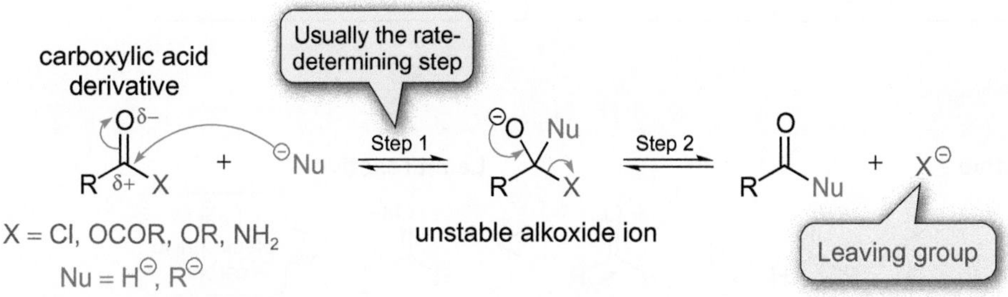

carboxylic acid derivative
Usually the rate-determining step
Step 1
Step 2
unstable alkoxide ion
Leaving group
X = Cl, OCOR, OR, NH2
Nu = H⊖, R⊖

Figure 24.3 General mechanism for nucleophilic acyl substitution with a negatively charged nucleophile.

Nucleophilic acyl substitution reactions are discussed in detail in Section 24.2 (p.1107).

Sometimes the mechanism in Figure 24.3 is abbreviated by not drawing the intermediate alkoxide ion as shown below. Notice that curly arrows showing the pair of electrons moving to and from the carbonyl oxygen atom *must* be included.

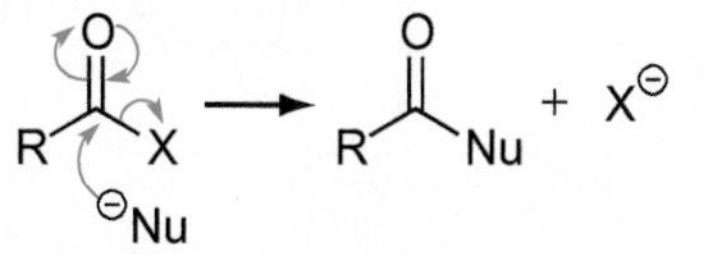

An acceptable abbreviated mechanism

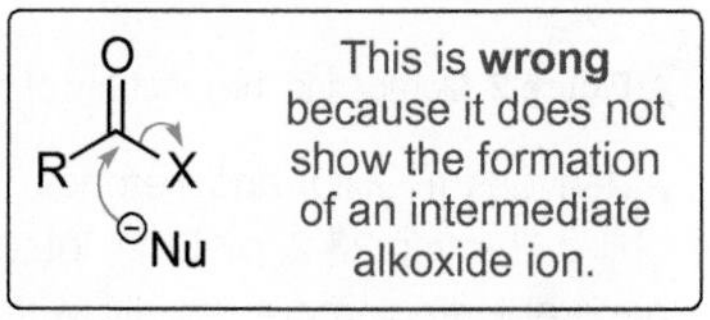

This is **wrong** because it does not show the formation of an intermediate alkoxide ion.

Leaving groups

The best leaving groups are neutral molecules or stable ions (Section 20.2, p.925). For an anion (A$^-$), a low pK_a value of the conjugate acid (HA) indicates that A$^-$ is relatively stable and a good leaving group.

Order of leaving groups (A$^-$):	Cl$^-$	>	$^-$OCOR	>	$^-$OR	>	$^-NH_2$
Order of pK_a of the conjugate acids (HA):	−7	<	4–5	<	16–17	<	38

Usually, the first step of nucleophilic acyl substitution is the rate-determining (slowest) step so the rate of the reaction depends on the concentrations of both the nucleophile and the carboxylic acid derivative. Fast nucleophilic acyl substitution reactions require strong nucleophiles and electrophilic carboxylic acid derivatives.

The rate-determining step is found by studying the kinetics of the reaction. See Section 9.6 (p.417).

Box 24.2 Relative reactivity of carboxylic acid derivatives

Carboxylic acid derivatives are not equally reactive to the same nucleophile. Acyl chlorides react most rapidly with nucleophiles, followed by acid anhydrides, esters, and, finally, amides (Figure 1). For example, the relative rates of hydrolysis (with water as the nucleophile) of an acyl chloride and an amide differ by a factor of 10^{13}. The differences in the rates of reaction are explained by the electronic effects of the substituents attached to the carbon atom in the C=O bonds.

For esters (RCO_2R) and amides (such as $RCONR_2$), the +M effects of the OR and NR_2 groups, respectively, are stronger than their −I effects (Section 19.1, p.875). For example, in an amide, delocalization of the lone pair of electrons on nitrogen weakens the C=O bond and strengthens the C–N bond. This is because delocalization produces a resonance form with a C–O single bond and C=N double bond ($RC(O^-)=NR_2^+$).

However, for acyl chlorides, the −I effect of Cl is stronger than the +M effect. Cl has a weak +M effect because the 3*p* orbital (which contains the lone pairs of electrons) is larger than the 2*p* orbital of carbon and they do not overlap efficiently.

→

→

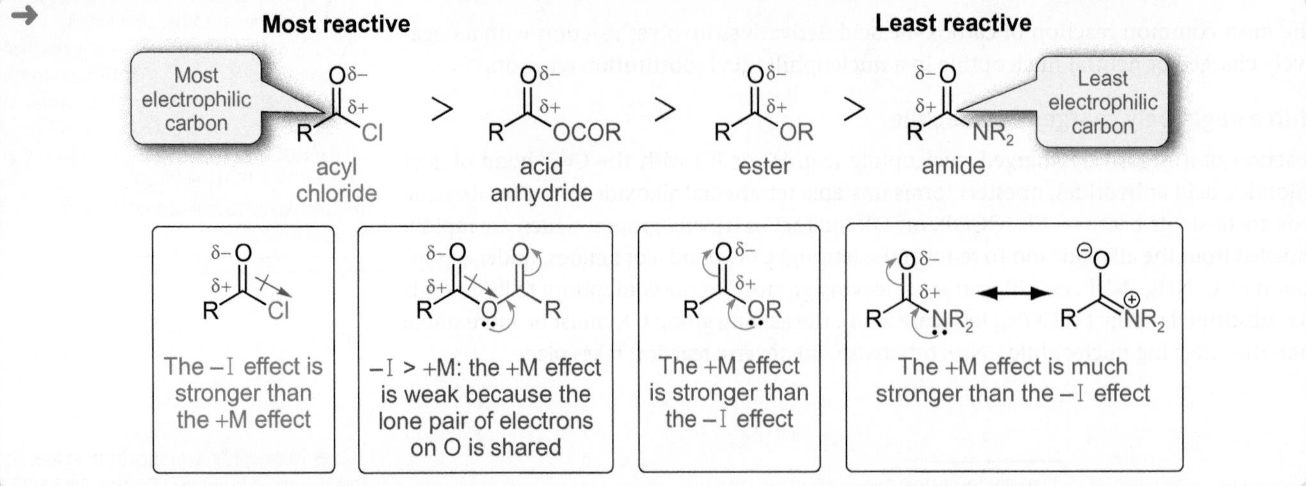
Most reactive
Least reactive
Most electrophilic carbon
Least electrophilic carbon
acyl chloride
acid anhydride
ester
amide
The −I effect is stronger than the +M effect
−I > +M: the +M effect is weak because the lone pair of electrons on O is shared
The +M effect is stronger than the −I effect
The +M effect is much stronger than the −I effect

Figure 1 Order of reactivity of carboxylic acid derivatives.

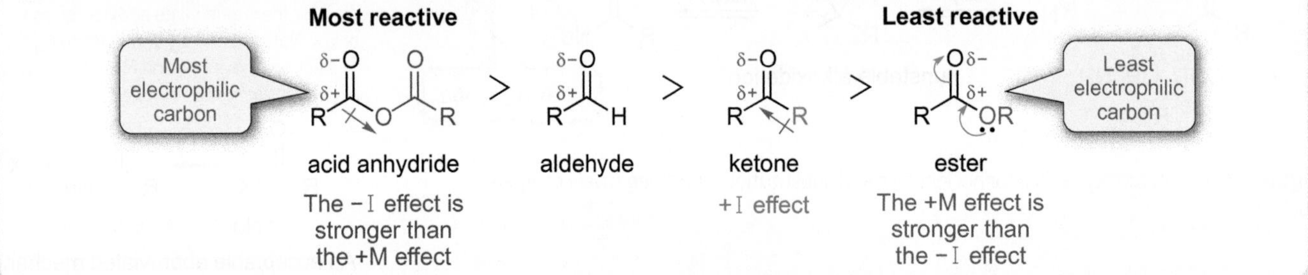
Most reactive
Least reactive
Most electrophilic carbon
Least electrophilic carbon
acid anhydride
aldehyde
ketone
ester
The −I effect is stronger than the +M effect
+I effect
The +M effect is stronger than the −I effect

Figure 2 Comparing the reactivity of carboxylic acid derivatives with that of aldehydes and ketones.

Aldehydes (RCHO) and ketones (RCOR) also react with nucleophiles (Section 23.2, p.1061). Interestingly, aldehydes and ketones generally react more rapidly with nucleophiles than esters and amides but they are less reactive than acyl chlorides or acid anhydrides (Figure 2). Again, you can explain the relative reactivities by considering the electronic effects of the substituents attached to the C=O bond. Acid anhydrides are more reactive than aldehydes or ketones because of the −I effect of the OCOR group, whereas esters are less reactive than aldehydes or ketones because of the +M effect of the OR group (Figure 3).

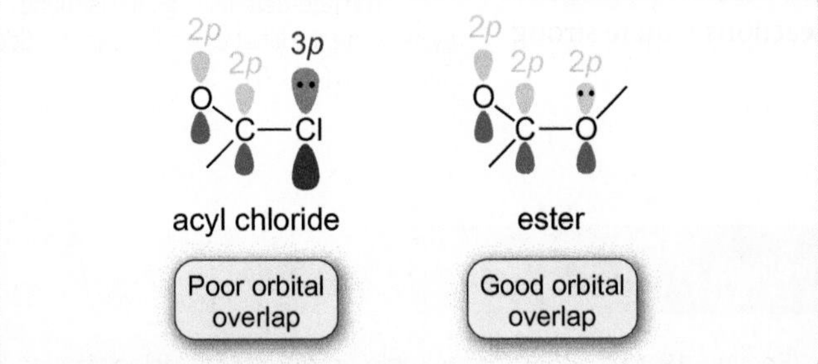
acyl chloride
Poor orbital overlap
ester
Good orbital overlap

Figure 3 For acyl chlorides (RCOCl) the −I effect is stronger than the +M effect because of poor orbital overlap.

Question

Place the five carbonyl compounds **A–E** in decreasing order of reactivity towards nucleophilic attack.

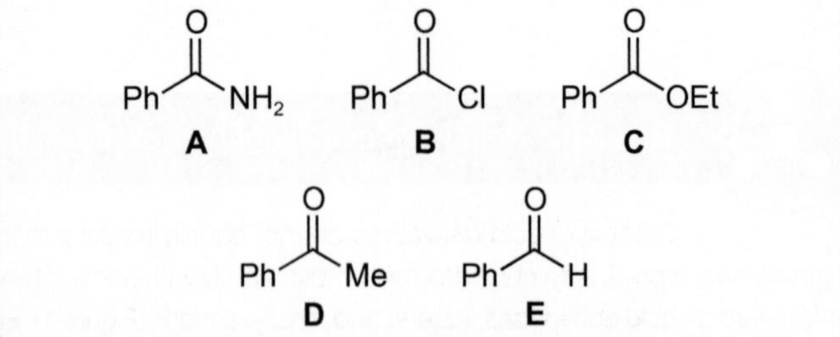
A
B
C
D
E

With a neutral nucleophile

Carboxylic acid derivatives also react with neutral nucleophiles (N̈uH), such as H_2O, ROH (an alcohol), RNH_2 or R_2NH (an amine), or RSH (a thiol). However, because carboxylic acid derivatives are not equally reactive to nucleophiles (Box 24.2), reactions with these relatively weak nucleophiles require different conditions.

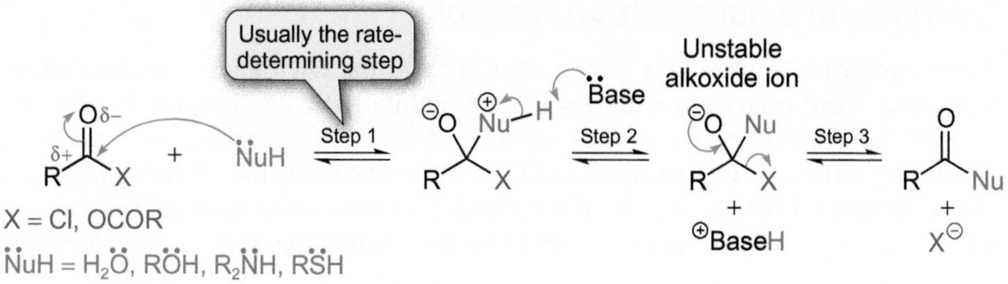
Usually the rate-determining step
Unstable alkoxide ion
Base
Step 1
Step 2
Step 3
X = Cl, OCOR
NuH = H2O, ROH, R2NH, RSH

Figure 24.4 General mechanism for nucleophilic acyl substitution of acyl chlorides and acid anhydrides with a neutral nucleophile.

Acyl chlorides (RCOCl) and acid anhydrides (RCOOCOR) are the most reactive carboxylic acid derivatives and these compounds generally react with neutral nucleophiles at room temperature and neutral pH (often in the presence of a weak base). The general mechanism is shown in Figure 24.4. The first step produces an intermediate containing both a negative charge and a positive charge, $RC(O^-)X(NuH^+)$ (this is called a **zwitterion** (Section 19.2, p.896)). In the second step, the zwitterion is deprotonated and, to facilitate this step, a weak base (Section 19.2, p.891) such as triethylamine (Et_3N; the pK_a value of the conjugate acid is 10.75) or pyridine (C_5H_5N; the pK_a value of the conjugate acid is 5) is often added. The mechanism now resembles that in Figure 24.3 on p.1103. The resulting tetrahedral alkoxide ion, $RC(O^-)XNu$, is unstable and, in the third step, ^-Cl or ^-OCOR is expelled from the alkoxide ion at the same time as the C=O bond is reformed. For the equilibrium to lie towards the substituted product, the leaving group ^-X must be more stable than ^-Nu; otherwise, the reverse reaction takes place.

Esters (RCO_2R) and amides (such as $RCONR_2$) are the least reactive carboxylic acid derivatives and reactions with neutral nucleophiles usually require an acid catalyst and heating (Figure 24.5). In the first step, the acid catalyst protonates the C=O bond of the ester or amide and this converts the carbon atom into a stronger electrophile ($RC(=OH^+)X$). The protonated carbonyl reacts with the nucleophile in step 2 to form a tetrahedral intermediate, $RC(OH)XNuH^+$. This is followed by transfer of a proton from the nucleophile to the OR or NR_2 group in steps 3 and 4. Protonation of the OR or NR_2 group converts it into a good leaving group (HOR or HNR_2), which is expelled in step 5 to reform a C=O bond. Finally, in step 6, deprotonation of the C=O bond forms the substituted product, RCONu. For the equilibrium to lie towards the substituted product, the leaving group XH must be more stable than NuH, otherwise, the reverse reaction takes place.

Proton transfer steps
Any base present in a reaction mixture can remove a proton in step 3. Any acid present in a reaction mixture can donate a proton to the tetrahedral intermediate in step 4. Sometimes, a solvent can facilitate proton transfer; for example, in an aqueous solvent, H_2O can act as a base and H_3O^+ can act as an acid.

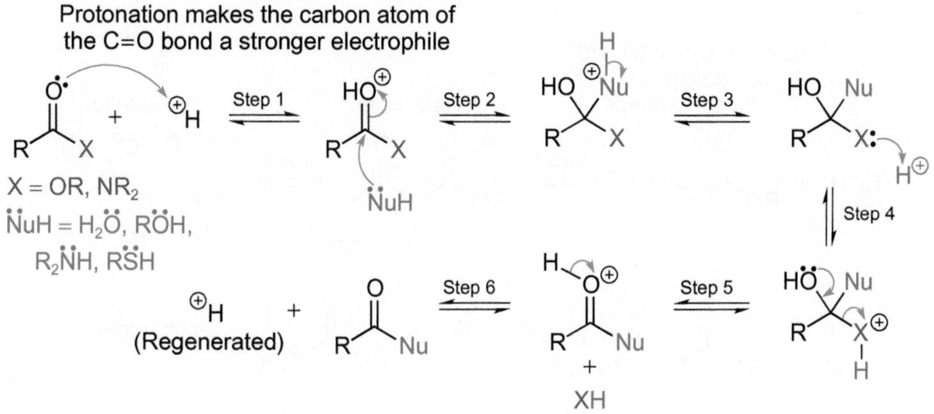
Protonation makes the carbon atom of the C=O bond a stronger electrophile
Step 1
Step 2
Step 3
Step 4
Step 5
Step 6
X = OR, NR2
NuH = H2O, ROH, R2NH, RSH
(Regenerated)

Figure 24.5 General mechanism for nucleophilic acyl substitution of esters and amides with a neutral nucleophile.

Acyl chlorides (RCOCl) need to be stored in the absence of moisture because they react with water, in nucleophilic acyl substitution reactions, to give carboxylic acids (RCO_2H) and HCl. For example, old bottles of benzoyl chloride (PhCOCl) usually have a crust of benzoic acid ($PhCO_2H$) around the screw lid.

Acyl chlorides (RCOCl) and acid anhydrides (RCO_2COR) generally react with neutral nucleophiles at room temperature and neutral pH (often in the presence of a weak base).

Reaction of esters (RCO_2R) or amides (such as $RCONR_2$) with neutral nucleophiles usually requires an acid catalyst and heating.

A protonated carbonyl group ($R_2C{=}OH^+$) has a very low pK_a value of around −7. Thus a 1.0 mol dm^{-3} solution of HCl (pH 0) will only protonate around 1 in 10^7 molecules of the carbonyl. Even at these low concentrations, the protonated carbonyls increase the rate of nucleophilic acyl substitutions because they are extremely reactive electrophiles.

In step 5, a very strong C=O bond (+742 kJ mol^{-1}) is formed at the expense of two weaker C–X bonds (for a C–O and a C–N bond, +358 and +286 kJ mol^{-1} is equal to +644 kJ mol^{-1}), so $RC(=OH^+)Nu$ will be favoured over $RC(OH)(Nu)XH^+$ at equilibrium.

In RC(OH)XNu, the OH, X and Nu groups are all basic so they can each be protonated. Protonation of any one of these groups is reversible.

Overview of α-substitution reactions

α-Substitution reactions are discussed in detail in Section 24.3 (p.1126).

Aldehydes and ketones undergo similar α-substitution reactions (Section 23.3, p.1082).

These reactions take place at a carbon atom next to the C=O bond—this is called the **α-position**. A hydrogen atom at the α-position is substituted by another group in a two-step reaction that occurs under either acidic or, more usually, basic conditions. For example, on reaction with a base, an ester (such as CH_3CO_2R) is deprotonated at the α-position to form an **ester enolate ion** (such as $^-CH_2CO_2R$). Ester enolate ions are strong nucleophiles that react with electrophiles to introduce a new substituent at the α-position, as shown in Figure 24.6.

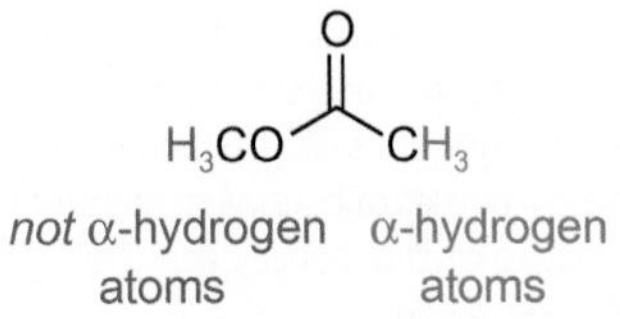
O
H₃CO
CH₃
not α-hydrogen atoms
α-hydrogen atoms

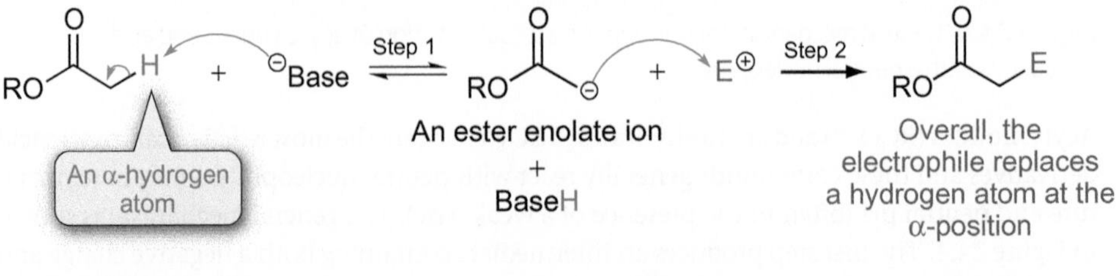
RO
H
+
Base
Step 1
An α-hydrogen atom
An ester enolate ion
+
BaseH
E⊕
Step 2
E
Overall, the electrophile replaces a hydrogen atom at the α-position

Figure 24.6 General mechanism for an α-substitution reaction of an ester under basic conditions.

Overview of carbonyl-carbonyl condensation reactions

Carbonyl–carbonyl condensation reactions are discussed in more detail in Section 24.3 (p.1128).

Aldehydes and ketones undergo similar carbonyl–carbonyl condensation reactions (Section 23.4, p.1089), called aldol condensations.

These reactions involve two carboxylic acid derivatives or one carboxylic acid derivative and an aldehyde (RCHO) or ketone (RCOR). For example, two carboxylic acid derivatives react with one another so that one of them undergoes an α-substitution reaction and the other undergoes a nucleophilic acyl substitution reaction.

This type of reaction is illustrated by the **Claisen condensation reaction** of an ester (such as CH_3CO_2R) shown in Figure 24.7. In the first step, a base removes a proton from the α-position of an ester to form an ester enolate ion ($^-CH_2CO_2R$). In step 2, the ester enolate ion reacts as a nucleophile and attacks the carbon atom of the C=O group in a molecule of the ester that has not been deprotonated. A new **carbon–carbon** bond forms to give an alkoxide ion ($RO_2CCH_2–C(O^-)(OR)CH_3$). The alkoxide ion is unstable and, in step 3, RO^- is expelled to re-form a C=O bond in a β-keto ester ($RO_2CCH_2COCH_3$). As soon

To convert the ester (CH_3CO_2R) into the β-keto ester ($CH_3COCH_2CO_2R$) requires one equivalent of the base. (One equivalent is a short way of saying 'one stoichiometric equivalent'; here, there are twice as many moles of ester as there are of base.) A molecule of ROH is eliminated when two ester molecules combine, so this is an example of a condensation reaction.

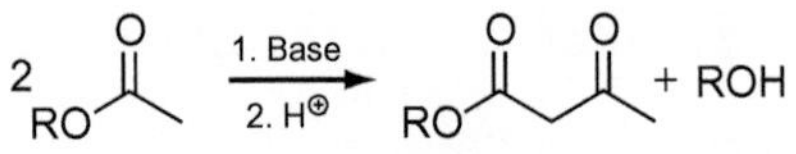
2
RO
1. Base
2. H⊕
+ ROH

Reactions like the Claisen condensation reaction, that form new carbon–carbon bonds, are very useful in synthesis as they convert reactants with smaller carbon chains into products with larger ones. A bit like stacking children's building bricks to make a vast array of super structures.

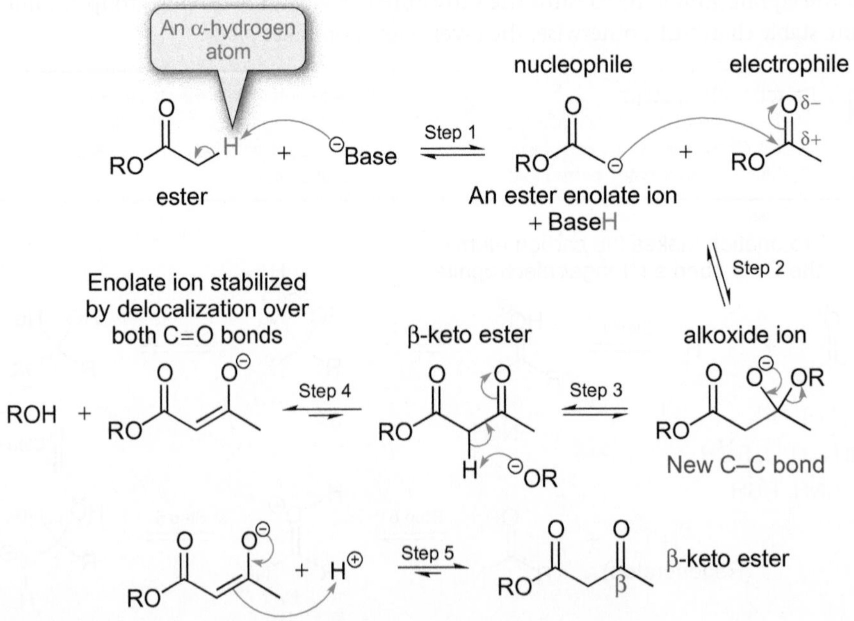
An α-hydrogen atom
ester
Base
Step 1
nucleophile
electrophile
An ester enolate ion
+ BaseH
δ−
δ+
Step 2
alkoxide ion
New C–C bond
Step 3
β-keto ester
Step 4
Enolate ion stabilized by delocalization over both C=O bonds
ROH +
OR
+ H⊕
Step 5
β
β-keto ester

Figure 24.7 General mechanism for a carbonyl–carbonyl condensation of an ester (Claisen condensation).

as the β-keto ester is formed, it is deprotonated by RO^- to form a resonance-stabilized enolate ion; $RO_2CCH^-COCH_3 \longleftrightarrow RO_2CCH{=}C(O^-)CH_3 \longleftrightarrow RO(O^-)C{=}CHCOCH_3$ (the negative charge is delocalized over both C=O bonds)—the β-keto ester is more acidic than the starting ester and so the equilibrium lies towards the deprotonated β-keto ester. Finally, on acid work-up (in step 5), the β-keto ester is reformed.

Summary

- Carboxylic acids and carboxylic acid derivatives (acyl chlorides, acid anhydrides, esters, and amides) have a C=O bond linked to an electronegative oxygen, nitrogen, or halogen atom.
- The C=O bond is polar and the electrons are attracted to the oxygen atom ($^{\delta+}C{=}O^{\delta-}$)—the oxygen atom reacts with electrophiles and the carbon atom reacts with nucleophiles.
- Deprotonation of a carboxylic acid (RCO_2H) forms a carboxylate ion (RCO_2^-), whereas deprotonation of an ester forms an ester enolate ion (H^+ is lost from the α-position).

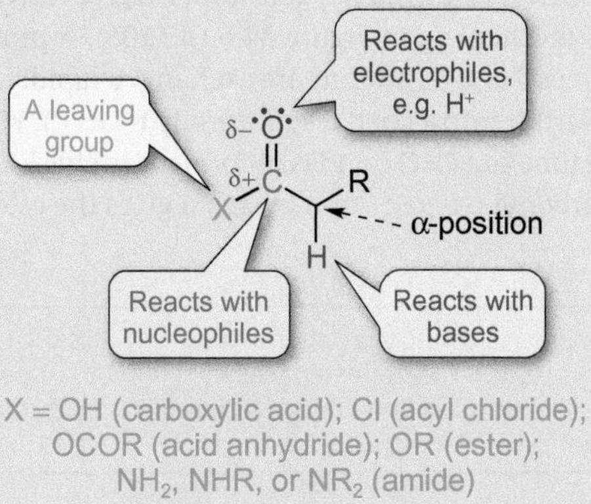
Reacts with electrophiles, e.g. H+
A leaving group
δ−
δ+
X
R
α-position
H
Reacts with nucleophiles
Reacts with bases
X = OH (carboxylic acid); Cl (acyl chloride); OCOR (acid anhydride); OR (ester); NH2, NHR, or NR2 (amide)

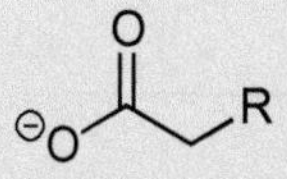
O
R
A carboxylate ion

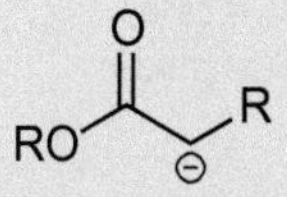
O
R
RO
An ester enolate ion

- Carboxylic acids undergo a limited number of nucleophilic substitution reactions—many nucleophiles act as bases and deprotonate the carboxylic acid, rather than attack the C=O bond.
- Carboxylic acid derivatives undergo nucleophilic acyl substitution reactions, α-substitution reactions, and carbonyl-carbonyl condensation reactions.
- In a nucleophilic acyl substitution reaction the Cl, OCOR, OR, or NR_2 group bonded to C=O acts as the leaving group. (Aldehydes and ketones do not undergo nucleophilic acyl substitution reactions because the R and/or H groups bonded to C=O are very poor leaving groups.)
- The order of reactivity of carboxylic acid derivatives in nucleophilic acyl substitutions is: acyl chloride (RCOCl) > acid anhydride (RCO_2COR) > ester (RCO_2R) > amide ($RCONR_2$).
- The order of reactivity of carbonyl compounds towards nucleophiles is: acyl chloride (RCOCl) > acid anhydride (RCO_2COR) > aldehyde (RCHO) > ketone (RCOR) > ester (RCO_2R) > amide ($RCONR_2$).

24.2 Nucleophilic acyl substitution reactions

In nucleophilic acyl substitution reactions, a carboxylic acid derivative is converted into another carbonyl compound. Often, a more reactive carboxylic acid derivative is converted into a less reactive carboxylic acid derivative. For example, reactive acyl chlorides (RCOCl) react with nucleophilic amines (RNH_2 or R_2NH) and alcohols (ROH) to form less reactive amides (RCONHR or $RCONR_2$) and esters (RCO_2R), respectively. As you may expect, the more reactive the carboxylic acid derivative, the greater the range of nucleophiles it reacts with. So, acyl chlorides react with a wider range of nucleophiles than amides.

Box 24.2 (p.1103) discusses the relative reactivity of carboxylic acid derivatives.

Carboxylic acids (RCO_2H) are exceptional because some nucleophiles act as bases and they deprotonate the carboxylic acid (to form RCO_2^-), rather than react with the C=O bond. For this reason, the number of important nucleophilic acyl substitution reactions of carboxylic acids is limited, the most important being esterification.

Nucleophilic acyl substitution reactions of carboxylic acids

Reaction of a carboxylic acid with an alcohol: preparation of esters

A nucleophilic acyl substitution reaction between a carboxylic acid (RCO_2H) and an alcohol (ROH) produces an ester, RCO_2R (in an **esterification** reaction). The mechanism is shown in Figure 24.8. The reaction requires an acid catalyst to protonate the carboxylic acid (on the oxygen atom in the C=O group, see margin) to convert it into a stronger electrophile, $RC(=OH^+)OH$—in the absence of an acid catalyst the reaction is very slow because a carboxylic acid is a weak electrophile and an alcohol is a weak nucleophile. Once the carboxylic acid is protonated in step 1, the alcohol (ROH) acts as a nucleophile and it attacks the carbon atom in the C=O bond in step 2; this forces a pair of electrons to move onto the positively charged oxygen atom (to give $RC(OH)_2ORH^+$). The mechanism follows the general mechanism in Figure 24.5 (p.1105). A proton is transferred to one of the OH groups in steps 3 and 4 (protons are exchanged rapidly between the –OR and –OH groups), which converts it into a good leaving group ($-OH_2^+$). Water is expelled as the leaving group at the same time as a C=O bond is re-formed (to give $RC(=OH^+)OR$). Finally, loss of a proton from the carbonyl oxygen atom in step 6 gives the ester (RCOOR).

To form an ester (RCO_2R), RCO_2H is commonly reacted with an excess of ROH, an acid catalyst, and a dehydrating agent (such as molecular sieves, Section 23.2, p.1075). Esterification is a **condensation reaction**, as a carboxylic acid and an alcohol combine to form an ester with loss of a small molecule; as this small molecule is water, esterification can also be described as a **dehydration reaction**.

Hydrolysis of an ester (e.g. using H^+/H_2O) forms a carboxylic acid and an alcohol (p.1105).

Reaction of an ester (RCO_2R^1) with an alcohol (R^2OH) and an acid catalyst, to produce a different ester (RCO_2R^2), is called a **transesterification reaction** (illustrated in Worked example 24.3, p.1122).

Interactive 3D animation of acid-catalysed esterification and ester hydrolysis.

Reaction of a carboxylic acid with an alcohol to form an ester is called an **esterification reaction**. Esterification is sometimes called **Fischer esterification** after Emil Fischer (1852–1919) who discovered the reaction. In 1902, Fischer was awarded the Nobel Prize in chemistry for his work on the synthesis of sugars and purines (purines are nitrogen bases found in the nucleic acids, DNA and RNA).

Carboxylic acids are protonated on the oxygen atom of the C=O group because this produces the most stable conjugate acid (Section 19.2, p.888).

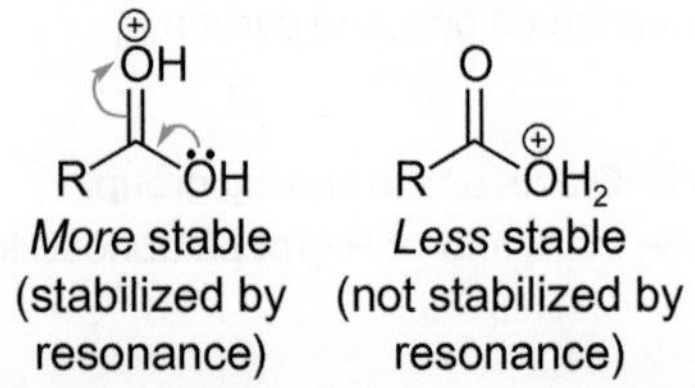
More stable
(stabilized by resonance)
Less stable
(not stabilized by resonance)

Note that all six steps in the esterification reaction are reversible (⇌). To force the equilibrium towards the ester (RCOOR) usually requires an excess of the alcohol (HOR) together with a dehydrating agent (such as anhydrous magnesium sulfate or molecular sieves), which removes the water from the reaction solution as soon as it is formed. Alternatively, the water can be removed from the reaction solution by distillation.

Anhydrous magnesium sulfate is commonly used as a drying agent in the laboratory, to remove traces of water from organic solutions. Solid magnesium sulfate is added to the organic solution, the flask is swirled gently, and the magnesium sulfate is then removed, typically by gravity filtration using a fluted filter paper.

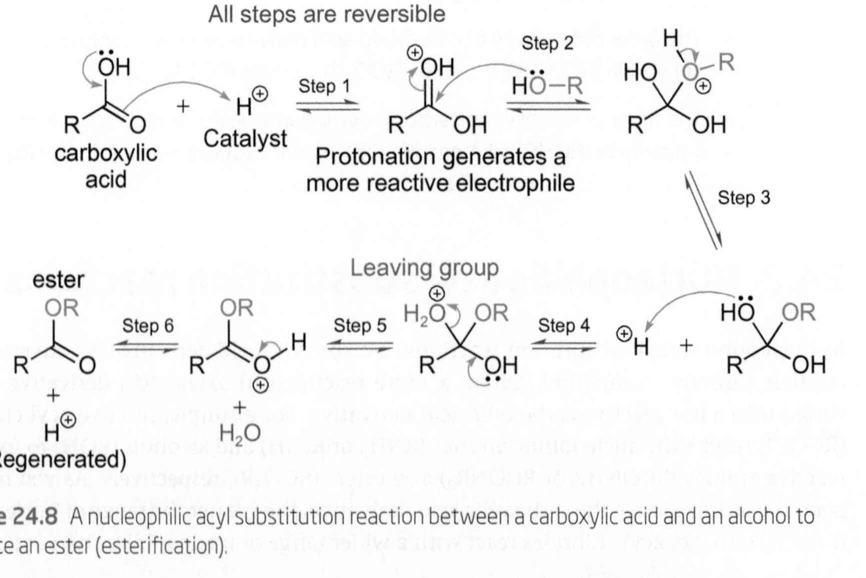
All steps are reversible
Step 1
Step 2
Step 3
Step 4
Step 5
Step 6
carboxylic acid
Catalyst
Protonation generates a more reactive electrophile
Leaving group
ester
(Regenerated)

Figure 24.8 A nucleophilic acyl substitution reaction between a carboxylic acid and an alcohol to produce an ester (esterification).

Box 24.3 Fragrant esters

Naturally occurring esters are present in both animals and plants. Small-sized esters, along with other volatile organic compounds, are responsible for the smells of fruits. Often, one compound plays a major role in producing the fruity smell. For example, the fragrance of pineapples is mainly due to ethyl butanoate, $CH_3CH_2CH_2CO_2CH_2CH_3$ (for how to name esters, see Section 2.7, p.102). Different esters have different fragrances and flavours and various synthetic esters are used as flavourings in the food industry.

A

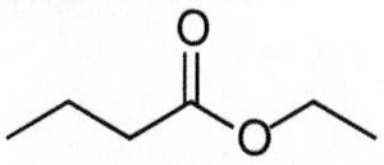
O
O

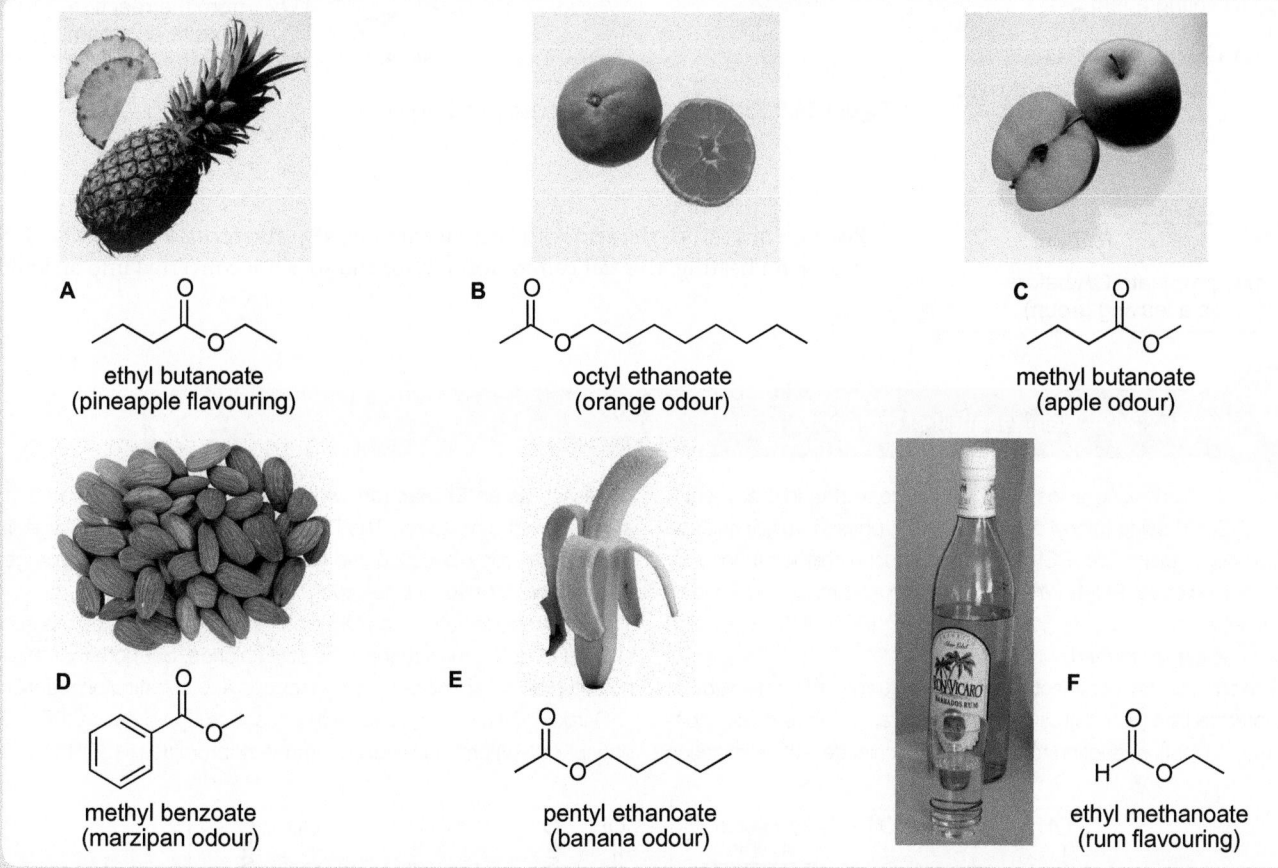

ethyl butanoate
(pineapple flavouring)

B

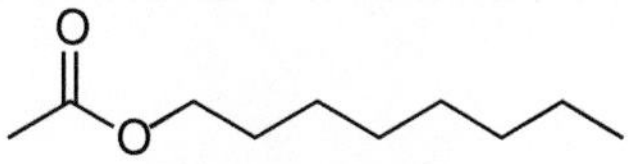
O
O

octyl ethanoate
(orange odour)

C

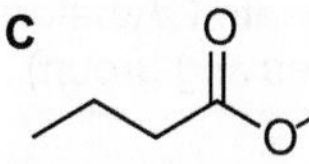
O
O

methyl butanoate
(apple odour)

D

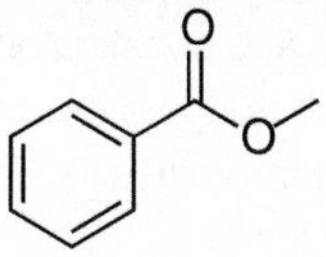
O
O

methyl benzoate
(marzipan odour)

E

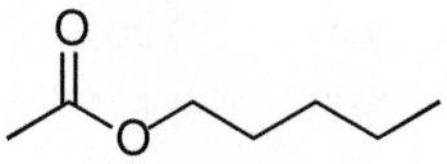
O
O

pentyl ethanoate
(banana odour)

F

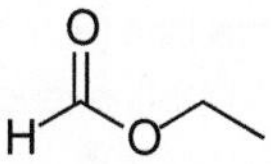
O
H
O

ethyl methanoate
(rum flavouring)

Question

All six esters (**A–F**) shown above are prepared by reaction of an alcohol with a carboxylic acid in an esterification reaction.

For each ester, draw the structures of the precursor alcohol and carboxylic acid.

Reaction of a carboxylic acid with ammonia and amines: preparation of amides

Carboxylic acids (RCO_2H) react with ammonia (NH_3) or primary or secondary amines (RNH_2 or R_2NH) at room temperature in acid–base reactions to form salts. When the salts are heated to high temperatures (typically 150 °C–200 °C) amides are formed in nucleophilic acyl substitution reactions (see Figure 24.9). At high temperatures, the small amount of ammonia (or primary or secondary amine) and carboxylic acid, which are in equilibrium with the salt, react to form the amide, $RCONH_2$ (or RCONHR or $RCONR_2$)—the equilibrium is shifted to the right because, at the high temperatures, as soon as water is formed, it is removed from the reaction mixture as steam. The nucleophilic acyl substitution reaction is reversible and the mechanism is similar to the esterification reaction in Figure 24.8—ammonia (or the primary or secondary amine) attacks the carbon atom in the C=O bond of the carboxylic acid and water is the leaving group.

Primary (RNH_2) and secondary (R_2NH) amines are stronger bases than ammonia (Section 192, p.891).

Whereas protonation of ammonia forms an ammonium ion, H_4N^+, protonation of an amine forms an aminium ion, commonly called an alkylammonium ion (e.g. $MeNH_3^+$ is the methanaminium ion, commonly called methylammonium).

The high temperatures required to make an amide from an amine and a carboxylic acid have limited the use of this reaction in synthesis. More commonly, a reagent (called an activating agent) is used to convert the OH group of the carboxylic acid into a good leaving group, so that it reacts with the amine, to form the amide, under mild conditions. (For activation of alcohols, see Section 20.2, p.925.)

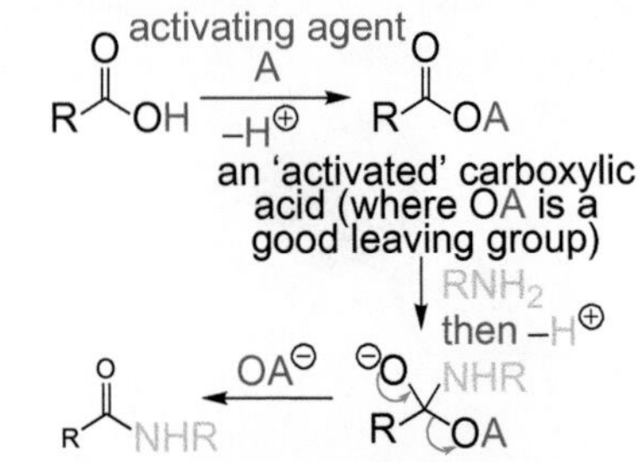
activating agent
A
an 'activated' carboxylic acid (where OA is a good leaving group)
RNH2
then –H⊕

(H+ may protonate OA before it acts as a leaving group)

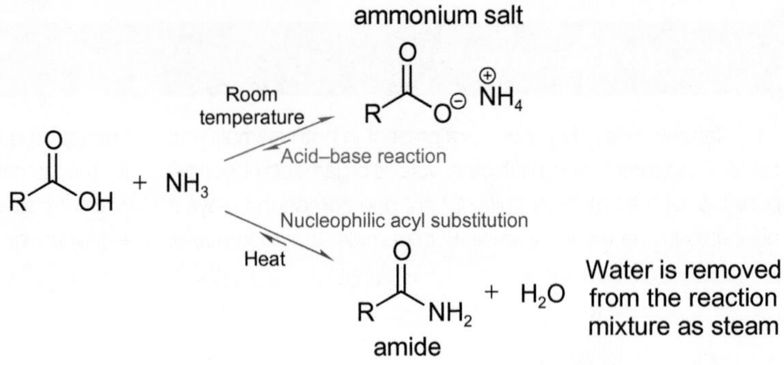
ammonium salt
Room temperature
Acid–base reaction
Nucleophilic acyl substitution
Heat
Water is removed from the reaction mixture as steam
amide

Figure 24.9 Reaction of a carboxylic acid with ammonia.

Reaction of a carboxylic acid with a tertiary amine (NR_3) also forms a salt, $RCO_2^{-}\,^{+}HNR_3$. However, on heating, this salt cannot form water and so is not converted into an amide.

Box 24.4 Halogenation of carboxylic acids

Carboxylic acids (RCO_2H) are converted into acyl chlorides (RCOCl) using thionyl chloride ($SOCl_2$), phosphorus(III) chloride (phosphorus trichloride, PCl_3), or phosphorus(V) chloride (phosphorus pentachloride, PCl_5). When a carboxylic acid reacts with thionyl chloride, a resonance-stabilized oxonium ion (RC(=OH$^+$)OSOCl) and a chloride ion are formed.

The chloride ion then reacts with the oxonium ion to form an unstable intermediate (RC(OH)(Cl)OSOCl), which rapidly undergoes a concerted elimination reaction to form the acyl chloride—the elimination is classed as an **Ei reaction**, where E stands for elimination and i stands for intramolecular. The Ei reaction is favoured by a large increase in entropy, because two gases (sulfur dioxide and hydrogen chloride) are formed as side-products (Section 14.4, p.669).

In this transformation, the OH group in the carboxylic acid is substituted by a Cl group to form the acyl chloride, but note that this is not a typical mechanism for a nucleophilic acyl substitution reaction.

Carboxylic acids also react with phosphorus(III) bromide (PBr_3) or phosphorus(V) bromide (PBr_5) to form acyl bromides (RCOBr).

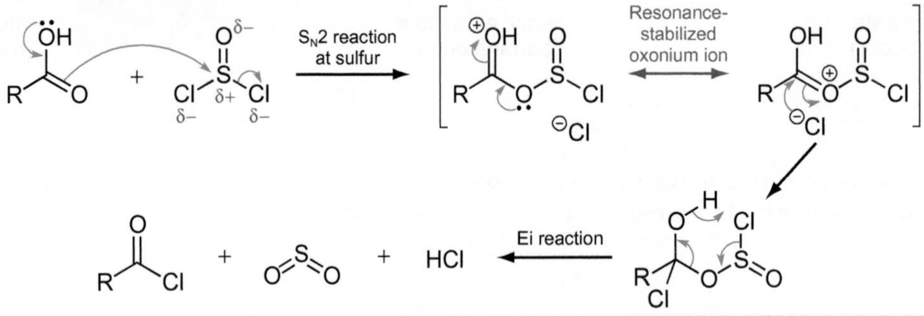
SN2 reaction at sulfur
Resonance-stabilized oxonium ion
Ei reaction

Question

The mechanisms of the reactions of carboxylic acids with PCl_5 and $SOCl_2$ are similar. As shown below, PCl_5 reacts with a carboxylic acid to form an intermediate phosphorus(V) compound (**A**), which breaks down to form the acyl chloride, $POCl_3$, and HCl.

Propose a mechanism to show how **A** is formed and then converted into the acyl chloride.

Interactive 3D animation of acyl chloride formation using thionyl chloride.

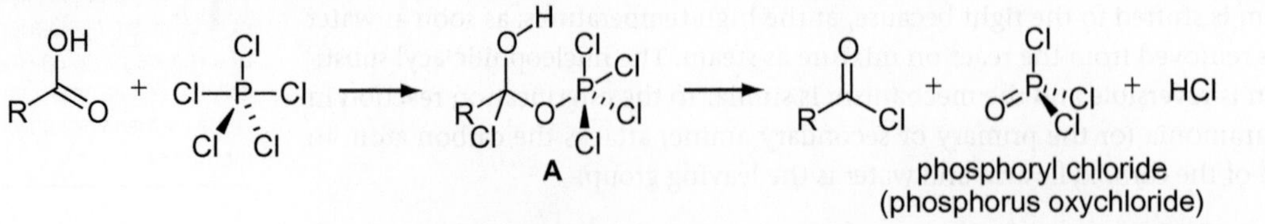
A
phosphoryl chloride
(phosphorus oxychloride)

Nucleophilic acyl substitution reactions of acyl chlorides

Of all the carboxylic acid derivatives, acyl chlorides (RCOCl) are the most reactive to nucleophiles (see Box 24.2, p.1103), so acyl chlorides react with the widest range of nucleophiles. These reactions are often used in synthesis to prepare other (less reactive) carboxylic acid derivatives. Some substitution reactions of acyl chlorides are summarized in Figure 24.10. In each reaction, the chlorine atom attached to the C=O bond acts as the leaving group.

The reaction of an acyl chloride to form an ester or an amide is commonly described as an **acylation reaction**, and the acyl chloride is the acylating agent. In an acylation reaction, an RC(=O) group is introduced into a compound (as in the Friedel–Crafts acylation reaction of benzene, Section 22.2, p.1023).

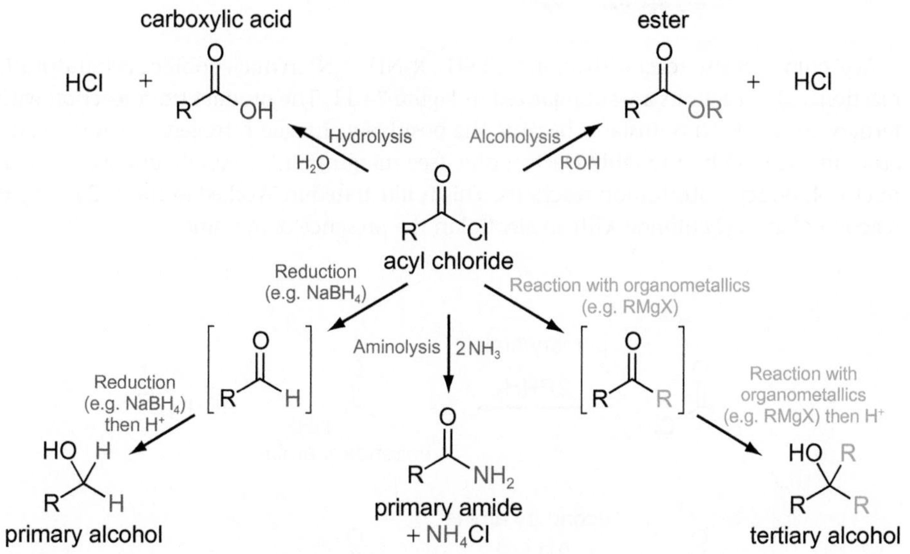
carboxylic acid
HCl +
R OH
Hydrolysis
H_2O
ester
+ HCl
R OR
Alcoholysis
ROH
R Cl
acyl chloride
Reduction
(e.g. $NaBH_4$)
Reaction with organometallics
(e.g. RMgX)
Aminolysis
2 NH_3
R H
R R
Reduction
(e.g. $NaBH_4$)
then H^+
HO H
R H
primary alcohol
R NH_2
primary amide
+ NH_4Cl
Reaction with
organometallics
(e.g. RMgX) then H^+
HO R
R R
tertiary alcohol

Figure 24.10 Nucleophilic acyl substitution reactions of acyl chlorides.

Nucleophilic acyl substitution in acyl chlorides is much faster than nucleophilic substitution in chloroalkanes. For example, compare the relative reaction rates of hydrolysis (in 80% ethanol : 20% water), where water is the nucleophile.

Compound	Relative rate
Benzoyl chloride, PhCOCl	1000
Benzyl chloride, $PhCH_2Cl$	1

It is easier for a nucleophile to approach the sp^2-hybridized carbon atom of an acyl chloride, as this is less hindered than the sp^3-hybridized carbon atom in a chloroalkane. Also, the carbonyl carbon atom in the acyl chloride is the stronger electrophile—the electronegative oxygen atom increases the electrophilicity of this carbon.

Acyl chlorides react with water at room temperature to form carboxylic acids in a hydrolysis reaction, following the general mechanism shown in Figure 24.4 (p.1105).

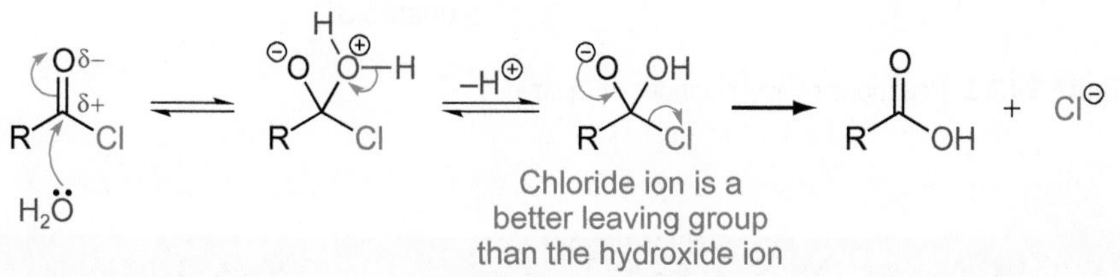
Oδ−
δ+
R Cl
$H_2\ddot{O}$
$-H^{\oplus}$
OH
R OH
+ $Cl^{\ominus}$
Chloride ion is a
better leaving group
than the hydroxide ion

The order of the steps is important. Deprotonation of the $-OH_2^+$ group (in the first tetrahedral intermediate) is required before the Cl^- ion is expelled (in the second tetrahedral intermediate). This is because, in the first tetrahedral intermediate, the better leaving group is H_2O, whereas in second tetrahedral intermediate, the better leaving group is Cl^-.

In the first step, water acts as a nucleophile and attacks the C=O bond of the acyl chloride (RCOCl) to form a tetrahedral intermediate ($RC(O^-)(Cl)OH_2^+$). The tetrahedral intermediate then loses a proton (often a base such as pyridine is added to the reaction mixture to remove the proton—see Worked example 24.1, p.1112) to form an unstable alkoxide ion ($RC(O^-)(Cl)OH$). In the final step, the alkoxide ion expels the chloride ion (Cl^-) to form the C=O bond of a carboxylic acid (RCOOH). Notice that the chloride ion (Cl^-) rather than the hydroxide ion (HO^-) acts as the leaving group.

Good leaving groups are stable anions that are weakly basic. The chloride ion (the pK_a value of HCl is −7) is a weaker base than the hydroxide ion (the pK_a value of H_2O is 15.7) so it is a better leaving group (Section 20.2, p.925).

Acyl chlorides also react with alcohols to form esters (RCOOR) in an alcoholysis reaction, and with ammonia to form primary amides ($RCONH_2$) in an aminolysis reaction. The mechanisms of these nucleophilic acyl substitution reactions are very similar to the reaction of an acyl chloride with water (for practice, try drawing the mechanism of reaction of an acyl chloride with an alcohol, ROH). Formation of a primary amide requires two equivalents of ammonia—one equivalent of ammonia acts as a nucleophile, whilst the second equivalent acts as a base (to form NH_4Cl).

In an **alcoholysis** reaction, a reactant is converted into two products by reaction with an alcohol. In an **aminolysis** reaction, a reactant is converted into two products by reaction with ammonia or an amine.

Interactive 3D animation of amide formation from an acyl chloride.

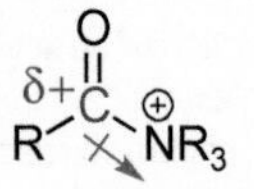

Strong −I effect increases the electrophilicity of the carbonyl carbon

Two equivalents is a short way of saying 'two stoichiometric equivalents', that is, twice as many moles of ammonia as there are of the acyl chloride.

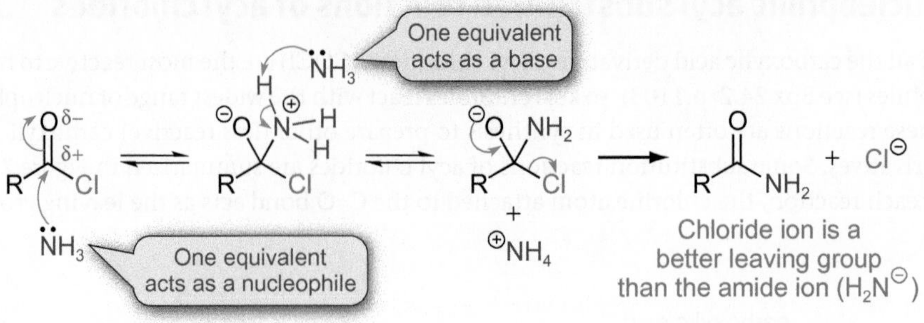
One equivalent acts as a base
One equivalent acts as a nucleophile
Chloride ion is a better leaving group than the amide ion ($H_2N^{\ominus}$)

Acyl chlorides also react with amines (RNH_2, R_2NH, R_3N) in nucleophilic acyl substitution reactions. The reactions are summarized in Figure 24.11. The product from reaction with a tertiary amine (R_3N) is unstable because the positively charged nitrogen makes the carbon atom in the C=O bond highly electrophilic (see margin) and it rapidly undergoes further nucleophilic acyl substitution reactions. This is illustrated in Worked example 24.1, by the reaction of an acyl chloride with an alcohol in the presence of pyridine.

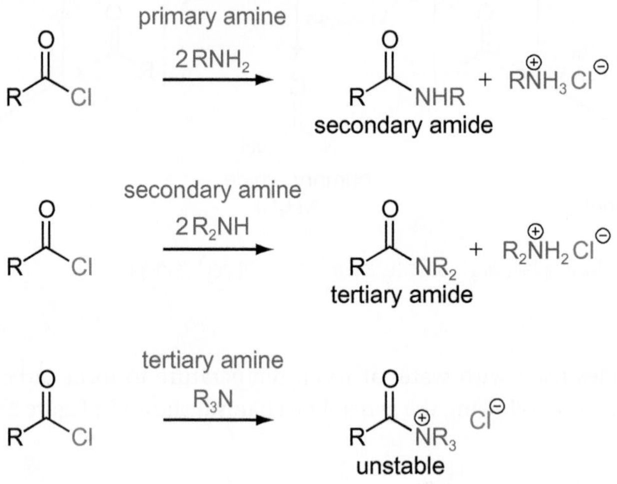
primary amine
2 RNH2
secondary amide
secondary amine
2 R2NH
tertiary amide
tertiary amine
R3N
unstable

Figure 24.11 Reactions of acyl chlorides with amines.

Worked example 24.1 Esters from acyl chlorides

Esters are prepared by reacting acyl chlorides with alcohols. Often a base such as pyridine is used to deprotonate the intermediate zwitterion. An outline of this reaction is shown here.

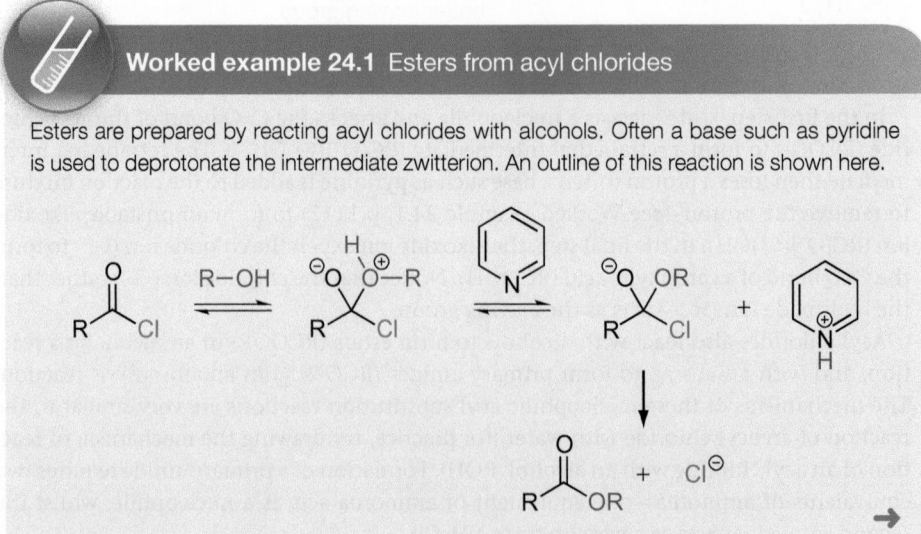

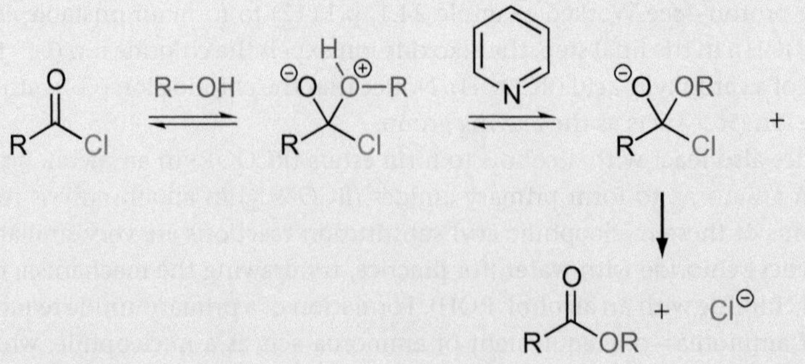
R−OH

→

(a) Use curly arrows to explain the mechanism of the reaction.

(b) In the presence of excess pyridine, pyridine is more likely than ROH to react with the acyl chloride to form the unstable intermediate **A**, which then reacts with ROH to form the ester.

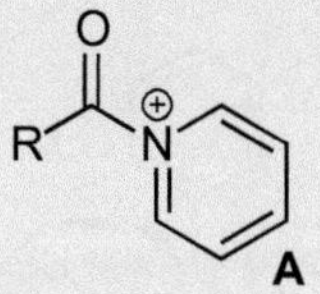
O
R
N
A

(i) Suggest a mechanism to explain the formation of **A** from an acyl chloride.

(ii) Would ROH react faster with **A** or with RCOCl? Give your reasons.

Strategy

(a) Recognize that this is a nucleophilic acyl substitution reaction in which the alcohol acts as the nucleophile and Cl^- is the leaving group.

(b)

(i) Recognize that this is a nucleophilic acyl substitution reaction in which pyridine acts as the nucleophile and Cl^- is the leaving group.

(ii) Identify the electronic effects of the substituents attached to the C=O bond in RCOCl and compound **A**. Use the electronic effects to decide whether RCOCl or compound **A** has the more electrophilic carbon atom in the C=O bond.

Solution

(a)

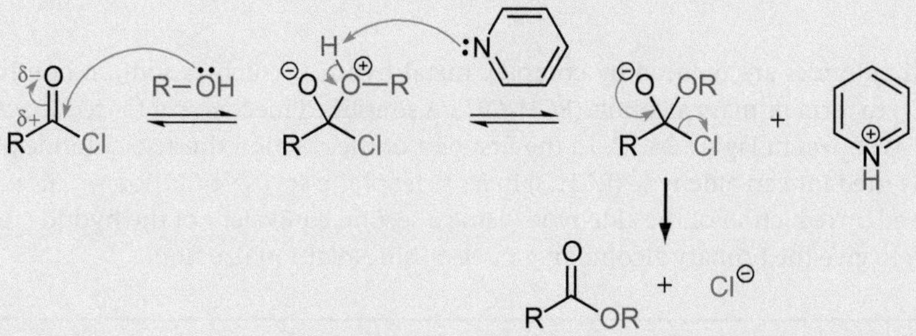
δ−O
δ+
R
Cl
R−ÖH
H
O
O−R
:N
O
OR
+
H
O
OR
+
Cl

(b) (i)

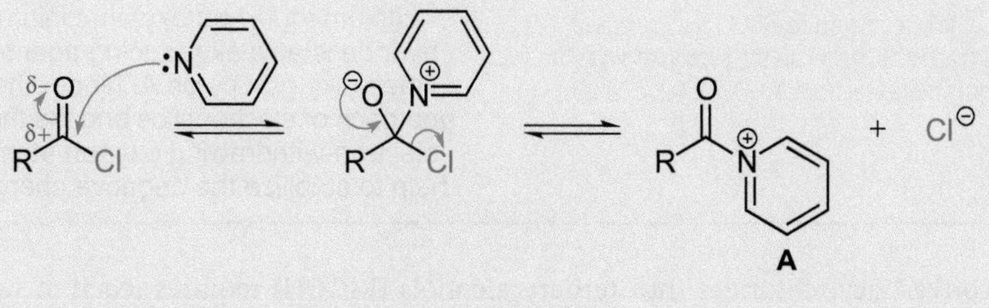
δ−O
δ+
R
Cl
:N
O
N
O
+
Cl
A

(ii) ROH reacts faster with **A** than with RCOCl. The carbon atom in the C=O bond of **A** is more electrophilic because the positively charged nitrogen in the pyridine ring has a strong −I effect (and no mesomeric effect). In RCOCl, although the chlorine atom has a strong −I effect, it also has a weak +M effect so the chlorine atom is not as electron-withdrawing as the positively charged pyridine group.

→

Sometimes, when forming an ester from an acyl chloride, a small quantity of pyridine is used together with one equivalent of a stronger base than pyridine, such as triethylamine (Et_3N). The triethylamine reacts with the HCl that is formed as a byproduct (giving $Et_3NH^+ Cl^-$), whereas the pyridine is used to speed up the acylation reaction by forming a highly electrophilic *N*-acylpyridinium intermediate (compound A). To form compound A pyridine must act as a nucleophile, and as pyridine speeds up the reaction but is recovered unchanged at the end of the reaction, this is called nucleophilic catalysis (and pyridine is called a nucleophilic catalyst).

Pyridine has a very unpleasant, distinctive odour—sour, putrid, and like rotting fish.

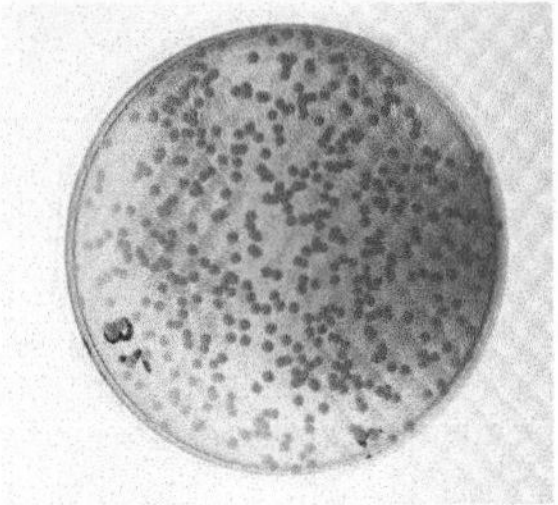

Researchers have recently re-engineered the biochemical pathway that poppies use to make opioids, such as morphine, and transplanted it into a species of yeast. The yeast uses 21 enzymes from plants, mammals, bacteria, and yeast to convert a sugar into thebaine, which has a similar structure to morphine. Opioids are used in medicine to relieve pain (Box 2.7, p.97), but currently, the amounts produced by the yeast are tiny compared with what is needed commercially.

➜

Question

Reaction of morphine with two equivalents of ethanoyl chloride and excess pyridine produces heroin. Give the structure of heroin.

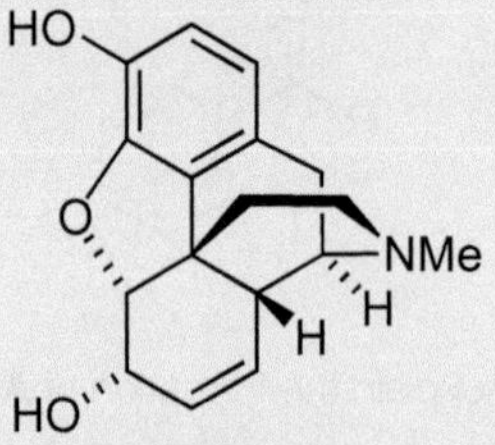
HO
O
NMe
H
H
HO

morphine

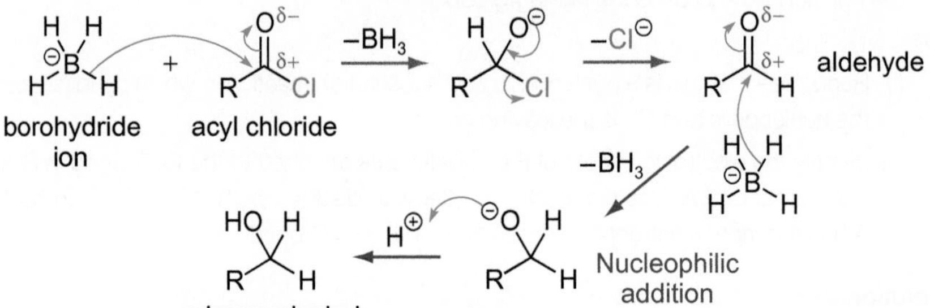
Nucleophilic acyl substitution
borohydride ion
acyl chloride
−BH3
−Cl⊖
aldehyde
−BH3
Nucleophilic addition
primary alcohol

Figure 24.12 Reduction of an acyl chloride with $[BH_4]^-$ to form a primary alcohol.

Acyl chlorides are reduced by complex metal hydrides (such as sodium borohydride, $NaBH_4$) to form primary alcohols (RCH_2OH). A simplified mechanism for reduction using $[BH_4]^-$ is shown in Figure 24.12. In the first part of the reaction the acyl chloride (RCOCl) is converted into an aldehyde (RCHO) in a nucleophilic acyl substitution reaction. This is followed by reduction of the aldehyde (using a second equivalent of the hydride reducing agent) to give the primary alcohol in a nucleophilic addition reaction.

ℹ Reaction of aldehydes (RCHO) and ketones (RCOR) with complex metal hydrides and Grignard reagents (RMgX) is discussed in Section 23.2 (pp.1062 and 1069). For clarity, the $[BH_4]^-$ reduction mechanism is simplified by assuming that BH_3 is a product of the reaction, whereas this hydride undergoes further reactions.

ℹ Reaction of RCOCl with $NaBH_4$ can be stopped at the aldehyde stage but, usually, the aldehyde is made by reacting RCOCl with the weaker reducing agent, lithium tri(*tert*-butoxy)-aluminium hydride, at low temperature. RCOCl is more easily reduced than RCHO because the carbon atom in the C=O bond is more electrophilic (Box 24.2, p.1103).

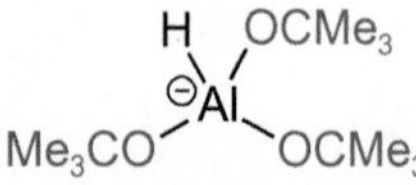
H OCMe3
Al
Me3CO OCMe3

Lithium tri(*tert*-butoxy)aluminium hydride is a weak reducing agent—the bulky groups on Al hinder the approach of electrophiles and the three electron-withdrawing oxygen atoms help to stabilize the negative charge.

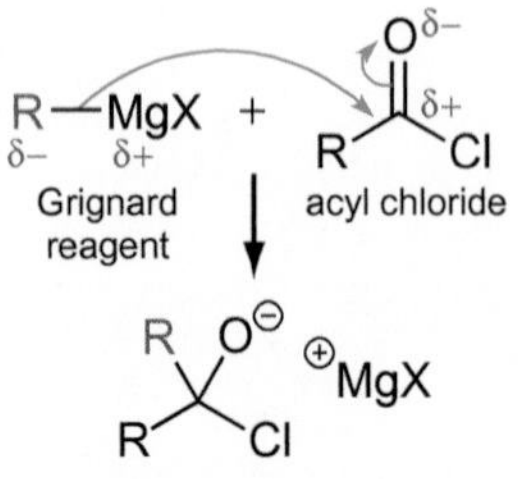
Grignard reagent
acyl chloride
MgX

The first step in the reaction of a Grignard reagent with an acyl chloride. Remember that, for the Grignard reagent, the tail of the curly arrow starts from the centre of the C–Mg bond.

To convert acyl chlorides into tertiary alcohols (R_3COH) requires reaction with two equivalents of an organometallic reagent such as a Grignard reagent (RMgX), as shown in Figure 24.10 (p.1111). The mechanism of the reaction is very similar to the reaction of acyl chlorides with complex metal hydrides shown in Figure 24.12.

In the first part of the reaction the acyl chloride is converted into a ketone (RCOR) in a nucleophilic acyl substitution reaction. This is followed by reaction of the ketone with a second equivalent of the Grignard reagent to give a tertiary alcohol (after protonation) in a nucleophilic addition reaction. For practice, try drawing a mechanism for the conversion of an acyl chloride into a tertiary alcohol. To help you, the first step is given in the margin.

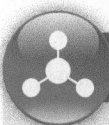

Box 24.5 Combinatorial chemistry

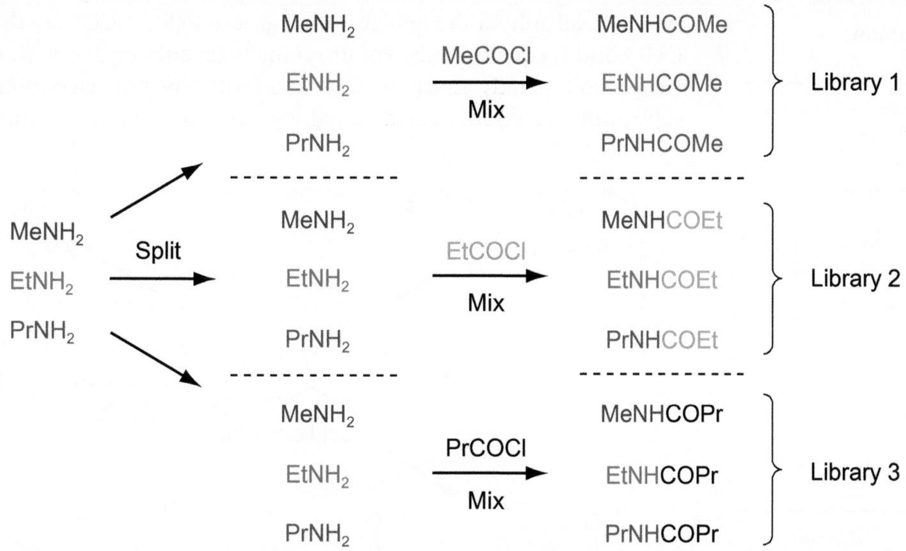
MeNH2
EtNH2
PrNH2
Split
MeNH2
EtNH2
PrNH2
MeCOCl
Mix
MeNHCOMe
EtNHCOMe
PrNHCOMe
Library 1
MeNH2
EtNH2
PrNH2
EtCOCl
Mix
MeNHCOEt
EtNHCOEt
PrNHCOEt
Library 2
MeNH2
EtNH2
PrNH2
PrCOCl
Mix
MeNHCOPr
EtNHCOPr
PrNHCOPr
Library 3

Combinatorial chemistry is a modern research method that involves the rapid preparation of large numbers of compounds. Several compounds are prepared simultaneously in one reaction mixture and the mixture is then tested, usually for biological activity. Most pharmaceutical companies use this technique because it has the potential to identify biologically active compounds much more quickly than the traditional approach of preparing, purifying, and then testing individual compounds for biological activity.

In an approach called 'split and mix', chemists prepare mixtures (known as libraries) of compounds with related chemical structures. Medicinal chemists often prepare libraries of amides (typically RCONHR or $RCONR_2$) because they are quickly and efficiently prepared under mild conditions, and the amide group is present in many biologically active compounds such as peptides and proteins.

To illustrate the principle of 'split and mix', consider the synthesis of mixtures of secondary amides (RCONHR) starting from three different primary amines (RNH_2) and three different acyl chlorides (RCOCl). Initially, equal amounts of methanamine ($MeNH_2$), ethanamine ($EtNH_2$), and propanamine ($PrNH_2$) are mixed together. The mixture is then split into three equal portions—each mixture is made up of 1/3 of methanamine, 1/3 of ethanamine, and 1/3 of propanamine.

Each of the three portions is then reacted with a different acyl chloride, ethanoyl chloride (MeCOCl), propanoyl chloride (EtCOCl), or butanoyl chloride (PrCOCl), to form three different mixtures. In library 1 all three amides contain a COMe group (from ethanoyl chloride), in library 2 all three amides contain a COEt group (from propanoyl chloride), while the amides in library 3 all contain a COPr group (from butanoyl chloride).

The three libraries are then tested separately to establish quickly whether changing the acyl group (from COMe to COEt or COPr) in a series of molecules affects their biological activity.

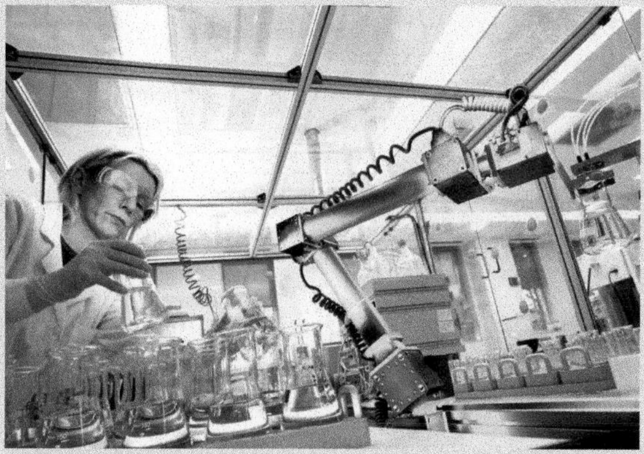

▲ Robots have greatly increased the speed and reduced the cost of making large libraries of compounds. In just one week, an automated laboratory can produce and test the biological activity of libraries containing a million compounds.

Combinatorial chemistry dramatically speeds up the number of compounds that can be prepared and tested, but a biological hit may not be due to just one compound in the mixture. It could be due to the combined properties of all or some of the compounds in the library. The random nature of the process has been likened to a chemical lottery, where biologically active compounds are found by chance rather than by design. It may also be difficult to determine which compound in the library is the most active. This generally requires the re-synthesis of all of the individual compounds in the library, which are then tested individually.

Question

Show how libraries of esters are constructed by reacting methanol, ethanol, and propanol with ethanoyl chloride, propanoyl chloride, and butanoyl chloride.

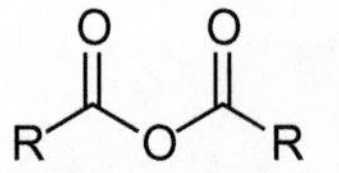

RCO_2COR
(R can be the same or different)
acid anhydride

The carboxylate ion (RCO_2^-) is a weak base (the pK_a value of RCO_2H is 4–5) and a good leaving group because the anion is stabilized by resonance (the negative charge is spread over both electronegative oxygen atoms). In comparison, the amide ion ($^-NH_2$) is a much stronger base (the pK_a value of NH_3 is 38) and it is a very poor leaving group.

To make aspirin (Box 19.7, p.900), acylation of salicylic acid is usually carried out in the laboratory using ethanoic anhydride ($CH_3CO_2COCH_3$), rather than ethanoyl chloride (CH_3COCl). As ethanoic anhydride is less reactive than ethanoyl chloride, this means it is easier to handle (it is less reactive to moisture in the air), less corrosive (hydrolysis of ethanoyl chloride forms fumes of HCl) and the reaction with salicylic acid is less exothermic (which is particularly important on a large scale). The anhydride is also cheaper to purchase.

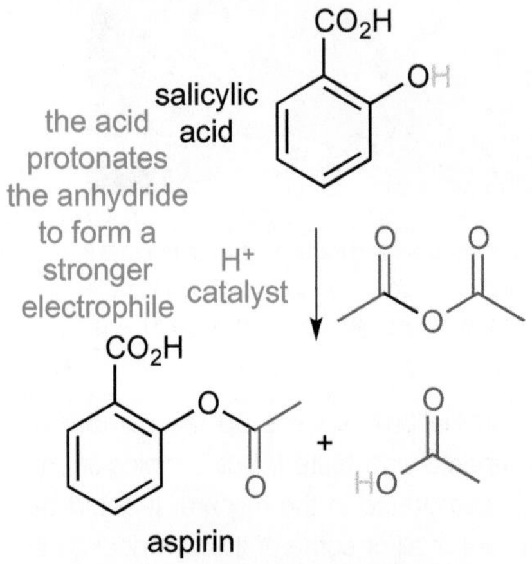

CO_2H
OH
salicylic acid
the acid protonates the anhydride to form a stronger electrophile
H^+ catalyst
CO_2H
O
HO
aspirin

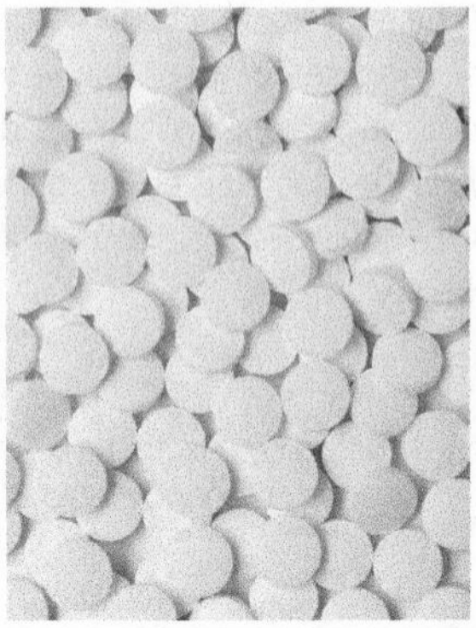

Nucleophilic acyl substitution reactions of acid anhydrides

Acid anhydrides (RCO_2COR) react with nucleophiles such as water, alcohols (ROH), and ammonia in nucleophilic acyl substitution reactions. In these reactions, the nucleophile attacks one of the C=O bonds and a carboxylate ion (RCO_2^-) acts as the leaving group. For symmetrical anhydrides (where both R groups in RCO_2COR are the same), attack at either C=O bond is equally likely. For unsymmetrical anhydrides (with different R groups), nucleophiles generally attack the C=O bond with the more electrophilic carbon atom. Some substitution reactions of acid anhydrides are summarized in Figure 24.13.

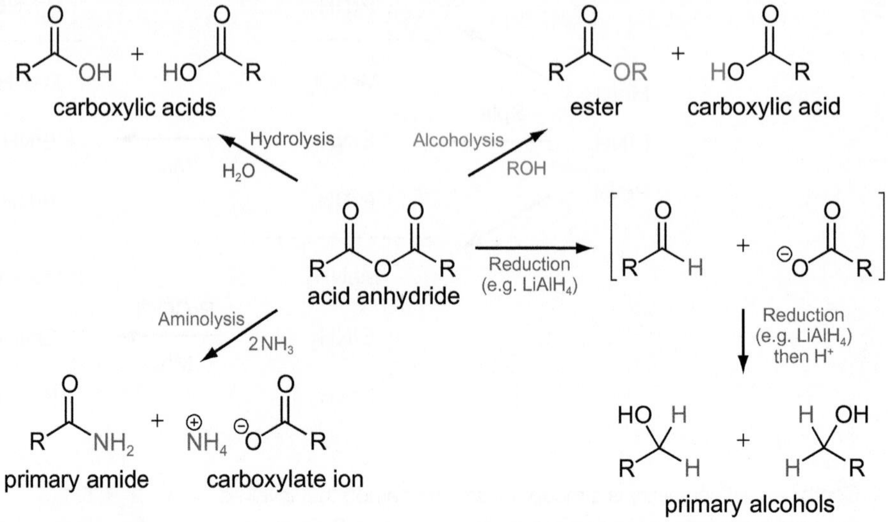

carboxylic acids
ester
carboxylic acid
Hydrolysis
H_2O
Alcoholysis
ROH
acid anhydride
Reduction (e.g. $LiAlH_4$)
Reduction (e.g. $LiAlH_4$) then H^+
Aminolysis
2 NH_3
primary amide
carboxylate ion
primary alcohols

Figure 24.13 Nucleophilic acyl substitution reactions of acid anhydrides. (The R groups can be the same or different.)

The mechanisms of these reactions follow the general mechanism in Figure 24.4 (p.1105), as illustrated below, by the reaction of an acid anhydride with ammonia (aminolysis).

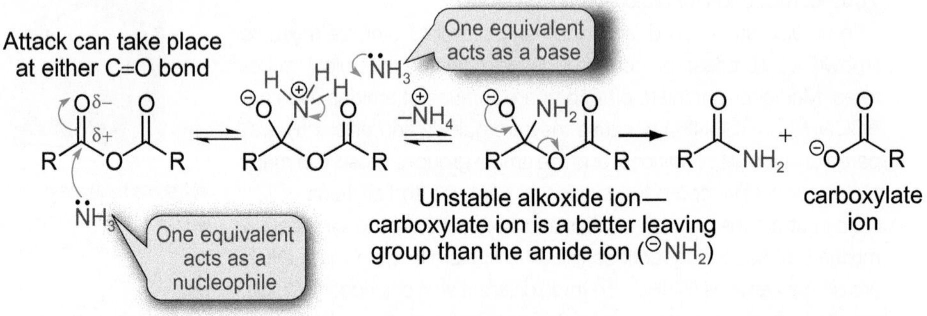

Attack can take place at either C=O bond
One equivalent acts as a base
One equivalent acts as a nucleophile
$-\overset{\oplus}{N}H_4$
Unstable alkoxide ion—carboxylate ion is a better leaving group than the amide ion ($^{\ominus}NH_2$)
carboxylate ion

Ammonia attacks one of the C=O bonds of the acid anhydride to form a tetrahedral intermediate, $RC(O^-)(NH_3^+)OCOR$. Deprotonation of the tetrahedral intermediate by a second equivalent of ammonia then produces an unstable alkoxide ion, $RC(O^-)(NH_2)OCOR$. In the final step, the alkoxide ion expels the carboxylate ion (RCO_2^-) to form the C=O bond of a primary amide ($RCONH_2$).

Primary and secondary amines (RNH_2 and R_2NH) react with anhydrides by a similar mechanism to that shown for ammonia, to form secondary and tertiary amides (RNHCOR and R_2NCOR), respectively. Worked example 24.2 discusses how reaction with an anhydride can be used to protect primary and secondary amines.

The mechanism for the reaction of acid anhydrides with complex metal hydrides, such as $LiAlH_4$, is more complicated because the two products of the nucleophilic acyl substitution reaction (an aldehyde, RCHO and a carboxylate ion, RCO_2^-) each react further, as shown in Figure 24.14. Overall, the acid anhydride is reduced to form two primary alcohol (RCH_2OH) molecules. The mechanism involves two nucleophilic acyl substitutions and two nucleophilic additions.

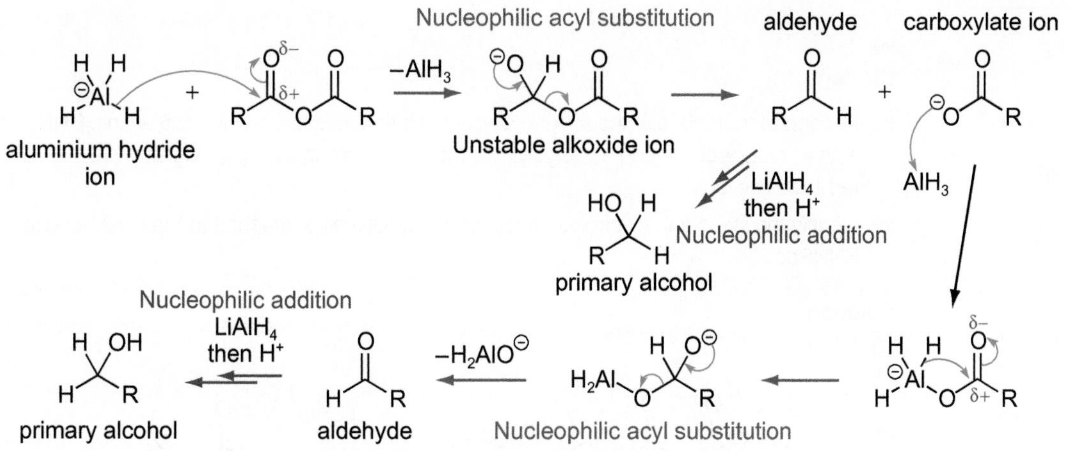

Nucleophilic acyl substitution
aldehyde
carboxylate ion
aluminium hydride ion
−AlH3
Unstable alkoxide ion
LiAlH4 then H+
Nucleophilic addition
primary alcohol
AlH3
Nucleophilic addition
LiAlH4 then H+
−H2AlO
primary alcohol
aldehyde
Nucleophilic acyl substitution

Figure 24.14 Reduction of an acid anhydride with $LiAlH_4$ to form a primary alcohol.

Initially, $LiAlH_4$ donates a hydride ion (H^-) to one of the C=O bonds of the acid anhydride to form an unstable alkoxide ion, $RC(O^-)(H)OCOR$. The alkoxide ion then expels the carboxylate ion (RCO_2^-) to form the C=O bond of an aldehyde in the first nucleophilic acyl substitution. $LiAlH_4$ subsequently reduces the aldehyde in a nucleophilic addition reaction (Section 23.2, p.1063), and it also reduces the carboxylate ion (although the exact mechanism for this step is less well understood). Reduction of the carboxylate ion may start by addition of the anion to AlH_3 (AlH_3 is produced in an earlier step) to form an unstable intermediate, $RCO_2AlH_3^-$, that could transfer a hydride ion from the aluminium atom to the C=O bond. This produces an alkoxide ion, $RC(O^-)H(OAlH_2)$, that expels H_2AlO^- to form the C=O bond of an aldehyde in a second nucleophilic acyl substitution. Finally, $LiAlH_4$ reduces the aldehyde in a second nucleophilic addition step.

$LiAlH_4$ reduces carboxylic acids to form primary alcohols. The reaction mechanism starts by deprotonation of the carboxylic acid (RCO_2H) to form a carboxylate ion (RCO_2^-). This is then reduced, possibly by the mechanism shown for reduction of the acid anhydride.

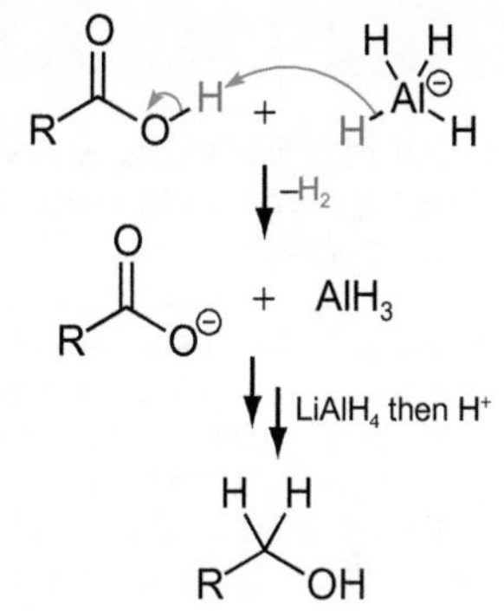

−H2
+ AlH3
LiAlH4 then H+

During the reduction, an Al–H bond (around +285 kJ mol^{-1}) in AlH_3 is replaced by a stronger Al–O bond (around +511 kJ mol^{-1}) in H_2AlO^-.

Visit the Online Resource Centre to view screencast 24.1 which explains why, in synthesis, amines often need to be protected as carbamates. It walks you through how an anhydride (called di-*tert*-butyl dicarbonate, $^tBuOC(=O)OC(=O)O^tBu$) converts an amine (RNH_2 or R_2NH) into a carbamate group (called a Boc group, $RNHCO_2{}^tBu$ or $R_2NCO_2{}^tBu$).

Worked example 24.2 Using an anhydride to protect amine groups

In multistep syntheses, nucleophilic amines (e.g. RNH_2) are often temporarily protected from reacting with electrophiles by conversion into carbamates (e.g. $RNHCO_2R$, see Section 2.7, p.105). One of the most common of this type of protecting group is the *tert*-**butyloxycarbonyl (Boc)** group. The Boc group is introduced by reaction of an amine with the anhydride Boc_2O ($^tBuOC(=O)OC(=O)O^tBu$) in the presence of Et_3N as shown below—Boc_2O is called an anhydride because of the COOCO group in the centre of the molecule.

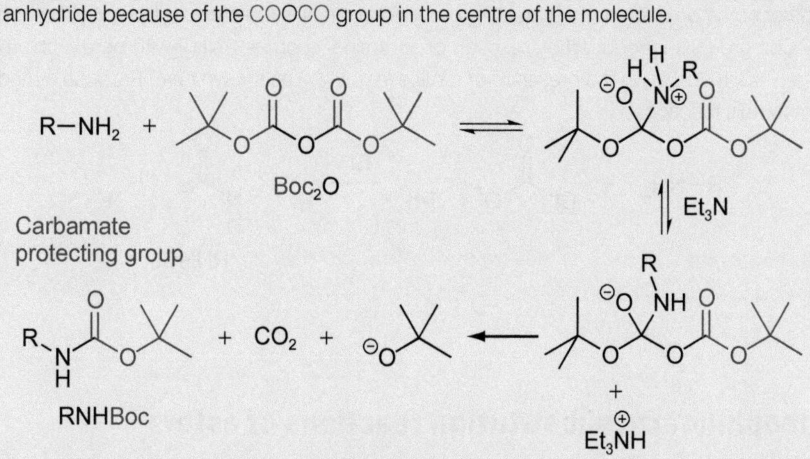

R−NH2 +
Boc2O
Et3N
Carbamate protecting group
+ CO2 +
RNHBoc
+
Et3NH

(a) Use curly arrows to explain the mechanism of the reaction shown here.

(b) One of the reasons the Boc group is widely used in synthesis is that this group is unreactive to most electrophiles and nucleophiles. Explain why the C=O bond in the Boc protecting group is relatively unreactive to nucleophiles.

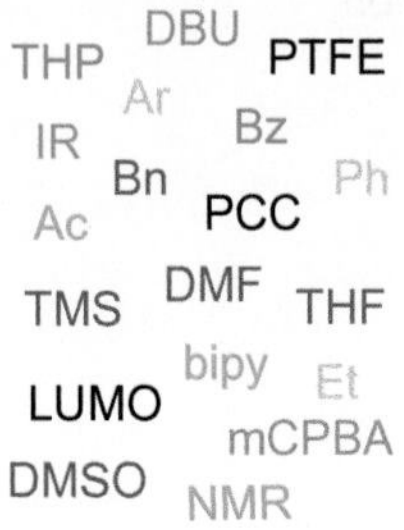
DBU
THP
PTFE
Ar
Bz
IR
Bn
Ph
Ac
PCC
DMF
TMS
THF
bipy
Et
LUMO
mCPBA
DMSO
NMR

The use of terms such as Boc and Cbz (sometimes shown in capitals, BOC and CBZ), for amine protecting groups, further illustrates how much chemists like to use abbreviations! For a list of commonly used abbreviations see Appendix 1 (p.1331).

CO2

The formation of RNHBoc is favoured by an increase in entropy as carbon dioxide gas is formed as a side-product.

The general formula of chloroformates is ROC(=O)Cl, where R can be an alkyl or aryl group. For example, ethyl chloroformate is EtOC(=O)Cl. Notice that the C=O group in a chloroformate is bonded to two possible leaving groups—reaction with a nucleophile leads to loss of Cl^- as this is a better leaving group than RO^-.

Strategy

(a) Recognize that this is a nucleophilic acyl substitution reaction in which the amine, RNH_2, acts as a nucleophile and $^-OCO_2CMe_3$ is the leaving group (forming CO_2 and $^-OCMe_3$, a *tert*-butoxide ion).

(b) Identify the steric and electronic effects of the substituents attached to the C=O bond in RNHBoc.

Solution

(a)

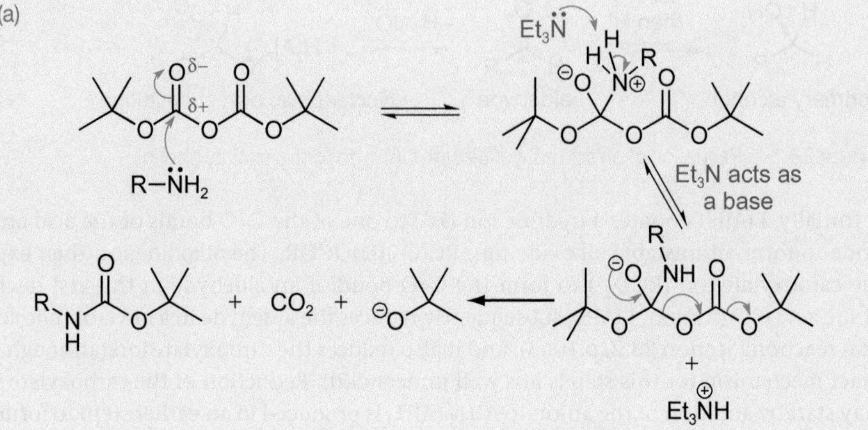
Et3N
δ−
δ+
R−NH2
Et3N acts as a base
CO2
Et3NH

(b) The carbon atom in the C=O bond of RNHBoc is only a weak electrophile. This is because of the +M effects of the adjacent nitrogen and oxygen atoms. Donation of the pair of electrons on nitrogen and a pair of electrons on oxygen towards the carbon atom in the C=O bond reduces its electrophilicity. Also, because the *tert*-butyl group is bulky, it hinders the approach of nucleophiles that could attack the C=O bond.

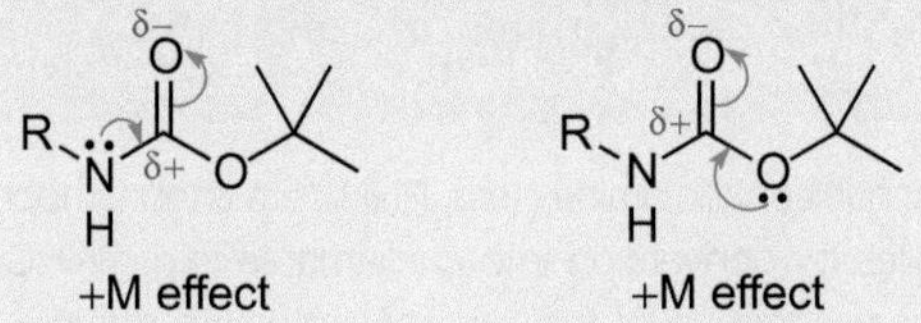
+M effect
+M effect

Question

Another common carbamate protecting group for amines is the **carboxybenzyl (Cbz)** group. The Cbz group is introduced by reaction of an amine (such as RNH_2) with benzyl chloroformate ($PhCH_2OCOCl$) in the presence of a base (e.g. Et_3N) as shown here. Propose a mechanism for this reaction.

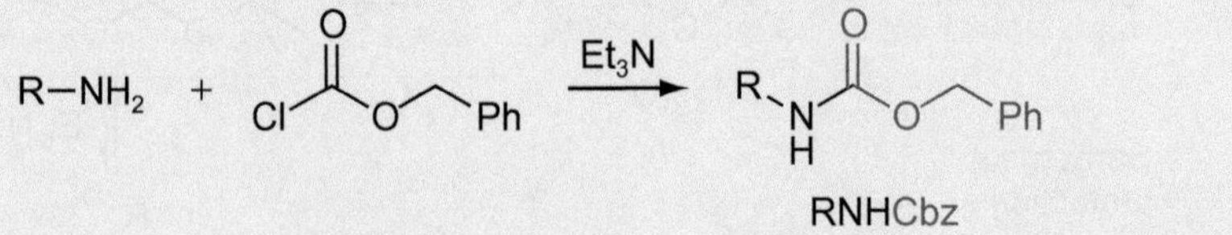
R−NH2
Et3N
RNHCbz

Nucleophilic acyl substitution reactions of esters

Esters (RCO_2R) react with nucleophiles such as water and ammonia in nucleophilic acyl substitution reactions. In these reactions, summarized in Figure 24.15, the nucleophile attacks the C=O bond and an alkoxide ion (RO^-) or, under acidic conditions, an alkanol (ROH) acts as the leaving group, following the general mechanisms in Figures 24.3 (p.1103) and 24.5 (p.1105).

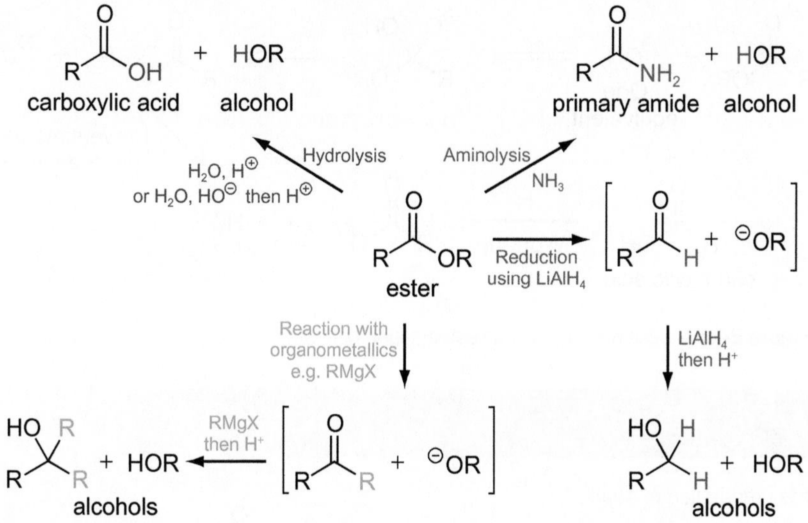
carboxylic acid
alcohol
primary amide
alcohol
Hydrolysis
Aminolysis
NH3
or H2O, HO⊖ then H⊕
ester
Reduction
using LiAlH4
Reaction with
organometallics
e.g. RMgX
LiAlH4
then H+
RMgX
then H+
alcohols
alcohols

Figure 24.15 Nucleophilic acyl substitution reactions of esters.

Notice than $LiAlH_4$ can reduce an ester, but $NaBH_4$ cannot.

Esters are protonated on the oxygen atom of the C=O group because this produces the most stable conjugate acid.

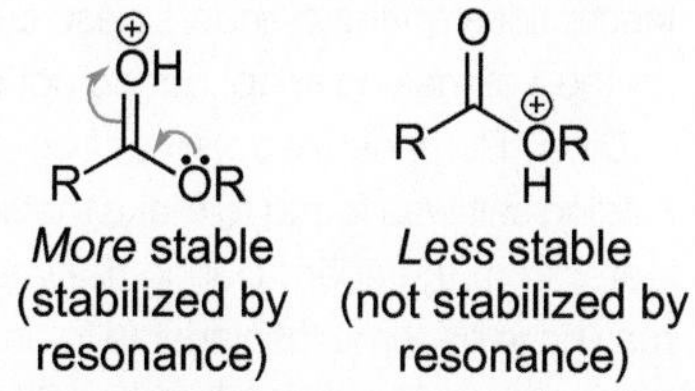
More stable
(stabilized by
resonance)
Less stable
(not stabilized by
resonance)

Hydrolysis of an ester to form a carboxylic acid (RCOOH) and an alcohol (ROH) requires either acidic or basic conditions. In the absence of H^+ or HO^-, the hydrolysis is slow because water is a relatively weak nucleophile and an ester is a relatively weak electrophile.

In an ester, the OR group has a strong +M effect and it donates a pair of electrons towards the carbon atom in the C=O bond, which reduces its electrophilicity (see Box 24.2, p.1103).

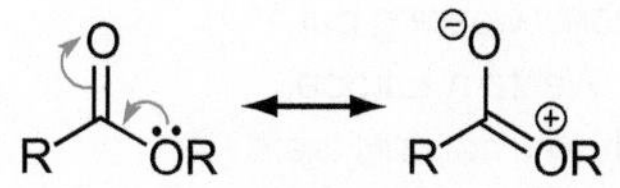

Interactive 3D animation of acid-catalysed hydrolysis of an ester.

Under acidic conditions, the rate of hydrolysis of an ester is increased because the ester is protonated, to form $RC(=OH^+)OR$ (Figure 24.16). Protonation of the ester converts it into a stronger electrophile, which reacts more rapidly with water (HOH). Notice that H^+ acts as a catalyst for the hydrolysis and that every step in the reaction is reversible. The reverse reaction, to form an ester from a carboxylic acid and an alcohol, is called **esterification** (p.1108).

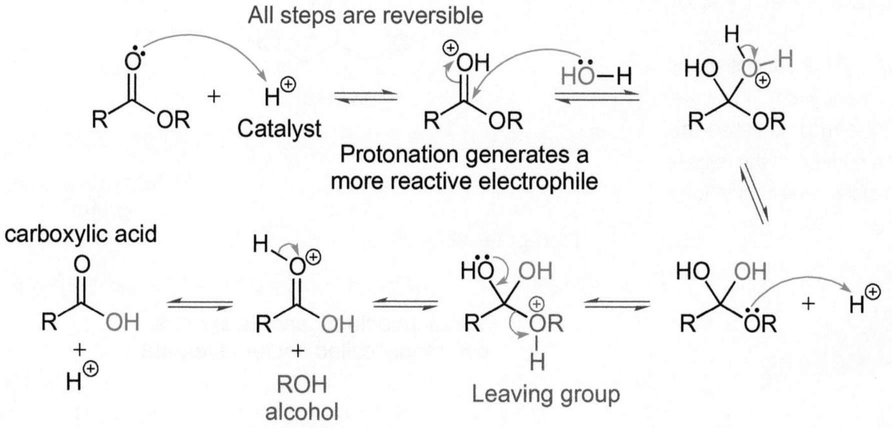
All steps are reversible
Catalyst
Protonation generates a
more reactive electrophile
carboxylic acid
ROH
alcohol
Leaving group

Figure 24.16 Acid-catalysed hydrolysis of an ester.

Heating methyl benzoate ($PhCO_2CH_3$) with aqueous sodium hydroxide forms methanol (CH_3OH) and sodium benzoate ($PhCO_2^-\ Na^+$). Adding HCl to the reaction mixture leads to formation of a white precipitate of benzoic acid ($PhCO_2H$), which can be isolated from the reaction solution by filtration using, for example, a Büchner funnel.

Under basic conditions, the rate of hydrolysis of an ester is increased because HO^- acts as the nucleophile (Figure 24.17). The hydroxide ion is a stronger nucleophile than water so it reacts more rapidly with the C=O bond in the ester. Notice that hydrolysis of an ester under basic conditions is irreversible (because a resonance-stabilized carboxylate ion, RCO_2^-, is formed) and that one equivalent of HO^- is required for complete hydrolysis of the ester (cf. the acid-catalysed hydrolysis). Hydrolysis of an ester using HO^-/H_2O is called a **saponification** reaction (see Box 24.6, p.1120).

Hydrolysis of an ester in alkaline solution is named saponification after *sapo*, the Latin word for soap.

> The hydroxide ion promotes the hydrolysis of an ester but it is *not* a catalyst, because HO^- is not regenerated.

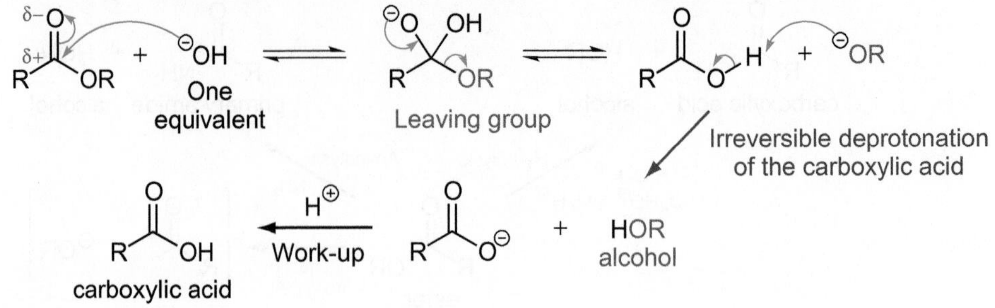
One equivalent
Leaving group
Irreversible deprotonation of the carboxylic acid
Work-up
carboxylic acid
HOR
alcohol

Figure 24.17 Base hydrolysis of an ester (saponification).

Box 24.6 Making soap

The earliest known evidence for the preparation of soap comes from a soap-like material found in clay cylinders, dating from 2800 BC, which were excavated from the ruins of ancient Babylon. Inscriptions on the cylinders describe the boiling of fats with ashes (a method of making soap) but do not describe the use of the material.

Later, Romans made soap from goat fat, wood ashes, and salt. Adding salt was found to make the soap hard, as the salt absorbs the water from the soap. During the later centuries of the Roman era, it may have become fashionable to use soap for personal washing but, immediately after the fall of the Roman Empire in Western Europe, there was little soap making or use of it in Europe. It was not until the fourteenth century that soap making started in England.

Today, soaps are made by heating animal fats or vegetable oils with sodium hydroxide or potassium hydroxide. Natural fats and oils contain triesters (compounds with three ester groups) that, when hydrolysed with sodium hydroxide or potassium hydroxide, form glycerol (propane-1,2,3-triol, $HOCH_2CHOHCH_2OH$) and sodium salts or potassium salts of long unbranched carbon chain carboxylic acids, called **fatty acids** (Box 18.4, p.836). It is the salts of the fatty acids ($RCO_2^-\ Na^+$ or $RCO_2^-\ K^+$) that are soaps.

Salts such as sodium stearate, $CH_3(CH_2)_{16}CO_2^-\ Na^+$, act as soaps by dissolving non-polar oil molecules, which carry dirt. In water, sodium stearate molecules come together to form balls called **micelles**. The polar carboxylate groups form the surface of the micelle because of their attraction to water. The non-polar alkyl chains form the interior to minimize their contact with water.

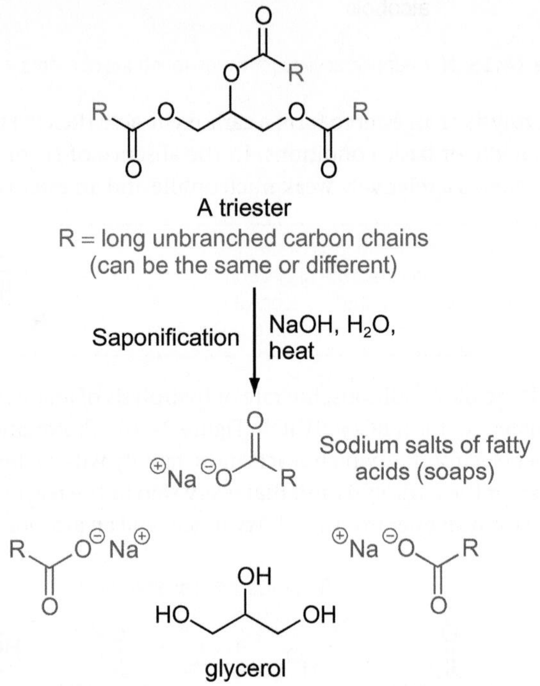
A triester
R = long unbranched carbon chains
(can be the same or different)
Saponification
NaOH, H2O, heat
Sodium salts of fatty acids (soaps)
glycerol

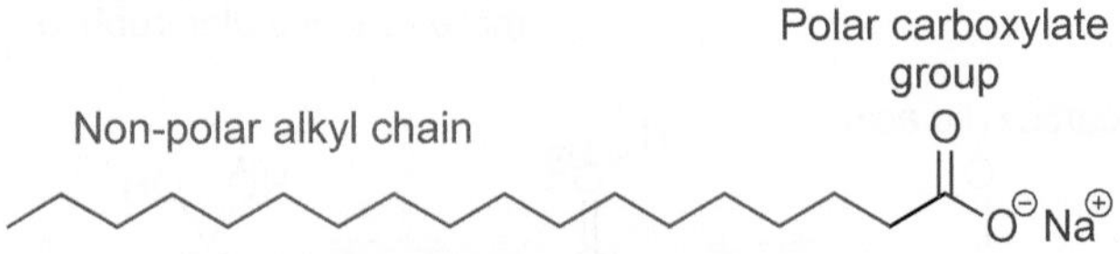
Polar carboxylate group
Non-polar alkyl chain
Sodium octadecanoate is more commonly called sodium stearate

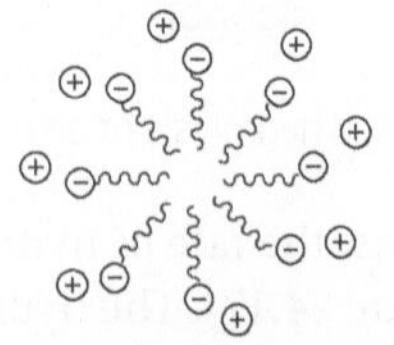

A micelle
When dissolved in water:
the non-polar alkyl chains are in the interior
the polar carboxylate groups are on the surface →

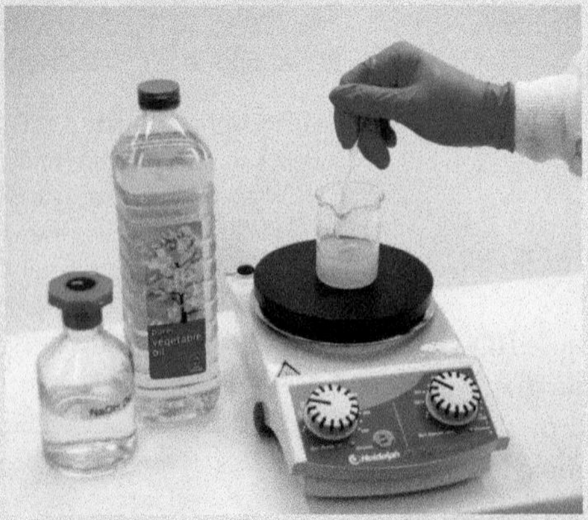

▲ Making soap by heating vegetable oil with sodium hydroxide.

→ When soap and non-polar oil molecules are mixed, the micelles break up and re-form to position the non-polar oil molecules inside the micelles. The non-polar oil molecules are now held within the micelles and are washed away with rinsing.

Question

Glyceryl tristearate (tristearin) is the main saturated fat in beef. Saponification of a molecule of glyceryl tristearate using an excess of aqueous sodium hydroxide produces three molecules of sodium stearate. Draw the structure of glyceryl tristearate and explain why it is a saturated fat.

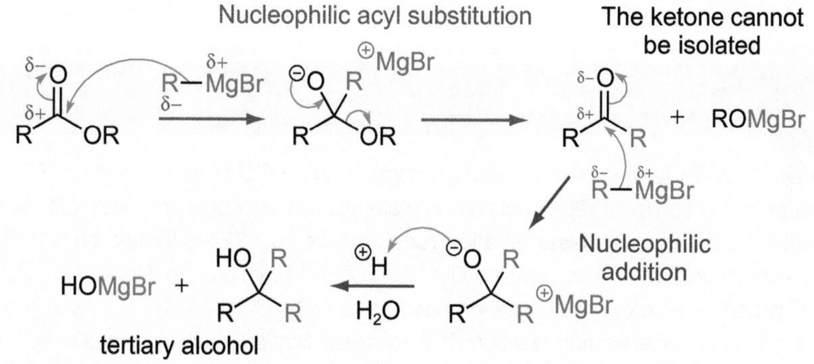

Nucleophilic acyl substitution
The ketone cannot be isolated
+ ROMgBr
Nucleophilic addition
HOMgBr +
H₂O
tertiary alcohol

Figure 24.18 Reaction of an ester with two equivalents of a Grignard reagent (RMgX) to give a tertiary alcohol.

Visit the Online Resource Centre to view screencast 24.2 which walks you through the mechanism of the reaction of an ester with a Grignard reagent and gives examples of the use of this reaction in synthesis.

Esters also react with other nucleophiles, such as ammonia, Grignard reagents (RMgX), and lithium aluminium hydride, $LiAlH_4$ (see Figure 24.15, p.1119). Heating an ester with ammonia forms a primary amide ($RCONH_2$), while reaction of an ester with two equivalents of a Grignard reagent, as shown in Figure 24.18, forms a tertiary alcohol ($R_2C(OH)R$). The tertiary alcohol is formed by an initial nucleophilic acyl substitution reaction to give a ketone (RCOR), followed by a nucleophilic addition reaction. The ketone cannot be isolated because it is a stronger electrophile than the ester so it reacts more rapidly with a Grignard reagent than the ester (Box 24.2, p.1103).

Reaction of esters with lithium aluminium hydride ($LiAlH_4$) forms a primary alcohol, RCH_2OH (Figure 24.19). An initial nucleophilic acyl substitution reaction gives an aldehyde (RCHO) that reacts with a second equivalent of hydride ion in a nucleophilic addition reaction. The aldehyde cannot be isolated because it is a stronger electrophile than the ester so it reacts more rapidly with lithium aluminium hydride than the ester (Box 24.2, p.1103).

The different reactivity of esters and aldehydes or ketones to nucleophiles is clearly illustrated in reactions with sodium borohydride ($NaBH_4$). Sodium borohydride is a weaker reducing agent than lithium aluminium hydride and, although it reduces aldehydes and ketones to alcohols, it cannot reduce esters. Consequently, in molecules containing both a ketone and an ester group, sodium borohydride selectively reduces the ketone group.

It is possible to reduce an ester to an aldehyde using an alternative reducing agent, called diisobutylaluminium hydride or DIBAL, $HAl(CH_2CHMe_2)_2$. As long as the reaction temperature is kept low, typically −70 °C, the use of one equivalent of this bulky reducing agent can afford good yields of aldehydes. This occurs because reaction of DIBAL with an ester forms a tetrahedral intermediate ($RCH(OR)OAl(CH_2CHMe_2)_2$) that is stable at low temperature—it only breaks down to the aldehyde on aqueous work-up.

$NaBH_4$ is a weaker reducing agent than $LiAlH_4$ (Section 23.2, p.1063).

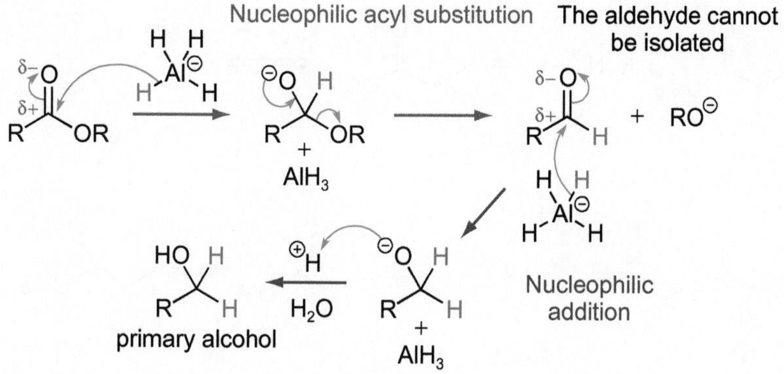

Nucleophilic acyl substitution
The aldehyde cannot be isolated
+ RO⊖
Nucleophilic addition
H₂O
primary alcohol

Figure 24.19 Reduction of an ester with $LiAlH_4$ to form a primary alcohol.

For clarity, the $LiAlH_4$ reduction mechanism is simplified by assuming that AlH_3 is a product of the reaction, whereas this hydride undergoes further reactions.

A chemoselective reduction is a reaction in which one functional group is reduced in preference to another functional group or groups (Section 19.3, p.909).

Notice that the reduction is not stereoselective. The secondary alcohol is formed as a racemate because $NaBH_4$ can react equally well with either face of the planar C=O bond.

Worked example 24.3 discusses how an ester reacts with an alcohol in a nucleophilic acyl substitution reaction.

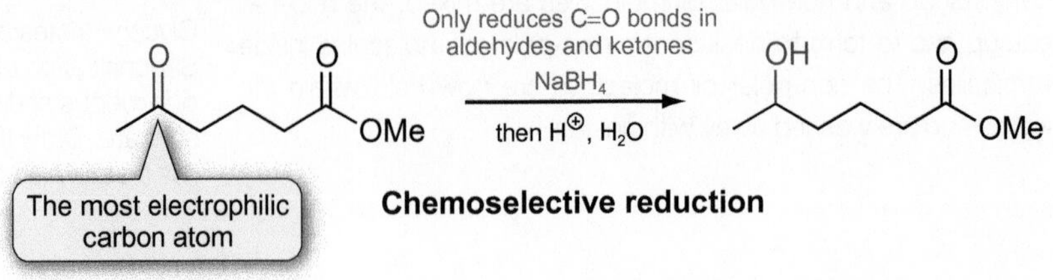
Only reduces C=O bonds in aldehydes and ketones
NaBH4
then H⊕, H2O
OMe
OH
OMe
The most electrophilic carbon atom
Chemoselective reduction

Worked example 24.3 Transesterification and aspirin

An ester (RCO_2R) is formed by reaction of a carboxylic acid (RCO_2H) with an alcohol (ROH) in an esterification reaction (p.1119). Reaction of an ester with an alcohol to produce a different ester and alcohol is called a **transesterification reaction** (see how PET is formed on p.1099). For example, reaction of methyl propanoate ($CH_3CH_2COOCH_3$) with propan-1-ol ($CH_3CH_2CH_2OH$) and H^+ produces a mixture of propyl propanoate ($CH_3CH_2COOCH_2CH_2CH_3$) and methanol (CH_3OH) in a transesterification reaction. The transesterification reaction comes to equilibrium so, to drive the reaction to the right, an excess of propan-1-ol is used and methanol (which has a lower boiling point than methyl propanoate, propan-1-ol, or propyl propanoate) is selectively distilled from the reaction mixture.

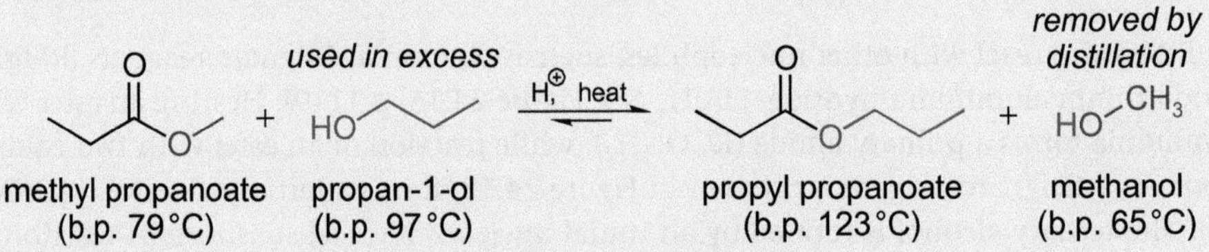
used in excess
removed by distillation
H⊕, heat
HO
HO
CH3
methyl propanoate (b.p. 79 °C)
propan-1-ol (b.p. 97 °C)
propyl propanoate (b.p. 123 °C)
methanol (b.p. 65 °C)

(a) Explain the role of H^+ in the transesterification reaction.

(b) Give a mechanism for the transesterification reaction.

Strategy

(a) Consider the effect of protonating the oxygen atom in the C=O bond of methyl propanoate.

(b) Recognize that this is an example of a nucleophilic acyl substitution reaction.

Solution

(a) Protonation of the C=O bond in methyl propanoate makes the ester a stronger electrophile and it reacts more rapidly with propan-1-ol.

(b)

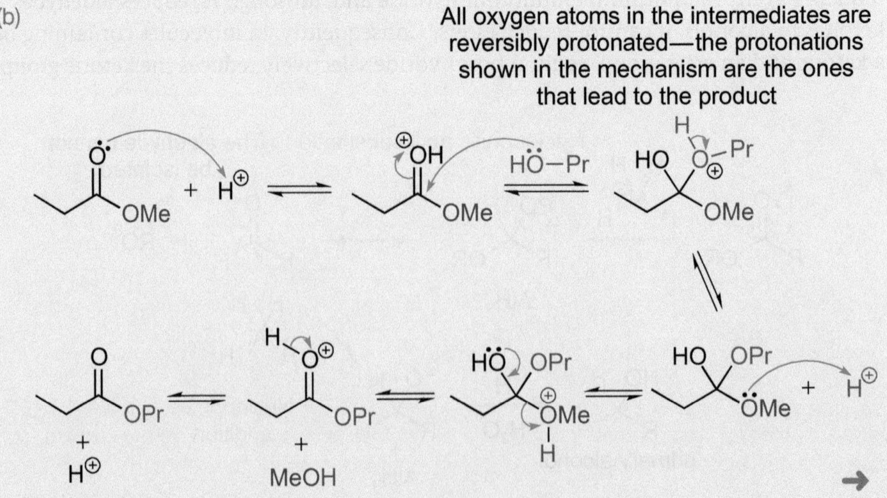
All oxygen atoms in the intermediates are reversibly protonated—the protonations shown in the mechanism are the ones that lead to the product
OMe
H⊕
OH
OMe
HO–Pr
HO
Pr
OMe
OPr
H⊕
OPr
MeOH
HO
OPr
OMe
H
HO
OPr
OMe
H⊕

Remember, when naming esters, the name of the alkyl or aryl group bonded to oxygen is given first (Section 2.7, p.103). This is followed by the name of the acid with the ending '-ic acid' replaced by '-ate'. So, propanoic acid ($CH_3CH_2CO_2H$) becomes propanoate.

Using distillation to separate liquids with different boiling points is discussed in Section 17.4 (p.805).

Tranesterification is used to convert naturally occurring vegetable oils into biodiesel. Typically methanol or ethanol reacts with the triesters in vegetable oils (Box 24.6, p.1120), often under basic conditions, to form methyl or ethyl esters of fatty acids (RCO_2CH_3 or $RCO_2CH_2CH_3$, where R is a long unbranched chain), called biodiesel. (Glycerol is also formed as a by-product.) Under basic conditions, methanol or ethanol is deprotonated to form more nucleophilic methoxide (CH_3O^-) or ethoxide ($CH_3CH_2O^-$) ions, which react with the triester in a series of nucleophilic acyl substitution reactions.

Question

In your body, compounds called prostaglandins induce inflammation, pain, and fever. To provide relief from pain, aspirin acts by inhibiting an enzyme that catalyses the formation of prostaglandins. Aspirin reacts with an OH group within the active site of the enzyme to convert it into an ester (in a transesterification reaction), which inhibits the production of prostaglandins.

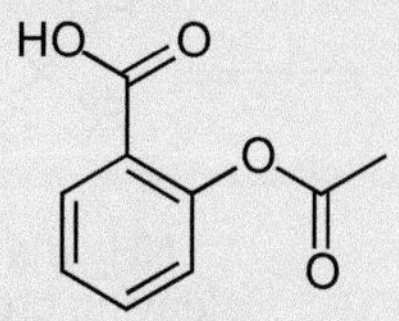
HO
O
O
O

aspirin

Representing the structure of the enzyme by Enzyme-OH, draw the structures of the products formed in the transesterification reaction.

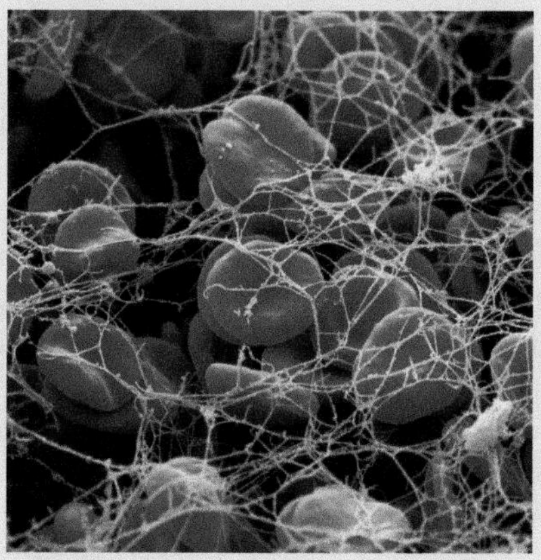

◀ Blood clots occur when red blood cells are trapped in a web of white threads made of an insoluble protein. Aspirin helps prevent blood clots by stopping the synthesis of prostaglandins, which initiate blood clotting.

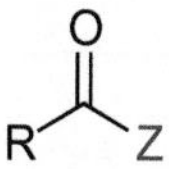
O
R
Z

Z = NH_2, primary amide ($RCONH_2$)
Z = NHR, secondary amide (RCONHR)
Z = NR_2, tertiary amide ($RCONR_2$)

The R groups can be the same or different.

Nucleophilic acyl substitution reactions of amides

Amides ($RCONH_2$, RCONHR, or $RCONR_2$) undergo relatively few nucleophilic acyl substitution reactions. The carbon atom in the C=O bond is not a good electrophile because of the electron-donating (+M) effect of the adjacent nitrogen atom (Box 24.2, p.1103). Electron donation from nitrogen gives the C–N bond some double-bond character so rotation about this bond can be slow—notice that a secondary amide (RCONHR) prefers to adopt a conformation in which the alkyl groups are *trans* to one another.

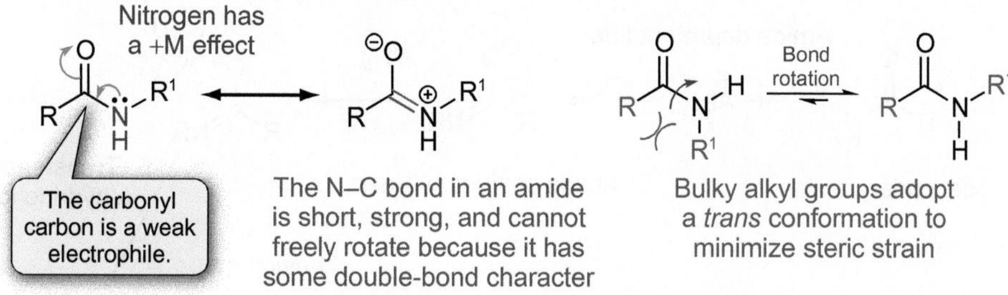
Nitrogen has a +M effect
The carbonyl carbon is a weak electrophile.
The N–C bond in an amide is short, strong, and cannot freely rotate because it has some double-bond character
Bond rotation
Bulky alkyl groups adopt a *trans* conformation to minimize steric strain

Amino acids are linked together by amide (peptide) bonds in peptides and proteins. To determine the number and types of amino acids in a peptide or protein, the amide bonds are hydrolysed to release the amino acids, which are then separated and identified.

Amide (peptide) bond in a dipeptide

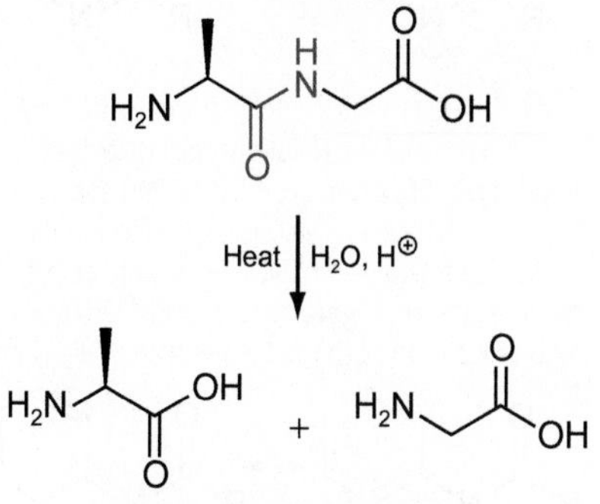
Heat
H_2O, $H^{\oplus}$

Constituent amino acids

To hydrolyse an amide bond requires heating in moderately concentrated aqueous acid (e.g. a 6 mol dm^{-3} solution of HCl) or in an aqueous solution of hydroxide ion (typically, a 2 mol dm^{-3} solution of NaOH). Hydrolysis converts the amide into a carboxylic acid and ammonia (from a primary amide, $RCONH_2$) or an amine (from a secondary

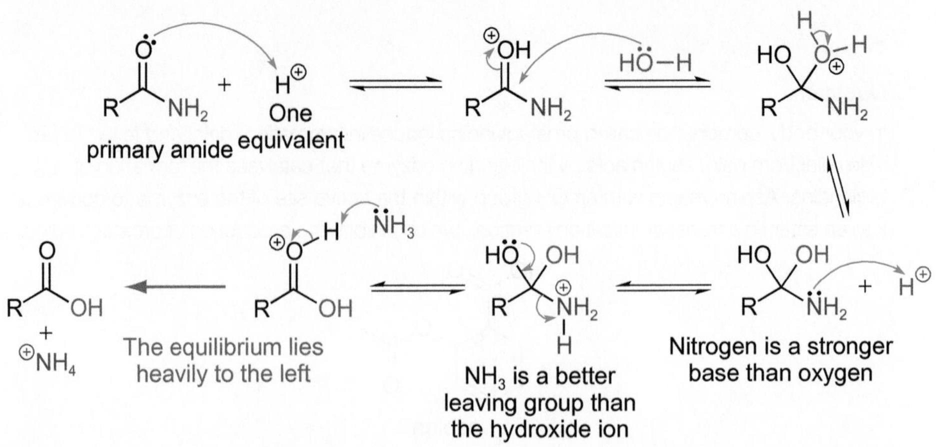
One equivalent
primary amide
The equilibrium lies heavily to the left
NH_3 is a better leaving group than the hydroxide ion
Nitrogen is a stronger base than oxygen

Figure 24.20 Acid-promoted hydrolysis of a primary amide (requires heat).

> Amides are protonated on the oxygen atom, rather than the nitrogen atom, because this produces the more stable conjugate acid.
>
> 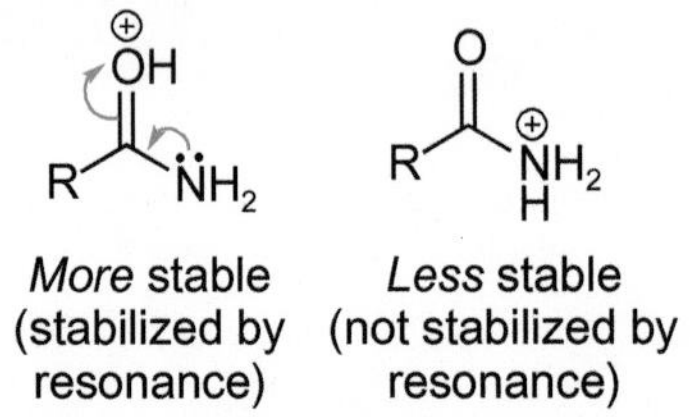
> *More* stable
> (stabilized by resonance)
> *Less* stable
> (not stabilized by resonance)

> As a guide, the weaker the basic strength of a group, the better leaving group it is. NH_3 is a better leaving group than ^-OH or $^-NH_2$ because NH_3 is the weakest base. A weak base has a conjugate acid with a relatively low pK_a value—the pK_a value of NH_4^+ is 9.2, for H_2O it is 16, and for NH_3 it is 38 (Section 19.2, p.891).

> Following deprotonation of a secondary amide (RCONHR), the conjugate base ($RCONR^-$) is stabilized by resonance.
>
> 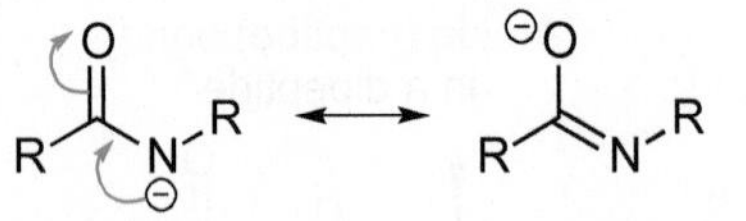
>
> The pK_a value of RCONHR is strongly affected by the nature of the R substituent attached to nitrogen. For example, MeCONHMe and MeCONHPh have pK_a values of around 26 and 21.5, respectively—MeCONHPh is more acidic because the conjugate base is stabilized by resonance with both the C=O and phenyl groups.
>
> 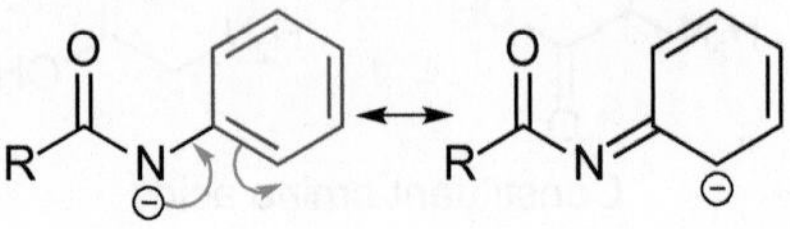

amide, RCONHR, or a tertiary amide, $RCONR_2$). In these reactions, water or hydroxide ion acts as the nucleophile and attacks the C=O bond of the amide following the general mechanisms in Figure 24.5 (p.1105) and Figure 24.3 (p.1103). For primary amides ($RCONH_2$), NH_3 or $^-NH_2$ act as the leaving group—$^-NH_2$ is an extremely poor leaving group (the pK_a value of NH_3 is 38) and in aqueous solutions it is immediately protonated to give NH_3.

For example, when a primary amide ($RCONH_2$) is heated in aqueous acid (Figure 24.20), the C=O bond is protonated and this converts it into a stronger electrophile, $RC(=OH^+)NH_2$. Attack by water forms a tetrahedral intermediate, $RC(OH_2^+)OH(NH_2)$, and a proton then moves from oxygen to nitrogen, giving $RC(OH)OH(NH_3^+)$, because the nitrogen atom is more basic than oxygen. A C=O bond is then formed by expelling NH_3 as a leaving group. Finally, NH_3 removes a proton from the protonated carboxylic acid, $RC(=OH^+)OH$, to form NH_4^+ and the carboxylic acid (RCOOH). For complete hydrolysis of the amide, one equivalent of H^+ is required.

Amides are reduced by lithium aluminium hydride, but these reactions are not usually nucleophilic acyl substitutions. For example, secondary amides (RCONHR) react with lithium aluminium hydride ($LiAlH_4$) to form secondary amines (RCH_2NHR), as shown in Figure 24.21. The first step of the mechanism probably involves deprotonation of the secondary amide by the $[AlH_4]^-$ ion to form the conjugate base ($RCONR^-$), which then adds to AlH_3 to give an unstable intermediate with a negative charge on aluminium, $RC(=NR)OAlH_3^-$. Transfer of a hydride ion from Al to the carbon atom in the C=N bond gives a tetrahedral intermediate that expels $^-OAlH_2$ to form the C=N bond of an imine (RCH=NR). The imine is reduced to the secondary amine (RCH_2NHR) in a nucleophilic addition reaction. Overall, the C=O bond in the amide is replaced by two C–H bonds in the amine.

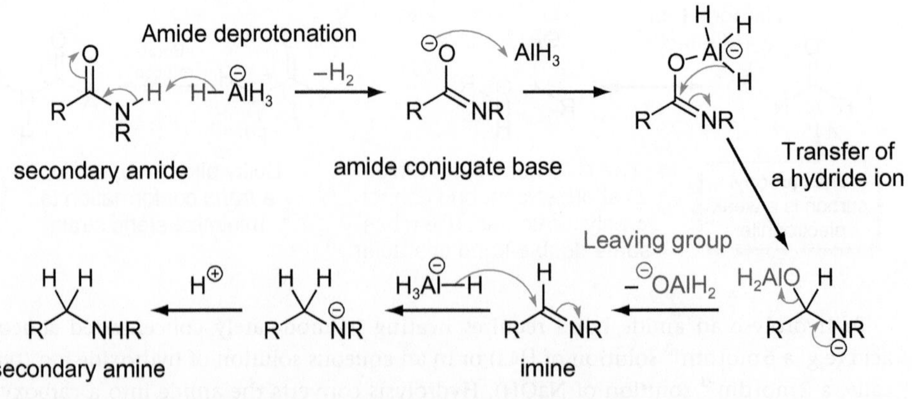
Amide deprotonation
$-H_2$
secondary amide
amide conjugate base
Transfer of a hydride ion
Leaving group
secondary amine
imine

Figure 24.21 Reduction of a secondary amide with $LiAlH_4$ to form a secondary amine.

Box 24.7 Hydrolysis of nitriles to form carboxylic acids

Nitriles (RC≡N) react with nucleophiles in nucleophilic addition reactions. Nucleophiles attack the partially positive carbon in the C≡N bond to give an intermediate with a negative charge on nitrogen, RC(=N$^-$)Nu, which reacts to give various products.

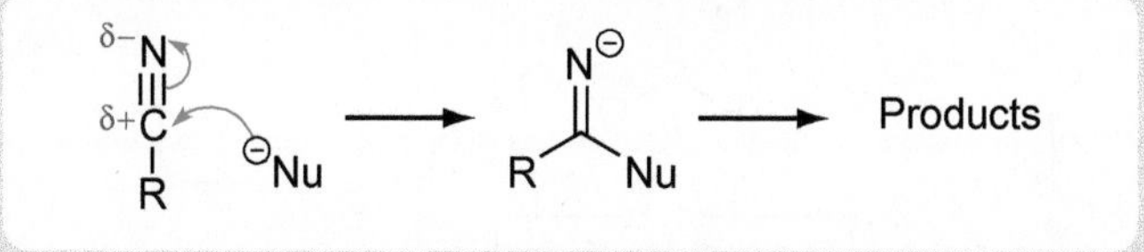
Products

For the nitrile to react with relatively weak nucleophiles, an acid is needed to protonate the nitrile to convert it into a stronger electrophile (RC≡NH$^+$).

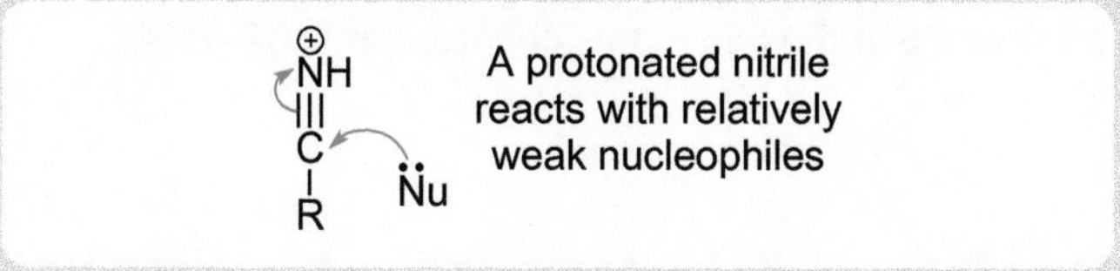
A protonated nitrile reacts with relatively weak nucleophiles

One of the most useful reactions of nitriles (RCN) is hydrolysis to form carboxylic acids (RCO_2H). Heating in aqueous acid slowly hydrolyses nitriles, as shown in Figure 1. Nucleophilic addition of water to the C≡N bond forms a primary amide ($RCONH_2$)—the conversion of RC(OH)=NH into $RCONH_2$ is an example of tautomerism (Section 24.3, p.1126). The primary amide is further hydrolysed to form a carboxylic acid (RCOOH), in a nucleophilic acyl substitution reaction, by the mechanism given in Figure 24.20. Note that the carbon atom in a C≡N bond is at the same oxidation level as the carbonyl carbon atom in an amide or carboxylic acid so their interconversion does not require a redox reagent (Section 19.2, p.899).

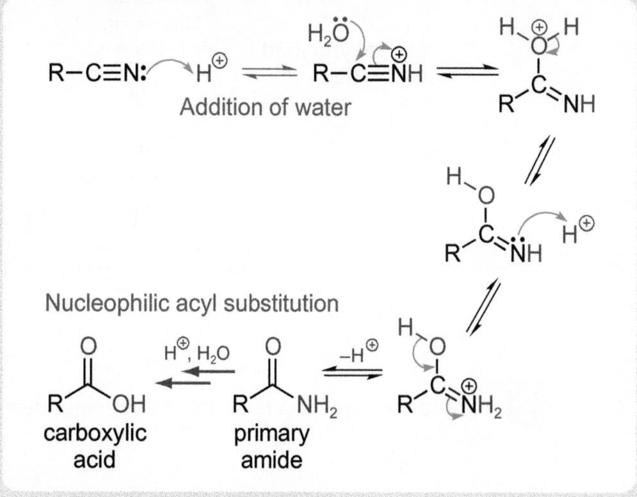
Addition of water
Nucleophilic acyl substitution
carboxylic acid
primary amide

▲ **Figure 1** Hydrolysis of a nitrile.

Question

Propose a mechanism for the following reaction, which involves an intermediate imine (see Section 2.7, p.100, for the structure of imines).

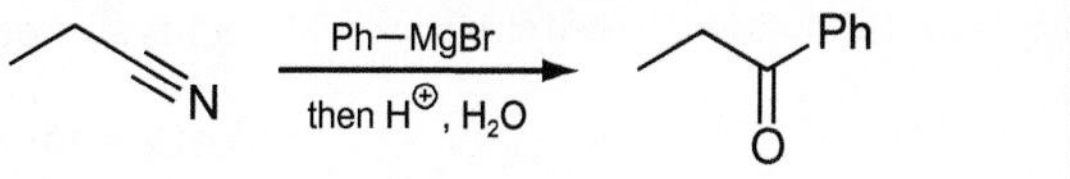
Ph−MgBr
then H⊕, H2O

Summary

- Acyl chlorides (RCOCl), acid anhydrides (RCO_2COR), esters (RCO_2R), and amides ($RCONH_2$, RCONHR, and $RCONR_2$) are called carboxylic acid derivatives because they can be prepared from a carboxylic acid (RCO_2H).
- Carboxylic acids and derivatives are useful compounds in synthesis because they can be converted into various functional groups. To convert one carboxylic acid derivative into another requires a nucleophilic acyl substitution reaction.

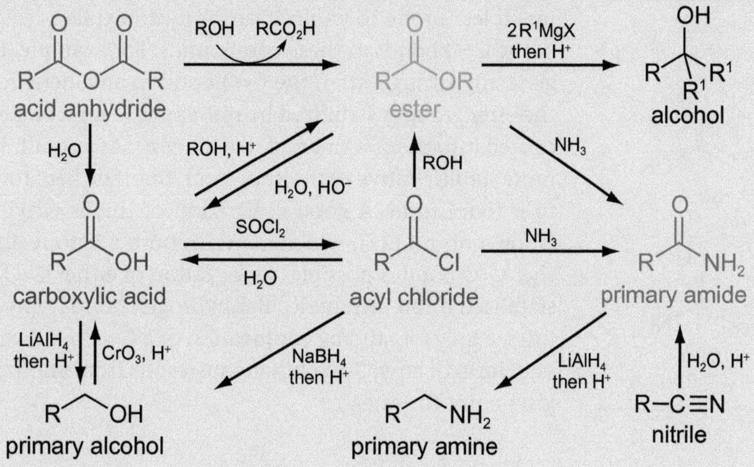
acid anhydride
ester
alcohol
carboxylic acid
acyl chloride
primary amide
primary alcohol
primary amine
nitrile

→

- Protonation of the C=O bond in a carboxylic acid or derivative converts it into a stronger electrophile, which reacts more rapidly with nucleophiles.

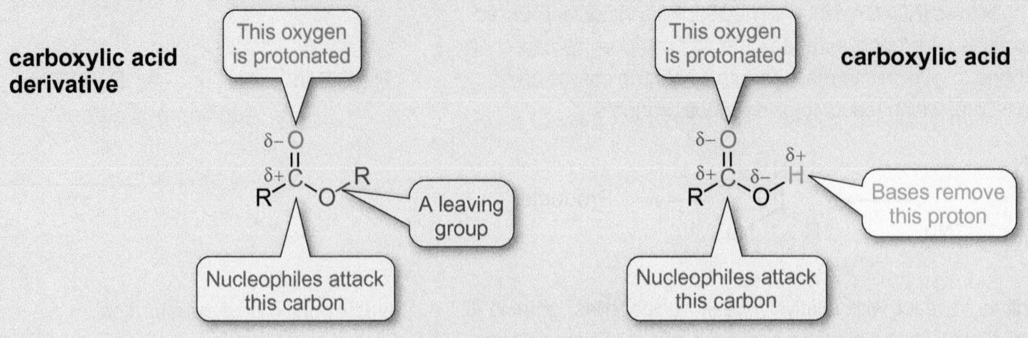
carboxylic acid derivative
This oxygen is protonated
A leaving group
Nucleophiles attack this carbon
This oxygen is protonated
carboxylic acid
Bases remove this proton
Nucleophiles attack this carbon

For practice questions on this topic, see questions 1–3 and 5 at the end of this chapter (pp.1134–1136).

24.3 α-Substitution and carbonyl-carbonyl condensation reactions

α-Substitution reactions

Enols and enolate ions formed from aldehydes and ketones are discussed in Section 23.3 (p.1082). Keto-enol tautomerism in aldehydes and ketones is described on p.1082.

α-Substitution reactions of carboxylic acids and derivatives involve substitution of a hydrogen atom by another group at the carbon atom next to the C=O bond (the α-position). As for aldehydes and ketones, the mechanisms of α-substitution reactions of carboxylic acids and derivatives involve intermediate enols or enolate ions.

Keto-enol tautomerism

Resonance stabilization of an ester:

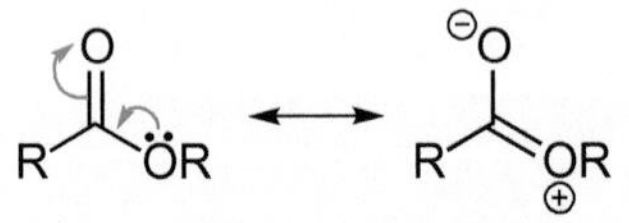

Carboxylic acids and derivatives with hydrogen atoms at the α-position are in equilibrium with structural isomers called enols. For most carboxylic acids and derivatives the more stable structural isomer is the keto form.

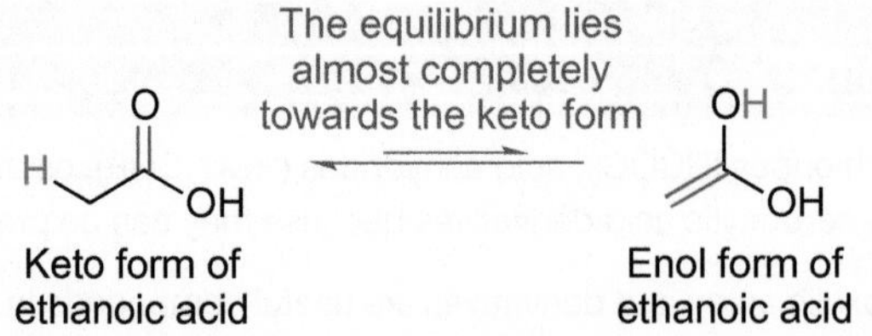
The equilibrium lies almost completely towards the keto form
Keto form of ethanoic acid
Enol form of ethanoic acid

Keto and enol forms of an ester (top) and a primary amide (bottom)—the keto forms are usually much more stable:

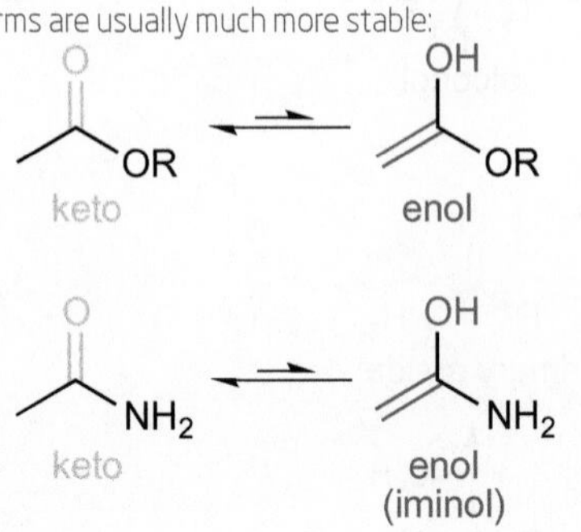
keto
enol
keto
enol (iminol)

Compared to aldehydes and ketones, carboxylic acids and derivatives are generally much less prone to enolization. You can explain this by comparing the bond enthalpies of the C=O bonds in these compounds. For example, the C=O bond in an ester (RCO_2R) is generally stronger than the C=O bond in an aldehyde (RCHO) or ketone (RCOR) because the ester group is stabilized by resonance (a lone pair of electrons on the OR group is delocalized into the C=O π bond, see margin). As a result, the keto form of an ester is generally more stable relative to the enol form, than the keto form of an aldehyde or ketone, relative to its enol form. A good illustration of this is ethyl 3-oxobutanoate, $CH_3COCH_2CO_2Et$ (shown on p.1127), which contains both a ketone and an ester group. Enolization of either C=O bond is possible—enolization of either C=O bond produces an enol form that is stabilized by an intramolecular hydrogen bond (represented by dotted, dashed, or hashed lines, e.g. IIIIIIIII) and by conjugation of a C=O bond with a C=C bond. However, the major enol form of ethyl 3-oxobutanoate results from enolization of the C=O bond in the ketone group, not the ester.

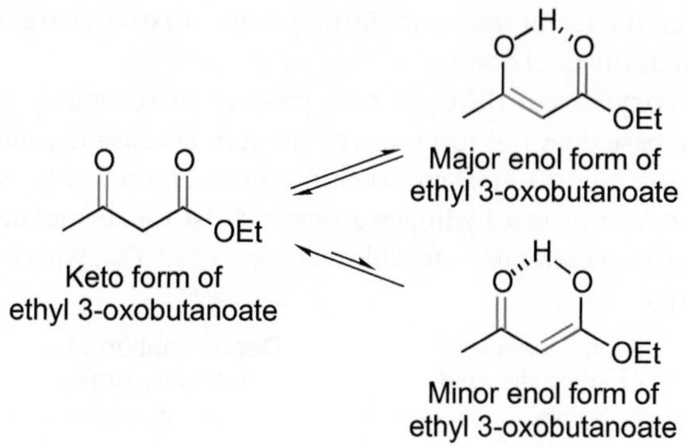

OEt
Major enol form of ethyl 3-oxobutanoate
OEt
Keto form of ethyl 3-oxobutanoate
OEt
Minor enol form of ethyl 3-oxobutanoate

Ethyl 3-oxobutanoate, commonly called ethyl acetoacetate, occurs naturally in strawberries. It is used a flavouring agent in foods and drinks.

Reactions of enols

Enols are nucleophiles and they react with electrophiles to introduce a new substituent at the α-position (see Figure 24.6, p.1106). For carboxylic acids and derivatives, a useful reaction of this type involves bromination of carboxylic acids using a mixture of PBr_3 and Br_2. This bromination reaction is called the **Hell–Volhard–Zelinsky (HVZ) reaction** (after the German chemists, Carl Magnus von Hell and Jacob Volhard, and the Russian chemist, Nikolai Dimitrievich Zelinsky).

The mechanism is shown in Figure 24.22. First, PBr_3 reacts with the carboxylic acid (RCOOH) to form an acyl bromide (RCOBr) and $HOPBr_2$. The $HOPBr_2$ then reacts with another molecule of the carboxylic acid. Overall, PBr_3 reacts with three equivalents of the carboxylic acid to form three equivalents of the acyl bromide and H_3PO_3 (phosphonic acid, more commonly called phosphorous acid).

In the presence of H_3PO_3, the acyl bromide undergoes acid-catalysed tautomerism to form the enol (MeCH=C(OH)Br). The enol form then acts as a nucleophile and attacks Br_2 to introduce a bromine atom at the α-position (giving MeCHBrC(=OH$^+$)Br). Only one bromine atom is introduced (when using one equivalent of Br_2) because the electron-withdrawing bromine atom (–I effect) in the brominated acyl bromide (MeCHBrCOBr) reduces the basicity of the oxygen atom. Because it is more difficult to protonate the C=O bond in a brominated acyl bromide, the rate of formation of the enol form is slower than for the acyl bromide. Finally, hydrolysis of the acyl bromide in a separate stage converts it into an α-brominated carboxylic acid (MeCHBrCOOH) in a nucleophilic acyl substitution reaction.

Halogenation of carboxylic acids to form acyl halides is discussed in Box 24.4 (p.1110).

Phosphorus tribromide (PBr_3) is also used to convert alcohols (ROH) into bromoalkanes (RBr), see Section 20.2 (p.926).

Reaction of the *E* or the *Z* isomer of the enol MeCH=C(OH)Br, with Br_2, gives the same product.

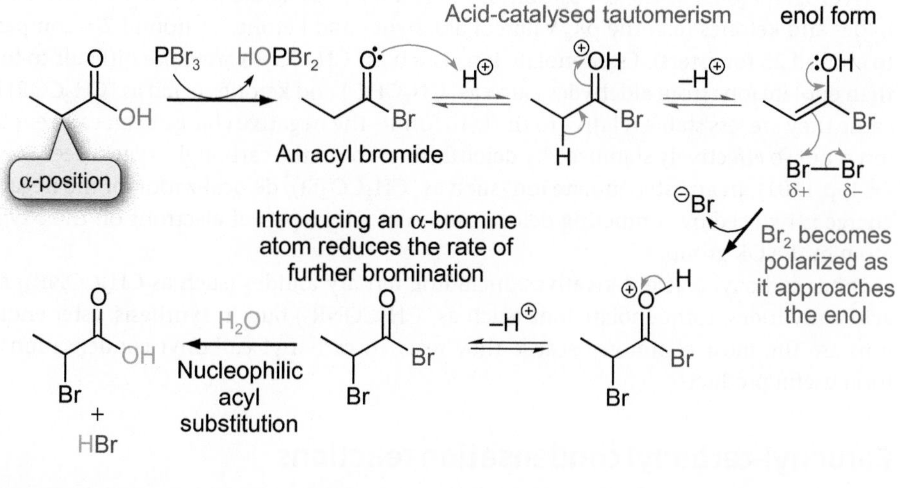

Acid-catalysed tautomerism
enol form
PBr3
HOPBr2
OH
Br
An acyl bromide
H⊕
–H⊕
Br
Br
H
α-position
Br—Br
δ+ δ–
⊖Br
Br2 becomes polarized as it approaches the enol
Introducing an α-bromine atom reduces the rate of further bromination
H2O
–H⊕
OH
Br
Br
Br
Br
Br
+
HBr
Nucleophilic acyl substitution

Figure 24.22 Bromination of a carboxylic acid using PBr_3 and Br_2 (the Hell–Volhard–Zelinsky reaction).

Notice that water reacts with the C=O bond of the acyl bromide, rather than the C–Br bond at the alpha position. This is because the carbonyl carbon atom is more electrophilic—the electronegative oxygen atom increases the electrophilicity of this carbon.

α-Bromocarboxylic acids are useful intermediates in synthesis as they react with various nucleophiles, in nucleophilic substitution reactions. For example, reaction of $RCHBrCO_2H$ with ammonia leads to substitution of the bromine atom and formation of an α-amino acid, $RCH(NH_2)CO_2H$.

Reactions of enolate ions

An enolate ion (often called simply an enolate) is formed by deprotonation of a carboxylic acid derivative at the α-position. The negatively charged enolate ion is stabilized by delo-

Compare the mechanism of α-bromination of a carboxylic acid with that of an aldehyde or ketone described in Section 23.3 (p.1085).

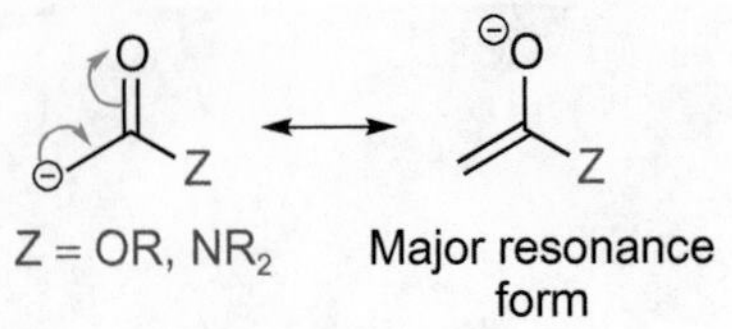
Z = OR, NR₂
Major resonance form

Resonance stabilization of the enolate ion of a carboxylic acid derivative.

Alkoxide ions (RO^-) act as nucleophiles and/or bases. Large alkoxide ions, such as the *tert*-butoxide ion Me_3CO^-, are better bases than nucleophiles (Section 20.5, p.954).

calization in which the major resonance form has the negative charge on the electronegative oxygen atom in the C=O bond.

Reacting a carboxylic acid (RCO_2H) or a primary or secondary amide ($RCONH_2$ or RCONHR) with a base does not form an enolate ion, because the most acidic hydrogen atoms are not at the α-position. Deprotonation forms anions, such as $CH_3CO_2^-$, that are stabilized by resonance. (The α-hydrogen atoms in $CH_3CO_2^-$ are not acidic because deprotonation would give an extremely unstable dianion, $^-CH_2CO_2^-$, with two negative charges in close proximity.)

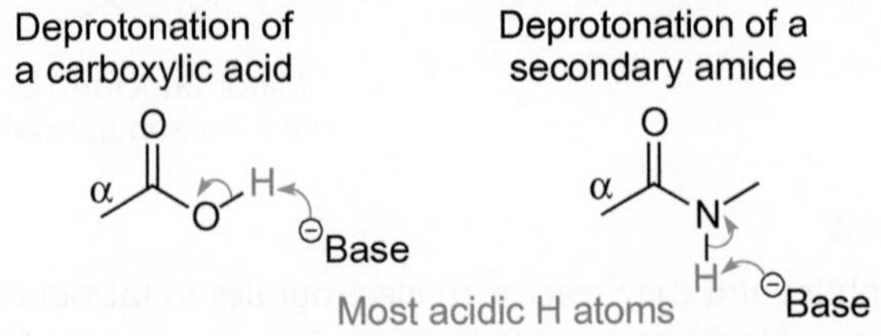
Deprotonation of a carboxylic acid
Deprotonation of a secondary amide
α
Base
Most acidic H atoms
Base

However, an ester (such as CH_3CO_2R) is converted into an ester enolate ion (such as $^-CH_2CO_2R$) by reaction with a base. Removal of an α-hydrogen atom by an alkoxide ion (RO^-) forms an equilibrium amount of the ester enolate ion. The choice of the alkoxide ion is determined by the structure of the ester side chain. For example, the ethoxide ion (^-OEt) is used to deprotonate ethyl esters (RCOOEt). Matching the alkoxide ion to the ester ensures that, if the ethoxide ion reacts with the ester group in a nucleophilic acyl substitution reaction, then an ethyl ester is regenerated (see margin).

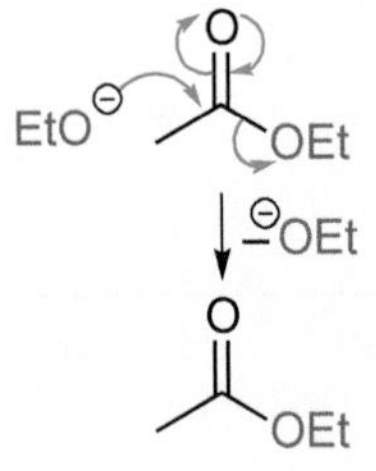
EtO
OEt
OEt
OEt

Regeneration of an ethyl ester.

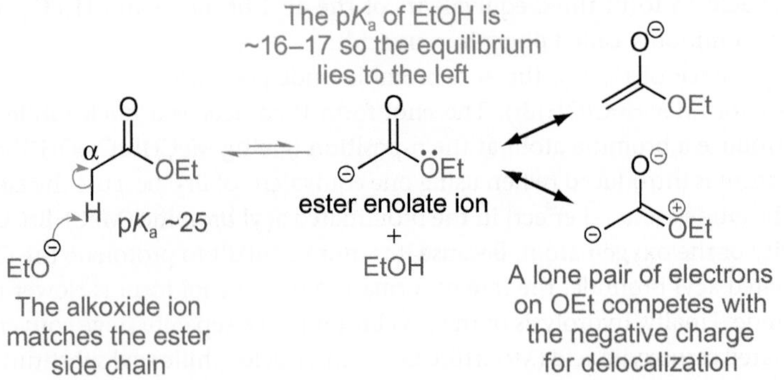
The pKa of EtOH is ~16–17 so the equilibrium lies to the left
OEt
α
OEt
OEt
H pKa ~25
EtO
ester enolate ion
+
EtOH
OEt
The alkoxide ion matches the ester side chain
A lone pair of electrons on OEt competes with the negative charge for delocalization

Acid–base reactions of organic compounds are discussed in Section 19.2 (p.888).

See Section 23.3 (p.1082) for the formation of enolate ions from aldehydes and ketones.

Hydrogen atoms at the α-position of esters are not as acidic as those in related aldehydes and ketones (e.g. the pK_a value of aldehydes and ketones is around 20, compared to around 25 for esters). Ester enolate ions (such as $^-CH_2CO_2R$) are more difficult to form than enolate ions from aldehydes (such as $^-CH_2CHO$) and ketones (such as $^-CH_2COR$) because they are less stable relative to the keto form—the negative charge in an ester enolate ion is not so effectively stabilized by delocalization onto the carbonyl oxygen (see Section 24.1, p.1101). In an ester enolate ion (such as $^-CH_2CO_2Et$), delocalization of the negative charge is reduced by competing delocalization of a lone pair of electrons on the oxygen atom in the OR group.

Other carboxylic acid derivatives, including tertiary amides (such as CH_3CONR_2) and acid anhydrides, form enolate ions (such as $^-CH_2CONR_2$) but, in synthesis, ester enolate ions are the most common because they react in carbonyl–carbonyl condensations to form useful products.

Carbonyl–carbonyl condensation reactions

In a condensation reaction, two molecules combine together to form a single organic product and a small molecule. The small molecule is often water but, in the carbonyl–carbonyl condensation of two esters (such as CH_3CO_2R), the small molecule formed is an alcohol (ROH).

In carbonyl–carbonyl condensation reactions, two carbonyl molecules react to form an organic product together with a small molecule. In the reaction of two esters (such as CH_3CO_2R), one molecule undergoes an α-substitution reaction and the other molecule undergoes a nucleophilic acyl substitution reaction. A molecule of an alcohol (ROH) is formed in the nucleophilic acyl substitution reaction.

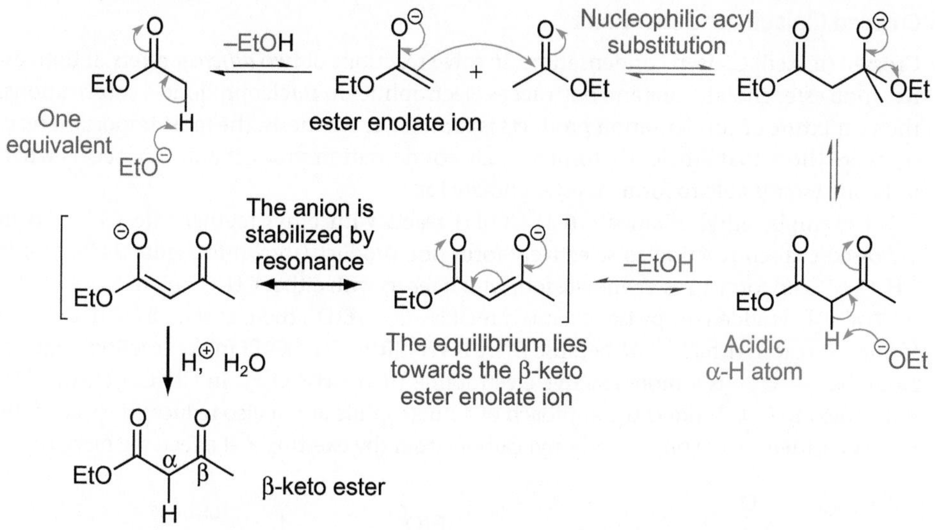
–EtOH
Nucleophilic acyl substitution
EtO
OEt
One equivalent
H
EtO
ester enolate ion
The anion is stabilized by resonance
–EtOH
The equilibrium lies towards the β-keto ester enolate ion
Acidic α-H atom
H, H2O
α
β
β-keto ester

Figure 24.23 Claisen condensation of ethyl ethanoate.

Reactions of two molecules of the same ester are called **Claisen condensations**. Reactions of two different ester molecules are called **crossed (mixed) Claisen condensations**. When a molecule containing two ester groups (a diester) reacts with itself in an **intramolecular Claisen condensation**, this is called a **Dieckmann reaction**. Like related condensations of aldehydes and ketones, ester condensations are some of the most useful reactions in synthesis because they make larger molecules by joining together small molecules or they can be used to convert linear molecules into rings.

For related carbonyl–carbonyl condensations of aldehydes and ketones, see Section 23.4 (p.1089).

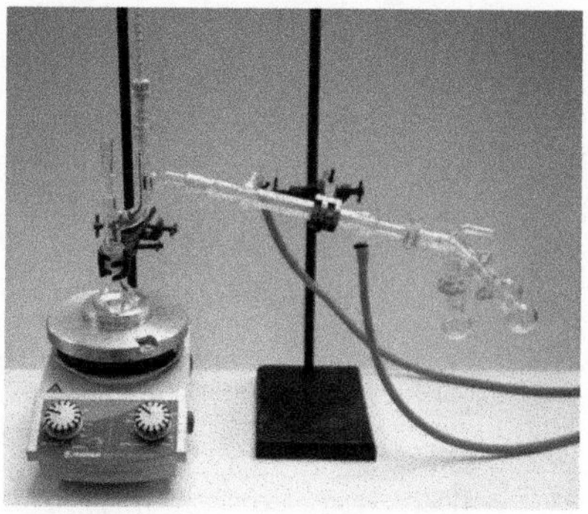

The Claisen condensation is named after the German chemist Rainer Ludwig Claisen (1851–1930), who invented the reaction at the University of Munich. He also designed a piece of laboratory glassware called the Claisen adapter shown above. The adapter allows reagents (usually liquids) to be introduced into a reaction mixture without taking apart the apparatus.

The Claisen condensation reaction

Reaction of two molecules of an ester with a base is called a Claisen condensation reaction. The mechanism of the Claisen condensation of ethyl ethanoate (CH_3CO_2Et) is shown in Figure 24.23. To start the reaction, a base (e.g. an alkoxide ion, EtO^-) removes an α-hydrogen from one molecule of the ester to form an ester enolate ion ($^-CH_2CO_2Et$). The ester enolate ion is a nucleophile and it attacks the C=O bond of a second molecule of the ester ion in a nucleophilic acyl substitution reaction to expel an alkoxide ion (^-OEt). A new bond is formed between the α-carbon atom of one ester molecule and the carbon atom in the C=O bond of a second ester molecule ($CH_3CO–CH_2CO_2Et$). The product is called a β-keto ester because it contains a ketone C=O bond, at the β-position, relative to the ester. Formation of the β-keto ester is reversible, but the equilibrium is pushed towards the condensation product by removal of an α-hydrogen atom from the β-keto ester ($CH_3COCH_2CO_2Et$). The α-hydrogen atoms in the β-keto ester are particularly acidic (pK_a ~11) because the resulting enolate ion ($CH_3COCH^-CO_2Et$) is stabilized by delocalization of the negative charge over two C=O bonds. Consequently, the β-keto ester is readily deprotonated by an alkoxide ion to form an enolate ion. On work-up with aqueous acid, the enolate ion is reprotonated to reform the β-keto ester ($CH_3COCH_2CO_2Et$).

The general mechanism for Claisen condensations is shown in Figure 24.7 (p.1106).

Notice that for both deprotonation steps (in Figure 24.23), the curly arrows are drawn to show the major resonance forms of the enolate ions (which have the negative charge on oxygen).

The pK_a values of $CH_3COCH_2CO_2Et$ and EtOH are ~11 and 16, respectively. The equilibrium constant for deprotonation of the β-keto ester is approximately 10^5 so the equilibrium lies well over towards the β-keto ester enolate ion. (To calculate the equilibrium constant from the pK_a values see Section 19.2, p.895.)

Note that, for an efficient Claisen condensation reaction, one equivalent of alkoxide ion (e.g. EtO^-) is required and the ester starting material must contain a minimum of two α-hydrogen atoms, so that the intermediate β-keto ester can be deprotonated.

It is interesting to see how the mechanism of the Claisen condensation reaction differs from that of the aldol reaction. In a Claisen condensation reaction the tetrahedral alkoxide ion expels RO^- to form a new C=O bond, whereas, in the aldol reaction, the tetrahedral alkoxide ion is protonated.

The aldol reaction is discussed in Section 23.4 (p.1089).

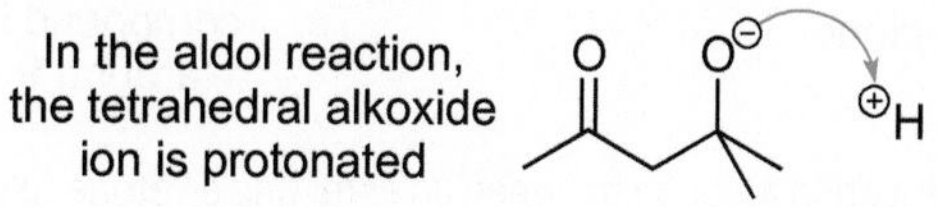
In the aldol reaction, the tetrahedral alkoxide ion is protonated
H

Interactive 3D animation of the Claisen condensation of ethyl ethanoate.

Polyketides
In nature, Claisen condensation reactions are used to form a large family of compounds called polyketides. Ethanoic acid molecules join together in the presence of enzymes to form carbon chains with a C=O bond on every other carbon. Modification of the chains (by cyclization, reduction, oxidation, and/or alkylation) produces a diverse range of structures.

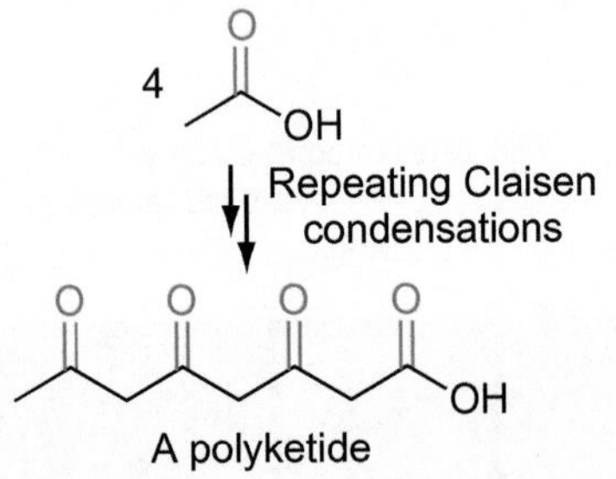
4
OH
Repeating Claisen condensations
OH
A polyketide

HCO_2Et is a stronger electrophile than CH_3CO_2Et.

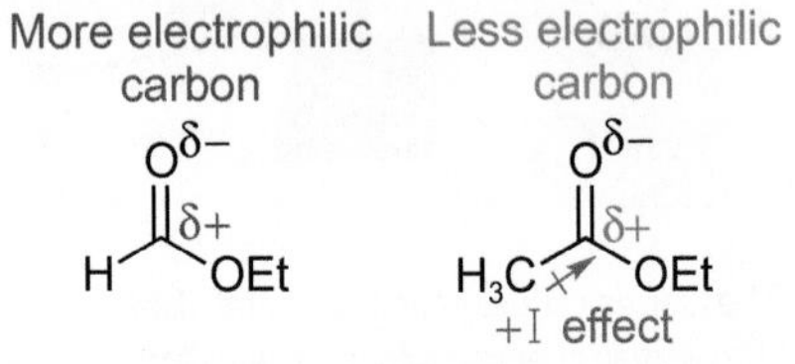
More electrophilic carbon
Less electrophilic carbon
δ−
δ+
H
OEt
δ−
δ+
H3C
OEt
+I effect

For related crossed (mixed) aldol reactions see Section 23.4 (p.1092).

Crossed Claisen condensations

Crossed (mixed) Claisen condensations involve reactions of two *different* esters. If both esters form ester enolate ions and both act as electrophiles in nucleophilic acyl substitutions, then a mixture of condensation products is formed. In synthesis, the most important reactions are those that efficiently form a single condensation product and this occurs when only one ester is able to form an ester enolate ion.

For example, ethyl ethanoate (CH_3CO_2Et) reacts with ethyl methanoate (HCO_2Et) in a crossed Claisen reaction to selectively form one product (shown in Figure 24.24). Only CH_3CO_2Et can form an ester enolate ion, which reacts with either CH_3CO_2Et or HCO_2Et. But if CH_3CO_2Et is added dropwise to a mixture of HCO_2Et/EtO^-, then, as soon as $^-CH_2CO_2Et$ is formed, it reacts with HCO_2Et because there is very little CH_3CO_2Et in the reaction mixture. Note that HCO_2Et is a more reactive electrophile than CH_3CO_2Et. In CH_3CO_2Et, the CH_3 group next to C=O hinders the approach of a nucleophile and it also reduces the size of the partial positive charge on the carbonyl carbon atom (by exerting a +I effect, see margin).

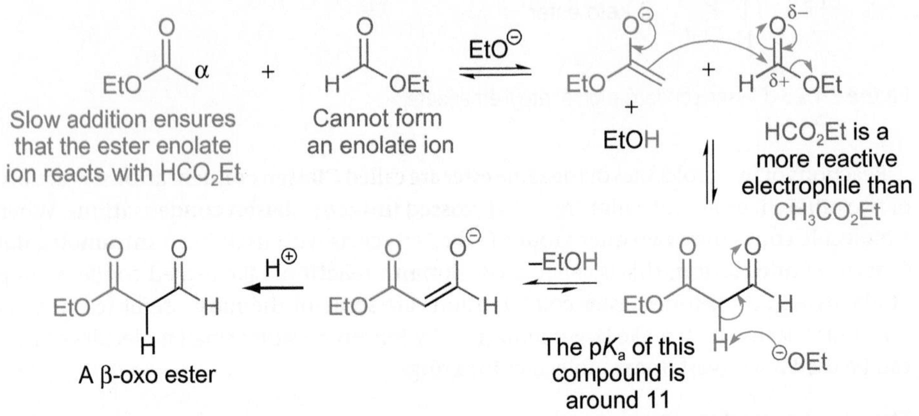
EtO
α
+
H
OEt
EtO⊖
EtO
+
H
δ+
OEt
δ−
Slow addition ensures that the ester enolate ion reacts with HCO2Et
Cannot form an enolate ion
+
EtOH
HCO2Et is a more reactive electrophile than CH3CO2Et
EtO
H
H
H⊕
EtO
H
−EtOH
EtO
H
A β-oxo ester
The pKa of this compound is around 11
H
⊖OEt

Figure 24.24 The crossed Claisen reaction between ethyl ethanoate (CH_3CO_2Et) and ethyl methanoate (HCO_2Et) gives a single product.

A related condensation reaction takes place between an ester and an aldehyde or ketone (see Figure 24.25). In this reaction, the ester acts as the electrophile and the aldehyde (such as CH_3CHO) or ketone (such as CH_3COR) is the nucleophile. Enolate ions from aldehydes (such as $^-CH_2CHO$) and ketones ($^-CH_2COR$) are more easily formed than those from esters because α-hydrogen atoms in aldehydes and ketones are more acidic (Section 24.1, p.1101). So, if both the ketone (or aldehyde) and ester have α-hydrogen atoms, it is possible to form mainly one condensation product by adding the ketone and the base dropwise to the ester. The use of this reaction to make a 1,3-dione is shown in Worked example 24.4.

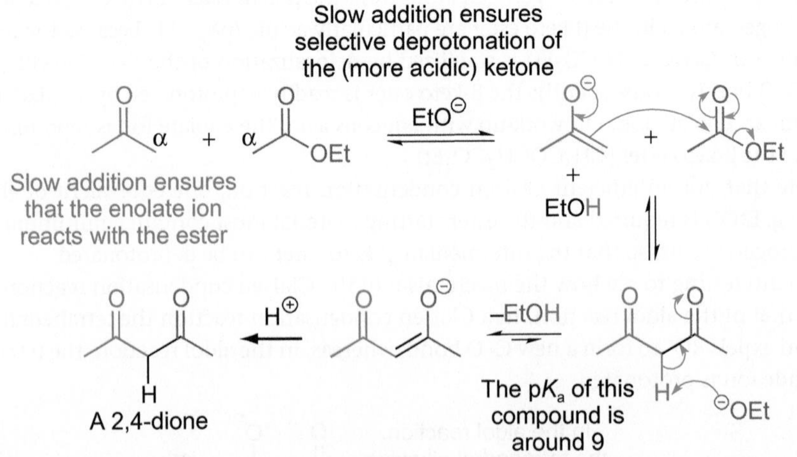
Slow addition ensures selective deprotonation of the (more acidic) ketone
α
+
α
OEt
EtO⊖
+
OEt
Slow addition ensures that the enolate ion reacts with the ester
+
EtOH
H⊕
−EtOH
H
A 2,4-dione
The pKa of this compound is around 9
H
⊖OEt

Figure 24.25 A condensation reaction between an ester and a ketone to give a single product.

The product from the condensation reaction between propanone and ethyl ethanoate is called pentane-2,4-dione, $CH_3COCH_2COCH_3$ (or commonly acetylacetone). The major enol form of pentane-2,4-dione is stabilized by conjugation and an intramolecular hydrogen bond (see Section 23.3, p.1084).

Worked example 24.4 Condensation reactions between an ester and a ketone

The following questions relate to the synthesis of 1-phenylbutane-1,3-dione.

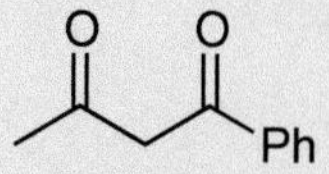

O
O
Ph

1-phenylbutane-1,3-dione

(a) Suggest two synthetic routes to 1-phenylbutane-1,3-dione that involve condensation reactions between an ester and a ketone.

(b) For each of your synthetic routes in part (a), highlight any competing carbonyl–carbonyl reactions that will lower the yield of 1-phenylbutane-1,3-dione.

Strategy

(a) Use a condensation reaction to make one of the C–C bonds in between the two C=O bonds in 1-phenylbutane-1,3-dione. Consider what ester and what ketone are required to make each C–C bond.

Ph
Ph

Either C–C or C–C is formed by a condensation reaction

(b) Identify the carbonyl compounds that are able to form enolate ions. Draw the different enolate ions and explain how they react with C=O bonds in competing carbonyl–carbonyl reactions, including aldol reactions (Section 23.4, p.1089).

Solution

(a)

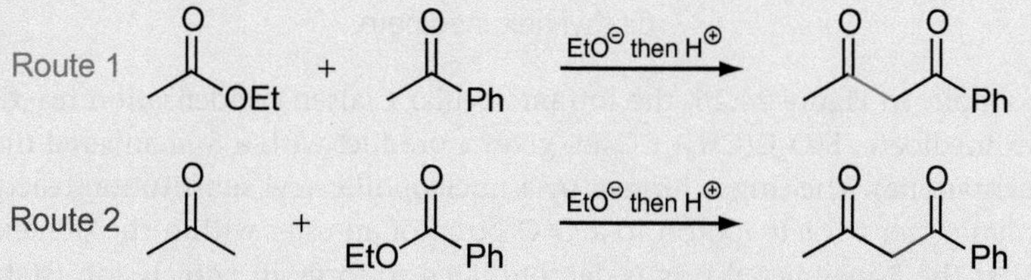

Route 1
OEt
Ph
EtO⊖ then H⊕
Ph
Route 2
EtO
Ph
EtO⊖ then H⊕
Ph

(b) • In route 1 both the ketone and ester can form enolate ions ($^{-}CH_2COPh$ and $^{-}CH_2CO_2Et$). Reaction of these enolate ions with the C=O bond of the ester and with the C=O bond of the ketone gives (after work-up with aqueous acid) 1-phenylbutane-1,3-dione and three by-products.

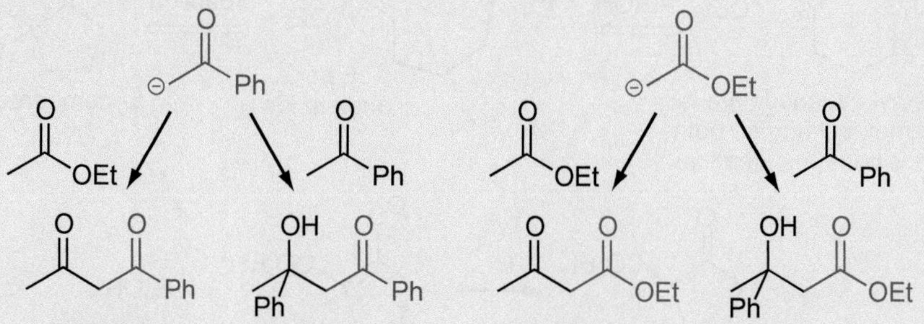

Ph
OEt
OEt
Ph
OEt
Ph
Ph
OH
Ph
Ph
OEt
OH
Ph
OEt
Desired condensation
Aldol reaction
Claisen condensation
Aldol reaction

Notice that the aldol reactions are not stereoselective. The β-hydroxy-ketones are formed as racemates because the enolate ions ($^{-}CH_2COPh$ and $^{-}CH_2CO_2Et$) can react equally well with either face of the planar C=O bond in CH_3COPh.

1-Phenylbutane-1,3-dione is commonly called benzoylacetone. The major enol forms of 1-phenylbutane-1,3-dione are stabilized by conjugation and an intramolecular hydrogen bond.

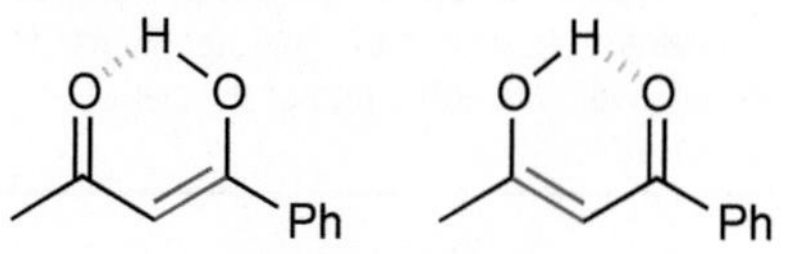

H
O
O
Ph
H
O
O
Ph

major enol forms

$PhCO_2Et$ is called ethyl benzoate. It can be prepared by the esterification of benzoic acid using ethanol and an acid catalyst. Like many esters (with low boiling points) it is has a pleasant, sweet odour, like cherries.

- In route 2, only the ketone is able to form an enolate ion ($^-CH_2COCH_3$), which reacts with the C=O bond of the ester or ketone to give (after work-up with aqueous acid) 1-phenylbutane-1,3-dione and a single by-product from the competing aldol reaction.

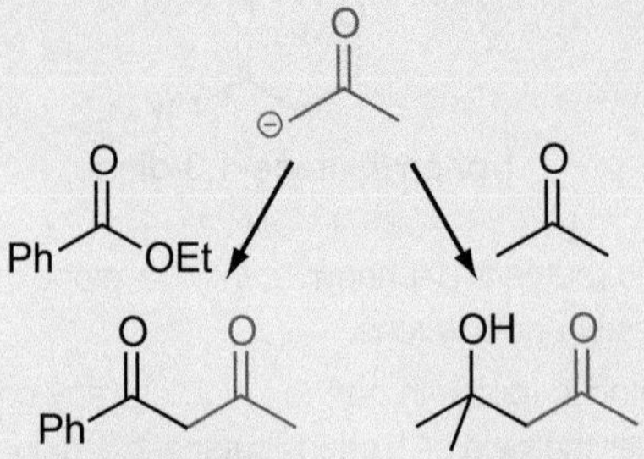
Ph
OEt
Ph
OH
Desired condensation
Aldol reaction

Question

For both synthetic approaches in (a), describe the experimental conditions that would maximize the yield of 1-phenylbutane-1,3-dione.

Propanone (CH_3COCH_3) is a more reactive electrophile than ethyl benzoate ($PhCO_2Et$). In $PhCO_2Et$ the Ph and OEt groups next to the C=O bond reduce the size of the partial positive charge on the carbonyl carbon by +M effects. In comparison, the methyl groups in propanone exert weaker +I effects.

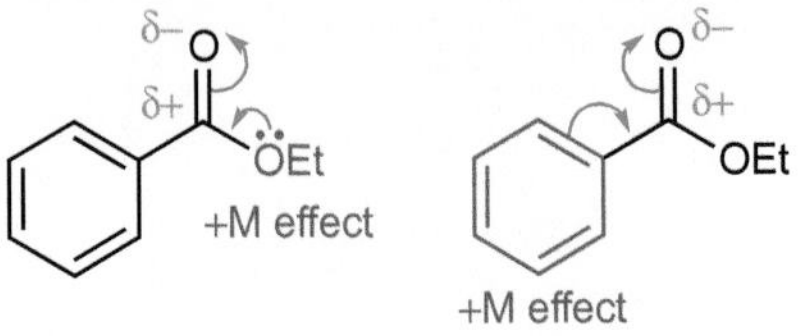
δ–
δ+
OEt
+M effect
δ–
δ+
OEt
+M effect

Intramolecular Claisen condensations: the Dieckmann reaction

In an **intramolecular Claisen condensation reaction (Dieckmann reaction)**, a starting material with two ester groups (a diester) is heated with a base to form a cyclic product, usually containing a 5- or a 6-membered ring. Cyclization of a 1,6-diester forms a 5-membered ring, while a 1,7-diester produces a 6-membered ring.

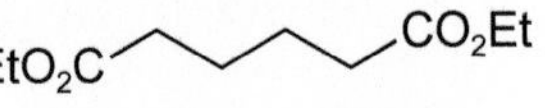
EtO2C
CO2Et

diethyl hexanedioate

Visit the Online Resource Centre to view screencast 24.3 walks you through the mechanism of some intramolecular Claisen condensations (Dieckmann reactions).

The Dieckmann reaction is named after the German chemist Walter Dieckmann (1869–1925), who invented the reaction at the University of Munich.

For example, in Figure 24.26, the intramolecular Claisen condensation reaction of diethyl hexanedioate, $EtO_2C(CH_2)_4CO_2Et$, gives a product with a 5-membered ring (called a cyclopentanone). The ring is formed by a nucleophilic acyl substitution reaction of an intermediate ester enolate ion on to a C=O bond of an ester within the same molecule. Notice that the 5-membered ring is deprotonated to form an enolate ion (stabilized by delocalization of the negative charge over the two C=O bonds), which is re-protonated on work-up with aqueous acid.

5- and 6-Membered rings are the most stable ring size and the easiest to form (see Section 18.2, p.827).

For related intramolecular aldol condensations see Section 23.4 (p.1093).

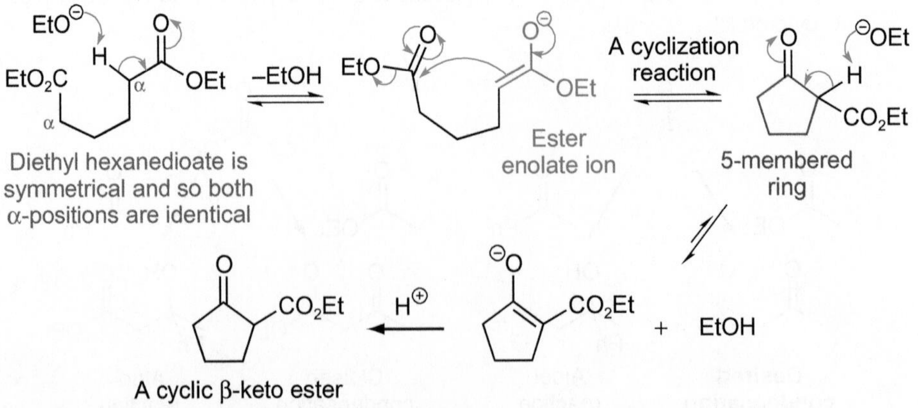
EtO
EtO2C
H
α
OEt
α
–EtOH
EtO
OEt
A cyclization reaction
OEt
H
CO2Et
Diethyl hexanedioate is symmetrical and so both α-positions are identical
Ester enolate ion
5-membered ring
CO2Et
H⊕
CO2Et
+ EtOH
A cyclic β-keto ester

Figure 24.26 Mechanism of the Dieckmann reaction.

The product of this Dieckmann reaction is called ethyl 2-oxocyclopentanecarboxylate. It is formed as a racemate because, in the final step, H^+ reacts equally well with either face of the planar enolate ion.

Summary

- For carboxylic acids and derivatives, the keto form is usually more stable than the enol form. The enol form and, particularly, the enolate ion are good nucleophiles that react with electrophiles in α-substitution reactions.

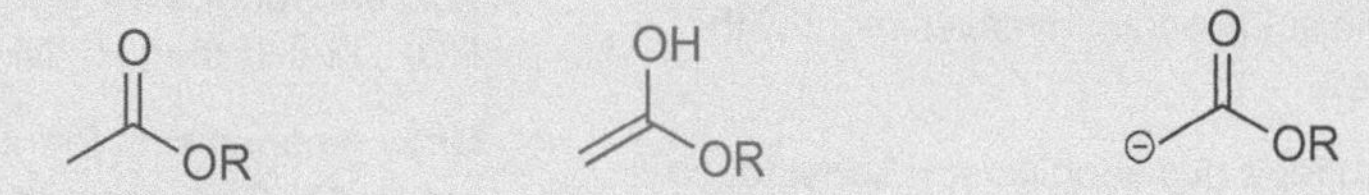
O
OR
OH
OR
O
OR

The **keto form** is generally the more stable

The **enol form** (An enol form is stabilized by conjugation and intramolecular H-bonding)

An ester enolate ion is a strong nucleophile

- α-Bromination of carboxylic acids using a mixture of PBr_3 and Br_2 is called the Hell–Volhard–Zelinsky (HVZ) reaction.
- Reaction of two molecules of the *same* ester to form a β-keto ester is called a Claisen condensation.

The ester enolate ion reacts with CO_2Et in a nucleophilic acyl substitution

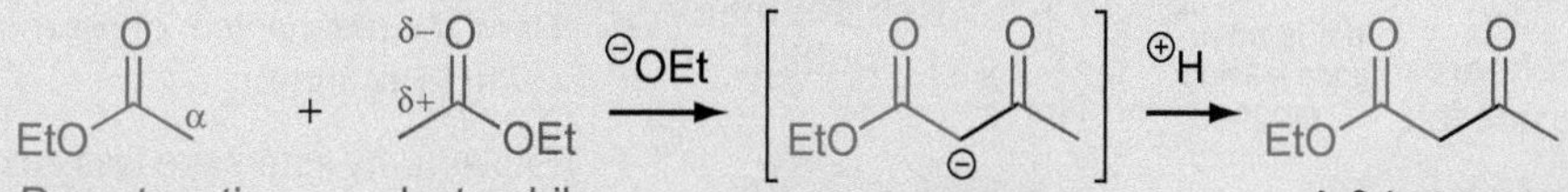
O
EtO
α
+
δ– O
δ+
OEt
OEt
O O
EtO
H
O O
EtO

α-Deprotonation forms an ester enolate ion

electrophile

Anion stabilized by delocalization over both C=O bonds

A β-keto ester

- Reactions of two *different* ester molecules are called crossed (mixed) Claisen condensations.
- Related condensation reactions take place between an ester and an aldehyde or ketone.

Slow addition of the base and ketone to the ester

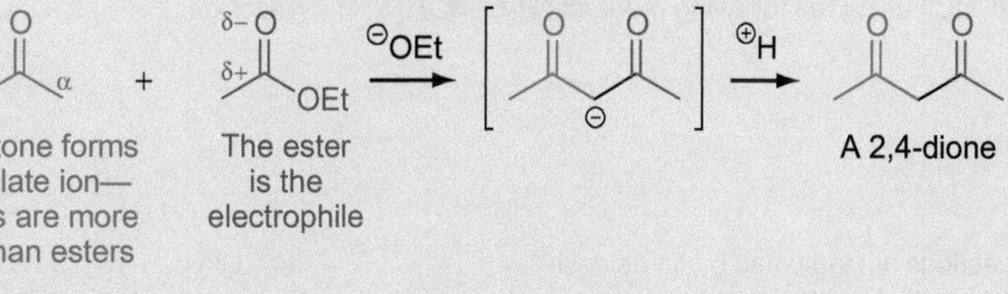
O
α
+
δ– O
δ+
OEt
OEt
O O
H
O O

The ketone forms an enolate ion—ketones are more acidic than esters

The ester is the electrophile

A 2,4-dione

- When a molecule containing two ester groups reacts with itself, this is called an intramolecular Claisen condensation or a Dieckmann reaction.

α-Deprotonation forms an ester enolate ion

Cyclization to give a 6-membered ring

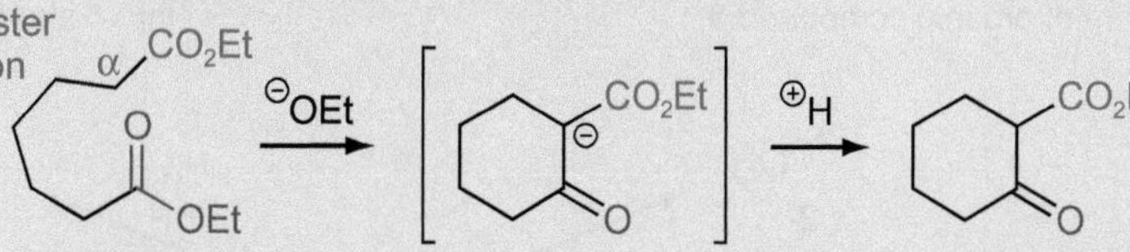
α
CO2Et
O
OEt
OEt
CO2Et
O
H
CO2Et
O

The electrophile

A cyclic β-keto ester

? For practice questions on these topics, see questions 1, 2, 4, and 6 at the end of this chapter (pp.1134–1137).

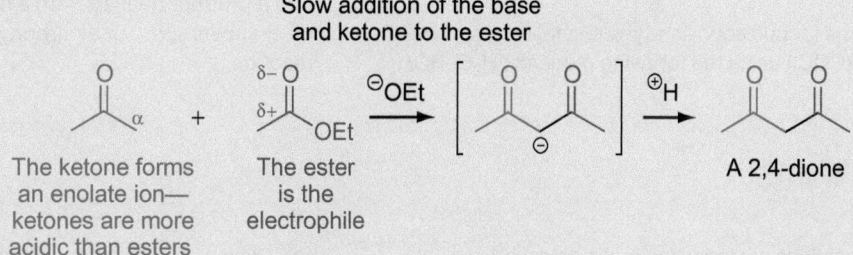

Concept review

By the end of this chapter, you should be able to do the following.

- Recognize carboxylic acids, acyl chlorides, acid anhydrides, esters, and amides.
- Understand how carboxylic acids and derivatives react with nucleophiles, H^+, or a base.
- Draw a general mechanism for a nucleophilic acyl substitution reaction and explain why carboxylic acid derivatives react with nucleophiles in substitution reactions rather than addition reactions.

Nucleophilic substitution versus nucleophilic addition

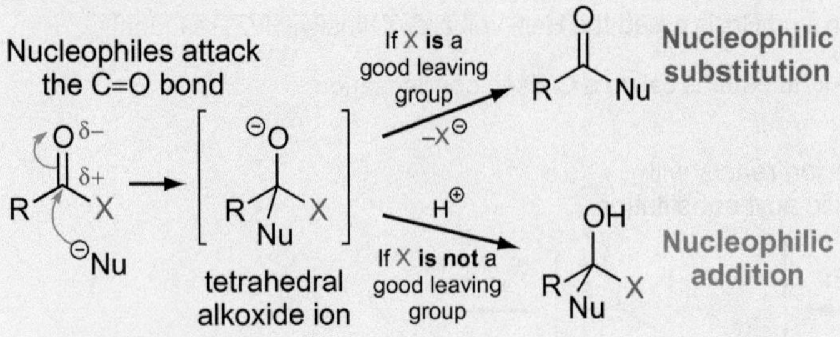
Nucleophiles attack the C=O bond
tetrahedral alkoxide ion
If X is a good leaving group
−X⊖
Nucleophilic substitution
H⊕
If X is not a good leaving group
Nucleophilic addition

- Explain why carboxylic acid derivatives are not equally reactive to nucleophiles.
- Draw general mechanisms for an α-substitution reaction and a carbonyl–carbonyl condensation reaction of a carboxylic acid derivative.
- Draw mechanisms for nucleophilic acyl substitution reactions of carboxylic acids (RCO_2H) using the following reagents: ROH, H^+ (catalyst); $SOCl_2$; NH_3 (heat).
- Draw mechanisms for nucleophilic acyl substitution reactions of acyl chlorides (RCOCl) using the following reagents: H_2O; ROH; $NaBH_4$ then H^+; NH_3 (two equivalents); RMgX (two equivalents) then H^+.
- Draw mechanisms for nucleophilic acyl substitution reactions of acid anhydrides (RCO_2COR) using the following reagents: H_2O; ROH; $LiAlH_4$ then H^+; NH_3 (two equivalents).
- Draw mechanisms for nucleophilic acyl substitution reactions of esters (RCO_2R) using the following reagents: H_2O, H^+; H_2O, HO^- then H^+; $LiAlH_4$ then H^+; NH_3; RMgX (two equivalents) then H^+.
- Draw mechanisms for nucleophilic acyl substitution reactions of amides (such as $RCONH_2$) using the following reagents: H_2O, H^+ (heat).
- Understand how nitriles (RCN) are hydrolysed to form carboxylic acids.
- Discuss keto–enol tautomerism in carboxylic acid derivatives and understand the factors that influence the stability of keto and enol forms.
- Draw the mechanism of α-bromination of a carboxylic acid using PBr_3 and Br_2.
- Explain why esters are less acidic than aldehydes or ketones.
- Draw mechanisms for Claisen condensations, crossed (mixed) Claisen condensations, and intramolecular Claisen condensations (Dieckmann reactions).
- Predict the structure of a product derived from a nucleophilic acyl substitution, an α-substitution, or a carbonyl–carbonyl condensation reaction of a carboxylic acid derivative.
- Propose reagents for converting a carboxylic acid derivative into a product derived from a nucleophilic acyl substitution, an α-substitution, or a carbonyl–carbonyl condensation reaction.

Questions

More challenging questions are indicated by an asterisk *.

1 The following questions are based on the reactions of ethyl ethanoate (**1**) shown below. (Sections 24.2 and 24.3)

(a) Give appropriate reagents for converting **1** into **2**. Is this an example of an oxidation or a reduction reaction?

(b) Give the structure of organic compound **3**.

(c) Oxygen-18 labelling is often used to establish the mechanism of a reaction. A ^{16}O oxygen in a starting material is replaced by an ^{18}O atom and the position of the ^{18}O atom at the end of the reaction recorded using mass spectrometry (Section 12.1, p.558).

(i) Draw a reaction mechanism to show how **1** is converted into **4** and EtOH.

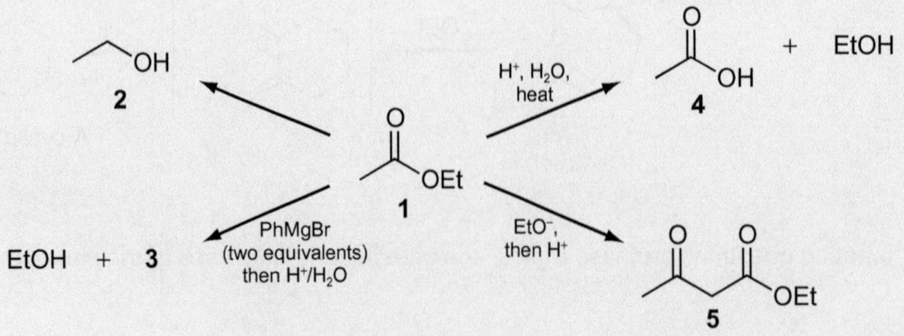
OH
2
H+, H2O, heat
OH
4
+ EtOH
OEt
1
EtOH + 3
PhMgBr (two equivalents) then H+/H2O
EtO−, then H+
OEt
5

(ii) If **1** is heated with $H_2{}^{18}O$ and H^+, where does the ^{18}O atom appear at the end of the reaction?

(iii) If ethyl ethanoate, labelled with ^{18}O as shown below, is heated with H_2O/H^+, where does the ^{18}O atom appear at the end of the reaction?

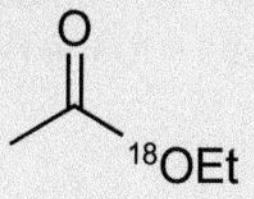

(d) Draw a reaction mechanism to show how **1** is converted into **5**.

(e) Draw the major enol form of compound **5**.

2 (a) The following questions relate to the formation of lactone **7** from keto-ester **6**. (Sections 24.2 and 24.3)

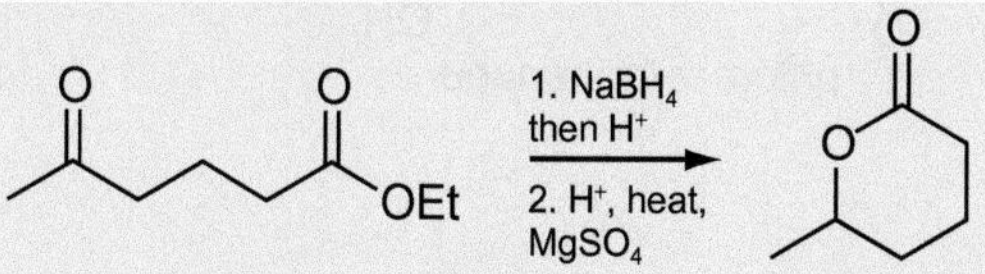

6 **7**

(i) Suggest a mechanism to explain the formation of **7**.

(ii) Explain why the reaction of **6** with $NaBH_4$ is chemoselective.

(iii) Give the structure of the organic product from reaction of **6** with $LiAlH_4$ then H^+.

(b) The following questions relate to the formation of compound **10**. (Sections 24.2 and 24.3)

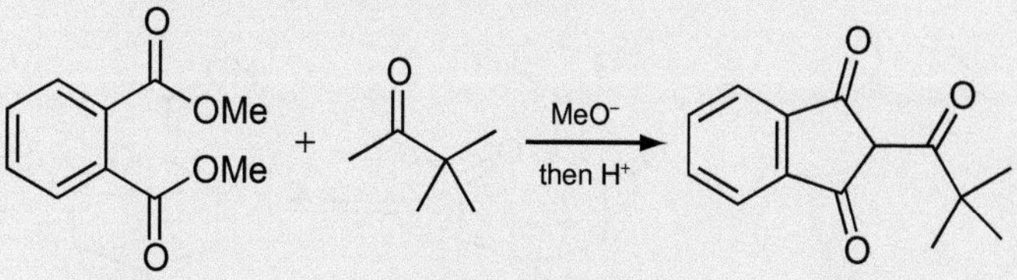

8 **9** **10**

(i) Suggest a mechanism to explain the formation of compound **10**.

(ii) Draw two different enol forms of compound **10**.

3 Part of a synthesis of the oral contraceptive norgestimate is shown below. (Section 24.2)

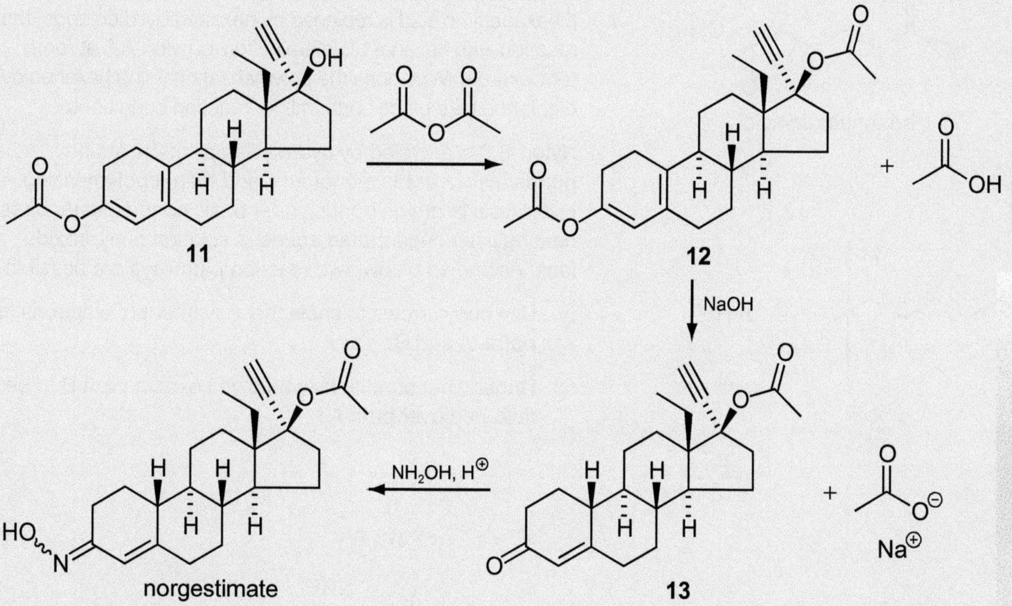
11
12
NaOH
NH2OH, H⊕
norgestimate
13

(a) To what class of compound does norgestimate belong—is it a peptide, a fatty acid, a steroid, or a carbohydrate?

(b) Propose a mechanism to explain the conversion of alcohol **11** into ester **12**.

(c) Give the structure of a reagent, other than ethanoic anhydride, that will convert **11** into **12**.

(d) Explain why NaOH reacts with only one of the ester groups in compound **12**. (Hint. Consider the relative electrophilicity of the carbonyl carbons.)

(e) What functional group is formed when compound **13** reacts with NH_2OH, H^+?

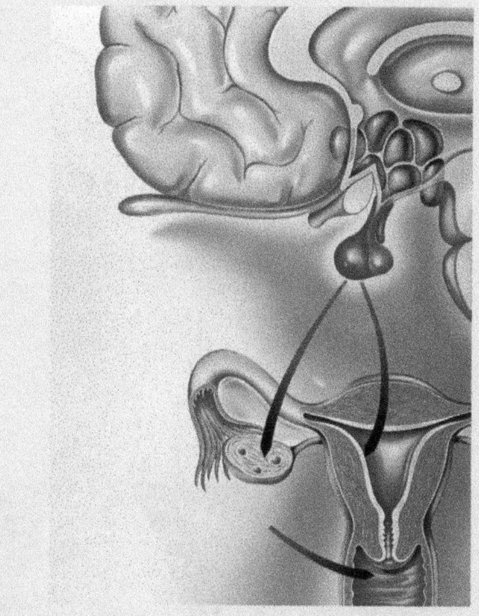

Hormones in female contraceptive pills target the hypothalamus of the brain (shown in blue), which stops the pituitary gland (red) releasing sex hormones, in turn preventing the monthly release of an egg from the ovaries (yellow ovals). The hormones also alter the lining of the uterus (lower right), preventing implantation should an egg be released.

4 In nature, aromatic compounds are formed by the cyclization of polyketide chains (containing alternating C=O and CH_2 groups, see p.1130) in the presence of enzymes. The polyketide chains have a thioester group (RCOSR) at the end of the chain, which, like esters, reacts in Claisen-type reactions. For example, thioester **14** is converted into phloracetophenone. Phloracetophenone, obtained from the plant *Curcuma comosa*, has been shown to lower cholesterol levels in animals. (Section 24.3)

(a) Propose a mechanism that explains how compound **15** is converted into phloracetophenone.

(b) For phloracetophenone, explain why the position of the equilibrium lies heavily to the right.

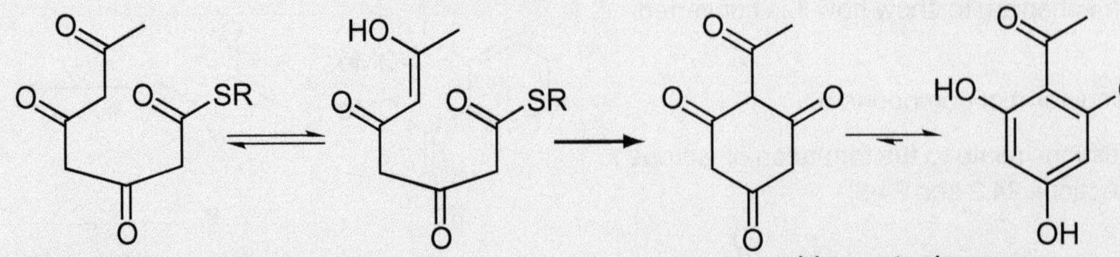

14 polyketide — **15** — phloracetophenone

5 Nylon-6,6 is a synthetic polyamide. It is formed by heating hexane-1,6-diamine with hexanedioic acid (adipic acid) in a condensation polymerization reaction. (Section 24.2)

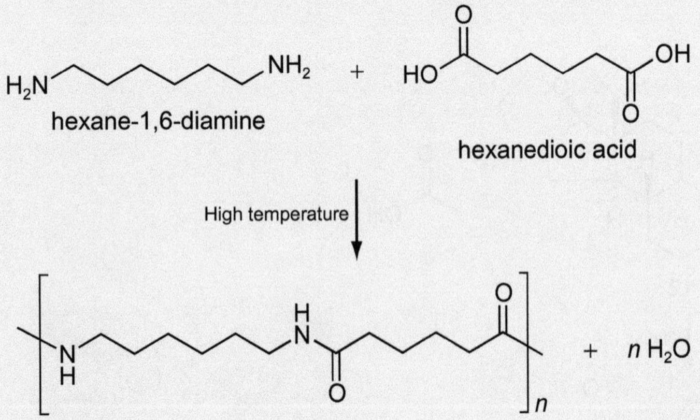
hexane-1,6-diamine
hexanedioic acid
High temperature
+ n H_2O

nylon-6,6

(a) Why is the formation of nylon-6,6 called a *condensation* polymerization?

(b) Why does the formation of nylon-6,6 from hexane-1,6-diamine and hexanedioic acid require heating?

(c) If hexanedioic acid is replaced by hexanedioyl dichloride, then reaction with hexane-1,6-diamine forms nylon-6,6 at room temperature. Why does the polymerization using hexanedioyl dichloride take place under milder reaction conditions?

(d) Nylon-6,6 is recycled by hydrolysing all of the amide bonds to reform the monomers, and then repolymerizing. Hydrolysis of amide bonds under basic conditions requires heating in a concentrated aqueous solution of hydroxide ions. As shown below, two reaction pathways are possible.

(i) Use curly arrows to show the movement of electrons in **paths A** and **B**.

(ii) Under what conditions would you expect **path B** to be favoured over **path A**?

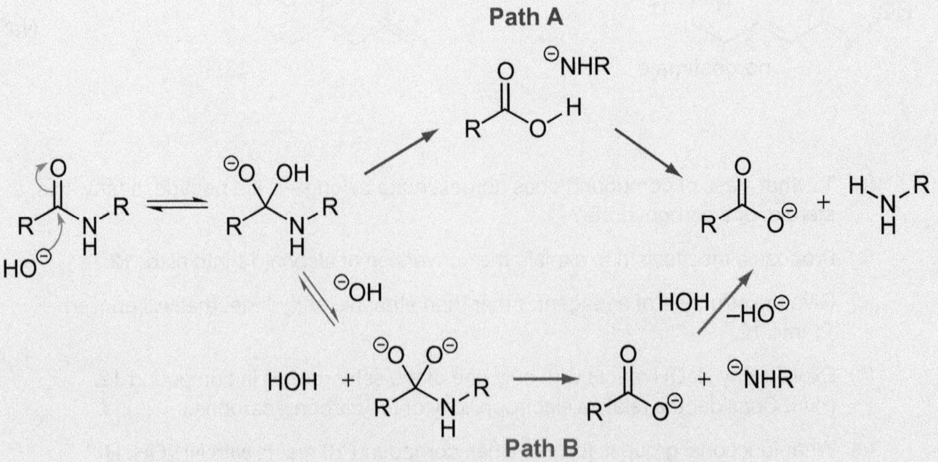
Path A
HOH
–HO⁻
HOH +
Path B

Artificial grass (or Astroturf) is made from nylon-6,6, which, when worn down, is often recycled.

6* Part of a synthesis of a naturally occurring steroid called equilenin is shown below. (Section 24.3)

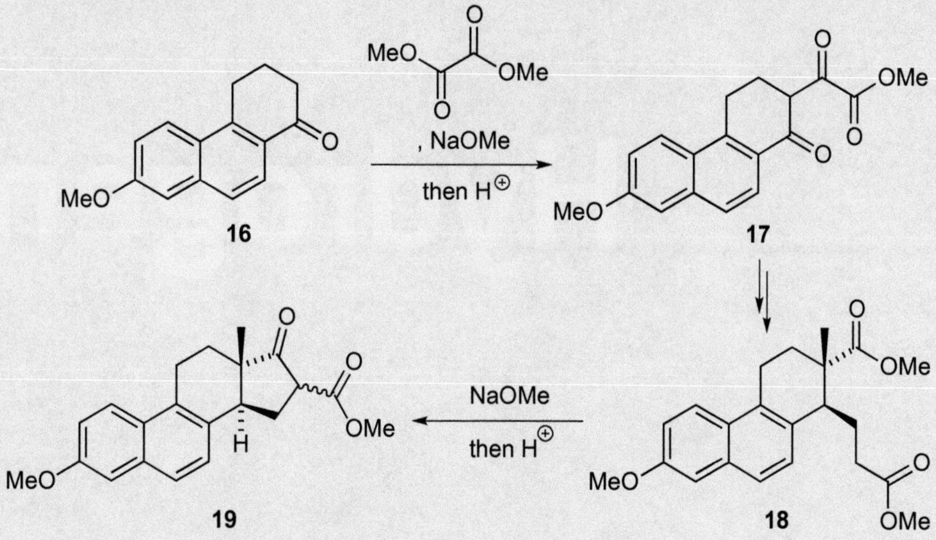
MeO
OMe
O
O
, NaOMe
then H⊕
MeO
16
17
18
19
NaOMe
then H⊕
OMe
H

(a) Propose a mechanism that explains how compound **16** is converted into compound **17**.

(b) Propose a mechanism that explains how compound **18** is converted into compound **19**.

Equilenin was first isolated in 1932 from the urine of pregnant mares.

25

Hydrogen

NASA

This chapter builds on the following topics:

- **Non-covalent interactions** Section 1.8, p.52
- **Polyatomic molecules** Chapter 5, p.216
- **Brønsted-Lowry acids and bases** Section 7.1, p.304

▲ The gas gun at the Lawrence Livermore National Laboratory, University of California, was used by Bill Nellis's team to recreate conditions on Jupiter.

◀ The Galileo spacecraft orbited Jupiter from 1995 to 2003.

The planet Jupiter

Jupiter is the largest planet in the solar system with a mass of over 2.5 times that of all the other planets combined. Jupiter is often called a *gas giant* as it is composed mainly of hydrogen and helium. This description is not very accurate, however, as most of the hydrogen on Jupiter is not gaseous.

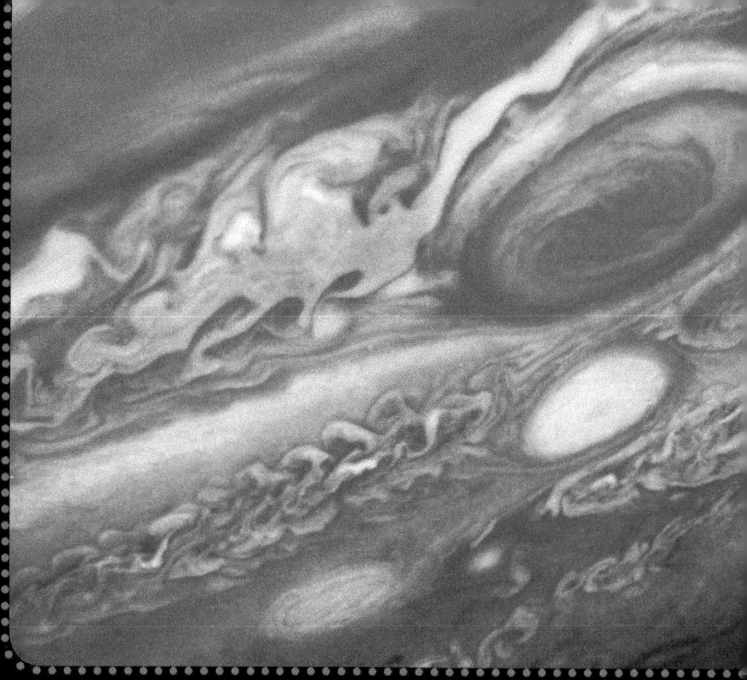

Experiments carried out in 1996 by Bill Nellis and his team at the University of California have shed some light on conditions inside Jupiter. These experiments involved subjecting samples of liquid hydrogen to enormous pressures. At 3000 K and a pressure of 1.8 million atmospheres, generated for a fraction of a second during the experiment, hydrogen behaves as a metal and conducts electricity. The existence of a metallic form of hydrogen was first proposed in the 1930s based on calculations from quantum mechanics, but Nellis's team was the first to provide direct evidence for this.

Under the conditions in the experiment, metallic hydrogen exists as a fluid rather than a solid. In this metallic fluid, about 90% of the hydrogen is in the form of H_2, with the remainder dissociated into hydrogen atoms. The electrical conduction is due to electrons hopping between H_2^+ ions. As the pressure increases, the percentage of atomic hydrogen also increases, and eventually the conductivity decreases.

We know that hydrogen exists as a metallic fluid on Jupiter because of the planet's magnetic field, which is about 10 times larger than that of the Earth. The Earth's magnetic field is a result of electric currents generated by convection within the liquid iron core of the planet. Jupiter's magnetic field is caused by similar convection in the metallic hydrogen fluid. The field is large because metallic hydrogen exists relatively close to the surface.

Trace elements on Jupiter include carbon, oxygen, nitrogen, and sulfur. These are present as their hydrides—methane, water, ammonia, and hydrogen sulfide—in the upper part of the planet. Methane exists as a gas, whereas H_2O, NH_3, and H_2S are present in the form of ice clouds. Scientists believe that these compounds exist in three separate cloud layers, the uppermost layer consisting of solid ammonia, the middle layer of ammonium hydrogen sulfide ($NH_4^+SH^-$), and the lowest layer of water ice.

Our understanding of Jupiter was improved dramatically by the Galileo spacecraft, which arrived at Jupiter after a 6 year journey from Earth. A probe released by this spacecraft descended into Jupiter's atmosphere in December 1995. The probe penetrated 200 km into the planet's atmosphere before it was destroyed by the massive pressure, but this was far enough to record and send back a wealth of information. One observation was that the abundances of helium and neon were both lower than anticipated from those known to be present in the early solar system when Jupiter was formed.

One explanation is that some of the helium condenses from the hydrogen fluid in the form of 'helium rain' which, being denser, falls towards the centre of the planet. Neon is very soluble in liquid helium, so it too is carried into the planet's interior. This would account for the low abundance of these elements in the atmosphere.

The Galileo spacecraft continued making observations and measurements on Jupiter and its moons until September 2003 when it too plunged into Jupiter's atmosphere. The spacecraft had discovered a possible ocean under the icy surface of the moon Europa, and it was deliberately destroyed to prevent it entering and contaminating this ocean.

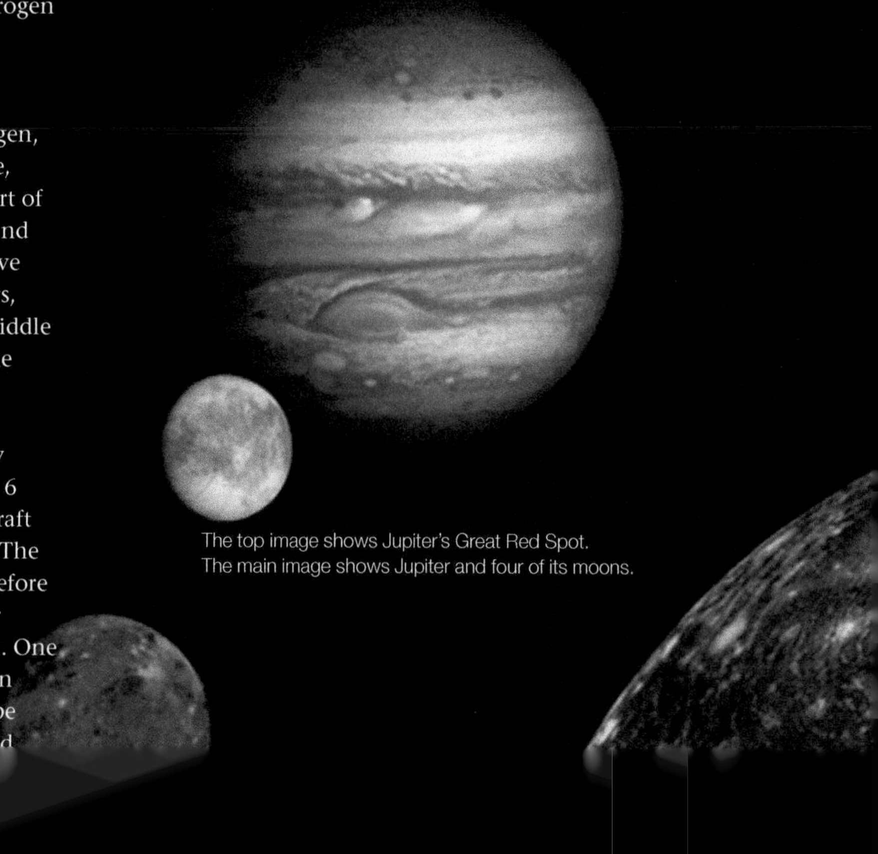

The top image shows Jupiter's Great Red Spot.
The main image shows Jupiter and four of its moons.

The atomic structure of hydrogen is described in Sections 3.3 (p.129) and 3.5 (p.134).

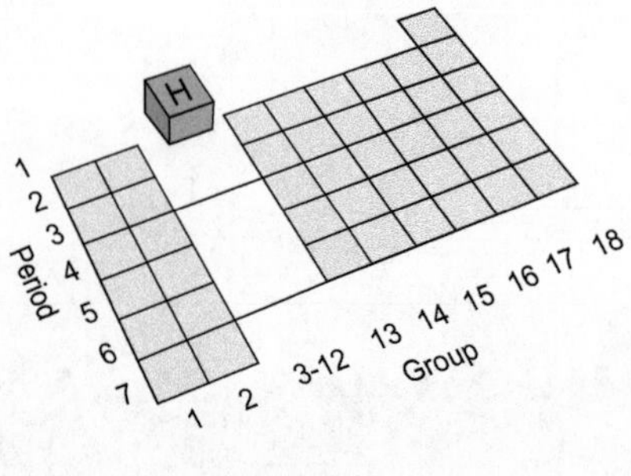
H
Period
1
2
3
4
5
6
7
1
2
3-12
13
14
15
16
17
18
Group
Hydrogen in the Periodic Table.

Hydrogen has the simplest atoms of all the elements. The nucleus of the most common isotope consists of a single proton, and the atom has just one electron. This simplicity means that the hydrogen atom is the only one for which the Schrödinger equation can be solved exactly.

Unlike the other elements, hydrogen doesn't fit neatly into the Periodic Table. Many Periodic Tables show hydrogen sitting at the top of Group 1 because of its electronic configuration ($1s^1$), which fits with the Group 1 valence shell configurations of ns^1. However, in chemical terms, hydrogen does not behave as a Group 1 element. It is a gas at room temperature and pressure, whereas the Group 1 elements are all reactive, electropositive metals. The chemistry of Group 1 is dominated by M^+ ions but, for hydrogen, H^+ and H^- ions are both common, as are covalent bonds.

Occasionally, hydrogen is shown sitting at the top of Group 17. This is because addition of one electron to hydrogen gives a Group 18 element configuration, just as it does for the halogens. Hydrogen also forms homonuclear diatomic molecules (H_2) like the halogens. In terms of reactivity though, hydrogen does not act like a halogen, and chemically it is just as out of place in Group 17 as it is in Group 1.

Since hydrogen does not fit neatly into any of the groups, it is often shown sitting isolated at the top of the Periodic Table, as on the inside front cover of this book. For the same reason, the chemistry of hydrogen is treated on its own in this chapter.

25.1 Elemental hydrogen

Hydrogen was first observed in 1671 by Robert Boyle. He observed the formation of a gas when iron filings were dissolved in hydrochloric acid and demonstrated that this gas was flammable. Despite this observation, the discovery of hydrogen is usually credited to Henry Cavendish, almost a hundred years after Boyle. Cavendish showed that the gas produced in this experiment combined with oxygen to form water. This was an important discovery, as it disproved the long-held view that water was one of the four fundamental elements of nature, along with fire, air, and earth.

Hydrogen is the most abundant element in the Universe, and the tenth most abundant element on Earth by mass. It does not, however, occur naturally on Earth in the elemental form. H_2 molecules are very light and can escape from the Earth's gravity into space. Most hydrogen on Earth occurs bonded to oxygen in the form of water or bonded to carbon in the form of hydrocarbons and other organic molecules.

Properties of hydrogen (H_2)

Hydrogen gas, $H_2(g)$, is colourless and odourless. It has a melting point of −259 °C (14 K) and a boiling point of −253 °C (20 K). These are both very low because of the weak intermolecular forces between H_2 molecules. The low molecular mass means that the density of hydrogen (0.082 g dm^{-3} at 25 °C) is the lowest of all the gases.

H_2 is generally unreactive at room temperature in the absence of a catalyst, largely because of the high H–H bond dissociation enthalpy (+436 kJ mol^{-1}). There are some exceptions to this, and hydrogen reacts explosively with O_2, F_2, and Cl_2 (see Section 25.2, p.1153).

Production of hydrogen

Over 60 million tonnes of hydrogen are produced annually throughout the world. Most is formed by heating natural gas or light crude oil fractions with steam at high temperature over a nickel oxide catalyst, a process that is called **steam reforming**. For methane

$$CH_4(g) + H_2O(g) \xrightleftharpoons{800\text{–}900\,°C} CO(g) + 3H_2(g) \quad \Delta H^{\ominus} = +205\ \text{kJ mol}^{-1}$$

The CO is oxidized to CO_2 by reaction with additional steam to generate more hydrogen. This process, typically using an iron catalyst, is known as the **shift reaction**

$$CO(g) + H_2O(g) \xrightleftharpoons{200-450\,°C} CO_2(g) + H_2(g) \quad \Delta H^{\ominus} = -42\,kJ\,mol^{-1}$$

Mixtures of H_2 and CO generated by steam reforming are known as synthesis gas (syngas). This is the starting point for the manufacture of other compounds, such as methanol. To prepare pure hydrogen from synthesis gas, the mixture is passed through a molecular sieve, usually a zeolite, to remove the CO and CO_2.

Examples of the structures and uses of zeolites are described in Chapter 6 (p.255).

Box 25.1 The hydrogen economy

The problems associated with fossil fuels such as oil, coal, and gas are well known. Their combustion leads to air pollution and emission of the greenhouse gas carbon dioxide, which is a major contributor to global warming. In addition, stocks of fossil fuels are limited, and there is only an estimated 50 years' supply of cheap, recoverable oil remaining. One of the biggest challenges facing civilization is finding a replacement for these fuels. Many people believe that this will be hydrogen.

Using hydrogen as a fuel would reduce pollution and greenhouse gas emissions, since the only product of hydrogen combustion is water. It would also reduce the dependency of Western countries on oil-producing states. A situation in which hydrogen has replaced fossil fuels as the main source of energy is known as the **hydrogen economy**.

Most hydrogen is currently manufactured from fossil fuels. This method of production loses many of the advantages of the hydrogen economy as the pollution and greenhouse gas emissions still occur, but in the production of the fuel rather than its combustion. For a pure hydrogen economy, the gas would need to be made without fossil fuels, for example by electrolysis of water using a pollution-free source of electricity such as solar power.

One of the important considerations for a fuel is its *energy density* (see Section 13.3, p.628). This is the amount of energy that is released for a given mass of the fuel. The energy density of hydrogen is 143 kJ g^{-1}, which is almost three times that of petrol (48 kJ g^{-1}) and almost five times that of coal (29 kJ g^{-1}). The high energy density of hydrogen is one of the reasons that liquid hydrogen was used as the fuel in the Space Shuttle (see Chapter 13, p.611).

The direct reaction between hydrogen and oxygen gases is explosive and difficult to control. For this reason, other ways of harnessing the energy produced by this reaction have been investigated. The most attractive of these is the combination of hydrogen and oxygen in a fuel cell to produce electricity.

Most car manufacturers are working on hydrogen fuel cell powered vehicles. While many are still experimental, Honda released the first commercial production hydrogen fuel cell cars in 2008. The Honda FCX Clarity stores the hydrogen under pressure and can be refuelled in a few minutes. The main drawbacks at the current time are the high cost of the vehicles and the relatively sparse distribution of refuelling stations. Another issue is the storage of hydrogen within these vehicles, and this topic is developed further in Box 25.3 (p.1147).

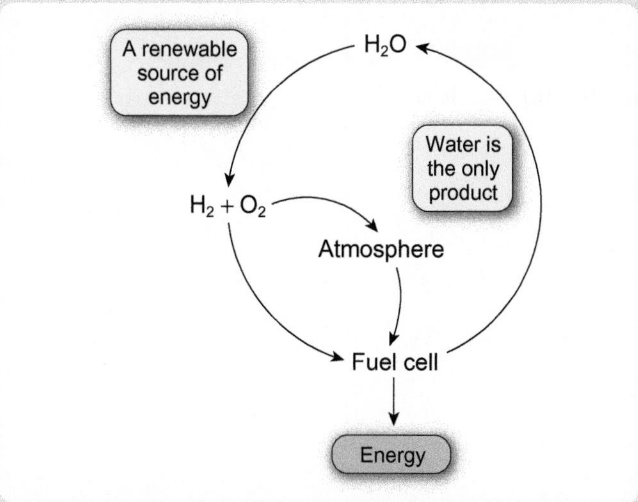
A renewable source of energy
H_2O
Water is the only product
$H_2 + O_2$
Atmosphere
Fuel cell
Energy

▲ The hydrogen economy requires a renewable source of H_2, which is then converted into water in fuel cells producing electricity.

▲ The Honda FCX Clarity is the world's first commercially produced car that is powered by hydrogen fuel cells.

Question

The enthalpy change of combustion for methane is −890 kJ mol^{-1} at 298 K. Use this to calculate the energy density for methane and compare it to the value given for hydrogen.

An alternative means of producing hydrogen is the decomposition of water. This is of great interest because it offers a way to produce hydrogen without also producing the greenhouse gas CO_2 (see Box 25.1). Although there is an almost limitless supply of water, this reaction involves breaking strong O–H bonds, so it requires a source of energy. Water decomposes to the elements on heating to above 2000 °C, but this temperature is far too high for the reaction to be commercially viable. The splitting of water into hydrogen and oxygen can be carried out using electrolysis

$$2H_2O(l) \xrightarrow{\text{electrolysis}} 2H_2(g) + O_2(g)$$

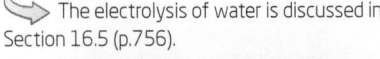

The electrolysis of water is discussed in Section 16.5 (p.756).

Electrolysis is currently a more expensive way to make H_2 than steam reforming, and it is only economically viable when electricity is cheap, such as in hydroelectric power plants.

Sunlight is another potential source of energy for the decomposition of water. This reaction occurs in plants as one of the key steps in photosynthesis, and scientists are trying to mimic this process in the laboratory. The photochemical splitting of water into oxygen and hydrogen has been achieved using titanium dioxide doped with platinum as a catalyst. Many scientists believe this will be an important source of H_2 in the future, but the process is not yet efficient enough to be used commercially.

The reaction between zinc and dilute hydrochloric acid is used to generate hydrogen.

In the laboratory, hydrogen is prepared from the reaction between an electropositive metal such as zinc and a dilute acid, in a similar reaction to that discovered by Boyle over 300 years ago

$$Zn(s) + 2H_3O^+(aq) \rightarrow Zn^{2+}(aq) + H_2(g) + 2H_2O(l)$$

In principle, any metal with a negative standard reduction potential will react with an acid to generate H_2. Some metals, such as aluminium, do not normally react with acids despite the negative standard reduction potential. Aluminium has a coating of unreactive oxide (Al_2O_3) on the surface and this physically prevents the reaction from occurring. The standard reduction potentials for a range of common metals are given in Table 25.1.

Standard reduction potentials, $E^{\ominus}$, are explained in Section 16.3, p.740.

Table 25.1 Standard reduction potentials for selected metals*

	$E^{\ominus}$/V	
$Au^{3+}(aq) + 3e^- \rightarrow Au(s)$	+1.50	These metal ions are reduced by $H_2(g)$ to give the metal.
$Hg^{2+}(aq) + 2e^- \rightarrow Hg(l)$	+0.86	
$Ag^+(aq) + e^- \rightarrow Ag(s)$	+0.80	
$Cu^{2+}(aq) + 2e^- \rightarrow Cu(s)$	+0.34	
$2H_3O^+(aq) + 2e^- \rightarrow H_2(g) + 2H_2O(l)$	0.00	
$Sn^{2+}(aq) + 2e^- \rightarrow Sn(s)$	−0.14	These metals are oxidized by acid to give the metal ion and $H_2(g)$. In some cases reaction is normally prevented by a coating of oxide on the metal surface.
$Ni^{2+}(aq) + 2e^- \rightarrow Ni(s)$	−0.25	
$Fe^{2+}(aq) + 2e^- \rightarrow Fe(s)$	−0.44	
$Zn^{2+}(aq) + 2e^- \rightarrow Zn(s)$	−0.76	
$Al^{3+}(aq) + 3e^- \rightarrow Al(s)$	−1.66	
$Mg^{2+}(aq) + 2e^- \rightarrow Mg(s)$	−2.37	
$Na^+(aq) + e^- \rightarrow Na(s)$	−2.71	
$Ca^{2+}(aq) + 2e^- \rightarrow Ca(s)$	−2.87	
$K^+(aq) + e^- \rightarrow K(s)$	−2.93	
$Li^+(aq) + e^- \rightarrow Li(s)$	−3.05	

* More standard reduction potentials are given in Table 16.2 (p.742).

Uses of hydrogen

Over half of the hydrogen produced worldwide is used in the fixation of atmospheric nitrogen. The Haber process involves the conversion of nitrogen and hydrogen into ammonia (NH_3).

The equilibrium reaction in the Haber process is discussed in Section 1.9 (p.57) and explored in more depth in Box 15.6 (p.720).

$$N_2(g) + 3H_2(g) \rightleftharpoons 2NH_3(g)$$

Another large-scale industrial use of hydrogen is the conversion of synthesis gas into methanol. This reaction is carried out at high pressure (250 atm) with a catalyst that consists of aluminium oxide pellets that contain small quantities of copper oxide and zinc oxide

$$2H_2(g) + CO(g) \xrightleftharpoons{330^\circ C} CH_3OH(g)$$

$$3H_2(g) + CO_2(g) \rightleftharpoons CH_3OH(g) + H_2O(g)$$

Hydrogen is widely used in the petrochemical industry for hydrogenation and desulfurization reactions. For example, benzene is converted into cyclohexane by hydrogenation, and this is used in the manufacture of nylon. Hydrogenation is a commonly used process in synthetic organic chemistry, and the reaction is usually catalysed by nickel or palladium. In the food industry, the hydrogenation of alkenes is used in the conversion of unsaturated fats and oils into saturated ones. Saturated fats are more stable to oxidation and have higher melting points than unsaturated fats, making them more convenient to use. An example is the conversion of vegetable oils into margarine and other spreads.

For more on the hydrogenation of fats see Box 2.4 (p.86).

Hydrogen is used to extract metals such as copper from their ores. This process involves dissolving ores containing copper(II) oxide, CuO, and copper(II) sulfide, CuS, in sulfuric acid, then passing hydrogen through the solution to reduce the Cu^{2+} ions to copper

$$Cu^{2+}(aq) + H_2(g) + 2H_2O(l) \rightarrow Cu(s) + 2H_3O^+(aq)$$

This reaction is the reverse of that between an electropositive metal and an acid to generate H_2, and is only possible for metals, such as copper, that have a positive standard reduction potential.

Summary

- Hydrogen does not occur native on Earth as H_2 gas. Hydrogen generally occurs bonded to oxygen in water, or bonded to carbon in organic compounds.
- The major industrial source of H_2 is from the reaction of hydrocarbons with steam (steam reforming), but in the future production from the electrolysis of water is likely to become more important.
- Hydrogen gas is used industrially to make ammonia and methanol, and in the reduction of unsaturated organic compounds and metal ions.

For practice questions on these topics, see questions 1 and 2 at the end of this chapter (p.1165)

25.2 Compounds of hydrogen

Hydrogen forms binary compounds with most of the other elements of the Periodic Table. These compounds, called **hydrides**, have a wide diversity of properties and structures. Lithium hydride (LiH) is an ionic solid that reacts violently with water to give an alkaline solution. In contrast, hydrogen chloride (HCl) is a gas at room temperature, and it dissolves in water to give an acidic solution.

A **binary hydride** contains hydrogen and one other element.

The key to understanding the wide differences in the chemistry of the binary hydrides lies in the electronegativity of hydrogen, which has a value of $\chi^P = 2.20$. When hydrogen forms a binary compound with a more electropositive element such as lithium, the hydrogen atom is described as **hydridic** and has an oxidation state of −1. The binary hydrides of Groups 1 and 2 are hydridic and are mostly ionic solids.

Ionization in solution only occurs with protic solvents, as a protic solvent is needed to solvate the anions. See Box 7.1 (p.305) for more details.

Ionization enthalpy and electron gain enthalpy are introduced in Section 3.7 (p.153). Electronegativity and bond polarity are discussed in Section 4.3 (p.177). The chemistry of acids is described in Chapter 7.

When hydrogen forms a binary compound to a more electronegative atom such as chlorine, the bond is polarized with the hydrogen atom δ+. In these cases the hydrogen atom is described as **protic** and has an oxidation state of +1. The hydrides of Groups 15, 16, and 17 are generally protic. The ionization enthalpy of hydrogen is very high (+1312 kJ mol^{-1}) and much larger than the electron gain enthalpy (–73 kJ mol^{-1}). This means that it is difficult to produce an H^+ ion, so these protic hydrides are covalent rather than ionic. Indeed, ions are only formed when the compound is dissolved in a solvent that is able to solvate the protons, and when this occurs the compounds act as acids. For example, when HCl dissolves in water it forms H_3O^+(aq) and Cl^-(aq)

$$HCl\,(aq) + H_2O\,(l) \rightarrow Cl^-\,(aq) + H_3O^+\,(aq)$$

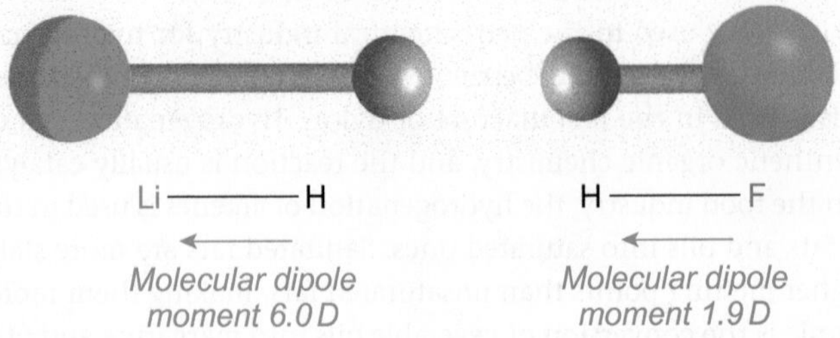
Li——H
H——F
Molecular dipole moment 6.0 D
Molecular dipole moment 1.9 D

The direction of the dipole moment is different in the molecules LiH and HF, due to the relative electronegativities of the two atoms. LiH forms molecules in the gas phase, but is an ionic solid in the solid state.

When hydrogen is bonded to an element with a similar electronegativity, the bonds are largely non-polar, but may have a small dipole in which the hydrogen atom can be either δ+ or δ–. In methane, for example, the hydrogen atoms carry small δ+ charges, whereas in diborane (B_2H_6) the hydrogen atoms carry small δ– charges.

The systematic names for hydride compounds depend on their protic or hydridic character. Hydridic compounds are simply called **hydrides**, for example, lithium hydride (LiH). When the hydrogen atom is protic or the X–H bond is almost non-polar, the systematic name for the compound ends in *-ane*. SiH_4 and GeH_4 are called silane and germane, respectively, and these are also their common names. For many of the hydrides, the older common names are in more general use than the systematic names, so you need to be familiar with them. For example, PH_3 is usually called phosphine rather than phosphane, whereas water and ammonia are hardly ever referred to by systematic names. The Group 17 hydrides and most Group 16 hydrides have two-word names with 'hydrogen' followed by the other element with an *-ide* ending, for example, hydrogen chloride and hydrogen sulfide.

Covalent hydride compounds can be divided into three classes, depending on the number of valence electrons present.

3-centre 2-electron bonds are discussed in Section 5.7 (p.250).

- In an **electron-precise compound**, all the valence electrons of the central atom are involved in forming bonds. The Group 14 hydrides are electron precise.
- In an **electron-deficient compound**, it is not possible to draw the structure using only 2-centre 2-electron bonds. 3-centre 2-electron bonds are present. BeH_2 and the Group 13 hydrides are electron deficient.
- In an **electron-rich compound**, not all of the electrons on the central atom are involved in bonding, so lone pairs are present. This means that electron-rich compounds can act as Lewis bases. The hydrides of Groups 15–17 are electron rich.

The simplest hydride formed by each *s*- and *p*-block element is shown in Figure 25.1, together with their classification as ionic, electron precise, electron deficient, or electron rich. Periodic trends in the behaviour of the binary hydrides are given at the end of this section (p.1154), whereas periodicity itself is described in Box 25.2.

Group 1	2	3–12	13	14	15	16	17
LiH lithium hydride	BeH_2 beryllium hydride		B_2H_6 diborane	CH_4 methane	NH_3 ammonia	H_2O water	HF hydrogen fluoride
NaH sodium hydride	MgH_2 magnesium hydride		AlH_3 aluminium hydride	SiH_4 silane	PH_3 phosphine	H_2S hydrogen sulfide	HCl hydrogen chloride
KH potassium hydride	CaH_2 calcium hydride	Sc–Zn	Ga_2H_6 gallium hydride	GeH_4 germane	AsH_3 arsine	H_2Se hydrogen selenide	HBr hydrogen bromide
RbH rubidium hydride	SrH_2 strontium hydride	Y–Cd	InH_3 indium hydride	SnH_4 stannane	SbH_3 stibine	H_2Te hydrogen telluride	HI hydrogen iodide
CsH caesium hydride	BaH_2 barium hydride	La–Hg	TlH_3 thallium hydride	PbH_4 plumbane	BiH_3 bismuthine		

- Ionic hydrides
- Electron-deficient covalent hydrides
- Electron-precise covalent hydrides
- Electron-rich covalent hydrides
- Hydride is not well characterized

Figure 25.1 The most common hydrides of the *s*- and *p*-block elements.

Box 25.2 Periodicity

Periodicity is a general term to describe the trends and recurrent variations that occur with increasing atomic number. Periodicity affects both physical and chemical properties, and the effects are observed for elements and their compounds. A good example in this chapter is the behaviour of the hydrides across a period (this is summarized on p.1154). Periodic trends arise from regular variations in atomic structure, which themselves lead to the form taken by the Periodic Table, as shown on the inside front cover. Regular changes in properties are seen going across the periods and down the groups of the Periodic Table.

Periodicity pervades inorganic chemistry, and for that reason it is not treated in a separate chapter in this book—instead, aspects of periodicity are integrated into all of the relevant chapters.

The table below lists the key sections in the book that deal with periodicity.

Section 3.7 (p.153)	Atomic properties and periodicity
Section 4.3 (p.177)	Electronegativity
Section 6.7 (p.298)	Predicting bond types
Section 7.7 (p.334)	Acidic and basic oxides
Section 25.2 (p.1143)	Compounds of hydrogen
Section 26.9 (p.1194)	Lithium and beryllium as exceptional elements
Section 26.11 (p.1196)	Diagonal relationships
Section 27.1 (p.1203)	General aspects and trends in the *p* block
Section 28.1 (p.1258)	The *d*-block elements
Section 28.2 (p.1263)	Chemistry of the first row *d*-block elements

s-block hydrides

Group 1 hydrides (MH)

The Group 1 hydrides have the general formula MH, and are formed by combination of the elements at high temperature. For example

$$2\,Na\,(s) + H_2(g) \rightarrow 2\,NaH\,(s)$$

These compounds are largely ionic, forming colourless, high-melting solids that have the NaCl structure.

Ionic compounds are discussed in Chapter 6. The NaCl structure is described in Section 6.4 (p.277).

The Group 1 hydrides dissolve in molten alkali halides and these solutions undergo electrolysis. For example, NaH dissolves in molten NaCl. When the liquid is electrolysed, sodium metal forms at the cathode and hydrogen gas is given off at the anode.

The polarizability of ions is explained in Section 6.6 (p.296). Lattice energies and the Born–Landé equation are introduced in Section 6.6 (p.293).

The H^- ion is very polarizable, so it is easily deformed by cations with high charge density. This means that the ionic radius of the hydride ion varies considerably with the cation and the combination of H^- with the small polarizing Li^+ cation leads to a high degree of covalent character in lithium hydride (LiH). As a result, the experimentally determined lattice energy of LiH is much greater than that predicted using the Born–Landé equation.

The Group 1 hydrides are strongly basic and react rapidly with water forming hydrogen and an alkaline solution

$$NaH(s) + H_2O(l) \rightarrow NaOH(aq) + H_2(g)$$

Group 1 hydrides: ionic structure (M^+H^-), though LiH has a high degree of covalent character.

The reactivity increases down the group. For example, NaH reacts more violently than LiH, and can ignite in moist air. This means that Group 1 hydrides need to be used under an inert atmosphere. The hydrides are usually sold as dispersions in oil, as these are easier to handle. You can obtain the pure hydride by washing the dispersion with pentane to remove the oil.

Group 2 hydrides (MH_2)

Group 2 chemistry is described in more detail in Chapter 26 (p.1183).

In Group 2, the chemistry of beryllium is often different from that of the other elements, and this is true of the synthesis and nature of the hydrides, MH_2.

Most Group 2 elements react with hydrogen to form the hydrides, for example

$$Mg(s) + H_2(g) \rightarrow MgH_2(s)$$

Beryllium does not react in this way, and BeH_2 is prepared from the decomposition of beryllium alkyl compounds. For example, when *t*-butylberyllium is heated to 210 °C it decomposes to BeH_2 and 2-methylpropene

The symbol Δ over the arrow stands for 'heat'.

$$Be\{C(CH_3)_3\}_2(s) \xrightarrow{\Delta} BeH_2(s) + 2(CH_3)_2C{=}CH_2(g)$$

In the gas phase, BeH_2 forms linear molecules. The bonding in these is described in Sections 5.4 (p.242) and 5.6 (p.246). 3-centre 2-electron bonding is explained in Section 5.7 (p.250).

BeH_2 has a three-dimensional covalent polymeric structure, a portion of which is shown in Figure 25.2. It is a colourless solid, which decomposes on heating above 250 °C. In this compound, the hydrogen atoms bridge between beryllium atoms using 3-centre 2-electron bonding.

The diagonal relationship between Li and Mg is described further and explained in Section 26.11 (p.1196).

The other Group 2 hydrides are crystalline, ionic solids, though the bonding in MgH_2 has a large degree of covalent character, similar to that in LiH. Similarities in the chemistry of lithium and magnesium are an illustration of the **diagonal relationship** between these elements.

BeH_2 is relatively stable in water, but the other Group 2 hydrides all react to give hydrogen gas and alkaline solutions. For example,

$$CaH_2(s) + 2H_2O(l) \rightarrow Ca(OH)_2(aq) + 2H_2(g)$$

Calcium hydride is used in laboratories as a drying agent for organic solvents such as ethanenitrile (acetonitrile, MeCN) and dichloromethane (CH_2Cl_2).

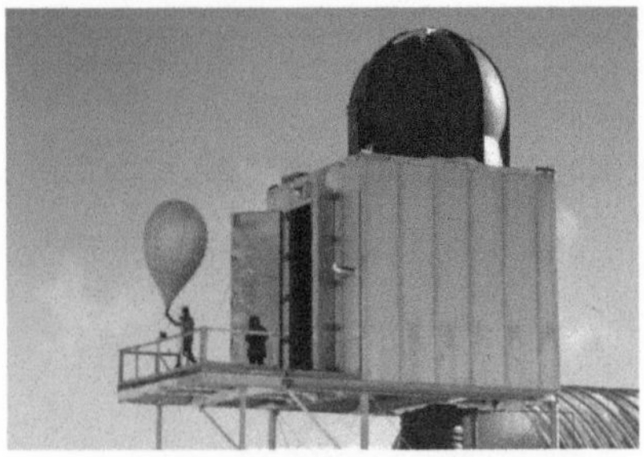

The reaction between CaH_2 and water is used at remote locations to generate the hydrogen required to fill weather balloons. This photo was taken at a weather station in Antarctica.

Group 2 hydrides: ionic structure ($M^{2+}(H^-)_2$) except for covalent BeH_2. MgH_2 has a high degree of covalent character. BeH_2 has a polymeric structure with bridging hydrogen atoms.

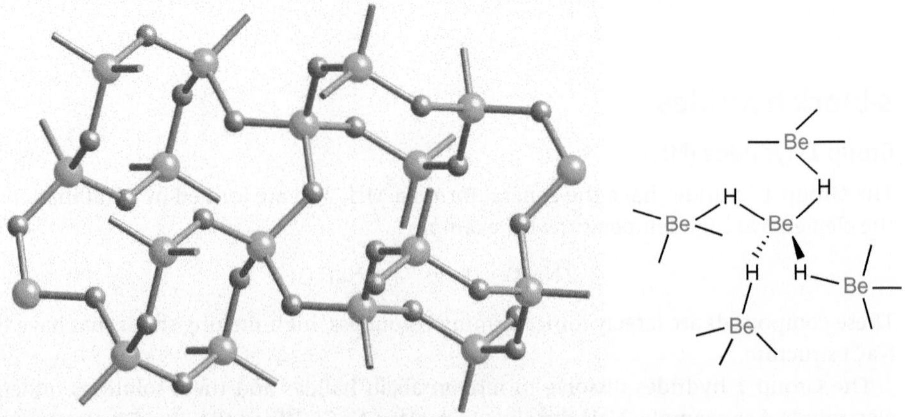
Be
H
H
Be
Be
H
H
Be
Be

Figure 25.2 Solid BeH_2 forms a three-dimensional polymeric structure with bridging hydrogen atoms.

Box 25.3 Hydrogen storage

The energy density, in kJ per gram, of hydrogen in high (see Box 25.1), but the low density of hydrogen gas means that its energy density in terms of *volume*, in kJ per dm^3, is low. This means that fuel tanks would need to be very large to store hydrogen gas. To get around this, vehicles such as the Honda FCX Clarity store hydrogen under pressure, though even at 340 atm the amount of hydrogen that can be stored is limited. An alternative way of storing hydrogen is within a stable compound that can react to release H_2 gas when it is needed.

One class of compounds that is being tested for hydrogen storage is the *metallic hydrides* of transition metals. Many *d*-block metals form non-stoichiometric compounds with hydrogen, and these are called metallic as they conduct electricity. Zirconium, for example, forms the hydride ZrH_x ($x = 1.3–1.8$). This has the fluorite (CaF_2) structure (see Section 6.4, p.278) with only some of the anion sites filled by hydrogen atoms.

In many metallic hydrides the hydrogen atoms move between the available sites in the compound. On heating, the compounds release H_2. Together with the variable composition, these properties make metallic hydrides attractive for hydrogen storage. These hydrides store hydrogen very efficiently: for example the compounds $FeTiH_2$ and $LaNi_5H_6$ both contain more hydrogen atoms per cm^3 than liquid hydrogen. Magnesium is also attractive as it can absorb 7.7% of its own mass in hydrogen to form MgH_2. This is higher than the US Department of Energy's ultimate target of 7.5% for hydrogen storage materials, which was set to provide the same driving distance per refill of fuel as in today's petrol or diesel vehicles. Unfortunately, the absorption and desorption of hydrogen in the interconversion of Mg and MgH_2 are too slow to be efficient, and the percentage does not include the mass of the container. These rates of absorption and desorption are improved by alloying the magnesium with nickel to give the metallic hydride Mg_2NiH_4 but this reduces the percentage mass of hydrogen that can be absorbed.

One of the major drawbacks of using metallic hydrides for storing hydrogen is their limited lifetime. The materials irreversibly absorb impurities in the hydrogen gas, and this reduces the number of sites available for hydrogen absorption. Other hydrides that are being looked at for hydrogen storage are borohydrides, such as $NaBH_4$ and $Mg(BH_4)_2$, and ammonia borane (NH_3BH_3).

Another approach is to develop porous materials that are capable of storing gases such as hydrogen within the pores. Coordination network structures are currently attracting research interest for this (see Box 28.4, p.1178).

Question

The alloys Mg_2Ni, FeTi, and $LaNi_5$ absorb hydrogen to form the metallic hydrides Mg_2NiH_4, $FeTiH_2$, and $LaNi_5H_6$, respectively. Work out the percentage of its own mass of hydrogen that each alloy absorbs. Do these compounds meet the US Department of Energy's target?

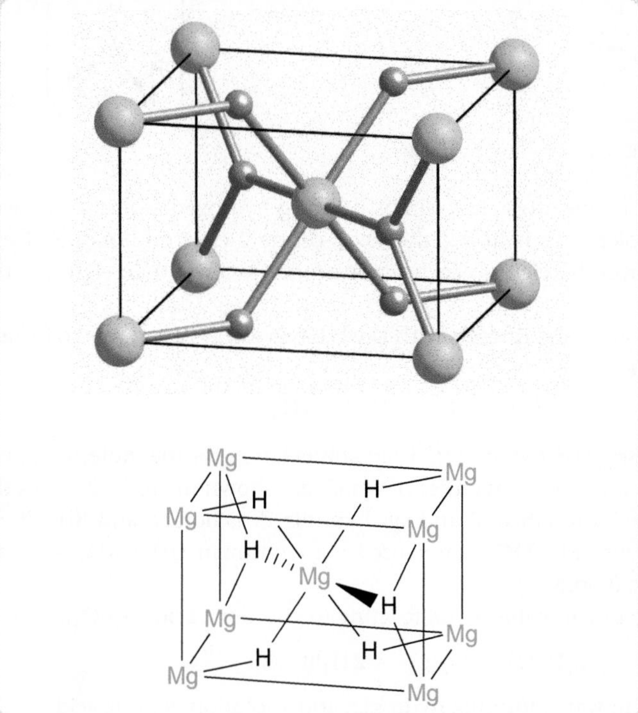
Mg
H
Mg
Mg
H
H
Mg
Mg
H
Mg
H
Mg
H
H
Mg
Mg

◄ Magnesium can absorb 7.7% of its own mass in hydrogen to form magnesium hydride, MgH_2, which has a similar structure to rutile (see Figure 6.26, p.283).

p-Block hydrides

Group 13 hydrides (XH_3 or X_2H_6)

The most common hydride known for each of the first three Group 13 elements is shown in Table 25.2 together with some properties.

The simplest boron hydride is borane (BH_3). The molecule has a trigonal planar structure in which the boron atom is sp^2 hybridized and has a vacant *p* orbital. Although BH_3 has been observed in the gas phase, it readily dimerizes and is normally isolated as diborane (B_2H_6).

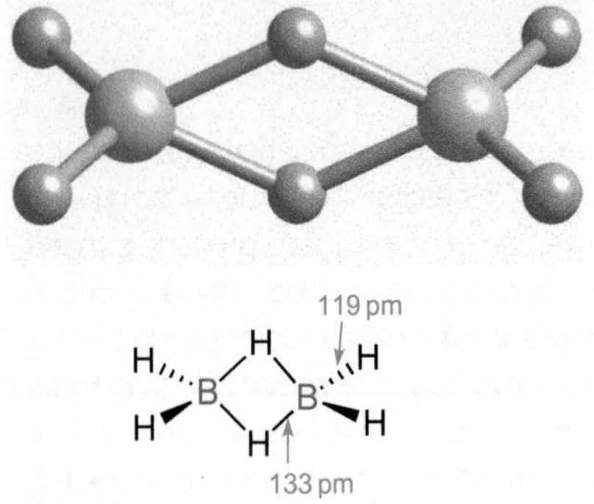
119 pm
133 pm

Figure 25.3 The structure of diborane (B_2H_6).

The bonding in B_2H_6 is described in Section 5.7 (p.250) and the use of B_2H_6 in the hydroboration of alkenes is described in Section 21.3 (p.982).

Lewis acids and bases are described in Section 7.8 (p.336).

Table 25.2 The Group 13 hydrides

Compound	m.p./°C	b.p./°C	$\Delta_f H^{\ominus}$/kJ mol^{-1}
Diborane (B_2H_6)	−165	−92	+35.6 (g)
Aluminium hydride (AlH_3)	150*		−46.0 (s)
Gallium hydride (Ga_2H_6)	−20	~0	

* AlH_3 decomposes at this temperature. The enthalpy change of formation, $\Delta_f H^{\ominus}$, is defined in Section 13.3 (p.624).

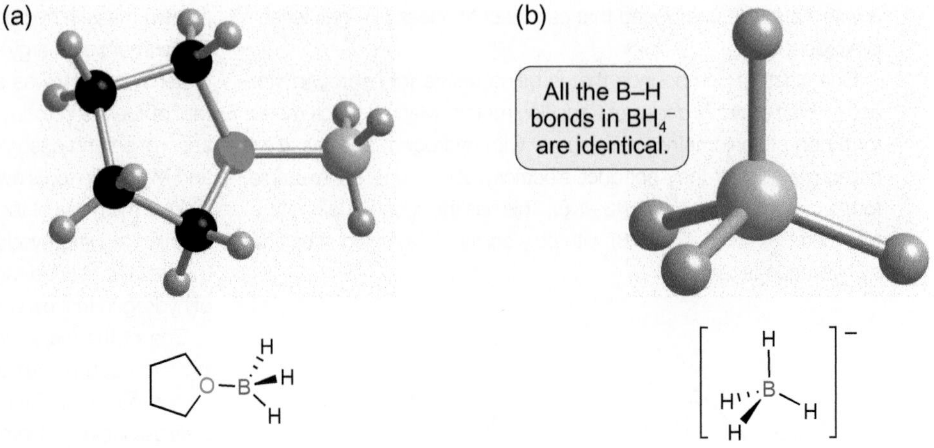
(a)
(b)
All the B–H bonds in BH_4^- are identical.

Figure 25.4 The structures of (a) $BH_3{\cdot}THF$ and (b) BH_4^-.

B_2H_6 is a gas at room temperature. It contains bridging hydrogen atoms, as shown in Figure 25.3, and the molecule is held together by 3-centre 2-electron bonding. B_2H_6 is often described as **electron deficient** as each boron atom has fewer than eight electrons in its valence orbitals.

B_2H_6 is a strong Lewis acid, and it reacts with Lewis bases such as ammonia to form stable adducts. For example,

$$B_2H_6 + 2NH_3 \rightarrow 2H_3\overset{-}{B}-\overset{+}{N}H_3$$

Other adducts include $BH_3{\cdot}THF$ (where THF is an abbreviation for the molecule tetrahydrofuran) and the BH_4^- anion, the structures of which are shown in Figure 25.4. In these adducts, the boron atom is tetrahedral and sp^3 hybridized. Diborane and $BH_3{\cdot}THF$ are both strong reducing agents. $BH_3{\cdot}THF$ is often used as a solution in THF, and this is a more convenient reagent than diborane gas.

B_2H_6 is spontaneously inflammable in air, reacting to give boric acid (H_3BO_3).

$$B_2H_6(g) + 3O_2(g) \rightarrow 2H_3BO_3(s)$$

B_2H_6 hydrolyses rapidly in water, forming hydrogen and a solution of boric acid

$$B_2H_6(g) + 6H_2O(l) \rightarrow 2H_3BO_3(aq) + 6H_2(g)$$

Boron forms a large number of other boranes with the general formulae B_nH_{n+4} and B_nH_{n+6}. For example, when B_2H_6 is heated to 100 °C, it forms decaborane ($B_{10}H_{14}$), the structure of which is shown in Figure 25.5.

Decaborane is a solid at room temperature and is stable in air. Generally, boranes become less flammable with increasing molecular mass.

The borohydride ion (BH_4^-) is isoelectronic with CH_4 though it is chemically very different. BH_4^- (see Figure 25.4) is hydridic and acts as a convenient source of H^- ions. The compound sodium borohydride, $NaBH_4$, is commonly used in organic chemistry to reduce aldehydes and ketones. $NaBH_4$ is more commonly used than simple hydrides such as NaH as it is more soluble in organic solvents and less basic, which reduces the side reactions.

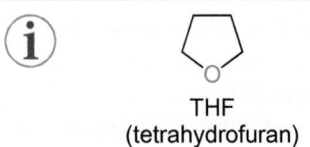

THF
(tetrahydrofuran)

Tetrahydrofuran (THF) is a cyclic ether (see Section 2.6, p.94).

The acidity of boric acid is described in Section 27.2 (p.1212) and its solid state structure is described in Section 25.3 (p.1159).

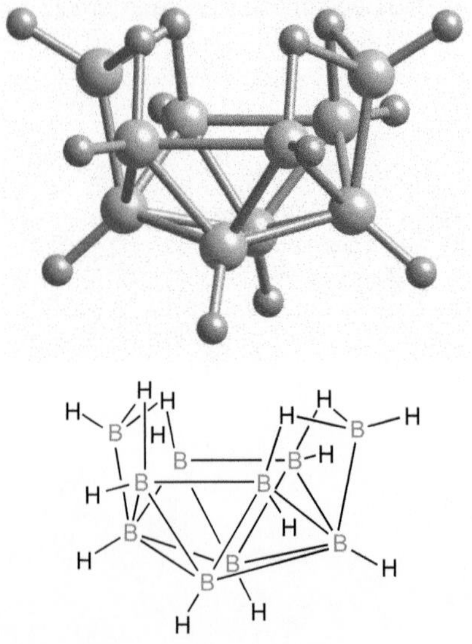

Figure 25.5 The structure of decaborane ($B_{10}H_{14}$).

Table 25.3 The Group 14 hydrides

Compound	m.p./°C	b.p./°C	$\Delta_f H^\ominus$/kJ mol^{-1}	X–H bond length/pm
Methane (CH_4)	−182	−164	−74.8 (g)	108.7
Silane (SiH_4)	−185	−112	+34.3 (g)	148.0
Germane (GeH_4)	−165	−88	+90.8 (g)	152.5
Stannane (SnH_4)	−146	−52	+162.8 (g)	171.1

The hydrides of the other Group 13 elements are not as stable, nor as structurally diverse as the boron hydrides. Aluminium hydride (AlH_3) is a polymer with bridging hydrogen atoms, and it is unstable above 150 °C. Gallium hydride was first prepared in the 1990s. Gallium hydride contains Ga_2H_6 molecules analogous to diborane, but it is unstable at room temperature. Indium and thallium hydrides have only been observed at very low temperatures, but have been characterized by infrared spectroscopy.

The reduction of carbonyl compounds with $NaBH_4$ is described in Section 23.2 (p.1062).

Generally, the decomposition temperatures of the hydrides decrease as the group is descended, and the compounds become more reactive. The trend in stability is related to the decrease in X–H bond enthalpies going down the group. This also affects the reactivity of the compounds and, for example, lithium aluminium hydride ($LiAlH_4$) is a much stronger reducing agent than $NaBH_4$.

Group 13 hydrides exist as covalently bonded dimers (B_2H_6, Ga_2H_6) or polymers (AlH_3). The stability of the compounds decreases going down the group.

Group 14 hydrides (XH_4)

Carbon forms an unlimited number of hydrides, the **hydrocarbons**, and these form the basis of organic chemistry. The structures of methane and ethane are described in Chapter 5, and the structure and chemistry of hydrocarbons are covered in Chapters 2 and 18.

Group 14 hydrides exist as XH_4 covalently bonded molecules. The stability of these hydrides decreases down the group.

The other Group 14 elements form tetrahydrides analogous to methane (Table 25.3). Silane (SiH_4), germane (GeH_4), and stannane (SnH_4) are all known, but plumbane (PbH_4) is less stable and consequently less well characterized. These compounds all have positive values of $\Delta_f H^\ominus$ and cannot be formed by direct combination of the elements.

Silane is prepared by reaction of silicon tetrachloride with a hydride source. The following reaction is carried out in diethyl ether

$$SiCl_4 + LiAlH_4 \rightarrow SiH_4 + AlCl_3 + LiCl$$

Unlike methane, silane spontaneously combusts in air to form silicon dioxide and water

$$SiH_4(g) + 2O_2(g) \rightarrow SiO_2(s) + 2H_2O(l)$$

Silane, germane, and stannane decompose to the elements on heating. This is particularly important for silane, as the reaction is used in the purification of silicon for the semiconductor industry

$$SiH_4(g) \xrightarrow{500\,°C} Si(s) + 2H_2(g)$$

Generally, the hydrides become less stable with respect to their elements as the group is descended. So, while SiH_4 decomposes at 500 °C, SnH_4 decomposes at room temperature.

Silane hydrolyses slowly in water, though more rapidly in the presence of acid. Higher silanes and germanes also exist, with chains of up to ten silicon or germanium atoms. Compounds such as disilane (Si_2H_6) and trigermane (Ge_3H_8) have similar structures to their carbon analogues ethane and propane. In line with the chemistry of their tetrahydrides, they are far more reactive than their carbon counterparts.

Group 15 hydrides (XH_3)

The Group 15 elements all form molecular trihydrides of the general formula XH_3. The compounds, together with some of their properties are given in Table 25.4. Ammonia, NH_3, is produced from the reaction between nitrogen and hydrogen in the Haber process (see Box 15.6, p.720). The annual worldwide production of NH_3 is over 140 million tonnes, the majority of which (~85%) is used to manufacture fertilizers, though nylon and nitric acid production are also significant.

Group 15 hydrides exist as XH_3 covalently bonded molecules. The stability and basicity of these hydrides decreases down the group.

Table 25.4 The Group 15 hydrides

Compound	m.p./°C	b.p./°C	$\Delta_f H^\ominus$/kJ mol^{-1}	X–H bond length/pm	H–X–H bond angle/°
Ammonia (NH_3)	−78	−33	−46.1 (g)	101.2	106.7
Phosphine (PH_3)	−133	−88	+5.4 (g)	142.0	93.3
Arsine (AsH_3)	−116	−55	+66.4 (g)	151.1	92.1
Stibine (SbH_3)	−92	−17	+145.1 (g)	170.4	91.6
Bismuthine (BiH_3)	−67	+17	+278 (g)	177.6	90.5

NH_3 is a polar covalent molecule. It acts as a Lewis base and, in aqueous solution, as a Brønsted–Lowry base

$$NH_3\,(aq) + H_2O\,(l) \rightleftharpoons NH_4^+\,(aq) + OH^-\,(aq)$$

Although it is normally regarded as a base, ammonia can also act as a Brønsted–Lowry acid. In liquid ammonia there is a degree of self-ionization into NH_4^+ and NH_2^- ions

$$2NH_3\,(l) \rightleftharpoons NH_4^+\,(am) + NH_2^-\,(am) \quad K_{am} = 1 \times 10^{-33}\,mol^2\,dm^{-6} \text{ at } -50\,°C$$

Liquid ammonia is a useful non-aqueous solvent, but because the boiling point of ammonia is −33 °C, it can only be handled below this temperature unless high pressure is used. Reactions are normally carried out using a propanone/dry ice bath to maintain the temperature below −33 °C. The Group 1 metals dissolve in liquid ammonia to form blue solutions containing solvated electrons.

Phosphine (PH_3), arsine (AsH_3), and stibine (SbH_3) all exist as colourless gases, with the stability of the molecules decreasing down the group. Bismuthine (BiH_3) is only stable below −45 °C. Phosphine is produced by the reaction of calcium phosphide (Ca_3P_2) with water

$$Ca_3P_2\,(s) + 6H_2O\,(l) \rightarrow 2PH_3\,(g) + 3Ca(OH)_2\,(aq)$$

Pure PH_3 ignites in air at 150 °C, but it is usually formed along with P_2H_4, which spontaneously ignites at room temperature.

In contrast to NH_3, the other Group 15 trihydrides are poor Lewis bases. This is related to the molecular structures of the compounds, and can be explained using valence bond theory. As the data in Table 25.4 show, the H–X–H angle decreases going down the group. The angle in NH_3 is consistent with sp^3 hybridization, which means that the lone pair is in a directional sp^3 hybrid orbital, and is available to donate to a Lewis acid. The angles in PH_3, AsH_3, and SbH_3 are much closer to 90°, which suggests that hybridization is less important for these compounds. This means that the *p* orbitals are used to form the bonds, and the lone pair is in an unhybridized *s* orbital. Since the *s* orbital is not directional, the lone pair is less available for donation to a Lewis acid. The difference in shape between NH_3 and PH_3 is illustrated in Figure 25.6.

Nitrogen, phosphorus, and arsenic also form compounds of the general formula X_2H_4, containing an X–X bond. N_2H_4 is known as **hydrazine**, and is a liquid at room temperature (m.p. 1.4 °C, b.p. 114 °C). The most favourable conformation of N_2H_4 is shown in Figure 5.21 (p.234).

Hydrazine is produced industrially in the Raschig reaction, which involves oxidation of ammonia with hypochlorite ions under alkaline conditions

$$2NH_3\,(aq) + ClO^-\,(aq) \xrightarrow{\text{base}} N_2H_4\,(aq) + Cl^-\,(aq) + H_2O\,(l)$$

Hydrazine has a high energy density, and reacts exothermically with oxygen

$$N_2H_4\,(l) + O_2\,(g) \rightarrow N_2\,(g) + 2H_2O\,(g) \quad \Delta H^\ominus = -622\,kJ\,mol^{-1}$$

This reaction in aqueous solution is used to remove dissolved oxygen from industrial water boilers. These boilers operate at high temperatures and pressures, and under these conditions oxygen is corrosive to the metal.

The ability of ammonia to act as a Brønsted–Lowry base is described in Section 7.1 (p.306) and as a Lewis base in Section 7.8 (p.336).

(am) indicates dissolved in liquid ammonia. It is analogous to (aq) which indicates dissolved in water. K_{am} is the self-ionization constant of ammonia, which is defined in an analogous manner to the self-ionization constant of water in Section 7.2 (p.313).

The reaction of Group 1 metals with liquid ammonia is described in more detail in Section 26.5 (p.1182).

FIFA

Smelling salts contain ammonium carbonate, which slowly decomposes to ammonia. The ammonia fumes irritate the membranes of the nose and lungs. This triggers a 'wake up' reflex that causes the muscles controlling breathing to work faster. In the 2006 FIFA World Cup Final, France's Thierry Henry is given smelling salts after receiving a head injury early in the match.

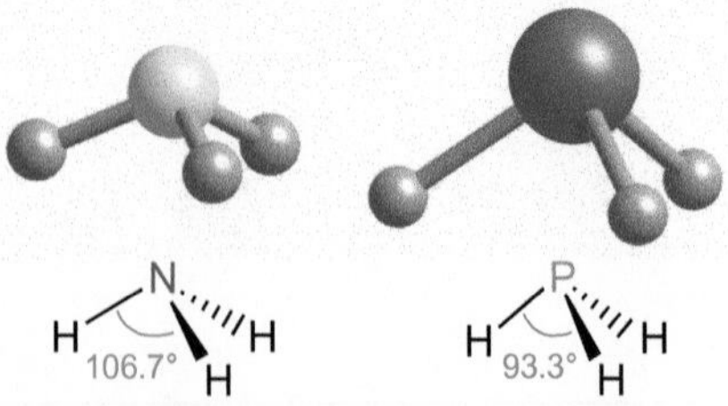
N
H
H
H
106.7°
P
H
H
H
93.3°

Figure 25.6 The structures of ammonia (NH_3) and phosphine (PH_3).

The trigonal pyramidal shape of the Group 15 hydrides is predicted by VSEPR theory (see Section 5.2, p.223). The valence bond approach to the bonding in ammonia is described in Section 5.4 (p.238).

Table 25.5 The Group 16 hydrides

Compound	m.p./°C	b.p./°C	$\Delta_f H^{\ominus}$/kJ mol^{-1}	X–H bond length/pm	H–X–H bond angle/°
Water (H_2O)	0	100	−285.8 (l)	95.8	104.5
Hydrogen sulfide (H_2S)	−85	−61	−20.6 (g)	133.6	92.1
Hydrogen selenide (H_2Se)	−66	−41	+29.7 (g)	146.0	90.9
Hydrogen telluride (H_2Te)	−49	−2	+99.6 (g)	169.0	90.0

Liquid hydrazine is used as a fuel in satellites and Earth orbiter vehicles, though in this case the oxidizing agent is N_2O_4, not O_2

$$2N_2H_4 + N_2O_4 \rightarrow 3N_2 + 4H_2O$$

Many satellites contain hydrazine fuel tanks to help maintain the correct altitude. The malfunctioning military satellite USA-193, shown being launched, was destroyed by a missile in 2008 to prevent the hydrazine fuel tank posing a danger on falling to Earth.

Group 16 hydrides (H_2X)

The Group 16 elements form dihydrides (H_2X). Some molecular and physical properties of these compounds are given in Table 25.5.

Water (H_2O) is a liquid at room temperature, whereas the other dihydrides are gases. The high melting point and boiling point of water and its unusually wide liquid range are a result of strong hydrogen bonding (see Section 25.3, p.1157). Water is amphoteric, being able to act as both an acid and a base. The self-ionization reaction occurs to a small extent in water generating H_3O^+(aq) and OH^-(aq)

$$2H_2O\,(l) \rightleftharpoons H_3O^+(aq) + OH^-(aq)$$

Water can act as a mild reducing agent and, on reaction with a strong oxidizing agent, the oxygen atom is oxidized to O_2. For example, water is oxidized by F_2

$$2F_2(g) + 6H_2O\,(l) \rightarrow O_2(g) + 4H_3O^+(aq) + 4F^-(aq)$$

Water can also act as an oxidizing agent and, on reaction with a reducing agent, the hydrogen atoms are reduced to H_2. An example is the reaction with a Group 1 metal such as sodium

$$2Na\,(s) + 2H_2O\,(l) \rightarrow 2NaOH\,(aq) + H_2(g)$$

The high relative permittivity (also called dielectric constant) of water makes it an excellent solvent for ionic compounds and for polar covalent molecules.

The self-ionization of water is described in Section 7.2 (p.313).

The relative permittivity of a solvent is defined in Section 17.3 (p.784). Values for different solvents are listed in Table 20.2 (p.937).

The Group 16 hydrides H_2S, H_2Se, and H_2Te are toxic gases at room temperature. They become increasingly less stable down the group, and H_2Se and H_2Te are both unstable with respect to decomposition into the elements. H_2S has a characteristic smell of rotten eggs, as it is formed from the decomposition of sulfur-containing egg proteins. The human nose can detect H_2S at concentrations as low as 0.5 ppb, but at higher concentrations it paralyses the olfactory nerve, so at lethal concentrations it is odourless.

Hydrogen sulfide is prepared in the laboratory by the reaction of iron(II) sulfide with acid

$$FeS\,(s) + 2H_3O^+(aq) \rightarrow H_2S\,(g) + Fe^{2+}(aq) + 2H_2O\,(l)$$

H_2S is a weak dibasic acid in water

$$H_2S\,(aq) + H_2O\,(l) \rightleftharpoons HS^-(aq) + H_3O^+(aq)$$

$$HS^-(aq) + H_2O\,(l) \rightleftharpoons S^{2-}(aq) + H_3O^+(aq)$$

A **dibasic acid** has two ionizable hydrogen atoms. See Section 7.6 (p.332) for details.

The acidity of the Group 16 hydrides increases down the group. For example, pK_{a1} for H_2S is 6.88, but pK_{a1} for H_2Se is 3.89. Compare these with pK_{a1} for H_2O, which is 15.7.

H_2S burns to produce sulfur in a restricted amount of air, or SO_2 in an excess of air

$$8H_2S\,(g) + 4O_2(g) \rightarrow S_8\,(s) + 8H_2O\,(g)$$

$$2H_2S\,(g) + 3O_2(g) \rightarrow 2SO_2(g) + 2H_2O\,(g)$$

As the data in Table 25.5 and Figure 25.7 show, the bond angles for H_2S, H_2Se, and H_2Te are all much less than the 104.5° bond angle observed in H_2O. This difference can be

Hydrogen sulfide is one of the gases dissolved in molten rock, and it is released by volcanoes such as Mount Rainier in the USA.

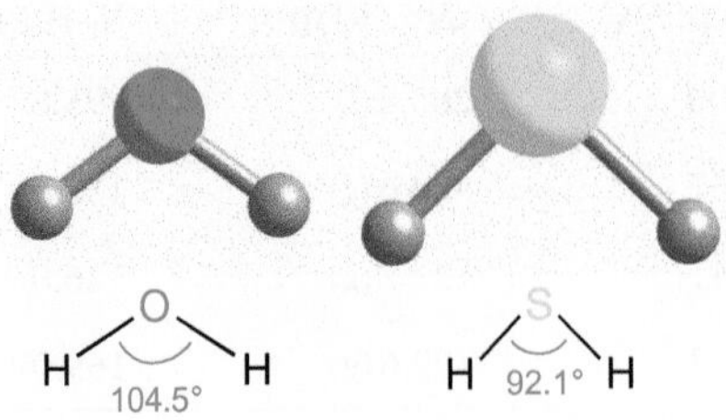
O
H
H
104.5°
S
H
H
92.1°

Figure 25.7 The molecular structures of water (H_2O) and hydrogen sulfide (H_2S).

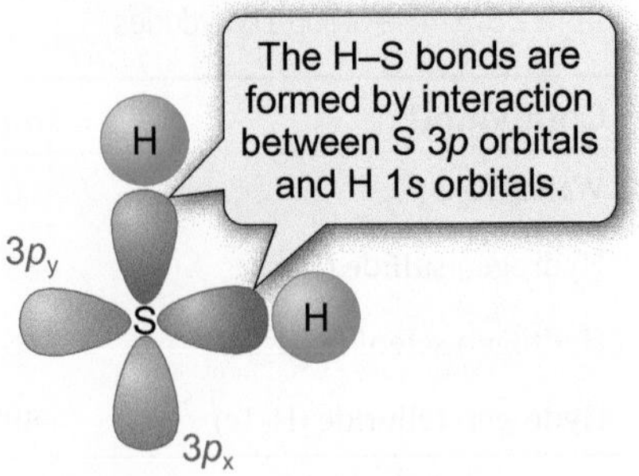
The H–S bonds are formed by interaction between S 3*p* orbitals and H 1*s* orbitals.
H
3*p*y
S
H
3*p*x

Figure 25.8 The lower bond angle in H_2S, compared to H_2O, can be explained by the lack of hybridization in the bonding.

The **torsion angle** in the four atom chain ABCD is the angle between the AB and DC bonds. This is equal to the angle between the ABC and BCD planes. This is illustrated for H_2O_2 in Figure 25.9.

explained in a similar way to the trend in bond angles for the Group 15 hydrides (p.1150) by using valence bond theory. In H_2O, the angle suggests the oxygen atom is sp^3 hybridized, with the slight decrease from the ideal tetrahedral angle due to repulsions between the lone pairs. The observed angle in H_2S suggests the sulfur atom is not hybridized.

The S–H bonds are formed instead by the interaction of *p* orbitals with hydrogen 1*s* orbitals, as seen in Figure 25.8.

Oxygen forms a second hydride, **hydrogen peroxide** (H_2O_2), which contains an O–O bond. Hydrogen peroxide is a pale blue viscous liquid when it is pure (m.p. –0.4 °C, b.p. 150 °C), but it is normally used as an aqueous solution. Hydrogen peroxide has the structure shown in Figure 25.9. The molecule is non-planar, and has a torsion angle of 90° in the solid state.

H_2O_2 is a very weak acid

$$H_2O_2\,(aq) + H_2O\,(l) \rightleftharpoons HO_2^-\,(aq) + H_3O^+\,(aq) \quad pK_{a1} = 11.8$$

but it is a much stronger oxidizing agent than water. For example, it oxidizes iron(II) to iron(III)

$$2\,Fe^{2+}(aq) + H_2O_2\,(aq) + 2\,H_3O^+(aq) \rightarrow 2\,Fe^{3+}(aq) + 4\,H_2O\,(l)$$

H_2O_2 is thermodynamically unstable with respect to disproportionation

$$\overset{-I}{2\,H_2O_2}\,(aq) \rightarrow \overset{-II}{2\,H_2O}\,(l) + \overset{0}{O_2}\,(g)$$

H_2O_2 is kinetically stable at room temperature, but its decomposition is catalysed by platinum, MnO_2, or catalase enzymes in living tissues.

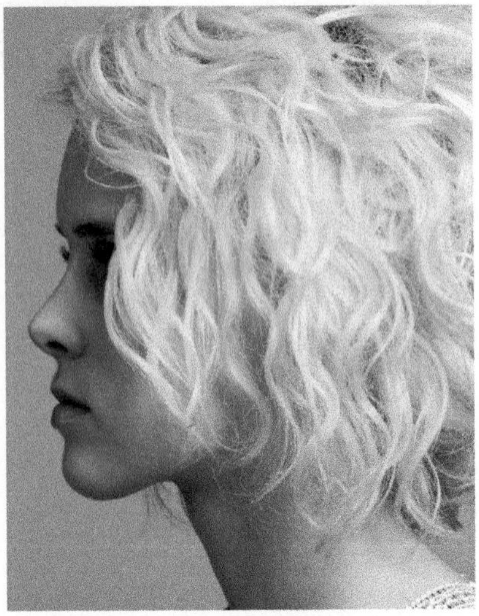

Aqueous solutions of hydrogen peroxide are used to bleach hair. H_2O_2 oxidizes eumelanin and phaeomelanin, the proteins that give hair its colour, into colourless forms.

A **disproportionation** reaction is one in which a compound undergoes oxidation and reduction simultaneously. You can think of it as self-oxidation and self-reduction.

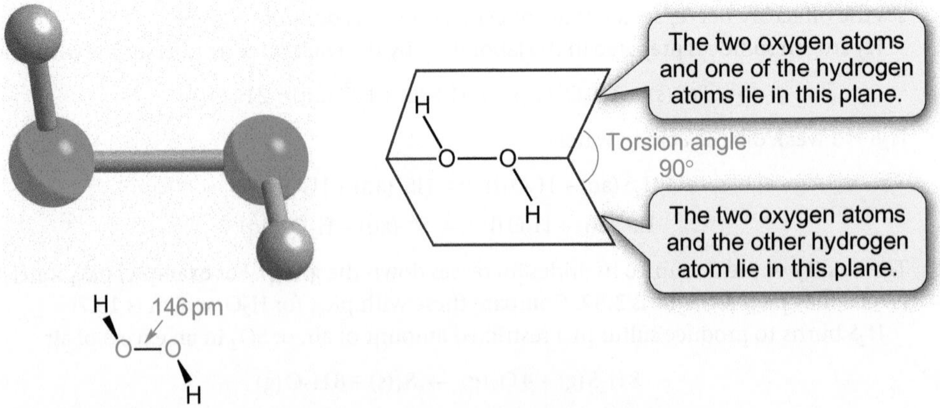
H
146 pm
O
O
H
The two oxygen atoms and one of the hydrogen atoms lie in this plane.
H
O
O
H
Torsion angle 90°
The two oxygen atoms and the other hydrogen atom lie in this plane.

Figure 25.9 The structure of hydrogen peroxide (H_2O_2). The torsion angle is the angle between the two OOH planes.

Group 16 hydrides exist as H_2X covalently bonded molecules. The stability of these hydrides decreases down the group.

Table 25.6 The Group 17 hydrides

	m.p./°C	b.p./°C	$\Delta_f H^{\ominus}$/kJ mol^{-1}	pK_a	X–H bond length/pm	D(H–X)/kJ mol^{-1}
Hydrogen fluoride (HF)	–83	20	–271.1 (g)	3.2	91.7	+570
Hydrogen chloride (HCl)	–115	–85	–92.3 (g)	–7	127.5	+431
Hydrogen bromide (HBr)	–88	–67	–36.4 (g)	–9	141.5	+366
Hydrogen iodide (HI)	–51	–35	+26.5 (g)	–10	160.9	+298

The sulfur analogue of hydrogen peroxide, H_2S_2, exists, as do other polysulfanes of the general formula H_2S_n (n = 3–8).

Group 17 hydrides (HX)

The Group 17 halides are gases with the general formula HX and are known as hydrogen halides. In aqueous solution they are referred to as hydrohalic acids. Some properties of these compounds are given in Table 25.6.

The hydrogen halides are formed by direct reaction of hydrogen and the appropriate halogen

$$H_2(g) + X_2(g) \rightarrow 2HX(g)$$

In the case of HF, the reaction is explosive. Mixtures of H_2 and Cl_2 are stable in the dark, but react explosively in the presence of light. Reactions of H_2 with Br_2 and I_2 are slower, requiring catalysts to get a reasonable rate, and for I_2 the reaction with H_2 does not go to completion.

The equilibrium between H_2, I_2, and HI is described in Section 1.9 (p.59).

In line with these observations, Table 25.6 shows that the enthalpy change of formation of HX becomes more positive as X increases in size. The main factor in this is the decrease in H–X bond dissociation enthalpies (D(H–X)) as the group is descended.

Section 13.3 (p.631) explains how to use bond dissociation enthalpies to calculate enthalpy changes of reaction.

The polarity of the H–X bond decreases down the group, so you might expect that the acidity would decrease in line with this. This is not the case, and the pK_a data in Table 25.6 shows that the acidity *increases* down the group. There are two factors leading to this trend. For the equilibrium

$$HX(aq) + H_2O(l) \rightleftharpoons X^-(aq) + H_3O^+(aq)$$

the decrease in H–X bond dissociation enthalpy makes the forward reaction from left to right *more favourable* as the group is descended. The larger the X^- anion, the weaker the attraction between X^- and H_3O^+. This means that the reverse reaction, from right to left, becomes *less favourable* as the group is descended. Both these factors contribute to the observed increase in the strength of the acids down the group, and together these outweigh the polarity effect. The large difference in pK_a between HCl and HF is due to very strong hydrogen bonding in hydrofluoric acid between the F^- and H_3O^+ ions, which reduces the concentration of free H_3O^+ ions in HF (aq). The hydrogen bonds in HF are particularly strong, and hydrogen bonding is even present in the gas phase, with the vapour containing a mixture of monomeric HF molecules plus hydrogen-bonded dimers and hexamers.

In liquid hydrogen halides there is a degree of self-ionization, in a similar manner to that in water and ammonia. In HF, both the cation and anion are solvated via a strong interaction with another HF molecule to give H_2F^+ and HF_2^- ions, respectively.

$$3HF(l) \rightleftharpoons H_2F^+(solv) + HF_2^-(solv) \quad K_{hf} = 2 \times 10^{11}\ mol^2\ dm^{-6} \text{ at } 0\ °C$$

K_{hf} is the self-ionization constant of hydrofluoric acid, which is defined in an analogous way to the self-ionization constant of water (see Section 7.2, p.313). The structure of HF_2^- is described in Section 25.3 (p.1156).

HCl is prepared by the reaction of sodium chloride with concentrated sulfuric acid

$$NaCl(s) + H_2SO_4(l) \rightarrow HCl(g) + NaHSO_4(s)$$

It is not possible to use analogous reactions to prepare HBr and HI. These hydrogen halides are oxidized by concentrated sulfuric acid, so phosphoric acid is used instead.

Group 17 hydrides exist as HX covalently bonded molecules. The stability of these hydrides decreases down the group and the acidity increases down the group.

Box 25.4 Hydrofluoric acid

HF may not be a strong acid, but it is an extremely reactive substance. Approximately 1 million tonnes of HF are prepared industrially each year worldwide by the reaction of concentrated sulfuric acid with calcium fluoride at 300 °C

$$CaF_2(s) + H_2SO_4(l) \rightarrow CaSO_4(s) + 2HF(g)$$

Although most HF is used in the manufacture of metal fluorides and organofluorine compounds, some is used in the etching of glass. When a glass surface is exposed to hydrofluoric acid it takes on a frosted appearance as the SiO_2 in the glass is converted into SiF_4 and H_2SiF_6

$$SiO_2(s) + 4HF(aq) \rightarrow SiF_4(g) + 2H_2O(l)$$

$$SiO_2(s) + 6HF(aq) \rightarrow H_2SiF_6(aq) + 2H_2O(l)$$

This process is used to frost the insides of light bulbs. Designs can be produced on glass surfaces by coating the object with wax, then inscribing a pattern through the wax layer. When the vessel is exposed to HF only the portion not covered by wax is etched.

Hydrofluoric acid must be handled with extreme care as it can cause severe burns. HF passes through the skin, so the resulting burns are deep and painful. In addition, there is usually a delayed onset to pain, which means that exposure may not be noticed at first. Once inside the body, the fluoride ions react with Mg^{2+} and Ca^{2+} ions, including those in bone, to form insoluble fluorides. The removal of Ca^{2+} ions from cell fluids results in a relative excess of K^+ ions, which causes nerve stimulation to heighten the pain.

▲ Hydrofluoric acid is used to etch glass. It has been used to create the frosted, semi-opaque appearance of these doors.

Question

Bearing in mind the reactivity of hydrofluoric acid, how would you expect it to be stored in the laboratory?

Periodic trends in the behaviour of the binary hydrides

Binary hydrides are often divided into classes based on their bonding. The Group 1 and 2 hydrides are known as **ionic hydrides** (or sometimes saline hydrides), the transition metal hydrides are known as **metallic hydrides**, and the Groups 13–17 hydrides are known as **covalent hydrides**, with the latter divided into electron-deficient, electron-precise, and electron-rich compounds depending on the number of electrons present (see Figure 25.1, p.1145). While there is some value in this approach, it is important to remember that there is a gradual variation in the properties across a row of the Periodic Table. As the electronegativity of the atom bonded to hydrogen increases, there is a change in nature of the hydride from hydridic to protic. The van Arkel–Ketelaar triangle in Figure 25.10 shows how the bonding character of the hydrides changes for the elements of the second period. The average electronegativity of the elements increases from the left to right, whereas the difference in electronegativity decreases to a minimum at carbon before increasing to HF, which shows considerable ionic character.

One effect of the changing nature from hydridic to protic across a row is that the hydrides become more acidic going from left to right across any row. For example, HCl is more acidic than H_2S, which in turn is more acidic than PH_3. Acidity of the hydrides also increases going down a group, as explained for the hydrohalic acids on p.1153.

The hydrides become less stable going down any group of the Periodic Table. In each group, the valence orbitals of the atoms get larger and more diffuse with increasing principal quantum number, *n*. This means that the interactions of these *ns* and *np* orbitals with the hydrogen 1*s* orbitals are reduced. As a consequence, the X–H bond enthalpies decrease down each group, as shown in Figure 25.11.

The data in Figure 25.11 also show that the X–H bond enthalpies tend to increase from left to right across each row of the Periodic Table. This is due to the increase in overlap between the orbitals caused by decreasing size of the X atom and an increase in ionic character of the bonds.

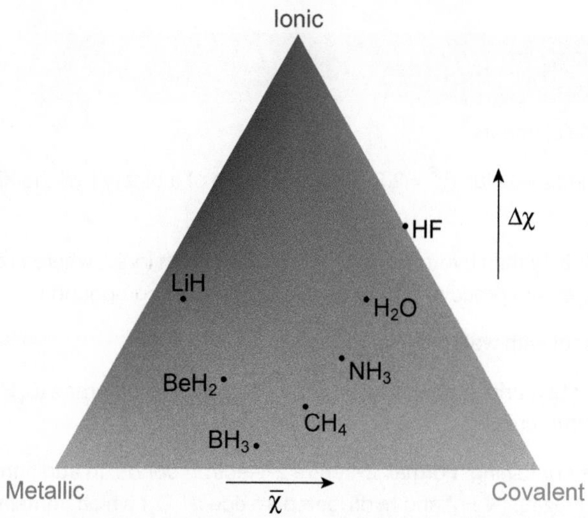
Ionic
HF
Δχ
LiH
H_2O
NH_3
BeH_2
CH_4
BH_3
Metallic
χ̄
Covalent

Figure 25.10 A van Arkel-Ketelaar triangle showing the bonding character of the hydrides for the second period elements. Although LiH is close to the metallic corner it does not have metallic properties, suggesting compounds are not metallic if the difference in electronegativity between the elements is quite high.

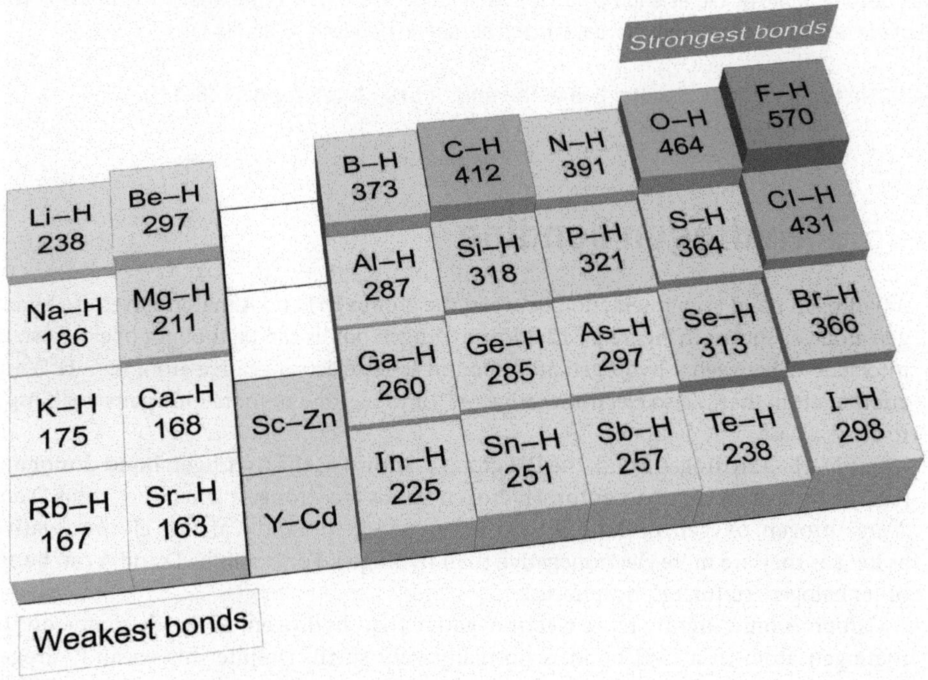
Strongest bonds
Li–H 238
Be–H 297
B–H 373
C–H 412
N–H 391
O–H 464
F–H 570
Na–H 186
Mg–H 211
Al–H 287
Si–H 318
P–H 321
S–H 364
Cl–H 431
K–H 175
Ca–H 168
Sc–Zn
Ga–H 260
Ge–H 285
As–H 297
Se–H 313
Br–H 366
Rb–H 167
Sr–H 163
Y–Cd
In–H 225
Sn–H 251
Sb–H 257
Te–H 238
I–H 298
Weakest bonds

Figure 25.11 Mean bond enthalpies for X-H bonds.

The elements of Groups 15, 16, and 17 in their high oxidation states do not form binary hydrides, despite the fact that equivalent halides such as SF_6 and ClF_3 are well established. One reason for this is the instability of the high oxidation state hydrides with respect to decomposition into a lower oxidation state hydride and H_2. For example, the reaction below shows the decomposition of the non-existent compound ClH_3 to form the lower oxidation state chloride HCl. This reaction is highly exothermic due to formation of very strong H–H bonds

$$\overset{-\text{III}}{ClH_3}(g) \rightarrow \overset{-\text{I}}{HCl}(g) + H_2(g)$$

The corresponding reaction for ClF_3 is less favourable as the F–F bond in the product F_2 is much weaker than the H–H bond.

The chemistry of ClF_3 is described in Box 5.3 (p.229).

Summary

- Hydrogen forms binary hydrides with most of the elements.
- The electronegativity of hydrogen has an intermediate value ($\chi^P = 2.20$), so the nature of a binary hydride XH_n depends on the electronegativity of the element X.
- Compounds with elements of lower electronegativity than hydrogen are hydridic and often ionic, whereas compounds with elements of higher electronegativity than hydrogen are protic and form covalently bonded compounds.
- Group 1 and 2 hydrides are mostly ionic, and react with water to give H_2.
- Group 13 hydrides are electron deficient and contain bridging hydrogen atoms, for example diborane (B_2H_6). 3-centre 2-electron bonding is required to explain their structures.
- Group 14–17 hydrides are covalent compounds containing 'normal' 2-centre 2-electron bonds. In addition to the simple hydrides XH_n, there are compounds such as hydrazine (N_2H_4) and hydrogen peroxide (H_2O_2) which contain X–X bonds. These compounds are electron precise (Group 14) or electron rich (Groups 15–17).
- For Groups 13–17, the stability of the hydrides decreases going down a group. This is caused by weakening of the X–H bonds, which is due to reduced orbital overlap as X gets larger.
- For Groups 15–17, the hydrides become more acidic as each group is descended. This is due to the weakening of the X–H bonds and the increasing size of the resulting anions, which weakens the attraction between X^- and H_3O^+.

? For practice questions on these topics, see questions 3–9 at the end of this chapter (pp.1165–1166).

Hydrogen bonds are introduced in Section 1.8 (p.54).

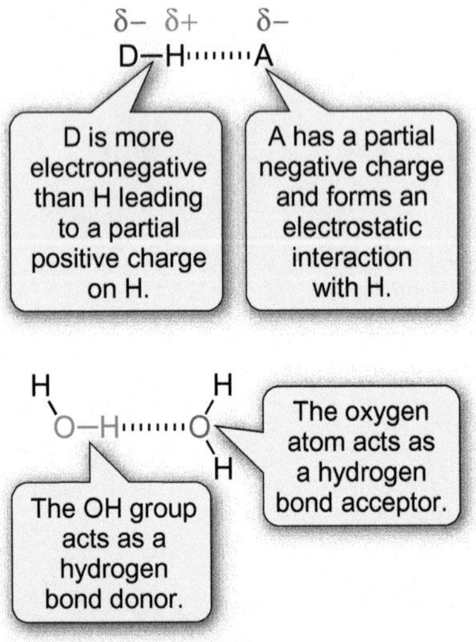
δ− δ+ δ−
D–H·······A
D is more electronegative than H leading to a partial positive charge on H.
A has a partial negative charge and forms an electrostatic interaction with H.
H
O–H·······O
H
H
The oxygen atom acts as a hydrogen bond acceptor.
The OH group acts as a hydrogen bond donor.

Figure 25.12 A hydrogen bond is an interaction between the groups DH and A that involves the hydrogen atom.

A hydrogen bond is defined by IUPAC as 'an attractive interaction between a hydrogen atom from a molecule or a molecular fragment X-H in which X is more electronegative than H, and an atom or a group of atoms in the same or a different molecule, in which there is evidence of bond formation.'

25.3 Hydrogen bonding

A hydrogen bond is an interaction between the groups DH and A that involves the hydrogen atom, as shown in Figure 25.12. Most hydrogen bonds can be thought of electrostatic interactions between a hydrogen atom bound to an electronegative atom ($D^{\delta-}$–$H^{\delta+}$) and another atom that is also electronegative ($A^{\delta-}$) and has one or more lone pairs enabling it to act as a base.

In a D–H···A hydrogen bond, the DH group is known as the **hydrogen bond donor** and A is the **hydrogen bond acceptor**. Hydrogen bonds are strongest when the atoms D and A are nitrogen, oxygen, or fluorine, but also occur when D and A are less electronegative, as long as they are more electronegative than hydrogen. For example, D and A can be another halogen, sulfur, or phosphorus.

Carbon is only slightly more electronegative than hydrogen, so the δ+ charge on the hydrogen atom in a C–H bond is normally very small. Despite this, weak hydrogen bonds in which the hydrogen bond donor is a C–H bond have been observed. This is most common when the carbon atom is also bonded to electronegative atoms, such as in $CHCl_3$, as this increases the size of δ+ on the hydrogen atom. Weak hydrogen bonds are also possible when the acceptor is not an electronegative atom with a lone pair as long as it has a filled orbital, such as a π-bonding orbital, that can interact with the hydrogen bond donor. There are C–H···π hydrogen bonds in the structure of solid ethyne, and these are responsible for the melting point of ethyne (−101 °C) being much higher than that of ethene (−169 °C).

In most hydrogen bonds the D–H distance is shorter than the A···H distance, but this is not true for the strongest hydrogen bonds. For example, hydrogen bonding between fluoride and HF gives HF_2^-, in which the hydrogen atom lies halfway between the two fluorine atoms. As the two H–F distances are the same, it is not possible to say which is the covalent H–F bond, and which is the H···F hydrogen bond. Unlike most hydrogen bonds, these very strong and symmetric hydrogen bonds have a high covalent character. The bonding in HF_2^- involves a 3-centre 4-electron interaction (Figure 25.13).

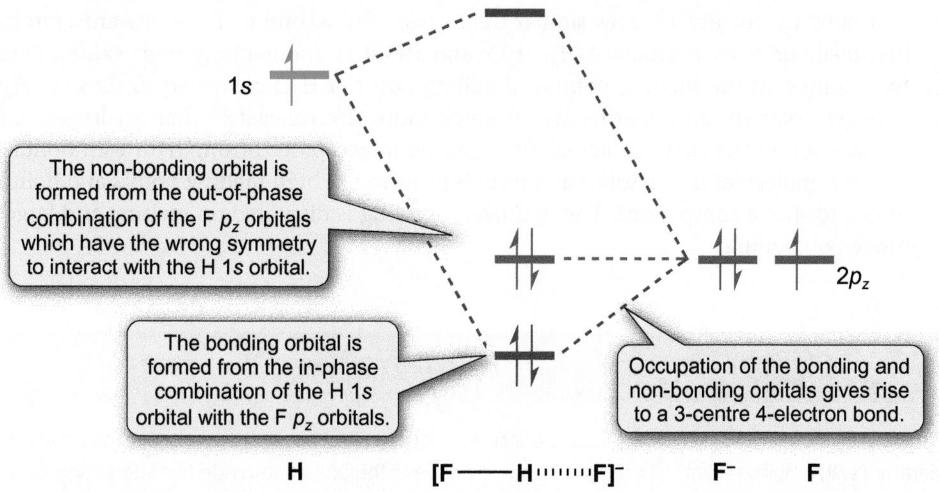
1s
The non-bonding orbital is formed from the out-of-phase combination of the F p_z orbitals which have the wrong symmetry to interact with the H 1s orbital.
The bonding orbital is formed from the in-phase combination of the H 1s orbital with the F p_z orbitals.
Occupation of the bonding and non-bonding orbitals gives rise to a 3-centre 4-electron bond.
$2p_z$
H
[F—H⋯F]⁻
F⁻
F

Figure 25.13 A partial molecular orbital energy level diagram showing the 3-centre 4-electron interaction in HF_2^-.

3-centre 4-electron interactions are discussed in Section 5.7 (p.249).

Hydrogen bonds vary considerably in strength, but a typical N–H⋯O or O–H⋯O hydrogen bond interaction has a bond enthalpy of between 20 kJ mol^{-1} and 30 kJ mol^{-1}, which is approximately one-tenth the strength of a typical covalent bond. This makes hydrogen bonds the strongest of the non-ionic intermolecular interactions.

The different non-covalent interactions are described in Section 1.8 (p.52).

Evidence for hydrogen bonds

Figure 25.14 shows the melting points and boiling points for the hydrides of Groups 14–17. Group 14 generally shows a steady increase for melting points and boiling points as the group is descended. This is due to the increase in the relative molecular mass of the molecules and the corresponding increase in London dispersion interactions between the molecules.

London dispersion interactions are introduced in Section 1.8 (p.52) and covered in more detail in Section 17.3 (p.786).

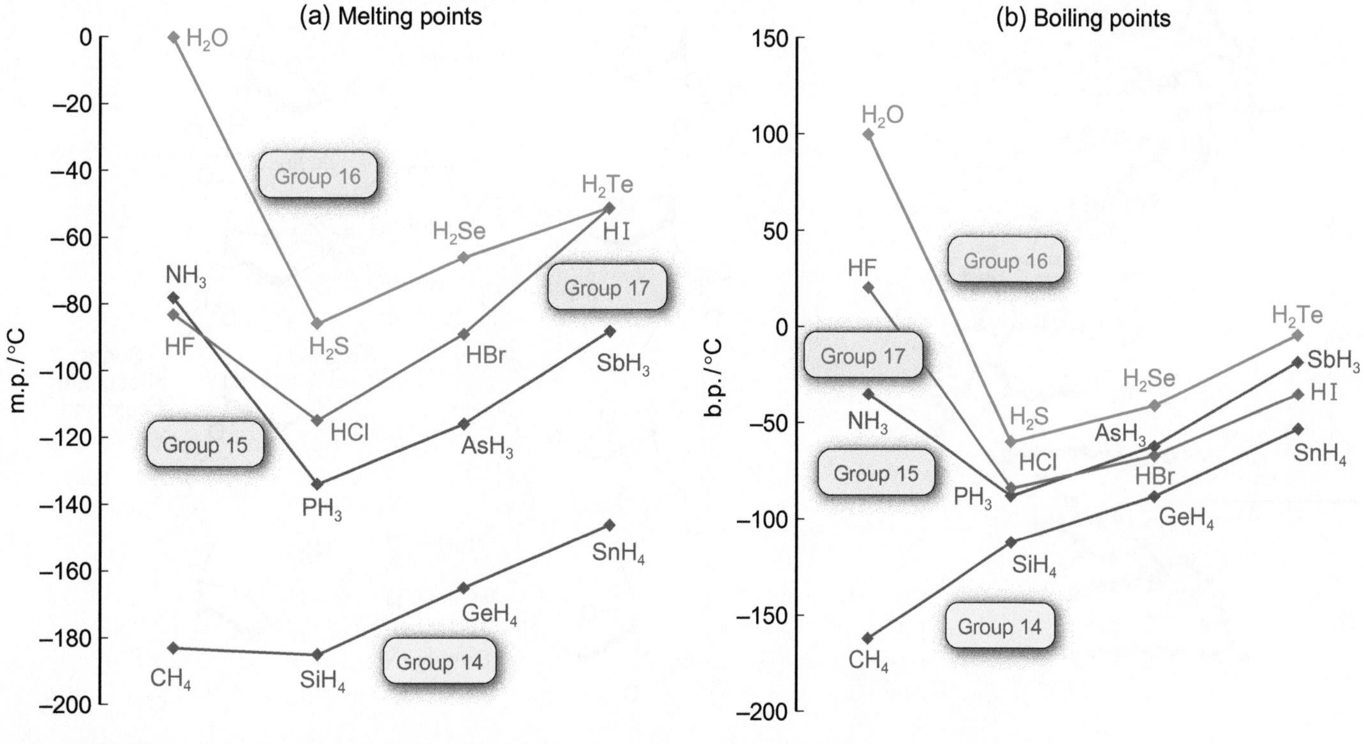
(a) Melting points
m.p./°C
0
−20
−40
−60
−80
−100
−120
−140
−160
−180
−200
H_2O
Group 16
H_2Te
H_2Se
HI
NH_3
Group 17
HF
H_2S
HBr
SbH_3
Group 15
HCl
AsH_3
PH_3
SnH_4
GeH_4
CH_4
SiH_4
Group 14
(b) Boiling points
b.p./°C
150
100
50
0
−50
−100
−150
−200
H_2O
HF
Group 16
Group 17
H_2Te
SbH_3
H_2Se
HI
NH_3
H_2S
AsH_3
SnH_4
HCl
Group 15
HBr
PH_3
GeH_4
SiH_4
Group 14
CH_4

Figure 25.14 (a) Melting points and (b) boiling points for the hydrides of Groups 14, 15, 16, and 17. The melting points and boiling points for NH_3, H_2O, and HF are anomalously high due to strong intermolecular hydrogen bonding.

The effect of hydrogen bonds on infrared spectra is described in Section 12.2 (p.572), and on NMR spectra in Section 12.3 (p.586).

Groups 15, 16, and 17 show similar trends from the second member onwards, but the first member of each group—NH_3, H_2O, and HF—has anomalously high values. Look, for example, at the melting point and boiling point of H_2O compared to those of H_2S. Nitrogen, oxygen, and fluorine are all much more electronegative than hydrogen, and all have lone pairs that can act as hydrogen bond acceptors. Strong hydrogen bonding between molecules in the solids and liquids leads to the high melting points and boiling points for these compounds. The hydrogen bonding is disrupted on melting and largely broken on boiling.

Box 25.5 Hydrogen bonding and life

You can see from Figure 25.14 that without hydrogen bonds water would be a gas at room temperature with a boiling point of about −75 °C. Life on Earth could not exist if this were the case. Life depends on hydrogen bonds.

Life has evolved to use hydrogen bonds in many ways. One of these is in the structure of deoxyribonucleic acid (DNA), which contains the genetic code of life (see Box 1.1, p.5). DNA molecules are long threads, built from a type of sugar molecule called deoxyribose and phosphate groups, which alternate along the chain. Each deoxyribose molecule in the chain is bonded to a nucleotide base. There are four different bases in DNA: adenine (A), cytosine (C), guanine (G), and thymine (T).

The structure of DNA consists of a right-handed double helix (see Box 19.4, p.883) in which two of the sugar–phosphate strands are wrapped around each other. The two strands are held together by hydrogen bonding between the bases, with adenine interacting with thymine, and cytosine interacting with guanine to form **base pairs**. There are approximately 3 billion base pairs in human DNA. The order of the bases in the chain provides the genetic code.

The structure of DNA was discovered in the 1950s. Rosalind Franklin and Maurice Wilkins were the first to determine from X-ray diffraction experiments that DNA has a helical structure, and subsequently James Watson and Francis Crick proposed the double helix and the existence of the base pairs.

▼ (a) The double helix of DNA. (b) A portion of DNA showing the hydrogen bonds between the base pairs.

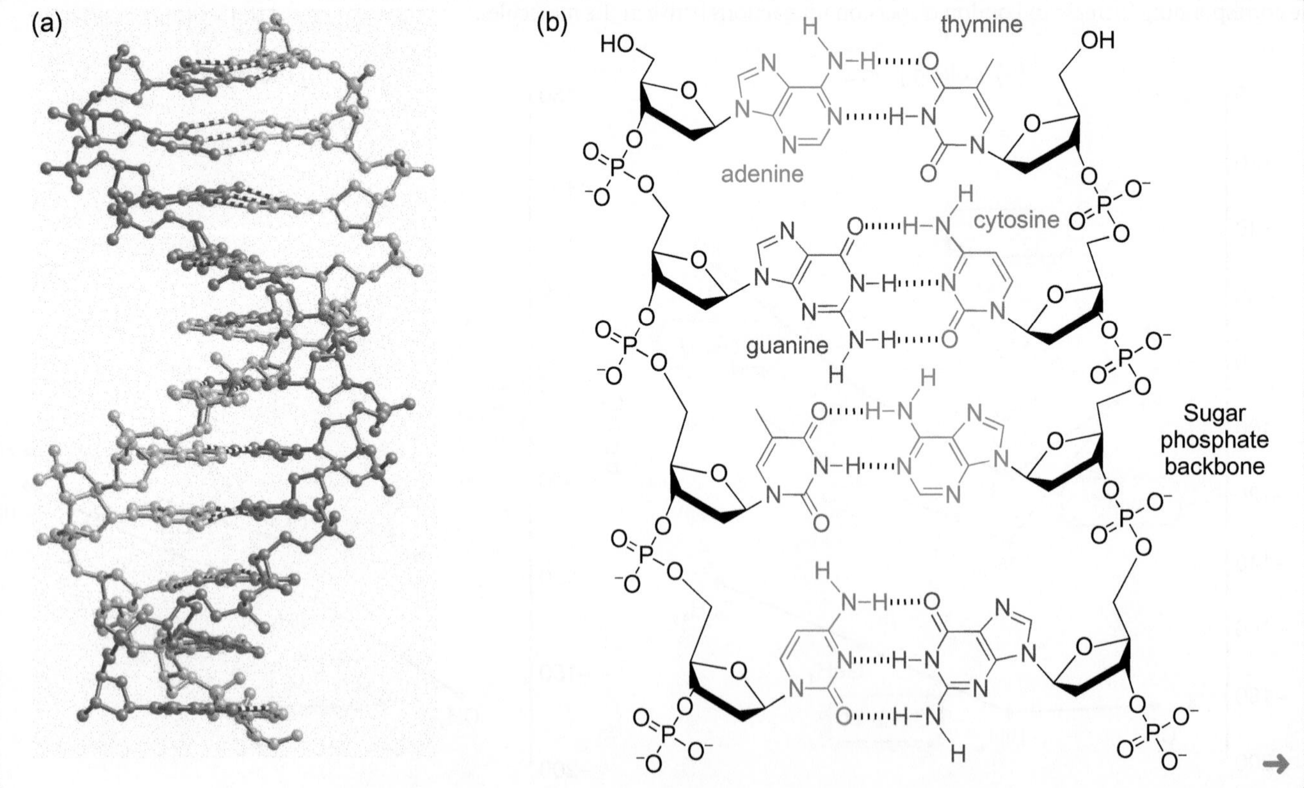
(a)
(b)
thymine
adenine
cytosine
guanine
Sugar phosphate backbone

→

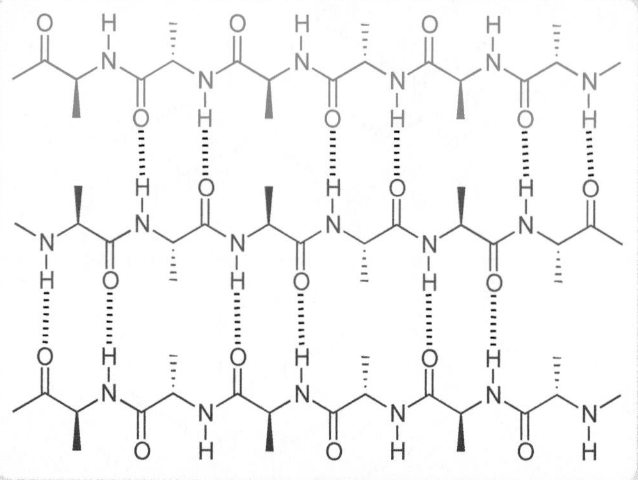

▲ The hydrogen bonding in the β-pleated sheet secondary structure of proteins.

▲ Spider silk contains a high proportion of β-pleated sheet regions which have a very regular structure. These sheets are separated by segments containing amino acids with bulky side chains, which are much less structured. The interplay between the hard, regular β-pleated sheet regions and the elastic, amorphous regions gives spider silk its strength and ability to stretch without breaking.

Consider a short segment of DNA, with the following sequence of bases

... ATCTGTAGCCTAAG ...
... TAGACATCGGATTC ...

The upper and lower strands are linked together by hydrogen bonds in the double helix so, for example, a C in one strand is always matched by a G in the other. In the process of replication, the two strands separate, and each acts as a template for the formation of a new strand, shown in red

... ATCTGTAGCCTAAG ... ... ATCTGTAGCCTAAG ...
... TAGACATCGGATTC ... ... TAGACATCGGATTC ...

After replication, there are two identical double helices.

Hydrogen bonds are also important in the three-dimensional structures of proteins. Proteins are polymers of amino acids, so they contain amide linkers between the monomers (see Chapter 14, p.655 and Section 24.2, p.1109). Hydrogen bonds occur between the N–H and carbonyl groups on different parts of the polymer backbone, and this leads to the secondary structure of the protein. The two most common hydrogen bond motifs are α-helices and β-pleated sheets. In β-pleated sheets, strands are joined by hydrogen bonds involving alternating amino acids on each strand.

Question

Why does cytosine form hydrogen bonds with guanine and not with adenine or thymine?

Hydrogen bonding in solids

Hydrogen bonding is important in the structures of many solids. HF forms hydrogen-bonded chains in the solid state, whereas boric acid (H_3BO_3) forms sheets. The compounds melamine and cyanuric acid crystallize together to form sheets in which the hydrogen bonding between the molecules is maximized. Melamine is used to test for cyanuric acid concentrations in swimming pools, as the hydrogen-bonded product is very insoluble. The turbidity of the solution after the melamine has been added is proportional to the cyanuric acid concentration. The structures of boric acid and melamine–cyanuric acid are shown in Figure 25.15.

The structure of solid HF is shown in Section 1.8 (p.54). The role of cyanuric acid in swimming pools is described in Box 7.4 (p.317).

Although H_2O is a simple molecule, the structure of ice is surprisingly complicated. Seventeen different crystalline ice structures are known, each of which is stable over a

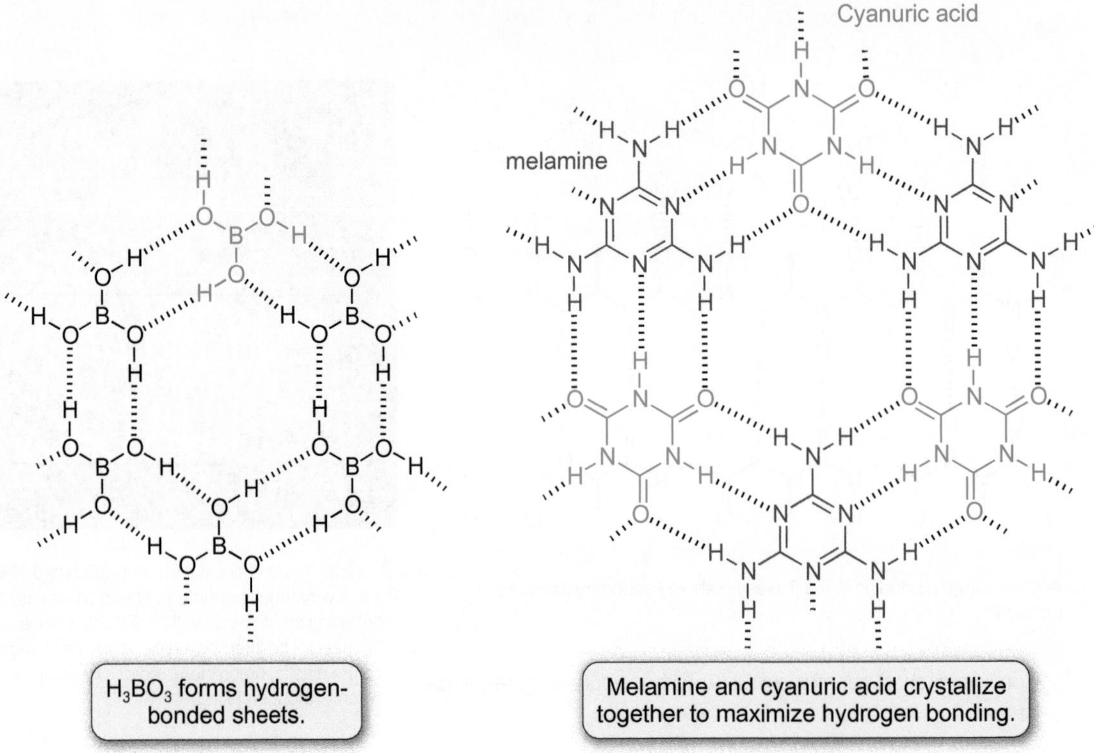
Cyanuric acid
melamine
H_3BO_3 forms hydrogen-bonded sheets.
Melamine and cyanuric acid crystallize together to maximize hydrogen bonding.

Figure 25.15 Hydrogen bonds in the solid state structures of H_3BO_3 and melamine–cyanuric acid.

The phase changes of water are described in Box 1.10 (p.50).

certain temperature and pressure range. Ordinary ice, which forms from water at 0 °C and atmospheric pressure, has a three-dimensional structure in which each oxygen atom is in an approximate tetrahedral environment, forming two covalent bonds and two hydrogen bonds. This gives rise to the very open network structure shown in Figure 25.16. As a result, ice is less dense than water so it floats on the surface. When ice melts, the density increases as some of the hydrogen bonds are broken and molecules move closer together.

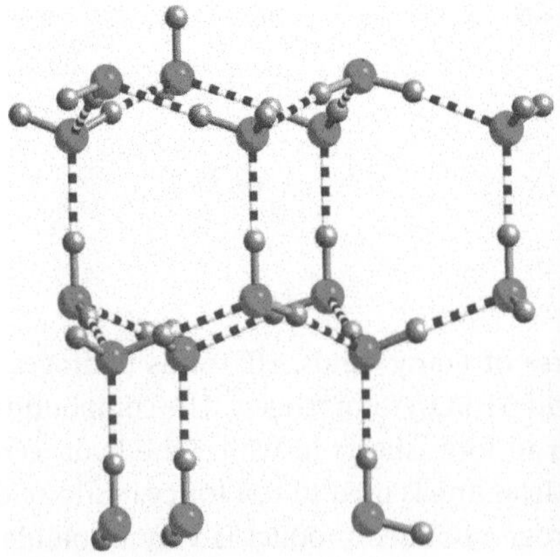

Figure 25.16 Ice has an open structure in which each water molecule contains two covalent OH bonds and accepts two hydrogen bonds from other water molecules. The hydrogen bonds are shown in blue and white.

The tip of an iceberg. Water is unusual in that the solid is less dense than the liquid. The difference in density is small so when ice floats, most of it is submerged.

Box 25.6 Burning ice

Imagine a form of ice that will burn when a lit match is held to it. This is methane hydrate, one of a class of compounds called **clathrate hydrates**. Clathrate hydrates are types of ice in which other molecules are trapped without being chemically bonded. Clathrate hydrates have larger void spaces than normal ice, and are only stable when the guest molecules occupy these voids. The structure of methane hydrate is made of linked polyhedra of water molecules. Each polyhedron contains 20 hydrogen-bonded water molecules forming a cage in which a methane molecule is trapped. If methane hydrate forms in a natural gas pipeline, it can lead to a dangerous blockage. To prevent this, pipelines in cold climates are insulated or chemical additives are used to stop methane hydrate from crystallizing.

Many different molecules can be included into clathrate hydrates. Cl_2 and SO_2 form clathrate hydrates at atmospheric pressure, but most gases only form hydrates under high pressure. Methane hydrate is stable at atmospheric pressure and low temperatures, though at high pressures it can exist far above the melting point of ice.

Deposits of methane clathrates occur naturally under the oceans at depths of over 300 m and also under permafrost. There is evidence that an abrupt warming of the Earth 55 million years ago was caused by the release of CH_4 from methane hydrate. Indeed, there is a fear amongst some environmentalists that global warming will lead to the release of methane from hydrates in the polar regions. Since methane is a greenhouse gas, rising concentrations in the atmosphere would cause an acceleration in global warming (see Chapter 9, p.383).

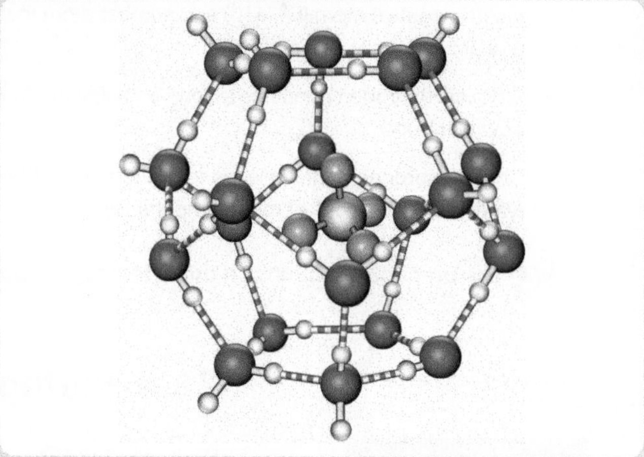

▲ Methane hydrate has a structure containing pentagonal dodecahedra of 20 water molecules. A CH_4 molecule, shown in grey and green, is present in each of the central cavities. Hydrogen bonds are shown in blue and white.

▲ In 1997, the biologist Charles Fisher discovered centipede-like worms living within mounds of methane hydrate on the floor of the Gulf of Mexico, at 700 m depth.

▲ Methane hydrate burning.

Question

Methane hydrate has the approximate formula $CH_4 \cdot 6.2\,H_2O$. What volume of methane could be released from 1 tonne of methane hydrate at 298 K? (1 mol of a gas has a volume of 24.5 dm^3 at this temperature and 1 atm pressure.)

Summary

- Most hydrogen bonds are best described as electrostatic interactions between a hydrogen atom bound to an electronegative atom ($D^{\delta-}$–$H^{\delta+}$) and another atom that is also electronegative ($A^{\delta-}$) and has one or more lone pairs enabling it to act as a base.
- A typical hydrogen bond has a bond enthalpy of 20–30 kJ mol^{-1}, which is approximately one-tenth the strength of a typical covalent bond.
- Strong hydrogen bonding leads to an increase in the melting point and boiling point. In the absence of hydrogen bonding, water would be a gas at room temperature.

For practice questions on these topics, see questions 10 and 11 at the end of this chapter (p.1166).

More details on isotopes are given in Section 1.3 (p.14).

D_2O is denser than H_2O. One consequence of this is that D_2O ice does not float in H_2O, but sinks. This photo shows H_2O ice at the top and D_2O ice at the bottom.

25.4 Isotope effects

Hydrogen has three isotopes. Although the vast majority of naturally occurring hydrogen is ^1H (sometimes called protium), a small fraction (0.0156%) is deuterium (^2H or D), which contains a neutron as well as a proton in its nucleus. The radioactive isotope tritium (^3H or T) occurs naturally in miniscule amounts, but it can be prepared artificially.

Since isotopes contain atoms of the same element with identical arrangements of electrons, it is normally assumed that they have identical chemical properties. This is not strictly true, as many physical and chemical properties depend on atomic mass. A deuterium atom has twice the mass of a normal hydrogen atom, so isotope effects are far more important in hydrogen compounds than they are for other elements, when the mass difference between isotopes is proportionately much smaller. For example, the mass difference between ^{13}C and ^{12}C is 8% and that between ^{238}U and ^{235}U is just over 1%.

Box 25.7 Tritium

Tritium is a radioactive isotope of hydrogen, with the nucleus of each atom containing two neutrons in addition to a proton. Tritium decays into an isotope of helium via β-emission (see Section 3.8, p.159) with a half-life of 12.3 years

$$^{3}_{1}H \rightarrow ^{3}_{2}He + e$$

Very small quantities of tritium are formed naturally in the upper atmosphere when cosmic rays strike nitrogen molecules, but most tritium is produced artificially in special reactors, by bombarding a ^6Li target with neutrons

$$^{6}_{3}Li + ^{1}_{0}n \rightarrow ^{4}_{2}He + ^{3}_{1}H$$

Tritium is used in self-powered light sources. These are present in many emergency exit signs, and are also used in some watches and night vision equipment. They typically have a useful life of 10–20 years, depending on the amount of tritium inside.

The illuminated face and hands of this watch use a tritium light source.

In the manufacture of a tritium light source, a small quantity of the gas is sealed in a borosilicate glass tube, the inside surfaces of which are coated with a phosphor. The electrons emitted from the β-decay cause an electron within the phosphor to be promoted to an excited state, and as the electron returns to the ground state it emits visible light. Tritium is particularly suited for use in light sources as the β-particles are of very low energy, so they do not penetrate the glass. This ensures that the lights are completely safe, provided the tubes are not broken. If you break a tritium light source and inhale the contents, you could theoretically receive a high dose of radiation. In practice, the tritium gas would quickly disperse, so overexposure is unlikely.

Question

Formation of tritium in the atmosphere involves the nuclear reaction of ^{14}N with a neutron. Write the nuclear equation for this process and identify the other product.

Bond dissociation enthalpies

The bond dissociation enthalpy of D_2 is +443 kJ mol^{-1}, whereas the corresponding value for H_2 is +436 kJ mol^{-1}. They are different because the zero point energy of a D–D bond is lower than that for a H–H bond, as shown in Figure 25.17. The bond dissociation enthalpy for any X–D bond is greater than that for the corresponding X–H bond, so more energy is needed to break a X–D bond than a X–H bond.

Zero point energy is introduced in Box 10.1 (p.454).

Infrared spectra

Substitution of hydrogen by deuterium leads to a decrease in the wavenumber of the X–H stretch. For example, the bond vibration of HCl gives an absorption band in the IR spectrum at 2990 cm^{-1}, whereas the bond vibration for DCl is observed at 2140 cm^{-1}.

The use of isotopic substitution in infrared spectroscopy and reduced mass are discussed in Section 10.5 (p.483).

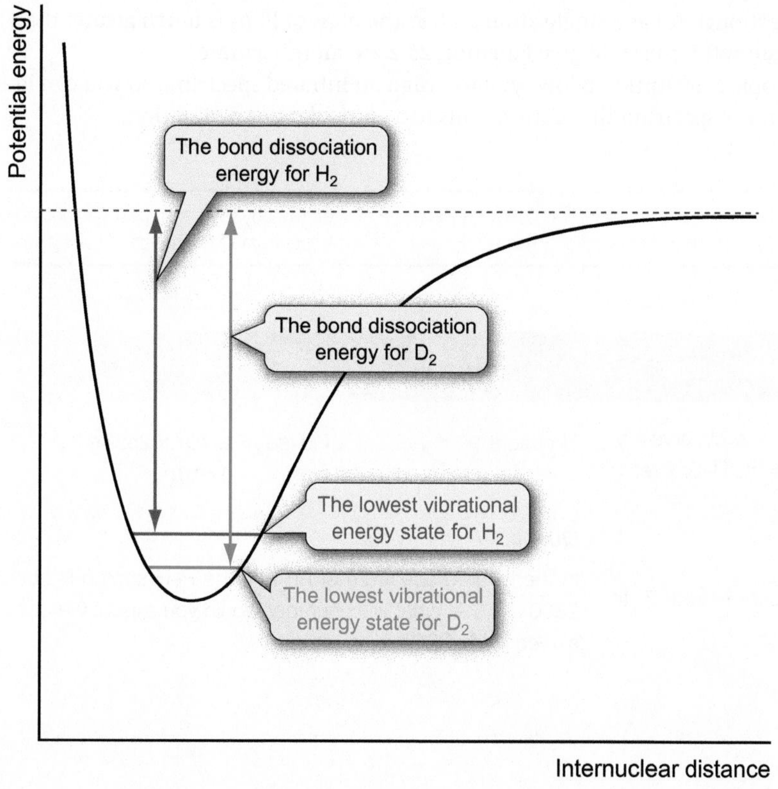
Potential energy
The bond dissociation energy for H2
The bond dissociation energy for D2
The lowest vibrational energy state for H2
The lowest vibrational energy state for D2
Internuclear distance

Figure 25.17 The zero point energy for a D–D bond is lower than that for a H–H bond. This means that D_2 has a higher bond dissociation energy than H_2.

The reduced mass, μ, for a diatomic molecule containing atoms of mass m_1 and m_2 is given by

$$\mu = \frac{m_1 m_2}{m_1 + m_2}$$

Equation 25.1 is the same as Equation 10.13 (p.467). If $m_1 >> m_2$, μ is approximately equal to m_2.

The bond-stretching frequency is related to the bond force constant, k, and the reduced mass of the system, μ, by Equation 25.1

$$\nu = \frac{1}{2\pi}\left(\frac{k}{\mu}\right)^{\frac{1}{2}} \qquad (25.1)$$

Therefore, if ν_H is the stretching frequency of the X–H bond and ν_D is the stretching frequency of the X–D bond

$$\frac{\nu_D}{\nu_H} = \frac{\left(\frac{1}{\cancel{2\pi}}\right)\left(\frac{\cancel{k}}{\mu_D}\right)^{\frac{1}{2}}}{\left(\frac{1}{\cancel{2\pi}}\right)\left(\frac{\cancel{k}}{\mu_H}\right)^{\frac{1}{2}}} = \frac{\left(\frac{1}{\mu_D}\right)^{\frac{1}{2}}}{\left(\frac{1}{\mu_H}\right)^{\frac{1}{2}}}$$

$$= \left(\frac{\mu_H}{\mu_D}\right)^{\frac{1}{2}}$$

The absorptions in infrared spectra are normally reported in units of wavenumber rather than frequency. The wavenumber ($\bar{\nu}$), normally given in units of cm^{-1}, is proportional to the frequency (ν), as shown in Equation 10.10b (p.466), so $\frac{\bar{\nu}_D}{\bar{\nu}_H} = \frac{\nu_D}{\nu_H}$.

where μ_H and μ_D are the reduced masses of the XH and XD systems. When the atom bonded to the hydrogen, X, is sufficiently large, the reduced mass of XD (μ_D) is approximately twice that of XH (μ_H). As a consequence

$$\frac{\nu_D}{\nu_H} \approx \left(\frac{1}{2}\right)^{\frac{1}{2}} = 0.707 \qquad (25.2)$$

Visit the Online Resource Centre to view screencast 25.1 which walks you through how deuteration affects an infrared spectrum.

The relationship between the X–H and X–D stretching frequencies in Equation 25.2 is useful for predicting the change in position of the X–H stretches when X–H is deuterated (i.e. H is replaced by D). The relationship holds for larger molecules too. For example, in Ph_2PH the P–H stretch involves only the phosphorus and hydrogen atoms, with the phenyl groups stationary relative to the phosphorus atom. This means that the PPh_2 group can be thought of as a single atom and, as the mass of PPh_2 is much greater than that of P, the assumptions used to give Equation 25.2 are more accurate.

Isotopic substitution allows you to assign an infrared spectrum, so you can identify the peak in the spectrum that corresponds to a particular bond vibration.

Deuterated solvents such as D_2O and $CDCl_3$ (*d*-chloroform) are commonly used in ^1H NMR spectroscopy so solvent peaks do not interfere with the spectrum of the compound being analysed.

Worked example 25.1 Isotopic substitution

In the infrared spectrum of HBr, the H–Br stretch is observed at 2559 cm^{-1}. At what wavenumber would you expect the D–Br stretch to be observed?

Strategy

Use Equation 25.2 and the relationship between ν and $\bar{\nu}$ to predict $\bar{\nu}$(D–Br).

Solution

Use Equation 25.2

$$\frac{\bar{\nu}_D}{\bar{\nu}_H} = \frac{\nu_D}{\nu_H} \approx \left(\frac{1}{2}\right)^{\frac{1}{2}} = 0.707$$

Therefore

$$\bar{\nu}(\text{D–Br}) = 0.707 \times 2559\ cm^{-1} = 1810\ cm^{-1}$$

Question

In the infrared spectrum of Ph_2PH, the P–H stretch is observed at 2300 cm^{-1}. At what wavenumber would you expect $\bar{\nu}$(P–D) to be observed in Ph_2PD?

Summary

- Bonds to deuterium are stronger than those to normal hydrogen because X–D bonds have lower zero point energies than X–H bonds.
- Deuterium labelling is useful in the assignment of infrared spectra. Large isotopic shifts are observed in infrared spectra due to the greater mass of ^{2}D over ^{1}H.

For practice questions on these topics, see questions 12 and 13 at the end of this chapter (p.1166).

Concept review

By the end of this chapter you should be able to do the following.

- Describe the means of production and the main uses of hydrogen gas.
- Understand why binary hydrides can be protic or hydridic, and predict the character of a particular hydride.
- Predict whether the bonding in a binary hydride is covalent or ionic, and whether a covalent hydride is likely to contain bridging hydrogen atoms.
- Describe how the stability and reactivity of hydride compounds changes across the Periodic Table and down each group.
- Explain why the hydrogen halides increase in acidity from HF to HI.
- Explain how hydrogen bonds are formed and why very strong hydrogen bonds can be described as covalent interactions.
- Predict the effect of hydrogen bonding on melting points and boiling points.
- Calculate the effects of replacing hydrogen by deuterium on infrared stretching frequencies.

Key equations

The effect of deuteration on IR spectra	$\frac{\nu_D}{\nu_H} \approx \left(\frac{1}{2}\right)^{\frac{1}{2}} = 0.707$	(25.2)

Questions

More challenging questions are indicated by an asterisk *.

1. What volume of hydrogen gas (at 298 K and 1 atm) is required for the extraction of 1 kg of copper from a Cu^{2+}(aq) solution? (1 mol gas occupies 24.5 dm^3 at this temperature and pressure.) (Section 25.1)
2. Using Table 25.1 (p.1145), which of the following metals would you expect to react with dilute acid to generate H_2: (a) Sn; (b) Ag; (c) Ba; (d) Ni; (e) Hg? Explain your answers. (Section 25.1)
3. Assign oxidation states to the elements in the following hydride compounds

   LiH, HI, NH_3, SiH_4, B_2H_6 (Section 25.2)
4. Predict the products and write balanced equations for the following reactions (Section 25.2):

   (a) CsH + water

   (b) B_2H_6 + pyridine

   (c) Si_2H_6 + oxygen

   (d) N_2H_4 + HNO_3(aq)

5. SF_6 is a stable compound. Suggest why the hydride analogue SH_6 does not exist. (Section 25.2)

6. Classify the following hydrides as ionic, covalent, or metallic: (a) BeH_2; (b) PH_3; (c) KH; (d) HCl; (e) $FeTiH_{1.8}$. For the covalent hydrides, state whether they exist as discrete molecules or are linked by bridging hydrogen atoms. (Section 25.2)

7. Suggest reasons why plumbane (PbH_4) is not well characterized. (Section 25.2)

8.* Calcium hydride is used at remote locations such as Antarctica as a source of hydrogen (see Section 25.2, p.1146). What mass of CaH_2 is required to fill a 500 dm^3 weather balloon at −50 °C and an atmospheric pressure of 95 500 Pa? (Section 25.2)

9. Use the following bond dissociation enthalpy data to calculate the enthalpy change of formation for HF(g) (Section 25.2).

$D_{(H–H)}$ = +436 kJ mol^{-1}

$D_{(F–F)}$ = +159 kJ mol^{-1}

$D_{(H–F)}$ = +570 kJ mol^{-1}

10. Explain the trend in hydrogen bond strengths:

F–H···F in HF_2^-	+220 kJ mol^{-1}
O–H···O in H_2O	+22 kJ mol^{-1}
S–H···S in H_2S	+7 kJ mol^{-1} (Section 25.3)

11. Account for the following observations. (Section 25.3)

(a) The boiling point of ammonia is higher than that of phosphine.

(b) The melting point of hydrogen selenide is higher than that for hydrogen sulfide.

(c) Water is denser than ice, whereas liquid methane is less dense than solid methane.

12. In the infrared spectrum of hydrogen iodide, $\bar{\nu}$(HI) is observed at 2310 cm^{-1}. Predict the value of $\bar{\nu}$(DI) in the deuterated analogue. (Section 25.4)

13.* In the IR spectrum of HCl, $\bar{\nu}$(HCl) is observed at 2990 cm^{-1}. Predict a value for $\bar{\nu}$(TCl) in the tritiated analogue. (Section 25.4)

14.* The energy density of hydrogen is 143 kJ g^{-1}. (Sections 25.1 and 25.2)

(a) How much energy is released on combustion of 1 m^3 hydrogen at 298 K and 1 atm?

(b) Given that the alloy Mg_2Ni reversibly absorbs hydrogen to form Mg_2NiH_4, what mass of Mg_2Ni would be required to absorb 1 m^3 hydrogen at 298 K and 1 atm?

(1 mol gas occupies 24.5 dm^3 at 298 K and 1 atm).

26

s-Block chemistry

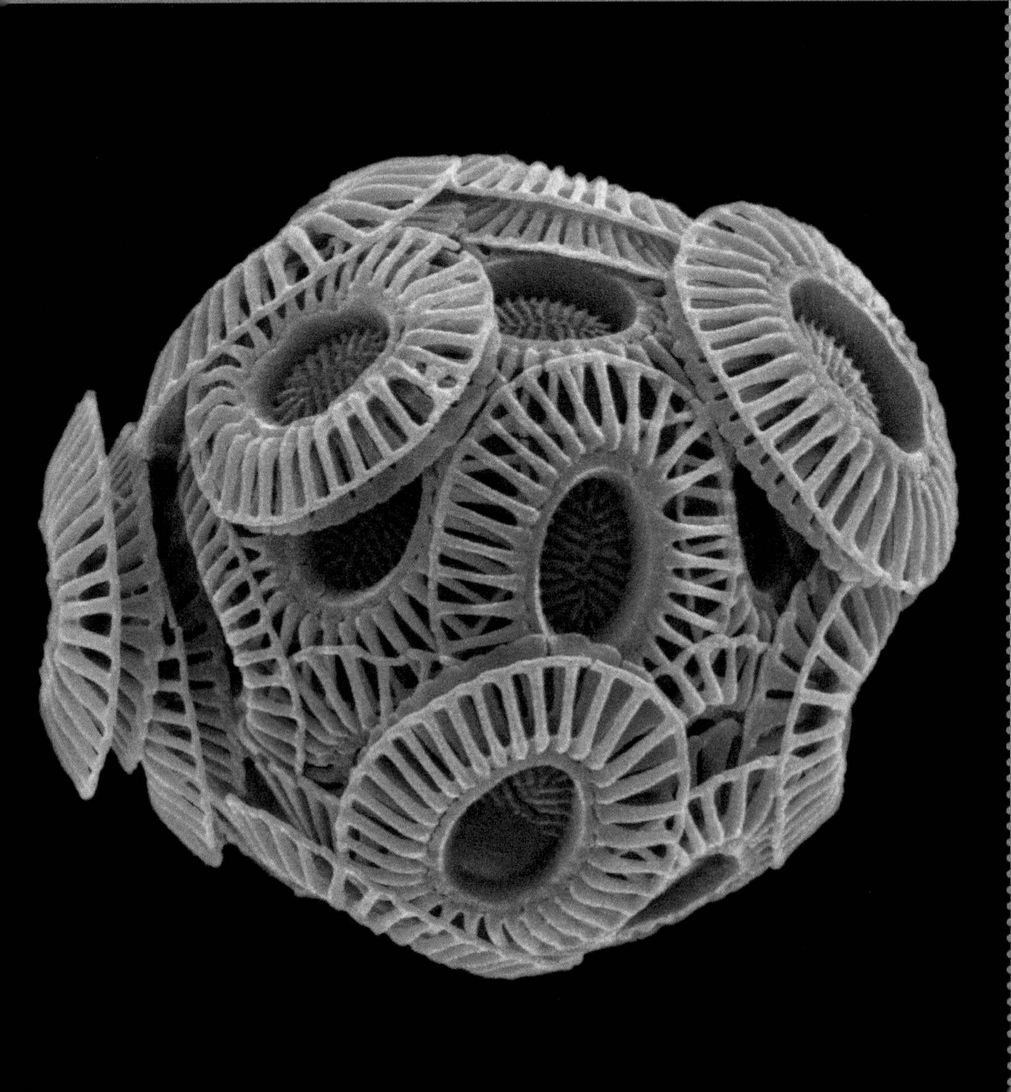

This chapter builds on the following topics:

- Atomic properties and periodicity Section 3.7, p.153
- Born–Haber cycles and lattice enthalpies Section 6.5, p.290
- Calculating lattice energy Section 6.6, p.293
- Lewis acids and bases Section 7.8, p.336
- Enthalpy changes in chemical reactions Section 13.3, p.622

◀ Scanning electron micrograph of a phytoplankton. The skeleton is a biomineral made of calcium carbonate.

Biominerals

In most environments, multicellular living organisms need strength and rigidity to function. Plants rely on cellulose and lignin to provide these properties, whereas animals use biominerals such as those found in shells and bone. Biominerals are *composite materials* consisting of an inorganic compound embedded in an organic matrix. This matrix is typically a protein and it controls the shape, size, and crystallinity of the inorganic compound so that the biomineral has different properties from the pure inorganic material. Many biominerals, including those found in shells and bone, are compounds of the Group 2 metal calcium. Bone is made from the calcium-phosphate mineral *hydroxyapatite* ($Ca_5(PO_4)_3(OH)$), crystals of which are embedded in fibres of the protein collagen.

Bone is not an inert substance in the living body. The hydroxyapatite is constantly being dissolved by cells known as *osteoclasts* and re-precipitated by different cells called *osteoblasts*. As a person ages, the ability of their body to replace dissolved bone decreases. This can lead to the condition *osteoporosis*, which literally means porous bones. Osteoporosis is characterized by abnormalities in the amount and structure of bone tissue, and it leads to a reduction in bone strength. This means that bones affected by osteoporosis are more likely to break than normal bones. Osteoporosis can affect people of all ages, but it is more common in the elderly and especially prevalent in women after the menopause.

Osteoporosis can be treated in two different ways. Hormonal treatments focus on replacement of the female hormone oestrogen in post-menopausal women. Oestrogen helps protect against osteoporosis as it reduces the amount of bone that is broken down. Non-hormonal treatments involve the use of bisphosphonate compounds such as *alendronate*, which is a salt of alendronic acid (shown right). Although the mode of action of bisphosphonates is not completely understood, scientists believe that anions such as that in alendronate bind to the surface of hydroxyapatite crystals where they are relatively stable to removal by the osteoclasts. This slows down the rate at which the bone is dissolved.

Hydroxyapatite is also the biomineral present in teeth. In the mouth, the surfaces of teeth are covered in a thin layer of saliva. Some bacteria have evolved to use the proteins in saliva as food. As they grow in the saliva film on a tooth, they form a deposit on the tooth known as plaque. This is removed by regular brushing, and toothpastes contain a mild abrasive such as silica or calcium carbonate to help this. One of the types of bacteria present in plaque—*Streptococcus mutans*—breaks down sugars into carboxylic acids such as ethanoic acid and butanoic acid. As these are formed, the pH in the mouth is reduced, and hydroxyapatite, which is basic, starts to dissolve. This eventually leads to tooth decay.

A simple way to help prevent tooth decay is the use of 'fluoride'. Most toothpastes contain 'fluoride', either in the form of sodium fluoride or sodium monofluorophosphate (Na_2PO_3F). The fluoride ions (F^-) react with hydroxyapatite by replacing the similarly sized OH^- ions to form *fluoroapatite* ($Ca_5(PO_4)_3F$) on the surfaces of the teeth. This has fewer defects in its lattice structure, which makes the teeth more resistant to decay.

▼ Alendronic acid.

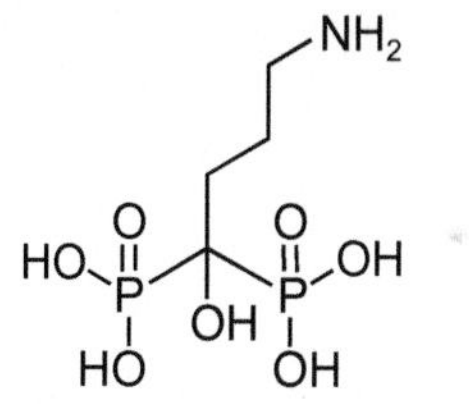

NH2
HO
P
O
OH
OH
P
O
OH
HO
OH

▼ Scanning electron micrographs of healthy bone (left) and bone affected by osteoporosis (right).

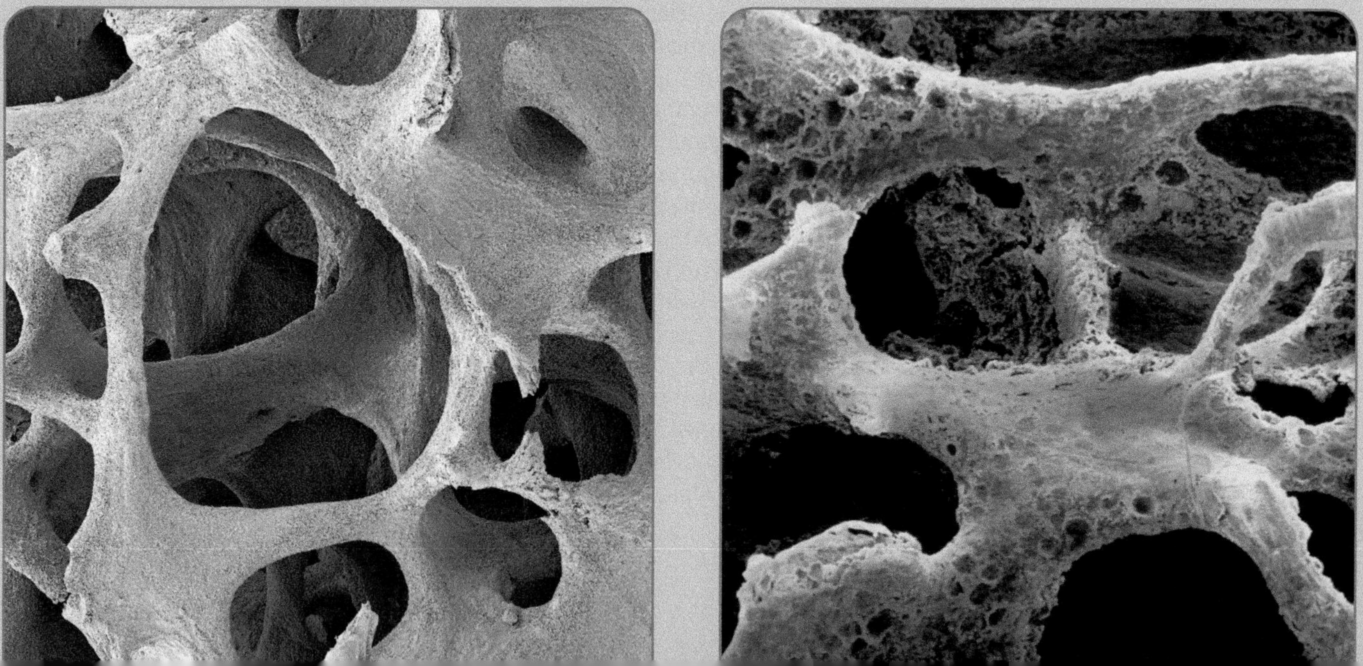

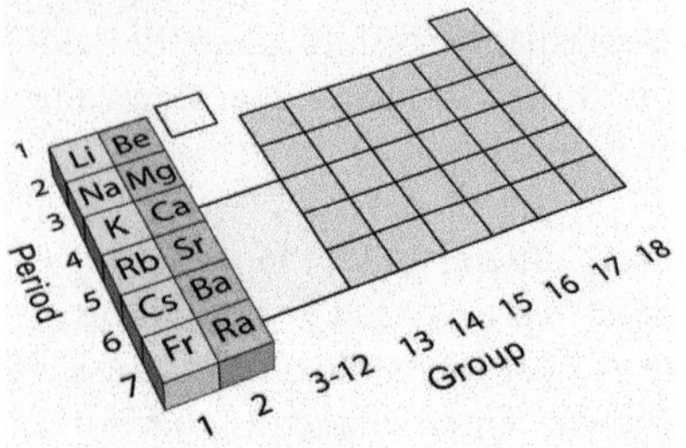
Li Be
Na Mg
K Ca
Rb Sr
Cs Ba
Fr Ra
Period
1 2 3 4 5 6 7
Group
1 2 3-12 13 14 15 16 17 18

Groups 1 and 2 in the Periodic Table.

Table 26.1 The 10 most abundant elements in the Earth's crust

	Element	Percentage (by mass)
1	Oxygen	46.6
2	Silicon	27.7
3	Aluminium	8.2
4	Iron	4.1
5	Calcium	4.1
6	Sodium	2.3
7	Potassium	2.1
8	Magnesium	2.1
9	Titanium	0.44
10	Hydrogen	0.14

The *s*-block is the collective name for the elements of Groups 1 and 2 of the Periodic Table. The name '*s*-block' comes from the valence electronic configurations of the elements: ns^1 and ns^2 for Group 1 and Group 2, respectively. Hydrogen has an electronic configuration of $1s^1$ and is sometimes included in Group 1. In this book hydrogen is treated on its own in Chapter 25 as its chemistry is very different from that of the Group 1 elements.

The Group 1 elements are often called the **alkali metals**, because they react with water to give alkaline solutions. The Group 2 elements are known as the **alkaline earth metals**. Earth is an old name for an oxide, so the name reflects the fact that the Group 2 oxides are basic.

Sodium, potassium, magnesium, and calcium are all among the ten most abundant elements in the Earth's crust (Table 26.1). The Group 1 and Group 2 elements are very reactive, so none exist naturally as a free metal. They always occur as compounds, with the oxidation state +1 for Group 1 and +2 for Group 2. The ions of sodium, potassium, magnesium, and calcium are all essential for life, and many Group 1 and 2 compounds are important industrially. You will see examples of the biological and industrial importance of Group 1 and Group 2 compounds throughout this chapter.

26.1 The Group 1 elements

Caesium is spelt 'cesium' in the USA.

Group 1 contains the elements lithium (Li), sodium (Na), potassium (K), rubidium (Rb), caesium (Cs), and francium (Fr). The first five members are metallic solids at room temperature and pressure, though the melting point of caesium is low enough for it to melt in hot weather. Francium is radioactive and has not been isolated as the pure element. It is found as a minor component in uranium minerals, and the longest-lived isotope has a half life of only 22 minutes.

The body-centred cubic structure is described in Section 6.2 (p.264) and metallic bonding is described in Section 6.3 (p.273).

Table 26.2 gives some physical properties of the Group 1 elements. The melting points, boiling points, and enthalpy changes of atomization ($\Delta_a H^{\ominus}$) are all low compared to other metallic elements. This is due to the relatively weak metallic bonding in the lattices. The elements have lower densities than other metals because of their large atomic radii and the relatively open body-centred cubic structure they all adopt.

Effective nuclear charge is introduced in Section 3.6 (p.150). Trends in ionization enthalpies are discussed in Section 3.7 (p.154).

The chemistry of the Group 1 elements is dominated by formation of M^+ cations. The outermost electron in a Group 1 atom lies in an *ns* orbital that is larger and more diffuse than the other occupied atomic orbitals and is well shielded from the nuclear charge. This means the effective nuclear charge felt by this electron is low, so these elements have low first ionization enthalpies. This is the main reason they are so reactive.

Table 26.2 Physical properties of the Group 1 elements

Element	A_r	Melting point / °C	Boiling point / °C	Density / $g\,cm^{-3}$	$\Delta_a H^{\ominus}$ / $kJ\,mol^{-1}$*
Lithium (Li)	6.94	181	1342	0.53	+159
Sodium (Na)	22.99	98	883	0.97	+108
Potassium (K)	39.10	64	759	0.89	+89
Rubidium (Rb)	85.47	39	688	1.53	+81
Caesium (Cs)	132.91	29	678	1.87	+76

* The enthalpy change of atomization ($\Delta_a H^{\ominus}$) is defined as the enthalpy change on conversion of a substance to a mole of gaseous atoms under standard conditions.

The Group 1 metals are very easily oxidized to M^+ ions, and they react with water and oxygen, as described in Section 26.2. Because of this reactivity, Group 1 metals are usually stored under oil.

The origins of flame test colours are described in Box 3.2 (p.128).

Group 1 metals have very distinctive visible spectra, and this is the basis of flame tests that allow compounds of these metals to be distinguished. The flame colours observed are:

- Li, red;
- Na, yellow-orange;
- K, lilac;
- Rb, blueish-purple;
- Cs, blue.

The sodium colour is particularly intense, and small quantities of sodium compounds can mask the presence of other metals.

The distinctive yellow colour of many street lights is a result of the sodium atomic spectrum. The lamp contains sodium vapour, and the emission is the same as that observed in the sodium flame test, involving an electronic transition between the 3*p* orbital and the 3*s* orbital. Sodium street lights are gradually being replaced with more energy efficient lighting.

Preparation and uses

Lithium and sodium metals are obtained industrially by electrolysis of their molten chlorides. For sodium, this is carried out in a Downs cell, as shown in Figure 26.1.

In this process, liquid sodium is produced at the cathode and chlorine gas is formed at the anode

$$2\,Na^+(melt) + 2\,e^- \rightarrow 2\,Na\,(l) \qquad \text{reduction at the cathode}$$

$$2\,Cl^-(melt) \rightarrow Cl_2\,(g) + 2\,e^- \qquad \text{oxidation at the anode}$$

Potassium, rubidium, and caesium metals are prepared by the reduction of their molten salts with sodium at high temperatures. For example,

$$KCl\,(melt) + Na\,(l) \rightarrow NaCl\,(melt) + K\,(l)$$

Potassium is more reducing than sodium, so it may seem odd that potassium can be produced in this way. The reaction is actually an equilibrium, and the more volatile potassium is obtained by fractional distillation. This displaces the equilibrium to the right-hand side and allows the reaction to proceed.

Electrolysis is described in Section 16.5 (p.755).

Sodium is the most important of the Group 1 metals industrially, followed by lithium and potassium. Sodium is used in the manufacture of compounds such as sodium borohydride and sodium hydride, the production of other electropositive metals such as uranium and zirconium, and in the manufacture of herbicides. However, the production of sodium metal has fallen by 80% since 1970. This is mainly due to the banning of lead in petrol. The main use of sodium in the past was in the manufacture of tetraethyllead ($PbEt_4$), which used a lead–sodium alloy

$$4\,PbNa + 4\,EtCl \rightarrow PbEt_4 + 4\,NaCl + 3\,Pb$$

'Et' is an abbreviation for the ethyl group, CH_3CH_2 (see Box 2.3, p.81).

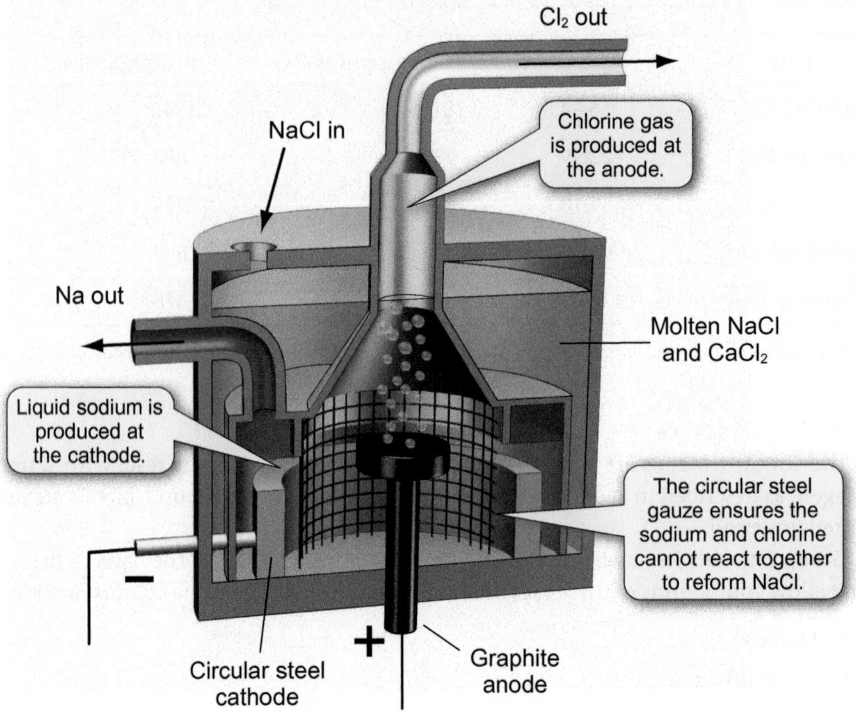

Cl_2 out
NaCl in
Chlorine gas is produced at the anode.
Na out
Molten NaCl and $CaCl_2$
Liquid sodium is produced at the cathode.
The circular steel gauze ensures the sodium and chlorine cannot react together to reform NaCl.
−
+
Circular steel cathode
Graphite anode

Figure 26.1 The Downs cell is used to obtain sodium from molten NaCl. $CaCl_2$ is added so that the cell can run at about 600 °C, which is lower than the melting point of NaCl (800 °C).

Tetraethyllead acts as an anti-knock agent in petrol. Environmental concerns about lead emissions led to a reduction in the use of leaded petrol, and leaded petrol was withdrawn from sale in the UK in 2000.

Summary

- The Group 1 elements are metallic solids with low melting points, low boiling points, and low enthalpy changes of atomization.
- The Group 1 metals are very reactive, and are readily oxidized to M^+ cations.
- Lithium and sodium are produced industrially by electrolysis of the molten chlorides, whereas potassium, rubidium, and caesium are obtained by reduction of their salts with sodium.

For a practice question on these topics, see question 1 at the end of this chapter (p.1198).

26.2 Group 1 compounds

Group 1 compounds are generally ionic and contain M^+ cations. Some lithium compounds, however, have a high degree of covalent character. This is especially true for the organolithium compounds that are described in Section 26.10 (p.1195).

Oxides

All of the Group 1 metals burn in air to form oxides, but the major product of combustion depends on the metal. Lithium burns to form lithium oxide (Li_2O), sodium gives the peroxide (Na_2O_2), whereas potassium and the heavier metals form superoxides (MO_2)

$$4\,Li\,(s) + O_2\,(g) \rightarrow 2\,Li_2O\,(s)$$
$$2\,Na\,(s) + O_2\,(g) \rightarrow Na_2O_2\,(s)$$
$$K\,(s) + O_2\,(g) \rightarrow KO_2\,(s)$$

Peroxides contain the O_2^{2-} ion and superoxides contain the O_2^- ion. Although Li_2O, Na_2O_2, KO_2, RbO_2, and CsO_2 are the normal products of combustion, all of the metals can form an oxide, peroxide, and superoxide under appropriate conditions. All of the Group 1 oxides are basic, and react with water to give the hydroxides. For example,

- O^{2-}, oxide ion
- O_2^{2-}, peroxide ion
- O_2^-, superoxide ion

$$Li_2O\,(s) + H_2O\,(l) \rightarrow 2\,LiOH\,(aq)$$
$$Na_2O_2\,(s) + 2\,H_2O\,(l) \rightarrow 2\,NaOH\,(aq) + H_2O_2\,(aq)$$

The bonding in the peroxide and superoxide ions is described in Section 4.10 (p.198).

The reaction between a superoxide and water also produces oxygen

$$4\,KO_2\,(s) + 2\,H_2O\,(l) \rightarrow 4\,KOH\,(s) + 3O_2\,(g)$$

This reaction is an example of a **disproportionation**, as the O_2^- ion is oxidized to O_2 and also reduced to OH^-. This reaction is used in some submarine breathing systems to generate oxygen. The potassium hydroxide produced in this reaction absorbs carbon dioxide and prevents harmful concentrations of this gas building up.

In a **disproportionation** reaction an intermediate oxidation state is converted to a higher and lower oxidation state.

The American Deep Submergence Rescue Vehicle 'Mystic' is shown here docked to a larger submarine. The Mystic contains a potassium superoxide breathing system.

Why does the type of oxide produced by combustion change going down the group? As a general rule *large anions are stabilized by large cations*. The ionic radius of the Group 1 cations increases down the group and the larger cations are better at stabilizing the large peroxide and superoxide ions with respect to decomposition into oxide and oxygen gas.

Since most *s*-block compounds are ionic, their lattice energies can be calculated using the Born–Landé equation (Equation 6.10, p.295) or, more simply, the Kapustinskii equation (Equation 6.11, p.297). As the magnitudes of the lattice energies are large, the differences between lattice energies and lattice enthalpies are small. This means that

$$\Delta_{latt}H^{\ominus} \approx \Delta_{latt}U^{\ominus}$$

Enthalpy cycles are used to calculate the enthalpy changes of reactions. For a general reaction, such as decomposition of Group 1 peroxides to oxides, you can calculate the enthalpy change of reaction for all of the compounds in the series. This enables you to see the general trend and to determine which are the most important enthalpy terms contributing to it. In this chapter you will see how this method is used to explain many of the properties of Group 1 and Group 2 compounds, and it is shown in Box 26.1 for the decomposition of Group 1 peroxides to oxides.

In most of the reactions in this chapter for which calculations are given for a series of related compounds, the entropy change of reaction, $\Delta_r S$, is largely independent of the identity of the metal. This is because the stoichiometry of a given reaction is generally the same for all the Group 1 metals. Since

$$\Delta_r G = \Delta_r H - T\Delta_r S \qquad (14.16)$$

if $\Delta_r S$ is constant, the trend in the change in the Gibbs energy of a reaction, $\Delta_r G$, in the series is largely due to differences in the change in enthalpy, $\Delta_r H$. This means that, as a good approximation, the overall trends in reactivity are given by variations in $\Delta_r H$. This approximation only holds if $\Delta_r H$ is reasonably large.

The relationship between enthalpy, entropy, and Gibbs energy is derived in Section 14.5 (p.675).

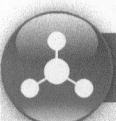

Box 26.1 Why is sodium peroxide more stable to heating than lithium peroxide?

When lithium peroxide is heated, it decomposes into lithium oxide and oxygen gas

$$2\,Li_2O_2(s) \rightarrow 2\,Li_2O(s) + O_2(g)$$

In contrast, sodium peroxide is stable at high temperatures. Why is there a difference in the thermal stabilities of these two peroxides?

Since both M_2O and M_2O_2 are ionic solids, you can construct an enthalpy cycle in which two of the terms are the lattice enthalpies for these compounds and the third relates to the disproportionation of O_2^{2-} into O_2 and O^{2-}. This third enthalpy change, $\Delta_x H^\ominus$, is independent of the metal and is equal to +850 kJ mol^{-1}. The enthalpy cycle is shown below.

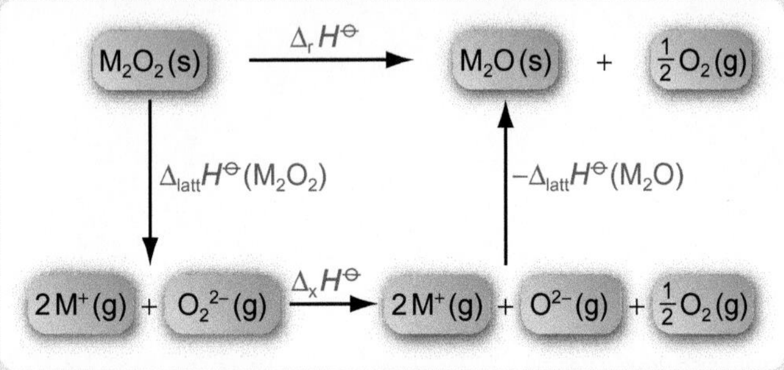

M2O2(s)
ΔrH⊖
M2O(s)
+
1/2 O2(g)
ΔlattH⊖(M2O2)
−ΔlattH⊖(M2O)
2M+(g) + O2 2−(g)
ΔxH⊖
2M+(g) + O2−(g) + 1/2 O2(g)

From the enthalpy cycle

$$\Delta_r H^\ominus = \Delta_{latt}H^\ominus(M_2O_2) + \Delta_x H^\ominus - \Delta_{latt}H^\ominus(M_2O)$$

$\Delta_x H^\ominus$ does not depend on the metal so it is the same for the decomposition of Li_2O_2 and Na_2O_2. You can obtain estimates for the lattice enthalpies using the Kapustinskii equation (Equation 6.11, p.297),

$$\Delta_{latt}U^\ominus = \frac{k\nu z_+ z_-}{r_+ + r_-} \qquad k = 107\,900\,\text{pm kJ mol}^{-1} \qquad (6.10)$$

z_+ and z_- are positive integers giving the charges on the ions, r_+ and r_- are the ionic radii, and ν is the number of ions in the formula unit. If you use the value of k in the non-standard units pm kJ mol^{-1} you can use r_+ and r_- in pm, and the lattice energy is given in kJ mol^{-1}. (The Kapustinskii equation gives a value for $\Delta_{latt}U^\ominus$, the lattice energy, rather than $\Delta_{latt}H^\ominus$, the lattice enthalpy. The difference between $\Delta_{latt}U^\ominus$ and $\Delta_{latt}H^\ominus$ is relatively small for this type of calculation so you can assume the two are the same.)

From Table 6.4 (p.286), the ionic radii for Li^+ and Na^+ are 76 pm and 102 pm, respectively. The ionic radius for O^{2-} is 140 pm (Table 6.5, p.286), whereas the thermochemical radius for O_2^{2-} is 167 pm (Table 6.9, p.297). Substituting these values into Equation 6.11 gives estimates for the lattice enthalpies of the oxides and peroxides of lithium and sodium

$$\Delta_{latt}H^\ominus(Li_2O) \approx \Delta_{latt}U^\ominus(Li_2O)$$
$$= \frac{(107\,900\,\cancel{\text{pm}}\,\text{kJ mol}^{-1}) \times 3 \times 1 \times 2}{(76+140)\,\cancel{\text{pm}}}$$
$$= +3000\,\text{kJ mol}^{-1}$$

$$\Delta_{latt}H^\ominus(Na_2O) \approx \Delta_{latt}U^\ominus(Na_2O)$$
$$= \frac{(107\,900\,\cancel{\text{pm}}\,\text{kJ mol}^{-1}) \times 3 \times 1 \times 2}{(102+140)\,\cancel{\text{pm}}}$$
$$= +2680\,\text{kJ mol}^{-1}$$

$$\Delta_{latt}H^\ominus(Li_2O_2) \approx \Delta_{latt}U^\ominus(Li_2O_2)$$
$$= \frac{(107\,900\,\cancel{\text{pm}}\,\text{kJ mol}^{-1}) \times 3 \times 1 \times 2}{(76+167)\,\cancel{\text{pm}}}$$
$$= +2\,660\,\text{kJ mol}^{-1}$$

$$\Delta_{latt}H^\ominus(Na_2O_2) \approx \Delta_{latt}U^\ominus(Na_2O_2)$$
$$= \frac{(107\,900\,\cancel{\text{pm}}\,\text{kJ mol}^{-1}) \times 3 \times 1 \times 2}{(102+167)\,\cancel{\text{pm}}}$$
$$= +2\,410\,\text{kJ mol}^{-1}$$

For a metal M, the lattice enthalpy for M_2O is always greater than that for M_2O_2. Both lattice enthalpies decrease with the increasing size of the cation, but crucially the lattice enthalpy for the oxide decreases at the faster rate because of the smaller size of the anion.

For lithium

$$\Delta_r H^\ominus = \Delta_{latt}H^\ominus(Li_2O_2) + \Delta_x H^\ominus - \Delta_{latt}H^\ominus(Li_2O)$$
$$= (2660\,\text{kJ mol}^{-1}) + (850\,\text{kJ mol}^{-1}) - (3000\,\text{kJ mol}^{-1})$$
$$= +510\,\text{kJ mol}^{-1}$$

For sodium

$$\Delta_r H^\ominus = \Delta_{latt}H^\ominus(Na_2O_2) + \Delta_x H^\ominus - \Delta_{latt}H^\ominus(Na_2O)$$
$$= (2410\,\text{kJ mol}^{-1}) + (850\,\text{kJ mol}^{-1}) - (2680\,\text{kJ mol}^{-1})$$
$$= +580\,\text{kJ mol}^{-1}$$

Remember, $\Delta_r H^\ominus$ is the enthalpy change for the decomposition of M_2O_2 to M_2O. For both Li and Na, $\Delta_r H^\ominus$ is endothermic, which makes sense since neither peroxide decomposes at room temperature. The decomposition temperature is when $\Delta_r G^\ominus$ is zero (see Section 14.5, p.676). Since $\Delta_r S^\ominus$ is likely to be similar for both metals, and

$$\Delta_r G^\ominus = \Delta_r H^\ominus - T\Delta_r S^\ominus \qquad (14.16)$$

the higher value of $\Delta_r H^\ominus$ for sodium leads to a higher decomposition temperature for Na_2O_2. Decomposition of the peroxide to the oxide therefore becomes less favourable going down the group.

Question

Estimate $\Delta_r H^\ominus$ for the decomposition of K_2O_2 to K_2O. The ionic radius of K^+ is 138 pm.

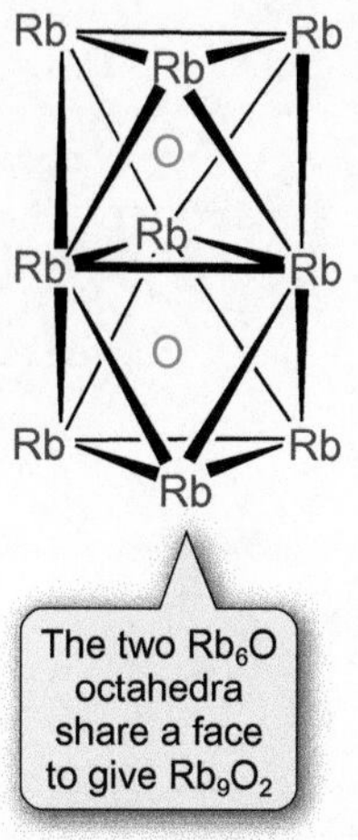
Rb
Rb
Rb
O
Rb
Rb
Rb
O
Rb
Rb
Rb
The two Rb_6O octahedra share a face to give Rb_9O_2

Figure 26.2 The structure of the rubidium suboxide Rb_9O_2 is made up from two Rb_6O octahedra that share a face.

Visit the Online Resource Centre to view screencast 26.1 which walks you through the calculation of $\Delta_r H^{\ominus}$ for the decomposition of Group 1 peroxides.

As well as forming oxides, peroxides, and superoxides, rubidium and caesium burn in limited amounts of oxygen to form a class of intensely coloured compounds called **suboxides**. Examples of suboxides include Rb_9O_2, $Cs_{11}O_3$, and $Cs_{21}O_3$. These compounds have structures based on octahedra of metal atoms with oxygen atoms at the centres of the octahedra. The structure of Rb_9O_2 is shown in Figure 26.2 and consists of two Rb_6O octahedra, sharing a face. Although the metal oxidation state in a suboxide appears to be less than +1, this is not really the case. The additional electrons are actually delocalized over the whole structure, so the formula for Rb_9O_2 could be written $(Rb^+)_9(O^{2-})_2(e^-)_5$ to emphasize this. These delocalized electrons give rise to metallic behaviour.

Dripping water on potassium leads to a violent reaction that forms potassium hydroxide and hydrogen gas. The gas is ignited by the heat produced by the reaction.

Hydroxides

All of the Group 1 metals react with water to give the hydroxide and hydrogen gas. For example,

$$2\,Na\,(s) + 2\,H_2O\,(l) \rightarrow 2\,NaOH\,(aq) + H_2\,(g)$$

These reactions are all very exothermic, and the violence of the reaction increases going down the group.

Lithium, sodium, and potassium are all less dense than water (Table 26.2, p.1171) so they react on the surface of the water. With the exception of lithium, the reactions are exothermic enough to melt the metals, and the reaction with potassium is sufficiently vigorous to ignite the hydrogen produced. Rubidium and caesium are denser than water, so they sink beneath the surface. They react extremely violently, and the energy released by ignition of the hydrogen can produce an explosion.

The Group 1 hydroxides are deliquescent crystalline solids, all of which are soluble in water. The solutions are strongly alkaline as the hydroxides are fully ionized into M^+ and OH^- ions.

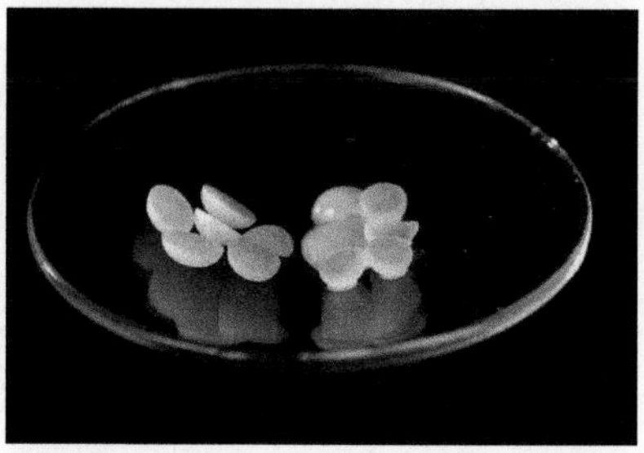

Potassium hydroxide is deliquescent. It absorbs moisture from the air and dissolves in it. The pellets on the left hand side are fresh whereas those on the right have been in air for 5 minutes.

A **deliquescent** substance absorbs water from the air and dissolves in the water it absorbs.

Sodium hydroxide is prepared industrially by the electrolysis of sodium chloride solution, and this is known as the **chloralkali process**. This reaction also produces hydrogen and chlorine gases. The overall equation for the process is

$$2\,NaCl\,(aq) + 2\,H_2O\,(l) \xrightarrow{\text{electrolysis}} 2\,NaOH\,(aq) + H_2\,(g) + Cl_2\,(g)$$

Approximately 66 million tonnes of sodium hydroxide are produced each year worldwide. It is used industrially as a base in the manufacture of organic and inorganic compounds, and in the production of paper, textiles, and detergents.

Bases are described in Section 7.2 (p.308).

The electrolysis of aqueous NaCl is discussed in Section 16.5 (p.757).

Naturally occurring deposits of sodium chloride are mined.

Lithium hydroxide is the only Group 1 hydroxide that decomposes readily on heating, and it forms lithium oxide and water vapour

$$2\,LiOH\,(s) \xrightarrow{\Delta} Li_2O\,(s) + H_2O\,(g)$$

The symbol Δ over the arrow stands for 'heat'.

Halides

The structures of the Group 1 halides are described in Section 6.4 (p.277).

The Group 1 halides are colourless ionic solids with high melting points. They are prepared in the laboratory by the reaction of the hydroxide or carbonate with the appropriate hydrohalic acid. For example,

$$Li_2CO_3\,(aq) + 2\,HBr\,(aq) \rightarrow 2\,LiBr\,(aq) + H_2O\,(l) + CO_2\,(g)$$

Sodium chloride is obtained either by mining naturally occurring deposits or by the evaporation of sea water, the composition of which is shown in Table 26.3. This is most feasible in hot countries where fast evaporation makes the process commercially viable or where the salt solution is already concentrated, such as Salt Lake in the USA.

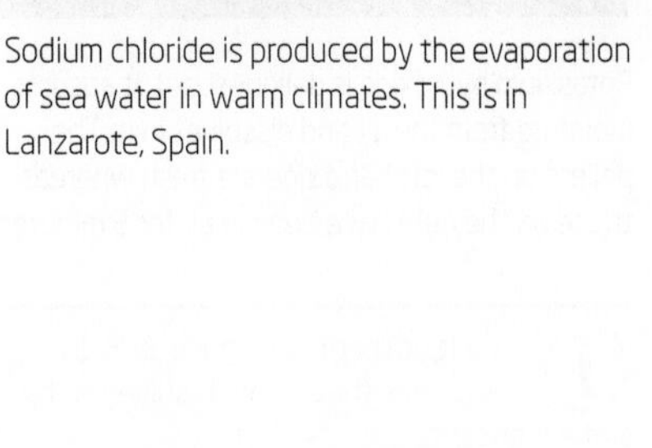

Sodium chloride is produced by the evaporation of sea water in warm climates. This is in Lanzarote, Spain.

Table 26.3 The major ions present in sea water

Ion	Average mass per kg of sea water / g
Chloride (Cl^-)	18.980
Sodium (Na^+)	10.556
Sulfate (SO_4^{2-})	2.649
Magnesium (Mg^{2+})	1.272
Calcium (Ca^{2+})	0.400
Potassium (K^+)	0.380
Hydrogencarbonate (HCO_3^-)	0.140

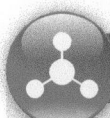

Box 26.2 Is salt bad for you?

Based on a number of studies, the UK Department of Health has concluded that reducing the average salt intake of the population would lead to a decrease in cardiovascular disease due to lower average blood pressure. It estimated in 2012 that the average person's salt intake was 8.1 g per day. This had dropped from 9.5 g per day in 2005, but was still above 6 g per day recommended by the Scientific Advisory Committee on Nutrition. About 75% of salt in the average British diet comes from processed foods, and the government is working with companies in the food industry to reduce this percentage.

While the debate on the amount of salt in the diet is set to continue, no-one doubts the importance of Na^+ ions in cell biology and that a certain concentration in the body is essential for life. The importance of sodium for nerve function is described in Box 16.5 (p.754).

This is true for all animals. Elephants, for example, need up to 100 g of salt a day. Salt occurs in plants and in the soil, but on Mount Elgon in Kenya heavy rainfalls cause much of this salt to leach away. The only natural salt source available is deep inside caves. Families of elephants enter these caves, moving as far as 150 m underground in total darkness. They excavate salt from seams in the rock, chiselling the salt from the walls with their tusks.

In the past, salt was less available than it is today, and consequently it had a much higher relative value. Indeed, salt was the only way of preserving meat over the winter. The Roman army made an allowance of salt to its soldiers, and this 'salarium' is the source of the word 'salary'.

▼ Elephants near Mount Elgon in Kenya get their salt from caves as far as 150 m underground.

Compounds with carbon or nitrogen

The Group 1 metals all react with ethyne (acetylene, C_2H_2) in liquid ammonia to form ethynides (also known as acetylides) which contain the HC_2^- monoanion or the C_2^{2-} dianion

$$2\,Li(s) + 2\,HC{\equiv}CH(g) \xrightarrow{\text{liquid NH}_3} 2\,Li^+\,{}^-C{\equiv}CH(s) + H_2(g)$$

$$2\,Li(s) + HC{\equiv}CH(g) \xrightarrow{\text{liquid NH}_3} Li^+\,{}^-C{\equiv}C^-\,Li^+(s) + H_2(g)$$

LiC_2H is used industrially in the production of vitamin A.

Lithium is the only Group 1 metal to form a stable binary nitride. Lithium nitride is prepared from the reaction of lithium with nitrogen at high temperature and pressure

$$6\,Li\,(l) + N_2\,(g) \xrightarrow{800\,°C,\ 10\,atm} 2\,Li_3N\,(s)$$

Visit the Online Resource Centre to view screencast 26.2 which walks you through the calculation of the enthalpy change of formation of Li_3N.

The relative stability of Li_3N is largely a consequence of its high lattice enthalpy. $\Delta_{latt}H^{\ominus}(Li_3N)$ is much higher than the equivalent values for the other Group 1 nitrides due to the small size of the Li^+ cation. Worked example 26.1 shows the importance of the $\Delta_{latt}H^{\ominus}(Li_3N)$ term in estimating a value for the enthalpy change of formation of Li_3N.

Group 1 ethynides and lithium nitride decompose in water to form LiOH together with ethyne and ammonia, respectively

$$Li_2C_2\,(s) + 2\,H_2O\,(l) \rightarrow 2\,LiOH\,(aq) + C_2H_2\,(g)$$

$$Li_3N\,(s) + 3\,H_2O\,(l) \rightarrow 3\,LiOH\,(aq) + NH_3\,(aq)$$

Worked example 26.1 Estimating the enthalpy change of formation of lithium nitride (Li_3N)

Use a Born–Haber cycle to estimate a value for the enthalpy change of formation of Li_3N.

$\Delta_a H^{\ominus}(Li) = +159\,kJ\,mol^{-1}$ $\Delta_i H(1)^{\ominus}(Li) = +520\,kJ\,mol^{-1}$

$\Delta_a H^{\ominus}(N) = +473\,kJ\,mol^{-1}$ $\Delta_{eg}H(1)^{\ominus}(N) + \Delta_{eg}H(2)^{\ominus}(N) + \Delta_{eg}H(3)^{\ominus}(N) = +2565\,kJ\,mol^{-1}$

Strategy

Construct a Born–Haber cycle for the formation of Li_3N from the elements (see Section 6.5, p.290).

Use the Kapustinskii equation (Equation 6.11, p.297) to estimate the lattice enthalpy of Li_3N.

Solution

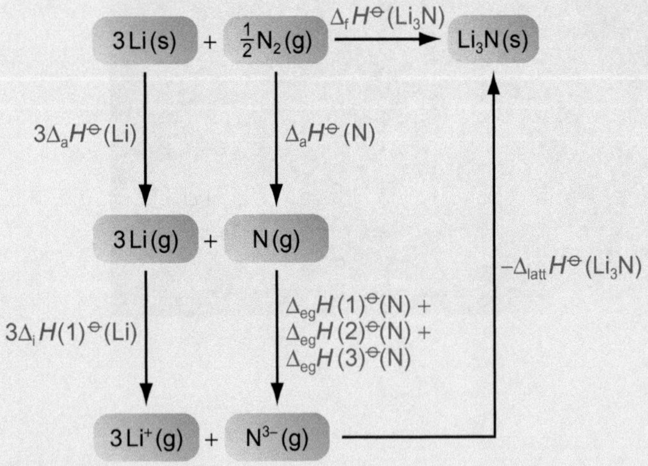

3Li(s) + ½N2(g)
ΔfH⊖(Li3N)
Li3N(s)
3ΔaH⊖(Li)
ΔaH⊖(N)
3Li(g) + N(g)
−ΔlattH⊖(Li3N)
3ΔiH(1)⊖(Li)
ΔegH(1)⊖(N) +
ΔegH(2)⊖(N) +
ΔegH(3)⊖(N)
3Li+(g) + N3−(g)

From the Born–Haber cycle

$$\Delta_f H^{\ominus}(Li_3N) = 3\Delta_a H^{\ominus}(Li) + 3\Delta_i H(1)^{\ominus}(Li) + \Delta_a H^{\ominus}(N) + \Delta_{eg}H(1)^{\ominus}(N) + \Delta_{eg}H(2)^{\ominus}(N) + \Delta_{eg}H(3)^{\ominus}(N) - \Delta_{latt}H^{\ominus}(Li_3N)$$

From the Kapustinskii equation,

$$\Delta_{latt}H^{\ominus}(Li_3N) \approx \Delta_{latt}U^{\ominus}(Li_3N) = \frac{kvz_+z_-}{r_+ + r_-} \quad (6.11)$$

$(k = 107\,900\,pm\,kJ\,mol^{-1})$

There are 4 ions in the formula unit, the charge on the cations, z_+, is 1, and the charge on the anions, z_-, is 3. The ionic radius for Li^+ is 76 pm and that for N^{3-} is 171 pm. Putting these values into the equation gives

$$\Delta_{latt}H^{\ominus}(Li_3N) = \frac{(107\,900\,\cancel{pm}\,kJ\,mol^{-1}) \times 4 \times 1 \times 3}{(76 + 171\,\cancel{pm})}$$

$$= +5240\,kJ\,mol^{-1}$$

Substituting this value into the expression for $\Delta_f H^{\ominus}(Li_3N)$ from the Born-Haber cycle gives

$$\Delta_f H^{\ominus}(Li_3N) = (3 \times +159\,kJ\,mol^{-1}) + (3 \times +520\,kJ\,mol^{-1}) + (+473\,kJ\,mol^{-1}) + (+2565\,kJ\,mol^{-1}) - (+5240\,kJ\,mol^{-1})$$

$$= -165\,kJ\,mol^{-1}$$

The negative value for $\Delta_f H^{\ominus}(Li_3N)$ confirms that the formation of Li_3N from the elements is exothermic. (Note the effect of the large value of $\Delta_{latt}H^{\ominus}(Li_3N)$ on determining the value of $\Delta_f H^{\ominus}(Li_3N)$.)

Question

Estimate a value for the enthalpy change of formation of Na_3N and comment on the stability of this compound with respect to the elements. Use the data in Tables 26.2 (p.1171), 3.6 (p.155), and 6.4 (p.286).

Compounds with oxoanions

The Group 1 metals form salts with oxoanions such as nitrates, carbonates, and sulfates.

Most Group 1 nitrates (MNO_3) decompose on heating to the nitrites (MNO_2)

$$2\,MNO_3\,(s) \xrightarrow{\Delta} 2\,MNO_2\,(s) + O_2\,(g)$$

but for $LiNO_3$ the decomposition leads to Li_2O

$$4\,LiNO_3\,(s) \xrightarrow{\Delta} 2\,Li_2O\,(s) + 4\,NO\,(g) + 3O_2\,(g)$$

These are both redox reactions, with the nitrogen of the nitrate being reduced and some of the oxygen atoms of the nitrate being oxidized.

In contrast to the nitrides, the nitrates become *more stable* with respect to decomposition as the group is descended. This can be explained using a similar argument to that used to explain the increasing stability of the peroxides with increasing cation size (Box 26.1, p.1174). The increasing stability of the nitrates down Group 1 is largely due to the decrease in the difference between the lattice enthalpies of the nitrate and nitrite, which makes decomposition less favoured.

Generally, compounds with small anions such as hydrides and nitrides tend to become *less stable* down the group. This is mainly due to the decrease in the lattice enthalpies of these compounds as the group is descended. When these compounds decompose on heating, they decompose to the elements. In contrast, compounds with large anions such as nitrates and peroxides tend to become *more stable* down the group. This is due to the decrease in the lattice enthalpies of the decomposition products.

The Group 1 carbonates are generally stable to heating, though Li_2CO_3 is an exception as it decomposes to Li_2O and CO_2. Sodium carbonate is manufactured by the Solvay process from sodium chloride and calcium carbonate, and about 50 million tonnes is produced globally every year. The biggest use of sodium carbonate is in glass manufacture.

Gunpowder contains carbon, sulfur, and potassium nitrate (saltpetre), which acts as an oxidizing agent. The introduction of gunpowder into Europe in the thirteenth century revolutionized warfare. Even the defences of castles were not strong enough to survive attack by cannons. Corfe Castle in Dorset, England was destroyed by gunpowder in 1646.

i NO_3^- nitrate; NO_2^- nitrite.

i Generally, compounds with small anions become less thermally stable down a group whereas compounds with large anions become more stable down a group.

i Lithium carbonate is used in the treatment of bipolar disorder, a debilitating illness characterized by periods of severe depression interspersed by bouts of euphoria. In 1998, researchers in Wisconsin showed that Li^+ regulates the neurotransmitter glutamate. Low concentrations of glutamate in the brain lead to depression, whereas high concentrations lead to mania. Treatment with lithium compounds stabilizes the glutamate levels.

The use of sodium carbonate in glass manufacture is described in Section 6.1 (p.257). Sodium hydrogencarbonate ($NaHCO_3$) is also known as bicarbonate of soda or baking soda. It is used in cooking as a raising agent. See Box 7.6 (p.334) for more details.

Summary

- The Group 1 metals burn in air to form Li_2O, Na_2O_2, KO_2, RbO_2, and CsO_2 as the major products.
- The Group 1 metals react violently with water to form H_2 gas and alkaline aqueous solutions of the hydroxides, MOH.
- The Group 1 metal halides are colourless ionic solids with the general formula, MX.
- The Group 1 metals react with ethyne in liquid ammonia to form ethynides MC_2H and M_2C_2, but only lithium forms a stable nitride, Li_3N.
- The ionic nature of the bonding in Group 1 compounds means that enthalpy cycles can be used to investigate trends in the stabilities of the compounds going down the group.
- The hydrides and nitrides become less stable down the group due to the decrease in their enthalpy changes of formation, which is largely a result of lattice enthalpies decreasing with increasing cation size.
- Peroxides, superoxides, nitrates, and carbonates become more stable down the group, as the decreasing difference between their lattice enthalpies and those of their decomposition products makes the decomposition reaction less favourable.
- Generally, large cations stabilize large anions.

? For practice questions on these topics, see questions 2–6 at the end of this chapter (p.1198).

26.3 Group 1 ions in solution

> The relatively small magnitude of the overall Gibbs energy change involved in dissolving a substance means that entropy changes cannot be ignored. Therefore the Gibbs energy change $\Delta_{sol}G^{\ominus}$ must be used instead of the enthalpy change $\Delta_{sol}H^{\ominus}$. See Section 13.3 (p.629) and Section 14.5 (p.675). The thermodynamics of dissolving is discussed in Section 17.4 (p.792).

> The **Gibbs energy change of hydration**, $\Delta_{hyd}G^{\ominus}$, of an ion is the Gibbs energy change for the process $A^+(g) \rightarrow A^+(aq)$ under standard conditions.

Most Group 1 salts are soluble in water. For salts containing large anions, such as chlorides, bromides, iodides, and nitrates, the solubility generally decreases down the group, and the lithium salts are the most soluble. For salts with small anions, such as fluorides and hydroxides, the solubility increases down the group, and the rubidium and caesium salts are the most soluble.

Whether an ionic compound of the general formula MX is soluble in water depends on the relative magnitudes of the lattice Gibbs energy, $\Delta_{latt}G^{\ominus}(MX)$, and the Gibbs energy changes of hydration of the ions, $\Delta_{hyd}G^{\ominus}(M^+)$ and $\Delta_{hyd}G^{\ominus}(X^-)$, as shown in Figure 26.3. This is another example of how an energy cycle can be used to explain trends down a group.

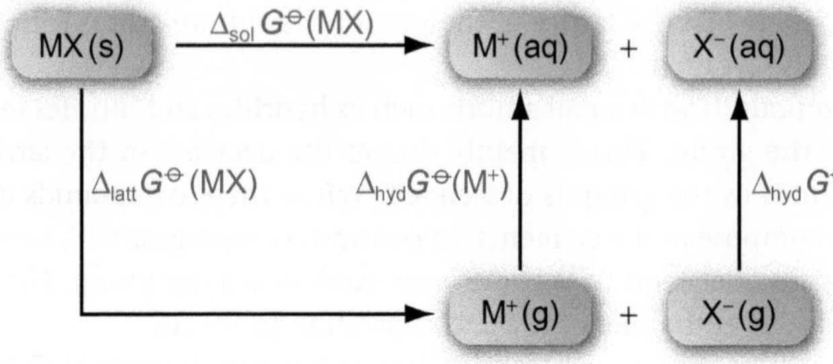

MX(s)
ΔsolG⊖(MX)
M+(aq)
+
X−(aq)
ΔlattG⊖(MX)
ΔhydG⊖(M+)
ΔhydG⊖(X−)
M+(g)
+
X−(g)

Figure 26.3 A Gibbs energy cycle for solubility.

> Visit the Online Resource Centre to view screencast 26.3 which walks you through the factors involved in the solubility of Group 1 salts.

The magnitudes of $\Delta_{latt}G^{\ominus}(MX)$, $\Delta_{hyd}G^{\ominus}(M^+)$, and $\Delta_{hyd}G^{\ominus}(X^-)$ are all large, whereas that for $\Delta_{sol}G^{\ominus}(MX)$ is small. This means that only small changes in the lattice Gibbs energies or the Gibbs energy changes of hydration are needed to cause a change from compounds being soluble to being insoluble. As a general rule, *compounds with large cations and small anions, or small cations and large anions are soluble*, whereas compounds with large cations and large anions, or small cations and small anions tend to be insoluble. The reasons for this are summarized in Figure 26.4.

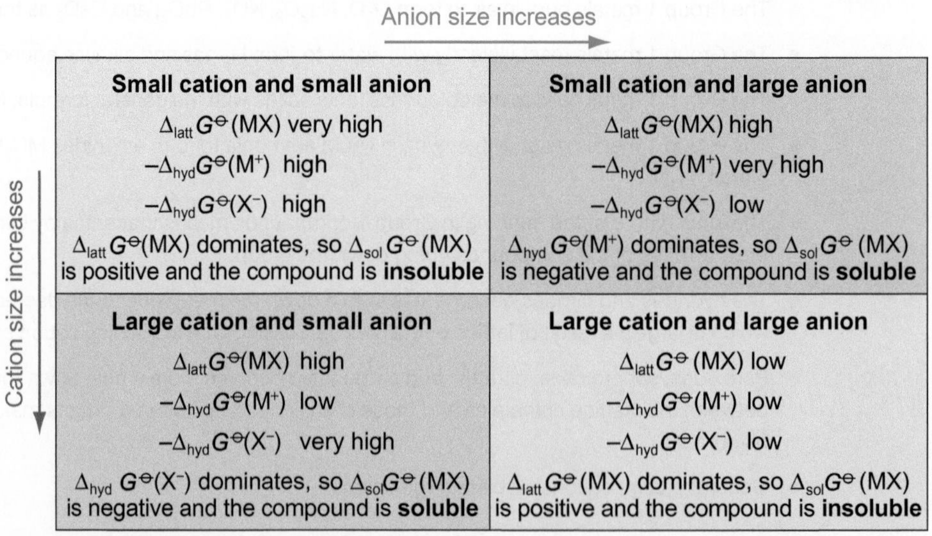

Anion size increases
Cation size increases
Small cation and small anion
ΔlattG⊖(MX) very high
−ΔhydG⊖(M+) high
−ΔhydG⊖(X−) high
ΔlattG⊖(MX) dominates, so ΔsolG⊖(MX) is positive and the compound is insoluble
Small cation and large anion
ΔlattG⊖(MX) high
−ΔhydG⊖(M+) very high
−ΔhydG⊖(X−) low
ΔhydG⊖(M+) dominates, so ΔsolG⊖(MX) is negative and the compound is soluble
Large cation and small anion
ΔlattG⊖(MX) high
−ΔhydG⊖(M+) low
−ΔhydG⊖(X−) very high
ΔhydG⊖(X−) dominates, so ΔsolG⊖(MX) is negative and the compound is soluble
Large cation and large anion
ΔlattG⊖(MX) low
−ΔhydG⊖(M+) low
−ΔhydG⊖(X−) low
ΔlattG⊖(MX) dominates, so ΔsolG⊖(MX) is positive and the compound is insoluble

Figure 26.4 The solubilities of Group 1 salts are affected by the sizes of the ions.

Summary

- For salts with large anions such as chlorides, bromides, and nitrates, the solubility in water decreases down Group 1.
- For salts with small anions such as fluorides and hydroxides, the solubility in water increases down Group 1.
- Generally, compounds are soluble if they have small cations and large anions, or large cations and small anions. In these cases the Gibbs energy change of hydration is greater than the lattice Gibbs energy.

For a practice question on these topics, see question 7 at the end of this chapter (p.1198).

26.4 Group 1 coordination chemistry

Metal ions are Lewis acids, so they are able to interact with Lewis bases. This type of interaction is the basis of **coordination chemistry**. A molecule or ion that interacts with a metal centre is known as a **ligand**, and a ligand is said to **coordinate** to a metal. A compound containing ligands coordinated to a metal is known as a coordination complex, or more simply a **complex**.

Lewis acids and bases are described in Section 7.8 (p.336). Coordination chemistry is particularly important for transition metals and is described in more detail in Chapter 28.

The Group 1 ions are singly charged and relatively large, so they have a low charge density. As a result, they are weak Lewis acids and, unlike many other metals, they coordinate only weakly to simple ligands such as water. This coordination becomes weaker down the group, as the charge density on the cation decreases.

For many years chemists believed that coordination chemistry was not important for Group 1 ions. In the late 1960s, however, Charles Pederson prepared a series of cyclic polyethers that were able to form stable complexes with the Group 1 metals in non-aqueous solvents. These ligands are known as **crown ethers**, and examples are shown in Figure 26.5. Crown ethers are examples of **macrocyclic ligands**.

In a macrocyclic ligand the donor atoms are all contained within a ring. Crown ethers, cryptands, and valinomycin are all examples of macrocyclic ligands.

The systematic names of crown ethers are long and complicated, so the common names [*x*]crown-*y* are normally used, where *x* is the number of atoms in the ring and *y* is the number of oxygen atoms. In the complexes, the metal ion sits in the centre of the ring, as shown in Figure 26.6, and interacts with lone pairs from all of the oxygen atoms.

The size of the ring is crucial in determining which Group 1 metal cation forms the most stable complex. The crown ethers are *size-selective*, as they form the most stable complexes with metal ions that fit best into the ring. For example, [18]crown-6 forms more stable complexes with K^+ than with either Na^+ or Cs^+. Na^+ is too small to coordinate to all of the oxygen atoms at the same time, whereas Cs^+ is too big to fit into the ring. The larger crown ether [21]crown-7 forms its most stable complexes with Cs^+.

Another class of ligands that binds Group 1 cations is the **cryptands**, and a potassium cryptand complex is shown in Figure 26.7. Cryptands form more stable complexes than crown ethers as less rearrangement of the ligand is needed to coordinate to the cation. Like crown ethers, cryptands are size-selective, and they form the most stable complexes with cations that fit best into the central cavity. Cryptand complexes are stable in water.

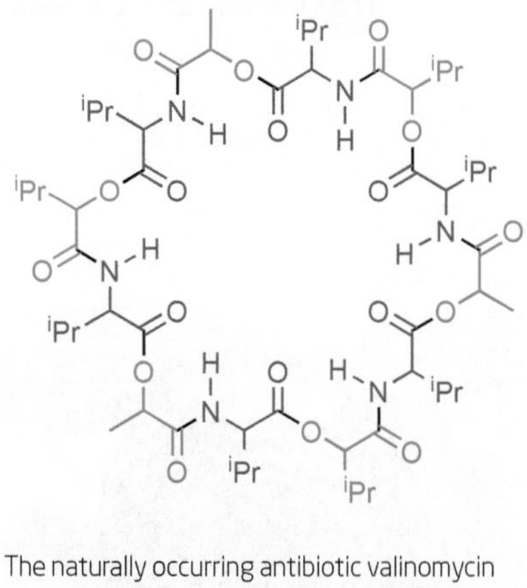

The naturally occurring antibiotic valinomycin is a cyclic molecule that coordinates K^+ in an analogous way to a crown ether. Valinomycin is a neutral molecule made up from 6 amino acids and 6 hydroxy acids alternating in a ring. It is very selective for K^+, as Na^+ is too small to interact with six oxygen atoms. The hydrophobic isopropyl groups allow valinomycin to transport K^+ ions across cell membranes, and this damages the bacteria.

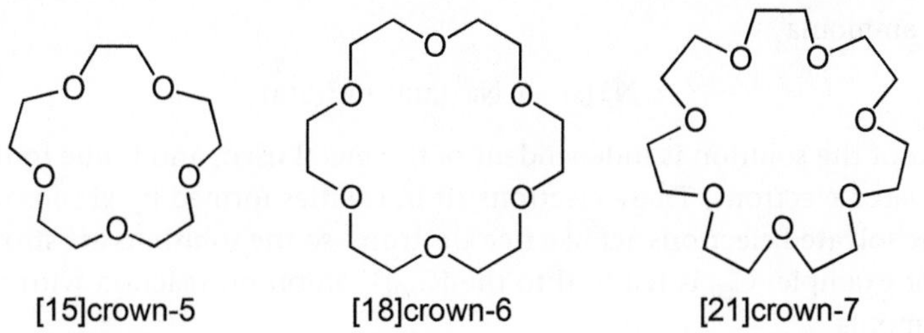
[15]crown-5
[18]crown-6
[21]crown-7

Figure 26.5 Examples of crown ethers.

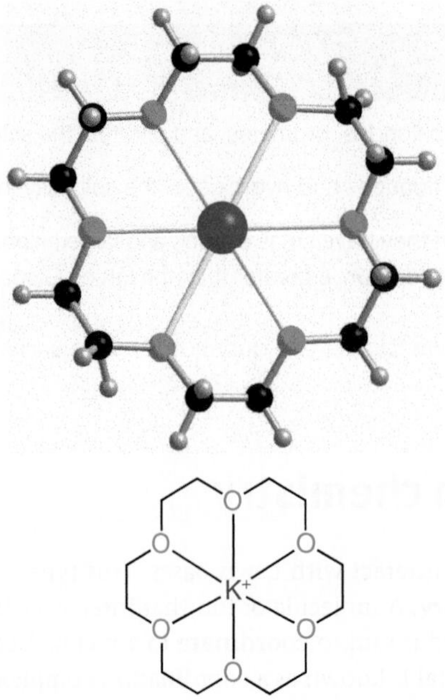
O
K+

Figure 26.6 In the complex between K^+ and [18]crown-6, the potassium ion sits in the centre of the crown ether ring and interacts with lone pairs on each of the oxygen atoms.

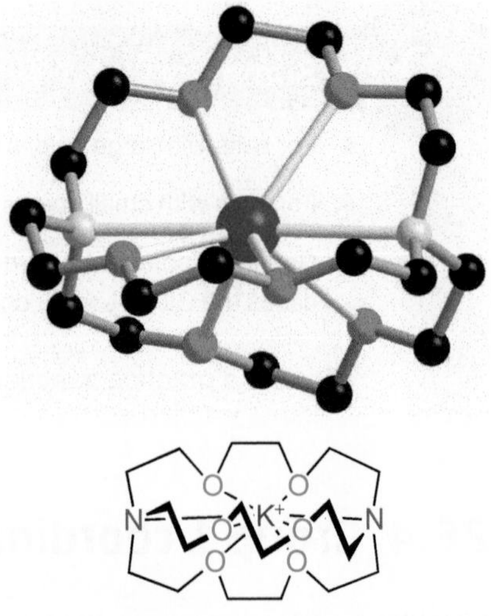
N
O
K+

Figure 26.7 Cryptands form more stable complexes with Group 1 ions than crown ethers. In the structure of the potassium complex shown, the hydrogen atoms have been removed to show the coordination more clearly.

Summary

- Group 1 ions have low charge densities so they bind only weakly to most ligands.
- Group 1 ions form stable complexes with macrocyclic ligands such as crown ethers and cryptands. These ligands are size-selective, and form the most stable complexes with the metal ions that fit best into the cavity.

? For a practice question on these topics, see question 8 at the end of this chapter (p.1198).

26.5 Reaction of Group 1 metals with liquid ammonia

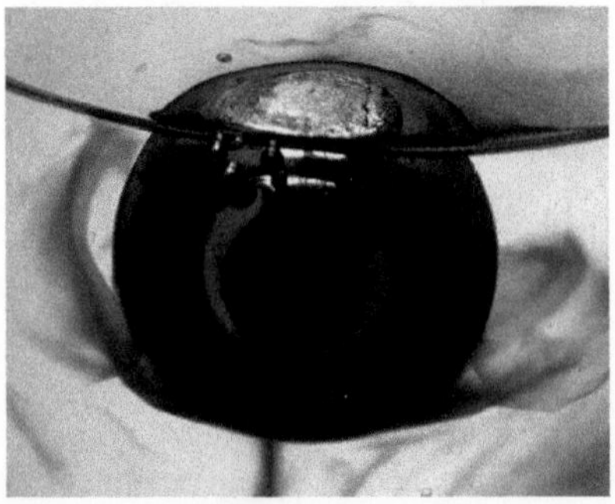

Figure 26.8 Group 1 metals, like this bead of sodium–potassium alloy, dissolve in liquid ammonia to give a dark blue solution. The colour is due to solvated electrons.

The Group 1 metals dissolve in liquid ammonia to give dark blue solutions, as shown in Figure 26.8. Unlike the reaction of sodium with water, the solvent is not being reduced. Instead the metal atoms ionize into cations and electrons, both of which are solvated by the liquid ammonia

$$Na\,(s) \rightarrow Na^+\,(am) + e^-\,(am)$$

The colour of the solution is independent of the metal used, and is due to the presence of the solvated electrons. These electrons sit in cavities formed by groups of NH_3 molecules. The solvated electrons act like free electrons, so the solutions are strong reducing agents. For example, C_{60} is reduced to the $[C_{60}]^{3-}$ anion on reaction with rubidium in liquid ammonia

$$C_{60} + 3\,Rb^+\,(am) + 3\,e^-\,(am) \xrightarrow{\text{liquid NH}_3} Rb_3[C_{60}]\,(am)$$

At high concentrations of the metal, the blue colour turns into a metallic bronze, and the liquid ammonia solution begins to conduct electricity. Spectroscopic studies show that under these conditions the electrons are delocalized throughout the solution in a similar manner to the delocalization of electrons in a solid metal.

The solvated electron solutions are thermodynamically unstable, and after several days ammonia is reduced to form sodium amide and hydrogen gas

$$2\,Na\,(s) + 2\,NH_3\,(l) \rightarrow 2\,NaNH_2\,(am) + H_2\,(g)$$

(am) denotes solution in liquid ammonia. Ammonia is a gas at room temperature, so liquid ammonia needs to be used at low temperatures, typically −60 °C.

Solvation is described in more detail in Box 7.1 (p.305).

Organic reductions using sodium in liquid ammonia are described in Section 21.2 (p.967).

Summary

- Group 1 metals dissolve in liquid ammonia to give blue solutions.
- These solutions contain solvated electrons and are very strong reducing agents.

26.6 The Group 2 elements

Group 2 consists of the elements beryllium (Be), magnesium (Mg), calcium (Ca), strontium (Sr), barium (Ba), and radium (Ra). All of the isotopes of radium are radioactive, and the longest lived is ^{226}Ra, which has a half life of 1599 years. The elements are metallic solids at room temperature and pressure, and some of their physical properties are given in Table 26.4.

The Group 2 metals have higher melting points, boiling points, and enthalpy changes of atomization than their Group 1 neighbours. This is an indication of stronger metallic bonding in the metal lattices due to the presence of two, rather than one, delocalized valence electrons per atom in the lattice.

The chemistry of the Group 2 elements is dominated by the +2 oxidation state, which is normally in the form of the M^{2+} ion. Ionic Group 2 compounds are stabilized by the relatively low values of the first and second ionization enthalpies of the metals and the high values of the lattice enthalpies for Group 2 salts. The first member of the group, beryllium, is exceptional as it forms a number of covalent compounds. The reasons behind this are explored in Section 26.9 (p.1194).

The Group 2 metals are much less reactive than Group 1 metals and, with the exception of barium, do not need to be stored under oil. Most Group 2 metals have a coating of oxide on their surfaces, which makes them less reactive than might be expected. They are

Table 26.4 Physical properties of the Group 2 elements

Element	A_r	Melting point / °C	Boiling point / °C	Density / g cm^{-3}	$\Delta_a H^{\ominus}$ / kJ mol^{-1}
Beryllium (Be)	9.01	1287	2471	1.85	+324
Magnesium (Mg)	24.31	650	1090	1.74	+147
Calcium (Ca)	40.08	842	1484	1.54	+178
Strontium (Sr)	87.62	777	1382	2.64	+164
Barium (Ba)	137.33	727	1897	3.62	+182
Radium (Ra)	226	696	1737	5	+159

sufficiently reactive, however, that the free metals do not occur naturally. Magnesium and calcium are the two most abundant Group 2 elements in the Earth's crust, and both have important biological roles. Magnesium is essential for both plants and animals, and is a component of many enzymes. It is also a component of chlorophyll, which is essential for photosynthesis (see Box 26.5, p.1192). Calcium compounds are used in nature to give rigidity, and are vital to the structures of bones and teeth.

The role of calcium in bones and teeth is described on p.1169.

The Group 2 metals are electropositive and react with acids to form M^{2+} (aq) and hydrogen gas. For example,

$$Mg(s) + 2H_3O^+(aq) \rightarrow Mg^{2+}(aq) + H_2(g) + 2H_2O(l)$$

For beryllium the reaction is slow because of the oxide coating on the surface. Beryllium is amphoteric, and also dissolves in strong alkalis to give $[Be(OH)_4]^{2-}$

A substance is **amphoteric** if it can react with both an acid and a base.

$$Be(s) + 4OH^-(aq) \rightarrow [Be(OH)_4]^{2-}(aq)$$

Preparation and uses

As well as being the most abundant Group 2 elements, magnesium and calcium are also the most important commercially. Magnesium is obtained from sea water, in which Mg^{2+} is the second most abundant cation (see Table 26.3, p.1176). In the first stage of the process, aqueous calcium hydroxide solution is added to the sea water, causing the less soluble $Mg(OH)_2$ to precipitate

$$Mg^{2+}(aq) + Ca(OH)_2(aq) \rightarrow Mg(OH)_2(s) + Ca^{2+}(aq)$$

The solid magnesium hydroxide is then treated with hydrochloric acid to produce $MgCl_2$, and the metal is liberated from molten $MgCl_2$ using electrolysis

$$Mg^{2+}(melt) + 2e^- \rightarrow Mg(l) \qquad \text{reduction at the cathode}$$

$$2Cl^-(melt) \rightarrow Cl_2(g) + 2e^- \qquad \text{oxidation at the anode}$$

Group 1 chlorides are added to the molten electrolyte to decrease the melting point of $MgCl_2$ and increase the energy efficiency of the electrolysis process.

Another source of magnesium is from minerals such as magnesite ($MgCO_3$) or dolomite ($MgCO_3{\cdot}CaCO_3$). The production of magnesium from dolomite is a two-step process. Firstly, the carbonates are converted into oxides by heating

$$MgCO_3{\cdot}CaCO_3(s) \xrightarrow{\Delta} MgO{\cdot}CaO(s) + 2CO_2(g)$$

The oxides are then treated with an alloy of iron and silicon known as ferrosilicon. The silicon in this alloy reduces the Mg^{2+} at a temperature of 1200 °C. This is above the boiling point of magnesium, so the metal vaporizes and is removed from the reaction mixture as a gas as soon as it is formed

$$2MgO{\cdot}CaO + Si \xrightarrow{\Delta} 2Mg + Ca_2SiO_4$$

The main use of metallic magnesium is in aluminium alloys, though it is also used as a desulfurization agent in the steel industry and as a reducing agent.

Calcium occurs naturally in many minerals, including limestone ($CaCO_3$), apatite [$Ca_3(PO_4)_2$], gypsum ($CaSO_4{\cdot}2H_2O$), and fluorspar (CaF_2). Limestone is the remains of the shells of ancient marine creatures. Many marine creatures use biominerals based on calcium carbonate to provide strength in a similar manner to that in which mammals use calcium phosphates such as hydroxyapatite.

The Seven Sisters cliffs in Sussex, England, are made of chalk, which is a form of $CaCO_3$. The chalk was formed about 80 million years ago from the bodies of marine micro-organisms.

Calcium metal is obtained from the reduction of CaO with aluminium

$$6CaO + 2Al \xrightarrow{\Delta} 3CaO{\cdot}Al_2O_3 + 3Ca$$

Under the high temperature conditions used, the calcium is formed as a vapour. Calcium metal does not have any large-scale industrial uses, but is used to prepare metals such as zirconium and in the manufacture of calcium hydride.

Box 26.3 Building materials

Just as nature uses calcium chemistry to provide strength for bones, teeth, and shells, humans have long used calcium compounds as construction materials. Limestone ($CaCO_3$) is a very common mineral that is used directly as a building material. When limestone is heated to 900 °C–1100 °C it loses CO_2 forming calcium oxide (CaO), known as 'quicklime'

$$CaCO_3(s) \xrightarrow{\Delta} CaO(s) + CO_2(g)$$

Kilns for making quicklime dating from the Stone Age have been discovered, and plaster based on CaO was used in the Egyptian pyramids, 4500 years ago. Calcium oxide reacts with water to form calcium hydroxide, called 'slaked lime', in an extremely exothermic reaction

$$CaO(s) + H_2O(l) \rightarrow Ca(OH)_2(s)$$

Calcium hydroxide has been used as a component of mortar in bricklaying for thousands of years. Mixed with water, $Ca(OH)_2$ slowly hardens by absorbing CO_2 from the air to reform $CaCO_3$. When fully hardened, this 'lime concrete' is water resistant.

Hydraulic cements rely on a reaction with water rather than CO_2 for their strength. **Cements** are normally based on CaO, Al_2O_3, and SiO_2. Tricalcium silicate (Ca_3SiO_5) is a typical component of cement. It hardens by absorbing water to form $Ca_3SiO_5{\cdot}2H_2O$. Long needle-like crystals of this hydrated calcium silicate are formed, which bond the cement particles together. Smaller needles then fill the gaps between the particles. This type of cement solidifies very quickly, which limits its usefulness. Adding gypsum ($CaSO_4{\cdot}2H_2O$) slows the solidification and this mixture, called Portland cement, is very commonly used. In the presence of gypsum, smaller hydrate crystals incorporating the sulfate ions form first, but these are not large enough to bridge between the cement particles. These small crystals recrystallize after 1–2 hours to give the larger crystals that lead to solidification.

Cement is used as a bonding agent in construction. When combined with sand it gives *mortar*, which is used to bond bricks or stones together. Adding gravel or other stone material gives *concrete*.

▲ Plaster based on CaO was used in the Egyptian pyramids.

Most modern *plasters* are based on calcium sulfate. The naturally occurring hydrated form of this compound, gypsum ($CaSO_4{\cdot}2H_2O$), is dehydrated by heating. *Plaster of Paris* is mainly a mixture of $CaSO_4{\cdot}0.5H_2O$ and anhydrous $CaSO_4$. When plaster sets, it absorbs water to reform the dihydrate. After an induction period, dihydrate crystals from the supersaturated solution nucleate around the particles. These needle-like crystals grow, forming a matted structure that gives the set plaster its strength. Like cement, plaster takes several days to harden to its full strength.

▲ Concrete is the most important modern construction material and is used in many buildings, roads, and bridges. 127 000 m^3 of concrete were used in the construction of the Millau Viaduct in France, which opened in 2004.

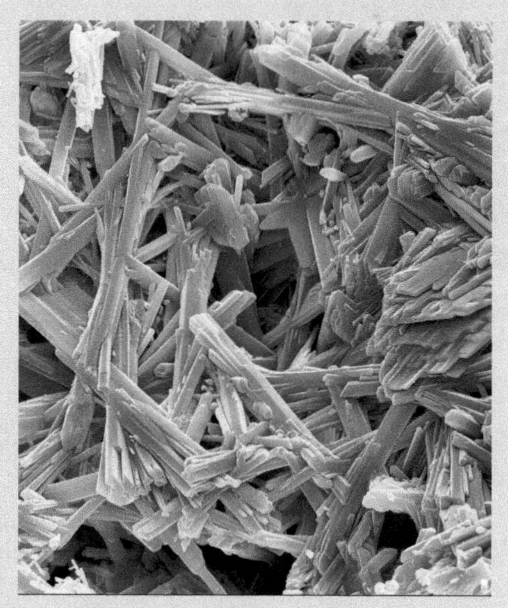

◀ $CaSO_4{\cdot}2H_2O$ crystals from set plaster of Paris.

Question

What advantages do hydraulic cements have over those based on calcium hydroxide?

Beryllium, strontium, and barium are produced on a much smaller scale than magnesium and calcium. The main sources of beryllium are the minerals beryl ($Be_3Al_2Si_6O_{18}$) and bertrandite ($Be_4Si_2O_8 \cdot H_2O$). Beryllium compounds are toxic as Be^{2+} ions can replace Mg^{2+} ions in enzymes, and disrupt their function. Beryllium metal is also extremely toxic, and exposure to beryllium dust or vapour causes a lung condition known as berylliosis. Beryllium is used in some alloys, but its use is declining due to this toxicity. Strontium occurs naturally as both the sulfate and the carbonate, whereas for barium the sulfate is the most important mineral.

Summary

- The Group 2 elements are metallic solids and typically form ionic compounds. Beryllium is exceptional in forming many covalent compounds.
- The Group 2 elements are less reactive than the Group 1 elements, but are readily oxidized to M^{2+} cations.
- The metals are produced by reduction of the oxides, but magnesium is also obtained from sea water.

? For a practice question on these topics, see question 9 at the end of this chapter (p.1198).

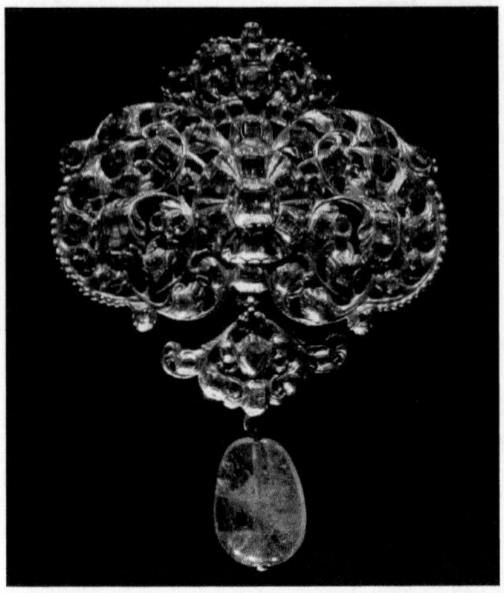

Emeralds are green varieties of the mineral beryl. The colour is caused by traces of chromium.

26.7 Group 2 compounds

Most Group 2 compounds are ionic, containing M^{2+} cations, though beryllium compounds tend to have high covalent character. As a result, the properties of many beryllium compounds are very different from those of the other Group 2 elements.

Oxides

The Group 2 metals burn in air, reacting with oxygen to give the oxides

$$2\,M(s) + O_2(g) \xrightarrow{\Delta} 2\,MO(s)$$

Magnesium burns with a bright white flame (see Figure 1.9, p.27). This reaction is used in fireworks, and calcium, strontium, and barium salts are often added to provide orange-red, crimson, and yellow-green colours, respectively. These are also the colours observed in flame tests for these elements. Neither beryllium salts nor magnesium salts give colours to flames.

The origin of the colours of fireworks is described in more detail in Box 3.2 (p.128).

Burning magnesium in air also produces some magnesium nitride (Mg_3N_2) from the reaction with nitrogen. Magnesium has been used in incendiary bombs as magnesium fires cannot be extinguished by water or CO_2. These react with the hot magnesium to form $Mg(OH)_2$ and $MgCO_3$, respectively.

The reaction of a Group 2 metal with oxygen becomes more exothermic as the group is descended, and barium can ignite in moist air. The greater reactivity of barium is also due to the lack of a protective oxide coating.

The heavier Group 2 metals are able to form peroxides. For example, BaO reacts with excess oxygen at 600 °C to form barium peroxide, BaO_2

$$2\,BaO(s) + O_2(g) \xrightarrow{\Delta} 2\,BaO_2(s)$$

See Section 6.4 (p.277 and p.280) for more details on the rock salt and wurtzite structures, respectively.

The Group 2 oxides are high melting solids. Most adopt the rock salt structure, though BeO forms the wurtzite structure.

The oxides are basic and, with the exception of BeO, dissolve in water to give alkaline solutions of the hydroxides. This reaction is very exothermic for CaO

$$CaO\,(s) + H_2O\,(l) \rightarrow Ca(OH)_2\,(aq)$$

but less so for MgO, which reacts only slowly.

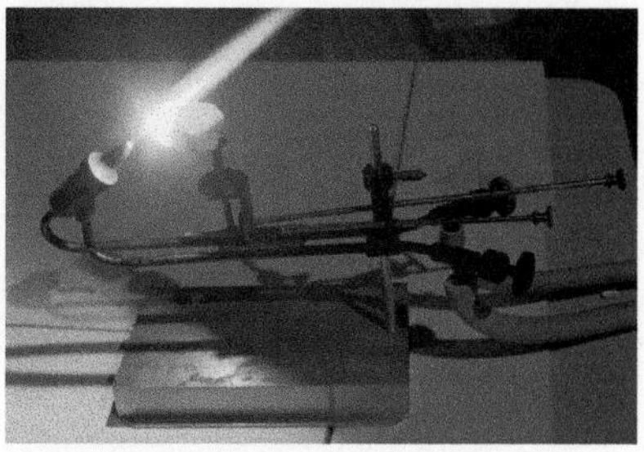

In the nineteenth century, theatre managers used calcium oxide (quicklime) to provide a spotlight. A *limelight*, such as that reconstructed above, burnt a mixture of oxygen and hydrogen over a chunk of calcium oxide, which caused the CaO to glow a brilliant white. The phrase *'in the limelight'* comes from actors becoming the centre of attention by being illuminated by these lamps.

Hydroxides

With the exception of beryllium, the Group 2 metals react with water to give the hydroxide, liberating hydrogen gas in the process. For example,

$$Ba\,(s) + 2\,H_2O\,(l) \rightarrow Ba(OH)_2\,(aq) + H_2\,(g)$$

As with Group 1, the reaction becomes more vigorous as the group is descended. Magnesium requires hot water or steam to react, whereas calcium reacts with cold water.

The Group 2 hydroxides are basic and sparingly soluble in water, though the solubility increases down the group. A suspension of $Mg(OH)_2$ in water is known as *milk of magnesia*, and is used as a stomach antacid. A solution of calcium hydroxide in water is known as *limewater*, and this is used as a test for carbon dioxide, which reacts to give a precipitate of insoluble $CaCO_3$

$$Ca(OH)_2\,(aq) + CO_2\,(g) \rightarrow CaCO_3\,(s) + H_2O\,(l)$$

The Group 2 hydroxides decompose on heating to give the oxides

$$M(OH)_2\,(s) \xrightarrow{\Delta} MO\,(s) + H_2O\,(g)$$

Beryllium hydroxide cannot be formed by reacting beryllium with water, but it can be prepared by reacting $BeCl_2$ with hydroxide ions

$$BeCl_2\,(s) + 2\,OH^-\,(aq) \rightarrow Be(OH)_2\,(s) + 2\,Cl^-\,(aq)$$

$Be(OH)_2$ is amphoteric. As well as reacting with acids in the manner expected for a hydroxide it dissolves in excess alkali to form a tetrahedral complex anion

$$Be(OH)_2\,(s) + 2\,H_3O^+\,(aq) \rightarrow Be^{2+}\,(aq) + 4\,H_2O\,(l)$$

$$Be(OH)_2\,(s) + 2\,OH^-\,(aq) \rightarrow [Be(OH)_4]^{2-}\,(aq)$$

Halides

The Group 2 metals react with the halogens to give the halides. For example, magnesium burns in chlorine gas to form magnesium chloride ($MgCl_2$)

$$Mg\,(s) + Cl_2\,(g) \xrightarrow{\Delta} MgCl_2\,(s)$$

Most of the Group 2 halides are ionic, but the beryllium compounds have polymeric structures with covalent bonds and bridging halides. The structure of beryllium chloride is shown in Figure 26.9.

Although both beryllium chloride and beryllium hydride form structures in which atoms bridge between beryllium atoms, the bonding in the two compounds is very different. Beryllium hydride is electron deficient, and the Be–H–Be bridges involve 3-centre 2-electron bonding. Beryllium chloride is not electron deficient because the bridging chloride donates three electrons, one through a covalent bond and two through a dative bond.

The halides are generally water soluble, though the fluorides tend to be less soluble than the other halides due to their higher lattice enthalpies.

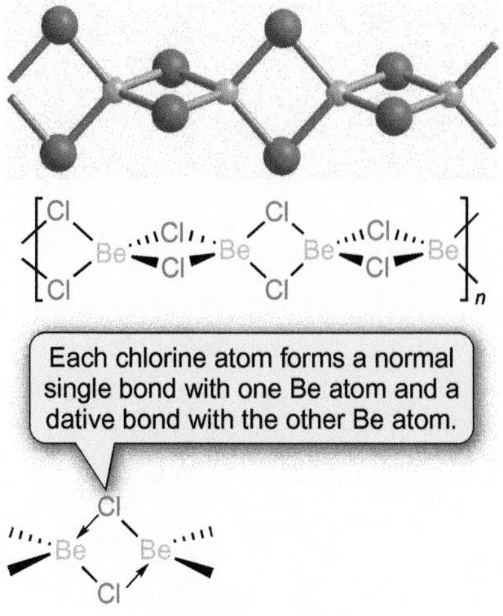
Cl
Be
Cl
Be
n
Each chlorine atom forms a normal single bond with one Be atom and a dative bond with the other Be atom.

Figure 26.9 The structure of $BeCl_2$ contains bridging chlorine atoms.

The structure of BeH_2 is described in Section 25.2 (p.1146).

Worked example 26.2 Why doesn't MgCl exist?

Use enthalpy cycles to explain why MgCl does not exist.

Strategy

To answer this, you first need to know whether MgCl is stable with respect to the elements. This can be found out by calculating its enthalpy change of formation using a Born–Haber cycle as in Worked example 26.1. Depending on the answer, you may then need to examine the stability of MgCl relative to $MgCl_2$.

Solution

The Born–Haber cycle below allows you to calculate $\Delta_f H^{\ominus}(\text{MgCl})$.

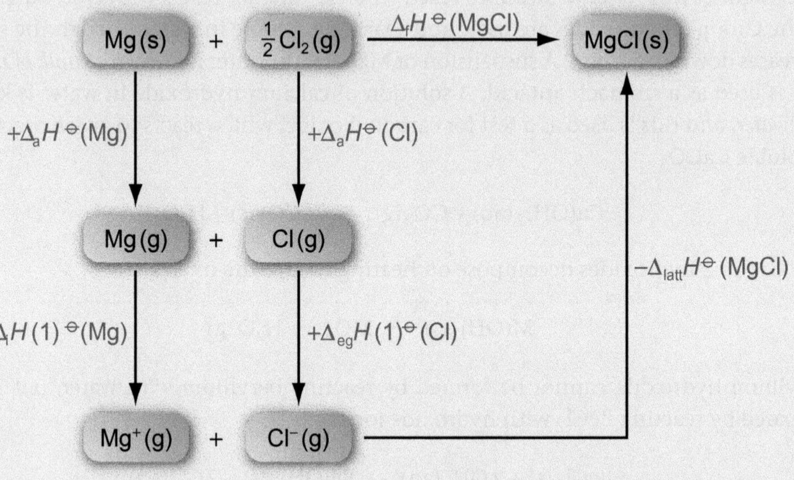

Mg (s)
+
½ Cl₂ (g)
ΔfH⦵ (MgCl)
MgCl (s)
+ΔaH⦵(Mg)
+ΔaH⦵ (Cl)
Mg (g)
+
Cl (g)
−ΔlattH⦵ (MgCl)
+ΔiH (1)⦵(Mg)
+ΔegH (1)⦵ (Cl)
Mg⁺ (g)
+
Cl⁻ (g)

$\Delta_a H^{\ominus}(\text{Mg}) = +147\,\text{kJ mol}^{-1}$ $\quad \Delta_a H^{\ominus}(\text{Cl}) = +121\,\text{kJ mol}^{-1}$

$\Delta_i H(1)^{\ominus}(\text{Mg}) = +738\,\text{kJ mol}^{-1}$ $\quad \Delta_{eg} H(1)^{\ominus}(\text{Cl}) = -349\,\text{kJ mol}^{-1}$

Use the Kapustinskii equation (Equation 6.11) to estimate a value for $\Delta_{latt} H^{\ominus}(\text{MgCl})$

$$\Delta_{latt} H^{\ominus} \approx \Delta_{latt} U^{\ominus} \frac{k\nu z_+ z_-}{r_+ + r_-} \qquad (k = 107\,900\,\text{pm kJ mol}^{-1}) \qquad (6.10)$$

The ionic radius for Cl^-, r_-, is 181 pm. Since Mg^+ compounds are unknown, the ionic radius of Mg^+ is also unknown. A reasonable assumption is that the ionic radius of Mg^+ is similar to that of the ionic radius of the neighbouring ion Na^+ (102 pm). Substituting these values,

$$\Delta_{latt} H^{\ominus}(\text{MgCl}) \approx \Delta_{latt} U^{\ominus}(\text{MgCl}) = \frac{(107\,900\,\cancel{\text{pm}}\,\text{kJ mol}^{-1}) \times 2 \times 1 \times 1}{(102 + 181)\,\cancel{\text{pm}}} = +763\,\text{kJ mol}^{-1}$$

From the Born–Haber cycle,

$$\begin{aligned}\Delta_f H^{\ominus}(\text{MgCl}) &= \Delta_a H^{\ominus}(\text{Mg}) + \Delta_i H(1)^{\ominus}(\text{Mg}) + \Delta_a H^{\ominus}(\text{Cl}) + \Delta_{eg} H(1)^{\ominus}(\text{Cl}) - \Delta_{latt} H^{\ominus}(\text{MgCl}) \\ &= (+147\,\text{kJ mol}^{-1}) + (+738\,\text{kJ mol}^{-1}) + (+121\,\text{kJ mol}^{-1}) + (-349\,\text{kJ mol}^{-1}) \\ &\quad - (+763\,\text{kJ mol}^{-1}) \\ &= -106\,\text{kJ mol}^{-1}\end{aligned}$$

This is a negative value, which means that MgCl is stable with respect to the elements. So why doesn't it exist? While a negative value of $\Delta_f S^{\ominus}(\text{MgCl})$ is a factor, another possibility is ➜

→ that MgCl is unstable with respect to disproportionation to Mg and $MgCl_2$. To assess this, construct a second enthalpy cycle, and calculate the enthalpy change of disproportionation, $\Delta_{dis}H^{\ominus}(MgCl)$.

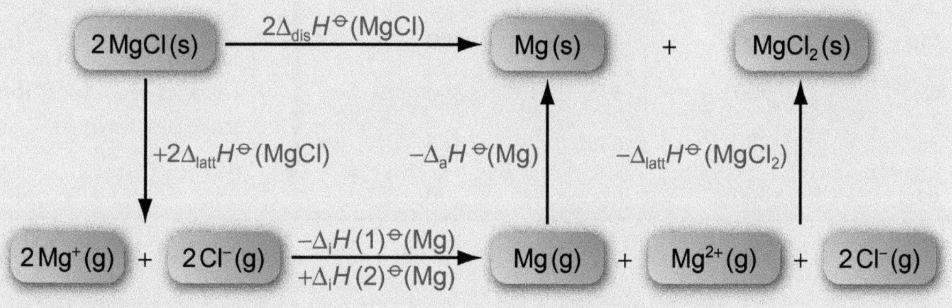
2 MgCl (s)
2Δdis H⊖(MgCl)
Mg (s)
+
MgCl2 (s)
+2Δlatt H⊖(MgCl)
−Δa H⊖(Mg)
−Δlatt H⊖(MgCl2)
2 Mg+ (g)
+
2 Cl− (g)
−Δi H (1)⊖(Mg)
+Δi H (2)⊖(Mg)
Mg (g)
+
Mg2+ (g)
+
2 Cl− (g)

In addition to the terms you used in the previous calculation, you need the second ionization enthalpy of magnesium, $\Delta_i H(2)^{\ominus}(Mg)$, which is +1451 kJ mol^{-1}, and the lattice enthalpy for $MgCl_2$. Again, estimate this using the Kapustinskii equation. The ionic radius for Mg^{2+} is 72 pm.

$$\Delta_{latt}H^{\ominus}(MgCl_2) \approx \Delta_{latt}U^{\ominus}(MgCl_2) = \frac{(107\,900\ \cancel{pm}\ kJ\,mol^{-1}) \times 3 \times 2 \times 1}{(72+181)\ \cancel{pm}} = +2560\ kJ\,mol^{-1}$$

From the enthalpy cycle

$2\Delta_{dis}H^{\ominus}(MgCl)$

$= 2\Delta_{latt}H^{\ominus}(MgCl) + \Delta_i H(2)^{\ominus}(Mg) - \Delta_i H(1)^{\ominus}(Mg) - \Delta_{latt}H^{\ominus}(MgCl_2) - \Delta_a H^{\ominus}(Mg)$

$= (2 \times +763\ kJ\,mol^{-1}) + (+1451\ kJ\,mol^{-1}) - (+738\ kJ\,mol^{-1}) - (+2560\ kJ\,mol^{-1}) - (+147\ kJ\,mol^{-1})$

$= -468\ kJ\,mol^{-1}$

So $\Delta_{dis}H^{\ominus}(MgCl) = -234\ kJ\,mol^{-1}$. The disproportionation reaction is strongly exothermic, so the reason for the non-existence of MgCl is its instability with respect to disproportionation. If MgCl ever formed, it would disproportionate into Mg and $MgCl_2$.

Question

Calculate the enthalpy change for the disproportionation of CaF to CaF_2 and Ca. Use your answer to comment on whether calcium(I) fluoride is known.

$\Delta_a H^{\ominus}(Ca) = +178\ kJ\,mol^{-1}$ $\quad \Delta_a H^{\ominus}(F) = +79\ kJ\,mol^{-1}$

$\Delta_i H(1)^{\ominus}(Ca) = +590\ kJ\,mol^{-1}$ $\quad \Delta_{eg}H(1)^{\ominus}(F) = -328\ kJ\,mol^{-1}$

$\Delta_i H(2)^{\ominus}(Ca) = +1145\ kJ\,mol^{-1}$

The ionic radii for Ca^{2+}, K^{+}, and F^{-} are 100 pm, 138 pm, and 133 pm, respectively.

Visit the Online Resource Centre to view screencast 26.4 which walks you through the explanation of why magnesium(I) chloride does not exist.

The reaction between calcium carbide and water has been used to provide light underground for caving and mining, and was once used in lighthouses. Water is dripped slowly onto CaC_2, and the ethyne released is burned to provide light.

Compounds with carbon or nitrogen

All of the Group 2 metals form binary compounds with carbon and nitrogen. Although commonly called carbides, the compounds MC_2 are really ethynides, since they contain the C_2^{2-} anion and react with water to give ethyne. For example,

$$CaC_2(s) + 2H_2O(l) \rightarrow Ca(OH)_2(aq) + C_2H_2(g)$$

The Group 2 metals all react with nitrogen on heating to give nitrides

$$3Mg(s) + N_2(g) \xrightarrow{\Delta} Mg_3N_2(s)$$

The nitrides react with water to give the metal hydroxide and ammonia

$$Mg_3N_2(s) + 6H_2O(l) \rightarrow 3Mg(OH)_2(aq) + 2NH_3(aq)$$

Calcium carbide is an ethynide ($Ca^{2+}\ {}^{-}C{\equiv}C^{-}$). The old name is calcium acetylide.

The chemistry of alkynes is described in Chapter 21. Group 1 ethynides are described in Section 26.2 (p.1177).

Table 26.5 Decomposition temperatures (°C) of the Group 2 carbonates, nitrates, and hydroxides

	MCO_3	$M(NO_3)_2$	$M(OH)_2$
Be		125	
Mg	400	450	300
Ca	900	575	390
Sr	1280	635	466
Ba	1360	675	700

The stability of MCO_3, $M(NO_3)_2$, and $M(OH)_2$ to thermal decomposition increases down the group.

Compounds with oxoanions

The temperature at which a reaction becomes spontaneous on heating is calculated in Worked example 14.7 (p.676).

Group 2 carbonates are not very soluble in water and decompose on heating to the oxides and CO_2

$$MCO_3(s) \xrightarrow{\Delta} MO(s) + CO_2(g)$$

The decomposition temperature of the carbonates increases as the group is descended, as shown in Table 26.5.

The decomposition temperature is related to the temperature at which the reaction becomes spontaneous, that is, when $\Delta_r G^\ominus$ becomes negative. For decomposition of a carbonate, $\Delta_r S^\ominus$ is positive and largely independent of M. This means that any changes in $\Delta_r G^\ominus$ are caused by changes in $\Delta_r H^\ominus$. $\Delta_r H^\ominus$ is related in turn to differences between the lattice enthalpies of MCO_3 and MO, in a similar manner to that described in Box 26.1 for the Group 1 peroxides and oxides.

The Group 2 nitrates also decompose on heating to give the oxides

$$2\,M(NO_3)_2(s) \xrightarrow{\Delta} 2\,MO(s) + 4\,NO(g) + 3\,O_2(g)$$

and the decomposition temperatures for these compounds are given in Table 26.5. As with the carbonates, the compounds become more stable down the group. The sulfates decompose on strong heating

$$MSO_4(s) \xrightarrow{\Delta} MO(s) + SO_3(g)$$

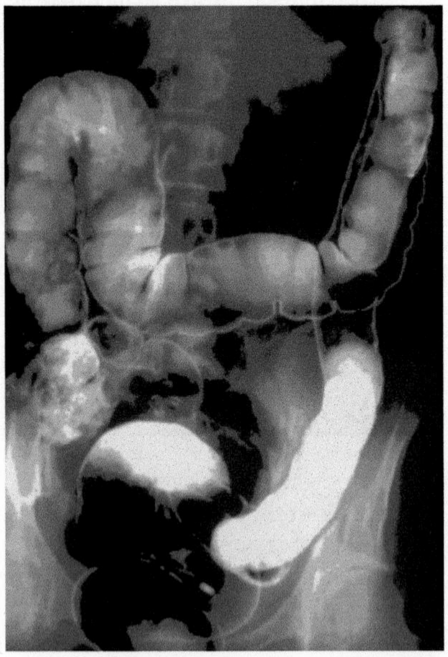

Barium sulfate is used as an X-ray contrast agent in images of the digestive tract. The patient is given a 'barium meal' containing barium sulfate. The high atomic number of barium means that it strongly absorbs X-rays, so its location is easy to see. In the photograph of a colon, the $BaSO_4$ shows up as the bright white area.

The trends in the solubility in water of Group 2 compounds are similar to those for Group 1 but, as the lattice enthalpies are higher, the compounds are generally less soluble. The solubility trends for the Group 2 sulfates are summarized in Figure 26.10.

Compounds with large cations and large anions tend to be insoluble. For example, $MgSO_4$ is very soluble, $CaSO_4$ much less so, and $BaSO_4$ very poorly soluble. Indeed, barium chloride solution is used to test for aqueous sulfate ions, as their presence leads to a dense white precipitate of $BaSO_4$.

Hydrated magnesium sulfate, $MgSO_4{\cdot}7H_2O$, is known medicinally as Epsom salts and is used as a relaxant and a laxative. The Mg^{2+} ions inhibit the absorption of water in the stomach, and the increased flow of water through the intestines helps flush them out.

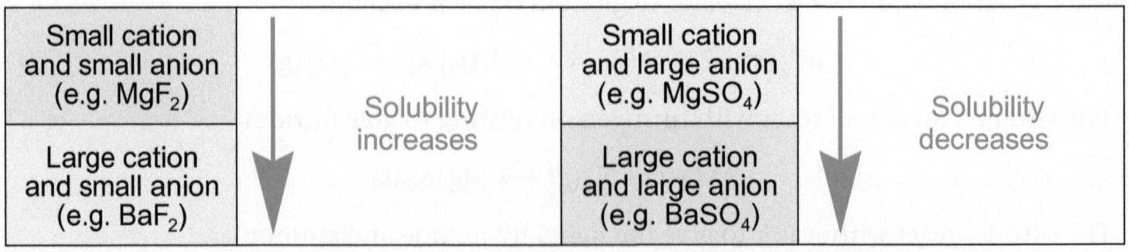
Small cation and small anion (e.g. MgF2)
Large cation and small anion (e.g. BaF2)
Solubility increases
Small cation and large anion (e.g. MgSO4)
Large cation and large anion (e.g. BaSO4)
Solubility decreases

Figure 26.10 The size of the cation affects the solubilities of Group 2 compounds.

Box 26.4 Hard water

Water that contains significant concentrations of ions such as Ca^{2+} is known as *hard water*. Soaps don't lather well in hard water and can form scums of insoluble calcium stearate (see Box 24.6 (p.1120)). If the main anion present is HCO_3^-, the hardness is described as *temporary*. When water containing HCO_3^- is heated, the anion decomposes

$$2\,HCO_3^-(aq) \xrightarrow{\Delta} CO_3^{2-}(aq) + CO_2(g) + H_2O(l)$$

If Group 2 cations are present, this reaction causes the insoluble Group 2 carbonates to precipitate. The limescale deposits observed in kettles and boilers in hard water areas are mainly calcium carbonate. *Permanent* hardness occurs when the main anion is sulfate. Permanent hardness cannot be removed by boiling.

Water softening involves removal of the Group 2 cations. Dishwashers contain ion exchange resins so that the clean utensils are not left with a coating of calcium carbonate. These resins are anionic polymer beads which, in the active form, contain Na^+ ions to balance the charges. When water containing Mg^{2+} or Ca^{2+} ions is in contact with the resin, the Group 2 cations bind to the resins and release the Na^+ ions into solution. This reduces the concentration of Group 2 cations in solution and prevents precipitation of the carbonates. You need to add salt at regular intervals to a dishwasher in order to regenerate the ion exchange resin, and replace the Group 2 cations with Na^+ cations.

While ion exchange works by relatively simple chemistry, there are many other products available for water softening for which the scientific background is less clear cut. The use of magnetic and electromagnetic fields for water softening has been around since the 1930s. How are these devices supposed to work? Many of the suppliers' websites mention pseudoscientific claims such as 'resonant energy forces'. A more reasonable theory is that magnetic fields change the form of the $CaCO_3$ crystals that are precipitated. There have been some scientific studies on this and, while a few support the idea, most have not reported a significant effect.

It would be unscientific to say simply that these magnetic treatments cannot work. However, if they do, it is surprising that after 70 years there is still no convincing evidence to support them.

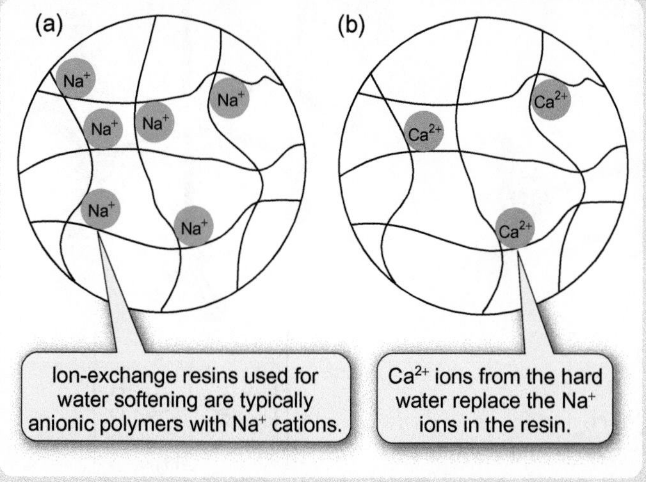
(a)
(b)
Na+
Ca2+
Ion-exchange resins used for water softening are typically anionic polymers with Na+ cations.
Ca2+ ions from the hard water replace the Na+ ions in the resin.

▲ In an ion-exchange resin, Ca^{2+} ions are removed from water and replaced by Na^+ ions.

▲ When temporary hard water is heated in a washing machine, calcium carbonate slowly precipitates and 'furs up' the heating element.

Question

Many limescale removers contain citric acid. Suggest how this works.

Summary

- The Group 2 metals burn in air to form the oxides, MO. For the heavier Group 2 metals, peroxides, MO_2, are also formed.
- With the exception of beryllium, the Group 2 elements react with water to give the hydroxide, $M(OH)_2$, and H_2 gas.
- With the exception of beryllium, the Group 2 elements form ionic halides, MX_2.
- The carbonates, nitrates, and hydroxides decompose on heating to oxides, with the decomposition temperature increasing down the group.

? For practice questions on these topics, see questions 10–16 at the end of this chapter (p.1198).

26.8 Group 2 coordination chemistry

The relationship between a metal ion and the type of ligands with which it forms the most stable complexes is discussed in Section 28.3 (p.1277).

The Group 2 M^{2+} ions are smaller than the Group 1 M^+ ions, and have twice the charge. This means that the charge densities on the Group 2 cations are much higher than those on Group 1 cations. As a result, Group 2 coordination chemistry is more extensive than that of Group 1. The most important ligands are those with oxygen and nitrogen donor atoms.

The increasing size of the M^{2+} ions down the group influences the coordination number, which is the number of atoms that are coordinated to the metal ion. Be^{2+} typically has a coordination number of 4, Mg^{2+} has a coordination number of 6, whereas the coordination numbers of Ca^{2+}, Sr^{2+}, and Ba^{2+} can be higher.

Group 2 cations are more strongly hydrated than Group 1 cations and, in aqueous solution, water molecules are coordinated to the cations, as shown in Figure 26.11. The charge density on the small Be^{2+} cation is so large that in solution the ion is always coordinated, and in aqueous solution it forms $[Be(H_2O)_4]^{2+}$ ions. Unlike the other Group 2 cations, solutions of Be^{2+} are acidic because of hydrolysis. The high charge density on the beryllium ion polarizes the O–H bonds of the coordinated water molecules, so that H^+ ions are more easily removed

The hydrolysis of $[Fe(H_2O)_6]^{3+}$ is described in Box 7.3 (p.312).

$$[Be(H_2O)_4]^{2+}(aq) + H_2O(l) \rightleftharpoons [Be(H_2O)_3(OH)]^+(aq) + H_3O^+(aq)$$

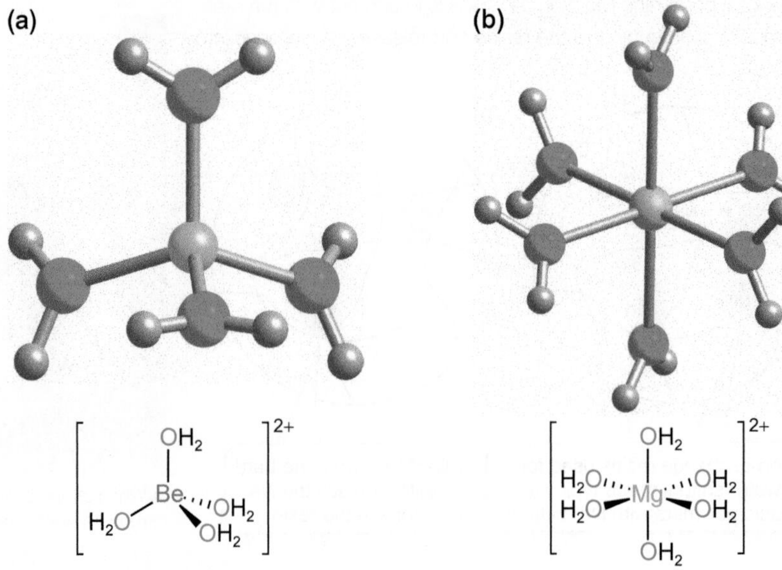
(a)
(b)
OH2
Be
H2O
OH2
OH2
2+
OH2
H2O
OH2
Mg
H2O
OH2
OH2
2+

Figure 26.11 The structures of the hydrated cations (a) $[Be(H_2O)_4]^{2+}$ and (b) $[Mg(H_2O)_6]^{2+}$.

Box 26.5 Chlorophylls

Chlorophylls are the pigment molecules that give leaves their green colour. They do this by absorbing light in both the red and violet parts of the visible light spectrum. Green light is not absorbed—it is reflected and is the colour you observe (see Box 10.7, p.491).

The absorption of light by chlorophylls is vitally important for life, as it provides the energy for photosynthesis, which is the process by which plants convert carbon dioxide and water into glucose. The overall change can be summarized as

$$6CO_2(g) + 6H_2O(l) \rightarrow C_6H_{12}O_6(aq) + 6O_2(g)$$

This is a redox reaction in which carbon dioxide is reduced by water.

Chlorophylls are magnesium complexes, and the structures of two of these compounds are shown on page 1193.

Both chlorophyll *a* and chlorophyll *b* contain a Mg^{2+} ion lying in the centre of a chlorin ring, and differ only in one of the side groups on the ring. The different side groups in chlorophyll *a* and chlorophyll *b* mean that they absorb light in slightly different parts of the visible spectrum, so the combination of the two molecules is more efficient at absorbing light than either on its own.

→

➜

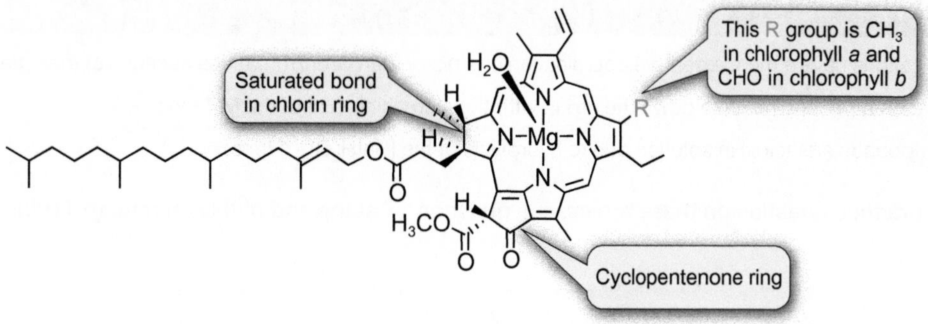
Saturated bond in chlorin ring
This R group is CH_3 in chlorophyll *a* and CHO in chlorophyll *b*
Cyclopentenone ring

▲ The structures of chlorophyll *a* and chlorophyll *b*.

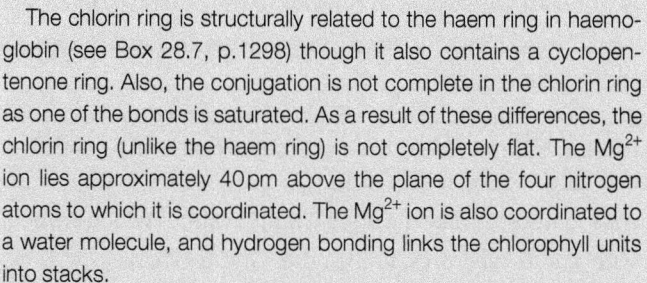

▲ The green colour of most plants is caused by the magnesium compound chlorophyll.

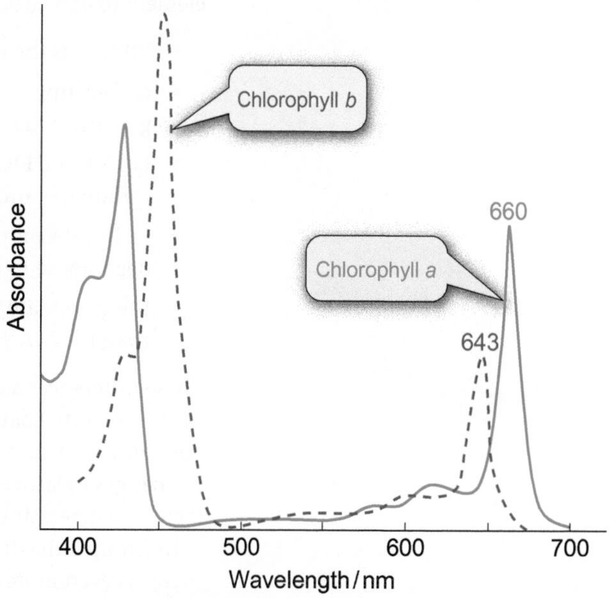
Chlorophyll *b*
Chlorophyll *a*
660
643
Absorbance
400
500
600
700
Wavelength/nm

▲ The absorption spectra of chlorophyll *a* and chlorophyll *b*.

The chlorin ring is structurally related to the haem ring in haemoglobin (see Box 28.7, p.1298) though it also contains a cyclopentenone ring. Also, the conjugation is not complete in the chlorin ring as one of the bonds is saturated. As a result of these differences, the chlorin ring (unlike the haem ring) is not completely flat. The Mg^{2+} ion lies approximately 40 pm above the plane of the four nitrogen atoms to which it is coordinated. The Mg^{2+} ion is also coordinated to a water molecule, and hydrogen bonding links the chlorophyll units into stacks.

When a chlorophyll molecule absorbs light, an electron is excited to a higher energy level. In this excited state, the electron is more readily transferred to another molecule. This starts a chain of electron transfer steps that ends with an electron being transferred to CO_2, which is the first step in its reduction. The chlorophyll that gave up an electron accepts an electron from another molecule at the end of a chain process that started with the removal of an electron from water. Overall, chlorophyll catalyses a photosynthetic redox reaction in which electrons are transferred from water to carbon dioxide

$$2H_2O \rightarrow O_2 + 4H^+ + 4e^- \quad \text{oxidation of water}$$

$$CO_2 + 4H^+ + 4e^- \rightarrow \tfrac{1}{6}C_6H_{12}O_6 + H_2O \quad \text{reduction of carbon dioxide}$$

The Mg^{2+} ion serves several purposes in chlorophyll—it keeps the chlorin rings rigid, so that energy is not lost as vibrations, it connects the molecules into stacks that facilitate electron transfer over distances of up to 2000 nm, and it increases the lifetime of the excited state so that the electron can be transferred.

Question

Suggest why the leaves on deciduous trees change colour and are lost in autumn.

Summary

- Group 2 cations have a more extensive coordination chemistry than Group 1 cations because of their greater charge density.
- Coordination numbers increase down the group with the increasing size of the M^{2+} ions.
- Be^{2+} compounds are acidic in solution due to hydrolysis of the $[Be(H_2O)_4]^{2+}$ ions.

For a practice question on these topics, see question 17 at the end of this chapter (p.1198).

26.9 Lithium and beryllium as exceptional elements

From the reactions described earlier in this chapter, you can see that the chemistry of the first element in each group is different from that of the other elements. For example, in Group 1:

- lithium is the only Group 1 metal that reacts with N_2 to give a nitride;
- on burning in air, lithium gives the oxide Li_2O, whereas the other Group 1 metals give peroxides or superoxides;
- Li_2CO_3 and LiOH both decompose on heating to give Li_2O, whereas the other Group 1 carbonates and hydroxides are generally stable to heating;
- $LiNO_3$ decomposes on heating to form Li_2O, whereas the other Group 1 nitrates decompose to the nitrites, MNO_2;
- the carbonate, fluoride, and hydroxide are all far less soluble in water for lithium than they are for the other Group 1 metals.

These differences suggest that lithium is somehow anomalous. This behaviour, however, fits the expected pattern of changes that occur with a decrease in the size of the cation and consequent increase in the charge density on the ion. The small cation size leads directly to changes in lattice enthalpies and Gibbs energy changes of hydration, which are the key factors in determining the thermal stabilities and solubilities of the Group 1 compounds.

In Group 2 the difference between the chemistry of beryllium and that of the other elements is even more dramatic:

- beryllium metal does not react with water, in contrast to the other Group 2 metals;
- beryllium is amphoteric—it reacts with acids as do the other Group 2 metals, but it also reacts with alkalis to form $[Be(OH)_4]^{2-}$;
- beryllium compounds have a much greater covalent character than those of the other Group 2 metals—for example, $BeCl_2$ has a polymeric covalently bonded structure, whereas the other chlorides are ionic solids;
- Be^{2+} is acidic in aqueous solution, unlike the other Group 2 cations which are neutral;
- BeO does not react with water, in contrast to the other Group 2 oxides.

Similar arguments apply here as for Group 1. The small size of Be^{2+} means that the charge density on this ion is very much higher than that on the other Group 2 cations and also makes lattice enthalpies very high. In addition, the high charge density means that the free Be^{2+} ion is strongly polarizing. It draws the electrons of neighbouring ions towards itself, which gives these bonds a much higher degree of covalent character than equivalent bonds of other Group 2 cations. The increase in electronegativity of the Group 2 element as the group is descended also contributes to the decrease in covalent character, as shown in the van Arkel–Ketelaar triangle for $BeCl_2$, $MgCl_2$, and $CaCl_2$ shown in Figure 26.12.

The polarizing power of the small Be^{2+} cation also explains the acidity of the aqueous solutions.

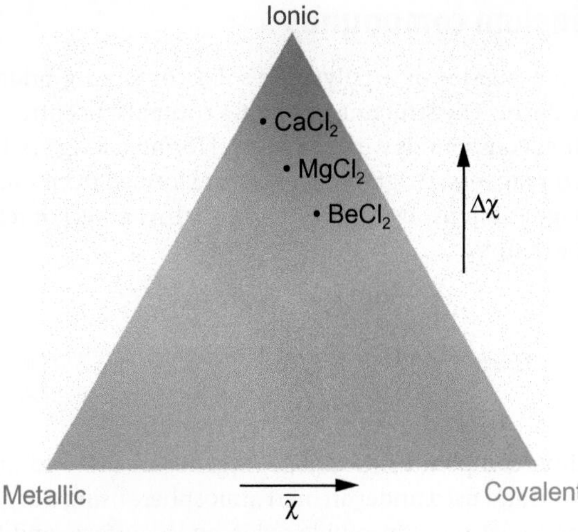
Ionic
• $CaCl_2$
• $MgCl_2$
• $BeCl_2$
$\Delta\chi$
Metallic
$\overline{\chi}$
Covalent

Figure 26.12 A van Arkel–Ketelaar triangle for the first three Group 2 chlorides. The compounds move towards the ionic corner as the group is descended.

van Arkel–Ketelaar triangles are introduced in Section 6.7, p.298.

Summary

- The anomalous nature of lithium and beryllium is due to the small size of the Li^+ and Be^{2+} ions, and their high charge densities. Although this leads to the chemistry of these ions being very different from the other members of their groups, this behaviour fits with predictions from the ionic model.

For a practice question on these topics, see question 18 at the end of this chapter (p.1198).

26.10 Organometallic compounds

The use of organolithium compounds in organic synthesis is described in Section 23.2 (p.1068).

Compounds containing one or more bonds between a metal and carbon are known as **organometallic**. Organometallic compounds are known for all of the Group 1 and Group 2 metals, but those of lithium and magnesium are the most common and most important synthetically. Since ions of these metals have high charge densities, there is a large degree of covalent character in the bonding of the organometallic compounds. The carbon atoms are polarized δ–, so the compounds are nucleophilic and used extensively in organic synthesis.

An **oligomer** is a polymer that contains only a small number of monomer units.

Organolithium compounds

Organolithium compounds of the general formula RLi (R = alkyl or aryl) are prepared by reacting lithium with a halogenoalkane

$$RI + 2\,Li \rightarrow RLi + LiI$$

Organolithium compounds are both air- and water-sensitive, so they must be prepared under an inert atmosphere (usually argon) using a solvent that has been rigorously dried. Two of the most important organolithium compounds are methyllithium (MeLi) and *n*-butyllithium (BuLi).

Although normally represented by the simple formula RLi, organolithium compounds exist as oligomers both in solution and the solid state. For example, solid methyllithium contains tetrameric $[MeLi]_4$ units as shown in Figure 26.13.

The 'cubic' Li_4C_4 cluster contains a tetrahedron of lithium atoms, with triply bridging methyl groups on the centre of each of the four faces. The bonding in $[MeLi]_4$ is mainly ionic, though with significant covalent character.

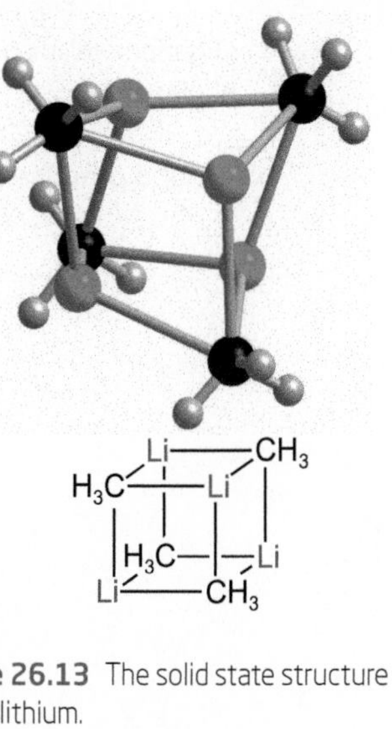
Li
CH_3
H_3C
Li
H_3C
Li
Li
CH_3

Figure 26.13 The solid state structure of methyllithium.

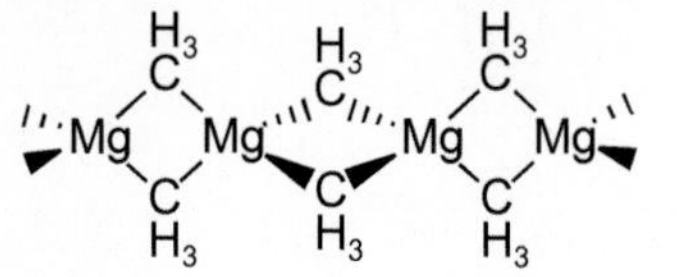

Figure 26.14 The solid state structure of $MgMe_2$ contains bridging methyl groups.

Organomagnesium compounds

Dimethylmagnesium, $MgMe_2$, is a polymeric solid containing bridging methyl groups, as shown in Figure 26.14. The structure of $MgMe_2$ contains 3-centre 2-electron bonding.

Organomagnesium compounds with the general formula RMgX, where X is a halide, are known as **Grignard reagents**. Grignard reagents are formed from the reaction of a halogenoalkane with magnesium using ethoxyethane (diethyl ether) or tetrahydrofuran (THF) as the solvent. For example,

$$\text{EtBr} + \text{Mg} \rightarrow \text{EtMgBr}$$

3-centre 2-electron bonding is described in Section 5.7 (p.250).

The use of Grignard reagents in organic synthesis is described in Section 23.2 (p.1068).

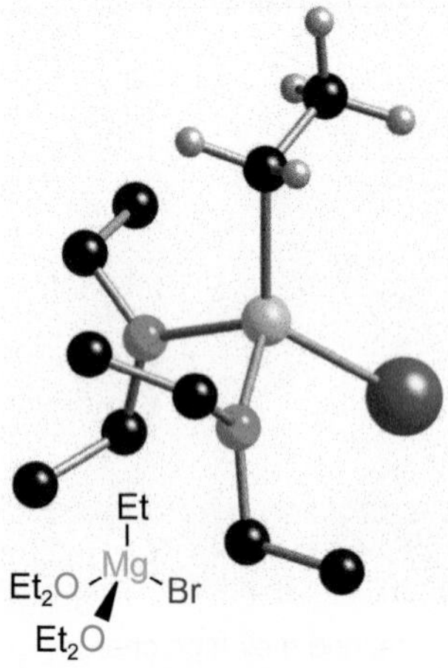
Et
Mg
Et_2O
Br
Et_2O

Figure 26.15 EtMgBr crystallizes with two coordinated molecules of ethoxyethane, which are shown without hydrogen atoms for clarity.

Like organolithium compounds, Grignard reagents are air- and moisture-sensitive, so they must be prepared and used under an inert atmosphere using a dry solvent. Magnesium metal normally has a coating of insoluble oxide on its surface, and this acts as a kinetic barrier to reaction. This problem is overcome by adding a trace amount of iodine or 1,2-dibromoethane, as these are able to penetrate the oxide layer and react with the metal to expose a more reactive surface.

Although normally represented by the simple formula RMgX, the actual structures of Grignard reagents are more complex. In the solid, the magnesium centres are normally tetrahedral, with solvent molecules occupying some of the coordination sites. When EtMgBr is crystallized from ethoxyethane, it forms a structure with two coordinated ethoxyethane molecules, as shown in Figure 26.15.

The compounds that are present in solution depend on a number of factors, including the nature of the solvent and the concentration. The following reaction, called the *Schlenk redistribution* reaction occurs

$$2\,\text{RMgX} \rightleftharpoons \text{MgR}_2 + \text{MgX}_2$$

and aggregates containing two or more magnesium atoms have also been observed.

Summary

- Lithium and magnesium form organometallic compounds that are useful reagents in organic synthesis.
- Organolithium compounds RLi exist as oligomers. For example, the structure of methyllithium contains $[MeLi]_4$ tetramers.
- Grignard reagents RMgX contain tetrahedral metal centres and often crystallize with coordinated solvent molecules.
- Organometallic compounds are important in organic synthesis due to the carbon atoms being able to act as nucleophiles.

? For a practice question on these topics, see question 19 at the end of this chapter (p.1198).

26.11 Diagonal relationships

Lithium and magnesium have many properties in common. For example:

- both metals react with N_2 to give nitrides;
- both metals burn in air to form the normal oxides and not peroxides;
- the carbonates and nitrates both decompose to the oxides on heating;
- both ions are more heavily hydrated than others in their respective group;
- both form an extensive number of organometallic compounds that have a large degree of covalent character in the M–C bonds.

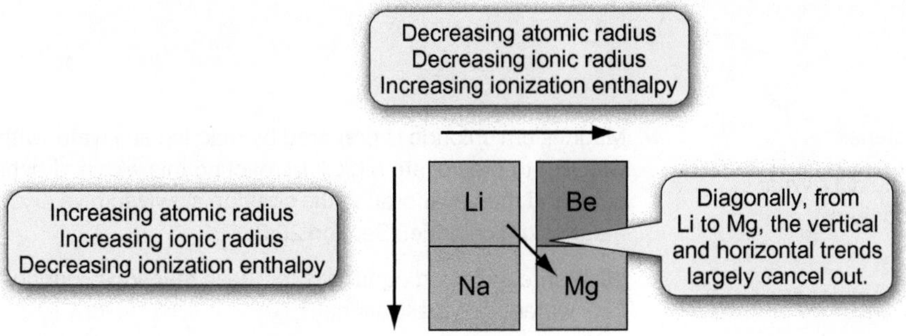
Decreasing atomic radius
Decreasing ionic radius
Increasing ionization enthalpy
Increasing atomic radius
Increasing ionic radius
Decreasing ionization enthalpy
Li
Be
Na
Mg
Diagonally, from Li to Mg, the vertical and horizontal trends largely cancel out.

Figure 26.16 The origin of diagonal relationships.

These similarities form the basis of a **diagonal relationship** between the two elements. Diagonal relationships occur because of the different ways in which many atomic properties vary down groups and across periods of the Periodic Table. For example, the atomic and ionic radii increase down Group 1 due to the increasing size of the outermost occupied atomic orbital, but decrease across the second period from lithium to neon due to ineffective shielding and the consequent increase in effective nuclear charge. Moving diagonally—one element down and one across to the right—these two trends largely cancel out, as shown in Figure 26.16. For example, the ionic radii of Li^+ and Mg^{2+} are similar, being 76 pm and 72 pm, respectively, for 6-coordinate structures.

Shielding is described in Section 3.6 (p.150).

The same principles apply to other atomic properties, and the consequences of this are similarities in the chemistry of the diagonally related elements. Lithium and magnesium often occur together in minerals, with the Li^+ ions substituted for Mg^{2+} ions.

The diagonal relationship between lithium and magnesium is the most important example of this principle, but it is not the only one. Diagonal relationships also exist between beryllium and aluminium, and between sodium and calcium. Other less strong diagonal relationships exist between boron and silicon, between carbon and phosphorus, and between nitrogen and sulfur.

Summary

- Since many trends work in different directions down groups and across periods, elements whose positions in the Periodic Table are diagonally related often have similar properties.
- The diagonal relationship between lithium and magnesium is particularly strong.

Concept review

By the end of this chapter, you should be able to do the following.

- Describe the preparation, uses, and reactivity of the *s*-block elements.
- Describe the synthesis and reactions of the *s*-block oxides, hydroxides, and halides.
- Explain why *s*-block compounds are largely ionic and explain the exceptions.
- Explain trends in reactivity and thermal stability using enthalpy cycles.
- Use enthalpy and Gibbs energy cycles to explain why some compounds don't exist.
- Describe changes in the solubility of *s*-block compounds and explain these using the lattice Gibbs energies and hydration Gibbs energies.
- Describe examples of *s*-block coordination complexes and explain why Group 2 coordination chemistry is more prevalent than Group 1 coordination chemistry.
- Justify why the chemistries of lithium and beryllium differ from those of the other elements in their groups.
- Describe examples of lithium and magnesium organometallic compounds.
- Explain the basis of the diagonal relationship between lithium and magnesium.

Questions

More challenging questions are indicated by an asterisk *.

1 Sodium can be used to prepare zirconium from $ZrCl_4$. What mass of sodium is required to yield 1 kg of zirconium? (Section 26.1)

2 Identify the products and write balanced equations for the following reactions (Section 26.2):

(a) adding caesium to water;

(b) burning rubidium in an excess of air;

(c) heating lithium nitrate.

3 Use a Born–Haber cycle and the data below to explain why $NaCl_2$ doesn't exist. Assume the ionic radius of Na^{2+} is the same as that for Mg^{2+}. (Section 26.2)

Ionic radii: Mg^{2+}, 72 pm; Cl^-, 181 pm

$\Delta_a H^\ominus(Na) = +108\,kJ\,mol^{-1}$ $\quad$ $\Delta_a H^\ominus(Cl) = +121\,kJ\,mol^{-1}$

$\Delta_i H(1)^\ominus(Na) = +496\,kJ\,mol^{-1}$ $\quad$ $\Delta_{eg} H(1)^\ominus(Cl) = -349\,kJ\,mol^{-1}$

$\Delta_i H(2)^\ominus(Na) = +4562\,kJ\,mol^{-1}$

4 Predict which compound in each of the following pairs has the higher decomposition temperature: (a) $NaNO_3$ and KNO_3; (b) LiH and KH; (c) Li_2CO_3 and $SrCO_3$. Give reasons for your answer. (Section 26.2)

5 A Group 1 metal, M, burns in air to produce X, which reacts with water to form Y. M also reacts with water to form Y, giving off a combustible gas Z. When Y is heated, it gives off steam to form X. Identify M, X, Y, and Z. (Section 26.2)

6* Construct an enthalpy cycle for the decomposition of a Group 1 nitrate to an oxide. Using the Kapustinskii equation to estimate values of the lattice energies, calculate values for the decomposition of lithium nitrate to lithium oxide and potassium nitrate to potassium oxide. Use the calculated values to predict which of these nitrates decomposes to the oxide on heating. (Section 26.2)

Values for ionic radii are given in Tables 6.4 (p.286), 6.5 (p.286), and 6.9 (p.297).

The enthalpy change for the reaction

$$2\,NO_3^-(g) \rightarrow O^{2-}(g) + 4\,NO(g) + 3\,O_2(g)$$

is $+2160\,kJ\,mol^{-1}$

7 Which of the following pairs of Group 1 compounds would you expect to be more soluble in water? (Section 26.3)

(a) $LiNO_3$ or $CsNO_3$

(b) LiOH or RbOH

(c) NaBr or CsBr.

8 Suggest how the crown ether below could be used to separate mixtures of Li^+ and K^+. (Section 26.4)

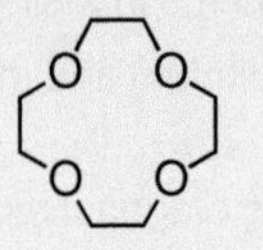
O
O
O
O

9* Magnesium chloride is prepared by reacting sea water with $Ca(OH)_2$ to precipitate $Mg(OH)_2$, reacting this with hydrochloric acid, and then evaporating the solution slowly to give crystals of magnesium chloride. (Section 26.6)

(a) Write balanced equations for these reactions, ignoring any water of crystallization.

(b) A sample of $MgCl_2$ obtained from this process contains 18.5% magnesium by mass. Assuming there are no impurities, how many molecules of water does the $MgCl_2$ crystallize with?

10 Identify the products and write balanced equations for the following reactions (Section 26.7):

(a) adding beryllium to sodium hydroxide solution;

(b) heating strontium carbonate;

(c) heating barium oxide in air.

11 When magnesium burns in air, both magnesium oxide and magnesium nitride are formed. Suggest how you could obtain a pure sample of magnesium oxide from this mixture. (Section 26.7)

12 The structure of $BeCl_2$ is shown in Figure 26.9 (p.1187). What is the geometry around the beryllium centres? Use this to predict a value for the Cl–Be–Cl bond angle. What is the most likely hybridization of the beryllium atom? (Section 26.7)

13 Calcium oxide is being used increasingly to reduce air pollution by the desulfurization of flue gases from power stations which contain SO_2. Suggest how SO_2 reacts with CaO. (Section 26.7)

14 Magnesium sulfate is more soluble in water than barium sulfate, but magnesium fluoride is less soluble than barium fluoride. Explain these observations in terms of the Gibbs energies involved. (Section 26.7)

15* Use a Born–Haber cycle to calculate the enthalpy change of formation of calcium(I) chloride. Does your answer help explain why this compound is unknown? If not, what else would help? (Section 26.7)

16 Use an enthalpy cycle to predict whether $Mg(OH)_2$ or $Ba(OH)_2$ has a higher decomposition temperature. (Section 26.7)

17 Explain why beryllium nitrate dissolves in water to give an acidic solution. (Section 26.8)

18 Beryllium compounds are mainly covalent, whereas other Group 2 compounds are predominantly ionic. Discuss the evidence for this statement, and suggest factors that contribute to it. (Section 26.9)

19* The solid state structure of dimethylmagnesium is polymeric, with bridging methyl groups. By considering the magnesium and carbon atoms as sp^3 hybridized, show how this bonding can be described as 3-centre 2-electron. (Section 26.10)

27

p-Block chemistry

This chapter builds on the following topics:

- **Oxidation and reduction** Section 1.4, p.27
- **Atomic properties and periodicity** Section 3.7, p.153
- **Electronegativity** Section 4.3, p.177
- **Molecular orbital energy level diagrams** Section 4.8–4.12, p.188
- **Valence shell electron pair repulsion theory** Section 5.2, p.223
- **Bond polarity and polar molecules** Section 5.3, p.235
- **Valence bond theory for polyatomic molecules** Section 5.4, p.236
- **Partial molecular orbital schemes** Section 5.7, p.249
- **Lewis acids and bases** Section 7.8, p.336
- **Enthalpy changes in chemical reactions** Section 13.3, p.622

◀ Photochemical smog over Los Angeles. The poor visibility is due to the presence of aerosols—small particles of liquid or solid containing a mixture of organic and inorganic compounds.

Photochemical smog

Photochemical smog, a yellow-brown haze with an unpleasant odour, is a major pollution problem in cities throughout the world. As well as being disagreeable, it can lead to respiratory problems, particularly in vulnerable groups such as babies, the elderly, and people with asthma. Photochemical smog was first reported in Los Angeles in the 1940s, but it is now an issue in many other places including Mexico City, Athens, Cairo, Beijing, Tokyo, and São Paulo. These cities have several features in common: they have high populations, a large number of vehicles, a warm, sunny climate, and restricted movement of air.

▲ Car exhausts are a major source of VOCs and NO_x, both components of photochemical smog.

The formation of photochemical smog involves some interesting *p*-block chemistry. The two main types of compound leading to photochemical smog are nitrogen oxides and volatile organic compounds (VOCs), which are mostly hydrocarbons and their oxidation products. NO and NO_2 are interconverted rapidly in the atmosphere, so they are collectively referred to as NO_x. Both NO_x and VOCs are emitted from vehicle exhausts and are called *primary pollutants*. The VOCs from vehicle exhausts are unburned and partially oxidized fuel. NO is formed when N_2 and O_2 react at high temperatures in vehicle engines.

In the presence of sunlight and oxygen, NO_x and the VOCs react to give a number of other compounds, known as *secondary pollutants*. The first step involves attack on the hydrocarbons by hydroxyl radicals, $^{\bullet}OH$, which occur naturally in the atmosphere (see Chapter 9, p.383)

$$RCH_3 + {}^{\bullet}OH \rightarrow RCH_2^{\bullet} + H_2O$$

$$RCH_2^{\bullet} + O_2 \rightarrow RCH_2O_2^{\bullet}$$

The peroxy radical ($RCH_2O_2^{\bullet}$ or $RO_2^{\bullet}$) reacts with NO to form NO_2, which is a source of the yellow-brown colour of the photochemical smog

$$RCH_2O_2^{\bullet} + NO \rightarrow RCH_2O^{\bullet} + NO_2$$

Once formed, NO_2 is photochemically decomposed back to NO in a process that also produces oxygen atoms. These react with O_2 molecules to form ozone (O_3)

$$NO_2 \xrightarrow{h\nu} NO + O$$

$$O_2 + O \longrightarrow O_3$$

Ozone is an important and beneficial component of the stratosphere due to its ability to absorb ultraviolet radiation (see Box 27.6, p.1233), but it creates problems closer to the ground. Ozone causes irritation to the respiratory system, leading to coughing, throat and nose irritations, shortness of breath, and chest pains.

Peroxy radicals also react with NO_2 to form peroxy nitrates

$$RO_2^{\bullet} + NO_2 \rightarrow RO_2NO_2$$

One of the most common peroxy nitrates in photochemical smog is peroxyacetyl nitrate (PAN, $CH_3CO_2ONO_2$). PAN is a lachrymator, which means it irritates the eyes.

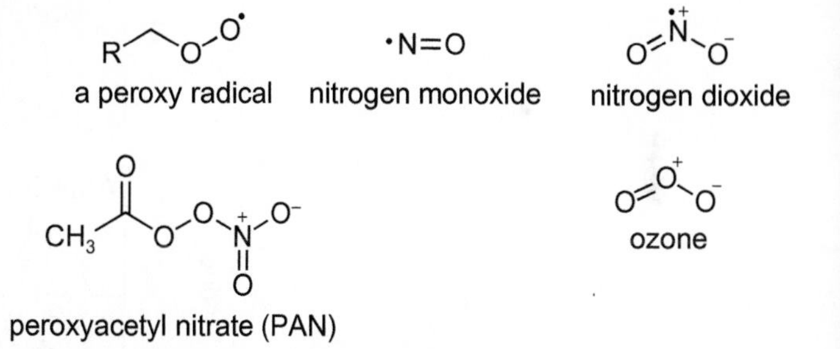
a peroxy radical
nitrogen monoxide
nitrogen dioxide
ozone
peroxyacetyl nitrate (PAN)

▲ Some *p*-block compounds present in smog.

How can photochemical smog be prevented? Emissions of NO_x and VOCs can be brought down by reducing the number of cars on the road. Many cities have attempted to do this by introducing regulations such as only allowing cars into the city centre on alternate days based on registration number. Emissions of NO_x and VOCs can be decreased by using catalytic converters. VOCs can also be decreased by changing the composition of the fuel used, reducing its volatility. Regulations in California have limited the use of hydrocarbon-containing products such as barbecue starter fluids, aerosol sprays, and oil-based paints, and these measures have improved air quality in Los Angeles.

The *p*-block elements are those with the valence electron configuration $ns^2\,np^x$, where x is between 1 and 6. The *p* block covers six groups of the Periodic Table. These are Groups 13, 14, 15, 16, 17, and 18, as shown in the Periodic Table on the inside front cover of the book. In older versions of the Periodic Table, the *p*-block groups are called Groups 3–8, but nowadays the group numbering normally includes the *d*-block elements, which form Groups 3–12.

The *p* block is the only block of the Periodic Table that contains non-metals as well as metals. Some *p*-block elements have properties that are intermediate between those of metals and non-metals, and these elements are often called **metalloids**. Figure 27.1 shows the *p*-block elements classified as metals, metalloids, or non-metals. The metallic character of the elements increases down each group, and decreases from left to right across each period. This means that the metal–non-metal boundary is diagonal, running roughly from the top left to the bottom right of the *p* block. Although the elements are conveniently described as metals, metalloids, and non-metals, the transitions are not exact and any element close to the boundary between the metals and non-metals will show some properties of both.

This chapter begins with a section on general aspects and the trends that occur going down the groups and across the periods. Sections 27.2–27.7 then cover the chemistry of the *p*-block elements and their compounds group by group. These sections concentrate

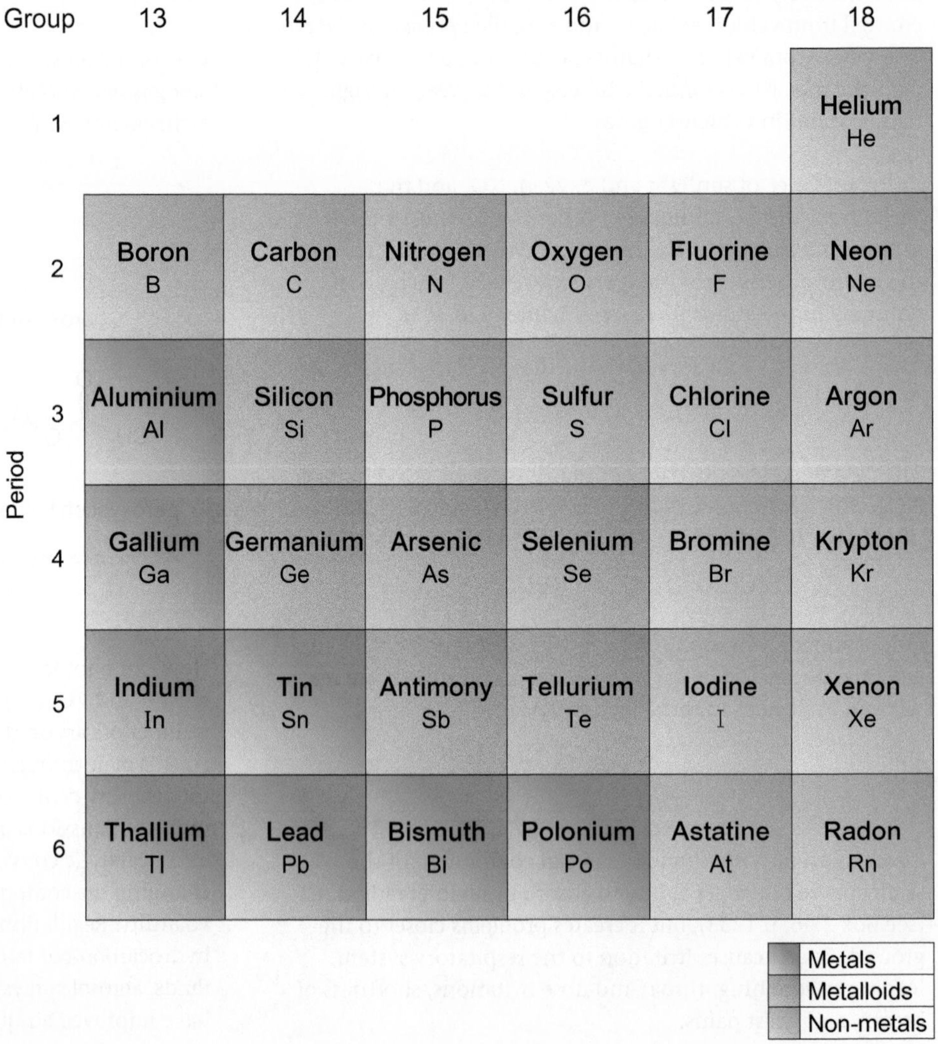
Group
13
14
15
16
17
18
Period
1
2
3
4
5
6
Helium He
Boron B
Carbon C
Nitrogen N
Oxygen O
Fluorine F
Neon Ne
Aluminium Al
Silicon Si
Phosphorus P
Sulfur S
Chlorine Cl
Argon Ar
Gallium Ga
Germanium Ge
Arsenic As
Selenium Se
Bromine Br
Krypton Kr
Indium In
Tin Sn
Antimony Sb
Tellurium Te
Iodine I
Xenon Xe
Thallium Tl
Lead Pb
Bismuth Bi
Polonium Po
Astatine At
Radon Rn
Metals
Metalloids
Non-metals

Figure 27.1 The elements of the *p* block and their classification as metals, metalloids, or non-metals.

on the elements of the second and third periods, with the heavier elements covered to a lesser extent in order to illustrate the changes that occur going down the groups. The descriptive chemistry focuses on the oxides, oxoacids, and halides, though other common compounds are also mentioned. Finally, Section 27.8 looks briefly at the organometallic chemistry of the *p*-block elements.

The *p*-block hydrides are described along with other hydrides in Section 25.2 (p.1147).

27.1 General aspects and trends in the *p* block

The changes that occur on going across a period or down a group are an aspect of **periodicity**. Many of these changes are related to trends in atomic properties such as ionization energy, atomic radius, and electronegativity, which are described in Chapters 3 and 4.

Trends in ionization energy, electron gain energy, and atomic radius are described in Section 3.7 (p.153). Trends in electronegativity are described in Section 4.3 (p.177).

There is a wide range of structural types possible for *p*-block compounds, with ionic lattices, covalent network structures, and discrete molecules containing covalent bonds all commonly observed. For the ionic compounds, the difference in electronegativity between the elements is less than that in analogous *s*-block compounds, so the bonding in *p*-block ionic compounds has considerable covalent character.

The geometries of discrete molecules can usually be predicted using VSEPR theory. In most of the covalently bonded compounds, the bonds are polar and the partial charges on the atoms are an important factor in determining the way in which a compound reacts.

VSEPR theory is explained in Section 5.2 (p.223).

Oxidation states

For a *p*-block compound, the oxidation states of the elements are assigned using their electronegativities. In silicon tetrafluoride (SiF_4), for example, fluorine is more electronegative than silicon, so silicon has the oxidation state +4 and fluorine has the oxidation state −1. One consequence of this approach is that the sign of the oxidation state of an element in a compound depends on the atoms it is bonded to. In NF_3, the nitrogen atom is in the +3 oxidation state as fluorine is in the −1 oxidation state. Contrast this with NH_3. Here, nitrogen is more electronegative than hydrogen, so the nitrogen atom is in the −3 oxidation state. Due to this variation with the substituents, oxidation states in covalently bonded compounds are often referred to as **formal oxidation states**.

Trends in oxidation states going across periods and down groups are covered later in this section (p.1204 and p.1207). The nitrogen oxides are described in Section 27.4 (p.1225).

Common formal oxidation states that are observed in the *p*-block groups are given in Table 27.1. Most *p*-block elements have more than one possible oxidation state. These tend to vary in units of 2, so in Group 15, for example, the most common oxidation states are +5 and +3. Compounds with the oxidation states of +2 and +4 are also known, but compounds with these oxidation states contain unpaired electrons, such as nitrogen monoxide [NO, N(II)] and nitrogen dioxide [NO_2, N(IV)]. Such radicals are only stable with electronegative substituents like oxygen or fluorine as this helps delocalize the unpaired electron and reduce the tendency of the radical to dimerize.

Table 27.1 Common formal oxidation states for the *p*-block elements

Group				
13	14	15	16	17
+3	+4	+5	+6	+7
+1	+2	+3	+4	+5
	−4	−3	+2	+3
			−2	+1
				−1

Lewis acidity and basicity

Many *p*-block compounds act as either Lewis acids or Lewis bases, and much *p*-block chemistry can be understood in these terms. The Group 13 compounds of the general formula AX_3 have only 6 valence electrons, so reaction with a Lewis base provides a means to achieve a filled octet. For example, BF_3 reacts with F^- to form BF_4^-.

A Lewis acid is a lone pair acceptor and a Lewis base is a lone pair donor. Lewis acids and bases are introduced in Section 7.8 (p.336).

In Groups 14 and 15, there is a difference between compounds of the second period element at the top of the group, carbon and nitrogen, respectively, and those of the heavier elements in each group. Carbon and nitrogen compounds of the general formulae CX_4 and NX_3 have 8 valence electrons and, since expansion of the octet is not possible for second period elements, these compounds are not Lewis acids. The heavier elements of these groups can be hypervalent so, for example, SiF_4 reacts with F^- to form SiF_6^{2-}. In this reaction F^- is behaving as a Lewis base and SiF_4 as a Lewis acid.

The Lewis acidity of the silicon halides is discussed in Section 27.3 (p.1221). Hypervalency is described in Section 5.1 (p.220).

Compounds of Groups 15, 16, and 17 contain one or more lone pairs of electrons on the central atom, so they have the potential to act as Lewis bases. Whether a lone pair is available

The difference in Lewis basicity between NH_3 and PH_3 is described in more detail in Section 25.2 (p.1150).

for donation to a Lewis acid depends on the type of orbital it is in. The reason why NH_3 is a stronger Lewis base than PH_3 can be understood using valence bond theory. For NH_3, the lone pair resides in a directional sp^3 hybrid orbital, so it is available for donation. In PH_3, the electron pair lies in an unhybridized spherical *s* orbital, so it is less available for donation.

Summary

- Most *p*-block elements can adopt more than one oxidation state, and the oxidation states normally vary in units of 2.
- Group 13 compounds act as Lewis acids, as can compounds of Groups 14 and 15 if they contain elements of the third period or lower.
- Groups 15, 16, and 17 compounds have lone pairs, so they can act as Lewis bases, but the strength of the base depends on the type of orbital the electron pair is in.

Trends across the periods of the *p* block

Metallic character of the elements

Effective nuclear charge is discussed in Section 3.6 (p.150). There are some irregularities in the general increase in ionization energies due to the electronic configurations of the atoms, and these are explained in Section 3.7 (p.153).

Generally, ionization energies increase from left to right across each period (see Figure 3.35, p.156) because of the increasing effective nuclear charge. In a metal, the valence electrons are not associated with a particular metal atom but are delocalized over the structure. This allows them to move when a potential difference is applied, so a metal conducts electricity. As the ionization energy increases, it becomes harder to remove the valence electrons from each atom. This leads to a decrease in metallic character from left to right across a period, and a decrease in conductivity. Consequently, metals occur on the left of each period and non-metals on the right.

Metallic bonding and semiconductors are described in Section 6.3 (p.273).

In some of the periods there are metalloids with intermediate properties between the metals and non-metals. In terms of conductivity, metalloids are often semiconductors. In the third period, aluminium is a metal, silicon is a metalloid, and phosphorus, sulfur, chlorine, and argon are non-metals. The classification of the *p*-block elements into metals, metalloids, and non-metals is shown in Figure 27.1 (p.1202).

Oxidation states

The oxidation states of the *p*-block elements in their oxides and halides are summarized in Figure 27.2.

For elements of the third and lower periods, the maximum oxidation state increases from left to right across the period. Look at the highest oxidation state oxides and fluorides of the third period elements

aluminium(III)	silicon(IV)	phosphorus(V)	sulfur(VI)	chlorine(VII)
Al_2O_3	SiO_2	P_4O_{10}	SO_3	Cl_2O_7
AlF_3	SiF_4	PF_5	SF_6	ClF_5 [Cl(V)]

In each case the maximum oxidation state oxide corresponds to involvement of all of the valence *s* and *p* electrons in bonding. There is a similar trend for the fluorides, with the exception of ClF_7, which is not known. This is due to size, and the difficulty in placing seven fluorine atoms around the relatively small central chlorine atom.

For the second period elements, expansion of the octet is not possible. In addition, the high electronegativity of oxygen and fluorine restricts the oxidation states that are possible for these elements. For the second period, the highest oxidation state oxides and fluorides formed are

boron(III)	carbon(IV)	nitrogen(V)	fluorine(–I)
B_2O_3	CO_2	N_2O_5	OF_2
boron(III)	carbon(IV)	nitrogen(III)	oxygen(II)
BF_3	CF_4	NF_3	OF_2

Group 13	14	15	16	17	18
					He
B +3 +3	C +2, +4 +4	N +1, +2, +3, +4, +5 +1, +2, +3	O +2, +1, −2	F −1 −1	Ne
Al +3 +3	Si +4 +4	P +3, +5 +3, +5	S +4, +6 +1, +4, +6	Cl +1, +4, +6, +7 +1, +3, +5	Ar
Ga +3 +1, +3	Ge +2, +4 +2, +4	As +3, +5 +3, +5	Se +4, +6 +4, +6	Br +4 +1, +3, +5	Kr +2
In +3 +1, +3	Sn +2, +4 +2, +4	Sb +3 +3, +5	Te +4, +6 +4, +6	I +5 +1, +3, +5, +7	Xe +6, +8 +2, +4, +6
Tl +1, +3 +1, +3	Pb +2, +4 +2, +4	Bi +3 +3, +5	Po +4	At	Rn

Figure 27.2 The oxidation states of the *p*-block elements in their main oxides (shown in blue) and halides (shown in red).

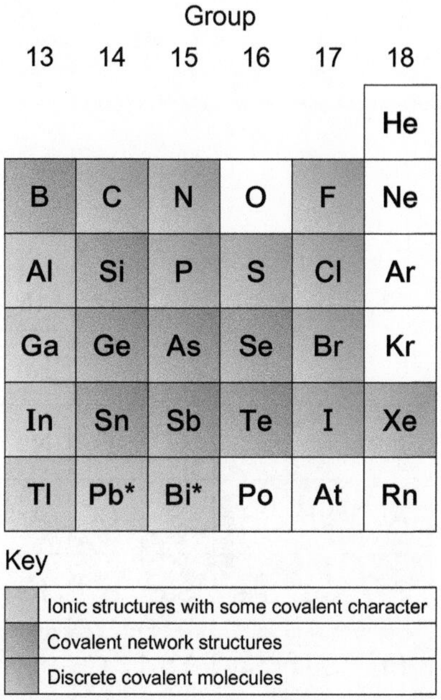
Group
13 14 15 16 17 18
He
B C N O F Ne
Al Si P S Cl Ar
Ga Ge As Se Br Kr
In Sn Sb Te I Xe
Tl Pb* Bi* Po At Rn
Key
Ionic structures with some covalent character
Covalent network structures
Discrete covalent molecules

Figure 27.3 The types of structure adopted by the *p*-block oxides.

* Pb(II), Bi(III).

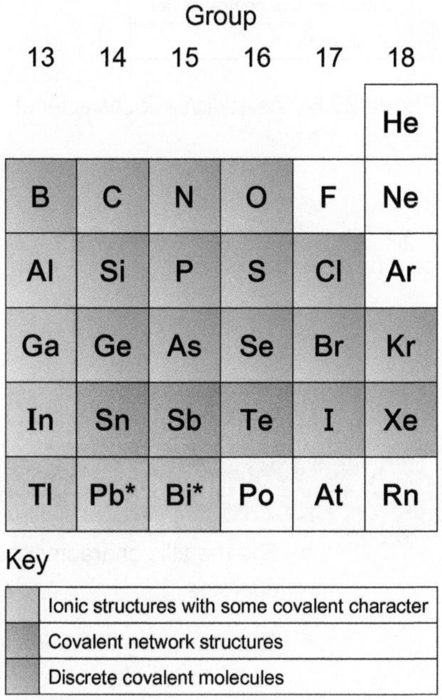
Group
13 14 15 16 17 18
He
B C N O F Ne
Al Si P S Cl Ar
Ga Ge As Se Br Kr
In Sn Sb Te I Xe
Tl Pb* Bi* Po At Rn
Key
Ionic structures with some covalent character
Covalent network structures
Discrete covalent molecules

Figure 27.4 The types of structure adopted by the *p*-block fluorides.

* Pb(II), Bi(III).

Structure and bonding in the oxides and halides

Figures 27.3 and 27.4 divide the structures of the *p*-block oxides and fluorides into ionic structures, covalent network structures, and discrete covalent molecules. The division into these structural types is not as clear cut as the figures suggest since an ionic lattice with significant covalent character is essentially the same as a covalent network structure with significant ionic character, but they help to show the trends.

Metals tend to form oxides and halides that have extended structures with a high degree of ionic character. The metal atom is less electronegative than oxygen or a halogen, and in the solid state electrons are transferred from the metal to the more electronegative atom. Moving across each period, the electronegativity of the elements increases. This means that the difference in electronegativity between the element and oxygen or a halogen decreases, so the ionic character of the bonding also decreases. Ionic lattices give way to covalent network structures or discrete covalent molecules. This is illustrated by the oxides that are formed by the third period elements.

Al_2O_3	SiO_2	P_4O_6/P_4O_{10}	SO_2/SO_3	Cl_2O/Cl_2O_7
Ionic solid with some covalent character	Covalent network structure with polar bonds	Covalent molecules	Covalent molecules	Covalent molecules

These compounds are shown on the van Arkel–Ketelaar triangle in Figure 27.5.

van Arkel–Ketelaar triangles are introduced in Section 6.7 (p.298).

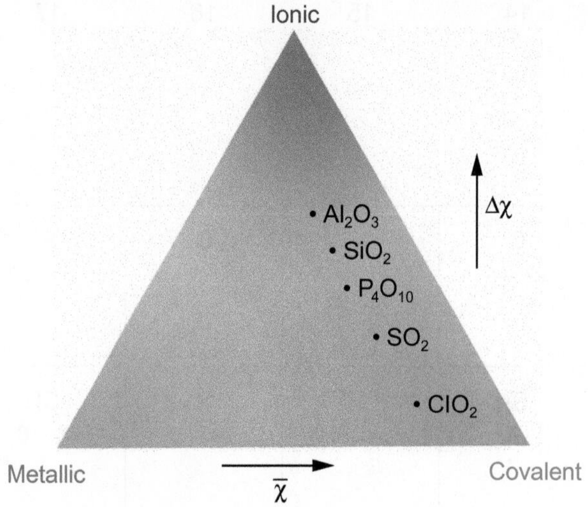
Ionic
• Al_2O_3
• SiO_2
• P_4O_{10}
• SO_2
• ClO_2
Δχ
Metallic
χ̄
Covalent

Figure 27.5 A van Arkel–Ketelaar triangle for the oxides of the third period elements.

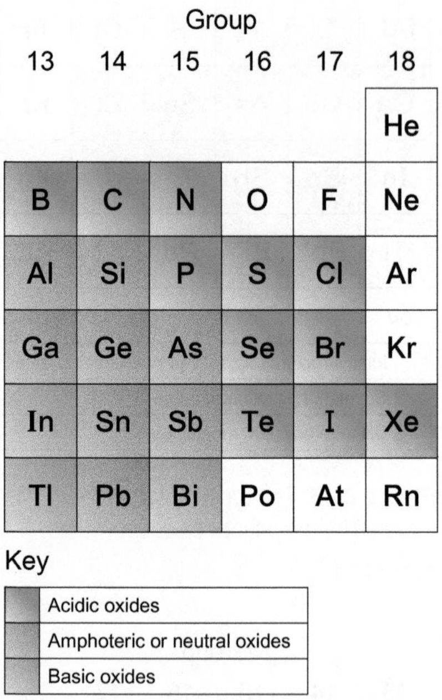
Group
13 14 15 16 17 18
He
B C N O F Ne
Al Si P S Cl Ar
Ga Ge As Se Br Kr
In Sn Sb Te I Xe
Tl Pb Bi Po At Rn
Key
Acidic oxides
Amphoteric or neutral oxides
Basic oxides

Figure 27.6 The acidic/basic character of the *p*-block oxides.

The acidity and basicity of oxides is described in Section 7.7 (p.334).

Acidity and basicity of the oxides

The oxides of electropositive metals are basic, and dissolve in water to form alkaline solutions. For example,

$$In_2O_3\,(s) + 3\,H_2O\,(l) \rightarrow 2\,In^{3+}\,(aq) + 6\,OH^-\,(aq)$$

Non-metal oxides tend to be acidic, reacting with water to form acidic solutions. For example,

$$Cl_2O_7\,(l) + 3\,H_2O\,(l) \rightarrow 2\,H_3O^+\,(aq) + 2\,ClO_4^-\,(aq)$$

In between these two extremes, the oxides can be amphoteric, which means they react as both acids and bases. Al_2O_3 is amphoteric, reacting with both acid and alkali

$$Al_2O_3\,(s) + 6\,H_3O^+\,(aq) + 3\,H_2O\,(l) \rightarrow 2\,[Al(H_2O)_6]^{3+}\,(aq)$$

$$Al_2O_3\,(s) + 2\,OH^-\,(aq) + 3\,H_2O\,(l) \rightarrow 2\,[Al(OH)_4]^-\,(aq)$$

The acidity and basicity of the *p*-block oxides is summarized in Figure 27.6.

Summary

- The metallic character of an element decreases from left to right across a period of the *p* block due to increasing ionization energies.
- The maximum oxidation state increases across a period, though expansion of the octet is not possible for the second period elements.
- The ionic character of the oxides and halides decreases across each period.
- The acidic character of the oxides increases across each period.

Trends down the groups of the *p* block

The largest differences going down a group are generally those between the elements of the second and third periods. For example, in Group 16, the difference in chemistry between oxygen (Period 2) and sulfur (Period 3) is greater than that between sulfur and selenium (Period 4). This can be understood by looking at the atomic radii (Figure 3.34, p.154). The atomic radius increases by 58% from oxygen (66 pm) to sulfur (104 pm), but by only 13% from sulfur to selenium (117 pm). This is caused by the presence of the *d*-block elements between calcium and gallium in the fourth period. The *d* orbitals have a poor shielding effect, so filling them results in a much larger increase in effective nuclear charge between calcium and gallium than that which occurs in the third period, between magnesium and aluminium.

Trends in atomic radii and ionization energies for the *d*-block elements are described in Section 28.1 (p.1259).

Pepto-Bismol
Heartburn
Indigestion
Nausea
Upset Stomach
Diarrhoea
PROTECTIVE COATING ACTION

Bismuth(III) salts are used in the treatment of heartburn and peptic ulcers.

Metallic character of the elements

The metallic character of the *p*-block elements is summarized in Figure 27.1 (p.1202). The ionization energies generally decrease down the groups of the Periodic Table. This is because the valence electrons lie in orbitals with successively higher principal quantum numbers, which have a greater average distance from the nucleus. The decrease in ionization energies means that the metallic character of the elements increases down a group. For example, in Group 14, carbon is a non-metal, and silicon and germanium are metalloids, whereas tin and lead are metals.

Oxidation states

The oxidation states of the *p*-block elements in their oxides and halides are summarized in Figure 27.2 (p.1205). For Groups 13–15, the maximum oxidation state (+3, +4, and +5, respectively) is relatively easy to obtain, and is observed for all of the elements. As these groups are descended, the oxidation state two lower than the maximum (+1, +2, and +3, respectively) becomes increasingly more stable and, for the last member, this is typically the most important oxidation state. So, for boron the +1 oxidation state is unimportant, but for thallium, at the bottom of the same group, it is as common as the +3 oxidation state. In a similar manner, at the bottom of Group 14, lead(II) is more common than lead(IV) and, at the bottom of Group 15, bismuth(III) is more common than bismuth(V).

This increase in importance of an oxidation state two less than the usual oxidation state for a group is called the **inert pair effect**. The name 'inert pair' comes from the fact that the lower oxidation state has the electronic configuration ns^2, and these two valence electrons are not used in bonding. The inert pair effect occurs because bond enthalpies decrease going down the group. Using the valence bond model of bonding, energy is required to promote the *s* electrons into hybrid orbitals so they can form bonds. For the lighter elements, this energy is compensated for by the energy released on forming two new bonds. The lower bond enthalpies for the heavier elements means that this is not the case, so formation of the higher oxidation state compound is less favoured.

For Groups 16 and 17, the maximum oxidation state (+6 and +7, respectively) is not shown by the first member of the group, oxygen and fluorine, respectively. This is due to the high electronegativity and high ionization energies of these elements. The maximum oxidation state becomes easier to obtain as the group is descended because the ionization energies and electronegativities decrease.

Trends in bond enthalpies

For an A–X bond, where A is a *p*-block element and X is a non-metal such as H, O, F, or Cl, bond enthalpies generally decrease down a group. If there are no lone pairs on X, the trend is smooth. This is shown by the bond enthalpies for A–H bonds, given in Table 27.2. For example, in Group 14, the bond enthalpies decrease in the order C–H > Si–H > Ge–H > Sn–H reflecting the poorer orbital overlap as the size of the Group 14 element increases.

If there are lone pairs on X, the trend down a group is different. The bond enthalpy for the bond involving the first element is generally lower than that for the second element. You can see this in Table 27.3 for the Group 14 A–Cl bonds. The bond enthalpies vary in

Table 27.2 Mean bond enthalpies (kJ mol^{-1}) for A–H bonds where A is a *p*-block element

Group			
14	15	16	17
C–H +412	N–H +391	O–H +464	F–H +570
Si–H +318	P–H +321	S–H +364	Cl–H +431
Ge–H +285	As–H +297	Se–H +313	Br–H +366
Sn–H +251	Sb–H +257	Te–H +238	I–H +298

Table 27.3 Mean bond enthalpies (kJ mol^{-1}) for A–F and A–Cl bonds where A is a *p*-block element

A–F bonds in Group				A–Cl bonds in Group			
14	15	16	17	14	15	16	17
C–F +467	N–F +278	O–F +214	F–F +159	C–Cl +346	N–Cl +190	O–Cl +206	F–Cl +256
Si–F +597	P–F +490	S–F +326	Cl–F +256	Si–Cl +400	P–Cl +326	S–Cl +271	Cl–Cl +242
Ge–F +452	As–F +484	Se–F +285	Br–F +250	Ge–Cl +349	As–Cl +322	Se–Cl +192	Br–Cl +218
Sn–F +414	Sb–F +440	Te–F +330	I–F +272	Sn–Cl +323	Sb–Cl +315	Te–Cl +311	I–Cl +211

the order C–Cl < Si–Cl > Ge–Cl > Sn–Cl. There are two reasons why Si–Cl bonds are stronger than C–Cl bonds. Firstly, the electron–electron repulsion between the electron pairs on the two atoms in a compound such as CCl_4 is greater than that in $SiCl_4$ as the atoms must be closer together to ensure good orbital overlap. Secondly, the electronegativity of silicon is less than that of carbon. This means the difference in electronegativity is greater between the elements in the Si–Cl bond than those in the C–Cl bond, so there is a greater ionic contribution to the bonding.

The trend is more exaggerated when A and X both have lone pairs, and in this case the A–X bond enthalpy for the bond involving the first element in the group is particularly low. This is shown by the Group 15 A–F bonds in Table 27.3. The data show that the N–F bond is very much weaker than the P–F bond. The major factor in this is the electron–electron repulsion between the non-bonded electrons (lone pairs) on the two atoms, which is much higher in the shorter N–F bond.

In a MO approach, the repulsion is between π-bonding and π*-antibonding electrons rather than between lone pairs.

Structure and bonding in the oxides, halides, and elements

The structure and bonding in the *p*-block oxides and fluorides is summarized in Figures 27.3 and 27.4 (p.1205). Going down a group, the difference in electronegativity between the element and oxygen increases. This means that the ionic character of the oxides increases as a group is descended. This is shown for the Group 15 oxides in the van Arkel–Ketelaar triangle in Figure 27.7. A similar process occurs for the halides, and helps explain why BiF_3 is an ionic solid whereas NF_3 is a gas with covalently bonded molecules.

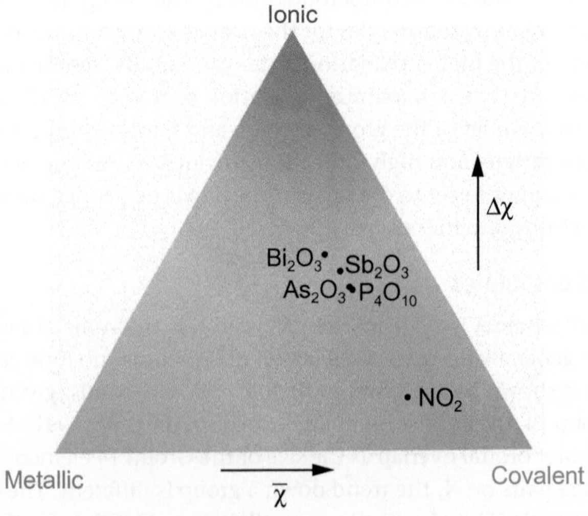
Ionic
Δχ
Bi2O3
Sb2O3
As2O3
P4O10
NO2
Metallic
χ
Covalent

Figure 27.7 A van Arkel–Ketelaar triangle for the Group 15 oxides.

The biggest differences in bonding are normally observed between the elements of the second and the third periods. This is largely due to the increased size in the atoms and the effects this has on bond strengths and coordination numbers. For example, oxygen forms discrete O_2 molecules with a double bond between the atoms, whereas sulfur forms a range of allotropes with S–S single bonds. This difference has two contributing factors. Firstly, π bonding decreases in strength with increasing size to a much greater extent than σ bonding. So, an S_2 double bond is much weaker than an O_2 double bond. Secondly, single bonds are particularly weak for N–N, O–O, and F–F because of the repulsions between the lone pairs, which are brought close together by the small size of the atoms.

The distance sensitivity of σ and π bonds is described in Section 4.9 (p.195).

The decreasing strength of multiple bonds down the groups is also a major factor in the difference in the structure of the oxides. Carbon forms discrete molecular oxides (CO_2, CO) in which multiple bonds are present. In contrast, SiO_2 forms covalent network structures containing Si–O single bonds.

See Section 27.3 (p.1217) for more details on the difference in structure between CO_2 and SiO_2.

Expansion of the octet is not possible for second period elements. This affects the nature of the halides, as those with maximum oxidation state (NF_5, OF_6) are not known for second period elements in Groups 15 and 16. The NF_4^+ cation, containing nitrogen(V), is known, but doesn't break the octet rule. The differences in size and electronegativity of the halogens influence the compounds they form. The smallest and most electronegative halogen, fluorine, forms high oxidation state compounds as a consequence of high A–F bond enthalpies. The lower the electronegativity, the lower the oxidation state favoured by the halogen.

The electronegativity of the halogen also affects the bond angles adopted in covalent molecules as described in Section 27.4 (p.1229).

Acidity and basicity of the oxides

The acidity and basicity of the *p*-block oxides is summarized in Figure 27.6 (p.1206). As the metallic character of the element increases, so does the basicity of its oxides. For example, in Group 15, the nitrogen and phosphorus oxides are acidic, reacting with water to form acidic solutions. Arsenic and antimony oxides are amphoteric and dissolve in both acids and alkalis. Bismuth(III) oxide (Bi_2O_3) is basic.

Summary

- The metallic character of an element increases down each group of the Periodic Table due to decreasing ionization energies.
- The '*Group −2*' oxidation state increases in importance down Groups 13–15. This is known as the inert pair effect.
- For an A–X bond, where A is a *p*-block element and X is a non-metal such as H, O, F, or Cl, bond enthalpies generally decrease down each group, but when the atoms have lone pairs the bond enthalpy is low for bonds involving the second period element due to high electron–electron repulsion.
- The ionic character of the oxides and halides increases down the groups and the oxides become more basic.
- π bonding becomes less important going down each group.

For practice questions on these topics, see questions 1–3 at the end of this chapter (p.1253). Questions 16–20 are on general *p*-block chemistry.

27.2 Group 13: boron, aluminium, gallium, indium, and thallium

Group 13 consists of the elements boron (B), aluminium (Al), gallium (Ga), indium (In), and thallium (Tl). The Group 13 elements have the electronic configuration $ns^2 np^1$, which means they have a maximum oxidation state of +3. This is the only important oxidation state for boron and aluminium, but the +1 oxidation state becomes more significant going down the group, a manifestation of the inert pair effect (Section 27.1, p.1207).

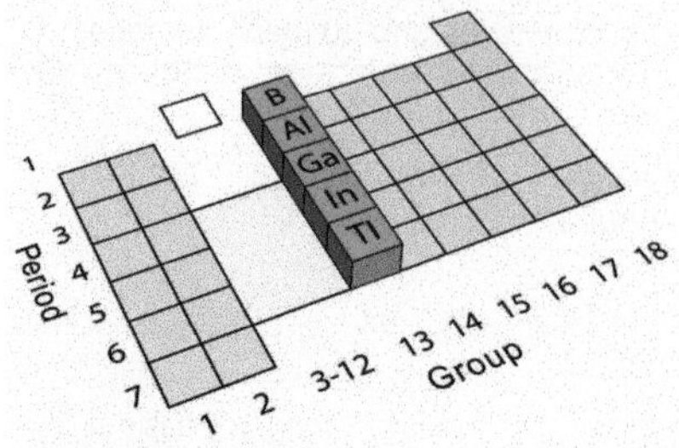
B
Al
Ga
In
Tl
Period
1
2
3
4
5
6
7
1 2 3-12 13 14 15 16 17 18
Group

Group 13 in the Periodic Table.

The first three ionization enthalpies for boron are very high, so ionic compounds containing the B^{3+} ion are not known. Instead boron compounds contain covalent bonds, whereas the other Group 13 elements form both ionic and covalent compounds. Group 13 elements have three valence electrons, so they commonly form three covalent bonds. These compounds of the general formula EX_3, where E is a Group 13 element and X is a monovalent atom or group, have only six valence electrons, two short of a filled octet. They possess a low energy empty *p* orbital and the compounds act as Lewis acids, accepting an electron pair to complete the octet. For example, BF_3 reacts with ethoxyethane to form the adduct $BF_3{\cdot}OEt_2$.

The structure and bonding in B_2H_6 are described in Section 5.7 (p.250), and the structures of other boron hydrides are described in Box 5.6 (p.252).

Many Group 13 compounds overcome their lack of valence electrons by forming 3-centre bonds. This leads to the formation of dimers and polymers containing bridging groups and, in the case of the boron hydrides, clusters such as B_5H_9. In some cases, such as B_2H_6, the structures cannot be explained by simple Lewis structures, and these compounds are called **electron deficient**.

The Group 13 elements

Selected physical properties of the Group 13 elements are given in Table 27.4. In terms of its chemical reactions, boron is best described as a non-metal, whereas aluminium, gallium, indium, and thallium are all metals.

Allotropes are different structural forms of the same element. See Section 6.1 (p.257).

Elemental boron is extracted from deposits of *borax* ($Na_2[B_4O_5(OH)_4]{\cdot}8H_2O$). There are several different allotropes of boron, and these all have structures containing B_{12} icosahedra interlinked into three-dimensional networks. Figure 27.8 shows the structure of one of the allotropes, called α-boron.

Electrolysis is described in more detail in Section 16.5 (p.755).

Aluminium metal is extracted commercially from the ore *bauxite*, which has the formula AlO(OH) or $Al(OH)_3$ depending on its source. The procedure involves two stages. The first of these, known as the *Bayer process*, is designed to remove insoluble impurities such as iron oxides and titanium oxide. The bauxite is dissolved in aqueous sodium hydroxide to convert it into aluminate, $[Al(OH)_4]^-$, ions. The insoluble impurities are removed by filtration, then the aluminate is cooled to precipitate $Al(OH)_3$. The solid hydroxide is finally heated to convert it into the oxide.

$$\mathrm{AlO(OH)\,(s) + OH^-\,(aq) + H_2O\,(l) \xrightarrow{140\,°C-250\,°C} [Al(OH)_4]^-\,(aq)}$$

$$\mathrm{[Al(OH)_4]^-\,(aq) \rightarrow Al(OH)_3\,(s) + OH^-\,(aq)}$$

$$\mathrm{2\,Al(OH)_3\,(s) \xrightarrow{1200\,°C} Al_2O_3\,(s) + 3\,H_2O\,(l)}$$

In the second stage, known as the *Hall–Héroult process*, aluminium metal is obtained from the purified aluminium oxide by electrolysis. Al_2O_3 is dissolved in *cryolite* ($Na_3[AlF_6]$) in order to reduce the melting point, and the electrolysis of the molten mixture is carried out at 940 °C–980 °C.

Aluminium has high tensile strength and low density, so it is widely used in the construction and manufacturing industries. Its low density coupled with a high electrical

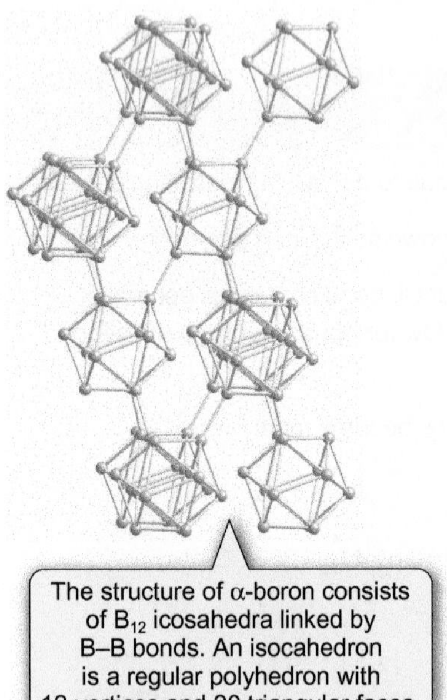
The structure of α-boron consists of B_{12} icosahedra linked by B–B bonds. An isocahedron is a regular polyhedron with 12 vertices and 20 triangular faces.

Figure 27.8 The structure of the allotrope α-boron.

Diagonal relationships are observed in the *p* block, with that between beryllium and aluminium particularly strong. The basis of diagonal relationships is described in Section 26.11 (p.1196).

Table 27.4 Physical properties and reduction potentials of the Group 13 elements

Element	A_r	m.p./°C	b.p./°C	Density/g cm^{-3}	$E^{\ominus}(M^{3+}\|M)$/V
Boron (B)	10.81	2075	4000	2.34	–
Aluminium (Al)	26.982	660	2519	2.70	−1.66
Gallium (Ga)	69.723	30	2204	5.90	−0.56
Indium (In)	114.82	157	2072	7.31	−0.34
Thallium (Tl)	204.38	304	1473	11.8	+1.26

The statue of Eros in Piccadilly Circus, London, erected in 1893, was one of the first statues to be made from aluminium. The large-scale use of aluminium metal only became practical with the sharp decrease in price following the development of the Hall–Héroult process.

The melting point of gallium is 30 °C, so body heat is sufficient to melt it.

conductivity makes it ideal for use in power lines. Pure aluminium is readily oxidized but, in practice, aluminium metal is resistant to corrosion because of a thin coating of the oxide that forms on the surface of the metal and acts as a physical barrier to reagents. The surface is said to be *passivated* by the oxide layer.

Box 27.1 Recycling aluminium

Almost every aluminium product, from cans and foil to window frames and cars, can be recycled at the end of its useful life without loss of metal quality. This recycling process is economically viable as it requires only 5% of the energy needed to produce aluminium from bauxite. Currently about one-third of the aluminium produced each year worldwide is recycled aluminium.

Three-quarters of the drinks cans sold in the UK are made of aluminium. What happens to one of these cans once it is thrown into a recycling bin? The cans are collected and flattened, then transported to a recycling plant. The flattened cans are shredded into pieces several centimetres long and these are passed through a magnetic separator to remove any steel. Lacquers on the inside and outside of the cans are burnt off by blowing hot air at a temperature of 500 °C through the shreds on a slow-moving conveyor belt. The exhaust gases are passed through a heat exchanger to heat up the incoming air and maximize energy efficiency. The pieces of shredded, de-coated aluminium are then loaded into a furnace and melted. The molten metal is processed into ingots, which are rolled into sheets, coiled, and transported to can manufacturers.

In countries with dedicated can collecting and recycling schemes, used drinks cans are back on supermarket shelves as new cans within 2 months. In 2012, 65% of aluminium cans in the UK were recycled. Finland and Belgium are Europe's top aluminium can recyclers, with recycling rates of 97%.

Question

Suggest why it takes so much more energy to extract aluminium from bauxite than to recycle it.

◄ Crushed aluminium cans awaiting recycling.

All of the Group 13 metals burn in air to form their oxides. For example

$$4\,Al(s) + 3\,O_2(g) \xrightarrow{\Delta} 2\,Al_2O_3(s)$$

Aluminium and gallium are both amphoteric, dissolving in acid and alkali to liberate hydrogen. For aluminium,

with acid: $2\,Al(s) + 6\,H_3O^+(aq) + 6\,H_2O(l) \rightarrow 2\,[Al(H_2O)_6]^{3+}(aq) + 3\,H_2(g)$

with alkali: $2\,Al(s) + 6\,H_2O(l) + 2\,OH^-(aq) \rightarrow 2\,[Al(OH)_4]^-(aq) + 3\,H_2(g)$

Indium and thallium both dissolve in dilute acids, but not alkalis, reflecting the increase in metallic character going down the group.

Summary

- Boron is a non-metal, whereas the other Group 13 elements are metals.
- The most common oxidation state for Group 13 is +3, but the +1 oxidation state becomes increasingly important going down the group.
- Aluminium and gallium are amphoteric, but indium and thallium show more metallic character and do not dissolve in alkalis.

The Group 13 oxides, hydroxides, and oxoanions

Boron forms a wide range of compounds containing B–O bonds. The simplest of these is boron oxide (B_2O_3), which is acidic. It reacts slowly with water to give boric acid ($B(OH)_3$, sometimes written H_3BO_3)

$$B_2O_3\,(s) + 3\,H_2O\,(l) \rightarrow 2\,B(OH)_3\,(aq)$$

B–O bonds are strong, with bond enthalpies in the range +560 kJ mol^{-1} to +790 kJ mol^{-1}. Anions containing B–O bonds are called **borates**, and there are many examples of these. Their structures include cyclic and linear polymers containing planar BO_3 and/or tetrahedral BO_4 units that are linked together by bridging oxygen atoms. Some examples of borates are shown in Figure 27.9.

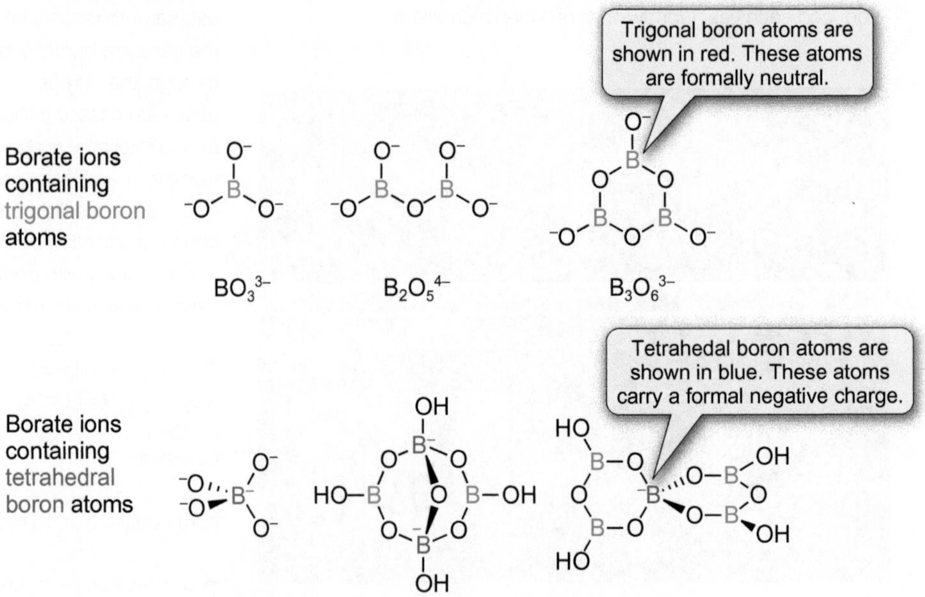
Trigonal boron atoms are shown in red. These atoms are formally neutral.
Borate ions containing trigonal boron atoms
BO3 3–
B2O5 4–
B3O6 3–
Tetrahedal boron atoms are shown in blue. These atoms carry a formal negative charge.
Borate ions containing tetrahedral boron atoms
BO4 5–
B4O5(OH)4 2–
B5O6(OH)4 –

Figure 27.9 Examples of borate anions. $B_4O_5(OH)_4^{2-}$ is the anion present in borax.

Lewis acids are introduced in Section 7.8 (p.336). The solid-state structure of boric acid is described in Section 25.3 (p.1160).

Boric acid has a layer structure in which the molecules within the layers are linked by hydrogen bonds. It is a weak acid but, in contrast to most oxoacids, the acidity is not due to loss of a proton from an O–H bond. Instead, the boron atom acts as a Lewis acid, interacting with a water molecule to form $B(OH)_3(OH_2)$, which loses a proton to form the tetrahedral anion $[B(OH)_4]^-$.

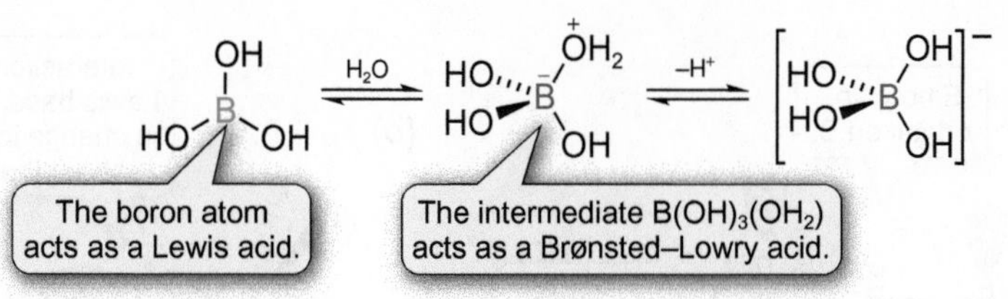
H$_2$O
−H$^+$
The boron atom acts as a Lewis acid.
The intermediate B(OH)$_3$(OH$_2$) acts as a Brønsted–Lowry acid.

On heating, boric acid loses water to form B_2O_3 in a stepwise manner, via metaboric acid (HBO_2)

$$\underset{\text{boric acid}}{2B(OH)_3\,(S)} \xrightarrow[-2H_2O]{\Delta} \underset{\text{metaboric acid}}{2HBO_2\,(S)} \xrightarrow[-H_2O]{\Delta} \underset{\text{boron oxide}}{B_2O_3\,(S)}$$

Metaboric acid exists in several forms, and the most common of these contains cyclic trimers (**1**).

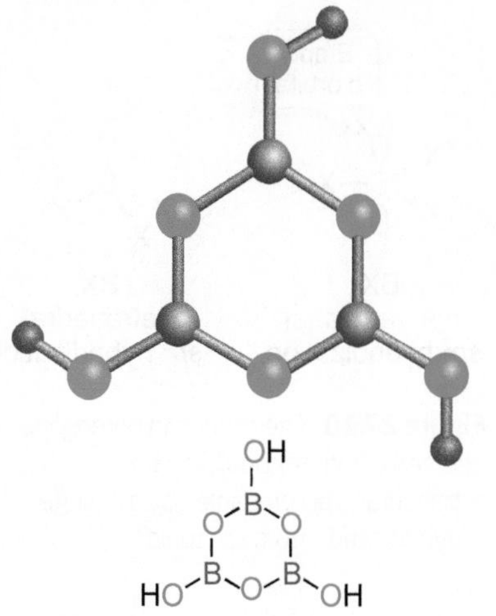

1 Metaboric acid trimer (($HBO_2)_3$).

Aluminium oxide (Al_2O_3) forms a number of polymorphs, which differ in their reactivity. α-Al_2O_3, also known as *corundum*, is a hard, unreactive solid that is used in ceramics and as an abrasive in toothpaste. γ-Al_2O_3 has a more open structure and as a consequence it is more reactive, showing amphoteric character.

Polymorphs are different structural forms of the same compound. See Section 6.1 (p.259). The reactions of Al_2O_3 with acid and alkali are described in Section 27.1 (p.1206).

Aluminium oxide is commonly called alumina.

In aqueous solution, the Group 13 metal cations form hexaaqua complexes with the general formula $[M(H_2O)_6]^{3+}$. The high charge density on the metal ion polarizes the O–H bonds of the coordinated water molecules. This means that the ions undergo hydrolysis and, as a result, the aqueous solutions are acidic. For the aluminium ions in aqueous solution

$$[Al(H_2O)_6]^{3+}(aq) + H_2O\,(l) \rightleftharpoons [Al(H_2O)_5(OH)]^{2+}(aq) + H_3O^+(aq)$$

$$[Al(H_2O)_5(OH)]^{2+}(aq) + H_2O\,(l) \rightleftharpoons [Al(H_2O)_4(OH)_2]^{+}(aq) + H_3O^+(aq)$$

$$[Al(H_2O)_4(OH)_2]^{+}(aq) + H_2O\,(l) \rightleftharpoons Al(OH)_3{\cdot}3\,H_2O\,(s) + H_3O^+(aq)$$

The related hydrolysis of hydrated Be^{2+} ions is described in Section 26.8 (p.1192) and that of hydrated *d*-block ions in Section 28.8 (p.1300).

Summary

- Boron forms the oxide B_2O_3 and a large number of borate anions containing trigonal planar BO_3 units and/or tetrahedral BO_4 units.
- Boric acid ($B(OH)_3$) is a monobasic acid. It acts as a Lewis acid, interacting with water to form $B(OH)_3(OH_2)$, which loses a proton to form $[B(OH)_4]^-$.
- Al_2O_3 is amphoteric, but some forms are unreactive.
- $[Al(H_2O)_6]^{3+}$ salts are acidic in aqueous solution due to hydrolysis.

The Group 13 halides

Boron halides

The boron halides BF_3 and BCl_3 are gases under standard conditions, whereas BBr_3 is a liquid and BI_3 is a solid. All have trigonal planar structures, which is consistent with sp^2 hybridization of the central boron atom. The compounds are all Lewis acids, and form adducts with Lewis bases.

The Lewis acidity increases in the order:

$$\underset{\text{least acidic}}{BF_3} < BCl_3 < BBr_3 < \underset{\text{most acidic}}{BI_3}$$

At first sight this order may be surprising. Fluorine is the most electronegative of the halogens, so the bonds in BF_3 are the most polar. This suggests that the boron atom in BF_3

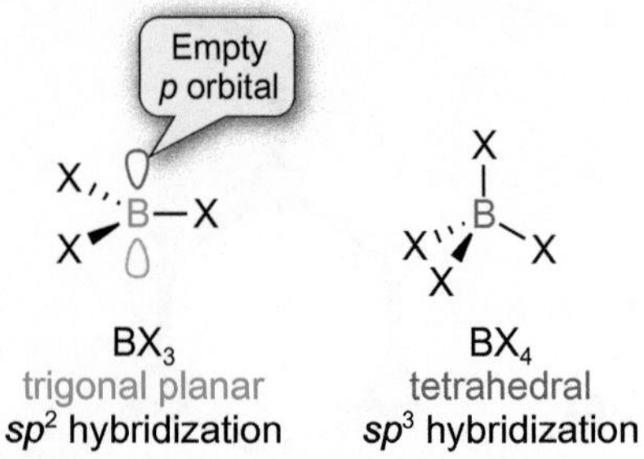
Empty
p orbital
BX$_3$
trigonal planar
sp^2 hybridization
BX$_4$
tetrahedral
sp^3 hybridization

Figure 27.10 The change in boron geometry from trigonal planar to tetrahedral is accompanied by a change in hybridization from sp^2 to sp^3.

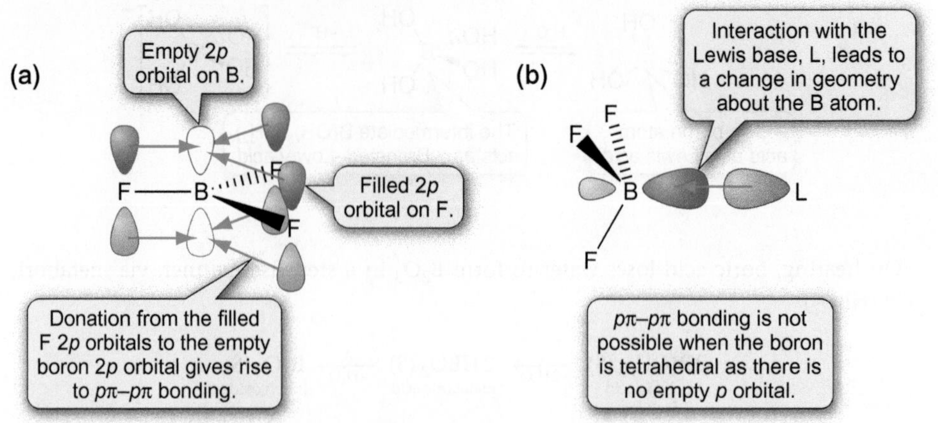
(a)
Empty 2*p* orbital on B.
Filled 2*p* orbital on F.
Donation from the filled F 2*p* orbitals to the empty boron 2*p* orbital gives rise to *pπ–pπ* bonding.
(b)
Interaction with the Lewis base, L, leads to a change in geometry about the B atom.
pπ–pπ bonding is not possible when the boron is tetrahedral as there is no empty *p* orbital.

Figure 27.11 (a) *pπ–pπ* bonding in trigonal planar boron halides. (b) For a tetrahedral adduct formed on reaction with a Lewis base, *pπ–pπ* bonding is not possible.

has the highest partial positive charge and so would interact most strongly with a Lewis base. Since this would give the opposite order to that observed, there must be another factor involved.

When a boron halide reacts with a Lewis base, the geometry around the boron atom changes from trigonal planar to tetrahedral. This is consistent with a change in the boron hybridization from sp^2 to sp^3 (Figure 27.10).

Why does this change in hybridization lead to the observed trend in Lewis acidities? The halogen atoms possess filled *p* orbitals that lie above and below the plane of the molecule, and these can donate electron density into the empty boron *p* orbital. This is known as ***pπ–pπ* bonding**, and is shown in Figure 27.11(a).

The distance sensitivity of π bonds is described in Section 4.9 (p.195).

π bonding is very distance-sensitive, so the *pπ–pπ* bonding is strongest in BF_3 and weakest in BI_3, decreasing in strength as the size of the halogen atoms increases. The *pπ–pπ* bonding can only occur with sp^2 hybridized boron, as there is no available orbital of the correct symmetry to interact with the halogen *p* orbitals on a sp^3 hybridized boron atom. The *pπ–pπ* bonding is therefore lost when the tetrahedral Lewis acid–Lewis base adduct forms (Figure 27.11(b)). Since *pπ–pπ* bonding is weaker in the halides with larger halogen atoms, the adducts form more readily for these.

Chemists are divided over the importance of *pπ–pπ* bonding in boron halides, and it is possible to explain the trend in Lewis acidities without this. The change in shape from trigonal planar to tetrahedral involves distortion of the bonds. The B–X bond strengths decrease with increasing size of the halogen and it takes less energy to distort a weaker bond.

In contrast to BH_3, the boron halides do not dimerize. Two factors in the stability of the BX_3 monomers are the presence of *pπ–pπ* bonding, which would be lost in B_2X_6 dimers, and the greater steric bulk that occurs with an increase in coordination number.

BF_3 reacts with metal fluorides to give salts that contain the tetrafluoroborate anion. In BF_4^-, all of the B–F bonds have the same bond length (145 pm), which is longer than the B–F bond in BF_3 (130 pm).

The bonds in many 3-coordinate boron compounds are shorter than would be expected on the basis of the atomic radii. In addition to possible *pπ–pπ* bonding, the other factors contributing to this are the large electronegativity difference between boron and other atoms, which leads to a high ionic component to the bonding, and the low amount of electron–electron repulsion, as there are only six valence electrons around the boron atom.

Aluminium halides

Aluminium fluoride (AlF_3) is a colourless, unreactive solid that sublimes at 1290 °C. It has an extended structure with a high degree of ionic character. In contrast, the other aluminium halides have considerably lower melting points (180 °C–260 °C), and they hydrolyse rapidly in moist air to give the hydrogen halide. For example,

$$AlCl_3\,(s) + 3\,H_2O\,(l) \rightarrow Al(OH)_3\,(s) + 3\,HCl\,(g)$$

3-centre 4-electron bonding is explained in Section 5.7 (p.249).

Aluminium chloride has a layer structure in the solid, but consists of covalently bonded Al_2Cl_6 dimers in the liquid and gas phases. These dimers are also present when aluminium chloride is dissolved in a non-polar solvent such as benzene. Aluminium bromide and aluminium iodide are dimeric as solids as well as in the liquid and gas phases, and in solution in nonpolar solvents. The dimers contain bridging halides, as shown in Figure 27.12 for Al_2Br_6. Each Al–Br–Al unit is held together by a 3-centre 4-electron interaction, with each bridging bromide donating three electrons, one in a normal covalent bond and the other two in a dative bond.

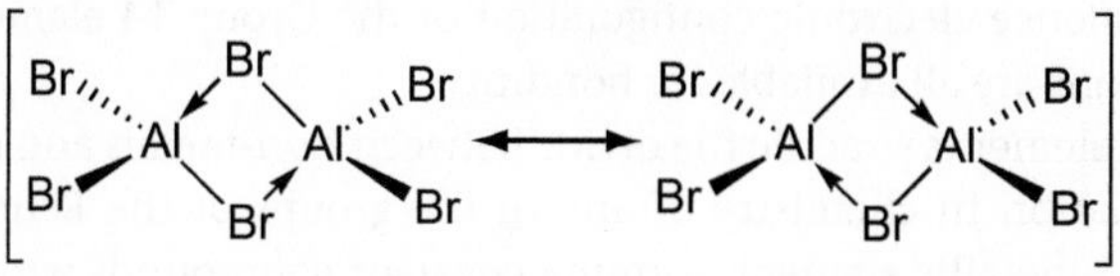
Br
Al
Br
Al
Br

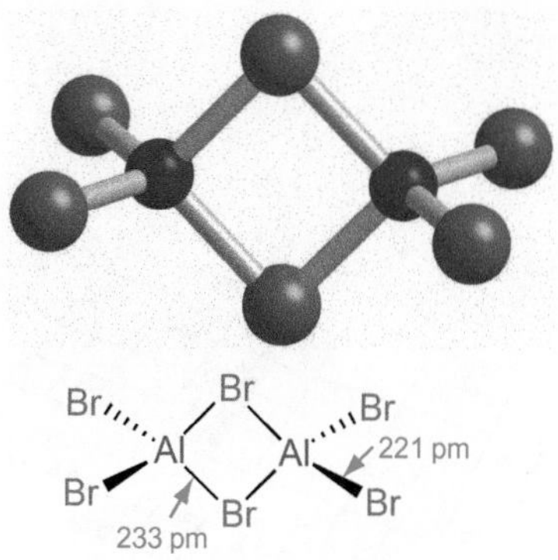
Br
Al
Al
221 pm
233 pm

Figure 27.12 The dimeric structure of aluminium bromide (Al_2Br_6).

The formation of the dimers is driven by the high Lewis acidity of the aluminium centre in AlX_3 and made possible by the larger size of aluminium compared to boron.

Summary

- The boron halides are monomeric and do not form dimers. *pπ–pπ* bonding in BX_3 would be lost on dimerization and there is greater steric repulsion with four large halogen atoms around a boron atom.
- The Lewis acidity of the boron halides increases in the order $BF_3 < BCl_3 < BBr_3 < BI_3$. This is because it takes less energy to distort the trigonal planar geometry with larger halogen atoms. This trend can be explained by weaker *pπ–pπ* bonding with increasing halogen size.
- Aluminium fluoride is a solid with high ionic character, but the other aluminium halides have structures containing covalent bonds. Al_2X_6 dimers with 3-centre 4-electron bonds are present in the liquid and gas phase or when the compounds are dissolved in nonpolar solvents.

Other Group 13 compounds

Many metals and non-metals form binary compounds with boron. For elements with a lower or similar electronegativity these are called **borides**. There are a wide range of borides, including those that are 'boron-poor' such as Mn_4B, and those that are 'boron-rich' such as CaB_6 and boron carbide ($B_{12}C_3$). Boron-poor borides contain isolated boron atoms, whereas boron-rich borides contain boron–boron bonds.

The Group 13 elements form 1:1 compounds with Group 15 elements. Boron nitride (BN) has several polymorphs, and the room temperature form has a layer structure similar to that of graphite, with the hexagonal layers containing 6-membered rings with alternating boron and nitrogen atoms, as shown in Figure 27.13. Unlike graphite, boron nitride does not conduct electricity.

Similar 6-membered rings are present in the molecule borazine ($B_3N_3H_6$, (2)). Borazine can be thought of as an inorganic analogue of benzene as it has a similar planar structure, though in terms of reactivity it has little aromatic character.

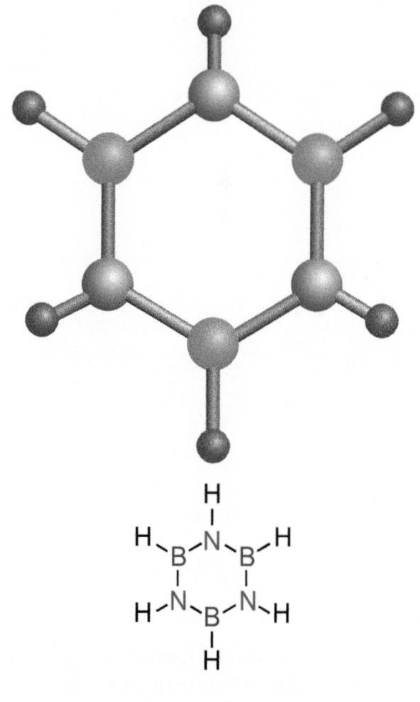
H
N
B

2 Borazine ($B_3N_3H_6$).

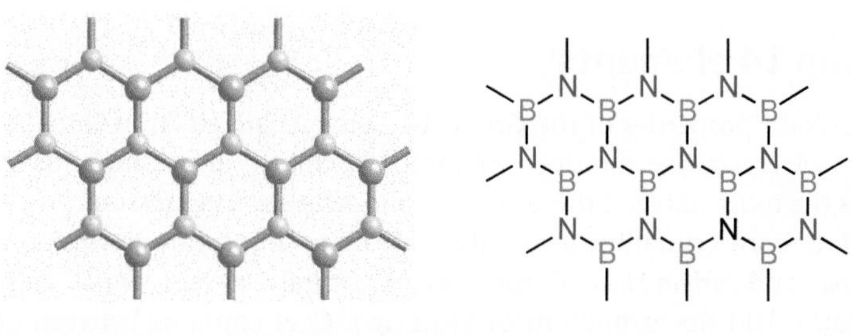
N
B

Figure 27.13 A layer in the structure of boron nitride (BN).

The aromatic nature of benzene is discussed in Section 22.1 (p.1004).

For practice questions on the topics covered in Section 27.2, see questions 4–6 at the end of this chapter (p.1253). Questions 16–20 are on general *p*-block chemistry.

27.3 Group 14: carbon, silicon, germanium, tin, and lead

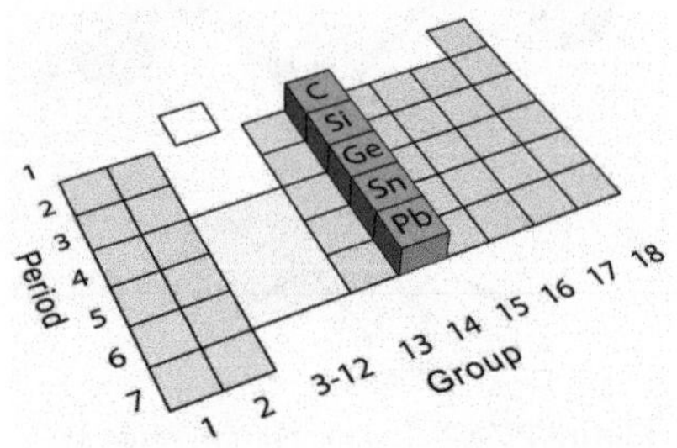
C
Si
Ge
Sn
Pb
Period
1
2
3
4
5
6
7
1 2 3-12 13 14 15 16 17 18
Group

Group 14 in the Periodic Table.

Group 14 consists of the elements carbon (C), silicon (Si), germanium (Ge), tin (Sn), and lead (Pb). The valence electronic configuration of the Group 14 elements is $ns^2\ np^2$, and these four electrons are all available for bonding.

The Group 14 elements straddle the divide between non-metals and metals and so show the greatest variation in chemistry of any of the groups of the Periodic Table. Carbon behaves as a non-metallic element, forming covalent compounds with other non-metals and structures with a high ionic character with electropositive metals. Silicon and germanium are typical metalloids, exhibiting both metallic and non-metallic properties. Tin and lead are both metallic, though tin shows some amphoteric behaviour, dissolving in both acid and alkali to generate hydrogen

$$\mathrm{Sn\,(s) + 2\,H_3O^+\,(aq) + 4\,H_2O\,(l) \rightarrow [Sn(H_2O)_6]^{2+}\,(aq) + H_2\,(g)}$$
$$\mathrm{Sn\,(s) + 2\,OH^-\,(aq) + 2\,H_2O\,(l) \rightarrow [Sn(OH)_4]^{2-}\,(aq) + H_2\,(g)}$$

The oxidation states of the Group 14 elements range from +4 in the dioxides and tetrahalides to −4 in the hydrocarbons. The +4 oxidation state becomes less important with respect to the +2 oxidation state as the group is descended. For example, carbon dioxide (CO_2) is the most stable oxide of carbon, whereas lead(IV) oxide (PbO_2) is a strong oxidizing agent, and is easily reduced to lead(II) oxide (PbO). The change in relative stability of the +4 and +2 oxidation states is an illustration of the inert pair effect.

The inert pair effect is explained in Section 27.1 (p.1207).

Catenation, the ability of an element to form covalent bonds with itself to give chains or rings, is particularly important for carbon, and carbon–carbon bonds lie at the heart of organic chemistry. Catenation becomes less important as the group is descended, partly due to the decreasing bond enthalpies of the X–X bonds (Table 27.5) and partly because the heavier atoms are larger and have orbitals that are available for reaction.

Double bonds and triple bonds are also more stable for carbon than for the other Group 14 elements (Table 27.5), and the chemistry of alkenes and alkynes is described in Chapter 21. For the heavier members of Group 14, multiply bonded compounds are much rarer, because of the distance sensitivity of π-overlap.

Table 27.5 Mean bond enthalpies for homonuclear single bonds and double bonds between Group 14 elements

Bond enthalpy / $kJ\,mol^{-1}$		Bond enthalpy / $kJ\,mol^{-1}$	
C–C	+347	C=C	+612
Si–Si	+226	Si=Si	+310
Ge–Ge	+188	Ge=Ge	+270
Sn–Sn	+152	Sn=Sn	+190

The Group 14 elements

The structures of graphite, diamond, and C_{60} are described in Section 6.1 (p.258).

Selected physical properties of the Group 14 elements are given in Table 27.6. There are three types of crystalline allotrope of carbon: graphite, diamond, and the fullerenes. Graphite is the most stable of these at room temperature and pressure.

In addition to the crystalline forms, there are many amorphous forms of carbon such as *coke*, *charcoal*, and *carbon black*. Carbon occurs naturally as coal, which was formed from organic matter laid down millions of years ago. Coal contains between 50% and 95% carbon and, when it is heated in the absence of air, volatile compounds such as water are driven off, leaving behind a material known as coke. Coke is a common industrial reducing agent and is used on a large scale to reduce iron oxide to iron metal in a blast furnace.

i **Amorphous** materials do not have long-range order in their structures. They are described in Section 6.1 (p.260).

Charcoal and carbon black are produced by heating organic matter and hydrocarbons respectively in the absence of air. Charcoal and carbon black can be produced in 'activated'

Table 27.6 Physical properties of the Group 14 elements

Element	A_r	m.p./°C	b.p./°C	Density/g cm^{-3}
Carbon (C)	12.011	3642*	–	2.2†/3.5‡
Silicon (Si)	28.085	1414	3265	2.33
Germanium (Ge)	72.60	938	2833	5.32
Tin (Sn)	118.71	232	2602	7.29§
Lead (Pb)	207.2	328	1749	11.3

*Sublimes. †Graphite. ‡Diamond. §White (β-tin).

Silicon is purified by slowly withdrawing a crystal seed from molten silicon. This grows a large, cylindrical silicon crystal from which wafers can be sliced.

forms by heating them to 800 °C–1000 °C in the presence of steam. These activated forms have very high surface areas so they can adsorb organic molecules. They are used in air purifiers, gas masks, and water purification plants.

Silicon is the second most abundant element in the Earth's crust after oxygen. Most rocks and sand contain silicon dioxide (SiO_2) or silicates. Elemental silicon is obtained by reduction of SiO_2 with coke in an electric arc furnace. Silicon is a semiconductor, and its conductivity is strongly affected by impurities. For this reason, the silicon that is used in the semiconductor industry needs to be extremely pure. In industry, silicon is purified by converting it into $SiHCl_3$, SiH_4, or $SiCl_4$, which is distilled and heated to decompose it back to silicon. For example,

$$Si(s) + 3HCl(g) \xrightarrow{300\,°C} SiHCl_3(g) + H_2(g)$$

$$2SiHCl_3(g) \xrightarrow[1000\,°C]{H_2} Si(s) + SiCl_4(g) + 2HCl(g)$$

This material is then recrystallized by dipping a crystal seed into molten silicon, then pulling the growing crystal upwards.

The extraction of silicon for use in silicon chips is discussed in Box 14.7 (p.682). Semiconductors are described in Section 6.3 (p.275).

Silicon dioxide is commonly called silica.

Summary

- Carbon is a non-metal, silicon and germanium are metalloids, and tin and lead are metals.
- Group 14 oxidation states vary from +4 to −4. As the group is descended the +2 oxidation state becomes increasingly important at the expense of the +4 oxidation state.

The Group 14 oxides

The large difference in behaviour between carbon and the other members of Group 14 is illustrated by the structures and properties of the oxides.

Carbon oxides

The structures of the main oxides of carbon are shown in Figure 27.14. Two of these compounds, carbon dioxide (CO_2) and carbon monoxide (CO), are very common. Both are colourless, odourless gases under standard conditions, and they are formed by the combustion of carbon or organic material

$$C(s) + O_2(g) \xrightarrow{\Delta} CO_2(g) \quad \text{in an excess of } O_2$$

$$2C(s) + O_2(g) \xrightarrow{\Delta} 2CO(g) \quad \text{in a limited supply of } O_2$$

Carbon dioxide is the thermodynamically favoured product of the oxidation reaction, and it is formed when combustion occurs in an excess of air. Carbon monoxide forms when carbon burns in a limited supply of air. CO is toxic because it bonds irreversibly to the iron atoms in haemoglobin, and this stops O_2 from being transported from the lungs to the cells where it is needed for respiration.

The structure and function of haemoglobin are described in Box 28.7 (p.1298). The bonding in CO is discussed in Section 4.12 (p.210).

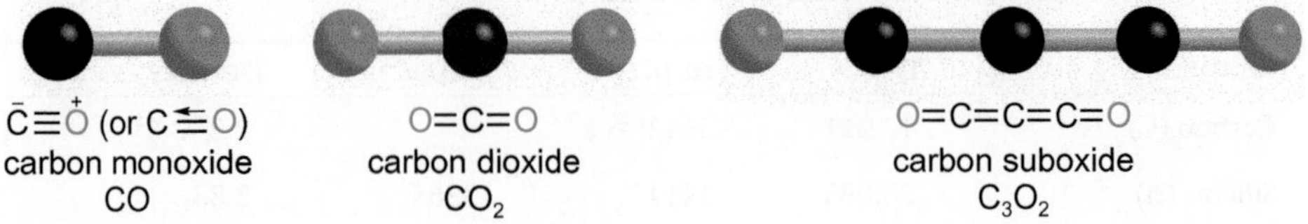
$\bar{C}\equiv\overset{+}{O}$ (or C≡O)
carbon monoxide
CO
O=C=O
carbon dioxide
CO_2
O=C=C=C=O
carbon suboxide
C_3O_2

Figure 27.14 The oxides of carbon. In the Lewis structure of CO, one of the bonds of the triple bond is formally a dative bond. CO_2 and C_3O_2 are linear molecules.

The CO_2/H_2CO_3 system occurs in many places. Its consequences for cave chemistry are described in Box 1.12 (p.58), for buffering in the blood in Box 7.5 (p.321), and for the oceans in Chapter 15 (p.695).

Carbon dioxide dissolves in water, and is present in solution in equilibrium with carbonic acid, H_2CO_3, and the H_3O^+, HCO_3^-, and CO_3^{2-} ions.

In addition to CO_2 and CO, carbon forms several other oxides of lower stability. The most common of these is carbon suboxide (C_3O_2; Figure 27.14).

Box 27.2 The greenhouse effect and global warming

Everyone has heard of the greenhouse effect, but not everyone realizes that it is a natural phenomenon that is essential for life on Earth.

Like any other warm body, the Earth emits energy in the form of electromagnetic radiation at a frequency that is related to its temperature. In the case of the Earth, the maximum intensity emissions are in the infrared part of the electromagnetic spectrum, and you feel them as heat. Some of this infrared radiation escapes into space, but part is absorbed by molecules in the atmosphere.

Some of this absorbed energy is converted to vibrational energy and translational energy and heats the lower atmosphere. The remainder is re-emitted. The energy is re-emitted in all directions—some is directed back towards the Earth and is re-absorbed, heating the Earth's surface and raising its temperature. This is the origin of the 'greenhouse effect'. Without it the average surface temperature of the Earth would be approximately −15 °C. The oceans would be frozen solid and the planet would be lifeless.

▼The mechanism of the greenhouse effect.

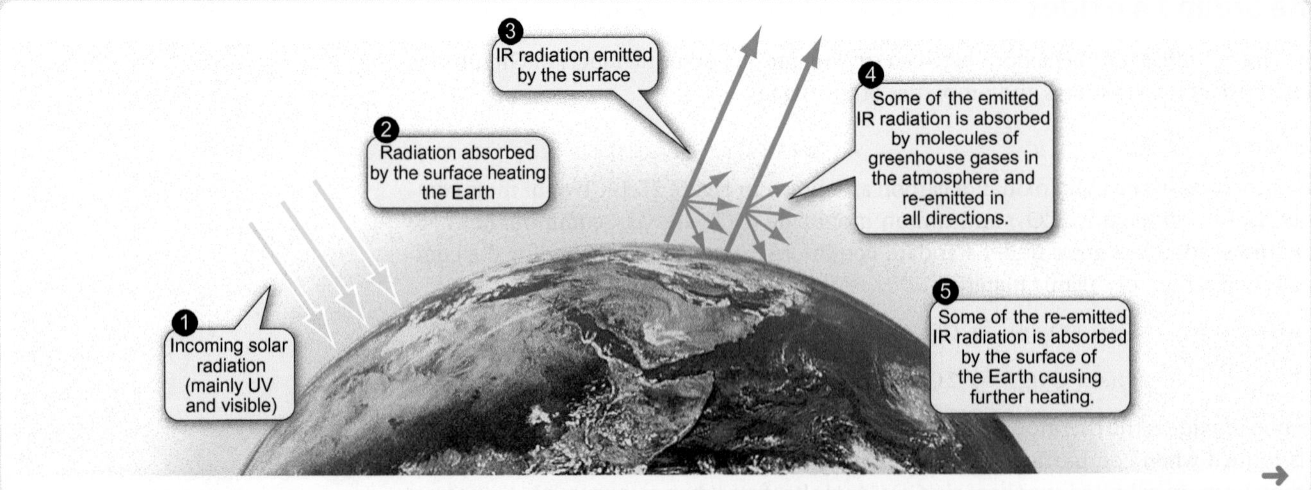
1
Incoming solar radiation (mainly UV and visible)
2
Radiation absorbed by the surface heating the Earth
3
IR radiation emitted by the surface
4
Some of the emitted IR radiation is absorbed by molecules of greenhouse gases in the atmosphere and re-emitted in all directions.
5
Some of the re-emitted IR radiation is absorbed by the surface of the Earth causing further heating.

The average temperature of the Earth has risen by over 0.5 °C during the last hundred years. There is now a broad consensus among scientists that this temperature increase is a result of human activity, particularly the increase in atmospheric CO_2 concentrations since the middle of the eighteenth century. The concentration of CO_2 in the atmosphere in 2015 reached 400 ppm by volume (0.0400%). Measurements of air trapped in ice core samples show that before around 1750, the start of the Industrial Revolution, the atmospheric concentration was 280 ppm. The large majority of experts agree that the main source of the additional CO_2 has been the burning of fossil fuels such as coal, oil, and gas.

Can we stop global warming and the resulting changes to the climate? The Paris accord, reached in 2015, witnessed 190 countries pledging to cut greenhouse gas emissions to a level that will limit global average temperature rise to 'well below 2 °C' compared to pre-industrial levels. Given each country made a voluntary, non-binding pledge, it remains to be seen how successful this will be. Many developed countries aim to use technology to solve the global warming problem. One possible strategy is to capture and store the

→

CO_2 rather than release it into the atmosphere. This is called *carbon sequestration*. One suggested method for carbon sequestration involves increasing the biomass of phytoplankton in the oceans, as the increased photosynthesis would remove CO_2 from the atmosphere.

Coal-fired power stations are the source of about 25% of the world's CO_2 emissions. In many of these power stations the flue gas is passed through an absorption tower where CO_2 reacts with an amine. Heating the product releases the CO_2 and regenerates the amine absorbent. The problem is then what can be done with the isolated CO_2. One possibility is to store it underground in the pores of sedimentary rocks. At depths of 800 m, the pressure is high enough for CO_2 to be a supercritical fluid (see Chapter 17, p.765), and this could replace the salt water that naturally occupies the pores. Old oil fields are being explored for this purpose. In oil fields, crude oil long ago displaced much of the water in the pores. Sequestration of CO_2 here is attractive to oil companies as it has the added advantage of boosting oil recovery.

Silicon oxides

Silicon dioxide (SiO_2) forms a range of covalent network structures all of which contain Si–O single bonds. Quartz is a common form of SiO_2, and sand consists of fragments of quartz that are often coloured yellow-brown by iron oxide impurities.

The covalent network structures of SiO_2 are very different from the discrete molecules formed by CO_2. This difference is related to the relative strengths of the single and double bonds to oxygen. The C=O bonds in CO_2 have a bond enthalpy of +805 kJ mol^{-1}, while the bond enthalpy of a C–O single bond is +358 kJ mol^{-1}. To convert from a network structure to a discrete molecule

$$CO_2\,(\text{network}) \rightarrow CO_2\,(g)$$

four C–O single bonds would need to be broken and two C=O double bonds formed. The enthalpy change for this reaction is

$$\Delta_r H^{\ominus} = (4 \times 358\,\text{kJ mol}^{-1}) + (-2 \times 805\,\text{kJ mol}^{-1}) = -178\,\text{kJ mol}^{-1}$$

The process is highly exothermic, so the silica-like structure for carbon dioxide is strongly unfavoured with respect to the gaseous molecules.

In contrast, Si=O bonds have a bond enthalpy of 638 kJ mol^{-1}, while the bond enthalpy of a Si–O single bond is 466 kJ mol^{-1}. This means that for the reaction

$$SiO_2\,(\text{network}) \rightarrow SiO_2\,(g)$$

$$\Delta_r H^{\ominus} = (4 \times 466\,\text{kJ mol}^{-1}) + (-2 \times 638\,\text{kJ mol}^{-1}) = +588\,\text{kJ mol}^{-1}$$

Here the process is strongly endothermic, so for SiO_2 the silica network is the more favoured structure.

The key factors in the difference in structure are, firstly the high strength of the C=O double bond compared to the Si=O bond, which arises from the distance sensitivity of π bonding, and, secondly the high strength of the Si–O single bond compared to the C–O bond, which arises largely as a consequence of the greater ionic character of the Si–O bond.

With the exception of hydrofluoric acid, SiO_2 does not react with acids. On heating with sodium carbonate, SiO_2 is converted to sodium silicate (Na_4SiO_4)

$$SiO_2\,(s) + 2\,Na_2CO_3\,(s) \xrightarrow{1500\,°C} Na_4SiO_4\,(s) + 2\,CO_2\,(g)$$

Silicates react with acids to form a gelatinous precipitate of hydrated silicon dioxide known as *silica gel*

$$SiO_4^{\,4-}\,(aq) + 4\,H_3O^+\,(aq) + (x-6)\,H_2O\,(l) \rightarrow SiO_2{\cdot}xH_2O\,(\text{gel})$$

When this gel is washed, dried, and granulated it has a high surface area and is used as a drying agent. Optical equipment such as cameras and binoculars are usually sold with small sachets of silica gel to keep the lenses dry.

Silicon forms many extended silicate anions that are based on SiO_4 tetrahedra, with oxygen atoms shared between adjacent tetrahedra. The tetrahedra can link together to form cyclic anions, extended chains, or extended sheets. Examples of silicates are shown in Figure 27.15. In aluminosilicates some of the silicon atoms have been exchanged for aluminium atoms.

The structures of two polymorphs of SiO_2 are described in Section 6.1 (p.259).

These calculations are simplifications as they use ΔH instead of ΔG. However, the changes in entropy, ΔS, would be similar as the two reactions both involve conversion of a solid to a gas. See Chapter 14 (p.675) for more details on the relationship between ΔH, ΔS, and ΔG.

Self-indicating silica gel contains cobalt(II) chloride. This turns from blue to pink as water is absorbed, so it is easy to judge whether the silica gel is still effective. The change in colour is due to a change in the Co^{2+} geometry from tetrahedral to octahedral (see Section 28.6, p.1291).

Asbestos is a name given to a range of fibrous materials, many of which are magnesium silicates with structures containing silicate chains linked by Si–O–Si bridges. In the past, asbestos was used extensively because of its fire-resistant properties. Inhalation of asbestos fibres can cause chronic lung disease, so its use is now restricted.

Zeolites are aluminosilicates that have channels running through the structure. The uses of these compounds are described in Chapter 6 (p.255).

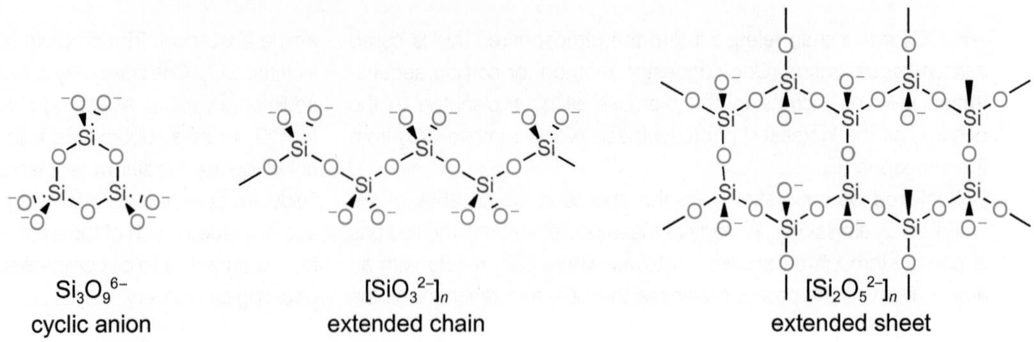
$Si_3O_9^{6-}$
cyclic anion
$[SiO_3^{2-}]_n$
extended chain
$[Si_2O_5^{2-}]_n$
extended sheet

Figure 27.15 Examples of silicate anions.

Box 27.3 Silicones—shiny hair and cosmetic surgery

Shampoos contain detergents to remove dirt and grease from hair. Unfortunately, these detergents also remove the natural oils that protect the hair. To prevent this, many people use a conditioner to coat their hair with a protective material and make it feel softer. Most conditioners are based on silicones, which are polymeric organosilicon compounds containing a silicon–oxygen backbone. Dimethylsiloxanes (**3**) with different chain lengths are the commonest silicones for conditioners, and these are known in the cosmetic industry as *dimethicone*.

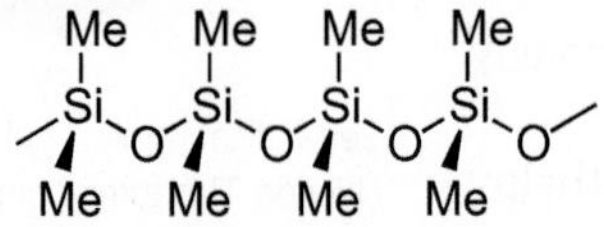

▲ **3** Polydimethylsiloxane, $(SiOMe_2)_n$

2-in-1 shampoos contain conditioners in the same bottle as the shampoo. These are difficult to formulate, as silicones are oils that are immiscible with water. One way to get around this problem is to use an emulsion of the silicone. Microdroplets of the silicone are suspended in the shampoo, and these remain inactive until the hair is rinsed. When the foam is removed, the microdroplets precipitate from solution and the silicone coats the hair.

NEW
PANTENE
2in1

▲ 2-in-1 shampoos are difficult to formulate.

Silicones are prepared by the hydrolysis of silicon chlorides such as Me_2SiCl_2. The hydrolysis initially forms $Me_2Si(OH)_2$, which then undergoes condensation polymerization to form the silicone. Silicone chains can contain the four different structural units shown below, which differ in the number of Si–O bonds they contain.

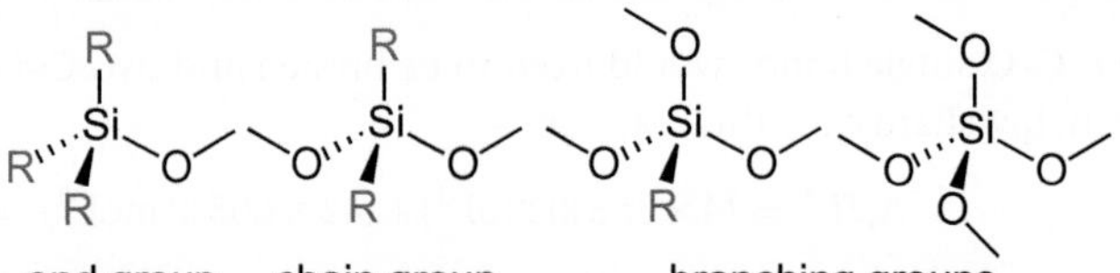
end group
chain group
branching groups

All silicones have a backbone containing strong Si–O bonds, and this gives them good thermal stability. The properties of the silicone depend on the nature of the organic substituents and the chain length. For example, the longer the chain, the more viscous the silicone. Silicone oils have many applications and are also used as lubricants, waterproof coatings, and insulators. Silicone polymers are used as stationary phases in chromatography, as described in Section 11.3 (p.535).

Cross-linking of the polymers gives silicone rubbers. These maintain their strength and flexibility over a wide temperature range and are used in adhesives and sealants, medical tubing, and heart-valve implants. Extensive cross-linking gives rise to a resin. Silicone resins are used in paints, to render plastics scratch-resistant, and for protecting building materials.

The most controversial use of silicones is in cosmetic surgery—in breast implants. Although silicones were chosen for use because of their non-toxicity, there have been reports of medical problems such as increased risk of cancer following implants. In some cases, the silicone has leaked from the implant. Although toxicity has not been confirmed, silicone implants are no longer available in some countries, and the controversy led to the bankruptcy of the biggest supplier of silicones in the USA.

Question

Suggest why carbon analogues of silicones, in which C atoms take the place of Si atoms, are unknown.

Summary

- The carbon oxides, CO_2 and CO, exist as discrete molecules containing multiple bonding.
- Silicon dioxide forms covalent network structures that contain Si–O single bonds. A wide range of silicate anions is known.

The Group 14 halides

All of the Group 14 tetrahalides are known, with one exception—PbI_4. Lead(IV) is too oxidizing to coexist with I^-, as it oxidizes it to iodine.

Carbon halides

The carbon tetrahalides are all covalent molecules with tetrahedral geometries. Although the individual C–X bonds are polar, these dipoles cancel out in the molecules. As a result, the carbon tetrahalides are non-polar and dissolve in non-polar solvents such as toluene. Under standard conditions, CF_4 is a gas, CCl_4 is a liquid, and CBr_4 and CI_4 are solids.

Bond polarities and polar molecules are described in Section 5.3 (p.235).

CCl_4 is known as both carbon tetrachloride and tetrachloromethane.

The stability of the carbon tetrahalides decreases as the halogen atom becomes larger. This is because the C–X bond enthalpy decreases in the order

C–F (highest bond enthalpy) > C–Cl > C–Br > C–I (lowest bond enthalpy)

This is consistent with the reduction in orbital overlap as the halogen atomic orbitals become larger and more diffuse. CI_4 decomposes in light or on heating to give tetraiodoethene (C_2I_4, (**4**))

$$2\,CI_4\,(s) \xrightarrow{\Delta} C_2I_4\,(s) + 2\,I_2\,(g)$$

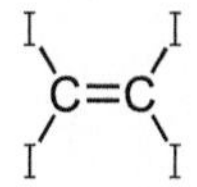

4 Tetraiodoethene (C_2I_4).

CF_4 and CCl_4 are very resistant to attack by acids and bases, oxidizing agents, and reducing agents. Hydrolysis to form CO_2 is thermodynamically favoured, but it does not occur. Carbon tetrachloride (CCl_4) is immiscible with water and was once used as a solvent in dry cleaning processes. Its use is now restricted as it is both a suspect carcinogen and a greenhouse gas. Chlorofluorocarbons such as $CFCl_3$ are much less toxic and were used extensively in the twentieth century as refrigerants and aerosol propellants but, following the recognition of their role in damaging the ozone layer, their use is now also restricted (see Box 27.8, p.1241).

Another carbon fluoride has an extended structure. Graphite reacts with fluorine gas at 400 °C to form graphite fluoride $(CF)_n$ (**5**), which is a colourless solid. This has a layer structure consisting of fused cyclohexane rings. The spacing between the layers is greater in graphite fluoride than it is in graphite, and graphite fluoride is an electrical insulator.

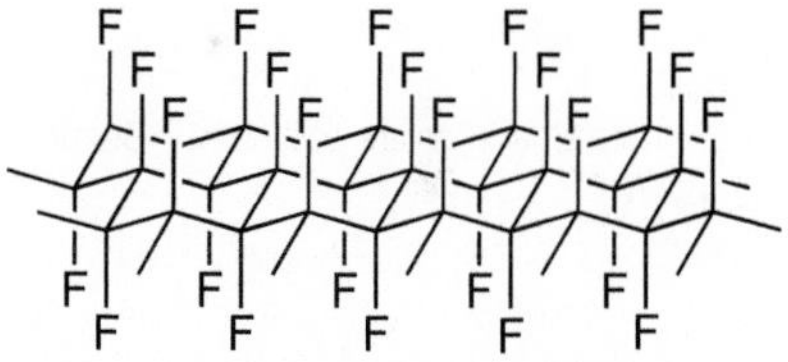

5 Graphite fluoride, $(CF)_n$.

The conformations of cyclohexane rings are described in Section 18.2 (p.827).

Silicon halides

At room temperature SiF_4 is a gas, $SiCl_4$ and $SiBr_4$ are liquids, and SiI_4 is a solid. The silicon tetrahalides have similar tetrahedral structures to their carbon analogues, but in contrast to the carbon compounds they are rapidly hydrolysed by water. For example, $SiCl_4$ reacts with water to form SiO_2 and hydrochloric acid

$$SiCl_4\,(l) + 2\,H_2O\,(l) \rightarrow SiO_2\,(s) + 4\,HCl\,(aq)$$

What is the cause of this difference in behaviour between CCl_4 and $SiCl_4$? Carbon is smaller than silicon, and an H_2O molecule cannot attack the sterically shielded carbon atom in CCl_4. The larger size of the silicon atom in $SiCl_4$ exposes it to nucleophilic attack.

Size is also a factor in why the silicon tetrahalides are able to act as Lewis acids, whereas the carbon tetrahalides do not. For the silicon tetrahalides, the Lewis acidity decreases in the order

Nucleophilic substitution in organic chemistry is described in Section 19.2 (p.903).

$$\underset{\text{most acidic}}{SiF_4} > SiCl_4 > SiBr_4 > \underset{\text{least acidic}}{SiI_4}$$

The trend in Lewis acidities of the boron halides is explained in Section 27.2 (p.1213).

This is the order expected on the basis of the electronegativities of the halogens, and is the opposite to that observed for the boron halides.

Summary

- CF_4 and CCl_4 are very unreactive compounds, but as the halogen increases in size the stability of the carbon tetrahalide towards thermal decomposition decreases.
- With the exception of the carbon compounds, the Group 14 tetrahalides are hydrolysed by water. The reaction is thermodynamically favoured for CX_4, but the carbon atom is sterically shielded by the halogen atoms.
- With the exception of the carbon compounds, the Group 14 tetrahalides are Lewis acids.

Other Group 14 compounds

Carbides

Ethynides are also known as ethynyl ions and acetylides.

Carbides are binary compounds of carbon with another element. Electropositive metals form **salt-like carbides** that contain either the C^{4-} ion (though with a high degree of covalent character) or the C_2^{2-} ion. These are known as methides and ethynides, respectively. Both types of carbide react with water, methides giving methane and ethynides giving ethyne (acetylene). For example,

$$Al_4C_3\,(s) + 12\,H_2O\,(l) \rightarrow 4\,Al(OH)_3\,(s) + 3\,CH_4\,(g)$$

$$CaC_2\,(s) + 2\,H_2O\,(l) \rightarrow Ca(OH)_2\,(s) + C_2H_2\,(g)$$

Non-metals form **covalent carbides**. Silicon carbide (SiC) has a covalent network structure similar to that of diamond and is used as an abrasive. It is formed from the reaction of SiO_2 with carbon at high temperature,

$$SiO_2\,(s) + 3\,C\,(s) \xrightarrow{2000\,°C} SiC\,(s) + 2\,CO\,(g)$$

Interstitial sites are described in Section 6.2 (p.270).

Many metals form **interstitial carbides**, in which the carbon atoms sit in interstitial sites within a metal lattice. This tends to increase the hardness of a metal as it makes it more difficult for the planes of atoms to slip past each other. Tungsten carbide (WC) is an interstitial carbide that is used for the cutting surfaces of drills. Several phases of steel also contain interstitial carbon atoms.

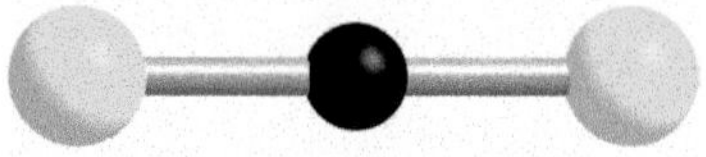

S=C=S

6 Carbon disulfide (CS_2).

N≡C−C≡N

7 Cyanogen ($(CN)_2$).

A **pseudohalogen** is a molecule that acts in an analogous manner to a halogen, and a **pseudohalide** is an anion that acts in an analogous manner to a halide ion.

Carbon-sulfur and carbon-nitrogen compounds

Carbon disulfide (CS_2, (**6**)) cannot be made by direct combination of the elements, but is formed from the gas-phase reaction of methane with sulfur over an SiO_2 or Al_2O_3 catalyst

$$CH_4\,(g) + 4\,S\,(g) \xrightarrow{600\,°C} CS_2\,(g) + 2\,H_2S\,(g)$$

CS_2 is a volatile, toxic liquid at room temperature. It typically has a characteristic, unpleasant smell of boiled cabbage, though this is actually caused by organic impurities. CS_2 is used in the manufacture of cellulose polymers such as cellophane.

The cyanide ion (CN^-) contains a carbon–nitrogen triple bond. In its reactions, the cyanide ion shows many similarities to halide ions, and is often described as a **pseudohalide**. Cyanide ions are, however, extremely toxic as they interact with the metal ions in metalloproteins, disrupting metabolic processes.

While cyanide is analogous to a halide ion, cyanogen ($(CN)_2$) is analogous to a dihalogen molecule and is an example of a **pseudohalogen**. Cyanogen (**7**) disproportionates in a base in a similar manner to a halogen (see Section 27.6, p.1243)

$$(CN)_2\,(g) + 2\,OH^-\,(aq) \rightarrow \underset{\text{cyanide ion}}{CN^-\,(aq)} + \underset{\text{cyanate ion}}{OCN^-\,(aq)} + H_2O\,(l)$$

Summary

- Most elements form binary compounds with carbon. These carbides can be ionic solids, covalent network structures, or the carbon atoms can occupy interstitial sites in metal lattices.
- Multiple bonds (C=C, C≡C, C=O, C≡N, C=N, C=S) are important for carbon, but less so for the other Group 14 elements.

For a practice question on the topic, see question 7 at the end of this chapter (p.1253). Questions 16–20 are on general *p*-block chemistry.

27.4 Group 15: nitrogen, phosphorus, arsenic, antimony, and bismuth

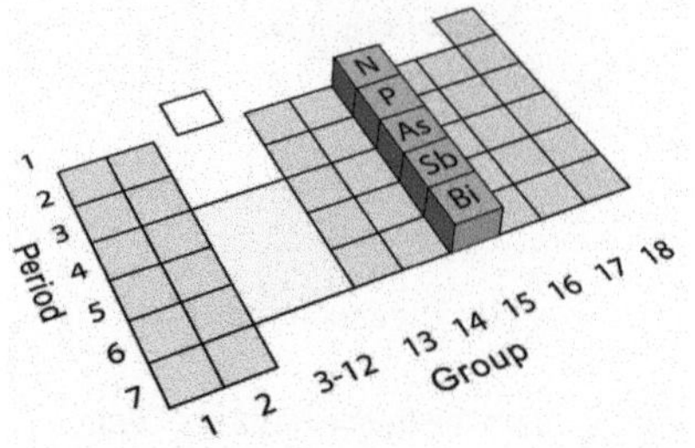
N
P
As
Sb
Bi
Period
1
2
3
4
5
6
7
1 2 3-12 13 14 15 16 17 18
Group

Group 15 in the Periodic Table.

Group 15 contains the elements nitrogen (N), phosphorus (P), arsenic (As), antimony (Sb), and bismuth (Bi). The valence electronic configuration of the Group 15 elements is $ns^2\ np^3$. Nitrogen and phosphorus can form the trianions nitride (N^{3-}) and phosphide (P^{3-}) for which the oxidation state is –3. All of the elements form covalent compounds with the most common oxidation states +3 and +5. Bismuth also forms trications (Bi^{3+}). With the exception of the NF_4^+ cation, nitrogen(V) is only observed in compounds containing multiple bonds such as the nitrate ion (**8**), and in these cases the nitrogen atom has a formal charge of +1.

Selected physical properties of the Group 15 elements are given in Table 27.7. Nitrogen and phosphorus are typical non-metals, and the metallic character of the elements increases down the group. Arsenic and antimony are often described as metalloids, whereas bismuth is a metal.

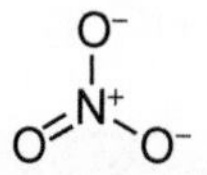

8 The nitrate anion (NO_3^-). One resonance form is shown.

Formal charge is explained in Section 5.1 (p.220).

Table 27.7 Physical properties of the Group 15 elements

Element	A_r	m.p./°C	b.p./°C	Density/g cm^{-3}
Nitrogen (N)	14.007	–210	–196	1.25×10^{-3}
Phosphorus (P)	30.974	44*	281*	1.82*
Arsenic (As)	74.922	614†	–	5.75‡
Antimony (Sb)	121.76	631	1587	6.68
Bismuth (Bi)	208.98	271	1564	9.79

*White. †Sublimes. ‡Grey.

The Group 15 elements

The bonding in N_2 is described in Sections 4.5 (p.180) and 4.11 (p.201).

Nitrogen occurs naturally in the elemental form as N_2 molecules, and is the main component (78%) of the atmosphere. It is a colourless, odourless gas that can be obtained pure by the fractional distillation of liquid air. The bond dissociation enthalpy of the N_2 molecule is very high (+945 kJ mol^{-1}) due to the strong triple bond, so N_2 is chemically unreactive. Other factors in the low reactivity are a large HOMO–LUMO gap, which prevents oxidation or reduction, and low polarizability, which prevents attack by electrophiles or nucleophiles.

The reaction with O_2

$$N_2(g) + O_2(g) \rightarrow 2NO(g) \qquad \Delta H^{\ominus} = +181\,\text{kJ mol}^{-1}$$

is highly endothermic, and only proceeds at high temperatures, for example in car engines and in power stations (see start of chapter, p.1201). The small amount of NO that occurs naturally in the atmosphere is formed from the reaction between N_2 and O_2 in lightning.

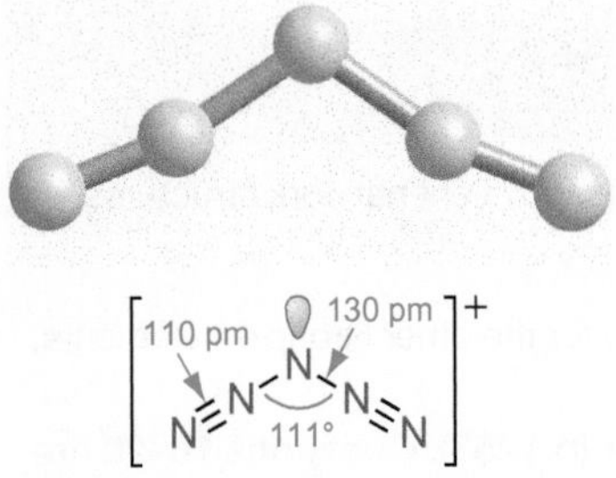
110 pm
130 pm
111°
+

9 The pentanitrogen cation (N_5^+).

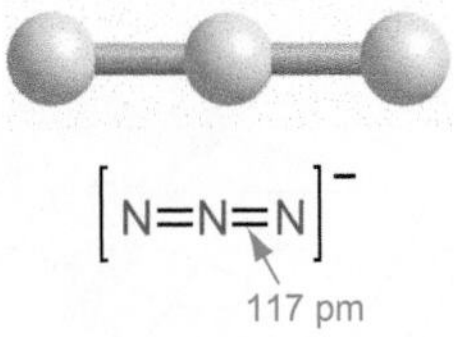
N=N=N
117 pm

10 The azide anion (N_3^-).

Visit the Online Resource Centre to view screencast 27.1 which walks you through the use of bond enthalpies to predict the structure of phosphorus.

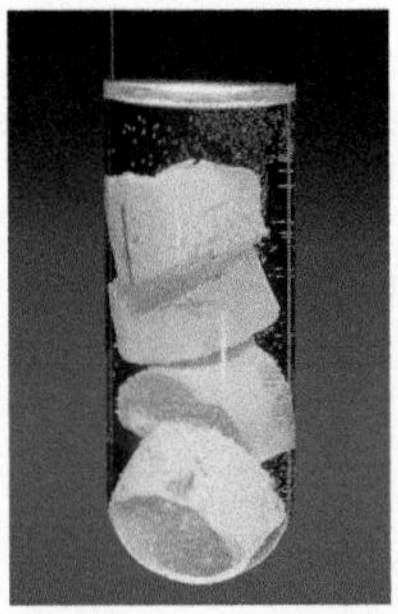

White phosphorus ignites spontaneously in air, so must be kept under water.

Both cations and anions that contain only nitrogen atoms are known. The pentanitrogen cation (N_5^+, (**9**)) was first prepared in 1999 and has a V-shaped structure. The azide anion (N_3^-, (**10**)) is linear.

Phosphorus occurs naturally in phosphate-containing minerals such as *apatites* (see Chapter 26, p.1169) and calcium phosphate. Elemental phosphorus is obtained by heating calcium phosphate with coke and SiO_2 in an electric furnace

$$2\,Ca_3(PO_4)_2\,(s) + 10\,C\,(s) + 6\,SiO_2\,(s) \xrightarrow{1400\,°C} 6\,CaSiO_3\,(l) + 10\,CO\,(g) + P_4\,(g)$$

The phosphorus vapour is collected and condensed to give a white, waxy solid that consists of P_4 tetrahedra. The solid is called *white phosphorus* and is one of several allotropes of phosphorus (Figure 27.16). At temperatures above 800 °C, P_4 partially dissociates into P_2 molecules, which contain triple bonds.

White phosphorus must be stored under water as it ignites spontaneously in air. The reactivity is due to the strain in the molecule, as all of the P–P–P bond angles in the tetrahedron are 60°. Because white phosphorus ignites so easily, it has been used in incendiary bombs. The earliest matches used white phosphorus in the match heads, but it is no longer used because it is dangerously flammable and toxic.

If white phosphorus is heated to 300 °C in the absence of air it is converted into *red phosphorus*. Red phosphorus has an amorphous structure, which is believed to consist of P_4 units linked into chains. Red phosphorus is much less reactive than white phosphorus because of the reduced strain, but it can be ignited by friction. Heating phosphorus at higher pressures leads to *black phosphorus*. There are several different forms of black phosphorus, all of which have more extended structures such as the layer structure shown in Figure 27.16. Black phosphorus is even less reactive than red phosphorus, so the reactivity of phosphorus decreases with the increasing connectivity within the structure. The energetics of the bonding in phosphorus is examined in Worked example 27.1.

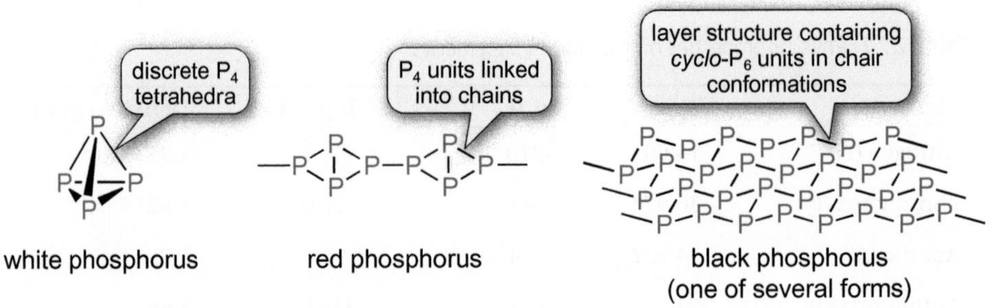
discrete P4 tetrahedra
P4 units linked into chains
layer structure containing cyclo-P6 units in chair conformations
white phosphorus
red phosphorus
black phosphorus
(one of several forms)

Figure 27.16 The allotropes of phosphorus.

Worked example 27.1 Bonding in the Group 15 elements

The mean bond enthalpy of a P–P bond is +198 kJ mol^{-1} while that of a P≡P bond is +489 kJ mol^{-1}. Calculate the enthalpy change for the reaction

$$P_4(g) \rightarrow 2P_2(g)$$

Use this value to explain why elemental phosphorus contains P_4 tetrahedra and not P_2 dimers as in N_2. Comment on any assumptions you have made.

Strategy

First, make sure you understand the structure of the tetrahedral P_4 molecule (Figure 27.16). Break down the reaction into two steps. In the first step, the six single bonds in the P_4 molecule break to give four phosphorus atoms. In the second step, two triple bonds form to give two P_2 molecules. Calculate the enthalpy change for the overall reaction using an enthalpy cycle.

Solution

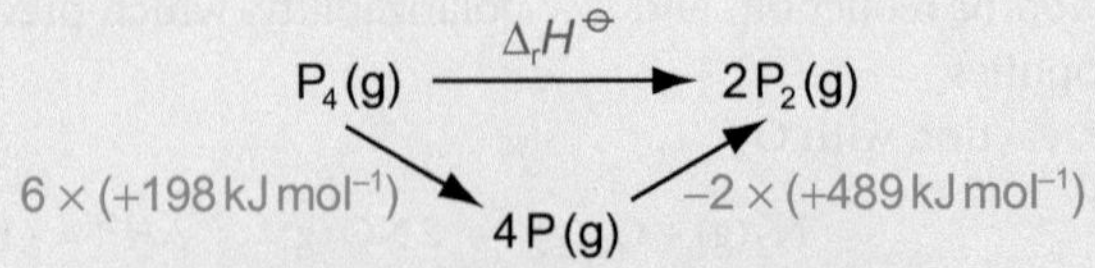
ΔrH⊖
P4(g)
2P2(g)
6 × (+198 kJ mol−1)
4P(g)
−2 × (+489 kJ mol−1)

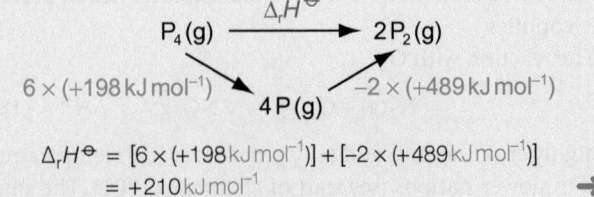

$$\Delta_r H^{\ominus} = [6 \times (+198\,\text{kJ mol}^{-1})] + [-2 \times (+489\,\text{kJ mol}^{-1})]$$
$$= +210\,\text{kJ mol}^{-1}$$

→

➜ The conversion of P_4 into 2 P_2 is highly endothermic so, under standard conditions, P_4 is the more stable form and phosphorus exists as P_4 molecules.

Strictly speaking, you need to use $\Delta G^{\ominus}$ not $\Delta H^{\ominus}$ to predict the feasibility of a reaction (see Section 14.5, p.675). In many reactions the change in entropy is small with respect to changes in enthalpy, so good predictions of the direction of a reaction at room temperature can be made just by considering $\Delta H^{\ominus}$. At high temperatures, $T\Delta S^{\ominus}$ becomes more significant. Converting P_4 to P_2 involves an increase in entropy, and this becomes more important at higher temperatures. This explains why P_4 decomposes to P_2 on heating to 800 °C.

Question

The mean bond enthalpies for the N–N and N≡N bonds are +158 kJ mol^{-1} and +945 kJ mol^{-1}, respectively. Calculate the enthalpy change for the conversion of tetrahedral N_4 into N_2 and use this value to comment on the molecular structure of nitrogen.

Summary

- Of the Group 15 elements, nitrogen and phosphorus are non-metals, arsenic and antimony are metalloids, and bismuth is a metal.
- Oxidation states vary from −3 to +5, with +3 and +5 the most common.
- Nitrogen occurs naturally as N_2 molecules in the atmosphere.
- Phosphorus forms several allotropes. White phosphorus is very reactive, igniting in air, but the reactivity of the allotropes decreases with increasing connectivity within the structure.

The Group 15 oxides and oxoacids

Nitrogen oxides

Nitrogen forms a range of oxides, the formulae and structures of which are shown in Figure 27.17. Of these oxides, NO and NO_2 are radicals and so contain unpaired electrons.

All the nitrogen oxides contain multiple bonds between nitrogen and oxygen, and none of the heavier Group 15 elements form oxides like them. With the exception of N_2O, the nitrogen oxides dissolve in water to form acidic solutions. All have positive enthalpy changes of formation.

The structure and properties of N_2O are described in more detail in Section 5.1 (p.221).

Dinitrogen oxide (nitrous oxide, N_2O) is prepared by carefully heating ammonium nitrate

$$\underset{\text{N(−III),N(V)}}{NH_4NO_3\,(s)} \xrightarrow{200\,°C} \underset{\text{N(I)}}{N_2O\,(g)} + 2\,H_2O\,(l)$$

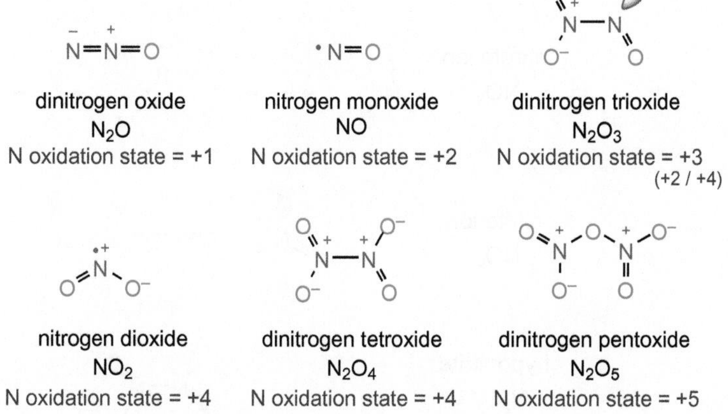

dinitrogen oxide
N₂O
N oxidation state = +1
nitrogen monoxide
NO
N oxidation state = +2
dinitrogen trioxide
N₂O₃
N oxidation state = +3
(+2 / +4)
nitrogen dioxide
NO₂
N oxidation state = +4
dinitrogen tetroxide
N₂O₄
N oxidation state = +4
dinitrogen pentoxide
N₂O₅
N oxidation state = +5

Figure 27.17 The structures of the most common nitrogen oxides. Only one resonance form of each is shown for simplicity.

A **comproportionation** reaction is one in which a high and low oxidation state of an element combine to give an intermediate oxidation state. It is the opposite of a disproportionation reaction (p.1152).

This is a **comproportionation** reaction, in which the NH_4^+ ion is oxidized and the NO_3^- ion is reduced.

Nitrogen monoxide (nitric oxide) is a radical, so you would expect it to dimerize by pairing up the unpaired electron on each of two molecules. However, it only dimerizes at very low temperatures. NO is prepared industrially on a large scale by the oxidation of ammonia. The reaction is carried out at 820 °C–950 °C and at pressures between 1 atm and 12 atm in the presence of a Pt/Rh catalyst

$$4\,NH_3\,(g) + 5O_2\,(g) \xrightarrow{\Delta,\,\text{Pt/Rh catalyst}} 4\,NO\,(g) + 6\,H_2O\,(l)$$

This reaction is the first step in the manufacture of nitric acid (HNO_3). In the manufacturing process, NO is then reacted with oxygen to give NO_2

$$2\,NO\,(g) + O_2\,(g) \rightarrow 2\,NO_2\,(g)$$

which is then treated with water to form HNO_3

$$3\,NO_2\,(g) + H_2O\,(l) \rightarrow 2\,HNO_3\,(aq) + NO\,(g)$$

The biological importance of NO is described in Box 4.11 (p.212).

NO_2 contributes to the odour and colour of photochemical smog (see chapter opener, p.1201).

The effect of change of temperature on the equilibrium between NO_2 and N_2O_4 is described in Section 1.9 (p.58) and Chapter 15 (p.696).

Nitrogen dioxide (NO_2) is a brown, paramagnetic gas with an unpleasant smell. NO_2 exists in equilibrium with its dimer, N_2O_4.

120 pm
134°
121 pm
135°
175 pm

$$2\,NO_2\,(g) \rightleftharpoons N_2O_4\,(g) \qquad \Delta H^{\ominus} = -57\,\text{kJ mol}^{-1}$$

In the solid state, nitrogen dioxide exists almost exclusively as N_2O_4.

When NO and NO_2 are mixed at low temperatures, they react to form N_2O_3. N_2O_3 only exists as the solid or liquid, both of which are blue. In the gas phase N_2O_3 dissociates into NO and NO_2.

In contrast, N_2O_5 exists as covalent molecules in the gas phase but as the ionic compound $NO_2^+NO_3^-$ in the solid. It is formed by dehydrating nitric acid with phosphorus(V) oxide.

Nitrogen oxoacids and oxoanions

There are two main nitrogen oxoacids, the structures of which are shown in Figure 27.18 together with their associated oxoanions.

Nitric acid (HNO_3) is important industrially as it is used in the manufacture of fertilizers and explosives. The nitrate ion (NO_3^-) has a trigonal planar structure with the three bond lengths identical, with some partial double bond character. Nitrate salts are very common, and some, such as KNO_3, are used as oxidizing agents. Most nitrates are soluble in water.

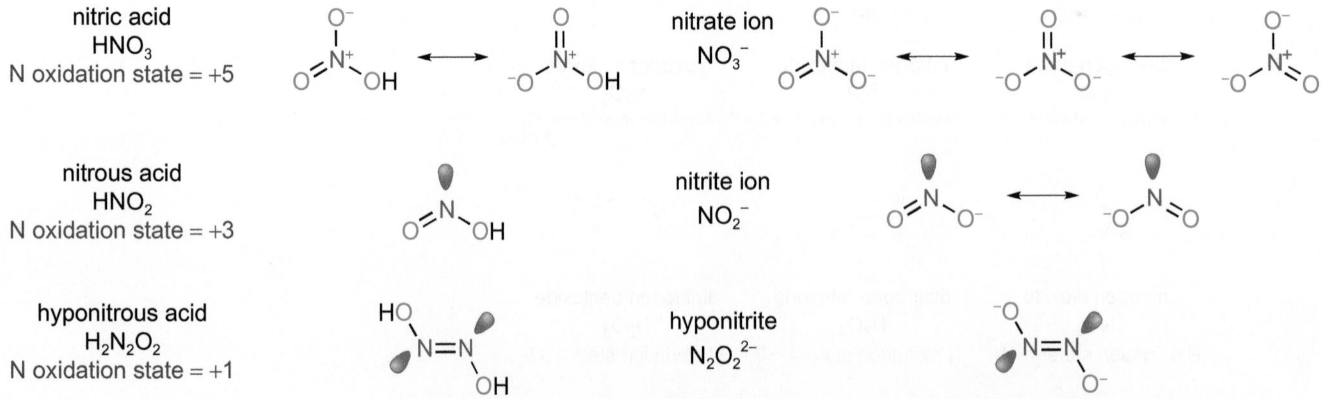
nitric acid
HNO3
N oxidation state = +5
nitrate ion
NO3-
nitrous acid
HNO2
N oxidation state = +3
nitrite ion
NO2-
hyponitrous acid
H2N2O2
N oxidation state = +1
hyponitrite
N2O2 2-

Figure 27.18 The structures of the nitrogen oxoacids and the anions derived from them.

Nitrous acid (HNO_2) is not stable as a pure compound, but is used as a reagent in aqueous solution. In contrast, the nitrite ion (NO_2^-) is stable and nitrite salts are used in the processing of foodstuffs.

Nitrous acid is used in the preparation of diazonium ions. See Section 22.4 (p.1041) for more details.

Nitrite (NO_2^-) forms a complex with haemoglobin, inhibiting its oxidation by air. Sodium nitrite is added to some cooked meat products such as ham, as the pink colour it gives is more attractive than the natural brown colour.

Phosphorus oxides

Phosphorus forms oxides in the +3 and +5 oxidation states. Phosphorus(III) oxide has the molecular formula P_4O_6 and phosphorus(V) oxide has the molecular formula P_4O_{10}. Both of the compounds have structures based on a P_4 tetrahedron, but with bridging oxygen atoms inserted between the phosphorus atoms, as shown in Figure 27.19. For P_4O_{10}, the phosphorus atoms are also bonded to terminal oxygen atoms. Mixed oxidation state compounds such as P_4O_7, in which only some of the phosphorus atoms are bonded to terminal oxygen atoms, are also known.

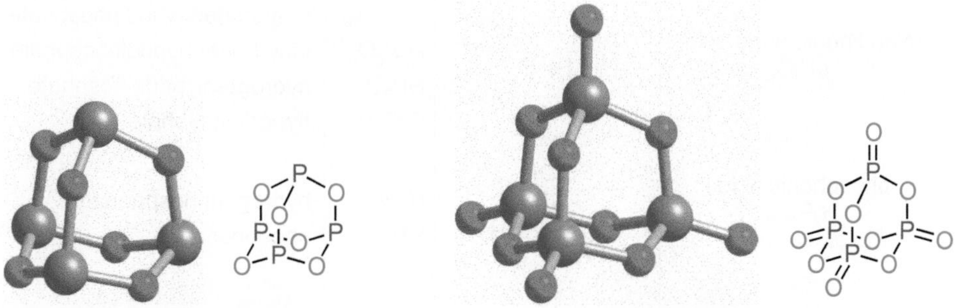

phosphorus(III) oxide P_4O_6 phosphorus(V) oxide P_4O_{10}

Figure 27.19 The structures of the main phosphorus oxides.

P_4O_6 and P_4O_{10} are formed by burning phosphorus in air. P_4O_6 is produced when the oxygen supply is limited, whereas P_4O_{10} is produced in the presence of excess oxygen. Both compounds are acidic. P_4O_{10} dissolves in water to give phosphoric acid, H_3PO_4, whereas P_4O_6 dissolves in water to give phosphonic acid, H_3PO_3

$$P_4O_{10}(s) + 6H_2O(l) \rightarrow \underset{\substack{\text{phosphoric acid} \\ \text{(phosphorus oxidation state +5)}}}{4H_3PO_4(aq)}$$

$$P_4O_6(s) + 6H_2O(l) \rightarrow \underset{\substack{\text{phosphonic acid} \\ \text{(phosphorus oxidation state +3)}}}{4H_3PO_3(aq)}$$

Phosphorus forms a wide range of oxoacids, some of which are shown in Figure 27.20 together with their oxoanions. Phosphoric acid is produced industrially on a large scale, and is used in the manufacture of fertilizers and detergents. Phosphoric acid is a tribasic acid, with all of the hydrogen atoms ionizable. In contrast, phosphonic acid (H_3PO_3) is dibasic whereas phosphinic acid (H_3PO_2) is monobasic. The basicity is related to the number of OH groups present.

For more details on the phosphorus oxoacids see Section 7.6 (p.330).

Diphosphoric acid and *cyclo*-triphosphoric acid are examples of condensed oxoacids. They are derived from condensation reactions between two or more molecules of phosphoric acid. For example,

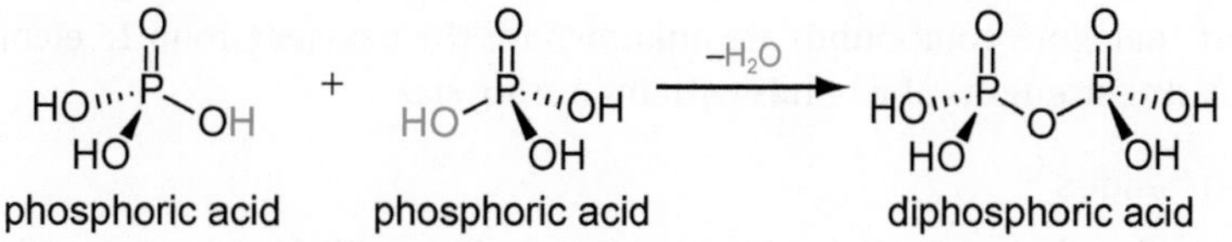

phosphoric acid phosphoric acid diphosphoric acid

Other phosphorus oxoacids such as hypodiphosphoric acid contain a P–P bond.

Phosphates are very important in living systems. In DNA, the bases are attached to a sugar-phosphate backbone (see Box 25.5, p.1158), whereas the interconversion between ATP and ADP is important is transferring energy within cells (see Box 14.8, p.683).

phosphoric acid H_3PO_4 P oxidation state = +5	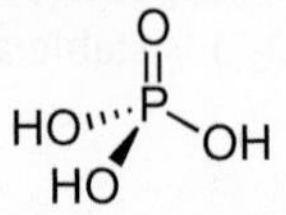	$H_2PO_4^-$ HPO_4^{2-} PO_4^{3-}	dihydrogenphosphate hydrogenphosphate phosphate
diphosphoric acid $H_4P_2O_7$ P oxidation state = +5	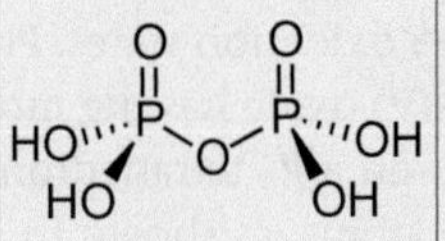	$H_3P_2O_7^-$ $H_2P_2O_7^{2-}$ $HP_2O_7^{3-}$ $P_2O_7^{4-}$	trihydrogendiphosphate dihydrogendiphosphate hydrogendiphosphate diphosphate
cyclo-triphosphoric acid $H_3P_3O_9$ P oxidation state = +5	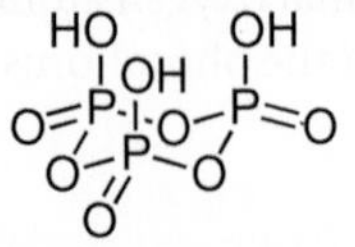	$H_2P_3O_9^-$ $HP_3O_9^{2-}$ $P_3O_9^{3-}$	dihydrogen-*cyclo*-triphosphate hydrogen-*cyclo*-triphosphate *cyclo*-triphosphate
hypodiphosphoric acid $H_4P_2O_6$ P oxidation state = +4	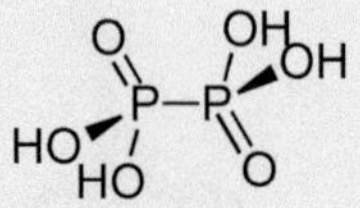	$H_3P_2O_6^-$ $H_2P_2O_6^{2-}$ $HP_2O_6^{3-}$ $P_2O_6^{4-}$	trihydrogenhypodiphosphate dihydrogenhypodiphosphate hydrogenhypodiphosphate hypodiphosphate
phosphonic acid H_3PO_3 P oxidation state = +3	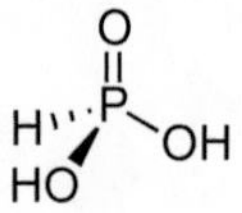	$H_2PO_3^-$ HPO_3^{2-}	hydrogenphosphonate phosphonate
phosphinic acid H_3PO_2 P oxidation state = +1	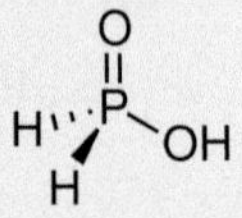	$H_2PO_2^-$	phosphinate

Figure 27.20 The structures of common phosphorus oxoacids and the anions derived from them.

Summary

- Nitrogen forms a range of oxides including N_2O, NO, and NO_2. These compounds all contain multiple bonds between nitrogen and oxygen.
- The phosphorus oxides have structures based on the P_4 tetrahedron with bridging oxygen atoms between the phosphorus atoms. For phosphorus(V) oxide, the phosphorus atoms are also bonded to terminal oxygen atoms.
- Nitrogen and phosphorus form many oxoacids, including HNO_3, HNO_2, and H_3PO_4.

The Group 15 halides

The Group 15 elements all form trihalides of the general formula AX_3, and, with the exception of nitrogen, pentahalides of the general formula AX_5. Nitrogen pentahalides are not known, though calculations suggest it may be possible to isolate NF_5, though this would involve accommodating 10 valence electrons around a small nitrogen atom. Dihalides of the general formula A_2X_4, containing an A–A bond between the Group 15 atoms, have been prepared. Nitrogen also forms the compound N_2F_2, which contains a N=N double bond, but analogous compounds are unknown for the heavier Group 15 elements due to the decreasing strengths of π bonds with increasing size.

Nitrogen halides

Some examples of nitrogen fluorides are shown in Figure 27.21. Nitrogen trifluoride (NF_3) is a colourless, thermodynamically stable gas and the molecules have a trigonal pyramidal

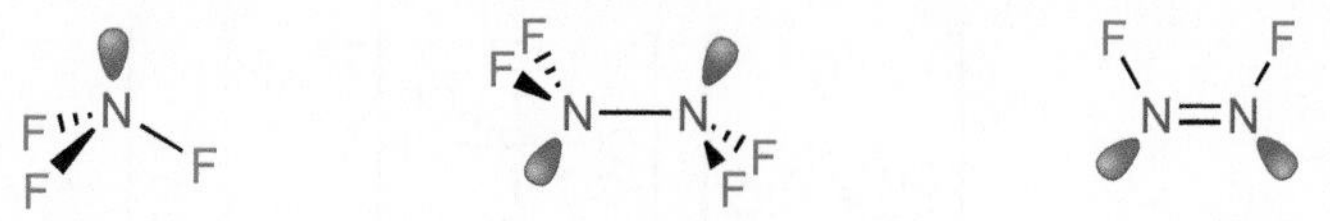

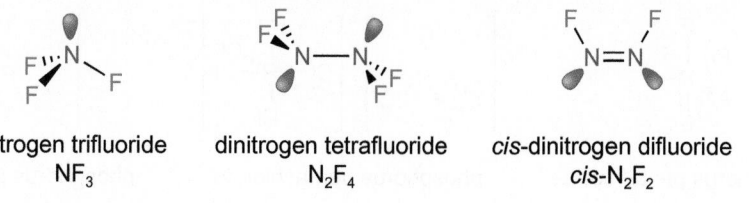

nitrogen trifluoride NF_3 | dinitrogen tetrafluoride N_2F_4 | *cis*-dinitrogen difluoride *cis*-N_2F_2

Figure 27.21 Examples of nitrogen fluorides.

geometry. NF_3 does not react with water, acids, or alkalis. It is less basic than ammonia due to the high polarity of the N–F bonds, which remove electron density from the nitrogen atom.

NF_3 is converted into dinitrogen tetrafluoride (N_2F_4) by heating with copper

$$2\,NF_3\,(g) + Cu\,(s) \xrightarrow{\Delta} N_2F_4\,(g) + CuF_2\,(s)$$

N_2F_4 exists in equilibrium with the paramagnetic monomer NF_2, which is blue

$$N_2F_4\,(g) \rightleftharpoons 2\,NF_2\,(g)$$

N_2F_4 reacts with $AlCl_3$ to give dinitrogen difluoride (N_2F_2). This has a double bond between the N atoms and exists as *cis* and *trans* isomers with the *cis* isomer being thermodynamically more stable.

Nitrogen trichloride (NCl_3) is far more reactive than NF_3. It is an explosive liquid that reacts rapidly with water to form ammonia and hypochlorous acid

$$NCl_3\,(l) + 3\,H_2O\,(l) \rightarrow NH_3\,(aq) + 3\,HOCl\,(aq)$$

In the past, this reaction was used to sterilize and bleach flour because of the HOCl produced. This process is now banned in the UK and USA due to the hazards associated with handling NCl_3. NBr_3 and NI_3 are even more reactive than NCl_3. When dry, NI_3 is very sensitive to touch, exploding violently to form N_2 and I_2.

Nitrogen triiodide is so shock-sensitive that it can be detonated with a feather. The colour of the vapour formed in the explosion is due to iodine.

The difference in reactivity between NF_3 and NCl_3 is discussed in Worked example 27.2 (p.1245).

Phosphorus halides

The phosphorus trihalides have trigonal pyramidal geometries. Phosphorus trifluoride (PF_3) is a gas under normal conditions. It hydrolyses only very slowly in water, though the reaction is faster in alkaline solutions

$$PF_3\,(g) + 3\,OH^-\,(aq) \rightarrow H_3PO_3\,(aq) + 3\,F^-\,(aq)$$

PF_3 is highly toxic as it bonds to the iron atoms in haemoglobin in a similar way to CO.

Phosphorus trichloride (PCl_3) is used industrially in the production of organophosphorus compounds that are used as flame retardants and pesticides. PCl_3 reacts vigorously with water to form phosphonic acid and HCl

$$PCl_3\,(l) + 3\,H_2O\,(l) \rightarrow H_3PO_3\,(aq) + 3\,HCl\,(g)$$

and with oxygen to form phosphorus oxytrichloride (**11**)

$$2\,PCl_3\,(l) + O_2\,(g) \rightarrow 2\,POCl_3\,(l)$$

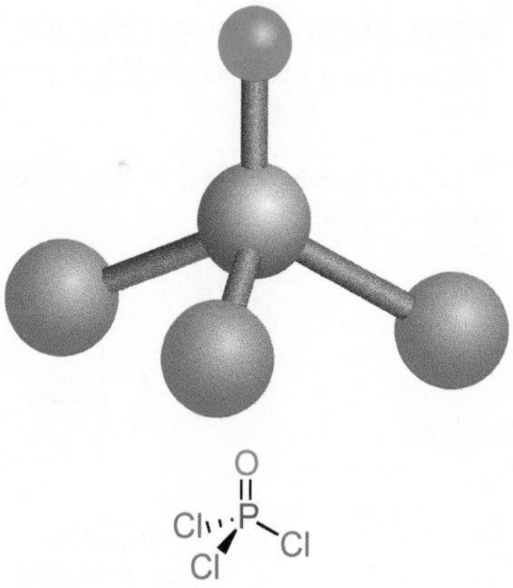

11 Phosphorus oxytrichloride ($POCl_3$).

The reaction of PCl_3 with alcohols is described in Section 20.2 (p.926).

The reaction of PCl_3 with water is very different from that of NCl_3. This is due to the difference in the polarity of the bonds. Nitrogen has a similar electronegativity to chlorine, and in NCl_3 it is the chlorine atoms that are attacked by the water molecules to form HOCl. In contrast, chlorine is more electronegative than phosphorus, so in PCl_3 the phosphorus atom carries a partial positive charge, and it is this that is attacked by the water.

Electronegativity also has a subtle effect on the bond angles in the phosphorus trihalides. The X–P–X angle increases from PF_3 (96°) to PCl_3 (100°), PBr_3 (101°), and PI_3 (102°). With the more electronegative fluorine atoms, the bonding pairs are located further from the phosphorus atom, so the steric effect of these electron pairs are reduced. The size of the halogen is another factor in the observed trend.

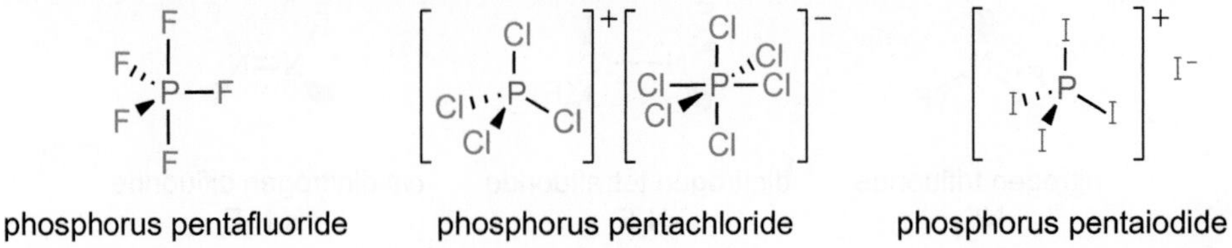
phosphorus pentafluoride
phosphorus pentachloride
phosphorus pentaiodide

Figure 27.22 The solid state structures of PF_5, PCl_5, and PI_5.

Visit the Online Resource Centre to view screencast 27.2 which walks you through the fluxionality of PF_5.

The phosphorus pentahalides have trigonal bipyramidal structures in the gas phase, but their solid state structures are more complicated, as shown in Figure 27.22. Solid PCl_5 has an ionic structure containing tetrahedral $[PCl_4]^+$ cations and octahedral $[PCl_6]^-$ anions. PBr_5 and PI_5 are also ionic but have the structures $[PBr_4]^+Br^-$ and $[PI_4]^+I^-$ respectively.

Box 27.4 Fluxional molecules

The axial and equatorial positions on a trigonal bipyramid are different, and a molecule such as PF_5 has two different P–F bond lengths. When solutions of PF_5 are studied using ^{19}F NMR spectroscopy (see Section 10.7 (p.498) and Worked example 10.8 (p.506)), only one signal is observed, and this appears as a doublet due to coupling with the ^{31}P nucleus. Nuclei in different chemical environments normally give different chemical shifts, so you would expect separate signals for the axial and equatorial F nuclei. Why is only one type of fluorine nucleus observed for PF_5?

Molecules are not static in solution and the bonds are constantly vibrating and rotating. The square pyramidal structure of PF_5 is only slightly higher in energy than the trigonal bipyramidal structure. As there is a low energy barrier between the two geometries, they interconvert in solution. This interconversion provides a means of scrambling the axial and equatorial fluorine atoms in a trigonal bipyramid, so that each fluorine atom spends two-fifths of its time in an axial position and three-fifths of its time in an equatorial position. A molecule that can rearrange its shape is known as fluxional, and the process involved in this particular type of fluxionality is called the **Berry pseudorotation**.

Every type of measurement has a characteristic timescale associated with it, which can be thought of as the minimum time required to make a measurement. The Berry pseudorotation process is fast on the NMR timescale, meaning that the atoms have interchanged positions many times before a signal can be recorded. The NMR spectrum therefore shows only the average positions of the nuclei.

Question

What would you expect to see in the ^{31}P NMR spectrum of PF_5?

▼ Berry pseudorotation scrambles the axial and equatorial fluorine atoms in PF_5.

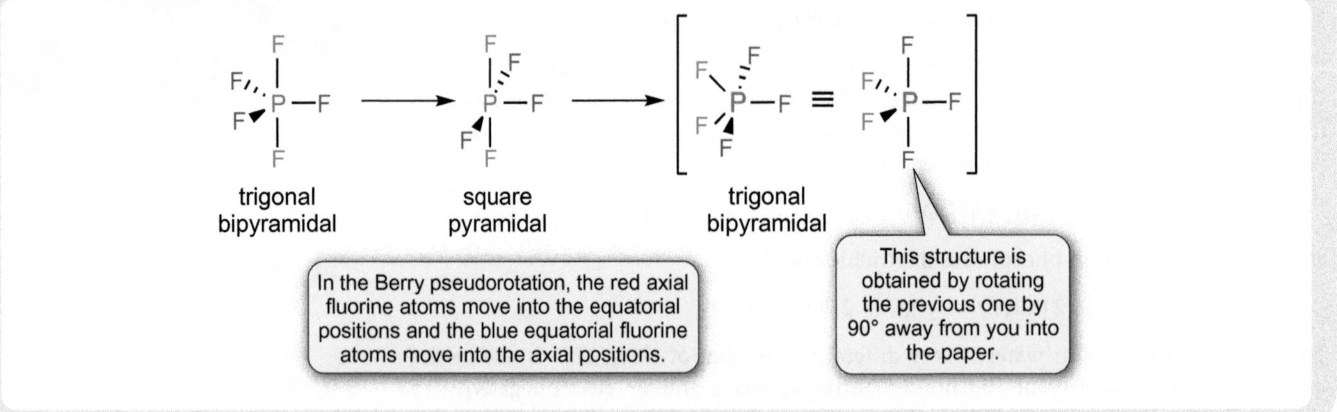
trigonal bipyramidal
square pyramidal
trigonal bipyramidal
In the Berry pseudorotation, the red axial fluorine atoms move into the equatorial positions and the blue equatorial fluorine atoms move into the axial positions.
This structure is obtained by rotating the previous one by 90° away from you into the paper.

Other Group 15 compounds

The reasons behind the stabilities of nitrides with small cations are discussed in Section 26.2 (p.1178).

Ionic nitride compounds, containing the N^{3-} ion, are only stable with small cations such as Li^+ and Mg^{2+}.

Nitrogen forms a range of sulfides. The reaction between SCl_2 and NH_3 leads to the compound S_4N_4 (**12**) which has a structure containing an 8-membered ring. S_4N_4 is stabilized by a degree of π bonding. It reacts with silver to form S_2N_2 (**13**) which also has a ring structure.

Phosphorus(V) sulfide (P_4S_{10}) has an analogous structure to P_4O_{10} (see p.1227). Other phosphorus sulfides are also known. For example, P_4S_3 (**14**) has a structure derived from the P_4 tetrahedron with three P–P bonds and three bridging sulfur atoms.

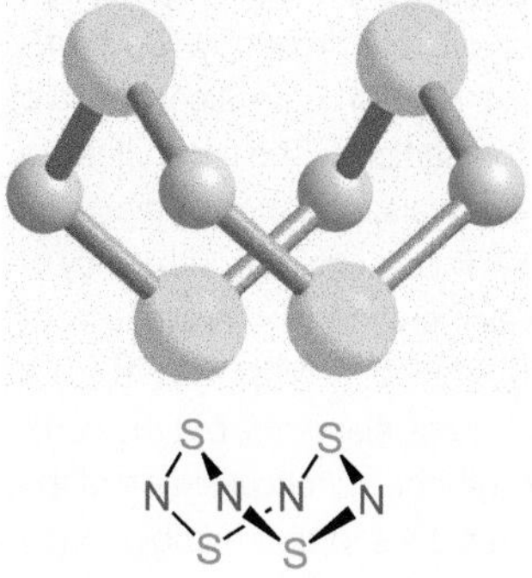

12 Tetrasulfur tetranitride (S_4N_4).

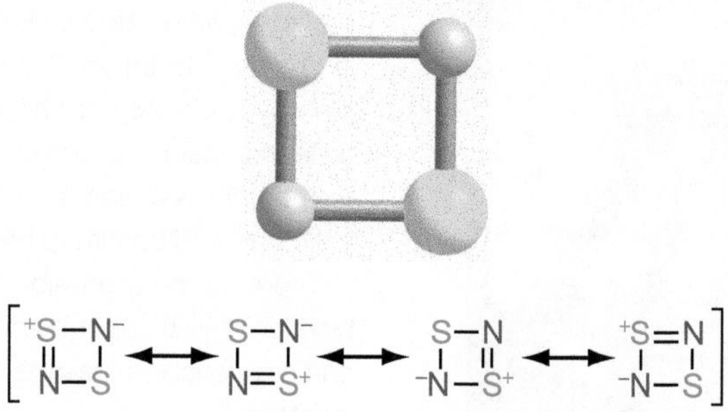

13 Disulfur dinitride (S_2N_2).

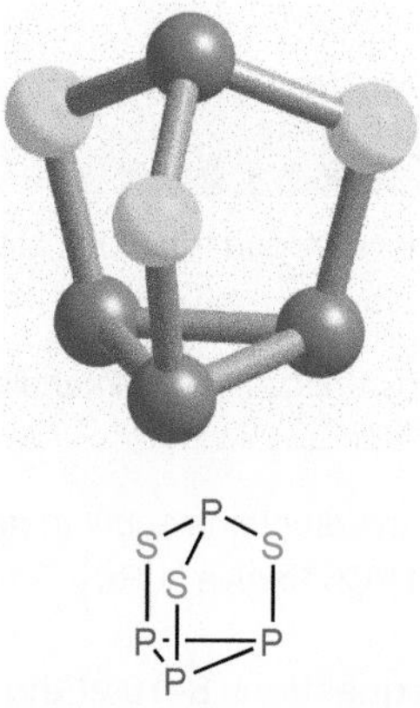

14 Tetraphosphorus trisulfide (P_4S_3).

Strike-anywhere matches contain P_4S_3, sulfur, and the oxidizing agent potassium chlorate ($KClO_3$). The fuel is ignited by friction, and the reaction produces P_4O_{10}, SO_2, and KCl. Safety matches such as that shown contain just sulfur and $KClO_3$. These matches only ignite in contact with the strip on the matchbox that contains red phosphorus.

Box 27.5 Firearms and forensic science

When a shooting has occurred, forensic scientists are able to help the police find out who fired the gun. They do this through studying materials known as *firearms discharge residues*, which are the compounds produced when a gun is fired.

A cartridge case is made up of three parts, a primer, a propellant, and a bullet. Pulling the trigger ignites a primer by friction. A typical primer contains a mixture of lead styphnate (lead 2,4,6-trinitro-1,3-benzenediolate), antimony sulfide, and barium nitrate. The primer burns rapidly, and its role is to ignite the propellant.

The propellant is normally either nitrocellulose or a mixture of nitrocellulose and nitroglycerine (see Box 13.7, p.643). These burn to generate a large volume of hot gas, which propels the bullet down the barrel of the gun.

Inorganic residues are formed by condensation of the vapours from the primer compounds, and these appear as hollow, spherical particles containing lead, antimony, and barium. These particles are collected by applying tapes to the hands of suspects and by vacuuming the crime scene.

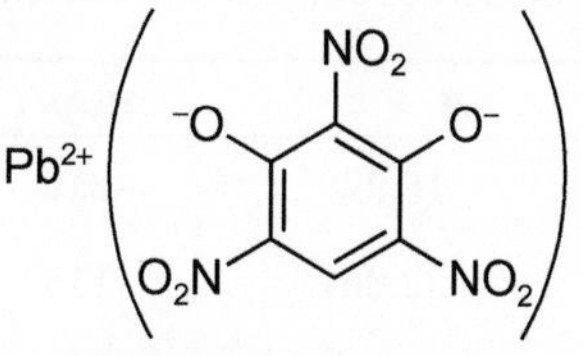

lead styphnate
explosive

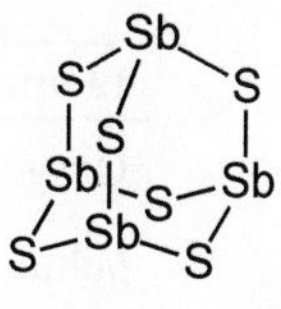

antimony(III) sulfide
fuel

$Ba(NO_3)_2$

barium nitrate
oxidizing agent

▲ The chemical components of the primer in a cartridge case.

→

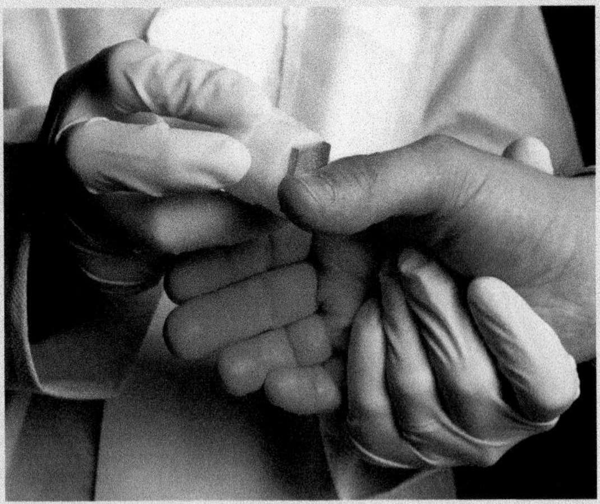

An adhesive surface is applied to the hand of a suspect in an attempt to find firearms discharge residues.

Forensic scientists analyse the residues using a scanning electron microscope (SEM) with energy dispersive spectroscopy. On bombardment with the electron beam of the SEM, core electrons are ionized from atoms on the surface of the particles. The vacancies are filled by electrons from higher energy orbitals, and these transitions are accompanied by the emission of X-ray photons. The frequency of the photons is characteristic for each element, and this enables lead, antimony, and barium to be identified.

There are many possible sources for these elements, but the combination of all three of them, together with the spherical shape of the particles, enables firearms discharge residues to be unambiguously identified.

Summary

- Nitrogen trifluoride is a stable, unreactive compound, but the other nitrogen trihalides are very reactive and the solids are explosive.
- Phosphorus forms trihalides, the molecules of which are trigonal pyramidal, and pentahalides, which are trigonal bipyramidal in the gas phase but have different structures in the solid states.
- NCl_3 and PCl_3 both react with water to give acidic solutions, but the reactions occur in different ways. Water attacks the chlorine atoms in NCl_3 to give HOCl and the phosphorus atom in PCl_3 to give H_3PO_3.

For practice questions on the topics in Section 27.4, see questions 8–10 at the end of this chapter (p.1253). Questions 16–20 are on general *p*-block chemistry.

27.5 Group 16: oxygen, sulfur, selenium, tellurium, and polonium

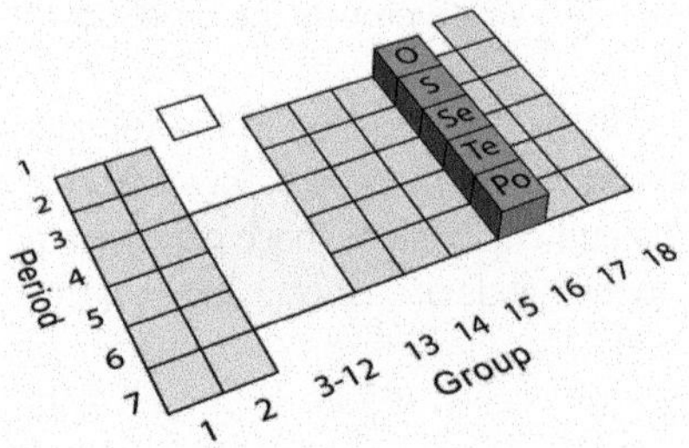
O
S
Se
Te
Po
Period
1
2
3
4
5
6
7
1 2 3-12 13 14 15 16 17 18
Group

Group 16 in the Periodic Table.

Group 16 contains the elements oxygen (O), sulfur (S), selenium (Se), tellurium (Te), and polonium (Po). These elements are collectively known as the **chalcogens**. As with the other groups, the metallic character of the elements increases down the group. Oxygen and sulfur are typical non-metals, selenium and tellurium have some metalloid properties, and polonium is normally regarded as a metal. Selected physical properties of the Group 16 elements are given in Table 27.8.

In older books sulfur is spelt 'sulphur'. The spelling with 'f' is recommended by IUPAC.

Table 27.8 Physical properties of the Group 16 elements

Element	A_r	m.p./°C	b.p./°C	Density/g cm^{-3}
Oxygen (O)	15.999	–219	–183	1.43×10^{-3}
Sulfur (S)	32.06	115	445	2.07
Selenium (Se)	78.971	221	685	4.79
Tellurium (Te)	127.60	450	988	6.23
Polonium (Po)	209	254	962	9.20

The valence electronic configuration of the Group 16 elements is $ns^2\ np^4$, though the group oxidation state of +6 is not observed for oxygen. The highest oxidation state possible for oxygen is +2, in the compound OF_2, but its most common oxidation state is −2. Oxygen forms compounds with virtually all other elements; the only exceptions are the Group 18 elements helium, neon, argon, and krypton. Oxidation states of +4 and +6 are possible for sulfur and the heavier chalcogens.

The Group 16 elements

Oxygen

The most common allotrope of oxygen is dioxygen (O_2) and this exists as a colourless, odourless gas at room temperature. Both the liquid and solid forms are pale blue in colour. O_2 is the second biggest component of the Earth's atmosphere, comprising 21% by volume. Industrially, O_2 is prepared by the fractional distillation of liquid air. In the laboratory, O_2 can be synthesized by the electrolysis of water or from decomposition of either potassium chlorate ($KClO_3$) or hydrogen peroxide (H_2O_2). Both of these decompositions are catalysed by MnO_2

The bonding in O_2 and its paramagnetism are described in Section 4.10 (p.196).

$$2\,KClO_3\,(s) \xrightarrow{\Delta} 2\,KCl\,(s) + 3\,O_2\,(g)$$

$$2\,H_2O_2\,(aq) \rightarrow 2\,H_2O\,(l) + O_2\,(g)$$

The chemistry of O_2 is dominated by its oxidizing ability. Combustion, corrosion, and respiration are all reactions in which O_2 acts as an oxidizing agent. Most of the industrial uses of O_2 also use this oxidizing power. For example, in the steel industry, O_2 is blown into molten steel to oxidize impurities, especially carbon.

Oxygen also forms a second allotrope, ozone (O_3, (**15**)). Ozone is a pale blue gas at room temperature and has a characteristic pungent odour. The formation of ozone from O_2 is extremely endothermic

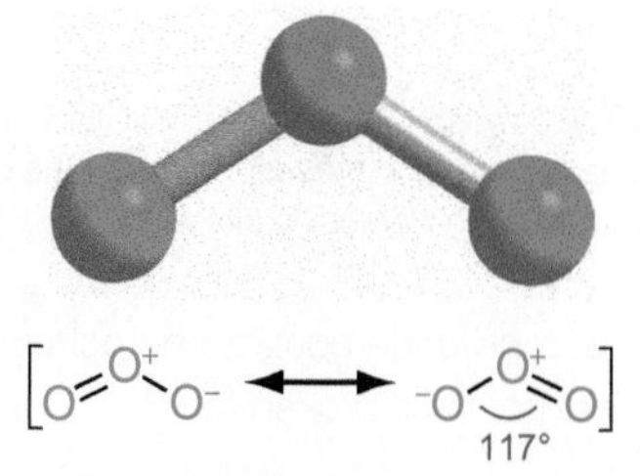

15 Ozone (O_3).

$$3\,O_2\,(g) \rightarrow 2\,O_3\,(g) \qquad \Delta H^{\ominus} = +285\ \text{kJ mol}^{-1}$$

and the reaction only occurs in an electrical discharge. This occurs naturally in lightning, and electric discharges are also used to prepare ozone artificially. You can sometimes detect the pungent smell of ozone near electric motors and other sources of electric sparks. Ozone is a powerful oxidizing agent and is used in water purification and the treatment of industrial waste. Ozone is useful for these purposes because it is converted only to harmless oxygen. The importance of ozone in the stratosphere is described in Box 27.6.

O_3 reacts with organic molecules containing C=C bonds to give ozonides as described in Section 21.4 (p.992). The role of ozone in photochemical smog is described on p.1201.

Box 27.6 The ozone layer

About 10% of the Sun's radiation energy is in the ultraviolet (UV) part of the electromagnetic spectrum (see Box 3.1, p.120). If all of the UV radiation that reached the atmosphere penetrated to the Earth's surface, life as we know it would not be possible. Fortunately, the majority of this UV radiation is absorbed in the upper atmosphere. Both dioxygen (O_2) and ozone (O_3) play roles in this absorption.

When an ultraviolet photon with a wavelength of 240 nm or lower (UV-C radiation) hits an O_2 molecule in the mesosphere or stratosphere, the energy absorbed is sufficient to break the bond between the atoms,

$$O_2\,(g) \xrightarrow{h\nu} 2\,O\,(g) \qquad \lambda \leq 240\,\text{nm}$$

Because of this photochemical reaction, most of the oxygen in the mesosphere exists as oxygen atoms. When two oxygen atoms collide, they re-combine to give O_2 molecules, which can then absorb more ultraviolet light. This process absorbs the majority of the UV-C radiation in the upper atmosphere.

Lower down in the stratosphere, in the absence of high-energy radiation with λ below 240 nm, most oxygen exists as O_2 molecules. The concentration of atomic oxygen is low, so it is unlikely that two oxygen atoms will collide to reform O_2. It is far more likely that an oxygen atom will collide with an O_2 molecule, leading to the formation of ozone

$$O\,(g) + O_2\,(g) \rightarrow O_3\,(g)$$

→

➜ This reaction is exothermic so it generates heat, and this is the reason for the increase in temperature with increasing height in the stratosphere.

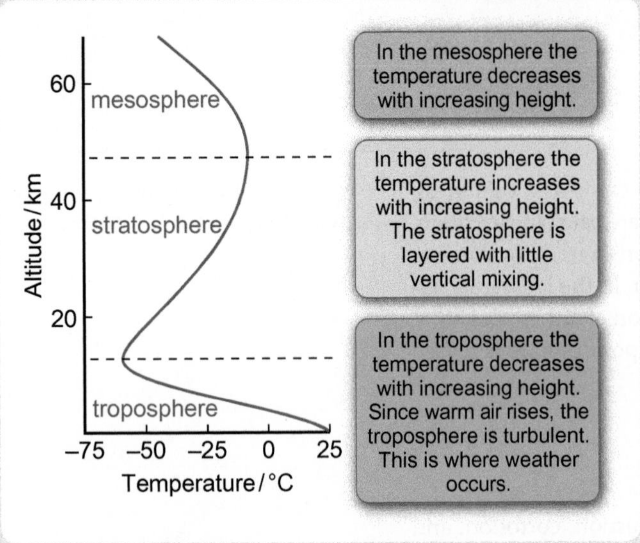

Altitude / km
60
40
20
mesosphere
stratosphere
troposphere
−75 −50 −25 0 25
Temperature / °C
In the mesosphere the temperature decreases with increasing height.
In the stratosphere the temperature increases with increasing height. The stratosphere is layered with little vertical mixing.
In the troposphere the temperature decreases with increasing height. Since warm air rises, the troposphere is turbulent. This is where weather occurs.

▲ The atmosphere is divided into three distinct regions—the troposphere, stratosphere, and mesosphere—on the basis of how the temperature changes with increasing altitude.

Ozone undergoes a photochemical reaction in which one of the O–O bonds is broken

$$O_3(g) \xrightarrow{h\nu} O_2(g) + O(g)$$

The bonds in ozone are weaker than those in O_2, so this reaction requires less energy than the splitting of O_2. It is responsible for absorbing UV-B radiation (λ = 320 nm–290 nm) and the longer wavelength UV-C radiation (λ = 290 nm–240 nm) that is not absorbed by O_2. The generated oxygen atom can then react with another O_2 to regenerate ozone or with an O_3 molecule to form O_2

$$O_2(g) + O(g) \rightarrow O_3(g)$$
$$O_3(g) + O(g) \rightarrow 2O_2(g)$$

Ozone is constantly forming, decomposing, and re-forming in the stratosphere during daylight hours. The reactions reach a steady state (see Section 9.6, p.419) so that the concentration of ozone remains constant, though the average lifetime of an ozone molecule is only 30 minutes. O_2 and O_3 are responsible for absorbing most of the UV-B and UV-C radiation that strikes the Earth's atmosphere and preventing it from reaching the surface.

There are other naturally occurring mechanisms that convert O_3 to O_2, and these involve radicals. For example, methyl chloride, which is produced by decaying vegetation, absorbs UV light in the upper atmosphere to produce chlorine atoms, which are often written $Cl^{\bullet}$ to emphasize the odd electron. These atoms react with ozone to form $ClO^{\bullet}$ radicals

$$Cl^{\bullet}(g) + O_3(g) \rightarrow ClO^{\bullet}(g) + O_2(g)$$

The $ClO^{\bullet}$ radical can react with either atomic oxygen in the upper stratosphere, or O_3 in the lower stratosphere giving O_2 and regenerating $Cl^{\bullet}$

$$ClO^{\bullet}(g) + O(g) \rightarrow Cl^{\bullet}(g) + O_2(g)$$

or

$$ClO^{\bullet}(g) + O_3(g) \rightarrow Cl^{\bullet}(g) + 2O_2(g)$$

In both cases, the overall effect is to catalytically convert O_3 into O_2. Since the chlorine atoms are regenerated, only a tiny concentration is required to remove a considerable amount of ozone.

Concerns were raised in the 1970s, when it was realized that unreactive chlorofluorocarbons (CFCs) were reaching the stratosphere and undergoing photodissociation to form $Cl^{\bullet}$ atoms. This upsets the natural steady state and leads to ozone depletion. Although $Cl^{\bullet}$ catalyses the conversion of O_3 to O_2, the reaction does not cycle endlessly. Chlorine is removed by the reaction of $ClO^{\bullet}$ with NO_2 to give chlorine nitrate ($ClONO_2$)

$$ClO^{\bullet}(g) + NO_2(g) \rightarrow ClONO_2(g)$$

and by reaction of $Cl^{\bullet}$ with methane to give HCl,

$$Cl^{\bullet}(g) + CH_4(g) \rightarrow HCl(g) + CH_3^{\bullet}(g)$$

Both $ClONO_2$ and HCl are regarded as inactive forms of chlorine, as they do not react with ozone. They are described as chlorine reservoirs and, as is shown in Box 27.8 (p.1241), under certain conditions they can be converted back into chlorine atoms with dramatic consequences.

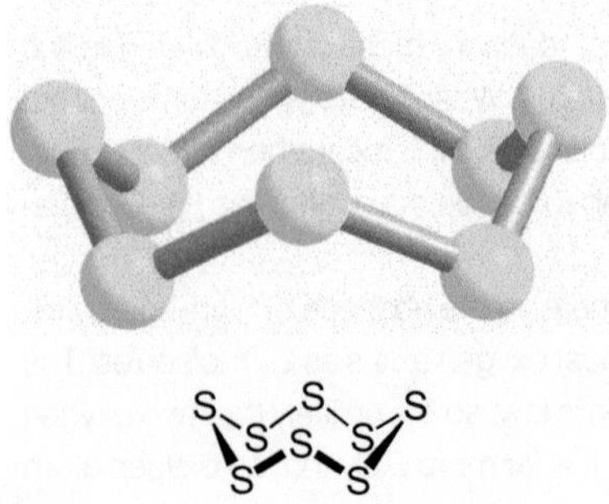

S S S S S S S S

16 Octasulfur (S_8). These molecules are present in both α-sulfur (rhombic sulfur) and β-sulfur (monoclinic sulfur).

Sulfur

Deposits of sulfur are found near to volcanoes and other areas of geological activity. Sulfur also occurs in deposits where it has been formed by the action of bacteria on H_2S. In addition, many minerals are sulfides, such as *galena* (PbS) and *sphalerite* (ZnS), or sulfates such as *gypsum* ($CaSO_4{\cdot}2H_2O$). Most sulfur is produced industrially as a by-product from the extraction of copper from sulfide ores and from cleaning natural gas and oil to produce low sulfur fuels.

There are many allotropes of sulfur, more than for any other element. The most stable at room temperature is α-sulfur (rhombic sulfur) which contains cyclic S_8 molecules (**16**). α-Sulfur can be converted by heating into other allotropes, as shown in Figure 27.23. β-Sulfur (monoclinic sulfur) also contains S_8 rings, but they are packed in a different way. In addition, rings containing between 6 and 20 sulfur atoms are also known. One of the

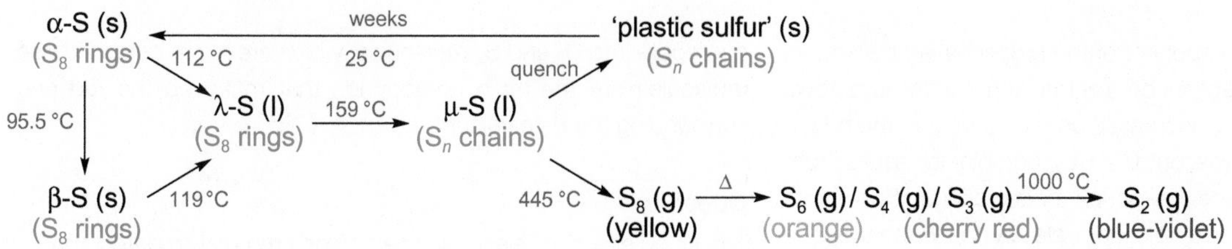
α-S (s)
(S8 rings)
weeks
'plastic sulfur' (s)
(Sn chains)
112 °C
25 °C
quench
95.5 °C
λ-S (l)
(S8 rings)
159 °C
μ-S (l)
(Sn chains)
β-S (s)
(S8 rings)
119°C
445 °C
S8 (g)
(yellow)
Δ
S6 (g)/ S4 (g)/ S3 (g)
(orange)
(cherry red)
1000 °C
S2 (g)
(blue-violet)

Figure 27.23 The allotropes of sulfur can be interconverted by changing the temperature. α-sulfur and β-sulfur are shown in Figure 1.21 (p.48).

reasons for the diversity of allotropes is the tendency of sulfur to **catenate**, forming chains and rings through formation of S–S bonds.

This ability to form S–S bonds is important biologically as these disulfide links are formed between the sulfur atoms of cysteine amino acids, and they help determine the three-dimensional shapes of proteins.

Sulfur is used industrially in the vulcanization of rubber. Disulfide links form bridges between polymer chains and make the rubber harder and more durable.

The thermodynamics of protein folding is discussed in Chapter 14 (p.655).

Box 27.7 Sulfur on Io

Io is the most volcanically active body known in the solar system, with volcanic plumes rising 300 km above its surface. Io is Jupiter's third largest moon, and its surface shows a wide variety of colours. The yellow, orange, red, and black colours are caused by deposits of sulfur at different temperatures. The white deposits are believed to be sulfur dioxide frost that has condensed out of Io's atmosphere onto the surface.

Our knowledge of Io is largely due to the success of the space probe Galileo (see Chapter 25, p.1139), which passed by Io several times between 1995 and 2003. Galileo detected more than 100 erupting volcanoes on the surface of Io.

▲ Jupiter's moon Io. The volcano Pele is on the bottom right, surrounded by a red ring formed from deposits of S_3 and S_4.

Why is Io so volcanically active? Io always points the same side towards Jupiter in its orbit around the planet, but Jupiter's other large moons, Ganymede, Europa, and Callisto, distort its orbit. The competing gravitational pulls subject Io to massive tidal forces that cause its surface to bulge by as much as 100 m. The constant friction generates enormous quantities of heat and high pressures in the surface layers, and these cause molten silicates and gases to rise through fractures in the crust and erupt onto the surface. The high level of volcanic activity means that the surface of Io constantly renews itself. The craters that are common on many planets and moons are absent from Io as they are buried by the frequent volcanic eruptions. Io's volcano Loki is the most powerful in the solar system, and emits more heat than all the volcanoes on Earth combined.

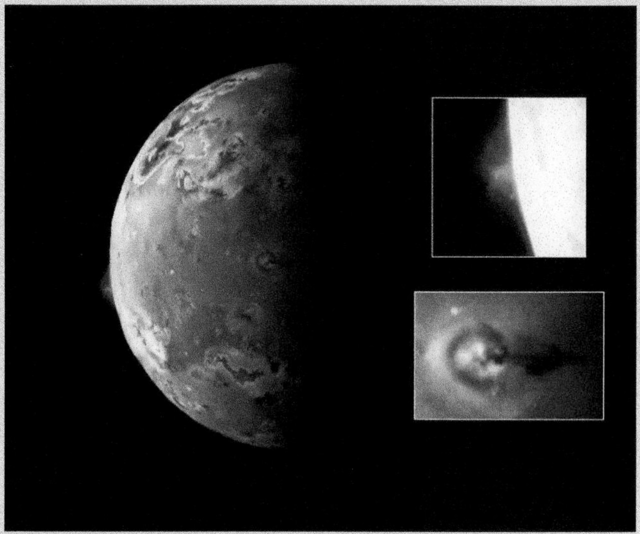

▲ A volcanic plume rises 140 km above the surface of Io.

→

→ Although the main component of the eruptions is now known to be molten silicates, scientists believe that some of the lava flows are primarily liquid sulfur. An increased understanding of the role of sulfur in Io's volcanism was obtained by combining results from Galileo with Hubble Space Telescope observations. Sulfur gas (S_2) was detected above Io's volcano, Pele, by the Hubble Space Telescope. S_2 is stable at the high temperatures inside the volcano but, once it is ejected and lands on the cold surface, the molecules aggregate into S_3 and S_4 molecules, which are red in colour. These molecules are the main compounds that make up the red ring surrounding the Pele volcano.

Question

The Hubble Space Telescope used a spectrometer to detect the presence of S_2. Would you expect this molecule to absorb infrared radiation (see Section 10.5, p.480)?

Anions of the Group 16 elements

The bonding in the O_2^- and O_2^{2-} ions is discussed in Section 4.10 (p.198). The occurrence of Group 1 peroxides and superoxides is described in Section 26.2 (p.1172).

The majority of metals in the Periodic Table form oxides, containing the O^{2-} ion, and sulfides, containing the S^{2-} ion. Ionic compounds containing the peroxide ion (O_2^{2-}) and the superoxide ion (O_2^-) are also well known. The X^{2-} ions become less stable going down Group 16. This is because the larger anions hold their electrons less tightly, so they are more easily oxidized.

Sulfur forms a range of polysulfides, containing S_n^{2-} ions (n = 2–6). These are anions of the catenated hydrides H_2S_n. They are prepared by the reaction between a sulfide and sulfur. For example,

$$2\,Cs_2S + S_8 \rightarrow 2\,Cs_2S_5$$

Summary

- Of the Group 16 elements, oxygen and sulfur are non-metals, selenium and tellurium are metalloids, and polonium is a metal.
- The group oxidation state of +6 is not possible for oxygen, which has a maximum oxidation state of +2 but is most commonly seen in the –2 oxidation state. The +4 and +6 oxidation states are observed for the other Group 16 elements.
- Oxygen forms two allotropes, dioxygen (O_2) and ozone (O_3). Sulfur forms a wide range of allotropes containing S–S single bonds.

The Group 16 oxides and oxoacids

Sulfur oxides

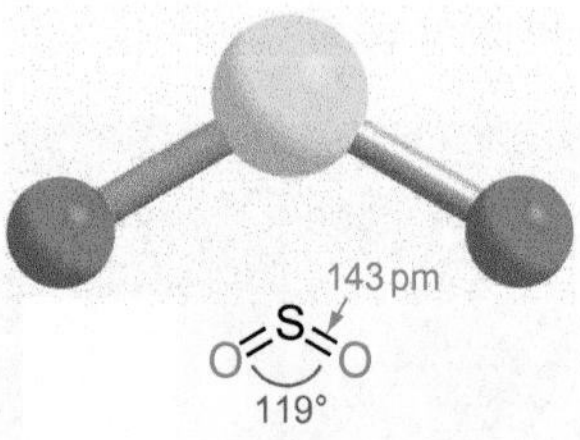

143 pm
119°

17 Sulfur dioxide (SO_2).

Sulfur forms two main oxides—sulfur dioxide (SO_2) and sulfur trioxide (SO_3). Sulfur dioxide (**17**) is a colourless, toxic gas that is formed when sulfur burns in air

$$S_8\,(s) + 8\,O_2\,(g) \rightarrow 8\,SO_2\,(g)$$

Sulfur dioxide is very soluble in water. The SO_2 (aq) molecules react with water to give an acidic solution containing ions derived from sulfurous acid (H_2SO_3)

$$SO_2\,(aq) + 2\,H_2O\,(l) \rightleftharpoons HSO_3^-\,(aq) + H_3O^+\,(aq)$$

$$HSO_3^-\,(aq) + H_2O\,(l) \rightleftharpoons SO_3^{2-}\,(aq) + H_3O^+\,(aq)$$

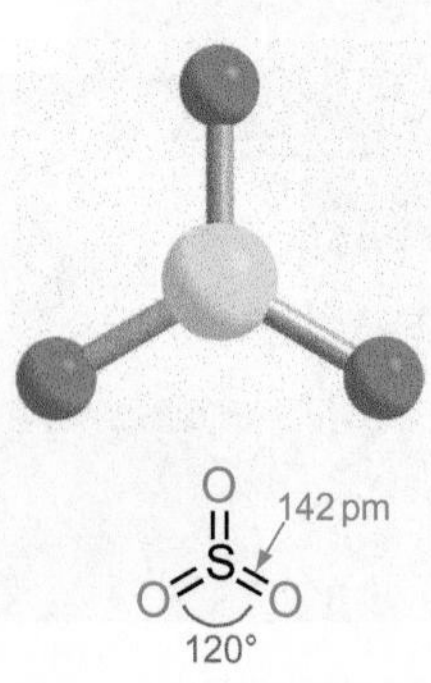

142 pm
120°

18 Sulfur trioxide (SO_3).

Sulfur dioxide reacts slowly with oxygen to form sulfur trioxide (SO_3). This reaction is important industrially in the manufacture of sulfuric acid, where it is catalysed by vanadium(V) oxide (V_2O_5)

$$2\,SO_2\,(g) + O_2\,(g) \rightarrow 2\,SO_3\,(g) \qquad \Delta H^{\ominus} = -191.2\,\text{kJ mol}^{-1}$$

The equilibrium constant decreases rapidly with increasing temperature, so the reaction is carried out at a relatively low temperature, between 435 °C and 635 °C. At lower temperatures, the reaction becomes too slow to be economical.

Sulfur trioxide is a volatile white solid at room temperature. The gaseous molecules are trigonal planar in shape, but in the solid, some of the SO_3 molecules (**18**) trimerize to form S_3O_9 rings (**19**). Sulfur trioxide reacts with water to form sulfuric acid (H_2SO_4)

$$SO_3\,(g) + H_2O\,(l) \rightarrow H_2SO_4\,(aq)$$

This reaction is strongly exothermic and causes a corrosive acid mist, so industrially SO_3 is dissolved in concentrated sulfuric acid rather than water. This forms *oleum*, which is a mixture of polysulfuric acids such as disulfuric acid ($H_2S_2O_7$). Water is carefully added to this to regenerate sulfuric acid.

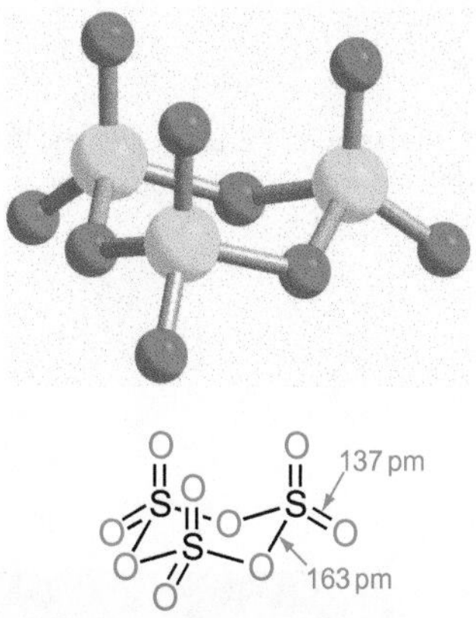
137 pm
163 pm

19 The sulfur trioxide trimer (S_3O_9).

Sulfur oxoacids and oxoanions

The structures of the main sulfur oxoacids are shown in Figure 27.24 together with their oxoanions.

sulfuric acid H_2SO_4 S oxidation state = +6	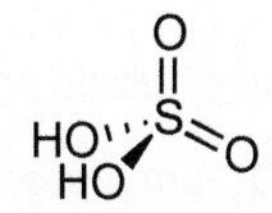	HSO_4^- hydrogensulfate SO_4^{2-} sulfate
disulfuric acid $H_2S_2O_7$ S oxidation state = +6	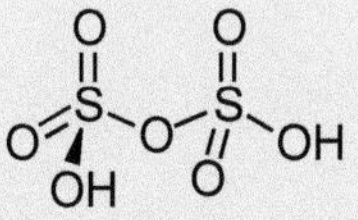	$HS_2O_7^-$ hydrogendisulfate $S_2O_7^{2-}$ disulfate
sulfurous acid H_2SO_3 S oxidation state = +4	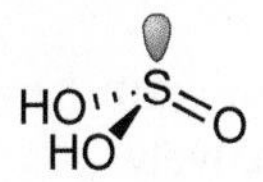	HSO_3^- hydrogensulfite SO_3^{2-} sulfite

Figure 27.24 The main oxoacids of sulfur and the anions derived from them.

The use of $CaSO_4{\cdot}2H_2O$ in the construction industry is described in Box 26.3 (p.1185), and the use of barium sulfate as an X-ray contrast agent is described in Section 26.7 (p.1190).

Sulfuric acid is produced worldwide on a larger scale than any other inorganic compound, with approximately 200 million tonnes produced annually. Sulfuric acid has many uses in synthetic chemistry, and is used to process metals as well as in the manufacture of fibres, paints and pigments, detergents, and fertilizers.

As well as being a strong acid, concentrated sulfuric acid is an oxidizing agent and a dehydrating agent. Its use as a dehydrating agent is shown by the reaction with sucrose ($C_{12}H_{22}O_{11}$), which produces carbon in a vigorous exothermic reaction

$$C_{12}H_{22}O_{11} \xrightarrow{\text{concentrated sulfuric acid}} 12\,C\,(s) + 11\,H_2O\,(l)$$

PYREX
125
100
150ml

Concentrated sulfuric acid is an excellent dehydrating agent, and it removes water from sucrose to form carbon.

A common chemical test for sulfate ions involves addition of barium chloride solution to an acidified solution. If sulfate is present, a dense white precipitate of barium sulfate is formed.

Sulfurous acid cannot be isolated as a pure compound, but its salts—sulfites, containing the SO_3^{2-} ion, and hydrogensulfites, containing the HSO_3^- ion—are stable compounds. Sulfites are formed by reacting dissolved SO_2 with a base such as sodium hydroxide

$$SO_2\,(aq) + 2\,OH^-\,(aq) \rightarrow SO_3^{2-}\,(aq) + H_2O\,(l)$$

Sulfur dioxide and sulfites are used in the food industry as preservatives. Sulfites are good reducing agents, as they are readily oxidized to sulfates.

Another oxoanion of sulfur is the thiosulfate ion ($S_2O_3^{2-}$), which is prepared by boiling sulfur in an alkaline solution of sodium sulfite

$$8\,SO_3^{2-}\,(aq) + S_8\,(s) \xrightarrow{\Delta} 8\,S_2O_3^{2-}\,(aq)$$

Thiosulfate ions are good reducing agents, and are oxidized to tetrathionate ions ($S_4O_6^{2-}$).

Many wines contain added sulfites as preservatives.

The prefix 'thio' is used when a sulfur atom has replaced an oxygen atom. See Section 1.4 (p.29).

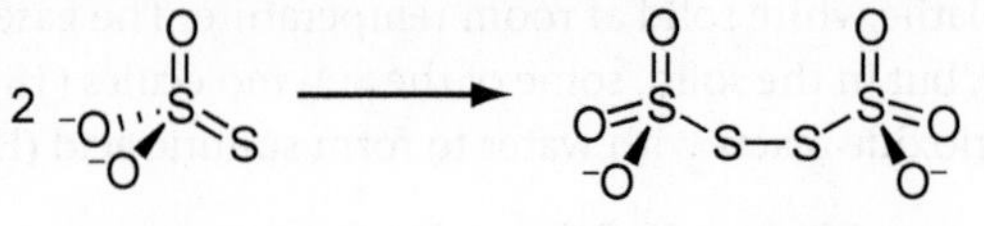

thiosulfate tetrathionate

$$2\,S_2O_3^{2-}\,(aq) \rightarrow S_4O_6^{2-}\,(aq) + 2\,e^-$$

This reaction is commonly used in redox titrations. The tetrathionate ion is one of a series of anions with the general formula $^-O_3S–S_n–SO_3^-$. These ions are known for $n = 0–22$.

Summary

- The two most common sulfur oxides are sulfur dioxide (SO_2) and sulfur trioxide (SO_3). SO_3 dissolves in water to form sulfuric acid.
- Sulfuric acid (H_2SO_4) is a strong acid, and when concentrated it is also an oxidizing agent and a dehydrating agent.
- Sulfurous acid (H_2SO_3) cannot be isolated, but its anions SO_3^{2-} and HSO_3^- are present in many salts.

The Group 16 halides

Oxygen halides

Oxygen forms the fluorides OF_2 and O_2F_2 (**20**). These compounds have similar structures to water and hydrogen peroxide, respectively.

OF_2 is a toxic, yellow gas formed in the reaction between fluorine and hydroxide ions

$$2\,OH^-(aq) + 2\,F_2\,(g) \rightarrow OF_2\,(g) + 2\,F^-(aq) + H_2O\,(l)$$

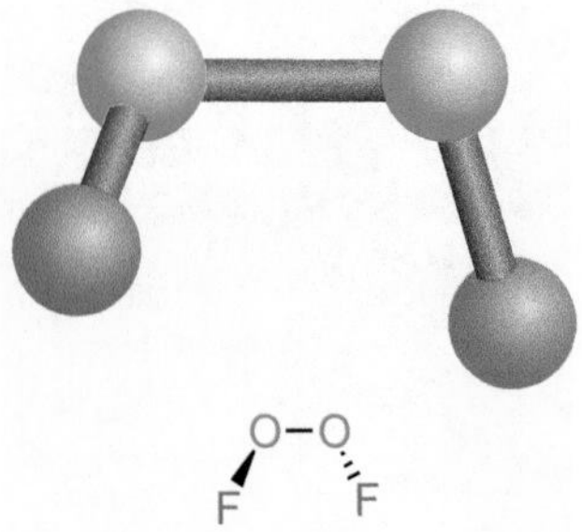

20 Dioxygen difluoride (O_2F_2).

The oxygen fluorides are the only compounds of oxygen in which the oxidation state is positive (+2 in OF_2 and +1 in O_2F_2), which means that the O–F bonds are polarized in the opposite direction from the O–Cl bonds in Cl_2O.

The stability of the oxygen dihalides decreases with increasing size of the halogen. Br_2O decomposes on melting at −18 °C, whereas I_2O is unknown. Cl_2O reacts with water to form hypochlorous acid (HOCl)

$$Cl_2O\,(g) + H_2O\,(l) \rightarrow 2\,HOCl\,(aq)$$

By convention, the more electronegative element is written last in a formula. This means that, in the oxygen halides, the fluoride is written OF_2 and the chloride is Cl_2O.

Chlorine, bromine, and iodine form oxides in higher oxidation states, and these are described in Section 27.6 (p.1243).

Sulfur halides

The main sulfur fluorides are S_2F_2, SF_4, and SF_6, all of which are gases under normal conditions. Disulfur difluoride (S_2F_2) exists as two isomers, one with a similar structure to O_2F_2 and the other containing a S=S double bond and sulfur atoms in two different oxidation states (Figure 27.25).

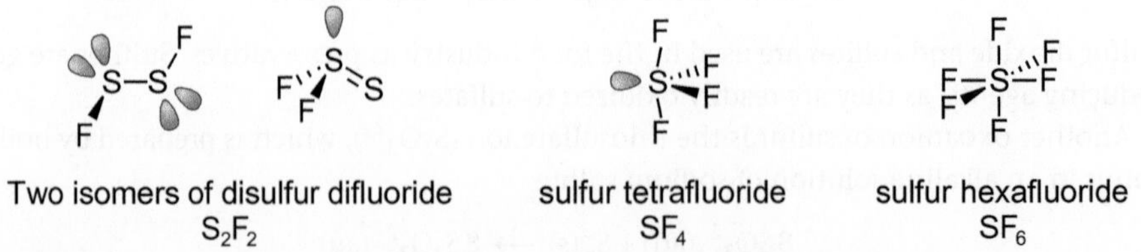
Two isomers of disulfur difluoride
S2F2
sulfur tetrafluoride
SF4
sulfur hexafluoride
SF6

Figure 27.25 The structures of the main sulfur fluorides.

Sulfur tetrafluoride is very reactive, and can act as both a Lewis acid and a Lewis base. In the reaction with fluoride

$$SF_4 + F^- \rightarrow SF_5^-$$

it reacts as a Lewis acid, accepting an electron pair, whereas in the reaction with BF_3 it acts as a Lewis base, transferring a fluoride rather than coordinating through the sulfur lone pair

$$SF_4 + BF_3 \rightarrow SF_3^+ BF_4^-$$

Sulfur hexafluoride is the main product of burning sulfur in fluorine gas. SF_6 is generally unreactive. The hydrolysis of SF_6 to form SO_3 is very favourable thermodynamically, but SF_6 does not react with water, or even with steam. This is because the fluorine atoms shield the central sulfur atom from attack. SF_6 is a good electrical insulator and is used as an insulating gas in switches on high voltage power lines.

The main sulfur chlorides are S_2Cl_2 and SCl_2, both of which are toxic liquids at room temperature. The higher oxidation state chlorides are much less stable than the fluorides. SCl_4 exists as $SCl_3^+Cl^-$ (**21**) but is only stable below −30 °C. Sulfur hexachloride (SCl_6) is unknown; fluorine is the only halogen capable of stabilizing sulfur in the +6 oxidation state.

The release and monitoring of small amounts of SF_6 has been used to model the effects of a poison gas release on the London Underground. SF_6 was chosen because it is non-toxic, inert, and can be detected in very low concentration.

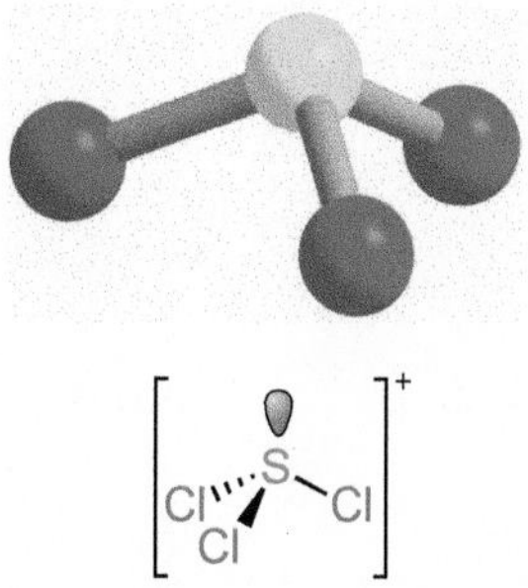
S
Cl
Cl
Cl
+

21 The trichlorosulfur cation (SCl_3^+).

The use of $SOCl_2$ as a chlorinating agent in organic chemistry is described in Section 20.2 (p.926).

Sulfur oxohalides

Sulfur forms two types of oxohalide, the sulfuryl halides SO_2X_2 (X = F, Cl) with oxidation state +6, and the thionyl halides, SOX_2 (X = F, Cl, Br) with oxidation state +4. The structures of the chlorides are shown in Figure 27.26.

Thionyl chloride is a colourless liquid at normal temperatures, and it reacts vigorously with water. It is used as a drying agent, especially in the preparation of anhydrous metal halides, and as a chlorinating agent in the manufacture of pesticides, pharmaceuticals, and dyes.

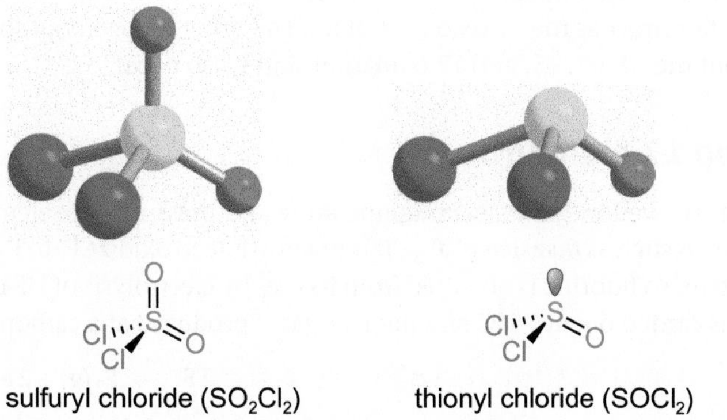
O
S
Cl
Cl
O
sulfuryl chloride (SO_2Cl_2)
S
Cl
Cl
O
thionyl chloride ($SOCl_2$)

Figure 27.26 The structures of sulfuryl chloride (SO_2Cl_2) and thionyl chloride ($SOCl_2$).

Summary

- Oxygen forms halides OF_2 and Cl_2O, which differ with respect to the polarization of the oxygen–halogen bond.
- The most common sulfur fluorides are S_2F_2, SF_4, and SF_6. Sulfur hexafluoride is very unreactive as the six fluorine atoms shield the central sulfur atom from attack.
- Chlorine is not oxidizing enough to stabilize the sulfur(VI) oxidation state, so sulfur tetrachloride is known, but not sulfur hexachloride.

For a practice question on the topics in Section 27.5, see question 11 at the end of this chapter (p.1253). Questions 16–20 are on general *p*-block chemistry.

27.6 Group 17: fluorine, chlorine, bromine, iodine, and astatine

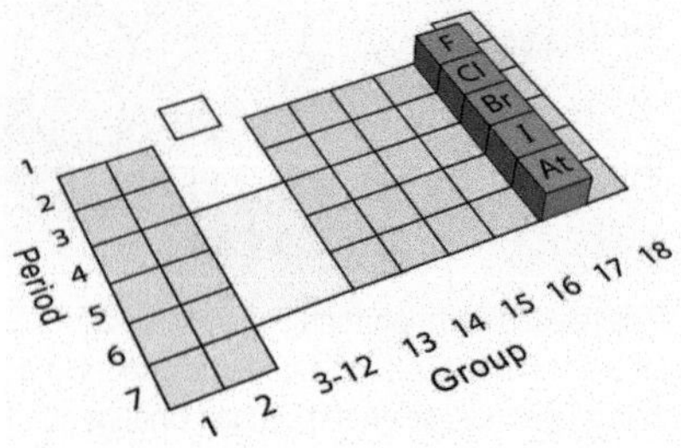
F
Cl
Br
I
At
Period
1
2
3
4
5
6
7
1 2 3-12 13 14 15 16 17 18
Group
Group 17 in the Periodic Table.

Group 17 contains the elements fluorine (F), chlorine (Cl), bromine (Br), iodine (I), and astatine (At). These elements are known collectively as the **halogens**. The halogens have the valence electron configuration $ns^2\ np^5$, so the atoms can readily gain an electron to form halide ions, X^-, or they can form covalent bonds. The electron gain enthalpies are all positive—the reactions are highly exothermic. All of the halogens are non-metals.

The halogens exist in the elemental form as X_2 molecules. Selected physical properties of the elements are given in Table 27.9. The halogens are all oxidizing agents, but the oxidizing power decreases going down the group, as is shown by the $E^\ominus$ values given in Table 27.9. The value for the iodine/iodide couple is sufficiently low that I^- can be used as a reducing agent.

Table 27.9 Physical properties and reduction potentials for the Group 17 elements

Element	A_r	m.p. / °C	b.p. / °C	Density / g cm^{-3}	$E^\ominus(X_2\|X^-)$ / V
Fluorine (F)	18.998	−220	−188	1.70×10^{-3}	+2.87
Chlorine (Cl)	35.45	−101	−34	3.21×10^{-3}	+1.36
Bromine (Br)	79.904	−7	59	3.12	+1.08
Iodine (I)	126.90	114	184	4.93	+0.54
Astatine (At)	210	300	–	–	–

Fluorine is the most oxidizing of all the elements in the Periodic Table, and in all of its compounds fluorine has the −1 oxidation state. The other halogens all show the −1 oxidation state, but the +1, +3, +5, and +7 oxidation states also occur.

The Group 17 elements

Fluorine is a pale yellow gas at room temperature and pressure. The element occurs naturally in minerals such as *fluorspar* (CaF_2). It is not possible to oxidize F^- to F_2 using a chemical oxidizing agent, so fluorine is obtained from its ores by electrolysis of HF in molten KF. The electrolysis is carried out at 75 °C and fluorine gas is produced at a carbon anode.

$$\text{oxidation at a carbon anode:} \quad 2F^- \rightarrow F_2(g) + 2e^-$$

$$\text{reduction at a steel cathode:} \quad 2H^+ + 2e^- \rightarrow H_2(g)$$

UF_6 is used in the enrichment of uranium. See Box 8.4 (p.367).

The main industrial uses of fluorine are in the production of uranium hexafluoride (UF_6) and sulfur hexafluoride (SF_6).

F_2 is extremely reactive, and reactions involving fluorine are generally highly exothermic. Values of the enthalpy changes of reaction are negative due to the low F–F bond dissociation enthalpy and high X–F bond enthalpies or high MF_n lattice enthalpies.

Sources of sodium chloride are described in Section 26.2 (p.1176). The chloralkali process is explained in Section 16.5 (p.757).

Chlorine is a yellow-green gas at room temperature and pressure. It is toxic and causes irritation of the lungs, and was used as a chemical weapon in the First World War (1914–1918). Cl_2 is obtained from sodium chloride by the electrolysis of aqueous sodium chloride (brine) in the chloralkali process, with about 56 million tonnes produced each year worldwide.

Much of the chlorine that is manufactured is used to prepare organochlorine compounds such as chloroethene, which are used to make polymers such as PVC. Chlorine is also used to prepare inorganic chlorides and bleach, and in the treatment of waste water.

In the laboratory, chlorine is usually prepared from the oxidation of hydrochloric acid by manganese(IV) oxide

$$\underset{\text{Mn(IV)}}{MnO_2(s)} + 4\underset{\text{Cl(–I)}}{HCl(aq)} \rightarrow \underset{\text{Mn(II)}}{MnCl_2(aq)} + 2H_2O(l) + \underset{\text{Cl(0)}}{Cl_2(g)}$$

Bromine is a red-brown liquid under normal conditions, whereas iodine is a solid that readily sublimes to form a violet vapour (see Figure 1.22, p.49). Both elements are obtained from sea water, where the anions are present in concentrations of up to 5000 ppm (bromide) and 100 ppm (iodide). Most of the world's production of bromine comes from the Dead Sea in Israel, where the concentration of halide ions is much higher than in normal sea water.

The sea water is concentrated, then treated with chlorine gas, exploiting the fact that chlorine is more oxidizing than either bromine or iodine. For example

$$2Br^-(aq) + Cl_2(g) \rightarrow Br_2(l) + 2Cl^-(aq)$$

The reaction between chlorine and iodide ions is described in Section 1.4 (p.30).

Astatine occurs in trace amounts in uranium and thorium ores. All of the isotopes of astatine are radioactive, and the longest-lived isotope, ^{210}At, has a half life of only 8.3 hours.

The bond dissociation enthalpies for the X_2 molecules are given in Table 27.10 together with enthalpy changes of formation for the gaseous halogen atoms. The bond dissociation enthalpies decrease down the group from chlorine to iodine because the overlap between the valence orbitals decreases as the atoms get larger. F_2 is an exception to this trend, and it has a much weaker bond dissociation enthalpy than might be expected. This is a consequence of the high electron–electron repulsion caused by the short F–F bond.

In the molecular orbital model, the repulsion is between fluorine π bonding and antibonding electrons, whereas, in the valence bond model, the repulsion is between the lone pairs on the two atoms, which are in sp^3 hybrid orbitals.

The enthalpy changes of formation for F(g) and Cl(g)

$$\tfrac{1}{2}F_2(g) \rightarrow F(g)$$

$$\tfrac{1}{2}Cl_2(g) \rightarrow Cl(g)$$

are equal to half of the bond dissociation enthalpy of F_2(g) and Cl_2(g), respectively. For Br(g) and I(g)

$$\tfrac{1}{2}Br_2(l) \rightarrow Br(g)$$

$$\tfrac{1}{2}I_2(g) \rightarrow I(g)$$

the enthalpy changes of formation also include a contribution from the enthalpy change of vaporization for Br_2(l) and the enthalpy change of sublimation for I_2(s).

Table 27.10 Bond dissociation enthalpies for Group 17 homonuclear bonds and the enthalpy changes of formation for X(g) [$\tfrac{1}{2}X_2(g) \rightarrow X(g)$]

	Bond dissociation enthalpy / kJ mol^{-1}	$\Delta H_f^\ominus$(X(g))/kJ mol^{-1}
F–F	+158.7	+79
Cl–Cl	+242.4	+121
Br–Br	+193.9	+112
I–I	+152.3	+107
At–At	+80	–

Box 27.8 Chlorofluorocarbons and the ozone hole

In 1985, the British geophysicist Joe Farman published a paper in the scientific journal *Nature* to report that the concentration of ozone in the stratosphere over the Antarctic was reduced by about 50% from September to early November (the Antarctic spring). This depletion became known as the ozone hole, and it was later shown to be an annual event.

What is happening? Over the sunless winter months the temperature in the stratosphere above the Antarctic drops to –80 °C. The temperature becomes cold enough for condensation to occur giving polar stratospheric clouds consisting of crystals of water, sulfuric acid, and nitric acid. There is a thin aqueous layer on the surface of these crystals, and chlorine nitrate, one of the chlorine reservoir molecules (see Box 27.6, p.1233), reacts on contact with this to form HOCl

$$ClONO_2(g) + H_2O(l) \rightarrow HOCl(aq) + HNO_3(aq)$$

HCl, another chlorine reservoir species, also dissolves in the aqueous layer to form H_3O^+ and Cl^- ions. HOCl and Cl^- comproportionate (see Section 27.4, p.1226) to give Cl_2, which escapes into the gas phase

$$HOCl(aq) + Cl^-(aq) \rightarrow Cl_2(g) + OH^-(aq)$$

During the dark winter months, concentrations of Cl_2 increase in the lower stratosphere. When the sunlight re-appears in the spring, the chlorine molecules decompose photochemically into atomic chlorine

$$Cl_2(g) \xrightarrow{h\nu} 2Cl^\bullet(g)$$

The large increase in $Cl^\bullet$ concentration speeds up the rate of ozone depletion leading to the observed drop in ozone concentration. There are not enough oxygen atoms around in the Antarctic spring for

➜ the catalytic cycle for ozone depletion described in Box 27.6 to be significant. Instead, ozone depletion takes place by the alternative mechanism outlined in Box 9.4 (p.403).

Some of the chlorine in the atmosphere is naturally occurring, but much comes from chlorofluorocarbons (CFCs; see Section 27.3, p.1221).

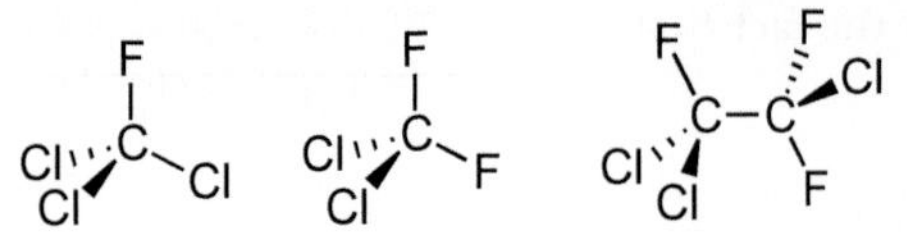

$CFCl_3$ CF_2Cl_2 $C_2F_3Cl_3$

▲ Examples of chlorofluorocarbons (CFCs).

The Montreal Protocol was signed in 1987. This is an international agreement to phase out CFC production and has been a success for global cooperation. As a result of this, all legal CFC production stopped in 1996 in developed countries, and in 2010 in developing countries. There have subsequently been a number of additions to the Montreal Protocol, and the production of other ozone-depleting chemicals such as tetrachloromethane (CCl_4), 1,1,1-trichloroethane (CH_3CCl_3), and the crop-fumigant bromomethane (CH_3Br) (see Chapter 20, p.917) are also being phased out.

The initial replacements for CFCs were hydrochlorofluorocarbons (HCFCs). These are similar to CFCs but also contain hydrogen. The presence of the hydrogen atom makes the molecule more reactive, so most HCFC molecules that are released do not reach the stratosphere. Those that do, however, release chlorine more quickly than CFCs. Due to this, and the global warming potential of these compounds in the troposphere, HCFCs are due to be phased out by 2030 in developed countries. Longer term replacements for CFCs were hydrofluorocarbons (HFCs) such as CH_2FCF_3, and these compounds are now used in most refrigerators. HFCs do not cause ozone depletion, but are of environmental concern due to their global warming potential (see Box 27.2, p.1218) and a global deal reached in 2016 will lead to an eventual phasing out of their use.

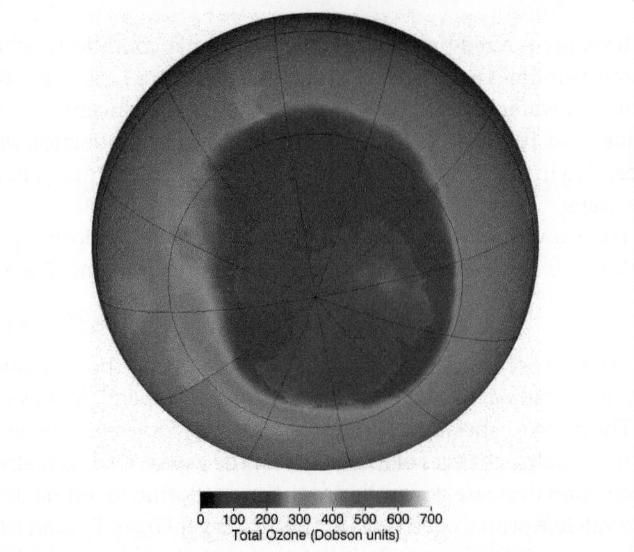
0 100 200 300 400 500 600 700
Total Ozone (Dobson units)

▲ Ozone concentration over the Antarctic showing the reduction in ozone, known as the ozone hole. This image was recorded on 15 September 2015. The size of the ozone hole varies from year to year with weather conditions. Scientists predict that the hole will not completely recover until about 2050 due to the long atmospheric lifetime of CFCs that were emitted before the ban. The concentration of ozone is measured in Dobson Units (DUs). Imagine if all of the ozone in a vertical column of air were compressed to standard temperature and pressure to form a slab. 1 Dobson Unit corresponds to a thickness of 0.01 mm.

Question

Suggest why compounds containing Cl or Br atoms cause ozone depletion in the stratosphere, but compounds containing F or H atoms do not.

Summary

- The Group 17 elements are known as the halogens. All are non-metals.
- The elements exist as diatomic molecules, X_2. Fluorine and chlorine are gases at room temperature, bromine is a liquid, and iodine is a solid.
- Astatine is radioactive, with the longest lived isotope having a half life of only several hours.
- Fluorine is the most electronegative element, and forms compounds only in the −1 oxidation state. For the other halogens, oxidation states up to +7 are also observed.
- Going down the group, the elements become less oxidizing.
- The X_2 bond dissociation enthalpies generally decrease going down the group, but the value for F_2 is anomalously low due to a high degree of electron–electron repulsion.

The Group 17 oxides

The halogen oxide compounds OX_2 and O_2X_2 are described in Section 27.6 (p.1238). This section concentrates on the higher oxidation state oxides, which exist for chlorine, bromine, and iodine. Since fluorine is more electronegative than oxygen, higher oxidation state oxides are unknown for fluorine. The structures of some of the chlorine oxides are shown in Figure 27.27.

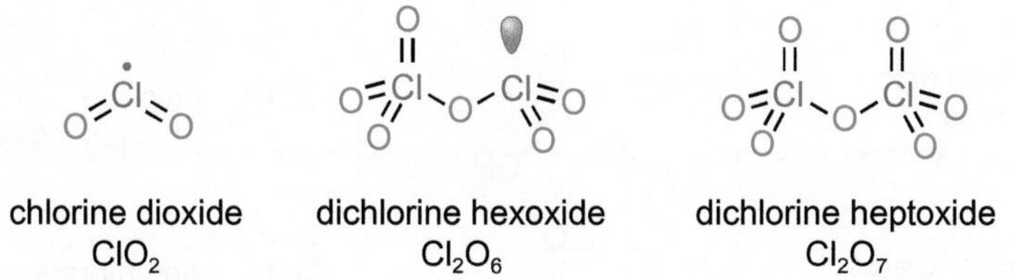
chlorine dioxide
ClO_2
dichlorine hexoxide
Cl_2O_6
dichlorine heptoxide
Cl_2O_7

Figure 27.27 The structures of the main chlorine oxides.

Chlorine dioxide (ClO_2) is a yellow, paramagnetic gas, and the molecules have a bent geometry. Despite having an unpaired electron, ClO_2 does not dimerize as the unpaired electron is delocalized over the whole molecule. ClO_2 is used industrially to bleach paper and to treat waste water.

ClO_2 reacts with ozone to form dichlorine hexoxide (Cl_2O_6) which is a dark red liquid that explodes in contact with organic compounds

$$2\,ClO_2\,(g) + 2\,O_3\,(g) \rightarrow Cl_2O_6\,(l) + 2\,O_2\,(g)$$

Cl_2O_6 reacts with water to form a mixture of oxoacids

$$Cl_2O_6\,(l) + H_2O\,(l) \rightarrow HClO_3\,(aq) + HClO_4\,(aq)$$

In the solid state, Cl_2O_6 exists as the ionic compound $ClO_2^+ClO_4^-$.

Dichlorine heptoxide (Cl_2O_7) is a colourless, oily liquid that is both shock-sensitive and explosive. It is formed by dehydrating perchloric acid ($HClO_4$) using concentrated phosphoric acid at –10 °C. The reverse reaction is also possible, and Cl_2O_7 reacts with water to regenerate perchloric acid.

Bromine oxides are less stable than chlorine oxides, and BrO_2 is unstable above –40 °C.

Iodine forms the most stable halogen oxides. I_2O_5 (**22**) is made by heating iodic acid (HIO_3) to 200 °C, and is a solid under normal conditions. It reacts with water to reform HIO_3

$$I_2O_5\,(s) + H_2O\,(l) \rightarrow 2\,HIO_3\,(aq)$$

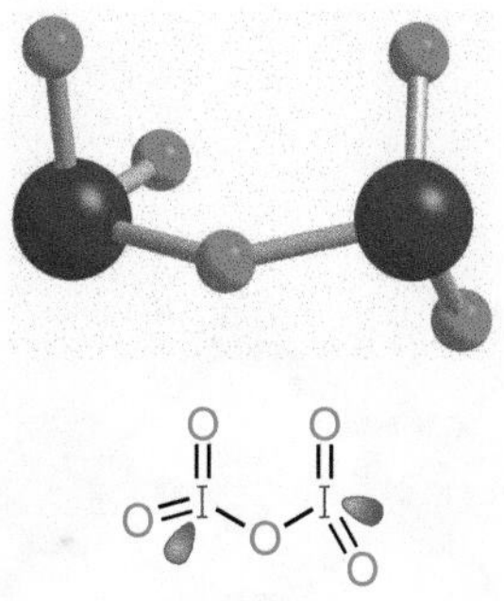

22 Diiodine pentoxide (I_2O_5).

The Group 17 oxoacids and oxoanions

The halogens form a wide range of oxoacids and oxoanions. Most of the acids cannot be isolated pure, but are used in aqueous solution. Fluorine forms only one oxoacid, HOF, as it is unable to form positive oxidation states. HOF is formed from the fluorination of ice at low temperature

$$F_2\,(g) + H_2O\,(s) \rightarrow HF\,(aq) + HOF\,(aq)$$

Solid HOF decomposes spontaneously to HF and O_2 at room temperature with a half life of 30 minutes.

The chlorine oxoacids are shown in Figure 27.28. Chlorine dissolves in water to form hypochlorous acid (HOCl)

The acidities of halogen oxoacids are described in Section 7.6 (p.330).

$$\underset{Cl(0)}{Cl_2}\,(g) + 2\,H_2O\,(l) \rightleftharpoons \underset{Cl(I)}{HOCl}\,(aq) + H_3O^+\,(aq) + \underset{Cl(-I)}{Cl^-}\,(aq)$$

When chlorine is dissolved in alkaline solution, HOCl is deprotonated, and the equilibrium above is pushed to the right-hand side. This allows compounds containing the hypochlorite anion (OCl^-) to be isolated

$$Cl_2\,(g) + 2\,OH^-\,(aq) \rightleftharpoons OCl^-\,(aq) + Cl^-\,(aq) + H_2O\,(l)$$

hypochlorous acid HOCl Cl oxidation state +1	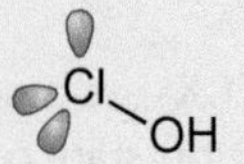	OCl^- (ClO^-)	hypochlorite [chloridooxygenate(1−)]
chlorous acid $HClO_2$ Cl oxidation state +3	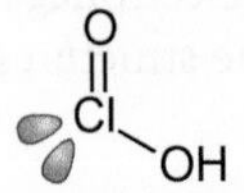	ClO_2^-	chlorite [dioxidochlorate(1−)]
chloric acid $HClO_3$ Cl oxidation state +5	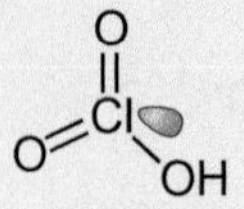	ClO_3^-	chlorate [trioxidochlorate(1−)]
perchloric acid $HClO_4$ Cl oxidation state +7	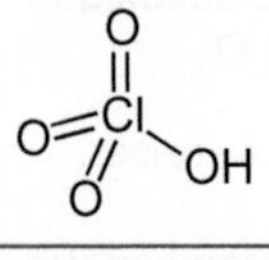	ClO_4^-	perchlorate [tetraoxidochlorate(1−)]

Figure 27.28 The structures of the chlorine oxoacids and the anions derived from them. The names in brackets are the systematic names approved by IUPAC.

The inorganic names used throughout this book are the IUPAC acceptable common names. The systematic names for the four chlorine oxyanions are given in Figure 27.28 alongside the common names.

The hypochlorite ion is also called the oxychloride ion or the chlorate(I) ion.

Bleach consists of an alkaline solution of sodium hypochlorite (NaOCl). The hypochlorite ion is thermodynamically unstable with respect to disproportionation to give chlorate ions and chloride ion, but this reaction occurs only very slowly

$$\underset{\text{Cl(I)}}{3\,OCl^-(aq)} \rightarrow \underset{\text{Cl(V)}}{ClO_3(aq)} + \underset{\text{Cl(−I)}}{2\,Cl^-(aq)}$$

The stability of the hypohalous acids (HOX) decreases down the group.

The chlorate ion, ClO_3^-, is formed in a disproportionation reaction when chlorine gas is dissolved in hot alkaline solution

$$\underset{\text{Cl(0)}}{3\,Cl_2(g)} + 6\,OH^-(aq) \rightarrow \underset{\text{Cl(−I)}}{5\,Cl^-(aq)} + \underset{\text{Cl(V)}}{ClO_3^-(aq)} + 3\,H_2O\,(l)$$

Chlorate compounds are strong oxidizing agents, and are used as the oxidants in safety matches (Section 27.4, p.1231). Chlorate ions can be oxidized to perchlorate ions (ClO_4^-) electrochemically.

Perchlorate ions are commonly used as anions in synthetic chemistry as they interact very poorly with metal ions. Perchlorates are strong oxidizing agents, but the reactions are usually very slow. Perchlorate compounds can be explosive so they should only be used with care.

The most stable of the bromine oxoacids is bromic acid ($HBrO_3$). Although periodate ions (IO_4^-) are known, iodine(VII) tends to form ions with higher coordination numbers such as $I_2O_9^{4-}$ (**23**).

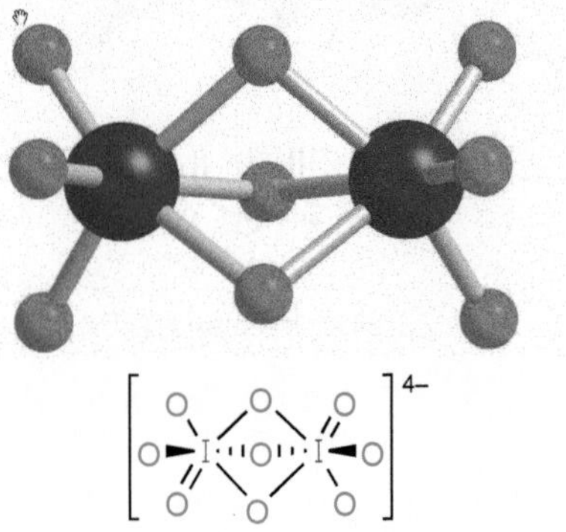

23 The $I_2O_9^{4-}$ anion

Summary

- Chlorine oxides include ClO_2, which is paramagnetic, and Cl_2O_7. All of the halogen oxides are acidic.
- Chlorine forms oxoacids in the +1 (HOCl), +3 ($HClO_2$), +5 ($HClO_3$), and +7 ($HClO_4$) oxidation states.

The *p*-block halides

The halide compounds of hydrogen and the *s*-block elements are described in Chapters 25 and 26, respectively, whilst the halides of the Groups 13–16 elements are described earlier in this chapter (pp.1213, 1221, 1228, and 1238). The halides have a wide variety of structures, stoichiometries, and properties, so any attempt to summarize these will inevitably lead to oversimplification. Nevertheless, some of the general trends are outlined here.

Fluorine reacts with all of the elements except helium, neon, and argon to form fluorides. Chlorine is less reactive, but still reacts with all of the elements with the exceptions of carbon, nitrogen, oxygen, and the Group 18 elements. The compounds of Group 17 elements with metals typically form extended structures with high ionic character when the metals are in low oxidation states. However, in high oxidation states, many metals form covalently bonded molecular halides. For example, $SnCl_4$, Al_2Br_6, and UF_6 are all molecular solids.

The compounds with non-metals contain covalent bonding. Fluorine is the most oxidizing of the halogens, and it stabilizes high oxidation states.

For the ionic compounds, the sizes of the halide ions increase down the group. As a consequence, the lattice enthalpies are highest for the fluorides, so the fluorides tend to be less soluble than the other halides. For the covalent compounds, the bond enthalpies decrease down the group due to the poorer overlap of the atomic orbitals.

Stability of the halides

For a *p*-block element, the stability of the halides generally decreases with increasing size of the halogen. So, the fluoride is more stable than the chloride, which in turn is more stable than the bromide. A thermodynamic cycle helps explain why this is so. A cycle for the enthalpy change of formation of AX_n, where A is a *p*-block element and X is a halogen, is shown in Figure 27.29. The cycle can be used to compare $\Delta_f H^\ominus(AX_n)$ for different halogens.

Of the three terms contributing to $\Delta_f H^\ominus(AX_n)$, the enthalpy change of formation of atom A does not depend on the halogen. The enthalpy changes of formation of the halogen atoms are given in Table 27.10 (p.1241). This term is notably lower for fluorine than it is for the other halogens due to the weakness of the F–F bond. The final term is the bond enthalpy of the A–X bonds, and examples of these are given in Table 27.3 (p.1208). The A–X bond enthalpies decrease with the increasing size of the halogens, due to both decreased orbital overlap and the decreasing ionic character of the bonds. The trends in $\Delta_f H^\ominus(X(g))$ and $\Delta_{be} H^\ominus(A–X)$ both act to make $\Delta_f H^\ominus(AX_n)$ less exothermic with increasing halogen size. Worked example 27.2 shows how enthalpy cycles can be used to calculate enthalpy changes of formation for *p*-block halides.

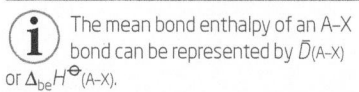

The mean bond enthalpy of an A–X bond can be represented by $\bar{D}$(A–X) or $\Delta_{be}H^\ominus$(A–X).

Visit the Online Resource Centre to view screencast 27.3 which walks you through calculations to compare the stability of NF_3 and NCl_3.

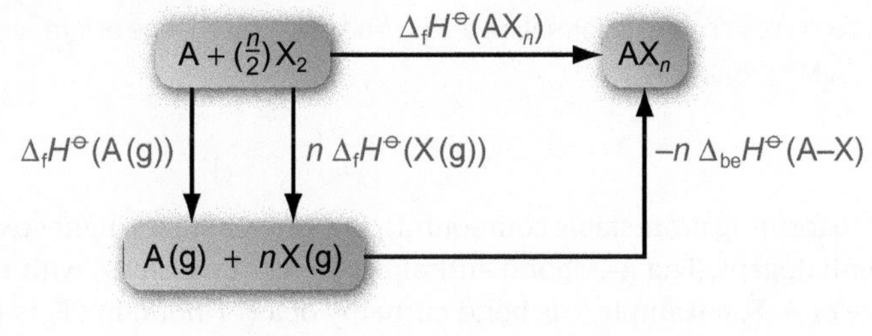

$\Delta_f H^\ominus(AX_n)$
$A + (\frac{n}{2})X_2$
AX_n
$\Delta_f H^\ominus(A(g))$
$n\,\Delta_f H^\ominus(X(g))$
$-n\,\Delta_{be} H^\ominus(A–X)$
$A(g) + nX(g)$

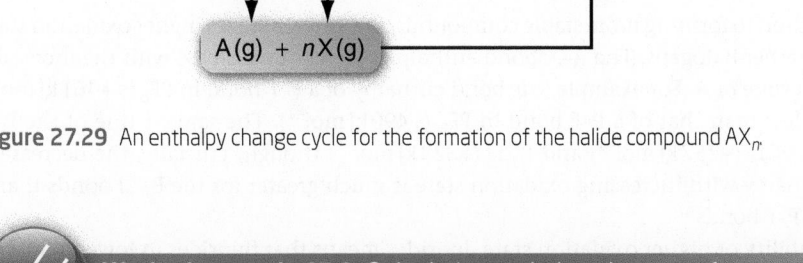

Figure 27.29 An enthalpy change cycle for the formation of the halide compound AX_n.

Worked example 27.2 Calculating enthalpy changes of formation of *p*-block halide compounds

Use the data in Table 27.3 (p.1208) and Table 27.10 (p.1241) to calculate $\Delta_f H^\ominus(NF_3)$ and $\Delta_f H^\ominus(NCl_3)$. The enthalpy change of formation of atomic nitrogen is equal to half of the N≡N bond dissociation enthalpy, that is, $\Delta_f H^\ominus(N) = +473\,kJ\,mol^{-1}$.

Strategy

Construct an enthalpy cycle similar to that in Figure 27.29 for NF_3 and for NCl_3.

Comment on the values obtained for the enthalpy changes of formation. →

➜ Solution

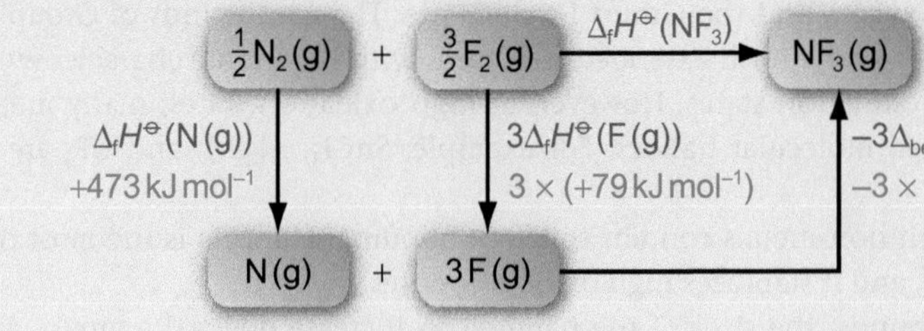

$\frac{1}{2}N_2(g)$ + $\frac{3}{2}F_2(g)$
$\Delta_f H^\ominus(NF_3)$
$NF_3(g)$
$\Delta_f H^\ominus(N(g))$
+473 kJ mol^{-1}
$3\Delta_f H^\ominus(F(g))$
3 × (+79 kJ mol^{-1})
$-3\Delta_{be}H^\ominus(N–F)$
−3 × (+278 kJ mol^{-1})
N(g) + 3F(g)

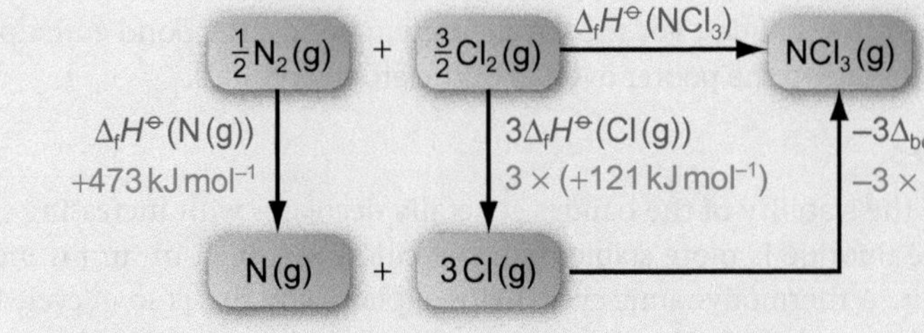

$\frac{1}{2}N_2(g)$ + $\frac{3}{2}Cl_2(g)$
$\Delta_f H^\ominus(NCl_3)$
$NCl_3(g)$
$\Delta_f H^\ominus(N(g))$
+473 kJ mol^{-1}
$3\Delta_f H^\ominus(Cl(g))$
3 × (+121 kJ mol^{-1})
$-3\Delta_{be}H^\ominus(N–Cl)$
−3 × (+190 kJ mol^{-1})
N(g) + 3Cl(g)

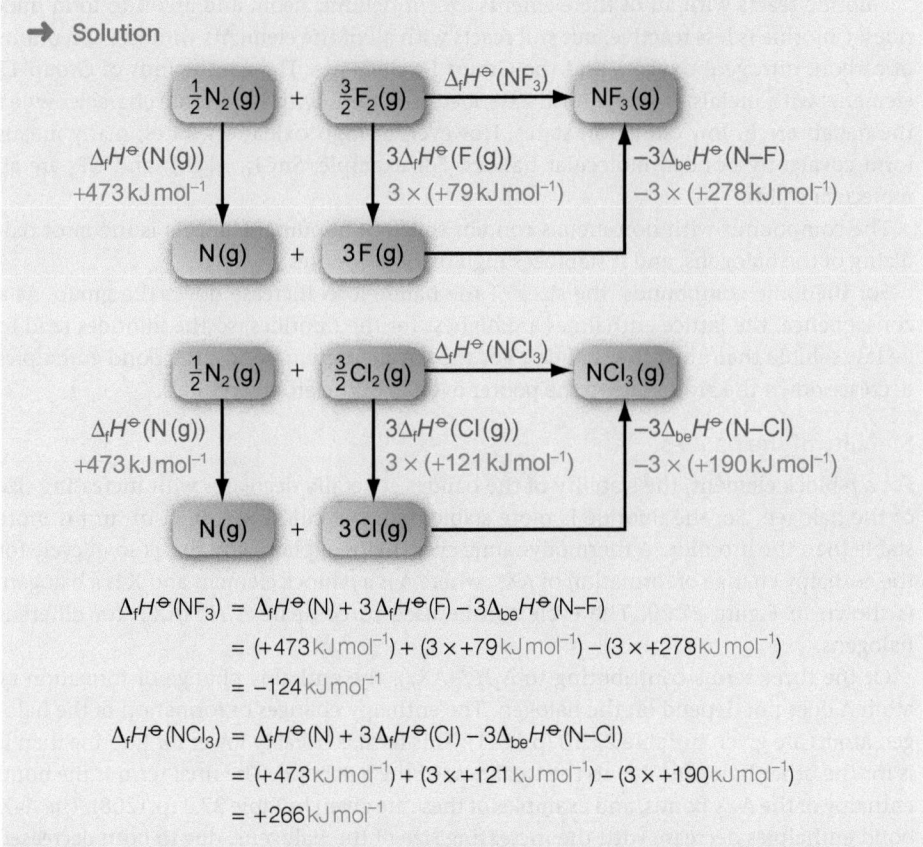

$$\begin{aligned}\Delta_f H^\ominus(NF_3) &= \Delta_f H^\ominus(N) + 3\Delta_f H^\ominus(F) - 3\Delta_{be}H^\ominus(N\text{–}F)\\ &= (+473\,\text{kJ mol}^{-1}) + (3\times{+79}\,\text{kJ mol}^{-1}) - (3\times{+278}\,\text{kJ mol}^{-1})\\ &= -124\,\text{kJ mol}^{-1}\end{aligned}$$

$$\begin{aligned}\Delta_f H^\ominus(NCl_3) &= \Delta_f H^\ominus(N) + 3\Delta_f H^\ominus(Cl) - 3\Delta_{be}H^\ominus(N\text{–}Cl)\\ &= (+473\,\text{kJ mol}^{-1}) + (3\times{+121}\,\text{kJ mol}^{-1}) - (3\times{+190}\,\text{kJ mol}^{-1})\\ &= +266\,\text{kJ mol}^{-1}\end{aligned}$$

The value of the enthalpy change of formation of NF_3 is negative (exothermic), whereas that of NCl_3 is positive (endothermic). This is consistent with the nature of the compounds, as NF_3 is a thermodynamically stable gas and NCl_3 is an explosive liquid (see Section 27.4, p.1228).

Question

Calculate the enthalpy change of formation of PF_3. The enthalpy change of formation of atomic phosphorus, $\Delta_f H^\ominus(P(g))$, is +315 kJ mol^{-1}.

In addition to forming more stable compounds, fluorine stabilizes higher oxidation states than the other halogens. The A–X bond enthalpies generally decrease with the increasing oxidation state of A. For example, the bond enthalpy of a P–F bond in PF_5 is +461 kJ mol^{-1}, which is less than that of a P–F bond in PF_3 (+490 kJ mol^{-1}). The same is true of the P–Cl bonds in PCl_5 (+257 kJ mol^{-1}) and PCl_3 (+323 kJ mol^{-1}) though, crucially, the decrease in bond enthalpy with increasing oxidation state is much greater for the P–Cl bonds than it is for the P–F bonds.

The stability of higher oxidation state fluorides means that fluorides in lower oxidation states are often unstable to disproportionation. For example, S_2F_2 disproportionates to form sulfur and SF_4

$$\underset{S(I)}{16\,S_2F_2(g)} \rightarrow \underset{S(0)}{3\,S_8(s)} + \underset{S(IV)}{8\,SF_4(g)}$$

Interhalogen compounds

The shapes of interhalogen molecules can be accurately predicted using VSEPR theory. See Section 5.2 (p.223) for details.

Compounds formed between the halogen elements are known as **interhalogens**. Interhalogens of the general formulae XY, XY_3, and XFY_5 are known, as is IF_7. Examples of these compounds are shown in Figure 27.30.

Interhalogens are prepared by direct reaction between the elements, with the products depending mainly on the stoichiometry. For example,

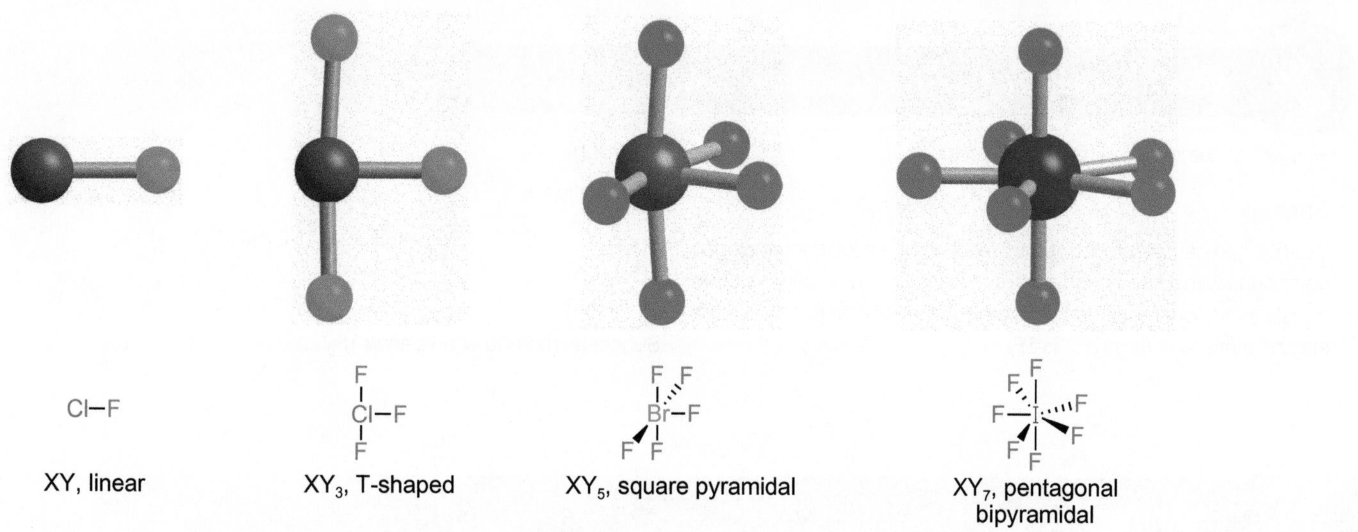
Cl–F
XY, linear
XY3, T-shaped
XY5, square pyramidal
XY7, pentagonal bipyramidal

Figure 27.30 Examples of interhalogen compounds.

$$Cl_2\,(g) + 3\,F_2\,(g) \xrightarrow{250^\circ C} 2\,ClF_3\,(g)$$
$$Cl_2\,(g) + 5\,F_2\,(g) \xrightarrow{350^\circ C} 2\,ClF_5\,(g)$$

Chlorine trifluoride (ClF_3) is extremely reactive and will even spontaneously ignite asbestos. It converts most oxides and chlorides to fluorides.

The uses of ClF_3 and BrF_3 are described in Box 5.3 (p. 229).

In addition to the neutral interhalogen compounds, many interhalogen cations and anions are known. The structures of some of these ions are shown in Figure 27.31.

The interhalogen cations are formed by removal of a fluoride anion from a neutral interhalogen compound by reacting it with a fluoride ion acceptor such as AsF_5. For example,

$$ClF_3 + AsF_5 \rightarrow ClF_2^{+}AsF_6^{-}$$

The interhalogen anions are formed by addition of a fluoride ion. For example,

$$ClF_3 + CsF \rightarrow Cs^{+}ClF_4^{-}$$

Polyiodide anions are related to interhalogen anions, and the most common polyiodide is the linear anion I_3^-. Iodine is not very soluble in water, but in the presence of iodide it forms the more soluble I_3^- ion

$$I_2\,(aq) + I^-\,(aq) \rightleftharpoons I_3^-\,(aq)$$

Larger polyiodide ions such as I_5^- and I_9^- are also known. Worked example 27.3 describes the synthesis of an ionic interhalogen compound.

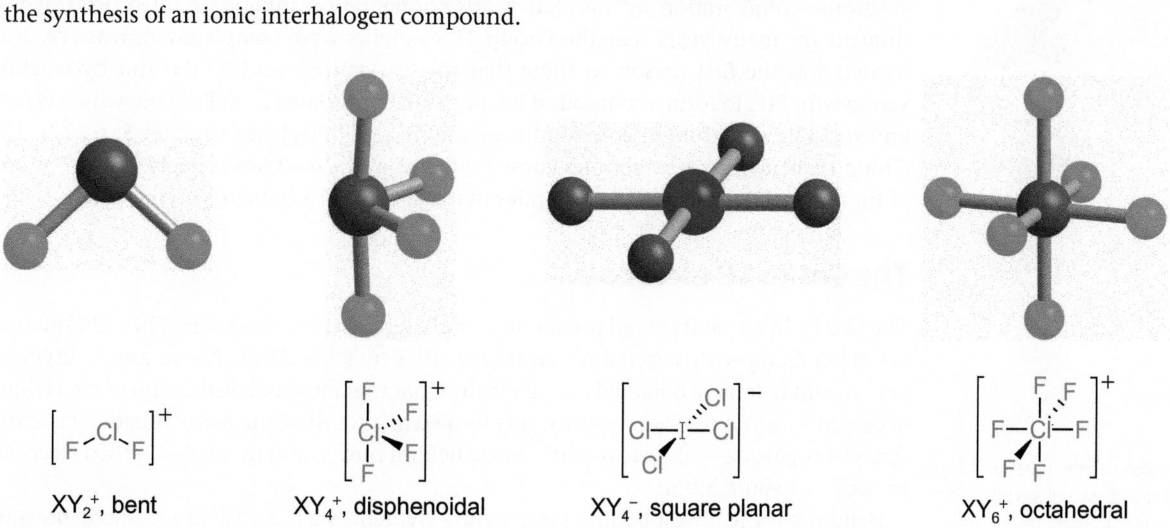
XY2+, bent
XY4+, disphenoidal
XY4−, square planar
XY6+, octahedral

Figure 27.31 Examples of interhalogen ions.

Worked example 27.3 Synthesis of interhalogen ions

Suggest a method of preparing IF_4^+ ions.

Strategy

Interhalogen cations are formed by reaction of a neutral interhalogen compound with a fluoride ion acceptor like AsF_5. The reaction involves transfer of a fluoride ion from the interhalogen molecule to AsF_5, so the starting material to form IF_4^+ is IF_5.

Solution

The reaction is

$$IF_5 + AsF_5 \rightarrow IF_4^+AsF_6^-$$

Question

Suggest a method of preparing IF_6^-.

Summary

- The stability of a non-metal halide AX_n, where A is a *p*-block element and X is a halogen, decreases with increasing size of X. The weakness of the F–F bond and the decreasing bond enthalpies of the A–X bonds with increasing size of X are the major factors behind this.
- Fluorine stabilizes high oxidation states. The A–F bond enthalpies decrease less with increasing coordination number than A–Cl bond enthalpies.
- Interhalogen compounds can be neutral, cationic, or anionic. The shapes of the molecules and ions can be predicted using VSEPR theory.

? For practice questions on the topics in Section 27.6, see questions 12–15 at the end of this chapter (p.1253). Questions 16–20 are on general *p*-block chemistry.

27.7 Group 18: helium, neon, argon, krypton, xenon, and radon

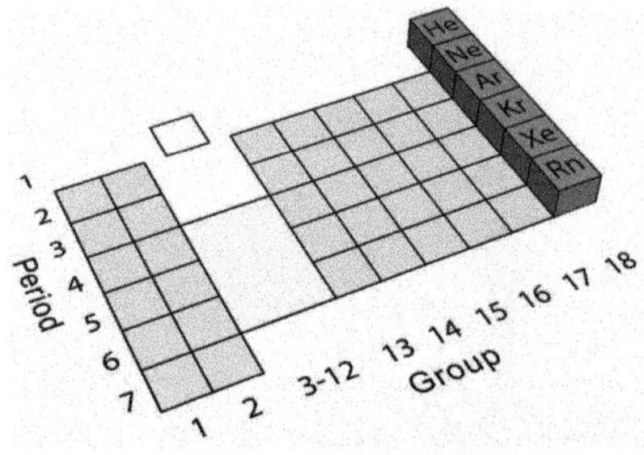
He
Ne
Ar
Kr
Xe
Rn
Period
1
2
3
4
5
6
7
1
2
3-12
13
14
15
16
17
18
Group

Group 18 in the Periodic Table.

Group 18 consists of the elements helium (He), neon (Ne), argon (Ar), krypton (Kr), xenon (Xe), and radon (Rn). All of these elements exist as colourless, monatomic gases with the electronic configuration $ns^2\ np^6$. Since this configuration involves a filled octet, it was thought for many years that the Group 18 elements were completely unreactive. Neil Bartlett was the first person to show that this is not true, and he did this by reacting xenon with PtF_6 to form a compound he originally formulated as $XePtF_6$, but which is now known to be a mixture of compounds including $[XeF][PtF_6]$ (see Chapter 5, p.217). The Group 18 elements are sometimes known together as the *inert gases* or *noble gases*. The first of these terms is misleading as it implies that the Group 18 elements do not react.

The Group 18 elements

The Group 18 elements are all present to some extent in the atmosphere. Their abundances are given along with selected physical properties in Table 27.11. Neon, argon, krypton, xenon, and radon are obtained commercially from the fractional distillation of air. Helium occurs in some natural gas deposits. It is formed as a result of the decay of radioactive isotopes by α-particle emission; α-particles are helium nuclei, and these pick up two electrons to become helium atoms.

Helium was discovered in the spectral lines of the Sun before it was discovered on Earth (see Box 3.3, p.131). Radioactivity is described in Section 3.8 (p.159).

Helium has the lowest boiling point of any element, 4.2 K, and at atmospheric pressure it does not form a solid, even when cooled to temperatures approaching absolute zero.

Table 27.11 Physical properties and atmospheric abundance of the Group 18 elements

Element	A_r	m.p. / °C	b.p. / °C	Density / 10^{-3} g dm^{-3}	Abundance in atmosphere / %
Helium (He)	4.0026	–272*	–269	0.18	5×10^{-4}
Neon (Ne)	20.180	–249	–246	0.90	1.8×10^{-3}
Argon (Ar)	39.948	–189	–186	1.78	0.94
Krypton (Kr)	83.798	–157	–153	3.73	1×10^{-4}
Xenon (Xe)	131.29	–112	–108	5.89	9×10^{-6}
Radon (Rn)	222	–71	–62	9.73	1×10^{-16}

* At 26 atm.

Neon lights, like these in Hong Kong, consist of narrow glass tubes, often shaped, containing neon gas at low pressure. When an electric current is passed through the gas, the excited molecules emit photons of characteristic wavelengths, which appear as red-orange light. The colour of the light emitted can be varied by adding small amounts of argon, or by coating the glass with phosphors.

Liquid helium is used as a coolant, enabling very low temperatures to be reached. Below 2.2K, liquid helium undergoes a transition to form a phase that exhibits the unusual property of *superfluidity* (see Chapter 17, p.766). At these temperatures, liquid helium has no viscosity, so it can move rapidly through any pore and can even flow upwards out of a container in an invisible film on the surface.

Box 27.9 Using argon to date rocks

Radiocarbon dating is used to date organic material (Box 3.8, p.161) but, since the half life of ^{14}C is 5730 years, it is only useful for samples up to about 100000 years old. It is possible to date rocks that are millions or billions of years old using radioactive isotopes with longer half lives.

^{40}K is a radioactive isotope of potassium that is present as a relatively low concentration (0.0117%) of naturally occurring potassium. This isotope has a half life of 1.277×10^9 years and two main decay pathways: β-emission produces ^{40}Ca, whereas electron-capture gives ^{40}Ar

β-emission $\quad {}^{40}_{19}K \rightarrow {}^{40}_{20}Ca + {}^{0}_{-1}e$

electron capture $\quad {}^{40}_{19}K + {}^{0}_{-1}e \rightarrow {}^{40}_{18}Ar$

Electron capture involves a reaction between the nucleus and a core electron in which a proton is converted to a neutron.

By comparing the ratio of ^{40}K to ^{40}Ar in a sample, geologists can determine the date that the rock was formed. The technique is useful to an archaeologist if a fossil is discovered between two layers of volcanic rock as it enables upper and lower limits for the date of the fossil to be determined. Potassium–argon dating has been used to date early hominid fossils in East Africa. This is possible because they occur in an area that was volcanically active when the fossils were deposited between one and five million years ago.

Argon–argon dating is a related but much more accurate dating technique. Samples of rock are irradiated with neutrons to convert ^{40}K into ^{39}Ar, which is not a naturally occurring isotope of argon. The ratio of ^{40}Ar to ^{39}Ar is then determined by mass spectrometry. This gives more accurate dates than potassium–argon dating and allows for the use of much smaller samples—milligrams rather than tens of grams. The Pompeii eruption of 79 AD has been dated to within a few years by argon–argon dating, which demonstrates that the technique can even be used accurately on samples less than 2000 years old.

▲ Samples from the Pompeii eruption of 79 AD have been accurately dated using Ar–Ar dating.

Question

^{40}Ca is also a product of the decay of ^{40}K. Suggest a reason why the ratio of ^{40}K to ^{40}Ca cannot be used to date rocks.

Group 18 compounds

The Group 18 elements have high enthalpy changes of ionization, but these decrease with increasing size of the atom. This means that the reactivity of the elements increases down the group, and this is witnessed by the range of compounds that the elements form. Xenon has the most extensive chemistry, and the majority of compounds are those with the most electronegative elements, fluorine and oxygen. No stable compounds of helium or neon are known, and the only compound of argon to have been prepared is HArF, which is only stable below –246 °C. This was reported in 2000 by Finnish researchers from the reaction of HF with solid argon. Krypton forms the stable difluoride KrF_2, and compounds have been isolated that contain the cations KrF^+ and $Kr_2F_3^+$.

Xenon fluorides

Xenon reacts with fluorine to form XeF_2, XeF_4, or XeF_6, depending on the conditions.

$$Xe(g) + F_2(g) \xrightarrow{5\ \text{atm},\ 400\ °C} XeF_2(g) \qquad (Xe:F_2\ \text{ratio} = 2:1)$$

$$Xe(g) + 2F_2(g) \xrightarrow{5\ \text{atm},\ 400\ °C} XeF_4(g) \qquad (Xe:F_2\ \text{ratio} = 1:5)$$

$$Xe(g) + 3F_2(g) \xrightarrow{50\ \text{atm},\ 400\ °C} XeF_6(g) \qquad (Xe:F_2\ \text{ratio} = 1:20)$$

A use of XeF_2 is described in Chapter 5 (p.217).

The XeF_2 molecules are linear, and the XeF_4 molecules are square planar. Both of these geometries are consistent with predictions from VSEPR theory.

Fluxionality is explained in Box 27.4 (p.1230).

XeF_6 does not form a structure based on a pentagonal bipyramid as might be expected from VSEPR theory. Instead, in the gas phase it has a distorted octahedral shape caused by the lone pair of electrons being directed through one of the triangular faces of the octahedron (Figure 27.32). The lone pair moves around the octahedron so XeF_6 is fluxional. In the solid state, xenon hexafluoride is ionic with the formula $XeF_5^+F^-$.

The xenon fluorides are all strong fluorinating agents, and act as fluoride ion donors and acceptors. For example, XeF_2 reacts with AsF_5 (a fluoride ion acceptor) to form an ionic compound containing the XeF^+ cation

$$XeF_2 + AsF_5 \rightarrow XeF^+AsF_6^-$$

Other xenon–fluorine ions such as $Xe_2F_3^+$, XeF_5^+, XeF_8^{2-}, and XeF_5^- are also known.

Xenon oxides and oxoanions

Xenon oxides are formed by hydrolysis of xenon fluorides. For example,

$$\underset{Xe(IV)}{6XeF_4(s)} + 12H_2O(l) \rightarrow \underset{Xe(VI)}{2XeO_3(aq)} + \underset{Xe(0)}{4Xe(g)} + 3O_2(g) + 24HF(aq)$$

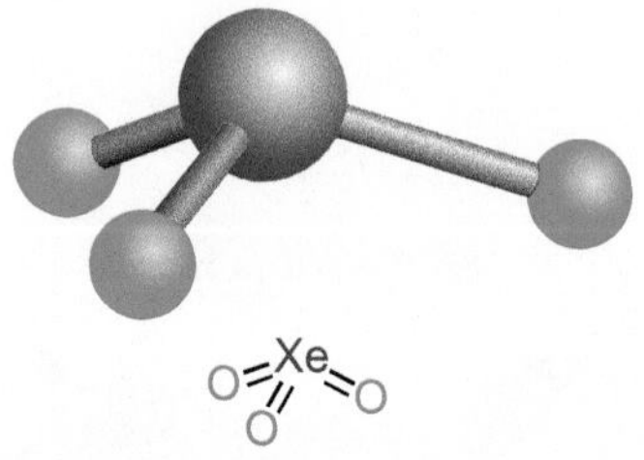

24 Xenon trioxide (XeO_3).

This is a disproportionation reaction, with xenon(IV) reacting to give xenon(VI) and xenon(0). Xenon trioxide (XeO_3, (**24**)) is an explosive white solid. It is stable in aqueous solution, but reacts with hydroxide ions to form $HXeO_4^-$

$$XeO_3(aq) + OH^-(aq) \rightarrow HXeO_4^-(aq)$$

The $HXeO_4^-$ ion is unstable with respect to disproportionation and decomposes to form XeO_6^{4-}, which contains xenon(VIII), and xenon gas

$$\underset{Xe(VI)}{HXeO_4^-(aq)} + 2OH^-(aq) \rightarrow \underset{Xe(VIII)}{XeO_6^{4-}(aq)} + \underset{Xe(0)}{4Xe(g)} + O_2(g) + 2H_2O(l)$$

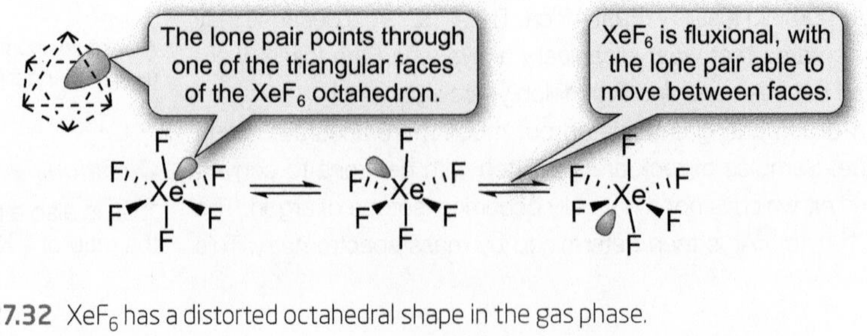
The lone pair points through one of the triangular faces of the XeF6 octahedron.
XeF6 is fluxional, with the lone pair able to move between faces.

Figure 27.32 XeF_6 has a distorted octahedral shape in the gas phase.

Xenon tetroxide (XeO_4) is also known, but it decomposes explosively to xenon and O_2. The compound HXeOXeH (**25**) has been prepared from the photolysis of water in solid xenon and shown to have a structure containing two xenon atoms.

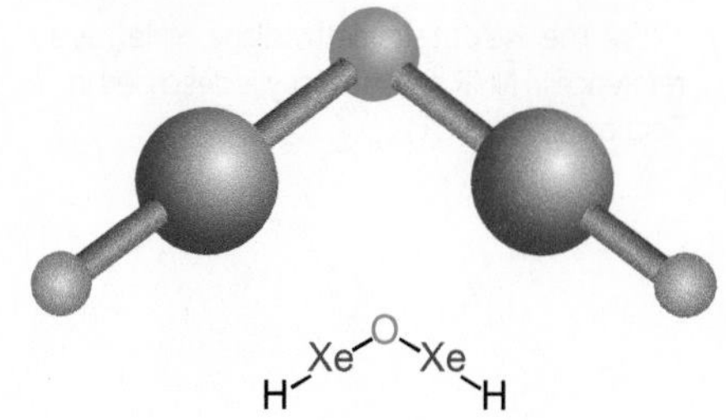
H
Xe
O
Xe
H

25 HXeOXeH.

Other xenon compounds

The known chemistry of xenon has grown remarkably over the past 25 years, and xenon compounds with bonds to carbon, nitrogen, chlorine, and metal atoms have all been characterized. Typically electron-withdrawing groups are needed on the carbon or nitrogen substituents in order to stabilize the compounds. For example, the phenyl cation $C_6H_5Xe^+$ is unknown, but the pentafluorophenyl cation $C_6F_5Xe^+$ exists.

Summary

- The Group 18 elements exist as monatomic gases. Radon is radioactive.
- Until the 1960s it was believed that compounds of the Group 18 elements were not possible, but compounds are now known for argon, krypton, and xenon, with xenon chemistry the most well developed.
- The xenon fluorides XeF_2, XeF_4, and XeF_6 are all known, and act as strong fluorinating agents.

27.8 *p*-Block organometallic chemistry

The metals and metalloids of the *p* block form organometallic compounds, many of which are useful in synthesis and in industry. In Group 13, the aluminium organometallic compounds are particularly useful, and triethylaluminium is used industrially together with $TiCl_4$ to form catalysts for the polymerization of ethene and propene. The development of these processes led to Karl Ziegler and Giulio Natta receiving the Nobel Prize for Chemistry in 1963.

Trimethylaluminium is a volatile, reactive liquid that is flammable in air and explodes on contact with water. It has a structure consisting of Al_2Me_6 dimers, as shown in Figure 27.33. Although the structure is superficially similar to that of Al_2Cl_6, Al_2Me_6 is electron deficient, whereas Al_2Cl_6 is not. This is because the methyl group does not have a lone pair or other filled orbital to donate to a metal centre. The bridging methyl groups in Al_2Me_6 are involved in 3-centre 2-electron bonds, similar to those involving the bridging hydrogen atoms in diborane (B_2H_6).

The dimeric structure of Al_2Me_6 is in contrast to the structure of BMe_3, which is monomeric. The greater size of the aluminium atom and the greater Lewis acidity of $AlMe_3$ are both factors in this difference.

Carbon forms covalent bonds to the other Group 14 atoms, and silyl groups such as $SiMe_3$ are useful in organic chemistry. The C–Si bond is relatively strong (mean bond

Tributyltin ($SnHBu_3$) was used extensively in the 1960s and 1970s in marine paints as it keeps the hulls of ships free of barnacles. Unfortunately, concentrations of less than 1 ng dm^{-3} in sea water cause infertility in oysters and mussels. The use of tributyltin was restricted in the 1980s and it is now banned in the European Union.

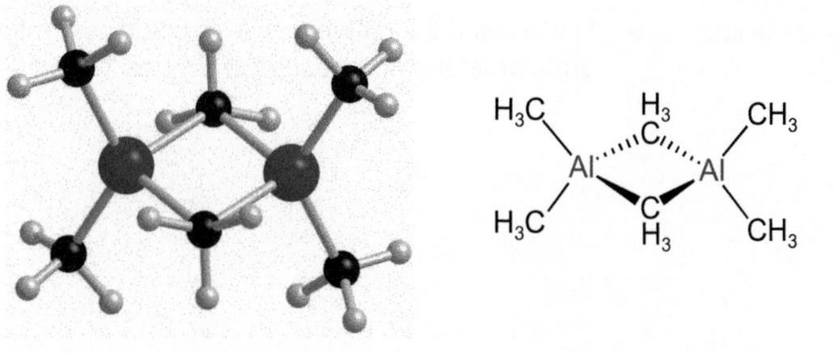
H3C
H3
C
CH3
Al
Al
H3C
C
H3
CH3

Figure 27.33 The dimeric structure of Al_2Me_6. In contrast to the aluminium halides, the bonding involves 3-centre 2-electron interactions in which the non-bonding orbitals are empty.

The use of tetramethylsilane, $SiMe_4$, as a reference in NMR spectroscopy is described in Section 10.7 (p.501).

enthalpy $+318\,kJ\,mol^{-1}$) and tetraalkylsilicon compounds have high thermal stabilities. The bonds decrease in strength with increasing size of the Group 14 element due to poorer orbital overlap, and the thermal stability of the compounds decreases in line with this. However, tetraalkyltin and tetraalkyllead compounds are more stable to reduction than their halide analogues. The organolead compound $PbEt_4$ has been used as an anti-knock agent in petrol. Lead emissions caused health problems, especially in children living near busy roads, and leaded petrol is now banned in the European Union and many other parts of the world.

As the electronegativity of the metal or metalloid increases, the bonding in its organometallic compounds becomes more covalent. As a consequence the compounds become less reactive. So, whereas *s*-block organometallics are highly reactive to air and water, Group 14 organometallics are air stable.

» Summary

- Trialkylaluminium compounds generally exist as electron-deficient dimers, in contrast to monomeric trialkylboron compounds.
- The thermal stability of Group 14 organometallics decreases with increasing size of the Group 14 element.
- As the electronegativity of the metal or metalloid increases, the bonding in its organometallic compounds becomes more covalent and the compounds become less reactive.

» Concept review

By the end of this chapter you should be able to do the following.

- Describe and explain trends in behaviour across the rows and down the groups of the *p* block.
- Understand why the metal/non-metal divide runs diagonally through the *p* block.
- Explain the basis of the inert pair effect and predict when it occurs.
- Describe the preparation, uses, and reactivity of the more important *p*-block elements.
- Describe the common oxidation states for the *p*-block elements.
- Describe the structures and key reactions of the *p*-block oxides, hydroxides, oxoacids, and halides.
- Explain the consequences of only having 6 valence electrons on the structures and reactions of Group 13 compounds.
- Use enthalpy cycles together with bond enthalpies to explain the differences in stabilities of compounds going down a group.
- Explain why fluorine is better at stabilizing high oxidation states than chlorine.
- Predict the shapes of simple *p*-block compounds using VSEPR theory.
- Understand the differences in structure and reactivity of organometallic compounds of Groups 13 and 14.

Questions

More challenging questions are indicated by an asterisk *.

1 Determine the oxidation state of the atom shown in red in the following molecules and ions: (a) $H_4P_2O_6$; (b) XeO_6^{4-}; (c) $[Sn(OH)_4]^{2-}$; (d) Se_4^{2+}; (e) S_3O_9; (f) S_2Cl_2. (Section 27.1)

2 Place the following elements in order of increasing first ionization energy (Section 27.1):

(a) antimony, arsenic, bismuth, nitrogen, phosphorus;

(b) carbon, fluorine, nitrogen, oxygen.

3 Are the following oxides acidic, basic, or amphoteric: (a) P_4O_{10}; (b) Bi_2O_3; (c) GeO_2? (Section 27.1)

4 Account for the difference in the B–F bond lengths observed in BF_3 (130 pm) and BF_4^- (145 pm). (Section 27.2)

5 The diagonal relationship between beryllium and aluminium is relatively strong. What evidence is there for this relationship? (Section 27.2)

6 Aluminium fluoride sublimes at over 1000 °C and is insoluble in benzene, whereas aluminium chloride melts at 192 °C and dissolves in benzene to give dimers. Account for these differences, and give reasons for any differences in the bonding in these compounds. (Section 27.2)

7 What is meant by the 'inert pair effect'? Why is the inert pair effect more important for lead than for silicon? (Sections 27.1, 27.3)

8 Draw the resonance structures for N_2O_5 and NO_2^-. (Section 27.4)

9 NF_2 exists in equilibrium with N_2F_4, whereas NH_2 dimerizes readily to form hydrazine, N_2H_4. Account for the difference in stability of NF_2 and NH_2 with respect to dimerization. (Section 27.4)

10 Use the data below to calculate the enthalpy changes of formation for $PCl_3(g)$ and $PI_3(g)$. Comment on the stabilities of these compounds. (Section 27.4)

$$\Delta_f H^\ominus(P(g)) = +315\ kJ\ mol^{-1}$$
$$\Delta_f H^\ominus(Cl(g)) = +121\ kJ\ mol^{-1}$$
$$\Delta_f H^\ominus(I(g)) = +107\ kJ\ mol^{-1}$$
$$\text{P–Cl bond enthalpy} = +326\ kJ\ mol^{-1}$$
$$\text{P–I bond enthalpy} = +184\ kJ\ mol^{-1}$$

11 Use the bond enthalpy data below to explain why oxygen exists as O_2 molecules and not O_8 rings, analogous to those formed by sulfur. (Section 27.5)

O–O +144 kJ mol^{-1}, O=O +498 kJ mol^{-1}

12 In an iodine–thiosulfate titration, thiosulfate is oxidized to tetrathionate ($S_4O_6^{2-}$). What are the average oxidation states of sulfur in thiosulfate and tetrathionate? (Section 27.6)

13 The bond angle in F_2O is 103.1°, which is less than the bond angle in water. Account for this difference. (Section 27.6)

14 Calculate the enthalpy change of formation of Cl_2O (g), given the bond enthalpies below. Would you expect to able to prepare Cl_2O from the elements? (Section 27.6)

Cl–Cl +242 kJ mol^{-1}, O=O +498 kJ mol^{-1}, Cl–O +205 kJ mol^{-1}

15* Iodine monofluoride (IF) disproportionates into IF_5 and I_2. Write a balanced equation for this reaction. Use the data below to calculate $\Delta H^\ominus$ for the disproportionation reaction. (Section 27.6)

$\Delta_f H^\ominus(IF(g)) = -95\ kJ\ mol^{-1}$, $\Delta_f H^\ominus(IF_5(g)) = -840\ kJ\ mol^{-1}$

16 Predict the products of each of the following reactions, and write a balanced equation in each case. (Sections 27.2, 27.4, 27.5, 27.7)

(a) $XeF_2 + AsF_5$

(b) $BI_3 + NH_3$

(c) $P_4O_{10} + H_2O$

(d) $SF_4 + BF_3$

17 Which of the following oxides are paramagnetic: NO_2; NO; N_2O_3; CO_2; CO; ClO_2; Cl_2O_7? (Sections 27.3, 27.4, 27.6)

18 Use VSEPR theory to predict the shapes of the following molecules and ions. (The background to VSEPR theory is given in Chapter 5.) (a) IF_5; (b) XeF_4; (c) IO_3^-; (d) $POCl_3$; (e) BrF_3; (f) XeO_2F_2. (Sections 27.4, 27.6, 27.7)

19 How would you prepare the following compounds: (a) $[IF_6]^+[SbF_6]^-$; (b) carbon suboxide; (c) XeO_3? (Sections 27.3, 27.6, 27.7)

20 H_3BO_3 is a monobasic acid whereas H_3PO_3 is a dibasic acid. Account for this difference. (Sections 27.2, 27.4)

28

d-Block chemistry

This chapter builds on the following topics:

- Oxidation and reduction Section 1.4, p.27
- Chemical equilibrium: how far has a reaction gone? Section 1.9, p.56
- Many-electron atoms Section 3.6, p.143
- Atomic properties and periodicity Section 3.7, p.153
- Lewis acids and bases Section 7.8, p.336

◀ Many pigments are *d*-block metal compounds.

Colouring with metal pigments

One of the characteristic features of *d*-block metal compounds is their wide range of colours. Many of these compounds have been used for centuries as the pigments in paints. Below are some examples.

Red	Red ochre	Fe_2O_3
	Vermillion	HgS
Green	Malachite	$CuCO_3{\cdot}Cu(OH)_2$
	Cobalt green	$CoO{\cdot}ZnO$
	Viridian	$Cr_2O_3{\cdot}2H_2O$
Yellow	Cobalt yellow	$K_3[Co(NO_2)_6]$
	Chrome yellow	$PbCrO_4$
	Yellow ochre	$Fe_2O_3{\cdot}H_2O$
Blue	Iron blue	$(NH_4)Fe[Fe(CN)_6]$
	Cerulean blue	$CoO{\cdot}n SnO_2$
	Cobalt blue	$CoO{\cdot}Al_2O_3$
	Azurite	$2CuCO_3{\cdot}Cu(OH)_2$

When a piece of art is being restored, it is important to identify the pigments used by the original artist, as this allows the restorer to return the artwork as closely as possible to its original appearance. There are several spectroscopic techniques that are useful for this, and the background to these is described in Chapter 10.

▼ A copper(II) phthalocyanine complex.

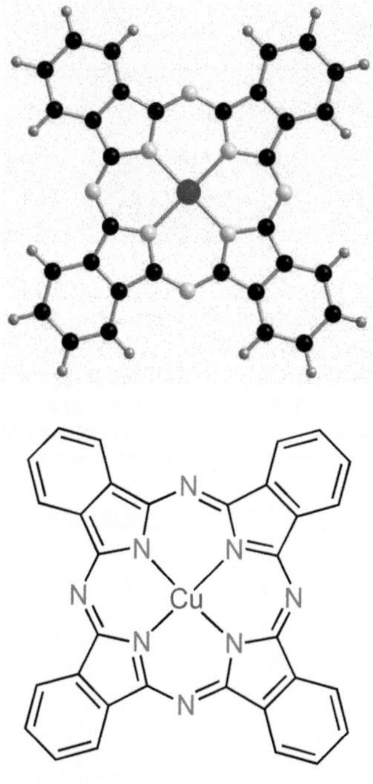
N
N
N
Cu
N
N
N
N
N

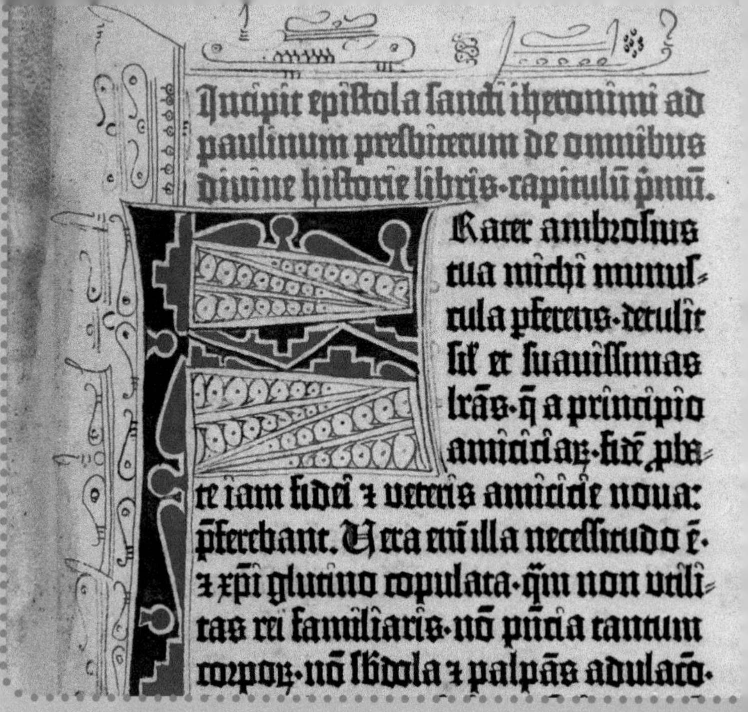

▲ A page from the Gutenberg bible.

In reflectance spectroscopy, visible light is shone on the artwork and the wavelengths of the reflected light are determined. Pigments absorb at different wavelengths, so each pigment has a characteristic reflectance spectrum. Another useful technique is Raman spectroscopy. This probes the vibrational structure of materials and, in the process, low energy laser light is shone on a sample of the artwork. This light is absorbed, and some is re-emitted at different wavelengths. As with reflectance spectroscopy, the nature of the pigments is determined from the emission spectrum. Raman spectroscopy was used to show that the blue colours in the fifteenth century Gutenberg bibles were due to the copper pigment azurite, not ultramarine as previously thought.

In the 1920s it was discovered by chance that the reaction between phthalic anhydride (the anhydride of benzene-1,2-dicarboxylic acid) and ammonia in the presence of iron gave a blue colour. This was caused by an iron phthalocyanine complex, and subsequent work showed the copper analogue was more versatile. For example, substitution of some of the hydrogen atoms by chlorine atoms changes the colour of the copper complex from blue to green. Copper phthalocyanines are now the most widespread blue and green pigments in use. Most blue and green car paints consist of a suspension of a copper phthalocyanine in a lacquer.

Insoluble pigments are converted into soluble dyes by attaching water-solubilizing groups such as sulfonate ($-SO_3^-$). Water-soluble copper phthalocyanine dyes are used to produce the blue and cyan colours in inkjet printers.

The origin of colour in *d*-block metal compounds is described in Section 28.6.

The *d*-block elements V, Cr, Mn, Fe, Co, Ni, Cu, Zn, and Mo are all essential for human life.

The *d*-block elements are all metals, and they are among the most useful of all the elements. Iron and copper have been known since prehistoric times, and their extraction and use have shaped human history. Nowadays, many of the *d*-block metals are used in construction materials. Nine of the elements are essential for life, and compounds of the *d*-block elements are used as catalysts, pigments, and medicines.

The *d*-block consists of the elements in Groups 3–12 of the Periodic Table. These elements are also known as the **transition metals**, as the elements provide a transition between the electropositive metals of the *s* block and the more electronegative metals of the *p* block. Generally, the *d*-block elements become more electronegative from left to right across a row. The terms '*d*-block metal' and 'transition metal' are not totally interchangeable. The IUPAC definition of a transition element is one whose atom has partially filled *d* orbitals, or that can give rise to cations with partially filled *d* orbitals. This definition excludes the elements in Group 12—zinc, cadmium, and mercury—which only form ions with a d^{10} configuration. These elements are therefore considered to be *d*-block metals but not transition metals.

The *d* block is shown in Figure 28.1, and the names of the elements together with their symbols are given in Table 28.1. The elements from scandium to zinc are in the fourth period of the Periodic Table, and these are generally referred to as the first row *d*-block elements. Similarly, the *d*-block elements of the fifth, sixth, and seventh periods of the Periodic Table are known as the second, third, and fourth row *d*-block elements, respectively.

The vertical groups in the *d* block are sometimes referred to as **triads**, despite the fact that there are four elements in all cases. The heaviest member at the bottom of each group is a synthetic element with a very short half life and has only been prepared on a miniscule scale. The chemistry of these elements is largely unknown. So, for example, the Group 8 triad consists of iron, ruthenium, and osmium, and the fourth member of the group, hassium, is normally ignored.

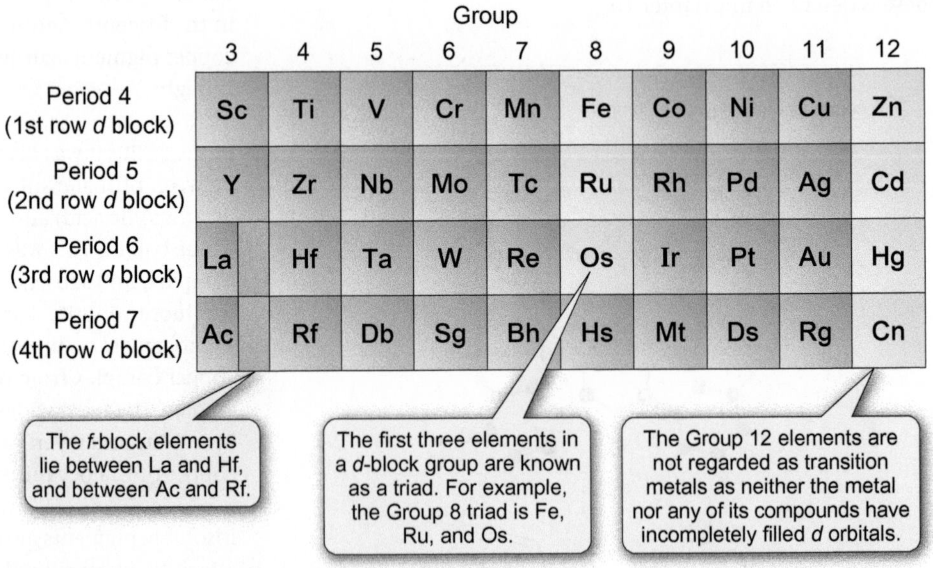
Group
3 4 5 6 7 8 9 10 11 12
Period 4 (1st row d block)
Sc Ti V Cr Mn Fe Co Ni Cu Zn
Period 5 (2nd row d block)
Y Zr Nb Mo Tc Ru Rh Pd Ag Cd
Period 6 (3rd row d block)
La Hf Ta W Re Os Ir Pt Au Hg
Period 7 (4th row d block)
Ac Rf Db Sg Bh Hs Mt Ds Rg Cn
The f-block elements lie between La and Hf, and between Ac and Rf.
The first three elements in a d-block group are known as a triad. For example, the Group 8 triad is Fe, Ru, and Os.
The Group 12 elements are not regarded as transition metals as neither the metal nor any of its compounds have incompletely filled d orbitals.

Figure 28.1 The *d*-block elements are those in Groups 3–12 of the Periodic Table.

Table 28.1 The *d*-block elements

1st row *d*-block elements	2nd row *d*-block elements	3rd row *d*-block elements	4th row *d*-block elements
Scandium, Sc	Yttrium, Y	Lanthanum, La	Actinium, Ac
Titanium, Ti	Zirconium, Zr	Hafnium, Hf	Rutherfordium, Rf
Vanadium, V	Niobium, Nb	Tantalum, Ta	Dubnium, Db
Chromium, Cr	Molybdenum, Mo	Tungsten, W	Seaborgium, Sg
Manganese, Mn	Technetium, Tc	Rhenium, Re	Bohrium, Bh
Iron, Fe	Ruthenium, Ru	Osmium, Os	Hassium, Hs
Cobalt, Co	Rhodium, Rh	Iridium, Ir	Meitnerium, Mt
Nickel, Ni	Palladium, Pd	Platinum, Pt	Darmstadtium, Ds
Copper, Cu	Silver, Ag	Gold, Au	Roentgenium, Rg
Zinc, Zn	Cadmium, Cd	Mercury, Hg	Copernicium, Cn

Box 28.1 Technetium and imaging the brain

Of the thirty *d*-block elements in Periods 4–6 of the Periodic Table, technetium is unique as it is the only one that does not occur naturally. Technetium was first prepared in 1937 by bombarding molybdenum with deuterium nuclei, although there is good evidence that minute amounts of technetium were detected in uranium ores over 10 years earlier. All of the isotopes of technetium are radioactive, and the longest living one, ^{98}Tc, has a half life of 4.2 million years. The most useful isotope is ^{99m}Tc, which is used for medical imaging. The 'm' stands for metastable, and indicates that the ^{99m}Tc nucleus exists in an excited state. This has a half life of 6 hours, and decays by emitting a γ-ray photon to form ^{99}Tc, in which the nucleus is in the normal ground state. ^{99m}Tc is formed by bombarding ^{98}Mo with neutrons to form ^{99}Mo, which then decays with a half life of 67 hours by β-emission to form ^{99m}Tc

$$^{98}_{42}Mo + ^{1}_{0}n \rightarrow ^{99}_{42}Mo \rightarrow ^{99m}_{43}Tc + ^{0}_{-1}e$$

This nuclear reaction takes place in a vessel called a 'technetium cow', which is 'milked' every 6 hours to remove soluble pertechnetate $[^{99m}TcO_4]^-$, leaving behind $[MoO_4]^{2-}$ which is adsorbed onto alumina.

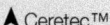

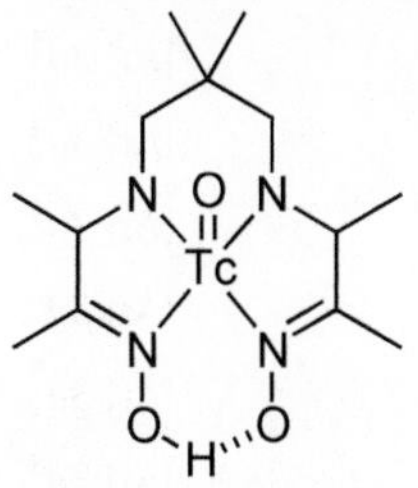

▲ Ceretec™

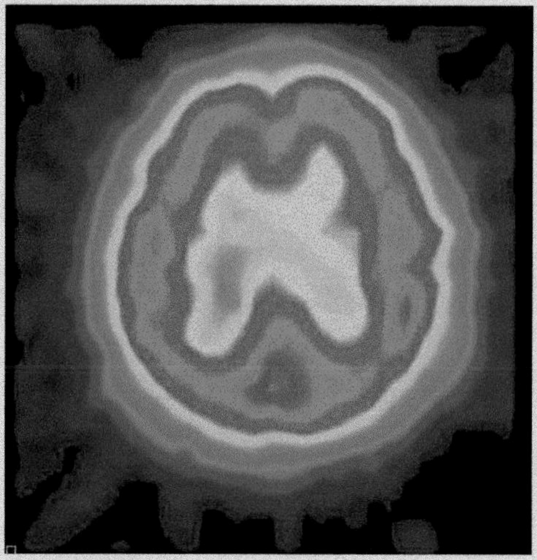

▲ A coloured SPECT (single photon emission computed tomography) scan of a healthy human brain. The red areas show the most active regions of the brain, which have taken up more ^{99m}Tc. This is detected by its emission of γ-ray photons.

The technetium compounds used for imaging are designed to target specific organs so the γ-rays they emit form a picture of that organ. Ceretec™ is a technetium imaging agent that is used to visualize blood flow in the brains of stroke patients. It also interacts with white blood cells, and this allows it to be used to locate abdominal infections.

Question

Some technetium compounds take several steps to prepare from the pertechnetate obtained from the 'technetium cow'. Suggest why such compounds are generally unsuitable for use as imaging agents.

Some books use the terms 'lanthanoids' and 'actinoids' for the two *f*-block series, as the -ide ending is usually associated in chemistry with anions. In practice, the terms lanthanide and actinide are more commonly used.

Although technically *d*-block elements, lanthanum and actinium in Group 3 are usually regarded as the first members of the two series of *f*-block elements known as the **lanthanides** and **actinides**.

The most abundant of the *d*-block elements in the Earth's crust are the first row *d*-block elements. The main focus of this chapter is on the chemistry of these elements and their compounds. Sections 28.1 and 28.2 deal with some general issues, whereas the remainder of the chapter concentrates on coordination chemistry. The aqueous chemistry of *d*-block metal compounds is described in Section 28.8, after the theories underlying coordination chemistry have been introduced.

28.1 The *d*-block elements

Electronic configuration

The electronic configurations of the first row *d*-block elements are given in Table 28.2, along with those for their M^{2+} and M^{3+} ions. Because the 3*d* orbitals are all of the same energy, they are each filled with a single electron first. Only after each of the five 3*d* orbitals is singly filled ($3d^5$), do the electrons start pairing up, from $3d^6$ onwards.

For a free atom of a first row *d*-block element, the 4*s* orbital is normally occupied, despite being higher in energy than the 3*d* orbitals. The 4*s* orbital is more diffuse than the 3*d* orbitals, so putting electrons in this orbital reduces electron–electron repulsions. This means that most first row *d*-block elements have the electronic configuration $3d^n\,4s^2$. Chromium ($3d^5\,4s^1$) and copper ($3d^{10}\,4s^1$) are exceptions to this. The reasons for this are detailed in Section 3.6 (p.148).

Since the 4*s* orbital is higher in energy than the 3*d* orbitals, when electrons are lost to form ions, it is the 4*s* electrons that are lost first. This is reflected in the electronic configurations of the *d*-block elements in ions and compounds, as these *never* contain 4*s* electrons unless the *d* orbitals are full. For example, the electronic configuration of V^{2+} is $3d^3$, *not* $3d^1\,4s^2$. Furthermore, when metal centres are involved in coordination complexes, the electron–electron repulsions are reduced. This also leads to electronic configurations in which the 4*s* orbitals are unoccupied. This means that the electronic configuration of Cr(0) in a compound is $3d^6$, not $3d^5\,4s^1$ as it is in atomic chromium.

In a compound of a first row *d*-block element, the 4*s* orbital is higher in energy than the 3*d* orbitals. For oxidation states of +2 and higher, the electronic configuration can be found by removing the 4*s* electrons, plus the appropriate number of 3*d* electrons, but for lower oxidation states the 4*s* electrons must be transferred into 3*d* orbitals before removing electrons. A simple way to work out the number of *d* electrons for a *d*-block metal in a compound is to subtract the metal oxidation state from its group number.

Visit the online resource centre to view screencast 28.1 which walks you through how to determine the electronic configuration of a metal ion.

Table 28.2 Electronic configurations of the first row *d*-block elements and their ions

Element (M)	Electronic configuration		
	M atom	M^{2+} ion	M^{3+} ion
Scandium, Sc	[Ar] $3d^1\,4s^2$	—	[Ar]
Titanium, Ti	[Ar] $3d^2\,4s^2$	[Ar] $3d^2$	[Ar] $3d^1$
Vanadium, V	[Ar] $3d^3\,4s^2$	[Ar] $3d^3$	[Ar] $3d^2$
Chromium, Cr	[Ar] $3d^5\,4s^1$	[Ar] $3d^4$	[Ar] $3d^3$
Manganese, Mn	[Ar] $3d^5\,4s^2$	[Ar] $3d^5$	[Ar] $3d^4$
Iron, Fe	[Ar] $3d^6\,4s^2$	[Ar] $3d^6$	[Ar] $3d^5$
Cobalt, Co	[Ar] $3d^7\,4s^2$	[Ar] $3d^7$	[Ar] $3d^6$
Nickel, Ni	[Ar] $3d^8\,4s^2$	[Ar] $3d^8$	[Ar] $3d^7$
Copper, Cu	[Ar] $3d^{10}\,4s^1$	[Ar] $3d^9$	[Ar] $3d^8$
Zinc, Zn	[Ar] $3d^{10}\,4s^2$	[Ar] $3d^{10}$	—

Worked example 28.1 Electronic configurations

What is the electronic configuration of Fe^{3+}?

Strategy

Write down the electronic configuration for an iron atom, then place all of the 4*s* electrons into 3*d* orbitals if they are available as the 3*d* orbitals are at lower energy in the ion or a compound.

Look at the charge on the ion, and remove the appropriate number of electrons.

(*Note*. This approach may be slightly longer than one based on removing the 4*s* electrons and the appropriate number of 3*d* electrons, but it gives the correct answer for all oxidation states.)

Solution

Fe has the electronic configuration [Ar] $3d^6\ 4s^2$.

In a *compound*, the electronic configuration for Fe(0) is [Ar] $3d^8$.

In Fe^{3+} there are three fewer electrons, so the electronic configuration for Fe^{3+} is [Ar] $3d^5$.

Question

Work out the electronic configurations of Co^{2+} and Co^+.

Trends in atomic properties across the first row of the *d* block

Atomic and ionic radii

Effective nuclear charge is discussed in Section 3.6 (p.150). Trends in atomic radii are described in Section 3.7 (p.153).

From lithium to fluorine, across the second period of the Periodic Table, the atomic radii of the elements decrease due to the increase in the effective nuclear charge, Z_{eff}. On moving one element to the right, the additional electron is not able to efficiently shield the increased nuclear charge, so the nuclear charge felt by the outermost electron is higher. For the *d*-block elements, the trend across a row is more complicated, and the difference between the left- and right-hand sides of the row is relatively small. The atomic radii of the first row *d*-block elements are given in Figure 28.2.

Moving from left to right, Z_{eff} increases, so you might expect the atomic radius to decrease across the series, as it does from lithium to fluorine. There is a decrease for the first four elements (Sc–Cr), but after this the trend changes. The value for manganese appears anomalously high, whereas the atomic radii for iron, cobalt, and nickel are very similar, and those for copper and zinc are slightly larger.

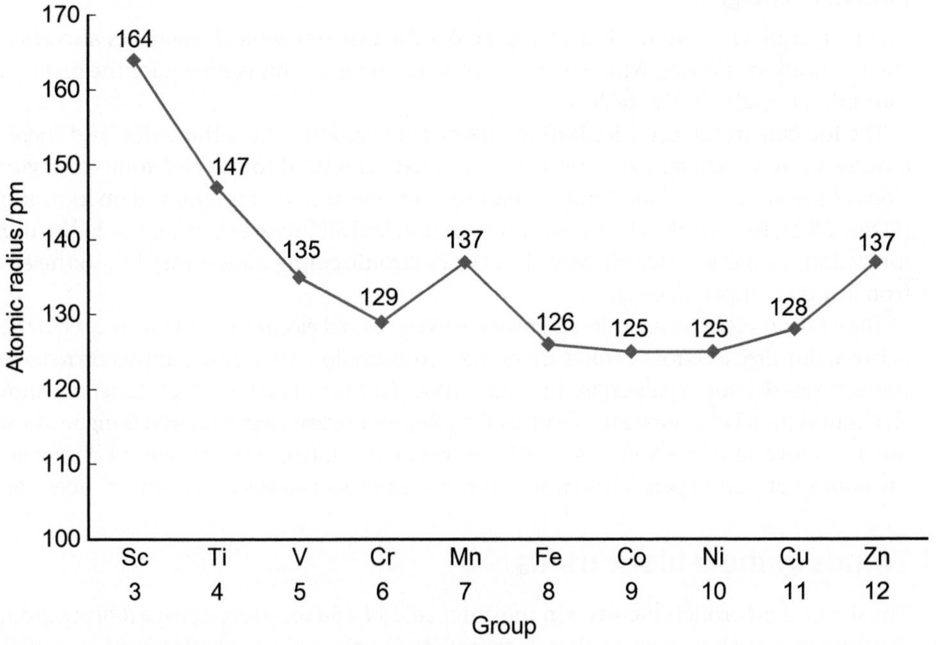
Atomic radius/pm
170
160
150
140
130
120
110
100
164
147
135
129
137
126
125
125
128
137
Sc
Ti
V
Cr
Mn
Fe
Co
Ni
Cu
Zn
3
4
5
6
7
8
9
10
11
12
Group

Figure 28.2 The atomic radii of the first row *d*-block elements.

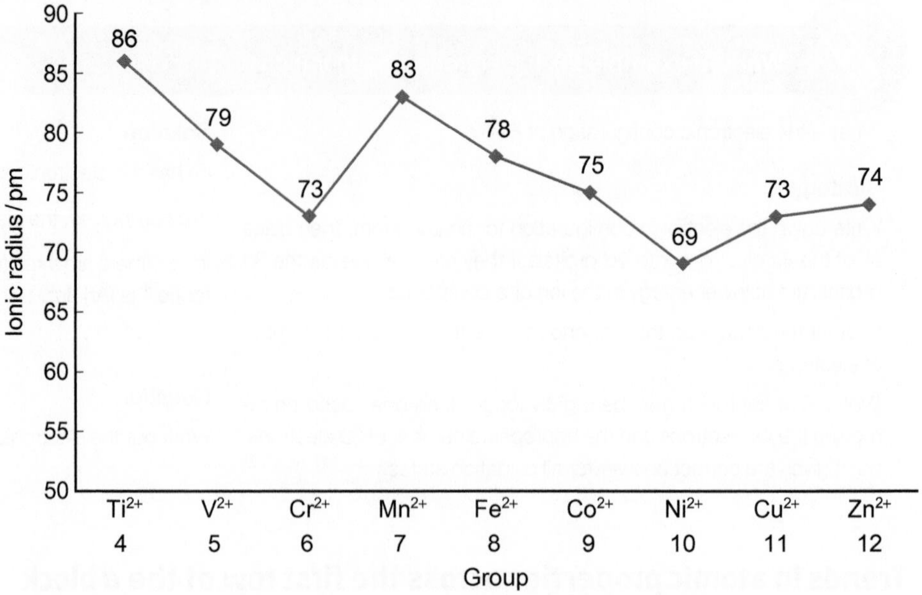
Ionic radius/pm
90
85
80
75
70
65
60
55
50
86
79
73
83
78
75
69
73
74
Ti2+ V2+ Cr2+ Mn2+ Fe2+ Co2+ Ni2+ Cu2+ Zn2+
4 5 6 7 8 9 10 11 12
Group

Values of ionic radii vary widely between different books due to the different ways in which they are measured. It is important to always use the same source when you are comparing values.

Figure 28.3 The ionic radii of the first row *d*-block M^{2+} ions with a coordination number of six. The values for Cr^{2+}, Mn^{2+}, Fe^{2+}, and Co^{2+} are for high spin ions (see Section 28.5, p.1284).

The deviation from the expected trend is caused by electron–electron repulsions, which become more important as the *d* orbitals are filled because of the increasing number of *d* electrons. For the elements on the right-hand side of the series, the increase in electron–electron repulsion on moving one element to the right is more significant than the increase in Z_{eff}, so the atomic radius increases.

Since there is little difference in the atomic radii of most of the first row transition metals, it is relatively easy for one to substitute for another in a lattice. This explains the widespread occurrence of alloys for these metals.

Alloys are discussed in Section 6.3 (p.275).

Ionic radii for the M^{2+} ions are given in Figure 28.3. These radii follow a similar trend to the atomic radii.

Ionization energies

The increasing effective nuclear charge across the first row *d*-block elements also affects their ionization energies, which increase from left to right. This is shown for the first three ionization energies in Figure 28.4.

The increase in the third ionization energy isn't smooth across the series, and there is a decrease from manganese to iron. This anomaly is related to the electronic configurations of the M^{2+} and M^{3+} ions and the nature of the electron that is removed on ionization (Figure 28.5). Fe^{2+} has the electronic configuration [Ar] $3d^6$, so the electron lost in the third ionization is a paired electron. Mn^{2+} has the electronic configuration [Ar] $3d^5$, so the electron lost is an unpaired electron.

The electron–electron repulsion is greater between paired electrons in the same orbital than between unpaired electrons in different orbitals, so ionization has a greater impact on reducing the electron–electron repulsion for Fe^{2+} than it does for Mn^{2+}. This means it is easier to remove an electron from Fe^{2+}, despite the fact that the effective nuclear charge on iron is higher. In addition, removal of an electron from Mn^{2+} involves a reduction in exchange energy as it reduces the number of pairs of parallel electrons, whereas removal of an electron from Fe^{2+} does not.

Exchange energy is described in Box 3.7 (p.147).

Trends in the *d*-block triads

The size of the *d* orbitals increases in the order $3d < 4d < 5d$ so, going down a *d*-block group, the valence orbitals become further away from the nucleus. On the basis of this, you might expect the atomic radii to increase smoothly down a group. As the data in Figure 28.6 show, the expected increase occurs from the first row *d*-block element to the second row

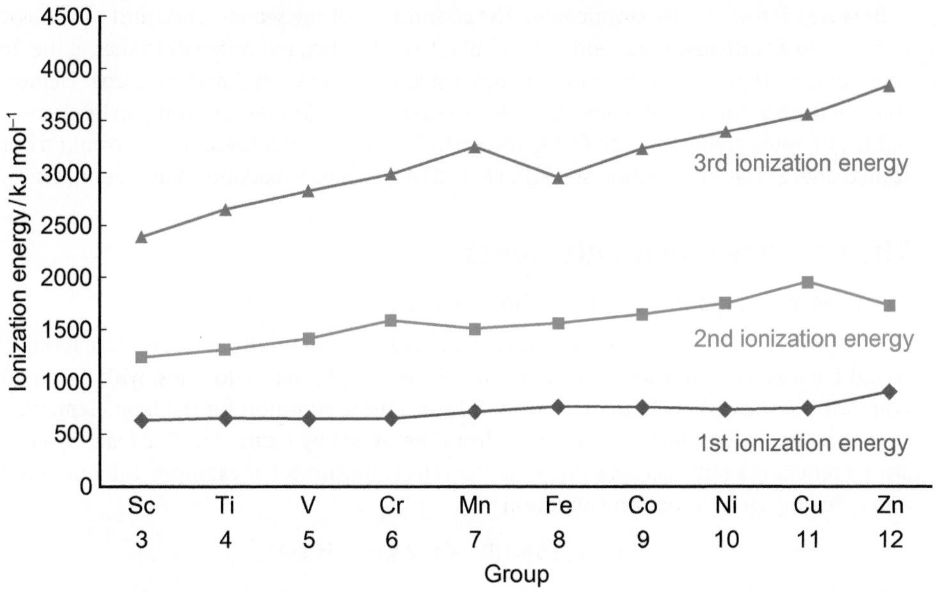
4500
4000
3500
3000
2500
2000
1500
1000
500
0
Ionization energy/kJ mol⁻¹
3rd ionization energy
2nd ionization energy
1st ionization energy
Sc Ti V Cr Mn Fe Co Ni Cu Zn
3 4 5 6 7 8 9 10 11 12
Group

Figure 28.4 The first, second, and third ionization energies for the first row *d*-block elements.

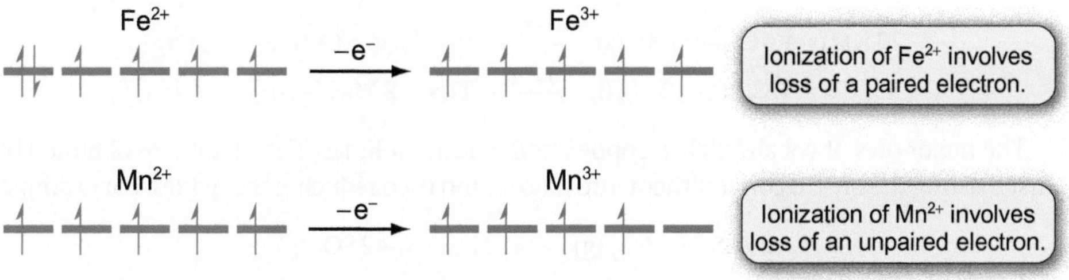
Fe2+
Fe3+
−e−
Ionization of Fe2+ involves loss of a paired electron.
Mn2+
Mn3+
−e−
Ionization of Mn2+ involves loss of an unpaired electron.

Figure 28.5 Removing an electron from Fe^{2+} and Mn^{2+}. For Fe^{2+} the electron–electron repulsion is reduced by a greater extent as a paired electron is removed. For Mn^{2+}, removal of an electron involves a loss of exchange energy.

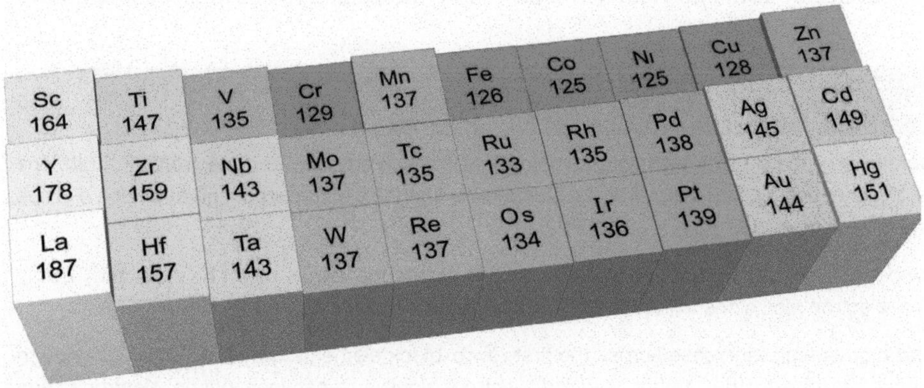
Sc 164
Ti 147
V 135
Cr 129
Mn 137
Fe 126
Co 125
Ni 125
Cu 128
Zn 137
Y 178
Zr 159
Nb 143
Mo 137
Tc 135
Ru 133
Rh 135
Pd 138
Ag 145
Cd 149
La 187
Hf 157
Ta 143
W 137
Re 137
Os 134
Ir 136
Pt 139
Au 144
Hg 151

Figure 28.6 Atomic radii of the first row, second row, and third row *d*-block elements.

element, but the atoms of the second and third row *d*-block elements are very similar in size. To understand why this is so, you need to take a broader look at the Periodic Table.

The Periodic Table is shown on the inside front cover.

The lanthanide elements lie before the start of the third row *d*-block elements, and this means that the 4*f* orbitals are filled before the 5*d* orbitals. The 4*f* orbitals are very poorly shielding, so there is a decrease in atomic radius from lanthanum (187 pm) at the start of the *f* block to lutetium (173 pm) at the end. This decrease is called the **lanthanide contraction**. As a result of the lanthanide contraction, there is a large decrease in atomic radius from the Group 2 element barium (224 pm) to the Group 4 metal hafnium (157 pm). It is the presence of the lanthanides that makes the atomic radii of the third row *d*-block metals smaller than expected.

Because of their similar atomic radii, the chemistries of the second and third row *d*-block elements in a particular triad tend to be similar to each other, but different to that of the first row element. High oxidation states are more stable for the second and third row elements than for the first row element. For example, in Group 8, iron forms oxides only in its +2 and +3 oxidation states, giving FeO and Fe_2O_3, respectively. In contrast, ruthenium and osmium both form oxides in the +4 oxidation state (RuO_2, OsO_2) and the +8 oxidation state (RuO_4, OsO_4).

OsO_4 is used in organic synthesis to form 1,2-diols from alkenes. See Section 21.4 (p.989).

The first row *d*-block elements

Occurrence and industrial preparation from ores

The Guggenheim museum in Bilbao, Spain, was one of the first buildings to be clad in titanium metal. This is guaranteed to resist corrosion for 100 years.

The first row *d*-block elements are all too reactive to occur naturally as free metals. From left to right across the row, there is a change in the nature of their major ores, with oxide ores common for the earlier elements, and sulfide ores more common for the later elements.

In most cases, the metals are obtained from the oxides by reduction. The reactions need strong reducing agents because the metals are electropositive. For example, calcium is used as a reducing agent to obtain vanadium

$$V_2O_5\,(s) + 5\,Ca\,(l) \xrightarrow{\Delta} 2\,V\,(s) + 5\,CaO\,(s)$$

Titanium is obtained from TiO_2 in a two-stage process. In the first step, the oxide is converted into the tetrachloride ($TiCl_4$). Gaseous $TiCl_4$ is then reduced by passing it through liquid magnesium under an argon atmosphere

$$2\,TiO_2\,(s) + 4\,Cl_2\,(g) + 3\,C\,(s) \xrightarrow{950\,°C} 2\,TiCl_4\,(g) + CO_2\,(g) + 2\,CO\,(g)$$
$$TiCl_4\,(g) + 2\,Mg\,(l) \xrightarrow{950\,°C} Ti\,(s) + 2\,MgCl_2\,(s)$$

See Section 28.3 (p.1277) for an explanation of the occurrence of oxide and sulfide ores using the concept of hard and soft acids and bases.

The major ores of cobalt, nickel, copper, and zinc are sulfides. The metals are obtained by first roasting the ores to convert them into oxides, and then reducing the oxides. For example,

$$2\,ZnS\,(s) + 3\,O_2\,(g) \xrightarrow{\Delta} 2\,ZnO\,(s) + 2\,SO_2\,(g)$$
$$ZnO\,(s) + C\,(s) \xrightarrow{\Delta} Zn\,(g) + CO\,(g)$$

Metal–carbon monoxide compounds are called **metal carbonyls**.

In the case of nickel, the crude metal obtained after reduction is purified in the *Mond process*. This involves converting the metal into the nickel(0) compound $[Ni(CO)_4]$, which is purified by distillation and decomposed into nickel metal by heating it above 200 °C.

Summary

- Although the electronic configurations of the *d*-block elements contain *s* electrons in addition to *d* electrons, in both ions and compounds the *s* orbital is of higher energy than the *d* orbitals. This means that the electronic configuration of a *d*-block element in a compound is *always* d^n and *never* $d^{n-2}s^2$.
- The atomic radii of the first row *d*-block elements do not decrease smoothly across the row as might be expected. This is because the increase in electron–electron repulsions across the series counteracts the effect of the increase in Z_{eff}.
- The first, second, and third ionization energies increase across the first row *d*-block elements with increasing Z_{eff}, though there are some anomalies. These are due to the greater electron–electron repulsions between paired electrons than between unpaired electrons and changes in exchange energy.
- In a particular triad, the atoms of the second and third row *d*-block elements are larger than the atoms of the first row element, though of similar size to each other. The similarity in sizes of the second and third row atoms is a consequence of the lanthanide contraction.
- Metals on the left of the *d* block exist naturally as oxides and are obtained from the ores by reduction. Metals on the right of the *d* block occur naturally as sulfides, and are obtained by conversion to the oxide followed by reduction.

For a practice question on these topics, see question 1 at the end of this chapter (p.1301).

28.2 Chemistry of the first row *d*-block elements

In this section, some of the key trends in *d*-block chemistry are described, along with the chemistry of the oxides and halides. Accounting for the chemistry of the metal ions in aqueous solution needs an understanding of coordination chemistry, so aqueous chemistry is described later in Section 28.8.

Trends in oxidation states

Most *d*-block elements have more than one possible oxidation state. For the first row, the two exceptions are scandium, which occurs only as scandium(III) in its compounds, and zinc, which except in special circumstances occurs only as zinc(II). The oxidation states adopted by the first row *d*-block elements are shown in Figure 28.7. The widest range of oxidation states occurs for the elements close to the centre of the *d* block. Manganese, for example, forms compounds in all 11 possible oxidation states, +7 (d^0) to −3 (d^{10}).

For the elements scandium to manganese, the maximum oxidation state occurs with the loss of all the 3*d* and 4*s* electrons. Scandium(III), titanium(IV), vanadium(V), chromium(VI), and manganese(VII) all have the electronic configuration [Ar] $3d^0$. This maximum oxidation state becomes more oxidizing from titanium(IV) to manganese(VII), and $[MnO_4]^-$ and $[Cr_2O_7]^{2-}$ are both powerful oxidizing agents. For the elements iron, cobalt, nickel, and copper, the maximum oxidation state contains 3*d* electrons and is very strongly oxidizing so it is relatively uncommon.

Redox processes involving the transition metals are described in Section 28.8 (p.1297).

At the start of the series (Sc → Cr), the +3 oxidation state is more important than the +2 oxidation state, which is strongly reducing. Moving from left to right across the row, the +2 oxidation state becomes more stable towards oxidation, and the +3 oxidation state becomes more readily reduced. For manganese, iron, and cobalt in the middle of the series, both oxidation states are important, but for nickel and copper the +3 oxidation state is much less common than the +2 oxidation state.

Sc	Ti	V	Cr	Mn	Fe	Co	Ni	Cu	Zn
				+7 (d^0)					
			+6 (d^0)	+6 (d^1)	+6 (d^2)				
		+5 (d^0)	+5 (d^1)	+5 (d^2)	+5 (d^3)	+5 (d^4)			
	+4 (d^0)	+4 (d^1)	+4 (d^2)	+4 (d^3)	+4 (d^4)	+4 (d^5)	+4 (d^6)		
+3 (d^0)	+3 (d^1)	+3 (d^2)	+3 (d^3)	+3 (d^4)	+3 (d^5)	+3 (d^6)	+3 (d^7)	+3 (d^8)	
	+2 (d^2)	+2 (d^3)	+2 (d^4)	+2 (d^5)	+2 (d^6)	+2 (d^7)	+2 (d^8)	+2 (d^9)	+2 (d^{10})
		+1 (d^4)	+1 (d^5)	+1 (d^6)	+1 (d^7)	+1 (d^8)	+1 (d^9)	+1 (d^{10})	
	0 (d^4)	0 (d^5)	0 (d^6)	0 (d^7)	0 (d^8)	0 (d^9)	0 (d^{10})		
		−1 (d^6)	−1 (d^7)	−1 (d^8)		−1 (d^{10})			
			−2 (d^8)	−2 (d^9)	−2 (d^{10})				
				−3 (d^{10})					
			−4 (d^{10})						

The high oxidation states shown in grey are rare, and their compounds are very strong oxidizing agents.

The most common oxidation states are highlighted in green.

Low oxidation states shown in red are not stable in aqueous solution, but are stabilized by ligands such as CO.

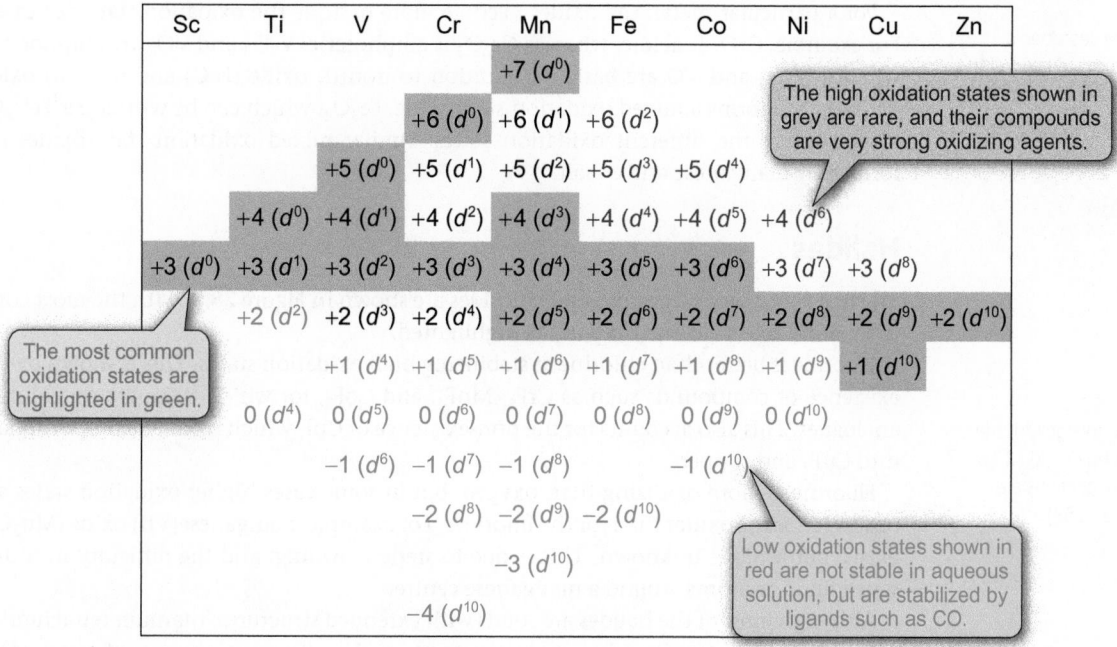

Figure 28.7 Oxidation states adopted by the first row *d*-block elements and their valence electronic configurations.

Oxidation state	Sc	Ti	V	Cr	Mn	Fe	Co	Ni	Cu	Zn
> +4			V_2O_5	CrO_3	Mn_2O_7					
+4		TiO_2	VO_2	CrO_2	MnO_2					
+3	Sc_2O_3	Ti_2O_3	V_2O_3	Cr_2O_3	Mn_2O_3	Fe_2O_3				
+2		TiO†	VO†		MnO	FeO‡	CoO	NiO	CuO	ZnO
+1									Cu_2O	
Mixed					Mn_3O_4	Fe_3O_4	Co_3O_4			

† Non-stoichiometric, with composition ranges from $MO_{0.7}$ to $MO_{1.3}$.
‡ Non-stoichiometric ($Fe_{0.95}O$).

Figure 28.8 The oxides of the first row *d*-block metals. The most stable compounds for each metal are highlighted in blue.

These variations in oxidation state stability are due to the increase in third ionization energy across the series (Figure 28.4, p.1261). As the third ionization energy increases, it becomes energetically less favourable to form the +3 oxidation state. The occurrence of more than one stable oxidation state for many *d*-block elements is a major factor in their use as catalysts.

The action of catalysts is discussed in Section 9.9 (p.437).

Oxides

The first row *d*-block oxides are summarized in Figure 28.8, with the most common and stable of these compounds highlighted.

Vanadium(V) oxide is a catalyst for the conversion of SO_2 to SO_3, which is used in the commercial production of sulfuric acid (see Section 27.5 (p.1236)).

As expected from the general trend in oxidation states described above, the maximum oxidation state oxide is possible up to manganese. These oxides become increasingly more oxidizing from TiO_2, through V_2O_5 and CrO_3, to Mn_2O_7. Mn_2O_7 has a molecular structure with covalent bonds, and it decomposes explosively on heating above 95 °C.

Amphoteric oxides are described in Section 7.7 (p.334). The coating of glass with titanium dioxide in self-cleaning windows is described in Box 6.5 (p.283).

For a particular metal, the oxides become more basic as the oxidation state decreases. For example, CrO_3 is acidic, whereas Cr_2O_3 is amphoteric. V_2O_5 and VO_2 are amphoteric, whereas V_2O_3 and VO are basic. In addition to iron(II) oxide (FeO) and iron(III) oxide (Fe_2O_3), iron forms a mixed oxidation state oxide, Fe_3O_4, which can be written $Fe^{II}Fe^{III}{}_2O_4$ to emphasize the different oxidation states. Similar mixed oxidation state oxides are formed by manganese and cobalt.

Halides

The first row *d*-block fluorides and chlorides are shown in Figure 28.9, with the most common and stable of these compounds highlighted.

Fluorine is better than chlorine at stabilizing high oxidation states. This is shown by the existence of compounds such as CrF_5, MnF_4, and CoF_3, for which there are no chloride analogues. This also accounts for the non-existence of CuF, which would disproportionate into CuF_2 and copper.

The ability of fluorine to stabilize high oxidation states is described in Section 27.6 (p.1245). Disproportionation reactions are introduced in Section 25.2 (p.1152).

Fluorine is more oxidizing than oxygen, but in some cases higher oxidation states are observed for the oxides than for the fluorides. For example, manganese(VII) oxide (Mn_2O_7) exists, but MnF_7 is unknown. This is due to steric crowding, and the difficulty in fitting seven fluorine atoms around a manganese centre.

Although most of the halides are solids with extended structures, titanium tetrachloride ($TiCl_4$) is a colourless liquid at room temperature. The structure consists of tetrahedral, covalently bonded molecules. $TiCl_4$ is susceptible to hydrolysis, and is similar chemically to Group 14 halides such as $SiCl_4$.

The chemistry of $SiCl_4$ is described in Section 27.3 (p.1221).

Oxidation state	Sc	Ti	V	Cr	Mn	Fe	Co	Ni	Cu	Zn
+5			VF_5	CrF_5						
+4		TiF_4	VF_4	CrF_4	MnF_4			NiF_4		
		$TiCl_4$	VCl_4	$CrCl_4$						
+3	ScF_3	TiF_3	VF_3	CrF_3	MnF_3	FeF_3	CoF_3	NiF_3		
	$ScCl_3$	$TiCl_3$	VCl_3	$CrCl_3$	$MnCl_3$	$FeCl_3$				
+2			VF_2	CrF_2	MnF_2	FeF_2	CoF_2	NiF_2	CuF_2	ZnF_2
		$TiCl_2$	VCl_2	$CrCl_2$	$MnCl_2$	$FeCl_2$	$CoCl_2$	$NiCl_2$	$CuCl_2$	$ZnCl_2$
+1										
									CuCl	

Figure 28.9 The fluorides and chlorides of the first row *d*-block metals. The most stable compounds for each metal are highlighted in blue for the fluorides and in pink for the chlorides.

Summary

- With the exceptions of Groups 3 and 12, *d*-block elements have more than one possible oxidation state, and this is reflected by the different oxides and halides the elements form.
- For the first row *d*-block elements the +3 oxidation state is more stable than the +2 oxidation state on the left-hand side, but the +2 oxidation state is more stable than the +3 oxidation state on the right-hand side.

For a practice question on these topics, see question 2 at the end of this chapter (p.1301).

28.3 Coordination chemistry

d-Block metal ions are Lewis acids and they form compounds with Lewis bases. The Lewis base donates an electron pair to the metal, and Lewis bases that are bonded to metal ions in this way are called **ligands**. Ligands are generally neutral, such as H_2O or NH_3, or anionic, such as Cl^-.

The ligands **coordinate** to the metal ion to give a **complex** (sometimes called a coordination complex), which if it is ionic is called a **complex ion** (Figure 28.10). The atom in a ligand that supplies the electron pair is called the **donor atom**. The ligands are in the **coordination sphere** of the metal. Although *s*- and *p*-block metals also form complexes, coordination chemistry is particularly widespread for *d*-block metal ions.

When cobalt(II) nitrate dissolves in water, six water molecules coordinate to the cobalt ion to form the $[Co(H_2O)_6]^{2+}$ complex ion. The coordination of the water molecules is shown in the formula by including them within square brackets. Crystallization of the solution gives the solid $[Co(H_2O)_6](NO_3)_2$, which is a **coordination compound** (Figure 28.11). The existence of complex ions was first recognized by the Swiss chemist Alfred Werner, who was awarded the Nobel Prize in Chemistry in 1913.

Some examples of simple ligands and the complexes that they form are shown in Table 28.3. When a molecule or ion coordinates to a metal ion, its name often changes.

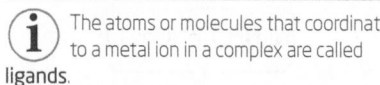

The atoms or molecules that coordinate to a metal ion in a complex are called ligands.

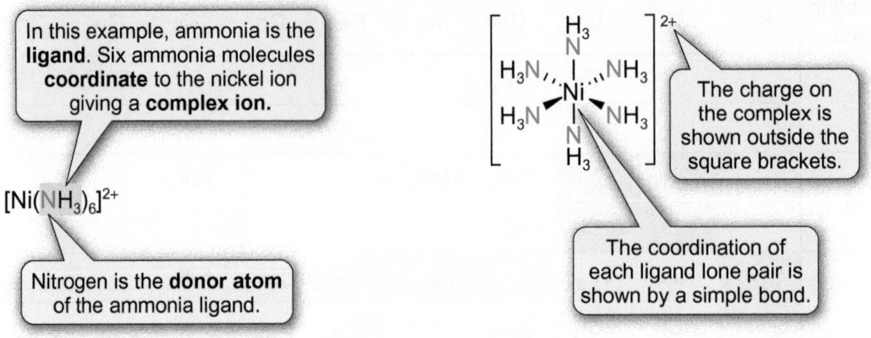

In this example, ammonia is the ligand. Six ammonia molecules coordinate to the nickel ion giving a complex ion.
[Ni(NH3)6]2+
Nitrogen is the donor atom of the ammonia ligand.
The charge on the complex is shown outside the square brackets.
The coordination of each ligand lone pair is shown by a simple bond.

Figure 28.10 An example of a complex ion, illustrating the key terms.

A coordinated water molecule is known as an *aqua* ligand, and the $[Co(H_2O)_6]^{2+}$ complex ion is called the hexaaquacobalt(II) ion. The rules for naming coordination compounds are examined on p.1270.

'Dentate' derives from the Greek for 'tooth', so polydentate ligands are 'many-toothed' ligands.

The ligands in Table 28.3 all contain one donor atom, and these ligands are known as **monodentate ligands**. Ligands with two donor atoms are known as **bidentate ligands**, and those with more than one donor atom are known generally as **polydentate ligands**. Examples of polydentate ligands are given in Table 28.4.

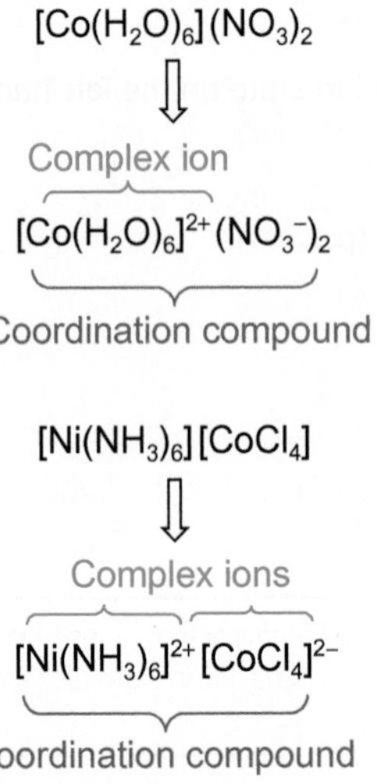

[Co(H2O)6](NO3)2
Complex ion
[Co(H2O)6]2+(NO3−)2
Coordination compound
[Ni(NH3)6][CoCl4]
Complex ions
[Ni(NH3)6]2+[CoCl4]2−
Coordination compound

Figure 28.11 A coordination compound contains one or more complex ions.

Me is an abbreviation for the methyl group, CH_3–. Similarly, Ph represents a phenyl group, C_6H_5–. See Section 2.5 (pp.81, 88).

The word chelate (pronounced 'key-late') is derived from the Greek word for 'claw'. The stability of chelates is described in Section 28.8 (p.1296).

Table 28.3 Examples of monodentate ligands and the complexes they form

Name of ion or molecule	Formula (donor atom in red)	Name of coordinated ligand*	Example of complex formed
Water	H_2O	aqua	$[Cr(H_2O)_6]^{3+}$
Ammonia	NH_3	ammine	$[Ni(NH_3)_6]^{2+}$
Fluoride	F^-	fluoro	$[TiF_6]^{2-}$
Chloride	Cl^-	chloro	$[CoCl_4]^{2-}$
Bromide	Br^-	bromo	$[NiBr_4]^{2-}$
Iodide	I^-	iodo	$[PtI_4]^{2-}$
Hydroxide	OH^-	hydroxo	$[Fe(OH)(H_2O)_5]^{2+}$
Methoxide	OMe^-	methoxo	$[Ti(OMe)_4]$
Cyanide	CN^-	cyano	$[Fe(CN)_6]^{4-}$
Thiocyanide	SCN^- or NCS^-	thiocyanato or isothiocyanato	$[Fe(SCN)(H_2O)_5]^{2+}$
Azide	N_3^-	azido	$[Co(N_3)(NH_3)_5]^{2+}$
Pyridine	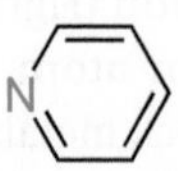	pyridine	$[NiCl_2(py)_2]$
Carbon monoxide	CO	carbonyl	$[Ni(CO)_4]$
Trimethylphosphine	PMe_3	trimethylphosphine	$[Pt(PMe_3)_4]$
Triphenylphosphine	PPh_3	triphenylphosphine	$[PdCl_2(PPh_3)_2]$
Dimethylsulfide	SMe_2	dimethylsulfide	$[AuCl(SMe_2)]$

* The current IUPAC recommendation is for anionic ligands to take the name of the free ligand with the ending changed from -e to -o. Using this rule, for example, chloride becomes chlorido on coordination, not chloro. In this book we have used the more commonly used familiar names.

Table 28.4 Examples of polydentate ligands and the complexes they form

Name of ion or molecule	Formula (donor atoms in red)	Name of coordinated ligand (old name from which abbreviation derives)	Abbreviation	Example of complex formed
Bidentate ligands				
Ethane-1,2-diamine	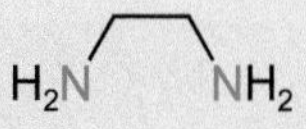	ethane-1,2-diamine (ethylenediamine)	en	$[Ni(en)_3]^{2+}$
Propane-1,3-diamine	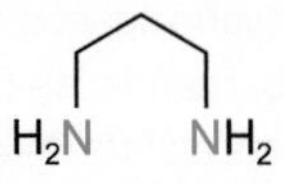	propane-1,3-diamine (1,3-propylenediamine)	pn	$[Co(pn)_3]^{3+}$
2,2′-Bipyridine	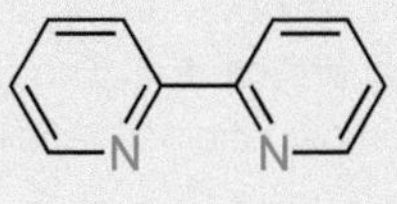	2,2′-bipyridine	bipy	$[Ru(bipy)_3]^{2+}$
1,10-Phenanthroline	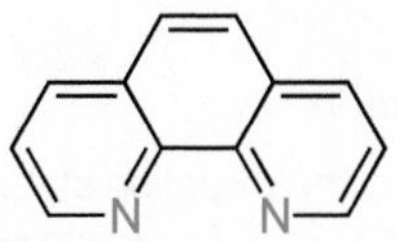	1,10-phenanthroline	phen	$[Fe(phen)_3]^{2+}$
Pentane-2,4-dionate ion (acetylacetonate ion)	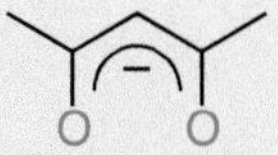	pentane-2,4-dionato (acetylacetonato)	$[acac]^-$	$[VO(acac)_2]$
Ethanedioate ion (oxalate ion)	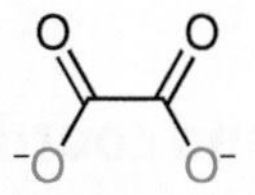	ethanedioato (oxalato)	$[ox]^{2-}$	$[Fe(ox)_3]^{3-}$
Tridentate ligands				
1,4,7-Triazaheptane (diethylenetriamine)	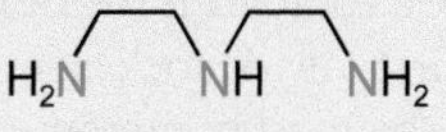	1,4,7-triazaheptane (diethylenetriamine)	dien	$[Ru(dien)_2]^{2+}$
2,2′:6′,2″-Terpyridine	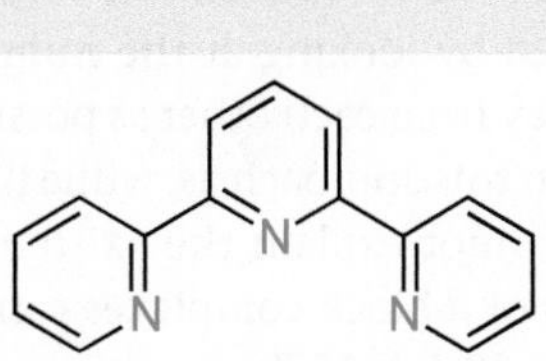	2,2′:6′,2″-terpyridine	terpy	$[Os(terpy)_2]^{2+}$
Hexadentate ligands				
Diamino-1,2-ethane-*N,N,N′,N′*-tetraethanoate ion (ethylenediaminetetraacetate ion)	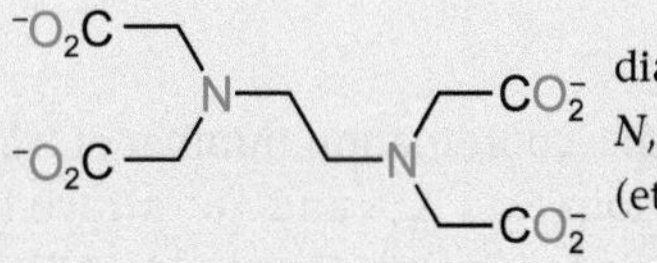	diamino-1,2-ethane-*N,N,N′,N′*-tetraethanoato (ethylenediaminetetraacetato)	$[edta]^{4-}$	$[Fe(edta)]^-$

The **denticity** of a ligand in a complex is the number of donor atoms that are coordinated to the metal ion. The following terms use this principle to describe ligands:

- 1 donor atom—monodentate;
- 2 donor atoms—bidentate;
- 3 donor atoms—tridentate;
- 4 donor atoms—tetradentate;
- 5 donor atoms—pentadentate;
- 6 donor atoms—hexadentate.

A complex containing a polydentate ligand is called a **chelate**, and polydentate ligands are known collectively as **chelating ligands**. When a bidentate ligand coordinates to a metal ion it forms a chelate ring (Figure 28.12). The most common bidentate ligands form 5- or 6-membered chelate rings.

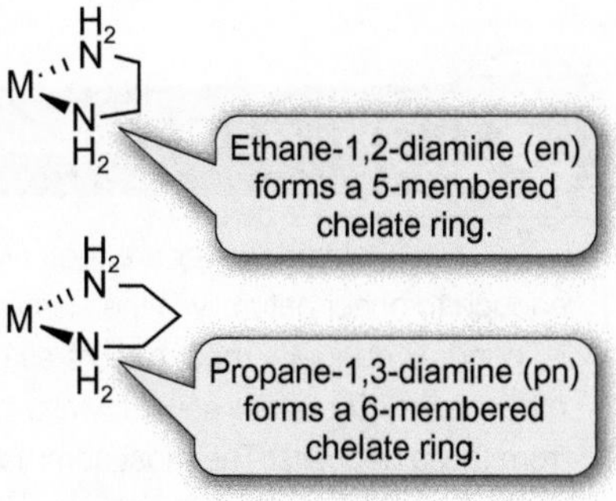
Ethane-1,2-diamine (en) forms a 5-membered chelate ring.
Propane-1,3-diamine (pn) forms a 6-membered chelate ring.

Figure 28.12 A metal ion (M) forms 5- and 6-membered chelate rings with the bidentate *N*-donor ligands en and pn (Table 28.4).

It is important to be able to work out the oxidation state of a metal in a complex from the formula. To do this, you need to know the charge on each of the ligands. The process is shown in Worked example 28.2.

Worked example 28.2 Oxidation states in coordination complexes

What is the oxidation state of the metal in the following complex ions:

(a) $[Ni(NH_3)_6]^{2+}$;

(b) $[Fe(CN)_6]^{4-}$;

(c) $[Fe(SCN)(H_2O)_5]^{2+}$?

Strategy

Look at the charge on each of the ligands. The sum of the ligand charges and the metal oxidation state is equal to the charge on the complex ion.

Solution

(a) The ammine ligands are neutral, so the total ligand charge is zero. In this case, the metal oxidation state is equal to the charge on the complex ion, so the nickel oxidation state is +2.

(b) The cyano ligands each carry a −1 charge, so the total charge on the six ligands is −6. The charge on the complex is −4, so the iron oxidation state is −4 − (−6) = +2. Iron is in the +2 oxidation state.

(c) The thiocyanato ligand carries a −1 charge, and the aqua ligands are neutral. The total charge on the six ligands is −1, so the iron oxidation state is + 2 − (−1) = +3. Iron is in the +3 oxidation state.

Question

What is the oxidation state of vanadium in the complex $[VO(acac)_2]$? (See Table 28.4 (p.1267) for the structure of the acac ligand.)

Coordination numbers and coordination geometry

The **coordination number** of a complex is the number of donor atoms that are bonded to the central metal ion. For a first row *d*-block element, the coordination number can vary from 2 to 8, but 4 and 6 are the most common values. The shapes of *d*-block complexes can sometimes be predicted by looking at the number of ligands involved and placing these in positions as far away from each other as possible in order to minimize steric repulsions. Care must be taken in this approach as, while it works for tetrahedral and octahedral complexes, this method cannot explain the existence of square planar complexes such as $[Ni(CN)_4]^{2-}$. The shapes of *d*-block complexes can be predicted more accurately using crystal field theory (Section 28.4, p.1279).

Coordination number 2

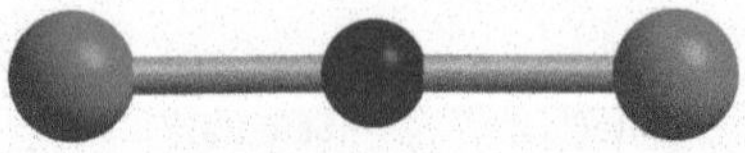

$[Cl{-}Cu{-}Cl]^-$

1 $[CuCl_2]^-$ (linear complex).

d-Block complexes with a coordination number of 2 are generally linear. This geometry is most common for the ions Cu^+, Ag^+, and Au^+, which have low charge densities. Examples of linear complexes include $[CuCl_2]^-$ (**1**) and $[Au(NH_3)_2]^+$.

Box 28.2 The extraction of gold

Gold and silver are the only *d*-block metals that are inert enough to occur naturally in the elemental form. Although some gold is found as nuggets, most occurs as grains or flakes of the metal in other rocks. There are several ways in which the gold is recovered from these deposits. The most common route, which was developed at the end of the nineteenth century, is known as the *cyanide process*. Cyanide ions react with gold in the presence of dissolved oxygen to form the water-soluble gold complex ion, $[Au(CN)_2]^-$

$$4\,Au(s) + 8\,CN^-(aq) + O_2(aq) + 2\,H_2O(l) \rightarrow 4\,[Au(CN)_2]^-(aq) + 4\,OH^-(aq)$$

This solution is separated from the undissolved rock. In the original process, the gold was then precipitated from the solution by reaction with zinc

$$2\,[Au(CN)_4]^-(aq) + Zn(s) \rightarrow [Zn(CN)_4]^{2-}(aq) + 2\,Au(s)$$

→

▲ Crystals of gold on a sample of sphalerite.

but nowadays a different process is used. The $[Au(CN)_2]^-$ ions are adsorbed onto charcoal, which is then washed with water that is heated to above 100 °C under pressure. This dissolves the $[Au(CN)_2]^-$ ions, separating these from other compounds, and the gold metal is obtained from the solution by electrolysis.

The cyanide process is controversial because of the high toxicity of CN^-. In the 1980s, leaks from Summitville mine in Colorado, USA, killed all aquatic life in 27 km of the Alamosa River.

A more recent process uses biotechnology to extract gold. When gold occurs in sulfide ores such as iron pyrites (FeS_2), as it does in parts of Australia, the cyanide route recovers less than 50% of the available gold. Bacteria such as *Sulfolobus acidocaldarius* and *Acidithiobacillus ferroxidans* are added to slurries containing the ore for several days at 50 °C. During this period, the bacteria dissolve the sulfide ores, catalysing the oxidation of FeS_2

$$4FeS_2\,(s) + 15O_2\,(g) + 6H_2O\,(l) \rightleftharpoons 4Fe^{3+}\,(aq) + 8SO_4^{2-}\,(aq) + 4H_3O^+\,(aq)$$

The Fe^{3+} ions generated in this reaction can also act as oxidizing agents, oxidizing more FeS_2, and the combined effect of these reactions is to dissolve the sulfide ores and liberate the gold.

Question

What is the oxidation state of gold in the $[Au(CN)_2]^-$ complex? What shape would you predict for the complex ion?

Coordination number 3

The coordination number 3 is not very common, but when it occurs the complexes are generally trigonal planar. Examples include the platinum(0) complex $[Pt(PPh_3)_3]$ and the mercury(II) complex ion $[HgI_3]^-$ (**2**).

Coordination number 4

There are two important geometries for a coordination number of 4—tetrahedral and square planar. Tetrachloro complexes can adopt either geometry, depending on the nature of the metal ion. So, $[CoCl_4]^{2-}$ (d^7) (**3**) is tetrahedral, whereas $[PtCl_4]^{2-}$ (d^8) (**4**) is square planar. Crystal field theory (Section 28.4) helps to explain this difference and can be used to predict which geometry a particular complex will adopt.

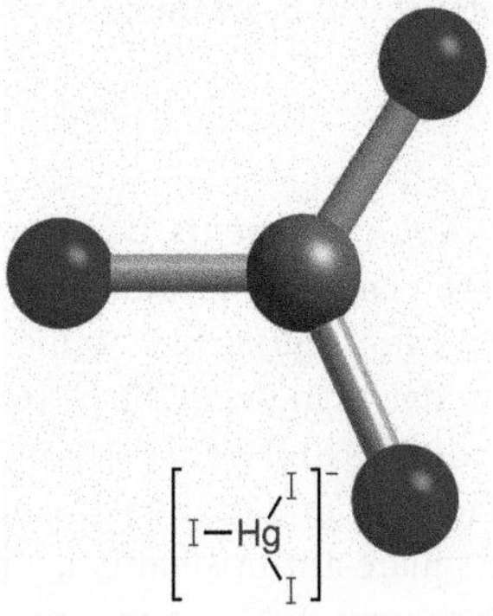
I
I–Hg
I
–

2 $[HgI_3]^-$ (trigonal planar).

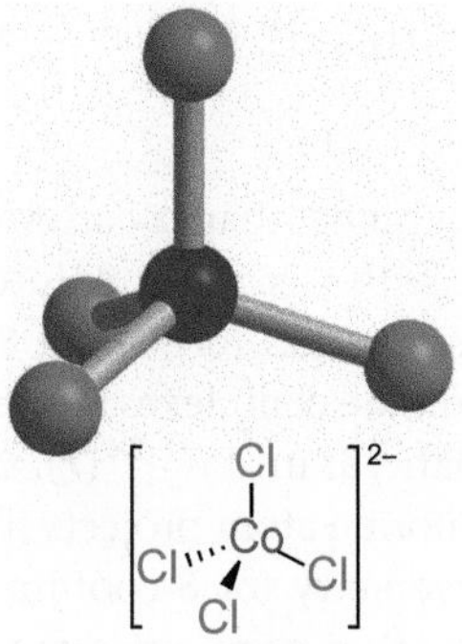
Cl
Cl Co Cl
Cl
2–

3 $[CoCl_4]^{2-}$ (tetrahedral).

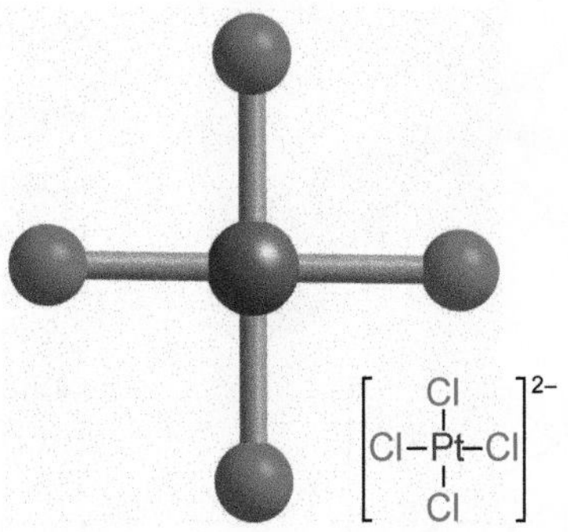
Cl
Cl–Pt–Cl
Cl
2–

4 $[PtCl_4]^{2-}$ (square planar).

Coordination number 5

The coordination number of 5 is much less common than either of the coordination numbers 4 or 6. 5-Coordinate complexes can be either trigonal bipyramidal or square pyramidal, and the difference in energy between these two geometries is relatively small. As shown in Figure 28.13, the ion $[Ni(CN)_5]^{3-}$ adopts both geometries, with the one that is present in a particular compound depending on the cation.

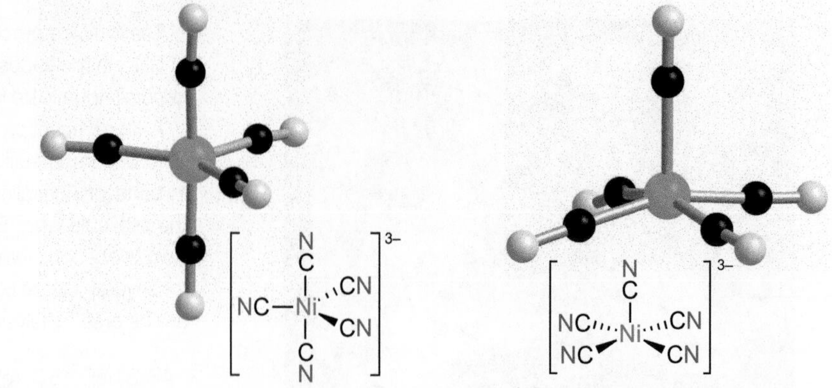

Figure 28.13 The 5-coordinate $[Ni(CN)_5]^{3-}$ complex can adopt either the trigonal bipyramidal geometry (left) or the square pyramidal geometry (right), depending on the cation in the compound.

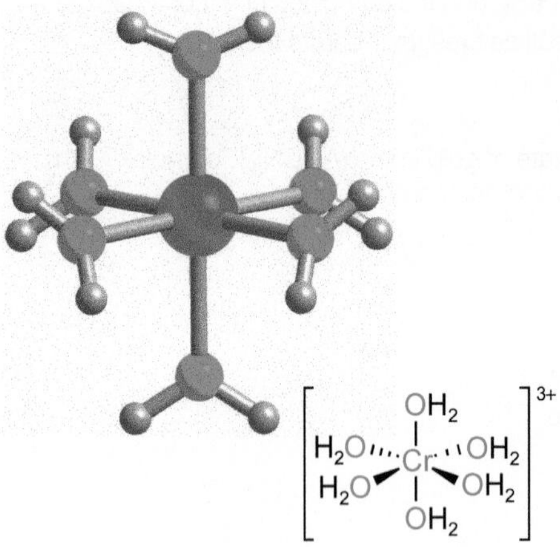

5 $[Cr(H_2O)_6]^{3+}$ (octahedral).

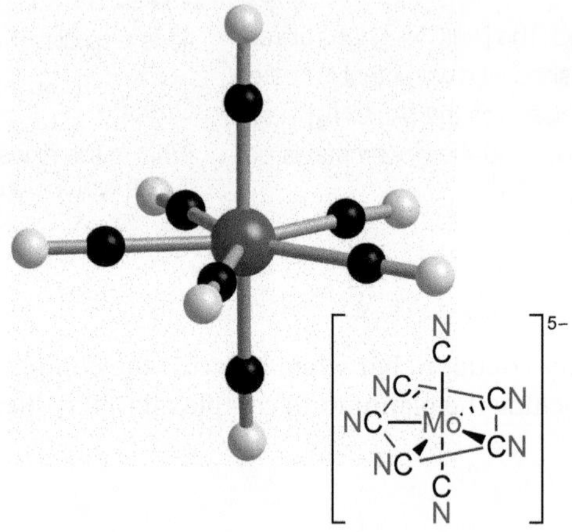

6 $[Mo(CN)_7]^{5-}$ (pentagonal bipyramidal).

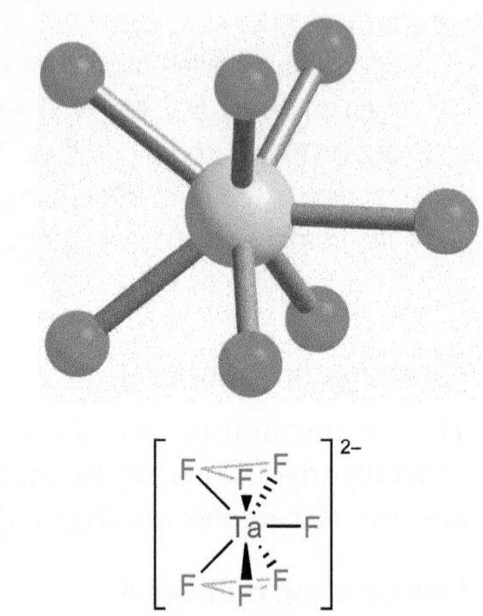

7 $[TaF_7]^{2-}$ (capped trigonal prismatic).

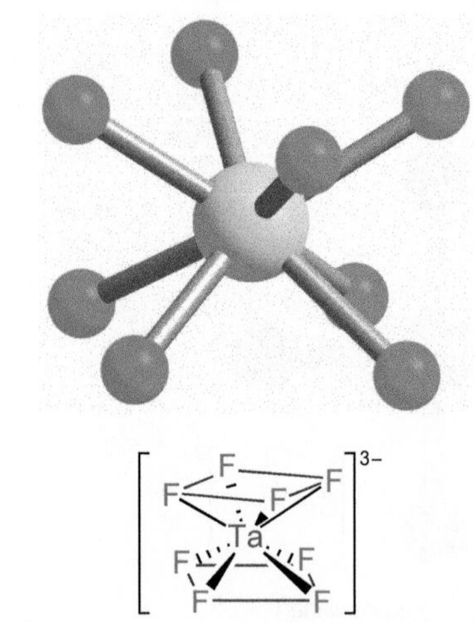

8 $[TaF_8]^{3-}$ (square-antiprismatic).

Coordination number 6

Six is the most common coordination number for first row *d*-block complexes. 6-Coordinate complexes are almost always octahedral. All of the first row *d*-block metal aqua complexes, such as $[Cr(H_2O)_6]^{3+}$ (**5**), have octahedral geometries.

Higher coordination numbers

Coordination numbers higher than six are rare for the first row *d*-block elements, but more common for elements in the second and third row because of their larger ionic radii. There are several possible geometries for 7- and 8-coordinate complexes. The most common geometries for 7-coordinate complexes are pentagonal bipyramidal, as in $[Mo(CN)_7]^{5-}$ (**6**), capped trigonal prismatic, as in $[TaF_7]^{2-}$ (**7**), and capped octahedral, as in $[Fe(edta)(H_2O)]^-$. For $[TaF_7]^{2-}$ (**7**), one fluorine atom projects through a square face of a TaF_6 trigonal prism. The most common geometry for 8-coordinate complexes is square-antiprismatic, as in $[TaF_8]^{3-}$ (**8**), although other geometries are also observed. Although cubic geometry is theoretically possible for an 8-coordinate complex, it is not observed. This is because the ligand–ligand repulsions in a cube are greater than those in a square-antiprism.

Naming coordination compounds

The names of ligands in complexes are given in Tables 28.3 (p.1266) and 28.4 (p.1267). In many cases the name of a neutral ligand is the same as that of the free molecule, but there are exceptions, such as aqua, ammine, and carbonyl, when the name of the molecule

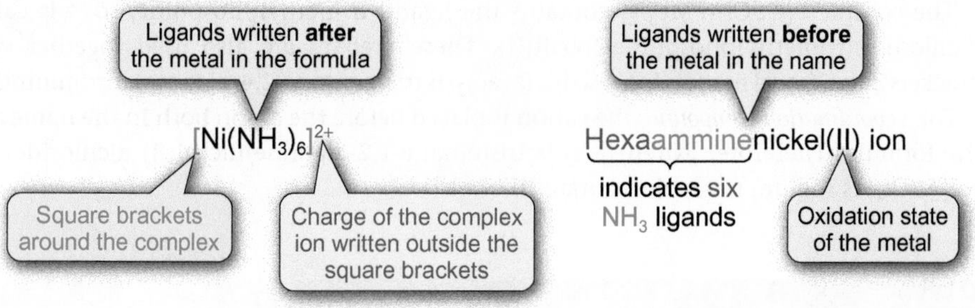
Ligands written **after** the metal in the formula
Ligands written **before** the metal in the name
[Ni(NH3)6]2+
Hexaamminenickel(II) ion
indicates six NH3 ligands
Square brackets around the complex
Charge of the complex ion written outside the square brackets
Oxidation state of the metal

Figure 28.14 Naming complexes.

changes on coordination. Names of negatively charged ligands end in -o, so chloride becomes chloro on coordination whereas sulfate becomes sulfato. When there is more than one of a particular type of ligand in a complex, a prefix is used before the ligand name to indicate this:

- di-, two;
- tri-, three;
- tetra-, four;
- penta-, five;
- hexa-, six.

The *name* of a complex is written as one word, with the ligands *before* the metal and its oxidation state, which is given as a Roman numeral in brackets (Figure 28.14). For example, $[Co(H_2O)_6]^{2+}$ is the hexaaquacobalt(II) ion and $[Ni(NH_3)_6]^{2+}$ is the hexaamminenickel(II) ion (Note that in the *formula* of the complex, the ligands are written *after* the metal).

When the complex is anionic, the ending *-ate* is added to the name of the metal ion. For example, $[NiCl_4]^{2-}$ is the tetrachloronickelate(II) ion. For some metals, shown in Table 28.5, the complex anion uses the Latin version of the name. For example, $[CuCl_4]^{2-}$ is the tetrachloro*cuprate*(II) ion *not* tetrachloro*copperate*(II).

The metals whose names change in anionic complexes are generally those for which the atomic symbol is derived from the Latin name. One exception is mercury (Hg), as its anionic complexes are called mercurates not hydrargyrates.

Table 28.5 *d*-Block metals that change their name in anionic complexes

Metal	Name in anionic complex
Copper, Cu	Cuprate
Gold, Au	Aurate
Iron, Fe	Ferrate
Silver, Ag	Argentate

When a complex contains more than one type of ligand, these are named in alphabetical order. The ion $[PtCl(NH_3)_3]^+$ is the triamminechloroplatinum(II) ion, because ammine comes before chloro in the alphabet. This is different from the order of the ligands in the formula, as here anionic ligands are put before neutral ligands. When there is more than one type of anionic or neutral ligand, they are ordered alphabetically in the formula of the complex based on the first letters of their formulae. For example, the diamminedibromod ichlorochromate(III) ion has the formula $[CrBr_2Cl_2(NH_3)_2]^-$.

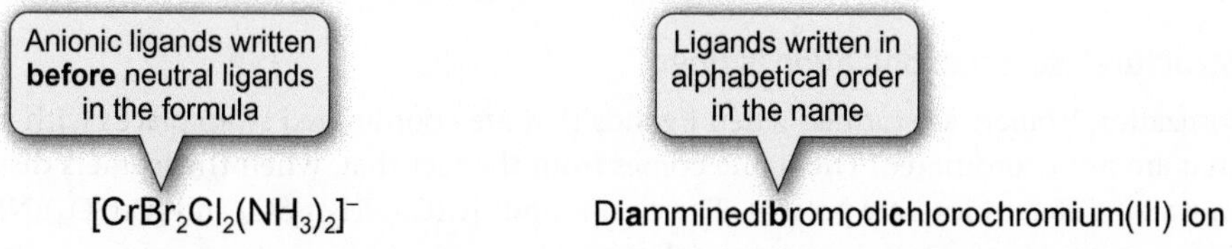
Anionic ligands written **before** neutral ligands in the formula
Ligands written in alphabetical order in the name
[CrBr2Cl2(NH3)2]−
Diamminedibromodichlorochromium(III) ion

If the *name* of the ligand already contains a numeric prefix (di-, tri, tetra-, etc.), for example, trimethylphosphine, the name is surrounded by brackets and, to avoid confusion, a different prefix is used to show how many of these ligands there are:

- bis-, two;
- tris-, three;
- tetrakis-, four;
- pentakis-, five;
- hexakis-, six.

The complex $[CoCl_2(PMe_3)_2]$ contains the ligand trimethylphosphine, so it is called dichlorobis(trimethylphosphine)cobalt(II). These prefixes are also used together with brackets if the ligand name is long, so $[Cr(acac)_3]$ is tris(pentane-2,4-dionato)chromium(III).

For a *coordination compound*, the cation is placed before the anion both in the name and the formula. Therefore, $[Ni(en)_3]Cl_2$ is tris(ethane-1,2-diamine)nickel(II) dichloride and $Na[AuCl_4]$ is sodium tetrachloroaurate(III).

Worked example 28.3 Naming coordination compounds

Name the following compounds:

(a) $K_3[Cr(CN)_6]$

(b) $[CrCl_2I(PEt_3)_3]$

(c) $[Fe(N_3)(H_2O)_5](NO_3)_2$.

Strategy

Use the rules given above to work out the names of the compounds.

Solution

(a) Potassium ions always carry a +1 charge, so the compound contains $[Cr(CN)_6]^{3-}$ anions. Since there are six cyano ions, all of which carry a −1 charge, the chromium has an oxidation state of +3. The complex is anionic, so the metal name ends in -ate. The anion is hexacyanochromate(III), so the compound is tripotassium hexacyanochromate(III) (often the tri- prefix is ignored, since there must be three potassium ions from the stoichiometry).

(b) The chromium ion is coordinated to two chloro ligands (Cl^-), one iodo ligand (I^-), and three neutral triethylphosphine ligands. This means that chromium has an oxidation state of +3. The ligands are named in alphabetical order, so chloro goes before iodo, which goes before triethylphosphine. Triethylphosphine has a numeric prefix in its name, so the name is written in brackets and the prefix tris is used to show that there are three of them. The compound is dichloroiodotris (triethylphosphine)chromium(III).

(c) Nitrate ions carry a −1 charge, so the compound contains $[Fe(N_3)(H_2O)_5]^{2+}$ cations. The iron is coordinated to an anionic azido ligand (N_3^-) and five neutral aqua ligands. This means that the iron atom has an oxidation state of +3. The complex ion is pentaaquaazidoiron(III). The compound is pentaaquaazidoiron(III) dinitrate.

Question

(a) What is the name of the compound $[CoI_2(en)_2]I$?

(b) What is the formula of the compound diaquadibromodi(methylamine)cobalt(III) nitrate?

Isomerism

Types of isomerism in organic chemistry are discussed in Section 18.1 (p.816).

There are two main classes of isomerism in coordination chemistry, just as there are in organic chemistry (Figure 28.15). **Structural isomers** are compounds in which the atoms are joined together in a different arrangement, whereas **stereoisomers** are compounds in which the atoms are joined in the same arrangement but the positions of the atoms in space are different.

Structural isomers: ionization isomers

Ionization isomers are formed when ligands that are coordinated swap places with those that are not coordinated. The name comes from the fact that, when the isomers dissolve in water, different ions are present. The compounds $[CrCl(NH_3)_5]SO_4$ and $[Cr(SO_4)(NH_3)_5]Cl$ are examples of ionization isomers (Figure 28.16).

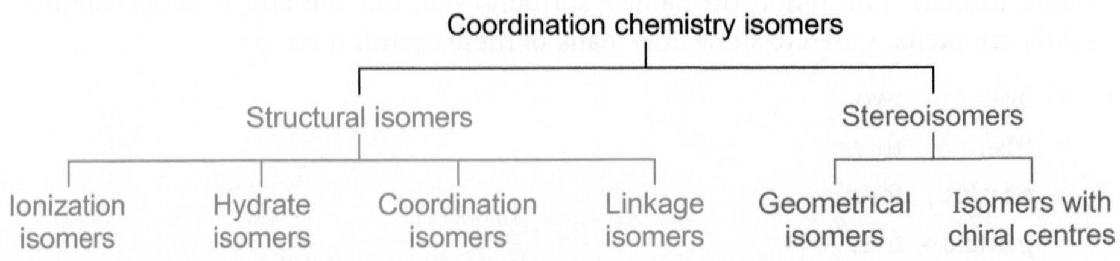
Coordination chemistry isomers
Structural isomers
Stereoisomers
Ionization isomers
Hydrate isomers
Coordination isomers
Linkage isomers
Geometrical isomers
Isomers with chiral centres

Figure 28.15 The different types of isomers observed in coordination chemistry.

Both contain a Cr^{3+} ion, five ammine ligands, a chloride, and a sulfate. In $[CrCl(NH_3)_5]SO_4$ the chloride anion is coordinated (it is a chloro ligand) and the sulfate anion is not coordinated (it is the **counter-ion**). Solutions of this compound contain $[CrCl(NH_3)_5]^{2+}$ cations and sulfate anions. In $[Cr(SO_4)(NH_3)_5]Cl$ the sulfate anion is coordinated (it is a sulfato ligand) and the chloride anion is not coordinated. Solutions of this compound contain $[Cr(SO_4)(NH_3)_5]^+$ cations and chloride anions.

You can distinguish the two isomers by adding silver nitrate solution to dilute solutions of the compounds, since only uncoordinated chloride is free to react. $[Cr(SO_4)(NH_3)_5]Cl$ gives a white precipitate of silver chloride, but $[CrCl(NH_3)_5]SO_4$ does not react

$$[Cr(SO_4)(NH_3)_5]Cl + Ag^+ \rightarrow [Cr(SO_4)(NH_3)_5]^+(aq) + AgCl(s)$$
$$[CrCl(NH_3)_5]SO_4 + Ag^+ \rightarrow \text{no reaction}$$

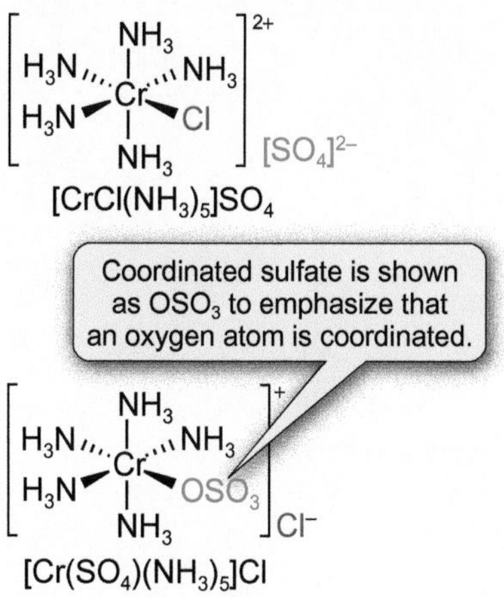

[CrCl(NH3)5]SO4
Coordinated sulfate is shown as OSO3 to emphasize that an oxygen atom is coordinated.
[Cr(SO4)(NH3)5]Cl

Figure 28.16 Ionization isomers.

Structural isomers: hydrate isomers

Hydrate isomers are a special case of ionization isomers, and these vary in the location of the water molecules, which may be coordinated as ligands or present as water of crystallization. The three chromium complexes below all have the empirical formula $CrCl_3{\cdot}6H_2O$, and are examples of hydrate isomers:

$[Cr(H_2O)_6]Cl_3$: hexaaquachromium(III) trichloride (violet);

$[CrCl(H_2O)_5]Cl_2{\cdot}H_2O$: pentaaquachlorochromium(III) dichloride hydrate (blue-green)

$[CrCl_2(H_2O)_4]Cl{\cdot}2H_2O$: tetraaquadichlorochromium(III) chloride dihydrate (green).

These isomers are distinguished using gravimetric analysis. A known mass of each isomer is dissolved in water and treated with $AgNO_3(aq)$. The mass of AgCl formed is measured and indicates the number of non-coordinated chloride ions that are present.

Gravimetric analysis is described in Section 1.5 (p.39).

Structural isomers: coordination isomers

When a coordination compound contains two complex ions, coordination isomers are possible. These differ in the distribution of the ligands between the two ions. The compounds $[Cr(NH_3)_6][CoF_6]$ and $[Co(NH_3)_6][CrF_6]$ are examples of coordination isomers. Both contain a hexaammine cation and a hexafluoro anion, but they differ in which ligands are coordinated to each metal ion.

Structural isomers: linkage isomers

Monodentate ligands that contain more than one possible donor atom are known as **ambidentate** ligands. For example, the thiocyanate ion (SCN^-) has lone pairs on both the nitrogen and the sulfur atoms.

The structure of the thiocyanate ion is described in Section 5.1 (p.222).

This means that the thiocyanate ion can coordinate through either the nitrogen atom or the sulfur atom. The donor atom is normally shown first in the formula and indicated in the name of the complex by being given in italics after the ligand name and the Greek letter κ:

> The Greek letter κ (kappa) is used to denote which atom of an ambidentate ligand coordinates in a complex.

M–SCN thiocyanato-κ*S* M–NCS thiocyanato-κ*N* (isothiocyanato)

Nitrite (NO_2^-) is another ambidentate ligand, as it can bind through a lone pair on either the nitrogen atom or one of the oxygen atoms

$M{-}NO_2$ nitrito-κ*N* M–ONO nitrito-κ*O*

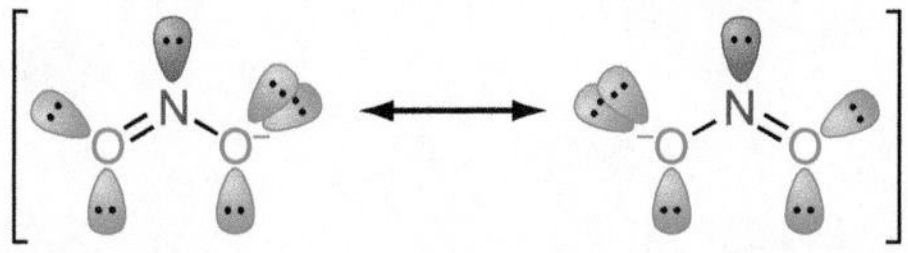

The compounds $[Co(NO_2)(NH_3)_5]Cl_2$ and $[Co(ONO)(NH_3)_5]Cl_2$, the cations of which are shown in Figure 28.17, were both identified by Werner in the early days of coordination chemistry. He assigned their structures on the basis of their colours. $[Co(NO_2)(NH_3)_5]Cl_2$ is yellow, similar to other cobalt(III) complexes containing six *N*-donors. In contrast,

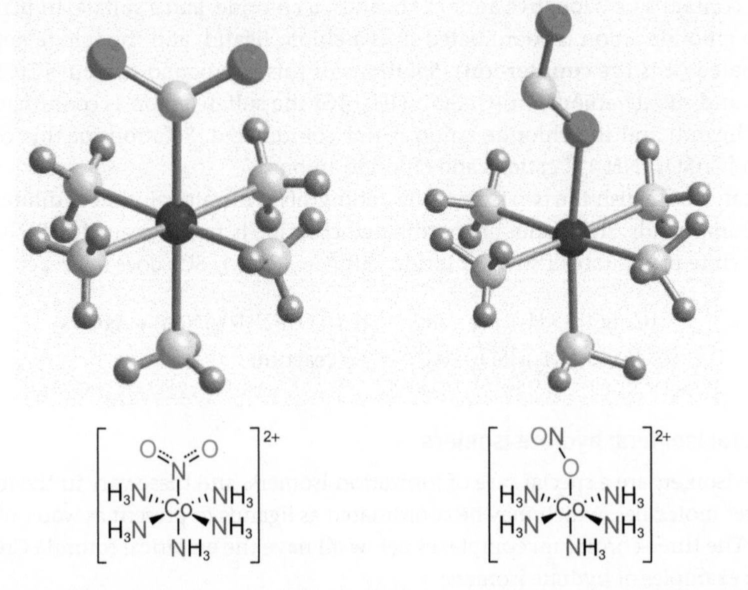

Figure 28.17 $[Co(NO_2)(NH_3)_5]^{2+}$ and $[Co(ONO)(NH_3)_5]^{2+}$ are linkage isomers. They differ with respect to the donor atom in the NO_2^- ligand.

$[Co(ONO)(NH_3)_5]Cl_2$ is red, similar to other cobalt(III) complexes containing five *N*-donors and one *O*-donor. Werner's interpretation was later shown to be correct by crystallography.

Stereoisomers: geometrical isomers

The simplest example of geometrical isomerism in coordination chemistry occurs when the complex can adopt two different geometries. For example, the $[Ni(CN)_5]^{3-}$ complexes shown in Figure 28.13 (p.1270) are geometrical isomers as are 4-coordinate complexes, which can be either tetrahedral or square planar.

Geometrical isomers are also possible in square-planar complexes of the general formula MX_2Y_2, when X and Y are different ligands. For example, the platinum complex $[PtCl_2(NH_3)_2]$ forms a *cis*-isomer (**9**) and a *trans*-isomer (**10**). This type of isomerism can have an important effect on the properties of the complexes. In this case, *cis*-$[PtCl_2(NH_3)_2]$ is an important anti-cancer agent used in chemotherapy (see Box 28.3), but *trans*-$[PtCl_2(NH_3)_2]$ is inactive.

Tetrahedral complexes of the general formula MX_2Y_2 do not form geometrical isomers as there is only one way in which the four ligands can be arranged.

Octahedral complexes of the general formula MX_2Y_4 can also form *cis* and *trans* isomers. For example, *cis*-$[CoCl_2(en)_2]^+$ (**11**) is violet, and *trans*-$[CoCl_2(en)_2]^+$ (**12**) is green.

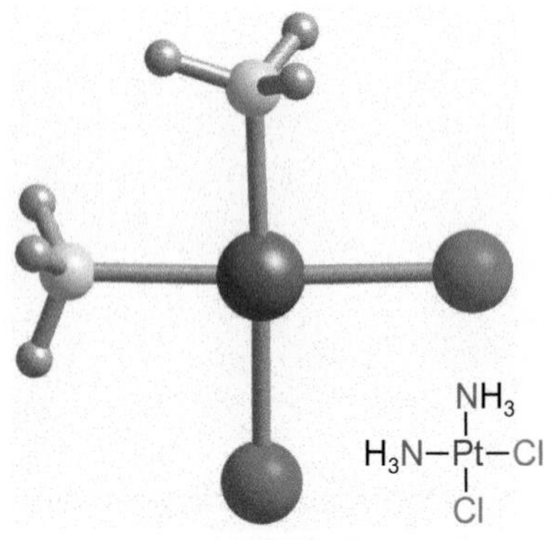

9 *cis*-$[PtCl_2(NH_3)_2]$.

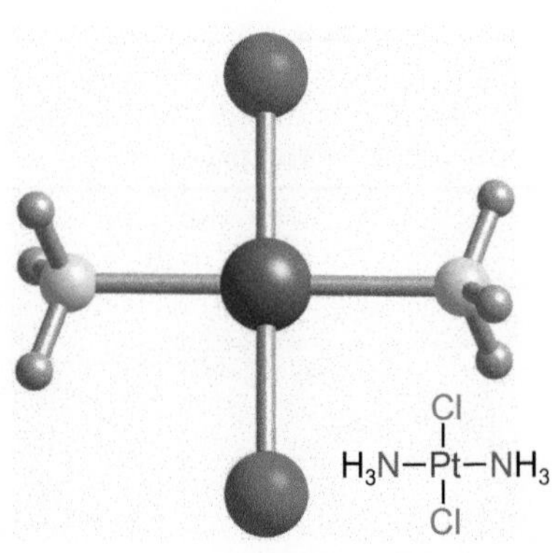

10 *trans*-$[PtCl_2(NH_3)_2]$.

The use of *cis* and *trans* in coordination chemistry is similar to that in organic chemistry. See Section 2.5 (p.85).

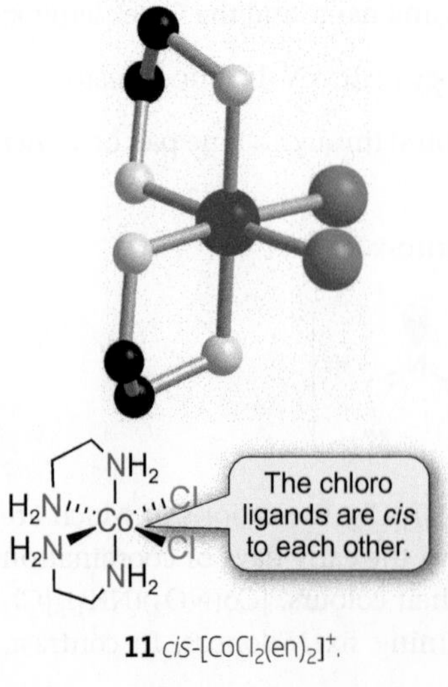
The chloro ligands are *cis* to each other.

11 *cis*-$[CoCl_2(en)_2]^+$.

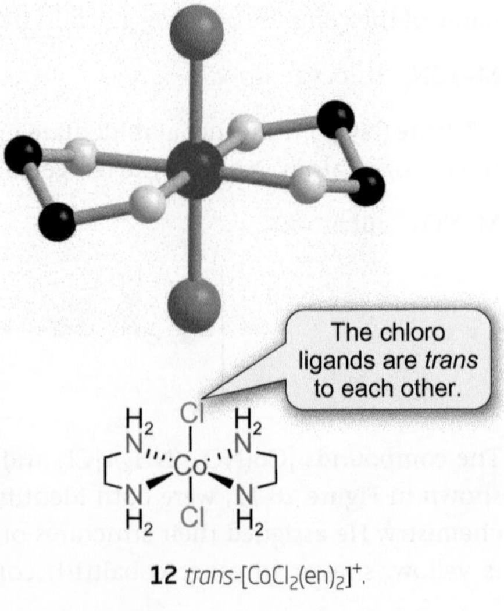
The chloro ligands are *trans* to each other.

12 *trans*-$[CoCl_2(en)_2]^+$.

Box 28.3 Platinum anti-cancer drugs

Cancers are diseases in which tumour cells replicate in an uncontrolled way. One of the best anti-cancer drugs is a platinum complex, *cis*-$[PtCl_2(NH_3)_2]$ (cisplatin), and its effectiveness was discovered by accident.

In the 1960s, the American scientist Barnett Rosenberg was studying the effect of electric fields on cell division. He discovered that, under the conditions of his experiment, the cells stopped dividing but they kept growing. This continued after the current was switched off, which suggested the electric field was not a direct cause of the behaviour. Eventually, Rosenberg realized that electrolysis was occurring at the platinum electrodes used to generate the electric field, and reaction of Pt^{2+} with the NH_4Cl buffer led to low concentrations of platinum ammine complexes. One of these complexes was cisplatin, which was shown to stop cell division. Cisplatin entered clinical trials in 1971 and was approved for use against cancers in 1978.

Although the chloro ligands are labile and easily substituted by other ligands, the concentration of chloride ions in body fluids is quite high so cisplatin reaches cells intact. The chloride ion concentration is much lower inside cells so, once cisplatin passes through a cell membrane, the chloro ligands are substituted by water molecules

$$cis\text{-}[PtCl_2(NH_3)_2](aq) + 2H_2O(l) \rightleftharpoons cis\text{-}[Pt(NH_3)_2(H_2O)_2]^{2+}(aq) + 2Cl^-(aq)$$

The aqua ligands are then replaced by nitrogen donors from two of the DNA base pairs (see Box 25.5, p.1158). These are normally two neighbouring guanine bases on the same strand of the DNA double helix. The formation of a chelate ring tilts the guanine rings from their normal stacked position, which disrupts the helix and interferes with cell division. Before platinum drugs, testicular cancer was resistant to all treatments, but nowadays most cases are curable if detected early enough.

Cisplatin has proved very successful, but there are problems with its use. It only affects a narrow range of tumours and some become resistant to it. It is also severely toxic and gives rise to many side-effects such as kidney damage, nausea, and hair loss. The toxicity occurs because cisplatin does not selectively target tumours, but affects all cells. The platinum coordinates to other molecules in the body, such as proteins, in addition to DNA.

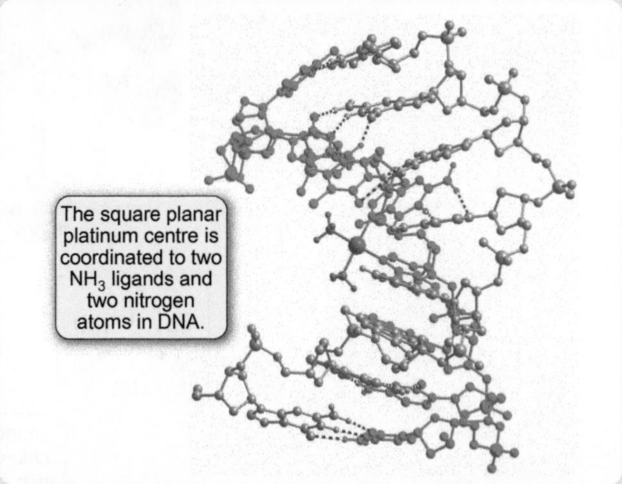
The square planar platinum centre is coordinated to two NH_3 ligands and two nitrogen atoms in DNA.

▲ Platinum coordinates to DNA causing it to twist, and this stops the DNA from replicating. In this figure the two strands of the DNA double helix are shown in different colours and the $Pt(NH_3)_2$ group is shown in green.

▲ Lance Armstrong won the Tour de France seven times in a row, from 1999 to 2005, but has now been stripped of his titles because he used banned substances. Remarkably though, he competed in these races after extensive chemotherapy. Platinum complexes were used to treat testicular cancer that had spread to his brain and lungs.

Since the anti-cancer properties of cisplatin were discovered, there has been a great effort to find other platinum compounds that work as well, or even better, but have fewer side-effects. Carboplatin and satraplatin are both examples of newer platinum anti-cancer drugs. Carboplatin is in clinical use, whereas satraplatin is in the final stages of clinical trials. Carboplatin contains a bidentate dicarboxylato ligand instead of the two chloro ligands. The dicarboxylato ligand is less labile, which makes carboplatin less toxic, so larger doses are possible. Satraplatin is a platinum(IV) complex, and is more soluble in water than cisplatin. This allows it to be taken orally rather than as an injection. Researchers believe satraplatin is reduced to platinum(II) in the body.

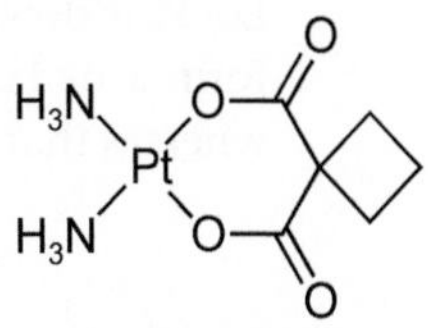

carboplatin

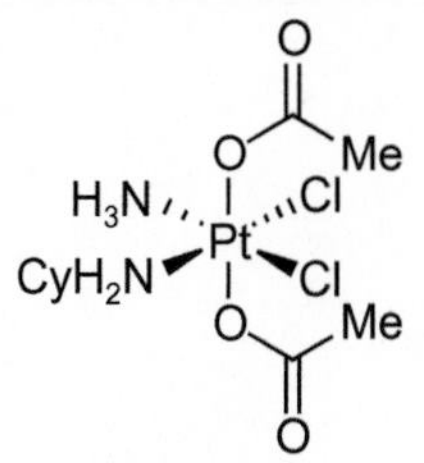

satraplatin
(Cy = cyclohexyl)

Question

What are the systematic names for cisplatin and satraplatin? (Ignore the stereochemistry for satraplatin.)

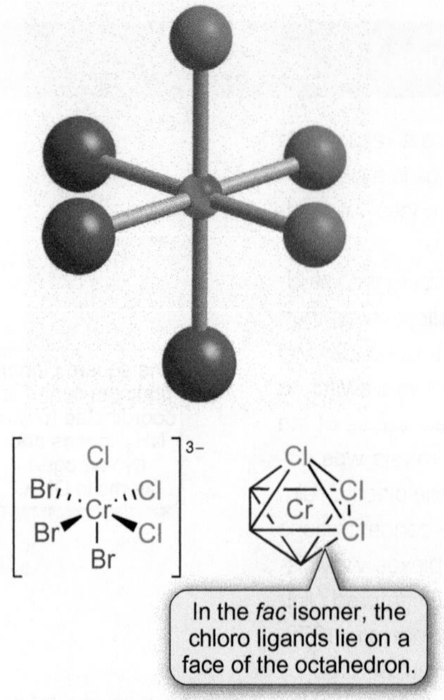
Cl
Br
Cr
Cl
Br
Cl
Br
3−
Cl
Cr
Cl
Cl
In the *fac* isomer, the chloro ligands lie on a face of the octahedron.

13 *fac*-$[CrBr_3Cl_3]^{3-}$.

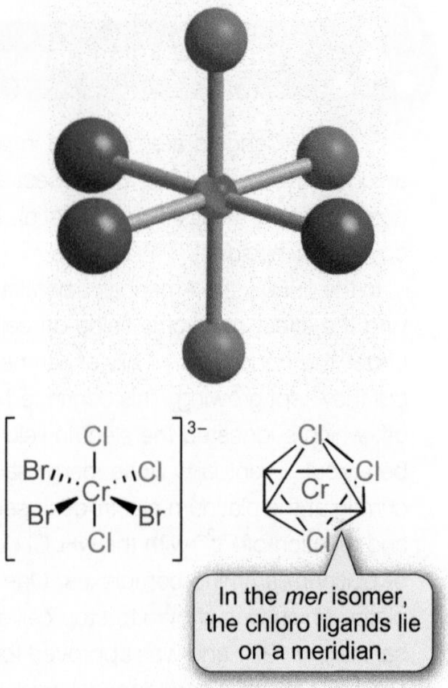
Cl
Br
Cr
Cl
Br
Br
Cl
3−
Cl
Cr
Cl
Cl
In the *mer* isomer, the chloro ligands lie on a meridian.

14 *mer*-$[CrBr_3Cl_3]^{3-}$.

In geography, a meridian is a circle on the Earth passing through the poles and any given point on the Earth's surface.

Octahedral complexes of the general formula MX_3Y_3 show a second type of geometrical isomerism. In *fac*-$[CrBr_3Cl_3]^{3-}$ (**13**) the three chloro ligands occupy a *face* of an octahedron, and in *mer*-$[CrBr_3Cl_3]^{3-}$ (**14**) the three chloro ligands occupy three positions on a *meridian*.

Stereoisomers: isomers with chiral centres

Chirality in organic chemistry is described in Section 18.4 (p.838). Isomers with chiral centres are sometimes called **optical isomers**.

Racemization is the conversion of a pure enantiomer to a racemic mixture. See Section 18.4 (p.841) for more details.

It helps to build models or use molecular modelling software to convince yourself that the two enantiomers of $[Cr(pn)_3]^{3+}$ or *cis*-$[CoCl_2(en)_2]^+$ are non-superimposable.

Isomers with chiral centres are **asymmetric molecules** that do *not* have a plane or centre of symmetry. Molecules or ions that are non-superimposable on their mirror images are called **enantiomers**.

In organic chemistry, enantiomers exist when a central atom, usually carbon, is bonded to four different groups. The same is true for tetrahedral metal complexes, so the tetrahedral cobalt(II) complex $[CoBrClFI]^{2-}$ (Figure 28.18) exists as a pair of enantiomers. In practice, this type of chirality is not very important for coordination complexes. This is because the metal–ligand bonds are **labile**. The ligands can break away and change places, so the enantiomers rapidly undergo racemization.

A more common type of chirality in coordination chemistry occurs for octahedral complexes that contain bidentate ligands. For example, the complex $[Cr(pn)_3]^{3+}$ (pn = propane-1,3-diamine) forms a pair of enantiomers, as shown in Figure 28.19. Bidentate ligands such as pn are much less labile than monodentate ligands such as Cl^-, so it is possible to isolate these enantiomers. These isomers are distinguished by Δ and Λ labels. Looking down a triangular face of the octahedron, as in Figure 28.20, the bidentate ligands form a helix. The enantiomer forming the right-handed, clockwise helix is labelled Δ, whereas that forming the left-handed, anticlockwise helix is labelled Λ.

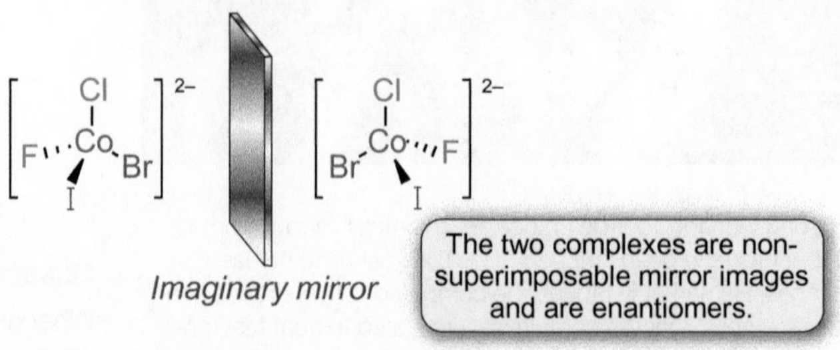
Cl
F
Co
Br
I
2−
Cl
Br
Co
F
I
2−
Imaginary mirror
The two complexes are non-superimposable mirror images and are enantiomers.

Figure 28.18 The tetrahedral complex $[CoBrClFI]^{2-}$ exists as a pair of enantiomers.

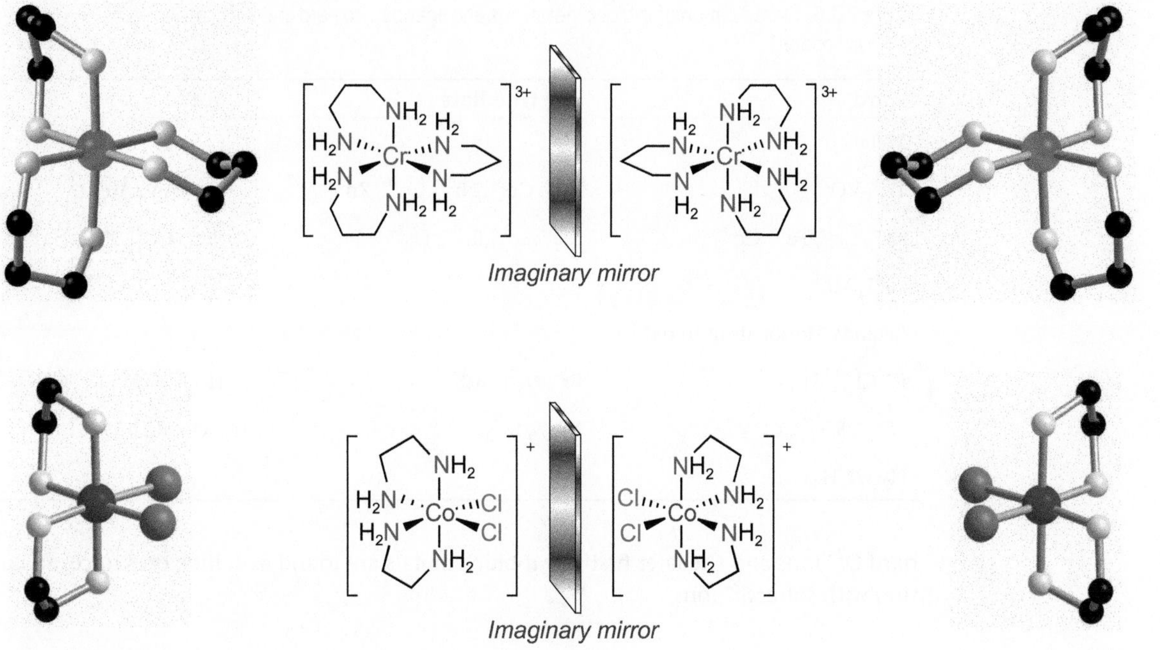
Imaginary mirror
Imaginary mirror

Figure 28.19 $[Cr(pn)_3]^{3+}$ and *cis*-$[CoCl_2(en)_2]^+$ exist as enantiomers.

Octahedral complexes with two bidentate ligands, such as *cis*-$[CoCl_2(en)_2]^+$ (**11**), also have chiral centres and exist as pairs of enantiomers (Figure 28.19). However, *trans*-$[CoCl_2(en)_2]^+$ (**12**) is not chiral as it contains a mirror plane.

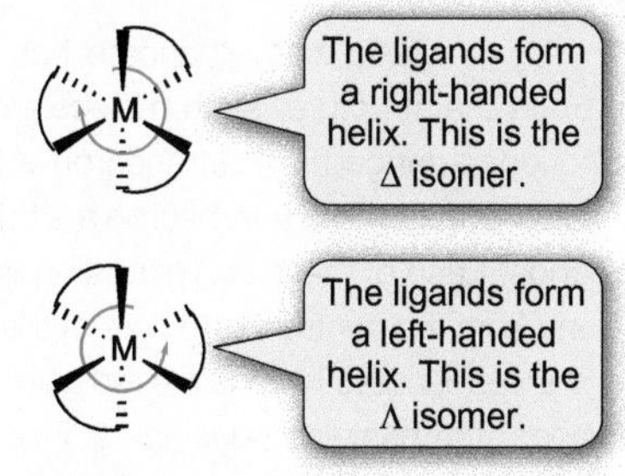
The ligands form a right-handed helix. This is the Δ isomer.
The ligands form a left-handed helix. This is the Λ isomer.

Figure 28.20 Δ and Λ isomers in tris(bidentate) complexes.

Hard and soft metals and ligands

Ti^{4+} ions form strong interactions with oxygen-donor ligands like methoxide (OMe^-) and water, but much weaker interactions with triphenylphosphine (PPh_3) and dimethylsulfide (SMe_2). In contrast, Pt^{2+} ions form strong bonds to PPh_3 and SMe_2, and much weaker bonds to OMe^- and H_2O. This preference of particular metals for particular ligands can be understood using the concept of **hard and soft acids and bases.**

The metal centres, which are Lewis acids, and the ligands, which are Lewis bases, are divided into classes.

- **Hard metals** carry a high charge or have a high charge density. These metals favour forming bonds to ligands with small, electronegative donor atoms, such as oxygen, nitrogen, or fluorine. Such ligands are known as **hard ligands**. The bonding between hard metals and hard ligands has a high degree of ionic character.
- **Soft metals** have a lower charge, and tend to be larger and more polarizable. These metals favour forming bonds with ligands that have larger, more polarizable donor atoms. Such ligands are known as **soft ligands**. The bonding between soft metals and soft ligands has a high degree of covalent character.

> ℹ Though useful for *d*-block metals, the concept of hard and soft acids and bases is used more widely in chemistry.

Examples of hard and soft metals and ligands are given in Table 28.6. Using this information, you can predict that Au^+ will form stronger bonds and, therefore, more stable complexes with I^- than with F^-, but that Cr^{3+} will form stronger bonds with F^- than with I^-. Some metals are intermediate, and can form strong bonds to both hard and soft ligands. There are also examples of intermediate ligands that form strong bonds to both hard metals and soft metals.

The most common ions formed by the first row *d*-block elements become softer moving from left to right across the period. For example, Ti^{4+} and Cr^{3+} are hard ions, whereas Ni^{2+} and Cu^{2+} are softer. This change in hardness is reflected in the natural occurrence of the metals, as the earlier first row *d*-block metals are found as oxide ores in combination with

Table 28.6 Classification of *d*-block metal ions and ligands into hard and soft categories. (R = alkyl or aryl.)

Hard	Intermediate	Soft
Metals		
Ti^{4+}, VO^{2+}	Fe^{2+}, Co^{2+}, Ni^{2+}, Cu^{2+}, Zn^{2+}	Cu^{+}, Ag^{+}, Au^{+}
Sc^{3+}, Cr^{3+}, Fe^{3+}, Co^{3+}	Rh^{3+}, Ir^{3+}, Ru^{3+}, Os^{3+}	Pd^{2+}, Cd^{2+}, Pt^{2+}, Hg^{2+}
Cr^{2+}, Mn^{2+}		
Ligands (donor atom in red)		
F^{-}, Cl^{-}, OH^{-}, O^{2-}	Br^{-}, N_3^{-}, NCS^{-}	H^{-}, R^{-}, CN^{-}, I^{-}, SCN^{-}
NO_3^{-}, SO_4^{2-}, CO_3^{2-}	NO_2^{-}, SO_3^{2-}	CO, PR_3, SR_2
H_2O, NH_3	py	

hard O^{2-} ions and the later first row *d*-block metals are found as sulfide ores in combination with softer S^{2-} ions.

Box 28.4 Coordination network structures

Basic zinc ethanoate has the formula $[Zn_4O(O_2CMe)_6]$ (**15**). In the structure, each molecule of the complex contains four zinc ions arranged in a tetrahedron with an oxide ion at the centre. The six edges of the tetrahedron are bridged by bidentate ethanoato ligands. Each of these six ligands is directed at 90° from the others, away from the central Zn_4O core of the complex.

If the ethanoato ligands are replaced by 1,4-benzenedicarboxylato ligands, the benzene rings link adjacent complexes together into a coordination network structure. Chemists are very interested in these structures because many of them are porous, in a similar way to zeolites (see Chapter 6, p.255). Coordination networks that have the potential for porosity are also called **metal–organic frameworks** (MOFs).

Coordination network structures typically crystallize from solution with solvent molecules in the pores, but these can often be removed by heating. Many bridging ligands have been used to make coordination networks, but the networks formed with dicarboxylates are particularly robust and often stable to over 300 °C. Once the solvent molecules have been removed, the pores can be used to store gas molecules. Hydrogen uptake into MOFs has been studied to see whether these materials can be used to safely store hydrogen. Initial results are promising, though more work is needed to find materials that will store enough hydrogen for them to be commercially viable. The storage of hydrogen for use in fuel cells is described in Box 25.3 (p.1147).

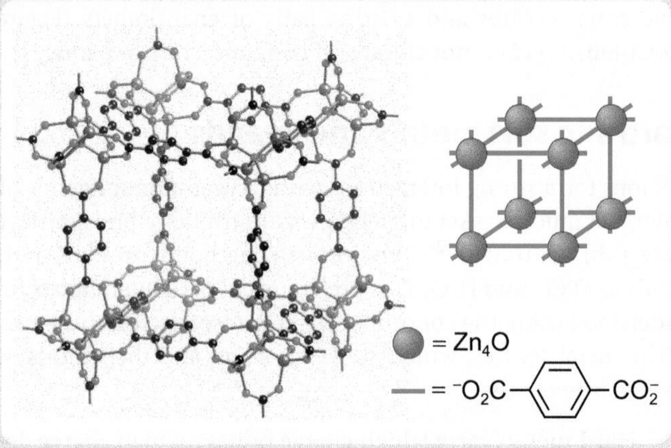
= Zn_4O
= $^{-}O_2C$–C_6H_4–CO_2^{-}

▲ Part of a metal–organic framework (MOF). The Zn_4O units are connected by the 1,4-benzenedicarboxylato ligands into a coordination network structure.

▼ **15** Basic zinc ethanoate ($[Zn_4O(O_2CMe)_6]$).

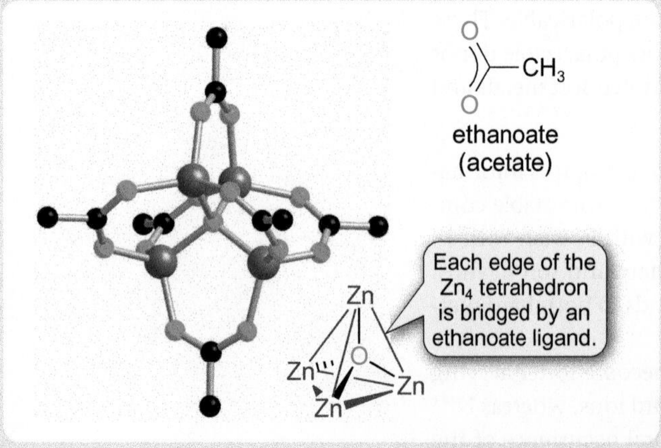
O
CH_3
O
ethanoate
(acetate)
Each edge of the Zn_4 tetrahedron is bridged by an ethanoate ligand.
Zn
O
Zn
Zn
Zn

Question

What is the formula of the coordination network structure formed when ethanoato ligands in (**15**) are replaced by 1,4-benzenedicarboxylato ligands?

Summary

- In a coordination complex, ligands are generally coordinated to a metal centre through electron pairs on their donor atoms.
- Monodentate ligands coordinate through one donor atom, whereas polydentate ligands coordinate through more than one donor atom, giving rise to chelate rings.
- The total number of donor atoms coordinated to a metal ion is known as the coordination number. For a first row *d*-block complex, this can vary from 2 to 8, but 4 and 6 are the most common.
- 4-Coordinate complexes can be tetrahedral or square planar, and 6-coordinate complexes are almost always octahedral.
- Structural isomers and stereoisomers are possible for coordination complexes. Ionization isomers, hydrate isomers, coordination isomers, and linkage isomers are types of structural isomers. Geometrical isomers and enantiomers are both types of stereoisomers.
- Metal ions and ligands can be divided into hard and soft classes. Hard metals have high charges or high charge densities, whereas soft metals are larger, with lower charges, and more polarizable. Hard ligands have small electronegative donor atoms such as N, O, or F, whereas soft ligands have larger, more polarizable donor atoms. Hard metals form more stable complexes with hard ligands and soft metals form more stable complexes with soft ligands.

? For practice questions on these topics, see questions 3–9 at the end of this chapter (p.1301).

28.4 Crystal field theory

Although the Lewis acid–Lewis base concept helps to explain the existence of coordination complexes, it cannot explain why certain geometries are favoured, nor can it explain the characteristic colours and magnetic properties of coordination compounds. For this, other approaches are needed. **Crystal field theory** is the simplest of these and it successfully explains many of the properties of coordination complexes.

This theory uses an electrostatic approach to the bonding, and treats the ligands as simple point charges. The electrostatic attraction between the metal ion and the negative point charges holds the complex together. Crystal field theory considers how the interactions between the ligand point charges and the metal *d* orbitals affect the energies of the *d* orbitals. The shapes of the five 3*d* orbitals are shown in Figure 28.21.

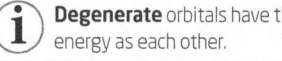

The origins of the 3*d* orbital shapes are described in Section 3.5 (p.142).

Octahedral complexes

In an isolated metal ion, the five *d* orbitals all have identical energies and are said to be **degenerate**. If this metal ion were surrounded by a spherical shell of negative charge, called a **spherical field**, there would be an electrostatic interaction between this field and the *d* electrons. This electrostatic repulsion leads to all of the *d* orbitals increasing in energy.

Degenerate orbitals have the same energy as each other.

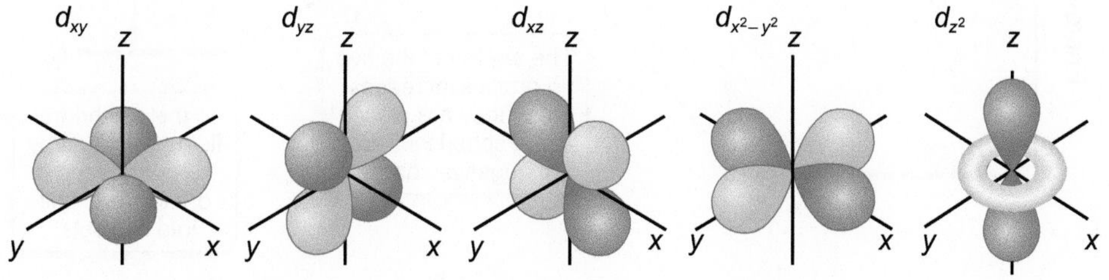
d_{xy}
d_{yz}
d_{xz}
$d_{x^2-y^2}$
d_{z^2}
z
y
x

Figure 28.21 The shapes of the five 3*d* orbitals.

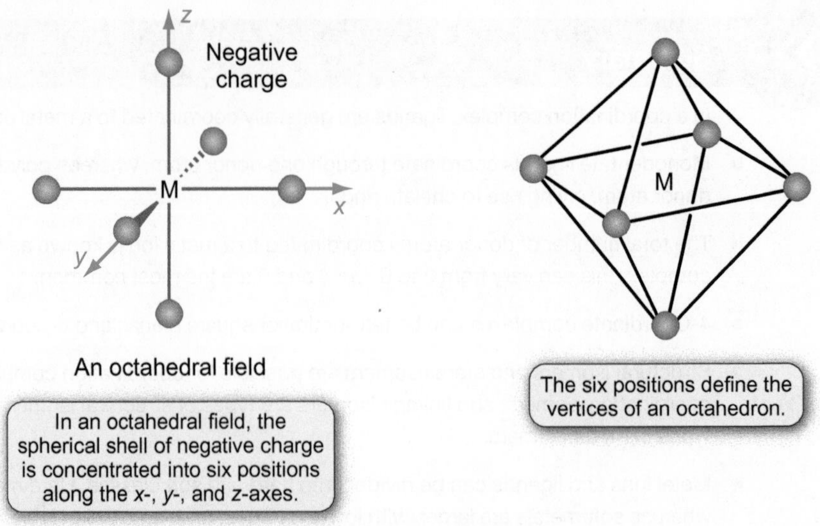
z
Negative charge
M
x
y
An octahedral field
In an octahedral field, the spherical shell of negative charge is concentrated into six positions along the x-, y-, and z-axes.
M
The six positions define the vertices of an octahedron.

Figure 28.22 In an octahedral field, six point charges interact with the metal ion. These charges are located on the *x*-, *y*-, and *z*-axes.

Imagine rearranging the negative charge of the spherical shell to concentrate it at six points, arranged along the *x*-, *y*-, and *z*-axes at the vertices of an octahedron, as shown in Figure 28.22. This is called an **octahedral field**. Since the total amount of charge has not altered, the average energy of the five *d* orbitals is the same in the octahedral field as it was in the spherical field. The five orbitals are, however, no longer of identical energy. Have a look at Figure 28.21 to make sure you understand the directions of the different *d* orbitals. Two of the orbitals, the $d_{x^2-y^2}$ and d_{z^2}, point directly along the axes towards the negative charges. These orbitals will be at a higher energy than the d_{xy}, d_{xz}, and d_{yz} orbitals, all of which point *between* the axes and so are less influenced by the negative charge. This leads to a splitting of the *d* orbitals into two sets, with three orbitals stabilized and two orbitals destabilized relative to their positions in a spherical field. This is shown in Figure 28.23.

This is how the *d* orbitals are split in an octahedral complex. In crystal field theory, degenerate orbitals are given group labels. In an octahedral complex, the d_{xy}, d_{xz}, and d_{yz} orbitals are collectively called the t_{2g} orbitals, and the $d_{x^2-y^2}$ and d_{z^2} orbitals are collectively called the e_g orbitals. The *g* part of the label reflects that fact that the orbitals are symmetric with respect to inversion.

g and *u* labels are described in Section 4.7 (p.186). The other parts of the labels contain additional information about the symmetry of the orbitals. The *t* label occurs when there are three orbitals of the same energy (triply degenerate) and the *e* label occurs when there are two orbitals of the same energy (doubly degenerate).

The energy gap between the two sets of orbitals is known as the **crystal field splitting energy**, and it is given the symbol Δ. The magnitude of Δ depends on the metal ion and the ligands. For an octahedral complex, the splitting is Δ_o (sometimes written Δ_{oct}). Since the average energy of the five *d* orbitals is unchanged on going from the spherical to the octahedral field, the three t_{2g} orbitals are *stabilized* by the same amount as the two *eg* orbitals are

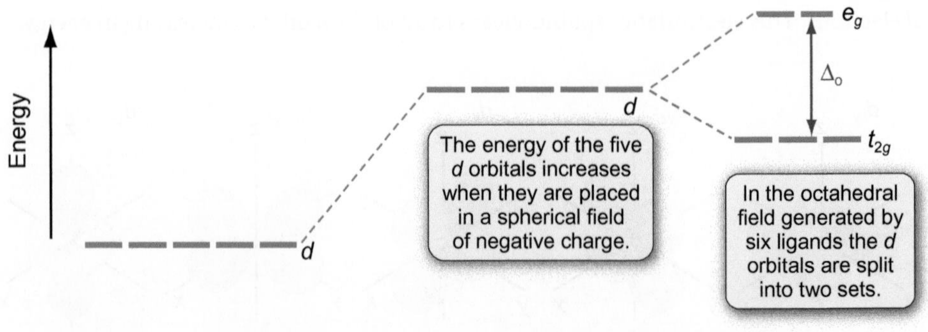
Energy
d
d
The energy of the five d orbitals increases when they are placed in a spherical field of negative charge.
e_g
Δ_o
t_2g
In the octahedral field generated by six ligands the d orbitals are split into two sets.

Figure 28.23 In an octahedral field, the *d* orbitals are split into two sets.

destabilized. Each t_{2g} orbital is stabilized by $-0.4\Delta_o$ $(-\frac{2}{5}\Delta_o)$, and each e_g orbital is destabilized by $+0.6\Delta_o$ $(+\frac{3}{5}\Delta_o)$.

$$\underset{\text{total stabilization}}{3 \times (-0.4\Delta_o)} + \underset{\text{total destabilization}}{2 \times (+0.6\Delta_o)} = 0$$

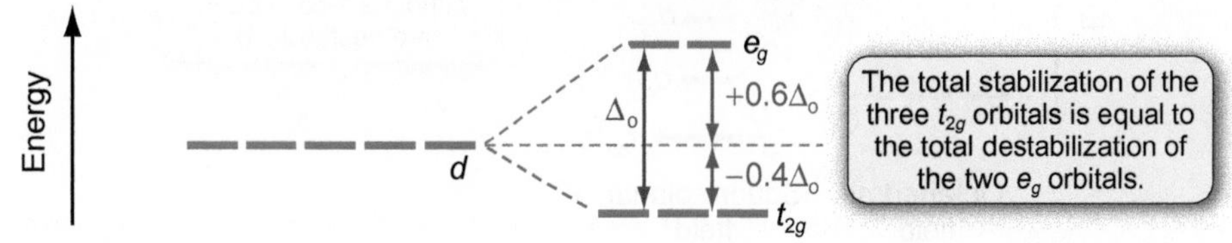
Energy
d
Δ_o
e_g
$+0.6\Delta_o$
$-0.4\Delta_o$
t_{2g}
The total stabilization of the three t_{2g} orbitals is equal to the total destabilization of the two e_g orbitals.

Tetrahedral complexes

You can use the same logic to determine how the energies of the *d* orbitals are split in other geometries. When the spherical field is rearranged into a tetrahedral field, none of the *d* orbitals is directed exactly towards the point charges (Figure 28.24). The $d_{x^2-y^2}$ and d_{z^2} orbitals are directed halfway between the point charges, whereas the d_{xy}, d_{xz}, and d_{yz} orbitals point more towards the charges. Therefore the d_{xy}, d_{xz}, and d_{yz} orbitals are more influenced by the charges than the $d_{x^2-y^2}$ and d_{z^2} orbitals, so they are destabilized relative to them. This leads to the splitting pattern shown in Figure 28.25.

The sets of orbitals in a tetrahedral complex are labelled *e* and t_2. The splitting between the *e* and t_2 orbitals, Δ_t (Δ_{tet}) is smaller than Δ_o. For the same metal ion and ligands, Δ_o and Δ_t are related by Equation 28.1.

$$\Delta_t = \tfrac{4}{9}\Delta_o \qquad (28.1)$$

The *g* subscript is not used for tetrahedral complexes because a tetrahedron does not have a centre of symmetry.

Square planar complexes

Although it is possible to derive the splitting pattern for a square planar complex in a similar way, starting from the spherical field, it is easier to start from an octahedral field and

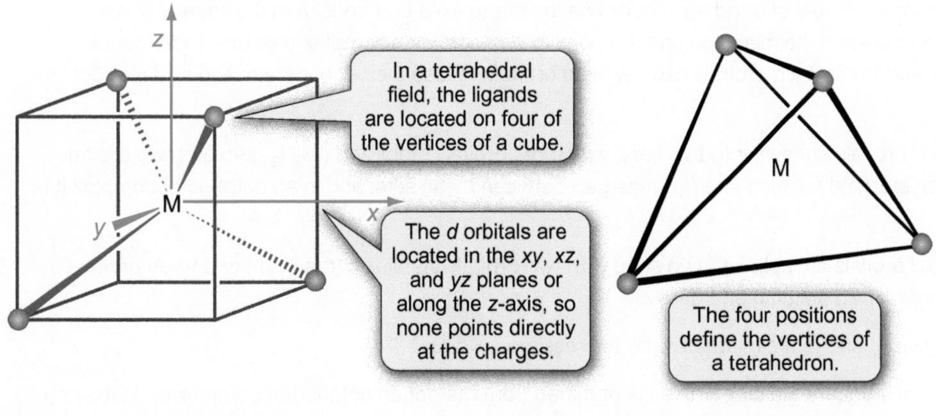
z
y
x
M
In a tetrahedral field, the ligands are located on four of the vertices of a cube.
The d orbitals are located in the xy, xz, and yz planes or along the z-axis, so none points directly at the charges.
M
The four positions define the vertices of a tetrahedron.

Figure 28.24 In a tetrahedral field all of the *d* orbitals are directed between the point charges.

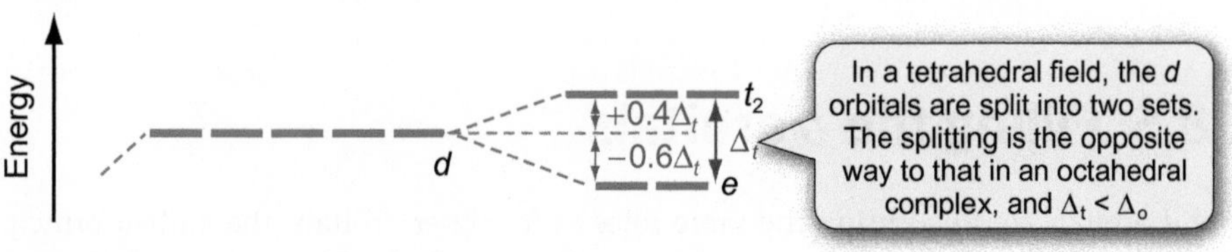
Energy
d
t_2
$+0.4\Delta_t$
$-0.6\Delta_t$
Δ_t
e
In a tetrahedral field, the d orbitals are split into two sets. The splitting is the opposite way to that in an octahedral complex, and $\Delta_t < \Delta_o$

Figure 28.25 In a tetrahedral field, the *d* orbitals are split into two sets.

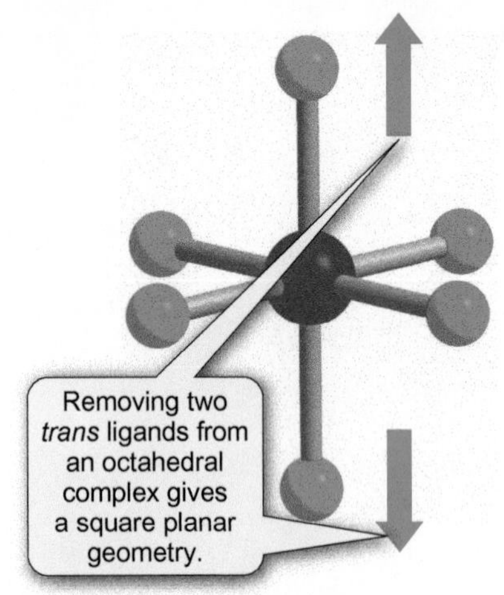
Removing two *trans* ligands from an octahedral complex gives a square planar geometry.

Figure 28.26 A square planar geometry is obtained by removing two ligands from an octahedral complex.

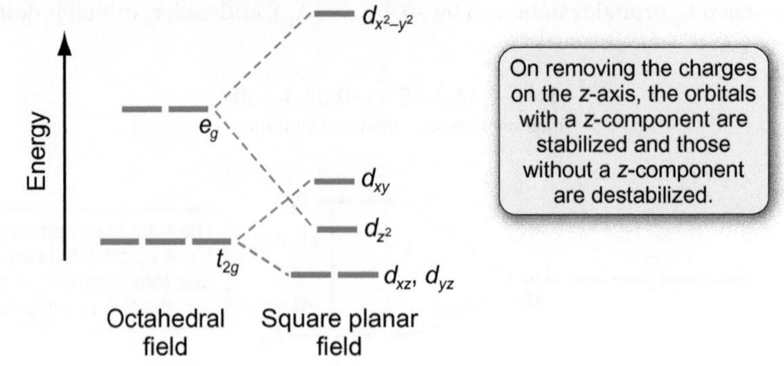
Energy
$d_{x^2-y^2}$
e_g
d_{xy}
d_{z^2}
t_{2g}
d_{xz}, d_{yz}
Octahedral field
Square planar field
On removing the charges on the z-axis, the orbitals with a z-component are stabilized and those without a z-component are destabilized.

Figure 28.27 The splitting of the *d* orbitals in a square planar field.

form a square planar complex by removing two of the ligands, as shown in Figure 28.26. If the ligands on the *z*-axis are removed, any orbital that points along this axis will experience less electrostatic repulsion, so it will be stabilized. This means that the d_{z^2} orbital is stabilized. The d_{xz} and d_{yz} orbitals are also stabilized, but to a lesser degree as these orbitals are not pointing directly along the *z*-axis. The $d_{x^2-y^2}$ and d_{xy} orbitals do not have a *z*-component, so they are not stabilized. Indeed, to maintain the same average energy of the five *d* orbitals as in a spherical field, the $d_{x^2-y^2}$ and d_{xy} orbitals must increase in energy. This leads to the orbital splitting pattern shown in Figure 28.27.

The occurrence of square planar complexes is described in Section 28.5 (p.1287).

The actual order of the *d* orbital energies in a square planar complex depends on the nature of the metal and the ligands, and in some cases may differ from that shown in Figure 28.27. Despite this, there is always a large energy gap between the highest energy orbital (the $d_{x^2-y^2}$) and the next highest orbital.

Summary

- Crystal field theory is an electrostatic model of bonding in coordination complexes that considers the ligands as point charges. Electrostatic attraction between the metal ion and the point charges holds a complex together. Electrostatic repulsion between the ligands and the metal *d* orbitals causes the *d* orbitals to lose their degeneracy and split into sets of different energies.
- In an octahedral complex, the *d* orbitals are split into two sets, with three orbitals stabilized (the t_{2g} set) and two orbitals destabilized (the e_g set). The crystal field splitting energy is the gap between these sets, and in an octahedral complex it is given the symbol Δ_o.
- In an tetrahedral complex, the *d* orbitals are split into two sets, with two orbitals stabilized (the e set) and three orbitals destabilized (the t_2 set). The crystal field splitting energy is Δ_t.
- The crystal field splitting energies Δ_o and Δ_t are related by the expression $\Delta_t = \frac{4}{9}\Delta_o$.
- The *d*-orbital splitting diagram for a square planar complex is obtained from that for an octahedral complex by stabilizing all of the orbitals with a *z*-component and destabilizing those without a *z*-component.

28.5 Filling the *d* orbitals

The aufbau principle (p.145), the Pauli exclusion principle (p.145), and Hund's rule (p.147) are explained in Section 3.6.

The *d* orbitals are filled using the same rules as for other orbitals: the aufbau principle along with the Pauli exclusion principle and Hund's rule. A Ti^{3+} ion has the electronic configuration [Ar] $3d^1$, so for the octahedral complex $[Ti(H_2O)_6]^{3+}$ there is one electron to

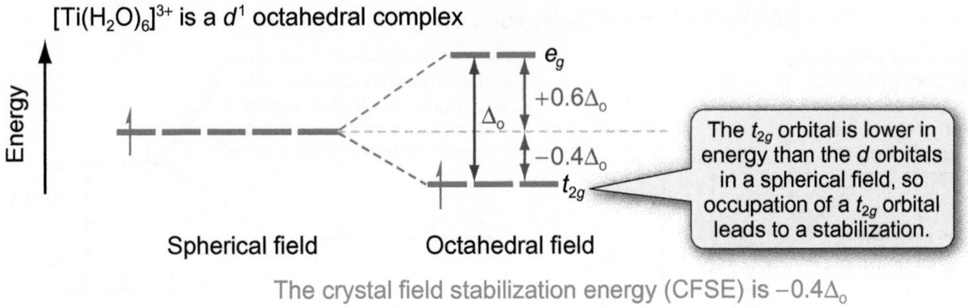
[Ti(H2O)6]3+ is a d1 octahedral complex
Energy
eg
+0.6Δo
Δo
−0.4Δo
t2g
The t2g orbital is lower in energy than the d orbitals in a spherical field, so occupation of a t2g orbital leads to a stabilization.
Spherical field
Octahedral field
The crystal field stabilization energy (CFSE) is −0.4Δo

Figure 28.28 The crystal field splitting diagram for $[Ti(H_2O)_6]^{3+}$, a d^1 complex.

place into the *d* orbitals. This will go into one of the t_{2g} orbitals, since these are of lower energy than the e_g orbitals (Figure 28.28).

As the t_{2g} orbitals are lower in energy than the *d* orbitals in a free ion in a spherical field, there is a stabilization associated with forming the octahedral complex. This is called the **crystal field stabilization energy (CFSE)**. For $[Ti(H_2O)_6]^{3+}$ the CFSE is equal to $-0.4\Delta_o$ because the electron occupies an orbital that is $0.4\Delta_o$ more stable than it would be in the free ion in a spherical field. The more negative the CFSE value is, the greater the stabilization on forming an octahedral complex.

V^{3+} and Cr^{3+} ions have valence electronic configurations $3d^2$ and $3d^3$, respectively. For octahedral $[V(H_2O)_6]^{3+}$ and $[Cr(H_2O)_6]^{3+}$ complexes, the electrons go into the t_{2g} orbitals and maximize the number of parallel electron spins, as expected from Hund's rule (Figure 28.29). The electronic configurations of the V^{3+} and Cr^{3+} ions in the complexes $[V(H_2O)_6]^{3+}$ and $[Cr(H_2O)_6]^{3+}$ are written $(t_{2g})^2$ and $(t_{2g})^3$, respectively.

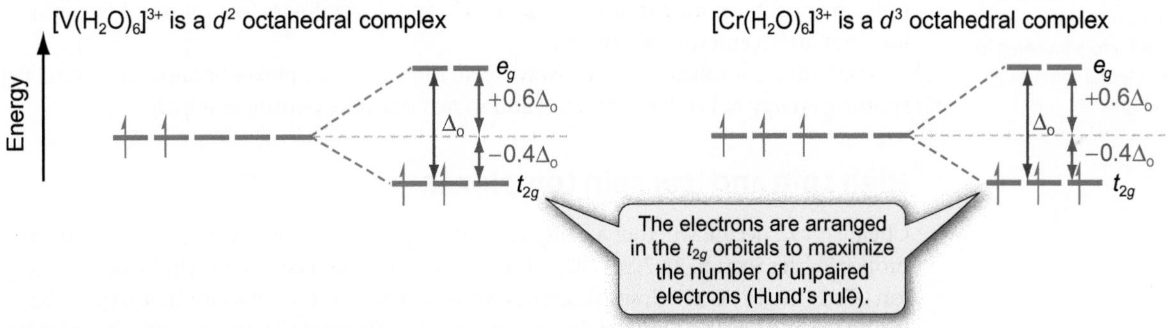
[V(H2O)6]3+ is a d2 octahedral complex
[Cr(H2O)6]3+ is a d3 octahedral complex
Energy
eg
+0.6Δo
Δo
−0.4Δo
t2g
eg
+0.6Δo
Δo
−0.4Δo
t2g
The electrons are arranged in the t2g orbitals to maximize the number of unpaired electrons (Hund's rule).

Figure 28.29 The crystal field splitting diagrams for $[V(H_2O)_6]^{3+}$ and $[Cr(H_2O)_6]^{3+}$.

Worked example 28.4 Crystal field stabilization energy

Calculate the crystal field stabilization energy (CFSE) for $[V(H_2O)_6]^{3+}$ in terms of Δ_o.

Strategy

Write down the valence electronic configuration for the vanadium ion in the complex.

Place the electrons into the lowest energy orbitals, obeying Hund's rule. Each electron in a t_{2g} orbital contributes $-0.4\Delta_o$, and each electron in an e_g orbital contributes $+0.6\Delta_o$ to the crystal field stabilization energy.

Solution

$[V(H_2O)_6]^{3+}$ is an octahedral complex of vanadium(III). The valence electronic configuration of the vanadium ion is $3d^2$. In an octahedral complex, both of these electrons are in t_{2g} orbitals, so the crystal field stabilization energy is given by

$$\text{CFSE} = 2 \times (-0.4\Delta_o) + 0 \times (+0.6\Delta_o) = -0.8\Delta_o$$

Question

Calculate the crystal field stabilization energy for $[Cr(H_2O)_6]^{3+}$.

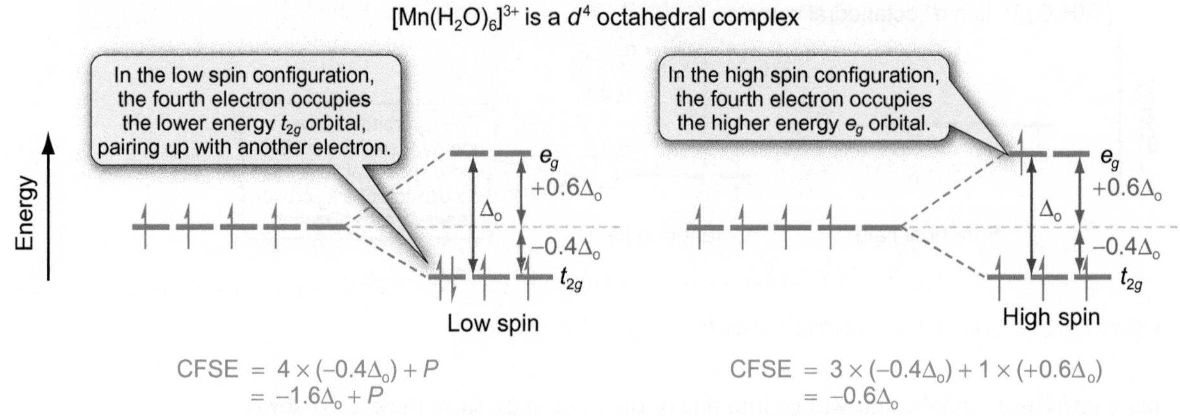
[Mn(H2O)6]3+ is a d4 octahedral complex
In the low spin configuration, the fourth electron occupies the lower energy t2g orbital, pairing up with another electron.
In the high spin configuration, the fourth electron occupies the higher energy eg orbital.
Energy
eg
+0.6Δo
Δo
−0.4Δo
t2g
Low spin
High spin
CFSE = 4 × (−0.4Δo) + P
= −1.6Δo + P
CFSE = 3 × (−0.4Δo) + 1 × (+0.6Δo)
= −0.6Δo

Figure 28.30 $[Mn(H_2O)_6]^{3+}$ is a d^4 complex. There are two possible ways that four electrons can be put into the t_{2g} and e_g orbitals, leading to low spin and high spin arrangements.

P is the **pairing energy**. It results from the greater electron–electron repulsion that occurs when two electrons are in the same orbital rather than being in different orbitals with parallel spins.

For the $[Mn(H_2O)_6]^{3+}$ ion ($3d^4$) there are two possible ways that the *d* orbitals can be occupied, as shown in Figure 28.30. The fourth electron can pair up with an electron in a t_{2g} orbital, or it can go into an e_g orbital. Putting the electron in the t_{2g} orbital leads to an increase in electron–electron repulsion, but putting the electron in the e_g orbital involves occupying a higher energy orbital. Which of these possibilities is favoured depends on the relative magnitudes of Δ_o and the pairing energy, *P*. If $\Delta_o > P$, the fourth electron will go into the t_{2g} orbital giving a **low spin complex**. If $\Delta_o < P$, the electron will go into the e_g orbital giving a **high spin complex**.

CFSEs for the electronic configurations d^1 to d^9 in both octahedral and tetrahedral geometries are summarized in Table 28.7. For d^1, d^2, d^3, d^8, and d^9 octahedral complexes there is only one possible configuration. For d^4, d^5, d^6, and d^7 octahedral complexes, high spin and low spin arrangements are possible.

Visit the Online Resource Centre to view screencast 28.2 which walks you through the factors involved in whether an octahedral complex is high spin or low spin.

Tetrahedral complexes almost always form high spin complexes because the crystal field splitting energy Δ_t is relatively small and so less than the pairing energy *P*.

High spin and low spin complexes

Chemists can decide whether a complex is high spin or low spin by measuring its magnetic properties (see Section 28.7, p.1292). Magnetic measurements show that the hexaaquairon(II) ion ($[Fe(H_2O)_6]^{2+}$) is a high spin complex, whereas the hexacyanoferrate(II) ion ($[Fe(CN)_6]^{4-}$) is a low spin complex. What is the reason for this difference? There are a number of factors that influence whether a compound is high spin or low spin. The key ones are the following.

- *The charge on the metal ion.* The higher the charge on the central ion, the larger the electrostatic interactions and hence the larger Δ. Increasing Δ makes it more likely that the complex will be low spin. For example, $[Co(H_2O)_6]^{3+}$ (d^6) is a low spin complex, whereas $[Co(H_2O)_6]^{2+}$ (d^7) is a high spin complex.
- *The nature of the metal.* Δ is generally larger for second and third row *d*-block ions than it is for first row *d*-block ions. In addition, the pairing energy *P* decreases down each triad as the orbitals get larger. Taken together, these factors mean that complexes of the second and third row metals are more likely to be low spin. For example, in Group 8, $[Ru(H_2O)_6]^{2+}$ is a low spin complex, but $[Fe(H_2O)_6]^{2+}$ is a high spin complex.
- *The nature of the ligands.* Some ligands produce larger values of Δ than others.

The information to create the spectrochemical series comes from UV–visible spectroscopy, hence the name.

The ranking of ligands in order of decreasing Δ is known as the **spectrochemical series**, and this is shown below for some common ligands

$$CO > CN^- > en > NH_3 > py > NCS^- > H_2O > OH^- > F^- > N_3^- > Cl^- > SCN^- > Br^- > I^-$$

Large Δ — Strong field ligands; Small Δ — Weak field ligands

Table 28.7 *d*-Electron arrangements and crystal field stabilization energies for d^1 to d^9 octahedral and tetrahedral complexes*

Valence electronic configuration of metal ion	Octahedral complex		Tetrahedral complex
d^1	e_g: — — t_{2g}: ↑ — — $-0.4\Delta_o$		t_2: — — — e: ↑ — $-0.6\Delta_t$
d^2	e_g: — — t_{2g}: ↑ ↑ — $-0.8\Delta_o$		t_2: — — — e: ↑ ↑ $-1.2\Delta_t$
d^3	e_g: — — t_{2g}: ↑ ↑ ↑ $-1.2\Delta_o$		t_2: ↑ — — e: ↑ ↑ $-0.8\Delta_t$
d^4	e_g: ↑ — t_{2g}: ↑ ↑ ↑ $-0.6\Delta_o$ (high spin)	e_g: — — t_{2g}: ↑↓ ↑ ↑ $-1.6\Delta_o + P$ (low spin)	t_2: ↑ ↑ — e: ↑ ↑ $-0.4\Delta_t$
d^5	e_g: ↑ ↑ t_{2g}: ↑ ↑ ↑ 0 (high spin)	e_g: — — t_{2g}: ↑↓ ↑↓ ↑ $-2.0\Delta_o + 2P$ (low spin)	t_2: ↑ ↑ ↑ e: ↑ ↑ 0
d^6	e_g: ↑ ↑ t_{2g}: ↑↓ ↑ ↑ $-0.4\Delta_o$ (high spin)	e_g: — — t_{2g}: ↑↓ ↑↓ ↑↓ $-2.4\Delta_o + 2P$ (low spin)	t_2: ↑ ↑ ↑ e: ↑↓ ↑ $-0.6\Delta_t$
d^7	e_g: ↑ ↑ t_{2g}: ↑↓ ↑↓ ↑ $-0.8\Delta_o$ (high spin)	e_g: ↑ — t_{2g}: ↑↓ ↑↓ ↑↓ $-1.8\Delta_o + P$ (low spin)	t_2: ↑ ↑ ↑ e: ↑↓ ↑↓ $-1.2\Delta_t$
d^8	e_g: ↑ ↑ t_{2g}: ↑↓ ↑↓ ↑↓ $-1.2\Delta_o$		t_2: ↑↓ ↑ ↑ e: ↑↓ ↑↓ $-0.8\Delta_t$
d^9	e_g: ↑↓ ↑ t_{2g}: ↑↓ ↑↓ ↑↓ $-0.6\Delta_o$		t_2: ↑↓ ↑↓ ↑ e: ↑↓ ↑↓ $-0.4\Delta_t$

* The number of pairing energies, *P*, given in the table for the low spin complexes is the *difference* between the number of paired electrons in the low spin and free atom cases.

Ligands that give rise to large values of Δ are known as **strong field ligands**, whereas ligands that give rise to small values of Δ are called **weak field ligands**.

To see how the nature of the ligands can affect whether a complex is high spin or low spin, consider the chromium(II) complex ion $[Cr(H_2O)_6]^{2+}$. Spectroscopic measurements

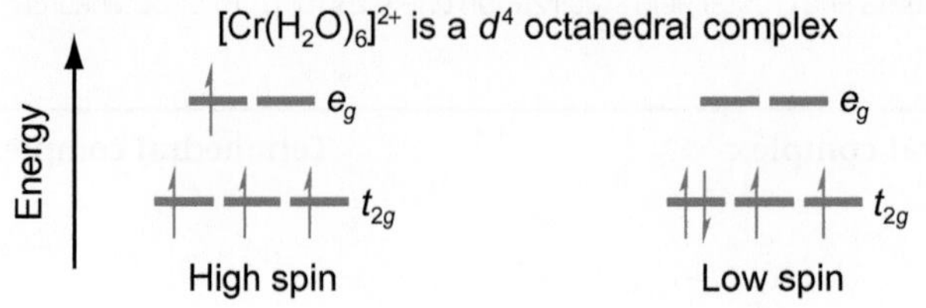
[Cr(H2O)6]2+ is a d4 octahedral complex
Energy
eg
t2g
High spin
eg
t2g
Low spin

High spin: CFSE $= -0.6\Delta_o = -0.6 \times 170\,\text{kJ mol}^{-1} = -102\,\text{kJ mol}^{-1}$

Low spin: CFSE $= -1.6\Delta_o + P = (-1.6 \times 170\,\text{kJ mol}^{-1}) + 245\,\text{kJ mol}^{-1} = -27\,\text{kJ mol}^{-1}$

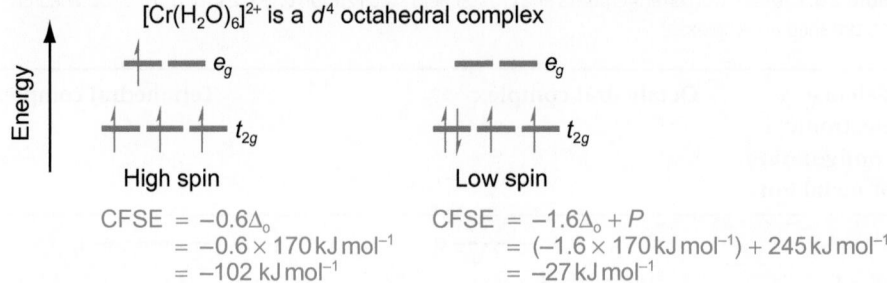

Figure 28.31 The high spin and low spin possibilities for $[Cr(H_2O)_6]^{2+}$.

show that Δ_o is 170 kJ mol^{-1} and that for Cr^{2+} the pairing energy, *P*, is 245 kJ mol^{-1}. The two possible arrangements of the *d* electrons are shown in Figure 28.31. $[Cr(H_2O)_6]^{2+}$ forms the high spin complex as this has the lower energy (i.e. it is more stable). Worked example 28.5 provides another illustration of this principle.

Worked example 28.5 High spin or low spin?

For the complex $[Cr(CN)_6]^{4-}$, spectroscopic measurements show that Δ_o is 380 kJ mol^{-1}. The pairing energy, *P*, for Cr^{2+} is 245 kJ mol^{-1}. Calculate the CFSE for high spin and low spin $[Cr(CN)_6]^{4-}$ and use these values to predict whether $[Cr(CN)_6]^{4-}$ is a high spin or a low spin complex.

Strategy

Use Table 28.7 to draw the electronic arrangements for the high spin and low spin complexes. Work out the CFSE for both possibilities. The complex will adopt the arrangement with the lower energy.

Solution

$[Cr(CN)_6]^{4-}$ is a low spin complex as this has the lower energy.

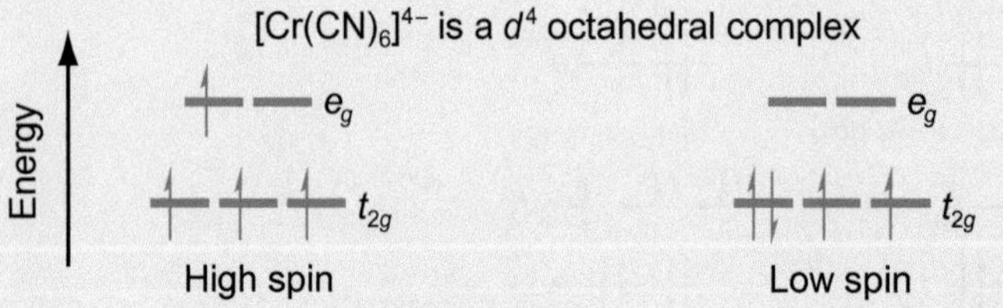
[Cr(CN)6]4- is a d4 octahedral complex
Energy
eg
t2g
High spin
eg
t2g
Low spin

High spin: CFSE $= -0.6\Delta_o = -0.6 \times 380\,\text{kJ mol}^{-1} = -228\,\text{kJ mol}^{-1}$

Low spin: CFSE $= -1.6\Delta_o + P = (-1.6 \times 380\,\text{kJ mol}^{-1}) + 245\,\text{kJ mol}^{-1} = -363\,\text{kJ mol}^{-1}$

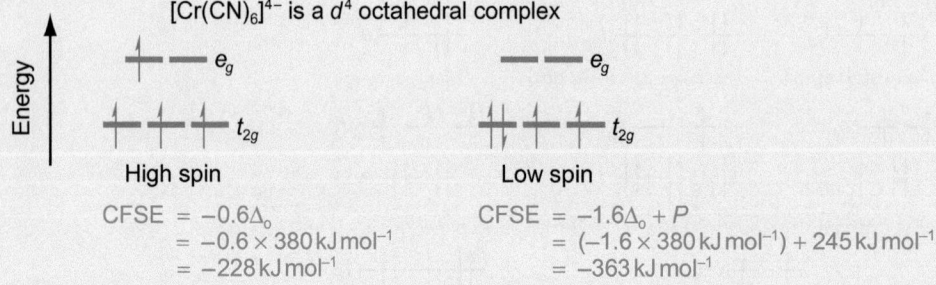

Question

For $[Fe(H_2O)_6]^{2+}$, Δ_o is 120 kJ mol^{-1}. Assume Fe^{2+} has the same pairing energy as Cr^{2+}. Calculate the CFSE for high spin and low spin $[Fe(H_2O)_6]^{2+}$, and predict whether this complex is high spin or low spin.

Although it is useful to be able to work out the value of the CFSE for the high spin and low spin possibilities, it is not necessary to calculate this every time you want to predict whether a particular complex is high spin or low spin. A complex will be high spin if $\Delta_o < P$, and low spin if $\Delta_o > P$. First row *d*-block hexaaqua complexes of the general formula $[M(H_2O)_6]^{2+}$ are normally high spin complexes, whereas hexacyano complexes are always low spin complexes.

The difference in size of high spin and low spin iron complexes is important in the function of haemoglobin. This is described in Box 28.7 (p.1298).

High spin ions are normally larger than low spin ions. For example, the ionic radius of high spin Fe^{2+} is 78 pm (see Figure 28.3, p.1260), whereas the ionic radius of low spin Fe^{2+} is only 61 pm. From Table 28.7, the arrangements of the *d* electrons are

Fe^{2+} high spin: $(t_{2g})^4\,(e_g)^2$

Fe^{2+} low spin: $(t_{2g})^6\,(e_g)^0$

Since the t_{2g} orbitals point between the ligands, and the e_g orbitals point at the ligands, occupation of the e_g orbitals leads to a greater degree of electron–electron repulsion between the *d* electrons and the ligands. This lengthens the bonds in the high spin complexes.

CFSE and geometry

Chromium(III) complexes are always octahedral, but cobalt(II) complexes can be octahedral or tetrahedral. This observation can be explained by looking at the crystal field stabilization energies for the ions in the two possible geometries.

Chromium(III) has the electronic configuration [Ar] $3d^3$. From Table 28.7, the CFSE for an octahedral d^3 complex is $-1.2\Delta_o$, whereas that for a tetrahedral d^3 complex is $-0.8\Delta_t$. Since $\Delta_t = \frac{4}{9}\Delta_o$ (Equation 28.1) you can substitute for Δ_t in terms of Δ_o,

$$\text{CFSE for a tetrahedral } d^3 \text{ complex} = -0.8\Delta_t = -0.8 \times \tfrac{4}{9}\Delta_o = -0.36\Delta_o$$

So

$$\text{CFSE for the octahedral geometry} = -1.2\Delta_o$$

$$\text{CFSE for the tetrahedral geometry} = -0.36\Delta_o$$

Therefore, octahedral complexes of chromium(III) complexes are considerably more stable than tetrahedral complexes, so it follows that chromium(III) complexes are always octahedral.

You can carry out the same comparison for a d^7 complex such as cobalt(II). From Table 28.7, the CFSE for a high spin octahedral d^7 complex is $-0.8\Delta_o$, whereas that for a tetrahedral d^7 complex is $-1.2\Delta_t$. Substituting for Δ_t,

$$\text{CFSE for a tetrahedral } d^7 \text{ complex} = -1.2\Delta_t = -1.2 \times \tfrac{4}{9}\Delta_o = -0.53\Delta_o$$

In this case, the difference in CFSE between the two geometries is relatively small so both geometries are possible.

Square planar geometry is based on octahedral geometry, with the two axial ligands removed. Square planar geometry only occurs when the additional CFSE is greater than the electrostatic energy lost on losing the two axial ligands. This is only the case for strong field ligands, for example, in $[Ni(CN)_4]^{2-}$, or for second and third row *d*-block metals, for example, in $[PtCl_4]^{2-}$. Square planar geometry is generally only observed for the electronic configurations d^8 and d^9.

The Jahn-Teller effect

In the structure of $[Cu(H_2O)_6]^{2+}$ (**16**), the measured copper–oxygen bond lengths are not all the same, and the bonds to two of the oxygen atoms are longer and weaker than those to the other four. This type of distortion from a regular octahedron is called a **tetragonal distortion**.

A **tetragonal distortion** is a distortion from a regular octahedral geometry in which the bonds to two *trans* ligands (the axial ligands) become either longer or shorter than the bonds to the other four ligands (the equatorial ligands).

The tetragonal distortion in $[Cu(H_2O)_6]^{2+}$ is an example of the **Jahn–Teller effect**. The electronic configuration of Cu^{2+} is $3d^9$, so there are three electrons to place in the two *eg* orbitals. This means there are two possible electronic arrangements, both of equal energy.

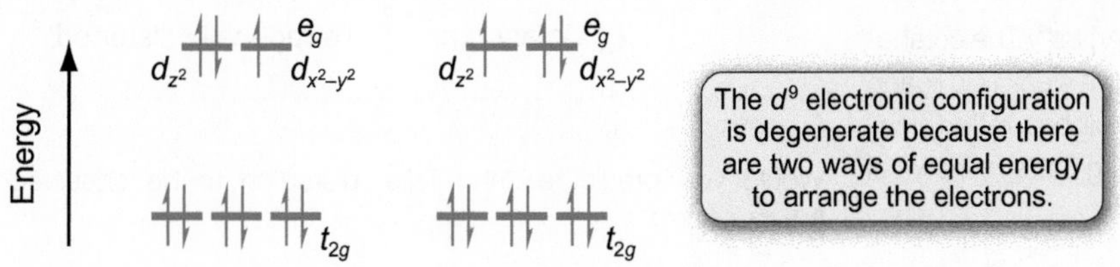

Energy
e_g
d_{z^2}
$d_{x^2-y^2}$
t_{2g}
The d^9 electronic configuration is degenerate because there are two ways of equal energy to arrange the electrons.

This ion has a tetragonal distortion. The two bonds to the *axial* ligands are longer than those to the *equatorial* ligands.

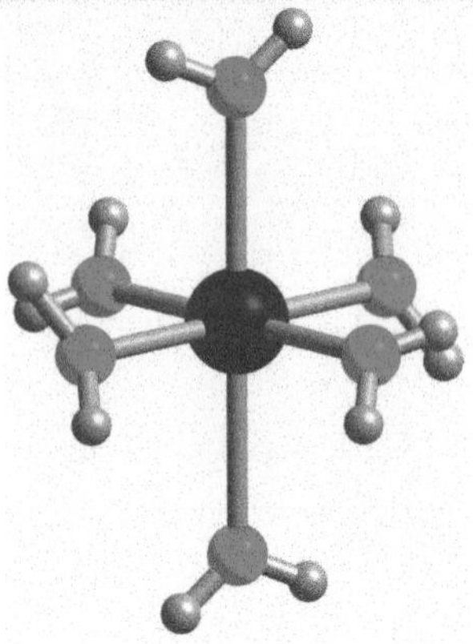

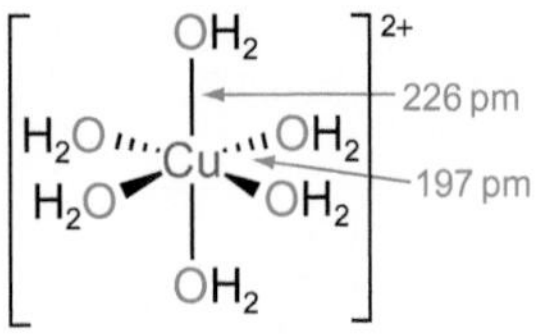

OH2
2+
226 pm
H2O
Cu
OH2
H2O
OH2
197 pm
OH2

16 $[Cu(H_2O)_6]^{2+}$.

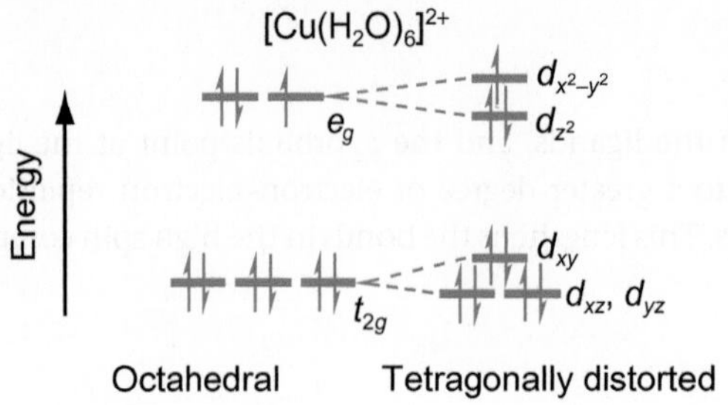

[Cu(H2O)6]2+
Energy
eg
dx2–y2
dz2
t2g
dxy
dxz, dyz
Octahedral
Tetragonally distorted

The Jahn–Teller distortion occurs when the e_g orbitals are unevenly occupied. The orbital energies are split and the complex undergoes a tetragonal distortion.

Figure 28.32 The Jahn–Teller effect in $[Cu(H_2O)_6]^{2+}$ leads to a splitting of the *d*-orbital energies from those in a regular octahedral geometry.

When there are two or more arrangements of electrons that are of equal energy, the state is described as degenerate.

A Jahn–Teller distortion occurs whenever a complex has a degenerate ground state. This is because the distortion causes one of the orbitals to have a lower energy than the others, and occupation of these leads to a lower overall energy for the complex. In the Cu^{2+} case, the Jahn–Teller distortion causes *d* orbitals with a *z*-component to be stabilized, whereas those without a *z*-component are destabilized. The Jahn–Teller effect is the cause of the tetragonal distortion in $[Cu(H_2O)_6]^{2+}$. The effect on the *d* orbital energies is shown in Figure 28.32.

Jahn–Teller distortions are particularly important for octahedral complexes with the valence electronic configurations d^4 and d^9, as these involve uneven occupation of the e_g orbitals. The Jahn–Teller effect is large because the $d_{x^2-y^2}$ and d_{z^2} orbitals have different shapes and are both directed at the ligands.

Jahn–Teller distortions also occur for other electronic configurations. For example, in d^1 octahedral complexes the electron can go into any one of the three t_{2g} orbitals. These three arrangements are degenerate, so d^1 complexes undergo a distortion that involves one of the *d* orbitals becoming lower in energy than the other two. Jahn–Teller distortions occur for many of the electronic configurations, but when they involve the t_{2g} orbitals they are relatively small, and often ignored.

Worked example 28.6 The Jahn–Teller effect

Would you predict a Jahn–Teller distortion in the complex ion $[Cr(H_2O)_6]^{2+}$?

Strategy

Determine the oxidation state of chromium in the complex and write down the valence electronic configuration of the chromium ion. Use the factors on p.1284 to decide whether the complex is likely to be high spin or low spin. If there is more than one way to arrange the electrons while obeying Hund's rule, the complex will undergo a Jahn–Teller distortion.

Solution

Chromium(II) has an electronic configuration of [Ar] $3d^4$. The octahedral complex is likely to be high spin, since water is a weak field ligand and the charge on the metal ion is only +2, so Δ_o will be relatively low. The complex has the high spin crystal field splitting pattern shown in Table 28.7 (p.1285).

There is one electron in each of the three t_{2g} orbitals, leaving one electron to go into one of the two e_g orbitals. This gives two possible arrangements, which are degenerate, so $[Cr(H_2O)_6]^{2+}$ will undergo a Jahn–Teller distortion. This will be relatively large, since it is the e_g orbitals that are unevenly filled.

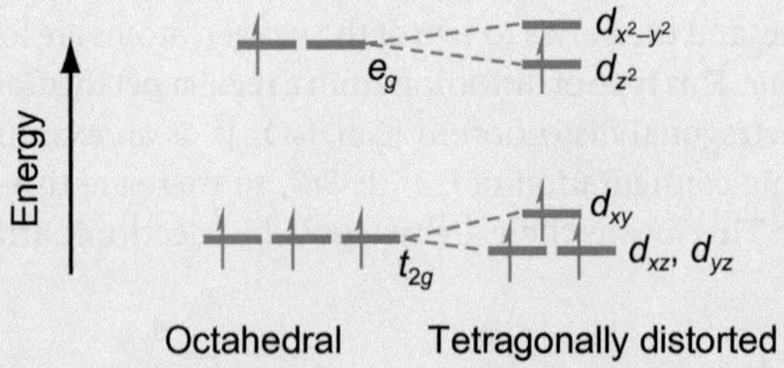

Energy
eg
dx2–y2
dz2
t2g
dxy
dxz, dyz
Octahedral
Tetragonally distorted

Question

Would you predict a Jahn–Teller distortion to be observed for $[Mn(H_2O)_6]^{2+}$?

Summary

- Uneven occupation of the *d* orbitals means that the electrons in a transition metal complex are generally at a lower energy than in the free ion in a spherical field. The difference is called the crystal field stabilization energy.
- Octahedral complexes with the electronic configurations d^4–d^7 have two possible arrangements, and the one that is adopted depends on the values of Δ_o and *P*, the pairing energy. If $\Delta_o < P$, a high spin complex is formed and, if $\Delta_o > P$, a low spin complex is formed.
- The ranking of ligands in order of decreasing Δ is called the spectrochemical series. Ligands with large values of Δ are called strong field ligands, and those with small values of Δ are called weak field ligands.
- Low spin complexes are favoured by a high charge on the metal ion and the presence of strong field ligands. Low spin complexes are more prevalent for second and third row *d*-block metals.
- Jahn–Teller distortions occur when there is a degenerate ground state. This is particularly significant for octahedral complexes in which the e_g orbitals are unevenly filled, which occurs in d^4 and d^9 configurations.

? For practice questions on these topics, see questions 10–14 at the end of this chapter (pp.1301–1302).

28.6 Colour in coordination compounds

One of the characteristic features of *d*-block metals is that many of their complex ions are coloured. Figure 28.33 shows the colours of a range of hexaaqua ions.

The colour can be explained using crystal field theory. The crystal field splitting energy, Δ, for most *d*-block metal complexes corresponds to the energy of a visible light photon. When white light passes through a solution containing $[Ti(H_2O)_6]^{3+}$ ions, photons of yellow and green light are absorbed. The energy of the light absorbed corresponds to excitation of an electron from a t_{2g} orbital to an e_g orbital (Figure 28.34). This gives rise to the visible absorption spectrum shown in Figure 28.35. Visible spectra of *d*-block ions are normally broad. This is because the transitions occur into one of a number of vibrational states of the electronically excited state, so the single band observed is really several overlapping peaks.

The violet colour observed for $[Ti(H_2O)_6]^{3+}$ is the colour of the light *transmitted* through the solution. It represents white light *minus* the light that has been absorbed. The observed colour is the complementary colour of that absorbed. The colour wheel in Figure 28.36 shows the relationship between absorbed and complementary colours.

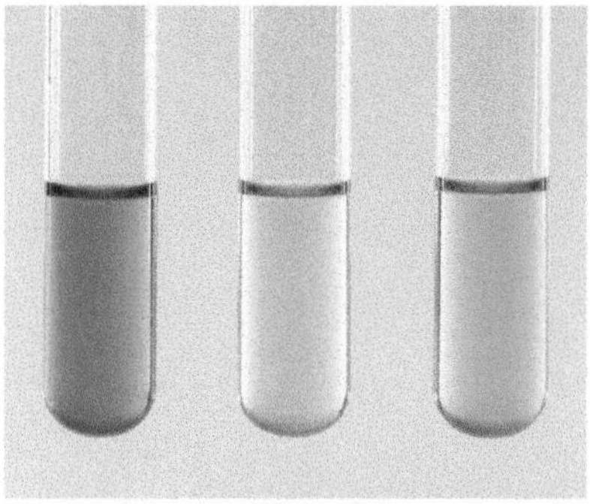

Figure 28.33 Complex ions are usually coloured. Some examples of hexaaqua complexes in aqueous solutions. From left to right: $[Co(H_2O)_6]^{2+}$, $[Ni(H_2O)_6]^{2+}$, and $[Cu(H_2O)_6]^{2+}$.

The reasons why visible spectra are broad are described in Section 10.6 (p.492).

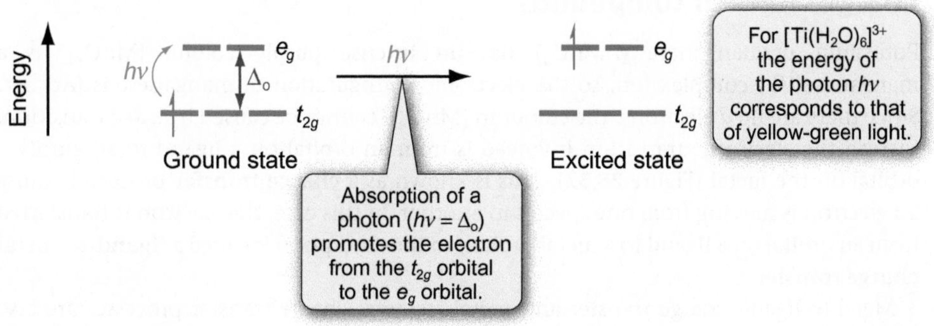
Energy
hν
e_g
Δ_o
t_{2g}
Ground state
hν
e_g
t_{2g}
Excited state
For $[Ti(H_2O)_6]^{3+}$ the energy of the photon hν corresponds to that of yellow-green light.
Absorption of a photon ($h\nu = \Delta_o$) promotes the electron from the t_{2g} orbital to the e_g orbital.

Figure 28.34 The origin of colour in $[Ti(H_2O)_6]^{3+}$. Absorption of a photon of the correct energy causes the electronic transition $t_{2g} \rightarrow e_g$.

The colours of many gemstones are caused by trace quantities of *d*-block ions in an ionic network or covalent network structure. Ruby and sapphire are impure forms of corundum (Al_2O_3). The red colour of rubies is due to Cr^{3+} and the blue colour of sapphires is due to Fe^{2+} and Ti^{4+}. Emerald is based on beryl ($Be_3Al_2(SiO_3)_6$) containing traces of Cr^{3+}.

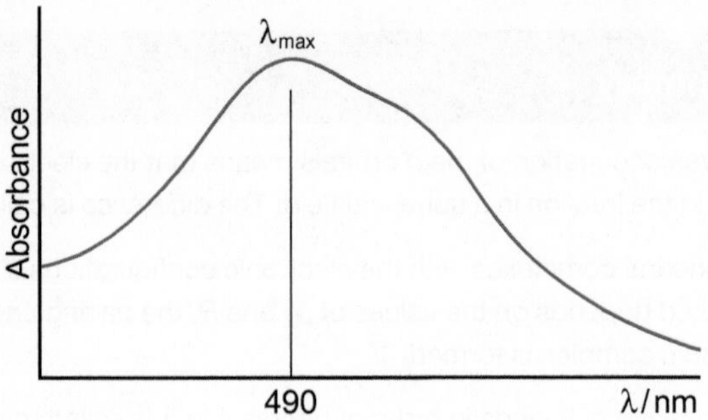

λ_{max}
Absorbance
490
λ/nm

Figure 28.35 The visible light absorption spectrum for $[Ti(H_2O)_6]^{3+}$.

Purple is opposite yellow-green on the colour wheel, so compounds that absorb yellow-green light, such as $[Ti(H_2O)_6]^{3+}$, appear purple.

Figure 28.36 In a colour wheel, the colour observed for a complex is opposite the colour of the light absorbed.

Crystal field theory helps to explain why different complexes of the same *d*-block ion have different colours. For example, solutions of $[Cr(H_2O)_6]^{3+}$ are violet, whereas those of $[Cr(NH_3)_6[^{3+}$ are yellow. Ammonia is higher in the spectrochemical series than water, so Δ_o is greater for the hexaammine complex than it is for the hexaaqua complex. For $[Cr(H_2O)_6]^{3+}$, green light is absorbed to give the violet complex and, for $[Cr(NH_3)_6]^{3+}$, higher energy violet light is absorbed to give the yellow complex.

Charge transfer compounds

Potassium permanganate ($KMnO_4$) has an intense purple colour. $[MnO_4]^-$ is a manganese(VII) complex ion, so the electronic configuration of manganese is [Ar] $3d^0$. Since there are no *d* electrons, the colour in $[MnO_4]^-$ cannot be caused by a *d–d* transition. Instead the electronic transition involved is from an orbital on a ligand to an empty *d* orbital on the metal (Figure 28.37). This is known as a **charge transfer** process, because an electron is moving from one species to another. In this case, the electron is transferred from an orbital on a ligand to a metal orbital, so the absorption is called a **ligand-to-metal charge transfer**.

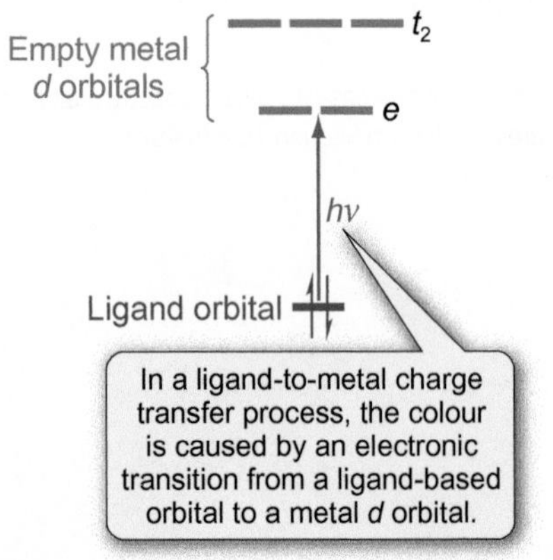

Empty metal *d* orbitals
t_2
e
$h\nu$
Ligand orbital
In a ligand-to-metal charge transfer process, the colour is caused by an electronic transition from a ligand-based orbital to a metal *d* orbital.

Figure 28.37 The purple colour of $[MnO_4]^-$ ions is due to a charge transfer process in which an electron is promoted from a ligand orbital to an empty *d* orbital on the metal.

Metal-to-ligand charge transfer and metal-to-metal charge transfer processes are also possible. For example, mixing the iron(II) complex $[Fe(CN)_6]^{4-}$ with an excess of Fe^{3+} (aq) gives an intense dark blue precipitate. This solid is known as *Prussian blue* and has the formula $Fe_4[Fe(CN)_6]_3$ (**17**). The compound has a structure in which cyanide ligands bridge between iron(II) and iron(III) centres. Visible light has the appropriate energy to transfer an electron from an orbital on the iron(II) centre to an orbital on iron(III), through the cyanide bridge.

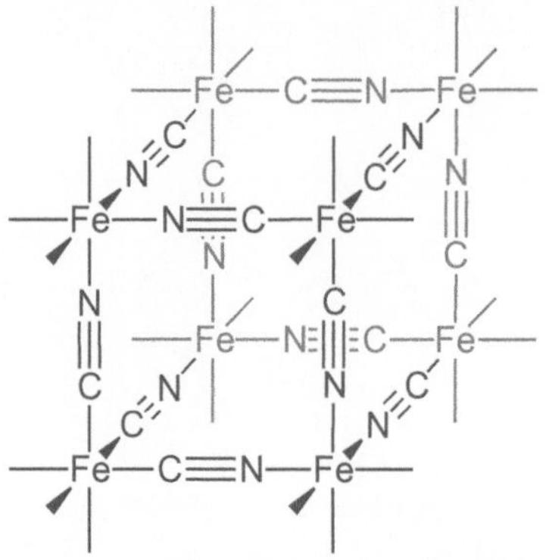

17 Prussian blue ($Fe_4[Fe(CN)_6]_3$).

Intensities

Figure 28.38 shows four solutions, all of the same molar concentration. Manganese(II) nitrate contains $[Mn(H_2O)_6]^{2+}$ ions, which are very pale pink. Cobalt(II) nitrate contains $[Co(H_2O)_6]^{2+}$ ions, which are a more intense pink. Sodium tetrachlorocobaltate(II) is an intense blue colour due to the presence of the $[CoCl_4]^{2-}$ ions, and potassium permanganate is a very intense purple. The colour of a compound is related to the energy gap between the orbitals involved in the electronic transition, as described on p.1289. The wavelength of maximum absorption, λ_{max}, is determined by measuring the visible spectrum.

The intensity of the colour is not related to λ_{max}. Instead, intensity is related to the absorbance, *A*, which is given by the Beer–Lambert law

$$A = \varepsilon c l \qquad (10.8)$$

For more on the Beer–Lambert law and selection rules see Section 10.3 (p.460).

where *c* is the concentration of the solution, *l* is the cell length, and ε is the molar absorption coefficient. The larger ε is, the more intense the colour. The value of ε for an electronic transition is influenced by selection rules. If a transition is allowed by a selection rule, it has a high probability of occurring, and ε will be large. If a transition is forbidden by a selection rule, it has a low probability of occurring, and ε will be small. For a *d*-block complex, there are three selection rules to consider.

- The **Laporte selection rule** states that, for an electronic transition to be allowed, the quantum number *l* must change by +1 or −1. This means that $s \rightarrow p$ transitions ($l = 0$ to $l = 1$) and $p \rightarrow d$ transitions ($l = 1$ to $l = 2$) are allowed, but $d \rightarrow d$ transitions ($l = 2$ to $l = 2$) are forbidden. This explains why $[MnO_4]^-$ is more intensely coloured than the other complexes shown in Figure 28.38. The colour in $[MnO_4]^-$ arises from a charge transfer process in which the electron moves from a ligand-based *p* orbital to a metal-based *d* orbital. This is not forbidden by the Laporte selection rule, whereas the colours of the other complexes all involve $d \rightarrow d$ transitions.
- The **parity selection rule** states that $g \rightarrow g$ and $u \rightarrow u$ transitions are forbidden, but $g \rightarrow u$ transitions are allowed. For an octahedral complex, the transition $t_{2g} \rightarrow e_g$ breaks the parity selection rule, so it is forbidden. For a tetrahedral complex, the transition $e \rightarrow t_2$ is not affected by the parity selection rule as there are no *g* and *u* labels. This means that tetrahedral complexes such as $[CoCl_4]^{2-}$ are more intensely coloured than octahedral complexes such as $[Co(H_2O)_6]^{2+}$.
- The **spin selection rule** states that the spin of an electron cannot change during an electronic transition. This is particularly important for d^5 complexes, as shown in Figure 28.39.

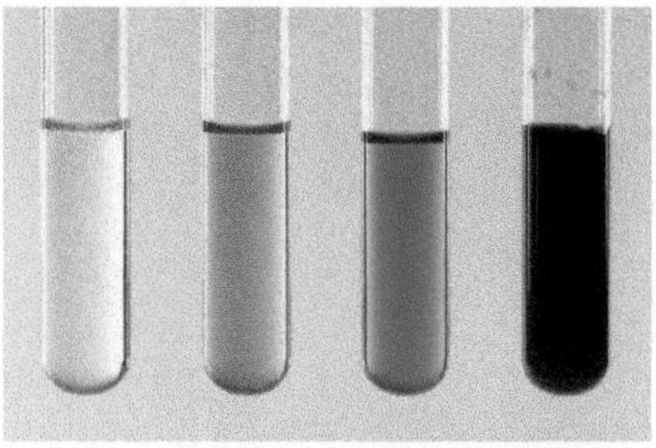

Figure 28.38 Solutions of $[Mn(H_2O)_6]^{2+}$, $[Co(H_2O)_6]^{2+}$, $[CoCl_4]^{2-}$, and $[MnO_4]^-$. Each solution contains the same concentration of the ions.

The electronic transition in $[Mn(H_2O)_6]^{2+}$ breaks all three selection rules, which explains why the colour is so weak.

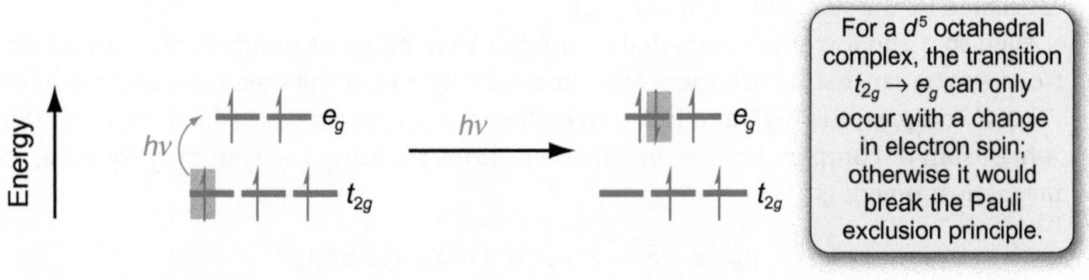
Energy
hν
e_g
t_{2g}
hν
e_g
t_{2g}
For a d^5 octahedral complex, the transition $t_{2g} \rightarrow e_g$ can only occur with a change in electron spin; otherwise it would break the Pauli exclusion principle.

Figure 28.39 The electronic transitions in d^5 complexes such as $[Mn(H_2O)_6]^{2+}$ require a change in electron spin, which breaks the spin selection rule.

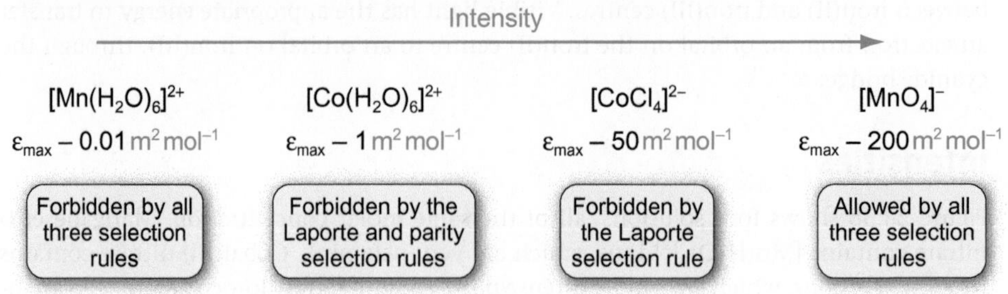
Intensity
[Mn(H2O)6]2+
εmax – 0.01 m2 mol–1
Forbidden by all three selection rules
[Co(H2O)6]2+
εmax – 1 m2 mol–1
Forbidden by the Laporte and parity selection rules
[CoCl4]2–
εmax – 50 m2 mol–1
Forbidden by the Laporte selection rule
[MnO4]–
εmax – 200 m2 mol–1
Allowed by all three selection rules

Summary

- Many *d*-block complexes are coloured because photons of visible light give rise to electronic transitions between the *d* orbitals.
- The colour observed for a complex ion is complementary to the colour of the light absorbed.
- The intensity of an absorption is affected by three selection rules—the Laporte, parity, and spin selection rules. The more selection rules that are broken, the less intense the absorption.
- Charge-transfer absorptions do not involve $d \rightarrow d$ transitions, so they generally give rise to intense colours.
- As a result of the parity selection rule, tetrahedral complexes are generally more intensely coloured than octahedral complexes.

For practice questions on these topics, see questions 13–15 at the end of this chapter (p.1302).

28.7 Magnetic properties

Paramagnetism is described in Box 4.2 (p.176).

μ_{so} means the magnetic moment calculated from the spin-only formula. The Bohr magneton is the unit of magnetic moments. It is a non-SI unit that is equal to $9.274 \times 10^{-24}\,\mathrm{J\,T^{-1}}$ where T is the symbol for a tesla, the unit of magnetic flux density.

In addition to the electron spin, the orbital angular momentum also contributes to paramagnetism. For most first row *d*-block compounds the orbital contribution can be ignored, and the experimental values are normally close to those predicted by only taking into account the contribution from the electron spin—hence the 'spin-only' formula. For second and third row compounds the orbital contribution needs to be included.

Many *d*-block complexes have unpaired electrons, and so are paramagnetic. For most first row *d*-block complexes, the magnetic moment of a complex (μ) is related to the number of unpaired electrons present (n) through the **spin-only formula**, given in Equation 28.2

$$\mu_{so} = \sqrt{n(n+2)}\,\mu_B \qquad (28.2)$$

where μ_B is a collection of fundamental constants ($eh/4\pi m_e$) known as the Bohr magneton.

For $[Ti(H_2O)_6]^{3+}$ there is one unpaired electron. From the spin-only formula

$$\mu_{so} = \sqrt{1(1+2)}\,\mu_B = \sqrt{3}\,\mu_B = 1.73\,\mu_B$$

The measured value of magnetic moment is known as the **effective magnetic moment** (μ_{eff}). For $[Ti(H_2O)_6](NO_3)_3$, μ_{eff} is $1.7\,\mu_B$, so this is within experimental error of the value calculated by the spin-only formula (μ_{so}).

High spin and low spin octahedral complexes have different numbers of unpaired electrons, so you can distinguish them experimentally by measuring their magnetic moments. For a d^5 complex, the high and low spin configurations are shown in Table 28.7 (p.1285). A high spin d^5 complex has five unpaired electrons so, using the spin-only formula, the magnetic moment is

$$\mu_{so} = \sqrt{5(5+2)}\,\mu_B = \sqrt{35}\,\mu_B = 5.92\,\mu_B$$

A low spin d^5 complex has only one unpaired electron. In this case, $\mu_{so} = 1.73\,\mu_B$, which is similar to that for d^1 complexes.

Worked example 28.7 Magnetic moments

$[Cr(NH_3)_6]Cl_2$ has a measured magnetic moment, μ_{eff}, of 4.85 μ_B. Is it a high spin or a low spin complex?

Strategy

Determine the oxidation state of the chromium in the complex and write down its valence electronic configuration. Use Table 28.7 (p.1285) to determine the number of unpaired *d* electrons in a high spin complex and in a low spin complex. Use the spin-only formula (Equation 28.2) to work out the expected values (μ_{so}) for both arrangements. Whichever is closer to the experimental value will be the configuration of the complex.

Solution

$[Cr(NH_3)_6]Cl_2$ is a chromium(II) complex, so the free ion has a valence electronic configuration of $3d^4$. From Table 28.7, the high spin complex has a configuration of $(t_{2g})^3\ (e_g)^1$ in which there are four unpaired electrons. Using the spin-only formula

$$\mu_{so} = \sqrt{n(n+2)}\,\mu_B = \sqrt{4(4+2)}\,\mu_B = \sqrt{24}\mu_B = 4.90\,\mu_B$$

The low spin complex has the configuration of $(t_{2g})^4\ (e_g)^0$ in which there are two unpaired electrons. Using the spin-only formula

$$\mu_{so} = \sqrt{2(2+2)}\,\mu_B = \sqrt{8}\,\mu_B = 2.83\,\mu_B$$

The measured magnetic moment of 4.85 μ_B is very close to the value calculated for the high spin complex, so $[Cr(NH_3)_6]Cl_2$ is a high spin complex.

Question

The magnetic moment of $K_3[Fe(ox)_3]$ is measured to be 5.95 μ_B. Is this a high spin or a low spin complex? (ox = ethanedioato. See Table 28.4, p.1267.)

Measurements of magnetic moments are used to determine whether 4-coordinate complexes are tetrahedral or square planar. $[NiCl_2(PPh_3)_2]$ is a nickel(II) complex, so the nickel atom has a valence electron configuration of $3d^8$. Figure 28.40 shows the *d* orbital splitting diagrams for the tetrahedral and square planar arrangements. A tetrahedral d^8 complex has two unpaired electrons so it is paramagnetic with μ_{so} equal to 2.83 μ_B, whereas a square planar d^8 complex has no unpaired electrons so it is diamagnetic, not paramagnetic ($\mu_{so} = 0\,\mu_B$). Experimental measurements show that $[NiCl_2(PPh_3)_2]$ is paramagnetic with μ_{eff} very close to that predicted from the spin-only formula. This is good evidence that the geometry of the complex is tetrahedral.

Visit the Online Resource Centre to view screencast 28.3 which walks you through the use of magnetic data to determine whether a complex is high spin or low spin.

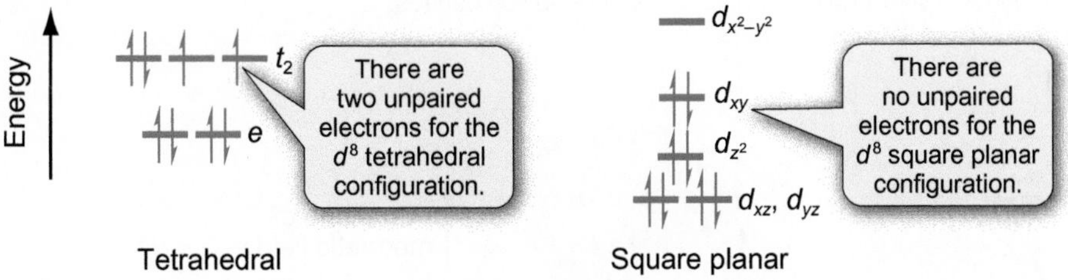
Energy
t_2
e
There are two unpaired electrons for the d^8 tetrahedral configuration.
Tetrahedral
$d_{x^2-y^2}$
d_{xy}
d_{z^2}
d_{xz}, d_{yz}
There are no unpaired electrons for the d^8 square planar configuration.
Square planar

Figure 28.40 The distribution of electrons in tetrahedral and square planar nickel(II) complexes.

Box 28.5 Ferromagnetism and recording information

Atoms and molecules with unpaired electrons are paramagnetic, and are weakly attracted to a magnet. In this case the **magnetic susceptibility**, χ, varies inversely with temperature, following the Curie law. The magnetic susceptibility is the degree of magnetization of a material in response to an applied magnetic field. The molar magnetic susceptibility, χ_m, is related to the effective magnetic moment (μ_{eff}) by the following equation

$$\mu_{eff} = 2.828\sqrt{\chi_m T}$$

Some materials, most notably the metals iron, nickel, and cobalt, have a much greater degree of magnetism. This is a result of a cooperative process called **ferromagnetism**. In a ferromagnetic material, the spins on the atoms or molecules spontaneously align in the same direction. This ferromagnetic ordering occurs below a critical temperature. Above this temperature, the thermal energy is high enough so that the spins are aligned randomly, and the material is paramagnetic but not ferromagnetic.

→

→

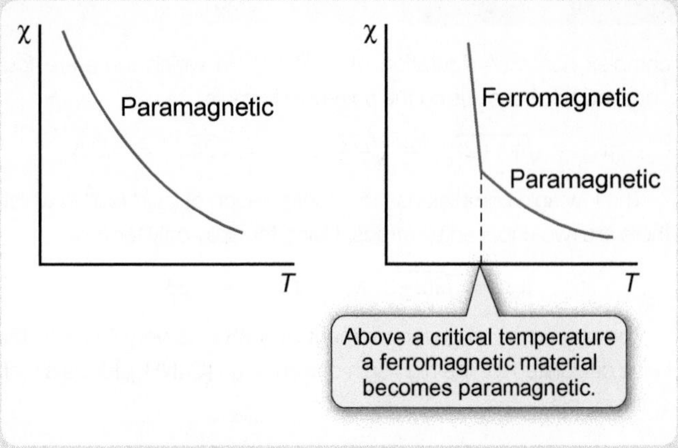
χ
Paramagnetic
T
χ
Ferromagnetic
Paramagnetic
T
Above a critical temperature a ferromagnetic material becomes paramagnetic.

▲ Variation of magnetic susceptibility, χ, with temperature, T, for paramagnetic and ferromagnetic materials.

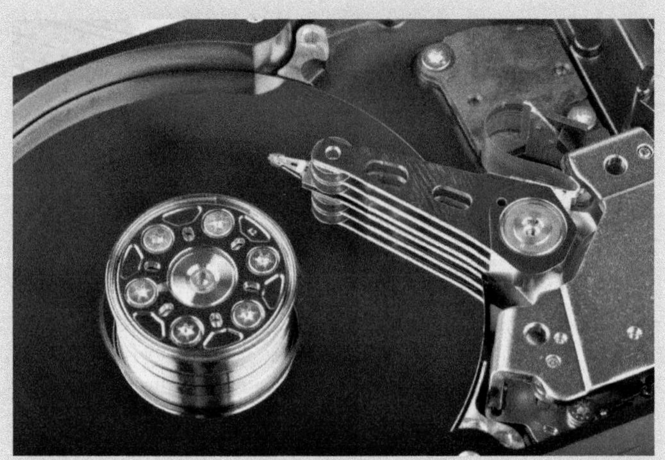

▲ A computer hard disk uses an electromagnet on the end of the arm to align domains of spins on the surface of the circular plate, in order to store information.

When a paramagnetic material is placed in a magnetic field, the magnetic moments on the individual molecules or atoms align in the direction of the field. When the field is removed, this alignment is lost. When a ferromagnetic material is placed in a magnetic field, the atoms or molecules also align in the direction of the field. When the field is removed, the orientation of the magnetic moments is retained, and the solid remains magnetized. The atomic distances need to be just right for paramagnetism to become ferromagnetism. In a ferromagnetic substance, large numbers of spins are aligned and 'locked together' over extensive regions of the solid, known as *domains*.

Ferromagnetism is not restricted to metals. Chromium(IV) oxide and iron(III) oxide are ferromagnetic and are used to record information. A typical magnetic tape consists of a plastic tape coated with Fe_2O_3 powder. The recording heads are electromagnets that align domains of spins as the tape passes beneath. These spin alignments remain in place for many years. When the tape is 'read', the magnetized material induces a voltage in the electromagnetic head which is held just above the tape. A similar process is used in computer hard disks, though in this case the magnetic material—Fe_2O_3 doped with cobalt—is layered onto an aluminium or glass disk. The strongest ferromagnets in production are based on NIB (neodymium–iron–boron, $Nd_2Fe_{14}B$). NIB magnets are used in mobile phones, microphones, and loudspeakers.

Question

The structure of Fe_2O_3 consists of a hexagonal close-packed array of oxide ions, with the iron ions in some of the octahedral sites. Given that oxide is a weak field ligand, determine the occupation of the t_{2g} and e_g orbitals on the iron centres.

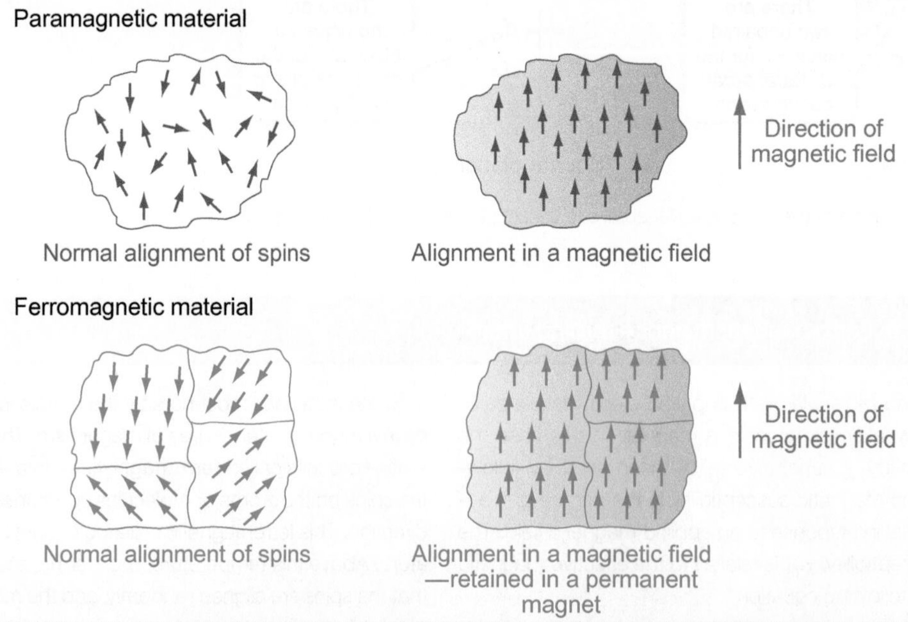
Paramagnetic material
Normal alignment of spins
Alignment in a magnetic field
Direction of magnetic field
Ferromagnetic material
Normal alignment of spins
Alignment in a magnetic field —retained in a permanent magnet
Direction of magnetic field

▲ The effects on paramagnetic and ferromagnetic materials of putting them into a magnetic field.

Summary

- The magnetic moment of a first row transition metal complex is related to the number of unpaired electrons through the spin-only formula.
- By comparing the measured magnetic moment with the value calculated using the spin-only formula you can determine whether an octahedral complex is high spin or low spin and whether a 4-coordinate complex is tetrahedral or square planar.

For practice questions on these topics, see questions 16–17 at the end of this chapter (p.1302).

28.8 Aqueous chemistry of the first row *d*-block ions

Equilibria and substitution reactions

The hexaamminenickel(II) complex is prepared by adding aqueous ammonia to a solution of a nickel salt in water. In this reaction, the aqua ligands coordinated to the nickel ion are replaced by ammine ligands

$$[Ni(H_2O)_6]^{2+}(aq) + 6NH_3(aq) \rightarrow [Ni(NH_3)_6]^{2+}(aq) + 6H_2O(l) \quad (28.3)$$

This is an example of a **ligand substitution** reaction.

Ligand substitution reactions occur stepwise, and for each of these steps there is an associated equilibrium constant, K_1, K_2, etc.

Equilibrium constants are described in Section 1.9 (p.56) and, in more detail, in Section 15.1 (p.697). Equilibrium constants for ligand substitution reactions are often called **stability constants**.

$$[Ni(H_2O)_6]^{2+}(aq) + NH_3(aq) \rightleftharpoons [Ni(NH_3)(H_2O)_5]^{2+}(aq) + H_2O(l)$$

$$K_1 = \frac{[Ni(NH_3)(H_2O)_5{}^{2+}][H_2O]}{[Ni(H_2O)_6{}^{2+}][NH_3]} = 630$$

$$[Ni(NH_3)(H_2O)_5]^{2+}(aq) + NH_3(aq) \rightleftharpoons [Ni(NH_3)_2(H_2O)_4]^{2+}(aq) + H_2O(l)$$

$$K_2 = \frac{[Ni(NH_3)_2(H_2O)_4{}^{2+}][H_2O]}{[Ni(NH_3)(H_2O)_5{}^{2+}][NH_3]} = 174$$

$$[Ni(NH_3)_2(H_2O)_4]^{2+}(aq) + NH_3(aq) \rightleftharpoons [Ni(NH_3)_3(H_2O)_3]^{2+}(aq) + H_2O(l)$$

$$K_3 = \frac{[Ni(NH_3)_3(H_2O)_3{}^{2+}][H_2O]}{[Ni(NH_3)_2(H_2O)_4{}^{2+}][NH_3]} = 53.7$$

$$[Ni(NH_3)_3(H_2O)_3]^{2+}(aq) + NH_3(aq) \rightleftharpoons [Ni(NH_3)_4(H_2O)_2]^{2+}(aq) + H_2O(l)$$

$$K_4 = \frac{[Ni(NH_3)_4(H_2O)_2{}^{2+}][H_2O]}{[Ni(NH_3)_3(H_2O)_3{}^{2+}][NH_3]} = 15.5$$

$$[Ni(NH_3)_4(H_2O)_2]^{2+}(aq) + NH_3(aq) \rightleftharpoons [Ni(NH_3)_5(H_2O)]^{2+}(aq) + H_2O(l)$$

$$K_5 = \frac{[Ni(NH_3)_5(H_2O)^{2+}][H_2O]}{[Ni(NH_3)_4(H_2O)_2{}^{2+}][NH_3]} = 5.62$$

$$[Ni(NH_3)_5(H_2O)]^{2+}(aq) + NH_3(aq) \rightleftharpoons [Ni(NH_3)_6]^{2+}(aq) + H_2O(l)$$

$$K_6 = \frac{[Ni(NH_3)_6{}^{2+}][H_2O]}{[Ni(NH_3)_5(H_2O)^{2+}][NH_3]} = 1.07$$

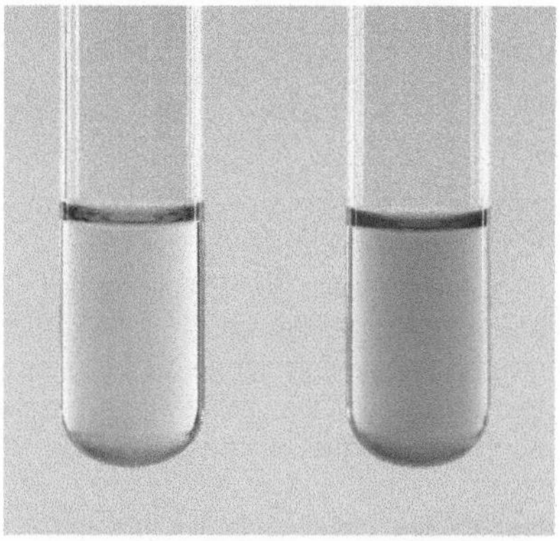

Aqueous solutions containing the ions $[Ni(H_2O)_6]^{2+}$ (left) and $[Ni(NH_3)_6]^{2+}$ (right).

Each successive equilibrium constant is lower than the previous one, which means that the position of the equilibrium lies less in favour of the products. There is a statistical explanation for this. In $[Ni(NH_3)(H_2O)_5]^{2+}$ there are five water molecules, and any one of these can be substituted by NH_3 to form $[Ni(NH_3)_2(H_2O)_4]^{2+}$. In $[Ni(NH_3)_2(H_2O)_4]^{2+}$ there are only four water molecules, so the probability of substituting one to form $[Ni(NH_3)_3(H_2O)_3]^{2+}$ is reduced.

The Jahn–Teller effect is described in Section 28.5 (p.1287).

For copper(II), the values of K_5 and K_6 for the reaction with ammonia are less than one, and the most abundant complex ion in solution when aqueous ammonia is added to Cu^{2+} (aq) is $[Cu(NH_3)_4(H_2O)_2]^{2+}$. This is due to the Jahn–Teller effect. Copper(II) has the valence electronic configuration $3d^9$, so octahedral complexes are tetragonally distorted with the axial ligands only very weakly bound. These axial ligands exchange very rapidly with the solvent (water), which is present in a much higher concentration than ammonia.

Often chemists are more interested in the equilibrium constant for full substitution than the stepwise equilibrium constants. This is given by the **formation constant** β_n, where n is the number of ligands. For an octahedral complex

$$\beta_6 = K_1K_2K_3K_4K_5K_6$$

For the substitution of the six aqua ligands by ammonia molecules in $[Ni(H_2O)_6]^{2+}$ (aq),

$$\begin{aligned}\beta_6 &= 630 \times 174 \times 53.7 \times 15.5 \times 5.62 \times 1.07 \\ &= 5.5 \times 10^8\end{aligned}$$

The very high value of β_6 for this reaction is due to a larger crystal field stabilization energy (CFSE) for $[Ni(NH_3)_6]^{2+}$ than for $[Ni(H_2O)_6]^{2+}$. This is a result of NH_3 being a stronger field ligand than H_2O.

The chelate effect

The structure of en is given in Table 28.4 (p.1267).

For the reaction between $[Ni(H_2O)_6]^{2+}$ and the bidentate ligand ethane-1,2-diamine (en)

$$[Ni(H_2O)_6]^{2+}(aq) + 3\,en(aq) \rightleftharpoons [Ni(en)_3]^{2+}(aq) + 6\,H_2O(l) \qquad (28.4)$$

$\beta_3 = 1.0 \times 10^{18}$, which is more than 1×10^9 higher than the value from the equivalent reaction with ammonia, even though both $[Ni(NH_3)_6]^{2+}$ and $[Ni(en)_3]^{2+}$ are octahedral complexes with six nitrogen donor atoms. The value of β is always higher for a chelate complex than it is for a complex with only monodentate ligands, even if the donor atoms are the same. This is called the **chelate effect**.

There are two factors contributing to the chelate effect. Once one end of a bidentate ligand is coordinated, the other end is forced to be close to the metal centre. Only a rotation of the ligand is required for coordination, so it is far more likely to coordinate than another molecule in solution.

The second factor arises from the entropy changes in the reactions. In the reaction between $[Ni(H_2O)_6]^{2+}$ and ammonia (Equation 28.3), there are seven species (molecules or ions) on the left-hand side of the equation, and seven on the right-hand side. This means that there is little change in entropy in going from left to right. In contrast, in the reaction between $[Ni(H_2O)_6]^{2+}$ and ethane-1,2-diamine (Equation 28.4), there are four species on the left-hand side of the equation, but seven on the right-hand side, which means there is an increase in entropy from left to right. The positive value for ΔS in Equation 28.4 is the major factor in making the ΔG for this reaction more negative than for Equation 28.3.

For a reaction to be spontaneous, the Gibbs energy change, ΔG, must be negative: $\Delta G = \Delta H - T\Delta S$. See Section 14.5 (p.675). The standard Gibbs energy change, $\Delta G^{\ominus}$, is related to the equilibrium constant, K, by the expression, $\Delta G^{\ominus} = -RT\ln K$. See Equation 15.4 (p.705).

Box 28.6 Trapping metal ions using the chelate effect

Ethane-1,2-diamine-*N,N,N′,N′*-tetraethanoate (ethylenediaminetetraacetate, normally abbreviated to $[edta]^{4-}$) is a hexadentate ligand (see Table 28.4, p.1267), and it coordinates to a metal ion through the two amine nitrogen donor atoms and four carboxylate oxygen donor atoms. This enables the ligand to wrap around a metal ion as in the cobalt(II) complex $[Co(edta)]^{2-}$ (**18**).

As a result of the chelate effect, the formation constants of $[edta]^{4-}$ complexes with *d*-block ions are very high. This means that compounds containing the $[edta]^{4-}$ ion, such as $CaNa_2[edta]$ and $Na_2[H_2edta]$, can be used to remove *d*-block metal ions from solutions. When used in this way, the compounds are known as *sequestrating agents*, and they are said to *sequester*, or isolate the *d*-block ions.

The sequestrating ability of $[edta]^{4-}$ is exploited in the food industry. $CaNa_2[edta]$ is often added to many processed foods such as mayonnaise, tinned beans, and soft drinks. These may contain trace amounts of *d*-block metal ions such as iron, copper, or zinc, which are present from the raw materials or the packaging. Beans absorb metal ions as they grow, and chick peas, for example, can contain over 10 ppm →

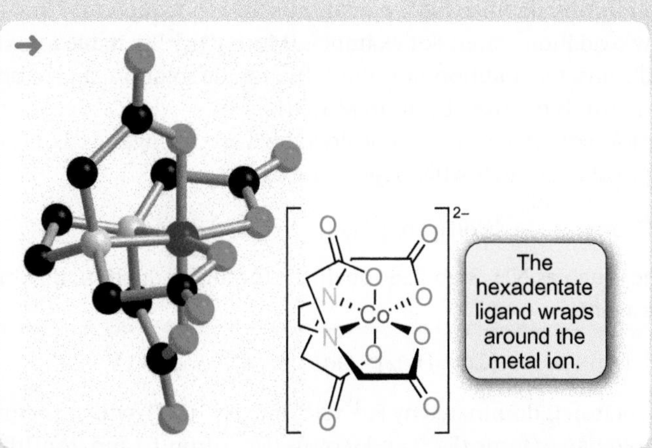
The hexadentate ligand wraps around the metal ion.

▲ 18 $[Co(edta)]^{2-}$.

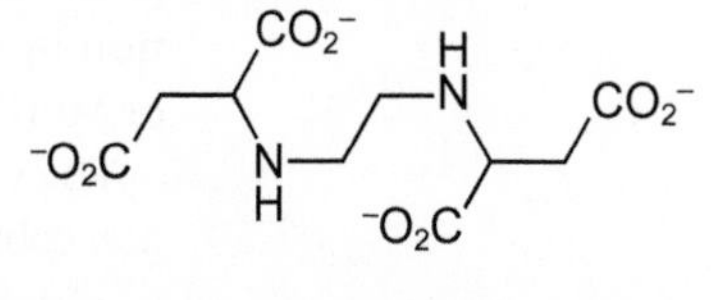

▲ 19 Ethane-1,2-diaminedisuccinate ($[edds]^{4-}$).

iron and 2 ppm copper. These concentrations would cause problems as the metal ions catalyse oxidation of the molecules that provide the flavour and colour. A metal ion concentration of less than 0.1 ppm would be enough to cause discoloration and degradation of vitamin C, but concentrations can be reduced to lower than this using edta compounds. $CaNa_2$[edta] is also used medicinally to treat heavy metal poisoning, using the same general principles.

$[Edta]^{4-}$ forms complexes with Group 2 metals as well as *d*-block metals. $Na_2[H_2edta]$ has been used in household products, such as washing powders and shampoos, in order to form complexes with the Ca^{2+} ions in hard water and prevent calcium salts from precipitating. This use of edta salts is now restricted because they are not biodegradable, and the structural isomer edds (ethane-1, 2-diaminedisuccinate) (**19**) is often used instead. $[Edds]^{4-}$ is also a hexadentate ligand so it forms complexes with very high formation constants, but in contrast to $[edta]^{4-}$ it is broken down during water treatment.

Question

Suggest how titrations with Na_4[edta] could be used in the quantitative analysis of metal ions.

Redox reactions

The important oxidation states of the first row *d*-block metals are shown in Figure 28.7 (p.1263). For many *d*-block metals, more than one oxidation state is common in aqueous solution, so redox reactions are possible. Vanadium(V) solutions are reduced by zinc in the presence of hydrochloric acid and the absence of air to violet V^{2+} (aq) ions. The reduction passes sequentially through vanadium(IV) (blue) and vanadium(III) (green) before reaching the final product.

Standard reduction potentials, are discussed in Section 16.3 (p.740).

For the early *d*-block metals, the reduction potential ($E^{\ominus}$) of the highest oxidation state increases with the oxidation state. So

$$[\overset{\text{IV}}{Ti}O]^{2+}(aq) + 2H^+(aq) + e^- \rightarrow Ti^{3+}(aq) + H_2O(l) \quad E^{\ominus} = +0.10\,V$$

$$[\overset{\text{V}}{V}O_2]^{+}(aq) + 2H^+(aq) + e^- \rightarrow [\overset{\text{IV}}{V}O]^{2+}(aq) + H_2O(l) \quad E^{\ominus} = +1.00\,V$$

$$[\overset{\text{VI}}{Cr_2}O_7]^{2-}(aq) + 14H^+(aq) + 6e^- \rightarrow 2Cr^{3+}(aq) + 7H_2O(l) \quad E^{\ominus} = +1.33\,V$$

$$[\overset{\text{VII}}{Mn}O_4]^{-}(aq) + 8H^+(aq) + 5e^- \rightarrow Mn^{2+}(aq) + 4H_2O(l) \quad E^{\ominus} = +1.51\,V$$

The use of $[Cr_2O_7]^{2-}$ and $[MnO_4]^-$ as oxidizing agents in organic chemistry is described in Box 23.2 (p.1064).

As a result, $[Cr_2O_7]^{2-}$ and $[MnO_4]^-$ are commonly used oxidizing agents.

The value of a reduction potential depends on the pH. Chromium(VI) exists as the orange dichromate ion ($[Cr_2O_7]^{2-}$) in acidic solution, and the yellow chromate ion ($[CrO_4]^{2-}$) in alkaline solution. The reduction potential is much lower for chromate

$$[\overset{\text{VI}}{Cr}O_4]^{2-}(aq) + 4H_2O(l) + 3e^- \rightarrow \overset{\text{III}}{Cr}(OH)_3(s) + 5OH^-(aq) \quad E^{\ominus} = -0.11\,V$$

so high pH stabilizes the +6 oxidation state towards reduction.

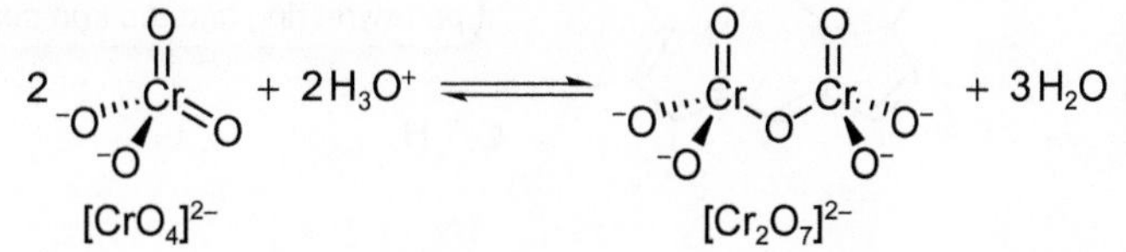
2
+ 2H3O+
+ 3H2O

$[CrO_4]^{2-}$ $[Cr_2O_7]^{2-}$

In general, alkaline conditions stabilize high oxidation states to reduction. They also promote oxidation of low oxidation states. For example, $[Mn(H_2O)_6]^{2+}$ is stable to oxidation by O_2 in acidic conditions. On addition of sodium hydroxide solution, a precipitate of $Mn(OH)_2$ is formed, and this is oxidized by air to $Mn(OH)_3$.

The value of the reduction potential can vary considerably with the ligands. For example, cobalt(III) is a strong oxidizing agent when H_2O is the ligand

$$[Co(H_2O)_6]^{3+}(aq) + e^- \rightarrow [Co(H_2O)_6]^{2+}(aq) \qquad E^\ominus = +1.92\,V$$

but nitrogen donor ligands such as NH_3 stabilize cobalt(III) towards reduction, making it a much weaker oxidizing agent

$$[Co(NH_3)_6]^{3+}(aq) + e^- \rightarrow [Co(NH_3)_6]^{2+}(aq) \qquad E^\ominus = +0.11\,V$$

The aqueous chemistry of iron is dominated by Fe^{3+} ($3d^5$) and Fe^{2+} ($3d^6$), which are sometimes referred to by their old names ferric (Fe^{3+}) and ferrous (Fe^{2+}). Iron(II) and iron(III) are readily interconverted

$$Fe^{3+}(aq) + e^- \rightarrow Fe^{2+}(aq) \qquad E^\ominus = +0.77\,V$$

Although the reduction potential for the conversion of Fe^{3+}(aq) to Fe^{2+}(aq) is positive, iron(II) compounds are slowly oxidized in aqueous solution due to the presence of dissolved oxygen

$$4\,Fe^{2+}(aq) + O_2(aq) + 4\,H_3O^+(aq) \rightarrow 4\,Fe^{3+}(aq) + 6\,H_2O(l)$$

Box 28.7 Haemoglobin and the transport of oxygen

In respiration, the reaction between glucose and oxygen to give water and carbon dioxide is used by the body as a source of energy. Humans rely on the iron compound haemoglobin in red blood cells to transport oxygen from the lungs, where it enters the body, to the tissues where it is needed for respiration. Every haemoglobin molecule contains four iron atoms, and each of these is coordinated to a dianionic *porphyrin ring*. Porphyrin acts as a tetradentate ligand through its nitrogen atoms, giving a planar *haem group*. Each haem group is bound on to a larger protein chain, and there are four such subunits in a haemoglobin molecule.

Understanding how haemoglobin works brings together many of the ideas discussed in this chapter, such as polydentate ligands, high spin and low spin complexes, and changes in ionic radius. The coordination geometry around the iron centres in haemoglobin is shown below.

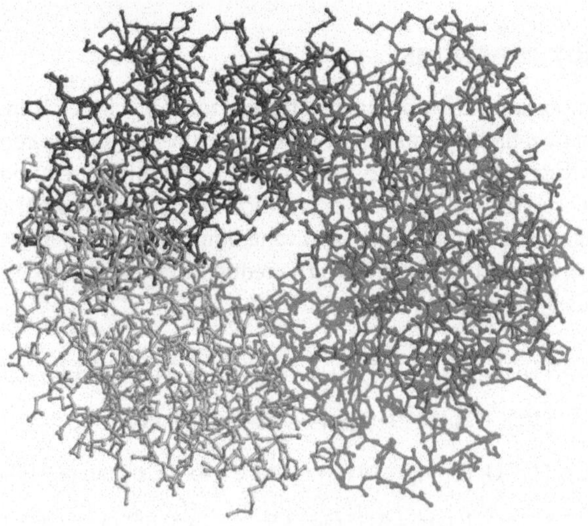

The structure of haemoglobin, with each of the four protein subunits shown in a different colour.

One of the four haem groups in haemoglobin.

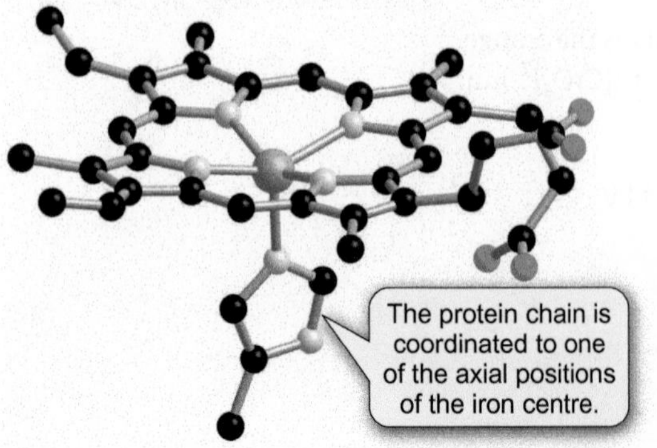
The protein chain is coordinated to one of the axial positions of the iron centre.

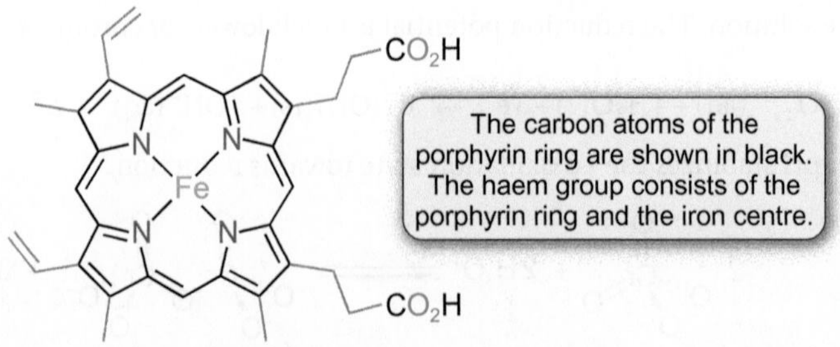
CO2H
N
N
Fe
N
N
CO2H
The carbon atoms of the porphyrin ring are shown in black. The haem group consists of the porphyrin ring and the iron centre.

→

→ The iron adopts a square pyramidal geometry, with the four porphyrin nitrogen donors in the equatorial plane and a histidine group from the protein chain coordinated in the axial position. This leaves one axial position unused.

These haem groups give blood its characteristic red colour. A molecule of O_2 can bond to the unused axial position of each iron, so each haemoglobin molecule is capable of bonding to four O_2 molecules, converting haemoglobin into oxyhaemoglobin. A pulse oximeter uses spectroscopy to measure the relative concentrations of haemoglobin and oxyhaemoglobin (see Box 11.7, p.546).

The iron atoms in haemoglobin are in the +2 oxidation state and are high spin, with four unpaired electrons. These atoms are too big to fit in the centre of the porphyrin ring so they sit approximately 40 pm to one side. O_2 is a relatively strong field ligand, so when an O_2 molecule enters the sixth coordination site of an iron atom there is an increase in Δ, and the electronic configuration changes to low spin. Since the e_g orbitals are not occupied, the iron atom decreases in size, and moves into the centre of the ring, as shown below.

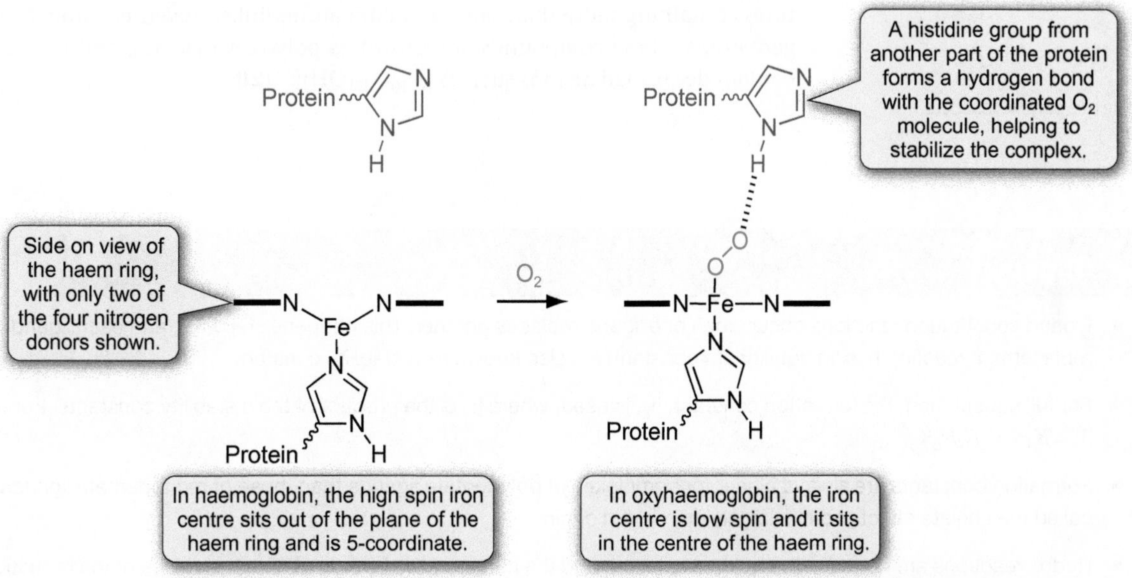
A histidine group from another part of the protein forms a hydrogen bond with the coordinated O_2 molecule, helping to stabilize the complex.
Protein
Protein
Side on view of the haem ring, with only two of the four nitrogen donors shown.
O_2
Fe
Protein
Protein
In haemoglobin, the high spin iron centre sits out of the plane of the haem ring and is 5-coordinate.
In oxyhaemoglobin, the iron centre is low spin and it sits in the centre of the haem ring.

▲ Coordination of oxygen to the iron centres causes a change from high spin to low spin, allowing the metal to move into the centre of the haem ring.

The four haem units in haemoglobin are widely separated but influence each other. When an iron atom coordinates to O_2 and moves into the haem ring, it pulls upon the protein chain. This causes conformational changes that increase the affinity of the other iron atoms for O_2. In places where the oxygen concentration is high, such as the lungs, haemoglobin is very efficient at binding O_2. This behaviour is different from what would be expected if the four haem groups acted independently.

When the oxyhaemoglobin reaches regions where the oxygen concentration is low, the O_2 molecules are released. The binding of O_2 is pH dependent, and O_2 is released more readily at low pH. This helps the release of O_2 in the tissues where metabolism is occurring, because the CO_2 produced dissolves in the blood and lowers the pH (see Box 7.5, p.321).

Carbon monoxide is toxic because it coordinates irreversibly to the haem iron atoms. This prevents haemoglobin from transporting oxygen from the lungs to the tissues, which in turn stops respiration from occurring.

▲ Octopi have blue rather than red blood. Unlike mammals, the octopus uses a copper compound called haemocyanin to transport oxygen around its body. Each haemocyanin molecule contains two copper atoms, and the O_2 molecule binds in a side-on coordination mode, bridging between the copper atoms.

Question

Attempts to make artificial blood using simple iron–porphyrin compounds, without the protein chain, fail because the resulting oxygen compounds dimerize with each O_2 molecule binding to two iron atoms. Suggest how this dimerization could be avoided.

The hydrolysis of Fe^{3+} (aq) in disused mines is described in Box 7.3 (p.312).

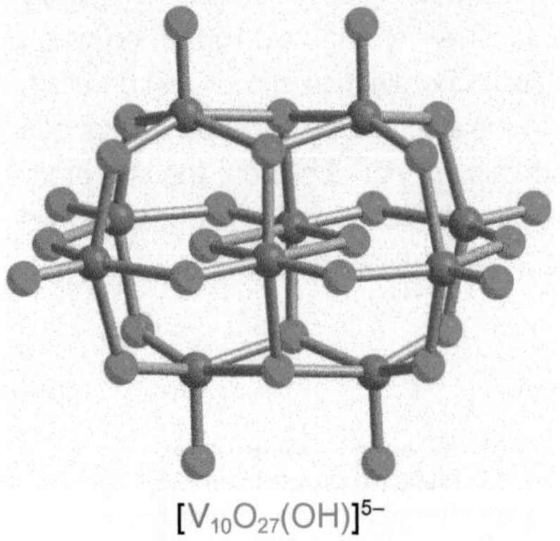
$[V_{10}O_{27}(OH)]^{5-}$

20 An example of a polyoxovanadate ($[V_{10}O_{27}(OH)]^{5-}$).

Hydrolysis

Ions with a charge of +3 or greater have a high charge density, which polarizes the O–H bonds of the coordinated water molecules, making the protons more easily lost. As a result, these ions hydrolyse in aqueous solution giving acidic solutions. For example

$$[Sc(H_2O)_6]^{3+} + H_2O \rightleftharpoons [Sc(OH)(H_2O)_5]^{2+} + H_3O^+$$

Hydrolysis affects the nature of the ions present in aqueous solution. Titanium(IV) and vanadium(V) are common oxidation states, but the $[Ti(H_2O)_6]^{4+}$ and $[V(H_2O)_6]^{5+}$ ions are unknown. For titanium(IV), the main species present in aqueous solutions is $[Ti(OH)_2(H_2O)_4]^{2+}$ though these ions are often written as TiO^{2+} (aq) for simplicity.

In strongly acidic solution, vanadium(V) is present as $[VO_2]^+$, and in strongly alkaline solutions it is vanadate ($[VO_4]^{3-}$). In between these extremes, there are a number of structures containing more than one vanadium atom, linked together through bridging oxygen atoms. These compounds are known as polyoxovanadates, and the most common contain decavanadate ions such as $[V_{10}O_{27}(OH)]^{5-}$ (**20**).

Summary

- Ligand substitution reactions occur when one ligand replaces another. This happens stepwise, and each ligand replacement reaction has an equilibrium constant, K_n, also known as a stability constant.
- For full substitution, the formation constant, β_n, is used, where β_n is the product of the *n* stability constants. For example, $\beta_6 = K_1K_2K_3K_4K_5K_6$.
- Formation constants are always higher for complexes of polydentate ligands than those of monodentate ligands. This is called the chelate effect and it is largely entropic in origin.
- Redox reactions are common in aqueous solution, and the nature of the ligands affects the values of the reduction potentials.
- Ions with a charge of +3 or more undergo hydrolysis, giving acidic solutions.

? For a practice question on these topics, see question 18 at the end of this chapter (p.1302).

Concept review

By the end of this chapter, you should be able to do the following.

- Work out the valence electronic configuration for any *d*-block ion.
- Describe the trends in atomic radii, ionic radii, and ionization energies across the first row of the *d* block and account for these.
- Explain why the second and third row elements in a *d*-block triad have similar chemistries, but are rather different from the first row element.
- Describe the general trend in the most stable oxidation states of the first row *d*-block elements, and explain its origin.
- Know the important geometries for coordination numbers between 2 and 8.
- Give the name of a coordination compound from its formula, and the formula of a coordination compound from its name.
- Work out the oxidation state of the metal in a coordination compound, its valence electronic configuration, and its coordination number.

- Work out if isomers are possible for a coordination compound and, if they are, describe the type of isomerism.
- Use the concept of hard and soft acids and bases to predict what type of ligands a particular ion will preferentially form complexes with.
- Work out the *d*-orbital splitting for octahedral, tetrahedral, and square planar complexes.
- Calculate the crystal field stabilization energy (CFSE) for an octahedral or tetrahedral complex.
- Predict whether a complex is high spin or low spin from values of the crystal field splitting energy, Δ, and the pairing energy, *P*.
- Predict whether a complex will undergo a Jahn–Teller distortion.
- Explain the basis of colour in coordination compounds.
- Calculate the spin-only magnetic moment, μ_{so}, for a *d*-block complex.
- Explain the origin of the chelate effect.

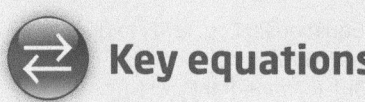

Key equations

Relationship between Δ_o and Δ_t	$\Delta_t = \frac{4}{9}\Delta_o$	(28.1)
The spin-only formula	$\mu_{so} = \sqrt{n(n+2)}\,\mu_B$	(28.2)

Questions

More challenging questions are indicated by an asterisk *.

You will find it useful to refer to Tables 28.3 and 28.4 for details of ligands and their abbreviations.

1 Give the electronic configurations of the following atoms and ions: (a) V^{3+}; (b) Fe; (c) Cr; (d) Ag^+; (e) Fe(0) (in $[Fe(CO)_5]$); (f) Cu^{2+}; (g) Ru^{2+}; (h) W^{4+}. (Section 28.1)

2 Explain why the +3 oxidation state becomes increasingly less stable with respect to the +2 oxidation state on moving from left to right across the first row of the d-block elements. (Section 28.2)

3 Give the oxidation state and coordination number of the metal in the following complexes: (a) $[Cr(CN)_6]^{3-}$; (b) $[PdCl_2(PBu_3)_2]$; (c) $[VO(acac)_2]$; (d) $[Ru(terpy)_2]^{2+}$; (e) $[Cr(ox)_3]^{3-}$. (Section 28.3)

4 Name the following coordination compounds (Section 28.3): (a) $[NiI(NH_3)_5]NO_3$; (b) $K[Co(ox)_2(H_2O)_2]$; (c) $[Fe(dien)_2]SO_4$; (d) $Na[AuBr_2Cl_2]$.

5 Determine the formulae for the following coordination compounds (Section 28.3):

(a) chloro(diethylsulfide)gold(I);

(b) tetraethylammonium tetrabromoferrate(III);

(c) bis(2,2′-bipyridine)dicarbonylmolybdenum(0);

(d) potassium hexanitrito-κO-cobaltate(III).

6 Compounds X and Y have the same molecular formula and contain 18.25% titanium, 40.54% chlorine, 36.59% oxygen, and 4.61% hydrogen by mass. When 0.5000 g of X was dissolved in water and treated with silver nitrate solution, 0.2731 g of silver chloride were precipitated. 0.5000 g of Y were dissolved in water and titrated against 0.2000 mol dm^{-3} silver nitrate solution, of which 28.2 cm^3 were required. Identify X and Y, and determine any possible isomers. (Section 28.3)

7 Which of the following pairs of compounds are isomers? Name the type of any isomerism present. (Section 28.3)

(a) $K_3[Fe(CN)_6]$ and $K_4[Fe(CN)_6]$

(b) $[CoCl(NH_3)_5]SeO_4$ and $[Co(SeO_4)(NH_3)_5]Cl$

(c) $[Pt(NH_3)_4][PtBr_4]$ and $[PtBr(NH_3)_3][PtBr_3(NH_3)]$

(d) $[Co(NO_2)(NH_3)_5]Cl_2$ and $[Co(ONO)(NH_3)_5]Cl_2$

8 Sketch the possible isomers for the following compounds and identify the type of isomerism involved: (a) $[Co(ox)_3]^{3-}$; (b) $[NiCl_2(en)_2]$; (c) $[NiCl_2(py)_2]$; (d) $[FeF_3(H_2O)_3]$. (Section 28.3)

9 Dimethyl sulfoxide (DMSO, Me_2SO) is an ambidentate ligand as it can act as an *S*-donor or an *O*-donor. (Sections 28.3, 5.2)

(a) Use VSEPR theory to show there is a lone pair on the sulfur atom.

(b) Use the hard and soft acids and bases theory to predict which atom will coordinate to (i) Pt^{2+} and (ii) Fe^{3+}.

10 For each of the following pairs of complexes, identify which has the higher value of Δ. Explain your answers. (Section 28.5)

(a) $[TiCl_6]^{3-}$ or $[TiCl_4]^-$

(b) $[NiCl_4]^{2-}$ or $[NiBr_4]^{2-}$

(c) $[Fe(H_2O)_6]^{2+}$ or $[Fe(H_2O)_6]^{3+}$

11 Explaining your reasoning, which of the following complex ions would you expect to be tetragonally distorted? (Section 28.5)

(a) $[Cu(NH_3)_6]^{2+}$

(b) $[Ni(NH_3)_6]^{2+}$

(c) $[Mn(H_2O)_6]^{2+}$

(d) $[Cr(H_2O)_6]^{3+}$

12 Write down expressions for the crystal field stabilization energy (CFSE) for a high spin and low spin d^4 complex in terms of Δ_o and *P*, the pairing energy. Use these expressions together with the data below to predict whether $[Mn(H_2O)_6]^{3+}$ and $[Re(H_2O)_6]^{3+}$ form high spin or low spin complexes. (Section 28.5)

	Δ_o/kJ mol^{-1}	P/kJ mol^{-1}
$[Mn(H_2O)_6]^{3+}$	250	300
$[Re(H_2O)_6]^{3+}$	400	180

13 (Sections 28.5, 28.6)

(a) Rank the following manganese complexes in order of increasing Δ. Explain your reasoning.

$[Mn(NH_3)_6]^{2+}$ $[MnCl_4]^{2-}$ $[Mn(CN)_6]^{4-}$ $[Mn(CN)_6]^{3-}$

(b) Rank the following manganese complexes in order of increasing intensity of colour. Explain your reasoning.

$[MnO_4]^-$ $[Mn(NH_3)_6]^{2+}$ $[MnCl_4]^-$ $[Mn(CN)_6]^{3-}$

14* The visible light absorption spectrum for $[Ti(H_2O)_6]^{3+}$ (aq) is shown in Figure 28.35 (p.1290). By considering the nature of the excited state, suggest why the absorption has a shoulder at a higher wavelength than λ_{max} at 490 nm, suggesting two electronic transitions at slightly different energies. (Sections 28.5, 28.6)

15 Copper(II) complexes are coloured but copper(I) complexes are generally colourless. Why is this? (Section 28.6)

16 $Na_2[Ni(CN)_4]$ is diamagnetic, not paramagnetic. Use this information to determine the geometry of the anion. (Section 28.7)

17 Use the spin-only formula (Equation 28.2, p.1292) to predict the magnetic moments of the following compounds: (a) $[Ni(en)_3]SO_4$; (b) $Na_2[CoCl_4]$; (c) *trans*-$[MnF_2(NH_3)_4]$; (d) $K_2[Cr_2O_7]$. (Section 28.7)

18 For the reaction between Ni^{2+}(aq) and 1,4,7-triazaheptane (dien), the equilibrium constants $K_1 = 6.5 \times 10^{10}$ and $K_2 = 1.4 \times 10^8$ have been determined. Write balanced equations for these reactions, and calculate the formation constant β_2. (Section 28.8)

Maths toolkit

The purpose of this toolkit is to give an overview of some of the basic mathematical techniques and relationships that are important in understanding the chemistry described in this book. It gives a brief description of all of the methods you need to study the material in this book and refers you to examples in the text where the mathematical technique is applied. The toolkit is not intended to teach you mathematics or to cover all of the mathematics needed to study a chemistry degree. For these you should consult one of the many 'mathematics for chemistry' textbooks.

MT1 Working with numbers

Other chemical contexts for integers are the quantum numbers used to define atomic orbitals (Section 3.5, p.134) and the quantum numbers used to label the energy levels between which transitions take place in molecular spectroscopy, for example, between rotational levels in Section 10.4 (p.466) and between vibrational levels in Section 10.5 (p.476).

The system of numbers that we use is based on whole numbers (called integer numbers), which may be positive, negative, or zero. For example, the atomic number of an element, Z, the number of protons in the nucleus, is an integer. For uranium, U, $Z = 92$.

Of course, not everything you encounter takes an integer value. Think about measuring a quantity of reactant needed for a reaction. It is unlikely that a whole number of kilograms or grams will be needed so it is necessary to subdivide integers. This is done by using fractions or, more commonly in scientific work, the decimal system.

By convention, the '–' sign before a negative number is *always* included but the '+' sign is often omitted before a positive number. No sign before a number means it is positive.

Fractions

In mathematical terminology, the number on top of the fraction is the **numerator**, the number on the bottom is the **denominator**.

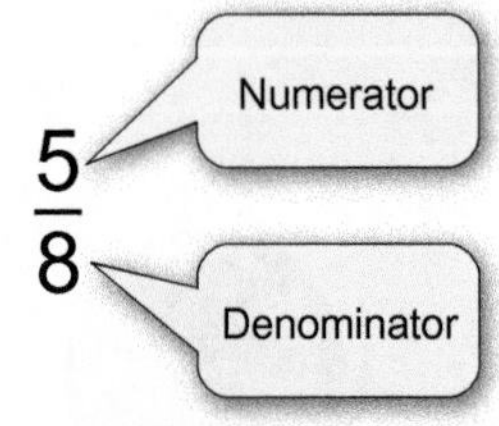
5
8
Numerator
Denominator

This is spoken as 'five eighths'.

Fractions represent the parts of a whole. For example, $\frac{1}{8}$ represents 1 part of a whole divided into 8; $\frac{5}{8}$ represents 5 parts of the whole divided into 8, as shown in Figure MT1.

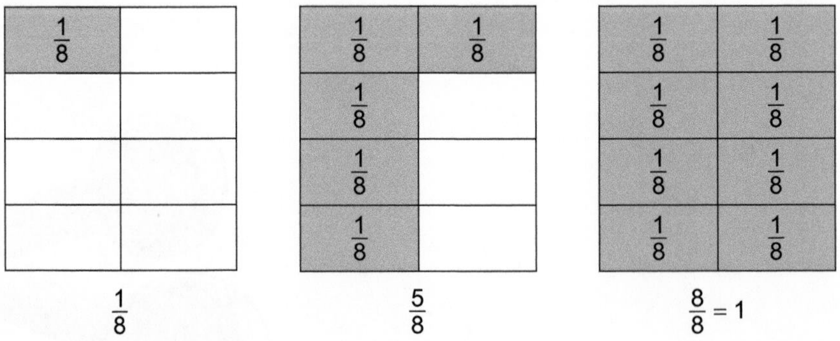
1/8
5/8
8/8 = 1

Figure MT1 A fraction indicates the number of parts of a whole.

Many fractions can be expressed in a number of equivalent ways. For example, $\frac{1}{4}$, $\frac{2}{8}$, $\frac{4}{16}$, $\frac{8}{32}$, $\frac{16}{64}$, and $\frac{32}{128}$ all represent the same fraction of a whole. It is usual to **cancel** the fractions so that they are expressed in their lowest (simplest) form. This involves dividing both the numerator and denominator by the same number until they cannot be divided further, as shown in Figure MT2.

Adding and subtracting fractions

The simplest case is where two fractions have the same denominator. Addition and subtraction are then straightforward, as in the three examples below.

$$\frac{1}{7}+\frac{3}{7}=\frac{4}{7} \qquad \frac{3}{5}-\frac{2}{5}=\frac{1}{5} \qquad \frac{7}{11}+\frac{5}{11}=1\frac{1}{11}$$

In the third example, the result of adding the fractions gives a number that is greater than 1, so it can also be written as a whole number and fraction

$$\frac{11}{11}=1, \text{ so } \frac{12}{11} \text{ can be written as } 1\frac{1}{11}$$

Another example is

$$\frac{3}{4}+\frac{3}{4}+\frac{3}{4}=\frac{9}{4}=2\frac{1}{4}$$

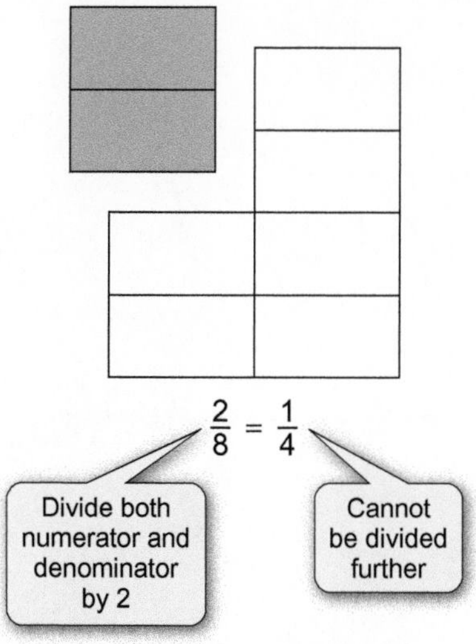
$\frac{2}{8}=\frac{1}{4}$
Divide both numerator and denominator by 2
Cannot be divided further

Figure MT2 Fractions are usually expressed in the lowest possible form.

If the denominators of two fractions are different, you need to find a **common denominator** before you can add or subtract the fractions. Both fractions must be expressed with the same denominator, that is, as fractions of the same whole. A common denominator is a number that can be divided by the denominators of the two fractions. The easiest way to find it is to multiply together the two denominators.

For example, to find the sum of $\frac{1}{4}+\frac{1}{3}$, multiply the denominators, $(4 \times 3) = 12$, to give the common denominator 12

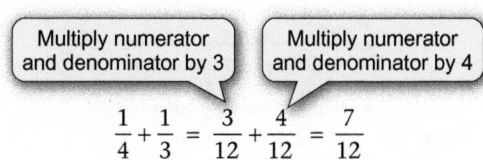

Multiply numerator and denominator by 3

Multiply numerator and denominator by 4

$$\frac{1}{4}+\frac{1}{3}=\frac{3}{12}+\frac{4}{12}=\frac{7}{12}$$

Sometimes you can cancel the resulting fraction to give a simpler fraction

Common denominator: $6 \times 8 = 48$

Divide numerator and denominator by 2 to simplify

$$\frac{1}{6}+\frac{1}{8}=\frac{8}{48}+\frac{6}{48}=\frac{14}{48}=\frac{7}{24}$$

In the above case, 24 is also a common denominator. It doesn't matter at what stage you cancel to simplify the fractions.

Multiplying fractions

From Figure MT1 you can see that $8\times\frac{1}{8}=1$ and that $4\times\frac{1}{8}=\frac{1}{2}$. Multiplying fractions uses this principle

$$8\times\frac{1}{8}=\frac{8}{1}\times\frac{1}{8}=\frac{8}{8}=1$$

To multiply fractions: multiply the numerators and then multiply the denominators

(Note that the integer 8 can be expressed as the fraction $\frac{8}{1}$.)

So, multiplying two fractions involves multiplying both numerators together to give a new numerator and both denominators together to give a new denominator, followed by cancelling, if necessary, to simplify the resulting fraction. For example

Divide numerator and denominator by 2 to simplify

Cannot be simplified further

$$\frac{2}{5}\times\frac{3}{4} = \frac{2\times 3}{5\times 4} = \frac{6}{20} = \frac{3}{10}$$

Another example is

Cannot be simplified further

$$\frac{3}{5}\times\frac{7}{10} = \frac{21}{50}$$

The process works for any number of fractions. For example

Divide numerator and denominator by 10 to simplify

$$\frac{2}{5}\times\frac{3}{4}\times\frac{5}{8} = \frac{2\times 3\times 5}{5\times 4\times 8} = \frac{30}{160} = \frac{3}{16}$$

Another example is

Divide numerator and denominator by 120 to simplify

$$\frac{1}{2}\times\frac{2}{3}\times\frac{3}{4}\times\frac{4}{5}\times\frac{5}{6} = \frac{120}{720} = \frac{1}{6}$$

Dividing fractions

Division is the reverse of multiplication. Dividing by 2 is the same as multiplying by $\frac{1}{2}$; dividing by 5 is the same as multiplying by $\frac{1}{5}$; dividing by $\frac{3}{4}$ is the same as multiplying by $\frac{4}{3}$. This forms the basis for dividing fractions. As an example, consider $\frac{1}{3}$ divided by $\frac{2}{5}$

$$\frac{1}{3}\div\frac{2}{5}$$

This is the same as $\frac{1}{3}$ multiplied by $\frac{5}{2}$

$$\frac{1}{3}\div\frac{2}{5} = \frac{1}{3}\times\frac{5}{2}$$

You can then carry out the multiplication as described above

$$\frac{1}{3}\div\frac{2}{5} = \frac{1}{3}\times\frac{5}{2} = \frac{1\times 5}{3\times 2} = \frac{5}{6}$$

To divide fractions, invert the second fraction and multiply

Two further examples are shown below

$$\frac{2}{5}\div\frac{3}{4} = \frac{2}{5}\times\frac{4}{3} = \frac{8}{15} \qquad \frac{3}{5}\div\frac{7}{10} = \frac{3}{5}\times\frac{30}{35} = \frac{6}{7}$$

Decimals

In science, non-integer numbers are usually represented by **decimal** numbers, rather than in terms of fractions. Decimals express numbers in terms of multiples and divisions of 10. For example, 123.4567 means '1 hundred + 2 tens + 3 units + 4 tenths + 5 hundredths + 6 thousandths + 7 ten-thousandths (see Figure MT3).

The **decimal point** shows where the fractional part of the number (i.e. the part that is less than 1) begins. The number of digits to the right of the decimal point is referred to as the number of decimal places in the number. The number in Figure MT3 has four **decimal places**, written as 4 d.p.

Rational numbers have a definite number of figures after the decimal point (e.g. 0.25 or 1.375) or have an infinite repeating pattern of digits (e.g. 2.173173173). They can also be written as a fraction; for example, 0.25 is $\frac{1}{4}$ and 0.3333 . . . (repeating infinitely) is $\frac{1}{3}$.

Irrational numbers are those that have an infinite number of decimal places, but with no repeating pattern. An irrational number cannot be written as a fraction. A common example of an irrational number is π (pi), the ratio of the circumference to the diameter of a circle (Figure MT4). Its value is often quoted to four decimal places, π = 3.1416, but there is an infinite number of decimal places. Another irrational number used in chemistry is e = 2.7183, the base of natural logarithms (see Section MT3, p.1316).

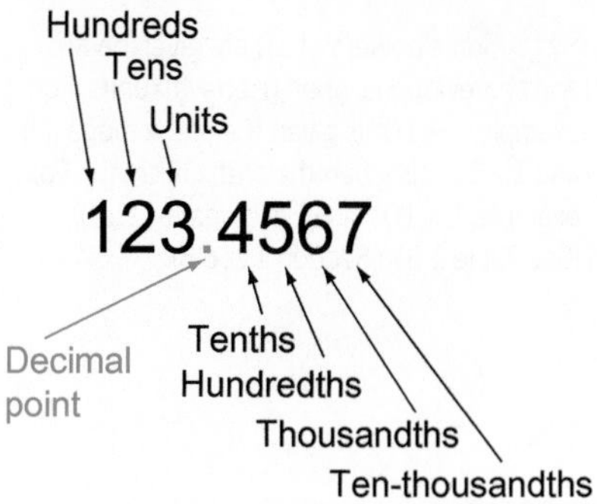
Hundreds
Tens
Units
123.4567
Decimal point
Tenths
Hundredths
Thousandths
Ten-thousandths

Figure MT3 Decimal numbers are based on multiples and divisions of ten.

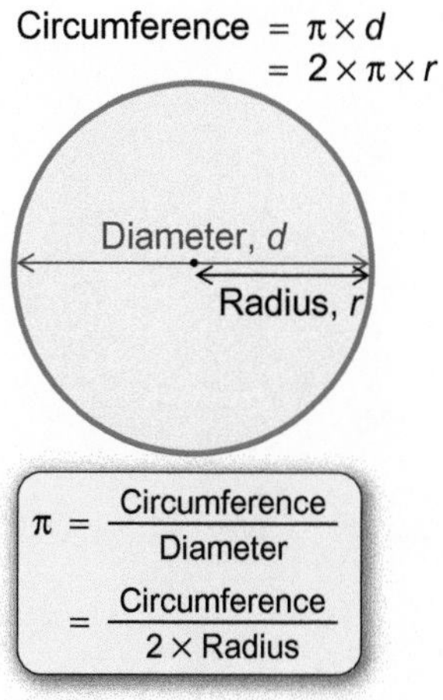
Circumference = π × d
= 2 × π × r
Diameter, d
Radius, r
π = Circumference / Diameter
= Circumference / 2 × Radius

Figure MT4 The value of π is the ratio of the circumference of a circle to its diameter.

> ℹ The value of π has been calculated to billions of digits, the first 22 of which are π = 3.141592653589793238462. In normal work, you are unlikely to need the value to more than 4 or 5 decimal places.

Large and small numbers: scientific notation

The number of atoms in one mole of an element is 602214000000000000000000 (602 sextillion!). This is the Avogadro constant. On the other hand, the C–H bond length in methane (CH_4) is 0.000000000109 m. Writing these numbers with the large number of zeros is inconvenient and can easily lead to errors.

To make dealing with large and small numbers more convenient, they are written in **scientific notation**, sometimes called **standard form**. The number is written as a decimal number between 1 and 10, multiplied by an appropriate power of 10. Any number can be written as digit × power of 10.

> ℹ The names given to large numbers vary and there is no definite system. For example, in Europe, 1 billion can sometimes mean 1×10^{12}; in the US, 1 billion = 1×10^{9}. For this reason it is better not to use names but to give large numbers in standard form.

For example, 100 can be written as $10 \times 10 = 10^2$. More completely, this is written as 1×10^2. 200 can be written as $(2 \times 100) = 2 \times 10^2$. Following this pattern, 5000 can be written as $(5 \times 10 \times 10 \times 10) = 5 \times 10^3$ and 7500 as 7.5×10^3. Using this notation, the Avogadro constant is written as 6.02214×10^{23}.

What about small numbers less than 1? 0.1 can be written as $1 \div 10 = 1 \times 0.1 = 1 \times 10^{-1}$. 0.003 is equivalent to $3 \div (10 \times 10 \times 10) = 3 \times 10^{-3}$. Thus, the C–H bond length in methane can be written as 1.09×10^{-10} m.

> ℹ A negative power of 10 indicates a number less than 1
>
> $1 \times 10^{-1} = 0.1$; $1 \times 10^{-2} = 0.01$

Combining numbers in scientific notation

Numbers written in scientific notation can only be *added or subtracted* if they are expressed to the same power of 10. For example, to add

$$(5 \times 10^3) + (2 \times 10^2)$$

> ℹ The power to which 10 is raised is sometimes called the **exponent**. Some calculators and computers use the notation that 3×10^8 is written 3E8, so 1.6×10^{-19} would be 1.6E–19. However, be careful not to get this mixed up with exponential functions (see Section MT3, p.1316).

Some powers of 10 are given names and abbreviations when used with units. For example, 1×10^6 is given the prefix mega (M) and 1×10^{-6} is given the prefix micro (μ). For example, $1 \times 10^{-6}\,g = 1$ microgram (1 μg). (See Table 1.3 in Section 1.2, p.6).

each number must be expressed as a number × 10^3 (or a number × 10^2)

$$(5+10^3)+(0.2\times10^3) = 5.2\times10^3$$

$$or\ (50\times10^2)+(2\times10^2) = 52\times10^2 = 5.2\times10^3$$

To *multiply* two numbers written in scientific notation, you need to carry out the process in two parts. You *multiply* the two decimal numbers and *add* the powers of 10 to give a new exponent. Here are two examples

Multiply the two decimal numbers

Add the powers of 10

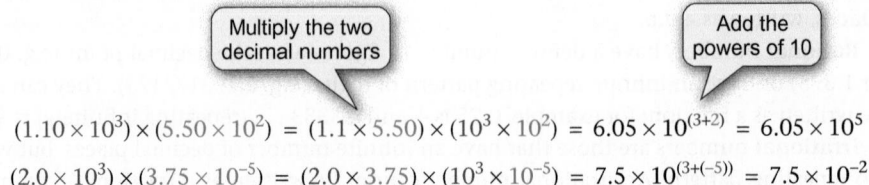

$$(1.10\times10^3)\times(5.50\times10^2) = (1.1\times5.50)\times(10^3\times10^2) = 6.05\times10^{(3+2)} = 6.05\times10^5$$

$$(2.0\times10^3)\times(3.75\times10^{-5}) = (2.0\times3.75)\times(10^3\times10^{-5}) = 7.5\times10^{(3+(-5))} = 7.5\times10^{-2}$$

To *divide* two numbers written in scientific notation, you *divide* the first decimal number by the second and *subtract* the powers of 10 to give the new exponent. For example

Divide the first decimal number by the second

Subtract the powers of 10

$$(5.5\times10^5)\div(1.1\times10^2) = (5.5\times1.1)\times(10^5\div10^2) = 5.0\times10^{(5-2)} = 5.0\times10^3$$

Significant figures and rounding of decimals

Consider the value of π. Using an increasing number of decimal places, you can write this as

$\pi = 3$	0 d.p.
$\pi = 3.1$	1 d.p.
$\pi = 3.14$	2 d.p.
$\pi = 3.142$	3 d.p.
$\pi = 3.1416$	4 d.p.
$\pi = 3.14159$	5 d.p.

The rules in **rounding** decimal places are:

- if the final digit is less than 5, then the preceding digit is rounded down so 3.142 becomes 3.14 to 2 d.p.;
- if the final digit is 5 or larger then rounding up occurs so 3.1416 becomes 3.142 to 3 d.p.

Another feature of numbers, especially important when you are quoting measurements, is the number of **significant figures (s.f.)**. Imagine you are weighing out 2 g of a reagent for a chemical reaction and the balance reads to the nearest whole number of grams. When the balance reads 2 g, the amount on the balance could be anywhere between 1.5 g and 2.5 g. You cannot quote the amount weighed to a greater accuracy than 2 g. The mass of 2 g is being weighed to 1 significant figure. A different balance reads to 0.1 g. Now, when the balance reads 2 g, you know that the amount is between 1.9 and 2.1 g so you can report it as 2.0 g, to 2 s.f. A third balance weighs to the nearest 0.001 g. Now, you can be confident that you have weighed out between 1.999 g and 2.001 g. The mass is now 2.000 g, to 4 s.f.

The number of significant figures is an indication of the accuracy of, or confidence in, a measurement. Values should be reported to the number of significant figures that leaves uncertainty only in the last digit. In rounding numbers to obtain the correct number of s.f., use the same rules as for rounding decimal places. Thus, the number 243.159 would be reported as

243.16	to 5 s.f.
243.2	to 4 s.f.
243	to 3 s.f.
240	to 2 s.f.
200	to 1 s.f.

A mass of 2 g can be expressed as 0.002 kg or 2000 mg. In both cases, the mass is expressed to 1 s.f.

Even though the units and the numbers of decimal places have changed, the number of significant figures has not. This is why, in quoting values, it is important to consider the correct number of significant figures, rather than looking at the number of decimal places.

Using the appropriate number of significant figures is also important when performing calculations. How do you decide how many significant figures to quote? Calculators often give answers with 8 or 10 digits and you need to decide how many of these are significant. Your final answer in a calculation should always be quoted to the smallest number of significant figures in the data you have.

For example, the energy, E, of a photon of electromagnetic radiation is related to its frequency, ν, and wavelength, λ, by Equations 3.2 and 3.1 in Section 3.2 (pp.118–120). Combining these two equations gives

$$E = \frac{hc}{\lambda}$$

where h is the Planck constant, 6.626×10^{-34} J s, and c is the speed of light, 2.998×10^{8} m s^{-1}. A photon of red light has a wavelength of 620 nm (6.2×10^{-7} m). Substituting these values into the equation above

$$\begin{aligned} E &= \frac{(6.626\times10^{-34}\ \text{J}\ \cancel{\text{s}})\times(2.998\times10^{8}\ \cancel{\text{m}}\ \cancel{\text{s}^{-1}})}{(6.2\times10^{-7}\ \cancel{\text{m}})} \\ &= 3.203991613\times10^{-19}\ \text{J} \\ &= 3.2\times10^{-19}\ \text{J} \end{aligned}$$

You can read more on how to set out calculations and deal with units in Section 1.2 (p.6).

A calculator gives the answer as $3.203991613 \times 10^{-19}$ J, but, of course, quoting this number of significant figures is nonsense. In the data provided, the values of h and c are both given to 4 s.f., but the wavelength is only given to 2 s.f. Therefore, the final answer can only be quoted to 2 s.f., so that $E = 3.2 \times 10^{-19}$ J.

Powers and indices

The **power** (sometimes called the **index**) notation is used where a number is multiplied by itself. For example, $4 \times 4 = 4^2$ is described as '4 to the power 2' or '4 *squared*'. Similarly, $4 \times 4 \times 4 = 4^3$ is '4 to the power 3' or '4 *cubed*'. $4 \times 4 \times 4 \times 4 \times 4 \times 4 \times 4 = 4^7$ is '4 to the power 7' and so on. From this definition, it follows that

The superscript in 4^7 indicates how many multiplications are performed and is called the **power** or the **index** (plural: indices).

$$\begin{aligned} 4^4 \times 4^3 &= (4\times4\times4\times4)\times(4\times4\times4) \\ &= 4^7 \end{aligned}$$

For any number, a

When multiplying numbers, add the indices

$$a^x \times a^y = a^{(x+y)} \qquad \text{(MT1)}$$

Following a similar argument

$$4^4 \div 4^3 = \frac{4^4}{4^3} = \frac{(4\times4\times4\times4)}{(4\times4\times4)} = 4^1 = 4$$

So

When dividing numbers, subtract the indices

$$a^x \div a^y = a^{(x-y)} \qquad \text{(MT2)}$$

Any number raised to the power 1 is simply the number itself, so

$$4^1 = 4 \quad \text{and} \quad 17^1 = 17$$

A special case is that any number raised to the power zero is 1, so

$$4^0 = 17^0 = 1$$

Negative indices indicate the **reciprocal** of a number, that is, 1 divided by the number

An index of –1 indicates the reciprocal of a number

$$a^{-1} = \frac{1}{a} \qquad \text{(MT3)}$$

Other negative indices are defined as follows

$$a^{-2} = \frac{1}{a^2}; \qquad a^{-3} = \frac{1}{a^3}; \qquad a^{-4} = \frac{1}{a^4}$$

So, in general

$$a^{-n} = \frac{1}{a^n} \qquad \text{(MT4)}$$

Here are three examples

$$4^{-1} = \frac{1}{4} \qquad 4^{-2} = \frac{1}{4^2} = \frac{1}{4 \times 4} = \frac{1}{16} \qquad 4^{-3} = \frac{1}{4^3} = \frac{1}{4 \times 4 \times 4} = \frac{1}{64}$$

Indices do not have to be whole numbers. Think about $4^{\frac{1}{2}}$. From the rules for adding indices

$$4^{\frac{1}{2}} \times 4^{\frac{1}{2}} = 4^{(\frac{1}{2}+\frac{1}{2})} = 4^1 = 4$$

The **square root** of a number is the number that when multiplied by itself gives the original number. The cube root is the number that when multiplied by itself twice gives the original number.

So, $4^{\frac{1}{2}}$ multiplied by itself gives 4. $4^{\frac{1}{2}}$ is the **square root** of 4, also written as $\sqrt{4}$. Similar reasoning shows that $4^{\frac{1}{3}}$ is the **cube root** of 4, that is, $4^{\frac{1}{3}} \times 4^{\frac{1}{3}} \times 4^{\frac{1}{3}} = 4^1 = 4$.

$$a^{\frac{1}{2}} = \sqrt{a} \qquad \text{(MT5a)}$$

$$a^{\frac{1}{3}} = \sqrt[3]{a} \qquad \text{(MT5b)}$$

Combining numbers in the correct order: BODMAS rules

Numbers can be combined using the familiar operations of addition, subtraction, multiplication, and division.

For addition or multiplication, the *order* in which the operation is carried out *is not* important. Addition and multiplication are said to be **commutative**.

$$25 + 59 \text{ is the same as } 59 + 25 \qquad 43 \times 59 \text{ is the same as } 59 \times 43$$

$$25 + 59 = 59 + 25 = 84 \qquad 43 \times 59 = 59 \times 43 = 2537$$

Non-commutative operations include subtraction and division. Here, the *order* of the operation *is* important

$$25-59=-34 \neq 59-25=+34 \qquad 43\div 59=0.729 \neq 59\div 43=1.372$$

> ≠ means 'is not equal to'.

Suppose you need to find the mass of 3 mol of water. Water has the formula H_2O so the mass of 1 mol in grams is the 'relative molecular mass, M_r, weighed in grams'. The M_r is calculated from the relative atomic masses, A_r, for hydrogen (1.0079) and oxygen (15.999).

It would be incorrect to write the mass of 3 mol of water as

$$\text{mass of water in g} = 3\times 15.999+2\times 1.0079$$

> Satisfy yourself that the order of carrying out the operations gives different answers. For example, $3\times 15.999=47.997$; $47.997+2=49.997$; $49.997\times 1.0079=50.392$, whereas $2\times 1.0079=2.0158$; $2.0158+15.999=18.0148$; $18.0148\times 3=54.044$.

because this is ambiguous and gives different answers depending on the order in which you perform the operations. To overcome this problem, the parts of a calculation that should be kept together are enclosed in brackets. In this case, you need to find three times the relative molecular mass of water, so

$$\text{mass of water in g} = 3\times(15.999+2\times 1.0079)$$

The order in which operations should be performed is given by the rules of precedence. A useful way of remembering these involves the mnemonic BODMAS. Expressions in brackets (B) have the highest priority and are worked out first; subtraction (S) has the lowest priority and is carried out last.

B	Brackets
O	pOwers
D	Division
M	Multiplication
A	Addition
S	Subtraction

The mass of 3 mol of water is therefore found by working out the expression in brackets first. Multiplication (M) is performed before the addition (A), so the order is

$$\begin{aligned}\text{mass of water in g} &= 3\times(15.999+2\times 1.0079)\\ &= 3\times(15.999+2.0158)\\ &= 3\times(18.0148)\\ &= 54.0444\end{aligned}$$

Here's another example. In molecular spectroscopy, the energy of vibration of the chemical bond in a diatomic molecule is approximated by Equation 10.21 (see Section 10.5, p.476):

$$E_v = \left(v+\frac{1}{2}\right)\frac{h}{2\pi}\left(\frac{k}{\mu}\right)^{\frac{1}{2}} \qquad (10.21)$$

where v is the vibrational quantum number, k is the force constant of the bond, and μ is the reduced mass of the atoms.

To use this expression, you must carry out the mathematical operations in the correct order. To apply the BODMAS rules to this, it helps to break down the equation to show all the operations involved

$$E_v = \left(v+\frac{1}{2}\right)\times\left(\frac{h}{2\times\pi}\right)\times\left(\frac{k}{\mu}\right)^{\frac{1}{2}}$$

For the first energy level, v = 0. An $^{1}H^{35}Cl$ molecule (see Worked example 10.7, p.482) has a force constant, $k = 511\,\text{N m}^{-1}$ and a reduced mass $\mu_{HCl} = 1.61\times 10^{-27}\,\text{kg}$. The energy of the ground state vibrational energy level is

> The value of $h = 6.626\times 10^{-34}$ J s is given on the back inside cover.

$$E_v = \left(0+\frac{1}{2}\right)\times\left(\frac{6.626\times 10^{-34}\,\text{J s}}{2\times 3.142}\right)\times\left(\frac{511\,\text{N m}^{-1}}{1.61\times 10^{-27}\,\text{kg}}\right)^{\frac{1}{2}}$$

The BODMAS rules give the highest priority to brackets (B), so the first step is to work out each of the brackets

$$E_v = \frac{1}{2}\times 1.054\times 10^{-34}\,\text{J s}\times\left(3.174\times 10^{29}\,\text{s}^{-2}\right)^{\frac{1}{2}}$$

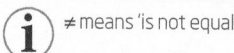

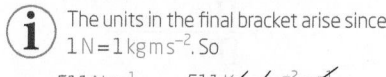

> The units in the final bracket arise since $1\,\text{N} = 1\,\text{kg m s}^{-2}$. So
>
> $$\frac{511\,\text{N m}^{-1}}{1.61\times 10^{-27}\,\text{kg}} = \frac{511\,\text{kg m s}^{-2}\,\text{m}^{-1}}{1.61\times 10^{-27}\,\text{kg}} = 3.174\times 10^{29}\,\text{s}^{-2}$$
>
> The manipulation of units in calculations is discussed in Section 1.2 (pp.6–11).

The next BODMAS priority is given to powers (O), so the next step is to work out the square root in the final bracket

$$E_v = \frac{1}{2} \times 1.054 \times 10^{-34}\ \text{J}\cancel{\text{s}} \times 5.634 \times 10^{14}\ \cancel{\text{s}^{-1}}$$

The final step is to perform the multiplications

$$E_v = 2.970 \times 10^{-20}\ \text{J}$$

Proportionality

> Note that proportionality refers to multiplication and division of the two quantities, *not* to addition and subtraction. For example, if you add 5 to one quantity, the other *does not* increase by 5.

Two quantities are said to be **proportional** if they are related so that, if one changes (e.g. doubles, triples, or halves), the other changes by a corresponding amount (i.e. doubles, triples, or halves).

For example, for a fixed amount of an ideal gas at constant volume

> If the temperature, T (measured in kelvin), of an ideal gas is doubled at constant volume, then the pressure, p, also doubles (see Section 8.1, p.345).

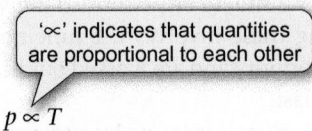

'∝' indicates that quantities are proportional to each other

$$p \propto T$$

> In a first order reaction, the rate of reaction ∝ concentration of the reactant (see Section 9.4, p.389).

Another way of representing proportionality is

$$p = c \times T \quad \text{or} \quad \frac{p}{T} = c$$

where c has a constant value and is called the **proportionality constant**.

For a fixed amount of an ideal gas at constant temperature, the volume goes down by a factor of 2 (halves) when the pressure doubles. This is written as

$$p \propto \frac{1}{V} \quad \text{or} \quad p \propto V^{-1} \quad \text{or} \quad p = \frac{c}{V}$$

Here, volume is *inversely* proportional to pressure.

Sums and products

> Σ (sum of) is upper-case Greek sigma; Π (product of) is upper-case Greek pi.

Two useful pieces of mathematical shorthand involve using the symbols Σ and Π, which stand for 'the sum of' and 'the product of', respectively.

For example, the sum of the first five integers, n_i, ($n_1 = 1$, $n_2 = 2$ ... $n_5 = 5$) is written as

$$\sum_{i=1}^{i=5} n_i = 1+2+3+4+5 = 15$$

> Examples of the use of Σ include summing the enthalpy changes of formation of the reactants and products to calculate the enthalpy change of a reaction using Equation 13.6 (p.626).
>
> $$\Delta_r H^{\ominus}_{298} = \sum \nu_i \Delta_f H^{\ominus}_{298\,(\text{products})} - \sum \nu_i \Delta_f H^{\ominus}_{298\,(\text{reactants})}$$
>
> An example of using Π is in the expression for the thermodynamic equilibrium constant, K, in Equation 15.2 (p.702)
>
> $$k = \frac{\prod (a_{(\text{products})})^{\nu_P}_{\text{eqm}}}{\prod (a_{(\text{reaxtants})})^{\nu R}_{\text{eqm}}}$$

The result of multiplying numbers together is called the **product**. The product of the first five integers is written as:

$$\prod_{i=1}^{i=5} n_i = 1 \times 2 \times 3 \times 4 \times 5 = 120$$

MT2 Algebra

Much of the algebra used in chemistry consists of devising and manipulating a formula or an equation that shows how quantities depend on each other. For example, the ideal gas equation, Equation 8.5 (p.349), shows how the pressure, p, volume, V, amount in moles, n, and the absolute temperature, T, of an ideal gas are related

$$pV = nRT \tag{8.5}$$

p, V, n, and T are known as **variables** because they can take different values. Equations can also contain **constants** whose values do not change. The ideal gas constant, R, is a proportionality constant ($R = 8.314\,\mathrm{J\,K^{-1}\,mol^{-1}}$). Equation 8.5 summarizes all the interrelationships between the variables.

Functions

For a first order reaction (see Section 9.4, p.389), the rate of reaction depends on the concentration of the reactant, A

$$\text{rate of reaction} = k[\mathrm{A}]$$

where k is a constant called the rate constant. The rate of reaction is a **dependent** variable, whereas [A] is an **independent** variable. In words, the equation says 'to find the value for the rate of reaction, multiply the value of the variable, [A], by the constant, k'. We say that 'the rate of reaction *is a* **function** of [A]'. The function tells you what mathematical procedures need to be performed on the variables. In this case, the function is 'multiply by k'. This is written as

The independent variable is usually the property you control and vary during an experiment, for example, the concentration of reactants, the time, or the temperature. The dependent variable is usually the property that you measure, for example, the rate of a reaction.

A function of [A]

$$\text{rate of reaction} = f([\mathrm{A}]) = k[\mathrm{A}]$$

In this case, the rate of reaction is a *linear* function of [A] (i.e. a plot of rate of reaction against [A] is a straight line). For a second order reaction, the rate of reaction depends on a different function of [A]

$$\text{rate of reaction} = f([\mathrm{A}]) = k[\mathrm{A}]^2 \qquad \text{(second order reaction)}$$

Functions may depend on several variables. For an ideal gas, p is a function of n, V, and T

A function of n, V, and T

The equation defines the nature of the function

$$p = f(n, V, T) = \frac{nRT}{V}$$

Rearranging equations

To obtain the expression for the pressure of an ideal gas above, Equation 8.5 had to be changed around so that it had the form '$p = \ldots$'. Rearranging equations like this is a common requirement in chemistry.

In arithmetic, adding the same quantity to both sides of an equation does not change anything. The same is true if both sides are multiplied by the same number or if any other operation is performed—as long as the same thing is done to both sides.

The same principle applies in algebra, so, to rearrange Equation 8.5 to give an expression in terms of p

$$pV = nRT \qquad (8.5)$$

The '=' sign can be thought of as the balance point of an equation. Adding or subtracting the same thing from both sides does not change the balance.

1 Divide both sides by V: $\dfrac{pV}{V} = \dfrac{nRT}{V}$

2 Cancel V on the left-hand side: $\dfrac{p\cancel{V}}{\cancel{V}} = \dfrac{nRT}{V}$ to give $p = \dfrac{nRT}{V}$

Similarly, Equation 8.5 can be rearranged to give an expression for *R*.

1 Divide both sides by nT: $\frac{pV}{nT} = \frac{nRT}{nT}$

2 Cancel n and T on the right-hand side: $\frac{pV}{nT} = \frac{\not{n}R\not{T}}{\not{n}\not{T}}$ to give $\frac{pV}{nT} = R$,

which is the same as $R = \frac{pV}{nT}$

As another example, the change in Gibbs energy during a reaction, ΔG, is given by Equation 14.16 (see Section 14.5, p.675)

$$\Delta G = \Delta H - T\Delta S \tag{14.16}$$

An expression for the entropy change, ΔS, can be found by rearranging the equation. You can do this in three stages.

1 Add $T\Delta S$ to each side: $\Delta G + T\Delta S = \Delta H - T\Delta S + T\Delta S$
On the right, $-T\Delta S + T\Delta S = 0$, so $\Delta G + T\Delta S = \Delta H$

2 Subtract ΔG from each side: $\Delta G + T\Delta S - \Delta G = \Delta H - \Delta G$
On the left, $\Delta G - \Delta G = 0$, so $T\Delta S = \Delta H - \Delta G$

3 Divide both sides by T and cancel: $\frac{\not{T}\Delta S}{\not{T}} = \frac{\Delta H - \Delta G}{T}$, so $\Delta S = \frac{\Delta H - \Delta G}{T}$

Inserting numbers into an equation

There are many examples throughout the book showing how to manipulate and cancel units in a calculation. See, for example, Worked examples 8.3 (p.352), 9.9 (p.428), 10.4 (p.470), 14.7 (p.676), and 17.2 (p.778).

It is particularly important when you insert numbers for quantities into an equation that you include the units for each quantity—and keep the units with the number for each quantity at every stage of the calculation. This ensures that the final result has the correct units.

Algebraic manipulation

The rules for rearranging symbols in equations are the same as the rules of arithmetic—after all, algebra is just using letters in an equation to represent numbers. Letters can be added, subtracted, multiplied, divided, squared, etc. just as numbers can. Some of the rules, such as using indices, are described in Section MT1. However, there are some rules that need to be used rather differently when algebraic terms are involved.

Think about the formula

Often in algebra, the multiplication sign is omitted so $ab = a \times b$.

$$a \times (b + c) \quad \text{(which is often written as } a(b+c))$$

The BODMAS rules are discussed in Section MT1 (p.1304).

Using the BODMAS rules, the term in brackets should be evaluated first—but, in this case, the values of *b* and *c* are not known. The multiplication is done by multiplying each term inside the brackets by the term outside

$$a \times (b + c) = (a \times b) + (a \times c) = ab + ac \tag{MT6}$$

Fractions are treated in the same way

$$\frac{a+b}{c} = \frac{1}{c}(a+b) = \frac{a}{c} + \frac{b}{c} \tag{MT7}$$

If you are unsure about rearrangements that you have made, it often helps to put numbers into the equations to check that you have the right answer. For Equation MT6, suppose $a = 3$, $b = 2$, and $c = 4$.
Compare $3 \times (2 + 4) = 3 \times (6) = 18$ with $3 \times (2 + 4) = (3 \times 2) + (3 \times 4) = (6 + 12) = 18$. Both methods give the same answer.

Suppose you want to multiply two bracketed terms, for example, $(a + b) \times (c + d)$. The two bracketed terms are treated by multiplying each pair of terms, as shown in Figure MT5

$$(a + b) \times (c + d) = (a + b)(c + d) = (ac + ad + bc + bd) \tag{MT8}$$

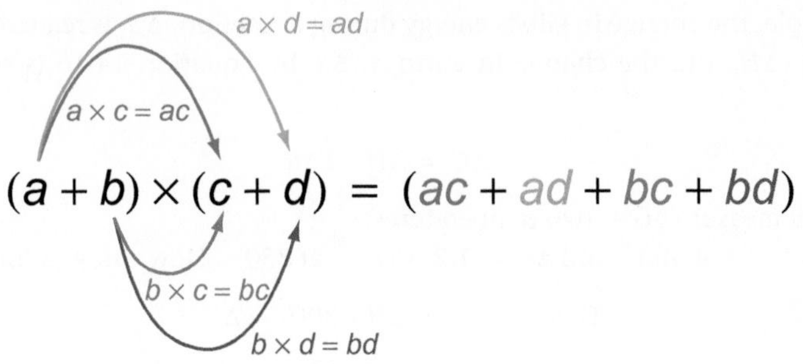
a × d = ad
a × c = ac
(a + b) × (c + d) = (ac + ad + bc + bd)
b × c = bc
b × d = bd

Figure MT5 Multiplication of terms in brackets.

Three useful examples of applying this method to bracketed terms are shown below

1 $(a+b)^2 = (a+b)\times(a+b) = (a\times a)+(a\times b)+(b\times a)+(b\times b)$

$$(a+b)^2 = a^2+2ab+b^2 \qquad \text{(MT9)}$$

2 $(a-b)^2 = (a-b)\times(a-b) = (a\times a)+(a\times -b)+(-b\times a)+(-b\times -b)$

$$(a-b)^2 = a^2-2ab+b^2 \qquad \text{(MT10)}$$

3 $(a+b)\times(a-b) = (a\times a)+(a\times -b)+(b\times a)+(b\times -b) = a^2-ab+ba-b^2$

$$(a+b)\times(a-b) = a^2-b^2 \qquad \text{(MT11)}$$

The square of a negative number is always positive,

so $(-b\times -b) = b^2$,

but $(b\times -b) = -b^2$.

When multiplying out brackets, it is usual to collect terms in the same power together to simplify the expression. For example

Multiply out the brackets: $(2x+2)\times(x-4) = 2x^2-8x+2x-8$

Collect terms in the same power together: $= 2x^2-6x-8$

Polynomial equations

The equation: $x^2-3x-4=0$ is an example of a **polynomial** equation, one containing a variable raised to various powers. The highest power is 2, so it is a **quadratic** equation. Equations where the highest power is 3, such as $4x^3+5x^2+3x+1=0$, are called **cubic equations**.

Solving simultaneous equations

Simultaneous equations are equations containing several variables, each of which has the same value in all the equations. For example, in the two equations below, x has the same value in both equations and y has the same value in both equations.

$$x+2y = 5 \qquad \text{(Equation 1)}$$

$$x+y = 3 \qquad \text{(Equation 2)}$$

To solve the equations to find values for x and y, Equation 2 is subtracted from Equation 1.

$$\text{(Equation 1)} - \text{(Equation 2)}$$

$$(x+2y)-(x+y) = 5-3$$

This can be simplified to

$$2y-y = 2, \quad \text{so} \quad y = 2$$

You can now substitute this value for y into Equation 1, which gives $x+4=5$, so $x=1$. You should then check that these values for x and y give the correct result when substituted into Equation 2. The values of x and y work in both equations—hence the equations are simultaneous.

For example, the change in Gibbs energy during a reaction, ΔG, is related to the change in enthalpy, ΔH, and the change in entropy, ΔS, by Equation 14.16 (see Section 14.5, p.675)

$$\Delta G = \Delta H - T\Delta S \qquad (14.16)$$

Suppose you measure ΔG at two temperatures:
$\Delta G = 17.5\,\mathrm{kJ\,mol^{-1}}$ at 300 K and $\Delta G = 11.2\,\mathrm{kJ\,mol^{-1}}$ at 350 K. How can you find ΔH and ΔS?

$$17.5\,\mathrm{kJ\,mol^{-1}} = \Delta H - 300\,\mathrm{K} \times \Delta S \qquad \text{(Equation 3)}$$

$$11.2\,\mathrm{kJ\,mol^{-1}} = \Delta H - 350\,\mathrm{K} \times \Delta S \qquad \text{(Equation 4)}$$

Subtracting Equation 4 from Equation 3

$$\begin{aligned} 17.5\,\mathrm{kJ\,mol^{-1}} - 11.2\,\mathrm{kJ\,mol^{-1}} - 11.2\,\mathrm{kJ\,mol^{-1}} &= (\Delta H - 300\,\mathrm{K} \times \Delta) - (\Delta H - 350\,\mathrm{K} \times \Delta S) \\ 6.3\,\mathrm{kJ\,mol^{-1}} &= \Delta H - \Delta H - 300\,\mathrm{K} \times \Delta S + 350\,\mathrm{K} \times \Delta S \\ &= +50\,\mathrm{K} \times \Delta S \\ \Delta S &= +0.126\,\mathrm{kJ\,K^{-1}\,mol^{-1}} \end{aligned}$$

Other examples where this technique can be used include using the Clausius–Clapeyron equation when vapour pressure is known at two temperatures (see Section 17.2, p.775) or when using the Arrhenius equation to find the value of a rate constant at a different temperature (see Section 9.7, p.425).

Now you can find ΔH by substituting this value for ΔS into Equation 3 (or Equation 4)

$$\begin{aligned} 17.5\,\mathrm{kJ\,mol^{-1}} &= \Delta H - 300\,\cancel{\mathrm{K}}\,(+0.126\,\mathrm{kJ}\,\cancel{\mathrm{K}}^{-1}\,\mathrm{mol^{-1}}) \\ &= \Delta H - 37.8\,\mathrm{kJ\,mol^{-1}} \\ \Delta H &= +55.3\,\mathrm{kJ\,mol^{-1}} \end{aligned}$$

MT3 Logarithms and exponentials

Logarithms were invented to provide an easier way of dealing with equations that contain exponents. They also provide a convenient way of dealing with numbers that vary widely in magnitude.

The numbers 10, 100, 1000, 10000, and so on, can be written in terms of powers of 10: 10^1, 10^2, 10^3, 10^4 . . .. In fact, *any* number—not just powers of 10—can be written in this way: $5 = 10^{0.699}$, $2 = 10^{0.3010}$, $125 = 10^{2.097}$. The exponent in these numbers is called the **logarithm to base 10** (written as $\log_{10}$) of the number.

In general,

$$\text{if } a = 10^x, \text{ then } x = \log_{10} a \qquad \text{(MT12)}$$

The exponent is the power to which 10 is raised. Scientific notation and exponents are discussed in Section MT1 (p.1304).

The relationship between the logarithm and the exponent when writing a number in scientific notation is shown in the following table.

100	1×10^2	$\log_{10}(1 \times 10^2)$	= 2
10	1×10^1	$\log_{10}(1 \times 10^1)$	= 1
1	1×10^0	$\log_{10}(1 \times 10^0)$	= 0
0.1	1×10^{-1}	$\log_{10}(1 \times 10^{-1})$	= −1
0.01	1×10^{-2}	$\log_{10}(1 \times 10^{-2})$	= −2

Logarithms to the base 10 are sometimes known as common logarithms.

One use of logarithms in chemistry is to define the 'p' scales, such as the pH scale for acidity and the pK_a scale for acidity constants (see Section 7.2, p.308). The pH scale is a logarithmic scale, where pH = $-\log_{10}[H_3O^+(aq)]$, so pH = 0 corresponds to $[H_3O^+(aq)]$ = 1 mol dm^{-3}, pH = 7 to $[H_3O^+(aq)] = 1 \times 10^{-7}$ mol dm^{-3}, and pH = 14 to $[H_3O^+(aq)]$ = 1×10^{-14} mol dm^{-3}.

Logarithms are not confined to base 10. In principle, any number can be written as a power of any other number. For example, $4 = 2^2$, $8 = 2^3$, $10 = 2^{3.322}$, $100 = 2^{6.644}$, so that

$$\log_2 8 = 3 \quad \text{and} \quad \log_2 100 = 6.664$$

A particularly useful form of logarithm is the **natural logarithm**. The base of natural logarithms is a fundamental constant of mathematics, e, which has a value of 2.7183. Natural logarithms are often written as 'ln'.

$$10 = e^{2.303} \quad \text{so,} \quad \log_e 10 = \ln 10 = 2.303 \qquad \text{(MT13)}$$

In general, for a logarithmic function

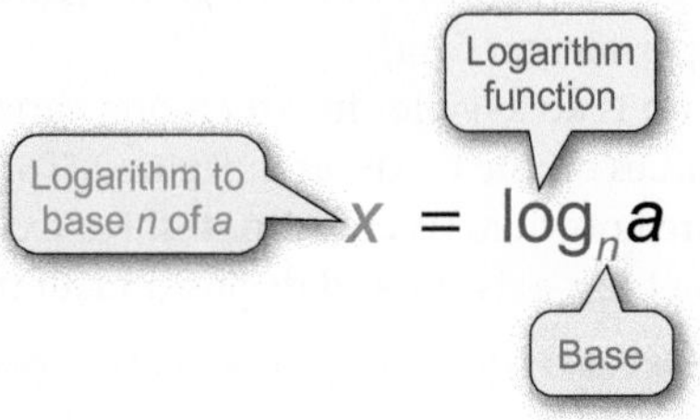
Logarithm function
Logarithm to base n of a
x = log_n a
Base

Like π, e is an irrational number, whose value cannot be expressed accurately as a fraction (see Section MT1, p.1304). To 5 significant figures, e = 2.7183.

Using logarithms

Below are some useful expressions that will help you use and manipulate logarithms.

Since $10 = 10^1$, it follows that $\log_{10}10 = 1$. In general

$$\log_n n = 1 \qquad \text{(MT14)}$$

so that $\ln_e = 1$. Since $10^0 = e^0 = 1$, it follows that $\log_{10}1 = \log_e 1 = \ln 1 = 0$. In general

$$\log_n 1 = 0 \qquad \text{(MT15)}$$

The rules for multiplying and dividing logarithms follow from the rules for indices described in Section MT1 (pp.1304).

$$\ln(a \times b) = \ln a + \ln b \qquad \text{(MT16)}$$

$$\ln\frac{a}{b} = \ln a - \ln b \qquad \text{(MT17)}$$

$$\ln\frac{1}{b} = \ln 1 - \ln b = 0 - \ln b = -\ln b \qquad \text{(MT18)}$$

$$\ln a^b = b \times \ln a \qquad \text{(MT19)}$$

Antilogarithms and exponentials

From Equation MT12, if $x = \log_{10}a$ then it follows that $a = 10^x$. a is known as the **antilogarithm** (antilog) of x. It is essentially the reverse of the logarithm function.

The antilog of a natural logarithm is known as the **exponential** function. It is usually written as e^x, or sometimes as $\exp(x)$.

$$e^x = \exp(x) = (2.7183)^x \qquad \text{(MT20)}$$

The number 2.7183 may seem a strange number on which to base a mathematical function. However, e has a series of properties that make it useful (see, for example, Section MT6, pp.1324). It crops up wherever systems grow or decay exponentially, and it often occurs in the analysis of chemical systems. For example, in a first order reaction, A → products, the concentration of A, [A], decreases as an exponential function of time, t

$$[A] = [A]_0\, e^{-kt} \qquad \text{(9.6c)}$$

where $[A]_0$ is the concentration of A at time $t = 0$, and k is the rate constant for the reaction at a particular temperature (see Section 9.4, p.389).

Other examples where exponentials occur in chemistry are the Arrhenius equation (Equation 9.24a), which describes the effect of temperature on the rate constant for a reaction (see Section 9.7, p.425), and the Boltzmann distribution (Equation 10.9), which describes the populations of energy levels in molecules (Section 10.3, p.458).

Most calculators have an e^x button, often labelled as the 'inverse ln'. Make sure that you don't confuse it with the 'EXP' button, which you use to enter powers of 10.

You often hear people describing anything that keeps on increasing as 'increasing exponentially'. However, in strict mathematical terms, the saying should be reserved for functions of the form $y = e^x$. A sketch of this function is shown in Figure MT8 (p.1319).

MT4 Graphical representations of functions

It is often very useful to have an idea of how a function depends on some variables and graphs are a useful way of picturing this. Plotting a graph is also a good way of looking for a relationship between experimental data.

The usual form of a graph is an x–y plot in which one variable is plotted on the x-axis with the corresponding values on the y-axis, as in Figure MT6. For a point on the plot, the positions along the axes are known as its **coordinates**. For example, (1, 3) means that the position of the point along the x-axis is 1 and its position along the y-axis is 3.

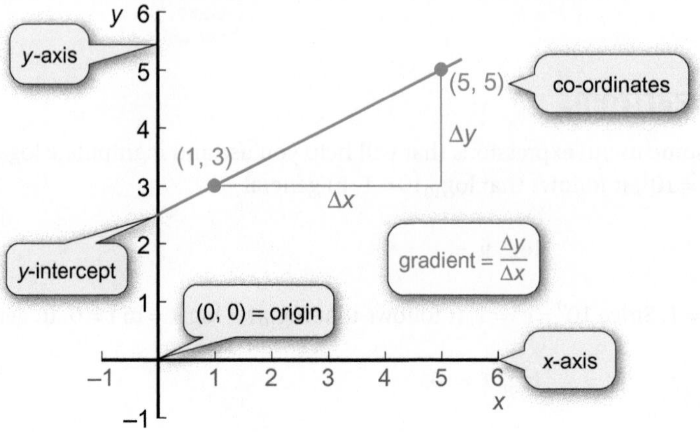
y
x
y-axis
co-ordinates
(5, 5)
(1, 3)
Δy
Δx
gradient = Δy/Δx
y-intercept
(0, 0) = origin
x-axis
6
5
4
3
2
1
–1
–1
1
2
3
4
5
6

Figure MT6 A straight line graph and calculation of the gradient.

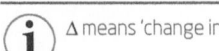

When plotting experimental data, the independent variable is plotted on the x-axis and the dependent variable on the y-axis. For example, if you measure the pressure of a gas at a number of different temperatures, the temperature is the independent variable (since you control and vary this); you measure the pressure so this is the dependent variable. You plot T on the x-axis and p on the y-axis.

Δ means 'change in'.

The simplest type of graph is a **linear** (straight line) plot as in Figure MT6. The **gradient** (slope) is a measure of the steepness of the line. It is found from the change in y (Δy) divided by the corresponding change in x (Δx). From the plot in Figure MT6

$$\text{gradient} = \frac{\Delta y}{\Delta x} = \frac{(5-3)}{(5-1)} = \frac{2}{4} = 0.5$$

Graphs where the value of y increases as x increases have positive gradients; those where the value of y decreases as x increases have negative gradients. The **intercept** is the value of y when $x = 0$.

The equation of a straight line

All straight lines can be represented by an equation of the form

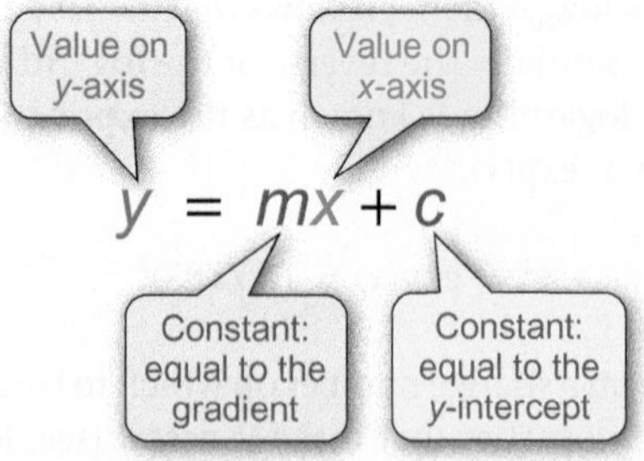
Value on y-axis
Value on x-axis
Constant: equal to the gradient
Constant: equal to the y-intercept

$$y = mx + c \qquad \text{(MT21)}$$

where x and y are variables and m and c are constants that correspond to the gradient and the y-intercept of the graph, respectively. Equation MT21 can be rewritten as

$$y = (\text{gradient} \times x) + (\text{intercept})$$

In Figure MT6, the intercept occurs at $y = 2.5$ and the gradient is 0.5. So the equation of the straight line is $y = 0.5x + 2.5$.

You can confirm that this is correct by substituting a value for x in the equation and checking that the value the equation gives for y corresponds to the value on the graph. For example, what is the value of y at $x = 3$? Using the equation, $y = (0.5 \times 3) + 2.5 = 4$. Reading off the value of y from the graph in Figure MT6 shows this to be correct.

If a line passes through the origin, the y-intercept is zero, so the equation has the form $y = mx$.

Note that Equation MT21 can also be written in the form $y = c + mx$ and you need to be able to recognize that an equation in this form corresponds to a straight line. (For example, see the variation of the Gibbs energy change with temperature (Section 14.6, p.686).)

The Beer–Lambert law, which describes the absorption of radiation by a substance (see Section 10.3, p.458), has the form $A = \varepsilon cl$. A plot of absorbance, A, against concentration, c, is a straight line passing through the origin with gradient $= \varepsilon \times l$. (ε and l are constants; ε is the molar absorption coefficient of the substance and l is the path length of the radiation in the solution.

Graphs of non-linear functions

Of course, not all functions encountered in chemistry are linear and give straight line plots. It is useful to have an idea of the shapes of some other functions.

Quadratics

Expressions in which the highest power of x is 2 (i.e. x^2) are called **quadratic** expressions. Figure MT7 shows a plot of $y = x^2$ contrasted with that of the linear plot of $y = x$. The shape of the $y = x^2$ plot is called a **parabola**. Plots of other quadratic functions (e.g. $y = 3x^2$, $y = 5x^2 + 3$, $y = 2x^2 - 4x + 6$, etc.) have the same general shape but intersect the y-axis at different points and have different rates of increase.

An example of a quadratic function is the dependence of potential energy on the extension, x, as a spring undergoes simple harmonic motion; $E_{PE} = \frac{1}{2}kx^2$. A plot of E_{PE} against x has the shape of a parabola (see Box 10.4, p.479).

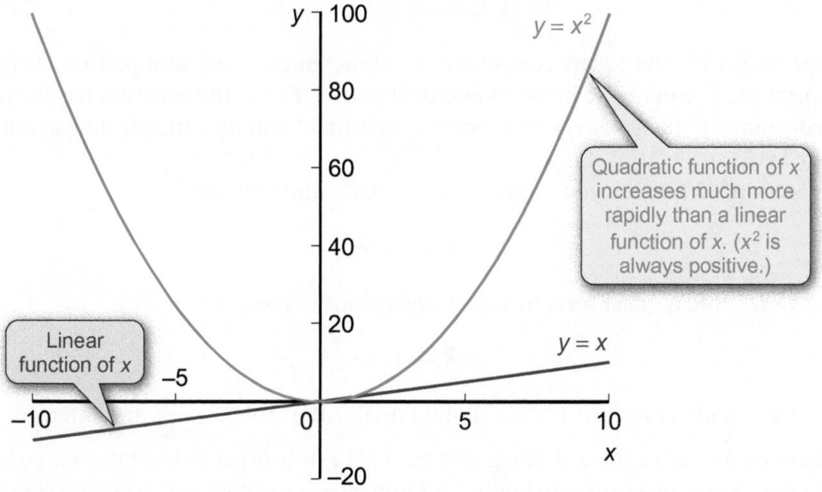
y
100
80
60
40
20
−20
y = x²
Quadratic function of x increases much more rapidly than a linear function of x. (x² is always positive.)
y = x
Linear function of x
−10
−5
0
5
10
x

Figure MT7 Plots of $y = x$ and $y = x^2$.

Logarithms and exponentials

Logarithmic and exponential functions occur frequently in chemistry. They have the shapes shown in Figure MT8.

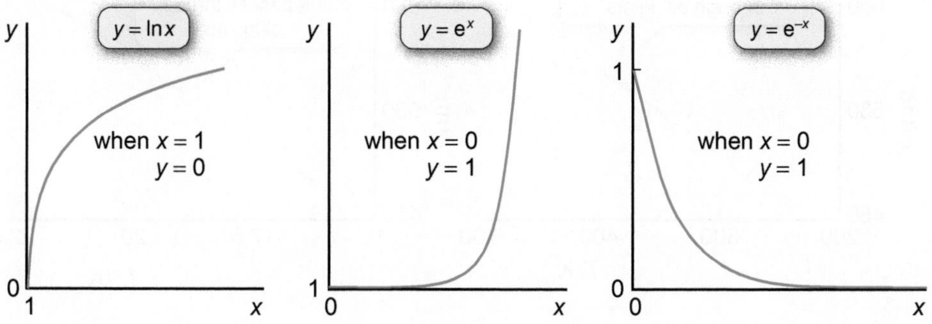
y
y = ln x
when x = 1
y = 0
0
1
x
y
y = eˣ
when x = 0
y = 1
1
0
x
y
y = e⁻ˣ
1
when x = 0
y = 1
0
0
x

Figure MT8 Plots of logarithmic and exponential functions of x.

The shape of $y = e^{-x}$ is particularly important. The value of e^{-x} falls as x increases, but the rate of change gradually decreases as e^{-x} gets closer to zero; the value of e^{-x} never actually reaches zero. An example is a plot of the concentration of the reactant, [A], against time, t, during a first order reaction (see Section 9.4, p.389)

$$[A] = [A]_0 \, e^{-kt} \tag{9.6c}$$

where $[A]_0$ is the concentration of A at time $t = 0$ and k is the rate constant. A plot of [A] against t is called an **exponential decay curve** (see Figure 9.9, p.394).

Converting equations to linear equations

While many relationships in chemistry are not linear, it is often useful to present the data in a form such that a straight line graph can be plotted.

The root mean square speed, c, of molecules in a gas is given by Equation 8.19 (Section 8.5, p.362)

$$c = \left(\frac{3RT}{M}\right)^{\frac{1}{2}} \tag{8.19}$$

R is the gas constant and, for a particular gas, the molar mass, M, is constant, so Equation 8.19 can be rewritten as

$$c = \left(\frac{3R}{M}\right)^{\frac{1}{2}} T^{\frac{1}{2}} = \text{constant} \times T^{\frac{1}{2}}$$

c depends on $T^{\frac{1}{2}}$, the square root of the absolute temperature, so a plot of c versus the temperature, T, will not be linear. However, if you let $T^{\frac{1}{2}} = x$, the equation has the form of $y = \text{constant} \times x$ (i.e. $y = mx$) and a plot of c against $T^{\frac{1}{2}}$ will be a straight line as shown in Figure MT9.

In Section 9.7 (p.425), you can see how the Arrhenius equation

Several other equations, such as the van 't Hoff equation (Section 15.5, p.713) and the Clausius–Clapeyron equation (Section 17.2, p.775) have similar forms to the Arrhenius equation and can be treated in the same way to give linear equations.

$$k = Ae^{-\frac{E_a}{RT}} \tag{9.24a}$$

is converted into a linear form by taking logarithms of both sides

$$\ln k = \ln A - \frac{E_a}{R}\left(\frac{1}{T}\right) \tag{9.24b}$$

so that a plot of $\ln k$ against $1/T$ is a straight line, with gradient $-\frac{E_a}{R}$ and intercept $\ln A$.

Note the format of the axis labels in Figure MT9. It is usual to label the axes of graphs using the format: measurement/units. This indicates the units used to anyone viewing the data and means that only pure numbers appear on the graph.

The same format is used when presenting data in tables. The table headings are written as measurement/units so that only numbers appear in the table.

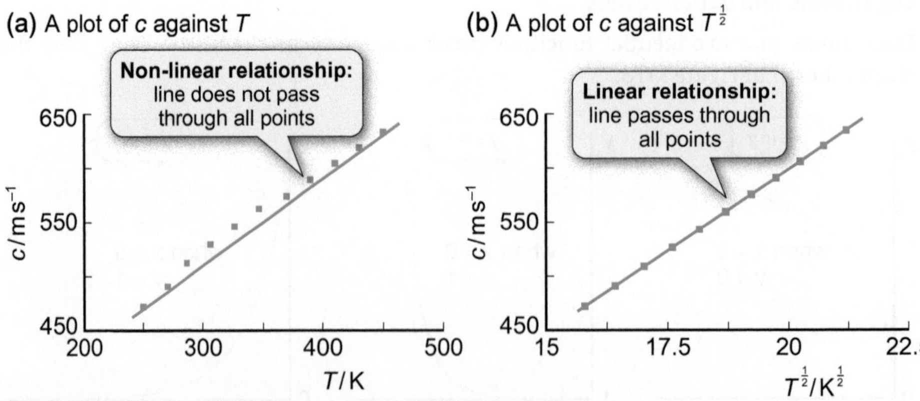

(a) A plot of c against T
Non-linear relationship: line does not pass through all points
650
550
450
c/ms⁻¹
200
300
400
500
T/K
(b) A plot of c against T^½
Linear relationship: line passes through all points
650
550
450
c/ms⁻¹
15
17.5
20
22.5
T^½/K^½

Figure MT9 Converting equations to give linear relationships is often useful. Here, plotting the average speed of gas molecules as a function of $T^{\frac{1}{2}}$ gives a straight line plot.

MT5 Geometry and trigonometry

Areas and volumes

Some topics in chemistry require an understanding of shapes in two and three dimensions and their areas and volumes. These are summarized in Figure MT10.

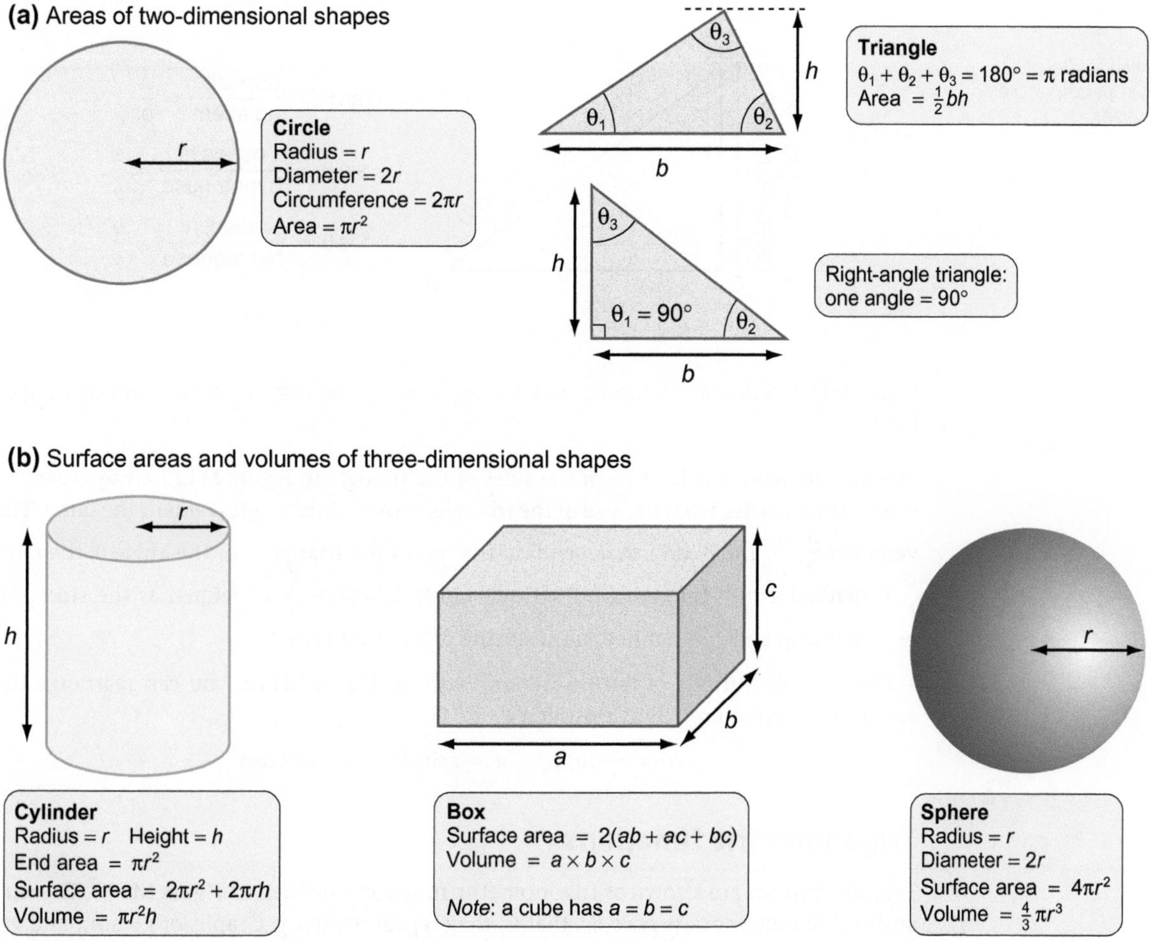
(a) Areas of two-dimensional shapes
r
Circle
Radius = r
Diameter = 2r
Circumference = 2πr
Area = πr²
θ3
θ1
θ2
h
b
Triangle
θ1 + θ2 + θ3 = 180° = π radians
Area = ½bh
θ1 = 90°
Right-angle triangle: one angle = 90°
(b) Surface areas and volumes of three-dimensional shapes
Cylinder
Radius = r Height = h
End area = πr²
Surface area = 2πr² + 2πrh
Volume = πr²h
a
c
Box
Surface area = 2(ab + ac + bc)
Volume = a × b × c
Note: a cube has a = b = c
Sphere
Radius = r
Diameter = 2r
Surface area = 4πr²
Volume = 4/3 πr³

Figure MT10 Areas and volumes. (a) Areas of two-dimensional shapes. (b) Surface areas and volumes of three-dimensional shapes.

Radians

Most commonly, angles are measured in degrees, symbol °; a complete circle is equivalent to turning through an angle of 360°. However, for some applications it is convenient to measure angles in other units, called radians. These are defined as in Figure MT11 as the ratio of the length of an arc of a circle to the radius of the circle. An angle of 1 radian has an arc length equivalent to the radius of the circle, so that $s = r$. An angle of 1 radian is 57.3°. Since the circumference of the whole circle is $2\pi r$, then

$$360° \text{ is equivalent to } \frac{2\pi r}{r} = 2\pi \text{ radians}$$

From this, it follows that $180° = \pi$ radians and $90° = \frac{\pi}{2}$ radians.

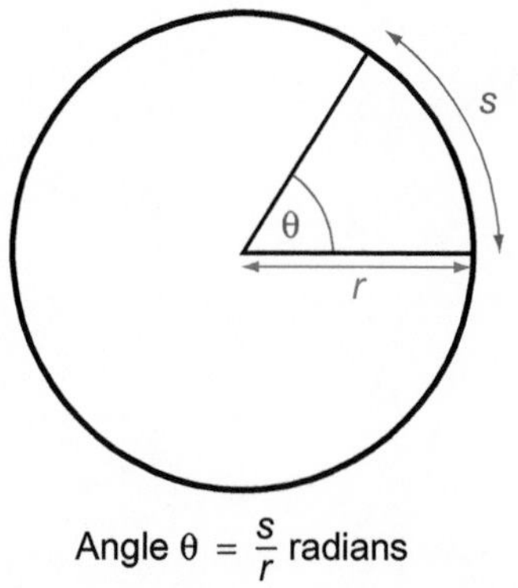
s
θ
r
Angle θ = s/r radians

Figure MT11 Angles can be measured in radians, defined by the angle subtended by an arc of length, s, on a circle of radius, r.

Right-angled triangles

Pythagoras's theorem states that *the square of the hypotenuse equals the sum of the squares of the other two sides.*

Consider the right-angled triangle shown in Figure MT12. Angle ACB is the right angle and is 90°. The side opposite the right angle is called the **hypotenuse**. The angles in any triangle add up to 180°, so angle ABC = 90° – θ. The lengths of the sides of a right-angled triangle are linked by **Pythagoras's theorem**

$$a^2 + b^2 = c^2 \quad \text{(MT22)}$$

Pythagoras's theorem is used to find the distance between ions in an ionic lattice structure (see Section 6.2, p.262).

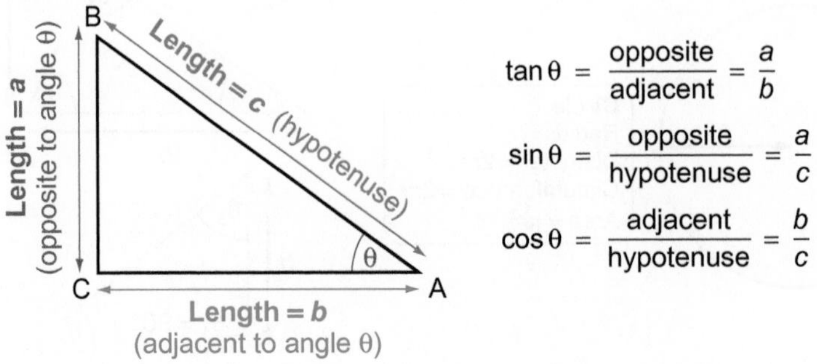
B
Length = c (hypotenuse)
Length = a
(opposite to angle θ)
θ
C
A
Length = b
(adjacent to angle θ)
tan θ = opposite/adjacent = a/b
sin θ = opposite/hypotenuse = a/c
cos θ = adjacent/hypotenuse = b/c

Figure MT12 Trigonometric functions are defined in terms of the lengths and angles in a right-angled triangle.

The *ratio* between the lengths of the sides of the triangle in Figure MT12 always remains the same no matter what the size of the triangle, provided the angles remain the same. The values of $\frac{a}{b}$, $\frac{a}{c}$, and $\frac{b}{c}$ do not depend on the size of the triangle. For the angle θ, the ratio $\frac{a}{c}$ is defined as the **tangent** of θ (written tan θ). Likewise, $\frac{b}{c}$ is defined as the **sine** of θ (written sin θ) and $\frac{b}{c}$ is defined as the **cosine** of θ (written cos θ).

From the definitions of tan θ, sin θ, and cos θ in Figure MT12, you can rearrange the equations to obtain relationships such as

$$a = b\tan\theta; \qquad a = c\sin\theta; \qquad b = c\cos\theta$$

Trigonometric functions

Tan, sin, and cos are known as trigonometric functions, defined in Figure MT12. The sine and cosine functions are *periodic*, that is, they repeat regularly. Graphs of $y = \sin x$ and $y = \cos x$ (Figure MT13) show how $\sin x$ and $\cos x$ change as the angle x changes.

The two plots have the same form but are shifted along the x-axis by $\frac{\pi}{2}$ radians (90°). Both repeat, that is, they have the same value, at intervals of 2π radians (360°), so that $\sin x = \sin(x + 2\pi)$ and $\cos x = \cos(x + 2\pi)$. In Figure MT13, the value of y oscillates smoothly between +1 and –1.

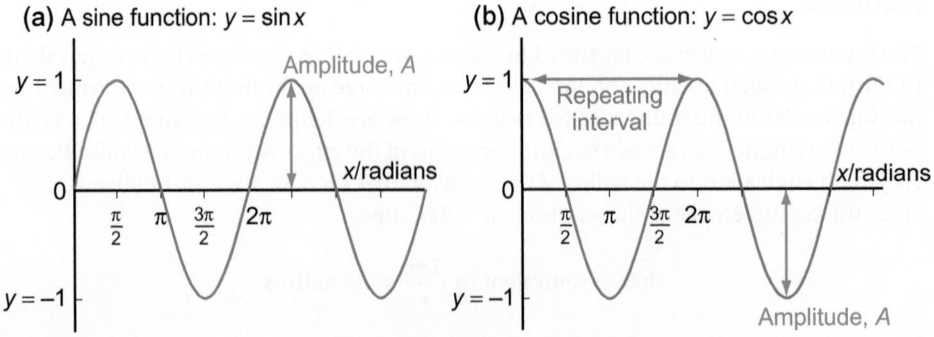
(a) A sine function: y = sin x
Amplitude, A
y = 1
0
y = –1
x/radians
π/2
π
3π/2
2π
(b) A cosine function: y = cos x
y = 1
Repeating interval
0
y = –1
x/radians
π/2
π
3π/2
2π
Amplitude, A

Figure MT13 The (a) sin and (b) cos functions are periodic with a repeating interval of 2π radians (360°).

A more general form of the sine function is

$$y = A \sin ax \qquad \text{(MT23)}$$

where A and a are constants. A is the maximum value of y (known as the **amplitude**); a defines the frequency with which waves oscillate between positive and negative values.

The variation in the electric field in electromagnetic radiation, such as visible light, is a sine function where $ax = 2\pi\nu t$. Here, y varies with time and ν is the frequency of the wave, that is, the number of waves that pass in 1 s. Since $c = \lambda\nu$ (Equation 3.1, p.118), the sine function can also be expressed in terms of the distance over which the light is propagated, and a is then related to the wavelength of the wave, λ. Figure 3.4 (p.118) shows the frequency, wavelength, and amplitude for a wave travelling with speed, c.

Another example of a sine function is the wavefunction that describes a particle in a box (see Box 10.1 in Section 10.2, p.454).

Polar and spherical coordinates

A point in a plane (such as on a graph) is usually represented by the distances along the x- and y-axes as in Figure MT14(a). This is known as a Cartesian representation. However, the position of the point could equally well be represented by its distance from the origin, r, and the angle, θ, made with one of the axes as in Figure MT14(b). The position of a point in these terms is defined by polar coordinates.

The same principle can be used in three dimensions. Three pieces of information are now needed to define the position of the point, as shown in Figure MT15. The Cartesian

Spherical coordinates are used in chemistry to define the wavefunctions of an electron, as described in Section 3.5 (p.134).

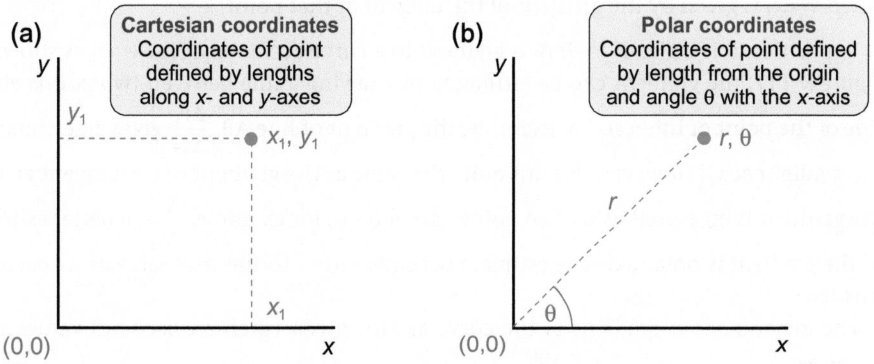

(a)
Cartesian coordinates
Coordinates of point defined by lengths along x- and y-axes
y
y1
x1, y1
x1
(0,0)
x
(b)
Polar coordinates
Coordinates of point defined by length from the origin and angle θ with the x-axis
y
r, θ
r
θ
(0,0)
x

Figure MT14 In two dimensions, the position of a point can be defined in terms of (a) its position along two axes (Cartesian coordinates) or (b) by its distance from the origin and angle with one of the axes (polar coordinates).

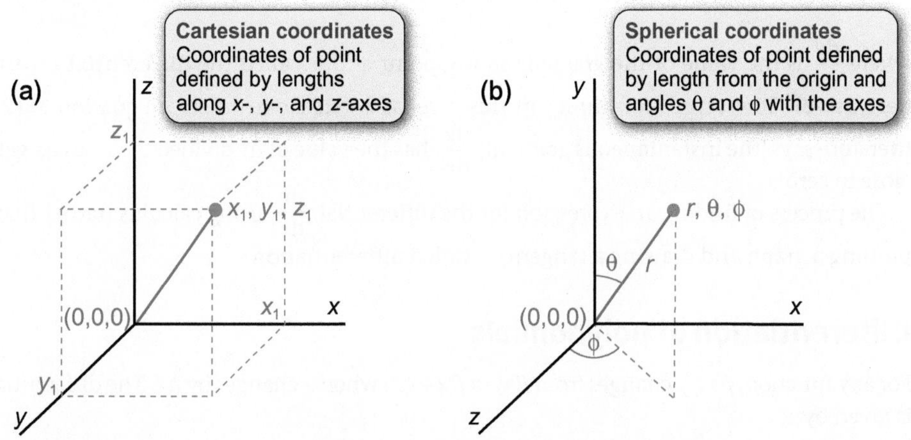

(a)
Cartesian coordinates
Coordinates of point defined by lengths along x-, y-, and z-axes
z
z1
x1, y1, z1
(0,0,0)
x1
x
y1
y
(b)
Spherical coordinates
Coordinates of point defined by length from the origin and angles θ and ϕ with the axes
y
r, θ, ϕ
θ
r
(0,0,0)
x
ϕ
z

Figure MT15 In three dimensions, the position of a point can be defined in terms of (a) its position along three axes (Cartesian coordinates) or (b) its distance from the origin and angles with two axes (spherical coordinates).

coordinates of the point are the distances along the x-, y-, and z-axes. Alternatively, you can use spherical coordinates (sometimes called spherical polar coordinates), which require the distance from the origin, r, together with the angles with two of the axes, θ and ϕ.

MT6 Differential calculus

Much of chemistry is concerned with systems that change. In kinetic studies, for example, the concentrations of reactants and products change with time; in thermodynamics, Gibbs energy changes with temperature; in quantum mechanics, electron density changes as a function of distance from the nucleus; in studies of phase equilibria, melting temperatures and boiling temperatures change with pressure. Calculus, and in particular **differential** calculus, provides a series of methods for mathematically describing changing systems.

Rate of change: the differential

The gradient of a straight line (as in Figure MT6, p.1324) is constant and does not change. The same value is obtained wherever on the line the gradient is measured. However, for a curve, the gradient continually changes and its value is different at different points along the curve. The gradient at any point is called the **instantaneous gradient** and is equal to the gradient of a tangent drawn to the curve at that point.

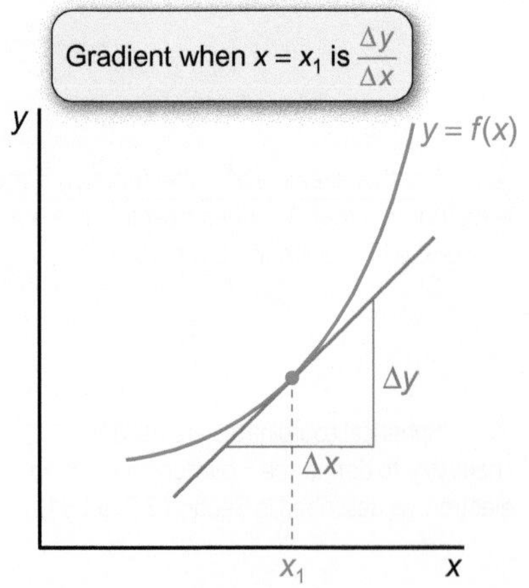
Gradient when $x = x_1$ is $\frac{\Delta y}{\Delta x}$
y
y = f(x)
Δy
Δx
x_1
x

Figure MT16 The tangent to a curve gives the instantaneous gradient.

$y = f(x)$ means that y is a function of x; see Section MT2 (p.1312).

In a kinetics experiment, y might be the concentration of product and x the time since the start of a reaction. The gradient of the curve at any point represents the rate of formation of the product; see Figure 9.3 (p.386).

Consider a general function $y = f(x)$ as shown in Figure MT16. The gradient of the curve when $x = x_1$ is given by the gradient of the tangent at that point, $\frac{\Delta x}{\Delta y}$.

In practice, it is difficult to draw a gradient to a curve accurately. However, as shown in Figure MT17, the gradient can be estimated by drawing a line between two points either side of the point of interest. For example, the gradient of line AB, $\frac{\Delta y_1}{\Delta x_2}$ gives an estimate of the gradient at x_1. However, it is not quite the same as the gradient of the tangent at x_1. If the gradient is measured using two points closer to x_1, for example, $\frac{\Delta y_2}{\Delta x_2}$, a better estimate of the gradient is obtained. The estimate becomes closer to the true value as Δx becomes smaller.

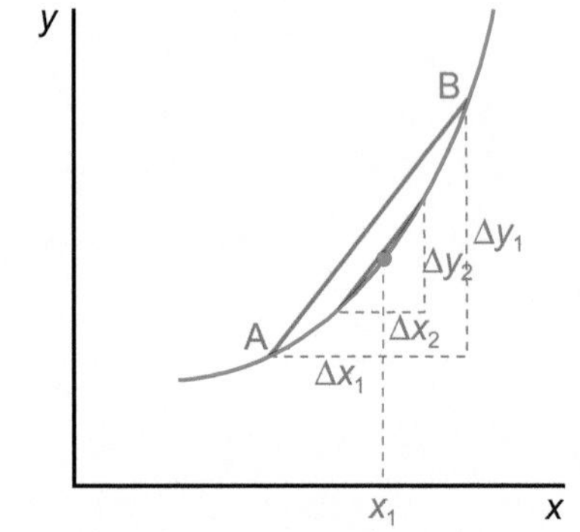
y
B
Δy_1
Δy_2
A
Δx_2
Δx_1
x_1
x

Figure MT17 The measured gradient becomes closer to the instantaneous value as the interval between A and B gets smaller.

The instantaneous gradient of the curve at x_1 is given when Δx becomes vanishingly small. This is given the symbol $\frac{dy}{dx}$.

$$\frac{dy}{dx} = \lim_{\Delta x \to 0} \frac{\Delta y}{\Delta x} \qquad \text{(MT24)}$$

where $\frac{dy}{dx}$ is the value of the gradient at any point and is called the **differential** (or the **derivative**). 'lim' stands for 'limit', in this case, as x gets closer to zero. Equation MT24 therefore says 'the instantaneous gradient, $\frac{dy}{dx}$, has the value of Δy divided by Δx, as Δx gets close to zero'.

The process of finding an expression for the differential, $\frac{dy}{dx}$, using calculus (rather than plotting a graph and drawing a tangent) is called differentiation.

The symbol $\frac{dy}{dx}$ (pronounced dee y by dee x) indicates rate of change of a function $y = f(x)$. Don't think of it as a ratio of dy divided by dx. The 'd's don't cancel out.

Differentiation of polynomials

For any function, $f(x)$, y changes from $f(x)$ to $f(x + \Delta x)$ when x changes by Δx. The differential is given by

$$\frac{dy}{dx} = \lim_{\Delta x \to 0} \frac{\Delta y}{\Delta x} = \lim_{\Delta x \to 0} \frac{f(x+\Delta x) - f(x)}{\Delta x} \qquad \text{(MT25)}$$

To illustrate how this can be used, consider the function $y = x^2$. The differential is

$$\frac{\mathrm{d}y}{\mathrm{d}x} = \lim_{\Delta x \to 0} \frac{f(x+\Delta x) - f(x)}{\Delta x} = \lim_{\Delta x \to 0} \frac{(x+\Delta x)^2 - x^2}{\Delta x}$$

A polynomial function is one in which various powers of the variable appear, for example, $y = 5x^3 - 2x^2 + 6x - 1$ (Section MT2, p.1312).

Working out the terms in brackets and simplifying

$x^2 - x^2 = 0$

Divide numerator and denominator by Δx

$$\frac{\mathrm{d}y}{\mathrm{d}x} = \lim_{\Delta x \to 0} \frac{(x^2 + 2x\Delta x + (\Delta x)^2) - x^2}{\Delta x} = \lim_{\Delta x \to 0} \frac{2x\Delta x + (\Delta x)^2}{\Delta x} = \lim_{\Delta x \to 0} (2x + \Delta x)$$

So, what happens to the value of $(2x + \Delta x)$ as Δx approaches zero? The value of Δx becomes so small that it is insignificant and

The differential of x^2 is $2x$

$$\frac{\mathrm{d}y}{\mathrm{d}x} = \frac{\mathrm{d}x^2}{\mathrm{d}x} = 2x$$

The differential can be written in several ways. For the function $y = f(x)$,

$$\frac{\mathrm{d}y}{\mathrm{d}x} = \frac{\mathrm{d}}{\mathrm{d}x}(y) = \frac{\mathrm{d}}{\mathrm{d}x} f(x).$$

The second way emphasizes that $\frac{\mathrm{d}}{\mathrm{d}x}$ is a mathematical function applied to y.

Equation MT25 can be applied to any function of x. For powers of x, the differentials are

$$\frac{\mathrm{d}x^3}{\mathrm{d}x} 3x^2; \quad \frac{\mathrm{d}x^4}{\mathrm{d}x} = 4x^3; \quad \frac{\mathrm{d}x^5}{\mathrm{d}x} = 5x^4$$

In general

The differential of x^n is nx^{n-1}

$$\frac{\mathrm{d}x^n}{\mathrm{d}x} = nx^{n-1} \tag{MT26}$$

Equation MT26 works for all values of n, including fractional and negative values, except when $n = 0$

$$\frac{\mathrm{d}x}{\mathrm{d}x} = \frac{\mathrm{d}x^1}{\mathrm{d}x} = 1x^0 = 1; \quad \frac{\mathrm{d}x^{\frac{1}{2}}}{\mathrm{d}x} = \frac{1}{2}x^{-\frac{1}{2}} = \frac{0.5}{\sqrt{x}}; \quad \frac{\mathrm{d}x^{-3}}{\mathrm{d}x} = -3x^{-2} = \frac{-3}{x^2}$$

When $n = 0$, $x^0 = 1$ and x has a constant value. The differential of a constant (i.e. a value that does not depend on x) is zero.

If the function of x is multiplied by a factor, for example, $y = 3x^2$, the factor simply multiplies the differential

$$\frac{\mathrm{d}}{\mathrm{d}x}(3x^2) = 3\frac{\mathrm{d}x^2}{\mathrm{d}x} = 3 \times (2x) = 6x; \quad \frac{\mathrm{d}}{\mathrm{d}x}(5x^4) = 5\frac{\mathrm{d}x^4}{\mathrm{d}x} = 5 \times (4x^3) = 20x^3$$

Where functions contain two or more terms, each is differentiated separately

$$\frac{\mathrm{d}}{\mathrm{d}x}(3x^2 + 2x + 3) = 6x + 2 + 0 = 6x + 2; \quad \frac{\mathrm{d}}{\mathrm{d}x}(4x^3 - 2x^2 - 5x) = 12x^2 - 4x - 5$$

The expressions here are of the form $y = f(x)$. However, you need to get used to seeing differentials in terms of other variables in chemistry, for example:

- $\frac{\mathrm{d}[A]}{\mathrm{d}t}$ variation of concentration of a substance with time (Section 9.4, p.389).
- $\frac{\mathrm{d}E_{PE}}{\mathrm{d}x}$ variation of potential energy with distance in the simple harmonic oscillator model for vibration of a diatomic molecule (Box 10.4, p.479).
- $\frac{\mathrm{d}p}{\mathrm{d}T}$ variation of pressure with the transition temperature in phase transitions (Section 17.2, p.775).

Differentiation of other functions

The differentials of several types of function are listed in Table MT1. You can work out the differentials for polynomials using Equation MT26. The differentials for other functions are more complex to derive and you need to learn the expressions for these. The exponential function, $y = \mathrm{e}^x$, is unique in that the differential has the same value as the function.

Table MT1 Differentials

$y = f(x)$	$\frac{d}{dx}f(x)$	Example
Constant	0	$\frac{d}{dx}(4) = 0$
x^n	nx^{n-1}	$\frac{d}{dx}(x^{-3}) = -3x^{-4}$
ax^n	anx^{n-1}	$\frac{d}{dx}(3x^2) = 3\times 2x = 6x$
e^x	e^x	
e^{ax}	ae^{ax}	$\frac{d}{dx}(e^{2x}) = 2e^{2x}$
$\ln x$	$\frac{1}{x}$	
$\ln ax$	$\frac{1}{x}$	
$B\ln ax$	$\frac{b}{x}$	$\frac{d}{dx}(5\ln 2x) = \frac{5}{x}$
$\sin x$	$\cos x$	
$\sin ax$	$a\cos ax$	$\frac{d}{dx}(\sin 2x) = 2\cos 2x$
$\cos x$	$-\sin x$	
$\cos ax$	$-a\sin ax$	$\frac{d}{dx}(\cos 5x) = -5\sin 5x$

The exponential function, $y = e^x$, is discussed in Section MT3 (p.1316) and Section MT4 (p.1318).

This means that the gradient of a plot of $y = e^x$ (see Figure MT8, p.1319) is always the same as the value of e^x.

Higher differentials

The differentials discussed so far have been **first differentials**—the functions have been differentiated once. The first differentials can be differentiated again to give the **second differential**. The second differential of y with respect to x is written as $\frac{d^2y}{dx^2}$.

For $y = 3x^3 - 4x^2 + 2x + 6$, the first differential is given by

$$\frac{dy}{dx} = 9x^2 - 8x + 2$$

The second differential comes from differentiating this expression

$$\frac{d^2y}{dx^2} = 18x - 8$$

Maxima and minima

The differential of a function, x, is equivalent to the gradient of a plot of $y = f(x)$ against x. The value of $\frac{dy}{dx}$ gives information on the shape of the plot, as shown in Figure MT18. The

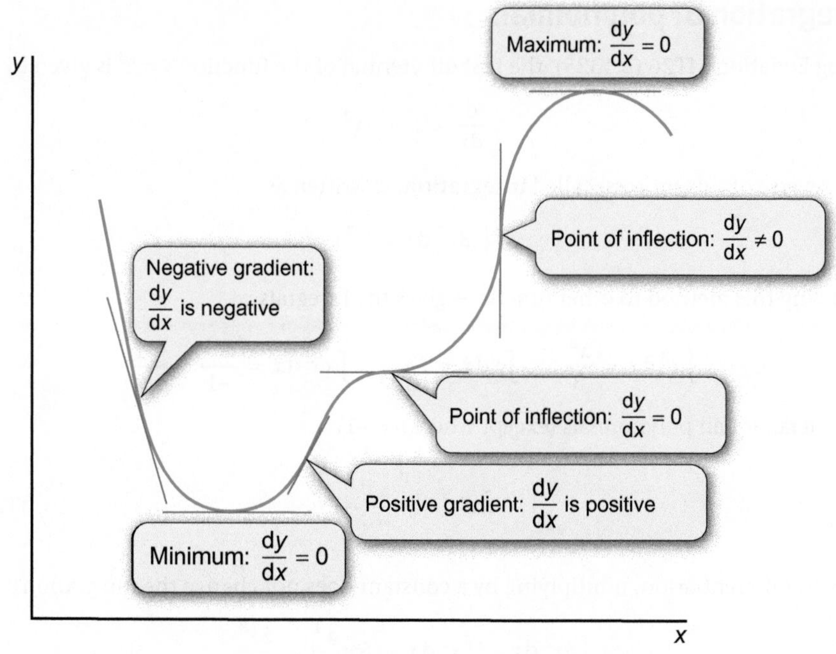

Maximum: $\frac{dy}{dx} = 0$
y
Point of inflection: $\frac{dy}{dx} \neq 0$
Negative gradient: $\frac{dy}{dx}$ is negative
Point of inflection: $\frac{dy}{dx} = 0$
Positive gradient: $\frac{dy}{dx}$ is positive
Minimum: $\frac{dy}{dx} = 0$
x

Figure MT18 The sign and value of the differential depend on the shape of the curve.

Maxima and minima represent **turning points** of a plot. Before a maximum, the gradient is positive and afterwards it is negative. Before a minimum, the gradient is negative and afterwards it is positive.

sign of $\frac{dy}{dx}$ shows which way the plot slopes. In particular, when the plot goes through a **turning point**, that is, a maximum or a minimum, the gradient, and hence the differential, must be zero. A major use of differential calculus in science is in finding the maxima and minima of functions.

Whether a point is a maximum or a minimum can be found from the sign of the second differential.

The variation of the intermolecular potential energy with the distance between two molecules, *r*, is given by the Lennard-Jones equation (see Section 17.3, p.783). Differentiation of the Lennard-Jones equation is used to find the equilibrium intermolecular distance, r_0, at which the potential energy is a minimum.

Type of turning point	$\frac{dy}{dx}$	$\frac{d^2y}{dx^2}$
Maximum	0	Negative
Minimum	0	Positive

At a **point of inflection,** $\frac{d^2y}{dx^2} = 0$ and the tangent to the curve *crosses* the curve. In Figure MT18, the gradient, $\frac{dy}{dx}$, is zero at one of the points of inflection but not at the other.

MT7 Integral calculus

Integration can be thought of as the reverse of differentiation. An **integral** is represented by the sign ∫. Integration is useful for finding out the value of a variable (e.g. distance travelled) when you know its rate of change (e.g. speed).

If the differential of *y* with respect to *x* is $\frac{dy}{dx}$, then the integral of $\frac{dy}{dx}$ with respect to *x* is *y*. This is written as

$$\int \frac{dy}{dx} dx = y \qquad \text{(MT27)}$$

The function being integrated is called the **integrand**.

Integration of polynomials

Using Equation MT26 (p.1325), the first differential of the function $y = x^3$ is given by

$$\frac{d}{dx}(x^3) = 3x^2$$

A polynomial function is one in which various powers of the variable appear, for example $y = 5x^3 - 2x^2 + 6x - 1$ (Section MT2, p.1312).

The reverse of this process, called **integration**, is written as

$$\int(3x^2\,dx = x^3)$$

Applying this method to other functions gives the integrals

$$\int x^2\,dx = \frac{x^3}{3} \qquad \int x\,dx = \frac{x^2}{2} \qquad \int x^{-2}\,dx = \frac{x^{-1}}{-1} = -\frac{1}{x}$$

In general, for all polynomials (except where $n = -1$)

Compare Equation MT28 with the general formula for differentiating a polynomial in Equation MT26 (p.1325).

$$\int x^n\,dx = \frac{x^{n+1}}{n+1} \qquad \text{(MT28)}$$

As with differentiation, multiplying by a constant does not change the integration

$$\int 5x^2\,dx = 5\int x^2\,dx = 5\times\frac{x^3}{3} = \frac{5x^3}{3}$$

Also, integrating the sum of a number of functions is the same as the sum of integrating them separately

$$\int(5x^3 + 2x^2 - x)\,dx = \frac{5x^4}{4} + 2\frac{x^3}{3} - \frac{x^2}{2}$$

Indefinite integration: the constant of integration

Think about the values of $\frac{dy}{dx}$ for the following functions

$$y = 3x^2 + 4; \qquad y = 3x^2 - 3; \qquad y = 3x^2 + 10; \qquad y = 3x^2 - 1000$$

The integrals in this section have been obtained by simply reversing the process of differentiation. They are indefinite integrals and should all include a *constant of integration* as explained below.

In each case, $\frac{dy}{dx} = 6x$ because differentiating the constants gives zero. So, what is the value of $\int 6x\,dx$? The integral of x can be evaluated, but not the constant. So, the integral is written as

$$\int 6x\,dx = 6\frac{x^2}{2} + \text{constant} = 3x^2 + c$$

where c is known as the **constant of integration**. Part of the process of applying integration to problems in chemistry is finding the value of the integration constant. Expressions containing constants of integration are known as **indefinite integrals**.

The integration of logarithmic and trigonometric functions can be found by applying the reverse of the differentials listed in Table MT1. These are summarized in Table MT2.

Definite integrals: integrating between limits

The expressions in Table MT2 are general formulae for the integrals. If you know the value of the constant, c, the integral is called a **definite integral**.

One way to eliminate the need for a constant is to integrate between two values of the variable. This is a definite integral and is called **integrating between limits**. For example, the integral $\int 6x\,dx$ between the values of $x = 2$ and $x = 4$ is written as

$$\int_{x=2}^{x=4} 6x\,dx$$

Table MT2 Integrals

$y = f(x)$	$\int f(x)dx$	Example
Constant	$ax + c$	$\int (4)dx = 4x + c$
x^n ($n \neq -1$)	$\frac{x^{n+1}}{n+1} + c$	$\int x^2 dx = \frac{x^3}{3} + c$
ax^n	$a\frac{x^{n+1}}{n+1} + c$	$\int 4x^2 dx = 4\frac{x^3}{3} + c$
e^x	$e^x + c$	
e^{ax}	$\frac{e^{ax}}{a} + c$	$\int e^{2x} dx = \frac{e^{2x}}{2} + c$
$x^{-1} = \frac{1}{x}$	$\ln x + c$	
$bx^{-1} = \frac{b}{x}$	$b \ln x + c$	$\int \frac{5}{x} dx = 5 \ln x + c$
$\sin x$	$-\cos x + c$	
$\sin ax$	$-\frac{1}{a}\cos ax + c$	$\int \sin 2x dx = -\frac{1}{2}\cos 2x + c$
$\cos x$	$\sin x + c$	
$\cos ax$	$-\frac{1}{a}\sin ax + c$	$\int \cos 5x dx = -\frac{1}{5}\sin 5x + c$

To work out the value of the integral, first integrate in the usual way

$$\int_{x=2}^{x=4} 6x\,dx = \left[3x^2 + c\right]_{x=2}^{x=4}$$

Then, work out the value of the integral for each value of x, and subtract the value for $x = 2$ from the value for $x = 4$

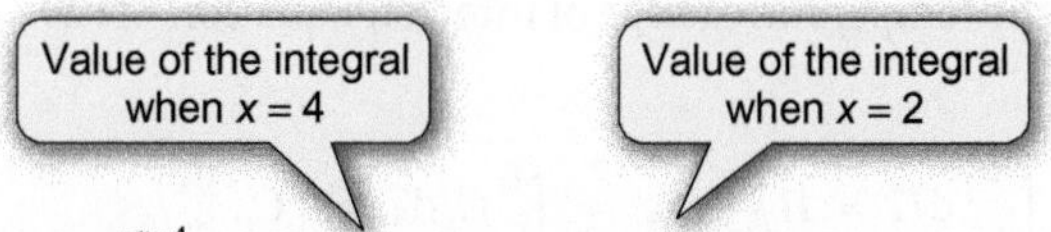
Value of the integral when x = 4
Value of the integral when x = 2

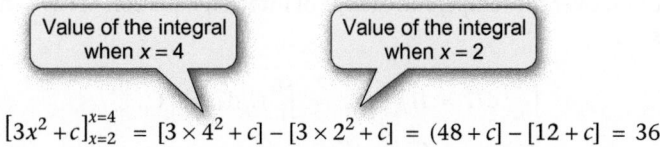

$$\left[3x^2 + c\right]_{x=2}^{x=4} = [3 \times 4^2 + c] - [3 \times 2^2 + c] = (48 + c] - [12 + c] = 36$$

In this way, the two constants cancel out, so they are not usually included in the working.

Using integration to measure areas

Integrating between limits is equivalent to finding the area under the curve between the two limits. For example, the molar heat capacity of a substance at constant pressure, C_p, is defined as

$$C_p = \frac{dH}{dT} \quad \text{where } H \text{ is enthalpy and } T \text{ is the temperature in kelvin.}$$

The enthalpy of a substance increases by an amount, dH, when the temperature increases by dT, so that

$$dH = C_p dT$$

To find the change in enthalpy on increasing the temperature from T_1 to T_2, you need to integrate each side of the equation between the limits T_1 and T_2

$$\int_{H_{T_1}}^{H_{T_2}} dH = \int_{T_1}^{T_2} C_p \, dT \qquad \text{(MT29)}$$

Integrating the left-hand side

$$\int_{H_{T_1}}^{H_{T_2}} dH = [H]_{H_{T_1}}^{H_{T_2}} = H_{T_2} - H_{T_1}$$

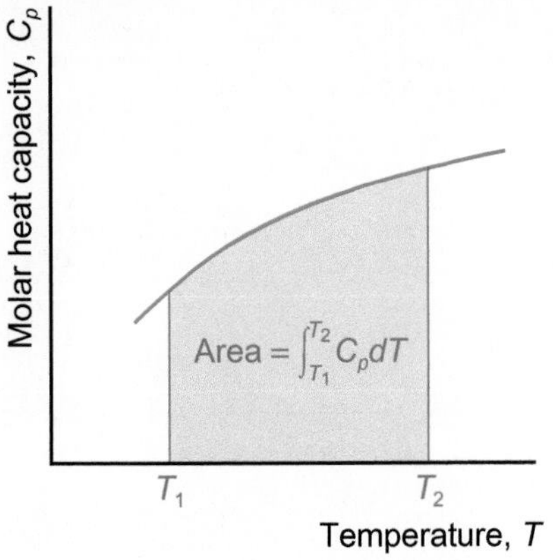
Molar heat capacity, C_p
Area = $\int_{T_1}^{T_2} C_p dT$
T_1
T_2
Temperature, T

Figure MT19 The enthalpy of a substance increases when heated from T_1 to T_2 by an amount equivalent to the shaded area. This is found by integration between the limits, $T = T_1$ and $T = T_2$.

Integrating the right-hand side is not so straightforward since C_p depends on temperature (see Figure MT19). However, the integral $\int_{T_1}^{T_2} C_p \, dT$ is equivalent to finding the shaded area under the curve in Figure MT19 and this can be worked out from the plot.

To show that the area under the curve is equivalent to the integral, divide the shaded region into a number of vertical strips, each with a width ΔT as shown in Figure MT20. If ΔT is small, each strip can be approximated by a rectangle of height $C_p(T)$ (the molar heat capacity at temperature T) and width ΔT. The total area is then the sum of the areas of all the strips.

The area of the first strip is $C_p(T_1) \times \Delta T$. The area of the second strip is $C_p(T_1 + \Delta T) \times \Delta T$. The total area is the sum of the areas of all the strips.

$$\text{Total area} = \sum_{1}^{n} C_p(T + n\Delta T) \times \Delta T$$

where n is the number of strips between T_1 and T_2.

The approximation comes closer to the real value as the number of strips is increased. As ΔT becomes infinitesimally small, the sum can be written as an integral between the limits $T = T_1$ and $T = T_2$.

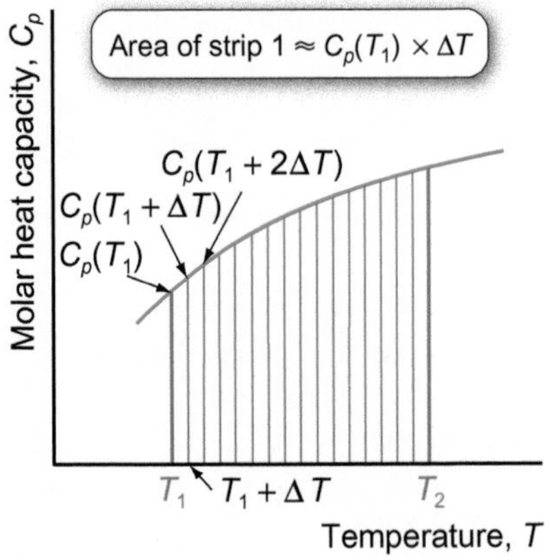
Area of strip 1 ≈ $C_p(T_1) \times \Delta T$
Molar heat capacity, C_p
$C_p(T_1 + 2\Delta T)$
$C_p(T_1 + \Delta T)$
$C_p(T_1)$
T_1
$T_1 + \Delta T$
T_2
Temperature, T

Figure MT20 Integrating is equivalent to adding the areas of all the strips.

The symbol Σ indicates that a summation is being performed (see MT1, p.1304).

This says 'integrate the function $C_p(T)$ as a function of temperature between the limits T_1 and T_2'. It is mathematically equivalent to saying 'find the area under a curve of C_p against T between T_1 and T_2'

$$\text{Total area} = \int_{T_1}^{T_2} C_p(T) \, dT$$

If C_p is constant over the temperature range of interest, Equation MT29 can be simplified and becomes

$$\int_{H_{T_1}}^{H_{T_2}} dH = H_{T_2} - H_{T_1} = \int_{T_1}^{T_2} C_p dT = \left[C_p\right]_{T_1}^{T_2}$$
$$= C_p(T_2 - T_1)$$

$H_{T_2} - H_{T_1}$ is the change in enthalpy, ΔH, and $(T_2 - T_1)$ is the change in temperature, ΔT, so provided C_p can be assumed to be constant

$$\Delta H = C_p \Delta T \qquad (13.9)$$

This corresponds to Equation 13.9 in Section 13.4 (p.634).

You can see another example of integrating between limits in Box 14.1 (p.663) to find the difference in entropy of a substance at two temperatures. Integration between limits is also used when deriving integrated rate equations to describe reaction kinetics (see Section 9.4, p.389).

Appendix 1

Abbreviations and symbols for quantities and units

Unavoidably, some symbols are used for more than one thing. The meaning is usually clear from the context, but always check that you understand what a symbol is representing. Note that symbols representing *physical quantities* are printed in *italics*. Symbols for units are *not* in italics. The reference (in brackets) given for each entry is to chapters where the abbreviation or symbol is used.

A	ampere (unit of electrical current) (1, 16)
A	absorbance (10, 11, 28)
A	area (8, 13, Maths Toolkit MT5)
A	frequency factor (in Arrhenius equation) (9)
A	Madelung constant (6)
A_r	relative atomic mass (1)
Å	angstrom (non-SI unit used for bond lengths) (1)
a	acceleration (1, 8)
a	thermodynamic activity (14, 15, 16, 17)
a	van der Waals constant for gases (8)
a_0	Bohr radius (3)
AAS	atomic absorption spectrometry (11)
AES	atomic emission spectrometry (11)
AO	atomic orbital (4)
Ac	acetyl (2)
Ar	general aryl group (2)
$[acac]^-$	pentane-2,4-dionato (acetylacetonato) (28)
atm	atmosphere (non-SI unit of pressure) (1, 8)
B	magnetic field strength (10)
B	rotational constant (10)
b	van der Waals constant for gases (8)
Bn	benzyl (2)
Boc	*tert*-butyloxycarbonyl group (24)
Bz	benzoyl (2)
bar	bar (unit of pressure) (1, 8)
bipy	2,2′-bipyridine (28)
bp	bonding pair (5)
b.p.	normal boiling point (usually quoted in °C) (1)
br	broad (IR absorption) (12)
C	coulomb (unit of electric charge) (1, 16)
C	Curie constant (in Curie's law for magnetic susceptibility) (28)
C_p	molar heat capacity at constant pressure (13, 14)
C_V	molar heat capacity at constant volume (13, 14)
°C	degrees Celsius (1)
C_s	specific heat capacity (13)
c	concentration of a solution (1)
c	speed of light (2, 10)
c	root mean square speed (of molecules) (8)
$\bar{c}$	mean speed (of molecules) (8)
CFSE	crystal field stabilization energy (28)
CI	chemical ionization (in mass spectrometry) (12)
CIP	Cahn–Ingold–Prelog (sequence rules for *R/S* nomenclature of chiral centres) (18)
Cbz	carboxybenzyl group (24)
cm	centimetre (unit of length) (1)
cm^3	cubic centimetre (unit of volume) (1)
cm^{-1}	reciprocal centimetre (unit of wavenumber) (10, 12)
D	debye (non-SI unit of electric dipole moment) (5, 17)
D	bond dissociation enthalpy (4, 13)
$\overline{D}$	mean bond enthalpy (13)
D	centrifugal distortion constant (in rotational spectroscopy) (10)
d	distance (1)
d	doublet (in NMR spectrum) (12)
DBU	1,8-diazabicyclo[5.4.0]undec-7-ene (19)
DMAP	4-(dimethylamino)pyridine (19)
DMF	*N,N*-dimethylformamide (2, 20)
DMSO	dimethyl sulfoxide (20)
Da	dalton (atomic mass unit) (1)
dien	1,4,7-triazaheptane (diethylenetriamine) (28)
dm^3	cubic decimetre (unit of volume) (1)
E	electrophile (19)
E	energy (3, 10)
E_{KE}	kinetic energy (1, 10)
E_{PE}	potential energy (1, 10, 17)

E_a	activation energy (9)
E_{cell}	zero-current cell potential (electromotive force, emf) (16)
$E^{\ominus}$	standard reduction potential (standard electrode potential) (16)
E-	*entgegen* (configuration of a substituted alkene using sequence rules) (18)
e	charge on the electron (6)
e^-	electron (1)
E1	elimination unimolecular (20, 21)
E2	elimination bimolecular (20, 21)
Ei	elimination intramolecular (24)
EI	electron impact (in mass spectrometry) (12)
EM	electromagnetic (radiation) (10)
ESI	electrospray ionization (in mass spectrometry) (12)
ESR	electron spin resonance (3, 10)
Et	ethyl (2)
$[edta]^{4-}$	diamino-1,2-ethane-*N,N,N′,N′*-tetraacetato (ethylenediamine tetraacetato) (28)
ee	enantiomeric excess (18)
emf	electromotive force (zero-current cell potential, E_{cell}) (16)
en	ethane-1,2-diamine (ethylenediamine) (28)
eV	electron volt (unit of energy) (4)
F	Faraday constant (16)
F	force (1, 13, 17)
FGI	functional group interconversion (22)
FT	Fourier transform (10, 12)
G	Gibbs energy (14, 15, 16, 17)
g	gram (unit of mass) (1)
g	acceleration due to gravity (1)
g	degeneracy of an energy state (10)
g	*gerade* (parity label for molecular orbital) (4)
GC–MS	gas chromatograph linked to a mass spectrometer (11)
GLC	gas–liquid chromatography (11)
GSC	gas–solid chromatography (11)
H	enthalpy (1,13)
h	height (1, 8)
h	Planck constant (3, 10)
hν	radiation energy (often used in an equation to indicate a photochemical reaction) (9, 27)
HOMO	highest occupied molecular orbital (4)
HPLC	high-performance liquid chromatography (11)
Hz	hertz (unit of frequency, s^{-1}) (3)
+I and –I	inductive effect (+ and – effects) (19)
I	current (16)
I	moment of inertia (10)
I	nuclear spin quantum number (10)
I	intensity of radiation (10)
I	ionization energy (3)
ICP	inductively coupled plasma (11)
IR	infrared (10, 12)
ISE	ion selective electrode (11)
J	joule (unit of energy) (1)
J	spin–spin coupling constant (10, 12)
J	rotational quantum number (10)
K	kelvin (unit of temperature) (1)
K ($K^{\ominus}$)	thermodynamic equilibrium constant (15)
K_a	acidity constant (acid dissociation constant) (7)
K_b	basicity constant (base dissociation constant) (7)
K_c	equilibrium constant in terms of concentrations (1, 15)
K_M	Michaelis constant (9)
K_p	equilibrium constant in terms of partial pressures of gases (1, 15)
K_{sp}	solubility constant (solubility product) (15)
K_w	self-ionization constant (ionic product) of water (7)
k	bond force constant (10)
k	Kapustinskii constant (6)
k	rate constant (9)
k_B	Boltzmann constant (9, 10)
L	litre (unit of volume; $1\ L = 1\ dm^3$) (1)
l	length, path length (1, 10)
l	secondary quantum number (angular, azimuthal, or orbital quantum number) (3)
LCAO	linear combination of atomic orbitals (4)
LCD	liquid crystal display (17)
LUMO	lowest unoccupied molecular orbital (4)
lp	lone pair (5)
M	molar mass (1)
+M and –M	mesomeric effect (+ and – effects) (19)
M_r	relative formula (molecular) mass (1)
M and $M^{\bullet+}$	molecular ion (in mass spectrometry) (12)
m	medium (intensity of IR absorption) (12)
m	metre (unit of length) (1)
m^3	cubic metre (unit of volume) (8, 16)
m	multiplet (in NMR spectrum) (12)
m	mass (1)
m-	*meta*- (3-position in a substituted benzene) (2, 22)
m_e	mass of the electron (3)
m_I	nuclear spin magnetic quantum number (10)
m_l	magnetic quantum number (3)
m_s	spin magnetic quantum number (3, 10)
m/z	mass:charge ratio (1, 12)
mCPBA	*meta*-chloroperbenzoic acid (21)
Me	methyl (2)
MALDI	matrix-assisted laser desorption/ionization (in mass spectrometry) (12)
MO	molecular orbital (4)
MRI	magnetic resonance imaging (10)
MS	mass spectrometry (1, 12)
mol	mole (unit of amount of substance) (1)
m.p.	normal melting point (usually quoted in °C) (1)

N	newton (unit of force) (1, 8, 13, 17)
N	number of molecules or nuclei (8, 10)
N_A	Avogadro constant (1)
n	amount in moles (1)
n	Born exponent (Born–Landé equation) (6)
n	number of (4)
n	principal quantum number (3)
n	translational quantum number (10)
NMR	nuclear magnetic resonance (10, 12)
Nu	nucleophile (19)
nm	nanometre (unit of length) (1, 3, 10)
o-	*ortho*- (2-position in a substituted benzene) (2, 22)
$[ox]^{2-}$	ethanedioato (oxalato) (28)
P	pairing energy (28)
p	momentum (3)
p	pressure (1, 8)
p_A	partial pressure of component A (8)
p-	*para*- (4-position in a substituted benzene) (2, 22)
PCB	polychlorinated biphenyl (11)
PCC	pyridinium chlorochromate (23)
Pa	pascal (unit of pressure) (1, 8)
Ph	phenyl (2)
pK_a	measure of the tendency of a molecule or ion to lose H^+ (7, 19)
Pr, ^{i}Pr	propyl, isopropyl (2)
phen	1,10-phenanthroline (28)
pm	picometre (unit of length) (1)
pn	propane-1,3-diamine (1,3-propylenediamine) (28)
ppm	parts per million (1)
ppb	parts per billion (1 billion = 1×10^9) (1)
Q	reaction quotient (15)
Q	total electric charge (16)
q	electric charge (17)
q	heat (13)
quin	quintet (in NMR spectrum) (12)
R	substituent group (often an alkyl group) (2)
R	gas constant (8)
R	resistance (16)
R_H	Rydberg constant (3)
R-	*rectus* (nomenclature for assigning the configuration of a chiral centre) (18)
r	internuclear distance (6, 17)
r	radius (6)
r_0	equilibrium bond length (10, 17)
r_+ and r_-	radius of cation and anion (6)
RASM	readily available starting material (22)
RDS	rate-determining step (9)
RSA	retrosynthetic analysis (22)
R(*r*)	radial wavefunction (3)
S	siemen (reciprocal ohm) (unit of electrical conductance) (16)
S	entropy (14)
S	shielding constant (Slater's rules) (3)
S-	*sinister* (nomenclature for assigning the configuration of a chiral centre) (18)
s	second (unit of time) (1)
s^{-1}	unit of frequency (Hz) (3, 10, 12)
s	singlet (in NMR spectrum) (12)
s	strong (intensity of IR absorption) (12)
s	electron spin quantum number (2, 10)
s	speed of individual molecules (8)
s	molar solubility (15)
$S^{\ominus}$	standard entropy (14)
SATP	standard ambient temperature and pressure (8)
SHE	standard hydrogen electrode (16)
S_N1	substitution nucleophilic unimolecular (20)
S_N2	substitution nucleophilic bimolecular (20)
S_Ni	substitution nucleophilic internal (20)
sep	septet (in NMR spectrum) (12)
sext	sextet (in NMR spectrum) (12)
T	tesla (unit of magnetic flux density) (10, 12)
T	transmittance (10)
T	temperature (in kelvin) (1, 8)
T_C	critical temperature (17)
T_b	normal boiling point (in kelvin) (1, 13, 17)
T_m	normal melting point (in kelvin) (1, 13, 17)
t	time (1, 9)
t_r	retention time (11)
$t_{1/2}$	half life (9)
t	triplet (in NMR spectroscopy) (12)
TLC	thin layer chromatography (11)
THF	tetrahydrofuran (2)
TM	target molecule (22)
TMS	tetramethylsilane (10, 12)
$TS^{\ddagger}$	transition state (9, 19)
terpy	2,2′:6′2″-terpyridine (28)
torr	torr (non-SI unit of pressure) (8)
U	internal energy (6, 13)
u	atomic mass unit (1)
u	*ungerade* (parity label for molecular orbital) (4)
UV	ultraviolet (3, 10)
UV/VIS	ultraviolet–visible (spectroscopy) (10)
V	volt (unit of potential difference) (16)
V	potential difference (voltage) (16)
V	volume (1, 8)
V_m	molar volume (8, 17)
v	vibrational quantum number (10)
v	velocity (8)
VB	valence bond (4)
VSEPR	valence-shell electron-pair repulsion (5)
W	watt (unit of power) (13)
W	number of ways of arranging molecules and their energies in a system (14)
w	weak (intensity of IR absorption) (12)
w	work done (13)

x_A	mole fraction of component A (8, 17)
x_e	anharmonicity correction (in vibrational spectroscopy) (10)
$Y(\theta, \phi)$	angular wavefunction (3)
Z	atomic number (1)
Z	collision frequency (8)
Z_{eff}	effective nuclear charge (3, 28)
Z-	*zusammen* (configuration of a substituted alkene using sequence rules) (18)
z_+ and z_-	charge on cation and anion (6)
α-carbon	C atom adjacent to the one in question (often the C atom adjacent to a C=O group) (2, 23)
$[\alpha]_D$	specific rotation measured using the sodium D-line (for a chiral compound in solution) (18)
α	degree of dissociation (15)
α	polarizability of a molecule (17)
β	stability constant of a complex (28)
β-carbon	C atom two along from the one in question (20, 24)
γ	magnetogyric ratio (10)
γ_A	activity coefficient of component A (17)
Δ	change in (1, 13, 14)
Δ	heat (when written above an equation arrow) (26)
Δ_o	octahedral crystal field splitting energy (28)
Δ_t	tetrahedral crystal field splitting energy (28)
$\Delta_a H$	enthalpy change of atomization (6, 26)
$\Delta_c H$	enthalpy change of combustion (13)
$\Delta_{cond} H$	enthalpy change of condensation (1, 17)
$\Delta_{diss} H$	bond dissociation enthalpy (4, 13)
$\Delta_{eg} H$	electron gain enthalpy (3, 6, 26)
$\Delta_f H$	enthalpy change of formation (6, 13)
$\Delta_{freezing} H$	enthalpy change of freezing (1)
$\Delta_{fus} H$	enthalpy change of fusion (melting) (1, 13, 17)
$\Delta_i H$	enthalpy change of ionization (ionization enthalpy) (6)
$\Delta_{latt} H$	lattice enthalpy (enthalpy change for the breakdown of an ionic lattice) (6)
$\Delta_r H$	enthalpy change of reaction (1, 13)
$\Delta_{reverse\ sub} H$	enthalpy change of reverse sublimation (1)
$\Delta_{sol} H$	enthalpy change of solution (13)
$\Delta_{solv} H$	enthalpy change of solvation (17)
$\Delta_{sub} H$	enthalpy change of sublimation (1, 13)
$\Delta_{vap} H$	enthalpy change of vaporization (1, 13, 17)
$\Delta_r G$	Gibbs energy change of reaction (for other subscripts see ΔH above) (14)
$\Delta^{\ddagger} G$	Gibbs energy of activation (9, 19)
$\Delta_{hyd} G$	Gibbs energy change of hydration of ions (26)
$\Delta_r S$	Entropy change of reaction (for other subscripts see ΔH above) (14)
$\Delta_{latt} U$	lattice energy (6)
δ	chemical shift (10, 12)

δ+ and δ–	partial positive and partial negative charge (5, 17, 19)
ε	molar absorption coefficient (10, 11)
ε_{max}	molar absorption coefficient corresponding to an absorption maximum (in an electronic spectrum) (10, 28)
ε	permittivity of a medium (17)
ε_0	permittivity of a vacuum (3, 6, 17)
θ	angle (used in polar and spherical coordinates) (3, Maths Toolkit MT5)
θ	temperature (in degrees Celsius) (1)
κ	conductivity (16)
Λ_m	molar conductivity (16)
Λ_m^o	limiting molar conductivity (at infinite dilution) (16)
λ	ionic conductivity (16)
λ	mean free path (8)
λ	wavelength (3)
λ_{max}	wavelength corresponding to an absorption maximum (in an electronic spectrum) (10, 28)
μ	electric dipole moment (17)
μ	magnetic moment of a complex (28)
μ	reduced mass (10)
μ_B	Bohr magneton (unit of magnetic moment) (28)
μ_{eff}	measured magnetic moment of a complex (28)
μ_{so}	spin-only magnetic moment (28)
ν	frequency (3)
ν	total number of ions in a formula unit (6)
$\tilde{\nu}$	wavenumber (11)
ν_i	stoichiometric coefficient in a balanced chemical equation (13, 14, 15)
ν_+ and ν_-	number of cations and anions, respectively, per formula unit (16)
π	osmotic pressure (17)
π	symmetry label for molecular orbital (4)
ρ	density (1, 8)
σ	collision cross-section (8)
σ	symmetry label for molecular orbital (4)
τ	volume (3)
Φ	electron work function (photoelectric effect) (3)
ϕ	angle (used in spherical coordinates) (3, Maths Toolkit MT5)
χ	magnetic susceptibility (28)
χ	electronegativity (4)
χ^P	Pauling electronegativity (4)
Ψ	wavefunction for an *N*-electron atom (4)
ψ	single electron wavefunction (3, 4)
Ω	ohm (unit of electrical resistance) (16)
$\ominus$	denotes standard state or standard conditions (in thermodynamics) (1, 13)
$\ominus'$	denotes biochemical standard state at pH7 (14)
1°, 2°, 3°, 4°	primary, secondary, tertiary, quaternary C atoms (2)
[A]	concentration of A in mol dm^{-3} (1)
$\Rightarrow$	indicates you are working backwards in a synthesis

Appendix 2

Representing reactions

You have probably represented most chemical reactions by balanced equations (see Section 1.4, p.21), which tell you the reactants, the products, and the ratios in which they react. A balanced equation is essential if you are calculating quantities.

In this book, you will find other ways of representing reactions, particularly when the reactions are part of a reaction scheme involving several stages. In general, reaction schemes show the reactants, products, and sometimes the reaction conditions, but chemists choose the representation that suits their purpose and focuses on what is important in the situation. For example, for reactions that give more than one product it is sometimes useful to draw a reaction scheme that clearly shows the formation of only one of the products (the one you are interested in). Also, on some occasions, it is important to give a detailed reaction mechanism, which shows every step of the reaction (curly arrows are usually drawn to show the movement of electrons; see Section 19.1, p.863).

Some general ways of representing reactions that will help you become familiar with the different representations used in this book are shown below.

(a) Conversion of reactants (A + B) into products (C + D) is commonly represented in one of seven different ways. A specific example of each representation is found on the pages highlighted.

(i) $A + B \rightarrow C + D$ — See p.24 and p.101.

(ii) $A + B \xrightarrow{h\nu \text{ or } \Delta} C + D$ — Reaction conditions are often shown above or below the arrow (e.g. $h\nu$ = light; Δ = heat). See p.921, 1186 and 1201.

(iii) $A + B \xrightarrow{-D} C$ — Focuses on the formation of C. See p.967 and p.1101.

(iv) 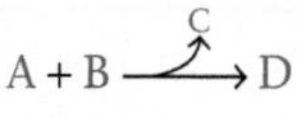 — Focuses on the formation of D. See p.960 and p.965.

(v) $A \xrightarrow{B} C + D$ — Focuses on the conversion of A into C and D (B is often a reagent). See p.86 and p.1020.

(vi) 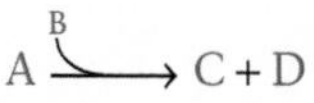 — Focuses on the conversion of A into C and D (B is often a reagent). It tends to be used in a reaction cycle. See part (c) and p.960.

(vii) 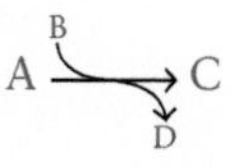 — Focuses on the conversion of A into C. See p.939.

(b) Often, reaction intermediates or transition states are included in the reaction scheme. Reaction intermediates are sometimes drawn inside brackets, especially when different resonance forms of a reaction intermediate are shown (p.874). Transition states are usually drawn inside brackets with the double dagger sign (‡) shown at the top right.

(i) $A + B \rightarrow E \rightarrow C + D$ — Shows the formation of a reaction intermediate (E). See p.903 and p.1080.

(ii) $A + B \rightarrow [E^1 \leftrightarrow E^2] \rightarrow C + D$ — Shows different resonance forms of a reaction intermediate (E). See p.923 and p.1083.

(iii) $A + B \rightarrow [F]^{\ddagger} \rightarrow C + D$ — Shows the transition state (F). See p.930.

(c) Reactions involving catalysts (cat) are commonly drawn to show catalytic cycles, which clearly illustrate how a catalyst is consumed and then regenerated in the course of a catalytic reaction sequence.

$A + B \xrightarrow{\text{cat}} C + D$ — See p.1019.

Appendix 3

Preparation and reactions of alcohols and amines

Chapter 2 describes the language of organic chemistry, including how to identify functional groups such as alcohols and amines (p.92 and p.95). In subsequent organic chemistry chapters, we have chosen to arrange the reactions of organic molecules by type of reaction mechanism, rather than by functional group. The chemistry of functional groups, such as halogenoalkanes, alkenes, and aldehydes and ketones, is closely linked to a particular type of reaction mechanism and is easy to locate. However, the chemistry of alcohols (RCH_2OH, R_2CHOH, and R_3COH) and amines (RNH_2, R_2NH, and R_3N) is spread over several chapters. The tables below give you a guide to where you can find information on the preparation and reactions of alcohols and amines.

Alcohols

Preparation of alcohols

Type of reaction	Reference
Hydrolysis of halogenoalkanes	Section 20.3 (p.936)
Hydration of an alkene	Section 21.3 (p.980)
Reduction of aldehydes and ketones	Section 23.2 (p.1062)
Reduction of acyl chlorides	Section 24.2 (p.1111)
Reduction of acid anhydrides	Section 24.2 (p.1116–1117)
Reduction of esters	Section 24.2 (p.1119, 1121–1122)

Reactions of alcohols

Type of reaction	Reference
Reaction of alcohols with bases	Section 19.2 (p.895)
Halogenation of alcohols	
Reaction of alcohols with hydrogen halides	Section 20.2 (p.925)
Reaction of alcohols with phosphorus trihalides	Section 20.2 (p.926)
Reaction of alcohols with 4-toluenesulfonyl chloride	Section 20.2 (p.927)
Reaction of alcohols with thionyl chloride	Section 20.2 (pp.926–927)
Deprotonation of alcohols followed by reaction with halogenoalkanes to form ethers	Worked example 19.3 (p.883), Section 20.3 (p.931)
Elimination reaction of alcohols to form alkenes (dehydration)	Section 21.2 (p.967)
Oxidation of primary and secondary alcohols to form aldehydes, ketones, and carboxylic acids	Section 19.2 (p.900), Box 23.2 (p.1064)
Reaction of alcohols with aldehydes and ketones to form hemiacetals and acetals	Section 23.2 (p.1073)
Acylation of alcohols to form esters (esterification)	
Reaction of alcohols with carboxylic acids	Section 24.2 (p.1108)
Reaction of alcohols with acyl chlorides	Section 24.2 (p.1111)
Reaction of alcohols with acid anhydrides	Section 24.2 (p.1116)
Reaction of alcohols with esters (transesterification)	Worked example 24.3 (p.1122)

Amines

Preparation of amines

Type of reaction	Reference
Reaction of ammonia with halogenoalkanes	Section 19.1 (p.880), Worked example 20.3 (p.939)
Reduction of aromatic nitro compounds	Section 22.4 (p.1040)
Reduction of imines	Box 23.6 (p.1079)
Hydrolysis of amides	Section 22.4 (p.1044), Section 24.2 (p.1124)
Reduction of amides	Section 24.2 (p.1124)

Reactions of amines

Type of reaction	Reference
Reaction of amines with acids	
Formation of salts	Section 19.2 (p.891)
Formation of amides	Section 24.2 (p.1109–1110)
Alkylation of amines by reaction with halogenoalkanes	Box 19.4 (p.883–884), Worked example 20.3 (p.939)
Reaction of amines with epoxides to form amino-alcohols	Box 20.4 (p.940–941)
Diazotization of aromatic amines	Section 22.4 (p.1041)
Reaction of primary amines with aldehydes and ketones to form imines	Section 23.2 (p.1077)
Reaction of secondary amines with aldehydes and ketones to form enamines	Section 23.2 (p.1078)
Acylation of primary and secondary amines by reaction with acyl chlorides to form amides	Section 22.4 (p.1043–1044; aromatic amine), Section 24.2 (p.1112), Box 24.5 (p.1115)
Acylation of a primary amine to form a carbamate	Worked example 24.2 (p.1117)

Appendix 4

Organic oxidation and reduction reactions

Organic functional groups can be interconverted by oxidation and reduction reactions. An introduction to redox reactions of organic compounds is given in Section 19.2 (p.899). In subsequent organic chemistry chapters, a variety of redox reactions are discussed. The tables below give you a guide to where you can find information on the oxidation and reduction of organic functional groups.

Oxidation

Starting material	Product	Reagent	Reference
Trialkylborane	Trialkylborate	H_2O_2, HO^-	Section 21.3 (p.982–983)
Alkene	Epoxide	RCO_3H	Section 21.3 (p.989)
Alkene	1,2-Diol	$KMnO_4$, aq. HO^- (below r.t.), or aq. OsO_4	Section 21.4 (p.989–990)
Alkene	Aldehyde and/or ketone	$KMnO_4$, aq. HO^- (r.t. or above), OsO_4 then HIO_4, O_3 then Me_2S, PPh_3, or Zn	Section 21.4 (p.991–992, 992–994)
Alkene	Carboxylic acid	O_3 then H_2O_2	Section 21.4 (p.992–994)
Toluene ($PhCH_3$)	Benzoic acid ($PhCO_2H$)	$KMnO_4$	Section 22.4 (p.1041)
Primary alcohol	Aldehyde	$KMnO_4$, $Na_2Cr_2O_7$, CrO_3, or PCC (pyridinium chlorochromate)	Box 23.2 (p.1064)
Primary alcohol	Carboxylic acid	$KMnO_4$, $Na_2Cr_2O_7$, or CrO_3	Box 23.2 (p.1064)
Secondary alcohol	Ketone	$KMnO_4$, $Na_2Cr_2O_7$, or CrO_3	Box 23.2 (p.1064)

Reduction

Starting material	Product	Reagent	Reference
Alkene	Alkane	H_2, Pd/C	Section 21.2 (p.967)
Alkyne	Alkane	H_2, Pd/C	Section 21.2 (p.967)
Alkyne	*Z*-Alkene	H_2, Pd/$CaCO_3$/PbO (Lindlar catalyst)	Section 21.2 (p.967)
Alkyne	*E*-Alkyne	Na, NH_3	Section 21.2 (p.967)
Aromatic nitro compound	Aromatic amine	Sn, HCl	Section 22.4 (p.1040)
Aromatic ketone	Alkylbenzene	Zn/Hg, HCl	Section 22.4 (p.1040)
Aldehyde	Primary alcohol	$NaBH_4$, $LiAlH_4$, $Na(CN)BH_3$, or HCHO/HO^-	Section 23.2 (p.1062–1065)
Ketone	Secondary alcohol	$NaBH_4$, $LiAlH_4$, or $Na(CN)BH_3$	Section 23.2 (p.1062–1064)
Thioacetal	Alkane	H_2, Raney Ni	Section 23.2 (p.1077)
Imine	Primary or secondary amine	$Na(CN)BH_3$	Box 23.6 (p.1079)
Acyl chloride	Primary alcohol	$NaBH_4$	Section 24.2 (p.1111 and 1114)
Acyl chloride	Aldehyde	$LiAlH(O^tBu)_3$	Section 24.2 (p.1114)
Acid anhydride	Primary alcohol	$LiAlH_4$	Section 24.2 (p.1116–1117)
Carboxylic acid	Primary alcohol	$LiAlH_4$	Section 24.2 (p.1117)
Ester	Primary alcohol	$LiAlH_4$	Section 24.2 (p.1119 and 1121)
Secondary amide	Secondary amine	$LiAlH_4$	Section 24.2 (p.1124)

Appendix 5

Properties of the elements

Name	Symbol	Atomic number	Relative atomic mass*	State	Density/g cm^{-3}	T_m/K	T_b/K
Actinium	Ac	89	[227]	solid	10.07	1323	3471
Aluminium	Al	13	26.982	solid	2.70	933	2792
Americium	Am	95	[243]	solid	12	1449	2284
Antimony	Sb	51	121.76	solid	6.68	904	1860
Argon	Ar	18	39.948	gas	1.78×10^{-3}	84	87
Arsenic (grey)	As	33	74.922	solid	5.75	889 (sub)	
Astatine	At	85	[210]	solid		575	
Barium	Ba	56	137.33	solid	3.62	1000	2170
Berkelium	Bk	97	[247]	solid	14.78 (α)	1269	
Beryllium	Be	4	9.0122	solid	1.85	1560	2744
Bismuth	Bi	83	208.98	solid	9.79	545	1837
Bohrium	Bh	107	[267]	solid			
Boron	B	5	10.81	solid	2.34	2348	4270
Bromine	Br	35	79.904	liquid	3.12	266	332
Cadmium	Cd	48	112.41	solid	8.69	594	1040
Caesium	Cs	55	132.91	solid	1.87	302	944
Calcium	Ca	20	40.078	solid	1.54	1115	1757
Californium	Cf	98	[251]	solid	15.1	1170	
Carbon (graphite)	C	6	12.011	solid	2.2	4098 (sub)	
Cerium	Ce	58	140.12	solid	6.77	1072	3716
Chlorine	Cl	17	35.45	gas	3.21×10^{-3}	172	239
Chromium	Cr	24	51.996	solid	7.15	2180	2944
Cobalt	Co	27	58.933	solid	8.9	1768	3200
Copernicium	Cn	112	[285]	solid			
Copper	Cu	29	63.546	solid	8.96	1358	2835
Curium	Cm	96	[247]	solid	13.51	1618	3370
Darmstadtium	Ds	110	[281]	solid			
Dubnium	Db	105	[268]	solid			

T_m and T_b are the melting and boiling points at 1 bar pressure. $\Delta_a H^\ominus$ is the enthalpy change of atomization. χ^P is the Pauling electronegativity. I_1, I_2, and I_3 are, respectively, the first, second, and third ionization energies. E_{eg} is the electron gain energy. For oxygen and sulfur, the values of E_{eg} given in brackets are the second election gain energies. Densities of gases are given at 273 K and 1 atm.

$\Delta_a H^\ominus$/kJ mol^{-1}	χ^P	I_1/kJ mol^{-1}	I_2/kJ mol^{-1}	I_3/ kJ mol^{-1}	E_{eg}/ kJ mol^{-1}	Oxidation states (main ones in bold)
406	1.1	499	1134		−34	+3
331	1.61	578	1817	2745	−42	+3
284		576				**+3** +4 +5 +6
264	2.05	831	1605	2440	−101	−3 **+3** +5
0		1521	2666	3931	> 0	0
303	2.18	944	1794	2735	−78	−3 **+3** +5
	2.2	920			−270	−1
179	0.89	503	965	3600	−14	+2
310		598				+3 +4
324	1.57	900	1757	14 849	> 0	+2
210	1.9	703	1612	2466	−91	**+3** +5
565	2.04	801	2427	3660	−27	+3
111.9	2.96	1140	2083	3500	−325	**−1** +1 +3 +4 +5 +6 +7
112	1.69	868	1631	3616	> 0	+2
77	0.79	376	2234	3400	−46	+1
178	1.00	590	1145	4912	−2	+2
196		606	1140			+3
717	2.55	1086	2353	4620	−122	−4 +2 **+4**
420	1.12	534	1047	1949	−63	**+3** **+4**
121.3	3.16	1251	2298	3822	−349	**−1** +1 +3 +4 +5 +6 +7
397	1.66	653	1591	2987	−64	+2 **+3** +4 +5 **+6**
427	1.88	760	1648	3232	−64	**+2** **+3** +4 +5
337	1.90	745	1958	3555	−119	+1 **+2**
386		578				**+3** +4

Name	Symbol	Atomic number	Relative atomic mass*	State	Density / $g\,cm^{-3}$	T_m/K	T_b/K
Dysprosium	Dy	66	162.50	solid	8.55	1685	2840
Einsteinium	Es	99	[252]	solid		1130	
Erbium	Er	68	167.26	solid	9.07	1802	3141
Europium	Eu	63	151.96	solid	5.24	1095	1802
Fermium	Fm	100	[257]	solid		1800	
Flevorium	Fl	114	[289]	solid			
Fluorine	F	9	18.998	gas	1.70×10^{-3}	53	85
Francium	Fr	87	[223]	solid		300	
Gadolinium	Gd	64	157.25	solid	7.90	1586	3546
Gallium	Ga	31	69.723	solid	5.90	303	2477
Germanium	Ge	32	72.63	solid	5.32	1211	3106
Gold	Au	79	196.97	solid	19.3	1337	3129
Hafnium	Hf	72	178.49	solid	13.3	2506	4876
Hassium	Hs	108	[270]	solid			
Helium	He	2	4.0026	gas	0.179×10^{-3}		4.2
Holmium	Ho	67	164.93	solid	8.80	1745	2970
Hydrogen	H	1	1.008	gas	0.0899×10^{-3}	14	20
Indium	In	49	114.82	solid	7.31	430	2345
Iodine	I	53	126.90	solid	4.93	387	458
Iridium	Ir	77	192.22	solid	22.6	2719	4701
Iron	Fe	26	55.845	solid	7.87	1811	3134
Krypton	Kr	36	83.798	gas	3.73×10^{-3}	116	120
Lanthanum	La	57	138.91	solid	6.15	1193	3737
Lawrencium	Lr	103	[262]	solid		1900	
Lead	Pb	82	207.2	solid	11.3	601	2022
Lithium	Li	3	6.94	solid	0.53	454	1615
Livermorium	Lv	116	[293]	solid			
Lutetium	Lu	71	174.97	solid	9.84	1936	3675
Magnesium	Mg	12	24.305	solid	1.74	923	1363
Manganese	Mn	25	54.938	solid	7.3	1519	2334
Meitnerium	Mt	109	[278]	solid			
Mendelevium	Md	101	[258]	solid		1100	
Mercury	Hg	80	200.59	liquid	13.55	234	630
Molybdenum	Mo	42	95.95	solid	10.2	2896	4912

$\Delta_a H^{\ominus}$/kJ mol^{-1}	χ^P	I_1/kJ mol^{-1}	I_2/kJ mol^{-1}	I_3/kJ mol^{-1}	E_{eg}/kJ mol^{-1}	Oxidation states (main ones in bold)
290	1.22	573	1126	2200	> 0	**+3**
133		619	1160			**+3**
316	1.24	589	1151	2194		**+3**
177		547	1085	2404	–83	+2 **+3**
		627				+2 **+3**
79.4	3.98	1681	3374	6050	–328	**–1**
	0.7	393			–47	+1
398	1.20	593	1167	1990		**+3**
272	1.81	579	1979	2965	–41	+1 **+3**
372	2.01	762	1537	3302	–119	+2 **+4**
368	2.4	890	1949		–223	+1 **+3**
618	1.3	659	1400	2250	+1	**+4**
0		2372	5251		> 0	**0**
301	1.23	581	1139	2204		**+3**
218.0	2.20	1312			–73	**–1 +1**
243	1.78	558	1821	2704	–29	**+1 +3**
106.8	2.66	1008	1846	3200	–295	**–1** +1 +3 +5 +7
669	2.2	865	1600		–151	**+3** +4
416	1.83	762	1562	2957	–15	+1 **+2 +3** +4 +6
0		1351	2350	3565	> 0	**0** +2
431	1.10	538	1067	1850	–45	**+3**
		470				+3
195	1.8	716	1450	3081	–35	**+2** +4
159	0.98	520	7298	11 815	–60	**+1**
428	1.0	524	1340	2022	–33	+2 **+3**
147	1.31	738	1451	7733	> 0	**+2**
283	1.55	717	1509	3248	> 0	**+2** +3 **+4 +7**
		635				+3
61	1.9	1007	1810	3300	> 0	**+1 +2**
659	2.16	684	1559	2618	–72	+2 +4 **+6**

Name	Symbol	Atomic number	Relative atomic mass*	State	Density / $g\,cm^{-3}$	T_m/K	T_b/K
Neodymium	Nd	60	144.24	solid	7.01	1289	3347
Neon	Ne	10	20.180	gas	0.900×10^{-3}	25	27
Neptunium	Np	93	[237]	solid	20.3	917	
Nickel	Ni	28	58.693	solid	8.90	1728	3186
Niobium	Nb	41	92.906	solid	8.57	2750	5017
Nitrogen	N	7	14.007	gas	1.25×10^{-3}	63	77
Nobelium	No	102	[259]	solid		1100	
Osmium	Os	76	190.23	solid	22.59	3306	5285
Oxygen	O	8	15.999	gas	1.43×10^{-3}	54	90
Palladium	Pd	46	106.42	solid	12.0	1828	3236
Phosphorus (white)	P	15	30.974	solid	1.82	317	554
Platinum	Pt	78	195.08	solid	21.5	2041	4098
Plutonium	Pu	94	[244]	solid	19.8	913	3501
Polonium	Po	84	[209]	solid	9.20	527	1235
Potassium	K	19	39.098	solid	0.89	337	1032
Praseodymium	Pr	59	140.91	solid	6.77	1204	3793
Promethium	Pm	61	[145]	solid	7.26	1315	3270
Protactinium	Pa	91	231.04	solid	15.4	1845	
Radium	Ra	88	[226]	solid	5	969	
Radon	Rn	86	[222]	gas	9.73×10^{-3}	202	211
Rhenium	Re	75	186.21	solid	20.8	3458	5869
Rhodium	Rh	45	102.91	solid	12.4	2237	3968
Roentgenium	Rg	111	[281]	solid			
Rubidium	Rb	37	85.468	solid	1.53	312	961
Ruthenium	Ru	44	101.07	solid	12.1	2607	4423
Rutherfordium	Rf	104	[263]	solid			
Samarium	Sm	62	150.36	solid	7.52	1345	2067
Scandium	Sc	21	44.956	solid	2.99	1814	3109
Seaborgium	Sg	106	[271]	solid			
Selenium (grey)	Se	34	78.971	solid	4.79	494	958
Silicon	Si	14	28.085	solid	2.33	1687	3538
Silver	Ag	47	107.87	solid	10.50	1235	2435
Sodium	Na	11	22.990	solid	0.97	371	1156
Strontium	Sr	38	87.62	solid	2.64	1050	1655

$\Delta_a H^{\ominus}/\text{kJ mol}^{-1}$	χ^P	$I_1/\text{kJ mol}^{-1}$	$I_2/\text{kJ mol}^{-1}$	$I_3/\text{kJ mol}^{-1}$	$E_{eg}/\text{kJ mol}^{-1}$	Oxidation states (main ones in bold)
327	1.14	533	1034	2130	−185	+3
0		2081	3952	6122	>0	0
465	1.3	605				+3 +4 **+5** +6
430	1.91	737	1753	3395	−112	+1 **+2** +3 +4
733	1.6	652	1350	2416	−88	+3 +4 **+5**
472.7	3.04	1402	2856	4578	>0	−3 +1 +2 +3 +4 **+5**
		642				**+2** +3
787	2.2	814	1600		−106	+2 +3 +4 +6 +8
249.2	3.44	1314	3389	5300	−141 (+798)	−1 **−2** +2
377	2.20	804	1875	3177	−54	+2
317	2.19	1012	1907	2914	−72	−3 +3 **+5**
566	2.2	864	1791		−205	**+2** +4
345	1.3	581	1080			**+3** +4
142	2.0	812			−183	−2 +2 +4 +6
89	0.82	419	3052	4420	−48	+1
357	1.13	528	1018	2086	−93	+3
		539	1052	2150		+3
563	1.5	568				+3 +4 **+5**
159	0.9	509	979		−10	+2
0		1037			>0	0
774	1.9	756	1260	2510	−14	**+3** **+4** +5 +7
556	2.28	720	1744	2997	−110	**+1** +3
81	0.82	403	2633	3900	−47	+1
651	2.2	710	1617	2747	−101	**+3** **+4**
		580				
207	1.17	545	1068	2260		+3
378	1.36	633	1235	2389	−18	+3
227	2.55	941	2045	2974	−195	−2 +2 +4
450	1.90	787	1577	3232	−134	+2 **+4**
285	1.93	731	2072	3361	−126	+1
108	0.93	496	4562	6910	−53	+1
164	0.95	549	1064	4138	−5	+2

Name	Symbol	Atomic number	Relative atomic mass*	State	Density / $g cm^{-3}$	T_m/K	T_b/K
Sulfur (rhombic)	S	16	32.06	solid	2.07	388	718
Tantalum	Ta	73	180.95	solid	16.4	3290	5731
Technetium	Tc	43	[98]	solid	11.5	2430	4538
Tellurium	Te	52	127.60	solid	6.23	723	1261
Terbium	Tb	65	158.93	solid	8.23	1629	3503
Thallium	Tl	81	204.38	solid	11.8	577	1746
Thorium	Th	90	232.04	solid	11.72	2023	5061
Thulium	Tm	69	168.93	solid	9.32	1818	2223
Tin (white)	Sn	50	118.71	solid	7.29	505	2875
Titanium	Ti	22	47.867	solid	4.51	1941	3560
Tungsten	W	74	183.84	solid	19.3	3695	5828
Uranium	U	92	238.03	solid	19.1	1408	4404
Vanadium	V	23	50.942	solid	6.0	2183	3680
Xenon	Xe	54	131.29	gas	5.89×10^{-3}	161	165
Ytterbium	Yb	70	173.05	solid	6.90	1097	1469
Yttrium	Y	39	88.906	solid	4.47	1795	3618
Zinc	Zn	30	65.38	solid	7.13	693	1180
Zirconium	Zr	40	91.224	solid	6.52	2128	4682

* [] denotes a radioactive element. The number given is the mass number of the most stable isotope.

Sources

Haynes, W. M. (ed.) (2015–16) *CRC Handbook of Chemistry and Physics*, 96th edn., CRC Press, Boca Raton, Florida.

Wieser, M. E. and Coplen, T. B. (2011) *Pure Appl. Chem.*, **83**, 359.

Elements from the Royal Society of Chemistry Data Book Website (http://www.rsc.org/education/teachers/learnnet/data/index_databases.htm). Reproduced by permission of the Royal Society of Chemistry.

www.webelements.com.

$\Delta_a H^{\ominus}$/kJ mol^{-1}	χ^P	I_1/kJ mol^{-1}	I_2/kJ mol^{-1}	I_3/kJ mol^{-1}	E_{eg}/kJ mol^{-1}	Oxidation states (main ones in bold)
277	2.58	1000	2252	3357	−200 (+640)	**−2** +1 +2 +4 +6
782	1.5	728	1500		−31	+3 +4 **+5**
678	2.10	702	1472	2850	−53	+4 +5 +7
197	2.1	869	1790	2698	−190	**−2** +2 +4
389		566	1112	2114	−112	+3
182	1.8	589	1971	2878	−36	+1 **+3**
602	1.3	609	1150	1930		+4
232	1.25	597	1163	2285	−99	+3
301	1.96	709	1412	2943	−107	**+2 +4**
473	1.54	659	1310	2653	−8	+1 +2 +3 **+4**
851	1.7	759	1550		−79	+2 +4 **+6**
533	1.7	598	1020			+2 +4 **+6**
516	1.63	651	1410	2828	−51	+2 +3 **+4 +5**
0	2.60	1170	2024	3099	> 0	**0** +2 +4 +6 +8
156		603	1175	2417	+2	**+2 +3**
425	1.22	600	1179	1980	−30	+3
130	1.65	906	1733	3833	> 0	+2
610	1.33	640	1260	2218	−41	**+3 +4**

Appendix 6

Electronic configurations of the elements and some important ions

Name	Symbol	Elemental configuration	Ionic configuration
Actinium	Ac	[Rn] $6d^1\ 7s^2$	
Aluminium	Al	[Ne] $3s^2\ 3p^1$	Al^{3+} [Ne]
Americium	Am	[Rn] $5f^7\ 7s^2$	
Antimony	Sb	[Kr] $4d^{10}\ 5s^2\ 5p^3$	
Argon	Ar	[Ne] $3s^2\ 3p^6$	
Arsenic	As	[Ar] $3d^{10}\ 4s^2\ 4p^3$	
Astatine	At	[Xe] $4f^{14}\ 5d^{10}\ 6s^2\ 6p^5$	
Barium	Ba	[Xe] $6s^2$	Ba^{2+} [Xe]
Berkelium	Bk	[Rn] $5f^9\ 7s^2$	
Beryllium	Be	[He] $2s^2$	
Bismuth	Bi	[Xe] $4f^{14}\ 5d^{10}\ 6s^2\ 6p^3$	Bi^{3+} [Xe] $4f^{14}\ 5d^{10}\ 6s^2$
Bohrium*	Bh	[Rn] $5f^{14}\ 6d^5\ 7s^2$	
Boron	B	[He] $2s^2\ 2p^1$	
Bromine	Br	[Ar] $3d^{10}\ 4s^2\ 4p^5$	Br^- [Kr]
Cadmium	Cd	[Kr] $4d^{10}\ 5s^2$	
Caesium	Cs	[Xe] $6s^1$	Cs^+ [Xe]
Calcium	Ca	[Ar] $4s^2$	Ca^{2+} [Ar]
Californium	Cf	[Rn] $5f^{10}\ 7s^2$	
Carbon	C	[He] $2s^2\ 2p^2$	
Cerium	Ce	[Xe] $4f^1\ 5d^1\ 6s^2$	
Chlorine	Cl	[Ne] $3s^2\ 3p^5$	Cl^- [Ar]
Chromium	Cr	[Ar] $3d^5\ 4s^1$	Cr^{3+} [Ar] $3d^3$
Cobalt	Co	[Ar] $3d^7\ 4s^2$	Co^{2+} [Ar] $3d^7$, Co^{3+} [Ar] $3d^6$
Copernicium*	Cn	[Rn] $5f^{14}\ 6d^{10}\ 7s^2$	
Copper	Cu	[Ar] $3d^{10}\ 4s^1$	Cu^+ [Ar] $3d^{10}$, Cu^{2+} [Ar] $3d^9$
Curium	Cm	[Rn] $5f^7\ 6d^1\ 7s^2$	

Name	Symbol	Elemental configuration	Ionic configuration
Darmstadtium*	Ds	[Rn] $5f^{14}$ $6d^8$ $7s^2$	
Dubnium*	Db	[Rn] $5f^{14}$ $6d^3$ $7s^2$	
Dysprosium	Dy	[Xe] $4f^{10}$ $6s^2$	
Einsteinium	Es	[Rn] $5f^{11}$ $7s^2$	
Erbium	Er	[Xe] $4f^{12}$ $6s^2$	
Europium	Eu	[Xe] $4f^7$ $6s^2$	
Fermium	Fm	[Rn] $5f^{12}$ $7s^2$	
Fluorine	F	[He] $2s^2$ $2p^5$	F^- [Ne]
Francium	Fr	[Rn] $7s^1$	Fr^+ [Rn]
Gadolinium	Gd	[Xe] $4f^7$ $5d^1$ $6s^2$	
Gallium	Ga	[Ar] $3d^{10}$ $4s^2$ $4p^1$	Ga^{3+} [Ar] $3d^{10}$
Germanium	Ge	[Ar] $3d^{10}$ $4s^2$ $4p^2$	
Gold	Au	[Xe] $4f^{14}$ $5d^{10}$ $6s^1$	
Hafnium	Hf	[Xe] $4f^{14}$ $5d^2$ $6s^2$	
Hassium*	Hs	[Rn] $5f^{14}$ $6d^6$ $7s^2$	
Helium	He	$1s^2$	
Holmium	Ho	[Xe] $4f^{11}$ $6s^2$	
Hydrogen	H	$1s^1$	H^- [He]
Indium	In	[Kr] $4d^{10}$ $5s^2$ $5p^1$	
Iodine	I	[Kr] $4d^{10}$ $5s^2$ $5p^5$	I^- [Xe]
Iridium	Ir	[Xe] $4f^{14}$ $5d^7$ $6s^2$	
Iron	Fe	[Ar] $3d^6$ $4s^2$	Fe^{2+} [Ar] $3d^6$, Fe^{3+} [Ar] $3d^5$
Krypton	Kr	[Ar] $3d^{10}$ $4s^2$ $4p^6$	
Lanthanum	La	[Xe] $5d^1$ $6s^2$	
Lawrencium*	Lr	[Rn] $5f^{14}$ $7s^2$ $7p^1$	
Lead	Pb	[Xe] $4f^{14}$ $5d^{10}$ $6s^2$ $6p^2$	Pb^{2+} [Xe] $4f^{14}$ $5d^{10}$ $6s^2$
Lithium	Li	[He] $2s^1$	Li^+ [He]
Lutetium	Lu	[Xe] $4f^{14}$ $5d^1$ $6s^2$	
Magnesium	Mg	[Ne] $3s^2$	Mg^{2+} [Ne]
Manganese	Mn	[Ar] $3d^5$ $4s^2$	Mn^{2+} [Ar] $3d^5$, Mn^{3+} [Ar] $3d^4$
Meitnerium*	Mt	[Rn] $5f^{14}$ $6d^7$ $7s^2$	
Mendelevium	Md	[Rn] $5f^{13}$ $7s^2$	
Mercury	Hg	[Xe] $4f^{14}$ $5d^{10}$ $6s^2$	
Molybdenum	Mo	[Kr] $4d^5$ $5s^1$	
Neodymium	Nd	[Xe] $4f^4$ $6s^2$	
Neon	Ne	[He] $2s^2$ $2p^6$	

Name	Symbol	Elemental configuration	Ionic configuration
Neptunium	Np	[Rn] $5f^4\,6d^1\,7s^2$	
Nickel	Ni	[Ar] $3d^8\,4s^2$	Ni^{2+} [Ar] $3d^8$
Niobium	Nb	[Kr] $4d^4\,5s^1$	
Nitrogen	N	[He] $2s^2\,2p^3$	N^{3-} [Ne]
Nobelium	No	[Rn] $5f^{14}\,7s^2$	
Osmium	Os	[Xe] $4f^{14}\,5d^6\,6s^2$	
Oxygen	O	[He] $2s^2\,2p^4$	O^{2-} [Ne]
Palladium	Pd	[Kr] $4d^{10}$	
Phosphorus	P	[Ne] $3s^2\,3p^3$	P^{3-} [Ar]
Platinum	Pt	[Xe] $4f^{14}\,5d^9\,6s^1$	
Plutonium	Pu	[Rn] $5f^6\,7s^2$	
Polonium	Po	[Xe] $4f^{14}\,5d^{10}\,6s^2\,6p^4$	
Potassium	K	[Ar] $4s^1$	K^+ [Ar]
Praseodymium	Pr	[Xe] $4f^3\,6s^2$	
Promethium	Pm	[Xe] $4f^5\,6s^2$	
Protactinium	Pa	[Rn] $5f^2\,6d^1\,7s^2$	
Radium	Ra	[Rn] $7s^2$	Ra^{2+} [Rn]
Radon	Rn	[Xe] $4f^{14}\,5d^{10}\,6s^2\,6p^6$	
Rhenium	Re	[Xe] $4f^{14}\,5d^5\,6s^2$	
Rhodium	Rh	[Kr] $4d^8\,5s^1$	
Roentgenium*	Rg	[Rn] $5f^{14}\,6d^{10}\,7s^1$	
Rubidium	Rb	[Kr] $5s^1$	Rb^+ [Kr]
Ruthenium	Ru	[Kr] $4d^7\,5s^1$	
Rutherfordium*	Rf	[Rn] $5f^{14}\,6d^2\,7s^2$	
Samarium	Sm	[Xe] $4f^6\,6s^2$	
Scandium	Sc	[Ar] $3d^1\,4s^2$	Sc^{3+} [Ar]
Seaborgium*	Sg	[Rn] $5f^{14}\,6d^4\,7s^2$	
Selenium	Se	[Ar] $3d^{10}\,4s^2\,4p^4$	Se^{2-} [Kr]
Silicon	Si	[Ne] $3s^2\,3p^2$	
Silver	Ag	[Kr] $4d^{10}\,5s^1$	
Sodium	Na	[Ne] $3s^1$	Na^+ [Ne]
Strontium	Sr	[Kr] $5s^2$	Sr^{2+} [Kr]
Sulfur	S	[Ne] $3s^2\,3p^4$	S^{2-} [Ar]
Tantalum	Ta	[Xe] $4f^{14}\,5d^3\,6s^2$	
Technetium	Tc	[Kr] $4d^5\,5s^2$	
Tellurium	Te	[Kr] $4d^{10}\,5s^2\,5p^4$	Te^{2-} [Xe]

Name	Symbol	Elemental configuration	Ionic configuration
Terbium	Tb	[Xe] $4f^9$ $6s^2$	
Thallium	Tl	[Xe] $4f^{14}$ $5d^{10}$ $6s^2$ $6p^1$	Tl^+ [Xe] $4f^{14}$ $5d^{10}$ $6s^2$, Tl^{3+} [Xe] $4f^{14}$ $5d^{10}$
Thorium	Th	[Rn] $6d^2$ $7s^2$	
Thulium	Tm	[Xe] $4f^{13}$ $6s^2$	
Tin	Sn	[Kr] $4d^{10}$ $5s^2$ $5p^2$	Sn^{2+} [Kr] $4d^{10}$ $5s^2$, Sn^{4+} [Kr] $4d^{10}$
Titanium	Ti	[Ar] $3d^2$ $4s^2$	Ti^{3+} [Ar] $3d^1$, Ti^{4+} [Ar]
Tungsten	W	[Xe] $4f^{14}$ $5d^4$ $6s^2$	
Uranium	U	[Rn] $5f^3$ $6d^1$ $7s^2$	
Vanadium	V	[Ar] $3d^3$ $4s^2$	V^{3+} [Ar] $3d^2$
Xenon	Xe	[Kr] $4d^{10}$ $5s^2$ $5p^6$	
Ytterbium	Yb	[Xe] $4f^{14}$ $6s^2$	
Yttrium	Y	[Kr] $4d^1$ $5s^2$	
Zinc	Zn	[Ar] $3d^{10}$ $4s^2$	Zn^{2+} [Ar] $3d^{10}$
Zirconium	Zr	[Kr] $4d^2$ $5s^2$	

* Tentative.

Source

Haynes, W.M. (ed.) (2015–16). *CRC handbook of chemistry and physics*, 96th edn. CRC Press, Boca Raton, Florida.

Appendix 7

Thermodynamic data for organic and inorganic compounds

The thermodynamic data in these tables refer to 298.15 K and 1 bar pressure. T_m is the melting point at 1 bar pressure, T_b is the boiling point at 1 bar pressure, M is the molar mass, $\Delta_f H^\ominus$ is the standard enthalpy change of formation, $\Delta_f G^\ominus$ is the standard Gibbs energy change of formation, $S^\ominus$ is the standard entropy, and C_p is the molar constant pressure heat capacity.

Organic compounds

Compound	Formula	State	T_m/K	T_b/K	M/g mol^{-1}	$\Delta_f H^{\ominus}$/ kJ mol^{-1}	$\Delta_f G^{\ominus}$/ kJ mol^{-1}	$S^{\ominus}$/ J K^{-1} mol^{-1}	C_p/ J K^{-1} mol^{-1}
Methane	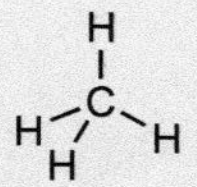	gas	91.1	109.1	16.04	−74.8	−50.8	186.3	35.7
Ethane	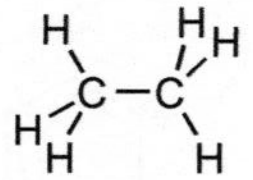	gas	89.8	184.5	30.07	−83.8	−32.8	229.6	52.5
Ethene	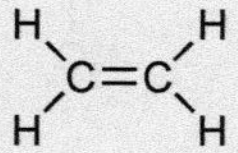	gas	104.1	169.4	28.05	+52.5	+68.2	219.3	42.9
Ethyne	H−C≡C−H	gas	171.7	189.1	26.04	+226.7	+209.2	200.9	44.0
Propane		gas	83.4	231	42.10	−103.9	−23.5	269.9	73.6
Cyclopropane		gas	145.5	240.4	42.08	+53.3	+104.5	237.6	55.6
Propene	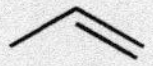	gas	87.9	225.7	42.08	+20.4	+62.8	267.1	64.3
Butane		gas	134.7	272.6	58.13	−125.6	−15.6	310.2	98.5
But–1–ene		gas	87.8	266.8	56.11	−0.6	+71.4	305.7	85.7
(*Z*)–but–1–ene	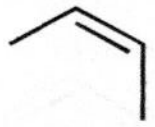	gas	134.2	276.8	56.11	−7.7	+66.0	300.9	80.2
(*E*)–2–but–2–ene		gas	167.6	274.0	56.11	−10.8	+63.1	296.6	87.7
Pentane		gas	143.1	309.2	72.15	−146.8	−8.20	347.8	120.1
Hexane		liquid	178.1	342.1	86.18	−198.7	−4.2	296.0	195.9
Cyclohexane		liquid	279.6	353.8	84.16	−156.4	+26.8	204.4	156.0
Heptane		liquid	182.5	371.5	100.21	−224.4	+1.3	328.6	224.6
Octane		liquid	216.3	398.8	114.23	−250.0	+6.4	361.2	255.7

Compound	Formula	State	T_m/K	T_b/K	M/g mol^{-1}	$\Delta_f H^{\ominus}$/ kJ mol^{-1}	$\Delta_f G^{\ominus}$/ kJ mol^{-1}	$S^{\ominus}$/ J K^{-1} mol^{-1}	C_p/ J K^{-1} mol^{-1}
Benzene	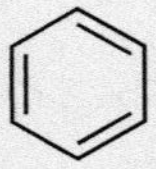	liquid	278.6	353.2	78.12	+49.0	+124.3	173.3	135.7
		gas				+82.9	+129.7	269.3	82.4
Methylbenzene (toluene)	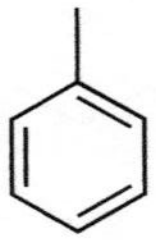	liquid	178.1	383.7	92.14	+12.0	+110.6	221.0	156.0
Naphthalene	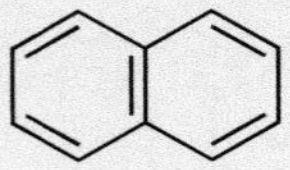	solid	353.6	491.1	128.18	+77.0		167.4	165.7
Methanol	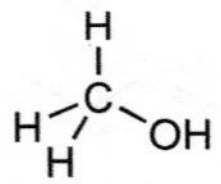	liquid	179.2	338.1	32.04	−238.4	−166.3	126.8	81.6
		gas				−201.0	−162.0	239.8	44.1
Ethanol	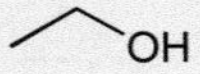	liquid	155.8	351.6	46.07	−277.6	−174.8	159.9	111.5
		gas				−235.3	−168.5	282.7	65.2
Propan-1-ol		liquid	146.6	370.5	60.10	−302.5	−171.3	192.8	144.4
Butan-1-ol		liquid	183.6	390.3	74.13	−328.0	−168.9	225.7	176.7
Phenol	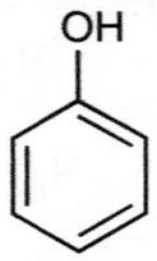	solid	316.1	454.8	94.12	−165.0	−47.5	144.0	127.2
Methanoic acid (formic acid)	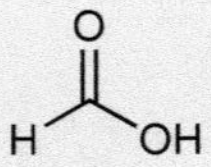	liquid	281.5	373.7	46.03	−425.1	−361.4	131.8	99.0
Ethanoic acid (acetic acid)	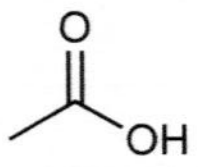	liquid	289.7	391.0	60.05	−484.5	−389.9	158.0	123.1

Benzoic acid	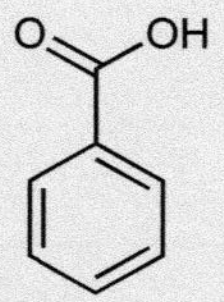	solid	393.5	522.0	122.13	–385.0	–245.3	167.6	146.7
(*RS*)–lactic acid	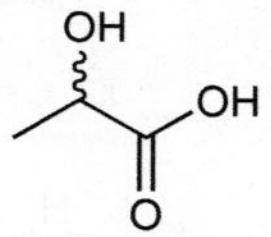	solid	326.1	376.1	90.08	–694.0			
Methanal (formaldehyde)	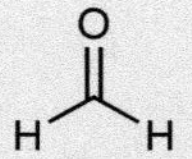	gas	181.1	252.2	30.03	–108.57	–113.0	218.8	35.4
Ethanal (acetaldehyde)	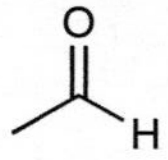	liquid	152.1	293.3	44.05	–196.40	–128.12	117.3	89.1
Propanone (acetone)	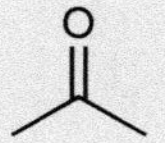	liquid	177.8	329.3	58.08	–249.4	–155.4	200.4	123.9
Ethyl ethanoate (ethyl acetate)	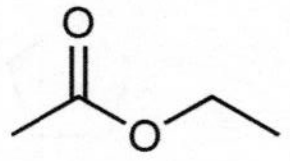	liquid	189.6	350.3	88.11	–479.9	–332.7	259.4	169.6
Ethanoic anhydride (acetic anhydride)	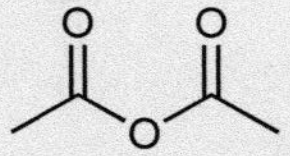	liquid	200	412.7	102.1	–625.0			168.2
Ethanoyl chloride (acetyl chloride)	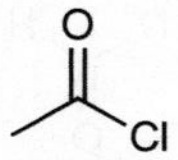	liquid	161.1	324.0	78.5	–272		200.8	117.2
α–D–glucose	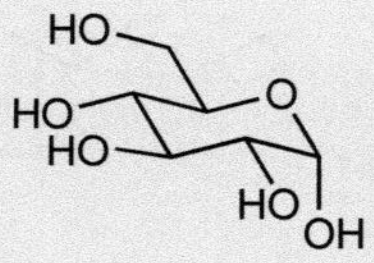	solid	423.1		180.16	–1274			
β–D–glucose	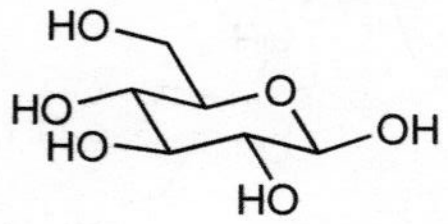	solid			180.16	–1268	–910	212	

Compound	Formula	State	T_m/K	T_b/K	M/g mol^{-1}	$\Delta_f H^\ominus$/ kJ mol^{-1}	$\Delta_f G^\ominus$/ kJ mol^{-1}	$S^\ominus$/ J K^{-1} mol^{-1}	C_p/ J K^{-1} mol^{-1}
Sucrose	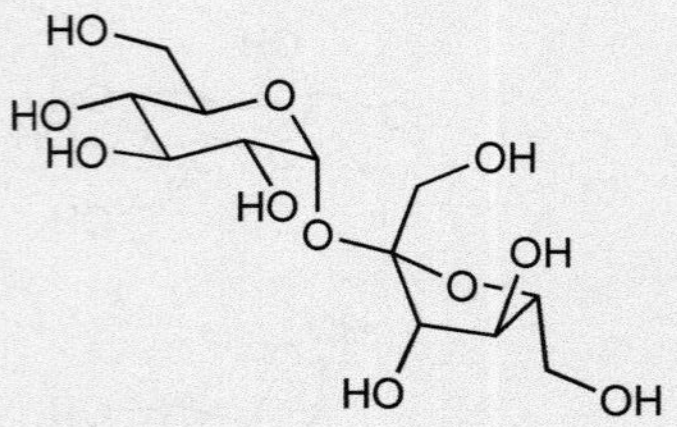 HO HO HO O HO O OH OH O HO OH	solid	462		342.30	–2222	–1543	360.2	425.5
Ethoxyethane (diethyl ether)	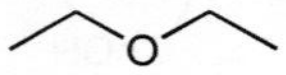 O	liquid	156.9	307.6		–271.2	–122.7	253.5	172.5
N,N–dimethylmethanamide (*N,N*–dimethylformamide, DMF)	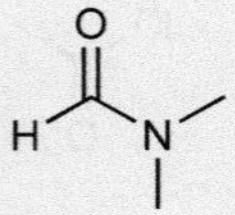 O H N	liquid	212.7	426		–231.2			148.2
Dimethyl sulfoxide (DMSO)	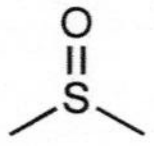 O S	liquid	291.6	464	78.13	–203.4		188.8	149.4
Tetrahydrofuran (THF)	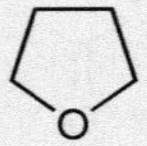 O	liquid	164.6	339	72.11	–211.4		203.9	124.1
Tetrachloromethane (carbon tetrachloride)	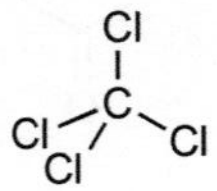 Cl C Cl Cl Cl	liquid	250.1	349.6	153.82	–128.4	–65.2	214.40	131.3
Trichloromethane (chloroform)	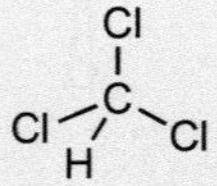 Cl C Cl H Cl	liquid	209.6	334.3	119.4	–134.3	–71.4	295.6	114.3
Dichloromethane	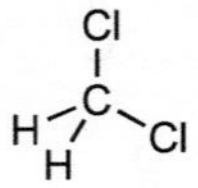 Cl C H H Cl	liquid	198.1	313		–124.3		174.5	102.3
Chloromethane	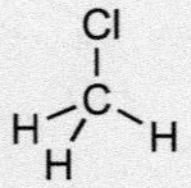 Cl C H H H	gas	176.0	248.9	50.5	–83.7	–57.4	234.4	

Inorganic elements and compounds

Formula	State	T_m/K	T_b/K	M/g mol^{-1}	$\Delta_f H^\ominus$/kJ mol^{-1}	$\Delta_f G^\ominus$/kJ mol^{-1}	$S^\ominus$/J K^{-1} mol^{-1}	C_p/J K^{-1} mol^{-1}
Aluminium								
Al	solid	933	2792	26.98	0	0	28.3	24.4
Al	gas				+326.4	+285.7	164.5	21.4
Al^{3+}	gas				+5483.17			
Al_2O_3	solid, α	2345	3253	101.96	−1675.7	−1582.3	50.9	79.0
$AlCl_3$	solid	sublimes	451	133.24	−704.2	−628.8	110.7	91.8
AlF_3	solid	1564			−1504.1	−1425.1	66.5	
Argon								
Ar	gas	84	87.5	39.95	0	0	154.8	20.8
Antimony								
Sb	solid	903	1650	121.75	0	0	45.7	25.2
SbH_3	gas	1815	256	124.8	+145.1	+147.8	232.8	41.1
Arsenic								
As	solid, α	1090	886	74.92	0	0	35.1	24.64
As	gas			74.92	+302.5	+261.0	174.2	20.8
AsH_3	gas	157	218	77.95	+66.4	+68.9	222.8	38.1
Barium								
Ba	solid	1000	2179	137.34	0	0	62.8	28.1
Ba	gas			137.34	+180	+146	170.2	20.8
BaO	solid	2191	2273	153.34	−553.5	−525.1	70.4	47.8
$BaCl_2$	solid	1236	1833	208.25	−858.6	−810.4	123.7	75.1
Beryllium								
Be	solid	1551	3243	9.01	0	0	9.5	16.4
Be	gas			9.01	+324.3	+286.6	136.3	20.8
Bismuth								
Bi	solid	544	1833	208.98	0	0	56.7	25.5
Bi	gas			208.98	+207.1	+168.2	187.0	20.8
Boron								
B	solid	2348	4273	10.81	0	0	5.9	
B	gas				+565	153.4		
BF_3	gas	129	173	67.8	−1136.0	−1120.3	254.4	
BCl_3	liquid	166	266	117.2	−427.2	−387.4	206.3	
B_2O_3	solid	723		69.6	−1254.5	−1182.4	77.8	
B_2H_6	gas	108	181	27.7	−135.6	+86.6	232.0	
Bromine								
Br_2	liquid	266	331	159.82	0	0	152.2	75.7
Br_2	gas			159.82	+30.9	+3.1	245.5	36.0
Br	gas			79.91	+111.9	+82.4	175.0	20.8
Br^-	gas			79.91	−219.07			
HBr	gas	185	206	90.92	−36.4	−53.5	198.7	29.1
Caesium								
Cs	solid	307	944	132.91	0	0	85.2	32.2
Cs	gas			132.91	+76.1	+49.1	175.6	20.8

Formula	State	T_m/K	T_b/K	M/g mol^{-1}	$\Delta_f H^\ominus$/kJ mol^{-1}	$\Delta_f G^\ominus$/kJ mol^{-1}	$S^\ominus$/J K^{-1} mol^{-1}	C_p/J K^{-1} mol^{-1}
Calcium								
Ca	solid	1115	1757	40.08	0	0	41.4	25.3
Ca	gas				+178.2	+144.3	154.9	20.8
CaO(s)	solid	2887	3123	56.08	−635.1	−604.0	39.8	42.8
$CaCO_3$ (calcite)	solid	1612		100.09	−1206.9	−1128.8	92.9	81.9
$CaCO_3$ (aragonite)	solid	793	1098	100.09	−1207.1	−1127.8	88.7	81.3
CaF_2	solid	1696	2773	78.08	1219.6	−1167.3	68.9	67.0
$CaCl_2$	solid	1055	1873	110.99	−795.8	−748.1	104.6	72.6
$CaBr_2$	solid	1003	1083	199.90	−682.8	−663.6	130.0	
$Ca(OH)_2$	solid	853		74.1	−986.1	−898.6	83.4	
Carbon								
C (graphite)	solid	3925		12.011	0	0	5.7	8.5
C (diamond)	solid	3832		12.011	+1.9	+2.9	2.4	6.1
C	gas			12.011	+716.7	+671.3	158.1	20.8
CO	gas	74	82	28.011	−110.5	−137.2	197.7	29.1
CO_2	gas	sublimes	195	44.010	−393.5	−394.4	213.7	37.1
CS_2	liquid	162	319	76.14	+89.70	+65.27	151.34	75.7
HCN	liquid	259	299	27.03	+108.9	+124.9	112.9	35.9
Chlorine								
Cl_2	gas	172	239	70.91	0	0	223.1	33.9
Cl	gas			35.45	+121.7	+105.7	165.2	21.8
Cl^-	gas			35.45	−233.13			
HCl	gas	158	188	36.46	−92.3	−95.3	186.9	29.1
Chromium								
Cr	solid	2130	2943	52.00	0	0	23.8	23.4
Cr	gas			52.00	+396.6	+351.8	174.5	20.8
Copper								
Cu	solid	1357	2835	63.54	0	0	33.1	24.4
Cu	gas			63.54	+338.3	+298.6	166.4	20.8
Cu_2O	solid	1508	2073	143.08	−168.6	−146.0	93.1	63.6
CuO	solid	1599	923	79.54	−157.3	−129.7	42.6	42.3
$CuSO_4$	solid	473	423	159.60	−771.4	−661.8	109.0	100.0
$CuSO_4 \cdot 5H_2O$	solid	383		249.68	−2279.7	−1879.7	300.4	280
Fluorine								
F_2	gas	53	85	38.00	0	0	202.8	31.3
F	gas			19.00	+79.0	+61.9	158.8	22.7
HF	gas	190	293	20.01	−271.1	−273.2	173.8	29.1
Gold								
Au	solid	1340	3130	196.97	0	0	47.4	25.4
Au	gas			196.97	+366.1	+326.3	180.5	20.8
Helium								
He	gas	1	4	4.003	0	0	126.2	20.8
Hydrogen								
H_2	gas	14	20.3	2.016	0	0	130.7	28.8
H	gas			1.008	+218.0	+203.3	114.7	20.8
H_2O	liquid	273.2	373.2	18.015	−285.8	−237.1	69.9	75.3
H_2O	gas			18.015	−241.8	−228.6	188.8	33.6
H_2O_2	liquid	273	323	34.015	−187.8	−120.4	109.6	89.1

Formula	State	T_m/K	T_b/K	M/g mol^{-1}	$\Delta_f H^\ominus$/kJ mol^{-1}	$\Delta_f G^\ominus$/kJ mol^{-1}	$S^\ominus$/J K^{-1} mol^{-1}	C_p/J K^{-1} mol^{-1}
Iodine								
I_2	solid	387	457	253.81	0	0	116.1	54.4
I_2	gas			253.81	+62.4	+19.3	260.7	36.9
I	gas			126.90	+106.8	+70.3	180.8	20.8
HI	gas	222	238	127.91	+26.5	+1.7	206.6	29.2
Iron								
Fe	solid	1810	3134	55.85	0	0	27.3	25.1
Fe	gas			55.85	+416.3	+370.7	180.5	25.7
FeO	solid	1642		71.85	−266.3	−244.1	60.8	50.0
Fe_3O_4 (magnetite)	solid	1867		231.54	−1018.4	−1015.4	146.4	143.4
Fe_2O_3 (haematite)	solid	1838		159.69	−824.2	−742.2	87.4	103.9
FeS	solid	1468		87.91	−100.0	−100.4	60.3	50.5
Krypton								
Kr		116	120	83.80	0	0	164.1	20.8
Lead								
Pb	solid	600	2020	207.19	0	0	64.8	26.4
Pb	gas			207.19	+195.0	+161.9	175.37	20.79
$PbCl_2$	solid	774	1223	278.1	−359.4	−314.1	136.0	
PbO (yellow)	solid			223.19	−217.3	−187.9	68.7	45.8
PbO (red)	solid			223.19	−219.0	−188.9	66.5	45.8
PbO_2	solid	563		239.19	−277.4	−217.3	68.6	64.6
Lithium								
Li	solid	454	1615	6.94	0	0	29.1	24.8
Li	gas			6.94	+159.3	+126.7	138.8	20.8
Li_2O	solid			29.9	−597.9	−561.2	37.6	
Magnesium								
Mg	solid	923	1363	24.31	0	0	32.7	24.9
Mg	gas			24.31	+147.7	+113.1	148.7	20.8
MgO	solid			40.31	−601.7	−569.4	26.9	37.1
$MgCO_3$	solid	623	1173	84.32	−1095.8	−1012.1	65.7	75.5
$MgCl_2$	solid	987	1685	95.22	−641.3	−591.8	89.6	71.4
$MgBr_2$	solid	973	1503	184.13	−524.3	−503.8	117.2	
Mercury								
Hg(l)	liquid	234	630	200.59	0	0	76.0	28.0
Hg(g)	gas			200.59	+61.3	+31.8	175.0	20.8
$HgCl_2$	solid	550	575	271.5	−224.3	−178.6	146.0	
HgO(s)	solid	773		216.59	−90.8	−58.5	70.3	44.1
Neon								
Ne	gas	25	27	20.18	0	0	146.3	20.8

Formula	State	T_m/K	T_b/K	M/g mol^{-1}	$\Delta_f H^\ominus$/kJ mol^{-1}	$\Delta_f G^\ominus$/kJ mol^{-1}	$S^\ominus$/J K^{-1} mol^{-1}	C_p/J K^{-1} mol^{-1}
Nitrogen								
N_2	gas	63	77	28.013	0	0	191.6	29.1
N	gas			14.007	+472.7	+455.6	153.3	20.8
NO	gas	110	121	30.01	+90.3	+86.6	210.8	29.8
N_2O	gas	182	185	44.01	+82.1	+104.2	219.9	38.5
NO_2	gas	262	294	46.01	+33.2	+51.3	240.1	37.2
N_2O_4	gas	262	294	92.01	+9.2	+97.9	304.3	77.3
N_2O_5	solid	303	320	108.01	–43.1	+113.9	178.2	143.1
N_2O_5	gas			108.01	+11.3	+115.1	355.7	84.5
HNO_3	liquid	231	356	63.01	–174.1	–80.7	155.6	109.9
NH_3	gas	195	240	17.03	–46.1	–16.5	192.5	35.1
NF_3	gas	67	144	71.0	–124.7	–83.3	260.6	
NCl_3	liquid	233	344	120.4	+230.1			
N_2H_4	liquid	275	387	32.05	+50.6	+149.4	121.2	139.3
NH_4NO_3	solid	443	483	80.04	–365.6	–183.9	151.1	84.1
NH_4Cl	solid	613	793	53.49	–314.4	–202.9	94.6	
Oxygen								
O_2	gas	55	90	31.999	0	0	205.1	29.4
O	gas			15.999	+249.2	+231.7	161.1	21.9
O_3	gas	81	161	47.998	+142.7	+163.2	238.9	39.2
Phosphorus								
P (white)	solid	317	553	30.97	0	0	41.1	23.8
P	gas			30.97	+314.6	+278.3	163.2	20.8
P_2	gas			61.95	+144.3	+103.7	218.1	32.1
P_4	gas			123.90	+58.9	+24.4	280.0	67.2
PH_3	gas	140		34.00	+5.4	+13.4	210.2	37.1
PCl_3	gas	161	349	137.33	–287.0	–267.8	311.8	71.8
PCl_3	liquid			137.33	–319.7	–272.3	217.1	
PCl_5	gas	435	440	208.24	–374.9	–305.0	364.6	112.8
PCl_5	solid			208.24	–443.5		166.5	
P_4O_{10}	solid	sublimes	573	283.89	–2984.0	–2697.0	228.9	211.7
P_4O_6	solid	297	448	219.89	–1640.1			
Potassium								
K	solid	336	1033	39.10	0	0	64.2	29.6
K	gas			39.10	+89.2	+60.6	160.3	20.8
K^+	gas			39.10	+514.3			
KOH	solid	633	1593	56.11	–424.8	–379.1	78.9	64.9
KF	solid	1131	1778	58.10	–576.3	–537.8	66.6	49.0
KCl	solid	1043	1773	74.56	–436.8	–409.1	82.6	51.3
KBr	solid	1007	1708	119.01	–393.8	–380.7	95.9	52.3
KI	solid	954	1603	166.01	–327.9	–324.9	106.3	52.9
K_2O	solid	623		94.2	–361.5	–322.8	102.0	77.4
KO_2	solid	653		71.1	–284.9	–239.5	116.7	
Silicon								
Si	solid	1685	2628	28.09	0	0	18.8	20.0
Si	gas			28.09	+455.6	+411.3	168.0	22.3
SiO_2 (α)	solid	1883	2503	60.09	–910.9	–856.6	41.8	44.4
$SiCl_4$	liquid	203	331	169.70	–687.0	–619.8	239.7	145.3

Formula	State	T_m/K	T_b/K	M/g mol^{-1}	$\Delta_f H^\ominus$/kJ mol^{-1}	$\Delta_f G^\ominus$/kJ mol^{-1}	$S^\ominus$/J K^{-1} mol^{-1}	C_p/J K^{-1} mol^{-1}
Silver								
Ag	solid	1235	2485	107.87	0	0	42.6	25.4
Ag	gas			107.87	+284.6	+245.7	173.0	20.8
Ag^+	gas			107.87	+105.8		73.45	
AgBr	solid	705	1573	187.78	−100.4	−96.9	107.1	52.4
AgCl	solid	728	1823	143.32	−127.1	−109.8	96.2	50.8
Ag_2O	solid	503		231.74	−31.1	−11.2	121.3	65.9
$AgNO_3$	solid	485	717	169.88	−124.4	−33.4	140.9	93.1
Sodium								
Na	solid	371	1156	22.99	0	0	51.2	28.2
Na	gas			22.99	+107.32	+76.76	153.71	20.79
NaOH	solid	592	1663	40.00	−425.6	−379.5	64.5	59.5
NaCl	solid	1074	1686	58.44	−411.2	−384.1	72.1	50.5
NaBr	solid	1020	1663	102.90	−361.1	349.0	86.8	51.4
NaI	solid	934	1577	149.89	−287.8	−286.1	98.5	52.1
Na_2O	solid	1548		62.0	−414.2	−375.5	75.1	
Na_2O_2	solid	733	930	78.0	−510.9	−447.7	95.0	
Sulfur								
S(α) (rhombic)	solid	386	718	32.06	0	0	31.8	22.6
S(β) (monoclinic)	solid	392	718	32.06	+0.3	0.1	32.6	23.6
S	gas			32.06	+278.8	+238.3	167.8	23.7
S_2	gas			64.13	+128.4	+79.3	228.2	32.5
SO_2	gas	200	263	64.06	−296.8	−300.2	248.2	39.9
SO_3	gas	290	318	80.06	−395.7	−371.1	256.7	50.7
H_2SO_4	liquid	283	611	98.08	−814.0	−690.0	156.9	138.9
H_2S	gas	188	212	34.08	−20.6	−33.6	205.8	34.2
SF_6	gas	sublimes	209	146.05	−1209	−1105.3	291.8	97.3
Tin								
Sn(α)	solid	505	2543	118.69	0	0	51.6	27.0
Sn	gas			118.69	+302.1	+267.3	168.5	20.3
SnO	solid	1353		134.69	−285.8	−256.9	56.5	44.3
SnO_2	solid	1903	2123	150.69	−580.7	+519.6	52.3	52.6
$SnCl_2$	solid	519	925	189.6	−325.1			
$SnCl_4$	liquid	240	387	260.5	−511.3	−440.2	258.6	
Xenon								
Xe	gas	161	165	131.30	0	0	169.7	20.8
Zinc								
Zn	solid	693	1180	65.37	0	0	41.6	25.4
Zn	gas			65.37	+130.7	+95.1	161.0	20.8
ZnO	solid	2248		81.37	−348.3	−318.3	43.6	40.3

Sources

Linstrom, P.J. and Mallard, W.G. (Eds.) (June 2005). *NIST Chemistry WebBook*, NIST Standard Reference Database Number 69. National Institute of Standards and Technology, Gaithersburg, Maryland, 20899. *http://webbook.nist.gov*

The Royal Society of Chemistry Data Book website. *http://www.chemsoc.org/networks/learnnet/data/index_databases.htm*

International Council for Science: Committee on Data for Science and Technology, CODATA key values for thermodynamics. *http://www.codata.org/resources/databases/key1.html*

Pedley, J.B., Naylor, B.R., and Kirby, S.P. (1977). *Thermochemical data of organic compounds.* Chapman and Hall, London.

Lide, D.R. (Ed.) (2002). *CRC handbook of chemistry and physics*, 83rd edn. CRC Press, Boca Raton, Florida.

Appendix 8

Ionic, atomic, and van der Waals radii for selected elements

Name	Cationic radius (r_+)/pm (coordination number in brackets)	Anionic radius (r_-)/pm (coordination number in brackets)	Atomic radius/pm	van der Waals radius/pm
Aluminium	Al^{3+} 39 (4), 54 (6)		143	184
Antimony	Sb^{3+} 76 (6)		141	206
Argon				188
Arsenic	As^{3+} 58 (6)	As^{3-} 220 (6)	121	185
Barium	Ba^{2+} 135 (6), 142 (8)		224	268
Beryllium	Be^{2+} 27 (4), 45 (6)		112	153
Bismuth	Bi^{3+} 103 (6), 117 (8)		182	207
Boron			88	192
Bromine		Br^- 196 (6)	114	185
Caesium	Cs^+ 167 (6), 174 (8)		272	343
Calcium	Ca^{2+} 100 (6), 112 (8)		197	231
Carbon			77	170
Chlorine		Cl^- 181 (6)	99	175
Chromium	Cr^{3+} 62 (6) Cr^{2+} 73 (6)		129	206
Cobalt	Co^{3+} 55 (6) Co^{2+} (LS) 56 (4), 65 (6) Co^{2+} (HS) 75 (6)		125	200
Copper	Cu^{2+} 57 (4), 73 (6) Cu^+ 60 (4), 77 (6)		128	196
Fluorine		F^- 131 (4), 133 (6)	64	147
Gallium	Ga^{3+} 47 (4), 62 (6)		153	187
Germanium			122	211
Helium				140

Name	Cationic radius (r_+)/pm (coordination number in brackets)	Anionic radius (r_-)/pm (coordination number in brackets)	Atomic radius/pm	van der Waals radius/pm
Hydrogen		H^- 208 (6)	37	110
Indium	In^{3+} 62 (4), 80 (6)		167	193
Iodine		I^- 220 (6)	133	198
Iron	Fe^{3+} 49 (4), 55 (6) Fe^{2+} (LS) 63 (4), 61 (6) Fe^{2+} (HS) 78 (6)		126	204
Krypton				202
Lead	Pb^{4+} 65 (4), 78 (6), 94 (8) Pb^{2+} 119 (6), 129 (8)		175	202
Lithium	Li^+ 59 (4), 76 (6), 92 (8)		157	182
Magnesium	Mg^{2+} 57 (4), 72 (6), 89 (8)		160	173
Manganese	Mn^{4+} 39 (4), 53 (6) Mn^{3+} 58 (6) Mn^{2+} 83 (6)		137	205
Neon				154
Nickel	Ni^{2+} 49 (4), 69 (6)		125	197
Nitrogen		N^{3-} 171 (6)	74	155
Oxygen		O^{2-} 138 (4), 140 (6), 142 (8)	66	152
Phosphorus		P^{3-} 190 (6)	110	180
Potassium	K^+ 137 (4), 138 (6), 151 (8)		235	275
Rubidium	Rb^+ 152 (6), 161 (8)		250	303
Scandium	Sc^{3+} 75 (6), 87 (8)		164	215
Selenium		Se^{2-} 198 (6)	117	190
Silicon			118	210
Sodium	Na^+ 99 (4), 102 (6), 118 (8)		191	227
Strontium	Sr^{2+} 118 (6), 126 (8)		215	249
Sulfur		S^{2-} 184 (6)	104	180
Tellurium	Te^{4+} 66 (4), 97 (6)	Te^{2-} 221 (6)	137	206
Thallium	Tl^{3+} 75 (4), 89 (6), 98 (8) Tl^+ 150 (6), 159 (8)		171	196
Tin	Sn^{4+} 55 (4), 69 (6), 81 (8)		158	217
Titanium	Ti^{4+} 42 (4), 61 (6), 74 (8) Ti^{3+} 67 (6) Ti^{2+} 86 (6)		147	211
Vanadium	V^{3+} 64 (6) V^{2+} 79 (6)		135	207
Xenon				216
Zinc	Zn^{2+} 60 (4), 74 (6), 90 (8)		137	201

Sources

Ionic and van der Waals radii: Haynes, W.M. (ed.), (2015–16). *CRC handbook of chemistry and physics*, 96th edn. CRC Press, Boca Raton, Florida.

Atomic radii: Wells, A.F. (1984). *Structural inorganic chemistry*, 5th edn. Clarendon Press, Oxford.

Appendix 9

Acidity constants

Values of pK_a quoted at 298 K in water except when otherwise noted.

Acid	Formula	pK_a*
Hydriodic acid	HI	–10
Perchloric acid	$HClO_4$	–10
Hydrobromic acid	HBr	–9
Hydrochloric acid	HCl	–7
Sulfuric acid	H_2SO_4	–3
Nitric acid	HNO_3	–1.4
Trichloroethanoic acid	CCl_3CO_2H	0.66†
Iodic acid	HIO_3	0.78
Oxalic acid	$(CO_2H)_2$	1.25
Phosphonic acid (phosphorous acid)	H_3PO_3	1.3†
Dichloroethanoic acid	Cl_2CHCO_2H	1.35
Sulfurous acid	H_2SO_3	1.85
Chlorous acid	$HClO_2$	1.94
Hydrogensulfate ion	HSO_4^-	1.99
Phosphoric acid	H_3PO_4	2.16
Chloroethanoic acid	$ClCH_2CO_2H$	2.87
Bromoethanoic acid	$BrCH_2CO_2H$	2.90
Hydrofluoric acid	HF	3.20
Nitrous acid	HNO_2	3.25
Methanoic acid	HCO_2H	3.75
Hydrogenoxalate ion	$HO_2CCO_2^-$	3.81
Benzoic acid	$C_6H_5CO_2H$	4.20
Ethanoic acid	CH_3CO_2H	4.76
Phenylammonium ion	$PhNH_3^+$	4.87
Propanoic acid	$CH_3CH_2CO_2H$	4.87
Pyridinium ion	$C_5H_5NH^+$	5.23

Acid	Formula	pK_a*
Carbonic acid	H_2CO_3	6.35
Hydrogen sulfide	H_2S	7.05
Hydrogensulfite ion	HSO_3^-	7.2
Dihydrogenphosphate ion	$H_2PO_4^-$	7.21
Hypochlorous acid	HClO (or HOCl)	7.40
Hydrazinium ion	$NH_2NH_3^+$	8.1
Hypobromous acid	HBrO (or HOBr)	8.55
Pentane-2,4-dione	$MeCOCH_2COMe$	9.0
Hydrocyanic acid	HCN	9.21
Ammonium ion	NH_4^+	9.25
Boric acid	H_3BO_3 (or $B(OH)_3$)	9.27†
Trimethylammonium ion	Me_3NH^+	9.80
Silicic acid	H_4SiO_4	9.9‡
Phenol	C_6H_5OH	9.99
Hydrogencarbonate ion	HCO_3^-	10.33
Ethylammonium ion	$EtNH_3^+$	10.65
Methylammonium ion	$MeNH_3^+$	10.66
Triethylammonium ion	Et_3NH^+	10.75
Hydrogen peroxide	H_2O_2	11.62
Hydrogenphosphate ion	HPO_4^{2-}	12.32
Water	H_2O	14.00
Methanol	MeOH	15.5
Hydrogensulfide ion	HS^-	19
Propan-2-one	MeCOMe	20
Ethyne	C_2H_2	25
Hydrogen	H_2	35
Ammonia	NH_3	38
Benzene	C_6H_6	43
Ethene	C_2H_4	44
Ethane	C_2H_6	50

* Values below −2 and above 18 are approximations.
† 293 K.
‡ 303 K.

Sources

Haynes, W.M. (ed.) (2015–16). *CRC handbook of chemistry and physics*, 96th edn. CRC Press, Boca Raton, Florida.

Smith, M.B. and March, J. (2007). *March's advanced organic chemistry: reactions, mechanisms, and structure*, 6th edn. Wiley-Interscience, New York.

Appendix 10

Bond lengths and bond enthalpies

Bond lengths and bond dissociation enthalpies *D*(A-B) for diatomic molecules

Bond	Bond length / pm	*D*(A–B) / kJ mol^{-1}
H–H	74.1	+435.8
F–F	141.2	+158.7
Cl–Cl	198.8	+242.4
Br–Br	228.1	+193.9
I–I	266.6	+152.3
H–F	91.7	+569.7
H–Cl	127.5	+431.4
H–Br	141.5	+366.2
H–I	160.9	+298.3
Li–Li	267.3	+105.0
Na–Na	307.9	+74.8
K–K	390.5	+57.0
O=O	120.7	+498.5
S=S	188.9	+430.0
N≡N	109.8	+944.9
P≡P	189.3	+489.1
C≡O	112.8	+1076.6
N≡O	115.1	+630.6

Bond lengths, bond angles, and mean bond enthalpies for polyatomic molecules

Bond	Compound	Bond length / pm	Bond angle /	Mean bond enthalpy / kJ mol^{-1}
H–C	CH_4	108.7	109.5	+412
H–Si	SiH_4	148.0	109.5	+318
H–Ge	GeH_4	152.5	109.5	+285
H–Sn	SnH_4	171.1	109.5	+251
H–N	NH_3	101.2	106.7	+391
H–P	PH_3	142.0	93.3	+321
H–As	AsH_3	151.1	92.1	+297
H–Sb	SbH_3	170.4	91.6	+257
H–O	H_2O	95.8	104.5	+464
H–S	H_2S	133.6	92.1	+364
H–Se	H_2Se	146.0	90.9	+313
H–Te	H_2Te	169.0	90.0	+238
N–N	N_2H_4	144.9		+158
O–O	H_2O_2	147.5		+144
O–O	O_3	127.2	117.5	+302
S–S	S_8	207	105	+266
O=S	SO_2	143.1	119.5	+552
Si–Si	Si_2H_6	233.1		+226
P–P	P_4	221		+198
C–F	CF_4	132.3	109.5	+485
C–Cl	CCl_4	176.7	109.5	+327
C–Br	CBr_4	193.5	109.5	+285
C–I	Cl_4	215	109.5	+234
C=O	CO_2	116.0	180	+805

Mean bond lengths and bond enthalpies

Mean bond enthalpies are given for A–H bonds in Table 27.2 (p.1207), and for A–F and A–Cl bonds in Table 27.3 (p.1208), where A is a Group 14 to Group 17 element. Oxidation states, coordination numbers, and substituents all affect bond lengths and mean bond enthalpies, so the experimentally determined values for a particular compound may differ from those given below. For more information on how oxidation states affect the mean bond enthalpies in *p*-block halides see Section 27.6 (p.1244).

Bond	Mean bond length / pm	Mean bond enthalpy / kJ mol^{-1}
C–C	153	+347
C=C	134	+612
C≡C	120	+838
C–H	109	+412
C–F	139	+467
C–Cl	179	+346
C–Br	194	+290
C–I	213	+228
C–N	146	+286
C=N	121	+615
C≡N	116	+887
C–O	142	+358
C=O	121	+742
C–B	160	+395
P–O	166	+380
P=O	149	+510
P–Cl	201	+326
P–Br	237	+270
S–O	158	+365
S=O	144	+520
S–Cl	207	+271
B–O	147	+515

Sources

Haynes, W.M. (ed.) (2015–16). *CRC handbook of chemistry and physics*, 96th edn. CRC Press, Boca Raton, Florida.

The Royal Society of Chemistry Data Book website. *http://www.chemsoc.org/networks/learnnet/data/index_databases.htm*

Appendix 11

The electromagnetic spectrum

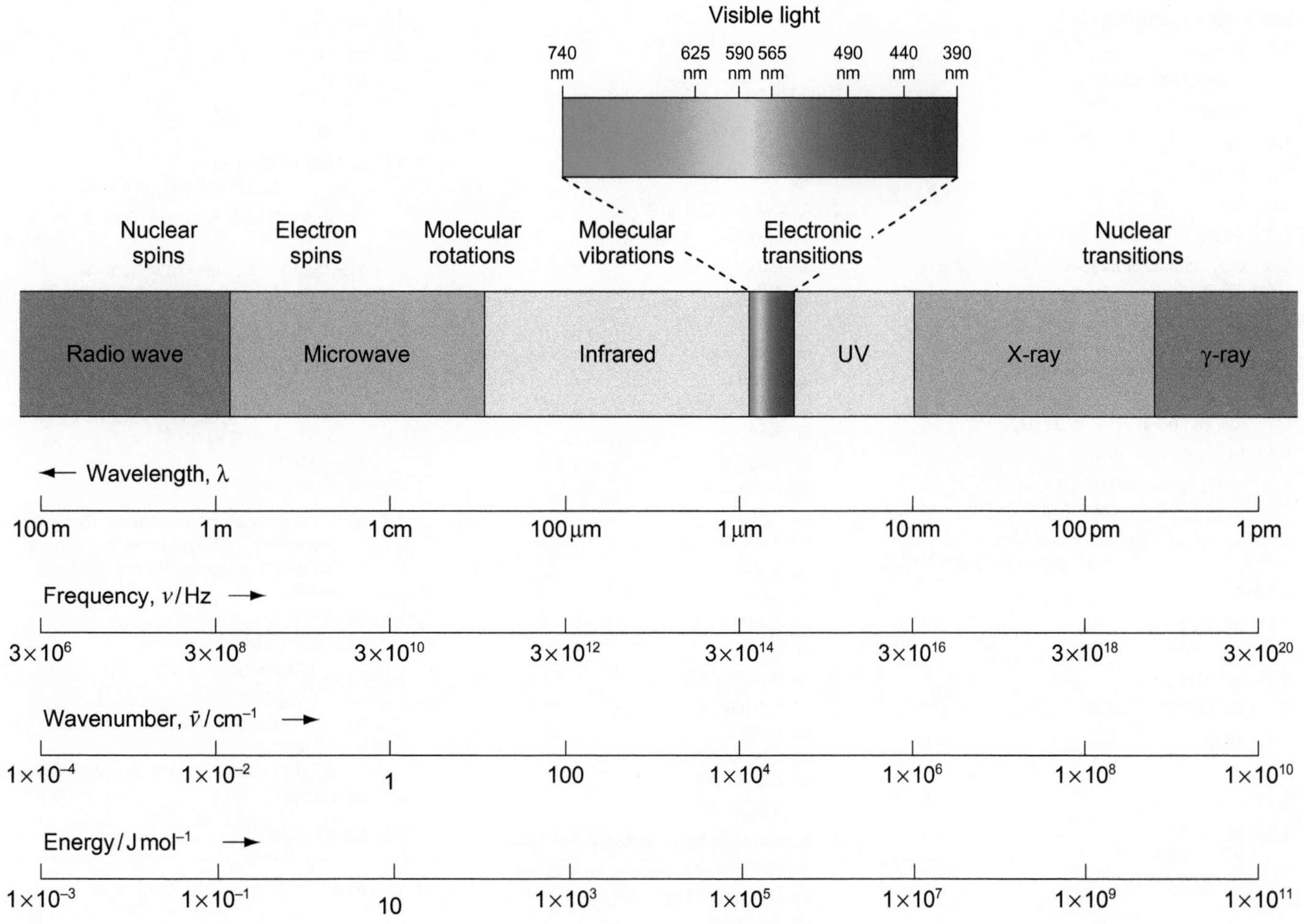

Visible light
740 nm
625 nm
590 nm
565 nm
490 nm
440 nm
390 nm
Nuclear spins
Electron spins
Molecular rotations
Molecular vibrations
Electronic transitions
Nuclear transitions
Radio wave
Microwave
Infrared
UV
X-ray
γ-ray
Wavelength, λ
100 m
1 m
1 cm
100 μm
1 μm
10 nm
100 pm
1 pm
Frequency, ν/Hz
3×10^6
3×10^8
3×10^10
3×10^12
3×10^14
3×10^16
3×10^18
3×10^20
Wavenumber, ν̄/cm^−1
1×10^−4
1×10^−2
1
100
1×10^4
1×10^6
1×10^8
1×10^10
Energy/J mol^−1
1×10^−3
1×10^−1
10
1×10^3
1×10^5
1×10^7
1×10^9
1×10^11

Answers

Chapter 1 Fundamentals

Worked examples

1.1 $kg\,mol^{-1}$

1.2 Thorium (Th), 144

1.3 79.99

1.4 340 g

1.5 $HClO_3$

1.6 52 g

1.7 ~25%

1.8 $CO_3^{2-}(aq) + 2H^+(aq) \rightarrow CO_2(g) + H_2O(l)$

1.9 (a) Selenium tetrafluoride
(b) Iron(III) perchlorate (or iron(III) chlorate(VII)). Using IUPAC nomenclature, it is iron(III) tetraoxidochlorate(1−)

1.10 Na is oxidized (0 in 2 Na to +1 in 2 NaOH). H is reduced (+1 in $2H_2O$ to 0 in H_2).

1.11 $Fe^{2+}(aq) \rightarrow Fe^{3+}(aq) + e^-$ oxidation
$MnO_4^-(aq) + 8H^+(aq) + 5e^- \rightarrow Mn^{2+}(aq) + 4H_2O(l)$ reduction
$MnO_4^-(aq) + 8H^+(aq) + 5Fe^{2+}(aq) \rightarrow 5Fe^{3+}(aq) + Mn^{2+}(aq) + 4H_2O(l)$

1.12 $0.137\ mol\,dm^{-3}$

1.13 (a) $60.1\ cm^3$
(b) 4.65 g

1.14 (a) $7.89 \times 10^{-4}\ mol\,dm^{-3}$
(b) $0.0585\ g\,dm^{-3}$

1.15 16.6%

1.16 −24 600 kJ

1.17 $1.64\ mol\,dm^{-3}$

1.18 (a) 23.7 atm
(b) (i) 98.5 atm
(ii) 24%

Boxes

1.2 ^{24}Mg has 12 neutrons, ^{25}Mg has 13 neutrons, ^{26}Mg has 14 neutrons. All have 12 protons.

1.3 ^{16}O has 8 neutrons, ^{17}O has 9 neutrons, ^{18}O has 10 neutrons. All three isotopes have 8 protons and 8 electrons.

1.4 72.8%

1.5 (a) 27 ppm, 0.0027%
(b) $6.7 \times 10^{-4}\ mol\,dm^{-3}$

1.6 (a) $6.51\ mg\,dm^{-3}$

1.7 (a) 22.3 kJ
(b) 5.81×10^{-21} J

1.9 (a) 695 kJ

1.10 (c) (i) $+51.1\ kJ\,mol^{-1}$
(ii) $-4.5\ kJ\,mol^{-1}$
(iii) $-6.01\ kJ\,mol^{-1}$

Questions

1 (a) 1.54×10^{-10} m
(b) 154 pm
(c) 1.54 Å

2 $T_m = 54.8$ K, $T_b = 90.2$ K

3 (a) $10000\ cm^3$ ($1 \times 10^4\ cm^3$)
(b) $0.01\ m^3$
(c) 10 L

4 $m\,s^{-1}$

5 (a) 0.905 mol
(b) 1.22 mol
(c) 6.40×10^{-4} mol
(d) 4.99 mol

6 (a) 2.03 mol
(b) 0.609 mol
(c) 6.48 mol
(d) 1.4×10^6 mol

7 (a) 434 g
(b) 16.9 g
(c) 0.131 g
(d) 5.27×10^{-5} g

8 (a) $C_4H_6O_4$ $118.1\ g\,mol^{-1}$
(b) $C_2H_3O_2$
(c) 40.7%
(d) 1.06×10^{-3} mol
(e) 6.38×10^{20}
(f) 2.55×10^{21}

9 Sodium azide NaN_3; sodium nitride Na_3N

10 (a) $NH_3(g) + HNO_3(aq) \rightarrow NH_4NO_3(aq)$
(b) $NH_3(g) + H^+(aq) \rightarrow NH_4^+(aq)$
(c) Acid–base
(d) $4NH_3 + 5O_2 \rightarrow 4NO + 6H_2O$
(e) NO: 1.76 kg, O_2: 2.35 kg
(f) 82%

11 (a) 37%
(b) 10.2 g

12 (a) $Fe(s) + CuSO_4(aq) \rightarrow FeSO_4(aq) + Cu(s)$
(b) $Fe(s) + Cu^{2+}(aq) \rightarrow Fe^{2+}(aq) + Cu(s)$
(c) Fe is oxidized (0 to +2).
Cu is reduced (+2 to 0).

13 (a) (i) carbon disulfide; (ii) dichlorine heptoxide; (iii) xenon hexafluoride; (iv) ammonium sulfate; (v) chromium(III) chloride; (vi) potassium periodate
(b) (i) Na_2SO_3; (ii) $BaCO_3$; (iii) $FeCl_2$; (iv) $Na_2S_2O_3$; (v) I_2O_5; (vi) N_2O

14 (a) $-\frac{1}{2}$
(b) −1
(c) −1
(d) −1
(e) +2

15* $Cr^{3+}(aq) + 8OH^-(aq) \rightarrow CrO_4^{2-}(aq) + 4H_2O(l) + 3e^-$ oxidation
$O_2^{2-}(aq) + 2H_2O(l) + 2e^- \rightarrow 4OH^-(aq)$ reduction
$2Cr^{3+}(aq) + 4OH^-(aq) + 3O_2^{2-}(aq) \rightarrow 2CrO_4^{2-}(aq) + 2H_2O(l)$

16* (a) 0.0294 mg
(b) $3.71 \times 10^{-7}\ mol\,dm^{-3}$

17* 0.0135 g

18 (a) $C_7H_{16}(l) + 11O_2(g) \rightarrow 7CO_2(g) + 8H_2O(l)$
$\Delta_c H^{\ominus} = -4817\ kJ\,mol^{-1}$
(b) −2408.5 kJ
(c) 2.08 kg

19* (a) London dispersion interactions; dipole–dipole interactions; hydrogen bonding. Hydrogen bonding is responsible for the strongest interactions.
(c) 3×10^{16} J
(d) 175 days (about 6 months)

20* (a) $K_c = \dfrac{[NO]^2[O_2]}{[NO_2]^2}$
(b) (i) The position of equilibrium moves to the right.
(ii) The position of equilibrium moves to the left.
(c) $0.0094\ mol\,dm^{-3}$

21 (a) $K_p = \dfrac{p_{CH_3CH}}{p_{CO}\,p_{H_2}^{\ 2}}$
(b) $8.8 \times 10^{-3}\ atm^{-2}$

Chapter 2 The language of organic chemistry

Worked examples

2.1 4-isopropyl-2-methylheptane

2.4 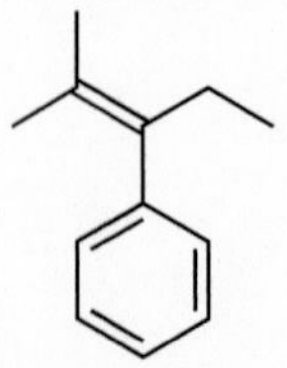

2.5 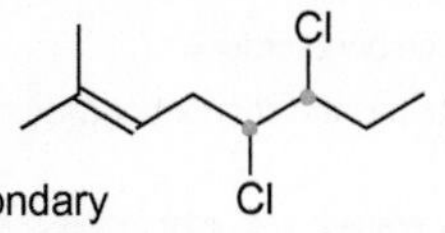
Cl
Cl

• = secondary carbon atom

2.6 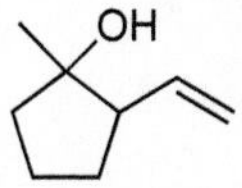
OH
Tertiary alcohol

2.7 *N-tert*-butyl-3-ethylhexan-2-amine (A secondary amine)

2.8 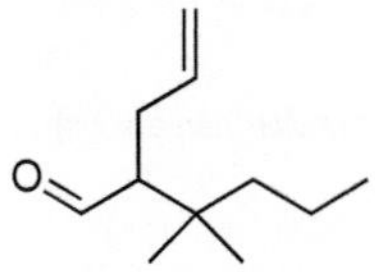
O

2.9 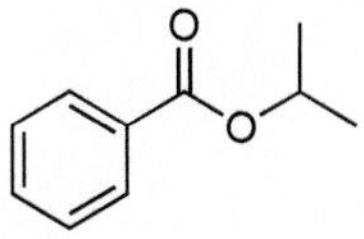
O
O

Questions

5 (a) 1-Bromo-4-methylpentane
(b) Prop-2-en-1-ol
(c) Methyl-3-aminopropanoate

6 (a) *Meta*-disubstituted benzene
(b) **1** 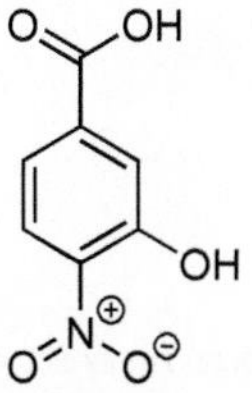
O
OH
OH
O
N
O
3 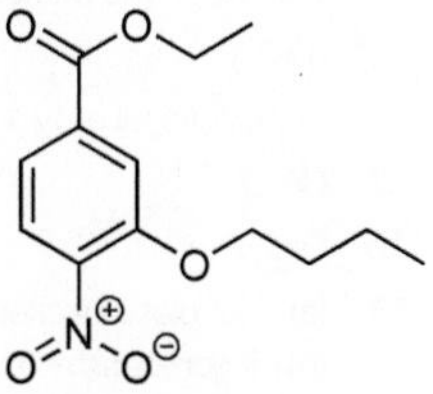
O
O
O
O
N
O
(c) **2**: Ethyl 3-hydroxy-4-nitrobenzoate,
4: 3-Butoxy-4-nitrobenzoyl chloride
(d) **5** 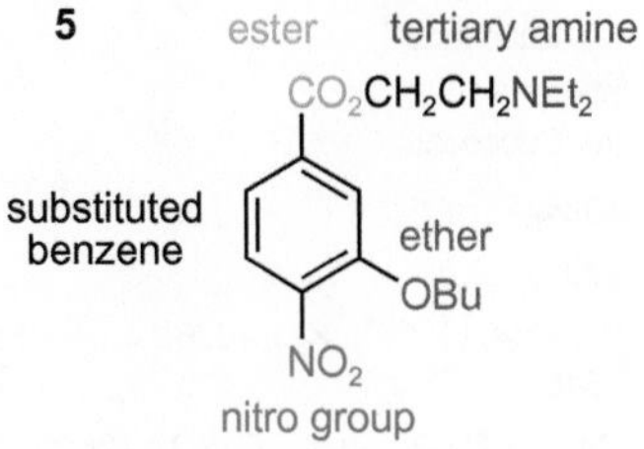
ester
tertiary amine
CO2CH2CH2NEt2
substituted benzene
ether
OBu
NO2
nitro group
(e) 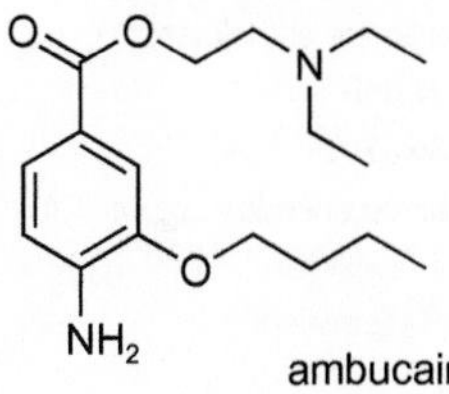
O
O
N
O
NH2
ambucaine

7 (a) 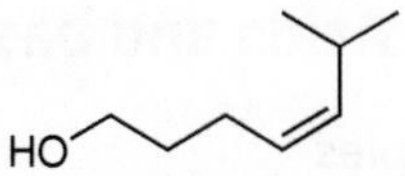
HO
(b) Primary carbon atom
(c) Diethyl propanedioate
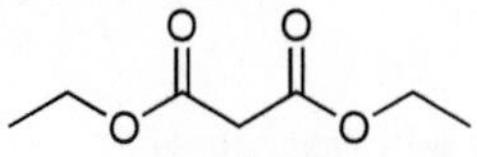
O
O
O
O

(d) **8** 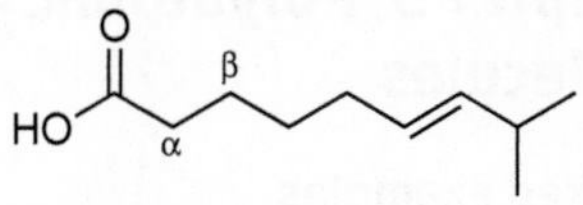
O
β
HO
α
(e) 4-(Aminomethyl)-2-methoxyphenol
(f) 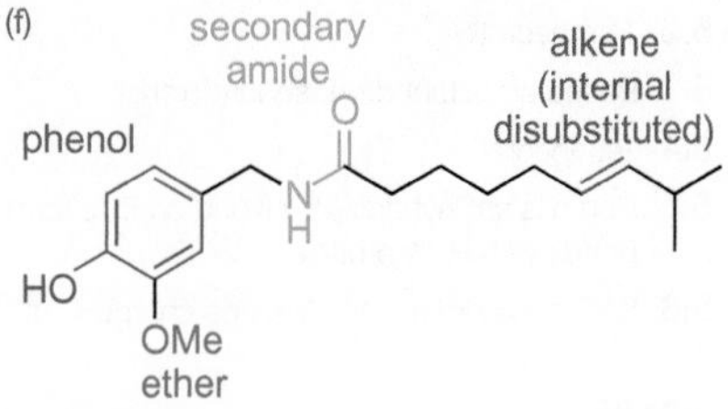
secondary amide
alkene (internal disubstituted)
phenol
O
N
H
HO
OMe
ether

8* (a) Primary
(b) 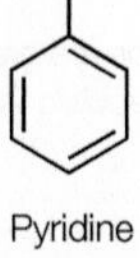
(c) Pyridine
(d) 2-(Chloromethyl)-4-methoxy-3,5-dimethylpyridine
(e) Thiol
(f) The positions of the two substituents on the benzene ring—it is a 1,3-disubstituted benzene
(g) 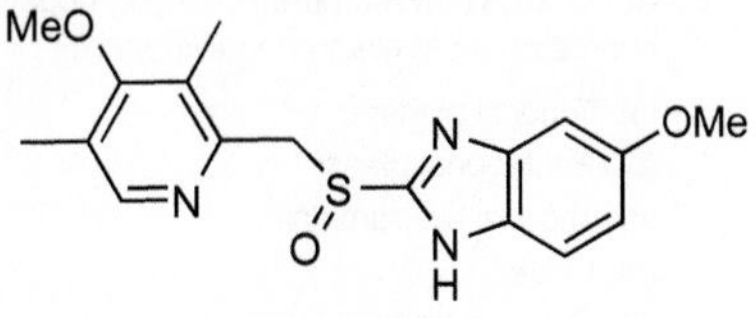
MeO
N
S
O
N
N
H
OMe
omeprazole

Chapter 3 Atomic structure and properties

Worked examples

3.1 330 m
3.2 3.63×10^{-4} J mol^{-1}
3.3 44 kJ mol^{-1}
3.4 First = 7.40×10^{13} Hz, second = 1.14×10^{14} Hz, third = 1.38×10^{14} Hz
3.5 328 kJ mol^{-1}
3.6 1.320×10^{-14} m
3.7 ≥57.9 m
3.8 6*d* orbital
3.9 6 radial nodes
3.10 $1s^2\,2s^2\,2p^6\,3s^2\,3p^6\,3d^{10}\,4s^2\,4p^6\,4d^{10}\,5s^2\,5p^4$, [Kr] $4d^{10}\,5s^2\,5p^4$
3.11 (a) 6.1
(b) 7.6
3.12 Two α-emissions and three β-emissions

Boxes

3.3 Lines at shorter wavelengths (blue-shifted)
3.7 6*K*
3.8 1.282×10^{-15} J
3.9 Year 3430
3.10 ${}^{241}_{95}\mathrm{Am} \rightarrow {}^{237}_{93}\mathrm{Np} + {}^{4}_{2}\mathrm{He}$

Questions

1 1.20×10^6 kJ mol^{-1}
2 6.06×10^{14} s^{-1} (Hz), violet end of the visible region
3 486 nm, visible
4 First line = 121 nm, second line = 103 nm, third line = 97 nm
5 98.9 pm
6 (a) Not allowed
(b) Allowed, 5*f* orbital
(c) Not allowed
(d) Allowed, 2*p* orbital
(e) Not allowed
7* 25 orbitals in total; one 5*s* orbital, three 5*p* orbitals, five 5*d* orbitals, seven 5*f* orbitals, nine 5*g* orbitals
8 $n = 3, l = 2$
9 (a) $n = 2, l = 0, m_l = 0$
(b) $n = 5, l = 3$, the possible values for m_l are +3, +2, +1, 0, −1, −2, and −3
(c) $n = 6, l = 1$, the possible values for m_l are +1, 0, and −1
12 (a) Excited state
(b) Ground state
(c) Excited state
(d) Impossible
13 (a) [Ar] $3d^{10}\,4s^2\,4p^3$ (3 unpaired electrons)
(b) [Ar] $3d^7\,4s^2$ (2 unpaired electrons)
(c) [Xe] $4f^{11}\,6s^2$ (3 unpaired electrons)
(d) [Kr] (0 unpaired electrons)
14* (a) Calcium
(b) Chlorine
(c) Rhodium
15 d^5, 10*K*; f^7, 21*K*
16 Phosphorus $Z_{eff} = 4.8$; arsenic $Z_{eff} = 6.3$
17 (a) Magnesium
(b) Magnesium
(c) Magnesium
18* (a) Due to the greater value for Z_{eff} for fluorine
(b) Due to the larger, more diffuse orbitals for phosphorus
(c) Removal of a paired electron for sulfur (p^4) compared to an unpaired electron for phosphorus (p^3) leads to a greater reduction in electron–electron repulsion. Removal of the paired electron leads to a reduction in exchange energy.
19 Bismuth-210, polonium-210
20 (a) X is ^{243}Bk
(b) Y is ^{271}Ds
(c) Z is ^{263}Sg

Chapter 4 Diatomic molecules

Worked examples

4.1 (a) Triple bond
(b) Single bond
4.2 Cl^- I^+
4.3 *u*
4.4 Yes (bond order 0.5)
4.5 Paramagnetic: O_2^+ and O_2^-
4.6 $\psi_{\text{in phase}} = N[\lambda\phi_{Li}(2s) + \phi_F(2p_z)]$, where $\lambda < 1$
4.7 CN^-

Boxes

4.1 X-rays have similar wavelengths to the distances between atoms in crystalline solids.

4.2 Paramagnetic: Na.

4.3 The electrons are closer together when they are in the same 1*s* orbital in H^-.

4.4 Although strong N≡N bonds and strong H–H bonds are broken, strong N–H bonds are formed. The net enthalpy change is negative.

4.7 Hydrogen molecules are light enough to escape the Earth's gravity.

4.8 $1510\ \text{kJ mol}^{-1}$

4.9 The red photons emitted from N_2 have a higher wavelength than the blue and purple photons from N_2^+, so their energy is lower.

4.10 $\psi = N[\phi_{Li}(2s) + \lambda\ \phi_{Na}(3s)]$

4.11 Formation of NO from N_2 and O_2 requires a high temperature, such as that in lightning.

Questions

1 | | |
|---|---|
| Cl_2 | $:\ddot{\underset{..}{\text{Cl}}}-\ddot{\underset{..}{\text{Cl}}}:$, bond order 1 |
| Se_2 | $\vdots\text{Se}=\text{Se}\vdots$, bond order 2 |
| HBr | $\text{H}-\ddot{\underset{..}{\text{Br}}}:$, bond order 1 |
| ClO^- | $:\ddot{\underset{..}{\text{O}}}-\ddot{\underset{..}{\text{Cl}}}:$, bond order 1 |

2 $\chi^P(\text{I}) = 2.41$

3 $D_{(\text{Cl–Br})} = 222\ \text{kJ mol}^{-1}$

4 $\chi^M(\text{F}) = 1005$, $\chi^M(\text{Cl}) = 800$, $\chi^M(\text{Br}) = 733$, $\chi^M(\text{I}) = 652$. χ^M decreases down the group in a similar manner to χ^P.

5 $\text{Cl}-\text{F} \leftrightarrow \text{Cl}^+\ \text{F}^- \leftrightarrow \text{Cl}^-\ \text{F}^+$. The ionic form $\text{Cl}^+\ \text{F}^-$ is more important than $\text{Cl}^-\ \text{F}^+$ as fluorine is more electronegative than chlorine.

6 Carbon and nitrogen are *sp* hybridized. One of these hybrid orbitals from each of the atoms interact to form a σ bond. The other points at 180° from the other atom and carries a lone pair. The unhybridized p_x and p_y orbitals on the two atoms interact to form π bonds (giving a bond order of 3).

7* $1\sigma_g \rightarrow 1\sigma_u$. The excited form of H_2 has a bond order of 0.

8 Bond orders: $Li_2 = 1$ [i.e. 1 – 0]; $Li_2^+ = \frac{1}{2}$ [i.e. $\frac{1}{2} - 0$]; $Li_2^- = \frac{1}{2}$ [i.e. $1 - \frac{1}{2}$]. Li_2^+ and Li_2^- are paramagnetic.

9 $\psi_{\text{in phase}} = 0.5^{1/2} \times [\phi_{2s}(\text{Li}_A) + \phi_{2s}(\text{Li}_B)]$; corresponds to the $2\sigma_g$ bonding orbital.

$\psi_{\text{out of phase}} = 0.5^{1/2} \times [\phi_{2s}(\text{Li}_A) - \phi_{2s}(\text{Li}_B)]$; corresponds to the $2\sigma_u{}^*$ antibonding orbital.

10 Peak at *m/z* 9 = Be^+, peak at *m/z* 18 = Be_2^+

12 Would not be expected to exist.

13 (a) Not isoelectronic
(b) Isoelectronic
(c) Isoelectronic

15 N_2, CO

17* $\lambda < 1$, bond order in both OH and OH^- is 1.

18 Bond order for CN^- is 3. CN would have a longer bond and be paramagnetic.

19 (a) π_u
(b) σ_g
(c) σ, B is more electronegative
(d) π*, B is more electronegative

Chapter 5 Polyatomic molecules

Worked examples

5.1 FOF

5.2 Tetrahedral

5.3 Based on octahedron, square planar

5.4 Tetrahedral

5.5 Boron is sp^2 hybridized. Hybrid orbitals form σ bonds with H 1*s* orbitals.

5.6 S–O bond order = $1\frac{1}{3}$ Average charge = $-\frac{2}{3}$

Boxes

5.2 36.4% oxygen by mass; greater than the percentage in air

5.3 Lone pair–bonding pair repulsions are greater than the bonding pair–bonding pair repulsions.

5.4 Bent

5.5 Gashes stress the figs, inducing them to release ethene. Ethene speeds up ripening.

Questions

1 BF_3 is electron poor; SiH_4 obeys the octet rule; CS_2 obeys the octet rule; PCl_3 obeys the octet rule; PF_5 is hypervalent; KrF_2 is hypervalent

2 Structure A (lowest formal charges, negative charge on most electronegative atom.)

3 (a) Trigonal planar
(b) Pentagonal planar
(c) Trigonal bipyramidal
(d) Linear
(e) Square planar
(f) Octahedral

4 (a) Bent
(b) Trigonal planar
(c) Distorted tetrahedral
(d) Bent
(e) Square pyramidal
(f) Bent around each nitrogen atom
(g) Disphenoidal

5 (a) Trigonal pyramidal
(b) Tetrahedral
(c) Bent
(d) Distorted tetrahedral

6 Bond angle: $NO_2^- < NO_2 < NO_2^+$

7 None

8 The left and middle molecules are polar.

9* PF_5 is non-polar. PF_3 and PCl_3 are polar with the phosphorus δ+ and the halogens δ–. Fluorine is more electronegative than chlorine.

10 (a) Methyl carbon is sp^3 hybridized, triply bonded carbon and nitrogen are *sp* hybridized.
(b) Methyl carbons are sp^3 hybridized, doubly bonded carbon and oxygen atoms are sp^2 hybridized.
(c) sp^2 hybridized

11 The ring carbons are sp^2 hybridized. The methyl carbons are sp^3 hybridized. The oxygen atoms are sp^2 hybridized. The nitrogen atoms bonded to the methyl group are sp^3 hybridized. The other nitrogen atom is sp^2 hybridized.

12* H_2S is a weaker base than H_2O (lone pair is s^2).

13 Average charge = $-\frac{1}{4}$

14 Average bond order = $1\frac{1}{2}$.

15 Sulfate has the shorter bonds. (bond order $1\frac{1}{2}$, vs. $1\frac{1}{3}$)

16 The average bond order for each Xe–F bond is $\frac{1}{2}$, which is lower than the value in XeF^+. (bond order 1)

17* Bond order: $\frac{1}{2}$

Chapter 6 Solids

Worked examples

6.1 Yes

6.4 ReO_3

6.5 0.155 (cations much smaller than anions)

6.6 Yes

6.7 $+2253\ \text{kJ mol}^{-1}$

6.8 $+636.4\ \text{kJ mol}^{-1}$

6.9 $+4070\ \text{kJ mol}^{-1}$

Boxes

6.3 $AgSbTe_2$

6.6 $$\Delta U_4 = 6 \times \frac{+(1\times 1)e^2}{4\pi\varepsilon_o(2)r} = \frac{2}{6} \times \frac{e^2}{4\pi\varepsilon_o r}$$

Questions

1 Boron has a covalent network structure; white phosphorus has a molecular structure; lead is a metal.

2 2 atoms in the unit cell

4* $8.82\ \text{g cm}^{-3}$

5 Substitutional alloy

7 Cu_2O

10 Fe_3O_4

11 (a) Not close packed
(b) Trigonal prismatic (Mo) and trigonal pyramidal (S)
(c) Weak interactions between A and B layers

12 (a) Fails (but close to borderline)
(b) Succeeds
(c) Succeeds
(d) Fails
(e) Succeeds

13 $+3844\ \text{kJ mol}^{-1}$

14 $+791.6\ \text{kJ mol}^{-1}$

15 KF: $+796\ \text{kJ mol}^{-1}$, KCl: $+676\ \text{kJ mol}^{-1}$, KBr: $+646\ \text{kJ mol}^{-1}$

16 $CaCl_2$: $+2300\ \text{kJ mol}^{-1}$, $CaSO_4$: $+2620\ \text{kJ mol}^{-1}$

17* Experimental lattice enthalpies:
$\Delta_{\text{latt}}H^{\ominus}(\text{LiCl}) = 859\ \text{kJ mol}^{-1}$;
$\Delta_{\text{latt}}H^{\ominus}(\text{AgCl}) = 915\ \text{kJ mol}^{-1}$;
Calculated lattice energies: $\Delta_{\text{latt}}U(\text{LiCl}) = 840\ \text{kJ mol}^{-1}$; $\Delta_{\text{latt}}U(\text{AgCl}) = 696\ \text{kJ mol}^{-1}$

18 KF: ionic, PbI_2: covalent

Chapter 7 Acids and bases

Worked examples

7.1 Conjugate acid pairs: $CH_3CO_2H/CH_3CO_2^-$ and NH_4^+/NH_3

7.2 pH 0.61

7.3 $0.023\ \text{mol dm}^{-3}$

7.4 F^- ions are a stronger base.

7.5 pH 4.35

7.6 pH 2.90

7.7 BI_3 (Lewis acid) + NMe_3 (Lewis base) → $I_3\overset{-}{B}-\overset{+}{N}Me_3$ (Adduct)

Boxes

7.2 98.1 g

7.3 4000 mol dm^{-3}

7.4 pH 4.2

7.6 Malic acid is a weak dibasic acid and sodium hydroxide is a strong base. The titration curve will be similar to Figure 7.12.

7.7 SbF_5 is trigonal bipyramidal, SbF_6^- is octahedral.

Questions

1

(a) $HCO_2H\,(aq)$ (acid) + $H_2O\,(l)$ (base) ⇌ $HCO_2^-(aq)$ (conjugate base) + $H_3O^+(aq)$ (conjugate acid)

(b) $EtNH_2\,(aq)$ (base) + $H_2O\,(l)$ (acid) ⇌ $EtNH_3^+(aq)$ (conjugate acid) + $OH^-(aq)$ (conjugate base)

(c) H_2SO_4 (acid) + $EtOH$ (base) ⇌ HSO_4^- (conjugate base) + $EtOH_2^+$ (conjugate acid)

2 (a) pH 2.25
(b) Basic (hydrolysis)

3 Basic

4 (a) 3.2×10^{-2} mol dm^{-3}
(b) 6.3×10^{-2} mol dm^{-3}

5 (a) $K_b = 1.3 \times 10^{-8}$ mol dm^{-3}, $pK_b = 7.9$
(b) $pK_a = 6.1$

6 Conjugate acid is $CH_3NH_3^+$. $pK_a = 10.66$

7* (a) $K_a = \dfrac{[H_3O^+(aq)][HPO_4^{2-}(aq)]}{[H_2PO_4^-(aq)]}$
(b) 1.6
(c) $[HPO_4^{2-}(aq)] = 1.2 \times 10^{-2}$ mol dm^{-3}, $[H_2PO_4^-(aq)] = 7.7 \times 10^{-3}$ mol dm^{-3}

8 (a) pH 1.0
(b) pH 1.2
(c) pH 1.6
(d) pH 7.0
(e) pH 12.3

9 (a) Phenolphthalein
(b) Bromophenol blue

10 (a) Between −3 and 0.8
(b) Between 1.8 and 3.3
(c) Between 1.8 and 3.3

12 $pK_{a1} = 3.1$, $pK_{a2} = 4.8$, $pK_{a3} = 6.4$.

13 (a) Lewis acid: I_2; Lewis base: I^-
(b) Lewis acid: HBr; Lewis base: NH_3
(c) Lewis acid: SiF_4; Lewis base: Pyridine
(d) Lewis acid: CO_2; Lewis base: OH^-

The reaction between NH_3 and HBr is also a Brønsted–Lowry acid–base reaction.

14 (a) Acts as a Lewis acid. Product is the adduct $O_2S \leftarrow NMe_3$
(b) Acts as a Lewis base. Product is the adduct $F_5Sb \leftarrow OSO$

Chapter 8 Gases

Worked examples

8.1 0.66 atm

8.2 0.304 m

8.3 0.38 m^3

8.4 2 atm

8.5 0.0224 m^3

8.6 0.3 atm

8.7 $c_{He} = 1340$ m s^{-1}, $c_{CO2} = 400$ m s^{-1}

8.8 O_2:CO_2:N_2 = 0.94:0.80:1.00

8.9 1.7×10^{10} s^{-1}

8.10 6.8×10^{-7} bar

8.11 % Reduction in V = 32.5%; % increase in p = 46.8%. The $a(n/V)^2$ correction factor makes the larger contribution.

Boxes

8.1 (a) 2.5 mol
(b) 1.6 mol
(c) 106 g

8.3 Make θ smaller

8.4 1.004

Questions

1 160 g mol^{-1}

2 Molar mass = 30.0 g mol^{-1}; ethane

3 286 cm^3

4 8.6×10^{12} molecules

5 0.129 g

6 (i) $p_{Ar} = 1.135$ bar $p_{N_2} = 0.545$ bar
(ii) 1.68 bar
(iii) 7.3 mol

7 0.260 g

8 (a) 1 bar
(b) $p_A = 1$ bar $p_B = 1.42$ bar
(c) 1.2 bar

9) 186 cm^3

10 $x_{N_2} = 0.495$, $p_{N_2} = 60200$ Pa, $p_{CO_2} = 61400$ Pa

11 3.3 times faster

12 1.16 times faster

13 Less, most molecules in air are heavier than water.

14 0.031 m^3

15 497 m s^{-1}

16 It will burst!

17 $Z = 3.9 \times 10^8$ s^{-1}, $\lambda = 0.98$ μm

18 2.2×10^7 Pa

20 Mass of O_2 = 926 g Mass of He = 328 g

22 (a) $p_{total} = p_{N_2} + p_{O_2} + p_{Ar} = 2$ atm
(b) 21%
(c) (i) 0.21 atm
(ii) 0.42 atm
(d) Twice as many molecules at 10 m compared to sea level.

23 (a) 125 mol
(b) 285 atm

24 (a) 1.37×10^{-7} m
(b) 4.60×10^9 m
(c) 2.0×10^7 s (about 230 days)

25 (a)

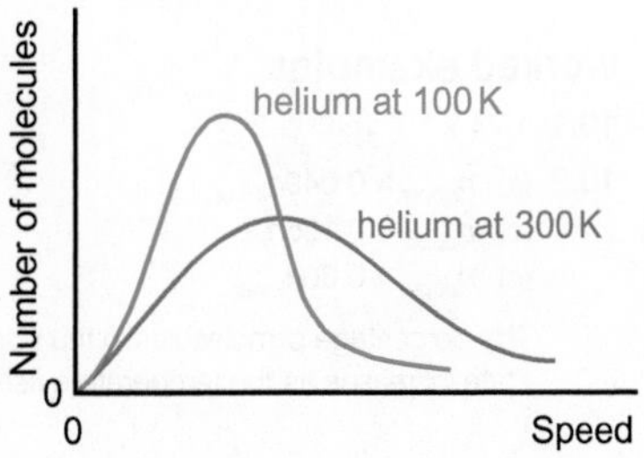
Number of molecules
helium at 100 K
helium at 300 K
0
0
Speed

(b)

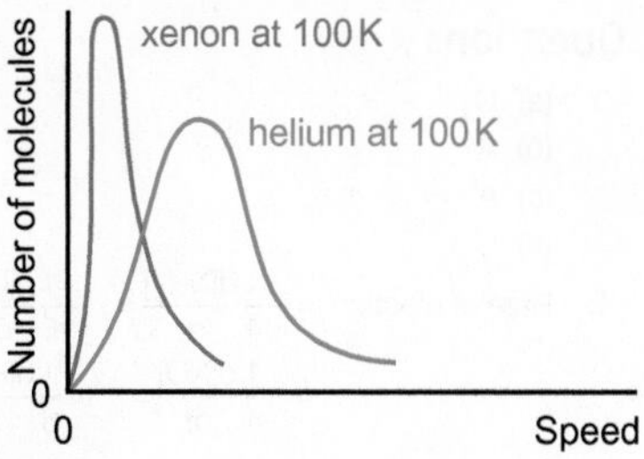
Number of molecules
xenon at 100 K
helium at 100 K
0
0
Speed

27 (a) helium
(b) hydrogen chloride
(c) helium has the highest root mean square speed; carbon dioxide has the lowest root mean square speed
(d) carbon dioxide

28 (a) 100 m s^{-1}
(b) 102 m s^{-1}
(c) 105 m s^{-1}

29 (a) 517 m s^{-1}
(b) 1200 K

30* (a) 146 m s^{-1}
(b) 5.8×10^{-4} m

32 (i) NO_2
(ii) He

33 ideal gas eqn: 732 K I.c. van der Waals: 679 K

Chapter 9 Reaction kinetics

Worked examples

9.1 15 mmol dm^{-3} s^{-1}

9.2 (a) $-\dfrac{d[CH_3Br]}{dt} = k[CH_3Br][OH^-]$
(b) Overall order 2

9.3 $\dfrac{1}{[CH_3^\bullet]_t} = \dfrac{1}{[CH_3^\bullet]_0} + 2kt$ A plot of $\dfrac{1}{[CH_3^\bullet]_t}$ against t is a straight line with gradient = $+2k$.

9.4 3.7×10^{10} dm^3 mol^{-1} s^{-1}

9.5 Br_2: zero order, H^+: first order, CH_3COCH_3: first order

9.6 Use the first plot of $[I_2]$ against t to measure at least three half lives in different areas of the curve. If these are *not* constant, the reaction is *not* first order.

9.7 (b) $2NO \rightarrow N_2O_2$: bimolecular, $N_2O_2 \rightarrow 2NO$: unimolecular, $N_2O_2 + O_2 \rightarrow 2NO_2$: bimolecular
(c) A simultaneous collision between three molecules is statistically very unlikely.

9.8 (a) 5×10^{-6} s^{-1}
(b) 40 hours

9.9 1.00×10^{-3} dm^3 mol^{-1} s^{-1}

Boxes

9.3 6 years

9.4 Differential rate equation:

$$\frac{d[ClO^\bullet]}{dt} = -2k[ClO^\bullet]^2$$

Integrated rate equation:

$$\frac{1}{[ClO^\bullet]_t} = \frac{1}{[ClO^\bullet]_0} + 2kt$$

9.7 3.49×10^{-5} mol dm^{-3}

9.8 $(rate)_{max} = 1.7$ μmol^{-1} dm^{-3} s^{-1}; $K_M = 1.1$ μmol dm^{-3}

Questions

1 (a) C
(b) A
(c) A
(d) C

2 Rate of reaction $= -\frac{1}{4}\frac{d[NH_3]}{dt} = -\frac{1}{5}\frac{d[O_2]}{dt}$
$= -\frac{1}{4}\frac{d[NO]}{dt} = -\frac{1}{6}\frac{d[H_2O]}{dt}$

3 $\frac{d[N_2O]}{dt} = -1.17 \times 10^{-5}\,\text{mol dm}^{-3}\,\text{s}^{-1}$,
$\frac{d[N_2]}{dt} = 1.17 \times 10^{-5}\,\text{mol dm}^{-3}\,\text{s}^{-1}$,
$\frac{d[O_2]}{dt} = 5.86 \times 10^{-6}\,\text{mol dm}^{-3}\,\text{s}^{-1}$

4 (a) Rate of reaction $= k[\text{Cl}\bullet][O_3]$; bimolecular
(b) Rate of reaction $= k[CH_3N_2CH_3]$; unimolecular
(c) Rate of reaction $= k[\text{Cl}\bullet]^2$; bimolecular
(d) Rate of reaction $= k[NO_2][F_2]$; bimolecular

5 (a) NO: 2; H_2: 1
(b) Overall order is 3
(c) (i) Double
(ii) Halve
(iii) Quadruple
(iv) Increase by a factor of 9.

6 $2.74 \times 10^{-2}\,\text{s}^{-1}$

7 $0.34\,\text{dm}^3\,\text{mol}^{-1}\,\text{s}^{-1}$

8 Rate $= k[CH_3COCH_3][H^+]$
$k = 4.0 \times 10^{-3}\,\text{dm}^3\,\text{mol}^{-1}\,\text{s}^{-1}$

9 $\frac{1}{16}$

10 $k = 0.00012\,\text{s}^{-1}$ $t_{1/2} = 5776\,\text{s}$

11 $0.018\,\text{mol dm}^{-3}$

12 $[CH_3\bullet] = 7.5 \times 10^{-9}\,\text{mol dm}^{-3}$

13 $k = 6.1 \times 10^{-5}\,\text{s}^{-1}$

14 (a) Second order with respect to OH^-(aq)

15 $4500\,\text{dm}^3\,\text{mol}^{-1}\,\text{s}^{-1}$

16* (a) Rate of reaction $= k[CH_3CHO][\bullet OH]$
(c) $3.0 \times 10^3\,\text{s}^{-1}$

17* $9.0 \times 10^9\,\text{dm}^3\,\text{mol}^{-1}\,\text{s}^{-1}$

18 Zero order, $1.5 \times 10^{-6}\,\text{mol dm}^{-3}\,\text{s}^{-1}$

19 $+220\,\text{kJ mol}^{-1}$

20 First order
69.3 s
$0.018\,\text{s}^{-1}$

21 $42.2\,\text{kJ mol}^{-1}$

22 (a) $k = 5.0 \times 10^5\,\text{dm}^3\,\text{mol}^{-1}\,\text{s}^{-1}$

24 (a) Strand A + strand B → stable double helix
(b) Rate of reaction = k[strand A][strand B]
(c) $k = \frac{k_2 k_1}{k_{-1}}$

25 $8.1 \times 10^6\,\text{dm}^3\,\text{mol}^{-1}\,\text{s}^{-1}$

26* (a) Initial step is second order (units of k_d are $\text{dm}^3\,\text{mol}^{-1}\,\text{s}^{-1}$). Other two steps are first order (units of k_{-d} and k_r are s^{-1}).
(b) $k = \frac{k_d k_r}{(k_{-d} + k_r)}$
(c) (i) if $k_r \gg k_{-d}$ $k = \frac{k_d k_r}{k_r} = k_d$
Rate-determining step is the diffusion of A and B to form the complex.
(ii) if $k_{-d} \gg k_r$ $k = \frac{k_d k_r}{k_{-d}} = K_d k_r$
Rate-determining step is the reaction of AB to form the products.

27 (b) The third step would have the highest Gibbs energy of activation.

28 (a) First order
(b) $k = 0.0076\,\text{min}^{-1}$ ($= 1.27 \times 10^{-4}\,\text{s}^{-1}$), $t_{1/2} = 91\,\text{min}$

29* (a) $[I^*] = \left(\frac{k_1}{k_{-1}}[I_2]\right)^{1/2}$
(b) $[H_2I^*] = \frac{k_2}{k_{-2}}[H_2][I^*]$
(c) Rate of reaction $= k_3 \frac{k_2}{k_{-2}}\frac{k_1}{k_{-1}}[H_2][I_2]$
(d) $k = k_3 \frac{k_2}{k_{-2}}\frac{k_1}{k_{-1}}$

30* (a) In a unimolecular reaction, the transition state involves just one molecule of the reactant in a high energy state.
(d) (i) Second stage (dissociation of $C_3H_8^*$)
(ii) First stage (collision of C_3H_8 and M)

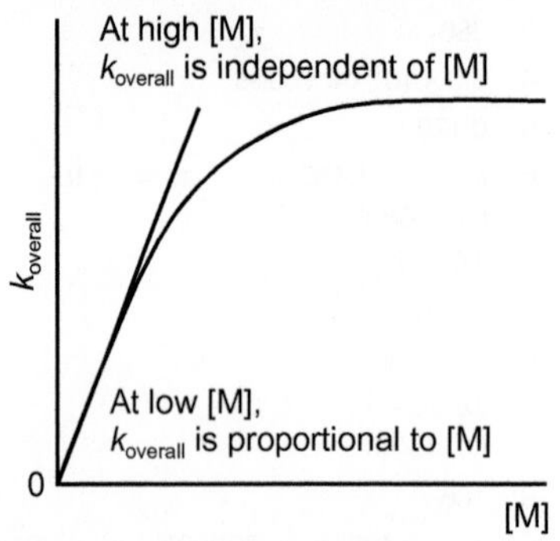

At high [M], $k_{overall}$ is independent of [M]
$k_{overall}$
At low [M], $k_{overall}$ is proportional to [M]
0
[M]

(e) Gradient $= \frac{1}{k_1}$; intercept is $\frac{k_{-1}}{k_1 k_2}$

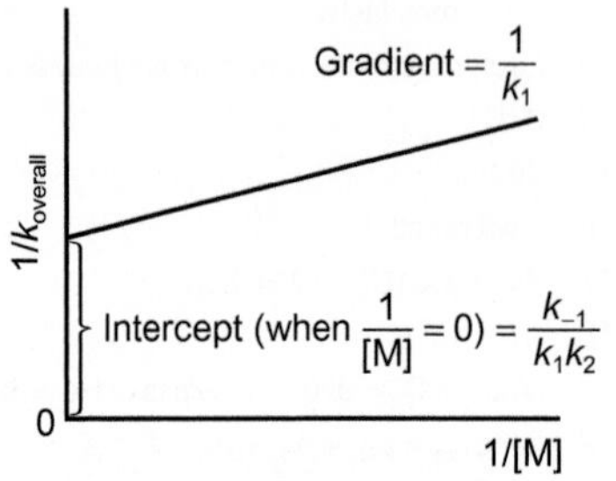

Gradient $= \frac{1}{k_1}$
$1/k_{overall}$
Intercept (when $\frac{1}{[M]} = 0$) $= \frac{k_{-1}}{k_1 k_2}$
0
1/[M]

31 (b) Rate of reaction = k[S][E]. The rate determining step is the reaction of E + S to form the complex ES.
(c) Rate of reaction = constant. The rate determining step is the conversion of ES to form E + P.

Chapter 10 Molecular spectroscopy

Worked examples

10.1 $1.24 \times 10^{-3}\,\text{mol dm}^{-3}$

10.2 (a) $n_{upper} = 0.049 n_{lower}$
(b) $n_{upper} = 0.135 n_{lower}$
(c) $n_{upper} = 0.30 n_{lower}$
The percentage of molecules in the upper energy state increases as the temperature rises.

10.3 0.128 nm

10.4 $E_{J=7} = 2.14 \times 10^{-21}\,\text{J}$, $E_{J=8} = 2.75 \times 10^{-21}\,\text{J}$, $30.4\,\text{cm}^{-1}$

10.5 HCl, ClO, CN

10.6 $B = 312 \times 10^9\,\text{s}^{-1}$, $r_0 = 0.129\,\text{nm}$

10.7 $2168\,\text{cm}^{-1}$

10.8 ^{19}F spectrum: doublet of peaks with almost equal intensity. ^{31}P spectrum: a quartet of peaks with intensities 1:3:3:1.

Boxes

10.1 $E = \frac{9h^2}{2ml^2}$, 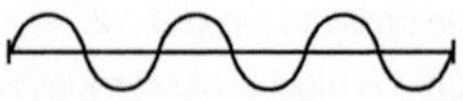

10.2 (a) Infrared
(b) (i) $3.14 \times 10^{-19}\,\text{J}$
(ii) $189\,\text{kJ mol}^{-1}$

10.3 0.1569 nm

10.6 (a) 15
(b) 9

10.7 (b) (i) $230\,\text{kJ mol}^{-1}$
(ii) $262\,\text{kJ mol}^{-1}$
(iii) $438\,\text{kJ mol}^{-1}$

Questions

1 (a) $2.22 \times 10^4\,\text{cm}^{-1}$
(b) $6.66 \times 10^{14}\,\text{Hz}$
(c) $4.4 \times 10^{-19}\,\text{J}$

2 $3.4 \times 10^{-22}\,\text{J}$

3 272 nm

4 Energy difference $= 1.13 \times 10^{-18}\,\text{J}$; wavelength $= 1.76 \times 10^{-7}\,\text{m}$

5* $E_1 = 1.07 \times 10^{-19}\,\text{J}$, $E_2 = 4.28 \times 10^{-19}\,\text{J}$, $E_3 = 9.64 \times 10^{-19}\,\text{J}$, $E_4 = 1.71 \times 10^{-18}\,\text{J}$; 266 nm

6 (a) $3.97 \times 10^{-20}\,\text{J}$, $n_{upper} = 6.46 \times 10^{-5} n_{lower}$
(b) $1.33 \times 10^{-23}\,\text{J}$, $n_{upper} = 0.997 n_{lower}$
(c) $3.97 \times 10^{-19}\,\text{J}$, $n_{upper} = 1.3 \times 10^{-42} n_{lower}$
(d) $5.0 \times 10^{-17}\,\text{J}$, $n_{upper} = 0$
(e) $6.63 \times 10^{-27}\,\text{J}$, $n_{upper} = n_{lower}$

7 (a) $T = 0.9$; $A = 0.046$
(b) $T = 0.1$; $A = 1.0$
(c) $T = 0.01$; $A = 2.0$

8 $44\,\text{m}^2\,\text{mol}^{-1}$

9 $\varepsilon = 29\,\text{m}^2\,\text{mol}^{-1}$; $c = 3.45 \times 10^{-3}\,\text{mol dm}^{-3}$

10 Absorbance = 0.717; molar absorption coefficient $= 143.3\,\text{m}^2\,\text{mol}^{-1}$; percentage transmitted = 3.69%

11* Concentration of Ni^{2+}(aq) $= 1.78 \times 10^{-5}\,\text{mol}^{-1}\,\text{dm}^{-3}$
Concentration of Co^{2+}(aq) $= 9.53 \times 10^{-6}\,\text{mol}^{-1}\,\text{dm}^{-3}$

12 (a) $n_{upper} = 9.8 \times 10^{-4} n_{lower}$, so the majority of molecules are in the ground state.
(b) $n_{upper} = 2.99 \times 10^{-83} n_{lower}$, so effectively all molecules are in the ground state.
(c) $n_{upper} = 0.976 n_{lower}$, so the molecules are equally distributed between the ground and excited states.

13 I_2 and Cl_2 are homonuclear and so do not possess a dipole; they do not adhere to the selection rule of rotational spectrum.

14 $10\,\text{cm}^{-1}$; $B = 1.5 \times 10^{11}\,\text{s}^{-1}$

15 (a) 0.143 nm
(b) $n_{upper} = 2.77 n_{lower}$ (remember degeneracy)

16 $B = 3.12 \times 10^{11}\,\text{Hz}$; Separation $= 20.8\,\text{cm}^{-1}$ or $6.24 \times 10^{11}\,\text{Hz}$

17* Bond length = 9.22×10^{-11} m. Separation between lines in DF = 21.9 cm^{-1}.

18 NO and HBr

19 (a) 30
(b) 48

21 409 $N m^{-1}$

22 39 vibrational modes. Some modes will not result in a change in polarizability so they do not appear in the spectrum.

23* Frequency = 8.7×10^{12} Hz

24* (a) Ground energy = 12.96 $kJ mol^{-1}$
(b) $n_{v=1} = 2.92 \times 10^{-5} n_{v=0}$
(c) Force constant = 1899 $kg s^{-2}$
(d) Peak shifts from 2168 cm^{-1} to 2120 cm^{-1}

26 ^2H: 1; ^3H: $\frac{1}{2}$ ^{14}N: 1; ^{15}N: $\frac{1}{2}$; ^{16}O: 0

29 Magnetic field strength = 9.39 T

Chapter 11 Analytical chemistry

Worked examples

11.1 (a) Analytes are the iron compounds (such as oxides and sulfides). Matrix is all the other material that is dug out of the ground.
(b) Analyte is the pesticide. Matrix is the remainder of the fruit.

11.2 14

11.3 1.34

11.4 365

11.5 1.83

Boxes

11.1 (a) 0.005 $mg g^{-1}$
(b) 5 $\mu g g^{-1}$
(c) 5 ppm

11.2 H^+, Na^+, K^+, Ca^{2+}: measurement of voltage generated by an electrochemical cell. O_2 and CO_2: measurement of current generated by an electrolysis cell (amperometric method).

11.3 GC–MS

11.8 (a) 0.023
(b) 7180 K
(c) 1.80×10^{-4}

Questions

1 (a) 0.004 $mol dm^{-3}$
(b) 4 $mol m^{-3}$
(c) 234 $mg dm^{-3}$
(d) 234 ppm (by mass)

2 7.0×10^{-7} $mol dm^{-3}$

3 Systematic or determinate error

4

	Method 1	Method 2
(a) Mean/ppm	10.09	10.094
(b) Standard deviation/ppm	0.324	0.037
(c) Range/ppm	0.98	0.12
(d) Coefficient of variation/%	3.21	0.367

5 Method 1 is less accurate, but more precise. Method 2 is more accurate, but less precise.

6 pH of carbonated drink = 2.8; pH of water sample = 6.4.

7 Benzoate concentration = 1.6×10^{-5} $mol dm^{-3}$

8* 1.03×10^{-6} $mol dm^{-3}$, pF = 5.99

9 1.57×10^{-3} $mol dm^{-3}$

10 0.14 $mol dm^{-3}$

11 1,2-isomer = 90%, 1,3-isomer = 6%, 1,4-isomer = 4%

12 (a) Gas chromatography.
(b) Extract in solvent, use gas chromatography employing an electron capture detector.
(c) HPLC with a UV/VIS or a refractive index detector and a chiral phase.

14 $\varepsilon = 2.49$ $m^2 mol^{-1}$

15 Fe^{2+} ions: 3.33×10^{-3} $mol dm^{-3}$, $[Fe(C_{10}H_8N_2)_3]^{2+}$ ions: 9.1×10^{-7} $mol dm^{-3}$

16* 0.55 ppm

17 Percentage of drug = 6.2%.

18 99.3% pure, by mass.

19 Concentration of contaminant = 0.0043 ppm

21 0.037 ppm

22 5 ppm

23 0.34 $ng g^{-1}$

24 (a) HPLC
(b) UV/VIS spectrophotometry
(c) TLC
(d) Gas chromatography
(e) Atomic absorption spectrophotometry

Chapter 12 Molecular characterization

Worked examples

12.1 Major product = $PhCHClCH_3$, minor product = $PhCH_2CH_2Cl$.

12.3 3400–3190 cm^{-1}: N–H stretch, 3000–2800 cm^{-1}: C–H stretch (from **A** and/or Nujol), 1650 cm^{-1}: C=O stretch of the $CONH_2$ group. Compound **A** is $CH_3CH_2CONH_2$

12.4 Me_2CHCHO

12.5 1-Phenylbutan-1-one

12.6 1,2-Dibromobenzene

12.7 2-Phenylethyl ethanoate

Boxes

12.4 900 MHz

12.6 (a) 5
(b) 2

Questions

1 (a) Both **A** and **B** contain a chlorine atom. Chlorine is usually a mixture of two isotopes, ^{35}Cl and ^{37}Cl, in a ratio of 3 : 1.
(b) Compound **C** contains a single bromine atom and will give two peaks, in a 1 : 1 ratio, because the natural abundances of ^{79}Br (50.7%) and ^{81}Br (49.3%) are about the same.
(c) Chlorine-containing fragment ions produced by cleavage of the CH_3–CO bond in the ketone groups of **A** and **B**.
(d) The C=O stretching vibration for **A–C** is lower because of the electron-donating (+M) effects of the benzene rings.
(e) Three types of hydrogen atoms in different electronic environments.
(f) 5
(g) The ^{13}C NMR spectrum of **A** has six signals, whereas **B** has eight signals.

2 (a) 1-Chlorobutane
(b) 2-Chlorobutane
(c) 1-Chloro-2-methylpropane

4 2-Methyl-1-propanol

5 4-Methoxybenzaldehyde (anisaldehyde)

Chapter 13 Energy and thermochemistry

Worked examples

13.2 6.58 K

13.3 12.5 minutes

13.4 2300 kJ

13.5 −87.9 $kJ mol^{-1}$

13.6 −123.9 $kJ mol^{-1}$

13.7 −74.8 $kJ mol^{-1}$

13.8 −1064 $kJ mol^{-1}$

13.9 −100.9 $kJ mol^{-1}$

13.10 −100 J

13.11 +769 J

13.12 −885 $kJ mol^{-1}$

13.13 −5134 $kJ mol^{-1}$

13.14 −2245 $kJ mol^{-1}$

Boxes

13.1 214 drops

13.2 (a) The water in both dishes evaporates as $H_2O(g)$ and cools the water by removing the heat of vaporization. In dish B evaporation is faster because the temperature is higher, so heat is removed faster, so it freezes first even though it starts at a higher temperature.
(b) If the dishes have lids on, $H_2O(g)$ cannot escape and heat will not be removed by evaporation. Dish B will no longer freeze first.

13.3 (a) Petrol: 48.0 $kJ g^{-1}$; diesel: 28.7 $kJ g^{-1}$; natural gas: 55.6 $kJ g^{-1}$; hydrogen: 143.0 $kJ g^{-1}$; methanol: 22.7 $kJ g^{-1}$; ethanol: 29.7 $kJ g^{-1}$; MTBE: 38.3 $kJ g^{-1}$. Hydrogen has the highest energy density and so is the most efficient in terms of energy per g of fuel
(b) Formula 1 racing cars in terms of energy per kg of fuel.
(c) Petrol + MTBE: 47.0 $kJ g^{-1}$; petrol (only): 48.0 $kJ g^{-1}$.

13.4 (a) $Mg(s) + 2H_2O(l) \rightarrow Mg(OH)_2(aq) + H_2(g)$, $\Delta_r H^{\ominus}_{298} = -356.6$ $kJ mol^{-1}$
(b) 75.2 kJ, 5.2 g

13.7 (a) (i) −5650 $kJ mol^{-1}$
(ii) −1413 $kJ mol^{-1}$
(b) (i) −17.96 kJ
(ii) −1431 $kJ mol^{-1}$

Questions

1 Intensive: density; temperature; pressure. Extensive: amount in moles; length

2 (a) 655 kJ
(b) 10.9 min

3 294.7 K

4 −33.4 kJ

5 +84 800 kJ (to 3 s.f.)

6 +30.1 kJ

7 +15.9 $kJ\,mol^{-1}$

8 −174.1 $kJ\,mol^{-1}$

9 −1703.1 kJ

10 −239 $kJ\,mol^{-1}$

11 +278.2 $kJ\,mol^{-1}$

12 −0.2 $kJ\,mol^{-1}$

13 (a) +5349 $kJ\,mol^{-1}$
(b) +5525 $kJ\,mol^{-1}$

14 −115.1 $kJ\,mol^{-1}$

15 +1.25 J

16 +5 kJ

17 +315 J

18 +57 700 J (to 3 s.f.)

19 −248 J

20 5.56 g

21 (a) 367 kJ
(b) 87.7 cal

22 (a) +21.7 $kJ\,mol^{-1}$
(b) −34 720 J
(c) 795 K

23 +40.36 $kJ\,mol^{-1}$

24* +491.7 $kJ\,mol^{-1}$

25 −56.8 $kJ\,mol^{-1}$

26 For the process butane → methylpropane, $\Delta_{isomerization}H = +277\ kJ\,mol^{-1}$

27* −91.1 $kJ\,mol^{-1}$

28* (a) −1268 $kJ\,mol^{-1}$
(b) −2806 $kJ\,mol^{-1}$

29* −1005 $kJ\,min^{-1}$

30. +205.7 $kJ\,mol^{-1}$

31. −238.6 $kJ\,mol^{-1}$

32. 0.440 g

33. $\Delta H = q = +203.5\ kJ$, $w = +15.51\ kJ$, $\Delta U = +188\ kJ$

34. −479.6 $kJ\,mol^{-1}$

35. +49.6 $kJ\,mol^{-1}$

36. −886.6 $kJ\,mol^{-1}$

37. −40.16 $kJ\,mol^{-1}$

Chapter 14 Entropy and Gibbs energy

Worked examples

14.1 (a) Negative
(b) Negative
(c) Positive

14.2 −22.1 $J\,K^{-1}\,mol^{-1}$

14.3 7.04 $J\,K^{-1}$

14.4

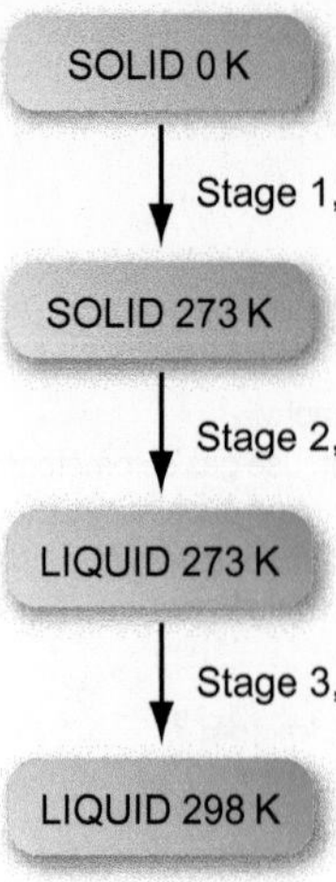

SOLID 0 K
Stage 1, ΔS_1
SOLID 273 K
Stage 2, ΔS_2
LIQUID 273 K
Stage 3, ΔS_3
LIQUID 298 K

14.5 (a) +160.6 $J\,K^{-1}\,mol^{-1}$
(b) +175.9 $J\,K^{-1}\,mol^{-1}$

14.6 (a) +160.2 $J\,K^{-1}\,mol^{-1}$
(b) +175.2 $J\,K^{-1}\,mol^{-1}$

14.7 1075 K

14.8 −93.1 $kJ\,mol^{-1}$

14.9 −1010 $kJ\,mol^{-1}$

14.10 −866.8 $kJ\,mol^{-1}$

Boxes

14.4 (a) Negative

14.6 (a) Styrene: −37.2 $kJ\,mol^{-1}$, 652 K.
α-methylstyrene: −3.8 $kJ\,mol^{-1}$, 334 K

14.7 (b) $\Delta_r H^{\ominus}_{298} = +910.9\ kJ\,mol^{-1}$,
$\Delta_r S^{\ominus}_{298} = +182.1\ J\,K^{-1}\,mol^{-1}$
(c) +856.6 $kJ\,mol^{-1}$
(d) 5002 K
(e) +582.3 $kJ\,mol^{-1}$
(f) 1911 K
(g) +462.3 $kJ\,mol^{-1}$, 2797 K

14.8 (a) +33.9 $J\,K^{-1}\,mol^{-1}$
(b) −17.1 $kJ\,mol^{-1}$

14.9 9.8 g

14.10 (a) Reaction 1: negative, Reaction 2: positive, Reaction 3: no change
(c) 1300 K

Questions

1 Spontaneous processes are those that, once started, will continue without any outside intervention until they come to equilibrium.

2 (a) Negative
(b) Negative
(c) Negative
(d) Positive
(e) Negative
(f) Negative
(g) Negative (positive if solution is dilute)
(h) Positive
(i) Negative

3 (a) The two gases gradually diffuse into each other until the concentrations of each gas are the same in both bulbs.
(b) $\Delta H = 0$, ΔS is positive, ΔG is negative
(c) Yes

4 +28.8 $J\,K^{-1}\,mol^{-1}$

5 (a) Negative
(b) Near zero
(c) Positive
(d) Near zero
(e) Positive

6 The entropy change in the system is positive because a solid is being converted to aqueous ions which have much higher entropy. This is more than enough to compensate for the negative entropy change in the surroundings caused by absorption of heat as the solid dissolves.

7 $\Delta S_{total}^{\ominus} = 32.38\ J\,K^{-1}\,mol^{-1}$

8 (a) $\Delta S(\text{water}) = -37.5\ J\,K^{-1}$
(b) $\Delta S(\text{refrigerator}) = +39.2\ J\,K^{-1}$
The overall entropy change is +1.7 $J\,K^{-1}$, so the Second Law is obeyed.

9

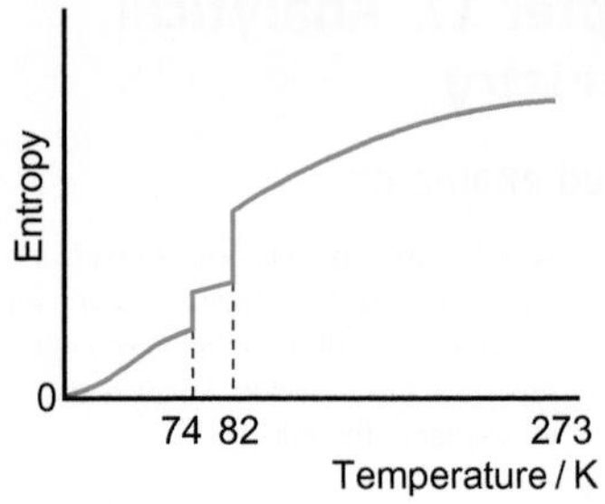

Entropy
0
74 82
273
Temperature / K

10 ΔH is negative, ΔS is negative, ΔG will depend on temperature. However, if condensation takes place spontaneously, ΔG must be negative.

11 $\Delta S_{ice} = +22\ J\,K^{-1}\,mol^{-1}$, $\Delta S_{water} = -22\ J\,K^{-1}\,mol^{-1}$

12 +6.5 $J\,K^{-1}\,mol^{-1}$

13 (a) −198.7 $J\,K^{-1}\,mol^{-1}$
(b) −153.1 $J\,K^{-1}\,mol^{-1}$
(c) +972.4 $J\,K^{-1}\,mol^{-1}$

14 $\Delta S(\text{system}) = +28.1\ J\,K^{-1}\,mol^{-1}$,
$\Delta S(\text{surroundings}) = -27.5\ J\,K^{-1}\,mol^{-1}$,
$\Delta_{fus}G = -710\ J\,mol^{-1}$, $T_m = 1075\ K$

15 1106 K

16 6.27 g

17 (a) −817.6 $kJ\,mol^{-1}$
(b) −72.4 $kJ\,mol^{-1}$
(c) 3415 m

18 (a) (i) −89.71 $kJ\,mol^{-1}$
(ii) +25.99 $kJ\,mol^{-1}$
(b) (i) −451.8 $kJ\,mol^{-1}$
(ii) −401.8 $kJ\,mol^{-1}$
(c) (i) +55.9 $kJ\,mol^{-1}$
(ii) +157.7 $kJ\,mol^{-1}$

19 $\Delta_r H^{\ominus}_{298} = -440.9\ kJ\,mol^{-1}$. The maximum work available is given by the Gibbs energy change, $\Delta_r G^{\ominus}_{298} = 395.9\ kJ$

20 349.2 K

21* 980.6 K

22* +114.6 $kJ\,mol^{-1}$

23* −32.7 $J\,K^{-1}\,mol^{-1}$

24* $\Delta_r H^{\ominus}_{263} = -5631\ J\,mol^{-1}$
$\Delta_r S^{\ominus}_{263} = -20.5\ J\,K^{-1}\,mol^{-1}$
$\Delta_r G^{\ominus}_{263} = -239.5\ J\,mol^{-1}$

25 (a) $-70.5\,\text{kJ mol}^{-1}$

(b) Yes, it is spontaneous since $\Delta_r G^\circ_{298} < 0$. $\Delta_r H^\circ_{298}$ is negative since strong N=O bonds are formed while $\Delta_r S^\circ_{298} < 0$ due to the increase in order as the reduction in the number of moles of gas during the reaction.

26 At 80 °C, $\Delta_r G^\circ_{298} = -16.7\,\text{kJ mol}^{-1}$ but at 200 °C, $\Delta_r G^\circ_{298} = +31.9\,\text{kJ mol}^{-1}$. The reaction changes from being spontaneous at 80 °C to non-spontaneous (i.e. the reverse reactions becomes spontaneous) at higher temperatures.

27 $-79.8\,\text{kJ mol}^{-1}$

28 The values of $\Delta_r G^\circ_{298}$ are $+55.3\,\text{kJ mol}^{-1}$ and $+163.9\,\text{kJ mol}^{-1}$. Neither reaction is spontaneous at 298 K (and given the values, unlikely to become so at any reasonable temperature) so the chemists should go back to the drawing board!

29 $+5.8\,\text{J K}^{-1}\,\text{mol}^{-1}$.

Chapter 15 Chemical equilibrium

Worked examples

15.1 (a) $K = \dfrac{(p_{NO}/p^\ominus)^4_{eqm}}{(p_{NH_3}/p^\ominus)^4_{eqm}(p_{O_2}/p^\ominus)^5_{eqm}}$

(b) $K = \dfrac{([Ca^{2+}(aq)]_{eqm}/1\,\text{mol dm}^{-3} \times [OH^-(aq)]^2_{eqm}/1\,\text{mol dm}^{-3})^2}{{}^{a}Ca(OH)_2(s)}$

(c) $K = \dfrac{(p_{CO}/p^\ominus)^4_{eqm}}{(p_{Ni(CO)_4}/p^\ominus)_{eqm}}$

Remember activities of pure liquids and solids ~1

15.2 0.0167 bar

15.3 More HI will form, Q is smaller than K.

15.4 $-8.50\,\text{kJ mol}^{-1}$

15.5 $\Delta_r G^\ominus_{473} = 0$ For a reaction with $K = 1$, $\Delta_r G^\ominus$ is zero at *any* temperature.

15.6 1.47 bar

15.7 At equilibrium: 0.28 mol of N_2O_4(g), 1.44 mol of NO_2(g)

15.8 At higher pressure, more CO_2(g) present in equilibrium mixture

15.9 (a) $\Delta_r H^\ominus = +40.3\,\text{kJ mol}^{-1}$, $\Delta_r S^\ominus = +62.8\,\text{J K}^{-1}\,\text{mol}^{-1}$

(b) $\Delta_r G^\ominus = +8.90\,\text{kJ mol}^{-1}$

Boxes

15.1 (a) $K = (a_{Mg^{2+}})_{eqm} \times (a_{F^-})^2_{eqm}$

$K = (a_{Ag^+})^2_{eqm} \times (a_{CO_3^{2-}})_{eqm}$

$K = (a_{Al^{3+}})_{eqm} \times (a_{OH^-})^3_{eqm}$

$K = (a_{Bi^{3+}})^2_{eqm} \times (a_{S^{2-}})^3_{eqm}$

(b) $K = (a_{Mg^{2+}}) \times (a_{F^-})^2$

$= 4s^3/\text{mol}^3\,\text{dm}^{-9}$

$K = (a_{Ag^+})^2_{eqm} \times (a_{CO_3^{2-}})_{eqm}$

$= 4s^3/\text{mol}^3\,\text{dm}^{-9}$

$K = (a_{Al^{3+}})_{eqm} \times (a_{OH^-})^3_{eqm}$

$= 27s^4/\text{mol}^4\,\text{dm}^{-12}$

$K = (a_{Bi^{3+}})^2_{eqm} \times (a_{S^{2-}})^3_{eqm}$

$= 108s^5/\text{mol}^5\,\text{dm}^{-15}$

(c) MgF_2: $1.2 \times 10^{-3}\,\text{mol dm}^{-3}$, $Al(OH)_3$: $2.5 \times 10^{-9}\,\text{mol dm}^{-3}$

(d) (i) $6.4 \times 10^{-7}\,\text{mol dm}^{-3}$

(ii) $1.3 \times 10^{-28}\,\text{mol dm}^{-3}$

15.2 (a) (i) *K* does not change.

(ii) *Q* increases.

(b) Solubility doubles.

15.4 (a) $K = \dfrac{(p_{\cdot HOSO_2}/p^\ominus)_{eqm}}{(p_{SO_2}/p^\ominus)_{eqm}(p_{\cdot OH}/p^\ominus)_{eqm}}$

(b) $\Delta_r G^\ominus = -71.0\,\text{kJ mol}^{-1}$, $K = 2.77 \times 10^{12}$

(c) At 500 K: 2.63×10^4, At 1100 K: 9.18×10^{-3}

15.5 (a) $K = \dfrac{(p_{CO_2}/p^\ominus) \times (1)}{1} = (p_{CO_2}/p^\ominus)$

(b) It remains the same.

(d) At 298 K: $\Delta_r G^\ominus_{298} = +130.4\,\text{kJ mol}^{-1}$, $K = 1.4 \times 10^{-23}$. At 1273 K: $\Delta_r G^\ominus_{1273} = -26.14\,\text{kJ mol}^{-1}$, $K = 11.8$

(e) 1110 K

(f) 11.8 bar

15.6 (a) (i) $-92.2\,\text{kJ mol}^{-1}$

(ii) $-113.7\,\text{kJ mol}^{-1}$

(iii) $-124.0\,\text{kJ mol}^{-1}$

(b) (i) $-33.0\,\text{kJ mol}^{-1}$

(ii) $+73.3\,\text{kJ mol}^{-1}$

(iii) $+129.6\,\text{kJ mol}^{-1}$

(e) (i) 6.1×10^5

(ii) 1.1×10^{-5}

(iii) 1.7×10^{-7}

(f) Forward reaction

Questions

1 (a) $K = \dfrac{1 \times (p_{NO_2}/p^\ominus)^4_{eqm}}{(p_{NH_3}/p^\ominus)^4_{eqm}(p_{O_2}/p^\ominus)^7_{eqm}}$

where $p^\ominus$ = 1 bar

(b) $K = \dfrac{([H_3O^+]_{eqm}/1\,\text{mol dm}^{-3}) \times ([CN^-]_{eqm}/1\,\text{mol dm}^{-3})}{([HCN]_{eqm}/1\,\text{mol dm}^{-3}) \times 1}$

(c) $K = \dfrac{(p_{PCl_3}/1\,\text{bar})_{eqm}(p_{Cl_2}/1\,\text{bar})_{eqm}}{(p_{PCl_5}/1\,\text{bar})_{eqm}}$

(d) $K = \dfrac{(p_{O_3}/1\,\text{bar})^2_{eqm}}{(p_{O_2}/1\,\text{bar})^3_{eqm}}$

(e) $K = \dfrac{([H_3O^+]_{eqm}/1\,\text{mol dm}^{-3}) \times ([OH^-]_{eqm}/1\,\text{mol dm}^{-3})}{1}$

(f) $K = \dfrac{([Zn^{2+}]_{eqm}/1\,\text{mol dm}^{-3})^3}{([Fe^{3+}]_{eqm}/1\,\text{mol dm}^{-3})^2}$

2 (a) $1.61 \times 10^{-10}\,\text{mol}^2\,\text{dm}^{-6}$

(b) $1.61 \times 10^{-8}\,\text{mol dm}^{-3}$

3 $K = 1.28$

4 (a) $Q = 5.00 \times 10^5$

(b) Since $Q > K$, the reaction will move towards the reactants as it comes to equilibrium.

5 $Q = 2.05 \times 10^5$

Since $K >> Q$, the reaction will move towards the products as it comes to equilibrium.

6 2.6×10^{-42}

7 $\Delta_r G^\ominus_{298} = -28.65\,\text{kJ mol}^{-1}$, $K = 1.05 \times 10^5$

8 Forward reaction

9 1.7 mol CS_2, 3.1 mol Cl_2, 0.30 mol S_2Cl_2, 0.30 mol CCl_4

10 (a) 1.39×10^{-4}

(b) Forward direction

11 Increasing the pressure and reducing the temperature.

12 (a) (i) No effect

(ii) Increases CO_2

(b) (i) Reduces *K*

(ii) Decreases CO_2

(c) (i) No effect

(ii) No effect

13 7.94×10^{-9}

14 0.3 mol

15 $0.0019\,\text{mol dm}^{-3}$

16 (a) $K = 4.40$

(b) $x_{cis} = 0.185$; $x_{trans} = 0.815$

17* $+52.9\,\text{kJ mol}^{-1}$

18* (a) 8.12×10^{-6}

(b) $\Delta_r H^\ominus = +626\,\text{J mol}^{-1}\,\text{k}^{-1}$, $\Delta_r S^\ominus = +338\,\text{J mol}^{-1}\,\text{k}^{-1}$

19 <Non-numerical answer. (It's a long derivation but is in a different file.)>

20 $K = 0.019$. Only limited decomposition would occur.

21* (a) No

(b) $0.33\,\text{mol dm}^{-3}$

22 $\Delta_r H^\ominus = +88.4\,\text{kJ mol}^{-1}$, $\Delta_r S^\ominus = +82.8\,\text{J K}^{-1}\,\text{mol}^{-1}$

23 $\Delta_r H^\ominus = +8.01\,\text{kJ mol}^{-1}$

24 If p_{H_2} increases, then to keep *K* constant, the concentrations of CO and H_2O would increase and the concentration of CO_2 would decrease.

25* (a) $K = \dfrac{[\text{protein G (unfolded)}]/1\,\text{mol dm}^3}{[\text{protein G(folded)}]/1\,\text{mol dm}^3}$

(b) $K = 1$

(c) $\Delta_r G^\ominus = 0$

(d) $19.7\,\text{kJ mol}^{-1}$

(e) 4.75×10^{-4}

(f) 9.68×10^{-4}

26* (a) $\Delta_r G^\ominus_{798} = -8.56\,\text{kJ mol}^{-1}$, $K = 3.63$

(b) 0.656 moles

27 0.0015 mol

28 (i) $K = 0.25$

(ii) the amount of HI would increase

29 (i) $\Delta_r H^\ominus_{298} = -114.2\,\text{kJ mol}^{-1}$

$\Delta_r S^\ominus_{298} = -146.5\,\text{J K}^{-1}\,\text{mol}^{-1}$

$\Delta_r G^\ominus_{298} = -43.5\,\text{kJ mol}^{-1}$

(ii) $K = 4.18 \times 10^7$

30 (i) All the $CaCO_3$(s) reacts to give a pressure of 0.18 bar. For (ii), (iii) and (iv), sufficient $CaCO_3$(s) reacts to bring the reaction to equilibrium, hence the pressure is 0.22 bar with some solid remaining in the container.

31 (i) $\Delta_r H^\ominus = +157.7\,\text{kJ mol}^{-1}$

$\Delta_r S^\ominus = +109.3\,\text{J K}^{-1}\,\text{mol}^{-1}$

(ii) $\Delta_r G^\ominus = +48.3\,\text{kJ mol}^{-1}$

(iii) Increasing the temperature would (a) increase the entropic contribution which favours dissociation (b) it is an endothermic reaction so Le Chatelier's principle predicts that the forward reaction is favoured.

Chapter 16 Electrochemistry

Worked examples

16.1 $Pt|H_2(g)|H^+(aq)||Cl^-|Cl_2(g)|Pt$

16.2 $V^{2+}(aq) + Fe^{3+}(aq) \rightarrow V^{3+}(aq) + Fe^{2+}(aq)$

16.3 The $E^{\ominus}$ value for the reduction of $MnO_4^-(aq)$ is more positive than that for the reduction of Fe^{3+}, so acidified $KMnO_4$ solution will oxidize $Fe^{2+}(aq)$ to $Fe^{3+}(aq)$.

16.4 +1.21 V

16.5 −210 kJ mol^{-1} (to 2 s.f.)

16.6 3.6×10^{15}

16.7 (a) $Cu^{2+}(aq) + H_2(g) \rightarrow Cu(s) + 2H^+(aq)$
(b) +0.34 V
(c) 9.0×10^{-4} mol dm^{-3}

16.8 603 s (approx. 10 min)

Boxes

16.1 (a) 1×10^{-4} mol m^{-3}
(b) 5×10^{-2} S m^2 mol^{-1}
(c) 0.055 μS cm^{-1}
(d) 7.9×10^{-4} mol dm^{-3}

16.3 (a) For zinc cell, $E^{\ominus} = -0.98$ V; for copper cell, $E^{\ominus} = +0.12$ V
(b) $E^{\ominus} = +1.10$ V

16.4 (a) The greater conductivity of solutions containing electrolytes allows a higher rate of reaction.
(b) Zinc.

16.5 (a) 120 mmol dm^{-3}
(b) $\Delta G_{K^+} = +8700$ J mol^{-1} (to 2 s.f.), $\Delta G_{Na^+} = -6000$ J mol^{-1}

16.6 (a) (i) $Pb(s), PbO_2(s)|H_2SO_4(aq)|PbSO_4(s), Pb(s)$
(ii) $Ni(s), NiO(OH)(s), Ni(OH)_2(s)|OH^-(aq)|MH(s), M(s)$
(b) $NiO(OH)(s) + MH \rightleftharpoons Ni(OH)_2(s) + M(s)$

Questions

1 (a) $MgSO_4$: $\Lambda_m° = 24.5$ mS m^2 mol^{-1}, Na_2CO_3 $\Lambda_m° = 27.0$ mS m^2 mol^{-1}
(b) 0.245 S m^{-1}

2* 1.88×10^{-10}

3 (a) Gold electrode (cathode)
(b) Aluminium electrode (anode)
(c) Aluminium electrode (anode)
(d) Gold electrode (cathode)
(e) Al (s)
(f) $Au^{3+}(aq) + 3e^- \rightarrow Au(s)$
(g) +3.16 V

4 (a) $Cd(s)|Cd^{2+}(aq)||Sn^{2+}(aq)|Sn(s)$
(b) $Pt|H^+(aq), H_2O_2(aq)|H_2(g)|Pt$
(c) $Pt|Sn^{2+}(aq), Sn^{4+}(aq)||Br_2(aq), Br^-(aq)|Pt$
(d) $Pt|Ag(s)|AgBr(s)|Br^-(aq)||Cu^{2+}(aq)|Cu(s)|Pt$
(e) $Pt|Fe^{2+}(aq), Fe^{3+}(aq)||MnO_4^-(aq), H^+(aq), Mn^{2+}(aq)|Pt$

5 (a) (i) $Fe(s) \rightarrow Fe^{2+}(aq) + 2e^-$; $Zn^{2+}(aq) + 2e^- \rightarrow Zn(s)$
(ii) −0.32 V
(b) (i) $H_2(g) \rightarrow 2H^+(aq) + 2e^-$; $2AgCl(s) + 2e^- \rightarrow 2Ag(s) + 2Cl^-(aq)$
(ii) +0.22 V
(c) (i) $2Hg(l) + 2Cl^-(aq) \rightarrow Hg_2Cl_2(s) + 2e^-$; $2AgCl(s) + 2e^- \rightarrow 2Ag(s) + 2Cl^-(aq)$
(ii) −0.05 V
(d) (i) $2Fe^{2+}(aq) \rightarrow 2Fe^{3+}(aq) + 2e^-$; $Sn^{4+}(aq) + 2e^- \rightarrow Sn^{2+}(aq)$
(ii) −0.62 V

6 H_2O_2 in acid

7 Under these conditions Cu (s) has a higher standard reduction potential than H^+ (hydrogen half cell), but a lower standard reduction potential than NO_3^-, therefore it can react with HNO_3 by reduction of NO_3^- to NO gas.

8 (a) Not spontaneous
(b) Spontaneous
(c) Spontaneous
(d) Spontaneous

9 The silver and aluminium establish an electrochemical cell with the NaCl solution as the salt bridge.

10 (a) $\Delta_r G^{\ominus} = -310.7$ kJ mol^{-1}; $K = 2.9 \times 10^{54}$
(b) $\Delta_r G^{\ominus} = -158.2$ kJ mol^{-1}; $K = 5.5 \times 10^{27}$
(c) $\Delta_r G^{\ominus} = -1024.7$ kJ mol^{-1}; $K = 1 \times 10^{100}$
(d) $\Delta_r G^{\ominus} = -54.03$ kJ mol^{-1}; $K = 3.0 \times 10^9$

11 (a) 0.06 V
(b) −0.03 V

12* 4.5×10^{-13}

13 (a) −0.46 V; +88.8 kJ mol^{-1}
(b) −0.43 V; +83.0 kJ mol^{-1}
(c) −0.49 V; +94.6 kJ mol^{-1}

14 2.9×10^9

15 +0.236 V

16* (a) RHS: $Cu^{2+}(aq) + 2e^- \rightarrow Cu(s)$; LHS: $Cu(s) \rightarrow Cu^{2+} + 2e^-$
(b) +0.030 V
(d) RHS: 0.49 mol dm^{-3}; LHS: 0.06 mol dm^{-3}

17 3+

18 2.8×10^6 s (32 days)

19 1.24×10^8 A

Chapter 17 Phase equilibrium and solutions

Worked examples

17.2 268 atm

17.3 41.0 kPa

17.4 62.8% benzene, 37.2% methylbenzene

17.5 $\gamma_{C_2H_5OH} = 1.06$, $\gamma_{CHCl_3} = 1.63$

17.6 8.4 mg dm^{-3} to 9.3 mg dm^{-3}

17.7 3000 Pa

Boxes

17.4 $p_{O_2} = 0.21$ atm = 21.3 kPa; 50% emulsion

17.5 29.1 bar

Questions

1 (a) W = solid, X = solid and liquid, Y = liquid and vapour, Z = vapour
(b) X = normal melting point, Y = normal boiling point, T = triple point, C = critical point
(c) (i) liquid → vapour
(ii) liquid → solid

2 (a) B
(b) Gas
(c) Gas to liquid to solid
(d) Sublimation
(e) Enthalpy change negative; entropy change negative
(f) Triple point at about 12 bar and 120 K
(g) Melting point 135 K; boiling point 230 K
(h) Supercritical phase
(i) Solid has higher density.

3 (a) 60 atm
(b) Monoclinic form of solid sulfur
(c) Ice I
(d) (i) Vaporize (sublimate)
(ii) Condense to a liquid
(iii) Solidify

5 1.61×10^7 Pa

6 114.9 °C

8 347.2 K

9* Melting temperature = −0.005 °C, boiling temperature = 71.0 °C

10* Melting point at 500 bar = 269.4 K

11* Enthalpy change = +28.2 kJ mol^{-1}; entropy change = +88.3 J K^{-1} mol^{-1}

12 i) $p = 1450$ Pa, $T_{triple} = 195.8$ K
ii) $\Delta_{fus}H = +8.52$ kJ mol^{-1}, $\Delta_{fus}S = 43.5$ J K^{-1} mol^{-1}

13 i) $\Delta_{vap}H = 42.9$ kJ mol^{-1}
ii) $T_B = 374.4$ K
iii) $\Delta_{vap}S = 114.6$ J K^{-1} mol^{-1}. It has a larger value because of significant intermolecular interactions.

14 281.1 K

15 1389 kJ mol^{-1}

16 Energy of interaction is in the ratio 1 : 1/78 : 1/2 (vacuum:water:oil), i.e. 1 : 0.0123 : 0.5

18 (a) Ethanol
(b) Methoxymethane
(c) Propanone
(d) (*E*)-1,2-dichloroethene

19 He, Ar, N_2, HCl, H_2O, NaCl

21 (a) 1 atm: 57.5 mg, 5 atm: 287 mg
(b) 250 cm^3

22 29 kPa

23 (a) $p_{ethanol} = 2431$ Pa, $p_{methanol} = 6980$ Pa
(b) $p_{total} = 9411$ Pa

24* $\gamma_{C_2H_5OH} = 1.06$, $\gamma_{C_7H_{14}} = 13.9$

25 718.1 Pa

26 (a) 332 K
(b) 0.82
(c) 326 K
(d) (i) Pure dichloromethane
(ii) Pure butanone

27 $\gamma_P = 1.14$; $\gamma_I = 1.10$

29 Zero degrees of freedom

30 $p_{hexane} = 12.4$ kPa $p_{octane} = 4.0$ kPa

31 $x_{benzene} = 0.77$ $x_{toluene} = 0.23$

32 $\pi = 2.18$ bar

Chapter 18 Isomerism and stereochemistry

Worked examples

18.1 Position isomers: **G** and **I**. Functional group isomers: **I** and **J**, **G** and **J**.

18.2 Eclipsed conformation is higher in energy.

18.3

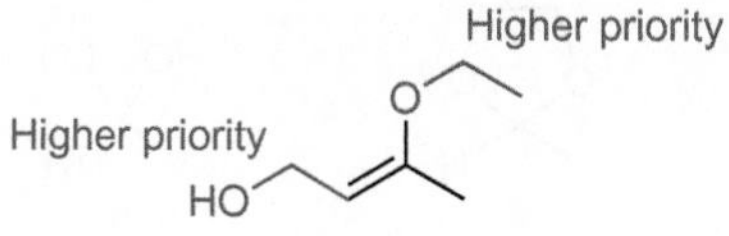
Higher priority
Higher priority
HO
O

(Z)-3-ethoxybut-2-en-1-ol

18.4 Observed $[\alpha]_D = +83$

Boxes

18.1 (a) *trans*-Isomer

18.6 L-Phenylalanine

Questions

1 (b) Syn-periplanar conformation

2 (a)

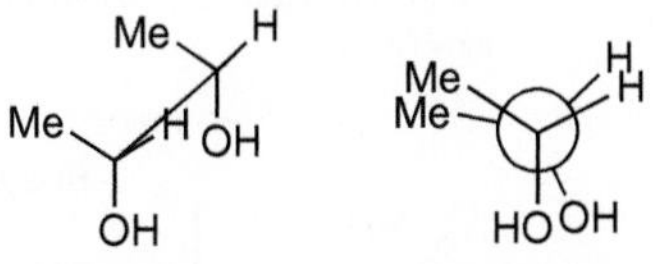
Me
H
Me
H
OH
OH
Me
Me
H
H
HO
OH

Sawhorse projection   Newman projection

(c) It contains two chiral centres but has an internal plane of symmetry (and so is achiral).

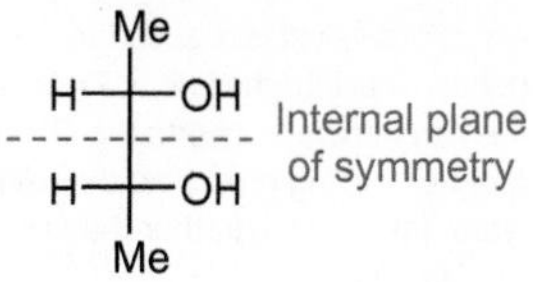
Me
H
OH
Internal plane of symmetry
H
OH
Me

(d)

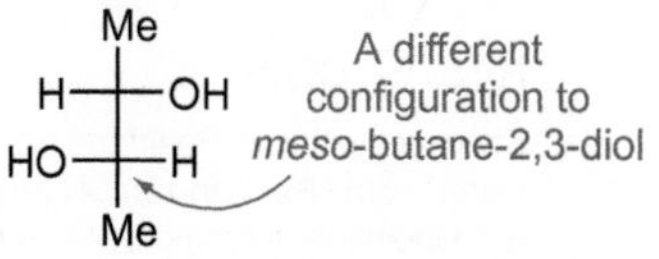
Me
H
OH
HO
H
Me
A different configuration to meso-butane-2,3-diol

3 (a) *E*-Configuration

(b)

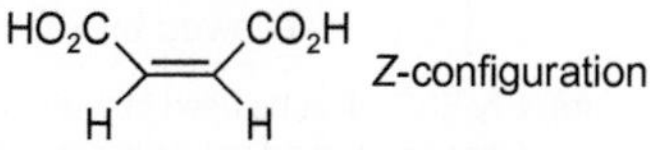
HO2C
CO2H
H
H
Z-configuration

(c) Because (–)-malic acid is formed in preference to (+)-malic acid.

7* (a) *S*

(b)

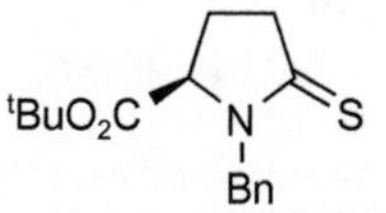
tBuO2C
N
S
Bn

(c) *E*

(d)

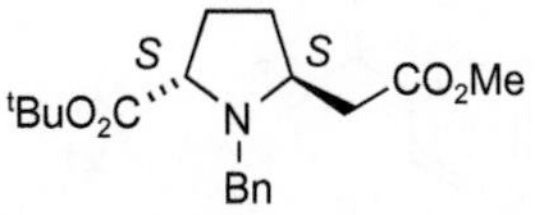
S
S
tBuO2C
N
CO2Me
Bn

4

(minor isomer)

(e) Diastereomers

(f)

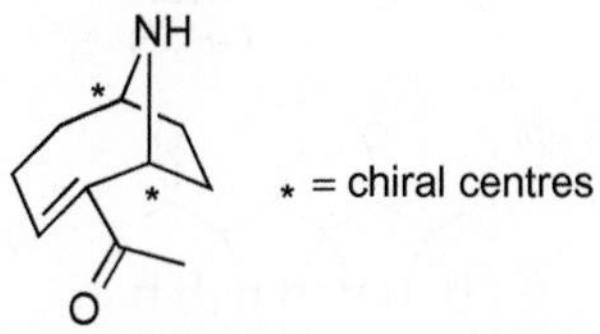
NH
O
* = chiral centres

(+)-anatoxin-*a*

Chapter 19 Organic reaction mechanisms

Worked examples

19.2 More resonance forms can be drawn for $PhCH{=}CHCH_2^+$ because of delocalization of the positive charge around the benzene ring.

19.5 **A** → **B**: Reduction,
B → **C**: Neither oxidation or reduction

19.6 (a) An electrophilic substitution reaction
(b) Step 1: nucleophilic addition. Step 2: nucleophilic acyl substitution.

19.7 (a) Step 1: Because it leads to the selective formation of an alkene with *Z* rather than *E* configuration. Step 2: Because it leads to the selective formation of a *cis*-epoxide rather than a *trans*-epoxide.
(b) Because the *Z*-alkene selectively forms the *cis*-epoxide; the corresponding *E*-alkene selectively forms the *trans*-epoxide.
(c) A diastereoselective reaction

Boxes

19.6 Steps 1 and 2 involve oxidation.

19.8 (b) Carbocation **A** (a tertiary carbocation) is more stable because it has three +I groups that delocalize the charge. A primary carbocation only has one +I group that can delocalize the charge.

Questions

1 (a) +I
(b) –I, –M
(c) –I, –M
(d) –I, +M
(e) –I, +M
(f) –I, –M

2 Primary carbocations without resonance (especially the methyl carbocation), are extremely unstable cations, so it is unlikely that they will exist as intermediates in reactions if other pathways are possible.

3 The allyl radical is stabilized by resonance.

4 (Least stable) **B** < **A** < **C** (Most stable)

6 (a) Three nucleophilic substitution reactions
(b) An elimination reaction (called a Hofmann elimination reaction)
(c) Because it leads to the selective formation of one structural isomer over another; but-1-ene (a terminal alkene) is formed in much higher yield than but-2-ene (an internal alkene).
(d) Because it leads to the selective formation of the *E*-isomer, rather than the *Z*-isomer, of but-2-ene.

7* (a) Oxidation level 0 in the starting material and 2 in the product. Step 1 is an oxidation reaction.
(b) Nucleophilic addition
(c) Oxidation level 3 in the starting material and 3 in the product. Step 3 is neither oxidation nor reduction.
(d) Nucleophilic acyl substitution
(e) Chemoselective
(f) Nucleophilic substitution
(g) Nucleophilic substitution

(h) The arrow indicates the configuration is *R*, but, the lowest priority group (H) is pointing towards you, so the configuration is *S*.

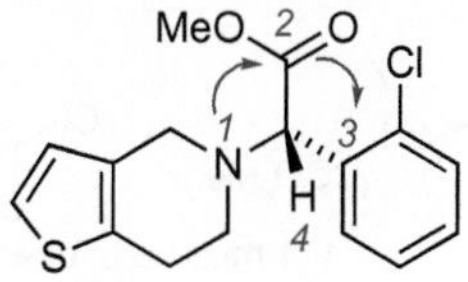
MeO
2
O
Cl
1
N
3
H
4
S

(i)

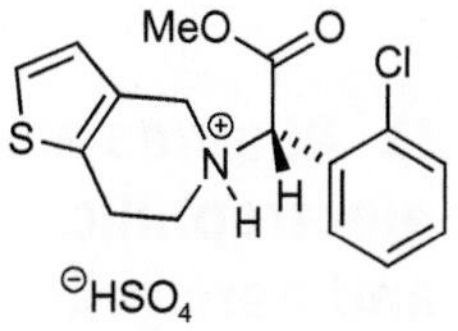
MeO
O
Cl
S
N
H
H
HSO4

Chapter 20 Halogenoalkanes: substitution and elimination reactions

Boxes

20.1 (a) Secondary carbon atom

20.2 (a) The RO• radical adds to the least hindered carbon atom of the C=C bond.
(b) The RO• radical adds to the C=C bond because this leads to the formation of the strongest bond (a C–O bond, which is stronger than an O–Cl bond).
(c)

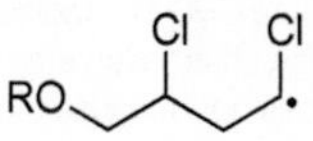
Cl
Cl
RO

Questions

4 (a)

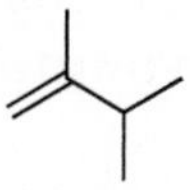

The tertiary bromoalkane undergoes an E2 reaction.

(b)

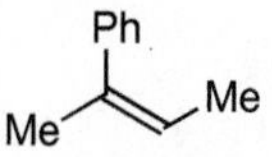
Ph
Me
Me

The secondary bromoalkane undergoes an E2 reaction.

(c)

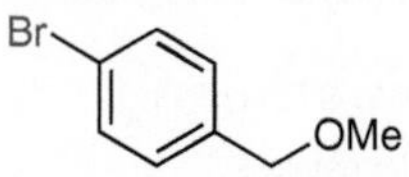
Br
OMe

The primary benzylic bromide undergoes an S_N1 reaction.

6 (a) By an S_N2 mechanism
(c) Slow addition of the substituted phenol and base to 1,4-dibromobutane. A low reaction temperature.

7* (b) The •CCl_3 radical adds to the less substituted carbon atom of the C=C bond, for steric reasons, so as to form the more stable radical (i.e. a secondary carbon-centred radical is formed, whereas attack at the more substituted carbon gives a primary carbon-centred radical).
(c) Addition of the •CCl_3 radical to the C=C bond forms a secondary carbon-centred radical which has a planar shape. Consequently, CCl_4 reacts equally with either face of the radical to form an equal mixture of *R* and *S* enantiomers.

(e) The *trans* isomer is the major diastereomer. It is the more stable isomer as the large ester and alkene substituents are attached to the opposite sides of the cyclopropane ring, which minimizes steric strain.

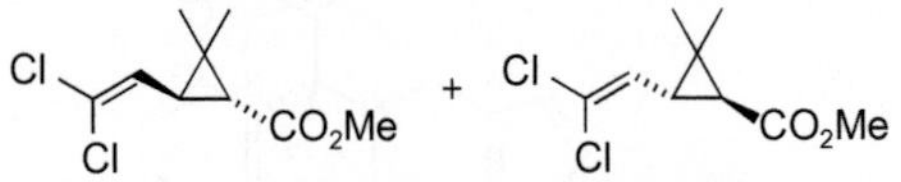
Cl
Cl
CO2Me
+
Cl
Cl
CO2Me

1 : 1 mixture of enantiomers (racemate)

Chapter 21 Alkenes and alkynes: electrophilic addition and pericyclic reactions

Worked examples

21.3 (a)

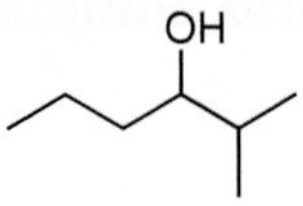
OH

21.4

O
OH
+
O
OH

21.5 H_2, Lindlar catalyst

Boxes

21.2 HCl does not add to a C=C in a radical chain reaction because the abstraction step is endothermic. HI does not because the addition step is endothermic.

Questions

2 (a) OsO_4, H_2O or $KMnO_4$/HO^-/H_2O at low temperature
(b) H^+, H_2O
(c) Br_2, H_2O
(d) BH_3 then H_2O_2, HO^-
(e) RCO_3H then H^+, H_2O (or HO^-, H_2O)
(f) O_3 then H_2O_2 (or $KMnO_4$, heat)
(g) RCO_3H
(h) O_3 then Me_2S (Zn or PPh_3)

3 (a) H_2, Lindlar catalyst
(b) (*Z*)-But-2-ene-1,4-diol
(c) RCO_3H
(d)

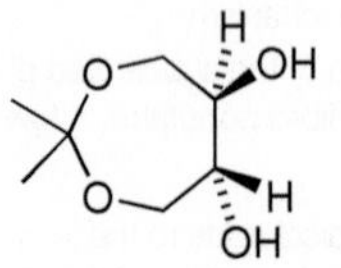
H
OH
O
O
H
OH

(e) An acetal

4 (a)

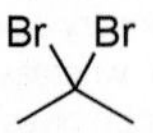
Br
Br

(b)

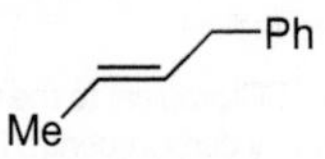
Ph
Me

(c)

5 (a) Two C=C bonds

(b)

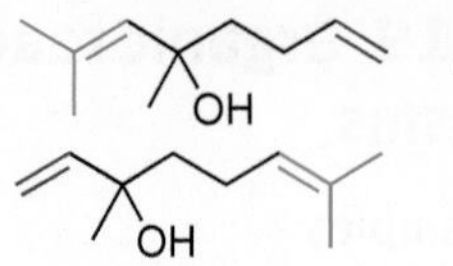
OH
&
OH

(c)

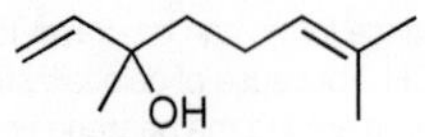
OH

6* (b)

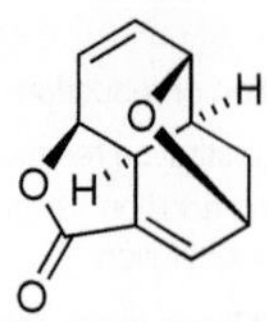
O
H
O
H
O

Chapter 22 Benzene and other aromatic compounds: electrophilic substitution reactions

Worked examples

22.1 Because the carbocations (formed on heterolytic cleavage of the C–Br bonds) have different stabilities.

22.3 (a) Compound **4**: faster, Compound **5**: slower, Compound **6**: slower.

Boxes

22.2 (a) H^+ adds to the terminal carbon atom of the C=C bond to form the more stable carbocation.
(b) Because the tertiary C–H bond is the weakest bond.

Questions

1 Compound **3** is aromatic.

2 (a)

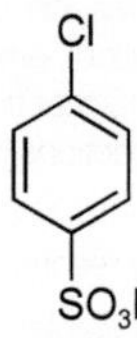
Cl
SO3H

(b)

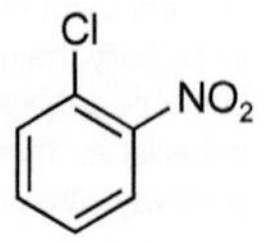
Cl
NO2

(c) Attack at the 2- and 4-positions produces the most stable carbocations.

3 (c) Benzocaine **5** reacts more rapidly with electrophiles.

Chapter 23 Aldehydes and ketones: nucleophilic addition and α-substitution reactions

Boxes

23.2 (a) H^+ adds to CrO_3 and this converts it into a stronger electrophile.

Questions

1 (a) $NaBH_4$ or $LiAlH_4$ then H^+. Reduction reaction.
(c) PhBr and Mg in dry diethyl ether

(d)

O
O
H
HO
CN
H

4 **5**

2 (a)

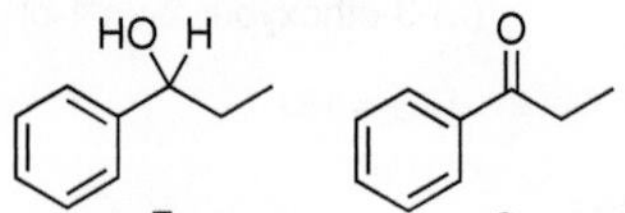
HO H
O

7 **8**

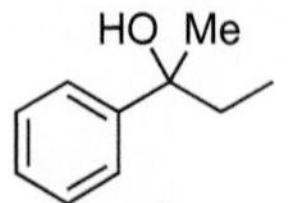
HO Me

9

(b) **6** into **10**: ethane-1,2-dithiol ($HSCH_2CH_2SH$), H^+, anhydrous $MgSO_4$; **10** into **11**: Raney Ni, H_2

3 (a)

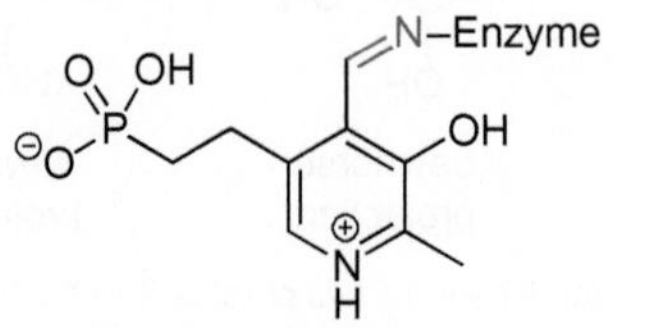
An imine
N–Enzyme
O
OH
O
P
OH
N
H

(b) Addition–elimination reaction
(c) An acid catalyst is required to convert the hemiaminal into the imine. Protonation of the hemiaminal converts the OH group into a better leaving group, so that elimination of water takes place (rather than HO^-).
(d)

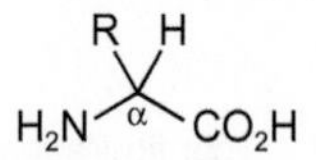
R H
H2N
α
CO2H

(e) Strecker synthesis. React the aldehyde or ketone with NH_4Cl and NaCN (to give an α-amino nitrile) followed by H^+/H_2O.

4 (a)

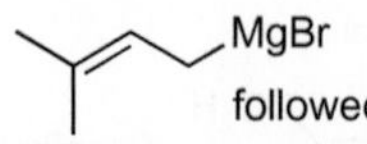
MgBr

followed by $H^{\oplus}$ H_2O

(b) Dry Et_2O must be used otherwise the Grignard reagent will react with water to form $Me_2C{=}CHCH_2CH_3$ and HOMgBr.

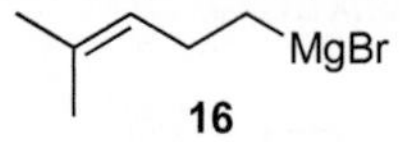
MgBr

16

(c) Step 1: CH_3CHO then H^+, H_2O; Step 2: CrO_3, H^+
(e) Aldehyde **14** would react more rapidly with $BrZnCH_2CO_2Et$ than ketone **17**.

5 (a)

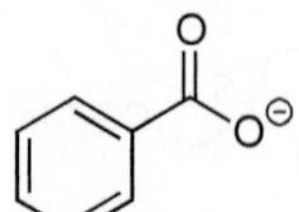
O
O

20

CHI_3 **21**

(c) HCHO, HO^-, heat

6* (a)

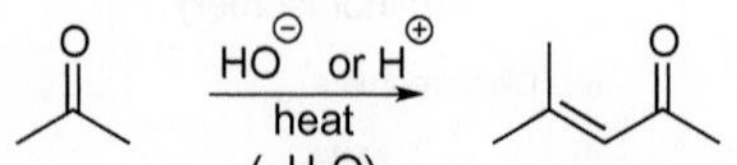
O
HO or H
heat
(–H2O)
O

24

(b)

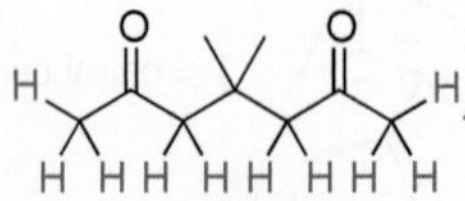
O
O
H
H
H H H H H H H H

25

There are 10 α-hydrogen atoms

(c) 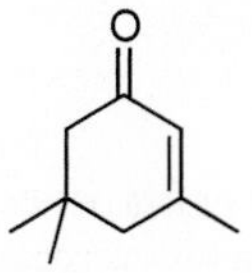
O
isophorone

Chapter 24 Carboxylic acids and derivatives: nucleophilic acyl substitution and α-substitution reactions

Worked examples

24.4 Route 1: Slow addition of a mixture of ethoxide ion and $PhCOCH_3$.

Route 2: Slow addition of a mixture of ethoxide ion and CH_3COCH_3.

Boxes

24.2 (Most reactive) **B** > **E** > **D** > **C** > **A** (Least reactive)

24.6 Sodium stearate is a saturated fat because the carbon chain does not contain any C=C bonds; it is fully saturated with hydrogen atoms.

Questions

1 (a) $LiAlH_4$ then H^+. Reduction.

(b) 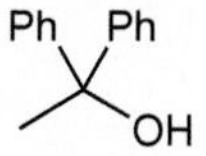
Ph
Ph
OH

(c) (ii) 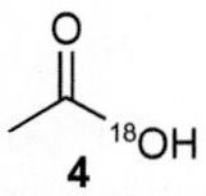
O
18OH
4

(c) (iii) $EtO^{18}H$

(e) 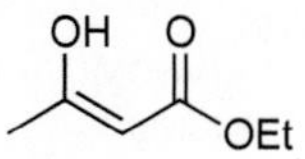
OH
O
OEt

3 (a) A steroid

(c) 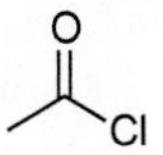
O
Cl

(e) An oxime

4 (b) Enolization of the three C=O bonds produces a stable aromatic ring.

5 (a) Two molecules of the monomer combine with the elimination of a molecule of water.

(b) At room temperature, the amine and carboxylic acid react to form a salt, not an amide bond.

(c) Acyl chlorides have a particularly electrophilic carbon atom and react rapidly with amines in nucleophilic acyl substitution reactions.

(d) (ii) Under a particularly high concentration of hydroxide ion.

Chapter 25 Hydrogen

Worked examples

25.1 1630 cm^{-1}

Boxes

25.1 55 $kJ\,g^{-1}$; this is much lower than the value for hydrogen (143 $kJ\,g^{-1}$)

25.3 Mg_2Ni: 3.8%, FeTi: 2.0%, $LaNi_5$: 1.4%. None of these compounds meet the US Department of Energy's target.

25.4 It cannot be kept in glass bottles and is generally stored in containers made of PTFE, poly (tetrafluoroethene).

25.5 Complementary hydrogen bonding groups.

25.6 192 dm^3

25.7 ${}^{14}_{7}N + {}^{1}_{0}n \rightarrow {}^{12}_{6}C + {}^{3}_{1}H$. The other product is ^{12}C.

Questions

1 386 dm^3

2 (a) Sn, (c) Ba, and (d) Ni have negative standard electrode potentials, so would be expected to displace hydrogen from dilute acids.

3 LiH: H = −1, Li = +1
HI: H = +1, I = −1
NH_3: H = +1, N = −3
SiH_4: H = −1, Si = +4
B_2H_6: H = −1, B = +3

4 (a) $CsH(s) + H_2O(l) \rightarrow CsOH(aq) + H_2(g)$

(b) $B_2H_6 + 2py \rightarrow 2H_3B \leftarrow py$

(c) $2Si_2H_6(g) + 7O_2(g) \rightarrow 4SiO_2(s) + 6H_2O(g)$

(d) $NH_2NH_2(aq) + H_3O^+(aq) \rightleftharpoons NH_2NH_3^+(aq) + H_2O(l)$

5 SH_6 would decompose exothermically to give H_2S and H_2. The reaction is highly exothermic because of the formation of strong H–H bonds.

6 (a) BeH_2: covalent hydride with a polymeric structure (it is electron deficient).

(b) PH_3: covalent hydride with a molecular structure (it is electron rich).

(c) KH: ionic hydride

(d) HCl: covalent hydride with a molecular structure (it is electron precise).

(e) $FeTiH_{1.8}$ is a metallic hydride.

8* 543 g

9 −273 $kJ\,mol^{-1}$

11 (a) Ammonia (NH_3) contains strong hydrogen bonds.

(b) The melting point is determined by London dispersion interactions, meaning the melting point increases with increasing relative molecular mass.

(c) Ice has a very open hydrogen bonding structure that partially collapses on converting the solid to the liquid.

12 1630 cm^{-1}

13* 1730 cm^{-1}

14* (a) 11 800 kJ (= 11.8 MJ)

(b) 2.19 kg

Chapter 26 *s*-Block chemistry

Worked examples

26.1 +107 $kJ\,mol^{-1}$

26.2 −406 $kJ\,mol^{-1}$; CaF will not exist.

Boxes

26.1 $\Delta_r H^{\ominus} = +635\,kJ\,mol^{-1}$

26.3 Hydraulic cements harden in air and water, not just air. Hydraulic cements are more durable, and less soluble in water.

26.4 Citric acid is a weak acid. It reacts with insoluble $CaCO_3$ to form soluble calcium citrate, CO_2, and water. Citrate is also a chelating ligand.

Questions

1 1.01 kg

2 (a) $2Cs(s) + 2H_2O(l) \rightarrow 2CsOH(aq) + H_2(g)$

(b) $Rb(s) + O_2(g) \rightarrow RbO_2(s)$

(c) $4LiNO_3(s) \rightarrow 2Li_2O(s) + 4NO(g) + 3O_2(g)$

3 The enthalpy change of formation of $NaCl_2$ is strongly endothermic (+2150 $kJ\,mol^{-1}$), so it is unstable with respect to the elements.

4 (a) KNO_3

(b) LiH

(c) $SrCO_3$

5 M is Li, Y is LiOH, X is Li_2O, Z is H_2

6* $LiNO_3$

7 (a) $LiNO_3$

(b) RbOH

(c) NaBr

8 The lithium ion fits in the crown ether cavity whereas the potassium ion is too big.

9* (a) $Mg^{2+}(aq) + Ca(OH)_2(aq) \rightarrow Mg(OH)_2(s) + Ca^{2+}(aq)$

$Mg(OH)_2(s) + 2H_3O^+(aq) \rightarrow Mg^{2+}(aq) + 4H_2O(l)$

$Mg^{2+}(aq) + 2Cl^-(aq) \xrightarrow{evaporation} MgCl_2(s)$

(b) 2

10 (a) $Be(s) + 4OH^-(aq) \rightarrow [Be(OH)_4]^{2-}(aq)$

(b) $SrCO_3(s) \rightarrow SrO(s) + CO_2(g)$

(c) $2BaO(s) + O_2(g) \rightarrow BaO_2(s)$

11 Add water to the mixture to convert both the nitride and the oxide into the hydroxide. Then heat the hydroxide to get a pure sample of MgO.

12 tetrahedral, 109.5°, sp^3 hybridization

13 $CaO(s) + SO_2(g) \rightarrow CaSO_3(s)$

15* −136 $kJ\,mol^{-1}$. Need to consider stability with respect to disproportionation.

16 $Ba(OH)_2$

17 The high charge density of Be^{2+} ensures it is hydrated in solution and that the O–H bonds are sufficiently polar for hydrolysis to occur.

Chapter 27 *p*-Block chemistry

Worked examples

27.1 −942 $kJ\,mol^{-1}$. The conversion of N_4 to N_2 is highly exothermic, so N_2 is more stable.

27.2 −918 $kJ\,mol^{-1}$

27.3 Addition of a fluoride ion to IF_5

Boxes

27.1 Extraction of aluminium from bauxite requires a higher temperature and a high electrical energy input for electrolysis.

27.3 The C–O bond is weaker than the Si–O bond. Carbon analogues of silicones would be thermodynamically unfavourable with respect to CO_2.

27.4 There is only one phosphorus environment, so only one signal would be seen. This would appear as a sextet as the phosphorus nucleus is coupled to five ^{19}F nuclei which are all equivalent on the NMR timescale.

27.7 No

27.8 The C–F and C–H bonds are too strong to be broken by ultraviolet radiation from the Sun.

27.9 ^{40}Ca is naturally present in rocks. It is impossible to distinguish between the ^{40}Ca created by radioactive decay and that already present.

Questions

1 (a) +4
(b) +8
(c) +2
(d) +0.5
(e) +6
(f) +1

2 (a) Bi < Sb < As < P < N
(b) C < O < N < F

3 (a) Acidic
(b) Basic
(c) Amphoteric

6 AlF_3 has an ionic structure whereas Al_2Cl_6 is covalent.

10 $\Delta_f H^\ominus (PCl_3, g) = -300\ kJ\,mol^{-1}$, $\Delta_f H^\ominus (PI_3, g) = +84\ kJ\,mol^{-1}$. PCl_3 is a stable molecule, but PI_3 is thermally unstable with respect to the elements.

11 The conversion of O_8 to four O_2 would be highly exothermic ($-840\ kJ\,mol^{-1}$), so O_2 is the more stable form of oxygen.

12 Thiosulfate, +2
Tetrathionate, +2.5

13 Fluorine is more electronegative than hydrogen. As a consequence, the electrons in the bond are pulled further form the central oxygen atom, and hence repel the lone pairs less.

14 $+81\ kJ\,mol^{-1}$. No.

15* $5IF \rightarrow IF_5 + 2I_2$; $-365\ kJ\,mol^{-1}$

16 (a) $XeF_2 + AsF_5 \rightarrow [XeF]^+[AsF_6]^-$
(b) $BI_3 + NH_3 \rightarrow I_3B \leftarrow NH_3$
(c) $P_4O_{10} + 6H_2O \rightarrow 4H_3PO_4$
(d) $SF_4 + BF_3 \rightarrow [SF_3]^+[BF_4]^-$

17 Paramagnetic: NO_2, NO, and ClO_2

18 (a) Square pyramidal
(b) Square planar
(c) Trigonal pyramidal
(d) Distorted tetrahedral
(e) T-shaped
(f) Disphenoidal

19 (a) $IF_7 + SbF_5 \rightarrow [IF_6]^+[SbF_6]^-$
(b) Dehydration of propanedioic acid with P_4O_{10}
(c) Reaction of XeF_4 with water

Chapter 28 *d*-Block chemistry

Worked examples

28.1 Co^{2+}: [Ar] $3d^7$, Co^+: [Ar] $3d^8$

28.2 +4

28.3 (a) Bis(ethane-1,2-diamine)diiodocobalt(III) iodide
(b) $[CoBr_2(H_2O)_2(MeNH_2)_2]NO_3$

28.4 $-1.2\Delta_o$

28.5 High spin: $-48\ kJ\,mol^{-1}$, low spin: $+202\ kJ\,mol^{-1}$. $[Fe(H_2O)_6]^{2+}$ is high spin.

28.6 No

28.7 High spin

Boxes

28.1 The half life of ^{99m}Tc is only 6 hours. If the compound cannot be made quickly the ^{99m}Tc will have decayed before the compound is used.

28.2 +1, linear

28.3 *cis*-diamminedichloroplatinum(II), amminedichloro(cyclohexylamine)-diethanoatoplatinum(IV)

28.4 $[Zn_4O(O_2CC_6H_4CO_2)_3]$

28.5 $(t_{2g})^3\ (e_g)^2$

Questions

1 (a) [Ar] $3d^2$
(b) [Ar] $3d^6\ 4s^2$
(c) [Ar] $3d^5\ 4s^1$
(d) [Kr] $4d^{10}$
(e) [Ar] $3d^8$
(f) [Ar] $3d^9$
(g) [Kr] $4d^6$
(h) [Xe] $4f^{14}\ 5d^2$

3 (a) +3, 6
(b) +2, 4
(c) +4, 5
(d) +2, 6
(e) +3, 6

4 (a) Pentaammineiodonickel(II) nitrate
(b) Potassium diaquadioxalatocobaltate(III)
(c) Bis(1,4,7-triazaheptane)iron(II) sulfate
(d) Sodium dibromodichloroaurate(III)

5 (a) $[AuCl(SEt_2)]$
(b) $[NEt_4][FeBr_4]$
(c) $[Mo(bipy)_2(CO)_2]$
(d) $K_3[Co(ONO)_6]$

6 X is $[TiCl_2(H_2O)_4]Cl{\cdot}2H_2O$. *Cis*- and *trans*-isomers of X. Y is $[Ti(H_2O)_6]Cl_3$. No isomers of Y.

7 (a) Not isomers
(b) Ionization isomers
(c) Coordination isomers
(d) Linkage isomers

8 (a) Enantiomers
(b) Geometrical isomers. In addition, the *cis*-isomer forms enantiomers.
(c) square planar *cis* and *trans* geometrical isomers or tetrahedral
(d) *fac* and *mer* geometrical isomers

9 (a)

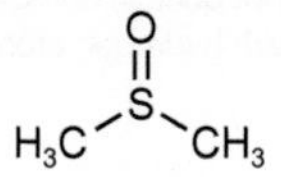
O
S
H3C
CH3

DMSO

(b) (i) *S*-donor
(ii) *O*-donor

10 (a) $[TiCl_6]^{3-}$
(b) $[NiCl_4]^{2-}$
(c) $[Fe(H_2O)_6]^{3+}$

11 Tetragonally distorted: $[Cu(NH_3)_6]^{2+}$ only

12 $[Mn(H_2O)_6]^{3+}$ = high spin complex, $[Re(H_2O)_6]^{3+}$ = low spin complex

13 (a) $[MnCl_4]^{2-} < [Mn(NH_3)_6]^{2+} < [Mn(CN)_6]^{4-} < [Mn(CN)_6]^{3-}$
(b) $[Mn(NH_3)_6]^{2+} < [Mn(CN)_6]^{3-} < [MnCl_4]^- < [MnO_4]^-$

16 Square planar

17 (a) $2.83\ \mu_B$
(b) $3.87\ \mu_B$
(c) $5.92\ \mu_B$
(d) $0\ \mu_B$

18 $[Ni(H_2O)_6]^{2+} + dien \rightleftharpoons [Ni(dien)(H_2O)_3]^{2+} + 3H_2O$, $[Ni(dien)(H_2O)_3]^{2+} + dien \rightleftharpoons [Ni(dien)_2]^{2+} + 3H_2O$
$\beta_2 = 9.1 \times 10^{18}$

Index

A

B

C

D

E

H

I

P

S

Figure Acknowledgements

Chapter 1

p.2 (background) CDC/SCIENCE PHOTO LIBRARY; **p.3 (top)** Image courtesy of Prof. Ally Lewis, Department of Chemistry, University of York; **p.3 (bottom)** Courtesy of Fundación CEAM, Valencia, Spain; **p.4 (left/top)** Janaka Dharmasena; **p.4 (left/middle)** Courtesy of Michael Graetzel. Photo credit: Alain Herzog; **p.4 (left/bottom)** © Professor Alexander M. Seifalian; **p.5 (Box 1.1 bottom/right)** Logo from Human Genone Project; **p.6 (Box 1.1)** Kansas State University; **p.7 (right)** Reproduced by permission of The International Union of Pure and Applied Chemistry (IUPAC); **p.11 (Fig. 1.5)** Crulina 98/ CC-BY-SA; **p.12 (Fig. 1.6)** Charles D Winters/Science Photo Library and IDPS, University of Bath; **p.13 (top/right)** Adrian Hughes/iStock; **p.15 (Box 1.2, Fig. 3)** Courtesy of Bruker Daltonik ; **p.17 (Box 1.3)** Wessex archaeology; **p.21 (right)** Paul Cowan/iStock; **p.25 (Box 1.4)** Marco Maccarini/ iStock; **p.26 (Fig. 1.8)** IDPS, University of Bath; **p.27 (Fig. 1.9)** Charles D Winters/Science Photo Library; **p.31 (Worked Example 1.10)** IDPS, University of Bath; **p.32 (Worked Example 1.11)** IDPS, University of Bath; **p.35 (Box 1.5)** NOAA; **p.37 (Fig. 1.10)** IDPS, University of Bath; **p.37 (Fig. 1.11)** Courtesy of Metrohm AG/Brinkmann Instruments Inc.; **p.38 (Box 1.6)** Courtesy of David Preston, Environment Agency; **p.39 (Fig. 1.12)** IDPS, University of Bath; **p.41 (Fig. 1.14)** Oxford University Press; **p.42 (Box 1.7, Fig. 1)** Matej Michelizza/ iStock; **p.45 (Box 1.9)** Image courtesy of Braun; **p.48 (Fig. 1.21)** Melissa Caroll/ iStock and Clive Streeter/Dorling Kindersley; **p.49 (Fig. 1.22)** IDPS, University of Bath; **p.51 (Box 1.10, Fig. 4)** Digital Vision; **p.54 (left)** Courtesy of Chen Ho at Pittsburgh Supercomputing Centre; **p.55 (Box 1.11 top/right)** simonkr/ iStock; **p.56 (left)** Courtesy of Nestlé Waters ; **p.59 (Box 1.12)** Ken Rygh Creative Art & Design/iStock

Chapter 2

p.69 TBC; **p.69 (bottom/right)** Oxford University Press; **p.70 (left)** Adam W. Brown/Origins of Life Experiment #1.x (July 2010)/By Adam Brown in collaboration with Robert Root-Bernstein; **p.71 (Box 2.1 right)** Courtesy of Maiken Naylor; Arts and Sciences Library, University at Buffalo; **p.72 (Box 2.2)** Mosquito courtesy of Michael Pettigrew/iStock; Algal bloom courtesy of Miriam Godfrey/NIWA; **p.73 (top/right)** Anna Bryukhanova/iStock; **p.74 (top/left)** TBC; **p.75 (top/right)** Martin Diebel/fStop Photos; **p.77 (top/right)** Dawn Roberts/iStock; **p.79 (right/top)** Oxford University Press; **p.79 (right/bottom)** Jon Candy/CC-BY-SA; **p.80 (bottom/left)** Andy Burrows; **p.82 (bottom/left)** Fancy Images ; **p.86 (Box 2.4 left)** Oxford University Press; **p.86 (Box 2.4right)** Courtesy of Raisio Group, Finland; **p.87 (right)** PhotoDisc; **p.88 (top/left)** Glowimages; **p.92 (Box 2.5)** Bettmann / Contributor; **p.93 (top/right)** Damir Spanic/iStock; **p.94 (Box 2.6)** Andrew Lambert Photography/Science Photo Library; **p.95 (top/right)** Geoff Kuchera/iStock; **p.97 (Box 2.7 right)** Anthony Hall/iStock; **p.100 (top/left)** Corbis Photos; **p.101 (right/top)** Henrik Larsson/ iStock; **p.101 (right/middle)** Oxford University Press; **p.102 (Box 2.8 right)** Prof. Dr. med. Otfried Müller, Institute of Anatomy, University of Bern, Switzerland; **p.103 (top/right)** Ben Heys/iStock; **p.104 (left)** Oxford University Press; **p.105 (Box 2.9 right)** Phil Jackson/iStock; **p.109 (Concept Review top/ right)** Oxford University Press; **p.110 (Questions left/top)** Stephen J Krasemann; **p.110 (Questions left/bottom)** Oxford University Press; **p.110 (Questions left/ bottom)** TBC; **p.111 (Questions top/right)** Oxford University Press

Chapter 3

p.112 (bottom/left) Image originally created by IBM Corporation (Don Eigler, IBM Almaden Research Center); **p.113 (top/right)** Peter Nellist; University of Oxford; **p.113 (background)** Image originally created by IBM Corporation. Originally published in: M.F.; Crommie, C.P. Lutz, D.M. Eigler. Confinement of electrons to quantum corrals; on a metal surface. Science 262, 218–220 (1993)"; **p.114 (top/left)** Courtesy of Edward Redish, University of Maryland;; **p.120 (Box 3.1 top/right)** Atmospheric Change: An Earth System Perspective by Thomas E. Graedel and Paul J. Crutzen. © 1993 by W H Freeman and company. Used with permission.; **p.120 (Box 3.1 bottom/left)** Courtesy NASA/JPL-Caltech; **p.122 (top/left)** travelpixpro/iStock; **p.123 (Fig. 3.7 right)** Dietrich Zawischa, Institut für Theoretische Physik, Universität Hannover; **p.124 (bottom/left)** Andrew Burrows; **p.129 (Box 3.2 bottom)** 1995 C Richard Megna - Fundamental Photographs; **p.129 (Box 3.2 top/right)** Caterina Foti; Foti International Fireworks; **p.130 (left)** Image 101; **p.131 (Box 3.3 bottom/ left)** National Optical Astronomy Observatories/Science Photo Library; **p.131 (Box 3.3 top/right)** NASA/JPL Caltech; **p.132 (Fig. 3.14)** Andrew Lambert Photography/Science Photo Library; **p.145 (Box 3.6)** Cordelia Molloy/ Science Photo Library; **p.149 (Box 3.8)** University of Pennslyvania; **p.161 (Box 3.9)** Courtesy of Jon Bodsworth; **p.162 (top/left)** Achim Zschau, GSI Helmholtzzentrum für Schwerionenforschung; **p.163 (Box 3.10 bottom)** Gwen Pilling

Chapter 4

p.168 (bottom/left) European Organisation for Astronomical Research in the Southern Hemisphere (ESO); **p.168 (bottom/right)** NASA/JPL Caltech; **p.169 (top/right)** NASA; **p.169 (background)** NASA; **p.170 (left)** NOAA; **p.173 (left/middle)** © 1916, American Chemical Society; **p.173 (right/ middle)** The Bancroft Library, University of California; **p.175 (Fig. 4.5)** © 1992 Richard Megna/Fundamental Photographs; **p.175 (centre)** Sean O'Riordon/ iStock; **p.176 (Box 4.2 top/right)** © High Field Magnet Laboratory, Radboud University ; **p.182 (Box 4.4 bottom/right)** Malcolm Fielding/Johnson Matthey PLC/ Science Photo Library; **p.183 (top/right)** Mallivan/istock; **p.193 (Worked Example 4.4)** NASA; **p.199 (Box 4.7 right)** NOAA; **p.200 (Box 4.8)** biosdi/ iStock; **p.203 (Box 4.9 right)** Courtesy of Sarah D'Angelo/ CC BY-SA 2.0; **p.211 (Worked Example 4.7)** OceanFishing/iStock

Chapter 5

p.216 (bottom/left) Hubert Raguet Eurelios/ Science Photo Library; **p.216 (bottom/right)** Courtesy of Neil Bartlett; **p.217 (top/left)** Courtesy of Neil Bartlett; **p.221 (Box 5.2)** Science Photo Library; **p.222 (Box 5.2)** Robert Young/ iStock; **p.229 (Box 5.3)** © Michel Chammas at Chammas cutters Inc; **p.232 (Box 5.4)** Heike Kampe/iStock; **p.233 (Box 5.4)** Kelly Robert McAllister, Washington Department of Fish and Wildlife; **p.235 (right)** Charles D Winters/ Science Photo Library; **p.240 (Box 5.5)** Mehmet Hilmi Barcin/iStock; **p.243 (bottom)** Richard Goerg/iStock

Chapter 6

p.254 Thomas Mounsey/iStock; **p.255 (background)** Sinclair Stammers/ Science Photo Library; **p.257 (right)** Andy Burrows; **p.258 (Fig. 6.2)** naddi/ iStock; **p.258 (left)** Courtesy of Sir Harry Kroto; **p.259 (Box 6.1right)** Roman Krochuk/iStock; **p.259 (Box 6.1 left)** Dr. Peter Harris/Science Photo Library; **p.259 (bottom/left)** Dr. Heinz Linke/iStock; **p.260 (left/top)** lantapix/iStock; **p.260 (left/middle)** rasoulali/iStock; **p.260 (left/bottom)** PhotoDisc; **p.261 (Box 6.2 right/top)** David Parker/IMI/Science Photo Library; **p.261 (Box 6.2 right/middle)** Yosemite/CC-BY-SA; **p.262 (left)** Tiffany Orpana/iStock; **p.276 (Box 6.3, Fig. 1 top)** CD: Blue Stocking/iStock ; **p.276 (Box 6.3, Fig. 1 bottom)** CD cross-section: Alberto Perez Veiga/iStock; **p.278 (left)** Rainer Walter Schmied/iStock; **p.279 (Box 6.4 bottom/left)** Philippe Plailly/Eurelios/ Science Photo Library; **p.280 (left/top)** Rob Lavinski, iRocks.com; **p.280 (left/ middle)** Rainer Walter Schmied/iStock; **p.281 (top/right)** Alexofdodd at the English language Wikipedia CC BY-SA 3.0; **p.283 (Box 6.5 bottom/right)** Courtesy of Pilkington Group Limited; **p.284 (Box 6.5)** Courtesy of Pilkington Group Limited; **p.286 (Fig. 6.28)** Adapted from Smart and Moore (2012) Solid state chemistry: an introduction, 4th ed. CRC press, with permission from Routledge

Chapter 7

p.302 (bottom/left) Ever/ iStock; **p.302 (bottom/right)** BigWest1/iStock; **p.305 (top/right)** RMAX/iStock; **p.306 (Box 7.1)** Fadil Kamarudin/iStock; **p.307 (Box 7.2 right/top)** Courtesy of David Briffet, author of The Acid Bath Murders; **p.307 (Box 7.2 right/bottom)** Alex Raths/iStock; **p.313 (Box 7.3 left)** D. Kirk Nordstrom, US Geological Survey; **p.313 (Box 7.3 right)** Ian Hamilton/iStock; **p.317 (Box 7.4)** Hans Caluwaerts/iStock; **p.319 (top/right)** John Christian Lonningdal/iStock; **p.321 (Box 7.5)** Courtesy of DtheZombie/ CC BY-SA 3.0; **p.322 (top/left)** TBC; **p.329 (Fig. 7.10)** Andrew Burrows; **p.331 (top/right)** Courtesy of Simon Cousins/CC BY 2.0; **p.335 (Box 7.6 top)** Left to right: Oxford University Press, Oxford University Press, Mykola Velychko/ iStock; **p.335 (Box 7.6 bottom/right)** Royal Society of Chemistry

Chapter 8

p.342 TBC; **p.343 (background)** TBC; **p.345 (top/right)** Jose Antonio Sanchez Reyes/iStock; **p.353 (Box 8.1 left)** uatp2/iStock; **p.357 (right)** PhotoDisc; **p.366 (top/left)** Saturated/iStock; **p.367 (top/right)** Andrew Lambert Photography/ Science Photo Library; **p.367 (Box 8.4, Fig. 1)** Courtesy of Oak Ridge National Laboratory; **p.371 (top/right)** PhotoDisc; **p.373 (right)** TBC

Chapter 9

p.382 NOAA; **p.383 (background)** George Burba/ 123RF; **p.384 (bottom/ left)** Peter Rejcek, National Science Foundation; **p.385 (top/left)** Brendan Hunter/iStock; **p.385 (top/middle)** Amy Kuck/iStock; **p.391 (right)** CREATIVE COMMONS; **p.397 (right/top)** Courtesy of Transitions Optical Inc; **p.397 (right/middle)** Courtesy of Transitions Optical Inc; **p.397 (right/bottom)** Courtesy of Transitions Optical Inc; **p.400 (Box 9.3 right)** anzeletti/iStock; **p.403 (Box 9.4, Fig. 1)** Reproduced from Discussions of the Faraday Society, 'Studies of free radical reactivity by the methods of flash photolysis: the photo chemical reaction between chlorine and oxygen', George Porter and Franklin J Wright, 14, 1953. With permission of The Royal Society of Chemistry; **p.403 (top/left)** Ooyoo/iStock; **p.403 (top/middle)** With kind permission from Ahmed Zewail, California Institute of Technology; **p.404 (Box 9.4, Fig. 2)** Courtesy of Robert Salter, School of Chemistry, University of Leeds; **p.411 (top/right)** Science Photo Library; **p.412 (Box 9.7)** miodrag ignjatovic/ iStock; **p.420 (bottom/left)** vovan13/iStock; **p.423** TBC; **p.424 (bottom/left)** Courtesy of Nathan Pitt and Peter Wothers, University of Cambridge; **p.425 (bottom)** TBC; **p.426 (top/left)** Paul Morton/iStock; **p.437 (bottom/right)** Courtesy of Thomas Ihle/ CC BY-SA 3.0

Chapter 10

p.449 (background) NASA; **p.451 (Fig. 10.3)** George Burba/123rf; **p.453 (Fig. 10.5)** Adapted from Burton et al (2000) Salters Advanced Chemistry: Chemical Ideas, Second edition, Heinemann, reprinted with permission from Pearson; **p.459 (bottom/right)** Charles D Winters/Science Photo Library; **p.464 (Box 10.2, Fig. 1)** NREL/US Dept of Energy/Science Photo Library; **p.466 (left)** Gusto images/Science photo library; **p.474 (left)** PhotoDisc; **p.477 (Fig. 10.21)** Oxford University Press; **p.483 (top/right)** LovLeah/iStock ; **p.489 (bottom/right)** TBC; **p.491 (top/right)** PhotoDisc; **p.499 (Fig. 10.40)** Alek K/ iStock; **p.500 (top)** Oxford University Press; **p.507 (Box 10.9 bottom/left)** Milanovic/iStock; **p.507 (Box 10.9 top/right)** Gusto images/Science photo library

Chapter 11

p.512 LEFTERIS PITARAKIS/AP/Press Association Images; **p.513 (bottom)** TBC; **p.514 (bottom/left)** Press Association Images; **p.514 (bottom/right)** Pidjoe/iStock; **p.517 (top/left)** "A man embracing a woman" by Dosso Dossi, 1524 © The National Gallery, London; **p.522 (Box 11.1)** Kelly Cline/iStock; **p.527 (Box 11.2)** Courtesy of Abbott Point of Care; **p.527 (bottom)** NASA/ JPL-Caltech/ University of Arizona; **p.529 (Fig. 11.10)** IDPS, University of Bath; **p.532 (left)** Maximillian Stock Ltd/ Science Photo Library ; **p.535 (Fig. 11.17)** Dr Tomasz Ganicz. Published under the GNU Free Documentation License. Photographed using equipment that is property of the centre of Molecular and Macromolecular Studies, Polish Academy of Science.; **p.536 (Fig. 11.18)** Dr Tomasz Ganicz. Published under the GNU Free Documentation License. Photographed using equipment that is property of the centre of Molecular and Macromolecular Studies, Polish Academy of Science.; **p.537 (Box 11.3)** Simon Fraser/Science Photo Library; **p.539 (Box 11.4)** Logo reproduced with permission from The Organisation for the Prohibition of Chemical Weapons (the OPCW) ; **p.540 (Fig. 11.21 bottom)** Oxford University Press; **p.543 (Box 11.6)** TBC; **p.544 (Fig. 11.23)** Oxford University Press; **p.546 (Box 11.7, Fig. 1)** Photomick/iStock; **p.547 (Fig. 11.25)** Soren Wedel Nielsen/ CC-BY-SA

Chapter 12

p.556 Colin Cuthbert/Science Photo Library; **p.557 (top/right)** dogayusufdokdok/iStock; **p.558 (right/middle)** Oxford University Press; **p.558 (left/middle)** Oxford University Press; **p.559 (bottom/right)** Oxford University Press; **p.560 (Box 12.1 left)** Jakub Niezabitowski/iStock photo; **p.560 (Box 12.1 right)** Institut für Anorganische und Analytische Chemie Universität Münster; **p.562 (top/left)** Jurc/iStock; **p.562 (bottom/left)** Oxford University Press; **p.568 (Box 12.2 bottom/left)** US Airforce Photo/Staff Sgt Eric T Sheler; **p.570** Stefan Klein/iStock; **p.571 (right/top)** Akinshin/iStock; **p.571 (right/middle)** PhotoDisc; **p.572 (bottom/left)** Oxford University Press; **p.573 (top/right)** Leezsnow/iStock; **p.578 (top/left)** Courtesy of Kurt Wüthrich; **p.579 (Box 12.3)** Oxford University Press; **p.583 (top/right)** etienne voss/ iStock; **p.587 (top/right)** Oxford University Press; **p.588 (top/left)** Logan Buell/iStock; **p.591 (top/right)** Oxford University Press; **p.593 (top/right)** Oxford University Press; **p.607 (Questions top/right)** Scott Bauer/USDA; **p.608 (Questions left/middle)** Copyright 2007, David Monniaux; **p.608 (Questions left/bottom)** www.fly53.com

Chapter 13

p.610 NASA; **p.611 (background)** NASA; **p.612 (top/left)** Courtesy of Alice Mumford; **p.612 (middle)** Stock Up Image Source; **p.615 (Worked Example 13.1left)** Courtesy of BREITLING ; **p.619 (Fig. 13.3)** Andrew Lambert Photography/Science Photo Library; **p.619 (Fig. 13.4)** Colin Soutari/iStock; **p.620** Helene Vallee/iStock; **p.621 (Box 13.2)** Jami Garrison/iStock; **p.623** AndreasKerrmann/iStock; **p.628** 1001Slide/iStock; **p.630 (Box 13.4 left)** Courtesy of Heater Meals which is a Registered Trademark of Innotech Products Ltd.; **p.630 (Box 13.4 right)** Courtesy of Koolpak; **p.644 (Box 13.7)** Courtesy of Bradford Landmark Society, Bradford, PA, USA; **p.645** Courtesy of Parr Instrument Company, Moline, IL; **p.648 (Box 13.8 left)** Courtesy of the Rowett Institute of Nutrition & Health; **p.649 (right)** Courtesy of Parr Instrument Company, Moline, IL

Chapter 14

p.654 Christian Anthony/iStock; **p.657 (Fig. 14.2)** Courtesy of Nathan Pitt and Peter Wothers; **p.657 (bottom/right)** Mass Ave 975/Wiki commons; **p.659 (top/right)** Courtesy of Daderot/ CC-BY-SA 3.0; **p.673 (Fig. 14.7)** Courtesy of Nathan Pitt and Peter Wothers; **p.674 (Box 14.4)** Courtesy of Art Davis USDA; **p.674 (Box 14.4, Fig.2)** Courtesy of Prof Sheena Radford, Astbury Centre for Structural Molecular Biology, University of Leeds; **p.679 (Box 14.6 bottom/ right)** Alexander Shalamov/iStock; **p.680 (bottom)** TBC; **p.682 (Box 14.7)**

Nicolas Loran/iStock; **p.685 (Box 14.9)** Courtesy of Diego Delso/ CC-BY-SA 3.0; **p.689 (Box 14.10)** The Yorck Project; **p.TBC (TBC TBCTBC)** TBC – *UNFIG11*

Chapter 15

p.694 Martin Strmko/iStock; **p.696 (centre)** © 1990 Richard Megna - Fundamental Photographs; **p.698 (top/left)** Courtesy of Mathias Krumbholz/ CC BY-SA 3.0; **p.701 (Box 15.1)** Courtesy of Henna/ CC BY-SA 1.0; **p.703 (top/right)** © Stefan K. Hetz; **p.704 (Box 15.2)** TBC; **p.707 (Box 15.4)** Al Braunworth/iStock; **p.713 (Box 15.5)** Angel-media/iStock; **p.715 (bottom/right)** Courtesy of Michael Murphy/ GFDL/CC-BY-SA. Diet Coke® is a registered trademark of the Coca-Cola company. All rights reserved.; **p.720 (Box 15.6)** TBC

Chapter 16

p.726 TBC; **p.727 (background)** Thomas Lammeyer/iStock; **p.729 (Fig. 16.1)** Richard Megna, Fundamental Photographs/ Science photo library; **p.729** GeorgiosArt/iStock; **p.732 (top/right)** Yoo Chung, CC-BY-SA; **p.732 (top/left)** Copyright © 2016 OTT Hydromet; **p.734 (Box 16.1)** Courtesy of Millipore Corporation; **p.736 (Fig. 16.5)** Andrew Lambert Photography/Science Photo Library; **p.738 (Box 16.2)** Achim Prill/iStock (left) Dmitry Kutalev/iStock (right); **p.741** Nissan leaf battery pack courtesy of Nissan Motor Limited; **p.744 (Worked Example 16.4)** © Ingram; **p.748** Copyright RS fabrications. Recycling Box manufactured by RS fabrications; **p.756** Courtesy of Xstrata Technology/ CC BY-SA 3.0; **p.757** Courtesy of Psaciemat/ CC BY-SA 3.0; **p.758** Uhde GmbH; **p.760 (Box 16.6)** Reproduced with kind permission from Energizer Group Ltd.

Chapter 17

p.764 PhotoDisc; **p.765 (background)** DNY59/iStock; **p.765 (bottom)** Courtesy of A.A. Clifford, K.D. Bartle and C.M. Rayner, University of Leeds; **p.766** Ruth Hughes; **p.768** Image courtesy of Izmaelt/ CC BY-SA 3.0; **p.770** Rowan Butler/iStock photo; **p.771** John Holman; **p.774 (Box 17.1)** Courtesy of Duncan W. Bruce, University of York; **p.777 (right/top)** Courtesy of Ansgar Walk/ CC BY-SA 2.5; **p.779 (bottom/right)** Design Pics Inc / Alamy Stock Photo; **p.802 (top/left)** Ruth Hughes; **p.803** © Dennis Kunkel Microscopy, Inc.; **p.805** From the GlenDronach Distillery - courtesy of the BenRiach Distillery Co. Ltd.

Chapter 18

p.814 Bombaert/iStock; **p.815 (top/right)** Peter Kneffel/apa/Corbis; **p.815 (bottom/right)** Oxford University Press; **p.816 (left/top)** Tom Grill/iStock; **p.816 (left/middle)** Oxford University Press; **p.817 (right)** Oxford University Press; **p.819 (right/middle)** Oxford University Press; **p.819 (right/bottom)** Geoff Tompkinson/Science Photo Library; **p.820 (bottom/left)** Peter Finnie/ iStock; **p.821** Oxford University Press; **p.823 (Box 18.1 left/top)** Marco Tenback/iStock; **p.823 (Box 18.1 left/middle)** PhotoDisc; **p.825 (Worked Example 18.2)** Oxford University Press; **p.826 (left/top)** Oxford University Press; **p.826 (left/middle)** En Tien Ou/iStock; **p.827 (right)** Kemie/iStock; **p.828** Skystorm/iStock; **p.829 (bottom/right)** Oxford University Press; **p.830 (Box 18.3left)** Simon Fraser/Science Photo Library; **p.830 (Box 18.3 right/top)** Chris Elwell/iStock; **p.835 (right)** Oxford University Press; **p.836 (Box 18.4)** Janice Haney Carr, Centers for Disease Control and Prevention/Public Health Image library; **p.838** Digital Vision; **p.839 (Box 18.5)** Adam Korzekwa/iStock (DNA) and Image of Van 't Hoff paper models courtesy of Museum Boerhaave Leiden; **p.839 (top/right)** Supermimicry/iStock; **p.840 (left/bottom)** Courtesy of Bellingham and Stanley ; **p.841 (right/middle)** HungryJenny; **p.843 (right)** Florida Stock/iStock; **p.844** Aldra/iStock; **p.844 (Worked Example 18.4)** Oxford University Press; **p.845 (right)** Oxford University Press; **p.846 (left)** JGI/iStock; **p.847 (Worked Example 18.5 bottom/right)** Juice Images; **p.848 (Box 18.7 top/right)** Larry Dunstan, Wellcome images/ CC BY-NC-ND 4.0; **p.854 (bottom/left)** Oxford University Press; **p.858 (Questions)** Matka Wariatka/iStock; **p.859 (Questions left/top)** Ingram; **p.859 (Questions right/top)** Oxford University Press; **p.859 (Questions right/middle)** Oxford University Press

Chapter 19

p.860 Issei Kato/Reuters/Corbis; **p.862** Oxford University Press; **p.865 (Box 19.2 right/top)** Oxford University Press; **p.865 (Box 19.2 right/bottom)** Oxford University Press; **p.871** Oxford University Press; **p.872 (top/left)** Oxford University Press; **p.876 (left)** Oxford University Press; **p.883** Oxford University Press; **p.887 (right)** Oxford University Press; **p.888 (bottom/left)** Oxford University Press; **p.889 (right)** Oxford University Press; **p.893 (top/right)** Oxford University Press; **p.894 (Box 19.5)** Rhino courtesy of Joris Van Ostaeyen/iStock. Dentist courtesy of Carmen Martinez Banus/iStock. Sea slug courtesy of Photo2222/CC BY-SA 3.0. Runners courtesy of PhotoDisc.; **p.901 (Box 19.7)** jcarroll-images/iStock; **p.904 (Box 19.8)** Courtesy of Remotec Inc., a Northrop Grumman Corporation Subsidiary ; **p.909** Oxford University Press; **p.912 (Worked Example 19.7bottom)** Scott Bauer/ USDA; **p.914 (Questions right)** Silva Merino/iStock; **p.915 (Questions right/top)** Oxford University Press; **p.915 (Questions right/middle)** © Sean Gladwell – Fotolia.com

Chapter 20

p.916 Olga LIS/iStock; **p.917 (top/right)** Courtesy of Trical Inc; **p.918 (Box 20.1right)** J Fontes ImagDOP/UAc; **p.919 (Box 20.1 left/top)** B. Hadlow, Australian National Botanic Gardens; **p.919 (Box 20.1 left/bottom)** Courtesy of Tom Murray ; **p.923 (Box 20.2 right)** NASA; **p.923** Oxford University Press; **p.928 (left)** Oxford University Press; **p.930 (left/middle)** Oxford University Press; **p.930 (left/bottom)** Shelly Perry/iStock Photo ; **p.934 (top/left)** Oxford University Press; **p.937 (top/right)** Courtesy of CEM Microwave Technology Ltd; **p.940 (Worked Example 20.3 right)** Courtesy of Montreal Biotech Inc; **p.942 (Box 20.4 left)** Dmitry Valberg/CC BY 2.0; **p.946 (bottom/left)** Oxford University Press; **p.947 (Box 20.5 top/right)** Jim Clare, Nature PL; **p.948 (bottom/left)** Oxford University Press; **p.953 (right)** Oxford University Press; **p.954 (bottom/left)** © 2005-2012, Asynt Ltd; **p.956** Oxford University Press; **p.958 (Questions top/left)** Soverign ISM Science photo library; **p.958 (Questions bottom/left)** Soverign ISM Science photo library; **p.959 (Questions right)** Stanislav Bokach/iStock; ;

Chapter 21

p.961 (top/right) Thomas Deerinck, NCMIR/Science Photo Library; **p.961 (background)** Ismael Montero Verdu/iStock; **p.962 (top/left)** Martin F Chillmaid/Science Photo Library; **p.965 (top)** Oxford University Press; **p.966 (Box 21.1 left)** Scott Bauer/USDA; **p.967 (top/right)** Oxford University Press; **p.968 (top/left)** Courtesy of Parr Instrument Company; **p.978 (top)** IDPS, University of Bath; **p.981 (Box 21.3 right/top)** Stephen Vaughn/USDA; **p.981 (Box 21.3 right/middle)** Chemical Design/Science Photo Library; **p.984 (left/middle)** Tom Merton; **p.990 (top/left)** Oxford University Press; **p.991 (top)** Oxford University Press; **p.993 (top/right)** Image Source / Alamy Stock Photo; **p.993 (bottom/right)** Ria Novotsi/Science Photo Library; **p.994 (top/left)** Oxford University Press; **p.996 (left/middle)** Reproduced with permission from Patrick Clement; **p.996 (left/bottom)** Oxford University Press; **p.1000 (Questions left/top)** PhotoDisc; **p.1000 (Questions left/bottom)** Oxford University Press; **p.1001 (Questions bottom)** TingTing/Picasa

Chapter 22

p.1002 Kateryna Govorushchenko/iStock; **p.1003 (background)** jussi Santaniemi/iStock; **p.1003 (top/right)** Victoria and Albert Museum; **p.1004 (left/middle)** busypix/iStock; **p.1004 (left/bottom)** Oxford University Press; **p.1005 (right)** Oxford University Press; **p.1008 (Box 22.2)** Courtesy of Joe Haupt; Vintage Radio Advertising/ CC BY-SA 2.0; **p.1010 (top/left)** Oxford University Press; **p.1012 (Box 22.3 left)** Science Museum; **p.1012 (Box 22.3 right)** Courtesy of Bayer Healthcare AG; **p.1021 (Box 22.4 left)** © britishcycling.org.uk; **p.1021 (Box 22.4 right)** Bill Grove/iStock Photo; **p.1024 (bottom/left)** Vassiliy Vassilenko/iStock; **p.1026 (Box 22.5 top/right)** Oxford University Press; **p.1030 (left)** Oxford University Press; **p.1036 (bottom/left)** Oxford University Press; **p.1037 (top/right)** Oxford University Press; **p.1038 (left/top)** Science Photo Library; **p.1038 (left/bottom)** Looney tunes characters, names and all related indicia are trademark and copyright of Warner Bros Entertainment Inc. all rights reserved.; **p.1039 (Box 22.6 left)** William R Sallaz,/New Sport/Corbis; **p.1041 (bottom/right)** Oxford University Press; **p.1042 (middle)** PhotoDisc; **p.1043 (left)** IDPS, University of Bath;

p.1045 (right) AFP; **p.1047 (Box 22.7 right)** AP Images/ Elise Amendola; **p.1049 (Box 22.8 top/right)** Joshua Sowin/iStock; **p.1052 (Questions right)** Miodrag Gajic I stock; **p.1053 (Questions right)** subtik iStock photo; **p.**TBC (TBC TBCTBC) Courtesy of Dharmadhyaksha/ CC BY-SA 3.0

Chapter 23

p.1054 pinobarile/iStock; **p.1056 (left/top)** Oxford University Press; **p.1056 (left/middle)** Oxford University Press; **p.1057 (top/right)** Oxford University Press; **p.1060 (Box 23.1 right/top)** Oxford University Press; **p.1060 (Box 23.1 right/bottom)** Kiyoshi Takahase Segundo/iStock; **p.1062 (left/middle)** Kaan Ates/iStock; **p.1063 (right/top)** Courtesy of DaimlerChrysler; **p.1063 (right/bottom)** LiAIH reacts violently; **p.1067 (right)** Image by Peggy Greb for the Agricultural Research Service USDA; **p.1068 (top/left)** Bob Walker/CC_BY-SA; **p.1069 (right)** Oxford University Press; **p.1070 (Box 23.4 top/right)** Oxford University Press; **p.1071 (top/right)** Paul R Blakemore, Department of Chemistry, Oregon State University; **p.1073 (top/right)** The kobal collection; **p.1074 (Box 23.5 left)** FreezeFrame studio/iStock; **p.1075 (top/right)** Courtesy of Jane Domingquez, CLAS News publications, University of Florida; **p.1076 (left/middle)** Neal Warren/iStock; **p.1076 (left/bottom)** Oxford University Press; **p.1078 (left/middle)** Courtesy of Montreal Biotech Inc.; **p.1078 (left/bottom)** Oxford University Press; **p.1083 (top/right)** Oxford University Press; **p.1084 (left)** Mark Wragg/iStock; **p.1087 (right)** Andrew Lambert/Science Photo Library; **p.1089 (right/bottom)** Gustaf Brundin/iStock; **p.1089 (right/middle)** Ruth Hughes; **p.1091 (right)** Oxford University Press; **p.1092 (left)** Oxford University Press; **p.1096 (Questions right/top)** Oxford University Press; **p.1096 (Questions right/bottom)** Kevin Herrin/iStock; **p.1097 (Questions right)** Keith Weller, USDA- ARS;

Chapter 24

p.1099 (top/right) Dave Logan/iStock; **p.1100 (Box 24.1)** Oxford University Press; **p.1105** Christian Biewer, University of York; **p.1106 (bottom/left)** Oxford University Press; **p.1108 (bottom/left)** AFP; **p.1109 (Box 24.3 bottom/left)** almonds: stockbyte; **p.1109 (Box 24.3 bottom/middle)** banana: PhotoDisc; **p.1109 (Box 24.3 bottom/right)** rum: john pozniak; **p.1109 (Box 24.3 top/left)** Oxford University Press; **p.1109 (Box 24.3 top/middle)** Oxford University Press; **p.1109 (Box 24.3 top/right)** Oxford University Press; **p.1114 (left/middle)** Oxford University Press; **p.1114 (left/top)** Oxford University Press; **p.1115 (Box 24.5 right)** Courtesy of Bayer Healthcare; **p.1116 (bottom/left)** Oxford University Press; **p.1118 (left)** Oxford University Press; **p.1119 (right)** Freundchen/CC-BY-SA; **p.1120 (Box 24.6 left)** Oxford University Press; **p.1122 (left)** Oxford University Press; **p.1123 (Worked Example 24.3)** Eye of Science/Science Photo Library; **p.1127 (top/right)** Oxford University Press; **p.1132 (top/left)** Oxford University Press

Chapter 25

p.1138 (right) NASA; **p.1138 (left)** NASA; **p.1139 (top/right)** NASA; **p.1139 (background)** NASA; **p.1141 (Box 25.1 right)** Honda Motor Company; **p.1142 (left)** University of Bath; **p.1146 (left)** NOAA; **p.1150 (left)** Action images/ Tony O'Brien; **p.1151 (right)** NASA JPL CALtech; **p.1152 (top/left)** YenWen Lu/iStock; **p.1152 (left/middle)** joseasreeves/iStock; **p.1154 (Box 25.4)** Matelux acid-etched glass, reproduced with permission of AGC flat glass UK; **p.1159 (Box 25.5 right)** leCajun/iStock; **p.1160 (bottom/right)** Corbis Digital/iStock; **p.1161 (Box 25.6 left)** Courtesy of the National Research Council of Canada; **p.1161 (Box 25.6 right/bottom)** Charles Fisher, Penn State University; **p.1162 (Box 25.7)** Tim La Duke of LaDuke Studios Holland, Michigan USA and provided by Cammenga Corporation; **p.1162 (left)** aimy27feb/iStock

Chapter 26

p.1168 Steve Gschmeissner/Science Photo Library; **p.1169 (bottom/left)** Steve Gschmeissner/Science Photo Library; **p.1169 (bottom/right)** Prof P Motta, Department of Anatomy,/University La Sapienza, Rome/Science Photo Library; **p.1171 (right)** Leadinglights/iStock; **p.1173 (left)** U.S. Navy Journalist 3rd class Wes Eplen.; **p.1175 (right/bottom)** IDPS, University of Bath; **p.1175 (right/top)** Charles D Winters/Science Photo Library ; **p.1176 (top)** Thomas Raupach/Still Pictures; **p.1176 (left)** Norbert Rohr/iStock; **p.1177 (Box 26.2)** Ian Redmond/Nature PL; **p.1179 (top/right)** PhotoDisc; **p.1182 (Fig. 26.8)** Peter Edwards, Department of Chemistry, University of Oxford; **p.1184 (bottom/left)** Wolfgang Glock/CC BY 3.0; **p.1185 (Box 26.3 left)** PhotoDisc; **p.1185 (Box 26.3 right/top)** Simon Cole/CC-BY-3.0; **p.1185 (Box 26.3 right/bottom)** Power and Syred/Science Photo Library; **p.1186 (left)** Junho Jung/ CC_BY-SA; **p.1187 (top/right)** Adam Hart Davis/Science Photo Library; **p.1189 (right)** Spydr/iStock; **p.1190 (left)** Gca/Science Photo Library; **p.1191 (Box 26.4 right)** Djordie Korovljevic/iStock; **p.1193 (Box 26.5 top)** nejmlez/ wikicommons

Chapter 27

p.1200 Daniel Stein/iStock; **p.1201 (top/right)** Olivier Blondeau/iStock; **p.1207 (top/right)** Studiomode/Alamy; **p.1210 (top/left)** Russ Lappa/ Science Photo Library; **p.1211 (top/left)** PhotoDisc; **p.1211 (Box 27.1)** Nancy Nehring/iStock; **p.1217 (top/right)** Michael Fielding, the boc group plc/ Science Photo Library; **p.1219 (right/top)** IDPS, University of Bath; **p.1219 (right/bottom)** Seraficus/iStock; **p.1220 (Box 27.3)** Alice Roberts neeeee Mumford; **p.1224 (left)** Charles D Winters/Science Photo Library ; **p.1227 (top/right)** Alnia Pavlova/iStock; **p.1229 (top/right)** Charles D Winters/ Science Photo Library ; **p.1231 (top/right)** ileha/istock; **p.1231 (Box 27.5)** Nikita Rogul/iStock; **p.1232 (Box 27.5)** Mauro Fermarielo/Science Photo Library ; **p.1235 (Box 27.7 left)** NASA/JPL/University of Arizona ; **p.1235 (Box 27.7 right)** NASA/JPL/University of Arizona ; **p.1237 (right/middle)** David Taylor/Science Photo Lib; **p.1237 (right/bottom)** Stockbyte; **p.1239 (top/right)** bostb/iStock; **p.1242 (Box 27.8)** Goddard Space Flight Centre, NASA; **p.1249 (top/right)** Corbis/Digital Stock; **p.1249 (Box 27.9)** Dhuss/ iStock; **p.1251 (bottom/right)** At Wag/istock

Chapter 28

p.1254 Leontura/iStock; **p.1255 (top/right)** Courtesy of the Bodleian Library; **p.1257 (Box 28.1 right)** ISM/Science photo library; **p.1262 (top/left)** MykReeve/CC-BY-SA; **p.1264 (left)** Materialscientist/CC-BY-SA; **p.1269 (Box 28.2)** USGS; **p.1275 (Box 28.3 bottom/left)** de:Benutzer Hase: CC_BY-SA; **p.1290 (top/left)** Guillermo Lobo/iStock; **p.1291 (Fig. 28.38)** IDPS, University of Bath; **p.1294 (Box 28.5 top/right)** Jim Jurica/iStock; **p.1295** IDPS, University of Bath; **p.1299 (Box 28.7 bottom/right)** photonimo/istock

The Greek alphabet

Α, α	alpha
Β, β	beta
Γ, γ	gamma
Δ, δ	delta
Ε, ε	epsilon
Ζ, ζ	zeta
Η, η	eta
Θ, θ	theta
Ι, ι	iota
Κ, κ	kappa
Λ, λ	lambda
Μ, μ	mu
Ν, ν	nu
Ξ, ξ	xi
Ο, ο	omicron
Π, π	pi
Ρ, ρ	rho
Σ, σ	sigma
Τ, τ	tau
ϒ, υ	upsilon
Φ, ϕ	phi
Χ, χ	chi
Ψ, ψ	psi
Ω, ω	omega

SI multiplication prefixes

femto	f	10^{-15}
pico	p	10^{-12}
nano	n	10^{-9}
micro	μ	10^{-6}
milli	m	10^{-3}
centi	c	10^{-2}
deci	d	10^{-1}
deca	da	10^{1}
kilo	k	10^{3}
mega	M	10^{6}
giga	G	10^{9}
tera	T	10^{12}

Physical constants

Quantity	Symbol	Value and units
Speed of light (in a vacuum)	c	$2.997\,92 \times 10^{8}\,\mathrm{m\,s^{-1}}$
Permittivity of a vacuum	ε_0	$8.854\,19 \times 10^{-12}\,\mathrm{C^2\,J^{-1}\,m^{-1}}$
Charge of an electron	e	$1.602\,18 \times 10^{-19}\,\mathrm{C}$
Electron mass	m_e	$9.109\,39 \times 10^{-31}\,\mathrm{kg}$
Proton mass	m_p	$1.672\,62 \times 10^{-27}\,\mathrm{kg}$
Neutron mass	m_n	$1.674\,93 \times 10^{-27}\,\mathrm{kg}$
Planck constant	h	$6.626\,08 \times 10^{-34}\,\mathrm{J\,s}$
Rydberg constant	R_H	$3.289\,84 \times 10^{15}\,\mathrm{s^{-1}}$ (Hz)
Avogadro constant	N_A	$6.022\,14 \times 10^{23}\,\mathrm{mol^{-1}}$
Faraday constant	$F\ (= N_A e)$	$9.648\,53 \times 10^{4}\,\mathrm{C\,mol^{-1}}$
Boltzmann constant	k_B	$1.380\,66 \times 10^{-23}\,\mathrm{J\,K^{-1}}$
Gas constant	$R\ (= N_A k_B)$	$8.314\,47\,\mathrm{J\,K^{-1}\,mol^{-1}}$
		$8.205\,74 \times 10^{-2}\,\mathrm{dm^3\,atm\,K^{-1}\,mol^{-1}}$ (non SI units)
Acceleration due to gravity	g	$9.806\,65\,\mathrm{m\,s^{-2}}$
Ratio of circumference to diameter of circle	π	3.141 59